SUPPLEMENTS AND N

For the Instructor

Instructor's Toolbox (0-13-092364-8)
This easy-to-use lecture organizer helps you integrate the entire Audesirk package by chapter. For qualified adopters, each toolbox includes:

- Instructor's Guide to Print and Media Resources
- Transparency Acetates—250 four-color selections from the text
- Transparency Masters—the balance of the text line art available as black-and-white masters
- Lecture Outlines
- Answers to end-of-chapter questions
- Group Activity Guidelines

Instructor Resource CD-ROM
Dual Platform (0-13-092662-0)
All of the instructor's print and media resources in one place for ease of use and convenience. The CD-ROM contains:

- Instructor's Guide to Print and Media Resources—electronically linked to take you immediately to the resources you want to use.
- Image Gallery—over 1000 illustrations, both labeled and unlabeled, formatted for large lecture-hall presentations; many photos are also included.
- ***MediaTutor Student CD-ROM***—the complete contents ofthe student CD-ROM designed for instructor use in the classroom. Instructors can select animations, images, and activities to use in lecture.
- PowerPoint Gallery—PowerPoint presentations for each chapter have been created for the instructor's convenience. Presentations are in two forms: images only, and images with lecture outline. Special PowerPoint and Flash animations by David Huffman of *Southwest Texas State University* are also included.
- Test Item File—all the test questions from the Test Item File are available and can be easily edited.

Test Item File (0-13-093446-1)
More than 2400 questions compiled by a carefully selected team of educators: Lewis Deaton, *University of Southwestern Louisiana*; Gail Gasparich, *Towson University*; Charles Good, *Ohio State University, Lima*; and Kate Lajtha, *Oregon State University*. Questions have a variety of styles to meet a range of needs. Users can select from conceptual, applied, and fact-based questions; users can also select questions according to level of difficulty.

PH Custom Test
Windows (0-13-092365-6)
Macintosh (0-13-092366-4)
Prentice Hall Custom Test allows instructors to create and edit exams electronically. With the On-Line Testing component, exams can be administered on-line and data can be automatically transferred for evaluation and grading.

ABCNEWS

ABC News/Prentice Hall Video Library for Biology (0-13-514191-5)
This unique video library contains brief segments (5 to 20 minutes long) from award-winning ABC News programs such as 20/20 and World News Tonight. This innovative resource provides a connection between biology and what students see in the news.

The Prentice Hall Laser Disc for Biology
(0-13-378100-3)

The Prentice Hall Laser Disc for Biology is a visual encyclopedia created for use with *Biology: Life on Earth.* The disc features more than 1000 still images, 30 minutes of video segments, and 25 minutes of three-dimensional animations. The animations focus on topics that are challenging to students and are particularly difficult to visualize (such as cellular respiration, photosynthesis, and enzyme function). The video segments focus on applications of biology to students' everyday world (such as DNA testing and CT and fMRI technologies).

Chapter 26

Activity 26.1 Feedback Loops and Homeostasis (active art)
Web Investigation: The Limits of Endurance
Issues in Biology: How Do Animals Cope with Environmental Change?
Bizarre Facts: Frog-cycles and Super-cool Squirrels?

Chapter 27

Activity 27.1 Heart Structure and Function (active art)
Activity 27.2 Cardiac Cycle (animation)
Web Investigation: Xenotransplants
Issues in Biology: Is Congestive Heart Failure an Epidemic?

Chapter 28

Activity 28.1 Human Respiratory Anatomy (active art)
Activity 28.2 Gas Exchange (animation)
Web Investigation: Lives up in Smoke
Issues in Biology: What Are the Physiological Stresses of Deep Diving?

Chapter 29

Activity 29.1 The Digestive System (active art)
Web Investigation: Fat in the Family
Issues in Biology: What Are the Side Effects of Eating Chocolate?

Chapter 30

Activity 30.1 Urinary System Anatomy (active art)
Activity 30.2 Urine Formation (animation)
Web Investigation: Hemodialysis and Hope
Issues in Biology: What Was the Penicillin Project?

Chapter 31

Activity 31.1 Inflammation (animation)
Activity 31.2 Clonal Selection (animation)
Web Investigation: Fighting the Flu
Issues in Biology: What Is an Allergy?

Chapter 32

Activity 32.1 Endocrine System Anatomy (active art)
Activity 32.2 Modes of Action of Hormones (animation)
Web Investigation: Losing on Steroids
Issues in Biology: What Were the Aftereffects of the Chernobyl Disaster?
Bizarre Facts: Sex Reversal in Fish

Chapter 33

Activity 33.1 Ions and Electrical Signals (animation)
Activity 33.2 Synapses (animation)
Web Investigation: From Tragedy to Triumph
Issues in Biology: What Is Alzheimer's Disease?
Bizarre Facts: A Real Shocker
Bizarre Facts: Stem Cells: Whither Goest Thou?

Chapter 34

Activity 34.1 Muscle Structure (active art)
Activity 34.2 Muscle Movement (animation)
Web Investigation: Healing Broken Bones
Issues in Biology: Part-Time Athletes; Full-Time Injuries?

Chapter 35

Activity 35.1 Human Reproductive Anatomy (active Art)
Activity 35.2 Hormonal Control of the Menstrual Cycle. (simulation)
Web Investigation: Mr. Mom?
Issues in Biology: Why Are Sexually Transmitted Diseases So Harmful?
Bizarre Facts: The Journey Ends In Lovers Meeting

Chapter 36

Activity 36.1 Stages of Animal Development (animation)
Activity 36.2 Human Development (active art)
Web Investigation: Far-Reaching Choices
Issues in Biology: What's Happening to Me?

Chapter 37

Activity 37.1 Innate Animal Behavior (simulation)
Web Investigation: Sex and Symmetry
Issues in Biology: What Is New About the Nature vs. Nurture Debate?
Bizarre Facts: A Trip of A Lifetime . . . on Gossamer Wings
Bizarre Facts: Light from Life

Chapter 38

Activity 38.1 Population Growth and Regulation (simulation)
Activity 38.2 Population Growth and Regulation in Human Populations (simulation)
Web Investigation: Acorns, Mice, Moths, Deer, and Disease
Issues in Biology: Are Zebra Mussels Really Invading?

Chapter 39

Activity 39.1 Effects of Keystone Species Extinction (simulation)
Activity 39.2 Primary succession (animation)
Web Investigation: Invasion of the Zebra Mussels
Issues in Biology: What Is Happening to the Red-Cockaded Woodpecker?
Bizarre Facts: Gut and Run!
Bizarre Facts: Twelve-Legged Frogs: A Serious Warning Or Nothing New?

Chapter 40

Activity 40.1 Building a Food Web (active art)
Web Investigation: Flight from Extinction
Issues in Biology: What Are Bioaccumulation and Endocrine-disrupting Chemicals?
Bizarre Facts: Giant Worms

Chapter 41

Activity 41.1 Climate Builds Biomes (animation)
Activity 41.2 Biomes of the World (active art)
Web Investigation: Wings of Hope
Issues in Biology: Why Should We Be Concerned about Biodiversity Loss and Deforestation?

Web Investigation: Sunshine Perils
Issues in Biology: Is There a Genetic Basis to Cancer?

Chapter 10

Activity 10.1 Transcription (animation)
Activity 10.2 Translation (animation)
Web Investigation: Boy or Girl?
Issues in Biology: What Can Flies with Eyes on Their Legs Tell Us about Gene Regulation?
Bizarre Facts: INDYan'a Fly

Chapter 11

Activity 11.1 Mitosis (animation)
Activity 11.2 Meiosis (animation)
Web Investigation: Rattlesnake Surprise
Issues in Biology: Why Can a Pause in Mitosis Be Crucial for a Cell's Health?

Chapter 12

Activity 12.1 Monohybrid Crosses (animation, simulation)
Activity 12.2 Dihybrid Crosses (simulation)
Web Investigation: Sleepy Genes
Issues in Biology: What Is the Most Common Genetic Disease in the US?
Bizarre Facts: Temperature Sex Determination
Bizarre Facts: A Typical Human: Brown Hair, Color Vision, Freckles, and Six Fingers!

Chapter 13

Activity 13.1 Genetic Recombination (animation)
Activity 13.2 Building a DNA Library (simulation)
Web Investigation: Teaching an Old Grain New Tricks
Issues in Biology: Were Neanderthals Contemporary to Modern Humans?
Bizarre Facts: Monkeyshine

Chapter 14

Activity 14.1 Analogous and Homologous Structures (active art)
Activity 14.2 Natural Selection for Antibiotic Resistance (simulation)
Web Investigation: A Missing Link Unearthed
Issues in Biology: What Is the Evolution Debate?

Chapter 15

Activity 15.1 The Bottleneck Effect (active art)
Activity 15.2 Natural Selection at Work: Alpine Skypilots (simulation)
Activity 15.3 Three Modes of Natural Selection (active art)
Web Investigation: Cause of Death: Evolution
Issues in Biology: What Is the Difference between Innate and Learned Behavior?

Chapter 16

Activity 16.1 Allopatric Speciation (active art)
Activity 16.2 Sympatric Speciation (active art)
Web Investigation: Lost World
Issues in Biology: What Is a Species?

Chapter 17

Activity 17.1 Endosymbiosis (animation)
Activity 17.2 Plate Tectonics (animation)
Web Investigation: Life on a Frozen Moon?
Issues in Biology: Who Were Our Early Human Ancestors?
Bizarre Facts: Out of Africa, or Not

Chapter 18

Activity 18.1 Taxonomic Classification (active art)
Web Investigation: Origin of a Killer
Issues in Biology: Classification Conundrums—What Is a Kingdom, Anyway?
Bizarre Facts: Branches of the Tree of Life

Chapter 19

Activity 19.1 Retrovirus Replication (animation)
Activity 19.2 Herpes Virus Replication (animation)
Activity 19.3 Bacterial Conjugation (animation)
Web Investigation: Agents of Death
Issues in Biology: Bacteria That Prevent Food Poisoning?
Bizarre Facts: Life on the Extreme Edge

Chapter 20

Activity 20.1 Fungi Structure (active art)
Activity 20.2 The Fungi (active art)
Web Investigation: Three Outings
Issues in Biology: You Said the Dog Fixed My Pizza?
Bizarre Facts: Just When You Thought It Was Safe...

Chapter 21

Activity 21.1 Evolution of Plant Structures (active art)
Activity 21.2 Fern Life Cycle (animation)
Web Investigation: Hunting for Medical Treasures
Issues in Biology: Where Have All the Food Plants Gone?

Chapter 22

Activity 22.1 Cnidocytes (animation)
Activity 22.2 Insect Metamorphosis (animation)
Web Investigation: The Search for a Sea Monster
Issues in Biology: What Is Happening to the Frogs?

Chapter 23

Activity 23.1 Plant Anatomy (active art)
Activity 23.2 Plant Transport Mechanisms (simulation, animation)
Activity 23.3 Primary and Secondary Growth (animation)
Web Investigation: A Beautiful Death Trap
Issues in Biology: How Is Hydroponics Possible?

Chapter 24

Activity 24.1 Flower Structure and Function (active art)
Activity 24.2 Fruit and Seed Structure and Development (animation)
Web Investigation: Walk Through a Meadow
Issues in Biology: How Are Plants Vegetatively Propagated?
Bizarre Facts: Frankenplants?

Chapter 25

Activity 25.1 Auxin Action (active art)
Activity 25.2 Is it Time to Grow? (simulation)
Web Investigation: A Chemical Cry for Help
Issues in Biology: What Is the Commercial Use of Plant Growth Regulators?

SUPPLEMENTS AND NEW MEDIA
For the Student

MediaTutor Student CD-ROM (0-13-092367-2)
(Free when shrink-wrapped with text)
The MediaTutor Student CD-ROM is the perfect companion to the text. Activities on the CD-ROM help students through difficult concepts found in the book: Students interact with animations, videos, and simulations to answer questions and solve problems. In the MediaTutor section at the end of each chapter in the book, the activities on the student CD-ROM are described. The description of each activity includes the topic, the overall goal, and the estimated time required to complete the activity. Extensive quizzing before, during, and after each CD-ROM activity ensures that students have the prerequisite knowledge needed to understand a new concept, that they understand the parts of a new concept as it is presented to them, and that they are able to put it all together.

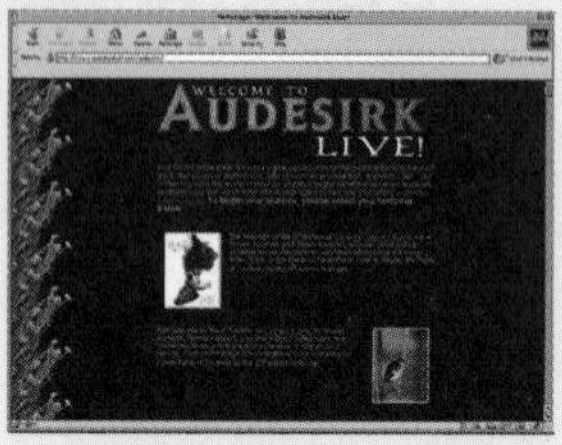

Audesirk Live!
WWW
(http://www.prenhall.com/audesirk)
Launch your exploration of biology on the Web with Audesirk Live! This innovative on-line resource center is designed specifically to support and enhance *Biology: Life on Earth, Sixth Edition.* Web Investigations found in the MediaTutor page of the book extend each chapter's Case Study; Issues in Biology address specific current topics mentioned in the text; and Concept Challenge allows students to test their knowledge with extensive self-quizzes for each chapter. Many other tools are provided to help instructors and students interact on-line.

Science on the Internet : A Student's Guide, 2000-2001 (0-13-028253-7)
By Andrew T. Stull
The perfect tool to help your students take advantage of Audesirk Live! on the Web and open up the world of the Internet for exploration. This practical resource gives clear steps for accessing our regularly updated biology resource area, as well as an overview of general investigation strategies.

Student Study Guide (0-13-092363-X)
By Joseph Chinnici, *Virginia Commonwealth University*, and Susan M. Wadkowski, *Lakeland Community College.*
This essential study tool will help students think through the biological concepts and reinforce key concepts presented in the text. It offers a wide range of study exercises and self-tests.

The New York Times **Themes of the Times**
Coordinated by Colleen Lee
This newspaper-format supplement brings together recent articles on the hottest issues in biology taken directly from the pages of the New York Times. This free supplement, available in quantity through your local representative, helps students make connections between the classroom and the world around them.

AIDS Update 2001 (0-13-090962-9)
By Gerald J. Stine, *University of North Florida, Jacksonville*
The paperback reader, updated annually, answers many of the questions students have about today's AIDS crisis.

For the Laboratory:

Exploration in Basic Biology. Ninth Edition (0-13-093031-8)
By Stanley E. Gunstream,
Pasadena City College
This best-selling laboratory manual can be used with *Biology: Life on Earth.* It includes 41 self-contained, easy-to-understand experiments that blend traditional experiments with new investigative exercises.

Instructor's Manual to Explorations in Basic Biology, Ninth Edition (0-13-061959-0)

Gunstream, Biological Explorations: A Human Approach, Fourth Edition
(0-13-089446-X)
By Stanley E. Gunstream,
Pasadena City College
Specifically designed for course in general biology where the human organism is emphasized – and for a growing number of courses in human biology– this lab manual contains 32 outstanding exercises by the author of our successful basic biology lab manual, *Explorations in Basic Biology.*

Instructors Manual to Biological Explorations. (0-13-090774-X)

Thinking About Biology: An Introductory Biology Manual
(0-13-633033-9) Mimi Bres and Arnold Weisshaar, both of *Prince George's Community College.*
The unique inquiry-based format of this lab manual promotes an active learning style. Extensively class-tested by more than 7000 students, it emphasizes problem solving.

Instructors Manual to Thinking About Biology
(0-13-952319-7)

Biology
LIFE ON EARTH

Biology

Sixth Edition

LIFE ON EARTH

Teresa Audesirk
Gerald Audesirk
University of Colorado at Denver

Bruce E. Byers
University of Massachusetts, Amherst

Prentice Hall

Upper Saddle River, New Jersey 07458

Library of Congress Cataloging-in-Publication Data

Audesirk, Teresa.
Biology : life on Earth / Teresa Audesirk, Gerald Audesirk, Bruce E. Byers.-- 6th ed.
p. cm.
Includes bibliographical references.
ISBN 0-13-089941-0
1. Biology. I. Audesirk, Gerald. II. Byers, Bruce E. III. Title.
QH308.2.A93 2001
570--dc21

2001021854

Executive Editor: Teresa Ryu
Editor in Chief, Biology: Sheri L. Snavely
Senior Development Editors: Shana Ederer; Karen Karlin
Production Editor: Tim Flem/PublishWare
Art Director: Jonathan Boylan
Managing Editor, Audio/Video Assets: Grace Hazeldine
Project Manager: Travis Moses-Westphal
Executive Marketing Manager: Jennifer Welchans
Marketing Director: John Tweeddale
Vice President of Production & Manufacturing: David W. Riccardi
Executive Managing Editor: Kathleen Schiaparelli
Director of Creative Services: Paul Belfanti
Director of Design: Carole Anson
Page Composition: PublishWare
Production Support: William Johnson; Elizabeth Gschwind
Manager of Formatting: Jim Sullivan
Manufacturing Manager: Trudy Pisciotti
Buyer: Michael Bell
Editor in Chief of Development: Carol Trueheart
Media Editors: Travis Moses-Westphal
Project Manager, Companion Web site: Elizabeth Wright
Assistant Managing Editor, Science Media: Alison Lorber
Media Production Editor: Rich Barnes
Supplements Production Editor: Dinah Thong
Editorial Assistants: Colleen Lee; Lisa Tarabokjia
Marketing Assistant: Anke Braun
Cover Designer: Tom Nery; John Christiana
Interior Designer: Tom Nery
Illustrators: Imagineering; Rolando Corujo; Hudson River Studios; Howard S. Friedman; David Mascaro; Edmund Alexander; Roberto Osti
Photo Research: Linda Sykes
Photo Research Administrator: Beth Boyd
Cover Photograph: "Weaver bird building nest, Maasai Mara National Reserve, Kanya/Manoj Shah"

Prentice Hall

Printed in the United States of America

College: ISBN 0-13-089941-0

10 9 8 7 6 5 4 3

High School: ISBN 0-13-093655-3

10 9 8 7 6 5 4 3 2

Prentice-Hall International (UK) Limited, *London*
Prentice-Hall of Australia Pty. Limited, *Sydney*
Prentice-Hall Canada Inc., *Toronto*
Prentice-Hall Hispanoamericana, S.A., *Mexico*
Prentice-Hall of India Private Limited, *New Delhi*
Prentice-Hall of Japan, Inc., *Tokyo*
Pearson Education Asia Pte. Ltd.
Editora Prentice-Hall do Brasil, Ltda., *Rio de Janeiro*

To Heather, Jack, and Lori and in memory of Eve and Joe

T. A. & G. A.

To Maija, Varis, and Ivars

B. E. B.

About the Authors

Both **Terry and Gerry Audesirk** grew up in New Jersey, where they met as undergraduates. After marrying in 1970, they moved to California, where Terry earned her doctorate in marine ecology at the University of Southern California and Gerry earned his doctorate in neurobiology at the California Institute of Technology. As postdoctoral students at the University of Washington's marine laboratories, they worked together on the neural bases of behavior, using a marine mollusk as a model system.

Terry and Gerry are now professors of biology at the University of Colorado at Denver, where they have taught introductory biology and neurobiology since 1982. In their research lab, funded by the National Institutes of Health, they investigate the mechanisms by which neurons are harmed by low levels of environmental pollutants.

Terry and Gerry share a deep appreciation of nature and of the outdoors. They enjoy hiking in the Rockies, running near their home in the foothills west of Denver, and attempting to garden at 7000 feet in the presence of hungry deer and elk. They are long-time members of many conservation organizations. Their daughter, Heather, has added another focus to their lives.

Bruce E. Byers, a midwesterner transplanted to the hills of western Massachusetts, is a professor in the biology department at the University of Massachusetts, Amherst. He's been a member of the faculty at UMass (where he also completed his doctoral degree) since 1993. Bruce teaches introductory biology courses for both nonmajors and majors; he also teaches courses in ornithology and animal behavior.

A lifelong fascination with birds ultimately led Bruce to scientific exploration of avian biology. His current research focuses on the behavioral ecology of birds, especially on the function and evolution of the vocal signals that birds use to communicate. The pursuit of vocalizations often takes Bruce outdoors, where he can be found before dawn, tape recorder in hand, awaiting the first songs of a new day.

Panel of Biology Educators
Biology: Life on Earth, Sixth Edition

We express sincere gratitude to the contributors who worked closely with the authors to ensure *Biology: Life on Earth's* continuing tradition of readability, accuracy, and relevance.

Rita Farrar
Louisiana State University
Research Areas: Immunoparasitology and immunopathology

I enjoy watching "non-science" majors lose their fear of science and find that they really can understand and learn biology. It is very gratifying to see them discover that they truly are a part of a fragile "web of life" on this planet and that even though they are not scientists, this web needs their understanding and diligent care.

James A. Hewlett
Finger Lakes Community College
Research Areas: Biotechnology and molecular ecology

My true love is in the teaching of a mixed-majors' undergraduate biology class. The mention of recombinant DNA brings out clamors for detailed protocols from majors, and concerns over the ethical and social aspects of such a practice from a liberal arts student. Art students become scientific illustrators as part of a class project while a business major is enthused over the marriage of biology and the NASDAQ. These combinations make teaching undergraduates enjoyable.

Edward Levri
Indiana University of Pennsylvania
Research Areas: Evolutionary ecology: host–parasite and predator–prey relationships

Teaching nonmajors' biology is one of the most rewarding aspects of my career. Nonmajors bring many different views about the world into the course. These different perspectives make each class unique and make the discovery process more enjoyable. I learn something new every time I teach it. It's always special to find some students who realize for the first time that they actually like science.

Kenneth A. Mason
University of Kansas
Research Area: Genetics of pigmentation in lower vertebrates

I enjoy teaching nonmajors' biology because I feel like I am doing something of benefit to a larger community than just the scientific community. If I can do my small part to increase science literacy and make biology more accessible to non-scientists, then I have made a genuine contribution to society.

Timothy Metz
Campbell University
Research Area: Plant biotechnology

I enjoy the challenge of helping students, especially those with limited science backgrounds, develop an appreciation of scientific investigation and an understanding of how biological knowledge is relevant to our individual and corporate lives. It is exciting to see students turned on to the critical-thinking process involved in science; to know that when the course is over, they leave better equipped to evaluate issues, ideas, and sources of information encountered throughout a lifetime.

Rhoda E. Perozzi
Virginia Commonwealth University
Research Areas: Plant ecology and biology education

A former student recently told me that my nonmajors' biology class was the only class in which she learned things that she uses every day. It is enabling students to grasp this relevance of biology to daily life that makes teaching a joy.

Susan M. Wadkowski
Lakeland Community College
Research Areas: Physiological plant biology and ecology

I enjoy teaching nonmajors' biology courses because they are a challenge. Students come into the course having "failed biology in high school" or "hated biology in high school," or "didn't do well." By the end of my course, most realize that biology is "not all that bad," "interesting," or even "fun." I love it when I know the students have learned, I've learned from them, and we had fun doing it.

Robin Wright
University of Washington
Research Areas: Cell biology and genetics

Teaching undergraduates is exciting to me because they bring fresh ideas and questions to problems that I frequently take for granted. These new perspectives are often valuable in helping to reveal preconceived notions in my own research approaches. In addition, since most of us AREN'T biologists, by bringing the excitement and wonder of biology to a more general audience, I hope to make a larger impact on how we, as a society, view science and research.

Brief Contents

1 An Introduction to Life on Earth 1

UNIT ONE

THE LIFE OF A CELL 19

2 Atoms, Molecules, and Life 20
3 Biological Molecules 36
4 Cell Membrane Structure and Function 56
5 Cell Structure and Function 74
6 Energy Flow in the Life of a Cell 98
7 Capturing Solar Energy: Photosynthesis 114
8 Harvesting Energy: Glycolysis and Cellular Respiration 130

UNIT TWO

INHERITANCE 147

9 DNA: The Molecule of Heredity 148
10 Gene Expression and Regulation 162
11 The Continuity of Life: Cellular Reproduction 184
12 Patterns of Inheritance 210
13 Biotechnology 242

UNIT THREE

EVOLUTION 267

14 Principles of Evolution 268
15 How Organisms Evolve 286
16 The Origin of Species 306
17 The History of Life on Earth 322
18 Systematics: Seeking Order Amidst Diversity 350
19 The Hidden World of Microbes 364
20 The Fungi 390
21 The Plant Kingdom 408
22 The Animal Kingdom 428

UNIT FOUR

PLANT ANATOMY AND PHYSIOLOGY 465

23 Plant Form and Function 466
24 Plant Reproduction and Development 496
25 Plant Responses to the Environment 518

UNIT FIVE

ANIMAL ANATOMY AND PHYSIOLOGY 533

26 Homeostasis and the Organization of the Animal Body 534
27 Circulation 548
28 Respiration 568
29 Nutrition and Digestion 582
30 The Urinary System 604
31 Defenses Against Disease: The Immune Response 618
32 Chemical Control of the Animal Body: The Endocrine System 642
33 The Nervous System and the Senses 662
34 Action and Support: The Muscles and Skeleton 698
35 Animal Reproduction 716
36 Animal Development 740
37 Animal Behavior 764

UNIT SIX

ECOLOGY 791

38 Population Growth and Regulation 792
39 Community Interactions 814
40 How Do Ecosystems Work? 834
41 Earth's Diverse Ecosystems 856

Essays

Earth Watch

Why Preserve Biodiversity? 15
Crops, Livestock, and Wild Genes 233
Do Recombinants Hold Environmental Promise or Peril? 254
Endangered Species: From Gene Pools to Gene Puddles 295
The Case of the Disappearing Mushrooms 403
Frogs in Peril 456
Plants Help Regulate the Distribution of Water 486
On Dodos, Bats, and Disrupted Ecosystems 506
Dams, Deforestation, and Diseases 635
Endocrine Deception 647
Have We Exceeded Earth's Carrying Capacity? 809
Exotic Invaders 830
Food Chains Magnify Toxic Substances 841
The Ozone Hole—A Puncture in Our Protective Shield 858
Humans and Ecosystems 884

Health Watch

Cholesterol—Friend and Foe 46
Why Can You Get Fat by Eating Sugar? 134
Sex, Aging, and Mutations 176
Prenatal Genetic Screening 260
Matters of the Heart 562
Smoking—A Life and Breath Decision 578
Eating Disorders—Betrayal of the Body 589
Ulcers: Digesting the Digestive Tract 597
When the Kidneys Collapse 612
Can We Beat the Flu Bug? 632
Drugs, Diseases, and Neurotransmitters 669
Healing the Spinal Cord 675
How Bones Heal 709
Osteoporosis—When Bones Become Brittle 710
Sexually Transmitted Diseases 731
The Placenta Provides Only Partial Protection 758

A Closer Look*

Protein Structure—A Hairy Subject 48
Chemiosmosis—ATP Synthesis in Chloroplasts 121
Glycolysis 136
The Mitochondrial Matrix Reactions 141
Chemiosmosis in Mitochondria 142
Introns, Exons, and Splicing 179
Viruses—How the Nonliving Replicate 368
The Nephron and Urine Formation 610
Ions and Electrical Signals 666
Hormonal Control of the Menstrual Cycle 728

Scientific Inquiry

Does Spoiled Meat Produce Maggots? 10
Radioactivity in Research 23
The Search for the Cell 78
The Discovery of the Double Helix 154
Cracking the Genetic Code 171
Much Ado About Dolly 192
In Search of the Huntington Gene 231
DNA Fingerprinting: A Tool for Medical Detectives 256
Charles Darwin—Nature Was His Laboratory 274
How Do We Know How Old a Fossil Is? 329
Molecular Genetics Reveals Evolutionary Relationships 356
How Were Plant Hormones Discovered? 522
Artificial Blood? Artificial Vessels? 556
Monoclonal Antibodies—Designer Drugs Fight Disease 629
Peering Into the "Black Box" 681
High-Tech Reproduction 734
The Promise of Stem Cells 748
Cycles in Predator and Prey Populations 798
Ants and Acacias—An Unlikely Match 825

Evolutionary Connections

Caribou Legs and Membrane Diversity 69
Knowing Your Relatives: Kin Selection and Altruism 300
Scientists Don't Doubt Evolution 318
Are Reptiles for Real? 359
Our Unicellular Ancestors 386
Fungal Ingenuity—Pigs, Shotguns, and Nooses 402
Are Humans a Biological Success? 460
What Are Some Special Adaptations of Roots, Stems, and Leaves 489
Adaptations for Pollination and Seed Dispersal 508
Rapid-Fire Plant Responses 528
The Evolution of Hormones 657
Uncommon Senses 691
Why Do Animals Play? 786

*See our Web site at http://www.prenhall.com/audesirk6 for additional "A Closer Look" essays:

Chapter 5 The Essentially Foreign Creatures Within Us: The Evolution of Chloroplasts and Mitochondria
Chapter 10 Rube Goldberg Genetics: Making New Proteins from Old Parts
Chapter 15 The Hardy-Weinberg Equilibrium Population
Chapter 26 Types of Epithelial Tissue
Chapter 27 Human Blood Cells
Chapter 28 Gills and Gas Exchange: Countercurrent Flow
Chapter 29 The Fate of Fats
Chapter 31 Cellular Communication During the Immune Response
Chapter 32 The Chemical Diversity of Vertebrate Hormones
Chapter 36 Stages of Development
Chapter 36 Homeobox Genes and the Control of Body Form
Chapter 38 The Mathematics of Population Growth

Contents

Chapter 1 An Introduction to Life on Earth 1

CASE STUDY The Life Around Us 1

1) **What Are the Characteristics of Living Things? 2**

Living Things Are Both Complex and Organized 2
Living Things Respond to Stimuli 4
Living Things Maintain Relatively Constant Internal Conditions Through Homeostasis 4
Living Things Acquire and Use Materials and Energy 5
Living Things Grow 5
Living Things Reproduce Themselves 6
Living Things As a Whole Have the Capacity to Evolve 6

2) **How Do Scientists Categorize the Diversity of Life? 6**

The Domains Bacteria and Archaea Consist of Prokaryotic Cells; the Domain Eukarya Is Composed of Eukaryotic Cells 8
Bacteria, Archaea, and Members of the Kingdom Protista Are Mostly Unicellular; Members of the Kingdoms Fungi, Plantae, and Animalia Are Primarily Multicellular 8
Members of the Different Kingdoms Have Different Ways of Acquiring Energy 8

3) **What Is the Science of Biology? 8**

Scientific Principles Underlie All Scientific Inquiry 8
The Scientific Method Is the Basis for Scientific Inquiry 9
Science Is a Human Endeavor 11
Scientific Theories Have Been Thoroughly Tested 11

4) **Evolution: The Unifying Theory of Biology 12**

Three Natural Processes Underlie Evolution 12

5) **How Does Knowledge of Biology Illuminate Everyday Life? 13**

CASE STUDY REVISITED The Life Around Us 16

Review Section 16

UNIT ONE

THE LIFE OF A CELL 19

Chapter 2 Atoms, Molecules, and Life 20

CASE STUDY Health Food? 21

1) **What Are Atoms? 21**

Atoms, the Basic Structural Units of Matter, Are Composed of Still Smaller Particles 21

2) **How Do Atoms Interact to Form Molecules? 24**

Atoms Interact with Other Atoms When There Are Vacancies in Their Outermost Electron Shells 24
Charged Atoms Called *Ions* Interact to Form Ionic Bonds 25
Uncharged Atoms Can Become Stable by Sharing Electrons, Forming Covalent Bonds 25
Hydrogen Bonds Are Weaker Electrical Attractions Between or Within Molecules with Polar Covalent Bonds 27

3) **Why Is Water So Important to Life? 27**

Water Interacts with Many Other Molecules 28
Water Molecules Tend to Stick Together 29
Water Can Form H^+ and OH^- Ions 30
Water Moderates the Effects of Temperature Changes 31
Water Forms an Unusual Solid: Ice 32

CASE STUDY REVISITED Health Food? 32

Review Section 32

Chapter 3 Biological Molecules 36

CASE STUDY Improving on Nature? 37

1) **Why Is Carbon So Important in Biological Molecules? 38**

2) **How Are Organic Molecules Synthesized? 39**

Biological Molecules Are Joined Together or Broken Apart by Adding or Removing Water 39

3) **What Are Carbohydrates? 39**

There Are a Variety of Monosaccharides with Slightly Different Structures 40
Disaccharides Consist of Two Single Sugars Linked by Dehydration Synthesis 41
Polysaccharides Are Chains of Single Sugars 41

4) **What Are Lipids? 43**

Oils, Fats, and Waxes Are Lipids Containing Only Carbon, Hydrogen, and Oxygen 43
Phospholipids Have Water-Soluble "Heads" and Water-Insoluble "Tails" 45
Steroids Consist of Four Carbon Rings Fused Together 45

5) **What Are Proteins? 45**

Proteins Are Formed from Chains of Amino Acids 46
Amino Acids Are Joined to Form Chains by Dehydration Synthesis 47
A Protein Can Have Up to Four Levels of Structure 49
The Functions of Proteins Are Linked to Their Three-Dimensional Structures 51

6) **What Are Nucleic Acids? 51**

DNA and RNA, the Molecules of Heredity, Are Nucleic Acids 52
Other Nucleotides Act as Intracellular Messengers, Energy Carriers, or Coenzymes 52

CASE STUDY REVISITED Improving on Nature? 53

Review Section 53

Chapter 4 Cell Membrane Structure and Function 56

CASE STUDY Vicious Venoms 57

1) How Is the Structure of a Membrane Related to Its Function? 57

The Plasma Membrane Isolates the Cell While Allowing Communication with Its Surroundings 57
Membranes Are "Fluid Mosaics" in Which Proteins Move Within Layers of Lipids 58
The Phospholipid Bilayer Is the Fluid Portion of the Membrane 58
A Mosaic of Proteins Is Embedded in the Membrane 59

2) How Do Substances Move Across Membranes? 60

Molecules in Fluids Move in Response to Gradients 60
Movement Across Membranes Occurs by Both Passive and Active Transport 61
Passive Transport Includes Simple Diffusion, Facilitated Diffusion, and Osmosis 62
Active Transport Uses Energy to Move Molecules Against Their Concentration Gradients 65
Cells Engulf Particles or Fluids by Endocytosis 65
Exocytosis Moves Material Out of the Cell 67

3) How Are Cell Surfaces Specialized? 68

A Variety of Junctions Allow Cells to Connect and Communicate 68
Some Cells Are Supported by Cell Walls 69
EVOLUTIONARY CONNECTIONS: Caribou Legs and Membrane Diversity 69

CASE STUDY REVISITED Vicious Venoms 71

Review Section 71

Chapter 5 Cell Structure and Function 74

CASE STUDY Microscopic Stowaways 75

1) What Are the Basic Features of Cells? 75

All Living Things Are Composed of One or More Cells 75
All Cells Share Certain General Features 76
There Are Two Basic Types of Cells: Prokaryotic and Eukaryotic 80

2) What Are the Features of Prokaryotic Cells? 80

3) What Are the Features of Eukaryotic Cells? 81

Eukaryotic Cells Contain Organelles 81
The Nucleus Is the Control Center of the Eukaryotic Cell 82
Eukaryotic Cells Contain a Complex System of Membranes 84
Vacuoles Serve Many Functions, Including Water Regulation, Support, and Storage 88
Mitochondria Extract Energy from Food Molecules, and Chloroplasts Capture Solar Energy 90
Plants Use Plastids for Storage 91
The Cytoskeleton Provides Shape, Support, and Movement 91
Cilia and Flagella Move the Cell or Move Fluid Past the Cell 93

CASE STUDY REVISITED Microscopic Stowaways 95

Review Section 95

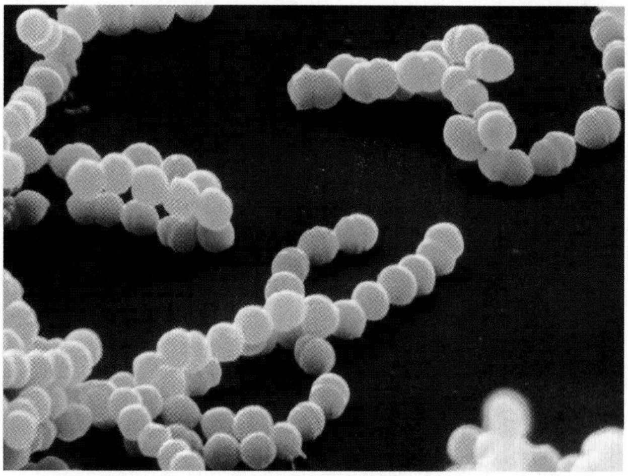

Chapter 6 Energy Flow in the Life of a Cell 98

CASE STUDY Energy Unleashed 99

1) What Is Energy? 100

The Laws of Thermodynamics Describe the Basic Properties of Energy 100
Living Things Use the Energy of Sunlight to Create the Low-Entropy Conditions Characteristic of Life 101

2) How Does Energy Flow in Chemical Reactions? 101

Exergonic Reactions Release Energy 101
Endergonic Reactions Require an Input of Energy 102
Coupled Reactions Link Exergonic with Endergonic Reactions 102

3) How Is Cellular Energy Carried Between Coupled Reactions? 103

ATP Is the Principal Energy Carrier in Cells 103
Electron Carriers Also Transport Energy Within Cells 105

4) How Do Cells Control Their Metabolic Reactions? 105

At Body Temperatures, Spontaneous Reactions Proceed Too Slowly to Sustain Life 106
Catalysts Reduce Activation Energy 106
Enzymes Are Biological Catalysts 107
The Structure of Enzymes Allows Them to Catalyze Specific Reactions 107
Cells Regulate the Amount and Activity of Their Enzymes 108
The Activity of Enzymes Is Influenced by Their Environment 108

CASE STUDY REVISITED Energy Unleashed 110

Review Section 110

Chapter 7 Capturing Solar Energy: Photosynthesis 114

CASE STUDY Did Dinosaurs Die from Lack of Sunlight? 115

1) What Is Photosynthesis? 116

Leaves and Chloroplasts Are Adaptations for Photosynthesis 116

Photosynthesis Consists of Light-Dependent and Light-Independent Reactions 117

2) **Light-Dependent Reactions: How Is Light Energy Converted to Chemical Energy? 118**

During Photosynthesis, Light Is First Captured by Pigments in Chloroplasts 118
The Light-Dependent Reactions Occur in Photosystems Within the Thylakoid Membranes 119

3) **Light-Independent Reactions: How Is Chemical Energy Stored in Glucose Molecules? 122**

The C_3 Cycle Captures Carbon Dioxide 122
Carbon Fixed During the C_3 Cycle Is Used to Synthesize Glucose 122

4) **What Is the Relationship Between Light-Dependent and Light-Independent Reactions? 123**

5) **Water, CO_2, and the C_4 Pathway 123**

When Stomata Are Closed to Conserve Water, Wasteful Photorespiration Occurs 124
C_4 Plants Reduce Photorespiration by Means of a Two-Stage Carbon-Fixation Process 124
C_3 and C_4 Plants Are Each Adapted to Different Environmental Conditions 124

CASE STUDY REVISITED Did the Dinosaurs Die from Lack of Sunlight? 126

Review Section 126

Chapter 8 Harvesting Energy: Glycolysis and Cellular Respiration 130

CASE STUDY The Flight of the Hummingbird 131

1) **How Is Glucose Metabolized? 132**

2) **How Is the Energy in Glucose Captured During Glycolysis? 132**

Glycolysis Breaks Down Glucose to Pyruvate, Releasing Chemical Energy 132
Some Cells Ferment Pyruvate to Form Lactate 135
Other Cells Ferment Pyruvate to Alcohol 137

3) **How Does Cellular Respiration Generate Still More Energy from Glucose? 137**

Pyruvate Is Transported to the Mitochondrial Matrix, Where It Is Broken Down via the Krebs Cycle 139

Energetic Electrons Produced by the Krebs Cycle Are Carried to Electron Transport Systems in the Inner Mitochondrial Membrane 140
Chemiosmosis Captures Energy Stored in a Hydrogen Ion Gradient and Produces ATP 141
Glycolysis and Cellular Respiration Influence the Way Entire Organisms Function 143

CASE STUDY REVISITED The Flight of the Hummingbird 143

Review Section 143

UNIT TWO

INHERITANCE 147

Chapter 9 DNA: The Molecule of Heredity 148

CASE STUDY Sunshine Perils 149

1) **How Did Scientists Discover That Genes Are Made of DNA? 149**

Transformed Bacteria Revealed the Link Between Genes and DNA 150

2) **What Is the Structure of DNA? 151**

DNA Is Composed of Four Nucleotides 151
DNA Is a Double Helix of Two Nucleotide Strands 152
Hydrogen Bonds Between Complementary Bases Hold the Two Strands Together 152
The Order of Nucleotides in DNA Can Encode Vast Amounts of Information 153

3) **How Does DNA Replication Ensure Genetic Constancy? 154**

The Replication of DNA Is a Critical Event in a Cell's Life 154
DNA Replication Produces Two DNA Double Helices, Each with One Old Strand and One New Strand 155
DNA Helicase Separates the Parental DNA Strands 155
DNA Polymerase Synthesizes New DNA Strands 156
One DNA Strand Is Synthesized in Short Segments That Are Connected by DNA Ligase 156
Proofreading Produces Almost Error-Free Replication of DNA 158
But Mistakes Do Happen 158

CASE STUDY REVISITED Sunshine Perils 158

Review Section 159

Chapter 10 Gene Expression and Regulation 162

CASE STUDY Boy or Girl? 163

1) **How Are Genes and Proteins Related? 164**

Genetic Studies by Beadle and Tatum Using a Common Mold Revealed the Relationship Between Genes and Enzymes 164
Most Genes Contain the Information for the Synthesis of a Single Protein 164
DNA Provides Instructions for Protein Synthesis via RNA Intermediaries 165
Overview: Genetic Information Is Transcribed into RNA and Translated into Protein 166

In the Genetic Code, a Sequence of Three Bases Specifies an Amino Acid or a "Stop" 166

2) **How Is Information in a Gene Transcribed into RNA? 167**

Initiation of Transcription Begins When RNA Polymerase Binds to the Promoter of a Gene 168
Elongation Proceeds Until RNA Polymerase Reaches a Termination Signal 168
Transcription Is Selective 168

3) **How Is the Sequence of a Messenger RNA Molecule Translated into Protein? 170**

Messenger RNA Carries the Code for Protein Synthesis from the Nucleus to the Cytoplasm 170
Ribosomal RNA Forms an Important Part of the Protein-Synthesizing Machinery of a Ribosome 170
Transfer RNA Molecules Decode the Sequence of Bases in mRNA into the Amino Acid Sequence of a Protein 170
During Translation, mRNA, tRNA, and Ribosomes Cooperate to Synthesize Proteins 170
Recap: Decoding the Sequence of Bases in DNA into the Sequence of Amino Acids in Protein Requires Transcription and Translation 173

4) **How Do Mutations in DNA Affect the Function of Genes? 174**

Mutations Result from Nucleotide Substitutions, Insertions, or Deletions 174
Mutations Have Different Effects on Protein Structure and Function 174
Mutations Provide the Raw Material for Evolution 175

5) **How Are Genes Regulated? 175**

Proper Regulation of Gene Expression Is Critical for an Organism's Development and Health 175
Eukaryotic Cells May Regulate the Transcription of Individual Genes, Regions of Chromosomes, or Entire Chromosomes 178

CASE STUDY REVISITED Boy or Girl? 181

Review Section 181

Chapter 11 The Continuity of Life: Cellular Reproduction 184

CASE STUDY Rattlesnake Surprise 185

1) **What Are the Functions of Cellular Reproduction? 186**

Binary Fission Is the Cell Division Process in Prokaryotes 186
Mitotic Cell Division Allows Development, Growth, Maintenance, and Repair of Body Tissues in Eukaryotes 186
Mitotic Cell Division Forms the Basis of Asexual Reproduction in Eukaryotes 187
Meiotic Cell Division Forms the Basis of Sexual Reproduction in Eukaryotes 188

2) **How Is DNA in Eukaryotic Cells Organized into Chromosomes? 188**

Cell Division Enables Accurate Passage of Chromosomes from One Generation to the Next 188
The Eukaryotic Chromosome Consists of a Linear DNA Double Helix Bound to Proteins 189
Eukaryotic Chromosomes Usually Occur in Homologous Pairs with Similar Genetic Information 190

3) **What Are the Events of the Eukaryotic Cell Cycle? 191**

Cell Division Is One Part of the Eukaryotic Cell Cycle 191
During Interphase, the Cell Grows in Size and Replicates Its DNA 194
Mitotic Cell Division Consists of Nuclear Division and Cytoplasmic Division 194

4) **What Are the Phases of Mitosis? 194**

During Prophase, the Chromosomes Condense and the Spindle Microtubules Form and Attach to the Chromosomes 194
During Metaphase, the Chromosomes Become Aligned Along the Equator of the Cell 196
During Anaphase, Sister Chromatids Separate and Are Pulled to Opposite Poles of the Cell 196
During Telophase, Nuclear Envelopes Form Around Both Groups of Chromosomes 197

5) **What Are the Events of Cytokinesis? 197**

6) **What Are Some Advantages of Sexual Reproduction? 198**

Mutations in DNA Are the Ultimate Source of Genetic Variability 198
Reshuffling Genes May Combine Different Alleles in Beneficial Ways 199
Meiotic Cell Division Produces Haploid Cells That Can Merge to Combine Genetic Material from Two Parents 199

7) **What Are the Events of Meiosis? 199**

Meiosis Separates Homologous Chromosomes in a Diploid Nucleus, Producing Haploid Daughter Nuclei 199
Meiosis I Separates Homologous Chromosomes into Two Daughter Nuclei 200
Meiosis II Separates Sister Chromatids into Four Daughter Nuclei 203
The Life Cycles of Organisms on Earth Usually Include Meiosis and Mitosis 203

8) **How Do Meiosis and Sexual Reproduction Produce Genetic Variability? 203**

Shuffling of Homologues Creates Novel Combinations of Chromosomes 203
Crossing Over Creates Chromosomes with Novel Combinations of Genes 205
Fusion of Gametes Adds Further Genetic Variability to the Offspring 205

CASE STUDY REVISITED Rattlesnake Surprise 205

Review Section 206

Chapter 12 Patterns of Inheritance 210

CASE STUDY Sleepy Genes 211

1) **How Did Gregor Mendel Lay the Foundations for Modern Genetics? 212**

Doing It Right: The Secrets of Mendel's Success 213

2) **How Are Single Traits Inherited? 213**

The Inheritance of Dominant and Recessive Alleles on Homologous Chromosomes Can Explain the Results of Mendel's Crosses 214
Mendel's Hypothesis Can Be Used to Predict the Outcome of New Types of Single-Trait Crosses 216

3) **How Are Multiple Traits on Different Chromosomes Inherited? 217**

Mendel Hypothesized That Genes on Different Chromosomes Are Inherited Independently 217
In an Unprepared World, Genius May Go Unrecognized 218

4) **How Are Genes Located on the Same Chromosome Inherited? 218**

Genes on the Same Chromosome Tend to Be Inherited Together 219
Recombination Can Create New Combinations of Linked Alleles 220

5) **How Is Sex Determined, and How Are Sex-Linked Genes Inherited? 220**

Sex-Linked Genes Are Found Only on the X or Only on the Y Chromosome 221

6) **What Are Some Variations on the Mendelian Theme? 222**

Alleles May Display Incomplete Dominance in Which the Phenotype of Heterozygotes Is Intermediate Between the Phenotypes of the Homozygotes 223
A Single Gene May Have Multiple Alleles 223
Many Traits Are Influenced by Several Genes 224
Single Genes Typically Have Multiple Effects on Phenotype 225
The Environment Influences the Expression of Genes 225

7) **How Are Human Genetic Disorders Investigated? 226**

8) **How Are Human Disorders Caused by Single Genes Inherited? 227**

Many Human Genetic Disorders Are Caused by Recessive Alleles 227

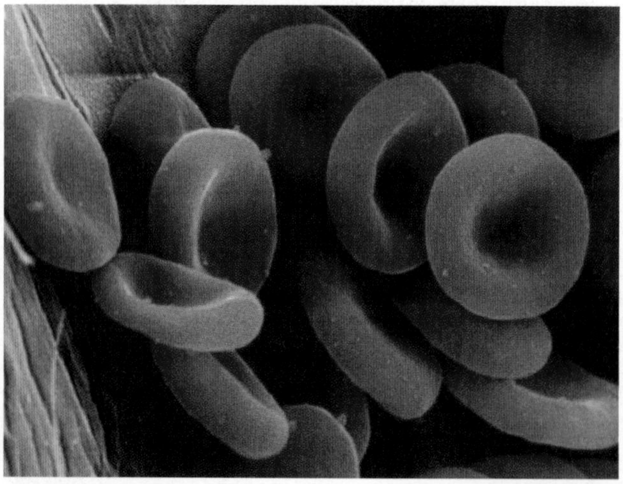

Many Human Genetic Disorders Are Caused by Dominant Alleles 228
Some Human Disorders Are Sex-Linked 229

9) **How Do Errors in Chromosome Number Affect Humans? 230**

Some Genetic Disorders Are Caused by Abnormal Numbers of Sex Chromosomes 230
Some Genetic Disorders Are Caused by Abnormal Numbers of Autosomes 234

CASE STUDY REVISITED Sleepy Genes 235

Review Section 235

Chapter 13 Biotechnology 242

CASE STUDY Teaching an Old Grain New Tricks 243

1) **What Is Biotechnology? 244**

2) **How Does DNA Recombination Occur in Nature? 244**

DNA Recombination Occurs Naturally Through Processes Such as Sexual Reproduction, Bacterial Transformation, and Viral Infection 244

3) **How Does DNA Recombination Occur in Genetic Engineering Laboratories? 246**

Restriction Enzymes Cut DNA at Specific Nucleotide Sequences 246
Insertion of Foreign DNA into a Vector Can Produce a Recombinant DNA Library 246

4) **How Can Researchers Identify Specific Genes? 248**

Restriction Fragment Length Polymorphisms Can Be Used to Identify Genes 248
Genes from One Organism Can Be Identified Based on Similarity to Related Genes in Other Organisms 249
Genes Can Be Identified Based on Their Protein Product 249

5) **What Are Some Applications of Biotechnology? 249**

Cloned Genes Provide Enough DNA for Gene Sequencing 249
DNA Fingerprinting Facilitates Genetic Detection on Many Fronts 251
Genetic Engineering Is Revolutionizing Agriculture 251
Will Biotechnology Create a Real Jurassic Park? 255

6) **What Are Some Medical Uses of Biotechnology? 255**

Knock-out Mice Provide Models of Human Genetic Diseases 255
Genetic Engineering Allows Production of Therapeutic Proteins 255
Human Gene Therapy Is Just Beginning 257
The Human Genome Project Has Completed a Working Draft of the Entire Human Genome 259

7) **What Are Some Ethical Implications of Human Biotechnology? 259**

Genetic Tests for Cystic Fibrosis and Inherited Forms of Breast Cancer Illustrate Potential Problems 259
Sickle-Cell Anemia and Tay-Sachs Disease Illustrate the Hazards and Benefits of Genetic Screening Programs 260
The Potential to Clone Humans Raises Further Ethical Issues 261

CASE STUDY REVISITED Teaching an Old Grain New Tricks 263

Review Section 263

UNIT THREE

EVOLUTION 267

Chapter 14 Principles of Evolution 268

CASE STUDY A Missing Link Unearthed 269

1) How Did Evolutionary Thought Evolve? 270
- Evidence Supporting Evolution Emerged Even Before Darwin's Time 270
- Both Darwin and Wallace Proposed That Evolution Occurs by Natural Selection 273
- Evolutionary Theory Arises from Scientific Observations and Conclusions Based on Them 274

2) How Do We Know That Evolution Has Occurred? 276
- The Fossil Record Provides Evidence of Evolutionary Change over Time 277
- Comparative Anatomy Provides Structural Evidence of Evolution 277
- Embryological Stages of Animals Can Provide Evidence of Common Ancestry 279
- Modern Biochemical and Genetic Analyses Reveal Relatedness Among Diverse Organisms 279

3) What Is the Evidence That Populations Evolve by Natural Selection? 279
- Artificial Selection Demonstrates That Organisms May Be Modified by Controlled Breeding 280
- Evolution by Natural Selection Occurs Today 280

4) A Postscript by Charles Darwin 282

CASE STUDY REVISITED A Missing Link Unearthed 282

Review Section 282

Chapter 15 How Organisms Evolve 286

CASE STUDY Cause of Death: Evolution 287

1) How Are Populations, Genes, and Evolution Related? 288
- Genes and the Environment Interact to Determine the Traits of Each Individual 288
- The Gene Pool Is the Sum of All the Genes in a Population 288
- Evolution Is the Change of Gene Frequencies Within a Population 288
- The Equilibrium Population Is a Hypothetical Population in Which Evolution Does Not Occur 289

2) What Causes Evolution? 289
- Mutations Are the Ultimate Source of Genetic Variability 289
- Gene Flow Between Populations Changes Allele Frequencies 289
- Small Populations Are Subject to Random Changes in Allele Frequencies 290
- Mating Within a Population Is Almost Never Random 293
- All Genotypes Are Not Equally Adaptive 294

3) How Does Natural Selection Work? 296
- Natural Selection Acts on the Phenotype, Which Reflects the Underlying Genotype 296
- Natural Selection Can Influence Populations in Three Major Ways 296
- A Variety of Processes Can Cause Natural Selection 298

- EVOLUTIONARY CONNECTIONS: Knowing Your Relatives: Kin Selection and Altruism 300

CASE STUDY REVISITED Cause of Death: Evolution 302

Review Section 302

Chapter 16 The Origin of Species 306

CASE STUDY Lost World 307

1) What Is a Species? 308

2) How Do New Species Form? 308
- Allopatric Speciation Can Occur in Populations That Are Physically Separated 309
- Sympatric Speciation Can Occur in Populations That Live in the Same Area 309
- Change over Time Within a Species Can Cause Apparent "Speciation" in the Fossil Record 313
- During Adaptive Radiation, One Species Gives Rise to Many 313

3) How Is Reproductive Isolation Between Species Maintained? 314
- Premating Isolating Mechanisms Prevent Mating Between Species 314
- Postmating Isolating Mechanisms Prevent Production of Vigorous, Fertile Offspring 316

4) What Causes Extinction? 316
- Localized Distribution and Overspecialization Make Species Vulnerable in Changing Environments 316
- Interactions with Other Organisms May Drive a Species to Extinction 317
- Habitat Change and Destruction Are the Leading Causes of Extinction 317
- EVOLUTIONARY CONNECTIONS: Scientists Don't Doubt Evolution 318

CASE STUDY REVISITED Lost World 318

Review Section 318

Chapter 17 The History of Life on Earth 322

CASE STUDY Life on a Frozen Moon? 323

1) How Did Life Begin? 324
- Prebiotic Evolution Was Controlled by the Early Atmosphere and Climate 324

2) **What Were the Earliest Organisms Like? 327**
The First Organisms Were Anaerobic Prokaryotes 327
Some Organisms Evolved the Ability to Capture Energy from the Sun 327
Aerobic Metabolism Arose in Response to the Oxygen Crisis 329
Eukaryotes Developed Membrane-Enclosed Organelles and a Nucleus 329

3) **How Did Multicellularity Arise? 331**
Multicellular Algae Developed Specialized Structures That Facilitated Their Invasion of Diverse Habitats 331
Multicellular Animals Developed Specializations That Allowed Them to Capture Prey, Feed, and Escape More Efficiently 331

4) **How Did Life Invade the Land? 333**
Some Plants Developed Specialized Structures That Adapted Them to Life on Dry Land 333
Some Animals Evolved Specialized Structures That Adapted Them to Life on Dry Land 334

5) **What Role Has Extinction Played in the History of Life? 336**
The Upward Trend in Species Diversity Has Been Interrupted by Periodic Mass Extinctions 336

6) **How Did Humans Evolve? 338**
Primate Evolution Has Been Linked to Grasping Hands, Binocular Vision, and a Large Brain 338
Hominids Evolved from Dryopithecine Primates 338
The Earliest Hominids Could Stand and Walk Upright 339
The Evolution of Human Behavior Is Highly Speculative 343

CASE STUDY REVISITED Life on a Frozen Moon? 345

Review Section 346

Chapter 18 Systematics: Seeking Order Amidst Diversity 350

CASE STUDY Origin of a Killer 351

1) **How Are Organisms Named and Classified? 351**
Taxonomy Originated as a Hierarchy of Categories 352
Modern Systematists Use Numerous Criteria for Classification 353

2) **What Are the Kingdoms of Life? 354**
The Five-Kingdom System Supplanted Earlier Classification Schemes 354
A Three-Domain System More Accurately Reflects Life's History 355
Kingdom-level Classification Remains Unsettled 355

3) **Why Do Taxonomies Change? 358**
Species Designations Change When New Information Is Discovered 358
The Biological Species Definition Can Be Difficult or Impossible to Apply 358
The Phylogenetic Species Concept Offers an Alternative Definition 358

4) **Exploring Biodiversity: How Many Species Exist? 358**
EVOLUTIONARY CONNECTIONS: Are Reptiles for Real? 359

CASE STUDY REVISITED Origin of a Killer 360

Review Section 361

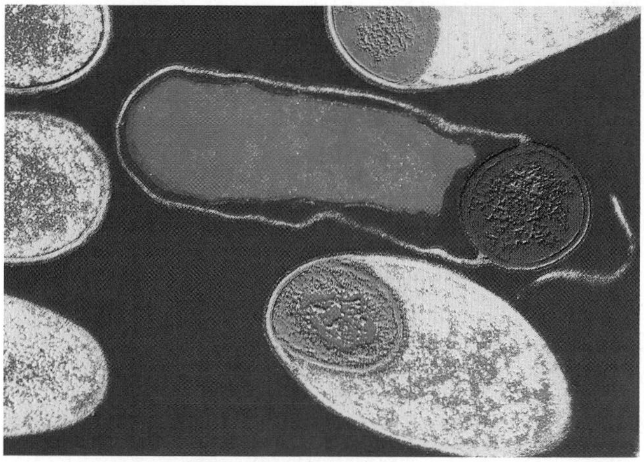

Chapter 19 The Hidden World of Microbes 364

CASE STUDY Agents of Death 365

1) **What Are Viruses, Viroids, and Prions? 366**
A Virus Consists of a Molecule of DNA or RNA Surrounded by a Protein Coat 366
Viral Infections Cause Diseases That Are Difficult to Treat 366
Some Infectious Agents Are Even Simpler Than Viruses 369
No One Is Certain How These Infectious Particles Originated 370

2) **Which Organisms Make Up the Prokaryotic Domains—Bacteria and Archaea? 370**
Bacteria and Archaea Are Fundamentally Different 371
Prokaryotes Are Difficult to Classify 371
Prokaryotes Exhibit a Variety of Shapes and Structures 371
Prokaryotes Reproduce by Binary Fission 373
Prokaryotes Are Specialized for Specific Habitats 374
Prokaryotes Exhibit Diverse Metabolisms 374
Prokaryotes Perform Many Functions That Are Important to Other Forms of Life 375
Some Bacteria Pose a Threat to Human Health 376

3) **Which Organisms Make Up the Kingdom Protista? 377**
Protists Are a Diverse Group Including Funguslike, Plantlike, and Animal-like Forms 378
The Water Molds and Slime Molds Are Funguslike Protists 378
The Algae Are Plantlike Protists 380
The Protozoa Are Animal-like Protists 383
EVOLUTIONARY CONNECTIONS: Our Unicellular Ancestors 386

CASE STUDY REVISITED Agents of Death 387

Review Section 387

Chapter 20 The Fungi 390

CASE STUDY Three Outings 391

1) **What Are the Main Adaptations of Fungi? 392**
Most Fungi Have Filamentous Bodies 392
Fungi Obtain Their Nutrients from Other Organisms 392
Most Fungi Can Reproduce Both Sexually and Asexually 392

2) **How Are Fungi Classified? 393**

The Chytrids Produce Swimming Spores 393
The Zygote Fungi Can Reproduce by Forming Diploid Zygospores 394
The Sac Fungi Form Spores in a Saclike Case Called an Ascus 394
The Club Fungi Produce Club-Shaped Reproductive Structures Called Basidia 396
The Imperfect Fungi Are Species in Which Sexual Structures Have Not Been Observed 396
Some Fungi Form Symbiotic Relationships 397

3) **How Do Fungi Affect Humans? 399**

Fungi Attack Plants That Are Important to People 400
Fungi Cause Human Diseases 400
Fungi Make Important Contributions to Gastronomy 401
Fungi Play a Crucial Ecological Role 402
EVOLUTIONARY CONNECTIONS: Fungal Ingenuity—Pigs, Shotguns, and Nooses 402

CASE STUDY REVISITED Three Outings 404

Review Section 404

Chapter 21 The Plant Kingdom 408

CASE STUDY Hunting for Medical Treasures 409

1) **What Are the Key Features of Plants? 410**

Plants Have Both a Sporophyte and a Gametophyte Generation 410

2) **What Is the Evolutionary Origin of Plants? 410**

Green Algae Gave Rise to Land Plants 411

3) **How Did Plants Invade and Flourish on Land? 411**

The Plant Body Increased in Complexity as Plants Made the Evolutionary Transition from Water to Dry Land 412
The Invasion of Land Required Protection and a Means of Dispersal for Sex Cells and Developing Plants 413
The Liverworts and Mosses Are Adapted to Moist Environments 413
The Vascular Plants, or Tracheophytes, Have Conducting Vessels That Also Provide Support 414
The Seedless Vascular Plants Include the Club Mosses, Horsetails, and Ferns 414
The Seed Plants Dominate the Land, Aided by Two Important Adaptations: Pollen and Seeds 414

CASE STUDY REVISITED Hunting for Medical Treasures 424

Review Section 424

Chapter 22 The Animal Kingdom 428

CASE STUDY The Search for a Sea Monster 429

1) **What Characteristics Define an Animal? 430**

2) **Which Anatomical Features Mark Branch Points on the Animal Evolutionary Tree? 430**

Lack of Tissues Separates Sponges from All Other Animals 430
Animals with Tissues Exhibit Either Radial or Bilateral Symmetry 430
Body Cavities Arose in Some Bilaterally Symmetrical Animals 431
Coelomates Include Two Distinct Evolutionary Lines 432

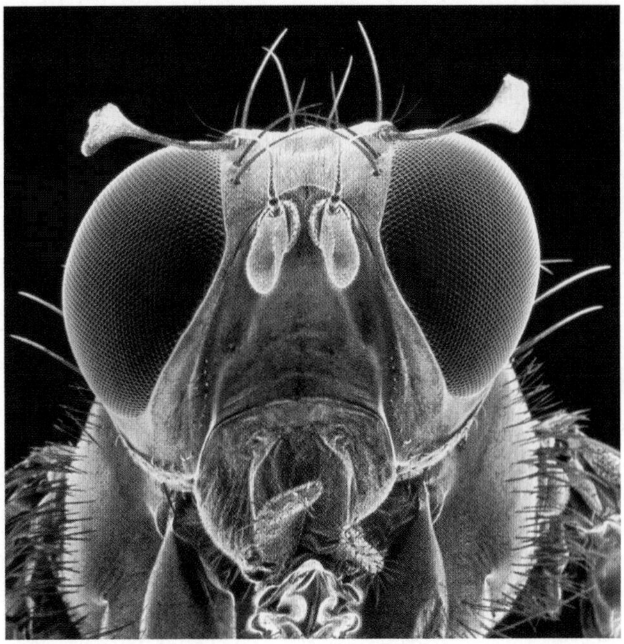

3) **What Are the Major Animal Phyla? 432**

The Sponges 433
The Hydra, Anemones, and Jellyfish 435
The Flatworms 437
The Roundworms 440
The Segmented Worms 441
The Insects, Arachnids, and Crustaceans 441
The Snails, Clams, and Squid 445
The Sea Stars, Sea Urchins, and Sea Cucumbers 448
The Tunicates, Lancelets, and Vertebrates 450
EVOLUTIONARY CONNECTIONS: Are Humans a Biological Success? 460

CASE STUDY REVISITED The Search for a Sea Monster 461

Review Section 461

UNIT FOUR

PLANT ANATOMY AND PHYSIOLOGY 465

Chapter 23 Plant Form and Function 466

CASE STUDY A Beautiful Death Trap 467

1) **How Are Plant Bodies Organized, and How Do They Grow? 468**

Flowering Plants Consist of a Root System and a Shoot System 468
During Plant Growth, Meristem Cells Give Rise to Differentiated Cells 468

2) **What Are the Tissues and Cell Types of Plants? 470**

The Dermal Tissue System Forms the Covering of the Plant Body 470
The Ground Tissue System Makes Up Most of the Young Plant Body 471

The Vascular Tissue System Consists of Xylem and Phloem 471

3) **Roots: Anchorage, Absorption, and Storage 473**

Primary Growth Causes Roots to Elongate 473
The Epidermis of the Root Is Very Permeable to Water 474
Cortex Makes Up Much of the Interior of a Young Root 475
The Vascular Cylinder Contains Conducting Tissues 475

4) **Stems: Reaching for the Light 476**

The Stem Includes Four Types of Tissue 476
Stem Branches Form from Lateral Buds Consisting of Meristem Cells 477
Secondary Growth Produces Thicker, Stronger Stems 477

5) **Leaves: Nature's Solar Collectors 479**

Leaves Have Two Major Parts: Blades and Petioles 479

6) **How Do Plants Acquire Nutrients? 481**

Roots Acquire Minerals by a Four-Step Process 481
Symbiotic Relationships Help Plants Acquire Nutrients 482

7) **How Do Plants Acquire Water and Transport Water and Minerals? 484**

Water Movement in Xylem Is Explained by the Cohesion–Tension Theory 484
Water Enters Roots Mainly by Pressure Differences Created by Transpiration 485
Adjustable Stomata Control the Rate of Transpiration 486

8) **How Do Plants Transport Sugars? 487**

The Pressure-Flow Theory Explains Sugar Movement in Phloem 488
EVOLUTIONARY CONNECTIONS: What Are Some Special Adaptations of Roots, Stems, and Leaves? 489
Some Specialized Roots Store Food; Others Photosynthesize 489
Some Specialized Stems Produce New Plants, Store Water or Food, or Produce Thorns or Climbing Tendrils 489
Specialized Leaves May Conserve and Store Water, Store Food, or Even Capture Insects 490

CASE STUDY REVISITED A Beautiful Death Trap 492

Review Section 492

Chapter 24 Plant Reproduction and Development 496

CASE STUDY Walk Through a Meadow 497

1) **What Are the Features of Plant Life Cycles? 497**

2) **How Did Flowers Evolve? 499**

Complete Flowers Have Four Major Parts 500

3) **How Do Gametophytes Develop in Flowering Plants? 500**

Pollen Is the Male Gametophyte 501
The Embryo Sac Is the Female Gametophyte 502

4) **How Does Pollination Lead to Fertilization? 503**

5) **How Do Seeds and Fruits Develop? 504**

The Seed Develops from the Ovule and Embryo Sac 504
The Fruit Develops from the Ovary Wall 506
Seed Dormancy Helps Ensure Germination at an Appropriate Time 506

6) **How Do Seeds Germinate and Grow? 507**

The Shoot Tip Must Be Protected 507
Cotyledons Nourish the Sprouting Seed 507
Controlling the Development of the Seedling 508
EVOLUTIONARY CONNECTIONS: Adaptations for Pollination and Seed Dispersal 508
Coevolution Matches Plants and Pollinators 508
Fruits Help Disperse Seeds 512

CASE STUDY REVISITED Walk Through a Meadow 514

Review Section 514

Chapter 25 Plant Responses to the Environment 518

CASE STUDY A Chemical Cry for Help 519

1) **What Are Plant Hormones, and How Do They Act? 519**

2) **How Do Hormones Regulate the Plant Life Cycle? 520**

Abscisic Acid Maintains Seed Dormancy; Gibberellin Stimulates Germination 520
Auxin Controls the Orientation of the Sprouting Seedling 520
The Genetically Determined Shape of the Mature Plant Is the Result of Interactions Among Hormones 524
Daylength Controls Flowering 525
Hormones Coordinate the Development of Seeds and Fruit 527
Senescence and Dormancy Prepare the Plant for Winter 528
EVOLUTIONARY CONNECTIONS: Rapid-Fire Plant Responses 528

CASE STUDY REVISITED A Chemical Cry for Help 530

Review Section 530

UNIT FIVE

ANIMAL ANATOMY AND PHYSIOLOGY 533

Chapter 26 Homeostasis and the Organization of the Animal Body 534

CASE STUDY The Limits of Endurance 535

1) Homeostasis: How Do Animals Maintain Internal Constancy? 536
- Negative Feedback Reverses the Effects of Changes 536
- Positive Feedback Drives Events to a Conclusion 536
- The Body's Internal Systems Act in Concert 537

2) How Is the Animal Body Organized? 538
- Animal Tissues Are Composed of Similar Cells That Perform a Specific Function 538
- Organs Include Two or More Interacting Tissue Types 541
- Organ Systems Consist of Two or More Interacting Organs 543

CASE STUDY REVISITED The Limits of Endurance 543

Review Section 543

Chapter 27 Circulation 548

CASE STUDY Xenotransplants 549

1) What Are the Major Features and Functions of Circulatory Systems? 550
- Animals Have Two Types of Circulatory Systems 550
- The Vertebrate Circulatory System Has Many Diverse Functions 550

2) How Does the Vertebrate Heart Work? 551
- Increasingly Complex and Efficient Hearts Have Arisen During Vertebrate Evolution 551
- The Vertebrate Heart Consists of Muscular Chambers Whose Contraction Is Controlled by Electrical Impulses 552

3) What Is Blood? 554
- Plasma Is Primarily Water in Which Proteins, Salts, Nutrients, and Wastes Are Dissolved 555
- Red Blood Cells Carry Oxygen from the Lungs to the Tissues 555
- White Blood Cells Help Defend the Body Against Disease 557
- Platelets Are Cell Fragments That Aid in Blood Clotting 557

4) What Are the Types and Functions of Blood Vessels? 558
- Arteries and Arterioles Are Thick-Walled Vessels That Carry Blood Away from the Heart 559
- Capillaries Are Microscopic Vessels That Allow the Blood and Body Cells to Exchange Nutrients and Wastes 559
- Veins and Venules Carry Blood Back to the Heart 560
- Arterioles Control the Distribution of Blood Flow 561

5) How Does the Lymphatic System Work with the Circulatory System? 561
- Lymphatic Vessels Resemble the Veins and Capillaries of the Circulatory System 562
- The Lymphatic System Returns Fluids to the Blood 562
- The Lymphatic System Transports Fats from the Small Intestine to the Blood 563
- The Lymphatic System Helps Defend the Body Against Disease 563

CASE STUDY REVISITED Xenotransplants 565

Review Section 565

Chapter 28 Respiration 568

CASE STUDY Lives Up in Smoke 569

1) Why Exchange Gases? 569

2) What Are Some Evolutionary Adaptations for Gas Exchange? 570
- Some Animals in Moist Environments Lack Specialized Respiratory Structures 570
- Respiratory Systems Facilitate Gas Exchange by Diffusion 571
- Gills Facilitate Gas Exchange in Aquatic Environments 571
- Terrestrial Animals Have Internal Respiratory Structures 572

3) How Does the Human Respiratory System Work? 573
- The Conducting Portion of the Respiratory System Carries Air to the Lungs 574
- Gas Exchange Occurs in the Alveoli 575
- Oxygen and Carbon Dioxide Are Transported Using Different Mechanisms 575
- Air Is Inhaled Actively and Exhaled Passively 576
- Breathing Rate Is Controlled by the Respiratory Center of the Brain 576

CASE STUDY REVISITED Lives Up in Smoke 577

Review Section 578

Chapter 29 Nutrition and Digestion 582

CASE STUDY Fat in the Family? 583

1) What Nutrients Do Animals Need? 584
- The Primary Sources of Energy Are Carbohydrates and Fats 584
- Lipids Include Fats, Phospholipids, and Cholesterol 584

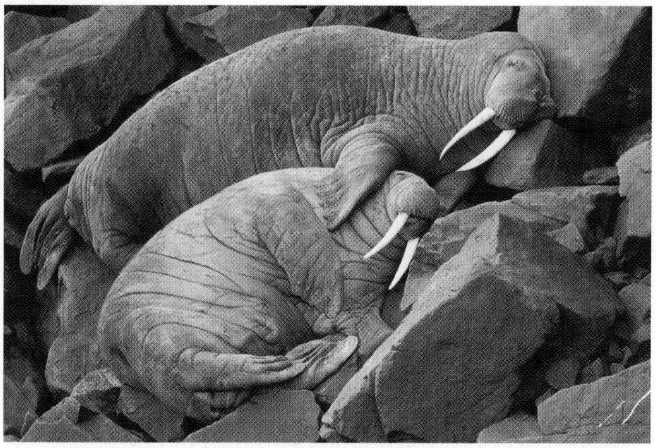

Carbohydrates, Including Sugars and Starches, Are a Source of Quick Energy 585
Proteins, Composed of Amino Acids, Perform a Wide Range of Functions Within the Body 585
Minerals Are Elements and Small Inorganic Molecules Required by the Body 586
Vitamins Are Required in Small Amounts and Play Many Roles in Metabolism 586
Nutritional Guidelines Help People Obtain a Balanced Diet 588
Are You Too Heavy? 589

2) How Is Digestion Accomplished? 589

An Overview of Digestion 589
Digestive Systems Are Adapted to the Lifestyle of Each Animal 590

3) How Do Humans and Other Mammals Digest Food? 593

The Mechanical and Chemical Breakdown of Food Begins in the Mouth 594
The Esophagus Conducts Food to the Stomach 595
Most Digestion Occurs in the Small Intestine 596
Most Absorption Occurs in the Small Intestine 597
Water Is Absorbed and Feces Are Formed in the Large Intestine 599
Digestion Is Controlled by the Nervous System and Hormones 599

CASE STUDY REVISITED Fat in the Family? 600

Review Section 600

Chapter 30 The Urinary System 604

CASE STUDY Hemodialysis and Hope 605

1) How Does Excretion Occur in Invertebrates? 605

Flame Cells Filter Fluids in Flatworms 606
Nephridia in Earthworms Resemble Parts of the Vertebrate Kidney 606

2) What Are the Functions of Vertebrate Urinary Systems? 606

3) How Does the Human Urinary System Function? 607

Urine Is Formed in the Kidneys 608
Blood Is Filtered by the Glomerulus 609
The Filtrate Is Converted to Urine in the Nephron 610
The Loop of Henle Allows Urine to Become Concentrated 610
The Kidneys Are Important Organs of Homeostasis 612

CASE STUDY REVISITED Hemodialysis and Hope 614

Review Section 615

Chapter 31 Defenses Against Disease: The Immune Response 618

CASE STUDY Fighting the Flu 619

1) How Does the Body Defend Against Invasion? 620

First, the Skin and Mucous Membranes Form Barriers to Invasion 620
Second, Nonspecific Internal Defenses Combat Microbes 621
Third, the Body Mounts an Immune Response Against Specific Microbes 623

2) What Are the Key Characteristics of the Immune Response? 623

First, the Immune System Must Recognize the Invader 624
Second, the Immune System Must Launch an Attack 627
Third, the Immune System Must Remember Its Past Victories 629

3) How Does Medical Care Augment the Immune Response? 631

Antibiotics Slow Down Microbial Reproduction 631
Vaccinations Stimulate the Development of Memory Cells 631

4) What Happens When the Immune System Malfunctions? 632

Allergies Are Misdirected Immune Responses 632
An Autoimmune Disease Is an Immune Response Against Some of the Body's Own Molecules 633
An Immune Deficiency Disease Results from the Inability to Mount an Effective Immune Response to Infection 634
AIDS Is a Devastating Immune Deficiency Disease 634
Cancer Can Evade or Overwhelm the Immune Response 637

CASE STUDY REVISITED Fighting the Flu 638

Review Section 639

Chapter 32 Chemical Control of the Animal Body: The Endocrine System 642

CASE STUDY Losing on Steroids 643

1) What Are the Characteristics of Animal Hormones? 643

Hormones Bind to Specific Receptors on Target Cells 644
Hormone Release Is Regulated by Feedback Mechanisms 647

2) What Are the Structures and Hormones of the Mammalian Endocrine System? 648

Mammals Have Both Exocrine and Endocrine Glands 648
The Hypothalamus Controls the Secretions of the Pituitary Gland 649
The Thyroid and Parathyroid Glands Influence Metabolism and Calcium Levels 652
The Pancreas Is Both an Exocrine and an Endocrine Gland 654
The Sex Organs Secrete Steroid Hormones 654
The Adrenal Glands Have Two Parts That Secrete Different Hormones 655
Many Types of Cells Produce Prostaglandins 656
Other Sources of Hormones Include the Pineal Gland, Thymus, Kidneys, Heart, Digestive Tract, and Fat Cells 656
EVOLUTIONARY CONNECTIONS: The Evolution of Hormones 657

CASE STUDY REVISITED Losing on Steroids 658

Review Section 659

Chapter 33 The Nervous System and the Senses 662

CASE STUDY From Tragedy to Triumph 663

1) What Are the Structures and Functions of Neurons? 664

2) How Is Neural Activity Produced and Transmitted? 665

Neurons Create Electrical Signals Across Their Membranes 665
Neurons Communicate at Synapses 665
Excitatory or Inhibitory Potentials Are Produced at Synapses and Integrated in the Cell Body 666
The Nervous System Uses Many Neurotransmitters 667

3) What Are Some General Features of Nervous Systems? 670

Information Processing Requires Four Basic Operations 670
Neural Pathways Direct Behavior 671
Increasingly Complex Nervous Systems Are Increasingly Centralized 671

4) How Is The Human Nervous System Organized? 672

The Peripheral Nervous System Links the Central Nervous System to the Body 672
The Central Nervous System Consists of the Spinal Cord and Brain 673
The Spinal Cord Is a Cable of Axons Protected by the Backbone 673
The Brain Consists of Many Parts Specialized for Specific Functions 676

5) How Does the Brain Produce the Mind? 680

The "Left Brain" and "Right Brain" Are Specialized for Different Functions 680
The Mechanisms of Learning and Memory Are Poorly Understood 682
Insights on How the Brain Creates the Mind Come from Diverse Sources 682

6) How Do Sensory Receptors Work? 683

7) How Is Sound Sensed? 683

The Ear Captures, Transmits, and Converts Sound into Electrical Signals 683
Sound Is Converted into Electrical Signals in the Cochlea 685

8) How Is Light Sensed? 686

The Compound Eyes of Arthropods Produce a Mosaic Image 686

The Mammalian Eye Collects, Focuses, and Transduces Light Waves 686
Binocular Vision Allows Depth Perception 689

9) How Are Chemicals Sensed? 690

The Ability to Smell Arises from Olfactory Receptors 690
Taste Receptors Are Located in Clusters on the Tongue 690
Pain Is a Specialized Chemical Sense 691
EVOLUTIONARY CONNECTIONS: Uncommon Senses 691

CASE STUDY REVISITED From Tragedy to Triumph 693

Review Section 693

Chapter 34 Action and Support: The Muscles and Skeleton 698

CASE STUDY Healing Broken Bones 699

1) How Do Muscles Work? 700

Skeletal Muscle Cell Structure and Function Are Closely Linked 701
Muscle Contraction Results from Thick and Thin Filaments Sliding Past One Another 702
Cardiac Muscle Powers the Heart 704
Smooth Muscle Produces Slow, Involuntary Contractions 704

2) What Does the Skeleton Do? 704

The Vertebrate Skeleton Serves Many Functions 705

3) Which Tissues Compose the Vertebrate Skeleton? 706

Cartilage Provides Flexible Support and Connections 706
Bone Provides a Strong, Rigid Framework for the Body 707

4) How Does the Body Move? 708

Muscles Move the Skeleton Around Flexible Joints 710

CASE STUDY REVISITED Healing Broken Bones 711

Review Section 712

Chapter 35 Animal Reproduction 716

CASE STUDY Mr. Mom? 717

1) How Do Animals Reproduce? 717

Asexual Reproduction Does Not Involve the Fusion of Sperm and Egg 718
Sexual Reproduction Requires the Union of Sperm and Egg 718

2) How Does the Human Reproductive System Work? 722

The Male Reproductive Tract Includes the Testes and Accessory Structures 722
The Female Reproductive Tract Includes the Ovaries and Accessory Structures 725
Copulation Allows Internal Fertilization 729

3) How Can People Limit Fertility? 732

Permanent Contraception Can Be Achieved Through Sterilization 732
There Are Three General Approaches to Temporary Contraception 733

CASE STUDY REVISITED Mr. Mom? 736

Review Section 736

Chapter 36 Animal Development 740

CASE STUDY Far-reaching Choices 741

1) How Do Indirect and Direct Development Differ? 742
- During Indirect Development, Animals Undergo a Radical Change in Body Form 742
- Newborn Animals That Undergo Direct Development Resemble Miniature Adults 743

2) How Does Animal Development Proceed? 744
- Cleavage Begins the Process 745
- Gastrulation Forms Three Tissue Layers 745
- Adult Structures Develop During Organogenesis 746
- Sexual Maturation Is Controlled by Genes and the Environment 746

3) How Is Development Controlled? 746
- Each Cell Contains the Entire Genetic Blueprint for the Organism 747
- Gene Transcription Is Precisely Regulated During Development 747

4) How Do Humans Develop? 750
- During the First 2 Months, Rapid Differentiation and Growth Occur 752
- The Placenta Secretes Hormones and Exchanges Materials Between Mother and Embryo 753
- Growth and Development Continue During the Last 7 Months 754
- Development Culminates in Labor and Delivery 756
- Milk Secretion Is Stimulated by Hormones of Pregnancy 757
- Aging Seems to Be Genetically Programmed 757

CASE STUDY REVISITED Far-reaching Choices 760

Review Section 760

Chapter 37 Animal Behavior 764

CASE STUDY Sex and Symmetry 765

1) How Do Innate and Learned Behaviors Differ? 766
- Innate Behaviors Can Be Performed Without Prior Experience 766
- Learned Behaviors Are Modified by Experience 766
- There Is No Sharp Distinction Between Innate and Learned Behaviors 768

2) How Do Animals Communicate? 771
- Visual Communication Is Most Effective over Short Distances 771
- Communication by Sound Is Effective over Longer Distances 772
- Chemical Messages Persist Longer But Are Hard to Vary 773
- Communication by Touch Helps Establish Social Bonds 773

3) How Do Animals Compete for Resources? 774
- Aggressive Behavior Helps Secure Resources 774
- Dominance Hierarchies Help Manage Aggressive Interactions 774
- Animals May Defend Territories That Contain Resources 776

4) How Do Animals Find Mates? 777
- Vocal and Visual Signals Encode Sex, Species, and Individual Quality 777
- Chemical Signals Bring Mates Together 778

5) What Kinds of Societies Do Animals Form? 779
- Group Living Has Advantages and Disadvantages 780
- Honeybees Form Complex Insect Societies 781
- Bullhead Catfish Form a Simple Vertebrate Society 782
- Naked Mole Rats Form a Complex Vertebrate Society 783

6) Can Biology Explain Human Behavior? 783
- The Behavior of Newborn Infants Has a Large Innate Component 784
- Behaviors Shared by Diverse Cultures May Be Innate 784
- People May Respond to Pheromones 785
- Comparisons of Identical and Fraternal Twins Reveal Genetic Components of Behavior 785
- EVOLUTIONARY CONNECTIONS: Why Do Animals Play? 786

CASE STUDY REVISITED Sex and Symmetry 787

Review Section 788

UNIT SIX
ECOLOGY 791

Chapter 38 Population Growth and Regulation 792

CASE STUDY Acorns, Mice, Moths, Deer, and Disease 793

1) How Do Populations Grow? 794
- Biotic Potential Can Produce Exponential Growth 794

2) How Is Population Growth Regulated? 796
- Exponential Growth Cannot Continue Indefinitely 796
- Environmental Resistance Limits Population Growth 797

3) How Are Populations Distributed in Space and Time? 802
- Populations Exhibit Differing Spatial Distributions 802
- Survivorship in Populations Follows Three Basic Patterns 803

4) How Is the Human Population Changing? 804
- The Human Population Is Growing Exponentially 804
- Technological Advances Have Increased Earth's Carrying Capacity for Humans 805

The Age Structure of a Population Predicts Its Future Growth 805
The U.S. Population Is Growing Rapidly 808

CASE STUDY REVISITED Acorns, Mice, Moths, Deer, and Disease 810

Review Section 810

Chapter 39 Community Interactions 814

CASE STUDY Invasion of the Zebra Mussels 815

1) Why Are Community Interactions Important? 816

2) What Are the Effects of Competition Among Species? 816
The Ecological Niche Defines the Place and Role of Each Species in Its Ecosystem 816
Adaptations Reduce the Overlap of Ecological Niches Among Coexisting Species 816
Competition Helps Control Population Size and Distribution 818

3) What Are the Results of Interactions Between Predators and Their Prey? 818
Predator–Prey Interactions Shape Evolutionary Adaptations 818

4) What Is Symbiosis? 823
Parasitism Harms, But Does Not Immediately Kill, the Host 824
In Mutualistic Interactions, Both Species Benefit 824

5) How Do Keystone Species Influence Community Structure? 824

6) Succession: How Does a Community Change over Time? 826
There Are Two Major Forms of Succession: Primary and Secondary 826
Succession Also Occurs in Ponds and Lakes 829
Succession Culminates in the Climax Community 829
Some Ecosystems Are Maintained in a Subclimax State 831

CASE STUDY REVISITED Invasion of the Zebra Mussels 831

Review Section 831

Chapter 40 How Do Ecosystems Work? 834

CASE STUDY Flight from Extinction 835

1) What Are the Pathways of Energy and Nutrients? 835

2) How Does Energy Flow Through Communities? 836
Energy Enters Communities Through Photosynthesis 836
Energy Is Passed from One Trophic Level to Another 838
Energy Transfer Through Trophic Levels Is Inefficient 838

3) How Do Nutrients Move Within and Among Ecosystems? 843
Carbon Cycles Through the Atmosphere, Oceans, and Communities 843
The Major Reservoir for Nitrogen Is the Atmosphere 844
The Major Reservoir for Phosphorus Is Rock 846
Most Water Remains Chemically Unchanged During the Water Cycle 846

4) What Is Causing Acid Rain and Global Warming? 847
Overloading the Nitrogen and Sulfur Cycles Causes Acid Rain 848
Interfering with the Carbon Cycle Contributes to Global Warming 849

CASE STUDY REVISITED Flight from Extinction 852

Review Section 852

Chapter 41 Earth's Diverse Ecosystems 856

CASE STUDY Wings of Hope 857

1) What Factors Influence Earth's Climate? 857
Both Climate and Weather Are Driven by the Sun 857
Many Physical Factors Also Influence Climate 858

2) What Conditions Does Life Require? 862

3) How Is Life on Land Distributed? 863
Terrestrial Biomes Support Characteristic Plant Communities 863
Rainfall and Temperature Determine the Vegetation a Biome Can Support 875

4) How Is Life in Water Distributed? 876
Freshwater Lakes Have Distinct Regions of Life 876
Freshwater Lakes Are Classified According to Their Nutrient Content 876
Marine Ecosystems Cover Much of Earth 878
Coastal Waters Support the Most Abundant Marine Life 878

CASE STUDY REVISITED Wings of Hope 886

Review Section 886

Appendix I 891

Appendix II 892

Glossary G-1

Photo Credits P-1

Index I-1

Preface

Will scientists clone a person in the foreseeable future? Are genetically engineered crops safe? Are people causing climate change? Is AIDS still spreading? Will physicians soon be transplanting pig hearts into people? The need for citizens to understand the basic concepts and issues of biology has never been more urgent. We have met with biology educators throughout the United States to discuss everything from the "big picture" to the details of how specific topics should be presented. These people combine exceptional teaching skills with expertise in diverse areas of biology. Such collaboration has been at the heart of this revision and of our effort to create a text that is responsive to the needs of today's students. Many of you are teaching a class that will be your students' final exposure to biology before they go out into the world. You told us that they should emerge from your course able to scrutinize science articles in the popular press with an educated and critical eye. Likewise, the course should prepare students to ask intelligent questions and make informed choices, both as voters and as consumers. You want your students to understand and appreciate the working of their own bodies. They should know about the other organisms with which we share Earth, the evolutionary forces that molded all life-forms, and how the complex interactions within ecosystems sustain us and all other life on Earth. Finally, you told us that you want a text with a relevancy that will leave students with a fascination for life that will inspire them to keep learning. This revision is our response. Now in its sixth edition, *Biology: Life on Earth. . . .*

. . . Actively Engages Students

Each chapter opens with a strikingly illustrated "Case Study." Our Case Studies are based on recent news items, on situations in which students might find themselves, or on particularly fascinating biological topics. For example, your students will witness a male seahorse giving birth (p. 717), learn about a remarkable butterfly industry that sustains a crucial tract of rain forest (p. 857), see cloned, genetically engineered pigs that are being raised as potential organ donors for humans (p. 549), and explore the questions and dilemmas that face a young woman during her first pregnancy (p. 741). Each Case Study is revisited at the end of the chapter, allowing students to explore the topic a bit further in light of what they have learned and, often, to find answers to questions raised in the initial study.

Throughout each chapter, our major headings pose important questions that encourage students to seek answers as they read. The full-sentence, conceptual subheadings both suggest answers to these questions and help students focus on the key points in each subsection.

We have placed icons next to key illustrations, tables, and concepts. These icons direct the student to a new section called the "MediaTutor" at the end of each chapter. Here, students will find descriptions of related topics that they can pursue (1) in our chapter Web site, *Audesirk Live!*, or (2) on the MediaTutor student CD-ROM, which contains interactive exercises and animations that engage students in learning and help them visualize biological concepts.

Finally, particularly where processes are illustrated, we have added annotations to figures (see, for example, Fig. 2-3, Fig. 3-11, Fig. 4-3, Fig. 23-22, and Fig. E36-01). These annotations place descriptions of each process at the points where they are most needed for clarity and reduce the need for lengthy, multi-part captions.

. . . Conveys That Biology Is Everywhere

Throughout the text, we relate key biological concepts both to everyday experiences and to important issues facing society. These references are woven into the text, introduced in end-of-chapter critical-thinking questions ("Applying the Concepts"), and highlighted in updated boxed essays. These essays cover a wide range of current topics in biology, from environmental issues ("Earth Watch"), to clinical discussions ("Health Watch"), to in-depth views of specific processes ("A Closer Look"), to the procedures biologists use in doing their work ("Scientific Inquiry"). For example:

Earth Watch: These environmental essays explore pressing issues such as the loss of biodiversity, the ozone hole, and invasions of exotic species.

Health Watch: These clinical essays investigate topics such as sexually transmitted diseases, the dangers of artificial steroids, and how smoking damages the lungs.

. . . Is Flexible and Easy to Use

At the start of each chapter, the conceptual questions and their responsive subheadings are brought together in an effective summary, "At a Glance." A consistent numbering system identifies the major conceptual headings—all the way from At a Glance, through the body of the chapter in major headings, and into the "Summary of Key Concepts." A "Key Terms" list identifies all the boldface terms that appear in the chapter, along with the page on which each term is introduced; key terms are defined in the Glossary at the back of the book.

The chapters in *Biology: Life on Earth* have been written to allow instructors flexibility in the order of use. Chapter 1 sets the stage for the coverage of topics in different sequence by providing an overview of the diversity of life and the principles of evolution. Cross-referencing among chapters allows students to seek additional information on specific topics in other parts of the book.

. . . Has Supporting Technology That Enhances Learning and Is Integrated into the Textbook

Biology: Life on Earth is fully supported by an array of both electronic and print ancillary materials. The Prentice Hall team has worked closely with biology educators and revised the text's supplements to ensure that they meet the needs both of students and instructors.

Instructional support in the classroom is the focus of the ancillary program. The Instructor CD-ROM acts as both a warehouse of presentation materials and as a master-of-ceremonies that allows the instructor to access all the available tools in one place at one time. Driven by a powerful database that allows instructors to search easily by table of contents, term, subject, media type, or figure number, it contains all the line art from the book, in both labeled and unlabeled styles, and many of the photos. The entire Student CD-ROM, along with guidelines for integrating the animations and exercises into lectures, is fully accessible from the Instructor CD-ROM. As an instructor, you will have the *Instructor's Guide to Print and Media Resources* on the CD-ROM and all of the questions from the Test Item File. The instructor's package includes 250 full-color acetates, supplemented by transparency masters that cover all the line art in the sixth edition. Prentice Hall's Custom Test provides a selection of 2400 sample test questions in a variety of styles, combined with custom testing software, allowing you to create and edit exams electronically.

An entirely new student CD-ROM accompanies each text and provides interactive exercises and animations. For easy reference and tight integration with our multimedia supplements, icons have been placed next to key illustrations, tables, and concepts throughout each chapter that direct students to the MediaTutor at the end of the chapter. Each MediaTutor is a resource for students that describes the specific content and objectives on the student CD-ROM and the "Web Investigation" for that particular chapter. By informing students about what to expect, MediaTutor allows them to follow their interests and needs, thus to make the best use of these resources.

The companion Web site, *Audesirk Live!*, has been completely updated with new questions, activities, and links to provide students with extensive opportunities for study and to provide instructors with additional sources for assignments. Our chapter Web sites are updated frequently, allowing students to explore topics relevant to each chapter on the World Wide Web. In *Audesirk Live!*, students will discover that biology is everywhere by exploring our "Issues in Biology" and "Bizarre Facts" sections. *The New York Times Themes of the Times*, our free newspaper-format supplement, provides a collection of recent science articles written by world-renowned science writers for this outstanding newspaper. Instructors can receive our *ABC NEWS/Prentice Hall Video Library for Biology*, containing short news segments from award-winning news programs. These effective teaching enhancements connect biology in the classroom to what's happening in the world today.

Finally, we have moved our popular "Group Activities" to the Web site for each chapter. These exercises, used successfully by our Panel of Biology Educators, encourage students to work out interesting problems in small groups and thereby become active participants in the learning process. By placing Group Activities on the Web, we allow our Web master to update them as new ideas arise. The *Instructor's Guide to Print and Media* includes additional Group Activities and provides suggestions about how instructors can incorporate these exercises into the classroom.

The Fundamental Philosophy of *Biology: Life on Earth*

Although our textbook continues to evolve in response to the changing needs of our audience, there are fundamental issues in the teaching of biology that don't change. *Biology: Life on Earth* still does the following:

Focuses on Concepts

Our conceptual questions and subheadings cast as sentences, At a Glance chapter-opening outlines, and end-of-chapter Summary of Key Concepts sections keep students focused on the important themes in each chapter. Figure captions include caption titles, which give the theme of each image and then provide more specific information. Because it is easy for students to lose sight of the underlying concepts in a welter of technical detail, we provide a general overview of complex subjects in the text itself and offer the details of such topics in A Closer Look essays:

A Closer Look: These essays focus on the more challenging details of topics such as chemiosmosis, cellular respiration, and urine formation in the nephron.

Communicates the Scientific Process

Biology is not just a compendium of facts and ideas; rather, it is the outgrowth of a dynamic process of inquiry and human endeavor. In many instances, we describe how scientists discovered specific facts. The scientific process is further highlighted in Scientific Inquiry essays:

Scientific Inquiry: With these essays, students will learn how fossils are dated and how PET scans are performed. They follow the development of the science of genetics from the laboratory of Watson and Crick to the Roslin Center in Scotland, where Ian Wilmut cloned a sheep from adult DNA, and even to Lake Maracaibo in Venezuela, where Nancy Wexler began to unlock the secrets of Huntington disease.

Emphasizes Unifying Themes

As Theodosius Dobzhansky so aptly put it, "Nothing in biology makes sense, except in the light of evolution." Throughout the text, students will find examples of how natural selection has produced organisms that are adapted to specific environments. In addition, many chapters end with an "Evolutionary Connections" section:

Evolutionary Connections: These lively discussions link chapter concepts into the broader perspective of evolution.

Our own concern for the environment can be found interwoven throughout this text and highlighted in Earth Watch essays as well. Whenever appropriate, we have tried to present students with the biological rationale for making sound environmental decisions in their daily lives.

Strives for Accuracy

A text is useless if it does not convey accurate information. To this end, we utilize multiple, high-level sources for our basic facts. Each unit of each new edition is carefully scrutinized by several individuals who are both talented educators and experts in the unit subject areas. They help us in our quest to present the material in a way that is both accurate and comprehensible. In between editions, our users serve as informal reviewers. We never fail to follow up on your queries, and the book is better as a result. As always, however, the buck stops with the authors. We take responsibility for the accuracy of the text material, and we take this responsibility seriously.

A Final Word

A course in introductory biology may be a student's first—and sometimes last—in-depth exposure to the fascinating complexity of life. As teachers, we recognize how easily a student can become mired in a plethora of new facts and unfamiliar terms while losing sight of the underlying concepts of biology. We have carefully revised *Biology: Life on Earth* to reduce unnecessary detail and excess terminology and to emphasize the ways in which an understanding of biology can enrich and enlighten day-to-day living. This text can serve the student in many ways—from an "owner's manual" of the human body to a "user's guide" for the environment. Why study biology? Maybe we're biased, but what can be more fascinating than learning about Life on Earth?

Acknowledgments

Biology: Life on Earth is truly a group effort. We are fortunate that Robin Wright, a member of our fifth edition Panel of Biology Educators, agreed to take on the revision of Unit Two. In addition to her teaching skills, she brings specialized knowledge of the rapidly changing applications of molecular biology to this task. Her careful revision has made the text more accurate and timely.

To meet the dauntingly complex challenge of putting together a text and supplement package of this magnitude, Prentice Hall has assembled an experienced and skilled development team. The text benefited considerably from the thoughtful suggestions of Developmental Editor Shana Ederer. She not only helped us keep the text clear, consistent, and student-friendly, she has also contributed substantially to the clarity of the revised art. Karen Karlin helped us tie up loose ends with her unique and valuable blend of experience, conscientiousness, and attention to detail. Tim Flem, our Production Editor, coordinated the efforts of the photo researcher, copy editor, art studio, and authors. He skillfully brought the art, photos, and manuscript together into a seamless whole while dealing good-naturedly with last minute improvements. Formatting this book is not an easy endeavor, but Tim applied his expertise with great attention to detail. Photo Researcher Linda Sykes tracked down excellent photos. Roberta Dempsey tackled the job of copyediting with exceptional skill. Proofreader Margaret Buresch read the final manuscript with scrupulous care to help ensure accuracy.

We also wish to thank Art Director Jonathan Boylan and Director of Design Carole Anson for guiding the text and cover design with flair and talent, and Managing Editor Grace Hazeldine for coordinating such an immense art program.

Travis Moses-Westphal, our Project Manager, played multiple roles from the beginning with much creativity and gusto. He deftly carried the vision for the media program and oversaw the seamless integration of the text, media, and print supplements. Media Editors Andy Stull and Kate Flickinger helped Travis move the media program forward. Editorial Assistant Colleen Lee was always around just when we needed her with good cheer and brought exceptional skills to traffic the manuscript.

Colleagues at other institutions have helped us enormously. Many, listed in the following pages, have stimulated us to rethink our presentation with careful, thoughtful reviews. Our new Panel of Biology Educators, listed on page vii, has met with us personally and made a special contribution to the sixth edition.

The expansive and crucial supplements package has many valued contributors. Joseph Chinnici and Susan Wadkowski prepared the Study Guide; Gail Gasparich, Charles Good, Kate Lajtha, and Lewis Deaton prepared the Test Item File; Timothy Metz was responsible for the Instructor's Guide to Print and Media Resources; Timothy Metz worked with David Huffman to organize the

Instructor CD-ROM; Joseph Coehlo, Jim Hewlett, Ken Mason, Paul Ramp, Diana Wheat, and Cal Young spent countless hours to deliver the content for the Student CD-ROM; and Kelly Johnson, Jeff Kenton, Chris Romero, Ina Pour-el, William Hayes, and Lydia Daniels were responsible for updating the Web site.

Jennifer Welchans, our Executive Marketing Manager, oversees a large and dedicated sales force with energy, talent, and enthusiasm. Jen provides inspired marketing concepts, shares success stories, and makes sure that user's comments always get through to the authors. We thank Paul Corey, now president of the Engineering, Science, and Mathematics Division of Prentice Hall, for his confidence and support through this and the past three editions. Finally, but most importantly, our editors: Editor in Chief Sheri Snavely has supported us now through four editions. Executive Editor Teresa Ryu has assumed the leadership of the team with talent and zeal, combined with a clear sense of where the project should be going and how to get it there without killing the authors. Her total commitment to the project, her organizational ability, and her sensitivity to all the people involved have been crucial to its success.

So here we acknowledge, with deep appreciation, our "coach" and all our teammates!

Terry and Gerry Audesirk
Bruce E. Byers

Panel of Multimedia Advisors and Reviewers

Media Contributors
Joseph Chinnici, *Virginia Commonwealth University*
Joseph Coelho, *Culver Stockton College*
Lydia Daniels, *University of Pittsburgh*
Lewis Deaton, *University of Southwestern Louisiana*
Gail Gasparich, *Towson State University*
William Hayes, *Delta State University*
James Hewlett, *Finger Lakes Community College*
David Huffman, *Southwest Texas State University*
J. Kelly Johnson, *University of Kansas*
Jeff Kenton, *Iowa State University*
Kate Lajtha, *Oregon State University*
Timothy Metz, *Campbell University*
Ina Pour-el, *DMACC–Boone Campus*
Paul Ramp, *Pellissippi State Technical College*
Chris Romero, *Front Range Community College*
Cal Young, *Fullerton College*

Media Reviewers
J. Gregory Burg, *University of Kansas*
Jerry Button, *Portland Community College*
Walter J. Conley, *State University of New York at Potsdam*
Jerry Cook, *Sam Houston State University*
David M. Demers, *University of Hartford*
Susannah Feldman, *Towson University*
Timothy L. Henry, *University of Texas Arlington*
James Hewlett, *Finger Lakes Community College*
Kelly Johnson, *University of Kansas*
Jeffrey Kiggins, *Blue Ridge Community College*
Harry Kurtz, *Sam Houston State University*
Kenneth A. Mason, *University of Kansas*
Timothy Metz, *Campbell University*
Marvin Price, *Cedar Valley College*
Chris Romero, *Front Range Community College*
Patricia Shields, *George Mason University*
Susan M. Wadkowski, *Lakeland Community College*
Stacy Wolfe, *Art Institutes International*
Robin Wright, *University of Washington*
Cal Young, *Fullerton College*

Sixth Edition Reviewers

Sara Chambers, *Long Island University*
Karen Dalton, *Community College of Baltimore County–Catonsville Campus*
Lewis Deaton, *University of Southwestern Louisiana*
Rosemarie Elizondo, *Reedley College*
Charles Good, *Ohio State University*
Lonnie J. Guralnick, *Western Oregon University*
Georgia Ann Hammond, *Radford University*
James Hewlett, *Finger Lakes Community College*
Leland N. Holland, *Paso-Hernando Community College*
Rebecca M. Jessen, *Bowling Green State University*
Patricia Lee-Robinson, *Chaminade University of Honolulu*
Edward Levri, *Indiana University of Pennsylvania*
Ann S. Lumsden, *Florida State University*
Linda Martin-Morris, *University of Washington*
Kenneth A. Mason, *University of Kansas*
Joseph R. Mendelson III, *Utah State University*
Timothy Metz, *Campbell University*
John W. Moon, *Harding University*
Jane Noble-Harvey, *University of Delaware*
David J. O'Neill, *Community College of Baltimore County, Dundalk Campus*
Rhoda E. Perozzi, *Virginia Commonwealth University*
Elsa C. Price, *Wallace State Community College*
Christopher F. Sacchi, *Kutztown University*
Anu Singh-Cundy, *Western Washington University*
Dan Tallman, *Northern State University*
Susan M. Wadkowski, *Lakeland Community College*
Brenda L. Young, *Daemen College*
Cal Young, *Fullerton College*

Previous Editions Reviewers

W. Sylvester Allred, *Northern Arizona University*
Judith Keller Amand, *Delaware County Community College*
William Anderson, *Abraham Baldwin Agriculture College*
Steve Arch, *Reed College*
Kerri Lynn Armstrong, *Community College of Philadelphia*
G. D. Aumann, *University of Houston*
Vernon Avila, *San Diego State University*
J. Wesley Bahorik, *Kutztown University of Pennsylvania*
Bill Barstow, *University of Georgia, Athens*
Colleen Belk, *University of Minnesota, Duluth*
Michael C. Bell, *Richland College*
Gerald Bergtrom, *University of Wisconsin*
Arlene Billock, *University of Southwestern Louisiana*
Brenda C. Blackwelder, *Central Piedmont Community College*
Raymond Bower, *University of Arkansas*
Marilyn Brady, *Centennial College of Applied Arts & Technology*
Virginia Buckner, *Johnson County Community College*
Arthur L. Buikema, Jr., *Virginia Polytechnic Institute*
William F. Burke, *University of Hawaii*
Robert Burkholter, *Louisiana State University*
Kathleen Burt-Utley, *University of New Orleans*
Linda Butler, *University of Texas, Austin*
W. Barkley Butler, *Indiana University of Pennsylvania*
Bruce E. Byers, *University of Massachusetts, Amherst*
Nora L. Chee, *Chaminade University*
Joseph P. Chinnici, *Virginia Commonwealth University*
Dan Chiras, *University of Colorado, Denver*
Bob Coburn, *Middlesex Community College*
Martin Cohen, *University of Hartford*
Mary U. Connell, *Appalachian State University*
Joyce Corban, *Wright State University*
Ethel Cornforth, *San Jacinto College, South*
David J. Cotter, *Georgia College*
Lee Couch, *Albuquerque Technical Vocational Institute*
Donald C. Cox, *Miami University of Ohio*
Patricia B. Cox, *University of Tennessee*
Peter Crowcroft, *University of Texas, Austin*
Carol Crowder, *North Harris Montgomery College*
Donald E. Culwell, *University of Central Arkansas*
Robert A. Cunningham, *Erie Community College, North*
David H. Davis, *Asheville-Buncombe Technical Community College*
Jerry Davis, *University of Wisconsin, LaCrosse*
Douglas M. Deardon, *University of Minnesota*
Lewis Deaton, *University of Southwestern Louisiana*
Fred Delcomyn, *University of Illinois, Urbana*
Lorren Denney, *Southwest Missouri State University*
Katherine J. Denniston, *Towson State University*
Charles F. Denny, *University of South Carolina, Sumter*
Jean DeSaix, *University of North Carolina, Chapel Hill*
Ed DeWalt, *Louisiana State University*
Daniel F. Doak, *University of California, Santa Cruz*
Matthew M. Douglas, *University of Kansas*
Ronald J. Downey, *Ohio University*
Ernest Dubrul, *University of Toledo*
Michael Dufresne, *University of Windsor*
Susan A. Dunford, *University of Cincinnati*
Mary Durant, *North Harris College*
Ronald Edwards, *University of Florida*
George Ellmore, *Tufts University*
Joanne T. Ellzey, *University of Texas, El Paso*
Wayne Elmore, *Marshall University*
Carl Estrella, *Merced College*
Nancy Eyster-Smith, *Bentley College*
Gerald Farr, *Southwest Texas State University*
Rita Farrar, *Louisiana State University*
Marianne Feaver, *North Carolina State University*
Linnea Fletcher, *Austin Community College, Northridge*
Charles V. Foltz, *Rhode Island College*
Douglas Fratianne, *Ohio State University*
Scott Freeman, *University of Washington*
Donald P. French, *Oklahoma State University*
Don Fritsch, *Virginia Commonwealth University*
Teresa Lane Fulcher, *Pellissippi State Technical Community College*
Michael Gaines, *University of Kansas*
Irja Galvan, *Western Oregon University*
Gail E. Gasparich, *Towson University*
Farooka Gauhari, *University of Nebraska, Omaha*
George W. Gilchrist, *University of Washington*
David Glenn-Lewin, *Iowa State University*
Elmer Gless, *Montana College of Mineral Sciences*
Charles W. Good, *Ohio State University, Lima*
Margaret Green, *Broward Community College*
Martin E. Hahn, *William Paterson College*
Madeline Hall, *Cleveland State University*
Blanche C. Haning, *North Carolina State University*
Helen B. Hanten, *University of Minnesota*
John P. Harley, *Eastern Kentucky University*
Stephen Hedman, *University of Minnesota*
Jean Helgeson, *Collins County Community College*
Alexander Henderson, *Millersville University*
Alison G. Hoffman, *University of Tennessee, Chattanooga*
Laura Mays Hoopes, *Occidental College*
Michael D. Hudgins, *Alabama State University*
Donald A. Ingold, *East Texas State University*
Jon W. Jacklet, *State University of New York, Albany*
Florence Juillerat, *Indiana University–Purdue University at Indianapolis*
Thomas W. Jurik, *Iowa State University*
Arnold Karpoff, *University of Louisville*
L. Kavaljian, *California State University*
Hendrick J. Ketellapper, *University of California, Davis*
Kate Lajtha, *Oregon State University*
William H. Leonard, *Clemson University*
Graeme Lindbeck, *University of Central Florida*
Jerri K. Lindsey, *Tarrant County Junior College, Northeast*
John Logue, *University of South Carolina, Sumter*
William Lowen, *Suffolk Community College*
Ann S. Lumsden, *Florida State University*
Steele R. Lunt, *University of Nebraska, Omaha*
Daniel D. Magoulick, *The University of Central Arkansas*
Paul Mangum, *Midland College*
Michael Martin, *University of Michigan*
Margaret May, *Virginia Commonwealth University*
D. J. McWhinnie, *De Paul University*
Gary L. Meeker, *California State University, Sacramento*
Thoyd Melton, *North Carolina State University*
Karen E. Messley, *Rockvalley College*
Glendon R. Miller, *Wichita State University*

Neil Miller, *Memphis State University*
Jack E. Mobley, *University of Central Arkansas*
Richard Mortenson, *Albion College*
Gisele Muller-Parker, *Western Washington University*
Kathleen Murray, *University of Maine*
Robert Neill, *University of Texas*
Harry Nickla, *Creighton University*
Daniel Nickrent, *Southern Illinois University*
Jane Noble-Harvey, *University of Delaware*
James T. Oris, *Miami University, Ohio*
Marcy Osgood, *University of Michigan*
C. O. Patterson, *Texas A & M University*
Fred Peabody, *University of South Dakota*
Harry Peery, *Tompkins–Cortland Community College*
Rhoda E. Perozzi, *Virginia Commonwealth University*
Bill Pfitsch, *Hamilton College*
Ronald Pfohl, *Miami University, Ohio*
Bernard Possident, *Skidmore College*
James A. Raines, *North Harris College*
Mark Richter, *University of Kansas*
Robert Robbins, *Michigan State University*
Paul Rosenbloom, *Southwest Texas State University*
K. Ross, *University of Delaware*
Mary Lou Rottman, *University of Colorado, Denver*
Albert Ruesink, *Indiana University*
Alan Schoenherr, *Fullerton College*
Edna Seaman, *University of Massachusetts, Boston*
Linda Simpson, *University of North Carolina, Charlotte*
Russel V. Skavaril, *Ohio State University*
John Smarelli, *Loyola University*
Shari Snitovsky, *Skyline College*
Jim Sorenson, *Radford University*
Mary Spratt, *University of Missouri, Kansas City*
Benjamin Stark, *Illinois Institute of Technology*
William Stark, *Saint Louis University*
Kathleen M. Steinert, *Bellevue Community College*
Barbara Stotler, *Southern Illinois University*
Gerald Summers, *University of Missouri, Columbia*
Marshall Sundberg, *Louisiana State University*
Bill Surver, *Clemson University*
Eldon Sutton, *University of Texas, Austin*
Dan Tallman, *Northern State University*
David Thorndill, *Essex Community College*
William Thwaites, *San Diego State University*
Professor Tobiessen, *Union College*
Richard Tolman, *Brigham Young University*
Dennis Trelka, *Washington & Jefferson College*
Sharon Tucker, *University of Delaware*
Gail Turner, *Virginia Commonwealth University*
Glyn Turnipseed, *Arkansas Technical University*
Lloyd W. Turtinen, *University of Wisconsin, Eau Claire*
Robert Tyser, *University of Wisconsin, La Crosse*
Robin W. Tyser, *University of Wisconsin, LaCrosse*
Kristin Uthus, *Virginia Commonwealth University*
F. Daniel Vogt, *State University of New York, Plattsburgh*
Nancy Wade, *Old Dominion University*
Jyoti R. Wagle, *Houston Community College, Central*
Michael Weis, *University of Windsor*
DeLoris Wenzel, *University of Georgia*
Jerry Wermuth, *Purdue University, Calumet*
Jacob Wiebers, *Purdue University*
Carolyn Wilczynski, *Binghamton University*
P. Kelly Williams, *University of Dayton*
Roberta Williams, *University of Nevada, Las Vegas*
Sandra Winicur, *Indiana University, South Bend*
Bill Wischusen, *Louisiana State University*
Chris Wolfe, *North Virginia Community College*
Colleen Wong, *Wilbur Wright College*
Wade Worthen, *Furman University*
Robin Wright, *University of Washington*
Tim Young, *Mercer University*

"Viewed from the distance of the moon, the astonishing thing about the earth, catching the breath, is that it is alive. The photographs show the dry, pounded surface of the moon in the foreground, dead as an old bone. Aloft, floating free beneath the moist, gleaming surface of bright blue sky, is the rising earth, the only exuberant thing in this part of the cosmos."

Lewis Thomas in* The Lives of a Cell *(1974)

Life on Earth is confined to a thin film encompassing Earth's surface: the biosphere. Earth, seen from the Moon, is an oasis of life in our solar system.

1 An Introduction to Life on Earth

AT A GLANCE

Case Study: The Life Around Us

1) What Are the Characteristics of Living Things?

Living Things Are Both Complex and Organized

Living Things Respond to Stimuli

Living Things Maintain Relatively Constant Internal Conditions Through Homeostasis

Living Things Acquire and Use Materials and Energy

Living Things Grow

Living Things Reproduce Themselves

Living Things As a Whole Have the Capacity to Evolve

2) How Do Scientists Categorize the Diversity of Life?

The Domains Bacteria and Archaea Consist of Prokaryotic Cells; the Domain Eukarya Is Composed of Eukaryotic Cells

Bacteria, Archaea, and Members of the Kingdom Protista Are Mostly Unicellular; Members of the Kingdoms Fungi, Plantae, and Animalia Are Primarily Multicellular

Members of the Different Kingdoms Have Different Ways of Acquiring Energy

3) What Is the Science of Biology?

Scientific Principles Underlie All Scientific Inquiry

The Scientific Method Is the Basis for Scientific Inquiry

Science Is a Human Endeavor

Scientific Theories Have Been Thoroughly Tested

4) Evolution: The Unifying Theory of Biology

Three Natural Processes Underlie Evolution

5) How Does Knowledge of Biology Illuminate Everyday Life?

Case Study Revisited: The Life Around Us

CASE STUDY

The Life Around Us

Next time you walk across campus, look at the astonishing array of creatures thriving in a place as domesticated as a college campus. Sparrows and squirrels communicate with others of their species through chirps and chatters. Within the trees, bushes, grass, and moss that blanket the campus with green, you may see honeybees or a butterfly flitting from flower to flower, gathering the sweet nectar that powers their flight and their own reproduction. Meanwhile, a spider, using several types of protein fibers, spins a web to trap these insects and harvest the energy in their bodies for its own uses. Mushrooms hiding in the grass are only the reproductive tips of a large underground network of fungal fibers. In addition to these obvious life-forms, countless microscopic organisms swim in the puddles left by the sprinklers or rain and thrive in the soil. And living on, in, and around all these organisms, and the humans observing them, are billions of bacteria—simple, single-celled organisms that have survived with little change for billions of years.

How did such an astounding variety of organisms evolve? How do they interact? How are these bacteria, fungi, plants, and animals alike, and how do they differ? What processes must occur for each organism to survive and reproduce? And how do living things differ from nonliving things? Questions such as these form the basis of the science of biology. They also relate to the very survival of the human species, because we too are part of the web of life. We evolved in response to the same sorts of survival needs and consist of chemicals found in all other living creatures. Similar processes allow us to survive and reproduce. As we journey through this book, we hope you will begin to sense the excitement and awe that comes with understanding life on Earth. ■

1 What Are the Characteristics of Living Things?

What is life? If you look up *life* in a dictionary, you will find definitions such as "the quality that distinguishes a vital and functioning being from a dead body," but you won't find out what that "quality" is. All of us have an intuitive understanding of what it means to be alive. Nevertheless, defining *life* is difficult, partly because living things are so diverse and nonliving matter in some cases seems so lifelike. A more fundamental difficulty in defining life is that living things cannot be described as simply the sum of their parts. The quality of life emerges as a result of the incredibly complex, ordered interactions among these parts. Because it is based on these **emergent properties**, life is a fundamentally intangible quality that defies any simple definition. We can, however, describe some of the characteristics of living things that, taken together, are not shared by nonliving objects. These characteristics are as follows:

1. Living things have a complex, organized structure that consists largely of organic molecules.
2. Living things respond to stimuli from their environment.
3. Living things actively maintain their complex structure and their internal environment, a process called *homeostasis*.
4. Living things acquire and use materials and energy from their environment and convert them into different forms.
5. Living things grow.
6. Living things reproduce themselves, using a molecular blueprint called *DNA*.
7. Living things, as a whole, have the capacity to evolve.

Let's explore these characteristics in more detail.

Living Things Are Both Complex and Organized

Compared with nonliving matter of similar size, living things are highly complex and organized. A crystal of table salt (Fig. 1-1a), for example, consists of just two chemical elements, sodium and chlorine, arranged in a precise cubical arrangement: The salt crystal is organized but simple. The oceans (Fig. 1-1b) contain some atoms of all the naturally occurring elements, but these atoms are randomly distributed: The oceans are complex but not organized. In contrast, even the tiny waterflea (Fig. 1-1c) contains dozens of different elements linked together in thousands of specific combinations that are further organized into ever larger and more complex assemblies to form structures such as eyes, legs, a digestive tract, and even a small brain.

Life on Earth consists of a hierarchy of structures, with each level of the hierarchy based on the one below it and providing the foundation for the one above it (Fig. 1-2). All of life is built on a chemical foundation

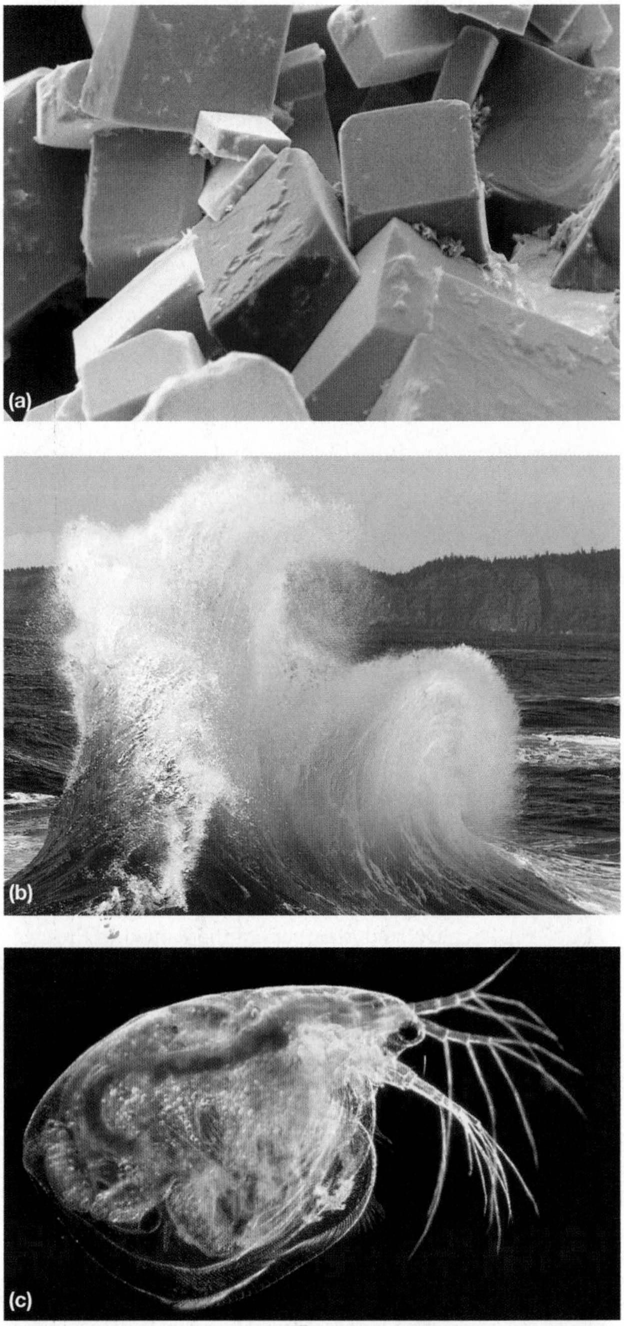

Figure 1-1 Life is both complex and organized
(a) Each crystal of table salt, sodium chloride, is a cube, showing great organization but minimal complexity. ***(b)*** The water and dissolved materials in the ocean represent complexity but very little organization. ***(c)*** Living things have both complexity and organization. The waterflea, *Daphnia pulex*, is only 1 millimeter long (1/1000 meter; smaller than the letter *i*), yet it has legs, a mouth, a digestive tract, reproductive organs, light-sensing eyes, and even a rather impressive brain in relation to its size.

based on substances called **elements**, each of which is a unique type of *matter*. An **atom** is the smallest particle of an element that retains the properties of that element. For example, a diamond is made of the element carbon. The smallest possible unit of the diamond is an individual carbon atom; any further division would

MediaTutor 1.1 What Is Life?

Biosphere	That part of Earth inhabited by living organisms; includes both the living and nonliving components
Ecosystem	A community together with its nonliving surroundings
Community	Two or more populations of different species living and interacting in the same area
Population	Members of one species inhabiting the same area
Species	Very similar, potentially interbreeding organisms
Multicellular Organism	An individual living thing composed of many cells
Organ System	Two or more organs working together in the execution of a specific bodily function
Organ	A structure usually composed of several tissue types that form a functional unit
Tissue	A group of similar cells that perform a specific function
Cell	The smallest unit of life
Organelle	A structure within a cell that performs a specific function
Molecule	A combination of atoms
Atom	The smallest particle of an element that retains the properties of that element
Subatomic Particle	Particles that make up an atom

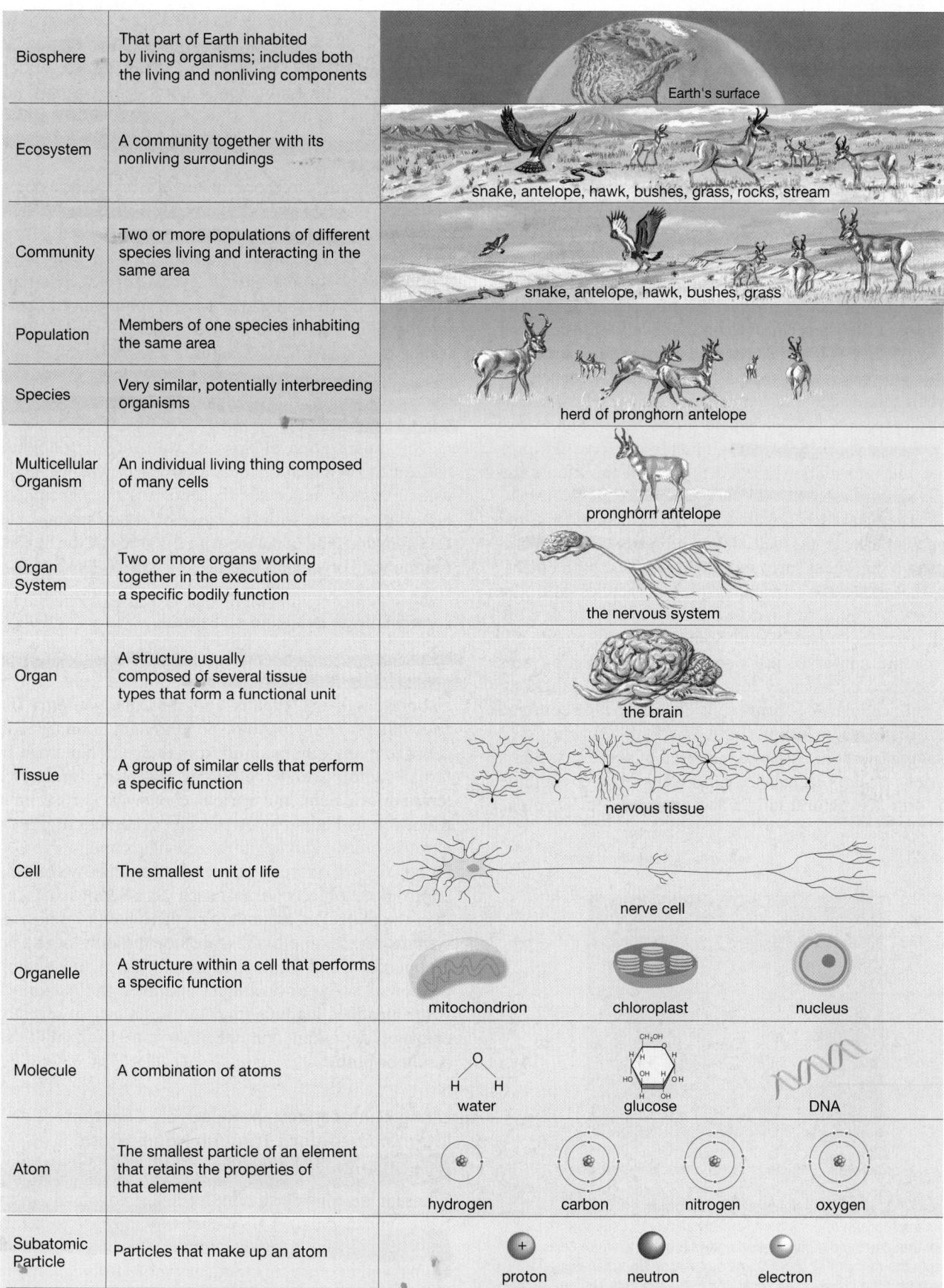

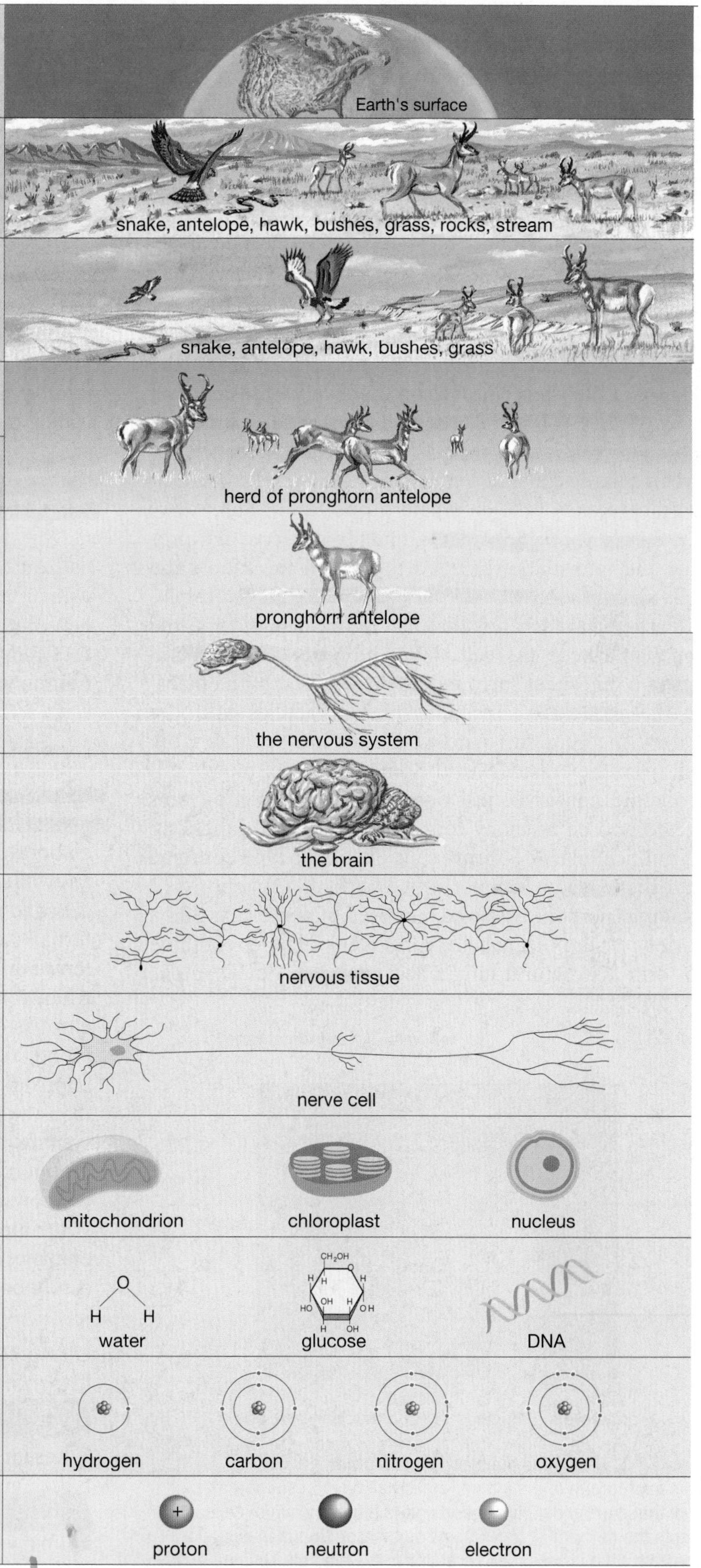

Figure 1-2 Levels of organization of matter
All life has a chemical basis, but the quality of life itself emerges on the cellular level. Interactions among the components of each level and the levels below it allow the development of the next-higher level of organization.

produce isolated **subatomic particles** that would no longer be carbon. Atoms may combine in specific ways to form assemblies called **molecules**; for example, one carbon atom can combine with two oxygen atoms to form a molecule of carbon dioxide. Although many simple molecules form spontaneously, only living things manufacture extremely large and complex molecules. The bodies of living things are composed primarily of complex molecules. The molecules of life are called **organic molecules**, meaning that they contain a framework of carbon, with at least some hydrogen bound to it. Although the chemical arrangement and interaction of atoms and molecules are the building blocks of life, the quality of life itself emerges on the level of the cell. Just as an atom is the smallest unit of an element, so the **cell** is the smallest unit of life (Fig. 1-3). The difference between a living cell and a conglomeration of chemicals illustrates some of the emergent properties of life.

All cells contain (1) **genes**, units of heredity that provide the information needed to control the life of the cell; (2) subcellular structures called **organelles**, miniature chemical factories that use the information in the genes and keep the cell alive; and (3) a **plasma membrane**, a thin sheet surrounding the cell that both encloses the **cytoplasm** (including the organelles and the watery medium surrounding them) and separates the cell from the outside world. Some life-forms, mostly microscopic, consist of just one cell, but larger life-forms are composed of many cells with specialized functions. In multicellular life-forms, cells of similar type combine to form **tissues**, which perform a particular function. For example, nervous tissue is composed of nerve cells and a variety of supporting cells. Various tissue types combine to form a structural unit called an **organ** (for example, the brain, which contains nervous tissue, connective tissue, and blood). Several organs that collectively perform a single function are called an **organ system**; for example, together the brain, spinal cord, sense organs, and nerves form the nervous system. All the organ systems functioning cooperatively make up an individual living thing, the **organism**.

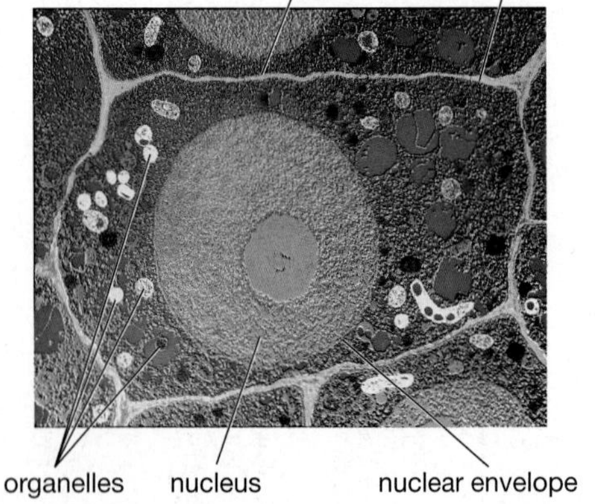

Figure 1-3 The cell is the smallest unit of life
This micrograph of a plant cell clearly shows the supporting cell wall that surrounds and supports plant (but not animal) cells. Just inside the cell wall, a thin plasma membrane (found in all cells) controls what substances enter and leave the cell. The nucleus, surrounded by a membrane called the *nuclear envelope*, contains the cell's DNA. The cell also contains several types of specialized organelles. Some store food, some break down food to provide usable energy, and, in plants, some organelles capure light energy.

Beyond individual organisms are broader levels of organization. A group of very similar, potentially interbreeding organisms constitutes a **species**. Members of the same species that live in a given area are considered a **population**. Populations of several species living and interacting in the same area form a **community**. A community plus its nonliving environment, including land, water, and atmosphere, constitute an **ecosystem**. Finally, the entire surface region of Earth inhabited by living things (but including the nonliving components) is called the **biosphere**.

The organization of this text will roughly follow the pattern of organization of life on Earth. We will begin with atoms and molecules, move to cells and principles of heredity, continue with the range of organisms and how they function, and conclude with the study of the interactions among organisms that are the focus of ecology.

Living Things Respond to Stimuli

Organisms perceive and respond to stimuli in their internal and external environments. Animals have evolved elaborate sensory organs and muscular systems that allow them to detect and respond to light, sound, chemicals, and many other stimuli from their surroundings. Internal stimuli are perceived by receptors for stretch, temperature, pain, and various chemicals. For example, when you feel hungry, you perceive contractions of your empty stomach and low levels of sugars and fats in your blood. You then respond to external stimuli by choosing appropriate objects to eat, such as a sandwich rather than the plate. Yet, animals, with their elaborate nervous systems and motile bodies, are not the only organisms that perceive and respond to stimuli. The plants on your windowsill grow toward light, and even the bacteria in your intestines manufacture a different set of digestive enzymes depending on whether you drink milk, eat candy, or both.

Living Things Maintain Relatively Constant Internal Conditions Through Homeostasis

Complex, organized structures are not easy to maintain. Whether we consider the molecules of your body or the books and papers on your desk, organization tends to disintegrate into chaos unless energy is used to sustain it. (We will explore this tendency more fully in Chapter 6.) To stay alive and function effectively, organisms must keep the conditions within their bodies fairly constant, a process called **homeostasis** (derived from Greek words meaning "to stay the same"). One of the many

Figure 1-4 Living things maintain homeostasis
Sweating helps cool former Olympic and World Heavyweight boxing champion George Foreman.

conditions that organisms regulate is body temperature. Among warm-blooded animals, for example, vital organs such as the brain and heart are kept at a warm, constant temperature despite wide fluctuations in environmental temperature.

Maintaining homeostasis is accomplished by a variety of automatic mechanisms. In the case of temperature regulation, these mechanisms include sweating during hot weather (Fig. 1-4), metabolizing more food in cold weather, and behaviors such as basking in the sun or even adjusting the thermostat in the room. Of course, not everything stays the same throughout an organism's life. Major changes, such as growth and reproduction, occur, but these are not failures of homeostasis. Rather, they are specific, genetically programmed parts of the organism's life cycle.

Living Things Acquire and Use Materials and Energy

Organisms need materials and energy to maintain their high level of complexity and organization, to grow, and to reproduce (Fig. 1-5). Organisms acquire the atoms and molecules they need from air, water, or soil or from other living things. These materials, called **nutrients**, are extracted from the environment and incorporated into the molecules of the organisms' bodies. The sum total of all the chemical reactions needed to sustain life is called **metabolism**.

Organisms obtain **energy**—the ability to do work, including carrying out chemical reactions, growing leaves in the spring, or contracting a muscle—in one of two basic ways. (1) Plants and some single-celled organisms capture the energy of sunlight and store it in energy-rich sugar molecules, a process called **photosynthesis**. (2) In contrast, neither fungi nor animals can photosynthesize, nor can most bacteria; these organisms must consume the energy-rich molecules contained in the bodies of other organisms. In either case, acquired energy is converted into a form that the organism can use or store for future use.

Ultimately, the energy that sustains nearly all life comes from sunlight, captured by photosynthetic organisms and incorporated into energy-rich molecules. Organisms that cannot photosynthesize depend on photosynthetic life-forms for food, either directly or indirectly. Thus, energy flows from the sun through nearly all forms of life. That energy is eventually released again as heat.

Living Things Grow

At some time in its life cycle, every organism becomes larger—that is, it *grows*. This characteristic is obvious for plants, birds, and mammals, all of which start out very small and undergo tremendous growth during their lives. Even single-celled bacteria, however, grow to about double their original size before they divide. In all cases, growth involves the conversion of materials acquired from the environment into the specific molecules of the organism's body.

Figure 1-5 Living things acquire energy and nutrients from the environment
The green plants seen here capture energy from the sun and nutrients from the air, water, and soil. The insect gains both energy and nutrients from the plants, while the toad will extract energy and nutrients from its insect prey. Without a continuous supply of solar energy, nearly all life would cease.

Figure 1-6 Living things reproduce
As they grow, these polar bear cubs will resemble, but not be identical to, their parents. The similarity and variability of offspring are crucial to the evolution of life.

Living Things Reproduce Themselves

The *continuity of life* occurs because organisms reproduce, giving rise to offspring of the same type (Fig. 1-6). The processes for producing offspring vary, but the result—the perpetuation of the parents' genetic material—is the same. The *diversity of life* occurs in part because offspring, though arising from the genetic material provided by their parents, are normally somewhat different from their parents, as explained briefly below and in Unit III. The mechanism by which traits are passed from one generation to the next, through a "genetic blueprint," produces these variable offspring.

DNA Is the Molecule of Heredity

All known forms of life use a molecule called **deoxyribonucleic acid**, or **DNA**, as the repository of hereditary information (Fig. 1-7). Genes are segments of the DNA molecule. Much of Unit II will be devoted to exploring the structure and function of this remarkable molecule. For now, we will simply note that an organism's DNA is its genetic blueprint or molecular instruction manual, a guide to both the construction and, at least in part, the operation of its body. When an organism reproduces, it passes a copy of its DNA to its offspring. The accuracy of the DNA copying process is astonishingly high: Only about one mistake occurs for every billion bits of information contained in the DNA molecule. But chance accidents to the genetic material also bring about changes in the DNA. The occasional errors and accidental changes, called **mutations**, produce variety. Without mutations, all life-forms might be identical. Indeed, there is reason to hypothesize that, without mutations, there would be no life. Variations, caused by mutations and superimposed on a background of overall genetic fidelity, make possible the final property of life, the capacity to evolve.

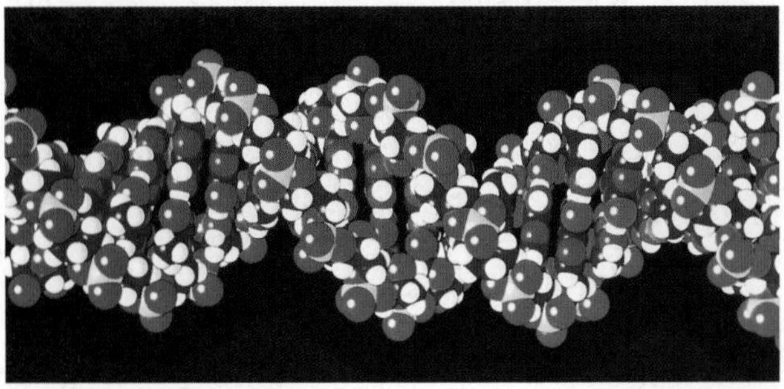

Figure 1-7 DNA
A computer-generated model of DNA, the molecule of heredity.

Living Things As a Whole Have the Capacity to Evolve

Although the genetic makeup of a single organism remains essentially the same over its lifetime, the genetic composition of a species as a whole changes over many generations. Over time, mutations and variable offspring provide diversity in the genetic material of a species. In other words, the species *evolves*. The scientific theory of **evolution** states that modern organisms descended, with modification, from preexisting life-forms, and that ultimately, all forms of life on Earth share a common ancestor. The most important force in evolution is **natural selection**, the process by which organisms with *adaptations* (characteristics that help them cope with the rigors of their environment) survive and reproduce more successfully than do others that lack those traits. Adaptive traits arising from genetic mutation are passed on to the next generation.

2 How Do Scientists Categorize the Diversity of Life?

Although all living things share the general characteristics discussed earlier, evolution has produced an amazing variety of life-forms. In the following brief description of the features used to classify organisms, we introduce you to the diversity of life on Earth. We describe the classification and structures of organisms in detail in Chapters 18 through 22.

Organisms can be grouped into three major categories, called **domains**: (1) Bacteria, (2) Archaea, and (3) Eukarya. This classification reflects fundamental differences among the cell types that comprise these organisms. Members of both the Bacteria and the Archaea normally consist of single, simple cells. Members of the Eukarya have bodies composed of one or more highly complex cells and are subdivided into four **kingdoms**: Protista, Fungi, Plantae, and Animalia (Fig. 1-8). There are exceptions to any simple set of criteria used to characterize the domains and kingdoms, but three characteristics are particularly useful: cell type, the number of cells in each organism, and the mode of nutrition—that is, energy acquisition (Table 1-1, p. 8).

(a)

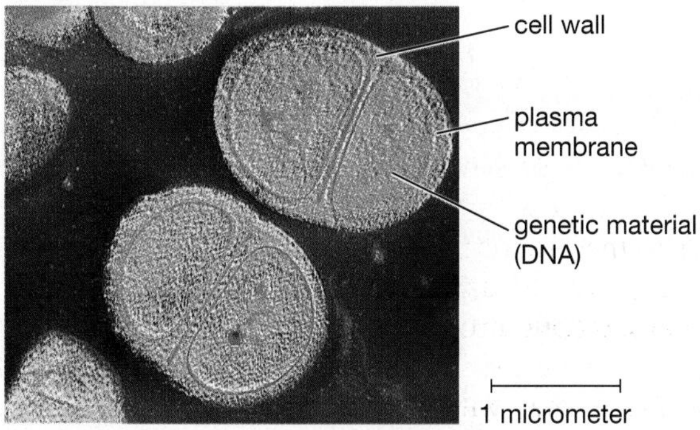

The domain Bacteria. A color-enhanced electron micrograph of a dividing bacterium. Bacteria are unicellular and prokaryotic; most are surrounded by a thick cell wall. Some bacteria photosynthesize, but most absorb food from their surroundings.

(b)

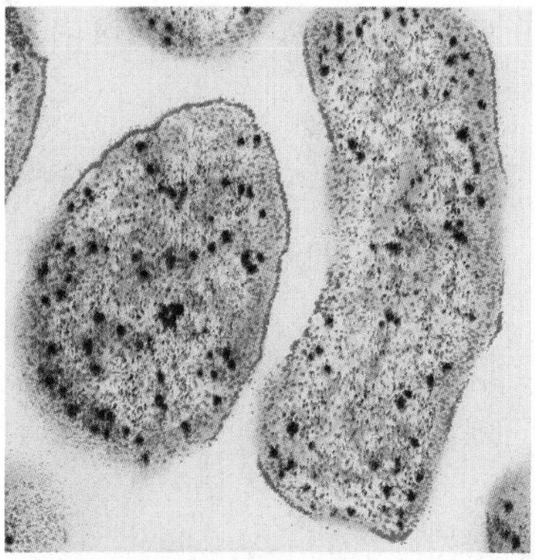

The domain Archaea. A color-enhanced electron micrograph of an archaean. The cell wall appears red, and DNA is scattered inside. Many archaeans can survive extreme conditions. This Antarctic species lives at temperatures as low as –2.5°C.

(c)

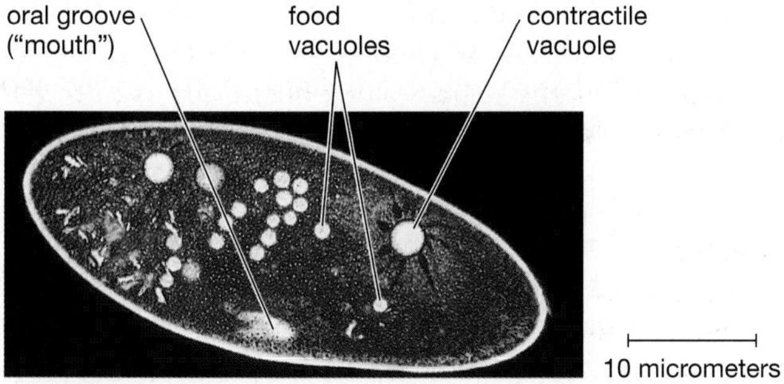

The kingdom Protista (domain Eukarya). This light micrograph of a *Paramecium* illustrates the complexity of these large, normally single, eukaryotic cells. Some protists photosynthesize, but others ingest or absorb their food. Many, including *Paramecium*, are mobile, moving with cilia or flagella.

(d)

The kingdom Fungi (domain Eukarya). An exotic mushroom found in Peru. Most fungi are multicellular. Fungi generally absorb their food, which is usually the dead bodies or wastes of plants and animals. The food is digested by enzymes secreted outside the fungal body. Most fungi cannot move.

(e)

The kingdom Plantae (domain Eukarya). This butterfly weed represents the flowering plants, the dominant members of the kingdom Plantae. Flowering plants owe much of their success to mutually beneficial relationships with animals, such as these pearl crescent butterflies, in which the flower provides food and the insect carries pollen from flower to flower, fertilizing them. Plants are multicellular, nonmotile eukaryotes that acquire nutrients by photosynthesis.

(f)

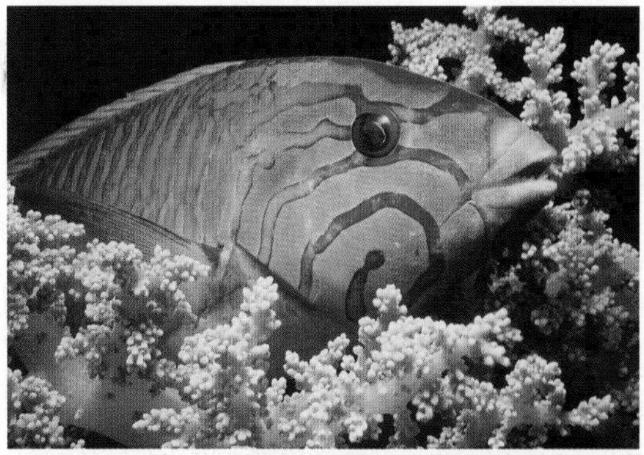

The kingdom Animalia (domain Eukarya). A wrasse rests on a soft coral. Animals are multicellular; animal bodies consist of a wide assortment of tissues and organs composed of specialized cell types. Most animals can move and respond rapidly to stimuli. The coral is a member of the largest group of animals: the invertebrates, which lack a backbone. This group also includes insects and mollusks. The wrasse is a vertebrate; like humans, it has a backbone.

Figure 1-8 The domains and kingdoms of life

Table 1-1 Some Characteristics Used in Classification of Organisms

Domain	Kingdom	Cell Type	Cell Number	Major Mode of Nutrition
Bacteria	(Under discussion)	Prokaryotic	Unicellular	Absorption, photosynthesis
Archaea	(Under discussion)	Prokaryotic	Unicellular	Absorption
Eukarya	Protista	Eukaryotic	Unicellular	Absorption, ingestion, or photosynthesis
	Fungi	Eukaryotic	Multicellular	Absorption
	Plantae	Eukaryotic	Multicellular	Photosynthesis
	Animalia	Eukaryotic	Multicellular	Ingestion

The Domains Bacteria and Archaea Consist of Prokaryotic Cells; the Domain Eukarya Is Composed of Eukaryotic Cells

There are two fundamentally different types of cells: (1) **prokaryotic** and (2) **eukaryotic**. *Karyotic* refers to the **nucleus** of a cell: a membrane-enclosed sac containing the cell's genetic material (see Fig. 1-3). *Eu* means "true" in Greek; eukaryotic cells possess a "true," membrane-enclosed nucleus. Eukaryotic cells are generally larger than prokaryotic cells and contain a variety of other organelles, many surrounded by membranes. Prokaryotic cells do not have a nucleus; their genetic material resides in their cytoplasm. They are usually small—only 1 or 2 micrometers long—and lack membrane-bound organelles. *Pro* means "before" in Greek; prokaryotic cells almost certainly evolved before eukaryotic cells (and, as we will see in Chapter 17, eukaryotic cells almost certainly evolved from prokaryotic cells). Bacteria and Archaea consist of prokaryotic cells; the cells of all the kingdoms of Eukarya are eukaryotic.

Bacteria, Archaea, and Members of the Kingdom Protista Are Mostly Unicellular; Members of the Kingdoms Fungi, Plantae, and Animalia Are Primarily Multicellular

Most members of the domains Bacteria and Archaea and members of the kingdom Protista from the domain Eukarya are single-celled, or **unicellular**, although a few live in strands or mats of cells with little communication, cooperation, or organization among them. Most members of the kingdoms Fungi, Plantae, and Animalia are many-celled, or **multicellular**; their lives depend on intimate communication and cooperation among specialized cells.

Members of the Different Kingdoms Have Different Ways of Acquiring Energy

All organisms need energy to live. Photosynthetic organisms capture energy from sunlight and store it in molecules such as sugars and fats. These organisms, including plants, some bacteria, and some protists, are therefore called **autotrophs**, meaning "self-feeders." Organisms that cannot photosynthesize must acquire energy prepackaged in the molecules of the bodies of other organisms; hence, these organisms are called **heterotrophs**, meaning "other-feeders." Many archaea, bacteria, and protists and all fungi and animals are heterotrophs. Heterotrophs differ in the size of the food they eat. Some, such as bacteria and fungi, absorb individual food molecules; others, including most animals, eat whole chunks of food and break them down to molecules in their digestive tracts (*ingestion*).

3 What Is the Science of Biology?

Biology is a science, and its principles and methods are the same as those of any other science. In fact, a basic tenet of modern biology is that living things obey the same laws of physics and chemistry that govern nonliving matter.

Scientific Principles Underlie All Scientific Inquiry

All scientific inquiry, including biology, is based on a small set of assumptions. Although we can never absolutely prove these assumptions, they have been so thoroughly tested and validated that we might call them scientific principles. These are the principles of *natural causality*, *uniformity in space and time*, and *common perception*.

Natural Causality Is the Principle That All Events Can Be Traced to Natural Causes

The first principle of science is natural causality. Over the course of human history, two approaches have been taken to the study of life and other natural phenomena. The first assumes that some events happen through the intervention of supernatural forces beyond our understanding. The ancient Greeks believed that the god Zeus hurled thunderbolts from the sky and that the god Poseidon caused earthquakes and storms at sea. In contrast, science adheres to the principle of **natural causality**: All events can be traced to natural causes that are potentially within our ability to comprehend. For example, epilepsy was once thought to be the result of a visitation from the gods. Today we realize that epilepsy is a disease of the brain in which groups of nerve cells are uncontrollably activated.

The principle of natural causality has an important corollary: The evidence we gather about the causes of natural events has not been deliberately distorted to fool us. This corollary may seem obvious, yet not so very long ago some people argued that fossils are not evidence of evolution but were placed on Earth by God as a test of our faith. If we cannot trust the evidence provided by nature, then the entire enterprise of science is futile.

The Natural Laws That Govern Events Apply Everywhere and for All Time

A second fundamental principle of science is that natural laws—laws derived from the study of nature—are uniform in space and time and do not change with distance or time. The laws of gravity, the behavior of light, and the interactions of atoms, for example, are the same today as they were a billion years ago and will hold true in Moscow just as well as in New York, or even on Mars. Uniformity in space and time is especially vital to biology, because many events of great importance to biology, such as the evolution of today's diversity of living things, happened before humans were around to observe them. Some people believe that each of the different types of organisms was individually created at one time in the past by the direct intervention of God, a philosophy called **creationism**. As scientists, we freely admit that we cannot disprove this idea. Creationism, however, is contrary to both natural causality and uniformity in time. The overwhelming success of science in explaining natural events through natural causes has led nearly all scientists to reject creationism.

Scientific Inquiry Is Based on the Assumption That People Perceive Natural Events in Similar Ways

A third basic assumption of science is that, as a general rule, all human beings perceive natural events in fundamentally the same way and that these perceptions provide us with reliable information about the natural world. Common perception is, to some extent, a principle peculiar to science. Value systems, such as those involved in the appreciation of art, poetry, and music, do not assume common perception. We may perceive the colors in a painting in a similar way (the scientific aspect of art), but we do not perceive the aesthetic value of the painting identically (the humanistic aspect of art). Values also differ radically among people, commonly owing to their culture or religious beliefs. Because value systems are subjective, not objective, science cannot solve certain types of philosophical or moral problems, such as the morality of abortion.

The Scientific Method Is the Basis for Scientific Inquiry

Given these assumptions, how do biologists study the workings of life? Scientific inquiry is a rigorous method for making observations of specific phenomena and searching for the order underlying those phenomena. Ideally, biology and the other sciences use the **scientific method**, which consists of four interrelated operations: (1) *observation*; (2) *hypothesis*; (3) *experiment*; and (4) *conclusion*. All scientific inquiry begins with an **observation** of a specific phenomenon. The observation, in turn, leads to questions, such as "How did this come about?" Then in a flash of insight, or more typically after long, hard thought, a hypothesis is formulated. A **hypothesis** is a supposition based on previous observations that is offered as an explanation for the observed phenomenon. To be useful, the hypothesis must lead to predictions that can be tested by additional controlled observations, or **experiments**. These experiments produce results that either support or refute the hypothesis, and a **conclusion** is drawn about its validity. A single experiment is never an adequate basis for a conclusion; the results must be repeatable not only by the original researcher but also by others.

Simple experiments test the assertion that a single factor, or **variable**, is the cause of a single observation. To be scientifically valid, the experiment must rule out a variety of other possible variables as the cause of the observation. So scientists design **controls** into their experiments, in which all the variables remain constant. Controls are then compared with the experimental situation, in which only the variable being tested is changed. In the 1600s, Francesco Redi used the scientific method to test the hypothesis that flies do not arise spontaneously from rotting meat (see "Scientific Inquiry: Does Spoiled Meat Produce Maggots?").

The scientific method can be used not only to generate new knowledge, but also to solve everyday problems. Let us consider an everyday situation in which you can apply the scientific method. Late to an appointment, you rush to your car and make the *observation* that it won't start. Immediately, you form a *hypothesis*: The battery is dead. Quickly you design an *experiment*: You replace your battery with the battery from your roommate's new car and try to start your car again. The result seems to confirm your hypothesis, because your car starts immediately.

But wait! You haven't provided controls for several variables. Perhaps your battery was fine all along, and you just needed to try to start the car again. Or perhaps the battery cable was loose and simply needed to be tightened. Realizing the need for a good *control*, you replace your old battery, making sure the cables are secured tightly, and attempt to restart the car. If your car repeatedly refuses to start with the old battery but then starts immediately when you put in your roommate's

Scientific Inquiry
Does Spoiled Meat Produce Maggots?

The critical experiments of Italian physician Francesco Redi (1621–1697) beautifully demonstrate the scientific method and also help illustrate the principle of natural causality, on which modern science is based. Redi investigated why maggots appear on spoiled meat. Before Redi, the appearance of maggots was considered to be evidence of **spontaneous generation**, the production of living things from nonliving matter.

Redi *observed* that flies swarm around fresh meat and that maggots appear on meat left out for a few days. He formed a testable *hypothesis*: The flies produce the maggots. In his *experiment*, Redi wanted to test just one variable: the access of flies to the meat. Therefore, he took two clean jars and filled them with similar pieces of meat. He left one jar open (the *control* jar) and covered the other with gauze to keep out flies (the *experimental* jar). He did his best to keep all the other variables the same (for example, the type of jar, the type of meat, and the temperature). After a few days, he observed that maggots swarmed over the meat in the open jar, but no maggots appeared on the meat in the covered jar. Redi *concluded* that his hypothesis was correct and that maggots are produced by flies, not by the meat itself (Fig. E1-1). Only through controlled experiments could the age-old hypothesis of spontaneous generation be laid to rest.

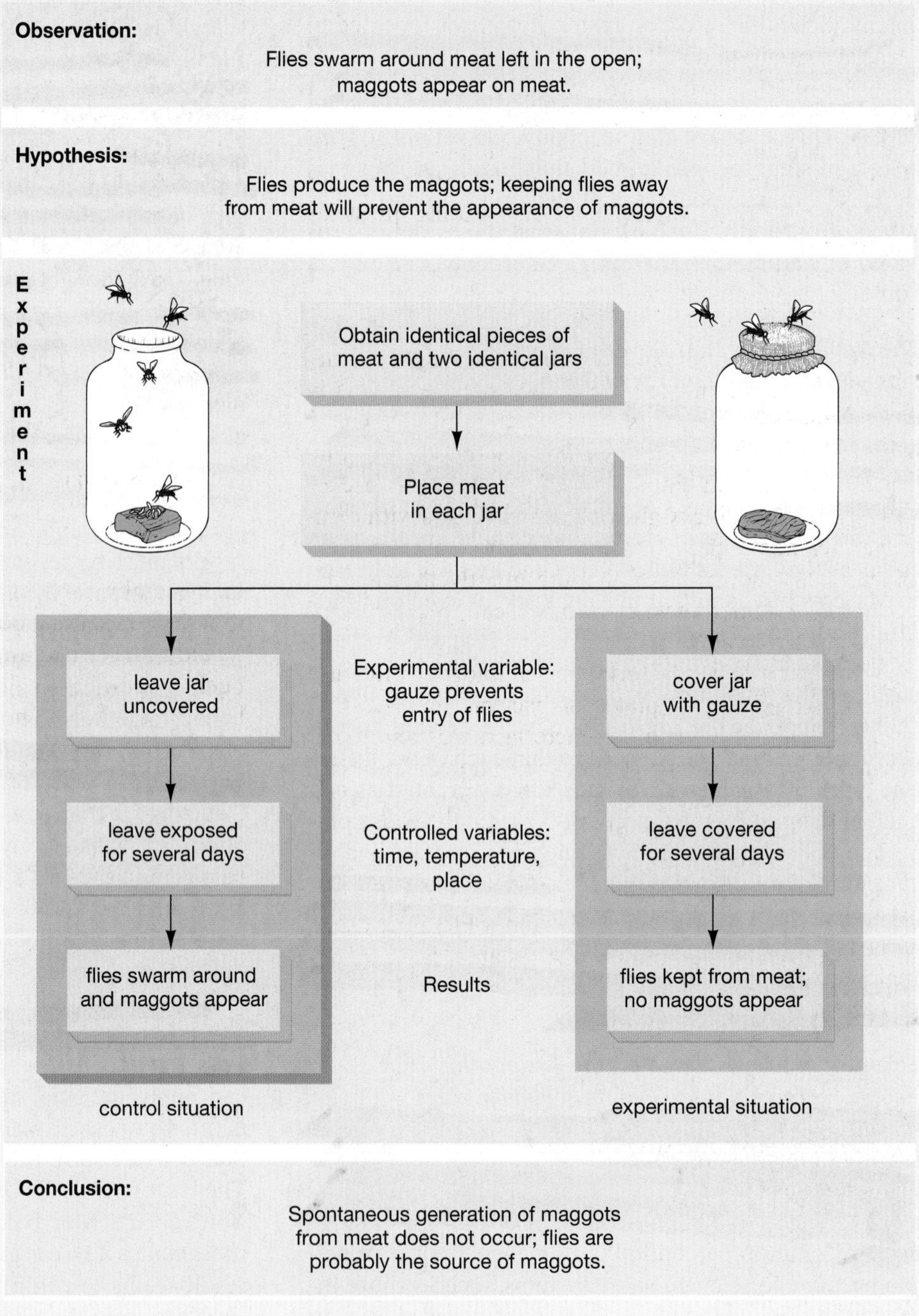
Observation:
Flies swarm around meat left in the open; maggots appear on meat.
Hypothesis:
Flies produce the maggots; keeping flies away from meat will prevent the appearance of maggots.
Experiment
Obtain identical pieces of meat and two identical jars
Place meat in each jar
leave jar uncovered
leave exposed for several days
flies swarm around and maggots appear
Experimental variable: gauze prevents entry of flies
Controlled variables: time, temperature, place
Results
cover jar with gauze
leave covered for several days
flies kept from meat; no maggots appear
control situation
experimental situation
Conclusion:
Spontaneous generation of maggots from meat does not occur; flies are probably the source of maggots.

Figure E1-1 The experiments of Francesco Redi

new battery, you have isolated a single *variable*, the battery. And (although you may have missed your appointment) you can now safely draw the *conclusion* that your old battery is dead.

The scientific method is powerful, but it is important to recognize its limitations. In particular, scientists can seldom be sure that they have controlled *all* the variables other than the one they are trying to study. Therefore, scientific conclusions must always remain tentative and are subject to revision if new observations or experiments demand it.

A final important element of science is *communication*. No matter how well designed an experiment is, it is useless if it is not communicated. Redi's experimental design and conclusions survive today because he recorded his methods and observations. If experiments are not communicated to other scientists in enough detail to be repeated, their conclusions cannot be verified. Without verification, scientific findings cannot be safely used as the basis for new hypotheses and further experiments. Scientific effort is wasted if it is not reported in clear and accurate detail.

A wonderful aspect of scientific inquiry is that whenever a scientist reaches a conclusion, the conclusion immediately raises further questions that lead to further hypotheses and more experiments (why did your battery die?). Science is a never-ending quest for knowledge.

Science Is a Human Endeavor

Scientists are real people. They are driven by the same ambitions, pride, and fears as other people, and they sometimes make mistakes. As you will read in Chapter 9, ambition played an important role in the discovery by James Watson and Francis Crick of the structure of DNA. Accidents, lucky guesses, controversies with competing scientists, and, of course, the intellectual powers of individual scientists contribute greatly to scientific advances. To illustrate what we might call "real science," let's consider an actual case.

When they study bacteria, microbiologists must use pure *cultures*—that is, plates of bacteria that are free from contamination by other bacteria, molds, and so on. Only by studying a single type at a time can we learn about the properties of that particular bacterium. Consequently, at the first sign of contamination, a culture is normally thrown out, often with mutterings about sloppy technique. On one such occasion, however, in the late 1920s, Scottish bacteriologist Alexander Fleming turned a ruined culture into one of the greatest medical advances in history.

One of Fleming's bacterial cultures became contaminated with a patch of a mold called *Penicillium*. Before throwing out the culture dish, Fleming observed that *no bacteria were growing near the mold* (Fig. 1-9). Why not? Fleming hypothesized that perhaps *Penicillium* releases a substance that kills off bacteria growing nearby. To test this hypothesis, Fleming grew some pure *Penicillium* in a liquid nutrient broth. He then filtered out the *Penicillium* mold and applied the liquid in which the mold had grown to an uncontaminated bacterial culture. Sure enough, something in the liquid killed the bacteria. Further research into these mold extracts resulted in the production of the first *antibiotic*: penicillin, a bacteria-killing substance that has since saved millions of lives. Fleming's experiments are a classic example of the use of scientific methodology. They began with an observation and proceeded to a hypothesis, followed by experimental tests of the hypothesis, which led to a conclusion. But the scientific method alone would have been useless without the

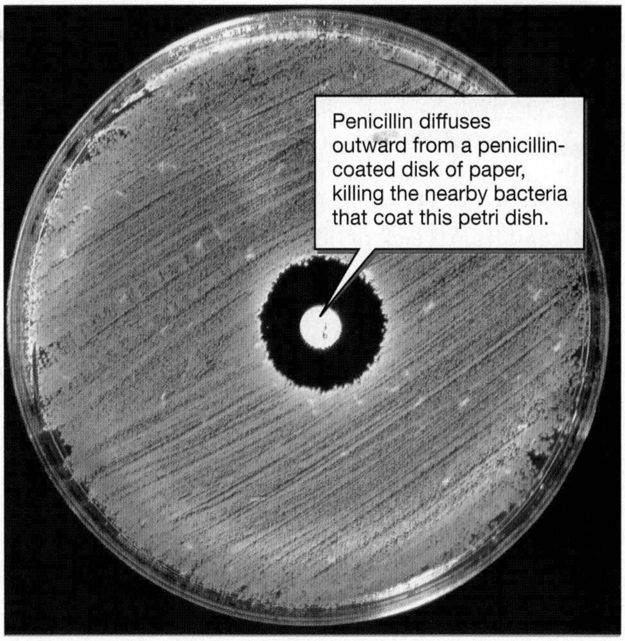

Penicillin diffuses outward from a penicillin-coated disk of paper, killing the nearby bacteria that coat this petri dish.

Figure 1-9 Penicillin kills bacteria

lucky combination of accident and a brilliant scientific mind. Had Fleming been a "perfect" microbiologist, he wouldn't have had any contaminated cultures. Had he been less observant, the contamination would have been just another spoiled culture dish. Instead, it was the beginning of antibiotic therapy for bacterial diseases. As French microbiologist Louis Pasteur said, "Chance favors the prepared mind."

Scientific Theories Have Been Thoroughly Tested

Scientists use the word *theory* in a way that is different from its everyday usage. If Dr. Watson were to ask Sherlock Holmes, "Do you have a theory as to the perpetrator of this foul deed?" in scientific terms, he would be asking Holmes for a hypothesis—an educated guess based on observable evidence, or clues. A **scientific theory** is far more general and more reliable than a hypothesis. Far from being an educated guess, a scientific theory is a general explanation of important natural phenomena, developed through extensive and reproducible observations. It is more like a *principle* or a *natural law* in common English usage. For example, scientific theories such as the atomic theory (that all matter is composed of atoms) and the theory of gravitation (that objects exert attraction for one another) are fundamental to the science of physics. Likewise, the *cell theory* (that all living things are composed of cells) and the theory of evolution are fundamental to the study of biology. Scientists describe fundamental principles as "theories" because a basic premise of scientific inquiry is that it must be performed with an open mind. If compelling evidence arises, a theory will be modified.

Scientific theories arise through *inductive reasoning.* **Inductive reasoning** is the process of creating a generalization based on many specific observations that support the generalization, coupled with an absence of observations that contradict it. Simplistically, the theory that Earth exerts gravitational forces on objects arises from repeated observations of objects falling down toward Earth and from a lack of observations of objects "falling up" that would contradict the theory. Likewise, the cell theory arises from the observation that all organisms that have the attributes of life are composed of one or more cells, and that nothing that is not composed of cells shares all these attributes. Once a scientific theory has been formulated, it can be used to support *deductive reasoning*. In science, **deductive reasoning** is the process of generating hypotheses about how a specific experiment or observation will turn out, based on a well-supported generalization, such as a scientific theory. For example, based on the cell theory, if a new organism is found that shares all the attributes of life, scientists can confidently deduce or hypothesize that it will be composed of cells. Of course, the new organism should be carefully scrutinized under the microscope for its cellular structure because, as we said earlier, if compelling new evidence arises, a theory can be modified.

One of the most important theories in biology is evolution. Ever since its formulation in the mid-1800s by two English naturalists, Charles Darwin and Alfred Russel Wallace, the theory of evolution has been supported by fossil finds, geological studies, radioactive dating of rocks, genetics, molecular biology, biochemistry, and breeding experiments. People who refer to evolution as "just a theory" profoundly misunderstand what scientists mean by the word *theory*.

4 Evolution: The Unifying Theory of Biology

Evolution is the unifying theory that explains the origin of diverse forms of life as a result of changes in their genetic makeup. As we noted earlier, the theory of evolution states that modern organisms descended, with modification, from preexisting life-forms. In the words of biologist Theodosius Dobzhansky, "Nothing in biology makes sense, except in the light of evolution." Why don't snakes have legs? Why are there dinosaur fossils but no living dinosaurs? Why are monkeys so like us, not only in appearance, but also in the structure of their genes and proteins? How are the diverse life-forms you saw on your walk across campus related? The answers to those questions, and thousands more, lie in the processes of evolution (which we shall examine in detail in Chapters 14 through 17). Evolution is so vital to our understanding and appreciation of biology that we must briefly review its important principles before going further.

Three Natural Processes Underlie Evolution

In the mid-1800s, Darwin and Wallace formulated the theory of evolution that is still the basis of our modern understanding. Evolution arises as a consequence of three natural processes: (1) *genetic variation* among members of a population; (2) *inheritance* of those variations by offspring of parents who carry the variation; and (3) *natural selection*, the survival and enhanced reproduction of organisms with favorable variations.

Much of the Variability Among Organisms Is Inherited

Look around at your classmates and notice how different they are. Although some of this variation is due to differences in environment and lifestyles, it is mainly influenced by our genes. Most of us could pump iron for the rest of our lives and never develop a body like Arnold Schwarzenegger's. From where does genetic variation arise? The genetic instructions—the genes—of all organisms are segments of molecules of deoxyribonucleic acid, or DNA. Occasionally, the DNA suffers an accident; perhaps radiation strikes the DNA molecule just so, causing a mutation and thereby altering its information content. Mistakes in the copying of DNA during reproduction, although rare, also occur. Such mutations, or changes in a gene, may affect the organism's appearance or ability to function. Many mutations have no effect or are harmless; some make the organism less able to function. But in rare cases mutations may improve an organism's ability to function. As a result of mutations, many of which occurred millions of years ago and have been passed from parent to offspring through countless generations, members of the same species tend to be slightly different from one another.

Natural Selection Tends to Preserve Genes That Help an Organism Survive and Reproduce

On the average, organisms that best meet the challenges of their environment will leave the most offspring. The offspring will inherit the genes that made their parents successful. Natural selection thus preserves genes that help organisms flourish in their environment. For example, a mutated gene providing instructions for larger teeth in beavers will be passed from parent to offspring. These offspring will be able to chew down trees more efficiently, build bigger dams and lodges, and eat more bark than can "ordinary" beavers. Because these big-toothed beavers will obtain more food and better shelter than their smaller-toothed relatives, they will probably raise more offspring. The offspring will inherit their parents' genes for larger teeth. Over time, less-successful, smaller-toothed beavers will become increasingly scarce; after many generations, all beavers will have large teeth.

Structures, physiological processes, or behaviors that aid in survival and reproduction in a particular environment are called **adaptations**. Most of the features that we admire so much in our fellow life-forms, such as the long limbs of deer, the wings of eagles, and the mighty columns of redwood trunks, are adaptations molded by millions of years of mutation and natural selection.

In the long run, however, what helps an organism survive today can become a liability tomorrow. If environments change—for example, as ice ages come and go—then the genetic makeup that best adapts organisms to their environment will also change over time. When new mutations happen to occur that increase the fitness of an organism in the altered environment, these mutations in their turn will spread throughout the population.

Over millennia, the interplay of environment, genetic variation, and natural selection results in evolution: the modification of the genetic makeup of species. In environments that are reasonably constant through time, such as the oceans, some well-adapted forms persist relatively unchanged and are sometimes called "living fossils." For example, sharks (Fig. 1-10) have retained essentially the same body form for tens of millions of years, as their sleek shape, powerful tail, acute sense of smell, and formidable teeth have made them superb predators.

In changing environments, some species do not experience the genetic changes that allow them to adapt. The rate of environmental change outstrips the rate of genetic changes, and those species go *extinct*—that is, no members of those species remain. The dinosaurs (Fig. 1-11) were mighty reptiles that could not withstand changing conditions 65 million years ago. Other species experience chance mutations that adapt them to meet new challenges. The mutation that produced the first stout, fleshy fins in what came to be known as lobefin fishes, for instance, enabled those fish to crawl on the bottom of shallow waters and ultimately to make the "jump" to life on land.

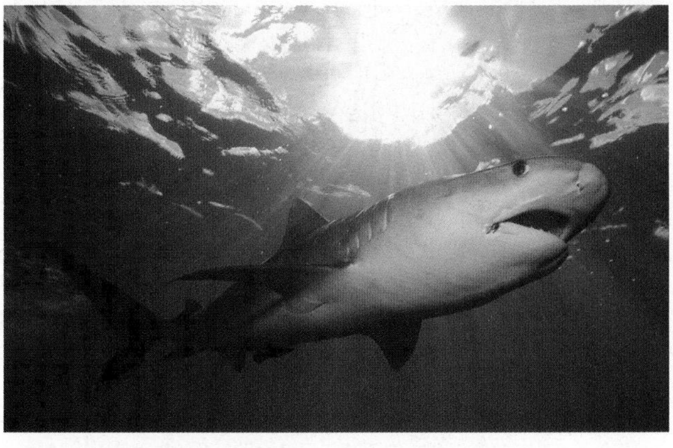

Figure 1-10 Ancient adaptations
This tiger shark, photographed in the clear waters of the Bahamas, possesses features that have characterized sharks for tens of millions of years: streamlined shape; a long, powerful tail; an acute sense of smell; and rows of sharp teeth.

Figure 1-11 A fossil of *Triceratops*
This *Triceratops* died in what is now Montana about 70 million years ago. No one is certain what caused the extinction of the dinosaurs, but we do know that the evolution of new adaptations in dinosaurs was unable to keep pace with changes in their habitat.

The result of evolution is a tremendous variety of species. Within particular habitats, these species have evolved complex interrelationships with one another and with their nonliving surroundings. The diversity of species and the complex interrelationships that sustain them are encompassed by the term **biodiversity**. In recent decades, the rate of environmental change has been drastically accelerated by a single species, *Homo sapiens* (modern humans). Few species are able to adapt to this rapid change. In habitats most affected by humans, many species are being driven to extinction. This concept is explored further in "Earth Watch: Why Preserve Biodiversity?"

5 How Does Knowledge of Biology Illuminate Everyday Life?

Some people regard science as a "dehumanizing" activity, feeling that too deep an understanding of the world robs us of vision and awe. Nothing could be further from the truth, as we repeatedly discover anew in our own lives. Years ago, we watched a bee foraging at a spike of lupine flowers. Members of the pea family, lupines have a complicated structure, with two petals on the lower half of the flower enclosing the pollen-laden male reproductive structures (*stamens*) and sticky pollen-capturing female reproductive structures (*stigma*). We had recently learned that in young lupine flowers (Fig. 1-12a), the weight of a bee pushing on these petals compresses the stamens, pushing pollen onto the bee's abdomen (Fig. 1-12b). In older flowers, the stigma protrudes through the lower petals; when a pollen-dusted bee visits, it usually leaves behind a few grains of pollen.

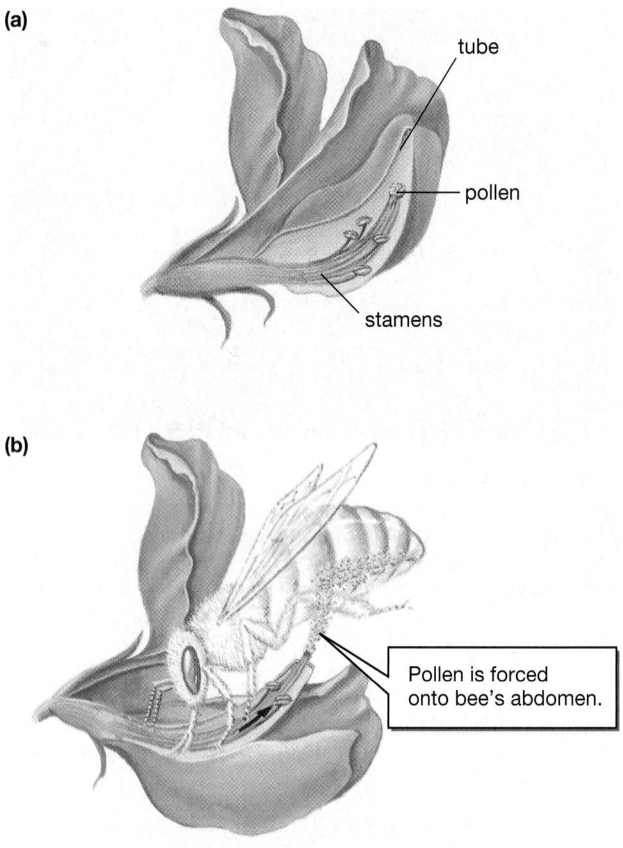

Figure 1-12 Complex adaptations help ensure pollination Lupines, like many other members of the pea family, have complex flowers. ***(a)*** The reproductive structures are enclosed within the lower petals. In young lupine flowers, the lower petals form a tube within which the stamens fit snugly. The stamens shed pollen within the tube. ***(b)*** When the weight of a foraging bee pushes on the lower petals, the stamens are thrust forward, and pollen is forced out the tube's end onto the bee's abdomen. Some pollen adheres to the abdomen and may come off on the sticky, pollen-receiving stigma of the next flower that the bee visits.

Did our newfound insights into the functioning of lupine flowers detract from our appreciation of them? Far from it. Rather, we now looked on lupines with new delight, understanding something of the interplay of form and function, bee and flower, which shaped the evolution of the lupine. A few months later we ventured atop Hurricane Ridge in Olympic National Park in Washington State, where the alpine meadows burst with color in August (Fig. 1-13). As we crouched beside a wild lupine, an elderly man stopped to ask what we were looking at so intently. He listened with interest as we explained the structure to him; he then went off to another patch of lupines to watch the bees foraging. He too felt the increased sense of wonder that comes with understanding.

We try to convey to you that dual sense of understanding and wonder throughout this text. We also emphasize that biology is not a completed work but an exploration that we have really just begun. As Lewis Thomas, a physician and natural philosopher, eloquently stated: "The only solid piece of scientific truth about which I feel totally confident is that we are profoundly ignorant about nature. Indeed, I regard this as the major discovery of the past hundred years of biology . . . but we are making a beginning."

We cannot urge you strongly enough, even if you are not contemplating a career in biology, to join in the journey of biological discoveries throughout your life. Don't think of biology as just another course to take, just another set of facts to memorize. Biology can be much more than that. It can be a pathway to a new understanding of yourself and of the life on Earth around you.

Figure 1-13 Wild lupines and subalpine fir trees Thousands of people visit Hurricane Ridge in Washington State's Olympic National Park each summer to gaze in awe at Mt. Olympus, but few bother to investigate the wonders at their feet.

Earth Watch
Why Preserve Biodiversity?

"The loss of species is the folly our descendants are least likely to forgive us."
E. O. Wilson, Professor, Harvard University

Ever since the United Nations' 1992 "Earth Summit" in Rio de Janeiro, Brazil, the word *biodiversity* has jumped out at us from magazines and news articles. What is biodiversity, and why should we be concerned with preserving it? *Biodiversity* refers to the total number of species within an ecosystem and to the resulting complexity of interactions among them; in short, it defines the "richness" of an ecological community.

Over the 3.5-billion-year history of life on Earth, evolution has produced an estimated 8 million to 10 million unique and irreplaceable species. Of these, scientists have named only about 1.4 million, and only a tiny fraction of this number has been studied. Evolution has not, however, merely been churning out millions of independent species. Over thousands of years, organisms in a given area have been molded by forces of natural selection exerted by other living species as well as by the nonliving environment in which they live. The outcome is the community, a highly complex web of interdependent life-forms whose interactions sustain one another. By participating in the natural cycling of water, oxygen, and other nutrients, and by producing rich soil and purifying wastes, these communities contribute to the sustenance of human life as well. The concept of biodiversity has emerged as a result of our increasing concern over the loss of countless forms of life and the habitat that sustains them.

The Tropics are home to the vast majority of all the species on Earth, perhaps 7 million to 8 million of them, living in complex communities. The rapid destruction of habitats in the Tropics, from rain forests to coral reefs, as a result of human activities is producing high rates of extinction of many species (Fig. E1-2). Most of these species have never been named, and others never even discovered. Aside from ethical concerns over eradicating irreplaceable forms of life, as we drive unknown organisms to extinction, we lose potential sources of medicine, food, and raw materials for industry.

For example, a wild relative of corn that is not only very disease-resistant but also *perennial* (that is, lasts more than one growing season) was found growing only on a 25-acre plot of land in Mexico that was scheduled to be cut and burned within a week of the discovery. The genes of this plant might one day enhance the disease-resistance of corn or create a perennial corn plant. The rosy periwinkle, a flowering plant found in the tropical forest of the island of Madagascar (off the eastern coast of Africa) produces two substances that are now widely marketed for the treatment of leukemia and Hodgkin's disease, a cancer of the lymphatic organs. Only about 3% of the world's flowering plants have been examined for substances that might fight cancer or other diseases. Closer to home, loggers of the Pacific Northwest frequently cut and burned the Pacific yew tree as a "nuisance species" until the active ingredient that has since gone into making the anticancer drug Taxol® was discovered in its bark. Animals too have proven useful in fighting cancer: In 1997 researchers isolated a potent anticancer compound from a species of coral that dwells in the Indian Ocean.

Figure E1-2 Biodiversity threatened
Destruction of tropical rain forests by indiscriminate logging threatens Earth's greatest storehouse of biological diversity. Interrelationships such as those that have evolved between this *Heliconia* flower and its hummingbird pollinator sustain these diverse communities and are threatened by human activities.

Many conservationists are also concerned that as species are eliminated, either locally or through total extinction, the communities of which they were a part might change and become less stable and more vulnerable to damage by diseases or adverse environmental conditions. Some experimental evidence supports this viewpoint, but the interactions within communities are so complex that these hypotheses are difficult to test. Clearly, some species have a much larger role than do others in preserving the stability of a given ecosystem. Which species are most crucial in each ecosystem? No one knows. Human activities have increased the natural rate of extinction by a factor of at least 100 and possibly by as much as 1000 times the pre-human rate. By reducing biodiversity to support increasing numbers of humans and wasteful standards of living, we have ignorantly embarked on an uncontrolled global experiment, using planet Earth as our laboratory. In their book *Extinction* (1981), Stanford ecologists Paul and Anne Ehrlich compare the loss of biodiversity to the removal of rivets from the wing of an airplane. The rivet-removers continue to assume that there are far more rivets than needed, until one day, upon takeoff, they are proven tragically wrong. As human activities drive species to extinction while we have little knowledge of the role each plays in the complex web of life, we run the risk of removing "one rivet too many."

CASESTUDYREVISITED

The Life Around Us

As you walk to class observing life, think about the "why" behind the "what" that you see. The plants on campus are green: This color allows a unique molecule, chlorophyll, to trap specific wavelengths of solar energy and use them to power the life of the plant. Bees and butterflies are gathering nectar containing sugar; the sugar was synthesized using energy trapped by chlorophyll and the simple ingredients of carbon dioxide and water. Why have plants evolved in ways that place this energy-rich nectar where it can be "stolen" by insects? Look more carefully at a bee. You may see yellow pollen clinging to its legs or to the hairs coating its body. The plants are "using" the insects to fertilize each other, and both are benefiting. Although the campus groundskeepers planted the flowers for you to enjoy, showy flowers evolved as a signal to insects saying "nectar is here—come and get it." With each breath you take, you are inhaling a "waste product" from the plants around you: oxygen. Wherever you look, if you look in the right way, you'll see evidence of the interdependence of living things, and you'll never take life on Earth for granted.

Identify two different types of organisms that you have seen interacting, such as bees and flowers. Now form a simple hypothesis about this interaction. Use the scientific method and your imagination to design an experiment that tests this hypothesis. Be sure to identify variables and control for them.

Summary of Key Concepts

1) What Are the Characteristics of Living Things?

Organisms possess the following characteristics: Their structure is complex and organized; they maintain homeostasis; they grow; they acquire energy and materials from the environment; they respond to stimuli; they reproduce; and they have the capacity to evolve.

2) How Do Scientists Categorize the Diversity of Life?

Organisms can be grouped into three major categories, called domains: Archaea, Bacteria, and Eukarya. Within the Eukarya are four kingdoms: Protista, Fungi, Plantae, and Animalia. Some features used to classify organisms are the type of cell(s) the organism possesses, the number of cells in each organism, and the mode of acquiring energy: *Cell Type*: The genetic material of eukaryotic cells is enclosed within a membrane-bound nucleus. Prokaryotic cells do not have a nucleus. *Cell Number*: Organisms may consist of a single cell (unicellular) or of many cells bound together and working cooperatively (multicellular). *Energy Acquisition*: Most autotrophic organisms obtain energy by capturing and storing the energy of sunlight in energy-rich molecules by means of photosynthesis. Heterotrophic organisms normally obtain energy by eating energy-rich molecules (food) synthesized in the bodies of other organisms. The food may be eaten in large chunks and broken down (ingestion) or may be absorbed molecule by molecule from the environment (absorption). The features of the domains and kingdoms are summarized in Table 1-1.

3) What Is the Science of Biology?

Biology is based on the scientific principles of natural causality, uniformity in space and time, and common perception. These principles are assumptions that cannot be directly proved but that are validated by experience. Knowledge in biology is acquired through the application of the scientific method. First, an observation is made. Then a hypothesis is formulated that suggests a natural cause for the observation. The hypothesis is used to predict the outcome of further observations or experiments. A conclusion is then drawn about the validity of the hypothesis. A scientific theory is a general explanation of natural phenomena, developed through extensive and reproducible experiments and observations.

4) Evolution: The Unifying Theory of Biology

Evolution is the theory that modern organisms descended, with modification, from preexisting life-forms. Evolution occurs as a consequence of (1) genetic variation among members of a population, caused by mutation; (2) inheritance of those variations by offspring; and (3) natural selection of the variations that best adapt an organism to its environment.

5) How Does Knowledge of Biology Illuminate Everyday Life?

The more you know about living things, the more fascinating they become!

Key Terms

adaptation *p. 13*
atom *p. 2*
autotroph *p. 8*
biodiversity *p. 13*
biosphere *p. 4*
cell *p. 4*
community *p. 4*
conclusion *p. 9*
control *p. 9*
creationism *p. 9*
cytoplasm *p. 4*
deductive reasoning *p. 12*
deoxyribonucleic acid (DNA) *p. 6*
domain *p. 6*
ecosystem *p. 4*
element *p. 2*
emergent property *p. 2*
energy *p. 5*
eukaryotic *p. 8*

evolution *p. 6*
experiment *p. 9*
gene *p. 4*
heterotroph *p. 8*
homeostasis *p. 4*
hypothesis *p. 9*
inductive reasoning *p. 12*
kingdom *p. 6*
metabolism *p. 5*
molecule *p. 4*
multicellular *p. 8*
mutation *p. 6*
natural causality *p. 8*
natural selection *p. 6*
nucleus *p. 8*
nutrient *p. 5*
observation *p. 9*
organ *p. 4*
organelle *p. 4*
organic molecule *p. 4*
organism *p. 4*
organ system *p. 4*
photosynthesis *p. 5*
plasma membrane *p. 4*
population *p. 4*
prokaryotic *p. 8*
scientific method *p. 9*
scientific theory *p. 11*
species *p. 4*
spontaneous generation *p. 10*
subatomic particle *p. 4*
tissue *p. 4*
unicellular *p. 8*
variable *p. 9*

Thinking Through the Concepts

Multiple Choice

1. *Which of the following is paired incorrectly?*
   a. organ—a structure formed of cells of similar types
   b. cell—the smallest unit of life
   c. genes—units of heredity
   d. cytoplasm—watery substance within cells that contains organelles
   e. plasma membrane—surrounds each cell
2. *A scientist examines an organism and finds that it is eukaryotic, heterotrophic, and multicellular and that it absorbs nutrients. She concludes that the organism is a member of the kingdom*
   a. Bacteria
   b. Protista
   c. Plantae
   d. Fungi
   e. Animalia
3. *Which statement is correct?*
   a. Eukaryotic cells are simpler than prokaryotic cells.
   b. *Heterotroph* means "self-feeder."
   c. Mutations are accidental changes in genes.
   d. A scientific theory is similar to an educated guess.
   e. Genes are proteins that produce DNA.
4. *Choose the answer that best describes the scientific method.*
   a. observation, hypothesis, experiment, absolute proof
   b. guess, hypothesis, experiment, conclusion
   c. observation, hypothesis, experiment, conclusion
   d. hypothesis, experiment, observation, conclusion
   e. experiment, observation, hypothesis, conclusion
5. *Which of the following are characteristics of living things?*
   a. They reproduce.
   b. They respond to stimuli.
   c. They are complex and organized.
   d. They acquire energy.
   e. all of the above
6. *The three natural processes that form the basis for evolution are*
   a. adaptation, natural selection, and inheritance
   b. predation, genetic variation, and natural selection
   c. mutation, genetic variation, and adaptation
   d. fossils, natural selection, and adaptation
   e. genetic variation, inheritance, and natural selection

? Review Questions

1. What are the differences between a salt crystal and a tree? Which is living? How do you know? How would you test your "knowledge"? What controls would you use?
2. What is the difference between a scientific theory and a hypothesis? Explain how each is used by scientists.
3. Define and explain the terms *natural selection*, *evolution*, *mutation*, *creationism*, and *population*.
4. Starting with the cell, list the hierarchy of organization of life, briefly explaining each level.
5. Define *homeostasis*. Why must organisms continuously acquire energy and materials from the external environment to maintain homeostasis?
6. Describe the scientific method. In what ways do you use the scientific method in everyday life?
7. What is evolution? Briefly describe how evolution occurs.

Applying the Concepts

1. In the heavily populated state of California, natural occurrences, including earthquakes, heavy rains, and grass and forest wildfires, have reduced the quality of life for many Californians. Look at it another way for a moment. What ecosystems are found in this state? What impacts have humans had on these natural ecosystems? Are changes in these areas the reason for fires and floods and mudslides? What needs to be done to alter the balance of "humans and nature"?

2. Design an experiment to test the effects of a new dog food, "Super Dog," on the thickness and water-shedding properties of the coats of golden retrievers. Include all the parts of a scientific experiment. Design objective methods to assess coat thickness and water-shedding ability.
3. Science is based on principles, including uniformity in space and time and common perception. Assume that humans one day encounter intelligent beings from a planet in another galaxy who evolved under very different conditions. Discuss the two principles mentioned above and how they would affect (1) the nature of scientific observations on the different planets and (2) communications about these observations.

For More Information

Attenborough, D. *Life on Earth*. Boston: Little, Brown, 1979. Gorgeously illustrated and beautifully written introduction to the diversity of life on Earth; the inspiration for our title.

Dawkins, R. *The Blind Watchmaker*. New York: W. W. Norton & Co., 1986. An engagingly written description of the process of evolution, which Dawkins compares to a blind watchmaker.

Ehrlich, P. R. *The Machinery of Nature*. New York: Simon & Schuster, 1986. Using layperson's terms, a foremost ecologist and author explains the science of ecology and the biological rationale for environmental concern.

Leopold, A. *A Sand County Almanac*. New York: Oxford University Press, 1949 (reprinted in 1989). A classic by a natural philosopher; provides an eloquent foundation for the conservation ethic.

Swain, R. B. *Earthly Pleasures*. New York: Charles Scribner's Sons, 1981. Insightful essays stress the interrelatedness and diversity of life.

Thomas, L. *The Medusa and the Snail*. New York: Bantam Books, 1980. The late physician, researcher, and philosopher Lewis Thomas shares his awe of the living world in a series of delightful essays.

Wilson, E. O. *The Diversity of Life*. New York: W. W. Norton & Co., 1992. A celebration of the diversity of life, how it evolved, and how humans are impacting it. Wilson's writings have won two Pulitzer prizes.

Answers to Multiple-Choice Questions

1. a 2. d 3. c 4. c 5. e 6. e

MEDIATUTOR
An Introduction to Life on Earth

CD Activities

Activity 1.1: Defining Life

Estimated time: 15 minutes

In this simulation you will explore the characteristics that define life and the way that we use the scientific method to ask questions about life. You will play the part of the scientist provided with some material that may contain living organisms—or not. It is up to you to make the determination.

Start the MediaTutor Student CD-ROM and enter the activity number in the Quick Search box to be taken directly to that activity.

Web Investigations

Case Study: The Life Around Us

Estimated time: 10 minutes

Most scientific discoveries are made by people who are just plain curious. The Internet can be a wonderful source of information for the novice scientist. Unfortunately, not all of the information is correct. Good scientists evaluate not only their own experimental data but any other information resources they may access. In this exercise you will learn how to evaluate a Web site.

Go to http://www.prenhall.com/audesirk6, the Audesirk Companion Web site. Select Chapter 1 and the Web Investigation to begin.

UNIT ONE

The Life of a Cell

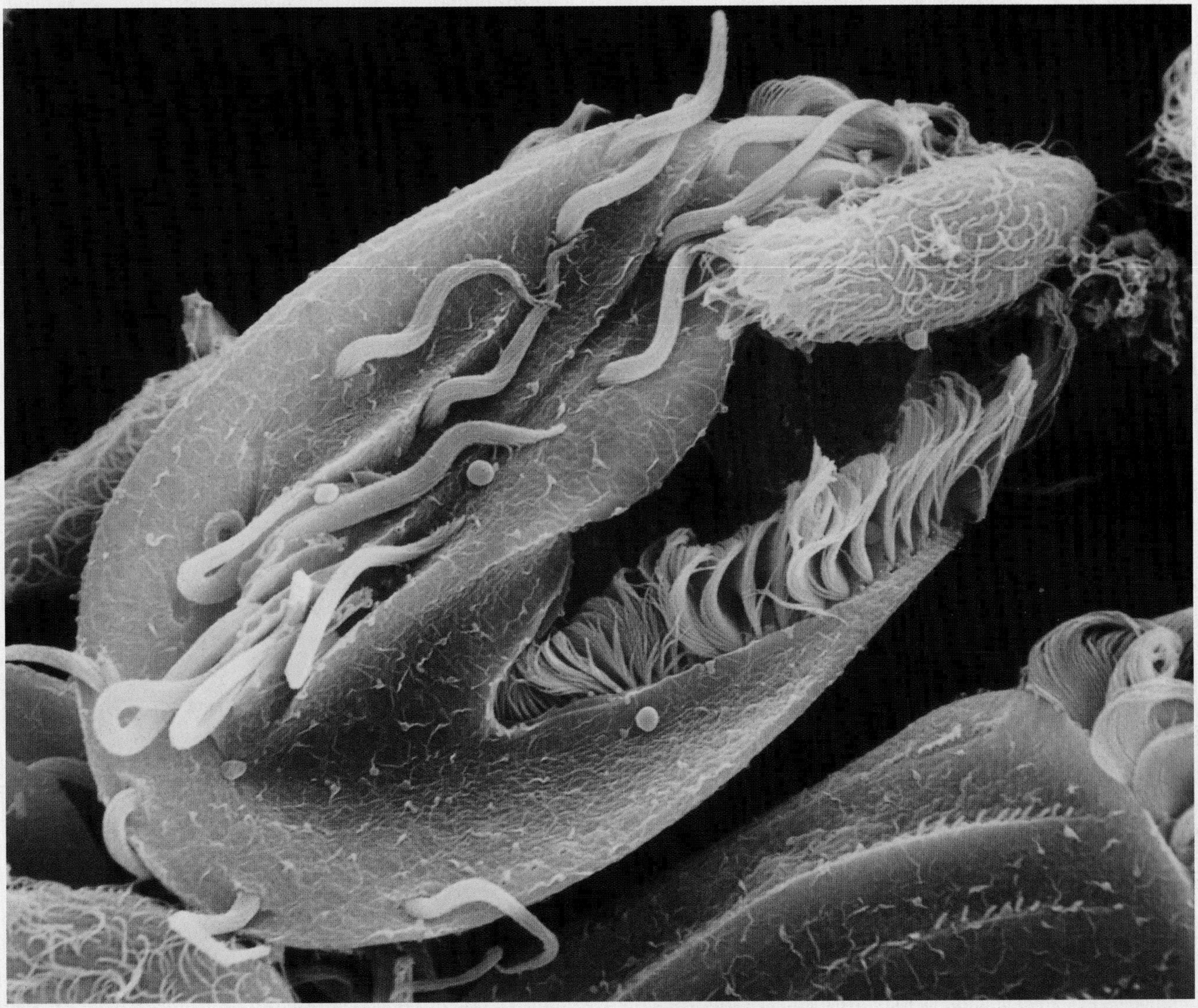

Single cells can be complex, independent organisms, such as these two ciliates, of the kingdom Protista. A large Euplotes *(about 300 μm in length) prepares to eat a much smaller* Paramecium. *Both are covered with cilia, short, beating structures used to move and ingest prey.*

"If there is magic on this planet, it is contained in water."

Loren Eiseley in* The Immense Journey *(1957)

Could something this delicious possibly be good for you? Chocolate, made from seeds found inside cacao pods (inset), contains high levels of antioxidants.

2 Atoms, Molecules, and Life

AT A GLANCE

Case Study: Health Food?

1) What Are Atoms?

Atoms, the Basic Structural Units of Matter, Are Composed of Still Smaller Particles

2) How Do Atoms Interact to Form Molecules?

Atoms Will Interact with Other Atoms Only When There Are Vacancies in Their Outermost Electron Shells

Charged Atoms Called *Ions* Interact to Form Ionic Bonds

Uncharged Atoms Can Become Stable by Sharing Electrons, Forming Covalent Bonds

Hydrogen Bonds Are Weaker Electrical Attractions Between or Within Molecules with Polar Covalent Bonds

3) Why Is Water So Important to Life?

Water Interacts with Many Other Molecules

Water Molecules Tend to Stick Together

Water Can Form H^+ and OH^- Ions

Water Moderates the Effects of Temperature Changes

Water Forms an Unusual Solid: Ice

Case Study Revisited: Health Food?

CASESTUDY

Health Food?

When 35 million people in the United States gave their loved ones boxes of chocolates last Valentine's Day, they knew they were giving sweet comfort—but health food? Sometimes described as "sinfully delicious," chocolate has often been a source of guilt for those who indulge (or overindulge) in it. Chocolate candy is certainly a significant source of fat and sugar calories, but recent research suggests that chocolate itself—the dark, bitter powder made from the seeds within cacao pods (see inset) may also be a significant source of protective molecules. Medical scientists have known for some time that many things that go wrong with our bodies can be traced to destructive molecules called *free radicals.* Many free radicals contain oxygen in a form that reacts strongly with, and damages, various biological molecules and their cellular structures. This process is called *oxidative stress.* Oxidative stress tears up cell membranes, breaks down DNA, and destroys enzymes, resulting in many aspects of aging, cancer, heart disease, and nervous system disorders. Unfortunately, oxidative stress is a fact of life, because as our cells use energy, they naturally produce free radicals. So what's a person to do? Well, maybe eat chocolate! ■

1) What Are Atoms?

Atoms, the Basic Structural Units of Matter, Are Composed of Still Smaller Particles

If you took a diamond (a form of carbon) and cut it into pieces, each piece would still be carbon. If you could make finer and finer divisions, you would eventually produce a pile of carbon atoms. **Atoms** are the fundamental structural units of matter. Atoms themselves, however, are composed of a central **atomic nucleus** (often called simply the *nucleus*; plural, *nuclei*, but don't confuse it with the nucleus of a cell). The nucleus contains two types of subatomic particles of equal weight: positively charged **protons** and uncharged **neutrons**. Subatomic particles called **electrons** orbit the atomic nucleus (Fig. 2-1). Electrons are lighter, negatively charged particles. An atom by itself has an equal number of electrons and protons and is therefore electrically neutral.

There are 92 types of atoms that occur naturally. Each type of atom forms the structural unit of a different element. An **element** is a substance that can neither be broken down nor converted to other substances by ordinary chemical means. The number of protons in the nucleus, called the **atomic number**, is a characteristic of each element. For example, every hydrogen atom has one proton in its nucleus, every carbon atom has six protons, and

MediaTutor 2.1 Interactive Atoms

every oxygen atom has eight. Each element has unique chemical properties based on the number and configuration of its subatomic particles. Some elements, such as oxygen and hydrogen, are gases at room temperature; others, such as lead, are extremely dense solids. Most elements are quite rare, and relatively few are essential to life on Earth. Table 2-1 lists the most common elements in the universe, on Earth, and in the human body. Notice how differently these elements are distributed.

Atoms of the same element may have different numbers of neutrons; when this occurs, the atoms are called **isotopes** of each other. Some, but not all, isotopes are **radioactive**; that is, they spontaneously break apart, forming different types of atoms and releasing energy in the process. Radioactive isotopes are extremely useful as "labels" in studying biological processes (see "Scientific Inquiry: Radioactivity in Research").

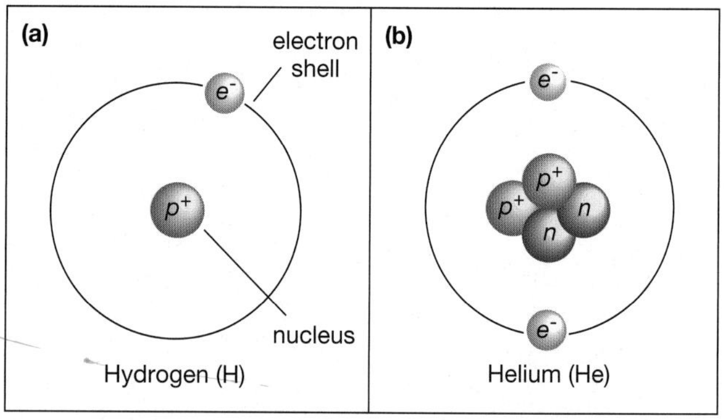

Figure 2-1 Atomic models
Structural representations of the two smallest atoms, ***(a)*** hydrogen and ***(b)*** helium. In these simplified models, the electrons are represented as miniature planets, circling in specific orbits around a nucleus that contains protons and neutrons.

Electrons Orbit the Nucleus at Fixed Distances, Forming Electron Shells That Correspond to Different Energy Levels

As you may know from experimenting with a magnet, like poles repel each other and opposite poles attract each other. In a similar way, electrons repel one another, owing to their negative electrical charge, and they are drawn to the positively charged protons of the nucleus. However, because of their mutual repulsion, only limited numbers of electrons can occupy the space closest to the nucleus. Large atoms can accommodate many electrons because their electrons orbit at increasing distances from their nucleus. The electrons orbit through a three-dimensional space; the orbits, which correspond to different energy levels, are called **electron shells** (Figs. 2-1 and 2-2, p. 24).

The electron shell closest to the atomic nucleus is the smallest and can hold only two electrons. The second shell can hold up to eight electrons. The electrons in an atom normally fill the shell closest to the nucleus and then begin to occupy the next shell. Thus, a carbon atom, with six electrons, has two electrons in the first shell, closest to the nucleus, and four electrons in its second shell (see Fig. 2-2). Nuclei and electron shells play complementary roles in atoms. Nuclei (assuming they are not radioactive) provide stability, while the electron shells allow interactions, or *bonds*, with other atoms. Nuclei resist disturbance by outside forces. Ordinary

Table 2-1 Common Elements Important in Living Organisms

Element	Symbol	Atomic Number[a]	Percent in Universe[b]	Percent in Earth[b]	Percent in Human Body[b]
Hydrogen	H	1	91	0.14	9.5
Helium	He	2	9	Trace	Trace
Carbon	C	6	0.02	0.03	18.5
Nitrogen	N	7	0.04	Trace	3.3
Oxygen	O	8	0.06	47	65
Sodium	Na	11	Trace	2.8	0.2
Magnesium	Mg	12	Trace	2.1	0.1
Phosphorus	P	15	Trace	0.07	1
Sulfur	S	16	Trace	0.03	0.3
Chlorine	Cl	17	Trace	0.01	0.2
Potassium	K	19	Trace	2.6	0.4
Calcium	Ca	20	Trace	3.6	1.5
Iron	Fe	26	Trace	5	Trace

[a]Atomic number = number of protons in the atomic nucleus.

[b]Approximate percentage of atoms of this element, by weight, in the universe, in Earth's crust, and in the human body.

Scientific Inquiry
Radioactivity in Research

As you read this text, you will encounter many statements that may cause you to wonder, How do they know *that*? How do biologists know that DNA is the genetic material of cells (Chapter 9)? How do paleontologists measure the ages of fossils (Chapter 17)? How do botanists know that sugars made in plant leaves during photosynthesis are transported to other parts of the plant in the sieve tubes of phloem (Chapter 23)? These discoveries, and many more, have been possible only through the use of radioactive isotopes.

Although all atoms of a particular element have the same number of protons, the number of neutrons may vary. Neutrons don't affect the chemical reactivity of an atom very much, but they add to the atom's mass, which can be detected by sophisticated instruments, such as mass spectrometers. Nuclei with "too many" neutrons break apart spontaneously, or *decay*, often emitting radioactive particles in the process. Those particles can also be detected—for example, with Geiger counters. The process by which a radioactive isotope spontaneously breaks apart is called *radioactive decay*.

A particularly fascinating and medically important use of radioactive isotopes is *positron emission tomography*, more commonly known as *PET scans* (Fig. E2-1a). In one common application of PET scans, a subject is given the sugar glucose that has been labeled with (that is, attached to) a harmless radioactive isotope of fluorine. When the nucleus of radioactive fluorine decays, it emits two bursts of energy that travel in opposite directions along the same line. Energy detectors arranged in a ring around the subject look at a "slice" of the brain, recording the nearly simultaneous arrival of the two energy bursts (Fig. E2-1b). A powerful computer then calculates the location within the subject at which the decay must have occurred and generates a color-coded map of the frequency of decays. Because the radioactive fluorine is attached to glucose molecules, this map reflects the glucose concentrations within the subject's brain. Since the brain uses this sugar for energy, the more active a brain cell is, the more glucose it uses, and the more radioactivity is concentrated there. For example, tumor cells (Fig. E2-1c) or brain regions in which epileptic seizures originate generally have excessively high glucose utilization and show up in PET scans as "hot spots." Normal brain cells activated by a specific mental task will also have higher glucose demands, which can be detected by PET scans.

Biology and medicine have profited immensely from close interactions with the other sciences, especially chemistry and physics. The development of PET scans required close cooperation with chemists (developing and synthesizing the radioactive probes), physicists (understanding interactions between electrons and short-lived, positively charged particles called *positrons,* as well as the geometry of the resulting energy emissions), and engineers (designing and building the electronic apparatus). Continued teamwork among scientists promises further advances in both the fundamental understanding of biological processes and applications in medicine and agriculture.

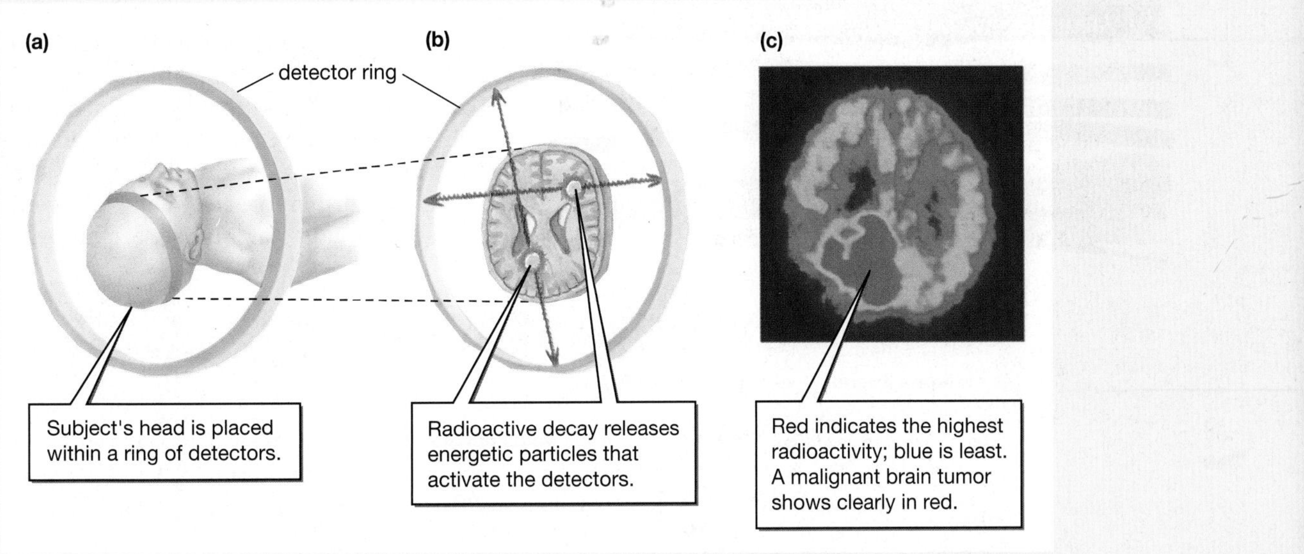
(a)
(b)
(c)
detector ring
Subject's head is placed within a ring of detectors.
Radioactive decay releases energetic particles that activate the detectors.
Red indicates the highest radioactivity; blue is least. A malignant brain tumor shows clearly in red.

Figure E2-1 How positron emission tomography works

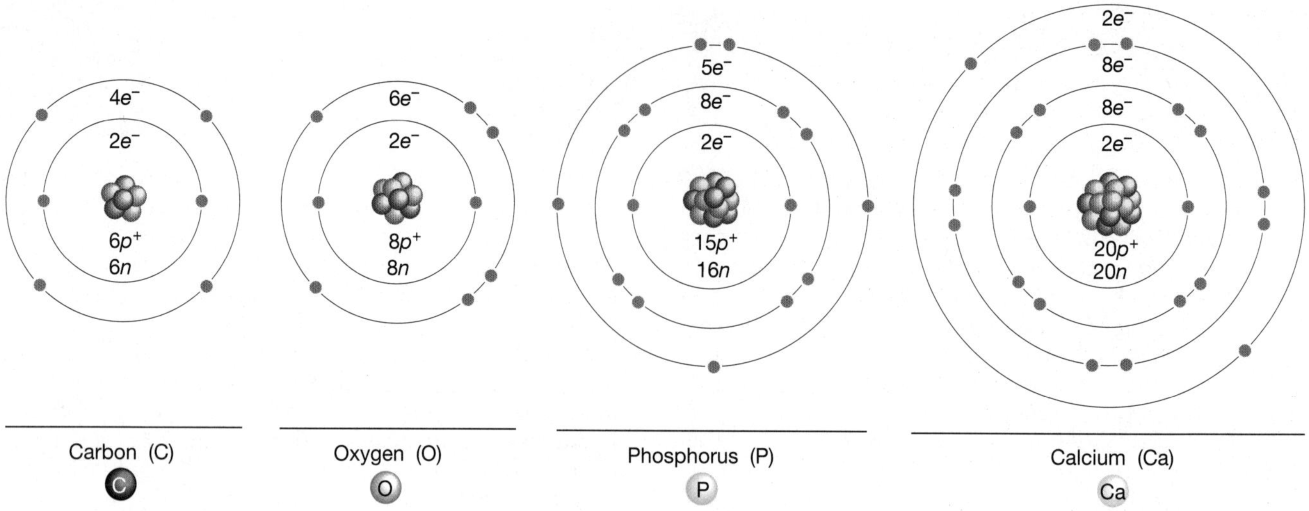

Figure 2-2 Electron shells in atoms
Most biologically important atoms have at least two shells of electrons. The first shell, closest to the nucleus, can hold two electrons; the next shell can hold a maximum of eight electrons. More distant shells can also hold eight electrons each.

sources of energy, such as heat, electricity, and light, hardly affect them at all. Because its nucleus is stable, a carbon atom remains carbon whether it is part of a diamond, carbon dioxide, or sugar. Electron shells, however, are dynamic; as you will soon see, atoms bond with one another by gaining, losing, or sharing electrons.

2 How Do Atoms Interact to Form Molecules?

Atoms Interact with Other Atoms When There Are Vacancies in Their Outermost Electron Shells

A **molecule** consists of two or more atoms, of either the same or of different elements, held together by interactions among their outermost electron shells. A substance whose molecules are formed of different types of atoms is called a **compound**. Atoms interact with one another according to two basic principles:

1. An atom will not react with other atoms when its outermost electron shell is completely full or empty. Such an atom is described as being *inert*.
2. An atom will react with other atoms when its outermost electron shell is only partially full. Such atoms are described as *reactive*.

To demonstrate these principles, consider three atoms: hydrogen, oxygen, and helium (see Fig. 2-1). Hydrogen (the smallest atom) has one proton in its nucleus and one electron in its single (and therefore outermost) electron shell, which can hold up to two electrons. The oxygen atom has six electrons in its outer shell, which can hold eight. In contrast, helium has two protons in its nucleus, and two electrons fill its single electron shell. Therefore, we predict that hydrogen and oxygen atoms, with partially empty outer shells, should be reactive, while helium atoms, with a full shell, should be stable. We might further predict that hydrogen and oxygen atoms could gain stability by reacting with each other.

Table 2-2 Chemical Bonds

Type of Bond	Bond Forms:
Weak Bonds: allow interactions between individual atoms or molecules	
Ionic bonds	Between positive and negative ions
Hydrogen bonds	Between a hydrogen atom involved in a polar covalent bond and another atom involved in a polar covalent bond
Hydrophobic interactions	Because interactions between water molecules exclude hydrophobic molecules
Strong Bonds: hold atoms together within molecules	
Covalent bonds	By the sharing of electron pairs; equal sharing produces nonpolar covalent bonds; unequal sharing produces polar covalent bonds

The single electrons from each of the two hydrogen atoms would fill the outer shell of oxygen, forming H_2O: water (see Fig. 2-4c). As predicted, hydrogen can react readily with oxygen—in fact, the reaction is an explosive one. The space shuttle and many other rockets use liquid hydrogen as fuel to power liftoff. The hydrogen fuel reacts with oxygen, releasing water as a by-product. In contrast, helium, with a full outer shell, is almost completely inert. Our case study mentioned free radicals. *Free radicals* are atoms or molecules that lack one or more electrons in their outer shells, making them highly reactive.

An atom with an outermost electron shell that is partially full can gain stability by losing electrons (emptying the shell completely), by gaining electrons (filling the shell), or by sharing electrons with another atom (with both atoms behaving as though they had full outer shells). The results of losing, gaining, and sharing electrons are **chemical bonds**; *attractive forces* that hold atoms together in molecules. Each element has chemical bonding properties that arise from the configuration of electrons in its outer shell (Table 2-2). **Chemical reactions**, the making and breaking of chemical bonds to form new substances, are essential for the maintenance of life and for the working of modern society. Whether they occur in a plant cell as it captures solar energy, your brain as it forms new memories, or your car's engine as it guzzles gas, chemical reactions consist of making new chemical bonds and/or breaking existing ones.

Charged Atoms Called *Ions* Interact to Form Ionic Bonds

Both atoms that have an almost empty outermost electron shell and atoms that have an almost full outermost shell can become stable by losing electrons (emptying their outermost shell) or by gaining electrons (filling their outermost shell). The formation of table salt (sodium chloride) demonstrates this principle. Sodium (Na) has only one electron in its outermost electron shell, and chlorine (Cl) has seven electrons in its outer shell—one electron short of being full (Fig. 2-3a). Sodium, therefore, can become stable by losing the electron to chlorine from its outer shell, leaving that shell empty; chlorine can fill its outer shell by gaining the electron. Atoms that have lost or gained electrons, altering the balance between protons and electrons, are *charged*. These charged atoms are called **ions**. To form sodium chloride, sodium loses an electron and thereby becomes a positively charged sodium ion (Na^+); chlorine picks up an electron and becomes a negatively charged chloride ion (Cl^-) (Fig. 2-3b,c).

Opposite charges attract; therefore, sodium ions and chloride ions tend to stay near one another. They form crystals that contain repeating orderly arrangements of the two ions (Fig. 2-3c). The electrical attraction between oppositely charged ions that holds them together in crystals is called an **ionic bond**. Ionic bonds are weak and easily broken, as occurs when salt is dissolved in water (see Table 2-2).

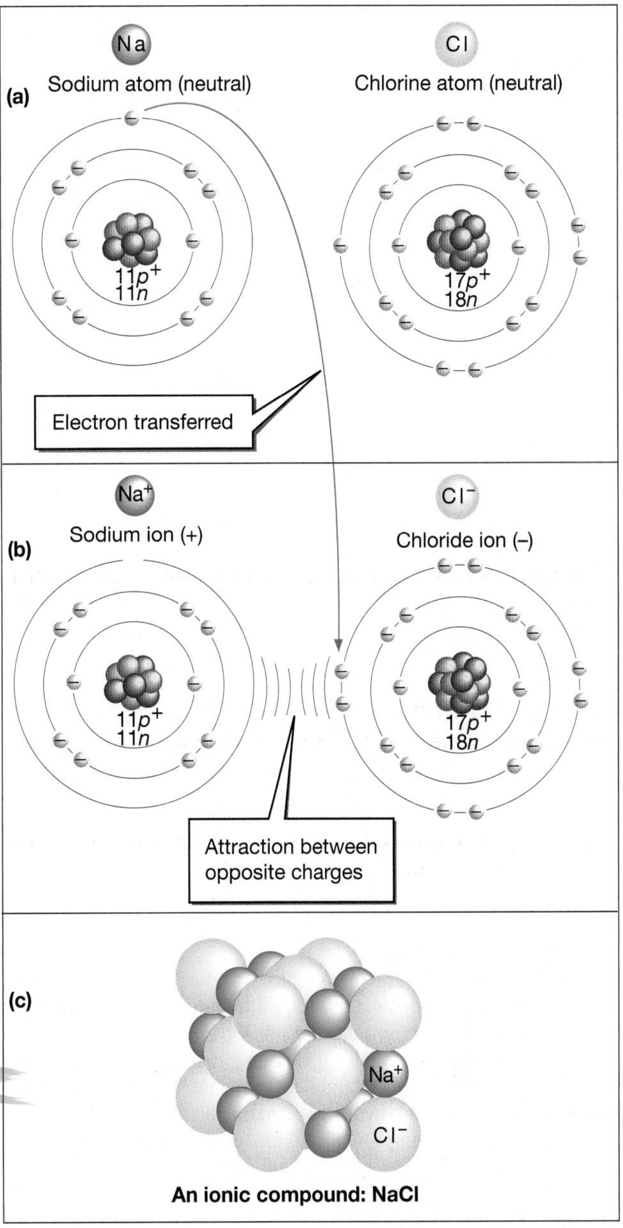

Figure 2-3 The formation of ions and ionic bonds ***(a)*** Sodium has only one electron in its outer electron shell; chlorine has seven. ***(b)*** Sodium can become stable by losing an electron, and chlorine can become stable by gaining an electron. Sodium becomes a positively charged ion and chlorine a negatively charged ion. ***(c)*** Because oppositely charged particles attract one another, the resulting sodium ions (Na^+) and chloride ions (Cl^-) nestle closely together in a crystal of salt, NaCl.

Uncharged Atoms Can Become Stable by Sharing Electrons, Forming Covalent Bonds

An atom with a partially full outermost electron shell can also become stable by sharing electrons with another atom, forming a **covalent bond**. Consider the

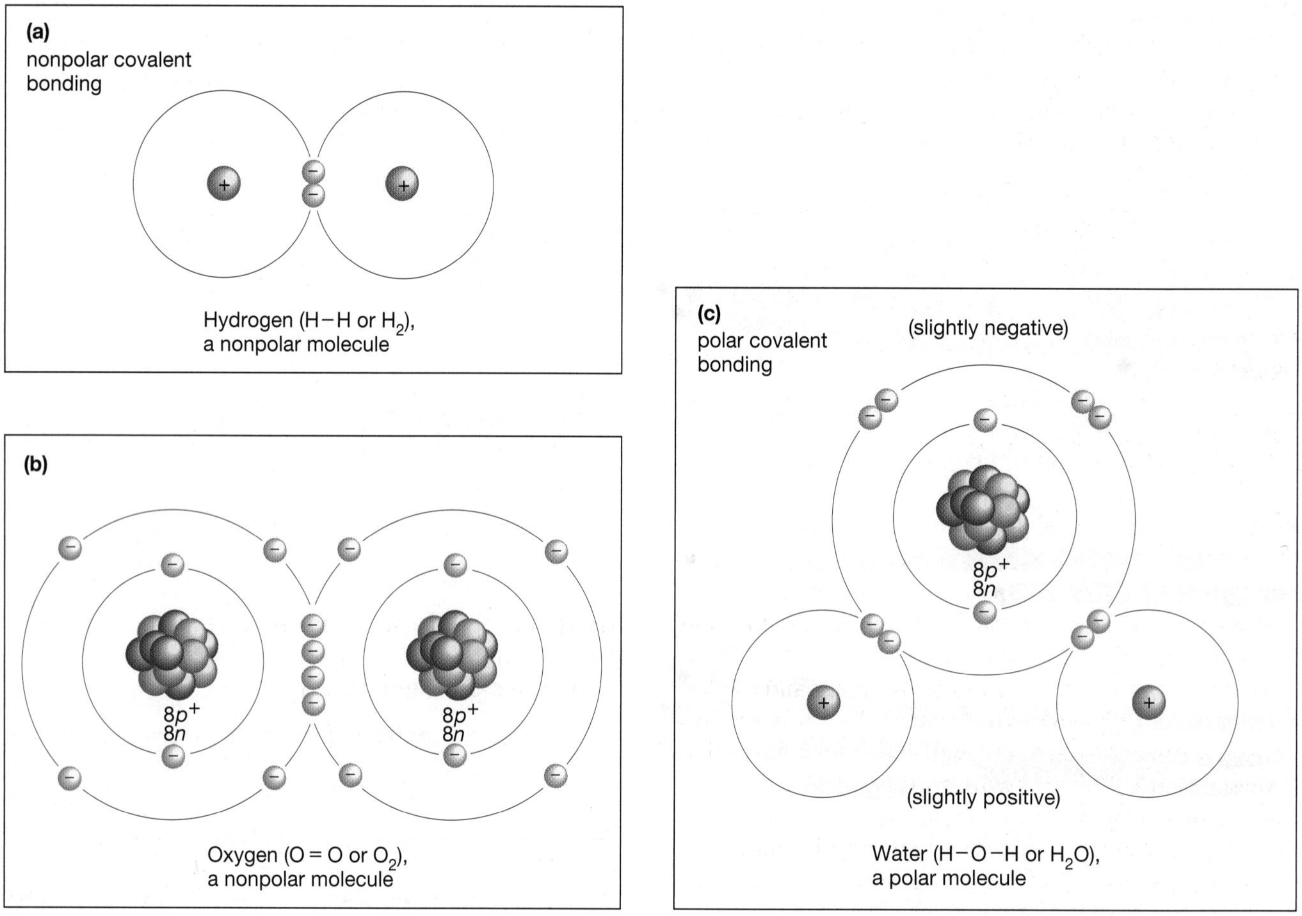

Figure 2-4 Covalent bonds
Electrons are shared between atoms to form covalent bonds. ***(a)*** In hydrogen gas, one electron from each hydrogen atom is shared, forming a single covalent bond. The resulting molecule of hydrogen gas is represented as H–H or H_2. ***(b)*** In oxygen gas, two oxygen atoms share four electrons, forming a double bond (O=O or O_2). ***(c)*** Oxygen lacks two electrons to fill its outer shell, so oxygen can make one bond with each of two hydrogen atoms to form water (H–O–H or H_2O). Oxygen exerts a greater pull on the electrons than does hydrogen, so the oxygen end of the molecule has a slight negative charge and the hydrogen end has a slight positive charge. This is an example of polar covalent bonding. The water molecule with its slightly charged ends is called a *polar molecule.*

hydrogen atom, which has one electron in a shell built for two. A hydrogen atom can become reasonably stable if it shares its single electron with another hydrogen atom, forming a molecule of hydrogen gas, H_2 (Fig. 2-4a). Because the two hydrogen atoms are identical, neither nucleus can exert more attraction and capture the other's electron. So the two electrons orbit around both nuclei for equal amounts of time, forming a single covalent bond; each hydrogen atom behaves almost as if it had two electrons in its shell. Two oxygen atoms also share electrons equally, producing a molecule of oxygen gas, O_2, with a double covalent bond (Fig. 2-4b). In such a bond, each atom contributes two electrons. If atoms share three pairs of electrons, a triple covalent bond is formed, as in nitrogen gas, N_2.

All covalent bonds are strong compared with ionic bonds, but some are stronger than others, depending on the atoms involved (see Table 2-2). Some covalent bonds, such as those in water (H_2O; Fig. 2-4c) and carbon dioxide (CO_2), are extremely stable—that is, it takes a lot of energy to break the bonds. Other bonds, such as those in oxygen gas (Fig. 2-4b) or gasoline, are less stable, and come apart more easily. When a chemical reaction occurs in which less stable bonds are broken and more stable bonds are formed (such as burning gasoline with oxygen to form carbon dioxide and water), energy is released, as we shall describe in Chapter 6.

Most Biological Molecules Utilize Covalent Bonding Covalent bonds are crucial to life because the atoms in most biological molecules are joined by covalent bonds. The molecules in proteins, sugars, bone, and cellulose are formed of atoms held together by covalent bonds. Hydrogen, carbon, oxygen, nitrogen, phosphorus, and sulfur are the most common atoms found in biological molecules. Except for hydrogen, each of these atoms needs at least two electrons to fill its outermost electron shell and can share electrons with two or

Table 2-3 Bonding Patterns of Atoms Commonly Found in Biological Molecules

Atom	Capacity of Outer Electron Shell	Electrons in Outer Shell	Number of Covalent Bonds Normally Formed	Common Bonding Patterns
Hydrogen	2	1	1	
Carbon	8	4	4	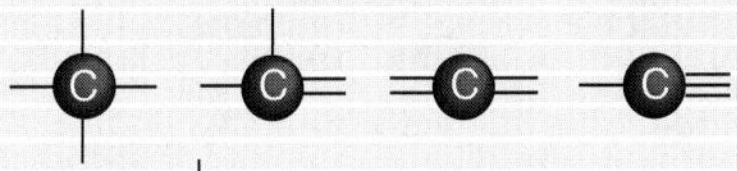
Nitrogen	8	5	3	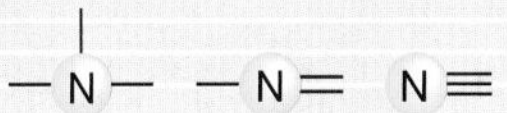
Oxygen	8	6	2	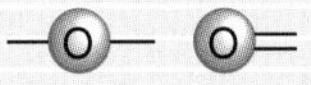
Phosphorus	8	5	5	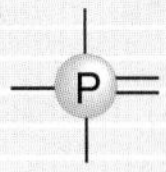
Sulfur	8	6	2	

more other atoms. Hydrogen can form a covalent bond with one other atom; oxygen and sulfur with two other atoms; nitrogen with three; and phosphorus and carbon with up to four. (Phosphorus is unusual; although it has only three spaces in its outer shell, it can form up to five covalent bonds with up to four other atoms.) This diversity of bonding arrangements permits biological molecules to be constructed in an almost infinite variety and complexity. Double and triple bonds increase the variety of shapes and functions of biological molecules. Table 2-3 summarizes bonding patterns in biological molecules.

Polar Covalent Bonds Form When Atoms Share Electrons Unequally

In hydrogen gas the two nuclei are identical, and the shared electrons spend equal time near each nucleus. Therefore, not only is the molecule as a whole electrically neutral, but each end, or *pole*, of the molecule is also electrically neutral. Such an electrically symmetrical bond is called a **nonpolar covalent bond** and the compound formed with such nonpolar bonds is a nonpolar molecule, such as hydrogen (H_2) or oxygen (O_2) (see Fig. 2-4a,b). But electron sharing in covalent bonds is not always equal. In many molecules, one nucleus may initially have a larger positive charge, and therefore attract the electrons more strongly, than does the other nucleus. This situation produces a **polar covalent bond.** Although the molecule as a whole is electrically neutral, it has charged parts: The atom that attracts the electrons more strongly then picks up a slightly negative charge (the negative pole of the molecule), and the other atom has a slightly positive charge (the positive pole). In water, for example, oxygen attracts electrons more strongly than does hydrogen, so the oxygen end of a water molecule is negative and each hydrogen is positive (see Fig. 2-4c). Water with its charged ends is a polar molecule.

Hydrogen Bonds Are Weaker Electrical Attractions Between or Within Molecules with Polar Covalent Bonds

Because of the polar nature of their covalent bonds, nearby water molecules attract one another. The partially negatively charged oxygens of some water molecules attract the partially positively charged hydrogens of other water molecules. This electrical attraction is called a **hydrogen bond** (Fig. 2-5; see Table 2-2). As we shall see shortly, hydrogen bonds between molecules give water several unusual properties that are essential to life on Earth.

Hydrogen bonds are common and important in biological molecules, as well as in water. They may occur whenever polar covalent bonds produce slightly negative and slightly positive charges that then attract one another. Both nitrogen and oxygen atoms attract electrons more strongly than do hydrogen atoms. Therefore, the nitrogen or oxygen pole of a nitrogen–hydrogen or oxygen–hydrogen bond is slightly negative, and the hydrogen pole is slightly positive. The resulting polar parts of the molecules can form hydrogen bonds with water, with other biological molecules, or with polar parts of the same molecule. Although individual hydrogen bonds are quite weak, many of them working together are quite strong. As we shall see in Chapter 3, hydrogen bonds play crucial roles in shaping the three-dimensional structures of proteins. In Chapter 9 you'll discover their importance in DNA .

3 Why Is Water So Important to Life?

Water is extraordinarily abundant on Earth, has unusual properties, and is so essential to life that it merits special consideration. Life is very likely to have arisen in

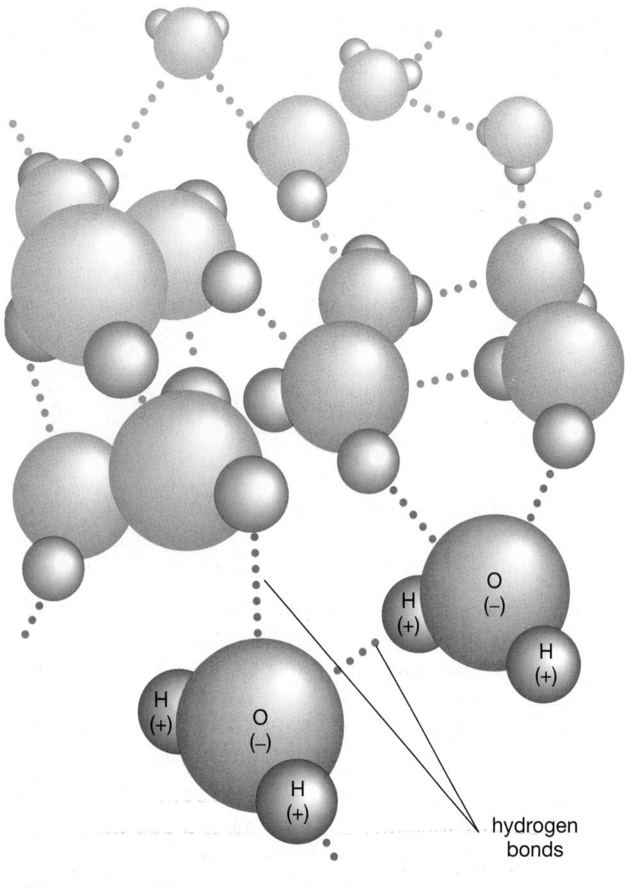

Figure 2-5 Hydrogen bonds
The partial charges on different parts of water molecules produce weak attractive forces called *hydrogen bonds* (dotted lines) between the hydrogens of one water molecule and the oxygens of other molecules.

MEDIATUTOR 2.2 Water and Life

the waters of the primeval Earth. Living organisms still contain about 60% to 90% water, and all life depends intimately on the properties of water. Why is water so crucial to life?

Water Interacts with Many Other Molecules

Water enters into many of the chemical reactions that occur in living cells. The oxygen that green plants release into the air is derived from water during photosynthesis. In manufacturing a protein, fat, nucleic acid, or sugar, your body produces water in the process; conversely, when you digest proteins, fats, and sugars in the foods you eat, water is used in the reactions. Why is water so important in biological chemical reactions?

Water is an extremely good **solvent**—that is, it is capable of dissolving a wide range of substances, including protein, salts, and sugars. Water or other solvents containing dissolved substances are called *solutions.* Recall that a crystal of table salt is held together by the electrical attraction between positively charged sodium ions and negatively charged chloride ions (see Fig. 2-3c). Because water is a polar molecule, it has positive and negative poles. If a salt crystal is dropped into water, the positively charged hydrogen ends of water molecules will be attracted to and will surround the negatively charged chloride ions, and the negatively charged oxygen poles of water molecules will surround the positively charged sodium ions. As water molecules enclose the sodium and chloride ions and shield them from interacting with each other, the ions separate from the crystal and drift away in the water—and the salt dissolves (Fig. 2-6).

Water also dissolves molecules held together by polar covalent bonds. Its positive and negative poles are attracted to oppositely charged regions of dissolving molecules. Ions and polar molecules are termed **hydrophilic** (Greek for "water-loving") because of their electrical attraction for water molecules. Many biological molecules, including sugars and amino acids, are hydrophilic and dissolve readily in water (Fig. 2-7). Water also dissolves gases such as oxygen and carbon dioxide. The fish swimming below the ice on a frozen lake rely on oxygen that dissolved before the ice formed, and they release CO_2 into solution in the water. By dissolving such a wide variety of molecules, the watery substance inside a cell provides a suitable environment for the countless chemical reactions essential to life on Earth.

Molecules that are uncharged and nonpolar, such as fats and oils, usually do not dissolve in water and hence are called **hydrophobic** ("water-fearing"). Nevertheless, water has an important effect on such molecules. Oils, for example, form globules when spilled into water. Oil

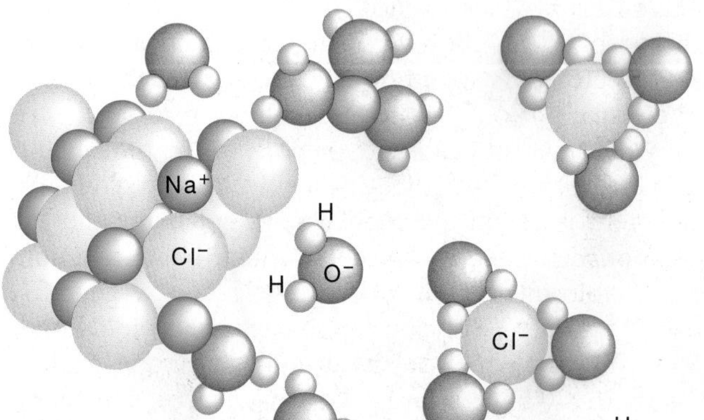

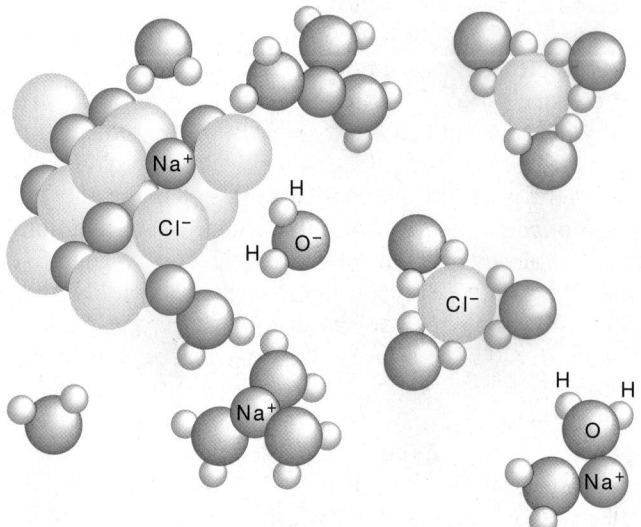

Figure 2-6 Water as a solvent
The polarity of water molecules allows water to dissolve polar and charged substances. When a salt crystal is dropped into water, the water surrounds the sodium and chloride ions with oppositely charged poles of the water molecules. Thus insulated from the attractiveness of other molecules of salt, the ions disperse, and the whole crystal gradually dissolves.

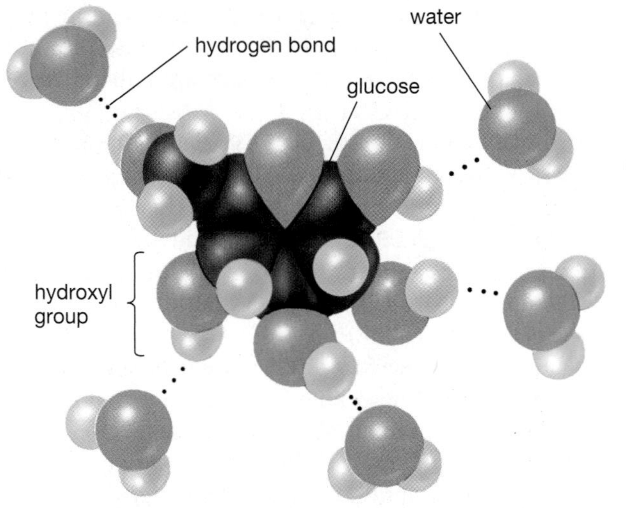

Figure 2-7 Water dissolves many biological molecules
Many biological molecules dissolve in water because they have polar parts—for example, OH^- (hydroxyl) groups—that can form hydrogen bonds with water molecules. As shown, hydrogen bonds can form between the hydroxyl groups on a glucose molecule (a simple sugar) and surrounding water molecules.

molecules in water disrupt the hydrogen bonding among adjacent water molecules. When oil molecules encounter one another in water, their nonpolar surfaces nestle closely together, surrounded by water molecules that form hydrogen bonds with one another but not with the oil. To separate again, the oil molecules would have to break apart the hydrogen bonds that link surrounding water molecules. Thus, the oil molecules remain together, forming a glistening droplet that floats on the water's surface. The tendency of oil molecules to clump together is described as a **hydrophobic interaction** (see Table 2-2). As we shall discuss in Chapter 4, the membranes of living cells owe much of their structure to hydrophobic interactions.

Water Molecules Tend to Stick Together

In addition to interacting with other molecules, water molecules interact with each other. Because hydrogen bonds interconnect individual water molecules, liquid water has high **cohesion**—that is, water molecules have a tendency to stick together. Cohesion among water molecules at the water's surface produces **surface tension**, the tendency for the water surface to resist being broken. If you've ever experienced the slap and sting of a belly flop into a swimming pool, you've discovered firsthand the power of surface tension. Surface tension can support fallen leaves, some spiders and water insects, and even a running lizard (Fig. 2-8a).

A more important role of cohesion in water occurs in the life of land plants. Since a plant absorbs water through its roots, how does the water reach the aboveground parts, especially if the plant is a 100-meter-tall redwood (Fig. 2-8b)? As we shall see in Chapter 23, water molecules are pulled up by the leaves. Water fills tiny tubes that connect the leaves, stem, and roots. Water molecules that evaporate from the leaves pull

(a) (b)

Figure 2-8 Cohesion among water molecules
(a) With webbed feet bearing specialized scales, the basilisk lizard of South America makes use of surface tension, caused by cohesion, to support its weight as it races across the surface of a pond. ***(b)*** In giant redwoods, cohesion holds water molecules together in continuous strands from the roots to the topmost leaves even 300 feet (about 100 meters) above the ground.

water up the tubes, much like a chain being pulled up from the top. The system works because the hydrogen bonds interconnecting water molecules are stronger than the weight of the water in the tubes, even 100 meters' worth; thus, the water "chain" doesn't break. Without the cohesion of water, there would be no land plants as we know them, and terrestrial life would undoubtedly have evolved quite differently. You may have realized by now that the "common bond" producing the sting of a belly flop, the ability of a lizard to run on water, and the movement of water up a tree is actually the hydrogen bond between water molecules.

Water exhibits another property, *adhesion*, a word that describes its tendency to stick to polar surfaces having slight charges that attract polar water molecules. Adhesion helps water move within small spaces, such as the thin tubes in plants that carry water from roots to leaves. If you stick the end of a narrow glass tube into water, the water will move a short distance up the tube. Put some water in a narrow glass bud vase or test tube and you'll see that the upper surface is curved; water pulls itself up the sides of the glass by its adhesion to the surface of the glass and by the cohesion among water molecules.

Water Can Form H^+ and OH^- Ions

Although water is generally regarded as a stable compound, individual water molecules constantly gain, lose, and swap hydrogen atoms. As a result, at any given time about two of every billion water molecules are ionized—that is, broken apart into hydrogen ions (H^+) and hydroxide ions (OH^-):

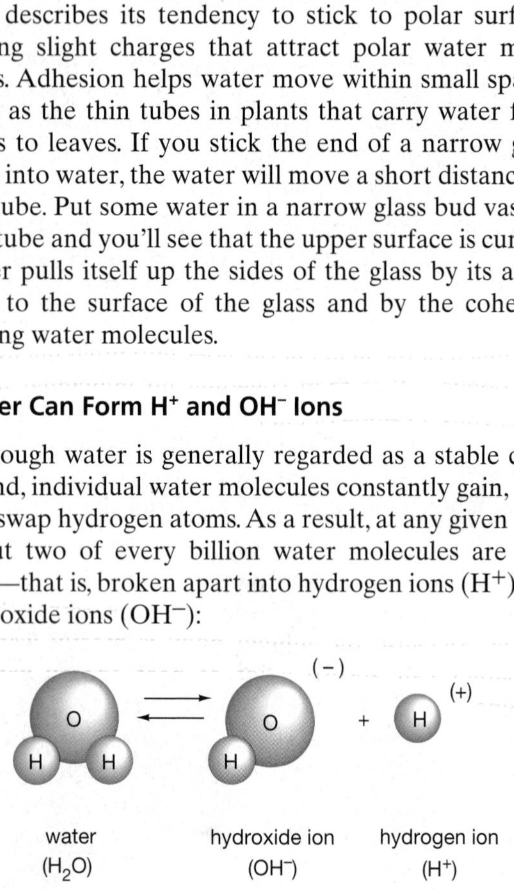

A hydroxide ion has gained an electron from the hydrogen ion, giving it a negative charge, while the hydrogen ion, which has lost its electron, now has a positive charge. Pure water contains equal concentrations of hydrogen ions and hydroxide ions.

In many solutions, however, the concentrations of H^+ and OH^- are not the same. If the concentration of H^+ exceeds the concentration of OH^-, the solution is **acidic**. An **acid** is a substance that releases hydrogen ions when it is dissolved in water. When hydrochloric acid (HCl), for example, is added to pure water, almost all of the HCl molecules separate into H^+ and Cl^-. Therefore, the concentration of H^+ greatly exceeds the concentration of OH^-, and the resulting solution is acidic. (Many acidic substances, such as lemon juice and vinegar, have a sour taste. This is because the sour-taste receptors on your tongue are specialized to respond to the excess of H^+.)

If the concentration of OH^- is greater, the solution is **basic**. A **base** is a substance that combines with hydrogen ions, reducing their number. If, for instance, sodium hydroxide (NaOH) is added to water, the NaOH molecules separate into Na^+ and OH^-. The OH^- combine with H^+, reducing their number. The solution is then basic.

The degree of acidity is expressed on the **pH scale** (Fig. 2-9), in which neutrality (equal numbers of H^+ and OH^-) is assigned the number 7. Acids have a pH below 7; bases have a pH above 7. Pure water, with equal concentrations of H^+ and OH^-, has a pH of 7. Each unit on the pH scale represents a tenfold change in the concentration of H^+. Thus, a cola drink (pH = 3) has a concentration of H^+ 10,000 times that of water (pH = 7)—no wonder it is bad for your teeth!

A Buffer Helps Maintain a Solution at a Relatively Constant pH

In most mammals, including humans, both the cell interior (cytoplasm) and the fluids that bathe the cells are nearly neutral (pH about 7.3 to 7.4). Small increases or decreases in pH may cause drastic changes in both the structure and function of biological molecules, leading to the death of cells or entire organisms. Nevertheless, living cells seethe with chemical reactions that take up or give off H^+. How, then, does the pH remain constant overall? The answer lies in the many buffers found in living organisms. A **buffer** is a compound that tends to maintain a solution at a constant pH by accepting or releasing H^+ in response to small changes in H^+ concentration. If the H^+ concentration rises, buffers combine with them; if the H^+ concentration falls, buffers release H^+. The result is that the concentration of H^+ is restored to its original level. Common buffers in living organisms include bicarbonate (HCO_3^-) and phosphate ($H_2PO_4^-$ and HPO_4^{2-}), both of which can accept or release H^+, depending on the circumstances. If the blood becomes too acidic, for example, bicarbonate accepts H^+ to form carbonic acid:

$$\underset{\text{(bicarbonate)}}{HCO_3^-} + \underset{\text{(hydrogen ion)}}{H^+} \rightarrow \underset{\text{(carbonic acid)}}{H_2CO_3}$$

If the blood becomes too basic, carbonic acid liberates hydrogen ions, which combine with the excess hydroxide ions, forming water:

$$\underset{\text{(carbonic acid)}}{H_2CO_3} + \underset{\text{(hydroxide ion)}}{OH^-} \rightarrow \underset{\text{(bicarbonate)}}{HCO_3^-} + \underset{\text{(water)}}{H_2O}$$

In either case, the result is that the blood pH remains near its normal value.

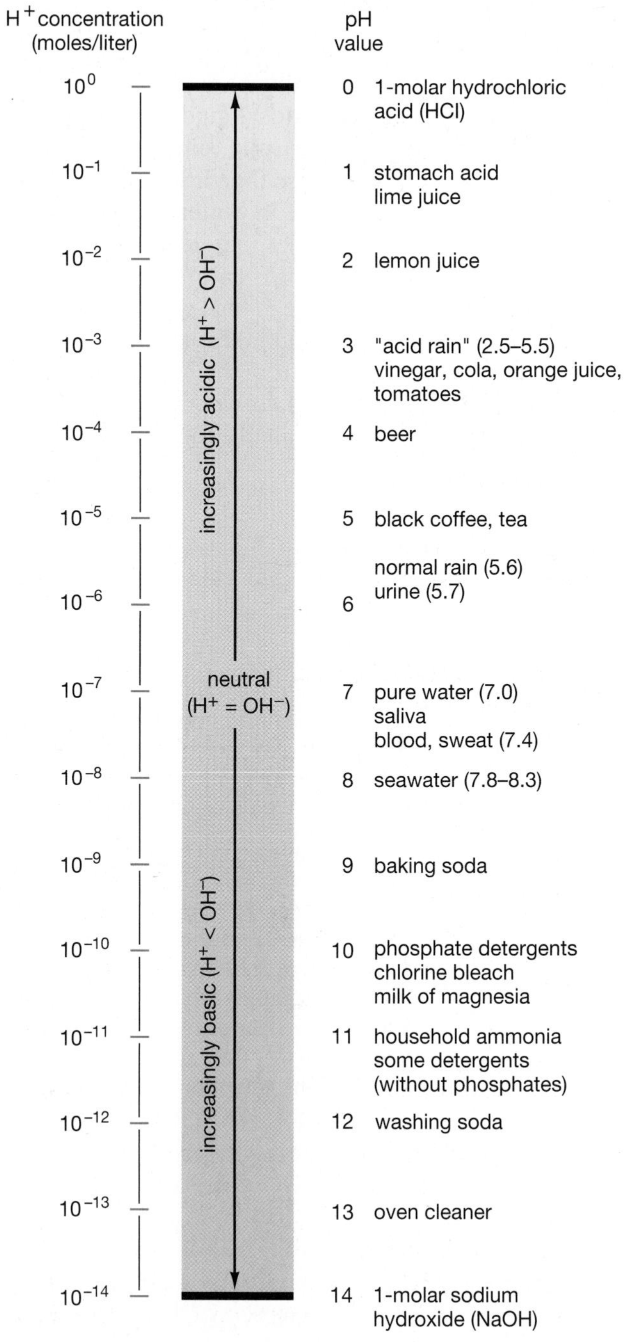

Figure 2-9 The pH scale
The pH scale expresses the concentration of hydrogen ions in a solution on a scale of 0 (very acidic) to 14 (very basic). Each unit of change in pH on the pH scale represents a tenfold change in the concentration of hydrogen ions. Lemon juice, for example, is about 10 times more acidic than orange juice, and the most severe acid rains in the northeastern United States are almost 1000 times more acidic than normal rainfall. Except for the inside of your stomach, nearly all the fluids in your body are finely adjusted to a pH of 7.4. The color coding corresponds to a common pH indicator dye, bromthymol blue, widely used by aquarium owners to monitor the pH of water for their fish.

Water Moderates the Effects of Temperature Changes

Your body and the bodies of other organisms can survive only within a limited temperature range. As we shall see in Chapter 6, high temperatures may damage enzymes that guide the chemical reactions essential to life. Low temperatures are also dangerous, because enzyme action slows as the temperature drops. Subfreezing temperatures within the body are usually lethal, because spearlike ice crystals can rupture cells.

Water has important properties that moderate the effects of temperature changes. These properties help keep the bodies of organisms within tolerable temperature limits. Also, large lakes and the oceans have a moderating effect on the climate of nearby land, making it warmer in winter and cooler in summer. First, some background: Temperature reflects the speed of molecules; the higher the temperature, the greater their average speed. Generally speaking, if heat energy enters a system, the molecules of that system move more rapidly, and the temperature of the system rises. Individual water molecules, however, are weakly linked to one another by hydrogen bonds (see Fig. 2-5). When heat enters a watery system such as a lake or a living cell, much of the heat energy goes into breaking hydrogen bonds rather than speeding up individual molecules. Thus, 1 **calorie** of energy will heat 1 gram of water 1 °C, whereas it takes only 0.6 calorie per gram to heat alcohol 1 °C, 0.2 calorie for table salt, and 0.02 calorie for common rocks such as granite or marble. So the energy required to heat a pound (a pint) of water only 1 °C would raise a pound of rock by 50 °C. If a lizard wants to warm up, it will seek out a rock, rather than a puddle. Because the human body is mostly water, a sunbather can absorb a lot of heat energy without sending his or her body temperature soaring—and many hot sunbathers can jump into a swimming pool to cool off without raising the temperature of the water very much. (The energy required to heat a gram of a substance by 1 °C is called its *specific heat*—water has a high specific heat.)

Second, water moderates the effects of high temperatures because it takes a great deal of heat, 539 calories per gram, to convert liquid water to water vapor. This, too, is due to the hydrogen bonds that interconnect individual water molecules. For a water molecule to evaporate, it must move quickly enough to break all the hydrogen bonds that hold it to the other water molecules in the solution. Only the fastest-moving water molecules, carrying the most energy, can break their hydrogen bonds and escape into the air as water vapor. The remaining liquid is cooled by the loss of these high-energy molecules. As children romp through a sprinkler on a hot summer day, water coats their bodies. Heat energy is transferred from their skin to the water and from the water to the vapor as the water evaporates. Evaporating just 1 gram of water cools 539 grams of a person's body 1 °C, so water is a very effective coolant. This also means that evaporating perspiration produces a great loss of heat without much loss of water. (The heat required to vaporize water is called its *heat of vaporization*—water's heat of vaporization is one of the highest known!)

Third, water moderates the effects of low temperatures because an unusually large amount of energy must be removed from molecules of liquid water before they form the precise crystal arrangement of ice. As a result, water freezes more slowly than do many other liquids at a given temperature and loses more heat to the environment in the process. (This property of water is called its *heat of fusion*, which is very high.)

Water Forms an Unusual Solid: Ice

Water, of course, will become a solid after prolonged exposure to temperatures below its freezing point. But even solid water is unusual. Most liquids become denser when they solidify, and the solid sinks. Ice is rather unique because it is less dense than liquid water. When a pond or lake starts to freeze in winter, the ice stays on top, forming an insulating layer that delays the freezing of the rest of the water. This insulation allows fish and other lake residents to survive in the liquid water below. If ice were to sink, ponds and lakes in much of North America would freeze solid, from the bottom up, during the winter, killing fish and underwater plants and making drinking water far less available to animals.

REVISITED CASE STUDY REVISITED CASESTUDYREVISITEDCA

Health Food?

You now know that a free radical is an atom or molecule containing an unfilled outer electron shell, which reacts vigorously with other molecules to fill its outer shell. Normal cellular activities produce a variety of molecules that contain an oxygen atom with an unfilled outer shell. Such molecules react vigorously with other molecules, damaging them in the process. If you've seen iron turn to rust (iron oxide), you've witnessed oxygen's reactive power. Oxidative stress refers to a kind of "biological rusting" in which free radicals that contain oxygen damage cells. Substances that react with these oxygen-containing free radicals and render them harmless are called *antioxidants*. Although cells contain some of their own antioxidants, your eating habits can also make a difference. Fruits and vegetables, particularly those with yellow, orange, and red colors, are good sources of antioxidant compounds; vitamins C and E are antioxidants also. Now, amazingly, researchers have given us an excuse to eat chocolate and feel good about it: cocoa powder contains high concentrations of *flavenoids*, which are powerful antioxidants. This research is in its early stages, and no studies have been done to determine whether high consumption of chocolate actually reduces the risk of cancer or heart disease, but there will certainly be no shortage of volunteers for this research. Although becoming fat by eating too much chocolate candy could counteract any positive effects of the cocoa powder itself, slim "chocoholics" have reason to relax and enjoy.

As you will learn in Chapter 8, oxygen is important for harvesting the maximum amount of energy from molecules such as sugar. Ross Hardison, a researcher at Pennsylvania State University, stated eloquently: "Keeping oxygen under control while using it in energy production has been one of the great compromises struck in the evolution of life on Earth." What did he mean by this?

Summary of Key Concepts

1) What Are Atoms?

An element is a substance that can neither be broken down nor converted to different substances by ordinary chemical means. The smallest possible particle of an element is the atom, which is itself composed of a central nucleus, containing protons and neutrons, and electrons outside the nucleus. All atoms of a given element have the same number of protons, which is different from the number of protons in the atoms of every other element. Electrons orbit the nucleus in electron shells, at specific distances from the nucleus. Each shell can contain a fixed maximum number of electrons. The chemical reactivity of an atom depends on the number of electrons in its outermost electron shell: An atom is most stable, and therefore least reactive, when its outermost shell is either completely full or empty.

2) How Do Atoms Interact to Form Molecules?

Atoms may combine to form molecules. The forces holding atoms together in molecules are called *chemical bonds*. Atoms that have lost or gained electrons are negatively or positively charged particles called *ions*. Ionic bonds are electrical attractions between charged ions, holding them together in crystals. When two atoms share electrons, covalent bonds form. In a nonpolar covalent bond, the two atoms share electrons equally. In a polar covalent bond, one atom may attract the electron more strongly than the other atom does; in this case, the strongly attracting atom bears a slightly negative charge, and the weakly attracting atom bears a slightly positive charge. Some polar covalent bonds give rise to hydrogen bonding, the attraction between charged regions of individual polar molecules or distant parts of a large polar molecule.

3) Why Is Water So Important to Life?

Properties of the water molecule that are important to living organisms include its ability to interact with many other molecules and to dissolve many polar and charged substances; to force nonpolar substances, such as fat, to assume certain types of physical organization; to participate in chemical reactions; to cohere to itself by using hydrogen bonds between water molecules; and to maintain a fairly stable temperature in the face of wide temperature fluctuations in the environment.

Key Terms

acid *p. 30*
acidic *p. 30*
atom *p. 21*
atomic nucleus *p. 21*
atomic number *p. 21*
base *p. 30*
basic *p. 30*
buffer *p. 30*
calorie *p. 31*
chemical bond *p. 25*
chemical reaction *p. 25*
cohesion *p. 29*
compound *p. 24*
covalent bond *p. 25*
electron *p. 21*
electron shell *p. 22*
element *p. 21*
hydrogen bond *p. 27*
hydrophilic *p. 28*
hydrophobic *p. 28*
hydrophobic interaction *p. 29*
ion *p. 25*
ionic bond *p. 25*
isotope *p. 22*
molecule *p. 24*
neutron *p. 21*
nonpolar covalent bond *p. 27*
pH scale *p. 30*
polar covalent bond *p. 27*
proton *p. 21*
radioactive *p. 22*
solvent *p. 28*
surface tension *p. 29*

Thinking Through the Concepts

Multiple Choice

1. *What is the purest form of matter that cannot be separated into different substances by chemical means?*
   a. compounds
   b. molecules
   c. atoms
   d. elements
   e. electrons
2. *Which phrase best describes chemical bonds?*
   a. physical bridges
   b. attractive forces
   c. shared protons
   d. atomic reactions
   e. all of these phrases are equally descriptive
3. *When an atom ionizes, what happens?*
   a. It shares one or more electrons with another atom.
   b. It emits energy as it loses extra neutrons.
   c. It gives up or takes up one or more electrons.
   d. It shares a hydrogen atom with another atom.
   e. none of the above
4. *If electrons in water molecules were equally attracted to hydrogen nuclei and oxygen nuclei, water molecules would be*
   a. more polar
   b. less polar
   c. unchanged
   d. triple bonded
   e. unable to form
5. *A covalent bond forms*
   a. when two ions are attracted to one another
   b. between adjacent water molecules, producing surface tension
   c. when one atom gives up its electron to another atom
   d. when two atoms share electrons
   e. between water molecules and fat globules
6. *What is the defining characteristic of an acid?*
   a. It donates hydrogen ions.
   b. It accepts hydrogen ions.
   c. It will donate or accept hydrogen ions, depending on the pH.
   d. It has an excess of hydroxide ions.
   e. It has a pH greater than 7.

? Review Questions

1. What are the six most abundant elements that occur in living organisms?
2. Distinguish among atoms and molecules; elements and compounds; and protons, neutrons, and electrons.
3. Compare and contrast covalent bonds and ionic bonds.
4. Why can water absorb a great amount of heat with little increase in its temperature?
5. Describe how water dissolves a salt. How does this phenomenon compare with the effect of water on a hydrophobic substance such as corn oil?
6. Define *acid*, *base*, and *buffer*. How do buffers reduce changes in pH when hydrogen ions or hydroxide ions are added to a solution? Why is this phenomenon important in organisms?

Applying the Concepts

1. Many "over-the-counter" substances are sold to bring relief from "acid stomach" or "heartburn." What is the chemical basis for these compounds? Why do they work?
2. Fats and oils do not dissolve in water; polar and ionic molecules dissolve easily in water. Detergents and soaps help clean by dispersing fats and oils in water so that they can be rinsed away. From your knowledge of the structure of water and the hydrophobic nature of fats, what general chemical structures (for example, polar or nonpolar parts) must a soap or detergent have, and why?
3. What would the effects be for aquatic life if the density of ice were greater than that of liquid water? What would be the impact on terrestrial organisms?
4. How does sweating help you regulate your body temperature? Why do you feel hotter and more uncomfortable on a hot, humid day than on a hot, dry day?

For More Information

Atkins, P. W. *Molecules*. New York: Scientific American Library, 1987. A layperson's introduction to atoms and molecules, with superb illustrations.

Glasheen, J. W., and McMahon, T. A. "Running on Water." *Scientific American*, September 1997. Answers the question, "How does the basilisk lizard run on water?"

Morrison, P., and Morrison, P. *Powers of Ten*. New York: W. H. Freeman, 1982. A fascinating journey from the universe to the nucleus of an atom.

Raloff, J. "Chocolate Hearts." *Science News,* March 18, 2000. Describes recent research indicating that chocolate is high in antioxidants.

Storey, K. B., and Storey, J. M. "Frozen and Alive." *Scientific American*, December 1990. By triggering ice formation here, suppressing it there, and loading up their cells with antifreeze molecules, some animals, including certain lizards and frogs, can survive with 60% of their body water frozen solid.

Answers to Multiple-Choice Questions

1.d 2.b 3.c 4.b 5.d 6.a

MediaTutor
Atoms, Molecules, and Life

CD Activities

Activity 2.1: Interactive Atoms

Estimated time: 5 minutes

This tutorial provides an introduction to atomic structure and chemical bonding. An understanding of simple chemistry is critical to understanding biology. This tutorial will provide the background to learn about more complex biological molecules. The bonds that hold together large, complex, biologically important molecules like DNA are the same as the bonds we will learn about here.

Activity 2.2: Water and Life

Estimated time: 5 minutes

Water is essential for life. In fact, water may be the only absolute essential for living systems. This tutorial explores the properties of water and how they relate to living systems.

Start the MediaTutor Student CD-ROM and enter the activity number in the Quick Search box to be taken directly to that activity.

Web Investigations

Case Study: Health Food?

Estimated time: 10 minutes

Ah, creamy, sweet, delicious chocolate! Most people love it. Some even think it's addictive. This exercise takes a closer, scientific look at an old favorite.

Go to http://www.prenhall.com/audesirk6, the Audesirk Companion Web site. Select Chapter 2 and the Web Investigation to begin.

Products from potato chips to diet sodas may contain artificially modified molecules that substitute for fats and sugar.

3 Biological Molecules

AT A GLANCE

Case Study: Improving on Nature?

1) **Why Is Carbon So Important in Biological Molecules?**

2) **How Are Organic Molecules Synthesized?**

Biological Molecules Are Joined Together or Broken Apart by Adding or Removing Water

3) **What Are Carbohydrates?**

There Are a Variety of Monosaccharides with Slightly Different Structures

Disaccharides Consist of Two Single Sugars Linked by Dehydration Synthesis

Polysaccharides Are Chains of Single Sugars

4) **What Are Lipids?**

Oils, Fats, and Waxes Are Lipids Containing Only Carbon, Hydrogen, and Oxygen

Phospholipids Have Water-Soluble "Heads" and Water-Insoluble "Tails"

Steroids Consist of Four Carbon Rings Fused Together

5) **What Are Proteins?**

Proteins Are Formed from Chains of Amino Acids

Amino Acids Are Joined to Form Chains by Dehydration Synthesis

A Protein Can Have Up to Four Levels of Structure

The Functions of Proteins Are Linked to Their Three-Dimensional Structures

6) **What Are Nucleic Acids?**

DNA and RNA, the Molecules of Heredity, Are Nucleic Acids

Other Nucleotides Act as Intracellular Messengers, Energy Carriers, or Coenzymes

Case Study Revisited: Improving on Nature?

CASESTUDY

Improving on Nature?

Protein, carbohydrate, fat, sugar: Most people know that these are the very building blocks of our bodies (and all other organisms on Earth). The special properties of biological molecules are part of what makes living bodies different from nonliving objects. As our understanding of these molecules increases, so do efforts to modify them. Why are we so busy trying to improve on nature's handiwork?

In societies blessed with an overabundance of food, obesity is a serious health problem. One goal of food scientists is to modify biological molecules to make them noncaloric. Sugar is a prime candidate; several artificial sweeteners such as saccharin, aspartame (Nutrasweet™), and the newly approved sucralose (Splenda™) add a sweet taste to foods, while providing few or no calories.

Government agencies such as the U.S. Food and Drug Administration (FDA) require exhaustive testing to demonstrate their safety of synthetic food products. Olestra (Olean™), an artificial oil substitute, was approved by the FDA in 1996 for use in snack foods. Olestra mimics the culinary properties of natural oils but is completely indigestible, ensuring that potato chips made with olestra have no fat calories and far fewer total calories than normal chips. It also means that some people's digestive systems might have difficulty coping with the product. To convince the FDA of its safety, olestra's developer, Procter & Gamble, submitted results from more than 150 studies on humans and animals.

How are these "nonbiological molecules" made? How can you, as an educated consumer, decide whether you want to add those tasty "nonfat" potato chips or artificially sweetened diet cola to your shopping cart? For starters, you need to understand real biological molecules, the building blocks of life. ■

1 Why Is Carbon So Important in Biological Molecules?

You have probably seen "organic" fruits and vegetables in your grocery store. To a chemist, this phrase is redundant; all produce is organic, because it is formed from biological molecules. In chemistry, the term **organic** is used to describe molecules that have a carbon skeleton and also contain some hydrogen atoms. The term *organic* is derived from the ability of living *organisms* to synthesize and use these molecules. **Inorganic** molecules include carbon dioxide and all molecules without carbon, such as water.

Although the common structure and function of the types of organic molecules among organisms afford unity, the tremendous range of organic molecules accounts for the diversity of living organisms and for the diversity of structures within single organisms and even within individual cells. This vast array of organic molecules, in turn, is possible because the carbon atom is so versatile. A carbon atom has four electrons in its outermost shell, with room for eight. Therefore, carbon atoms are able to form many bonds. They become stable by sharing four electrons with other atoms, forming up to four single covalent bonds or fewer double or triple covalent bonds. Molecules with many carbon atoms can assume complex shapes, including chains, branches, and rings, the basis for an amazing diversity of molecules.

Organic molecules are much more than just complicated skeletons of carbon atoms, however. Attached to the carbon backbone are groups of atoms, called **functional groups**, that determine the characteristics and chemical reactivity of the molecules. These functional groups are far less stable than the carbon backbone and are more likely to participate in chemical reactions. The common functional groups found in organic molecules are shown in Table 3-1.

The similarity among organic molecules from all forms of life is a consequence of two main features: (1) the use of the same set of functional groups in virtually all organic molecules in all types of organisms and (2) the use of the "modular approach" to synthesizing large organic molecules.

Table 3-1 Important Functional Groups in Biological Molecules

Group	Structure	Properties	Types of Molecules
Hydrogen (–H)	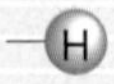	Polar or nonpolar, depending on which atom hydrogen is bonded to; involved in condensation and hydrolysis	Almost all organic molecules
Hydroxyl (–OH)	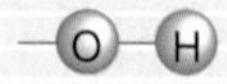	Polar; involved in condensation and hydrolysis	Carbohydrates, nucleic acids, alcohols, some acids, and steroids
Carboxyl (–COOH)	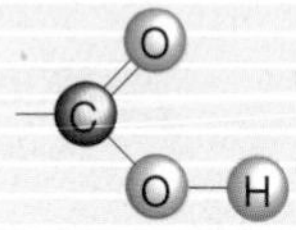	Acidic; negatively charged when H^+ dissociates; involved in peptide bonds	Amino acids, fatty acids
Amino ($-NH_2$)	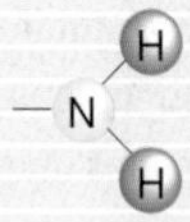	Basic; may bond an additional H^+, becoming positively charged; involved in peptide bonds	Amino acids, nucleic acids
Phosphate ($-H_2PO_4$)	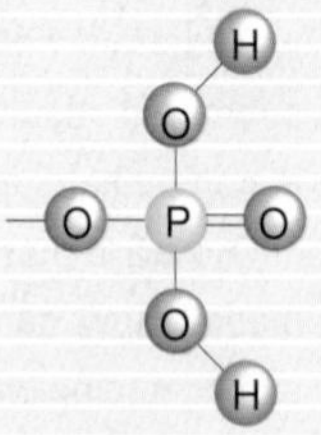	Acidic; up to two negative charges when H^+ dissociates; links nucleotides in nucleic acids; energy-carrier group in ATP	Nucleic acids, phospholipids
Methyl ($-CH_3$)	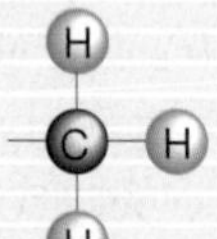	Nonpolar; tends to make molecules hydrophobic	Many organic molecules; especially common in lipids

MEDIATUTOR 3.1 Structure of Biological Molecules

2 How Are Organic Molecules Synthesized?

In principle, there are two ways to manufacture a large, complex molecule: by combining atom after atom following an extremely detailed blueprint or by preassembling smaller molecules and hooking them together. Just as trains are made by coupling engines to various train cars, so too life takes the modular approach. Small organic molecules (for example, *sugars*) are used as **subunits** to synthesize longer molecules (for example, *starches*), like cars in a train. The individual subunits are often called **monomers** (from Greek words meaning "one part"); long chains of monomers are called **polymers** ("many parts"). Most organic molecules in our bodies are in the form of either monomers or polymers.

Biological Molecules Are Joined Together or Broken Apart by Adding or Removing Water

You already learned some of the reasons why water is so important to life; here you will learn that water plays a central role in reactions that break down biological molecules to liberate subunits that your body can use. In addition, when complex biological molecules are synthesized in the body, water is often produced as a by-product.

The subunits that make up large biological molecules are almost always linked together by a chemical reaction called a **dehydration synthesis** (literally, "to form by removing water"). In a dehydration synthesis, a hydrogen atom (–H) is removed from one subunit and a hydroxyl group (–OH) is removed from a second subunit, creating openings in the outer electron shells of the two subunits. These openings are filled by sharing electrons between the subunits, creating a covalent bond that links them. The free hydrogen and hydroxyl ions then combine to form a molecule of water (H_2O):

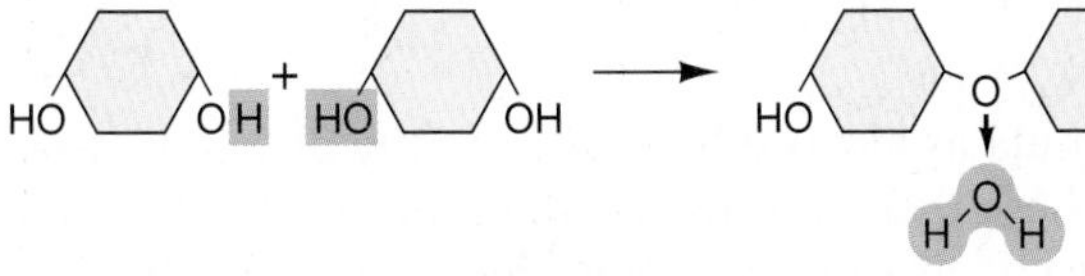

Dehydration synthesis

The reverse reaction, **hydrolysis** ("to break apart with water"), can split the molecule into individual subunits:

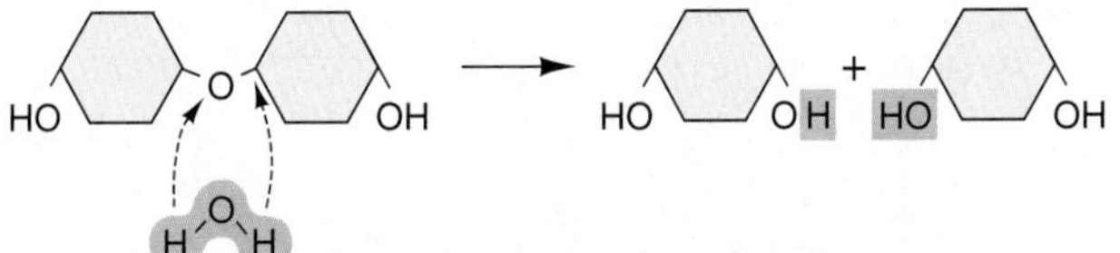

Hydrolysis

Considering how complicated living things are, it might surprise you that nearly all biological molecules fall into one of only four general categories: carbohydrates, lipids, proteins, or nucleic acids (Table 3-2).

3 What Are Carbohydrates?

Carbohydrates are molecules composed of carbon, hydrogen, and oxygen in the approximate ratio of 1:2:1. All carbohydrates are either small, water-soluble **sugars** (*glucose* or *fructose*, for example) or chains, such as starch or *cellulose*, that are made by stringing sugar subunits together. If a carbohydrate consists of just one sugar molecule, it is called a **monosaccharide** (Greek for "single sugar"). When two or more monosaccharides are linked, they form a **disaccharide** ("two sugars") or a **polysaccharide** ("many sugars").

Carbohydrates such as sugars and starches are important energy sources for most organisms. Consider a breakfast that includes blueberry pancakes, syrup, and orange juice. Your pancakes consist mainly of carbohydrates stored in the seeds of wheat or other grains. The sugar that sweetens your syrup, blueberries, and orange juice is also stored by plants as an energy source. Other carbohydrates, such as cellulose and similar molecules, provide structural support for individual cells or even for the entire bodies of organisms as diverse as plants, fungi, bacteria, and insects.

Most single sugars (monosaccharides) have a backbone of three to seven carbon atoms. Most of the carbon atoms have both a hydrogen (–H) and a hydroxyl group (–OH) attached to them, so carbohydrates generally have the approximate chemical formula $(CH_2O)_n$. (Here n is the number of carbons in the backbone.) This formula explains the origin of the name *carbohydrate*, which literally means "carbon plus water." When dissolved in water, such as in the cytoplasm of a cell, the carbon backbone of a sugar usually "circles up" into a ring. Here are a linear representation and two ways the circular (or "ring") structure of a carbohydrate may be drawn, all for the sugar glucose:

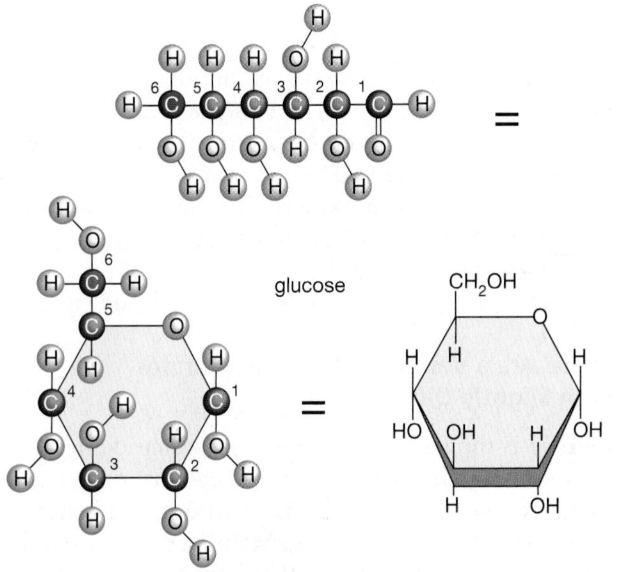

Table 3-2 The Principal Biological Molecules

Class of Molecule	Principal Subtypes (subunits in parentheses)	Example	Function
Carbohydrate: Normally contains carbon, oxygen, and hydrogen, in the approximate formula $(CH_2O)_n$	(*Monosaccharide*: Simple sugar)	Glucose	Important energy source for cells; subunit of which most polysaccharides are made
	Disaccharide: Two monosaccharides bonded together	Sucrose	Principal sugar transported throughout bodies of land plants
	Polysaccharide: Many monosaccharides (normally glucose) bonded together	Starch Glycogen Cellulose	Energy storage in plants Energy storage in animals Structural material in plants
Lipid: Contains high proportion of carbon and hydrogen; usually nonpolar and insoluble in water	*Triglyceride*: Three fatty acids bonded to glycerol	Oil, fat	Energy storage in animals, some plants
	Wax: Variable numbers of fatty acids bonded to long-chain alcohol	Waxes in plant cuticle	Waterproof covering on leaves and stems of land plants
	Phospholipid: Polar phosphate group and two fatty acids bonded to glycerol	Phosphatidylcholine	Common component of membranes in cells
	Steroid: Four fused rings of carbon atoms with functional groups attached	Cholesterol	Common component of membranes of eukaryotic cells; precursor for other steroids such as testosterone, bile salts
Protein: Chains of amino acids; contain carbon, hydrogen, oxygen, nitrogen, and sulfur	(*amino acids*)	Keratin	Helical protein, principal component of hair
		Silk	Protein produced by silk moths and spiders
		Hemoglobin	Globular protein composed of four subunit peptides; transport of oxygen in vertebrate blood
Nucleic acid: Made of nucleotide subunits; may consist of a single nucleotide or long chain of nucleotides	*Long-chain nucleic acids*	Deoxyribonucleic acid (DNA)	Genetic material of all living cells
		Ribonucleic acid (RNA)	Genetic material of some viruses; in living cells, essential in transfer of genetic information from DNA to protein
	(Single nucleotides)	Adenosine triphosphate (ATP)	Principal short-term energy-carrier molecule in cells
		Cyclic adenosine monophosphate (cyclic AMP)	Intracellular messenger

It is in the ring form that sugars link together to make disaccharides (see Fig. 3-1) and polysaccharides.

Most small carbohydrates are soluble in water. As in water molecules, the O–H bond in a hydroxyl group is polar, because oxygen attracts electrons more strongly than does hydrogen. Hydrogen bonds between water molecules and the polar hydroxyl groups of the carbohydrate keep the carbohydrate in solution (see Fig. 2-7).

There Are a Variety of Monosaccharides with Slightly Different Structures

Glucose is the most common monosaccharide in living organisms and is the subunit of which most polysaccharides are made. Glucose has six carbons, and hence its chemical formula is $C_6H_{12}O_6$. Many organisms synthesize other monosaccharides that have the same chemical formula as glucose but have slightly different structures. These include *fructose* (the "corn sugar" found in corn syrup, also the molecule that makes your orange juice taste sweet) and *galactose* (part of lactose, or "milk sugar"), shown below:

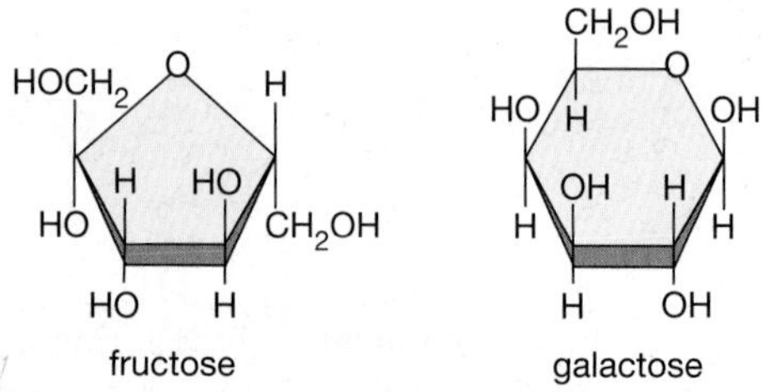

Some other common monosaccharides, such as *ribose* and *deoxyribose*, have five carbons:

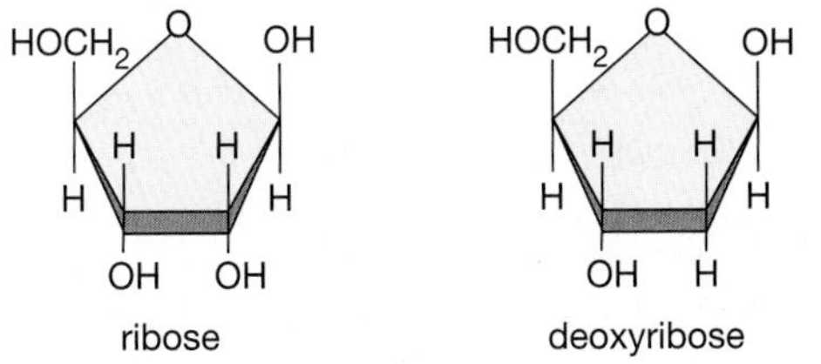

Ribose and deoxyribose are parts of the genetic molecules *ribonucleic acid* (RNA) and *deoxyribonucleic acid* (DNA), respectively.

Disaccharides Consist of Two Single Sugars Linked by Dehydration Synthesis

Monosaccharides, especially glucose and its relatives, have a short life span in a cell. Most are either broken down to free their chemical energy for use in various cellular activities or are linked by dehydration synthesis to form disaccharides or polysaccharides (Fig. 3-1). Disaccharides are often used for short-term energy storage, especially in plants. Perhaps you had coffee with cream and sugar at breakfast. Common disaccharides include **sucrose** (glucose plus fructose), which you stirred into your coffee; **lactose** (milk sugar: glucose plus galactose), found in the milk you poured in your coffee; and **maltose** (glucose plus glucose, which will form in your digestive tract as you break down the starch in your pancakes). When energy is required, the disaccharides are broken apart into their monosaccharide subunits by hydrolysis.

Polysaccharides Are Chains of Single Sugars

Try chewing a cracker for a long time to allow enzymes in your saliva to break it down into its component sugars. Does it taste sweeter the longer you chew? It should; monosaccharides (usually glucose) are joined together into polysaccharides to form **starch** (in plants; Fig. 3-2) or **glycogen** (in animals) for long-term energy storage. Starch is commonly formed in roots and seeds—in the case of your cracker, from the seeds of wheat. Starch may occur as coiled, unbranched chains of up to 1000 glucose subunits or, more commonly, as huge branched chains of up to half a million glucose monomers. Glycogen, stored as an energy source in the liver and muscles of animals such as humans, is generally much smaller than starch and has branches every 10 to 12 glucose subunits. Having many small branches probably makes it easier to split off the glucose subunits for quick energy release.

Many organisms also use polysaccharides as structural materials. One of the most important structural polysaccharides is **cellulose**, which makes up most of the cell walls of plants and about half the bulk of a tree trunk (Fig. 3-3). When you picture the vast fields and forests that blanket much of our planet, you will not be surprised to learn that there is probably more cellulose on Earth than all other organic molecules put together. Ecologists estimate that about a trillion tons of cellulose are synthesized each year!

Like starch, cellulose consists of glucose subunits bonded together. However, whereas most animals can easily digest starch, only a few microbes, such as those in the digestive tracts of cows or termites, can digest cellulose. Why is this the case, given that both starch and cellulose consist of glucose? The orientation of the bonds between subunits is different in the two polysaccharides. In cellulose, every other glucose is "upside down" (compare Fig. 3-2c with Fig. 3-3). This bond orientation prevents the digestive enzymes of animals from attacking the bonds between glucose subunits. Enzymes synthesized by certain microbes, though, can break these bonds. As a result, cellulose is food for these microbes. But for most animals, cellulose is *roughage* or *fiber*, material that passes undigested through the digestive tract.

The hard outer coverings (exoskeletons) of insects, crabs, and spiders are made of **chitin**, a polysaccharide in which the glucose subunits have been chemically modified by the addition of a nitrogen-containing functional group (Fig. 3-4, p. 43). Interestingly, chitin also stiffens the cell walls of many fungi. Bacterial cell walls contain other types of modified polysaccharides, as do the lubricating fluids in our joints and the transparent corneas of our eyes.

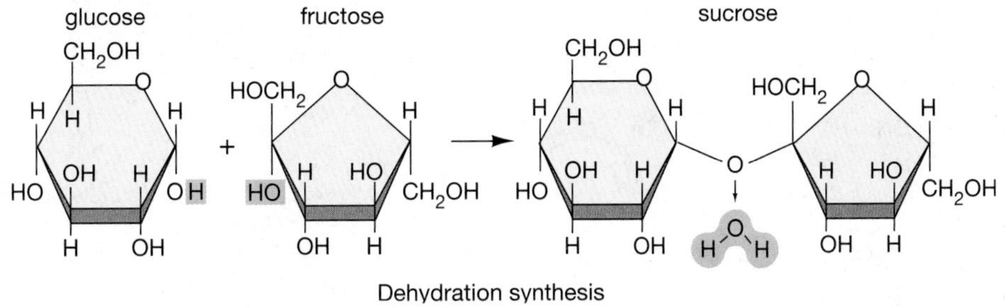

Figure 3-1 Synthesis and breakdown of a disaccharide
The disaccharide sucrose is synthesized by a dehydration synthesis reaction in which a hydrogen (–H) is removed from glucose and a hydroxyl group (–OH) is removed from fructose. A water molecule is formed in the process, leaving the two monosaccharide rings joined by single bonds to the remaining oxygen atom. Hydrolysis of sucrose is just the reverse of its synthesis, as water is split and added to the monosaccharides.

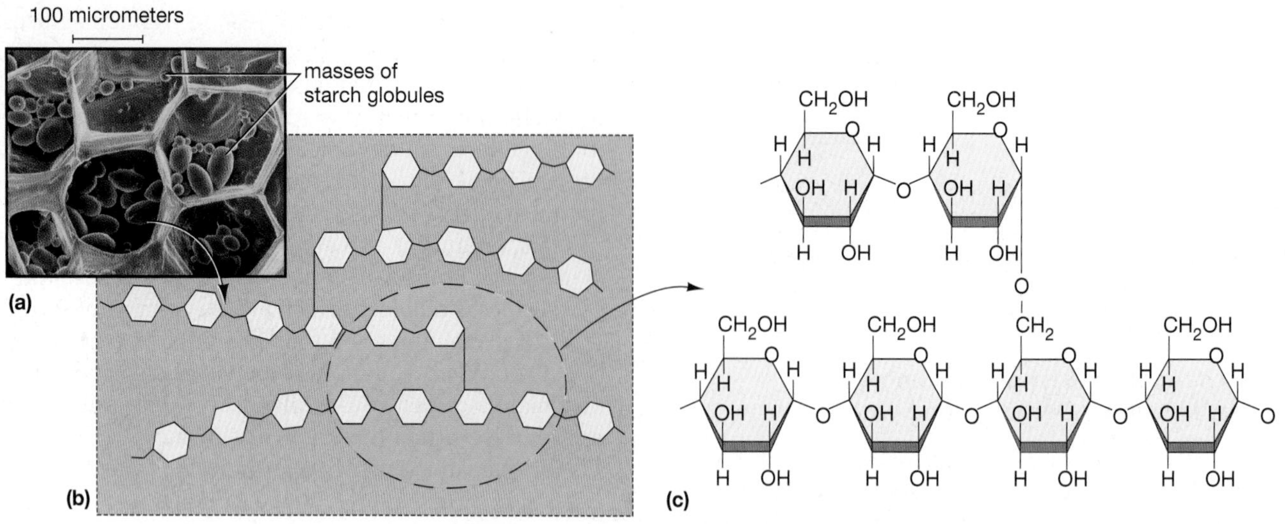

Figure 3-2 Starch is an energy-storage polysaccharide made of glucose subunits
(a) Starch globules inside individual potato cells. Most plants synthesize starch, which forms water-insoluble globules consisting of many starch molecules, which make up the bulk of this potato. *(b)* A small portion of a single starch molecule. Starch commonly occurs as branched chains of up to half a million glucose subunits. *(c)* The precise structure of the blue highlighted portion of the starch molecule in (b). Note the linkage between the individual glucose subunits for comparison with cellulose (Fig. 3-3).

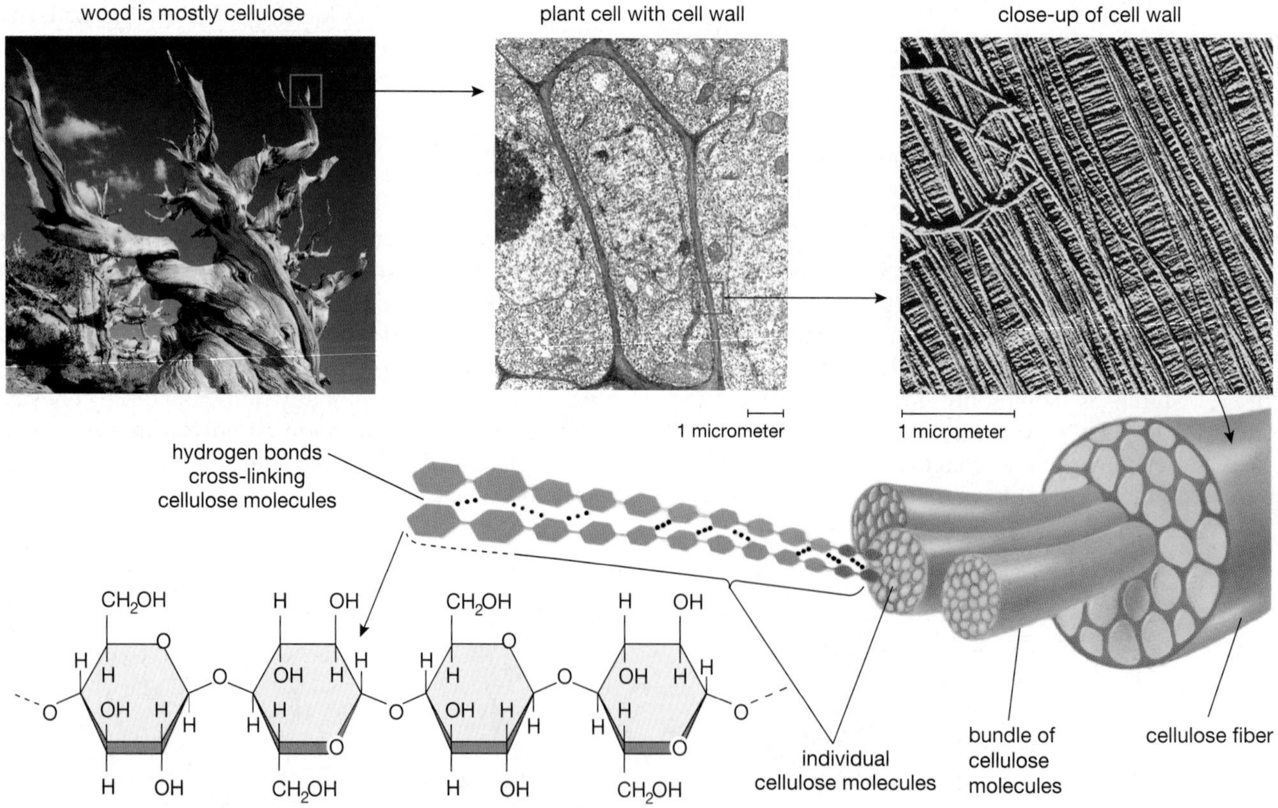

Figure 3-3 Cellulose structure and function
Cellulose, like starch, is composed of glucose subunits, but the orientation of the bond between subunits in cellulose is different (compare with Fig. 3-2c) such that every other glucose molecule is "upside down." Unlike starch, cellulose has great structural strength, due partly to the difference in bonding and partly to the arrangement of parallel molecules of cellulose into long, cross-linked fibers. Plant cells often lay down cellulose fibers in layers that run at angles to each other, resulting in resistance to tearing in both directions. The final product can be incredibly tough, as testified by this 3000-year-old bristlecone pine in California's White Mountains.

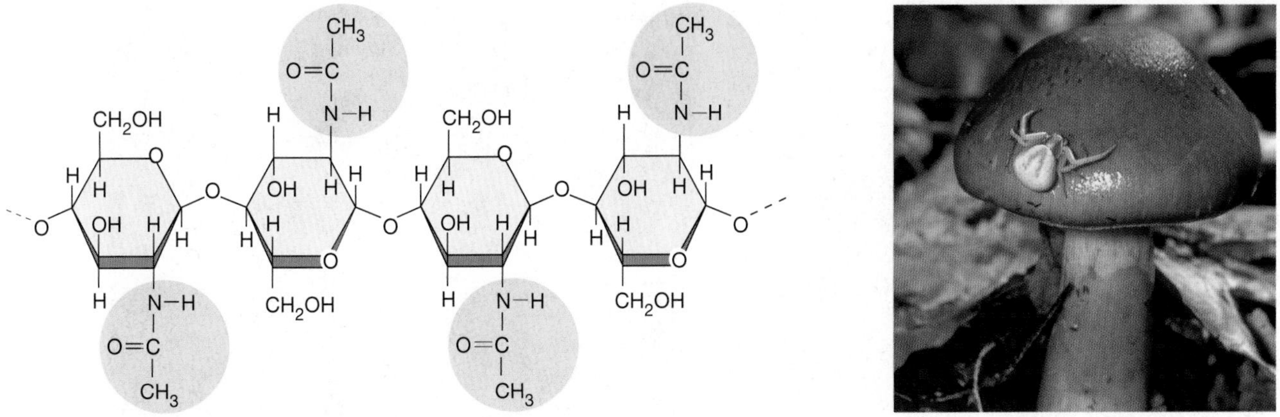

Figure 3-4 Chitin: A unique polysaccharide
Chitin has the same "alternating upside down" bonding of glucose molecules as cellulose has, but in chitin the glucose subunits are modified by the replacement of one of the hydroxyl groups with a nitrogen-containing functional group (yellow). Tough, slightly flexible chitin supports the otherwise soft bodies of arthropods (insects, spiders, and their relatives) and fungi.

Many other molecules, including *mucus*, some chemical messengers called *hormones*, and many molecules in the plasma membrane, consist in part of carbohydrate. Perhaps the most interesting of these molecules are the nucleic acids (discussed later in this chapter), which carry hereditary information.

4 What Are Lipids?

Lipids are a diverse assortment of molecules, all of which share two important features. *First*, lipids contain large regions composed almost entirely of hydrogen and carbon, with nonpolar carbon-carbon or carbon-hydrogen bonds. *Second*, these nonpolar regions make lipids hydrophobic and insoluble in water. The various types of lipids serve a wide variety of functions. Some lipids are energy-storage molecules; some form waterproof coverings on plant or animal bodies; some make up the bulk of all the membranes of a cell; still others are hormones.

Lipids are classified into three major groups: (1) oils, fats, and waxes, which are similar in structure and contain only carbon, hydrogen, and oxygen; (2) phospholipids, structurally similar to oils but also containing phosphorus and nitrogen; and (3) the fused-ring family of steroids.

Oils, Fats, and Waxes Are Lipids Containing Only Carbon, Hydrogen, and Oxygen

Oils, fats, and waxes are related in three ways: (1) They contain only carbon, hydrogen, and oxygen; (2) they contain one or more **fatty acid** subunits, which are long chains of carbon and hydrogen with a *carboxyl group* (–COOH) at one end; and (3) they usually do not have ring structures. **Fats** and **oils** are formed by dehydration synthesis from three fatty acid subunits and one molecule of **glycerol**, a short, three-carbon molecule with one hydroxyl group (–OH) per carbon:

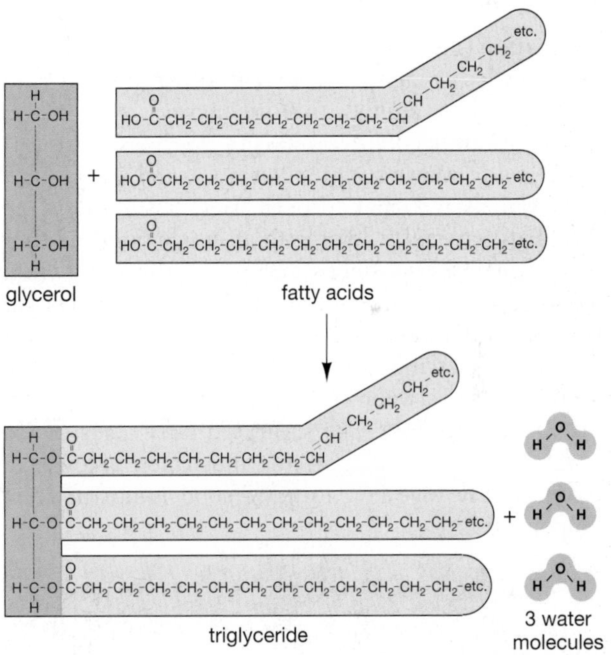

This structure of three fatty acids joined to one glycerol molecule gives fats and oils their chemical name, **triglycerides**. Notice that a double bond between two carbons in the fatty acid subunit forms a kink in the chain.

Fats and oils have a high concentration of chemical energy, about 9.3 Calories per gram, compared with 4.1 for sugars and proteins. (A *Calorie* with a capital *C* equals 1000 calories; the Calorie is used in measuring the energy content of foods.) Because fats are so "dense" in calories, fat substitutes such as Olestra may be especially appealing to dieters. Fats and oils are used for long-term energy storage in both plants and animals. For example, bears that feast during summer and fall put on fat to tide them over during their winter

(a)

(b)

Figure 3-5 Lipids
(a) A European brown bear ready to hibernate. Fat is an efficient way to store energy. If this bear stored the same amount of energy in carbohydrates instead of fat, she probably would be unable to walk! ***(b)*** Wax is a highly saturated lipid that remains very firm at normal outdoor temperatures. Its rigidity allows it to be used to form the strong but thin-walled hexagons of this honeycomb.

hibernation (Fig. 3-5a). Because fats store the same energy with less weight than do carbohydrates, fat is an efficient way for animals to store energy.

The difference between a fat (such as beef fat), which is a solid at room temperature, and an oil (such as the one used to make french fries) lies in their fatty acids. Fats have fatty acids with all single bonds in their carbon chains. Hydrogens occupy all the other bond positions on the carbons. The resulting fatty acid is said to be **saturated**, because it is "saturated" with hydrogens—it has as many hydrogens as possible. Lacking double bonds between carbons, the carbon chain of the fatty acid is straight. The saturated fatty acids of fats (such as the beef fat molecule illustrated below) can nestle closely together, forming solid lumps at room temperature:

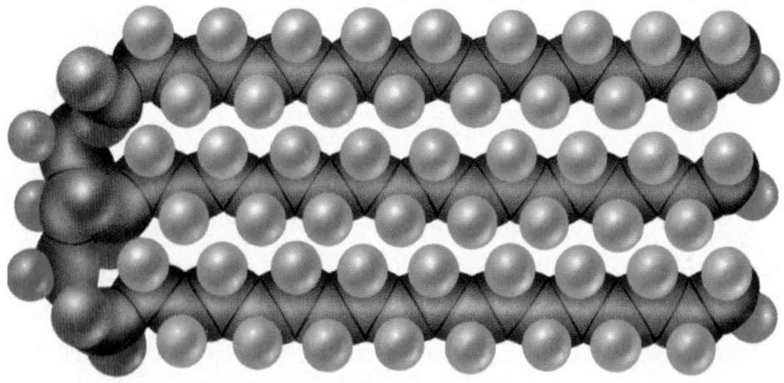

Beef fat (saturated)

If there are double bonds between some of the carbons, and consequently fewer hydrogens, the fatty acid is said to be **unsaturated.** Oils have mostly unsaturated fatty acids. The double bonds in the unsaturated fatty acids of oils produce kinks in the fatty acid chains, as illustrated by the linseed oil molecule (top of next column):

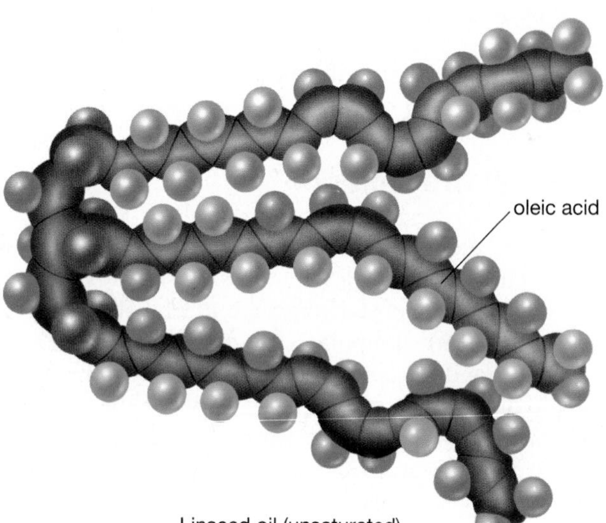

Linseed oil (unsaturated)

The kinks keep oil molecules apart; as a result, oil is liquid at room temperature. An oil can be converted to a fat by breaking the double bonds between carbons, replacing them with single bonds, and adding hydrogens to the remaining bond positions. This is the "hydrogenated oil" listed in the ingredients on a box of margarine, which allows the margarine to be solid at room temperature. Interestingly, most of the saturated fat in the human diet comes from animals; butter, bacon fat, and the fat on steak are examples. By contrast, we get most of our unsaturated oils from the seeds of plants, where it is used by their developing embyos. Corn oil, peanut oil, and canola oil are all examples.

Waxes are chemically similar to fats. They are highly saturated, making them solid at normal outdoor

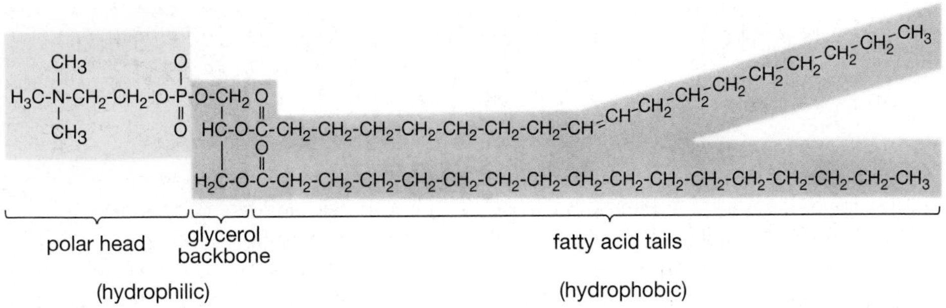

Figure 3-6 Phospholipids
Phospholipids are similar to fats and oils, except that only two fatty acid tails are attached to the glycerol backbone. The third position on the glycerol is occupied by a polar head composed of a phosphate group ($-PO_4^-$) to which a second, typically nitrogen-containing, functional group is attached. The phosphate group is negatively charged, and the nitrogen-containing group is positively charged.

temperatures. Waxes form a waterproof coating over the leaves and stems of land plants. Animals synthesize waxes as waterproofing for mammalian fur and insect exoskeletons and, in a few cases, to build elaborate structures such as beehives (Fig. 3-5b).

Phospholipids Have Water-Soluble "Heads" and Water-Insoluble "Tails"

The plasma membrane that separates the inside of a cell from the outside world contains several types of **phospholipids.** A phospholipid is similar to an oil, except that one of the three fatty acids is replaced by a phosphate group with a short, polar functional group, typically a nitrogen-containing group, attached to the end (Fig. 3-6). Unlike the two fatty acid "tails," which are insoluble in water, the phosphate-nitrogen "head" is polar, or charged, and is water soluble. Thus, a phospholipid has two dissimilar ends: a hydrophilic head attached to hydrophobic tails. As you will see in Chapter 4, this dual nature of phospholipids is crucial to the structure and function of the plasma membrane.

Steroids Consist of Four Carbon Rings Fused Together

Steroids are structurally different from all the other lipids. All steroids are composed of four rings of carbon fused together with various functional groups protruding from them (Fig. 3-7); note the basic steroid "skeleton" in color). One type of steroid is *cholesterol*; an egg yolk supplies more than half your recommended daily allotment. Cholesterol is a vital component of the membranes of animal cells and is also used by cells to synthesize other steroids. Other steroids the body synthesizes from cholesterol include male and female sex hormones, hormones that regulate salt levels, and bile that assists in fat digestion. Why, then, has cholesterol gotten so much bad publicity? Find out in "Health Watch: Cholesterol—Friend and Foe."

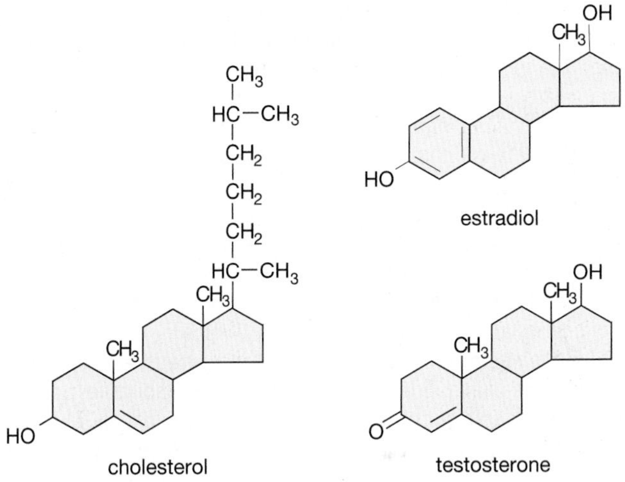

Figure 3-7 Steroids
Steroids are synthesized from cholesterol. All steroids have almost the same molecular structure (colored rings). Great differences in steroid function can be a result of even minor differences in functional groups attached to the rings. Notice the similarity in structure between the male sex hormone testosterone and the female sex hormone estradiol (a type of estrogen).

5 What Are Proteins?

Proteins are molecules composed of one or more chains of *amino acids*. Proteins perform many functions; this diversity of function is made possible by the diversity of protein structures (Table 3-3). **Enzymes** are important proteins that guide almost all the chemical reactions that occur inside cells, as you will learn in Chapter 6. Because each enzyme assists only one or a few specific reactions, most cells contain hundreds of different enzymes. Other types of proteins are used for structural purposes, such as *elastin*, which gives skin its elasticity; *keratin*, the principal protein of hair, horns, and claws; and the silk of spider webs and silk moth cocoons

Health Watch
Cholesterol—Friend and Foe

Cholesterol is a steroid with a bad reputation. Why are so many products now advertising themselves as "cholesterol free" or "low in cholesterol"? After all, cholesterol is a crucial component of cell membranes. It is also the raw material for the production of bile (which helps us digest fats), vitamin D, and both male and female sex hormones.

Although this steroid is crucial to life, medical researchers have found that individuals with excessively high levels of cholesterol in their blood are at increased risk for heart attacks and strokes. Unfortunately, the cholesterol builds up "silently" and gives no warning signs. A person may not know that anything is wrong until he or she actually suffers a heart attack. Cholesterol contributes to the formation of obstructions in arteries, called *plaques,* which in turn can promote the formation of blood clots. These clots can break loose and block an artery carrying blood to the heart, causing a heart attack, or to the brain, causing a stroke. (Plaque formation is described in more detail in "Health Watch: Matters of the Heart" in Chapter 27.) Where does cholesterol come from? Cholesterol comes from animal-derived foods; it is essentially nonexistent in plants. There are several sources in typical breakfast foods. For example, egg yolks are a particularly rich source; sausages, bacon, whole milk, and butter contain it as well. Have you ever heard of "good" and "bad" cholesterol? Because cholesterol molecules are nonpolar, they do not dissolve in blood (which is mostly water). Thus, the cholesterol molecules are transported through blood in packets surrounded by special carrier molecules called *lipoproteins* (phospholipids plus proteins). Cholesterol in high-density lipoproteins packets ("HDL cholesterol," which has more protein and less lipid) is the "good" kind; these packets transport cholesterol to the liver, where it is removed from circulation and further metabolized (used in bile synthesis, for example). Cholesterol in low-density lipoprotein packets ("LDL cholesterol" with less protein and more lipid) is the "bad" kind; this is the form in which cholesterol circulates to cells throughout the body and can be deposited on artery walls. A high ratio of HDL ("good") to LDL ("bad") is correlated with reduced risk of heart disease. A complete cholesterol screening test will distinguish between these two forms in your blood.

Perhaps you've also heard about "trans" fatty acids as dietary villains. Margarine, particularly the type molded into sticks, is made from unsaturated vegetable oil, which is artificially hardened by saturating it. This process creates "trans" fatty acids (*trans* refers to the location of the double bonds between carbons), which do not occur in nature. Research suggests that these trans fatty acids are not metabolized normally and can both increase LDL cholesterol and decrease HDL cholesterol.

Another source of cholesterol is your own body, which can synthesize cholesterol from other lipids. Because of genetic differences, some people's bodies manufacture more than others. People with high cholesterol (about 25% of all adults in the United States) can often reduce their levels by eating a diet low in both cholesterol and saturated fats. For people with excessively high cholesterol who are unable to reduce it adequately by changing their diets, doctors often prescribe cholesterol-reducing drugs.

(Fig. 3-8). Still other types of proteins are used for energy and material storage (*albumin* in egg white, *casein* in milk), transport (*hemoglobin* to carry oxygen in the blood), and cell movement (contractile proteins in muscle). Some hormones (insulin, growth hormone), *antibodies* (which help fight disease and infection), and many poisons (rattlesnake venom) are also proteins.

Table 3-3 The Many Functions of Proteins

Function	Example
Structure	Collagen in skin; keratin in hair, nails, horns
Movement	Actin and myosin in muscle
Defense	Antibodies in bloodstream
Storage	Zeatin in corn seeds
Signals	Growth hormone in bloodstream
Catalysis	Enzymes: catalyze nearly every chemical reaction in our cells; DNA polymerase (makes DNA); pepsin (digests protein); amylase (digests carbohydrates); ATP synthase (makes ATP)

Proteins Are Formed from Chains of Amino Acids

Proteins are polymers of **amino acids**. All amino acids have the same fundamental structure, consisting of a central carbon bonded to four different functional groups: a nitrogen-containing *amino group* ($-NH_2$); a carboxyl, or carboxylic acid, group ($-COOH$); a hydrogen; and a variable group (represented by the letter R):

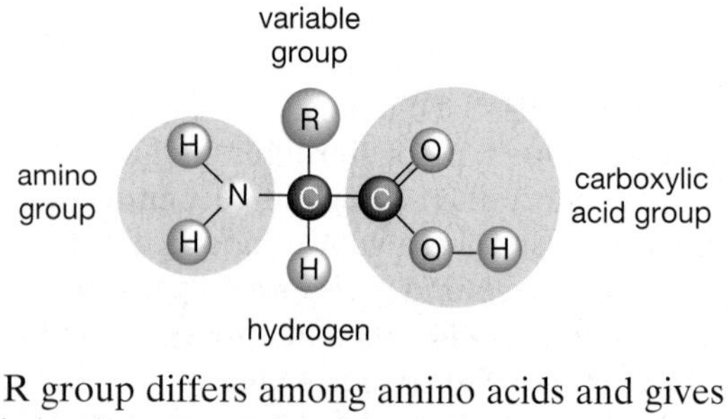

The R group differs among amino acids and gives each its distinctive properties (Fig. 3-9). Twenty amino acids are commonly found in the proteins of organisms.

Some amino acids are hydrophilic; their R groups are polar and soluble in water. Others are hydrophobic, with nonpolar R groups that are insoluble in

(a)

(b)

(c)

Figure 3-8 Structural proteins
Common structural proteins include those of *(a)* hair, *(b)* horn, and *(c)* spider web silk.

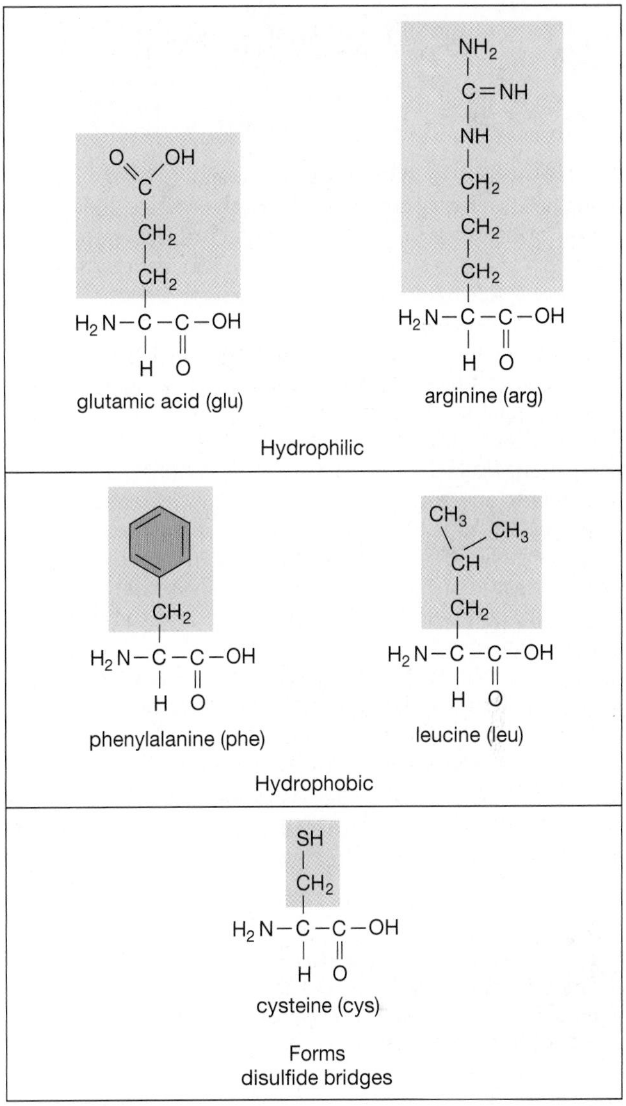

Figure 3-9 Amino acid diversity
The diversity of amino acid structures is a consequence of differences in the variable R group (colored blue). Some amino acids are classified according to the variable R group as hydrophilic, others as hydrophobic. Cysteine stands in a class by itself. Two cysteines in distant parts of a protein molecule can form a covalent bond between their sulfur atoms, creating a disulfide bridge that brings the cysteines very close together and bends the protein chain.

water. Another type of amino acid, cysteine, has sulfur in its R group and can form bonds with other cysteines, linking protein chains together. These bonds between the R groups of cysteine are called **disulfide bridges.**

Amino acids differ in their chemical and physical properties—size, water solubility, electrical charge—because of their different R groups. Therefore, the exact sequence of amino acids dictates the function of each protein: whether it is water soluble or not, whether it is an enzyme or a hormone or a structural protein. Scrambled sequences of amino acids are useless. In some cases, just one wrong amino acid can cause a protein to function incorrectly.

Amino Acids Are Joined to Form Chains by Dehydration Synthesis

Like lipids and polysaccharides, proteins are formed by dehydration synthesis. The nitrogen of the amino group ($-NH_2$) of one amino acid is joined to the carbon of the

A Closer Look
Protein Structure—A Hairy Subject

A single strand of human hair, thin and not even alive, is nonetheless a highly organized, complex structure. Hair is composed mostly of a single, helical protein called *keratin*. If we look closely at the structure of hair, we can learn a great deal about biological molecules and chemical bonds and why human hair behaves as it does.

A single hair consists of a hierarchy of structures (Fig. E3-1). The outermost layer is a set of overlapping shingle-like scales that protect the hair and keep it from drying out. Inside the hair lie closely packed, cylindrical dead cells, each filled with long strands called *microfibrils*. Each microfibril is a bundle of *protofibrils*, and each protofibril contains helical keratin molecules twisted together. A strand of hair grows because living cells in the hair follicle embedded in the skin whip out new keratin at the rate of 10 turns of the protein helix every second.

Pull on both ends of a detached hair, and you will notice that it is rather strong. Hair gets its strength from three types of chemical bonds. First, the individual molecules of keratin are held in their helical shape by many hydrogen bonds. Before a hair will break, all the hydrogen bonds of all the keratin molecules in one cross-sectional plane of the strand must break to allow the helix to be stretched to its maximal extent. Second, each molecule is cross-linked to neighboring keratin molecules

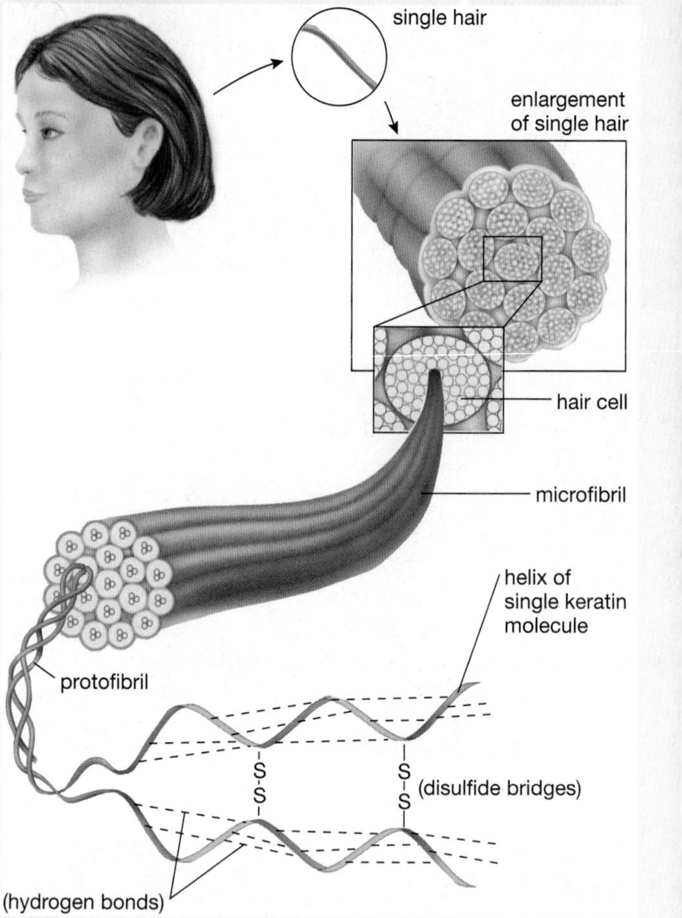

Figure E3-1 The organization of hair
At the microscopic level, a single hair is organized into bundles of fibers within further bundles of fibers. Hydrogen bonds and disulfide bridges between cysteines impart strength and elasticity to individual hairs.

carboxyl group (–COOH) of another amino acid by a single covalent bond (Fig. 3-10). This bond is called a **peptide bond**, and the resulting chain of two amino acids is called a **peptide**. More amino acids are added, one by one, until the protein is complete. Amino acid chains in living cells vary in length from three to thousands of amino acids. Often, the word *protein* or *polypeptide* is reserved for long chains—say, 50 or more amino acids in length—and *peptide* is used for shorter chains.

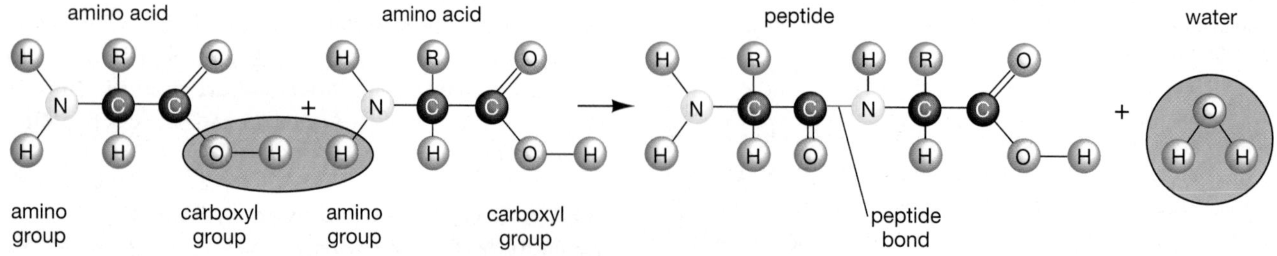

Figure 3-10 Protein synthesis
In protein synthesis, a dehydration synthesis joins the carbon of the carboxyl acid group of one amino acid to the nitrogen of the amino group of a second amino acid. The resulting covalent bond is called a *peptide bond.*

by disulfide bridges between cysteines (particular amino acids). Some of these bridges must break as the hair stretches. Finally, at least one peptide bond in each keratin molecule must break before the strand as a whole breaks.

Hair is also fairly stiff. The stiffness arises from hydrogen bonds within the individual helices of keratin and from disulfide bridges that hold neighboring keratin molecules together. When hair gets wet, however, the hydrogen bonds between turns of the helices are replaced by hydrogen bonds between the amino acids and the water molecules surrounding them, so the helices collapse. Wet hair is therefore very limp. If wet hair is rolled onto curlers and allowed to dry, the hydrogen bonds re-form in slightly different places, holding the hair in a curve. However, the slightest moisture, even humid air, allows these hydrogen bonds to rearrange into their natural configuration, and normally straight hair straightens out.

Pull gently, and you will discover still another property of hair. It stretches and then springs back into shape when you release the tension. When hair stretches, many of the hydrogen bonds within each keratin helix are broken, allowing the helix to be extended. Most of the covalent disulfide bonds between different levels of the helices, in contrast, are distorted by stretching but do not break. When tension is released, these disulfide bridges contract, returning the hair to its normal length.

Finally, each hair has a characteristic shape: It may be straight, wavy, or curly. The curliness of hair is genetically specified and is determined biochemically by the arrangement of disulfide bridges (see Fig. E3-1). Curly hair has disulfide bridges cross-linking the various keratin molecules at *different levels*, whereas straight hair has bridges mostly at the *same level*. When straight hair is given a "permanent," two lotions are applied. The first lotion breaks disulfide bonds between neighboring helices. The hair is then rolled tightly onto curlers, and a second solution, which re-forms the bridges, is applied. The new disulfide bridges connect helices at different levels, holding the strands of hair in a curl. These new bridges are more or less permanent, and genetically straight hair can be transformed into biochemically curly hair. As new hair grows in, it will have the genetically determined arrangement of bridges and will not be curly.

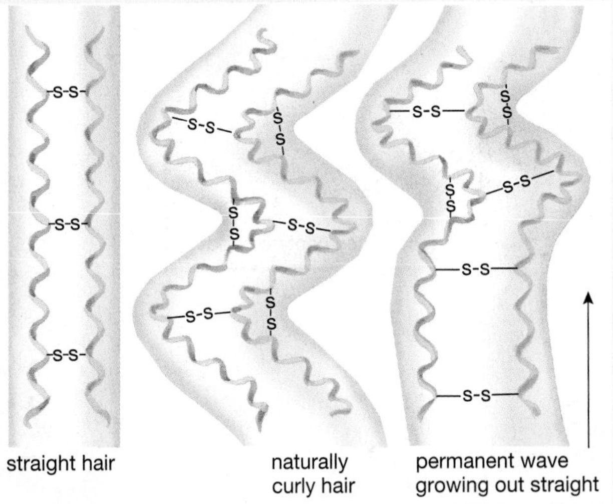

A Protein Can Have Up to Four Levels of Structure

The phrase "amino acid chains" may evoke images of proteins as floppy, featureless structures, but they are not. Proteins are, instead, highly organized molecules that come in a variety of shapes. Biologists recognize four levels of organization in protein structure. A single molecule of hemoglobin, the oxygen-carrying protein in red blood cells, exhibits all four structural levels (Fig. 3-11). The **primary structure** is the sequence of amino acids that make up the protein (Fig. 3-11a). This sequence is coded by the genes. Different types of proteins have different sequences of amino acids.

You may recall that hydrogen bonds can form between parts of molecules that have slight negative and slight positive charges, which attract one another (Ch. 2). Hydrogen bonds between parts of amino acids cause many protein chains to form one of two simple, repeating **secondary structures**. Many proteins, such as the hair protein keratin, and subunits of the hemoglobin molecule (Fig. 3-11b) have a coiled, springlike secondary structure called a **helix**. Hydrogen bonds between the relatively negative oxygen of the –C=O and the relatively positive hydrogen of the –N–H groups of amino acids hold the turns of the coils together. Some proteins, such as silk, consist of many protein chains lying side by side, with hydrogen bonds holding adjacent chains together in a **pleated sheet** arrangement, another type of secondary structure (Fig. 3-12, p. 51).

In addition to their secondary structures, most proteins assume complex, three-dimensional **tertiary structures** (Fig. 3-11c). Disulfide bridges formed between cysteine amino acids may bring otherwise distant parts of a single peptide close together. In keratin, helices are held together in various ways by disulfide bonds, depending on whether the hair is straight or curly (see "A Closer Look: Protein Structure—A Hairy Subject").

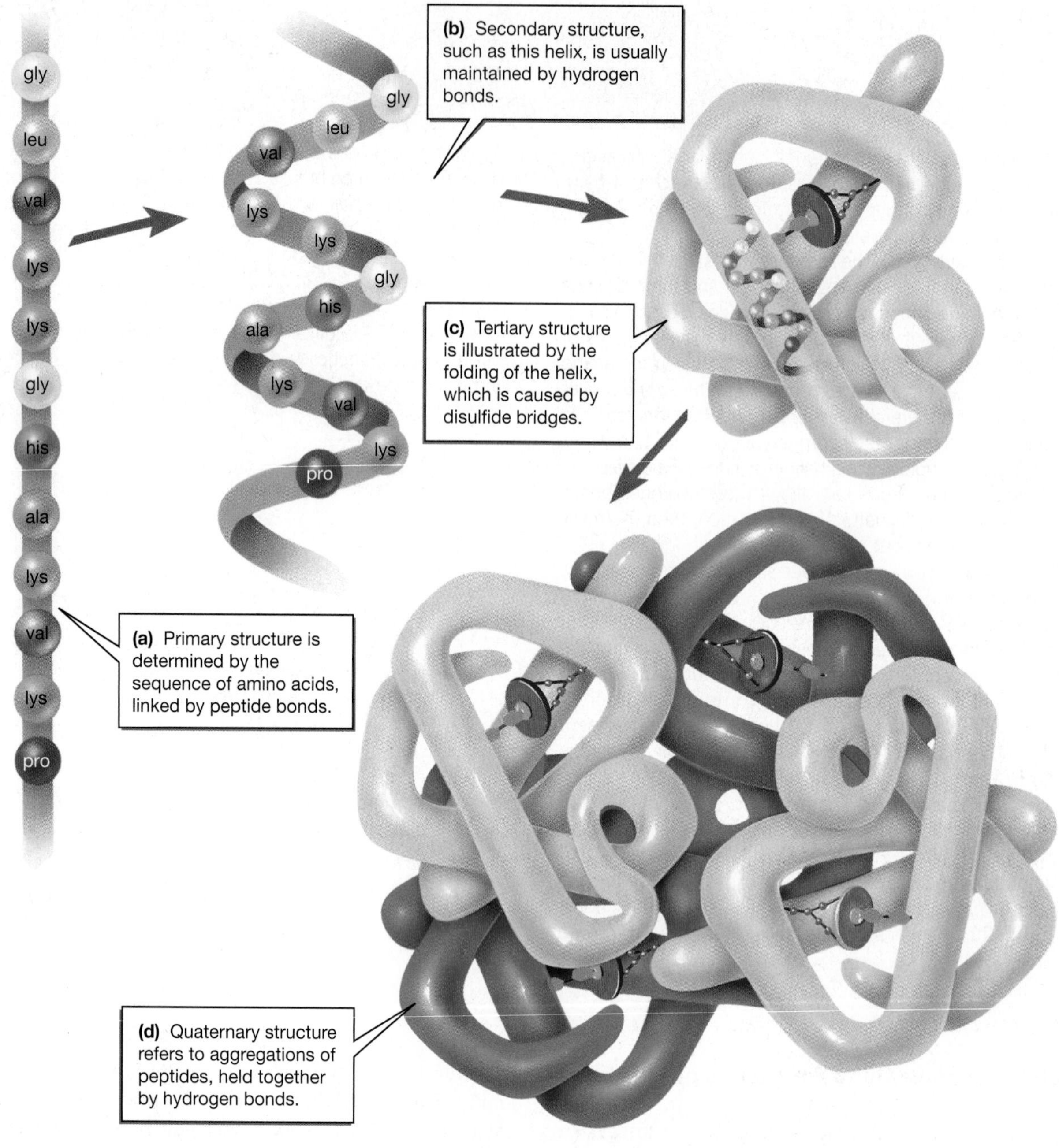

gly
leu
val
lys
lys
gly
his
ala
lys
val
lys
pro
(b) Secondary structure, such as this helix, is usually maintained by hydrogen bonds.
(c) Tertiary structure is illustrated by the folding of the helix, which is caused by disulfide bridges.
(a) Primary structure is determined by the sequence of amino acids, linked by peptide bonds.
(d) Quaternary structure refers to aggregations of peptides, held together by hydrogen bonds.

Figure 3-11 The four levels of protein structure
Levels of protein structure are represented here by hemoglobin, the oxygen-carrying protein in red blood cells. All levels of protein structure are determined by the amino acid sequence of the protein, interactions among the R groups of the amino acids (primarily hydrogen bonds and disulfide bridges between cysteines), and interactions between the R groups and their surroundings (generally water or lipids).

Probably the most important influence on the tertiary structure of a protein is its cellular environment—specifically, whether the protein is dissolved in the water of the cytoplasm, in the lipids of the membranes, or straddling the two environments. Hydrophilic amino acids can form hydrogen bonds with nearby water molecules, whereas hydrophobic amino acids cannot. Therefore, a protein dissolved in water folds into an irregular glob, with its hydrophilic amino acids facing the outside watery environment and its hydrophobic ones clustered in the center of the molecule.

Peptides may sometimes join into aggregations to form the fourth level of protein organization, called **quaternary structure.** Hemoglobin consists of two pairs of very similar peptides, held together by hydrogen bonds (Fig. 3-11d). Each peptide holds an iron-containing organic molecule called a *heme* (the red disks in Fig. 3-11c, d), which can bind one molecule of oxygen.

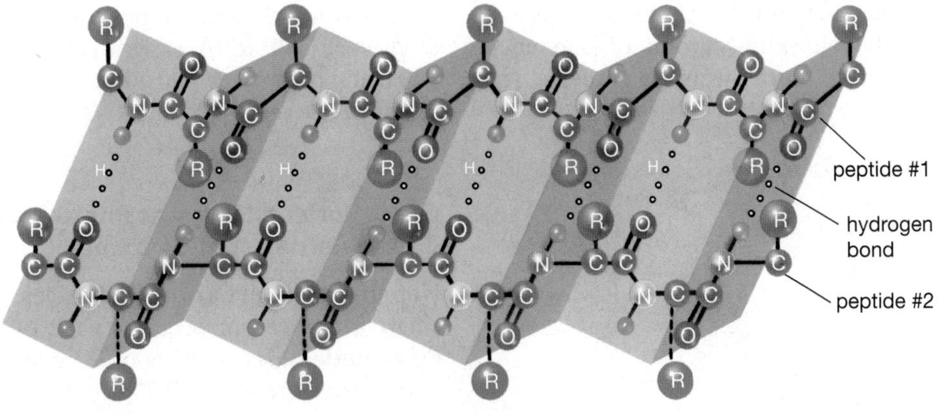

Figure 3-12 The pleated sheet is an example of protein secondary structure
In a pleated sheet, several peptide chains lie side by side (here shown horizontally). Hydrogen bonds between peptides (vertical dotted lines) hold the peptide chains together. The R groups (green) project alternately above and below the sheet. Despite its accordion-pleated appearance, each peptide chain is in a fully extended state and cannot easily be stretched farther. For this reason, pleated sheet proteins such as silk are strong, but not elastic.

The Functions of Proteins Are Linked to Their Three-Dimensional Structures

Within a protein, the exact type, position, and number of amino acids bearing specific R groups determine both the structure of the protein and its biological function. In any given protein, however, some amino acids are more important than others. In hemoglobin, for example, certain amino acids bearing specific R groups must be present in precisely the right places to hold the iron-containing heme group that binds oxygen. Some of the other amino acids are interchangeable if they are functionally equivalent. For instance, the amino acids on the outside of a hemoglobin molecule serve mostly to keep it dissolved in the cytoplasm of a red blood cell. Therefore, as long as they are hydrophilic, the sequence of the amino acids is unimportant. As we will see in Chapter 12, however, replacing a hydrophilic amino acid with a hydrophobic amino acid can have catastrophic effects on the solubility of the hemoglobin molecule. In fact, such a substitution is the molecular cause of a painful and sometimes life-threatening disorder called sickle-cell anemia.

For an amino acid to be in the proper location within a protein, the amino acids must be in their proper sequence. Likewise, the protein must have the correct secondary and tertiary structures so that the amino acid is correctly positioned within the protein. When the secondary and tertiary structures of a protein are altered (leaving the peptide bonds between amino acids intact), the protein is said to be **denatured**, and it will no longer perform its function. There are many ways to denature protein. If you had an egg for breakfast, the heat of the frying pan denatured the albumin protein in the "egg white," causing its appearance to change from clear to white and its texture to change from liquid to solid. Sterilization using heat or ultraviolet rays denatures the proteins of bacteria or viruses and causes them to lose their function. Salty, acidic solutions also denature proteins—dill pickles are preserved in this way.

6 What Are Nucleic Acids?

Nucleic acids are long chains of similar but not identical subunits called **nucleotides**. All nucleotides have a three-part structure: (1) a five-carbon sugar (*ribose* or *deoxyribose*), (2) a phosphate group, and (3) a nitrogen-containing base that differs among nucleotides:

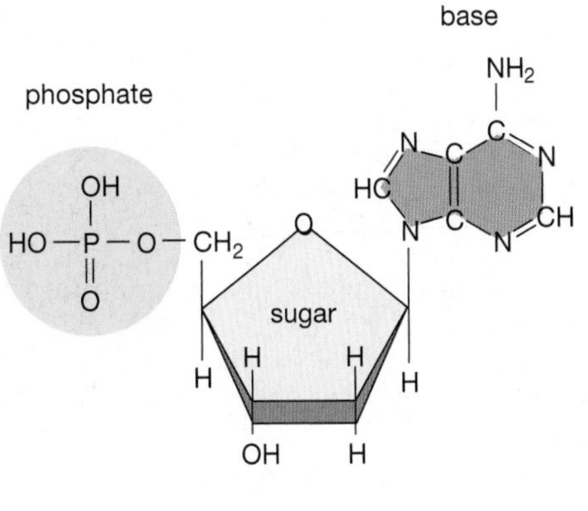

Deoxyribose nucleotide

There are two types of nucleotides, the ribose nucleotides (containing the sugar ribose) and the deoxyribose nucleotides (containing the sugar deoxyribose). Deoxyribose nucleotides bond to four types of nitrogen-containing bases—adenine, guanine, cytosine, and thymine. Similarly, ribose nucleotides bond to four types of bases—adenine, guanine, cytosine, and uracil instead of thymine.

Nucleotides may be strung together in long chains as nucleic acids, with the phosphate group of one nucleotide covalently bonded to the sugar of another:

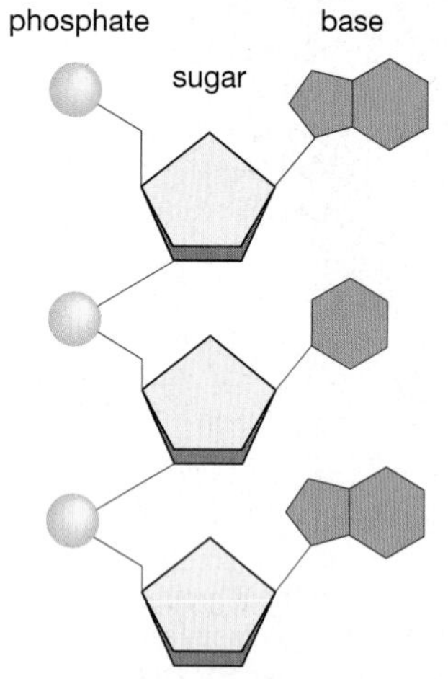

Nucleotide chain

There are two types of nucleic acids: (1) *deoxyribonucleic acid* and (2) *ribonucleic acid*.

DNA and RNA, the Molecules of Heredity, Are Nucleic Acids

Deoxyribose nucleotides form chains millions of units long called **deoxyribonucleic acid**, or **DNA**. DNA is found in the chromosomes of all living things. Its sequence of nucleotides, like the dots and dashes of a biological Morse code, spells out the genetic information needed to construct the proteins of each organism. Chains of ribose nucleotides, called **ribonucleic acid**, or **RNA**, are copied from the central repository of DNA in the nucleus of each cell. RNA carries DNA's genetic code into the cell's cytoplasm and directs the synthesis of proteins. (We cover DNA and RNA in detail in Chapters 9 and 10.)

Other Nucleotides Act as Intracellular Messengers, Energy Carriers, or Coenzymes

Not all nucleotides are part of nucleic acids. Some exist singly in the cell or occur as parts of other molecules. **Cyclic nucleotides**, such as *cyclic adenosine monophosphate* (*cyclic AMP*; Fig. 3-13a), are intracellular messengers that carry information from the plasma membrane to other molecules in the cell. Cyclic AMP is synthesized when certain hormones come in contact with the plasma membrane. Cyclic AMP then stimulates essential reactions in the cytoplasm or nucleus.

Some nucleotides have extra phosphate groups. These diphosphate and triphosphate nucleotides, such as **adenosine triphosphate** (*ATP*; Fig. 3-13b), are unstable molecules that carry energy from place to place within a cell. They capture energy where it is produced (during photosynthesis, for example) and give it up to drive energy-demanding reactions elsewhere (say, to synthesize a protein).

Finally, certain nucleotides assist enzymes in their role of promoting and guiding chemical reactions. These nucleotides are called **coenzymes**. Most coenzymes consist of a nucleotide combined with a vitamin (Fig. 3-13c). You will learn more about energy-carrier nucleotides, enzymes, and coenzymes in Chapter 6, where we discuss energy production and use in the cell.

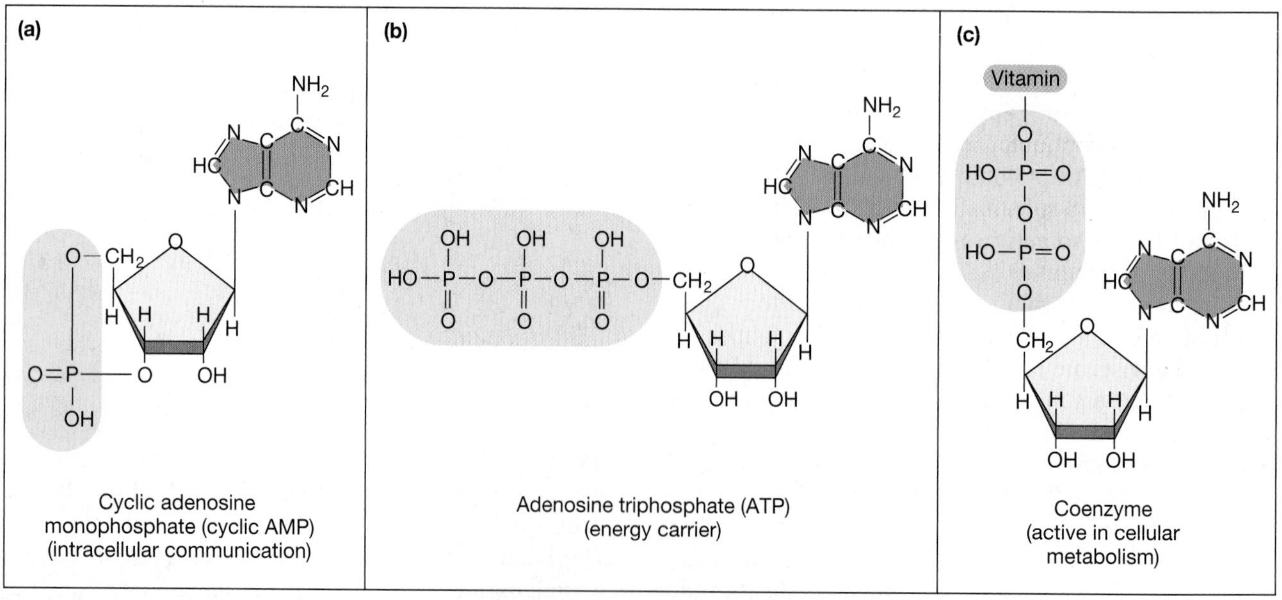

Figure 3-13 A sampling of the diversity of nucleotides
Individual nucleotides, constructed of a sugar, a phosphate group, and a nitrogen-containing base, are often modified by the addition of different functional groups and serve a variety of cellular functions.

CASE STUDY REVISITED

Improving on Nature?

The artificial sweetener aspartame carries an important warning: "Phenylketonurics: contains phenylalanine." Aspartame is composed of two amino acids: aspartic acid and phenylalanine. You now know that amino acids are the building blocks of proteins (p. 45). (Researchers aren't sure why this combination excites the same receptors on our tongues as does sugar.) Phenylketonurics lack an enzyme that converts phenylalanine into tyrosine, another amino acid. Thus, phenyketonurics have abnormally high levels of phenylalanine, which is toxic to the developing nervous system, causing brain damage and mental retardation. A pregnant woman with phenylketonuria must restrict her intake of this amino acid to avoid harming her developing child. Newborns are immediately tested for phenylalanine levels in urine, and they are put on a diet low in phenylalanine if they have this metabolic disorder.

The newly approved sucralose is made from sucrose that has been modified so that three of its hydroxyl groups (p. 39) are replaced with chlorine atoms. Sucralose activates our sweet receptors 600 times as effectively as sucrose, but our enzymes cannot digest it, so it provides no calories. Look for more sucralose-sweetened products in the near future, since it is more stable than other artificial sweeteners.

To understand olestra, recall that oils (p. 43) combine a glycerol backbone with three fatty acid chains. Olestra, however, contains a sucrose backbone with six to eight fatty acids attached. Apparently, the large number of fatty acid chains prevents digestive enzymes from reaching the digestible sucrose backbone of the olestra molecule. Since the molecule is not broken into absorbable fragments, it is excreted unchanged. Results of exhaustive tests on olestra left two concerns. First, since the oil passes undigested through the intestine, it acts as an intestinal lubricant and can have a laxative effect on people who consume large quantities. Second, olestra can reduce absorption of fat-soluble vitamins in food by dissolving the vitamins and carrying them out of the body. Upon approving olestra, the FDA required that foods containing it be supplemented with fat-soluble vitamins and be identified with a warning label that reads: "This product contains olestra. Olestra may cause abdominal cramping and loose stools. Olestra inhibits the absorption of some vitamins and nutrients. Vitamins A, D, E, and K have been added."

Some experts argue that the advantages of being able to eat appealing food while limiting fat and sugar consumption makes artificial sweeteners and fake oils and fats worthwhile. Others contend that people should select natural, nutritious foods that are already low in sugar and fats and shun these artificial dieting aids. Think of arguments on both sides of this issue. You might want to look at the case study and its revisitation in Chapter 29 for more about obesity.

Summary of Key Concepts

1) Why Is Carbon So Important in Biological Molecules?

Organic molecules are so diverse because the carbon atom is able to form many types of bonds. This ability, in turn, allows organic molecules (molecules with a backbone of carbon and hydrogen atoms) to form many complex shapes, including chains, branches, and rings.

2) How Are Organic Molecules Synthesized?

Most large biological molecules are polymers synthesized by linking together many smaller subunits, or monomers. Chains of subunits are connected by covalent bonds through dehydration synthesis; the chains may be broken apart by hydrolysis reactions. The most important organic molecules fall into one of four classes: carbohydrates, lipids, proteins, and nucleic acids. Their major characteristics are summarized in Table 3-2.

3) What Are Carbohydrates?

Carbohydrates include sugars, starches, chitin, and cellulose. Sugars (monosaccharides and disaccharides) are used for temporary storage of energy and for the construction of other molecules. Starches and glycogen are polysaccharides that serve for longer-term energy storage in plants and animals, respectively. Cellulose and related polysaccharides form the cell walls of bacteria, fungi, plants, and some microorganisms.

4) What Are Lipids?

Lipids are nonpolar, water-insoluble molecules of diverse chemical structure that include oils, fats, waxes, phospholipids, and steroids. Lipids are used for energy storage (oils and fats), as waterproofing for the outside of many plants and animals (waxes), as the principal component of cellular membranes (phospholipids), and as hormones (steroids).

5) What Are Proteins?

Proteins are chains of amino acids. Both the structure and the function of a protein are determined by the sequence of amino acids in the chain. Proteins can be enzymes (which guide chemical reactions), structural molecules (hair, horn), hormones (insulin), or transport molecules (hemoglobin).

6) What Are Nucleic Acids?

Nucleic acid molecules are chains of nucleotides. Each nucleotide is composed of a phosphate group, a sugar group, and a nitrogen-containing base. The two types of nucleic acids are deoxyribonucleic acid (DNA) and ribonucleic acid (RNA). Other nucleotides include intracellular messengers (cyclic AMP), energy-carrier molecules (ATP), and coenzymes.

Key Terms

adenosine triphosphate *p. 52*
amino acid *p. 46*
carbohydrate *p. 39*
cellulose *p. 41*
chitin *p. 41*
coenzyme *p. 52*
cyclic nucleotide *p. 52*
dehydration synthesis *p. 39*
denatured *p. 51*
deoxyribonucleic acid (DNA) *p. 52*
disaccharide *p. 39*
disulfide bridge *p. 47*
enzyme *p. 45*
fat *p. 43*
fatty acid *p. 43*
functional group *p. 38*
glucose *p. 40*
glycerol *p. 43*
glycogen *p. 41*
helix *p. 49*
hydrolysis *p. 39*
inorganic *p. 38*
lactose *p. 41*
lipid *p. 43*
maltose *p. 41*
monomer *p. 39*
monosaccharide *p. 39*
nucleic acid *p. 51*
nucleotide *p. 51*
oil *p. 43*
organic *p. 38*
peptide *p. 48*
peptide bond *p. 48*
phospholipid *p. 45*
pleated sheet *p. 49*
polymer *p. 39*
polysaccharide *p. 39*
primary structure *p. 49*
protein *p. 45*
quaternary structure *p. 50*
ribonucleic acid (RNA) *p. 52*
saturated *p. 44*
secondary structure *p. 49*
starch *p. 41*
steroid *p. 45*
subunit *p. 39*
sucrose *p. 41*
sugar *p. 39*
tertiary structure *p. 49*
triglyceride *p. 43*
unsaturated *p. 44*
wax *p. 44*

Thinking Through the Concepts

Multiple Choice

1. *Which of the following is not a function of polysaccharides in organisms?*
   a. energy storage
   b. storage of hereditary information
   c. formation of cell walls
   d. structural support
   e. formation of exoskeletons

2. *Characteristics of carbon that contribute to its ability to form an immense diversity of organic molecules include its*
   a. tendency to form covalent bonds
   b. ability to bond with up to four other atoms
   c. capacity to form single and double bonds
   d. ability to bond together to form extensive, branched or unbranched carbon skeletons
   e. all of the above

3. *Foods that are high in fiber are most likely to be derived from*
   a. plants
   b. dairy products
   c. meat
   d. fish
   e. all of the above

4. *Proteins differ from one another because*
   a. the peptide bonds linking amino acids differ from protein to protein
   b. the sequence of amino acids in the polypeptide chain differs from protein to protein
   c. each protein molecule contains its own unique sequence of sugar molecules
   d. the number of nucleotides in each protein varies from molecule to molecule
   e. the number of nitrogen atoms in each amino acid differs from the number in all others

5. *Which, if any, of the following choices does not properly pair an organic compound with one of its building blocks (subunits)?*
   a. polysaccharide–monosaccharide
   b. fat–fatty acid
   c. nucleic acid–glycerol
   d. protein–amino acid
   e. all are paired correctly

6. *Which of the following statements about lipids is false?*
   a. A wax is a lipid.
   b. Unsaturated fats are liquid at room temperature.
   c. The body doesn't need any cholesterol.
   d. Both male and female sex hormones are steroids.
   e. Beef fat is highly saturated.

? Review Questions

1. Which elements are common components of biological molecules?
2. List the four principal types of biological molecules, and give an example of each.
3. What roles do nucleotides play in living organisms?
4. One way to convert corn oil to margarine (solid at room temperature) is to add hydrogen atoms, decreasing the number of double bonds in the molecules of oil. What is this process called? Why does it work?
5. Describe and compare dehydration synthesis and hydrolysis. Give an example of a substance formed by each chemical reaction, and describe the specific reaction in each instance.
6. Distinguish among the following: monosaccharide, disaccharide, and polysaccharide. Give two examples of each and their functions.
7. Describe the synthesis of a protein from amino acids. Then describe primary, secondary, tertiary, and quaternary structures of a protein.
8. Most structurally supportive materials in plants and animals are polymers of special sorts. Where would we find cellulose? Chitin? In what way(s) are these two polymers similar? Different?
9. Which kinds of bonds or bridges between keratin molecules are altered when hair is (a) wet and allowed to dry on curlers and (b) given a permanent wave?

Applying the Concepts

1. A preview question for Chapter 4: In Chapter 2 you learned that hydrophobic molecules tend to cluster when immersed in water. In this chapter, you discovered that a phospholipid has a hydrophilic head and hydrophobic tails. What do you think would be the configuration of phospholipids that are immersed in water?
2. Many birds must store large amounts of energy to power flight during migration. Which type of organic molecule would be the most advantageous for energy storage? Why?
3. Remember the nuclear accident at Chernobyl in 1986? A scientist suspects that the food in a nearby ecosystem may have been contaminated with radioactive nitrogen over a period of months. Which substances in plants and animals could be examined for radioactivity to test his hypothesis?
4. Fat contains twice as many calories per unit weight as carbohydrate does, so fat is an efficient way for animals, who must move about, to store energy. Compare the way fat and carbohydrates interact with water, and explain why this interaction also gives fat an advantage for weight-efficient energy storage.

For More Information

Doolittle, R. F. "Proteins." *Scientific American*, October 1985. Proteins are the molecular tools that, directly or indirectly, carry out almost all the functions of a cell.

Goodsell, D. S. *The Machinery of Life*. New York: Springer, 1993. Goodsell depicts the molecules of a cell in all their three-dimensional, interactive glory. A great way to gain a feel for the beauty and intricacy of the organic molecules of life.

Hill, J. W., and Kolb, D. K. *Chemistry for Changing Times*. 9th ed. Englewood Cliffs, NJ: Prentice Hall, 2001. A chemistry textbook for nonscience majors that is both clearly readable and thoroughly enjoyable.

Radetsky, P. "Kim's Coils." *Discover*, June 1995. Biochemist Peter Kim is uncovering the relationship between structure and function in the elaborate coiling of proteins.

Sharon, N., and Lis, H. "Carbohydrates in Cell Recognition." *Scientific American*, January 1993. Sugar-protein complexes on the surfaces of cells regulate cell identification and interaction between cells. See also Sharon's earlier article, "Carbohydrates," *Scientific American*, November 1980.

Answers to Multiple-Choice Questions

1. b 2. e 3. a 4. b 5. c 6. c

MediaTutor
Biological Molecules

CD Activities

Activity 3.1: Structure of Biological Molecules

Estimated time: 5 minutes

Most important biological molecules are both organic compounds and macromolecules. In this tutorial you will explore the various classes of important biological macromolecules.

Activity 3.2: Functions of Macromolecules

Estimated time: 5 minutes

The important macromolecules in cells are necessary for most cellular functions. This tutorial introduces the major functions of the four main types of macromolecules: carbohydrates, lipids, proteins, and nucleic acids.

Start the MediaTutor Student CD-ROM and enter the activity number in the Quick Search box to be taken directly to that activity.

Web Investigations

Case Study: Improving on Nature?

Estimated time: 10 minutes

Let's take a closer look at olestra. While most artificial sweeteners, food colors, and fat replacers can be used for any food product, olestra is currently approved by the U.S. Food and Drug Administration (FDA) for use only in "savory snacks," such as potato chips. Some experts say that olestra is safe. Others think the side effects outweigh the benefits. In this exercise we will examine the pros and cons of the issue. Visit the sites below and answer the questions.

Go to http://www.prenhall.com/audesirk6, the Audesirk Companion Web site. Select Chapter 3 and the Web Investigation to begin.

"To stay alive, you have to be able to hold out against equilibrium, maintain imbalance, bank against entropy, and you can only transact this business with membranes in our kind of world."

Lewis Thomas in* The Lives of a Cell *(1974)

A rattlesnake prepares to strike.

4 Cell Membrane Structure and Function

AT A GLANCE

Case Study: Vicious Venoms

1) **How Is the Structure of a Membrane Related to Its Function?**

The Plasma Membrane Isolates the Cell While Allowing Communication with Its Surroundings

Membranes Are "Fluid Mosaics" in Which Proteins Move Within Layers of Lipids

The Phospholipid Bilayer Is the Fluid Portion of the Membrane

A Mosaic of Proteins Is Embedded in the Membrane

2) **How Do Substances Move Across Membranes?**

Molecules in Fluids Move in Response to Gradients

Movement Across Membranes Occurs by Both Passive and Active Transport

Passive Transport Includes Simple Diffusion, Facilitated Diffusion, and Osmosis

Active Transport Uses Energy to Move Molecules Against Their Concentration Gradients

Cells Engulf Particles or Fluids by Endocytosis

Exocytosis Moves Material Out of the Cell

3) **How Are Cell Surfaces Specialized?**

A Variety of Junctions Allow Cells to Connect and Communicate

Some Cells Are Supported by Cell Walls

Evolutionary Connections: Caribou Legs and Membrane Diversity

Case Study Revisited: Vicious Venoms

CASESTUDY

Vicious Venoms

Freshmen roommates at a university in southern California, Karl and Mark, eager to explore their new environs, drove to a trailhead in the Mojave Desert. Karl kidded Mark about carrying his cellular phone—what kind of a wilderness experience could you have with a phone along? Mark joked about the large field guide, "Desert Flora and Fauna," weighing down Karl's pack. Competitive and athletic, after several miles of hiking they spotted a rocky bluff and raced each other to the top. The sudden exertion in the hot sun made Karl feel momentarily light-headed, and he reached for a rocky outcropping to steady himself. Expecting solid stone, he gasped at the feel of thick, scaly coils writhing under his hand. A sudden unmistakable warning rattle was followed almost immediately by an intense burning pain at the base of his thumb. Karl's yell brought Mark running. Seeing the huge snake slithering back into the protection of a crevice, Mark helped Karl, who was now feeling dizzy and nauseated, to lie down and then quickly rummaged in his pack for his cellular phone. The 911 dispatcher told them to wait quietly for a medical evacuation helicopter. By the time they heard the chopper, they had used Karl's field guide to identify the rattler as a Western Diamondback. Before he reached the hospital, a large bruised-looking area was spreading over Karl's hand, his blood pressure had dropped, and the paramedics were administering oxygen because he was gasping for air. Why was his hand bruising rapidly? Why did Karl feel short of breath? What treatment will Karl receive? ■

1) How Is the Structure of a Membrane Related to Its Function?

The Plasma Membrane Isolates the Cell While Allowing Communication with Its Surroundings

In Chapter 1 we defined the *cell* as the smallest unit of life. Each cell is surrounded by a thin **plasma membrane**, which can be considered a gatekeeper, allowing only specific substances in or out and passing chemical messages from the external environment to the cell's interior. As gatekeeper, the plasma membrane must perform three general functions:

1. Selectively isolate the cell's contents from the external environment.
2. Regulate the exchange of essential substances between the cell's contents and the external environment.
3. Communicate with other cells.

These are formidable tasks for a structure so thin that 10,000 plasma membranes stacked atop one another would scarcely equal the thickness of this page. The key to membrane function lies in membrane structure. Membranes are not simply homogeneous sheets: They are complex, heterogeneous structures with different parts performing very distinct functions, and they change dynamically in response to their surroundings.

Most cells have internal membranes as well as a plasma membrane that surrounds the cell. These internal membranes form compartments in which specialized biochemical activities can occur. All the membranes of a cell have a similar basic structure: proteins floating in a double layer of lipids. Lipids are responsible for the isolating function of membranes, whereas proteins regulate the exchange of substances and communication with the environment. Although this chapter focuses on the plasma membrane, much of the information presented here also applies to other cellular membranes.

Membranes Are "Fluid Mosaics" in Which Proteins Move Within Layers of Lipids

The **fluid mosaic model** of cellular membranes was developed by cell biologists S. J. Singer and G. L. Nicolson in 1972. According to this model, a membrane, when viewed from above, looks something like a lumpy, constantly shifting mosaic of tiles (Fig. 4-1). A double layer of phospholipids forms a viscous, fluid "grout" for the mosaic; an assortment of proteins are the "tiles," which can move about within the phospholipid layers. Thus, although the components within the plasma membrane remain relatively constant, the overall distribution of proteins and various types of phospholipids can change over time. As strange as this model may seem, it captures something of the dynamic quality of real membranes. Now let's look more closely at the structure of membranes.

The Phospholipid Bilayer Is the Fluid Portion of the Membrane

As you learned in Chapter 3, a phospholipid consists of two very different parts: (1) a polar, hydrophilic head (attracted to water) and (2) a pair of nonpolar, hydrophobic tails (repelled by water). Notice that the double bond (making the lipid unsaturated) introduces a kink in the tail that will help keep the membrane fluid at lower temperatures:

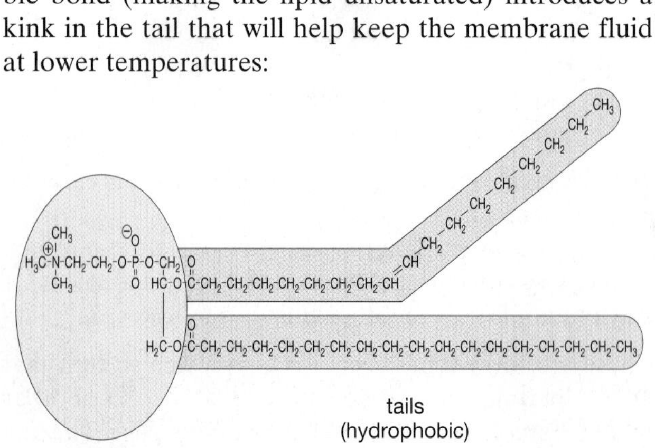

All cells are surrounded by a watery medium. Single-celled organisms may live in fresh water or in the ocean, while animal cells are bathed in a weakly salty *extracellular fluid* that filters out of the blood. The **cytoplasm** consists of all of a cell's internal contents (including all the organelles except the nucleus, in eukaryotes); cytoplasm is mostly water. Plasma membranes separate a watery cytoplasm from a watery external environment, and similar membranes surround watery compartments within the cell. Under these conditions, phospholipids spontaneously arrange themselves into a double layer called a **phospholipid bilayer**, in which the hydrophilic heads form the outer borders and the hydrophobic tails "hide" inside:

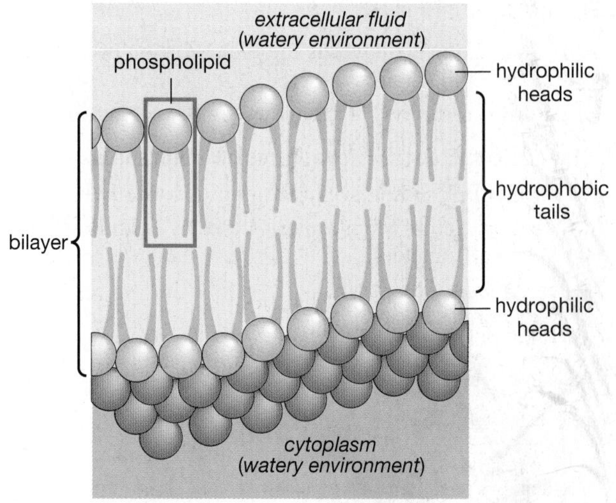

Hydrogen bonds can form between water and the phospholipid heads, so the hydrophilic heads face the cytoplasm or the extracellular fluid, forming the outer layer of the bilayer. Hydrophobic interactions (see Chapter 2) cause the phospholipid tails to hide inside the bilayer. Because individual phospholipid molecules are not bonded to one another and the lipid tails contain saturated bonds, this double layer is quite fluid; individual phospholipids move about easily within each layer.

Most biological molecules, including salts, amino acids, and sugars, are polar and water soluble: hydrophilic. In fact, most substances that contact a cell are water soluble, hydrophilic, and so cannot easily pass through the nonpolar, hydrophobic fatty acid tails of the phospholipid bilayer. The phospholipid bilayer is largely responsible for the first of the three membrane functions listed earlier—selectively isolating the cell's contents from the external environment. The isolation is not complete, however. As we will describe later, very small molecules, such as water, and uncharged, lipid-soluble molecules can pass relatively freely through the lipid bilayer.

In most cells, the phospholipid bilayer of membranes also contains cholesterol. Some cellular membranes have just a few cholesterol molecules; others have as many cholesterol molecules as they do phospholipids. Cholesterol affects membrane structure and function in several ways: It makes the bilayer stronger, more flexible

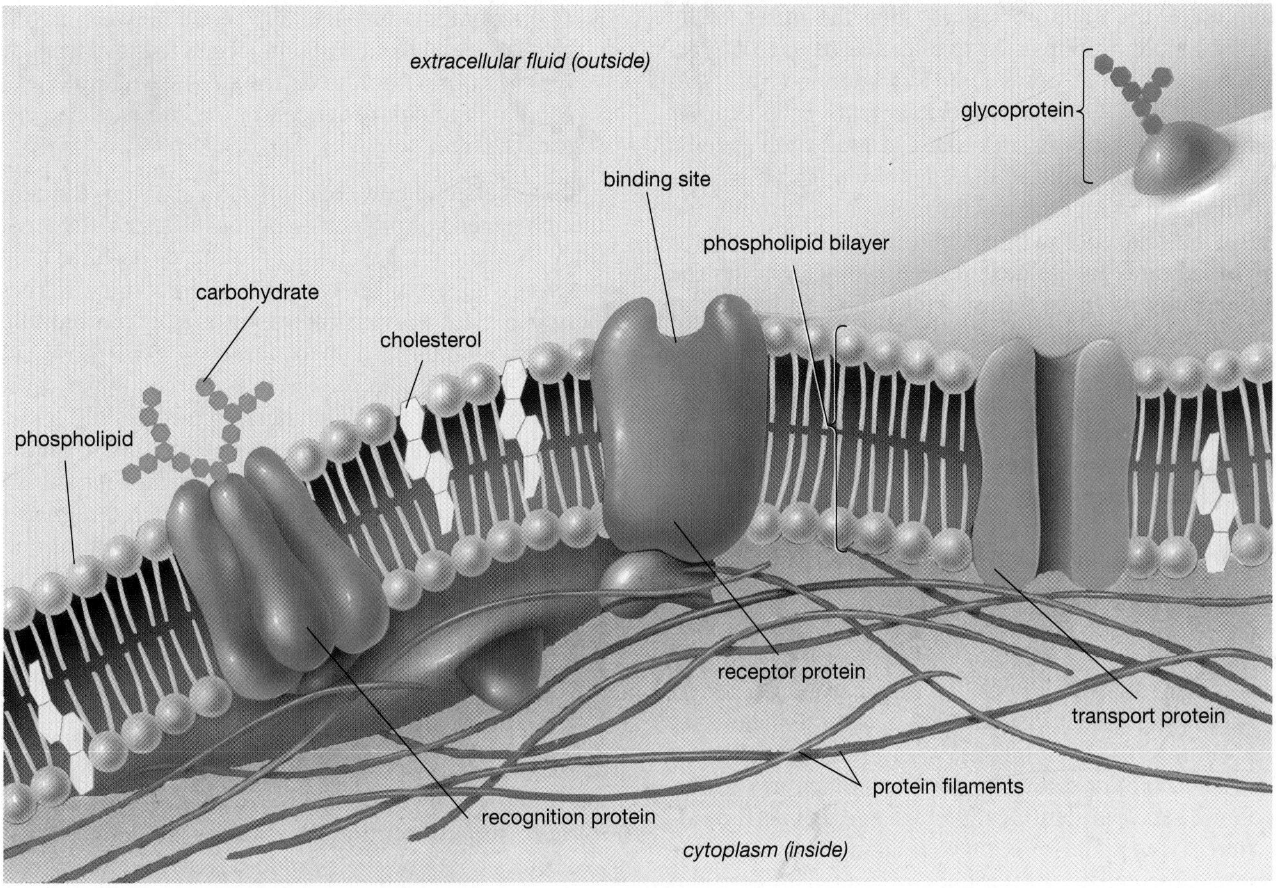

Figure 4-1 The plasma membrane is a fluid mosaic
The plasma membrane is a bilayer of phospholipids in which various proteins are embedded. Many proteins have carbohydrates attached to them, forming glycoproteins. The wide variety of membrane proteins fall mostly into three categories: transport proteins, receptor proteins, and recognition proteins.

MediaTutor
4.1 Membrane Structure and Transport

but less fluid, and less permeable to water-soluble substances such as ions or monosaccharides.

The flexible, somewhat fluid nature of the bilayer is very important for membrane function. As you breathe, move your eyes, and turn the pages of this book, cells in your body change shape. If their plasma membranes were stiff instead of flexible, cells would break open and die. Further, as you will learn in Chapter 5, membranes within eukaryotic cells are in constant motion. Membrane-enclosed compartments ferry substances into the cell, carry materials within the cell, and expel them to the outside, merging membranes in the process. This flow and merger of membranes is made possible by the fluid nature of the lipid bilayer.

A Mosaic of Proteins Is Embedded in the Membrane

Thousands of proteins are embedded within or attached to the surface of a membrane's phospholipid bilayer. Collectively, these proteins regulate the movement of substances through the membrane and communicate with the environment. Many of the proteins in plasma membranes have carbohydrate groups attached to them, especially to the parts that stick outside the cell. These membrane proteins and their attached carbohydrates are called **glycoproteins**.

Many membrane proteins can move about within the relatively fluid phospholipid bilayer. Others, however, are anchored in place to a network of protein filaments within the cytoplasm. The attachments between plasma membrane proteins and the underlying protein filaments produce the characteristic shapes of animal cells, from the dimpled discs of red blood cells to the elaborate branching of nerve cells.

There are three major categories of membrane proteins, each of which serves a different function: (1) *transport proteins,* (2) *receptor proteins,* and (3) *recognition proteins* (see Fig. 4-1).

(1) **Transport proteins** regulate the movement of hydrophilic (water-soluble) molecules through the plasma membrane. Some transport proteins, called **channel proteins**, form pores or channels that allow small water-soluble molecules to pass through the membrane (see Fig. 4-1). Every plasma membrane bears a large assortment of channel proteins. The size of their pores, and the

charges on the amino acids that line the pores, make these channel proteins selective for the passage of particular ions such as potassium (K^+), sodium (Na^+), and calcium (Ca^{2+}). Other transport proteins, called **carrier proteins**, have binding sites that can grab onto specific molecules on one side of the membrane. The transport protein then changes shape, in some cases through the use of cellular energy, and moves the molecule across the membrane. In the next section, we will discuss the mechanisms whereby transport proteins move molecules across the membrane.

(2) **Receptor proteins** trigger cellular responses when specific molecules in the extracellular fluid, such as hormones or nutrients, bind to them. Most cells bear dozens of types of receptors on their plasma membranes. When activated by the appropriate molecule, some receptors set off elaborate sequences of cellular changes, such as increased metabolic rate, cell division, movement toward a nutrient source, or secretion of hormones. Other receptors act like gates on channel proteins; activating the receptor opens the gates, allowing ions to flow through the channels. For example, receptors allow nerve cells in your brain to communicate with one another (see Chapter 33).

(3) **Recognition proteins**, many of which are glycoproteins, serve as identification tags and cell-surface attachment sites. The cells of your immune system, for example, recognize a bacterium as a foreign invader and target it for destruction. These same immune cells ignore the trillions of cells in your own body because your body cells have different identification glycoproteins on their surfaces. During development, the growth of nerve fibers from your spinal cord down to the muscles in your feet is guided by attachments between recognition proteins on the nerve cell and the other cells it traverses on its way to the muscle.

As you can see from these brief descriptions, membrane proteins are largely responsible for moving substances across the membrane and for communicating with other cells.

2 How Do Substances Move Across Membranes?

Molecules in Fluids Move in Response to Gradients

Because the plasma membrane separates the fluid in the cell's cytoplasm from its fluid extracellular environment, let's begin our study of membrane transport with a brief look at the characteristics of fluids. We must start with a few definitions:

1. A **fluid** is a liquid or a gas—that is, any substance that can move or change shape in response to external forces without breaking apart.
2. The **concentration** of molecules in a fluid is the number of molecules in a given unit of volume.
3. A **gradient** is a physical difference between two regions of space that causes molecules to move from one region to the other. Cells frequently generate or encounter gradients of concentration, pressure, and electrical charge.

To understand how concentration gradients influence the movement of molecules or ions within a fluid, consider a sugar cube dissolving in coffee, or perfume molecules moving from an open bottle into the air. These substances are moving in response to a **concentration gradient**, a difference in concentration of those substances between one region and another. What causes this movement? The individual molecules in a fluid move continuously, bouncing off one another in random directions. Over time, these random movements will produce a net movement of molecules from regions of high concentration to regions of low concentration, a process called **diffusion**. By analogy with gravity, we will refer to such movements as going "down" the concentration gradient. If there are no factors opposing this movement, such as electrical charge or pressure differences or physical barriers, the movement of molecules from regions of high to low concentration will continue until the substance is evenly dispersed throughout the fluid or the air. In this evenly dispersed state, called a *dynamic equilibrium*, the concentration gradient no longer exists. Molecules continue their random movements and collisions (the *dynamic* aspect), but there is no longer any change in concentration occurring; the substance has reached an *equilibrium* with its surroundings.

To watch diffusion in action, place a drop of food coloring in a glass of water, and check its progress every few minutes. With time, the drop will seem to spread out and become paler, until eventually, even without stirring, the entire glass of water will be uniformly faintly colored. Molecules of dye, simply owing to random motion, move out into the water from the region of high dye concentration into the surrounding water where the dye concentration is low (Fig. 4-2). Simultaneously, random motion causes some water molecules to enter the dye droplet, and the net movement of water is from the high water concentration outside the drop into the lower water concentration inside the drop.

At first, there is a very steep concentration gradient, and the dye diffuses rapidly. As the concentration differences lessen, the dye diffuses more and more slowly. In other words, the greater the concentration gradient, the faster the rate of diffusion. However, as long as the concentration of dye within the expanding drop is greater than the concentration of dye in the rest of the glass, the net movement of dye will continue until it becomes uniformly dispersed in the water. Then, with no concentration gradient of either dye or water, diffusion stops. Individual molecules still move about randomly within the glass, but no changes occur in concentration of either water or dye. A dynamic equilibrium has been established.

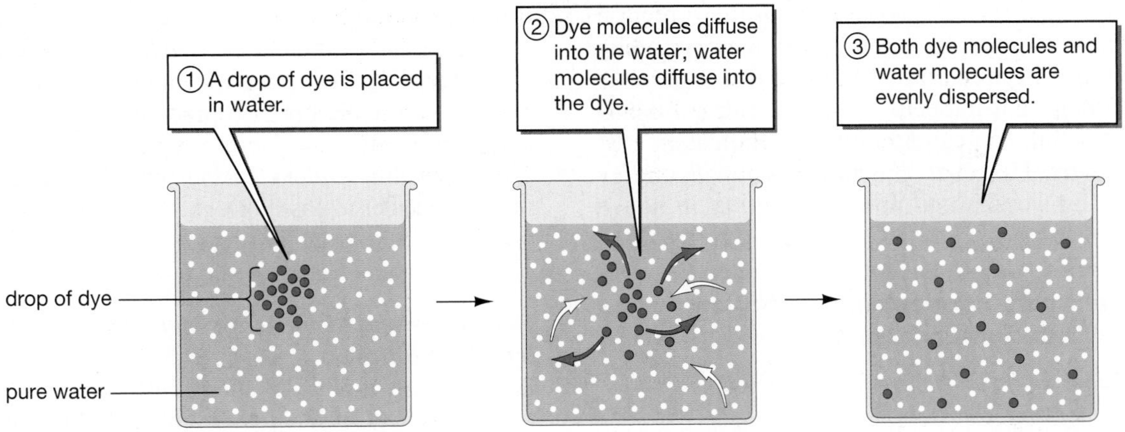

① A drop of dye is placed in water.
② Dye molecules diffuse into the water; water molecules diffuse into the dye.
③ Both dye molecules and water molecules are evenly dispersed.
drop of dye
pure water

Figure 4-2 Diffusion of a dye in water

As you can appreciate from this simple experiment, diffusion cannot move molecules rapidly over long distances. Although the drop of dye immediately begins to diffuse into the water, it will take many minutes or hours for the dye to disperse uniformly. As you will learn in Chapter 5, the slow rate of diffusion over long distances is one of the reasons that cells are small.

Summing Up
The Principles of Diffusion

1. Diffusion is the net movement of molecules down a gradient from high to low concentration.
2. The greater the concentration gradient, the faster the rate of diffusion.
3. If no other processes intervene, diffusion will continue until the concentration gradient is eliminated.
4. Diffusion cannot move molecules rapidly over long distances.

Movement Across Membranes Occurs by Both Passive and Active Transport

There are significant concentration gradients of ions and molecules across the plasma membrane of each cell because the cytoplasm of a cell is very different from the extracellular fluid. In its role as gatekeeper of the cell, the plasma membrane provides for two types of movement: (1) *passive transport* and (2) *active transport* (Table 4-1).

During **passive transport**, substances move into or out of cells down concentration gradients. This movement by itself requires no expenditure of energy, since the concentration gradients provide the potential energy that drives the movement and controls the direction of movement, into or out of the cell. The lipids and protein pores of the plasma membrane regulate which molecules can cross, but they do not influence the direction of movement.

During **active transport**, the cell uses energy to move substances *against* a concentration gradient. In

Table 4-1 Transport Across Membranes

Passive transport	Movement of substances across a membrane, going down a gradient of concentration, pressure, or electrical charge. Does not require the cell to expend energy.
Simple diffusion	Diffusion of water, dissolved gases, or lipid-soluble molecules through the phospholipid bilayer of a membrane.
Facilitated diffusion	Diffusion of (normally water-soluble) molecules through a channel or carrier protein.
Osmosis	Diffusion of water across a differentially permeable membrane—that is, a membrane that is more permeable to water than to dissolved molecules.
Energy-requiring transport	Movement of substances across a membrane, usually against a concentration gradient, using cellular energy.
Active transport	Movement of individual small molecules or ions through membrane-spanning proteins, using cellular energy, normally ATP.
Endocytosis	Movement of large particles, including large molecules or entire microorganisms, into a cell by engulfing extracellular material, as the plasma membrane forms membrane-bound sacs that enter the cytoplasm.
Exocytosis	Movement of materials out of a cell by enclosing the material in a membranous sac that moves to the cell surface, fuses with the plasma membrane, and opens to the outside, allowing its contents to diffuse away.

this case, transport proteins do control the direction of movement. A helpful analogy for understanding the difference between passive and active transports is to consider what happens when you ride a bike. If you don't pedal, you can go only downhill, as in passive transport. However, if you put enough energy into pedaling, you can go uphill as well, as in active transport.

Passive Transport Includes Simple Diffusion, Facilitated Diffusion, and Osmosis

Plasma Membranes Are Differentially Permeable to Diffusion of Molecules

Diffusion can occur from one part of a fluid to another or across a membrane separating two fluid compartments. Many molecules cross plasma membranes by diffusion, driven by differences between their concentration in the cytoplasm and in the external environment. Because of the properties of the plasma membrane, different molecules cross the plasma membrane at different locations and at different rates. Therefore, plasma membranes are said to be **differentially permeable**—that is, they allow some molecules to pass through, or *permeate*, but prevent other molecules from passing. A barrier that prevents the passage of all molecules is said to be *impermeable.*

Some Molecules Move Across Membranes by Simple Diffusion

Lipid-soluble molecules such as ethyl alcohol and vitamin A easily diffuse across the phospholipid bilayer, as do very small molecules, including water and dissolved gases such as oxygen and carbon dioxide. This process is called **simple diffusion** (Fig. 4-3a). Generally, the rate

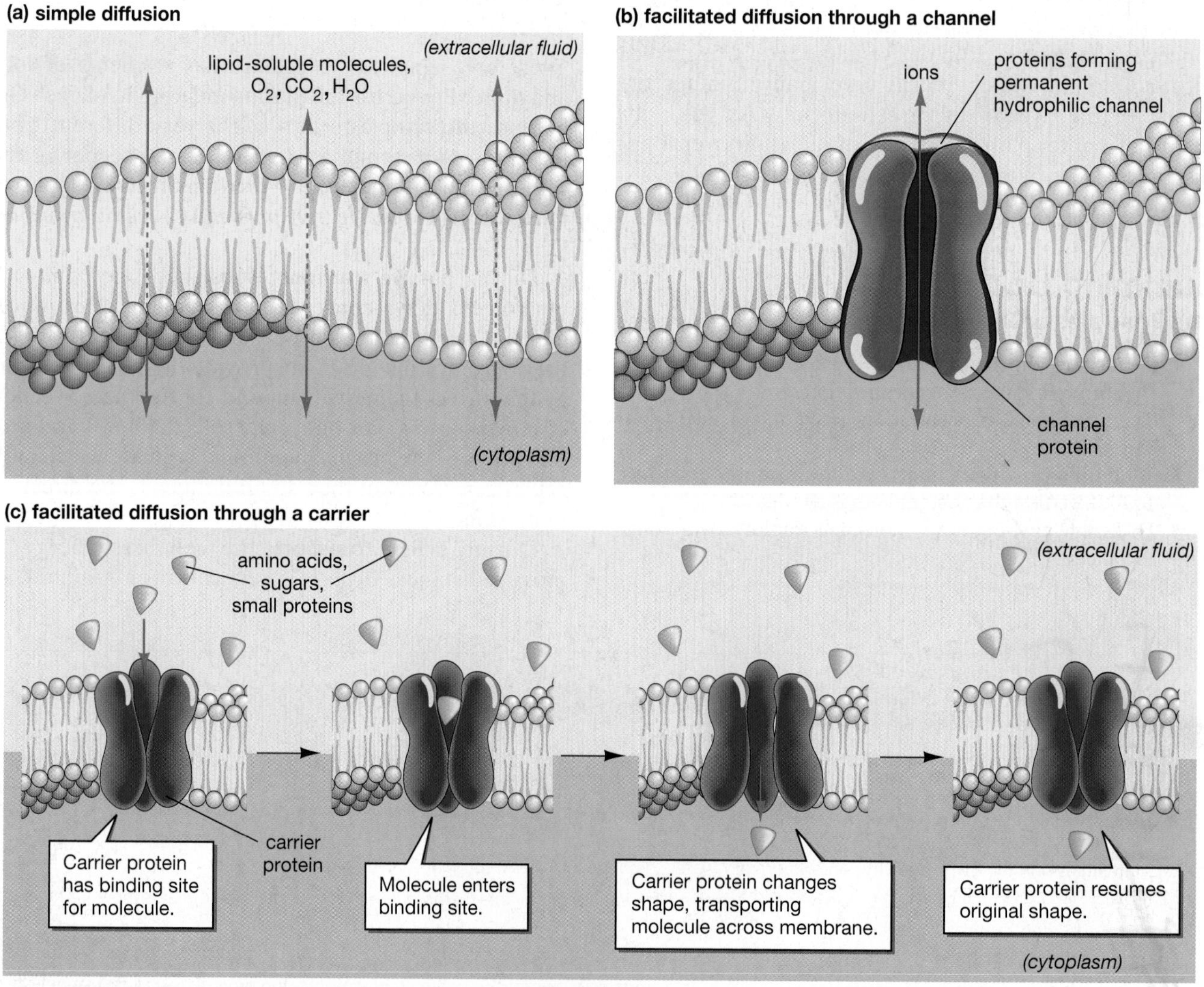

Figure 4-3 Diffusion through the plasma membrane
(a) Simple diffusion through the phospholipid bilayer: Gases such as oxygen and carbon dioxide and lipid-soluble molecules can diffuse directly through the phospholipids. ***(b)*** Facilitated diffusion through a channel: Some molecules cannot pass through the bilayer on their own. Protein channels (pores) allow some water-soluble molecules, principally ions, to enter or exit the cell. ***(c)*** Facilitated diffusion through a carrier: Carrier proteins may bind a specific molecule and, as a result, change their own shape, passing the molecule through the middle of the protein to the other side of the membrane.

of simple diffusion is a function of the concentration gradient across the membrane, the size of the molecule, and how easily it dissolves in lipids (its lipid *solubility*): Large concentration gradients, small molecule size, and high lipid solubility all increase the rate of simple diffusion.

Other Molecules Cross the Membrane by Facilitated Diffusion, with the Help of Membrane Transport Proteins

Most water-soluble molecules, such as ions (K^+, Na^+, Ca^{2+}), amino acids, and monosaccharides, cannot move through the phospholipid bilayer on their own. These molecules can diffuse across only with the aid of one of two types of transport proteins: *channel proteins* and *carrier proteins*. This process is called **facilitated diffusion**.

Channel proteins form pores, or channels, in the lipid bilayer through which certain ions can cross the membrane (Fig. 4-3b). Most channel proteins have a specific interior diameter and distribution of electrical charges that allow only particular ions to pass through. Nerve cells, for example, have separate channels for sodium ions, potassium ions, and calcium ions.

Carrier proteins bind specific molecules, such as particular amino acids, sugars, or small proteins, from the cytoplasm or extracellular fluid. Binding triggers a change in the shape of the carrier that allows the molecules to pass through the protein and across the plasma membrane. Facilitated diffusion occurs through carrier proteins that do not use cellular energy. These carrier proteins move molecules only down their concentration gradients (Fig. 4-3c).

Because they must rely on transport proteins, molecules that cross the membrane by facilitated diffusion usually do so more slowly than do those that cross by simple diffusion through the lipid bilayer.

Osmosis Is the Diffusion of Water Across Membranes

Water also diffuses from regions of high water concentration to regions of low water concentration. However, the diffusion of water across differentially permeable membranes has such dramatic and important effects on cells that we refer to it by a special name: **osmosis**.

What do we mean when we describe a solution as having a "high water concentration" or a "low water concentration"? The answer is simple: Pure water has the highest water concentration. Any substance added to pure water displaces some of the water molecules; the resulting solution will have a lower water content than pure water. Dissolved substances may form weak bonds with some of the water molecules, making these water molecules unavailable to diffuse across the membrane (Fig. 4-4a). The higher the concentration of dissolved substances, the lower the concentration of water. A very simple differentially permeable membrane might have pores just large enough for water to pass through but small enough to be impermeable to sugar molecules.

Consider a bag made of a special plastic that is permeable to water but not to sugar. What will happen if we placed a sugar solution in the bag and then immerse

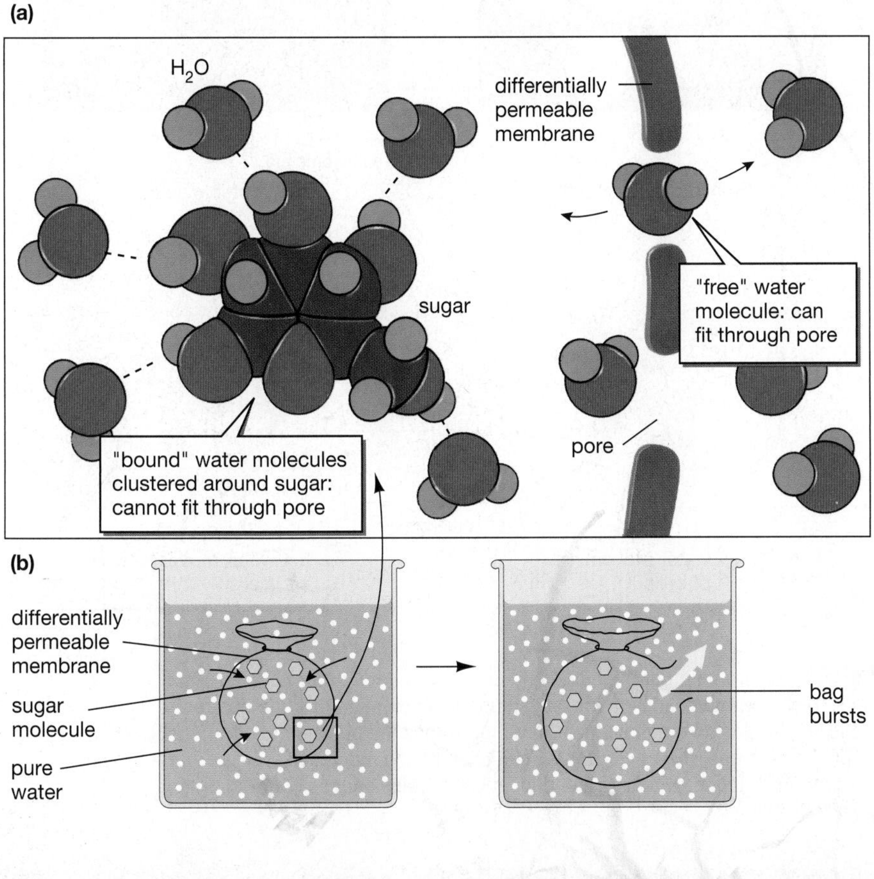

Figure 4-4 Osmosis
(a) Membrane pores allow "free" water molecules to pass through, but sugar molecules are too large. "Bound" water molecules, attracted to the sugars by hydrogen bonds, are also kept from passing through the pore. ***(b)*** A membrane is differentially permeable to free water molecules (white dots) but not to larger molecules such as sugar (yellow hexagons) or water molecules held to the sugars by hydrogen bonds. If a bag made of such a membrane is filled with a sugar solution and suspended in pure water, free water molecules will diffuse down their concentration gradient from the high concentration of water outside the bag to the lower concentration of water inside the bag. The bag swells up as water enters. If the bag is weak enough, the increasing water pressure will cause it to burst.

MediaTutor 4.2 Osmosis

the sealed bag in pure water? The principles of osmosis tell us that the bag will swell. If it is weak enough, it will burst (Fig. 4-4b).

Summing Up
The Principles of Osmosis

1. Osmosis is the diffusion of water across a differentially permeable membrane.
2. Water moves across a membrane down its concentration gradient from a high concentration of free water molecules to a low concentration of free water molecules (or down a pressure gradient from high pressure to low pressure).
3. Dissolved substances reduce the concentration of free water molecules in a solution.

Osmosis Across the Plasma Membrane Plays an Important Role in the Lives of Cells

Most plasma membranes are highly permeable to water. Because all cells contain dissolved salts, proteins, sugars, and so on, the flow of water across the plasma membrane depends on the concentration of water in the liquid that bathes the cells. The extracellular fluid of animals is usually **isotonic** ("having the same strength") to the cytoplasmic fluid within each cell; that is, the concentration of water inside is the same as that outside the cells, so there is no net tendency for water to enter or leave the cells (Fig. 4-5a). Note that the *types of dissolved particles* are seldom the same inside and outside the cells, but the *total concentration of all dissolved particles* is equal, with the result that the water concentration inside is equal to that outside the cells.

For example, if red blood cells are taken out of the body and immersed in salt solutions of varying concentrations, the effects of the differential permeability of the plasma membrane to water and dissolved particles become dramatically apparent. If the solution has a higher salt concentration than the cytoplasm of the red blood cell (that is, if the solution has a lower water concentration), water will leave the cells by osmosis. The cells will shrivel up until the concentrations of water inside and outside become equal (Fig. 4-5b). Solutions that have a higher concentration of dissolved particles than does a cell's cytoplasm, and thus cause water to leave the cell by osmosis, are termed **hypertonic** ("having greater strength").

Conversely, if the solution has little or no salt, water will enter the cells, causing them to swell (Fig. 4-5c). If

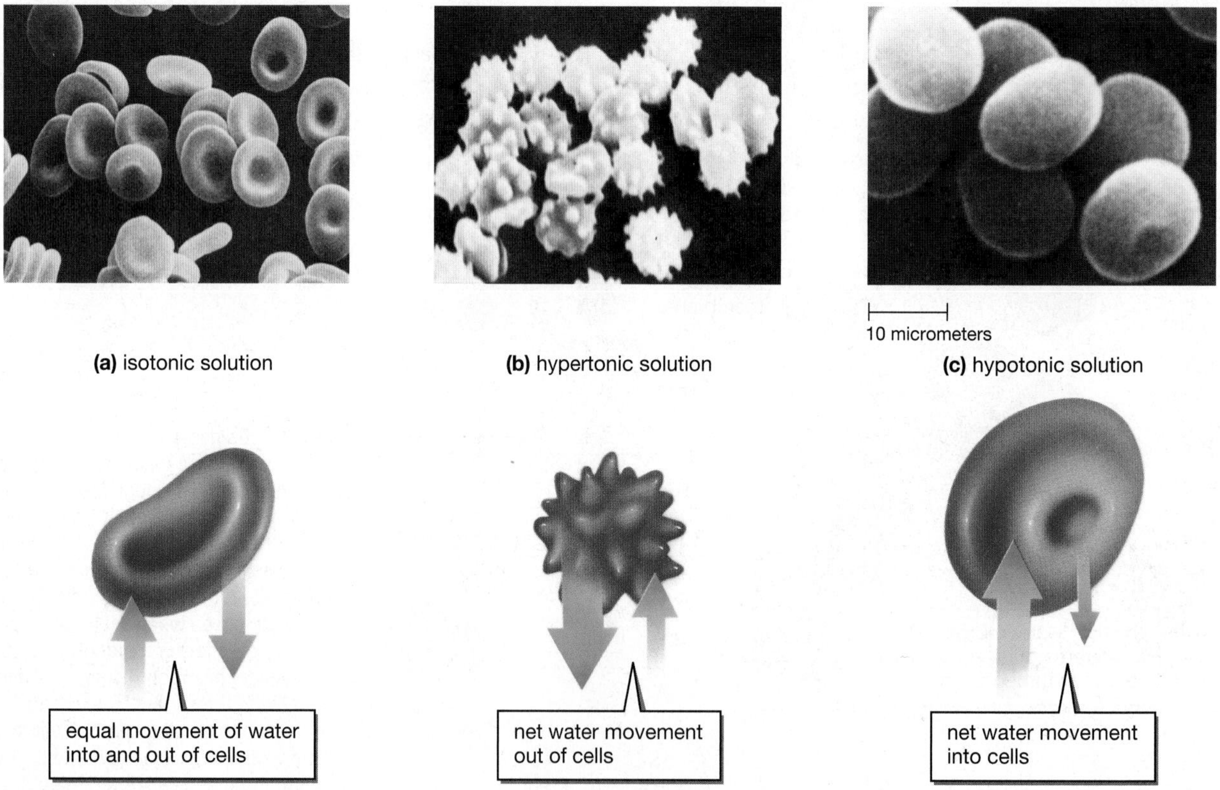

Figure 4-5 The effects of osmosis
Red blood cells are normally suspended in the fluid environment of the blood and cannot regulate water flow across their plasma membranes. ***(a)*** If red blood cells are immersed in an isotonic salt solution, which has the same concentration of dissolved substances as the blood cells do, there is no net movement of water across the plasma membrane. The red blood cells keep their characteristic dimpled disk shape. ***(b)*** A hypertonic solution, with too much salt, causes water to leave the cells, shriveling them up. ***(c)*** A hypotonic solution, with less salt than is in the cells, causes water to enter, and the cells swell.

red blood cells are placed in pure water, they will swell and eventually burst. Solutions that have a lower concentration of dissolved particles than a cell's cytoplasm, and thus cause water to enter the cell by osmosis, are called **hypotonic** ("having lesser strength"). This process is why your fingers wrinkle up after a long bath. It may seem like your fingers are shrinking, but they're not. Instead, water is diffusing into the outer skin cells of your fingers, swelling them more rapidly than the cells underneath and causing the wrinkling.

The swelling caused by osmosis can have considerable effects on cells. As we shall see in Chapter 5, protists such as *Paramecium* that live in fresh water have special structures called *contractile vacuoles* to eliminate the water that continuously leaks in. In contrast, water entry into *central vacuoles* of plant cells helps support the plant. Osmosis across plasma membranes is crucial to the functioning of many biological systems, including water uptake by plant roots (Chapter 23), absorption of dietary water from the intestine (Chapter 29), and reabsorption of water and minerals in kidneys (Chapter 30).

Active Transport Uses Energy to Move Molecules Against Their Concentration Gradients

All cells need to move some materials "uphill" across their plasma membranes, against concentration gradients. For example, every cell requires some nutrients that are less concentrated in the environment than in the cell's cytoplasm. Therefore, diffusion would cause the cell to lose, not gain, these nutrients. Other substances, such as sodium and calcium ions in your brain cells, must be maintained at much lower concentrations inside the cells than in the extracellular fluid. When these ions diffuse into the cells, they must be pumped out again against their concentration gradients.

In active transport, membrane proteins use cellular energy to move individual molecules across the plasma membrane, usually against their concentration gradient (Fig. 4-6). Active-transport proteins span the width of the membrane and have two active sites. One active site (which may be either on the face of the plasma membrane in contact with the cytoplasm or on the face in contact with the extracellular fluid, depending on the transport protein) binds a particular molecule, say a calcium ion. The second site (always on the inside of the membrane) binds an energy-carrier molecule, normally adenosine triphosphate (ATP). The ATP donates energy to the protein, causing it to change shape and move the calcium ion across the membrane. Active-transport proteins are often called *pumps*, in analogy to water pumps, because they use energy to move molecules "uphill" against a concentration gradient. We will see that plasma membrane pumps are vital in mineral uptake by plants (Chapter 23), mineral absorption in your intestines (Chapter 29), and maintaining concentration gradients essential to nerve cell functioning (Chapter 33).

Cells Engulf Particles or Fluids by Endocytosis

Many cells acquire or expel particles or substances that are too large to diffuse across a membrane regardless of concentration gradients. Cells have evolved several processes that use cellular energy to move materials

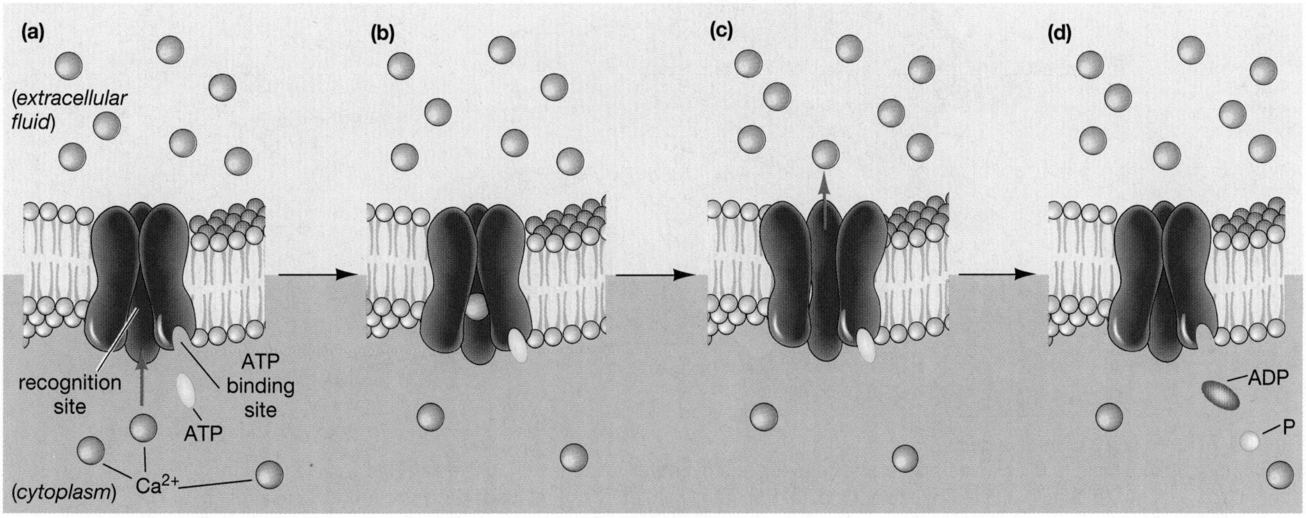

Figure 4-6 Active transport
Active transport uses cellular energy to move molecules across the plasma membrane, often against a concentration gradient. ***(a)*** A transport protein (blue) has an ATP binding site and a recognition site for the molecules to be transported, in this case calcium ions (Ca^{2+}). ***(b)*** The transport protein binds ATP and Ca^{2+}. ***(c)*** Energy from ATP changes the shape of the transport protein and moves the ion across the membrane. ***(d)*** The carrier releases the ion and the remnants of the ATP (ADP and P) and resumes its original configuration.

into or out of the cell. Cells can acquire fluids or particles from their extracellular environment, especially large proteins or entire microorganisms such as bacteria, by a process called **endocytosis** (Greek for "into the cell"). During endocytosis, the plasma membrane engulfs the fluid droplet or particle and pinches off a membranous sac called a **vesicle** with the fluid or particle inside, into the cytoplasm (Fig. 4-7). We can distinguish three types of endocytosis on the basis of the size of the particle acquired and the method of acquisition: (1) *pinocytosis*, (2) *receptor-mediated endocytosis*, and (3) *phagocytosis*.

Pinocytosis Moves Liquids into the Cell

In **pinocytosis** ("cell drinking"), a very small patch of plasma membrane dimples inward as it surrounds extracellular fluid and buds off into the cytoplasm as a tiny vesicle (Fig. 4-7a). Pinocytosis moves a droplet of extracellular fluid, contained within the dimpling patch of membrane, into the cell. Therefore, the cell acquires materials in the same concentration as in the extracellular fluid.

Receptor-Mediated Endocytosis Moves Specific Molecules into the Cell

Cells can take up certain molecules (cholesterol, for example) most efficiently by a process known as **receptor-mediated endocytosis** (Fig. 4-7b), a mechanism that can selectively concentrate specific molecules inside a cell. Most plasma membranes bear many receptor proteins on their outside surfaces, each with a binding site for a particular nutrient molecule. In most cases, these receptors move through the phospholipid bilayer and accumulate in depressions of the plasma membrane called *coated pits* (Fig. 4-8). If the right molecule contacts a receptor protein in one of these coated pits, that molecule attaches to the binding site. The coated pit deepens into a U-shaped pocket that eventually pinches off into the

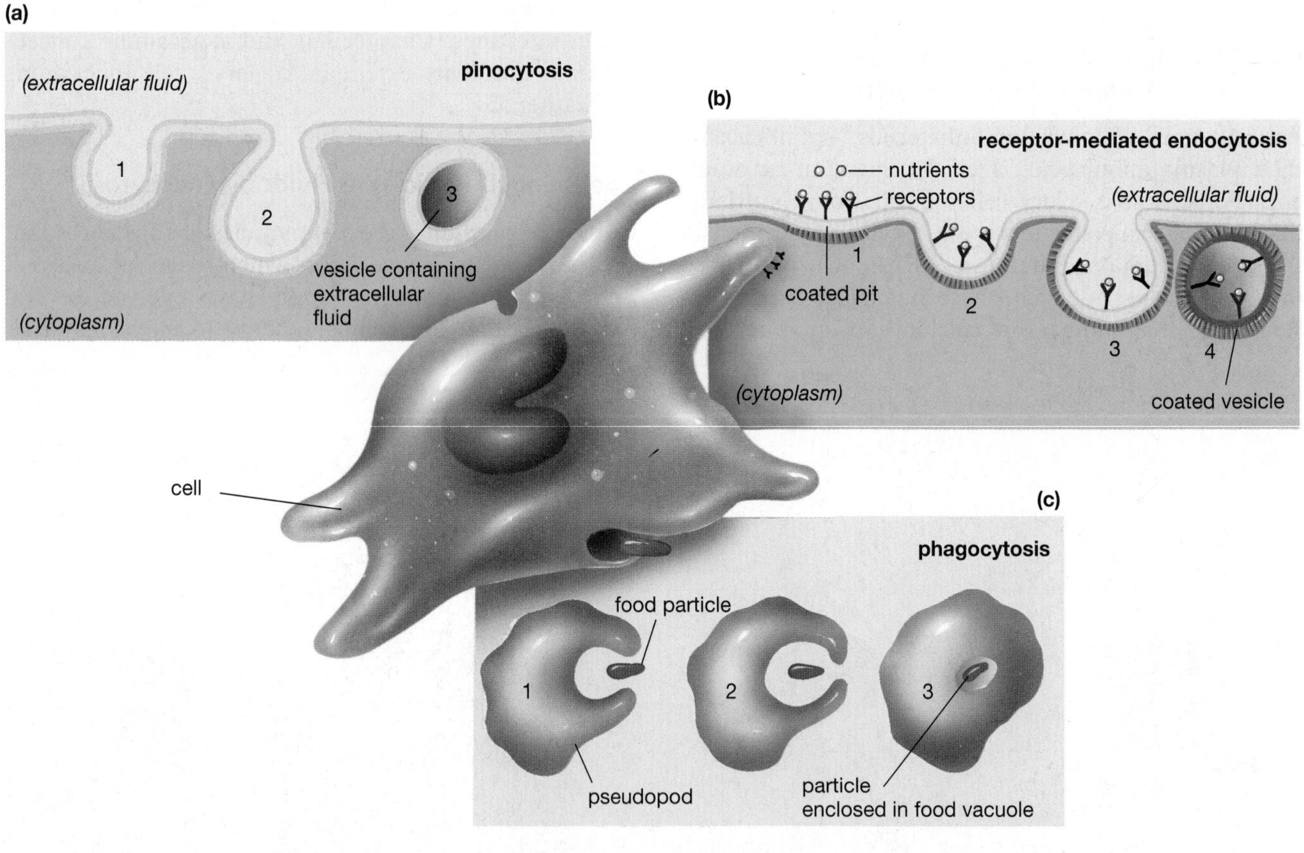

Figure 4-7 Three types of endocytosis
(a) Pinocytosis: A dimple in the plasma membrane deepens and eventually pinches off as a fluid-filled vesicle, which contains a random sampling of the extracellular fluid. ***(b)*** Receptor-mediated endocytosis: Receptor proteins selectively bind molecules (for example, nutrients) in the extracellular fluid. The receptors migrate along the plasma membrane to dimpling sites (coated pit). The membrane dimples inward, carrying the receptor-captured molecule with it. The end of the coated pit buds off a coated vesicle into the cell's cytoplasm. The vesicle contains both extracellular fluid and a high concentration of the molecules that bind to the receptors. ***(c)*** Phagocytosis: Extensions of the plasma membrane, called *pseudopodia*, encircle an extracellular particle (for example, food). The ends of the pseudopodia fuse, forming a large vesicle (a food vacuole) containing the engulfed particle.

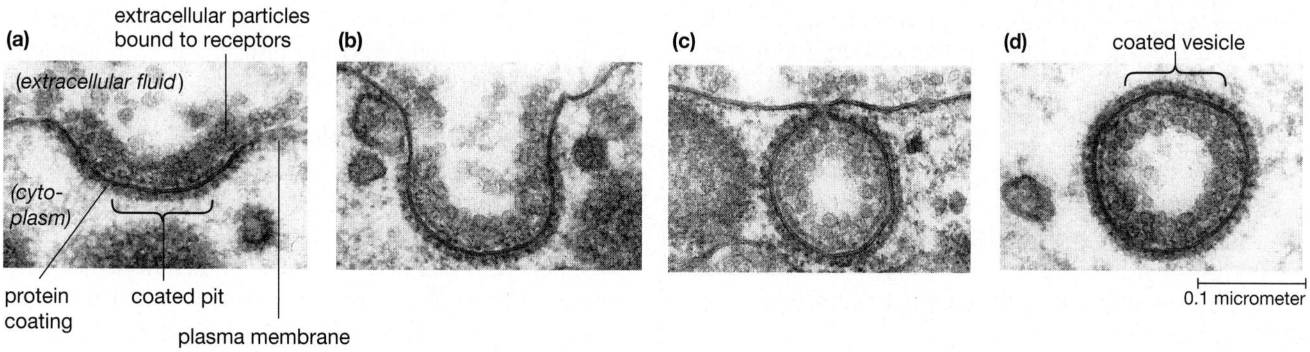

Figure 4-8 Receptor-mediated endocytosis
These electron micrographs illustrate the sequence of events in receptor-mediated endocytosis. ***(a)*** This type of endocytosis begins with a shallow depression in the plasma membrane, coated on the inside with a protein (dark, fuzzy substance in the micrographs) and bearing receptor proteins on the outside (not visible). ***(b,c)*** The pit deepens and ***(d)*** eventually pinches off as a coated vesicle. The protein coating is eventually recycled back to the plasma membrane.

cytoplasm as a *coated vesicle.* Both the receptor–nutrient complex and a bit of extracellular fluid move into the cell in the coated vesicle.

Phagocytosis Moves Large Particles into the Cell

Phagocytosis (which means "cell eating") is used by the cell to pick up large particles, including whole microorganisms (Fig. 4-7c). When the freshwater protist *Amoeba*, for example, senses a tasty *Paramecium*, *Amoeba* extends parts of its surface membrane. These membrane extensions are called **pseudopodia** (Latin for "false foot"; singular, **pseudopod**). The pseudopodia ends fuse around the luckless *Paramecium*, and the prey is carried into the interior of the *Amoeba* inside a vesicle, called a *food vacuole,* for digestion. Like *Amoeba*, white blood cells also use phagocytosis and intracellular digestion to engulf and destroy bacteria that have invaded your body.

Exocytosis Moves Material Out of the Cell

Cells often use the reverse of endocytosis, a process called **exocytosis** (Greek for "out of the cell"), to dispose of unwanted materials, such as the waste products of digestion, or to secrete materials, such as hormones, into the extracellular fluid (Fig. 4-9). During exocytosis, a membrane-enclosed vesicle carrying material to be expelled moves to the cell surface, where the vesicle's

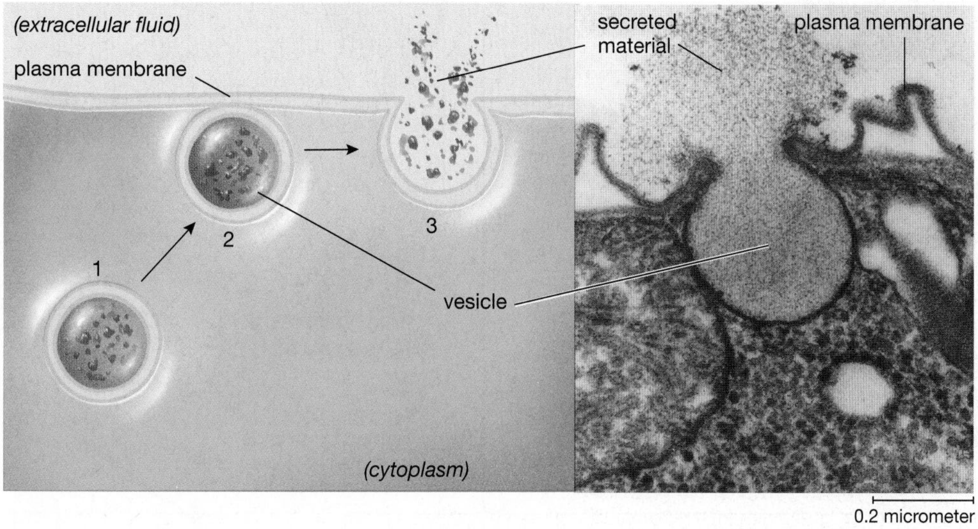

Figure 4-9 Exocytosis
Exocytosis is functionally the reverse of endocytosis. The material to be ejected from the cell is encapsulated into a membrane-bound vesicle that moves to the plasma membrane and fuses with it. As the vesicle opens to the outside, its contents leave by diffusion.

membrane fuses with the cell's plasma membrane. The vesicle then opens to the extracellular fluid, and its contents diffuse out.

3 How Are Cell Surfaces Specialized?

A Variety of Junctions Allow Cells to Connect and Communicate

In multicellular organisms, plasma membranes hold together clusters of cells and provide avenues through which cells communicate with their neighbors. Depending on the organism and the cell type, four types of connection may occur between cells: (1) *desmosomes*, (2) *tight junctions*, (3) *gap junctions*, and (4) *plasmodesmata.*

Desmosomes Attach Cells Together

Animals, as you know, tend to be flexible, mobile organisms. Many of an animal's tissues are stretched, compressed, and bent as the animal moves. Cells in the skin, intestine, urinary bladder, and other organs must adhere firmly to one another to avoid tearing under the stresses of movement. Such animal tissues have junctions called **desmosomes**, which hold adjacent cells together (Fig. 4-10a). In a desmosome the membranes of adjacent cells are glued together by proteins and carbohydrates. Protein filaments attached to the insides of the desmosomes extend into the interior of each cell, further strengthening the attachment.

Tight Junctions Make the Cell Leakproof

The animal body contains many tubes or sacs that must hold their contents without leaking; a leaky urinary bladder would spell disaster for the rest of the body. Spaces between the cells that line such tubes or sacs are sealed with strands of protein to form **tight junctions** (Fig. 4-10b). The membranes of adjacent cells nearly fuse along a series of ridges, effectively forming leakproof gaskets between cells. Continuous tight junctions sealing each cell to its neighbors prevent molecules from escaping between cells.

Gap Junctions and Plasmodesmata Allow Communication Between Cells

Multicellular organisms must coordinate the actions of their component cells. In animals, many cells, including heart muscle cells, most gland cells, some brain cells, and every cell of very young embryos, communicate through protein channels that directly connect the insides of adjacent cells. These cell-to-cell channels are called **gap junctions** (Fig. 4-11a). Hormones, nutrients, ions, and even electrical signals can pass through the channels at gap junctions.

Virtually all of the living cells of plants are connected to one another by **plasmodesmata** (singular, **plasmodesma**). Plasmodesmata are openings in the walls of

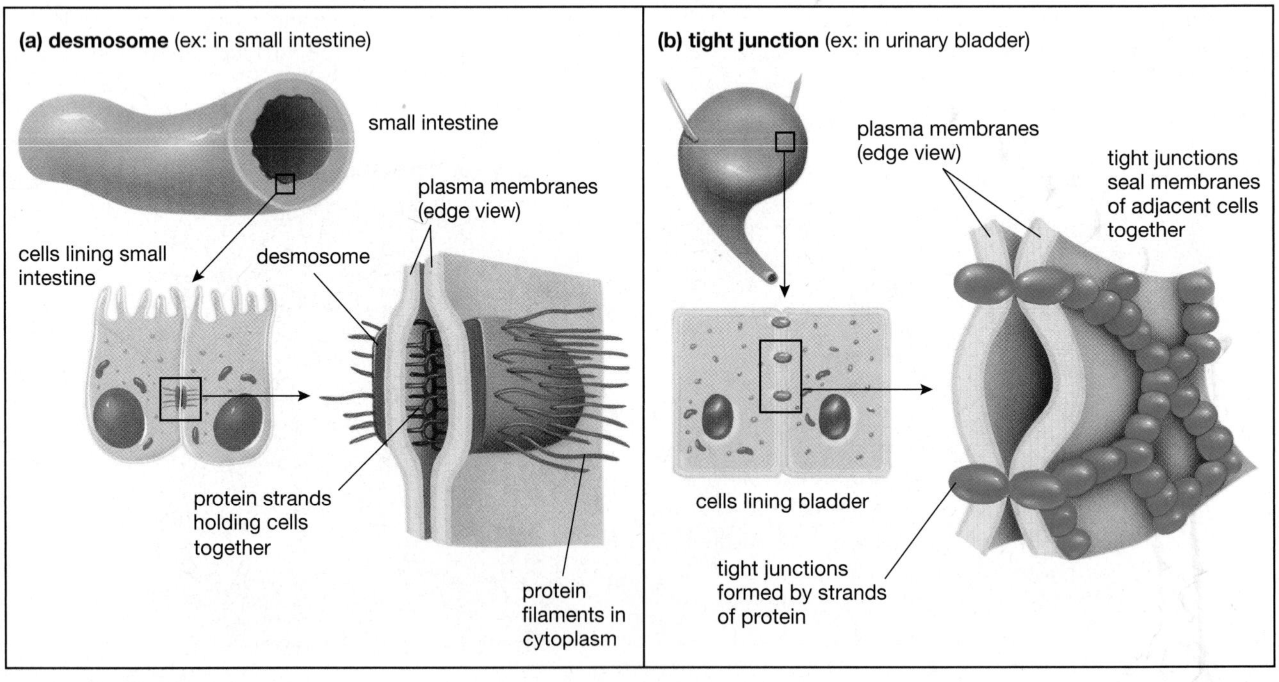

Figure 4-10 Cell attachment structures
(a) Cells lining the small intestine are firmly attached to one another by desmosomes. Protein filaments bound to the inside surface of each desmosome extend into the cytoplasm and attach to other filaments inside the cell, strengthening the connection between cells. ***(b)*** Leakage between cells of the urinary bladder is prevented by close-fitting tight junctions.

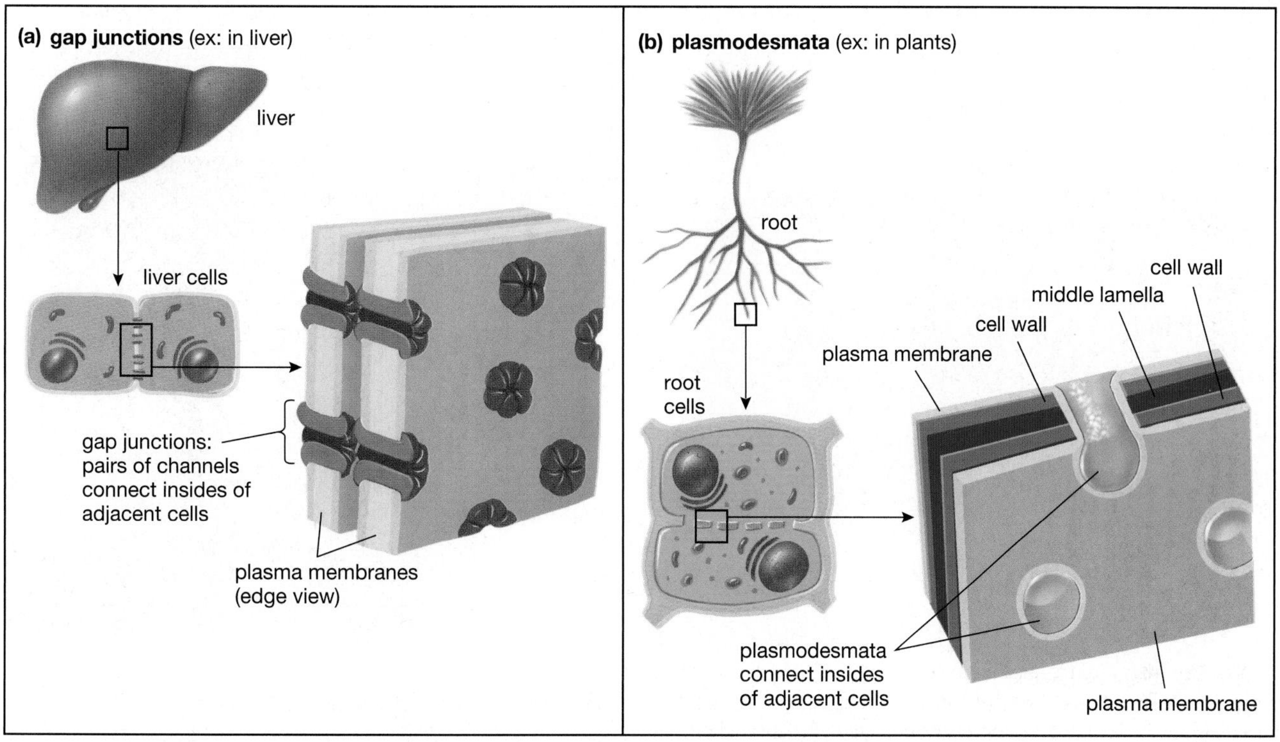

Figure 4-11 Cell communication structures
(a) Gap junctions, such as those between cells of the liver, contain cell-to-cell channels that interconnect the cytoplasm of adjacent cells. *(b)* Plant cells are interconnected by plasmodesmata, which pass through openings in the walls of adjacent plant cells.

adjacent plant cells, lined with plasma membrane and filled with cytoplasm. Plasmodesmata create continuous cytoplasmic bridges between the insides of adjacent cells (Fig. 4-11b). Many plant cells have thousands of plasmodesmata, allowing water, nutrients, and hormones to pass quite freely from one cell to another.

Some Cells Are Supported by Cell Walls

The outer surfaces of the cells of bacteria, plants, fungi, and some protists are covered with nonliving, typically stiff coatings called **cell walls**. Single-celled organisms that live in the ocean, for example, may have outer coverings of cellulose, protein, or glassy silica that protect the delicate plasma membrane. Plant cell walls are composed of cellulose and other *polysaccharides*, whereas fungal cell walls are made of polysaccharides and the modified polysaccharide *chitin*. (We described both polysaccharides and chitin in Chapter 3.) Bacterial cell walls have a chitin-like framework to which short chains of amino acids and other molecules are attached.

Cell walls are produced by the cells they surround. Plant cells secrete cellulose through their plasma membranes, forming the **primary cell wall**. Many plant cells later secrete more cellulose and other polysaccharides beneath the primary wall to form a thick **secondary cell wall**, pushing the primary cell wall away from the plasma membrane. In some plant cells, the secondary wall becomes thicker than the cell inside it. The primary cell walls of adjacent cells are joined by the **middle lamella**, a layer made primarily of *pectin* (Fig. 4-12). Pectin is the polysaccharide that makes jelly solidify.

Cell walls support and protect otherwise fragile cells. For example, cell walls allow plants and mushrooms to resist the forces of gravity and wind and to stand erect on land. Tree trunks, composed almost entirely of cellulose and other materials laid down over the years and capable of supporting impressive loads, are the ultimate proof of the strength of cell walls.

Although strong, cell walls are normally porous, permitting easy passage of small molecules such as minerals, water, oxygen, carbon dioxide, amino acids, and sugars. (Otherwise, the cell within would soon die.) However, the structure that really governs the interactions that occur between a cell and its external environment is the plasma membrane.

Evolutionary Connections

Caribou Legs and Membrane Diversity

The membranes of all cells are similar in structure, reflecting the common evolutionary heritage of all life on Earth. Membrane function varies tremendously from organism to organism, however, and even from cell to

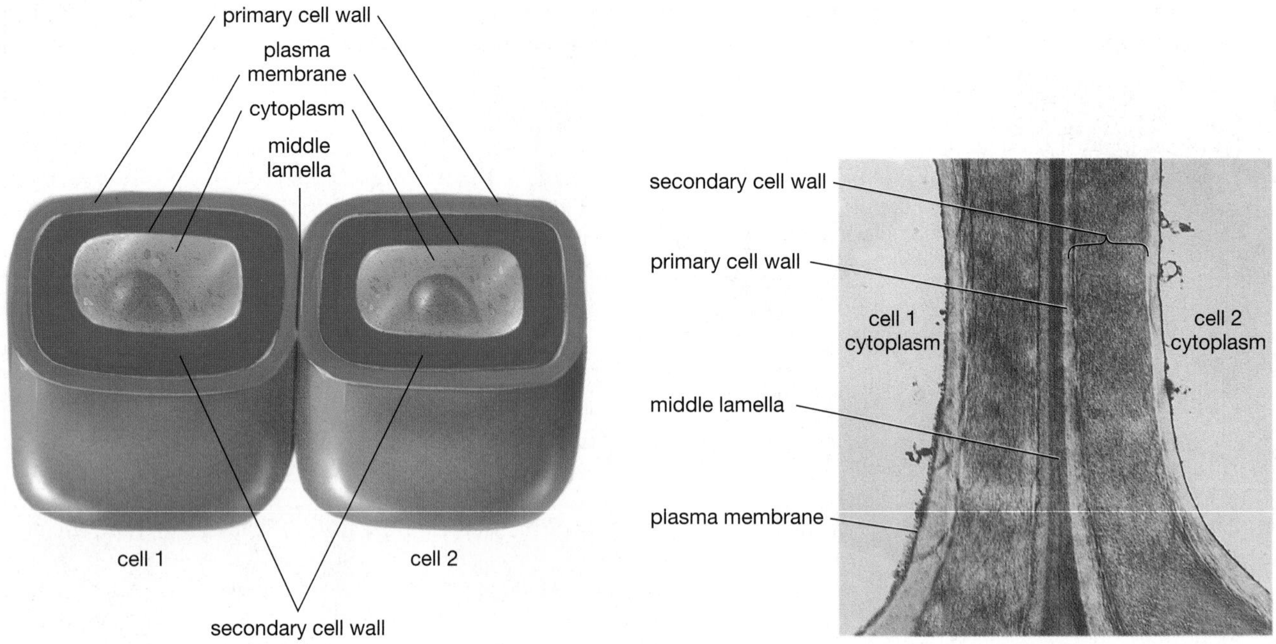

Figure 4-12 Plant cell walls
Each plant cell secretes cellulose and other carbohydrates to form its primary cell wall, just outside the plasma membrane. Many cells may then secrete additional cellulose and other polysaccharides beneath the primary wall, forming a secondary cell wall, which pushes the primary cell wall away from the plasma membrane. The middle lamella separates adjacent plant cells.

cell within a single organism. This diversity arises largely from the different proteins and phospholipids in the membrane, which have evolved under differing environmental pressures.

Our discussion of membranes emphasized the unique functions of membrane proteins. Consequently, you may think that the phospholipids are just a waterproof place for the proteins to reside. This isn't quite true, as we can see by examining the plasma membrane phospholipids in cells of the legs of caribou, animals that live in very cold regions of North America (Fig. 4-13). During the long arctic winters of these regions, temperatures plummet far below freezing. For caribou to keep their legs and feet really warm would waste precious energy. These conditions have favored the evolutions of specialized arrangements of arteries and veins in caribou legs that allow the temperature of their lower legs to drop almost to freezing (0 °C), thus conserving body heat. The upper legs and main trunk of the body, in contrast, remain at about 37 °C. To remain fluid at these radically different temperatures requires the phospholipids in the membranes of cells in the upper legs of caribou to be very different from those near the hooves.

Remember, the membrane of a cell needs to be somewhat fluid to allow the membrane proteins to move to sites where they are needed. The fluidity of a membrane is a function of the fatty acid tails of its phospholipids: Unsaturated fatty acids (with some double bonds between carbons) remain more fluid at lower temperatures than do saturated fatty acids (see Chapter 3). Caribou have a range of fatty acids in the phospholipids of the cells in their legs. The membranes of cells near the chilly hoof have lots of unsaturated fatty acids, whereas the membranes of cells near the warmer trunk have more saturated ones. This arrangement gives the plasma membranes throughout the leg the proper fluidity despite great differences in temperature.

As important as the phospholipids are, the membrane proteins play the major roles in determining cell function and in governing the interactions between a cell and its neighbors. Protein molecules can be altered as a result of mutations that change their amino acid composition, shapes, and functions. Over billions of

Figure 4-13 Caribou browse on the frozen Alaskan tundra
The lipid composition of the membranes in the cells of a caribou's legs varies with distance from the animal's trunk. Unsaturated phospholipids predominate in the lower leg; more saturated phospholipids are found in the upper leg.

years, an incredible diversity of proteins has evolved. Every nerve cell in your body, for instance, has membrane proteins essential for producing electrical signals and conducting them along the nerves to various parts of your body. Other membrane proteins receive chemical messages from neighboring nerve cells or from hormones and other chemicals in the blood. Each cell in the brain has a specific set of membrane proteins, allowing it to respond to some stimuli while ignoring others. In fact, your ability to read this page depends on the proteins that reside in the membranes of your light-detecting cells and in the membranes of your brain cells.

As we progress through this book, we shall return many times to the concepts of membrane structure presented in this chapter. Understanding the diversity of membrane lipids and proteins is the key to understanding not just the isolated cell, but also entire organs, which function as they do largely because of the properties of the membranes of their component cells.

CASE STUDY REVISITED

Vicious Venoms

Snake venoms are complex brews of poisonous proteins. The venom of the Western Diamondback rattlesnake is rich in enzymes called *phospholipases*, which break down the phospholipids of the plasma membranes of cells, causing the cells to rupture and die. Cell death blackens the tissue around the bite. In the bloodstream, the phospholipases attack red blood cells, reducing the oxygen-carrying capacity of the blood and causing the victim to become short of breath. Once carried to muscles, phospholipases also attack muscle cell membranes, exposing the complex contractile proteins within the muscle cells to attack by protein-digesting enzymes in the venom. Proteins that strengthen blood vessels also come under attack, and hemorrhaging may result if the vessels rupture. The venom also contains enzymes that attack proteins that allow the blood to clot, making the hemorrhaging more dangerous. When internal bleeding is combined with loss of the oxygen-carrying red blood cells, it is no wonder the victim is short of breath and may go into shock.

Karl was extremely lucky that Mark had brought his cell phone. Trying to hike back to the car would have quickly spread the venom throughout his body, and the added delay would have decreased his chances of survival. Fortunately, Mark was able to get expert advice immediately: keep Karl lying down—since he was suffering from shock—and as immobile as possible until help arrives. Because they had identified the snake, the hospital had the correct antivenin waiting. Antivenin contains proteins that bind to and neutralize the various toxins in the snake venom. The Western Diamondback, with its half-inch-long fangs and large stores of venom, is responsible for more U.S. snakebite deaths than any other species. Of the 7000 to 8000 people bitten by poisonous snakes in the United States each year, only 10 to 15 people are killed by the venom.

Phospholipases and other digestive enzymes are found in animal (including human) digestive tracts as well as in snake venom. Why do we have phospholipases in our digestive tracts? Since snakes swallow their prey whole, what are two very different roles for the phospholipases and other enzymes in snake venom?

Summary of Key Concepts

1) How Is the Structure of a Membrane Related to Its Function?

The plasma membrane has three major functions: (1) it selectively isolates the cytoplasm from the external environment, (2) it regulates the flow of materials into and out of the cell, and (3) it communicates with other cells. The membrane consists of a bilayer of phospholipids in which a variety of proteins are embedded. There are three major categories of membrane proteins: (1) transport proteins, which regulate the movement of most water-soluble substances through the membrane; (2) receptor proteins, which bind molecules in the external environment, triggering changes in the metabolism of the cell; and (3) recognition proteins, which serve as identification tags and attachment sites.

2) How Do Substances Move Across Membranes?

Diffusion is the movement of particles from regions of higher concentration to regions of lower concentration. In simple diffusion, water, dissolved gases, and lipid-soluble molecules diffuse through the phospholipid bilayer. In facilitated diffusion, water-soluble molecules cross the membrane through protein channels or with the assistance of protein carriers. In both cases, molecules move down their concentration gradients, and cellular energy is not required.

Osmosis is the diffusion of water across a differentially permeable membrane and down its concentration gradient. Dissolved substances decrease the concentration of free water molecules. Osmosis does not require cellular energy.

In active transport, protein carriers in the membrane use cellular energy (ATP) to drive the movement of molecules across the plasma membrane, usually against concentration gradients. Large molecules (for example, proteins), particles of food, microorganisms, and extracellular fluid may be acquired by endocytosis, either by pinocytosis, receptor-mediated endocytosis, or phagocytosis. The secretion of substances such as hormones and the excretion of wastes from a cell are accomplished by exocytosis.

3 How Are Cell Surfaces Specialized?

Cells may be connected to one another by a variety of junctions. Desmosomes attach cells firmly to one another, preventing the tearing of a tissue during movement or stress. Tight junctions seal off the spaces between adjacent cells, leakproofing organs such as the urinary bladder. Gap junctions in animals and plasmodesmata in plants are the locations at which the cytoplasms of two adjacent cells are interconnected.

The outer surfaces of some protist cells and of each bacterial, plant, and fungal cell are surrounded by a rigid cell wall outside the plasma membrane. The cell wall, produced by the cell that it surrounds, protects and supports that cell.

Key Terms

active transport *p. 61*
carrier protein *p. 60*
cell wall *p. 69*
channel protein *p. 59*
concentration *p. 60*
concentration gradient *p. 60*
cytoplasm *p. 58*
desmosome *p. 68*
differentially permeable *p. 62*
diffusion *p. 60*
endocytosis *p. 66*
exocytosis *p. 67*
facilitated diffusion *p. 63*
fluid *p. 60*
fluid mosaic model *p. 58*
gap junction *p. 68*
glycoprotein *p. 59*
gradient *p. 60*
hypertonic *p. 64*
hypotonic *p. 65*
isotonic *p. 64*
middle lamella *p. 69*
osmosis *p. 63*
passive transport *p. 61*
phagocytosis *p. 67*
phospholipid bilayer *p. 58*
pinocytosis *p. 66*
plasma membrane *p. 57*
plasmodesma *p. 68*
primary cell wall *p. 69*
pseudopod *p. 67*
receptor-mediated endocytosis *p. 66*
receptor protein *p. 60*
recognition protein *p. 60*
secondary cell wall *p. 69*
simple diffusion *p. 62*
tight junction *p. 68*
transport protein *p. 59*
vesicle *p. 66*

Thinking Through the Concepts

Multiple Choice

1. *Active transport through the plasma membrane occurs through the action of*
   a. diffusion
   b. membrane proteins
   c. DNA
   d. water
   e. osmosis

2. *The following is a characteristic of a plasma membrane:*
   a. It separates the cell contents from its environment.
   b. It is permeable to certain substances.
   c. It is a lipid bilayer with embedded proteins.
   d. It contains pumps for moving molecules against their concentration gradient.
   e. all of the above

3. *If an animal cell is placed into a solution whose concentration of dissolved substances is higher than that inside the cell,*
   a. the cell will swell
   b. the cell will shrivel
   c. the cell will remain the same size
   d. the solution is described as hypertonic
   e. both (b) and (d) are correct

4. *Small, nonpolar hydrophobic molecules such as fatty acids*
   a. pass readily through a membrane's lipid bilayer
   b. diffuse very slowly through the lipid bilayer
   c. require special channels to enter a cell
   d. are actively transported across cell membranes
   e. must enter the cell via endocytosis

5. *Which of the following would be least likely to diffuse through a lipid bilayer?*
   a. water
   b. oxygen
   c. carbon dioxide
   d. sodium ions
   e. the small, nonpolar molecule butane

6. *Which of the following processes causes substances to move across membranes without the expenditure of cellular energy?*
   a. endocytosis
   b. exocytosis
   c. active transport
   d. diffusion
   e. pinocytosis

? Review Questions

1. Describe and diagram the structure of a plasma membrane. What are the two principal types of molecules in plasma membranes? What are the four principal functions of plasma membranes?
2. What are the three categories of proteins commonly found in plasma membranes, and what is the function of each?
3. Define *diffusion*, and compare that process to osmosis. How do these two processes help plant leaves remain firm?
4. Define *hypotonic*, *hypertonic*, and *isotonic*. What would be the fate of an animal cell immersed in each of the three types of solution?
5. Describe the following types of transport processes: simple diffusion, facilitated diffusion, active transport, pinocytosis, receptor-mediated endocytosis, phagocytosis, and exocytosis.
6. Name four types of cell-to-cell junctions and the function of each. Which junctions allow communication between the interiors of adjacent cells?

Applying the Concepts

1. Different cells have somewhat different plasma membranes. The plasma membrane of a *Paramecium*, for example, is only about 1% as permeable to water as the plasma membrane of a human red blood cell. Referring back to our discussion of the effects of osmosis on red blood cells and the role of contractile vacuoles in *Paramecium,* what do you think is the function of the low water permeability of *Paramecium*? What molecular differences do you think might account for this low water permeability?
2. A preview question for Chapter 31: The integrity of the plasma membrane is essential for cellular survival. Could the immune system utilize this fact to destroy foreign cells that have invaded the body? How might cells of the immune system disrupt membranes of foreign cells? (Two hints: Virtually all cells can secrete proteins, and some proteins form pores in membranes.)
3. A preview question for Chapter 23: Plant roots take up minerals (inorganic ions such as potassium) that are dissolved in the water of the soil. The concentration of such ions is usually much lower in the soil water than in the cytoplasm of root cells. Design the plasma membrane of a hypothetical mineral-absorbing cell, with special reference to mineral-permeable channel proteins and mineral-transporting active transport proteins. Justify your choice of channels and active-transport proteins.
4. Red blood cells will swell up and burst when placed in a hypotonic solution such as pure water. Why don't we swell up and burst when we swim in water that is hypotonic to our cells and body fluids?

For More Information

McNeil, P. L. "Cell Wounding and Healing." *American Scientist*, May–June 1991. The fluidity of plasma membrane phospholipids makes cells able to withstand minor damage.

Rothman, J. E., and Orci, L. "Budding Vesicles in Living Cells." *Scientific American*, March 1996. Membranes within cells form small containers called *vesicles*, which transport materials inside the cell. Researchers are discovering the mechanisms by which these containers are formed.

Sharon, N., and Lis, H. "Carbohydrates in Cell Recognition." *Scientific American*, January 1993. Carbohydrates, normally attached to proteins as part of glycoproteins, identify cells, serve as parts of receptors in hormone binding, and regulate the attachment and movement of cells.

Answers to Multiple-Choice Questions

1. b 2. e 3. e 4. a 5. d 6. d

MEDIATUTOR
Cell Membrane Structure and Function

CD Activities

Activity 4.1: Membrane Structure and Transport

Estimated time: 10 minutes

Cell membranes control what is allowed in or kept out of the cell. In this tutorial you will interact with a virtual membrane to control the transport of various molecules into and out of the cell.

Activity 4.2: Osmosis

Estimated time: 5 minutes

Cells are always both filled with and surrounded by water. In this tutorial you will explore how osmosis, the movement of water in response to a concentration gradient, affects the cell.

Start the MediaTutor Student CD-ROM and enter the activity number in the Quick Search box to be taken directly to that activity.

Web Investigations

Case Study: Vicious Venoms

Estimated time: 10 minutes

Many people are afraid of snakes. A person bitten by a poisonous snake might die without anti-venom treatment, yet venom components can also be lifesaving drugs. This exercise takes a closer look at snake bites and venom.

Go to http://www.prenhall.com/audesirk6, the Audesirk Companion Web site. Select Chapter 4 and the Web Investigation to begin.

"It is all too easy now to underestimate cells. We have known about them for such large fractions of our lives that, for the most part, we cease being aware of how remarkable they really are."

Bruce Alberts, Dennis Bray, Julian Lewis, Martin Raff, Keith Roberts, and James Watson from the Prologue to* Molecular Biology of the Cell *(1983)

David R. Scott, an *Apollo 12* astronaut, stands on the moon in 1969. A breach in his suit would cause almost immediate death for his body and the individual cells that it comprises. In a remarkable display of resilience, cells of the bacterium *Streptococcus* (inset), which normally live in the human nose and throat, survived for more than 2 years on the moon.

5 Cell Structure and Function

AT A GLANCE

Case Study: Microscopic Stowaways

1) **What Are the Basic Features of Cells?**
All Living Things Are Composed of One or More Cells
All Cells Share Certain General Features
There Are Two Basic Types of Cells: Prokaryotic and Eukaryotic

2) **What Are the Features of Prokaryotic Cells?**

3) **What Are the Features of Eukaryotic Cells?**
Eukaryotic Cells Contain Organelles
The Nucleus Is the Control Center of the Eukaryotic Cell
Eukaryotic Cells Contain a Complex System of Membranes
Vacuoles Serve Many Functions, Including Water Regulation, Support, and Storage
Mitochondria Extract Energy from Food Molecules, and Chloroplasts Capture Solar Energy
Plants Use Plastids for Storage
The Cytoskeleton Provides Shape, Support, and Movement
Cilia and Flagella Move the Cell or Move Fluid Past the Cell

Case Study Revisited: Microscopic Stowaways

CASESTUDY

Microscopic Stowaways

In 1967 the unmanned *Surveyor 3* spacecraft landed on the moon, carrying a camera to photograph the bleak lunar landscape. Two years and 7 months later, in November 1969, the *Apollo 12* astronauts followed. As part of their mission, they visited the *Surveyor 3* craft, opened the camera case, and collected a small sample of the insulating polyurethane foam inside. Placing the sample in a sterile container, the astronauts continued their lunar tasks. Later, back on Earth, scientists painstakingly opened the container under sterile conditions, stroked it with a sterile swab, removed some of the sample and placed it in a sterile plate with growth medium, and incubated it at human body temperature (37 °C). A few days later, they gazed with excitement at the results: dozens of colonies of bacteria dotted the dishes. What could still be alive after more than two and a half years on the lunar surface? Scientists at the U.S. Communicable Disease Center in Atlanta, Georgia, made the positive identification: *Streptococcus mitis*, a harmless resident of the human mouth, nose, and throat, resembling those shown in the photo inset. Perhaps one of the NASA scientists had sneezed while assembling the camera, unwittingly adding a group of stowaway space travelers to the cargo of the *Surveyor* craft.

Our bodies are made of far more elaborate and fragile cells. Molded by the forces of natural selection, these complex cells have specialized and become interdependent. Human cells, working together, have allowed us to travel to the moon and to gaze in awe at the bacteria that survived there. In this chapter we explore the two fundamentally different types of cells that make up all living things. ■

1) What Are the Basic Features of Cells?

All Living Things Are Composed of One or More Cells

In the late 1850s Austrian pathologist Rudolf Virchow wrote, "Every animal appears as a sum of vital units, each of which bears in itself the complete characteristics of life." Furthermore, Virchow predicted, "All cells come from cells." The three principles of modern cell theory evolved directly from Virchow's statements:

1. Every living organism is made up of one or more cells.
2. The smallest living organisms are single cells, and cells are the functional units of multicellular organisms.
3. All cells arise from preexisting cells.

All living things, from the *Streptococcus* bacteria to your own body, are composed of cells. Whereas each bacterium consists of a single, relatively simple cell, your body consists of trillions of complex cells, each specialized to perform a specific function. To survive, all cells must obtain energy and nutrients from their environment. They must synthesize a variety of proteins and other molecules necessary for their growth and repair, and they eliminate wastes. Many cells need to interact with other cells. To ensure the continuity of life, cells must also reproduce. These activities are accomplished by specialized parts of each cell, which are described in the following sections.

All Cells Share Certain General Features

In spite of their diversity, cells share certain common features summarized in Table 5-1 and described below.

Table 5-1 Common Features of All Cells

Molecular components	Protein, amino acids, lipids, carbohydrates, sugars, nucleotides, DNA, RNA
Structural components	Plasma membrane, cytoplasm, ribosomes
Metabolism	Extracts energy and nutrients from the environment; uses energy and nutrients to build, repair, and replace cellular parts

The Plasma Membrane Encloses the Cell and Mediates Interactions Between the Cell and Its Environment

As we saw in Chapter 4, the **plasma membrane** consists of a phospholipid bilayer in which a wide variety of proteins are embedded. The plasma membrane performs three major functions:

1. It isolates the cell's contents from the external environment.
2. It regulates the flow of materials into and out of the cell, for example, acquiring nutrients and expelling wastes.
3. It allows interaction with other cells.

Cells Use DNA as a Hereditary Blueprint

Each cell has genetic material, an inherited hereditary blueprint that stores the instructions for making all the other parts of the cell and for producing new cells. The genetic material in all cells is **deoxyribonucleic acid (DNA)**. In *eukaryotic* cells (plants, animals, fungi, and protists), the DNA is contained within a separate, membrane-bound structure called the **nucleus**. In *prokaryotic* cells (bacteria and archaeans), the DNA, although localized to a particular region within the cell, is not separated by membranes from the rest of the cell's interior.

All Cells Contain Cytoplasm

The **cytoplasm** consists of all the material inside the plasma membrane and outside the DNA-containing region. The fluid portion of the cytoplasm in prokaryotic and eukaryotic cells contains water, salts, and an assortment of organic molecules; in each case it is a thick soup of proteins, lipids, carbohydrates, salts, sugars, amino acids, and nucleotides (see Chapter 3). Most of the cell's *metabolic activities*—that is, the sum of all the biochemical reactions that underlie life—occur in the cell cytoplasm. Protein synthesis is one example. This complex process takes place on special structures called **ribosomes**, located in the cytoplasm of all cells. One of the many types of proteins that cells synthesize are *enzymes* that allow metabolic reactions to occur, as we will see in the following chapter.

All Cells Obtain Energy and Nutrients from Their Environment

To maintain their incredible complexity, all cells must continuously acquire and expend energy. As we explain in Chapters 6, 7, and 8, essentially all of the energy powering life on Earth originates in sunlight. Cells that can harness this energy directly and incorporate it into high-energy molecules provide the source of energy for nearly all other life-forms. The building blocks of biological molecules, including carbon, nitrogen, oxygen, and a variety of minerals, ultimately comes from the environment—the air, water, rocks, and other forms of life. All cells obtain the materials to generate the molecules of life, and the energy to power this synthesis, from their living and nonliving environment.

Cell Function Limits Cell Size

Most cells are small, ranging from about 1 to 100 micrometers (millionths of a meter) in diameter (Fig. 5-1). Because cells are so small, their discovery awaited the invention of the microscope. Ever since the first cells were seen in the late 1600s, scientists have devised increasingly sophisticated ways of observing them, as described in "Scientific Inquiry: The Search for the Cell."

Why are most cells small, and why do large organisms consist of many cells rather than one large cell? The answer lies in the need for cells to exchange nutrients and wastes with their external environment through the plasma membrane. As you learned in Chapter 4, many nutrients and wastes move into, through, and out of cells by *diffusion*, the movement of molecules from places of high concentration of those molecules to places of low concentration. As a roughly spherical cell becomes larger, its innermost regions become farther away from the membrane. Diffusion is quite a slow process; for example, in a giant cell 8.5 inches (20 centimeters) in diameter, oxygen molecules would take more than *200 days* to diffuse to the center of the cell! Further, as a cell enlarges, its volume increases more rapidly than its surface area does. For example, a cell that triples in radius becomes 27 times

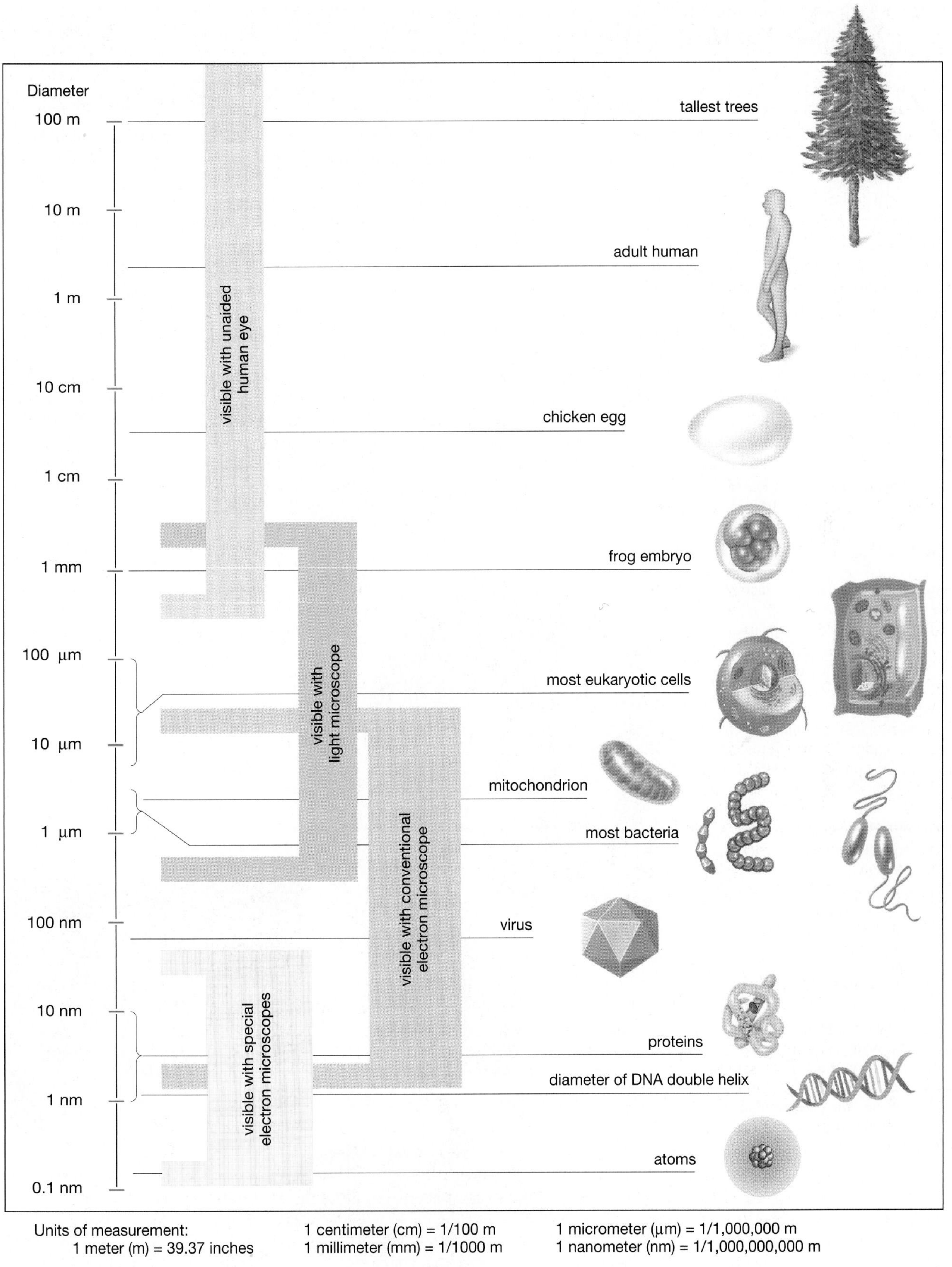

Figure 5-1 Relative sizes
Dimensions commonly encountered in biology range from about 100 meters (the height of the tallest redwoods) through a few micrometers (the diameter of most cells) to a few nanometers (the diameter of many large molecules). Note that in the metric system (used almost exclusively in science and in many regions of the world), separate names are given to dimensions that differ by factors of 10, 100, and 1000.

Scientific Inquiry
The Search for the Cell

Human understanding of the cellular nature of life came slowly. In 1665 English scientist and inventor Robert Hooke reported observations with a primitive microscope. He aimed his instrument at an "exceeding thin . . . piece of Cork" and saw "a great many little Boxes" (Fig. E5-1a). Hooke called the boxes "cells," because he thought they resembled the tiny rooms, or cells, occupied by monks. Cork comes from the dry outer bark of the Mediterranean oak. Hooke wrote that in the living oak and other plants, "These cells [are] fill'd with juices."

In the 1670s Dutch microscopist Anton van Leeuwenhoek was constructing his own simple microscopes and observing a previously unknown world. His descriptions of myriad "animalcules" (his term for protists) in rain, pond water, and well water caused quite an uproar, because water was consumed without treatment in those days. He made careful observations of an enormous range of microscopic specimens, including red blood cells, sperm, and the eggs of small insects such as weevils, aphids, and fleas. His discoveries struck a blow to the then-common belief in spontaneous generation; at that time, fleas were believed to emerge spontaneously from sand or dust, as were weevils from grain!

More than a century passed before biologists began to understand the role of cells in life on Earth. Microscopists first noted that many plants consist entirely of cells. The thick wall surrounding all plant cells, first observed by Hooke, made their observations easier. Animal cells, however, escaped notice until the 1830s, when German zoologist Theodor Schwann saw that cartilage contains cells that "exactly resemble [the cells of] plants." In 1839, after studying cells for years, Schwann was confident enough to publish his *cell theory*, calling cells the elementary particles of both plants and animals. By the mid-1800s, German botanist Matthias Schleiden had further refined science's view of cells when he wrote: "It is . . . easy to perceive that the vital process of the individual cells must form the first, absolutely indispensable fundamental basis" of life.

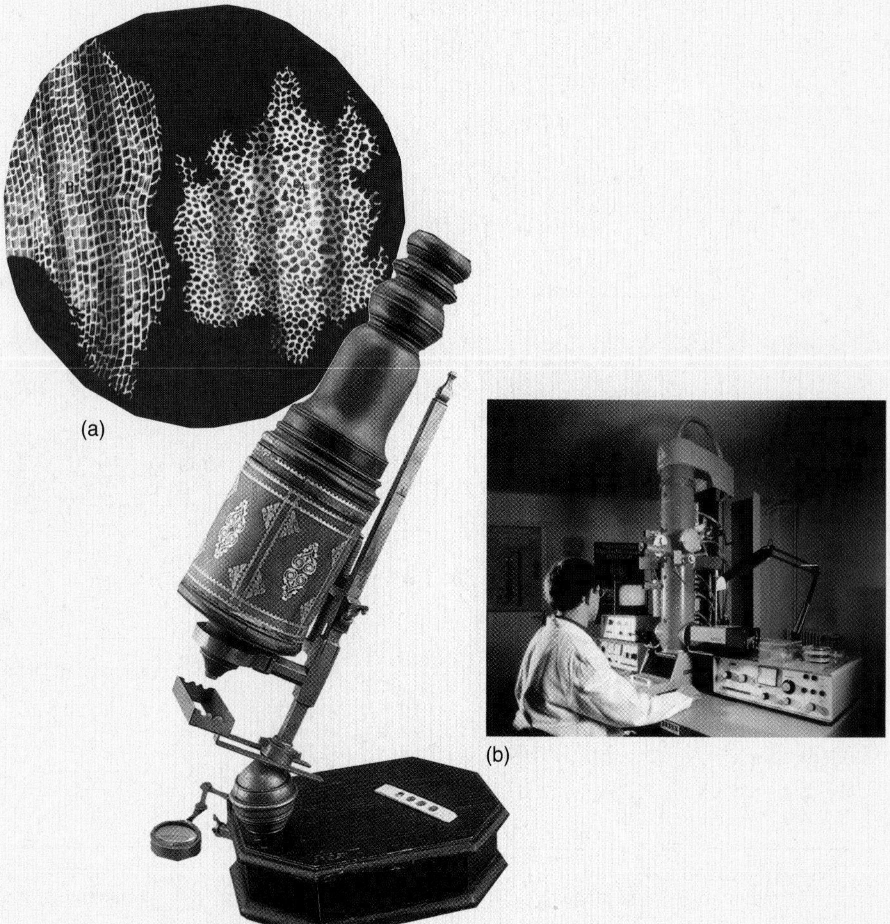

Figure E5-1 Microscopes yesterday and today
(a) Robert Hooke's drawings of the cells of cork, as he viewed them with an early light microscope similar to the one shown here. The cells of cork are not living, and only the cell walls remain to outline the cell. ***(b)*** A scanning electron microscope. This allows a three-dimensional image of the surfaces of specimens. The specimen is prepared by coating it with a thin layer of heavy metal (such as platinum), which is vaporized and deposited on the surface of the object to be viewed. The microscope directs an electron beam over the coated surface, and the metal ions on the surface release electrons in response. The number of electrons released depends on the angle of the electron beam relative to the surface of the specimen, which in turn is determined by the contours of the specimen. Thus, the final image, formed by detection of the emitted electrons, has a three-dimensional appearance.

Ever since the pioneering efforts of Robert Hooke and Anton van Leeuwenhoek, biologists, physicists, and engineers have collaborated in the development of a variety of advanced microscopes to view the cell and its components:

Light microscopes use lenses, usually made of glass, to focus and magnify light rays that either pass through or bounce off a specimen. Light microscopes provide a wide range of images, depending on how the specimen is illuminated and whether it has been stained (Fig. E5-2a). The *resolving power* of light microscopes—that is, the smallest structure that can be seen—is about 1 micrometer (a millionth of a meter).

Electron microscopes use beams of electrons instead of light. The electrons are focused by magnetic fields rather than by lenses. Some types of electron microscopes can resolve structures as small as a few nanometers (billionths of a meter). *Transmission electron microscopes* (TEMs) pass electrons through a thin specimen and can reveal the details of interior cell structure, including organelles and plasma membranes (Fig. E5-2b). *Scanning electron microscopes* (SEMs; see Figure E5-1b) bounce electrons off specimens that have been coated with metals and provide three-dimensional images. SEMs can be used to view the surface details of structures that range in size from entire insects down to cells and even organelles (Fig. E5-2c,d).

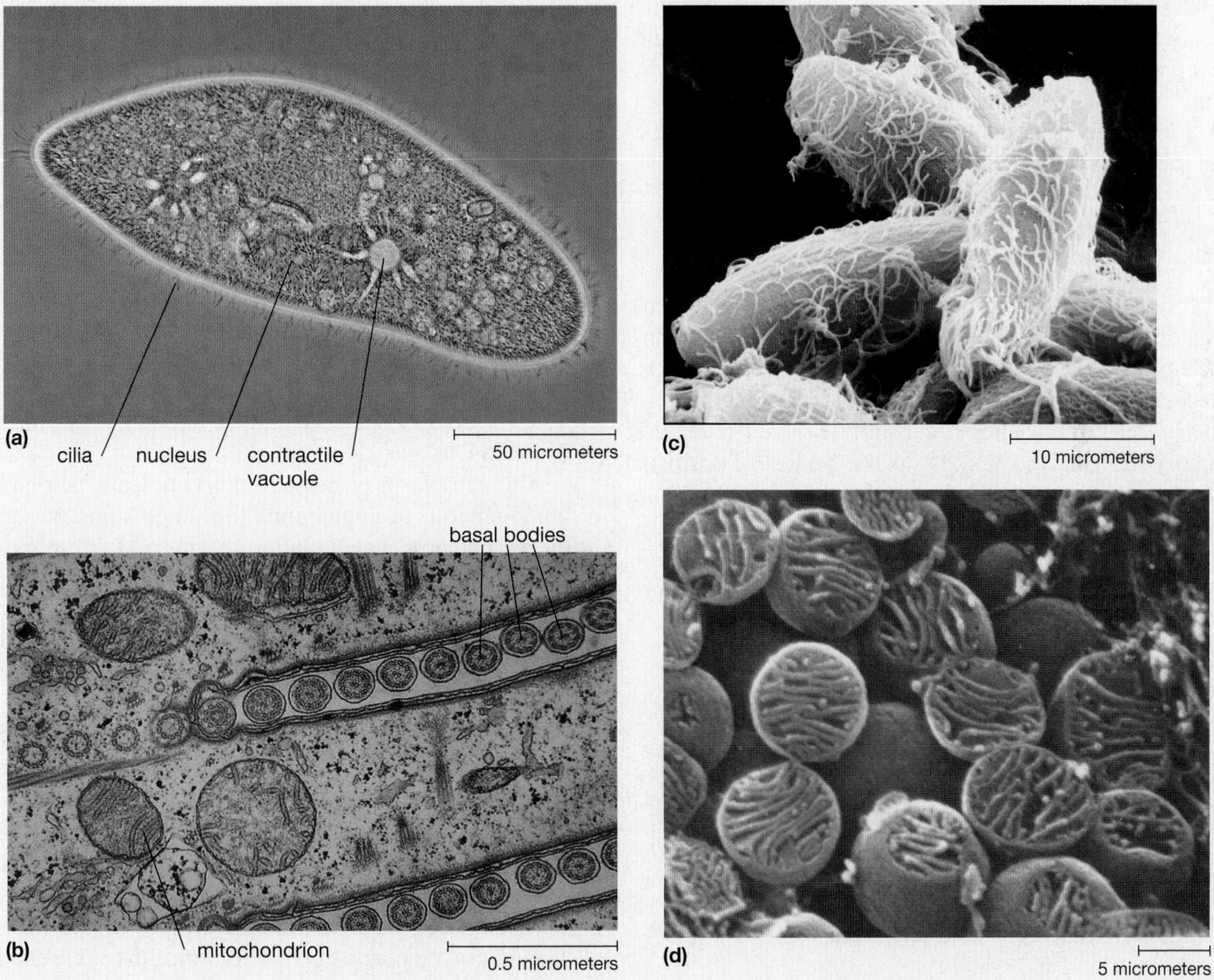

Figure E5-2 A comparison of microscope images
(a) A living *Paramecium* viewed through a light microscope. ***(b)*** A false-color TEM photo showing the basal bodies at the bases of the cilia that cover *Paramecium.* In this single-celled freshwater protist, mitochondria are also visible. ***(c)*** A false-color SEM photo of a *Paramecium.* ***(d)*** An SEM photo at much higher magnification, showing mitochondria, many of which are sliced open.

greater in volume but only nine times greater in surface area.

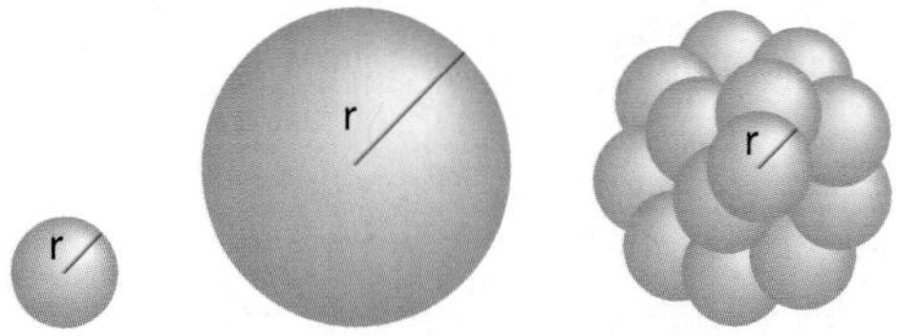

distance to center (r)	1.0	3.0	1.0
surface area $(4\pi r^2)$	12.6	113.1	339.4
volume $(4/3\ \pi r^3)$	4.2	113.1	113.1
area/volume	3.0	1.0	3.0

So, a larger cell has a greater need for the exchange of nutrients and wastes with the environment but a relatively smaller expanse of plasma membrane through which to make these exchanges. In a very large, roughly spherical cell, the surface area of the membrane would be too small to keep up with the cell's metabolic needs. Some cells, such as your neurons and muscle cells, can get very large because they have an elongated shape that increases their membrane surface area.

There Are Two Basic Types of Cells: Prokaryotic and Eukaryotic

All forms of life are composed of only two fundamentally different types of cells. The first type, which include the bacteria and archaeans, is called **prokaryotic**, Greek for "before the nucleus" (Fig. 5-2). The second type of cell, which almost certainly evolved from the prokaryotic cell and makes up the bodies of protists, plants, fungi, and animals, is called **eukaryotic**, for "true nucleus" (see Figs. 5-3 and 5-4). As their names imply, one striking difference between prokaryotic cells and eukaryotic cells is that the genetic material of eukaryotic cells is contained within a membrane-enclosed nucleus. In contrast, the genetic material of prokaryotic cells is not enclosed within a membrane. In the following sections, we discuss the features of prokaryotic and eukaryotic cells.

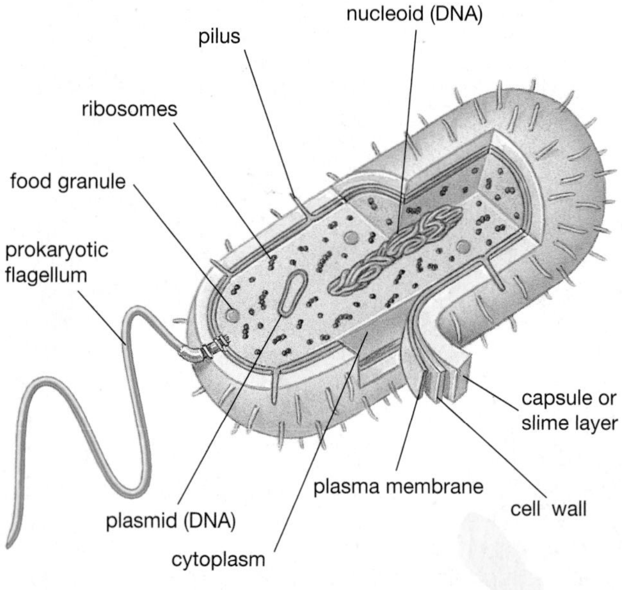

Figure 5-2 A generalized prokaryotic cell

2 What Are the Features of Prokaryotic Cells?

Most prokaryotic cells are very small (less than 5 micrometers long), with a relatively simple internal structure (Fig. 5-2). Most are surrounded by a relatively stiff cell wall, which confers shape and protects the bacterial cell. Antibiotics, such as penicillin, fight bacterial infections by interfering with cell wall synthesis, causing the bacteria to rupture. Some bacteria can move, propelled by flagella that are different in structure from those found in eukaryotic cells. Surface projections made of protein, called **pili** (singular, **pilus**), are used to attach some types of bacteria to surfaces or to exchange genetic material. **Capsules** or **slime layers** are polysaccharide or protein coatings that some disease-causing bacteria secrete outside their cell wall. These coatings help the bacteria attach to their hosts and may allow them to evade attack by immune cells. In its normal encapsulated form, *Streptococcus pneumoniae*, a close relative of the bacteria that survived on the moon, will readily kill mice; a mutant lacking the capsule, however, is harmless. We discuss bacterial shapes and cell walls in more detail in Chapter 19.

The cytoplasm of most prokaryotic cells is relatively homogeneous in appearance (although some photosynthetic bacteria have elaborate internal membranes). Prokaryotic cells each have a single, circular strand of DNA. It is usually coiled, attached to the plasma membrane, and concentrated in the region of the cell called the **nucleoid**. The nucleoid is not, however, physically separated from the rest of the cytoplasm by a membrane.

Prokaryotic cells lack nuclei as well as the other membrane-enclosed *organelles* that eukaryotic cells possess. Nonetheless, some prokaryotic cells use membranes to organize molecules responsible for a series of biochemical reactions. Photosynthetic bacteria possess internal membranes, which contain the light-capturing proteins and enzymes that catalyze the synthesis of high-energy molecules. In prokaryotic cells, reactions that harvest energy from the breakdown of sugars are catalyzed by enzymes that may be localized along the inner plasma membrane or that may be free in the cytoplasm. Bacterial cytoplasm contains ribosomes, structures composed of proteins and **ribonucleic acid (RNA)**, on which proteins are synthesized. The cytoplasm also may contain *food granules* that store energy-rich materials, such as glycogen. Prokaryotic cells are compared

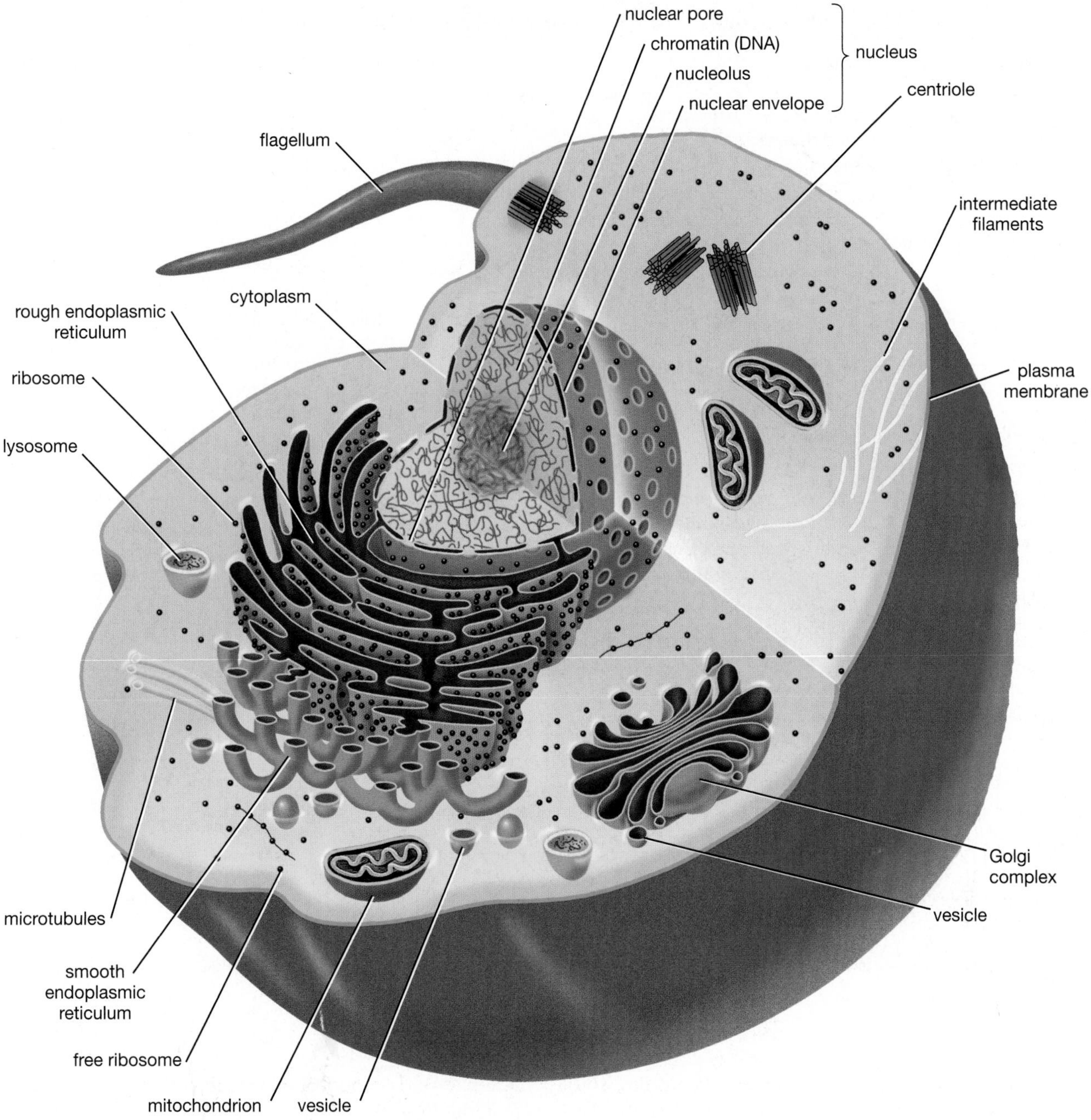

Figure 5-3 A generalized animal cell

with the eukaryotic cells of plants and animals in Table 5-2 . The diversity and specialized structures of bacteria are covered in more detail in Chapter 19.

3 What Are the Features of Eukaryotic Cells?

Eukaryotic Cells Contain Organelles

Eukaryotic cells differ from prokaryotic cells in many ways. For one thing, they are usually larger than prokaryotic cells—typically more than 10 micrometers in diameter. However, the cytoplasm of eukaryotic cells also houses a variety of membrane-enclosed structures called **organelles** that perform specific functions within the cell. A network of protein fibers, the **cytoskeleton** gives shape and organization to the cytoplasm of eukaryotic cells. Many of the organelles are attached to the cytoskeleton.

Eukaryotic cells, however, are not all alike. Figures 5-3 and 5-4 (p. 83) illustrate the structures that are found in animal and plant cells, respectively, although few individual cells possess all the features shown in either drawing. Each type of cell has a few unique organelles not found in the other. Plant cells, for example, contain chloroplasts, plastids, and a central vacuole not

Table 5-2 Cell Structures, Their Functions, and Their Distribution in Living Cells

Structure	Function	Prokaryotes	Plants	Animals
Cell surface				
Cell wall	Protects, supports cell	Present	Present	Absent
Plasma membrane	Isolates cell contents from environment; regulates movement of materials into and out of cell; communicates with other cells	Present	Present	Present
Organization of genetic material				
Genetic material	Encodes information needed to construct cell and control cellular activity	DNA	DNA	DNA
Chromosomes	Contain and control use of DNA	Single, circular, no proteins	Many, linear, with proteins	Many, linear, with proteins
Nucleus	Membrane-bound container for chromosomes	Absent	Present	Present
Nuclear envelope	Encloses nucleus; regulates movement of materials into and out of nucleus	Absent	Present	Present
Nucleolus	Synthesizes ribosomes	Absent	Present	Present
Cytoplasmic structures				
Mitochondria	Produce energy by aerobic metabolism	Absent	Present	Present
Chloroplasts	Perform photosynthesis	Absent	Present	Absent
Ribosomes	Provide site of protein synthesis	Present	Present	Present
Endoplasmic reticulum	Synthesizes membrane components and lipids	Absent	Present	Present
Golgi complex	Modifies and packages proteins and lipids; synthesizes carbohydrates	Absent	Present	Present
Lysosomes	Contain intracellular digestive enzymes	Absent	Present	Present
Plastids	Store food, pigments	Absent	Present	Absent
Central vacuole	Contains water and wastes; provides turgor pressure to support cell	Absent	Present	Absent
Other vesicles and vacuoles	Contain food obtained through phagocytosis; contain secretory products	Absent	Present (some)	Present
Cytoskeleton	Gives shape and support to cell; positions and moves cell parts	Absent	Present	Present
Centrioles	Synthesize microtubules of cilia and flagella; may produce spindle in animal cells	Absent	Absent (in most)	Present
Cilia and flagella	Move cell through fluid or move fluid past cell surface	Present[a]	Absent (in most)	Present

[a]Many prokaryotes have structures called *flagella,* but these are not made of microtubules and move in a fundamentally different way than eukaryotic cilia or flagella do.

found in animal cells. Animal cells possess centrioles. You may want to refer to these illustrations as we describe the structures of the cell in more detail. The major components of eukaryotic cells are listed in Table 5-2 and explained in more detail in the following sections.

The Nucleus Is the Control Center of the Eukaryotic Cell

Deoxyribonucleic acid (DNA) is the genetic material of all living cells. A cell's DNA is the repository of information needed to construct the cell and direct the countless chemical reactions necessary for life and reproduction. The genetic information in DNA is used selectively by the cell, depending on its stage of development and its environmental conditions. In eukaryotic cells the DNA is housed within the nucleus.

The nucleus is an organelle, usually the largest in the cell, consisting of three readily distinguishable parts (Fig. 5-5, p. 84). The **nuclear envelope** separates the nuclear material from the cytoplasm. Inside the nuclear envelope, the nucleus contains a granular-looking material called **chromatin** and a darker region called the **nucleolus**, described in the following sections.

The Nuclear Envelope Allows Selective Exchange of Materials

The nucleus is isolated from the rest of the cell by a nuclear envelope that consists of a double membrane. The membrane is perforated with tiny membrane-lined channels called *pores*. Water, ions, and small molecules such as ATP can pass freely through the pores, but the passage of large molecules, particularly proteins, pieces of ribosomes, and RNA, is regulated by specialized "gatekeeper proteins" that line each nuclear pore. Ribosomes stud the outer nuclear membrane, which is continuous with membranes of the rough endoplasmic reticulum, described later (see Figs. 5-3 and 5-4).

MediaTutor 5.1 Cell Structure

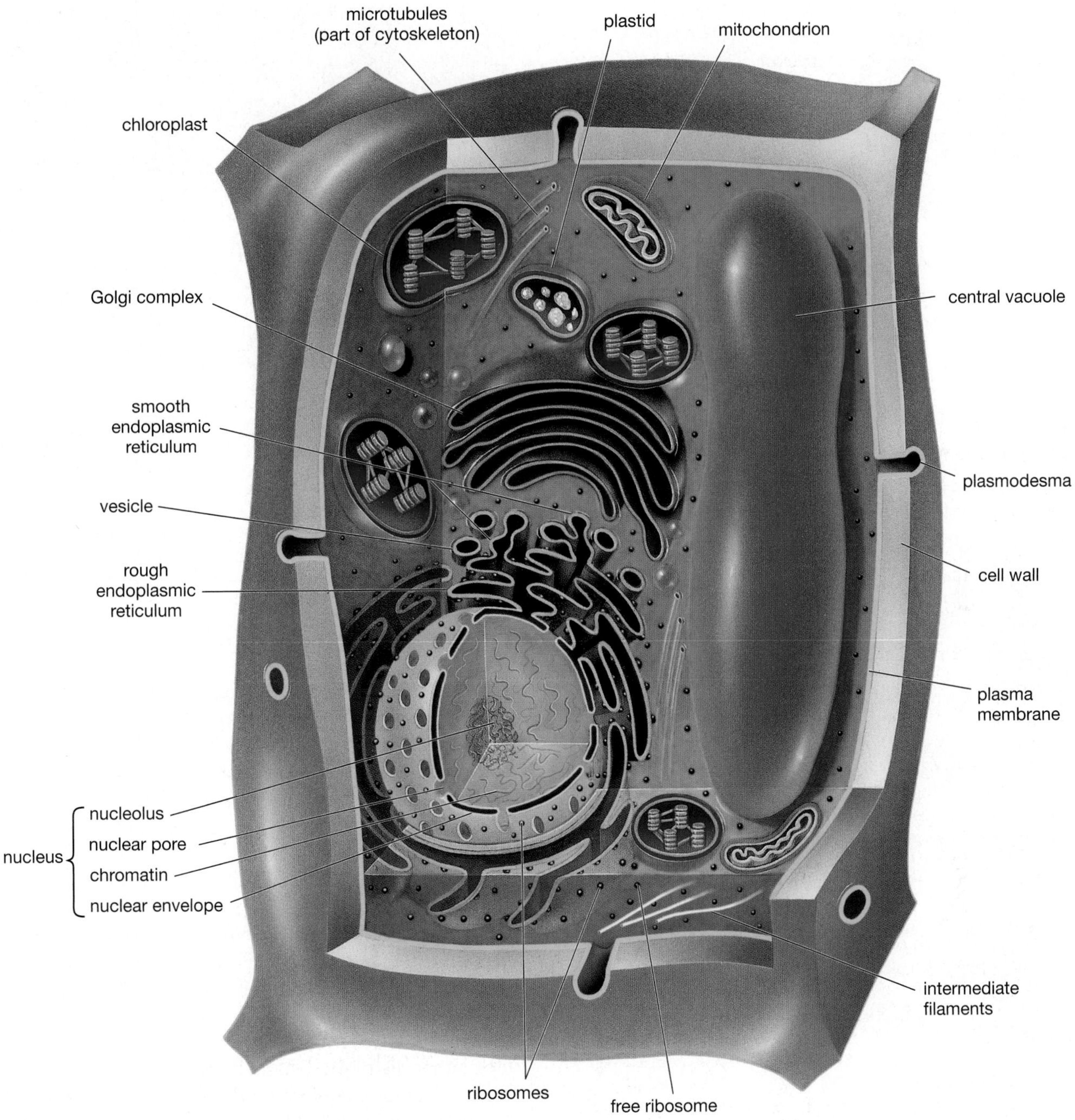

Figure 5-4 A generalized plant cell

Chromatin Consists of DNA and Its Associated Proteins

Because the nucleus is highly colored by common stains used in light microscopy, early microscopists named the nuclear material *chromatin*, meaning "colored substance." Biologists have since learned that chromatin consists of DNA associated with proteins. Eukaryotic DNA and its associated proteins form long strands called **chromosomes** ("colored bodies"). When cells divide, each chromosome coils upon itself, becoming thicker and shorter. The resulting "condensed" chromosomes are easily visible even with light microscopes (Fig. 5-6).

Chemical reactions within the cell that are responsible for growth and repair, nutrient and energy acquisition and use, and reproduction are governed by the information encoded in DNA. Because the DNA stays in the nucleus, whereas most of the chemical reactions that it controls occur in the cytoplasm, information molecules must be exchanged between the nucleus and the cytoplasm. Genetic information is copied from DNA into molecules of RNA, which move through the pores of the nuclear envelope into the cytoplasm. This information is then used to direct the synthesis of cellular proteins. These proteins include enzymes, which catalyze

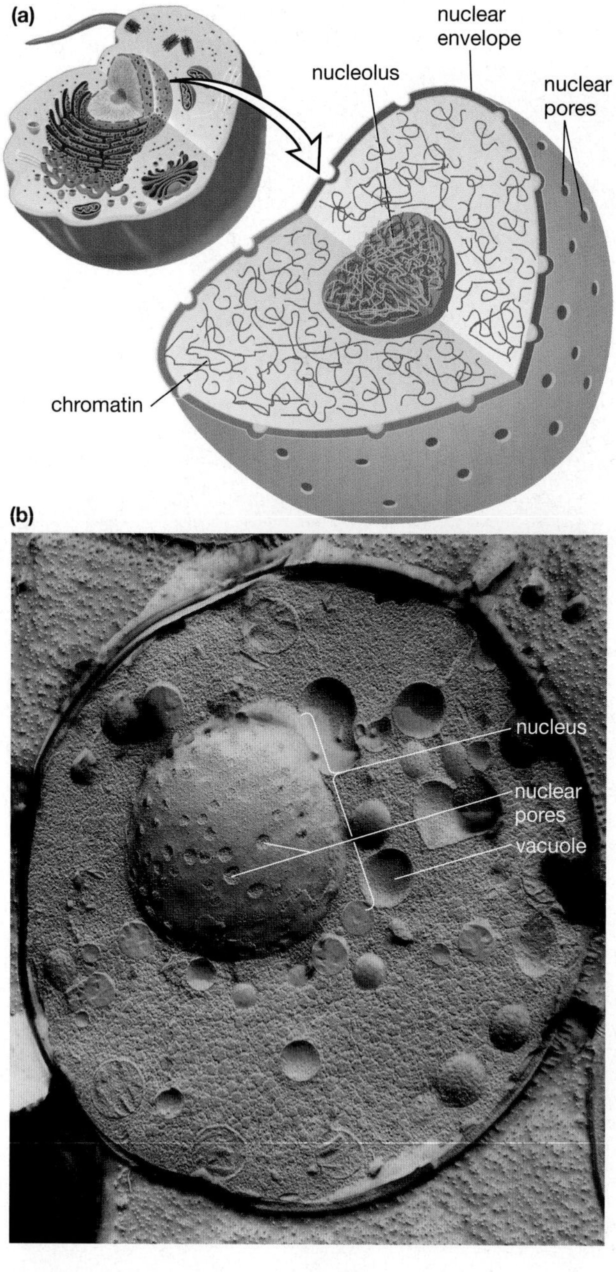

Figure 5-5 The nucleus
(a) The nucleus is bounded by a double outer membrane. Inside are chromatin (chromosomes in an uncondensed state) and a nucleolus, which contains DNA coding for ribosomal RNA, ribosomes in various stages of synthesis, and associated proteins. ***(b)*** An electron micrograph of a yeast cell that has been frozen and broken open to reveal its internal structures. The large nucleus, with nuclear pores penetrating its nuclear membrane, is clearly visible.

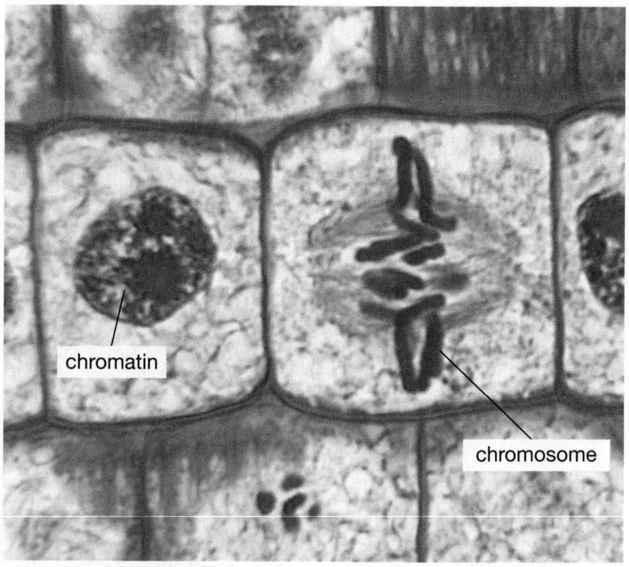

Figure 5-6 Chromosomes
Chromosomes, seen here in a light micrograph of a dividing cell (on the right) in an onion root tip, are the same material (DNA and proteins) as the chromatin seen in nondividing cells adjacent to it, but in a more compact state.

and regulate chemical reactions; membrane proteins, which govern interactions between the cell and its environment; and a variety of structural proteins. Some of these proteins pass from the cytoplasm into the nucleus and regulate the transfer of information from DNA to RNA, depending on what is happening in the cytoplasm and in the extracellular environment. We take a closer look at these processes in Chapter 10.

The Nucleolus Is the Site of Ribosome Assembly

Most eukaryotic nuclei have one or more darkly staining regions called *nucleoli* ("little nuclei"; one nucleolus is shown in Fig. 5-5a). The nucleolus consists of *ribosomal RNA*, proteins, ribosomes in various stages of synthesis, and DNA (bearing genes that specify the blueprint for ribosomal RNA).

Nucleoli are the sites of ribosome synthesis. A ribosome is a small particle composed of RNA and proteins that serves as a kind of "workbench" for the synthesis of proteins. Just as a workbench can be used to construct many different objects, any ribosome can be used to synthesize any of the thousands of proteins made by a cell. In electron micrographs, ribosomes appear as dark granules, either distributed in the cytoplasm (Fig. 5-7) or clustered along the membranes of the nuclear envelope and the endoplasmic reticulum (see Figs. 5-3, 5-4, and 5-8).

Eukaryotic Cells Contain a Complex System of Membranes

All eukaryotic cells have an elaborate system of membranes that enclose the cell and create internal compartments that allow a huge variety of processes to occur within the cytoplasm. This membrane system is composed of the *plasma membrane* and several organelles, including the endoplasmic reticulum, nuclear envelope, Golgi complex, and a variety of membrane-enclosed sacs such as lysosomes. The various parts of the membrane system of the cells can exchange membrane material with one another and also transfer materials surrounded by membrane to different compartments for different types of processing (see Figs. 5-3 and 5-4).

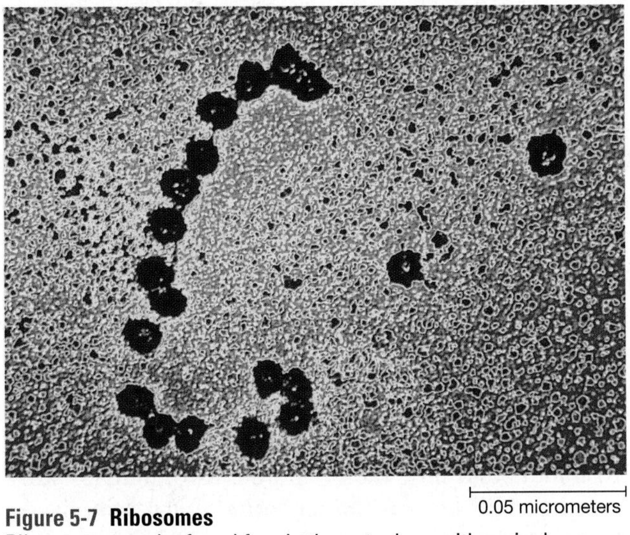

Figure 5-7 Ribosomes
Ribosomes may be found free in the cytoplasm either singly or strung along messenger RNA molecules as they participate in protein synthesis. Ribosomes also stud the rough endoplasmic reticulum, giving it a rough appearance and allowing the synthesis of proteins within the ER.

The Plasma Membrane Both Isolates the Cell and Allows Selective Interactions Between the Cell and Its Environment

The plasma membrane forms the outer boundary of the living part of a cell, enclosing the cytoplasm. It is a marvelously complex structure that must perform the seemingly contradictory functions of separating the cytoplasm of the cell from the outside environment while providing for the transport of selected substances into or out of the cell. In plants, fungi, and some protists, a *cell wall* is secreted through the plasma membrane and forms an outer, protective coating. Both plasma membrane and eukaryotic cell wall structure and function were discussed in Chapter 4.

The Endoplasmic Reticulum Forms Membrane-Enclosed Channels Within the Cytoplasm

The **endoplasmic reticulum (ER)** is a series of interconnected membrane-enclosed tubes and channels in the cytoplasm (Fig. 5-8); the ER membrane is continuous with the nuclear membrane. Eukaryotic cells have two forms of ER: rough and smooth. Numerous ribosomes stud the outside of the **rough endoplasmic reticulum**; in contrast, **smooth endoplasmic reticulum** lacks ribosomes.

The different structures of smooth and rough ER reflect different functions. Enzymes embedded in the membranes of the smooth ER are the major site of lipid synthesis, including the phospholipids and cholesterol used in membrane formation. Smooth ER in liver cells contains enzymes that detoxify harmful drugs and metabolic by-products. In some cells the smooth ER synthesizes other types of lipids as well, such as the steroid hormones testosterone and estrogen, which are produced in the reproductive organs of mammals.

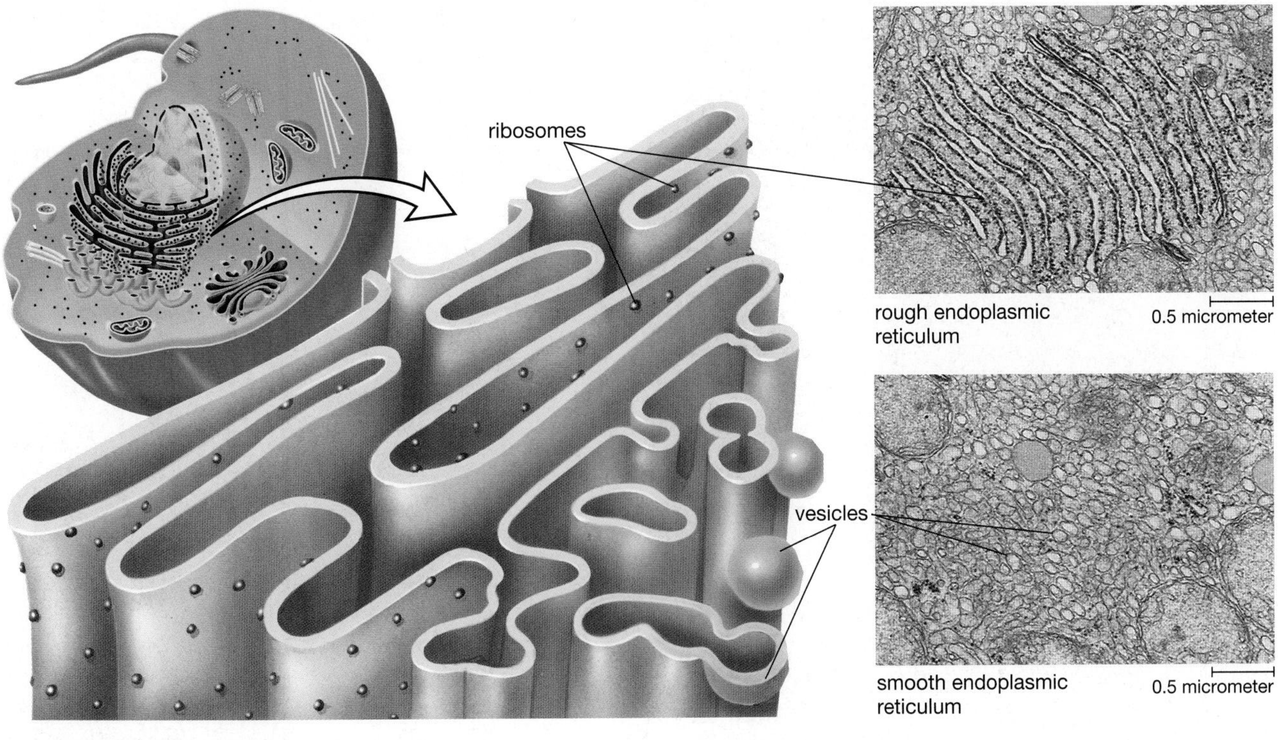

Figure 5-8 Endoplasmic reticulum
There are two types of endoplasmic reticulum: rough ER, coated with ribosomes, and smooth ER, without ribosomes. Although in electron micrographs the ER looks like a series of tubes and sacs, it is actually a maze of folded sheets and interlocking channels. In many cells the rough and smooth ER are thought to be continuous, as depicted in the drawing. Ribosomes (black) stud the cytoplasmic face of the rough ER membrane.

The ribosomes on the outside of rough ER synthesize proteins, including membrane proteins. Therefore, the ER can synthesize itself, both lipid and protein components. While some most of the membrane synthesized in the ER forms new or replacement ER membrane, some of it moves inward to replace nuclear membrane or outward to maintain the Golgi complex, lysosomes, and the plasma membrane.

Ribosomes on rough ER also manufacture the proteins such as digestive enzymes and protein hormones (for example, insulin) that some secretory cells export into their surroundings. These proteins are synthesized by the ribosomes attached to the outside of the ER, and as they are synthesized, they are inserted through the ER membrane into the ER interior. The proteins synthesized for secretion or use within the cell then move through the ER channels and accumulate in pockets. These pockets then bud off, forming membrane-bound sacs called **vesicles** that carry their protein cargo to the Golgi complex.

The Golgi Complex Sorts, Chemically Alters, and Packages Important Molecules

The **Golgi complex** is a specialized set of membranes derived from the endoplasmic reticulum that looks very much like a stack of flattened sacs (Fig. 5-9). Vesicles from the ER fuse with one face of the Golgi complex, adding their membrane to the Golgi complex and emptying their contents into the Golgi sacs. Other vesicles bud off the Golgi complex on the opposite face of the stack, carrying away specific proteins, lipids, and other complex molecules. The Golgi complex performs the following three major functions:

1. The Golgi separates proteins and lipids received from the ER according to their destinations; for example, it separates digestive enzymes that are bound for lysosomes from hormones that the cell will secrete.
2. The Golgi modifies some molecules—for instance, it adds sugars to proteins to make glycoproteins.
3. The Golgi packages these materials into vesicles that are then transported to other parts of the cell or to the plasma membrane for export.

The Travels of a Secreted Protein

To understand how the membranous organelles work together, let's look at the secretion of an antibody. An antibody is a protein, secreted by a type of white blood cell, that binds to foreign invaders (such as bacteria) and helps destroy them. Antibody proteins are synthesized on ribosomes of the rough ER and then packaged into vesicles formed from the ER membrane. These vesicles travel to the Golgi, where the membranes fuse, releasing the protein into the Golgi complex. Here, carbohydrates are attached to the protein, which is then repackaged into vesicles formed from Golgi membrane. The vesicle containing the completed antibody travels to the plasma membrane and fuses with it, releasing the antibody outside the cell, where it will make its way into the bloodstream to help defend your body against infection.

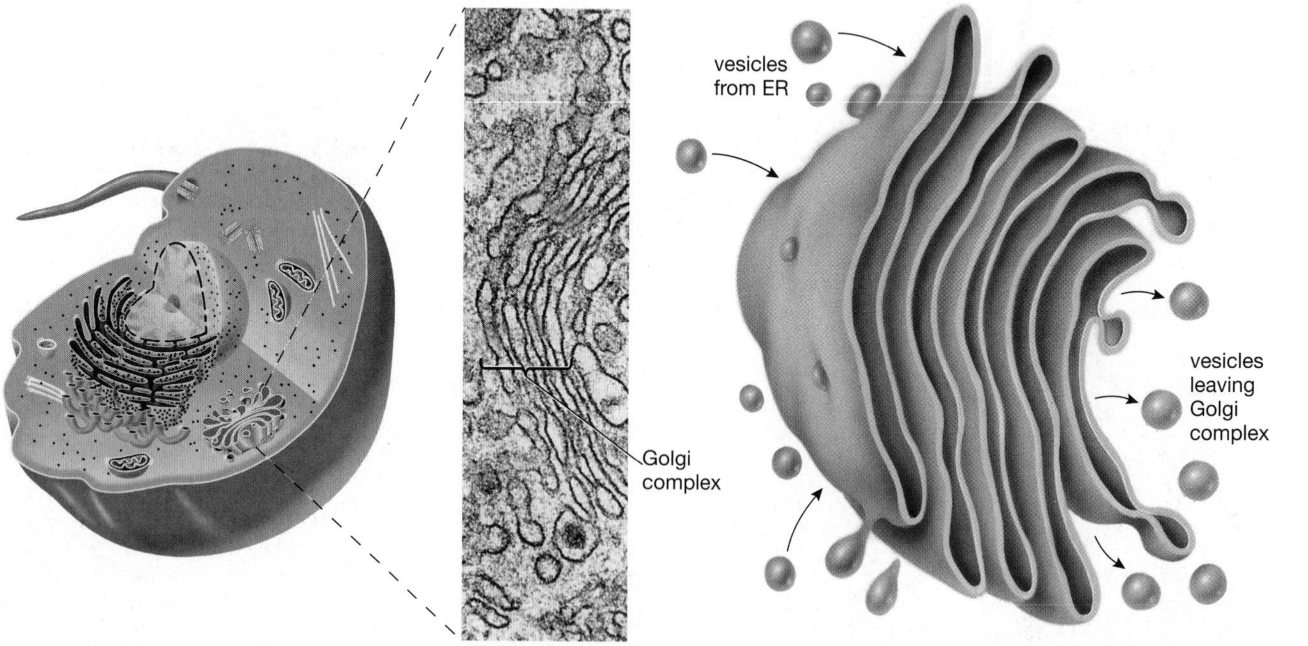

Figure 5-9 The Golgi complex
The Golgi complex is a stack of flat membranous sacs derived from the endoplasmic reticulum. Vesicles constantly bud off from and fuse with the Golgi and ER, transporting material from the ER to the Golgi and back again, and from the Golgi to plasma membrane, lysosomes, and vesicles. Vesicles join the Golgi on one face and leave the Golgi from its opposite face.

Lysosomes Serve as the Cell's Digestive System

Some of the proteins manufactured in the ER and sent to the Golgi complex are intracellular digestive enzymes that can break down proteins, fats, and carbohydrates into their component subunits. In the Golgi these enzymes are packaged in membranous vesicles called **lysosomes** (Fig. 5-10). One major function of lysosomes is to digest food particles, which range from individual proteins to complete microorganisms.

As you have seen in Chapter 4, many cells "eat" by *phagocytosis*—that is, by engulfing extracellular particles with extensions of the plasma membrane. The food particles are then moved into the cytoplasm, enclosed within membranous sacs, now called **food vacuoles**. Lysosomes recognize these food vacuoles and fuse with them. The contents of the two vesicles mix, and the lysosomal enzymes digest the food into amino acids, monosaccharides, fatty acids, and other small molecules. These simple molecules then diffuse out of the lysosome and into the cytoplasm to nourish the cell. Cell biologists continue to search for the key to how lysosomes recognize these food vacuoles.

Lysosomes also digest excess cellular membranes, or defective or malfunctioning organelles, such as *mitochondria* or *chloroplasts*. After identifying these organelles, the cell encloses them in vesicles made of membrane from the ER. These vesicles fuse with lysosomes, and digestive enzymes within the lysosome enable the cell to recycle valuable materials from the defunct organelles. Scientists are still researching the problem of how the cell identifies organelles that have outlived their usefulness.

Membrane Synthesized in the Endoplasmic Reticulum Flows Through the Membrane System of the Cell in an Orderly Way

The nuclear envelope, rough and smooth ER, Golgi complex, lysosomes, food vacuoles, and plasma membrane all form an integrated membrane system. Membrane is

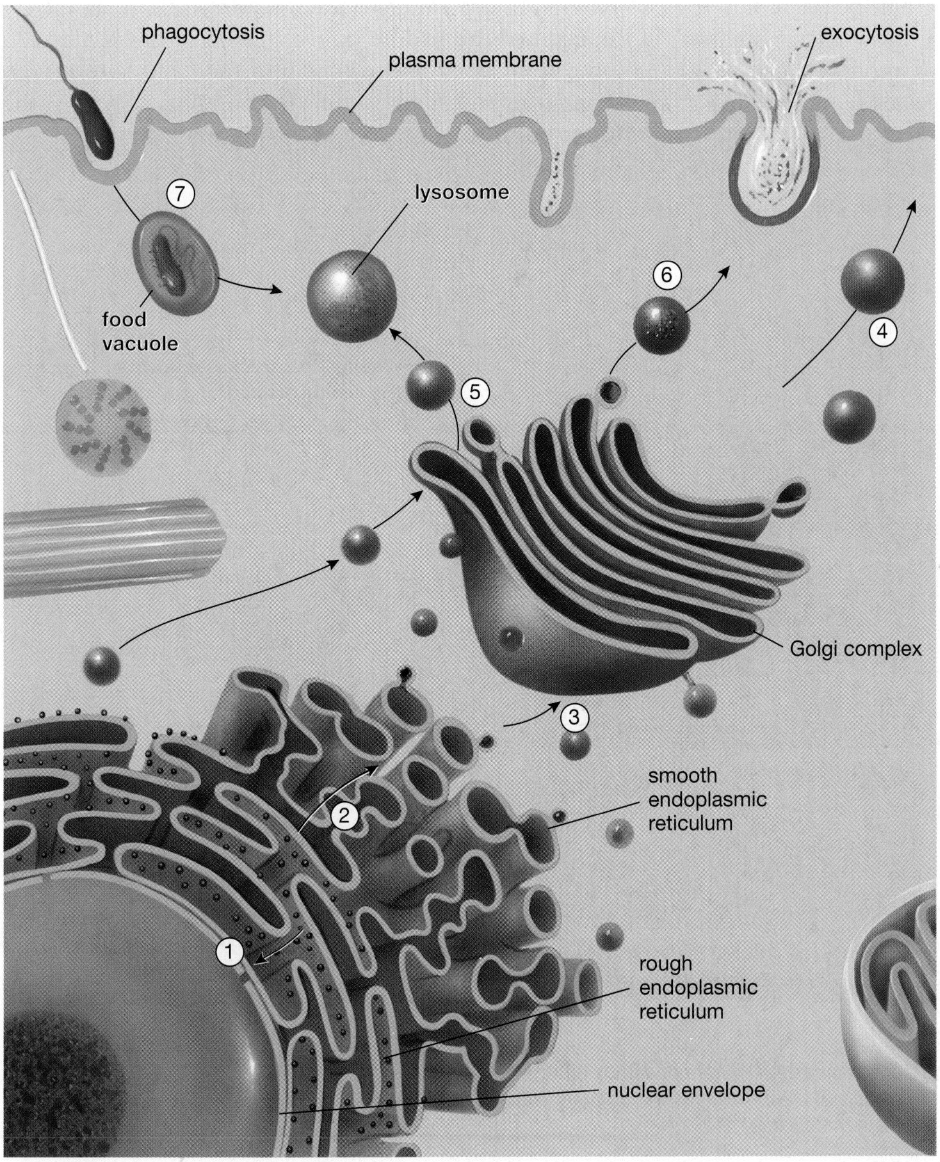

Figure 5-10 The flow of membrane within the cell
Membrane is synthesized by the endoplasmic reticulum. (1) Some of the membrane moves inward to form new nuclear envelope, some moves outward to form (2) smooth ER and (3) Golgi membrane. From the Golgi, membrane moves to form (4) new plasma membrane and (5) membrane surrounding other organelles such as lysosomes. Some proteins synthesized in the rough ER are modified in the smooth ER and travel in vesicles to the Golgi, where they are further modified and sorted. Some of these proteins are packaged in vesicles bound for the plasma membrane, where they will be (6) secreted from the cell, and some are packaged in lysosomes surrounded by membrane from the Golgi. Lysosomes may fuse with food vacuoles (7), allowing intracellular digestion of food particles.

MediaTutor
5.2 Membrane Traffic

synthesized in the ER and flows back and forth between these structures in an orderly way. As an example, let's look at the movement of materials destined for inclusion in the plasma membrane (see Fig. 5-10). The ER synthesizes the phospholipids and proteins that make up the plasma membrane and buds off a vesicle whose membrane includes these plasma membrane components. The vesicle fuses with the Golgi complex. Plasma membrane material continues on through the Golgi, where it may be modified—for example, by adding sugars to make glycoproteins or *glycolipids* (lipids to which a sugar is attached). Eventually the plasma membrane material becomes an "outward-bound" vesicle that buds off the far side of the Golgi and moves to the cell surface. The vesicle fuses with the plasma membrane, replenishing and enlarging the membrane.

The Golgi complex processes and packages all membrane-enclosed materials produced by the cell. Many of the vesicles pinched off from the Golgi contain secretory products (for example, hormones) that are released outside the cell. Lysosomes contain digestive enzymes and often fuse with food vacuoles to carry out intracellular digestion. How these diverse materials are recognized, purified, modified properly, separated out, and individually packaged remains a challenge to cell biologists.

Vacuoles Serve Many Functions, Including Water Regulation, Support, and Storage

Most cells contain one or more **vacuoles**, which are fluid-filled sacs surrounded by a single membrane. Some, such as the food vacuoles that form during phagocytosis, are temporary features of cells. However, many cells contain permanent vacuoles that have important roles in maintaining the integrity of the cell, most notably by regulating the cell's water content.

Freshwater Microorganisms Have Contractile Vacuoles

Freshwater protists such as *Paramecium* consist of a single, eukaryotic cell. Many of these organisms possess complex **contractile vacuoles** composed of collecting ducts, a central reservoir, and a tube leading to a pore in the plasma membrane (Fig. 5-11). Because fresh water is hypotonic to the cytoplasm of these organisms, water constantly enters the cell by osmosis. The increasing volume of incoming water might soon burst the fragile creature if it did not have a mechanism to excrete the water. Cellular energy is used to pump salts from the cytoplasm of the protist into collecting ducts. Water follows by osmosis and drains into the central reservoir. When the reservoir is full, it contracts, squirting the water out through a pore in the plasma membrane.

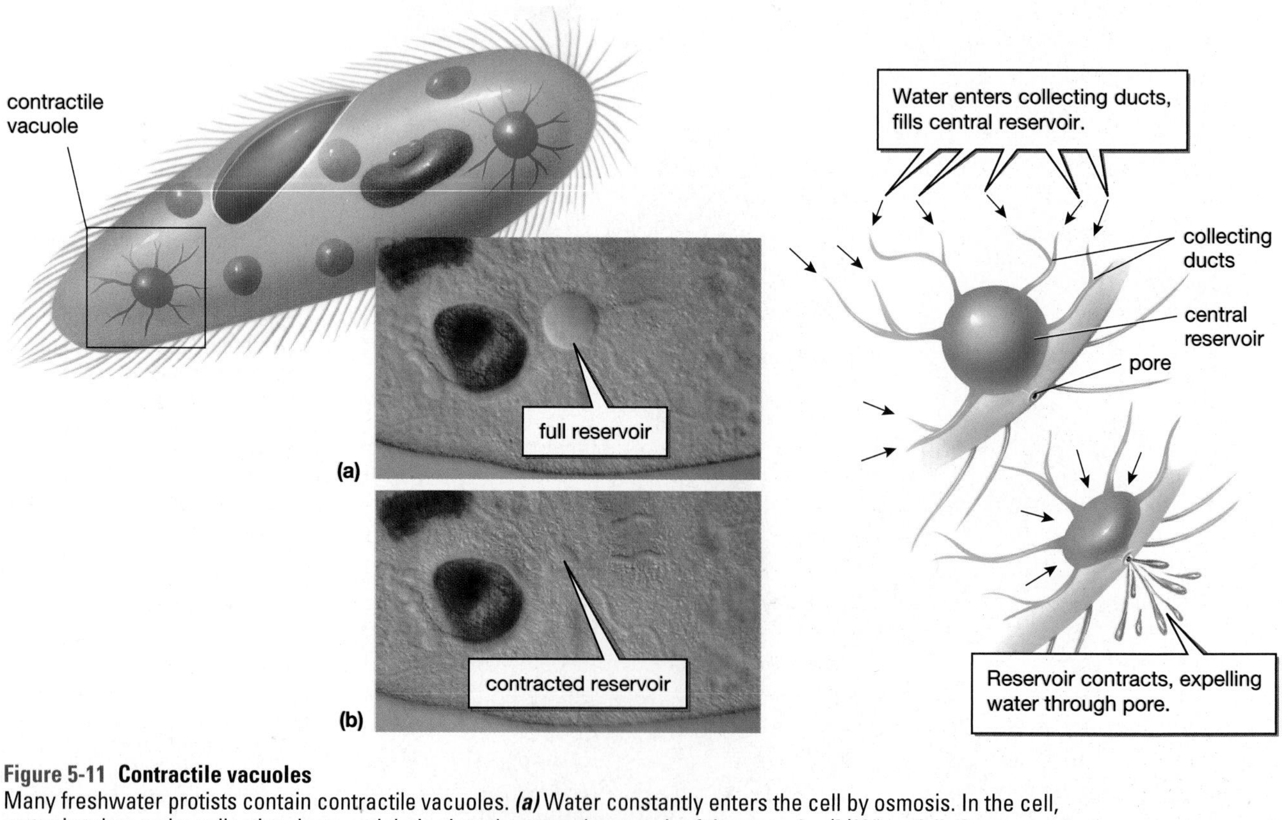

Figure 5-11 Contractile vacuoles
Many freshwater protists contain contractile vacuoles. ***(a)*** Water constantly enters the cell by osmosis. In the cell, water is taken up by collecting ducts and drains into the central reservoir of the vacuole. ***(b)*** When full, the reservoir contracts, expelling the water through a pore in the plasma membrane.

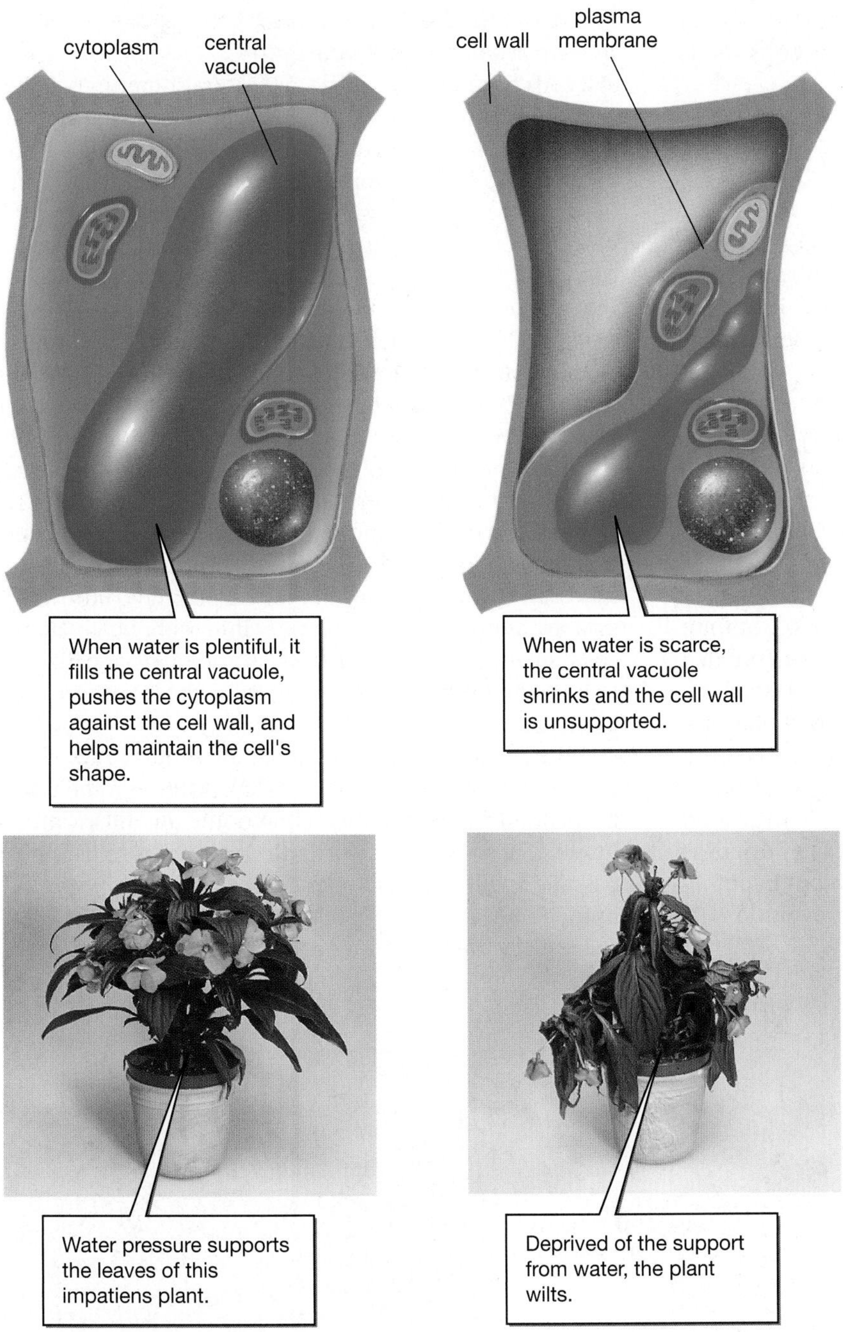

cytoplasm
central vacuole
cell wall
plasma membrane
When water is plentiful, it fills the central vacuole, pushes the cytoplasm against the cell wall, and helps maintain the cell's shape.
When water is scarce, the central vacuole shrinks and the cell wall is unsupported.
Water pressure supports the leaves of this impatiens plant.
Deprived of the support from water, the plant wilts.

Figure 5-12 The central vacuole and turgor pressure in plant cells

Plant Cells Have Central Vacuoles

Three-quarters or more of the volume of many plant cells is occupied by a large **central vacuole** [Fig. 5-12 (top); see also Fig. 5-4]. The central vacuole has several functions. Filled mostly with water, the central vacuole is involved in the cell's water balance. It also provides a dump site for hazardous wastes, which plant cells often cannot excrete. Some plant cells store extremely poisonous substances, such as sulfuric acid, in their vacuoles, which deter animals from munching on the otherwise tasty leaves. Vacuoles may also store sugars and amino acids not immediately needed by the cell. Blue or purple pigments stored in central vacuoles are responsible for the colors of many flowers.

These dissolved substances make the vacuole contents hypertonic to the cell cytoplasm, which in turn is usually hypertonic to the extracellular fluid that bathes the cells. Water therefore enters the vacuole by osmosis, which tends to make it swell. The pressure of the water within the vacuole, called **turgor pressure**, pushes the fluid portion of the cytoplasm up against the cell wall with considerable force (see Fig. 5-12). Cell walls are usually somewhat flexible, so both the overall shape and the rigidity of the cell depend on turgor pressure within the cell. Turgor pressure thus provides support for the nonwood parts of plants. If you forget to water your houseplants, the central vacuoles and cytoplasm lose water and the cells shrink away from their cell walls. Just as a balloon goes limp when its air leaks out, so too the plant droops as its cells lose turgor pressure.

Mitochondria Extract Energy from Food Molecules, and Chloroplasts Capture Solar Energy

Every cell requires a continuous supply of energy to manufacture complex molecules and structures, to acquire nutrients from the environment and excrete waste materials, to move, and to reproduce. All eukaryotic cells have **mitochondria** that convert energy stored in sugar to ATP. Plant cells also have **chloroplasts**, which can capture energy directly from sunlight and store it in sugar molecules.

Biologists believe that both mitochondria and chloroplasts evolved from prokaryotic bacteria that took up residence long ago within the cytoplasm of cells ancestral to eukaryotic cells, a process called *endosymbiosis* (literally, "living together inside." Mitochondria and chloroplasts are similar to each other in many ways. Both are about 1 to 5 micrometers in diameter and are surrounded by a double membrane. Both have assemblies of enzymes that synthesize ATP, although the assemblies are used in a very different manner in chloroplasts than in mitochondria. Finally, both have characteristics, including their own DNA and ribosomes, as well as their general size and shape, that seem to be remnants of their origin as free-living prokaryotic cells. The **endosymbiont hypothesis** of mitochondrial and chloroplast evolution is discussed in more detail in Chapter 17.

Mitochondria Use Energy Stored in Food Molecules to Produce ATP

All cells have mitochondria, which are sometimes called the "powerhouses of the cell" because they extract energy from food molecules and store it in the high-energy bonds of ATP. As you shall see in Chapter 8, different amounts of energy can be released from a food molecule, depending on how it is metabolized. The breakdown of food molecules begins without the use of oxygen by enzymes in the fluid portion of the cytoplasm. This **anaerobic** (without oxygen) metabolism does not convert very much food energy into ATP energy. Mitochondria enable a eukaryotic cell to utilize oxygen to break down high-energy molecules. These **aerobic** (with oxygen) reactions are much more effective in generating energy than are the anaerobic reactions; 18 or 19 times more ATP is generated by aerobic metabolism in the mitochondria than by anaerobic metabolism in the cytoplasm. Not surprisingly, mitochondria are found in large numbers in metabolically active cells, such as muscle, and are less abundant in cells that are less metabolically active, such as those of bone and cartilage.

Mitochondria are round, oval, or tubular sacs made of a pair of membranes (Fig. 5-13). Although the outer mitochondrial membrane is smooth, the inner membrane loops back and forth to form deep folds called **cristae** (singular, **crista**, meaning "crest"). As a result, the mitochondrial membranes enclose two fluid-filled spaces: the **intermembrane compartment** between the

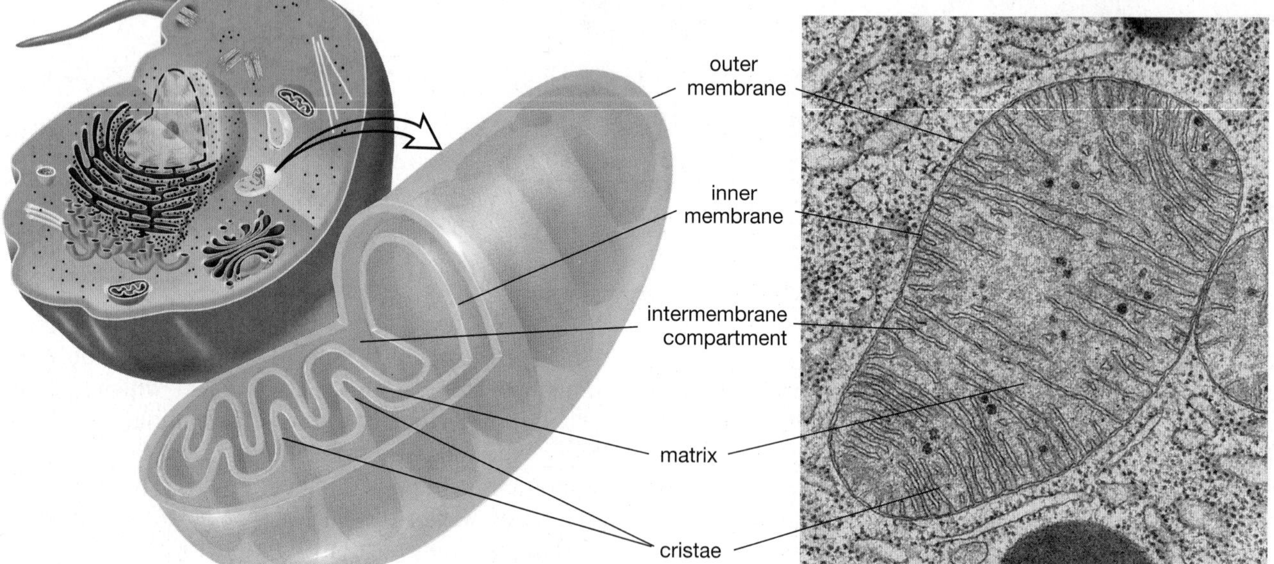

Figure 5-13 A mitochondrion
Mitochondria consist of a pair of membranes enclosing two fluid compartments, the intermembrane compartment between the outer and inner membranes and the matrix within the inner membrane. The outer membrane is smooth, but the inner membrane forms deep folds called *cristae*. Mitochondria are the site of aerobic metabolism.

inner and outer membranes and the **matrix**, or inner compartment, within the inner membrane. Some of the reactions of food metabolism occur in the fluid matrix contained within the inner membrane; the rest are conducted by a series of enzymes attached to the membranes of the cristae within the intermembrane compartment. The role of mitochondria in energy production is described in detail in Chapter 8.

Chloroplasts Are the Site of Photosynthesis

If chloroplasts didn't exist, there would be none of the eukaryotic life-forms that dominate Earth today, and you wouldn't be reading this. Chloroplasts (Fig. 5-14) are specialized organelles, surrounded by a double membrane; photosynthesis in the eukaryotic cells of plants and in photosynthetic protists occurs in the chloroplasts. The inner membrane of the chloroplast encloses a fluid called the **stroma**. Within the stroma are interconnected stacks of hollow membranous sacs. The individual sacs are called **thylakoids**, and a stack of sacs is a **granum** (plural, **grana**).

The thylakoid membranes contain the green pigment molecule **chlorophyll** (which gives plants their green color) as well as other pigment molecules. During photosynthesis, chlorophyll captures the energy of sunlight and transfers it to other molecules in the thylakoid membranes. These molecules in turn transfer the energy to ATP and other energy-carrier molecules. The energy carriers diffuse into the stroma, where their energy is used to drive the synthesis of sugar from carbon dioxide and water. Photosynthesis is described in more detail in Chapter 7.

Plants Use Plastids for Storage

Chloroplasts are highly specialized **plastids.** Plastids are organelles found only in plants and photosynthetic protists. Plastids are surrounded by a double membrane and serve a variety of functions. Plants and photosynthetic protists use nonchloroplast types of plastids as storage containers for various molecules, including pigments that give ripe fruits their yellow, orange, or red colors. In plants that continue growing from one year to the next, plastids store photosynthetic products from the summer for use during the following winter and spring. Most plants convert the sugars made during photosynthesis into starch, which is stored in plastids (Fig. 5-15). Potatoes, for example, are masses of cells, each stuffed with starch-filled plastids.

The Cytoskeleton Provides Shape, Support, and Movement

Organelles do not drift about the cytoplasm haphazardly; most are attached to a network of protein fibers called the *cytoskeleton* (Fig. 5-16). Even individual enzymes, which are often parts of complex metabolic pathways, may be fastened in sequence to the cytoskeleton, so that molecules can be passed from one

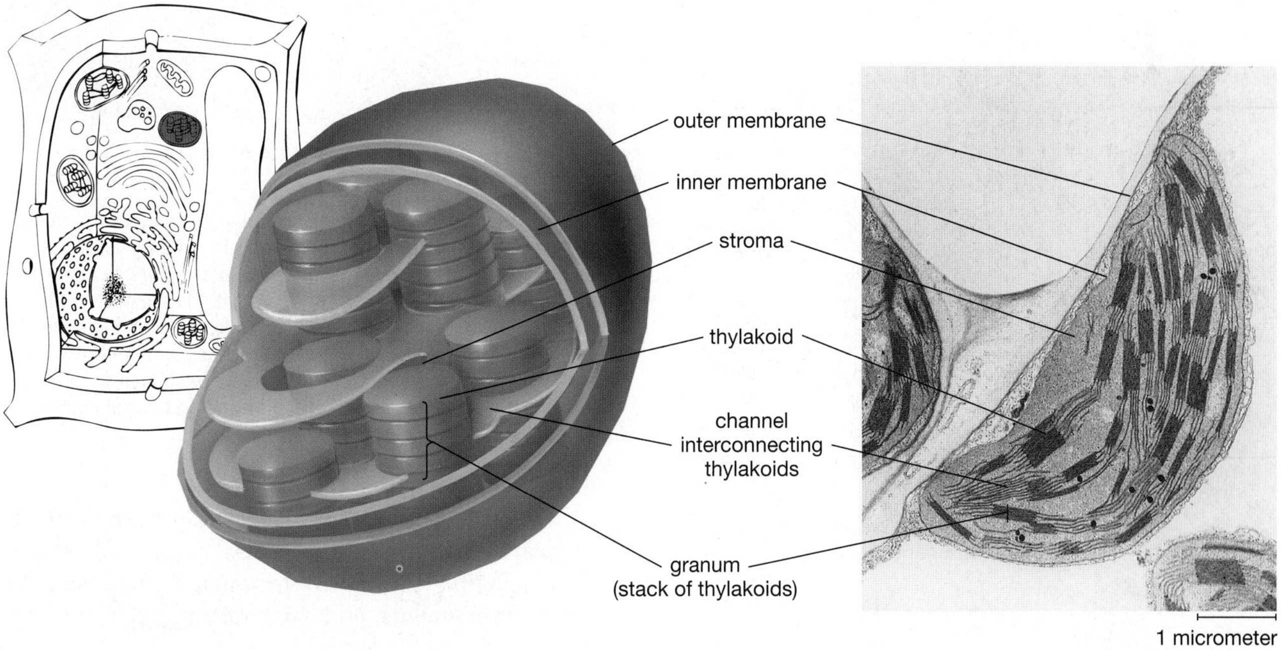

Figure 5-14 A chloroplast
Chloroplasts are surrounded by a double membrane, although the inner membrane is not usually visible in electron micrographs. Semifluid stroma are enclosed by the inner membrane; stroma contain embedded stacks of sacs, collectively referred to as *grana*. The individual sacs of the grana are called *thylakoids*. Chlorophyll is embedded in the membranes of the thylakoids.

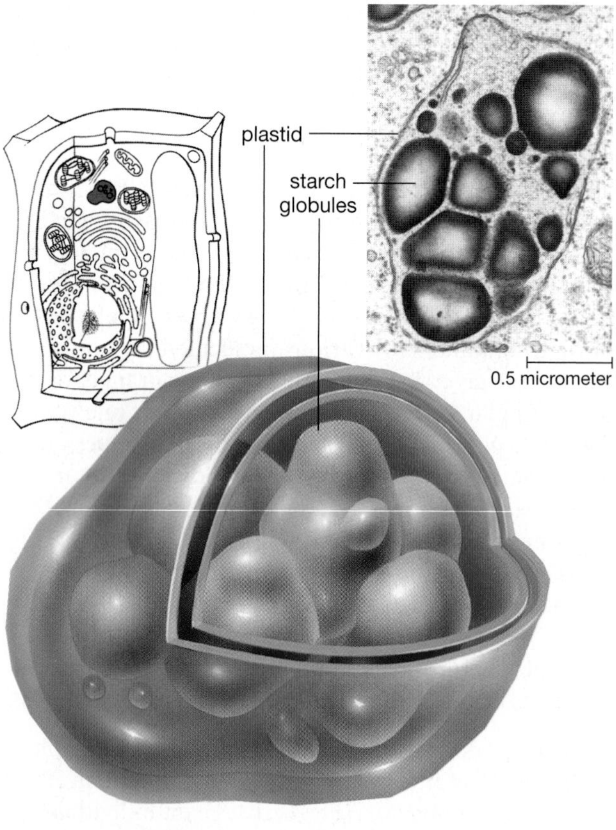

Figure 5-15 A plastid
Plastids, found in the cells of plants and photosynthetic protists, are organelles surrounded by a double outer membrane. Chloroplasts are the most familiar plastids; other types store various materials, such as the starch filling these plastids in potato cells.

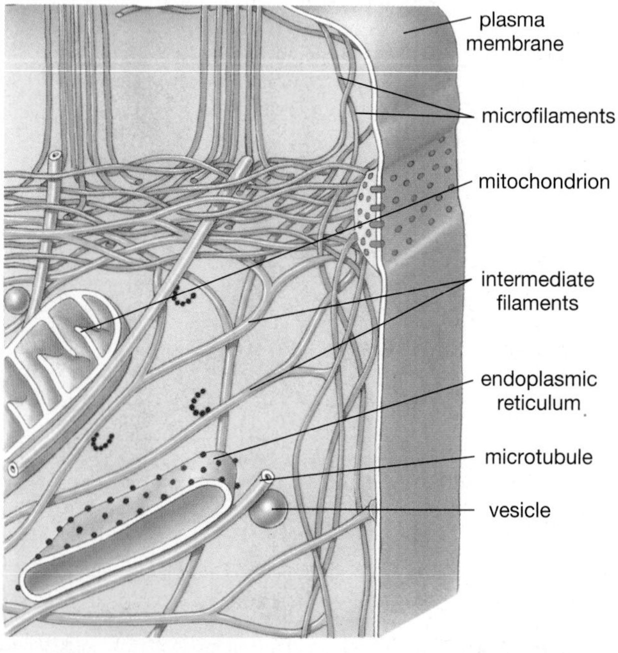

Figure 5-16 The cytoskeleton
Eukaryotic cells are given shape and organization by the cytoskeleton, which consists of three types of proteins: microtubules, intermediate filaments, and microfilaments.

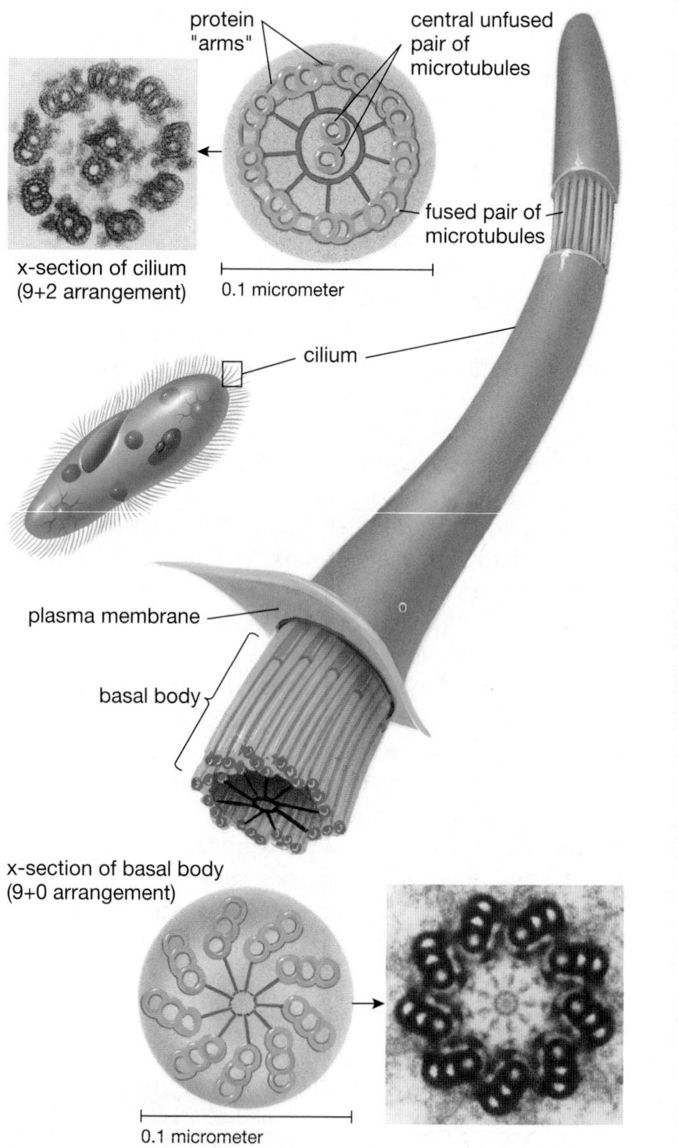

Figure 5-17 Cilia and flagella
Both cilia and flagella contain microtubules arranged in an outer ring of nine fused pairs of microtubules surrounding a central unfused pair (a 9+2 arrangement). The nine outer pairs have "arms" made of protein that interact with adjacent pairs to provide the force for bending. Cilia and flagella arise from basal bodies formed from centrioles located just beneath the plasma membrane. The fused pairs of microtubles arise from these basal bodies, which have nine fused triplets of microtubules and no central microtubules ("9+0").

enzyme to the next in the correct order for a particular chemical transformation. Several types of protein fibers, including thin **microfilaments**, medium-sized **intermediate filaments**, and thick **microtubules**, make up the cytoskeleton (Table 5-3).

The cytoskeleton performs the following important functions:

1. *Cell shape*. In cells without cell walls, the cytoskeleton, especially networks of intermediate filaments, determines the shape of the cell.

Table 5-3 Components of the Cytoskeleton

	Structure	Protein Structure	Function
Microfilaments	Twisted double strands, each consisting of a string of protein subunits; about 7 nm in diameter and up to several centimeters long (in muscle cells)	Actin 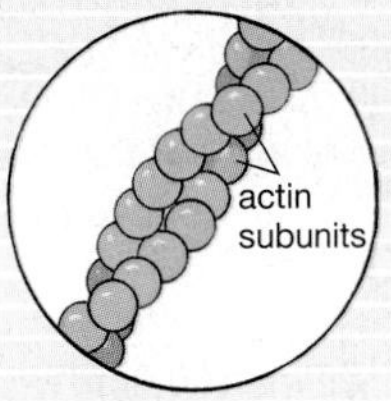	Muscle contraction; changes in cell shape, including cytoplasmic division in animal cells; cytoplasmic movement; movement of pseudopodia
Intermediate filaments	Consist of eight subunits composed of ropelike protein strands; 8–12 nm in diameter and 10–100 μm in length	Protein varies with tissue type 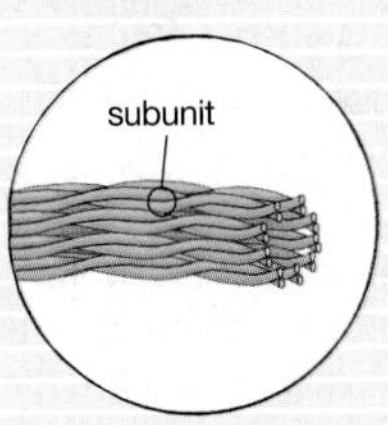	Maintenance of cell shape; attachments of microfilaments in muscle cells; support of nerve cell extensions; attach cells together (desmosomes)
Microtubules	Tubes consisting of spiraling two-part protein subunits; about 25 nm in diameter and can be 50 μm in length	Tubulin 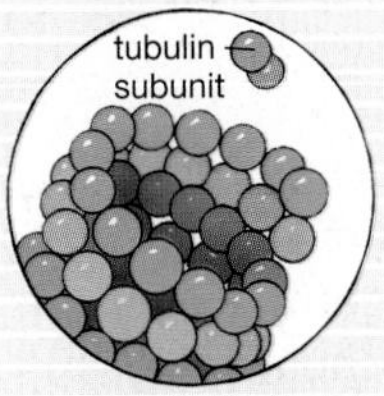	Movement of chromosomes during cell division; movement of organelles within cytoplasm; movement of cilia and flagella

2. *Cell movement.* The assembly, disassembly, and sliding of microfilaments and microtubules cause cell movement. Cell movement includes both the "crawling" of white blood cells, the contraction of muscle cells, and the migration and shape changes that occur during the development of multicellular organisms.
3. *Organelle movement.* Microtubules and microfilaments move organelles from place to place within a cell. For example, microfilaments attach to vesicles formed during *endocytosis* (see Chapter 4), when large particles are engulfed by the plasma membrane and pull the vesicles into the cell. Vesicles budded off the ER and Golgi complex are probably guided by the cytoskeleton as well.
4. *Cell division.* Microtubules and microfilaments are essential to cell division in eukaryotic cells. First, when eukaryotic nuclei divide, microtubules move the chromosomes into the daughter nuclei. Second, in animal cells, division of the cytoplasm of a single parent cell into two new daughter cells results from the contraction of a ring of microfilaments that pinch the "waist" of the parent cell around the middle. In addition, *centrioles* (see Fig. 5-3), which play a role in animal cell division, are composed of microtubules. Cell division will be covered in detail in Chapter 11.

Cilia and Flagella Move the Cell or Move Fluid Past the Cell

Both **cilia** (Latin for "eyelash") and **flagella** ("whip") are slender extensions of the plasma membrane. Each cilium and flagellum contains a ring of nine fused pairs of microtubules, with an unfused pair of microtubules in the center of the ring (forming what is called a *"9+2" arrangement*; Fig. 5-17). The microtubules span the length of the cilium or flagellum. This pattern of microtubules is produced by a centriole located in the cytoplasm just beneath the plasma membrane. A **centriole** is a short, barrel-shaped ring consisting of nine microtubule triplets, with no microtubules in the center (forming a *"9+0" arrangement*). Centrioles move to the plasma membrane and provide a center for the formation of cilia or flagella; in this situation, two of the members of each triplet give rise to pairs of microtubules in the cilium or flagellum (see Fig. 5-17). After it begins forming the cilium or flagellum, the centriole is referred to as a **basal body**, because it is located at the base of these structures, anchoring them to the plasma membrane.

Tiny "arms" of protein attach neighboring pairs of microtubules in cilia and flagella. When these arms flex, they slide one pair of microtubules past the adjacent pairs, causing the cilia or flagellum to move. The energy

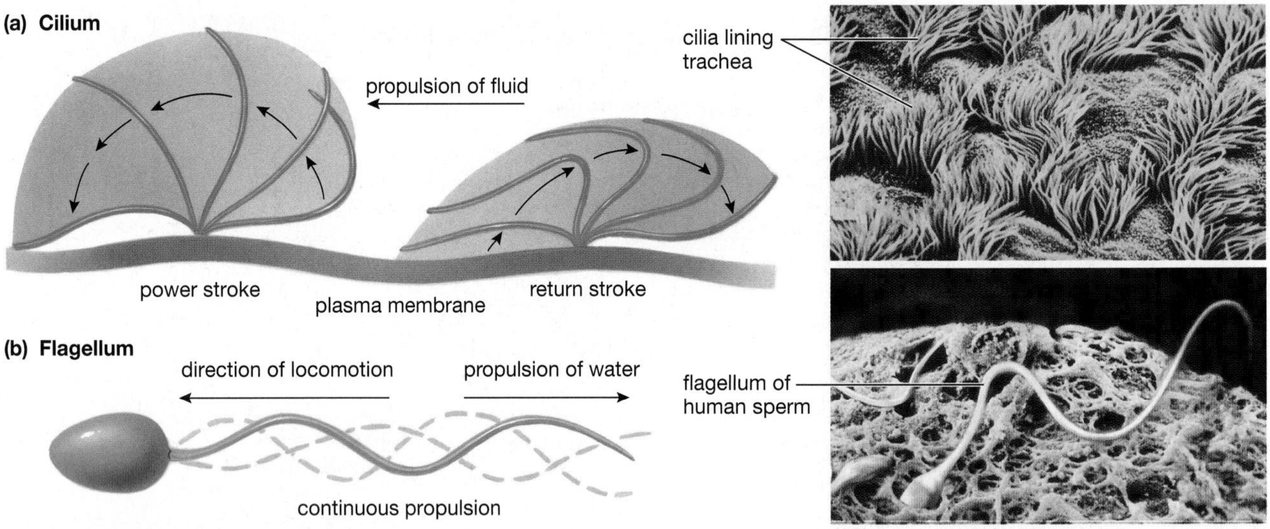

Figure 5-18 How cilia and flagella move
(a) (left) Cilia usually "row" along, providing a force of movement parallel to the plasma membrane, just as oars provide movement parallel to the sides of a rowboat. Their movement resembles the arms of a swimmer doing the breast stroke. (right) SEM photo of cilia lining the trachea (which conducts air to the lungs); these cilia sweep out mucus and trapped particles. ***(b)*** (left) Flagella move in a wavelike motion, providing continuous propulsion perpendicular to the plasma membrane. In this way a flagellum attached to a sperm can move the sperm straight ahead. (right) A human sperm cell on the surface of a human egg cell.

of ATP powers the movement of the protein arms during microtubule sliding. Cilia and flagella often move almost continuously and, consequently, require enormous supplies of ATP, which are generated by mitochondria that are usually found in abundance near the basal bodies.

The main differences between cilia and flagella lie in their length, number, and the direction of the force they generate. In general, cilia are shorter (about 10 to 25 micrometers long) and more numerous than flagella. They provide force in a direction parallel to the plasma membrane, like the oars in a canoe. This is accomplished through a "rowing" motion (Fig. 5-18a). Flagella are longer (50 to 75 micrometers), usually fewer in number, and provide force perpendicular to the plasma membrane, like the engine on a motorboat (Fig. 5-18b).

Some unicellular organisms, such as *Paramecium* and *Euglena*, use cilia or flagella to move about. Most animal sperm also rely on flagella for movement (see Fig. 5-18b). In small multicellular animals, cilia are occasionally used for locomotion, moving the animal through water. Many small aquatic invertebrates, for example, swim by the coordinated beating of rows of cilia, like the oars on a Roman galley ship. More commonly, however, cilia move fluids and suspended particles past a surface. Ciliated cells line such diverse structures as the gills of oysters (moving food- and oxygen-rich water), the oviducts of female mammals (moving the eggs along from the ovary to the uterus), and the respiratory tracts of most land vertebrates (clearing mucus that carries debris and microorganisms from the windpipe and lungs; see the photograph in Fig 5-18a).

Prokaryotic cells may also bear slender protrusions that undulate or spin, thereby enabling the cell to move about. However, the "flagella" of prokaryotic cells do not contain microtubules and have no evolutionary relationship to eukaryotic flagella or cilia.

CASESTUDYREVISITED

Microscopic Stowaways

The amazing bacteria, *Streptococcus mitis*, which thrive in the warm moisture of the human body, had survived the pressures of take-off, the complete vacuum of outer space, temperatures averaging only 20 degrees above absolute zero (–273 °C), without nutrients, energy, or water. They had not reproduced under these freeze-dried conditions on the moon, but they had survived—and this particular bacterium is not even noted for hardiness. Bacteria that form sturdy resting structures called *endospores* have been cultured from the intestines of 2000-year-old human mummies. In fact, scientists claim that they have cultured endospores from the gut of a bee that died and was preserved in amber (tree sap) at least 25 million years ago. Bacteria thrive in the near-boiling temperatures of hot springs in Yellowstone National Park, have been found miles below Earth's surface, and are embedded deep within polar ice. Bacteria make most eukaryotic cells seem fragile by comparison. Why? One reasonable hypothesis is that today's bacteria have descended with relatively little change from the first forms of life to colonize the planet about 3.5 billion years ago, under conditions far less hospitable than those that exist today. Their very simplicity confers a hardiness unmatched by the far more complex eukaryotic cells.

Bacteria almost never form any sort of grouping that resembles a multicelled organism with an internal division of labor among its cells. Think about the attributes of prokaryotic bacterial cells, and the eukaryotic cells of which all multicellular organisms are comprised. Develop some hypotheses explaining why only eukaryotic cells have evolved into multicellular organisms.

Summary of Key Concepts

1) What Are the Basic Features of Cells?

The principles of the cell theory are:

1. Every living organism is made up of one or more cells.
2. The smallest living organisms are single cells, and cells are the functional units of multicellular organisms.
3. All cells arise from preexisting cells.

Cells are limited in size because they must exchange materials with their surroundings by diffusion. Because diffusion is relatively slow, the interior of the cell must never be too far from the plasma membrane, and the plasma membrane must have a large surface area through which materials can diffuse relative to the volume of its cytoplasm. Both of these constraints limit the size of cells.

All cells are either prokaryotic or eukaryotic. Prokaryotic cells, or bacteria, are small and relatively simple in structure. More complex eukaryotic cells make up all other forms of life: protists, plants, fungi, and animals.

2) What Are the Features of Prokaryotic Cells?

Prokaryotic cells are generally very small with a relatively simple internal structure. Most are surrounded by a relatively stiff cell wall. The cytoplasm of prokaryotic cells lacks membrane-enclosed organelles (although some photosynthetic bacteria have elaborate internal membranes). A single, circular strand of DNA is found in the nucleoid. Table 5-2 compares prokaryotic cells to the eukaryotic cells of plants and animals.

3) What Are the Features of Eukaryotic Cells?

Genetic material (DNA) is contained within the nucleus, which is bounded by the double membrane of the nuclear envelope. Pores in the nuclear envelope regulate the movement of molecules between nucleus and cytoplasm. The genetic material of eukaryotic cells is organized into linear strands called *chromosomes*, which consist of DNA and proteins. The nucleolus consists of ribosomal RNA and ribosomal proteins, as well as the genes that code for ribosome synthesis. Ribosomes are particles of ribosomal RNA and protein that are the sites of protein synthesis.

The membrane system of a cell consists of the plasma membrane, endoplasmic reticulum (ER), Golgi complex, and vesicles derived from these membranes. Endoplasmic reticulum with ribosomes, called rough ER, manufactures many cellular proteins. Endoplasmic reticulum without ribosomes, called smooth ER, manufactures lipids. The ER is the site of all membrane synthesis within the cell. The Golgi complex is a series of membranous sacs derived from the ER. The Golgi complex processes and modifies materials synthesized in the rough and smooth ER. Some substances in the Golgi are packaged into vesicles for transport elsewhere in the cell. Lysosomes are vesicles that contain digestive enzymes, which digest food particles and defective organelles.

All eukaryotic cells contain mitochondria, organelles that use oxygen to complete the metabolism of food molecules, capturing much of their energy as ATP. Cells of plants and some protists contain plastids, including chloroplasts, which capture the energy of sunlight during photosynthesis, enabling the cells to manufacture organic molecules, particularly sugars, from simple inorganic molecules. Both mitochondria and chloroplasts probably originated from bacteria. Storage plastids store pigments or starch.

Many eukaryotic cells contain sacs, called *vacuoles*, that are bounded by a single membrane and that store food or wastes, excrete water, or support the cell. Some protists have contractile vacuoles, which collect and expel water. Plants use central vacuoles to support the cell as well as to store wastes and toxic materials.

The cytoskeleton organizes and gives shape to eukaryotic cells and moves and anchors organelles. The cytoskeleton is composed of microfilaments, intermediate filaments, and microtubules (see Table 5-3). Cilia and flagella are whiplike extensions of the plasma membrane that contain microtubules in a characteristic pattern. These structures move fluids past the cell or move the cell through its fluid environment.

Study Note

Figures 5-3 and 5-4 illustrate the overall structure of animal and plant cells, respectively. Table 5-2 lists the principal organelles, their functions, and their occurrence in prokaryotic cells, animal cells, and plant cells.

Key Terms

aerobic *p. 90*
anaerobic *p. 90*
basal body *p. 93*
capsule *p. 80*
central vacuole *p. 89*
centriole *p. 93*
chlorophyll *p. 91*
chloroplast *p. 90*
chromatin *p. 82*
chromosome *p. 83*
cilium *p. 93*
contractile vacuole *p. 88*
crista *p. 90*
cytoplasm *p. 76*
cytoskeleton *p. 81*
deoxyribonucleic acid (DNA) *p. 76*
endoplasmic reticulum (ER) *p. 85*
endosymbiont hypothesis *p. 90*
eukaryotic *p. 80*
flagellum *p. 93*
food vacuole *p. 87*
Golgi complex *p. 86*
granum *p. 91*
intermediate filament *p. 92*
intermembrane compartment *p. 90*
lysosome *p. 87*
matrix *p. 91*
microfilament *p. 92*
microtubule *p. 92*
mitochondrion *p. 90*
nuclear envelope *p. 82*
nucleoid *p. 80*
nucleolus *p. 82*
nucleus *p. 76*
organelle *p. 81*
pilus *p. 80*
plasma membrane *p. 76*
plastid *p. 91*
prokaryotic *p. 80*
ribonucleic acid (RNA) *p. 80*
ribosome *p. 76*
rough endoplasmic reticulum *p. 85*
slime layer *p. 80*
smooth endoplasmic reticulum *p. 85*
stroma *p. 91*
thylakoid *p. 91*
turgor pressure *p. 89*
vacuole *p. 88*
vesicle *p. 86*

Thinking Through the Concepts

Multiple Choice

1. *The outermost boundary of an animal cell is the*
   a. plasma membrane b. nucleus
   c. cytoplasm d. cytoskeleton
   e. cell wall
2. *Which organelle contains a eukaryotic cell's chromosomes?*
   a. Golgi complex b. ribosomes
   c. nucleus d. mitochondria
   e. chloroplast
3. *Most of the cell's ATP is synthesized in the*
   a. Golgi complex b. ribosomes
   c. nucleus d. mitochondria
   e. chloroplast
4. *Which organelle sorts, chemically modifies, and packages newly synthesized protein?*
   a. Golgi complex
   b. ribosomes
   c. nucleus
   d. mitochondria
   e. chloroplast
5. *Membrane-enclosed digestive organelles that contain enzymes are called*
   a. lysosomes
   b. smooth endoplasmic reticulum
   c. cilia
   d. Golgi complex
   e. mitochondria
6. *A series of membrane-enclosed channels studded with ribosomes are called*
   a. lysosomes
   b. Golgi complex
   c. rough endoplasmic reticulum
   d. mitochondria
   e. smooth endoplasmic reticulum

? Review Questions

1. Diagram "typical" prokaryotic and eukaryotic cells, and describe their important similarities and differences.
2. Which organelles are common to both plant and animal cells, and which are unique to each?
3. Define *stroma* and *matrix*.
4. Describe the nucleus, including the nuclear envelope, chromatin, chromosomes, DNA, and the nucleolus.
5. What are the functions of mitochondria and chloroplasts? Why do scientists believe that these organelles arose from prokaryotic cells?
6. What is the function of ribosomes? Where in the cell are they typically found?
7. Describe the structure and function of the endoplasmic reticulum and Golgi complex.
8. How are lysosomes formed? What is their function?
9. Diagram the structure of cilia and flagella.

Applying the Concepts

1. If muscle biopsies (samples of tissue) were taken from the legs of a world-class marathon runner and a typical couch potato, which would you expect to have a higher density of mitochondria? Why? What about a muscle biopsy from the biceps of a weight lifter?
2. One of the functions of the cytoskeleton in animal cells is to give shape to the cell. Plant cells have a fairly rigid cell wall surrounding the plasma membrane. Does this mean that a cytoskeleton is superfluous for a plant cell? Defend your answer in terms of other functions of the cytoskeleton.
3. Most cells are very small. What physical and metabolic constraints limit cell size? What problems would an enormous cell encounter? What adaptations might help a very large cell survive?

For More Information

de Duve, C. "The Birth of Complex Cells." *Scientific American*, April 1996. Describes the mechanisms by which the first eukaryotic cells were produced from prokaryotic ancestors.

Ford, B. J. "The Earliest Views." *Scientific American*, April 1998. The author used the original microscopes of Anton van Leeuwenhoek to see the microscopic world as Leeuwenhoek saw it. Photographic images taken through these early and very primitive instruments reveal remarkable detail.

Goodsell, D. S. *The Machinery of Life*. New York: Springer, 1993. Wonderful, drawn-to-scale images of the organelles and molecules of the cell.

Ingber, D. E. "The Architecture of Life." *Scientific American*, January 1998. Counteracting forces stabilize the design of organic structures, from carbon compounds to the cytoskeleton-reinforced architecture of the cell.

Kiester, E., Jr. "A Bug in the System." *Discover*, February 1991. Mitochondria may be the descendants of bacteria that live within our cells. Because mitochondria are essential for human life, defects in mitochondrial genes can cause some devastating diseases.

Murray, M. "Life on the Move." *Discover*, March 1991. Many cells, including *Amoeba* and white blood cells, crawl about by means of tiny protein "motors" to manipulate their cytoskeleton.

Rothman, J. E., and Orci, L. "Budding Vesicles in Living Cells." *Scientific American*, March 1996. Researchers are uncovering the mechanisms by which cells form vesicles.

Answers to Multiple-Choice Questions

1. a 2. c 3. d 4. a 5. a 6. c

MEDIATUTOR
Cell Structure and Function

CD Activities

Activity 5.1: Cell Structure

Estimated time: 5 minutes

The basic unit of biological structure is the cell. All living things consist of cells. In this tutorial you will explore the structure of the basic cell types.

Activity 5.2: Membrane Traffic

Estimated time: 5 minutes

This tutorial presents the details of membrane traffic in the cell. This includes where new membrane is synthesized and how proteins that are synthesized on the rough endoplasmic reticulum are sent to their proper locations. The complementary processes of phagocytosis and exocytosis are also explored. These allow the import and export of large molecules and larger complexes that could never pass through even the largest transport proteins in the cell.

Start the MediaTutor Student CD-ROM and enter the activity number in the Quick Search box to be taken directly to that activity.

Web Investigations

Case Study: Microscopic Stowaways

Estimated time: 10 minutes

Bugs in space. Sounds like a science fiction thriller. Bacteria that survive on the moon may get all the headlines, but conditions can be harsh on Earth, too. This exercise looks at extremophiles, organisms that not only survive but thrive under some pretty unusual conditions. Could such organisms survive in space? Could they be dangerous?

Go to http://www.prenhall.com/audesirk6, the Audesirk Companion Web site. Select Chapter 5 and the Web Investigation to begin.

"Disorder spreads through the universe, and life alone battles against it."

G. Evelyn Hutchinson

The bodies of these runners are efficiently converting energy stored in fats and carbohydrates to the energy of movement and heat. Their pounding footsteps noticeably shake the Verrazano Bridge during the New York Marathon.

6 Energy Flow in the Life of a Cell

AT A GLANCE

Case Study: Energy Unleashed

1) **What Is Energy?**

The Laws of Thermodynamics Describe the Basic Properties of Energy

Living Things Use the Energy of Sunlight to Create the Low-Entropy Conditions Characteristic of Life

2) **How Does Energy Flow in Chemical Reactions?**

Exergonic Reactions Release Energy

Endergonic Reactions Require an Input of Energy

Coupled Reactions Link Exergonic with Endergonic Reactions

3) **How Is Cellular Energy Carried Between Coupled Reactions?**

ATP Is the Principal Energy Carrier in Cells

Electron Carriers Also Transport Energy Within Cells

4) **How Do Cells Control Their Metabolic Reactions?**

At Body Temperatures, Spontaneous Reactions Proceed Too Slowly to Sustain Life

Catalysts Reduce Activation Energy

Enzymes Are Biological Catalysts

The Structure of Enzymes Allows Them to Catalyze Specific Reactions

Cells Regulate the Amount and the Activity of Their Enzymes

The Activity of Enzymes Is Influenced by Their Environment

Case Study Revisited: Energy Unleashed

CASESTUDY

Energy Unleashed

Lungs heaving and legs pumping, runners push themselves to the limits of their endurance—not only running 26 miles, but doing it fast. The 20,000 runners in the New York Marathon collectively convert more than 50 million Calories of energy into moving their bodies, shaking the Verrazano Bridge, and heating the air around them. Once finished, they douse their overheated bodies with water and refuel on high-energy snacks, since they've "burned up" the equivalent of 30,000 pounds of chocolate cake! Finally, cars, buses, and airplanes—burning vast quantities of fuel and releasing enormous amounts of heat—carry the runners back to their homes throughout the world.

What exactly is energy? Do the same principles that govern energy use in the engines of cars and airplanes govern its use by our bodies and other organisms? Why do our bodies generate heat, and why do we give off more heat when exercising than when we do watching TV? How do the runners store energy and make it available to their muscles when they need it? We often talk about "burning" Calories. Setting a spoonful of sugar on fire or eating it and allowing your body to "burn" it involve similar reactions. In both cases, oxygen combines with sugar to produce carbon dioxide, water, and heat. However, our bodies aren't roasted when we metabolize food, and some of the energy is captured in other molecules that can be used to power muscle movement or a huge variety of processes within our cells. How do we control the breakdown of high-energy molecules to produce useful energy? In this chapter we discuss the physical laws that govern energy flow in the universe, how energy flow in turn governs chemical reactions, and how the chemical reactions within living cells are controlled by the cell's own molecules. ■

1 What Is Energy?

Energy can be defined simply as the *capacity to do work*, including synthesizing molecules, moving objects, and generating heat and light. There are two types of energy: *kinetic energy* and *potential energy*. Both of these, in turn, exist in many different forms. **Kinetic energy**, or *energy of movement*, includes light (movement of photons), heat (movement of molecules), electricity (movement of electrically charged particles), and movement of large objects. **Potential energy**, or *stored energy*, includes chemical energy stored in the bonds that hold atoms together in molecules, electrical energy stored in a battery, and positional energy stored in a diver poised to spring (Fig. 6-1). Under the right conditions, kinetic energy can be transformed into potential energy, and vice versa. For example, the diver converted kinetic energy of movement into potential energy of position when he climbed up to the platform; when he eventually jumped off, the potential energy was converted back into kinetic energy.

Figure 6-1 From potential to kinetic energy
The body of a diver perched atop the platform has potential energy, because the heights of the platform and the pool are different. As he dives, the potential energy is converted to the kinetic energy of motion of the diver's body. Finally, some of this kinetic energy is transferred to the water, which itself is set in motion.

MediaTutor 6.1 Energy and Chemical Reactions

To understand energy flow, we need to know two things: (1) the quantity of available energy and (2) the usefulness of the energy. These are the subjects of the laws of thermodynamics, which we will now explore.

The Laws of Thermodynamics Describe the Basic Properties of Energy

The **laws of thermodynamics** define the basic properties and behavior of energy. The **first law of thermodynamics** states that, assuming there is no influx of energy, the total amount of energy remains constant. Energy can neither be created nor destroyed by ordinary processes, but it can be changed in form (for example, from chemical energy to heat energy). The first law is therefore often called the *law of conservation of energy*. When you drive a car, you convert the potential chemical energy of gasoline into the kinetic energy of movement and heat. The total amount of energy remains the same, although it has changed in form. Likewise, a runner is converting the potential chemical energy from food into the same total amount of kinetic energy of movement plus heat.

The **second law of thermodynamics** states that when energy is converted from one form to another, the amount of useful energy decreases. Put another way, the second law states that all spontaneous changes result in a more uniform distribution of energy, reducing the energy differences that are essential for doing work; in other words, energy is spontaneously converted from more useful into less useful forms.

To illustrate the second law, let's examine a car engine burning gasoline. The kinetic energy of the moving vehicle is much less than the chemical energy originally contained in the gasoline. According to the first law, the total amount of energy remains constant; where is the "missing" energy? The burning gas not only moved the car but also heated up the engine, the exhaust system, and the air around the car. The friction of tires on the pavement slightly heated the road as well, so, as the first law dictates, no energy is missing. However, the energy released as heat is in a less usable form; it merely increased the random movement of molecules in the engine, the road, and the air.

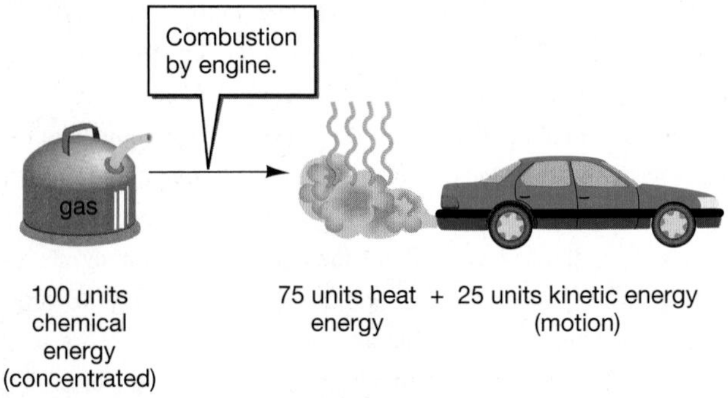

So, too, the energy of heat that runners liberate to the air when food "burns" in their cells cannot be harnessed to allow them to run farther or faster. Thus, the second law tells us that no process—including processes carried on in the body—is 100% efficient.

The second law of thermodynamics also tells us something about the organization of matter. Regions of concentrated energy tend to be regions of great orderliness. The eight carbon atoms in a single molecule of gasoline have a much more orderly arrangement than do the carbon atoms of the eight separate, randomly moving molecules of carbon dioxide that are formed when the gasoline burns. Therefore, we can also phrase the second law in terms of the organization of matter: Unless energy is added to the system, processes that proceed spontaneously result in an increase in randomness and disorder. This tendency toward loss of orderliness and high-level energy and an increase in randomness, disorder, and low-level energy is called **entropy**. We all experience the tendency toward entropy in our homes. Frequent inputs of energy are required to keep debris confined to the trash can, newspapers to folded stacks, books to their shelves, and clothes to drawers and closets. Without our energetic cleaning and organizing efforts, these items tend to end up in their lowest-energy state—a state of disorder. When G. Evelyn Hutchinson said, "Disorder spreads through the universe, and life alone battles against it," he was making an eloquent reference to entropy and the second law of thermodynamics.

Living Things Use the Energy of Sunlight to Create the Low-Entropy Conditions Characteristic of Life

If you think about the second law of thermodynamics, you may wonder how life can exist at all. If chemical reactions, including those inside living cells, cause the amount of unusable energy to increase, and if matter tends toward increasing randomness and disorder, how can organisms accumulate the concentrated energy and precisely ordered molecules that characterize living things? The answer is that nuclear reactions in the sun produce energy in the form of sunlight, a process that also produces vast increases in entropy. Living things on Earth use a continuous input of solar energy to synthesize complex molecules and maintain orderly structures—to "battle against disorder." The highly organized, low-entropy systems that characterize life do not violate the second law; they are achieved at the expense of an enormous loss of usable energy from the sun. The entropy of the solar system as a whole constantly increases.

2 How Does Energy Flow in Chemical Reactions?

A **chemical reaction** is a process that forms and breaks chemical bonds that hold atoms together. Chemical reactions convert one set of chemical substances, the **reactants**, into another set, the **products**. All chemical reactions are either exergonic or endergonic. A reaction is **exergonic** (Greek for "energy out") if the reactants contain more energy than the products. Consequently, the reaction releases energy:

Exergonic reaction

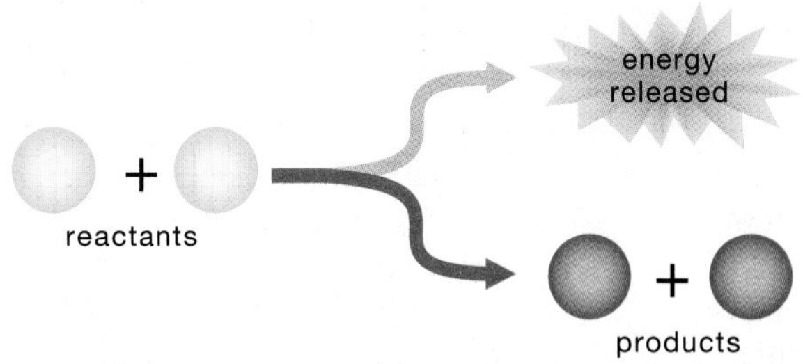

Conversely, a reaction is **endergonic** (Greek for "energy in") if the products contain more energy than the reactants. According to the second law of thermodynamics, endergonic reactions require an influx of energy from some outside source:

Endergonic reaction

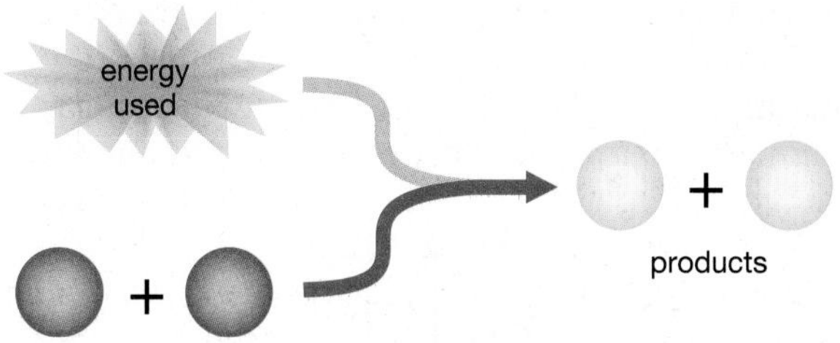

Let's look at two processes that illustrate these types of reactions: burning sugar and photosynthesis.

Exergonic Reactions Release Energy

In an exergonic reaction, the reactants contain more energy than the products. The sugar that the runners' bodies use as fuel contains more energy than the carbon dioxide and water that are produced when that sugar breaks down. The extra energy is liberated as muscular movement and heat. Sugar can also be burned, as any cook can tell you. When sugar is burned by a flame, similar to the way it is burned in the body, it reacts with oxygen (O_2) to produce carbon dioxide (CO_2) and water (H_2O) and energy, as described by this equation:

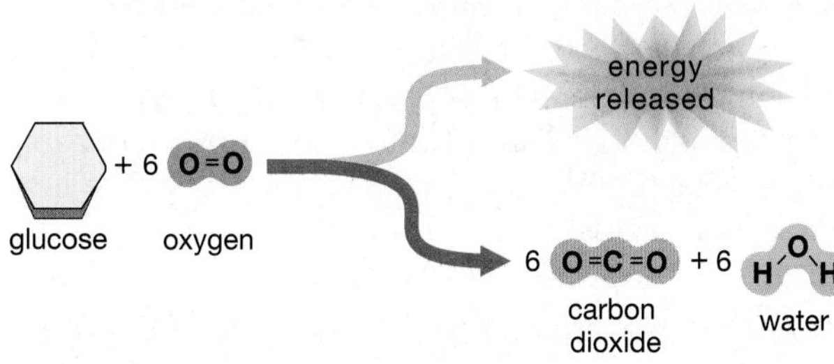

This reaction illustrates two important concepts, diagrammed in Figure 6-2a. First, molecules of sugar and

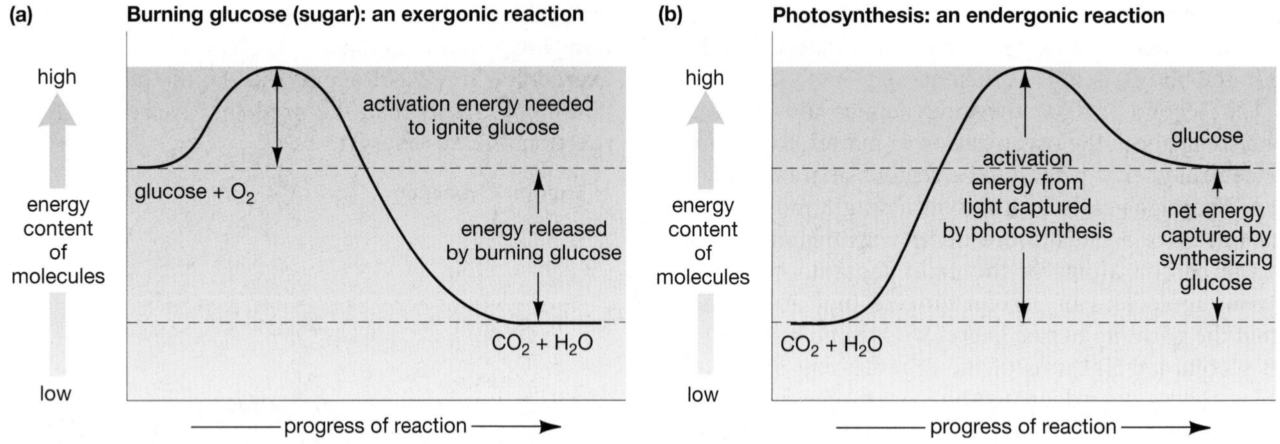

Figure 6-2 Energy relations in exergonic and endergonic reactions
(a) An exergonic ("downhill") reaction, such as the burning of sugar, proceeds from high-energy reactants (here, glucose) to low-energy products (CO_2 and H_2O). The energy difference between the chemical bonds of the reactants and products is released as heat. To start the reaction, however, an initial input of energy—the activation energy—is required. ***(b)*** An endergonic ("uphill") reaction, such as photosynthesis, proceeds from low-energy reactants (CO_2 and H_2O) to high-energy products (glucose) and therefore requires a net input of energy, in this case from sunlight.

oxygen contain much more energy than molecules of carbon dioxide and water, so the reaction releases energy. Energy release allows exergonic reactions to occur without a net input of energy. Once ignited, sugar will continue to burn spontaneously. It may be helpful to think of exergonic reactions as running "downhill," from high energy to low energy.

Although burning sugar releases energy, a spoonful of sugar doesn't burst into flames by itself. This observation leads to the second important concept: All chemical reactions require an initial input of energy to get started, something like giving a rock poised at the top of a hill a push to start it rolling down. In chemical reactions, this initial energy input or "push" is called the **activation energy**. Chemical reactions require activation energy to get started because a shell of negatively charged electrons surrounds atoms and molecules. For two molecules to react with each other, their electron shells must be forced together, despite their mutual electrical repulsion. That forcing requires energy. The usual source of activation energy is the kinetic energy of movement. Molecules moving with sufficient speed collide hard enough to force their electron shells to mingle and react. Because molecules move faster as the temperature increases, most chemical reactions occur more readily at high temperatures. The initial heat provided by a match setting sugar on fire allows these reactions to begin. The combination of sugar with oxygen then releases enough of its own heat to sustain the reaction. Now, think about lighting a match; where does the heat to start that reaction come from?

Endergonic Reactions Require an Input of Energy

In contrast to what happens when sugar or a match is burned, many reactions in living systems result in products that contain *more energy* than the reactants. Sugar, produced in photosynthetic organisms, contains far more energy than the carbon dioxide and water from which it is formed. The protein in a muscle cell contains more energy than the individual amino acids that were joined together to synthesize it. In other words, synthesizing complex biological molecules requires an input of energy (Fig. 6-2b). As we shall see in Chapter 7, photosynthesis in green plants takes low-energy water and carbon dioxide and produces oxygen and high-energy sugar from them:

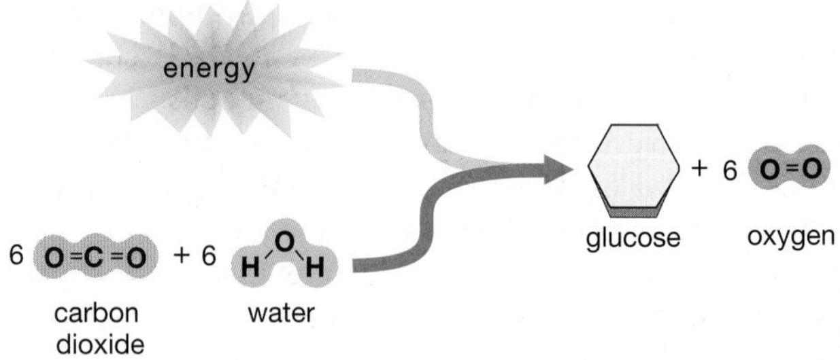

We might call endergonic reactions "uphill" reactions, going from low energy to high energy, like pushing our rock to the top of the hill in the first place. Photosynthesis requires energy, which plants (and algae and some types of bacteria) obtain from sunlight. But where do we get the energy to synthesize muscle protein and other complex biological molecules?

Coupled Reactions Link Exergonic with Endergonic Reactions

Endergonic reactions require energy from other sources; they obtain this energy from exergonic, or energy-releasing, reactions. In a **coupled reaction** (Fig. 6-3), an exergonic reaction provides the energy needed to drive an endergonic reaction. When you drive a car, the exergonic reaction of burning gasoline provides the energy for the

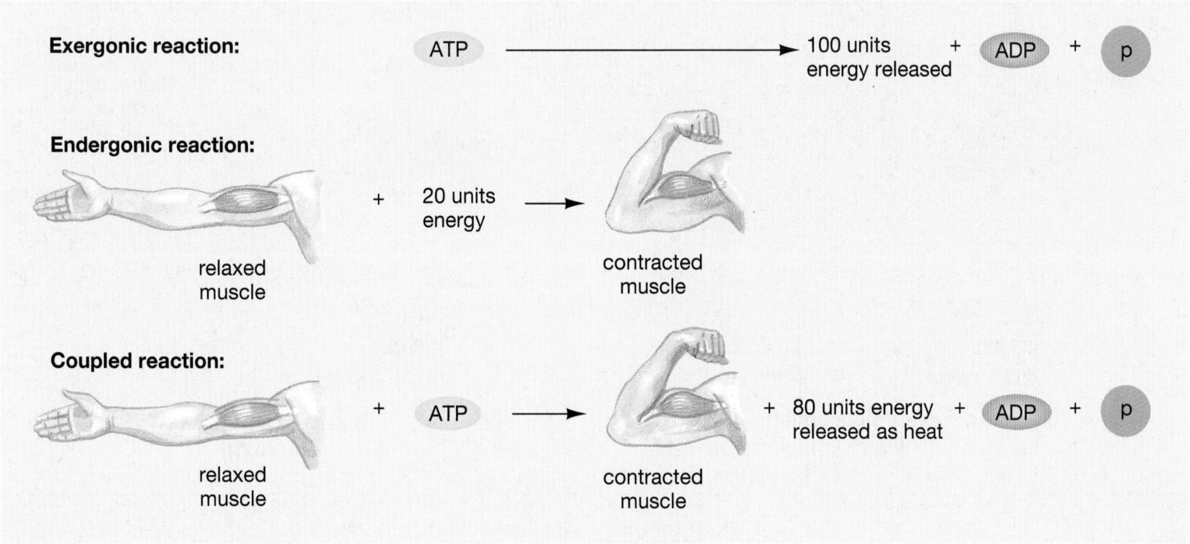

Figure 6-3 A coupled reaction
Muscle movement is an endergonic reaction coupled to the exergonic reaction of ATP breakdown. The energy released by ATP as it is broken down into ADP plus phosphate (P) exceeds the energy put into muscle contraction, so the overall reaction is exergonic (energy units are arbitrary).

endergonic reaction of starting a stationary car into motion and keeping it going; in the process, much energy is lost as heat. Photosynthesis is another coupled reaction. In photosynthesis, the exergonic reaction occurs in the sun, and the endergonic reaction occurs in the plant. Most of the energy liberated by the sun is lost as heat, so the second law of thermodynamics still applies: usable energy decreases.

Living organisms constantly use the energy given off by exergonic reactions (such as the chemical breakdown of sugar) to drive essential endergonic reactions (such as brain activity, muscular contraction or other types of movement, or the synthesis of complex molecules), as shown in Figure 6-3. Endergonic reactions cannot occur unless, somewhere in the body, an exergonic reaction has already happened to provide the energy to drive it. Further, since some energy is lost as heat every time it is transformed, the energy provided by exergonic reactions must exceed that needed to drive endergonic reactions. The exergonic and endergonic parts of coupled reactions often occur in different places, so there also must be some way to transfer the energy from the exergonic reaction that releases energy to the endergonic reaction that consumes it. In coupled reactions occurring within cells, energy is usually transferred from place to place by *energy-carrier* molecules, of which the most common is ATP.

3 How Is Cellular Energy Carried Between Coupled Reactions?

As we saw earlier, cells couple reactions so that the energy released by exergonic reactions is used to drive endergonic reactions. In the case of a runner, the breakdown of a sugar (glucose) releases energy; this energy release is coupled to energy-consuming reactions that cause muscle contraction. But glucose cannot be used directly for muscle contraction. Instead, the energy from glucose must be transferred to an **energy-carrier molecule**, which provides the muscle protein with energy to contract. Energy carriers work something like rechargeable batteries, picking up an energy charge at an exergonic reaction, moving to another location within the cell, and releasing the energy to drive an endergonic reaction. Because energy-carrier molecules are unstable, they are used only for temporary energy transfer within cells. They are not used to transfer energy from cell to cell, nor are they used for long-term energy storage. Muscles store energy in glycogen, a stable molecule that consists of chains of glucose molecules. When energy is needed, for example, when the marathon begins, the glycogen in the body is broken down by enzymes first to glucose, then to carbon dioxide and water. The energy is then captured and transferred to muscle protein molecules by energy-carrier molecules such as ATP.

ATP Is the Principal Energy Carrier in Cells

Several exergonic reactions in cells produce **adenosine triphosphate**, or **ATP**, the most common energy-carrier molecule in cells. By providing energy for a wide variety of endergonic reactions, ATP serves as a common currency of energy transfer. For this reason, it is sometimes called the "energy currency" of living cells. As you learned in Chapter 3, ATP is a nucleotide composed of the nitrogen-containing base adenine, the sugar ribose, and three phosphate groups (Fig. 6-4). Energy released in cells through glucose breakdown is

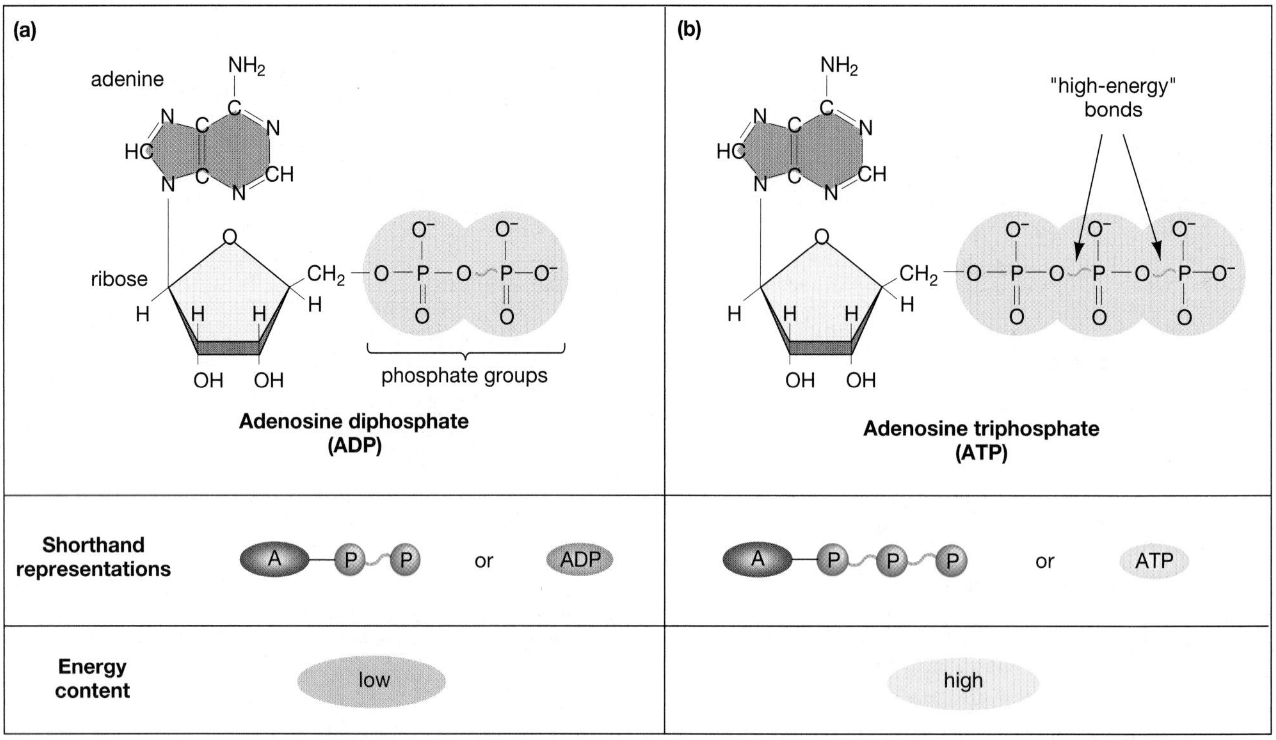

Figure 6-4 ADP and ATP
A phosphate group is added to ***(a)*** ADP (adenosine diphosphate) to make ***(b)*** ATP (adenosine triphosphate). In most cases, only the last phosphate group and its high-energy bond are used to carry energy and transfer it to endergonic reactions within a cell. In much of the remainder of the text, we will employ the shorthand representations of ATP and ADP.

MediaTutor
6.2 Energy and Life

used to synthesize ATP from **adenosine diphosphate (ADP)** and phosphate:

ATP synthesis: Energy is stored in ATP

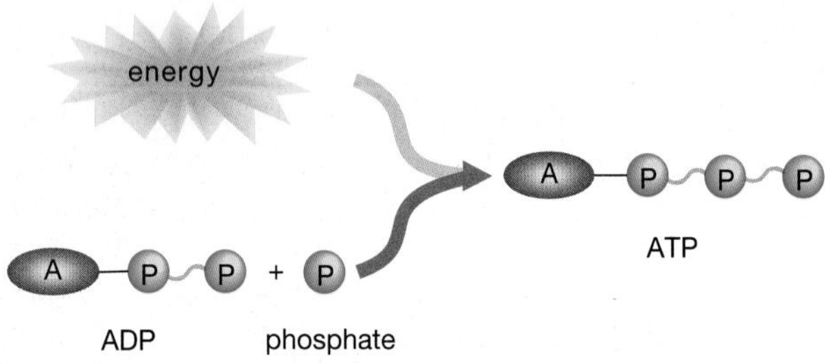

ATP stores this energy within its chemical bonds and can carry the energy to sites in the cell that perform energy-requiring reactions, such as the synthesis of proteins or muscle contraction (see Fig. 6-3). The ATP is then broken down to form ADP and phosphate:

ATP breakdown: Energy of ATP is released

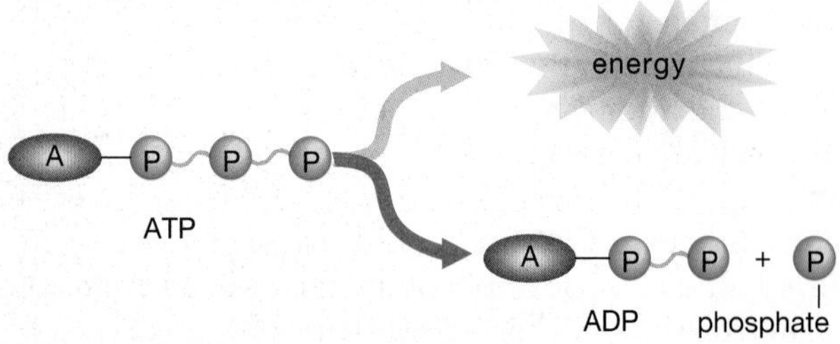

During these energy transfers, some heat is given off at each stage, and there is an overall loss of usable energy (Fig. 6-5).

ATP is admirably suited to carry energy within cells. The bonds joining the last two phosphate groups of ATP to the rest of the molecule (sometimes called *high-energy bonds*) require a large amount of energy to form, so considerable energy can be trapped from exergonic reactions by synthesizing ATP molecules. ATP is also unstable; it readily releases its energy in the presence of appropriate enzymes. Under most circumstances, only the bond joining the last phosphate group (the one joining phosphate to ADP to form ATP) carries energy from exergonic to endergonic reactions.

The life span of an ATP molecule in a living cell is very short, because this energy carrier is continuously formed, broken down to ADP and phosphate, and resynthesized. If the molecules of ATP that you use just sitting at your desk all day could be captured (instead of recycled), they would weigh 40 kg—nearly 90 pounds of ATP! A marathon runner may use a pound of ATP every minute. (The ADP must be quickly converted back to ATP, or it would be a very brief run.) As you can see, ATP is *not* a long-term energy-storage molecule. More stable molecules, such as sucrose, glycogen, starch, or fat, store energy in your body for hours, days, or months.

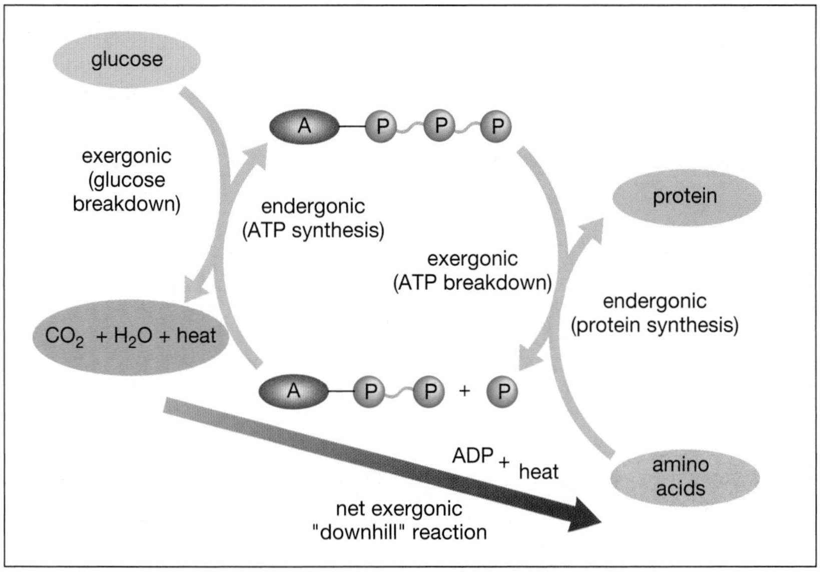

Figure 6-5 Coupled reactions within living cells Exergonic reactions (such as glucose metabolism) drive the endergonic reaction of ATP synthesis from ADP. The ATP molecule moves to a part of the cell where the breakdown of ATP liberates some of this energy to drive an essential endergonic reaction (such as protein synthesis). The ADP and phosphate are recycled to the exergonic reactions; they will be converted back to ATP. The overall reaction is "downhill": More energy is produced by the exergonic reaction than is needed to drive the endergonic reaction. The extra energy is released as heat.

Electron Carriers Also Transport Energy Within Cells

In addition to ATP, other carrier molecules can also transport energy within a cell. In some exergonic reactions, including both glucose metabolism and the light-capturing stage of photosynthesis, some energy is transferred to electrons. These energetic electrons (in some cases, along with hydrogen atoms) are captured by **electron carriers** (Fig. 6-6). Common electron carriers include *nicotinamide adenine dinucleotide* (NAD^+) and its relative *flavin adenine dinucleotide* (FAD). Loaded electron carriers then donate the electrons, along with their energy, to other molecules. You will learn more about electron carriers and their role in cellular metabolism in Chapters 7 and 8.

4 How Do Cells Control Their Metabolic Reactions?

Cells are miniature, incredibly complex chemical factories. The **metabolism** of a cell is the sum of all its chemical reactions. Many of these reactions are linked in sequences called **metabolic pathways** (Fig. 6-7). Photosynthesis (Chapter 7) is one such pathway. *Glycolysis*, the series of reactions that begin the digestion of glucose (Chapter 8), is another example. Different metabolic pathways may utilize the same molecules; as a result, all the reactions and all the molecules of a cell are interconnected in a single, enormously complicated metabolic pathway.

The chemical reactions in a cell are governed by the same laws of thermodynamics that control other

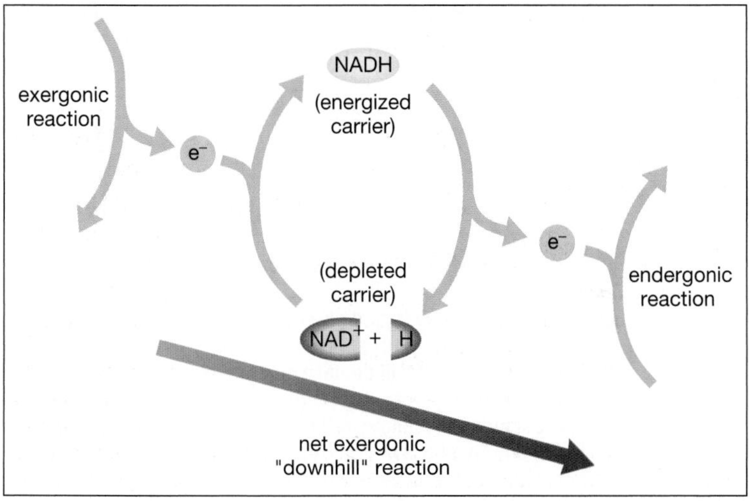

Figure 6-6 Electron carriers Electron-carrier molecules such as nicotinamide adenine dinucleotide (NAD^+) pick up electrons generated by exergonic reactions and hold them in high-energy outer electron shells. Hydrogen atoms are often picked up simultaneously. The electron is then deposited, energy and all, with another molecule to drive an endergonic reaction, typically the synthesis of ATP.

Figure 6-7 Simplified view of metabolic pathways
The original reactant molecule, A, undergoes a series of reactions, each catalyzed by a specific enzyme. The product of each reaction serves as the reactant for the next reaction in the pathway. Metabolic pathways are commonly interconnected such that the product of a step in one pathway may serve as a reactant for the next reaction in that pathway or for a reaction in another pathway.

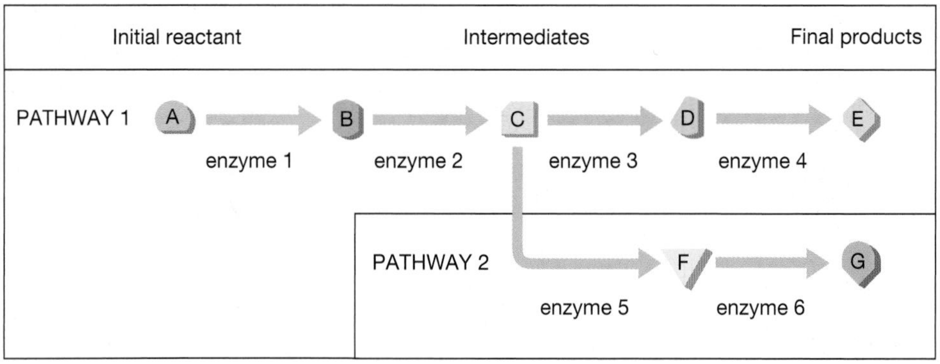

reactions. How, then, do orderly metabolic pathways arise? The biochemistry of cells is finely tuned in three ways:

1. Cells regulate chemical reactions through the use of proteins called *enzymes*.
2. Cells couple reactions together, powering energy-requiring endergonic reactions with the energy released by exergonic reactions.
3. Cells synthesize energy-carrier molecules that capture energy from exergonic reactions and transport it to endergonic reactions.

At Body Temperatures, Spontaneous Reactions Proceed Too Slowly to Sustain Life

In general, the speed at which a reaction occurs is determined by its activation energy, that is, how much energy is required to start the reaction (see Fig. 6-2). Reactions with low activation energies can proceed swiftly at body temperature, whereas reactions with high activation energies, such as gasoline combining with oxygen, are practically nonexistent at similar temperatures. Most reactions can be accelerated by raising the temperature, thereby increasing the speed of molecules.

The reaction of sugar with oxygen to yield carbon dioxide and water is exergonic, but it has an enormous activation energy. At the temperatures found in living organisms, sugar and many other energetic molecules would almost never break down spontaneously and give up their energy. It is the enzymes produced by living cells that allow sugar to be an important energy source for life on Earth. Let's see how enzymes and other *catalysts* influence chemical reactions.

Catalysts Reduce Activation Energy

Catalysts are molecules that speed up the rate of a reaction without themselves being used up or permanently altered. Catalysts speed up reactions by reducing the activation energy (Fig. 6-8). As an example of catalytic action, let's consider the catalytic converters in many automobile engines. When gasoline is burned completely, the final products are carbon dioxide and water:

$$2\,C_8H_{18} + 25\,O_2 \rightarrow 16\,CO_2 + 18\,H_2O + \text{energy}$$
(octane)

However, flaws in the combustion process generate other substances, including poisonous carbon monoxide (CO). Carbon monoxide reacts spontaneously but slowly with oxygen in the air to form carbon dioxide:

$$2\,CO + O_2 \rightarrow 2\,CO_2 + \text{energy}$$

In large cities, so many vehicles emit so much CO that the spontaneous reaction of CO with O_2 can't keep pace, and unhealthy levels of carbon monoxide accumulate.

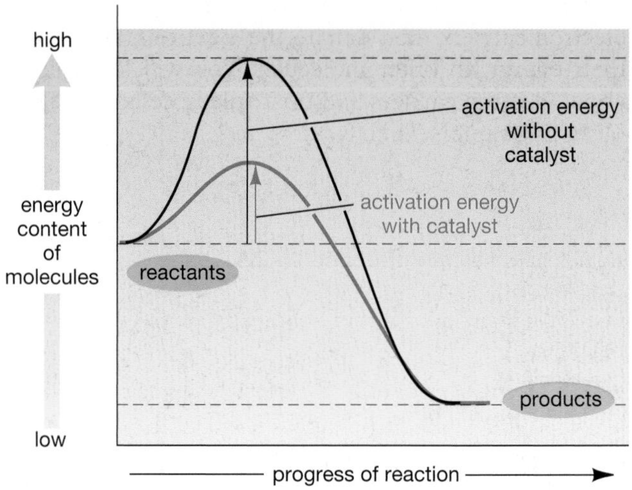

Figure 6-8 Activation energy controls the rate of chemical reactions
High activation energy (black curve) means that reactant molecules must collide very forcefully in order to react. Only very fast-moving molecules will collide hard enough to react, so reactions with high activation energies proceed slowly at low temperatures, where most molecules move relatively slowly. Catalysts lower the activation energy of a reaction (red curve), so a much higher proportion of molecules move fast enough to react when they collide. Therefore, the reaction proceeds much more rapidly.

Enter the catalytic converter. Platinum catalysts in the converter hasten the conversion of CO to CO_2, thereby reducing air pollution.

Note three important principles about all catalysts:

1. Catalysts speed up reactions.
2. Catalysts can speed up only those reactions that would occur spontaneously anyway, but at a much slower rate.
3. Catalysts are not consumed in the reactions they promote. No matter how many reactions they accelerate, the catalysts themselves are not permanently changed.

Enzymes Are Biological Catalysts

Enzymes are biological catalysts, usually proteins, synthesized by living organisms. Enzymes possess the characteristics of catalysts that we just described. But enzymes have two additional attributes that set them apart from nonbiological catalysts:

1. Enzymes are usually very specific, catalyzing at most only a few types of chemical reactions. In most cases, an enzyme catalyzes a single reaction that involves one or two specific molecules but leaves similar molecules untouched. (You may recall from Chapter 3, for example, that animals have enzymes that break apart starch molecules but leave cellulose intact, despite the fact that both starch and cellulose are composed of glucose subunits.)
2. Enzyme activity is regulated—that is, enhanced or suppressed—in many cases by the very molecules whose reactions they catalyze.

The Structure of Enzymes Allows Them to Catalyze Specific Reactions

Why are enzymes specific, and how are they regulated? Enzyme function is intimately related to enzyme structure. Enzymes are proteins with complex three-dimensional shapes. Each enzyme has a "pocket," called the **active site**, into which reactant molecules, called **substrates**, can enter.

The active site of each enzyme has a distinctive shape and distribution of electrical charge that is complementary to those of its substrate. Because the enzyme and its substrate must fit together, only certain molecules can enter the active site. For example, several enzymes are required to completely digest all the proteins we eat, because each enzyme breaks apart only a specific sequence of amino acids.

How does an enzyme catalyze a reaction? First, both the shape and the charge of the active site force substrates to enter the enzyme in specific orientations (Fig. 6-9, step ①). Second, when substrates enter the active site, both substrate and active site change shape (step ②). Certain amino acids within the part of the protein

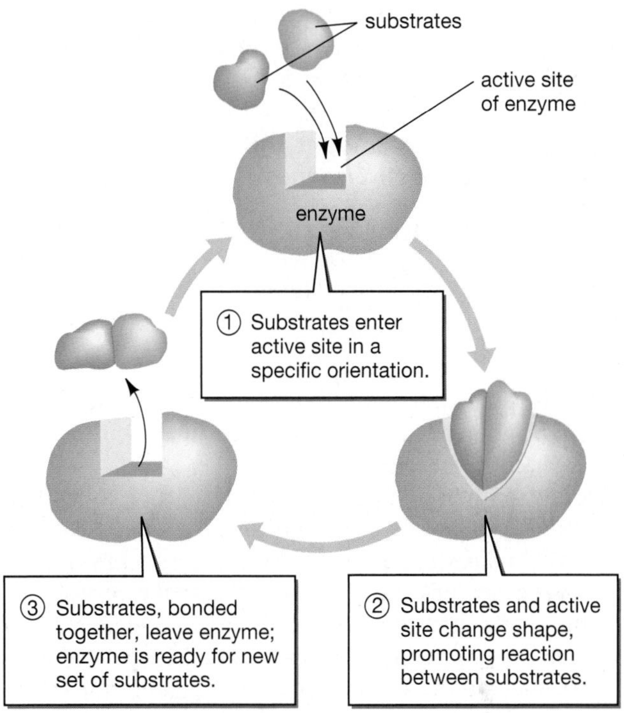

Figure 6-9 The cycle of enzyme–substrate interactions

that forms the active site may temporarily bond with atoms of the substrates, or electrical interactions between the amino acids in the active site and substrates may distort the chemical bonds within the substrates. The combination of substrate selectivity, substrate orientation, temporary chemical bonds, and bond distortion promotes the specific chemical reaction catalyzed by a particular enzyme. When the final reaction between the substrates is finished, the product(s) no longer fit properly into the active site and are expelled (step ③). The temporary changes in shape, charge, and bonding patterns within the enzyme revert to their original configuration, and the enzyme is ready to accept another set of substrates (back to step ①).

How do enzymes speed up the rate of chemical reactions? Enzymes are precisely regulated and promote only very specific reactions. The breakdown or synthesis of a molecule within a cell normally occurs in many discrete steps, each catalyzed by a different enzyme (see Fig. 6-7), which lowers the activation energy for its particular reaction (see Fig. 6-8). Just as carving a staircase into a cliff allows the cliff to be climbed—one small step at a time—this series of low activation energy steps allow the overall reaction to surmount its otherwise high activation energy "cliff," so the reaction can proceed at body temperature. The reason we are not roasted by burning sugar in our bodies is that, unlike the situation in which sugar is set on fire, our cells break down sugar in many small steps, each liberating small quantities of energy. As you will learn in Chapter 8, the breakdown of sugar by enzymes actually takes about 16 steps. Along the way, energy is inevitably lost as heat, but much is also captured in energy-carrier molecules, as described previously.

Cells Regulate the Amount and the Activity of Their Enzymes

To be useful, the metabolic reactions within cells must be carefully controlled; they must occur at the proper rate and with the proper timing. Cells have evolved many ways of regulating enzyme activity, including the following:

1. Cells regulate the synthesis of enzymes to meet their changing needs. Reactions can occur only if the necessary enzymes are available.
2. Cells synthesize some enzymes in inactive forms and activate them only when needed. An example is the protein-digesting enzyme pepsin described in the following section. Pepsin doesn't digest the proteins of the cells that produce it, because it is secreted in an inactive form with its active site blocked. In the acidic environment of the stomach, the blocking portion of the protein is removed, and the enzyme becomes active.
3. Cells inhibit enzymes when adequate amounts of the enzyme's product are available. In **feedback inhibition**, the activity of an enzyme is inhibited by its own product or by a subsequent product produced further along in a metabolic pathway (Fig. 6-10). For example, assume that an enzyme catalyzes the conversion of one amino acid to another. The concentration is regulated automatically by feedback inhibition if the end-product amino acid inhibits the enzyme once it reaches sufficiently high concentrations.
4. Certain enzymes are subject to **allosteric regulation**. In allosteric regulation, enzyme action is enhanced or inhibited by small organic molecules that act as regulators. The regulator is neither the substrate nor the product of the enzyme, but may be the end product of a series of reactions in which the enzyme is involved. A regulator molecule will bind to a special *allosteric regulatory site* on the enzyme (Fig. 6-11a). As a result of this binding, the active site of the enzyme is changed (*allosteric* literally means "other shape"), and the enzyme may become either more or less able to bind its substrates (Fig. 6-11b). The specific enzyme and specific regulator molecule determine whether allosteric regulation increases or decreases enzyme activity. Allosteric regulation is one mechanism of feedback inhibition (see Fig. 6-10).
5. In some cases, two or more molecules that are somewhat similar in structure compete for the active site of an enzyme, a situation called **competitive inhibition** (Fig. 6-11c). Some poisons are competitive inhibitors that keep an enzyme from breaking down its normal substrate. In another scenario two types of alcohol, ethanol (the normal substrate, found in alcoholic beverages) and methanol (a poison), compete for the active site of the enzyme *alcohol dehydrogenase*. Methanol breakdown by this enzyme produces formaldehyde, which can cause blindness. Taking advantage of competitive inhibition, doctors administer ethanol to methanol-poisoning victims. By competing with methanol for the active site, ethanol blocks formaldehyde production.

The Activity of Enzymes Is Influenced by Their Environment

Enzymes, which are proteins, have very complex three-dimensional structures that are required for their proper function but that are also sensitive to environmental conditions. Recall from Chapter 3 that much of the three-dimensional structure of proteins is the result of hydrogen bonds between partially charged amino acids. These bonds can be altered by their chemical and

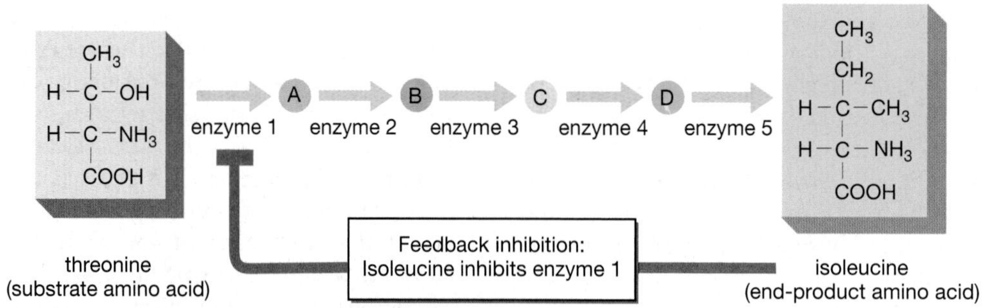

Figure 6-10 Enzyme regulation by feedback inhibition
In this example, the first enzyme in the metabolic pathway that converts threonine (an amino acid substrate) to isoleucine (an amino acid product) is inhibited by high concentrations of isoleucine. If a cell lacks isoleucine, the first enzyme is not inhibited, and the pathway proceeds rapidly. As isoleucine concentrations build up, the isoleucine binds to the first enzyme and gradually shuts down the pathway. When concentrations of isoleucine drop and fewer molecules are available to inhibit the enzyme, the pathway resumes production.

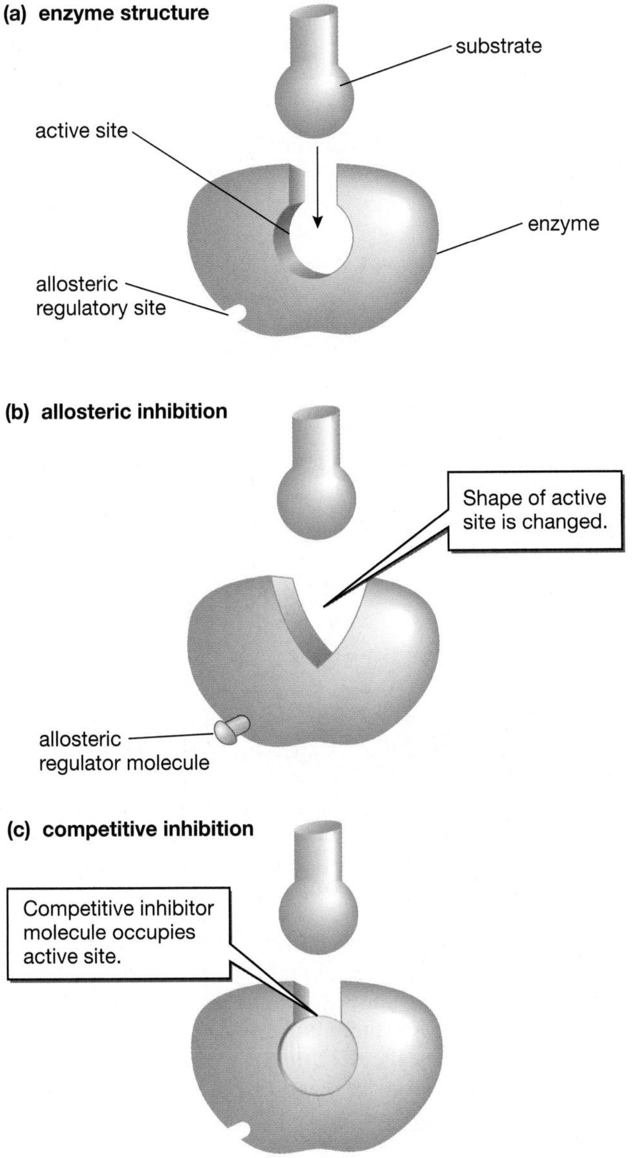

Figure 6-11 Enzyme regulation by allosteric regulation and competitive inhibition
(a) Many enzymes have an active site and an allosteric regulatory site on different parts of the molecule. ***(b)*** When enzymes are inhibited by allosteric regulation, binding by a regulator molecule alters the active site so the enzyme is less compatible with its substrate. ***(c)*** During competitive inhibition, a molecule somewhat similiar to the substrate fits into the active site and blocks entry of the substrate.

physical surroundings. Each enzyme has evolved to function optimally at a particular pH, salt concentration, and temperature. Some also require the presence of other molecules called *coenzymes*, typically derived from water-soluble vitamins, in order to function.

Most enzymes function optimally at a pH between 6 and 8, the level found in most body fluids and maintained within cells. An exception is the protein-digesting enzyme *pepsin*. Pepsin is converted from an inactive to an active form by the highly acidic pH (2) conditions within the stomach. At this pH, the excess of hydrogen ions causes hydrogen to attach to certain locations on the enzyme, altering its configuration and exposing the active site. In proteins that function best at neutral pH (7), the acidic environment would distort the enzyme's structure and destroy its normal function.

Before the advent of refrigeration, foods such as meat were commonly preserved by using concentrated salt solutions. These solutions kill most bacteria, partly by interfering with enzyme function. Salts dissociate into ions, which form bonds with enzymes that interfere with the enzymes' normal three-dimensional structure, thus destroying enzyme activity. Dill pickles are very well preserved in a vinegar-salt solution, which combines both salty and acidic conditions. Organisms that live in highly salty environments, as you might predict, have enzymes whose configuration *depends* on the presence of salt ions.

Temperature also affects the rate of enzyme-catalyzed reactions. Because molecules move more rapidly at higher temperatures, their random movements are then more likely to bring them into contact with the active site of an appropriate enzyme. Thus, these reactions are accelerated by moderately higher temperatures and slowed by lower temperatures. Stories abound of people who have fallen into frozen lakes for extended periods and survived. In one true story, a boy who fell through the ice was rescued and survived unharmed after 20 minutes under water. Although at normal body temperature the brain would die after about 4 minutes without oxygen, the boy's body temperature was lowered by the icy water, slowing all his metabolic reactions and drastically reducing his need for oxygen. However, when temperatures rise too high, the hydrogen bonds that regulate enzyme shape may be broken apart by the excessive molecular motion. Think of the protein in egg white and how it is completely altered in color and texture by cooking. Far lower temperatures than those required to fry an egg can still be too hot to allow enzymes to function properly.

Some enzymes require helper molecules called **coenzymes** to function. These organic molecules bind to the enzyme and interact with the substrate molecule. Coenzymes help weaken the bonds of the substrate, allowing it to react with an enzyme. Many water-soluble vitamins (such as the B vitamins) are essential to humans because they are used by the body to synthesize coenzymes.

In summary, the ability of an enzyme to catalyze reactions is controlled by many factors, including the amount of active enzyme, levels of allosteric regulator molecules, the concentration of inhibitor molecules, the concentration of substrates, pH, temperature, ionic environment, and, in some cases, the presence of coenzymes. In a healthy cell the interactions among these molecules and precisely regulated environmental conditions maintain suitable concentrations of both substrates and products.

CASE STUDY REVISITED

Energy Unleashed

The bodies of marathon runners generate large amounts of heat: as each ATP is broken down to power muscle contraction, some of the energy is converted to the kinetic form as movement, and some is lost as heat in keeping with the laws of thermodynamics. Rapidly contracting muscle cells also warm up from friction as they slide past one another. Although runners and all other organisms use sugar as fuel, we life-forms burn it in a controlled fashion, using enzymes rather than flames. While the heat of a lit match will cause sugar and oxygen molecules to collide violently enough to cause a vigorous and ungoverned reaction, enzymes use chemical means to orient, distort, and reconfigure molecules into new combinations that release energy in small, discrete steps that eventually result in ATP production. The ATP that doesn't power muscle contraction releases heat as it is utilized in endergonic reactions that produce the wide range of biological molecules that your body comprises. Heat generated as a by-product of every biochemical transformation is used by warm-blooded animals to maintain a high body temperature that allows us to move faster and respond more quickly to stimuli than if our body temperatures were lower.

Although runners do not go up in flames, their bodies do generate enough heat that—if they did not have efficient ways to get rid of it—their body temperatures would rise to levels at which enzyme function would be impaired. First, explain why proper enzyme function is disrupted by high temperatures. Then, using your own personal experience and your knowledge of the properties of water from Chapter 3, discuss various ways that runners dissipate the excess heat generated by their "burning" of energy-rich molecules.

Summary of Key Concepts

1) What Is Energy?

Energy is the capacity to do work. Kinetic energy is the energy of movement (light, heat, electricity, movement of large particles). Potential energy is stored energy (chemical energy, positional energy). The flow of energy among atoms and molecules obeys the laws of thermodynamics. The first law of thermodynamics states that, assuming there is no influx of energy, the total amount of energy remains constant, although it may change in form. The second law of thermodynamics states that any use of energy causes a decrease in the quantity of concentrated, useful energy and an increase in the randomness and disorder of matter. Entropy is a measure of disorder within a system.

2) How Does Energy Flow in Chemical Reactions?

Chemical reactions fall into two categories. In exergonic reactions, the product molecules have less energy than do the reactant molecules, so the reaction releases energy. In endergonic reactions, the products have more energy than do the reactants, so the reaction requires an input of energy. Exergonic reactions can occur spontaneously, but all reactions, including exergonic ones, require an initial input of energy (the activation energy) to overcome electrical repulsions between reactant molecules. Exergonic and endergonic reactions may be coupled such that the energy liberated by an exergonic reaction drives the endergonic reaction. Organisms couple exergonic reactions, such as light-energy capture or sugar metabolism, with endergonic reactions, such as the synthesis of organic molecules.

3) How Is Cellular Energy Carried Between Coupled Reactions?

Energy released by chemical reactions within a cell is captured and transported within the cell by energy-carrier molecules, such as ATP and electron carriers. These molecules are the major means by which cells couple exergonic and endergonic reactions that occur at different places in the cell.

4) How Do Cells Control Their Metabolic Reactions?

Cellular reactions are linked in interconnected sequences called *metabolic pathways*. The biochemistry of cells is regulated in three ways: first, through the use of protein catalysts called *enzymes*; second, by coupling endergonic with exergonic reactions; and third, through the use of energy-carrier molecules that transfer energy within cells.

High activation energies slow many reactions, even exergonic ones, to an imperceptible rate under normal environmental conditions. Catalysts lower the activation energy and thereby speed up chemical reactions without being permanently changed themselves. Organisms synthesize protein catalysts called *enzymes* that promote one or a few specific reactions. The reactants temporarily bind to the active site of the enzyme, making it easier to form the new chemical bonds of the products. Enzyme action is regulated in many ways, including: (1) by altering the rate of enzyme synthesis, (2) by activating previously inactive enzymes, (3) by feedback inhibition, (4) by allosteric regulation, and (5) by competitive inhibition. Environmental conditions including pH, salt concentration, and temperature can promote or inhibit enzyme function, by altering their three-dimensional structure.

Key Terms

activation energy *p. 102*
active site *p. 107*
adenosine diphosphate (ADP) *p. 104*
adenosine triphosphate (ATP) *p. 103*
allosteric regulation *p. 108*
catalyst *p. 106*
chemical reaction *p. 101*
coenzyme *p. 109*
competitive inhibition *p. 108*
coupled reaction *p. 102*
electron carrier *p. 105*
endergonic *p. 101*
energy *p. 100*
energy-carrier molecule *p. 103*
entropy *p. 101*
enzyme *p. 107*
exergonic *p. 101*
feedback inhibition *p. 108*
first law of thermodynamics *p. 100*
kinetic energy *p. 100*
laws of thermodynamics *p. 100*
metabolic pathway *p. 105*
metabolism *p. 105*
potential energy *p. 100*
product *p. 101*
reactant *p. 101*
second law of thermodynamics *p. 100*
substrate *p. 107*

Thinking Through the Concepts

Multiple Choice

1. *According to the first law of thermodynamics, the total amount of energy in the universe*
a. is always increasing
b. is always decreasing
c. varies up and down
d. is constant
e. cannot be determined

2. *What is predicted by the second law of thermodynamics?*
a. Energy is always decreasing.
b. Disorder cannot be created or destroyed.
c. Systems always tend toward greater states of disorder.
d. All potential energy exists as chemical energy.
e. all of the above

3. *Which statement about exergonic reactions is true?*
a. The products have more energy than do the reactants.
b. The reactants have more energy than do the products.
c. They will not proceed spontaneously.
d. Energy input reverses entropy.
e. none of the above

4. *ATP is important in cells because*
a. it transfers energy from exergonic reactions to endergonic reactions
b. it is assembled into long chains that make up cell membranes
c. it acts as an enzyme
d. it accelerates diffusion
e. all of the above

5. *How does an enzyme increase the speed of a reaction?*
a. by changing an endergonic to an exergonic reaction
b. by providing activation energy
c. by lowering activation energy requirements
d. by decreasing the concentration of reactants
e. by increasing the concentration of products

6. *Which of the following statements about enzymes is (are) true?*
a. They interact with specific reactants (substrates).
b. Their three-dimensional shapes are closely related to their activities.
c. They change the shape of the reactants.
d. They have active sites.
e. all of the above

? Review Questions

1. Explain why organisms do not violate the second law of thermodynamics. What is the ultimate energy source for most forms of life on Earth?

2. Define *metabolism*, and explain how reactions can be coupled to one another.

3. What is activation energy? How do catalysts affect activation energy? How does this change the rate of reactions?

4. Describe some exergonic and endergonic reactions that occur in plants and animals very regularly.

5. Describe the structure and function of enzymes. How is enzyme activity regulated?

Applying the Concepts

1. A preview question for ecology: When a brown bear eats a salmon, does the bear acquire all the energy contained in the body of the fish? Why or why not? What implications do you think this answer would have for the relative abundance (by weight) of predators and their prey? Does the second law of thermodynamics help explain the title of the book *Why Big Fierce Animals Are Rare*?
2. Many people in sub-Saharan Africa have experienced the effects of malnutrition and starvation, but the very young are most severely affected. Some individuals suffer permanent disability even if food is provided. How could a lack of food intake interfere with functions of individual cells and tissues? Which tissues are likely to suffer the most irreversible damage?
3. As you learned in Chapter 3, the subunits of virtually all organic molecules are joined by condensation reactions and can be broken apart by hydrolysis reactions. Why, then, does your digestive system produce separate enzymes to digest proteins, fats, and carbohydrates—in fact, several of each type?
4. Suppose someone tried to disprove the existence of evolution with the following argument: "According to evolutionary theory, organisms have increased in complexity through time. However, evolution of increased biological complexity contradicts the second law of thermodynamics. Therefore, evolution is impossible." How would you respond to this argument in support of evolution?

For More Information

Baker, J. J. W., and Allen, G. E. *Matter, Energy, and Life*. 4th ed. Reading, MA: Addison-Wesley, 1981. An excellent introduction to chemical-energy principles for those interested in biology.

Farid, R. S. "Enzymes Heat Up." *Science News*, May 9, 1998. Scientists explore new ways to synthesize enzymes that will function at high temperatures.

Fenn, J. *Engines, Energy, and Entropy*. New York: W. H. Freeman, 1982. Elegantly simple introduction to the laws of thermodynamics and their relationship to everyday life.

Answers to Multiple-Choice Questions

1.d 2.c 3.b 4.a 5.c 6.e

MEDIATUTOR
Energy Flow in the Life of a Cell

CD Activities

Activity 6.1: Energy and Chemical Reactions:

Estimated time: 5 minutes

The cell is a chemical factory; to understand it, we must understand its basic chemistry. This tutorial introduces the concept of energy and explores the basic concepts of thermodynamics.

Activity 6.2: Energy and Life

Estimated time: 10 minutes

The concepts of thermodynamics introduced in the last tutorial are applied to living systems in this tutorial. The central importance of ATP as an energy carrier is detailed, electron carriers are introduced, and the importance of enzymes is described.

Start the MediaTutor Student CD-ROM and enter the activity number in the Quick Search box to be taken directly to that activity.

Web Investigations

Case Study: Energy Unleashed

Estimated time: 15 minutes

While biologists study the sophisticated biochemical processes organisms use to maintain relatively constant energy levels, dieters focus on two simple rules. Take in too few calories, and weight loss, disability, and eventually death will result. Take in too much energy—that is, too many calories—and you gain weight (or mass). But how much is too much? In this exercise we will use online calorie calculators to explore how different factors effect energy needs. Remember that this is only an exercise. The data obtained should not be used as the basis for any dietary or exercise program.

Go to http://www.prenhall.com/audesirk6, the Audesirk Companion Web site. Select Chapter 6 and the Web Investigation to begin.

". . . the cells of life bound themselves into clans, a multitude of cells
To make one being—as the molecules before
Had made of many one cell. Meanwhile they had invented
Chlorophyll and ate sunlight, cradled in peace
On the warm waves."

Robinson Jeffers in* The Beginning and the End *(1963)

As *Tyrannosaurus* threatens *Triceratops*, a giant meteor and its fragments streak toward Earth. Both hunter and hunted will die in this cataclysmic event. Such an event may have driven 70% of all life-forms to extinction 65 million years ago.

7 Capturing Solar Energy: Photosynthesis

AT A GLANCE

Case Study: Did the Dinosaurs Die from Lack of Sunlight?

1) **What Is Photosynthesis?**

Leaves and Chloroplasts Are Adaptations for Photosynthesis

Photosynthesis Consists of Light-Dependent and Light-Independent Reactions

2) **Light-Dependent Reactions: How Is Light Energy Converted to Chemical Energy?**

During Photosynthesis, Light Is First Captured by Pigments in Chloroplasts

The Light-Dependent Reactions Occur in Photosystems Within the Thylakoid Membranes

3) **Light-Independent Reactions: How Is Chemical Energy Stored in Glucose Molecules?**

The C_3 Cycle Captures Carbon Dioxide

Carbon Fixed During the C_3 Cycle Is Used to Synthesize Glucose

4) **What Is the Relationship Between Light-Dependent and Light-Independent Reactions?**

5 **Water, CO_2, and the C_4 Pathway**

When Stomata Are Closed to Conserve Water, Wasteful Photorespiration Occurs

C_4 Plants Reduce Photorespiration by Means of a Two-Stage Carbon Fixation Process

C_3 and C_4 Plants Are Each Adapted to Different Environmental Conditions

Case Study Revisited: Did the Dinosaurs Die from Lack of Sunlight?

CASESTUDY

Did the Dinosaurs Die from Lack of Sunlight?

It is summer and the year is 65,000,000 B.C. The Cretaceous period is about to reach an abrupt and catastrophic end. On an Earth where the continents we know as the Americas are largely submerged by shallow seas, a 20-foot *Triceratops* grazes in lush, tropical vegetation in what is now southern California. From the cover of dense forest nearby, a 45-foot, 8-ton *Tyrannosaurus rex* watches and waits. Suddenly, a deafening roar startles the animals, and they gaze upward to see a fireball obliterating the blue sky. A meteorite 6 miles in diameter has entered the atmosphere and is about to irrevocably alter life on Earth. Although any creatures that witnessed the event were immediately incinerated by the blast wave from the impact, plants and animals all over the world would experience the aftereffects. As it plowed into the ocean at the tip of the Yucatan Peninsula, the meteorite dug a crater a mile deep and 120 miles wide. The force of its impact sent trillions of tons of debris from Earth's crust and from the meteorite itself rocketing into the atmosphere. The heat of the blast would have caused fires that may have charred 25% of all the vegetation on land. Ashes, smoke, and dust obliterated the sun. Earth was plunged into a night that lasted for months. What would happen today if the sun were obscured for months on end? Did a meteorite really destroy the dinosaurs? ■

1 What Is Photosynthesis?

At least 2 billion years ago, some cells, through chance mistakes (*mutations*) in their genetic machinery, acquired the ability to harness the energy in sunlight. These cells combined simple inorganic molecules—carbon dioxide and water—into more complex organic molecules such as glucose. In the process of *photosynthesis*, these cells captured a small fraction of the sunlight's energy, storing it as chemical energy in these complex organic molecules. Exploiting this new source of energy without competition, early photosynthetic cells filled the seas, releasing oxygen as a by-product. A new element in the atmosphere, free oxygen was harmful to many organisms. But the endless variation produced by random genetic mistakes eventually gave rise to some cells that could survive in oxygen and, later, to cells that made use of oxygen to break down glucose in a new, more efficient process: *cellular respiration*. Today, most forms of life on Earth—including you—depend on the sugar produced by photosynthetic organisms as an energy source and release the energy from that sugar through cellular respiration, using the photosynthetic by-product oxygen (Fig. 7-1). Sunlight powers life on Earth and is captured only by photosynthesis, which we explore in this chapter. In Chapter 8 we examine the processes by which all living things break down the sugary energy-storage molecules produced by photosynthesis and reclaim the energy for powering other metabolic reactions.

Starting with the simple molecules of carbon dioxide (CO_2) and water, **photosynthesis** converts the energy of sunlight into chemical energy stored in the bonds of glucose ($C_6H_{12}O_6$) and releases oxygen. The simplest overall chemical reaction for photosynthesis is:

$$6\,CO_2 + 6\,H_2O + \text{light energy} \rightarrow C_6H_{12}O_6 + 6\,O_2$$

Photosynthesis occurs in eukaryotic plants and algae and certain types of prokaryotes, or bacteria, all of which are *autotrophs* (literally, "self-feeders"). In this chapter we will restrict our discussion of photosynthesis to plants, particularly land plants. Photosynthesis in plants takes place within chloroplasts, most of which are contained in leaf cells. Let us begin, then, with a brief look at the structures of leaves and chloroplasts. Chapter 23 examines leaf structure and function more thoroughly.

Leaves and Chloroplasts Are Adaptations for Photosynthesis

The leaves of most land plants are only a few cells thick; their structure is elegantly adapted to the demands of photosynthesis (Fig. 7-2). The flattened shape of leaves exposes a large surface area to the sun, and their thinness ensures that sunlight can penetrate them, reaching the light-trapping chloroplasts inside. Both the upper and lower surfaces of a leaf consist of a layer of transparent cells, the *epidermis*. The outer surface of both epidermal layers is covered by the *cuticle*, a waxy, waterproof covering that reduces the evaporation of water from the leaf. A leaf obtains CO_2 for photosynthesis from the air; adjustable pores in the epidermis, called **stomata** (singular, **stoma**; Greek for "mouth"), open and

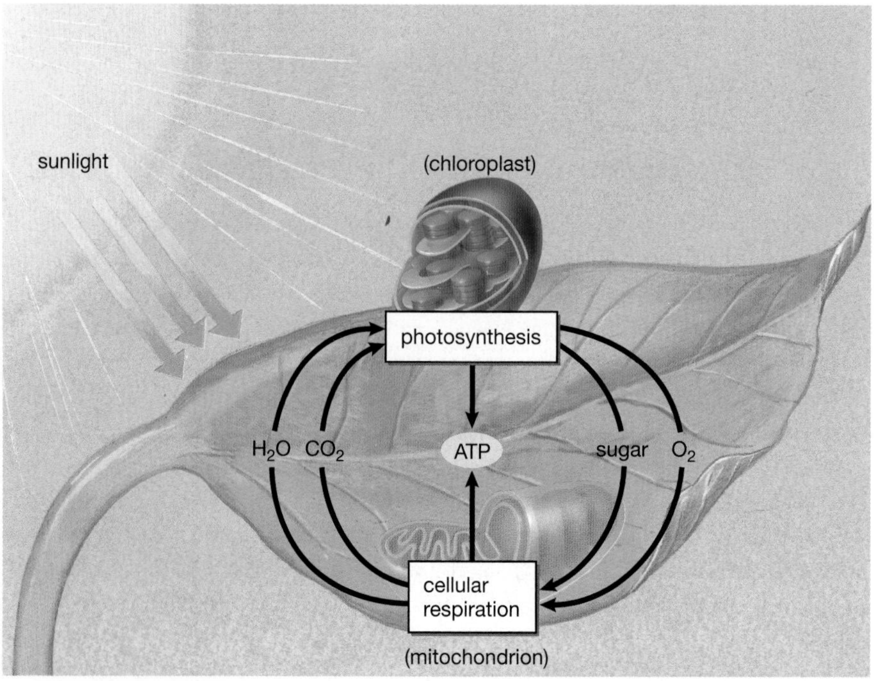

Figure 7-1 Interconnections between photosynthesis and cellular respiration Chloroplasts in green plants use the energy of sunlight to synthesize high-energy carbon compounds such as glucose from low-energy molecules of water and carbon dioxide. Plants themselves, and other organisms that eat plants or one another, extract energy from these organic molecules by cellular respiration, yielding water and carbon dioxide once again. This energy in turn drives all the reactions of life.

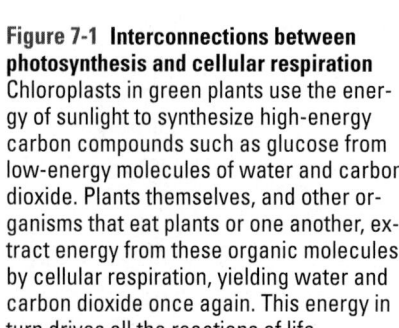

MediaTutor 7.1 Photosynthesis

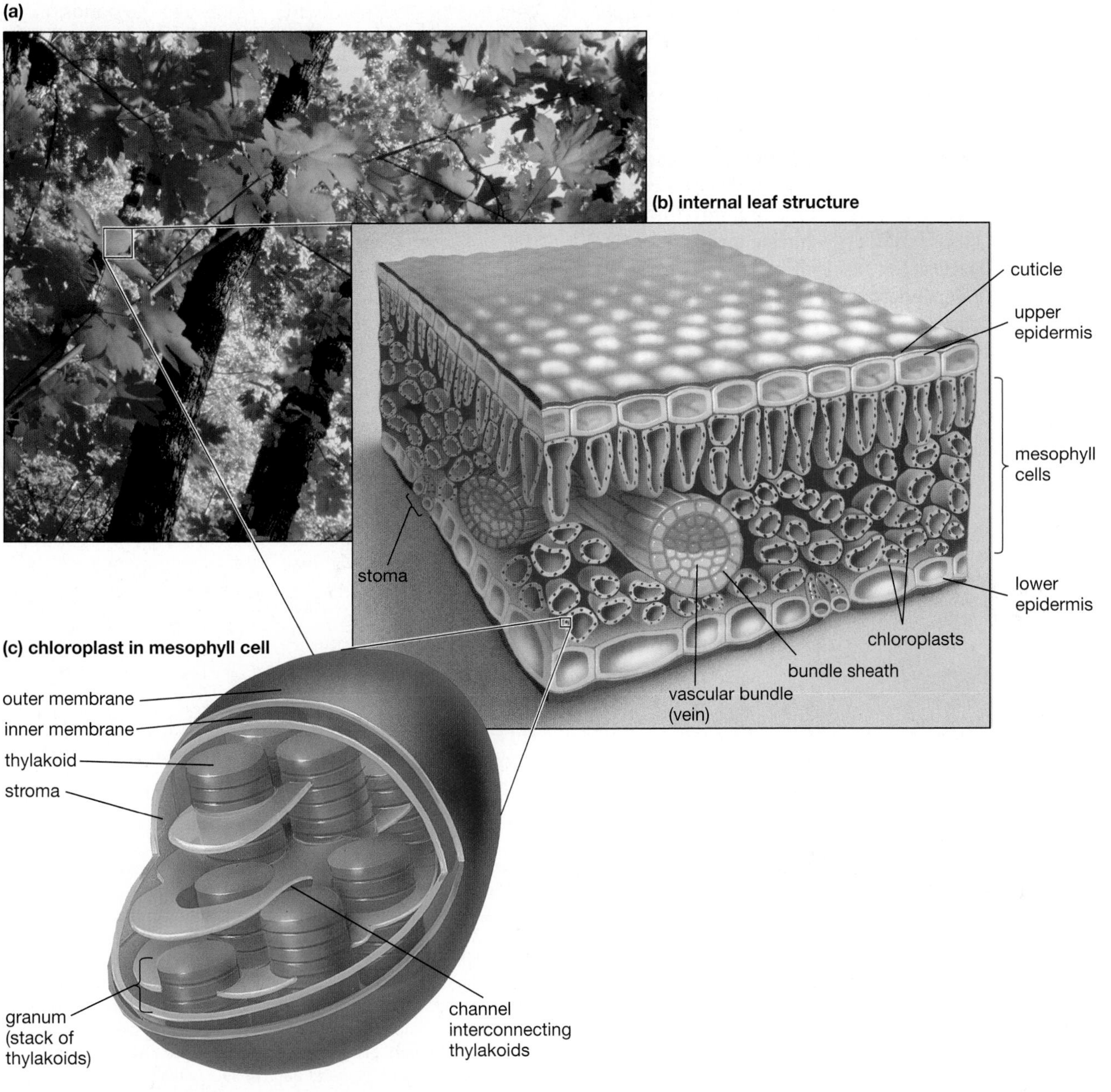

Figure 7-2 An overview of photosynthetic structures
(a) Photosynthesis occurs in chloroplasts, which are located primarily in the leaves of land plants. ***(b)*** A section of a leaf is shown ***(c)*** with a single chloroplast isolated and enlarged.

close at appropriate times to admit CO_2. Inside the leaf are a few layers of cells collectively called **mesophyll** (which means "middle of the leaf"). The mesophyll cells contain the vast majority of a leaf's chloroplasts (see Fig. 7-2), and consequently photosynthesis occurs principally in these cells. *Vascular bundles*, or veins, supply water and minerals to the mesophyll cells and carry the sugars they produce to other parts of the plant.

As we discussed in Chapter 5, chloroplasts are organelles that consist of a double outer membrane enclosing a semifluid medium, the **stroma** (see Fig. 7-2). Embedded in the stroma are disk-shaped, interconnected membranous sacs called **thylakoids**. In most chloroplasts the thylakoids are piled atop one another in stacks called **grana** (singular, **granum**). The light-dependent reactions of photosynthesis occur within the membranes of the thylakoids.

Photosynthesis Consists of Light-Dependent and Light-Independent Reactions

The simple summary of the overall reaction of photosynthesis obscures the fact that photosynthesis actually involves dozens of enzymes catalyzing dozens of individual reactions. The reactions can be divided into (1) *light-dependent reactions* and (2) *light-independent reactions*. Each reaction occurs in a different site in the chloroplast, but the two reactions are coupled by energy-carrier molecules.

1. In the **light-dependent reactions**, chlorophyll and other molecules in the membranes of the thylakoids capture sunlight energy and convert some of it into the chemical energy stored in energy-carrier molecules (ATP and NADPH). Oxygen gas is released as a by-product.
2. In the **light-independent reactions**, enzymes in the stroma use the chemical energy of the carrier molecules to drive the synthesis of glucose or other organic molecules.

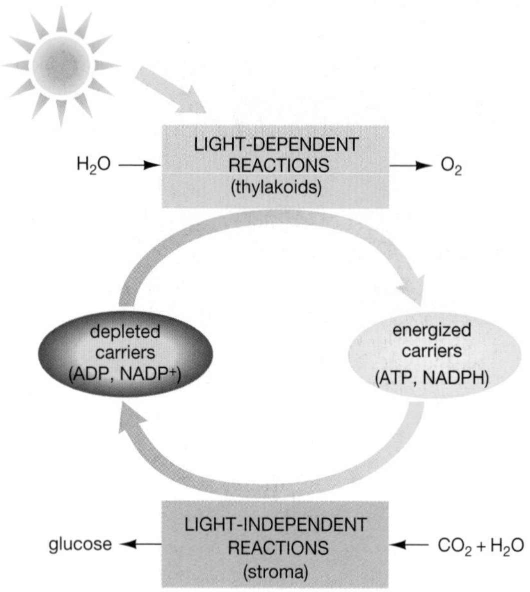

2 Light-Dependent Reactions: How Is Light Energy Converted to Chemical Energy?

The light-dependent reactions convert the energy of sunlight into the chemical energy of two different carrier molecules: the familiar energy carrier ATP and the electron carrier *NADPH* (*nicotinamide adenine dinucleotide phosphate*). These molecules will then be used to power the synthesis of high-energy storage molecules, such as glucose, during the light-independent reactions.

During Photosynthesis, Light Is First Captured by Pigments in Chloroplasts

The sun emits energy in a broad spectrum of electromagnetic radiation. The *electromagnetic spectrum* ranges from short-wavelength gamma rays, through ultraviolet, visible, and infrared light, to very long wavelength radio waves (Fig. 7-3a). Light and the other types of radiation are composed of individual packets of energy called **photons**. The energy of a photon corresponds

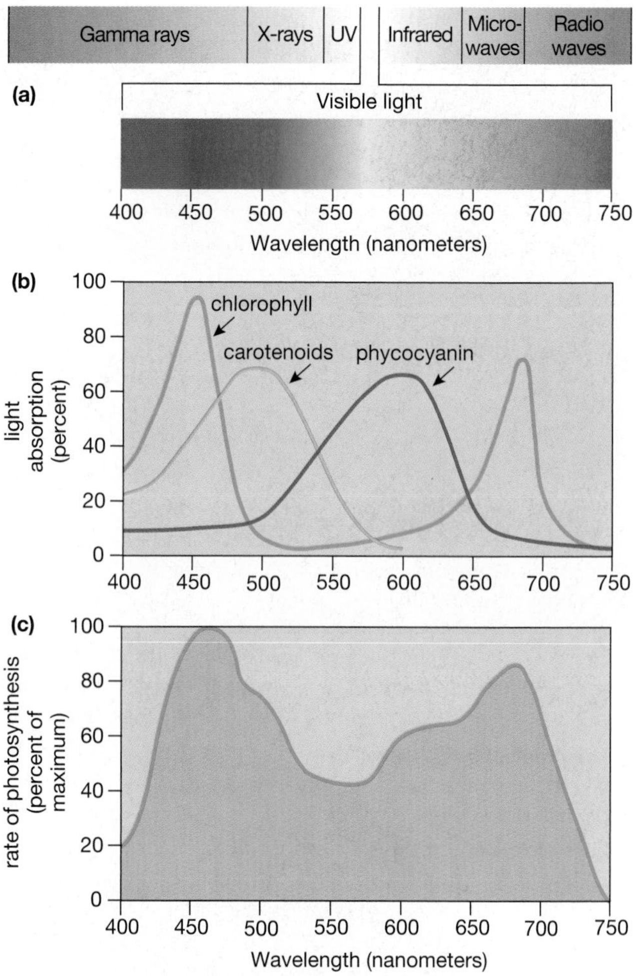

Figure 7-3 Light, chloroplast pigments, and photosynthesis ***(a)*** Visible light, a small part of the electromagnetic spectrum (top line), consists of wavelengths that correspond to the colors of the rainbow. ***(b)*** Different pigments selectively absorb certain colors; the height of the curves represents the ability of each pigment to absorb light of each color. Chlorophyll (green curve) strongly absorbs violet, blue, and red light. The other pigments in chloroplasts absorb other colors of light. ***(c)*** Photosynthesis is driven to some extent by all colors of light, because collectively, the pigments in chloroplasts absorb some of each visible wavelength.

to its wavelength: Short-wavelength photons are very energetic, whereas longer-wavelength photons have lower energies. Visible light consists of wavelengths with energies that are strong enough to alter the shape of certain pigment molecules but weak enough not to damage crucial molecules such as DNA. It is no coincidence that these wavelengths, with "just the right amount" of energy, also stimulate the pigments in our eyes and allow us to see the world around us.

One of three processes occurs when light strikes an object such as a leaf: The light is either (1) *absorbed*, (2) *reflected* (bounced back again), or (3) *transmitted* (passed through). Light that is absorbed can heat up the object or drive biological processes, such as photosynthesis. Light that is reflected or transmitted reaches our eyes and gives an object its color.

Chloroplasts contain several types of molecules that absorb different wavelengths of light. **Chlorophyll**, the key light-capturing molecule in thylakoid membranes, strongly absorbs violet, blue, and red light but reflects green. Since green wavelengths reach our eyes, we see chlorophyll as green (Fig. 7-3b). Most leaves appear green because they are rich in chlorophyll. Thylakoids also contain other molecules, called **accessory pigments**, that capture light energy and transfer it to chlorophyll. **Carotenoids** absorb blue and green light and reflect mostly yellow, orange, or red, which we see; **phycocyanins** absorb green and yellow and thus we see the blue or purple wavelengths they reflect. Because all wavelengths of light are absorbed to some degree by either chlorophyll, carotenoids, or phycocyanins, all wavelengths can drive photosynthesis to some extent in plants with those pigments (Fig. 7-3c).

The Light-Dependent Reactions Occur in Photosystems Within the Thylakoid Membranes

The thylakoid membranes contain highly organized assemblies of proteins, chlorophyll, accessory pigment molecules, and electron-carrier molecules. These assemblies are called **photosystems.** Each thylakoid contains thousands of copies of two kinds of photosystems, named *photosystem I* and *photosystem II.* Each photosystem consists of two major parts: (1) a *light-harvesting complex* and (2) an *electron transport system.* Each **light-harvesting complex** contains about 300 chlorophyll and accessory pigment molecules. These molecules absorb light and pass the energy to a specific chlorophyll molecule called the **reaction center.** As an analogy with television reception, the light-absorbing pigments are called *antenna molecules*, because they gather energy and transfer it to the energy-processing reaction center. The reaction-center chlorophyll is located next to the second part of the photosystem: the **electron transport system**, a series of electron carrier molecules also embedded in the thylakoid membrane.

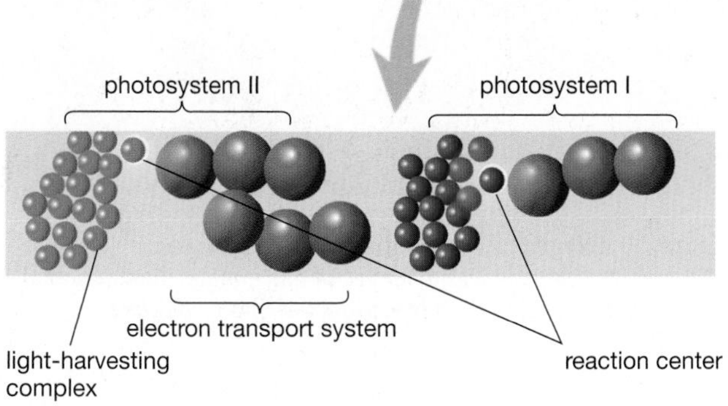

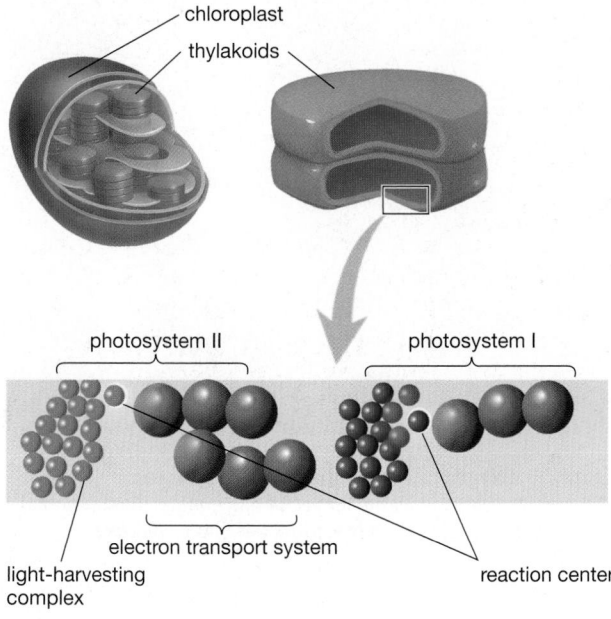

When the reaction-center chlorophyll receives energy from the antenna molecules, one of the reaction center electrons absorbs the energy, leaves the chlorophyll, and jumps over to the electron transport system. This energetic electron moves from one carrier to the next. At some of the transfers, the electron releases energy. The energy drives reactions that result in the synthesis of ATP from ADP plus phosphate or NADPH from $NADP^+$. (*NADP* is the electron carrier NAD, described in Chapter 6, plus a phosphate group.) With this overall scheme in mind, let's look more closely at the actual sequence of events in the light-dependent reactions (Fig. 7-4).

Photosystem II Generates ATP

For historical reasons, the photosystems are numbered "backward": The usual process of light-energy capture is most easily understood by starting with photosystem II and following the events caused by the capture of two photons of light. The light-dependent reactions begin when photons of light are absorbed by the light-harvesting complex of photosystem II (step ① in Fig. 7-4). Each photon's energy passes from molecule to molecule until it reaches the reaction center, where it boosts an electron out of the chlorophyll molecule (step ②). The first electron carrier of the adjacent electron transport system instantly accepts these energized electrons (step ③). The electrons move from one carrier molecule to the next, releasing energy as they go. Some of the energy is used to pump hydrogen ions (H^+) across the thylakoid membrane, generating a H^+ ion gradient within the thylakoid. This gradient drives the synthesis of ATP by a process known as **chemiosmosis** (step ④). "A Closer Look: Chemiosmosis—ATP Synthesis in Chloroplasts" describes chemiosmosis in more detail.

Photosystem I Generates NADPH

Meanwhile, light rays have also been striking the light-harvesting complex of photosystem I (step ⑤ in Fig. 7-4). Each photon striking photosystem I ejects an electron from its reaction-center chlorophyll (step ⑥). These electrons jump to the electron transport system of photosystem I (step ⑦). Photosystem I's reaction-center chlorophyll immediately obtains replacements for its lost electrons from the last electron carrier in photosystem II's electron transport system. Photosystem I's high-energy electrons move through its electron transport system to the electron carrier $NADP^+$. Each $NADP^+$ molecule picks up two energetic electrons and one hydrogen ion, forming NADPH (step ⑧). Both $NADP^+$ and NADPH are water-soluble molecules dissolved in the chloroplast stroma.

Splitting Water Maintains the Flow of Electrons Through the Photosystems

Overall, electrons flow from the reaction center of photosystem II, through the photosystem II electron transport system, to the reaction center of photosystem I,

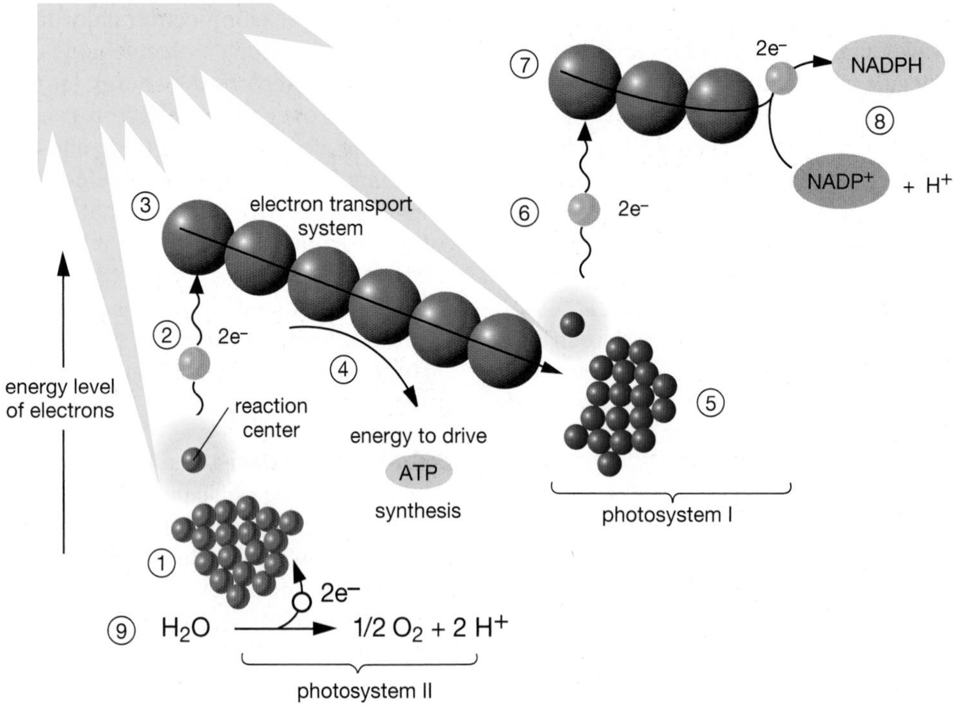

Figure 7-4 Thylakoid structure and the light-dependent reactions of photosynthesis
A summary of the light-dependent reactions. ① Light is absorbed by the light-harvesting complex of photosystem II, and the energy is passed to the reaction-center chlorophyll molecule. ② This energy ejects electrons out of the reaction center. ③ The electrons pass to the adjacent electron transport system. ④ The transport system passes the energetic electrons along, and some of their energy is used to pump hydrogen ions into the thylakoid interior. The hydrogen ion gradient thus generated can drive ATP synthesis. ⑤ Light strikes photosystem I, ⑥ causing it to emit electrons. ⑦ The electrons are captured by the photosystem I electron transport system. The electrons lost from the reaction center of photosystem I are replaced by those from the transport system of photosystem II. ⑧ The energetic electrons from photosystem I are captured in molecules of NADPH. ⑨ The electrons lost from the reaction center of photosystem II are replaced by electrons obtained from splitting water. This reaction also releases oxygen.

through the photosystem I electron transport system, and on to form NADPH. To sustain this one-way flow of electrons, photosystem II's reaction center must be continuously supplied with new electrons to replace the ones it gives up. These replacement electrons come from water (step ⑨ in Fig. 7-4). In a series of reactions that scientists are still deciphering, photosystem II's reaction-center chlorophyll attracts electrons from water molecules within the thylakoid compartment, causing the water molecules to split apart:

$$H_2O \rightarrow \frac{1}{2}\,O_2 + 2\,H^+ + 2\,e^-$$

For every two photons captured by photosystem II, two electrons are boosted out of the reaction-center chlorophyll and are replaced by the two electrons obtained by splitting one water molecule. As water molecules are split, their oxygen atoms combine to form molecules of oxygen gas, O_2. The oxygen may be used directly by the plant in its own respiration or given off to the atmosphere (Fig. 7-5).

Figure 7-5 Oxygen is a by-product of photosynthesis
The bubbles released by the leaves of this aquatic plant (*Elodea*) are composed of oxygen, a by-product of photosynthesis.

A Closer Look

Chemiosmosis—ATP Synthesis in Chloroplasts

In the electron transport system of photosystem II, energetic electrons move from carrier to carrier. Until about 1960, it was thought that ATP synthesis was directly coupled to these electron transfers: Where the exergonic steps (reactions in which there is a net release of energy) were large enough, the energy given up by the electrons drove ATP synthesis. However, it turns out that ATP synthesis in chloroplasts is *not* a simple coupled reaction. The electron transfers do not directly drive ATP synthesis; rather, the energy released during the transfers is used to generate a concentration gradient of hydrogen ions across the thylakoid membrane. In a separate reaction, the energy stored in this gradient then powers ATP synthesis. Let us follow these reactions.

In the first step of the light-dependent reactions of photosynthesis, a photon strikes the light-harvesting complex of photosystem II; the energy is absorbed and passed to the reaction-center chlorophyll. There, an electron absorbs the energy and is ejected out of the chlorophyll molecule. Within a billionth of a second, the electron is captured by the first electron carrier of the adjacent electron transport system.

As the electron moves from one carrier molecule to the next, it loses energy at each transfer. The energy released from the exergonic reaction of electron movement is used to transport hydrogen ions actively across the thylakoid membrane from the stroma into the thylakoid interior:

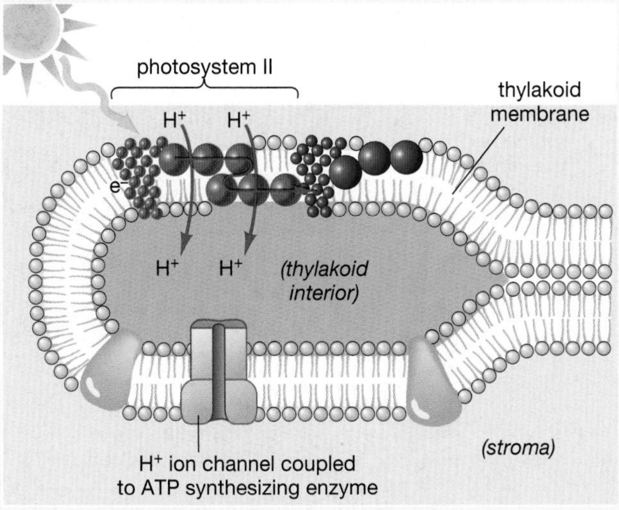

This transport raises the concentration of hydrogen ions (and therefore also the positive charge) inside the thylakoid, creating a gradient of both hydrogen ions and positive charge across the thylakoid membrane. Using energy to create a gradient of charge and ion concentration is very much like charging a rechargeable battery. The thylakoid membrane does not allow hydrogen ions to leak out, except at specific protein channels that are coupled to ATP-synthesizing enzymes. When hydrogen ions flow through these channels, down their gradients of charge and concentration, the energy released drives the synthesis of ATP:

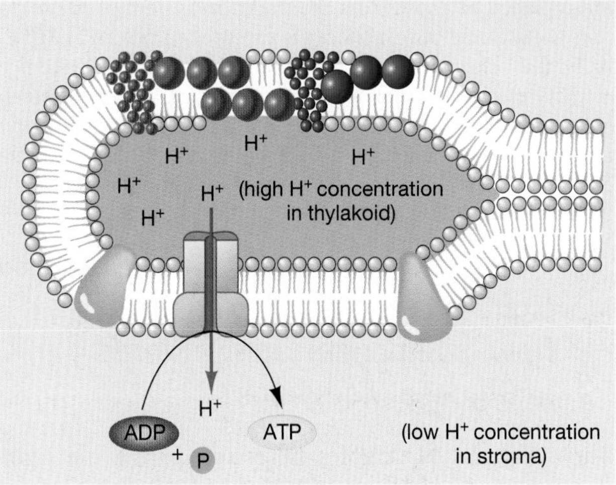

How does a gradient of hydrogen ions generate energy to drive ATP synthesis? Compare the hydrogen ion gradient across the thylakoid membrane to water stored behind a dam at a hydroelectric plant. When gates within the dam are opened, the rapid flow of water is directed through turbines that are turned by the rushing water and convert this energy of motion into electrical energy—they generate electricity. The thylakoid operates in an analogous way. Hydrogen ions in the thylakoid interior can move down their gradients into the stroma only through the ATP-synthesizing hydrogen ion channels. Something like flowing water, this flow of hydrogen ions can do work—namely, drive ATP synthesis from ADP + phosphate. Apparently, about one ATP molecule is synthesized for every three protons that pass through the channel.

Scientists are still investigating the exact way that the ATP-synthesizing proton channel works. However, this general mechanism of ATP synthesis was first proposed in 1961 by British biochemist Peter Mitchell, who called it *chemiosmosis*. Chemiosmosis has been shown to be the mechanism of ATP generation in chloroplasts, mitochondria (as we shall see in Chapter 8), and bacteria. For his brilliant hypothesis, Mitchell was awarded the Nobel Prize for Chemistry in 1978.

Summing Up
Light-Dependent Reactions

① The light-dependent reactions begin with the absorption of light by the light-harvesting complex of photosystem II. ② The light energy energizes electrons from the reaction center of the complex, causing them to be ejected. ③ The electrons are transferred to photosystem II's electron transport system. As the electrons pass through the transport system, they release energy. ④ Some of the energy is used to create a hydrogen ion gradient that drives ATP synthesis. ⑤ Meanwhile, light is absorbed by the light-harvesting complex of photosystem I. ⑥ The light energy ejects electrons from the reaction center that ⑦ are picked up by photosystem I's electron transport system. Electrons lost from the reaction center are replaced by those from the transport system of photosystem II. ⑧ Some of this energy is captured as NADPH. ⑨ Finally, the "electron-deprived" chlorophyll of photosystem II attracts electrons from water molecules. A water molecule splits, donating electrons to the photosystem II chlorophyll and generating oxygen as a by-product. The products of the light-dependent reactions are NADPH, ATP, and O_2.

3 Light-Independent Reactions: How Is Chemical Energy Stored in Glucose Molecules?

The ATP and NADPH synthesized during the light-dependent reactions are dissolved in the fluid stroma that surrounds the thylakoids. There, they provide the energy to power the synthesis of glucose from carbon dioxide and water, a process that requires enzymes that are also dissolved in the stroma. The reactions that eventually produce glucose are called the light-independent reactions because they can occur independently of light as long as ATP and NADPH are available.

The C_3 Cycle Captures Carbon Dioxide

Carbon dioxide capture occurs in a set of reactions known as the **Calvin-Benson cycle** (after its discoverers), or the **C_3 cycle** (three-carbon cycle), because some of the important molecules in the cycle have three carbon atoms in them (Fig. 7-6). The C_3 cycle requires 1) CO_2 (normally from the air); 2) a CO_2-capturing sugar, *ribulose bisphosphate* (RuBP); 3) enzymes to catalyze all the reactions; and 4) energy in the form of ATP and NADPH (normally from the light-dependent reactions).

The C_3 cycle is best understood if we mentally divide it into ① carbon fixation, ② the synthesis of *glyceraldehyde-3-phosphate* (G3P), and ③ the regeneration of RuBP. Keep track of the carbons as you follow Figure 7-6.

1. *Carbon fixation*. The C_3 cycle begins (and ends) with RuBP, a five-carbon sugar. Enzymes combine six RuBP molecules with CO_2 from the atmosphere to form six molecules of an extremely unstable six-carbon compound. This compound spontaneously reacts with water to form 12 three-carbon molecules of *phosphoglyceric acid* (PGA), whose three carbons give the C_3 cycle its name. The capturing of CO_2 is called **carbon fixation**, because it "fixes" gaseous CO_2 into a relatively stable organic molecule, PGA (step ① in Fig. 7-6).
2. *Synthesis of G3P*. In a series of enzyme-catalyzed reactions, energy donated by ATP and NADPH (generated during the light-dependent reactions) is used to convert PGA to G3P (step ②).
3. *Regeneration of RuBP*. Through complex reactions requiring ATP energy, 10 of the 12 molecules of G3P (10×3 carbons) regenerate the six molecules of RuBP (6×5 carbons; step ③) used at the start of carbon fixation. The remaining two molecules of G3P will be used to synthesize glucose.

Carbon Fixed During the C_3 Cycle Is Used to Synthesize Glucose

If the C_3 cycle starts with RuBP, adds carbon from CO_2, and ends one "cycle" with RuBP again, there is carbon left over from the captured CO_2. Using the simplest

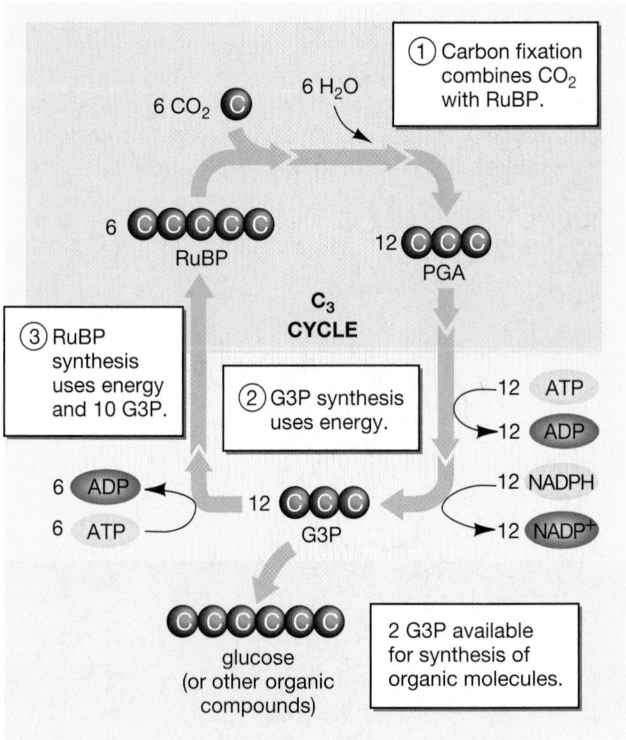

Figure 7-6 The C_3 cycle of carbon fixation
① Six molecules of RuBP react with 6 molecules of CO_2 and 6 molecules of H_2O to form 12 molecules of PGA. This reaction is carbon fixation, the capture of carbon from CO_2 into organic molecules. ② The energy of 12 ATPs and the electrons and hydrogens of 12 NADPHs are used to convert the 12 PGA molecules to 12 G3Ps. ③ Energy from 6 ATPs is used to rearrange 10 G3Ps into 6 RuBPs, completing one turn of the C_3 cycle. The remaining 2 G3P molecules are further processed into glucose or other organic molecules such as glycerol, fatty acids, or the carbon skeleton of amino acids, depending on the needs of the plant.

"carbon accounting" numbers shown in Figure 7-6, if you start and end one passage through the cycle with six molecules of RuBP, two molecules of G3P are left over. These two G3P molecules (three carbons each) are combined to form one molecule of glucose (six carbons). Later, glucose may be broken down during cellular respiration or linked together in chains to form starch (a storage molecule) or cellulose (a major component of cell walls), or modified further into amino acids, lipids, and other cellular constituents.

Summing Up
Light-Independent Reactions

For the synthesis of one molecule of glucose via one passage through the C_3 cycle: Six molecules of CO_2 are captured by six molecules of RuBP. A series of reactions driven by energy from ATP and NADPH produces 12 molecules of G3P. Two molecules join to become one molecule of glucose. Further ATP energy is used to regenerate the 6 RuBP molecules from 10 G3P molecules. Light-independent reactions generate glucose and depleted energy carriers (ADP and $NADP^+$) that will be recharged during light-dependent reactions.

4 What is the Relationship Between Light-Dependent and Light-Independent Reactions?

At this point, it would be useful for you to go back to page 118 and review the summary and in-text figure of the relationship between light-dependent and light-independent reactions. Figure 7-7 illustrates these in the context of where each occurs in the chloroplast. Each illustration stresses the interdependence of these two sets of reactions in the overall process of photosynthesis. Stated simply, the "photo" part of photosynthesis refers to the capture of light energy by the light-dependent reactions. The "synthesis" part of photosynthesis refers to the synthesis of glucose that occurs during the light-independent reactions using the energy captured by the light-dependent reactions. Stated in more detail, the light-dependent reactions in the membranes of the thylakoids use light energy to "charge up" the energy-carrier molecules ADP and $NADP^+$ to form ATP and NADPH. These energized carriers move to the stroma, where their energy is used to drive glucose synthesis by means of the light-independent reactions, catalyzed by enzymes dissolved in the stroma. The depleted carriers, ADP and $NADP^+$, then return to the light-dependent reactions for recharging to ATP and NADPH.

5 Water, CO_2, and the C_4 Pathway

Photosynthesis requires light and carbon dioxide. Therefore, you might think that an ideal leaf should have a large surface area to intercept lots of sunlight and be very porous to allow lots of CO_2 to enter the leaf from the air. For land plants, however, being porous to CO_2 also allows water to evaporate easily from the leaf. Loss of water from leaves is a prime cause of stress to land plants and may even be fatal.

Many plants have evolved leaves that are a compromise between obtaining adequate CO_2 supplies and reducing water loss: large, waterproof leaves with adjustable pores, the stomata, that readily admit carbon dioxide (see Fig. 7-2; Chapter 23 describes more details

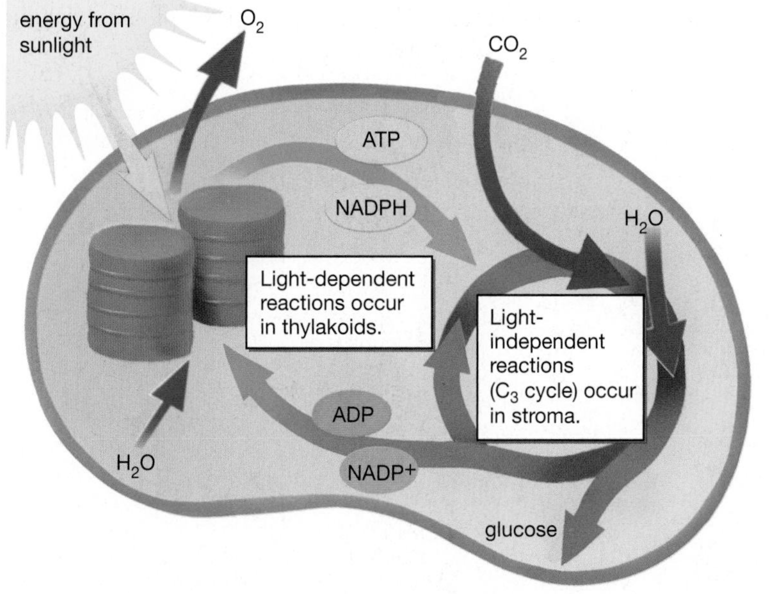

Figure 7-7 A summary diagram of photosynthesis The light-dependent reactions in the thylakoids convert the energy of sunlight into the chemical energy of ATP and NADPH. Part of the sunlight energy is also used to split H_2O, forming O_2. In the stroma, the light-independent reactions (C_3 cycle) use the energy of ATP and NADPH to convert CO_2 and H_2O to glucose. The depleted carriers, ADP and $NADP^+$, return to the thylakoids to be recharged by the light-dependent reactions.

of leaf structure). When water supplies are adequate, the stomata open, letting in CO_2. If the plant is in danger of drying out, the stomata close. Closing the stomata reduces evaporation but has two disadvantages: It reduces CO_2 intake and restricts the release of O_2, which is produced during photosynthesis.

When Stomata Are Closed to Conserve Water, Wasteful Photorespiration Occurs

What happens to carbon fixation when the stomata close, CO_2 levels drop, and O_2 levels rise? Unfortunately, the enzyme that catalyzes the reaction of RuBP with CO_2 is not very selective: It can cause *either* CO_2 *or* O_2 to combine with RuBP (Fig. 7-8a). When O_2, rather than CO_2, is combined with RuBP, a wasteful process occurs called **photorespiration**. During photorespiration (as in cellular respiration), O_2 is used up and CO_2 is generated. But unlike cellular respiration, photorespiration does not produce any useful cellular energy, and it prevents the C_3 pathway from synthesizing glucose. Thus, photorespiration undermines the plant's ability to fix carbon.

Some photorespiration occurs all the time, even under the best of conditions. Recent research suggests that this process, in moderation, actually helps protect C_3 plants (plants that utilize the C_3 pathway) from the damaging effects of too much light. During hot, dry weather, the stomata seldom open; CO_2 from the air can't get in, and the O_2 generated by photosynthesis can't get out. With CO_2 levels low and O_2 levels very high, photorespiration dominates (see Fig. 7-8a). Plants, especially seedlings, may die during hot, dry weather because they are unable to capture enough energy in glucose to meet their metabolic needs.

C_4 Plants Reduce Photorespiration by Means of a Two-Stage Carbon-Fixation Process

Some plants, described as *C_4 plants*, have evolved a way to reduce photorespiration and boost photosynthesis during dry weather. In the leaves of "regular" (C_3) plants, almost all the chloroplasts reside in the mesophyll cells. In plants such as corn and crabgrass, which thrive in relatively hot and dry conditions, both mesophyll cells and **bundle-sheath cells** (which surround the vascular bundles) contain chloroplasts (Fig. 7-8b). These plants use a two-stage carbon-fixation pathway, the **C_4 pathway**.

In C_4 plants, the mesophyll cells contain a three-carbon molecule called *phosphoenolpyruvate* (PEP) instead of RuBP. The CO_2 reacts with PEP to form the four-carbon molecule *oxaloacetate* (for which C_4 plants are named). This reaction is highly specific for CO_2 and is not hindered by high O_2 concentrations. Oxaloacetate is used as a shuttle to transport carbon from mesophyll cells to bundle-sheath cells. There it breaks down, releasing CO_2 again. The high CO_2 concentration created in the bundle-sheath cells now allows the regular C_3 cycle to proceed with less competition from oxygen. The remnant of the shuttle molecule returns to the mesophyll cells, where ATP energy is used to regenerate the PEP molecule.

C_3 and C_4 Plants Are Each Adapted to Different Environmental Conditions

Plants using the C_4 process to fix carbon are locked into this pathway, which uses up more energy to produce glucose than the C_3 pathway. C_4 plants have an advantage when light energy is abundant but water is not. However, if water is plentiful (so the stomata of C_3 plants can stay open and let in lots of CO_2) or if light levels are low, the C_3 carbon fixation pathway is more efficient.

Consequently, C_4 plants thrive in deserts and in midsummer in temperate climates, when light energy is plentiful but water is scarce. The C_3 plants have the advantage, however, in cool, wet, cloudy climates, because the C_3 pathway is more energy-efficient. This is why your lawn of lush Kentucky bluegrass (a C_3 plant) may be taken over by spiky crabgrass (a C_4 plant) during a long, hot, dry summer.

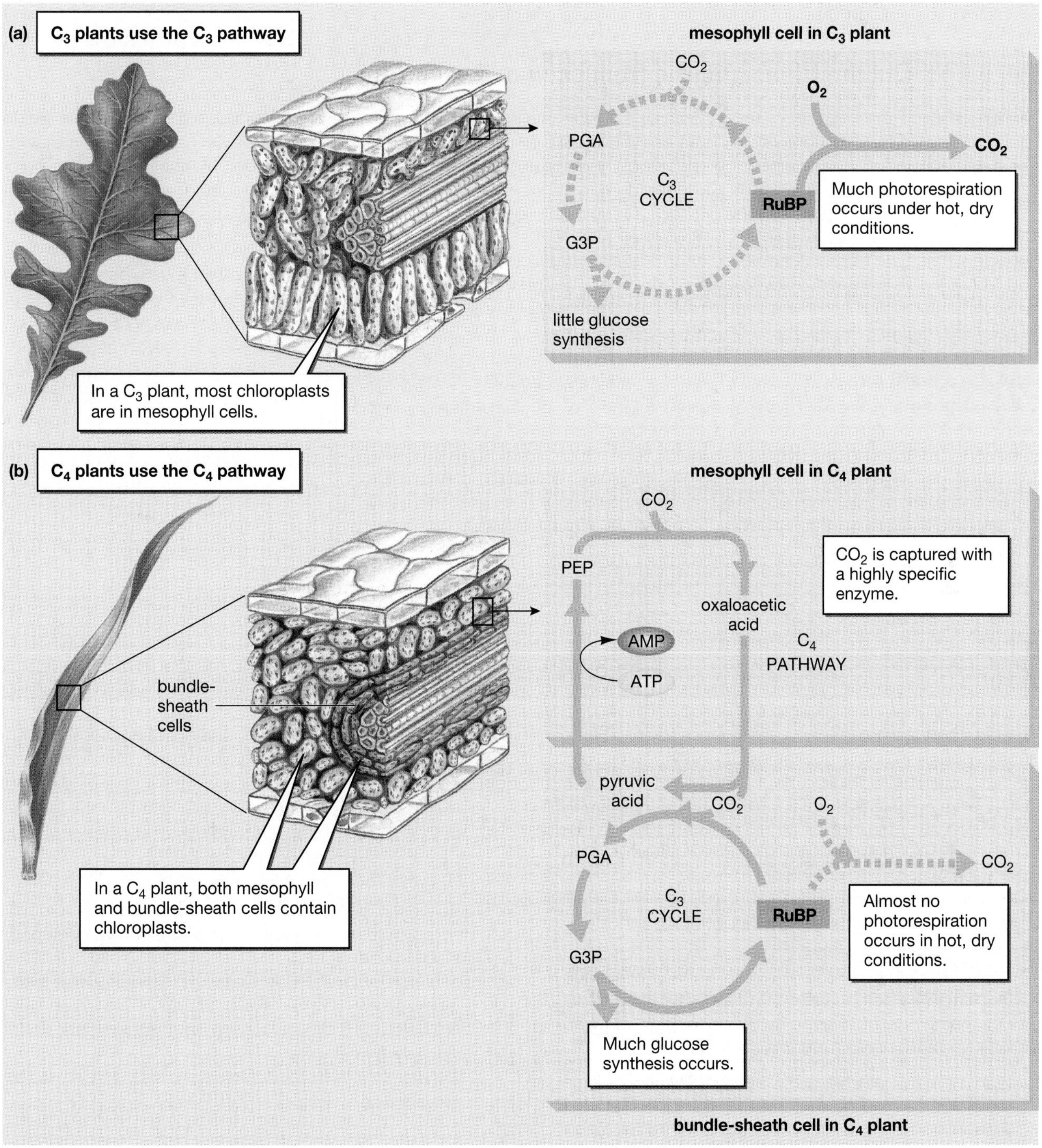
(a) C_3 plants use the C_3 pathway
mesophyll cell in C_3 plant
CO_2
O_2
PGA
CO_2
C_3 CYCLE
RuBP
Much photorespiration occurs under hot, dry conditions.
G3P
little glucose synthesis
In a C_3 plant, most chloroplasts are in mesophyll cells.
(b) C_4 plants use the C_4 pathway
mesophyll cell in C_4 plant
CO_2
PEP
CO_2 is captured with a highly specific enzyme.
oxaloacetic acid
AMP
ATP
C_4 PATHWAY
bundle-sheath cells
pyruvic acid
CO_2
O_2
PGA
CO_2
C_3 CYCLE
RuBP
Almost no photorespiration occurs in hot, dry conditions.
G3P
In a C_4 plant, both mesophyll and bundle-sheath cells contain chloroplasts.
Much glucose synthesis occurs.
bundle-sheath cell in C_4 plant

Figure 7-8 Comparison of C_3 and C_4 plants
(a) In C_3 plants, only the mesophyll cells carry out photosynthesis. All carbon fixation occurs by the C_3 pathway. With low CO_2 and high O_2 levels, photorespiration dominates in C_3 plants, because the enzyme that should catalyze the RuBP + CO_2 reaction catalyzes the RuBP + O_2 reaction instead. ***(b)*** In C_4 plants, both the mesophyll cells and bundle-sheath cells contain chloroplasts and participate in photosynthesis. The initial carbon-fixation step in the mesophyll cells is a reaction between phosphoenolpyruvic acid (PEP) and CO_2, with which O_2 does not compete. A four-carbon molecule of oxaloacetate is produced and releases CO_2 in the bundle-sheath cells, thus maintaining a high CO_2 concentration in their chloroplasts. Higher CO_2 levels allow efficient carbon fixation in the C_3 pathway of the bundle-sheath cells with little photorespiration. Notice that the regeneration of PEP requires energy: Two phosphates are removed from ATP to produce AMP (adenosine monophosphate).

CASESTUDYREVISITED

Did the Dinosaurs Die from Lack of Sunlight?

Paleontologists (scientists who study fossils) have chronicled the extinction of approximately 70% of all living species through the disappearance of their fossils at the end of the Cretaceous period. In sites from around the globe, researchers have found a thin layer of clay deposited around 65 million years ago; the clay has about 30 times the typical levels of a rare element called *iridium*, which is found in high concentrations in some meteorites. The clay also contains soot, such as would have been deposited in the aftermath of massive fires. Did a meteorite wipe out the dinosaurs? Many scientists believe it did. Certainly the evidence of an enormous meteorite impact, dated to 65 million years ago, is clear on the Yucatan Peninsula. Other scientists believe that more gradual climate changes, possibly triggered by intense volcanic activity, produced conditions that would no longer support the enormous reptiles. Volcanoes also spew out soot and ash, and iridium is found in higher levels in Earth's molten mantle than on its surface, so furious volcanic activity could also explain the iridium layer. Either scenario would significantly reduce the amount of sunlight and immediately impact the rate of photosynthesis. Large herbivores (plant-eaters), such as *Triceratops,* which might have needed to consume hundreds of pounds of vegetation daily, would suffer if plant growth slowed significantly. Predators such as *Tyrannosaurus,* which fed on herbivores, would also suffer. In the Cretaceous as now, sunlight captured by photosynthesis powers all the dominant forms of life on Earth—interrupting this vital flow of energy would be catastrophic.

We remain vulnerable to an impact such as might have wiped out much of life at the end of the Cretaceous period. If such an impact were to occur, and if we assume that humanity had 6 months notice, speculate on its effects on civilization. What would be the most constructive measures that humanity could take (from a biological standpoint) to reduce the damage?

Summary of Key Concepts

1) What Is Photosynthesis?

Photosynthesis captures the energy of sunlight to convert the inorganic molecules of carbon dioxide and water into high-energy organic molecules such as glucose. In plants, photosynthesis takes place in the chloroplasts, in two major steps: the light-dependent and the light-independent reactions.

2) Light-Dependent Reactions: How Is Light Energy Converted to Chemical Energy?

The light-dependent reactions occur in the thylakoids. Light excites electrons in chlorophyll molecules and transfers the energetic electrons to electron transport systems. The energy of these electrons drives three processes:

1. *Photosystem II generates ATP.* Some of the energy from the electrons is used to pump hydrogen ions into the thylakoids. The hydrogen ion concentration is therefore higher inside the thylakoids than in the stroma outside. Hydrogen ions move down this concentration gradient through ATP-synthesizing enzymes in the thylakoid membranes, providing the energy to drive ATP synthesis.
2. *Photosystem I generates NADPH.* Some of the energy, in the form of energetic electrons, is added to electron-carrier molecules of $NADP^+$ to make the highly energetic carrier NADPH.
3. *Splitting water maintains the flow of electrons through the photosystems.* Some of the energy is used to split water, generating electrons, hydrogen ions, and oxygen.

3) Light-Independent Reactions: How Is Chemical Energy Stored in Glucose Molecules?

In the stroma of the chloroplasts, both ATP and NADPH provide the energy that drives the synthesis of glucose from CO_2 and H_2O. The light-independent reactions occur in a cycle of chemical reactions called the *Calvin-Benson,* or C_3, cycle. The C_3 cycle has three major parts:

1. *Carbon fixation.* Carbon dioxide and water combine with ribulose bisphosphate (RuBP) to form phosphoglyceric acid (PGA).
2. *Synthesis of G3P.* PGA is converted to glyceraldehyde-3-phosphate (G3P), using energy from ATP and NADPH. G3P may be used to synthesize organic molecules, such as glucose.
3. *Regeneration of RuBP.* Ten molecules of G3P are used to regenerate 6 molecules of RuBP, again using ATP energy.

4) What Is the Relationship Between Light-Dependent and Light-Independent Reactions?

The light-dependent reactions produce the energy carrier ATP and the electron carrier NADPH. Energy from these carriers is used in the synthesis of organic molecules during the light-independent reactions. The depleted carriers, ADP and $NADP^+$, return to the light-dependent reactions for recharging.

5) Water, CO_2, and the C_4 Pathway

The enzyme that catalyzes the reaction between RuBP and CO_2 may also catalyze a reaction, called *photorespiration,* between RuBP and O_2. Photorespiration prevents carbon fixation and does not generate ATP. If CO_2 concentrations drop too low or if O_2 concentrations rise too high, photorespiration may exceed carbon fixation. Some plants have evolved an additional step for carbon fixation that minimizes photorespiration. In the mesophyll cells of these C_4 plants, CO_2 combines with phosphoenolpyruvic acid (PEP) to form the four-carbon molecule oxaloacetate. Oxaloacetate is transported into adjacent bundle-sheath cells, where it releases CO_2, thereby maintaining a high CO_2 concentration in those cells. This CO_2 is then fixed in the C_3 cycle.

Key Terms

accessory pigments *p. 119*
bundle-sheath cell *p. 124*
C_3 cycle *p. 122*
C_4 pathway *p. 124*
Calvin-Benson cycle *p. 122*
carbon fixation *p. 122*
carotenoid *p. 119*
chemiosmosis *p.119*
chlorophyll *p. 119*
electron transport system *p. 119*
granum *p. 117*
light-dependent reactions *p. 118*
light-harvesting complex *p. 119*
light-independent reactions *p. 118*
mesophyll *p. 117*
photon *p. 118*
photorespiration *p. 124*
photosynthesis *p. 116*
photosystems *p. 119*
phycocyanin *p. 119*
reaction center *p. 119*
stoma *p. 116*
stroma *p. 117*
thylakoid *p. 117*

Thinking Through the Concepts

Multiple Choice

1. *Photosynthesis is measured in the leaf of a green plant exposed to different wavelengths of light. Photosynthesis is*
   a. highest in green light
   b. highest in red light
   c. highest in blue light
   d. highest in red and blue light
   e. the same at all wavelengths

2. *Where do the light-dependent reactions of photosynthesis occur?*
   a. in the stomata
   b. in the chloroplast stroma
   c. within the thylakoid membranes of the chloroplast
   d. in the leaf cell cytoplasm
   e. in leaf cell mitochondria

3. *The oxygen produced during photosynthesis comes from*
   a. the breakdown of CO_2
   b. the breakdown of H_2O
   c. the breakdown of both CO_2 and H_2O
   d. the breakdown of oxaloacetate
   e. photorespiration

4. *The role of accessory pigments is to*
   a. provide an additional photosystem to generate more ATP
   b. allow photosynthesis to occur in the dark
   c. prevent photorespiration
   d. donate electrons to chlorophyll reaction centers
   e. capture additional light energy and transfer it to the chlorophyll reaction centers

5. *The generation of ATP by electron transport in photosynthesis and cellular respiration depends on*
   a. a proton gradient across a membrane
   b. proton pumps driven by electron transport chains
   c. an ATP-synthesizing enzyme complex
   d. a, b, and c are all required for ATP generation
   e. none of the above are required for ATP generation

6. *Where do the light-independent, carbon-fixing reactions occur?*
   a. in the guard cell cytoplasm
   b. in the chloroplast stroma
   c. within the thylakoid membranes
   d. at night in the thylakoids
   e. in mitochondria

? Review Questions

1. Write the overall equation for photosynthesis. Does the overall equation differ between C_3 and C_4 plants?
2. Draw a diagram of a chloroplast, and label it. Explain specifically how chloroplast structure is related to its function.
3. Briefly describe the light-dependent and light-independent reactions. In what part of the chloroplast does each occur?
4. What is the difference between carbon fixation in C_3 and in C_4 plants? Under what conditions does each mechanism of carbon fixation work most effectively?
5. Describe the process of chemiosmosis in chloroplasts, tracing the flow of energy from sunlight to ATP.

Applying the Concepts

1. Many lawns and golf courses are planted with bluegrass, a C_3 plant. In the spring, the bluegrass grows luxuriously. In the summer, crabgrass, a weed and a C_4 plant, often appears and spreads rapidly. Explain this sequence of events, given the normal weather conditions of spring and summer and the characteristics of C_3 versus C_4 plants.
2. Suppose an experiment is performed in which plant I is supplied with normal carbon dioxide but with water that contains radioactive oxygen atoms. Plant II is supplied with normal water but with carbon dioxide that contains radioactive oxygen atoms. Each plant is allowed to perform photosynthesis, and the oxygen gas and sugars produced are tested for radioactivity. Which plant would you expect to produce radioactive sugars, and which plant would you expect to produce radioactive oxygen gas? Why?
3. You continuously monitor the photosynthetic oxygen production from the leaf of a plant illuminated by white light. Explain what will happen (and why) if you place (a) red, (b) blue, and (c) green filters between the light source and the leaf.
4. A plant is placed in a CO_2-free atmosphere in bright light. Will the light-dependent reactions continue to generate ATP and NADPH indefinitely? Explain how you reached your conclusion.
5. You are called before the Ways and Means Committee of the House of Representatives to explain why the U.S. Department of Agriculture should continue to fund photosynthesis research. How would you justify the expense of producing, by genetic engineering, the enzyme that catalyzes the reaction of RuBP with CO_2 and prevents RuBP from reacting with oxygen as well as CO_2? What are the potential applied benefits of this research?

For More Information

Bazzazz, F. A., and Fajer, E. D. "Plant Life in a CO_2-Rich World." *Scientific American*, January 1992. Burning fossil fuels is increasing CO_2 levels in the atmosphere (see Chapter 40). This increase could tip the balance between C_3 and C_4 plants.

Govindjee, and Coleman, W. J. "How Plants Make Oxygen." *Scientific American*, February 1990. The generation of oxygen during photosynthesis is just beginning to be understood.

Grodzinski, B. "Plant Nutrition and Growth Regulation by CO_2 Enrichment." *BioScience*, 1992. How higher CO_2 levels influence plant metabolism.

Hall, D. O., and Rao, K. K. *Photosynthesis*. 5th ed. New York: Cambridge University Press, 1994. An excellent short book recommended to any student interested in finding out more about photosynthesis.

Hinkle, P. C., and McCarthy, R. E. "How Cells Make ATP." *Scientific American*, March 1978. A good explanation of chemiosmosis, which is a difficult concept for many students.

Monastersky, R. "Children of the C_4 World." *Science News*, January 3, 1998. What role did a shift in global vegetation toward C_4 photosynthesis play in the evolution of humans?

Mooney, H. A., Drake, B. G., Luxmoore, R. J., Oechel, W. C., and Pitelka, L. F. "Predicting Ecosystems' Responses to Elevated CO_2 Concentrations." *BioScience*, 1994. What effects will CO_2 enrichment of the atmosphere due to human activities have on ecosystems?

Zimmer, C. "The Processing Plant." *Discover*, September 1995. Describes organisms inhabiting the watery digestive chamber of the pitcher plant, which is both photosynthetic and carnivorous.

Answers to Multiple-Choice Questions

1. d 2. c 3. b 4. e 5. d 6. b

MEDIATUTOR
Capturing Solar Energy: Photosynthesis

CD Activities

Activity 7.1: Photosynthesis

Estimated Time: 15 minutes

The energy that powers the biosphere comes from the sun. The process that converts radiant energy into useful energy for living systems is photosynthesis. This process also evolves oxygen, which animals breathe and use during cellular respiration. The process of photosynthesis is explored in depth, relating it to the overall energy flow in the biosphere and to the process of respiration.

Start the MediaTutor Student CD-ROM and enter the activity number in the Quick Search box to be taken directly to that activity.

Web Investigations

Case Study: Did the Dinosaurs Die from Lack of Sunlight?

Estimated time: 10 minutes

While most people view the death of the dinosaurs as a unique biological disaster, geological records suggest that global mass extinctions have occurred several times. In the past two hundred years more species have become extinct than in many previous millennia. This exercise will examine the causes of some modern extinctions in the hope that they might shed some light on the other mass extinctions, including the one that ended the age of dinosaurs.

Go to http://www.prenhall.com/audesirk6, the Audesirk Companion Web site. Select Chapter 7 and the Web Investigation to begin.

With wings beating 60 times per second, the ruby-throated hummingbird has a metabolic rate 50 times that of a human. The muscles of its wings are packed with mitochondria, which supply the ATP needed to meet the bird's energy demands.

8 Harvesting Energy: Glycolysis and Cellular Respiration

AT A GLANCE

Case Study: The Flight of the Hummingbird

1) **How Is Glucose Metabolized?**

2) **How Is the Energy in Glucose Harvested During Glycolysis?**

Glycolysis Breaks Down Glucose to Pyruvate, Releasing Chemical Energy

Some Cells Ferment Pyruvate to Form Lactate

Other Cells Ferment Pyruvate to Alcohol

3) **How Does Cellular Respiration Generate Still More Energy from Glucose?**

Pyruvate Is Transported to the Mitochondrial Matrix, Where It Is Broken Down via the Krebs Cycle

Energetic Electrons Produced by the Krebs Cycle Are Carried to Electron Transport Systems in the Inner Mitochondrial Membrane

Chemiosmosis Captures Energy Stored in a Hydrogen Ion Gradient and Produces ATP

Glycolysis and Cellular Respiration Influence the Way Entire Organisms Function

Case Study Revisited: The Flight of the Hummingbird

CASESTUDY

The Flight of the Hummingbird

When a broadtailed hummingbird crashed into the glass door of Susan Heriford's home in Colorado, the unfortunate event had a happy ending. She picked up the bird and (unaware that hummingbirds can't walk) placed it in the grass. When it recovered enough to fly, it flew to her shoulder. The bird adopted the Heriford family, including their pet dog and rabbit, as companions. It perched on Susan's hand to eat, followed her family on hikes, and slept on a branch outside their bedroom window. In September the bird migrated south into Mexico and Susan wondered if she would ever see it again. The following summer, "Buddy" returned, as friendly as ever, and to the Herifords' delight, fathered a new generation of hummingbirds. Hummingbirds have extraordinary energy demands. Their wings are a blur, beating 60 times per second, "burning" calories at a rate 50 times that of an average human. Hummingbirds must eat frequently, sipping the sugar-rich nectar of flowers for energy and munching on small flying insects for protein. The rufous hummingbird makes the longest migration, from the Alaskan coast to central Mexico. As it passes through Colorado, it brazenly chases broadtails like Buddy from their favorite flowers and feeders. For some hummers, migration is a very dangerous undertaking. For example, ruby-throated hummingbirds fly continuously over 620 miles of open sea, across the Gulf of Mexico from the southeastern United States to Mexico and Central America.

In Chapter 7 we described how plants trap the energy of sunlight and store it in sugar, some of which becomes nectar that feeds the hummingbird. How does the hummingbird extract the energy and convert it to muscular movement that powers flight? How does the ruby-throated hummingbird store enough energy from the sugar it eats to make it across the Gulf of Mexico? You'll find answers as you read this chapter and as we revisit this topic later. ■

1 How Is Glucose Metabolized?

Most cells can metabolize a variety of organic molecules to produce ATP. We focus here on the metabolism of glucose for three reasons. First, virtually all cells metabolize glucose for energy at least part of the time. Some, such as nerve cells in your brain, rely predominantly on glucose as a source of energy. Second, glucose metabolism is less complex than the metabolism of most other organic molecules. Finally, when using other organic molecules as energy sources, cells usually convert the molecules to glucose or other compounds that enter the pathways of glucose metabolism (see "Health Watch: Why Can You Get Fat by Eating Sugar?").

As you learned in Chapter 7, photosynthetic organisms capture and store the energy of sunlight in glucose. During glucose breakdown, that energy is released and used to make ATP. The chemical equations for glucose formation by photosynthesis and for the complete metabolism of glucose back to CO_2 and H_2O (the original reactants in photosynthesis) are almost perfectly symmetrical:

Photosynthesis:

$$6\ CO_2 + 6\ H_2O + \text{sunlight energy} \rightarrow C_6H_{12}O_6 + 6\ O_2$$

Complete Glucose Metabolism:

$$C_6H_{12}O_6 + 6\ O_2 \rightarrow 6\ CO_2 + 6\ H_2O + \text{chemical and heat energies}$$

This symmetry might lead you to believe that a cell can convert all of the chemical energy contained in a glucose molecule to high-energy bonds of ATP. Unfortunately, according to the second law of thermodynamics, "you can't break even"; in other words, the conversion of energy into different forms always results in the decrease of the amount of concentrated, useful energy. In fact, most of the energy listed on the right-hand side of the glucose metabolism equation is heat energy, not the chemical energy of ATP. Nevertheless, a cell can extract a great deal of energy, in the form of ATP, from glucose if the glucose molecule is completely broken down to CO_2 and H_2O.

Figure 8-1 summarizes the major steps of glucose metabolism in eukaryotic cells. The first stage, *glycolysis*, does not require oxygen and proceeds in exactly the same way under both aerobic (with oxygen) and anaerobic (without oxygen) conditions. Glycolysis splits apart a single glucose molecule (a six-carbon sugar) into two three-carbon molecules of **pyruvate.** This splitting releases a small fraction of the chemical energy stored in the glucose, some of which is used to generate two ATP molecules. The presence of oxygen becomes an issue only in the processes that follow glycolysis. Under anaerobic conditions, the pyruvate is usually converted by fermentation into lactate or ethanol. Fermentation does not produce more ATP energy. Both glycolysis and fermentation occur in the fluid portion of the cytoplasm.

The pyruvate produced by glycolysis may also enter the mitochondria. There, if oxygen is available, cellular respiration uses oxygen to break pyruvate down completely to carbon dioxide and water, generating an additional 34 or 36 ATP molecules (the amount differs from cell to cell). The extra ATP produced by cellular respiration is so important to most organisms that anything that interferes with its production, such as the poison cyanide or lack of oxygen, quickly results in death.

2 How Is the Energy in Glucose Captured During Glycolysis?

The initial reactions that break down glucose without the use of oxygen are collectively called **glycolysis** (in Greek, "to break apart a sweet"). Glycolysis is believed to be one of the most ancient of all biochemical pathways, because it is used by every living creature on the planet. This sequence of reactions occurs in the fluid portion of the cytoplasm and results in a molecule of glucose being cleaved into two molecules of pyruvate. Glycolysis produces relatively little energy, only two molecules of ATP and two molecules of the electron carrier NADH. But without it, life would rapidly be extinguished. Reduced to its essentials, glycolysis consists of two major parts (each with several steps): (1) glucose activation and (2) energy harvest (Fig. 8-2).

Glycolysis Breaks Down Glucose to Pyruvate, Releasing Chemical Energy

Before glucose is broken down to release its energy, it must be activated—a process that actually uses up energy. During glucose activation, a molecule of glucose undergoes two enzyme-catalyzed reactions, each of which uses ATP energy (Fig. 8-2a). These reactions convert a relatively stable glucose molecule into a highly unstable "activated" molecule of *fructose bisphosphate*. Fructose is a sugar molecule similar to glucose; *bisphosphate* refers to the two phosphate groups acquired from the ATP molecules. Forming fructose bisphosphate costs the cell two ATP molecules. But this initial consumption of energy is necessary to produce much greater energy returns in the long run.

In the energy harvest steps, fructose bisphosphate splits apart into two three-carbon molecules of glyceraldehyde 3-phosphate (G3P; Fig. 8-2b). (In Chapter 7 we encountered G3P in the C_3 cycle of photosynthesis.) Each G3P molecule then goes through a series of reactions that convert it to pyruvate. During these reactions, two ATP are generated for each G3P, for a total of four

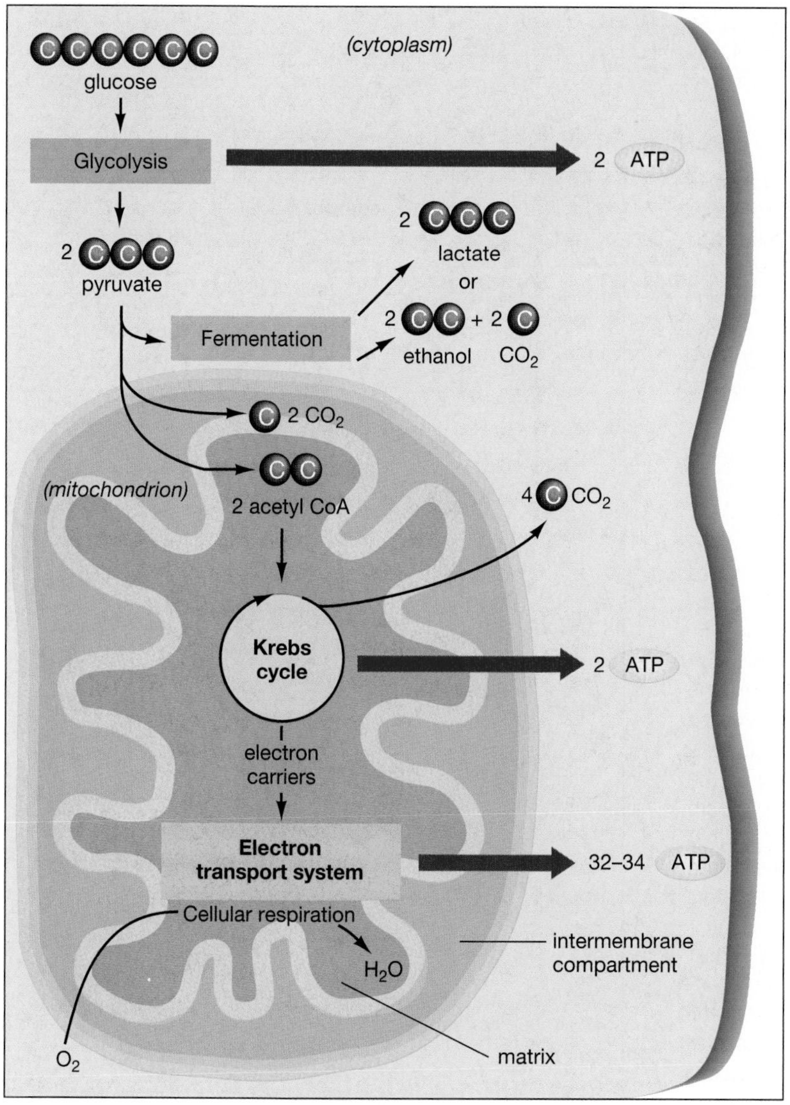

Figure 8-1 A summary of glucose metabolism
Refer to this diagram as we progress through the reactions of glycolysis (in the fluid portion of the cytoplasm) and cellular respiration (in the mitochondria). The breakdown of glucose occurs in stages, with various amounts of energy harvested as ATP along the way. The vast majority of the ATP is produced in the mitochondria, justifying their nickname, "powerhouse of the cell."

MEDIATUTOR 8.1 Glucose Metabolism

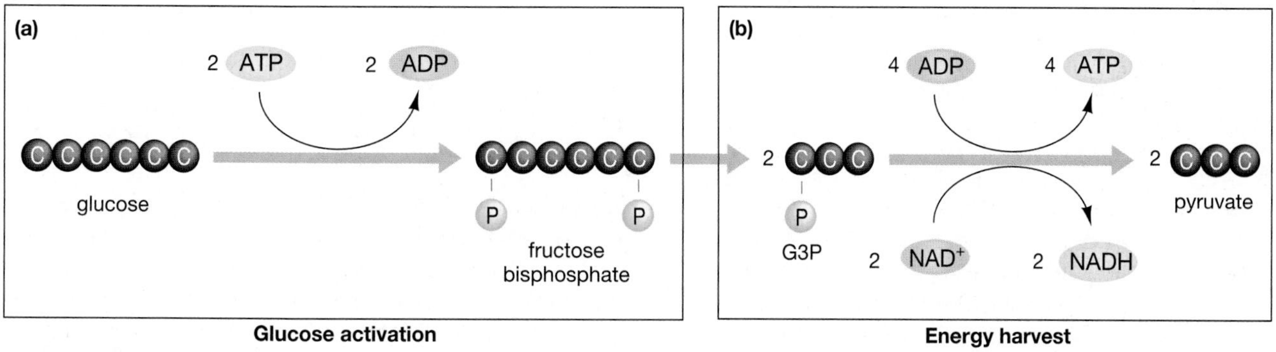

Figure 8-2 The essentials of glycolysis
(a) Glucose activation: The energy of two ATP molecules is used to convert glucose to the highly reactive fructose bisphosphate. Fructose bisphosphate splits into two smaller, but still reactive, molecules of glyceraldehyde-3-phosphate (G3P). ***(b)*** Energy harvest: The two G3P molecules undergo a series of reactions that generate four ATP and two NADH molecules. Therefore, glycolysis results in a net harvest of two ATP and two NADH molecules per glucose molecule.

Health Watch
Why Can You Get Fat by Eating Sugar?

As you know, humans do not live by glucose alone. Nor does the typical diet contain exactly the required amounts of each nutrient. Accordingly, the cells of the human body seethe with biochemical reactions, synthesizing one amino acid from another, making fats from carbohydrates, and channeling surplus organic molecules of all types into energy storage or release. Let us look at two examples of these metabolic transformations: the production of ATP from fats and proteins and the synthesis of fats from sugars.

How Are Fats and Proteins Metabolized? Even the leanest people have some fat in their bodies. During fasting or starvation, the body mobilizes these fat reserves for ATP synthesis, because even the bare maintenance of life requires a continuous supply of ATP, and seeking out new food sources demands even more energy. Fat metabolism flows directly into the pathways of glucose metabolism.

Chapter 3 illustrated the structure of a fat: three fatty acids connected to a glycerol backbone. In fat metabolism, the bonds between the fatty acids and glycerol are hydrolyzed (broken into subunits by the addition of water). The glycerol part of a fat, after activation by ATP, feeds directly into the middle of the glycolysis pathway (Fig. E8-1). The fatty acids are transported into the mitochondria, where enzymes in the inner membrane and matrix chop them up into acetyl groups. These groups attach to coenzyme A to form acetyl CoA, which enters the Krebs cycle.

In individuals with severe starvation or those feeding almost exclusively on protein, amino acids can be used to produce energy. First, the amino acids are converted to pyruvate, acetyl CoA, or the compounds of the Krebs cycle. These molecules then proceed through the remaining stages of cellular respiration, yielding amounts of ATP that vary with their point of entry into the pathway.

How Is Fat Synthesized from Sugar? The body, in addition to having developed ways of coping with fasting or starvation, has also evolved strategies for coping with situations in which food intake exceeds current energy needs. The sugars and starches in corn flakes, candy bars, or the nectar of flowers can be converted into fats for energy storage. Complex sugars, such as starches and sucrose, are first hydrolyzed into their monosaccharide subunits (see Chapter 3). The monosaccharides are broken down to pyruvate and converted to acetyl CoA. If the cell needs ATP, the acetyl CoA will enter the Krebs cycle. If the cell has plenty of ATP, acetyl CoA will be used to make fatty acids by a series of reactions that are essentially the reverse of fatty-acid breakdown. Thus, the hummingbird can double its body weight by eating sugar and getting fat before it flies south. In humans the liver synthesizes fatty acids, but fat storage is relegated to fat cells, with their all-too-familiar distribution in the body, particularly around the waist and hips.

Energy use, fat storage, and nutrient intake are usually precisely balanced. Where the balance point lies, however, varies from person to person. Some people seem able to eat nearly continuously without ever storing much fat; other people crave high-calorie foods even when they have a lot of stored fat. From an evolutionary perspective, overeating during times of easy food availability is highly adaptive behavior. If hard times come, heavier people may survive while the more trim succumb to starvation. Only recently (over evolutionary time) have people in societies such as ours had continuous access to inexpensive, high-calorie food. Under these conditions, the drive to eat and the adaptation of storing excess food as fat leads to obesity, a growing health problem in the United States.

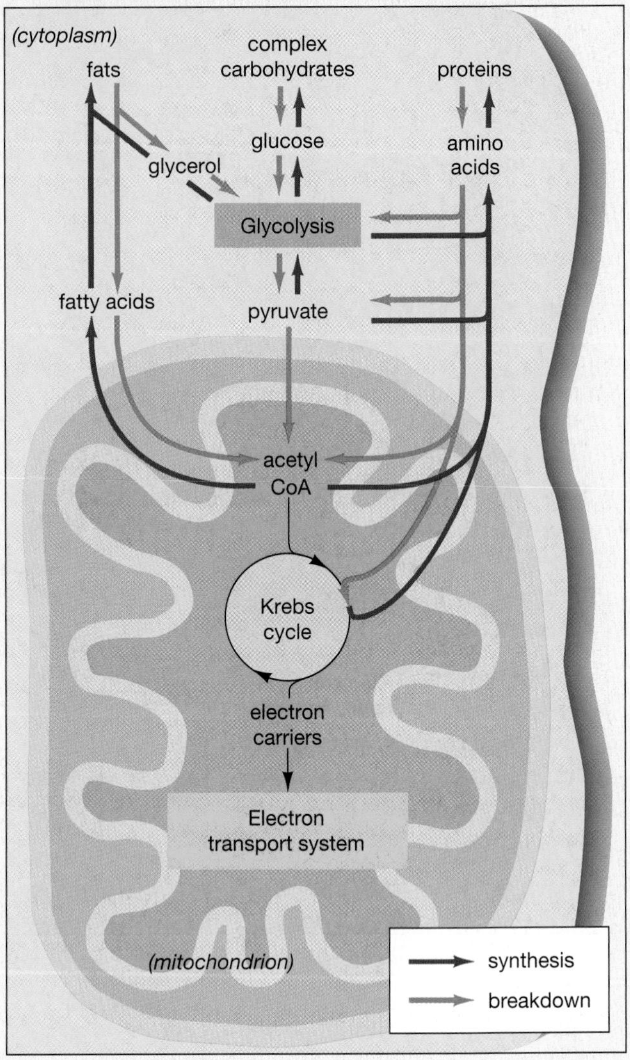

Figure E8-1 How various nutrients yield energy
Fats, carbohydrates such as starches, and proteins can all be broken down metabolically into molecules that enter glycolysis or the Krebs cycle, where they are used to generate ATP.

ATPs. Because two ATPs were used to activate the glucose molecule in the first place, *there is a net gain of only two ATPs per glucose molecule.* At another step along the way from G3P to pyruvate, two high-energy electrons and a hydrogen ion are added to the "empty" electron carrier NAD^+ to make the "energized" carrier NADH (this is a slightly different electron carrier from the $NADP^+$ used in photosynthesis). Two G3P molecules are produced per glucose molecule, so two NADH carrier molecules are formed when those G3P molecules are converted to pyruvate. For a complete description of glycolysis, see "A Closer Look: Glycolysis."

Summing Up
Glycolysis

Each molecule of glucose is broken down to two molecules of pyruvate. During these reactions, two ATP molecules and two NADH electron carriers are formed.

Carrier molecules such as NAD^+ capture energy by accepting high-energy electrons. Carriers can transport these electrons to sites where their energy is used to form ATP. One major difference between anaerobic (without oxygen) and aerobic (with oxygen) glucose breakdown is how these high-energy electrons are used. In the absence of oxygen, pyruvate acts as the electron acceptor from NADH, producing ethanol or lactate; this process is called **fermentation**. During cellular respiration, which occurs in the presence of oxygen, oxygen becomes the electron acceptor, allowing the pyruvate to be fully broken down and its energy harvested as ATP.

Some Cells Ferment Pyruvate to Form Lactate

Many organisms (particularly microorganisms) thrive in places where oxygen is rare or absent, such as in the stomach and intestine of animals, deep in soil, in sediments underlying lakes and oceans, or in bogs and marshes. Even some of our own body cells must cope without oxygen for brief periods. In anaerobic conditions (the conditions in which life, and probably glycolysis, evolved), NADH production is *not* used as a method of energy capture; it is actually a way of getting rid of hydrogen ions and electrons produced during the breakdown of glucose to pyruvate. But this disposal method poses a problem for the cell, because NAD^+ is used up as it accepts electrons and hydrogen ions in becoming NADH. Without a way to regenerate NAD^+ and to dispose of the electrons and hydrogen ions, glycolysis would have to stop once the supply of NAD^+ was exhausted.

Fermentation solves this problem by enabling pyruvate to act as the final acceptor of electrons and hydrogen ions from NADH. Thus, NAD^+ is regenerated for use in further glycolysis. There are two main types of fermentation: one type converts pyruvate to lactate and the other converts pyruvate to carbon dioxide and ethanol.

Fermentation to lactate occurs in your muscles when you exercise vigorously, such as when you race to class after you've overslept, or in the muscles of a runner sprinting through the finish line (Fig. 8-3a, p. 137). (You may hear this compound called "lactic acid"; lactate is the ionized form of lactic acid that is in solution in the cytoplasm.) Even though working muscles need lots of ATP and cellular respiration generates much more ATP than does glycolysis, cellular respiration is limited by the organism's ability to provide oxygen (by breathing, for example). While you exercise vigorously, you may not be able to get enough air into your lungs and enough oxygen into your blood to supply your muscles with sufficient oxygen to allow cellular respiration to meet all their energy needs.

When deprived of adequate oxygen, your muscles do not immediately stop working. Instead, glycolysis continues for a while, providing its meager two ATP molecules per glucose and generating both pyruvate and NADH. Then, to regenerate NAD^+, muscle cells ferment pyruvate molecules to lactate, using electrons and hydrogen ions from NADH:

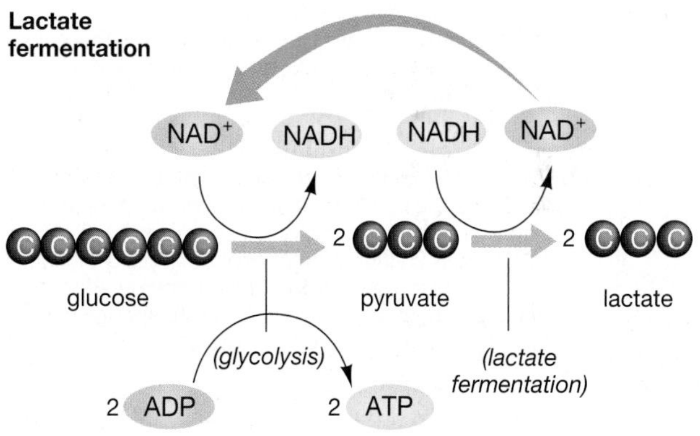

In high concentrations, however, lactate is toxic to your cells. Soon, it causes intense discomfort and fatigue, forcing you to stop or at least slow down. As you rest, breathing rapidly after your sprint, oxygen once more becomes available and the lactate is converted back to pyruvate. Interestingly, the conversion from lactate to pyruvate occurs not in the muscle cells, which lack the necessary enzymes, but in the liver. This pyruvate is then broken down by cellular respiration into carbon dioxide and water.

Various microorganisms, including the bacteria that produce yogurt, sour cream, and cheese, also use lactate fermentation. As you may know, acids taste sour, and lactate (lactic acid) contributes to the distinctive tastes of these foods. Some microorganisms ferment even when oxygen is present; others are poisoned by oxygen.

A Closer Look
Glycolysis

Glycolysis is a series of enzyme-catalyzed reactions that break down a single molecule of glucose into two molecules of pyruvate. To help you follow the reactions, we show only the "carbon skeletons" of glucose and the molecules produced during glycolysis. Each blue arrow represents a reaction catalyzed by at least one enzyme.

1. A glucose molecule is energized by the addition of a high-energy phosphate from ATP.
2. The molecule is slightly rearranged, forming fructose.
3. Then a second phosphate is added from another ATP.
4. The resulting molecule, fructose-1,6-bisphosphate, is split into two three-carbon molecules, one DHAP (dihydroxacetone phosphate) and one G3P. Each has one phosphate attached.
5. DHAP rearranges into G3P. From now on, there are two molecules of G3P going through the identical reactions.
6. Each G3P undergoes two almost-simultaneous reactions. Two electrons and a hydrogen ion are donated to NAD^+ to make the energized carrier NADH, and an inorganic phosphate (P) is attached to the carbon skeleton with a high-energy bond. The resulting molecules of 1,3-bisphosphoglycerate have two high-energy phosphates.
7. One phosphate from each bisphosphoglycerate is transferred to ADP to form ATP, for a net of two ATPs. This transfer compensates for the initial two ATPs used in glucose activation.
8. After another rearrangement, the second phosphate from each phosphoenolpyruvate is transferred to ADP to form ATP, leaving pyruvate as the final product of glycolysis. There is a net profit of two ATPs from each glucose molecule.

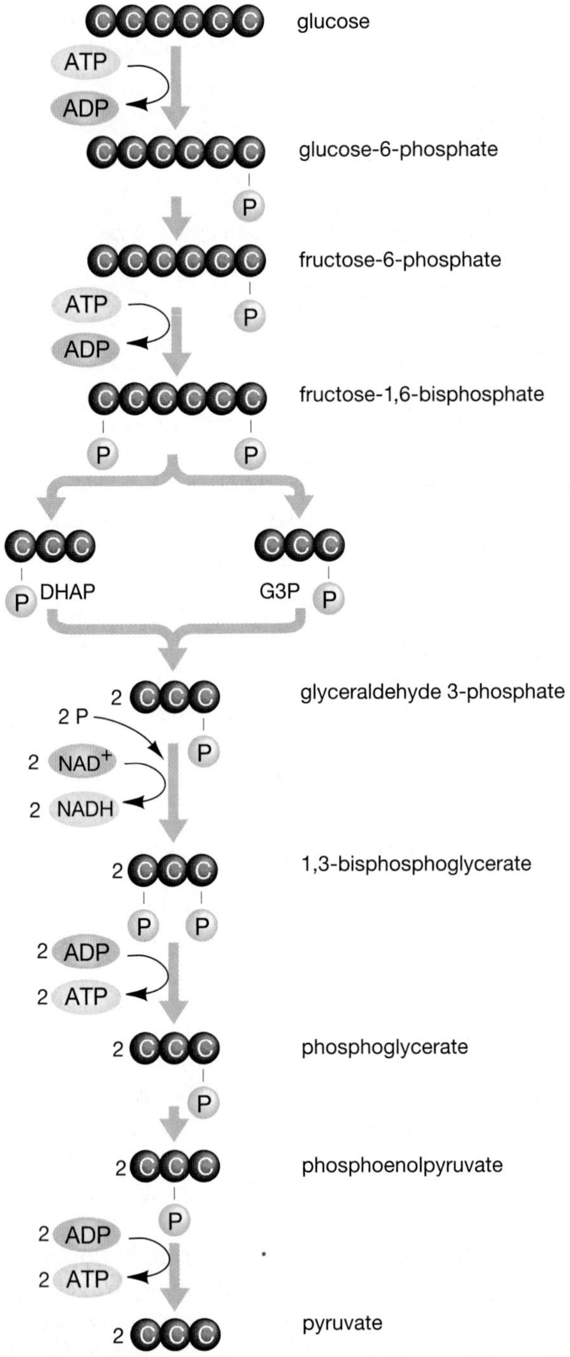

(a)

(b)

Figure 8-3 Fermentation
(a) Sprinters at the end of a race. A runner's respiratory and circulatory systems cannot supply oxygen to her leg muscles fast enough to keep up with the demand for energy, so glycolysis and lactate fermentation must provide the ATP. Panting after the race brings in the oxygen needed to remove the lactate through cellular respiration. ***(b)*** Bread rises as CO_2 is liberated by fermenting yeast, which converts glucose to ethanol via the alcoholic fermentation pathway.

Other Cells Ferment Pyruvate to Alcohol

Many microorganisms use another process to regenerate NAD^+ under anaerobic conditions: *alcoholic fermentation*. These reactions produce ethanol and CO_2 (rather than lactate) from pyruvate, using hydrogen ions and electrons from NADH:

Alcoholic fermentation

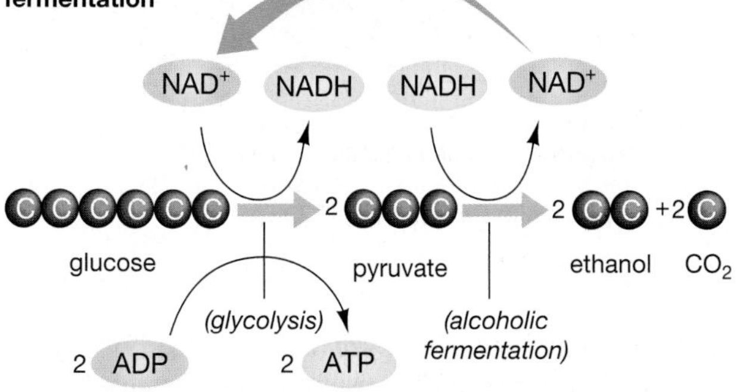

Sparkling wines, such as champagne, are bottled while the yeasts are still alive and fermenting, trapping both the alcohol and the CO_2. When the cork is removed, the pressurized CO_2 is released, sometimes explosively. Baker's yeast in bread dough produces CO_2, making the bread rise; the alcohol generated by the yeast evaporates while the bread is baking (Fig. 8-3b).

3 How Does Cellular Respiration Generate Still More Energy from Glucose?

Cellular respiration is a series of reactions, occurring under aerobic conditions, in which large amounts of ATP are produced. During cellular respiration, the pyruvate produced by glycolysis is broken down to carbon dioxide and water. The final reactions of cellular respiration require oxygen because oxygen acts as the final acceptor of electrons.

In eukaryotic cells, cellular respiration occurs in the mitochondria. Recall from Chapter 5 that a mitochondrion has two membranes that produce two compartments: an inner compartment that is enclosed by the inner membrane and contains the fluid **matrix**, and an **intermembrane compartment** between the two membranes (Fig. 8-4). Most of the ATP produced during cellular respiration is

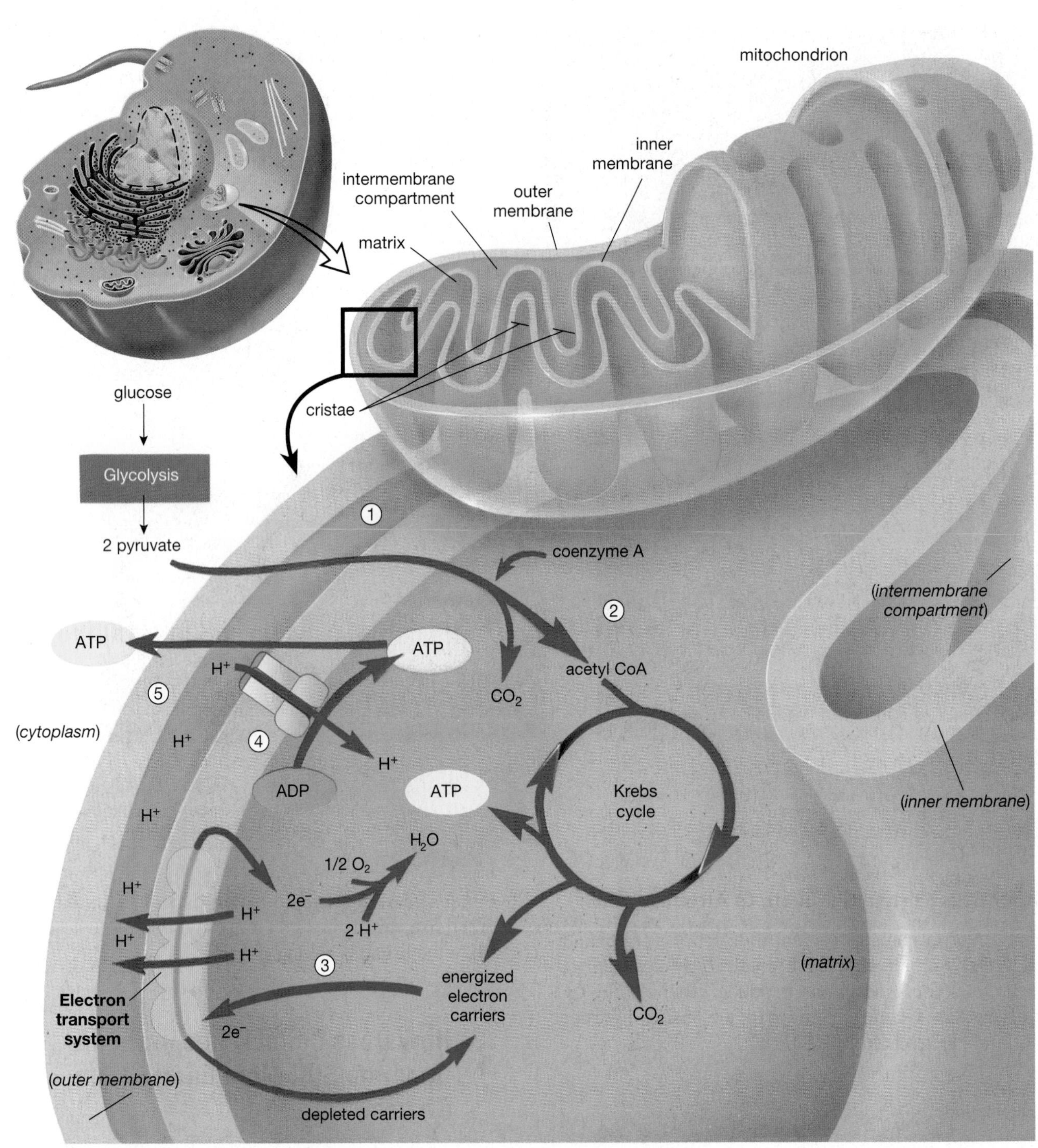

Figure 8-4 Cellular respiration
Cellular respiration takes place in the mitochondria, whose structure, like that of a chloroplast, reflects the compartmentalized reactions that occur there. The inner membrane separates the inner compartment, containing the soluble enzymes of the matrix, from the intermembrane compartment (between the inner and outer membranes). The diagram summarizes the essential steps of glucose metabolism, from glycolysis in the cytoplasm to the transport of ATP out of the mitochondrion and back into the cytoplasm.

generated by enzyme-catalyzed reactions in the matrix, by electron transport proteins in the inner membrane, and by the movement of hydrogen ions through ATP-synthesizing proteins in the inner membrane.

Figure 8-4 summarizes the main events of cellular respiration:

① The two molecules of pyruvate produced by glycolysis are transported across both mitochondrial membranes and into the matrix.

② Each pyruvate is split into CO_2 and a two-carbon acetyl group, which enters the *Krebs cycle* (discussed in the next section). The Krebs cycle releases the remaining carbons as CO_2, produces one ATP from each pyruvate, and donates energetic electrons to several electron-carrier molecules.

③ The electron carriers donate their energetic electrons to the electron transport system of the inner membrane. There the energy of the electrons is used to transport H^+ from the matrix to the intermembrane compartment. At the end of the system, the electrons combine with O_2 and H^+ to form H_2O. Depleted carriers are reused in the Krebs cycle.

④ In chemiosmosis, the hydrogen ion gradient created by the electron transport system discharges through ATP-synthesizing enzymes in the inner membrane, and the energy is used to produce ATP.

⑤ ATP is transported out of the mitochondrion into the fluid of the cytoplasm, where it provides energy for cellular activities.

We have already discussed glycolysis; now let's look a little more closely at the processes of cellular respiration in the mitochondria. These two processes are summarized in Table 8-1.

Pyruvate Is Transported to the Mitochondrial Matrix, Where It Is Broken Down via the Krebs Cycle

Recall that pyruvate is the end product of glycolysis and that it is synthesized in the fluid portion of the cytoplasm. The pyruvate diffuses down its concentration gradient into the mitochondria, through pores in the mitochondrial membranes, until it reaches the mitochondrial matrix, where it is used in cellular respiration.

In the matrix, pyruvate reacts with a molecule called *coenzyme A* (Fig. 8-5, step 1). Each pyruvate is split into CO_2 and a two-carbon molecule called an *acetyl group*, which immediately attaches to coenzyme A (CoA), forming an acetyl–coenzyme A complex (*acetyl CoA* for short). During this reaction, two energetic electrons and a hydrogen ion are transferred to NAD^+, forming NADH.

The next stages of the reaction form a cyclic pathway known as the **Krebs cycle**, named after its discoverer, Hans Krebs, a German-born biochemist who immigrated to Britain and won the Nobel Prize for this work in 1953. The Krebs cycle is also called the *citric acid cycle* because citrate (the ionized form of citric acid) is the first molecule produced in the cycle. During the Krebs cycle (Fig. 8-5, step 2), each acetyl CoA briefly combines with a molecule of oxaloacetate. The two-carbon acetyl group is donated to the four-carbon *oxaloacetate* (which, as you may remember from Chapter 7, also functions in the carbon fixation stage of the C_4 pathway) to form the six-carbon *citrate*. Coenzyme A is released once again. (Coenzyme A, like an enzyme, is not permanently altered during these reactions and is reused many times.) Mitochondrial enzymes then lead each citrate through a number of rearrangements that regenerate the oxaloacetate, give off two CO_2 molecules, and capture most of the energy of the acetyl group as one ATP and four electron carriers—one $FADH_2$ (flavin adenine dinucleotide) and three NADH.

Table 8-1 Summary of Glycolysis and Cellular Respiration

Process	Location	Reactions	Electron Carriers Formed	ATP Yield (per glucose molecule)
Glycolysis	Fluid cytoplasm	Glucose broken down to the two pyruvates	2 NADH	2 ATP
Cellular Respiration				
Acetyl CoA formation	Matrix of mitochondrion	Each pyruvate combined with coenzyme A to form acetyl CoA and CO_2	2 NADH	
Krebs cycle	Matrix of mitochondrion	Acetyl group of acetyl CoA metabolized to two CO_2	6 NADH, 2 $FADH_2$	2 ATP
Electron transport	Inner membrane, intermembrane compartment	Energy of electrons from NADH, two $FADH_2$ used to pump H^+ into intermembrane compartment, H^+ gradient used to synthesize ATP: three ATP per NADH, two ATP per $FADH_2$		32–34 ATP*

*Glycolysis produces two NADH molecules in the fluid portion of the cytoplasm of the cell. Unlike the NADH and $FADH_2$ molecules generated in the matrix of the mitochondrion, the electrons from these two NADH molecules must be transported into the matrix before they can enter the electron transport system of the inner membrane. In most eukaryotic cells, the energy of one ATP molecule is used to transport the electrons from one NADH molecule into the matrix. Thus, the two "glycolytic NADH" molecules net only two ATPs, not the usual three, during electron transport. The heart and liver cells of mammals, however, use a different transport system, one that does not consume ATP. In these cells, then, the "glycolytic NADH" molecules net three ATPs apiece, just as the "mitochondrial NADH" molecules do.

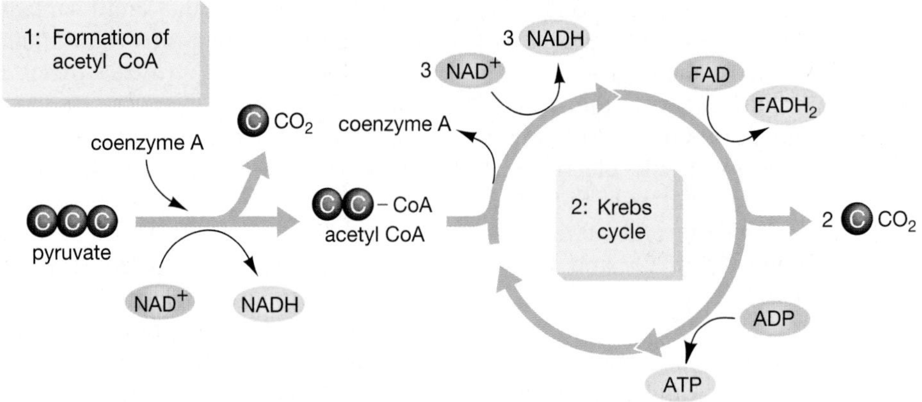

Figure 8-5 The essential reactions in the mitochondrial matrix
Pyruvate reacts with coenzyme A to form CO_2 and acetyl CoA. During this reaction, an energetic electron is added to NAD^+ to form NADH. When acetyl CoA enters the Krebs cycle, the acetyl group combines with oxaloacetate to form citrate, and coenzyme A is released. One turn through the reactions of the cycle produces three molecules of NADH, one molecule of $FADH_2$, two of CO_2, and one molecule of ATP for each acetyl CoA. Because each glucose molecule yields two pyruvates, the total energy harvest per glucose molecule in the matrix is two ATP, eight NADH (two from the synthesis of acetyl CoA, six from the Krebs cycle), and two $FADH_2$.

The essay "A Closer Look: The Mitochondrial Matrix Reactions" shows the complete set of reactions that occur in the mitochondrial matrix, from acetyl CoA formation through the Krebs cycle.

Summing Up
The Mitochondrial Matrix Reactions

The synthesis of acetyl CoA produces one CO_2 and one NADH per pyruvate. The Krebs cycle produces two CO_2, one ATP, three NADH, and one $FADH_2$ per acetyl CoA. Therefore at the conclusion of the matrix reactions, the two pyruvates that are produced from a single glucose molecule have been completely broken down by the addition of oxygen to form six CO_2 molecules. In the process, two ATPs, eight NADH, and two $FADH_2$ electron carriers have been produced.

Energetic Electrons Produced by the Krebs Cycle Are Carried to Electron Transport Systems in the Inner Mitochondrial Membrane

At this point the cell has gained only four ATP molecules from the original glucose molecule: two during glycolysis and two during the Krebs cycle. The cell has, however, captured many energetic electrons in carrier molecules: 2 NADH during glycolysis plus 8 more NADH and 2 $FADH_2$ from the matrix reactions, for a total of 10 NADH and 2 $FADH_2$. The carriers deposit their electrons in **electron transport systems** located in the inner mitochondrial membrane (Fig. 8-6). These electron transport systems are similar in function to those embedded in the thylakoid membrane of chloroplasts. The energetic electrons move from molecule to molecule along the transport systems. Energy released by the electrons during these transfers is used to pump hydrogen ions from the matrix, across the inner membrane, and into the intermembrane compartment during *chemiosmosis* (discussed in the next section).

Finally, at the end of the electron transport system, oxygen and hydrogen ions accept the energetically depleted electrons: Two electrons, one oxygen atom, and two hydrogen ions combine to form water. This step clears out the transport system, leaving it ready to run through more electrons. Without oxygen, the electrons would "pile up" in the transport system; the hydrogen

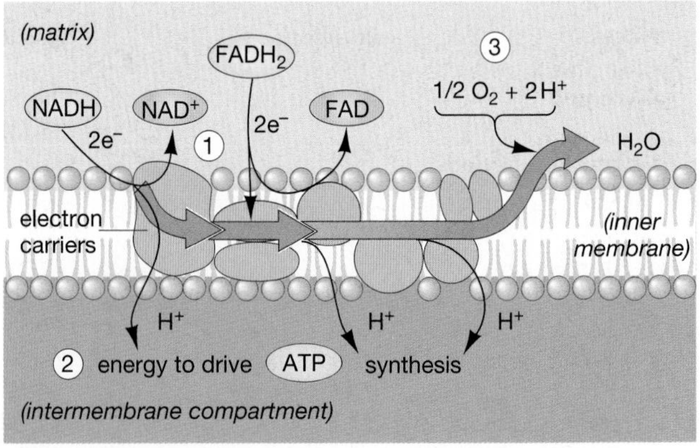

Figure 8-6 The electron transport system of mitochondria
① The electron carrier molecules NADH and $FADH_2$ deposit their energetic electrons with the carriers of the transport system located in the inner membrane. ② The electrons move from carrier to carrier within the transport system. Some of their energy is used to pump hydrogen ions across the inner membrane from the matrix into the intermembrane compartment. This creates a hydrogen ion gradient that is used to drive ATP synthesis. ③ At the end of the electron transport system, the energy-depleted electrons combine with oxygen and hydrogen ions in the matrix to form water.

A Closer Look
The Mitochondrial Matrix Reactions

Mitochondrial matrix reactions occur in two stages: (1) the formation of acetyl coenzyme A and (2) the Krebs cycle. Recall that glycolysis produces two pyruvates from each glucose molecule, so each set of matrix reactions occurs twice during the metabolism of a single glucose molecule.

First Stage: Formation of Acetyl Coenzyme A
Pyruvate is split to CO_2 and an acetyl group. The acetyl group attaches to coenzyme A to form acetyl CoA. Simultaneously, NAD^+ receives two electrons and a hydrogen ion to make NADH. The acetyl CoA enters the second stage of the matrix reactions.

Second Stage: The Krebs Cycle

① Acetyl CoA donates its acetyl group to oxaloacetate to make citrate.

② Citrate is rearranged to form isocitrate.

③ Isocitrate loses a carbon to CO_2, forming α-ketoglutarate; NADH is formed from NAD^+.

④ Alpha-ketoglutarate loses a carbon to CO_2, forming succinate; NADH is formed from NAD^+, and additional energy is stored in ATP. By this point, two molecules of CO_2 have been given off. (These two CO_2 molecules, along with the one that was released during the formation of acetyl CoA, account for the three carbons of the original pyruvate.)

⑤ Succinate is converted to fumarate, and the electron carrier FAD is charged up to $FADH_2$.

⑥ Fumarate is converted to malate.

⑦ Malate is converted to oxaloacetate, and NADH is formed from NAD^+.

The Krebs cycle produces three molecules of CO_2 and NADH, one $FADH_2$, and one ATP per acetyl CoA. NADH and $FADH_2$ will donate their electrons to the electron transport system of the inner membrane, where the energy of the electrons will be used to synthesize ATP by chemiosmosis.

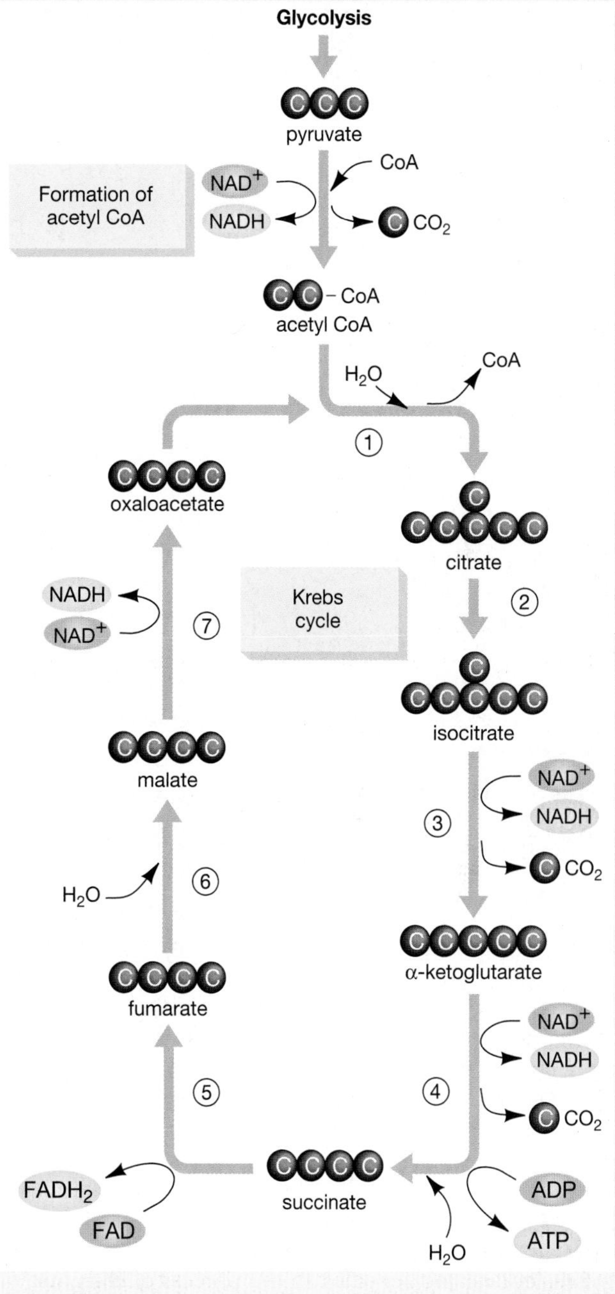

ions would not be pumped across the inner membrane. The hydrogen ion gradient would soon dissipate, and ATP synthesis would stop.

Chemiosmosis Captures Energy Stored in a Hydrogen Ion Gradient and Produces ATP

Why pump hydrogen ions across a membrane? Hydrogen ion pumping across the inner membrane generates a large H^+ concentration gradient—that is, a high concentration of hydrogen ions in the intermembrane compartment and a low concentration in the matrix. Recall from Chapter 6 that, according to the second law of thermodynamics, energy must be expended to produce this nonuniform distribution of hydrogen ions, sort of like pumping water up into an elevated storage tank. Energy is released when the hydrogen ions are allowed to move down their concentration gradient—like opening the valves of the storage tank and allowing the water to rush out. This energy can be captured because the inner membrane is impermeable to

A Closer Look
Chemiosmosis in Mitochondria

ATP synthesis in mitochondria is similar to the process of chemiosmosis described for chloroplasts in Chapter 7. The inner membrane of a mitochondrion has an electron transport system that functions similarly to the one in the thylakoids. Further, the intermembrane compartment between the outer and inner membranes of a mitochondrion is analogous to the interior of a thylakoid.

Anatomically, the arrangement in mitochondria looks like this:

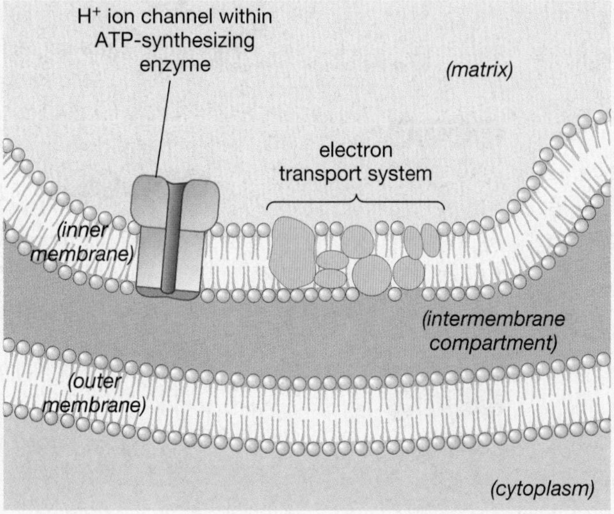

The electron carriers formed during glycolysis and the Krebs cycle—NADH and $FADH_2$—deposit their electrons with the electron transport system of the inner membrane. (For clarity, $FADH_2$ is not shown in the illustration.) As they pass through the electron transport system, the electrons provide the energy to pump hydrogen ions (H^+) across the inner membrane, from the matrix to the intermembrane compartment:

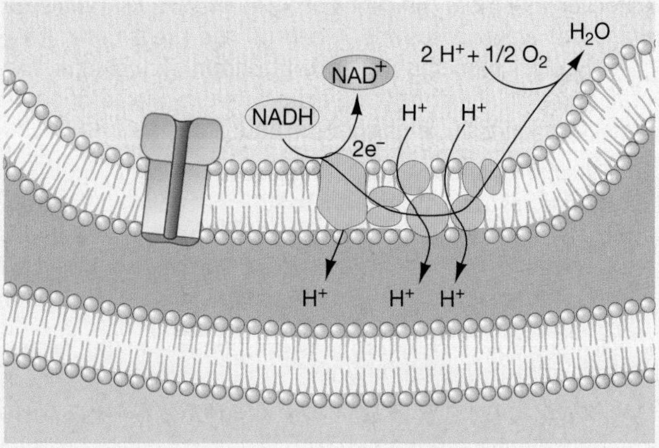

This pumping process increases the H^+ concentration in the intermembrane compartment and decreases the H^+ concentration in the matrix; therefore, a H^+ gradient is produced across the inner membrane. Like the thylakoid membrane of a chloroplast, the inner membrane of a mitochondrion is permeable to H^+ only at channels that are coupled with ATP-synthesizing enzymes. The movement of hydrogen ions down their concentration gradient through these channels drives ATP synthesis:

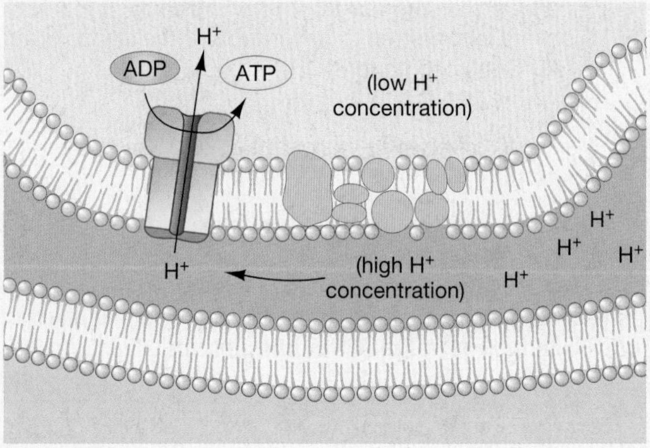

hydrogen ions except at protein channels that are part of ATP-synthesizing enzymes. In the process of **chemiosmosis**, hydrogen ions move down their concentration gradient from the intermembrane compartment to the matrix through these ATP-synthesizing enzymes. The flow of hydrogen ions provides the energy to synthesize 32 to 34 molecules of ATP by combining ADP (adenosine diphosphate) and phosphate for each molecule of glucose that is broken down. "A Closer Look: Chemiosmosis in Mitochondria" examines chemiosmosis in more detail.

The ATP that was synthesized in the matrix during chemiosmosis is transported across the inner membrane from the matrix to the intermembrane compartment. It then diffuses out of the mitochondrion to the surrounding cytoplasm through the outer membrane, which is very permeable to ATP. These ATP molecules provide most of the energy needed by the cell. ADP simultaneously diffuses from the fluid of the cytoplasm across the outer membrane and is transported across the inner membrane to the matrix, replenishing the supply of ADP.

Summing Up
Electron Transport and Chemiosmosis

Electrons from the electron carriers NADH and $FADH_2$ enter the electron transport system of the inner mitochondrial membrane. Here their energy is used to generate a hydrogen ion gradient across the inner membrane. The movement of hydrogen ions down their gradient through the pores of ATP-synthesizing enzymes drives the synthesis of 32 to 34 molecules of ATP. At the end of the electron transport system, two electrons combine with one oxygen atom and two hydrogen ions to form water.

Glycolysis and Cellular Respiration Influence the Way Entire Organisms Function

Many students believe that the details of glycolysis and cellular respiration are hard to learn and don't really help them understand the living world around them. Have you ever read a murder mystery and wondered how cyanide could kill a person almost instantly? Cyanide reacts with one of the proteins in the electron transport system, immediately blocking the movement of electrons through the system and bringing cellular respiration to a screeching halt. Even under normal conditions, metabolic processes within individual cells have enormous impacts on the functioning of entire organisms. For a hummingbird to beat its wings, for your brain to process this information you are reading, for your hand to turn the pages of this book, cells require a continuous supply of energy. For an extreme example, let's consider Olympic track events.

Humans, like hummingbirds, must regulate energy reserves and energy use. Why is the average speed of the 5000-meter run in the Olympics slower than that of the 100-meter dash? During the dash, or during the sprint across the finish line of a marathon, runners' leg muscles use more ATP than cellular respiration can supply, because their bodies cannot deliver enough oxygen to keep up with the demand. Glycolysis and lactate fermentation can keep the muscles supplied with ATP for a short time, but soon the toxic effects of lactate buildup cause fatigue and cramps. Although runners can do a 100-meter dash anaerobically, distance runners must pace themselves, using cellular respiration to power their muscles for most of the race and saving the anaerobic sprint for the finish.

Marathon runners face somewhat the same dilemma that migrating hummingbirds do. A marathon may require 3000 kilocalories of stored energy, with cellular respiration supplying nearly all the ATP. Marathoners train by running 50 or 100 miles a week, not so much to build up their leg muscles as to build up the capacity of their respiratory and circulatory systems to deliver enough oxygen to their muscles. An efficient transport of oxygen to the cells is necessary to support the cellular respiration that such vigorous exercise demands.

As you can see, sustaining life depends on efficiently obtaining, storing, and using energy. By gaining an understanding of the principles of cellular respiration, you can more fully appreciate the energy-related adaptations of living organisms.

REVISITEDCASESTUDYREVISITEDCASESTUDYREVISITEDCASE

The Flight of the Hummingbird

To fly more than 600 miles over open water, the ruby-throated hummingbird must store a great deal of energy. Hummers store the highest-energy molecules possible—fat—and extract the maximum usable energy during flight. A ruby-throated hummingbird weighs 0.11 to 0.16 ounces (2 to 3 grams, about as much as a penny) before it puts on weight for migration. It adds as much as 2 grams of fat in late summer, nearly doubling its weight. Recall from Chapter 3 that fats contain more than twice as much energy per unit weight as do proteins or carbohydrates. If a hummer had to store glycogen or protein for energy, it would be too heavy to lift off. Even so, the hummer must still generate every ATP molecule possible out of each fat molecule. The hummer that just makes it to Guatemala on 2 grams of fat by using cellular respiration would collapse before reaching the Gulf Coast if it used lactate fermentation instead. The cells of a hummingbird's flight muscles are packed with mitochondria, so each cell is capable of producing large quantities of ATP. Furthermore, the hummingbird's respiratory system is exquisitely designed to extract oxygen from the air even while exhaling, enabling the lungs to provide a constant supply of oxygen to the cells. Even during strenuous flight, cellular respiration never falters for lack of oxygen.

Some flowers are adapted to attract hummingbirds. They store relatively large quantities of nectar at the end of long, tubular compartments that bees cannot reach. The flowers are often red or orange (colors that attract hummingbirds). Look at the opening photo and form a hypothesis about why some plants have evolved to "waste" so much energy by putting their sugar into nectar, and why flowers that attract hummingbirds with large amounts of nectar prevent bees from reaching the nectar by their shape. (HINT: If you are stumped by this, you may find some ideas in Chapter 24.)

Summary of Key Concepts

1) How Is Glucose Metabolized?

Cells produce usable energy by breaking down glucose to lower-energy compounds and capturing some of the released energy as ATP. In glycolysis, glucose is metabolized in the fluid portion of the cytoplasm to two molecules of pyruvate, generating two ATP molecules. In the absence of oxygen, pyruvate is converted by fermentation to lactate or ethanol and CO_2. If oxygen is available, the pyruvates are metabolized to CO_2 and H_2O through cellular respiration in the mitochondria, generating much more ATP than does fermentation.

Figure 8-7 and Table 8-1 summarize the locations, major mechanisms, and overall energy harvest for the complete metabolism of glucose from glycolysis through cellular respiration.

2) How Is the Energy in Glucose Harvested During Glycolysis?

During glycolysis, a molecule of glucose is activated by the addition of phosphates from two ATP molecules to form fructose bisphosphate. In a series of reactions, the fructose bisphosphate is broken down into two molecules of pyruvate. These reactions produce four ATP molecules and two NADH electron carriers. Because two ATPs were used in the activation steps, the net yield from glycolysis is two ATPs and two NADHs. Glycolysis, in addition to providing a small yield of ATP, uses up NAD^+ to produce NADH. Once the cell's supply of NAD^+ is consumed, glycolysis must stop. NADH may be regenerated by fermentation, with no additional ATP gain, or by cellular respiration, which also produces additional ATP.

3) How Does Cellular Respiration Generate Still More Energy from Glucose?

If oxygen is available, cellular respiration can occur. The pyruvates are transported into the matrix of the mitochondria. In the matrix each pyruvate reacts with coenzyme A to form acetyl CoA plus CO_2. One NADH is also formed at this step. The two-carbon acetyl group of acetyl CoA enters the Krebs cycle, which releases the remaining 2 carbons as CO_2. One ATP, 3 NADHs, and 1 $FADH_2$ are also formed for each acetyl group that goes through the cycle. At this point, each glucose molecule has produced 4 ATPs (2 from glycolysis and 1 from each acetyl CoA via the Krebs cycle), 10 NADHs (2 from glycolysis, 1 from each pyruvate during the formation of acetyl CoA, and 3 from each acetyl CoA during the Krebs cycle), and 2 $FADH_2$s (1 from each acetyl CoA during the Krebs cycle).

The NADHs and $FADH_2$s deliver their energetic electrons to the proteins of the electron transport system embedded in the inner mitochondrial membrane. The energy of the electrons is used to pump hydrogen ions across the inner membrane from the matrix to the intermembrane compartment. At the end of the electron transport system, the depleted electrons combine with hydrogen ions and oxygen to form water. This is the oxygen-requiring step of cellular respiration. During chemiosmosis, the hydrogen ion gradient created by the electron transport system is used to produce ATP, as the hydrogen ions diffuse back across the inner membrane through channels in ATP-synthesizing enzymes. Electron transport and chemiosmosis yield 32 to 34 additional ATPs, for a net yield of 36 to 38 ATPs per glucose molecule.

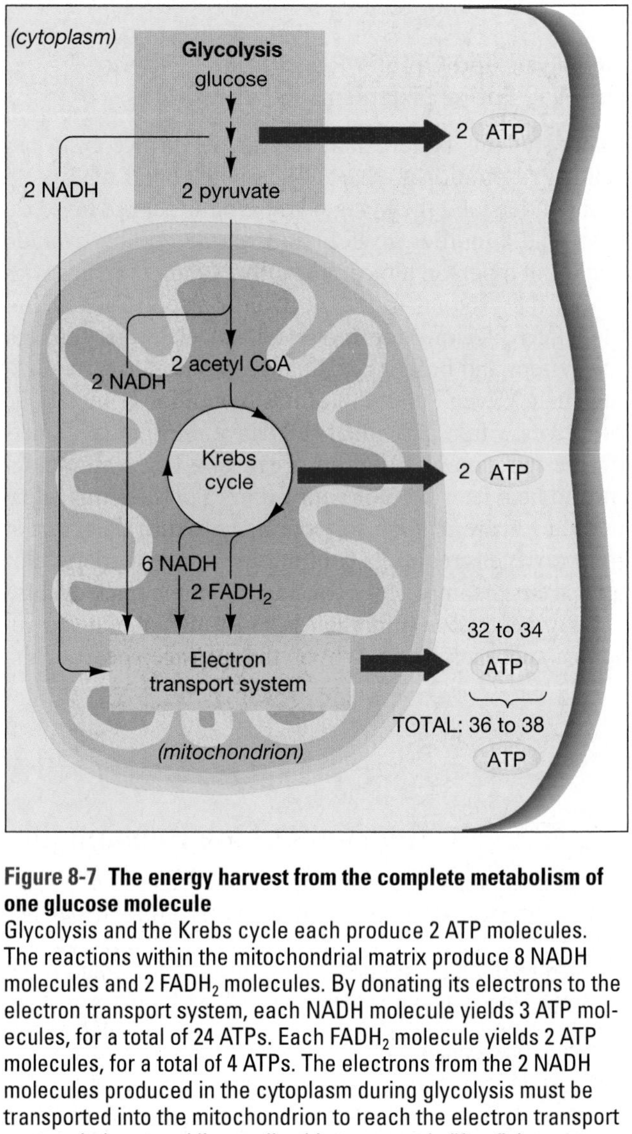

Figure 8-7 The energy harvest from the complete metabolism of one glucose molecule
Glycolysis and the Krebs cycle each produce 2 ATP molecules. The reactions within the mitochondrial matrix produce 8 NADH molecules and 2 $FADH_2$ molecules. By donating its electrons to the electron transport system, each NADH molecule yields 3 ATP molecules, for a total of 24 ATPs. Each $FADH_2$ molecule yields 2 ATP molecules, for a total of 4 ATPs. The electrons from the 2 NADH molecules produced in the cytoplasm during glycolysis must be transported into the mitochondrion to reach the electron transport system. In heart and liver cells, this transport is "free"; in most cells, transport costs 1 ATP per NADH. The 2 "glycolytic NADH" molecules therefore yield either 4 or 6 ATP molecules, depending on the cell type. Consequently, the energy harvest from electron transport is 32 to 34 ATPs. Including 2 ATPs from glycolysis and 2 ATPs from the Krebs cycle, the total energy yield from glucose metabolism is 36 to 38 ATPs.

Key Terms

cellular respiration *p. 137*
chemiosmosis *p. 142*
electron transport system *p. 140*
fermentation *p. 135*
glycolysis *p. 132*
intermembrane compartment *p. 137*
Krebs cycle *p. 139*
matrix *p. 137*
pyruvate *p. 132*

Thinking Through the Concepts

Multiple Choice

1. *Where does glycolysis occur?*
   a. cytoplasm
   b. matrix of mitochondria
   c. inner membrane of mitochondria
   d. outer membrane of mitochondria
   e. stroma of chloroplast

2. *Where does respiratory electron transport occur?*
   a. cytoplasm
   b. matrix of mitochondria
   c. inner membrane of mitochondria
   d. outer membrane of mitochondria
   e. stroma of chloroplast

3. *What is the product of the fermentation of sugar by yeast in bread dough that is essential for the rising of the dough?*
   a. lactate
   b. ATP
   c. ethanol
   d. CO_2
   e. O_2

4. *The majority of ATP produced in aerobic respiration comes from*
   a. glycolysis
   b. the Krebs cycle
   c. chemiosmosis
   d. fermentation
   e. photosynthesis

5. *The process that converts glucose into two molecules of pyruvate is*
   a. glycolysis
   b. fermentation
   c. the Krebs cycle
   d. respiratory electron transport
   e. the Calvin-Benson cycle

6. *The process that causes lactate buildup in muscles during strenuous exercise is*
   a. glycolysis
   b. fermentation
   c. the Krebs cycle
   d. respiratory electron transport
   e. the Calvin-Benson cycle

? Review Questions

1. Starting with glucose ($C_6H_{12}O_6$), write the overall reactions for (a) aerobic respiration and (b) fermentation in yeast.
2. Draw a labeled diagram of a mitochondrion, and explain how its structure is related to its function.
3. What role do the following play in respiratory metabolism: (a) glycolysis, (b) mitochondrial matrix, (c) inner membrane of mitochondria, (d) fermentation, and (e) NAD^+?
4. Outline the major steps in (a) aerobic and (b) anaerobic respiration, indicating the sites of ATP production. What is the overall energy harvest (in terms of ATP molecules generated per glucose molecule) for each?
5. Describe the Krebs cycle. In what form is most of the energy harvested?
6. Describe the mitochondrial electron transport system and the process of chemiosmosis.
7. Why is oxygen necessary for cellular respiration to occur?

Applying the Concepts

1. Some years ago a freight train overturned, spilling a load of grain. Because the grain was unusable, it was buried in the embankment. Yeasts are common in the soil. Although there is no shortage of other food locally, the local bear population has created a nuisance by continually uncovering the grain. What do you think has happened to the grain to make them do this, and how is it related to human cultural evolution?
2. In detective novels, "the odor of bitter almonds" is the telltale clue to murder by cyanide poisoning. Cyanide works by attacking the enzyme that transfers electrons from respiratory electron transport to O_2. Why is it not possible for the victim to survive by using anaerobic respiration? Why is cyanide poisoning almost immediately fatal?
3. More than a century ago, French biochemist Louis Pasteur described a phenomenon, now called "the Pasteur effect," in the wine-making process. He observed that in a sealed container of grape juice containing yeast, the yeast will consume the sugar very slowly as long as oxygen remains in the container. As soon as the oxygen is gone, however, the rate of sugar consumption by the yeast increases greatly and the alcohol content in the container rises. Discuss the Pasteur effect on the basis of what you know about aerobic and anaerobic cellular respiration.
4. Some species of bacteria that live at the surface of sediment on the bottom of lakes are facultative anaerobes; that is, they are capable of either aerobic or anaerobic respiration. How will their metabolism change during the summer when the deep water becomes anoxic (deoxygenated)? If the bacteria continue to grow at the same rate, will glycolysis increase, decrease, or remain the same after the lake becomes anoxic? Explain why.

5. The dumping of large amounts of raw sewage into rivers or lakes typically leads to massive fish kills, although sewage itself is not toxic to fish. Similar fish kills also occur in shallow lakes that become covered in ice during the winter. What kills the fish? How might you reduce fish mortality after raw sewage is accidentally released into a small pond containing large bass?
6. Although respiration occurs in all living cells, different cells respire at different rates. Explain why. How could you predict the relative respiratory rates of different tissues in a fish by microscopic examination of cells?
7. Imagine that a starving cell reached the stage where every bit of its ATP was depleted and converted to ADP plus phosphate. If that cell were placed in fresh nutrient broth at this point, will it recover and survive? Explain your answer based on what you know of glucose breakdown.

For More Information

Calder, W. A. "Red-Hot Hummers." *Nature Conservancy,* March/April 1998. Beautiful photos and lively writing describe the trials of the tiny rufous hummingbird as it fuels up for its long migration South.

McCarty, R. E. "H^+-ATPases in Oxidative and Photosynthetic Phosphorylation." *BioScience*, January 1985. A description of the structure and function of the ATP-synthesizing enzymes in mitochondria and chloroplasts.

Nelson, M., Burgess, T. L., Alling, A., Alverez-Romo, N., Dempster, W. F., Walford, R. L., and Allen, J. P. "Using a Closed Ecological System to Study Earth's Biosphere." *BioScience*, 1993. An artificial ecosystem allows scientists to learn more about how natural ecosystems function.

Answers to Multiple-Choice Questions

1. a 2. c 3. d 4. c 5. a 6. b

MEDIATUTOR
Glycolysis and Cellular Respiration

CD Activities

Activity 8.1: Glucose Metabolism

Estimated Time: 15 minutes

The mechanisms that cells use to derive energy from the sugar glucose are explored. The passage of glucose through glycolysis and the Krebs cycle are detailed, along with the fates of products of glycolysis under both anaerobic and aerobic conditions. The oxidative phosphorylation of ATP is also presented.

Start the MediaTutor Student CD-ROM and enter the activity number in the Quick Search box to be taken directly to that activity.

Web Investigations

Case Study: The Flight of the Hummingbird

Estimated time: 5 minutes

Myth #1: Hummingbirds live on a strict diet of sugar water (nectar). Myth #2: They obtain this nectar by sucking it out of flowers, using their beaks as straws. Fact: The feeding behaviors and diet of hummingbirds are perfectly adapted to their energy needs. What, how, and where do hummingbirds eat?

Go to http://www.prenhall.com/audesirk6, the Audesirk Companion Web site. Select Chapter 8 and the Web Investigation to begin.

UNIT TWO

Inheritance

Inheritance provides for both similarity and difference. These fruits are all apples and share many similarities because their genes are nearly identical. The differences in their color, shape, size, and taste result from tiny differences in their genes.

"A structure this pretty just had to exist."

James Watson in* The Double Helix *(1968)

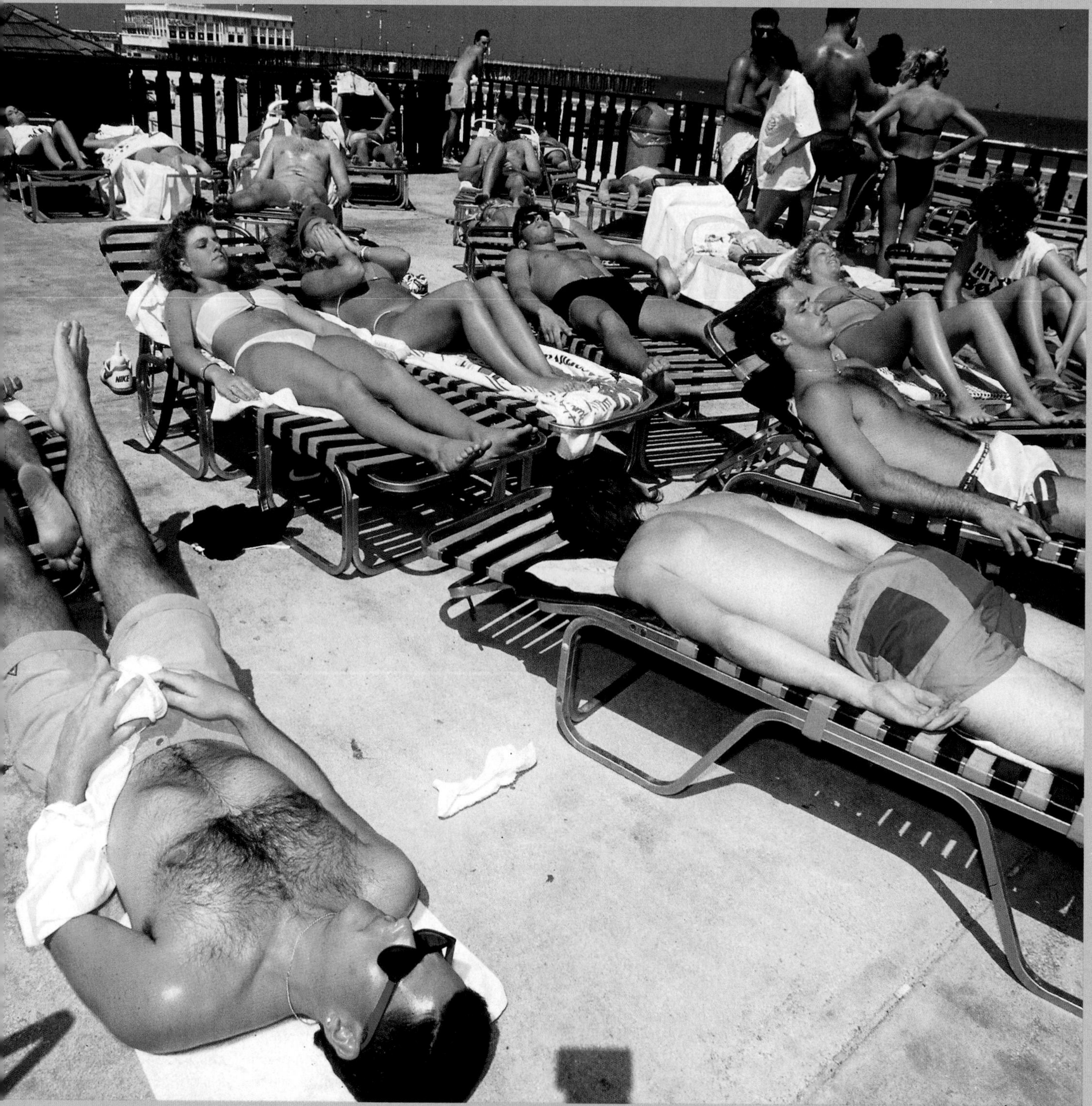

Within the famous double helix structure of DNA lies the explanation of the link between sunbathing and an increased risk of developing skin cancer.

9 DNA: The Molecule of Heredity

AT A GLANCE

Case Study: Sunshine Perils

1) **How Did Scientists Discover That Genes Are Made of DNA?**

Transformed Bacteria Revealed the Link Between Genes and DNA

2) **What Is the Structure of DNA?**

DNA Is Composed of Four Nucleotides

DNA Is a Double Helix of Two Nucleotide Strands

Hydrogen Bonds Between Complementary Bases Hold the Two DNA Strands Together

The Order of Nucleotides in DNA Can Encode Vast Amounts of Information

3) **How Does DNA Replication Ensure Genetic Constancy?**

The Replication of DNA Is a Critical Event in a Cell's Life

DNA Replication Produces Two DNA Double Helices, Each with One Old Strand and One New Strand

DNA Helicase Separates the Parental DNA Strands

DNA Polymerase Synthesizes New DNA Strands

One DNA Strand Is Synthesized in Short Segments That Are Connected by DNA Ligase

Proofreading Produces Almost Error-Free Replication of DNA

But Mistakes Do Happen

Case Study Revisited: Sunshine Perils

CASESTUDY

Sunshine Perils

Given Rachel's hectic schedule as a college junior, skin cancer was the last thing on her mind. Compared to varsity swimming, her studies, and her part-time job, the bumpy, black mole on her back didn't seem important. Rachel would have ignored it completely, but her swim coach asked her to have it checked by a physician. So she scheduled an appointment with her family doctor to have the mole removed. The doctor told her that he could remove the mole in his office and the wound should heal in time for the next swim meet, leaving almost no scar.

After the appointment, Rachel didn't think about the mole at all. However, her doctor called back three days later. Following his general policy, he had sent the tissue to a laboratory for examination. The diagnosis was a type of cancer called *melanoma.*

Melanoma is a relatively common skin cancer that usually begins in pigmented cells in the lower parts of the skin. The cancer can spread to other parts of the body, including internal organs. The resulting disease is challenging to treat and frequently deadly. The American Dermatology Association estimates that more than 44,000 people in the United States will be diagnosed with melanoma this year. Likewise, about 7000 people are expected to die as a result of melanoma. The incidence of melanoma appears to be rising. For example, between 1998 and 1999, dermatologists noted a 6% increase in its occurrence. Many physicians suspect that the cause may be increased exposure to sunlight.

How can sunlight, the carrier of life-giving energy for the planet, cause cancer? To answer this question, we need to understand the basic structure of DNA, the hereditary molecule. ■

1 How Did Scientists Discover That Genes Are Made of DNA?

Just 60 years ago, no one knew that *deoxyribonucleic acid,* or **DNA**, is the molecule that carries the blueprints for all forms of life on Earth. We now know that the molecular "instructions" in DNA direct the life of each cell in an organism, and confer on each cell its specialized characteristics. DNA also enables organisms, or cells within an organism, to transmit information accurately from one generation the next generation. The discovery of how DNA carries life's blueprints is one of the crowning achievements of 20th-century biology.

Like most scientific breakthroughs, discovering DNA's structure and how it worked required the incremental advances of dozens of scientists over a period of decades. By the late 1800s, scientists had learned that heritable information exists in discrete units called **genes**. However, they could not provide a precise definition of a gene. Scientists merely knew that genes determine many of the heritable differences among individuals within a species. For example, a gene for flower color determines whether an individual pea plant has white flowers or purple flowers. Studies of dividing cells provided strong evidence that genes are located in threadlike structures within cells, called **chromosomes**. Scientists also discovered that chromosomes contain both DNA and **protein**, indicating that genes are made of either DNA or protein. For the first half of the 20th century, most scientists thought that genes were made of the protein components of the chromosomes. However, experiments using bacteria eventually provided clear proof that genes are made of DNA.

Transformed Bacteria Revealed the Link Between Genes and DNA

In the 1920s a British researcher named Frederick Griffith was trying to make a vaccine to prevent pneumonia infections, a major cause of death at that time. Some vaccines are made using a weakened strain of the bacteria or virus that doesn't cause illness. Injecting this weakened but living strain into an animal can promote immunity against the disease-causing strains. Other vaccines use disease-causing bacteria or viruses that have been killed by exposure to heat or chemicals. Griffith was trying to make a vaccine using two strains of the *Streptococcus pneumonia* bacterium. One strain, R, did not cause pneumonia when injected into mice (Fig. 9-1a). The other strain, S, was deadly when injected, causing pneumonia and killing the mice in a day or two (Fig. 9-1b). When the S strain was heated to kill it and then injected into mice, it did not cause disease (Fig. 9-1c). So Griffith had two possible vaccines: the living R strain and the heat-killed S strain.

He then did a control experiment, expecting to confirm that these two bacterial strains did not cause pneumonia. He mixed living R-strain bacteria *together with*

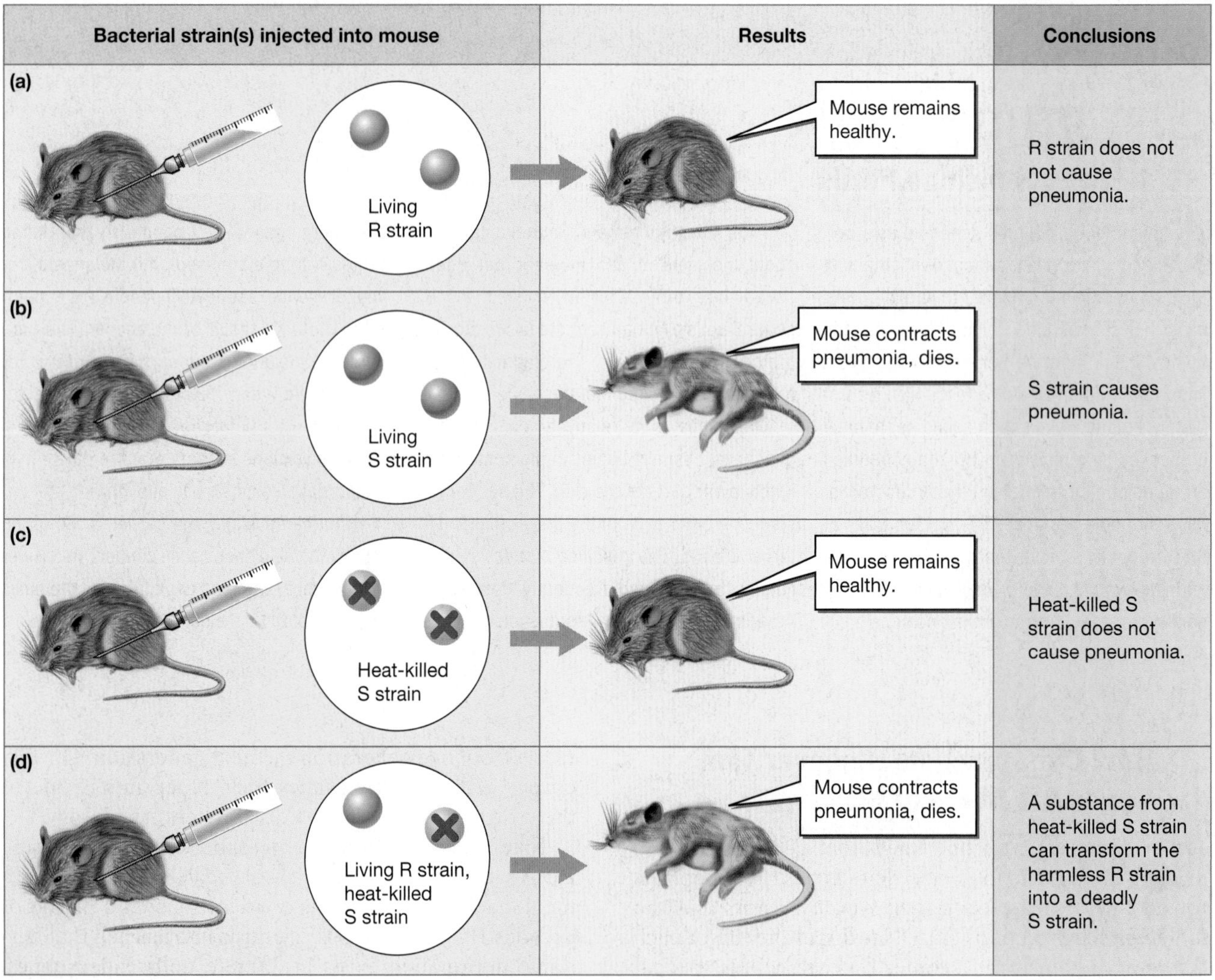

Figure 9-1 Transformed bacteria
Griffith's discovery that bacteria can be transformed from harmless to deadly laid the groundwork for the discovery that DNA contains genes.

heat-killed S-strain bacteria. Then he injected the mixture of strains into mice (Fig. 9-1d). Because neither of these bacterial strains causes pneumonia on its own, he expected the mice to be free of disease and to live. Imagine his surprise when they sickened and died. When he autopsied the mice, he recovered *living* S-strain bacteria from their organs. The simplest interpretation of these results is that some substance in the heat-killed S strains transformed the living but harmless R strain into a deadly S strain. These newly transformed S-strain cells could then multiply and kill the mice. Griffith never made an effective pneumonia vaccine. However, his surprising discovery of transformed bacteria was a critical step in the discovery that genes are made of DNA.

A little more than a decade after Griffith's experiments, three researchers at Rockefeller University began to search for the molecule that could transform bacteria. Oswald Avery, Colin MacLeod, and Maclyn McCarty were able to purify the molecules from the S-strain bacteria that could transform the R strain into a deadly S strain. They were surprised to discover that the molecules were DNA.

We can better interpret the results of Griffith's experiments based on this discovery. Heating the S-strain cells killed them but did not completely destroy their DNA. What happened when the killed S-strain bacteria were mixed in a test tube with living R-strain bacteria? Fragments of DNA from the dead S-strain cells entered into some of the R-strain cells. If these fragments of DNA contained the genes needed to cause disease, an R-strain cell was transformed into a deadly S-strain cell. Thus, these researchers concluded that genes, the heritable units that determine many of an organism's characteristics, are made of DNA. However, many years of additional debate and experimentation were needed to convince everyone of DNA's central role in heredity.

2 What Is the Structure of DNA?

Knowing that genes are made of DNA does not answer critical questions about inheritance: How does DNA encode genetic information? How is DNA duplicated so that information can be accurately passed from one cell to its daughter cells? The secrets of DNA function, and therefore of heredity itself, could be found only by understanding the three-dimensional structure of the DNA molecule.

DNA Is Composed of Four Nucleotides

At the same time that geneticists were studying transformed bacteria, biochemists were learning more about the molecular composition of DNA. Based on their studies, we know that the DNA of every organism on Earth, from bacteria to bison, is composed of four small subunits called **nucleotides**. Each nucleotide in DNA consists of three parts: (1) a phosphate group, (2) a sugar called *deoxyribose*, and (3) one of four possible nitrogen-containing **bases—adenine (A)**, **thymine (T)**, **guanine (G)**, or **cytosine (C)**.

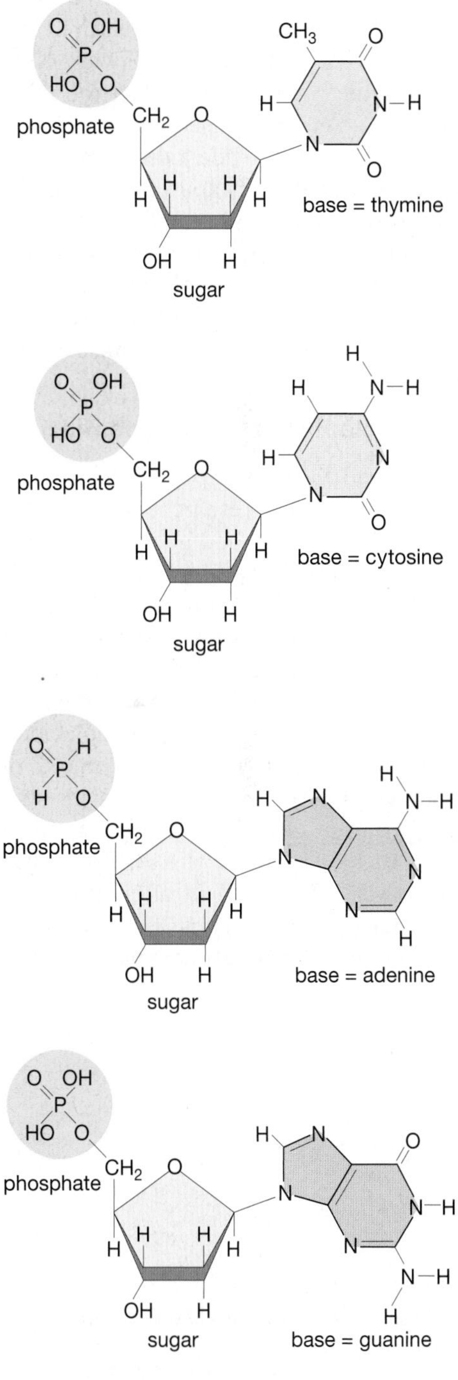

In the 1940s Chargaff analyzed the relative amounts of the four nucleotides in the DNA from various species and found a curious consistency. The DNA of any given species contains *equal amounts of adenine and thymine*, as well as *equal amounts of cytosine and guanine.* Chargaff's observation was certainly significant, if only someone could figure out what it meant about DNA structure.

Determining the structure of any biological molecule is no simple task, even for scientists today. Nevertheless, 50 years ago, several scientists had begun to study DNA in hopes of learning more about its structure. British scientists Maurice Wilkins and Rosalind Franklin used X-ray diffraction to study DNA structure. They bombarded crystals of purified DNA with X-rays and recorded how the X-rays bounced off the DNA molecules (Fig. 9-2a). As you can see, the resulting "diffraction" pattern does not provide a direct picture of DNA structure. However, Wilkins and Franklin used the pattern to determine much about the DNA molecule. First, the DNA molecule is helical: That is, it is twisted like a corkscrew. Second, the DNA molecule has a uniform diameter of 2 nanometers (2 billionths of a meter). Third, the DNA molecule consists of repeating subunits.

DNA Is a Double Helix of Two Nucleotide Strands

The chemical and X-ray diffraction data did not provide enough information for researchers to work out the structure of DNA; some good guesses were also needed. (See "Scientific Inquiry: The Discovery of the Double Helix.") Combining a knowledge of how complex organic molecules bond together with an intuition that "important biological objects come in pairs," James Watson and Francis Crick proposed a new model for DNA. They suggested that the DNA molecule consists of two separate DNA **strands**, or "polymers," of linked nucleotides (Fig. 9-3). Within each DNA strand, the phosphate group of one nucleotide bonds to the sugar of the next nucleotide in the strand. This bonding pattern produces a "backbone" of alternating, covalently bonded sugars and phosphates. The nucleotide bases protrude from the **sugar-phosphate backbone.** All of the nucleotides within a single DNA strand are oriented in the same direction. This consistent orientation makes the two ends of a single DNA strand different: One end of the DNA strand has a "free" or unbonded sugar and the other end of the DNA strand has a "free" or unbonded phosphate.

Hydrogen Bonds Between Complementary Bases Hold the Two DNA Strands Together

Watson and Crick proposed that two DNA strands are held together by hydrogen bonds that form between the protruding bases of the two separate DNA strands. These bonds give DNA a ladder-like structure, with the sugar-phosphate backbones on the outside and the nucleotide bases on the inside (Fig. 9-3b). However, the DNA strands are not straight. Instead, they are twisted about each other to form a **double helix**, much like a ladder twisted lengthwise into a circular staircase shape. In addition to twisting around each other in the double helix, the two DNA strands in a DNA double helix are oriented in opposite directions. At each end of the double helix, one DNA strand ends with a free phosphate and the other DNA strand ends with a free sugar (Fig. 9-3a).

Take a closer look at the pairs of hydrogen-bonded bases that form each rung of the double helix ladder. Notice that adenine forms hydrogen bonds only with thymine and that guanine forms hydrogen bonds only with cytosine (Fig. 9-3c). These A–T and G–C pairs are called **complementary base pairs**, and their presence explains Chargaff's results. Because an A in one DNA strand always pairs with a T in the other strand, the amount of A in DNA always equals the amount of T. Similarly, because a G in one DNA strand always pairs with a C in the other DNA strand, the amount of G always equals the amount of C.

The structure of DNA was solved. At The Eagle Pub in Cambridge, England, on March 7, 1953, Francis

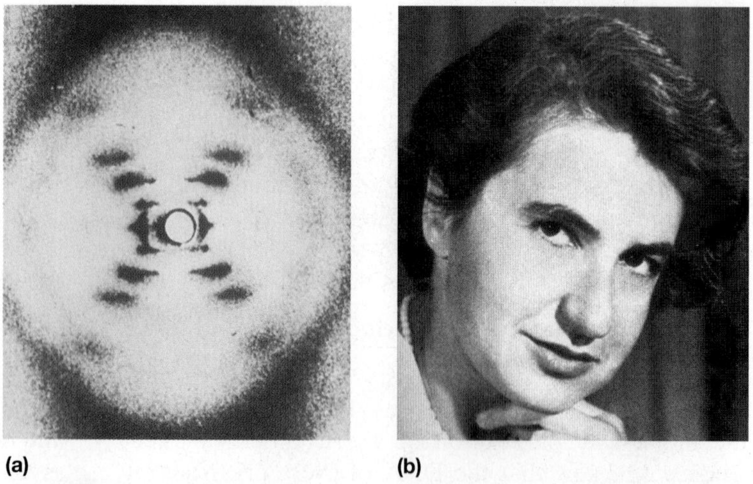

Figure 9-2 X-ray diffraction studies of DNA taken by Rosalind Franklin
(a) The X formed of dark spots is characteristic of helical molecules such as DNA. Measurements of various aspects of the pattern indicate the dimensions of the DNA helix; for example, the distance between the dark spots corresponds to the distance between turns of the helix. ***(b)*** Rosalind Franklin, before her untimely death at the age of 37 in 1958, published about 40 scientific papers.

(a) (b)

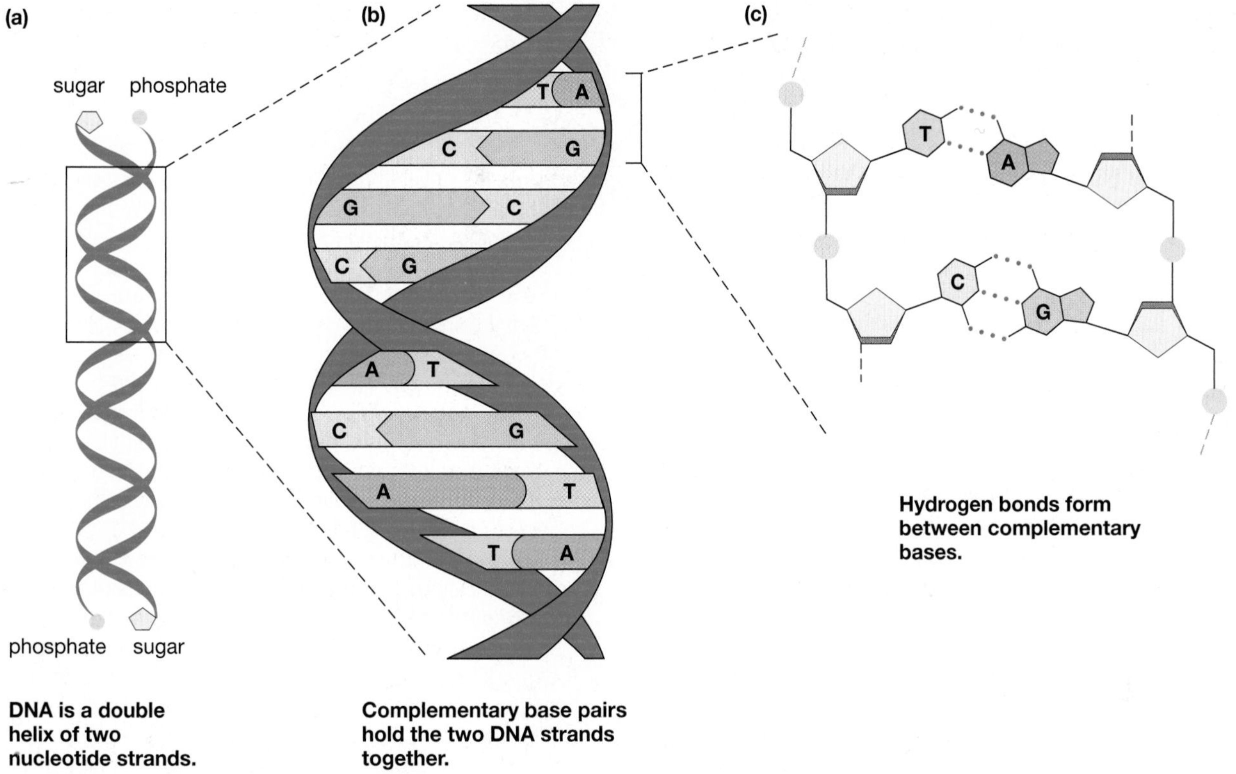

Figure 9-3 The Watson-Crick model of DNA structure
(a) Two strands of DNA wind about each other in a double helix, like a twisted ladder. The two DNA strands run in opposite directions. This directionality is especially clear at the ends of the double helix, where the terminal nucleotide of one strand has an unbonded ("free") sugar and the terminal nucleotide on the other strand has an unbonded ("free") phosphate. ***(b)*** Complementary base pairs (adenine and thymine, guanine and cytosine) hold the two DNA strands together. ***(c)*** Hydrogen bonding between specific base pairs in the center of the helix hold the two strands together. Three hydrogen bonds hold guanine to cytosine; two hydrogen bonds hold adenine to thymine.

MEDIATUTOR 9.1 DNA Structure

Crick proclaimed to the lunchtime crowd, "We have discovered the secret of life." This claim was not far from the truth. Although further data would be needed to confirm the details, within just a few years, the model revolutionized biology, from genetics to medicine. As we will later see, the revolution continues to change biology today in ways that neither Watson nor Crick imagined.

The Order of Nucleotides in DNA Can Encode Vast Amounts of Information

Look again at the elegant structure of DNA shown in Figure 9-3. Can you see why many scientists had trouble thinking of DNA as the carrier of genetic information? Consider the many characteristics of just one type of organism. How can the color of a bird's feathers, the size and shape of its beak, the ability to make appropriate nests for its chicks, the bird's song, and the bird's ability to navigate between Canada and Mexico every year all be determined by a molecule with just four simple parts?

Although it took many more experiments to answer this question, scientists eventually found evidence to support an equally simple explanation. It's not the *number* of different subunits, but their *order* that's important. Within a DNA strand, the four types of bases can be arranged in any linear order. Each sequence of bases represents a unique set of genetic instructions, like a biological Morse code. (The Morse code represents the entire alphabet, and therefore the entire English language, with only two symbols.) A stretch of DNA that is just 10 nucleotides long can have more than a million possible sequences of the four bases. Because an organism has millions (in bacteria) to billions (in plants or animals) of nucleotides, DNA molecules can encode a staggering amount of information.

So we know that genes are present on DNA and that DNA is a double helix of two nucleotide strands held together by complementary base pairs. We've come a long way, but much remains to be discovered. One major question is this: How are identical copies of DNA formed when cells divide?

Scientific Inquiry
The Discovery of the Double Helix

In the early 1950s many biologists realized that the key to understanding inheritance lay in the structure of DNA. They also knew that whoever deduced the correct structure of DNA would receive recognition, very possibly the Nobel Prize. Linus Pauling of Caltech was the person most likely to solve DNA structure. Pauling probably knew more about the chemistry of large organic molecules than did any person alive. Like Rosalind Franklin and Maurice Wilkins, Pauling was an expert in X-ray diffraction techniques. In 1950 he used these techniques to show that many proteins were coiled into single-stranded helices (see Chapter 3). Pauling, however, had two main handicaps. First, for years he had concentrated on protein research, and therefore he had little data about DNA. Second, he was active in the peace movement. At that time some government officials, including Senator Joseph McCarthy, considered such activity to be potentially subversive and possibly dangerous to national security. This latter handicap may have proved decisive.

The second most likely competitors were Wilkins and Franklin, the British scientists who had set out to determine the structure of DNA by using X-ray diffraction patterns. In fact, they were the only scientists who had very good data about the general shape of the DNA molecule. Unfortunately for them, their methodical approach was also slow.

The door was open for the eventual discoverers of the double helix, James Watson and Francis Crick, two scientists with neither Pauling's tremendous understanding of chemical bonds nor Franklin and Wilkins's expertise in X-ray analysis. Watson and Crick did no experiments in the ordinary sense of the word; instead, they spent their time thinking about DNA, trying to construct a molecular model that made sense and fit the data. Because they were working in England and because Wilkins was very open about his and Franklin's data, Watson and Crick were familiar with all the X-ray information relating to DNA. This information was just what Pauling lacked. Because of Pauling's presumed subversive tendencies, the U.S. State Department refused to issue him a passport to leave the United States, so he could neither attend meetings at which Wilkins presented the X-ray data nor visit England to talk with Franklin and Wilkins directly. Watson and Crick knew that Pauling was working on DNA structure and were terrified that he would beat them to it. In his book *The Double Helix,* Watson recounts his belief that, if Pauling could have seen the X-ray pictures, "in a week at most, Linus would have the structure."

You might be thinking by now, "But wait just a minute! That's not fair. If the goal of science is to advance knowledge, then everyone should have access to all the data. If Pauling was the best, he should have discovered the double helix first." Perhaps so. But science is an activity of scientists, who, after all, are people too. Although virtually all scientists want to see the advancement and benefit of humanity, each individual also wants to be the one responsible for that advancement and to receive the credit and the glory. Linus Pauling remained in the dark about the correct X-ray pictures of DNA and was beaten to the correct structure (Fig. E9-1). When Watson and Crick discovered the double helix structure of DNA, Watson described it in a letter to Max Delbruck, a friend and advisor at Caltech. He asked Delbruck not to reveal the contents of the letter to Pauling until their structure was formally published. Delbruck, perhaps more of a model scientist, firmly believed that scientific discoveries belong in the public domain and promptly told Pauling all about it. With the class of a great scientist and a great person, Pauling graciously congratulated Watson and Crick on their brilliant solution to the DNA structure. The race was over.

Figure E9-1 The discovery of DNA
James Watson and Francis Crick with a model of DNA structure.

3 How Does DNA Replication Ensure Genetic Constancy?

The Replication of DNA Is a Critical Event in a Cell's Life

Cells reproduce by a complex process of cell division that produces two daughter cells from a single parental cell. To produce healthy daughter cells, cell division must ensure that each daughter cell receives a copy of all of the parent cell's genetic information. Consequently, at an early stage of cell division, the parent cell must synthesize two exact copies of its DNA, a process known as **DNA replication**. Many cells in an adult human never divide at all and therefore do not replicate their DNA. In most of the thousands of cells that *do* divide, initiation of DNA replication irreversibly commits the cell to division. If a cell tries to replicate its DNA without stockpiling enough raw materials or energy to complete the process, it can die. Consequently, the timing of replication is exquisitely regulated. Such regulation ensures that DNA replication does not begin unless and until a cell is ready to divide. These controls also ensure that the cell's DNA is replicated exactly one time—no more, no less—prior to each cell division.

DNA Replication Produces Two DNA Double Helices, Each with One Old Strand and One New Strand

Once the "decision" is made to divide, the cell replicates its DNA (see overview in Fig. 9-4). Recall that the cell's DNA is contained within structures called *chromosomes*. Each chromosome contains a single DNA double helix. DNA replication produces two identical DNA double helices, each of which will be passed, within its chromosome, to one of the new daughter cells.

DNA replication begins when enzymes pull apart the parental DNA double helix, so that the bases of the two parental DNA strands no longer form base pairs with one another. Other enzymes move along each separated parental DNA strand, selecting **free nucleotides** with bases that are complementary to the parental DNA strand. The enzymes join these free nucleotides to form two new DNA strands, each complementary to one of the parental DNA strands.

When replication is complete, one parental DNA strand and its newly synthesized, complementary daughter DNA strand wind together into one double helix. At the same time, the other parental strand and its daughter strand wind together into a second double helix. In forming a new double helix, the process of DNA replication conserves one parental DNA strand and produces one newly synthesized strand. Hence the process is called **semiconservative replication**.

These new double helices are held together while the cell prepares for division. The chromosome is now called a *duplicated chromosome*. Each duplicated chromosome contains two chromatids (or sister chromatids). Within each chromatid is an identical double helix of DNA, a product of DNA replication. When sister chromatids separate during cell division, one chromatid is delivered to each daughter cell. Thus, both daughter cells have exactly the same genetic information as the original parent cell. In this way, the integrity of genetic information is maintained from cell division to cell division. We will take a closer look at the process of replication in the following sections.

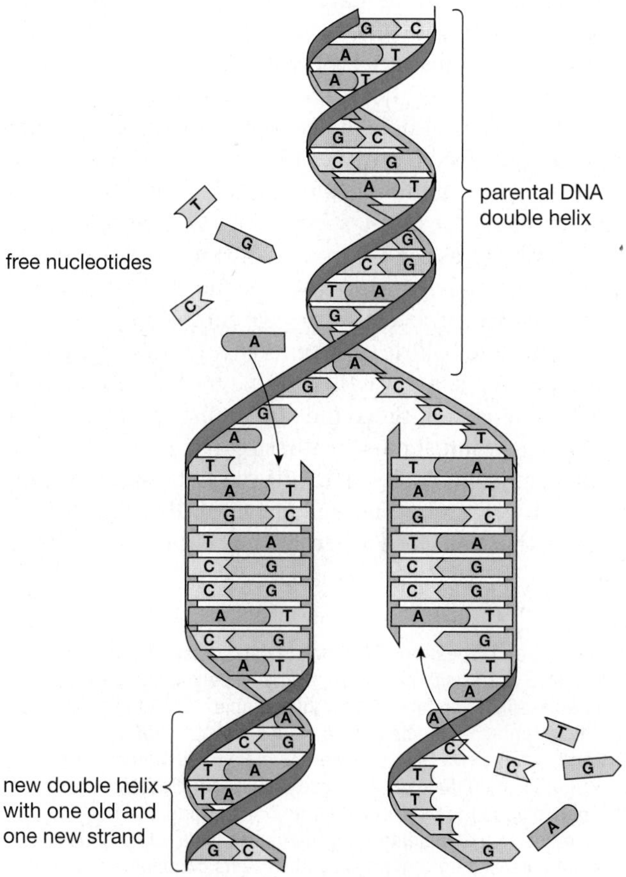

Figure 9-4 Basic features of DNA replication
During replication, enzymes separate the parental DNA double helix, breaking the hydrogen bonds between complementary bases. Other enzymes select complementary nucleotides and add them to the growing daughter strands. Each parental strand and its new daughter strand form a new double helix.

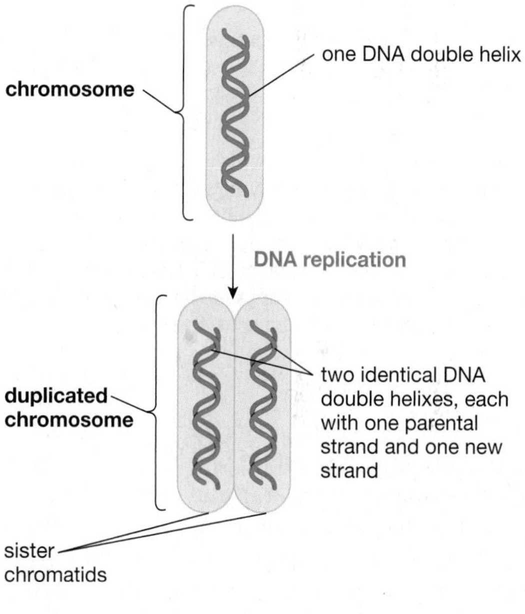

DNA Helicase Separates the Parental DNA Strands

DNA replication requires the action of dozens of enzymes. One key enzyme is **DNA helicase**, "an enzyme that breaks apart the helix." Together with related enzymes, DNA helicase uses energy from ATP to break the hydrogen bonds between complementary base pairs that hold the two parental DNA strands together. This activity separates and unwinds the parental DNA double helix, forming a **replication "bubble."** Within the replication bubble, the nucleotide bases of the parental DNA strands are no longer paired with one another. Each replication bubble contains two replication "forks" where the two parental DNA strands have not yet been unwound (Fig. 9-5). These replication forks can resemble a fork in a road. Imagine you are traveling down the parental DNA double helix on one side of a

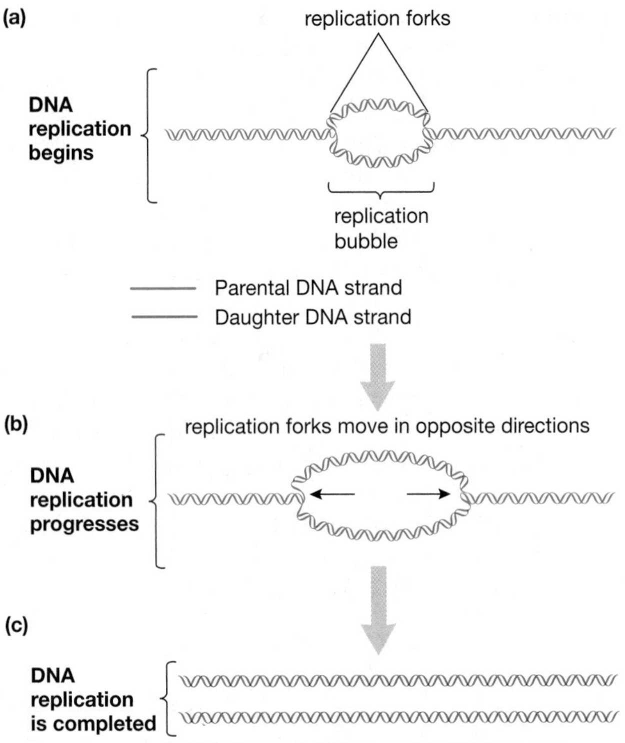

Figure 9-5 Replication bubble
DNA replication begins when DNA helicase and related enzymes unwind portions of the parental DNA double helix, creating a replication bubble. In complex cells, DNA replication occurs simultaneously at many locations on the parental DNA double helix, resulting in multiple replication bubbles. DNA replication is completed when all adjacent replication bubbles meet to form two complete and separate DNA double helices.

replication bubble. When you reach a replication bubble, the two DNA strands separate; you must choose whether to follow one parental DNA strand or the other.

In eukaryotic cells, many replication bubbles occur simultaneously on each chromosome so that all of the DNA can be replicated in time for cell division to occur properly. The bubbles grow as DNA replication progresses until the bubbles finally meet up as replication is completed.

DNA Polymerase Synthesizes New DNA Strands

DNA polymerase ("an enzyme that makes a DNA polymer") plays a critical role in the synthesis of new DNA strands. At each replication fork, DNA polymerase and other enzymes synthesize two new DNA strands that are complementary to the two parental strands. During this process, DNA polymerase recognizes an unpaired nucleotide base in the parental strand and matches it up with a free nucleotide that has the correct complementary base. For example, DNA polymerase pairs up an exposed adenine base in the parental strand with a thymine base in a free nucleotide. Then DNA polymerase catalyzes formation of new covalent bonds that link the phosphate of the incoming free nucleotide to the sugar of the previously added nucleotide in the growing daughter strand. In this way, DNA polymerase synthesizes the sugar-phosphate backbone of the daughter strand (Fig. 9-6).

One DNA Strand Is Synthesized in Short Segments That Are Connected by DNA Ligase

Like other enzymes, DNA polymerase is extremely specific. It can only add new nucleotides onto the free sugar end of the new DNA strand that it is making. Because the two strands of the parental DNA double helix are oriented in opposite directions, the new complementary DNA strands must also be synthesized in opposite directions. Thus, as DNA helicase moves along the parental DNA double helix, separating the parental strands, one DNA polymerase "moves" along in the same direction, adding nucleotides to form a long continuous daughter DNA strand (Fig. 9-6a). The second DNA polymerase must "move" in the opposite direction, adding nucleotides to a second daughter DNA strand. However, because the parental double helix is only unwound a little at a time, this DNA polymerase can only go a short way before it runs into a region of DNA that is not unwound and must stop (see Fig. 9-6a). As a result, it must synthesize a daughter strand in small segments. These segments are subsequently connected together by another enzyme, **DNA ligase** ("an enzyme that ties DNA together"). The joining process is

Figure 9-6 (opposite page) Details of DNA replication
(a) One DNA strand can be synthesized as a long, continuous strand. The other DNA strand must be synthesized as a series of short segments that are connected by DNA ligase. ***(b)*** Synthesis of new DNA strands involves complex enzyme machinery. ① A large complex of enzymes assembles at each replication fork. Within this complex, DNA helicase separates the two DNA strands, unwinding a small portion of the parental double helix. The complex also contains two DNA polymerase molecules, one attached to each parental strand. ② The two DNA polymerase molecules match up complementary free nucleotides with the parental DNA strands and join up their sugar-phosphates to form the backbones of the new DNA strands. Because nucleotides in DNA can only be added to the free sugar end of a strand, the DNA polymerase on one strand (upper strand in this diagram) can synthesize a complementary DNA strand in a continuous segment. The other DNA polymerase (on the lower, looping strand in this diagram) must synthesize its complementary DNA strand in small segments as the DNA replication machinery unwinds more and more of the original DNA double helix. ③ The replication machinery advances, unwinding more parental DNA and synthesizing more complementary daughter DNA. ④ Another enzyme, DNA ligase, joins together the segments of newly synthesized DNA, finally producing a continuous daughter strand that is complementary to the original strand. Replication produces two DNA double helices, each of which has one strand from the original DNA double helix and one strand that has just been synthesized.

MEDIATUTOR 9.2 DNA Replication

(a) Continuous versus segmented DNA replication

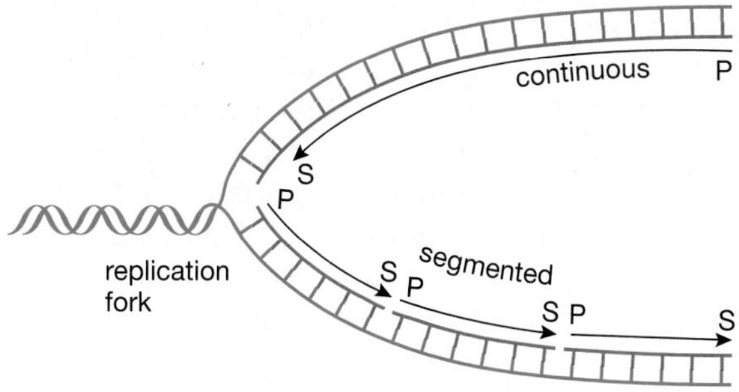
continuous
P
S
P
replication fork
segmented
S P
S P
S

(b) Detailed view of DNA replication

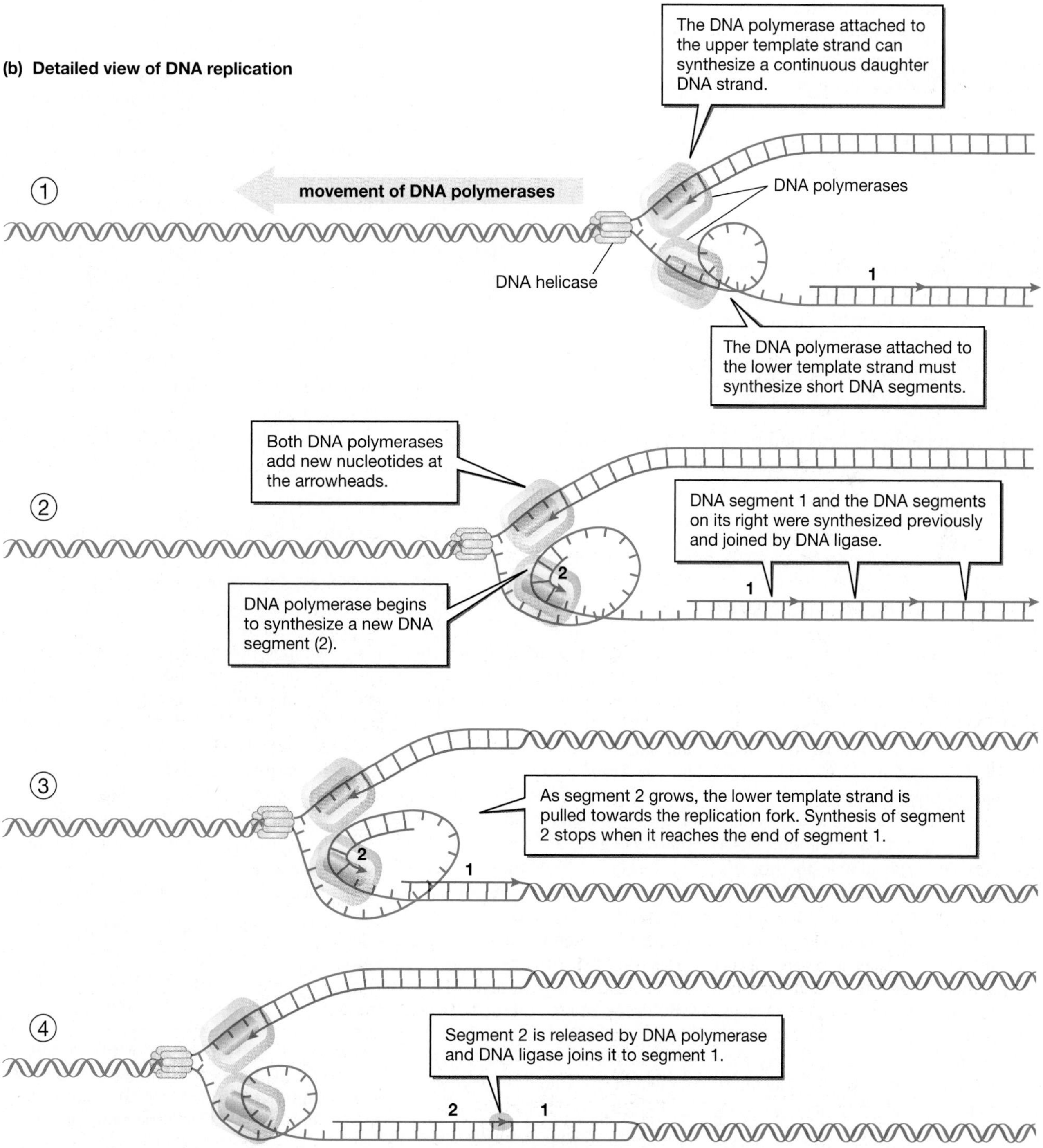
The DNA polymerase attached to the upper template strand can synthesize a continuous daughter DNA strand.
①
movement of DNA polymerases
DNA polymerases
DNA helicase
1
The DNA polymerase attached to the lower template strand must synthesize short DNA segments.
Both DNA polymerases add new nucleotides at the arrowheads.
②
DNA segment 1 and the DNA segments on its right were synthesized previously and joined by DNA ligase.
2
1
DNA polymerase begins to synthesize a new DNA segment (2).
③
As segment 2 grows, the lower template strand is pulled towards the replication fork. Synthesis of segment 2 stops when it reaches the end of segment 1.
2
1
④
Segment 2 is released by DNA polymerase and DNA ligase joins it to segment 1.
2
1

repeated as many as 10 million times for a single human chromosome until the daughter strand has been completely synthesized. DNA ligase also connects the DNA strands synthesized by adjacent replication forks and plays an important role in repairing DNA that has been damaged by sunlight.

Recent experiments provide new insights into the structure of DNA polymerase and its interaction with the parental DNA strands (Fig. 9-6b). DNA polymerase partly encircles the parental DNA strand, positioning the parental strand within a groove in the DNA polymerase structure. It appears that DNA polymerase pulls the parental strand through this groove as it synthesizes the new daughter DNA strand. Thus, the DNA polymerase appears to be stationary, while the parental DNA strand moves through it. Both DNA polymerase assemblies are oriented in the same direction. One of the parental DNA strands passes straight through the DNA polymerase. This parental strand can be used by DNA polymerase to synthesize a new complementary strand that is long and continuous. However, the other parental strand forms a loop to place it in the proper orientation for DNA polymerase. The new DNA strand that is complementary to this looping parental DNA strand must be synthesized in short segments (see Fig. 9-6b).

Proofreading Produces Almost Error-Free Replication of DNA

The specificity of hydrogen bonding between complementary base pairs makes DNA replication highly accurate. Nevertheless, no process in cells, not even DNA replication, is perfect. Partly because replication is so fast (up to about 700 nucleotides per second), DNA polymerase matches bases incorrectly about once in every 10,000 base pairs. However, the completed DNA strands contain only about one mistake in every billion base pairs. This phenomenal accuracy is ensured by a variety of DNA repair enzymes, including some forms of DNA polymerase. These enzymes "proofread" each daughter strand during and after its synthesis and make any necessary repairs.

But Mistakes Do Happen

Despite this amazing accuracy, neither we nor any other life-forms are free of DNA damage. In addition to mistakes made during DNA replication, the DNA in each cell in your body loses about 10,000 bases every day due to spontaneous chemical breakdown just because your temperature is about 98.6 °F! A variety of environmental conditions can also damage DNA. For example, whenever you go out in the sunshine, DNA in some of your skin cells is damaged by ultraviolet light. Although special DNA repair enzymes are continuously on call to repair this damage, inevitably some errors are not repaired. A cell with DNA damage may function normally (for example, if the damage occurs in a noncritical stretch of DNA), it may survive but not function as efficiently as before, or it may die. Deterioration in the accuracy of DNA replication as people get older may contribute to the aging process, a topic that is explored in Chapter 10 in "Health Watch: Sex, Aging, and Mutations."

REVISITED CASESTUDYREVISITED CASESTUDYREVISITEDC

Sunshine Perils

Even on a cloudy day, ultraviolet rays in sunlight penetrate our uncovered skin and sometimes directly hit nucleotides in our DNA. Thymine and cytosine are particularly prone to UV damage. For example, if the UV rays hit a DNA strand with two adjacent thymines, the two thymines can become improperly linked together. Unless the damage is repaired, DNA polymerase may be unable to properly replicate this region of DNA and may insert incorrect nucleotides into the daughter strand. If the damage occurs in one of the many genes involved in control of cell division, the progeny of the damaged cell may begin to undergo a series of changes that can ultimately form a cancer.

When you get a sunburn, many skin cells sustain more DNA damage than they can repair. These severely damaged cells commit a kind of cellular suicide. That's why you peel after a bad sunburn: The dead cells are being sloughed off. If less damage occurs, special repair enzymes can repair the damaged DNA strand. People who lack one or more of the enzymes needed to repair UV-induced damage can have a variety of disorders, including xeroderma pigmentosum, that make them very sensitive to sunlight and prone to getting cancers in general.

Rachel had frequently been sunburned as a child, and all that sun may have caused Rachel's cancer. Studies indicate that getting sunburned when we are kids raises our risk of developing melanoma about 3 times higher than that of those who were not sunburned. People with darker skin who burn less easily are 15 times less likely to develop melanomas, although they are not totally immune.

Because her melanoma was caught at a very early stage, Rachel's prognosis for total recovery is good. You, too, can learn to spot potential problems before they become deadly, since recognizing a possible melanoma is as easy as ABCD: Look at your moles for *A*symmetry, irregular *B*order, irregular *C*olor, or a *D*iameter larger than the eraser on the end of a pencil. Check out this text's Web site to learn some tips for recognizing potential problems. And be sure to visit your doctor promptly to have any suspicious spots checked out thoroughly.

Clearly, DNA damage can lead to cancers. Many scientists hypothesize that DNA damage may also be responsible for some aspects of aging. Why could DNA damage lead to certain characteristics of aging, such as wrinkles, thinning hair, and slower healing? Assuming this hypothesis is true, what actions might people take to slow the aging process?

Summary of Key Concepts

1) How Did Scientists Discover That Genes Are Made of DNA?

By the turn of the century, scientists knew that genes must be made of either protein or DNA. Studies by Griffith showed that genes can be transferred from one bacterial strain into another. This transfer could transform the bacterial strain from harmless to deadly. Avery, MacLeod, and McCarty showed that DNA was the molecule that could transform bacteria. Thus, genes must be made of DNA.

2) What Is the Structure of DNA?

DNA is composed of subunits called *nucleotides*, linked together into long strands. Each nucleotide consists of a phosphate group, the five-carbon sugar deoxyribose, and a nitrogen-containing base. Four types of bases occur in DNA: adenine, guanine, thymine, and cytosine. Within DNA, two nucleotide strands wind about one another to form a double helix. Within each strand, the sugars of one nucleotide are linked to the phosphate of the next nucleotide, forming a sugar-phosphate "backbone" on each side of the double helix. The nucleotide bases of each strand pair up in the middle of the helix, held together by hydrogen bonds. Only specific pairs of bases, called *complementary base pairs*, can bond together in the helix: adenine bonds with thymine, and guanine bonds with cytosine.

3) How Does DNA Replication Ensure Genetic Constancy?

When cells reproduce, they must replicate their DNA so that each daughter cell receives all of the original genetic information. During DNA replication, enzymes such as DNA helicase unwind the two parental DNA strands. The enzyme DNA polymerase binds to each parental DNA strand, selects free nucleotides with complementary bases, and links the nucleotides together to form new DNA strands. The sequence of nucleotides in each newly formed strand is complementary to the sequence of a parental strand. Replication is semiconservative because, when DNA replication is complete, both new DNA double helices consist of one parental DNA strand and one newly synthesized, complementary strand. Each newly synthesized DNA strand is complementary to one of the parental DNA strands and an exact copy of the other parental DNA strand. The two new DNA double helices are therefore duplicates of the parental DNA double helix.

Key Terms

adenine *p. 151*
bases *p. 151*
chromosomes *p. 150*
complementary base pairs *p. 152*
cytosine *p. 151*
DNA *p. 149*
DNA helicase *p. 155*
DNA ligase *p. 156*
DNA polymerase *p. 156*
DNA replication *p. 154*
double helix *p. 152*
free nucleotides *p. 155*
genes *p. 149*
guanine *p. 151*
nucleotides *p. 151*
protein *p. 150*
replication bubble *p. 155*
semiconservative replication *p. 155*
strands *p. 152*
sugar-phosphate backbone *p. 152*
thymine *p. 151*

Thinking Through the Concepts

Multiple Choice

1. *How many different possible base sequences are there in a nucleotide chain three nucleotides in length?*
   a. 1
   b. 3
   c. 9
   d. 64
   e. more than 64

2. *Because each base pairs with a complementary base, in every DNA molecule the amount of*
   a. cytosine equals that of guanine
   b. cytosine equals that of thymine
   c. cytosine equals that of adenine
   d. each nucleotide is unrelated to all others
   e. each nucleotide is equal to all others

3. *Semiconservative replication refers to the fact that*
   a. each new DNA molecule contains two new single DNA strands
   b. DNA polymerase uses free nucleotides to synthesize new DNA molecules
   c. certain bases pair with specific bases
   d. each parental DNA strand is joined with a new strand containing complementary base pairs
   e. mistakes are made during DNA replication

4. *DNA helicase*
   a. cleaves hydrogen bonds that join the two strands of DNA
   b. converts two single strands of DNA into a double helix
   c. is a unique form of DNA
   d. adds nucleotides to newly forming DNA molecules
   e. proofreads the newly formed DNA strand

5. *DNA polymerase*
   a. can advance in either direction along a single strand of DNA
   b. cleaves hydrogen bonds that join the two strands of DNA
   c. creates a polymer that consists of many molecules of DNA
   d. adds appropriate nucleotides to a newly forming DNA strand
   e. is the protein found in conjunction with DNA in eukaryotic chromosomes

6. *Which of the following are incorrectly matched?*
   a. complementary base pairs ↔ adenine and cytosine
   b. bases ↔ adenine, thymine, cytosine, and guanine
   c. nucleotide ↔ phosphate and sugar and base
   d. eukaryotic chromosome ↔ DNA and protein
   e. enzymes involved in DNA replication ↔ DNA polymerase and DNA helicase

? Review Questions

1. Draw the general structure of a nucleotide. Which parts are identical in all nucleotides, and which can vary?
2. Name the four types of nitrogen-containing bases found in DNA.
3. Which bases are complementary to one another? How are they held together in the double helix of DNA?
4. Describe the structure of DNA. Where are the bases, sugars, and phosphates in the structure?
5. Describe the process of DNA replication.

Applying the Concepts

1. As you learned in "Scientific Inquiry: The Discovery of the Double Helix," scientists in different laboratories often compete with one another to make new discoveries. Do you think this competition helps promote scientific discoveries? Sometimes, researchers in different laboratories collaborate with one another. What advantages does collaboration offer over competition? What factors might provide barriers to collaboration and lead to competition?

2. Today, scientific advances are being made at an astounding rate, and nowhere is this more evident than in our understanding of the biology of heredity. Using DNA as a starting point, do you believe there are limits to the knowledge people should acquire? Defend your answer.

For More Information

Crick, F. *What Mad Pursuit: A Personal View of Scientific Discovery*. New York: Basic Books, 1998. Another view of the race to determine the structure of DNA, by Francis Crick himself.

Gibbs, W. W. "Peeking and Poking at DNA." *Scientific American (Explorations),* March 31, 1997. An update of new techniques for studying the DNA molecule, such as atomic force microscopy.

Judson, H. F. *The Eighth Day of Creation.* Cold Spring Harbor, NY: Cold Spring Harbor Laboratory Press, 1993. A very readable historical perspective on the development of genetics.

Leutwyler, K. "Turning Back the Strands of Time." *Scientific American (Explorations),* February 2, 1998. A brief discussion of telomeres, the repeating DNA regions at the ends of chromosomes.

Mirsky, A. E. "The Discovery of DNA." *Scientific American,* June 1968. The early history of DNA research.

Olby, R. *The Path to the Double Helix.* Seattle: University of Washington Press, 1975. A historical perspective on the development of genetics in the twentieth century.

Radman, M., and Wagner, R. "The High Fidelity of DNA Duplication." *Scientific American,* August 1988. Faithful duplication of chromosomes requires both reasonably accurate initial replication of DNA sequences and final proofreading.

Rennie, J. "DNA's New Twists." *Scientific American,* March 1993. A reprise of new information on DNA structure and function.

Watson, J. D. *The Double Helix.* New York, Atheneum, 1968. If you still believe the Hollywood images that scientists are either maniacs or cold-blooded, logical machines, be sure to read this book. Although hardly models for the behavior of future scientists, Watson and Crick are certainly human enough!

Weinberg, R. "How Cancer Arises." *Scientific American,* September 1996. An overview of the molecular basis of cancer: mutations in DNA.

Answers to Multiple-Choice Questions

1.d 2.a 3.d 4.a 5.d 6.a

MEDIATUTOR
DNA: The Molecule of Heredity

CD Activities

Activity 9.1: DNA Structure

Estimated time: 5 minutes

The structure of the DNA helix is key to its ability to store and transmit genetic information. In this tutorial, you will explore the parts of the DNA helix.

Activity 9.2: DNA Replication

Estimated time: 10 minutes

Making exact copies of our genetic material (chromosomes) is crucial to life. In this tutorial, you will explore the mechanism of DNA replication.

Start the MediaTutor Student CD-ROM and enter the activity number in the Quick Search box to be taken directly to that activity.

Web Investigations

Case Study: Sunshine Perils

Estimated time: 10 minutes

Skin cancer is now the most common cancer in the United States. Who is most at risk? What are the treatment alternatives? How can it be prevented? This exercise explores the biochemical, medical, and psychological features of skin cancers.

Go to http://www.prenhall.com/audesirk6, the Audesirk Companion Web site. Select Chapter 9 and Web Investigation to begin.

Many of the basic differences in the body structure of males and females can ultimately be traced to differences in a single gene.

10 Gene Expression and Regulation

AT A GLANCE

Case Study: Boy or Girl?

1) How Are Genes and Proteins Related?

Genetic Studies by Beadle and Tatum Using a Common Mold Revealed the Relationship Between Genes and Enzymes

Most Genes Contain the Information for the Synthesis of a Single Protein

DNA Provides Instructions for Protein Synthesis via RNA Intermediaries

Overview: Genetic Information Is Transcribed into RNA and Translated into Protein

In the Genetic Code, a Sequence of Three Bases Specifies an Amino Acid or a "Stop"

2) How Is Information in a Gene Transcribed into RNA?

Initiation of Transcription Occurs When RNA Polymerase Binds to the Promoter of a Gene

Elongation Proceeds Until RNA Polymerase Reaches a Termination Signal

Transcription Is Selective

3) How Is the Sequence of a Messenger RNA Molecule Translated into Protein?

Messenger RNA Carries the Code for Protein Synthesis from the Nucleus to the Cytoplasm

Ribosomal RNA Forms an Important Part of the Protein-Synthesizing Machinery of a Ribosome

Transfer RNA Molecules Decode the Sequence of Bases in mRNA into the Amino Acid Sequence of a Protein

During Translation, mRNA, tRNA, and Ribosomes Cooperate to Synthesize Proteins

Recap: Decoding the Sequence of Bases in DNA into the Sequence of Amino Acids in Protein Requires Transcription and Translation

4) How Do Mutations in DNA Affect the Function of Genes?

Mutations Result from Nucleotide Substitutions, Insertions, or Deletions

Mutations Have Different Effects on Protein Structure and Function

Mutations Provide the Raw Material for Evolution

5) How Are Genes Regulated?

Proper Regulation of Gene Expression Is Critical for an Organism's Development and Health

Eukaryotic Cells May Regulate the Transcription of Individual Genes, Regions of Chromosomes, or Entire Chromosomes

Case Study Revisited: Boy or Girl?

CASESTUDY

Boy or Girl?

"Is it a boy or a girl?" When we first hear about the birth of a baby, the answer to this question is often the first thing we want to know. For most of us, the response will send us to the gift shop searching for very different items: frilly, pink, and cuddly for girls; solid, blue, and sporty for boys. Such differences in what is considered "appropriate" for boys and girls reflect societal ideals rather than biological facts. However, boys and girls clearly have many physical differences that are biologically determined. An obvious difference is the presence of a vagina in girls and a penis in boys. Additional physical differences become obvious as children grow into adults. Men are usually heavier, stronger, and grow beards. Women are usually smaller, less muscular, and grow breasts. However, the genes of men and women do not differ so dramatically: Human males and females have almost exactly the same genes. In fact, boys have all of the genes needed to make female genitalia and girls have all of the genes needed to make male genitalia. In boys, the action of a single gene activates the male developmental pathway and deactivates the female developmental pathway. Without this gene we would all be physically female. How can a single gene determine something as complex as the sex of a human being? To solve this mystery, you will need to learn more about how genes work. ■

1 How Are Genes and Proteins Related?

Information, by itself, never *does* anything. For example, a book may describe in detail how to build a house, but unless that information is translated into action, no house will ever be built. Likewise, although the information in our DNA contains an incredible amount of information, it is not capable of carrying out any action on its own. So how does DNA determine whether your eyes are brown or blue, or the nature of any other inherited characteristic?

Proteins are the "molecular workhorses" of the cell, responsible for building cellular components and carrying out necessary biochemical reactions. Each cell contains a particular set of proteins; the activities of these proteins determine the cell's shape, movements, function, and capacity to reproduce. Clearly, the information in DNA must be linked to proteins. But how? Answering this question required clever experiments, often involving surprisingly simple organisms.

Genetic Studies by Beadle and Tatum Using a Common Mold Revealed the Relationship Between Genes and Enzymes

By the 1940s scientists knew that cells synthesize molecules in a series of linked steps called *biochemical pathways.* Each step in a biochemical pathway is catalyzed by an enzyme. (Remember: Proteins that catalyze a chemical reaction are called *enzymes*.) Within a biochemical pathway, the product produced by one eynzyme becomes the substrate of the next enzyme in the pathway. In this way, a biochemical pathway resembles a molecular assembly line. Scientists also knew that many specific characteristics of an organism are determined by **genes**. Perhaps genes and enzymes are somehow related.

During World War II, U.S. geneticists George Beadle and Edward Tatum studied the genetics of *Neurospora* to test the relationship between enzymes and genes. *Neurospora* is a mold commonly found growing on bread. Unlike humans, *Neurospora* have all the enzymes needed to make their own vitamin B_6 and each of the 20 different amino acids. Thus, normal *Neuorspora* can grow on a simple medium (food source) that does not contain vitamin B_6 or supplemental amino acids. The researchers studied many strains of *Neurospora*, including some with mutations. A *mutation* is a change in the base sequence of DNA.

Beadle and Tatum hypothesized that genes might encode enzymes. If this hypothesis were true, a mutation (defect) in a specific gene would disrupt synthesis of a specific enzyme. Without this enzyme, one of the mold's molecular assembly lines would not function properly, making the mold unable to synthesize molecules needed for growth.These mutant *Neurospora* strains would not be able to grow unless the simple medium had been supplemented with additional components.

The scientists found evidence to support their hypothesis by identifying a large number of different mutant *Neurospora* strains. For example, one of the first mutants they described could grow only if the simple medium were enriched with vitamin B_6. This mutant was no longer able to synthesize vitamin B_6 but had to obtain it from the environment, just as we do by eating food. However, this mutant *Neurospora* could make all of its own amino acids. Thus, this mutation affected only the biochemical pathway for vitamin B_6 synthesis. Beadle and Tatum crossed the vitamin B_6-requiring mutant to normal *Neurospora*. The progeny from these crosses demonstrated that the mutant strain was defective in only a single gene; this single mutation affected only one specific biochemical pathway.

Subsequent experiments refined the initial observation. For example, they isolated several single-gene mutants that could grow if the simple growth medium were supplemented with the amino acid arginine but not with vitamin B_6. Arginine is synthesized from citrulline, which in turn is synthesized from ornithine (Fig. 10-1b). Some of the arginine-requiring mutants could grow only if supplemented with arginine, but not if supplemented with citrulline or ornithine (Fig. 10-1a). These strains had a defect in the enzyme that converts citrulline to arginine. Other mutant strains could grow if they were supplemented with either arginine or citrulline, but not if supplemented with ornithine (Fig. 10-1a). These mutants had a defect in the enzyme that converts ornithine to citrulline. Thus, a mutation in a single gene affected only a single enzyme within a single biochemical pathway. Beadle and Tatum were able to conclude that one gene encodes one enzyme. The importance of this observation was recognized in 1958 with a Nobel Prize, which was also shared by Joshua Lederberg, one of Tatum's students.

Most Genes Contain the Information for the Synthesis of a Single Protein

Many of the proteins within cells are not enzymes. For example, keratin is a structural protein in hair and nails, but it does not catalyze any chemical reaction. In addition, many enzymes are composed of more than one protein subunit. For example, DNA polymerase is composed of more than a dozen proteins. Thus, the "one gene, one enzyme" relationship proposed by Beadle and Tatum was later clarified to "one gene, one polypeptide." A **polypeptide** is simply a chain of amino acids, joined by covalent bonds called *peptide bonds*. Many biochemists informally use polypeptide and protein interchangeably, as we will do in this text. There are exceptions to the "one gene, one protein" rule, including several in which the final product of a gene isn't protein but a nucleic acid called *ribonucleic acid* (RNA), described below. Nevertheless, as a generalization, each gene encodes the information for a single protein.

(a) Growth characteristics of normal and mutant *Neurospora* on simple medium with different supplements show that defects in a single gene lead to defects in a single enzyme.

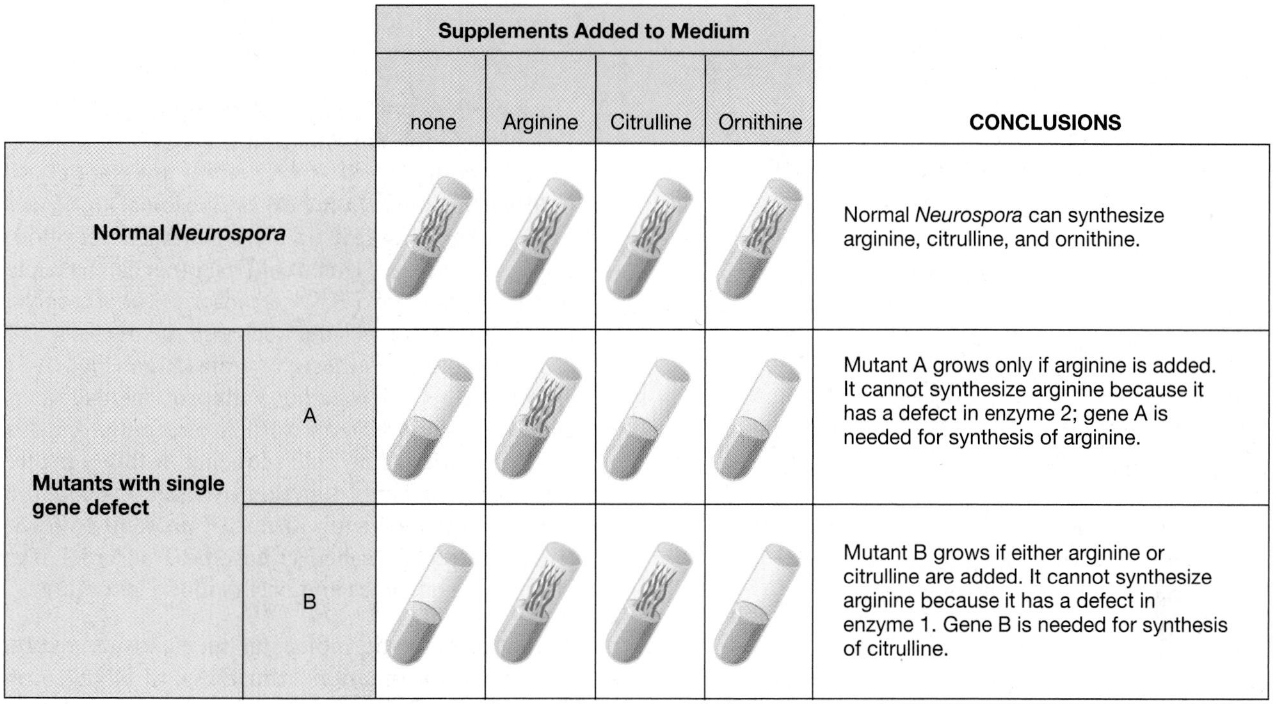
Supplements Added to Medium
none
Arginine
Citrulline
Ornithine
CONCLUSIONS
Normal Neurospora
Normal Neurospora can synthesize arginine, citrulline, and ornithine.
Mutants with single gene defect
A
Mutant A grows only if arginine is added. It cannot synthesize arginine because it has a defect in enzyme 2; gene A is needed for synthesis of arginine.
B
Mutant B grows if either arginine or citrulline are added. It cannot synthesize arginine because it has a defect in enzyme 1. Gene B is needed for synthesis of citrulline.

(b) The biochemical pathway for synthesis of the amino acid arginine involves two steps, each catalyzed by a different enzyme.

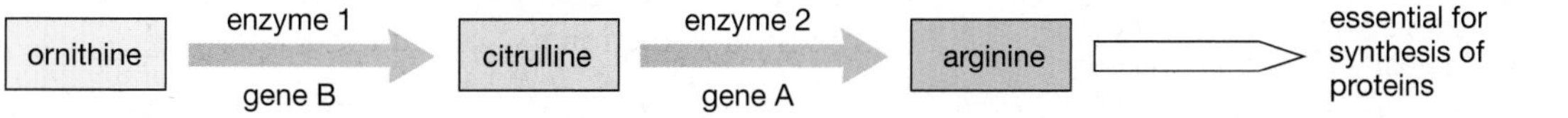
ornithine
enzyme 1
gene B
citrulline
enzyme 2
gene A
arginine
essential for synthesis of proteins

Figure 10-1 Beadle and Tatum's experiments with *Neurospora* mutants
Experiments with *Neurospora* led Beadle and Tatum to discover the relationship between an individual gene and an individual enzyme.

DNA Provides Instructions for Protein Synthesis via RNA Intermediaries

The DNA of a eukaryotic cell is located in the cell's nucleus, but protein synthesis occurs on ribosomes in the cytoplasm. Therefore, DNA cannot directly guide protein synthesis. There must be an intermediary, a molecule that carries the information from DNA in the nucleus to the ribosomes in the cytoplasm. This molecule is **ribonucleic acid**, or **RNA**.

RNA is similar to DNA but differs structurally in three respects: (1) RNA is normally single-stranded; (2) RNA has the sugar ribose instead of deoxyribose in its backbone; and (3) RNA has the base uracil instead of the base thymine, which is found in DNA (Table 10-1).

DNA codes for synthesis of three major types of RNA: (1) *messenger RNA* (mRNA), (2) *ribosomal RNA* (rRNA), and (3) *transfer RNA* (tRNA) (Fig. 10-2). All of these RNA molecules are involved in converting the nucleotide sequence of genes into the amino acid sequence of proteins. We will examine their functions in more detail shortly.

Table 10-1 A Comparison of DNA and RNA

	DNA	RNA	
Strands	2	1	
Sugar	deoxyribose	ribose	
Types of Bases	adenine (A), thymine (T) cytosine (C), guanine (G)	adenine (A), uracil (U) cytosine (C), guanine (G)	
Base Pairs	DNA:DNA A–T T–A C–G G–C	RNA:DNA A–T U–A C–G G–C	RNA:RNA A–U U–A C–G G–C
Function	Contains genes; sequence of bases in most genes determines the amino acid sequence of a protein	**Messenger RNA (mRNA):** carries the code for a protein-coding gene from DNA to ribosomes **Ribosomal RNA (rRNA):** combines with proteins to form ribosomes, the structures that link amino acids to form a protein **transfer RNA (tRNA):** carries the proper amino acid to the ribosome, per the code in mRNA	

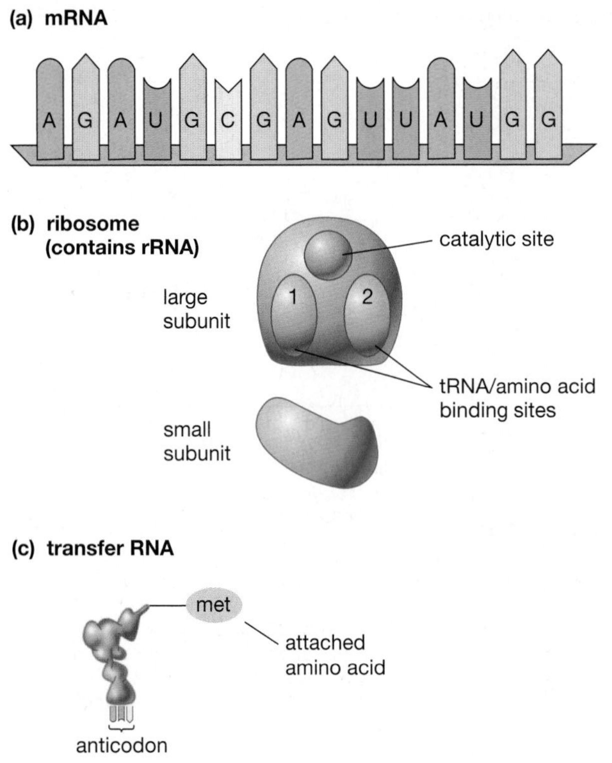

Figure 10-2 Cells synthesize three major types of RNA
RNA consists of a single nucleotide strand whose bases are complementary to the bases within the template strand of the gene. There are three major types of RNA: ***(a)*** Messenger RNA (mRNA) carries within its base sequence the information for the amino acid sequence of a protein. ***(b)*** Ribosomes contain both ribosomal RNA (rRNA) and proteins. The ribosome is divided into a small and large subunit that join together during protein synthesis. The small subunit binds the mRNA; the large subunit binds tRNA and catalyzes the formation of bonds between amino acids to form a protein. ***(c)*** One side of transfer RNA contains an anticodon, which is a sequence of three nucleotides that can form base pairs with a codon in mRNA. Enzymes within the cytoplasm attach a specific amino acid to the opposite side of the tRNA so that it can carry the proper amino acid to the ribosome for incorporation into a new protein.

Overview: Genetic Information Is Transcribed into RNA and Translated into Protein

Information from DNA is used to direct the synthesis of proteins in a two-step process (Table 10-2):

1. During RNA synthesis, or **transcription** (Fig. 10-3a), the information contained in the DNA of a specific gene is copied into RNA, either **messenger RNA (mRNA)**, **transfer RNA (tRNA)**, or **ribosomal RNA (rRNA)**. Thus, we can now devise a more general description of a gene: a gene is a segment of DNA that can be copied, or transcribed, into RNA. Transcription is catalyzed by RNA polymerase and occurs in the nucleus.
2. During protein synthesis, or **translation** (Fig. 10-3b), tRNA and rRNA, together with proteins, use the nucleotide sequence in an mRNA molecule to synthesize a specific amino acid sequence within a protein. The rRNA does not function by itself. Instead, it assembles together with dozens of proteins to form a complex structure called a **ribosome**. Translation is catalyzed by ribosomes and occurs in the cytoplasm.

To understand the molecular mechanisms involved in transfer of information from DNA to RNA to protein, geneticists first had to break the language barrier: How does the language of nucleotide sequences in DNA and messenger RNA translate into the language of amino acid sequences in proteins? This translation relies on a "dictionary" called the *genetic code*.

In the Genetic Code, a Sequence of Three Bases Specifies an Amino Acid or a "Stop"

We have used the word *code* to refer to the information that is stored in DNA and ultimately translated into the amino acid sequence of proteins. This **genetic code** is conceptually similar to Morse code: a set of symbols (bases in nucleic acids, dots and dashes in Morse code)

Table 10-2 Processes Involved in the Use and Inheritance of Genetic Information

Process	Information for Process	Product	Major Enzyme or Structure Involved in Process	Type of Base Pairing Required
Transcription (synthesis of RNA)	Short segment of one DNA strand (template strand within a gene)	One RNA molecule (mRNA, tRNA, rRNA)	RNA polymerase	DNA–RNA: DNA bases in template strand form base pairs with RNA bases in new RNA molecule.
Translation (synthesis of protein)	mRNA	One protein molecule	Ribosome (also requires tRNA)	RNA–RNA: Codon in mRNA forms base pairs with anti-codon in tRNA.
Replication (synthesis of DNA; occurs only before cells divide)	Entire length of both DNA strands	Two DNA double helices (each with one old and one new strand)	DNA polymerase	DNA-DNA: DNA bases of each parental strand base pair with DNA bases in the newly synthesized strands.

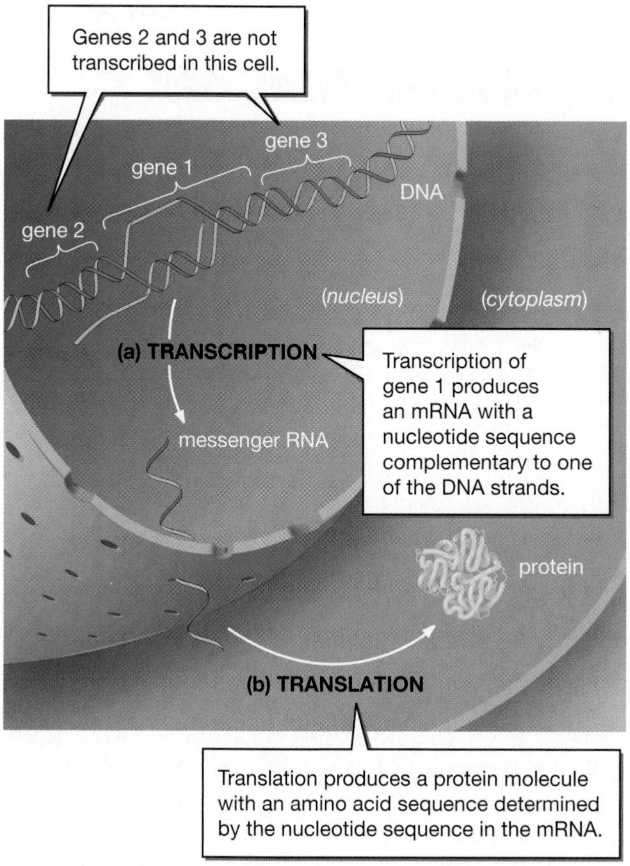
Genes 2 and 3 are not transcribed in this cell.
gene 3
gene 1
DNA
gene 2
(nucleus)
(cytoplasm)
(a) TRANSCRIPTION
Transcription of gene 1 produces an mRNA with a nucleotide sequence complementary to one of the DNA strands.
messenger RNA
protein
(b) TRANSLATION
Translation produces a protein molecule with an amino acid sequence determined by the nucleotide sequence in the mRNA.

Figure 10-3 Genetic information flows from DNA to RNA to protein Cellular information is stored within the base sequence of DNA. Transcription, the process of RNA synthesis, occurs in the nucleus. During transcription, the nucleotide sequence in a gene specifies the nucleotide sequence in a complementary RNA molecule. For protein-encoding genes, the product is an mRNA molecule that exits from the nucleus and enters the cytoplasm where translation occurs. During translation, the sequence in an mRNA molecule specifies the amino acid sequence in a protein.

that can be translated into another set of symbols (amino acids in proteins, letters of the alphabet). What combinations of bases stand for which amino acids?

Both DNA and RNA contain four different bases: A, T (or U in RNA), G, and C (see Table 10-1). However, proteins are made of 20 different amino acids. Therefore, one base cannot code for just one amino acid because there are simply not enough types of bases. The genetic code must rely on a short sequence of bases to encode each amino acid, just as Morse code relies on a short sequence of dots and dashes to encode each letter of the alphabet. If a sequence of two bases codes for an amino acid, there would be 16 possible combinations, not enough to account for all 20 amino acids. A three-base sequence, however, gives 64 possible combinations of bases, which is more than enough. Under the assumption that nature operates as economically as possible, biologists hypothesized that the genetic code must be a triplet code: Three bases specify a single amino acid. In 1961 Francis Crick and three co-workers demonstrated that this hypothesis is correct (see "Scientific Inquiry: Cracking the Genetic Code" on page 171).

For any language to be understood, the users must know what the words mean, where words start and stop, and where sentences begin and end. Crick's experiments demonstrated that all the "words" of the genetic code are three bases long and that a set of three bases signifies one amino acid. Shortly after this discovery, researchers began to decipher the genetic code. They ground up bacteria and isolated the components needed to synthesize proteins. To this mixture, they added artificial mRNA, which allowed them to control what "words" were to be transcribed. Researchers could then see which amino acids were incorporated into the resulting proteins. For example, an mRNA strand composed entirely of uracil (UUUUUUUU . . .) directed the mixture to synthesize a protein composed solely of the amino acid phenylalanine. Therefore, the triplet UUU must specify phenylalanine. Because the genetic code was deciphered by using these artificial mRNAs, the code is often written in terms of the base triplets in mRNA (rather than in DNA) that code for each amino acid (Table 10-3). These mRNA triplets are called **codons**.

What about punctuation? Given that one mRNA molecule may contain thousands of bases, how does the cell recognize where the code for a given protein starts and stops? Research showed that all proteins originally begin with the same amino acid, methionine, which is specified by the codon AUG. In eukaryotes, the first AUG in an mRNA sequence determines where protein synthesis will begin. In addition, three codons—UAG, UAA, and UGA—are **stop codons**, which signal the completion of protein synthesis. When the ribosome encounters a stop codon, it releases both the newly synthesized protein and the mRNA.

All 61 possible animo acid-specifying codons are used in the genetic code, but only 20 amino acids need to be encoded. Thus, except for methionine and tryptophan, each amino acid can be specified by several different codons in mRNA. For example, six different codons all code for leucine (see Table 10-3), so whether UUA or CUG is present in the mRNA sequence, ribosomes will insert the amino acid leucine (Leu) into the growing amino acid chain. However, each codon specifies one, and only one, amino acid.

2 How Is Information in a Gene Transcribed into RNA?

We can view transcription as a process consisting of (1) *initiation*, (2) *elongation*, and (3) *termination*. These three steps correspond to the three major parts of most genes in both eukaryotes and prokaryotes: (1) a *promoter* region at the beginning of the gene, where initiation of transcription begins; (2) the "body" of the gene

Table 10-3 The Genetic Code (Codons of mRNA)

First Base	Second Base: U	Second Base: C	Second Base: A	Second Base: G	Third Base
U	UUU Phenylalanine (Phe)	UCU Serine (Ser)	UAU Tyrosine (Tyr)	UGU Cysteine (Cys)	U
	UUC Phenylalanine	UUC Serine	UAC Tyrosine	UGC Cysteine	C
	UUA Leucine	UCA Serine	UAA Stop	UGA Stop	A
	UUG Leucine	UCG Serine	UAG Stop	UGG Tryptophan (Trp)	G
C	CUU Leucine (Leu)	CCU Proline (Pro)	CAU Histidine (His)	CGU Arginine (Arg)	U
	CUC Leucine	CCC Proline	CAC Histidine	CGC Arginine	C
	CUA Leucine	CCA Proline	CAA Glutamine (Gln)	CGA Arginine	A
	CUG Leucine	CCG Proline	CAG Glutamine	CGG Arginine	G
A	AUU Isoleucine (Ile)	ACU Threonine (Thr)	AAU Asparagine (Asp)	AGU Serine (Ser)	U
	AUC Isoleucine	ACC Threonine	AAC Asparagine	AGC Serine	C
	AUA Isoleucine	ACA Threonine	AAA Lysine (Lys)	AGA Arginine (Arg)	A
	AUG Methionine (Met)	ACG Threonine	AAG Lysine	AGG Arginine	G
G	GUU Valine (Val)	GCU Alanine (Ala)	GAU Aspartic acid (Asp)	GGU Glycine (Gly)	U
	GUC Valine	GCC Alanine	GAC Aspartic acid	GGC Glycine	C
	GUA Valine	GCA Alanine	GAA Glutamic acid (Glu)	GGA Glycine	A
	GUG Valine	GCG Alanine	GAG Glutamic acid	GGG Glycine	G

where elongation of the RNA strand occurs; and (3) a termination signal at the end of the gene, where RNA synthesis ceases, or terminates (Fig. 10-4).

Initiation of Transcription Begins When RNA Polymerase Binds to the Promoter of a Gene

A different version of the enzyme **RNA polymerase** synthesizes each major type of RNA: mRNA, tRNA, and rRNA. To initiate transcription, RNA polymerase must first locate the beginning of the genes that should be transcribed in that particular cell. This is no small task. We estimate that a human cell contains about 30,000 genes, most of which are only transcribed in certain cells, only at certain times, or only in certain individuals. So RNA polymerase must somehow select the appropriate genes to transcribe in each type of cell and at each stage of the cell's life. This regulation requires control regions within the gene, including the **promoter** region. The promoter is a non-transcribed sequence of DNA bases that marks the beginning of the gene. Depending on conditions inside and outside the cell, proteins bind to control regions of a gene, blocking or enhancing the binding of RNA polymerase to the promoter. Usually, if RNA polymerase binds to the promoter, transcription of that gene will begin (Fig. 10-4a).

Elongation Proceeds Until RNA Polymerase Reaches a Termination Signal

When it binds to the promoter of a gene, RNA polymerase changes shape, forcing the DNA double helix at the beginning of the gene to partially unwind (Fig. 10-4a). RNA polymerase then travels in one direction along one of the DNA strands, synthesizing a single strand of RNA that is complementary to that strand of DNA (Fig. 10-4b). We call this strand of DNA the **template strand.** During the process of transcription, RNA forms the same base pairs as DNA, except that in RNA, uracil, rather than thymine, pairs with adenine (see Table 10-1).

After about 10 nucleotides have been added to the growing RNA chain, the first nucleotides in the RNA molecule separate from the DNA template strand. This separation allows the two DNA strands to re-form the DNA–DNA base pairs of the original DNA double helix (Fig. 10-4b,c). Thus, as transcription continues to elongate the RNA molecule, one end of the RNA drifts away from the DNA; the other end remains attached to the DNA template strand by RNA polymerase (Fig. 10-5, p. 170).

RNA polymerase continues along the template strand until it reaches a sequence of DNA bases in the gene known as the *termination signal.* At this point, RNA polymerase releases the completed RNA molecule and detaches from the DNA. The RNA polymerase is then free to bind to another promoter and synthesize another RNA molecule (see Fig. 10-4d).

Transcription Is Selective

The transcription of genes into RNA is selective in two major ways. First, in any cell, transcription normally copies the DNA of only selected genes into RNA. Some of these genes are transcribed in all cells, since they encode proteins or RNA molecules that are essential for the life of any cell. For example, all cells need to synthesize proteins, so they transcribe all of the tRNA genes, rRNA genes, and genes for ribosomal proteins. Other genes are transcribed only in certain types of cells. For example, even though every cell in your body contains the gene for insulin, a protein hormone that regulates glucose uptake, that gene is transcribed only in certain cells in your pancreas. Thus, transcription is restricted to only those genes needed by that particular type of cell at that particular time of its life.

We have already encountered a second way in which transcription is selective. With only rare exceptions,

MediaTutor
10.1 Transcription

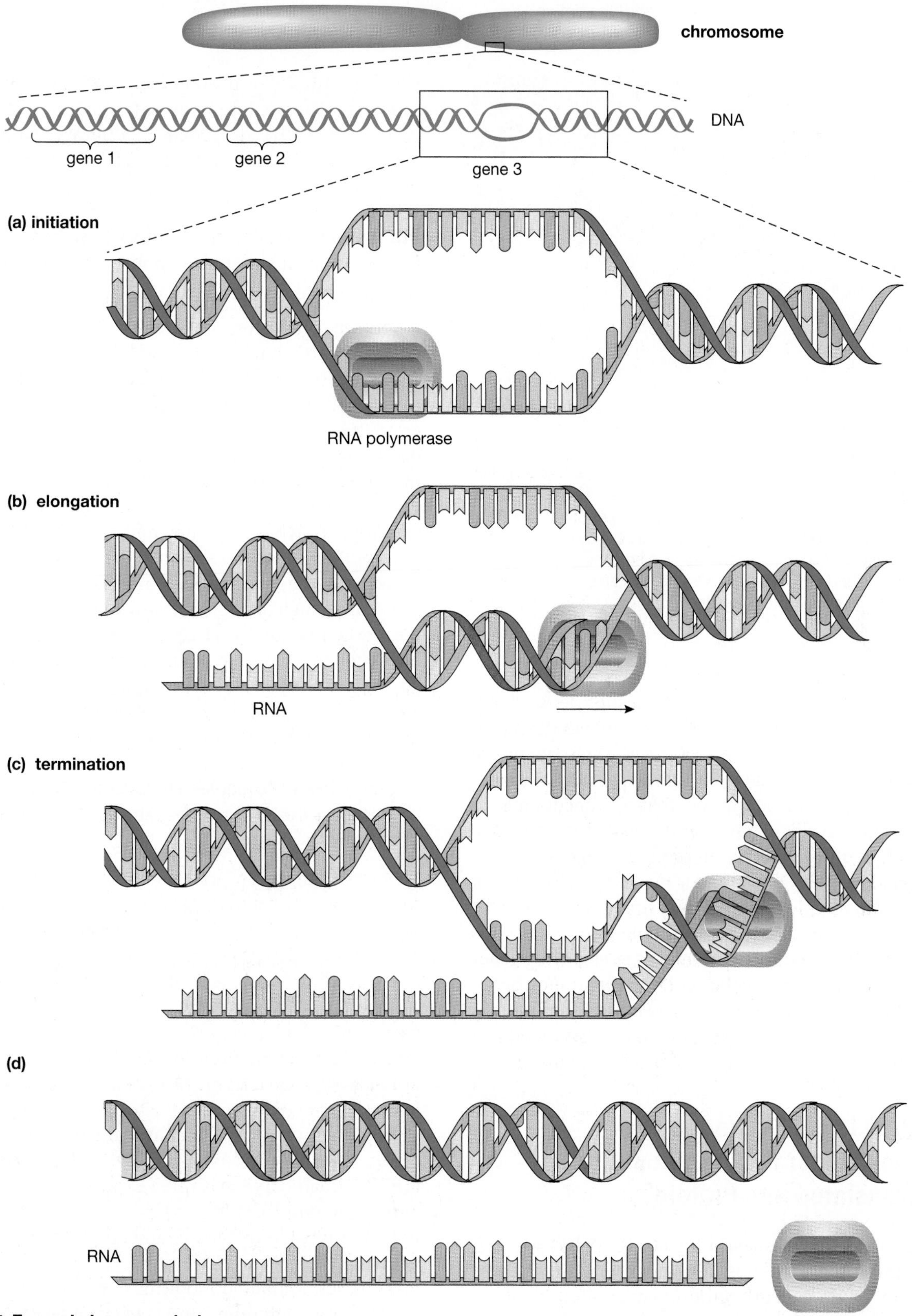

Figure 10-4 Transcription occurs in three steps
Most chromosomes contain hundreds or thousands of genes. Each gene occupies a particular position in the DNA of the chromosome. Within the gene, one of the DNA strands will serve as the template for RNA synthesis. ***(a)*** Initiation: The enzyme RNA polymerase binds to the promoter region of DNA near the beginning of a gene. In different cells, RNA polymerase binds to the promoters of different genes, depending on conditions inside and outside the cell. Binding of RNA polymerase forces the DNA double helix near the promoter to separate. ***(b)*** Elongation: RNA polymerase then travels along the DNA template strand, catalyzing the addition of nucleotides into an RNA molecule. The nucleotides in the RNA are complementary to the template strand of the DNA. The DNA double helix rewinds a short distance after RNA polymerase passes. The process continues until ***(c)*** termination: At the end of a gene, the RNA polymerase encounters a sequence of DNA called a *termination signal.* ***(d)*** At this point, RNA polymerase detaches from the DNA and releases the RNA molecule, allowing the DNA double helix to completely rewind. Initiation of another round of transcription can occur before the previous RNA polymerase has completed its RNA, so that many RNA molecules can be produced nearly simultaneously from a single gene.

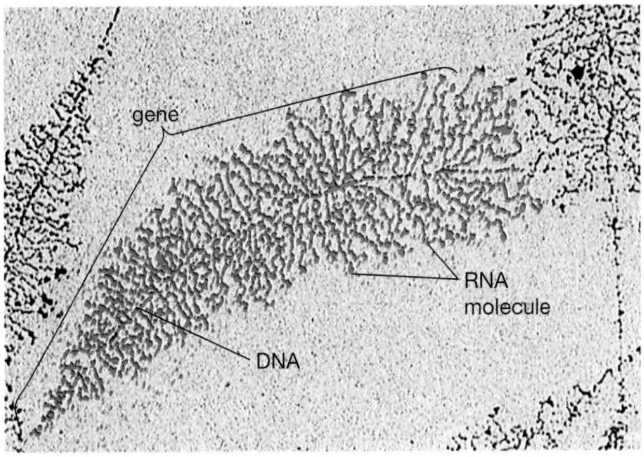

Figure 10-5 RNA transcription in action
This electron micrograph shows the progress of RNA transcription in the egg of an African clawed toad. In each treelike structure, the central "trunk" is DNA and the "branches" are RNA molecules. A series of RNA polymerase molecules are traveling down the DNA, synthesizing RNA as they go. The beginning of the gene is on the left. Therefore, the short RNA molecules on the left have just begun to be synthesized; the long RNA molecules on the right are almost finished.

transcription copies only one of the two strands of DNA into RNA. Why? In answering this question, let's consider a protein-coding gene. Remember, the two strands of DNA are *complementary*, not *identical*. The sequence of bases on one strand codes for synthesis of a complementary mRNA that can be translated into a functional protein. Recall that this DNA strand is called the *template strand:* It is the template, or pattern, from which the complementary mRNA strand is made. The sequence of bases on the complementary DNA strand has a different base sequence, one that is very unlikely to code for a useful protein. The template strand of neighboring genes within a chromosome can be on different strands of the DNA double helix (see Fig. 10-3). Thus, transcription is selective because only certain genes are transcribed at a given time and only one strand of a gene is transcribed.

3 How Is the Sequence of a Messenger RNA Molecule Translated into Protein?

As their names suggest, each type of RNA has a specific role in protein synthesis. This section describes these roles and the process of translation in greater detail.

Messenger RNA Carries the Code for Protein Synthesis from the Nucleus to the Cytoplasm

All RNA is produced by transcription of DNA, but only mRNA carries the code for the amino acid sequence of a protein (see Fig. 10-2a). In eukaryotic cells, mRNA molecules are synthesized in the nucleus and enter the cytoplasm through the pores in the nuclear envelope. In the cytoplasm, mRNA binds to ribosomes, which synthesize a protein based on the mRNA base sequence. The gene remains safely stored in the nucleus, like a valuable document in a library, while mRNA carries the information to the cytoplasm to be used in protein synthesis.

Ribosomal RNA Forms an Important Part of the Protein-Synthesizing Machinery of a Ribosome

Recall that ribosomes, the structures that carry out translation, are composites containing rRNA and many different proteins. Each ribosome is composed of two subunits—one small and one large. Unless they are actively synthesizing proteins, the two subunits remain separate (see Fig. 10-2b). However, when protein synthesis is occurring, the small ribosomal subunit and large ribosomal subunit come together with an mRNA molecule in between them. The large ribosomal subunit has binding sites for two tRNA molecules and a catalytic site for joining together the amino acids attached to the tRNA molecules. Ribosomes are critical for cell function since most cells have a high demand for protein synthesis. For example, certain white blood cells secrete antibody proteins into your bloodstream. One of these cells can produce 20,000 antibody proteins per second! As a result, most eukaryotic cells have tens of thousands of ribosomes to keep up with the cell's need for new proteins.

Transfer RNA Molecules Decode the Sequence of Bases in mRNA into the Amino Acid Sequence of a Protein

Delivery of the appropriate amino acids to the ribosome for incorporation into the growing protein chain depends on the activity of tRNA. Each cell synthesizes 61 different types of tRNA, one type for each different amino acid-encoding codon. (Recall that 3 of the 64 codons are stop codons.) Twenty enzymes in the cytoplasm, one for each amino acid, recognize the tRNA molecules and attach the correct amino acid to one end (see Fig. 10-2c). For example, one enzyme recognizes all of the the six different leucine tRNAs and attaches a leucine to each one.

The ability of tRNA to deliver the proper amino acid depends on specific base-pairing between tRNA and mRNA. Each tRNA has three exposed bases, called the **anticodon**, that form base pairs with the mRNA codon. For example, the mRNA codon AUG forms base pairs with the anticodon UAC of a tRNA that has an attached methionine amino acid. The ribosome can then incorporate methionine into a growing protein chain.

During Translation, mRNA, tRNA, and Ribosomes Cooperate to Synthesize Proteins

Now that we have introduced the major molecules involved in translation, let's look at the actual events (Fig. 10-6). Like transcription, translation has three steps: (1) *initiation* of protein synthesis, (2) *elongation* of the protein chain, and (3) *termination* of translation.

Scientific Inquiry
Cracking the Genetic Code

The hypothesis that three bases in DNA code for one amino acid in a protein is an attractive and logical possibility. But how could you prove that nature actually uses a triplet code?

Francis Crick and his co-workers exposed *bacteriophages* (or simply *phages*), viruses that multiply inside bacteria, to a chemical called acridine. Acridine causes one, two, or three nucleotides to be inserted into the viral DNA molecule at random places near the beginning of a particular gene. The researchers found that inserting one or two nucleotides into the DNA caused the synthesis of defective enzymes that prevented the viruses from multiplying inside their host bacteria. Inserting three nucleotides, however, sometimes produced phages that synthesized normal or nearly normal enzymes.

Crick concluded that during RNA synthesis, a gene's DNA is "read" in a linear order, in which each sequence of three bases makes up a "word"; that is, three bases in DNA encode a single amino acid. The code must somehow specify the beginning and end of a gene, but genes have no spaces or punctuation between words. How did he arrive at these conclusions?

To understand Crick's reasoning, let's suppose that we have a gene with the repetitive sequence of bases (TAG) as shown in Fig. E10-1a. If DNA always has three-letter words, then this base sequence spells out the word "TAG" over and over again. You don't need spaces to separate the words if they all have three and only three letters.

Now suppose that the acridine inserts another nucleotide with the base cytosine near the beginning of the gene (Fig. E10-1b). Most of the gene now reads GTA . . ., which is the code for a different amino acid and produces a nonfunctional enzyme. Inserting another nucleotide nearby again changes the amino acid specified, as most of the gene now reads AGT . . ., and results in another nonfunctional enzyme (Fig. E10-1c). A third insertion, however, results in most of the gene reading TAG . . . again (Fig. E10-1d). In this case, most of the enzyme would be synthesized correctly and might work well enough to allow the viruses to reproduce.

Crick concluded that only a triplet code could account for the fact that three insertions, and not one or two, restore near-normal enzyme function. Further, only a code without punctuation or spaces between words would be confused by any of the insertions. Finally, a code without punctuation between words could work only if the start points and stop points of protein synthesis are clearly marked. All three conclusions have been found to be correct, as we describe in the text.

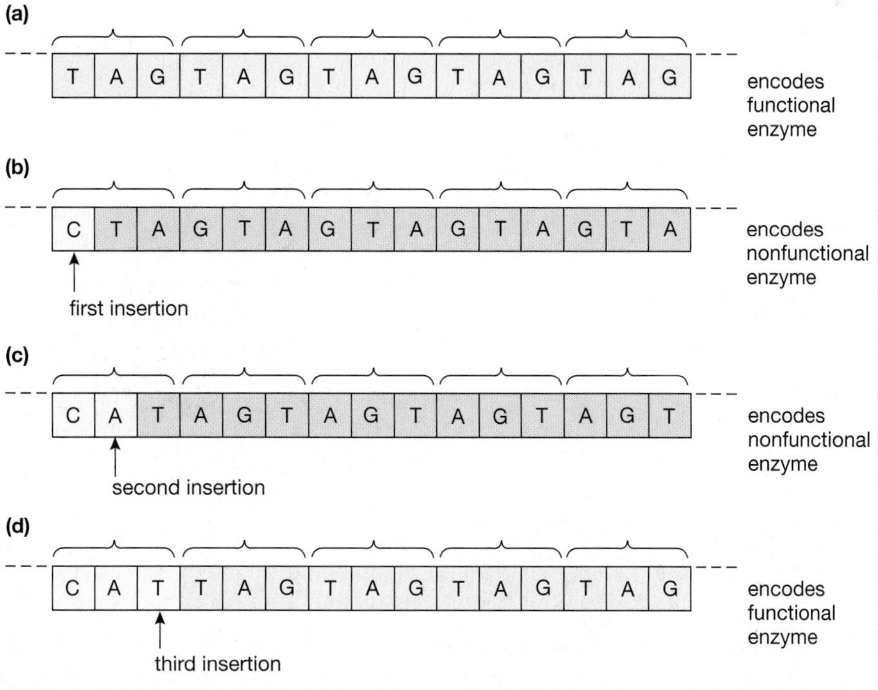

Figure E10-1 How insertion mutations helped crack the genetic code

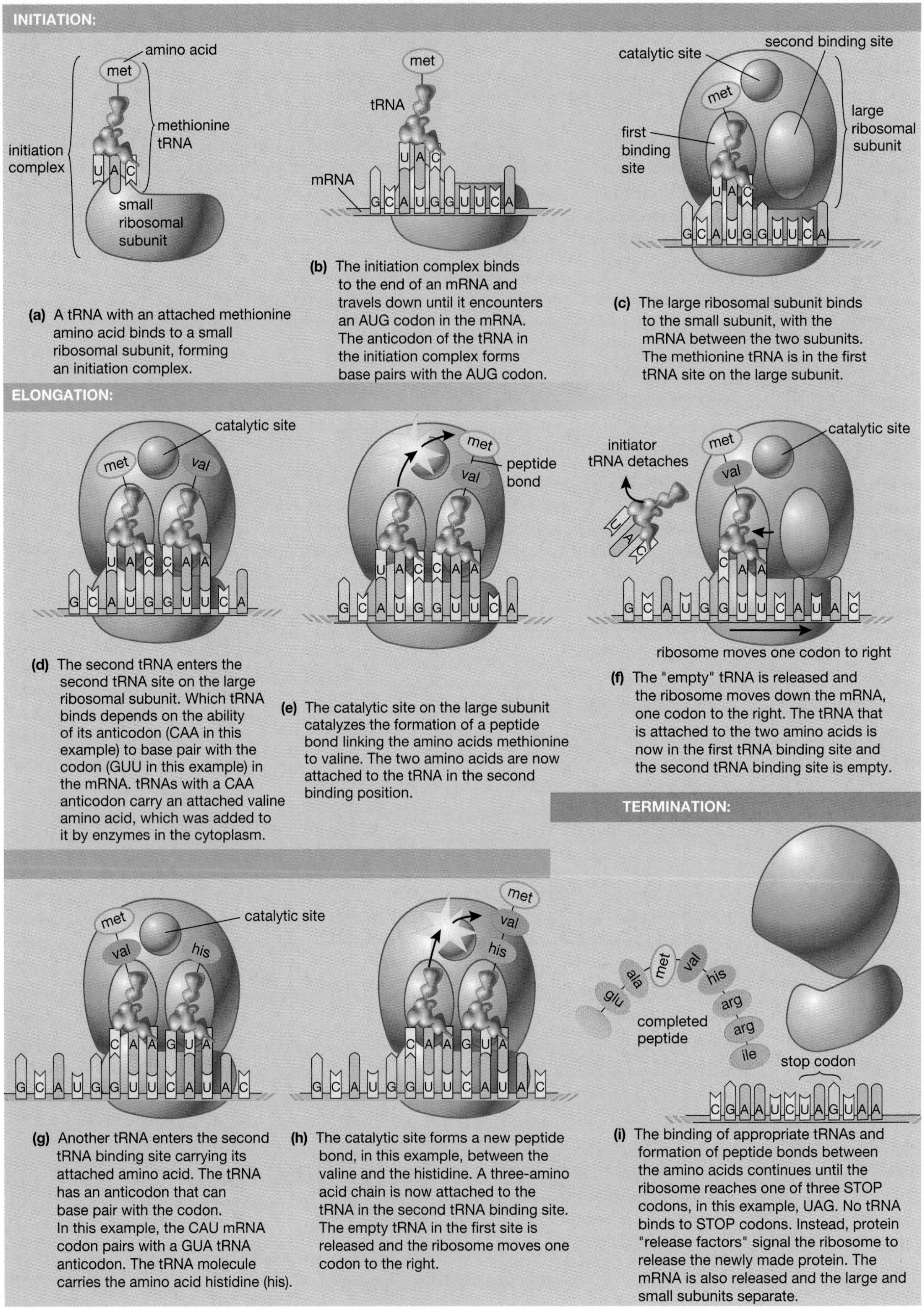

(a) A tRNA with an attached methionine amino acid binds to a small ribosomal subunit, forming an initiation complex.

(b) The initiation complex binds to the end of an mRNA and travels down until it encounters an AUG codon in the mRNA. The anticodon of the tRNA in the initiation complex forms base pairs with the AUG codon.

(c) The large ribosomal subunit binds to the small subunit, with the mRNA between the two subunits. The methionine tRNA is in the first tRNA site on the large subunit.

(d) The second tRNA enters the second tRNA site on the large ribosomal subunit. Which tRNA binds depends on the ability of its anticodon (CAA in this example) to base pair with the codon (GUU in this example) in the mRNA. tRNAs with a CAA anticodon carry an attached valine amino acid, which was added to it by enzymes in the cytoplasm.

(e) The catalytic site on the large subunit catalyzes the formation of a peptide bond linking the amino acids methionine to valine. The two amino acids are now attached to the tRNA in the second binding position.

(f) The "empty" tRNA is released and the ribosome moves down the mRNA, one codon to the right. The tRNA that is attached to the two amino acids is now in the first tRNA binding site and the second tRNA binding site is empty.

(g) Another tRNA enters the second tRNA binding site carrying its attached amino acid. The tRNA has an anticodon that can base pair with the codon. In this example, the CAU mRNA codon pairs with a GUA tRNA anticodon. The tRNA molecule carries the amino acid histidine (his).

(h) The catalytic site forms a new peptide bond, in this example, between the valine and the histidine. A three-amino acid chain is now attached to the tRNA in the second tRNA binding site. The empty tRNA in the first site is released and the ribosome moves one codon to the right.

(i) The binding of appropriate tRNAs and formation of peptide bonds between the amino acids continues until the ribosome reaches one of three STOP codons, in this example, UAG. No tRNA binds to STOP codons. Instead, protein "release factors" signal the ribosome to release the newly made protein. The mRNA is also released and the large and small subunits separate.

Figure 10-6 Translation is the process of protein synthesis
Protein synthesis, or translation, decodes the base sequence of an mRNA into the amino acid sequence of a protein.

Initiation: Protein Synthesis Begins When tRNA and mRNA Bind to a Ribosome

The amino acid methionine can be present at many locations within a protein's amino acid sequence. However, methionine also holds a unique place in protein synthesis. The first amino acid in all newly synthesized proteins is methionine. Only an AUG codon specifies methionine and only tRNAs with a UAC anticodon carry methionine (see Table 10-2). The first AUG codon in a eukaryotic mRNA sequence specifies where translation will begin. An "initiation complex," which contains the small ribosomal subunit, a methionine tRNA, and several other proteins, also plays a critical role in the start of translation (Fig. 10-6a). The initiation complex binds to the end of an mRNA molecule and moves along until it encounters the first AUG codon in the mRNA sequence. At this point, the AUG codon within the mRNA forms base pairs with the UAC anticodon of the methionine tRNA (Fig. 10-6b). The large ribosomal subunit then attaches to the small subunit and to the methionine tRNA, sandwiching the mRNA between the two ribosome subunits; the methionine tRNA is in the first tRNA binding site on the large subunit (Fig. 10-6c). The ribosome is now fully assembled and ready to begin translation.

Elongation and Termination: Protein Synthesis Proceeds One Amino Acid at a Time Until a Stop Codon Is Reached

The assembled ribosome covers about 30 nucleotides of the mRNA. It holds two mRNA codons in alignment with the two tRNA binding sites of the large subunit. With the help of certain proteins, a second tRNA, with the complementary anticodon and an attached amino acid moves into the second tRNA binding site on the large subunit (Fig. 10-6d). The two amino acids attached to the two tRNAs are now next to one another. The large subunit catalytic site breaks the bond holding the first amino acid to its tRNA and forms a peptide bond between the two amino acids (Fig. 10-6e). At the end of this step, the first tRNA is "empty," and the second tRNA carries a two-amino-acid chain.

With help from other proteins, the ribosome releases the empty tRNA and shifts to the next codon in the mRNA molecule (Fig. 10-6f). The tRNA holding the elongating chain of amino acids shifts too, moving from the second to the first binding site of the ribosome. A new complementary tRNA, carrying the next amino acid, binds to the empty second site (Fig. 10-6g). The catalytic site on the large subunit now links the third amino acid onto the growing protein chain (Fig. 10-6h). The "empty" tRNA leaves the ribosome, the ribosome shifts to another codon, and the process is repeated as the ribosome moves down the mRNA one codon at a time.

Recall that a termination signal in the DNA template strand signaled to RNA polymerase that transcription of the gene was complete. Ribosomes also require a signal to indicate that synthesis of the protein is complete. This signal is a stop codon within the mRNA molecule. Unlike the other 61 codons, stop codons do not bind to a tRNA. Instead, special proteins bind to the ribosome when it encounters a stop codon, forcing the ribosome to release the finished protein chain and the mRNA (Fig. 10-6i). The ribosome disassembles into large and small subunits, which can then be used to translate another mRNA.

How fast does translation occur? Under optimum conditions, a ribosome can synthesize 5 to 15 peptide bonds per second. Thus, translation of an mRNA encoding an antibody protein, which is approximately 500 amino acids long, takes less than 2 minutes. Most proteins are 100 to 200 amino acids long and can be synthesized in less than a minute.

Recap: Decoding the Sequence of Bases in DNA into the Sequence of Amino Acids in Protein Requires Transcription and Translation

We can now understand how a cell decodes the genetic information stored in its DNA to synthesize a protein. Each step involves the pairing of complementary bases and requires action of a variety of proteins and enzymes. Follow these steps in Figure 10-7:

(a) With exceptions such as the genes for tRNA and rRNA, each gene codes for a single protein.

(b) Transcription of a protein-coding gene produces an mRNA molecule that is complementary to one DNA

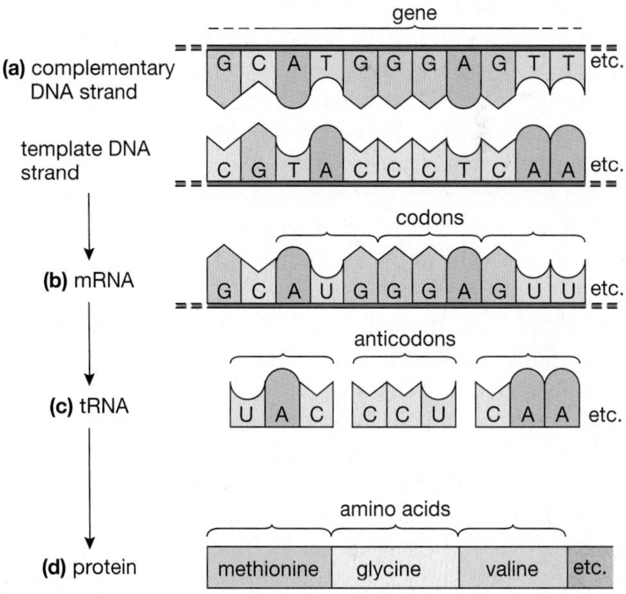

Figure 10-7 Complementary base pairing is critical at each step in decoding genetic information
(a) DNA contains two strands: the template strand is used by RNA polymerase to synthesize an RNA molecule; the other strand, which is complementary to the template strand, is needed for DNA replication. ***(b)*** Bases in the template strand of DNA are transcribed into a complementary mRNA. Codons are sequences of three bases that specify an amino or a stop during protein synthesis. ***(c)*** Unless it is a stop codon, each mRNA codon forms base pairs with the anticodon of a tRNA molecule that carries a specific amino acid. ***(d)*** The ribosome links the amino acids together, forming the protein.

strand of the gene. Starting from the first AUG, each codon within the mRNA is a sequence of three bases that specify a single amino acid or a "stop."

(c) Enzymes in the cytoplasm attach the appropriate amino acid to each tRNA based on the tRNA's anticodon.

(d) During translation, tRNAs carry their attached amino acids to the ribosome. The appropriate amino acid is selected based on the complementary base pairs formed between the bases in the mRNA codon and the bases in the tRNA anticodon. The ribosome then links the amino acids together in sequence to form a protein.

The decoding "chain of events," moving from DNA bases to mRNA codons to tRNA anticodons to amino acids, results in synthesis of a protein with a specific amino acid sequence. This amino acid sequence is ultimately determined by the base sequence within a single gene.

4 How Do Mutations in DNA Affect the Function of Genes?

Up to now, we have emphasized the precision and fidelity of DNA replication and its transcription, as well as the translation of mRNA into proteins. Of course, nothing alive is perfect. Mistakes can occur in any of these processes. A single faulty mRNA or a single defective protein normally doesn't affect a cell very much, because there are many correct copies of these molecules in the cell at the same time to carry out the proper cellular functions. However, a single faulty copy of a *gene* may be much more serious, since all of the proteins encoded by the mutant gene may be defective. We explore two cases in which a single mutation causes far-reaching results in the essay "Health Watch: Sex, Aging, and Mutations."

Changes in the sequence of bases in DNA are called **mutations**. How can a mutation occur? One way is through a mistake in base pairing during replication as a cell prepares for cell division. Despite the best efforts of proofreading enzymes, each DNA replication results in a few mistakes. Changes in bases can also occur spontaneously. In addition, certain chemicals (such as aflatoxins, synthesized by some molds that live on grain and peanuts) and some types of radiation (such as X rays and ultraviolet rays in sunlight) increase the frequency of base-pairing errors during replication, or even induce changes in DNA composition between replications.

Random changes in DNA sequence are unlikely to code for improvements in the functioning of the gene's products, much as randomly changing words in the middle of the play *Hamlet* would be unlikely to improve on Shakespeare's work. Some mutations, however, have no effect or (in very rare instances) are even beneficial, as you will learn later in this chapter.

Mutations Result from Nucleotide Substitutions, Insertions, or Deletions

In a **nucleotide substitution**, a pair of bases becomes incorrectly matched. Repair enzymes recognize the mismatch, cut out the incorrect nucleotide, and replace it with a nucleotide that bears a complementary base. Occasionally, however, the enzymes replace the *correct* nucleotide instead of the incorrect one. The resulting base pair is complementary, but it is the wrong pair of nucleotides. These mutations are also called **point mutations**, because they change individual nucleotides in the DNA sequence. An **insertion mutation** occurs when one or more new nucleotide pairs are inserted into a gene. A **deletion mutation** occurs when one or more nucleotide pairs are removed from a gene.

Mutations Have Different Effects on Protein Structure and Function

If a mutation occurs in cells whose offspring become gametes (sperm or eggs), that mutation may be passed on to future generations. But how does a change in base sequence affect the organism that inherits the mutated DNA? As the box "Scientific Inquiry: Cracking the Genetic Code" points out, deletions and insertions of one or two nucleotides can have particularly catastrophic effects on a gene, because all of the codons that follow the deletion or insertion will be altered. The protein synthesized from an mRNA containing such changed codons is almost certain to be nonfunctional.

Nucleotide substitutions within a protein-coding gene can produce at least four different outcomes (Table 10-4). As a concrete example, let's consider mutations that occur in the gene encoding beta-globin, one of the proteins in hemoglobin. Hemoglobin is the oxygen-binding protein found in red blood cells. In all but the last example, we will consider the results of mutations that occur in the sixth codon, GAG, which specifies the amino acid glutamic acid.

1. *The protein may be unchanged:* Recall that most amino acids can be encoded by several different codons. For example, if a mutation changes the beta-globin DNA sequence from GAG to GAA, the new codon still codes for glutamic acid. Therefore, the protein synthesized from the mutated gene remains the same even though the DNA sequence is different.
2. *The new protein may be equivalent to the original one:* Many proteins have regions whose exact amino acid sequence is relatively unimportant. For example, in beta-globin, the amino acids on the outside of the protein must be hydrophilic to keep the protein dissolved in the cytoplasm of red blood cells. Exactly *which* hydrophilic amino acids are on the outside doesn't matter much. For example, a family in the Japanese town of Machida was found to contain a mutation from GAG to CAG, replacing glutamic

Table 10-4 Effects of Mutations in the Hemoglobin Gene

	DNA (template strand)	mRNA	Amino Acid	Properties of Amino Acid	Functional Effect on Protein	Disease
Original codon 6	CTC	GAG	Glutamic acid	Hydrophilic	Normal protein function	None
Mutation 1	CTT	GAA	Glutamic acid	Hydrophilic	Neutral, normal protein function	None
Mutation 2	CTC	CAG	Glutamine	Hydrophilic	Neutral, normal protein function	None
Mutation 3	CAC	GUG	Valine	Hydrophobic	Lose water solubility; compromises protein function	Sickle-cell anemia
Original codon 17	TTC	AAG	Lysine	Hydrophilic	Normal protein function	None
Mutation 4	ATC	UAG	Stop codon	Ends translation after amino acid 16	Synthesizes only part of protein; eliminates protein function	Beta-thalassemia

acid (hydrophilic) with glutamine (also hydrophilic). Hemoglobin containing this mutant beta-globin protein is known as *Hemoglobin Machida* and appears to function well. Mutations such as that in Hemoglobin Machida and in the previous example are called **neutral mutations** because they do not detectably change the function of the encoded protein.

3. *Protein function may be changed by an altered amino acid sequence:* A mutation from GAG to GTG (GUG in mRNA) replaces glutamic acid (hydrophilic) with valine (hydrophobic). This substitution, which is the genetic defect in sickle-cell anemia (see Chapter 12), causes hemoglobin that contains the mutant beta-globin protein to stick to one another, clumping up and distorting the shape of the red blood cells. These changes can cause serious illness.
4. *Protein function may be destroyed by a premature stop codon:* To consider another type of mutation, we will shift our attention to the 17th codon in the beta-globin gene. This codon, AAG, specifies the amino acid lysine. A mutation from AAG to TAG (UAG in mRNA) changes an amino acid codon to a stop codon. This premature stop codon halts translation of beta-globin mRNA before the protein is completed. As a result, no beta-globin protein is made in individuals who inherit this mutant gene from both their mother and their father. Individuals with this type of mutation have a condition called beta-thalassemia, which can be fatal unless treated with blood transfusions.

Mutations Provide the Raw Material for Evolution

If mutations in gametes are not lethal, they may be passed on to future generations. In humans, mutation rates in genes studied thus far range from 1 in every 100,000 gametes to 1 in 1,000,000 gametes. For reference, a man releases about 300 to 400 million sperm per ejaculation, so on average each ejaculate contains about 600 sperm with new mutations. Although most mutations are neutral or potentially harmful, mutations are essential for evolution, because these random changes in DNA sequence are the ultimate source of all genetic variation. New base sequences undergo natural selection as organisms compete to survive and reproduce. Occasionally, a mutation proves beneficial in the organism's interactions with its environment. The mutant base sequence may spread throughout the population and become common as organisms that possess it outcompete rivals that bear the original, unmutated base sequence. This process is described in detail in Unit III.

5 How Are Genes Regulated?

Knowing how proteins are synthesized and the likely effects of gene mutations on protein function does not provide a complete understanding of gene function in an organism's life. For example, simply knowing the existence of a gene that determines the sex of mammals does not tell us why some mammals are male and some are female. To understand how genes affect an organism's characteristics, we must also consider how cells control the presence, concentration, and activity of the proteins encoded by genes.

Proper Regulation of Gene Expression Is Critical for an Organism's Development and Health

No one is sure exactly how many genes are encoded in the human genome, but recent estimates suggest that it is about 30,000 genes. Each of these genes is present in

Health Watch
Sex, Aging, and Mutations

Imagine that you are a girl in her mid-teens. Your breasts have filled out, but you have not begun to menstruate. Accompanied by a concerned parent, you visit your family doctor, who takes a small blood sample to do a chromosome test. When the results come in, you and your parents are called back to the office, where a genetic counselor gives you amazing news: Your sex chromosomes are XY, a combination that would normally give rise to a male. The reason you have not begun to menstruate is that you lack ovaries and a uterus but instead have testes that have remained inside your abdominal cavity. You have levels of male hormones (collectively called *androgens,* such as testosterone) circulating in your blood that are normal for a male. In fact, these male hormones, produced by the testes, have been present since early in your development. The problem is that your cells cannot respond to them—a rare condition called **androgen insensitivity**. This condition was a problem for Maria Jose Martinez Patino, an outstanding Spanish athlete who reached the Olympics, only to be barred from the hurdles competition because her chromosomes indicated that she was male. After three years of struggle, the fact that she had developed as a female was finally recognized, and she was allowed to compete against others of her gender.

Many male features, including the formation of a penis, the descent of the testes into sacs outside the body cavity, and sexual characteristics that develop at puberty, such as a beard and increased muscle mass, occur because various body cells are responding to male sex hormones produced by the testes. In normal males, many body cells have androgen receptor proteins in their cytoplasm. These proteins bind to male hormones such as testosterone and escort them into the nucleus, where the hormone receptor complex binds to DNA and influences the transcription of genes into mRNA. The mRNA molecules will serve as templates for the synthesis of specific proteins that contribute to maleness. In different groups of cells, the androgen receptor–testosterone complex influences gene transcription in different ways, giving rise to

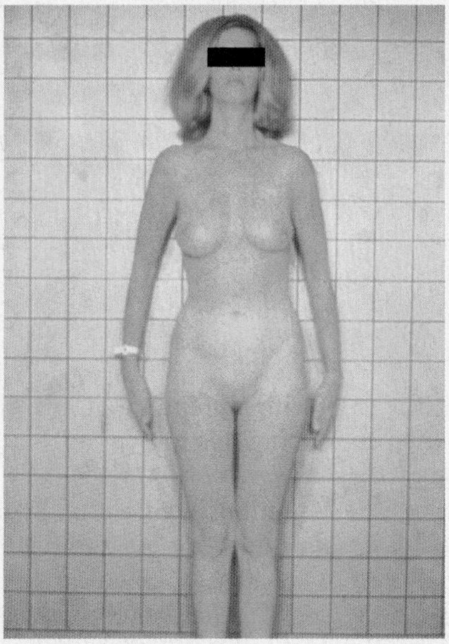

Figure E10-2 Androgen insensitivity leads to female features
This individual has an X and a Y chromosome. As a result, she has testes that produce testosterone, but these are located in her abdomen. In addition, she lacks ovaries and a uterus. A mutation in her androgen receptor genes makes her cells unable to respond to the testosterone her testes produce and produces the obviously female appearance.

most of your body cells, but individual cells *express* (transcribe and translate) only a small fraction of these genes—namely, those genes that are appropriate for the function of that particular cell type. Muscle cells, for example, synthesize large amounts of the contractile proteins actin and myosin but do not synthesize insulin or hair proteins.

Expression of a gene also changes over time, depending on the body's needs from moment to moment. For example, during pregnancy, milk-producing cells in a woman's breasts multiply tremendously. Immediately after the baby's birth, those cells begin expressing the gene encoding casein, the major protein in milk. This change in gene expression allows the mother to produce large amounts of protein-laden milk to feed her new baby. Finally, an individual contains genes that may never be expressed at any time during his or her entire lifetime. For example, a human male does not express the casein gene. However, he passes a copy of this gene to his daughters, who will express the gene if they become pregnant.

An organism's environment also helps to determine which genes are transcribed. For example, in birds living at temperate climates, increasing length of day stimulates the sex organs (testes or ovaries) to increase in size. The sex organs in turn produce sex hormones that influence a variety of physiological processes and behaviors. Under the influence of sex hormones, birds begin to produce eggs and sperm, to sing, to mate, and to build nests. The proliferation of cells in the sex organs, the production of hormones by these cells, and the effects of those hormones on other cells throughout the body all result, directly or indirectly, from changes in gene expression.

Expression of genetic information by a cell is a multistep process, beginning with transcription of DNA and commonly ending with a protein that performs a particular function. Regulation of gene expression can occur a wide range of male characteristics. As for all other proteins, androgen receptors are coded by specific genes. An individual who inherits a mutant androgen receptor gene from both her mother and father will be unable to make a functional androgen receptor protein. Her cells will be unable to respond to the testosterone that the testes produce. Thus, a change in the nucleotide sequence of a single gene, causing a single type of defective protein to be produced, can cause a person who is genetically male (XY) to look like and feel like a woman (Fig. E10-2).

A second type of mutation is providing clues to the mystery of why people age. Why will your hair whiten, skin wrinkle, joints ache, and eyes cloud as you become elderly? A small number of individuals carry a defective gene that causes **Werner syndrome**, which causes a type of premature aging (Fig. E10-3). People with this disorder die of age-related conditions, typically by age 50. Recent research has localized the mutations in most people with Werner syndrome to a gene that codes for an enzyme involved in DNA replication. As you have seen, the accurate replication of DNA is crucial to the production of normally functioning cells. If a mutation interferes with the ability of enzymes to promote accurate DNA replication and to proofread and repair errors in DNA, then mutations will accumulate progressively in cells throughout the body.

The fact that an overall increase in mutations caused by defective replication enzymes produces symptoms of old age provides support for one hypothesis about the way many of the symptoms of normal aging occur. During a typical long (say, 80-year) life span, mutations gradually accumulate because of mistakes in DNA replication and environmentally induced DNA damage. Eventually, these mutations interfere with nearly every aspect of body functioning and contribute to death from "old age."

Disorders such as androgen insensitivity and Werner syndrome provide profound insights into the impact of mutations, the function of specific genes and their protein products, the ways hormones regulate gene transcription, and even the mystery of aging.

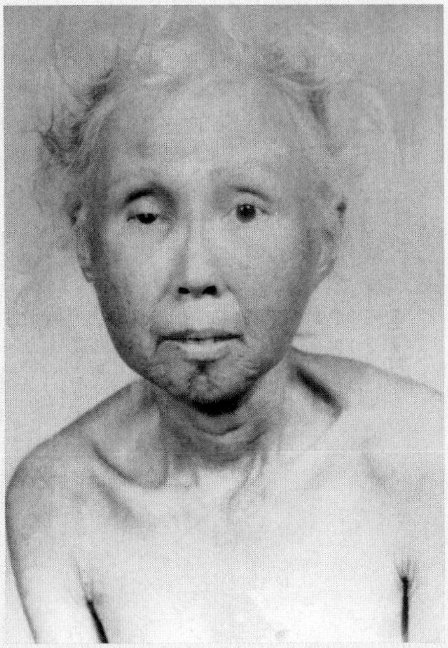

Figure E10-3 A 48-year-old woman with Werner syndrome This condition, most common among people of Japanese ancestry, is the result of a mutation that interferes with proper DNA replication, increasing the incidence of mutations throughout the body.

at any of these steps, four of which are illustrated in Figure 10-8, as follows:

① ***Cells can control the frequency at which an individual gene is transcribed.*** Transcription of genes is turned on or off, according to the demand for the protein (or RNA) product that they encode. Thus, whether a specific gene is transcribed and the frequency of its transcription depend on the type of cell and on the metabolic activity of both the cell and the organism as a whole. We consider this aspect of regulation in greater detail in the next section. In addition, transcription in eukaryotes is complicated by the fact that most eukaryotic protein-coding genes contain segments of nucleotides that do not code for protein; following transcription, these nucleotide segments must be removed from the newly synthesized RNA molecule to create a functional mRNA. This process, described in "A Closer Look: Introns, Exons, and Splicing," can also be regulated to control the amount of mRNA produced from a particular gene.

② ***Different messenger RNAs may be translated at different rates.*** Messenger RNAs vary in their stability and in the rate at which they are translated into protein. Some mRNAs are extremely stable, making them available for repeated translation so they can produce protein for long time periods. Other mRNAs rapidly degrade so that they only produce protein for a short time. Furthermore, depending on metabolic requirements, a cell may completely block the translation of certain mRNAs until a signal activates their translation. This block is especially evident in egg cells, which contain thousands of stored mRNAs that are not translated unless the egg is fertilized.

③ ***Proteins may require modification before they can carry out their functions.*** Many proteins must be modified before they become active. For instance,

Figure 10-8 An overview of "information flow" in a cell
This simplified diagram shows the major steps from DNA to protein to chemical reactions catalyzed by enzymes. Regulation of gene expression may occur at any or all steps.

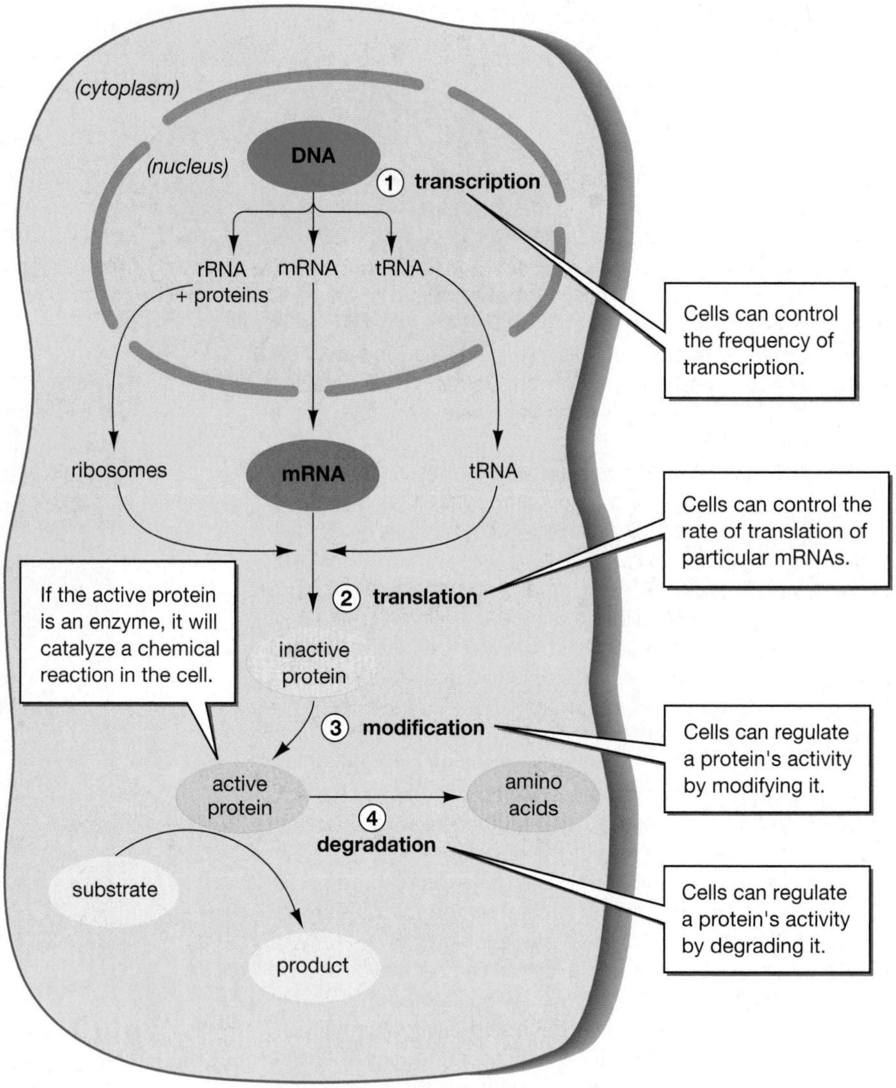

the protein-digesting enzymes produced by cells of your stomach wall and pancreas are initially synthesized in an inactive form. This initial inactivity prevents the enzymes from digesting the very cells that produce them. After these inactive forms are secreted into the digestive tract, portions of the enzymes are then snipped out to unveil the enzyme's active site. Other modifications can temporarily activate or inactivate a protein's function, allowing second-to-second control of the protein's activity.

④ ***The life span of a protein can be regulated.*** Most proteins have a limited life span within the cell and a variety of mechanisms exist in cells to continually degrade proteins and other cellular components. By preventing or promoting a protein's degradation, cells can rapidly adjust the amount of a particular protein within it.

Each of these methods of regulating gene activity is important and all are used to some extent by virtually all eukaryotic cells. In addition, an individual gene may be subject to regulation at many or all of the steps described above.

Eukaryotic Cells May Regulate the Transcription of Individual Genes, Regions of Chromosomes, or Entire Chromosomes

In eukaryotic cells, transcriptional regulation can operate on at least three levels: (1) the individual gene, (2) regions of chromosomes, or (3) entire chromosomes.

Regulatory Proteins That Bind to the Gene's Promoter Alter the Transcription of Individual Genes

Some of the best-known examples of transcriptional regulation are the cellular actions of steroid hormones, such as sex hormones. To explore this regulation in more detail, let's return to the example of sex hormone function in birds. The gene for the egg-white protein, albumin, is not transcribed in the winter when birds are not breeding and sex hormone levels are low. During the breeding season, the ovaries in female birds produce the sex hormone estrogen. Estrogen crosses the membrane of cells in the oviduct and binds to a receptor protein in the cytoplasm. The estrogen–receptor complex then enters the nucleus and attaches to a DNA control region near the promoter of the albumin gene.

A Closer Look
Introns, Exons, and Splicing

In the 1970s molecular geneticists discovered that most eukaryotic protein-coding genes have much more DNA than is needed to encode the amino acids of proteins. These genes consist of two or more segments with nucleotide sequences that encode for a protein, interrupted by other nucleotide sequences that are not translated into a protein. The researchers called the coding segments **exons**, because they are *ex*pressed in protein, and the noncoding segments **introns**, because they *int*ervene between the exons (Fig. E10-4a). Most eukaryotic genes—and even some genes in archaebacteria—have introns. In fact, one gene coding for a particular type of connective tissue in chickens has about 50 introns!

Transcription of a eukaryotic gene produces a very long RNA, starting before the first exon and ending after the last exon (Fig. E10-4b). The resulting RNA contains both introns and exons. To convert this RNA molecule into true mRNA that contains only the exons needed to code for the protein, enzymes in the nucleus precisely cut the RNA molecule apart at the junctions between intron and exon sequences, splice together the protein-coding segments (exons), and discard the rest.

Why are eukaryotic genes split up like this? Gene fragmentation appears to serve at least two functions. The first function is to allow a cell to produce multiple proteins from a single gene by splicing exons together in different ways. Rats, for example, have a gene that is transcribed in both the thyroid and the brain. In the thyroid, one splicing arrangement results in the synthesis of a hormone called *calcitonin*, which helps regulate calcium concentrations in the blood. In the brain, a different splicing arrangement results in the synthesis of a short protein that is probably used as a molecular messenger for communication among brain cells.

The second function of interrupted genes is more speculative but is supported by some good experimental evidence: Fragmented genes may provide a quick and efficient way for eukaryotes to evolve new proteins with new functions. Chromosomes sometimes break apart, and their parts may reattach to different chromosomes. If the breaks occur within the noncoding introns of genes, exons may be moved intact from one chromosome to another. Most such errors would be harmful. But some of these shuffled exons might code for a protein subunit that has a specific function (binding ATP, for example). In rare instances, adding this subunit to an existing gene may cause the gene to code for a new protein with useful functions. The accidental exchange of exons among genes produces new eukaryotic genes that will, on occasion, enhance survival and reproduction of the organism that carries them.

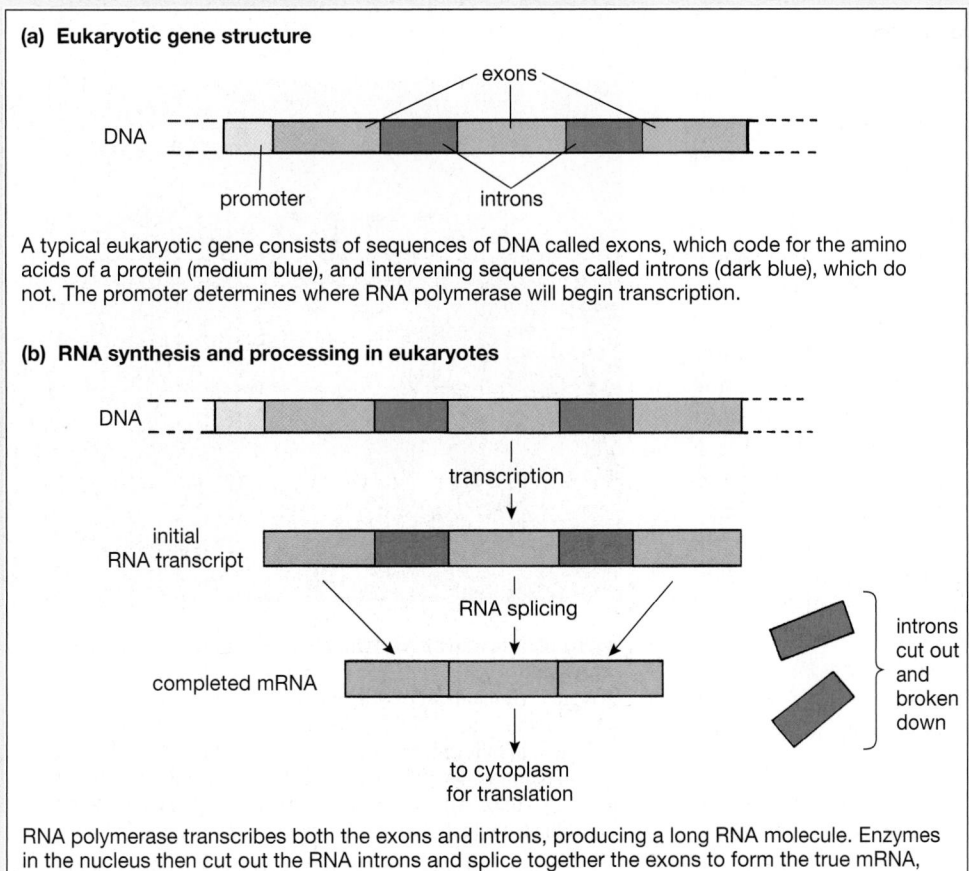

(a) Eukaryotic gene structure

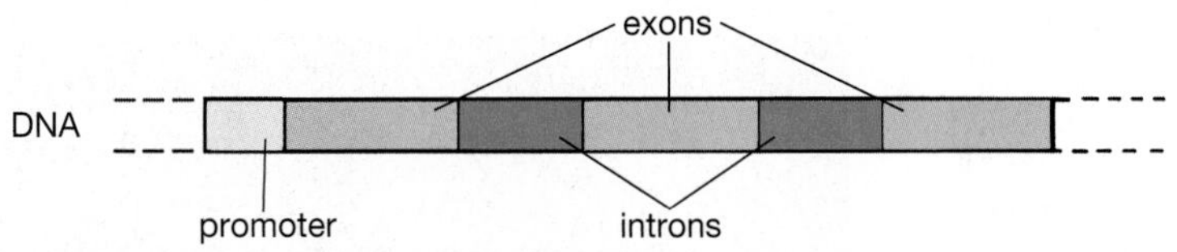

A typical eukaryotic gene consists of sequences of DNA called exons, which code for the amino acids of a protein (medium blue), and intervening sequences called introns (dark blue), which do not. The promoter determines where RNA polymerase will begin transcription.

(b) RNA synthesis and processing in eukaryotes

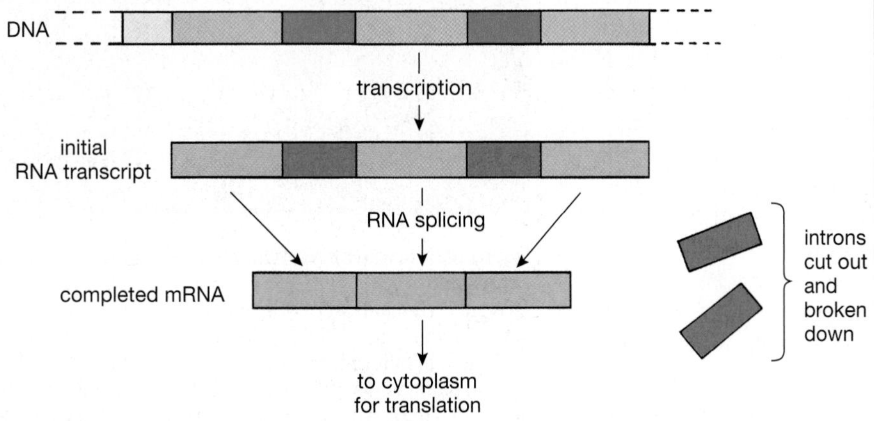

RNA polymerase transcribes both the exons and introns, producing a long RNA molecule. Enzymes in the nucleus then cut out the RNA introns and splice together the exons to form the true mRNA, which moves out of the nucleus and is translated on the ribosomes.

Figure E10-4 Eukaryotic genes contain introns and exons

This attachment makes it easier for RNA polymerase to bind to the promoter of the gene. As a result, oviduct cells transcribe large amounts of albumin mRNA that is translated into the large amounts of albumin protein needed to make eggs. Similar activation of gene transcription by steroid hormones occurs in other animals, including humans. The importance of hormonal regulation of transcription during development is illustrated by genetic defects in which receptors for sex hormones are nonfunctional (see "Health Watch: Sex, Aging, and Mutations"). In such cases, cells within the individual are unable to respond to the hormone, short-circuiting critical events in sexual development.

Some Regions of Chromosomes Are Condensed and Not Normally Transcribed

Certain parts of eukaryotic chromosomes are in a highly condensed, compact state in which most of the DNA seems to be inaccessible to RNA polymerase. Some of these regions are structural parts of chromosomes that don't contain genes. Other tightly condensed regions contain functional genes that are not currently being transcribed. When the product of a gene is needed, the portion of the chromosome containing that gene becomes "decondensed"—loosened so that the nucleotide sequence is accessible to RNA polymerase and transcription can occur.

Entire Chromosomes May Be Inactivated, Thereby Preventing Transcription

In some cases an entire chromosome may be condensed, making it largely inaccessible to RNA polymerase. An example occurs in the sex chromosomes of female mammals. Male mammals usually have an X and a Y chromosome (XY) and females usually have two X chromosomes (XX). As a consequence, females have the capacity to synthesize twice as much protein from genes on their two X chromosome as males with only one X chromosome. Such imbalance in gene number can be detrimental, as illustrated by Down Syndrome. Down Syndrome results from an extra copy of Chromosome 21, which encodes many fewer genes than the X chromosome. To balance expression of genes on the female's X chromosomes, one of her X chromosomes is condensed into a tight mass. In a light microscope, the inactivated, condensed X chromosome shows up in the nucleus as a dark spot called a **Barr body**, named after its discoverer, Murray Barr. As a result, the bodies of female mammals (including human females) are interspersed with patches of cells in which one of the X chromosomes is active and patches of cells in which the other X chromosome is active.

The results of this phenomenon are easily observed in calico cats (Fig. 10-9). The cat X chromosome contains a gene encoding an enzyme that produces fur pigment. This gene comes in two versions, one producing orange fur and the other producing black fur. If one X chromosome in a female cat has the orange version of the fur color gene and the other X chromosome has the black version of the fur color gene, the cat will have patches of orange and black fur. These patches represent areas of skin that developed from cells in the early embryo in which different X chromosomes were inactivated. Calico coloring is therefore almost exclusively found in female cats, since male cats have only one X chromosome, which is active in all of their cells. Male cats will therefore have black fur or orange fur, but not both.

To prevent males from competing against women unfairly at the Olympics, officials have attempted to verify that athletes who compete in women's events are truly female by performing a gene-based sex test. Women who "pass" the test are given a gender certification card, a requirement for a female athlete's participation in many athletic competitions. One type of sex test used as recently as the 1996 Olympics in Atlanta examines stained cells taken from the mouth of the athlete for the presence of Barr bodies. This test caused major problems for a female hurdler from Spain, Maria Jose Martinez Patino, when no Barr bodies were found in her cells. Learn why in "Health Watch: Sex, Aging, and Mutations."

Figure 10-9 Inactivation of the X chromosome regulates gene expression
This calico female cat carries a gene for orange fur on one X chromosome and a gene for black fur on her other X chromosome. Inactivation of different X chromosomes produces the black and orange patches. The white color is due to an entirely different gene, on a different chromosome, that prevents pigment formation, regardless of which X chromosome is present in the cells.

CASESTUDYREVISITED

Boy or Girl?

How does knowledge about transcription and translation help us understand the physical differences between males and females? By the 1930s biologists knew that one or more genes on the Y chromosome were essential for determining whether a mammal would develop into a male or a female. In 1990 a search for this gene lead to discovery of the SRY gene, for "*s*ex-determining *r*egion on the *Y* chromosome." The SRY gene is found in all mammals, including humans. Experiments with mice demonstrated its importance in sex determination. If a mouse embryo with two X chromosomes is given a copy of SRY, the embryo develops completely male characteristics: It has a penis and testes and behaves like a male mouse. (These XX male mice are sterile, however, because other genes located on the Y chromosome are apparently needed for production of functional sperm.) Thus, female (XX) mammals have all of the genes needed to be male but usually aren't, because they don't have an SRY gene. Likewise, mouse embryos that lack an SRY gene develop as females, regardless of whether they have two X chromosomes or an X and a Y. Thus, male (XY) mammals have all the genes needed to be female but usually aren't, because they have an SRY gene.

How does the SRY gene exert such a major effect on a mammal's characteristics? Even though it is critical for sex determination, the SRY gene is transcribed for only a short time in embryonic development, and only in cells that will become the testes. The SRY gene is then apparently permanently inactivated for the rest of the animal's life. In the short time that it is transcribed, the protein encoded by the SRY gene apparently initiates the expression of many other genes; the proteins produced by these genes are needed for development of testes. Once formed, the testes in the tiny embryo secrete testosterone, which activates other genes, leading to development of the penis and scrotum. Boy or girl? It depends on the carefully regulated expression of many genes, with a single gene, SRY, serving as the initial genetic switch to activate male development.

Many scientists are concerned about the presence of chemicals in the environment that behave somewhat like sex hormones. Animals exposed to these chemicals may have the right chromosomes to be males but nevertheless look and behave like females. How could interference with sexual development be detrimental to endangered species?

Summary of Key Concepts

1) How Are Genes and Proteins Related?

Genes are segments of DNA that can be transcribed into RNA and, for most genes, translated into protein. Transcription produces the three types of RNA needed for translation: (1) messenger RNA (mRNA), (2) transfer RNA (tRNA), and (3) ribosomal RNA (rRNA). During translation, tRNA and rRNA collaborate with enzymes and other proteins to decode the sequence of bases in mRNA and produce a protein with the amino acid sequence specified by the gene. The genetic code consists of codons, sequences of three bases in mRNA that specify either an amino acid in the protein chain or the end of protein synthesis (stop codons).

2) How Is Information in a Gene Transcribed into RNA?

Within an individual cell, only certain genes are transcribed. When the cell requires the product of a gene, RNA polymerase binds to the promoter region of the gene and synthesizes a single strand of RNA. This RNA is complementary to the template strand in the gene's DNA double helix.

3) How Is the Sequence in a Messenger RNA Molecule Translated into Protein?

In eukaryotes, mRNA carries the genetic information from the nucleus to the cytoplasm, where ribosomes can use this information to synthesize a protein. Ribosomes contain rRNA and proteins organized into large and small subunits. These subunits come together at the first AUG codon of the mRNA molecule to form the complete protein-synthesizing machine. tRNAs deliver the appropriate amino acids to the ribosome for incorporation into the growing protein. Which tRNA binds, and consequently which amino acid is delivered, depends on base pairing between the anticodon of the tRNA and the codon of the mRNA. Two tRNAs, each carrying an amino acid, bind simultaneously to the ribosome's large subunit, which then catalyzes the formation of peptide bonds between the amino acids. As each new amino acid is attached, one tRNA detaches, and the ribosome moves down one codon and binds to another tRNA that carries the next amino acid specified by mRNA. Addition of amino acids to the growing protein continues until a stop codon is reached, signaling the ribosome to disassemble and to release both the mRNA and the newly formed protein.

4) How Do Mutations in DNA Affect the Functions of Genes?

A mutation is a change in the nucleotide sequence of a gene. Mutations can be caused by mistakes in base pairing during replication, by chemical agents, and by environmental factors such as radiation. Common types of mutations include changes in a nucleotide base pair (point mutations) and insertions or deletions of nucleotide base pairs. Mutations may be neutral or harmful, but in rare cases a mutation will promote better adaptation to the environment and thus will be favored by natural selection.

5) How Are Genes Regulated?

The expression of a gene requires that it is transcribed and translated, and the resulting protein performs some action within the cell. Which genes are expressed in a cell at any given time is regulated by the function of the cell, the developmental stage of the organism, and the environment. Control of gene regulation can occur at many steps along the information flow. The amount of mRNA synthesized from a particular gene can be regulated by increasing or decreasing the rate of its transcription, as well as by the stability of the mRNA itself. Rates of translation of mRNAs can also be regulated. Regulation of transcription and translation affects how many protein molecules are produced from a particular gene. Even after they are synthesized, many proteins must be modified by enzymes before they can function. In addition to regulation of individual genes, cells can regulate transcription of groups of genes. For example, entire chromosomes or parts of chromosomes may be condensed and inaccessible to RNA polymerase, whereas other portions are expanded, allowing transcription to occur.

Key Terms

androgen insensitivity *p. 176*
anticodon *p. 170*
Barr body *p. 180*
codon *p. 167*
deletion mutation *p. 174*
exon *p. 179*
gene *p. 164*
genetic code *p. 166*
insertion mutation *p. 174*
intron *p. 179*
messenger RNA (mRNA) *p. 166*
mutation *p. 174*
neutral mutation *p. 175*
nucleotide substitution *p. 174*
point mutation *p. 174*
polypeptide *p. 164*
promoter *p. 168*
ribonucleic acid (RNA) *p. 165*
ribosomal RNA (rRNA) *p. 166*
ribosome *p. 166*
RNA polymerase *p. 168*
stop codon *p. 167*
template strand *p. 168*
transcription *p. 166*
transfer RNA (tRNA) *p. 166*
translation *p. 166*
Werner syndrome *p. 177*

Thinking Through the Concepts

Multiple Choice

1. *A gene*
   a. is synonymous with a chromosome
   b. is composed of mRNA
   c. is a specific segment of nucleotides in DNA
   d. contains only those nucleotides required to synthesize a protein
   e. specifies the sequence of nutrients required by the body
2. *Which of the following is a single-stranded molecule that contains the information for assembly of a specific protein?*
   a. transfer RNA
   b. messenger RNA
   c. exon DNA
   d. intron DNA
   e. ribosomal RNA
3. Anticodon *is the term applied to*
   a. the list of amino acids that corresponds to the genetic code
   b. the concept that multiple codons sometimes code for a single amino acid
   c. the part of the tRNA that interacts with the codon
   d. the several three-nucleotide stretches that code for "stop"
   e. the part of the tRNA that binds to an amino acid
4. *DNA*
   a. takes part directly in protein synthesis by leaving the nucleus and being translated on the ribosome
   b. takes part indirectly in protein synthesis; the DNA itself stays in the nucleus
   c. has nothing to do with protein synthesis; it is involved only in cell division
   d. is involved in protein synthesis within the nucleus
   e. codes for mRNA but not for tRNA or rRNA
5. *Synthesis of a protein based on the sequence of messenger RNA*
   a. is catalyzed by DNA polymerase
   b. is catalyzed by RNA polymerase
   c. is called *translation*
   d. is called *transcription*
   e. occurs in the nucleus
6. *A Barr body is*
   a. a condensed X chromosome
   b. an organelle involved in protein synthesis
   c. another term for chromosomes as they are seen during cell division
   d. visible in cells of both males and females
   e. present only in males, because their female chromosomes are inactivated

? Review Questions

1. How does RNA differ from DNA?
2. What are the three types of RNA? What is the function of each?
3. Define the following terms: *genetic code*; *codon*; *anticodon*. What is the relationship among the bases in DNA, the codons of mRNA, and the anticodons of tRNA?
4. How is mRNA formed from a eukaryotic gene?
5. Diagram and describe protein synthesis.
6. Explain how complementary base pairing is involved in both transcription and translation.
7. Describe some mechanisms of gene regulation.
8. Define *mutation*, and give one example of how a mutation might occur. Would you expect most mutations to be beneficial or harmful? Explain your answer.

Applying the Concepts

1. Some antibiotics, such as erythromycin and streptomycin, bind to the small ribosomal subunit in bacteria and inhibit translation. Why do these drugs kill bacteria? How might bacteria evolve so that they become resistant to such antibiotics?
2. Although they are rare, male calico cats do occur. In fact, about 1 in 3000 calico cats is a male. Can you come up with an explanation that explains the origin of male calico cats? Most male calico cats are infertile. Why? About 1 in 10,000 male calico cats is fertile. Considering chromosomes, genes, and mutations, can you explain the differences between infertile and fertile male calico cats?
3. As you have learned in this chapter, many factors influence gene expression, including hormones. The use of anabolic steroids and growth hormones among athletes has created controversy in recent years. Hormones certainly affect gene expression, but, in the broadest sense, so do vitamins and foods. What do you think are appropriate guidelines for the use of hormones? Should athletes take steroids or growth hormones? Should children at risk of being unusually short be given growth hormones? Should parents be allowed to request growth hormones for their children of normal height in the hope of producing a future basketball player?

For More Information

Grunstein, M. "Histones as Regulators of Genes." *Scientific American*, October 1992. Histones are proteins associated with DNA in eukaryotic chromosomes. Once thought to be merely a scaffold for DNA, they are actually important in gene regulation.

Nirenberg, M. W. "The Genetic Code: II." *Scientific American*, March 1963. Nirenberg describes some of the experiments in which he deciphered much of the genetic code.

Tjian, R. "Molecular Machines That Control Genes." *Scientific American*, February 1995. Complexes of proteins regulate which genes are transcribed in a cell and therefore help determine the cell's structure and function.

Travis, J. "Biology's Periodic Table." *Science News*, March 22, 1997. A brief synopsis of the Human Genome Project.

Answers to Multiple-Choice Questions

1. c 2. b 3. c 4. b 5. c 6. a

MEDIATUTOR
Gene Expression and Regulation

CD Activities

Activity 10.1: Transcription

Estimated time: 10 minutes

The genetic information that is stored in the DNA found in the nucleus of a cell must be conveyed to the cytoplasm, where it can be translated into a specific sequence of amino acids. The intermediate molecule is RNA, synthesized by the process of transcription. This tutorial investigates the steps involved in transcription.

Activity 10.2: Translation

Estimated time: 10 minutes

Transcription provides a messenger RNA transcript of the gene. Now it is the task of the cell to take the information encoded in this messenger RNA and translate it into a specific sequence of amino acids. This process occurs on the ribosomes and involves a number of steps illustrated in this tutorial.

Start the MediaTutor Student CD-ROM and enter the activity number in the Quick Search box to be taken directly to that activity.

Web Investigations

Case Study: Boy or Girl?

Estimated time: 15 minutes

Everyone knows that chromosomes determine gender. Baby girls have two X chromosomes (XX), while baby boys are XY. But what about individuals with XO, XXY, or XYY genomes? Why do some female babies have XY karyotypes (chromosome sets)? Gender biology is much more complicated than one might think.

Go to http://www.prenhall.com/audesirk6, the Audesirk Companion Web site. Select Chapter 10 and Web Investigation to begin.

"All cells come from cells."

Rudolf Virchow, German physician and early proponent of the cell theory (1858)

Reproduction of timber rattlesnakes is usually unremarkable: Males provide sperm; females provide eggs; and baby rattlesnakes develop from fertilized eggs. When female rattlesnakes produce offspring without any contribution from a male, however, there's some explaining to do!

11 The Continuity of Life: Cellular Reproduction

AT A GLANCE

Case Study: Rattlesnake Surprise

1) **What Are the Functions of Cellular Reproduction?**
Binary Fission Is the Cell Division Process in Prokaryotes
Mitotic Cell Division Allows Development, Growth, Maintenance, and Repair of Body Tissues in Eukaryotes
Mitotic Cell Division Forms the Basis of Asexual Reproduction in Eukaryotes
Meiotic Cell Division Forms the Basis of Sexual Reproduction in Eukaryotes

2) **How Is DNA in Eukaryotic Cells Organized into Chromosomes?**
Cell Division Enables Accurate Passage of Chromosomes from One Generation to the Next
The Eukaryotic Chromosome Consists of a Linear DNA Double Helix Bound to Proteins
Eukaryotic Chromosomes Usually Occur in Homologous Pairs with Similar Genetic Information

3) **What Are the Events of the Eukaryotic Cell Cycle?**
Cell Division Is One Part of the Eukaryotic Cell Cycle
During Interphase, the Eukaryotic Cell Grows in Size and Replicates Its DNA
Mitotic Cell Division Consists of Nuclear Division and Cytoplasmic Division

4) **What Are the Phases of Mitosis?**
During Prophase, the Chromosomes Condense and the Spindle Microtubules Form and Attach to the Chromosomes
During Metaphase, the Chromosomes Become Aligned Along the Equator of the Cell
During Anaphase, Sister Chromatids Separate and Are Pulled to Opposite Poles of the Cell
During Telophase, Nuclear Envelopes Form Around Both Groups of Chromosomes

5) **What Are the Events of Cytokinesis?**

6) **What Are Some Advantages of Sexual Reproduction?**
Mutations in DNA Are the Ultimate Source of Genetic Variability
Reshuffling Genes May Combine Different Alleles in Beneficial Ways
Meiotic Cell Division Produces Haploid Cells That Can Merge to Combine Genetic Material from Two Parents

7) **What Are the Events of Meiosis?**
Meiosis Separates Homologous Chromosomes in a Diploid Nucleus, Producing Haploid Daughter Nuclei
Meiosis I Separates Homologous Chromosomes into Two Daughter Nuclei
Meiosis II Separates Sister Chromatids into Four Daughter Nuclei
The Life Cycles of Organisms on Earth Usually Include Meiosis and Mitosis

8) **How Do Meiosis and Sexual Reproduction Produce Genetic Variability?**
Shuffling of Homologues Creates Novel Combinations of Chromosomes
Crossing Over Creates Chromosomes with Novel Combinations of Genes
Fusion of Gametes Adds Further Genetic Variability to the Offspring

Case Study Revisited: Rattlesnake Surprise

CASESTUDY

Rattlesnake Surprise

David Chiszar, a professor in the psychology department at the University of Colorado, got a big surprise when he walked into his laboratory in 1995. His laboratory shelves are lined with dozens of glass cases; behind the glass, the subjects of Dr. Chiszar's research coil a safe distance from the professor and any visitors brave enough to enter. Although most are eerily silent, an occasional research subject emits a dry, rattling sound when the professor passes by. Dr. Chiszar is a herpetologist—in particular, he studies snakes. The day of the surprise, he went to the glass case that held a 14-year-old female timber rattlesnake. Timber rattlesnakes are poisonous but relatively unaggressive snakes that are found only in the eastern half of the United States. Severe declines in timber rattlesnake populations have led several states, including New York and Massachusetts, to list it as a threatened or endangered species, making it illegal to harass, kill, or collect timber rattlesnakes.

Dr. Chiszar had raised this timber rattlesnake since she was a two-day-old baby, not much bigger than an earthworm. Now, as he peered into her cage, he was astonished to see a tiny rattlesnake baby next to the 2½-foot-long female snake. Even though she had never been in contact with a male snake, she had given birth to a *male* timber rattlesnake. Like other pit vipers, timber rattlesnakes don't lay eggs, but bear their young alive. Genetic studies showed that the baby snake really was the offspring of the 14-year-old rattler and that he didn't have a "father" in the usual sense of the word.

How can a female organism produce offspring without the help of a male? To understand, you need to learn more about the basic patterns of cellular reproduction in life on Earth. ■

1 What Are the Functions of Cellular Reproduction?

The critical steps in reproduction, including human reproduction, occur at the microscopic level of the cell. Cellular reproduction enables a parent cell to accurately distribute both genes and cell components to its daughter cells in a process called **cell division**. In prokaryotes, cell division occurs via binary fission. In eukaryotes, cell division can occur via either mitosis or meiosis.

Binary Fission Is the Cell Division Process in Prokaryotes

Prokaryotic cells undergo a simple method of cell division called **binary fission**, which means "splitting in two" (Figure 11-1). Binary fission produces two cells that are genetically identical. Binary fission begins only if sufficient nutrients are available and other conditions are favorable. At the beginning of the process, the cell replicates its DNA double helix. Prokaryotic DNA usually forms a closed circular structure and contains about 4000 genes. Replication produces two identical DNA double helices that attach to the plasma membrane at nearby, but separate, points (Fig. 11-1, ②). As the cell continues to synthesize additional cellular components and grow, the plasma membrane expands, pushing the DNA double helices apart (Fig. 11-1, ③). When the cell has approximately doubled in size, the plasma membrane around the middle of the cell grows inward between the two DNA attachment sites (Fig. 11-1, ④). Fusion of the plasma membrane completes binary fission and produces two daughter cells (Fig. 11-1, ⑤). In some bacteria, these individual cells may remain attached to each other, forming long chains or clusters.

Following binary fission, each daughter cell contains one DNA double helix and about half of the original cell's cytoplasm. If conditions remain suitable, the daughter cells will acquire nutrients, replicate their DNA, grow until their size roughly doubles, and undergo another binary fission. Under ideal conditions, binary fission in prokaryotes proceeds rapidly. For example, the common intestinal bacterium *Escherichia coli* can complete binary fission in about 20 minutes. Luckily, ideal conditions for bacterial growth are not frequently present in our intestines, or the bacteria would soon weigh more than the rest of us!

Mitotic Cell Division Allows Development, Growth, Maintenance, and Repair of Body Tissues in Eukaryotes

Mitotic cell division involves a process of nuclear division called *mitosis* followed by a single cell division. As a result, mitotic cell division produces two daughter cells that are genetically identical to the original cell. This type of cell division plays several roles in the lives of multicellular eukaryotic organisms. First, in conjunction

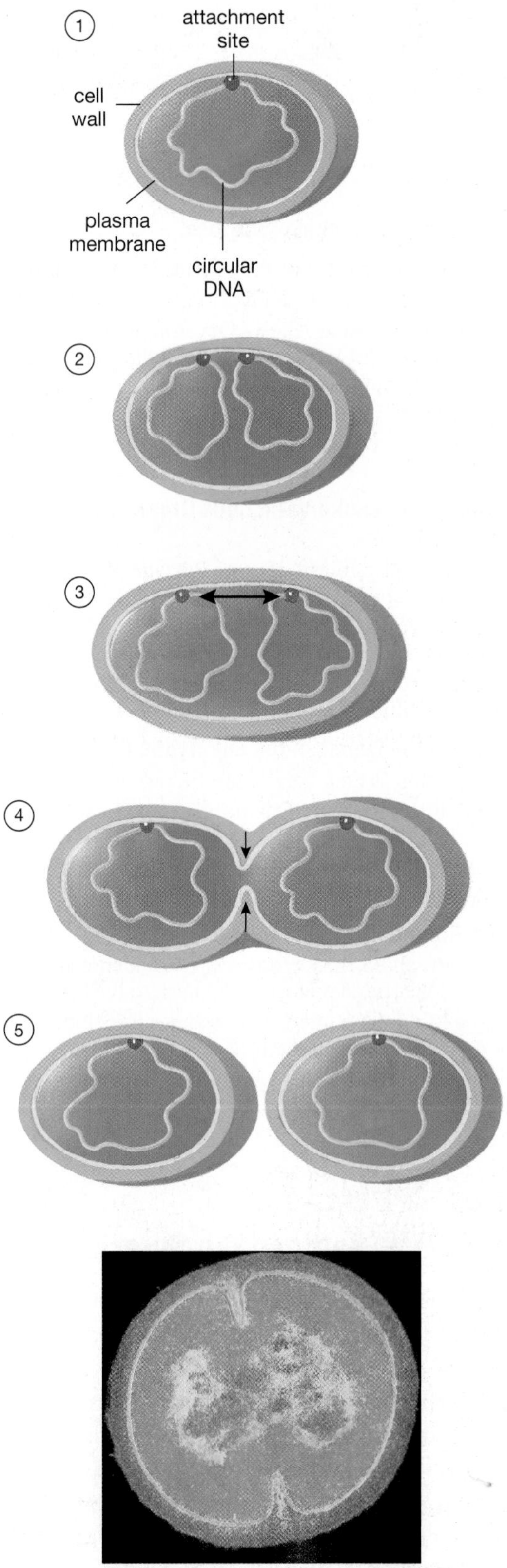

Figure 11-1 Prokaryotic cells divide by binary fission
① The circular DNA double helix attaches to the plasma membrane at one point. ② The DNA replicates and the two DNA double helices attach to the plasma membrane at nearby points. ③ As the cell grows, new plasma membrane is added between the attachment points, pushing them further apart. ④ The plasma membrane grows inward at the middle of the cell. ⑤ The parent cell has divided into two daughter cells.

with the differential expression of genes in different cells, it allows a fertilized egg to produce the cells needed in a newborn organism (Fig. 11-2a–c). Mitotic cell division then allows the organism to grow into an adult with perhaps trillions of individual cells.

Mitotic cell division also allows an organism to maintain its tissues, many of which require frequent replacement (Fig. 11-2d). For example, your skin cells live for only about 2 weeks. Your red blood cells become worn out after about 4 months. The cells of your stomach lining, exposed to acid and digestive enzymes, survive only about 3 days. Without mitotic cell division to replace these short-lived cells, your body would soon be unable to function properly. Mitotic cell divisions also allow the body to repair itself, or even regenerate following injury. Activation of mitotic cell division is partly responsible for the ability of the liver to regenerate—the only organ in our body with this capability. Even if half of the liver is lost, mitotic cell divisions will replace the lost part in just a few weeks, restoring normal liver function. The regenerative property of the liver makes it possible for a living individual to donate a part of his or her liver to someone who needs a liver transplant: Only the right lobe of the donor's liver is removed. In the donor, cells in the remaining portion of the liver undergo mitotic cell division until it is the same size as before. In the recipient, cells in the donated portion of the liver undergo mitosis cell division to form a full-sized liver.

Mitotic Cell Division Forms the Basis of Asexual Reproduction in Eukaryotes

Mitotic cell division also provides the basis of **asexual reproduction**, in which offspring are formed from a single parent without the uniting of male and female gametes. This mode of reproduction is normal for many unicellular organisms, such as *Tetrahymena* and yeasts (Fig. 11-3a,b). Many multicellular organisms can also reproduce asexually. In this process, small replicas of the parent grow by means of mitotic cell division. Like its relative the sea anemone, a *Hydra* can reproduce by growing a miniature replica of itself as a bud (Fig. 11-3c). Eventually the bud separates from its parent, going off to live independently. Because mitosis produces genetically identical cells, these offspring are genetically identical to their parents; they are called **clones.** Although it is a reasonable hypothesis, asexual reproduction is *not* responsible for the baby rattlesnake described in the Case Study.

Many plants and fungi reproduce both asexually and sexually. The beautiful aspen groves of Colorado, Utah, and New Mexico (Fig. 11-4) develop asexually from shoots growing up from the root system of a single parent tree. The entire grove, although seeming to be a population of separate trees to the admiring visitor, may be considered to be a single individual, with its multiple trunks interconnected by a common root system. These

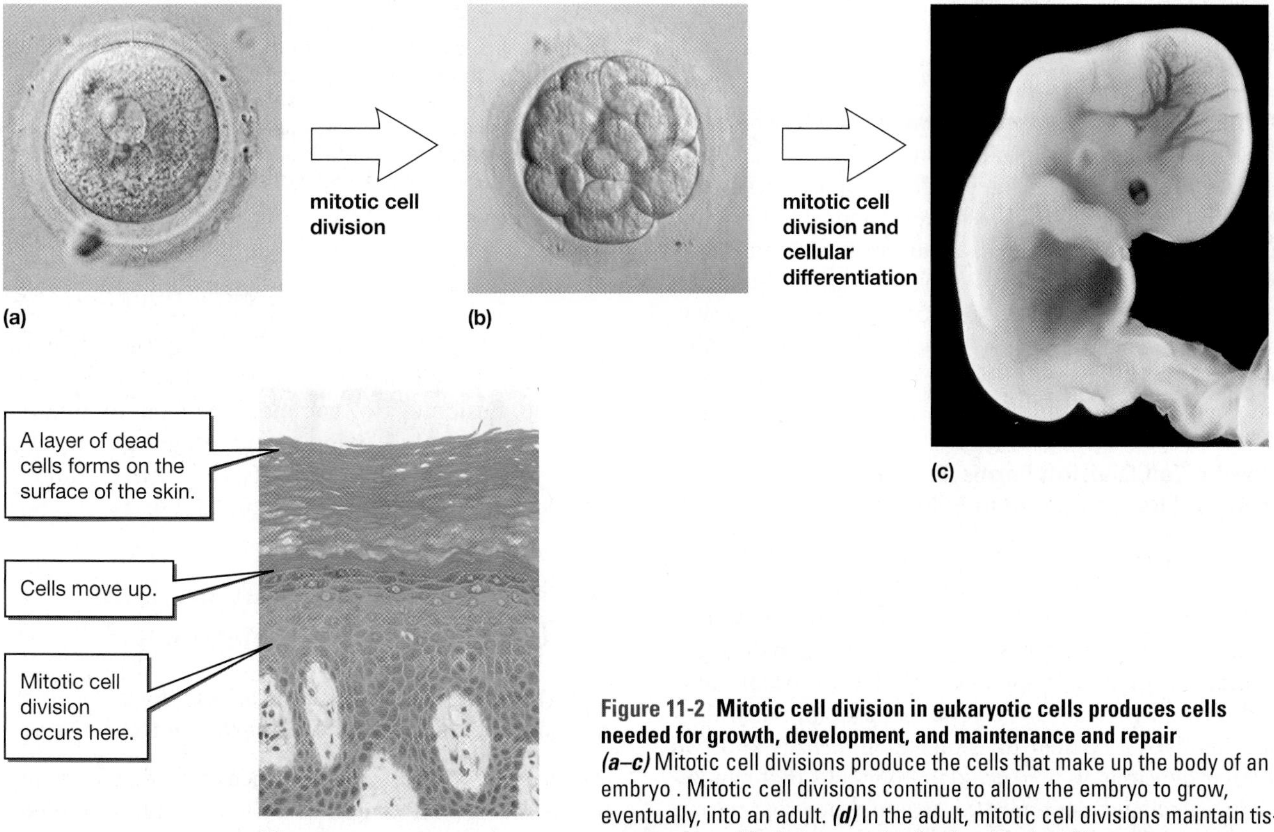

mitotic cell division
mitotic cell division and cellular differentiation
(a)
(b)
(c)
A layer of dead cells forms on the surface of the skin.
Cells move up.
Mitotic cell division occurs here.
(d)

Figure 11-2 Mitotic cell division in eukaryotic cells produces cells needed for growth, development, and maintenance and repair ***(a–c)*** Mitotic cell divisions produce the cells that make up the body of an embryo . Mitotic cell divisions continue to allow the embryo to grow, eventually, into an adult. ***(d)*** In the adult, mitotic cell divisions maintain tissues such as skin that are made of cells with short life spans.

Figure 11-3 Mitotic cell division in eukaryotes enables asexual reproduction
(a) In unicellular microorganisms, such as the protist *Tetrahymena*, mitotic cell division produces two new, independent organisms. ***(b)*** Yeast, a unicellular fungus, reproduces by mitotic cell division. ***(c)*** *Hydra*, a freshwater relative of jellyfish, grows a miniature replica of itself (a bud) protruding from its side. When fully developed, the bud breaks off and assumes independent life.

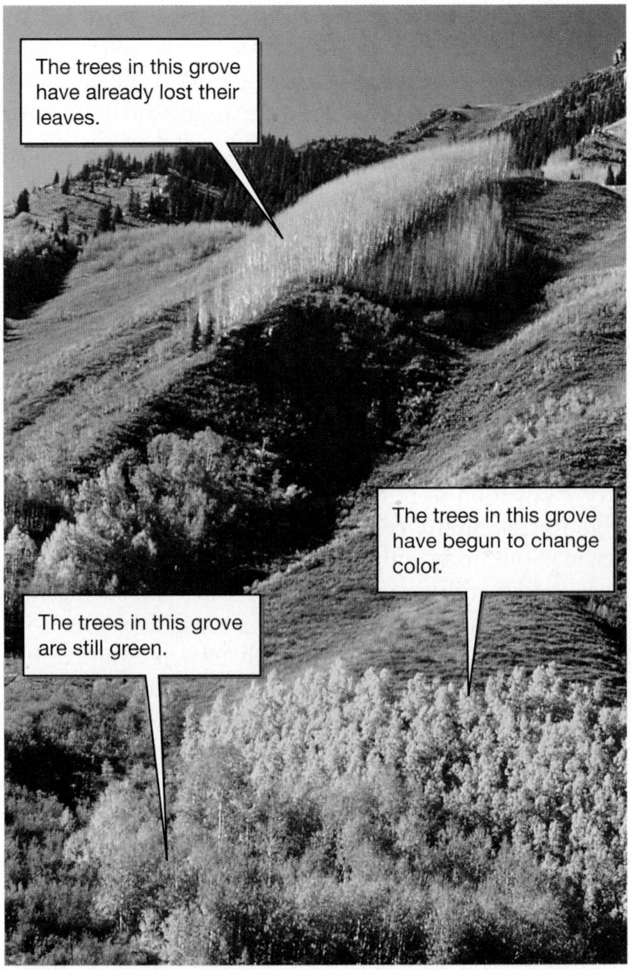

Figure 11-4 The trees in aspen groves are genetically identical
Each tree grows up from the roots of a single ancestral tree. This photo shows three separate groves near Aspen, Colorado. In fall the behavior of their leaves shows the genetic identity within a grove and the genetic difference between groves. One entire grove is still green (bottom), the second has turned bright gold (middle), and the third has already lost its leaves (top).

plants can also reproduce by making seeds, which are produced via sexual reproduction.

As you will discover later in this chapter and as you explore the diversity of life in later units, plants, many fungi, and some unicellular algae produce sex cells by mitotic cell division. Mitosis also gave rise to the nucleus used to produce Dolly, the cloned sheep, as you will learn in "Scientific Inquiry: Much Ado About Dolly" (page 192). This type of asexual reproduction in mammals can occur only in the laboratory.

Meiotic Cell Division Forms the Basis of Sexual Reproduction in Eukaryotes

Most eukaryotic organisms reproduce sexually, a process that produces offspring with a mixture of genetic material from two parents. Sexual reproduction is made possible by a process of division called **meiotic cell division**. In mammals, meiotic cell division occurs only in the ovary or testes. The process of meiotic cell division involves a nuclear division called *meiosis* and two cellular divisions to produce daughter cells that can become specialized **gametes** (for example, eggs or sperm). These gametes carry half of the genetic material of the parent. Thus, the cells produced by meiotic cell division are not genetically identical to each other *or* to the original cell. Fusion of two gametes, one from each parent, reconstitutes a full complement of genetic material, forming a genetically novel offspring that is similar to both parents, but identical to neither (Fig. 11-5).

2 How Is DNA in Eukaryotic Cells Organized into Chromosomes?

Cell Division Enables Accurate Passage of Chromosomes from One Generation to the Next

In all types of cell division, accurate distribution of genes to the new daughter cells is a major challenge. Eukaryotic cells meet this challenge by controlling

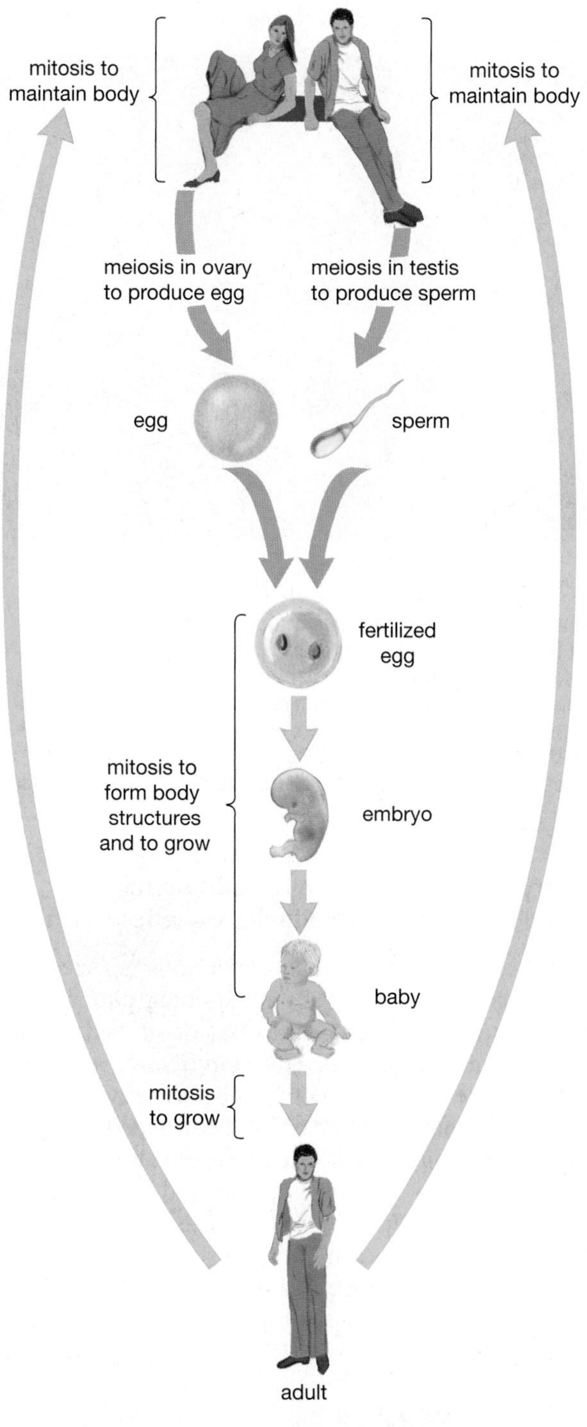

Figure 11-5 Mitotic and meiotic cell division in humans
Mitotic cell divisions maintain and repair the bodies of adults. Within ovaries, meiotic cell division produces eggs; within testes, meiotic cell division produces sperm. Fusion of an egg and a sperm produce a fertilized egg that undergoes hundreds of mitotic cell divisions to produce a baby that can grow by additional mitotic cell divisions into an adult, completing the life cycle.

the behavior of chromosomes. How are genes and chromosomes related? Recall that genes are sequences of DNA ranging from a few hundred to many thousands of nucleotides long. The sequences of nucleotides in genes encode information for the synthesis of the RNA and protein molecules needed to build the cell and carry out its activities. A single DNA double helix can contain hundreds or thousands of genes, arranged in a particular linear order along the DNA double helix.

DNA does not float free within the cell. Instead, each double helix is packaged into a **chromosome** (see Chapter 5). Every species has a particular number of chromosomes, and consequently, a specific number of DNA double helices. Human body cells have 46 chromosomes, each containing a DNA double helix. The largest human chromosome, chromosome 1, contains approximately 3000 genes, including the gene for GBA, an enzyme needed to degrade certain types of fats. Defects in the GBA gene cause Gaucher's disease, which can be lethal. One of the smallest human chromosomes, chromosome 22, contains about 600 genes, including a gene for one of the crystallins, which are major proteins in the lens of your eye.

The Eukaryotic Chromosome Consists of a Linear DNA Double Helix Bound to Proteins

Recall that the DNA of prokaryotes is a single circular double helix with no free ends. In contrast, the DNA double helices in eukaryotes are *linear* rather than circular. In addition, the DNA is bound to large numbers of proteins, including histones, that organize and compact the DNA so that it can fit into the nucleus. Fitting all the DNA into this tiny space is no small task. If it were laid end to end, the total DNA in a single cell in your body would be about 6 feet long, and this DNA must fit into a nucleus that is at least a million times smaller! The degree of DNA compaction, or **condensation**, varies with the stage of the cell cycle. During cell growth, the DNA is maximally dispersed, making it readily available to enzymes needed for transcription and replication. In this extended state, individual chromosomes are too thin to be visible in light microscopes. During cell division, however, the chromosomes must be sorted out and moved into two daughter nuclei. Just as thread is easier to organize when it is wound tightly onto spools, sorting and transporting chromosomes is easier when they are condensed and shortened (Fig. 11-6). In a process that is still not completely understood, proteins fold up the DNA of each chromosome into compact structures that can be seen in a light microscope.

At the time it condenses, the DNA within each chromosome has already replicated, forming two DNA double helices that remain attached to each other at a specialized region of DNA called the **centromere**. *Centromere* means "middle body," but a chromosome's centromere can be located almost anywhere along the DNA double helix. While the two chromosomes are attached at their centromeres, we refer to each attached chromosome as a sister **chromatid**. Thus, DNA

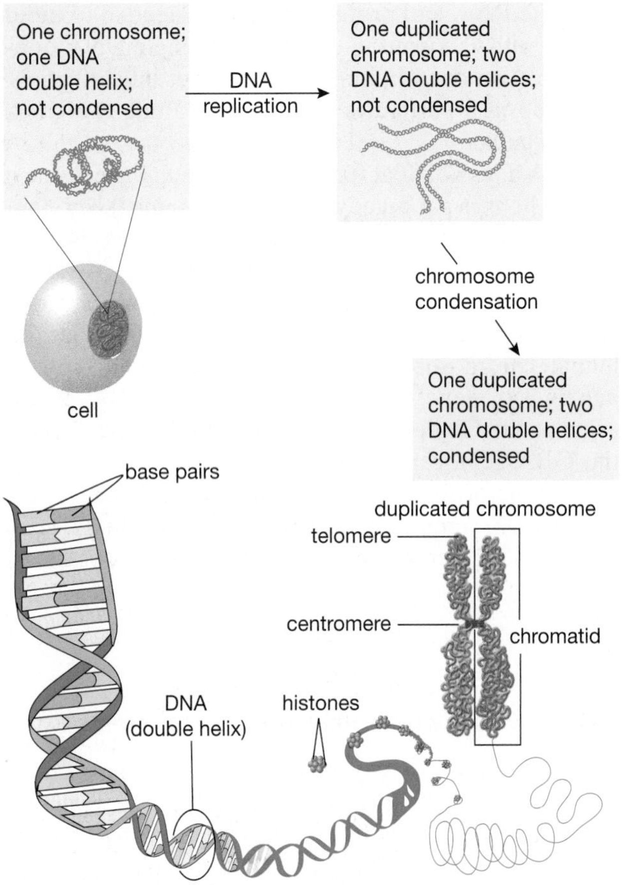

Figure 11-6 Chromosome condensation
Before cell division begins, the DNA double helix in each chromosome replicates. The two newly made double helices remain associated with each other. When cell division begins, the duplicate chromosomes condense and become visible. At this stage, each duplicated chromosome contains two identical DNA double helices. Each DNA double helix is contained within a sister chromatid. Condensation of the duplicated chromosome requires action of many proteins, including histones.

replication produces a **duplicated chromosome** with two identical sister chromatids (Fig. 11-7).

During mitotic cell division, the two sister chromatids separate, and each chromatid becomes an independent chromosome that is delivered to one of the two daughter cells.

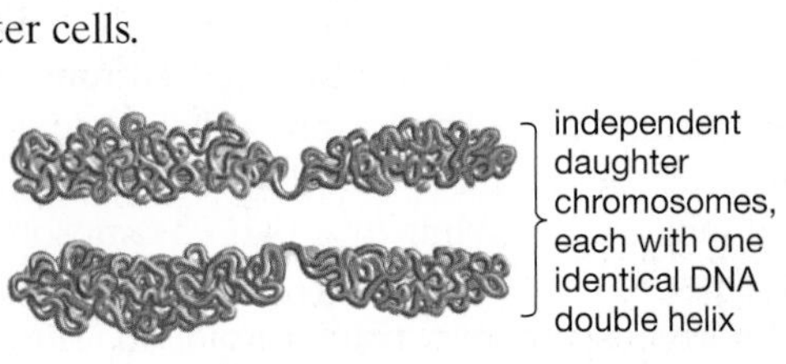

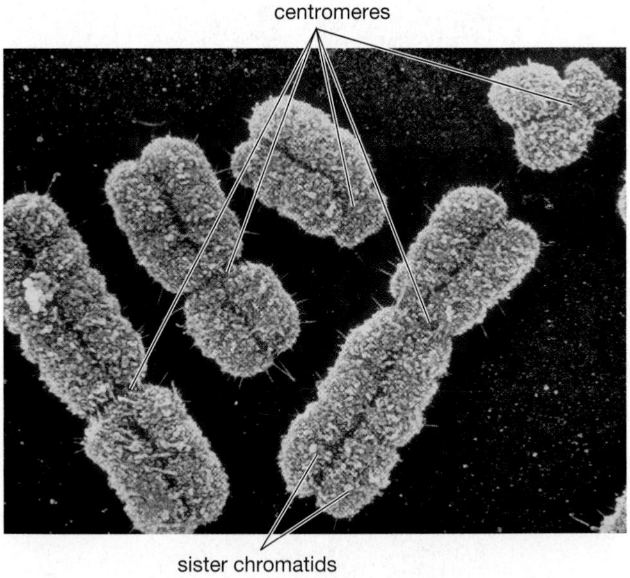

Figure 11-7 Human chromosomes during mitosis
The DNA and associated proteins in these duplicated human chromosomes have coiled up into the thick, short sister chromatids attached at the centromere. Each strand of "texture" is a loop of chromosome. During cell division, the condensed chromosomes are about 5 to 20 micrometers long. At other times, the chromosomes uncoil until they are about 10,000 to 40,000 micrometers long.

Eukaryotic Chromosomes Usually Occur in Homologous Pairs with Similar Genetic Information

The chromosomes of each eukaryotic species have characteristic shapes, sizes, and staining patterns (Fig. 11-8). When we view an entire set of stained chromosomes from a single cell (the **karyotype**), it becomes clear that the nonreproductive cells of many organisms, including humans, contain pairs of chromosomes. With a single exception we will discuss shortly, both members of each pair are the same length and have the same staining pattern. This similarity in size, shape, and staining results because each chromosome in a pair carries the same genes, arranged in the same order. Chromosomes that contain the same genes are called *homologous chromosomes* or **homologues**, meaning "to say the same thing." Cells with pairs of homologous chromosomes are called **diploid**, meaning "times two."

Let's consider a human skin cell. Although it has 46 chromosomes in all, it does not have 46 completely *different* chromosomes. Instead, it has 23 different types of chromosomes, each present as a pair of homologues. The cell has two copies of chromosome 1, two copies of chromosome 2, and so on, up through chromosome 22. The cell also has two **sex chromosomes:** two X chromosomes *or* an X and a Y chromosome. Thus, your skin cell has two sets of 23 chromosomes.

Most cells within our bodies are diploid. However, during sexual reproduction, cells in our ovaries or testes undergo meiotic cell division to produce gametes (sperm or eggs) with only one set of 23 chromosomes. Cells that

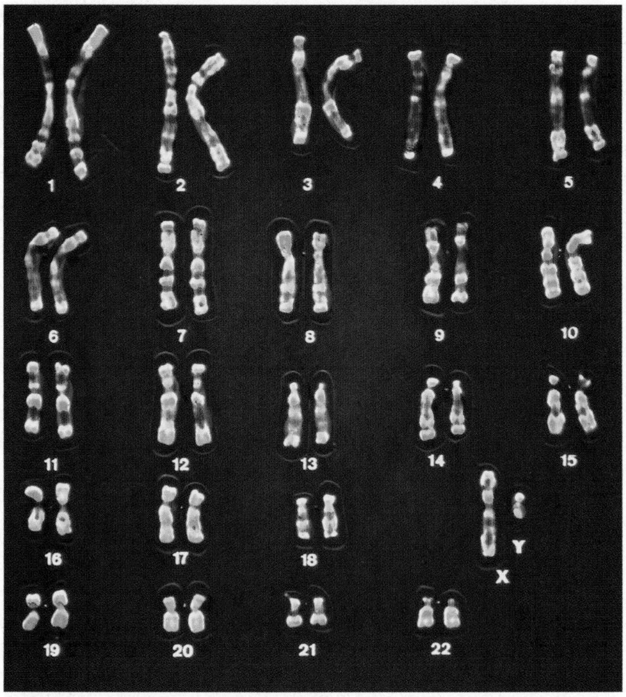

Figure 11-8 A human male karyotype
Staining and photographing the entire set of duplicated chromosomes within a single dividing cell produces a karyotype. Pictures of the individual chromosomes are cut out and arranged in descending order of size. The chromosomes occur in pairs (homologues) that are similar in both size and staining patterns and have similar genetic material. If these chromosomes were from a female, the cell would usually contain two X chromosomes. Notice that the Y chromosome is much smaller than the X chromosome.

contain only one of each type of chromosome are called **haploid**. Fusion of two haploid cells produces a diploid cell with two copies of each type of chromosome.

In biological shorthand, the number of different types of chromosomes in a species is called the *haploid number* and is designated n. For humans, $n = 23$ because we have 23 types of chromosomes (chromosome 1 to chromosome 22 and a sex chromosome). Diploid cells contain $2n$ chromosomes, two of each type of chromosome. Each human nonreproductive cell thus has 46 (2×23) chromosomes. Many plants may have $3n$, $4n$, or even more chromosomes!

Organism	n (haploid number)	$2n$ (diploid number)
Human	23	46
Gorilla, chimpanzee, bonobo, orangutan	24	48
Dog	39	78
Cat	19	38
Shrimp	127	254
Fruitfly	4	8
Pea	7	14
Potato	24	48
Sweet potato	45	90

An exception to the rule that homologous chromosomes contain the same genes occurs in male mammals and other species in which chromosomes carry the genes for sex determination. In male mammals, one pair of chromosomes, the X and Y, have very different sizes and staining patterns and carry different sets of genes. However, the X and Y chromosomes behave like homologues during the cell division process that produces eggs and sperm (meiotic cell division), and so the X and Y are considered as a pair in our chromosome bookkeeping.

3 What Are the Events of the Eukaryotic Cell Cycle?

Cell division is just a small part of the cycle of cell growth and asexual reproduction known as the **cell cycle**. The eukaryotic cell cycle is divided into two major phases: interphase (shown in yellow) and mitotic cell division (shown in blue) (Fig. 11-9).

Cell Division Is One Part of the Eukaryotic Cell Cycle

In spite of its importance, cell division is actually just a small part of a cycle of cell growth and reproduction called the *cell cycle*. Newly formed cells usually acquire nutrients from their environment, synthesize additional cellular components, and grow larger. After a variable amount of time, depending on the organism,

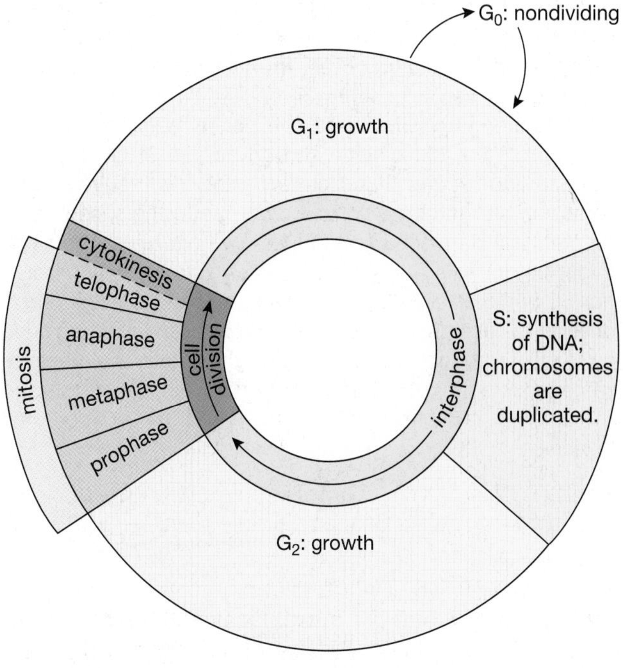

Figure 11-9 The eukaryotic cell cycle
The eukaryotic cell cycle consists of two major phases, interphase and cell division. Each is divided into subphases.

Scientific Inquiry
Much Ado About Dolly

"And what is the Scientific Community doing? THEY'RE CLONING SHEEP. Great! Just what we need! Sheep that look MORE ALIKE than they already do!"

Dave Barry, 1997

Early in 1997 a short article in the journal *Nature* caused headlines throughout the world. Dr. Ian Wilmut and associates of the Roslin Institute in Edinburgh, Scotland, had done what many scientists believed to be impossible: They cloned a mammal, using a nucleus taken from adult tissue. This type of **cloning** is the process of artificially creating a *clone*—a new individual that is genetically identical to an existing individual. As you are aware, every body cell has a complete set of genes. Researchers had already discovered that the nucleus of a cell in extremely early stages of embryonic development, before cells have specialized, could be used to replace the nucleus of an egg cell. The embryonic cell nucleus then directs the development of a new individual, with no need for a fusion of sperm and egg. Until Dr. Wilmut's breakthrough, attempts to make clones from nuclei taken from older tissues (from embryos that had more than 16 cells) had failed. Many researchers hypothesized that the inactivation of selected genes that make, say, neurons different from skin cells and liver cells different from fat cells, must be permanent. What Dr. Wilmut's research demonstrated is that differentiation is reversible under the right conditions (Fig. E11-1). They used cells taken from the udder of a 6-year-old ewe and grown in culture dishes. Depriving these cells of nutrients forced the cells into the non-dividing (G_0) stage of the cell cycle—a factor that seems crucial to allowing all the genes to be expressed when the nucleus was transplanted into an egg cell.

Cloning is a highly artificial form of asexual reproduction, because the nucleus that directs development of the offspring was produced mitotically from a diploid cell and not by the fusion of haploid gametes produced by meiosis. Like the hydra budding from its parent, the cloned mammal is literally "a chip off the old block." The advantage of cloning cells from an adult rather than from an embryo is that you know what you are getting. A prized wool-producing sheep could theoretically be cloned to produce a whole herd. The same might be done with a cow that has exceptional milk production. Cloning an adult allows scientists to use asexual reproduction to take advantage of the natural variability provided by sexual reproduction. By observing the adults, the most advantageous gene combinations produced during meiosis and the union of gametes can be selected and then perpetuated in clones.

The increasing ability of scientists to manipulate individual genes makes cloning even more advantageous. For example, let's say genetic engineering produces a cow that secretes large quantities of a specialized antibiotic in its milk. This antibiotic could be extracted inexpensively to fight hard-to-treat infections. Such a cow would represent an enormous investment in money, time, and ingenuity. But what if the cow were infertile? Or what if none of its offspring carried the gene to produce the antibiotic? Enter adult cloning. Tiny amounts of tissue from the engineered cow would yield millions of cells, each containing the full genetic "blueprint" for an antibiotic-producing cow.

The implications of Dr. Wilmut's research are far-reaching. Scientists now have proof that the long-inactivated genes of specialized adult cells can be made functional again. Perhaps neurons in the brain region destroyed by Parkinson's disease could be made to divide again, replacing those that have perished. Perhaps more genes for red blood cell production could be switched on in an anemia patient.

Dolly and other cloned mammals also promise to provide insights into aging. Do the genes of each species dictate a maximum life span? If so, Dolly may have been born middle-aged, because her genes came from the cell of a 6-year-old ewe. Recent results support this idea. The ends of eukaryotic chromosomes have special repetitive nucleotide sequences called *telomeres*. In most cells in an animal, each cell division leaves the telomeres a tiny bit shorter. Old animals, then, have shorter telomeres than young animals. Dolly's telomeres are shorter than usually seen in a 3-year-old sheep. Will her life be shorter as a result? It may take some time to know, since a sheep's average life span is 13 years. The scientific world is watching, but they are also keeping busy. Dr. Wilmut's research group has produced 5 additional sheep clones. Cows, mice, and pigs have been cloned; scientists have even produced clones of clones! Meanwhile, Dolly has been mated twice to a Welsh mountain ram named David. Her first pregnancy produced a ewe named Bonnie. Her second produced triplets, two rams and one ewe. Her offspring appear normal, but only time will tell.

What really caught the attention of people everywhere is that Dolly raises the possibility of human cloning. Although the technique required 277 attempts before one healthy lamb was born, the technology is relatively simple. There is no theoretical reason why the method could not be applied to humans. For example, adult cloning could produce a child who had exactly the same set of genes as its mother, a beloved grandparent, a famous sports figure, a Nobel Prize winner, or a terrorist. Think about it. We'll revisit this topic in Chapter 13.

the type of cell, and the nutrients available, the cell divides. Each daughter cell may then enter another cell cycle, producing additional cells. Alternatively, the newly formed cells may exit the cell cycle and never divide again, or they may divide only if they receive signals to do so. In our bodies, cells in our bone marrow and skin divide as frequently as once a day. Cells in our liver divide only once or twice a year. Most of our nerve cells and our muscle cells never divide after they mature; if one of these cells dies, it is not replaced.

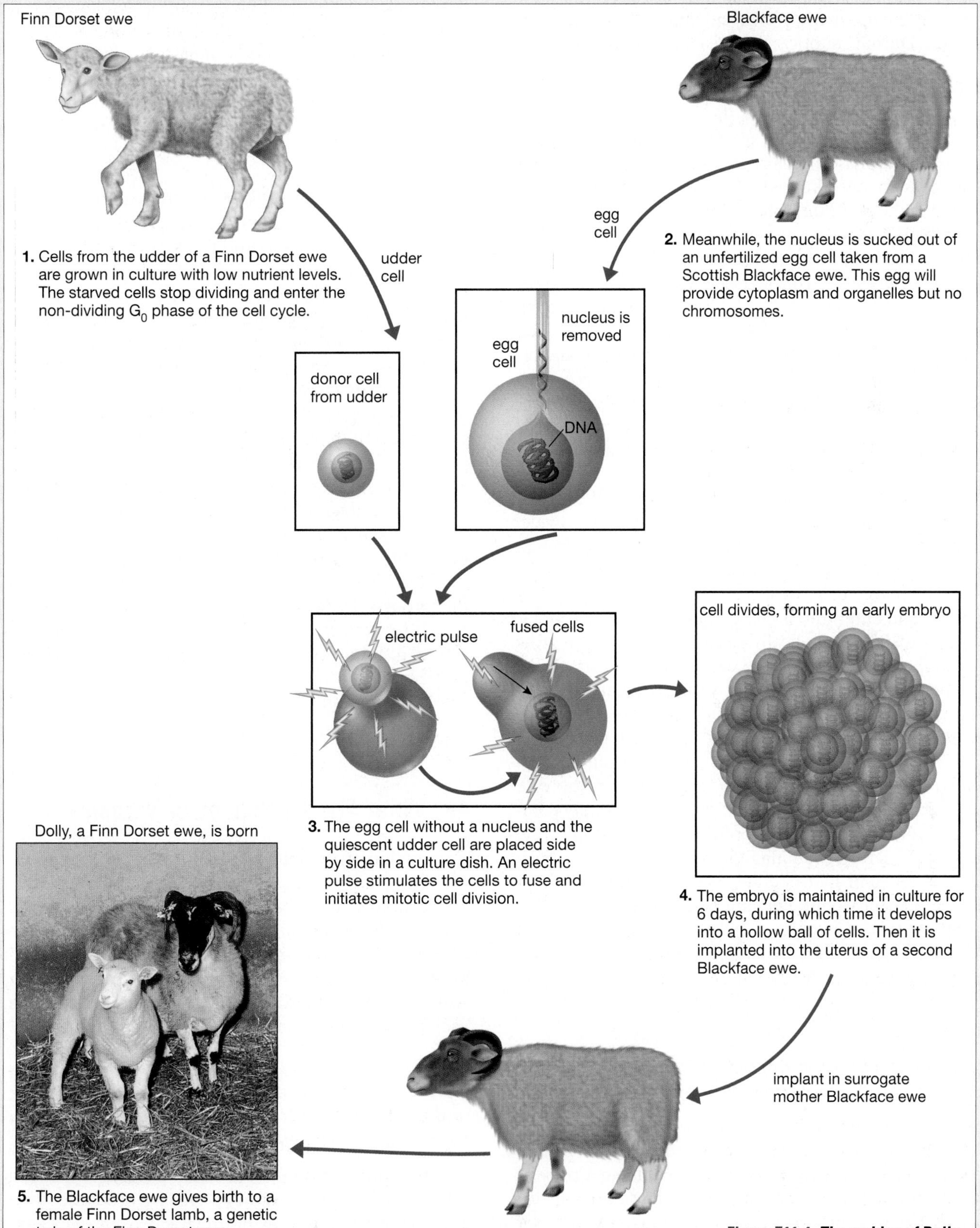
Finn Dorset ewe
Blackface ewe
1. Cells from the udder of a Finn Dorset ewe are grown in culture with low nutrient levels. The starved cells stop dividing and enter the non-dividing G_0 phase of the cell cycle.
udder cell
egg cell
2. Meanwhile, the nucleus is sucked out of an unfertilized egg cell taken from a Scottish Blackface ewe. This egg will provide cytoplasm and organelles but no chromosomes.
donor cell from udder
egg cell
nucleus is removed
DNA
electric pulse
fused cells
cell divides, forming an early embryo
Dolly, a Finn Dorset ewe, is born
3. The egg cell without a nucleus and the quiescent udder cell are placed side by side in a culture dish. An electric pulse stimulates the cells to fuse and initiates mitotic cell division.
4. The embryo is maintained in culture for 6 days, during which time it develops into a hollow ball of cells. Then it is implanted into the uterus of a second Blackface ewe.
implant in surrogate mother Blackface ewe
5. The Blackface ewe gives birth to a female Finn Dorset lamb, a genetic twin of the Finn Dorset ewe.

Figure E11-1 The making of Dolly

During Interphase, the Eukaryotic Cell Grows in Size and Replicates Its DNA

The eukaryotic cell cycle can be divided into two major phases: interphase and mitotic cell division (see Fig. 11-9). During **interphase**, the cell acquires nutrients from its environment, grows, and duplicates its chromosomes. During mitotic cell division, one copy of each chromosome and usually about half the cytoplasm (including mitochondria, ribosomes, and other organelles) are parceled out into each of the two daughter cells.

Most eukaryotic cells spend the majority of their time in interphase, preparing for cell division. For example, cells in our skin, which divide every day, spend about 22 hours in interphase. Interphase itself contains three subphases: G_1 (gap or growth phase 1), S (DNA synthesis), and G_2 (gap or growth phase 2). To explore these stages, let us consider a newly formed daughter cell. This cell enters the G_1 portion of interphase, during which it acquires or synthesizes the materials needed for cell division. If the cell grows to a proper size and receives the necessary signals, it replicates its DNA. This period is called *S phase* because it is when DNA *synthesis* occurs. Following S phase, the cell completes its growth in a G_2 phase before leaving interphase.

G_1 is an important time in the life of a cell. During this phase, the cell is sensitive to internal and external signals that help the cell "decide" whether to divide. Once that decision is made, the cell progresses through the entire cell cycle. Depending on the cell type and external signals, a cell can also exit from the cell cycle during G_1 and enter into a phase known as G_0. Cells in G_0 are alive and metabolically active. They may even grow in size, but they do not replicate their DNA or divide. Some cells, including those in your heart muscle, your eyes, and your brain, appear to be trapped in G_0 forever. If these cells are damaged, they cannot be replaced.

Progression through the cell cycle is carefully regulated. At key times, molecular signals within the cell, called *checkpoints*, ensure that the cell accurately completes all of the necessary processes in one step of the cell cycle before beginning the next step. For example, one checkpoint that operates in G_1 prevents cells that are too small from beginning to replicate their DNA. Without this checkpoint, a cell might begin replication but then run out of raw material before it could complete replication. Such a failure could be lethal to the cell or lead to formation of mutations.

Mitotic Cell Division Consists of Nuclear Division and Cytoplasmic Division

Mitotic cell division has two major parts: mitosis (nuclear division) and cytokinesis (cytoplasmic division). **Mitosis** produces two nuclei, each containing a copy of every chromosome that was present in the original nucleus. The term *mitosis* comes from the Greek word for "thread"; during mitosis, the duplicated chromosomes become condensed and visible as threadlike structures. In most cells, the cell divides following mitosis to form two daughter cells, each with one nucleus and about half of the cytoplasm. This process is called **cytokinesis** (from the Greek word for "cell movement"). Cytokinesis provides both daughter cells with all of the organelles, nutrients, enzymes, and other molecules they need to maintain life. Although mitosis and cytokinesis are usually coupled, they can be independent events. Some cells, including some tumor cells, undergo mitosis *without* cytokinesis. This process produces single cells with many nuclei.

The two daughter cells produced by mitotic cell division are genetically identical to each other and to the parent cell. This observation has profound implications for the development of multicellular organisms. If you trace the lineage of each cell currently in your body, you will ultimately arrive at a single fertilized egg cell. This cell underwent a mitotic cell division to produce two genetically identical daughter cells; these daughter cells underwent mitotic cell division to produce four genetically identical daughter cells. The divisions continued, ultimately producing the trillions of genetically identical cells that make up your body. Thus, as incredible as it seems, a cell in your brain, a cell in your liver, and a cell in your collar bone have the same genes. How can they have the same genes but be so different from one another in structure and in function? These cells have undergone **differentiation**, the process by which cells assume their specialized structures and functions. The differences in the cells reflect differences in the genes they are currently expressing and in the genes that they expressed as they were developing.

4 What Are the Phases of Mitosis?

Completion of the interphase processes, G_1, S, and G_2, prepares a cell to undergo mitosis. For convenience, we divide mitosis into four phases, based on the appearance and behavior of the chromosomes: (1) *prophase*, (2) *metaphase*, (3) *anaphase*, and (4) *telophase*. As with most biological processes, however, these phases are not really discrete events. Rather, they form a continuum, each phase merging into the next.

During Prophase, the Chromosomes Condense and the Spindle Microtubules Form and Attach to the Chromosomes

The first phase of mitosis is called **prophase** (meaning "the stage before" in Greek). During prophase, three major events occur: (1) the duplicated chromosomes

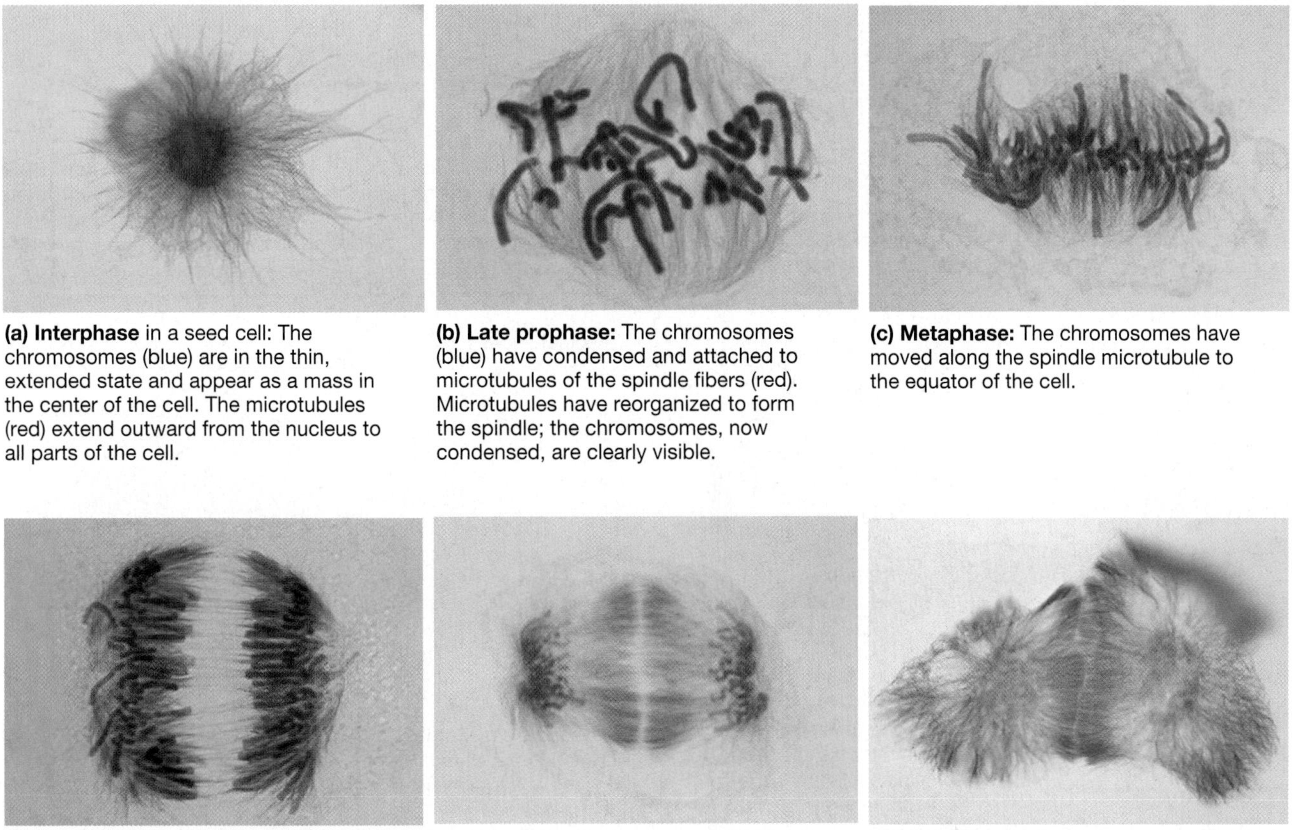

(a) Interphase in a seed cell: The chromosomes (blue) are in the thin, extended state and appear as a mass in the center of the cell. The microtubules (red) extend outward from the nucleus to all parts of the cell.

(b) Late prophase: The chromosomes (blue) have condensed and attached to microtubules of the spindle fibers (red). Microtubules have reorganized to form the spindle; the chromosomes, now condensed, are clearly visible.

(c) Metaphase: The chromosomes have moved along the spindle microtubule to the equator of the cell.

(d) Anaphase: Sister chromatids have separated, and one set of chromosomes moves along the spindle microtubule to each pole of the cell.

(e) Telophase: The chromosomes have gathered into two clusters, one at the site of each future nucleus.

(f) Resumption of interphase: The chromosomes are relaxing again into their extended state. The spindle microtubules are disappearing, and the microtubules of the two daughter cells are rearranging into the interphase pattern.

Figure 11-10 The cell cycle in a plant cell
Interphase and mitosis in the African blood lily. The chromosomes are stained bluish-purple, and the spindle microtubules are stained pink to red. Compare these micrographs with the drawings of mitosis in an animal cell shown in Figure 11-11. (Each micrograph is of a different single cell that has been fixed and stained at a particular stage of mitosis. Only chromosomes and microtubules have been stained.)

condense, (2) the spindle microtubules form, and (3) the chromosomes are captured by the spindle (Figs. 11-10b and 11-11b–c).

Recall that chromosome duplication (S phase) occurs during interphase. Therefore, when mitosis begins, each chromosome *already* consists of two sister chromatids attached to one another at the centromere. During prophase, the duplicated chromosomes coil up and condense (Figs. 11-10b and 11-11b–c). In addition, the nucleolus, a structure within the nucleus where ribosomes assemble, disappears.

After the duplicated chromosomes condense, the **spindle microtubules** begin to assemble. Proper movements of chromosomes during mitosis in all eukaryotic cells depend on these spindle microtubules. In animal cells (but not in plant cells) the spindle microtubules originate from a region that contains a pair of microtubule-containing **centrioles**. During interphase, a new pair of centrioles forms near the previously existing pair. During prophase, each centriole pair migrates to opposite sides of the nucleus. Each centriole pair serves as a central point from which the spindle microtubules radiate, both inward toward the nucleus and outward toward the plasma membrane. These points are called *spindle poles*.

As the spindle microtubules form into a complete basket around the nucleus, the nuclear envelope disintegrates, releasing the duplicated chromosomes and allowing them to interact with spindle microtubules. At the centromere, each sister chromatid has a protein-containing structure called a **kinetochore** that serves as an attachment site for the ends of spindle microtubules. The kinetochore of each sister chromatid binds to spindle microtubules that lead to opposite poles of the cell (Fig. 11-11b). Thus, one kinetochore of the duplicated chromosome is tethered to one spindle pole and the other kinetochore is tethered to the other spindle pole. When the sister chromatids separate later

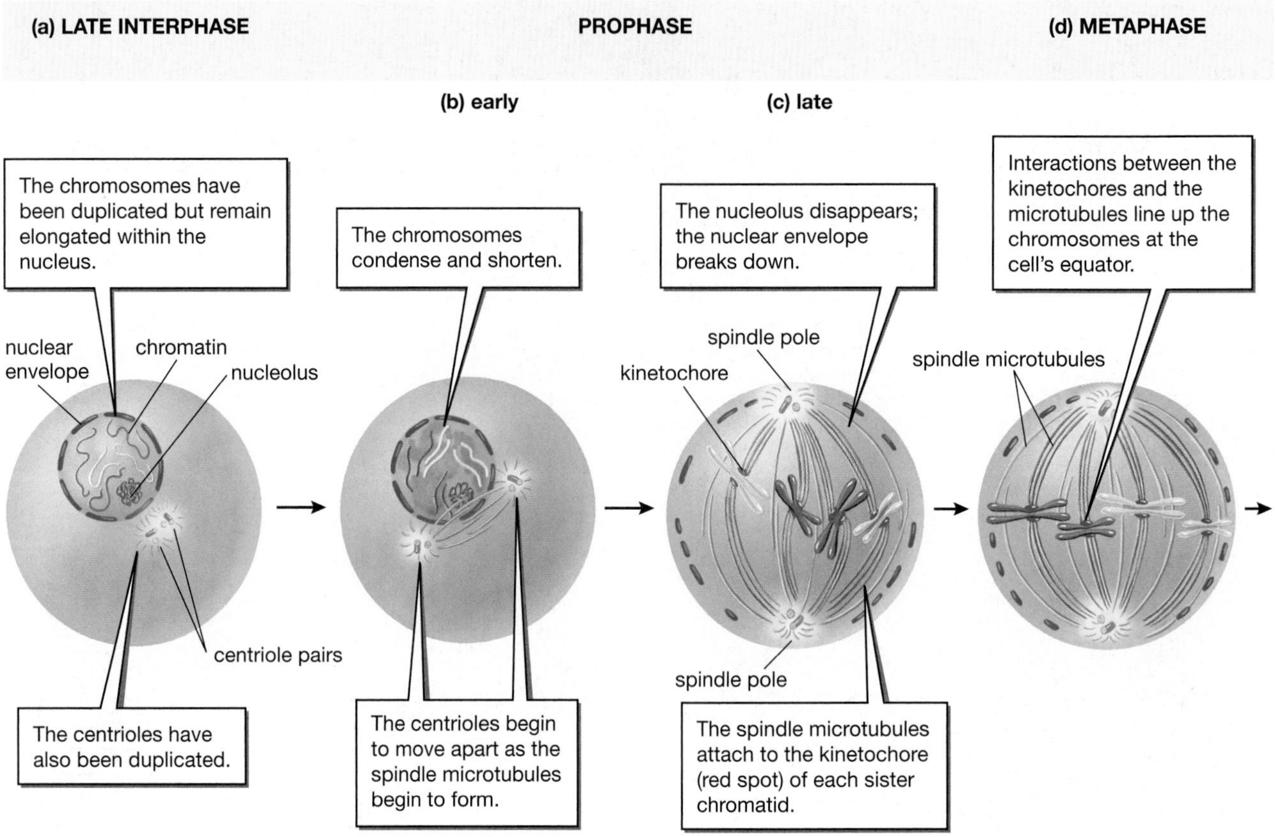
(a) LATE INTERPHASE
PROPHASE
(d) METAPHASE
(b) early
(c) late
The chromosomes have been duplicated but remain elongated within the nucleus.
The chromosomes condense and shorten.
The nucleolus disappears; the nuclear envelope breaks down.
Interactions between the kinetochores and the microtubules line up the chromosomes at the cell's equator.
nuclear envelope
chromatin
nucleolus
spindle pole
kinetochore
spindle microtubules
centriole pairs
spindle pole
The centrioles have also been duplicated.
The centrioles begin to move apart as the spindle microtubules begin to form.
The spindle microtubules attach to the kinetochore (red spot) of each sister chromatid.

Figure 11-11 The cell cycle in an animal cell
(a) Late interphase: The chromosomes have been duplicated but remain elongated and relaxed within the nucleus. The centrioles have also been duplicated. ***(b)*** Early prophase: The chromosomes condense, shortening and thickening. The centrioles begin to move apart, and the spindle microtubules begin to form between them. ***(c)*** Late prophase: The nucleolus disappears; the nuclear envelope breaks down, and the spindle microtubules attach to the kinetochore of each sister chromatid (red spot). ***(d)*** Metaphase: Interactions between the kinetochores and the microtubules have lined up the chromosomes at the cell's equator. ***(e)*** Anaphase: Chromatids separate at the centromere,
(continued on next page)

MEDIATUTOR 11.1 Mitosis

in mitosis, the newly independent chromosomes will move along the spindle microtubules to opposite poles. Some of the spindle microtubules do not attach to chromosomes; rather, they have free ends that overlap along the cell's equator. As we will see, these unattached spindle microtubules will push the two spindle poles apart later in mitosis.

During Metaphase, the Chromosomes Become Aligned Along the Equator of the Cell

At the end of prophase, the two kinetochores of each duplicated chromosome are connected to spindle microtubules from opposite poles. As a result, each duplicated chromosome is connected to both spindle poles. During **metaphase** (the "middle stage"), the two kinetochores on a duplicated chromosome engage in a "tug of war." During this process the kinetochore controls the length of the spindle microtubules. The microtubules lengthen and shorten, until each duplicated chromosome lines up properly along the equator of the cell, with one kinetochore facing each pole (Figs. 11-10c and 11-11d).

During Anaphase, Sister Chromatids Separate and Are Pulled to Opposite Poles of the Cell

At the beginning of **anaphase** (Figs. 11-10d and 11-11e), the sister chromatids separate and the protein "motors" in the kinetochores pull the kinetochores (and their attached chromosomes) poleward along the spindle microtubules. As the kinetochores tow their chromosomes toward the poles, the unattached spindle microtubules interact and lengthen to push the poles of the cell apart, forcing the cell into an oval shape (see Fig. 11-11e). Because the sister chromatids are identical copies of the original chromosomes, both clusters of chromosomes that form on opposite poles of the cell contain one copy of every chromosome that was in the original cell.

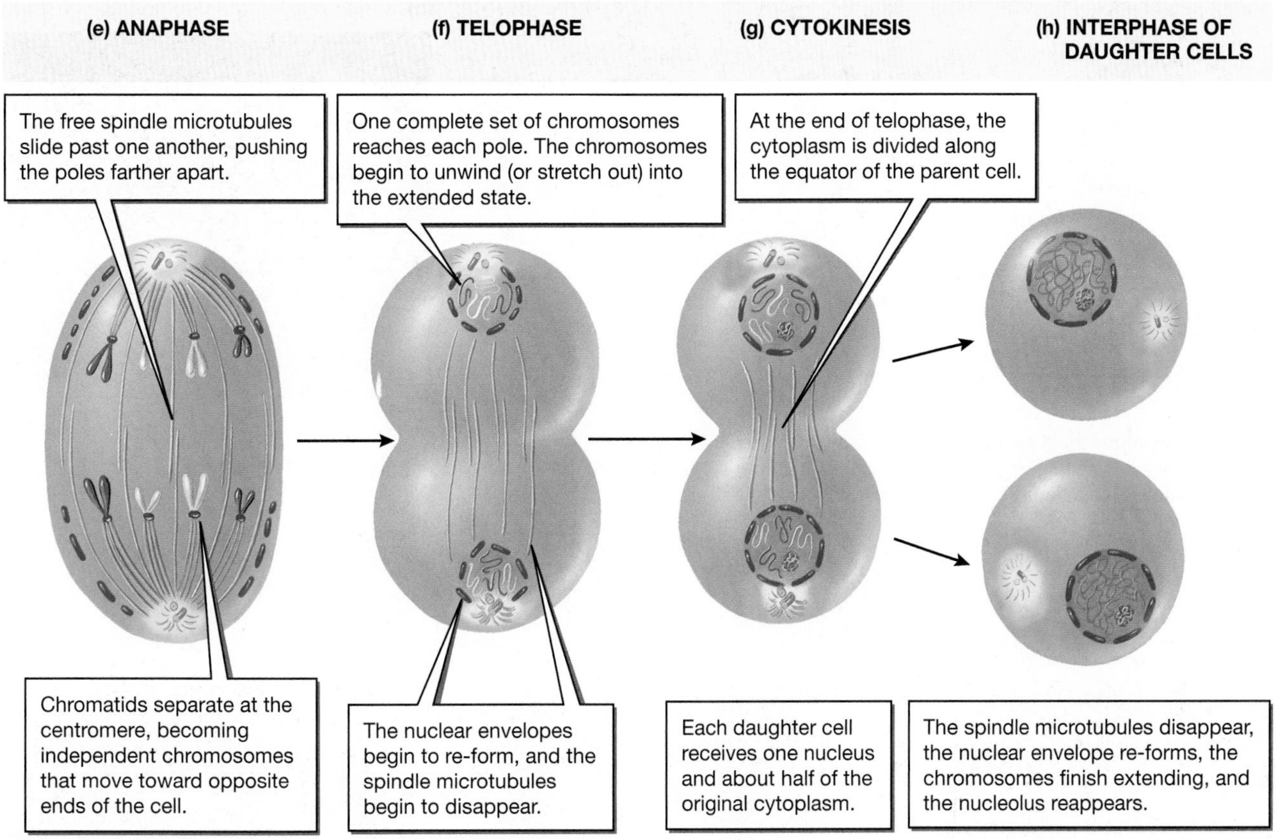
(e) ANAPHASE
(f) TELOPHASE
(g) CYTOKINESIS
(h) INTERPHASE OF DAUGHTER CELLS
The free spindle microtubules slide past one another, pushing the poles farther apart.
One complete set of chromosomes reaches each pole. The chromosomes begin to unwind (or stretch out) into the extended state.
At the end of telophase, the cytoplasm is divided along the equator of the parent cell.
Chromatids separate at the centromere, becoming independent chromosomes that move toward opposite ends of the cell.
The nuclear envelopes begin to re-form, and the spindle microtubules begin to disappear.
Each daughter cell receives one nucleus and about half of the original cytoplasm.
The spindle microtubules disappear, the nuclear envelope re-forms, the chromosomes finish extending, and the nucleolus reappears.

Figure 11-11 ***(continued)***
becoming independent chromosomes that move toward the opposite poles of the cell. The free spindle microtubules slide past one another, pushing the poles farther apart. ***(f)*** Telophase: One complete set of chromosomes reaches each pole. The chromosomes relax into their extended state, the spindle microtubules begin to disappear, and the nuclear envelopes begin to re-form. ***(g)*** Cytokinesis: At the end of telophase, the cytoplasm is divided along the equator of the parent cell, with each daughter cell receiving one nucleus and about half the original cytoplasm. ***(h)*** Interphase of daughter cells: The daughter cells enter interphase. The spindle microtubules disappear, the nuclear envelope re-forms, the chromosomes finish extending, and the nucleolus reappears.

During Telophase, Nuclear Envelopes Form Around Both Groups of Chromosomes

When the chromosomes reach the poles, **telophase** (the "end stage") has begun (Figs. 11-10e and 11-11f). The spindle microtubules disintegrate, and a nuclear envelope forms around each group of chromosomes. The chromosomes revert to their extended state, and the nucleoli reappear. In most cells, cytokinesis occurs during telophase, separating each daughter nucleus into a separate cell (Fig. 11-11f). The daughter cells enter interphase (Figs. 11-10f and 11-11g).

5 What Are the Events of Cytokinesis?

In animal cells (normally during telophase), microfilaments attached to the plasma membrane form a ring around the equator of the cell. During cytokinesis, the ring contracts and constricts the cell's equator, much like pulling the drawstring around the waist of a pair of sweatpants. Eventually the "waist" contracts down to nothing, dividing the cytoplasm into two new daughter cells (Fig. 11-12).

Cytokinesis in plant cells is quite different, perhaps because the stiff cell wall makes it impossible to divide one cell into two by pinching at the waist. Instead, the Golgi complex buds off carbohydrate-filled vesicles that line up along the cell's equator between the two nuclei (Fig. 11-13). The vesicles fuse, producing a structure called the **cell plate**, which is shaped like a flattened sac, surrounded by plasma membrane, and filled with sticky carbohydrates. When enough vesicles have fused, the edges of the cell plate merge with the original plasma membrane around the circumference of the cell. The carbohydrate formerly contained in the vesicles remains between the plasma membranes as part of the cell wall.

Following cytokinesis, the cell enters G_1 of interphase, thus completing the cell cycle (Figs. 11-10f and 11-11h).

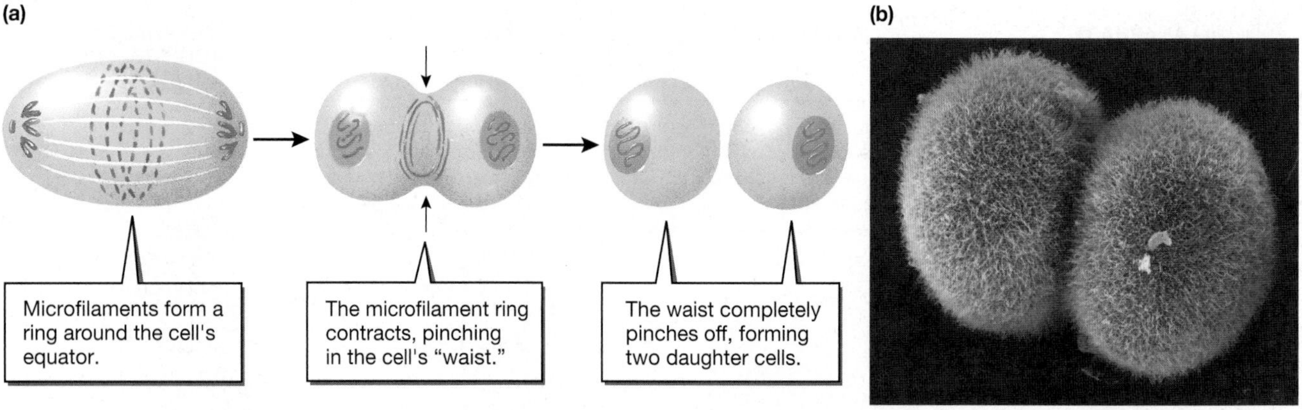
(a)
(b)
Microfilaments form a ring around the cell's equator.
The microfilament ring contracts, pinching in the cell's "waist."
The waist completely pinches off, forming two daughter cells.

Figure 11-12 Cytokinesis in an animal cell
(a) A ring of microfilaments just beneath the plasma membrane contracts around the equator of the cell, pinching it in two. ***(b)*** Cytokinesis has almost separated the two daughter cells.

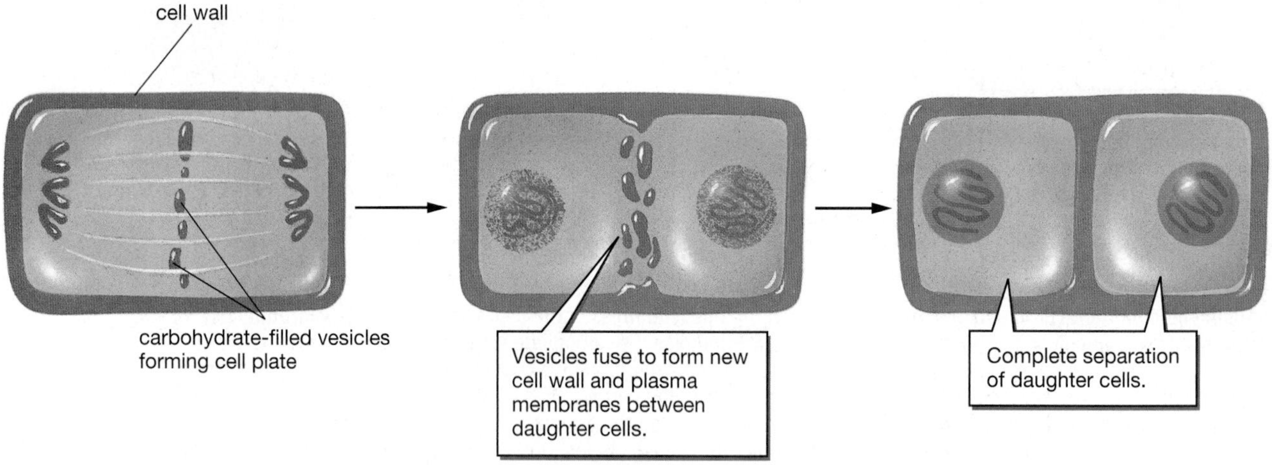
cell wall
carbohydrate-filled vesicles forming cell plate
Vesicles fuse to form new cell wall and plasma membranes between daughter cells.
Complete separation of daughter cells.

Figure 11-13 Cytokinesis in a plant cell
Carbohydrate-filled vesicles produced by the Golgi complex congregate at the equator of the cell, forming the cell plate. The membranes of the vesicles will fuse to form the two plasma membranes separating the daughter cells; their carbohydrate contents form part of the cell wall.

6 What Are Some Advantages of Sexual Reproduction?

The largest organism discovered on the planet is a mushroom, whose underground, branching filaments cover 2200 acres in eastern Oregon. Nearly all of this organism was produced by mitotic cell division. Clearly, asexual reproduction via mitotic cell division must work pretty well! Why, then, have nearly all known forms of life, even the simplest, evolved ways of sexual reproduction? Mitosis can only produce clones, genetically identical offspring. In contrast, **sexual reproduction** enables reshuffling of genes among individuals to produce genetically unique offspring. The nearly universal presence of sexual reproduction is evidence of the tremendous evolutionary advantage that DNA exchange among individuals confers on a species.

Mutations in DNA Are the Ultimate Source of Genetic Variability

As we saw in Chapter 10, the fidelity of DNA replication and proofreading minimizes the number of errors, but changes in DNA base sequences do occur, producing mutations. Although most mutations are neutral or harmful, they are also the raw material for evolution. Bacteria are different from bison, and you are different from your predecessors, because of differences in DNA sequence that originally arose as mutations.

Mutations are perpetuated by DNA replication. Unless they are lethal, mutations are passed to offspring and become a part of the genetic makeup of each species. Such mutations form **alleles**, alternate forms of a given gene that confer variability on individuals, such as black, brown, or blond hair. Shown below is an example of two alleles of a human hemoglobin gene, which encodes the oxygen-carrying protein in red blood cells:

Human Hemoglobin Gene

Normal allele: . . . tggtggtctacccttggacccag . . .
↓ ↓
North Chicago allele: . . . tggtggtctactcttggacccag . . .

The protein produced by this allele has higher oxygen affinity than that produced by the normal allele.

Reshuffling Genes May Combine Different Alleles in Beneficial Ways

As we saw earlier, most eukaryotic organisms that exist today are *diploid*, containing pairs of homologous chromosomes. Homologous chromosomes have the same genes, but each homologue may have different alleles of these genes. For example, one homologue may have the normal allele of the hemoglobin gene and the other homologue may have the North Chicago allele of the hemoglobin gene.

To illustrate the advantage of sexual reproduction, consider a simple hypothetical example of two individuals of the same species. Each individual possesses a particularly beneficial allele of a different gene. We'll call one of these genes "Freeze" and other "Coloration." Individual A possesses an allele of the Freeze gene that causes individual A to remain motionless when a predator is near. Individual B possesses an allele of the Coloration gene that gives individual B camouflage coloration. So, individual A is brightly colored but freezes when near a predator. Individual B has camouflage coloration but continues to move about when near a predator.

	Freeze Gene	**Coloration Gene**
Individual A	Motionless allele: freezes when predator approaches	Motion allele: remains active when predator approaches
Individual B	Bright allele: has bright coloration that stands out from surroundings	Camouflage allele: has coloration that blends into surroundings

Clearly it would be advantageous for a single individual to have both the motionless allele of the Freeze gene *and* the camouflage allele of the Coloration gene. How might this advantage be accomplished? Not through asexual reproduction, which produces offspring that are genetically identical to the single parent. These useful alleles cannot be combined in a single offspring unless two parent organisms combine their DNA, a process that requires sexual reproduction.

Meiotic Cell Division Produces Haploid Cells That Can Merge to Combine Genetic Material from Two Parents

The first eukaryotic cells to evolve, about 1 billion to 1.5 billion years ago, were probably haploid, with only one copy of each chromosome. Relatively early on, two events evolved in single-celled eukaryotic organisms that allowed the organisms to shuffle and recombine genetic information. First, two haploid (parental) cells fused, resulting in a diploid cell with two copies of each chromosome. This cell could reproduce by mitotic cell division, producing diploid daughter cells. Second, this diploid cell evolved a variation in the process of cell division, called *meiotic cell division* (discussed in the next section). Meiotic cell division produces haploid gametes, each containing one copy of each chromosome. A haploid gamete from individual A could contain a motionless allele, and a haploid gamete from individual B could contain a freeze allele. Fusion of these gametes would produce an individual with camouflage coloration that also becomes motionless when a predator approaches, a combination of traits that might help this individual survive longer than others.

7 What Are the Events of Meiosis?

Meiosis Separates Homologous Chromosomes in a Diploid Nucleus, Producing Haploid Daughter Nuclei

The key to sexual reproduction in eukaryotic cells is **meiosis**, the production of haploid nuclei with unpaired chromosomes from diploid parent nuclei with paired chromosomes. The word *meiosis* comes from a Greek word meaning "to diminish"; consistent with this meaning, meiosis reduces the number of chromosomes by half. In meiotic cell division (meiosis followed by cytokinesis), each daughter cell receives one member from each pair of homologous chromosomes. For example, each diploid cell in your body contains 23 *pairs* of chromosomes; meiotic cell division produces gametes (sperm or eggs) with 23 chromosomes, one of each type.

Because meiosis evolved from mitosis, many of the structures and events in meiosis are similar or identical to those of mitosis. However, meiotic cell division differs from mitotic cell division in a major way: during meiotic cell division, the cell undergoes one round of DNA replication followed by two nuclear divisions known as *meiosis I* and *meiosis II*. The behavior of chromosomes during meiosis I differs tremendously from mitosis: In meiosis I, homologous chromosomes pair up and exchange DNA and sister chromatids remain connected to each other. In meiosis II, the behavior of chromosomes appears identical to mitosis: sister chromatids separate and each is pulled to the opposite spindle pole.

MEIOSIS I

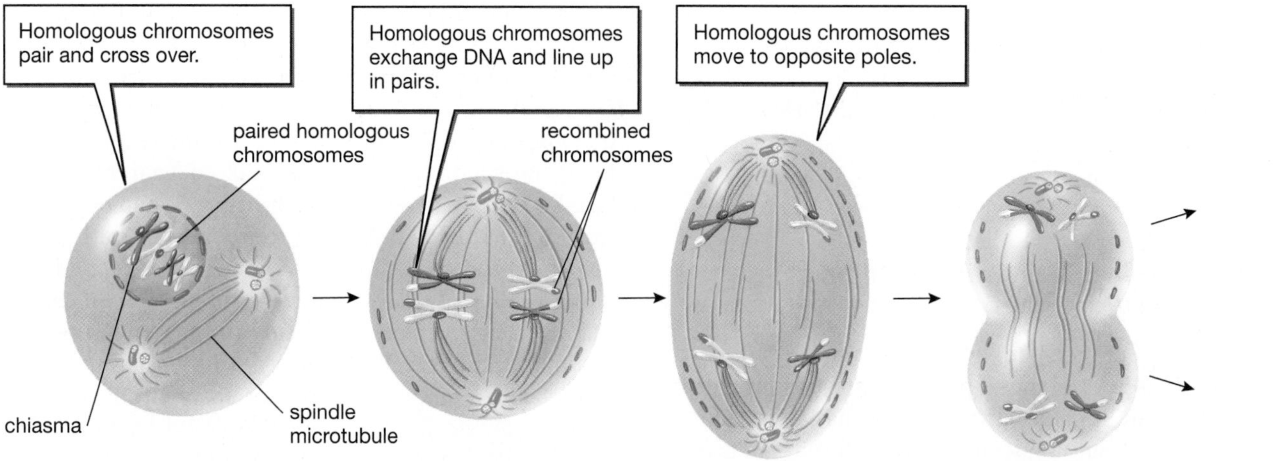
Homologous chromosomes pair and cross over.
Homologous chromosomes exchange DNA and line up in pairs.
Homologous chromosomes move to opposite poles.
paired homologous chromosomes
recombined chromosomes
chiasma
spindle microtubule

(a) Prophase I. Duplicated chromosomes condense. Homologous chromosomes pair up and chiasmata occur as chromatids of homologues exchange parts. The nuclear envelope disintegrates, and spindle microtubules form.

(b) Metaphase I. Paired homologous chromosomes line up along the equator of the cell. One homologue of each pair faces each pole of the cell and attaches to spindle microtubules via its kinetochore (red).

(c) Anaphase I. Homologues separate, one member of each pair going to each pole of the cell. Sister chromatids do not separate.

(d) Telophase I. Spindle microtubules disappear. Two clusters of chromosomes have formed, each containing one member of each pair of homologues. The daughter nuclei are therefore haploid. Cytokinesis commonly occurs at this stage. There is little or no interphase between meiosis I and meiosis II.

MEDIATUTOR 11.2 Meiosis

Figure 11-14 The details of meiotic cell division
In meiotic cell division (meiosis and cytokinesis), the homologous chromosomes of a diploid cell are separated, producing four haploid daughter cells. Each daughter cell contains one member of each pair of parental homologous chromosomes. In these diagrams, two pairs of homologous chromosomes are shown, large and small. The yellow chromosomes are from one parent (for example, the father), and the violet chromosomes are from the other parent. *(continued on next page)*

Because diploid cells have two homologous chromosomes, one round of DNA replication produces four chromatids for each type of chromosome, two chromatids in each duplicated chromosome. Meiosis separates each of these four chromatids into a different nucleus, thus reducing the number of chromosomes in each nucleus from diploid to haploid. Because each nucleus is usually separated by cell division into a different cell, meiosis produces four haploid cells from a single diploid parent cell.

The phases of meiosis have the same names as the roughly equivalent phases in mitosis, followed by a I or II to distinguish the two nuclear divisions that occur in meiosis. In the descriptions that follow, we assume that cytokinesis accompanies the nuclear divisions.

Meiosis I Separates Homologous Chromosomes into Two Daughter Nuclei

In multicellular organisms, meiosis occurs only in a relatively few, specialized cells in the reproductive organs, such as the testes and ovaries of animals. Meiosis begins in these cells with the duplication of the chromosomes. As in mitosis, the sister chromatids of each chromosome remain attached to one another at the centromere. From this point on, meiosis I differs from mitosis.

During Prophase I, Homologous Chromosomes Pair Up and Exchange DNA

During mitosis, homologous chromosomes behave as completely separate entities. In contrast, during *prophase I*, homologous chromosomes line up side by side and exchange segments of DNA (Fig. 11-14a and Fig. 11-15, step ①, p. 202). We'll call one homologue the "maternal chromosome" and the other the "paternal chromosome," because one was originally inherited from the organism's mother and the other from the organism's father. During prophase I, proteins bind the maternal and paternal homologues together so that they match up exactly along their entire length, much like closing a zipper (Fig. 11-15, step ②). In addition, large enzyme complexes assemble at several places along the paired chromosomes (Fig. 11-15, step ③). The enzymes cut through the DNA backbones within the chromosomes and graft the broken DNA ends together again (Fig. 11-15, step ④). Usually, the maternal DNA is joined to the paternal DNA and vice versa, forming

MEIOSIS II

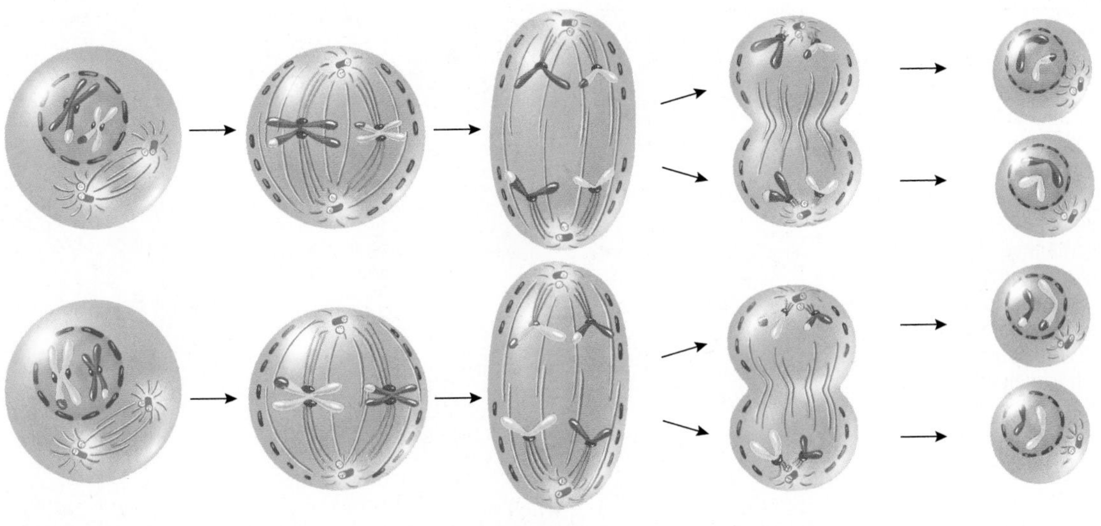

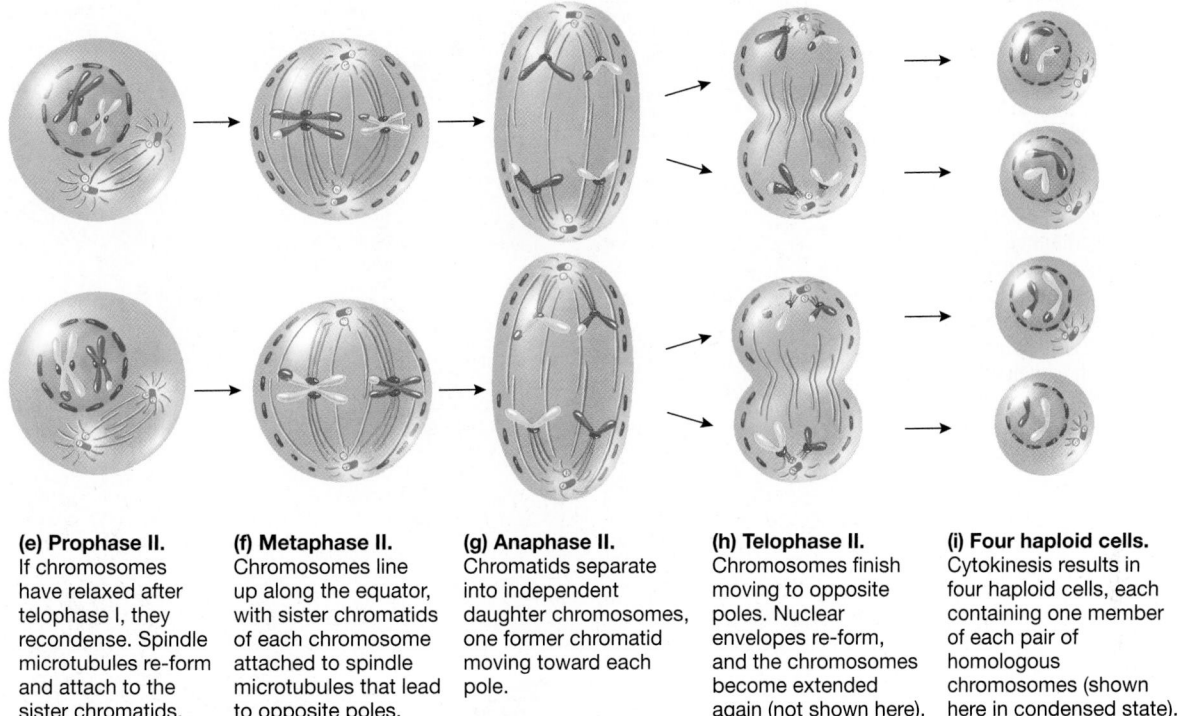

(e) Prophase II. If chromosomes have relaxed after telophase I, they recondense. Spindle microtubules re-form and attach to the sister chromatids.

(f) Metaphase II. Chromosomes line up along the equator, with sister chromatids of each chromosome attached to spindle microtubules that lead to opposite poles.

(g) Anaphase II. Chromatids separate into independent daughter chromosomes, one former chromatid moving toward each pole.

(h) Telophase II. Chromosomes finish moving to opposite poles. Nuclear envelopes re-form, and the chromosomes become extended again (not shown here).

(i) Four haploid cells. Cytokinesis results in four haploid cells, each containing one member of each pair of homologous chromosomes (shown here in condensed state).

Figure 11-14 ***(continued)***

crosses, or **chiasmata** (singular, **chiasma**) where the maternal and paternal chromosomes intertwine. In human cells, each pair of homologues in meiosis I will form two to three chiasmata. Eventually the enzyme complexes detach from the chromosomes, and the protein zippers that held homologues together tightly then disappear. However, the homologues remain associated with each other via the chiasmata (Fig. 11-15, step ⑤).

The formation of chiasmata results in exchange of DNA between maternal and paternal chromosomes, a process called **crossing over**. If (as is likely) the chromosomes had different alleles, then neither chromosome is quite the same as it was before the chiasmata formed. The result of crossing over, then, is genetic **recombination**: the formation of new combinations of alleles on a chromosome.

As in mitosis, the spindle microtubules begin to assemble outside the nucleus during prophase I. Near the end of prophase I, the nuclear envelope breaks down and the spindle microtubules capture the chromosomes by attaching to their kinetochores.

During Metaphase I, Paired Homologous Chromosomes Line Up at the Equator of the Cell

During *metaphase I*, interactions between the kinetochores and the spindle microtubules move the paired homologues to the equator of the cell (Fig. 11-14b). Unlike mitosis, in which *individual duplicated chromosomes* line up along the equator, during metaphase I of meiosis *homologous pairs of duplicated chromosomes* line up along the equator.

The key to understanding meiosis is to understand how the duplicated chromosomes line up at metaphase I. Before going further, then, let's look more closely at the differences between chromosome attachment to spindle microtubules in mitosis versus meiosis I (Fig. 11-16, p. 203). First, in mitosis the homologues do not interact with one another but instead attach independently to the spindle. In meiosis I the homologues remain associated with each other via chiasmata and attach to the spindle as a unit containing the maternal and paternal homologues. Second, in mitosis the duplicated chromosome has two functional kinetochores, one on each sister chromatid. Both kinetochores attach to spindle microtubules, so that each sister chromatid is attached to microtubules that pull toward opposite poles (Fig. 11-16a). In meiosis I the duplicated chromosome has only one functional kinetochore, so that each duplicated chromosome within a homologous pair attaches to spindle microtubules that pull toward opposite poles. These differences in attachment explain what will happen at anaphase: in mitosis, the *sister chromatids separate* and move to opposite poles. In contrast, in meiosis I the sister chromatids of each duplicated chromosome remain

Figure 11-15 The mechanism of crossing over
(1) Homologous chromosomes pair up side by side. (2) One end of each chromosome binds to the nuclear envelope. Protein strands "zip" homologous chromosomes together. (3) Homologous chromosomes are fully joined by protein strands. Recombination enzymes bind to the chromosomes. (4) Recombination enzymes snip chromatids apart and reattach the chromatids. Chiasmata are formed when one end of a chromatid of a paternal chromosome (yellow) is attached to the other end of a chromatid of a maternal chromosome (violet). (5) The protein strands and recombination enzymes leave as the chromosomes condense. The chiasmata remain as locations where homologous chromosomes are twisted around each other, helping to hold homologues together.

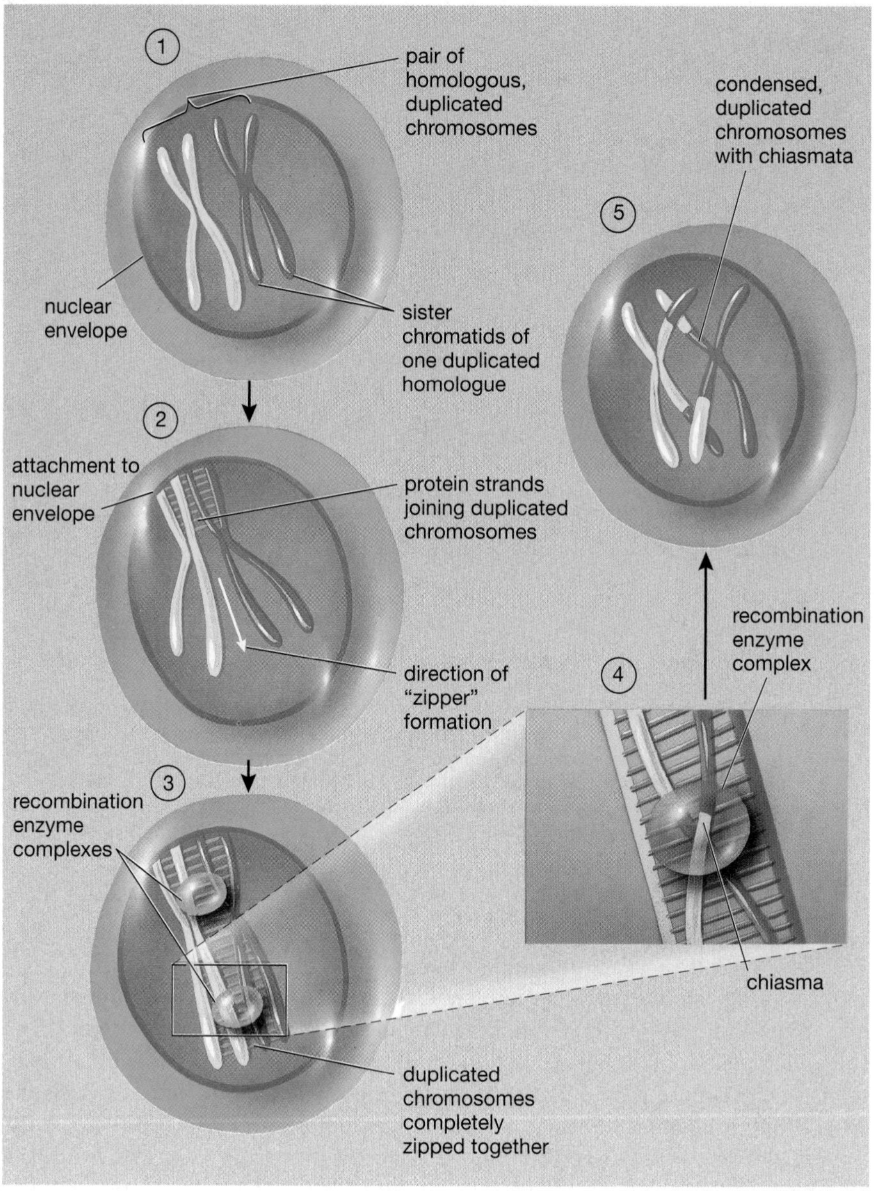

attached to each other; the *homologues separate* and move to opposite poles (Fig. 11-16b).

At meiosis I which member of a pair of chromosomes faces which pole of the cell is random. The maternal chromosome may face "north" for some pairs and "south" for other pairs. This randomness (also called *independent assortment*), together with genetic recombination caused by crossing over, is responsible for the genetic diversity within the haploid cells produced by meiosis.

During Anaphase I, Homologous Chromosomes Separate

In *anaphase I*, the homologues separate from one another and are towed by their kinetochore to opposite poles of the cell (see Fig. 11-14c). One duplicated chromosome of each homologous pair (still consisting of two sister chromatids) moves to each pole of the dividing cell. At the end of anaphase I, the cluster of chromosomes at each pole contains one member of each pair of homologous chromosomes. For example, in humans, one cluster will contain the maternal copy of chromosome 1 and the other cluster will contain the paternal copy of chromosome 1.

During Telophase I, Two Haploid Clusters of Duplicated Chromosomes Form

In *telophase I*, the spindle microtubules disappear. Cytokinesis commonly occurs at this phase and the nuclear envelope may reappear. Following meiosis I, cells have a diploid DNA *content* since they have two chromatids of each chromosome. However, they are sometimes considered haploid because they have only one of each type of chromosome, either the maternal or the paternal homologue. Here's a hint about that baby rattlesnake at the beginning of the chapter: What would happen to the cell if the meiotic process stopped at this point?

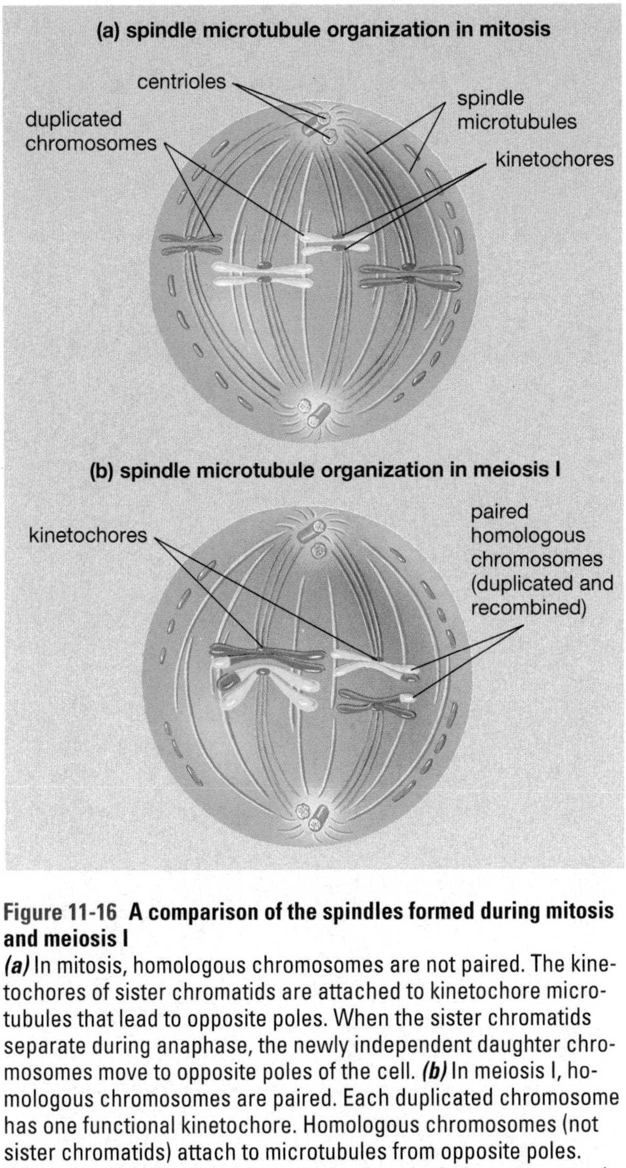

Figure 11-16 A comparison of the spindles formed during mitosis and meiosis I
(a) In mitosis, homologous chromosomes are not paired. The kinetochores of sister chromatids are attached to kinetochore microtubules that lead to opposite poles. When the sister chromatids separate during anaphase, the newly independent daughter chromosomes move to opposite poles of the cell. ***(b)*** In meiosis I, homologous chromosomes are paired. Each duplicated chromosome has one functional kinetochore. Homologous chromosomes (not sister chromatids) attach to microtubules from opposite poles. During anaphase I, sister chromatids of each chromosome remain together, moving to the same pole, but homologous chromosomes separate and move to opposite poles.

Meiosis II Separates Sister Chromatids into Four Daughter Nuclei

Completion of meiosis I is usually followed immediately by meiosis II, with little or no intervening interphase during which no DNA replication occurs. Typically, the chromosomes remain condensed. During meiosis II, the sister chromatids of each duplicated chromosome separate in a process that is virtually identical to mitosis. During *prophase II*, the spindle microtubules re-form (see Fig. 11-14e). The duplicated chromosomes attach individually to spindle microtubules as they did in mitosis. Each chromatid contains a functional kinetochore that attaches to spindle microtubules. Thus, each sister chromatid in a duplicated chromosome attaches to spindle microtubules that extend to opposite poles of the cell. During *metaphase II*, the duplicated chromosomes line up at cell's equator. During *anaphase II*, the sister chromatids separate and are towed to opposite poles. *Telophase II* and cytokinesis conclude meiosis II as nuclear envelopes re-form, the chromosomes relax into their extended state, and the cytoplasm divides. Commonly, both daughter cells produced in meiosis I undergo meiosis II, producing a total of four haploid cells from the original parental diploid cell.

Now that we have covered all of the processes in detail, examine Table 11-1 to review and compare mitotic and meiotic cell division.

The Life Cycles of Organisms on Earth Usually Include Meiosis and Mitosis

Since we are humans ourselves, it is usually pretty easy to understand where mitosis and meiosis fit into the life cycle of humans—and other animals (see Fig. 11-5). However, as you explore the diversity of life on Earth in later units, you will find a variety of life cycles among single-celled organisms, plants, and fungi.

Despite this variability, the life cycles of almost all eukaryotic organisms have a common overall pattern. First, two haploid cells fuse, bringing together genes from two parent organisms and endowing the resulting diploid cell with new gene combinations. Second, at some point in the life cycle, meiosis occurs, re-creating haploid cells. Third, at some point, mitosis of either haploid or diploid cells, or both, results in the growth of multicellular bodies and/or asexual reproduction. The seemingly vast differences among life cycles of, say, fungi, ferns, and humans are due to variations in the part of the life cycle in which mitosis and meiosis occur and in the relative proportions of the life cycle spent in the diploid and haploid states.

8 How Do Meiosis and Sexual Reproduction Produce Genetic Variability?

Shuffling of Homologues Creates Novel Combinations of Chromosomes

As we noted earlier, genetic variability among organisms is essential for survival and reproduction in a changing environment, and therefore for evolution. Mutations occurring randomly over millions of years are the ultimate source of genetic variability within populations of organisms that exist today. However, the genetic variability from one generation of a species to the next relies on meiosis.

How does meiosis produce genetic diversity? One mechanism is the random assortment of maternal and paternal homologues to the daughter cells at meiosis I.

Table 11-1 A Comparison of Mitotic and Meiotic Cell Divisions in Animal Cells

Feature	Mitotic Cell Division	Meiotic Cell Division
Cells in which it occurs	Body cells	Gamete-producing cells
Final chromosome number	Diploid—2*n* (46 for humans); two copies of each chromosome (homologous pairs)	Haploid—1*n* (23 for humans); one member of each homologous pair
Number of daughter cells	Two, identical to the parent cell and to each other	Four, containing recombined chromosomes due to crossing over
Number of cell divisions per DNA replication	One	Two
Function in animals	Development, growth, repair and maintenance of tissues, asexual reproduction	Gamete production for sexual reproduction

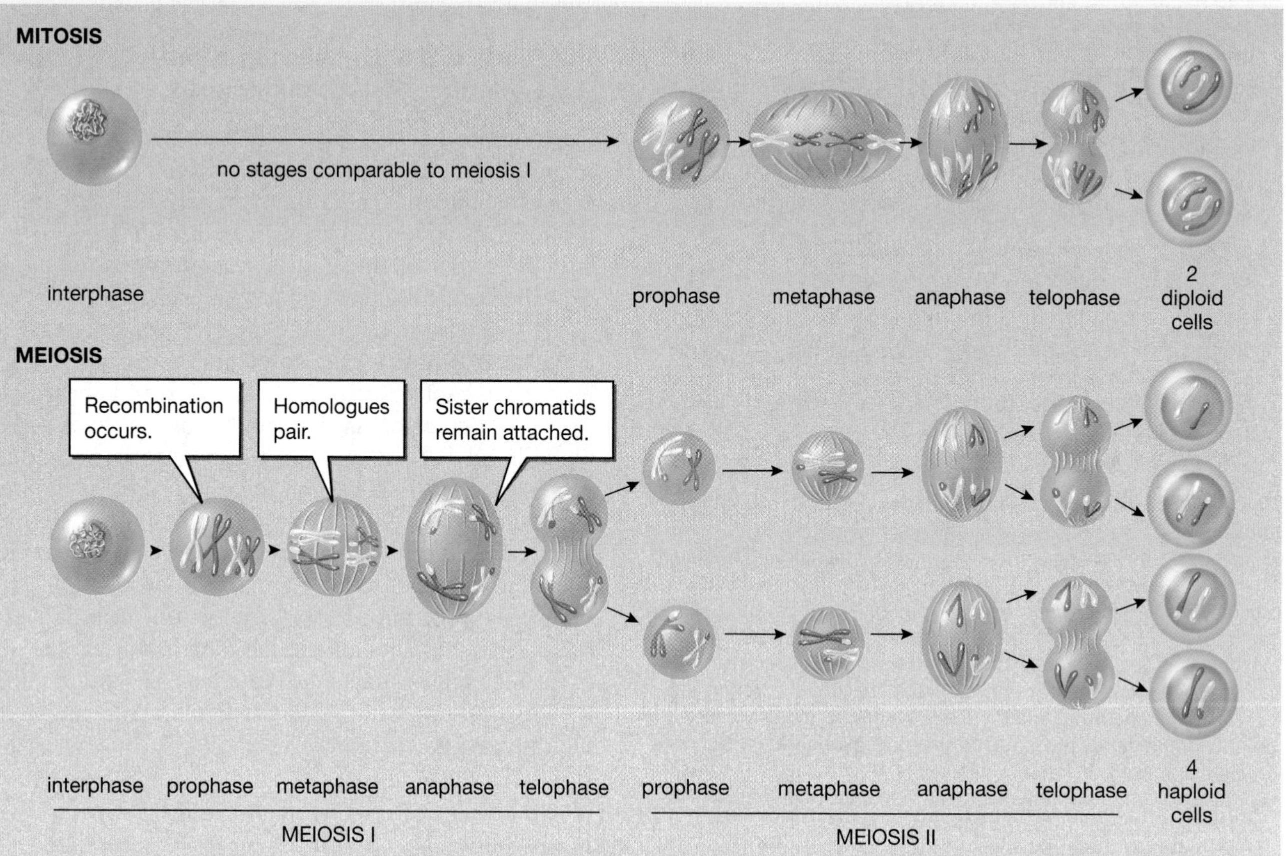

In these diagrams, comparable phases are aligned. In both mitosis and meiosis, chromosomes are replicated during interphase. Meiosis I, with the pairing of homologous chromosomes, formation of chiasmata, exchange of chromosome parts, and separation of homologues to form haploid daughter nuclei, has no counterpart in mitosis. Meiosis II, however, is similar to mitosis.

Remember, at metaphase I the paired homologues line up at the cell's equator. For each pair of homologues, the maternal chromosome faces one pole and the paternal chromosome faces the opposite pole, but which homologue faces which pole is random.

Let's consider meiosis in mosquitoes, which have three pairs of homologous chromosomes ($n = 3, 2n = 6$). For the sake of simplicity, we'll represent these chromosomes as large, medium, and small. To keep track of the homologues, let's color-code the maternal chromosomes yellow and the paternal chromosomes violet. At metaphase I, the chromosomes can align in four configurations:

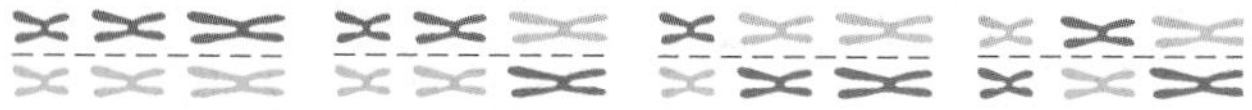

Anaphase I can therefore produce eight possible sets of chromosomes ($2^3 = 8$):

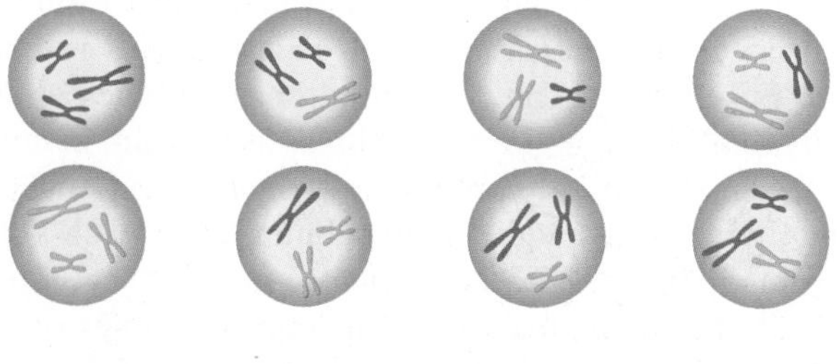

Each of these chromosome clusters will then undergo meiosis II to produce two gametes. Therefore, mosquitoes, with three pairs of homologous chromosomes, can produce gametes with eight different chromosome sets. For humans, with 23 pairs of homologous chromosomes, meiosis produces gametes with more than 8 million (2^{23}) different combinations of paternal and maternal chromosomes.

Crossing Over Creates Chromosomes with Novel Combinations of Genes

In addition to the genetic variation resulting from the random assortment of parental chromosomes, crossing over during meiosis produces chromosomes with combinations of alleles that differ from those of either parent. In fact, these new combinations may have *never before* existed. Because homologous chromosomes cross over in new and different places at each meiotic division, a parent is unlikely to produce two gametes that carry exactly the same combinations of alleles. In essence, every egg and every sperm is genetically unique.

Fusion of Gametes Adds Further Genetic Variability to the Offspring

The final mechanism by which sexual reproduction produces variety occurs at fertilization. Two gametes, each probably containing unique combinations of alleles, fuse to form a diploid offspring. Let's look at how fertilization contributes to genetic variability in human beings. As we saw, one person can produce 2^{23} (about 8 million) different gametes, based solely on the random separation of the homologues. Fusion of gametes from just two people could produce 8 million × 8 million, or more than 6 trillion genetically different children! This variation takes into account only random assortment and fertilization! When you also consider the almost endless variability produced by crossing over, is it any wonder that, with the exception of identical twins, there is truly no one just like you?

CASESTUDYREVISITED

Rattlesnake Surprise

With the information about mitosis and meiosis, you can understand how that female rattlesnake managed to produce a baby without the help of a male. Although it may surprise you, such events are not all that rare in the world of reptiles, some birds, and insects. In fact, some species of whiptail lizards and geckos lack males entirely. The females of these species routinely produce offspring without the requirement for fertilization. For reptiles and birds, these progeny appear to result when meiosis II is not properly completed in females. This abnormal meiotic division produces two cells that have two identical copies of each chromosome. The resulting egg can sometimes develop into a viable, diploid embryo in a process called *parthenogenesis*, Greek for "virgin birth." In the Beltsville Small White breed of turkeys, up to 40% of the eggs laid by the hens contain parthenogenetically produced embryos. Many hatch into apparently normal male turkeys. Are these offspring genetically identical to their mothers? If you consider the events of meiosis, you can see that the answer to the question is, "No." For example, recombination can still occur, producing new combinations of alleles on the chromosomes.

Scientists believe that a similar process of abnormal meiosis occurred in the timber rattler, allowing her to produce a viable offspring even though no sperm was involved. Although it is rare, parthenogenesis has been observed in other snakes, and some scientists think that this is an important mode of reproduction in nature, where finding a mate might be difficult.

But, wait a minute! How could these turkey and the rattler babies be male? Why aren't they female? Visit the Web site for this textbook if you want to discover the answer to this question.

Many scientists stress the importance of sexual reproduction for a species' long-term survival and evolution. However, the success of male-less species of lizards is not consistent with this view. Propose several hypotheses that might account for the surprising success of these parthenogenetic lizard species. Could your hypotheses be extended broadly to cover all other species?

Summary of Key Concepts

1) What Are the Functions of Cellular Reproduction?

Prokaryotes reproduce by a process of cell division called *binary fission* in which the cell grows, replicates its DNA, and divides. Eukaryotic cells can divide by mitosis or meiosis. Meiotic cell division produces gametes, which have only half of the parent's DNA. Fusion of gametes creates a fertilized egg that has a different genetic makeup from either parent. Mitotic cell division of a fertilized egg produces genetically identical cells that differentiate into an embryo. Additional mitotic cell divisions and differentiation allow the embryo to grow, eventually, into an adult. Mitotic cell division also maintains body tissues, such as skin and blood, and can be used to repair damage to some organs. Asexual reproduction is based on mitotic cell divisions, resulting in formation of clones that are genetically identical to the parent.

2) How Is DNA in Eukaryotic Cells Organized into Chromosomes?

Each chromosome in a eukaryotic cell consists of a single DNA double helix and proteins that organize the DNA. During cell growth, the chromosomes are extended and accessible for use in reading their genetic instructions. During cell division, the chromosomes condense into short, thick structures. Eukaryotic cells typically contain pairs of chromosomes called *homologues*. Homologues have virtually identical appearance because they carry the same genes with similar nucleotide sequences. Cells with pairs of homologous chromosomes are diploid. Cells with only a single member of each chromosome are haploid.

3) What Are the Events of the Eukaryotic Cell Cycle?

The eukaryotic cell cycle consists of interphase and cell division. During interphase, the cell grows and duplicates its chromosomes (that is, replicates its DNA). Interphase is divided into G_1 (growth phase I), S (DNA synthesis), and G_2 (growth phase II). During G_1, some cells may exit the cell cycle to enter a non-dividing state called G_0. Cells may remain permanently in G_0, or may be induced to reenter the cell cycle. Mitotic cell division consists of two processes: (1) mitosis (nuclear division) and (2) cytokinesis (cytoplasmic division). Mitosis parcels out one copy of every chromosome into two separate nuclei, and cytokinesis subsequently encloses each nucleus in a separate cell. Mitosis produces two genetically identical daughter cells.

4) What Are the Phases of Mitosis?

The chromosomes are duplicated during interphase, prior to mitosis. The two identical copies, called *chromatids*, remain attached to one another at the centromere during the early stages of mitosis. Mitosis consists of four phases (see Fig. 11-11):

1. **Prophase:** The chromosomes condense and their kinetochores attach to the spindle microtubules that form at this time.
2. **Metaphase:** The chromosomes move along their attached spindle microtubules to the equator of the cell.
3. **Anaphase:** The two chromatids of each duplicated chromosome separate and move along the spindle microtubules to opposite poles of the cell.
4. **Telophase:** The chromosomes relax into their extended state, and nuclear envelopes re-form around each new daughter nucleus.

5) What Are the Events of Cytokinesis?

Cytokinesis normally occurs at the end of telophase and divides the cytoplasm into approximately equal halves, each containing a nucleus. In animal cells, the plasma membrane is pinched in along the equator by a ring of microfilaments. In plant cells, new plasma membrane forms along the equator by the fusion of vesicles produced by the Golgi complex.

6) What Are Some Advantages of Sexual Reproduction?

Genetic differences among organisms originate as mutations. Mutations preserved within a species produce alternate forms of genes called alleles. Alleles in different individuals of a species may be combined in offspring through sexual reproduction.

7) What Are the Events of Meiosis?

Meiosis separates homologous chromosomes and produces haploid cells with only one homologue from each pair. These haploid cells or their descendants fuse to form diploid cells that receive one homologue from each parent, reestablishing pairs of homologous chromosomes. During interphase before meiosis, chromosomes are duplicated. The cell then undergoes two specialized cell divisions—meiosis I and meiosis II—to produce four haploid daughter cells.

Meiosis I: During prophase I, homologous duplicated chromosomes, each consisting of two chromatids, pair up and exchange parts by crossing over. During metaphase I, homologues move together as a pair to the cell's equator, one member of each pair facing opposite poles of the cell. Homologous chromosomes separate during anaphase I, and two nuclei form during telophase I. Each daughter nucleus receives only one member of each pair of homologues. The sister chromatids remain attached to each other throughout meiosis I.

Meiosis II: Meiosis II occurs in both daughter nuclei and resembles mitosis in a haploid cell. The duplicated chromosomes move to the cell's equator during metaphase II. The two chromatids of each chromosome separate and move to opposite poles of the cell during anaphase II. This second division produces four haploid nuclei. Cytokinesis normally occurs during or shortly after telophase II, producing four haploid cells.

8) How Do Meiosis and Sexual Reproduction Produce Genetic Variability?

The random shuffling of homologous maternal and paternal chromosomes creates new chromosome combinations. Crossing over creates chromosomes with allele combinations that may never before have occurred on single chromosomes. Because of crossing over, a parent probably never produces any two gametes that are completely identical. The fusion of two such genetically unique gametes adds further genetic variability to the offspring.

Key Terms

allele *p. 198*
anaphase *p. 196*
asexual reproduction *p. 187*
binary fission *p. 186*
cell cycle *p. 191*
cell division *p. 186*
cell plate *p. 197*
centriole *p. 195*
centromere *p. 189*
chiasma (chiasmata) *p. 201*
chromatid *p. 189*
chromosome *p. 189*
clone *p. 187*
cloning *p. 192*
condensation *p. 189*
crossing over *p. 201*
cytokinesis *p. 194*
differentiation *p. 194*
diploid *p. 190*
duplicated chromosome *p. 190*
gamete *p. 188*
haploid *p. 191*
homologue *p. 190*
interphase *p. 194*
karyotype *p. 190*
kinetochore *p. 195*
meiosis *p. 199*
meiotic cell division *p. 188*
metaphase *p. 196*
mitosis *p. 194*
mitotic cell division *p. 194*
prophase *p. 194*
recombination *p. 201*
sex chromosome *p. 190*
sexual reproduction *p. 198*
spindle microtubule *p. 195*
telophase *p. 197*

Thinking Through the Concepts

Multiple Choice

1. *At which stage of mitosis are chromosomes arranged along a plane at the midline of the cell?*
   a. anaphase
   b. telophase
   c. metaphase
   d. prophase
   e. interphase
2. *A diploid cell contains in its nucleus*
   a. an even number of chromosomes
   b. an odd number of chromosomes
   c. one copy of each homologue
   d. either an even or an odd number of chromosomes
   e. two sister chromatids of each chromosome during G_1
3. *Synthesis of new DNA occurs during*
   a. prophase
   b. interphase
   c. mitosis
   d. cytokinesis
   e. formation of the cell plate
4. *Which statement is most correct?*
   a. All mutations are harmful.
   b. Both mitosis and meiosis add to genetic diversity.
   c. Crossing over helps each gamete get a different set of alleles in meiosis.
   d. Mitosis always makes diploid daughter cells; meiosis always produces gametes.
   e. The only haploid cells are gametes.
5. *When do homologous chromosomes pair up?*
   a. only in mitosis
   b. only in meiosis I
   c. only in meiosis II
   d. in both mitosis and meiosis
   e. in neither mitosis nor meiosis
6. *Curiously, there is no crossing over of any chromosome in the male fruit fly* Drosophila, *which has four pairs of chromosomes. How many different combinations of maternal vs. paternal chromosomes are possible in a male fruit fly's sperm?*
   a. 2
   b. 4
   c. 8
   d. 16
   e. many more than the above

? Review Questions

1. Diagram and describe the eukaryotic cell cycle. Name the various phases, and briefly describe the events that occur during each. What is the role of the cell cycle in a human?
2. Define *mitosis* and *cytokinesis*. What changes in cell structure result when cytokinesis does not occur after mitosis?
3. Diagram the stages of mitosis. How does mitosis ensure that each daughter nucleus receives a full set of chromosomes?
4. Define the following terms: *homologous chromosome, centromere, kinetochore, chromatid, diploid, haploid.*
5. Describe and compare the process of cytokinesis in animal cells and in plant cells.
6. Diagram the events of meiosis. At which stage do homologous chromosomes separate?
7. Describe homologue pairing and crossing over. At which stage of meiosis do they occur? Name two functions of chiasmata.
8. In what ways are mitosis and meiosis similar? In what ways are they different?
9. Describe how meiosis provides for genetic variability. If an animal had a haploid number of 2 (no sex chromosomes), how many genetically different types of gametes could it produce? (Assume no crossing over.) If it had a haploid number of 5?

Applying the Concepts

1. Nerve cells in the adult human central nervous system, as well as heart muscle cells, remain in the G_0 portion of interphase. In contrast, cells lining the inside of the small intestine divide frequently. Discuss this difference in terms of why damage to the nervous system and heart muscle cells (such as caused by a stroke or heart attack) is so dangerous. What do you think might happen to tissues such as the intestinal lining if some disorder or drug blocked mitoses in all cells of the body?
2. Cancer cells divide out of control. Side effects of chemotherapy and radiation therapy that fight cancers include loss of hair and of the gastrointestinal lining, producing severe nausea. Note that cells in hair follicles and intestinal lining divide frequently. What can you infer about the mechanisms of these treatments? What would you look for in an improved cancer therapy?
3. Some animal species can reproduce either asexually or sexually, depending on the state of the environment. Asexual reproduction tends to occur in stable, favorable environments; sexual reproduction is more common in unstable and/or unfavorable circumstances. Discuss the advantages or disadvantages this behavior might have on survival of the species in an evolutionary sense or on survival of individuals.
4. Would you predict that a clone produced as Dolly was would be a carbon copy of the animal from which the nucleus was obtained? What factors might cause the two sheep to differ? What variables might cause a human clone to differ from the individual who donated the genetic material?

For More Information

Grant, M. C. "The Trembling Giant." *Discover*, October 1993. Aspen groves are really single individuals: huge, slowly spreading from the roots of the original parent tree, and potentially almost immortal.

Lanza, R. P., Dresser, B. L., and Damiani, P. "Cloning Noah's Ark." *Scientific American,* November 2000. Cloning rare and endangered species may offer hope to prevent extinction.

Nash, M. "The Age of Cloning." *Time*, March 10, 1997. Well-written and well-illustrated description of the technique and implications of using an adult cell to clone a sheep.

Travis, J. "A Fantastical Experiment." *Science News*, April 5, 1997. A clear description of the cloning of Dolly the sheep and some of its implications.

Wilmut, I. "Cloning for Medicine." *Scientific American,* December 1998. Explanation of why cloning experiments might have medical applications.

Answers to Multiple-Choice Questions

1.c 2.a 3.b 4.c 5.b 6.d

MEDIATUTOR
The Continuity of Life: Cellular Reproduction

CD Activities

Activity 11.1: Mitosis

Estimated time: 8 minutes

Each of us starts as a single cell—the fusion between sperm and egg. Yet by the time you are old enough to read this, you are made of trillions of cells, all descended from that original single cell. This is the result of repeated cell divisions. Every time one cell divides, it must accurately distribute copies of the chromosomes of the parental cell, producing two genetically identical cells. This process of chromosome distribution is mitosis. In this tutorial you will explore the steps involved in mitosis.

Activity 11.2: Meiosis

Estimated time: 8 minutes

"Variety is the spice of life." Sexual reproduction is designed to generate genetic variety, or diversity. In order for the fusion of sperm and egg to produce a cell with the correct number of chromosomes, the sperm and egg must each have one-half the normal number of chromosomes. To accomplish this, certain cells in the body undergo chromosome reduction division, meiosis. This tutorial will lead you through the steps of meiosis and ask you to determine how meiosis compares with mitosis.

Start the MediaTutor Student CD-ROM and enter the activity number in the Quick Search box to be taken directly to that activity.

Web Investigations

Case Study: Rattlesnake Surprise

Estimated time: 10 minutes

We can outline the essentials of mammalian reproduction in a relatively simple manner. A female must mate with a male. Each contributes a half set of chromosomes to the offspring, who end up with the same number of chromosomes as the parents. Is it the same for the birds, the bees, and the snakes? Find out by answering questions in the Web Investigation.

Go to http://www.prenhall.com/audesirk6, the Audesirk Companion Web site. Select Chapter 11 and Web Investigation to begin.

Narcoleptic dogs collapse if they become excited by a food treat or by roughhousing. Studies of these dogs have uncovered a gene that may help us understand narcolepsy in humans.

12 Patterns of Inheritance

AT A GLANCE

Case Study: Sleepy Genes

1) **How Did Gregor Mendel Lay the Foundations for Modern Genetics?**
Doing It Right: The Secrets of Mendel's Success

2) **How Are Single Traits Inherited?**
The Inheritance of Dominant and Recessive Alleles on Homologous Chromosomes Can Explain the Results of Mendel's Crosses
Mendel's Hypothesis Can Be Used to Predict the Outcome of New Types of Single-Trait Crosses

3) **How Are Multiple Traits on Different Chromosomes Inherited?**
Mendel Hypothesized That Genes on Different Chromosomes Are Inherited Independently
In an Unprepared World, Genius May Go Unrecognized

4) **How Are Genes Located on the Same Chromosome Inherited?**
Genes on the Same Chromosome Tend to Be Inherited Together
Recombination Can Create New Combinations of Linked Alleles

5) **How Is Sex Determined, and How Are Sex-Linked Genes Inherited?**
Sex-Linked Genes Are Found Only on the X or Only on the Y Chromosome

6) **What Are Some Variations on the Mendelian Theme?**
Alleles May Display Incomplete Dominance in Which the Phenotype of Heterozygotes Is Intermediate Between the Phenotypes of the Homozygotes
A Single Gene May Have Multiple Alleles
Many Traits Are Influenced by Several Genes
Single Genes Typically Have Multiple Effects on Phenotype
The Environment Influences the Expression of Genes

7) **How Are Human Genetic Disorders Investigated?**

8) **How Are Human Disorders Caused by Single Genes Inherited?**
Many Human Genetic Disorders Are Caused by Recessive Alleles
Many Human Genetic Disorders Are Caused by Dominant Alleles
Some Human Disorders Are Sex-Linked

9) **How Do Errors in Chromosome Number Affect Humans?**
Some Genetic Disorders Are Caused by Abnormal Numbers of Sex Chromosomes
Some Genetic Disorders Are Caused by Abnormal Numbers of Autosomes

Case Study Revisited: Sleepy Genes

CASESTUDY

Sleepy Genes

Jason was a good student throughout grade school and middle school. However, when he began high school, he started to have odd symptoms that made his academic performance plummet. He was unable to stay awake during class, even when he had gotten plenty of sleep the night before. Even worse were the periods of paralysis called *cataplexy.* If he was startled by a slamming locker door, he might collapse and be unable to move for a few minutes. In his freshman year he broke three pairs of glasses as a result of these bouts of cataplexy. The diagnosis finally came when Jason was a junior in high school. Along with about 200,000 other Americans, Jason suffers from a disorder of the central nervous system called *narcolepsy.* Learning the name of his disease didn't cure Jason, but it has helped him to manage the symptoms. For example, he now takes amphetamines to keep himself awake during his normal daily activities. Even with these stimulants, he still needs to nap throughout the day, which he can manage by carefully organizing his schedule. Antidepressants seem to help prevent the embarrassing and often dangerous instances of cataplexy.

Jason and the thousands of other narcoleptics share their plight with several colonies of dogs that are being studied at Stanford University. The excitement of getting a doggie biscuit can trigger cataplexy in these narcoleptic Daschunds, Dobermans, or Labrador retrievers. They regain muscle control a short time later, apparently none the worse for undergoing the uncontrollable collapse. In August 1999 the Stanford researchers reported the culmination of 36 years of study on these animals: They had discovered a gene that is defective in some of their narcoleptic dogs. The hope is that this information may help in developing new treatments for sleep disorders such as narcolepsy. How can scientists find genes? To answer this question, we must first go back in time about 140 years to a monastery garden in what is now the Czech Republic, where the science of genetics was born. ■

1 How Did Gregor Mendel Lay the Foundations for Modern Genetics?

Before settling down as a monk in the monastery of St. Thomas in Brünn (now Brno, in the Czech Republic), Gregor Mendel (Fig. 12-1) tried his hand at several pursuits, including health care and teaching. To earn his teaching certificate, Mendel attended the University of Vienna for 2 years, where he studied botany and mathematics, among other subjects. This training proved crucial to his later experiments, which were the foundation for the modern science of genetics. At the St. Thomas monastery in the mid-1800s, Mendel carried out both his monastic duties and a groundbreaking series of experiments on inheritance in the common edible pea. Although Mendel worked without knowledge of genes or chromosomes, we can more easily follow his experiments after a brief look at some modern genetic concepts.

A gene's specific physical location on a chromosome is called its **locus** (plural, *loci*) (Fig. 12-2). Homologous chromosomes carry the same **genes**, located at the same loci. Although the nucleotide sequence at a given gene locus is *similar* on homologous chromosomes, the sequence may not be *identical*. These differences in nucleotide sequences at the same gene locus on two homologous chromosomes produce alternate forms of the gene, called **alleles**. Human A, B, and O blood types, for example, are produced by the three different alleles of the blood type gene.

If both homologous chromosomes in an organism have the *same* allele at a given gene locus, the organism is said to be **homozygous** at that gene locus. (*Homozygous* comes from Greek words meaning "same pair.") For example, the chromosomes in Figure 12-2 are homozygous at the loci for both the M and D genes. If two homologous chromosomes have *different* alleles at a given gene locus, the organism is **heterozygous** ("different pair") at that locus and can be called a **hybrid**. The chromosomes in Figure 12-2 are heterozygous at the locus for the Bk gene. Recall from Chapter 11 that, during meiosis, homologous chromosomes are separated, so each gamete receives one member of each pair of homologous chromosomes. As a result, every gamete has only one allele for each gene. Therefore, all the gametes produced by an organism that is homozygous at a particular gene locus will contain the same allele. Gametes produced by an organism that is heterozygous at the same gene locus are of two kinds: Half of the gametes contain one allele, and half contain the other allele.

Interestingly, both the patterns of inheritance and many essential facts about genes, alleles, and the distrib-

Figure 12-1 Gregor Mendel
A portrait of Mendel, painted in about 1888, after he had completed his pioneering genetics experiments.

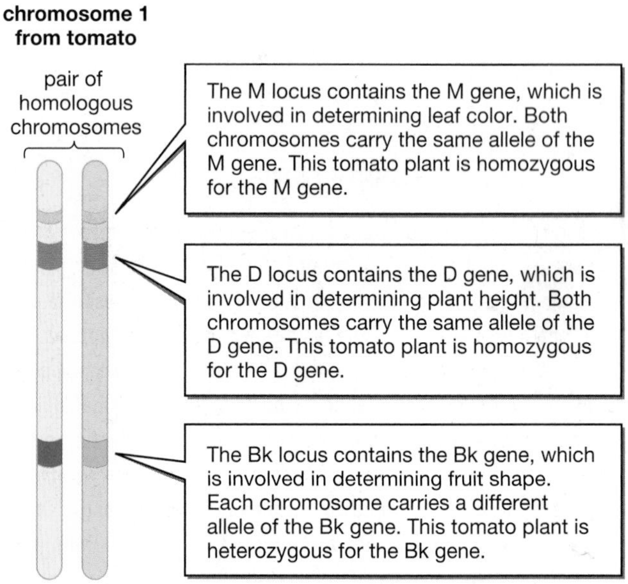

chromosome 1 from tomato
pair of homologous chromosomes
The M locus contains the M gene, which is involved in determining leaf color. Both chromosomes carry the same allele of the M gene. This tomato plant is homozygous for the M gene.
The D locus contains the D gene, which is involved in determining plant height. Both chromosomes carry the same allele of the D gene. This tomato plant is homozygous for the D gene.
The Bk locus contains the Bk gene, which is involved in determining fruit shape. Each chromosome carries a different allele of the Bk gene. This tomato plant is heterozygous for the Bk gene.

Figure 12-2 The relationships among genes, alleles, and chromosomes
Each homologous chromosome carries the same set of genes. Each gene is located at the same relative position, or locus, on its chromosome. Differences in nucleotide sequences at the same gene locus produce different alleles of the gene. Diploid organisms have two alleles of each gene.

ution of alleles in gametes and zygotes during sexual reproduction were deduced by Gregor Mendel long before DNA, chromosomes, or meiosis had been discovered. Because his experiments are a succinct, elegant example of science in action, let's follow Mendel's paths of discovery.

Doing It Right: The Secrets of Mendel's Success

There are three key steps to any successful experiment in biology: (1) choosing the right organism with which to work, (2) designing and performing the experiment correctly, and (3) analyzing the data properly. Mendel was the first geneticist to complete all three steps.

Mendel's choice of the edible pea (Fig. 12-3a) as an experimental subject was critical to the success of his experiments. The flowers of peas and other flowering plants contain structures that produce gametes. Each pollen grain contains a male gamete, which for simplicity we'll call *sperm*. Pollination allows sperm to fertilize an egg cell, located within the ovary in the flower's interior. The petals of a pea flower enclose all of the flower structures, normally preventing another flower's pollen from entering (Fig. 12-3b). Instead, each pea flower normally supplies its own pollen, so the egg cells in each flower are fertilized by sperm from the pollen of the same flower. This process is called **self-fertilization.** Consider a pea plant that is homozygous at the locus for flower color: All of its progeny will have the same flower color, which is the same as the parent plant. Such plants are called **true-breeding.** Even in Mendel's time, commercial seed dealers sold many types of true-breeding pea varieties.

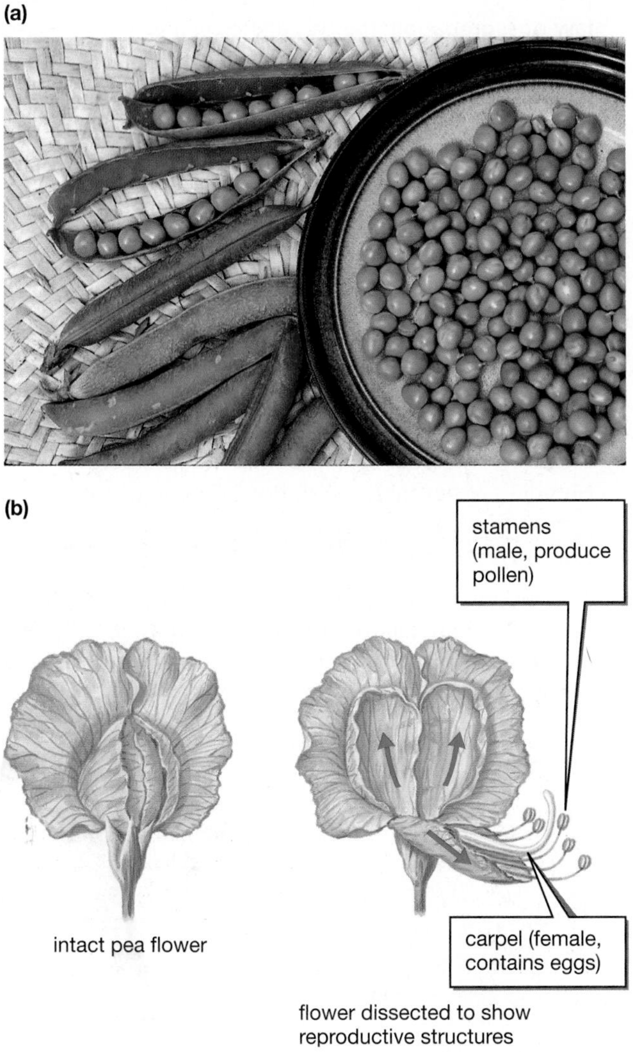

Figure 12-3 The seeds and flowers of the edible pea
(a) Edible or garden peas are important cool weather crops. ***(b)*** In the intact pea flower (left), the lower petals form a container enclosing the reproductive structures—the stamens (male) and carpel (female). Pollen normally cannot enter the flower from outside, so peas normally self-fertilize.

Although peas normally self-fertilize, plant breeders can also mate two plants by hand, a process called **cross-fertilization.** Breeders pull apart the petals and remove the pollen-producing structures (stamens), preventing self-fertilization (see Fig. 12-3b). By dusting the egg-producing structures of the flower with pollen from the plants they have selected, breeders can control fertilization. In this way, two plants can be mated to see what types of offspring they produce.

Before Mendel, scientists trying to understand heredity looked at the entire organism and its progeny. Instead, Mendel chose to study individual *traits*—heritable characteristics—that had unmistakably different forms, such as white flowers versus purple flowers. He also worked with one trait at a time. These factors allowed Mendel to see through to the underlying principles of inheritance. Mendel also followed the inheritance of those traits for several generations, counting the numbers of offspring with each type of trait. By analyzing these numbers using statistics, the basic patterns of inheritance became clear. The use of statistics as a tool to verify the validity of results has since become an extremely important practice in biology.

2 How Are Single Traits Inherited?

MEDIATUTOR
12.1 Monohybrid Crosses

Mendel started as simply as possible. He raised varieties of pea plants that were true-breeding for different forms of a single trait and cross-fertilized them. Mendel saved the resulting hybrid seeds and grew them the following year to determine their characteristics.

In one of these experiments, Mendel cross-fertilized a white-flowered pea with a purple-flowered one. This was the *parental generation*, denoted by the letter *P*. When he grew the resulting seeds, he found that all the first-generation offspring (the "first filial," or F_1, generation) produced purple flowers:

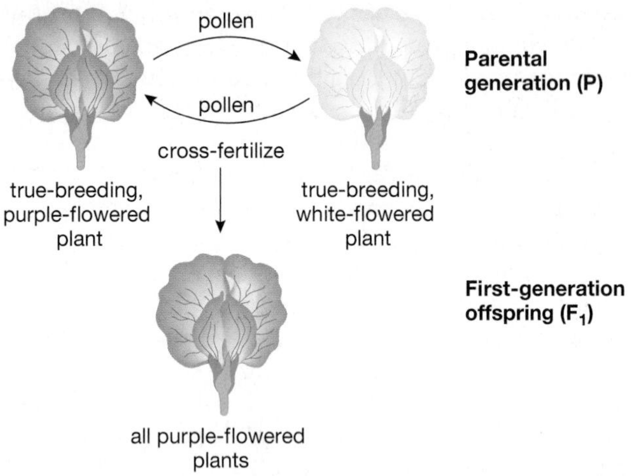

What had happened to the white color? The flowers of the hybrids were just as purple as their parent. The white color seemed to have disappeared in the F_1 offspring.

Mendel then allowed the F_1 flowers to self-fertilize, collected the seeds, and planted them the next spring. In the second generation (F_2), about three-fourths of the plants had purple flowers and one-fourth had white flowers:

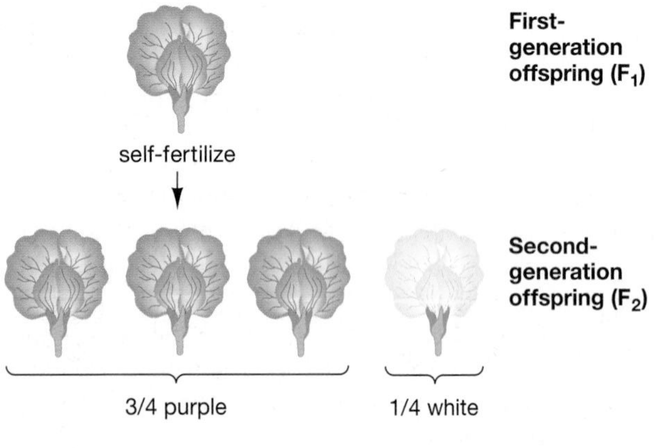

The exact numbers were 705 purple and 224 white, or a ratio of about 3 purple to 1 white. This result showed that the gene that produced white flowers had not disappeared but had only been "hidden."

Mendel allowed the F_2 plants to self-fertilize and produce yet a third (F_3) generation. He found that all the white-flowered F_2 plants produced white-flowered offspring; that is, they were true-breeding. For as many generations as he had time and patience to raise, white-flowered parents always gave rise to white-flowered offspring. The purple-flowered F_2 plants were of two types: About $\frac{1}{3}$ of these were true-breeding for purple; the remaining $\frac{2}{3}$ were hybrids that produced both purple- and white-flowered offspring, again in the ratio of 3 to 1. Therefore, the F_2 generation included $\frac{1}{4}$ true-breeding purple plants, $\frac{1}{2}$ hybrid purple, and $\frac{1}{4}$ true-breeding white.

The Inheritance of Dominant and Recessive Alleles on Homologous Chromosomes Can Explain the Results of Mendel's Crosses

Mendel's results, supplemented by our knowledge of genes and homologous chromosomes, allow us to develop a five-part hypothesis:

1. Each trait is determined by pairs of discrete physical units, which we now call *genes*. Each individual has two alleles for a given gene, such as the gene that determines flower color. One allele of the gene is present on each homologous chromosome. True-breeding peas with white flowers have different alleles of the "flower-color" gene than true-breeding purple-flowered peas.
2. The pairs of genes on homologous chromosomes separate from each other during gamete formation, so each gamete receives only one allele of an organism's pair of genes. This conclusion is known as Mendel's **law of segregation**: the two alleles of a gene segregate from one another at meiosis. When a sperm fertilizes an egg, the resulting offspring receives one allele from the father and one from the mother.
3. Which allele becomes included in a gamete is determined by chance. This randomness occurs because the separation of homologous chromosomes during meiosis is random.
4. When two different alleles are present in an organism, one (the **dominant** allele) may mask the expression of the other (the **recessive** allele). The dominant allele does not, however, alter the physical presence of the recessive allele. Each allele, whether dominant or recessive, is passed into the individual's gametes. In Mendel's experiments with flower color, the allele for purple flowers is dominant, and the allele for white flowers is recessive.
5. True-breeding organisms have two of the same alleles for a given gene; these organisms are homozygous. All of the gametes from a homozygous individual have the same allele for that gene.

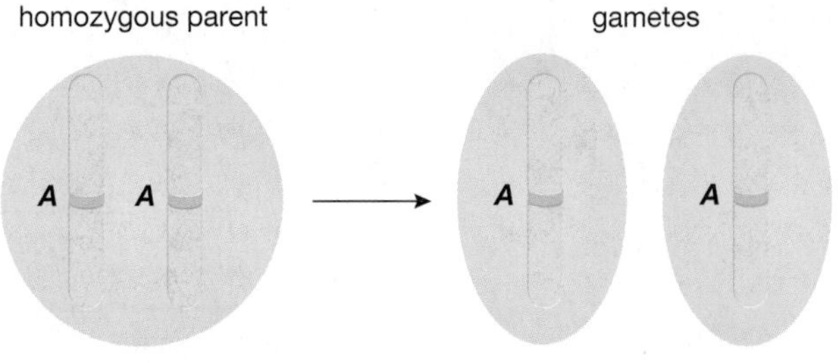

Hybrids have two different alleles for that gene; they are heterozygous. Half of the organism's gametes contain one allele for that gene; half contain the other allele. A heterozygous organism, with two different alleles of gene "A," produces equal numbers of gametes with each of the two alleles:

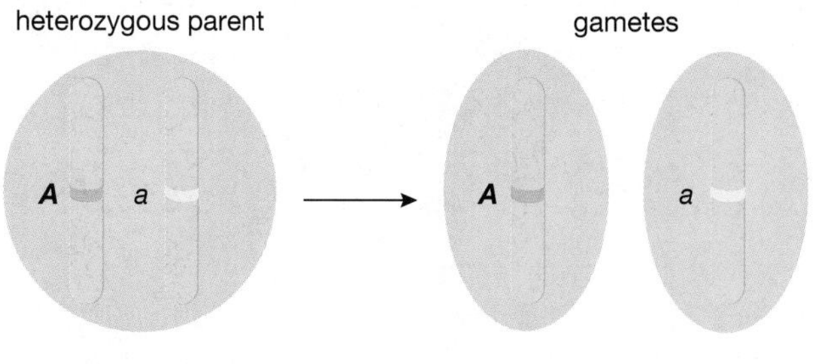

Let's see how Mendel's hypothesis explains the results of his experiments with flower color. Using letters to represent the different alleles, we will assign the uppercase letter *P* to the allele for purple (dominant) and the lowercase letter *p* to the allele for white (recessive). (By Mendel's convention, the dominant allele is represented by a capital letter.) A true-breeding (homozygous) purple-flowered plant has two alleles for purple flowers (*PP*), whereas a white-flowered plant has two alleles for white flowers (*pp*). All the sperm and eggs produced by a *PP* plant carry the *P* allele; all the sperm and eggs of a *pp* plant carry the *p* allele:

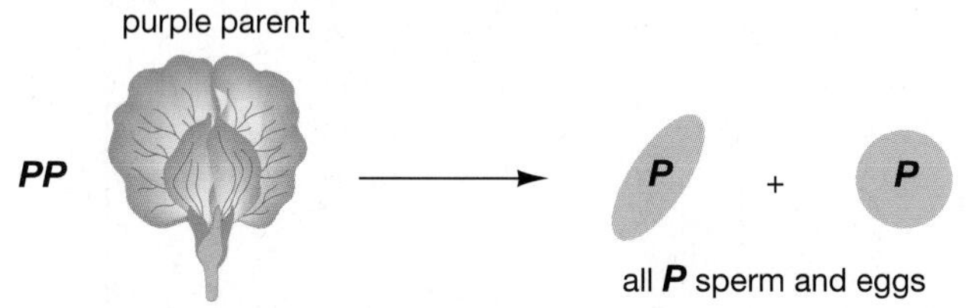

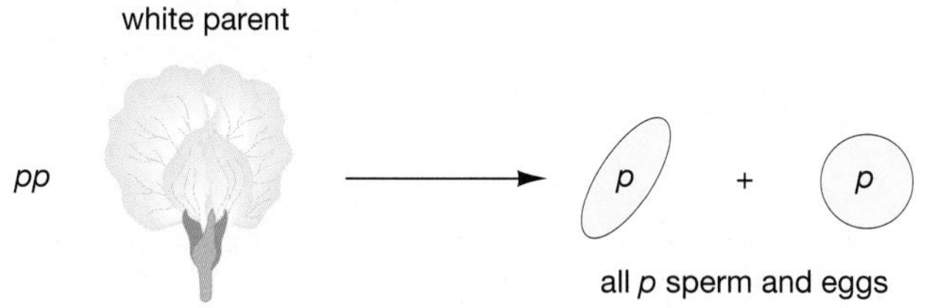

The F_1 hybrid offspring are produced when *P* sperm fertilize *p* eggs or when *p* sperm fertilize *P* eggs. In either case, the F_1 offspring are *Pp*. Because *P* is dominant to *p*, all the offspring are purple:

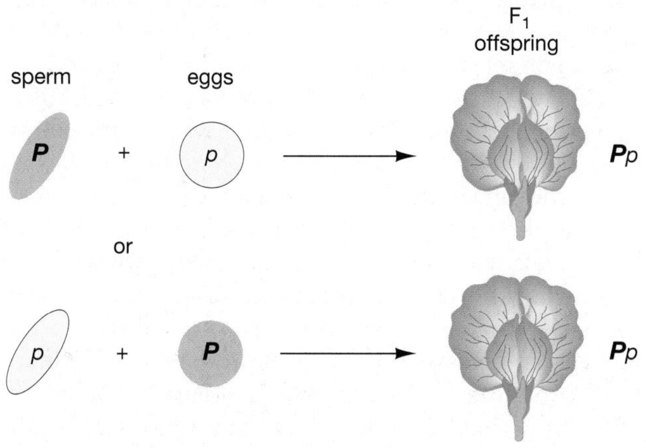

Each gamete produced by a heterozygous *Pp* plant has an equal chance of receiving either the *P* allele or the *p* allele. That is, the hybrid plant produces equal numbers of *P* and *p* sperm and equal numbers of *P* and *p* eggs. When a *Pp* plant self-fertilizes, each type of sperm has an equal chance of fertilizing each type of egg:

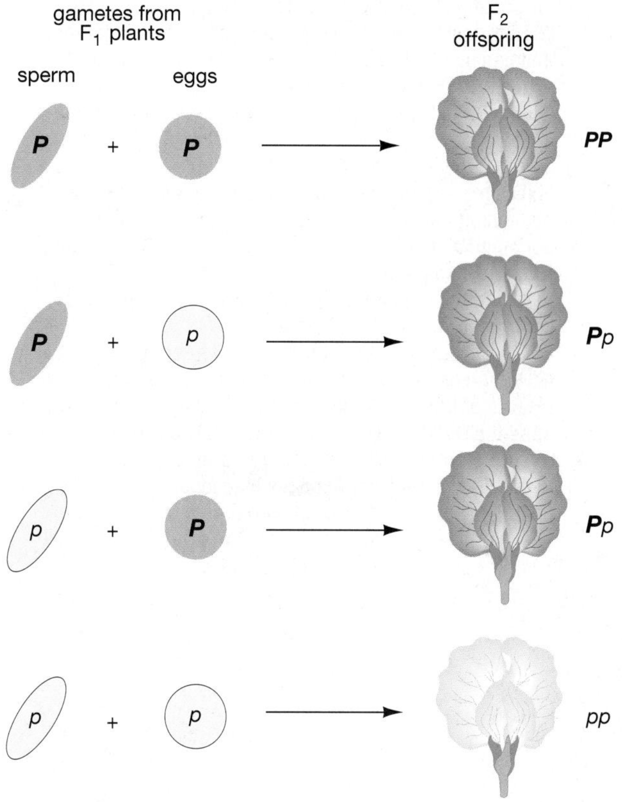

Therefore, three types of offspring can be produced: *PP*, *Pp*, and *pp*. The three types occur in the approximate proportions of ¼ *PP*, ½ *Pp*, and ¼ *pp*.

The actual combination of alleles carried by an organism (for example, *PP* or *Pp*) is its **genotype.** The organism's traits, including its outward appearance, behavior, digestive enzymes, blood type, or any other observable or measurable feature, make up its **phenotype.** As we have seen, plants with either the *PP* or the *Pp* genotype make purple flowers. Thus, even though they have different genotypes, they have the same phenotype. Therefore, the F_2 generation consists of three genotypes (¼ *PP*, ½ *Pp*, and ¼ *pp*) but only two phenotypes (¾ purple and ¼ white).

The **Punnett square method**, named after a famous geneticist of the early 1900s, R. C. Punnett, is an intuitive way to predict the genotypes and phenotypes of offspring. Figure 12-4 shows how to use a Punnett square to determine the proportion of offspring that arise from the self-fertilization of a flower that is heterozygous for color (or from any two gametes that are heterozygous for a single trait). This figure also provides the fractions that allow you to calculate the same outcomes based on probability theory. As you use these "genetic bookkeeping" techniques, keep in mind that, in a real experiment, the offspring will

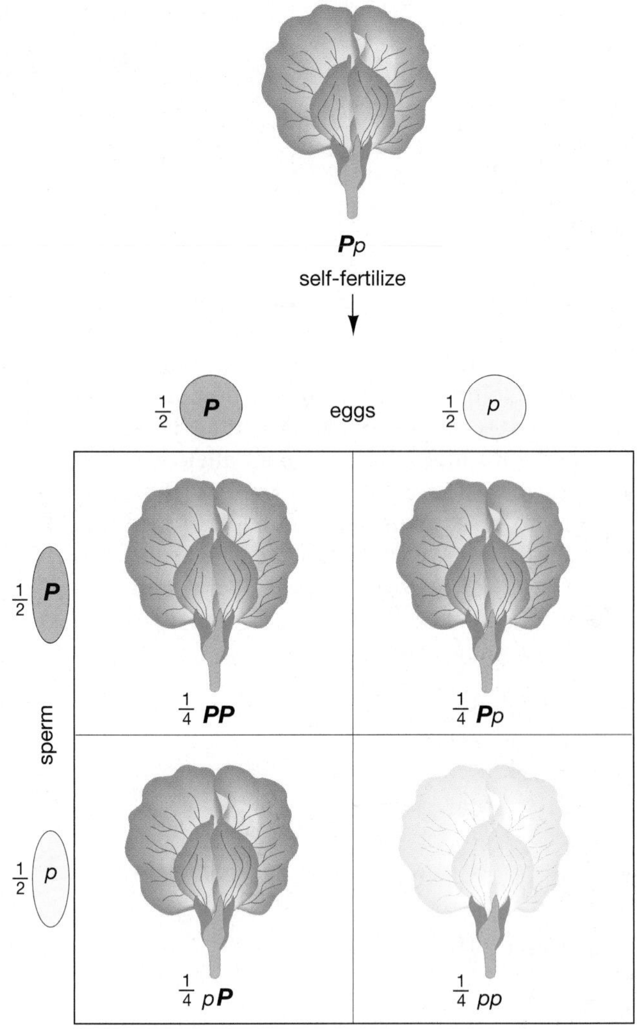

Figure 12-4 The Punnett square method
The Punnett square method allows you to predict both genotypes and phenotypes of specific crosses; here we use it for a cross between plants that are heterozygous for a single trait, flower color.
(1) Assign letters to the different alleles; use uppercase for dominant and lowercase for recessive.
(2) Determine all the types of genetically different gametes that can be produced by the male and female parents.
(3) Draw the Punnett square, with each row and column labeled with one of the possible genotypes of sperm and eggs, respectively. (We have included the fractions of these genotypes with each label.)
(4) Fill in the genotype of the offspring in each box by combining the genotype of sperm in its row with the genotype of the egg in its column. (We have placed the fractions in each box.)
(5) Count the number of offspring with each genotype. (Note that *Pp* is the same as *pP*.)
(6) Convert the number of offspring of each genotype to a fraction of the total number of offspring. In this example, out of four fertilizations, only one is predicted to produce the *pp* genotype, so $\frac{1}{4}$ of the total number of offspring produced by this cross is predicted to be white. To determine phenotypic fractions, add the fractions of genotypes that would produce a given phenotype. For example, purple flowers are produced by $\frac{1}{4}$ *PP* + $\frac{1}{4}$ *Pp* + $\frac{1}{4}$ *pP*, for a total of $\frac{3}{4}$ of the offspring.

occur only in *approximately* the predicted proportions. These fluctuations in the data reflect the fact that each egg is fertilized by only one sperm. So, some of the sperm that are produced do not get the chance to fertilize an egg. In addition, some eggs may not be fertilized. Let's consider an example. We know that each time a baby is conceived, it has a 50:50 chance of being a boy or a girl. So, in a family with six children, we expect to see three boys and three girls. However, because of the randomness of fertilization, families with six boys or six girls do occur. Only when we examine many families with six children do we discover the 50:50 ratio of girls to boys.

Mendel's Hypothesis Can Be Used to Predict the Outcome of New Types of Single-Trait Crosses

You have probably recognized that Mendel used the scientific method, observing results and formulating a hypothesis based on them. The scientific method has a third step: to use the hypothesis to predict the results of other experiments and to see if those experiments support or refute the hypothesis. For example, if the hybrid F_1 flowers have one allele for purple and one for white (*Pp*), then Mendel could predict the outcome of cross-fertilizing these *Pp* plants with homozygous recessive white plants (*pp*). Can you? Mendel predicted that there would be equal numbers of *Pp* (purple) and *pp* (white) offspring, and this is indeed what he found.

This type of experiment also has practical uses. Cross-fertilization of an individual with an unknown genotype but a dominant phenotype (in this case, a purple flower) with a homozygous recessive individual (a white flower) is called a **test cross**. The progeny of individuals with these genotypes can test whether the unknown genotype is homozygous or heterozygous (in this case, for purple color). When crossed with a homozygous recessive (*pp*), a homozygous dominant (*PP*) produces all phenotypically dominant offspring, whereas a heterozygous dominant (*Pp*) yields offspring with both dominant and recessive phenotypes in a 1:1 ratio:

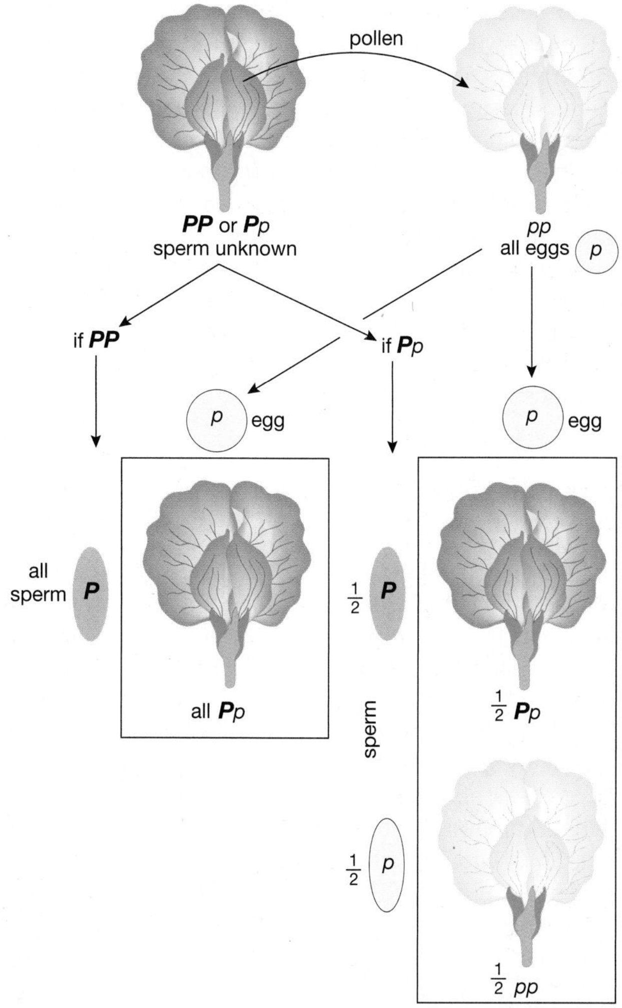

TRAIT	DOMINANT FORM	RECESSIVE FORM
Seed shape	smooth	wrinkled
Seed color	yellow	green
Pod shape	inflated	constricted
Pod color	green	yellow
Flower color	purple	white
Flower location	at leaf junctions	at tips of branches
Plant size	tall (1.8 to 2 meters)	dwarf (0.2 to 0.4 meters)

Figure 12-5 Traits of pea plants that Mendel studied

3 How Are Multiple Traits on Different Chromosomes Inherited?

Mendel Hypothesized That Genes on Different Chromosomes Are Inherited Independently

Having determined the **inheritance** modes for single traits, Mendel then turned to the more complex question of multiple traits. Again, he used a variety of traits in peas (Fig. 12-5). He began by crossbreeding plants that differed in two traits—for example, seed color (yellow or green) and seed shape (smooth or wrinkled). From other crosses of plants with these traits, Mendel already knew that the smooth allele of the seed shape gene (*S*) is dominant to the wrinkled allele (*s*). In addition, the yellow allele of the seed color gene (*Y*) is dominant to the green allele (*y*) (see Fig. 12-5). He crossed a true-breeding plant with smooth, yellow seeds (*SSYY*) to a true-breeding plant with wrinkled, green seeds (*ssyy*). All the F_1 offspring, therefore, were genotypically *SsYy*. They also all had the same phenotype: smooth, yellow seeds. Allowing these F_1 plants to self-fertilize, Mendel found that the F_2 generation consisted of 315 plants with smooth, yellow seeds, 101 with wrinkled, yellow seeds, 108 with smooth, green seeds, and 32 with wrinkled, green seeds, a ratio of about 9:3:3:1. The F_2 generations produced from other crosses of gametes that were heterozygous for two traits had similar phenotypic ratios.

Mendel realized that these results could be explained if the genes for seed color and seed shape are inherited independently of each other and do not influence each other during gamete formation. For each trait, ¾ of the offspring should show the dominant phenotype (*SS* and *Ss* genotypes) and ¼ should show the recessive trait (ss), producing a 3:1 ratio. Independent combination of two 3:1 ratios yields a 9:3:3:1 ratio (Fig. 12-6). We can see from Mendel's results that this is just what happened. There were 423 plants with smooth seeds (of either color) to 133 with wrinkled ones (about 3:1) and 416 plants with yellow seeds (of either shape) to 140 with green ones (about 3:1). Figure 12-6 shows how a Punnett square can be used to determine the outcome of a cross between gametes that are heterozygous for two traits and how two independent 3:1 ratios combine to form an overall 9:3:3:1 ratio.

The independent inheritance of two or more distinct traits is called the **law of independent assortment**. It

(a)

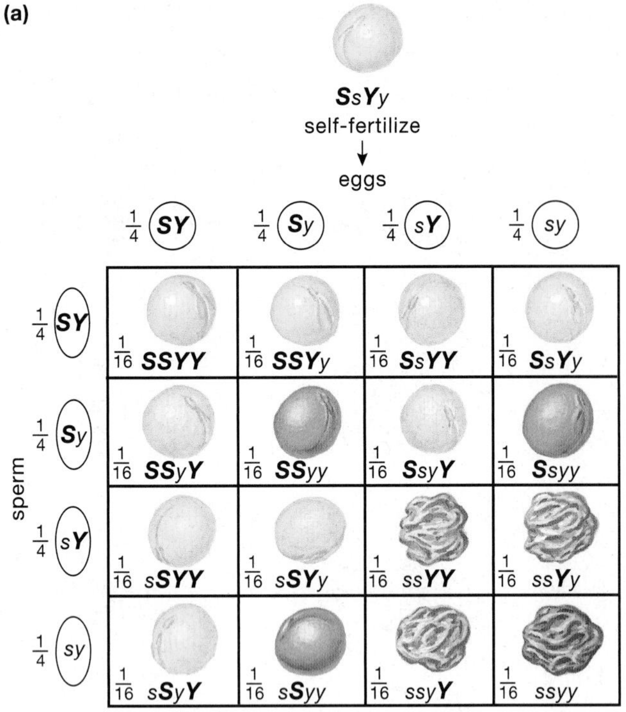

(b)

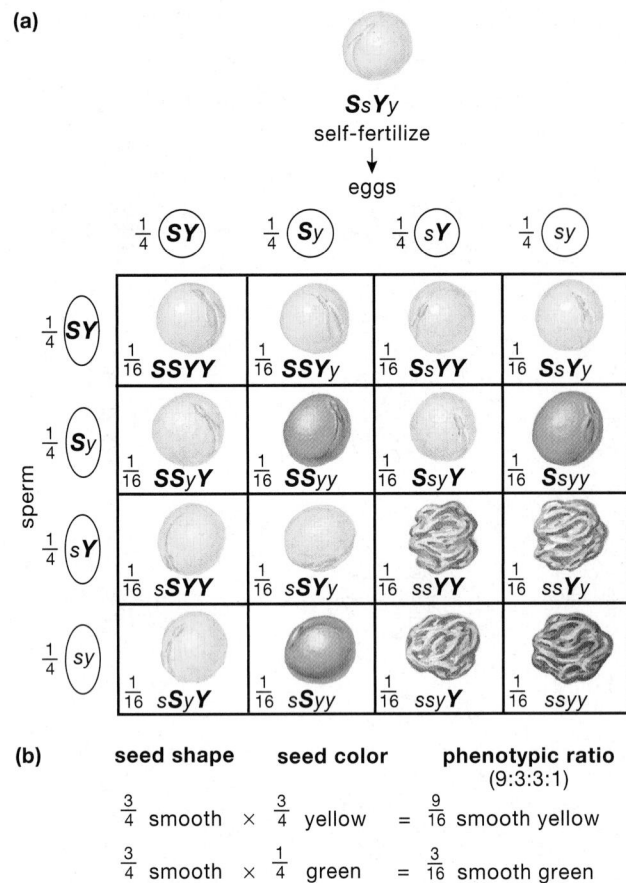

seed shape		seed color		phenotypic ratio (9:3:3:1)
$\frac{3}{4}$ smooth	×	$\frac{3}{4}$ yellow	=	$\frac{9}{16}$ smooth yellow
$\frac{3}{4}$ smooth	×	$\frac{1}{4}$ green	=	$\frac{3}{16}$ smooth green
$\frac{1}{4}$ wrinkled	×	$\frac{3}{4}$ yellow	=	$\frac{3}{16}$ wrinkled yellow
$\frac{1}{4}$ wrinkled	×	$\frac{1}{4}$ green	=	$\frac{1}{16}$ wrinkled green

Figure 12-6 Predicting genotypes and phenotypes for a cross between gametes that are heterozygous for two traits
Here we are working with both seed color and shape, with yellow (*Y*) dominant to green (*y*), and smooth (*S*) dominant to wrinkled (*s*). ***(a)*** Punnett square analysis. In this cross, both parents are heterozygous for each trait (or a single individual heterozygous for both traits self-fertilize). There are now 16 boxes in the Punnett square. In addition to predicting all the genotypic combinations, the Punnett square predicts ¾ yellow seeds, ¼ green seeds, ¾ smooth seeds, and ¼ wrinkled seeds, just as we would expect from crosses made of each trait separately. ***(b)*** Probability theory can be used to predict phenotypes that result from a cross between gametes that are heterozygous for two traits. The fraction of genotypes from each sperm and egg combination is illustrated within each box of the Punnett square. Adding the fractions for the same genotypes will give the genotypic ratios. Converting each genotype to a phenotype and then adding their numbers reveals that ¾ of the offspring will be smooth and ¼ will be wrinkled and that ¾ will be yellow and ¼ will be green. Multiplying these independent probabilities produces predictions for the phenotype of offspring. These ratios are identical to those generated by the Punnett square.

MEDIATUTOR 12.2 Dihybrid Crosses

states that the alleles of one gene may be distributed to the gametes independently of the alleles for other genes. Recall the events of meiosis in Chapter 11; how might independent assortment of alleles for two different genes occur? *Hint:* Peas have seven pairs of homologous chromosomes. Independent assortment will occur when the genes being studied are on different chromosomes. When paired homologous chromosomes line up during metaphase I, which homologue faces which pole of the cell is random, and the orientation of one homologous pair does not influence other pairs. Therefore, when the homologues separate during anaphase I, the alleles of genes on different chromosomes are distributed, or "assorted," independently (Fig. 12-7). Amazingly, each of the several traits Mendel had chosen to study happened to be controlled by a single gene located on a different chromosome.

In an Unprepared World, Genius May Go Unrecognized

In 1865 Gregor Mendel presented his theories of inheritance to the Brünn Society for the Study of Natural Science, and they were published the following year. His paper did not mark the beginning of genetics. In fact, it didn't make any impression at all on the study of biology during his lifetime. Mendel's experiments, which eventually spawned one of the most important scientific theories in all of biology, simply vanished from the scene. Apparently, very few biologists read his paper, and those who did failed to recognize its significance or discounted it because it contradicted prevailing ideas of inheritance.

It was not until 1900 that three biologists—Carl Correns, Hugo de Vries, and Erich Tschermak—working independently and knowing nothing of Mendel's work, rediscovered the principles of inheritance. No doubt to their intense disappointment, when they searched the scientific literature before publishing their results, they found that Mendel had scooped them more than 30 years earlier. To their credit, they graciously acknowledged the important work of the Augustinian monk, who had died in 1884.

4 How Are Genes Located on the Same Chromosome Inherited?

Gregor Mendel knew nothing about the physical nature of genes or chromosomes. Only later, after scientists had seen chromosomes in a microscope and figured out how they are distributed during mitosis and meiosis, did scientists learn that chromosomes are the vehicles of inheritance. It also became obvious that there are many more traits (and therefore many more genes) than there are chromosomes. Genes are parts of chromosomes and each chromosome contains many genes. This fact has important implications for inheritance that Mendel's experiments did not reveal.

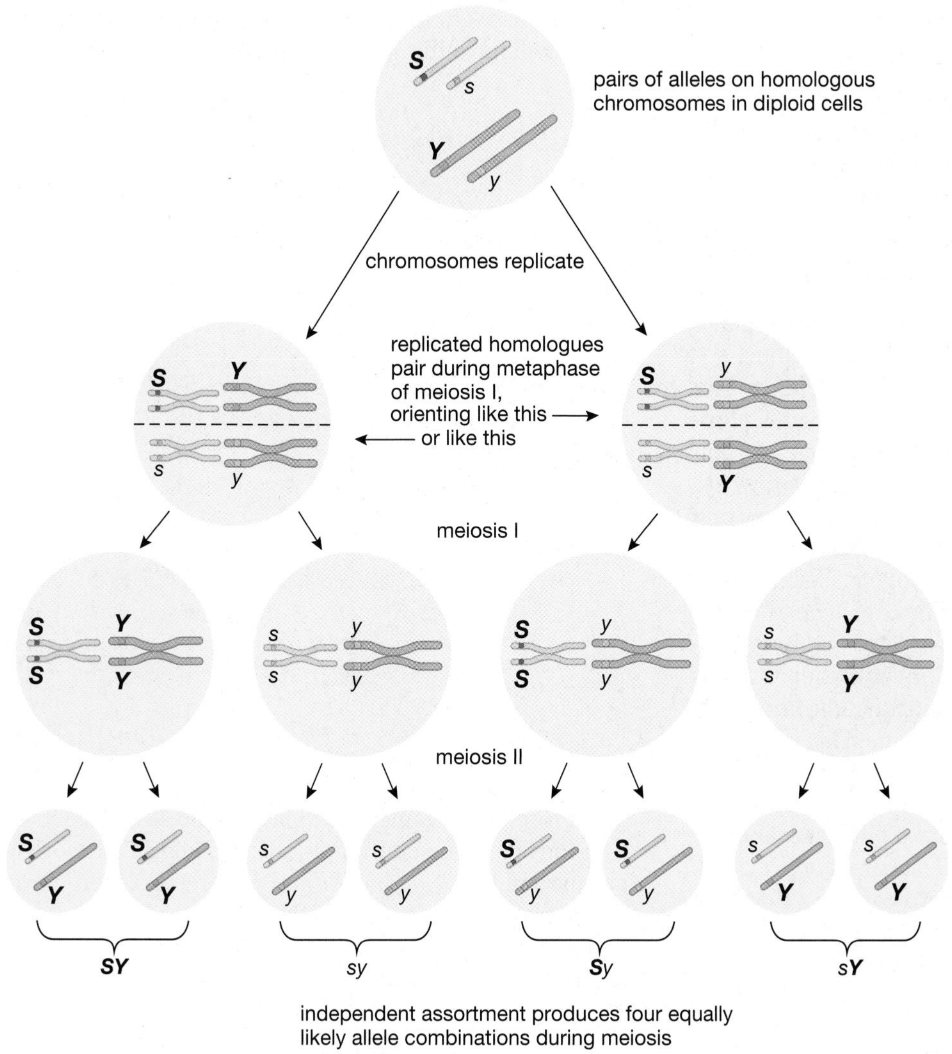

Figure 12-7 Independent assortment of alleles
Chromosome movements during meiosis produce independent assortment of alleles of two different genes. Because meiosis occurs in many reproductive cells in a plant, each combination is equally likely to occur. Therefore, an F_1 plant would produce gametes in the predicted proportions ¼ *SY*, ¼ *sy*, ¼ *sY*, and ¼ *Sy*.

Genes on the Same Chromosome Tend to Be Inherited Together

If chromosomes assort independently during meiosis I, then only genes located on *different chromosomes* will assort independently into gametes. In contrast, genes on the *same chromosome* tend to be inherited together. These genes are *linked* together, because they are encoded on the same DNA double helix within the chromosome. Genetic **linkage** is the inheritance of certain genes as a group because they are on the same chromosome. Alleles of linked genes tend to be inherited together and do not assort independently. One of the first pairs of linked genes to be discovered were found in the sweet pea, a different species from Mendel's garden pea. In sweet peas, the gene for flower color and the gene for pollen grain shape are carried on the same chromosome. Thus, the alleles for these genes normally assort *together* into gametes during meiosis and are inherited together as a result.

Consider a hybrid sweet pea plant with red flowers and long pollen that has the following chromosomes:

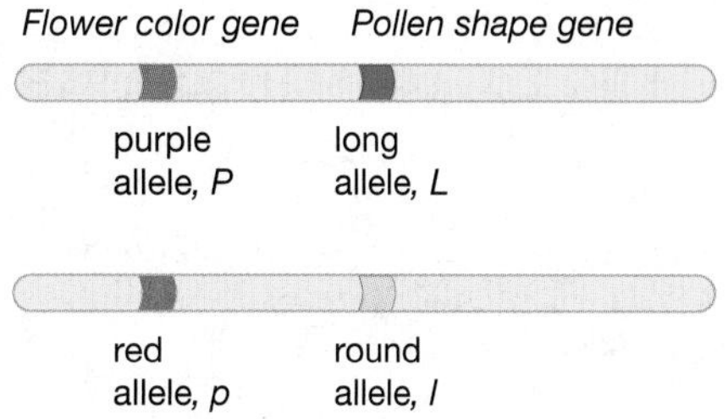

Note that the purple allele of the flower color gene and the long allele of the pollen shape gene are located on one homologous chromosome. The red allele of the flower color gene and the round allele of the pollen shape gene are located on the other homologue. The gametes produced by this sweet pea plant are likely to have *either* purple and long alleles *or* red and round alleles. This pattern of inheritance breaks the law of independent assortment, which states that the alleles for flower color and pollen shape should segregate independently of one another into the gametes. Instead, they tend to stick together through meiosis.

Recombination Can Create New Combinations of Linked Alleles

Although they tend to be inherited together, genes on the same chromosome do not always stay together. In the sweet pea cross just described, for example, the F_2 generation will commonly include a few plants in which the genes for flower color and pollen shape have been inherited as if they were not linked. That is, some of the gametes will have the red, round alleles and some will have the purple, long alleles. Why aren't the alleles for genes on the same chromosome always inherited together?

As you learned in Chapter 11, during prophase I of meiosis, the nonsister chromatids of homologous chromosomes intertwine at sites called *chiasmata*. At chiasmata, segments of homologous chromosomes are exchanged with each other, a process called **crossing over**. The resulting recombination is common; at least one exchange between each homologous chromosome pair occurs during each meiosis. The exchange of corresponding segments of DNA during crossing over produces new allele combinations on both homologous chromosomes. Then, when homologous chromosomes separate at anaphase I, the chromosomes that each haploid daughter cell receives will have different sets of alleles from those of the parent cell.

Crossing over during meiosis explains the appearance of new combinations of alleles that were previously linked. Let us revisit that sweet pea plant, this time during the early stages of meiosis I, when the chromosomes have duplicated and homologous chromosomes are paired up:

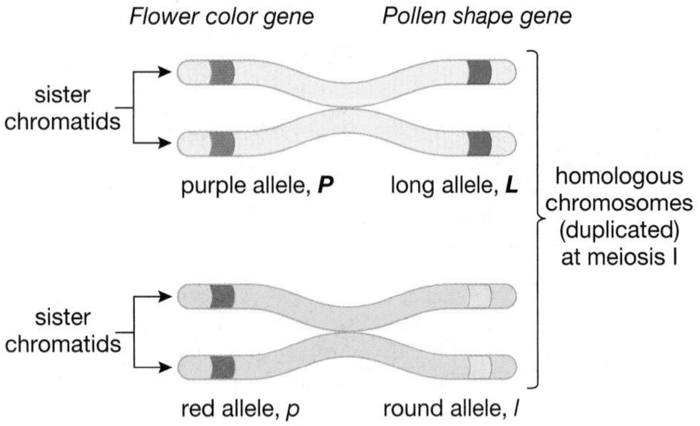

Each homologous chromosome will have one or more regions where crossing over will occur. Imagine that one cross-over occurs between the genes for flower color and pollen shape. Thus, nonsister chromatids of homologous chromosomes exchange alleles for flower color:

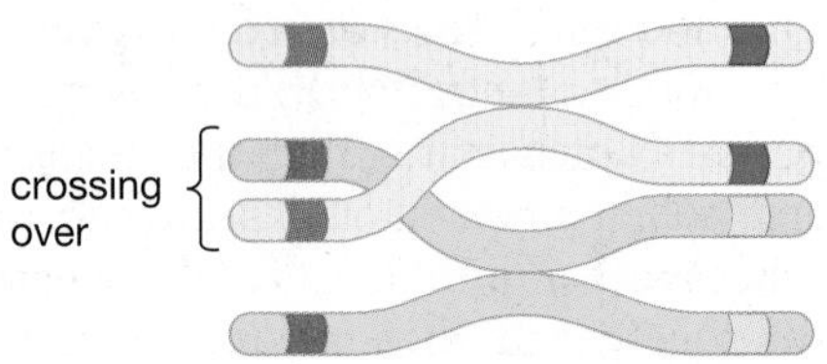

At anaphase I, the separated homologous chromosomes will have this gene composition:

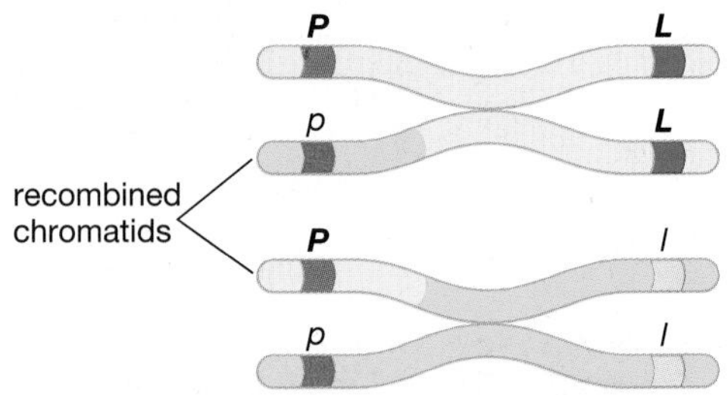

Four types of chromosomes are then distributed to the haploid daughter cells during meiosis II:

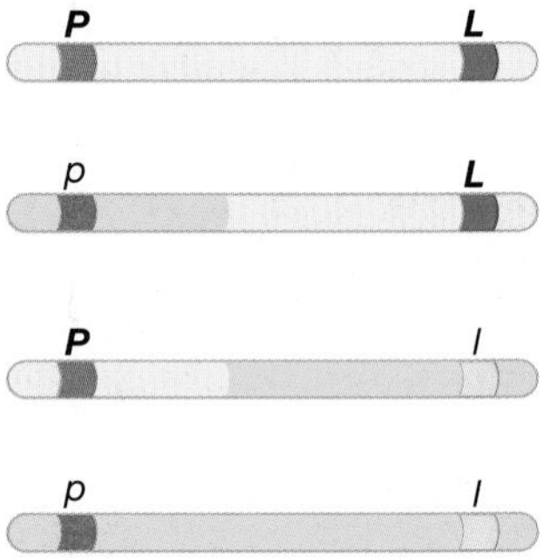

Therefore, some gametes will be produced with each of four chromosome configurations: *PL* and *pl* (the original parental types), and *Pl* and *pL* (recombined chromosomes). By exchanging DNA between homologous chromosomes, this **genetic recombination** creates new combinations of alleles.

5 How Is Sex Determined, and How Are Sex-Linked Genes Inherited?

In mammals and many insects, males have the same number of chromosomes as females do, but one "pair," the **sex chromosomes**, is very different in appearance and genetic composition. Females have two identical sex chromosomes, called *X chromosomes*, whereas males have one X chromosome and one *Y chromosome* (Fig. 12-8). Although the Y chromosome normally carries far fewer genes than does the X chromosome, a small part of both sex chromosomes is homologous. As a result, the X and Y chromosomes pair up during prophase of meiosis I and separate during anaphase I. All other chromosomes, which occur in pairs that have identical appearance in both males and females, are called **autosomes**. Numbers of chromosomes vary tremendously among species, but there is always only one pair of sex chromosomes. For example, the fruit fly *Drosophila* has four pairs of chromosomes (three pairs of autosomes and one pair of sex chromosomes), humans have 23 pairs (22 pairs of autosomes and one pair of sex chromosomes), and dogs have 39 pairs (38 pairs of autosomes and one pair of sex chromosomes).

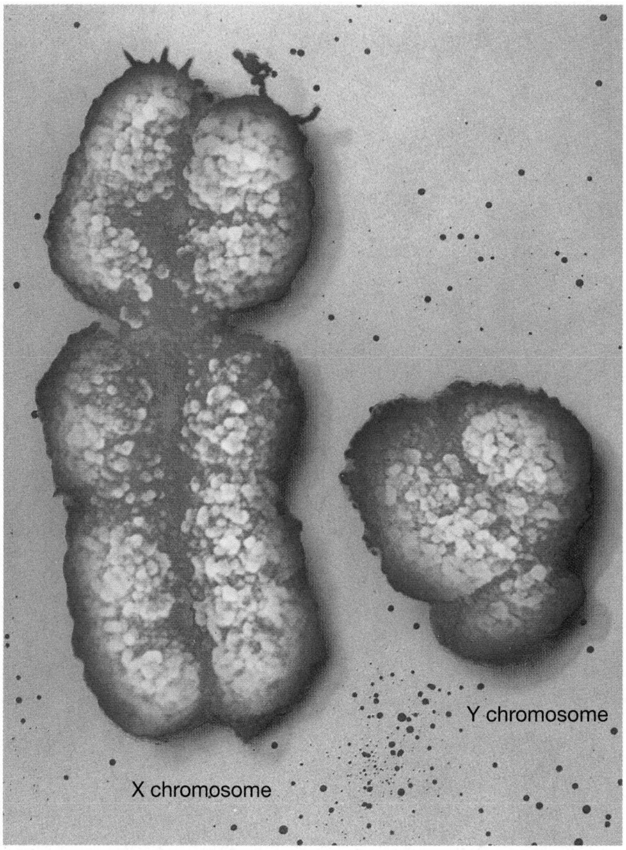

Figure 12-8 Photomicrograph of human sex chromosomes
Notice the small size of the Y chromosome, which carries relatively few genes.

For organisms in which males are XY and females are XX, the sex chromosome carried by the sperm determines the sex of the offspring (Fig. 12-9). During sperm formation, the sex chromosomes segregate, and each sperm receives either the X or the Y chromosome (plus one member of each pair of autosomes). The sex chromosomes also segregate during egg formation, but because females have two X chromosomes, every egg receives one X chromosome (along with one member of each pair of autosomes). An offspring is male if an egg is fertilized by a Y-bearing sperm or female if an egg is fertilized by an X-bearing sperm.

Sex-Linked Genes Are Found Only on the X or Only on the Y Chromosome

Genes that are on one sex chromosome but not on the other are said to be **sex-linked**. In many animals, the Y chromosome carries only a few genes. The human Y chromosome contains about 20 genes, most of which determine maleness or male fertility. In contrast, the X chromosome contains many genes that have nothing to do with specifically female traits. The human X chromosome contains about 1500 genes that have no counterpart on the Y chromosome, including the genes for color vision, blood clotting, and certain structural proteins in

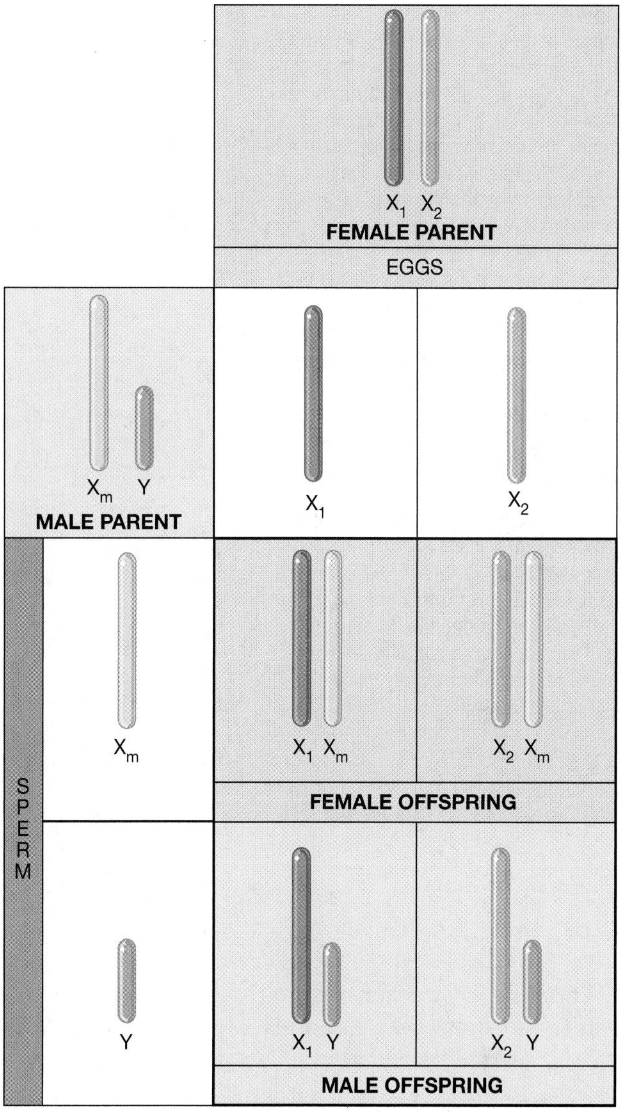

Figure 12-9 Sex determination in mammals
Male offspring receive their Y chromosome from the father; female offspring receive the father's X chromosome (labeled X_m). Both male and female offspring receive an X chromosome (either X_1 or X_2) from the mother.

muscles. Therefore, because they have two X chromosomes, females can be either homozygous or heterozygous for genes on the X chromosome. Normal dominant versus recessive relationships among alleles will be expressed. Males, in contrast, must fully express all alleles they have on their single X chromosome, whether those alleles are dominant or recessive. For this reason, in humans most cases of recessive traits encoded by genes on the X chromosome, such as color blindness, hemophilia, and certain types of muscular dystrophy, occur in males. We will return to this concept later in the chapter.

How does sex linkage affect inheritance? Let's look at the first example of sex linkage to be discovered, the inheritance of eye color in the fruit fly *Drosophila*. Because these flies are small, reproduce rapidly, are easy to grow in the laboratory, and have few chromosomes,

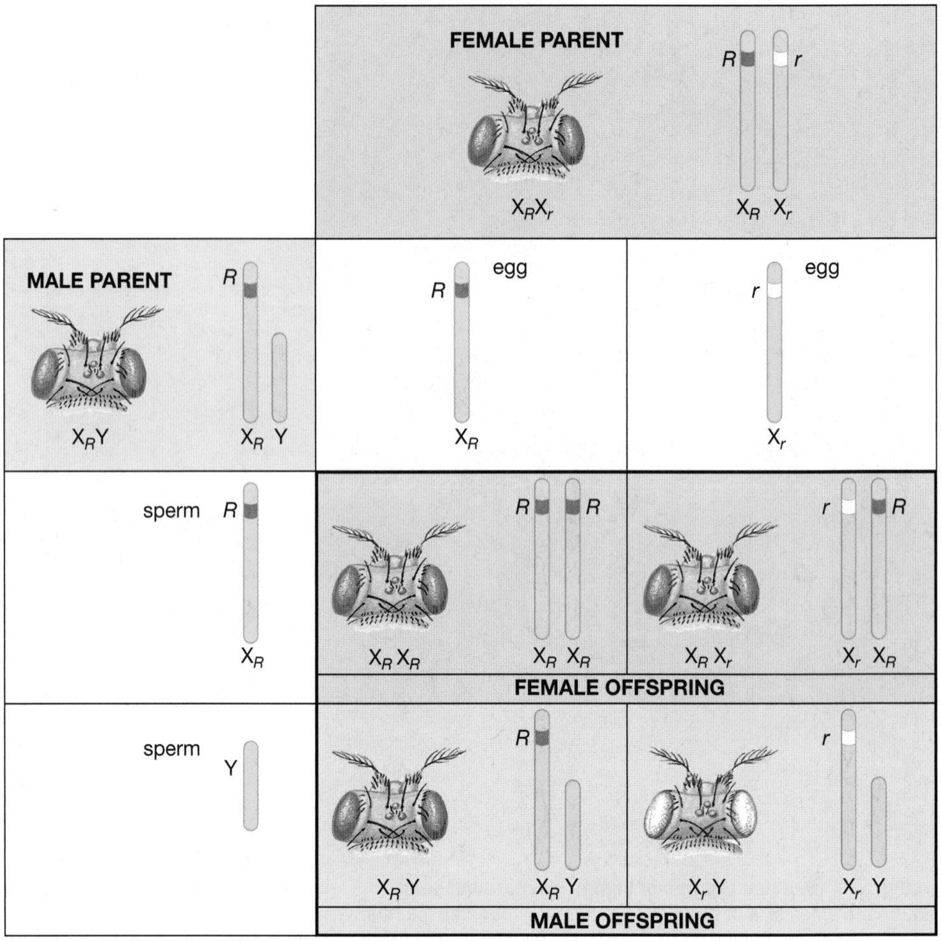

Figure 12-10 Sex-linked inheritance of eye color in fruit flies The gene for eye color is located on the X chromosome; the Y chromosome does *not* contain an eye color gene. Normal red eyes (*R*) is dominant to the mutant allele for white eyes (*r*). When a white-eyed male is mated to a homozygous red-eyed female, all members of the F_1 generation have red eyes, since all progeny received an *R* gene from their mother. All F_1 females are heterozygous, receiving the *r* allele from their father and the *R* allele from their mother. Male offspring receive only the *R* allele from their mother. Here is the result of crossing the F_1 progeny with each other. The single *R* allele of the F_1 male parent is passed on to his daughters. Thus, all the daughters will have red eyes. The father passes on his Y chromosome, which lacks an eye color gene, to his sons. Half the sons will have white eyes; the other half inherited an *R* allele from their F_1 mother.

Drosophila have been favored subjects for genetics studies for more than a century. Normally, *Drosophila* have red eyes. In the early 1900s, researchers in the laboratory of Thomas Hunt Morgan at Columbia University discovered a male fly with white eyes. This white-eyed male was mated to a true-breeding, red-eyed female. All the resulting offspring were red-eyed flies, suggesting that white eye color (*r*) is recessive to red (*R*). The F_2 generation, however, was a surprise: There were nearly equal numbers of red-eyed males and white-eyed males, but no females had white eyes! A test cross of the F_1 red-eyed females to true-breeding red-eyed females and to the original white-eyed male yielded roughly equal numbers of red-eyed and white-eyed males and females.

From these data, could you figure out how eye color is inherited? Morgan made the brilliant hypothesis that *the gene for eye color must be located on the X chromosome and that the Y chromosome has no corresponding gene* (Fig. 12-10). In the F_1 generation, both male and female offspring received an X chromosome, with its *R* allele for red eyes, from their mother. The F_1 males received a Y chromosome from their father with no allele for eye color, and so the males had an *R-* genotype, and red-eyed phenotype. (Here "-" indicates no gene for eye color on the Y chromosome.) The F_1 females received the father's X chromosome with its *r* allele, so the females had an *Rr* genotype and red-eyed phenotype. Thus, both male and female F_1 offspring had red eyes.

Crossing two F_1 flies, *R- X Rr,* resulted in an F_2 generation with the chromosome distribution shown in Figure 12-10. All the F_2 females received one X chromosome from their F_1 male parent, with its *R* allele, and therefore had red eyes. All the F_2 males inherited their single X chromosome from their F_1 female parent, heterozygous for eye color (*Rr*). So the F_2 males had a 50:50 chance of receiving an X chromosome with the *R* allele or one with the *r* allele. *With no corresponding gene on the Y chromosome, the F_2 males displayed the phenotype determined by the allele on the X chromosome.* Therefore, half the F_2 males had red eyes, and half had white eyes.

6 What Are Some Variations on the Mendelian Theme?

In our discussion of patterns of inheritance thus far, we have made some major simplifying assumptions: that each trait is completely controlled by a single gene, that there are only two possible alleles of each gene, and that one allele is completely dominant to the other, recessive, allele. Most traits, however, are influenced in more varied and subtle ways than this.

Alleles May Display Incomplete Dominance in Which the Phenotype of Heterozygotes Is Intermediate Between the Phenotypes of the Homozygotes

In his pea experiments, Mendel encountered a particularly simple situation—heterozygotes with one dominant allele had the same phenotype as homozygotes with two dominant alleles—but this is often not the case. In snapdragons, for example, crossing homozygous red-flowered plants (*RR*) with homozygous white-flowered ones (*R′R′*) produces F_1 hybrids that do not bear red flowers. Instead, the F_1 (*RR′*) flowers are pink. When the heterozygous phenotype is intermediate between the two homozygous phenotypes, the pattern of inheritance is called **incomplete dominance.** In genetic nomenclature, incompletely dominant alleles are often given uppercase symbols with a superscript to denote the different alleles. In the snapdragon example, the two alleles are denoted as *R* and *R′*. The apparent blending of the phenotype in cases of incomplete dominance is not the result of any change in alleles. In homozygous flowers of the F_2 generation, the red and white colors are as strong as ever (Fig. 12-11). The F_2 offspring include about ¼ red (*RR*), ½ pink (*RR′*), and ¼ white (*R′R′*) flowers.

What causes incomplete dominance? Recall that each gene encodes the information for production of a protein. The different alleles of a gene may encode proteins with different functional capacity. For example, in snapdragons, the *R* allele codes for an active enzyme that catalyzes the formation of red pigment; the *R′* allele codes for a defective, nonfunctional enzyme. When only one copy of the *R* allele is present, less red pigment is produced and the resulting flower appears pink. Plants with the *RR* genotype produce lots of red pigment and have red flowers, but those with the *R′R′* genotype produce no pigment and have white flowers.

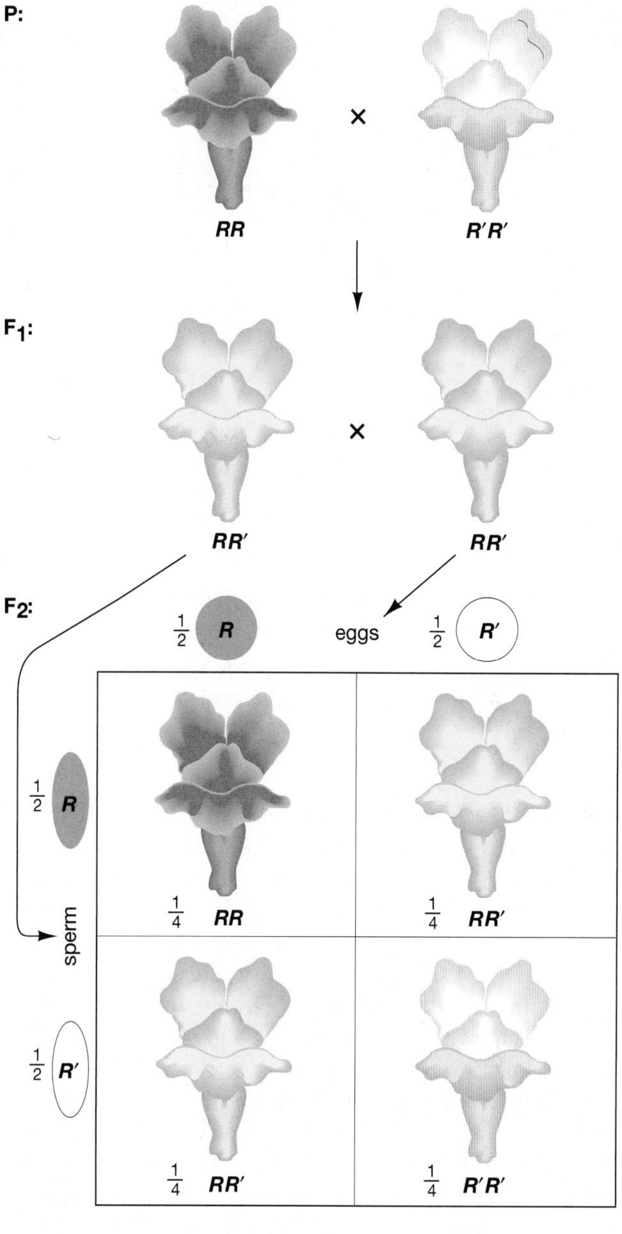

Figure 12-11 Incomplete dominance
(a) The inheritance of flower color in snapdragons is an example of incomplete dominance. (In such cases, we will use capital letters for both alleles, here *R* and *R′*.) Hybrids (*RR′*) have pink flowers, whereas the homozygotes are red (*RR*) or white (*R′R′*). Because heterozygotes can be distinguished from homozygous dominants, the distribution of phenotypes in the F_2 generation (¼ red: ½ pink: ¼ white) is the same as the distribution of genotypes (¼ *RR*: ½ *RR′*: ¼ *R′R′*).

A Single Gene May Have Multiple Alleles

Alleles arise through mutation, and the same gene in different individuals may have different mutations, each producing a new allele. If we could sample all the individuals of a species, we would typically find dozens of alleles for every gene. Thus, rather than each gene having only two alleles, most genes have **multiple alleles.** One eye-color gene in *Drosophila*, for example, has more than a thousand alleles. Depending on how they are combined, eye color in fruit flies can be white, yellow, orange, pink, brown, or red. Remember, however, that an *individual* diploid organism can have only two alleles for a given gene. If the alleles are the same, the organism is homozygous and if the alleles are different, the organism is heterozygous.

The blood types of humans are another example of multiple alleles. The blood types A, B, AB, and O arise as a result of three different alleles (for simplicity, we will designate them *A, B,* and *o*) of a single gene located on chromosome 9. This gene codes for an enzyme responsible for adding sugar molecules to the ends of glycoproteins that protrude from the surfaces of red blood cells. Alleles *A* and *B* code for enzymes that add different sugars to the glycoproteins (we will call the resulting molecules glycoproteins *A* and *B*, respectively). Allele *o* codes for a nonfunctional enzyme that

Table 12-1 Human Blood Group Characteristics

Blood Type	Genotype	Red Blood Cells	Has Plasma Antibodies to:	Can Receive Blood from:	Can Donate Blood to:	Frequency in U.S.
A	*AA* or *Ao*	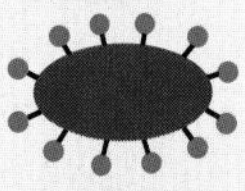 A glycoprotein	B glycoprotein	A or O (no blood with B glycoprotein)	A or AB	40%
B	*BB* or *Bo*	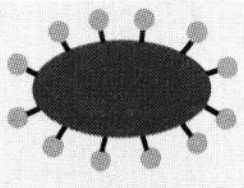 B glycoprotein	A glycoprotein	B or O (no blood with A glycoprotein)	B or AB	10%
AB	*AB*	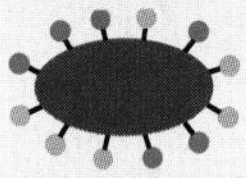 Both A and B glycoproteins	Neither A nor B glycoprotein	AB, A, B, O (universal recipient)	AB	4%
O	*oo*	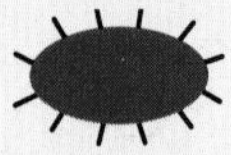 Neither A nor B glycoprotein	Both A and B glycoproteins	O (no blood with A or B glycoprotein)	O, AB, A, B (universal donor)	46%

doesn't add any sugar molecule. An individual may have one of six genotypes: *AA, BB, AB, Ao, Bo,* or *oo* (Table 12-1). Alleles *A* and *B* are dominant to *o*. Therefore, individuals with genotypes *AA* or *Ao* have only type A glycoproteins and have type A blood. Those with genotypes *BB* or *Bo* synthesize only type B glycoproteins and have type B blood. Homozygous recessive *oo* individuals lack both types of glycoproteins and have type O blood. In individuals with type AB blood, both enzymes are present, so that both A and B glycoproteins are present on red blood cells. Since the AB combination of alleles produces a different phenotype from AA or BB, *A* and *B* alleles are said to be *codominant* to one another. In **codominance**, *both phenotypes are expressed in heterozygotes*.

People make antibodies to the type of glycoprotein(s) that they lack. These antibodies are proteins in blood plasma that bind to foreign glycoproteins by recognizing different end-sugar molecules. The antibodies cause red blood cells that bear the foreign glycoproteins to clump together and also to rupture. The resulting clumps and fragments can clog small blood vessels and damage vital organs such as the brain, heart, lungs, or kidneys. This means that blood type must be determined and matched carefully before a blood transfusion is made.

Type O blood, lacking any end sugars, is not attacked by antibodies in A, B, or AB blood, so it can be transfused safely to all other blood types. (The antibodies present in transfused blood become too diluted to cause problems.) People with type O blood are called "universal donors." But O blood carries antibodies to both A and B glycoproteins, so type O individuals can receive transfusions of only type O blood. Can you predict the blood type of people called "universal recipients"? Table 12-1 summarizes blood types and transfusion characteristics.

Many Traits Are Influenced by Several Genes

If you look around your class, you are likely to see people of varied heights, skin colors, and body builds, to consider just a few obvious traits. Traits such as these are not governed by single genes but are influenced by interactions among two or more genes—as well as by interactions with the environment. Many traits, such as human skin color, may have several phenotypes or even seemingly continuous variation that cannot be split up into convenient, easily defined categories. This is an example of **polygenic inheritance**, a form of inheritance in which the interaction of two or more genes contribute to a single phenotype.

Eye color is another good example of polygenic inheritance. The color of the human iris varies from very pale blue through green to almost black, depending on the amount of pigment, yellowish brown melanin, in the outer layer of the iris. If there is little or no pigment here, the iris appears blue, owing to the same type of light scattering that makes the sky appear blue. People with more melanin in the outer iris may have eyes that appear green (blue plus yellowish brown), brown, or almost black. At least two (and probably more) genes direct the synthesis of melanin in the front of the iris, with each gene having two alleles that show incomplete

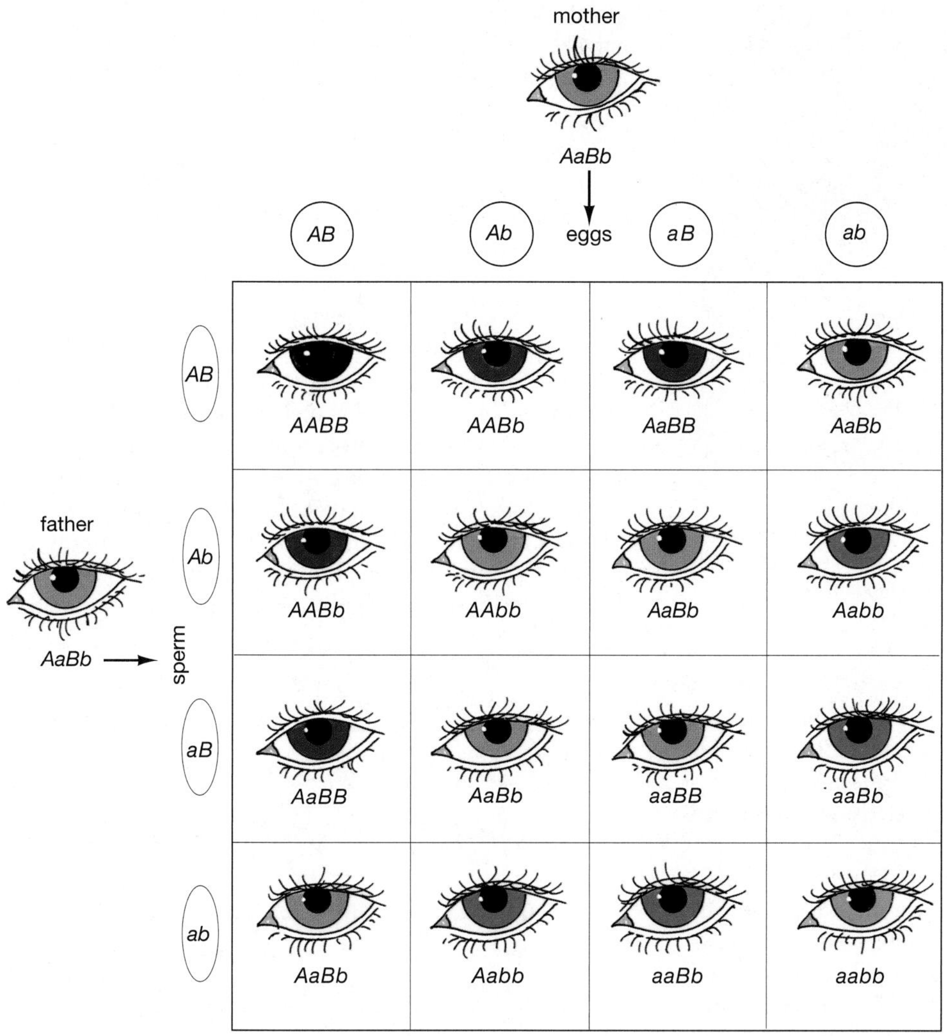

Figure 12-12 Human eye color At least two separate genes, each with two incompletely dominant alleles, determine human eye color. A brown-eyed man and a brown-eyed woman, each heterozygous for both genes, could have children with five different eye colors, ranging from light blue (no dominant alleles) through light brown (two dominant alleles) to almost black (all four dominant alleles).

dominance. The simplest scheme of two genes can create five shades of eye colors from one set of parents (Fig. 12-12).

As you might imagine, the more genes that contribute to a single trait, the greater the number of phenotypes and the finer the distinctions among them. When more than three pairs of genes contribute to a trait, differences between phenotypes are small, and it is extremely difficult to classify the phenotypes reliably. At least three or four genes control skin color, producing the observed variation from very dark to very light coloration.

Single Genes Typically Have Multiple Effects on Phenotype

We have just seen that a single phenotype may result from the interaction of several genes. The reverse is also true: Single genes commonly have multiple phenotypic effects, a phenomenon called **pleiotropy**. A good example is the SRY gene, discovered in 1990 on the Y chromosome. The SRY gene (short for "sex-determining region of the Y chromosome") codes for a protein that activates other genes; those genes in turn code for proteins that switch on male development in an embryo. Under the influence of the genes activated by the SRY protein, sex organs develop into testes. The testes, in turn, secrete sex hormones that stimulate the development of both internal and external male reproductive structures, such as the epididymis, seminal vesicles, prostate gland, penis, and scrotum. So, a single gene can affect many body structures.

The Environment Influences the Expression of Genes

An organism is not just the sum of its genes. In addition to the genotype, the environment in which an organism lives profoundly affects its phenotype. A striking example of environmental effects on gene action occurs in the Himalayan rabbit, which, like the Siamese cat, has pale body fur but black ears, nose, tail, and feet (Fig. 12-13). The Himalayan rabbit actually has the genotype for black fur all over its body. However, the enzyme that

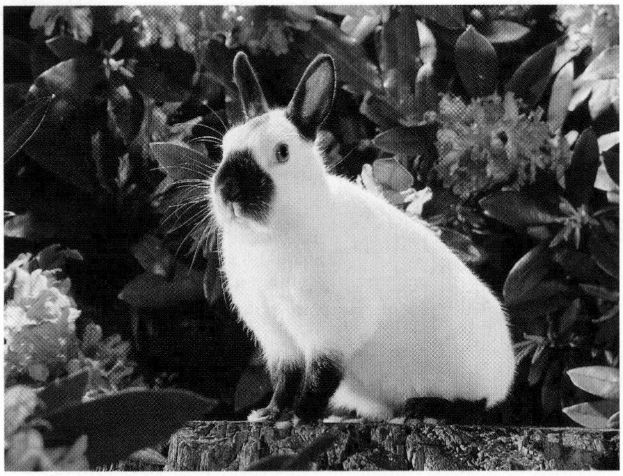

Figure 12-13 Environmental influence on phenotype
The expression of the gene for black fur in the Himalayan rabbit is a simple case of interaction between genotype and environment in producing a particular phenotype. Cool areas (nose, ears, and feet) allow expression of the gene for black fur.

produces the black pigment is inactive at temperatures above about 34 °C (93 °F). At typical ambient temperatures, extremities such as the ears and feet are cooler than the rest of the body, so black pigment can be produced there. The main body surface is warmer than 34 °C, so the fur there is pale.

Most environmental influences are more complicated and subtle. The complexity of environmental influences is particularly true of human characteristics. The polygenic trait of skin color is modified by the environmental effects of sun exposure, which stimulates melanin production. Height, another polygenic trait, can be reduced by poor nutrition, an environmental factor.

Intelligence, too, has both genetic and environmental components. Dozens of studies have compared IQ levels in people of varying family relationships. Even when separated at birth and raised in different environments, identical twins make similar scores on IQ tests, although not as similar as identical twins who have been raised together. Brothers and sisters who are not twins differ more than do twins but are still fairly similar. Thus, the more genetically related two people are, the more similar their IQ scores. These findings indicate that IQ testing capability has a genetic component. However, unrelated people who have been raised together as children (for example, adoptees) show more similarity on IQ tests than do unrelated people reared apart. Therefore, it is fair to say that *both heredity and environment play major roles in the development of various mental abilities* and almost certainly other personality traits as well.

The interactions between complex genetic systems and varied environmental conditions can create a continuum of phenotypes that defies analysis into genetic and environmental components. The human generation time is long, and the number of offspring per couple is small. Add to these factors the countless subtle ways in which people respond to their environments, and you can see why it may be exceedingly difficult to determine the precise genetic basis of complex human traits such as intelligence or musical or athletic ability.

7 How Are Human Genetic Disorders Investigated?

Because experimental crosses with humans are out of the question, human geneticists search medical, historical, and family records to study past crosses. Records extending across several generations can be arranged in the form of family **pedigrees**, diagrams that show the genetic relationships among a set of related individuals (Fig. 12-14). Careful analysis of pedigrees can reveal whether a particular trait is inherited in a dominant, recessive, or sex-linked pattern. Since the mid-1960s, analysis of human pedigrees, combined with molecular

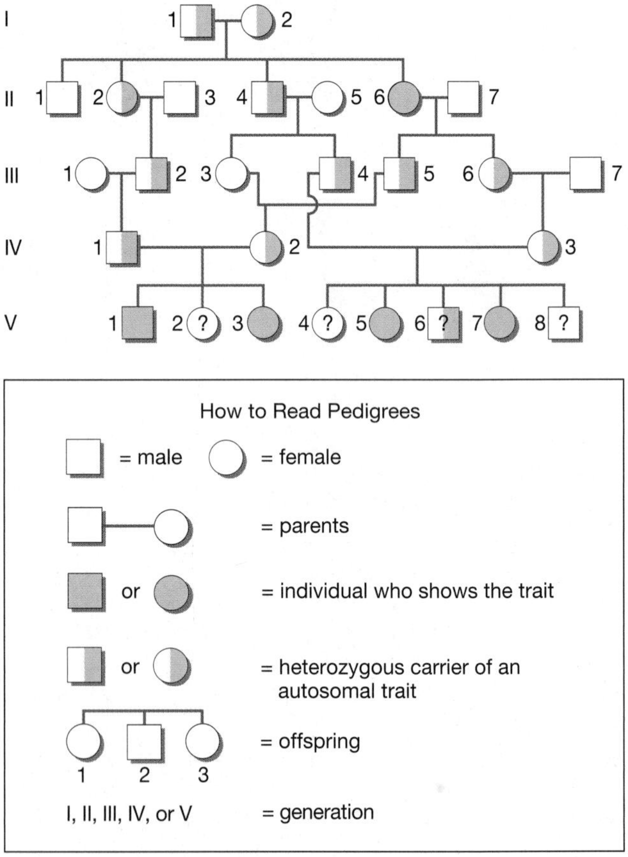

Figure 12-14 A family pedigree
This pedigree is for a recessive trait, such as albinism. Both of the original parents are carriers. Because the allele for albinism is rare, pairing between carriers is an unlikely event. However, the chance that each of two *related* people will carry a rare recessive allele (inherited from a common ancestor) is much higher than normal. As a result, pairings between cousins or even closer relations are the cause of a disproportionate number of recessive diseases. In this family, pairings between cousins occurred three times—between III 3 and III 5, III 4 and IV 3, and IV 1 and IV 2.

genetic technology, have produced great strides in understanding human genetic diseases. For instance, geneticists now know the genes responsible for dozens of inherited diseases, such as sickle-cell anemia, prostate cancer, and cystic fibrosis. Research in molecular genetics promises to increase our ability to predict genetic diseases and perhaps even to cure them, a topic we explore further in Chapter 13.

8 How Are Human Disorders Caused by Single Genes Inherited?

Many common human traits, such as freckles, long eyelashes, cleft chin, and widow's peak hairline, are inherited in a simple Mendelian fashion; that is, each trait appears to be controlled by a single gene with a dominant and a recessive allele. Here, we will concentrate on a few examples of medically important genetic disorders and the ways in which they are transmitted from one generation to the next.

Many Human Genetic Disorders Are Caused by Recessive Alleles

The human body depends on the integrated actions of thousands of enzymes and other proteins. A mutation in the gene coding for one of these enzymes can impair or destroy enzyme function. However, the presence of one normal allele may generate enough functional enzyme or other protein to enable heterozygotes to be phenotypically indistinguishable from homozygotes with two of the normal alleles.

Therefore, for many genes, a normal allele, which encodes a functional protein, is dominant to a mutant allele, which encodes a nonfunctional protein. The reverse is also true: For these genes, a mutant allele is recessive to a normal allele. In other words, the disease occurs only in individuals who inherit two copies of the mutant allele.

Heterozygous individuals are **carriers** of the recessive genetic trait: They are phenotypically normal but can pass on their defective recessive allele to their offspring. Geneticists estimate that each of us carry a recessive allele of 5 to 15 genes, each of which would cause a serious genetic defect if we were homozygous for the recessive allele. Every time we have a child, there is a 50:50 chance that we will pass on the defective allele. However, an unrelated man and woman are unlikely to possess a defective allele in the *same* gene. So they are unlikely to produce a child who is homozygous for a recessive allele for one of these genetic diseases. Related couples, however (especially first cousins or closer), have inherited some of their genes from recent common ancestors. Therefore, they are much more likely to carry a defective allele in the same gene and, if such couples bear children, they have a 1 in 4 chance of having a child affected by the genetic disease or disorder (see Fig. 12-14).

Albinism Results from a Defect in Melanin Production

An enzyme called *tyrosinase* is needed to produce melanin, the dark pigment in skin cells. The gene that encodes tyrosinase is called *TYR*. If an individual is homozygous for a mutant *TYR* allele that encodes a defective tyrosinase enzyme, albinism results (Fig. 12-15). *Albinism* in humans and other mammals is manifested

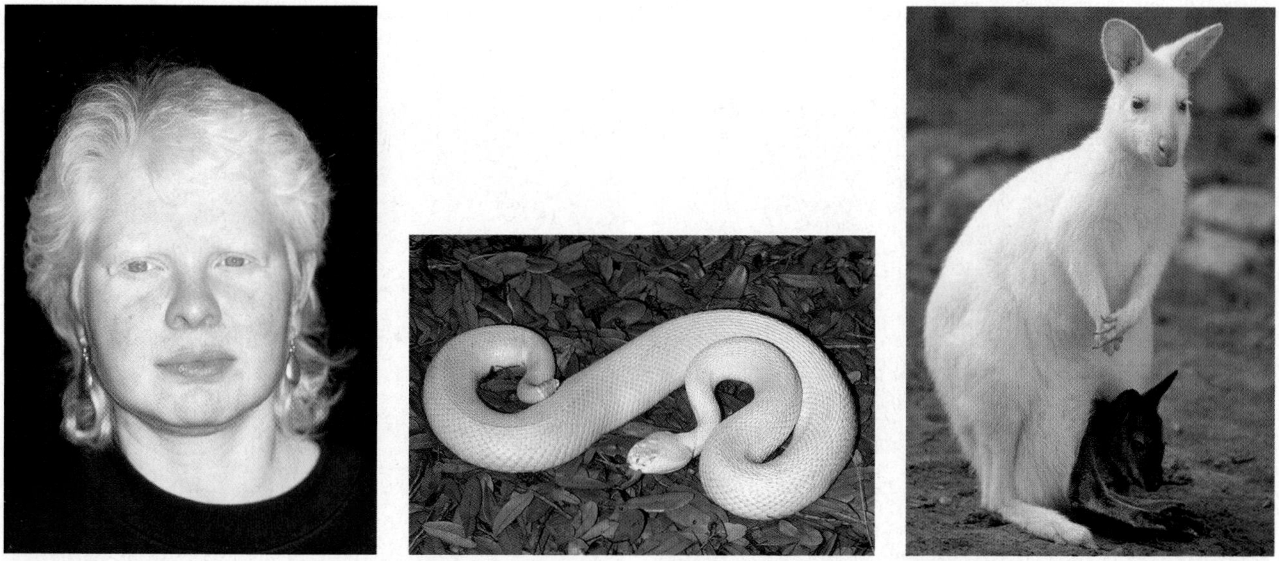

(a) (b) (c)

Figure 12-15 Albinism
Albinism is controlled by a single, recessive allele. Melanin is found throughout the animal kingdom, and albinos of many species have been observed: ***(a)*** human, ***(b)*** rattlesnake, and ***(c)*** wallaby. The female wallaby, having mated with a normally pigmented male, carries a normally colored offspring in her pouch.

as white skin and hair and pink eyes (because blood vessels in the retina are visible in the absence of masking melanin pigment).

Sickle-Cell Anemia Is Caused by a Defective Allele for Hemoglobin Synthesis

Sickle-cell anemia, a recessive disease in which defective hemoglobin is produced, results from a specific mutation in the hemoglobin gene. The hemoglobin protein, which gives red blood cells their color, transports oxygen in the blood. In sickle-cell anemia, the substitution of one nucleotide results in a single incorrect amino acid at a crucial position in hemoglobin, altering the properties of the hemoglobin molecule. Under conditions of low oxygen (such as in muscles during exercise), masses of hemoglobin molecules in each red blood cell clump together. The clumps force the red blood cell out of its normal disk shape (Fig. 12-16a) into a longer, sickle shape (Fig. 12-16b). The sickled cells are more fragile than normal red blood cells, making them likely to break; they also tend to aggregate, clogging capillaries. Tissues "downstream" of the block do not receive enough oxygen or have their wastes removed. This lack of bloodflow can cause pain, especially in joints. Paralyzing strokes can occur if blocks occur in blood vessels in the brain. Other symptoms include anemia, because so many red blood cells are destroyed, and reduced immunity to disease. Although heterozygotes have about half normal and half abnormal hemoglobin, they usually have few sickled cells and are not disabled by the disease—in fact, many world-class athletes are heterozygotes for the sickle-cell allele.

About 8% of the African-American population is heterozygous for sickle-cell anemia, reflecting a genetic legacy of African origins. In some regions of Africa, 15% to 20% of the population carries the allele. The prevalence of the sickle-cell allele in Africa is explained by the fact that heterozygotes have some resistance to the parasite that causes malaria. We will explore this benefit further in Chapter 13. Individuals of Mediterranean, Middle Eastern, Central or South American, and East Indian ancestry also have increased risk of carrying the allele responsible for sickle-cell anemia.

Heterozygous carriers of the sickle-cell allele can be detected by a blood test, but until fairly recently there was no way to learn whether a fetus is homozygous or heterozygous. If two heterozygous carriers have children, every conception will have a 1 in 4 chance of producing a child who is homozygous for the sickle-cell allele. This child will have sickle-cell anemia. DNA analysis techniques can distinguish the normal hemoglobin allele from the sickle-cell allele. Analysis of fetal cells allows medical geneticists to diagnose sickle-cell anemia in fetuses. In addition, analysis of a single cell from an in vitro fertilized embryo can identify embryos homozygous for the sickle-cell anemia allele (see "Health Watch: Prenatal Genetic Screening" in Chapter 13).

Many Human Genetic Disorders Are Caused by Dominant Alleles

Many normal physical traits, including cleft chin and freckles, are inherited as dominant traits. Many serious genetic diseases are also caused by dominant alleles. For example, the On-line Mendelian Inheritance in Man database lists about 1300 recessive disease genes and about 1300 dominant disease genes. For dominant diseases to be passed on to offspring, at least one parent must be suffering from the disease. Thus, dominant diseases must not affect the ability of the affected individual to survive into adulthood and to have children.

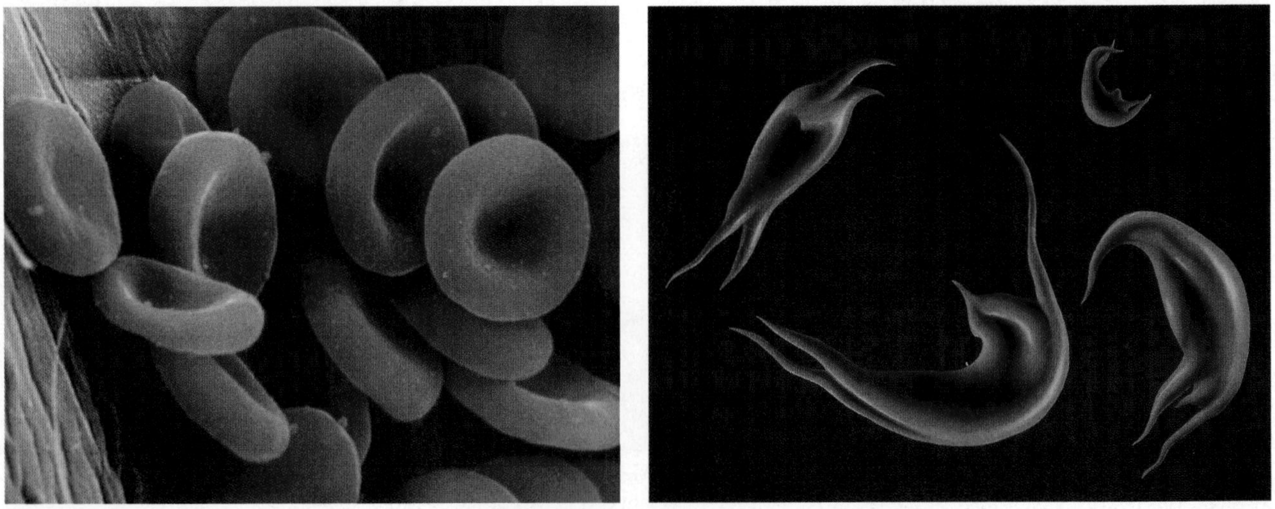

(a) (b)

Figure 12-16 Sickle-cell anemia
(a) Normal red blood cells are disc-shaped with indented centers. ***(b)*** Sickled red blood cells in a person with sickle-cell anemia occur when blood oxygen is low. In this shape they are fragile and tend to clump together, clogging capillaries.

Alternately, the dominant allele may result from a new mutation formed in a normal individual's eggs or sperm. In this case, neither parent suffers from the disease.

How can a mutant allele be dominant to the normal allele? Some dominant alleles produce an abnormal protein that interferes with the function of the normal one. For example, some proteins must link together into long chains in order to perform their function in the cell. The abnormal protein may enter into a chain but prevent addition of new protein "links." These shortened chain fragments may be unable to properly perform a needed function. Other dominant alleles may encode proteins that carry out new, toxic reactions. Finally, dominant alleles may encode a protein that is overactive, performing its function at inappropriate times and places.

One dominant genetic disease in humans is **Huntington disease**, which causes a slow, progressive deterioration of parts of the brain. This unstoppable deterioration results in the loss of motor coordination, flailing movements, personality disturbances, and eventual death. The symptoms of Huntington disease typically do not appear until 30 to 50 years of age. Therefore, many people pass the allele to their children before they suffer the first symptoms. In 1983 both molecular genetic technology and painstaking pedigree analysis were used to localize the Huntington gene to a small part of chromosome 4 (see "Scientific Inquiry: In Search of the Huntington Gene"). Geneticists finally isolated the Huntington gene in 1993 and a few years later identified the gene's product, a protein dubbed "huntingtin." The normal function of the protein remains unknown. The mutant protein forms large aggregates in nerve cells that appear to result, ultimately, in nerve cell death.

Some Human Disorders Are Sex-Linked

As we described earlier, the X chromosome contains many genes that have no counterpart on the Y chromosome. Because males have only one X chromosome, they have only one allele for each of these genes. This single allele will be expressed with no possibility of its activity being "hidden" by expression of another allele, a phenomenon called *sex-linked inheritance.*

A son receives his X chromosome from his mother and passes it only to his daughters. Thus, sex-linked disorders caused by a recessive allele have a unique pattern of inheritance. Such disorders appear far more frequently in males and typically skip generations: An affected male passes the trait on to a phenotypically normal, carrier daughter, who in turn bears affected sons. The most familiar genetic defects due to recessive alleles of X-chromosome genes are *red-green color blindness* (Fig. 12-17) and **hemophilia** (Fig. 12-18). Hemophilia is caused by a recessive allele on the X chromosome that results in a deficiency of one of the proteins needed for blood clotting. People with hemophilia bleed excessively from a wound or from mild damage to internal structures, and they bruise easily. Hemophiliacs often have anemia due to blood loss. But even before modern treatment with clotting factor derived from donor blood, some hemophiliac males

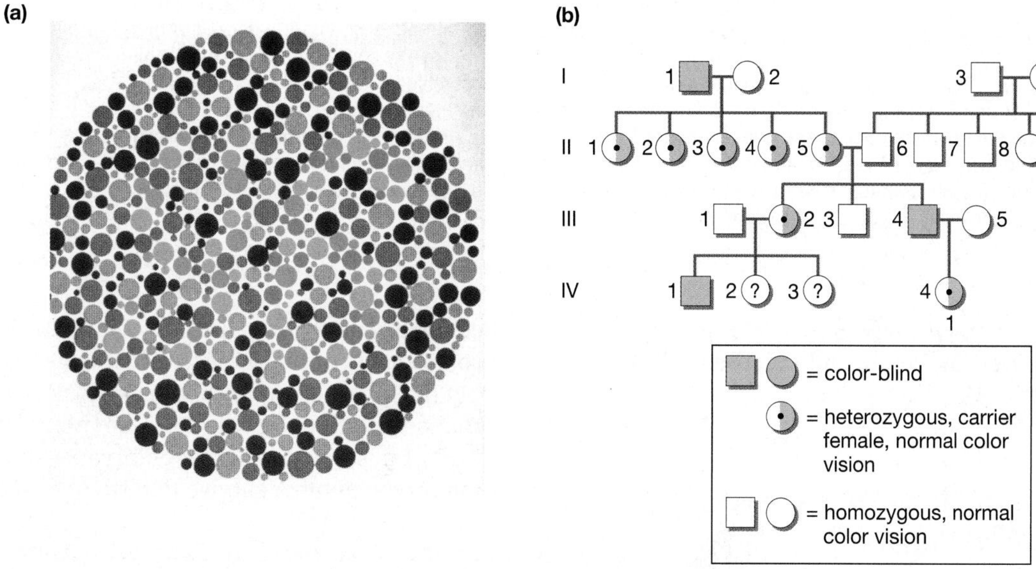

Figure 12-17 Color blindness, a sex-linked recessive trait
(a) This figure, called an *Ishihara chart* after its inventor, distinguishes color-vision defects. People with red-deficient vision see a 6, and those with green-deficient vision see a 9. People with normal color vision see 96. ***(b)*** Pedigree of one of the authors (G. Audesirk: III 4), showing sex-linked inheritance of red-green color blindness. Both the author and his maternal grandfather (I 1) are color deficient; his mother and her four sisters carry the trait but have normal color vision. This pattern of more-common occurrence in males and of transmission from affected male to carrier female to affected male is typical of sex-linked recessive traits.

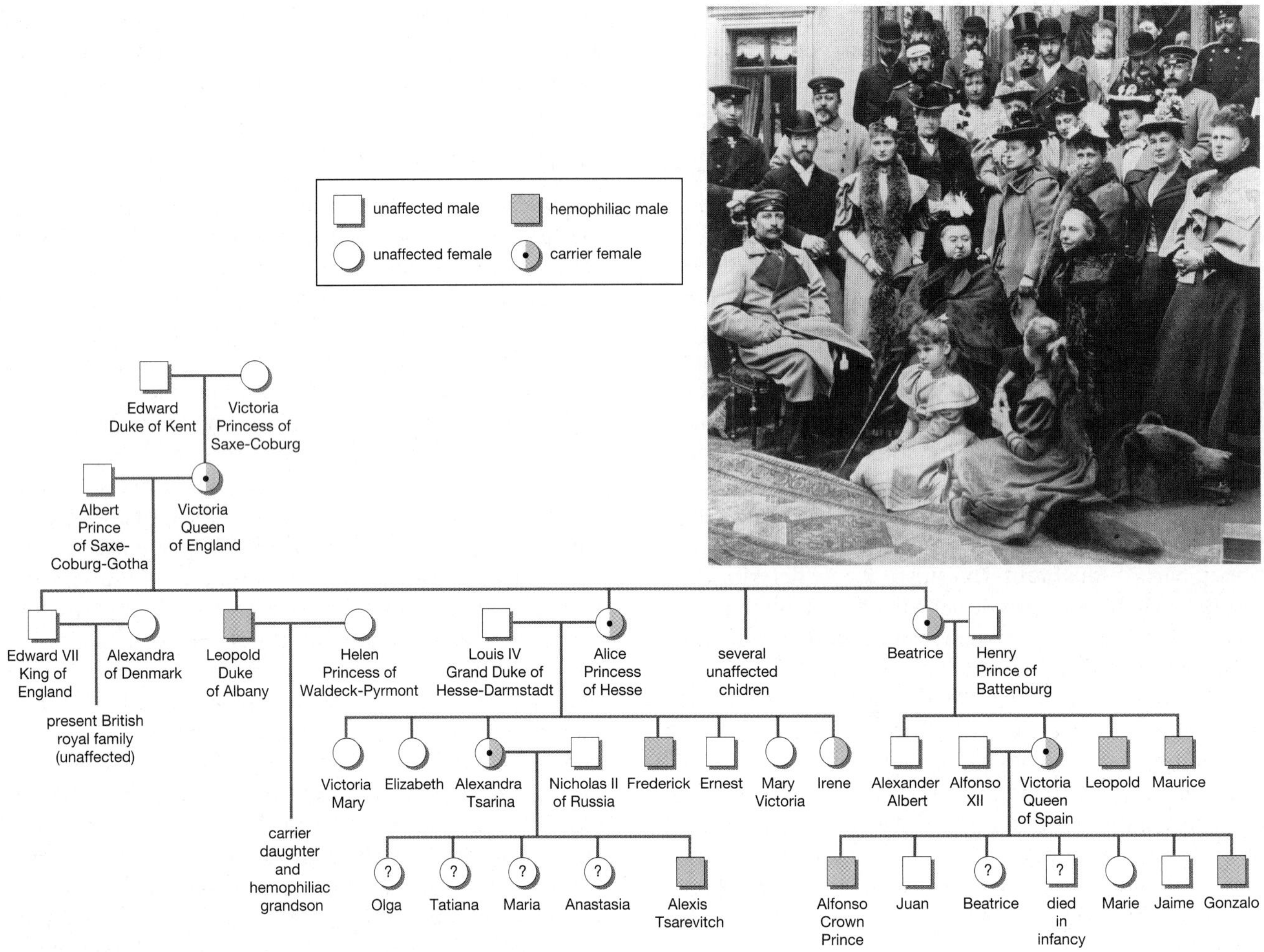

Figure 12-18 Hemophilia among the royal families of Europe
A famous genetic pedigree involves the transmission of sex-linked hemophilia from Queen Victoria of England (seated center front, with cane, 1885) to her offspring and eventually to virtually every royal house in Europe. Because Victoria's ancestors were free of hemophilia, the hemophilia allele must have arisen as a mutation either in Victoria herself or in one of her parents (or as a result of marital infidelity). Extensive intermarriage among royalty spread Victoria's hemophilia allele throughout Europe. Her most famous hemophiliac descendant was great-grandson Alexis, tsarevitch (crown prince) of Russia. The Tsarina Alexandra (Victoria's granddaughter) believed that the monk Rasputin, and no one else, could control Alexis's bleeding. Rasputin may actually have used hypnosis to cause Alexis to cut off circulation to bleeding areas by muscular contraction. The influence that Rasputin had over the imperial family may have contributed to the downfall of the tsar during the Russian Revolution. In any event, hemophilia was not the cause of Alexis's death; he was killed with the rest of this family by the Bolsheviks (Communists) in 1918.

survived to pass on their defective allele to their daughters, who carried the allele and could pass it to their sons.

9 How Do Errors in Chromosome Number Affect Humans?

In Chapter 11 we examined the intricate mechanisms of meiosis, which ensure that each sperm and egg receive only one chromosome from each homologous pair. Not surprisingly, this elaborate dance of the chromosomes occasionally misses a step, resulting in gametes that have too many or too few chromosomes. Such errors in meiosis, called **nondisjunction**, can affect the number of either sex chromosomes or autosomes. Most embryos that arise from the fusion of gametes with abnormal chromosome numbers spontaneously abort, accounting for 20% to 50% of all miscarriages. Some embryos with abnormal chromosome number survive to birth or beyond.

Some Genetic Disorders Are Caused by Abnormal Numbers of Sex Chromosomes

Because the X and Y chromosomes pair up during meiosis, sperm usually carry either an X or a Y chromosome. Nondisjunction of sex chromosomes during male meiosis produces sperm that have 22 autosomes and no sex chromosome. Such sperm are called are "O" (lack-

Scientific Inquiry
In Search of the Huntington Gene

As a young boy in 1860, George Huntington encountered a mother and daughter, both emaciated and uncontrollably twitching, grimacing, twisting, and bowing. "I stared in wonderment, almost in fear," he later wrote. Huntington went on to study medicine at Columbia University and spent the rest of his life investigating this disease, now called Huntington disease (HD) in his honor. Dr. Huntington showed that HD occurs in about half the children of an affected parent, evidence that it is inherited as a dominant allele of a gene.

More than 100 years later, in the late 1960s, Leonore Wexler was diagnosed with HD. In response, her husband, Milton Wexler, a clinical psychologist, founded the Hereditary Disease Foundation in 1968, hoping to find a cure or an effective treatment for the disease that eventually killed his wife. Nancy Wexler, a young woman of 22, was profoundly affected by her mother's suffering and especially by her mother's progressive loss of mental function. Trained as a clinical psychologist like her father, Dr. Wexler decided to devote her life to finding a cure for HD and now serves as president of the Hereditary Disease Foundation. She taught herself genetics by attending lectures, workshops, and seminars on the subject, by holding many informal discussions with geneticists, and by reading extensively. Dr. Wexler learned that in several villages near Lake Maracaibo in Venezuela, an unusually high percentage of the villagers died of HD. Visiting these villages, she found that about 1 in 5 people there were stricken with the disease (Fig. E12-1). All of them descended from a Portuguese sailor who married a local woman seven generations earlier, in the 1800s.

By 1981 Dr. Wexler began assembling a pedigree showing the relationships of this huge, genetically interconnected Venezuelan population, where families with 13 or more children are common. More than 10,000 individuals have been added to this extensive pedigree. Wexler also collected blood samples from the villagers with HD and shipped the samples to the laboratory of Dr. James Gusella at Massachusetts General Hospital in Boston. In 1993 Gusella's lab, using molecular genetic technology, identified a region of DNA near the tip of human chromosome 4 as the location of the gene for HD. Comparing the genetic information in the Venezuelan blood samples with the relationships depicted in the Venezuelan pedigree and others, Gusella's group devised a screening test. This test very accurately determines whether a person has inherited the gene for HD from an affected parent or is free of the "genetic time bomb," as Dr. Wexler describes the Huntington gene. Unfortunately, even though we know the specific gene responsible, no treatment has yet been discovered to slow or halt the progression of Huntington disease.

This lack of treatments may partly explain why Nancy Wexler has decided not to be tested to determine whether she has inherited her mother's Huntington gene. As Dr. Wexler explains, each individual must make a deeply personal decision as to whether the pain of knowing the gene is present outweighs the joy of knowing the gene is absent, or whether living with the anxiety of not knowing is preferable. The availability of this test is of special importance to an individual who may have inherited the Huntington gene but wants to have children who are unaffected by the disease. Genetic counseling can help the individual make informed, personal decisions about their reproduction options.

Figure E12-1 Nancy Wexler tracks the gene for Huntington disease
Nancy Wexler, right, with one of the many inhabitants of the Lake Maracaibo region of Venezuela who have Huntington disease.

ing any sex chromosome). Alternately, the sperm can have two sex chromosomes and be XX, YY, or XY. Nondisjunction of the X chromosome in female meiosis produces O or XX eggs instead of eggs with one X chromosome. When normal gametes fuse with these defective sperm or eggs, the zygotes have normal numbers of autosomes but abnormal numbers of sex chromosomes (Table 12-2). The most common abnormalities are XO, XXX, XXY, and XYY. (Genes on the X chromosome are essential to survival, and any embryo without at least one X chromosome spontaneously aborts very early in development.)

Turner Syndrome (XO)

About one in every 3000 phenotypically female babies has only one X chromosome, a condition known as **Turner syndrome**. Thus, approximately 47,000 individuals in the U.S. have Turner syndrome. At puberty, hormone deficiencies prevent XO females from menstruating or developing secondary sexual characteristics, such as enlarged breasts. Treatment with estrogen promotes physical development. However, because most individuals with Turner syndrome lack mature eggs, the hormone treatment does not induce menstruation or reverse the infertility typically seen in XO women. Additional characteristics of Turner syndrome

Table 12-2 Effects of Nondisjunction of the Sex Chromosomes During Meiosis

Nondisjunction in Father

Sex Chromosomes of Defective Sperm	Sex Chromosomes of Normal Egg	Sex Chromosomes of Offspring	Phenotype
0 (none)	X	XO	Female—Turner syndrome
XX	X	XXX	Female—Trisomy X
YY	X	XYY	Male—"XYY male"
XY	X	XXY	Male—Klinefelter syndrome

Nondisjunction in Mother

Sex Chromosomes of Normal Sperm	Sex Chromosomes of Defective Egg	Sex Chromosomes of Offspring	Phenotype
X	0 (none)	XO	Female—Turner syndrome
Y	0 (none)	YO	Dies as embryo
X	XX	XXX	Female—Trisomy X
Y	XX	XXY	Male—Klinefelter syndrome

include short stature, folds of skin around the neck, and increased risk of cardiovascular disease, kidney defects, and hearing loss. Since women with Turner syndrome have only one X chromosome, they display X-linked recessive disorders, such as hemophilia and color blindness, much more frequently than do XX women.

The differences between XO and XX women suggest that the inactivation of one X chromosome in XX females (Chapter 10), is not complete. Otherwise, XX, with only one "active" X chromosome and XO women, with only one X chromosome, should have identical features. Some genes on the inactivated X chromosome must be functional in XX females, preventing the Turner syndrome traits.

Trisomy X (XXX)

About one in every 1000 women have three X chromosomes. Thus, approximately 140,000 in the U.S. have three X chromosomes rather than two. Most such women have no detectable defects, except for a tendency to be tall and a higher incidence of below-normal intelligence. Unlike women with Turner syndrome, most **trisomy X** women are fertile and, interestingly enough, almost always bear normal XX and XY children. Some unknown mechanism must operate during meiosis to prevent the extra X chromosome from being included in the egg.

Klinefelter Syndrome (XXY)

About one male in every 1000 is born with two X chromosomes and one Y chromosome. Thus, approximately 134,000 individuals in the U.S. have XXY chromosomes, rather than XY (or XX). Most of these men go through life never realizing that they have an extra X chromosome. However, at puberty, some of these men show mixed secondary sexual characteristics, including partial breast development, broadening of the hips, and small testes. These symptoms are known as **Klinefelter syndrome**. XXY men are usually infertile because of low sperm count but are not impotent. They are usually diagnosed when the man and his partner seek medical help when they are unable to conceive a baby.

XYY Males

Another common type of sex chromosome abnormality is XYY, occurring in about one male in every 1000. Approximately 134,000 males in the U.S. have XYY chromosomes. You might expect that having an extra Y chromosome, which presumably has few genes, would not make very much difference, and this seems to be true in most cases. However, XYY males usually have high levels of testosterone, often have severe acne, and are tall (about two-thirds of XYY males are over 6 feet tall, compared with the average male height of 5 feet 9 inches). Some appear to have slightly lower scores on IQ tests than their brothers who have XY chromosome constitution. Debates continue about whether XYY males are genetically predisposed to violence. One estimate is that an XYY individual has a 24 times higher risk of behavioral problems or criminal activity. However, most studies only involve a few XYY individuals and many have used questionable statistics. In several countries men accused of murder have attempted to use their XYY makeup as a defense, like the insanity plea. They were not acquitted; only a tiny percentage of XYY males ever commit any sort of crime, so an extra Y chromosome certainly doesn't force anyone into a life of violence.

Earth Watch
Crops, Livestock, and Wild Genes

The science of genetics, which began in a monastery more than a century ago, is critically important to humankind today. Advances in the diagnosis, prevention, and cure of inherited diseases, advances in forensic science, and the creation of bioengineered drugs and crops are changing our society. Research into crop and livestock improvement represents some of the greatest successes, as well as challenges, for geneticists. Virtually all the corn grown in the U.S. today, for example, is hybrid corn, the result of crossing inferior parent strains. Faced with increasing populations and diminishing farmland, agricultural geneticists are constantly trying to develop strains of plants or animals that produce more food per unit input of energy, labor, money, and space.

Scientists take three principal approaches to improving crops and livestock. First, breeders can search for individual plants or animals with the desired characteristics. The characteristics selected for preservation through selective breeding could arise from mutations or from chance recombinations of preexisting alleles. This approach relies on "good" alleles appearing spontaneously and on our ability to recognize them. Second, molecular geneticists can try to create superior alleles in the laboratory or transplant them from one species to another, as we shall describe in Chapter 13. The third approach is to look for desirable traits in wild populations of the same or closely related species. Breeders might then cross-breed or otherwise incorporate the desired genetic information into crop plants or livestock. This last approach shows great promise and, in fact, is how many crops and livestock breeds were developed in the first place, thousands of years ago. Wheat, for example, was a cross between varieties of wild grass, aided by irregularities in meiosis that resulted in a plant with multiple sets of chromosomes, producing large, edible kernels.

Rare species and local varieties of widespread species have been important in crop development in the past and hold promise for the future. In the 1960s, for example, plant geneticists cross-bred commercial strawberries with a wild variety growing in Cottonwood Canyon, Utah. The result was strawberries that produce fruit year-round. More recently, Florida breeders have developed a heat-tolerant blueberry bush by cross-breeding domestic blueberries with the rabbit-eye blueberry of the South. Geneticists at Oregon State University are developing a wildflower called ***meadowfoam*** into a source of high-quality commercial oil for the pharmaceutical and electronics industries. One of the most useful meadowfoam species is found in a 6-square-mile area near Medford, Oregon, and nowhere else in the world.

Today, our reserves of wild genes are diminishing at a frightening rate, largely due to the pressures of human population and development. Most of the estimated 10 million species of organisms on Earth have never even been scientifically named or described, much less studied. Sadly, we are in danger of losing these species without ever having gotten to know them. With the destruction of rain forests and other previously undisturbed tracts of land all over the world (perhaps even in our own backyards), countless species and genetic varieties are disappearing rapidly. For example, the anticancer drug Taxol® was discovered in the Pacific yew tree, which grows in restricted areas of the Pacific Northwest. In 1987 a researcher brought back samples of a rainforest tree from Borneo. Scientists at the National Cancer Institute tested an extract from the tree, a member of the genus *Calophyllum*, against the AIDS virus and found the extract to be extremely effective in stopping viral replication in cultured cells. When the researcher returned to Borneo to collect more, the tree he had sampled had been cut down, and the other local *Calophyllum* trees were of a different species with only weak anti-viral activity. Fortunately, the same species of tree was protected in the Singapore Botanic Garden. Researchers were able to isolate and then synthesize the compound, which began clinical trials in human AIDS patients in 1997.

Preserving the genes of the resident plants and animals is an important, but little recognized, reason to preserve wilderness. This preservation is also an important goal of efforts to save endangered species, for once a species becomes extinct, its genes are lost forever. Genes, with all their various alleles, have evolved over hundreds of millions of years and represent one of our most valuable and irreplaceable natural resources.

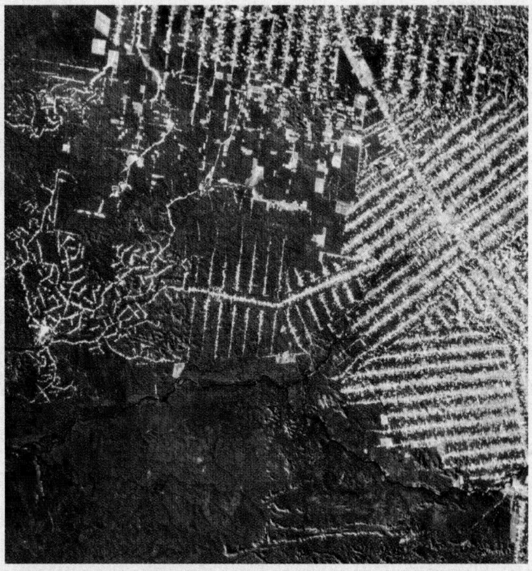

Figure E12-2 Deforestation
This *Landsat* image highlights the loss of rain forest in Brazil. The dark green color shows healthy forest; the pink colors show areas that have been clear-cut. Such deforestation can lead to extinction of species, removing their unique genetic composition from Earth's genetic repertoire.

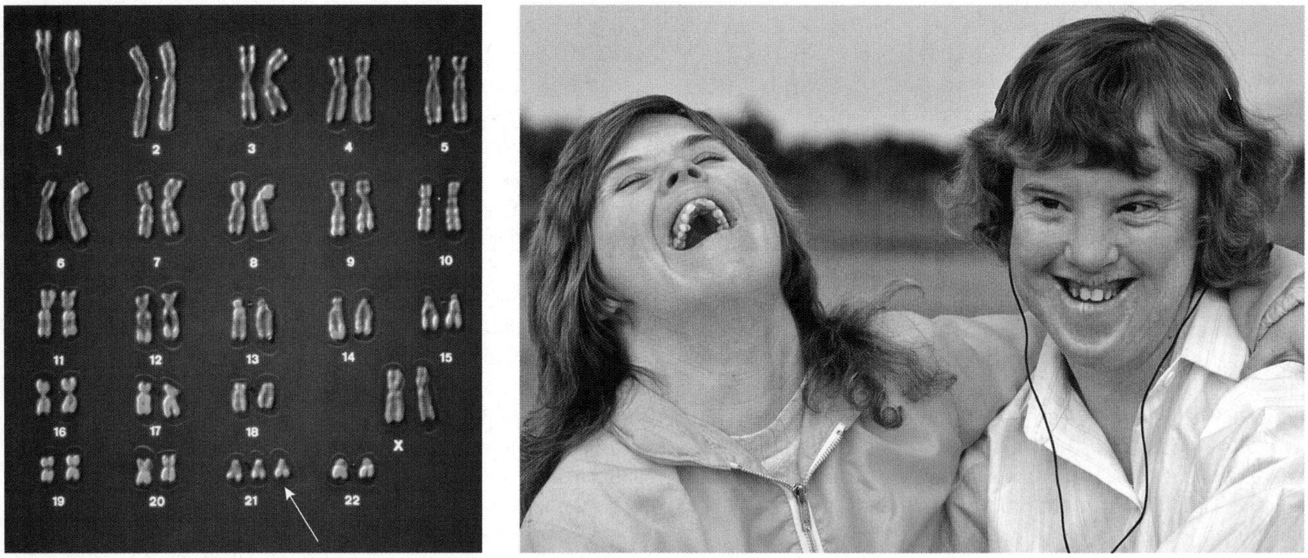

(a) (b)

Figure 12-19 Trisomy 21, or Down syndrome
(a) Karyotype of a Down syndrome child reveals three copies of chromosome 21. ***(b)*** These girls have the relaxed mouth and distinctively shaped eyes typical of Down syndrome.

Some Genetic Disorders Are Caused by Abnormal Numbers of Autosomes

Nondisjunction of the autosomes may also occur, producing eggs or sperm missing an autosome or with two copies of an autosome. Fusion with a normal gamete (bearing one copy of each autosome) leads to an embryo with either one or three copies of the affected autosome. Embryos that have only one copy of any of the autosomes aborts so early in development that the woman never knows she was pregnant. Embryos with three copies of an autosome (trisomy) also usually spontaneously abort. However, a small fraction of embryos with three copies of chromosomes 13, 18, or 21 can develop sufficiently to be born. In the case of trisomy 21, the baby can live into adulthood.

Trisomy 21, or Down Syndrome

In about one of every 900 births, the child inherits an extra copy of chromosome 21, a condition called **trisomy 21** or **Down syndrome.** Thus, approximately 304,000 individuals in the U.S. have Down syndrome. Children with Down syndrome have several distinctive physical characteristics, including weak muscle tone, a small mouth held partially open because it cannot accommodate the tongue, and distinctively shaped eyelids (Fig. 12-19). Much more serious defects include low resistance to infectious diseases, heart malformations, and mental retardation so severe that only about one in 25 ever learns to read and only one in 50 learns to write.

The frequency of nondisjunction increases with the age of the parents, especially that of the mother (Fig. 12-20). Nondisjunction in sperm accounts for about 25% of the cases of Down syndrome, and there is a small increase in such defective sperm with increasing age of the father. Since the 1970s, it has become more common for couples to delay having children, increasing the probability of trisomy 21. Fetal trisomy can be diagnosed by examining the chromosomes of fetal cells. However, because many fewer older women have children than younger ones, the majority of Down syndrome children are born to young mothers.

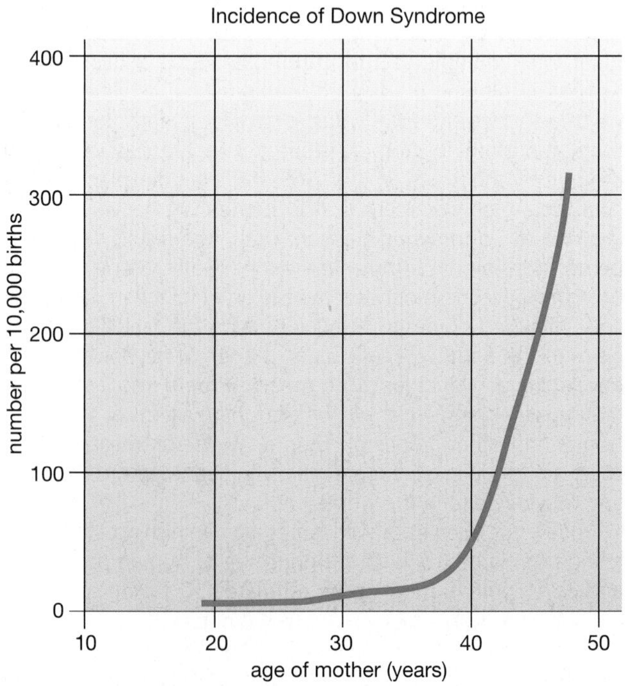

Figure 12-20 Down syndrome frequency increases with maternal age
The increase in frequency of Down syndrome after maternal age 35 is quite dramatic.

CASE STUDY REVISITED

Sleepy Genes

Following in the footsteps of Gregor Mendel, researchers at the Stanford Center for Narcolepsy crossed narcoleptic dogs to one another and analyzed the progeny. This analysis showed that narcolepsy in dogs results from a recessive allele of a single gene. But where is this gene and what kind of protein does it encode? Further studies by Dr. Mignot's research group revealed that the narcolepsy gene is on chromosome 12. After years of effort, they were eventually able to clone the gene—only to find that it had already been discovered! The gene is called *Hcrt2,* and it encodes a protein receptor that is present on the cell surfaces of some cells in the hypothalamus (a part of the brain). The protein encoded by *Hcrt2* binds to signaling molecules called *hypocretins*. In the narcoleptic dogs the receptor was defective, making their brain cells ignore the molecular signal delivered by the hypocretins. Narcoleptic mice, studied by another research group, appear to have normal *Hcrt2* genes, but they have a different mutation that prevents them from producing hypocretins at all. In a most interesting turn of events, hypocretins (also called *orexins*) had been under investigation because of their role in controlling feeding behavior. Thus, the molecular systems that control sleep and feeding may have some common features.

What about the human connection? In January 2000 Mignot's research group and their collaborators reported that seven out of nine patients with narcolepsy did not produce hypocretin. These individuals may have a genetic defect similar to that of the narcoleptic mice. Two of the nine narcoleptic humans did produce hypocretins. These individuals may have a mutation in the hypocretin receptor gene similar to that in the narcoleptic dogs. Intense investigation is under way to use this information to develop treatments for patients with narcolepsy. This knowledge might also allow scientists to develop better medicines to promote sleep in people suffering from insomnia.

Because genetics is important to so many aspects of human behavior, defense attorneys might consider using a defendant's genetic constitution as a strategy to excuse criminal behavior. First, take the side of the defense and present an argument about why a defendant's genes should be considered as a factor in the criminal behavior. Then take the prosecution's side and present an argument about why a defendant's genes do not excuse criminal behavior.

Summary of Key Concepts

1) How Did Gregor Mendel Lay the Foundations for Modern Genetics?

Homologous chromosomes carry the same genes located at the same loci, but the genes at a particular locus can exist in alternate forms called *alleles*. An organism whose homologous chromosomes carry the same allele at a locus is homozygous for that particular gene. If the alleles at a locus differ, the organism is heterozygous for that gene. Gregor Mendel deduced many principles of inheritance in the mid-1800s, before the discovery of DNA, genes, chromosomes, or meiosis. He did this by choosing an appropriate experimental subject, designing his experiments carefully, following progeny for several generations, and analyzing his data statistically.

2) How Are Single Traits Inherited?

A trait is an observable or measurable feature of the organism's phenotype, such as eye color or blood type. Traits are inherited in particular patterns that depend on the types of alleles that parents pass on to their offspring. Each parent provides its offspring with one copy of every gene, so that the offspring inherits a pair of alleles for every gene. The combination of alleles in the offspring determines whether or not it displays a particular trait. Dominant alleles mask the expression of recessive alleles. The masking of recessive alleles can result in organisms with the same phenotype that different genotypes. That is, organisms with two dominant alleles (homozygous dominant) have the same phenotype as do organisms with one dominant and one recessive allele (heterozygous). Because each allele segregates randomly during meiosis, we can use the laws of probability to predict the relative proportions of offspring with a particular trait.

3) How Are Multiple Traits on Different Chromosomes Inherited?

If the genes for two traits are located on separate chromosomes, they will be assorted independently of one another into the egg or sperm. Thus, crossing two organisms that are heterozygous at two loci on separate chromosomes produces offspring with 16 different genotypes. If the alleles are typical dominant and recessive alleles, these progeny will display only 4 different phenotypes.

4) How Are Genes Located on the Same Chromosome Inherited?

Genes located on the same chromosome are linked to one another (encoded on the same DNA double helix) and they tend to be inherited together. Unless the alleles are separated by chromosomal recombination, the two alleles will be passed together to the offspring.

5) How Is Sex Determined, and How Are Sex-Linked Genes Inherited?

In many animals, sex is determined by sex chromosomes, often designated X and Y. The rest of the chromosomes, identical in the two sexes, are called *autosomes*. In many animals, females have two X chromosomes, whereas males have one X and one Y chromosome. The Y chromosome has many fewer genes than the X chromosome. Because males have only one copy of most X chromosome genes, recessive traits on the X chromosome are more likely to be phenotypically expressed in males.

6) What Are Some Variations on the Mendelian Theme?

Not all inheritance follows the simple dominant–recessive pattern:

1. In incomplete dominance, heterozygotes have a phenotype that is intermediate between the two homozygous phenotypes.
2. Codominance occurs when two types of proteins, each encoded by a different allele at a single locus, both contribute to the phenotype.
3. Many traits are determined by several different genes at different loci that contribute to the phenotype, a phenomenon called *polygenic inheritance*.
4. Many genes have multiple effects on the organism's phenotype (pleiotropy).
5. The environment influences the phenotypic expression of most, if not all, traits.

7) How Are Human Genetic Disorders Investigated?

The genetics of humans is similar to the genetics of other animals, except that experimental crosses are not feasible. Analysis of family pedigrees and, more recently, molecular genetic techniques must be used to determine the mode of inheritance of human traits.

8) How Are Human Disorders Caused by Single Genes Inherited?

Many genetic disorders are inherited as recessive traits; therefore, only homozygous recessive persons show symptoms of the disease. Heterozygotes are called *carriers*, because they carry the recessive allele but do not express the trait. Many other diseases are inherited as simple dominant traits. In such cases, only one copy of the dominant allele is needed to cause the disease symptoms. The human Y chromosome bears few genes other than those that determine maleness; therefore, men phenotypically display whichever allele they carry on their single X chromosome, a phenomenon called *sex-linked inheritance.*

9) How Do Errors in Chromosome Number Affect Humans?

Errors in meiosis can result in gametes with abnormal numbers of sex chromosomes or autosomes. Many people with abnormal numbers of sex chromosomes have distinguishing physical characteristics. Abnormal numbers of autosomes typically lead to spontaneous abortion early in pregnancy. In rare instances, the fetus may survive to birth, but severe mental and physical deficiencies always occur, as is the case with Down syndrome, or trisomy 21. The likelihood of abnormal numbers of chromosomes increases with increasing age of the mother, and, to a lesser extent, the father.

Key Terms

allele *p. 212*
autosome *p. 220*
carrier *p. 227*
codominance *p. 224*
cross-fertilization *p. 213*
crossing over *p. 220*
dominant *p. 214*
Down syndrome *p. 234*
gene *p. 212*
genetic recombination *p. 220*
genotype *p. 215*
hemophilia *p. 229*
heterozygous *p. 212*
homozygous *p. 212*
Huntington disease *p. 229*
hybrid *p. 212*
incomplete dominance *p. 223*
inheritance *p. 217*
Klinefelter syndrome *p. 232*
law of independent assortment *p. 217*
law of segregation *p. 214*
linkage *p. 219*
locus *p. 212*
multiple alleles *p. 223*
nondisjunction *p. 230*
pedigree *p. 226*
phenotype *p. 215*
pleiotropy *p. 225*
polygenic inheritance *p. 224*
Punnett square method *p. 215*
recessive *p. 214*
self-fertilization *p. 213*
sex chromosome *p. 220*
sex-linked *p. 221*
sickle-cell anemia *p. 228*
test cross *p. 216*
trisomy 21 *p. 234*
trisomy X *p. 232*
true-breeding *p. 213*
Turner syndrome *p. 231*

Thinking Through the Concepts

Multiple Choice

1. *An organism is described as* Rr:*red. The* Rr *is the organism's [A]; red is the organism's [B]; and the organism is [C].*
   a. [A] phenotype; [B] genotype; [C] degenerate
   b. [A] karyotype; [B] hybrid; [C] recessive
   c. [A] genotype; [B] phenotype; [C] heterozygous
   d. [A] gamete; [B] linkage; [C] pleiotropic
   e. [A] zygote; [B] phenotype; [C] homozygous

2. *The 9:3:3:1 ratio is a ratio of*
   a. phenotypes in a test cross
   b. phenotypes in a cross of individuals that differ in one trait
   c. phenotypes in a cross of individuals that differ in two traits
   d. genotypes in a cross of individuals that differ in one trait
   e. genotypes in a cross of individuals that differ in two traits

3. *A lawyer tells a male client that blood type cannot be used to his advantage in a paternity suit against the client because the child could, in fact, be the client's, according to blood type. Which of the following is the only possible combination supporting this hypothetical circumstance? (Answers are in the order mother:father:child.)*
   a. A:B:O
   b. A:O:B
   c. AB:A:O
   d. AB:O:AB
   e. B:O:A
4. *A heterozygous red-eyed female* Drosophila *mated with a white-eyed male would produce*
   a. red-eyed females and white-eyed males in the F_1
   b. white-eyed females and red-eyed males in the F_1
   c. half red- and half white-eyed females and all white-eyed males in the F_1
   d. all white-eyed females and half red- and half white-eyed males in the F_1
   e. half red- and half white-eyed females as well as males in the F_1
5. *Which is NOT true of sickle-cell anemia?*
   a. It is most common in African Americans.
   b. It involves a one-amino-acid change in hemoglobin.
   c. It involves red blood cells.
   d. It is lethal in heterozygotes because it is dominant.
   e. It confers some resistance to malaria.
6. *Sex-linked disorders such as color blindness and hemophilia are*
   a. caused by genes on the X chromosome
   b. caused by genes on the autosome
   c. caused by genes on the Y chromosome
   d. expressed only in men
   e. expressed only when two chromosomes are homozygous recessive

? Review Questions

1. Define the following terms: *gene, allele, dominant, recessive, true-breeding, homozygous, heterozygous, cross-fertilization, self-fertilization.*
2. Explain why genes located on the same chromosome are said to be linked. Why do alleles of linked genes sometimes separate during meiosis?
3. Define *polygenic inheritance*. Why could polygenic inheritance allow parents to produce offspring that are notably different in eye or skin color than either parent?
4. What is sex linkage? In mammals, which sex would be most likely to show recessive sex-linked traits?
5. What is the difference between a phenotype and a genotype? Does knowledge of an organism's phenotype always allow you to determine the genotype? What type of experiment would you perform to determine the genotype of a phenotypically dominant individual?
6. If one (heterozygous) parent of a couple has Huntington disease, calculate the fraction of that couple's children that would be expected to develop the disease. What if both parents were heterozygous?
7. Why are most genetic diseases inherited as recessives rather than dominants?
8. Define *nondisjunction*, and describe the common syndromes caused by nondisjunction of sex chromosomes and autosomes.

Applying the Concepts

1. Sometimes the term *gene* is used rather casually. Compare and contrast use of the terms *allele* and *locus* as alternatives to *gene*.
2. Using the information in the chapter, explain why AB individuals are referred to as "universal recipients" in terms of blood transfusions and why people with type O blood are called "universal donors."
3. Mendel's numbers seemed almost too perfect to be real; some believe he may have cheated a bit on his data. Perhaps he continued to collect data until the numbers matched his predicted ratios, then stopped. Recently, there has been much publicity over violations of scientific ethics, including researchers' plagiarizing others' work, using other scientists' methods to develop lucrative patents, or just plain fabricating data. How important an issue is this for society? What are the boundary lines of ethical scientific behavior? How should the scientific community or society "police" scientists? What punishments would be appropriate for violations of scientific ethics?
4. Although American society has been described as a "melting pot," people often engage in "assortative mating," in which they marry others of similar height, socioeconomic status, race, and IQ. Discuss the consequences to society of assortative mating among humans. Would society be better off if people mated more randomly? Discuss why or why not.
5. *Eugenics* is the term applied to the notion that the human condition might be improved by improving the human genome. Do you think there are both good and bad sides to eugenics? What examples can you think of to back up your stand? What would a eugenicist think of the medical advances that have ameliorated the problems of hemophilia?
6. Think about some of the personal, religious, and economic issues related to prenatal counseling and diagnosis. Would you avoid having children if you knew that both you and your spouse were heterozygous for a recessive disorder that is fatal at an early age and may involve considerable

suffering? What would you do if you or your spouse were pregnant and learned that your offspring, if born, would be homozygous for such a disorder? Is the situation qualitatively different for Down syndrome? (Down syndrome children have a life expectancy of 20 to 30 years with mental retardation but can have productive lives.) If you were heterozygous for Huntington disease, would you want to avail yourself of the medical diagnostic tests now available to find out? Would your answer be different if you were planning to have a baby?

Genetics Problems

(Note: An extensive group of genetics problems, with answers, can be found in the Study Guide.)

1. In certain cattle, hair color can be red (homozygous *RR*), white (homozygous *R′R′*), or roan (a mixture of red and white hairs, heterozygous *RR′*).

a. When a red bull is mated to a white cow, what genotypes and phenotypes of offspring could be obtained?

b. If one of the offspring in (a) were mated to a white cow, what genotypes and phenotypes of offspring could be produced? In what proportion?

2. The palomino horse is golden in color. Unfortunately for horse fanciers, palominos do not breed true. In a series of matings between palominos, the following offspring were obtained:

65 palominos, 32 cream-colored,
34 chestnut (reddish brown)

What is the probable mode of inheritance of palomino coloration?

3. In the edible pea, tall (*T*) is dominant to short (*t*), and green pods (*G*) are dominant to yellow pods (*g*). List the types of gametes and offspring that would be produced in the following crosses:

a. *TtGg* × *TtGg* b. *TtGg* × *TTGG* c. *TtGg* × *Ttgg*

4. In tomatoes, round fruit (*R*) is dominant to long fruit (*r*), and smooth skin (*S*) is dominant to fuzzy skin (*s*). A true-breeding round, smooth tomato (*RRSS*) was cross-bred with a true-breeding long, fuzzy tomato (*rrss*). All the F_1 offspring were round and smooth (*RrSs*). When these F_1 plants were bred, the following F_2 generation was obtained:

Round, smooth: 43 Long, fuzzy: 13

Are the genes for skin texture and fruit shape likely to be on the same chromosome or on different chromosomes? Explain your answer.

5. In the tomatoes of problem 4, an F_1 offspring (*RrSs*) was mated with a homozygous recessive (*rrss*). The following offspring were obtained:

Round, smooth: 583	Round, fuzzy: 21
Long, fuzzy: 602	Long, smooth: 16

What is the most likely explanation for this distribution of phenotypes?

6. In humans, hair color is controlled by two interacting genes. The same pigment, melanin, is present in both brown-haired and blond-haired people, but brown hair has much more of it. Brown hair (*B*) is dominant to blond (*b*). Whether any melanin can be synthesized depends on another gene. The dominant form (*M*) allows melanin synthesis; the recessive form (*m*) prevents melanin synthesis. Homozygous recessives (*mm*) are albino. What will be the expected proportions of phenotypes in the children of the following parents?

a. *BBMM* × *BbMm*
b. *BbMm* × *BbMm*
c. *BbMm* × *bbmm*

7. In humans, one of the genes determining color vision is located on the X chromosome. The dominant form (*C*) produces normal color vision; red-green color blindness (*c*) is recessive. If a man with normal color vision marries a color-blind woman, what is the probability of their having a color-blind son? A color-blind daughter?

8. In the couple described in problem 7, the woman gives birth to a color-blind but otherwise normal daughter. The husband sues for a divorce on the grounds of adultery. Will his case stand up in court? Explain your answer.

Answers to Genetics Problems

1. a. A red bull (*RR*) is mated to a white cow (*R′R′*). The bull will produce all *R* sperm; the cow will produce all *R′* eggs. All the offspring will be *RR′* and will have roan hair (codominance).

b. A roan bull (*RR′*) is mated to a white cow (*R′R′*). The bull produces half *R* and half *R′* sperm; the cow produces *R′* eggs. Using the Punnett square method:

sperm \ eggs	*R′*
R	*R R′*
R′	*R′ R′*

Using probabilities:

sperm	**egg**	**offspring**
$\frac{1}{2}$ *R*	*R′*	$\frac{1}{2}$ *RR′*
$\frac{1}{2}$ *R′*	*R′*	$\frac{1}{2}$ *R′R′*

The predicted offspring will be $\frac{1}{2}$ *RR′* (roan) and $\frac{1}{2}$ *R′R′* (white).

2. The offspring occur in three types, classifiable as dark (chestnut), light (cream), and intermediate (palomino). This distribution suggests incomplete dominance, with the alleles for chestnut (*C*) combining with the allele for cream (*C′*) to produce palomino heterozygotes (*CC′*). We can test this hypothesis by examining the offspring numbers. There are approximately $\frac{1}{4}$ chestnut (*CC*), $\frac{1}{2}$ palomino (*CC′*), and $\frac{1}{4}$ cream (*C′C′*). If palominos are heterozygotes, we would expect the cross *CC′* × *CC′* to yield $\frac{1}{4}$ *CC*, $\frac{1}{2}$ *CC′*, and $\frac{1}{4}$ *C′C′*. Our hypothesis is supported.

3. a. *TtGg* × *TtGg*. This is a "standard" cross for differences in two traits. Both parents produce *TG, Tg, tG*, and *tg* gametes. The expected proportions of offspring are $\frac{9}{16}$ tall green, $\frac{3}{16}$ tall yellow, $\frac{3}{16}$ short green, $\frac{1}{16}$ short yellow.

b. *TtGg* × *TTGG*. In this cross, the heterozygous parent produces *TG, Tg, tG*, and *tg* gametes. However, the homozygous dominant parent can produce only *TG* gametes. Therefore, all offspring will receive at least one *T* allele for tallness and one *G* allele for green pods, and thus all the offspring will be tall with green pods.

c. *TtGg* × *Ttgg*. The second parent will produce two types of gametes, *Tg* and *tg*. Using a Punnett square:

sperm \ eggs	*Tg*	*tg*
TG	*TTGg*	*TtGg*
Tg	*TTgg*	*Ttgg*
tG	*TtGg*	*ttGg*
tg	*Ttgg*	*ttgg*

The expected proportions of offspring are $\frac{3}{8}$ tall green, $\frac{3}{8}$ tall yellow, $\frac{1}{8}$ short green, $\frac{1}{8}$ short yellow.

4. If the genes are on separate chromosomes—that is, assort independently—then this would be a typical two-trait cross with expected offspring of all four types (about $\frac{9}{16}$ round smooth, $\frac{3}{16}$ round fuzzy, $\frac{3}{16}$ long smooth, and $\frac{1}{16}$ long fuzzy). However, only the parental combinations show up in the F_2 offspring, indicating that the genes are on the same chromosome.

5. The genes are on the same chromosome and are quite close together. On rare occasions, crossing over occurs between the two genes, producing recombination of the alleles.

6. a. *BBMM* (brown) × *BbMm* (brown). The first parent can produce only *BM* gametes, so all offspring will receive at least one dominant allele for each gene. Therefore, all offspring will have brown hair.

b. *BbMm* (brown) × *BbMm* (brown). Both parents can produce four types of gametes: *BM, Bm, bM,* and *bm*. Filling in the Punnett square:

sperm \ eggs	*BM*	*Bm*	*bM*	*bm*
BM	*BBMM*	*BBMm*	*BbMM*	*BbMm*
Bm	*BBMm*	*BBmm*	*BbMm*	*Bbmm*
bM	*BbMM*	*BbMm*	*bbMM*	*bbMm*
bm	*BbMm*	*Bbmm*	*bbMm*	*bbmm*

All *mm* offspring are albino, so we get the expected proportions $\frac{9}{16}$ brown-haired, $\frac{3}{16}$ blond-haired, $\frac{4}{16}$ albino.

c. *BbMm* (brown) × *bbmm* (albino):

sperm \ eggs	*bm*
BM	*BbMm*
Bm	*Bbmm*
bM	*bbMm*
bm	*bbmm*

The expected proportions of offspring are $\frac{1}{4}$ brown-haired, $\frac{1}{4}$ blond-haired, $\frac{1}{2}$ albino.

7. A man with normal color vision is *CY* (remember, the Y chromosome does not have the gene for color vision). His color-blind wife is *cc*. Their expected offspring will be:

sperm \ eggs	*c*
C	*Cc*
Y	*cY*

We therefore expect that all the daughters will have normal color vision and all the sons will be color-blind.

8. The husband should win his case. All his daughters must receive one X chromosome, with the *C* allele, from him and therefore should have normal color vision. If his wife gives birth to a color-blind daughter, her husband cannot be the father (unless there was a new mutation for color blindness in his sperm line, which is very unlikely).

For More Information

Hoyt, E. "Wild Relatives." *Living Wilderness*, Summer 1990. Undisturbed ecosystems are a valuable gene bank with great potential for improving crop plants.

Horgan, J. "Eugenics Revisited." *Scientific American*, June 1993. Determining patterns of inheritance, let alone finding suspect genes, for behavioral disorders is notoriously difficult. What's more, it isn't at all clear what one should do with such data, if they become available.

Kahn, P. "Gene Hunters Close in on Elusive Prey." *Science*, March 8, 1996. Using pedigrees and biotechnology, researchers are uncovering the causes of diseases that involve complex interactions between genes and the environment.

McClintock, J. "Let Sleeping Dogs Arise." *Discover*, February 2000. Interesting article about Dr. Mignot's narcoleptic dogs at Stanford and how it is helping us understand narcolepsy in humans.

McGue, M. "The Democracy of the Genes." *Nature*, July 1997. The environment has a more important role in the development of intelligence than previously thought.

O'Brien, S. J. "A Role for Molecular Genetics in Biological Conservation." *Proceedings of the National Academy of Sciences*, June 1994, Vol. 91. Habitat destruction has reduced many wild populations to tiny remnants of their former size. Genetic analysis is used to determine how much loss of genetic diversity has occurred due to inbreeding.

Pines, M. "In the Shadow of Huntington's." *Science 84,* May 1984; Grady, D. "The Ticking of a Time Bomb in the Genes." *Discover,* June 1987; Roberts, L., "Huntington's Gene: So Near, Yet So Far." *Science,* February 1990; and Morell, V. "Huntington's Gene Finally Found." *Science*, April 1993. These articles focus on the story of the scientific detective work involved in the diagnosis of Huntington disease by using recombinant DNA techniques, the frustrations of searching for the gene, and, finally, finding it. Particularly poignant because one of the lead investigators may herself be a victim.

Sapienza, C. "Parental Imprinting of Genes." *Scientific American*, October 1990. It is not quite true that all genes are equal, regardless of whether they have been inherited from mother or father. In some cases, which parent a gene comes from greatly alters its expression in the offspring.

Siegel, J. M. "Narcolepsy." *Scientific American,* January 2000. This article summarizes the characteristics of narcolepsy and current research on its cause.

Stern, C., and Sherwood, E. R. *The Origin of Genetics: A Mendel Source Book*. San Francisco: W. H. Freeman, 1966. There is no substitute for the real thing, in this case a translation of Mendel's original paper to the Brünn Society.

Wivel, N. A., and Walters, L. "Germ-line Gene Modification and Disease Prevention: Some Medical and Ethical Perspectives." *Science*, 1993. If we can fix a defective allele in an egg or zygote, should we?

Answers to Multiple-Choice Questions

1. c 2. c 3. a 4. e 5. d 6. a

MEDIATUTOR
Patterns of Inheritance

CD Activities

Activity 12.1: Monohybrid Crosses

Estimated time: 10 minutes

When Gregor Mendel first examined crosses of pea plants, he analyzed single-trait differences and found some very interesting results. This tutorial will recreate one of Mendel's early experiments and show that a Punnett square can be used to determine how probability figures into genetic analysis.

Activity 12.2: Dihybrid Crosses

Estimated time: 10 minutes

If a Punnett square can be used to demonstrate the probability of genetics with monohybrid crosses, can it also be used for a dihybrid cross, tracking two separate traits at once? In this tutorial, we will examine a set of crosses, focusing on two separate genetic traits. We will see how probability can help us understand this situation also.

Start the MediaTutor Student CD-ROM and enter the activity number in the Quick Search box to be taken directly to that activity.

Web Investigations

Case Study: Sleepy Genes

Estimated time: 15 minutes

It is a wonderful time to be a medical geneticist. After years of diagnosing genetic diseases, predicting their clinical consequences, and, in many cases, being unable to offer effective treatments, the era of gene deconstruction has arrived. Narcolepsy is only one of many disorders now associated with a specific gene product. For some syndromes, targeted biochemical and gene therapies are already being developed. Read about new discoveries in human/medical genetics.

Go to http://www.prenhall.com/audesirk6, the Audesirk Companion Web site. Select Chapter 12 and Web Investigation to begin.

Farmers in the Philippines plant young rice sprouts in a flooded rice paddy. Can genetically modified rice and other crops help to feed a hungry world? Or will tinkering with genes unlock a Pandora's box of problems? It depends on whom you ask.

13 Biotechnology

AT A GLANCE

Case Study: Teaching an Old Grain New Tricks

1) **What Is Biotechnology?**

2) **How Does DNA Recombination Occur in Nature?**

DNA Recombination Occurs Naturally Through Processes Such as Sexual Reproduction, Bacterial Transformation, and Viral Infection

3) **How Does DNA Recombination Occur in Genetic Engineering Laboratories?**

Restriction Enzymes Cut DNA at Specific Nucleotide Sequences

Insertion of Foreign DNA into a Vector Can Produce a Recombinant DNA Library

4) **How Can Researchers Identify Specific Genes?**

Restriction Fragment Length Polymorphisms Can Be Used to Identify Genes

Genes from One Organism Can Be Identified Based on Similarity to Related Genes in Other Organisms

Genes Can Be Identified Based on Their Protein Product

5) **What Are Some Applications of Biotechnology?**

Cloned Genes Provide Enough DNA for Gene Sequencing

DNA Fingerprinting Facilitates Genetic Detection on Many Fronts

Genetic Engineering Is Revolutionizing Agriculture

Will Biotechnology Create a Real Jurassic Park?

6) **What Are Some Medical Uses of Biotechnology?**

Knock-out Mice Provide Models of Human Genetic Diseases

Genetic Engineering Allows Production of Therapeutic Proteins

Human Gene Therapy Is Just Beginning

The Human Genome Project Has Completed a Working Draft of the Entire Human Genome

7) **What Are Some Ethical Implications of Human Biotechnology?**

Genetic Tests for Cystic Fibrosis and Inherited Forms of Breast Cancer Illustrate Potential Problems

Sickle-Cell Anemia and Tay-Sachs Disease Illustrate the Hazards and Benefits of Genetic Screening Programs

The Potential to Clone Humans Raises Further Ethical Issues

Case Study Revisited: Teaching an Old Grain New Tricks

CASESTUDY

Teaching an Old Grain New Tricks

Rice is the major food for about two-thirds of the humans on Earth. A bowl of rice provides a good supply of carbohydrates and some protein but is a poor source of most vitamins, including vitamin A. Unless people eat sufficient fruits and vegetables along with the rice to provide this essential vitamin, a condition called *vitamin A deficiency* results. Although uncommon in the U.S., each year this condition causes the deaths of more than a million children in Asia, Africa, and Latin America. In addition, more than 250,000 children in these developing nations become blind each year as a result of vitamin A deficiency. In 1999 biotechnology provided a simple possible cure: genetically engineered rice.

This new rice has a golden-yellow color because it contains elevated levels of beta-carotene, the vitamin A precursor that gives carrots their bright color. Eating approximately 10 ounces of the bioengineered rice each day (a typical amount in Asian diets) would prevent vitamin A deficiency.

How do scientists create genetically modified plants and animals? What are some potential risks and benefits of this technology? In this chapter, we explore these questions and discuss the increasing role that biotechnology plays in life on Earth. ■

1 What Is Biotechnology?

In its broadest sense, **biotechnology** is any commercial use or alteration of organisms, cells, or biological molecules to achieve specific practical goals. By this definition, biotechnology is nearly as old as civilization itself. Archeological studies show that as long as 10,000 years ago, Neolithic cultures in Egypt and the Near East used yeast cells to produce beer and wine, much as we do today. Selective breeding of plants and animals in agriculture has a similarly lengthy history. Squash fragments preserved in a dry cave in Mexico were recently dated as 8000 to 10,000 years old. These squash have larger seeds and thicker rinds than those of wild varieties, providing evidence of selective breeding by humans—an early form of genetic manipulation. Prehistoric art and animal remains suggest that dogs, sheep, goats, pigs, and camels were also domesticated and selectively bred beginning about 10,000 years ago.

Selective breeding continues to be an important tool in biotechnology. However, modern biotechnology also commonly uses **genetic engineering**, the modification of genetic material to achieve specific goals. Genetically engineered cells or organisms may have genes deleted, added, or changed. Major goals of genetic engineering include:

1. to learn more about cellular processes, including inheritance and gene expression;
2. to provide better understanding and treatment of diseases, particularly genetic disorders; and
3. to generate economic and social benefits, including efficient production of valuable biological molecules and improved plants and animals for agriculture.

A key tool in genetic engineering is **recombinant DNA**. Recombinant DNA contains genes or portions of genes from different organisms, often from different species. Large amounts of recombinant DNA can be grown in bacteria, viruses, or yeast and then transferred into other species, including animals and plants. These plants and animals, which express DNA that has been modified or derived from another species, are called **transgenic**. Since its development in the 1970s, recombinant DNA technology has grown explosively, providing new methods, applications, and possibilities for genetic engineering. Today most research labs involved in analysis of cell structure, genetics, molecular basis of disease, and evolution routinely use recombinant DNA technology in their experiments. Many genetically engineered products are used in preference to previously available ones, including human insulin and enzymes used for making cheese. However, also growing are concerns about the wisdom and safety of some of these methods and products.

2 How Does DNA Recombination Occur in Nature?

Most of us tend to think of a species' genetic make-up as relatively stable and static. However, many natural processes can transfer DNA from one organism—or even one species—to another. In fact, many recombinant DNA technologies used in the laboratory are based on these naturally occurring DNA recombination processes.

DNA Recombination Occurs Naturally Through Processes Such as Sexual Reproduction, Bacterial Transformation, and Viral Infection

Regardless of how it occurs, recombination changes the genetic makeup of organisms. DNA recombination within many species occurs during sexual reproduction: Crossing over during meiosis I exchanges DNA from each maternal chromosome with the homologous DNA from the paternal chromosome. Following recombination, the chromosomes contain new combinations of alleles, different from either parent. Thus, each egg and sperm produced by meiosis can be considered "recombinant."

Many people tend to consider recombination within a species during sexual reproduction as "natural" and therefore good but think of recombinations performed in the laboratory between different species as "unnatural" and therefore intrinsically bad. However, recombinations between species also occur in nature, as described next.

Transformation May Combine DNA from Different Bacterial Species

Bacteria can undergo several types of recombination that allow gene transfer between unrelated species. A process called **transformation** enables bacteria to pick up free DNA from the environment. The free DNA may be part of the chromosome of another bacterium, including DNA from another bacterial species. Discovery of such chromosomal transformation was one of the first indications that DNA carries genes. Transformation may also occur when bacteria pick up tiny circular DNA molecules called **plasmids** (Fig. 13-1). Many types of bacteria contain plasmids, which can also be found in some fungi, algae, and protists. Plasmids range in size from about 1000 to 100,000 nucleotides long. For comparison, the *E. coli* chromosome is around 4,600,000 nucleotides long. A single bacterium may contain dozens or even hundreds of copies of a plasmid. When the bacterium dies, it releases these plasmids into the environment, where they can be picked up by bacteria of similar or different species. In addition, living bacteria can often pass a copy of their plasmid directly to other living bacteria. Passing of plasmids from living bacteria to living yeast has also been documented!

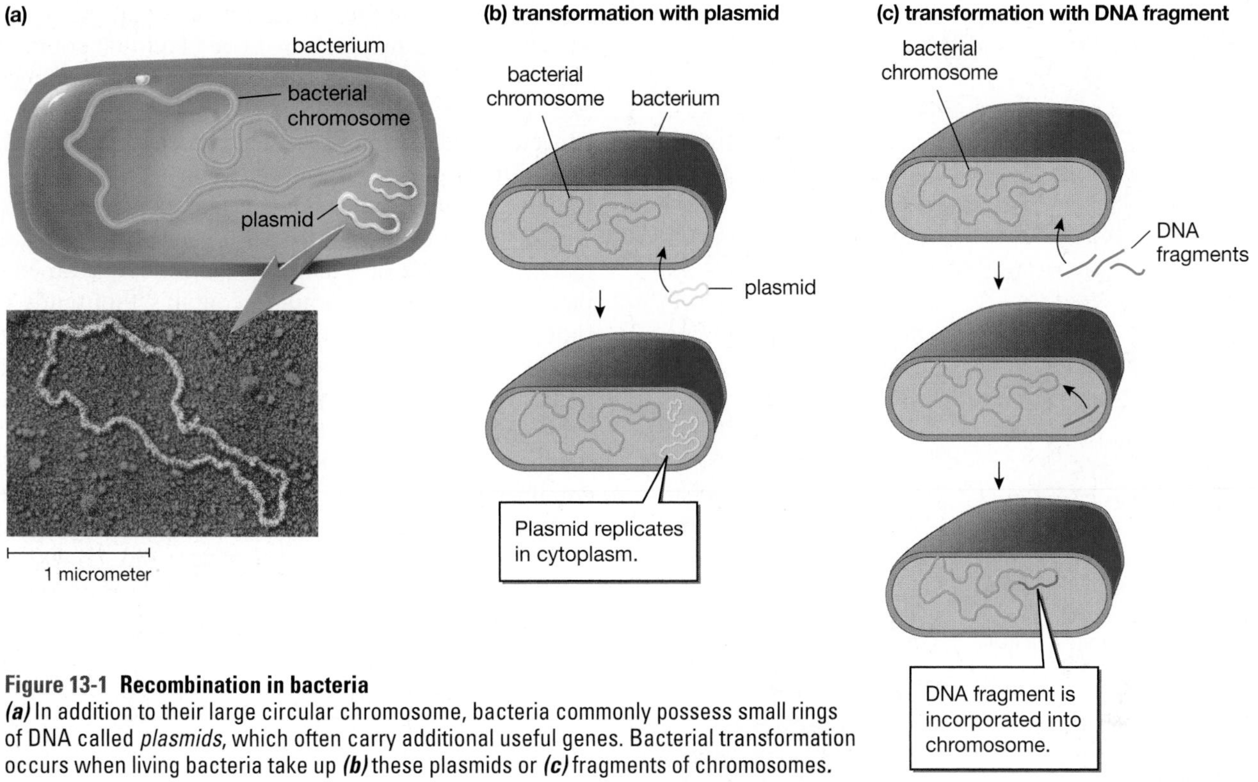

Figure 13-1 Recombination in bacteria
(a) In addition to their large circular chromosome, bacteria commonly possess small rings of DNA called *plasmids,* which often carry additional useful genes. Bacterial transformation occurs when living bacteria take up ***(b)*** these plasmids or ***(c)*** fragments of chromosomes.

MediaTutor 13.1 Genetic Recombination

What use are plasmids? The bacterium's chromosome contains all the genes the cell normally needs for basic survival. However, genes carried by plasmids often allow the bacteria that carry them to grow in novel environments. Some plasmids contain genes that allow bacteria to metabolize novel energy sources, such as petroleum or other hydrocarbons. Other plasmids carry genes that cause disease symptoms, such as diarrhea, in the animal or other organism that the bacterium infects. (From the bacterium's point of view, diarrhea in the infected animal may be beneficial in that it enables the bacterium to spread and infect new hosts.) Still other plasmids carry genes that enable the bacterium to grow in the presence of an antibiotic, such as penicillin. In environments where antibiotic use is high, particularly hospitals, bacteria carrying these antibiotic-resistance plasmids can quickly spread among patients and health care workers, making antibiotic-resistant infections a serious problem.

Viruses May Transfer DNA Between Bacteria and Between Eukaryotic Species

Viruses, which are little more than genetic material encased in a protein-containing coat, transfer their genetic material into cells during infection. Within the infected cell, viral genes replicate and direct the synthesis of viral proteins. The replicated genes and new viral proteins assemble inside the cell, forming new viruses that are released to infect new cells (Fig. 13-2). Usually, a particular virus infects and replicates only in the cells of certain bacterial, animal, or plant species. For example, the canine distemper virus, which causes a frequently fatal disease in dogs, usually only infects dogs, raccoons, ferrets, otters, and related species. People who care for these sick animals are not at risk for contracting this viral disease.

During many viral infections, viral DNA sequences become incorporated into one of the host cell's chromosomes. The viral DNA can remain there for days, months, or even years. The cell replicates the incorporated viral DNA as well as the rest of its DNA every time the cell divides. When new viruses are produced from the incorporated DNA, they can mistakenly incorporate human genes into the virus genome, creating a recombinant virus. When such viruses infect new cells, they also transfer a portion of the previous host cell's DNA. Occasionally, viruses cross species barriers to infect new species. For example, canine distemper, which usually doesn't infect cats, has killed thousands of lions in the Serengeti plain of Africa. During such cross-species infections, the new host may acquire genes that originally belonged to an unrelated species.

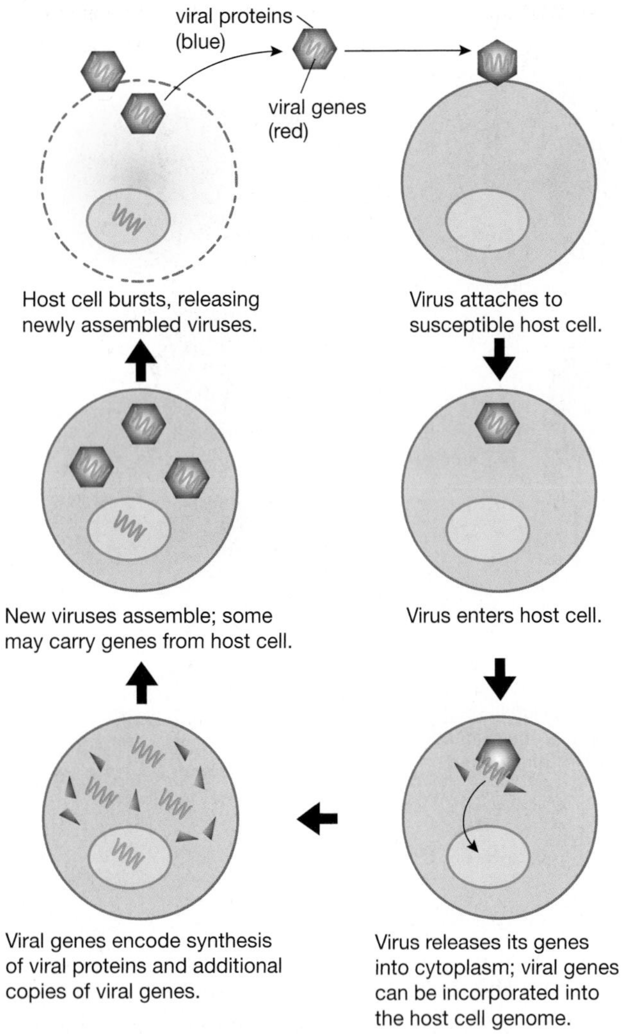

Figure 13-2 Viruses may transfer genes
A virus infecting a eukaryotic cell uses its own genetic material and the host's cellular machinery to copy the virus. When copies are released from the cell, they can infect new host cells. Segments of the host DNA may be incorporated into the viral genome and transferred to hosts of other species.

3 How Does DNA Recombination Occur in Genetic Engineering Laboratories?

The tools for creating and analyzing recombinant DNA molecules are constantly being improved and simplified. What once could be done only in specialized university research labs is now routinely done in high school or even middle school biology classes. Nevertheless, several key technologies continue to be cornerstones of laboratory DNA recombination.

Restriction Enzymes Cut DNA at Specific Nucleotide Sequences

In the Case Study, you learned that hundreds of thousands of children go blind each year because of insufficient vitamin A in their diets. Imagine that you wanted to make a vitamin A-enriched rice plant that could provide the daily requirement of vitamin A in one bowl of rice. How would you go about doing this? One approach would be to find the genes needed to make vitamin A from one organism and transfer these genes to the rice plant. How could you find these genes? You would probably begin with an organism that naturally made high levels of vitamin A. Daffodils are an example. But the daffodil genome contains thousands and thousands of genes; which ones are the genes you want?

An important tool for finding genes relies on the ability of **restriction enzymes** to cut up large DNA molecules into smaller, more manageable pieces. Many bacteria produce one or more restriction enzymes, which cut through the DNA backbones wherever they encounter a particular sequence of nucleotides. For example, the EcoRI restriction enzyme from *E. coli* cuts through any DNA double helix wherever it encounters the following sequence:

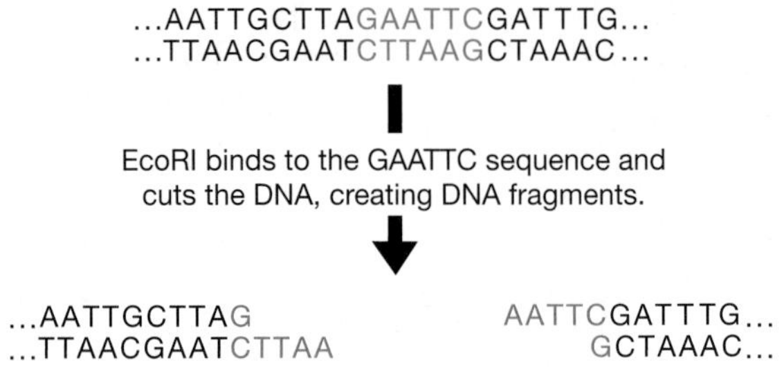

Some enzymes (such as EcoRI) produce a staggered cut through the DNA double helix, leaving a small region of single-stranded DNA at the cut ends. Other enzymes produce blunt, double-stranded ends. In nature, restriction enzymes defend bacteria against viral infections by cutting apart invading viral DNA. The bacteria protect their own DNA by attaching methyl ($-CH_3$) groups to some of the DNA nucleotides. This modified DNA cannot be recognized by restriction enzymes. Researchers have isolated dozens of restriction enzymes and use them to cut DNA at specific sites, producing shorter DNA fragments. Incubating DNA with a restriction enzyme produces smaller DNA fragments with known sequences at their ends. Such fragments can be used to generate a collection of recombinant DNA called a "library."

Insertion of Foreign DNA into a Vector Can Produce a Recombinant DNA Library

To create a transgenic rice plant, we need large amounts of pure daffodil DNA containing the genes we want. How can large amounts of DNA be produced? Scientists realized early on that they could harness plasmids and viruses to synthesize DNA. They added new features to the plasmids and viruses that made it easy to insert foreign DNA. Such specialized plasmids and viruses are called **vectors**. Biotechnology companies routinely develop and market new, improved vectors for research and other recombinant DNA uses. Researchers across the world also provide the vectors that they have developed to other researchers free of charge. When a recombinant plasmid is introduced into a bacterium, the bacterium replicates the recombinant plasmid whenever the bacterium divides. So, by simply growing the bacterium that contains the desired plasmid, the scientists can produce as much of the recombinant DNA as they need.

Inserting foreign DNA into a vector is simple (Fig. 13-3). For example, DNA can be isolated from a daffodil, or any other organism, and cut with EcoRI, producing hundreds of thousands of small DNA fragments. Each daffodil DNA fragment has the same nucleotide sequence at its ends: TTAA on one end of each fragment, and AATT on the other end. The vector DNA is also cut with EcoRI, so that the cut ends of the vector DNA are complementary to those of the daffodil DNA. Next, the vector DNA fragments are combined with the daffodil DNA fragments. DNA ligase, the enzyme used by cells during DNA replication to join DNA strands together, is added to the mixture. DNA ligase joins the vector DNA and daffodil DNA together, creating recombinant DNA molecules.

These daffodil/bacteria recombinant DNA molecules are then added to a culture of bacteria. Via the process of transformation, each bacterium can take up a different recombinant DNA molecule. In this way, each bacterium contains a different portion of the daffodil genome carried in a vector. Growing a culture of these bacteria produces billions of copies of the various recombinant DNA molecules.

The entire process is analogous to taking a set of encyclopedias, cutting the pages apart wherever you find the word *their* and then inserting the fragments into separate folders. Each folder contains just a small portion of the encyclopedia, but the collection of folders contains the entire encyclopedia. Similarly, each genetically modified bacterium contains just a small portion of the daffodil genome, but the collection of bacteria contains the entire daffodil genome. Such a collection is called a **DNA library**. This choice of names is unfortunate, since it brings to mind nicely organized and catalogued information. In contrast, a DNA library is typically a collection of billions of genetically modified bacteria in a single tiny test tube. Returning to the encyclopedia analogy, it's as if the folders were randomly stored in a box. To identify the daffodil genes needed to produce vitamin A, we would have to find the bacteria that have these genes and separate these bacteria from the millions of others that have other daffodil gene sequences. Isolation of a particular bacterial strain that carries a particular foreign gene from a DNA library is what scientists mean when they say that they have "cloned a gene."

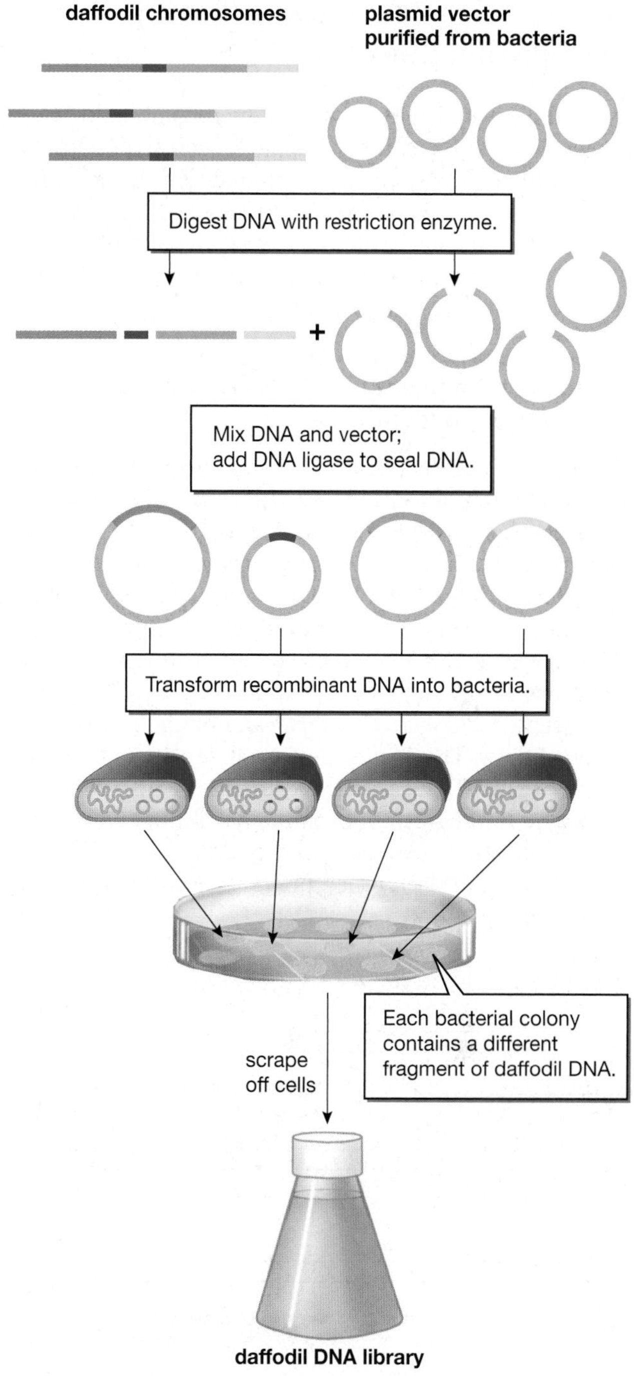

Figure 13-3 Building a DNA library
Chromosomes, purified from the organism of interest, are cut with a restriction enzyme and mixed together with copies of a bacterial plasmid that has been cut with the same restriction enzyme. When the severed DNA molecules are mixed, each plasmid DNA joins with a piece of foreign DNA and the DNA backbones are joined together by DNA ligase, forming recombinant DNA molecules. The new recombinant DNA molecules are taken up into bacteria by transformation. The bacteria grow in dishes containing nutrients until they form colonies. Within each colony, each bacterial cell contains the same recombinant plasmid that carries the same fragment of foreign DNA. However, different colonies contain different recombinant plasmids with different pieces of foreign DNA. The bacterial cells are scraped up together into a test tube, creating a "DNA library."

MediaTutor 13.2 Building a DNA Library

4 How Can Researchers Identify Specific Genes?

As you may realize by now, creating recombinant DNA molecules is not a major challenge for scientists. However, identifying the specific recombinant DNA molecules that contain the gene you are interested in can take years. The challenge of identifying a particular gene (that is, cloning the gene) is met in a variety of ways, three of which we describe next.

Restriction Fragment Length Polymorphisms Can Be Used to Identify Genes

By the early 1980s libraries of human DNA had been created. Scientists could use these libraries, together with restriction digestion and pedigrees, to identify specific genes using *restriction fragment length polymorphisms*. Recall that the DNA sequence of two individuals is very similar, but not identical. Some of these differences in DNA sequence create differences in the length of DNA fragments produced by digestion with a restriction enzyme. Two DNA double helices are depicted below. The location of each site recognized by EcoRI (GAATTC) is marked by an arrow. The DNA molecule in the bottom example has a single base pair difference that changes the middle GAATTC to CAATTC. This difference prevents EcoRI from cutting at that location.

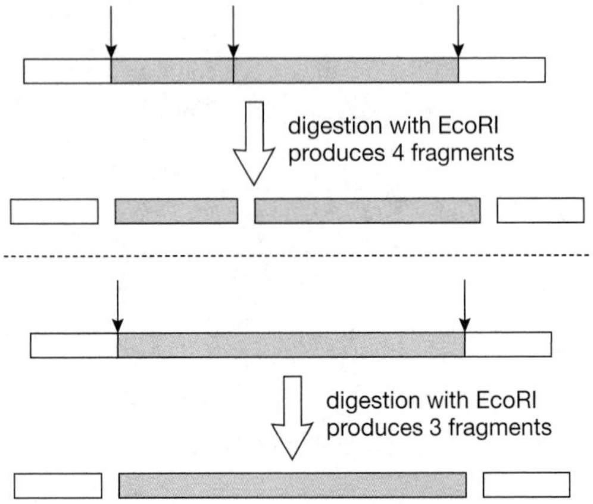
digestion with EcoRI produces 4 fragments
digestion with EcoRI produces 3 fragments

Following digestion with EcoRI, the DNA molecule in the top example will produce four smaller fragments; the one on the bottom will produce three smaller fragments. Differences in DNA sequence such as these that alter the size of restriction fragments are called **restriction fragment length polymorphisms**, or **RFLPs** for short.

RFLPs can be detected by separating restriction fragments via **gel electrophoresis**. In this process, the mixture of DNA restriction fragments is loaded into an indented area, or well, in a slab of agar. Agar is a carbohydrate produced by seaweed that is also used to prepare various human foods. For example, the shiny "jelly" that surrounds a beautiful fruit torte is often made of agar. The agar slab is called a *gel*. When an electrical current is applied to the gel, the negatively charged DNA fragments move toward the positively charged electrode. The smaller fragments can slip through the spaces between the agar molecules more easily than the larger fragments. So, the smaller fragments move more rapidly, leaving larger fragments behind. Eventually the DNA fragments are separated by size, forming distinct bands on the gel.

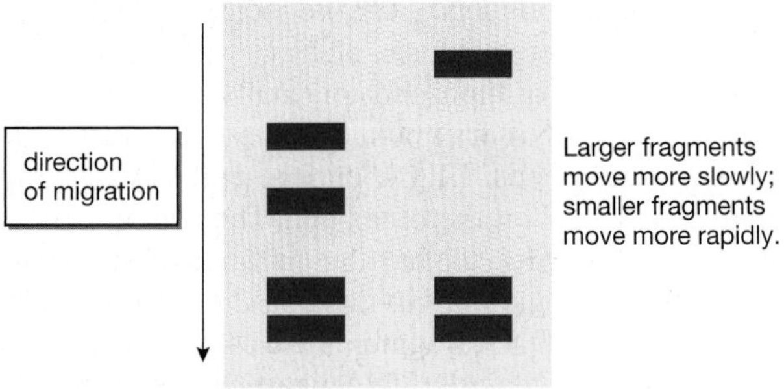
direction of migration
Larger fragments move more slowly; smaller fragments move more rapidly.

How are restriction fragment length polymorphisms useful to researchers? If an RFLP is inherited along with the gene for a particular trait, then the sequence of nucleotides in the RFLP must be close to the gene of interest. The RFLP may even be within the gene! All we need to do is identify the bacteria in the DNA library that contain the RFLP DNA.

One way to identify bacteria carrying a specific gene or other DNA sequence takes advantage of DNA probes. A **DNA probe** is a short sequence of single-stranded DNA that can form base pairs with the DNA that we seek. The DNA probe has been modified, often by adding a radioactive atom, so that we can easily detect its presence. The bacteria that comprise our daffodil DNA library are spread out on solid growth medium and allowed to grow into separate colonies. Each colony is formed from a single cell that divided multiple times until it formed a mound of identical cells. By placing a filter paper on top of the colonies, we can transfer some of the cells from each colony onto the filter. (The plate with the remaining cells is stored so that bacterial colonies in which we are interested can be recovered later.) The cells on the filter are broken open so that the DNA probe can enter. After bathing these broken cells with the DNA probe and washing away excess, the probe will remain only in some of the cells. Specifically, the probe will be present only in cells containing daffodil DNA that can form base pairs with the DNA probe. We can detect cells that contain a radioactive DNA probe by laying a piece of X-ray film on top of the filter. The radioactivity in the DNA probe exposes the film, producing a dark spot on top of the bacterial colonies that we seek. We can then go back to the original plate and pick the colonies of interest, grow the cells, and analyze the DNA that they contain.

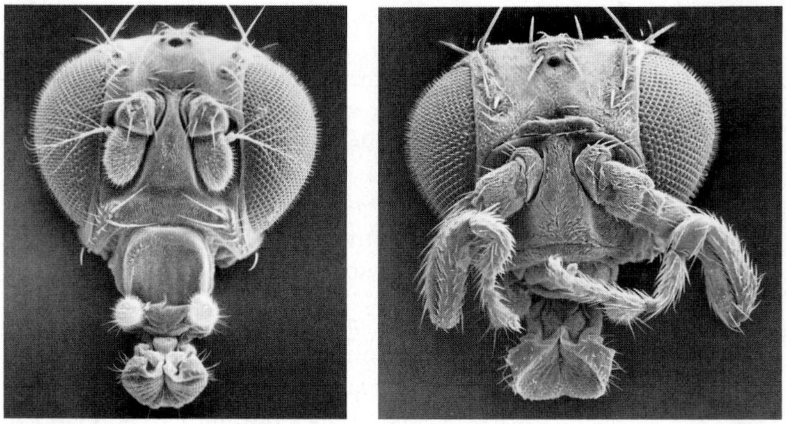

Figure 13-4 Cloned genes from one organism can be used to find similar genes in another organism
The fly on the left has normal antennae. The mutant fly on the right has leg structures instead of antenna. The mutated gene responsible for this defect is called *Antennapedia*. Using a cloned fly *Antennapedia* gene as a probe of a human DNA library, a human version of the *Antennapedia* gene was found. Both genes are expressed in similar regions of the developing embryos.

Genes from One Organism Can Be Identified Based on Similarity to Related Genes in Other Organisms

Once a gene has been cloned from any organism, it can be used to search for related genes in other organisms. Consider homeotic genes, which are extremely important for proper development of animal body structures. One of the first homeotic genes to be cloned was a fruit fly gene called *Antennapedia*. Mutations in this gene cause the antenna of fruit flies to develop abnormally (Fig. 13-4). Instead of antenna, these flies develop legs! Much to the surprise of fruit fly researchers, the use of the *Antennapedia* gene as a probe identified a large class of related genes in other species, including humans. All of these genes share some sequence similarities, and most appear to serve as "master control switches" in early development. Thus, studying flies with legs growing where their antenna should be led to a much more complete understanding of early development in all animals.

Genes Can Be Identified Based on Their Protein Product

Some vectors are designed so that the inserted foreign gene is transcribed and translated in the transformed cell. If researchers wish to clone the gene encoding a particular enzyme, they can use the *protein's* activity to identify cells that contain that gene. For example, the gene encoding a key enzyme in cholesterol synthesis was cloned because yeast cells with high levels of this protein become resistant to a drug that inhibits the enzyme's activity. Only those cells with high levels of the enzyme (and consequently extra copies of the gene) can survive in the presence of the drug.

5 What Are Some Applications of Biotechnology?

Once a gene has been cloned (isolated), it can be put to a variety of uses. Some applications of genes in biotechnology are described next.

Cloned Genes Provide Enough DNA for Gene Sequencing

After cloning, the exact nucleotide sequence of a gene can be determined. The most commonly used method relies on a variation of the polymerase chain reaction, a method that produces large quantities of specific segments of DNA in a test tube. This process, called the **polymerase chain reaction (PCR)** technique, does not require a living organism (Fig. 13-5). Developed in 1986 by Kary B. Mullis of the Cetus Corporation, PCR allows billions of copies of selected genes to be made faster and more cheaply than they can be grown in bacteria. PCR is so elegant and so crucial to continued advances in molecular biology that it earned Mullis a share in the Nobel Prize for Chemistry in 1993.

How Does the Polymerase Chain Reaction Provide Large Quantities of a Specific DNA Segment?

The PCR technique is based on the activity of DNA polymerase, the enzyme that synthesizes new DNA strands. Recall from Chapter 10 that during DNA replication, enzymes unwind the double-stranded helix and that each strand is used as a template to produce its complementary strand. DNA polymerase links up appropriate nucleotides to form each new single-stranded molecule of DNA. PCR requires three major components: DNA containing the nucleotide sequence that you want to synthesize, two short sequences of DNA called *primers*, and DNA polymerase. The primers form base pairs with the original DNA molecule, flanking the stretch of DNA to be synthesized: one primer binds at the "beginning" of the DNA sequence to be synthesized, and the other primer binds to the "end" of the DNA sequence to be synthesized. DNA polymerase will synthesize the DNA that lies between the positions where the primers bind.

PCR involves the following basic steps, which are repeated for as many cycles as needed to generate sufficient copies of the DNA segment (usually about 30 cycles):

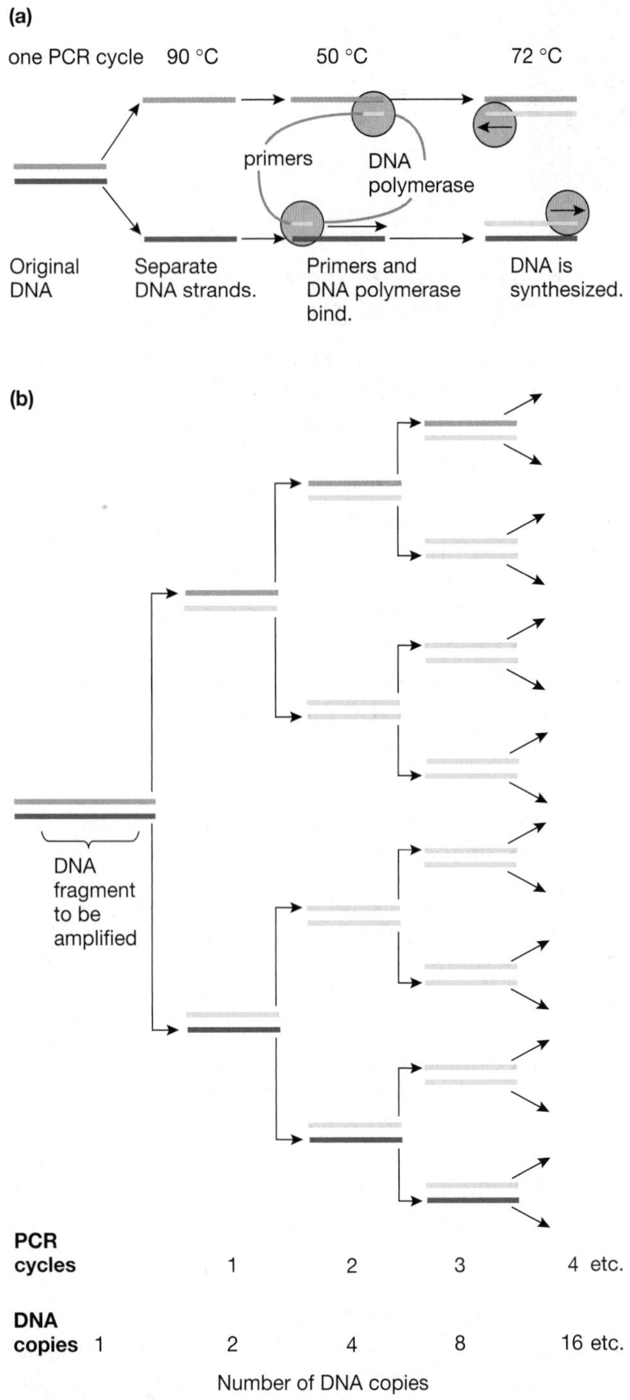

Figure 13-5 PCR copies a specific DNA sequence
The polymerase chain reaction consists of a series of 20 to 30 cycles. ***(a)*** During each cycle, ① the DNA strands are separated by heat, ② primers form base pairs with the target DNA that will be copied, and ③ complementary DNA molecules are synthesized by a heat resistant DNA polymerase enzyme. ***(b)*** After each cycle, the amount of the target DNA doubles.

1. DNA is separated into single strands by heating to about 90 °C.
2. When the temperature is lowered to about 50 °C, the two primers form complementary base pairs with the original DNA strands.
3. Heat-resistant DNA polymerase (isolated from bacteria that thrive in hot springs or hot ocean vents) joins up nucleotides, synthesizing new DNA strands that are complementary to those in the original DNA.

With appropriate mixtures of primers, free nucleotides, and DNA polymerase, a PCR machine automatically runs heating and cooling cycles over and over again. Each cycle takes only a few minutes, so PCR can produce billions of copies of a gene or DNA segment in a single afternoon, starting, if necessary, from a single molecule of DNA (see Fig. 13-5). The DNA produced by PCR can then be examined for RFLPs, used to generate new recombinant DNA molecules, or used for many other purposes.

How Do Researchers Use the Polymerase Chain Reaction to Reveal the Sequence of a Specific DNA Segment?

A variation on the standard PCR reaction can be used to determine the nucleotide sequence of DNA. First, instead of using a pair of primers, only one primer is added to the PCR reaction mixture. This modification allows DNA polymerase to synthesize only one DNA strand, specifically the DNA strand that is complementary to the primer. Second, a fraction of each nucleotide added to the reaction is labeled with a fluorescent molecule, or tag. Each type of nucleotide (A, T, G, and C) is labeled with a different-colored tag. During PCR, DNA polymerase incorporates complementary nucleotides into the growing DNA strand as usual. If DNA polymerase adds an unlabeled nucleotide to the DNA strand that it is synthesizing, it continues along and adds the next complementary nucleotide to the growing chain. However, when DNA polymerase adds one of the fluorescently labeled nucleotides, it stops synthesis of the new DNA strand. As a result, this modified PCR reaction produces a series of single-stranded DNA fragments, each fragment one nucleotide longer than the next. Whether the terminal nucleotide is A, T, G, or C can be determined by its color. When the fragments are separated by gel electrophoresis, an automated sensor records the color of the nucleotide at the end of each increasingly longer DNA fragment. In this way, the nucleotide sequence complementary to the original DNA strand can be determined.

Why is knowing the nucleotide sequence of a gene useful? For one thing, the nucleotide sequence determines the amino acid sequence of the encoded protein. Once the protein is known, its potential function in the cell can often be deduced. DNA sequences are also essential for diagnosing genetic diseases. For example, scientists determined the nucleotide sequence of one allele of the cystic fibrosis gene in 1989. Within a very short time, researchers had also determined the sequence of the alleles of this gene found in cystic fibrosis patients. Most of the patients had a defective allele that was missing three nucleotides. With this knowledge, researchers developed methods for rapidly determining whether an individual is a heterozygous carrier of this allele. Such genetic testing can also tell parents whether an unborn child has inherited two copies of the mutant cystic fibrosis allele and will thus develop the disease. Genetic tests are available for other inherited diseases, including sickle-cell anemia (see "Health Watch: Prenatal Genetic Screening"). In the U.S., state legislation mandates that all newborns be tested for several genetic diseases, including phenylketonuria and hypothyroidism. Most states also mandate testing for sickle-cell anemia and related hemoglobin defects. Our ability to diagnose infections ranging from tuberculosis to HIV also increasingly relies on genetic tests; these tests determine whether DNA from the disease-causing bacterium or virus is present in the patient.

In one of the most remarkable instances of cooperation and collaboration in modern biology, scientists worldwide have essentially completed the ultimate DNA sequencing task—to sequence the entire human genome, described in more detail later in this chapter.

DNA Fingerprinting Facilitates Genetic Detection on Many Fronts

DNA fingerprinting is a type of RFLP analysis that has rapidly emerged as a tool for genetic inquiry in many different contexts. Just as each person has a unique set of fingerprints, each person's DNA produces a unique set of restriction fragments. The pattern of these fragments creates a unique "DNA fingerprint" that can be used to identify an individual (Fig. 13-6). DNA fingerprinting has proven extremely useful for establishing the innocence of suspected criminals, both before and after trials. For example, in the last ten years, DNA fingerprinting has shown that at least ten death row inmates were actually innocent of the crimes for which they had received the death penalty. In 2000, several bills were introduced in Congress proposing legislation to guarantee a convicted criminal's right to such DNA tests.

DNA fingerprinting also has dozens of other uses; it enables us to test potential organ donors for compatibility with a particular patient, to monitor pedigree claims for livestock, to examine relationships among ancient human populations, and to determine whether endangered species laws are being complied with. For example, DNA fingerprinting of whale meat purchased in a Japanese market showed that some of the meat came from endangered whale species; the harvesting of these species was prohibited by international laws. DNA fingerprinting of a 5000-year-old frozen corpse found in the Alps in 1991 showed that this "Ice Man" was genetically related to people living today in northern Europe.

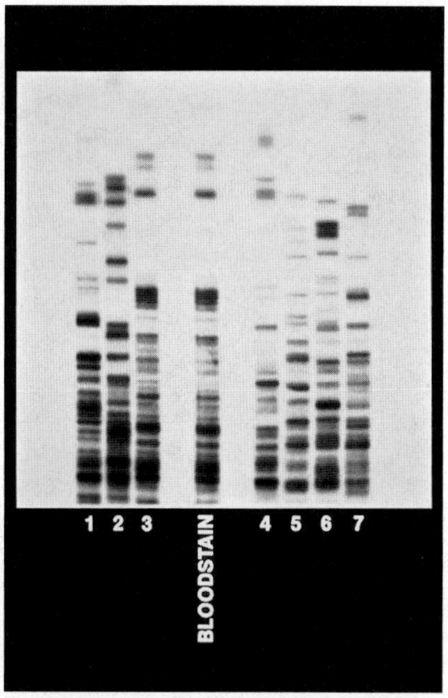

Figure 13-6 DNA fingerprinting in forensics
The differences in restriction fragment lengths provide a DNA fingerprint as individual and unique as a normal fingerprint. Here, DNA fingerprinting clearly shows that the blood of only one suspect matches that found at the crime scene. Which of the seven subjects can be eliminated as suspects? [Cellmark Diagnostics, Germantown, MD]

Genetic Engineering Is Revolutionizing Agriculture

Genetic engineering is rapidly replacing traditional breeding techniques to produce improved strains of crops. Transgenic plants can be produced in several ways. One commonly used technique involves inserting recombinant DNA via a plasmid (called the *Ti* plasmid) found in a certain soil bacterium that infects plants. Alternately, genes can be directly inserted into plant cells using a "gene gun" that blasts DNA-coated particles directly into cells. The transgenic tobacco plant in Figure 13-7 contains a firefly gene that encodes the enzyme luciferase. When this enzyme cleaves its substrate, it produces light. The plant's eerie glow clearly demonstrates that the firefly gene has been incorporated successfully into the plant genome.

Production of transgenic crops (or animals) requires approval by the U.S. Department of Agriculture, the Environmental Protection Agency, and/or the Food and Drug Administration. These agencies also monitor compliance with governmental regulation of transgenic organisms. As of November 2000, more than 5000 field trials of transgenic crops had taken place. Based on these tests, 73 applications for commercial production of transgenic crops had been examined by the USDA. Of these, 52 varieties have been deemed safe for commercial production; these crops can now be grown and crossed with no special government control or oversight. Transgenic crops currently approved for commercial growth are listed in Table 13-1.

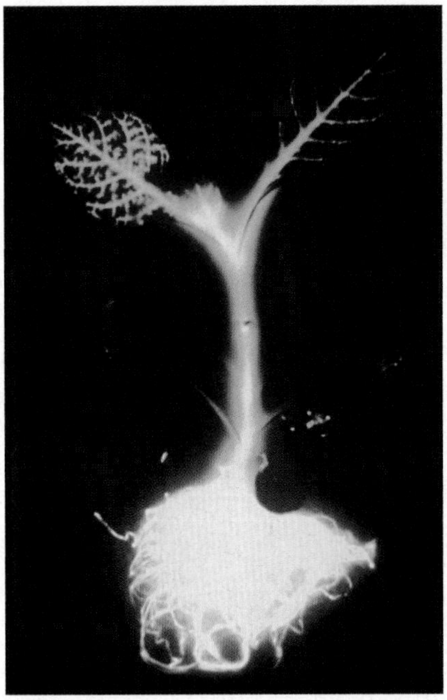

Figure 13-7 Inserting genes into plant cells
Genetic engineering techniques have been used to insert a firefly gene that codes for the enzyme luciferase into this tobacco plant. The enzyme breaks down the chemical luciferin, releasing light in the process. This plant has taken up water containing luciferin.

Genetic Engineering Can Make Plants Resistant to Diseases, Insects, and Weeds

At present, the largest agricultural application of genetic engineering lies in producing crops that are more resistant to diseases, insects, and weeds (Fig. 13-8). Scientists often incorporate genes into the crop's genome that allow crop plants to resist the effects of herbicides. Thus, herbicides can be applied to fields after the crop plants have sprouted to kill competing weeds. In January 2001 the U.S. Department of Agriculture estimated that 8% of the corn, 57% of the soybeans, and 46% of the cotton grown in the U.S. was an herbicide-resistant variety. To create resistance to insects, the most common strategy is to incorporate genes from a bacterium (*Bacillus thuringiensis*, or Bt) that code for the synthesis of a natural insecticide called Bt toxin. The Bt toxin gene has been introduced into more than 50 crop plants, including corn, soybeans, and potatoes. Genes that confer resistance to plant viruses have also being engineered into plants such as squash.

Genetic Engineering May Produce Plants with Therapeutic Benefits

In addition to producing plants that can resist herbicides and pests, some genetic engineers are trying to create plants that can serve as easy, cheap vaccines. Scientists have engineered potatoes that might be eaten raw, in small amounts, to produce resistance to hepatitis, type I diabetes, cholera, and a strain of *Escherichia coli* (*E. coli*) bacteria that causes serious diarrhea. Attempts are under way to use bananas instead of potatoes, since they are much tastier than raw potatoes and might be fed to infants. Such a banana vaccine against the dangerous *E. coli* strain would be particularly valuable in developing nations, where millions of infants and children die each year of diarrhea as a result of *E. coli* infections.

Genetic Engineering May Improve Domesticated Animals

To create transgenic animals, the cloned DNA is injected into a fertilized egg and, in the case of mammals, returned to a surrogate mother to allow it to develop. The

Table 13-1 Genetically Engineered Crops with USDA Approval

Genetically Engineered Trait	Potential Advantage	Examples of Bioengineered Crops Receiving USDA Approval Between 1992 and 2000
Resistance to herbicide	Application of herbicide kills weeds, but not crop plants, producing higher crop yields.	beet, canola, corn, cotton, flax, potato, rice, soybean, tomato
Resistance to pests	Crop plants suffer less damage from insects, producing higher crop yields.	corn, cotton, potato
Resistance to disease	Plants are less prone to infection by viruses, bacteria, or fungi, producing higher crop yields.	papaya, potato, squash
Sterile	Transgenic plants cannot cross with wild varieties, making them safer for the environment and more economically productive for the seed companies that produce them.	chicory, corn
Altered oil content	Oils can be made healthier for human consumption or can be made similar to more expensive oils (such as palm or coconut).	canola, soybean
Altered ripening	Fruits can be more easily shipped with less damage, producing higher returns for farmer.	tomato

(a) (b)

Figure 13-8 Transgenic crops
(a) This squash has been genetically engineered to resist two common plant viruses. ***(b)*** NewLeaf® potatoes (middle row) have been genetically engineered to resist the Colorado potato beetle. The genetically engineered strain clearly has an advantage over the nonengineered strains in the rows on either side.

resulting offspring are tested to see if they express the foreign gene, and those that do are bred with each other to produce animals homozygous for the foreign gene.

Progress in the genetic engineering of livestock has been much slower and less successful than that in crop plants. For example, in an attempt to produce faster-growing, leaner hogs, researchers transferred human and cow growth-hormone genes into pigs. On a high-protein diet, the pigs grew faster, but they developed arthritis, ulcers, and sterility and died prematurely. Growth-hormone genes from rainbow trout have been introduced into salmon, carp, and catfish, causing them to grow 40% faster than normal (Fig. 13-9). These fish are not yet available for human consumption,

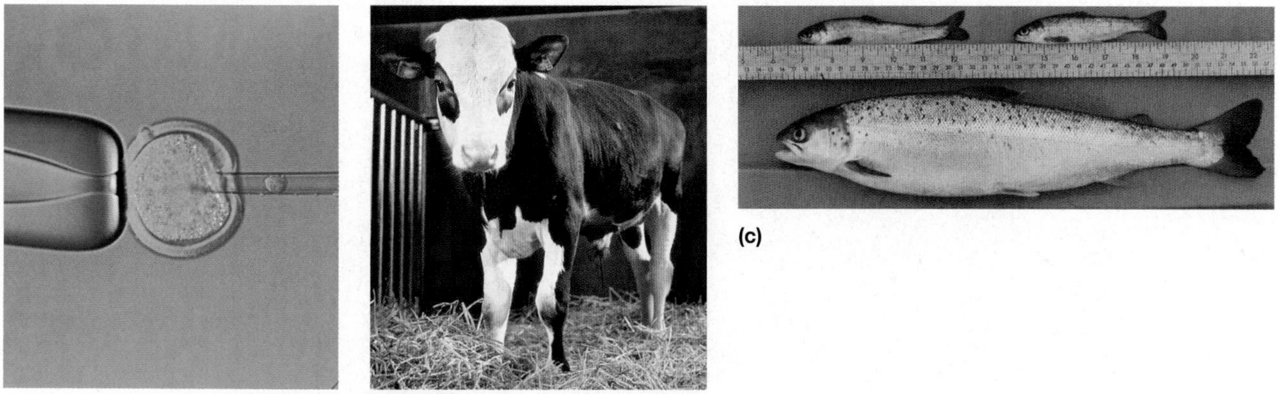

(a) (b) (c)

Figure 13-9 Transgenic animals
(a) A fertilized egg is injected with cloned DNA. Afterward, the egg it will be transferred into a surrogate mother (in the case of mammalian eggs) or allowed to develop (in the case of fish eggs). ***(b)*** This transgenic calf has been genetically engineered to secrete lactoferrin in its milk. Lactoferrin is a protein normally found in human breast milk, making this cow's milk better for human babies. ***(c)*** A gene for growth hormone has been genetically engineered into this Atlantic salmon (shown below two "normal" siblings), making it grow faster.

Earth Watch

Do Recombinants Hold Environmental Promise or Peril?

Scarcely a century ago, the towering American chestnut tree dominated many eastern hardwood forests (Fig. E13-1a). Today the splendor of the chestnuts is barely a memory; few notice the struggling sprouts emerging from the roots of the former giants. The American chestnut has been all but driven to extinction by the ravages of the chestnut blight fungus, inadvertently imported from Asia around 1900 on Asian chestnuts. However, there is a possibility that the American chestnut will reappear, thanks to the techniques of biotechnology. Scientists have engineered a weakened chestnut blight fungus that carries a fungal virus. They hope that after they infect the trees with the recombinant fungus, the debilitating virus will spread from it to the lethal fungus, allowing the chestnuts to recover (Fig. E13-1b). Because Asian chestnuts resist the blight, scientists are also trying to identify and transfer the gene conferring blight-resistance from the Asian chestnut to the American variety.

In 1975 Kurt Benirschke of the San Diego Zoo initiated a program to freeze tissue samples from a variety of endangered species. Although he never envisioned that cloning animals would be possible, the cloning of Dolly the sheep (see "Scientific Inquiry: Much Ado About Dolly" in Chapter 11) has raised the possibility of using such samples to clone other species. Could cloning be a way to increase the populations of endangered animals that may be nearly extinct in the wild and do not reproduce well in zoos? Surrogate mothers of related species might bear the young of endangered—or even recently extinct—species from which living tissue had been saved and frozen. The hurdles are enormous, but so are the possibilities.

What perils do genetically engineered species pose? One concern is that with crop plants now resistant to herbicides, farmers will use more of these chemicals, potentially jeopardizing human health and the environment. Further, plant genes are occasionally carried between species by plant viruses or bacteria, so genes for herbicide resistance might be transferred from crops to weeds. Even if genes for herbicide resistance are not transferred, increased herbicide use will select for weeds with their own mutations for herbicide-resistance. In either of these scenarios, herbicides might ultimately become much less effective. Similarly, widespread cultivation of plants containing the Bt toxin would favor the evolution of insects that could tolerate or inactivate the toxin, rendering it useless.

Successful experiments with domestic crops have led to research into the genetic engineering of species that provide wood, fiber, animal forage, ornamental landscaping, or food (such as faster-growing salmon, Fig. 13-9c). These types of genetically modified species are much more likely than crops (such as cotton or corn) to invade natural ecosystems, with unforeseen and unpredictable consequences. Will they spread and displace native species, disrupting the ecological balance?

Genetic engineers correctly assert that humans have practiced crude genetic engineering for millennia, breeding plants and animals with desired properties. Modern biotechnology is merely a faster, more precise version of standard agricultural practice. What's more, various forms of genetic recombination occur all the time in nature. The features of biotechnology that provide both vast promise and a potential threat are the specificity with which biotechnology can (or will soon be able to) direct genetic changes, the speed with which genetic changes can be made, and the ability to transfer genetic material between species in ways that could never occur in the wild.

(a)

(b)

Figure E13-1 Biotechnology may help the American chestnut
(a) A healthy American chestnut, which escaped the blight because it had been transplanted far west of the chestnut's normal range before the invasion of the blight fungus. ***(b)*** Chestnut blight fungus has produced a lesion on this chestnut tree. The lesioned area was treated with infected fungus, and the tree is walling off the infection with a growth of new tissue.

and many question the value of such animals. We discuss some of the environmental implications of biotechnology in "Earth Watch: Do Recombinants Hold Environmental Promise or Peril?"

Will Biotechnology Create a Real Jurassic Park?

Modern techniques for DNA analysis are yielding information long buried in the DNA of extinct, even fossilized animals and plants. Indeed, DNA has been isolated from a 15-million-year-old magnolia leaf, a 20- to 40-million-year-old bee, and a 120-million-year-old weevil. Isolation and analysis of DNA from Neanderthal skeletons have demonstrated that they are unlikely to have contributed genetic information to modern humans.

Given all these amazing reports, might dinosaur DNA also be retrieved and used to clone a dinosaur? To date, no one has succeeded in retrieving DNA from dinosaur remains (Fig. 13-10). Even if dinosaur DNA were recovered, it is likely that the DNA would be too fragmented to allow us to reconstruct a complete set of dinosaur genes. Recent attempts to clone a wooly mammoth have captured headlines and imaginations through the world, but, as yet, no one has revived a long-extinct species using genetic engineering. In contrast, it appears that ancient microbes have actually been revived without the cloning steps. For example, fungi from grass lining the 5000-year-old shoes of the "Ice Man" appear to have been successfully isolated and cultured. Researchers have even reported that they have cultured bacteria that had been trapped for 250 million years within a salt crystal taken from a New Mexico salt mine!

Figure 13-10 Ancient DNA
Insects (such as this biting midge) some 100 million years old have been found almost perfectly preserved in amber, which dehydrates them. Recent research has dashed hopes of finding intact DNA from these insects, from their stomach contents, or from the bones of fossilized dinosaurs.

6 What Are Some Medical Uses of Biotechnology?

Biotechnology applied to the human genome promises to have a profound and growing impact on human health. One increasingly available application is the use of this technology to screen for genetic defects. Potential parents can learn if they are carriers of a genetic disorder; an embryo can be diagnosed early in a pregnancy (see "Health Watch: Prenatal Genetic Screening"). In the following sections we discuss some other important medical applications of biotechnology for today and tomorrow.

Knock-out Mice Provide Models of Human Genetic Diseases

Once a gene has been cloned, scientists can use the information to make "knock-out" mice. Creation of knock-out mice combines recombinant DNA techniques, cell culturing, and manipulation of early-stage embryos, followed by selective breeding. The result is a mouse strain in which both alleles of a normal gene are replaced with nonfunctional copies. One important use of knock-out mice is to create animal models of human diseases, including cystic fibrosis, sickle-cell anemia, Huntington disease, cancers, Alzheimer disease, and mad cow disease. These genetically engineered animals enable researchers to test the safety and value of new drugs or other treatments before they are used on human patients. Hundreds of different knock-out mice strains are currently available for research.

Genetic Engineering Allows Production of Therapeutic Proteins

Using giant vats of bacteria or yeast that carry and express human genes, biotechnology firms generate large quantities of human proteins. The first human protein made by recombinant DNA technology was insulin. Prior to 1982, when recombinant human insulin was first licensed for use, diabetics used insulin extracted from the pancreas of cattle or pigs slaughtered for meat. Although these insulin proteins are very similar to human insulin, the slight differences caused an allergic reaction in about 5% of diabetics. This problem is avoided with the recombinant human insulin. Recombinant human growth hormone became available in 1985. Some of the other human proteins produced by recombinant DNA technology are listed in Table 13-2.

In addition to producing human proteins in bacteria and yeast, pharmaceutical companies also engineer cows, goats, pigs, rabbits, rats, and sheep to produce various therapeutic human proteins. For example, people who inherit a potentially fatal form of emphysema have a defective gene for the production of the protein alpha-1-antitrypsin (AAT), which they must take as a drug. The gene for the production of human AAT has now been introduced into sheep, who then secrete AAT into

Table 13-2 Examples of Currently Used Products Produced by Recombinant DNA Methods

Type of Product	Purpose	Example			
		Product	Year Approved	Genetic Engineering	Product Used as:
Human hormones	Used in treatment of diabetes, growth deficiency	Humulin™ (human insulin)	1982	Human gene inserted into bacteria	Purified protein
Human cytokines (regulate immune system function)	Used in bone marrow transplants and to treat cancers and viral infections, including hepatitis and genital warts	Leukine™ (granulocyte-macrophage colony stimulating factor)	1991	Human gene inserted into yeast	Purified protein
Antibodies (immune system proteins)	Used to fight infections, cancers, diabetes, organ rejection, and multiple sclerosis	Herceptin™ (antibodies to HER2 protein, expressed at high levels in some breast cancer cells)	1998	Recombinant antibody genes inserted into cultured hamster cell line	Purified protein
Viral proteins	Used to generate vaccines against viral diseases and for diagnosing viral infections	Energiz-B™ (Hepatitis B vaccine)	1989	Viral gene inserted into yeast	Purified protein
Enzymes	Used in treatment of heart attacks, cystic fibrosis, and other diseases, and production of cheeses and detergents	Activase™ (tissue plasminogen activator)	1987	Human gene inserted into cultured hamster cell line	Purified enzyme

Scientific Inquiry
DNA Fingerprinting: A Tool for Medical Detectives

The use of restriction fragment length polymorphisms (RFLPs) in DNA fingerprinting has caused a revolution in many fields of biology. Nowhere has the impact of this technique been felt more than in forensics, the collection of information to be used as evidence in court proceedings. DNA testing was first used in a court case in 1986 in which blood samples from males in an entire British village were tested to solve the rape and murder of two young women. Since then, DNA testing has since entered the mainstream of forensic science. The PCR (polymerase chain reaction) enables investigators to produce a DNA fingerprint from a semen stain, the few cells at the base of a hair shaft, tiny skin fragments found under a victim's fingernails after a struggle, or even a speck of dried blood. Forensic scientists have recently discovered that they can swab off objects regularly handled by a suspect, such as a telephone handset, a briefcase handle, or the inside of vinyl gloves, and use DNA fingerprinting on the collected "fingerprints" themselves! Even the back of a stamp that was licked to attach it to an envelope has enough DNA for analysis!

Some fascinating cases have used DNA from other species to solve murders. When a Canadian woman was murdered in 1994, suspicion fell on her estranged, common-law husband, who was living with his parents and a white cat named Snowball. When investigators found that a discarded man's jacket stained with the victim's blood also had a few white cat hairs clinging to it, they called on Marilyn Menotti-Raymond and co-workers at the National Cancer Institute, who had compiled an extensive map of domestic cat DNA. The researchers extracted DNA from the root of one of the hairs on the jacket, synthesized additional copies using PCR, and compared it to hair and blood samples from Snowball. The resulting perfect match was instrumental in convicting the suspect of murder in 1996.

In addition to providing pivotal evidence for obtaining convictions, the use of DNA fingerprinting has also been instrumental in proving the innocence of dozens of people already convicted of crimes in jury trials. At least 10 of these individuals were on death row, awaiting execution for crimes their milk—as much as 35 grams per liter (just over 1 ounce per quart). This protein may also help prevent lung damage in individuals with cystic fibrosis. In February 2000, clinical trials of this recombinant protein for treatment of patients with cystic fibrosis were under way.

Human Gene Therapy Is Just Beginning

More than 4000 diseases in humans, including cystic fibrosis, sickle-cell anemia, and Huntington disease, result from defects in a single gene. Many more diseases, such as cancer, heart disease, arthritis, and asthma, involve impaired functioning of multiple genes. Gene therapy offers potential treatments for these diseases, as well as infections such as AIDS. Thus, the promise of human gene therapy is enormous. The challenges are equally large.

One major challenge to successful gene therapy is to deliver the functional gene into a large number of the patient's cells. Several approaches for getting a patient's cells to take up DNA are used (Table 13-3). Even if the challenge of getting large numbers of cells to incorporate the gene is solved, gene therapy faces another major challenge: The newly incorporated gene must be efficiently expressed so that the gene will produce enough of its protein to help the patient overcome the disease.

In the U.S., more than 418 clinical gene therapy trials involving more than 2000 patients were under way or had been completed by early 2001. Gene therapy has been most successful for a few children with a rare genetic disorder called *SCID* (*severe combined immunodeficiency*). Such children do not have a functional ADA gene, which encodes an enzyme called adenosine deaminase (ADA). Without this enzyme, individuals cannot develop a functioning immune system. Until the 1980s, children with SCID died in infancy or early childhood, unable to fight off infections. At that time, scientists developed an injectable version of the ADA enzyme prepared from cattle. Bone marrow transplants were also made available for treatment, provided a compatible donor could be found. In 1990 the first test of human gene therapy was performed on a SCID patient, 4-year-old Ashanti DeSilva. Some of her white blood cells were removed, genetically altered with a retrovirus containing a functional ADA gene, and then returned to her bloodstream. By late in 2000, at least 18 children in five countries had received similar gene therapy for SCID (Fig. 13-11). French scientists are hopeful that two children have been permanently cured by this therapy. Other patients are not cured but continue to require injections of ADA.

The limited success of gene therapy has been disappointing. Even more sobering, at least one individual has died as a direct result of his participation in a gene therapy trial. In September 1999 Jesse Gelsinger, an 18-year-old from Arizona, suffered a fatal immune reaction to the virus used in a gene therapy trial. This fatality has caused the National Institutes of Health to increase their scrutiny of all gene therapy trials currently under way. Nevertheless, most researchers in the field remain optimisitic that gene therapy approaches to treating human diseases will eventually prove highly successful.

they did not commit. Currently, only New York and Illinois have laws that guarantee the right of convicted death row inmates to DNA testing. Some states actually have statute of limitation laws that prevent introduction of new evidence after a certain numbers of years have passed following the guilty verdict. In such cases, the convicted person can be denied access to new testing, even if it might prove his or her innocence.

Other areas of biology and medicine are being revolutionized by DNA fingerprinting methods. For example, in 1997 Oxford researchers using DNA fingerprinting announced the longest human lineage ever traced. They extracted DNA from the tooth of a 9000-year-old skeleton found in 1903 in a cave near the town of Cheddar, England. Wondering if descendants of the "Cheddar Man" might still live in the area, they analyzed DNA from local families; amazingly, they found a match in a Cheddar school teacher, separated from his ancestor by 300 generations. In 1998, DNA testing of living descendants indicated that Eston Hemings, the last son born to Sally Hemings, a slave of Thomas Jefferson, was fathered by Thomas Jefferson, the third president of the United States. However, the testing also showed that Jefferson could not have been the father of her oldest son, Thomas. Also in 1998, DNA testing identified remains initially interred in the "Tomb of the Unknowns," a mausoleum containing remains of unidentified war casualties. The tested remains were found to belong to First Lieutenant Michael Blassie; the remains were removed from the tomb and buried near his parents' home in St. Louis. The Pentagon announced that no casualties in future wars would be unidentified. In 2000 the preserved heart of a 10-year-old boy who died in 1795 was shown by DNA testing to be from the son of Marie Antoinette and King Louis XVI. These results ended more than 200 years of speculation that the son might been secretly rescued from his prison cell and thereby survived.

Table 13-3 Possible Vectors for Human Genes

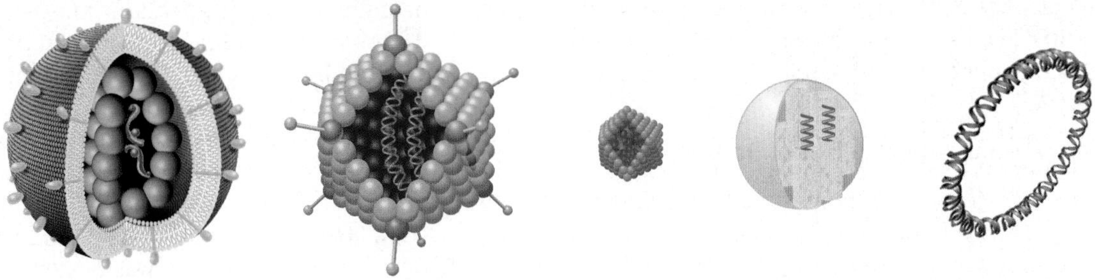

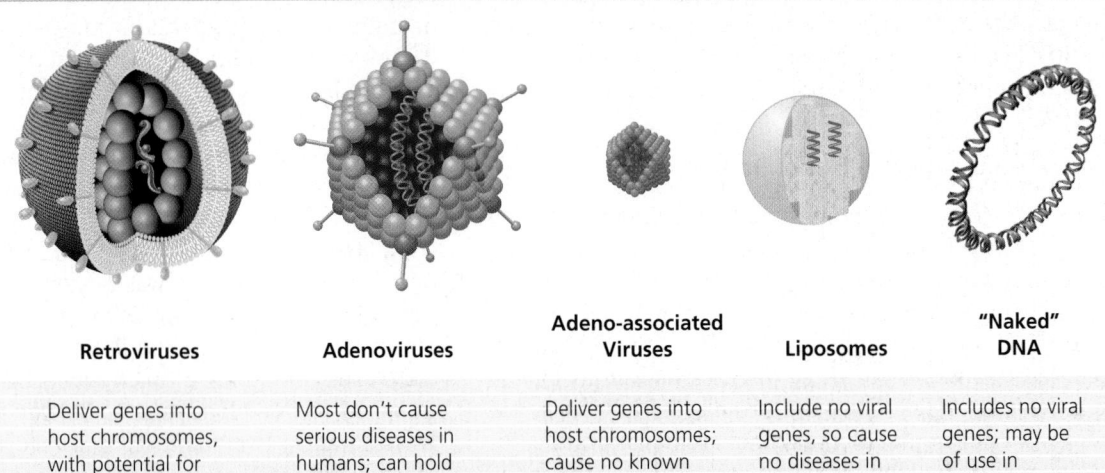

	Retroviruses	Adenoviruses	Adeno-associated Viruses	Liposomes	"Naked" DNA
Potential advantages	Deliver genes into host chromosomes, with potential for long-term cure	Most don't cause serious diseases in humans; can hold many foreign genes	Deliver genes into host chromosomes; cause no known diseases in humans	Include no viral genes, so cause no diseases in humans	Includes no viral genes; may be of use in vaccination
Drawbacks of vectors under study	Genes are delivered randomly, so could disrupt host genes; many infect only dividing cells	Genes may function inconsistently if delivery is incomplete or if attacked by immune system	Can hold few foreign genes	Not as effective as viruses at delivering genes into cells	Ineffective at gene delivery; not stable in most body tissues

(a)

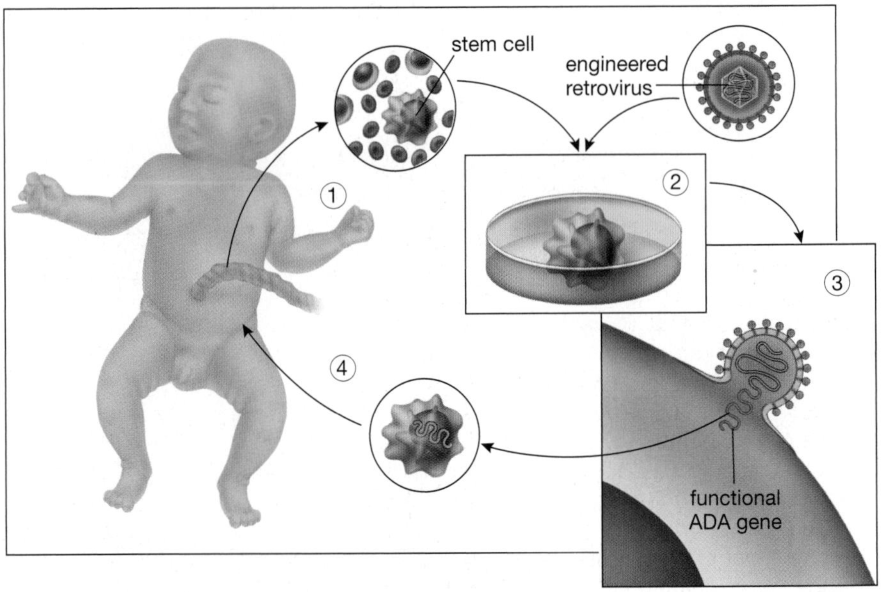

(b)

Figure 13-11 Hope through gene therapy
(a) Andrew Gobea (shown with his mother) received gene therapy on stem cells as a newborn. Eventually, he may produce adequate numbers of normal immune cells on his own. ***(b)*** Genetic engineering of stem cells may permanently replace the defective ADA gene. ① Stem cells are harvested from umbilical cord blood. ② A retrovirus engineered to contain normal human ADA genes is mixed with the stem cells in culture. ③ The retrovirus transmits its DNA, including the functional gene, into the stem cells. ④ The engineered stem cells are injected into the same newborn to take up residence in bone marrow and produce normal immune cells.

The Human Genome Project Has Completed a Working Draft of the Entire Human Genome

The Human Genome Project was launched in 1990 by the National Institutes of Health (NIH) and the Department of Energy (DOE). The project has become an international effort to determine the sequence of nucleotides in each of the roughly 35,000 human genes. Initially, the time, effort, and cost (an estimated $3 billion) of sequencing the human genome seemed daunting. Technological improvements made this goal attainable, much more rapidly than researchers originally imagined possible. In February 2001 two groups of scientists involved in the Human Genome Project published independent working drafts of the human genome sequence. Data from the federally funded Human Genome Project is made available to all researchers, free of charge, via Web sites maintained by the National Center for Biotechnology Information. The technology and information resulting from the Human Genome Project is hastening completion of similar projects under way to determine the DNA sequence of chickens, dogs, cattle, corn, rice, swine, barley, and other organisms. This information will join the already completed sequences of baker's yeast, fruit flies, a roundworm, and about 30 different bacterial species. Such vast amounts of information allow novel approaches in biology that were undreamed of just several years ago.

7 What Are Some Ethical Implications of Human Biotechnology?

As a direct or indirect result of the Human Genome Project, the ability to screen for human diseases is increasing at an exponential rate. One company has developed a technology that can simultaneously analyze DNA from hundreds of patients, looking for mutations in several genes at once. Physicians will soon have access to dozens of easily administered tests that will reveal whether a patient carries a specific gene. These new powers demand a new set of decisions, both ethical and economic, by individuals, physicians, and society.

Genetic Tests for Cystic Fibrosis and Inherited Forms of Breast Cancer Illustrate Potential Problems

Cystic fibrosis (CF) is the most common fatal genetic disease in the U.S. Approximately 850 babies are born in the U.S. each year with cystic fibrosis, most to families with no known history of this disease. To reduce the number of these babies, a 1997 advisory panel of the National Institutes of Health recommended that testing for CF be offered to all couples who consider having a baby. If this policy is adopted, many couples who have never heard of CF will learn that they both carry a copy of the CF allele and that their chances are 1 in 4 of producing a child with CF. Will they get the medical information and counseling they need to deal with the results? They will need to be educated about the symptoms, treatments, and monetary and emotional costs, about the "genetic lottery" they are about to enter, and about their options. Should they remain childless or adopt? Should they conceive and then use prenatal genetic diagnosis to determine whether their child will develop CF? If tests are positive, then what? Is abortion acceptable to the couple? If not, they will give birth to a child who will struggle for breath and be hospitalized repeatedly for respiratory infections despite daily doses of antibiotics. The child will need to lie down and be pounded thoroughly on the chest and back twice a day to dislodge the thick buildup of mucus and will require a special diet fortified with digestive enzymes. Despite these treatments, their child will have a 50% chance of dying before he or she reaches age 30. The costs will be staggering. Should society bear these costs for a family who knowingly takes this risk? Are insurance companies justified in denying coverage for CF treatment to the children of parents who both knowingly carry the CF gene?

Further dilemmas are posed by genetic tests that can indicate an individual has an elevated risk of developing Alzheimer's disease, breast cancer, or ovarian cancer. The presence of the gene does not condemn a woman to cancer, for example, but it does mean that her chances of developing cancer are very high. Some women find living with this "genetic time bomb," and the risk of passing it to their children, very distressing. Many have chosen not to have the test. Others, finding that they carry the gene, have made the painful decision to have both breasts and/or both ovaries surgically removed. Genetic counselors are concerned that women who find they have not inherited the defective gene might become lax about regular screening procedures; these women still have the same risks as others for non-inherited types of breast cancer. Many women who choose to be tested for the gene do so anonymously or under a false name to prevent the information from reaching their insurance companies.

The increasing availability and use of tests for genetic diseases raise major concerns about genetic discrimination by insurance companies and employers. Many people worry that, under pressure to hold down rising costs, insurance companies will require information about genetic diseases and will use this information as a basis for denying or restricting the availability of life insurance or health insurance. Cases are on record in which companies terminated or denied health insurance

Health Watch
Prenatal Genetic Screening

Prenatal diagnosis of a variety of genetic disorders, including cystic fibrosis, sickle-cell anemia, Tay-Sachs disease, and Down syndrome, requires samples of fetal cells or chemicals produced by the fetus. Presently, there are two main techniques used to obtain these samples—*amniocentesis* and *chorionic villus sampling*—and a technique for analyzing fetal cells from maternal blood is under development. Once samples are collected, several tests, including some that use recombinant DNA techniques, can be performed that allow prenatal diagnosis of many genetic disorders.

Amniocentesis The human fetus, like all animal embryos, develops in a watery environment. As you'll see in Chapter 36, a waterproof membrane called the *amnion* surrounds the fetus and holds the fluid. As the fetus develops, it sheds some of its own cells into the fluid, which is called *amniotic fluid*. When a fetus is 16 weeks or older, amniotic fluid can be collected safely by a procedure called **amniocentesis**. A physician determines the position of the fetus by ultrasound scanning, inserts a sterilized needle through the abdominal wall, the uterus, and amnion, and withdraws 10 to 20 milliliters of fluid (Fig. E13-2). Biochemical analysis may be performed on the fluid immediately, but there are very few cells in the fluid sample. For many analyses, such as karyotyping for Down syndrome, the cells must first be grown in culture, where they multiply. After a week or two, there are normally enough cells for karyotyping or other analyses.

Chorionic Villus Sampling As you'll see in Chapter 36, the *chorion* is a membrane that is produced by the fetus and becomes part of the placenta. The chorion produces many small projections, called *villi*; the loss of a few villi seems to cause no harm. In **chorionic villus sampling (CVS)**, a physician inserts a small tube into the uterus through the mother's vagina and suctions off a few fetal villi for analysis (see Fig. E13-2). CVS has two major advantages over amniocentesis. First, it can be done much earlier in pregnancy—as early as the eighth week. This factor is especially important if the woman is contemplating a therapeutic abortion in case the fetus has a major defect. Second, the sample contains a much higher concentration of fetal cells than amniocentesis can obtain, so analyses can be performed immediately rather than a week or two later. However, chorionic cells tend to be more likely to have abnormal numbers of chromosomes (even when the fetus is normal), which complicates karyotyping.

Fetal Cells from Maternal Blood Fetal cells become detectable in maternal blood between the 6th and 12th weeks of pregnancy. The goal of a new procedure is to harvest and analyze fetal cells collected from a 20-milliliter sample (just over a tablespoon) of maternal blood. If successful, this procedure might replace both amniocentesis and CVS as a means of obtaining fetal cells. By eliminating the need to penetrate the uterus, blood sampling greatly reduces the risk of injury and infection to both mother and fetus. But only about 1 in every 100,000 cells in maternal blood comes from the fetus, so isolating fetal cells efficiently is a major challenge. A large study is currently in progress to determine the feasibility of this technique.

Analyzing the Samples Several types of analyses can be performed on the fetal cells or on the amniotic fluid (see Fig. E13-2). Biochemical analysis is used to determine the concentration of chemicals in the amniotic fluid. For example, Tay-Sachs disease and many other metabolic disorders can be detected by the low concentration of the enzymes that normally catalyze specific metabolic pathways or by the abnormal accumulation of precursors or by-products. Analysis of the chromosomes of the fetal cells can show if all the chromosomes are present, if there are too many or too few of some, and if any chromosomes show structural abnormalities.

based on required genetic tests, or family history of genetic diseases, such as Huntington disease. As more genetic information becomes available and treatments become more costly, this trend is sure to intensify. State and federal legislatures are devising new laws to address these issues. By April 1999 at least 24 states had passed laws prohibiting genetic discrimination. Thus far, no federal laws dealing with genetic information have been passed, although several have been introduced in the House of Representatives or the Senate.

Sickle-Cell Anemia and Tay-Sachs Disease Illustrate the Hazards and Benefits of Genetic Screening Programs

An example of the hazards of widespread genetic screening without adequate education occurred in the 1970s when a major screening program for **sickle-cell anemia** was launched and, in some areas, even made mandatory. Although the intentions were good, the results were misused through ignorance and prejudice. Healthy carriers of the sickle-cell trait (most of whom are African American) were led to believe they were sick. Some African-American carriers were denied insurance; others were blocked from admission to the U.S. Air Force Academy or from jobs as flight attendants—on the inaccurate assumption that they were more likely to be affected by the lack of oxygen at high altitudes. Suggestions that carriers (who make up almost 7% of the African-American population) should avoid having children led to fears that the testing was intended as a means of reducing the numbers of this racial minority.

Testing for **Tay-Sachs disease**, which is prevalent among Jews of Eastern European descent, is a contrasting success story. Tay-Sachs is a fatal degenerative disease

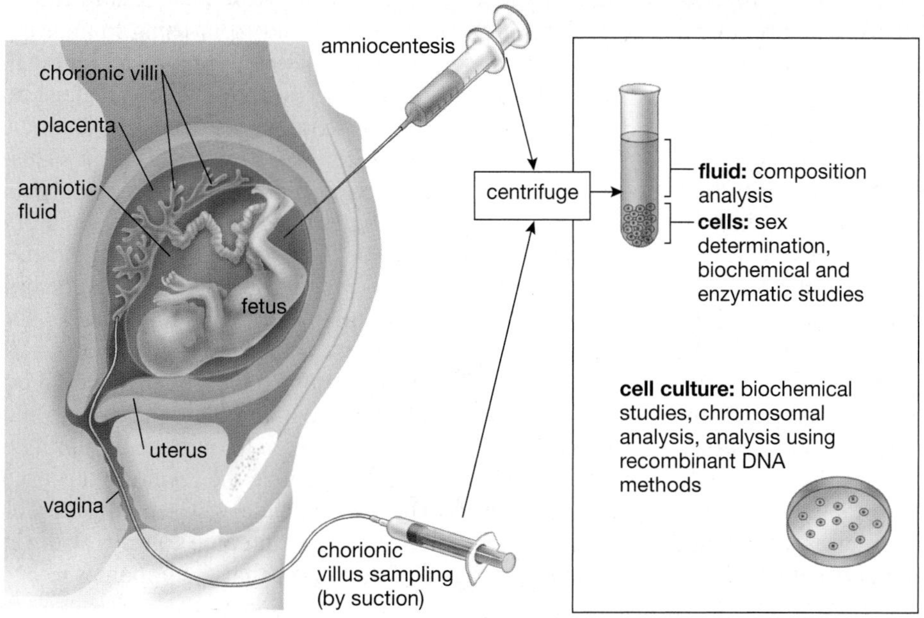

Figure E13-2 Prenatal cell sampling techniques
Two methods of obtaining fetal cell samples—amniocentesis and chorionic villus sampling—and some of the tests performed on the fetal cells.

Recombinant DNA techniques can be used to analyze the DNA of fetal cells to detect many defective alleles, such as those for cystic fibrosis, Tay-Sachs, or sickle-cell anemia. Prior to the development of PCR, fetal cells typically had to be grown in culture for as long as 2 weeks before cells had multiplied sufficiently. Now, the second step in prenatal diagnosis is to extract the DNA from a few cells and to use PCR to synthesize the region containing the gene in question. After a few hours, enough DNA is available for techniques such as RFLP analysis, which is used to detect the allele that causes sickle-cell anemia. Technicians cut the PCR-synthesized DNA with restriction enzymes. Because the single-nucleotide substitution that causes sickle-cell anemia happens to destroy a cutting site for a particular restriction enzyme, there will be a recognizable difference in the length of certain restriction fragments. This difference allows the hemoglobin genotype of the fetus to be easily determined. If the infant is homozygous for the sickle-cell allele, some therapeutic measures can be taken. In particular, regular doses of penicillin greatly reduce bacterial infections that otherwise kill about 15% of homozygous children. Further, knowing that a child has the disorder ensures correct diagnosis and rapid treatment, should "sickling crises" occur.

in which an initially normal infant gradually loses mental and physical abilities, becomes blind and deaf, and dies in early childhood (Fig. 13-12). This disease is recessive, like sickle-cell anemia, and results from deficiency of enzymes that regulate lipid breakdown in the brain. It is such a devastating illness that few people will knowingly give birth to such a child. Voluntary genetic testing for Tay-Sachs within the Jewish community, coupled with prenatal testing and careful genetic counseling, has dramatically reduced the incidence of this fatal disorder.

The Potential to Clone Humans Raises Further Ethical Issues

The cloning of Dolly the sheep from the nucleus of an adult cell introduced into the cytoplasm of an egg (see "Scientific Inquiry: Much Ado About Dolly" in Chapter 11) led to a flurry of debate and proposed legislation

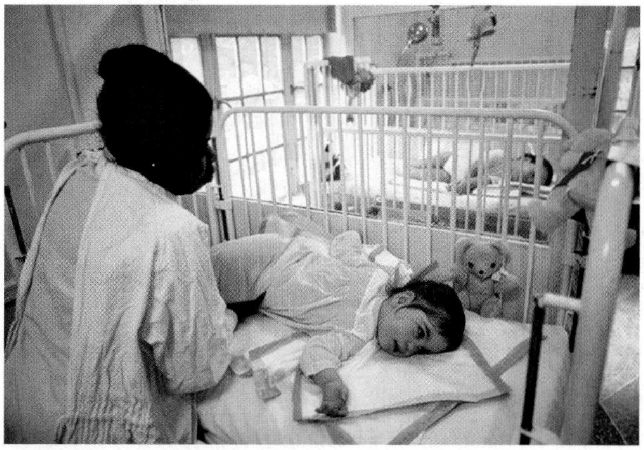

Figure 13-12 Tay-Sachs disease
Apparently bright, normal children with Tay-Sachs disease begin to deteriorate at about 6 months of age. As the nervous system is progressively destroyed, the children gradually lose all ability to function, and most die in early childhood. There is no treatment.

regarding human cloning. There are no known technical barriers to the eventual development of human cloning. Are there ethical barriers that should restrict this technology to nonhuman species? One argument is that there is no valid reason to clone humans. Perhaps a small number of infertile couples would want to perpetuate their genes in this way, but there are already so many reproductive options (including sperm or egg donors) that cloning would not meet a pressing need. On the other hand, such technology might allow permanent correction of genetic defects in the couple's children (Fig. 13-13). For some people, the possibility that cloning would be used by unscrupulous individuals in positions of power to perpetuate themselves or to create copies of certain people to fill specific functions raises the specter of a "Brave New World" type of society. Some argue that biological uniqueness is a fundamental human right and intrinsic to human dignity.

Countering these arguments are those who point out that identical twins are more similar to one another than clones would ever be, because twins share not only the same genes, but also cytoplasmic factors from the same original egg cell, the same uterine environment, and the same home environment and upbringing. An individual trying to produce a self-copy might be sorely disappointed at the differences these environmental variables (not to mention growing up in a different generation) might produce.

Human cloning is illegal in England and Norway, but not in the U.S. However, at present, federally funded U.S. researchers are prohibited from using human embryos in research, a necessary step in human cloning. In contrast, privately funded research in the U.S. does not operate under these restrictions and has used human embryos discarded after in-vitro fertilization. These studies have succeeded in isolating embryonic stem cells, the precursors for all types of adult tissues. These embryonic stem cells might be used to clone humans, but they also might be used to regenerate adult tissues, such as bone marrow, hearts, and lungs. For example, recent success in using embryonic stem cells to regrow nerve cells offers hope of treating paralysis. Clearly, advances in our understanding and ability to manipulate the genes of humans and other species threaten to outpace the ability of society to understand and utilize the resulting technologies effectively. The next decade will see considerable upheaval as society struggles to come to grips with this new knowledge and the power that accompanies it.

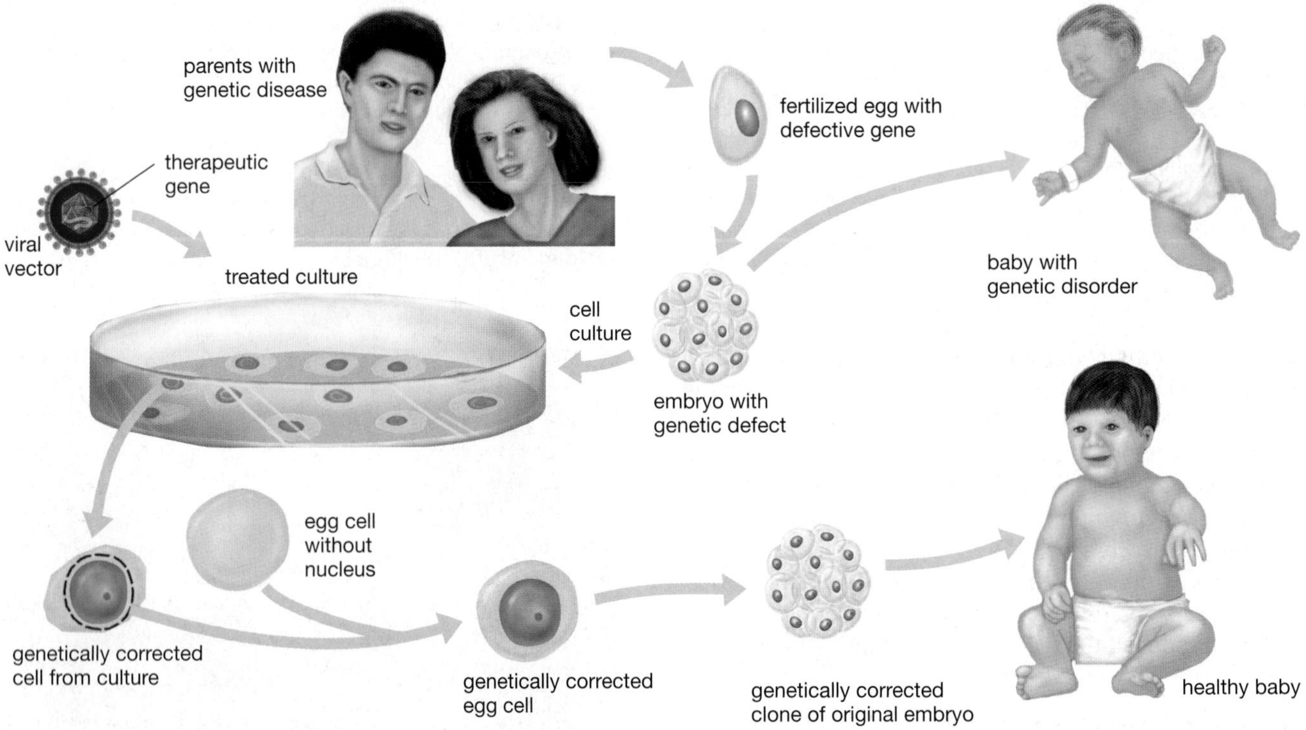

Figure 13-13 Human cloning technology might allow permanent correction of genetic defects
Researchers might derive human embryos from eggs fertilized in culture dishes, using sperm and eggs from the natural parents, one or both of whom have a genetic disorder. When an embryo containing a defective gene grows into a small cluster of cells, a single cell could be removed from the embryo and the defective gene replaced by means of an appropriate vector. Then the repaired nucleus could be implanted into another egg (taken from the mother) whose nucleus had been removed. The repaired, diploid egg cell could then be implanted in the mother's uterus for normal development.

CASESTUDYREVISITED

Teaching an Old Grain New Tricks

Creating a rice strain with high levels of vitamin A precursors in their seeds was not a simple task. In fact, many scientists were skeptical that it could be done at all. However, Ingo Potrykus, a Swiss researcher, and Peter Beyer, a German researcher, teamed up and, with their students, inserted three genes into the rice genome. Two genes came from daffodils and one from a bacterium. Regulatory DNA sequences were included with the genes to control their expression, so that the inserted genes would be turned on in the rice seeds. The resulting genetically modified rice plants produce enough beta-carotene to prevent vitamin A deficiency if three bowls of the rice are eaten each day. The research to produce the golden rice was funded by non-profit organizations, including The Rockefeller Institute, the European Community Biotech Program, and the Swiss Federal Office for Education and Science. Seeds for the bioengineered rice are being supplied free of charge to agricultural research centers in developing countries. Local scientists or farmers can cross the bioengineered strain with their local rice strains to produce more nutritious rice that will grow well in local conditions.

Many people herald this golden rice as the first in a wave of bioengineered crops that will have higher nutritional values. Some even argue that these crops will help decrease the suspicion with which some consumers view biotechnology. Up till now, most bioengineering of crops has focused on meeting the needs of the farmer and farming industry. For example, pest-resistant potatoes indirectly benefit the consumer by allowing decreased amounts of pesticides to be used in farming. These same potatoes directly benefit the farmer, who obtains increased yields with less cost, and the biotech company that provides the seed potatoes for planting. Availability of a bioengineered crop that might prevent death and blindness in hundreds of thousands of impoverished children would seem more difficult for consumers to oppose.

Many consumers are skeptical of the safety of genetically modified animals and plants. Take a scientific approach to this issue: Make a list of the potential negative consequences of genetic engineering of crops and livestock. State each potential negative consequence as a hypothesis, for example, "The introduction of genes from another species may. . . ." Describe an experiment that would test that hypothesis. What evidence would disprove the hypothesis for each potential negative consequence on your list?

Summary of Key Concepts

1) What Is Biotechnology?

Biotechnology is any industrial or commercial use or alteration of organisms, cells, or biological molecules to achieve specific practical goals. Modern biotechnology generates altered genetic material via genetic engineering. Genetic engineering frequently involves the production of recombinant DNA by combining DNA from different organisms. DNA is transferred between organisms by means of vectors, such as bacterial plasmids. The resulting organisms are called *transgenic*. Some major goals of genetic engineering are to increase our understanding of gene function, to treat disease, and to improve agriculture.

2) How Does DNA Recombination Occur in Nature?

DNA recombination occurs naturally through processes such as sexual reproduction (during crossing over), bacterial transformation, in which bacteria acquire DNA from plasmids or other bacteria, and viral infection, in which viruses incorporate fragments of DNA from their hosts and transfer the fragments to members of the same or other species.

3) How Does DNA Recombination Occur in Genetic Engineering Laboratories?

DNA can be cut into specific, reproducible fragments using restriction enzymes. Such fragments can then be inserted into a vector, which allows the DNA to be replicated within a host cell. The joining of a DNA fragment and a vector produces recombinant DNA molecules.

4) How Can Researchers Identify Specific Genes?

To identify a specific gene, researchers usually must isolate a recombinant DNA molecule that contains the gene of interest. Many methods are available to identify specific genes, including analysis of RFLP (restriction fragment length polymorphism), similarity to other previously cloned genes, the protein product produced by the gene, and rescue of a mutation. RFLP analysis is also used for DNA fingerprinting, in which restriction enzymes are used to break the DNA into fragments whose different lengths are characteristic of different individuals. DNA fingerprints can establish relatedness of individuals and is used in forensics to analyze biological material at a crime scene and link it to the suspect.

5) What Are Some Applications of Biotechnology?

Once a gene has been identified, sufficient quantities can be generated so that the nucleotide sequence of the gene can be determined. DNA sequencing typically uses a process called the *polymerase chain reaction* (*PCR*), which can also be used to synthesize specific fragments of DNA. Researchers can then synthesize and study the coded protein or devise tests for mutated forms of the gene. Biotechnology is revolutionizing agriculture by allowing scientists to engineer plants that resist pests and herbicides and that have superior qualities for shipping and eating, and to produce better fibers.

6 What Are Some Medical Uses of Biotechnology?

Knock-out mice, in which a specific gene has been rendered nonfunctional, provide laboratory animal models of human genetic diseases that can be studied to determine disease mechanisms and to test various treatments. Genetic engineering allows production of therapeutic proteins, such as human insulin, by culturing huge quantities of bacteria, yeast, or mammalian cultured cells that contain the recombinant gene and extracting the protein. Farm animals can also be genetically engineered to secrete proteins of interest in milk. Human gene therapy is in its infancy. Vectors such as viruses have been used to deliver functional genes to specific cells that lack them. This technique has been used successfully in treating children with a deficiency in adenosine deaminase, an enzyme essential in immune-system function. The sequences of many bacteria and several eukaryotic organisms has already been determined. In addition, an ambitious collaborative project to sequence the entire human genome is nearing completion.

7 What Are Some Ethical Implications of Human Biotechnology?

Dozens of tests for genetic diseases are already available, and many more are under development. To administer these tests appropriately, the physician or genetic counselor must decide who should receive the tests, and then educate the patients about the disorder and assist them in making a series of difficult decisions about their own health or that of potential offspring. Questions include whether to conceive or adopt a child, whether to have prenatal testing, and whether to abort a fetus with a fatal or permanently debilitating disorder. Should insurance companies pay the lifelong costs to treat children with genetic diseases born to parents who know they carry the defective gene? How can society protect individuals from various types of discrimination on the basis of their genetic makeup? Should research on human cloning techniques be permitted? Both the rapid acquisition of knowledge about genetics and the power to manipulate the genome of humans and other organisms threaten to outpace our ability to deal with the consequences.

Key Terms

amniocentesis *p. 260*
biotechnology *p. 244*
chorionic villus sampling (CVS) *p. 260*
DNA fingerprinting *p. 251*
DNA library *p. 247*
DNA probe *p. 248*
gel electrophoresis *p. 248*
genetic engineering *p. 244*
plasmid *p. 244*
polymerase chain reaction (PCR) *p. 249*
recombinant DNA *p. 244*
restriction enzyme *p. 246*
restriction fragment length polymorphism (RFLP) *p. 248*
sickle-cell anemia *p. 260*
Tay-Sachs disease *p. 260*
transformation *p. 244*
transgenic *p. 244*
vector *p. 247*

Thinking Through the Concepts

Multiple Choice

1. *Restriction enzymes are*
a. isolated from bacterial cells
b. used to produce DNA fingerprints
c. used to create recombinant plasmids
d. used to create a DNA library
e. all of the above

2. *Restriction fragment length polymorphisms*
a. all of the following
b. can be used to detect differences in DNA among individuals
c. have been used to identify human genes
d. can be used to analyze fetal cells to detect disorders prenatally
e. are produced through the use of restriction enzymes

3. *The polymerase chain reaction*
a. is a method of synthesizing human protein from human DNA
b. takes place naturally in bacteria
c. can produce billions of copies of a DNA fragment in several hours
d. uses restriction enzymes
e. is relatively slow and expensive compared with other types of DNA purification

4. *Knock-out mice*
a. are a particularly aggressive strain
b. have been genetically engineered to produce a human protein
c. produce human proteins
d. have been genetically engineered to lack a specific mouse gene
e. are naturally produced mutant mice

5. *DNA fingerprinting*
a. requires large amounts of DNA
b. is useful only for forensic analysis
c. can involve analysis of RFLPs
d. can only prove guilt, never innocence
e. is a constitutional right

6. *A child with a single allele for sickle-cell anemia*
a. is likely to die before age 30 from this disease
b. will have difficulty getting enough oxygen at high altitudes
c. will have difficulty participating in sports
d. will need immediate treatment at birth
e. will be able to lead a perfectly normal life

? Review Questions

1. Describe three natural forms of genetic recombination, and discuss the similarities and differences between recombinant DNA technology and these natural forms of genetic recombination.
2. What is a plasmid? How are plasmids involved in bacterial transformation?
3. What is a restriction enzyme? How can restriction enzymes be used to splice a piece of human DNA within a plasmid?
4. What is a DNA library? Briefly describe the steps involved in creating a DNA library of a mouse genome.
5. What is a restriction fragment length polymorphism (RFLP)? Describe how might be useful in proving innocence of a murder suspect or determining relatedness among people.
6. Describe the polymerase chain reaction.
7. Describe several uses of genetic engineering in agriculture.
8. Describe several uses of genetic engineering in human medicine.
9. Describe amniocentesis and chorionic villus sampling, including the advantages and disadvantages of each. What are their medical uses?

Applying the Concepts

1. Discuss the ethical issues that surround the release of bioengineered organisms (plants, animals, or bacteria) into the environment. What could go wrong? What precautions might prevent the problems you listed from occurring? What benefits do you think would justify the risks?
2. Do you think that using recombinant DNA technologies to change the genetic composition of a human egg cell is ever justified? If so, what restrictions should be placed on such a use? What about human cloning?
3. Are there any conditions under which insurance companies or employers should be allowed access to the results of tests for genetic defects? Discuss this issue from the standpoint of the company or employer, the individual who was tested, and society as a whole.
4. In what ways was the program to test for sickle-cell anemia in the 1970s flawed? What criteria should be met (education, privacy guarantees, etc.) before any genetic screening test is administered to the general public?
5. If you were contemplating having a child, would you want both yourself and your spouse tested for the cystic fibrosis gene? If both of you were carriers, how would you deal with this decision?

For More Information

Beardsley, T. "Vital Data." *Scientific American*, March 1996. Discusses the Human Genome Project and the problems and prospects of the tests for genetic diseases that are arising as a spin-off of this project.

Collins, F. S. and others. "New Goals for the U.S. Human Genome Project: 1998–2003." *Science*, October 1998. Explains goals of human genome project, including those involving bioethics.

Fackelmann, K. A. "DNA Dilemmas." *Science News*, December 17, 1994. Uses case histories to explore the ethical controversies surrounding human applications of genetic engineering.

Gibbs, N. "Baby, It's You! And You, and You. . . ." *Time,* February 19, 2001. How close are we to cloning a human?

Gibbs, W. W. "Plantibodies." *Scientific American*, November 1997. Clinical trials using human antibodies engineered into corn began in 1998; the target was tooth decay!

Gura, T. "New Genes Boost Rice Nutrients," *Science*, August 1999. Explanation of how rice was genetically engineered to produce vitamin A precursors.

Karapelou, J. "Gene Therapy: Special Delivery." *Discover*, January 1996. Describes several methods on trial for delivering therapeutic genes into human cells.

Langridge, W. H. R. "Edible Vaccines." *Scientific American*, September 2000. Current state of attempts to develop plants that produce vaccines or treatments for diseases.

Lanza, R. P., Dresser, B.L., and Damiani, P. "Cloning Noah's Ark." *Scientific American*, November 2000. Using domestic animals as surrogate mothers for endangered species provides some hope for staving off extinction.

Marvier, M. "Ecology of Transgenic Crops." *American Scientist,* March/April 2001. A thoughtful article that weighs the benefits, risks, and uncertainties of bioengineered crops.

Miller, R. V. "Bacterial Gene Swapping in Nature." *Scientific American*, January 1998. How likely is it that genes introduced into bioengineered organisms might be inadvertently transferred to wild organisms?

Mirsky, S., and Rennie, J. "What Cloning Means for Gene Therapy." *Scientific American*, June 1997. Could human cloning technology help permanently repair genetic defects?

Roberts, L. "To Test or Not to Test." *Science*, January 1990. Should the entire childbearing population be screened for carriers of cystic fibrosis?

Ronald, P. C. "Making Rice Disease-Resistant." *Scientific American*, November 1997. Rice, a staple crop for millions of people, has finally been engineered for disease resistance.

Scientific American, June 1997. A special issue devoted to the prospects of human gene therapy.

Travis, J. "Cystic Fibrosis Controversy." *Science News*, May 10, 1997. Might gene therapy in the womb help CF patients?

Answers to Multiple-Choice Questions

1.e 2.a 3.c 4.d 5.c 6.e

MediaTutor
Biotechnology

CD Activities

Activity 13.1: Genetic Recombination

Estimated time: 10 minutes

We often think of DNA as being static and unchanging. In fact, it is anything but static. This tutorial will demonstrate some of the ways that genetic recombination can occur in bacteria. As you will see, there are many opportunities for bacteria to exchange genetic material with other bacteria. What are the consequences of these exchanges? Does this happen in other organisms too?

Activity 13.2: Building a DNA Library

Estimated time: 10 minutes

Biotechnology uses modern technology to modify biological molecules. For example, biotechnology enables scientists to produce (or overproduce) certain molecules that are needed by humans. Often, this technique can be used to compensate for an individual's inability to produce these molecules. An important step in this technique is to find the particular gene in question. The traditional way to find the gene is to create a DNA library containing all of the DNA of an organism, but in small pieces. This tutorial will lead you through the steps involved in producing a DNA library from a human being.

Start the MediaTutor Student CD-ROM and enter the activity number in the Quick Search box to be taken directly to that activity.

Web Investigations

Case Study: Teaching an Old Grain New Tricks

Estimated time: 15 minutes

Improving nature is not a new idea. Most domestic animals are the products of highly selective breeding. Think about milk cattle, beef cattle, hunting dogs, herding dogs, draft horses, and even lab rats. The introduction of corn hybrids in the mid-twentieth century doubled acreage yields and reduced the genetic diversity of seed corn. To many people, "biotechnology" means the introduction of potentially dangerous weed- or pesticide-resistance genes into food crop genomes. Genetically-engineered golden rice is meant to help feed the world. When evaluating genetic engineering, should we take the intended result of the modification into account?

Go to http://www.prenhall.com/audesirk6, the Audesirk Companion Web site. Select Chapter 13 and Web Investigation to begin.

UNIT THREE

Evolution

The ghostly grandeur of ancient bones evokes images of a lost world. Fossil remnants of extinct creatures, such as this Triceratops *dinosaur skeleton, provide clues to the biologists who attempt to reconstruct the history of life.*

"Nothing in biology makes sense except in the light of evolution."

Theodosius Dobzhansky

Newly discovered fossils of feathered dinosaurs such as *Caudipteryx* (shown here in an artist's reconstruction) provide powerful evidence that today's birds descended from dinosaur ancestors.

14 Principles of Evolution

AT A GLANCE

Case Study: A Missing Link Unearthed

1) **How Did Evolutionary Thought Evolve?**

Evidence Supporting Evolution Emerged Even Before Darwin's Time

Both Darwin and Wallace Proposed That Evolution Occurs by Natural Selection

Evolutionary Theory Arises from Scientific Observations and Conclusions Based on Them

2) **How Do We Know That Evolution Has Occurred?**

The Fossil Record Provides Evidence of Evolutionary Change over Time

Comparative Anatomy Provides Structural Evidence of Evolution

Embryological Stages of Animals Can Provide Evidence of Common Ancestry

Modern Biochemical and Genetic Analyses Reveal Relatedness Among Diverse Organisms

3) **What Is the Evidence That Populations Evolve by Natural Selection?**

Artificial Selection Demonstrates That Organisms May Be Modified by Controlled Breeding

Evolution by Natural Selection Occurs Today

4) **A Postscript by Charles Darwin**

Case Study Revisited: A Missing Link Unearthed

CASESTUDY

A Missing Link Unearthed

As the twentieth century drew to an end, the little village of Sihetun in northeastern China became the unlikely setting for some of history's most breathtaking fossil discoveries. Working in dusty quarries around the village, Chinese fossil hunters extracted the exquisitely preserved remains of some previously undiscovered types of dinosaurs. New dinosaur discoveries are always cause for celebration among students of evolutionary biology, but the new Chinese specimens sported a distinctive feature that propelled them to superstar status: They had feathers. Plainly visible along the margin of these clearly dinosaurian fossil skeletons were impressions of what were unmistakably feathers. For the first time, scientists had hard evidence of the existence of feathered dinosaurs.

As descriptions of the growing collection of new fossils were published in 1998 and 1999, incredulity gave way to delight among many paleontologists (the scientists who study fossils). A flurry of news stories alerted the public to the exciting new developments. Why such a big fuss over feathers? Because the new discoveries seemed to finally confirm the controversial theory that dinosaurs, those icons of extinction and former rulers of Earth, were the ancestors of the birds that today flit about the trees in our backyards and flock to our feeders. The fossils are evidence that modern birds arose from a branch of the dinosaur family tree, a branch whose members had feathers. That is, today's birds arose through **evolution**, the process by which the characteristics of individuals in a population change over time. ■

1 How Did Evolutionary Thought Evolve?

Evidence Supporting Evolution Emerged Even Before Darwin's Time

Pre-Darwinian science, heavily influenced by theology, held that all organisms were created simultaneously by God and that each distinct life-form remained fixed, immutable, and unchanging from the moment of its creation. This explanation of how life's diversity came to be was elegantly expressed by the ancient Greek philosophers, especially Plato and Aristotle. Plato (427–347 B.C.) proposed that each object on Earth was merely a temporary reflection of its divinely inspired "ideal form." Plato's student Aristotle (384–322 B.C.) categorized all organisms into a linear hierarchy that he called the "ladder of Nature."

These ideas formed the intellectual basis of the view that the form of each type of organism is permanently fixed. This view reigned unchallenged for nearly 2000 years. By the eighteenth century, however, several lines of newly emerging evidence began to erode the dominance of this static view of Creation.

Exploration of New Lands Revealed a Staggering Diversity of Life

As early European naturalists explored the newly discovered lands of Africa, Asia, and America, they found that the number of **species**, or different types of organisms, was much greater than anyone had suspected. Naturalists also observed that some of these exotic species closely resembled one another, yet also differed in some characteristics. These observations led some naturalists to consider that perhaps species could change after all. Perhaps some of the similar species might have developed from a common ancestor.

Fossils in Rocks Resembled Parts of Living Organisms

As new lands were explored, excavations for roads, mines, and canals revealed that many rocks occur in layers (Fig. 14-1). In some cases, a few strangely shaped rocks or fragments were found embedded within one of these layers. These **fossils** (from the Latin, meaning "dug up") resembled parts of living organisms. At first, fossils were thought to be ordinary rocks that wind, water, or people had worked into lifelike forms. As more and more fossils were discovered, however, it became obvious that they were the remains of plants or animals that had died long ago and been changed into or in some way preserved in rock (Fig. 14-2). The rapidly accumulating fossil discoveries also revealed that fossils come in many forms. The classic image of a fossil is of bones or other hard parts (such as shells or wood) that have been transformed into rock by eons of geological processing. But fossils also include casts, molds, and other impressions that organisms left in ancient sediments before decomposing. Some of the most interesting and informative fossils are the trails, burrows, tracks, or droppings that organisms left behind. In fact, any tangible trace of an organism that is preserved in rock or sediments is a fossil.

Figure 14-1 The Grand Canyon of the Colorado River
Layer upon layer of sedimentary rock forms the walls of the Grand Canyon. The canyon strata (rock layers) cover more than a billion years of evolutionary history.

These windows into the past are fascinating in and of themselves, but the distribution of fossils in rock can also be revealing. After studying fossils carefully, the British surveyor William Smith (1769–1839) realized that certain fossils were always found in the same layers of rock. Further, the organization of fossils and rock layers was consistent: Fossil type A could always be found in a rock layer resting atop an older layer containing fossil type B, which in turn rested atop a still older layer containing fossil type C, and so on.

Fossil remains also showed a remarkable progression of form. Most fossils found in the lowest (and therefore oldest) rock layers were very different from modern forms; the resemblance to modern forms gradually increased upward toward younger rocks, as if there were indeed a ladder of Nature stretching back in time. Many of these fossils were the remains of plant or animal species that had gone *extinct*—that is, no members of the species still lived on Earth (Fig. 14-3, p. 272). Putting these facts together, scientists came to the inescapable conclusion that different types of organisms had lived at various times in the past.

But what did this newfound richness of organisms, both living and extinct, mean? Was each organism produced by a separate act of Creation? If so, why? And why bother to create so many types, letting thousands of types go extinct? The French naturalist Georges Louis LeClerc (1707–1788), known by the title Comte de Buffon, suggested that perhaps the

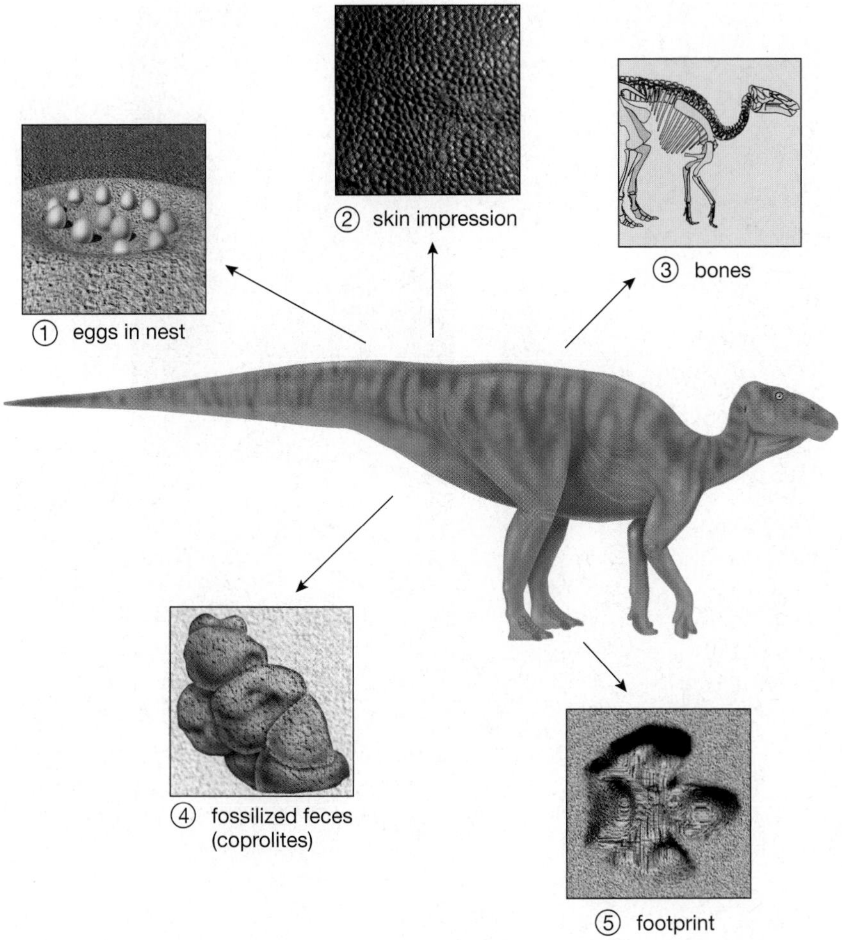

Figure 14-2 Types of fossils
Many kinds of evidence may be preserved in rock to form fossils, such as (1) eggs, (2) impressions of skin, (3) bones, (4) feces, or (5) footprints.

original Creation provided a relatively small number of founding species and that some of the modern species had been "conceived by Nature and produced by Time"—that is, they had evolved through natural processes. Most people were not convinced. First, Buffon could not provide a mechanism whereby Nature could "conceive" new species. Second, no one thought that Earth was old enough to allow sufficient time for the "production" of new species.

Geology Provided Evidence That Earth Is Exceedingly Old

In the early 1700s, few scientists suspected that Earth could be more than a few thousand years old. Counting generations in the Old Testament, for example, yields a maximum age of 4000 to 6000 years. From the descriptions of plants and animals from ancient writers such as Aristotle, it was clear that wolves, deer, lions, and other European organisms had not changed in more than 2000 years. How, then, could whole new species have arisen if Earth was created only a couple of thousand years before Aristotle's time?

To account for a multitude of species, both extinct and modern, while preserving the notion of Creation, Georges Cuvier (1769–1832) proposed the theory of **catastrophism**. Cuvier, a French paleontologist, hypothesized that a vast supply of species was created initially. Successive catastrophes (such as the Great Flood described in the Bible) produced the layers of rock and destroyed many species, fossilizing some of their remains in the process. The reduced flora and fauna of the modern world, he theorized, are the species that survived the catastrophes. However, if modern *species* have survived from an original Creation, then many *individuals* of those species should have died in the ancient catastrophes. Surely some of them would have been fossilized, and even the lowest and oldest rock layers should contain fossils of present-day species. Unfortunately for Cuvier's hypothesis, the vast majority of fossils are of extinct species. To account for this observation, the French geologist Louis Agassiz (1807–1873) proposed that there was a new creation after each catastrophe and that modern species result from the most recent creation. The fossil record forced Agassiz to hypothesize at least 50 separate catastrophes and creations!

Alternatively, perhaps Earth *is* old enough to allow for the production of new species. Geologists James Hutton (1726–1797) and Charles Lyell (1797–1875) contemplated the forces of wind, water, earthquakes, and

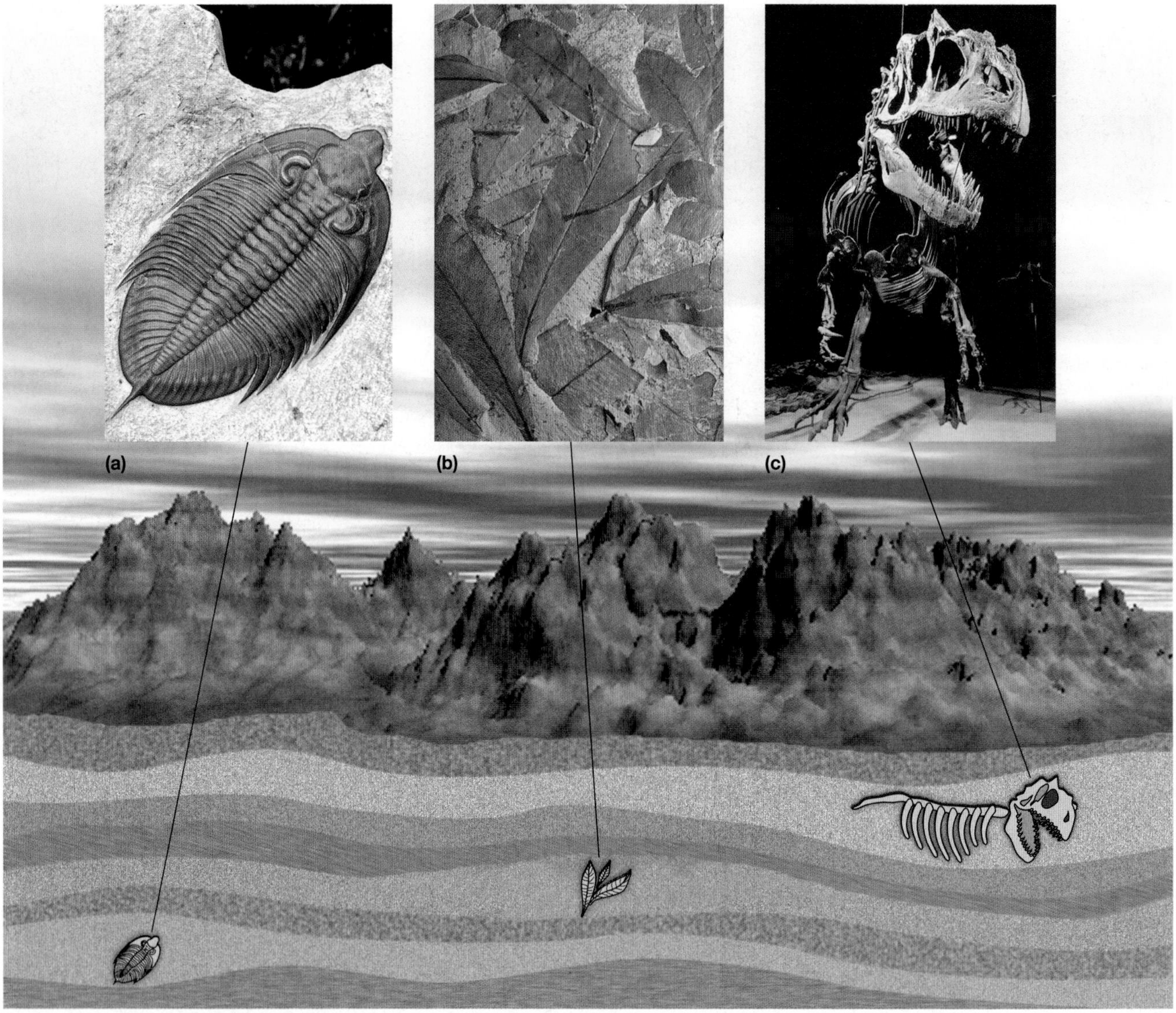

Figure 14-3 Fossils of extinct organisms
Fossils provide strong support for the idea that today's organisms were not created all at once but arose over time by the process of evolution. If all species were created simultaneously, we would not expect a fossil record in which ***(a)*** trilobites appear earlier than ***(b)*** seed ferns, which in turn appear before ***(c)*** dinosaurs, such as *Allosaurus*. Trilobites had gone extinct by about 230 million years ago, seed ferns by about 150 million years ago, and dinosaurs by 65 million years ago.

volcanism. They concluded that there was no need to invoke catastrophes to explain the findings of geology. Don't rivers in flood lay down layers of sediment? Don't lava flows produce layers of basalt? Why, then, should we assume that layers of rock are evidence of anything but ordinary natural processes, occurring repeatedly over long periods of time? This concept, called **uniformitarianism**, had profound implications. If slow, natural processes alone are able to produce layers of rock thousands of feet thick, then Earth must be old indeed, many millions of years old. Hutton and Lyell, in fact, concluded that Earth was eternal: "No Vestige of a Beginning, no Prospect of an End," in Hutton's words. (Modern geologists estimate that Earth is about 4.5 billion years old; see Chapter 17's feature, "Scientific Inquiry: How Do We Know How Old a Fossil Is?") Thus, Hutton and Lyell provided the time for evolution to occur. But there was still no convincing mechanism.

Early Biologists Proposed Mechanisms for Evolution

One of the first to propose a mechanism for evolution was French biologist Jean Baptiste Lamarck (1744–1829). Lamarck was impressed by the progression of forms in the fossil record. Older fossils tend to be simpler, whereas younger fossils tend to be more complex and more like existing organisms. In 1801 Lamarck hypothesized that organisms evolved through the **inheritance of acquired characteristics**, a process in which the bodies of living organisms are modified through the use or disuse of parts, and these modifications are inherited by offspring. (As it turns out, the first part of his hypothesis is correct to some extent, but the second part is not.) Why would bodies be modified? Lamarck proposed that all organisms possess an innate drive for perfection, an urge to climb the ladder of Nature. In his best-known example, Lamarck hypothesized that ancestral giraffes stretched their necks to feed on leaves that grow high up in trees, and as a result their necks became slightly longer. Their offspring would have inherited these longer necks and stretched even farther to reach still higher leaves. Eventually, this process might have produced modern giraffes with very long necks indeed.

Today, Lamarck's theory seems silly: The fact that a prospective father pumps iron doesn't mean that his children will look like Arnold Schwarzenegger. Remember, though, that in Lamarck's day no one had the foggiest idea how inheritance worked. Gregor Mendel wouldn't even be born for another 20 years, and his principles of inheritance weren't incorporated into mainstream biology until the early twentieth century.

Although Lamarck's theory fell by the wayside, by the mid-1800s some biologists were beginning to realize that the fossil record and the similarities between fossil forms and modern species could be best explained if present-day species had evolved from preexisting ones. The question remained: *But how?* In 1858 Charles Darwin and Alfred Russel Wallace, working separately, provided convincing evidence that the driving force behind evolutionary change was natural selection.

Both Darwin and Wallace Proposed That Evolution Occurs by Natural Selection

Although their social and educational backgrounds were very different, Darwin and Wallace were quite similar in some respects. Both had traveled extensively in the Tropics and had studied the staggering variety of plants and animals living there. Both found that some species differed only in a few fairly subtle, but ecologically important, features (Fig. 14-4). Both were familiar with the fossil record, which showed a trend of increasing complexity through time. Finally, both were aware of the studies of Hutton and Lyell, who had proposed that Earth is extremely ancient. These facts suggested to both Darwin and Wallace that species change over time; that is, species evolve. Both sought a mechanism that might direct evolutionary change over many generations.

(a) large ground finch, beak suited to large seeds

(b) small ground finch, beak suited to small seeds

(c) warbler finch, beak suited to insects

(d) vegetarian tree finch, beak suited to leaves

Figure 14-4 Darwin's finches, residents of the Galapagos Islands
The Galapagos Islands are home to a group of closely related species of finches, each of which specializes in eating a different type of food. Natural selection has favored those individuals best suited to exploit each food source efficiently. The result is a wide range of beak sizes and shapes among otherwise similar birds.

Scientific Inquiry
Charles Darwin—Nature Was His Laboratory

Charles Darwin's Voyage on the *Beagle* Sowed the Seeds for His Theory of Evolution

Like many modern students, Charles Darwin excelled only in subjects that intrigued him. Although his father was a physician, Darwin was uninterested in medicine and unable to stand the sight of surgery. He eventually obtained a degree in theology from Cambridge University, although theology too was of minor interest to him. What he really liked to do was to tramp over the hills, observing plants and animals, collecting new specimens, scrutinizing their structures, and categorizing them.

In 1831, when Darwin was only 22 years old (Fig. E14-1), the British government sent Her Majesty's Ship *Beagle* on a 5-year surveying expedition that went first along the coastline of South America and then around the world. The *Beagle* would carry along a naturalist to observe and collect geological and biological specimens encountered along the route. Thanks to a professor's recommendation, Darwin was offered the position of naturalist aboard the *Beagle*. The *Beagle* sailed to South America, making many stops along the coast. There Darwin observed the plants and animals of the Tropics and was stunned by the diversity of species compared with that of Europe. Although he boarded the *Beagle* convinced of the permanence of species, his experiences soon led him to doubt this. He discovered a snake with rudimentary hind limbs, calling it "the passage by which Nature joins the lizards to the snakes." Another snake he encountered vibrated its tail like a rattlesnake but had no rattles and therefore made no noise. Similarly, Darwin noticed that penguins used their wings to paddle through the water rather than fly through the air. If a Creator had individually created each animal in its present form, to suit its present environment, what could be the purpose behind these makeshift arrangements?

Perhaps the most significant stopover of the voyage was the month spent on the Galapagos Islands, off the northwestern coast of South America. There Darwin found huge tortoises (Fig. E14-2a); in fact, *galapagos* means "tortoise" in Spanish. Different islands were home to distinctively different types of tortoises. On islands without tortoises, prickly pear cacti grew with their juicy (though spiny) pads and fruits spread out over the ground. On islands where tortoises lived, the prickly pears grew substantial trunks, bearing the fleshy pads and fruits high above the reach of the voracious and tough-mouthed tortoises (Fig. E14-2b). Darwin also found several varieties of mockingbirds and finches; as with the tortoises, different islands supported slightly different forms. Could the differences in these organisms have arisen after they became isolated from one another on separate islands? The diversity of tortoises and birds "haunted" him for years afterward.

Figure E14-1 A painting of Charles Darwin as a young man

In 1836 Darwin returned to England after 5 years on the *Beagle* and became established as one of the foremost naturalists of his time. But constantly gnawing on his mind was the problem of the origin of species. Part of the solution came to him from an unlikely source: the writings of an English economist and clergyman, Thomas Malthus. In his *Essay on Population*, Malthus wrote, "It may safely be pronounced, therefore, that [human] population, when unchecked, goes on doubling itself every 25 years, or increases in a geometrical ratio." Darwin realized that a similar principle holds true for plant and an-

In 1858 Darwin and Wallace each described a mechanism for evolution in remarkably similar papers that were presented to the Linnaean Society in London. Like Gregor Mendel's manuscript on the principles of genetics, their papers had little impact. The secretary of the society, in fact, wrote in his annual report that nothing very interesting happened that year. Fortunately, the next year Darwin published his monumental *On the Origin of Species by Means of Natural Selection*, which attracted a great deal of attention to the new theory. (See "Scientific Inquiry: Charles Darwin—Nature Was His Laboratory.")

Evolutionary Theory Arises from Scientific Observations and Conclusions Based on Them

Darwin and Wallace concluded that life's huge variety of excellent designs arose by a process of descent with modification, in which the members of each generation differ slightly from the members of the preceding generation, and these small changes accumulate over long stretches of time to produce major transformations. The chain of logic leading to this powerful conclusion turns out to be surprisingly simple and straightforward. We summarize their theory in modern terms:

(a)

(b)

Figure E14-2 Tortoises act as agents of selection
(a) Galapagos tortoises feed on prickly pear cactuses. *(b)* On islands with tortoises, a young cactus quickly grows a tall trunk, which lifts the succulent pads beyond the reach of the tortoises.

imal populations. In fact, most organisms can reproduce much more rapidly than can humans (consider rabbits, dandelions, and houseflies) and consequently could produce overwhelming populations in short order. Nonetheless, the world is not chest-deep in rabbits, dandelions, or flies: Natural populations do not grow "unchecked" but tend to remain approximately constant in size. Clearly, vast numbers of individuals must die in each generation, and most must not reproduce.

From his experience as a naturalist, Darwin realized that the individual members of a species typically differ from one another in form and function. Further, which individuals die in each generation is not arbitrary but depends to some extent on the structures and abilities of the organisms. This observation was the source of the theory of evolution by natural selection. As Darwin's colleague Alfred Wallace put it, "Those which, year by year, survived this terrible destruction must be, on the whole, those which have some little superiority enabling them to escape each special form of death to which the great majority succumbed." Here you see the origin of the expression "survival of the fittest." That "little superiority" that confers greater fitness might be better resistance to cold, more efficient digestion, or any of hundreds of other advantages, some very subtle. Everything now fell into place. Darwin wrote, "It at once struck me that under these circumstances favorable variations would tend to be preserved, and unfavorable ones to be destroyed." If the favorable variations were inheritable, then the entire species would eventually consist of individuals possessing the favorable trait. With the continual appearance of new variations (due, as we now know, to mutations), which in turn are subject to further selection, "the result . . . would be the formation of new species. Here, then, I had at last got a theory by which to work."

When Darwin finally published *On the Origin of Species* in 1859, his evidence had become truly overwhelming. Although its full implications would not be realized for decades, Darwin's theory of evolution by natural selection has become a unifying concept for virtually all of biology.

Observation 1: A natural **population**, which consists of all the individuals of one species in a particular area, has the potential to grow rapidly, because organisms can produce far more offspring than are required merely to replace the parents.

Observation 2: Nevertheless, the number of individuals in a natural population tends to remain relatively constant over time.

Conclusion 1: Therefore, it must be the case that more organisms are born than survive and reproduce. If some individuals fail to survive, it must also be true that organisms compete to survive and reproduce. In each generation, many individuals must die young, fail to reproduce, produce few offspring, or produce less-fit offspring that fail to survive and reproduce in their turn.

Observation 3: Individual members of a population differ from one another in their ability to obtain resources, withstand environmental extremes, escape predators, and so on.

Conclusion 2: These differences among individuals help determine which individuals will survive and reproduce most successfully, thereby leaving the most offspring. This process, in which those individuals

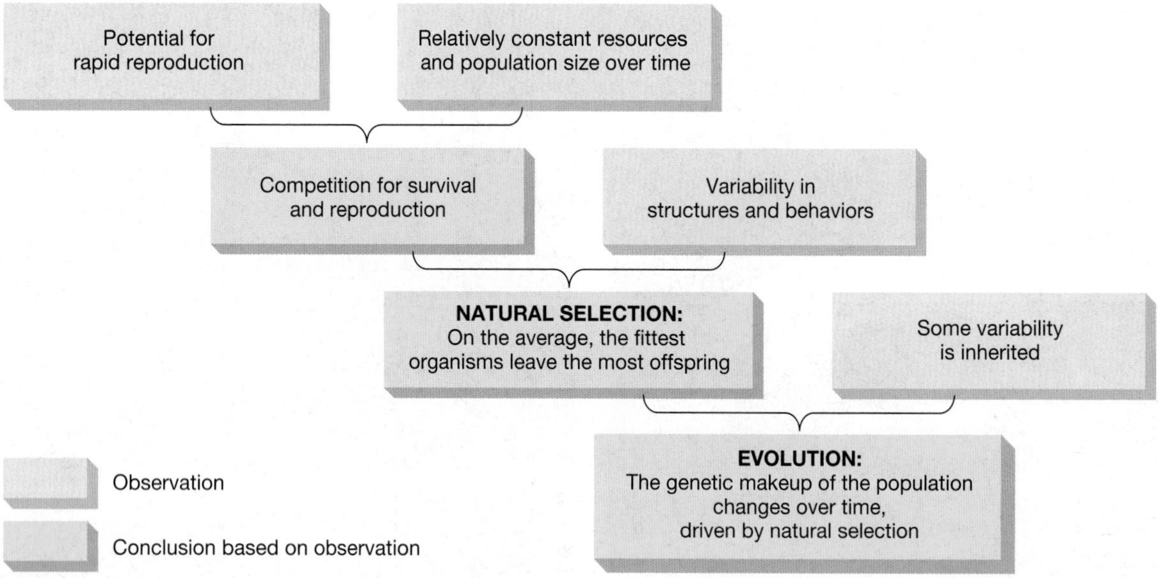

Figure 14-5 A flowchart of evolutionary reasoning
This chart is based on the hypotheses of Darwin and Wallace but incorporates ideas from modern genetics.

whose traits best adapt them to their environment leave a larger number of offspring, is known as **natural selection**.

Observation 4: At least some of the variation among individuals in traits that affect survival and/or reproduction is due to genetic differences that may be passed on from parent to offspring.

Conclusion 3: Because better-adapted individuals leave more offspring, the traits (and underlying genes) of the most well-adapted individuals will be passed to a larger proportion of the individuals in subsequent generations. Over many generations, this differential, or unequal, reproduction among individuals with different genetic makeup changes the overall genetic composition of the population. This process is evolution by natural selection.

Figure 14-5 charts these observations and conclusions.

As you know, the principles of genetics had not yet been discovered when Darwin published *On the Origin of Species*. Our observation 4, therefore, was an untested assumption for Darwin and a grave weakness in his theory. Although he could not explain how inheritance operated, Darwin's theory made an important prediction that we now know is correct. According to Darwin, the variations that appear in natural populations arise purely by chance. Unlike Lamarck, he proposed no internal drives for perfection or any other mechanisms that would ensure that variations would be favorable. Molecular genetics has shown that Darwin was correct: Variations arise as a result of chance mutations in DNA (see Chapters 9 and 10).

How might natural selection among chance variations change the makeup of a species? In *On the Origin of Species*, Darwin proposed the following example. "Let us take the case of a wolf, which preys on various animals, securing [them] by . . . fleetness. . . . The swiftest and slimmest wolves would have the best chance of surviving, and so be preserved or selected. . . . Now if any slight innate change of habit or structure benefited an individual wolf, it would have the best chance of surviving and of leaving offspring. Some of its young would probably inherit the same habits or structure, and by the repetition of this process, a new variety might be formed." The same argument would apply to the wolf's prey, in which the fastest or most alert would be the most likely to avoid predation and would pass on these traits to its offspring. Notice that natural selection is acting on individuals within a population. Over generations, the population would change as a greater percentage of its individuals acquired favorable adaptive traits. An individual cannot evolve, but a population can.

Although it is easiest to understand how natural selection would cause *changes within a species*, under the right circumstances, the same principles might produce *entirely new species*. In Chapter 16 we shall discuss the circumstances that give rise to new species. Here we shall briefly review some of the evidence for evolution.

2 How Do We Know That Evolution Has Occurred?

Virtually all biologists consider evolution to be a fact. Although scientists may still debate the relative importance of different *mechanisms* of evolutionary change, exceedingly few biologists dispute that evolution occurs. Why? Because an overwhelming body of evidence permits no other conclusion. The key lines of evidence include such sources as the fossil record, comparative anatomy (the study of anatomical structures for comparative purposes among species), embryology, and biochemistry and genetics.

The Fossil Record Provides Evidence of Evolutionary Change over Time

Because fossils are the remains of members of species ancestral to modern species, we would expect to find progressive series of fossils leading from an ancient, primitive organism, through several intermediate stages, and culminating in the modern forms. Probably the best known of such series are the fossil horses (Fig. 14-6), but giraffes, elephants, and several mollusks also show a roughly gradual evolution of body form over time. These fossil series suggest that new species evolved from, and replaced, previous species. Certain sequences of fossil snails have such slight gradations in form between successive fossils that paleontologists cannot easily decide where one species leaves off and the next one begins. The fossil record also documents larger-scale evolutionary transitions, such as the link between dinosaurs and birds that was confirmed by fossils of dinosaurs with feathers.

Comparative Anatomy Provides Structural Evidence of Evolution

Appearance has long been used as an indicator of the relatedness of organisms. Structure, inexorably tied to function, also provides evidence of descent with modification.

Unrelated Species in Similar Environments Have Evolved Similar Forms

Evolution by natural selection also predicts that, given similar environmental demands, unrelated species might independently evolve superficially similar structures, a process called **convergent evolution.** Such outwardly similar body parts in unrelated organisms are called **analogous structures.** The wings of flies and of birds are analogous structures that have arisen by convergent evolution; the fat-insulated, streamlined shapes

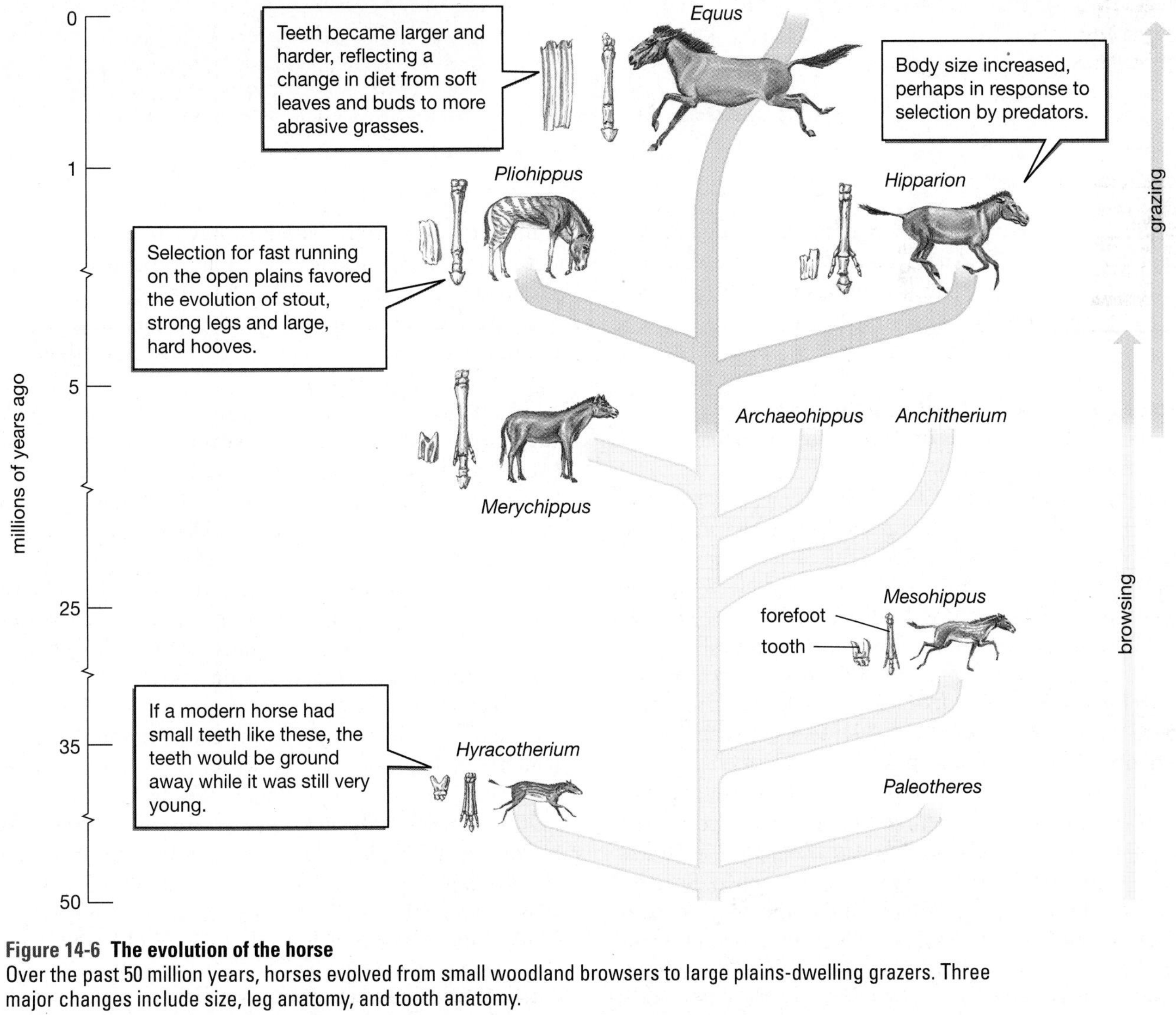

Figure 14-6 The evolution of the horse
Over the past 50 million years, horses evolved from small woodland browsers to large plains-dwelling grazers. Three major changes include size, leg anatomy, and tooth anatomy.

Figure 14-7 Analogous structures Similar environmental pressures acting on unrelated animals may result in the convergent evolution of outwardly similar structures. The wings of ***(a)*** insects and ***(b)*** birds and the sleek, streamlined shapes of ***(c)*** seals and ***(d)*** penguins are examples of such analogous structures.

(a) (b)

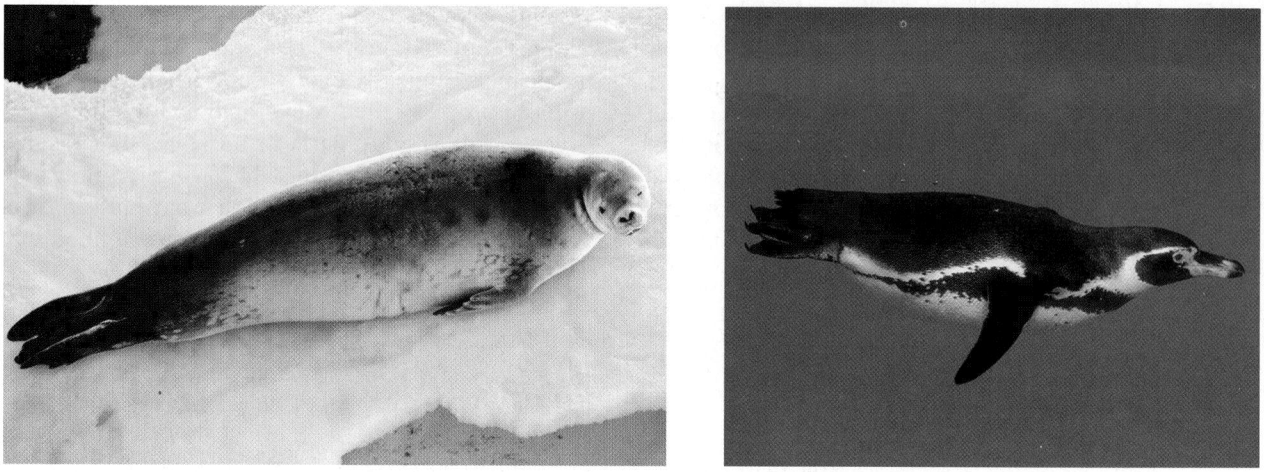

(c) (d)

MEDIATUTOR 14.1 Analogous and Homologous Structures

of seals (mammals) and of penguins (birds) are another example (Fig. 14-7). Analogous structures may be very different in internal anatomy, because the parts are not derived from common ancestral structures.

Homologous and Vestigial Structures Provide Evidence of Relatedness of Organisms Adapted to Different Environments

Modern organisms are adapted to a wide variety of habitats and lifestyles. The forelimbs of birds and mammals, for example, are variously used for flying, swimming, running over several types of terrain, and grasping objects, such as branches and tools. Despite this enormous diversity of function, the internal anatomy of all bird and mammal forelimbs is remarkably similar (Fig. 14-8). It seems inconceivable that the same bone arrangements would be used to serve such diverse functions if each animal had been created separately. Such similarity is exactly what we would expect, however, if bird and mammal forelimbs were derived from a common ancestor. Through natural selection, each has been modified to perform a particular function. Such internally similar structures are called **homologous structures**, meaning that they have the same evolutionary origin despite possible differences in current function. Studies of comparative anatomy have long been used to determine the relationships among organisms, on the grounds that the more similar the internal structures of two species, the more closely related the species must be; that is, the more recently they must have diverged from a common ancestor.

Evolution by natural selection also helps explain the curious circumstance of **vestigial structures** that serve no apparent purpose. Examples include such things as molar teeth in vampire bats (which live on a diet of blood and therefore don't chew their food) and pelvic bones in whales and certain snakes (Fig. 14-9, p. 280). Both of these vestigial structures are clearly homologous to structures that are found in—and used by—other vertebrates (animals with a backbone). Their continued existence in animals that have no use for them is best explained as a sort of "evolutionary baggage." For example, the ancestral mammals from which whales evolved had four legs and a well-developed set of pelvic bones. Whales do not have hind legs, yet they have small pelvic and leg bones embedded in their sides. During whale evolution, losing the hind legs provided a selective advantage, better streamlining the body for movement through water. The result is the modern whale with small, useless, and unused pelvic bones.

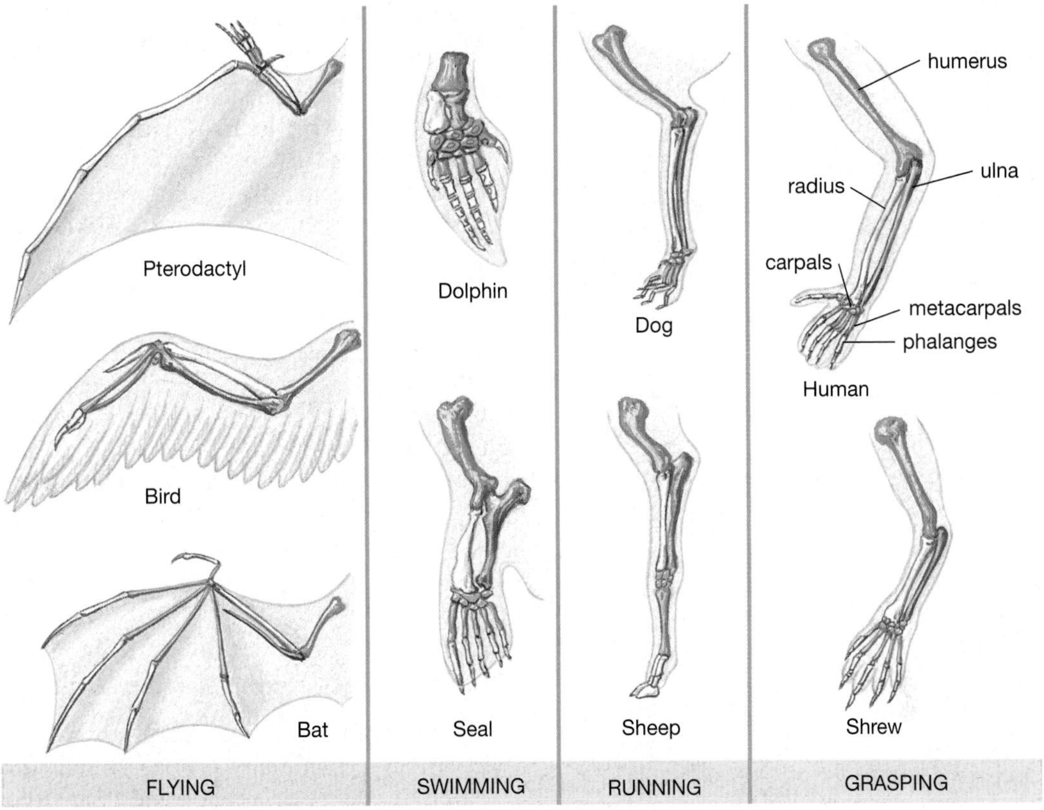

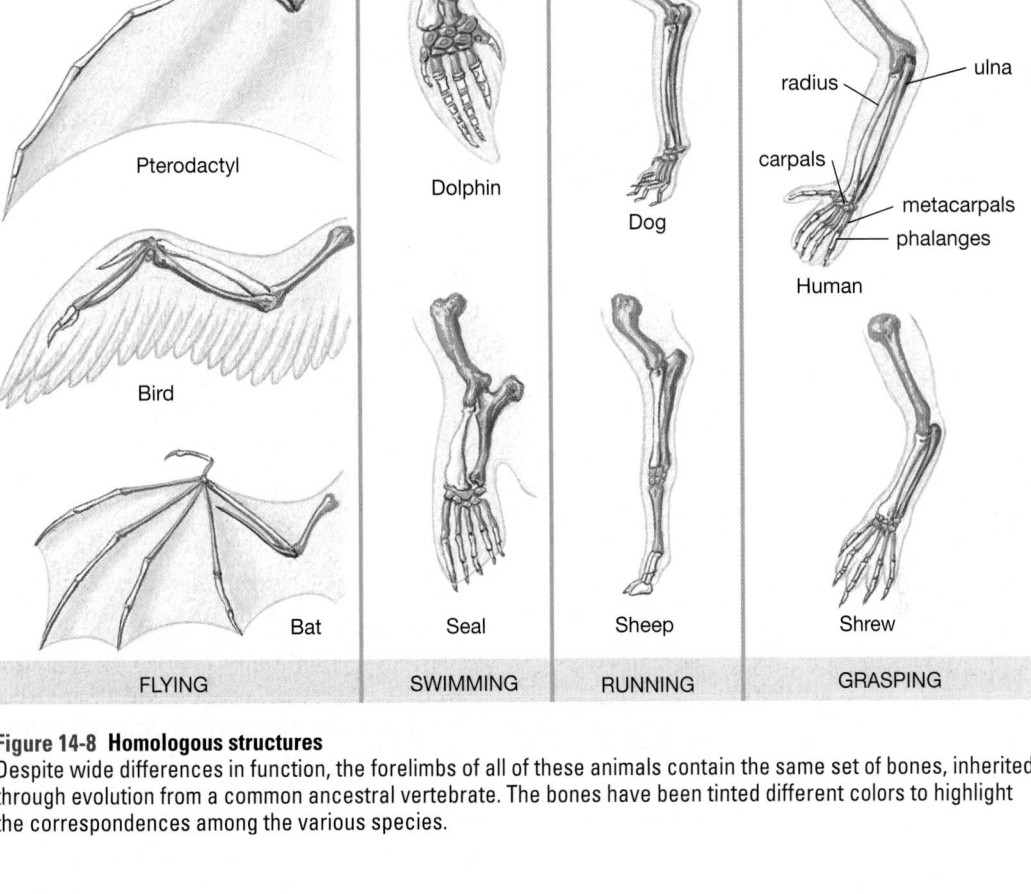

Figure 14-8 Homologous structures
Despite wide differences in function, the forelimbs of all of these animals contain the same set of bones, inherited through evolution from a common ancestral vertebrate. The bones have been tinted different colors to highlight the correspondences among the various species.

Embryological Stages of Animals Can Provide Evidence of Common Ancestry

In the early 1800s German embryologist Karl von Baer noted that all vertebrate embryos look quite similar to one another early in their development (Fig. 14-10). In their early embryonic stages, fish, turtles, chickens, mice, and humans all develop tails and gill slits. Only fish go on to develop gills, and only fish, turtles, and mice retain substantial tails. Why do such diverse vertebrates have similar developmental stages? The only plausible explanation is that ancestral vertebrates possessed genes that direct the development of gills and tails. All of their descendants still have those genes. In fish these genes are active throughout development, resulting in gill-bearing and tail-bearing adults. In humans and chickens these genes are active only during early developmental stages, and the structures are lost or inconspicuous in adults.

Modern Biochemical and Genetic Analyses Reveal Relatedness Among Diverse Organisms

Biochemistry and molecular biology provide striking evidence of the evolutionary relatedness of all living organisms. At the most fundamental biochemical levels, all living cells are very similar. For example, all cells have DNA as the carrier of genetic information; all use RNA, ribosomes, and approximately the same genetic code to translate that genetic information into proteins; all use roughly the same set of 20 amino acids to build proteins; and all use ATP as an intracellular energy carrier. Species share similarities in chromosome structure, in sequences of amino acids in proteins, and in DNA composition; each of these parameters is now used to investigate relatedness among organisms. We shall return to the methods of measuring evolutionary relatedness in Chapter 18.

3 What Is the Evidence That Populations Evolve by Natural Selection?

We have seen that evidence of evolution comes from several types of sources. But what is the evidence that evolution occurs by the mechanism of natural selection?

MEDIATUTOR
14.2 Natural Selection for Antibiotic Resistance

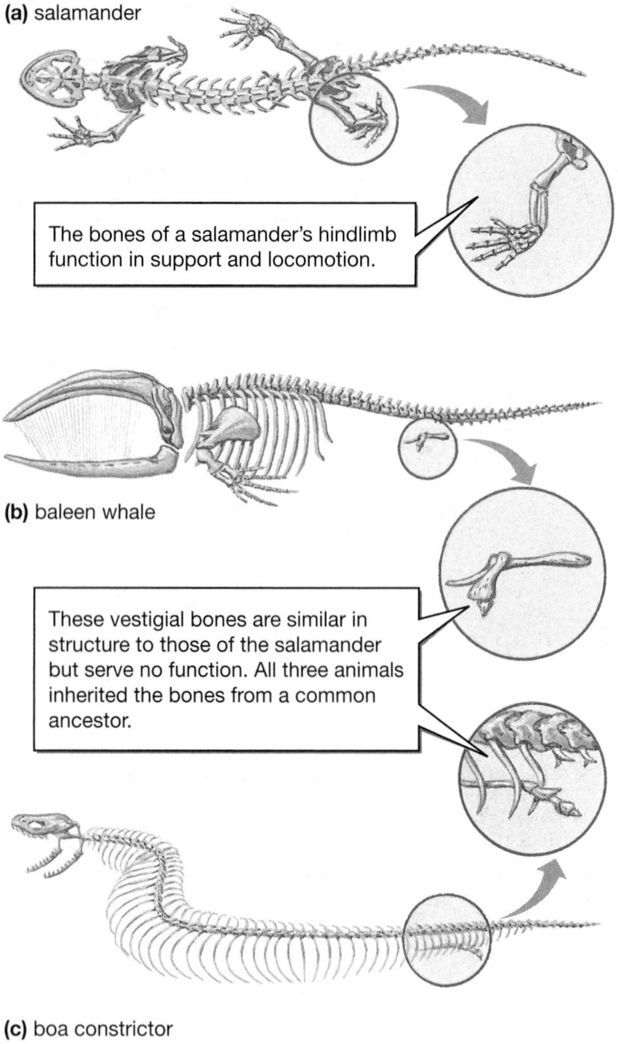

(a) salamander
The bones of a salamander's hindlimb function in support and locomotion.
(b) baleen whale
These vestigial bones are similar in structure to those of the salamander but serve no function. All three animals inherited the bones from a common ancestor.
(c) boa constrictor

Figure 14-9 Vestigial structures
Many organisms have vestigial structures that serve no apparent function. Such structures were most likely inherited, in modified form, from ancestors that did use them.

Artificial Selection Demonstrates That Organisms May Be Modified by Controlled Breeding

One line of evidence supporting evolution by natural selection that particularly impressed Charles Darwin was **artificial selection**: the breeding of domestic plants and animals to produce specific desirable features. The various breeds of dogs provide a striking example of artificial selection (Fig. 14-11). Dogs descended from wolves, and even today the two will readily cross-breed. With rare exceptions, however, few modern dogs resemble wolves. Some breeds are so different from one another that they would be considered separate species if they were found in the wild. Interbreeding would hardly be possible without a lot of human assistance. If humans could breed such radically different dogs in a few hundred to at most a few thousand years, Darwin reasoned, it seemed quite plausible that natural selection could produce the spectrum of living organisms in hundreds of millions of years.

Evolution by Natural Selection Occurs Today

The logic of natural selection gives us no reason to believe that evolutionary change is limited to the past. After all, inherited variation and competition for access to resources are certainly not limited to the past. If Darwin and Wallace were correct that those conditions lead inevitably to evolution by natural selection, then scientific observers and experimenters ought to be able to detect evolutionary change as it occurs. And they have. For example, researchers have confirmed that the coloration of male guppies (a species of tropical fish) can change when they move into a new environment.

On the island of Trinidad, guppies live in streams that are also inhabited by several species of larger, predatory fish that frequently dine on guppies. In upstream portions

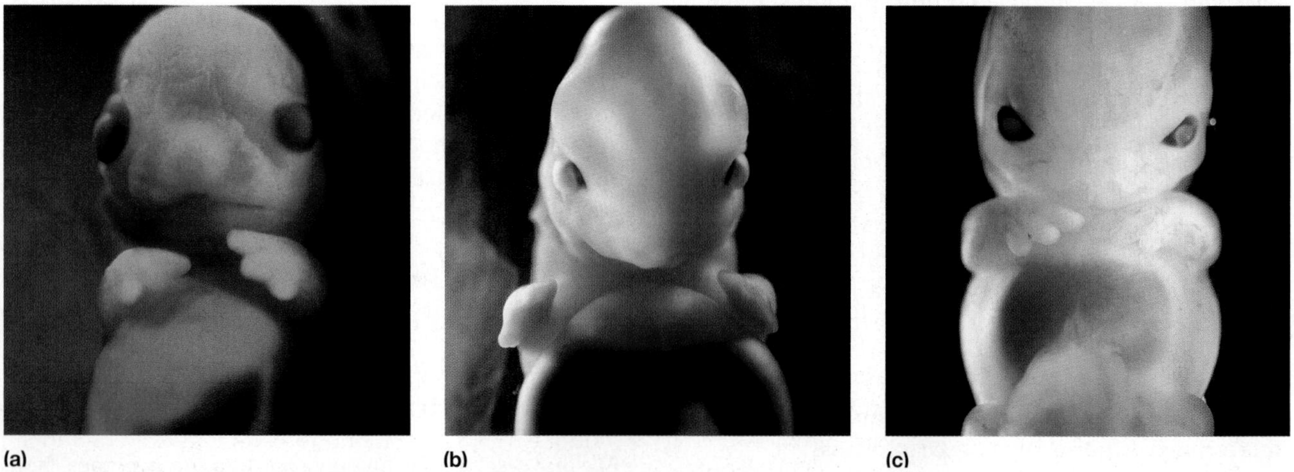

Figure 14-10 Embryological stages reveal evolutionary relationships
Early embryonic stages of a *(a)* lemur, *(b)* pig, and *(c)* human, showing strikingly similar anatomical features.

(a) (b)

Figure 14-11 Dog diversity illustrates artificial selection
A comparison of ***(a)*** the ancestral dog (the gray wolf, *Canis lupus*) and ***(b)*** various breeds of dog. Artificial selection by humans has caused a great divergence in size and shape of dogs in only a few thousand years.

of these streams, however, the water is too shallow for the predators, and any guppies that manage to find their way to these shallower waters are free of danger from predators. When scientists compared a group of male guppies that had colonized an upstream area to those that remained downstream, they found that the upstream guppies had become much more brightly colored than the downstream guppies. The explanation for this difference stems in part from the sexual preferences of female guppies. The females prefer to mate with the most brightly colored males, so the brightest males have a large advantage when it comes to reproduction. In predator-free areas, male guppies are free to evolve the bright colors that females prefer. Bright color, however, also makes guppies more conspicuous to predators, and therefore more likely to be eaten. Thus, where predators are common, they act as agents of natural selection. In predator-rich areas, the brightest males are less likely to survive, and duller males have the advantage. The color difference between the upstream and downstream guppy populations is a direct result of natural selection.

Natural selection is also evident in the unfortunately numerous instances of insect pests evolving resistance to the pesticides with which we try to control them. Florida homeowners were dismayed to realize that roaches were ignoring a formerly effective poison bait called Combat®. Researchers discovered that the bait had acted as an agent of natural selection. Roaches that liked it were consistently killed; those that survived had inherited a rare mutation that caused them to dislike glucose, a type of sugar found in the corn syrup used as bait in Combat. By the time researchers identified the problem in the early 1990s, the formerly rare mutation had become widespread in Florida's urban roach population.

In addition to observing natural selection in the wild, scientists have also devised numerous experiments that confirm the action of natural selection. For example, one group of evolutionary biologists released small groups of *Anolis sagrei* lizards onto 14 small Bahamian islands that were previously uninhabited by lizards. The original lizards came from a population on Staniel Cay, an island with tall vegetation, including plenty of trees. In contrast, the islands to which the small colonial groups were introduced had few or no trees and were covered mainly with small shrubs and other low-growing plants.

The biologists returned to those islands 14 years after releasing the colonists and found that the original small groups of lizards had given rise to thriving populations of hundreds of individuals. On all 14 of the experimental islands, lizards had legs that were shorter and thinner than lizards from the original source population on Staniel Cay. In just over a decade, it appeared, the lizard populations had in some measure adapted to new environments. Why had the new lizard populations evolved shorter, thinner legs? Long legs allow greater speed for escaping predators, but shorter legs allow for more agility and maneuverability on narrow surfaces. So natural selection favors legs that are as long and thick as possible while still allowing sufficient maneuverability. When the lizards were moved from an environment with thick-branched trees to an environment with only thin-branched bushes, the individuals with formerly favorable long legs were at a disadvantage. In the new environment more agile, shorter-legged individuals were better able to escape predators and survive to produce a greater number of offspring. Thus, members of subsequent generations would have shorter legs on average.

Note two important points about these examples:

1. *The variations on which natural selection works are produced by chance mutations.* The bright coloration in Trinidadian guppies, distaste for glucose in Florida cockroaches, and shorter legs in Bahamian lizards were not *produced* by the pollution, poisoned corn syrup, or the thinner branches. The mutations that produced each of these beneficial traits arose spontaneously.
2. *Evolution by natural selection selects for organisms that are best adapted to a particular environment.* Natural selection is *not* a mechanism for producing ever-greater degrees of perfection. Natural selection does not select for the "best" in any absolute sense, but only in the context of a particular environment, which varies from place to place and which may change over time. A trait that is advantageous under one set of conditions may become disadvantageous if conditions change. In the presence of poisoned corn syrup, a distaste for glucose yields an advantage to a cockroach, but under natural conditions avoiding glucose would cause the insect to bypass good sources of food.

4 A Postscript by Charles Darwin

"It is interesting to contemplate an entangled bank, clothed with many plants of many kinds, with birds singing on the bushes, with various insects flitting about, and with worms crawling through the damp earth, and to reflect that these elaborately constructed forms . . . have all been produced by laws acting around us. These laws, taken in the highest sense, being Growth with Reproduction; Inheritance [and] Variability; a Ratio of Increase so high as to lead to a Struggle for Life, and as a consequence to Natural Selection, entailing Divergence of Character and Extinction of less-improved forms. . . . There is grandeur in this view of life, with its several powers, having been originally breathed into a few forms or into one; and that, whilst this planet has gone cycling on according to the fixed law of gravity, from so simple a beginning endless forms most beautiful and most wonderful have been, and are being, evolved."

These are the concluding sentences of Darwin's *On the Origin of Species*.

CASESTUDYREVISITED

A Missing Link Unearthed

The feathered dinosaur bandwagon hit a bit of a bump in early 2000 when a much-publicized new fossil of a feathered dinosaur species turned out to be a fake. The fraudulent fossil had been purchased from a dealer in Utah, and someone had apparently doctored it by gluing a dinosaur's tail onto a fossil bird. The embarrassed scientists who fell for the hoax, however, maintain that the fraud in no way affects the validity of previously discovered feathered dinosaur fossils.

Fossil fakery notwithstanding, even well-accepted feathered dinosaurs have failed to completely resolve the controversy over the evolutionary origin of birds. A small group of skeptical paleontologists questions whether the new fossils really provide truly conclusive evidence that dinosaurs gave rise to birds. For one thing, say the skeptics, the fossils of the feathered dinosaurs are much younger (by about 30 million years) than the fossils of *Archaeopteryx*, the oldest known bird, so these feathered dinosaur species could not have been the ancestors of birds. To the skeptics, this anomaly suggests that the fossils represent not dinosaurs, but rather birds that had lost the power of flight (as have modern birds, such as the ostrich and penguin).

Most paleontologists, however, shrug off the question of the new fossils' age. To them, the comparatively young age of the fossils means nothing more than that some species of feathered dinosaurs persisted after the line leading to birds had branched off (much like apes and humans both persist today, millions of years after the human and ape lineages diverged). To proponents of the bird–dinosaur connection, feathered dinosaurs of any age demonstrate conclusively that pre-bird dinosaurs evolved feathers and that these feathered dinosaurs were the precursors to birds.

What kind of evidence would provide the strongest proof that birds are descended from dinosaurs?

Summary of Key Concepts

1 How Did Evolutionary Thought Evolve?

Historically, the most common explanation for the origin of species has been the divine Creation of each species in its present form, and species were believed not to have changed significantly since their creation. Evidence provided by the diversity of living things, by fossils, and by geology challenged this view, although no convincing mechanism for the evolution of present-day species from previous ones was proposed. Since the middle of the nineteenth century, scientists have concluded that species originate by the operation of natural laws, as a result of changes in the genetic makeup of the populations of organisms. This process is called *evolution*.

Charles Darwin and Alfred Russel Wallace independently proposed the theory of evolution by natural selection. Their theory can be concisely expressed as three conclusions based on four observations. These are summarized in modern biological terms in Figure 14-5.

2) How Do We Know That Evolution Has Occurred?

Many lines of evidence indicate that evolution has occurred, including the following:

1. Fossils of ancient species tend to be simpler in form than modern species. Sequences of fossils have been discovered that show a graded series of changes in form. Both of these facts would be expected if modern species evolved from older species.
2. Species thought to be related through evolution from a common ancestor show many similar anatomical structures. Examples include the limbs of amphibians, reptiles, birds, and mammals.
3. Stages in embryological development are quite similar among very different types of vertebrates.
4. Similarities in chromosome structure, sequences of amino acids in proteins, and DNA composition all support the notion of descent of related species through evolution from common ancestors.

3) What Is the Evidence That Populations Evolve by Natural Selection?

Similarly, many lines of evidence indicate that natural selection is the chief mechanism driving changes in the characteristics of species over time, including the following:

1. Rapid, heritable changes have been produced in domestic animals and plants by selectively breeding organisms with desired features (artificial selection). The immense variations in species produced in a few thousand years of artificial selection by humans makes it seem likely that much larger changes could be wrought by hundreds of millions of years of natural selection.
2. Evolution can be observed today. Both natural and human activities drastically change the environment over short periods of time. Significant changes in the characteristics of species have been observed in response to these environmental changes.

Key Terms

analogous structures *p. 277*
artificial selection *p. 280*
catastrophism *p. 271*
convergent evolution *p. 277*
evolution *p. 269*
fossil *p. 270*
homologous structures *p. 278*
inheritance of acquired characteristics *p. 272*
natural selection *p. 276*
population *p. 275*
species *p. 270*
uniformitarianism *p. 272*
vestigial structure *p. 278*

Thinking Through the Concepts

Multiple Choice

1. *Your arm is homologous with*
   a. a seal flipper
   b. an octopus tentacle
   c. a bird wing
   d. a sea star arm
   e. both a and c
2. *All organisms share the same genetic code. This commonality is evidence that*
   a. evolution is occurring now
   b. convergent evolution has occurred
   c. evolution occurs gradually
   d. all organisms are descended from a common ancestor
   e. life began a long time ago
3. *Which of the following are fossils?*
   a. pollen grains buried in the bottom of a peat bog
   b. the petrified cast of a clam's burrow
   c. the impression a clam shell made in mud, preserved in mudstone
   d. an insect leg sealed in plant resin
   e. all of the above
4. *In Africa, there is a species of bird called the yellow-throated longclaw. It looks almost exactly like the meadowlark found in North America, but they are not closely related. This is an example of*
   a. uniformitarianism
   b. artificial selection
   c. gradualism
   d. vestigial structures
   e. convergent evolution
5. *Which of the following are examples of vestigial structures?*
   a. your tailbone
   b. your ear lobes
   c. sixth fingers found in some humans
   d. your kneecap
   e. none of the above
6. *Which of the following would stop evolution by natural selection from occurring?*
   a. if humans became extinct because of a disease epidemic
   b. if a thermonuclear war killed most living organisms and changed the environment drastically
   c. if ozone depletion led to increased ultraviolet radiation, which caused many new mutations
   d. if all individuals in a population were genetically identical, and there was no genetic recombination, sexual reproduction, or mutation
   e. all of the above

? Review Questions

1. Selection acts on individuals, but only populations evolve. Explain why this is true.
2. Distinguish between catastrophism and uniformitarianism. How did these hypotheses contribute to the development of evolutionary theory?
3. Describe Lamarck's theory of inheritance of acquired characteristics. Why is it invalid?
4. What is natural selection? Describe how natural selection might have caused differential reproduction among the ancestors of a fast-swimming predatory fish, such as the barracuda.
5. Describe how evolution occurs through the interactions among the reproductive potential of a species, the normally constant size of natural populations, variation among individuals of a species, natural selection, and inheritance.
6. What is convergent evolution? Give an example.
7. How do biochemistry and molecular genetics contribute to the evidence that evolution occurred?

Applying the Concepts

1. Does evolution through natural selection produce "better" organisms in an absolute sense? Are we climbing the "ladder of Nature"? Defend your answer.
2. Both the study of fossils and the idea of special creation have had an impact on evolutionary thought. Discuss why one is considered scientific endeavor and the other not scientific.
3. In evolutionary terms, "success" can be defined in many different ways. What are the most successful organisms you can think of in terms of (a) persistence over time, (b) sheer numbers of individuals alive now, (c) numbers of species (for a lineage), and (d) geographical range?
4. In what sense are humans currently acting as "agents" of selection on other species? Name some organisms that are *favored* by the environmental changes humans cause.
5. Darwin and Wallace's discovery of natural selection is one of the great revolutions in scientific thought. Some scientific revolutions spill over and affect the development of philosophy and religion. Is this true of evolution? Does (or should) the idea of evolution by natural selection affect the way humans view their place in the world?
6. In your mind, what scientific question currently represents the "mystery of mysteries"?

For More Information

Altman, S. A. "The Monkey and the Fig." *American Scientist,* May–June 1989. An amusing, informative discussion of many evolutionary themes, cast as a Socratic dialogue.

Darwin, C. *On the Origin of Species by Means of Natural Selection.* Garden City, NY: Doubleday, 1960 (originally published in 1859). An impressive array of evidence amassed to convince a skeptical world.

Dawkins, R. "God's Utility Function." *Scientific American,* November 1995. One evolutionary biologist's view of the nature of evolutionary change and how it relates to the meaning of life. Dawkins is an articulate advocate of the "selfish gene" view of evolution, in which genes use organisms as tools in a contest for superiority in self-replication.

Dennet, D. *Darwin's Dangerous Idea.* New York: Simon & Schuster, 1995. A philosopher's view of Darwinian ideas and their application to the world outside biology. A thought-provoking book that seems to have inspired admiration and condemnation in roughly equal proportions.

Eiseley, L. C. "Charles Darwin." *Scientific American,* February 1956. An essay on the life of Darwin by one of his foremost American biographers. Even if you need no introduction to Darwin, read this anyway as an introduction to Eiseley, author of many marvelous essays.

Gould, S. J. *Ever Since Darwin,* 1977; *The Panda's Thumb,* 1980; and *The Flamingo's Smile,* 1985. New York: W. W. Norton. A series of witty, imaginative, and informative essays, mostly from *Natural History* magazine. Many deal with various aspects of evolution.

Grant, P. R. "Natural Selection and Darwin's Finches." *Scientific American,* October 1991. A drought in the Galapagos Islands provides dramatic evidence of natural selection as an agent of evolutionary change.

Weiner, J. "Evolution Made Visible." *Science,* January 6, 1995. A clear summary of modern evidence for evolution in action.

Answers to Multiple-Choice Questions

1. e 2. d 3. e 4. e 5. b 6. d

MEDIATUTOR
Principles of Evolution

CD Activities

Activity 14.1: Analogous and Homologous Structures

Estimated time: 5 minutes

Different animal species commonly share structures but use those structures in strikingly different ways. Such differences were part of the evidence that led Charles Darwin to conclude that all animals were not created in their current form. This exercise will help you to learn the difference between structures that are shared as a result of inheritance from a common ancestor and those that are similar as a result of common function.

Activity 14.2: Natural Selection for Antibiotic Resistance

Estimated time: 15 minutes

Rifampin is a drug that has been used to kill the bacterium that causes tuberculosis (TB) in humans. Many populations of this bacterium, *Mycobacterium tuberculosis*, have evolved resistance to Rifampin. After completing the tutorial, you will be able to describe the process of evolution by natural selection using the case study of *Mycobacterium tuberculosis*.

Start the MediaTutor Student CD-ROM and enter the activity number in the Quick Search box to be taken directly to that activity.

Web Investigations

Case Study: A Missing Link Unearthed

Estimated time: 10 minutes

Finding and analyzing the fossil record is hard work. Sometimes the results are quite unexpected. This exercise will explore the fossil record of birds.

Go to http://www.prenhall.com/audesirk6, the Audesirk Companion Web. Siteselect Chapter 14 and Web Investigation to begin.

"What but the wolf's tooth whittled so fine
The fleet limbs of the antelope?
What but fear winged the birds, and hunger
Jewelled with such eyes the great goshawk's head?"

Robinson Jeffers in **The Bloody Sire** ***(1941)***

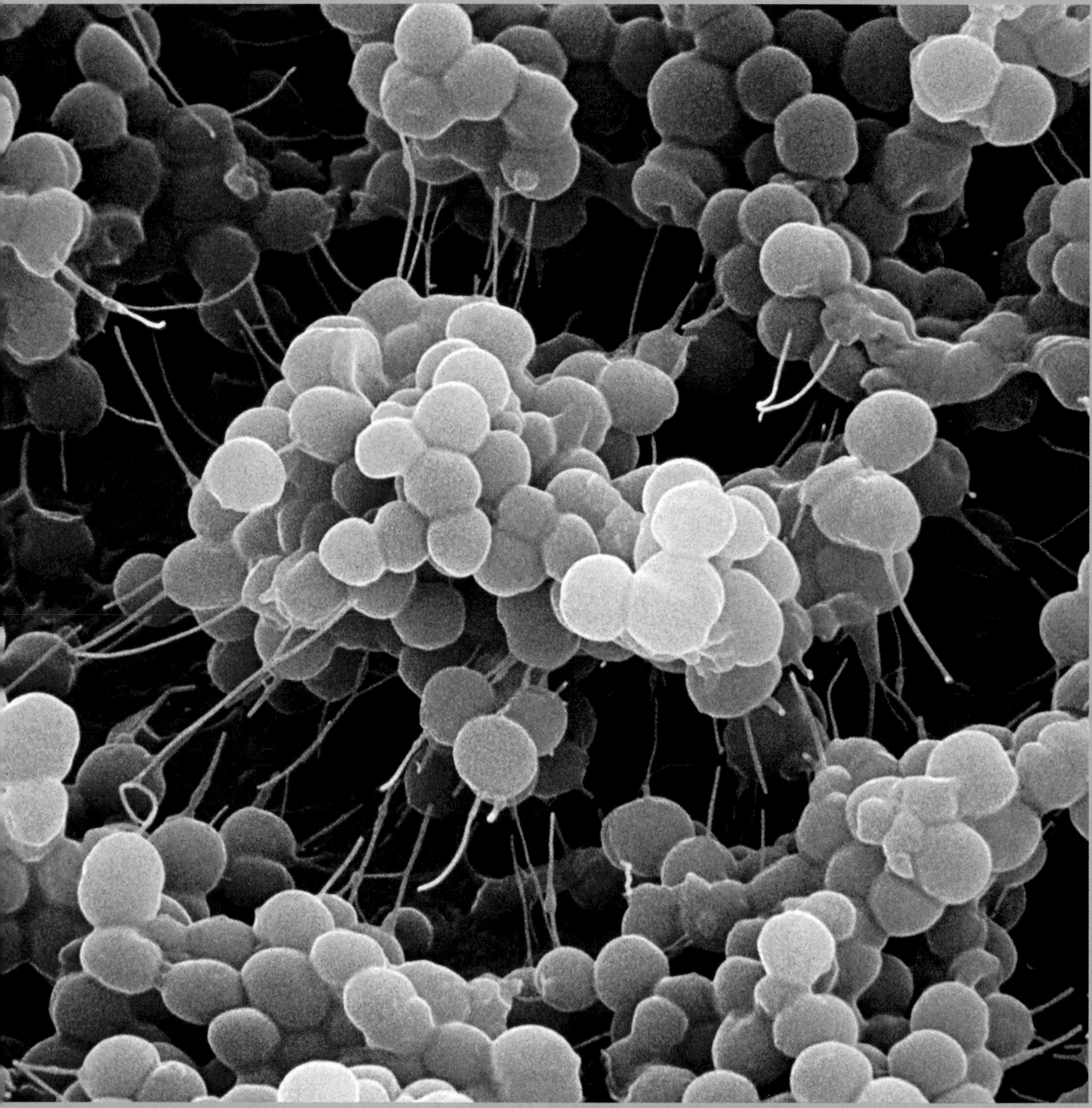

Staphylococcus aureus is among the many bacterial species that have evolved resistance to antibiotics. The evolution of such supergerms threatens to reverse our success in the battle against infectious disease.

15 How Organisms Evolve

AT A GLANCE

Case Study: Cause of Death: Evolution

1) **How Are Populations, Genes, and Evolution Related?**

Genes and the Environment Interact to Determine the Traits of Each Individual

The Gene Pool Is the Sum of All the Genes in a Population

Evolution Is the Change of Gene Frequencies Within a Population

The Equilibrium Population Is a Hypothetical Population in Which Evolution Does Not Occur

2) **What Causes Evolution?**

Mutations Are the Ultimate Source of Genetic Variability

Gene Flow Between Populations Changes Allele Frequencies

Small Populations Are Subject to Random Changes in Allele Frequencies

Mating Within a Population Is Almost Never Random

All Genotypes Are Not Equally Adaptive

3) **How Does Natural Selection Work?**

Natural Selection Acts on the Phenotype, Which Reflects the Underlying Genotype

Natural Selection Can Influence Populations in Three Major Ways

A Variety of Processes Can Cause Natural Selection

Evolutionary Connections: Knowing Your Relatives: Kin Selection and Altruism

Case Study Revisited: Cause of Death: Evolution

CASESTUDY

Cause of Death: Evolution

On a cold winter night in 1999, an 11-month-old infant arrived at a North Dakota emergency room. The child was ill with pneumonia and a blood infection, and doctors immediately began treatment with antibiotics. Despite this prompt treatment, however, the baby died within hours. The death of any child is a terrible tragedy, but this incident was especially chilling because it represented the escalation of a frightening trend: the rise of "supergerms," strains of disease-causing bacteria that defy destruction by antibiotic medicines. The young victim was the fourth area child to die of infection by antibiotic-resistant *Staphylococcus aureus* (commonly known as "staph"). The four deaths marked a disturbing new development in the story of drug-resistant staph, because the fatal infections were acquired not in a hospital (the usual source of drug-resistant infections) but in the children's homes and communities. As has long been feared, resistant staph bacteria are spreading beyond the hospital door.

The spread of drug-resistant staph is the latest manifestation of a trend that has been developing since antibiotics first came into widespread use in the 1940s. Antibiotics are losing their effectiveness against a host of diseases caused by bacteria, including food poisoning, tuberculosis, gonorrhea, meningitis, and urinary and respiratory tract infections. The rapid spread of resistance in bacterial populations raises the specter of incurable diseases, resistant to all known treatments.

In retrospect, this disturbing state of affairs is an unsurprising outcome of our unrelenting attack on disease-causing bacteria. The spread of a novel characteristic (such as antibiotic resistance) through a population in response to changing environmental conditions (such as the introduction of antibiotics) is precisely the sort of outcome that a biologist might predict, in accordance with our understanding of the mechanics of evolutionary change. ■

1 How Are Populations, Genes, and Evolution Related?

The changes that we see in an individual organism as it grows and develops are not evolutionary changes. Instead, evolutionary changes are those that occur from generation to generation, the changes that cause descendants to be different from their ancestors. Furthermore, we can't really detect evolutionary changes across generations by looking at single individuals. For example, if you observed a mature man who stood five feet tall, could you conclude that humans were evolving to become shorter? Obviously not. If you wanted to know about evolutionary change in human height, your first step would be to measure many humans across many generations to see if the average height changed over time. Clearly, evolution is a property not of individuals but of **populations**, which include all the individuals of a species living in a given area.

The recognition that evolution is a population-level phenomenon was one of Darwin's key insights. But populations are composed of individuals, and it is the actions and fates of individuals that determine which characteristics will be passed to descendant populations. In this fashion, inheritance provides the link between the lives of individual organisms and the evolution of populations. We will therefore begin our discussion of the processes of evolution by reviewing the principles of genetics as they apply to individuals and then extend those principles to the genetics of populations. You may want to refer to Unit II to refresh your memory on specific points.

Genes and the Environment Interact to Determine the Traits of Each Individual

Each cell of every organism contains a repository of genetic information encoded in the DNA of its chromosomes. Recall that a *gene* is a segment of DNA located at a particular place on a chromosome. The sequence of nucleotides in a gene encodes the sequence of amino acids of a protein, normally an enzyme that catalyzes one particular reaction in the cell. At a given gene's location, there can be slightly different nucleotide sequences, called *alleles*, as we saw in Chapter 12. Different alleles generate different forms of the same enzyme. In this way, various alleles of the gene that influences eye color in humans, for example, help produce eyes that are brown, or blue, or green, and so on. In any population of organisms, there are usually two or more alleles of each gene. An individual of a diploid or polyploid species whose alleles of a particular gene are all of the same type is *homozygous*, and an individual with alleles of different types for that gene is *heterozygous*. The specific alleles borne on an organism's chromosomes (its *genotype*) interact with the environment to influence development of its physical and behavioral traits (its *phenotype*).

Let's illustrate these principles with an example that should be familiar to you from Unit II. A pea flower is colored purple because a chemical reaction in its petals converts a colorless molecule to a purple pigment. When we say that a pea plant has the allele for purple flowers, we mean that a particular stretch of DNA on one of its chromosomes contains a sequence of nucleotides that codes for the enzyme that catalyzes this reaction. A pea with the allele for white flowers has a different sequence of nucleotides at the corresponding position on one of its chromosomes. The enzyme for which that different sequence codes cannot produce purple pigment. If a pea is homozygous for the white allele, its flowers produce no pigment and thus are white.

The Gene Pool Is the Sum of All the Genes in a Population

We can often deepen our understanding of a concept or phenomenon by looking at it from more than one perspective. In studying evolution, looking at the process from the point of view of a gene has proven to be an enormously effective tool. In particular, evolutionary biologists have made excellent use of the tools of a branch of genetics called **population genetics**, which deals with the frequency, distribution, and inheritance of alleles in populations. To take advantage of this potent aid to understanding the mechanisms of evolution, you will need to learn to use a few of the basic concepts of population genetics.

In population genetics, the **gene pool** is defined as the sum of all the genes in a population. In other words, the gene pool consists of all the alleles of all the genes in all the individuals of that population. Each particular gene can also be considered to have its own gene pool, which consists of all the alleles of that specific gene in a population. If we added up the alleles of that gene over all the individuals in a population, we could determine the relative proportions of the different alleles, a number called the **allele frequency**. For example, a population of 100 pea plants would contain 200 copies of the allele that controls flower color (because pea plants are diploid). If 50 of those 200 alleles were of the type that codes for white flowers, then we would say that the frequency of that allele in the population is 0.25 (or 25%), because 50/200 = 0.25.

Evolution Is the Change of Gene Frequencies Within a Population

A casual observer might choose to define evolution on the basis of changes in the outward appearance or behaviors of the members of a population. The population geneticist, however, looks at a population and sees a gene pool that just happens to be divided into the packages that we call individual organisms. So any phenotypic changes that we observe in the individuals that make up the population can also be viewed as the outward expression of underlying changes to the gene pool. A population geneticist therefore defines evolution as changes in allele frequencies that occur in a gene pool over time. *Evolution is nothing more or less than a change in the genetic makeup of populations over generations.*

The Equilibrium Population Is a Hypothetical Population in Which Evolution Does Not Occur

It will be easier to understand the forces that cause populations to evolve if we first consider the characteristics of a population that would *not* evolve. In 1908 English mathematician Godfrey H. Hardy and German physician Wilhelm Weinberg independently developed a simple mathematical model now known as the **Hardy-Weinberg principle**. This model showed that under certain conditions allele frequencies and genotype frequencies in a population will remain constant no matter how many generations pass. In other words, evolution will not occur in this population. Population geneticists call this idealized, evolution-free population an **equilibrium population**, which will remain in **genetic equilibrium** as long as several conditions are met:

1. *There must be no mutation.*
2. *There must be no* **gene flow** *between populations*; that is, there must be no net migration of alleles into the population (through immigration) or out of the population (through emigration).
3. *The population must be very large.*
4. *All mating must be random,* with no tendency for certain genotypes to mate with specific other genotypes.
5. *There must be no natural selection*; that is, all genotypes must be equally adaptive and reproduce with equal success.

Under these conditions, allele frequencies within a population will remain the same indefinitely. If one or more of these conditions are violated, then allele frequencies may change: Evolution may occur.

As you might expect, few if any natural populations are truly in genetic equilibrium. Then what is the importance of the Hardy-Weinberg principle? The Hardy-Weinberg conditions are useful starting points for studying the mechanisms of evolution. In the following sections, we will examine each condition, show that natural populations often fail to meet it, and illustrate the consequences of such failures. In this way, we can better understand both the inevitability of evolution and the forces that drive evolutionary change.

2 What Causes Evolution?

Population genetics theory predicts that the Hardy-Weinberg equilibrium can be disrupted by deviations from any of its five main underlying conditions. We might therefore predict five major causes of evolutionary change: (1) mutation, (2) gene flow, (3) small population size, (4) nonrandom mating, and (5) natural selection.

Mutations Are the Ultimate Source of Genetic Variability

A population will remain in genetic equilibrium only if no *mutations* (changes in DNA sequence) occur, but mutations are inevitable. Although cells have efficient mechanisms that protect the integrity of their genes, including enzymes that constantly scan the DNA and repair flaws caused by radiation, chemical damage, or mistakes in copying, some changes in nucleotide sequence nevertheless slip past the checking and repair systems. When one of these changes occurs in a cell that produces gametes, the mutation may be passed to an offspring and enter the gene pool of a population.

How significant is mutation in altering the gene pool of a population? Mutations are rare, occurring once in 100,000 to 1,000,000 genes per individual in each generation. Therefore, mutation by itself is not a major force in evolution. However, *mutations are the source of new alleles,* new heritable variations on which other evolutionary processes can work. As such, they are the foundation of evolutionary change. *Without mutations there would be no evolution and no diversity among life-forms.*

Mutations are not goal-directed. A mutation does not arise as a result of, or in anticipation of, environmental necessities (Fig. 15-1). A mutation simply happens and may in turn produce a change in the structure or function of the organism. Whether that change is helpful or harmful or neutral, now or in the future, depends on environmental conditions over which the organism has little or no control. The mutation provides *potential*; other forces, such as migration and especially natural selection, that act on that potential may favor the spread of a mutation through the population or eliminate it.

Gene Flow Between Populations Changes Allele Frequencies

When individuals move from one population to another and interbreed at the new location, alleles are transferred from one gene pool to another. This movement of alleles, or gene flow, between populations alters the distribution of alleles among populations. Baboons, for example, live in social groupings called *troops.* Within each troop, all the females mate with a handful of dominant males. Juvenile males usually leave the troop. If they are lucky, they join and perhaps even become dominant in another troop. Thus, the male offspring of one troop carry genes to the gene pools of other troops.

Movement of breeding organisms between populations has two significant effects:

1. *Gene flow spreads advantageous alleles throughout the species.* Suppose that a new allele arises in one

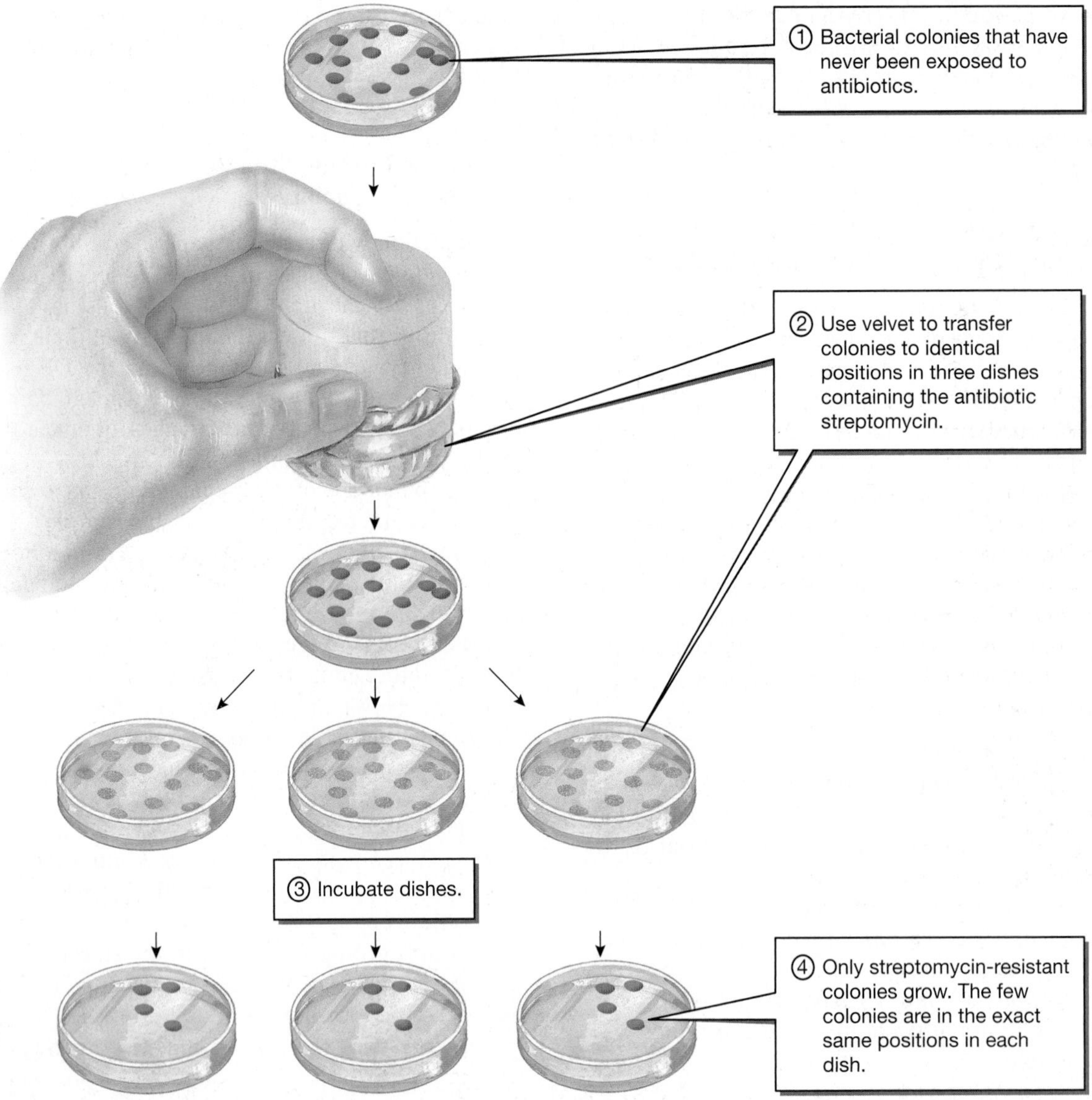
① Bacterial colonies that have never been exposed to antibiotics.
② Use velvet to transfer colonies to identical positions in three dishes containing the antibiotic streptomycin.
③ Incubate dishes.
④ Only streptomycin-resistant colonies grow. The few colonies are in the exact same positions in each dish.

Figure 15-1 Mutations occur spontaneously
This experiment demonstrates that mutations occur spontaneously and not in response to environmental pressures. The bacterial colonies in the original petri dish have never been exposed to antibiotics. When members of each colony are transferred to three new dishes that contain the antibiotic streptomycin, the exact positions of the "parent" colonies are duplicated. Only a few daughter colonies grow in the new dishes, and these surviving colonies grow in the exact same positions in all three dishes. The identical patterns show that mutations for resistance to streptomycin must have already been present in the original dish, before the appearance of any environmental pressure in the form of streptomycin.

population and that this new allele benefits the organisms that possess it. Migration can carry this new allele to other populations of the species.

2. *Gene flow helps maintain all the organisms over a large area as one species.* If migrants constantly carry genes back and forth among populations, then the populations can never develop large differences in allele frequencies. Isolation of a population, with no gene flow to or from other populations of the same species, is a key factor in the origin of new species, as we will discuss in Chapter 16.

Small Populations Are Subject to Random Changes in Allele Frequencies

To remain in genetic equilibrium, a population must be so large that chance events have no impact on its overall genetic makeup. Disaster may befall even the fittest organism. The maple seed that falls into a pond never sprouts; the deer and elk blasted away by the eruption of Mount St. Helens left no descendants. If a population is sufficiently large, chance events are unlikely to alter the overall gene frequency, because such events would be expected to interfere equally with the reproduction of organisms of all genotypes. In a small population, however, certain alleles may be carried by only a few organisms. Chance events could reduce or even eliminate such alleles from the population, altering its genetic makeup.

Genetic Drift Is an Example of Random Genetic Change in Small Populations

Chance events are much more likely to change allele frequencies in a small population than in a large population; this happens by a process called **genetic drift.** Imagine, for example, two populations of amoebas in which each individual amoeba is either red or blue, and cell color is controlled by alternate alleles of a single gene. Each of our two imaginary populations is evenly divided between red and blue individuals. One population, however, has only four individuals in it, while the other has 10,000.

Now let's picture reproduction in our imaginary amoeba populations. Let's allow only half of the individuals in each population to reproduce by binary fission. To do so, each reproducing amoeba splits in half to yield two amoebas, each of which is the same color as the "parent." In the large population, 5000 amoebas will reproduce, yielding a new generation of 10,000. What are the chances that all 10,000 members of the new generation will be red? Just about nil. In fact, it would be extremely unlikely for even 3000 amoebas to be red or for 7000 to be red. The most likely outcome is that about half will be red and half blue, just as in the original population. In this large population, then, we would not expect a major change in allele frequencies to occur from generation to generation (Fig. 15-2a).

In the small population, the situation is different. Only two individual amoebas will reproduce, and there is a 25% chance that both reproducers will be red. (This outcome is as likely as flipping two coins and having both come up heads.) If only red amoebas reproduce, then the next generation will consist entirely of red amoebas—a relatively likely outcome. Within a single generation, it is possible for the allele for blue color to disappear from the population.

Figure 15-2b illustrates two important points about genetic drift:

1. *Genetic drift tends to reduce genetic variability within a small population.* In extreme cases, all members of a population may become genetically identical. (The top blue line in Figure 15-2b represents a population in which only the *A* allele is present at generation six).
2. *Genetic drift tends to increase genetic variability between populations.* Purely as a result of chance, separate populations may evolve extremely different allele frequencies. (Compare the frequencies of allele *A* at generation six in the two populations represented by the top blue line and the bottom red line in Figure 15-2b).

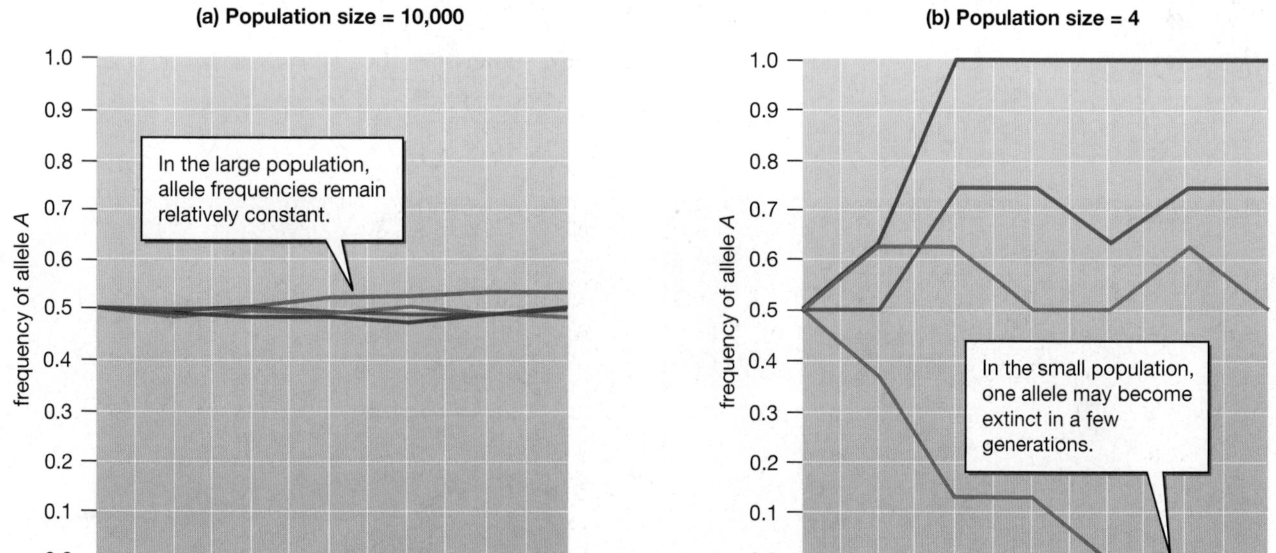

Figure 15-2 Genetic drift
Computer-generated graphs illustrating the effect of population size on genetic drift. Each colored line represents one simulation of a population in which two alleles, *A* and *a*, were initially present in equal proportions, and in which randomly chosen individuals reproduced.

Two special cases of genetic drift, called the *population bottleneck* and the *founder effect*, further illustrate the enormous consequences that small population size may have on the allele frequencies of a species.

A Population Bottleneck Is an Example of Genetic Drift

In a **population bottleneck**, a population undergoes a drastic reduction in size—as a result of a natural catastrophe or overhunting, for example. Then only a few individuals are available to contribute genes to the entire future population. As our amoeba example showed, population bottlenecks can cause both *changes in allele frequencies* and *reductions in genetic variability* (Fig. 15-3a). Even if the population later increases significantly in size, these genetic effects of the bottleneck may remain for hundreds or thousands of generations.

Loss of genetic variability has been documented in the northern elephant seal and the cheetah (Fig. 15-3b,c). The elephant seal was hunted almost to extinction in the 1800s; by the 1890s only about 20 had survived. Because dominant male elephant seals typically monopolize breeding, with a single male mating with a stable group of females, one male may have fathered all the offspring at this extreme bottleneck point. Since then, elephant seals have increased in number to about 30,000 individuals, but biochemical analysis shows that all northern elephant seals are genetically almost identical. Other species of seals, whose populations have historically always remained large, exhibit much more genetic variability. The rescue of the northern elephant seal from extinction is rightly regarded as a triumph of conservation. With very little genetic variation, however, the elephant seal has much less potential to evolve in response to environmental changes. No matter how many elephant seals there are, the species must be considered to be threatened with extinction. Cheetahs are also genetically uniform, although the reason for the bottleneck is unknown. Consequently, cheetahs too could be gravely threatened by small changes in their environment.

A special case of a population bottleneck is the **founder effect**, which occurs when isolated colonies are founded by a small number of organisms. A flock of birds, for instance, may become lost during migration or may be blown off course by a storm. (This is thought to have happened in the case of Darwin's finches in the Galapagos Islands.) Among humans, small groups may migrate for religious or political reasons. Such a small

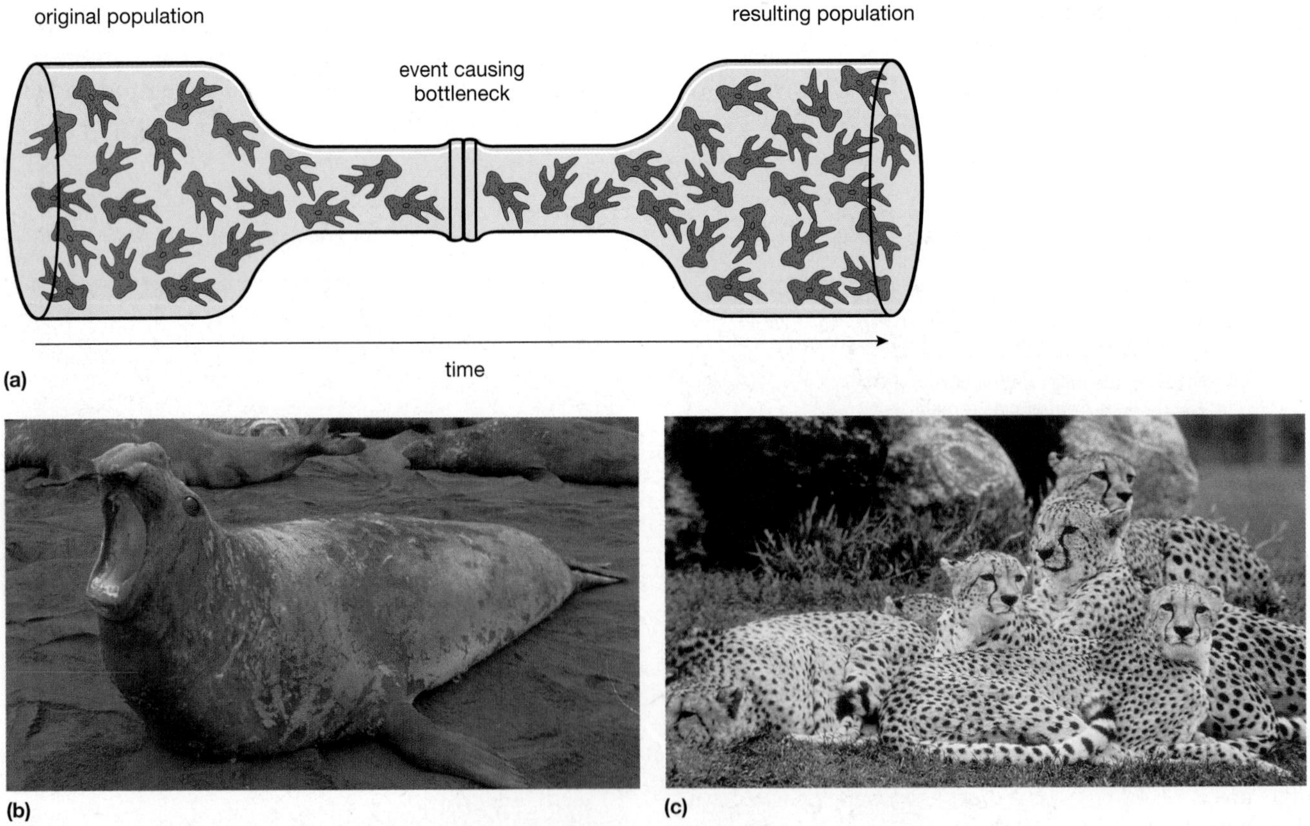

Figure 15-3 Genetic bottlenecks reduce variability
(a) If a population is reduced to a very small number of individuals, the gene pool is reduced and a population bottleneck occurs. The recovered population shows reduced genetic and phenotypic variability, because all are offspring of the few organisms that survived the bottleneck. Both ***(b)*** the northern elephant seal and ***(c)*** the cheetah passed through a population bottleneck in the recent past, resulting in an almost total loss of genetic diversity. As a result, the ability of these populations to adapt to changing environments is very limited.

MEDIATUTOR 15.1 The Bottleneck Effect

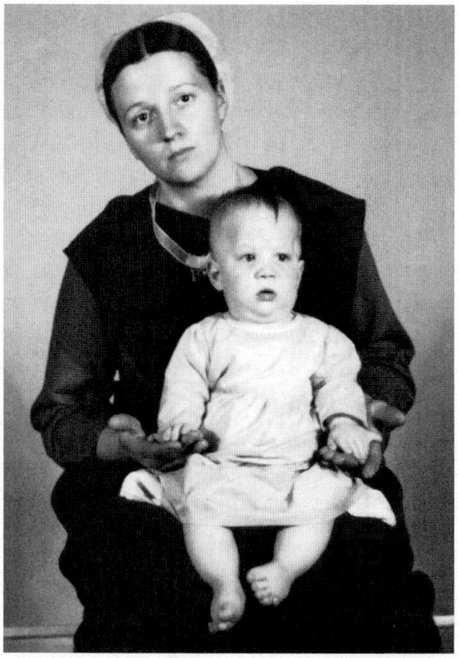

Figure 15-4 **A human example of the founder effect**
An Amish woman with her child, who suffers from a set of genetic defects known as Ellis–van Creveld syndrome (short arms and legs, extra fingers, and, in some cases, heart defects). The founder effect accounts for the prevalence of Ellis-van Creveld syndrome among the Amish residents of Lancaster County, Pennsylvania.

group may have allele frequencies that are very different from the frequencies of the parent population because of chance inclusion of disproportionate numbers of certain alleles in the founders. If the isolation of the founders is maintained for a long period of time, a sizable new population may arise that differs greatly from the original population. For example, a set of genetic defects known as Ellis–van Creveld syndrome (Fig. 15-4) is far more common among the Amish inhabitants of Lancaster County, Pennsylvania, than among the general population. Today's Lancaster County Amish are descended from only 200 or so eighteenth-century immigrants, and one couple among these immigrants is known to have carried the Ellis–van Creveld allele. In such a small founder population this single occurrence meant that the allele was carried by a comparatively high proportion of the Amish founder population (1 or 2 carriers out of 200, versus perhaps 1 in 1000 in the general population). This high initial allele frequency, combined with subsequent genetic drift, has led to extraordinarily high levels of Ellis–van Creveld syndrome among this Amish group.

How much does genetic drift contribute to evolution? No one really knows. Only rarely are natural populations extremely small or completely cut off from gene flow with other populations. These small and/or isolated populations, however, may make an important contribution to evolutionary change. Such populations are probably a major source of new species, as we shall see in Chapter 16.

Mating Within a Population Is Almost Never Random

Any form of nonrandom mating that is based on phenotype will alter allele frequencies in the population where it occurs, and organisms seldom mate strictly randomly. For example, most animals have limited mobility and are most likely to mate with nearby members of their species. Further, they may choose to mate with certain individuals of their species rather than with others. The snow goose is a case in point. Individuals of this species come in two "color phases"; some snow geese are white, and others are blue (Fig. 15-5). Although both white and blue geese are members of the same species, mate choice is not random with respect to color phase. The birds exhibit a strong tendency to mate with a partner of the same color. This kind of preference for mates that are similar to one's self is known as *assortative mating.*

Another common form of nonrandom mating occurs in animal species with mating systems in which only a few dominant males gain reproductive access to females. In these species, which include elephant seals, deer, elk, baboons, and bighorn sheep, all of the females in a population are fertilized by a small number of males. Fertilization typically follows some sort of contest among males, which may involve showing off with loud sounds or flashy colors, making threatening gestures, or actual combat (Fig. 15-6).

In many animal species, mating is not random because one sex, normally the female, controls mate selection and is quite picky about who qualifies for the privilege of breeding. Males display their virtues, such

Figure 15-5 **Nonrandom, assortative mating among snow geese**
Some snow geese have white plumage, and others have dark blue plumage. These geese are most likely to mate with another bird of the same color.

Figure 15-6 Male competition promotes nonrandom mating Two male bighorn sheep spar during the fall mating season. In many species, the losers of such contests are unlikely to mate.

as the bright plumage of a peacock (Fig. 15-7) or the rich territory of a songbird. A female evaluates the males and chooses her mate. We shall explore this phenomenon in more detail later in this chapter.

All Genotypes Are Not Equally Adaptive

Genetic equilibrium requires that all genotypes must be equally adaptive—that is, no one genotype has any advantage over the others. It is probably true that some alleles are adaptively neutral, so organisms possessing any of several alleles will be equally likely to survive and reproduce. However, this is clearly not true of all alleles in all environments. Any time an allele confers, in Alfred Russel Wallace's words, "some little superiority," natural selection will favor more successful reproduction by the individuals who possess it. This phenomenon is perhaps best illustrated with an example.

The antibiotic penicillin first came into widespread use during World War II, when it was used to combat infections in wounded soldiers. Suppose that an infantryman, brought to a field hospital after suffering a gunshot wound in his arm, develops a staph infection in that arm. A medic sizes up the situation and resolves to treat the wounded soldier with an intravenous drip of penicillin. As the antibiotic courses through the soldier's blood vessels, millions of staph bacteria die before they can reproduce. A few individual bacteria, however, carry a rare allele that codes for an enzyme that helps destroy any penicillin that comes in contact with the bacterial cell. (This allele is a variant of a gene that normally codes for an enzyme involved in breaking down the bacterium's waste products.) The individuals carrying this rare allele are able to survive and reproduce, and their offspring inherit the penicillin-destroying allele. After a few generations, the frequency of the penicillin-destroying allele has soared to nearly 100%, and the frequency of the normal, waste-processing allele has declined to near zero. As a result of the antibiotic's selective killing power, *evolution will have occurred in the population of staph bacteria within that soldier's body.* The gene pool of the staph population has changed, and natural selection, in the form of bacterial destruction by penicillin, has caused the change.

This example illustrates four important points about evolution:

1. *Natural selection does not cause genetic changes in individuals.* The allele for penicillin resistance arose spontaneously, long before penicillin was dripped into the soldier's vein. Penicillin did not cause resistance to appear; its presence merely favored the survival of individuals with penicillin-destroying alleles over that of individuals with waste-processing alleles.
2. *Natural selection acts on individuals, but evolution occurs in populations.* The agent of natural selection,

Figure 15-7 The male peacock's showy tail has evolved through sexual selection
If females are picky when deciding on a male with which to mate, they might tend to favor males with slightly longer or more colorful tails. Their sons would then inherit tails that are larger and more colorful than average. If females in the next generation also preferred long, colorful tails, the cycle would continue. Given enough cycles of female selectivity, the scene pictured above would become the species' norm.

Earth Watch

Endangered Species: From Gene Pools to Gene Puddles

Ever since the Endangered Species Act was passed in 1973, the United States has had an official policy of protecting rare species. In fact, the real goal of the act is not protection but recovery; as one U.S. Fish and Wildlife Service official said, "The goal is to get species *off* the list." Wildlife biologists try to determine how large a population a species needs to have before it is no longer in danger of extinction from unpredictable events, such as a couple of years of drought or an epidemic of parasites. If a species reaches this critical population size, it is no longer legally "endangered" by extinction.

Does a "large enough" population (which usually is still very small by historical standards) really ensure a species' survival? From our discussion of genetic drift and population bottlenecks, you probably realize that the answer is no. If the population of a species has been reduced to the point at which it is declared an endangered species, then it probably has lost much of its genetic diversity. As ecologist Thomas Foose aptly put it, loss of habitat and the consequent reduction in population size mean that "gene pools are being converted into gene puddles." Even if the species recovers in numbers, the variety of its original gene pool has been lost. When the forces of natural selection change at some future time, the species may not have the necessary genetic variability to produce individuals adapted to the new environment, and it may then go extinct.

What can be done? The best solution, of course, is to leave enough habitat of diverse types so that species never become endangered in the first place. The human population, however, has grown so large and appropriated so much of Earth's resources that this solution is not possible in many places (Fig. E15-1). For many species, the only solution is to preserve enough habitat so that the remaining population is large enough to retain all or most of the total genetic diversity of the species. If we are to be the caretakers of the planet and not merely its ultimate consumers, then protection of other life-forms and their genetic heritage will be our continuing responsibility.

Figure E15-1 Endangered by habitat destruction
This orangutan and its young, who live in the tropical rain forests of Borneo, are among the innumerable species whose continued survival is threatened by habitat destruction.

penicillin, acted on individual staph bacteria. As a result, some individuals reproduced and some did not. However, it was the population as a whole that evolved as its allele frequencies changed.

3. *Evolution is a change in the allele frequencies of a population, owing to unequal success at reproduction among organisms bearing different alleles.* In evolutionary terminology, the **fitness** of an organism is measured by its reproductive success: In our example, the penicillin-resistant bacteria had greater fitness than the normal bacteria did, because the resistant bacteria produced greater numbers of viable offspring.
4. *Evolutionary changes are not "good" or "progressive" in any absolute sense.* The resistance alleles were favored only because of the presence of penicillin in this particular soldier's body; in another environment, the resistance allele may well have placed its carriers at a disadvantage relative to other bacteria that could process waste more effectively.

Natural selection is not the *only* evolutionary force. As we have seen, mutation provides initial variability in heritable traits. The chance effects of genetic drift may change allele frequencies, even spawning new species. Further, evolutionary biologists are just now beginning to appreciate the power of random catastrophe in shaping the history of life on Earth—massively destructive events that may exterminate flourishing and floundering species alike. Nevertheless, it is natural selection that shapes the evolution of populations as they adapt to their changing environment. For this reason, we will examine the mechanisms of natural selection in some detail.

3 How Does Natural Selection Work?

To most people, the words **natural selection** are synonymous with the phrase "survival of the fittest." Natural selection evokes images of wolves chasing caribou, of lions snarling angrily in competition over a zebra carcass. Natural selection, however, is not really about *survival* alone; it is also about *reproduction*. It is certainly true that if an organism is to reproduce, it must survive long enough to do so. In some cases, it is also true that a longer-lived organism has more chances to reproduce. But no organism lives forever, and the only way that its genes continue into the future is through successful reproduction. When an organism that fails to reproduce dies, its genes die with it. The organism that reproduces lives on, in a sense, through the genes that it has passed on to its offspring. Therefore, although evolutionary biologists often discuss survival, partly because survival is usually easier to measure than reproduction, natural selection is really an issue of **differential reproduction**: Individuals bearing certain alleles leave more offspring (who inherit those alleles) than do other individuals with different alleles.

Natural Selection Acts on the Phenotype, Which Reflects the Underlying Genotype

Although we have defined evolution in terms of changes in the genetic composition of a population, it is important to recognize that natural selection cannot act directly on the genotypes of individual organisms. Rather, natural selection acts on phenotypes, the actual structures and behaviors displayed by the members of a population. Selection on phenotypes, however, inevitably affects the genotypes present in a population, because phenotypes and genotypes are closely tied. For example, we know that a pea plant's height is strongly influenced by the plant's alleles of certain genes. If a population of pea plants were to encounter environmental conditions that favored taller plants, then taller plants would leave more offspring. These offspring would carry the alleles that contributed to their parents' height. Thus, if natural selection favors a particular phenotype, it will necessarily also favor the underlying genotype.

Natural Selection Can Influence Populations in Three Major Ways

Biologists recognize three major categories of natural selection based on its effect on the population over time (Fig. 15-8):

1. **Directional selection** favors individuals who possess values for a trait at one end of the distribution range of a particular trait and selects against both average individuals and individuals at the opposite extreme of the distribution. For example, directional selection may favor small size and select against both average and large individuals in a population.
2. **Stabilizing selection** favors individuals who possess an average value for a trait (for example, intermediate body size) and selects against individuals with extreme values.
3. **Disruptive selection** favors individuals who possess relatively extreme values for a trait at the expense of individuals with average values. Disruptive selection favors organisms at both ends of the distribution of the trait (for example, both large and small body sizes).

Directional Selection Shifts Character Traits in a Specific Direction

If environmental conditions change in a consistent way—for example, if the climate becomes colder—then a species may evolve in a consistent direction in response—for example, with thicker fur. The evolution of long necks in giraffes was almost certainly due to directional selection (Fig. 15-8a): Ancestral giraffes with longer necks obtained more food and therefore reproduced more prolifically than their shorter-necked contemporaries did. Antibiotic resistance in bacteria is another example of directional selection.

Stabilizing Selection Acts Against Individuals Who Deviate Too Far from the Average

Directional selection can't go on forever. What happens once a species is well adapted to a particular environment? If the environment doesn't change, then most variations that appear through new mutations or recombination of old alleles during sexual reproduction will be harmful. Therefore, the species will typically undergo stabilizing selection, which favors the survival and reproduction of "average" individuals (Fig. 15-8b). Stabilizing selection commonly occurs when a single trait is under opposing environmental pressures from two different sources. Biologist M. K. Hecht, for example, studied lizards of the genus *Aristelliger*. He found that small lizards have a hard time defending territories but that large lizards are more likely to be preyed on by owls. Therefore, *Aristelliger* lizards are under stabilizing selection that favors an "average" body size.

It is widely assumed, although difficult to prove, that many traits are under stabilizing selection. We have already mentioned several. Although the lengths of legs and necks of giraffes probably originated under directional selection for feeding on leaves high up in trees, they are almost certainly now under stabilizing selection, as a compromise between the advantage of reaching higher leaves for food and the disadvantage of vulnerability while drinking water (Fig. 15-9). Similarly, female mate choice probably drove the evolution of elaborate sexual displays in many birds, but now increased vulnerability to predators may exert stabilizing

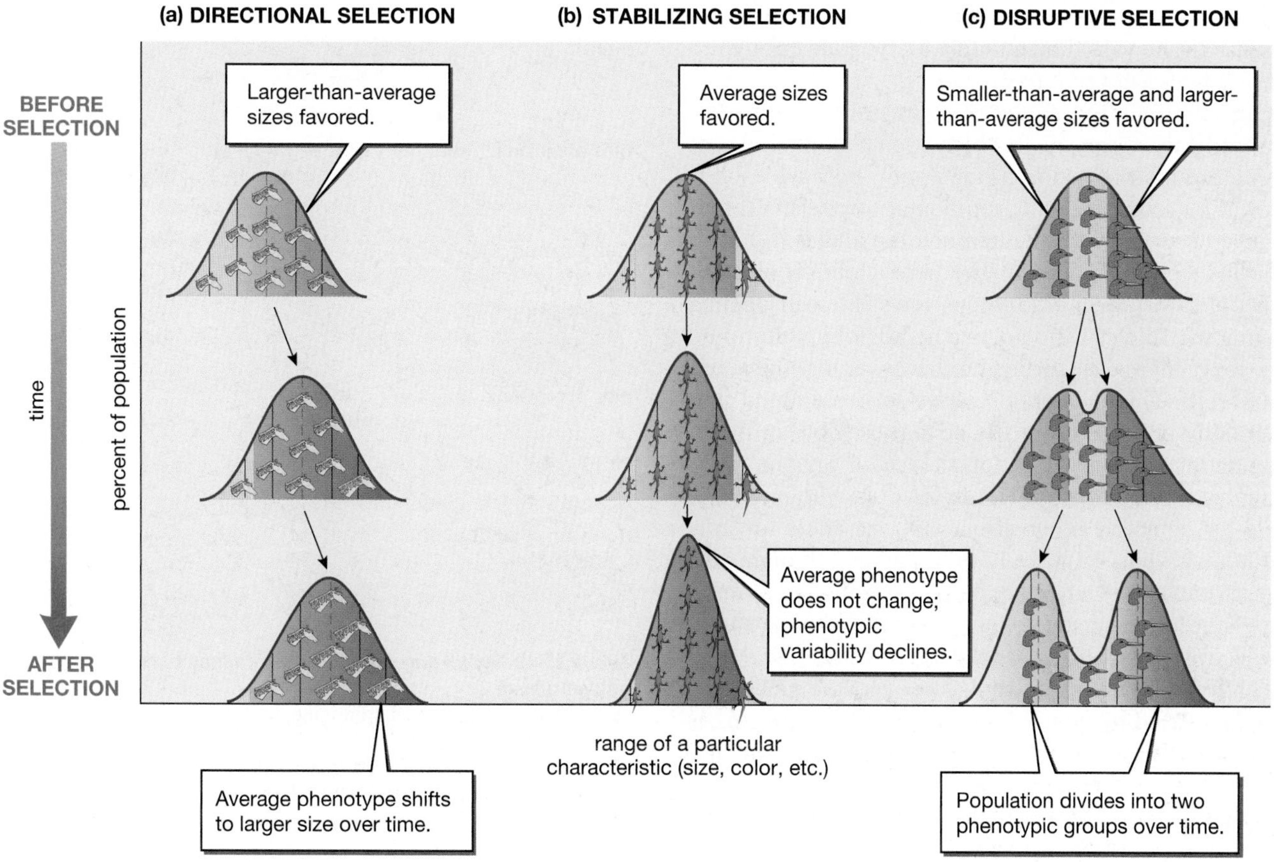
(a) DIRECTIONAL SELECTION
(b) STABILIZING SELECTION
(c) DISRUPTIVE SELECTION
BEFORE SELECTION
time
AFTER SELECTION
percent of population
Larger-than-average sizes favored.
Average sizes favored.
Smaller-than-average and larger-than-average sizes favored.
Average phenotype does not change; phenotypic variability declines.
range of a particular characteristic (size, color, etc.)
Average phenotype shifts to larger size over time.
Population divides into two phenotypic groups over time.

Figure 15-8 Three ways natural selection affects a population over time
A graphical illustration of three ways natural selection, acting on a normal distribution of phenotypes (in these examples, size), can affect a population over time. In all graphs, the beige areas represent individuals that are selected against—that is, do not reproduce as successfully as do the individuals in the purple range.

MEDIATUTOR 15.2 Three Modes of Natural Selection

(a) (b)

Figure 15-9 A compromise between opposing environmental pressures
(a) The long neck and legs of a giraffe are a definite advantage in feeding on acacia leaves high up in trees. ***(b)*** But a giraffe has to get into an extremely awkward and vulnerable position to drink. Feeding and drinking thus place opposing environmental pressures on the length of neck and legs.

MEDIATUTOR 15.3 Natural Selection at Work: Alpine Skypilots

selection: If a peacock's tail became so long that he couldn't fly, he would be unlikely to live long enough to woo a female with that flashy tail.

Opposing environmental pressures may give rise to **balanced polymorphism**, in which two or more alleles of a gene are maintained in a population because each is favored by a separate environmental force. This seems to have occurred with the hemoglobin alleles in native Africans (see Chapter 12). The hemoglobin molecules of people who are homozygous for sickle-cell anemia (having two alleles for defective hemoglobin) clump up into long chains, distorting and weakening their red blood cells. This distortion causes severe anemia and potentially death. Before the advent of modern medicine, people homozygous for sickle-cell anemia were strongly selected against. Heterozygotes, who have one allele for defective hemoglobin and one allele for normal hemoglobin, suffer only mild anemia, though they may be adversely affected during strenuous exercise. Under these circumstances, you might think that natural selection would eliminate the sickle-cell allele.

Far from being eliminated, however, the sickle-cell allele is carried by nearly half the people in some areas of Africa. This distribution seems to result from the counterbalancing effects of anemia and malaria, a disease that formerly caused high death rates in equatorial Africa. The parasites that cause malaria multiply rapidly within the red blood cells of homozygous normal individuals. Before effective medical treatments were discovered, homozygous normals commonly died of malaria. Heterozygotes, however, enjoy some protection against malaria. Malaria parasites inside a heterozygote's red blood cells use up oxygen, causing the sickle-cell hemoglobin to clump and the cells to become sickle shaped. Infected sickled cells are destroyed by the spleen before the parasites can complete their development. Heterozygotes, therefore, have mild anemia but do not succumb to malaria.

During the evolution of African populations, heterozygotes survived better than either type of homozygote did and reproduced the most successfully. As a result, both the normal hemoglobin allele and the sickle-cell allele have been preserved (Fig. 15-10).

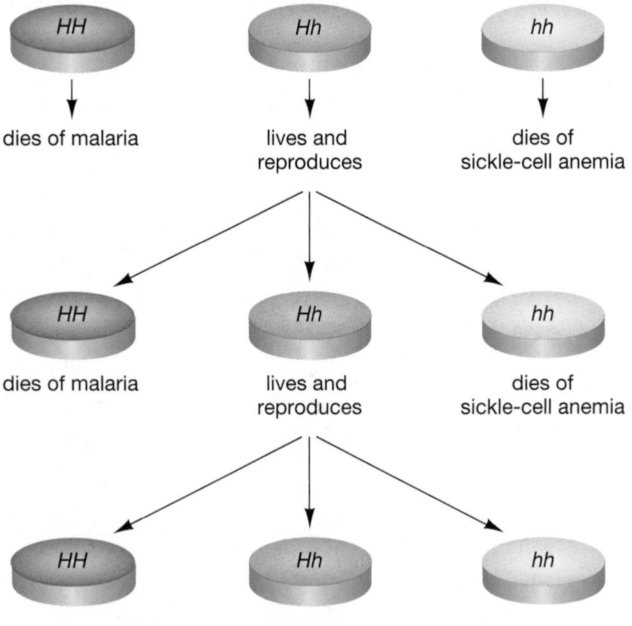

Figure 15-10 Stabilizing selection can produce balanced polymorphism
Two or more alleles, each producing a different phenotype, may be maintained in a population by opposing environmental pressures. The alleles for normal (*H*) and sickle-cell (*h*) hemoglobin are maintained by selection against both homozygotes. Heterozygotes (*Hh*) reproduce the most, thereby keeping both alleles in the population.

Disruptive Selection Adapts Individuals Within a Population to Different Habitats

Disruptive selection (Fig. 15-8c) may occur when a population occupies an area that provides different types of resources that can be used by the species. In this situation, different characteristics best adapt individuals to use each type of resource. For example, individual black-bellied seedcrackers (small, seed-eating birds found in the forests of Africa) have beaks that come in one of two sizes. A bird may have a large beak or small beak, but very few birds have a medium-sized beak. The species' food source includes both hard seeds and soft seeds, and each bird seems to specialize in eating only one type of seed. Cracking hard seeds requires a large, stout beak. However, a smaller, pointier beak is apparently a more efficient tool for processing soft seeds. Individuals with intermediate-sized beaks have a lower survival rate than do individuals with either large or small beaks. Disruptive selection in black-bellied seedcrackers thus favors birds with both large and small beaks, but not those with medium-sized beaks.

A Variety of Processes Can Cause Natural Selection

As we've just seen, natural selection consists of nothing more than the plain fact that some phenotypes reproduce more successfully than others do. What makes this simple process such a powerful agent of change is that only the "best" phenotypes pass traits to subsequent generations. But what makes a phenotype "best"? Successful phenotypes are those that have the best *adaptations* to their particular environment. **Adaptations** are characteristics that help an individual survive and reproduce in an environment that includes not only physical factors but also the other organisms with which the individual interacts. These other organisms include predators, prey, and same-species competitors.

The nonliving (*abiotic*) component of the environment includes such factors as climate, availability of water, and minerals in the soil. The abiotic environment establishes the "bottom line" requirements that an organism must have to survive and reproduce. However, many of the adaptations that we see in modern organisms have arisen because of interactions with other organisms—the living (*biotic*) component of the environment. As Darwin wrote, "The structure of every organic being is related . . . to that of all other organic beings, with which it comes into competition for food or residence, or from which it has to escape, or on which it preys." A simple example illustrates this concept.

A buffalo grass plant sprouts in a small patch of soil in the eastern Wyoming plains. Its roots must be able to take up enough water and minerals for growth and reproduction, and to that extent it must be adapted to its abiotic environment. Even in the dry prairies of Wyoming, this requirement is relatively trivial provided that the plant is alone and protected in its square meter of soil. In reality, however, many other plants—including other grasses, sagebrush bushes, and annual wildflowers—also sprout in that same patch of soil. If our buffalo grass is to survive, it must compete for resources with the other plants. Its long, deep roots and efficient mineral uptake processes have evolved not so much because the plains are dry, but because it must share the dry prairies with other plants. Further, cattle (and, in the past, bison) graze the prairies. Buffalo grass is extremely tough, with silica (glass) compounds reinforcing the blades, an adaptation that discourages grazing. Over millennia, tougher plants were harder to eat and so survived better and reproduced more than did less-tough plants—another adaptation to the biotic environment.

Competition for Scarce Resources Favors the Best-Adapted Individuals

One of the major agents of natural selection in the biotic environment is **competition** with other members of the same species. As Darwin wrote in *On the Origin of Species*, "The struggle almost invariably will be most severe between the individuals of the same species, for they frequent the same districts, require the same food, and are exposed to the same dangers." In other words, no competing organism has such similar requirements for survival as does another member of the same species. For example, both lazuli buntings and western bluebirds are brightly colored in blue, red, and white, and both nest and rear their young in the foothills of the Rocky Mountains in the summer. But they do not compete very much with each other, because they eat different foods: Bluebirds mostly catch insects, whereas buntings specialize in seeds. Each mosquito eaten by a bluebird makes little difference to a bunting but makes it harder for other bluebirds to find enough to eat.

Different species may also compete for the same resources, although generally to a lesser extent than do individuals within a species. As we will discuss more fully in Chapter 39, whether a particular plot of prairie is covered with grass, sagebrush, or trees is at least partly determined by competition among those plants for scarce soil moisture.

During Predation, Both Predator and Prey Act as Agents of Selection

When two species interact extensively, each exerts strong selection pressures on the other. When one evolves a new feature or modifies an old one, the other typically evolves new adaptations in response. As the Red Queen told Alice in Lewis Carroll's *Through the Looking Glass*, "Here, you see, it takes all the running you can do to keep in the same place." This constant, mutual feedback between two species is called **coevolution**. Perhaps the most familiar form of coevolution is found in *predator–prey* relationships.

Although we commonly think of predation as one animal preying upon another animal, **predation** includes any situation in which one organism eats another. In some instances, coevolution between *predators* (those who do the eating) and *prey* (those who are eaten) is a sort of "biological arms race," with each side evolving new adaptations in response to "escalations" by the other. Darwin used the example of wolves and deer: Wolf predation selects against slow or careless deer, thus leaving faster, more-alert deer to reproduce and continue the species. In their turn, alert, swift deer select against slow, clumsy wolves, because such predators cannot acquire enough food.

Symbiosis Produces Adaptations for Living in Intimate Association with Another Species

The term **symbiosis** describes a relationship in which individuals of different species live in direct contact with one another for prolonged periods of intimate interaction. Symbiosis incorporates several kinds of ecological relationships, including *parasitism*, in which one species lives and feeds on a larger species; *commensalism*, in which one species benefits and the other remains unharmed; and *mutualism*, in which both species benefit. The different types of symbiosis are described in Chapter 39. From an evolutionary perspective, of all the types of biotic interaction, symbiosis leads to the most-intricate coevolutionary adaptations. Whereas a given predator usually preys on several species and may interact with a particular species only occasionally, partners in symbiosis typically live together virtually their entire lives. Thus, each species must continually adjust to any evolutionary changes developed by the other.

Sexual Selection Favors Traits That Help an Organism Mate

In many species of animals, males compete for access to matings with females. This competition can take the form of contests for dominance among males, or it may involve behaviors or physical features that females find attractive. In the latter case, males may compete for the attention of females through song, elaborate displays, the defense of large territories, or even by building elaborate structures, such as those of the bowerbird (as we shall see in Chapter 37). Choosing a male with a good territory is obviously advantageous, because good territories provide adequate food and shelter to raise young. However, females often also prefer elaborate "fashions" in their mates, such as bright colors and long feathers or fins that may make the male more vulnerable to predation. Why? A popular hypothesis is that structures and colors that do not serve any clear adaptive purpose actually provide the females with an outward sign of the males' fitness. Only vigorous, energetic males can survive when burdened with conspicuous coloration or large tails. Similarly, males that are sick or under parasitic attack may be dull and frumpy compared with healthy males. Whatever the exact selective mechanisms, it is thought that many of the elaborate structures and behaviors found only in males have evolved through selection caused by female mate choice: Only the flashy males transmitted their genes to the next generation.

Both the conspicuous structures and bizarre courtship behaviors of some male animals, and the apparent willingness of females to base their mate choices on these features, frequently seem to be at odds with efficient survival and reproduction. As we've stated above, exaggerated ornamentation may make males more attractive to females, but it also makes the males more vulnerable to predators. Darwin was intrigued by this apparent contradiction. He coined the term **sexual selection** to describe the special kind of natural selection that acts on traits that help an animal mate.

Kin Selection Favors Altruistic Behaviors

Evolution is often portrayed as being a bloody and vicious fight for survival. This portrayal, however, is not the complete picture. Although it is true that competitive and predatory interactions influence the evolution of most species, cooperation and even self-sacrifice may be favored by selection. **Altruism** refers to any behavior that endangers an individual organism or reduces its reproductive success but benefits other members of its species. Altruistic behaviors are common in the animal kingdom. A male scrub jay helps feed and defend the nestlings of another mated pair of jays instead of seeking out a territory and mate of his own (Fig. 15-11); female worker bees forego reproduction and devote their lives to raising the offspring of the hive queen; and young male baboons scout around the edges of the troop, even though doing so increases their danger from leopards.

You might think that altruism runs counter to natural selection: If altruism is encoded in an organism's genes, those genes are placed at risk every time the altruist performs one of its brave behaviors. But natural selection can indeed select for altruistic genes, if the altruistic individual helps relatives who possess the same alleles. This special case of natural selection is called **kin selection** and is discussed at greater length next.

Figure 15-11 Altruism by a helper at the nest
A yearling Florida scrub jay helps feed and defend his younger siblings. The young bird chooses to help at his parents' nest rather than to have a nest of his own. This apparent self-sacrifice is actually selfish behavior that represents the young bird's best chance to preserve his evolutionary legacy.

Evolutionary Connections

Knowing Your Relatives: Kin Selection and Altruism

From an evolutionary viewpoint, how can we explain behavior that helps other individuals while reducing the fitness of the performing individual (that is, altruistic behavior)? Surely, if a mutation arose that caused altruistic behavior and the bearers of that mutation died or failed to reproduce because of their self-sacrificing behaviors, their "altruistic alleles" would disappear from the population. Maybe, or maybe not. To understand the evolution of altruism, we will need to introduce a new concept: inclusive fitness. The **inclusive fitness** of an individual is determined by the individual's success at contributing its own genes to the next generation. This contribution has two components: (1) a direct contribution from producing offspring *and* (2) an indirect contribution gained by helping relatives produce more offspring than would otherwise have been possible.

To see how altruism might increase the inclusive fitness of an individual, let's consider the Florida scrub jay. Year-old jays rarely mate and reproduce. Instead, these yearlings remain at their parents' nest and help feed and protect the parents' next brood.

Altruistic yearlings forego reproduction for at least one year; some die from predation or accidents and never reproduce at all. How, then, can this behavior be adaptive? It all has to do with the reproductive options available to young Florida scrub jays. In their particular ecological setting, suitable breeding habitat is in very short supply. Long-term studies have shown that a young jay that leaves its home territory is fairly unlikely to be successful at finding a territory and raising a brood. But if a young bird stays on home territory, it may eventually inherit the territory or carve out a smaller area for itself. For this reason, selection has favored the behavior of "staying home." Still, this doesn't explain why the stay-at-homes expend valuable energy to help raise their younger siblings. Answering *that* question requires us to recognize that, on average, a diploid, sexually reproducing animal shares 50% of its genes with each sibling. A scrub jay—or any other animal—is just as related to its siblings as it would be to its own offspring. Therefore, if a yearling jay's help at the nest enables the survival of a sibling who would not otherwise have survived, the helper has accomplished as much in fitness terms as if it had raised an offspring of its own. As long as the average extra (indirect) fitness that results from helping is greater than the average (direct) fitness that would have resulted from an attempt at independent breeding, then the altruistic helping behavior is the most adaptive behavior. This phenomenon, whereby the actions of an individual increase the survival or reproductive success of its relatives, is called *kin selection*.

As this example suggests, *kin selection can favor the evolution of altruism if the altruistic behavior benefits relatives that bear the alleles that help cause altruistic behavior.* In most cases, an animal will not know if another carries the altruism allele. A yearling jay that helped out at the nest of unrelated adult jays would probably waste its time and effort.

Identification of relatives isn't too hard to imagine in the case of jays and their parents. Many biologists, however, have objected to kin selection as a plausible explanation of other instances of altruistic behaviors, arguing that animals cannot evaluate degrees of relatedness. Two findings seem to address this objection. First, many social groups, including wolf packs and baboon troops, are actually family groups. Therefore, an animal would not have to identify relatives for its altruistic behaviors to benefit them the most. Second, many animals, including birds, monkeys, tadpoles (Fig. 15-12), bees, and even tunicate larvae, can indeed identify relatives. Given the choice between relatives and strangers, these animals preferentially associate with their relatives, even if they were separated at birth and have never seen those relatives before. If animals selectively form related groups, then altruistic behaviors will most likely benefit relatives. Although it is not the only mechanism of natural selection, kin selection has been a powerful environmental force in the evolution of altruism in many species, probably including humans.

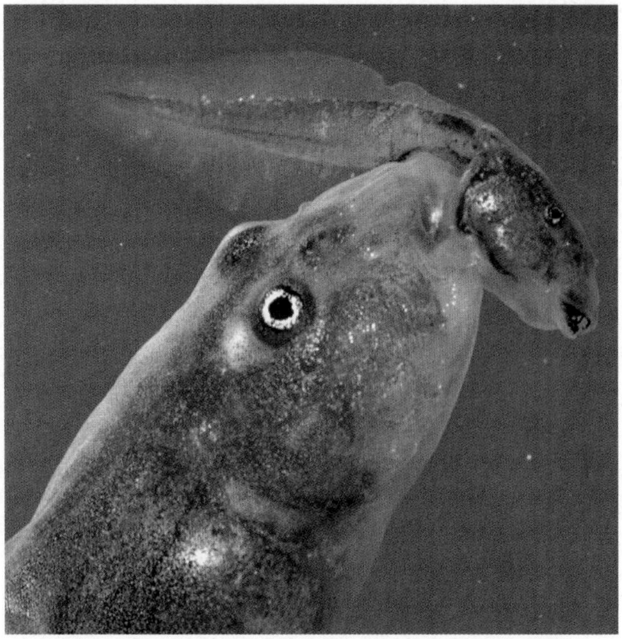

Figure 15-12 Cannibalistic animals don't eat close relatives Spadefoot toad tadpoles, found in transient water holes of the Arizona desert, are cannibalistic. Many of their prey, however, are released unharmed after being tasted briefly. Researchers have discovered that the tadpoles can indeed distinguish, and spit out, their own brothers and sisters, preferring to eat unrelated members of their own species.

CASE STUDY REVISITED

Cause of Death: Evolution

The evolution of antibiotic resistance in bacterial populations is a direct consequence of natural selection applied by widespread use of antibiotic drugs. When a new antibiotic is first introduced, it kills the vast majority of bacteria exposed to it. The surviving bacterial cells, however, may include individuals whose genomes happen to include a mutant gene that confers resistance. As Darwin understood, individuals carrying the resistance gene will leave behind a disproportionately large share of offspring, which inherit the gene. If the environment consistently contains an antibiotic, bacteria carrying the resistance gene will eventually come to predominate. Because bacteria reproduce so rapidly and have comparatively high rates of mutation, evolutionary change leading to resistant populations is often rapid.

We have accelerated the pace of the evolution of antibiotic resistance by introducing massive quantities of antibiotics into the bacteria's environment. Each year, U.S. physicians prescribe more than 100 million courses of antibiotics; the Centers for Disease Control estimates that about half of these prescriptions are unnecessary. An additional 20 million pounds of antibiotics are fed to farm animals annually. The use of antibacterial soaps and cleansers has become routine in many households. As a result of this massive alteration of the bacterial environment, resistant bacteria are now found not only in hospitals and the bodies of sick people but are also widespread in our food supply and in the environment. Our heavy use (many would say overuse) of antibiotics means that susceptible bacteria are under constant attack and that resistant strains have little competition. In our fight against disease, we rashly overlooked some basic principles of evolutionary biology and are now paying a heavy price.

How can further evolution of antibiotic resistance be prevented?

Summary of Key Concepts

1) How Are Populations, Genes, and Evolution Related?

The gene pool of a population is the total of all the alleles of all the genes carried by the members of that population. The sources of genetic variability within a population are mutation, which produces new genes and alleles, and recombination during sexual reproduction. In its broadest sense, evolution is a change in the frequencies of alleles in a population's gene pool due to enhanced reproduction by individuals who bear certain alleles.

Allele frequencies in a population will remain constant over generations only if the following conditions are met: (1) There is no mutation; (2) there is no gene flow—that is, no net migration of alleles into or out of the population; (3) the population is very large; (4) all mating is random; (5) all genotypes reproduce equally well (that is, no natural selection). These conditions are rarely, if ever, met in nature. Understanding why they are not met leads to an understanding of the mechanisms of evolution.

2) What Causes Evolution?

1. Mutations are random, undirected changes in DNA composition. Although most mutations are neutral or harmful to the organism, some prove advantageous in certain environments. Mutations are rare and do not change allele frequencies very much, but they provide the raw material for evolution.
2. In gene flow between populations, if the alleles that a migrant carries are different from those in the population from which the migrant comes or to which it migrates, allele frequencies will change.
3. In any population, chance events kill or prevent reproduction by some of the individuals. If the population is small, chance events may eliminate a disproportionate number of individuals who bear a particular allele, thereby greatly changing the allele frequency in the population. This change is termed *genetic drift*.
4. Many organisms do not mate randomly. If only certain members of a population can mate, then the next generation of organisms in the population will all be offspring of this select group, whose allele frequencies may differ from those of the population as a whole. Population bottleneck and founder effect, two types of genetic drift, illustrate the consequences of small population size on allele frequency.
5. The survival and reproduction of organisms are influenced by their phenotype. Because phenotype depends at least partly on genotype, natural selection tends to favor the reproduction of certain alleles at the expense of others.

3) How Does Natural Selection Work?

Natural selection is an issue of differential, or unequal, reproduction. Over time, natural selection can affect a population in three ways:

1. *Directional selection.* Individuals with characteristics that are different from average in one direction (for example, smaller) are favored both over average individuals and over individuals that differ from average in the opposite direction.
2. *Stabilizing selection.* Individuals with the average "value" for a characteristic are favored over individuals with extreme values.
3. *Disruptive selection.* Individuals with extreme values for a characteristic are favored over individuals with average values.

Natural selection occurs as a result of the interactions of organisms with both the biotic (living) and abiotic (nonliving) parts of their environments. Within a species, sexual selection and altruism are two types of natural selection. When two or more species interact extensively so as to exert mutual environmental pressures on each other for long periods of time, both of them evolve in response. Such coevolution can occur as a result of any type of relationship between organisms, including competition, predation, and symbiosis.

Key Terms

adaptation *p. 298*
allele frequency *p. 288*
altruism *p. 300*
balanced polymorphism *p. 298*
coevolution *p. 299*
competition *p. 299*
differential reproduction *p. 296*
directional selection *p. 296*
disruptive selection *p. 296*
equilibrium population *p. 289*
fitness *p. 295*
founder effect *p. 292*
gene flow *p. 289*
gene pool *p. 288*
genetic drift *p. 291*
genetic equilibrium *p. 289*
Hardy-Weinberg principle *p. 289*
inclusive fitness *p. 300*
kin selection *p. 300*
natural selection *p. 296*
population *p. 288*
population bottleneck *p. 292*
population genetics *p. 288*
predation *p. 299*
sexual selection *p. 300*
stabilizing selection *p. 296*
symbiosis *p. 299*

Thinking Through the Concepts

Multiple Choice

1. *Genetic drift is a ______________ process.*
   a. random
   b. directed
   c. selection-driven
   d. coevolutionary
   e. uniformitarian

2. *Most of the 700 species of fruit flies found in the Hawaiian archipelago are each restricted to a single island. One hypothesis to explain this pattern is that each species diverged after a small number of flies had colonized a new island. This mechanism is called*
   a. sexual selection
   b. genetic equilibrium
   c. disruptive selection
   d. the founder effect
   e. assortative mating

3. *You are studying leaf size in a natural population of plants. The second season is particularly dry, and the following year the average leaf size in the population is smaller than the year before. But the amount of overall variation is the same, and the population size hasn't changed. Also, you've done experiments that show that small leaves are better adapted to dry conditions than are large leaves. Which of the following has occurred?*
   a. genetic drift
   b. directional selection
   c. stabilizing selection
   d. disruptive selection
   e. the founder effect

4. *You have bacteria thriving in your gastrointestinal tract. This is an example of*
   a. inclusive fitness
   b. balanced polymorphism
   c. symbiosis
   d. kin selection
   e. altruism

5. *Lamarckian evolution, the inheritance of acquired characteristics, could occur*
   a. if each gene had only one allele
   b. if individuals had different phenotypes
   c. if the genotype was altered by the same environmental changes that altered the phenotype
   d. if the phenotype was altered by the environment
   e. under none of these conditions

6. *Of the following possibilities, the best way to estimate an organism's evolutionary fitness is to measure the*
   a. size of its offspring
   b. number of eggs it produces
   c. number of eggs it produces over its lifetime
   d. number of offspring it produces over its lifetime
   e. number of offspring it produces over its lifetime that survive to breed

? Review Questions

1. What is a gene pool? How would you determine the allele frequencies in a gene pool?
2. Define *equilibrium population*. Outline the conditions that must be met for a population to stay in genetic equilibrium.
3. How does population size affect the likelihood of changes in allele frequencies by chance alone? Can significant changes in allele frequencies (that is, evolution) occur as a result of genetic drift?
4. If you measured the allele frequencies of a gene and found large differences from the proportions predicted by the Hardy-Weinberg principle, would that prove that natural selection is occurring in the population you are studying? Review the conditions that lead to an equilibrium population, and explain your answer.
5. People like to say that "you can't prove a negative." Study the experiment in Figure 15-1 again, and comment on what it demonstrates.

6. Describe the three ways in which natural selection can affect a population over time. Which way(s) is (are) most likely to occur in stable environments, and which way(s) might occur in rapidly changing environments?
7. What is sexual selection? How is sexual selection similar to and different from other forms of natural selection?
8. Briefly describe competition, symbiosis, and altruism, and give an example of each.
9. Define *kin selection* and *inclusive fitness.* Can these concepts help explain the evolution of altruism?

Applying the Concepts

1. In North America the average height of adult humans has been increasing steadily for decades. Is directional selection occurring? What data would justify your answer?
2. Malaria is rare in North America. In populations of African Americans, what would you predict is happening to the frequency of the hemoglobin allele that leads to sickling in red blood cells? How would you go about determining if your prediction is true?
3. By the 1940s the whooping crane population had been reduced to fewer than 50 individuals. Thanks to conservation measures, their numbers are now increasing. What special evolutionary problems do whooping cranes have now that they have passed through a population bottleneck?
4. In many countries, conservationists are trying to design national park systems so that "islands" of natural area (the big parks) are connected by thin "corridors" of undisturbed habitat. The idea is that this arrangement will allow animals and plants to migrate between refuges. Why would such migration be important?
5. A preview question for Chapter 16: A species is all the populations of organisms that potentially interbreed with one another but that are reproductively isolated from (cannot interbreed with) other populations. Using the five assumptions of the Hardy-Weinberg principle as a starting point, what factors do you think would be important in the splitting of a single ancestral species into two modern species?

For More Information

Allison, A. C. "Sickle Cells and Evolution." *Scientific American,* August 1956. The story of the interaction between sickle-cell anemia and malaria in Africa.

Dawkins, R. *Climbing Mount Improbable.* New York: Norton, 1996. An eloquent book-length tribute to the power of natural selection to design intricate adaptations. The chapter on the evolution of the eye is an instant classic.

Fellman, B. "To Eat or Not to Eat." *National Wildlife,* February–March 1995. How animals with altruistic behaviors identify their relatives.

Gould, S. J. "The Evolution of Life on the Earth." *Scientific American,* October 1994. The importance of chance and catastrophe in shaping modern life.

Levy, S. B. "The Challenge of Antibiotic Resistance." *Scientific American,* March 1998. An excellent summary of the public health implications of antibiotic resistance. Also discusses some strategies for ameliorating the problem.

May, R. M. "The Evolution of Ecological Systems." *Scientific American,* September 1979. Coevolution accounts for much of the structure of natural communities of plants and animals.

O'Brien, S. J., Wildt, D. E., and Bush, M. "The Cheetah in Peril." *Scientific American,* May 1986. According to molecular and immunological techniques, a population bottleneck has reduced the genetic variability of the world's cheetahs almost to zero.

Ryan, M. J. "Signals, Species, and Sexual Selection." *American Scientist,* January–February 1990. Ryan explores a variety of experiments on sexual selection, including the genetic basis of male characteristics and female choice.

Stebbins, G. L., and Ayala, F. "The Evolution of Darwinism." *Scientific American,* July 1985. A synthesis of molecular and classical evolutionary methodologies.

Answers to Multiple-Choice Questions

1. a 2. d 3. b 4. c 5. c 6. e

MEDIATUTOR
How Organisms Evolve

CD Activities

Activity 15.1: The Bottleneck Effect

Estimated time: 5 minutes

Genetic drift is one of the major mechanisms of evolutionary change and has significant implications for human interactions with natural populations of organisms on Earth. This simple animation illustrates the bottleneck effect, a case of genetic drift, in a hypothetical population of beetles.

Activity 15.2: Three Modes of Natural Selection

Estimated time: 10 minutes

Natural selection can have various effects on a population. Three possible modes of natural selection, known as directional, stabilizing, and disruptive selection, are demonstrated in this simulation.

Activity 15.3: Natural Selection at Work: Alpine Skypilots

Estimated time: 15 minutes

The size of alpine skypilot flowers differs between the bare alpine and forested sub-alpine environments. What factors contribute to this difference? Is it natural selection or some other factor at work? This tutorial will allow you to walk through Candace Galen's experiments as she uncovered this mystery.

Start the MediaTutor Student CD-ROM and enter the activity number in the Quick Search box to be taken directly to that activity.

Web Investigations

Case Study: Cause of Death: Evolution

Estimated time: 15 minutes

Antibiotic resistance is reaching crisis proportions. Diseases that were all but eradicated only a few years ago are back and resistant to all known treatments. This exercise will take a brief look at the basic biology, medical repercussions, and social implications of bacterial antibiotic resistance.

Go to http://www.prenhall.com/audesirk6, the Audesirk Companion Web site. Select Chapter 15 and Web Investigation to begin.

". . . endless forms most beautiful and most wonderful have been, and are being, evolved."

Charles Darwin in* On the Origin of Species by Means of Natural Selection *(1859)

The saola, unknown to science until 1992, is one of a number of previously undiscovered species recently found in the mountains of Vietnam. The area's distinctive assemblage of species probably arose during a past period of geographic isolation.

16 The Origin of Species

AT A GLANCE

Case Study: Lost World

1) **What Is a Species?**

2) **How Do New Species Form?**

Allopatric Speciation Can Occur in Populations That Are Physically Separated

Sympatric Speciation Can Occur in Populations That Live in the Same Area

Change over Time Within a Species Can Cause Apparent "Speciation" in the Fossil Record

During Adaptive Radiation, One Species Gives Rise to Many

3) **How Is Reproductive Isolation Between Species Maintained?**

Premating Isolating Mechanisms Prevent Mating Between Species

Postmating Isolating Mechanisms Prevent Production of Vigorous, Fertile Offspring

4) **What Causes Extinction?**

Localized Distribution and Overspecialization Make Species Vulnerable in Changing Environments

Interactions with Other Organisms May Drive a Species to Extinction

Habitat Change and Destruction Are the Leading Causes of Extinction

Evolutionary Connections: Scientists Don't Doubt Evolution

Case Study Revisited: Lost World

CASESTUDY

Lost World

The steep, rain-drenched slopes of Vietnam's Annamite Mountains are remote and forbidding, cloaked in tropical mists that lend an air of mystery and concealment to the forested mountains. As it turns out, this remote redoubt did indeed conceal a most astonishing and pleasing biological surprise: the saola, a hoofed, horned mammal that was unknown to science until the early 1990s. The discovery of a new species of large mammal at this late date was a complete shock. After centuries of exploration and exploitation in every corner of the world's forests, deserts, and savannas, scientists were certain that no large-sized mammal species could have escaped detection. As long ago as 1812, French naturalist Georges Cuvier wrote that "there is little hope of discovering new species of large quadrupeds." And yet, the saola, 3 feet high at the shoulder, weighing up to 200 pounds and sporting 20-inch black horns, remained hidden in Annamite Mountain forests, outside the realm of scientific knowledge until 1992 (though local tribespeople had apparently been hunting the creature for some time).

Since the discovery of the saola, scientists have described several additional new (if smaller-sized) mammal species, including the giant muntjac (also known as the barking deer) and a strange rabbit that has short ears and a brown-striped coat. This spate of discoveries has revealed the Vietnamese mountains as a kind of lost world of animals. Isolated by inhospitable terrain and the wars fought in Vietnam during the last century, the animals of the Annamite Mountains remained unknown to scientists. In the face of increased scientific attention, however, this lost world is increasingly revealed, and the curious biologist may wonder why these wonderfully unfamiliar species are concentrated in this particular part of the world. Before we can fully consider that question, we will need to explore the evolutionary process by which new species arise. ■

1 What Is a Species?

Before we can study the origin of species, we must first clarify our definition of the term. Throughout most of human history, "species" was a poorly defined concept. To most Europeans, the word *species* simply referred to one of the "kinds" produced by the biblical Creation. In this view, humans could not possibly know the criteria of the Creator, but could only attempt to distinguish among species on the basis of visible differences in structure. In fact, *species* is Latin for "appearance."

On a coarse scale, it is easy to use quick visual comparisons to distinguish species. For example, warblers are clearly different from eagles, which are obviously different from ducks. But it is far more difficult to distinguish among different species of warblers or eagles or ducks. By what standard do scientists make these finer distinctions?

Today, biologists define **species** as "groups of actually or potentially interbreeding natural populations, which are reproductively isolated from other such groups." This definition, known as the *biological-species concept*, has at least one major limitation. It does not help us discern species boundaries among asexually reproducing organisms, which can't interbreed. Despite this limitation, most biologists accept the biological-species concept and its emphasis on reproductive community as the main criterion for identifying species of sexually reproducing organisms. (Scientists who study bacteria and other organisms whose reproduction is mainly asexual must use alternate species definitions; one such definition is discussed on p. 358.)

Biologists have found that differences in appearance do not always mean that two populations belong to different species. For example, field guides published in the 1970s listed the myrtle warbler and Audubon's warbler (Fig. 16-1) as distinct species; these birds differ in geographic range and in the color of their throat feathers. More recently, scientists decided that these birds are merely local varieties of the same species. The reason: Where their ranges overlap, these warblers interbreed, and the offspring are just as vigorous and fertile as the parents.

2 How Do New Species Form?

Despite his exhaustive exploration of the process of natural selection, Charles Darwin never proposed a complete mechanism of **speciation**, the process by which new species form. One scientist who did play a large role in describing the process of speciation was Ernst Mayr of Harvard University, an ornithologist (expert on birds) and a pivotal figure in the history of evolutionary biology. Mayr developed the species definition given above. He was also among the first to recognize that speciation depends on two factors: the (1) isolation and (2) genetic divergence of two populations.

1. *Isolation of populations.* If two populations are to become sufficiently distinct that interbreeding is difficult or impossible, then there must be relatively little gene flow (migration) between them. If there is a great deal of gene flow, then genetic changes in one population will soon become widespread in the other as well.
2. *Genetic divergence.* It is not sufficient for two populations simply to be isolated. They will become separate species only if, during the period of isolation, they

(a) (b)

Figure 16-1 Interbreeding blurs the distinction between species
(a) The myrtle warbler and ***(b)*** Audubon's warbler were formerly thought to be two separate species but are now considered to be merely local varieties of one widespread species.

evolve sufficiently large genetic differences. The differences must be large enough that, if the isolated populations are reunited, they can no longer interbreed and produce vigorous, fertile offspring. If isolated populations are small, chance events may generate significant genetic differences by genetic drift (see Chapter 15). In both small and large populations, different environmental pressures in separate environments may favor the evolution of large genetic differences.

Speciation has seldom been observed in the wild (with the exception of "instant" speciation in plants by *polyploidy*, described later in this chapter). Nevertheless, evolutionary biologists have synthesized theories, observations, and experiments to devise hypothetical mechanisms for the origin of new species. These mechanisms fall into two broad categories: (1) **allopatric speciation**, in which two populations are geographically separated from one another, and (2) **sympatric speciation**, in which two populations share the same geographical area (Fig. 16-2).

At first glance, you might think that sympatric speciation violates our first principle of speciation, isolation of populations, because the speciating populations live in the same locale. However, it is *isolation from gene flow* that is crucial to speciation. Although such isolation may be most commonly imposed by a physical barrier such as a river, two populations living in the same area can also experience restricted gene flow if they occupy different habitats within the area (for example, marshes as opposed to forests) or breed in different, non-overlapping time periods. Therefore, the principle still holds: *Isolation from gene flow is the key to both allopatric speciation and sympatric speciation.*

Allopatric Speciation Can Occur in Populations That Are Physically Separated

Allopatric speciation can occur when different parts of a population become physically separated by an impassible barrier. Physical separation could occur if, for example, some members of a population of land-dwelling organisms drifted, swam, or flew to a remote oceanic island. Populations of water-dwelling organisms might be split when geological processes such as volcanism or continental drift create new land barriers that subdivide previously continuous seas or lakes. Geological change can also divide terrestrial populations (Fig. 16-3, p. 311). Portions of populations can become stranded in patches of suitable habitat that become isolated by climate shifts. You can probably imagine myriad other scenarios that could lead to the geographic subdivision of a population.

If two or more populations become *geographically isolated* for any reason, little or no migration (and therefore little or no gene flow; see Chapter 15) can occur between them. If the pressures of natural selection differ in the separate locations, then the populations may accumulate genetic differences. Alternatively, genetic differences may arise if one or more of the separated populations is small enough for genetic drift to occur, which may be especially likely in the aftermath of a founder event (in which a few individuals become isolated from the main body of the species). In either case, genetic differences between the separated populations may eventually become large enough to make interbreeding impossible. At that point, the two populations will have become separate species. Most evolutionary biologists believe that geographic isolation followed by allopatric speciation has been the most common source of new species, especially among animals.

Sympatric Speciation Can Occur in Populations That Live in the Same Area

Sympatric speciation refers to speciation that occurs within a single geographical area. Sympatric speciation, like allopatric speciation, requires limited gene flow. There are two likely mechanisms whereby gene flow can be reduced between members of a single population in a given area: (1) ecological isolation and (2) chromosomal aberrations.

Ecological Isolation Restricts Different Populations to Different Habitats Within the Same Area

If the same geographical area contains two distinct types of habitats (for example, distinct food sources, nesting places, and so on), different members of a single species may begin to specialize in one habitat or the other. If conditions are right, natural selection for habitat specialization may cause the formerly single species to split into two species. Such a split seems to be occurring right before biologists' eyes, so to speak, in the case of the fruit fly *Rhagoletis pomonella* (Fig. 16-4, p. 311).

Rhagoletis is a parasite of the American hawthorn tree. This fly lays its eggs in the hawthorn's fruit; when the maggots hatch, they eat the fruit. About 150 years ago, entomologists (scientists who study insects) noticed that *Rhagoletis* had begun to infest apple trees, which were introduced into North America from Europe. It appears today that *Rhagoletis* is splitting into two species, one that breeds on apples and one that sticks to hawthorns. Substantial genetic differences exist between the two groups. At least some of these genetic differences, such as the ones that affect the timing of emergence of the adult flies, are important for survival on a particular host plant.

The two kinds of flies will become two species only if they maintain reproductive separation. Apple trees and hawthorns are typically quite close together, and flies, after all, can fly. So why don't apple-flies and hawthorn-flies interbreed and cancel out any incipient genetic differences? First, female flies usually lay their eggs in the same type of fruit in which they developed. Males also tend to rest on the same type of fruit in which they

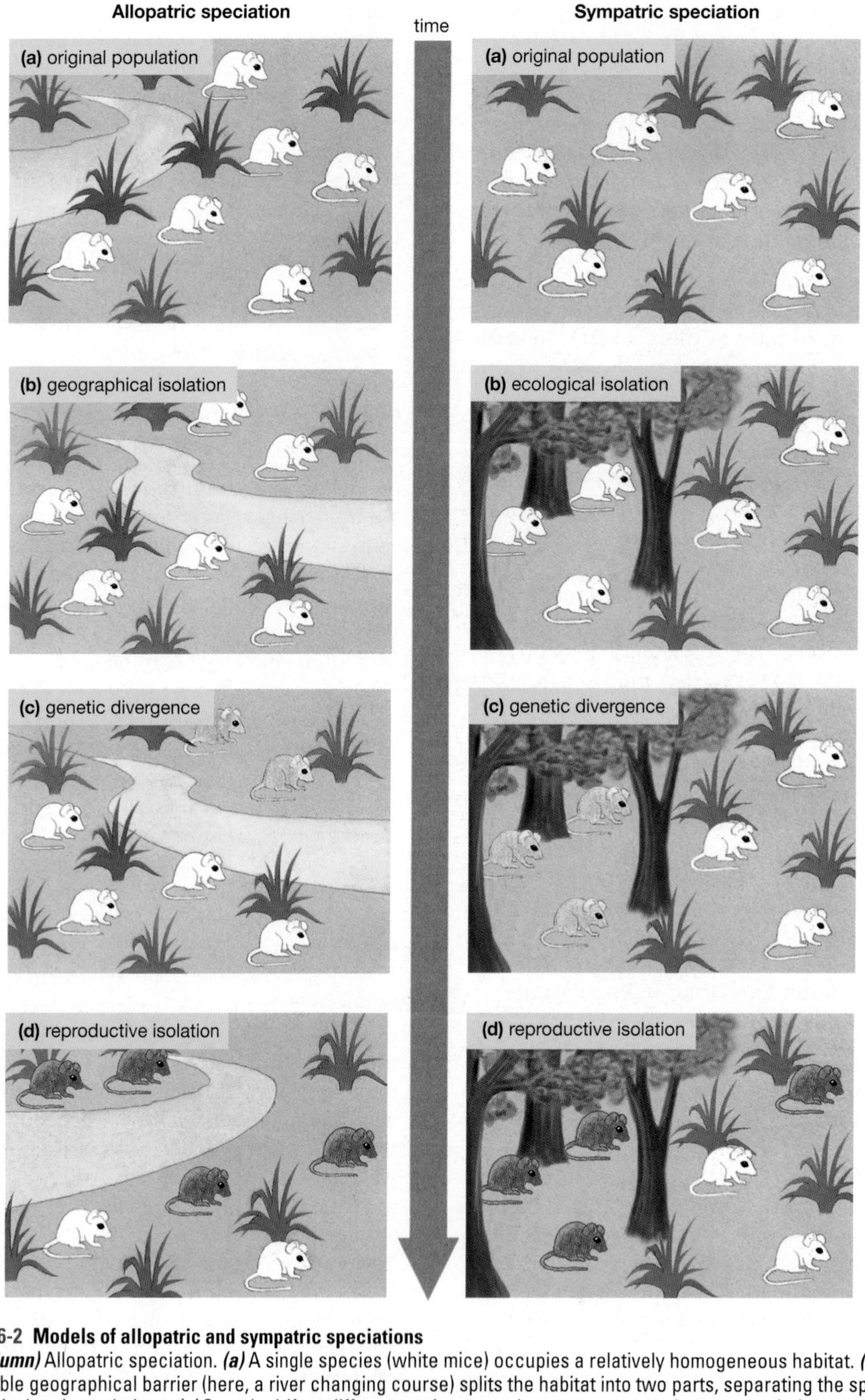

MEDIATUTOR
16.1 Allopatric Speciation
16.2 Sympatric Speciation

Figure 16-2 Models of allopatric and sympatric speciations
(Left column) Allopatric speciation. ***(a)*** A single species (white mice) occupies a relatively homogeneous habitat. ***(b)*** An impassable geographical barrier (here, a river changing course) splits the habitat into two parts, separating the species into two isolated populations. ***(c)*** Genetic drift or different environmental pressures cause the two populations to diverge genetically (tan vs. white mice). ***(d)*** The barrier is removed (the river changes course again), and the members of the two populations can share the same habitat. If the genetic differences between the two populations have become large enough that interbreeding cannot occur (that is, they are reproductively isolated from one another), then the two populations constitute separate species (brown vs. white mice). ***(Right column)*** Sympatric speciation. ***(a)*** A single species (white mice) occupies a homogeneous habitat. ***(b)*** Climate change or other factors form two distinctly different habitats that are still physically part of the same general region; that is, there are no barriers to movement between habitats. ***(c)*** Different environmental pressures in the two habitats lead to genetic divergence of organisms living in each (tan vs. white mice). ***(d)*** Sufficient genetic divergence causes reproductive isolation; former occupants of the two different habitats are now separate species (brown vs. white mice).

(a) (b)

Figure 16-3 Geographical isolation
To determine if these two squirrels are members of different species, we must know if they're "actually or potentially interbreeding." Unfortunately, it is hard to tell, because ***(a)*** the Kaibab squirrel lives only on the north rim of the Grand Canyon and ***(b)*** the Abert squirrel lives exclusively on the south rim. The two populations are geographically separated but still quite similar. Have they diverged enough after their separation to be considered separate species? On the basis of our current knowledge, it is impossible to say.

developed. Therefore, apple-liking males will encounter and mate with apple-liking females. Second, apples mature 2 or 3 weeks later than hawthorn fruits do, and the two types of flies emerge with a timing appropriate for their chosen host fruits. Thus, the two varieties of flies have very little chance of meeting. Although some interbreeding between the two types of flies occurs, it seems they are well on their way to speciation. Will they make it? Entomologist Guy Bush suggests, "Check back with me in a few thousand years."

Changes in Chromosome Number Can Cause Immediate Reproductive Isolation of a Population

In some instances, new species can arise nearly instantaneously through changes in chromosome number. A common speciation mechanism in plants is **polyploidy** (Fig. 16-5), the acquisition of multiple copies of each chromosome. As you know from Chapter 11, most plants and animals have paired chromosomes and are described as diploid. Occasionally, especially in plants, a fertilized egg duplicates its chromosomes but doesn't divide into two daughter cells. The resulting cell thus becomes *tetraploid*, with four copies of each chromosome. If all of the subsequent cell divisions are normal, this tetraploid zygote will develop into a plant that consists of tetraploid cells. Most tetraploid plants are vigorous and healthy, and many can successfully complete meiosis to form viable gametes. The gametes, however, are diploid (whereas meiosis normally produces haploid gametes from diploid cells). These diploid gametes can fuse with other diploid gametes to produce new tetraploid offspring, so it is no problem for tetraploids to interbreed with other tetraploids of that species or to self-fertilize (as many plants do).

If, however, a tetraploid interbreeds with a diploid individual from the "parental" species, the outcome is not so successful. For example, if a diploid sperm from a tetraploid plant fertilizes a haploid egg cell of the parental species, the offspring will be *triploid*, with three copies of each chromosome. Many triploid individuals experience problems during growth and development. Even if the triploid offspring develops normally, it will be sterile: When a triploid cell attempts to undergo meiosis, the odd number of chromosomes makes chromosome-pairing impossible. Meiosis fails, and viable gametes are not formed. Thus, the offspring of diploid–tetraploid matings are inevitably sterile, so tetraploid plants and their diploid parents form distinct reproductive communities that cannot interbreed successfully. A new species can form in a single generation.

Figure 16-4 Sympatric isolation
Two sympatric populations of the fruit fly species *Rhagoletis pomonella* may be evolving into two separate, reproductively isolated species.

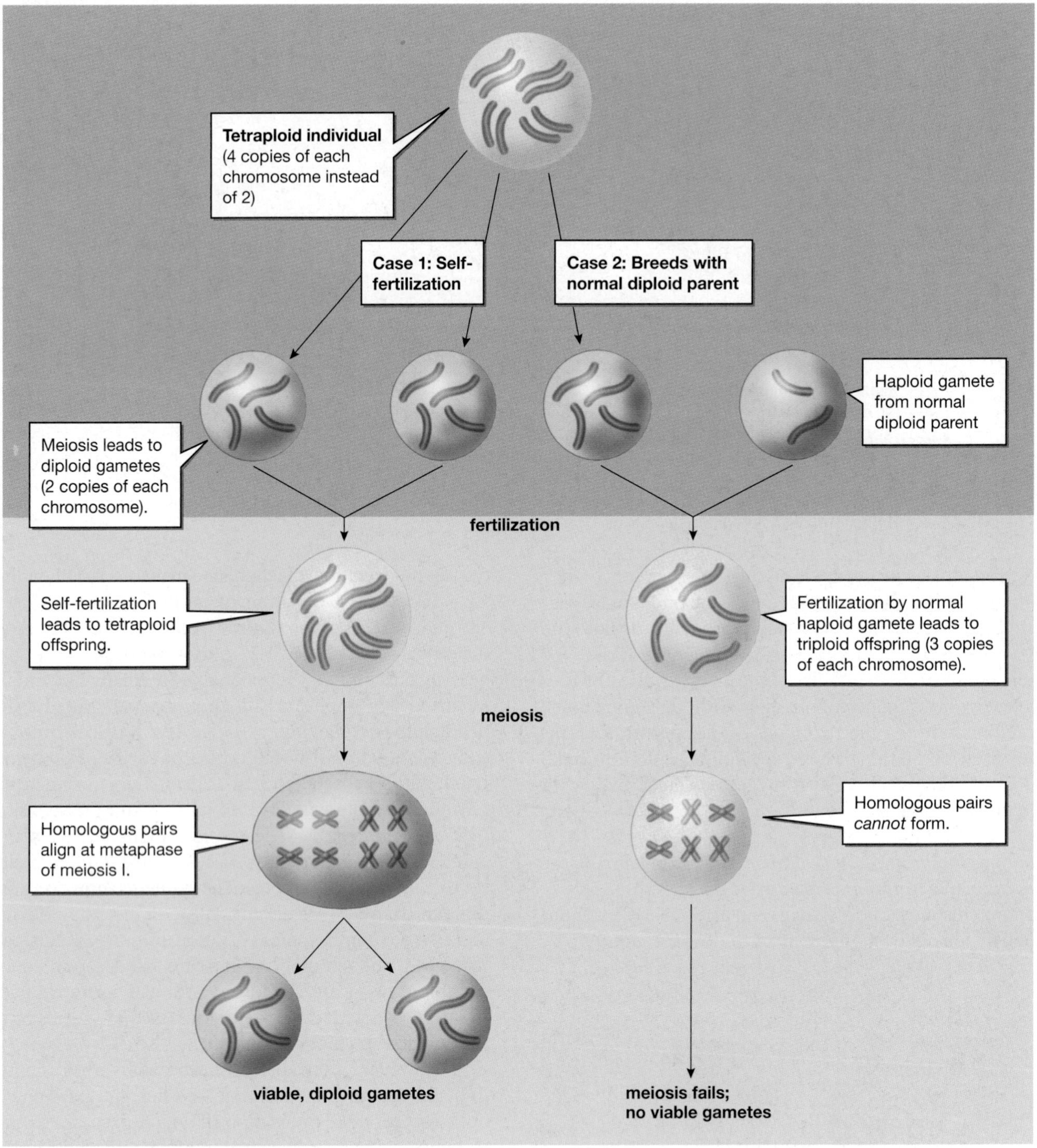

Figure 16-5 Speciation by polyploidy
Chromosomal mutations, especially in plants, can result in polyploid individuals with extra copies of each chromosome. This example shows that a tetraploid mutant can successfully self-fertilize (or can interbreed with other tetraploid individuals) to yield a new generation of tetraploids but that matings between tetraploids and normal diploid individuals will yield only sterile offspring. Tetraploid mutants are thus reproductively isolated from their diploid ancestors and may constitute a new species.

Why is speciation by polyploidy common in plants but not in animals? Many plants can either self-fertilize or reproduce asexually, or both. If a tetraploid plant self-fertilizes, then its offspring will also be tetraploid. Asexual offspring, of course, are genetically identical to the parent and are also tetraploid. In either case, the new tetraploid plant may perpetuate itself and form a new species. Most animals, however, cannot self-fertilize or reproduce asexually. Therefore, if an animal produced a tetraploid offspring, the offspring would have to mate with a member of the diploid parental species and would produce all triploid off-

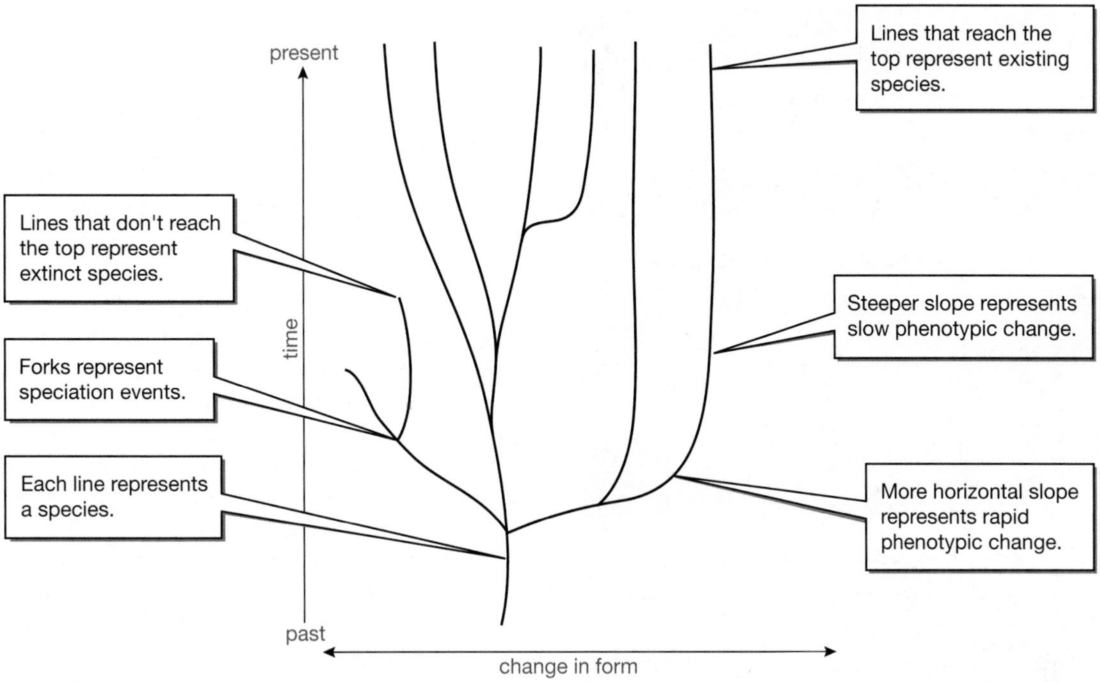

Figure 16-6 Interpreting an evolutionary tree
Evolutionary history is often represented by an evolutionary tree, a graph in which the vertical axis plots time and the horizontal axis stands for change in form (the greater the horizontal distance between points, the more phenotypic differences there are). Any evolutionary tree that can be drawn to depict the evolutionary relationships among a group of species is a portion of a much larger tree of life that connects all living things.

spring. The triploid offspring would almost certainly be sterile. Speciation by polyploidy is extremely common in plants; in fact, nearly half of all species of flowering plants are polyploid, and many of them are tetraploid.

Change over Time Within a Species Can Cause Apparent "Speciation" in the Fossil Record

The mechanisms of speciation and reproductive isolation that we have described lead to forking branches in the *evolutionary tree* of life (Fig. 16-6), as one species splits into two. This kind of branching is a key source of evolutionary change, but changes *within* a species over time are also important. As generations pass and evolutionary innovations accumulate, the members of a species may come to be very different from their distant ancestors, even if no speciation takes place—that is, even if a new species doesn't form.

When a biologist encounters two populations of living organisms, he or she can, if the populations live in the same place, devise a test to see if the two populations are reproductively isolated and, therefore, separate species. For a paleontologist (a scientist who studies fossils), however, things are not so simple. Fossils cannot breed, so it is difficult to determine whether they were reproductively isolated from other fossils. Furthermore, different fossil organisms may be found in different rock layers that were laid down thousands or even millions of years apart, and it is typically impossible to know if any branching, speciation events occurred between an older organism and a younger one. For these reasons, paleontologists must typically assign extinct organisms to species without reference to the biological-species concept. When comparing differences between extinct organisms, it is often impossible to tell if the differences arose due to branching evolution or to change within a single line of descent. Given the futility of applying the biological-species concept to fossils, many paleontologists choose to use a different system in which it is considered acceptable to assign different species names to anatomically different fossils, even if the two fossils may simply represent different time points on a single evolutionary branch.

During Adaptive Radiation, One Species Gives Rise to Many

In some cases, a species gives rise to many new species in a relatively short time. This process, called **adaptive radiation**, occurs when populations of a single species invade a variety of new habitats and evolve in response to the differing environmental pressures in those habitats. Adaptive radiation has occurred many times and in many groups of organisms. Adaptive radiation normally results from one of two causes. First, a species may encounter a wide variety of unoccupied habitats—for example, when the ancestors of Darwin's finches colonized the Galapagos Islands, or when marsupial

Figure 16-7 Adaptive radiation
More than 300 species of cichlid fishes inhabit Lake Malawi in East Africa. These species are found nowhere else, and all of them descended from a single ancestral population within a million years. This dramatic adaptive radiation has led to a collection of closely related species with an array of adaptations for exploiting the many different food sources in the lake.

mammals first invaded Australia, or when an ancestral cichlid fish species arrived at Lake Malawi (Fig. 16-7). With no competitors except other members of their own species, all the available habitats and food sources were rapidly utilized by new species that evolved from the original invaders.

3 How Is Reproductive Isolation Between Species Maintained?

The process of speciation is primarily a story about the evolution of mechanisms that prevent interbreeding. Genetic divergence during the period of isolation is a necessary condition for the origin of a new species, but it is not sufficient unless part of that genetic divergence happens to cause the development of something that ensures **reproductive isolation** (Table 16-1). The structural and/or behavioral modifications that prevent interbreeding are called **isolating mechanisms**.

Isolating mechanisms have a clear adaptive value. Any individual that mates with a member of another species will probably produce unfit or sterile offspring, thereby "wasting" its reproductive effort and contributing nothing to future generations. Thus, there is evolutionary pressure to avoid mating across species boundaries. Incompatibilities that prevent mating between species are called **premating isolating mechanisms**.

When premating isolation fails or has not yet evolved, members of different species may mate. If, however, the resulting hybrid offspring die during development, then naturally the two species are still reproductively isolated from one another. In some cases, however, viable hybrid offspring are produced. Even so, if the hybrids are less fit than their parents or are themselves infertile, the two species may still remain separate, with little or no gene flow between them. Incompatibilities that prevent the formation of vigorous, fertile hybrids between species are called **postmating isolating mechanisms**.

Premating Isolating Mechanisms Prevent Mating Between Species

Mechanisms that prevent mating between species include geographical isolation, ecological isolation, temporal isolation, behavioral isolation, and mechanical incompatibility.

Table 16-1 Mechanisms of Reproductive Isolation

Premating Isolating Mechanisms: any structure, physiological function, or behavior that prevents organisms of two populations from mating.

1. **Geographical isolation:** the separation of two populations by a physical barrier.
2. **Ecological isolation:** lack of interbreeding between populations that occupy distinct habitats within the same general area.
3. **Temporal isolation:** the inability of populations to interbreed if they have significantly different breeding seasons.
4. **Behavioral isolation:** lack of interbreeding between populations of animals that differ substantially in courtship and mating rituals.
5. **Mechanical incompatibility:** the inability of male and female organisms to exchange gametes, normally because the reproductive structures are incompatible.

Postmating Isolating Mechanisms: any structure, physiological function, or developmental abnormality that prevents organisms of two populations, once mating has occurred, from producing vigorous, fertile offspring.

1. **Gametic incompatibility:** the inability of sperm from one population to fertilize eggs of another population.
2. **Hybrid inviability:** the failure of a hybrid offspring of two different populations to survive to maturity.
3. **Hybrid infertility:** reduced fertility (or complete sterility) in hybrid offspring of two different populations.

Geographical Isolation Prevents Members of Different Species from Meeting One Another

Members of different species cannot mate if they never get near one another. As we have already seen, **geographical isolation** typically provides the conditions for speciation in the first place. However, we cannot determine if geographically separated populations necessarily constitute distinct species. Should the barrier separating the two populations disappear (an intervening river changes course, for example), the reunited populations might interbreed freely and not be separate species after all. If they cannot interbreed, then other mechanisms, such as different courtship rituals, must have developed during their isolation. Geographical isolation, therefore, is usually considered to be a mechanism that *allows new species to form* rather than a mechanism that *maintains reproductive isolation between species.*

Ecological Isolation Confines Members of Different Species to Different Habitats

Two populations that have different resource requirements may use different local habitats within the same general area and thus exhibit **ecological isolation.** White-crowned and white-throated sparrows, for example, have extensively overlapping ranges. The white-throated sparrow, however, frequents dense thickets, whereas the white-crowned sparrow inhabits fields and meadows, seldom penetrating far into dense growth. The two species may coexist within a few hundred yards of one another and yet seldom meet during the breeding season. A more dramatic example is provided by the more than 750 species of fig wasp (Fig. 16-8). Each species of fig wasp breeds in (and pollinates) the fruits of a particular species of fig, and each fig species hosts one and only one species of pollinating wasp. Although ecological isolation may slow down interbreeding, it seems unlikely that it could prevent gene flow entirely. Other mechanisms normally also contribute to reproductive isolation.

Figure 16-8 Ecological isolation
This tiny female fig wasp must find her way to a tree of the single fig species to which her reproductive cycle is bound. She must then enter the developing fruit through a pore so small that her passage through it can rip off her wings. Once inside, she will never depart. After laying eggs (and depositing pollen that she carried from the fig where she hatched), she will die.

Temporal Isolation Occurs Between Species That Breed at Different Times

Even if two species occupy similar habitats, they cannot mate if they have different breeding seasons, a phenomenon called **temporal** (time-related) **isolation**. Bishop pines and Monterey pines coexist near Monterey on the California coast (Fig. 16-9). Viable hybrids have been produced between these two species in the laboratory. In the wild, however, they release their pollen at different times: The Monterey pine releases pollen in early spring, the bishop pine in summer. Therefore, the two species never cross-breed under natural conditions. Hawthorn-liking and apple-liking *Rhagoletis* fruit flies are also partially isolated from one another because they emerge from their host fruits and breed at slightly different times of the year.

Different Courtship Rituals Create Behavioral Isolation

Among animals, the elaborate courtship colors and behaviors that so enthrall human observers have evolved not only as recognition and evaluation signals between male and female; they may also aid in distinguishing among species. These different signals and behaviors create **behavioral isolation.** The striking colors and calls of male songbirds, for example, may attract females of their own species, but females of other species treat them with the utmost indifference. Among frogs, males are often impressively indiscriminate, jumping on every female in sight, regardless of the species, when the spirit moves them. Females,

Figure 16-9 Temporal isolation
Bishop pines, such as these, and Monterey pines coexist in nature. In the laboratory they produce fertile hybrids. In the wild, however, they do not interbreed, because they release pollen at different times of the year.

however, approach only male frogs that croak the "ribbet" appropriate to their species. If they do find themselves in an unwanted embrace, they utter the "release call," which causes the male to let go. As a result, few hybrids are produced.

Mechanical Incompatibility Occurs When Physical Barriers Between Species Prevent Fertilization

In rare instances, ecological, temporal, and behavioral isolating mechanisms fail, and male and female of different species attempt to mate. Among animal species with internal fertilization (in which the sperm is deposited inside the female's reproductive tract), the male's and female's sexual organs simply may not fit together. Among plants, differences in flower size or structure may prevent pollen transfer between species. For example, different species may attract different pollinators (see Chapter 24 for a description of the interesting deceptions used by flowers to lure specific types of pollinators). Isolating mechanisms of this type are called **mechanical incompatibility**.

Postmating Isolating Mechanisms Prevent Production of Vigorous, Fertile Offspring

In some cases premating isolation fails, and mating occurs between members of different species. However, gene flow between the two species will still be restricted if the mating fails to produce vigorous, fertile hybrid offspring. Postmating isolating mechanisms include gametic incompatibility, hybrid inviability, and hybrid infertility.

Gametic Incompatibility Occurs When Sperm from One Species Are Unable to Fertilize Eggs of Another

Even if a male inseminates a female of a different species, his sperm may not fertilize her eggs, a situation called **gametic incompatibility**. For example, the fluids of the female reproductive tract may weaken or kill sperm of other species. Among plants, chemical incompatibility may prevent the germination of pollen from one species that lands on the stigma (pollen-catching structure) of the flower of another species.

Hybrid Inviability Occurs If Hybrid Offspring Survive Poorly

If cross-species fertilization does occur, the resulting hybrid may be weak or even unable to survive, a situation called **hybrid inviability**. The genetic programs directing development of the two species may be so different that hybrids abort early in development. Even if the hybrid survives, it may display behaviors that are mixtures of the two parental types. In attempting to do some things the way species *A* does them and other things the way species *B* does them, the hybrid may be hopelessly uncoordinated. Hybrids between certain species of lovebirds, for example, have great difficulty learning to carry nest materials during flight and probably could not reproduce in the wild.

Hybrid Infertility Occurs if Hybrid Offspring Are Unable to Produce Normal Sperm or Eggs

Most animal hybrids, such as the mule (a cross between a horse and a donkey) or the liger (a zoo-based cross between a male lion and a female tiger), are sterile. **Hybrid infertility** prevents hybrids from passing on their genetic material to offspring. A common reason is the failure of chromosomes to pair properly during meiosis, so eggs and sperm never develop. As we saw earlier, among plants that have speciated by polyploidy, any offspring produced by mating between the diploid "parent" species and the tetraploid "daughter" species will be triploid and sterile.

4 What Causes Extinction?

Every living organism must eventually die, and the same is true of species. Just like individuals, species are "born" (through the process of speciation), persist for some period of time, and then perish. The ultimate fate of any species is **extinction**, the death of all the members of a species. In fact, at least 99.9% of all the species that have ever existed are now extinct. The natural course of evolution, as revealed by the fossil record, is continual turnover of species as new ones arise and old ones go extinct.

The proximate cause of extinction is probably always environmental change, either in the living or the nonliving parts of the environment. Two characteristics seem to predispose a species to extinction when the environment changes: localized distribution and overspecialization. Three major environmental changes that drive a species to extinction are competition among species, the introduction of new predators or parasites, and habitat destruction.

Localized Distribution and Overspecialization Make Species Vulnerable in Changing Environments

Species vary widely in their range of distribution and, hence, in their susceptibility to extinction. Some species, such as herring gulls, white-tailed deer, and humans, inhabit entire continents or even the whole Earth; others, such as the Devil's Hole pupfish (Fig. 16-10), have extremely limited ranges. Obviously, if a

Figure 16-10 Very localized distribution can endanger a species The Devil's Hole pupfish is found in only one spring-fed water hole in the Nevada desert. During the last glacial period, the southwestern deserts received a great deal of rainfall, forming numerous lakes and rivers. As the rainfall decreased, pupfish populations were isolated in shrinking small springs and streams. Isolated small populations and differing environmental conditions caused the ancestral pupfish species to split into several very restricted modern species, all of which are on the brink of extinction.

species has a localized distribution (occurs in only a very small area), any disturbance of that area could easily result in extinction. If Devil's Hole dries up due to climatic change or well-drilling nearby, its pupfish will immediately vanish. Conversely, wide-ranging species normally do not succumb to local environmental catastrophes.

Another factor that may make a species vulnerable to extinction is overspecialization. Each species develops genetic adaptations in response to pressures from its environment. These adaptations may limit the organism to a very specialized set of environmental conditions. The Everglades kite, for example, is a bird of prey that feeds only on the apple snail, a freshwater snail (Fig. 16-11).

Figure 16-11 Extreme specialization places species at risk The Everglades kite feeds exclusively on the apple snail, found in swamps of the southeastern United States. Such behavioral specialization renders the kite extremely vulnerable to any environmental change that may exterminate its single species of prey.

As the swamps of the American Southeast are drained for farms and developments, the snail population shrinks. If the snail becomes extinct, the kite will surely go extinct along with it.

Interactions with Other Organisms May Drive a Species to Extinction

As described earlier, interactions such as competition, predation, and parasitism serve as forces of natural selection. In some cases, these same forces can lead to extinction rather than to adaptation.

Competition for limited resources occurs in all environments. If a species' competitors evolve superior adaptations and the species doesn't evolve fast enough to keep up, it may become extinct. A particularly striking example of extinction through competition occurred in South America, beginning about 2.5 million years ago. At that time, the isthmus of Panama rose above sea level and formed a land bridge between North America and South America. After the previously separated continents were connected, the mammal species that had evolved in isolation on each continent were able to mix. Many species did indeed expand their ranges, as North American mammals moved southward and South American species drifted northward. As they moved, each species encountered resident species that occupied the same kinds of habitats and exploited the same kinds of resources. The ultimate result of the ensuing competition was that the North American species diversified and underwent an adaptive radiation that displaced the vast majority of the South American species, many of which went extinct. Clearly, evolution had bestowed on the North American species some (as yet unknown) set of adaptations that enabled their descendants to exploit resources more efficiently and effectively than their South American counterparts could.

Habitat Change and Destruction Are the Leading Causes of Extinction

Habitat change, both contemporary and prehistoric, is the single greatest cause of extinctions. Present-day habitat destruction due to human activities is proceeding at a frightening pace. Many biologists believe that we are presently in the midst of the fastest-paced and most widespread episode of species extinction in the history of life. Loss of tropical forests is especially devastating to species diversity. As many as half the species presently on Earth may be lost over the next 50 years as the tropical forests that contain them are cut for timber and to clear land for cattle and crops. (See "Earth Watch: Endangered Species: From Gene Pools to Gene Puddles," p. 295.) We will discuss extinctions due to prehistoric habitat change in Chapter 17.

Evolutionary Connections

Scientists Don't Doubt Evolution

In the popular press, conflicts among evolutionary biologists are sometimes seen as conflicts about evolution itself. We occasionally read statements to the effect that new theories are overthrowing Darwin's and casting doubt on the reality of evolution. Nothing could be further from the truth. Despite some disagreements about the details of the evolutionary process, biologists unanimously agree that evolution occurred in the past and is still occurring today. The only argument is over the relative importance of the various mechanisms of evolutionary change in the history of life on Earth, their pace, and which forces were most important in shaping the evolution of particular species. Meanwhile, wolves still tend to catch the slowest caribou, small populations still undergo genetic drift, habitats still change or disappear, and perhaps somewhere in our galaxy another meteorite is swinging our way. Evolution continues, still generating, in Darwin's words, "endless forms most beautiful."

CASE STUDY REVISITED

Lost World

One possible explanation for the distinctive collection of species found in the Annamite Mountains of Vietnam lies in the geological history of the region. During the ice ages that have occurred repeatedly over the past million years or so, the area covered by tropical forests must have shrunk dramatically. Organisms that depended on the forests for survival would have been restricted to any remaining "islands" of forest and isolated from their fellows in other, distant patches of forest. What is now the Annamite Mountain region may well have been an isolated forest during periods of glacial advance. As we learned in this chapter, this kind of isolation can set the stage for allopatric speciation and may have created the conditions that gave rise to the saola, giant muntjac, striped rabbit, and other unique denizens of Vietnamese forests.

Ironically, we have discovered the lost world of Vietnamese animals at a moment when that world is in grave danger of disappearing. Economic development in Vietnam has brought logging and mining to ever more remote regions of the country, and Annamite Mountain forests are being cleared at an unprecedented rate. Increasing local human population means that local animals are hunted heavily; most of our knowledge of the saola comes from carcasses found in local markets. All of the newly discovered mammals of Vietnam are quite rare, seen only infrequently by even local hunters. Fortunately, the Vietnamese government has established a number of national parks and nature preserves in key areas. Only time will tell if these measures are sufficient to ensure the survival of the mysterious mammals of the Annamites.

Do you think the search for undiscovered species should continue? What value or benefit to humans does the search for new species provide?

Summary of Key Concepts

1) What Is a Species?

According to the biological-species concept, a species is defined as all the populations of organisms that are potentially capable of interbreeding under natural conditions and that are reproductively isolated from other populations.

2) How Do New Species Form?

Speciation, the development of new species, requires that two populations be isolated from gene flow between them and develop significant genetic divergence. Allopatric speciation occurs by geographical isolation and subsequent divergence of the separated populations through genetic drift or natural selection. Sympatric speciation occurs by ecological isolation and subsequent divergence or by rapid chromosomal changes, such as polyploidy. Fossil species must normally be defined on the basis of anatomy alone, as information on reproductive status cannot be retrieved from them.

3) How Is Reproductive Isolation Between Species Maintained?

Reproductive isolation between species may be maintained by one or more of several mechanisms, collectively known as *premating isolating mechanisms* and *postmating isolating mechanisms*. Premating isolating mechanisms include geographical isolation, ecological isolation, temporal isolation, behavioral isolation, and mechanical incompatibility. Postmating isolating mechanisms include gametic incompatibility, hybrid inviability, and hybrid infertility.

4) What Causes Extinction?

Two factors that increase the likelihood of extinction, the death of all the members of a species, are localized distribution and overspecialization. Factors that actually cause extinctions include competition among species, the introduction of new predators or parasites, and habitat destruction.

Key Terms

adaptive radiation *p. 313*
allopatric speciation *p. 309*
behavioral isolation *p. 315*
ecological isolation *p. 315*
extinction *p. 316*
gametic incompatibility *p. 316*
geographical isolation *p. 315*
hybrid infertility *p. 316*
hybrid inviability *p. 316*
isolating mechanism *p. 314*
mechanical incompatibility *p. 316*
polyploidy *p. 311*
postmating isolating mechanism *p. 314*
premating isolating mechanism *p. 314*
reproductive isolation *p. 314*
speciation *p. 308*
species *p. 308*
sympatric speciation *p. 309*
temporal isolation *p. 315*

Thinking Through the Concepts

Multiple Choice

1. *Many closely related species of marine invertebrates exist on either side of the isthmus of Panama. They probably resulted from*
a. premating isolation
b. isolation by distance
c. postmating isolation
d. gametic incompatibility
e. allopatric speciation

2. *Many hybrids are sterile because their chromosomes don't pair up correctly during meiosis. Why aren't polyploid plants sterile?*
a. They backcross to the parental generation.
b. Most are triploid.
c. They cross-pollinate.
d. They self-fertilize, using their diploid gametes.
e. Their eggs develop directly, without fertilization.

3. *In many species of fireflies, males flash to attract females. Each species has a different flashing pattern. This is probably an example of*
a. ecological isolation
b. temporal isolation
c. geographical isolation
d. premating isolation
e. postmating isolation

4. *Under the biological-species concept, the main criterion for identifying a species is*
a. anatomical distinctiveness
b. behavioral distinctiveness
c. geographic isolation
d. reproductive isolation
e. gametic incompatibility

5. *In terms of changes in gene frequencies, founder events result in*
a. gradual accumulation of many small changes
b. large, rapid changes
c. polyploidy
d. hybridization
e. mechanical incompatibility

6. *After the demise of the dinosaurs, mammals evolved rapidly into many new forms because of*
a. the founder effect
b. a genetic bottleneck
c. adaptive radiation
d. geological time
e. genetic drift

? Review Questions

1. Define the following terms: *species*, *speciation*, *allopatric speciation*, and *sympatric speciation*. Explain how allopatric and sympatric speciations might work, and give a hypothetical example of each.

2. Many of the oak tree species in central and eastern North America hybridize (interbreed). Are they "true species"?

3. Review the material on the possibility of sympatric speciation in *Rhagoletis* varieties that breed on apples or hawthorns. What types of genotypic, phenotypic, or behavioral data would convince you that the two forms have become separate species?

4. A drug called *colchicine* affects the mitotic spindle fibers and prevents cell division after the chromosomes have doubled at the start of meiosis. Describe how you would use colchicine to produce a new polyploid species of your favorite garden flower.

5. List and describe the different types of premating and postmating isolating mechanisms.

Applying the Concepts

1. The biological-species concept has no meaning with regard to asexual organisms, and it is difficult to apply to extinct organisms that we know only as fossils. Can you devise a meaningful, useful species definition that would apply in all situations?
2. Seedless varieties of fruits and vegetables, created by breeders, are triploid. Explain why they are seedless.
3. Why do you suppose there are so many *endemic* species—that is, species found nowhere else—on islands? And why have the overwhelming majority of recent extinctions occurred on islands?
4. A contrarian biologist you've met claims that the fact that humans are pushing other species into small, isolated populations is good for biodiversity because these are the conditions that lead to new speciation events. Comment.
5. Southern Wisconsin is home to several populations of gray squirrels (*Sciurus carolinensis*) with black fur. Design a study to determine if they are actually separate species.
6. It is difficult to gather data on speciation events in the past or to perform interesting experiments about the process of speciation. Does this difficulty make the study of speciation "unscientific"? Should we abandon the study of speciation?

For More Information

Cowie, R. H. "Variation in Species, Diversity and Shell Shape in Hawaiian Land Snails." *Evolution* (49), 1996. A discussion of the ecological relationships involved in the sympatric speciation of Hawaiian land snails.

Eldredge, N. *Fossils: The Evolution and Extinction of Species*. New York: Harry N. Abrams, 1991. A nicely illustrated exploration of a paleontologist's approach to examining and interpreting the past, including speciation events.

Grant, P. R. "Natural Selection and Darwin's Finches." *Scientific American*, October 1991. A temporary climate change altering the food supply created selection pressures on finches that altered their average beak shape within several generations.

Quammen, D. *The Song of the Dodo*. New York: Scribner, 1996. Beautifully written exposition of the biology of islands. Read this book to understand why islands are known as "natural laboratories of speciation."

Wilson, E. O. *The Diversity of Life*. New York: W. W. Norton, 1992. Elegant description of how species arise, how they disappear, and why we should preserve them.

Answers to Multiple-Choice Questions

1.e 2.d 3.d 4.d 5.b 6.c

MEDIATUTOR
The Origin of Species

CD Activities

Activity 16.1: Allopatric Speciation

Estimated time: 10 minutes

Allopatric speciation is the primary means by which new species evolve. This animation demonstrates how one species splits to form two that are separated in space. You will be challenged at various stages along the way to predict what will happen next.

Activity 16.2: Sympatric Speciation

Estimated time: 10 minutes

Sympatric speciation is a means by which new species evolve. This animation demonstrates how one species splits to form two without being separated in space. You will be challenged at various stages along the way to predict what will happen next.

Start the MediaTutor Student CD-ROM and enter the activity number in the Quick Search box to be taken directly to that activity.

Web Investigations

Case Study: Lost World

Estimated time: 15 minutes

Is the great age of exploration over? Think again. Thousands of new organisms, ecosystems, and other natural wonders remain to be discovered. This exercise looks at a few recently discovered animal species and how they were found.

Go to http://www.prenhall.com/audesirk6, the Audesirk Companion Web site. Select Chapter 16 and Web Investigation to begin.

"[T]hence life was born,
Its nitrogen from ammonia, carbon from methane,
Water from the cloud and salts from the young seas . . . the cells of life
Bound themselves together into clans, a multitude of cells
To make one being—as the molecules before
Had made of many one cell."

Robinson Jeffers in* The Beginning and the End *(1963)

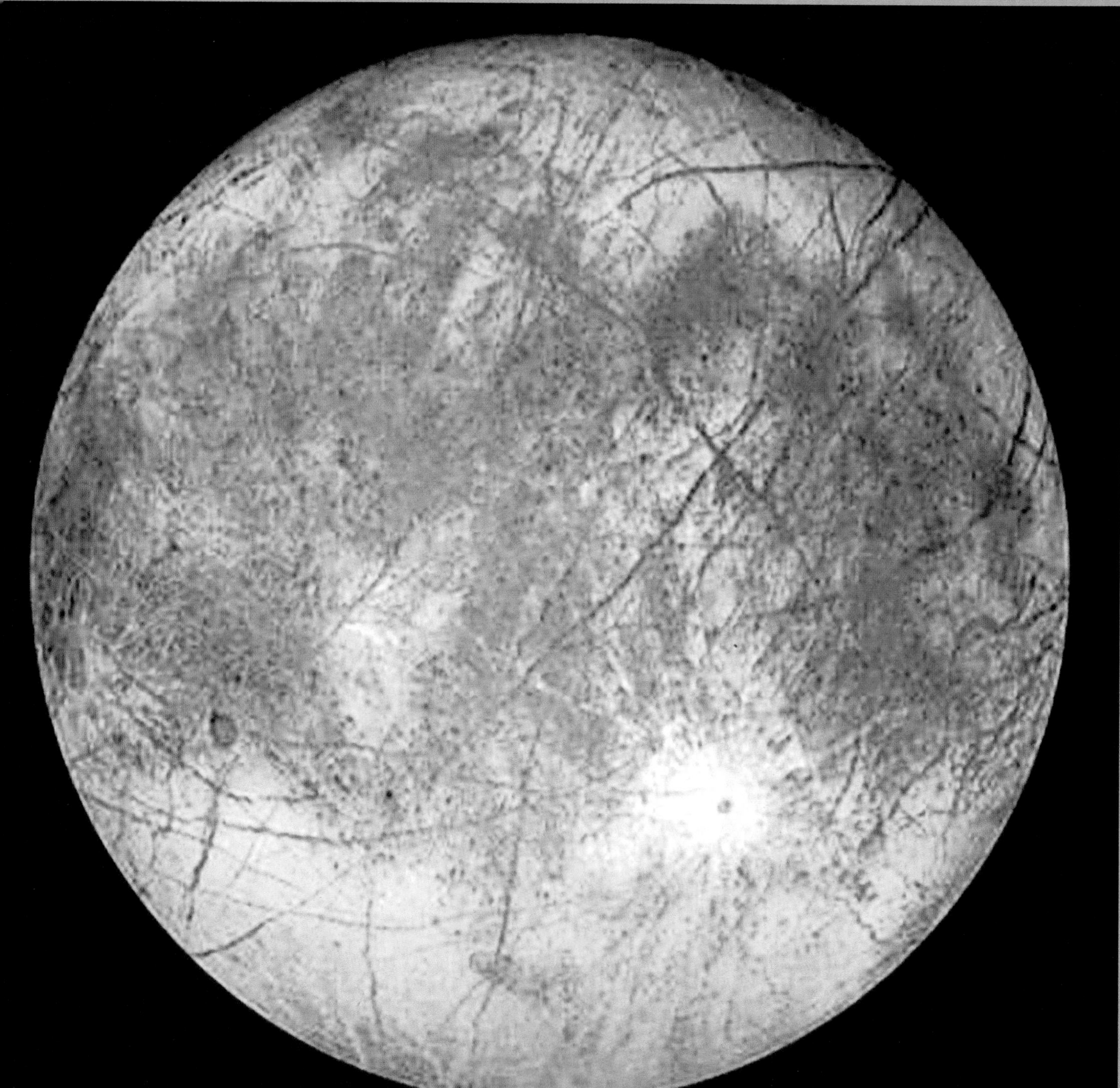

The icy surface of Jupiter's moon Europa may conceal a liquid ocean. Could this extraterrestrial ocean harbor life?

17 The History of Life on Earth

AT A GLANCE

Case Study: Life on a Frozen Moon?

1) **How Did Life Begin?**

Prebiotic Evolution Was Controlled by the Early Atmosphere and Climate

2) **What Were the Earliest Organisms Like?**

The First Organisms Were Anaerobic Prokaryotes

Some Organisms Evolved the Ability to Capture Energy from the Sun

Aerobic Metabolism Arose in Response to the Oxygen Crisis

Eukaryotes Developed Membrane-Enclosed Organelles and a Nucleus

3) **How Did Multicellularity Arise?**

Multicellular Plants Developed Specialized Structures That Facilitated Their Invasion of Diverse Habitats

Multicellular Animals Developed Specializations That Allowed Them to Capture Prey, Feed, and Escape More Efficiently

4) **How Did Life Invade the Land?**

Some Plants Developed Specialized Structures That Adapted Them to Life on Dry Land

Some Animals Developed Specialized Structures That Adapted Them to Life on Dry Land

5) **What Role Has Extinction Played in the History of Life?**

The Upward Trend in Species Diversity Has Been Interrupted by Periodic Mass Extinctions

6) **How Did Humans Evolve?**

Primate Evolution Has Been Linked to Grasping Hands, Binocular Vision, and a Large Brain

Hominids Evolved from Dryopithecine Primates

The Earliest Hominids Could Stand and Walk Upright

The Evolution of Human Behavior Is Highly Speculative

Case Study Revisited: Life on a Frozen Moon?

CASESTUDY

Life on a Frozen Moon?

On the cold, dark surface of Europa, all is silent. On the horizon, the huge orange and purple mass of Jupiter looms. High overhead, a spidery metal spacecraft speeds across the sky, its instruments whirring. Within moments it is gone, hurtling back into space.

Back on Earth, a group of NASA scientists gathers around a computer screen. They gaze intently at the screen, examining images of Europa that have just been beamed home by *Galileo,* an unmanned spacecraft. Why so much interest in pictures from one of Jupiter's 16 moons? Because Europa has been deemed to be among the celestial objects most likely to hold extraterrestrial life.

Europa's status as a possible home to extraterrestrial life is new. Data gathered back in the 1970s showed the moon's surface to be entirely covered with frozen water, and this icy world did not appear to be particularly hospitable to life. More recently, however, the *Galileo* spacecraft has flown quite close to Europa's surface and has produced a wealth of detailed pictures. These images, together with other data gathered by *Galileo,* have fostered a new outlook on the frozen moon.

The most recent images of Europa show that the moon's frozen surface is littered with huge cracks, humps, and crevices. These deformities indicate that the ice sheet is moving and shifting, and careful analyses by scientists suggest that the observed pattern of cracks is what would be expected if the ice was resting atop liquid water. This seemingly cold and lifeless world may actually conceal a huge, liquid ocean, warmed by heat from the moon's rocky core and hidden deep beneath a vast sheet of ice.

What is the significance of this finding? Water is essential to life; all forms of life on Earth require water to survive. According to our current understanding of the origin of life on Earth, water is a key component of the conditions under which life can arise. The discovery of Europa's shifting ice sheets is our first indication that liquid water may exist somewhere in the solar system other than Earth. If life could arise in the watery environment of early Earth, why not in the watery environment of Europa? ■

1 How Did Life Begin?

How and when did life first appear on Earth? Just a few centuries ago, this question would have been considered trivial. Although no one knew how life *first* arose, people thought that new living things appeared all the time, through **spontaneous generation** from both nonliving matter and other, unrelated forms of life. In 1609 a French botanist wrote, "There is a tree . . . frequently observed in Scotland. From this tree leaves are falling; upon one side they strike the water and slowly turn into fishes, upon the other they strike the land and turn into birds." Medieval writings abound with similar observations and delightful recipes for creating life—even human beings. Microorganisms were thought to arise spontaneously from broth, maggots from meat, and mice from mixtures of sweaty shirts and wheat.

In 1668 Italian physician Francesco Redi disproved the maggots-from-meat hypothesis simply by keeping flies (whose eggs hatch into maggots) away from uncontaminated meat (see Chapter 1). In the mid-1800s Louis Pasteur in France and John Tyndall in England disproved the broth-to-microorganism idea (Fig. 17-1). Although their work effectively demolished the notion of spontaneous generation, it did not address the question of how life on Earth originated in the first place. Or, as biochemist Stanley Miller put it, "Pasteur never proved it didn't happen once, he only showed that it doesn't happen all the time."

For almost half a century, the subject lay dormant. Eventually, biologists returned to the question of the origin of life. In the 1920s and 1930s, Alexander Oparin in Russia and John B. S. Haldane in England noted that the oxygen-rich atmosphere that we know would not have permitted the spontaneous formation of the complex organic molecules necessary for life. Oxygen reacts readily with other molecules, disrupting chemical bonds and thus tending to keep molecules simple. Oparin and Haldane speculated that the atmosphere of the young Earth was very low in oxygen and rich in hydrogen and that, under these atmospheric conditions, life could have arisen from nonliving matter through ordinary chemical reactions. This process of chemical evolution is called **prebiotic evolution**, or evolution before life existed.

Prebiotic Evolution Was Controlled by the Early Atmosphere and Climate

The primordial Earth differed greatly from the planet we now enjoy. When Earth first formed about 4.5 billion years ago, it was quite hot. A multitude of meteors smashed into the forming planet, and the kinetic energy of these extraterrestrial rocks was converted into heat on impact. Still more heat was released as radioactive atoms decayed. The rock composing Earth melted, and heavier elements such as iron and nickel sank to the center of the planet, where they remain molten today. Gradually Earth cooled, and elements combined to form compounds of many sorts. Virtually all of the oxygen combined with hydrogen to form water, with carbon to form carbon dioxide, or with heavier elements to form minerals. After millions of years, Earth cooled enough to allow water to exist as a liquid; it must have rained for thousands of years as water vapor condensed out of the cooling atmosphere. As the water struck the surface, it dissolved many minerals, forming a weakly salty ocean. Lightning from storms, heat from volcanoes, and intense ultraviolet light from the sun all poured energy into the young seas.

Judging from the chemical composition of the rocks that formed at this time, geochemists have deduced that the primitive atmosphere probably contained carbon dioxide, methane, ammonia, hydrogen, nitrogen, hydrochloric acid, hydrogen sulfide, and water vapor. But there was virtually no free oxygen in the early atmosphere, because oxygen atoms were bound up in water, carbon dioxide, and minerals.

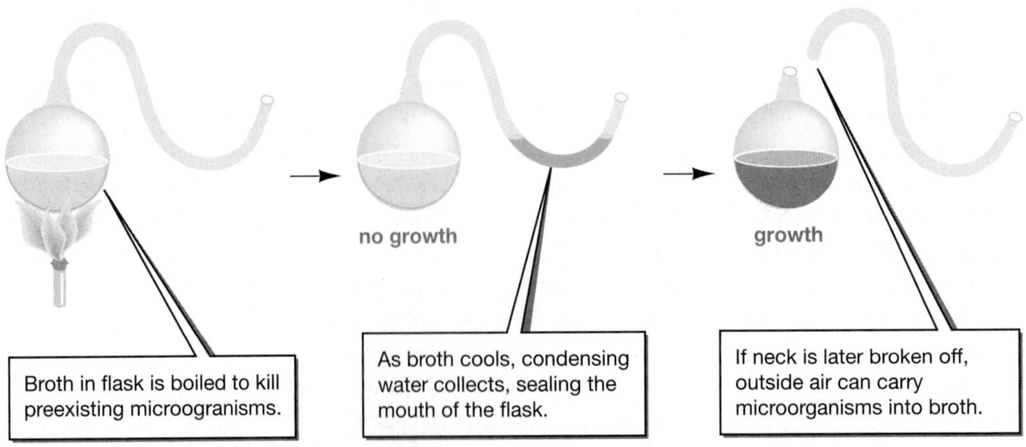

no growth
growth
Broth in flask is boiled to kill preexisting microorganisms.
As broth cools, condensing water collects, sealing the mouth of the flask.
If neck is later broken off, outside air can carry microorganisms into broth.

Figure 17-1 Spontaneous generation refuted
Louis Pasteur's experiment disproving the spontaneous generation of microorganisms in broth.

Organic Molecules Can Be Synthesized Spontaneously Under Prebiotic Conditions

Inspired by the ideas of Oparin and Haldane, Stanley Miller, then a graduate student, and his advisor, Harold Urey of The University of Chicago, set out in 1953 to simulate prebiotic evolution in the laboratory. They tried to reproduce the atmosphere of early Earth by mixing water vapor, ammonia, hydrogen, and methane in a flask. An electrical discharge mimicked the intense energy of early Earth's lightning storms. In this experimental microcosm, Miller and Urey found that simple organic molecules appeared after just a few days (Fig. 17-2). Similar experiments by Miller and others have produced amino acids, short proteins, nucleotides, adenosine triphosphate (ATP), and other molecules characteristic of living things. Interestingly, the exact composition of the "atmosphere" used in these experiments is unimportant, provided that hydrogen, carbon, and nitrogen are available and that free oxygen is excluded. Similarly, a variety of energy sources, including ultraviolet light, electrical discharge, and heat, are all about equally effective. Even though geochemists may never know exactly what the primordial atmosphere was like, organic molecules certainly were synthesized on ancient Earth. Additional organic molecules probably arrived from space when meteors and comets crashed into the surface of Earth. (Analysis of present-day meteors recovered from impact craters on Earth has revealed that some meteors contain relatively high concentrations of amino acids and other simple organic molecules similar to those generated in the Miller-Urey experiment.)

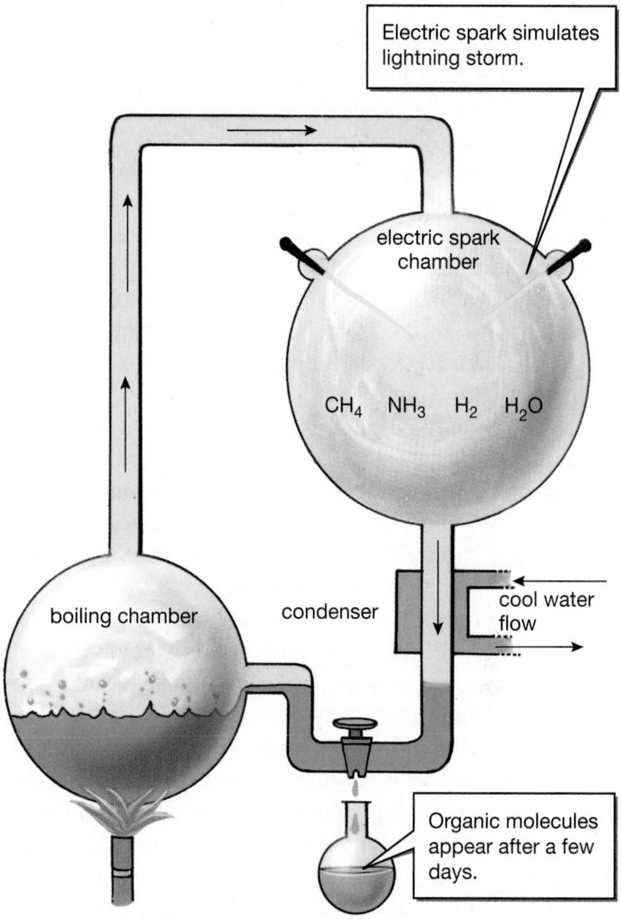

Figure 17-2 The experimental apparatus of Stanley Miller and Harold Urey
Life's very earliest stages left no fossils, so evolutionary historians have pursued a strategy of re-creating in the laboratory the conditions that may have prevailed on early Earth. The mixture of gases in the spark chamber simulates Earth's early atmosphere.

Prebiotic Conditions Would Allow Organic Molecules to Accumulate

Prebiotic synthesis would not have been very efficient or very fast. Nonetheless, in a few hundred million years, large quantities of organic molecules could accumulate, especially because they didn't break down nearly as fast back then. On Earth today, most organic molecules have a short life; either they are digested by living organisms or they react with atmospheric oxygen. Primeval Earth lacked both life and free oxygen, so these sources of degradation were absent. However, the primordial atmosphere also lacked an ozone layer, a region high in the atmosphere that is enriched with ozone (O_3) molecules, which absorb some of the sun's high-energy ultraviolet (UV) light before it reaches Earth. During the early history of Earth, before the ozone layer formed, UV bombardment, which can break apart organic molecules, must have been fierce. Some places, however, such as those beneath rock ledges or at the bottoms of even fairly shallow seas, would have been protected from UV radiation. In these locations, organic molecules may have accumulated to relatively high levels.

Even in areas protected from the sun, however, it is unlikely that molecules dissolved in a huge ocean could reach the concentrations necessary to form spontaneously the more-complex molecules that arose in the next stage of prebiotic evolution. The chemical reactions in which simple molecules combined to form larger molecules such as RNA and proteins required that the reacting molecules be packed closely together. Scientists have proposed several mechanisms by which the requisite high concentrations might have been achieved on early Earth. One possibility is that shallow pools at the ocean's edge were filled with water by waves crashing onto the shore. Afterward, some of the water in the pool might have evaporated, concentrating the dissolved substances. Given enough cycles of refilling and evaporation, the molecules in these pools could have become a concentrated "primordial soup" in which spontaneous chemical reactions could generate complex organic molecules. These molecules could then have become the building blocks of the first living organisms.

Was RNA the First Self-Reproducing Molecule?

One of the key features of living things is their capacity to reproduce. This capacity depends on the special ability of the DNA molecule to replicate itself. In modern cells, DNA encodes the information the cells need to synthesize the proteins that carry out most cellular functions. Self-replication of DNA ensures that this information is transferred to subsequent generations of cells. But how did this system of self-copying arise? It is far from obvious how any of the components of a primordial soup could have become endowed with the ability to make copies of themselves.

In the 1980s Thomas Cech of the University of Colorado and Sidney Altman of Yale University offered an intriguing solution to this question. They discovered that certain small RNA molecules, dubbed **ribozymes**, act as enzymes that catalyze cellular reactions, including the synthesis of more RNA molecules. It seems inevitable that, during hundreds of millions of years of prebiotic chemical synthesis, RNA nucleotides may have occasionally bonded together to form short RNA chains. Let us suppose that, purely by chance, one of these RNA chains was a ribozyme that could catalyze the synthesis of copies of itself from free ribonucleotides in the surrounding waters. This first ribozyme probably wasn't very good at its job and made lots of mistakes. These mistakes were the first mutations. Like modern mutations, most undoubtedly ruined the catalytic abilities of the "daughter molecules," but a few may have been improvements. Molecular evolution could begin, as ribozymes with increased speed and accuracy of replication reproduced faster, making more and more copies of themselves.

Even after small self-replicating molecules arose, the transition to the modern "DNA → RNA → protein" mechanism must have involved a complex series of intermediate steps. One particularly difficult problem is that, although small nucleotide chains can catalyze their own replication, longer nucleotide chains cannot replicate without the assistance of specific protein enzymes that are encoded by the long nucleotide chains themselves. So how could the earliest RNA molecules get long enough to encode the necessary proteins if those proteins couldn't exist until the RNA molecules were long enough? Fortunately, the laws of physics and chemistry allow a solution to this knotty problem, as well as to the similarly tricky problem of how RNA gradually receded into its present role as an intermediary between DNA and protein enzymes. The details are fascinating but too intricate to describe here. Readers interested in exploring this stage of chemical evolution in all its glory should read Manfred Eigen's book *Steps Towards Life* (see "For More Information" on page 348).

Protocells May Have Consisted of Ribozymes Within Microspheres

Self-replicating molecules alone do not constitute life; some kind of enclosing membrane is also required. The precursors of the earliest biological membranes may have been simple structures that formed spontaneously from purely physical, mechanical processes. For example, chemists have shown that if water containing proteins and lipids is agitated to simulate waves beating against ancient shores, hollow structures called **microspheres** form (Fig. 17-3). These hollow balls resemble living cells in several respects. They have a well-defined outer boundary that separates internal contents from the external solution. If the composition of the microsphere is right, a "membrane" forms that is remarkably similar in appearance to a real cell membrane. Under certain conditions microspheres can absorb material from the solution ("feed"), grow, and even divide.

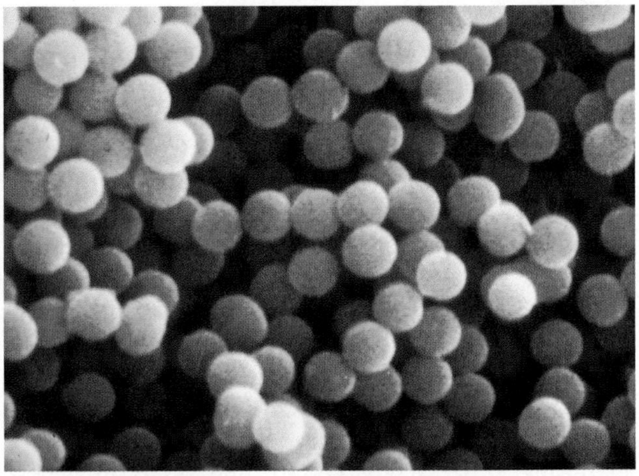

Figure 17-3 Do microspheres resemble the earliest cells? Cell-like microspheres can be formed by agitating proteins and lipids in a liquid medium. Such microspheres can take in material from the surrounding solution, grow, and even "reproduce," as these are doing.

If a microsphere happened to surround the right ribozymes, something a bit like a living cell would have been formed. We could call it a **protocell**, structurally similar to a cell but not a living thing. The ribozymes and their protein products would have been protected from free-roaming ribozymes in the primordial soup. Nucleotides and amino acids might have diffused across the membrane and have been used to synthesize new RNA and protein molecules. After sufficient growth, the microsphere may have divided, with a few copies of both ribozymes and proteins becoming incorporated into each daughter microsphere. If so, the path to the evolution of the first cells would be nearly at its end. Was there a particular moment when a nonliving protocell gave rise to something alive? Probably not. Like most evolutionary transitions, the change from protocell to living cell was a continuous process, with no sharp boundary between one state and the next.

But Did All This Happen?

The above scenario, although plausible and consistent with many research findings, is by no means certain. One of the most striking aspects of origin-of-life research is the great diversity of assumptions, experiments, and contradictory hypotheses (see the suggested reading by Iris Fry, *The Emergence of Life on Earth,* for a taste of the controversies). Researchers disagree as to whether life arose in quiet pools, in the sea, in moist films on the surfaces of clay or iron pyrite (fool's gold), or in furiously hot deep-sea vents. (Proponents of a deep-sea vent origin of life suggest that similar vents might keep Europa's seas thawed and provide the conditions for an independent origin of life there.) A few researchers even argue that life arrived on Earth from space. Can we draw any conclusions from the research done so far? No one really knows the answer to that question, but we can offer a few observations.

First, the experiments of Miller and others show that amino acids, nucleotides, and other organic molecules, along with simple membrane-like structures, would have formed in abundance on the primordial Earth. Second, early chemical evolution had long periods of time and huge areas of the Earth available to it. Given sufficient time and a sufficiently large pool of actors, even extremely rare events can occur many times. So, even if prebiotic evolution yielded only simple molecules, and the earliest catalysts were not very efficient, and the earliest membranes were unsophisticated, vast expanses of available time and space would have increased the likelihood of each small step on the path from primordial soup to living cell.

Most biologists accept that the origin of life was probably an inevitable consequence of the working of natural laws. We should emphasize, however, that this proposition is not proved and never will be. Biologists investigating the origin of life have neither millions of years nor trillions of gallons of reaction solutions with which to work!

2 What Were the Earliest Organisms Like?

Life began during the *Precambrian era*, an interval designated by geologists and paleontologists, who have devised a hierarchical naming system of eras, periods, and epochs to delineate the immense span of geologic time (Table 17-1). The oldest fossil organisms found so far are in Precambrian rocks that are about 3.5 billion years old, based on radiometric dating techniques (see "Scientific Inquiry: How Do We Know How Old a Fossil Is?"). Chemical traces in older rocks have led some paleontologists to believe that life is even older, perhaps as old as 3.9 billion years. What were the earliest cells like?

The First Organisms Were Anaerobic Prokaryotes

The first cells were prokaryotic; that is, their genetic material was not sequestered from the rest of the cell within a membrane-limited nucleus. These cells probably obtained nutrients and energy by absorbing organic molecules from their environment. There was no free oxygen in the atmosphere, so the cells must have metabolized the organic molecules anaerobically. You will recall from Chapter 8 that anaerobic metabolism yields only small amounts of energy.

As you probably have already recognized, the earliest cells were primitive anaerobic bacteria. As these ancestral bacteria multiplied, however, they must have eventually used up the organic molecules produced by prebiotic synthesis. Simpler molecules, such as carbon dioxide and water, were still very abundant, as was energy, in the form of sunlight. What was lacking, then, was not materials or energy itself, but *energetic molecules*—molecules in which energy is stored in chemical bonds.

Some Organisms Evolved the Ability to Capture Energy from the Sun

Eventually, some cells evolved the ability to use the energy of sunlight to drive the synthesis of complex, high-energy molecules from simpler molecules: In other words, photosynthesis appeared. Photosynthesis requires a source of hydrogen, and the very earliest photosynthetic bacteria probably used hydrogen sulfide gas for this purpose (much as today's purple photosynthetic bacteria do). Eventually, however, Earth's supply of hydrogen sulfide (which is produced mainly by volcanoes) must have run low. The shortage of hydrogen sulfide set the stage for evolution of photosynthetic bacteria that were able to use the planet's most abundant source of hydrogen: water (H_2O).

Water-based photosynthesis converts water and carbon dioxide to sugar, releasing oxygen as a by-product. The advent of this new method for capturing energy introduced significant amounts of free oxygen to the atmosphere for the first time. At first, the newly liberated oxygen was quickly consumed by reactions with other molecules in the atmosphere and in Earth's *crust*, or surface layer. One especially common reactive atom in the crust was iron, and much of the new oxygen combined with iron atoms to form huge deposits of iron oxide (also known as *rust*).

Once all the accessible iron had turned to rust, the concentration of free oxygen in the atmosphere began to increase. Chemical analysis of rocks suggests that significant amounts of free oxygen first appeared in the atmosphere about 2.2 billion years ago, produced by bacteria that were probably very similar to modern cyanobacteria (you will undoubtedly breathe in some oxygen molecules today that were expelled by one of those 2-billion-year-old cyanobacteria). Atmospheric oxygen levels increased steadily until reaching a stable level about 1.5 billion years ago. Since that time, the proportion of oxygen in the atmosphere has been nearly constant, as the amount of oxygen released by photosynthesis worldwide is neatly balanced by the amount that is consumed by aerobic respiration.

Table 17-1 The History of Life on Earth

Era	Period	Epoch	Years Ago* (millions)	Major Events
Precambrian			4600	Origin of solar system and Earth.
			4000–3900	Appearance of first rocks on Earth.
			3900–3500	First living cells (prokaryotes).
			3500	Origin of photosynthesis (in cyanobacteria).
			2200	Accumulation of free oxygen in atmosphere.
			2000–1700	First eukaryotes.
			By 1000	First multicellular organisms.
			About 1000	First animals (soft-bodied marine invertebrates).
Paleozoic	Cambrian		544–505	Primitive marine algae flourish; origin of most marine invertebrate types; first fish.
	Ordovician		505–440	Invertebrates, especially arthropods and mollusks, dominant in sea; first fungi.
	Silurian		440–410	Many fish, trilobites, mollusks in sea; first vascular plants; invasion of land by plants; invasion of land by arthropods.
	Devonian		410–360	Fishes and trilobites flourish in sea; first amphibians and insects; first seeds and pollen.
	Carboniferous		360–286	Swamp forests of tree ferns and club mosses; first conifers; dominance of amphibians; numerous insects; first reptiles.
	Permian		286–245	Massive marine extinctions, including last of trilobites; flourishing of reptiles and decline of amphibians; continents aggregated into one land mass, Pangaea.
Mesozoic	Triassic		245–208	First mammals and dinosaurs; forests of gymnosperms and tree ferns; breakup of Pangaea begins.
	Jurassic		208–146	Dominance of dinosaurs and conifers; first birds; continents partially separated.
	Cretaceous		146–65	Flowering plants appear and become dominant; mass extinctions of marine life and some terrestrial life, including last dinosaurs; modern continents well separated.
Cenozoic	Tertiary	Paleocene	65–54	Widespread flourishing of birds, mammals, insects, and flowering plants; continents are shifted into modern positions; mild climate at beginning of period, with extensive mountain building and cooling toward end.
		Eocene	54–38	
		Oligocene	38–23	
		Miocene	23–5	
		Pliocene	5–1.8	
	Quaternary	Pleistocene	1.8–0.01	Evolution of genus *Homo*; repeated glaciations in Northern Hemisphere; extinction of many giant mammals.
		Recent	0.01–present	

* From University of California Museum of Paleontology, April 2000.

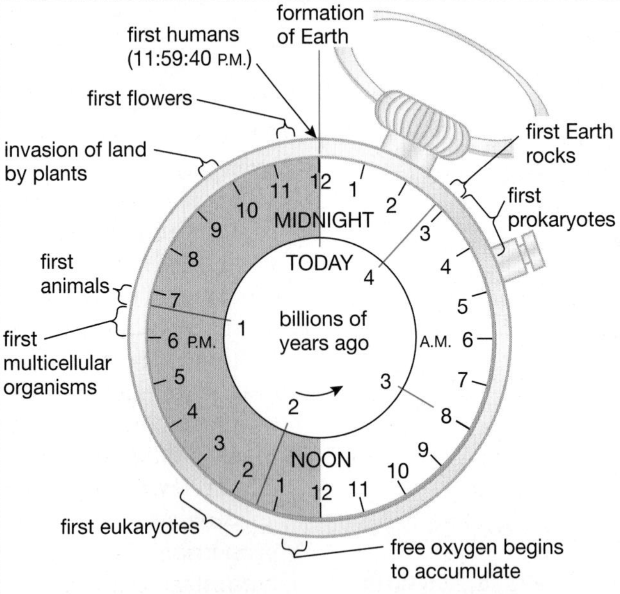

Earth's history projected on a 24-hour day

Scientific Inquiry
How Do We Know How Old a Fossil Is?

Early geologists could date rock layers and their accompanying fossils only in a *relative* way: Fossils found in deeper layers of rock were generally older than those found in shallower layers. With the discovery of radioactivity, it became possible to determine *absolute* dates, within certain limits of uncertainty. The nuclei of radioactive elements spontaneously break down, or *decay*, into other elements. For example, carbon-14 (usually written ^{14}C) emits an electron to become nitrogen-14 (^{14}N). Each radioactive element decays at a rate that is independent of temperature, pressure, or the chemical compound of which the element is a part. The time it takes for half the radioactive nuclei to decay at this characteristic rate is called the *half-life*. The half-life of ^{14}C, for example, is 5730 years.

How are radioactive elements used in determining the age of rocks? If we know the rate of decay and measure the proportion of decayed nuclei to undecayed nuclei, we can estimate how much time has passed since these radioactive elements were trapped in rock. This process is called *radiometric dating*. A particularly straightforward dating technique uses the decay of potassium-40 (^{40}K), which has a half-life of about 1.25 billion years, into argon-40 (^{40}Ar). Potassium is a very reactive element commonly found in volcanic rocks such as granite and basalt. Argon, however, is an unreactive gas. Let us suppose that a volcano erupts with a massive lava flow, covering the countryside. All the ^{40}Ar, being a gas, will bubble out of the molten lava, so when the lava cools and solidifies into rock, it will start out with no ^{40}Ar. Any ^{40}K present in the hardened lava will decay to ^{40}Ar, half the ^{40}K decaying every 1.25 billion years. The ^{40}Ar gas will be trapped in the rock. A geologist could take a sample of the rock and determine the proportion of ^{40}K to ^{40}Ar (Fig. E17-1). If the analysis finds equal amounts of the two elements, the geologist will conclude that the lava hardened 1.25 billion years ago. With appropriate care, such age estimates are quite reliable. If a fossil is found beneath a lava flow dated at, say, 500 million years, then we know that the fossil is at least that old.

Some radioactive elements, as they decay, can even give an estimate of the age of the solar system. Analysis of uranium, which decays to lead, has shown that the oldest meteorites and moon rocks collected by astronauts are about 4.6 billion years old.

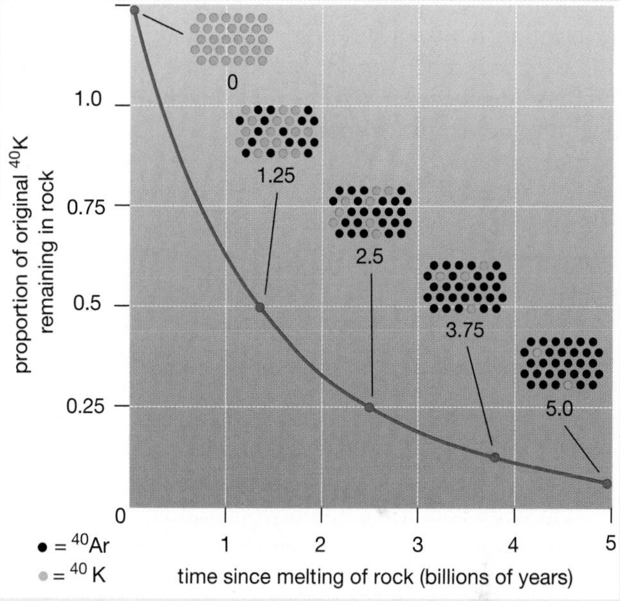

Figure E17-1 The relationship between time and the decay of radioactive ^{40}K to ^{40}Ar

Aerobic Metabolism Arose in Response to the Oxygen Crisis

Oxygen is potentially very dangerous to living things, because it reacts with organic molecules, destroying them. Many of today's anaerobic bacteria perish when exposed to what is for them a deadly poison, oxygen. The accumulation of oxygen in the atmosphere of early Earth probably exterminated many organisms and fostered the evolution of cellular mechanisms for detoxifying oxygen. The oxygen crisis also provided the environmental pressure for the next great advance in the Age of Microbes: the ability to use oxygen in metabolism. This ability not only provides a defense against the chemical action of oxygen, but actually channels its destructive power through aerobic respiration to generate useful energy for the cell. Because the amount of energy available to a cell is vastly increased when oxygen is used to metabolize food molecules, aerobic cells had a significant selective advantage.

Eukaryotes Developed Membrane-Enclosed Organelles and a Nucleus

Hordes of bacteria would offer a rich food supply to any organism that could eat them. There are no fossil records of the first predatory cells, but paleobiologists speculate that once a suitable prey population (such as these bacteria) appeared, predation would have evolved quickly. These predators would have been specialized prokaryotic cells that lacked cell walls and consequently were able to engulf whole bacteria as prey. According to the most widely accepted hypothesis, these predators were otherwise quite primitive, being capable of neither photosynthesis nor aerobic metabolism. Although they could capture large food particles, namely bacteria, they metabolized them inefficiently. By about 1.7 billion years ago, however, one predator probably gave rise to the first eukaryotic cell.

Eukaryotic cells differ from prokaryotic cells in many ways, but perhaps most fundamental is the presence in eukaryotes of a membrane-bound nucleus that contains the cell's genetic material. Other key eukaryote structures are the organelles used for energy metabolism: mitochondria and (in plants only) chloroplasts. How did these organelles evolve?

Mitochondria and Chloroplasts May Have Arisen from Engulfed Bacteria

The **endosymbiont hypothesis**, championed most forcefully by Lynn Margulis of the University of Massachusetts, proposes that primitive cells acquired the precursors of mitochondria and chloroplasts by engulfing certain types of bacteria. These cells and the bacteria trapped inside them (*endo* means "within") gradually entered into a *symbiotic* relationship, a close association between different types of organisms over an extended time. Let us suppose that an anaerobic predatory cell captured an aerobic bacterium for food, as it often did, but for some reason failed to digest this particular prey (Fig. 17-4). The aerobic bacterium remained alive and well. In fact, it was better off than ever, because the cytoplasm of its predator/host was chock-full of half-digested food molecules, the remnants of anaerobic metabolism. The aerobe absorbed these molecules and used oxygen to metabolize them, thereby gaining enormous amounts of energy. So abundant were the aerobe's food resources, and so bountiful its energy production, that the aerobe must have leaked energy, probably as ATP or similar molecules, back into its host's cytoplasm. The anaerobic predatory cell with its symbiotic bacteria could now metabolize food aerobically, gaining a great advantage over its anaerobic compatriots. Soon its progeny filled the seas. Eventually, the endosymbiotic bacterium lost its ability to live independently of its host, and the mitochondrion was born.

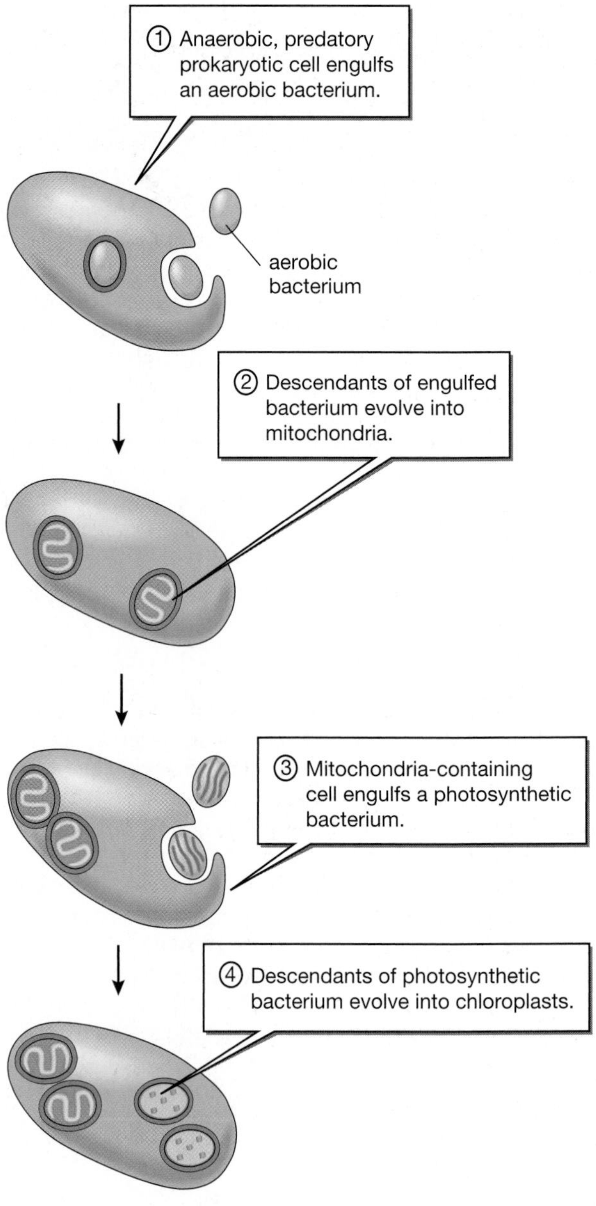

Figure 17-4 The probable origin of mitochondria and chloroplasts in eukaryotic cells

MEDIATUTOR 17.1 Endosymbiosis

One of these successful new cellular partnerships must have managed a second feat: It captured a photosynthetic cyanobacterium and similarly failed to digest its prey (see Fig. 17-4). The cyanobacterium flourished in its new host and gradually evolved into the first chloroplast. Other eukaryotic organelles may have also originated through endosymbiosis. Many paleobiologists believe that cilia, flagella, centrioles, and microtubules may all have evolved from a symbiosis between a spirilla-like bacterium (a form of bacteria with an elongated corkscrew shape) and a primitive eukaryotic cell.

Several lines of evidence support the endosymbiont hypothesis. For example, mitochondria, chloroplasts, and centrioles each contain their own minute supply of DNA, which some researchers interpret as a remnant of the DNA originally contained within the symbiotic bacterium. Another kind of support comes from *living intermediates*, organisms alive today that are similar to a hypothetical ancestral condition and thus help show that a proposed evolutionary pathway is plausible. For example, the amoeba *Pelomyxa palustris* lacks mitochondria but hosts a permanent population of aerobic bacteria that carry out much the same role. Similarly, a variety of corals, some clams, a few snails, and at least one species of *Paramecium* harbor a permanent collection of algae in their cells (Fig. 17-5). These examples of modern cells that host bacterial endosymbionts suggest that we have no reason to doubt that similar symbiotic associations could have occurred almost 2 billion years ago and led to the first eukaryotic cells.

The Origin of the Nucleus Is More Obscure

The evolution of the nucleus is more obscure. One possibility is that the plasma membrane folded inward, surrounding the DNA. This would create the nuclear

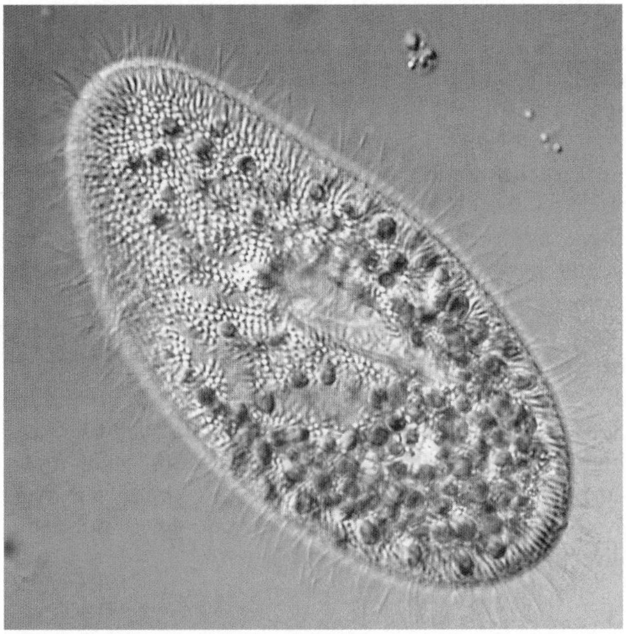

Figure 17-5 A modern intracellular symbiosis
Is an endosymbiotic origin of cellular organelles plausible? The plausibility of any scenario for evolutionary history is enhanced if we can point to living examples of the proposed intermediate stages. The ancestors of the chloroplasts in today's plant cells may have resembled *Chlorella*, the green, unicellular algae living symbiotically within the cytoplasm of the *Paramecium* pictured here.

membrane. Further infoldings could have produced the endoplasmic reticulum, which is continuous with the nuclear membrane. An alternative hypothesis is that, like many other eukaryotic organelles, the nucleus arose as a result of endosymbiosis. In this scenario, the engulfed bacterium took control of its host. But regardless of how the nucleus originated, having the DNA sequestered within the nucleus seems to have conferred great advantages, perhaps by allowing for finer regulation of the genetic material. Today, organisms composed of eukaryotic cells are by far the most visible forms of life on Earth.

3 How Did Multicellularity Arise?

Once predation had evolved, increased size became an advantage. A larger cell could more easily engulf a smaller cell while also being more difficult for other predatory cells to ingest. Most larger organisms can also move faster than small ones, making successful predation and escape more likely. But enormous single cells have problems. Oxygen and nutrients going into the cell and waste products going out must diffuse through the plasma membrane. As we pointed out in Chapter 6, the larger a cell becomes, the less surface membrane is available per unit volume of cytoplasm. There are only two ways that an organism larger than a millimeter or so in diameter can survive. First, it can have a low metabolic rate so that it doesn't need much oxygen or produce much carbon dioxide. This seems to work for certain very large unicellular algae. Alternatively, an organism may be multicellular—that is, it may consist of many small cells packaged into a larger, unified body.

The first multicellular organisms almost certainly evolved in the sea, but the fossil record reveals little about the origin of multicellularity. Unicellular eukaryotic fossils first appear in rocks that are about 1.7 billion years old, but the first fossil evidence of multicellular organisms is in rocks that are 700 million years younger. Some of these billion-year-old rocks have yielded fossil traces of animal tracks and burrows and of multicellular organisms that may have been the precursors of today's seaweeds. But because the earliest multicellular organisms almost certainly had no skeleton or other hard parts, they left few fossils. Consequently, we may never learn very much about them.

Multicellular Algae Developed Specialized Structures That Facilitated Their Invasion of Diverse Habitats

During the Precambrian era, unicellular eukaryotic cells containing chloroplasts gave rise to the first multicellular algae. Multicellularity would have provided at least two advantages for these seaweeds. First, large, many-celled algae would have been difficult for unicellular predators to swallow. Second, specialization of cells would have conferred the potential for staying in one place in the brightly lit waters of the shoreline, as rootlike structures burrowed in sand or clutched onto rocks, while leaflike structures floated above in the sunlight. The green, brown, and red algae lining our shores today—some, such as the brown kelp, more than 60 meters (200 feet) long—are the descendants of these early multicellular algae.

Multicellular Animals Developed Specializations That Allowed Them to Capture Prey, Feed, and Escape More Efficiently

A wide variety of Precambrian invertebrate animals (animals lacking backbones) appear in rocks laid down between 650 million and 544 million years ago. Many of these ancient animals are quite different in appearance from any animals that appear later in the fossil record. Some paleontologists believe that they represent a kind of failed evolutionary experiment in body plans that failed to survive the test of time. Most of these Precambrian animals appear rather suddenly in the fossil record and disappear just as suddenly, so it is difficult to determine either their ancestry or the fate of their descendants. Fossil ancestors of today's sponges, however, have been found in Precambrian rocks about 570 million years old, so at least some representatives of modern animal groups had arisen by that time. For the most part, though, the familiar forms of modern invertebrate animals do not appear in the fossil record until the Cambrian period, marking the beginning of the Paleozoic era, about 544 million years ago. Like the Precambrian animals, the first fossils of Cambrian animals reveal an adaptive radiation that had already yielded a diverse array of complex body plans. The evolutionary history that produced such an impressive range of different animal forms remains obscure.

For animals, one of the advantages of multicellularity is the potential for eating larger prey. Coevolution of predator and prey led to increased sophistication in many kinds of animals. By the Silurian period (440 million to 410 million years ago), mud-skimming, armored trilobites were preyed on by (now-extinct) ammonites and the chambered nautilus, which still survives almost unchanged in deep Pacific waters (Fig. 17-6). A major evolutionary trend of this era was toward greater mobility. Predators often need to travel over wide areas in search of suitable prey, and speedy escape is also an advantage for prey. Locomotion is normally accomplished by the contraction of muscles that move body parts through their attachments to some sort of skeleton. Most invertebrates at this time possessed either an internal hydrostatic skeleton, much like a water-filled tube (worms), or an external skeleton covering the body (arthropods such as trilobites). Greater sensory capabilities and more-sophisticated nervous systems evolved along with the ability to move more efficiently. Senses for detecting touch, chemicals, and light became highly developed. The senses were typically concentrated in the head end of the animal, along with a nervous system capable of handling the sensory information and directing appropriate behaviors.

(a)

(b)

(c)

(d)

Figure 17-6 Diversity of ocean life during the Silurian period
(a) Characteristic life of the oceans during the Silurian period, 440 million to 410 million years ago. Among the most common fossils from that time are ***(b)*** the trilobite and its predators, ***(c)*** the ammonites, and the nautiloids. Although ***(d)*** illustrates a living *Nautilus*, the Silurian nautiloids were very similar in structure, showing that a successful body plan may exist virtually unchanged for hundreds of millions of years.

About 500 million years ago, one group of animals developed a new form of support for the body: an internal skeleton. These early fishes were inconspicuous members of the ocean community, but by 400 million years ago fishes were a diverse and prominent group. By and large, the fishes proved to be faster than the invertebrates, with more-acute senses and larger brains. Eventually, they became the dominant predators of the open seas.

4 How Did Life Invade the Land?

One of the more thrilling subplots in the long tale of life's history is the story of life's invasion of land after more than 3 billion years of a strictly watery existence. In moving to solid ground, organisms had many obstacles to overcome. Life in the sea provides buoyant support against gravity, but on land an organism must bear its weight against the crushing force of gravity. The sea provides ready access to life-sustaining water, but a terrestrial organism must find adequate water. Sea-dwelling plants and animals can reproduce by means of mobile sperm and/or eggs that swim to each other through the water, but land-dwellers must ensure that their gametes be protected from drying out.

Despite the obstacles to life on land, the vast empty spaces of the Paleozoic landmass represented a tremendous evolutionary opportunity. The potential rewards of terrestrial life were especially great for plants. Water strongly absorbs light, so even in the clearest water, photosynthesis is limited to the upper few hundred meters, and usually much less. Out of the water, the sun is dazzlingly bright, permitting rapid photosynthesis. Furthermore, terrestrial soils are rich storehouses of nutrients, whereas seawater tends to be low in certain nutrients, particularly nitrogen and phosphorus. Finally, the Paleozoic sea swarmed with plant-eating animals, but the land was devoid of animal life. The plants that first colonized the land would have ample sunlight, untouched nutrient sources, and no predators.

Some Plants Developed Specialized Structures That Adapted Them to Life on Dry Land

In moist soils at water's edge, a few small green algae began to grow, taking advantage of the sunlight and nutrients. They didn't have large bodies to support against the force of gravity, and, living right in the film of water on the soil, they could easily obtain water. About 400 million years ago, some of these algae gave rise to the first multicellular land plants. Initially simple, low-growing forms, land plants rapidly developed solutions to two of the main difficulties of plant life on land: (1) obtaining and conserving water and (2) staying upright despite gravity and winds. Waterproof coatings on aboveground parts reduced water loss by evaporation, and rootlike structures delved into the soil, mining water and minerals. Specialized cells formed tubes called *vascular tissues* to conduct water from roots to leaves. Extra-thick walls surrounding certain cells enabled stems to stand erect.

Primitive Land Plants Retained Swimming Sperm and Required Water to Reproduce

Reproduction out of water presented greater challenges. Like animals, plants produce sperm and eggs, which must be able to meet if reproduction is to be accomplished. The first land plants had swimming sperm, presumably much like some of today's marine algae that have retained swimming sperm (and in some cases swimming eggs as well). Consequently, the earliest plants were restricted to swamps and marshes, where the sperm and eggs could be released into the water, or to areas with abundant rainfall, where the ground would occasionally be covered with water. Later, plants that required water for reproduction prospered when wet conditions were widespread. In particular, the warm and moist climate that prevailed during the Carboniferous period (360 million to 286 million years ago) permitted the formation of vast forests of giant tree ferns and club mosses that produced swimming sperm (Fig. 17-7). The coal we mine today is derived from the fossilized remains of those forests.

Seed Plants Encased Sperm in Pollen Grains, Allowing Them to Flourish in Dry Habitats

Meanwhile, some plants inhabiting drier regions had evolved reproductive strategies that no longer depended on films of water. The eggs of these plants were retained on the parent plant, and the sperm were encased in drought-resistant pollen grains that blew on the wind from plant to plant. When the pollen landed near an egg, it released sperm cells directly into living tissue, eliminating the need for a surface film of water. The fertilized egg remained on the parent plant, where it developed inside of a seed, which provided protection and nutrients for the developing embryo within.

The earliest seed-bearing plants appeared in the late Devonian (375 million years ago) and produced their seeds along branches, without any specialized structures to hold them. By the middle of the Carboniferous period, however, a new kind of seed-bearing plant had arisen. These plants, called **conifers**, protected their developing seeds inside cones. Conifers, which did not depend on water for reproduction, flourished and spread during the Permian period (286 to 245 million years ago), when mountains rose, swamps drained, and the climate became much drier. The good fortune of the conifers, however, was not shared by the tree ferns and giant club mosses, which, with their swimming sperm, largely went extinct.

Figure 17-7 The swamp forest of the Carboniferous period
The treelike plants are tree ferns and giant club mosses, both now mostly extinct.

Flowering Plants Enticed Animals to Carry Pollen

About 140 million years ago, during the Cretaceous period, the flowering plants appeared, having evolved from a group of conifer-like plants. Flowering plants are pollinated by insects, and this mode of reproduction seems to have conferred an evolutionary advantage. Flower pollination by insects is far more efficient than pollination by wind; wind-pollinated plants such as conifers must produce an enormous amount of pollen because the vast majority of pollen grains fail to reach their target. Flowering plants also evolved other advantages, including more-rapid reproduction and, in some cases, much more rapid growth. Today, flowering plants dominate the land, except in cold northern regions, where conifers still prevail.

Some Animals Evolved Specialized Structures That Adapted Them to Life on Dry Land

Soon after land plants evolved, providing potential food sources for other organisms, animals emerged from the sea. The first animals to move onto land were arthropods (the group that today includes insects, spiders, scorpions, centipedes, and crabs) Why arthropods? The answer seems to be that they were **preadapted** for land life—they already possessed structures that, purely by chance, were suited to life on land. Foremost among these preadaptations was the external skeleton, or **exoskeleton**, a hard covering surrounding the body, such as the shell of a lobster or crab. Exoskeletons are both waterproof and strong enough to support a small animal against the force of gravity.

Arthropods also solved another difficulty experienced by land animals: breathing. To allow gas exchange, respiratory surfaces must be kept moist, and doing so is difficult in the dry air. Some arthropods, such as land crabs and spiders, evolved what amounts to an internal gill, kept moist within a waterproof sac. Insects developed *tracheae*, small branching tubes directly penetrating the body, with adjustable openings in the exoskeleton that lead to the outside air.

For millions of years, arthropods had the land and its plants to themselves, and for tens of millions of years more, they were the dominant animals. Dragonflies with a wingspan of 28 inches (70 centimeters) flew among the Carboniferous tree ferns, while millipedes 6.5 feet (2 meters) long munched their way across the swampy forest floor. Eventually, however, the arthropods' splendid isolation came to an end.

Amphibians Evolved from Lobefin Fishes

About 400 million years ago, a group of Silurian fishes called the *lobefins* appeared, probably in fresh water. Lobefins had two important preadaptations to land-life: (1) stout, fleshy fins with which they crawled about on the bottoms of shallow, quiet waters, and (2) an outpouching of the digestive tract that could be filled with air, like a primitive lung. One group of lobefins colonized very shallow ponds and streams, which shrank during droughts and whose water often became oxygen-poor. By taking air into their lungs, these lobefins could obtain oxygen anyway. Some of their descendants began to use their fins to crawl from pond to pond in search of prey or water, as some modern fish can do today (Fig. 17-8).

The benefits of feeding on land and moving from pool to pool favored the evolution of fish that could stay out of water for longer periods and that could move about more effectively on land. With improvements in lungs and legs, the amphibians evolved from

Figure 17-8 A fish that walks on land
Some modern fish are able to move about on land, much like the ancient lobefin fish that gave rise to amphibians. Mudskippers use their strong pectoral fins to move across dry areas in their swampy habitat.

lobefins, first appearing in the fossil record about 350 million years ago. If an amphibian could have thought about such things, it would have thought that the Carboniferous swamp forests were heaven itself: no predators to speak of, abundant prey, and a warm, moist climate. As with the insects and millipedes, some amphibians evolved gigantic size, including salamanders more than 10 feet (3 meters) long.

Despite their success, the early amphibians were still not fully adapted to life on land. Their lungs were simple sacs without very much surface area, so they had to obtain some of their oxygen through their skin. Therefore, their skin had to be kept moist, which restricted them to swampy habitats where they wouldn't dry out. Further, amphibian sperm and eggs could not survive in dry surroundings and had to be deposited in watery environments. So, although amphibians could move about on land, they could not stray too far from the water's edge. As with the tree ferns and club mosses, when the climate turned dry at the beginning of the Permian period about 286 million years ago, amphibians were in trouble.

Reptiles, Which Evolved from Amphibians, Developed Several Adaptations to Dry Land

Just as the conifers had been evolving on the fringes of the swamp forests, so too was a group of amphibians evolving, developing adaptations to drier conditions. These amphibians ultimately gave rise to the reptiles, which achieved four great advances over their amphibian relatives. (1) They evolved *internal fertilization*, in which the reproductive tract of the female reptile provided a watery environment for the sperm. Thus, sperm transfer could occur on land, without the reptile having to venture back to the dangerous swamps full of fish and amphibian predators. (2) Reptiles evolved shelled, waterproof eggs that enclosed a supply of water for the developing embryo, again providing freedom from the swamps. (3) The ancestral reptiles evolved scaly, waterproof skin that helped prevent the loss of body water to the dry air. (4) Reptiles evolved improved lungs that were able to provide the entire oxygen supply for an active animal. As the climate dried during the Permian period, reptiles became the dominant land vertebrates, relegating amphibians to swampy backwaters, where most remain today.

A few tens of millions of years later, the climate returned to more moist and equable conditions, providing for lush plant growth. Once again, gigantic size was selected for, as certain families of reptiles evolved into the dinosaurs. The variety of dinosaur forms was enormous—from carnivores (Fig. 17-9) to herbivores; from those that dominated the land, to others that took to the air, to still others that returned to the sea. Dinosaurs were among the most successful animals ever, if we consider persistence as a measure of success. They flourished for more than a hundred million years, until about 65 million years ago, when the last dinosaurs went extinct. No one is certain why they died out, but a gigantic meteorite impact seems to have been the final blow (see below).

Even during the age of dinosaurs, many reptiles remained quite small. One major difficulty faced by small reptiles is the maintenance of a high body temperature. Being active on land seems to require a rather warm body to maximize the efficiency of the nervous system and muscles. But a warm body loses heat to the environment unless the air is also warm. Heat loss is a bigger problem for smaller animals, which have a larger surface area per unit of weight than do larger animals. Many species of small reptiles have retained slow metabolisms and have coped with the heat-loss problem by developing lifestyles in which they remain active only when the air is sufficiently warm. Two groups of small reptiles, however, independently followed a different evolutionary pathway: they developed insulation. One group evolved feathers, and another evolved hair.

Reptiles Gave Rise to Both Birds and Mammals

In the ancestral birds, insulating feathers helped retain body heat. Consequently, these animals could be active in cool habitats and during the night, when their scaly relatives became sluggish. Later, some ancestral birds evolved longer, stronger feathers on their forelimbs, perhaps under selection for better ability to glide from trees or to jump after insect prey. Ultimately, feathers evolved into structures capable of supporting powered flight. Fully developed, flight-capable feathers are present in 150-million-year-old fossils, so the earlier insulating structures that eventually developed into flight feathers must have been present well before that time.

The earliest mammals coexisted with the dinosaurs but were small creatures, probably living in trees and being active mostly at night. When the dinosaurs went extinct, surviving mammals colonized the habitats left

Figure 17-9 A reconstruction of a Cretaceous forest
By the Cretaceous era, flowering plants dominated terrestrial vegetation. Dinosaurs, such as the predatory pack of 6-foot-long *Velociraptors* shown here, were the preeminent land animals. Although small by dinosaur standards, *Velociraptor* was a formidable predator with great running speed, sharp teeth, and deadly, sickle-like claws on its hind feet.

empty by the extinction. Mammal species prospered, diversifying into the array of modern forms.

Unlike birds, which retained the reptilian habit of laying eggs, mammals evolved live birth and the ability to feed their young with secretions of the mammary (milk-producing) glands. Ancestral mammals also developed hair, which provided insulation. Because these structures do not fossilize, we may never know when the uterus, mammary glands, and hair first appeared, or what their intermediate forms looked like. Recently, however, a team of paleontologists found bits of fossil hair preserved in *coprolites*, which are fossilized animal feces. These coprolites, found in the Gobi Desert of China, were deposited by an anonymous predator 55 million years ago, so mammals have presumably had hair at least that long.

5 What Role Has Extinction Played in the History of Life?

If there is a moral to the great tale of life's history, it is that nothing lasts forever. The story of life can be read as long series of evolutionary dynasties, with each new dominant group rising, ruling the land or the seas for a period of time and, inevitably, falling into decline and extinction. Dinosaurs are the most famous of these collapsed dynasties, but the list of extinct groups known only from fossils is impressively long. Despite the inevitability of extinction, however, the overall trend has been for species to arise at a faster rate than they disappear, so the number of different species on Earth has tended to increase over time.

The Upward Trend in Species Diversity Has Been Interrupted by Periodic Mass Extinctions

Over much of life's history, the process of dynastic succession proceeded in a steady, inexorable manner. This slow and steady turnover of species, however, has been interrupted by episodes of **mass extinction** (Fig. 17-10). These mass extinctions are characterized by the relatively sudden disappearance of a wide variety of species over a large part of Earth. In the most catastrophic episodes of extinction, more than half of the planet's species disappeared. The worst episode of all, which occurred 245 million years ago at the end of the Permian period, wiped out more than 90% of the world's species, and life came perilously close to disappearing altogether.

Mass extinctions have had a profound impact on the course of life's history, repeatedly redrawing the picture of life's diversity. What could have caused such dramatic changes in the fortunes of so many species? Many evolutionary biologists believe that climate changes must have played an important role. When the climate changes, as it has done many times over the course of Earth's history, organisms that were adapted for survival under one set of environmental conditions may be unable to survive under a drastically different set of conditions. In particular, at times when warm climates gave way to drier, colder climates with more variable temperatures, species may have gone extinct after failing to adapt to the harsh new conditions.

One cause of climate change is *plate tectonics*. Earth's surface is divided into portions called *plates*, which include the continents and the seafloor (Fig. 17-11). The solid plates slowly move above a viscous

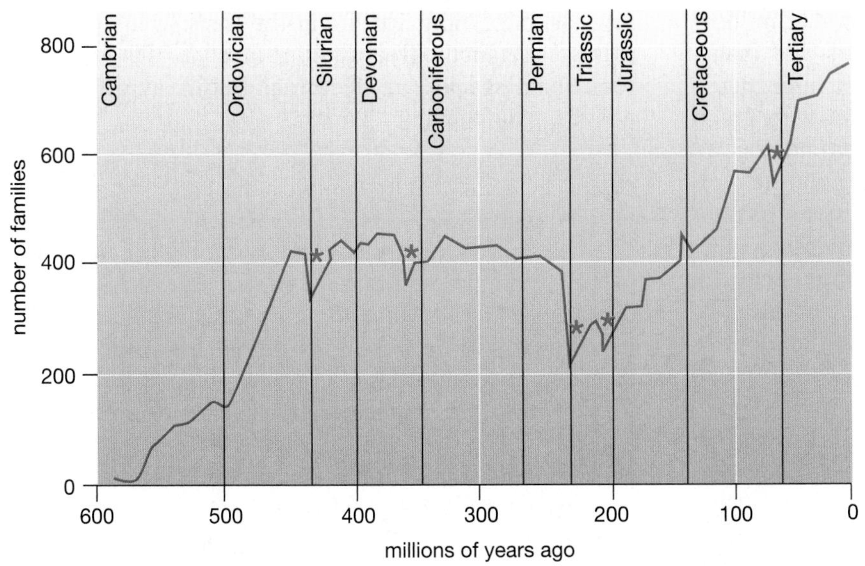

Figure 17-10 Episodes of mass extinction This graph plots diversity of marine animals over time, as reconstructed from the fossil record. Note the general trend toward increasing diversity, punctuated by periods of sometimes rapid extinction. Five of these declines in diversity, marked by asterisks, are so steep that they qualify as catastrophic mass extinctions.

but fluid layer. As the plates wander, their positions may change in latitude. For example, 350 million years ago much of North America was located at or near the equator, an area characterized by consistently warm and wet tropical weather. But plate tectonics carried the continent up into temperate and arctic regions. As a result, the tropical climate was replaced by a regime of seasonal changes, cooler temperatures, and less rainfall.

The geological record indicates that most mass extinction events coincided with periods of climatic change. To many scientists, however, the rapidity of mass extinctions suggests that the slow process of climate change could not, by itself, be responsible for such large-scale disappearances of species. Perhaps more sudden events must also play a role. For example, catastrophic geological events, such as massive volcanic eruptions, could have had a devastating effect. Geologists have found evidence of past volcanic eruptions so huge that they make the 1980 Mount St. Helens explosion look like a firecracker by comparison. Even such gigantic eruptions, however, would directly affect only a relatively small portion of Earth's surface.

The search for the causes of mass extinctions took a fascinating turn in the early 1980s when Luis and Walter

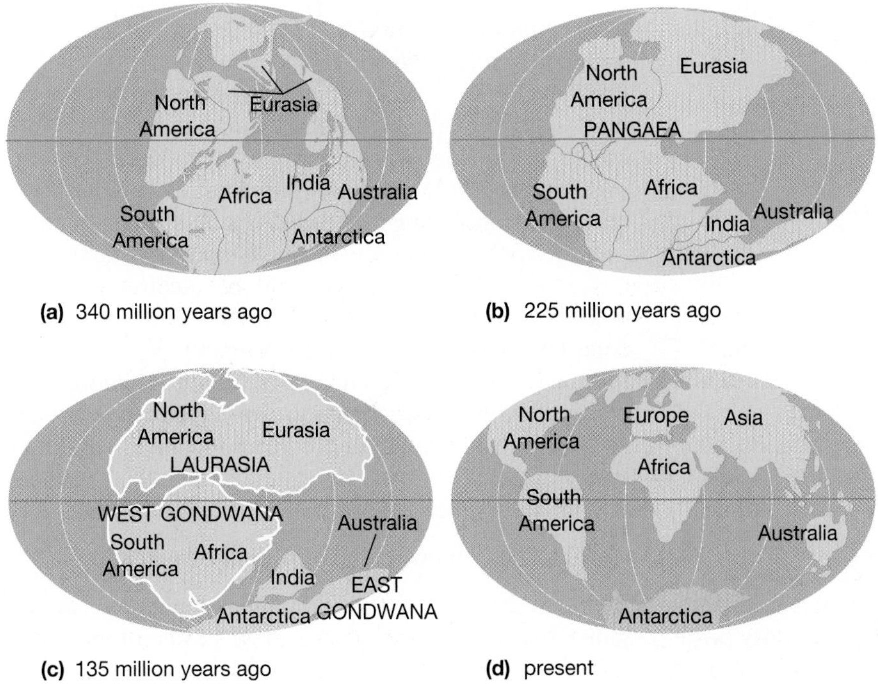

Figure 17-11 Plate tectonics is tied to climate change
Although slow, the drifting of continents—passengers on plates moving on Earth's surface as a result of plate tectonics—can cause tremendous environmental changes. The solid surfaces of the continents slide about over a viscous, but fluid, layer. ***(a)*** About 340 million years ago, much of what is now North America was positioned at the equator. ***(b)*** All the plates eventually fused together into one gigantic landmass, which geologists call Pangaea. ***(c)*** Gradually Pangaea broke up into Laurasia and Gondwanaland, which itself eventually broke up into West and East Gondwana. ***(d)*** Further plate motion eventually resulted in the modern positions of the continents. Plate tectonics continues today; the Atlantic Ocean, for example, widens by a few centimeters each year.

MEDIATUTOR 17.2 Plate Tectonics

Alvarez proposed that the extinction event of 65 million years ago, which wiped out the dinosaurs and many other species, was caused by the impact of a huge meteorite. The Alvarez brothers' idea was met with great skepticism when it was first introduced, but geological research since that time has generated a great deal of evidence that a massive impact did indeed occur 65 million years ago. In fact, researchers have identified the Chicxulub crater, a 100-mile-wide crater buried beneath the Yucatan Peninsula of Mexico, as the impact site of a giant meteorite, 10 miles in diameter, that collided with Earth just at the time that dinosaurs disappear from the fossil record.

Could this immense meteorite strike have caused the mass extinction that coincided with it? No one knows for sure, but scientists suggest that such a massive impact would have thrown so much debris into the atmosphere that the entire planet would have been plunged into darkness for a period of years. With little light reaching the planet, temperatures would have dropped precipitously and the photosynthetic capture of energy (upon which all life ultimately depends) would have declined drastically. The worldwide "impact winter" would have spelled doom for the dinosaurs and a host of other species.

6 How Did Humans Evolve?

Humans are intensely interested in their own origin and evolution, especially in trying to determine what conditions led to the evolution of the gigantic human brain. This obsession has triggered sweeping speculations despite an often skimpy fossil record. Therefore, although the outline of human evolution that we will present is a synthesis of current thought on the subject, it is by no means as well understood as, say, the genetic code. Paleontologists disagree about the interpretation of the fossil evidence, and many ideas may have to be revised as new fossils are found.

Primate Evolution Has Been Linked to Grasping Hands, Binocular Vision, and a Large Brain

Fossils of **primates** (lemurs, monkeys, apes, and humans) are relatively rare compared to those of many other animals. The primate fossil record is nonetheless sufficient to show that the most-likely primate ancestors were insect-eating tree shrews, whose fossils are found in rocks about 80 million years old. Nimble, probably nocturnal animals, tree shrews were smaller than all but the tiniest modern primates.

Over the next 50 million years, the descendants of the tree shrews evolved forms that are similar to modern tarsiers, lemurs, and monkeys (Fig. 17-12). Early primates, which probably fed on fruits and leaves, evolved several adaptations for life in the trees. Many modern primates retain the tree-dwelling lifestyle of their ancestors. Some species are nocturnal, but many others are active during the day.

Grasping Hands in Early Primates Allowed Both Powerful and Precise Manipulations

The tree shrews possessed handlike paws for holding on to small branches, and the primates further refined these appendages. Most primates are much larger than tree shrews, so only relatively large tree limbs could bear their weight. Long, grasping fingers that could wrap around and hold larger limbs made life in the trees a little safer. The primate line that led to humans evolved hands that could perform both delicate maneuvers, using a *precision grip* (used by modern humans for manipulating small objects, writing, and sewing), and powerful actions, using a *power grip* (swinging a club, thrusting with a spear).

Binocular Vision Provides Accurate Depth Perception

One of the earliest primate adaptations seems to have been large, forward-facing eyes (see Fig. 17-12). Jumping from branch to branch is risky business unless an animal can accurately judge where the next branch is located. Accurate depth perception was made possible by binocular vision, provided by forward-facing eyes with overlapping fields of view. Another key adaptation was color vision. We cannot, of course, tell if a fossil animal had color vision, but modern primates have excellent color vision, and it seems reasonable to assume that earlier primates did too. Many primates feed on fruit, and color vision helps in detecting ripe fruit among a bounty of green leaves.

A Large Brain Facilitated Hand-Eye Coordination and Complex Social Interactions

Primates have brains that are larger, relative to their body size, than almost all other animals. No one really knows for certain what environmental forces favored the evolution of large brains. It seems reasonable, however, that controlling and coordinating binocular, color vision; rapid locomotion through trees; and dexterous movements of the hands would be facilitated by increased brain power. Most primates also have fairly complex social systems, which probably would require relatively high intelligence. If sociality promoted increased survival and reproduction, then there would have been environmental pressures for the evolution of larger brains.

Hominids Evolved from Dryopithecine Primates

Between 20 million and 30 million years ago, in the moist tropical forests of Africa, a group of primates called the *dryopithecines* diverged from the monkey

(a) (b) (c)

Figure 17-12 Representative primates
(a) Tarsier, ***(b)*** lemur, and ***(c)*** lion-tail macaque monkey. Note that all have relatively flat faces, with forward-looking eyes providing binocular vision. All also have color vision and grasping hands. These features served as preadaptations for tool and weapon use by early humans.

line. The dryopithecines appear to have been ancestral to the **hominids** (humans and their fossil relatives) and *pongids* (the great apes). About 18 million years ago, global climatic cooling began to shrink the vast expanses of forest, splitting up the woodlands into isolated islands dotted on a sea of grasslands. Diversification of habitat and isolation of small populations led to the diversification of dryopithecines. One of these primate groups gave rise to the later hominids and pongids.

The Earliest Hominids Could Stand and Walk Upright

The hominid line is believed to have diverged from the ape lineage sometime between 5 million and 8 million years ago, but direct evidence of that transition is absent from the fossil record. The first trace of our hominid ancestors appears in 4.4-million-year-old rocks in Ethiopia, where excavations have yielded fragments of *Ardipithecus ramidus*. *Ardipithecus* was clearly a hominid, sharing several anatomical features with better-known, later members of the group. But this oldest known member of our family also exhibits other features (such as thin tooth enamel and thick arm bones) that are more characteristic of apes, so *Ardipithecus* may represent a point on our family tree that is fairly near the split from the apes.

Ardipithecus is known from only a few bones and teeth. A more extensive record of early hominid evolution does not really begin until about 4 million years ago. That date marks the beginning of the fossil record of the genus *Australopithecus* ("southern ape," after their original discovery site in southern Africa; Fig. 17-13). The brains of *australopithecines* (as the various species of *Australopithecus* are collectively known) were fairly large but still much smaller than those of modern humans. Although the earliest australopithecines had legs that were shorter, for their height, than those of modern humans, their knee joints allowed them to straighten their legs fully, permitting efficient bipedal (upright, two-legged) locomotion. Footprints almost 4 million years old, discovered in Tanzania by anthropologist Mary Leakey, show that even the earliest australopithecines could, and at least sometimes did, walk upright.

Upright posture is an extremely important feature of hominids, because it freed their hands from use in walking. Later hominids thus were able to carry weapons, manipulate tools, and eventually achieve the cultural revolutions produced by modern *Homo sapiens*. What environmental pressures led to the evolution of bipedalism? The short answer is that no one knows, but British physiologist Peter Wheeler has an intriguing hypothesis: Our ancestors developed the ability to walk upright because this posture exposes a minimum body surface to the blazing savanna sun. Wheeler constructed a small-scale model of an ancient australopithecine and photographed it in bipedal and quadrupedal (four-legged)

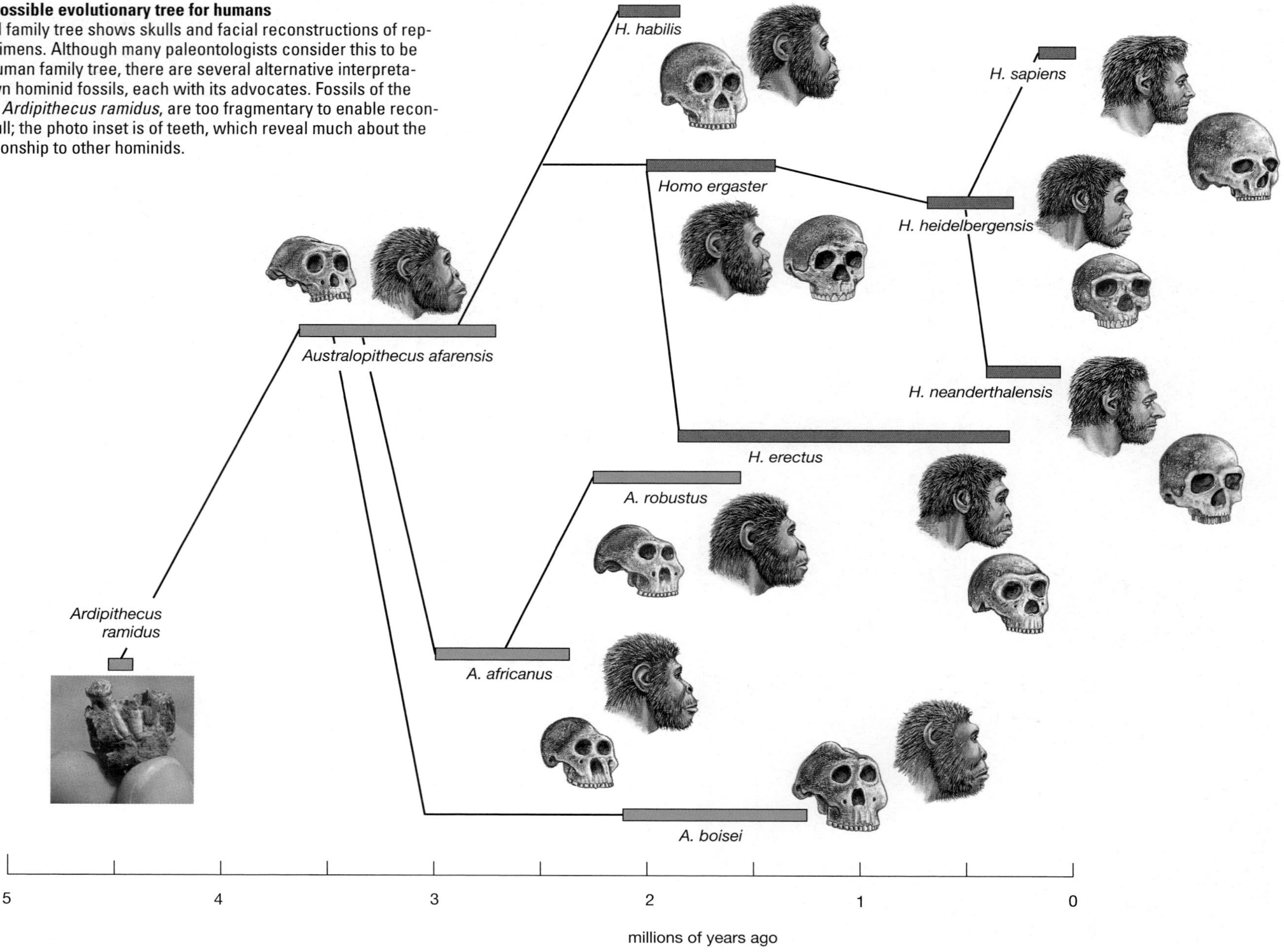

Figure 17-13 A possible evolutionary tree for humans
This hypothetical family tree shows skulls and facial reconstructions of representative specimens. Although many paleontologists consider this to be the most likely human family tree, there are several alternative interpretations of the known hominid fossils, each with its advocates. Fossils of the earliest hominid, *Ardipithecus ramidus*, are too fragmentary to enable reconstruction of a skull; the photo inset is of teeth, which reveal much about the specimen's relationship to other hominids.

postures from a series of overhead locations that imitated the pathway of the sun. He found that a four-legged posture exposed the model to 60% more solar radiation than did the upright posture. Further, says Wheeler, standing upright exposes more of the body to cooler breezes a few feet above the ground. Together, these factors would increase the time the animal could spend foraging in the sun, allowing it to increase its access to food.

The oldest australopithecine species, represented by fossilized teeth, skull fragments, and arm bones, was unearthed near an ancient lake bed in Kenya from sediments that were dated, by means of radioactive isotopes, as between 3.9 million and 4.1 million years old (see "Scientific Inquiry: How Do We Know How Old a Fossil Is?"). It was named *Australopithecus anamensis* by its discoverers (*anam* means "lake" in the local Ethiopian language). The second most ancient australopithecine, called *Australopithecus afarensis*, was discovered in the Afar region of Ethiopia in 1974. Fossil remains of this species as much as 3.9 million years old have been unearthed. Later, the *A. afarensis* line apparently gave rise to at least two distinct forms: the small, omnivorous *A. africanus* (which in size and eating habits was similar to *A. afarensis*), and the large, herbivorous *A. robustus* and *A. boisei*. All of the australopithecine species had apparently gone extinct by 1.2 million years ago, but one of them (probably either *A. afarensis* or *A. africanus*) first gave rise to a new branch of the hominid family tree, the genus *Homo*.

The Genus Homo *Diverged from the Australopithecines 2.5 Million Years Ago*

Hominids that are sufficiently similar to modern humans to be placed in the genus *Homo* first appear in the fossil record in Africa, beginning about 2.5 million years ago. The earliest African *Homo* fossils are assigned to one of two species, *H. rudolfensis* or *H. habilis* (see Fig. 17-13), whose body and brain were larger than those of the australopithecines but which retained the apelike long arms and short legs of their australopithecine forebearers. In contrast, the skeletal anatomy of *H. ergaster*, a species whose fossils first appear 2 million years ago, has limb proportions more like those of modern humans. This species is believed by many paleoanthropologists (scientists who study human origins) to be on the evolutionary branch that led ultimately to our own species, *H. sapiens*. In this view, *H. ergaster* was the common ancestor of two distinct branches of hominids. The first branch led to *H. erectus*, which was the first hominid species to leave Africa. The brain of the average *H. erectus* was as large as the smallest modern adult human brains (about 1000 cubic centimeters). The face, unlike that of modern humans (see Fig. 17-13), featured large brow ridges above the eyes, a slightly protruding face, and no chin.

The second branch from *H. ergaster* gave rise to *H. heidelbergensis*. After migrating to Europe, *H. heidelbergensis* gave rise to the Neanderthals, *H. neanderthalensis*. Meanwhile, back in Africa, another branch split off from the *H. heidelbergensis* lineage. This branch ultimately became *H. sapiens*, modern humans.

The Evolution of Homo *Was Accompanied by Advances in Tool Technology*

Hominid evolution is closely tied to the development of tools, a hallmark of hominid behavior. The oldest tools discovered so far were found in 2.5-million-year-old East African rocks, concurrent with the early emergence of the genus *Homo*. Early *Homo*, whose cheek teeth were much smaller than those of its australopithecine ancestors, might have first used stone tools to break and crush tough foods that were hard to chew. Hominids constructed their earliest tools by striking one rock with another to chip off fragments and leave a sharp edge behind. Over the next several hundred thousand years, tool-making techniques in Africa gradually became more advanced. By 1.7 million years ago, tools were more sophisticated, with flakes chipped symmetrically from both sides of a rock to form "biface" tools ranging from hand axes, used for cutting and chopping, to points, probably used on spears (Fig. 17-14a,b). *Homo ergaster* and other bearers of these weapons presumably ate meat, probably both hunting and scavenging for the remains of prey killed by other predators. Biface tools were carried to Europe at least 600,000 years ago by migrating populations of *H. heidelbergensis*, and the Neanderthal descendants of these emigrants took stone tool construction to new heights of skill and delicacy (Fig. 17-14c).

Neanderthals Had Large Brains and Ritualistic Behaviors

Neanderthals first appear in the European fossil record about 150,000 years ago; by about 70,000 years ago, they had spread throughout Europe and western Asia. By 30,000 years ago, however, Neanderthals were extinct. Contrary to the popular image of a hulking, stoop-shouldered "cave man," Neanderthals were quite similar to modern humans in many ways. Although more heavily muscled, Neanderthals walked fully erect, were dexterous enough to manufacture finely crafted stone tools, and had brains, on the average, slightly larger than those of modern humans. Although many of the European fossils show heavy brow ridges and a broad, flat skull, others, particularly from areas around the eastern shores of the Mediterranean Sea, were remarkably like ourselves. Neanderthal remains also show evidence of modern behaviors, particularly ritualistic burial ceremonies (Fig. 17-15). Neanderthal skeletons have been discovered in clearly marked burial sites that were surrounded with stones and typically included offerings of

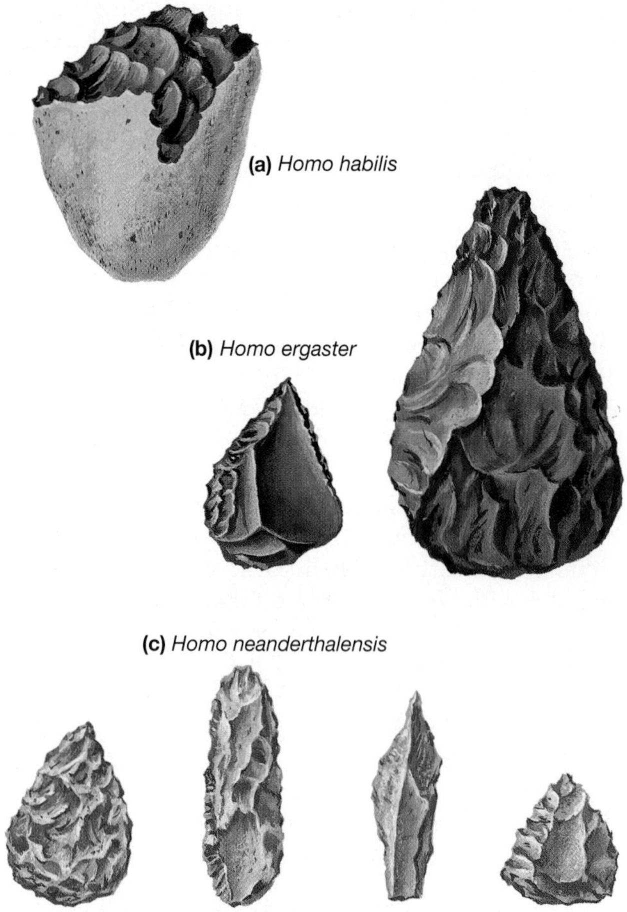

Figure 17-14 Representative hominid tools
(a) *Homo habilis* produced only fairly crude chopping tools called *hand axes,* usually unchipped on one end to hold in the hand. ***(b)*** *Homo ergaster* manufactured much finer tools. The tools were typically sharp all the way around the stone, so at least some of these blades were probably tied to spears rather than held in the hand. ***(c)*** Neanderthal tools were works of art, with extremely sharp edges made by flaking off tiny bits of stone. In comparing these weapons, note the progressive increase in the number of flakes taken off the blades and the corresponding decrease in flake size. Smaller, more numerous flakes produce a sharper blade and suggest more insight into tool making (perhaps passed down culturally from experienced tool makers to apprentices), more patience, finer control of hand movements, or perhaps all three.

flowers, bear skulls, and food. Altars found at other Neanderthal sites were probably used in "religious" rites associated with a bear cult.

Until recently most anthropologists believed that Neanderthals were simply an archaic variety of *H. sapiens*. It is becoming increasingly clear, however, that *H. neanderthalensis* was a separate species. Dramatic evidence in support of this hypothesis has come from two groups of researchers who were able to isolate and analyze ancient DNA from two different 30,000-year-old Neanderthal skeletons. The researchers determined the nucleotide sequence of a Neanderthal gene and compared it to the same gene from a large number of different modern humans from various parts of the

Figure 17-15 Neanderthal burial
Museum reconstruction of a Neanderthal burial site. Some Neanderthals were buried with elaborate ceremonies, with bear skulls placed on the grave of the deceased.

world. The Neanderthal gene sequence was much different from that of the modern humans and indicated that the evolutionary branch leading to *H. neanderthalensis* diverged from the ancestral human line hundreds of thousands of years prior to the emergence of modern *H. sapiens*.

Modern Humans Emerged Only 150,000 Years Ago

Finally, about 150,000 years ago, anatomically modern humans appear in the African fossil record. The location of these fossils suggests that *Homo sapiens* originated in Africa, but most of our knowledge about our own early history comes from European and Middle Eastern fossil humans collectively known as *Cro-Magnons* (after the district in France in which their remains were first discovered). Cro-Magnons, which appear in the fossil record beginning about 90,000 years ago, had domed heads, smooth brows, and prominent chins (just like us!). Their tools were precision instruments not very different from the stone tools used until recently in many parts of the world. Behaviorally, Cro-Magnons seem to have been similar to, but more sophisticated than, Neanderthals. Perhaps the most remarkable accomplishment of Cro-Magnons is the magnificent art left in caves such as Altamira in Spain and Lascaux in France (Fig. 17-16). No one knows exactly why these paintings were made, but they attest to minds fully as human as our own.

Cro-Magnons coexisted with Neanderthals in Europe and the Middle East for perhaps up to 50,000 years before the Neanderthals disappeared. Some researchers believe that Cro-Magnons interbred extensively with Neanderthals, so Neanderthals were essentially absorbed into the human genetic mainstream. Other scientists disagree, citing mounting evidence such as the fossil DNA described earlier, and suggest that later-arriving Cro-Magnons simply overran and displaced the less-well-adapted Neanderthals. Neither hypothesis, however, does a good job of explaining how the two kinds of hominids managed to occupy the same geographical areas for such a long time. The persistence in one area of two similar but distinct groups for tens of thousands of years seems inconsistent with both interbreeding *and* direct competition. Perhaps *H. neanderthalensis* and *H. sapiens* were simply different biological species, unable to interbreed successfully but able to coexist in the same habitat.

Africa is where the human family tree has its roots, but hominids found their way out of Africa on numerous occasions. For example, *H. erectus* reached tropical Asia almost 2 million years ago, apparently thrived there, and eventually spread across Asia. Similarly, *H. heidelbergensis* made it to Europe at least 780,000 years ago. It is increasingly clear that the genus *Homo* made repeated long-distance emigrations, beginning just as soon as sufficiently capable limb anatomy evolved. What is less clear is how all this wandering is related to the origin of modern *H. sapiens*. According to one hypothesis (the basis of the scenario outlined earlier), *H. sapiens* emerged in Africa and dispersed less than 150,000 years ago, spreading into the Near East, Europe, and Asia and supplanting all other hominids (Fig. 17-17a). But some paleoanthropologists believe that populations of *H. sapiens* evolved in many regions simultaneously from the already widespread populations of *H. erectus*. According to this hypothesis, continued migrations and interbreeding among *H. erectus* populations maintained them as a single species as they gradually evolved into *H. sapiens* (Fig. 17-17b). A steadily increasing number of comparative molecular studies of modern humans support the "single African origin" model of the genesis of our species, but both points of view are consistent with the fossil record, and the question remains unsettled.

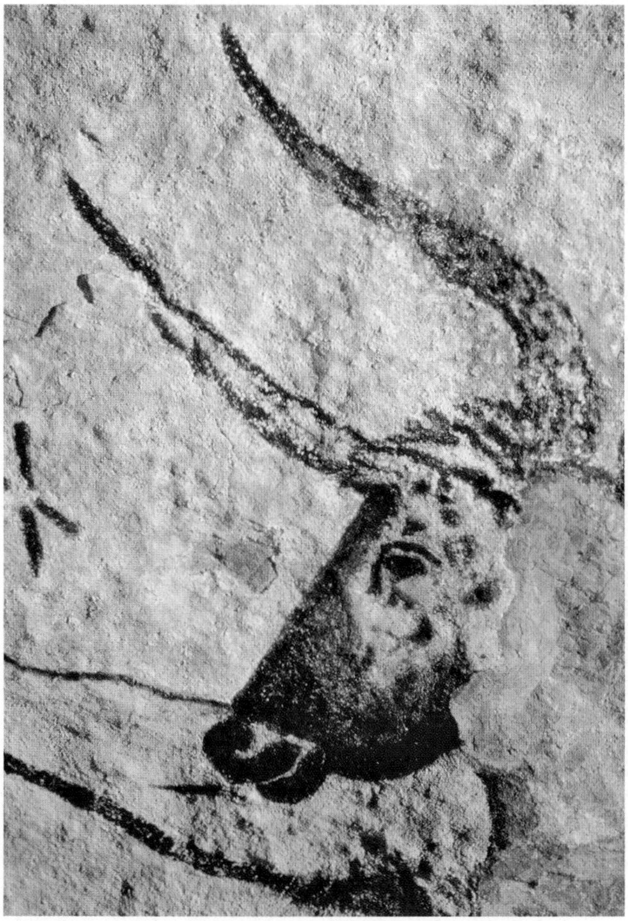

Figure 17-16 The sophistication of Cro-Magnon people
Cave paintings by Cro-Magnon people remarkably preserved by the relatively constant underground conditions of a cave in Lascaux, France.

The Evolution of Human Behavior Is Highly Speculative

Perhaps the most hotly debated subject in human evolution is the development of human behavior. Except in rare instances, such as the Neanderthal burials and the Cro-Magnon cave paintings, there is no direct evidence of the behavior of prehistoric hominids. Biologists and anthropologists have offered a few hypotheses in accordance with the fossil record and the behavior of modern humans and other animals.

The fossil record shows that the development of truly huge brains occurred within the last couple of million years. What were the selective advantages of large brains? This subject continues to stir controversy among paleoanthropologists. The enlarging brain may have been selected both because it provided for improved hand-eye coordination and because it facilitated complex social interactions.

As the early hominids such as *Australopithecus* descended from the trees into the savanna, they began to walk upright. Bipedal locomotion allowed them to carry things in their hands as they walked. Hominid fossils reveal shoulder joints capable of powerful throwing motions and an *opposable thumb* (a thumb that can be placed directly against each of the other fingers) that would enhance the ability to manipulate objects. These early hominids probably could see well and judge depth accurately with binocular vision. The brain expansion that occurred in the australopithecines may have been at least partly related to integration of visual input and control of hand and arm movements.

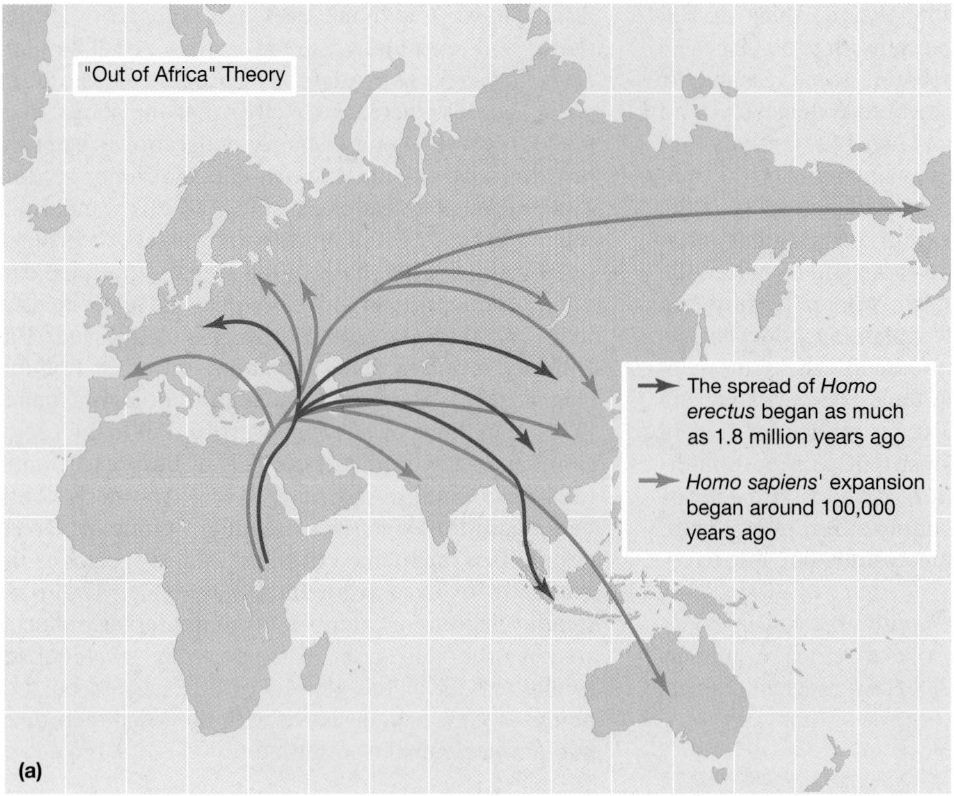

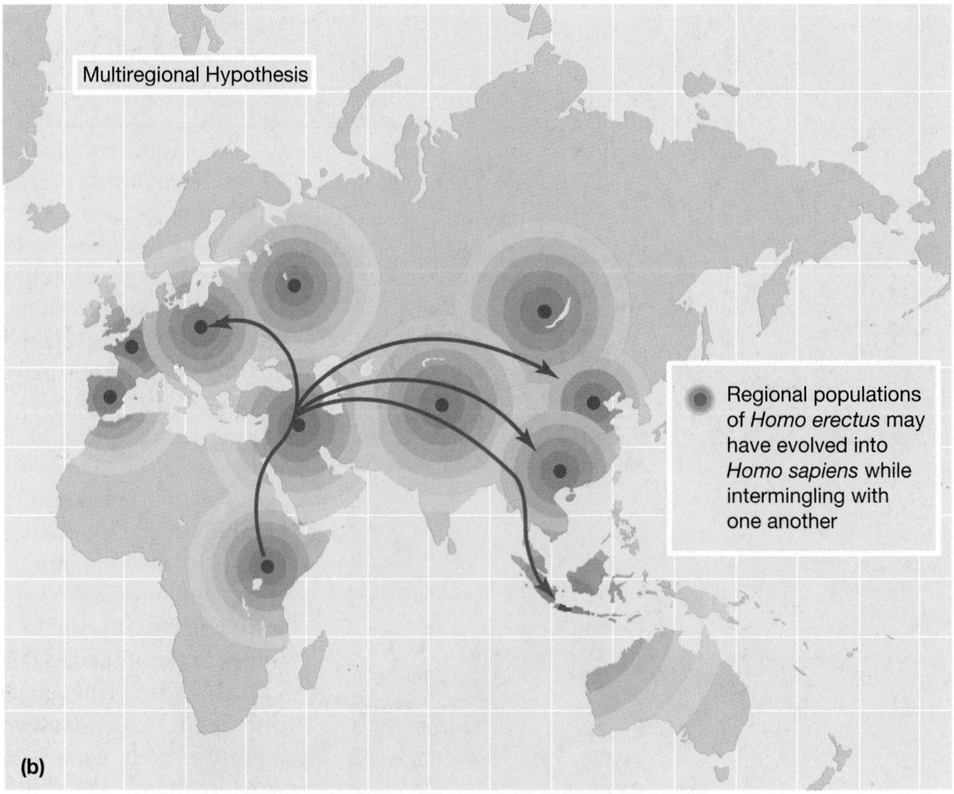

Figure 17-17 Competing hypotheses for the evolution of *Homo sapiens*
(a) The "out of Africa" advocates hypothesize that *H. sapiens* evolved in Africa, then migrated throughout the Near East, Europe, and Asia. ***(b)*** The "multiregional" hypothesis states that populations of *H. sapiens* evolved in many regions simultaneously from the already widespread populations of *H. erectus*.

Homo erectus, *H. ergaster*, and perhaps *H. habilis* and *H. rudolfensis* before them, were social. So, of course, are many monkeys and apes. The later hominids, however, seem to have engaged in a new type of social activity: cooperative scavenging and hunting. *Homo erectus*, in particular, seems to have hunted extremely large game, which calls for cooperation in all phases of the hunt. Some paleoanthropologists believe that selection favored larger and more-powerful brains as an adaptation for success in cooperative hunting. (Lest you become too impressed with such speculation, remember that lions and wolves also hunt cooperatively yet are not noticeably more intelligent than their relatives, such as leopards and coyotes, who do not.)

The Cultural Evolution of Humans Now Far Outpaces Biological Evolution

In recent millennia, human evolution has come to be dominated by **cultural evolution**, the transmission of learned behaviors from generation to generation. Our recent evolutionary success, for example, was engendered not so much by new physical adaptations as by a series of cultural and technological revolutions. The first such revolution was the development of tools, which began with the early hominids. Tools increased the efficiency with which food and shelter could be acquired and thus increased the number of individuals that could survive within a given ecosystem. About 10,000 years ago, human culture underwent a second revolution when people discovered how to grow crops and domesticate animals. This agricultural revolution dramatically increased the amount of food that could be extracted from the environment, and the human population surged, increasing from about 5 million at the dawn of agriculture to around 750 million by 1750. The subsequent industrial revolution gave rise to the modern economy and its attendant improvements in public health. Longer lives and lower infant mortality led to truly explosive population growth, and today we are more than 6 billion strong.

Human cultural evolution and the accompanying increases in human population have had profound effects on the continuing biological evolution of other life forms. Our agile hands and minds have transformed much of Earth's terrestrial and aquatic habitats. Humans have become the single overwhelming agent of natural selection. In the words of evolutionary biologist Stephen Jay Gould of Harvard University, "We have become, by the power of a glorious evolutionary accident called intelligence, the stewards of life's continuity on Earth. We did not ask for this role, but we cannot abjure it. We may not be suited for it, but here we are."

CASE STUDY REVISITED

Life on a Frozen Moon?

Is there really liquid water on Europa? Calculations that estimate the amount of heat produced by the gravitational push and pull exerted on Europa by Jupiter's other moons indicate that the moon's interior should be warm enough to sustain a liquid ocean. A more definitive answer, however, appears to be within reach. In 2003 (according to current plans) the *Europa Orbiter* spacecraft will be launched from Earth and, after a few years of travel to the vicinity of Jupiter, will enter into orbit around Europa. Its mission will include subsurface radar measurements that should definitively determine whether or not liquid is present.

If liquid water *is* present on Europa, has life arisen in it? If so what form has European life assumed? Answers to these questions do not appear to be close at hand, but some tantalizing clues might be found right here on Earth. The potential source of such clues is Lake Vostok, a huge lake (similar in size to Lake Ontario) that is buried more than two miles beneath the Antarctic ice pack. The cold waters of Lake Vostok, under extremely high pressure from the immense weight of the ice above it and isolated from Earth's atmosphere for perhaps hundreds of thousands of years, represent an environment analogous to that which might be found beneath Europa's ice sheet.

Biologists would very much like to know if Lake Vostok harbors any living organisms and, if so, how they survive. If, for example, bacteria are alive in the lake, what is their source of energy? How do they cope with a cold environment? Answers to questions such as these could help set the stage for investigation of Europa's subsurface seas. Recent drilling to within a few hundred feet of Lake Vostok's surface revealed some preserved bacteria (apparently not still alive), raising hopes that the lake itself will harbor life. Scientists have held off of drilling all the way to the lake for fear of contaminating the lake with microbes from the surface or of inadvertently releasing unknown organisms from the lake into the surface environment. Efforts are under way to devise methods to explore the lake without risking cross-contamination. Once such methods are devised, they could be adapted for use in future exploration of similar environments on Europa.

In addition to water, what other conditions would be necessary for life to have arisen on Europa? If we were somehow able to explore the entire universe, would we find that life is common, rare, or unique to Earth?

Summary of Key Concepts

1) How Did Life Begin?

Before life arose, lightning, ultraviolet light, and heat formed organic molecules from water and the components of primordial Earth's atmosphere. These molecules probably included nucleic acids, amino acids, short proteins, and lipids. By chance, some molecules of RNA may have had enzymatic properties, catalyzing the assembly of copies of themselves from nucleotides in Earth's waters. These may have been the forerunners of life. Protein-lipid microspheres enclosing these RNA molecules may have formed the first cell-like structures, protocells.

2) What Were the Earliest Organisms Like?

The first fossil cells are found in rocks about 3.5 billion years old. These cells were prokaryotes that fed by absorbing organic molecules that had been synthesized in the environment. Because there was no free oxygen in the atmosphere at that time, energy metabolism must have been anaerobic. As the cells multiplied, they depleted the organic molecules that had been formed by prebiotic synthesis. Some cells developed the ability to synthesize their own food molecules by using simple inorganic molecules and the energy of sunlight. These earliest photosynthetic cells were probably ancestors of today's cyanobacteria.

Photosynthesis releases oxygen as a by-product, and by about 2.2 billion years ago significant amounts of free oxygen had accumulated in the atmosphere. Aerobic metabolism, which generates more cellular energy than does anaerobic metabolism, probably arose about this time.

Eukaryotic cells evolved by about 1.7 billion years ago. The first eukaryotic cells probably arose as symbiotic associations between predatory prokaryotic cells and bacteria. Mitochondria may have evolved from aerobic bacteria engulfed by predatory cells. Similarly, chloroplasts may have evolved from photosynthetic cyanobacteria.

3) How Did Multicellularity Arise?

Multicellular organisms evolved from eukaryotic cells, first appearing about 1 billion years ago. Multicellularity offers several advantages, including increased speed of locomotion and greater size. The first multicellular organisms arose in the sea. In plants, increased size due to multicellularity offered some protection from predation. Specialization of cells allowed plants to anchor themselves in the nutrient-rich, well-lit waters of the shore. For animals, multicellularity allowed more-efficient predation and more-effective escape from predators. These in turn provided environmental pressures for faster locomotion, improved senses, and greater intelligence.

4) How Did Life Invade the Land?

The first land organisms were probably algae. The first multicellular land plants appeared about 400 million years ago. Although the land required special adaptations for support of the body, reproduction, and the acquisition, distribution, and retention of water, the land also offered abundant sunlight and protection from aquatic herbivores. Soon after land plants evolved, arthropods invaded the land. Absence of predators and abundant land plants for food probably facilitated the invasion of the land by animals.

The earliest land vertebrates evolved from lobefin fishes, which had leglike fins and a primitive lung. A group of lobefins evolved into the amphibians about 350 million years ago. Reptiles evolved from amphibians, with several further adaptations for life on land: internal fertilization, waterproof eggs that could be laid on land, waterproof skin, and better lungs. Birds and mammals evolved independently from separate groups of reptiles. Major advances included a high, constant body temperature and insulation over the body surface.

5) What Role Has Extinction Played in the History of Life?

The history of life has been characterized by constant turnover of species as species go extinct and are replaced by new ones. Mass extinctions, in which large numbers of species disappear within a relatively short time, have occurred periodically. Mass extinctions were probably caused by some combination of climate change and catastrophic events, such as volcanic eruptions and meteor impacts.

6) How Did Humans Evolve?

One group of mammals evolved into the tree-dwelling primates. Primates show several preadaptations for human evolution: forward-facing eyes for binocular vision, color vision, grasping hands, and relatively large brains. Between 20 million and 30 million years ago, some primates descended from the trees; these were the ancestors of apes and humans. The australopithecines arose in Africa about 4 million years ago. These hominids walked erect, had larger brains than did their forebearers, and fashioned primitive tools. One group of australopithecines gave rise to a line of hominids in the genus *Homo*, which in turn gave rise to modern humans.

Key Terms

conifer *p. 333*
cultural evolution *p. 345*
endosymbiont hypothesis *p. 330*
exoskeleton *p. 334*
hominid *p. 339*
mass extinction *p. 336*
microsphere *p. 326*
preadaptation *p. 334*
prebiotic evolution *p. 324*
primate *p. 338*
protocell *p. 326*
ribozyme *p. 326*
spontaneous generation *p. 324*

Thinking Through the Concepts

Multiple Choice

1. *There was no free oxygen in the early atmosphere because most of it was tied up in*
   a. water
   b. ammonia
   c. methane
   d. rock
   e. radioactive isotopes
2. *RNA became a candidate for the first information-carrying molecule when Tom Cech and Sidney Altman discovered that some RNA molecules can act as enzymes that*
   a. degrade proteins
   b. turn light into chemical energy
   c. split water and release oxygen gas
   d. synthesize copies of themselves
   e. synthesize amino acids
3. *The earliest living organisms were*
   a. multicellular
   b. eukaryotes
   c. prokaryotes
   d. photosynthesizers
   e. aerobic
4. *Which three of the following observations support Lynn Margulis's endosymbiont hypothesis for the origin of chloroplasts and mitochondria from ingested bacteria?*
   a. Aerobic respiration takes place in mitochondria.
   b. Mitochondria have their own DNA.
   c. Photosynthesis takes place in chloroplasts.
   d. Chloroplasts have their own DNA.
   e. Bacterial plasma membranes are strikingly similar to the inner membrane of mitochondria.
5. *The exoskeleton of early, marine-dwelling arthropods can be considered a preadaptation for life on land because that shell*
   a. can support an animal's weight against the pull of gravity
   b. allows a wide diversity of body types
   c. resists drying
   d. absorbs light
   e. both a and c
6. *The evolution of the shelled, waterproof egg was an important event in vertebrate evolution because it*
   a. led to the Cambrian explosion
   b. was the first example of parents caring for their young
   c. allowed the colonization of freshwater environments
   d. freed organisms from having to lay their eggs in water
   e. allowed internal fertilization of eggs

? Review Questions

1. What is the evidence that life might have originated from nonliving matter on primordial Earth? What kind of evidence would you like to see before you would accept this hypothesis?
2. If they were so much more efficient at producing energy, why didn't the first cells with aerobic metabolism extinguish cells with only anaerobic metabolism?
3. Explain the endosymbiont hypothesis for the origin of chloroplasts and mitochondria.
4. Name two advantages of multicellularity in both plants and animals.
5. What advantages and disadvantages would terrestrial existence have had for the first plants to invade the land? For the first land animals?
6. Outline the major adaptations that emerged during the evolution of vertebrates, from fish to amphibians to reptiles to birds and mammals. Explain how these adaptations increased the fitness of the various groups for life on land.
7. Outline the evolution of humans from early primates. Include in your discussion such features as binocular vision, grasping hands, bipedal locomotion, social living, tool making, and brain expansion.

Applying the Concepts

1. What is cultural evolution? Is cultural evolution more or less rapid than biological evolution? Why?
2. Do you think that studying our ancestors can shed light on the behavior of modern humans? Why or why not?
3. A biologist would probably answer the age-old question "What is life?" by saying "the ability to self-replicate." Do you agree with this definition? If so, why? If not, how would you define life in biological terms?
4. Traditional definitions of humans have emphasized "the uniqueness of humans" because we possess language and use tools. But most animals can communicate with other individuals in sophisticated ways, and many vertebrates use tools to accomplish tasks. Pretend that you are a biologist from Mars, and write a taxonomic description of the species *Homo sapiens*.
5. Extinctions have occurred throughout the history of life on Earth. Why should we care if humans are causing a mass extinction event now?
6. The "out of Africa" and "multiregional" hypotheses of *Homo sapiens'* evolution make contrasting predictions about the extent and nature of genetic divergence among human races. One predicts that races are old and highly diverged genetically; the other predicts that races are young and little diverged genetically. What data would help you determine which hypothesis is closer to the truth?
7. In biological terms, what do you think was the most significant event in the history of life? Explain your answer.

For More Information

de Duve, C. "The Birth of Complex Cells." *Scientific American*, April 1996. Narrative describing the origin of complex eukaryotic cells by repeated instances of endosymbiosis.

Diamond, J. "How to Speak Neanderthal." *Discover*, January 1990. Could the Neanderthals speak? How could we tell? Jared Diamond lucidly explains the arguments and methodology behind a scientific dispute.

Eigen, M. *Steps Towards Life*. New York: Oxford University Press, 1992. A Nobel laureate lays out a scenario for the origin of life and describes supporting experimental evidence. Challenging material, concisely and precisely explained.

Fenchel, T., and Finlay, B. J. "The Evolution of Life Without Oxygen." *American Scientist*, January/February 1994. Clues to the origin of the first eukaryotic cells are provided by symbiotic relationships of organisms in oxygen-free environments.

Fry, I. *The Emergence of Life on Earth: A Historical and Scientific Overview*. Brunswick, NJ: Rutgers University Press, 2000. A thorough review of research and hypotheses concerning the origin of life.

Hay, R. L., and Leakey, M. D. "The Fossil Footprints of Laetoli." *Scientific American*, February 1982. The actual footprints of a hominid family were discovered by Hay and Leakey in volcanic ash 3.5 million years old.

Maynard Smith, J., and Szathmary. E. *The Origins of Life: From the Birth of Life to the Origin of Language*. New York: Oxford University Press, 1999. A thought-provoking review of the major shifts that have occurred over the 3.5 billion year history of life.

Monastersky, R. "The Rise of Life on Earth." *National Geographic*, March 1998. A beautifully illustrated and engaging description of current ideas and evidence of how life arose.

Morell, V. "Announcing the Birth of a Heresy." *Discover*, March 1987. Paleontologists Robert Bakker and Jack Horner speculate that dinosaurs were not the plodding beasts of monster movies but warm-blooded, advanced animals that may even have cared for their young.

Pappalardo, R., Head, J., and Greeley, R. "The Hidden Ocean of Europa." *Scientific American,* October 1999. A summary of the evidence that Europa's icy surface conceals a liquid ocean; illustrated with photos from the *Galileo* mission.

Shreeve, J. "Find of the Century?" *Discover*, January 1997. A brief account of the evidence presented by NASA scientists in support of the proposition that life once existed on Mars.

Tattersall, I. "Out of Africa . . . Again and Again?" *Scientific American*, April 1997. A review of the fossil evidence of human evolutionary history.

Ward, P. D. *The End of Evolution: On Mass Extinctions and the Preservation of Biodiversity.* New York: Bantam Books, 1994. An engaging first-person account of a paleontologist's investigation of the causes of mass extinction.

Answers to Multiple-Choice Questions

1. a 2. b 3. c 4. b, d, e 5. e 6. d

MediaTutor
The History of Life on Earth

CD Activities

Activity 17.1: Endosymbiosis

Estimated time: 5 minutes

Endosymbiosis is thought to have generated one of the most important events in evolutionary history: the origin of eukaryotic organisms. Recent advances have uncovered a great deal of evidence in support of the endosymbiont theory. This animation demonstrates how symbiosis may have given rise to such organelles as mitochondria and chloroplasts.

Activity 17.2: Plate Tectonics

Estimated time: 5 minutes

The idea of continental drift was largely scoffed at when it was first proposed. Now, the theory of plate tectonics forms a major foundation of geophysics. Like the theory of evolution, plate tectonics represents a profound change in our understanding of the world. This animation demonstrates how the continents have moved over the course of time and points out some effects of continental drift on the distributions and evolutionary histories of organisms.

Start the MediaTutor Student CD-ROM and enter the activity number in the Quick Search box to be taken directly to that activity.

Web Investigations

Case Study: Life on a Frozen Moon?

Estimated time: 10 minutes

What are the origins of life? Since water is one of the basic elements of life as we know it, many astrobiologists believe that the best way to detect unknown life in space is to look for liquid water. Recent observations suggest that there may be liquid water on Europa, one of the moons of Jupiter. This exercise will take a closer look at the fourth biggest satellite of Jupiter.

Go to http://www.prenhall.com/audesirk6, the Audesirk Companion Web site. Select Chapter 17 and Web Investigation to begin.

"Systems of classification are not hat racks, objectively presented to us by nature. They are dynamic theories developed by us to express particular views about the history of organisms. Evolution has provided a set of unique species ordered by differing degrees of genealogical relationship. Taxonomy, the search for this natural order, is the fundamental science of history."

Stephen Jay Gould in* Natural History *(1987)

Biologists studying the evolutionary history of type 1 human immunodeficiency virus (HIV-1) discovered that the virus, which causes AIDS, probably originated in chimpanzees.

18 Systematics: Seeking Order Amidst Diversity

AT A GLANCE

Case Study: Origin of a Killer

1) **How Are Organisms Named and Classified?**
Taxonomy Originated as a Hierarchy of Categories
Modern Systematists Use Numerous Criteria for Classification

2) **What Are the Kingdoms of Life?**
The Five-Kingdom System Supplanted Earlier Classification Schemes
A Three-Domain System More Accurately Reflects Life's History
Kingdom-level Classification Remains Unsettled

3) **Why Do Taxonomies Change?**
Species Designations Change When New Information Is Discovered
The Biological Species Definition Can Be Difficult or Impossible to Apply
The Phylogenetic Species Concept Offers an Alternative Definition

4) **Exploring Biodiversity: How Many Species Exist?**

Evolutionary Connections: Are Reptiles for Real?

Case Study Revisited: Origin of a Killer

CASESTUDY

Origin of a Killer

One of the world's most frightening diseases is also one of its most mysterious. Acquired immune deficiency syndrome (AIDS) appeared seemingly out of nowhere, and when it was first recognized in the early 1980s, no one knew what caused it or where it came from. Scientists raced to solve the mystery and, within a few years, had identified human immunodeficiency virus (HIV) as the infectious agent that caused AIDS. Once HIV had been identified, attention turned to the question of its origin.

Finding the source of HIV required an evolutionary approach. To ask "Where did HIV come from?" is really to ask "What kind of virus was the ancestor of HIV?" Biologists who examine questions of ancestry are known as *systematists*. Systematists strive to categorize organisms according to their evolutionary history, to build classifications that accurately reflect the structure of the tree of life. When a systematist concludes that two species are closely related, it means that the two species share a recent common ancestor from which both species evolved.

The systematists who explored the ancestry of HIV discovered that the closest relatives of HIV are found not among other viruses that infect humans, but among those that infect monkeys and apes. In fact, the latest research on HIV's evolutionary history has concluded that the closest relative of HIV-1 (the type of HIV that is most responsible for the worldwide AIDS epidemic) is a virus strain that infects a particular chimp subspecies that inhabits a limited range in West Africa. So the ancestor of the virus that we now know as HIV-1 did not evolve from a preexisting human virus but must have somehow jumped from West African chimpanzees to humans. ■

1 How Are Organisms Named and Classified?

Systematics is the science of reconstructing **phylogeny**, or evolutionary history. A key part of systematics is **taxonomy** (from the Greek word *taxis*, meaning "arrangement"), the science of naming organisms and placing them into categories on the basis of their evolutionary relationships. There are eight major categories: (1) **domain**, (2) **kingdom**, (3) **phylum**, (4) **class**, (5) **order**, (6) **family**, (7) **genus**, and (8) **species**. These taxonomic categories form a nested hierarchy in which each level includes all of the other levels below it. Each domain contains many kingdoms, each kingdom contains many phyla, each phylum includes many classes, each class includes many orders, and so on. As we move down the hierarchy, smaller and smaller groups are included. Each category is increasingly narrow and specifies a group whose common ancestor is increasingly recent. Table 18-1 includes some examples of classifications of specific organisms.

MediaTutor 18.1 Taxonomic Classification

Table 18-1 Classification of Selected Organisms, Reflecting Their Degree of Relatedness*

	Human	Chimpanzee	Wolf	Fruit Fly	Sequoia Tree	Sunflower
Domain	**Eukarya**	**Eukarya**	**Eukarya**	**Eukarya**	**Eukarya**	**Eukarya**
Kingdom	**Animalia**	**Animalia**	**Animalia**	**Animalia**	**Plantae**	**Plantae**
Phylum	**Chordata**	**Chordata**	**Chordata**	Arthropoda	Coniferophyta	Anthophyta
Class	**Mammalia**	**Mammalia**	**Mammalia**	Insecta	Coniferosida	Dicotyledoneae
Order	**Primates**	**Primates**	Carnivora	Diptera	Coniferales	Asterales
Family	Hominidae	Pongidae	Canidae	Drosophilidae	Taxodiaceae	Asteraceae
Genus	*Homo*	*Pan*	*Canis*	*Drosophila*	*Sequoiadendron*	*Helianthus*
Species	*sapiens*	*troglodytes*	*lupus*	*melanogaster*	*giganteum*	*annuus*

*Boldface categories are those that are shared by more than one of the organisms classified. Genus and species names are always italicized or underlined.

The **scientific name** of an organism is formed from the two smallest taxonomic categories, the genus and the species. Each genus includes a group of very closely related species, and each species within a genus includes populations of organisms that can potentially interbreed under natural conditions. Thus, the genus *Sialia* (bluebirds) includes the eastern bluebird (*Sialia sialis*), the western bluebird (*Sialia mexicana*), and the mountain bluebird (*Sialia currucoides*)—very similar birds that normally do not interbreed (Fig. 18-1).

Each two-part scientific name is unique, so referring to an organism by its scientific name rules out any chance of ambiguity or confusion. For example, the bird *Gavia immer* is commonly known in North America as the common loon, in Great Britain as the northern diver, and by a variety of other names in non-English-speaking countries. But the Latin scientific name *Gavia immer* is recognized by biologists worldwide, overcoming language barriers and allowing precise communication.

Note that, by convention, scientific names are always underlined or *italicized*. The first letter of the genus name is always capitalized, and the first letter of the species name is always lowercased. The species name is never used alone but is always be paired with its genus name.

Taxonomy Originated as a Hierarchy of Categories

Aristotle (384–322 B.C.) was among the first to attempt to formulate a logical, standardized language for naming living things. Based on characteristics such as structural complexity, behavior, and degree of development at birth, he classified about 500 organisms into 11 categories. Aristotle's categories formed a hierarchical

Figure 18-1 Three species of bluebird
Despite their obvious similarity, these three species of bluebird remain distinct because they do not interbreed. The three species shown are (from left to right) the eastern bluebird *(Sialia sialis)*, the western bluebird *(Sialia mexicana)*, and the mountain bluebird *(Sialia currucoides)*.

structure, with each category more inclusive than the one beneath it, a concept that is still used today.

Building on this foundation more than 2000 years later, Swedish naturalist Carl von Linné (1707–1778)—who called himself Carolus Linnaeus, a latinized version—laid the groundwork for the modern classification system. He placed each organism into a series of hierarchically arranged categories on the basis of its resemblance to other life-forms, and he also introduced the scientific name composed of genus and species. Nearly 100 years later, Charles Darwin (1809–1882) published *On the Origin of Species*, which demonstrated that all organisms are connected by common ancestry. Taxonomists then began to recognize that taxonomic categories ought to reflect the pattern of evolutionary relatedness among organisms. The more categories two organisms share, the closer their evolutionary relationship.

Modern Systematists Use Numerous Criteria for Classification

Systematists seek to reconstruct the tree of life, but they must do so without much direct knowledge of evolutionary history. Because they can't see into the past, they must infer the past as best they can, based on similarities among living organisms. Not just any similarity will do, however. Some observed similarities stem from convergent evolution in organisms that are not closely related; such similarities are not useful to systematists. Instead, systematists value the similarities that arise when two kinds of organisms share a feature because both inherited it from a common ancestor. Therefore, one of a systematist's main tasks is to distinguish informative similarities caused by common ancestry from the less useful similarities that result from convergent evolution. In the search for informative similarities, systematists look at many different kinds of characteristics.

Historically, the most important and useful distinguishing characteristics have been anatomical. Taxonomists look carefully at similarities in external body structure (see Fig. 18-1) and at internal structures, such as skeletons and muscles. For example, homologous structures, such as the finger bones of dolphins, bats, seals, and humans (see Fig. 14-8), provide evidence of a common ancestor. To detect relationships between more closely related species, taxonomists may use microscopes to discern finer details—the number and shape of the "teeth" on the tonguelike *radula* of a snail, the shape and position of the bristles on a marine worm, or the fine structure of pollen grains of a flowering plant (Fig. 18-2).

Developmental stages also provide clues to common ancestry. For example, although the sea squirt, a type of tunicate, spends its adult life permanently attached to rocks on the ocean floor, its free-swimming larva has a nerve cord, tail, and gills, placing it in the same phylum as the vertebrates: phylum Chordata.

The developmental and anatomical characteristics shared by related organisms are expressions of underlying genetic similarities, so it stands to reason that evolutionary relationships among species must also be reflected in genetic similarities. Unfortunately, direct genetic comparisons were not possible for most of the history of biology. In the past two decades, however,

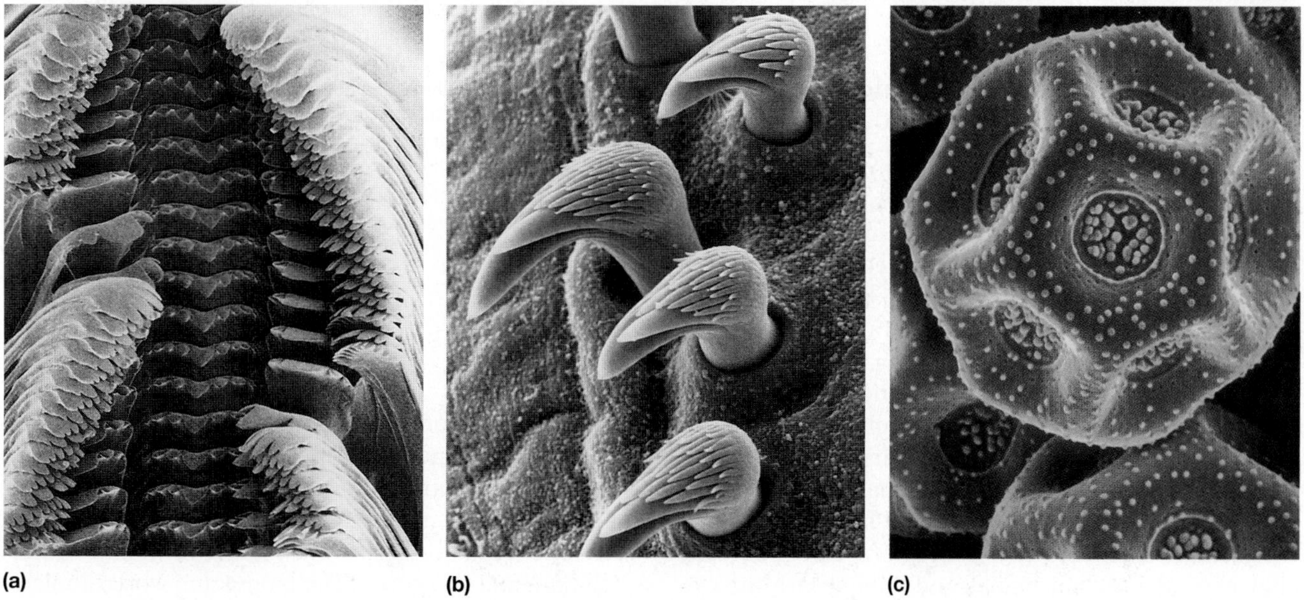

(a) (b) (c)

Figure 18-2 Microscopic structures help taxonomists classify organisms
(a) The "teeth" on a snail's tonguelike radula (a structure used in feeding) or ***(b)*** the bristles on a marine worm are potential taxonomic criteria. ***(c)*** The shape and surface features of pollen grains help identify plants, particularly from fossil deposits and sediments.

advances in the techniques of molecular genetics have triggered a revolution in studies of evolutionary relationships. For the first time, the nucleotide sequence of DNA (that is, genotype), rather than phenotypical features such as appearance, behavior, or even proteins, can be used to investigate relatedness among different types of organisms. Some of the key methods and findings of genetic analysis are explored in the essay "Scientific Inquiry: Molecular Genetics Reveals Evolutionary Relationships."

Despite its power, this new direct access to DNA has not yet completely eliminated other ways of making genetic comparisons. For example, relatedness among organisms can also be evaluated by examining the structure of their chromosomes. Among the findings derived from this technique is that the chromosomes of chimpanzees and humans are extremely similar, showing that these two species are very closely related (Fig. 18-3).

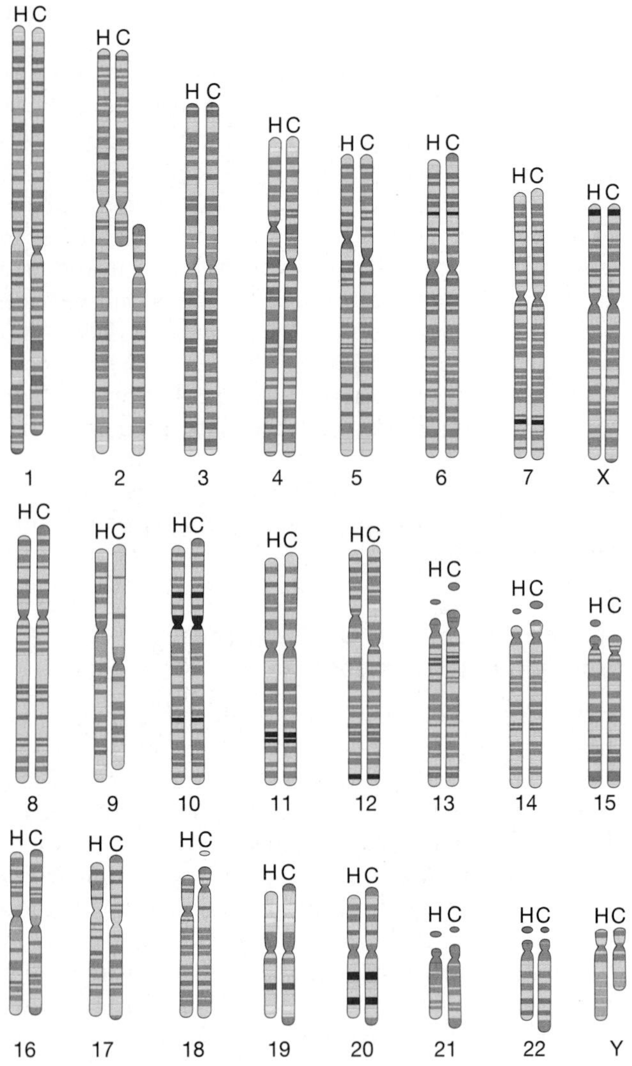

Figure 18-3 Human and chimp chromosomes are similar
Chromosomes from different species can be compared by means of banding patterns that are revealed by staining. The comparison illustrated here, between human chromosomes (left member of each pair; H) and chimpanzee chromosomes (C), reveals that the two species are genetically very similar. In fact, it has been estimated that 99% of the two genomes is identical. The numbering system shown is that used for human chromosomes; note that human chromosome 2 corresponds to a combination of two chimp chromosomes.

2 What Are the Kingdoms of Life?

Before 1970, taxonomists classified all forms of life into two kingdoms: Animalia and Plantae. All bacteria, fungi, and photosynthetic protists were considered plants, and the protozoa were classified as animals. As scientists learned more about fungi and microorganisms, however, it became apparent that the two-kingdom system oversimplified the true nature of evolutionary history. To help rectify this problem, in 1969 Robert H. Whittaker proposed a five-kingdom classification scheme that was eventually adopted by most systematists.

The Five-Kingdom System Supplanted Earlier Classification Schemes

Whittaker's five-kingdom system divides unicellular microorganisms into two kingdoms, based primarily on whether they show prokaryotic or eukaryotic cellular organization. The kingdom Monera consists of generally single-celled prokaryotic organisms, whereas the kingdom Protista consists of generally single-celled eukaryotic organisms. The remaining three kingdoms (Plantae, Fungi, and Animalia) include only eukaryotic organisms, most of which are multicellular. These three kingdoms of multicellular eukaryotes can be distinguished on the basis of their methods of acquiring nutrients. Members of the kingdom Plantae photosynthesize, and members of the kingdom Fungi secrete enzymes outside their bodies and then absorb the externally digested nutrients. In contrast, members of the kingdom Animalia ingest their food and then digest it, either within an internal cavity or within individual cells.

Because it more accurately reflected our understanding of evolutionary history, the five-kingdom system was an improvement over the old two-kingdom system. As our understanding has continued to grow, however, it has proved necessary to change our view of life's most fundamental categories. The pioneering work of microbiologist Carl Woese has shown that systematists have overlooked a fundamental event in the early history of life, one that demands a new and more-accurate classification of life.

A Three-Domain System More Accurately Reflects Life's History

Since the 1970s Woese and other biologists interested in the phylogeny of microorganisms have studied the biochemistry of prokaryotic organisms. These researchers, focusing on nucleotide sequences of the RNA that is found in ribosomes, determined that what had been considered the kingdom Monera actually consists of two very different kinds of organisms. Woese has dubbed these two groups the Bacteria and the Archaea (Fig. 18-4).

Despite superficial similarities in their appearance under the microscope, the Bacteria (sometimes called *eubacteria*) and the Archaea (also known as *archaebacteria*) are radically different. The members of these two groups are no more closely related to one another than either one is to any eukaryote (Fig 18-5). The tree of life split into three parts very early in the history of life, long before the eukaryotes gave rise to plants, animals, and fungi. So Woese has proposed, and many systematists agree, that we ought to classify life into three broad categories called *domains*: the (1) *Bacteria*, (2) *Archaea*, and (3) *Eukarya*.

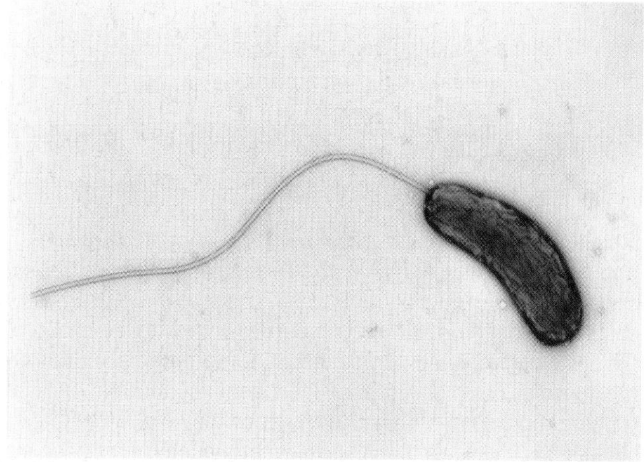

(a)

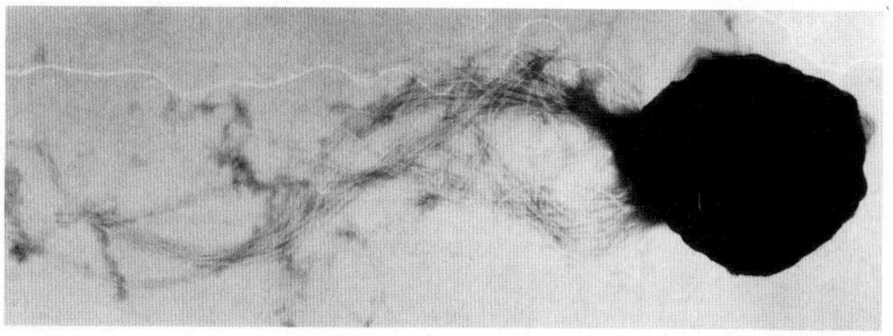

(b)

Figure 18-4 Two domains of prokaryotic organisms
Although superficially similar in appearance, ***(a)*** *Vibrio cholerae* and ***(b)*** *Methanococcus jannaschi* are less closely related than a mushroom and an elephant. Each prokaryote belongs to a fundamentally different domain of life. *Vibrio* is in the domain Bacteria, and *Methanococcus* is in the Archaea.

Kingdom-level Classification Remains Unsettled

If we accept the three-domain classification system, we're left with some taxonomic dilemmas with regard to the status of our five kingdoms. It does not make sense to lump all prokaryotes into the single kingdom Monera, because the Bacteria and the Archaea are only very distantly related. If life has three main branches, we can't justify lumping two of them (Archaea and Bacteria) into one kingdom while splitting the third branch (Eukarya) into four kingdoms. A related problem is that if the domain Eukarya can contain four kingdoms, then the other two domains may also contain more than one kingdom. In fact, there were a number of very early splits both within the Archaea and within the Bacteria that long predated, for example, the split between plants and animals. If the plant–animal split is ancient enough to warrant kingdom status for each group, then each of the even-older splits within the Bacteria and Archaea should also merit kingdom status. Thus, we would need a lot more than five kingdoms to construct a phylogenetically accurate classification of life. Some systematists have even suggested that plants, animals, and fungi don't deserve kingdom status. After all, they represent only tiny twigs way down at the end of the eukaryote branch of the tree of life.

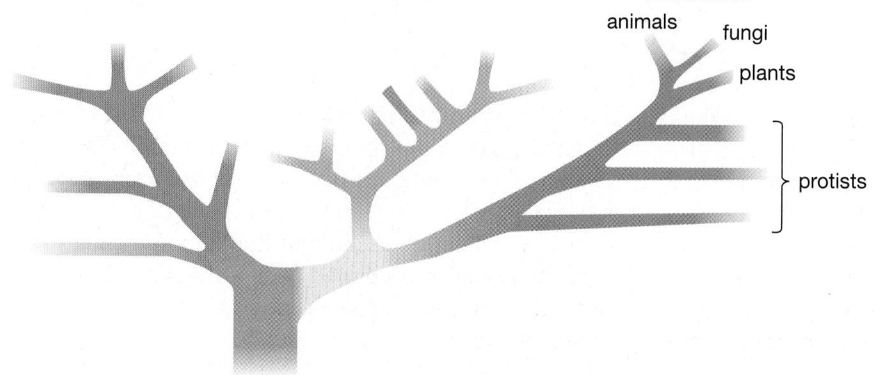

Figure 18-5 The tree of life
The three domains of life represent the earliest branches in evolutionary history.

Scientific Inquiry
Molecular Genetics Reveals Evolutionary Relationships

Evolution results from the accumulation of inherited changes in populations. Because DNA is the molecule of heredity, evolutionary changes must be reflected in changes in DNA. Systematists have long known that comparing DNA within a group of species would be a powerful method for inferring evolutionary relationships, but direct access to genetic information was nothing more than a dream for most of the history of systematics. Today, however, **DNA sequencing**—determining the sequence of nucleotides in segments of DNA—is comparatively cheap, easy, and widely available. The *polymerase chain reaction* (PCR, see Chapter 13) allows systematists to accumulate large samples of DNA from organisms easily, and automated machinery makes sequence determination a comparatively simple task. Sequencing has rapidly become one of the primary tools for uncovering phylogeny.

The logic underlying molecular systematics is straightforward. It is based on the observation that when a single species divides into two species, the gene pool of each resulting species begins to accumulate mutations. The particular mutations in each species, however, will differ because the species are now evolving independently, with no gene flow between them. As time passes, more and more genetic differences accumulate. So, a systematist who has obtained DNA sequences from representatives of both species can compare the two species' nucleotide sequences at any given location in the genome. Fewer differences indicate more closely related organisms.

Putting the simple principles outlined above into practice usually involves some more-sophisticated thinking. For example, sequence comparisons become far more complex when a researcher wishes to assess relationships among, say, 20 or 30 species. Fortunately, mathematicians and computer programmers have devised some very clever computer-assisted methods for comparing large numbers of sequences and deriving the phylogeny that is most likely to account for the observed sequence differences.

Molecular systematists must also use care in choosing which segment of DNA to sequence. Different parts of the genome evolve at different rates, and it is crucial to sequence a DNA segment whose rate of change is well matched to the phylogenetic question at hand. In general, slowly evolving genes work best for comparing distantly related organisms, and rapidly changing portions of the genome are best for analyzing closer relationships. It is sometimes difficult to find any single gene that will yield sufficient information to provide an accurate picture of evolutionary change across the genome, so sequences from several different genes are often needed to construct reliable phylogenies (Fig. E18-1).

Today, sequence data are piling up with unprecedented rapidity, and systematists have access to sequences from an ever-increasing number of species. For several species, including some bacteria, an archaean, and a yeast, the entire genome has been sequenced. The Human Genome Project is scheduled to fully complete its work by the year 2003, and then our own DNA sequences will be a matter of public record. The revolution in molecular biology has fostered a great leap forward in our understanding of evolutionary history.

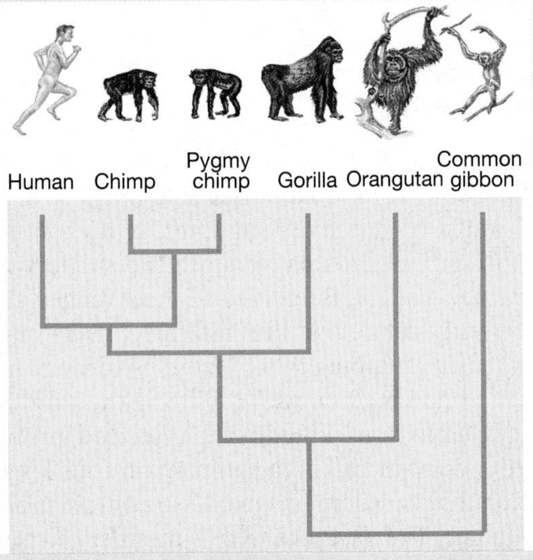

Figure E18-1 Relatedness can be determined by comparing DNA sequences
This evolutionary tree was derived from the nucleotide sequences of several different genes that are common to humans and apes.

Overall, we're now at a rather awkward moment in the evolution of high-level classification. The systematic data strongly support the three-domain system, but the thorny issues of kingdom-level classification have yet to be definitively worked out. At the same time, it is unsatisfying to downplay the seemingly fundamental differences among protists, plants, animals, and fungi. In this text, our imperfect solution will be to reserve judgment on kingdom-level classification within the two prokaryotic domains and to retain the four kingdoms within the Eukarya. Table 18-2 compares the characteristics within those four kingdoms, and Figure 18-6 shows their evolutionary relationships. This arrangement is useful, but it is important for you to understand that it is derived from a five-kingdom system that may be replaced some time in the future by a more phylogenetically accurate system.

Table 18-2 Some Characteristics of the Eukaryotic Kingdoms

Kingdom	Cell Type	Cell Number	Major Mode of Nutrition	Motility (movement)	Cell Wall	Reproduction
Protista	Eukaryotic	Unicellular	Absorb, ingest, or photosynthesize	Both motile and nonmotile	Present in algal forms: varies	Both sexual and asexual
Animalia	Eukaryotic	Multicellular	Ingest	All motile at some stage	Absent	Both sexual and asexual
Fungi	Eukaryotic	Most multicellular	Absorb	Generally nonmotile	Present: chitin	Both sexual and asexual
Plantae	Eukaryotic	Multicellular	Photosynthesize	Generally nonmotile	Present: cellulose	Both sexual and asexual

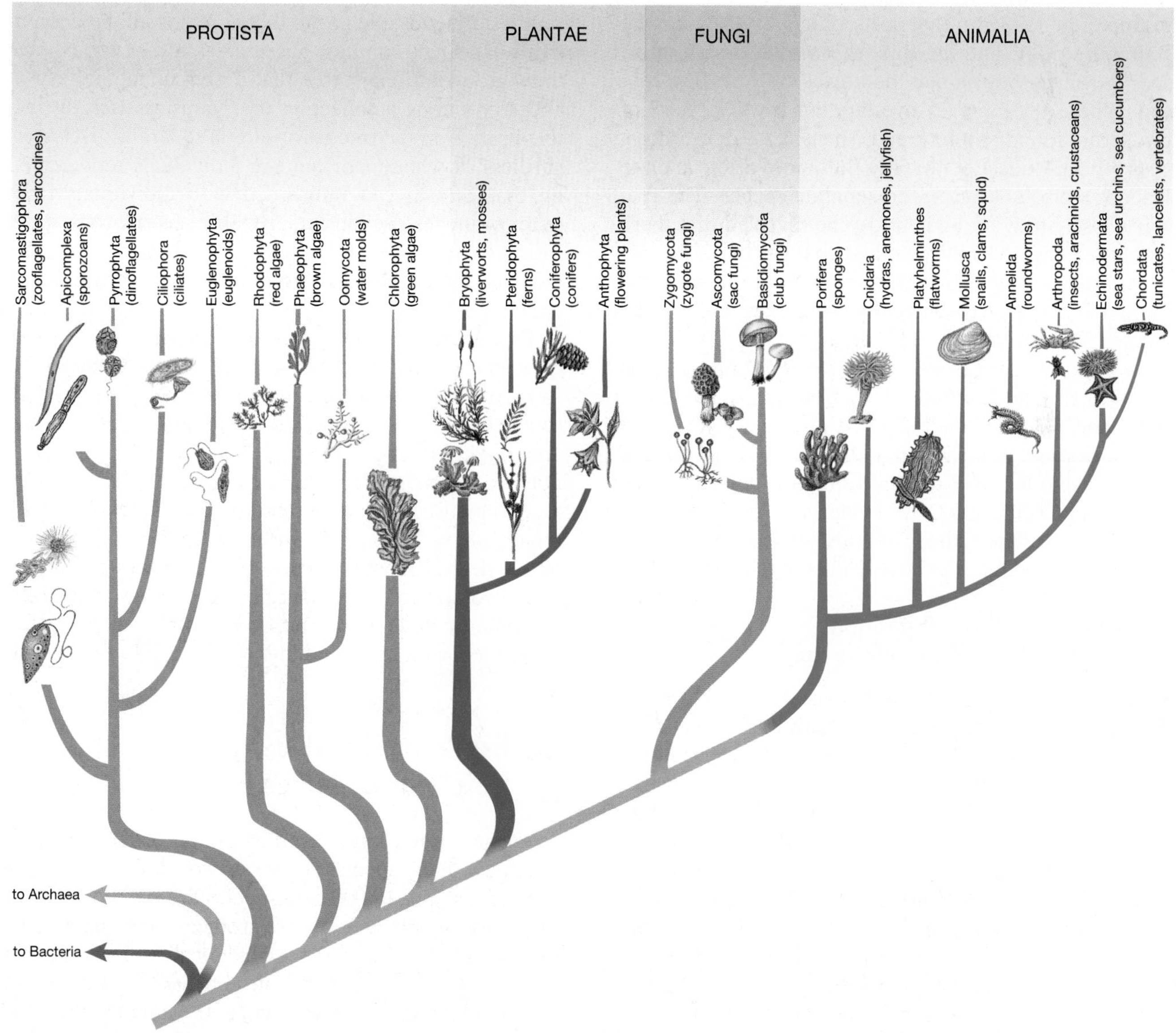

Figure 18-6 A closer look at the eukaryotic tree of life
The four kingdoms within the domain Eukarya and some of the major phyla or divisions within them are illustrated.

3 Why Do Taxonomies Change?

As the emergence of the three-domain system shows, the hypotheses of evolutionary relationships on which taxonomy is based are subject to revision as new data emerge. Even the five kingdoms, representing deep, old branchings of the tree of life and the most fundamental categories of taxonomy, will have to be modified in light of recent advances in systematics. Such changes at the kingdom level occur only rarely, but at the other end of the taxonomic spectrum, among species designations, revisions are more frequent.

Species Designations Change When New Information Is Discovered

Systematists regularly propose changes in species-level taxonomy, as researchers uncover new information. For example, in 1973 ornithologists officially declared the Baltimore oriole and the Bullock's oriole to be a single species, which they named the northern oriole. The reason: Where their ranges overlap, the two "species" of birds were found to interbreed. Then, in 1997, ornithologists officially declared that the Baltimore oriole and the Bullock's oriole were indeed separate species. The reason: Closer study of the hybrid zone revealed that there was little gene flow between the two types of oriole.

Another example is the red wolf from the southern United States, which is currently listed as an endangered species. Researchers recently analyzed the DNA from mitochondria taken from red wolves and found it to be identical to mitochondrial DNA from coyotes in some cases and to mitochondrial DNA from gray wolves in others. Because mitochondrial DNA is inherited directly from the mother (via the mitochondria present in the original egg cell), this DNA evidence strongly suggests that red wolves are actually hybrids between gray wolves and coyotes and may not be a distinct species after all.

The Biological Species Definition Can Be Difficult or Impossible to Apply

In some cases, systematists find themselves unable to say with certainty where one species ends and another begins. As we noted in Chapter 16, asexually reproducing organisms pose a particular challenge to taxonomists, because the criterion of interbreeding (the basis of the "biological species" definition that we have used in this text) cannot be used to distinguish among species. The irrelevance of this criterion in studies of asexual organisms leaves plenty of room for investigators to disagree about which asexual populations constitute a species, especially when comparing groups that have similar phenotypes. For instance, some taxonomists recognize 200 species of the British blackberry (a plant that can produce seeds *parthenogenetically*—that is, without fertilization), but others recognize only 20 species.

The difficulty of applying the biological species definition to asexual organisms is a serious problem for systematists. After all, a significant portion of Earth's organisms reproduces without sex. Most bacteria, archaea, and protists, for example, reproduce asexually most of the time. Some systematists argue that we need a more universally applicable definition of species, one that won't exclude asexual organisms and that doesn't depend on the criterion of reproductive isolation.

The Phylogenetic Species Concept Offers an Alternative Definition

A number of alternative species definitions have been proposed over the history of evolutionary biology, but none has been sufficiently compelling to displace the biological species definition. One alternative definition, however, has been gaining adherents in recent years. The *phylogenetic species concept* defines a species as "the smallest diagnosable group that contains all the descendants of a single common ancestor." In other words, if we draw an evolutionary tree that describes the pattern of ancestry among a collection of organisms, each distinctive branch on the tree constitutes a separate species, regardless of whether or not the individuals represented by that branch can interbreed with individuals from other branches. As you can probably see, rigorous application of the phylogenetic species concept would vastly increase the number of different species recognized by systematists.

Proponents and opponents of the phylogenetic species concept are currently engaged in a vigorous debate about its merits. Perhaps one day the phylogenetic species concept will replace the biological species concept as the "textbook definition" of species. In the meantime, taxonomic categories will continue to be debated and revised as systematists learn more and more about evolutionary relationships, particularly with the application of techniques derived from molecular biology. Although the precise evolutionary relationships of many organisms continue to elude us, taxonomy is enormously helpful in ordering our thoughts and investigations into the diversity of life on Earth.

4 Exploring Biodiversity: How Many Species Exist?

Scientists do not know even within an order of magnitude how many species share our world. Each year, between 7000 and 10,000 new species are named, most of them insects, many from the tropical rain forests. The total number of named species is currently about 1.4 million. However, many scientists believe that 7 million to 10 million species may exist, and estimates range as high as 100 million. This total range of species diversity is known as **biodiversity**. Of all the species that have been identified thus far, about 5% are prokaryotes and protists. An additional 22% are plants and fungi, and the rest are animals. This distribution has little to do with the actual abundance of these organisms and a lot to do with the size of the organisms, how easy they are to classify, how accessible they are, and the number of scientists studying them. Historically, systematists have chiefly focused on large or conspicuous organisms in temperate regions, but biodiversity is greatest among small, inconspicuous organisms in the Tropics. In addition to the overlooked species on land and in shallow waters, an entire new "continent" of species lies largely unexplored on the deep-sea floor. From the limited samples available, scientists estimate that hundreds of thousands of unknown species may reside there.

Although about 5000 species of prokaryotes have been described and named, prokaryotic diversity remains largely unexplored. Consider a study by Norwegian scientists who analyzed DNA to count the number of different bacteria species present in a small sample of forest soil. To distinguish among species, the researchers arbitrarily defined bacterial DNA as coming from separate species if it differed by at least 30% from that of any other bacterial DNA in the sample. Using this criterion, they reported more than 4000 types of bacteria in their soil sample and an equal number of forms in a sample of shallow marine sediment.

Our ignorance of the full extent of life's diversity adds a new dimension to the tragedy of the destruction of the tropical rain forests, discussed in Chapter 41. Although these forests cover only about 6% of Earth's land area, they are believed to be home to two-thirds of the world's existing species, most of which have never been studied or named. Because these forests are being destroyed so rapidly, Earth is losing many species that we will never even know existed! For example, in 1990 a new species of primate, the black-faced lion tamarin, was discovered in a small patch of dense rain forest on an island just off the east coast of Brazil (Fig. 18-7). Researchers estimate that no more than 260 individuals of this squirrel-sized monkey remain in the wild, and captive breeding may be its only hope for survival. At current rates of deforestation, most of the tropical rain forests, with their undescribed wealth of life, will be gone within the next century.

Figure 18-7 The black-faced lion tamarin
This rare species of monkey was discovered in 1990 in the vanishing Brazilian rain forest.

Evolutionary Connections

Are Reptiles for Real?

The ever-changing nature of taxonomy sometimes means that even the most entrenched and familiar categories must give way to new understandings of evolutionary history. Take reptiles, for example. Even schoolchildren are familiar with this well-known group of scaly animals and know that it includes snakes, lizards, turtles, alligators, and crocodiles. But many systematists would like to do away with the reptiles. Why? Because reptiles do not form what systematists call a *natural group.*

The goal of modern systematics is to devise classifications that accurately reflect evolutionary history. One characteristic of such classifications is that each designated group should contain only organisms that are more closely related to one another than to any organisms outside the group. So, for example, the members of family Canidae (which includes dogs, wolves, foxes, and coyotes) are more closely related to each other than to any member of any other family. Another way to state this principle is to say that each designated group should contain *all* of the living descendants of a common ancestor (Fig. 18-8a). In the lingo of systematics, such groups are said to be **monophyletic**.

The reptiles, alas, are not a monophyletic group (Fig. 18-8b). As historically defined, the reptiles exclude the birds, which are now known to fall squarely within the reptile family tree. The reptiles, therefore, do not include all living descendants of the common ancestor that gave rise to snakes, lizards, turtles, crocodilians, and birds. So systematists would prefer to dispense with the former Class Reptilia in favor of a scheme that names only monophyletic groups. The word *reptiles*, however, will most likely be with us for some time to come, if only because so many people (including systematists) are simply accustomed to using it. And the word does, after all, provide a convenient way to describe a group of animals that share some interesting adaptations, even if that group isn't monophyletic.

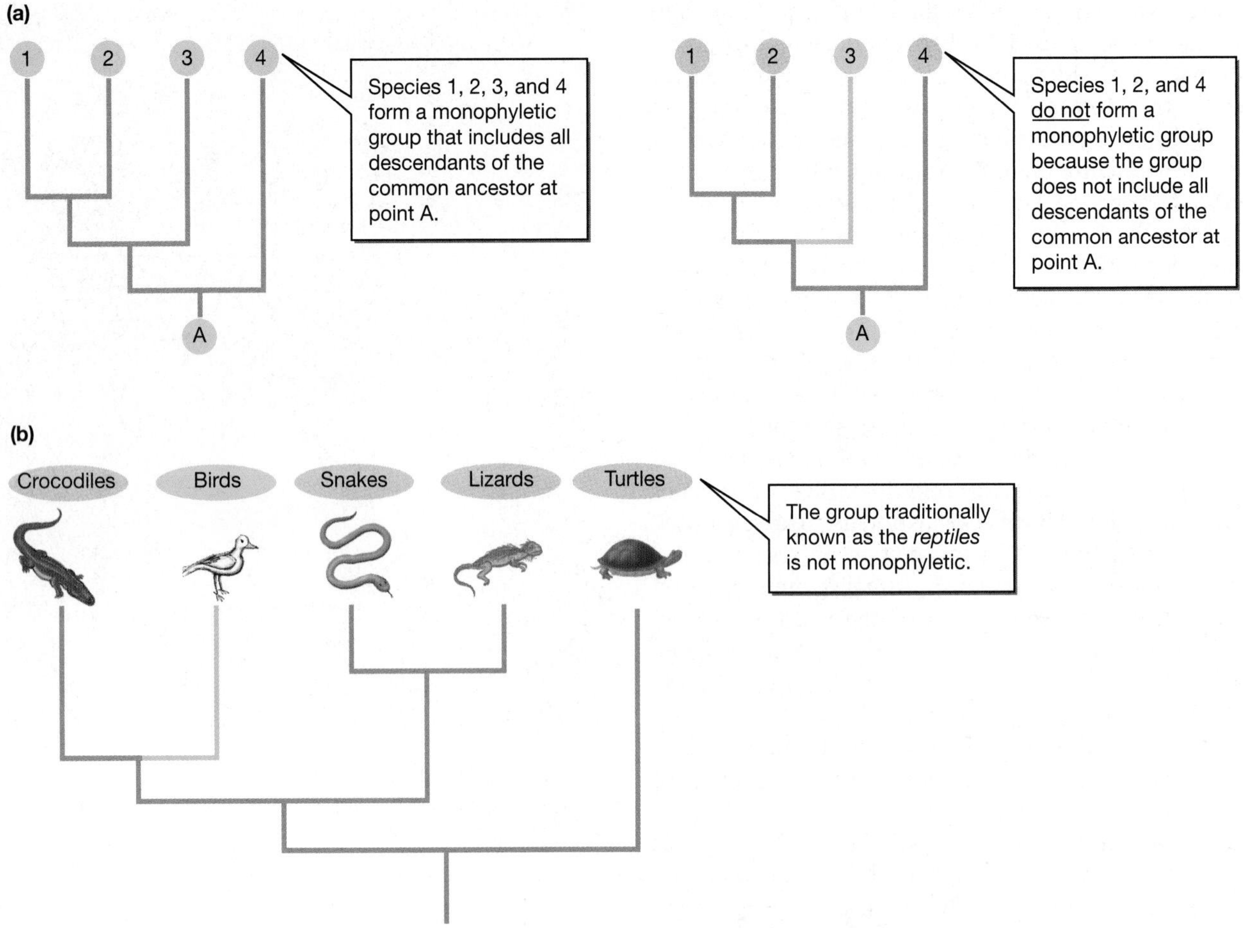

Figure 18-8 Reptiles are not a monophyletic group
Only groups that contain all of the descendants of a common ancestor are considered to be monophyletic groups.

CASESTUDYREVISITED

Origin of a Killer

What evidence has persuaded evolutionary biologists that HIV originated in apes and monkeys? To understand the evolutionary thinking behind this conclusion, examine the evolutionary tree shown in Figure 18-9. This tree illustrates the phylogeny of HIV and its close relatives, as revealed by a comparison of RNA sequences among different viruses. Note the positions on the tree of the four human viruses (two strains of HIV-1 and two of HIV-2). One strain of HIV-1 is more closely related to a chimpanzee virus than to the other strain of HIV-1. Similarly, one strain of HIV-2 is more closely related to pig-tailed macaque SIV than to the other strain of HIV-2. Both HIV-1 and HIV-2 are more closely related to ape or monkey viruses than to one another.

The only way for the evolutionary history shown in the tree to have emerged is if viruses jumped between host species. If HIV had evolved strictly within human hosts, the human viruses would have been each other's closest relatives. Because the human viruses do not cluster together on the phylogenetic tree, we can infer that cross-species infection has occurred, probably on multiple occasions. The mainstream view is that transmission occurred in connection with human consumption of monkeys (HIV-2) and chimpanzees (HIV-1), but not all investigators agree with that assessment. Some feel that it is more likely that HIV was transferred via contaminated vaccines. Edward Hooper, for example, has gathered evidence in support of the hypothesis that HIV-1 jumped to humans in an oral polio vaccine that was administered to more than a million people in Africa between 1957 and 1960. The vaccine may have been prepared using chimpanzee kidneys.

Can understanding the origin of HIV help researchers devise better ways to treat and control the spread of AIDS? In what way?

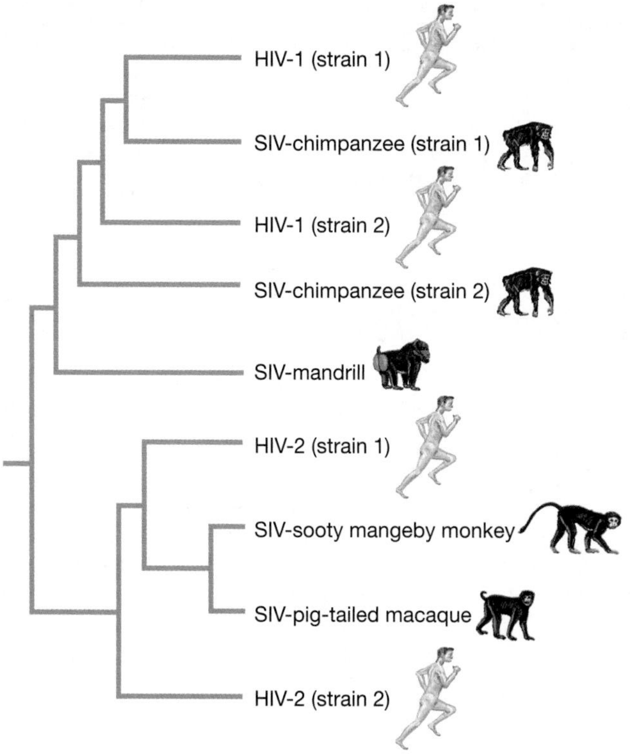

Figure 18-9 Evolutionary analysis helps reveal the origin of HIV
In this phylogeny of some immunodeficiency viruses, the viruses with human hosts do not cluster together. This lack of congruence between the evolutionary histories of the viruses and their host species suggests that the viruses must have jumped between host species. (Note that SIV stands for *simian immunodeficiency virus.*)

Summary of Key Concepts

1) How Are Organisms Named and Classified?

Taxonomy is the science by which organisms are classified and placed into hierarchical categories that reflect their evolutionary relationships. The eight major categories, in order of decreasing inclusiveness, are (1) domain, (2) kingdom, (3) division or phylum, (4) class, (5) order, (6) family, (7) genus, and (8) species. The scientific name of an organism is composed of its genus name and species name. A hierarchical concept was first used by Aristotle, but in the mid-1700s Linnaeus laid the foundation for modern taxonomy. In the 1860s, evolutionary theory proposed by Charles Darwin provided an explanation for the observed similarities and differences among organisms, and modern taxonomists attempt to classify organisms according to their evolutionary relationships. Today, taxonomists use features such as anatomy, developmental stages, and biochemical similarities to categorize organisms. Molecular biological techniques are used to determine the sequences of nucleotides in DNA and RNA and of amino acids in proteins. Biochemical similarities among organisms are a measure of evolutionary relatedness.

2) What Are the Kingdoms of Life?

A five-kingdom classification system consisting of the kingdoms Monera, Protista, Animalia, Fungi, and Plantae is widely used, but a new and increasingly accepted three-domain system will require revisions at the kingdom level of classification. The three domains, representing the three main branches of life, are Bacteria, Archaea, and Eukarya. (See Table 18-2 for characteristics of the four eukaryotic kingdoms.)

3) Why Do Taxonomies Change?

Taxonomic categories are subject to revision as new information is discovered. Species boundaries may be hard to define, particularly in the case of asexually reproducing species. However, taxonomy is essential for precise communication and contributes to our understanding of the evolutionary history of life.

4) Exploring Biodiversity: How Many Species Exist?

Although only about 1.4 million species have been named, estimates of the total number of species range up to 100 million. New species are being identified at the rate of 7000 to 10,000 annually, mostly in tropical rain forests.

Key Terms

biodiversity *p. 358*
class *p. 351*
domain *p. 351*
DNA sequencing *p. 356*
family *p. 351*
genus *p. 351*
kingdom *p. 351*
monophyletic *p. 359*
order *p. 351*
phylogeny *p. 351*
phylum *p. 351*
scientific name *p. 352*
species *p. 351*
systematics *p. 351*
taxonomy *p. 351*

Thinking Through the Concepts

Multiple Choice

1. *It is possible to imagine the various levels of taxonomic classification as a kind of "family tree" for an organism. If the kingdom is analogous to the trunk of the tree, which taxonomic category would be analogous to the large limbs coming off that trunk?*
   a. class b. family
   c. order d. phylum
   e. subfamily
2. *The more taxonomic categories two organisms share, the more closely related those organisms are in an evolutionary sense. Which scientist's work led to this insight?*
   a. Aristotle b. Darwin
   c. Linnaeus d. Whittaker
   e. Woese
3. *Which of the following criteria could not be used to determine how closely related two types of organisms are?*
   a. similarities in the presence and relative abundance of specific molecules
   b. DNA sequence
   c. the presence of homologous structures
   d. developmental stages
   e. occurrence of both organisms in the same habitat
4. *Which one of the following habitats appears to have the greatest number of species?*
   a. the seafloor b. deserts
   c. tropical rain forests d. grasslands
   e. mountaintops
5. *Which of the following pairs is the most distantly related?*
   a. archaea and bacteria b. protists and fungi
   c. plants and animals d. fish and starfish
   e. fungi and plants
6. *An organism is described to you as having many nuclei-containing cells, each surrounded by a cell wall of chitin and absorbing its food. In which kingdom or domain would you place it?*
   a. Plantae b. Protista
   c. Animalia d. Fungi
   e. Archaea

? Review Questions

1. What contributions did Aristotle, Linnaeus, and Darwin each make to modern taxonomy?
2. What features would you study to determine whether a dolphin is more closely related to a fish or to a bear?
3. What techniques might you use to determine whether the extinct cave bear is more closely related to a grizzly bear or to a black bear?
4. Only a small fraction of the total number of species on Earth has been scientifically described. Why?
5. In England, "daddy long-legs" refers to a long-legged fly, but the same name refers to a spider-like animal in the United States. How do scientists attempt to avoid such confusion?

Applying the Concepts

1. There are many areas of disagreement among taxonomists. For example, there is no consensus about whether the red wolf is a distinct species or about how many kingdoms are within the domain Bacteria. What difference does it make whether biologists consider the red wolf a species, or into which kingdom a bacterial species falls? As Shakespeare put it, "What's in a name?"
2. The pressures created by human population growth and economic expansion place storehouses of biological diversity such as the Tropics in peril. The seriousness of the situation is clear when we consider that probably only 1 out of every 20 tropical species is known to science at present. What arguments can you make for preserving biological diversity in poor and developing countries? Does such preservation require that these countries sacrifice economic development? Suggest some solutions to the conflict between the growing demand for resources and the importance of conserving biodiversity.
3. During major floods, only the topmost branches of submerged trees may be visible above the water. If you were asked to sketch the branches below the surface of the water solely on the basis of the positions of the exposed tips, you would be attempting a reconstruction somewhat similar to the "family tree" by which taxonomists link various organisms according to their common ancestors (analogous to branching points). What sources of error do both exercises share? What advantages do modern taxonomists have?

4. The Florida panther, found only in the Florida Everglades, is currently classified as an endangered species, protecting it from human activities that could lead to its extinction. It has long been considered a subspecies of cougar (mountain lion), but recent mitochondrial DNA studies have shown that the Florida panther may actually be a hybrid between American and South American cougars. Should the Florida panther be protected by the Endangered Species Act?

For More Information

Avise, J. C. "Nature's Family Archives." *Natural History*, March 1989. Shows how evolutionary relationships can be determined by analyzing differences in DNA contained in mitochondria.

Gould, S. J. "What Is a Species?" *Discover*, December 1992. Discusses the difficulties of distinguishing separate species.

Mann, C., and Plummer, M. *Noah's Choice: The Future of Endangered Species*. New York: Knopf, 1995. A thought-provoking look at the hard choices we must make with regard to protecting biodiversity. Which species will we choose to preserve? What price are we willing to pay?

Margulis, L., and Sagan, D. *What Is Life?* London: Weidenfeld & Nicolson, 1995. A lavishly illustrated survey of life's diversity. Also includes an account of life's history and a meditation on the question posed by the title.

Margulis, L., and Schwartz, K. *Five Kingdoms*. 2nd ed. New York: W. H. Freeman, 1988. An illustrated paperback guide to the diversity of life.

May, R. M. "How Many Species Inhabit the Earth?" *Scientific American*, October 1992. Although no one knows the precise answer to this question, an effective estimate is crucial to our effort to manage our biological resources.

Moffett, M. W. *The High Frontier: Exploring the Tropical Rainforest Canopy*. Cambridge, MA: Harvard University Press, 1994. The tremendous diversity of life in the rainforest treetops is only now becoming known to us. This book documents the unexpected and spectacular diversity of animals in the upper reaches of this endangered habitat.

Wilson, E. O. *The Diversity of Life*. Cambridge, MA: Harvard University Press, 1992. An outline of the processes that created the diversity of life and a discussion of the threats to that diversity and the steps required to preserve it.

Answers to Multiple-Choice Questions

1. d 2. b 3. e 4. c 5. a 6. d

MEDIATUTOR

Systematics: Seeking Order Amidst Diversity

CD Activities

Activity 18.1: Taxonomic Classification

Estimated time: 5 minutes

Taxonomic classification is a means of categorizing all species of organisms on Earth. This exercise will help you understand the hierarchy of taxonomic groups used by biologists.

Start the MediaTutor Student CD-ROM and enter the activity number in the Quick Search box to be taken directly to that activity.

Web Investigations

Case Study: Origin of a Killer

Estimated time: 10 minutes

The search for the origins of the HIV viruses involves the often conflicting activities of evolutionary biologists, epidemiologists, clinicians, and social activists. Alternate theories, intricate data analyses, and contradictory conclusions are common. This exercise will explore popular theories and recent developments in HIV studies.

Go to http://www.prenhall.com/audesirk6, the Audesirk Companion Web site. Select Chapter 18 and Web Investigation to begin.

This soldier, clad in gear designed to protect him from assault by biological weapons, serves as a vivid reminder that not all applications of biological knowledge are beneficial.

19 The Hidden World of Microbes

AT A GLANCE

Case Study: Agents of Death

1) **What Are Viruses, Viroids, and Prions?**

A Virus Consists of a Molecule of DNA or RNA Surrounded by a Protein Coat

Viral Infections Cause Diseases That Are Difficult to Treat

Some Infectious Agents Are Even Simpler Than Viruses

No One Is Certain How These Infectious Particles Originated

2) **Which Organisms Make Up the Prokaryotic Domains—Bacteria and Archaea?**

Archaea and Bacteria Are Fundamentally Different

Prokaryotes Are Difficult to Classify

Prokaryotes Exhibit a Variety of Shapes and Structures

Prokaryotes Reproduce by Binary Fission

Prokaryotes Are Specialized for Specific Habitats

Prokaryotes Exhibit Diverse Metabolisms

Prokaryotes Perform Many Functions That Are Important to Other Forms of Life

Some Bacteria Pose a Threat to Human Health

3) **Which Organisms Make Up the Kingdom Protista?**

Protists Are a Diverse Group Including Funguslike, Plantlike, and Animal-like Forms

The Water Molds and Slime Molds Are Funguslike Protists

The Algae Are Plantlike Protists

The Protozoa Are Animal-like Protists

Evolutionary Connections: Our Unicellular Ancestors

Case Study Revisited: Agents of Death

CASE STUDY

Agents of Death

One night in April 1979 the sound of a small explosion rolled through small Russian city of Sverdlovsk. Those residents who noticed it at all briefly wondered about its source and went back to sleep. The next day, however, many residents in the vicinity of the nondescript industrial building in which the explosion had occurred began to feel ill. At first the illness seemed like nothing more than a cold, but victims quickly developed a high fever and began to find breathing difficult as their lungs filled with fluid. Within a few days, hundreds, perhaps thousands, of residents of Sverdlovsk had died.

We may never know exactly how many perished during that hellish week in Sverdlovsk, because authorities in what was then the Soviet Union hushed up the incident. Why such secrecy? Because the source of the disease was a top-secret facility for developing agents of biological warfare. An accident there had released a cloud of disease-causing bacteria in the air.

The bacteria that spread through Sverdlovsk that night were *Bacillus anthracis*, which cause a disease called *anthrax*. Anthrax, which normally infects domestic animals such as goats and sheep, is fatal to 90% of its human victims. It has emerged as the disease of choice among developers of biological weapons, which do their damage by intentionally spreading disease-causing *microbes* (as single-celled organisms are collectively known). Anthrax bacteria are easily isolated from infected animals, are cheap and easy to culture in large quantities and, once produced, can be dried into a powder that remains viable for years. The powder is easily "weaponized" by packing it into a missile warhead or other delivery device, and a small volume of bacteria can infect a very large number of people. The specter of anthrax and other potential biological weapons reminds us that biological knowledge can be tapped for destructive purposes.

The harmful microbes that cause human diseases (and that are sometimes recruited as potential biological weapons) represent only a small portion of the world's diverse array of single-celled life. In the following pages, we explore some of that microbial diversity. We also discuss viruses and other particles that interact with cells, even though many biologists do not consider these entities to be living organisms. ■

1 What Are Viruses, Viroids, and Prions?

Viruses possess no membranes of their own, no ribosomes on which to make proteins, no cytoplasm, and no source of energy. They cannot move or grow on their own, and they can reproduce only inside a **host** cell—the cell a virus or other infectious agent infects. The simplicity of viruses makes it impossible to call them cells and, indeed, seems to place them outside the realm of living things.

A Virus Consists of a Molecule of DNA or RNA Surrounded by a Protein Coat

A virus particle is so small (0.05–0.2 micrometer [μm: 1/1000 of a millimeter] in diameter) that seeing it requires the enormous magnification of an electron microscope (Fig. 19-1). Viruses consist of two major parts: (1) a molecule of hereditary material and (2) a coat of protein surrounding the molecule. The hereditary molecule may be either DNA or RNA and may be single-stranded or double-stranded, linear or circular. The protein coat may be surrounded by an envelope formed from the plasma membrane of the host cell (Fig. 19-2a).

Viruses are unable to grow or reproduce on their own, even if placed in a rich broth of nutrients at optimal temperature. They lack the complex cellular organization that these activities require. A virus's protein coat, however, is specialized to enable the virus to penetrate the cells of a specific host. After a virus enters a host cell, the viral genetic material takes command. The hijacked host cell is forced to "read" the viral genes and to use the instructions encoded there to produce the components of new viruses. The pieces are rapidly assembled, and an army of new viruses bursts forth to invade and conquer neighboring cells (Fig. 19-2b; see "A Closer Look: Viruses—How the Nonliving Replicate").

Viral Infections Cause Diseases That Are Difficult to Treat

Each type of virus is specialized to attack a specific host cell (Fig. 19-3). As far as we know, no organism is immune to all viruses. Even bacteria fall victim to viral invaders; viruses that infect bacteria are called **bacteriophages** (Fig. 19-4).

Within a particular organism, viruses specialize in attacking particular cell types. Viruses responsible for the common cold, for example, attack the membranes of the respiratory tract; those causing measles infect the skin; and the rabies virus attacks nerve cells. One type of herpes virus specializes in the mucous membranes of the mouth and lips, causing cold sores; a second type, transmitted through sexual contact, produces similar sores on or near the genitals. Herpes viruses take up permanent residence in the body, erupting periodically (typically during times of stress) as infectious sores. The devastating disease AIDS (acquired immune deficiency syndrome), which cripples the body's immune system, is caused by a virus that attacks a specific type of white blood cell that controls the body's immune response (as

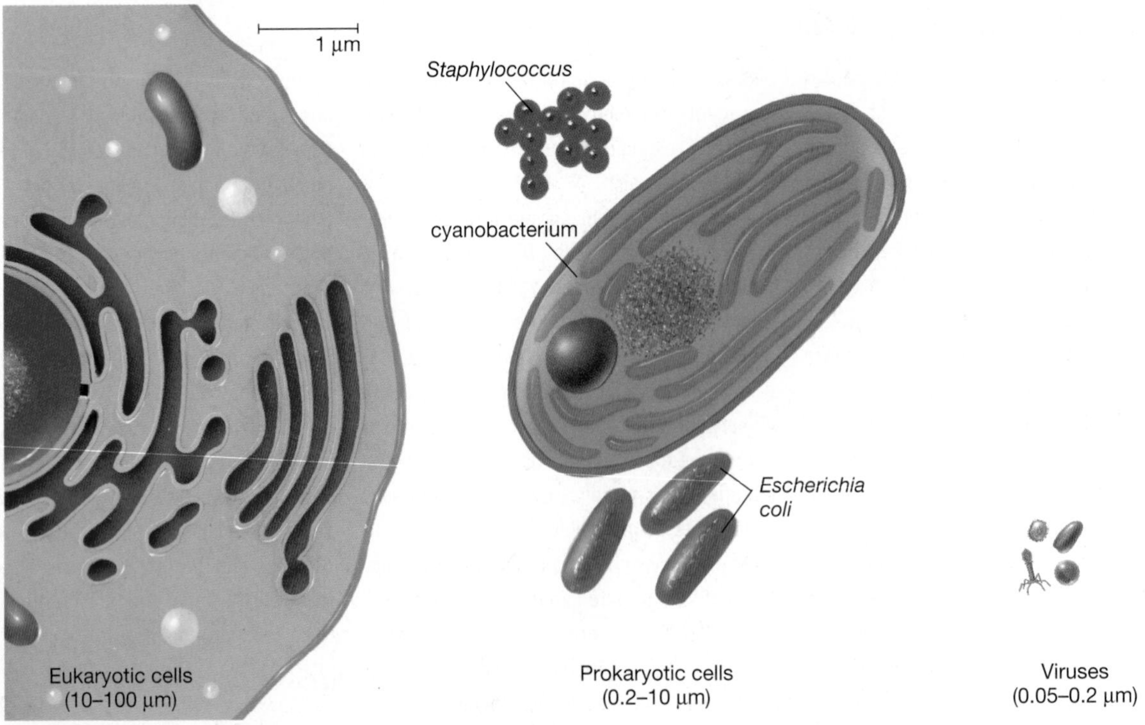

Figure 19-1 The sizes of microorganisms
The relative sizes of eukaryotic cells, prokaryotic cells, and viruses (1 μm = 1/1000 millimeter).

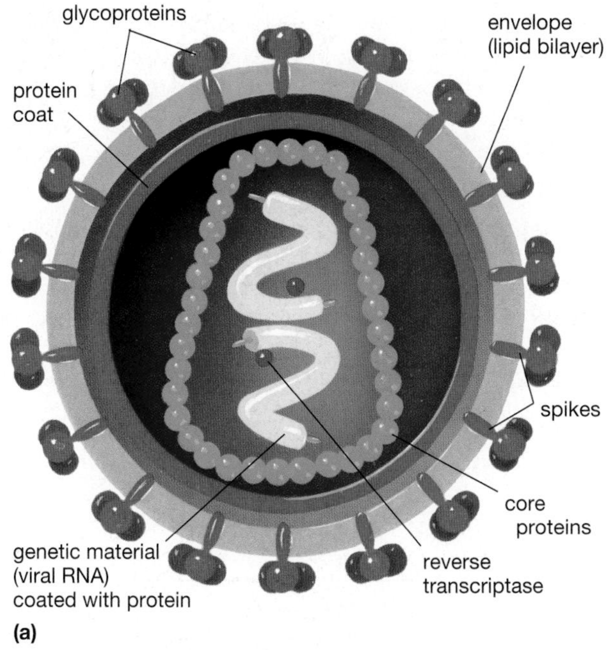

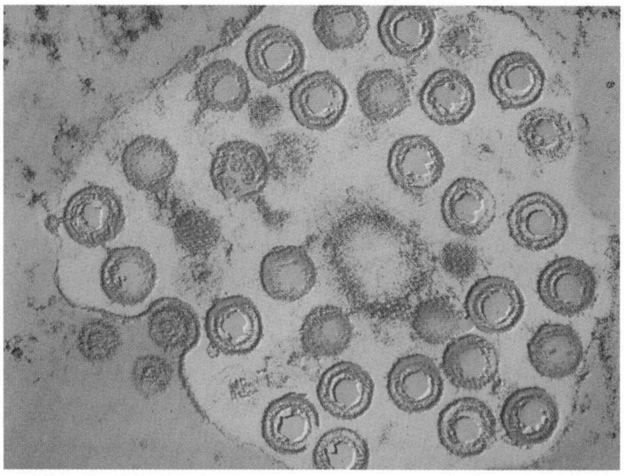

Figure 19-2 Viral structure and replication
(a) Cross section of the virus that causes AIDS. Inside is genetic material surrounded by a protein coat and molecules of reverse transcriptase, an enzyme that catalyzes the transcription of DNA from the viral RNA template after the virus enters the host cell. The virus that causes AIDS is among those that also have an outer envelope that is formed from the host cell's plasma membrane. Spikes made of glycoprotein (protein and carbohydrate) project from the envelope and help the virus attach to its host cell. ***(b)*** In this electron micrograph, herpes viruses are seen packed into an infected cell.

we shall see in Chapter 31). Viruses have also been linked to some types of cancer, such as T-cell leukemia, a cancer of the white blood cells. The recent identification of the papilloma virus, which causes genital warts, in 90% of cervical cancers sampled suggests that this virus may cause these cancers.

Because viruses are intracellular infectious agents that require the cellular machinery of their host, the illnesses they cause are difficult to treat: Antiviral agents may destroy host cells as well as viruses. The antibiotics that are so effective against many bacterial infections are useless against viruses. Viral diseases are thus intriguing to developers of biological weapons, who are especially interested in the Ebola virus. The virus causes Ebola hemorrhagic fever, a serious disease that kills more than 90% of its victims, most of whom

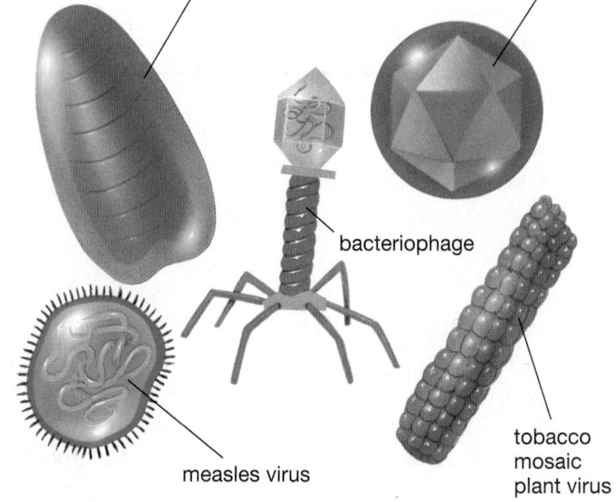

Figure 19-3 Viruses come in a variety of shapes
Viral shape is determined by the nature of the virus's protein coat. Viruses such as the rabies and herpes viruses are surrounded by an extra envelope derived from membranes of the host cell.

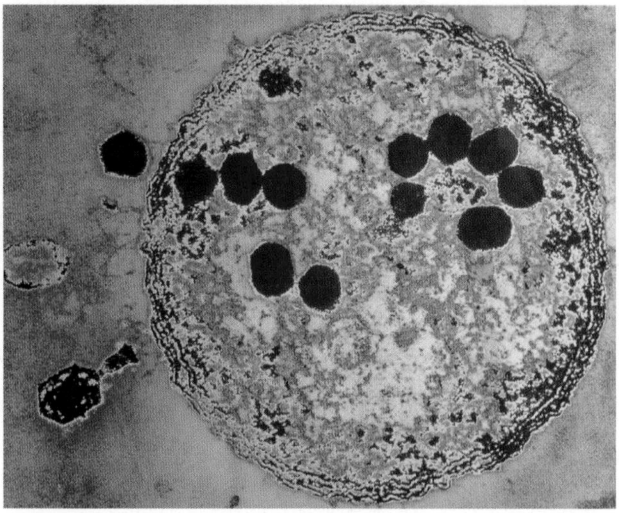

Figure 19-4 Some viruses infect bacteria
In this electron micrograph, bacteriophages are seen attacking a bacterium. They have injected their genetic material inside, leaving their protein coats clinging to the bacterial cell wall. The black objects inside the bacterium are newly forming viruses.

A Closer Look
Viruses—How the Nonliving Replicate

Viruses multiply, or "replicate," by using their own genetic material, which consists of single-stranded or double-stranded RNA or DNA, depending on the virus. This material serves as a template (a "blueprint") for the viral proteins and genetic material required to make new viruses. Viral enzymes may participate in replication as well, but the overall process depends on the biochemical machinery that the host cell uses to make its own proteins.

Viral replication follows a general sequence:

1. **Penetration.** Viruses may be engulfed by their host cell (endocytosis). Some viruses have surface proteins that bind to receptors on the host cell's plasma membrane and stimulate endocytosis. Other viruses are coated with an envelope that can fuse with the host's membrane. The viral genetic material is then released into the cytoplasm.

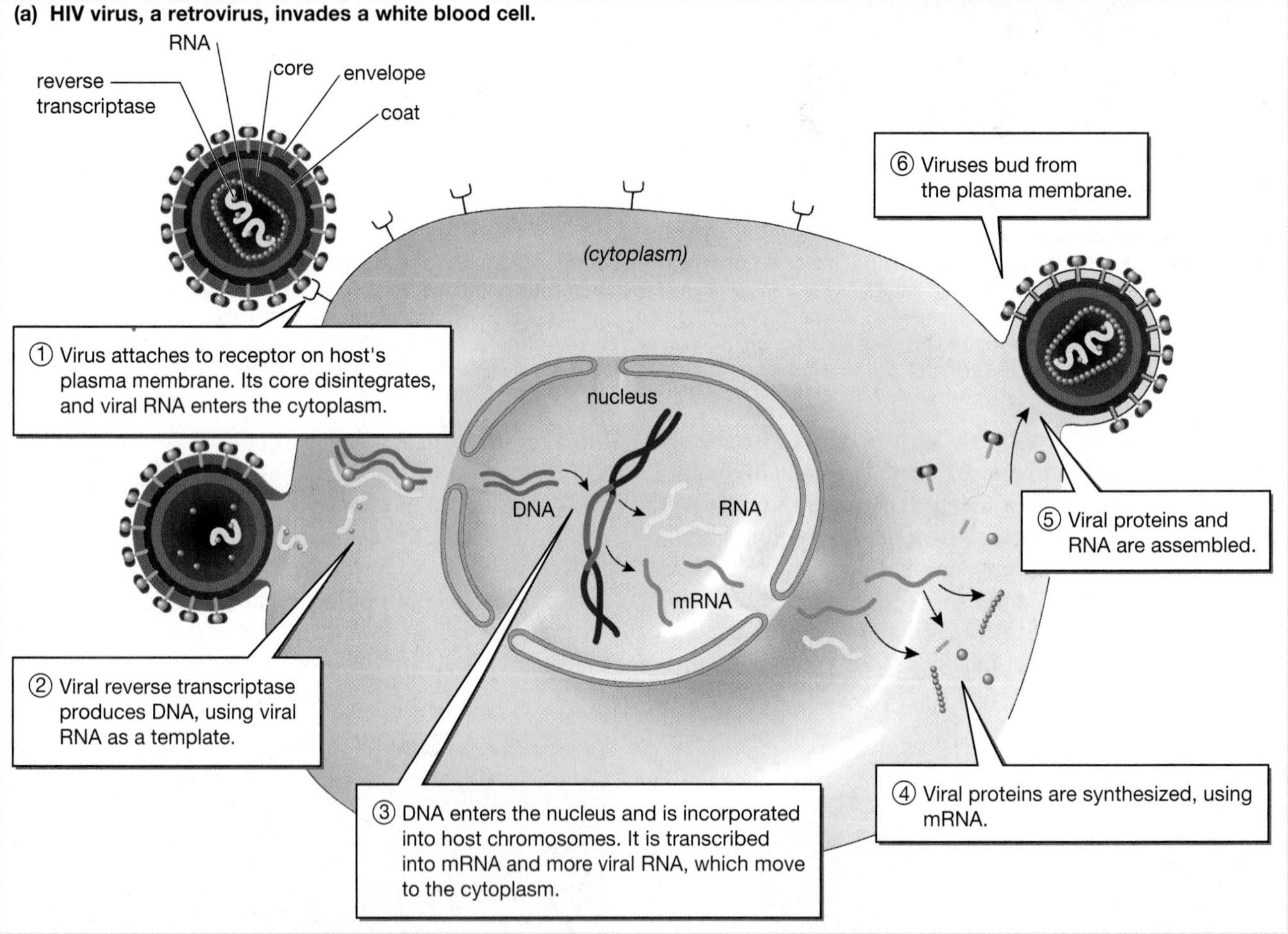
(a) HIV virus, a retrovirus, invades a white blood cell.
RNA
reverse transcriptase
core
envelope
coat
(cytoplasm)
⑥ Viruses bud from the plasma membrane.
① Virus attaches to receptor on host's plasma membrane. Its core disintegrates, and viral RNA enters the cytoplasm.
nucleus
DNA
RNA
mRNA
⑤ Viral proteins and RNA are assembled.
② Viral reverse transcriptase produces DNA, using viral RNA as a template.
③ DNA enters the nucleus and is incorporated into host chromosomes. It is transcribed into mRNA and more viral RNA, which move to the cytoplasm.
④ Viral proteins are synthesized, using mRNA.

Figure E19-1 How viruses replicate

are in Africa. Ebola is especially worrisome, both as an emerging infectious disease and as a potenial biological weapon, because there is currently no effective treatment and no vaccine to prevent infections.

Despite the difficulty of attacking viruses as they "hide" within cells, a number of antiviral drugs have been developed. Many of these drugs destroy or block the functioning of enzymes that the target virus requires for replication. In almost all cases, however, the benefits of antiviral drugs are limited, because viruses quickly evolve resistance to the drugs. Mutation rates are very high in viruses, in part because viruses lack mechanisms for correcting errors that occur during replication of DNA or RNA. It is thus almost inevitable that when a population of viruses is under attack by an antiviral drug, a mutation that confers resistance to the drug will arise. The resistant viruses prosper and replicate in great numbers, eventually spreading to new human hosts. Ultimately, resistant viruses come to predominate, and a formerly helpful antiviral drug is rendered ineffective.

2. **Replication.** The viral genetic material is copied many times.
3. **Transcription.** Viral genetic material is used as a blueprint to make messenger RNA (mRNA).
4. **Protein synthesis.** In the host cytoplasm, viral mRNA is used to synthesize viral proteins.
5. **Viral assembly.** The viral genetic material and enzymes are surrounded by their protein coat.
6. **Release.** Viruses emerge from the host cell by "budding" from the cell membrane or by bursting the cell.

Here we depict two types of viral life cycle. In Figure E19-1a, the *human immunodeficiency virus (HIV)*, which causes AIDS, is a *retrovirus*. Retroviruses use single-stranded RNA as a template to make double-stranded DNA by using a viral enzyme called *reverse transcriptase*. Many other retroviruses exist, and several cause cancers or tumors. In Figure E19-1b, the *herpes virus* contains double-stranded DNA that is transcribed into mRNA.

(b) Herpes virus, a double-stranded DNA virus, invades a skin cell.

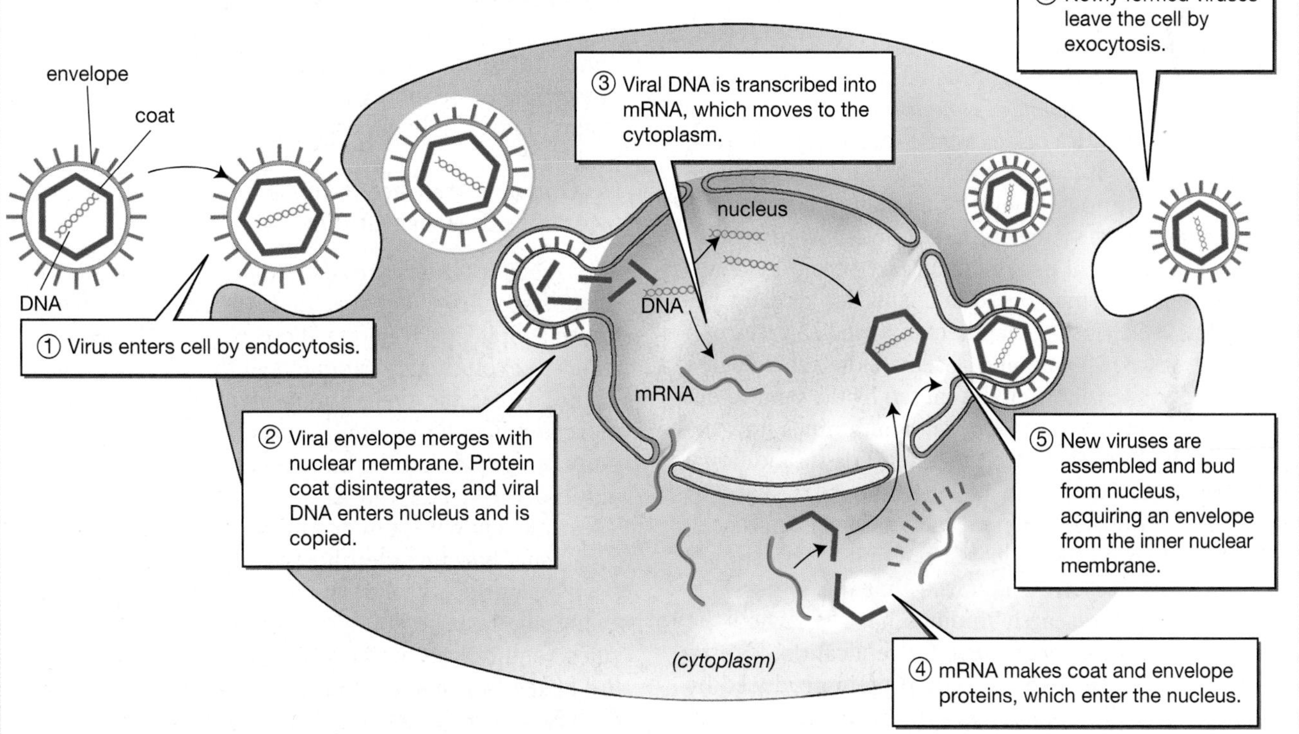
envelope
coat
DNA
① Virus enters cell by endocytosis.
② Viral envelope merges with nuclear membrane. Protein coat disintegrates, and viral DNA enters nucleus and is copied.
③ Viral DNA is transcribed into mRNA, which moves to the cytoplasm.
nucleus
DNA
mRNA
(cytoplasm)
④ mRNA makes coat and envelope proteins, which enter the nucleus.
⑤ New viruses are assembled and bud from nucleus, acquiring an envelope from the inner nuclear membrane.
⑥ Newly formed viruses leave the cell by exocytosis.

MEDIATUTOR 19.2 Herpes Virus Replication

Some Infectious Agents Are Even Simpler Than Viruses

In 1971 plant pathologist T. O. Diener discovered that some plant diseases are caused by particles only one-tenth the size of normal plant viruses. Called **viroids**, these particles are merely short strands of RNA that lack even a protein coat. Viroids apparently enter the nucleus of the infected cell, where they direct the synthesis of new viroids. About a dozen crop diseases, including cucumber pale fruit disease, avocado sunblotch, and potato spindle tuber disease, have been attributed to viroids.

Prions are even more puzzling than viroids. In the 1950s, physicians studying the Fore, a primitive tribe in New Guinea, were puzzled to observe numerous cases of a fatal degenerative disease of the nervous system, which the Fore called *kuru*. The symptoms of **kuru**—loss of coordination, dementia, and ultimately death—were similar to those of the rare but more widespread

Creutzfeldt-Jakob disease in humans and of *scrapie*, a disease of domestic livestock. Each of these diseases typically results in brain tissue that is spongy—riddled with holes. The researchers in New Guinea eventually determined that kuru was transmitted by ritual cannibalism; members of the Fore tribe honored their dead by consuming their brains. This practice has since stopped, and kuru has virtually disappeared. Clearly, kuru was caused by an infectious agent transmitted by infected brain tissue—but what was that agent?

In 1982 Nobel-Prize-winning neurologist Stanley Prusiner published evidence that scrapie (and, by extension, kuru, Creutzfeldt-Jakob disease, and a number of other, similar afflictions) is caused by an infectious agent that consists of nothing but protein. This idea seemed preposterous at the time, because most scientists believed that infectious agents must contain genetic material such as DNA or RNA in order to replicate. But Prusiner and his colleagues were able to isolate the infectious agent from scrapie-infected hamsters and to demonstrate that it contained no nucleic acids. The researchers called these protein-only, infectious particles *prions* (Fig. 19-5).

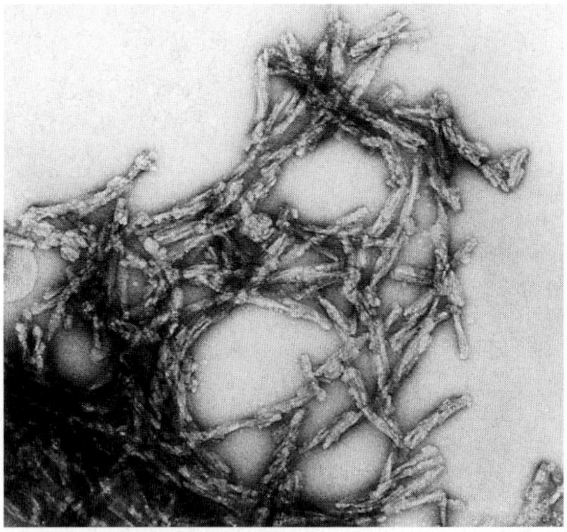

Figure 19-5 Prions: puzzling proteins
A section from the brain of a hamster infected with scrapie (a degenerative disease of the nervous system) contains fibrous clusters of prion proteins.

Of current concern is the possibility that humans can be infected with *bovine spongiform encephalopathy* ("mad cow disease"), a prion disease that affects cattle, by eating beef from infected animals. Laboratory research indicates that prions could potentially infect across species boundaries, so the spread of mad cow disease to humans is a very real possibility. In fact, recent evidence suggests that transmission to humans has already occurred. In 1996 British researchers reported 15 cases of a previously unknown variant of Creutzfeldt-Jakob disease that has since been shown to be caused by the same agent that causes mad cow disease. Health authorities now believe that these cases, as well as the 60 or so additional cases that have since been reported, are very likely to have been transmitted when the victims ate infected beef.

How can a protein replicate itself and be infectious? Not all researchers are convinced that this is possible. However, recent research findings have sketched the outline of a possible mechanism for replication. It turns out that prions consist of a single protein produced by normal nerve cells. Some copies of this normal protein molecule, for reasons still poorly understood, become folded into the wrong shape and are thus transformed into infectious prions. Once present, prions can apparently induce other, normal copies of the protein molecule to become transformed into prions. Eventually, the concentration of prions in nerve tissue may get high enough to cause cell damage and degeneration. Why would a slight alteration to a normally benign protein turn it into a dangerous cell-killer? No one knows.

Another peculiarity of the prion diseases is that they can be inherited as well as transmitted by infection. Recent research has shown that certain small mutations in the gene that codes for the "normal" prion protein increase the likelihood that the protein will fold into its abnormal form. If one of these mutations is genetically passed on to offspring, a tendency to develop a prion disease may also be inherited.

No One Is Certain How These Infectious Particles Originated

The origin of viruses, viroids, and prions is obscure. Some scientists believe that the huge variety of mechanisms for self-replication among these particles reflects their status as evolutionary remnants of the very early history of life, before the main line of evolution settled on the more familiar large, double-stranded DNA molecules. Another possibility is that viruses, viroids, and prions may be the degenerate descendants of parasitic cells. These ancient parasites (organisms that live in or on host organisms, harming their hosts in the process) may have been so successful at exploiting their hosts that they eventually lost the ability to synthesize all of the molecules required for survival and became dependent on the host's biochemical machinery. Whatever the origin of these infectious particles, their success poses a continuing challenge to living things.

2 Which Organisms Make Up the Prokaryotic Domains—Bacteria and Archaea?

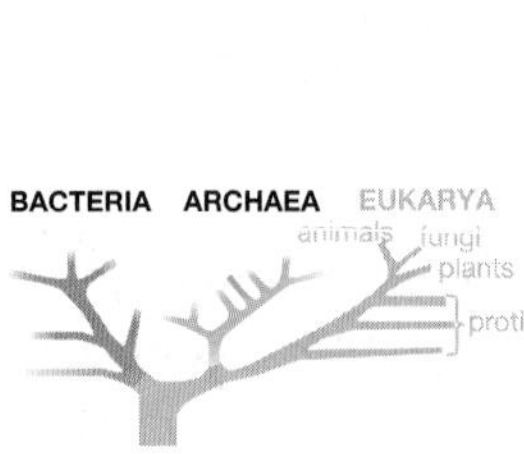

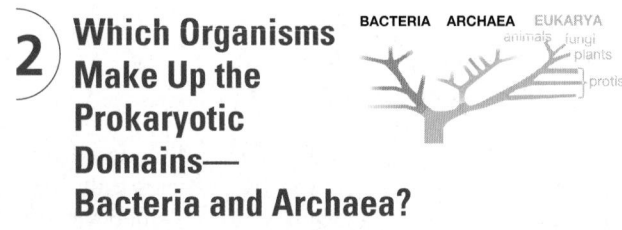

Two of life's three domains (see Chapter 18) consist entirely of prokaryotes, single-celled microbes that lack organelles such as the nucleus, chloroplasts, and mitochondria. (See Chapter 5 for a comparison of prokaryotic and eukaryotic cells.) In the past, all prokaryotes were categorized as bacteria. On the basis of an improved understanding of evolutionary history, however, we now recognize two fundamentally different prokaryotic domains, **Bacteria** and **Archaea**. Both bacteria and archaea are normally very small, ranging from about 0.2 to 10 micrometers in diameter, compared with eukaryotic cells, whose diameters range from about 10 to 100 micrometers. About 250,000 average-sized bacteria or archaea could congregate on the period at the end of this sentence.

Bacteria and Archaea Are Fundamentally Different

Until recently, the archaea were thought to be part of the same evolutionary lineage as bacteria and were therefore inaccurately called "archaebacteria." Since the 1970s, however, it has become increasingly clear that the archaea diverged from the bacteria very early in the history of life and that the two groups are only very distantly related. Recent analysis of RNA nucleotide sequences has revealed that the archaea are more closely related to eukaryotes than are the bacteria. It thus appears that life split into archaeal and bacterial lineages very early in its history and that eukaryotes are derived from a later branch of the archaeal line.

Bacteria and archaea are superficially similar in appearance under the microscope, but the ancient evolutionary separation between the two domains is revealed by striking differences in the most basic structural and biochemical features. For example, the rigid **cell wall** that encases bacterial cells contains **peptidoglycan**, but the cell walls of archaea lack this uniquely bacterial substance. The structure and composition of other cellular components, such as plasma membranes, ribosomes, and RNA polymerases, also differ between the two domains, as do key features of basic processes such as transcription and translation.

Prokaryotes Are Difficult to Classify

The sharp biochemical differences between archaea and bacteria make distinguishing the two domains a straightforward matter, but classification within each domain poses particular challenges to taxonomists. Prokaryotes are tiny and structurally simple and simply do not exhibit the huge array of anatomical and developmental differences that are used to infer the evolutionary history of plants, animals, and other eukaryotes. Consequently, prokaryotes have historically been classified on the basis of such features as shape, means of locomotion, pigments, nutrient requirements, the appearance of *colonies* (groups of bacteria that descend from a single cell), and staining properties. For example, the **Gram stain**, a staining technique, distinguishes two types of cell-wall construction in bacteria, enabling us to classify them as either *gram-positive* or *gram-negative.*

In recent years our understanding of the evolutionary history of the prokaryotic domains has been greatly expanded by comparisons of DNA and RNA nucleotide sequences. On the basis of this new information, taxonomists now classify bacteria into about 13 to 15 kingdoms and archaea into 3 kingdoms. As mentioned in Chapter 18, however, prokaryote taxonomy is a rapidly changing field, and consensus on kingdom-level classification has thus far proved elusive. With new DNA sequence data being generated at a furious pace, and with new and distinctive types of bacteria and archaea being discovered and described on a regular basis, prokaryote classification schemes will likely continue to be dynamic and frequently revised for some time to come.

Prokaryote taxonomy is also complicated by the difficulty of defining species boundaries in prokaryotes. Because their reproduction is usually asexual, species definitions based on the ability of populations to interbreed are meaningless. One increasingly popular alternative approach is to define prokaryotic species strictly on the basis of genetic differences. Under this approach, two types of prokaryote are declared to be separate species if the number of DNA sequence differences between them exceeds some threshold level. Some researchers have proposed that a 3% difference be adopted as the threshold level for species identification but, again, there is no consensus among investigators.

Prokaryotes Exhibit a Variety of Shapes and Structures

A Cell Wall Gives Prokaryotes Protection and Their Characteristic Shapes

The cell wall that surrounds prokaryotic cells protects them from osmotic rupture in watery environments and gives characteristic shapes to different types of bacteria and archaea. The most common shapes are rodlike, spherical, and corkscrew-shaped (Fig. 19-6). The cell walls of some bacterial species are surrounded by sticky **capsules** or **slime layers**, composed of polysaccharide or protein. Capsules help certain disease-causing bacteria escape detection by their host animal's immune system. Slime layers allow the bacteria that cause tooth decay to adhere in masses to the smooth surface of a tooth. This slime forms the basis of dental plaque (Fig. 19-7). Other bacteria cover themselves with a fuzz of hairlike projections called **pili** (singular, **pilus**). Pili are made of protein and generally serve to attach the bacterium to other cells. The pili of some infectious bacteria, such as those causing the sexually transmitted disease gonorrhea, attach to the cell membranes of their host, facilitating infection. Some bacteria produce special sex pili, described later.

Some Prokaryotes Can Move by Using Simple Flagella

Some bacteria and archaea are equipped with **flagella** (singular, **flagellum**). These are simpler in structure than the flagella seen in some eukaryotic cells, as discussed in

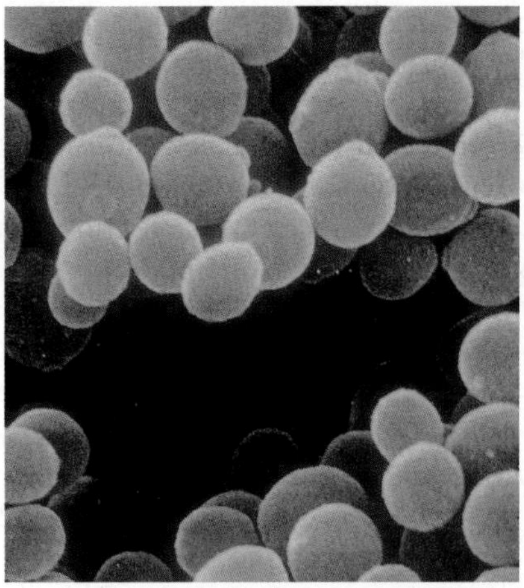

(a) 0.10 μm

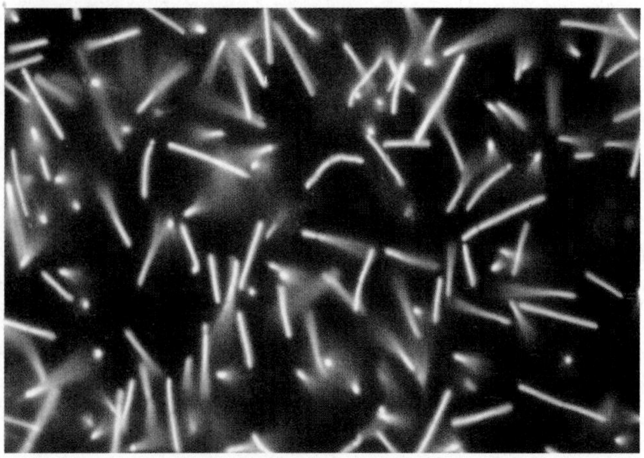

(b)

Figure 19-6 Three common prokaryote shapes
(a) Spherical bacteria of the genus *Micrococcus*; ***(b)*** rod-shaped archaean of the genus *Methanopyrus*; and ***(c)*** corkscrew-shaped bacteria of the species *Leptospirosis interrogans*.

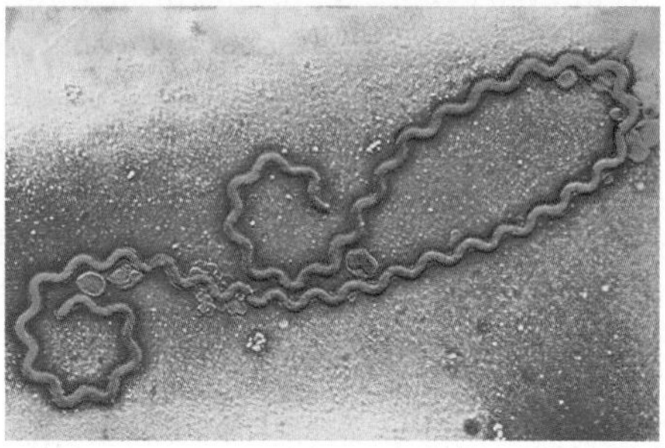

(c) 0.25 μm

Chapter 6. Prokaryote flagella may appear singly at one end of a cell, in pairs (one at each end of the cell), as a tuft at one end of the cell (Fig. 19-8a), or scattered over the entire cell surface. Flagella can rotate rapidly, propelling the organism through its liquid environment. In bacteria, a unique wheel-like structure embedded in the bacterial membrane and cell wall allows the flagellum to rotate (Fig. 19-8b).

Flagella allow prokaryotes to disperse into new habitats, to migrate toward nutrients, and to leave unfavorable environments. Flagellated bacteria may orient toward various stimuli, a behavior called a **taxis**. Some bacteria are **chemotactic**, moving toward chemicals given off by food or away from toxic chemicals. Some are **phototactic**, moving toward or away from light, depending on the habitat they require. Other flagellated bacteria are **magnetotactic**. These forms detect Earth's magnetic field by means of tiny magnets formed from iron crystals within their cytoplasm. They use this unique sensory system to direct their beating flagella to move themselves downward into aquatic sediments.

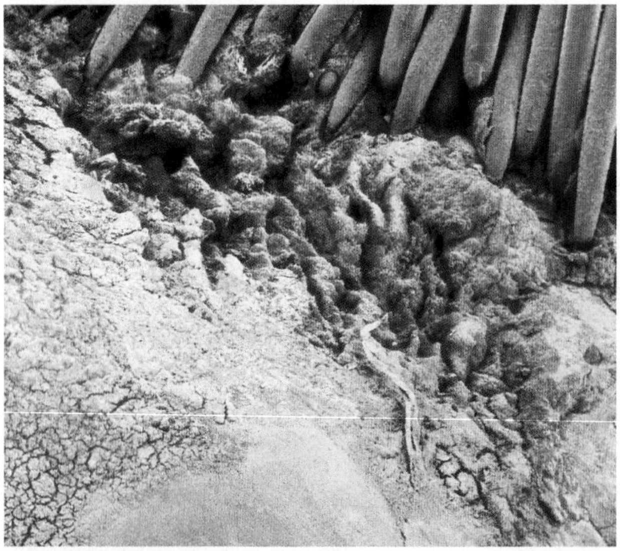

Figure 19-7 The cause of tooth decay
The human body provides a safe environment for a diverse assortment of bacteria, including these bacteria that inhabit the mouth. These bacteria possess a slime layer that allows them to cling to tooth enamel, where they can cause tooth decay unless they are removed by their chief antagonist, a toothbrush (seen here as green bristles).

Protective Endospores Allow Some Bacteria to Withstand Adverse Conditions

When environmental conditions become inhospitable, many rod-shaped bacteria form protective resting structures called **endospores**. The endospore (literally, "inside spore") forms inside the bacterium (Fig. 19-9). It contains genetic material and a few enzymes encased within a thick protective coat. Metabolic activity ceases. Endospores are resistant structures that can survive extremely unfavorable conditions. Some can withstand boiling for an hour or more. Others can survive for extraordinarily long periods. In the most extreme example of such longevity, scientists recently discovered bacterial spores that had been sealed inside rock for 250 million years. After being carefully extracted from their rocky "tomb," the spores were incubated in test tubes. Amazingly, live bacteria developed from the an-

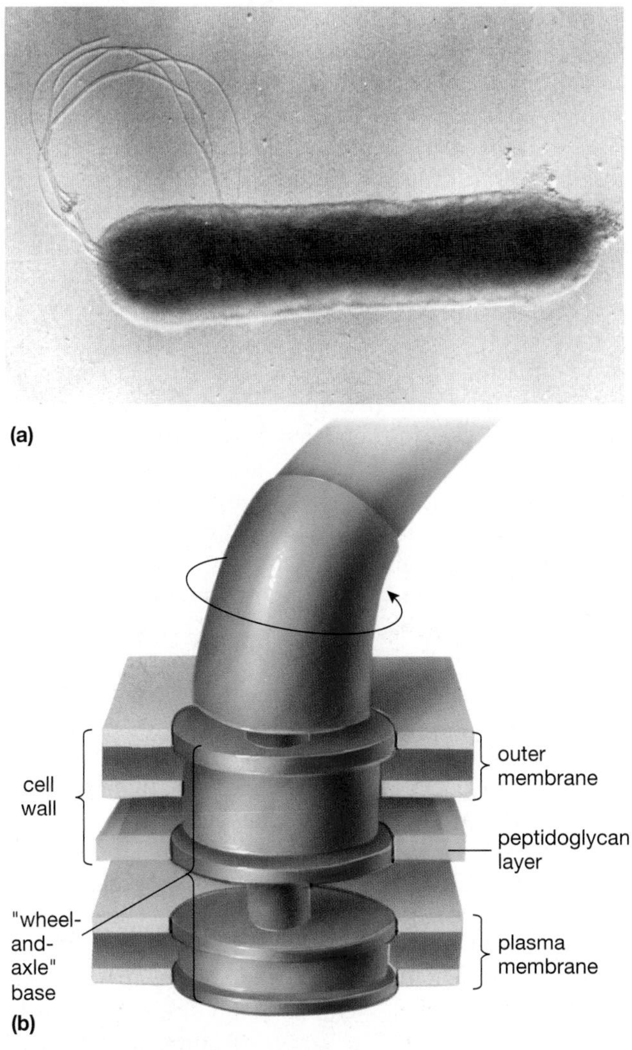

Figure 19-8 The prokaryote flagellum
(a) A flagellated archaean of the genus *Aquifex* uses its flagella to move toward favorable environments. ***(b)*** In bacteria, a unique "wheel-and-axle" arrangement anchors the flagellum within the cell wall and plasma membrane, so the flagellum can rotate rapidly.

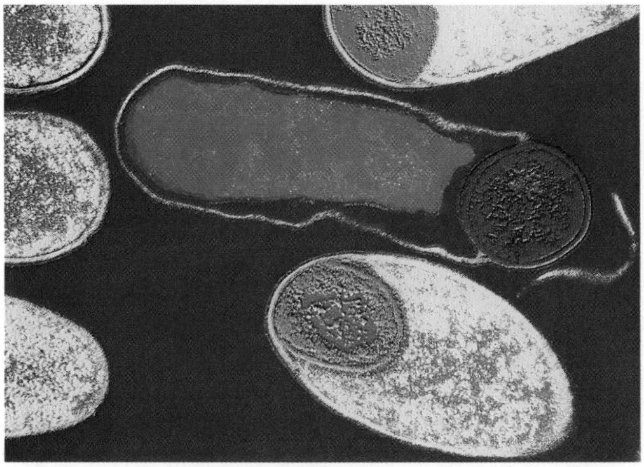

Figure 19-9 Spores protect some bacteria
Resistant endospores, here colored red, have formed inside bacteria of the genus *Clostridium*, which causes the potentially fatal food poisoning called *botulism.*

cient spores, which were older than the oldest dinosaur fossils.

Endospores are one of the main reasons that the bacterial disease anthrax has become the most likely agent of biological warfare or terrorism. The bacterium that causes anthrax forms endospores, and the spores provide the means by which the designers of biological weapons intend to disperse the the bacteria. The spores can be stored indefinitely and can survive the harsh conditions of a missile launch and high-altitude travel. At their target, the spores can survive dispersal into the atmosphere and can remain viable until inhaled by a potential victim.

Prokaryotes Reproduce by Binary Fission

Most prokaryotes reproduce asexually by a simple form of cell division called **binary fission** (see Chapter 11), which produces genetically identical copies of the original cell (Fig. 19-10). Under ideal conditions, a prokaryotic cell can divide about once every 20 minutes, potentially giving rise to sextillions (10^{21}) of offspring in a single day! This rapid reproduction allows bacteria to exploit temporary habitats, such as a mud puddle or warm pudding. Recall that many mutations, the source of genetic variability, occur as a result of mistakes in DNA replication during cell division (see Chapter 10). Thus, the rapid reproductive rate of bacteria provides ample opportunity for new forms to arise and also allows mutations that enhance survival to spread quickly.

Some bacteria transfer genetic material from a donor bacterium to a recipient bacterium during a process called **bacterial conjugation**. Some conjugating bacteria

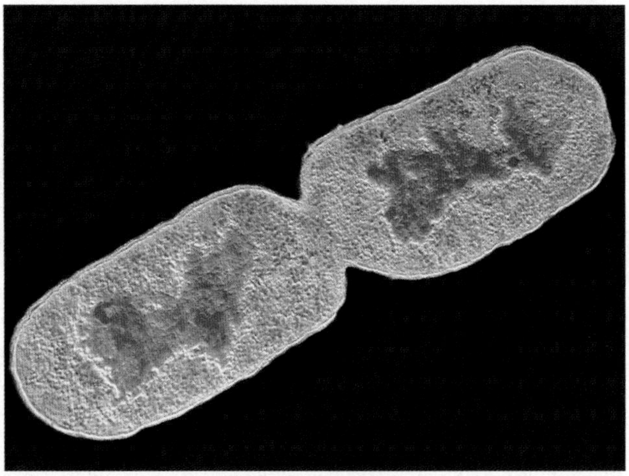

Figure 19-10 Reproduction in prokaryotes
Prokaryotic cells reproduce by a simple form of cell division called *binary fission.* In this color-enhanced electron micrograph, *Escherichia coli*, a normal component of the human intestine, is dividing. Red areas are genetic material.

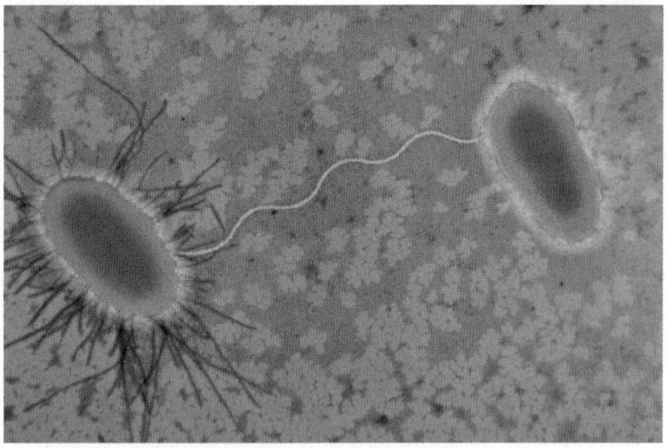

MEDIATUTOR 19.3 Bacterial Conjugation

Figure 19-11 Conjugation: prokaryotic "mating"
Conjugation—the transfer of genetic material—occurs through a special large, hollow sex pilus, shown here connecting a pair of *Escherichia coli.* One bacterium (at top right) acts as a donor, transferring DNA to the recipient. In this photo, the donor bacterium is bristling with nonsex pili that probably help it attach to surfaces.

use specialized hollow structures called *sex pili* to transfer genetic material (Fig. 19-11). The genetic material that is transferred during bacterial conjugation is located outside the single, circular bacterial chromosome, in a structure called a **plasmid**. A plasmid is a small, circular DNA molecule that may carry genes for antibiotic resistance or even alleles of genes also found on the main bacterial chromosome. Researchers in molecular genetics have made extensive use of bacterial plasmids, as described in Chapter 13. Conjugation produces new genetic combinations that may allow the resulting bacteria to survive under a greater variety of conditions.

Prokaryotes Are Specialized for Specific Habitats

Prokaryotes occupy virtually every habitat, including those where extreme conditions prevent occupation by other forms of life. For example, some bacteria thrive in near-boiling environments, such as the hot springs of Yellowstone National Park (Fig. 19-12). Many archaea live in even hotter environments, including springs where the water actually boils or deep-ocean vents, where superheated water is spewed through cracks in Earth's crust at temperatures of up to 230 °F (110 °C). It is also pretty warm 1.7 miles (2.8 kilometers) below Earth's surface, which is where scientists recently discovered a new bacterial species. Bacteria and archaea are also found in very cold environments, such as Antarctic sea ice that remains frozen nearly year-round. Even extreme chemical conditions fail to impede invasion by prokaryotes. Thriving colonies of bacteria and archaea live in the Dead Sea, where a salt concentration seven times that of the oceans precludes all other life, and in waters that are as acidic as vinegar or as alkaline as household ammonia. Of course, rich bacterial communities also reside in a full range of more-moderate habitats, including in and on the healthy human body. An animal need not remain healthy to harbor bacteria, though. Recently, a colony of bacteria was found dormant within the intestinal contents of a mammoth that had lain in a peat bog for 11,000 years.

No single species of prokaryote, however, is as versatile as these examples may suggest. In fact, most prokaryotes are specialists. One species of archaea that inhabits deep-sea vents, for example, grows optimally at 223 °F (106 °C) and stops growing altogether at temperatures below 194 °F (90 °C). Clearly, this species could not survive in a less extreme habitat. Bacteria that live on the human body are also specialized; different species colonize the skin, the mouth, the respiratory tract, the large intestine, and the urogenital tract.

Prokaryotes Exhibit Diverse Metabolisms

Prokaryotes are able to colonize such diverse habitats partly because they are able to use such a wide variety of nutrient sources. Some types of bacteria, such as the

Figure 19-12 Some prokaryotes thrive in extreme conditions
Hot springs harbor bacteria and archaea that are both heat-tolerant and mineral-tolerant. Several species of cyanobacteria paint these hot springs in Yellowstone National Park with vivid colors, and each is confined to a specific area determined by temperature range. The bacterial pigments aid in photosynthesis.

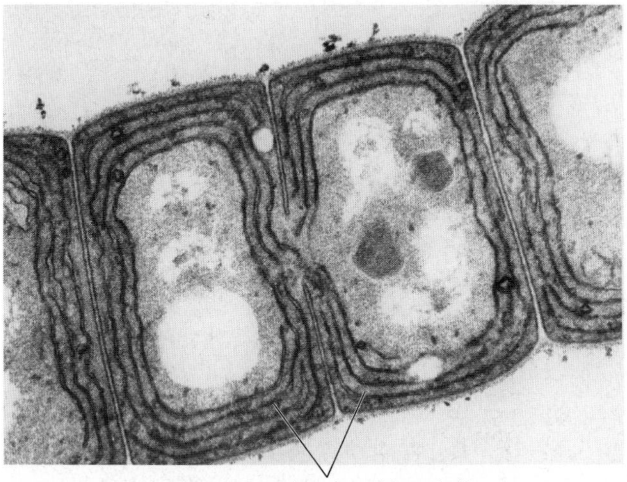

Figure 19-13 Cyanobacteria
Electron micrograph of a section through a cyanobacterial filament (genus *Oscillatoria*). Chlorophyll is located on the membranes visible within the cells.

cyanobacteria (Fig. 19-13), engage in plantlike photosynthesis. Like green plants, cyanobacteria possess chlorophyll, produce oxygen as a by-product of photosynthesis, and exist only where light and oxygen are available. Other photosynthetic bacteria, known as the sulfur bacteria, use hydrogen sulfide (H_2S) instead of water (H_2O) in photosynthesis, releasing sulfur instead of oxygen. No photosynthetic archaea are known.

Many species of bacteria and archaea are **chemosynthetic**, deriving energy through reactions that combine oxygen with inorganic molecules such as sulfur, ammonia, or nitrite. In the process they may release *sulfates* or *nitrates*, crucial plant nutrients, into the soil. Some archaea are **methanogens**, organisms that convert carbon dioxide to methane (sometimes called "swamp gas"). Many prokaryotes are **anaerobes**, which means that they do not require oxygen to extract energy. Some anaerobes, such as many of the archaea found in hot springs and the bacterium that causes *tetanus*, are actually poisoned by oxygen. Others are opportunists, engaging in fermentation when oxygen is lacking and switching to aerobic respiration (a more efficient process) when oxygen becomes available.

Prokaryotes Perform Many Functions That Are Important to Other Forms of Life

Did you know that if it weren't for bacteria, we would not be eating steaks and hamburgers? Cows, along with sheep and some nondomestic mammals such as deer, are members of a group known as *ruminants*. Ruminants subsist by eating leaves, but they can't actually digest that plant material themselves. Instead, they depend on certain bacteria that have the unusual ability to break down cellulose, the principal component of plant cell walls. Some of these bacteria have entered into a **symbiotic** (literally, "living together") relationship with the ruminants, and live in the animals' digestive tracts, where they help liberate nutrients from plant fodder that the animals are unable to break down themselves. Without the symbiotic bacteria, cows could not survive. No cows, no hamburgers.

Prokaryotes also have other important impacts on human nutrition. Many foods, including cheese, yogurt, and sauerkraut, are produced by the action of bacteria. Symbiotic bacteria also inhabit your intestines. These bacteria feed on undigested food and synthesize such nutrients as vitamin K and vitamin B_{12}, which the human body absorbs.

Humans could not live without plants, and plants are entirely dependent on bacteria. In particular, plants are unable to capture nitrogen from that element's most abundant reservoir, the atmosphere. Plants need nitrogen to grow. To acquire it, they depend on **nitrogen-fixing bacteria**, which live both in soil and in specialized *nodules*, small, rounded lumps on the roots of certain plants (**legumes**, which include alfalfa, soybeans, lupines, and clover; Fig. 19-14). The nitrogen-fixing bacteria capture nitrogen gas (N_2) from air trapped in the soil and combine it with hydrogen to produce ammonium (NH_4^+), a nitrogen-containing nutrient that plants can use directly.

Prokaryotes also play a crucial role in recycling waste. Many substances that we consider to be waste can serve as food to archaea and bacteria, most species of which are heterotrophic, obtaining energy by breaking down complex organic (carbon-containing) molecules. The range of compounds attacked by prokaryotes is staggering. Nearly anything that human beings can synthesize, including detergents and the poisonous solvent benzene, some prokaryote can destroy. The term *biodegradable* (meaning "broken down by living things") refers largely to the work of prokaryotes. Even oil is biodegradable. Soon after the tanker *Exxon Valdez* dumped 11 million gallons of crude oil into Prince William Sound, Alaska, researchers from Exxon sprayed oil-soaked beaches with a fertilizer that encouraged the growth of natural populations of oil-eating bacteria. Within 15 days the oil deposits were noticeably reduced compared with unsprayed areas. Prokaryotes are much less successful, however, at degrading most kinds of plastic. The search for useful, economical plastics that are also biodegradable is therefore a major focus of industrial research.

The appetite of some prokaryotes for nearly any organic compound is the key to their important role as decomposers in ecosystems. While feeding themselves, bacteria break down the waste products and dead bodies of plants and animals, freeing nutrients for reuse. The recycling of nutrients provides the basis for continued life on Earth, as we shall see in Chapter 40.

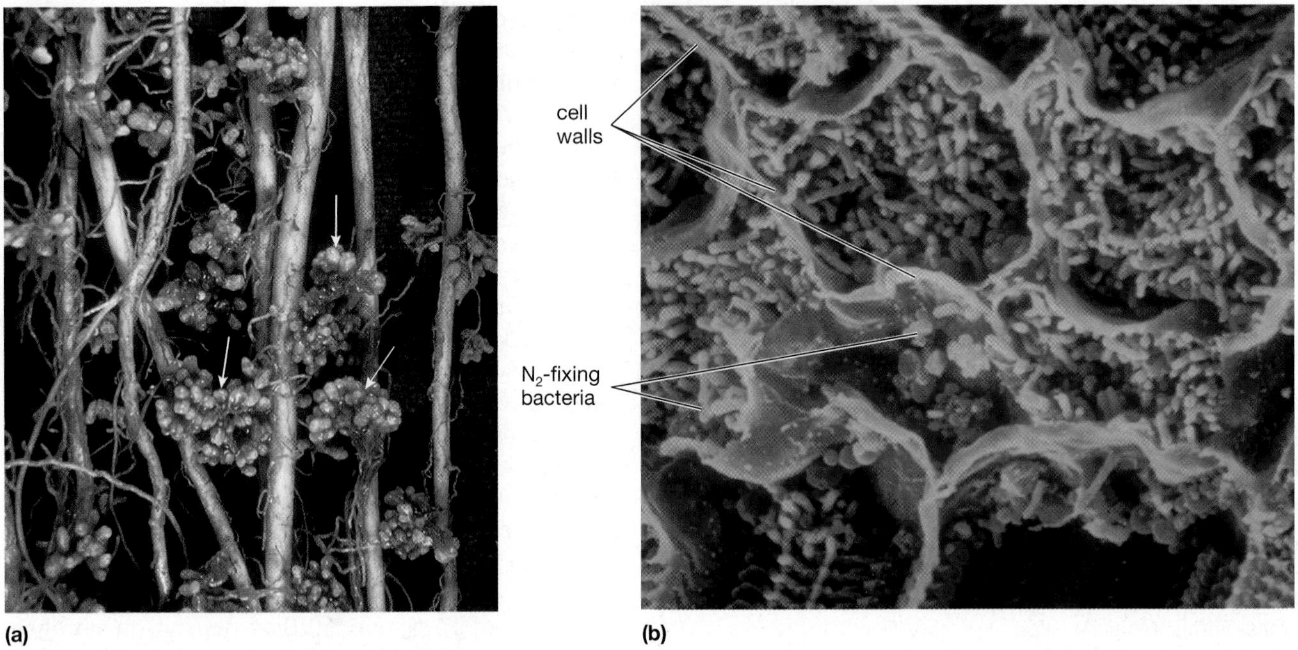

Figure 19-14 Nitrogen-fixing bacteria in root nodules
(a) Special chambers called nodules on the roots of a legume (alfalfa) provide a protected and constant environment for nitrogen-fixing bacteria. ***(b)*** This scanning electron micrograph shows the nitrogen-fixing bacteria inside cells within the nodules.

Some Bacteria Pose a Threat to Human Health

Despite the benefits some bacteria provide, the feeding habits of certain bacteria threaten our health and well-being. These **pathogenic** ("disease-producing") bacteria synthesize toxic substances that cause disease symptoms. (No pathogenic archaea have been identified yet.)

Some bacteria produce toxins that attack the nervous system. Examples of such pathogens include *Clostridium tetani*, which causes tetanus, and *C. botulinum*, which causes *botulism* (a lethal form of food poisoning). Both of these bacterial species are anaerobes that survive as spores until introduced into a favorable, oxygen-free environment. For example, a deep puncture wound may provide a means through which tetanus bacteria can penetrate a human body and reach a place where they will be protected from contact with oxygen. As they multiply, the bacteria release their paralyzing poison into the bloodstream. For botulism bacteria, a sealed container of canned food that has been improperly sterilized may provide a haven. Thriving on the nutrients so thoughtfully provided by their human benefactors, these anaerobes produce a toxin so potent that a single gram could kill 15 million people. Perhaps inevitably, this potent poison has caught the attention of biological weapon designers, who are presumed to have added it to their arsenals.

Bacterial diseases have had a significant impact on human history. Perhaps the most infamous example is the *plague*, or "Black Death," which killed 100 million people during the mid-fourteenth century. In many parts of the world, one-third or more of the population died. Plague is caused by the highly infectious bacteria *Yersinia pestis*, which is spread by fleas that feed on infected rats and then move to human hosts. Although plague has not reemerged as a large-scale epidemic, about 2000 to 3000 people worldwide are diagnosed with the disease each year.

Some bacterial pathogens seem to emerge suddenly. *Lyme disease*, for example, was unknown until 1975. This disease, named after the town of Old Lyme, Connecticut, where it was first described, is caused by the spiral-shaped bacterium *Borrelia burgdorferi*. The bacterium is carried by the deer tick and transmitted to humans who are bitten by the tick. At first, the symptoms resemble flu, with chills, fever, and body aches. If untreated, weeks or months later the victim may experience rashes, bouts of arthritis, and in some cases abnormalities of the heart and nervous system. Both physicians and the general public are becoming more familiar with the disease, so more victims are receiving treatment before serious symptoms develop.

Perhaps the most frustrating pathogens are those that come back to haunt us long after we believed that we had them under control. Tuberculosis, a bacterial disease once almost vanquished in developed countries, is again on the rise in the United States and elsewhere. Two sexually transmitted bacterial diseases, *gonorrhea* and *syphilis*, have reached epidemic proportions around the globe. Cholera, a water-transmitted disease that flourishes when raw sewage contaminates drinking water or fishing areas, is under control in developed countries but remains a major killer in poorer parts of the world.

Some pathogenic bacteria are so widespread and ubiquitous that we cannot expect to ever be totally free of their damaging effects. For example, different forms of the abundant streptococcus bacterium produce a variety of diseases. One type of streptococcus causes strep throat. Another, *Streptococcus pneumoniae*, causes pneumonia by stimulating an allergic reaction that clogs the lungs with fluid. Yet another form of strep has gained fame as the "flesh-eating bacterium." A small percentage of people who become infected with this strain of strep experience severe symptoms, described luridly in tabloid newspapers with such headlines as "Killer Bug Ate My Face." About 800 Americans each year are victims of *necrotizing fasciitis* (as the "flesh-eating" infection is more properly known), and about 15% of these victims die. The streptococci enter through broken skin and spew out toxins that either destroy flesh directly or stimulate an overwhelming and misdirected attack by the immune system against the body's own cells. A limb can be destroyed in hours, and in some cases only amputation can halt the rapid tissue destruction. In other cases, these rare strep infections sweep through the body, causing death within a matter of days; Jim Henson, creator of the Muppets, died as a result of such an infection.

Although some bacteria assault the human body, most of the bacteria with which we share our bodies are harmless and many are beneficial. For example, the normal bacterial community in the female vagina creates an environment that is hostile to infections by parasites such as yeasts. Bacteria harmlessly inhabiting our intestines are an important source of vitamin K. As the late physician, researcher, and author Lewis Thomas so aptly put it, "Pathogenicity is, in a sense, a highly skilled trade, and only a tiny minority of all the numberless tons of microbes on the Earth has ever been involved in it; most bacteria are busy with their own business, browsing and recycling the rest of life."

3 Which Organisms Make Up the Kingdom Protista?

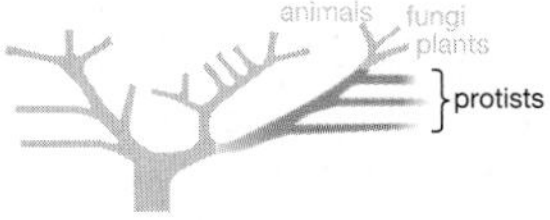

The third domain, Eukarya, includes one group of microbes, the kingdom **Protista**, as well as multicellular organisms (the kingdoms Fungi, Plantae, and Animalia, which we shall discuss in Chapters 20 through 22, respectively). To appreciate the single-celled eukaryotes of the kingdom Protista, we have to overcome our size bias. Protists are largely invisible to us as we go about our daily lives. If we could somehow inhabit their microscopic scale, however, we might be more impressed with their spectacular and beautiful forms, their varied and active lifestyles, their astonishingly diverse modes of reproduction, and the variety of structural and physiological innovations that are possible within the limits of a single cell.

Eukaryotic cells contain many membrane-bound organelles that prokaryotic cells lack. Organelles such as mitochondria and chloroplasts are similar in size to typical bacteria and indeed probably evolved from bacteria (see Chapter 17). Within the kingdom Protista are some of the most complex eukaryotic cells in existence, with organelles taking on functions served by organs in multicellular organisms.

Since Anton van Leeuwenhoek first observed protists through his simple homemade microscope in 1674, at least 50,000 species have been described. Although some protists form colonies, most consist of a single eukaryotic cell. Most protists reproduce asexually by mitotic cell division, but many are also capable of a form of sexual reproduction called *conjugation* (Fig. 19-15). All three major modes of nutrition are represented in

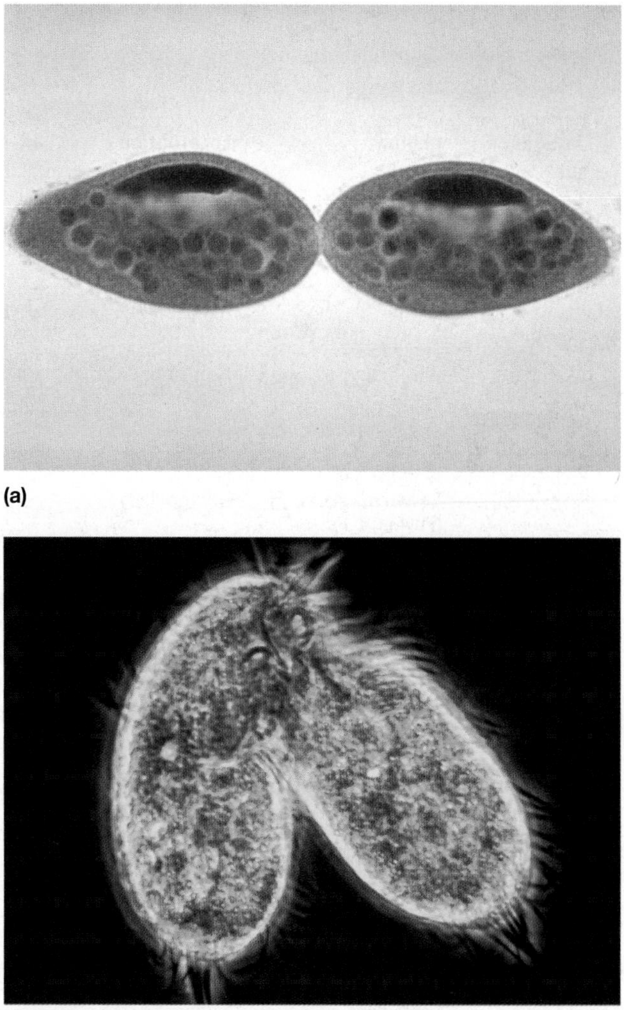

(a)

(b)

Figure 19-15 Two modes of protistan reproduction
(a) *Paramecium*, a ciliate, reproduces asexually by cell division that results in two daughters identical to the original parent.
(b) Mating in a ciliate, *Euplotes*. Genetic material is exchanged across a cytoplasmic bridge. After the exchange occurs, new individuals formed by cell division will have gene combinations different from those of either parent cell.

this kingdom: For example, algae capture solar energy through photosynthesis; predatory protists ingest their food; and parasitic forms can absorb nutrients from their surroundings.

Protists Are a Diverse Group Including Funguslike, Plantlike, and Animal-like Forms

The kingdom Protista is extremely diverse. Historically, the mostly single-celled organisms that compose this kingdom have been grouped on the basis of apparent similarities to the cells of fungi, plants, or animals. We now recognize that these similarities are for the most part superficial and do not generally reflect shared evolutionary history with members of the other eukaryotic kingdoms. Nonetheless, categorizing protists as funguslike, plantlike, or animal-like (Table 19-1) is both traditional and convenient, and we will follow that convention in the descriptions that follow. Keep in mind, however, that most of the protist groups in each of these categories of convenience represent branches of the tree of life that split off long before the branches that led to fungi, plants, and animals. Thus, the members of each category (funguslike, plantlike, animal-like) are not closely related to fungi, plants or animals, and are not necessarily closely related to one another.

The Water Molds and Slime Molds Are Funguslike Protists

Some protists are characterized by physical similarity to the filaments or fruiting bodies of fungi and have a funguslike mode of nutrition. Like fungi, they absorb nutrients from the soil, water, or tissues of other organisms and typically help decompose dead organisms. The funguslike protists form three groups: (1) the water molds, (2) the acellular slime molds, and (3) the cellular slime molds.

Table 19-1 The Major Groups of Protists

General Category	Phylum	Locomotion	Nutrition	Representative Features	Representative Genus
Plantlike protists: algae	Dinoflagellates (Phylum Pyrrophyta)	Swim with two flagella	Autotrophic; photosynthetic	Many bioluminescent; often have cellulose wall; most marine	*Gonyaulax* (causes red tide)
	Diatoms (Phylum Chrysophyta)	Glide along surfaces	Autotrophic; photosynthetic	Have silica shells; most marine	*Navicula* (glides toward light)
	Euglenoids (Phylum Euglenophyta)	Swim with one flagellum	Autotrophic; photosynthetic	Have an eyespot; all freshwater	*Euglena* (common pond-dweller)
	Red algae (Phylum Rhodophyta)	Nonmotile	Autotrophic; photosynthetic	Some deposit calcium carbonate; most marine	*Porphyra* (used as food in Japan)
	Brown algae (Phylum Phaeophyta)	Nonmotile	Autotrophic; photosynthetic	"Seaweeds" of temperate oceans	*Macrocystis* (forms kelp forests)
	Green algae (Phylum Chlorophyta)	Swim with flagella (some species)	Autotrophic; photosynthetic	Closest relatives of land plants	*Ulva* (sea lettuce)
Funguslike protists: water molds and slime molds	Water molds (Phylum Oomycota)	Swim with flagella (gametes)	Heterotrophic	Filamentous bodies	*Plasmopara* (causes downy mildew)
	Acellular (plasmodial) slime molds (Phylum Myxomycota)	Sluglike mass oozes over surfaces	Heterotrophic	Form multinucleate plasmodium	*Physarum* (forms a large bright orange mass)
	Cellular slime molds (Phylum Acrasiomycota)	Amoeboid cells extend pseudopodia; sluglike mass crawls over surfaces	Heterotrophic	Form pseudoplasmodium with individual amoeboid cells	*Dictyostelium* (often used in laboratory studies)
Animal-like protists: protozoa	Zooflagellates (Phylum Sarcomastigophora)	Swim with flagella	Heterotrophic	Inhabit soil or water or may be parasitic	*Trypanosoma* (causes African sleeping sickness)
	Sarcodines (Phylum Sarcomastigophora)	Extend pseudopodia	Heterotrophic	Both naked and shelled forms exist	*Amoeba* (common pond-dweller)
	Sporozoans (Phylum Apicomplexa)	Nonmotile	Heterotrophic; all parasitic	Form infectious spores	*Plasmodium* (causes malaria)
	Ciliates (Phylum Ciliophora)	Swim with cilia	Heterotrophic	Most complex single cells	*Paramecium* (fast-moving pond-dweller)

The Water Molds Have Had Important Impacts on Humans

The **water molds**, or *oomycetes*, form a small phylum of filamentous protists that includes both inoffensive species that live in water and damp soil and some species of profound economic importance. For example, a water mold causes the disease known as *downy mildew* of grapes (Fig. 19-16). Its inadvertent introduction into France from the United States in the late 1870s nearly destroyed the French wine industry. Another oomycete has destroyed millions of avocado trees in California; still another is responsible for *late blight*, a devastating disease of potatoes. When accidentally introduced into Ireland about 1845, this protist destroyed nearly the entire potato crop, causing the devastating potato famine during which a million people in Ireland starved and many more emigrated to the United States.

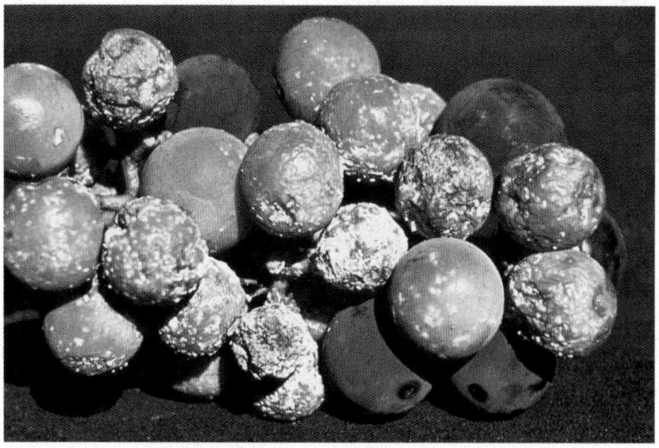

Figure 19-16 A parasitic water mold
Downy mildew, a plant disease caused by the water mold *Plasmopara*, nearly destroyed the French wine industry in the 1870s.

The Acellular Slime Molds Form a Multinucleate Mass of Cytoplasm Called a Plasmodium

The life cycle of the *slime mold* consists of two phases: a mobile feeding stage and a stationary reproductive stage called a **fruiting body**. The **acellular**, or **plasmodial, slime molds** consist of a mass of cytoplasm that may spread thinly over an area of several square meters. Although the mass contains thousands of diploid nuclei, the nuclei are not confined in discrete cells surrounded by plasma membranes, as in most multicellular organisms. This structure, called a **plasmodium**, explains why these protists are described as "acellular" (without cells). The plasmodium oozes through decaying leaves and rotting logs, engulfing food such as bacteria and particles of organic material. The mass may be bright yellow or orange—a large plasmodium can be rather startling (Fig. 19-17a). Dry conditions or starvation stimulate the plasmodium to form a fruiting body, on which haploid spores are produced (Fig. 19-17b). The spores are dispersed and germinate under favorable conditions, eventually giving rise to a new plasmodium.

The Cellular Slime Molds Live as Independent Cells but Aggregate into a Pseudoplasmodium When Food Is Scarce

The **cellular slime molds** live in soil as independent haploid cells that move and feed by producing extensions called **pseudopods** (literally, "false feet"). These extensions surround and engulf food such as bacteria. In the best-studied genus, *Dictyostelium*, individual cells release a chemical signal when food becomes scarce. This signal attracts nearby cells into a dense aggregation that forms a sluglike mass called a **pseudoplasmodium** ("false plasmodium") because it actually consists of individual cells (Fig. 19-18). The pseudoplasmodium then behaves like a multicellular organism. After crawling toward a source of light, the cells in the aggregation take on specific roles, forming

(a) (b)

Figure 19-17 The acellular slime mold *Physarum*
(a) *Physarum* oozes over a stone on the damp forest floor. ***(b)*** When food becomes scarce, the mass differentiates into black fruiting bodies in which spores are formed.

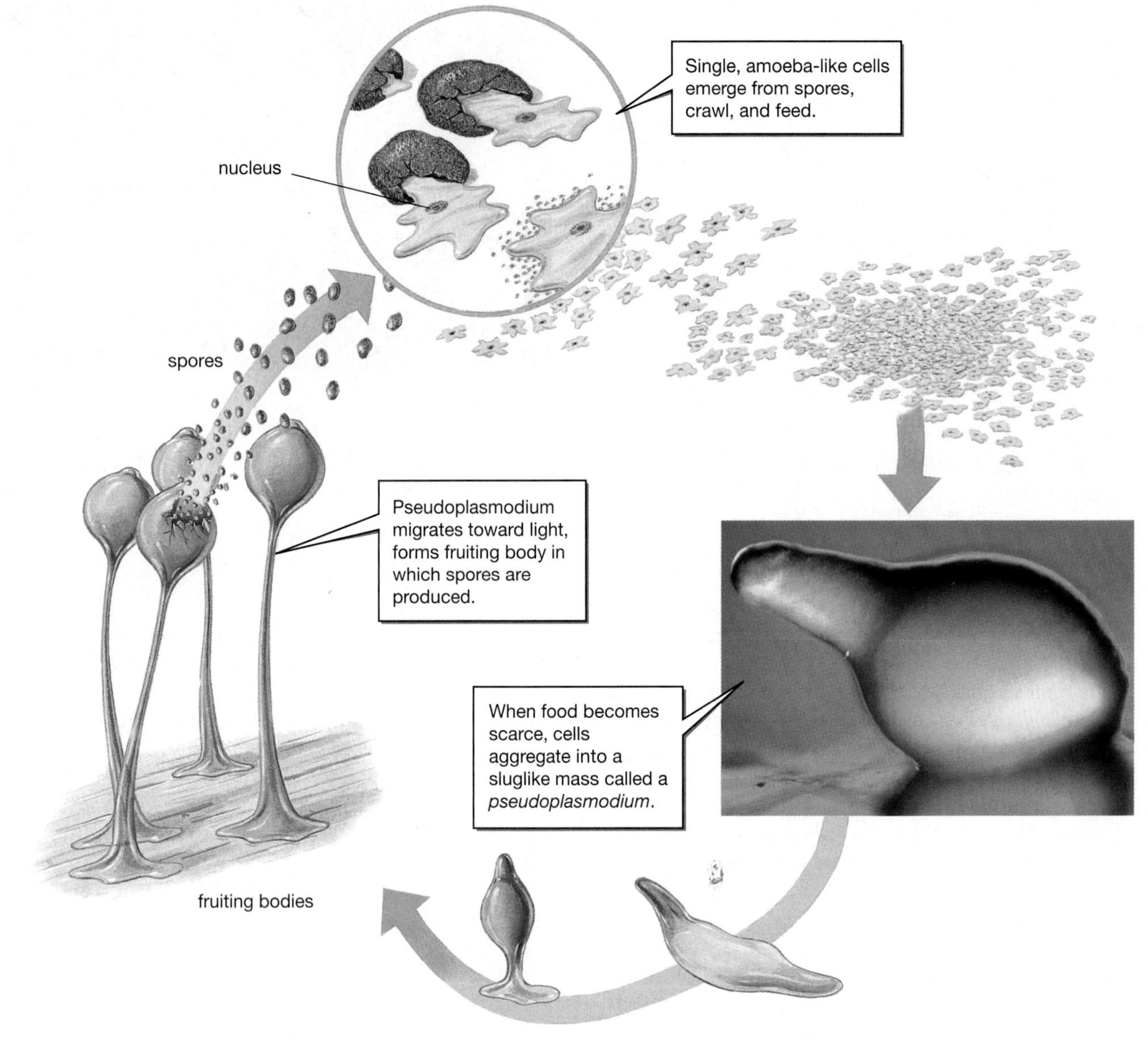
Single, amoeba-like cells emerge from spores, crawl, and feed.
nucleus
spores
Pseudoplasmodium migrates toward light, forms fruiting body in which spores are produced.
When food becomes scarce, cells aggregate into a sluglike mass called a *pseudoplasmodium*.
fruiting bodies

Figure 19-18 The life cycle of a cellular slime mold

a fruiting body. Haploid spores formed within the fruiting body are dispersed by wind and germinate directly into new amoeboid individuals.

The Algae Are Plantlike Protists

Photosynthetic protists, collectively known as **algae** (singular, **alga**), are widely distributed in oceans and lakes. Most species are single-celled, but some form multicellular aggregations that are commonly known as *seaweeds*. The single-celled algal species are collectively known as **phytoplankton**. Although they are microscopic, phytoplankton are immensely important. Marine phytoplankton account for nearly 70% of all the photosynthetic activity on Earth, absorbing carbon dioxide, recharging the atmosphere with oxygen, and supporting the complex web of aquatic life.

There are six major phyla of plantlike protists: (1) the dinoflagellates, (2) the diatoms, (3) the euglenoids, (4) the red algae, (5) the brown algae, and (6) the green algae.

Dinoflagellates Swim by Means of Two Whiplike Flagella

Dinoflagellates are named for the motion created by their two whiplike flagella (*dino* is Greek for "whirlpool"). One flagellum encircles the cell, and the second projects behind it. Some dinoflagellates are covered only by a cell membrane; others have cellulose walls that resemble armor plates (Fig. 19-19). Although some types live in fresh water, dinoflagellates are especially abundant in the ocean, where they are an important food source for larger organisms. Many dinoflagellates are bioluminescent, producing a brilliant blue-green light when disturbed.

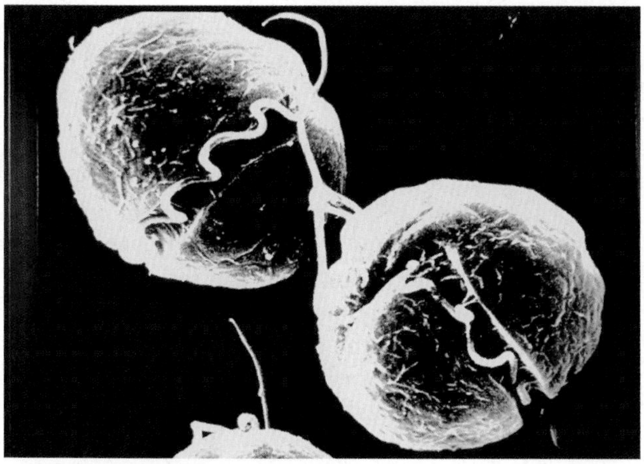

Figure 19-19 Dinoflagellates
Two dinoflagellates covered with protective cellulose armor. Two flagella lie within the grooves that encircle the body.

Specialized dinoflagellates known as *zooxanthellae* live within the tissues of corals, some clams, and even other protists, where the algae provide photosynthetic nutrients and remove carbon dioxide. Reef-building coral are found only in the shallow, well-lit waters in which their zooxanthellae can survive.

When the water is warm and rich in nutrients, a dinoflagellate population explosion may occur. Dinoflagellates can become so numerous that the water is dyed red by the color of their bodies, causing a "red tide" (Fig. 19-20). During red tides, fish die by the thousands, suffocated by clogged gills or by the oxygen depletion that results from the decay of billions of dinoflagellates. But oysters, mussels, and clams have a feast, filtering millions from the water for food. In the process, however, they concentrate a nerve poison the dinoflagellates produce in their bodies. Humans who eat these mollusks may be stricken with potentially lethal paralytic shellfish poisoning.

Figure 19-20 A red tide
The explosive reproductive rate of certain dinoflagellates under the right environmental conditions can produce dinoflagellate concentrations so great that their microscopic bodies dye the seawater red or brown, as in this bay in Mexico.

Diatoms Encase Themselves Within Glassy Walls

The **diatoms**, photosynthesizers found in both fresh and salt water, are so important to marine food webs that they have been called the "pastures of the sea." They produce protective shells of *silica* (glass), some of exceptional beauty (Fig. 19-21). These shells consist of top and bottom halves that fit together like a pillbox or petri dish. Accumulations of the glassy walls of diatoms over millions of years have produced fossil deposits of "diatomaceous earth" that may be hundreds of meters thick. This slightly abrasive substance is widely used in products such as toothpaste and metal polish.

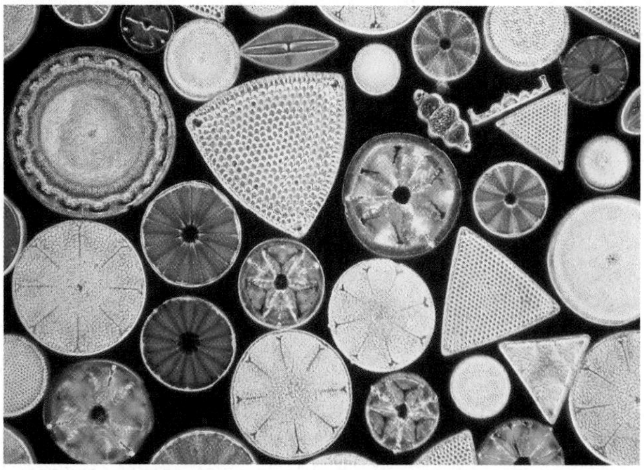

Figure 19-21 Some representative diatoms
This photomicrograph illustrates the intricate, microscopic beauty and variety of the glassy walls of diatoms.

Euglenoids Lack a Rigid Covering and Swim by Means of Flagella

The **euglenoids** live mostly in fresh water and are named after the group's best-known representative, *Euglena* (Fig. 19-22), a complex single cell that moves

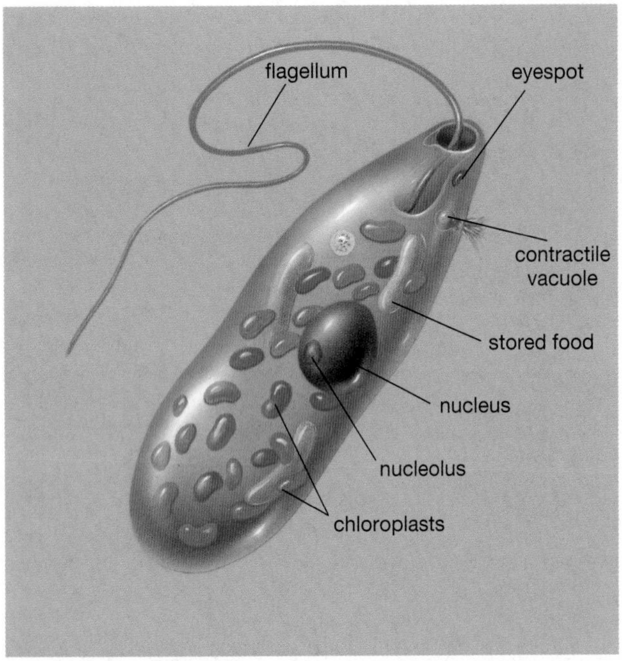

Figure 19-22 ***Euglena*****, a representative euglenoid**
Euglena's elaborate, single cell is packed with green chloroplasts, which will disappear if the protist is kept in darkness.

about by whipping its flagellum through water. Unlike diatoms or dinoflagellates, euglenoids lack a rigid outer covering, so some can move by wriggling as well as by whipping the flagellum. Some euglenoids also possess simple light-sensing organelles consisting of a photoreceptor, called an *eyespot*, and an adjacent patch of pigment. The pigment shades the photoreceptor only when light strikes from certain directions, enabling the organism to determine the direction of the light source. Using information from the photoreceptor, the flagellum propels the protist toward light levels appropriate for photosynthesis.

The Red Algae Live Primarily in Clear Tropical Oceans

The red algae, which range in color from bright red to nearly black, derive their color from red pigments that mask their green chlorophyll. Red algae are mostly marine and strictly multicellular (Fig. 19-23). They dominate in deep, clear tropical waters, where their red pigments absorb the deeply penetrating blue-green light and transfer this light energy to chlorophyll, where it is used in photosynthesis.

Some species of red algae deposit calcium carbonate, which forms limestone, in their tissues and contribute to the formation of reefs (see Fig. 19-23). Other species are harvested for food in Asia. Red algae also provide certain gelatinous substances with commercial uses, including carrageenan (used as a stabilizing agent in products such as paints, cosmetics, and ice cream) and agar (a substrate for growing bacterial colonies in laboratories). However, the major importance of these and other algae lies in their photosynthetic ability: The energy they capture and the food they synthesize help support the heterotrophs in marine ecosystems.

The Brown Algae Dominate in Cool Coastal Waters

The brown algae increase their light-gathering ability by means of brownish yellow pigments that (in combination with green chlorophyll) produce the brown to olive-green color of these algae. Like the red algae, brown algae are almost entirely marine and strictly multicellular. This group includes the dominant "seaweed" species that dwell along rocky shores in the temperate (cooler) oceans of the world, including the eastern and western coasts of the United States. Brown algae live in habitats ranging from nearshore, where they cling to rocks that are exposed at low tide, to far offshore. Several species use gas-filled floats to support their bodies (Fig. 19-24a). Some of the giant kelp found along the Pacific coast reach heights of 325 feet (100 meters) and may grow more than 6 inches (15 centimeters) in a single day. With their dense growth and towering height (Fig. 19-24b), kelp form undersea forests that provide food, shelter, and breeding areas for a variety of marine animals.

The Green Algae Live Mostly in Ponds and Lakes and Probably Gave Rise to the Land Plants

The green algae, a large and diverse group of species, include both multicellular and unicellular forms. Some green algae, such as *Spirogyra*, form thin filaments

Figure 19-23 Red algae
Red coralline algae from the Pacific Ocean off California provide an anchoring site for bright yellow hydroids (phylum Cnidaria). Coralline algae, which deposit calcium carbonate within their bodies, contribute to coral reefs in tropical waters.

(a)

(b)

Figure 19-24 Diverse brown algae
(a) *Fucus*, a genus found near shores, is shown here exposed at low tide. Notice the gas-filled floats, which provide buoyancy in water. ***(b)*** The giant kelp *Macrocystis* forms underwater forests off southern California.

from long chains of cells (Fig. 19-25). Other species of green algae form colonies containing clusters of cells that are somewhat interdependent and that constitute a structure intermediate between unicellular and multicellular forms. These colonies range from a few cells to a few thousand cells, as in species of *Volvox*. Most green algae are small, but some large species inhabit

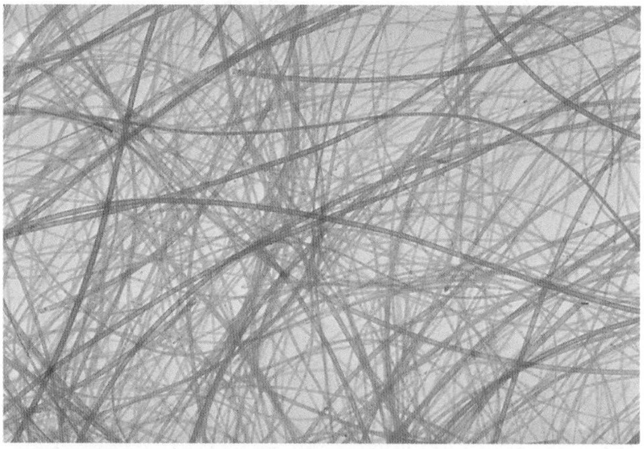

Figure 19-25 A form of green algae
Spirogyra is a filamentous green alga composed of strands only one cell thick.

the sea. For example, the green alga *Ulva*, or sea lettuce, is similar in size to the leaves of its namesake.

The green algae are of special interest because, unlike the other groups of plantlike protists, green algae are actually closely related to plants. Plants share a common ancestor with some types of green algae, and many researchers believe that the very earliest plants were similar to today's multicellular green algae.

The Protozoa Are Animal-like Protists

The **protozoa** are described as animal-like because they can move and they obtain their food from other organisms. They are placed into three major phyla: (1) the zooflagellates and sarcodines (amoebae), (2) the sporozoans, and (3) the ciliates. All are unicellular, eukaryotic, and heterotrophic; they differ in methods of locomotion.

Zooflagellates Possess Flagella

All **zooflagellates** possess at least one flagellum, which may propel the organism, sense the environment, or ensnare food. Many zooflagellates are free-living, inhabiting soil and water; others are symbiotic, living inside other organisms in a relationship that may be either mutually beneficial or parasitic. One symbiotic form can digest cellulose and lives in the gut of termites, where it helps them extract energy from wood. A more dangerous symbiotic zooflagellate in the genus *Trypanosoma* is responsible for African sleeping sickness, a potentially fatal disease (Fig. 19-26). Like many parasites, this organism has a complex life cycle, part of which is spent in the tsetse fly. While feeding on the blood of a mammal, the fly transmits the trypanosome to the mammal. The parasite then develops in the new host (which may be a human) and enters the bloodstream. It may then be ingested by another tsetse fly that bites the host, thus beginning a new cycle of infection.

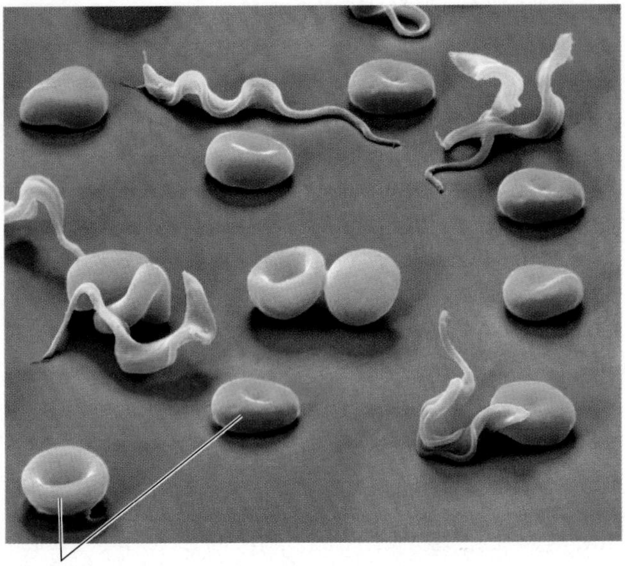

Figure 19-26 A disease-causing zooflagellate
This photomicrograph shows human blood that is heavily infested with the corkscrew-shaped, parasitic zooflagellate *Trypanosoma*, which causes African sleeping sickness.

Another parasitic zooflagellate, *Giardia*, is an increasing problem in the United States, particularly to hikers who drink from apparently pure mountain streams. *Cysts* (tough resting structures) of this flagellate are released in the feces of infected humans, dogs, or other animals (a single gram of feces may contain 300 million cysts) and enter freshwater streams and even community reservoirs. Cysts develop into the adult form (Fig. 19-27) in the small intestine of their mammalian host. In humans, infections can cause severe diarrhea, dehydration, nausea, vomiting, and cramps. Fortunately, these infections can be cured with drugs, and deaths from *Giardia* infections are uncommon.

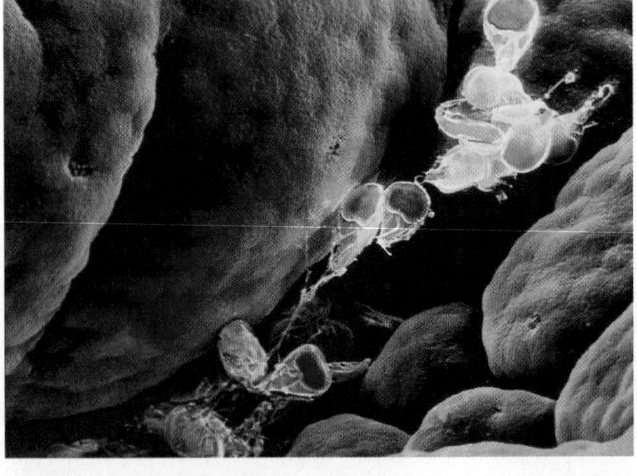

Figure 19-27 *Giardia:* the curse of campers
A zooflagellate (genus *Giardia*) that may infect drinking water, causing gastrointestinal disorders, is shown here in the human small intestine.

Sarcodines, Including the Amoebae, Move by Means of Pseudopods

Sarcodines possess flexible plasma membranes that they can extend in any direction to form pseudopodia, which are used for locomotion and for engulfing food (Fig. 19-28). The group includes amoebae, heliozoans, foraminiferans, and radiolarians.

Amoebae are common in freshwater lakes and ponds. Many amoebae are predators that stalk and engulf prey, but some species are parasites. One parasitic form causes amoebic dysentery, a disease that is prevalent in warm climates. The dysentery-causing amoeba multiplies in the intestinal wall, triggering severe diarrhea.

Heliozoans ("sun animals"), a striking form of freshwater sarcodine, may be found floating in ponds or attached by stalks to an underwater plant or rock (Fig. 19-29). They have stiff, needlelike pseudopodia, each of which is supported internally by a bundle of microtubules. Some heliozoans cover themselves with intricate and delicate shells of silica.

The **foraminiferans** and **radiolarians** are primarily marine sarcodines that also produce beautiful shells. The shells of foraminiferans are constructed mostly of calcium carbonate (chalk; Fig. 19-30a); those of radiolarians are of silica (Fig. 19-30b). These elaborate shells are pierced by myriad openings through which pseudopods extend. The chalky shells of foraminiferans, accumulating over millions of years, have resulted in immense deposits of limestone such as those that form the famous White Cliffs of Dover, England.

Sporozoans Are Parasitic and Have No Means of Locomotion

All **sporozoans** are parasitic, living inside the bodies and sometimes inside the individual cells of their hosts. They are named after their ability to form infectious spores, resistant structures transmitted from one host to another through food, water, or the bite of an infected insect. As adults, sporozoans have no means of locomotion.

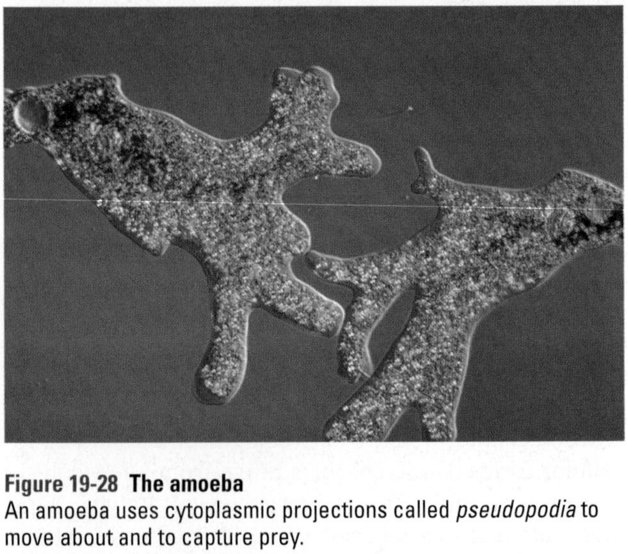

Figure 19-28 The amoeba
An amoeba uses cytoplasmic projections called *pseudopodia* to move about and to capture prey.

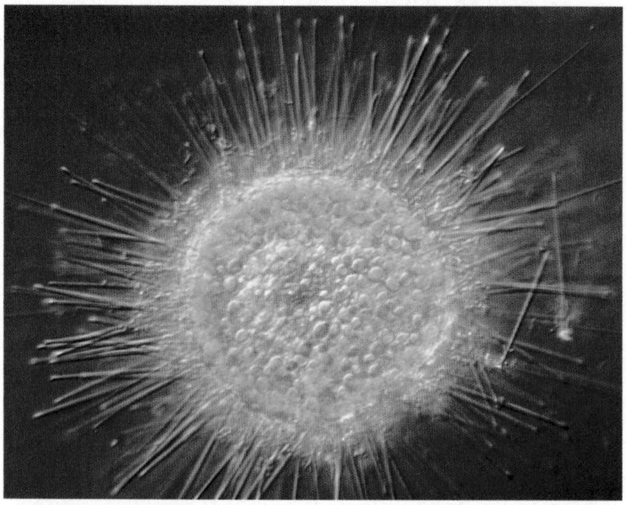

Figure 19-29 Heliozoans
Heliozoans are beautiful freshwater sarcodines; the needlelike pseudopodia are clearly visible in this specimen (genus *Acanthocystis*).

Many have complex life cycles, a common feature of parasites. A well-known example is the malarial parasite *Plasmodium*. Parts of its life cycle are spent in the stomach, and later the salivary glands, of the female *Anopheles* mosquito. When the mosquito bites a human, it passes the *Plasmodium* to the unfortunate victim. The sporozoan develops in the liver, then enters the blood, where it reproduces rapidly in red blood cells. The release of large quantities of spores through the rupture of the blood cells causes the recurrent fever of malaria. Uninfected mosquitoes may acquire the parasite by feeding on the blood of a malaria victim, spreading the parasite when it bites another person.

Although the drug chloroquine kills the malarial parasite, drug-resistant populations of *Plasmodium* are, unfortunately, spreading rapidly throughout Africa, where the disease is prevalent. Programs to eradicate mosquitoes have failed because the mosquitoes rapidly evolve resistance to pesticides.

Ciliates, Named for Their Locomotory Cilia, Are the Most Complex of the Protozoa

Ciliates, which inhabit fresh or salt water, represent the peak of unicellular complexity. They possess many specialized organelles, including the **cilia** (singular, **cilium**), the short hairlike outgrowths after which they are named. The cilia may cover the cell, or they may be localized. In the well-known freshwater genus *Paramecium* (Fig. 19-31), rows of cilia cover the entire body surface. Their coordinated beating propels the cell through the water at a protistan speed record of a millimeter per second. Although only a single cell, *Paramecium* responds to its environment as if it had a well-developed nervous system. Confronted with some noxious chemical or with a physical barrier, the cell immediately backs up by reversing the beating of its cilia and then proceeds in a new direction.

Some ciliates, such as *Didinium*, are accomplished predators (Fig. 19-32). Prey items are subdued and moved to a mouthlike opening, the *oral groove*. From there, the food item moves into the cytoplasm, where it is surrounded by a food vacuole, which forms a temporary "stomach." Digestive enzymes in the food vacuole break down the meal. Excess water is accumulated in a contractile vacuole, which periodically contracts, emptying the fluid to the outside through a hole called the *anal pore*.

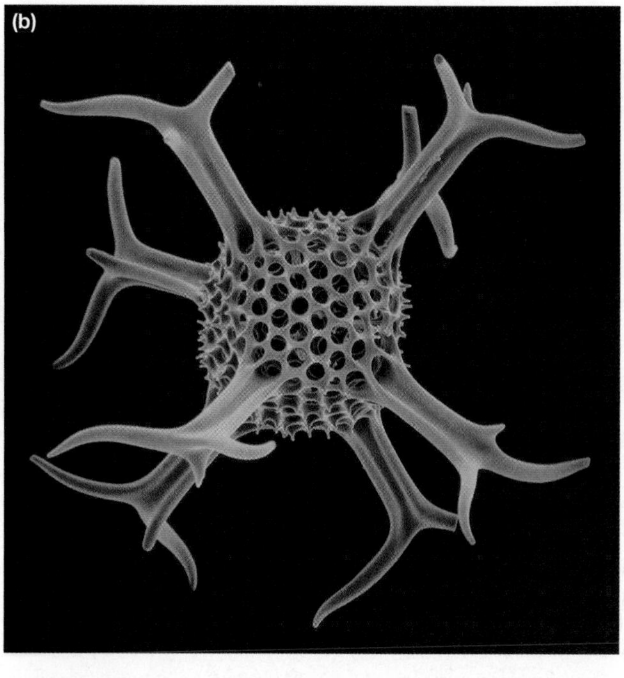

Figure 19-30 Foraminiferans and radiolarians
(a) The chalky shells of foraminiferans show numerous interior chambers. ***(b)*** The delicate, glassy shell of a radiolarian. Pseudopodia, which sense the environment and capture food, extend out through the openings in the shell.

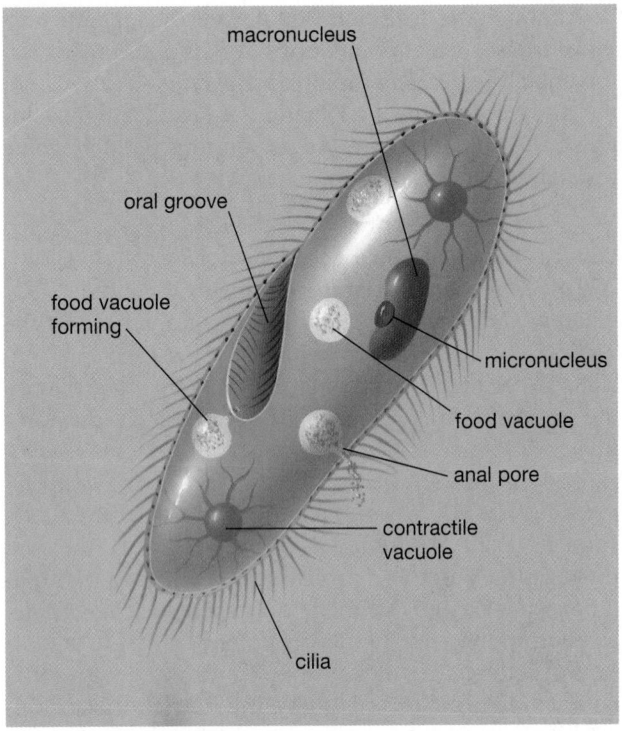

Figure 19-31 The complexity of ciliates
The ciliate *Paramecium* illustrates some important ciliate organelles. The oral groove acts as a mouth, and food vacuoles, miniature digestive systems, form at its apex; waste is expelled by exocytosis through an anal pore. The contractile vacuoles regulate water balance.

Figure 19-32 Microscopic predator
In this scanning electron micrograph, the predatory ciliate *Didinium* attacks a *Paramecium*. Note that the cilia of *Didinium* are confined to two bands, whereas *Paramecium* has cilia over its entire body. Ultimately, the predator will engulf and consume its prey. This microscopic drama could occur on a pinpoint with room to spare.

Evolutionary Connections

Our Unicellular Ancestors

Today's microbes include organisms that are probably quite similar to the ancient species that ultimately gave rise to the complex multicellular organisms that are the most conspicuous inhabitants of modern Earth. For example, the external appearance of many modern prokaryotes is basically indistinguishable from that of 3.5-billion-year-old fossilized cells. Similarly, the metabolism of today's anaerobic, heat-loving archaea is probably similar to the methods of energy acquisition that were used by Earth's earliest inhabitants, long before any oxygen entered the atmosphere. And modern purple sulfur bacteria and cyanobacteria are probably not too different from the first photosynthetic organisms that appeared more than 2 billion years ago.

Life might still consist solely of prokaryotic single cells if the protists, with their radical eukaryotic design, had not appeared on the scene nearly 2 billion years ago. As you learned in the discussion of the endosymbiont theory in Chapter 17, eukaryotic cells may have originated when one prokaryote, perhaps a bacterium capable of aerobic respiration, took up residence inside a partner, forming the first "mitochondrion." A separate but equally crucial merger may have occurred when a photosynthetic bacterium (probably resembling a cyanobacterium) took up residence within a nonphotosynthetic partner and became the first "chloroplast." The foundations of multicellularity were laid with the eukaryotic cell, whose intricacy allowed specialization of entire cells for specific functions within a multicellular aggregation. Thus, primitive protists, some absorbing nutrients from the environment, some photosynthesizing, and others consuming their food in chunks, almost certainly followed divergent evolutionary paths that led to the three multicellular kingdoms—the fungi, plants, and animals—which are the subjects of the following three chapters.

REVISITEDCASESTUDYREVISITEDCASESTUDYREVISITEDCASE

Agents of Death

How real is the threat of biological weapons? The U.S. government is certainly taking it seriously. After Iraq admitted that during the 1991 Gulf War it possessed missiles tipped with anthrax-containing warheads, the United States began a program (now under way) to immunize all members of the armed forces against anthrax. The immunization program focuses on anthrax, which is believed to be the biological agent most likely to be used, but many other infectious agents are believed to be potential weapons. Among the other likely candidates: the viruses that cause smallpox and Ebola hemorrhagic fever and the bacteria that cause plague. An additional worry comes from evidence that some countries are trying to use genetic engineering to "improve" pathogens, for example by adding antibiotic-resistance genes to plague bacteria so that victims of an attack would be more difficult to treat and a higher proportion would die.

Perhaps even more frightening than the thought that some nations might be tempted to make use of biological weapons is the possibility that terrorist groups might gain access to them. Little expertise is required to culture pathogenic bacteria or viruses, and the necessary supplies and equipment are easily acquired. The United States has already experienced one incident of "bio-terrorism": in 1984 disgruntled members of a religious group spread *Salmonella* bacteria on salad bars in a small town in Oregon. More than 700 people fell ill as a result, though none died. Still, the incident shows how simple it would be to mount a biological attack. It is not difficult to imagine the destruction that could be wrought by a more determined group with a more dangerous pathogen.

Defending a civilian population against biological attacks is difficult. An attack cannot be readily detected, as the pathogens themselves are invisible and symptoms may take hours or days to appear after an attack. Few people are vaccinated against anthrax or plague or even smallpox, and no vaccination exists for some potential weapons (such as the Ebola virus). Even treating victims after an attack would probably be ineffective. Inhaled anthrax, for example, is almost always fatal because antibiotics are effective against it only if administered soon after exposure to the bacteria; by the time recognizable symptoms arise, it is usually too late for treatment. In any case, the logistics of treating the victims of a mass attack would probably overwhelm our public health system. In the end, we must rely on politics, diplomacy, and humankind's widespread revulsion at the concept of biological warfare to protect us from its terrifying destructive potential.

What kinds of research might lead to new methods for protecting humanity from the threat of biological weapons?

Summary of Key Concepts

Microorganisms are classified in three domains: Archaea, Bacteria, and Eukarya. Archaea and Bacteria consist of tiny prokaryotic cells that lack organelles such as nuclei, mitochondria, and chloroplasts. Eukarya contains organisms with eukaryotic cells that possess the full range of organelles and resemble the cells of multicellular organisms. Within Eukarya, a diverse assemblage of single-celled species forms the kingdom Protista. Without the photosynthetic, nitrogen-trapping, and decomposing abilities of the prokaryotes and the photosynthetic activities of protists, life as we know it would grind to a halt.

1) What Are Viruses, Viroids, and Prions?

Viruses are parasites consisting of a protein coat that surrounds genetic material. They are noncellular and unable to move, grow, or reproduce outside a living cell. They invade cells of a specific host and use the host cell's energy, enzymes, and ribosomes to produce more virus particles, which are liberated when the cell ruptures. Many viruses are pathogenic to humans, including those causing colds and flu, herpes, AIDS, and certain forms of cancer.

Viroids are short strands of RNA that can invade a host cell's nucleus and direct the synthesis of new viroids. To date, viroids are known to cause only certain diseases of plants.

Prions have been implicated in diseases of the nervous system, such as kuru, Creutzfeldt-Jakob disease, and scrapie. Prions are unique in that they lack genetic material. They are composed solely of mutated prion protein, which may act as an enzyme, catalyzing the formation of more prions from normal prion protein.

2) Which Organisms Make Up the Prokaryotic Domains—Bacteria and Archaea?

Members of the domains Bacteria and Archaea—the bacteria and archaea—are unicellular and prokaryotic. Archaea and bacteria are not closely related and differ in several fundamental features, including cell wall composition, ribosomal RNA sequence, and membrane lipid structure. A cell wall determines the characteristic shapes of prokaryotes: round, rodlike, or spiral. Certain types of bacteria can form spores that disperse widely and withstand inhospitable environmental conditions. Prokaryotes obtain energy in a variety of ways. Some, including the cyanobacteria, rely on photosynthesis. Others are chemosynthetic, breaking down inorganic molecules to obtain energy. Heterotrophic forms are capable of consuming a wide variety of organic compounds. Many are anaerobic, able to obtain energy from fermentation when oxygen is not available.

Some bacteria are pathogenic, causing disorders such as pneumonia, tetanus, botulism, and the venereal diseases gonorrhea and syphilis. Most bacteria, however, are harmless to humans and play important roles in natural ecosystems. Bacteria and archaea have colonized nearly every habitat on Earth, including hot, acidic, very salty, and anaerobic environments. Some live in the digestive tracts of larger organisms, such as cows and sheep, and break down cellulose. Nitrogen-fixing bacteria enrich the soil and aid in plant growth; many others live off the dead bodies and wastes of other organisms, liberating nutrients for reuse.

3) Which Organisms Make Up the Kingdom Protista?

The kingdom Protista consists of organisms composed of single, highly complex eukaryotic cells. They are classified as funguslike, plantlike, and animal-like.

The funguslike acellular (plasmodial) slime molds form a multinucleate plasmodium that crawls in amoeboid fashion, ingesting decaying organic matter. Drought or starvation stimulates the formation of a fruiting body on which spores are formed. Cellular slime molds exist as independent amoeboid cells. Under adverse conditions, these cells aggregate in response to a chemical signal and form a pseudoplasmodium that differentiates into a spore-forming fruiting body.

The plantlike algae are important photosynthetic organisms in marine and freshwater ecosystems. They include dinoflagellates, diatoms, euglenoids, red algae, brown algae, and green algae.

Protozoa are nonphotosynthetic protists that absorb or ingest their food. They are widely distributed in soil and water; some are parasitic. They include the zooflagellates, the amoeboid sarcodines, the parasitic sporozoans, and the predatory ciliates.

Key Terms

acellular slime mold *p. 379*
alga *p. 380*
amoeba *p. 384*
anaerobe *p. 375*
Archaea *p. 371*
Bacteria *p. 371*
bacterial conjugation *p. 373*
bacteriophage *p. 366*
binary fission *p. 373*
capsule *p. 371*
cellular slime mold *p. 379*
cell wall *p. 371*
chemosynthetic *p. 375*
chemotactic *p. 372*
ciliate *p. 385*
cilium *p. 385*
cyanobacterium *p. 375*
diatom *p. 381*
dinoflagellate *p. 380*
endospore *p. 372*
euglenoid *p. 381*
flagellum *p. 371*
foraminiferan *p. 384*
fruiting body *p. 379*
Gram stain *p. 371*
heliozoan *p. 384*
host *p. 366*
kuru *p. 369*
legume *p. 375*
magnetotactic *p. 372*
methanogen *p. 375*
nitrogen-fixing bacterium *p. 375*
pathogenic *p. 376*
peptidoglycan *p. 371*
phototactic *p. 372*
phytoplankton *p. 380*
pilus *p. 371*
plasmid *p. 374*
plasmodial slime mold *p. 379*
plasmodium *p. 379*
prion *p. 369*
Protista *p. 377*
protozoan *p. 383*
pseudoplasmodium *p. 379*
pseudopod *p. 379*
radiolarian *p. 384*
sarcodine *p. 384*
slime layer *p. 371*
sporozoan *p. 384*
symbiotic *p. 375*
taxis *p. 372*
viroid *p. 369*
virus *p. 366*
water mold *p. 379*
zooflagellate *p. 383*

Thinking Through the Concepts

Multiple Choice

1. *Which of the following is true?*
   a. Viruses cannot reproduce outside a host cell.
   b. Prions are infectious proteins.
   c. Viroids lack a protein coat.
   d. Some viruses cause cancer.
   e. all of the above

2. *Which structure is used to transfer genetic material between bacteria?*
   a. flagellum b. pilus
   c. peptidoglycan d. spore
   e. capsule

3. *Most pathogenic bacteria cause disease by*
   a. directly destroying individual cells of the host
   b. fixing nitrogen and depriving the host of this nutrient
   c. producing toxins that disrupt normal functions
   d. depleting the energy supply of the host
   e. depriving the host of oxygen

4. *Cyanobacteria*
   a. have chlorophyll b. are chemosynthetic
   c. can live without oxygen d. are archaea
   e. all of the above

5. *Which organisms are sometimes called the "pastures of the sea"?*
   a. dinoflagellates b. foraminiferans
   c. diatoms d. radiolarians
   e. amoebae

6. *Which of the following pairs of organism and disease is incorrect?*
   a. sporozoan: malaria b. prion: kuru
   c. bacterium: syphilis d. archaean: gonorrhea
   e. virus: AIDS

? Review Questions

1. Describe the structure of a typical virus. How do viruses replicate?
2. List the major differences between prokaryotes and protists.
3. Describe some of the ways in which bacteria obtain energy and nutrients.
4. What are nitrogen-fixing bacteria, and what role do they play in ecosystems?
5. What is an endospore? What is its function?
6. Describe some examples of bacterial symbiosis.
7. What is the importance of dinoflagellates in marine ecosystems? What happens when they reproduce rapidly?
8. What is the major ecological role played by unicellular algae?
9. What protozoan group consists entirely of parasitic forms?
10. Describe the life cycle and mode of transmission of the malarial parasite.

Applying the Concepts

1. In some developing countries, antibiotics can be purchased without a prescription. Why do you think this is done? What biological consequences would you predict?
2. Before the discovery of prions, many (perhaps most) biologists would have agreed with the statement, "It is a fact that no infectious organism or particle can exist that lacks nucleic acid (such as DNA or RNA)." What lessons do prions have to teach us about nature, science, and scientific inquiry? You may wish to review Chapter 1 to help answer this question.
3. Recent research shows that ocean water off southern California has warmed by 2–3 °F (1–1.5 °C) over the past four decades, possibly due to the greenhouse effect. This warming has led indirectly to a depletion of nutrients in the water and thus a decline in photosynthetic protists such as diatoms. What effects is this warming likely to have for the life in the oceans?
4. Argue for and against the statement, "Viruses are alive."
5. The internal structure of many protists is much more complex than that of cells of multicellular organisms. Does this mean that the protist is engaged in more complex activities than the multicellular organism? If not, why should the protistan cell be much more complicated?
6. Why would the lives of multicellular animals be impossible if prokaryotic and protistan organisms did not exist?

For More Information

Cole, L. A. "The Specter of Biological Weapons." *Scientific American*, December 1996. Assesses the likelihood of germ warfare, the politics of preventing it, and the biological agents that make it possible.

Madigan, M., and Marrs, B. "Extremophiles." *Scientific American*, April 1997. Prokaryotes that prosper under extreme conditions, and potential industrial uses of the enzymes that they use to do so.

Prusiner, S. "The Prion Diseases." *Scientific American*, January 1995. A description of prions and the research that led to their discovery, from the point of view of the most influential scientist in the field.

Answers to Multiple-Choice Questions

1. e 2. b 3. c 4. a 5. c 6. d

MediaTutor
The Hidden World of Microbes

CD Activities

Activities 19.1 and 19.2: Virus Replication

Viruses take over our cellular machinery and use it to produce more viruses. Understanding how viruses replicate may provide a way to prevent or cure many diseases. In **Activity 19.1: Retrovirus Replication** *(Estimated time: 5 minutes)*, you will look at how viruses such as HIV multiply in human cells. In **Activity 19.2: Herpes Virus Replication** *(Estimated time: 5 minutes)*, you will look at how a herpes virus uses a different mechanism to accomplish the same task.

Activity 19.3 Bacterial Conjugation

Estimated Time: 5 minutes

Bacterial conjugation is a mechanism by which some prokaryotes are able to exchange genes. It has a significant impact on humans because it can allow antibiotic resistance to spread. This animation shows how a donor bacterium gives plasmid DNA to a recipient.

Start the MediaTutor Student CD-ROM and enter the activity number in the Quick Search box to be taken directly to that activity.

Web Investigations

Case Study: Agents of Death

Estimated time: 10 minutes

While biological weapons are not new, the bacterial strains, delivery mechanisms, and toxins have become much more sophisticated. In this exercise we will study biological weapons from a biological, functional, and social perspective. Have biological weapons been effective in the past? How dangerous are current designs, and what can be done to defend against them?

Go to http://www.prenhall.com/audesirk6, the Audesirk Companion Web site. Select Chapter 19 and Web Investigation to begin.

Despite its graceful appearance, the death cap, *Amanita phalloides,* contains one of the world's deadliest poisons.

20 The Fungi

AT A GLANCE

Case Study: Three Outings

1) **What Are the Main Adaptations of Fungi?**

Most Fungi Have Filamentous Bodies

Fungi Obtain Their Nutrients from Other Organisms

Most Fungi Can Reproduce Both Sexually and Asexually

2) **How Are Fungi Classified?**

The Chytrids Produce Swimming Spores

The Zygote Fungi Can Reproduce by Forming Diploid Zygospores

The Sac Fungi Form Spores in a Saclike Case Called an Ascus

The Club Fungi Produce Club-Shaped Reproductive Structures Called Basidia

The Imperfect Fungi Are Species in Which Sexual Structures Have Not Been Observed

Some Fungi Form Symbiotic Relationships

3) **How Do Fungi Affect Humans?**

Fungi Attack Plants That Are Important to People

Fungi Cause Human Diseases

Fungi Make Important Contributions to Gastronomy

Fungi Play a Crucial Ecological Role

Evolutionary Connections: Fungal Ingenuity—Pigs, Shotguns, and Nooses

Case Study Revisited: Three Outings

CASESTUDY

Three Outings

As the work week drew to a close, Anita was discussing her weekend plans with her co-workers, Roger and Whitney. They were mildly surprised to discover that each of them had separately planned to go mushroom hunting, but the coincidence was not too great, as collecting mushrooms was a popular activity in their small Oregon town. After all, the surrounding moist Northwest forests provided ideal growing conditions for an abundant mushroom crop.

The following Saturday, Anita rose early, gathered up her collecting baskets and some newspapers in which to wrap her prizes, and drove to her favorite collecting spot. She spent a pleasant day roaming the mountainsides, field guide in hand, digging edible mushrooms, and carefully placing them in her basket.

Later that day, Anita had a delicious, satisfying supper. She owed her tasty meal almost entirely to members of the kingdom Fungi. The centerpiece was a casserole she prepared with the choice king bolete and chanterelle mushrooms that she'd gathered herself. This wild fare was augmented by a slice of home-baked bread (which owed its airy texture to the activity of yeast, a type of fungus) and a chunk of bleu cheese (which derives its flavor from a mold, another type of fungus). As Anita ate her fungal feast, she sipped a glass of wine (fermented courtesy of a yeast).

On Sunday afternoon, Anita decided to call Roger and Whitney to compare notes. She phoned Roger first, and received shocking news. Roger was in the intensive care unit of the hospital, gravely ill and awaiting a liver transplant. Apparently, Roger had mistakenly consumed some specimens of the highly poisonous death cap mushroom, confusing them with an edible species. Shaken, Anita called Whitney to share the bad news. But Whitney could not come to the phone because he was in the county jail, awaiting trial on felony charges of possession of a Schedule 1 controlled substance. Whitney had spent his Saturday collecting mushrooms, but in his case the collection consisted of hallucinogenic *Psilocybe* mushrooms, possession of which is illegal in all 50 states.

The tale of Anita, Roger, and Whitney happens to be an invention. Each element of the story, however, is quite plausible, as fungi do indeed affect human lives as gourmet treats, everyday foods, deadly poisons, and mind-altering drugs. In each of these roles, and many others, fungi have played a fascinating role in human affairs. Read on to find out more about the inconspicuous but often potent members of kingdom Fungi. ■

1 What Are the Main Adaptations of Fungi?

Most Fungi Have Filamentous Bodies

When you think of a fungus, you probably picture a mushroom. Mushrooms, however, are just temporary reproductive structures that extend from the main bodies of certain kinds of fungus. The body of almost all fungi is a **mycelium** (Fig. 20-1a), which is an interwoven mass of one-cell-thick, threadlike filaments called **hyphae** (singular, **hypha**; Fig. 20-1b). Depending on the species, hyphae either consist of single elongated cells with numerous nuclei or are subdivided by partitions called **septa** (singular, **septum**) into many cells, each containing from one to many nuclei. Pores in the septa allow cytoplasm to stream between cells, distributing nutrients. Like plant cells, fungal cells are surrounded by cell walls. Unlike plant cells, however, fungal cell walls are strengthened by chitin, the same substance found in the exoskeletons of arthropods (as we shall see in Chapter 22).

Fungi are not able to move about. They compensate for this lack of mobility with filaments that can grow rapidly in any direction within a suitable environment. The fungal mycelium quickly infuses itself into aging bread or cheese, beneath the bark of decaying logs, or into the soil. Periodically, the hyphae grow together and differentiate into reproductive structures that project above the surface beneath which the mycelium grows. These structures, including mushrooms, puffballs, and the powdery molds on unrefrigerated food, represent only a fraction of the complete fungal body, but are typically the only part of the fungus that we can easily see.

Fungi Obtain Their Nutrients from Other Organisms

Like animals, fungi are heterotrophic, surviving by breaking down nutrients stored in the bodies or wastes of other organisms. Some fungi are **saprobes**, which digest the bodies of dead organisms. Others are parasitic, feeding on living organisms and causing disease. Others, including *lichens* and *mycorrhizae*, live in symbiotic relationships with other organisms. There are even a few predatory fungi, which attack tiny worms in soil.

Unlike animals, fungi do not ingest food. Instead, they secrete enzymes that digest complex molecules outside their bodies, breaking down the molecules into smaller subunits that can be absorbed. Fungal filaments can penetrate deeply into a source of nutrients and are only one cell thick, presenting an enormous surface area through which to secrete enzymes and absorb nutrients. This mode of securing nutrition has served the fungi well. Almost every biological material can be consumed by one or another fungal species, so nutritional support for fungi is likely to be at hand in nearly every terrestrial habitat.

Most Fungi Can Reproduce Both Sexually and Asexually

Unlike plants and animals, fungi form no embryos. Instead, fungi propagate by means of tiny, lightweight reproductive packages known as **spores**, which are extraordinarily mobile, even though most lack a means for self-propulsion. Spores are distributed far and wide as hitchhikers on the outside of animal bodies, as passengers inside the digestive systems of animals that have eaten them, or as airborne drifters, cast aloft by chance or shot into the atmosphere by elaborate reproductive

MEDIATUTOR 20.1 Fungi Structure

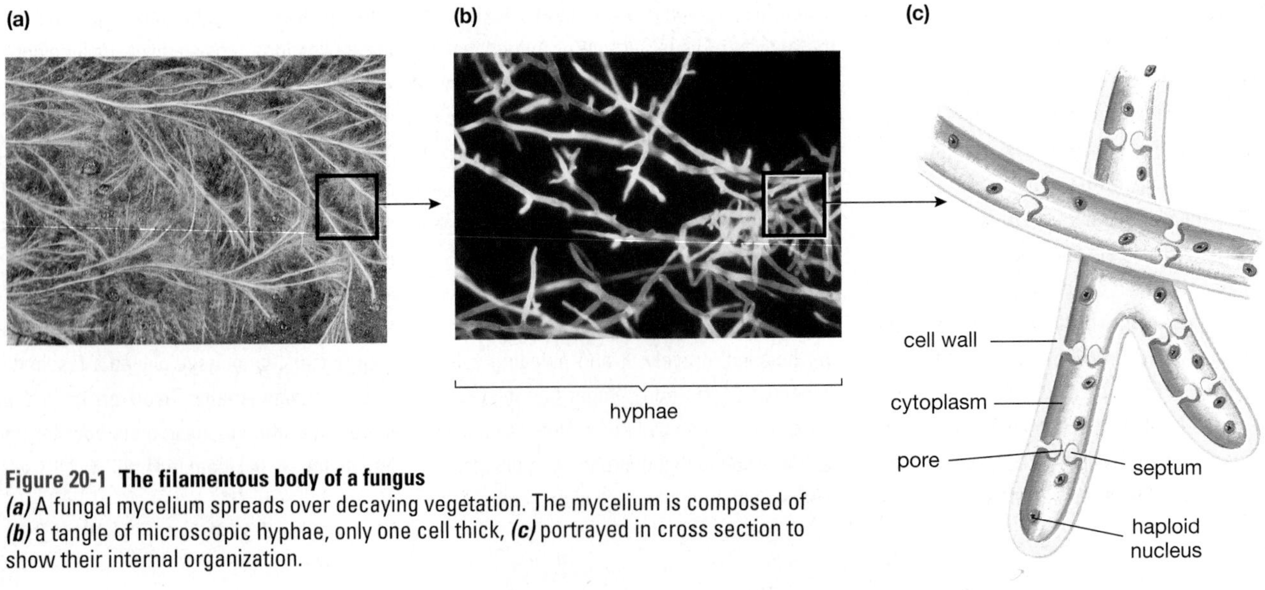

Figure 20-1 The filamentous body of a fungus
(a) A fungal mycelium spreads over decaying vegetation. The mycelium is composed of ***(b)*** a tangle of microscopic hyphae, only one cell thick, ***(c)*** portrayed in cross section to show their internal organization.

Figure 20-2 Some fungi can eject spores
A ripe earthstar mushroom, struck by a drop of water, releases a cloud of spores that will be dispersed by air currents.

structures (Fig. 20-2). In addition, spores are often produced in great numbers (a single giant puffball may contain 5 trillion sexual spores; see Fig. 20-7a). The fungal combination of prodigious reproductive capacity and highly mobile spores ensures that fungi are ubiquitous in terrestrial environments and accounts for the inevitable growth of fungi on every uneaten sandwich and container of leftovers.

In general, fungi are capable of both asexual and sexual reproduction. For the most part, asexual reproduction is the default mode under stable conditions, with sexual reproduction occurring mainly under conditions of environmental change or stress. Both asexual and sexual reproduction ordinarily involve the production of spores within special fruiting bodies that project above the mycelium (though asexual spores in some species may not be contained in a fruiting body).

The bodies and spores of fungi are haploid (contain only a single copy of each chromosome). Diploid structures form only during a brief period during the sexual portion of the fungal life cycle. Thus, a haploid mycelium produces haploid asexual spores by mitosis. If an asexual spore is deposited in a favorable location, it will begin mitotic divisions and develop into a new mycelium. This simple reproductive cycle results in the rapid production of genetically identical clones of the original mycelium.

Sexual reproduction begins when a filament of one mycelium comes into contact with a filament from a second mycelium that is of a different, compatible mating type (the different mating types of fungi are analogous to the different sexes of animals, except that there are often more than two mating types). If conditions are suitable, the two hyphae may fuse, so that nuclei from the two different hyphae share a common cell. This merger of hyphae is followed (immediately in some species, after some delay in others) by fusion of the different haploid nuclei to form a diploid zygote. The zygote then undergoes meiosis to form haploid sexual spores. These spores are dispersed, germinate, and divide by mitosis to form new haploid mycelia. Unlike the cloned offspring of asexual spores, these sexually produced fungal bodes are genetically distinct from either parent.

2 How Are Fungi Classified?

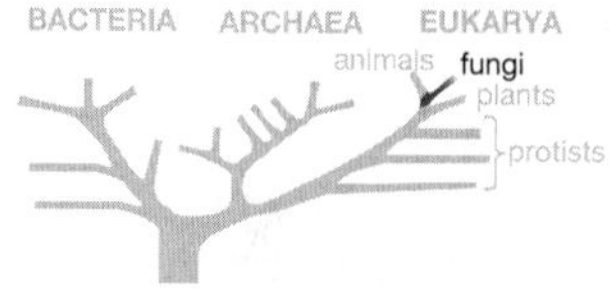

MediaTutor 20.2 The Fungi

Although nearly 100,000 species of modern fungi have been described, biologists have only begun to comprehend the diversity of these organisms—at least 1000 additional species are described each year. The phyla of fungi are the Chytridiomycota (chytrids), Zygomycota (zygote fungi), Ascomycota (sac fungi), and Basidiomycota (club fungi) (Table 20-1). In addition, species that cannot be readily classified are placed for convenience in a group known as the *deuteromycetes* (imperfect fungi).

The Chytrids Produce Swimming Spores

Unlike other types of fungi, most chytrids live in water. The chytrids (Fig. 20-3) are further distinguished from other fungi by their swimming spores, which require water for dispersal (so even soil-dwelling chytrids require a film of water for reproduction). A chytrid spore propels itself through the water by means of a single flagellum located on one end of the spore. No other group of fungi has flagella.

Research by fungal systematists suggests that the chytrids form an ancient group that predates and gave rise to the other groups of modern fungi. This conclusion is bolstered by the fossil record, as the oldest known fossil fungi are chytrids that were found in rocks more than 600 million years old. Ancestral fungi may well have been similar in habit to today's aquatic and marine chytrids, so fungi (like plants and animals) probably originated in a watery environment before colonizing land.

Most chytrid species feed on dead aquatic plants or other detritus in watery environments, but some species are parasites of plants or animals. One such parasitic chytrid is believed to be a major cause of the current worldwide die-off of frogs, which threatens many species and has apparently caused the extinction of several

Table 20-1 The Major Divisions of Fungi

Common Name (Phylum)	Reproductive Structures	Cellular Characteristics	Economic and Health Impacts	Representative Genera
Chytrids (Chytridiomycota)	Flagellated spores	Cell walls contain chitin; septa are absent	Contribute to decline of frog populations	*Batrachochytrium* (frog pathogen)
Zygote fungi (Zygomycota)	Produce sexual diploid zygospores	Cell walls contain chitin; septa are absent	Cause soft fruit rot and black bread mold	*Rhizopus* (causes black bread mold); *Pilobolus* (dung fungus)
Sac fungi (Ascomycota)	Sexual spores formed in saclike ascus	Cell walls contain chitin; septa are present	Cause molds on fruit; can damage textiles; cause Dutch elm disease and chestnut blight; include yeasts and morels	*Saccharomyces* (yeast); *Ophiostoma* (causes Dutch elm disease)
Club fungi (Basidiomycota)	Sexual reproduction involves production of haploid basidiospores on club-shaped basidia	Cell walls contain chitin; septa are present	Cause smuts and rusts on crops; include some edible mushrooms	*Amanita* (poisonous mushroom); *Polyporus* (shelf fungus)

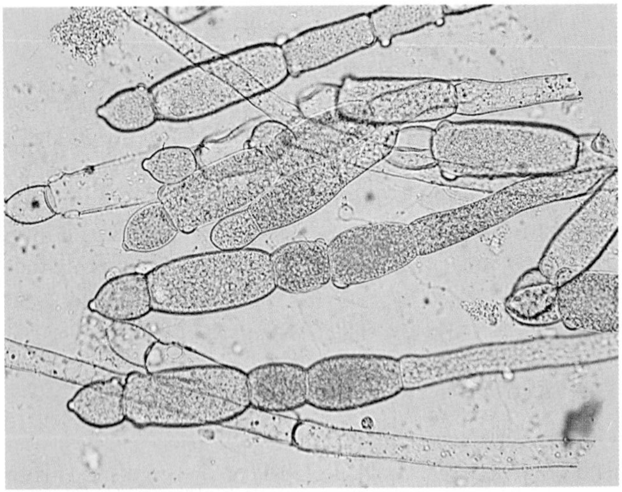

Figure 20-3 Chytrid filaments
These filaments of the chytrid fungus *Allomyces* are in the midst of sexual reproduction. The orange structures visible on many of the filaments will release male gametes; the clear structures will release female gametes. Chytrid gametes are flagellated, and these swimming reproductive structures aid dispersal of members of this mostly aquatic phylum.

species already. No one yet understands exactly why this fungal disease emerged as a major cause of death in frogs. One hypothesis is that frog populations under stress from pollution and other environmental challenges might be more susceptible to infection by chytrids.

The Zygote Fungi Can Reproduce by Forming Diploid Zygospores

The *zygomycetes*, also called the **zygote fungi**, generally live in soil or on decaying plant or animal material. Among the zygomycetes are species that belong to the genus *Rhizopus*, which cause the familiar annoyances of soft fruit rot and black bread mold. The life cycle of the black bread mold, which reproduces both asexually and sexually, is depicted in Figure 20-4a. Asexual reproduction in zygote fungi is initiated by the formation of haploid spores in black spore cases called **sporangia** (Fig. 20-4b). These spores disperse through the air and, if they land on a suitable substrate (such as a piece of bread), germinate to form new haploid hyphae.

If two hyphae of different mating types of zygote fungi come into contact, sexual reproduction may ensue. The two hyphae "mate sexually," and their nuclei fuse to produce diploid **zygospores** (Fig. 20-4c), tough, resistant structures that give this group its name. Zygospores can remain dormant for long periods until environmental conditions are favorable for growth. Like asexually produced spores, zygospores disperse and germinate, but instead of producing new hyphae directly, they undergo meiosis. As a result, they form structures that bear haploid spores, which develop into new hyphae.

The Sac Fungi Form Spores in a Saclike Case Called an Ascus

The *ascomycetes*, or **sac fungi**, also reproduce both asexually and sexually. Asexual spores of sac fungi are produced at the tips of specialized hyphae. During sexual reproduction, spores are produced by a complex sequence of events that begins when hyphae of two different mating types fuse. This sequence culminates in the formation of **asci** (singular, **ascus**), which are saclike cases that contain several spores and that give this phylum its name.

Some ascomycetes live in decaying forest vegetation and form either beautiful cup-shaped reproductive

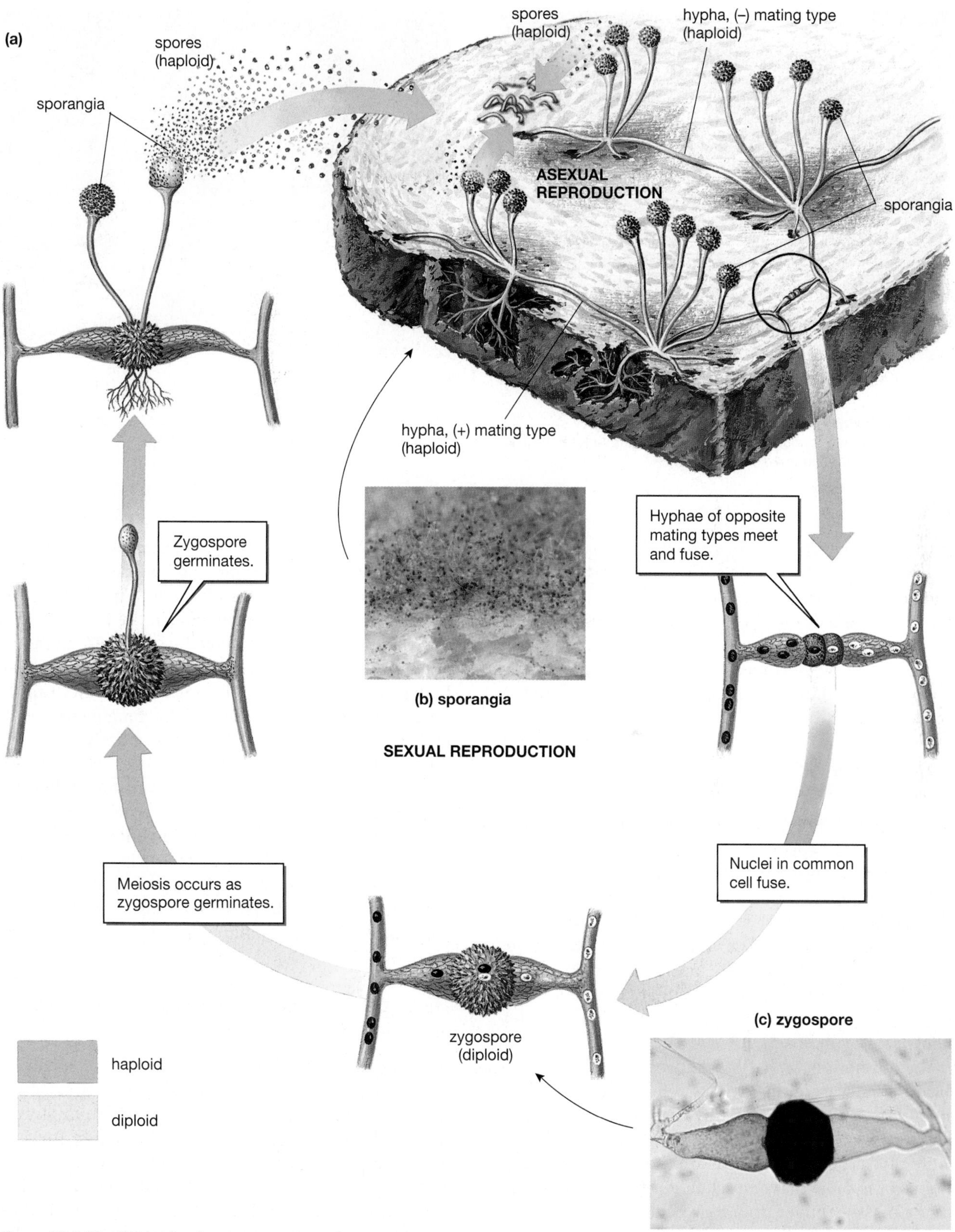

Figure 20-4 The life cycle of a zygomycete
(a) Top: During asexual reproduction in the black bread mold (genus *Rhizopus*), haploid spores, produced within ***(b)*** sporangia, disperse and germinate on food such as bread. The resulting haploid hyphae may complete the asexual cycle by producing sporangia and spores. Bottom: During sexual reproduction, hyphae of different mating types (designated + and – on the bread) contact one another and fuse, producing ***(c)*** a diploid zygospore. The zygospore undergoes meiosis and germinates, producing sporangia. The sporangia liberate haploid spores, which germinate into hyphae that can enter either the asexual or sexual cycle.

(a)

(b)

Figure 20-5 Diverse ascomycetes
(a) The cup-shaped fruiting body of the scarlet cup fungus. *(b)* The morel, an edible delicacy. (Consult an expert before sampling any wild fungus—some are deadly!)

structures (Fig. 20-5a) or corrugated, mushroomlike fruiting bodies called *morels* (Fig. 20-5b). This phylum also includes many of the colorful molds that attack stored food and destroy fruit and grain crops and other plants. Among the ascomycetes we also find the yeasts (some of the few unicellular fungi) and the species that produces penicillin, the first antibiotic and still an important medicine.

The Club Fungi Produce Club-Shaped Reproductive Structures Called Basidia

Basidiomycetes are called the **club fungi** because they produce club-shaped reproductive structures. Members of this phylum typically reproduce sexually (Fig. 20-6), with hyphae of different mating types fusing to form filaments in which each cell contains two nuclei, one from each parent. The nuclei do not themselves fuse until the formation of specialized, club-shaped diploid cells called **basidia** (singular, **basidium**). Basidia in turn give rise to haploid reproductive **basidiospores** by meiosis.

The formation of basidia and basidiospores takes place in special fruiting bodies, which are familiar to most of us as mushrooms, puffballs (Fig. 20-7a, p. 398), shelf fungi (Fig. 20-7b), and stinkhorns (Fig. 20-7c). These reproductive structures are actually dense aggregations of hyphae that emerge under proper conditions from a massive underground mycelium. On the undersides of mushrooms are leaflike gills on which basidia are produced. Basidiospores are released by the billions from the gills of mushrooms or through openings in the tops of puffballs and are dispersed by wind and water.

Falling on fertile ground, a mushroom basidiospore may germinate and form haploid hyphae. These hyphae grow outward from the original spore in a roughly circular pattern as the older hyphae in the center die. The subterranean body periodically sends up numerous mushrooms, which emerge in a ringlike pattern called a **fairy ring** (Fig. 20-8). The diameter of the fairy ring can reveal the approximate age of the fungus—the wider the ring diameter, the older the underlying fungus. Some fairy rings are estimated to be 700 years old. Basidomycete mycelia can grow even older than that and can expand over huge areas. Recently, researchers using genetic analysis have documented a mycelium that covers 38 acres in northern Michigan and is estimated to be at least 1500 years old. Another, on the forested slopes of Mt. Adams in Washington State, encompasses 1500 acres, although its age is "only" 400 to 1000 years.

The Imperfect Fungi Are Species in Which Sexual Structures Have Not Been Observed

In many fungus species, researchers have never observed sexual reproduction or the special structures that result from sexual reproduction. These mysterious species cannot be readily classified, because fungi are categorized on the basis of their sexual structures (flagellated spores, zygospores, asci, and basidia). Systematists place these unclassifiable forms into a group known as the *deuteromycetes,* or **imperfect fungi**. Unlike the chytrids, zygomycetes, ascomycetes, and basidiomycetes, the deuteromycetes do not form a natural evolutionary group of closely related organisms linked by descent from a common ancestor. Instead, the

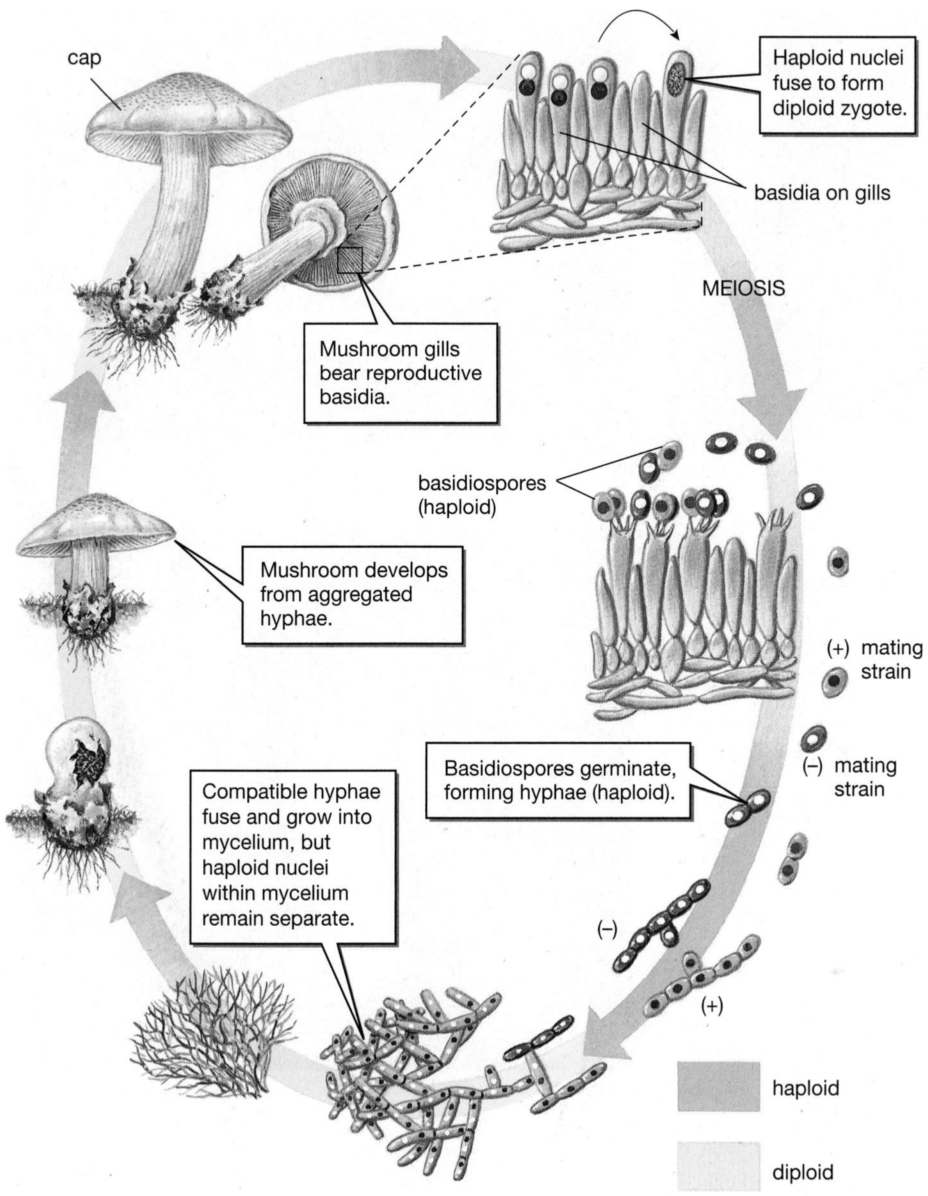
cap
Haploid nuclei fuse to form diploid zygote.
basidia on gills
MEIOSIS
Mushroom gills bear reproductive basidia.
basidiospores (haploid)
Mushroom develops from aggregated hyphae.
(+) mating strain
(–) mating strain
Basidiospores germinate, forming hyphae (haploid).
Compatible hyphae fuse and grow into mycelium, but haploid nuclei within mycelium remain separate.
(–)
(+)
haploid
diploid

Figure 20-6 The life cycle of a "typical" basidiomycete
The mushroom (top left) is a reproductive structure formed from aggregated hyphae made up of cells that each contain two haploid nuclei. Within the cap, leaflike gills bear numerous basidia (top right). Within each basidium, the two haploid nuclei fuse, producing a diploid zygote. The zygote then undergoes meiosis, forming haploid basidiospores that are released from the basidia (right). After dispersal by wind or water, the basidiospores germinate, forming haploid hyphae. When hyphae of different mating types meet, some of the cells fuse. These cells, each containing two haploid nuclei, produce an extensive underground mycelium (bottom). Under appropriate conditions, portions of the mycelium aggregate, swell, and differentiate, poking up through the soil as mushrooms and completing the cycle.

deuteromycetes are a category of convenience, a kind of holding pen for species that cannot be fit into one of the four fungal phyla. It is possible that some deuteromycetes truly lack a sexual stage, but most are believed to actually belong to one of the four main phyla. In fact, many former deuteromycetes have been placed into the appropriate phylum once sexual reproduction was observed.

Some Fungi Form Symbiotic Relationships

Lichens Are Formed by Fungi That Live with Photosynthetic Algae or Bacteria

Lichens are **symbiotic** (literally, "living together") associations between fungi and unicellular green algae or cyanobacteria. Lichens are sometimes described as fungi that have learned to garden, because the fungal member of the partnership "tends" the photosynthetic algal or bacterial partner by providing shelter and protection from harsh conditions. In this protected environment, the photosynthetic member of the partnership uses solar energy to manufacture simple sugars, producing food for itself but also some excess food that is consumed by the fungus. In fact, the fungus seems to consume the lion's share of the photosynthetic product (up to 90% in some species), leading some researchers to conclude that the symbiotic relationship in lichens is really much more one-sided than it is usually portrayed. This view was bolstered by the discovery that, in lichens that include algal symbionts, fungal hyphae actually

(a)

(b)

(c)

Figure 20-7 Diverse basidiomycetes
(a) The giant puffball *Lycopedon giganteum* may produce up to 5 trillion spores. ***(b)*** Shelf fungi, the size of dessert plates, are conspicuous on trees. ***(c)*** The spores of stinkhorns are carried on the outside of a slimy cap that smells terrible to humans, but appeals to flies. The flies lay their eggs on the stinkhorn, and inadvertently disperse the spores that stick to their bodies.

Figure 20-8 A mushroom fairy ring
Mushrooms emerge in a fairy ring from an underground fungal mycelium, growing outward from a central point where a single spore germinated, perhaps centuries ago.

penetrate the cell walls of the algae, in much the same manner as those of fungi that parasitize plants (Fig. 20-9).

Thousands of different fungal species (mostly ascomycetes) form lichens (Fig. 20-10), each in combination with one of a much smaller number of algal or bacterial species. Together, these organisms form a unit so tough and undemanding of nutrients that it is among the first living things to colonize newly formed volcanic islands. Brightly colored lichens have also invaded other inhospitable habitats, ranging from deserts to the Arctic, and can even grow on bare rock. Understandably, lichens in extreme environments grow very slowly, with Arctic colonies, for example, expanding at rates as low as 1 to 2 inches per 1000 years. Despite their slow growth, lichens can persist for long periods of time; some lichens in the Arctic are more than 4000 years old.

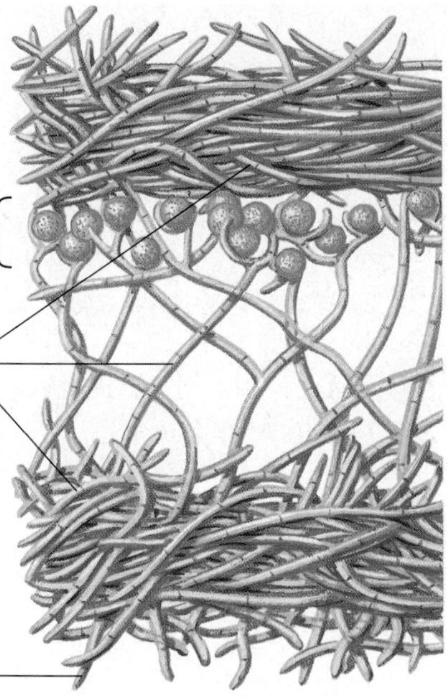

Figure 20-9 The lichen: a symbiotic partnership
Most lichens have a layered structure bounded on the top and bottom by an outer layer formed from fungal hyphae. Attachments formed from fungal hyphae emerge from the lower layer and anchor the lichen to a surface, such as a rock or a tree. An algal layer in which the alga and fungus grow in close association lies beneath the upper layer of hyphae.

Mycorrhizae Are Fungi Associated with the Roots of Many Plants

Mycorrhizae (singular, **mycorrhiza**) are important symbiotic associations between fungi and plant roots. More than 5000 species of mycorrhizal fungi (including representatives of all the major groups of fungi) are known to grow in intimate association with the roots of about 80% of all plants that have roots, including most trees. These associations benefit both the plant and its fungal partner. The hyphae of mycorrhizal fungi surround the plant root and commonly invade the root cells (Fig. 20-11). The fungus digests and absorbs minerals and organic nutrients from the soil, passing some of them directly into the root cells. The fungus also absorbs water and passes it to the plant—an advantage in dry, sandy soils. In return, sugar produced photosynthetically by the plant is passed from the root to the fungus. Plants that participate in this unique relationship, especially those in poor soils, tend to grow larger and more vigorously than do those deprived of the fungus.

Some scientists believe that mycorrhizal associations may have been important in the invasion of land by plants more than 400 million years ago. Such a relationship between an aquatic fungus and a green alga (ancestral to terrestrial plants) could have helped the alga acquire the water and mineral nutrients it needed to survive out of water.

3 How Do Fungi Affect Humans?

The average person gives little thought to fungi, except perhaps for an occasional, momentary appreciation for the white button mushrooms on a pizza. This inattention is unwarranted, however, as fungi play a large part in human affairs.

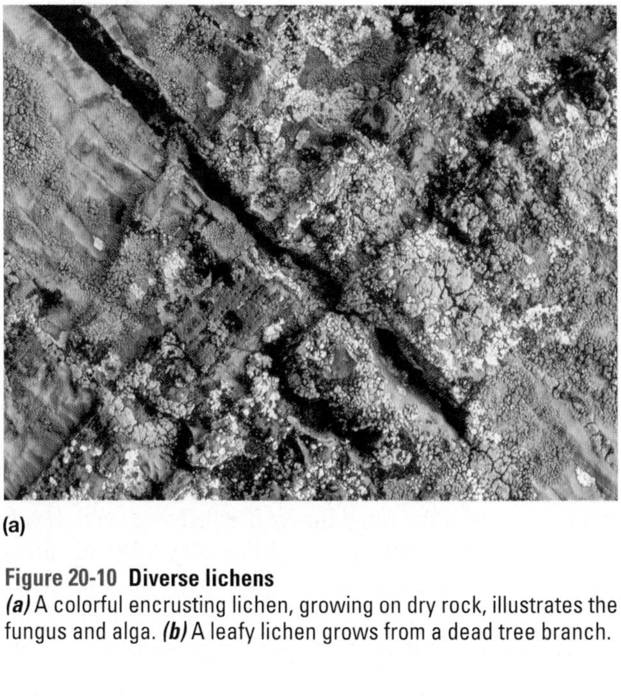

(a)

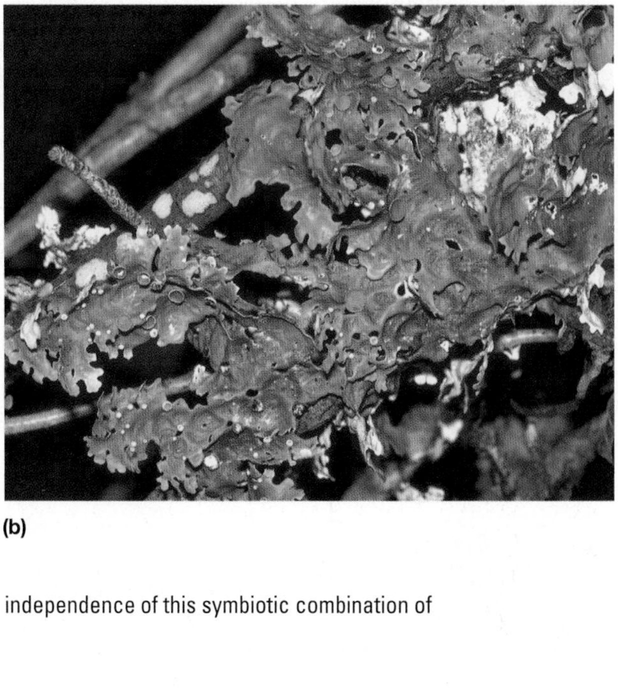

(b)

Figure 20-10 Diverse lichens
(a) A colorful encrusting lichen, growing on dry rock, illustrates the tough independence of this symbiotic combination of fungus and alga. ***(b)*** A leafy lichen grows from a dead tree branch.

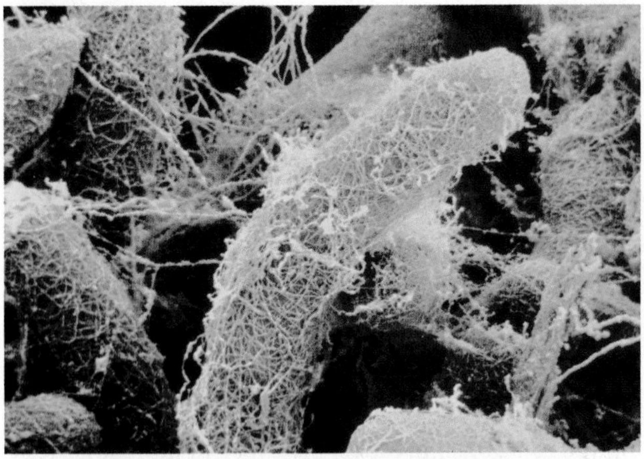

Figure 20-11 Mycorrhizae enhance plant growth
Hyphae of mycorrhizae entwining about the root of an aspen tree. Plants grow significantly better in a symbiotic association with these fungi, which help make nutrients and water available to the roots.

Figure 20-13 A helpful fungal parasite
Although some parasitic fungi attack crops, others are used by farmers to control insect pests. Here, a *Cordyceps* species has killed a grasshopper.

Fungi Attack Plants That Are Important to People

Fungi cause the majority of plant diseases, and some of the plants that they infect are important to humans. For example, fungal pathogens have a devastating effect on the world's food supply. Especially damaging are the basidiomycete plant pests descriptively called *rusts* and *smuts*, which cause billions of dollars' worth of damage to grain crops annually (Fig. 20-12). Fungal diseases also affect the appearance of our landscape. The American elm and the American chestnut, two tree species that were once prominent in many of America's parks, yards, and forests, were destroyed on a massive scale by the ascomycetes that cause Dutch elm disease and chestnut blight. Today, few people can recall the graceful forms of large elms and chestnuts, as these are almost entirely absent from the landscape.

Figure 20-12 Corn smut
This basidiomycete pathogen destroys millions of dollars' worth of corn each year. Even a pest like corn smut has its admirers, though. In Mexico this fungus is known as *huitlacoche* and is considered to be a great delicacy.

Fungi continue to attack plant tissues long after they have been harvested for human use. To the dismay of homeowners, a host of different fungal species attack wood, causing it to rot. Some ascomycete molds secrete the enzymes cellulase and protease, which can cause significant damage to cotton and wool textiles, especially in warm, humid climates where molds flourish.

The fungal impact on agriculture and forestry is not entirely negative, however. Fungal parasites that attack insects and other arthropod pests (Fig. 20-13) can be an important ally in pest control. Farmers who wish to reduce their dependence on toxic and expensive chemical pesticides are increasingly turning to biological methods of pest control, including the application of "fungal pesticides." Fungal pathogens are currently used to control a variety of pests, including termites, rice weevils, tent caterpillars, aphids, and citrus mites.

Fungi Cause Human Diseases

As if the damage that fungi do to our food and forest products were not enough, the kingdom Fungi also includes parasitic species that attack humans directly. Among the most familiar fungal diseases are those caused by ascomycetes that attack the skin: athlete's foot, jock itch, and ringworm. These diseases, though unpleasant, are not life threatening and can usually be treated with antifungal ointments. Prompt treatment can also usually control another common fungal disease, vaginal infections caused by the yeast *Candida albicans* (Fig. 20-14). Fungi can also infect the lungs when victims inhale spores of the fungi that cause diseases such as valley fever and histoplasmosis. Like other fungal infections, these diseases can, if promptly and properly diagnosed, be controlled with antifungal drugs. If untreated, however, they can develop into serious, systemic

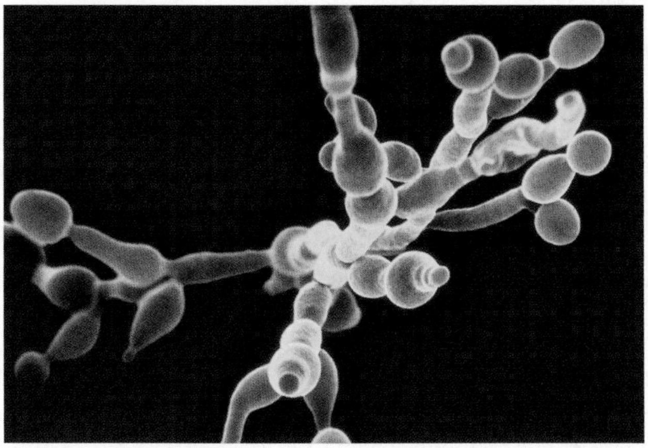

Figure 20-14 The unusual yeast
Yeasts are ununusal, normally nonfilamentous ascomycetes that reproduce most commonly by budding. The yeast shown here is *Candida*, a common cause of vaginal infections.

Figure 20-15 *Penicillium*
Penicillium growing on an orange. Reproductive structures, which coat the fruit's surface, are visible, while hyphae beneath draw nourishment from inside. The antibiotic penicillin was first isolated from this fungus.

infections. Singer Bob Dylan, for instance, became gravely ill with histoplasmosis when the fungus infected the pericardial membrane surrounding his heart.

In addition to their role as agents of infectious disease, some fungi produce toxins that are dangerous to humans. Of particular concern are toxins produced by fungi that grow on grains and other foodstuffs that have been stored in too-moist conditions. For example, molds of the genus *Aspergillus* produce highly toxic, carcinogenic compounds known as aflatoxins. Aflatoxins can contaminate a variety of foods, but some, such as peanuts, seem especially susceptible. Since aflatoxins were discovered in the 1960s, food growers and processors have developed methods for reducing the growth of *Aspergillus* in stored crops, so the amount of aflatoxin in the nation's peanut butter supply has been greatly reduced.

One infamous toxin-producing fungus is the ascomycete *Claviceps purpurea,* which infects rye plants and causes a disease known as ergot. This fungus produces several toxins, which can affect humans if infected rye is ground into flour and consumed. This happened frequently in northern Europe in the Middle Ages, to devastating effect. Then, ergot poisoning was typically fatal, but before dying, victims experienced terrible symptoms. One of the ergot toxins is a vasoconstrictor, which means that it constricts blood vessels and reduces blood flow. The effect can be so extreme that gangrene develops and limbs actually shrivel and fall off. Other ergot toxins cause symptoms that include a burning sensation, vomiting, convulsive twitching, and vivid hallucinations. Today, new agricultural techniques have effectively eliminated ergot poisoning, but a legacy remains as the hallucinogenic drug LSD, which is derived from a component of the ergot toxins.

As with our account of plant diseases, we must conclude this section by noting that fungi have also had positive impacts on human health. The modern era of life-saving antibiotic medicines was ushered in by the discovery of penicillin, which is produced by an ascomycete mold (Fig. 20-15). Penicillin was the first antibiotic and is still used, along with other fungi-derived antibiotics, to combat bacterial diseases. Other important drugs are also derived from fungi, among them cyclosporin, which is used to suppress the immune response during organ transplants so that the body is less likely to reject the foreign organ.

Fungi Make Important Contributions to Gastronomy

Fungi make an important contribution to human nutrition. The most obvious components of this contribution are the fungi that we consume directly: wild and cultivated basidiomycete mushrooms and ascomycetes such as morels and the rare and prized truffle (see "Evolutionary Connections: Fungal Ingenuity—Pigs, Shotguns, and Nooses"). The role of fungi in cuisine, however, also has less visible manifestations. For example, some of the world's most famous cheeses, including Roquefort, Camembert, Stilton, and Gorgonzola, gain their distinctive flavors from ascomycete molds that grow on them as they ripen. Perhaps the most important and pervasive fungal contributors to the food supply, however, are the single-celled ascomycetes (a few species are basidiomycetes) known as *yeasts.*

The discovery that we could harness yeasts to enliven our culinary experience is surely a key event in human history. Among the many foods and beverages that depend on yeasts for their production are bread, wine, and beer, which are consumed so widely that it is difficult to imagine a world without them. All derive their special qualities from fermentation by yeasts. Fermentation occurs when yeasts extract energy from sugar and, as by-products of the metabolic process, emit carbon dioxide and ethyl alcohol. So, as yeasts consume the fruit sugars in grape juice, the sugars are replaced by alcohol, and wine is the result. Eventually, the increasing concentration of alcohol kills the yeasts, ending fermentation. If this occurs before all available grape sugar is consumed, the wine will be sweet; if the sugar is exhausted, the wine will be dry.

Beer is brewed from grain (usually barley), but yeasts are not very effective at consuming the carbohydrates that compose grain kernels. For the yeasts to do their work, the barley grains must first be sprouted (recall that grains are actually seeds). Germination converts carbohydrates to sugar, so the sprouted barley provides an excellent food source for the yeasts. As with wine, fermentation converts sugars to alcohol, but beer brewers capture the carbon dioxide by-product as well, giving the beer its characteristic bubbly carbonation.

In bread making, carbon dioxide is the crucial fermentation product. The yeasts added to bread dough do produce alcohol as well as carbon dioxide, but the alcohol evaporates during baking. In contrast, the carbon dioxide is trapped in the dough, where it forms the bubbles that give bread its light, airy texture (and saves us from a life of eating sandwiches made of crackers). So the next time you're enjoying a slice of French bread with Camembert cheese and a nice glass of Chardonnay, or a slice of pizza and a cold bottle of your favorite brew, you might want to quietly give thanks to the yeasts. Our diets would certainly be a lot duller without the help we get from fungal partners.

Fungi Play a Crucial Ecological Role

No account of the fungi would be complete without mention of their fundamental ecological importance. The fungi are Earth's undertakers, consuming the dead of all kingdoms and returning their component substances to the ecosystems from which they came. The extracellular digestive activities of many fungi liberate nutrients such as carbon, nitrogen, and phosphorus compounds and minerals that can be used by plants. If fungi and bacteria were suddenly to disappear, the consequences would be disastrous. Nutrients would remain locked in the bodies of dead plants and animals, the recycling of nutrients would grind to a halt, soil fertility would rapidly decline, and waste and organic debris would accumulate. In short, ecosystems would collapse.

Evolutionary Connections

Fungal Ingenuity—Pigs, Shotguns, and Nooses

Natural selection, operating over millions of years on the diverse forms of fungi, has produced some remarkable adaptations by which fungi disperse their spores and obtain nutrients. A few of these adaptations are highlighted here.

The Rare, Sexy Truffle

Although many fungi are prized as food, none are as avidly sought as the truffle (Fig. 20-16). The finest Italian truffles may sell for as much as $1500 per pound. Truffles are the spore-containing structures of an ascomycete that forms a mycorrhizal association with the roots of oak trees. Although it develops underground, the truffle has evolved a fascinating mechanism to entice animals, especially wild pigs, to dig it up. The truffle releases a chemical that closely resembles the pig's sex attractant. As aroused pigs dig up and devour the truffle, millions of spores are scattered to the winds. Truffle collectors use muzzled pigs to hunt their quarry; a good truffle-pig can smell an underground truffle 150 feet (about 50 meters) away!

The Shotgun Approach to Spore Dispersal

If you get close enough to a pile of horse manure to scrutinize it, you may observe the beautiful, delicate reproductive structures of the zygomycete *Pilobolus* (Fig. 20-17). Despite their daintiness, these are actually fungal shotguns. The clear bulbs, capped with sticky black spore cases, extend from hyphae that penetrate the dung. As the bulbs mature, the sugar concentration inside them increases, and water is drawn in by osmosis.

Figure 20-16 The truffle
Truffles, rare ascomycetes (each about the size of a small apple), are a gastronomic delicacy.

Figure 20-17 An explosive zygomycete
The delicate, translucent reproductive structures of the zygomycete *Pilobolus* will literally blow their tops when ripe, dispersing the black caps with their payload of spores.

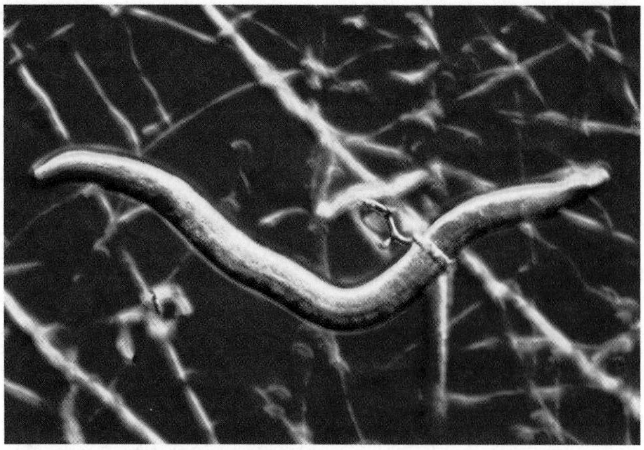

Figure 20-18 Nemesis of nematodes
Arthrobotrys, the nematode (roundworm) strangler, traps its nematode prey in a nooselike modified hypha that swells when the inside of the loop is contacted.

Meanwhile, the bulb begins to weaken just below its cap. Suddenly, like an overinflated balloon, the bulb bursts and blows its spore-carrying top up to 3 feet (1 meter) away.

The airborne spores may well land in some leaves of grass, because *Pilobolus* bends toward the light as it grows, thereby increasing the chances that its spores will be directed toward open pasture. Spores that adhere to grass remain there until consumed by a grazing herbivore, perhaps a horse. Later (some distance away) the horse will deposit a fresh pile of manure containing *Pilobolus* spores that have passed unharmed through its digestive tract. The spores germinate, and growing hyphae penetrate the manure (a rich source of nutrients), ultimately sending up new projectiles to continue this ingenious cycle.

The Nematode Nemesis

Microscopic *nematodes* (roundworms) abound in rich soil, and fungi have evolved several fascinating forms of nematode-nabbing hyphae that allow them to exploit this rich source of protein. Some fungi produce sticky pods that adhere to passing nematodes and penetrate the worm body with hyphae, which then begin digesting the worm from within. One species shoots a microscopic harpoonlike spore into passing nematodes; the spore develops into a new fungus inside the worm. The fungal strangler *Arthrobotrys* produces nooses formed from three hyphal cells. When a nematode wanders into the noose, its contact with the inner parts of the noose stimulates noose cells to swell with water (Fig. 20-18). In a fraction of a second, the hole constricts, trapping the worm. Fungal hyphae then penetrate and feast on their prey.

Earth Watch
The Case of the Disappearing Mushrooms

Mycologists (scientists who study fungi) and gourmet cooks may seem to have little in common, but recently they have become united in a common concern: European mushrooms are rapidly declining in numbers, in average size, and in species diversity. Although the problem is most easily recognized in Europe, where people have been gathering wild mushrooms for centuries, American mycologists are also alarmed; the same decline may be occurring here as well. Why are mushrooms disappearing? Overhunting of edible mushrooms is not the culprit, because poisonous forms are equally affected. The loss is evident in all types of mature forests, so changing forest management practices could not be the cause. The most likely cause is air pollution, because the loss of mushrooms is greatest where the air contains the highest levels of ozone, sulfur, and nitrogen. Although mycologists have not yet determined exactly how air pollution harms mushrooms, the evidence is clear. In Holland, for example, the average number of fungal species per 1000 square meters has dropped from 37 down to 12 over the past several decades. Twenty out of 60 fungal species surveyed in England are declining. Concern is intensified by the fact that the mushrooms most affected are those whose hyphae form mycorrhizal associations with tree roots. Trees with diminished mycorrhizae may have less resistance to periodic droughts or extreme cold spells. Because air pollution is also harming forests directly, the additional loss of mycorrhizae could be devastating.

CASE STUDY REVISITED

Three Outings

As our intrepid mushroom hunter Anita contemplated the fates of her colleagues Roger (in a coma) and Whitney (in jail), she began to wonder about the complex chemistry of fungi. It is not clear, from an evolutionary standpoint, why fungi should contain such an array of molecules that act as medicines, flavors, toxins, and intoxicants for humans. Nonetheless, some fungi do contain chemicals that have profound impacts on human physiology. Anita's friends simply ran afoul of some especially potent examples.

Some of the deadliest poisons known to humankind are found in mushrooms. Especially notable in this regard are certain species in the genus *Amanita*, which have evocative common names such as death cap and destroying angel. These names are apt, as even a single bite of one can be lethal. The destructive agents in *Amanita* mushrooms are molecules known as *amatoxins*. Amatoxins inhibit the action of one of the RNA polymerases in human cells. As a result, messenger RNA is not made, proteins are not synthesized, and cell growth slows or stops. Damage is most severe in the liver, where toxins tend to accumulate and cellular activity is especially intense. Often, a liver transplant is the only way to save victims of *Amanita* poisoning. Most such victims are small children and inexperienced collectors like Roger who make serious errors in mushroom identification.

Some collectors are searching not for the makings of a gourmet meal but for the psychoactive properties of certain mushrooms. Mind-altering mushrooms are found in a number of fungal genera, but those in the genus *Psilocybe* are perhaps the most infamous. *Psilocybe* mushrooms have a long history of ritual and ceremonial use in native cultures of the Americas, but today the possession and sale of such mushrooms is illegal. Some people nonetheless persist in consuming them, even though such individuals risk lengthy prison terms.

The punch that these mushrooms pack derives from the alkaloid chemicals psilocybin and psilocin. Users report that ingestion of mushrooms containing these compounds is followed by intoxication, euphoria, and intense hallucinations. An unknown number of people collect *Psilocybe* for personal consumption and economic gain. As Whitney discovered, the threat of legal jeopardy is not an idle one; arrests are frequent in regions where *Psilocybe* are common.

The experiences of Anita, Roger, and Whitney illustrate some of the rewards and risks of mushroom collecting. Growing interest in gourmet foods and outdoor activity and growing markets for wild mushrooms (legal and illegal) have greatly increased the number of collectors in the nation's forests and fields. Accordingly, we can expect to see more unusual meals, emergency liver transplants, and mushroom-related arrests.

Fungi seem especially likely to manufacture compounds that are flavorful, poisonous, or mind-altering. Which aspects of fungi's way of life might account for the prevalence of these compounds?

Summary of Key Concepts

1) What Are the Main Adaptations of Fungi?

Fungal bodies generally consist of filamentous hyphae, which are either multicellular or multinucleate and form large, intertwined networks called *mycelia.* Fungal nuclei are generally haploid. A cell wall of chitin surrounds fungal cells.

All fungi are heterotrophic, either parasitic or saprobic. They secrete digestive enzymes outside their bodies and absorb the liberated nutrients.

Fungal reproduction is varied and complex. Asexual reproduction can occur either through fragmentation of the mycelium or through asexual spore formation. Sexual spores form after compatible haploid nuclei fuse to form a diploid zygote, which undergoes meiosis to form haploid sexual spores. Both asexual and sexual spores produce haploid mycelia through mitosis.

2) How Are Fungi Classified?

The major phyla of fungi and their characteristics are summarized in Table 20-1. A lichen is a symbiotic association between a fungus and algae or cyanobacteria. This self-sufficient combination can colonize bare rock. Mycorrhizae are associations between fungi and the roots of most vascular plants. The fungus derives photosynthetic nutrients from the plant roots and in return carries water and nutrients into the root from the surrounding soil.

3) How Do Fungi Affect Humans?

The majority of plant diseases are caused by parasitic fungi. Some parasitic fungi can help control insect crop pests. Others can cause human diseases, including ringworm, athlete's foot, and common vaginal infections. Some fungi produce toxins that can harm humans. Nonetheless, fungi add variety to the human food supply, and fermentation by fungi helps make wine, beer, and bread.

Fungi are extremely important decomposers in ecosystems. Their filamentous bodies penetrate rich soil and decaying organic material, liberating nutrients through extracellular digestion.

Key Terms

ascus *p. 394*
basidiospore *p. 396*
basidium *p. 396*
club fungus *p. 396*
fairy ring *p. 396*
hypha *p. 392*
imperfect fungus *p. 396*
lichen *p. 397*
mycelium *p. 392*
mycorrhiza *p. 399*
sac fungus *p. 394*
saprobe *p. 392*
septum *p. 392*
sporangium *p. 394*
spore *p. 392*
symbiotic *p. 397*
zygospore *p. 394*
zygote fungus *p. 394*

Thinking Through the Concepts

Multiple Choice

1. *There are no fungi that are*
   a. predators
   b. photosynthetic
   c. decomposers
   d. parasites
   e. symbiotic
2. *With what plant organ do mycorrhizae interact?*
   a. roots
   b. leaves
   c. stems
   d. fruits
   e. all parts of the plant
3. *What term refers to the mass of threads that forms the body of most fungi?*
   a. hyphae
   b. mycelium
   c. sporangia
   d. ascus
   e. basidium
4. *Which of the following pairs is incorrect?*
   a. fruit rot: zygote fungus
   b. edible mushroom: club fungus
   c. black bread mold: sac fungus
   d. shelf fungus: club fungus
   e. yeast: sac fungus
5. *Which of the following statements is true both of fungi and of animals?*
   a. Both photosynthesize.
   b. Both form embryos.
   c. They interact to form lichens.
   d. Both fix nitrogen.
   e. Both are heterotrophic.
6. *Which of the following structures would you expect to find in the corn smut fungus?*
   a. ascospores
   b. basidiospores
   c. asci
   d. zygospores
   e. fairy rings

? Review Questions

1. Describe the structure of the fungal body. How do fungal cells differ from most plant and animal cells?
2. What portion of the fungal body is represented by mushrooms, puffballs, and similar structures? Why are these structures elevated above the ground?
3. What two plant diseases, caused by parasitic fungi, have had an enormous impact on forests in the United States? In which division are these fungi found?
4. List some fungi that attack crops. In which phyla is each?
5. Describe asexual reproduction in fungi.
6. What is the major structural ingredient in fungal cell walls?
7. List the major phyla of fungi, describe the feature that gives each its name, and give one example of each.
8. Describe how a fairy ring of mushrooms is produced. Why is the diameter related to its age?
9. Describe two symbiotic associations between fungi and organisms from other kingdoms. In each case, explain how each partner in these associations is affected.

Applying the Concepts

1. Dutch elm disease in the United States is caused by an *exotic*—that is, an organism (in this case, a fungus) introduced from another part of the world. What damage has this introduction done? What other fungal pests fall into this category? Why are parasitic fungi particularly likely to be transported out of their natural habitat? What can governments do to limit this importation?
2. The discovery of penicillin revolutionized the treatment of bacterial diseases. However, penicillin is now rarely prescribed. Why is this? HINT: Refer back to Chapter 15.

3. The discovery of penicillin was the result of a chance observation by an observant microbiologist, Alexander Fleming. How would you search systematically for new antibiotics produced by fungi? Where would you look for these fungi?
4. Fossil evidence indicates that mycorrhizal associations between fungi and plant roots existed in the late Paleozoic era, when the invasion of land by plants began. This evidence suggests an important link between mycorrhizae and the successful invasion of land by plants. Why might mycorrhizae have been important fungi in the colonization of terrestrial habitats by plants?
5. General biology texts in the 1960s included fungi in the plant kingdom. Why do biologists no longer consider fungi as legitimate members of the plant kingdom?
6. What ecological consequences would occur if humans, using a new and deadly fungicide, destroyed all fungi on Earth?

For More Information

Angier, N. "A Stupid Cell with All the Answers." *Discover,* November 1986. Fascinating description of the uses of yeasts in molecular biology, including beautiful illustrations.

Barron, G. "Jekyll-Hyde Mushrooms." *Natural History*, March 1992. Fungi include a variety of forms; some absorb decaying plant material, and others prey on microscopic worms.

Dix, N. J., and Webster, J. *Fungal Ecology*. London: Chapman & Hall, 1995. A rather technical but comprehensive and readable account of fungal diversity that focuses on the roles of fungi in different ecological communities.

Hudler, G. W. *Magical Mushrooms, Mischievous Molds*. Princeton, NJ: Princeton University Press, 1998. Engaging treatment of the fungi, focusing on their importance in human affairs.

Kiester, E. "Prophets of Gloom." *Discover*, November 1991. Lichens can be used as indicators of air quality and provide evidence of deteriorating environmental conditions.

Radetsky, P. "The Yeast Within." *Discover*, March 1994. Describes the newly discovered ability of baker's yeast to form filaments and the possible implications of this for the study of yeasts that cause vaginal infections.

Schaechter, E. *In the Company of Mushrooms*. Cambridge: Harvard University Press, 1997. An accessible account of the world of mushrooms, written in a warm, personal style.

Strobel, G. A., and Lanier, G. N. "Dutch Elm Disease." *Scientific American*, August 1981. A description of the life cycle of this parasitic fungus and of techniques that limit its spread.

Vogel, S. "Taming the Wild Morel." *Discover*, May 1988. Describes the research that has allowed these rare delicacies to be cultivated in the laboratory.

Answers to Multiple-Choice Questions

1. b 2. a 3. b 4. c 5. e 6. b

MEDIATUTOR
The Fungi

CD Activities

Activity 20.1: Fungi Structure

Estimated time: 5 minutes

In this tutorial students are asked to explore the major fungi structures and to complete a series of labeling activities.

Activity 20.2: The Fungi

Estimated time: 5 minutes

In this tutorial you are shown examples of each of the major divisions of fungi and their distinguishing characteristics. Afterward you will be quizzed on your understanding of those characteristics.

Start the MediaTutor Student CD-ROM and enter the activity number in the Quick Search box to be taken directly to that activity.

Web Investigations

Case Study: Three Outings

Estimated time: 5 minutes

Wild mushrooms can be delicious, debilitating, or deadly. Can they be collected safely? This exercise will take a closer look at the dangers and difficulties of mushroom hunting.

Go to http://www.prenhall.com/audesirk6, the Audesirk Companion Web site. Select Chapter 20 and Web Investigation to begin.

The bark of the Pacific yew is the source of a potent anticancer drug. Plants hold many such medicinal treasures, most of them still undiscovered.

21 The Plant Kingdom

AT A GLANCE

Case Study: Hunting for Medical Treasures

1) **What Are the Key Features of Plants?**

Plants Have Both a Sporophyte and a Gametophyte Generation

2) **What Is the Evolutionary Origin of Plants?**

Green Algae Gave Rise to Land Plants

3) **How Did Plants Invade and Flourish on Land?**

The Plant Body Increased in Complexity as Plants Made the Evolutionary Transition from Water to Dry Land

The Invasion of Land Required Protection and a Means of Dispersal for Sex Cells and Developing Plants

The Liverworts and Mosses Are Adapted to Moist Environments

The Vascular Plants, or Tracheophytes, Have Conducting Vessels That Also Provide Support

The Seedless Vascular Plants Include the Club Mosses, Horsetails, and Ferns

The Seed Plants Dominate the Land, Aided by Two Important Adaptations: Pollen and Seeds

Case Study Revisited: Hunting for Medical Treasures

CASESTUDY

Hunting for Medical Treasures

When Vicki M. was diagnosed with ovarian cancer, her first reaction was shock, followed by despair. The diagnosis felt like a death sentence. However, after learning more about the disease from her doctors, Vicki began to feel more optimistic. Improved treatment regimens have bettered patients' chances of surviving the disease, and Vicki, whose tumor was identified early in its development, was a good candidate for aggressive treatment. Her treatment would be multifaceted, and a key component would be the tumor-suppressing drug Taxol, which is especially effective against ovarian cancer.

Like many other important medicines, Taxol was a gift from nature. This powerful drug was not concocted in a research lab, but occurs naturally in the bark of the American yew tree (*Taxus brevifolia*). Since its approval by the FDA in 1992, Taxol has helped relieve the suffering of thousands of people with ovarian and other cancers and has grown into a pharmaceutical titan that now generates more than one billion dollars in annual sales.

The medical and financial success of Taxol and other plant-derived medicines serves to inspire a continuing search for the additional treasures that are no doubt hidden in plants. This search for biological riches is sometimes known as *bioprospecting*. With millions of dollars to be made and millions of human lives to be saved, one might think that the planet's forests would be filled with armies of bioprospectors, filling sample bags with leaves, roots, and bark. Bioprospecting activity, however, has yet to become the important and lucrative activity that many observers predicted. What's the holdup?

One problem is that successful prospecting tends to create intense demand for the species that produce valuable medicines, and existing plant populations may not be able to meet the demand. For example, a 100-year-old American yew tree yields only a single dose of Taxol, so supplying the demand for the drug could threaten the yew with extinction. What's more, widespread collecting of yews inflicts considerable damage on the old-growth Pacific forests in which the yew is found.

Can we find ways to garner the benefits of bioprospecting without sacrificing the very ecosystems that we will need for future bioprospecting? We will return to this question following a survey of the "green kingdom," whose chemical treasures tantalize the medical profession and the pharmaceutical industry. ■

1 What Are the Key Features of Plants?

Plants are multicellular and able to use photosynthesis to convert water and carbon dioxide to sugar. Neither multicellularity nor the ability to photosynthesize is unique to plants, but the simultaneous appearance of those traits in a single organism is quite rare outside the green kingdom. The most distinctive feature of plants, however, is their reproductive cycle.

Plants Have Both a Sporophyte and a Gametophyte Generation

The plant life cycle is characterized by **alternation of generations** (Fig. 21-1), in which separate diploid and haploid generations alternate with one another. (Recall that a diploid organism has two sets of chromosomes; a haploid organism, one set.) In the diploid generation, the plant body consists of diploid cells and is known as the **sporophyte**. Certain cells of sporophytes undergo meiosis to produce haploid spores. These spores divide by mitosis and develop into multicellular, haploid plants called **gametophytes**. The gametophytes ultimately produce male and female haploid gametes by mitosis. The gametes fuse to form diploid **zygotes**, which develop into a diploid sporophyte, and the cycle begins again.

The evolutionary history of plants has been marked by a tendency for the sporophyte generation to become increasingly prominent as the longevity and size of the gametophyte generation shrinks (Table 21-1). Thus, the earliest plants are believed to have been similar to today's *nonvascular plants*, such as mosses. Nonvascular plants have a sporophyte that is smaller than the gametophyte and remains attached to it. In contrast, plants whose evolutionary origin is somewhat later, such as ferns and other *seedless vascular plants*, feature a life cycle in which the sporophyte is dominant, and the gametophyte is a much smaller, independent plant. Finally, in the most recently evolved group of plants, the *seed plants*, gametophytes are microscopic and barely recognizable as an alternate generation. These tiny gametophytes, however, still produce the eggs and sperm that unite to form the zygote that develops into the diploid sporophyte.

2 What Is the Evolutionary Origin of Plants?

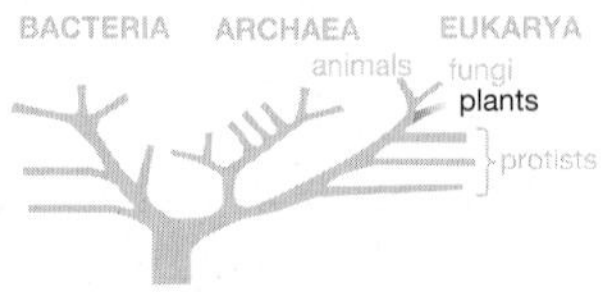

The ancestors of plants were photosynthetic protists, probably similar to today's algae. Like modern algae, the organisms that gave rise to plants presumably lacked true roots, stems, leaves, and complex reproduc-

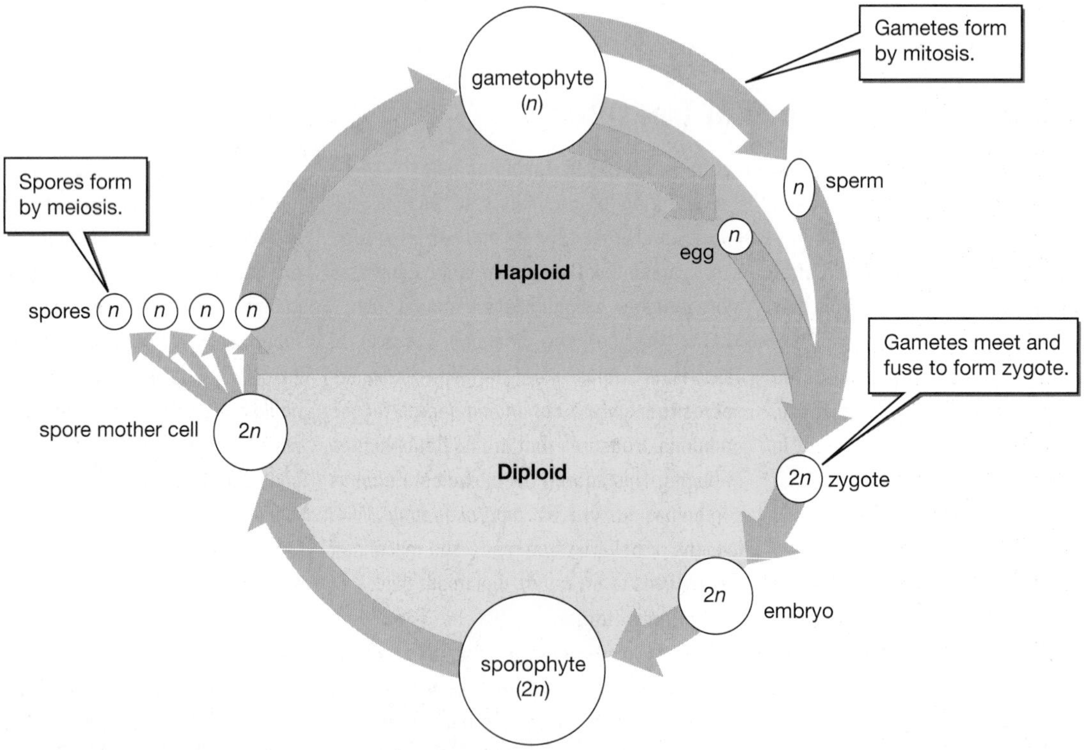

Figure 21-1 Alternation of generations in plants
As shown in this generalized depiction of a plant life cycle, a diploid sporophyte generation produces haploid spores through meiosis. The spores develop into a haploid gametophyte generation that produces haploid gametes by mitosis. The fusion of these gametes results in a diploid zygote that develops into the sporophyte plant.

Table 21-1 Features of the Major Plant Groups

Phylum	Relationship of Sporophyte and Gametophyte	Transfer of Reproductive Cells	Early Embryonic Development	Dispersal	Water and Nutrient Transport Structures	Typical Habitat
Liverworts and mosses (Bryophyta)	Gametophyte dominant—sporophyte develops from zygote retained on gametophyte	Motile sperm swims to stationary egg retained on gametophyte	Occurs within archegonium of gametophyte	Haploid spores carried by wind	Present	Moist terrestrial
Ferns (Pteridophyta)	Sporophyte dominant—develops from zygote retained on gametophyte	Motile sperm swims to stationary egg retained on gametophyte	Occurs within archegonium of gametophyte	Haploid spores carried by wind	Present	Moist terrestrial
Conifers (Coniferophyta)	Sporophyte dominant—microscopic gametophyte develops within sporophyte	Wind-dispersed pollen carries sperm to stationary egg in cone	Occurs within a protective seed containing a food supply	Seeds containing diploid sporophyte embryo dispersed by wind or animals	Present	Varied terrestrial habitats—dominate in dry, cold climates
Flowering plants (Anthophyta)	Sporophyte dominant—microscopic gametophyte develops within sporophyte	Pollen, dispersed by wind or animals, carries sperm to stationary egg within flower	Occurs within a protective seed containing a food supply; seed encased in fruit	Fruit, carrying seeds, dispersed by animals, wind, or water	Present	Varied terrestrial habitats—dominant terrestrial plant

tive structures such as flowers or cones. All of these features appeared later in the evolutionary history of plants (Fig. 21-2).

Green Algae Gave Rise to Land Plants

Of today's different groups of algae, green algae are probably most similar to ancestral plants. This supposition stems from the close phylogenetic relationship between the two groups. DNA comparisons have shown that green algae are plants' closest living relatives. In addition, other lines of evidence support the hypothesis that land plants evolved from green algal ancestors:

1. Green algae use the same type of chlorophyll and accessory pigments in photosynthesis as do land plants. In contrast, the photosynthetic pigments of other algal groups, such as the red algae and the brown algae, differ from those of plants.
2. Green algae store food as starch, as do land plants, and have cell walls made of cellulose, similar in composition to those of land plants. Again, the food storage and cell wall molecules of red and brown algae are different.

Today's green algae live mainly in fresh water, which suggests that the early evolutionary history of the group may have occurred in freshwater habitats. If so, the green algae would have been subjected to environmental pressures that resulted in adaptations that enhanced their potential to give rise to land-dwelling organisms.

Freshwater habitats, in contrast to the nearly constant environmental conditions of the ocean, are highly variable. Water temperature can fluctuate seasonally or even daily, and changing levels of rainfall can lead to fluctuations in the concentration of chemicals in the water or even to periods in which the aquatic habitat dries up. Ancient freshwater green algae must have evolved features that enabled them to withstand extremes of temperature and periods of dryness. These adaptations served their descendants well as they invaded land.

3 How Did Plants Invade and Flourish on Land?

The terrestrial world may be green now, but it didn't start out that way. When plants first made the transition ashore more than 400 million years ago, the land was

MEDIATUTOR
21.1 Evolution of Plant Structure

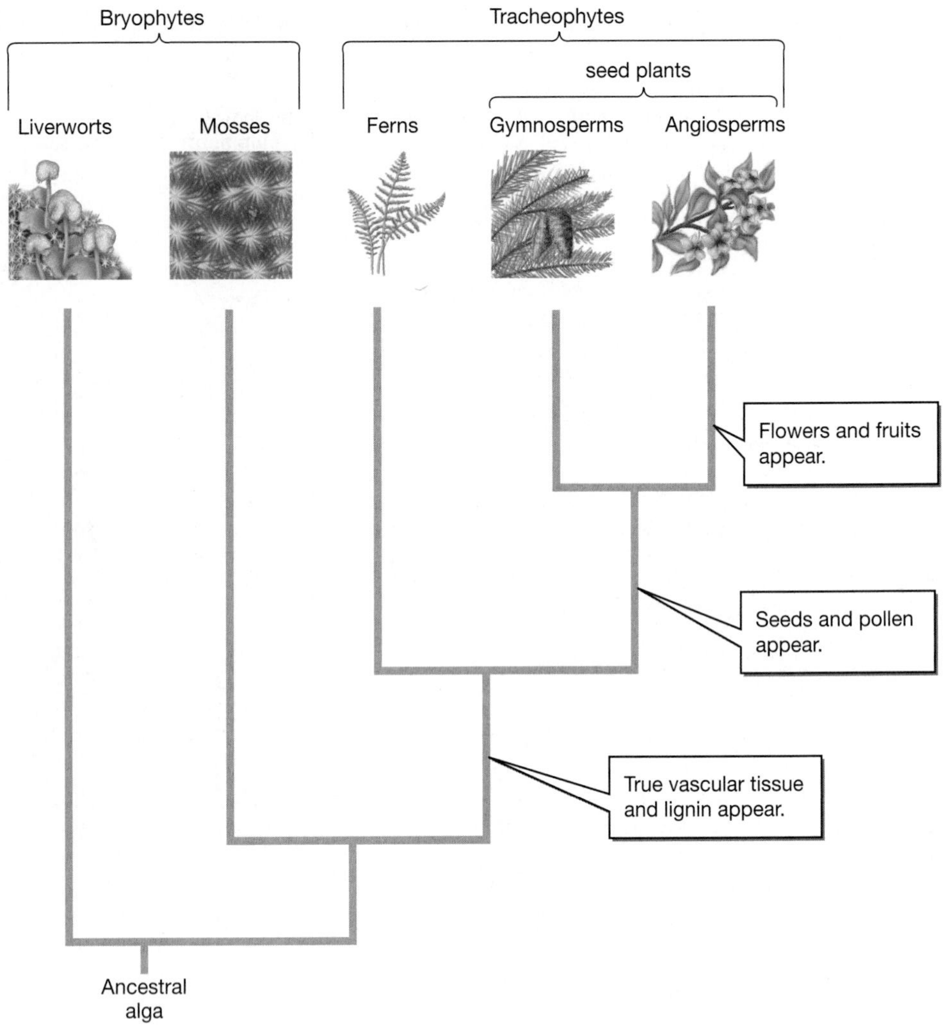

Figure 21-2 Evolutionary tree of some major plant groups

barren and desolate, inhospitable to life. From a plant's evolutionary viewpoint, however, it was also a land of opportunity, free of competitors and predators, and full of carbon dioxide and sunlight (the raw materials for photosynthesis, which are present in far higher concentrations in air than in water). So once natural selection had shaped the adaptations that helped plants overcome the obstacles to terrestrial living, plants prospered and diversified.

The Plant Body Increased in Complexity as Plants Made the Evolutionary Transition from Water to Dry Land

When plants pioneered the land, they faced a range of challenges posed by terrestrial environments. On land, the supportive buoyancy of water is missing, the plant body is no longer bathed in a nutrient solution, and the air tends to dry things out. These conditions favored the evolution of structures that support the body, vessels that transport water and nutrients to all parts of the plant, and structures that conserve water. The resulting adaptations to dry land include some structural features that arose early in plant evolution; now these features are common to virtually all land plants. They include:

1. Roots or rootlike structures that anchor the plant and/or absorb water and nutrients from the soil
2. A waxy **cuticle** that covers the surfaces of leaves and stems and that limits the evaporation of water
3. Pores called **stomata** (singular, **stoma**) in the leaves and stems that open to allow gas exchange but close when water is scarce, thus reducing the amount of water lost to evaporation

Other key adapations occurred somewhat later in the transition to terrestrial life, and are now widespread but not universal among plants (most non-vascular plants, a group decribed below, lack them):

4. Conducting vessels that transport water and minerals upward from the roots and that move photosynthetic products from the leaves to the rest of the plant body
5. The stiffening substance **lignin**, which is a rigid polymer that impregnates the conducting vessels and supports the plant body, helping the plant expose maximum surface area to sunlight

The Invasion of Land Required Protection and a Means of Dispersal for Sex Cells and Developing Plants

The adaptations described above allowed an increasing diversity of plant forms to exploit dry land. Life on land, however, also required new methods of transporting sperm to eggs. Unlike aquatic and marine forms, whose gametes (sex cells) and zygotes (fertilized sex cells) can be carried passively by water currents or can actively swim using flagella, land plants cannot always rely on water to carry their sex cells and disperse their fertilized eggs. So the most successful groups of land plants are those that evolved (1) methods of gamete and zygote dispersal that are independent of water and (2) structures that protect developing embryos from drying out.

Protected embryos and waterless dispersal of sex cells were achieved with the origin of the seed plants and the key evolutionary innovations that they introduced: *pollen*, *seeds*, and, later, *flowers* and *fruits*. Early seed plants produced dry, microscopic pollen grains that allowed wind, instead of water, to carry the male gametes. Seeds provided protection and nourishment for developing embryos and the potential for enhanced dispersal in terrestrial environments. Later came the evolution of flowers, which enticed animal pollinators that delivered pollen more precisely than did wind. Fruits also attracted animal foragers, which consumed them and dispersed the indigestible seeds in their feces.

The Liverworts and Mosses Are Adapted to Moist Environments

Two major groups of land plants arose from the ancient algal ancestors. One group, the *nonvascular plants*, or **bryophytes**, requires a moist environment to reproduce and thus straddles the boundary between aquatic and terrestrial life, much like the amphibians of the animal kingdom. The other group, the *vascular plants*, or **tracheophytes**, has been able to colonize drier habitats.

Bryophytes retain some characteristics of their algal ancestors. They lack true roots, leaves, and stems. They do possess rootlike anchoring structures called **rhizoids** that bring water and nutrients into the plant body, but bryophytes are *nonvascular*—they lack well-developed structures for conducting water and nutrients. They must instead rely on slow diffusion or poorly developed conducting tissues to distribute water and other nutrients. As a result, their body size is limited. Size is also limited by the absence of lignin; without this stiffening agent, bryophyte bodies cannot achieve much upward growth. Most bryophytes are less than 1 inch (less than 2.5 centimeters) tall.

Among the bryophyte features that do represent adaptations to terrestrial existence are their enclosed reproductive structures. The **archegonia** (singular, **archegonium**), in which eggs develop, and the **antheridia** (singular, **antheridium**), where sperm are formed, prevent the gametes from drying out. In some bryophyte species both archegonia and antheridia are located on the same plant; in other species each individual plant is either male or female. In all bryophytes the sperm must swim to the egg (which emits a chemical attractant) through a film of water. Bryophytes that live in drier areas must time their reproduction to coincide with rains.

The major bryophyte representatives are the liverworts and mosses (Fig. 21-3). Liverworts and most

Figure 21-3 Bryophytes
(a) Liverworts grow in moist, shaded areas. This is the female gametophyte plant, bearing umbrella-like archegonia, which hold the eggs. Sperm must swim up the stalks through a film of water to fertilize the eggs. ***(b)*** Moss plants showing both stages in the life cycle. The short, leafy green plants are the haploid gametophytes; the stalks are the diploid sporophyte generation. The sporophytes are less than a half inch (about 1 centimeter) in height.

mosses are most abundant in areas where moisture is plentiful. Some mosses, however, possess a waterproof cuticle that retains moisture and stomata that can be closed, preventing water loss. These mosses can survive in deserts, on bare rock, and in far northern and southern latitudes where humidity is low and liquid water is scarce for much of the year.

The life cycle of a moss is shown in Figure 21-4. The larger "leafy" plant body is the haploid gametophyte, which forms sperm and eggs by mitosis. (The gametophyte is found at the lower right in Fig. 21-4.) The fertilized egg is retained in the archegonium, where the embryo grows and matures into a small diploid sporophyte that remains attached to the parent gametophyte plant. At maturity, the sporophyte produces haploid spores by meiosis within a capsule. When the capsule is opened, spores are released and dispersed by the wind. If a spore lands in a suitable environment, it may develop into another haploid gametophyte plant.

The Vascular Plants, or Tracheophytes, Have Conducting Vessels That Also Provide Support

Although the early bryophytes were quite successful at colonizing land, their growth habit left a prime evolutionary opportunity available. Even after moister regions were covered with the short green fuzz of bryophytes, any plant that could stand taller than an inch or two would benefit by basking in sunlight while shading its short competitors. Natural selection thus favored two types of adaptations that allowed some plants to become taller: (1) structures that provided support for the body and (2) vessels that conducted water and nutrients absorbed by the roots into the upper portions of the plant. In the **vascular** (from the Latin word for "vessel-bearing") plants, specialized groups of conducting cells (which we will call **vessels**) impregnated with the stiffening substance lignin served both supportive and conducting functions.

The Seedless Vascular Plants Include the Club Mosses, Horsetails, and Ferns

Seedless vascular plants first appear in the fossil record of the Devonian period (410 to 360 million years ago), and by the end of that period had become quite diverse, including plants of treelike proportions. During the subsequent Carboniferous period (286 million to 360 million years ago), seedless vascular plants dominated the landscape. Their bodies—transformed by heat, pressure, and time—are burned today as coal. Some descendants of the seedless vascular plant groups of that era have survived until modern times, but present-day club mosses, horsetails, and ferns are much diminished in size, and their dominant role has been largely assumed by the more-versatile seed plants.

The club mosses are now limited to representatives a few inches in height (Fig. 21-5a, p. 416). Their leaves are small and scalelike, resembling the leaflike structures of mosses. Club mosses of the genus *Lycopodium*, commonly known as ground pine, form a beautiful ground cover in some temperate coniferous and deciduous forests.

Modern horsetails form a single genus, *Equisetum*, that contains only 15 species, most less than 1 meter tall (Fig. 21-5b). The bushy branches of some species lend them the common name horsetails; the leaves are reduced to tiny scales on the branches. They are also called "scouring rushes," because they were used by early European settlers of North America to scour pots and floors. All species of *Equisetum* deposit large amounts of silica (glass) in their outer layer of cells, giving them an abrasive texture.

The ferns, with 12,000 species, are far more successful (Fig. 21-5c). In the Tropics, "tree ferns" still reach heights reminiscent of their ancestors in the Carboniferous period. Ferns are the only seedless vascular plants that have broad leaves. Broad leaves can capture more sunlight, and this advantage over the small-leaved club mosses and horsetails may account for the relative success of modern ferns.

The life cycle of a fern is depicted in Figure 21-6, p. 417. A major difference between vascular plants and bryophytes is that the diploid sporophyte is dominant in vascular plants. On special leaves of club mosses and ferns, and on conelike structures of horsetails, haploid spores are produced in structures called *sporangia*. The spores are dispersed by the wind and give rise to tiny, haploid gametophyte plants, which produce sperm and eggs. The gametophyte generation retains two traits that are reminiscent of the bryophytes. First, the small gametophytes lack conducting vessels. Second, as in bryophytes, the sperm must swim through water to reach the egg, so these plants still depend on the presence of water for sexual reproduction.

The Seed Plants Dominate the Land, Aided by Two Important Adaptations: Pollen and Seeds

As mentioned above, the invasion of terrestrial habitats by plants hinged in large measure on crucial evolutionary innovations. In particular, (1) pollen made it possible to achieve fertilization without the need for gametes to travel through water, and (2) seeds made it possible for embryos to develop without being immersed in water. Together, these two key advances have helped seed plants maintain their preeminence on land for the past 250 million years.

Pollen (also called a *pollen grain*) is all that remains of the male gametophyte of seed plants. In seed plants, both male and female gametophytes (which produce the sex cells) are greatly reduced in size,

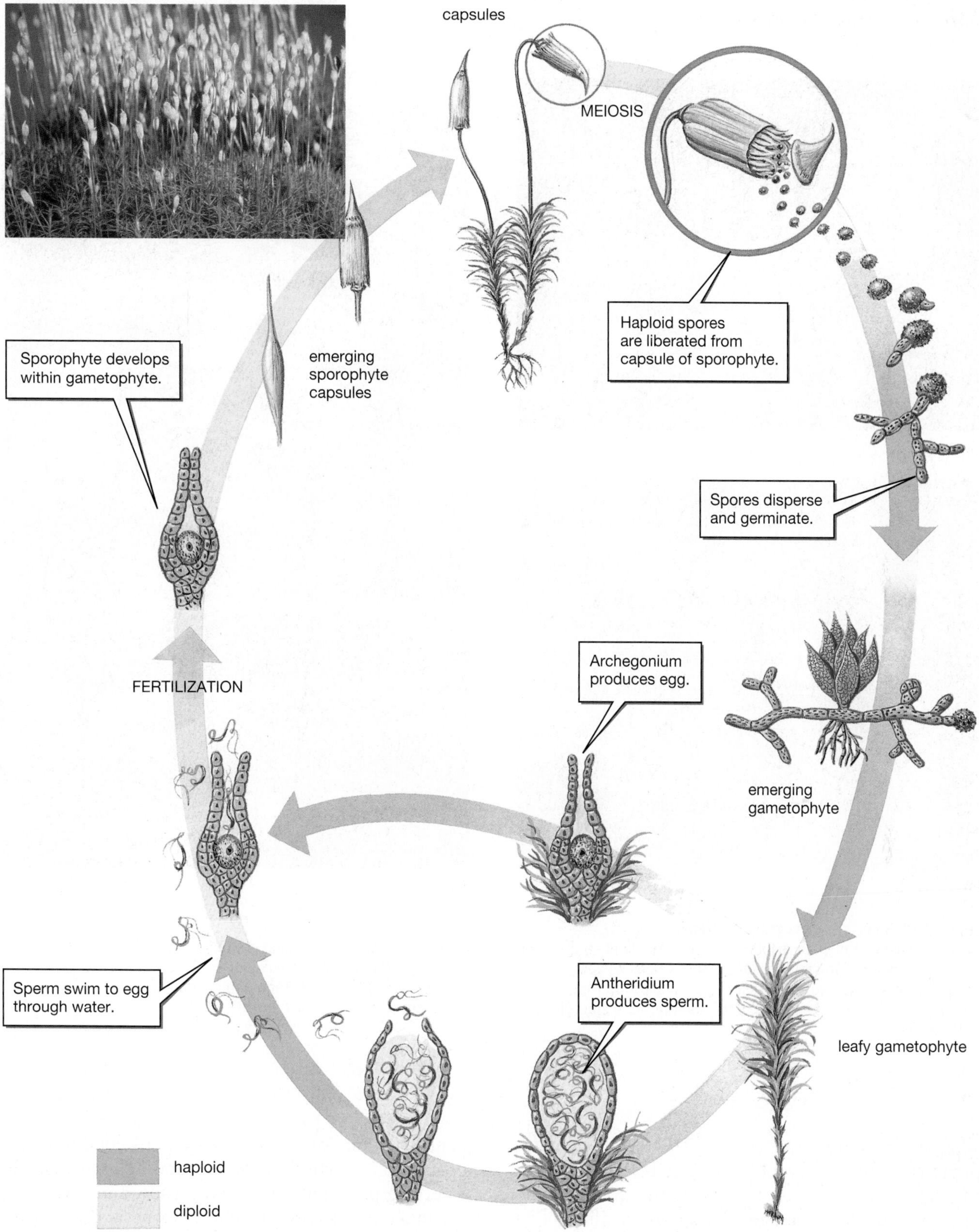
capsules
MEIOSIS
Haploid spores are liberated from capsule of sporophyte.
Sporophyte develops within gametophyte.
emerging sporophyte capsules
Spores disperse and germinate.
Archegonium produces egg.
FERTILIZATION
emerging gametophyte
Sperm swim to egg through water.
Antheridium produces sperm.
leafy gametophyte
haploid
diploid

Figure 21-4 Life cycle of a moss
The life cycle of a moss shows alternation of diploid and haploid generations. The leafy green gametophyte (lower right), which typically grows in patches resembling a cushion, is the haploid generation that produces sperm and eggs. The sperm develop in the antheridium and must swim through a film of water to the egg (which remains in the archegonium where it is formed). The zygote develops into a stalked, diploid sporophyte that emerges from the gametophyte plant. The sporophyte is topped by a brown capsule in which haploid spores are produced by meiosis. These are dispersed and germinate, producing another green gametophyte generation. **(Inset)** Moss plants. The short, leafy green plants are haploid gametophytes; the reddish brown stalks are diploid sporophytes, less than a half inch (about 1 centimeter) tall.

(a)

(c)

(b)

Figure 21-5 Some seedless vascular plants
Seedless vascular plants are found in moist woodland habitats. ***(a)*** The club mosses (sometimes called ground pines) grow in temperate forests. This specimen is releasing spores. ***(b)*** The giant horsetail (genus *Equisetum*) extends long, narrow branches in a series of rosettes. Its leaves are reduced to insignificant scales. At right is a cone-shaped spore-forming structure. ***(c)*** The leaves of this deer fern are emerging from coiled fiddleheads.

whereas the sporophyte is large. The female gametophyte is a small group of haploid cells that produces the egg. The male gametophyte is the pollen grain. Sperm-producing cells are carried within the pollen grains, which are dispersed by wind or by animal pollinators, such as bees. Thus, the distribution of seed plants is not limited by the need for water in which sperm can swim to the egg; seed plants are fully adapted to life on dry land.

The second reproductive adaptation is the seed itself. Analogous to the eggs of birds and reptiles, **seeds** consist of an embryonic plant, a supply of food for the embryo, and a protective outer coat (Fig. 21-7, p. 418). The *seed coat* maintains the embryo in a state of suspended animation or dormancy until conditions are proper for growth. The stored food helps sustain the emerging plant until it develops roots and leaves and can make its own food by photosynthesis. (Seed structure is discussed in detail in Chapter 24.) Some seeds possess elaborate adaptations that make possible dispersal by wind, water, and animals. These adaptations have helped them invade nearly every habitat on Earth.

Seed plants are grouped into two general types: (1) *gymnosperms*, which lack flowers, and (2) *angiosperms*, the flowering plants. Although these groupings are not official taxonomic categories, they are useful in organizing our discussion of the seed plants.

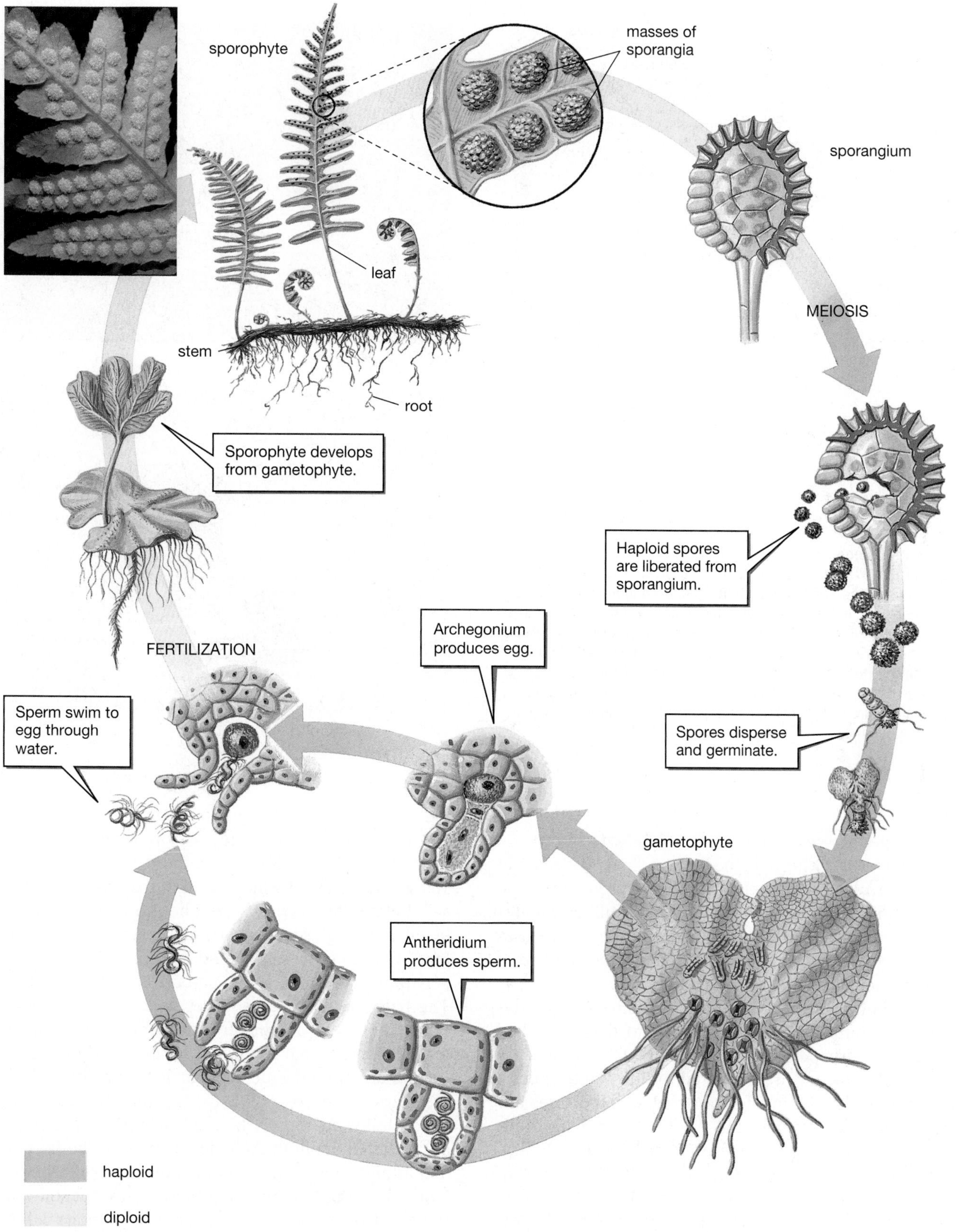
sporophyte
masses of sporangia
sporangium
leaf
stem
root
MEIOSIS
Sporophyte develops from gametophyte.
Haploid spores are liberated from sporangium.
Archegonium produces egg.
FERTILIZATION
Sperm swim to egg through water.
Spores disperse and germinate.
gametophyte
Antheridium produces sperm.
haploid
diploid

Figure 21-6 Life cycle of a fern
The fern's life cycle shows alternation of generations. The dominant plant body (upper left) is the diploid sporophyte. Haploid spores, formed in sporangia located on the underside of certain leaves, are dispersed by the wind to germinate on the moist forest floor into inconspicuous haploid gametophyte plants. On the lower surface of these small, sheetlike gametophytes, male antheridia and female archegonia produce sperm and eggs. The sperm must swim to the egg, which remains in the archegonium. The zygote develops into the large sporophyte plant. **(Inset)** Underside of a fern leaf, showing clusters of sporangia.

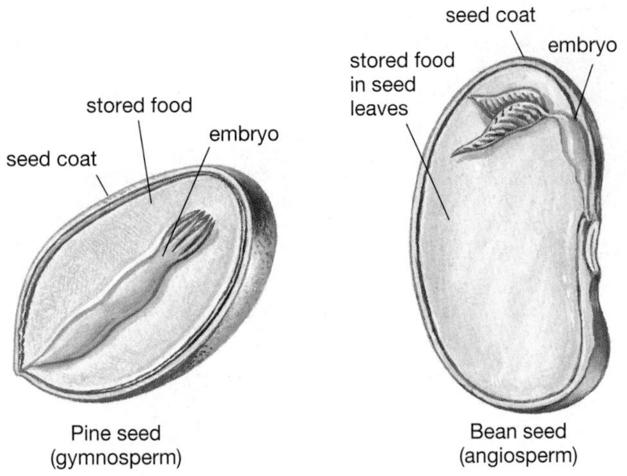

(a) (b)

(c)

(d)

Figure 21-7 Seeds
Seeds from ***(a)*** a gymnosperm (pine) and ***(b)*** an angiosperm (bean). Both consist of an embryonic plant and stored food confined within a seed coat. In the angiosperm seed, food is stored within large seed leaves, which take up most of the volume of the seed. Seeds exhibit diverse adaptations for dispersal, including ***(c)*** the dandelion's tiny, tufted seeds that float in the air and ***(d)*** the massive, armored seeds (protected inside the fruit) of the coconut palm, which can survive prolonged immersion in seawater as they traverse oceans.

Gymnosperms Are Nonflowering Seed Plants

Gymnosperms (whose name means "naked seed" in Greek) evolved earlier than the flowering plants. One group, the **conifers**, with 500 species, still dominates large areas of our planet. Other gymnosperms, such as the ginkgos and cycads, have declined to a small remnant of their former range and abundance.

The ginkgos were probably the first of the surviving seed plants to evolve, becoming widespread during the Jurassic period, which began 208 million years ago. Today they are represented by the single species *Ginkgo biloba*, the maidenhair tree. Ginkgo trees are either male or female; the female trees bear foul-smelling, fleshy seeds the size of cherries (Fig. 21-8a). Ginkgos have been maintained by cultivation, particularly in Asia; if not for this cultivation, they might be extinct today. Because they are more resistant to pollution than are most other trees, ginkgos (normally, the male trees) have been extensively planted in U.S. cities. Recently, the leaves of the ginkgo have gained notoriety as an herbal remedy that purportedly improves memory.

Cycads look like large ferns, from which they probably evolved (Fig. 21-8b). Today there are approximately 160 species, most of which dwell in tropical or subtropical climates. Most cycads are about 3 feet (1 meter) in height, although some species can reach 65 feet (20 meters). Cycads grow slowly and live for a long time; one Australian specimen is estimated to be 5000 years old. The fleshy seeds of cycads were once a staple food in Guam, but they contain a toxin that may cause a neurological disorder that resembles Parkinson's disease.

Conifers spread widely as Earth became drier during the Permian period (286 to 245 million years ago), which followed the Carboniferous period. Today they are most abundant in the cold latitudes of the far north and at high elevations where conditions are rather dry. Not only is rainfall limited in these areas, but soil water remains frozen and unavailable during the long winters. Conifers, including pines, firs, spruce, hemlocks, and cypresses, are adapted to dry, cold conditions in several ways. First, conifers retain green leaves throughout the year, enabling these plants to continue photosynthesizing and growing slowly during times when most other

(a)

(b)

Figure 21-8 Two uncommon gymnosperms
(a) The ginkgo, or maidenhair tree, has been kept alive by cultivation in China and Japan. Relatively resistant to pollution, these trees have become popular in U.S. cities. This ginkgo is female and bears fleshy seeds the size of large cherries, which are noted for their foul smell when ripe. ***(b)*** A cycad. Common in the age of dinosaurs, these are now limited to about 160 species living in warm, moist climates. Like ginkgos, cycads have separate sexes.

plants become dormant. Conifers are often called **evergreens** for this reason. Second, conifer leaves are actually thin needles covered with a thick cuticle whose small waterproof surface minimizes evaporation. Finally, conifers produce an "antifreeze" in their sap that enables them to continue transporting nutrients in subfreezing temperatures. This substance gives them their fragrant "piney" scent.

Reproduction is similar in all conifers, with pines serving as a good illustration (Fig. 21-9). The tree itself is the diploid sporophyte. It develops both male and female cones. The male cones are relatively small (normally 3/4 of an inch, about 2 centimeters, or less) and delicate structures. They release clouds of pollen during the reproductive season and then disintegrate (Fig. 21-9, top). Each pollen grain is a male gametophyte, consisting of several specialized haploid cells, some of which form tiny winglike structures that allow the pollen to be carried by the wind for long distances. Immense clouds of pollen are released by the male cones; inevitably some pollen grains land by chance on a female cone.

Each female cone consists of a series of woody scales arranged in a spiral around a central axis (Fig. 21-9, top). At the base of each scale are two **ovules** (immature seeds), within which diploid spore cells form and undergo meiosis to produce haploid female gametophytes. These gametophytes develop and produce egg cells. A pollen grain that lands nearby sends out a pollen tube that slowly burrows into the female gametophyte. After nearly 14 months, the tube finally reaches the egg cell and releases the sperm that fertilize it. The fertilized egg becomes enclosed in a seed as it develops into a tiny embryonic plant. The seed is liberated when the cone matures and its scales separate.

Angiosperms Are Flowering Seed Plants

Modern flowering plants, or **angiosperms**, have dominated Earth for more than 100 million years. The group is incredibly diverse, with more than 230,000 species. Angiosperms range in size from the diminutive duckweed (Fig. 21-10a, p. 421), a few millimeters in diameter, to the mighty eucalyptus tree (Fig. 21-10b), more than 325 feet (100 meters) tall. From desert cactus to tropical orchids to grasses to parasitic mistletoe, angiosperms rule over the plant kingdom.

Three major adaptations have contributed to the enormous success of angiosperms: (1) flowers, (2) fruits, and (3) broad leaves. **Flowers**, the structures in which both male and female gametophytes are formed, may have evolved when a gymnosperm ancestor formed an association with animals (most likely insects), which carried their pollen from plant to plant. According to this scenario, the relationship between these ancient gymnosperms and their animal pollinators was so beneficial that natural selection favored the evolution of showy flowers that advertised the presence of pollen to insects and other animals. The animals benefited by eating some of the protein-rich pollen, whereas the plant benefited from the animals' unwitting transportation of pollen from plant to plant. With this animal assistance, flowering plants no longer needed to produce prodigious quantities of pollen and send it flying on the fickle winds to ensure fertilization.

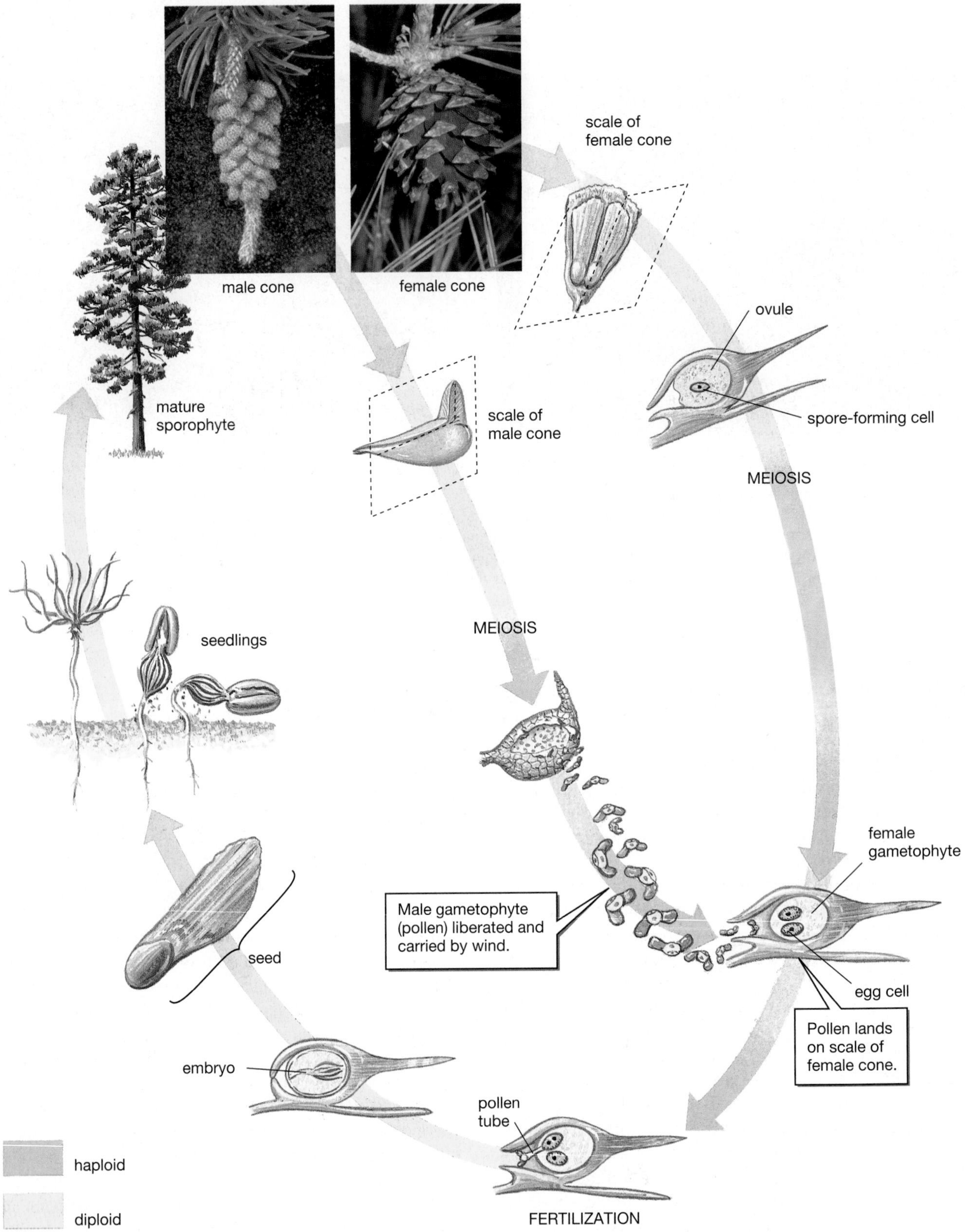

Figure 21-9 Life cycle of the pine

The pine tree is the sporophyte generation (upper left), bearing both male and female cones. Within the ovules at the base of each scale of the female cone, diploid spore-forming cells undergo meiosis, producing haploid spore cells, one of which develops into the female gametophyte. The female gametophyte in turn produces egg cells. In the male cones, meiosis produces the male gametophytes: the pollen. Pollen grains, carrying sperm, are dispersed by the wind and land on the scales of the female cone. The pollen grows a pollen tube that penetrates the female gametophyte and conducts the sperm to the egg. The fertilized egg then develops into an embryonic plant enclosed in a seed formed from the ovule. The seed is eventually released from the cone, germinates, and grows into a sporophyte tree.

Figure 21-10 Diverse angiosperms
(a) The smallest angiosperm is the duckweed, found floating on ponds. These specimens are about 1/8 inch (3 millimeters) in diameter. ***(b)*** The largest angiosperms are eucalyptus trees, which can reach 150 feet (45 meters) in height. Both ***(c)*** grasses and many trees, such as ***(d)*** this birch, in which flowers are shown as buds (green) and blossom (brown), have inconspicuous flowers and rely on wind for pollination. ***(e)*** Flowers, such as those shown on this butterfly weed and eucalyptus tree (inset in part b), entice insects and other animals that carry pollen between individual plants.

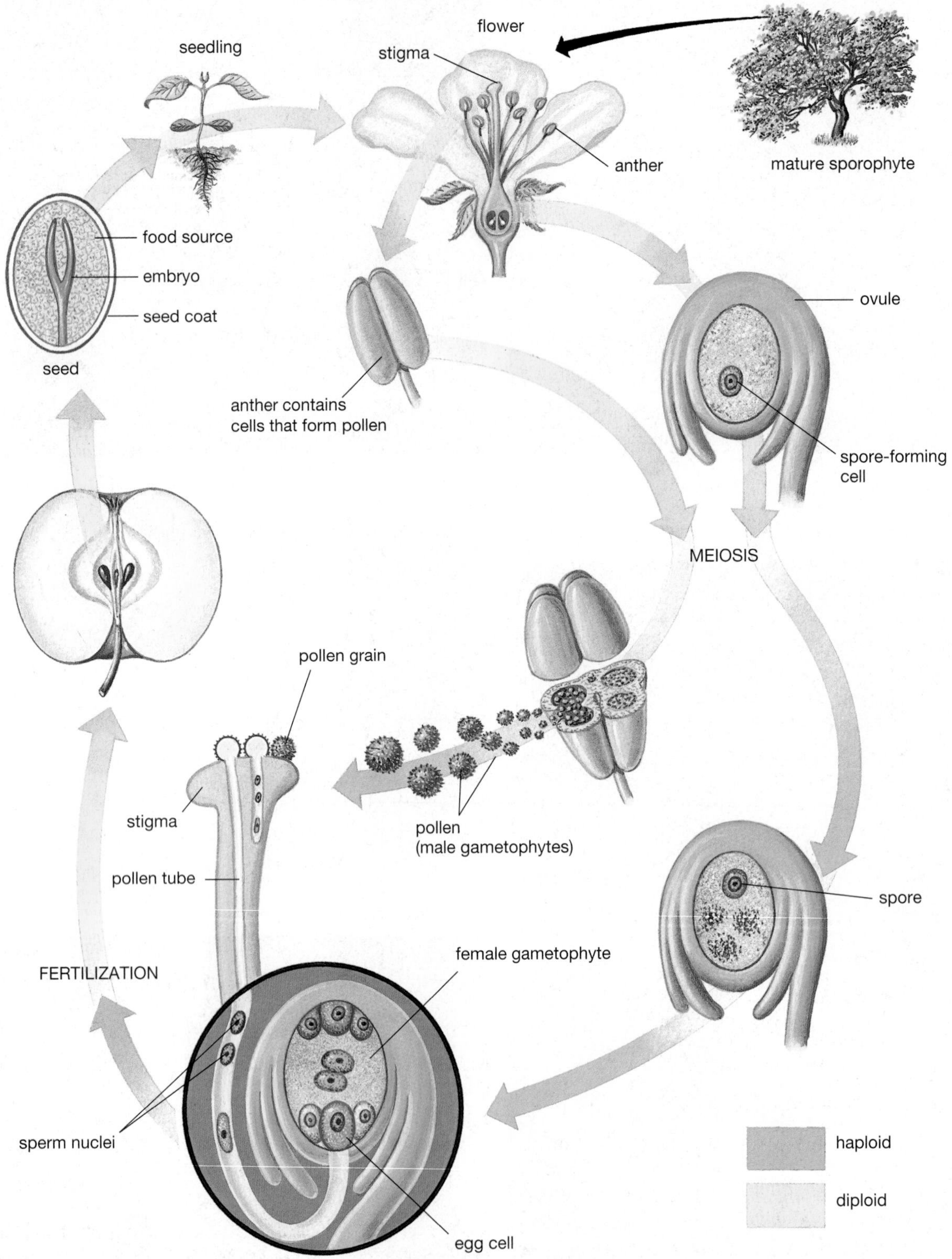

Figure 21-11 Life cycle of a flowering plant
The dominant plant (upper right) is the diploid sporophyte, whose flowers normally produce both male and female gametophytes. Male gametophytes (pollen grains) are produced within the anthers, where diploid spore-forming cells undergo meiosis, producing haploid spore cells. These spores divide mitotically to produce pollen, in which two sperm are formed. The female gametophyte develops within the ovule of the ovary. There diploid spore-forming cells undergo meiosis, producing a haploid spore cell. The spore divides mitotically to produce the female gametophyte, whose contents include one egg cell. After landing on the stigma, a pollen grain produces a pollen tube that burrows down to the ovary and into the female gametophyte. There it releases its sperm. One sperm fuses with the egg to form a zygote. The second eventually forms a source of food for the developing embryo. The ovule gives rise to the seed, which contains the endosperm and an embryo that develops from the zygote. The seed is dispersed, germinates, and develops into a mature sporophyte.

In the angiosperm life cycle (Fig. 21-11), flowers develop on the dominant sporophyte plant. (We shall discuss this process in more detail in Chapter 24.) Male gametophytes (pollen) are formed inside a structure called the *anther*; the female gametophyte develops from an ovule within a part of the flower called the *ovary*. The egg in turn develops within the female gametophyte. Fertilization occurs when the pollen forms a tube through the *stigma*, a sticky pollen-catching structure of the flower, and bores into the ovule. There the zygote develops into an embryo enclosed in a seed formed from the ovule.

The ovary surrounding the seed matures into a **fruit**, the second adaptation that has contributed to the success of angiosperms. Just as flowers encourage animals to transport pollen, so, too, many fruits entice animals to disperse seeds. If an animal eats a fruit, many of the enclosed seeds will pass through animal digestive tracts unharmed, perhaps to fall at a suitable location for germination. Not all fuits, however, depend on edibility for dispersal. Dog owners are well aware, for example, that some fruits (called *burs*) disperse by clinging to animal fur. Others, such as the fruits of maples, form wings that carry the seed through the air. The variety of dispersal mechanisms made possible by various fruits have helped the angiosperms invade nearly all possible terrestrial habitats.

A third feature that gives angiosperms an adaptive advantage in warmer, wetter climates is broad leaves. When water is plentiful, as it is during the warm growing season of temperate and tropical climates, broad leaves give trees an advantage by collecting more sunlight for photosynthesis. In regions with seasonal variation in growing conditions, the extra energy gained during favorable periods allows trees and shrubs to drop their leaves when conditions deteriorate, which reduces evaporative water loss during periods when water is in short supply (Fig. 21-12). In temperate climates, such periods occur during the fall and winter, when virtually all temperate angiosperm trees and shrubs drop their leaves. In the Tropics and subtropics, most angiosperms are evergreen, but species that inhabit certain tropical climates where periods of drought are common may drop their leaves to conserve water during the dry season.

The advantages of broad leaves are offset by some evolutionary costs. In particular, broad, tender leaves are much more appealing to herbivores than are the tough, waxy needles of conifers. As a result, angiosperms have developed a range of defenses against mammalian and insect herbivores. These adaptations include physical defenses such as thorns, spines, and resins that toughen the leaves. But the evolutionary struggle for survival has also led to a host of chemical defenses—compounds that make plant tissue poisonous or distasteful to potential predators. Many of the compounds responsible for chemical defense have properties that humans have exploited for medicinal and culinary uses. Medicines such as Taxol and aspirin, stimulants such as nicotine and caffeine, and spicy flavors such as mustard and peppermint are all derived from angiosperm plants.

Figure 21-12 Two ways of coping with the dryness of winter The evergreen (a conifer) retains its needles throughout the year. The small surface area and heavy cuticle of the needles slow the loss of water through evaporation. In contrast, the aspen (an angiosperm) sheds its leaves each fall. The dying leaves turn brilliant shades of gold as pigments used to capture light energy for photosynthesis are exposed when the chlorophyll disintegrates.

CASE STUDY REVISITED

Hunting for Medical Treasures

The progress of bioprospecting has been hampered by concerns about its effects on biodiversity conservation. Commercial success of a plant-derived drug may threaten the source species' survival, and in such cases the best solution may be to find alternative means of producing the drug. In the case of Taxol, researchers have been working hard to find ways to synthesize the drug without using American yews. A great deal of Taxol is now made by synthetic modification of a similar molecule found in a related species, the European yew (*Taxus baccata*). Unlike the American yew, the European yew is easily cultivated and grows relatively rapidly. What's more, the useful molecule is found in the wood and leaves, instead of just in the bark. So European yews can be grown on plantations where, five years after planting, the entire plant is harvested, ground up, and used for synthesis of Taxol. A large amount of plant tissue is still required for this process, so Taxol remains quite expensive. Ideally, drug manufacturers would like to be able to make completely synthetic Taxol that requires no plant inputs. Three different methods for synthesizing Taxol have already been developed, but these methods have thus far proved to be unsuitable for commercial use.

Why do so many plant species contain substances that have medicinal properties for humans? What evolutionary forces may have led to the presence of these substances?

Summary of Key Concepts

1) What Are the Key Features of Plants?

The kingdom Plantae consists of eukaryotic, photosynthetic, multicellular organisms. The ability of plants and other photosynthetic organisms to capture the energy of sunlight in high-energy molecules provides nearly all other forms of life on Earth with a source of usable energy. Plants exhibit alternation of generations in which a haploid gametophyte generation alternates with a diploid sporophyte generation. There has been a general evolutionary trend toward reduction of the haploid gametophyte, which is dominant in bryophytes but microscopic in seed plants.

2) What Is the Evolutionary Origin of Plants?

Photosynthetic protists, probably green algae, gave rise to the first plants. Ancestral plants were probably similar to modern multicellular green algae, which have photosynthetic pigments, starch molecules, and cell wall components that are similar to those of plants. The freshwater heritage of green algae may have endowed them with qualities that enabled their descendants to invade land.

3) How Did Plants Invade and Flourish on Land?

As plants became increasingly adapted to a terrestrial existence, they developed (1) rootlike structures for anchorage and for absorption of water and nutrients; (2) a waxy cuticle to slow the loss of water through evaporation; (3) stomata that can open, allowing gas exchange, and that can also close, preventing water loss; (4) conducting vessels to transport water and nutrients throughout the plant; and (5) a stiffening substance, called *lignin*, to impregnate the vessels and support the plant body.

As plants invaded the land, new reproductive structures and strategies evolved. Reduction of the male gametophyte to pollen allowed wind to replace water in carrying sperm to eggs. Flowers attracted animals, who carry pollen more precisely and efficiently than wind can, and fruit enticed animals to disperse seeds. Seeds nourish, protect, and help disperse developing embryos.

Two major groups of plants, bryophytes and tracheophytes, arose from the ancient algal ancestors. Bryophytes, including the liverworts and mosses, are small, simple land plants that lack conducting vessels. Although some have adapted to dry areas, most live in moist habitats. Reproduction in bryophytes requires water through which the sperm swims to the egg.

In tracheophytes, or vascular plants, a system of vessels, stiffened by lignin, conducts water and nutrients absorbed by the roots into the upper portions of the plant and supports the body as well. Owing to this support system, seedless vascular plants, including the club mosses, horsetails, and ferns, can grow larger than bryophytes. As in bryophytes, the sperm of seedless tracheophytes must swim to the egg for sexual reproduction to occur, and the gametophyte lacks conducting vessels.

Vascular plants with seeds have two major additional adaptive features: pollen and seeds. Seed plants are often classified into two categories: gymnosperms and angiosperms. Gymnosperms include ginkgos, cycads, and the highly successful conifers. These plants were the first fully terrestrial plants to evolve. Their success on dry land is partially due to the evolution of the male gametophyte into the pollen grain. Pollen protects and transports the male gamete, eliminating the need for the sperm to swim to the egg. The seed, a protective resting structure containing an embryo and a supply of food, is a second important adaptation contributing to the success of seed plants.

Angiosperms, the flowering plants, dominate much of the land today. In addition to pollen and seeds, angiosperms also produce flowers and fruits. The flower allows angiosperms to utilize animals as pollinators. In contrast to wind, animals can in some cases carry pollen farther and with greater accuracy and less waste. Fruits may attract animal consumers, which incidentally disperse the seeds in their feces.

Key Terms

alternation of generations *p. 410*
angiosperm *p. 419*
antheridium *p. 413*
archegonium *p. 413*
bryophyte *p. 413*
conifer *p. 418*
cuticle *p. 412*
evergreen *p. 419*
flower *p. 419*
fruit *p. 423*
gametophyte *p. 410*
gymnosperm *p. 418*
lignin *p. 412*
ovule *p. 419*
pollen *p. 414*
rhizoid *p. 413*
seed *p. 416*
sporophyte *p. 410*
stoma *p. 412*
tracheophyte *p. 413*
vascular *p. 414*
vessel *p. 414*
zygote *p. 410*

Thinking Through the Concepts

Multiple Choice

1. *Which of the following organisms bear fruit?*
   a. mosses b. pine trees
   c. maple trees d. liverworts
   e. ferns

2. *In which of the following plants is the gametophyte the dominant generation?*
   a. mosses b. ferns
   c. pine trees d. sunflowers
   e. The sporophyte is dominant in all of the above.

3. *What is the function of a fruit?*
   a. It attracts pollinators.
   b. It provides food for the developing embryo.
   c. It stores excess food produced by photosynthesis.
   d. It helps ensure seed dispersal from the parent plant.
   e. It evolved so that people would cultivate the plant, ensuring its survival.

4. *What is the function of lignin?*
   a. It provides support for the plant.
   b. It waterproofs plant surfaces.
   c. It stores food.
   d. It promotes gas exchange in plant leaves.
   e. It transports dissolved nutrients.

5. *Which of the following plants produces sperm that swim to the egg?*
   a. walnut tree
   b. Douglas fir tree, a conifer
   c. rattlesnake fern
   d. common dandelion
   e. none of the above

6. *Which of the following is the correct sequence during alternation of generations?*
   a. sporophyte—diploid spores—gametophyte—haploid gametes
   b. sporophyte—haploid spores—gametophyte—haploid gametes
   c. sporophyte—haploid gametes—gametophyte—haploid spores
   d. sporophyte—haploid gametes—gametophyte—diploid spores
   e. sporophyte—diploid gametes—gametophyte—diploid spores

? Review Questions

1. What is meant by "alternation of generations"? What two generations are involved? How does each reproduce?
2. Explain the evolutionary changes in plant reproduction that adapted plants to increasingly dry environments.
3. Describe evolutionary trends in the life cycles of plants. Emphasize the relative sizes of the gametophyte and sporophyte.
4. From which algal phylum did green plants probably arise? Explain the evidence that supports this hypothesis.
5. List the structural adaptations necessary for the invasion of dry land by plants. Which of these adaptations are possessed by bryophytes? By ferns? By gymnosperms and angiosperms?
6. The number of species of flowering plants is greater than the number of species in the rest of the plant kingdom. What feature(s) are responsible for the enormous success of angiosperms? Explain why.
7. List the adaptations of gymnosperms that have helped them become the dominant tree in dry, cold climates.
8. What is a pollen grain? What role has it played in helping plants colonize dry land?
9. The majority of all plants are seed plants. What is the advantage of a seed? How do plants that lack seeds meet the needs served by seeds?

Applying the Concepts

1. You are a geneticist working for a firm that specializes in plant biotechnology. Explain what *specific* parts (fruit, seeds, stems, roots, etc.) of the following plants you would try to alter by genetic engineering, what changes you would try to make, and why, on (a) corn, (b) tomatoes, (c) wheat, and (d) avocados.

2. Prior to the development of synthetic drugs, more than 80% of all medicines were of plant origin. Even today, indigenous tribes in remote Amazonian rain forests can provide a plant product to treat virtually any ailment. Herbal medicine is also widely and successfully practiced in China. Most of these drugs are unknown to the Western world. But the forests from which much of this plant material is obtained are being converted to agriculture. We are in danger of losing many of these potential drugs before they can be discovered. What steps can you suggest to preserve these natural resources while also allowing nations to direct their own economic development?

3. Only a few hundred of the hundreds of thousands of species in the plant kingdom have been domesticated for human use. One example is the almond. The domestic almond is nutritious and harmless, but its wild precursor can cause cyanide poisoning. The oak makes potentially nutritious seeds (acorns) that contain very bitter-tasting tannins. If we could breed the tannin out of acorns, they might become a delicacy. Why do you suppose we have failed to domesticate oaks?

For More Information

Cox, P. A., and Balick, M. J. "The Ethnobotanical Approach to Drug Discovery." *Scientific American*, June 1994. Biologists seek new pharmaceutical compounds by analyzing the plants used as drugs by indigenous cultures.

Diamond, J. "How to Tame a Wild Plant." *Discover*, September 1994. Cultivated plants have ecological and genetic properties that make them well suited for agriculture.

Doyle, J. "DNA, Phylogeny, and the Flowering of Plant Systematics." *BioScience*, June 1993. Chloroplast DNA is used in reconstructing evolutionary relationships, but there are difficulties with each of the methods in use.

Grant, M. C. "The Trembling Giant." *Discover*, 1993. The largest living thing is a clone of 47,000 trees covering 106 acres in the Wasatch Mountains of Utah.

Joyce, C. *Earthly Goods: Medicine-Hunting in the Rainforest.* Boston: Little, Brown, 1994. Science and adventure combine in this account of prospecting for new medicines and the people who do it.

Kaufman, P. B. *Plants—Their Biology and Importance.* New York: Harper & Row, 1989. Complete, readable coverage of all aspects of plant taxonomy, physiology, and evolution.

Milot, V. "Blueprint for Conserving Plant Diversity." *BioScience*, June 1989. Points out the importance of preserving genetic diversity in endangered plant species.

Nicholson, R. "Death and Taxus." *Natural History*, September 1992. The bark of the yew tree contains compounds that are used to treat cancer.

Answers to Multiple-Choice Questions

1. c 2. a 3. d 4. a 5. c 6. b

MEDIATUTOR
The Plant Kingdom

CD Activities

Activity 21.1: Evolution of Plant Structure

Estimated time: 10 minutes

This tutorial illustrates the evolutionary changes that have occurred in plant structures and how the changes have allowed the plants to exploit different habitats. This tutorial does not illustrate evolution's path but rather the evolutionary end points of related structures in selected plant groups.

Activity 21.2: Fern Life Cycle

Estimated time: 5 minutes

In this tutorial you will explore the different stages and structures in the life cycle of the fern.

Start the MediaTutor Student CD-ROM and enter the activity number in the Quick Search box to be taken directly to that activity.

Web Investigations

Case Study: Hunting for Medical Treasures

Estimated time: 10 minutes

For every new plant-derived miracle drug, thousands of plants are tested that have no therapeutic function. In this exercise we will look at some important plant-derived pharmaceuticals, how they were discovered, and what lessons can be learned from current discovery processes.

Go to http://www.prenhall.com/audesirk6, the Audesirk Companion Web site. Select Chapter 21 and Web Investigation to begin.

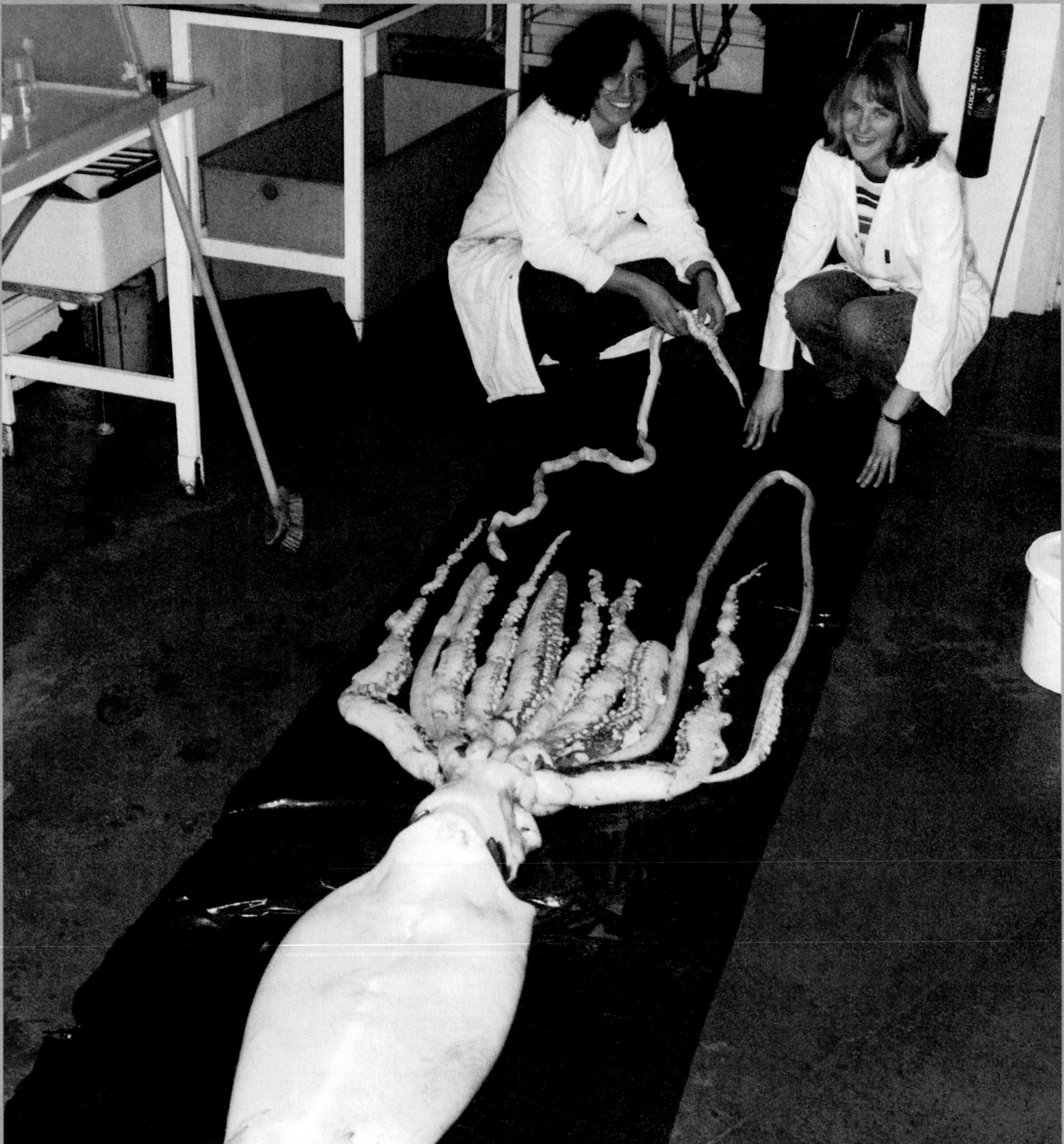

The giant squid is Earth's largest invertebrate animal, but no human has ever observed a living giant squid in its natural habitat.

22 The Animal Kingdom

AT A GLANCE

Case Study: The Search for a Sea Monster

1) **What Characteristics Define an Animal?**

2) **Which Anatomical Features Mark Branch Points on the Animal Evolutionary Tree?**

Lack of Tissues Separates Sponges from All Other Animals

Animals with Tissues Exhibit Either Radial or Bilateral Symmetry

Body Cavities Arose in Some Bilaterally Symmetrical Animals

Coelomates Include Two Distinct Evolutionary Lines

3) **What Are the Major Animal Phyla?**

The Sponges

The Hydra, Anemones, and Jellyfish

The Flatworms

The Roundworms

The Segmented Worms

The Insects, Arachnids, and Crustaceans

The Snails, Clams, and Squid

The Sea Stars, Sea Urchins, and Sea Cucumbers

The Tunicates, Lancelets, and Vertebrates

Evolutionary Connections: Are Humans a Biological Success?

Case Study Revisited: The Search for a Sea Monster

CASESTUDY

The Search for a Sea Monster

Everyone loves a good mystery, and a mystery involving a gigantic, fearsome predator is even better. Consider the puzzle of *Architeuthis*, the giant squid. The giant squid is the world's largest invertebrate animal, reaching lengths of 60 feet (18 meters) or more. Each of its huge eyes, the largest in the animal kingdom, can be as large as a human head. The squid's 10 tentacles, two of them longer than the others, are covered with powerful suckers. The suckers contain sharp, claw-like hooks to better grasp the prey that is then pulled toward the mouth, where a heavily muscled beak tears food items apart. The giant squid is one of the most imposing organisms on Earth, yet we know nothing of its habits and lifestyle. What does it eat? Does it swim with its head elevated or pointed downward? How does it mate? Does it live alone or in groups? Even these basic questions about the behavior of the giant squid remain unanswered, because no one has ever seen a giant squid alive in its natural habitat, or even in an aquarium.

Our limited scientific knowledge of the giant squid comes entirely from specimens that have been found dead or dying, washed up on beaches, caught in fishermen's nets, or contained in the stomachs of sperm whales (which consume vast quantities of squid, including the occasional giant squid). More than 200 such specimens have been reported over the past century, and written accounts of squid corpses go back to the sixteenth century. Live squids, however, remain elusive because they inhabit deep ocean waters, beyond the reach of human divers.

Clyde Roper, a biologist at the Smithsonian Institution, has devoted much of his professional life to a quest to view and study live giant squids. We will describe some of Dr. Roper's undersea search methods and his latest effort to find the giant squid at the close of this chapter, following a survey of the extraordinary diversity of animal life. ■

1 What Characteristics Define an Animal?

It is difficult to devise a concise definition of the term *animal.* No single feature fully characterizes animals, so we define the group with a list of characteristics. None of these characteristics is unique to animals, but, taken together, they distinguish animals from members of other kingdoms:

1. Animals are multicellular.
2. Animals are heterotrophic—they obtain their energy by consuming the bodies of other organisms.
3. Animals typically reproduce sexually. Although animal species exhibit a tremendous diversity of reproductive styles, most are capable of sexual reproduction.
4. Animal cells lack a cell wall.
5. Animals are motile (able to move about) during some stage of their life. Even the stationary sponges have a free-swimming *larval* stage (a juvenile form).
6. Most animals are able to respond rapidly to external stimuli as a result of the activity of nerve cells, muscle tissue, or both.

2 Which Anatomical Features Mark Branch Points on the Animal Evolutionary Tree?

By the Cambrian period, which began 544 million years ago, most of the animal phyla that currently populate Earth were already present. Unfortunately, the pre-Cambrian fossil record contains no representatives of the direct ancestors of today's animals, so we cannot rely on fossils to reveal the sequence in which the animal phyla arose. Instead, animal systematists have historically looked to features of animal anatomy and embryological development for clues about the evolutionary history of animals. Some of these features appear to mark major branching points on the animal evolutionary tree and represent milestones in the evolution of the different body plans of modern animals. In the following sections we describe these evolutionary milestones and their legacies in the bodies of modern animals.

Lack of Tissues Separates Sponges from All Other Animals

One of the earliest major innovations in animal evolution was the appearance of **tissues**—groups of similar cells integrated into a functional unit, such as a muscle. Today, almost all animals have bodies that include tissues. The only animals that have retained the ancestral lack of tissues are the sponges. In sponges, individual cells may have specialized functions but act more or less independently and are not organized into true tissues. This unique feature of sponges suggests that the split between sponges and the evolutionary branch leading to all other animal phyla must have occurred very early in the history of animals. An ancient common ancestor without tissues gave rise to both the sponges and the remaining tissue-containing phyla.

Animals with Tissues Exhibit Either Radial or Bilateral Symmetry

The evolutionary advent of tissues coincided with the first appearance of body symmetry; all animals with true tissues also have symmetrical bodies. An animal is said to be symmetrical if it can be bisected along at least one plane such that the resulting halves are mirror images of one another. Note that, unlike the asymmetrical sponges, any symmetrical animal has an upper, or **dorsal**, surface and a lower, or **ventral**, surface.

The symmetrical, tissue-bearing animals can be divided into two groups, one that contains animals that exhibit **radial symmetry** (Fig. 22-1a) and one with animals that exhibit **bilateral symmetry** (Fig. 22-1b). In radial symmetry, any plane through a central axis divides the object into roughly equal halves. In contrast, a bilaterally symmetrical animal can be divided into roughly mirror-image halves only along a particular, single plane through the central axis.

The difference between radially and bilaterally symmetrical animals reflects another major branching point in the animal evolutionary tree. This split separated the ancestors of the radially symmetrical cnidarians (jellyfish, anemones, and corals) and ctenophores (comb jellies) from the ancestors of the remaining animal phyla, all of which are bilaterally symmetrical.

Radially Symmetrical Animals Have Two Embryonic Tissue Layers; Bilaterally Symmetrical Animals Have Three

The distinction between radial and bilateral symmetry in animals is closely tied to a corresponding difference in the number of tissue layers, called **germ layers**, that arise during embryonic development. Animals with radial symmetry have two germ layers: an inner layer of **endoderm** (forming the lining of most hollow organs) and an outer layer of **ectoderm** (forming the tissue that covers the body and lines its inner cavities and nerve tissues). Bilaterally symmetrical animals add a third germ layer. Between the endoderm and ectoderm is a layer of **mesoderm** (forming muscle and, when present, the circulatory and skeletal systems).

The parallel evolution of symmetry type and number of germ layers helps us make sense of the potentially puzzling case of the echinoderms (starfish, sea cucumbers, and sea urchins). Adult echinoderms are radially symmetrical, yet our evolutionary tree places them squarely within the bilaterally symmetrical group. It turns out, however, that echinoderms have three germ layers, as well as several other characteristics (some described below) that unite them with the bilaterally symmetrical animals. So, the immediate ancestors of

UNIT FOUR

Plant Anatomy and Physiology

A prairie blanketed with Texas bluebonnets in the spring. Aside from providing a beautiful vista, flowers such as these are adapted to attract pollinators. They are actually elaborately modified leaves.

A sundew *(Drosera rotundifolia)* grasps a lacewing fly with sticky, enzyme-laden hairs. Nutrients from the insect's body help sundews to thrive in nitrogen-poor soils. The genus name *Drosera* comes from the Greek, meaning "dewy plant," referring to its glistening droplets. The species name *rotundifolia* appropriately means "round leaves."

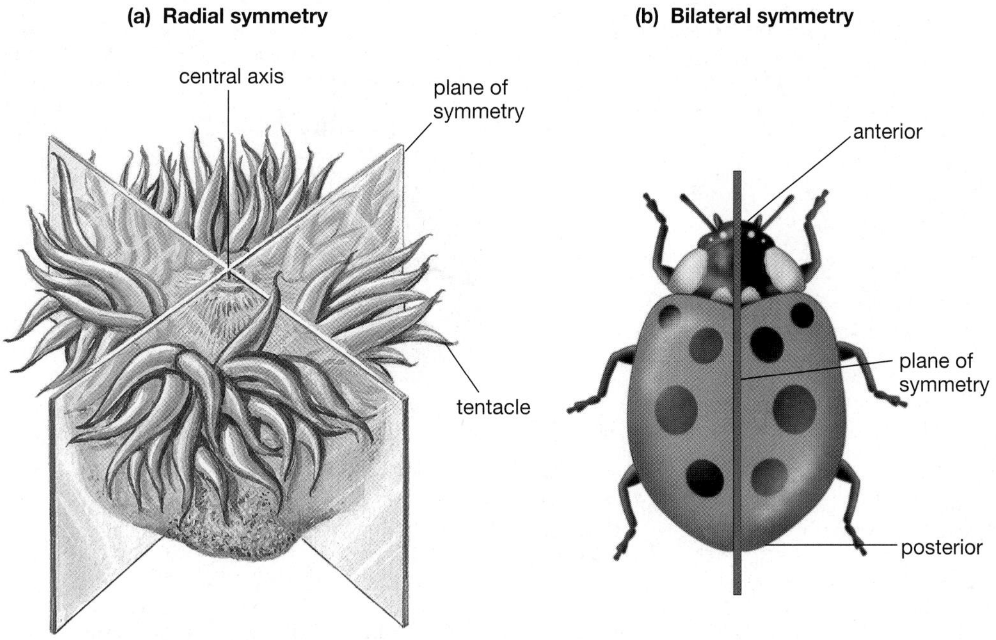

Figure 22-1 Body symmetry and cephalization
(a) A radially symmetrical body. Any plane that passes through the central axis divides the body into mirror-image halves. Animals in these groups lack a well-defined head. ***(b)*** Bilateral symmetry. The body can be split into two mirror-image halves only along a particular plane that runs down the midline. Animals with bilateral symmetry have an anterior head end and a posterior tail end.

echinoderms must have been bilaterally symmetrical, and the group subsequently evolved radial symmetry (a case of convergent evolution; see page 277). Even now, however, larval echinoderms retain bilateral symmetry.

The Evolution of Bilateral Symmetry Was Accompanied by Cephalization

Radially symmetrical animals tend either to be **sessile** (fixed to one spot, as in sea anemones) or to drift around on currents (as in jellyfish). Such animals may encounter food or threats from any direction, so a body that is essentially "facing" all directions at once is advantageous. In contrast, most bilaterally symmetrical animals move under their own power in a particular direction. Resources are most likely to be encountered by the part of the animal that is closest to the direction of movement. The evolution of bilateral symmetry was therefore accompanied by **cephalization,** the concentration of sensory organs and a brain in a defined head region. Cephalization produces an **anterior** (head) end, where sensory cells, sensory organs, clusters of nerve cells, and organs for ingesting food are concentrated. The other end of a cephalized animal is designated **posterior** and typically features a tail (Fig. 22-1b).

Body Cavities Arose in Some Bilaterally Symmetrical Animals

After the origin of bilateral symmetry, the next major anatomical feature to arise during animal evolution was a fluid-filled cavity between the digestive tube (or gut, where food is digested and absorbed) and the outer body wall. This feature appeared some time after the split between radially and bilaterally symmetrical animals. Some phyla of bilateral animals retain the ancestral lack of a body cavity. For example, flatworms have no cavity between their gut and body wall; the space is filled with solid tissue (Fig. 22-2a).

In the phyla that arose after the origin of body cavities, the gut and body wall were separated by a space, which created a "tube-within-a-tube" body plan. This body plan freed the gut from the constraints imposed by direct attachment to the body wall and created a space in which new organs could evolve.

Animals with body cavities fall into two groups, based on the structure of the cavity. Members of some phyla have a body cavity known as a **pseudocoelom** and are collectively known as *pseudocoelomates* (Fig. 22-2b). A pseudocoelom is a body cavity that is *not* completely surrounded by mesoderm-derived tissue. The roundworms (nematodes) are the largest pseudocoelomate group. Phyla whose members have a type of body cavity known as a **coelom** (Fig. 22-2c) are called *coelomates.* A coelom is a fluid-filled body cavity that *is* completely lined with a thin layer of tissue that develops from mesoderm. The annelids (segmented worms), arthropods (insects, spiders, crustaceans), mollusks (clams and snails), echinoderms, and chordates (which include humans) are coelomate phyla. Systematists have long believed that the distinction between pseudocoelomates and coelomates represents another major fork in the evolutionary

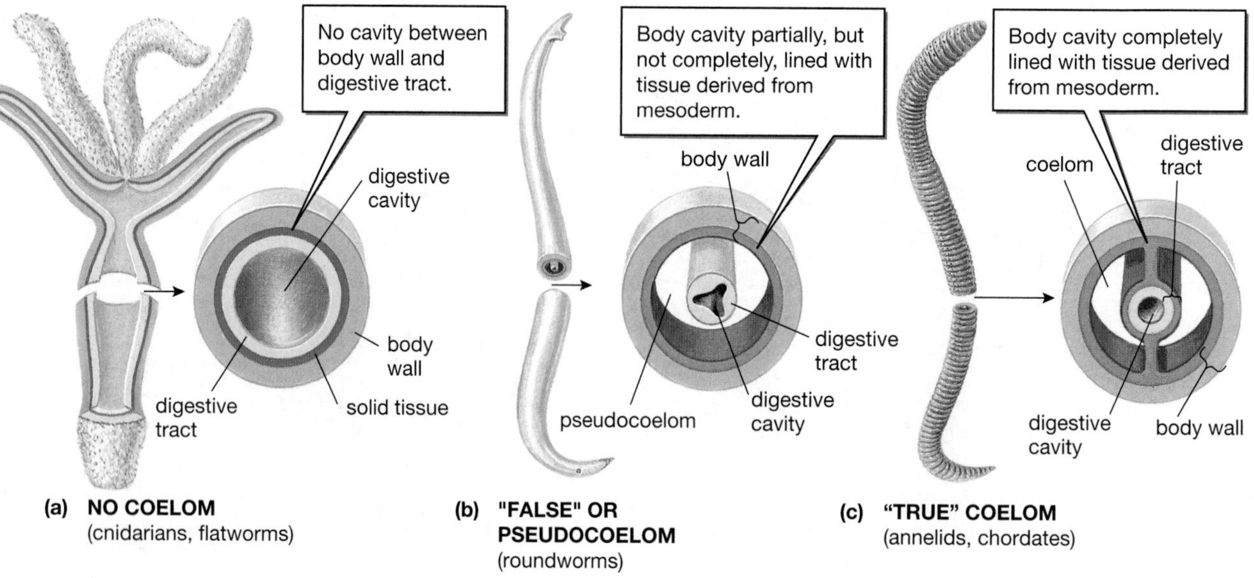

Figure 22-2 Body cavities
(a) Cnidarians and flatworms have no cavity between the body wall and digestive tract. ***(b)*** Roundworms are pseudocoelomates. ***(c)*** Annelids have a true coelom.

tree of animals, but recent molecular data (comparisons of DNA among different phyla) suggest that the pseudocoelomate phyla may instead represent different branches within the coelomate group. If that assessment is correct, then the pseudocoelom is not a precursor of the coelom but a modification of it.

A body cavity can serve a variety of functions. In the earthworm the coelom acts as a kind of skeleton, providing support for the body and a framework against which muscles can act. In other coelomate animals, internal organs are suspended in the coelomic fluid, which serves as a protective buffer between them and the outside world. The coelom has also allowed the internal organs, such as the heart and digestive tract, to move independently of the body wall. Thanks to your coelom, you may remain externally inactive after a meal even though your digestive tract is churning energetically.

Coelomates Include Two Distinct Evolutionary Lines

Among the animal phyla whose members have a true coelom, embryological development follows a variety of pathways. These diverse developmental pathways, however, can be grouped into two categories, known as **protostome** and **deuterostome** development. In protostome development the coelom forms within the space between the body wall and the digestive cavity. In deuterostome development the coelom forms as outgrowths of the digestive cavity. The two types of development also differ in the pattern of cell division immediately after fertilization and the method by which the mouth and anus are formed. Protostomes and deuterostomes represent distinct evolutionary branches within the coelomate animals. Annelids, arthropods, and mollusks exhibit protostome development; echinoderms and chordates are deuterostomes.

3 What Are the Major Animal Phyla?

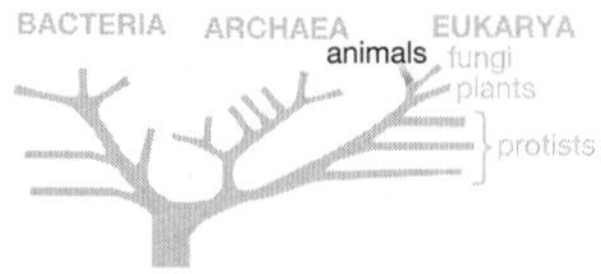

It's easy to overlook the differences among the multitude of small, boneless animals in the world. Even Carolus Linnaeus, the originator of modern taxonomy, recognized only two phyla of animals without backbones (insects and "worms"). Since the time of Linnaeus, however, systematists have come to appreciate that the animal lineage underwent many ancient splits. Some of the major branches of the animal evolutionary tree are summarized in Figure 22-3 and Table 22-1 (p. 434).

For convenience, biologists often place animals in one of two major categories: (1) **vertebrates**, those with a backbone, or vertebral column, and (2) **invertebrates**, those lacking a backbone. The vertebrates—fish, amphibians, reptiles, birds, and mammals—are perhaps the most conspicuous animals from a human point of view, but less than 3% of all known animal species on Earth are vertebrates. The vast majority of animals are invertebrates.

The earliest animals probably originated from colonies of protists whose members had become specialized to perform distinct roles within the colonial body. In our survey of the kingdom Animalia, we will begin with the sponges, whose body plan most closely resembles the probable ancestral protozoan colonies. Our discussion will proceed through a series of representative invertebrate phyla and will end with the vertebrates in the phylum Chordata.

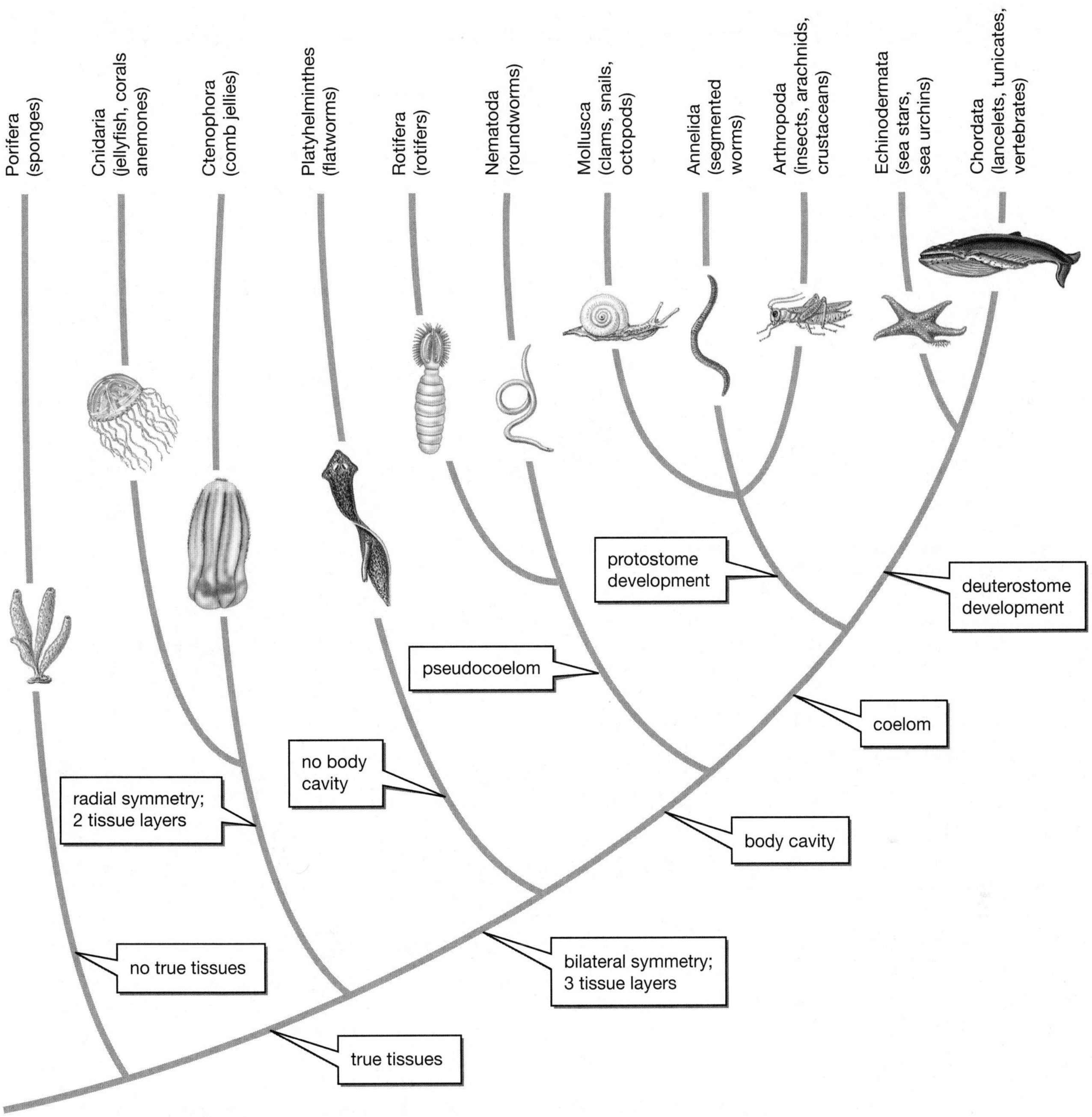

Figure 22-3 An evolutionary tree of some major animal phyla

The Sponges

Sponges (phylum Porifera) lack true tissues and organs. In some ways, a sponge resembles a colony of single-celled organisms, and a few systematists believe that sponges should be classified as the protist group most closely related to animals, rather than within the animal kingdom. The colony-like properties of sponges were revealed in an experiment performed by embryologist H. V. Wilson in 1907. Wilson mashed a sponge through a piece of silk, thereby breaking it apart into single cells and cell clusters. He then placed these tiny bits of sponge into seawater and waited for 3 weeks. By the end of the experiment, the cells had reaggregated into a functional sponge, demonstrating that individual sponge cells had been able to survive and function independently.

All sponges have a similar general body plan. The body is perforated by numerous tiny pores, through which water enters, and by fewer, large openings (called *oscula*), through which it is expelled. Within the sponge, water travels through canals. During its passage, oxygen is extracted, microorganisms are filtered out and taken

Table 22-1 Comparison of the Major Animal Phyla

Common name (Phylum)		Sponges (Porifera)	Hydra, Anemones, Jellyfish (Cnidaria)	Flatworms (Platyhelminthes)	Roundworms (Nematoda)
Body Plan	**Level of organization**	Cellular—lack tissues and organs	Tissue—lack organs	Organ system	Organ system
	Germ layers	Absent	Two	Three	Three
	Symmetry	Absent	Radial	Bilateral	Bilateral
	Cephalization	Absent	Absent	Present	Present
	Body cavity	Absent	Absent	Absent	Pseudocoel
	Segmentation	Absent	Absent	Absent	Absent
Internal Systems	**Digestive system**	Intracellular	Gastrovascular cavity; some intracellular	Gastrovascular cavity	Separate mouth and anus
	Circulatory system	Absent	Absent	Absent	Absent
	Respiratory system	Absent	Absent	Absent	Absent
	Excretory system (fluid regulation)	Absent	Absent	Canals with ciliated cells	Excretory gland cells
	Nervous system	Absent	Nerve net	Head ganglia with longitudinal nerve cords	Head ganglia with dorsal and ventral nerve cords
	Reproduction	Sexual; asexual (budding)	Sexual; asexual (budding)	Sexual (some hermaphroditic); asexual (body splits)	Sexual (some hermaphroditic)
	Support	Endoskeleton of spicules	Hydrostatic skeleton	Absent	Hydrostatic skeleton
	Number of known species	5000	9000	12,000	12,000

into individual cells where they are digested, and wastes are released (Fig. 22-4, p. 436).

Sponges have three major cell types (see Fig. 22-4), each with a specialized role. Flattened **epithelial cells** cover their outer body surfaces. Some epithelial cells are modified into *pore cells,* which surround pores, controlling their size and regulating the flow of water. The pores are closed when harmful substances are present. **Collar cells** maintain a flow of water through the sponge by beating a flagellum that extends into the inner canal. The collar that surrounds the flagellum acts as a fine sieve, filtering out microorganisms that are then ingested by the cell. Some of the food is passed to the **amoeboid cells.** These cells roam freely between the epithelial and collar cells, digesting and distributing nutrients, producing reproductive cells, and secreting small skeletal projections called **spicules.**

Sponges come in a variety of shapes and sizes. Some species have a well-defined shape, but others grow free-form over underwater rocks (Fig. 22-5, p. 436). The largest sponges can grow to more than 3 feet (1 meter) in height. An internal skeleton composed of calcium carbonate (chalk), silica (glass), or protein spicules provides support for the body (see Fig. 22-4). The natural bath sponge, now largely replaced by factory-made cellulose imitations, is actually a proteinaceous sponge skeleton.

All sponges live in water, mostly in saltwater environments. Adult sponges are generally sessile, attaching themselves to rocks or other underwater surfaces. Sponges may reproduce asexually by **budding**, in which the adult produces miniature versions of itself that drop off and assume an independent existence, or may reproduce sexually through the fusion of sperm and eggs. Fertilized eggs develop inside the adult into active larvae

Segmented Worms *(Annelida)*	Insects, Arachnids, Crustaceans *(Arthropoda)*	Snails, Clams, Squid *(Mollusca)*	Sea Stars, Sea Urchins *(Echinodermata)*	Vertebrates *(Chordata)*
Organ system	Organ system	Organ system	Organ system	Organ system
Three	Three	Three	Three	Three
Bilateral	Bilateral	Bilateral	Bilateral larvae, radial adults	Bilateral
Present	Present	Present	Absent	Present
Coelom	Coelom	Coelom	Coelom	Coelom
Present	Present	Absent	Absent	Present (but reduced)
Separate mouth and anus	Separate mouth and anus	Separate mouth and anus	Separate mouth and anus (normally)	Separate mouth and anus
Closed	Open	Open	Absent	Closed
Absent	Tracheae, gills, or book lungs	Gills, lungs	Tube feet, skin gills, respiratory tree	Gills, lungs
Nephridia	Excretory glands resembling nephridia	Nephridia	Absent	Kidneys
Head ganglia with paired ventral cords; ganglia in each segment	Head ganglia with paired ventral nerve cords; ganglia in segments, some fused	Well-developed brain in some cephalopods; several paired ganglia, most in the head; nerve network in body wall	Head ganglia absent; nerve ring and radial nerves; nerve network in skin	Well-developed brain; dorsal nerve cord
Sexual (some hermaphroditic)	Normally sexual	Sexual (some hermaphroditic)	Sexual (some hermaphroditic); asexual by regeneration (rare)	Sexual
Hydrostatic skeleton	Exoskeleton	Hydrostatic skeleton	Endoskeleton of plates beneath outer skin	Endoskeleton of cartilage or bone
9000	1,000,000	50,000	6500	40,000

that escape through the oscula. Water currents disperse the larvae to new areas, where they settle and develop into adult sponges.

The Hydra, Anemones, and Jellyfish

Cnidarians (phylum Cnidaria) come in a bewildering and beautiful variety of forms (Fig. 22-6, p. 437), all of which are actually variations on two basic body plans: the **polyp** (Fig. 22-7a, p. 437) and the **medusa** (Fig. 22-7b). The generally tubular polyp is adapted to a life spent quietly attached to rocks. The polyp has **tentacles**, extensions that reach upward for grasping, stinging, and immobilizing prey. Although the medusa ("jellyfish") swims weakly by contracting its bell-shaped body, it is primarily carried by ocean currents, trailing its tentacles like multiple fishing lines. Both polyp and medusa develop from just two germ layers—the interior endoderm and the exterior mesoderm; between those layers is a jellylike **mesoglea**. Polyps and medusae are radially symmetrical, with body parts arranged in a circle around the mouth and digestive cavity (see Fig. 22-1a). This arrangement of parts is well suited to these animals, which are either sessile or carried randomly by water currents, because they are prepared to respond to prey or threats from any direction.

The cells of cnidarians are organized into distinct tissues, including contractile tissue that acts like muscle. The nerve cells are organized into tissue called a **nerve net**, which branches through the body and controls the contractile tissue to bring about movement and feeding behavior. Most cnidarians lack true organs, however, and they have no brain.

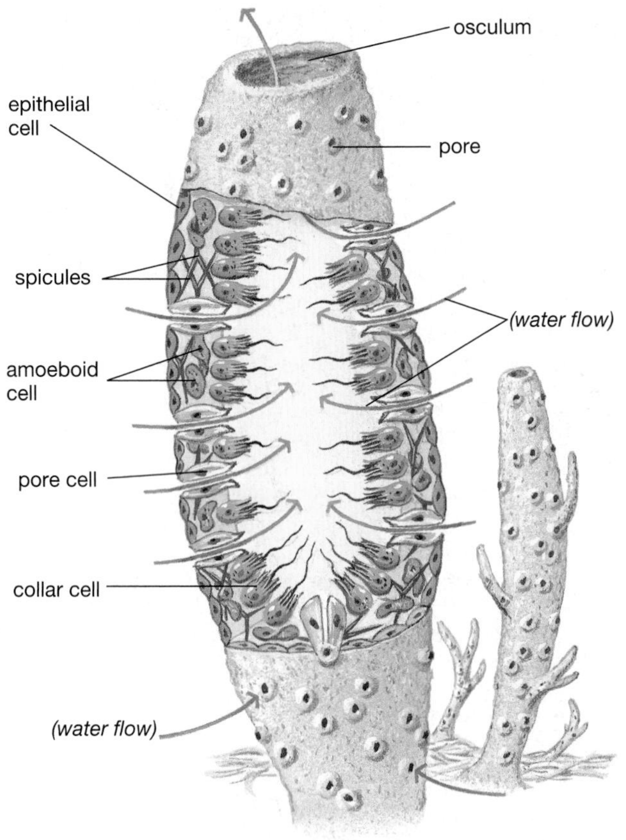

Figure 22-4 The body plan of sponges
All sponges have a similar body plan. Currents created by collar cells draw in water through numerous tiny pores. Microscopic food particles are filtered out by collar cells and shared among the various cell types. Water exits through larger pores, the oscula. Spicules form a supportive internal skeleton.

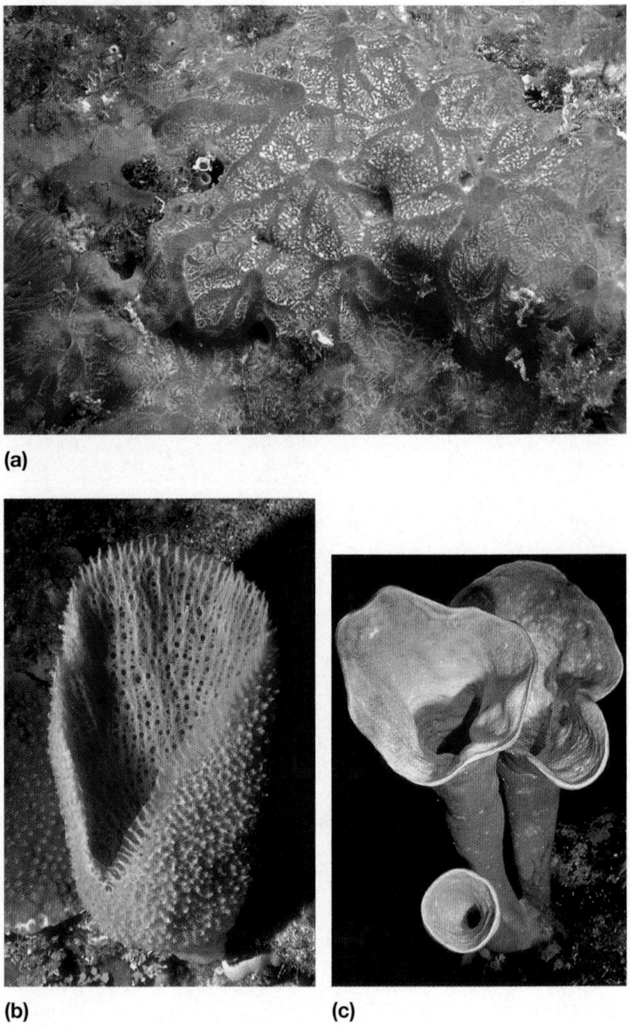

Figure 22-5 The diversity of sponges
Sponges come in a wide variety of sizes, shapes, and colors. Some are more than a meter tall; others, such as ***(a)*** this fire sponge, grow in free-form pattern over undersea rocks. ***(b)*** Tiny appendages attach this tubular sponge to rocks, whereas ***(c)*** this reef sponge with flared tubular openings attaches to a coral reef.

Cnidarian tentacles are armed with **cnidocytes**, cells containing structures that, when stimulated by contact, explosively inject poisonous or sticky filaments into prey (Fig. 22-8, p. 438). The venom of some cnidarians can cause painful stings in humans unfortunate enough to come into contact with them, and the stings of a few jellyfish species can even be life-threatening. The most deadly of these species is a "sea wasp," *Chironex fleckeri*, which is found in the waters off northern Australia and southeast Asia. The amount of venom found in a single sea wasp could kill up to 60 people, and the victim of a serious stinging may die within minutes of being stung.

The function of cnidocytes, however, in not to sting human swimmers but to capture prey. Although all cnidarians are predatory, none hunt actively. Instead, they rely on their victims' blundering by chance into the grasp of their enveloping tentacles. Stung and firmly grasped, the prey is forced through an expansible mouth into a digestive sac, the **gastrovascular cavity**. Digestive enzymes secreted into this cavity break down some of the food, and further digestion occurs within the cells lining the cavity. Because the gastrovascular cavity has only a single opening, when digestion is completed, undigested material is expelled through the mouth. Although this two-way traffic prevents continuous feeding, it is adequate to support the low energy demands of these animals.

Like sponges, cnidarians are confined to watery habitats, and most species are marine. One group of cnidarians, the corals, is of particular ecological importance (see Fig. 22-6c). Coral polyps form large colonies, and each member of the colony secretes a hard skeleton of calcium carbonate. The skeletons persist long after the corals' death, serving as a base to which others may attach themselves. The cycle continues until, after thousands of years, massive coral reefs are formed. Corals are restricted to the warm, clear waters of the Tropics, where their reefs form undersea habitats that are the basis of an ecosystem of stunning diversity and unparalleled beauty (see Chapter 41).

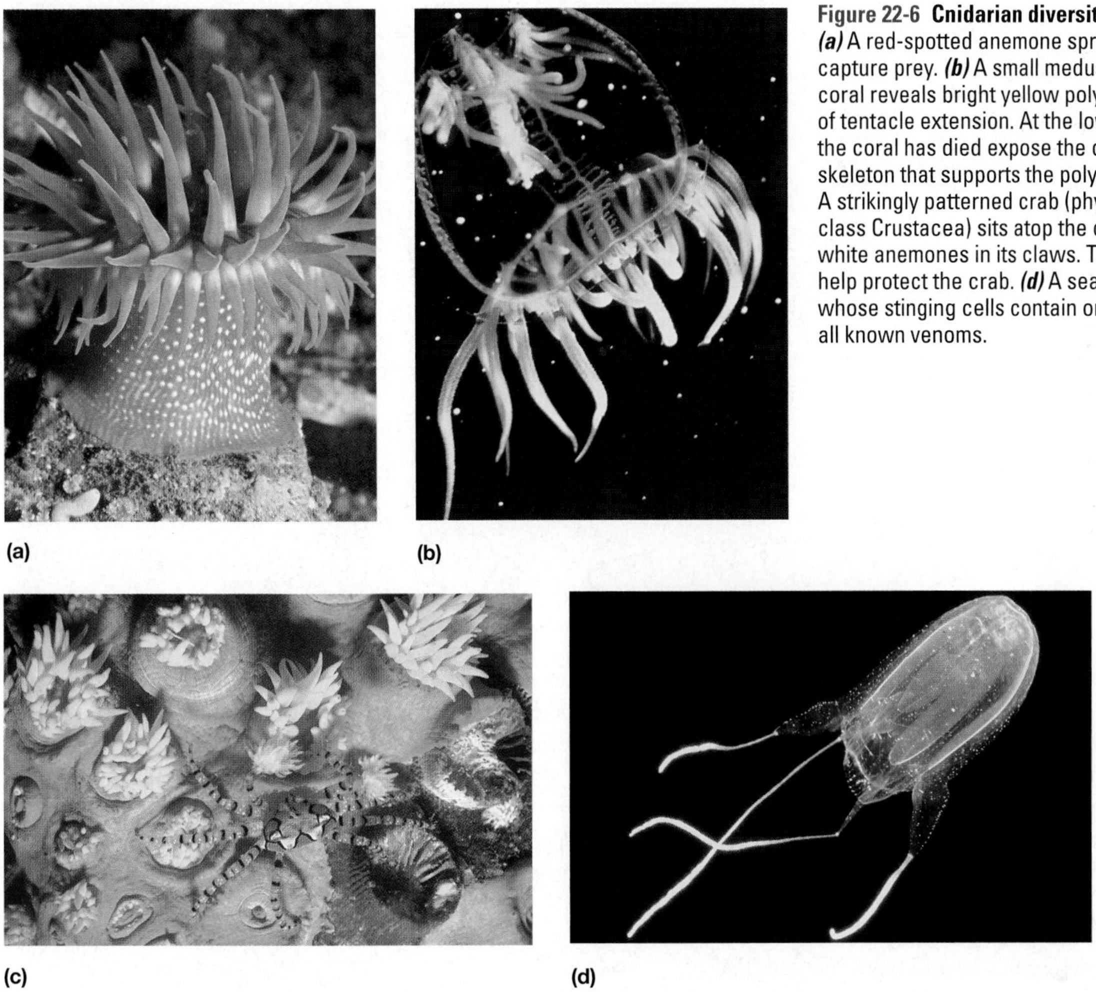

Figure 22-6 Cnidarian diversity
(a) A red-spotted anemone spreads its tentacles to capture prey. ***(b)*** A small medusa. ***(c)*** A close-up of coral reveals bright yellow polyps in various stages of tentacle extension. At the lower right, areas where the coral has died expose the calcium carbonate skeleton that supports the polyps and forms the reef. A strikingly patterned crab (phylum Arthropoda, class Crustacea) sits atop the coral, holding tiny white anemones in its claws. Their stinging tentacles help protect the crab. ***(d)*** A sea wasp, a cnidarian whose stinging cells contain one of the most toxic of all known venoms.

The Flatworms

The phylum Platyhelminthes (the *flatworms*) forms one of the earliest-branching lineages of bilaterally symmetrical, rather than radially symmetrical, animals (see Fig. 22-1). This body plan and its accompanying cephalization are adaptations that foster active movement. Cephalized, bilaterally symmetrical animals possess an anterior end, which is the first part of a moving animal to encounter the environment ahead. Sense organs are concentrated in the anterior portion of the body. This arrangement enhances the animal's ability to make an appropriate response to the stimuli it encounters (for example, eating food items and retreating from obstacles).

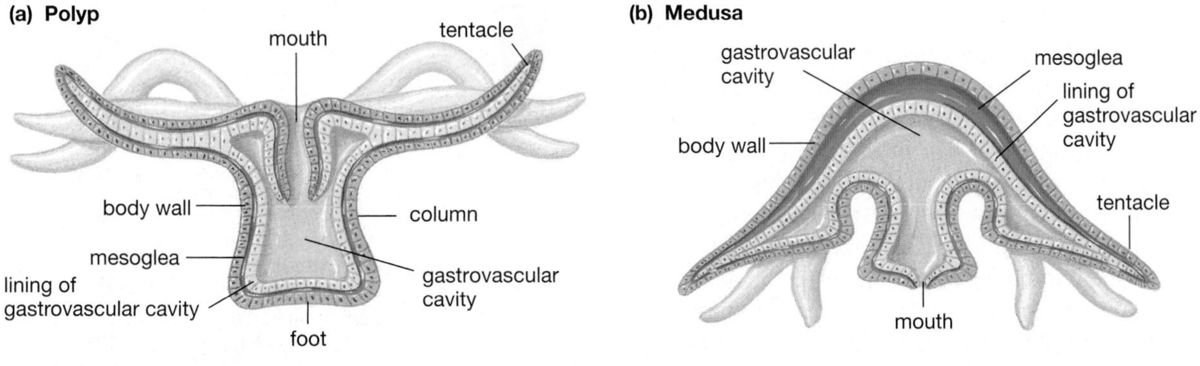

Figure 22-7 Polyp and medusa
The two basic body forms of cnidarians are actually variations on a single, simple theme. ***(a)*** The polyp form is seen in hydra (see Fig. 22-8), sea anemones (Fig. 22-6a), and the individual polyps within a coral (Fig. 22-6c). ***(b)*** The medusa form, seen in the jellyfish (Fig. 22-6b), resembles an inverted polyp. Both forms exhibit radial symmetry, with body parts arranged in a circle around a central axis.

MEDIATUTOR 22.1 Cnidocytes

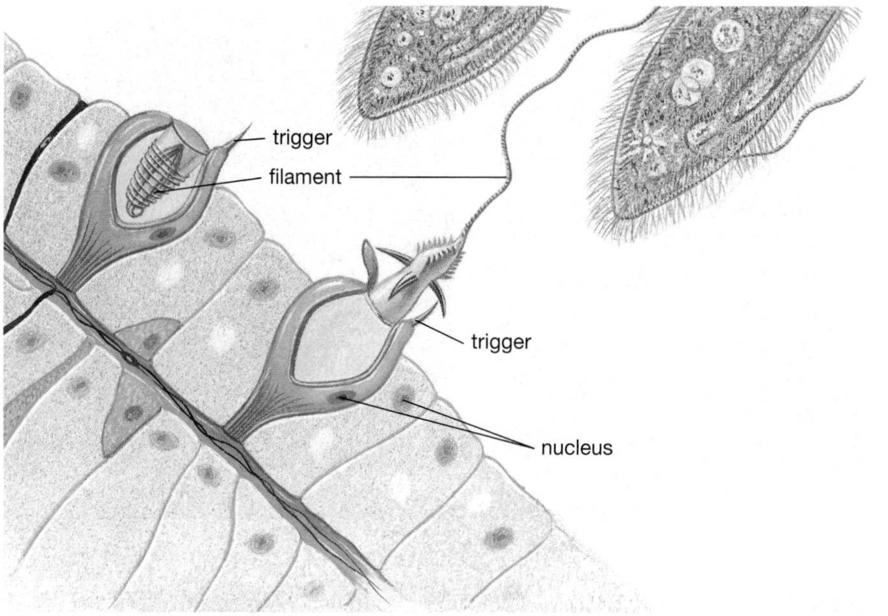

Figure 22-8 Cnidarian weaponry: the cnidocyte
At the slightest touch to the trigger of a special structure in their cnidocytes, cnidarians, such as this hydra, violently expel a poisoned filament. The hollow filament turns inside out, impaling the prey and injecting a paralyzing venom. These structures are microscopic. Only a few species inject enough venom to harm a human.

Unlike members of the radially symmetrical phyla, flatworms have well-developed organs, in which tissues are grouped into functional units. For example, flatworms such as the freshwater *planarians* (Fig. 22-9) have sense organs, including eyespots for detecting light and dark, and cells responsive to chemical and tactile stimuli. To process information, flatworms have clusters of nerve cells called **ganglia** (singular, **ganglion**) in the head that form a simple brain. Paired neural structures called **nerve cords** conduct nervous signals to and from the ganglia.

Despite the presence of some organs, flatworms lack respiratory and circulatory systems. In the absence of a respiratory system, gas exchange is accomplished by direct diffusion between body cells and the environment. This mode of respiration is possible because the small size and flat shape of flatworm bodies ensure that no body cell is very far from the surrounding environment. In the absence of a circulatory system, nutrients move directly from the digestive tract to body cells. The digestive cavity has a branching structure (see Fig. 22-9a) that reaches all parts of the body and allows digested nutrients to diffuse into nearby cells. The digestive cavity has only one opening to the environment, so undigested waste must pass out through the same opening that also serves as a mouth.

Flatworms may be either parasitic or **free-living**. **Parasites** are organisms that live in or on the body of another organism, called a *host*, which is harmed as a result of the relationship, whereas free-living organisms do not live in such intimate association with members of another species. Some parasitic flatworms can infect humans. For example, *tapeworms* can infect people who eat improperly cooked beef, pork, or fish that has been infected by the worms. Worm larvae form encapsulated resting structures, called **cysts**, in the muscles of these animals. The cysts hatch in the human digestive tract, where the young tapeworms attach themselves to the lining of the intestine. There they may grow to a length of more than 20 feet (7 meters), absorbing digested nutrients directly through their outer surface, and eventually releasing packets of eggs that are shed in the host's feces. If pigs, cows, or fish eat food contaminated with

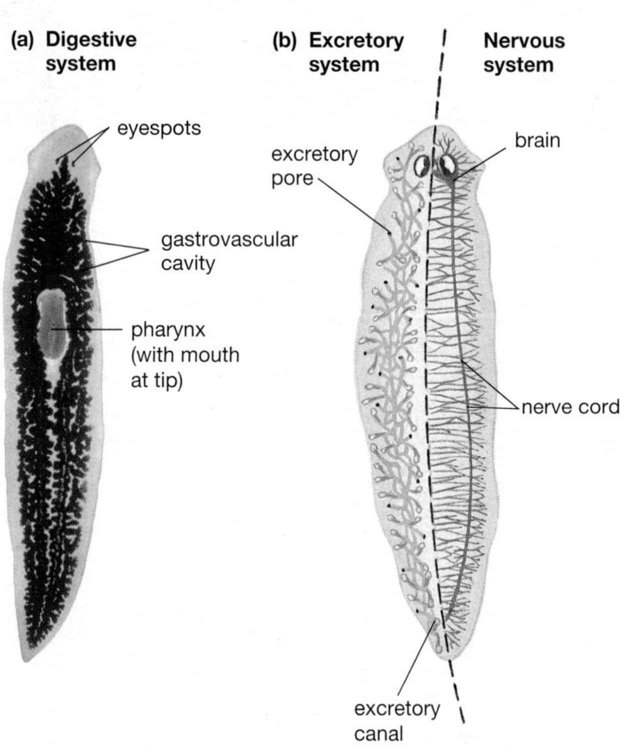

Figure 22-9 Flatworm organ systems
Flatworms such as planarians have well-developed organ systems. ***(a)*** The elaborately branched digestive system, the centrally located ventral pharynx, and eyespots in the head are clearly visible. ***(b)*** (Left) The excretory system consists of branching tubes that conduct excess fluid to the outside through numerous pores. Cilia keep the fluid moving. (Right) The nervous system of flatworms shows clear cephalization, with eyes and a brain composed of ganglia cells in a well-defined head. Ladderlike nerve cords carry signals through the rest of the body.

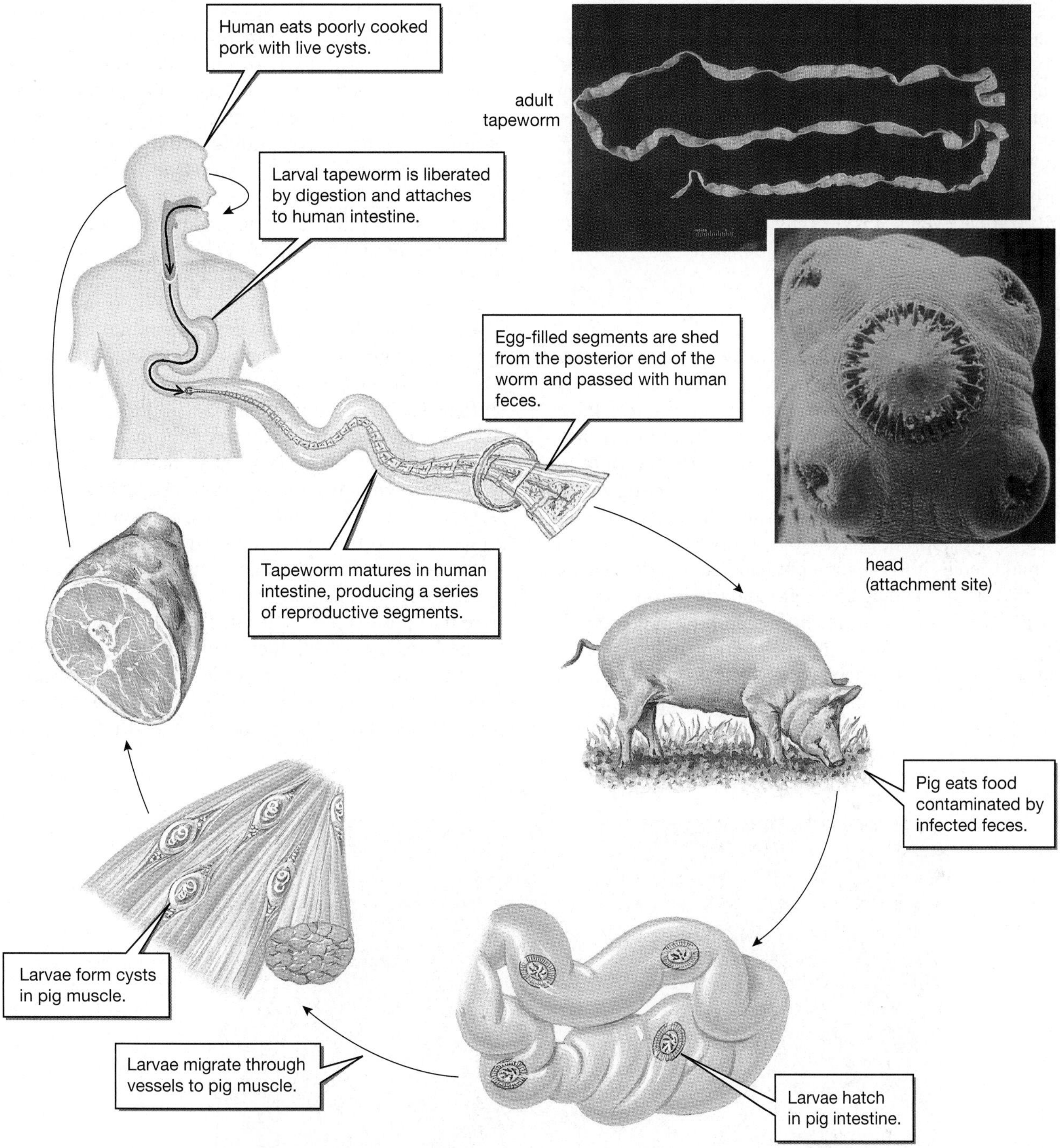

Human eats poorly cooked pork with live cysts.
adult tapeworm
Larval tapeworm is liberated by digestion and attaches to human intestine.
Egg-filled segments are shed from the posterior end of the worm and passed with human feces.
Tapeworm matures in human intestine, producing a series of reproductive segments.
head (attachment site)
Pig eats food contaminated by infected feces.
Larvae form cysts in pig muscle.
Larvae migrate through vessels to pig muscle.
Larvae hatch in pig intestine.

Figure 22-10 The life cycle of the human pork tapeworm
Each reproductive unit, or proglottid, is a self-contained reproductive factory that includes both male and female sex organs.

infected human feces, the eggs hatch in the animal's digestive tract, releasing larvae that burrow into its muscles and form cysts, thereby continuing the infective cycle (Fig. 22-10).

Another group of parasitic flatworms is the *flukes.* Of these, the most devastating are liver flukes (common in Asia) and blood flukes, such as those of the genus *Schistosoma,* which cause the disease schistosomiasis. Like most parasites, flukes have a complex life cycle that includes an intermediate host (a snail, in the case of *Schistosoma*). Prevalent in Africa and parts of South America, schistosomiasis affects an estimated 200 million people worldwide. Its symptoms include diarrhea, anemia, and possible brain damage. Efforts to control the spread of schistosomiasis are sometimes impeded by the unintentional consequences of other human activities. For example, in Egypt, the irrigation ditches filled by the Aswan Dam, completed in 1968, have contributed to the spread of schistosomiasis by creating an extensive new habitat for the snail host.

Flatworms can reproduce both sexually and asexually. Free-living forms may reproduce by cinching themselves around the middle until they separate into two halves, each of which regenerates its missing parts. All forms can reproduce sexually; most are **hermaphroditic**—that is, they possess both male and female sexual organs (see Fig. 22-10). This feature is a great advantage to parasitic forms, because each worm can reproduce through self-fertilization, even if it is the only individual present in its host.

The Roundworms

Although you may be blissfully unaware of their presence, *roundworms* (phylum Nematoda) are nearly everywhere. Roundworms, also called *nematodes*, have colonized nearly every habitat on Earth, and they play an important role in breaking down organic matter. They are extremely numerous; a single rotting apple may contain 100,000 roundworms. Billions thrive in each acre of topsoil. In addition, almost every plant and animal species hosts several parasitic nematode species.

In addition to being abundant and ubiquitous, roundworms are diverse. Although only about 12,000 roundworm species have been named, there may be as many as 500,000. Most are microscopic, such as the one shown in Figure 22-11, but some parasitic forms reach a meter in length.

Nematodes have a rather simple body plan, featuring a tubular gut that runs from mouth to anus. This kind of digestive tract, with two openings and a one-way digestive path, is much more efficient than the single-opening digestive systems of cnidarians and flatworms. Roundworms also have a fluid-filled pseudocoelom that surrounds the organs and forms a **hydrostatic skeleton**, a framework against which muscles can act. A tough, flexible, nonliving cuticle encloses and protects the thin, elongated body (see Fig. 22-11). Sensory organs in the head transmit information to a simple "brain," composed of a nerve ring.

Like flatworms, nematodes lack circulatory and respiratory systems. Because most nematodes are extremely thin and have low energy requirements, diffusion suffices for gas exchange and distribution of nutrients. Most nematodes reproduce sexually, and the sexes are separate, with the male (who is normally smaller) fertilizing the female by placing sperm inside her body.

During your life, you may be parasitized by one of the 50 species of roundworms that infect humans. Most such worms are relatively harmless, but there are important exceptions. For example, hookworm larvae in soil may bore into human feet, enter the bloodstream, and travel to the intestine, where they cause continuous bleeding. Another dangerous nematode parasite, *Trichinella*, causes the disease trichinosis. *Trichinella* worms can infect people who eat improperly cooked infected pork,

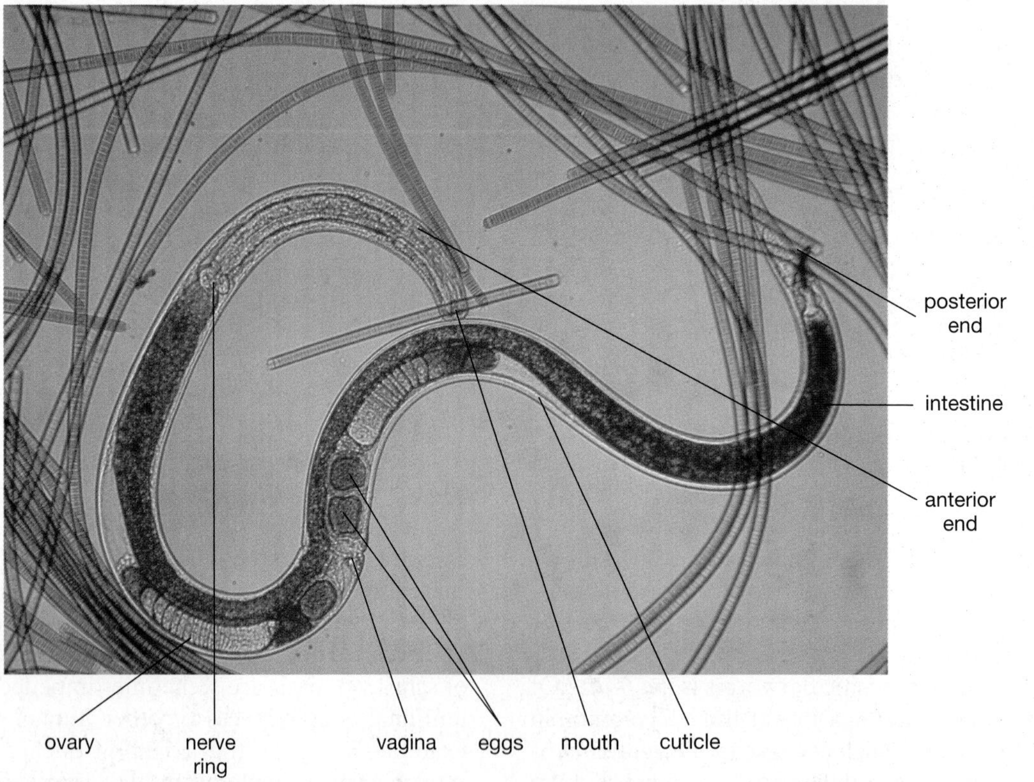

Figure 22-11 A freshwater nematode
Eggs can be seen inside this female freshwater nematode, which feeds on algae.

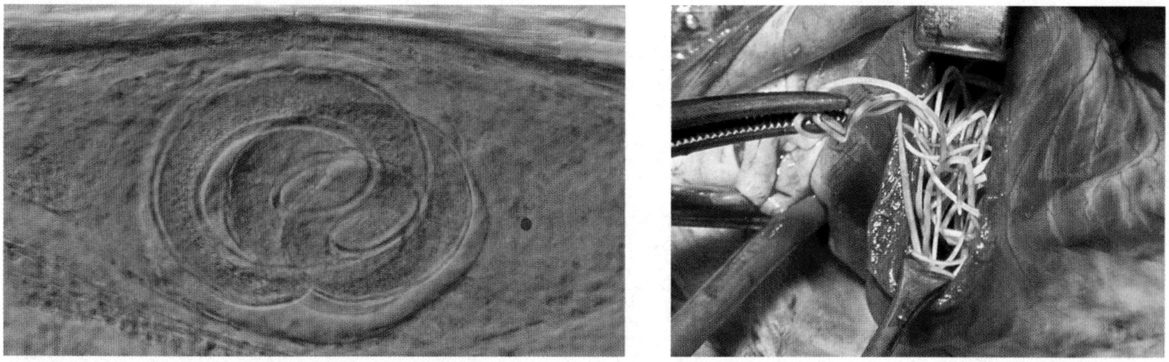

(a) (b)

Figure 22-12 Some parasitic nematodes
(a) Encysted larva of the *Trichinella* worm in the muscle tissue of a pig, where it may live for up to 20 years. ***(b)*** Adult heartworms in the heart of a dog. The juveniles are released into the bloodstream, where they may be ingested by mosquitoes and passed to another dog by the bite of an infected mosquito.

which can contain up to 15,000 larval cysts per gram (Fig. 22-12a). The cysts hatch in the human digestive tract and invade blood vessels and muscles, causing bleeding and muscle damage.

Parasitic nematodes can also endanger domestic animals. Dogs, for example, are susceptible to heartworm, which is transmitted by mosquitoes (Fig. 22-12b). In the southern United States, and increasingly in other parts of the country, heartworm poses a severe threat to the health of unprotected pets.

The Segmented Worms

A prominent feature of the segmented worms, or *annelids* (phylum Annelida), is the division of the body into a series of repeating segments. Externally, these segments appear as ringlike depressions on the surface. Internally, most of the segments contain identical copies of nerve ganglia, excretory structures, and muscles. **Segmentation** is advantageous for locomotion, because the body compartments, each of which is controlled by separate muscles, collectively are capable of more complex movement than is seen in the nonsegmented worms. Another feature that distinguishes annelids from flatworms and roundworms is a fluid-filled true coelom between the body wall and the digestive tract (see Fig. 22-2c). The incompressible fluid in the coelom of many annelids is confined by the partitions between the segments and serves as a hydrostatic skeleton, making possible such feats as burrowing through soil.

In contrast to the nematodes, the annelids have a well-developed **closed circulatory system** that distributes gases and nutrients throughout the body (see Chapter 27). In closed circulatory systems (including yours), blood remains confined to the heart and blood vessels. In the earthworm, for example, blood with oxygen-carrying hemoglobin is pumped through well-developed vessels by five pairs of "hearts" (Fig. 22-13). These hearts are actually short segments of specialized blood vessels that contract rhythmically. The blood is filtered and wastes are removed by excretory organs called **nephridia** (singular, **nephridium**), which are found in many of the segments. Nephridia resemble the individual tubules of the vertebrate kidney (see Chapter 30). The annelid nervous system consists of a simple ganglionic brain in the head and a series of repeating paired segmental ganglia joined by a pair of ventral nerve cords that pass along the length of the body.

Digestion in annelids occurs in a series of compartments, each specialized for a different phase of food processing (see Fig. 22-13). For example, in the earthworm, a muscular **pharynx** draws in food, consisting of bits of decaying plant and animal debris in soil. The food is conducted through the esophagus to a storage chamber, the *crop*, and then released slowly into the muscular *gizzard*, where it is ground into tiny particles by muscular contractions of the gizzard and the sharp-edged sand grains it contains. The food then passes into the intestine, where it is digested and nutrients are absorbed. Undigested food and soil exit through the anus.

The phylum Annelida includes three main subgroups, the *oligochaetes*, the *polychaetes*, and the *leeches*. The oligochaetes include the familiar earthworm and its relatives. Polychaetes live primarily in the ocean. Some polychaetes have paired fleshy paddles on most of their segments, used in locomotion. Others live in tubes from which they project feathery gills that both exchange gases and sift the water for microscopic food (Fig. 22-14a,b). Leeches (Fig. 22-14c) live in freshwater or moist terrestrial habitats, and are either carnivorous or parasitic—some preying on smaller invertebrates, others sucking the blood of larger animals.

The Insects, Arachnids, and Crustaceans

Arthropods are the dominant animals on Earth. In terms of both number of individuals and number of species, no other phylum comes close to the phylum

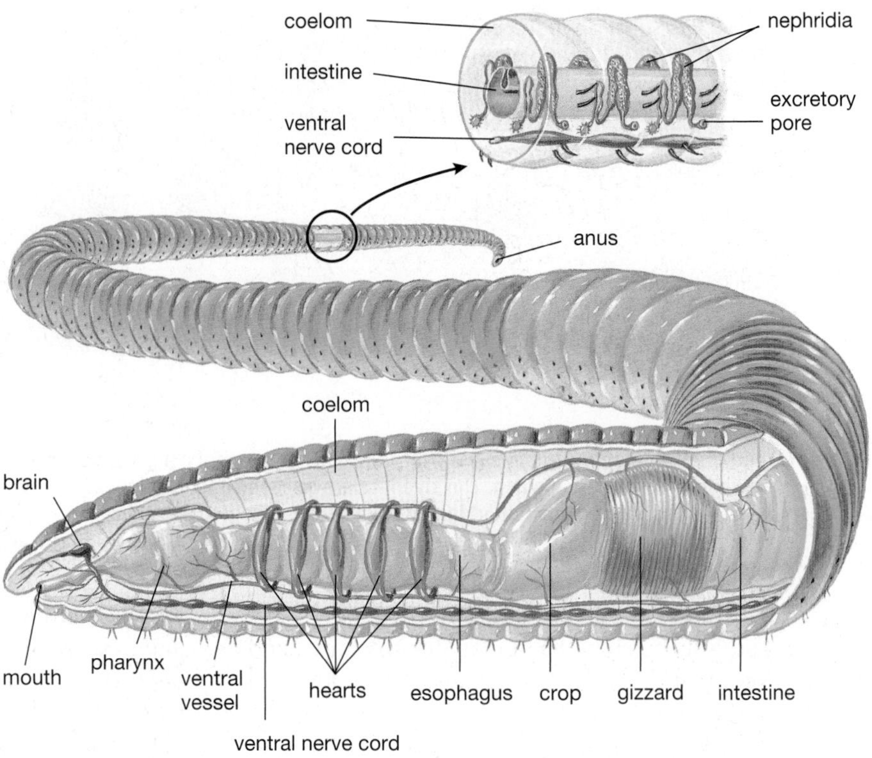

Figure 22-13 An annelid, the earthworm
This diagram shows an enlargement of segments, many of which are repeating similar units separated by partitions. The digestive system, which has both a mouth and an anus, is divided into a series of compartments specialized to process food in an orderly sequence.

(a)

(b)

(c)

Figure 22-14 Diverse annelids
(a) A polychaete annelid projects brightly spiraling gills from a tube attached to rock. When the gills retract, the tube is covered by the trap door visible on the lower right. ***(b)*** The "fireworm" polychaete swims by using paddles on each segment. The bristles on each paddle can deliver a fiery sting. ***(c)*** This leech, a freshwater annelid, shows numerous segments. The sucker encircles its mouth, allowing it to attach to its prey. Doctors used leeches medicinally up until the 1800s to suck "tainted" blood from patients; even today leeches are used by some surgeons to prevent blood from building up around healing wounds.

Arthropoda. About 1 million arthropod species have been discovered, and scientists estimate that up to 9 million remain undescribed. The phylum includes a huge array of forms, such as the insects (class Insecta); the spiders and their relatives (class Arachnida); and the crabs, shrimp, and their relatives (class Crustacea).

The success of the arthropods can be attributed to several important adaptations that have allowed them to exploit nearly every habitat on Earth. These adaptations include an exoskeleton, segmentation, efficient gas-exchange mechanisms, and well-developed circulatory, sensory, and nervous systems.

The **exoskeleton** (*exo* means "outside" in Greek) is an external skeleton that encloses the arthropod body like a suit of armor. In places it is thin and flexible, allowing movement of the paired, *jointed appendages*. The exoskeleton is secreted by the *epidermis* (the outer layer of skin) and is composed chiefly of protein and a polysaccharide called **chitin**. This external skeleton protects against predators and is responsible for arthropods' greatly increased agility relative to their wormlike ancestors. By providing stiff but flexible appendages and rigid attachment sites for muscles, the exoskeleton makes possible the flight of the bumblebee and the intricate, delicate manipulations of the spider as it weaves its web (Fig. 22-15). The exoskeleton also contributed enormously to the arthropod invasion of dry terrestrial habitats (arthropods were the earliest terrestrial animals, see Chapter 17) by providing a watertight covering for delicate, moist tissues such as those used for gas exchange.

Like a suit of armor, the arthropod exoskeleton poses some unique problems. First, because it cannot expand as the animal grows, the exoskeleton must be shed, or **molted**, periodically and replaced with a larger one (Fig. 22-16). Molting uses energy and leaves the animal temporarily vulnerable until the new skeleton hardens. ("Soft-shelled" crabs are simply regular "hard-shelled" crabs that are caught and eaten during this delicate period.) The exoskeleton is also heavy; its weight increases exponentially as the animal grows. It is no coincidence that the largest arthropods are crustaceans (crabs and lobsters), whose watery habitat supports much of their weight.

Figure 22-15 The exoskeleton allows precision movements A garden spider, having immobilized its prey with a paralyzing venom, rapidly encases the prey in its web. Such dexterous manipulations are made possible by the exoskeleton and jointed appendages characteristic of arthropods.

Figure 22-16 The exoskeleton must be molted periodically A newly emerged praying mantis (a predatory insect) hangs beside its outgrown exoskeleton (left).

Segmentation in arthropods is evidence that they share a common ancestor with annelids. Arthropod segments, however, tend to be fewer, less distinct from one another, and specialized for different functions such as sensing the environment, feeding, and movement (Fig. 22-17). For example, in insects sensory and feeding structures are concentrated on the anterior-most segment, known as the **head**, and digestive structures are largely confined to the **abdomen**, which is the segment at the animal's posterior end. Between the head and the abdomen is the **thorax**, the segment to which structures used in locomotion, such as wings and walking legs, are attached.

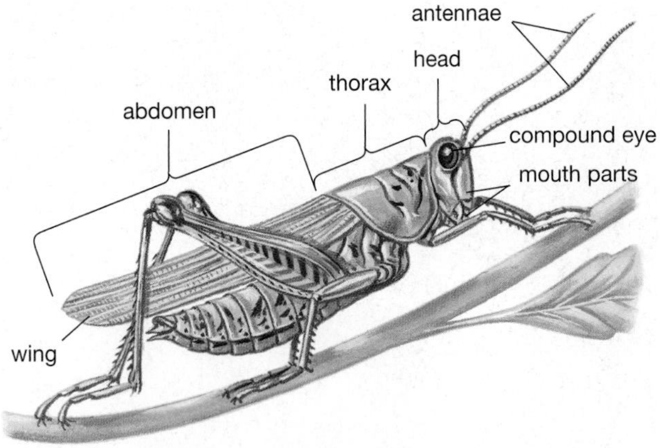

Figure 22-17 Segments are fused and specialized in insects Insects such as this grasshopper show fusion and specialization of body segments into a distinct head, thorax, and abdomen. Segments are visible beneath the wings on the abdomen.

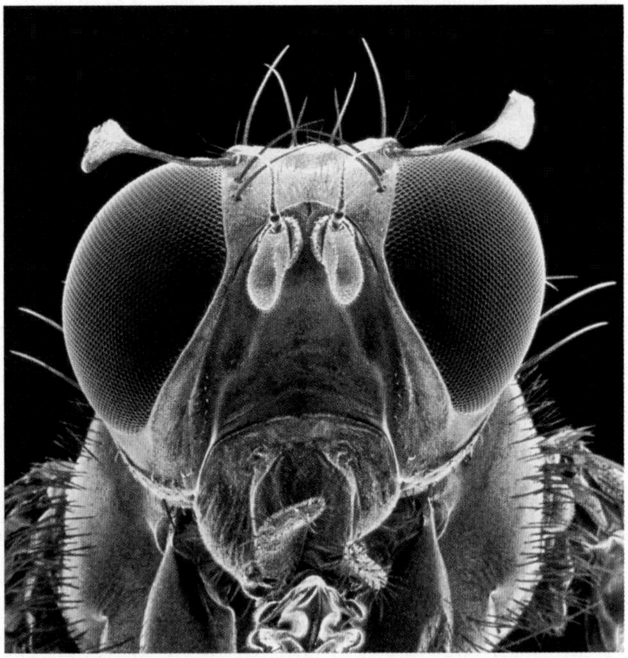

Figure 22-18 Arthropods possess compound eyes This scanning electron micrograph shows the compound eye of a fruit fly. Compound eyes consist of an array of similar light-gathering and sensing elements whose orientation gives the arthropod a panoramic view of the world. Insects have reasonably good image-forming ability and good color discrimination.

Efficient gas exchange is required to supply adequate oxygen to the muscles that allow the rapid flight, swimming, or running displayed by many arthropods. Gas exchange is accomplished by **gills** (thin, external respiratory membranes) in aquatic forms such as the crustaceans, and by either **tracheae** (singular, **trachea**; a network of narrow, branching respiratory tubes) or **book lungs** (specialized respiratory structures of arachnids) in terrestrial forms. Arthropods also have well-developed circulatory systems with a feature not seen in annelids: the **hemocoel**, or blood cavity. Blood not only travels through vessels but also empties into the hemocoel, where it bathes the internal organs directly. This arrangement, known as an **open circulatory system**, is also present in most mollusks.

Most arthropods possess a well-developed sensory system, including **compound eyes**, which have multiple light detectors (Fig. 22-18), and acute chemical and tactile senses. The arthropod nervous system is similar in plan to that of annelids, but more complex. It consists of a brain composed of fused ganglia in the head, connected to a series of ganglia along the length of the body and linked by a ventral nerve cord. The capacity for finely coordinated movement combined with sophisticated sensory abilities and a well-developed nervous system has allowed the evolution of complex behavior. In fact, the social behavior of some insect species, such as the honeybee, is as intricate and complex as any vertebrate behavior. Although many people associate communication and learning with vertebrates, both play important roles in insect societies.

Insects

The number of described *insect* species (members of the class Insecta; Fig. 22-19) is about 850,000, roughly three times the total number of known species in all other classes of animals combined. Insects have three pairs of legs, normally supplemented by two pairs of wings. Insects' capacity for flight distinguishes them from all other invertebrates and has contributed to their enormous success (Fig. 22-19c). As anyone who has unsuccessfully pursued a fly can testify, flight helps in escaping from predators. It also allows the insect to find widely dispersed food. Swarms of locusts (Fig. 22-19d) have been traced all the way from Saskatchewan, Canada, to Texas on the trail of food. Flight requires rapid and efficient gas exchange, which insects accomplish by means of tracheae. The network of tracheae conducts air to all parts of the body.

During their development, insects undergo **metamorphosis**, which commonly involves a radical change in body form from juvenile to adult. In insects with complete metamorphosis, the immature form, called a **larva**, is worm-shaped (for example, the maggot of a housefly or the caterpillar of a moth or butterfly; see Fig. 22-19e). The larva hatches from an egg, grows by eating voraciously and shedding its exoskeleton several times, then forms a nonfeeding form called a **pupa**. Encased in an outer covering, the pupa undergoes a radical change in body form, emerging in its adult winged form. The adults mate and lay eggs, continuing the cycle. Metamorphosis may include a change in diet as well as in shape, thereby eliminating competition for food between adults and juveniles and in some cases allowing the insect to exploit different foods when they are most available. For instance, a caterpillar that feeds on new green shoots in springtime metamorphoses into a butterfly that drinks nectar from the summer's blooming flowers. Some insects undergo a more

Figure 22-19 The diversity of insects
(a) The rose aphid sucks sugar-rich juice from plants. ***(b)*** A mating pair of Hercules beetles. The large "horns" are found only on the male. ***(c)*** A June beetle displays its two pairs of wings as it comes in for a landing. The outer wings protect the abdomen and the inner wings, which are relatively thin and fragile. ***(d)*** Insects such as this locust can cause devastation of both crops and natural vegetation. ***(e)*** Caterpillars are larval forms of moths or butterflies. This caterpillar larva of the Australian fruit-sucking moth displays large eyespot patterns that may frighten potential predators, who mistake them for eyes of a large animal.

MEDIATUTOR 22.2 Insect Metamorphosis

gradual metamorphosis (called *incomplete metamorphosis*), hatching as young that bear some resemblance to the adult, then gradually acquiring more-adult features as they grow and molt.

Arachnids

The *arachnids* include spiders, mites, ticks, and scorpions (Fig. 22-20). All members of the class Arachnida have eight walking legs, and most are carnivorous; many subsist on a liquid diet of blood or predigested prey. For example, spiders, the most numerous arachnids, first immobilize their prey with a paralyzing venom. They then inject digestive enzymes into the helpless victim (typically an insect) and suck in the resulting soup. Arachnids breathe by using either tracheae, book lungs, which are unique to arachnids, or both. Arachnids have simple eyes, each with a single lens, in contrast to the compound eyes of insects and crustaceans. The eyes are particularly sensitive to movement, and in some species they probably can form images. Most spiders have eight eyes placed in such a way as to give them a panoramic view of predators and prey.

Crustaceans

The *crustaceans*, including crabs, crayfish, lobster, shrimp, and barnacles, make up the Crustacea, the only class of arthropods whose members live primarily in the water (Fig. 22-21). Crustaceans range in size from microscopic maxillopods that live in the spaces between grains of sand to the largest of all arthropods, the Japanese crab, with legs spanning nearly 12 feet (4 meters). Crustaceans have two pairs of sensory *antennae*, but the rest of their appendages are highly variable in form and number, depending on the habitat and lifestyle of the species. Most crustaceans have compound eyes similar to those of insects, and nearly all respire by means of gills.

The Snails, Clams, and Squid

Mollusks share a common ancestry with their fellow protostomes, the annelids and arthropods. In terms of species diversity, they are second (albeit a distant second) only to the arthropods. Members of the phylum Mollusca have a moist, muscular body supported by a hydrostatic skeleton. Some protect their bodies with a

(a) (b) (c)

Figure 22-20 The diversity of arachnids
(a) The tarantula is among the largest spiders but is relatively harmless. ***(b)*** Scorpions, found in warm climates, including deserts of the southwestern United States, paralyze their prey with venom from a stinger at the tip of the abdomen. A few species can harm humans. ***(c)*** Ticks before (left) and after feeding on blood. The uninflated exoskeleton is flexible and folded, allowing the animal to become grotesquely bloated while feeding.

Figure 22-21 The diversity of crustaceans
(a) The microscopic waterflea, *Daphnia,* is common in freshwater ponds. Notice the eggs developing within the body. ***(b)*** The sowbug, found in dark, moist places such as under rocks, leaves, and decaying logs, is one of the few crustaceans to invade the land successfully. ***(c)*** The hermit crab protects its soft abdomen by inhabiting an abandoned snail shell. ***(d)*** The gooseneck barnacle uses a tough, flexible stalk to anchor itself to rocks, boats, or even animals such as whales. Other types of barnacles attach with shells that resemble miniature volcanoes (see Fig. 22-24b). Early naturalists thought barnacles were mollusks until the jointed legs, seen extending into the water, were observed.

(a) (b) (c) (d)

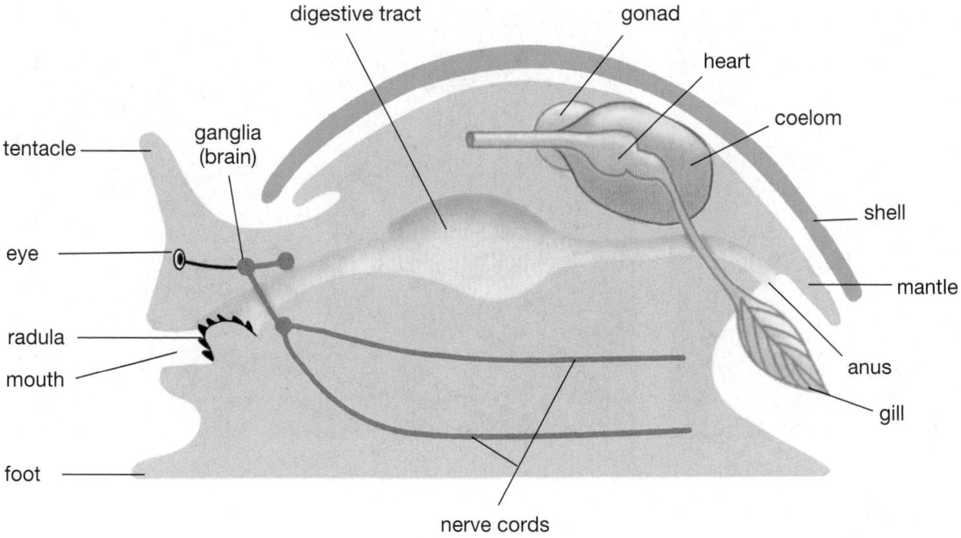

Figure 22-22 A generalized mollusk
The general body plan of a mollusk, showing the mantle, foot, gills, shell, radula, and other features that are seen in most (but not all) species of mollusk.

shell of calcium carbonate; others escape predation by moving swiftly or, if caught, by tasting terrible. Mollusks have a **mantle**, an extension of the body wall that forms a chamber for the gills and secretes the shell in shelled species. Most mollusks have an open circulatory system, like that of arthropods, with blood directly bathing the organs in a hemocoel. The nervous system, like that of annelids and arthropods, consists of ganglia connected by nerves, but many more of the ganglia are concentrated in the brain. Reproduction is sexual; some species have separate sexes, and others are hermaphroditic. Although mollusks are enormously diverse, a simplified diagram of the body plan of a mollusk is shown in Figure 22-22.

Among the many classes of mollusks, we will discuss three in more detail: snails and their relatives (the class Gastropoda), clams and their relatives (the class Bivalvia), and octopuses and their relatives (the class Cephalopoda).

Snails and Slugs

Snails and slugs, collectively known as *gastropods*, crawl on a muscular *foot*, and many are protected by shells that vary widely in form and color (Fig. 22-23a). Not all gastropods are shelled, however. Sea slugs, for example, lack shells, but their brilliant colors warn potential predators that they are poisonous, or at least taste bad (Fig. 22-23b).

Gastropods feed with a **radula**, a flexible ribbon of tissue studded with spines that is used to scrape algae from rocks or to grasp larger plants or prey (see Fig. 22-22). Most snails use gills, typically enclosed in a cavity beneath the shell, for respiration. Gases can also diffuse readily through the skin of most gastropods, and most sea slugs rely on this mode of gas exchange. The few gastropod species (including the destructive garden snails and slugs) that live in terrestrial habitats use a simple lung for breathing.

(a) **(b)**

Figure 22-23 The diversity of gastropod mollusks
(a) A Florida tree snail displays a brightly striped shell and eyes at the tip of stalks that are retracted instantly if touched.
(b) Spanish shawl sea slugs prepare to mate. The brilliant colors of many sea slugs warn potential predators that they are distasteful.

(a) (b)

Figure 22-24 The diversity of bivalve mollusks
(a) This swimming scallop parts its hinged shells. The upper shell is covered with an encrusting sponge. ***(b)*** Mussels attach to rocks in dense aggregations exposed at low tide. White barnacles are attached to the mussel shells and surrounding rock.

Scallops, Clams, Mussels, and Oysters

Included among the *bivalves* (class Bivalvia) are scallops, oysters, mussels, and clams (Fig. 22-24). Not only do members of the class lend exotic variety to the human diet, they are also important members of the nearshore marine community. Bivalves possess two shells connected by a flexible hinge. A strong muscle clamps the shells closed in response to danger; this muscle is what you are served when you order scallops in a restaurant.

Clams use a muscular foot for burrowing in sand or mud. In mussels, which are sessile, the foot is smaller and is used to help secrete threads that anchor the animal to rocks. Scallops lack a foot and move by a sort of whimsical jet propulsion achieved by flapping their shells together. Bivalves are filter feeders, using their gills as both respiratory and feeding structures. Water is circulated over the gills, which are covered with a thin layer of mucus that traps microscopic food particles. Food is conveyed to the mouth by the beating of cilia on the gills. Probably because they filter-feed and do not move extensively, bivalves have "lost their heads" during the course of their evolution.

Octopuses, Squid, and Cuttlefish

The *cephalopods* are a fascinating group that contains octopuses, nautiluses, cuttlefish, and squids (Fig. 22-25), including the largest invertebrate, the giant squid. All cephalopods are predatory carnivores, and all are marine. In these mollusks, the foot has evolved into tentacles with well-developed chemosensory abilities and suction disks for detecting and grasping prey. Prey grasped by tentacles may be immobilized by a paralyzing venom in the saliva before being torn apart by beaklike jaws.

Cephalopods move rapidly by jet propulsion, which is accomplished by the forceful expulsion of water from the mantle cavity. Octopuses may also travel along the seafloor by using their tentacles like multiple undulating legs. The rapid movements and generally active lifestyles of cephalopods are made possible in part by their closed circulatory systems. Cephalopods are the only mollusks with closed circulation, which transports oxygen and nutrients more efficiently than open circulatory systems do.

Cephalopods have highly developed brains and sensory systems. The cephalopod eye rivals our own in complexity and exceeds it in efficiency of design. The cephalopod brain, especially that of the octopus, is exceptionally large and complex for an invertebrate brain. It is enclosed in a skull-like case of cartilage and endows the octopus with highly developed capabilities to learn and remember. In the laboratory, octopuses can rapidly learn to associate certain symbols with food and to open a screw-cap jar to obtain food.

The Sea Stars, Sea Urchins, and Sea Cucumbers

Echinoderms are found only in marine environments, and their common names tend to evoke their saltwater habitats. Sand dollars, sea urchins, sea stars (or starfish), sea cucumbers, and sea lilies are among the members of the phylum Echinodermata (Fig. 22-26). The name *echinoderm* (Greek, "hedgehog skin") stems from the bumps or spines that extend from the skin of most echinoderms. These spines are especially well developed in sea urchins and much reduced in sea stars and sea cucumbers. Echinoderm bumps and spines are actually extensions of an **endoskeleton** (internal skeleton) composed of plates of calcium carbonate that lie beneath the outer skin.

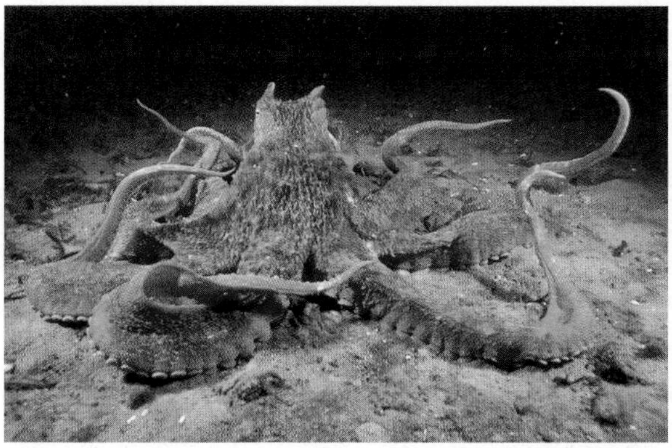

(a)

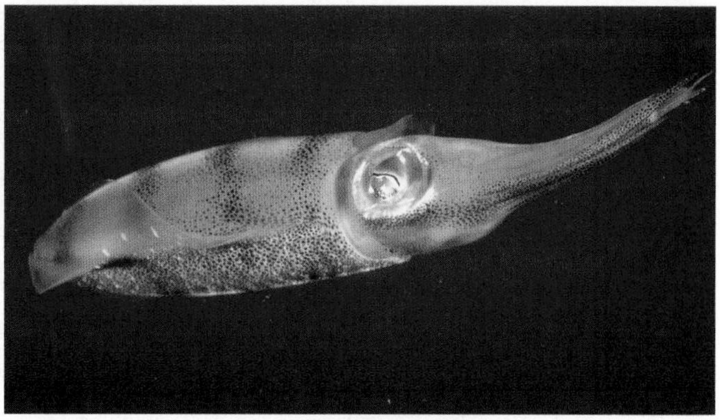

(b)

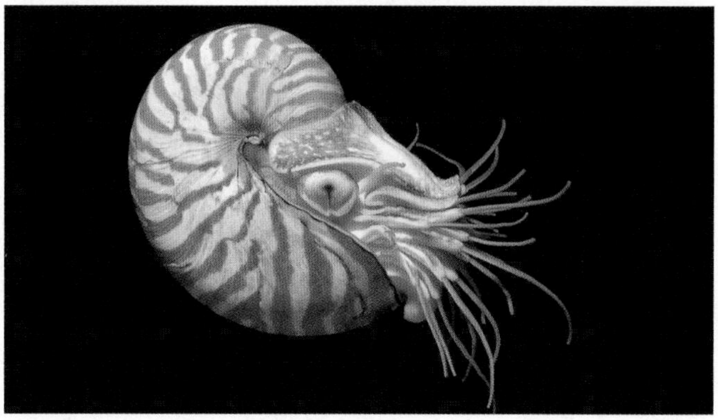

(c)

Figure 22-25 The diversity of cephalopod mollusks
(a) An octopus can crawl rapidly by using its eight suckered tentacles. It can alter its color and skin texture to blend with its surroundings. In emergencies this mollusk can jet backward by vigorously contracting its mantle. Octopuses and squid can emit clouds of dark purple ink to confuse pursuing predators. ***(b)*** The squid moves entirely by contracting its mantle to generate jet propulsion, which pushes the animal backward through the water. The giant squid is the largest invertebrate, reaching a length of 15 meters (almost 50 feet), including tentacles. ***(c)*** The chambered nautilus secretes a shell with internal, gas-filled chambers that provide buoyancy. Note the well-developed eyes and the tentacles used to capture prey.

Echinoderms exhibit deuterostome development and are linked by common ancestry with the other deuterostome phyla, including the chordates (described below). Deuterostomes form a group of branches on the larger evolutionary tree of bilaterally symmetrical animals, but in echinoderms bilateral symmetry is expressed only in embryos and free-swimming larvae. An adult echinoderm, in contrast, is radially symmetrical and lacks a head. This absence of cephalization is consistent with the sluggish or (in some forms) sessile existence of echinoderms. Most echinoderms move only very slowly as they feed on algae or small particles sifted from sand or water. Some echinoderms are slow-motion predators. Sea stars, for example, slowly pursue even slower-moving prey, such as bivalve mollusks.

Echinoderms move on numerous tiny **tube feet**, delicate cylindrical projections that extend from the lower surface of the body and terminate in a suction cup. Tube feet are part of a unique echinoderm feature, the **water-vascular system**, which functions in locomotion,

(a) (b) (c)

Figure 22-26 The diversity of echinoderms
(a) A sea cucumber feeds on debris in the sand. ***(b)*** The sea urchin's spines are actually projections of the internal skeleton. ***(c)*** The sea star has reduced spines and typically has five arms.

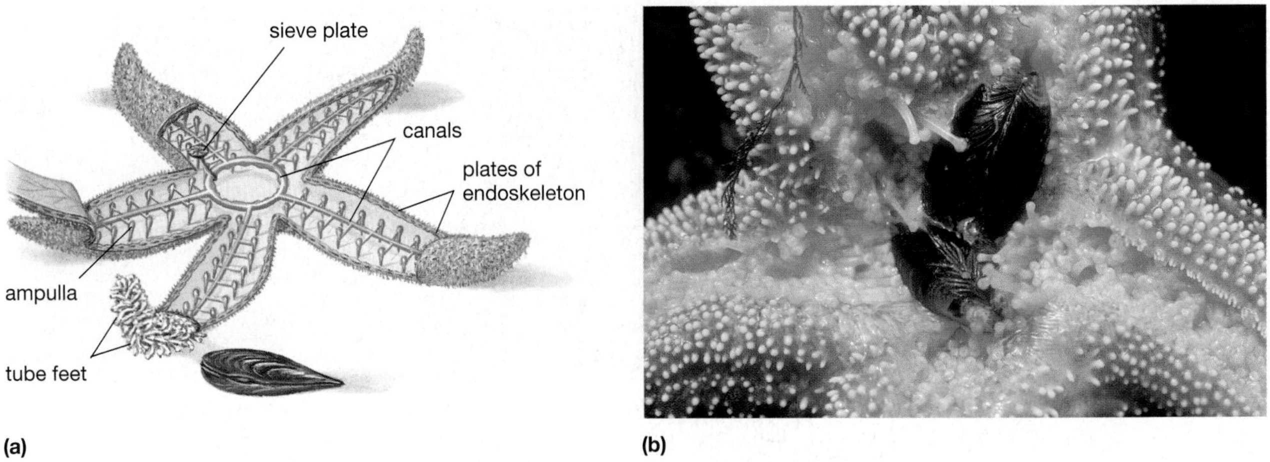

Figure 22-27 The water-vascular system of echinoderms
(a) Seawater enters through the sieve plate and is transported into the circular central canal, branches of which distribute water to each of the arms. The water inflates squeeze-bulb-like ampullae, which expand and contract to extend or retract the tube feet. The plates of the endoskeleton are embedded in the body wall. ***(b)*** The sea star often feeds on mollusks such as this mussel. A feeding sea star attaches numerous tube feet to the mussel's shells, exerting a relentless pull. Then, the sea star turns the delicate tissue of its stomach inside out, extending it through its centrally located ventral mouth. The stomach can fit through an opening in the bivalve shells of less than 1 millimeter. Once insinuated between the shells, the stomach tissue secretes digestive enzymes that weaken the mollusk, causing it to open further. Partially digested food is transported to the upper portion of the stomach, where digestion is completed.

respiration, and food capture (Fig. 22-27). Seawater enters through an opening (the **sieve plate**) on the animal's upper surface and is conducted through a circular central canal, from which branch a number of radial canals. These canals conduct water to the tube feet, each of which is controlled by a muscular squeeze bulb (**ampulla**). Contraction of the bulb forces water into the tube foot, causing it to extend. The suction cup may be pressed against the seafloor or a food object, to which it adheres tightly until pressure is released.

Echinoderms have a relatively simple nervous system with no distinct brain. Movements are loosely coordinated by a system consisting of a nerve ring that encircles the esophagus, radial nerves to the rest of the body, and a nerve network through the epidermis. In sea stars, simple receptors for light and chemicals are concentrated on the arm tips, and sensory cells are scattered over the skin. Echinoderms lack a circulatory system, although movement of the fluid in their well-developed coelom serves this function. Gas exchange occurs through the tube feet and, in some forms, through numerous tiny "skin gills" that project through the epidermis. Most species have separate sexes and reproduce by shedding sperm and eggs into the water, where fertilization occurs.

Many echinoderms are able to regenerate lost body parts, and these regenerative powers are especially potent in sea stars. In fact, a single arm of a sea star is capable of developing into a whole animal, provided that part of the central body is attached to it. Before this ability was widely appreciated, a group of mussel fisherman tried to rid mussel beds of predatory sea stars by hacking them into pieces and throwing the pieces back. Needless to say, the strategy backfired.

The Tunicates, Lancelets, and Vertebrates

It's hard to believe that humans have anything in common with a sea squirt. Nonetheless, we share the phylum Chordata not only with the birds and the apes but also with the tunicates (sea squirts) and little fishy-looking creatures called *lancelets*. What characteristics unite us with these creatures that seem so different? All chordates share deuterostome development (which is also characteristic of echinoderms) and are further united by four features that all possess at some stage of their lives:

1. A **notochord**. A stiff but flexible rod that extends the length of the body and provides an attachment site for muscles.
2. A **dorsal, hollow nerve cord**. Lying dorsal to the digestive tract, this hollow neural structure develops a thickening at its anterior end that becomes a brain.
3. **Pharyngeal gill slits**. Located in the pharynx (the cavity behind the mouth), these may form functional respiratory openings or may appear only as grooves during an early stage of development.
4. A **post-anal tail**. An extension of the body past the anus.

This list may seem puzzling because, although humans are chordates, at first glance we seem to lack every feature except the nerve cord. But evolutionary relationships are sometimes seen most clearly during early stages of development; it is during our embryonic life that we develop, and lose, our notochord, our gill slits, and our tails (Fig. 22-28). Humans share these chordate features with all other vertebrates and with two invertebrate chordate groups, the lancelets and the tunicates.

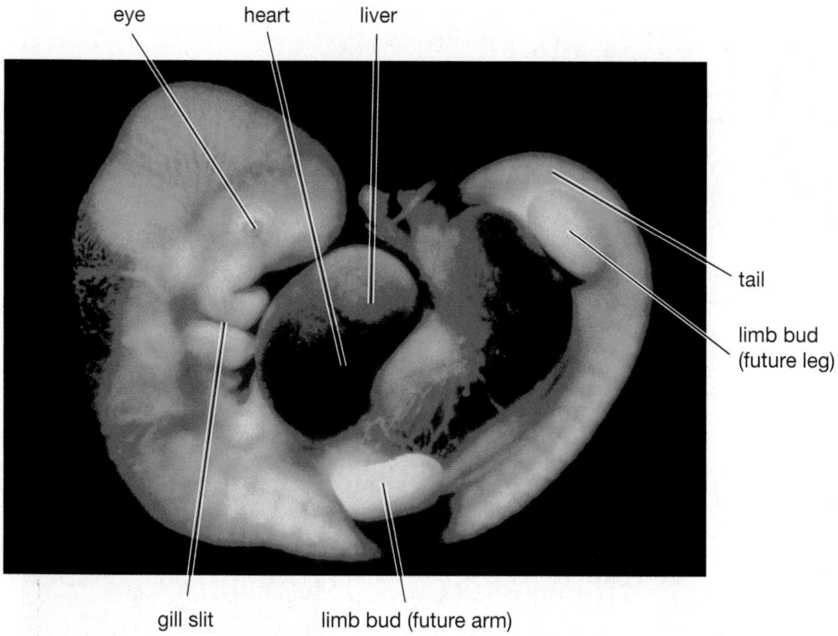

Figure 22-28 Chordate features in the human embryo
This 5-week-old human embryo is about 1 centimeter long and clearly shows external gill slits (more properly called grooves, since they do not penetrate the body wall) and a tail. Although the tail will disappear completely, the gill grooves contribute to the formation of the lower jaw.

Invertebrate Chordates

The invertebrate chordates lack the backbone that is the defining feature of vertebrates. These chordates comprise two groups of organisms, the *lancelets* and the *tunicates.* The small (2 inches, or about 5 centimeters, long) fishlike lancelet spends most of its time half-buried in the sandy sea bottom, filtering tiny food particles from the water. As can be seen in Figure 22-29a, all the typical chordate features are present in the adult lancelet.

The tunicates form a larger group of marine invertebrate chordates, which includes the sea squirts. It is difficult to imagine a less likely relative of humans than the sessile, filter-feeding, vaselike sea squirt (Fig. 22-29b). Its ability to move is limited to forceful contractions of its saclike body, which can send a jet of seawater into the face of anyone who plucks it from its undersea home; hence the common name sea squirt. Adult sea squirts may be immobile, but their larvae swim actively and possess all the diagnostic chordate features (see Fig. 22-29b).

Vertebrates

In vertebrates the embryonic notochord is normally replaced during development by a backbone, or **vertebral column**. The vertebral column is composed of bone or **cartilage**, a tissue that resembles bone but is less brittle and more flexible. This column supports the body, offers attachment sites for muscles, and protects the delicate nerve cord and brain. It is also part of a living endoskeleton that can grow and repair itself. Because this internal skeleton provides support without the armorlike weight of the arthropod exoskeleton, it has allowed vertebrates to achieve great size and mobility and has contributed to their invasion of the land and the air.

Vertebrates show other adaptations that have contributed to their successful invasion of most habitats. One is the presence of paired appendages. These first appeared as fins in fish and served as stabilizers for swimming. Over millions of years, some fins were modified by natural selection into legs that allowed animals to crawl onto dry land, and later into wings that allowed some to take to the air. Another adaptation that has contributed to the success of vertebrates is an increase in the size and complexity of their brains and sensory structures, which allow vertebrates to perceive their environment in detail and to respond to it in a great variety of ways.

The sequence of evolutionary events that led from invertebrate chordates to the first, fishlike vertebrates remains shrouded in mystery, because fossils of intermediate forms have never been discovered. Today, vertebrates are represented by jawless fishes, cartilaginous fishes, bony fishes, amphibians, reptiles, birds, and mammals.

Jawless Fishes The earliest vertebrate fossils are those of strange *jawless fishes* protected by bony armor plates. Today, two groups of jawless fishes survive: the *hagfishes* (class Myxini) and the *lampreys* (class Petromyzontiformes). Although both hagfishes and lampreys have eel-shaped bodies and smooth, unscaled skin, the two groups represent distinct, early branches of the chordate evolutionary tree. The branch leading to modern hagfishes is the more ancient of the two. The hagfish body is stiffened by a notochord, but the "skeleton" is limited to a few small cartilage elements, one of which forms a rudimentary braincase. Because hagfishes lack skeletal elements that surround and protect the nerve chord, most systematists do not consider them to be vertebrates. Instead, they represent the chordate group that is most closely related to the vertebrates.

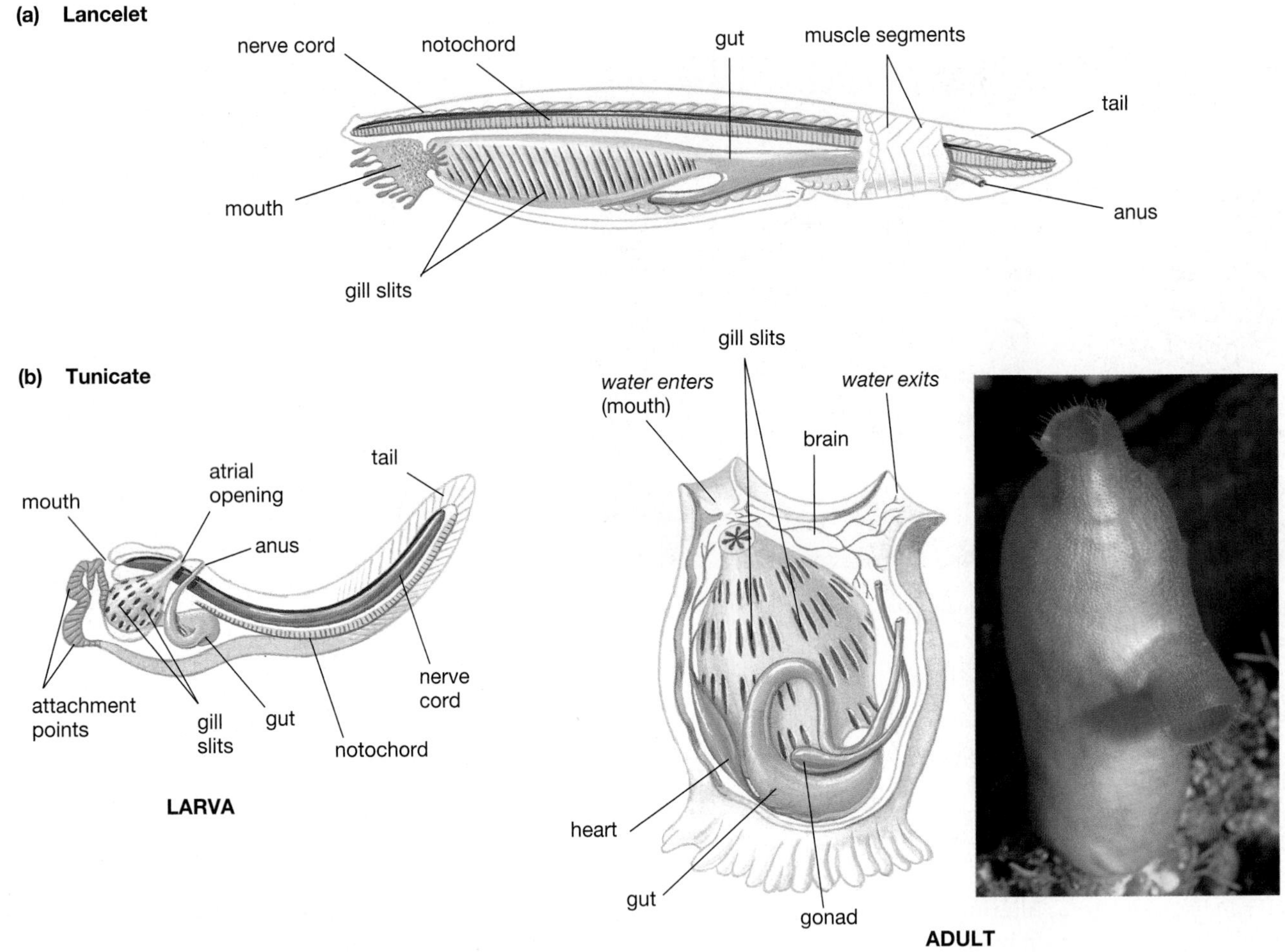

Figure 22-29 Invertebrate chordates
(a) A lancelet, a fishlike invertebrate chordate. The adult organism exhibits all the diagnostic features of chordates.
(b) The sea squirt larva (left) also exhibits all the chordate features. The adult sea squirt (a type of tunicate, middle) has lost its tail and notochord and has assumed a sedentary life, as shown in the photo (right).

Hagfishes are exclusively marine (Fig. 22-30a). They live near the ocean floor, often burrowing in the mud, and feed primarily on polychaete worms. They will, however, also eagerly attack dead and dying fish, using pincerlike teeth to burrow into their prey's body and consume the soft internal organs. Hagfish are regarded with great disgust by fishermen because they secrete massive quantities of slime as a defense against predators. Despite their well-deserved reputation as "slime balls of the sea," hagfishes are avidly pursued by many commercial fishermen, because the leather industry in some parts of the world provides a market for hagfish skin. Most leather items that purport to be "eel skin" are in fact made from tanned hagfish skin.

The spinal cord of lampreys is protected by segments of cartilage, so lampreys are considered to be true vertebrates. They live in both fresh and salt waters, but the marine forms must return to fresh water to spawn. Some lamprey species are parasitic. Parasitic lampreys have suckerlike mouths lined with teeth, which they use to attach themselves to larger fish (Fig. 22-30b). Using rasping teeth on its tongue, the lamprey excavates a hole in the host's body wall, through which it sucks blood and body fluids. Beginning in the 1920s, lampreys spread into the Great Lakes, where, in the absence of effective predators, they have multiplied prodigiously and greatly reduced commercial fish populations, including the lake trout. Vigorous measures to control the lamprey population have allowed some recovery of the other fish populations of the Great Lakes.

About 425 million years ago, in the mid-Silurian period, jawless fishes that were ancestors of the lampreys and hagfishes gave rise to a group of fish that possessed an important new structure: jaws. Jaws allowed fish to grasp and chew their food, permitting them to exploit a much wider range of food sources than could jawless fish. Although the earliest forms of jawed fishes have been extinct for 230 million years, they gave rise to the two major classes of jawed fishes that survive today: the cartilaginous fishes (class Chondrichthyes) and the bony fishes (class Osteichthyes).

(a)

(b)

Figure 22-30 Jawless fishes
(a) The colorful hagfishes live in communal burrows in mud, feeding on polychaete worms. ***(b)*** Some lampreys are parasitic, attaching to fish (such as this carp) with suckerlike mouths lined with rasping teeth (inset).

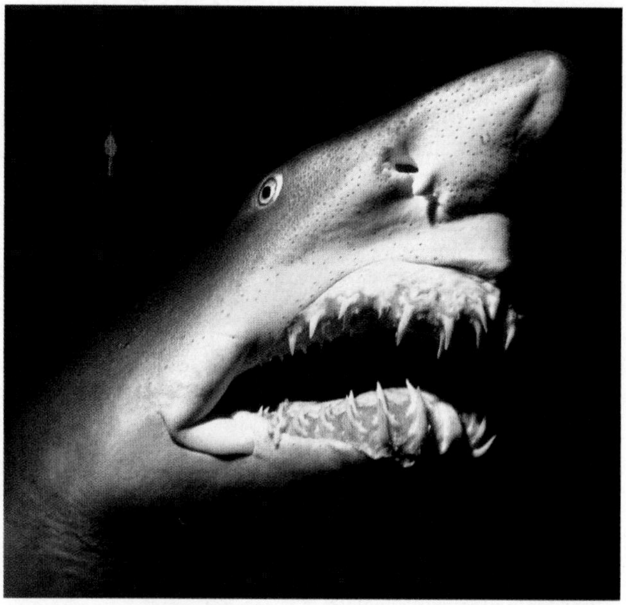

(a)

(b)

Figure 22-31 Cartilaginous fishes
(a) A sand tiger shark displaying several rows of teeth. As outer teeth are lost, they are replaced by new ones formed behind them. Both sharks and rays lack a swim bladder and tend to sink toward the bottom when they stop swimming. ***(b)*** The tropical blue-spotted sting ray swims by graceful undulations of lateral extensions of the body.

Cartilaginous Fishes The class Chondrichthyes, whose name is derived from Greek words meaning "cartilage fishes," includes 625 marine species, among them the sharks, skates, and rays (Fig. 22-31). The *cartilaginous fishes* are graceful predators that lack any bone in their skeleton, which is formed entirely of cartilage. The body is protected by a leathery skin roughened by tiny scales. Members of this group respire via gills. Although some must swim to circulate water through their gills, most can pump water across their gills. Like all fish, the cartilaginous fishes have a two-chambered heart.

Many shark species sport several rows of razor-sharp teeth, the back rows moving forward as the front teeth are lost to age and use (Fig. 22-31a). Although a few species consider us potential prey, most sharks are shy of humans. Skates and rays (Fig. 22-31b) are also retiring creatures, although some can inflict dangerous wounds with a spine near their tail, and others produce a powerful electrical shock that can stun their prey. Some cartilaginous fishes are very large. The whale shark, for example, can grow to more than 45 feet (15 meters) in length.

Bony Fishes Just as our size bias makes us overlook the most-diverse invertebrate groups, our habitat bias makes us overlook the most-diverse vertebrates. The most

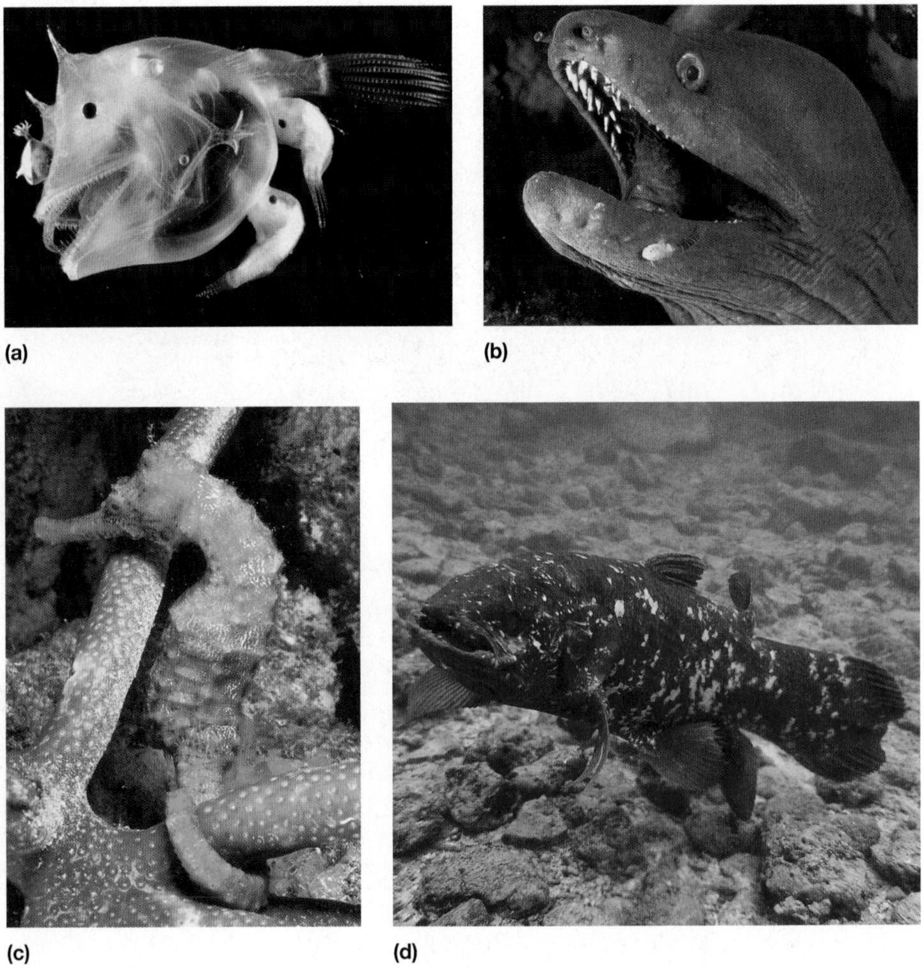

Figure 22-32 The diversity of the bony fishes
Bony fishes have colonized nearly every aquatic habitat. ***(a)*** This female deep-sea angler fish attracts prey with a living lure that projects just above her mouth. The fish is ghostly white because in the 2000-meter depth where anglers live, no light penetrates and thus colors are superfluous. Male deep-sea angler fish are extremely small and remain as permanent parasites attached to the female, always available to fertilize her eggs. Two parasitic males can be seen attached to this female. ***(b)*** This tropical green moray eel lives in rocky crevices. A small fish (a banded cleaner goby) on its lower jaw eats parasites that cling to the moray's skin. ***(c)*** The tropical seahorse may anchor itself with its prehensile tail (adapted for grasping) while the animal feeds on small crustaceans. ***(d)*** A rare photo of a coelacanth, a fish once believed to be long extinct, in its natural habitat off the coast of South Africa.

diverse and abundant vertebrates are not the birds or predominantly terrestrial mammals. The vertebrate diversity crown belongs instead to the lords of the oceans and fresh water, the *bony fishes* of the class Osteichthyes, whose skeletons are composed of bone rather than cartilage. From the snakelike moray eel to the bizarre, luminescent deep-sea forms to the streamlined tuna, this enormously successful group has spread to nearly every possible watery habitat, both freshwater and marine (Fig. 22-32). Although about 17,000 species have been identified, scientists predict that perhaps nearly twice this number may exist if the undescribed species from deep waters and remote areas are considered. The ocean's potential to conceal species was illustrated in dramatic fashion by the case of the *coelacanth*, a bony fish that had been known only from 75-million-year-old fossils. In 1939 coelacanths were found to be alive and well in deep water off the coast of South Africa (Fig. 22-32d). Since that initial discovery, coelacanths have also been found in other locations, including waters near Madagascar and Indonesia.

The ancestors of modern bony fish probably had lungs in addition to gills (the *swim bladder*, a sort of internal balloon that allows most bony fishes to float effortlessly at any level, probably evolved from these ancestral lungs). In addition, some groups of early bony fishes developed modified fleshy fins that could be used (in an emergency) as legs, dragging the fish from a drying puddle to a deeper pool. We know from fossils that at least one species even evolved actual limbs, although the function of limbs in a water-dwelling organism is not well understood. Such ancestors ultimately gave rise to the first vertebrates to make the first tentative invasion of the land: the amphibians.

Amphibians To those animals that first crawled from the water, life on land offered many advantages, including abundant food and freedom from aquatic predators. But the price was high. Deprived of water's support, the body was heavy and clumsy to drag along on modified fins or weak, poorly developed legs. Unsupported by water, gills collapsed and became useless. The dry air and relentless sun sucked vital water from unprotected skin and eggs, and terrestrial temperatures fluctuated dramatically compared with the relatively steady environmental conditions of the sea. Much of the evolutionary change that accompanied the transition to terrestrial life consisted of modifications that helped organisms overcome these challenges.

(a)

(b)

(c)

Figure 22-33 Amphibian means "double life"
The double life of amphibians is illustrated by the transition from ***(a)*** the completely aquatic larval tadpole to ***(b)*** the adult bullfrog leading a semiterrestrial life. ***(c)*** The red salamander is restricted to moist habitats in the eastern United States. Salamanders hatch in a form that closely resembles the adult.

The species of the class Amphibia straddle the boundary between aquatic and terrestrial existence (Fig. 22-33). The limbs of amphibians show varying degrees of adaptation to movement on land, from the belly-dragging crawl of salamanders to the long leaps of frogs and toads. A three-chambered heart (in contrast to the two-chambered heart of fishes) circulates blood more efficiently (as we shall see in Chapter 27), and lungs replace gills in most adult forms. Amphibian lungs, however, are poorly developed and must be supplemented by the skin, which serves as an additional respiratory organ. This respiratory function requires that the skin remains moist, a constraint that greatly restricts the range of amphibian habitats on land.

Amphibians are also tied to moist habitats by their breeding behavior, which requires water. Their fertilization is normally external and must therefore occur in water so that the sperm can swim to the eggs. The eggs must remain moist, as they are protected only by a jelly-like coating that leaves them vulnerable to loss of water by evaporation. The mechanics of keeping the eggs moist varies considerably among different amphibian species, but in many species the requisite moisture is secured by the simple expedient of laying the eggs in water. In some amphibian species, fertilized eggs develop into aquatic larvae such as the tadpoles of some frogs and toads. The dramatic transformation of these aquatic larvae into semiterrestrial adults gives the class Amphibia its name, which means "double life." Their double life and their thin, permeable skin have made amphibians particularly vulnerable to pollutants and to environmental degradation, as described in "Earth Watch: Frogs in Peril."

Reptiles and Birds The reptiles include the lizards and snakes (by far the most successful of the modern groups) and the turtles, alligators, and crocodiles (Fig. 22-34). Reptiles evolved from an amphibian ancestor about 250 million years ago. Early reptiles—the dinosaurs—ruled the land for nearly 150 million years.

Some reptiles, particularly desert dwellers such as tortoises and lizards, are completely independent of their aquatic origins. This independence was achieved through a series of adaptations, of which three are outstanding: (1) Reptiles evolved a tough, scaly skin that resists water loss and protects the body. (2) Reptiles evolved internal fertilization, in which the male deposits sperm within the female's body. (3) Reptiles evolved a shelled **amniote egg**, which can be buried in sand or dirt, far from water with its hungry predators. The shell prevents the egg from drying out on land. An internal membrane, the **amnion**, encloses the embryo in the watery environment that all developing animals require (Fig. 22-35, p. 457).

In addition to these features, reptiles have more-efficient lungs than do earlier vertebrates, dispensing with the skin as a respiratory organ. The three-chambered heart became modified, allowing better separation of oxygenated and deoxygenated blood, and the limbs and skeleton evolved features that provided better support and more-efficient movement on land.

Earth Watch
Frogs in Peril

Frogs and toads have frequented Earth's ponds and swamps for nearly 150 million years and somehow survived the Cretaceous catastrophe that extinguished the dinosaurs and so many other species about 65 million years ago. Their evolutionary longevity, however, appears to offer inadequate defense against the environmental changes wrought by human activities. Over the past decade, *herpetologists* (biologists who study reptiles and amphibians) from around the world have documented an alarming decline in amphibian populations. Thousands of species of frogs, toads, and salamanders are experiencing a dramatic decrease in numbers, and many have apparently gone extinct.

The phenomenon is not a local one; population crashes have been reported from every part of the globe. Yosemite toads and yellow-legged frogs are disappearing from the mountains of California; tiger salamanders have been nearly wiped out in the Colorado Rockies. Leopard frogs, eagerly chased by rural children, are becoming rare in the United States. Logging destroys the habitats of amphibians from the Pacific Northwest to the Tropics (Fig. E22-1), but even amphibians in protected areas are dying. In the Monteverde Cloud Forest Preserve in Costa Rica, the golden toad (see Fig. 35-6) was common in the early 1980s but has not been seen since 1989. The gastric brooding frog of Australia fascinated biologists by swallowing its eggs, brooding them in its stomach, and later regurgitating fully formed offspring. This species was abundant and seemed safe in a national park. Suddenly, in 1980, the gastric brooding frog disappeared and hasn't been seen since.

The causes of the worldwide decline in amphibian diversity are poorly understood, but researchers have recently discovered that frogs and toads in many places are succumbing to infection by a pathogenic fungus. The fungus has been found in the skin of dead and dying frogs in widespread locations, including Australia, Central America, and the western United States. In those places, discovery of the fungus has coincided with massive frog and toad die-offs, and most herpetologists agree that the fungus is causing the deaths. It seems unlikely, however, that the fungus alone is responsible for the worldwide decline of amphibians. For one thing, die-offs have occurred in many places where the fungus has not been found. In addition, many herpetologists believe that the fungal epidemic would not have arisen if the frogs and toads had not

Figure E22-1 Amphibians in danger
The corroboree toad, shown here with its eggs, is rapidly declining in its native Australia. Tadpoles are developing within the eggs. The thin water-permeable and gas-permeable skin of the adult and the jellylike coating around the eggs make them vulnerable to both air and water pollutants.

(a) (b) (c)

Figure 22-34 The diversity of reptiles
(a) The mountain king snake has a color pattern very similar to that of the poisonous coral snake, which potential predators avoid. This mimicry helps the harmless king snake elude predation. ***(b)*** The outward appearance of the American alligator, found in swampy areas of the South, is almost identical to that of 150-million-year-old fossil alligators. ***(c)*** The tortoises of the Galapagos Islands, Ecuador, may live to be more than 100 years old.

first been weakened by other stresses. So, if the fungus is not doing all of the damage on its own, what are the other possible causes of amphibian decline? All of the most likely causes stem from human modification of the biosphere—the portion of Earth that sustains life.

Habitat destruction, especially the draining of wetland habitats that are especially hospitable to amphibian life, is one major cause of the decline. The unique biology of amphibians also makes them especially vulnerable to toxic substances in the environment. In particular, amphibian bodies at all stages of life are protected only by a thin, permeable skin through which pollutants can easily penetrate. To make matters worse, the double life of many amphibians exposes their permeable skin to a wide range of aquatic and terrestrial habitats and to a correspondingly wide range of environmental toxins. For example, many temperate amphibians' eggs and larvae develop in ponds and streams during the spring, a time when the melting of snow and ice formed from acid rain causes a pulse of intense acidity in freshwater ecosystems. In addition, many amphibians are herbivorous as larvae and insectivorous as adults, so they're exposed to both herbicides and insecticides in their food as their diet shifts.

Amphibian eggs can also be damaged by ultraviolet (UV) light, according to recent research by Andrew Blaustein, an ecologist at Oregon State University. Blaustein demonstrated that the eggs of some species of frogs in the Pacific Northwest are sensitive to damage from UV light and that the most sensitive species are experiencing the most drastic declines. Unfortunately, many parts of Earth are subject to increasingly intense UV radiation levels, because atmospheric pollutants have caused a thinning of the protective ozone layer (as we shall see in Chapter 40).

Increased UV radiation may also be tied to another disturbing trend among frogs and toads, the increasing incidence of grotesquely deformed individuals. The deformities seem to result from environmental rather than genetic causes, as they occur in a variety of species and in widely separated geographic areas. Some researchers think the deformities may be caused by newly developed pesticides that have been applied on a large scale in recent years. Other biologists, including Stanley Sessions of Hartwick College, cite evidence suggesting that some deformities—especially the most common one, extra limbs—are caused by parasitic infections during embryonic development. Sessions has found that many frogs with extra legs are infested with a parasitic flatworm, and other researchers have shown that tadpoles experimentally infected with the flatworm developed into deformed adults. What's the tie-in with UV light? Well, UV radiation may harm frogs by suppressing the immune response. In Sessions's view, a compromised immune system could leave a developing tadpole more vulnerable to parasitic infection.

Many scientists believe that the troubles of amphibians signal an overall deterioration of Earth's ability to support life. According to this line of reasoning, the highly sensitive amphibians are providing an early warning of environmental degradation that will eventually affect more-resistant organisms as well. Equally worrisome is the observation that amphibians are not just sensitive indicators of the health of the biosphere; they are also crucial components of many ecosystems. They may keep insect populations in check, in turn serving as food for larger carnivores. Their decline will further disrupt the balance of these delicate communities. Margaret Stewart, an ecologist at the State University of New York, Albany, aptly summarized the problem: "There's a famous saying among ecologists and environmentalists: 'Everything is related to everything else. '. . . You can't wipe out one large component of the system and not see dramatic changes in other parts of the system."

Figure 22-35 The amniote egg
An anole lizard struggles free of its egg. The amniote egg encapsulates the developing embryo in a liquid-filled membrane (the amnion), ensuring that development occurs in a watery environment, even if the egg is far from water.

(a) (b) (c)

Figure 22-36 The diversity of birds
(a) The delicate hummingbird beats its wings about 60 times per second and weighs about 0.15 ounce (4 grams). ***(b)*** This young frigate bird, a fish-eater from the Galapagos Islands, has nearly outgrown its nest. ***(c)*** The ostrich is the largest of all birds, weighing more than 300 pounds (136 kilograms); its eggs weigh more than 3 pounds (1500 grams).

One very distinctive group of "reptiles" comprises the birds (Fig. 22-36). Although birds have traditionally been classified as a group separate from reptiles (class Aves *versus* class Reptilia), modern systematists have shown that birds are really a subset of an evolutionary group that includes both birds and the groups that have been traditionally designated as reptiles (see page 359 in Chapter 18 for a fuller explanation). The first birds appear in the fossil record roughly 150 million years ago (Fig. 22-37) and are distinguished from other reptiles by the presence of feathers, which are essentially a highly specialized version of reptilian body scales. Modern birds retain scales on their legs—a testimony to the ancestry they share with the rest of the reptiles.

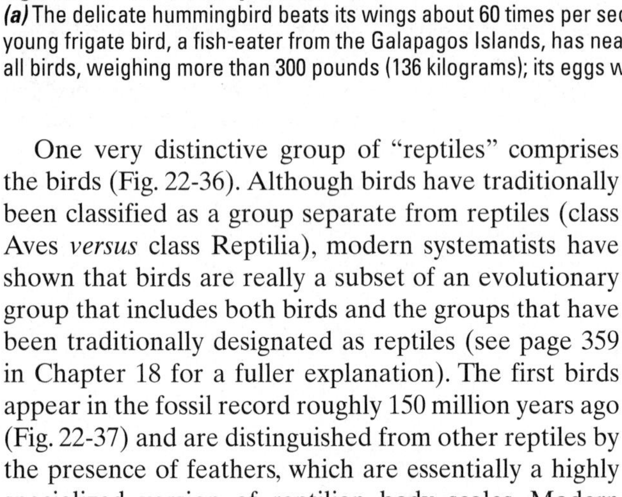

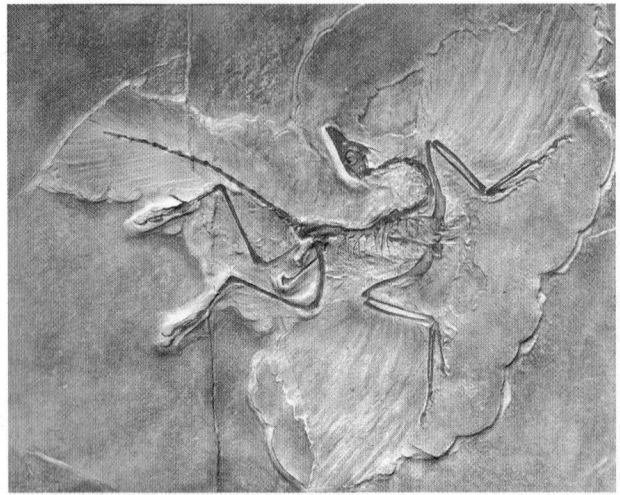

Figure 22-37 *Archaeopteryx,* the "missing link" between reptiles and birds
The earliest known bird is *Archaeopteryx,* preserved here in 150-million-year-old limestone. Feathers, a feature unique to birds, are clearly visible, but the reptilian ancestry of birds is also apparent: Like a modern reptile (but unlike a modern bird), *Archaeopteryx* had teeth, a tail, and claws.

Bird anatomy and physiology are dominated by adaptations that help meet the rigorous demands of flight. In particular, birds are exceptionally light for their size. Hollow bones reduce the weight of the skeleton to a fraction of that of other vertebrates, and many bones present in other reptiles are lost or fused with other bones in birds. Reproductive organs are considerably reduced in size during nonbreeding periods, and female birds possess only a single ovary, further minimizing weight. The shelled egg that contributed to the reptiles' success on land frees the mother bird from carrying her developing offspring internally. Feathers form lightweight extensions to the wings and the tail for the lift and control required for flight, and they also provide lightweight protection and insulation for the body. The nervous system of birds accommodates the special demands of flight with extraordinary coordination and balance combined with acute eyesight.

Birds are also able to maintain body temperatures high enough to allow their muscles and metabolic processes to operate at peak efficiency, supplying the power necessary to fly regardless of the outside temperature. This physiological ability to maintain an internal temperature that is usually higher than that of the surrounding environment is characteristic of both birds and mammals, which are therefore sometimes described as *warm-blooded.* In contrast, the body temperature of invertebrates, fish, amphibians, and reptiles varies with the temperature of their environment, though these animals may exert some control of their body temperature by behavioral means (such as basking in the sun or seeking shade).

Warm-blooded animals such as birds have a high metabolic rate, which increases their demand for energy and requires efficient oxygenation of tissues. Birds thus must eat frequently and possess circulatory and respiratory adaptations that help meet the need for efficiency. The heart of a bird has four chambers, serving to prevent the mixing of oxygenated and deoxygenated blood (alligators and crocodiles also have four-chambered hearts). The respiratory system of birds is supplemented by air sacs that provide a continuous supply of oxygenated air to the lungs, even while the bird exhales.

Mammals One branch of the reptile evolutionary tree gave rise to a group that evolved hair and diverged to form the mammals. The mammals first appeared approximately 250 million years ago but did not diversify and assume terrestrial prominence until after the dinosaurs went extinct roughly 65 million years ago. Like birds, mammals are warm-blooded and have high metabolic rates. In most mammals, fur protects and insulates the warm body. Like birds, alligators, and crocodiles, mammals have four-chambered hearts that increase the amount of oxygen delivered to the tissues. Legs designed for running rather than crawling make many mammals fast and agile. In contrast to birds, whose bodies are almost uniformly molded to the requirements of flight, mammals have evolved a remarkable diversity of form. The bat, mole, impala, whale, seal, monkey, and cheetah exemplify the radiation of mammals into nearly all habitats, with bodies finely adapted to their varied lifestyles (Fig. 22-38).

Mammals are named for the milk-producing **mammary glands** used by all female members of this class to suckle their young (Fig. 22-38c). In addition to these unique glands, the mammalian body is arrayed with sweat, scent, and sebaceous (oil-producing) glands, none of which are found in reptiles. With the exception of the egg-laying **monotremes**, such as the platypus and spiny anteater (Fig. 22-39a), mammalian embryos develop in the *uterus*, a muscular organ in the female reproductive tract. In one group of mammals, the **marsupials** (including opossums, koalas, and kangaroos), the period of uterine development is short and the young are born at a very immature stage of development. Immediately after birth they crawl to a nipple, firmly grasp it, and, nourished by milk, complete their development. In most, but not all, marsupial species, this post-birth development takes place in a protective pouch (Fig. 22-39b). Most

Figure 22-38 The diversity of mammals
(a) A humpback whale gives its offspring a boost. ***(b)*** A bat, the only mammal capable of true flight, navigates at night by using a kind of sonar. Large ears aid in detecting echoes as its high-pitched cries bounce off nearby objects. ***(c)*** Mammals are named after the mammary glands with which females nurse their young, as illustrated by this mother cheetah. ***(d)*** The male orangutan can reach 75 kilograms (165 pounds). These gentle, intelligent apes occupy swamp forests in limited areas of the Tropics. They are endangered by hunting and habitat destruction.

(a) (b)

Figure 22-39 Nonplacental mammals
(a) Monotremes, such as this platypus from Australia, lay leathery eggs resembling those of reptiles. The newly hatched young obtain milk from slitlike openings in the mother's abdomen. ***(b)*** Marsupials, such as the wallaby, give birth to extremely immature young who immediately grasp a nipple and develop within the mother's protective pouch (inset).

mammal species are **placental** mammals and retain their young in the uterus for a much longer period. *Placental* refers to the **placenta**, the uterine structure that functions in gas, nutrient, and waste exchange between the circulatory systems of the mother and embryo.

The mammalian nervous system has contributed significantly to the success of the mammals by making possible behavioral adaptation to changing and varied environments. The brain is more highly developed than in any other class, endowing mammals with unparalleled curiosity and learning ability. The highly developed brain allows mammals to alter their behavior on the basis of experience and helps them survive in a changing environment. Relatively long periods of parental care after birth allow some mammals to learn extensively under parental guidance; humans and other primates are good examples. In fact, the large brains of humans have been the major factor leading to human domination of Earth, as explored in the following section.

Evolutionary Connections

Are Humans a Biological Success?

Physically, human beings are fairly unimpressive biological specimens. For such large animals, we are not very strong or very fast, and we lack the natural weapons of fang and claw. It is the human brain, with its tremendously developed cerebral cortex, that truly sets us apart from other animals. Our brains give rise to our minds, which, in bursts of solitary brilliance and in the collective pursuit of common goals, have created wonders. No other animal could sculpt the graceful columns of the Parthenon, much less reflect on the beauty of this ancient Greek temple. We alone can eradicate smallpox and polio, domesticate other life-forms, penetrate space with rockets, and fly to the stars in our imaginations.

And yet, are we, as it appears at first glance, the most successful of all living things? The duration of human existence is a mere instant in the 3.5-billion-year span of life on Earth. But during the last 300 years, the human population has increased from 0.5 billion to 6 billion and now grows by a million people every 4 days. Is this a measure of our success? As we have expanded our range over the globe, we have driven at least 300 major species to extinction. Within your lifetime, the rapid destruction of tropical rain forests and other diverse habitats may wipe out millions of species of plants, invertebrates, and vertebrates, most of which we will never know. Many of our activities have altered the environment in ways that are hostile to life, including our own. Acid from power plants and automobiles rains down on the land, threatening our forests and lakes—

and eroding the marble of the Parthenon. Deserts spread as land is stripped by overgrazing and the demand for firewood. Our aggressive tendencies, spurred by pressures of expanding wants and needs, their scope magnified by our technological prowess, have given us the capacity to destroy ourselves and most other lifeforms as well.

The human mind is the source of our most pressing problems—and our greatest hope for solving them. Will we now devote our mental powers to reducing our impact, controlling our numbers, and preserving the ecosystems that sustain us and other forms of life? Are we a phenomenal biological success—or a brilliant catastrophe? Perhaps the next few centuries will tell.

CASE STUDY REVISITED

The Search for a Sea Monster

Clyde Roper's search for the giant squid has led him to organize three major expeditions over the past few years. The first of these began in 1996, off the Azores Islands in the Atlantic Ocean. Because sperm whales are known to prey upon giant squids, Roper believed that the whales might lead him to the squids. To test this idea, he and his team affixed video cameras to sperm whales, so that the scientists could see what the whales were seeing. The "crittercams" revealed a great deal of new information about sperm whale behavior but, alas, no footage of giant squids.

The next Roper-led expedition took place the following year in the Kaikoura Canyon, an area of very deep water (3300 feet, or 1000 meters) off the coast of New Zealand. This spot was chosen because deep-sea fishing boats had recently captured several giant squid in the vicinity. Cameras were again deployed on sperm whales, but this time the mobile cameras were supplemented by a stationary, baited camera and an unmanned, remote-controlled submarine. Again, however, a large investment of time, money, and equipment yielded no squid sightings.

In 1999 Roper assembled a team of scientists for a return to Kaikoura Canyon. This time, the group was able to use Deep Rover, a one-person submarine that could carry an observer to depths of 2200 feet. The scientists used Deep Rover to explore the canyon, following sperm whales in hopes that the huge mammals would lead them to giant squid.

Despite 30 days of trying, the expedition team once again failed to glimpse a giant squid. They did amass a wealth of data on a virtually unknown deep-sea ecosystem, but gathered no new clues about the mysterious "sea monster." The life of the giant squid remains a mystery. Says Dr. Roper, "We probably know more about the dinosaurs than about the giant squid."

If you were in charge of the giant squid project, what method would you suggest for finding and observing the species in its natural habitat?

Summary of Key Concepts

1) What Characteristics Define an Animal?

Animals are multicellular, sexually reproducing, heterotrophic organisms. Most can perceive and react rapidly to environmental stimuli and are motile at some stage in their lives. Their cells lack a cell wall.

2) Which Anatomical Features Mark Branch Points on the Animal Evolutionary Tree?

The earliest animals had no tissues, a feature retained by modern sponges. All other modern animals have tissues. Animals with tissues can be divided into radially symmetrical and bilaterally symmetrical groups. During embryonic development, radially symmetrical animals have two germ layers, bilaterally symmetrical animals have three. Bilaterally symmetrical animals also tend to have sense organs and clusters of neurons concentrated in the head, a process called *cephalization*. Some phyla of bilaterally symmetrical animals lack body cavities, but most have pseudocoeloms or true coeloms. Coelomate animals can be divided into two main groups, one of which undergoes protostome development, the other of which undergoes deuterostome development.

3) What Are the Major Animal Phyla?

The bodies of sponges (phylum Porifera) are typically free-form in shape and are sessile. Sponges have relatively few types of cells. Despite the division of labor among the cell types, there is little coordination of activity. Sponges lack the muscles and nerves required for coordinated movement, and digestion occurs exclusively within the individual cells.

The hydra, anemones, and jellyfish (phylum Cnidaria) have tissues. A simple network of nerve cells directs the activity of contractile cells, allowing loosely coordinated movements. Digestion is extracellular, occurring in a central gastrovascular cavity with a single opening. Cnidarians exhibit radial symmetry, an adaptation to both the free-floating lifestyle of the medusa and the sedentary existence of the polyp.

Flatworms (phylum Platyhelminthes) have a distinct head with sensory organs and a simple brain. A system of canals forming a network through the body aids in excretion. They lack a body cavity.

The pseudocoelomate roundworms (phylum Nematoda) possess a separate mouth and anus.

The segmented worms (phylum Annelida) are the most complex of the worms, with a well-developed closed circulatory system and excretory organs that resemble the basic unit of the vertebrate kidney. The segmented worms have a compartmentalized digestive system, like that of vertebrates, which processes food in a sequence. Annelids also have a true coelom, a fluid-filled space between the body wall and the internal organs.

Arthropods, the insects, arachnids, and crustaceans (phylum Arthropoda), are the most diverse and abundant organisms on Earth. They have invaded nearly every available terrestrial and aquatic habitat. Jointed appendages and well-developed nervous systems make possible complex, finely coordinated behavior. The exoskeleton (which conserves water and provides support) and specialized respiratory structures (which remain moist and protected) enable the insects and arachnids to inhabit dry land. The diversification of insects has been enhanced by their ability to fly. Crustaceans, which include the largest arthropods, are restricted to moist, usually aquatic habitats and respire by using gills.

The snails, clams, and squid (phylum Mollusca) lack a skeleton; some forms protect the soft, moist, muscular body with a single shell (many gastropods and a few cephalopods) or a pair of hinged shells (the bivalves). The lack of a waterproof external covering limits this phylum to aquatic and moist terrestrial habitats. Although the body plan of gastropods and bivalves limits the complexity of their behavior, the cephalopod's tentacles are capable of precisely controlled movements. The octopus has the most complex brain and the best-developed learning capacity of any invertebrate.

The sea stars, sea urchins, and sea cucumbers (phylum Echinodermata) are an exclusively marine group. Like other complex invertebrates and chordates, echinoderm larvae are bilaterally symmetrical; however, the adults show radial symmetry. This, in addition to a primitive nervous system that lacks a definite brain, adapts them to a relatively sedentary existence. Echinoderm bodies are supported by a nonliving internal skeleton that sends projections through the skin. The water-vascular system, which functions in locomotion, feeding, and respiration, is a unique echinoderm feature.

The phylum Chordata includes two invertebrate groups, the lancelets and tunicates, as well as the familiar vertebrates. All chordates possess a notochord, a dorsal hollow nerve cord, pharyngeal gill slits, and a post-anal tail at some stage in their development. Vertebrates are a subphylum of chordates that have a backbone, which is part of a living endoskeleton. Vertebrate evolution is believed to have proceeded from the fishes to amphibians to reptiles, which gave rise to both birds and mammals. The heart increases in complexity from the two chambers in fishes, to three in amphibians and most reptiles, to four in the warm-blooded birds and mammals. The evolution of efficient transport of oxygen supported active lifestyles, high metabolic rates, and constant body temperatures.

During the progression from fishes to amphibians to reptiles, a series of adaptations evolved that helped vertebrates colonize dry land. All amphibians have legs, and most have simple lungs for breathing in air rather than in water. Most are confined to relatively damp terrestrial habitats by their need to keep their skin moist, use of external fertilization, and requirement that their eggs and larvae develop in water. Reptiles, with well-developed lungs, dry skin covered with relatively waterproof scales, internal fertilization, and the amniote egg with its own water supply, are well adapted to the driest terrestrial habitats. Birds and mammals are also fully terrestrial and have additional adaptations, such as an elevated body temperature, which allows the muscles to respond rapidly regardless of the temperature of the environment. The bird body is molded for flight, with feathers, hollow bones, efficient circulatory and respiratory systems, and well-developed eyes. Mammals have insulating hair and give birth to live young that are nourished with milk. The mammalian nervous system is the most complex in the animal kingdom, providing mammals with enhanced learning ability that helps them adapt to changing environments.

Key Terms

abdomen *p. 443*
amnion *p. 455*
amniote egg *p. 455*
amoeboid cell *p. 434*
ampulla *p. 450*
anterior *p. 431*
bilateral symmetry *p. 430*
book lung *p. 444*
budding *p. 434*
cartilage *p. 451*
cephalization *p. 431*
chitin *p. 443*
closed circulatory system *p. 441*
cnidocyte *p. 436*
coelom *p. 431*
collar cell *p. 434*
compound eye *p. 444*
cyst *p. 438*
deuterostome *p. 432*
dorsal *p. 430*
ectoderm *p. 430*
endoderm *p. 430*
endoskeleton *p. 448*
epithelial cell *p. 434*
exoskeleton *p. 443*
free-living *p. 438*
ganglion *p. 438*
gastrovascular cavity *p. 436*
germ layer *p. 430*
gill *p. 444*
head *p. 443*
hemocoel *p. 444*
hermaphroditic *p. 440*
hydrostatic skeleton *p. 440*
invertebrate *p. 432*
larva *p. 444*
mammary gland *p. 459*
mantle *p. 447*
marsupial *p. 459*
medusa *p. 435*
mesoderm *p. 430*
mesoglea *p. 435*
metamorphosis *p. 444*
molt *p. 443*
monotreme *p. 459*
nephridium *p. 441*
nerve cord *p. 438*
nerve net *p. 435*
notochord *p. 450*
open circulatory system *p. 444*
parasite *p. 438*
pharyngeal gill slit *p. 450*
pharynx *p. 441*
placenta *p. 460*
placental *p. 460*
polyp *p. 435*
post-anal tail *p. 450*
posterior *p. 431*
protostome *p. 432*
pseudocoelom *p. 431*
pupa *p. 444*
radial symmetry *p. 430*
radula *p. 447*
segmentation *p. 441*
sessile *p. 431*
sieve plate *p. 450*
spicule *p. 434*
tentacle *p. 435*
thorax *p. 443*
tissue *p. 430*
trachea *p. 444*
tube foot *p. 449*
ventral *p. 430*
vertebral column *p. 451*
vertebrate *p. 432*
water-vascular system *p. 449*

Thinking Through the Concepts

Multiple Choice

1. *Which of the following animals have radial symmetry?*
   a. jellyfish
   b. sea star
   c. sea anemone
   d. sea urchin
   e. all of the above
2. *Animals in which of the following phyla have collar cells?*
   a. Porifera
   b. Cnidaria
   c. Annelida
   d. Gastropoda
   e. Chordata
3. *What is a radula?*
   a. a flexible supportive rod on the dorsal surface of chordates
   b. a stinging cell used by sea anemones to capture prey
   c. a gas-exchange structure in insects
   d. a spiny ribbon of tissue used for feeding in snails
   e. a locomotory structure of sea stars
4. *Which of the following groups of animals includes the first vertebrates to appear on Earth?*
   a. the jawless fishes
   b. the cartilaginous fishes
   c. the bony fishes
   d. the cephalopods
   e. none of the above
5. *Which of the following animals molts its exoskeleton, allowing the animal to grow larger?*
   a. the blue crab
   b. the bat sea star
   c. the scallop
   d. the sea urchin
   e. the Venus clam
6. *Which of the following has a digestive cavity with a single opening?*
   a. sponges
   b. sea stars
   c. nematodes
   d. flatworms
   e. earthworms

? Review Questions

1. List the distinguishing characteristics of each of the phyla discussed in this chapter, and give an example of each.
2. Briefly describe each of the following adaptations, and explain the adaptive significance of each: amniote egg, bilateral symmetry, cephalization, closed circulatory system, coelom, placenta, radial symmetry, segmentation.
3. Describe and compare respiratory systems in the three major arthropod classes.
4. Describe the advantages and disadvantages of the arthropod exoskeleton.
5. State in which of the three major mollusk classes each of the following characteristics is found:
   a. two hinged shells
   b. a radula
   c. tentacles
   d. some sessile members
   e. the best-developed brains
   f. numerous eyes
6. Give three functions of the water-vascular system of echinoderms.
7. To what lifestyle is radial symmetry an adaptation? Bilateral symmetry?
8. List the vertebrate groups that have each of the following:
   a. a skeleton of cartilage
   b. a two-chambered heart
   c. the amniote egg
   d. warm-bloodedness
   e. a four-chambered heart
   f. a placenta
   g. lungs supplemented by air sacs
9. Distinguish between vertebrates and invertebrates. List the major phyla found in each broad grouping.
10. List four distinguishing features of chordates.
11. Describe the ways in which amphibians are adapted to life on land. In what ways are amphibians still restricted to a watery or moist environment?
12. List the adaptations that distinguish reptiles from amphibians and help reptiles adapt to life in dry terrestrial environments.
13. List the adaptations of birds that contribute to their ability to fly.
14. How do mammals differ from birds, and what adaptations do they share?
15. How has the mammalian nervous system contributed to the success of mammals?

Applying the Concepts

1. The class Insecta is the largest taxon of animals on Earth. Its greatest diversity is in the Tropics, where habitat destruction and species extinction are occurring at an alarming rate. What biological, economic, and ethical arguments can you advance to persuade people and governments to preserve this biological diversity?
2. Animals, particularly laboratory rats, are used extensively in medical research. What advantages and disadvantages do lab rats have as research subjects in medicine? What are the advantages and disadvantages of using animals more closely related to humans?
3. Discuss at least three ways in which the ability to fly has contributed to the success and diversity of insects.
4. Discuss and defend what attributes you would use to define biological success among animals. Are humans a biological success by these standards? Why or why not?

For More Information

Blaustein, A. "Amphibians in a Bad Light." *Natural History*, October 1994. Recent declines in amphibian population size and overall diversity are linked to possible harm from ultraviolet light that is penetrating a depleted ozone layer.

Brusca, R. C., and Brusca, G. J. *Invertebrates*. Sunderland, MA: Sinauer, 1990. A thorough survey of the invertebrate animals in textbook format but readable and filled with beautiful and informative drawings.

Chadwick, D. H. "Planet of the Beetles." *National Geographic*, March 1998. The beauty and diversity of beetles, which comprise one-third of the world's insects, are described in text and photographs.

Diamond, J. "Stinking Birds and Burning Books." *Natural History*, October 1994. A description of a recently described species of bird (the pitohuis) and its peculiar chemical ecology.

Duellman, W. E. "Reproductive Strategies of Frogs." *Scientific American*, July 1992. Free-living tadpoles are only one way in which these amphibians progress from egg to adult.

Hamner, W. "A Killer Down Under." *National Geographic*, August 1994. Among the most poisonous animals in the world is the box jellyfish, which lives off the coast of northern Australia.

Horridge, G. A. "The Compound Eye of Insects." *Scientific American*, July 1977. A close look at the multifaceted eyes of insects.

Luoma, J. R. "Vanishing Frogs." *Audubon*, May–June 1997. Brief summary of declining amphibian populations, accompanied by striking photographs of some of the affected frog species.

McMenamin, M. A. S., and McMenamin, D. L. S. *The Emergence of Animals: The Cambrian Breakthrough.* New York: Columbia University Press, 1990. A look at the adaptive radiation that led to an explosion of animal forms at the beginning of the Cambrian period.

Montgomery, S. "New Terror of the Deep." *International Wildlife*, July–August 1992. A description of the threat that overfishing by humans poses to shark populations.

Morell, V. "Life on a Grain of Sand." *Discover*, April 1995. The sand beneath shallow waters is home to an incredible range of microscopic creatures.

Rahn, H., Ar, A., and Paganelli, C. V. "How Bird Eggs Breathe." *Scientific American*, February 1979. (Offprint No. 1420). A description of the remarkable adaptations of the amniote egg for gas exchange.

Rennie, J. "Living Together." *Scientific American*, January 1992. Most parasites do not kill their hosts, and the interaction of parasite and host provides some fascinating insights into the evolution of life on Earth.

Answers to Multiple-Choice Questions

1. e 2. a 3. d 4. a 5. a 6. d

MEDIATUTOR
The Animal Kingdom

CD Activities

Activity 22.1: Cnidocytes

Estimated time: 5 minutes

Cnidocytes are specialized cells unique to and characteristic of all cnidarians—jellies, sea anemones, corals, and their relatives. Cnidocytes function in diverse ways, such as obtaining food, defense, locomotion, and attachment. The mechanism by which cnidocytes operate provides an interesting example of how a common mechanism—osmosis—is used in an uncommon way, to deploy a physical and chemical defense and predation system.

Activity 22.2: Insect Metamorphosis

Estimated time: 5 minutes

Insects are perhaps the most successful group of organisms on Earth in terms of species diversity and sheer numbers. One key to the success of insects may be their ability to undergo metamorphosis, a profound change in body form from the juvenile to the adult stage. In this simple animation, the metamorphosis of a periodical cicada is demonstrated as a nymph, having lived underground for 17 years while feeding on tree roots, changes into a winged adult, which will fly, feed on tree sap, mate, and die within a few weeks.

Start the MediaTutor Student CD-ROM and enter the activity number in the Quick Search box to be taken directly to that activity.

Web Investigations

Case Study: The Search for a Sea Monster

Estimated time: 10 minutes

Many fascinating creatures can be found in the oceans' depths. This exercise looks at the relatively limited information available about a few, better-known species.

Go to http://www.prenhall.com/audesirk6, the Audesirk Companion Web site. Select Chapter 22 and Web Investigation to begin.

23 Plant Form and Function

AT A GLANCE

Case Study: A Beautiful Death Trap

1) **How Are Plant Bodies Organized, and How Do They Grow?**

Flowering Plants Consist of a Root System and a Shoot System

During Plant Growth, Meristem Cells Give Rise to Differentiated Cells

2) **What Are the Tissues and Cell Types of Plants?**

The Dermal Tissue System Forms the Covering of the Plant Body

The Ground Tissue System Makes Up Most of the Young Plant Body

The Vascular Tissue System Consists of Xylem and Phloem

3) **Roots: Anchorage, Absorption, and Storage**

Primary Growth Causes Roots to Elongate

The Epidermis of the Root Is Very Permeable to Water

Cortex Makes Up Much of the Interior of a Young Root

The Vascular Cylinder Contains Conducting Tissues

4) **Stems: Reaching for the Light**

The Stem Includes Four Types of Tissue

Stem Branches Form from Lateral Buds Consisting of Meristem Cells

Secondary Growth Produces Thicker, Stronger Stems

5) **Leaves: Nature's Solar Collectors**

Leaves Have Two Major Parts: Blades and Petioles

6) **How Do Plants Acquire Nutrients?**

Roots Acquire Minerals by a Four-Step Process

Symbiotic Relationships Help Plants Acquire Nutrients

7) **How Do Plants Acquire Water and Transport Water and Minerals?**

Water Movement in Xylem Is Explained by the Cohesion–Tension Theory

Water Enters Roots Mainly by Pressure Differences Created by Transpiration

Adjustable Stomata Control the Rate of Transpiration

8) **How Do Plants Transport Sugars?**

The Pressure-Flow Theory Explains Sugar Movement in Phloem

Evolutionary Connections: Special Adaptations of Roots, Stems, and Leaves

Case Study Revisited: A Beautiful Death Trap

CASE STUDY

A Beautiful Death Trap

It is high summer in a water-saturated bog in Georgia. A lacewing fly hovers silently as it watches for insect prey, its iridescent blue body glinting in the sunlight. Tiring, it swoops downward to rest on some innocent-looking vegetation, but immediately begins a frantic attempt to escape as it feels its legs grasped by sticky hairs projecting from the leaves. The lacewing has landed on a sundew plant. As the hapless insect struggles to free itself, the plant's tendrils slowly and insidiously bend toward the fly's body, which becomes mired in sticky red droplets. But the droplets are more than just sticky—they are packed with protein-digesting enzymes, as well as enzymes that digest the insect's chitinous exoskeleton. The exoskeleton-digesting enzymes are derived, not from the plant, but from bacteria that thrive in the droplets. Over the next few days the lacewing's body will become food for the plant. What forces of natural selection have driven this small bog-dwelling plant to act as a meat-eater? How can the glandular hairs of the plant actually move toward the insect? You'll find answers at the end of this chapter. ■

1 How Are Plant Bodies Organized, and How Do They Grow?

Flowering Plants Consist of a Root System and a Shoot System

Flowering plants, or *angiosperms*, consist of two major regions, the *root system* and the *shoot system* (Fig. 23-1). The **root system** consists of all the roots of a plant. **Roots** are branched portions of the plant body that are usually embedded in soil. Plant roots serve six major functions: They (1) anchor the plant in the ground; (2) absorb water and minerals from the soil; (3) store surplus sugars manufactured during photosynthesis; (4) transport water, minerals, sugars, and hormones to and from the shoot; (5) produce some hormones; and (6) interact with soil fungi and microorganisms that provide nutrients to the plant.

The rest of the plant is the **shoot system**, usually located above ground. The shoot system consists of *leaves*, *buds*, and (in season) *flowers* and *fruits*, all borne on **stems**, which are typically branched. The functions of shoots include (1) photosynthesis, mainly in leaves and young green stems; (2) transport of materials among leaves, flowers, fruits, and roots; (3) reproduction; and (4) hormone synthesis.

Figure 23-2 illustrates the two broad groups of flowering plants, or angiosperms: the **monocots**, which include grasses, lilies, palms, and orchids, and the **dicots**, which include deciduous trees (those that drop their leaves in winter), bushes, and many garden flowers. As we shall see in Chapter 24, the first leaflike structures an angiosperm embryo produces are called *cotyledons*, or *seed leaves*, for which these two broad groups are named. Monocots and dicots differ in the structure of their flowers, leaves, vascular tissue, root pattern, and seeds. Don't worry about terms that are not yet familiar to you; for now, look over Figure 23-2, and refer to it later as we examine the parts of the flowering plants in more detail.

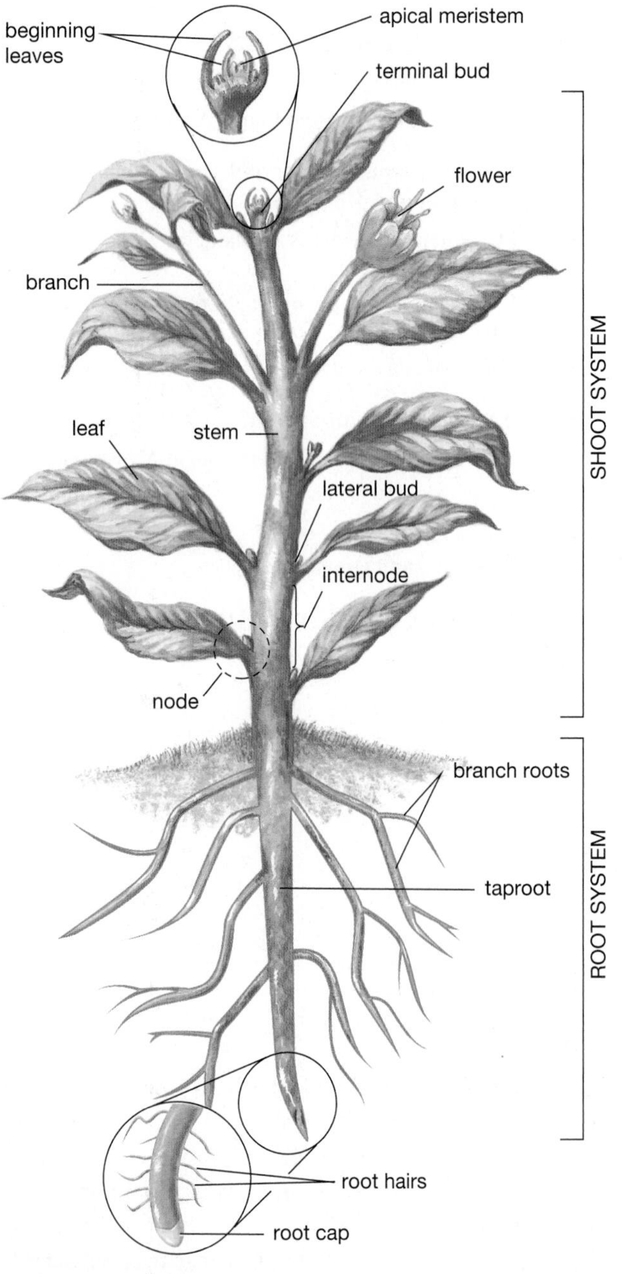

Figure 23-1 Flowering plant structure
A flowering plant consists of the root system and the shoot system. The root system is normally highly branched. Root hairs, typically microscopic, increase surface area for nutrient absorption. The shoot system includes stems (from which branches grow), with buds and leaves. In the appropriate season, the shoot may bear flowers and fruit. Our model plant is a dicot.

MEDIATUTOR 23.1 Plant Anatomy

During Plant Growth, Meristem Cells Give Rise to Differentiated Cells

Animals and plants develop in dramatically different ways. One difference is the timing and distribution of growth. As you grew from a baby to an adult, all parts of your body became larger. When you reached your adult height, you stopped growing (upward, at least!). In contrast, flowering plants grow throughout their lives, never reaching a stable "adult" body form. Moreover, most plants grow longer only at the tips of their branches and roots, and structures that developed earlier remain in exactly the same place; a swing tied to a tree branch does not move farther from the ground each year. Why do plants grow this way?

From the moment they sprout, plants are composites of two fundamentally different categories of cells: meristem cells and differentiated cells. Embryonic, undifferentiated **meristem cells** are capable of mitotic cell division. Some of the offspring, or daughter cells, of the dividing meristem cells lose the ability to divide and become part of the nongrowing portions of the plant body. These **differentiated cells** are specialized in structure and function. Continued divisions of meristem cells, then, keep a plant growing throughout its life, whereas

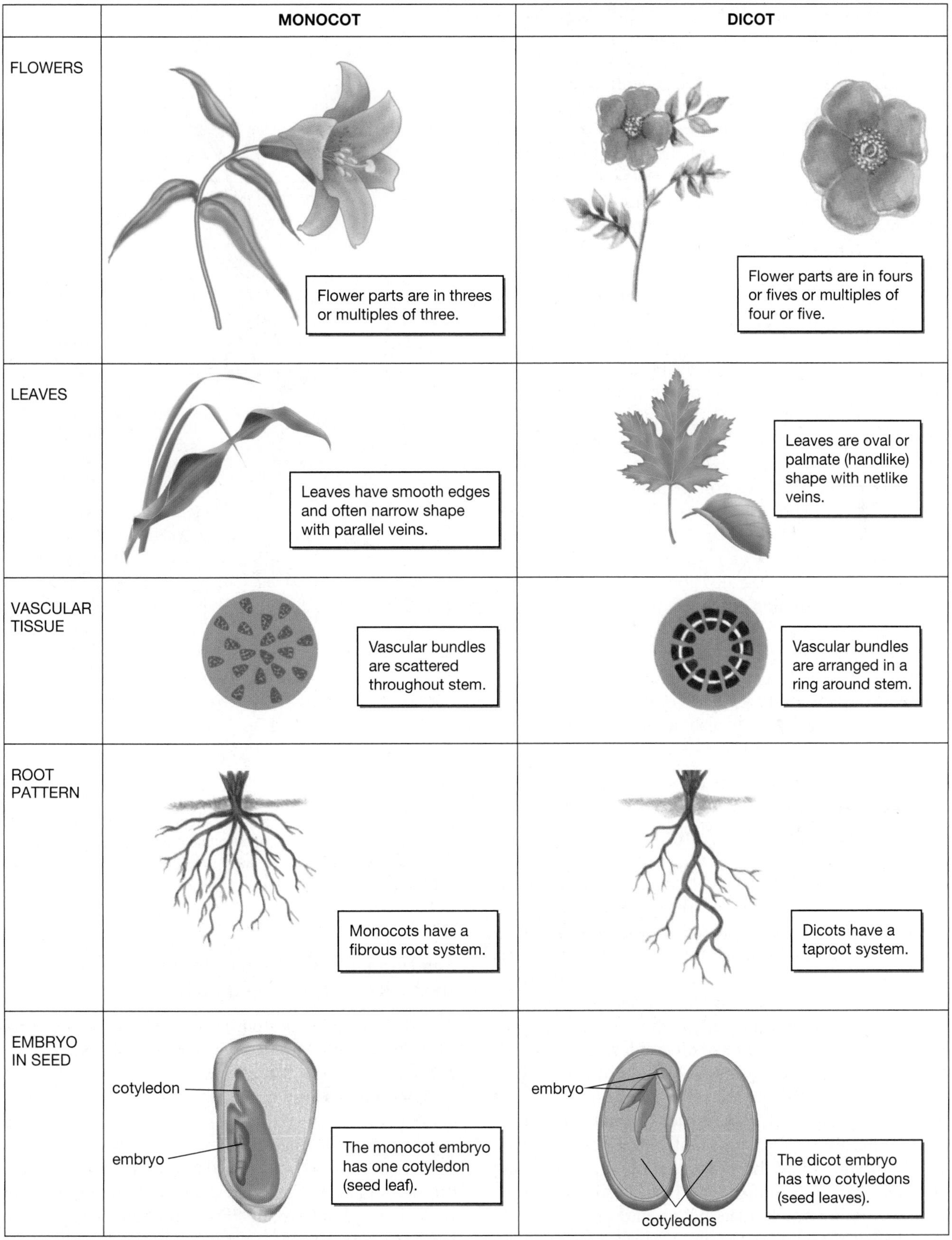
MONOCOT
DICOT
FLOWERS
Flower parts are in threes or multiples of three.
Flower parts are in fours or fives or multiples of four or five.
LEAVES
Leaves have smooth edges and often narrow shape with parallel veins.
Leaves are oval or palmate (handlike) shape with netlike veins.
VASCULAR TISSUE
Vascular bundles are scattered throughout stem.
Vascular bundles are arranged in a ring around stem.
ROOT PATTERN
Monocots have a fibrous root system.
Dicots have a taproot system.
EMBRYO IN SEED
cotyledon
embryo
The monocot embryo has one cotyledon (seed leaf).
embryo
cotyledons
The dicot embryo has two cotyledons (seed leaves).

Figure 23-2 Monocots and dicots compared
Distinguishing traits of the two major classes of flowering plants, the monocots and the dicots. The word *monocot* refers to the fact that the embryo in the seed has a single ("mono") cotyledon, or seed leaf; *dicot* means "two cotyledons."

their differentiated daughter cells form more stable or permanent parts of the plant, such as mature leaves or the trunks of trees.

Plants grow through the division and differentiation of two major types of meristem cells: apical meristems and lateral meristems. **Apical meristems** ("tip meristems") are located at the ends of roots and shoots, including main stems and branches (see Figs. 23-1, 23-9). **Lateral meristems** ("side meristems"), also called **cambia** (singular, **cambium**), form cylinders that run parallel to the long axis of roots and stems (see Fig. 23-15).

Plant growth takes two forms: (1) *primary growth* and (2) *secondary growth*. **Primary growth** occurs by the mitotic cell division of apical meristem cells followed by differentiation of the resulting daughter cells. This type of growth takes place in the growing tips of roots and shoots of plants. Primary growth is responsible both for increase in length and for the development of the specialized plant structures. The elongation of roots and shoots through primary growth allows them to enter new space from which to collect light, nutrients, and water. It also explains why your swing never gets any higher off the ground.

Secondary growth, which is responsible for increases in diameter, occurs by the division of lateral meristem cells and differentiation of their daughter cells. Secondary growth causes the stems and roots of most conifers (cone-bearing trees, or evergreens) and dicots to become thicker and woodier as they age. Although we later discuss secondary growth only in stems, keep in mind that secondary growth also occurs in roots.

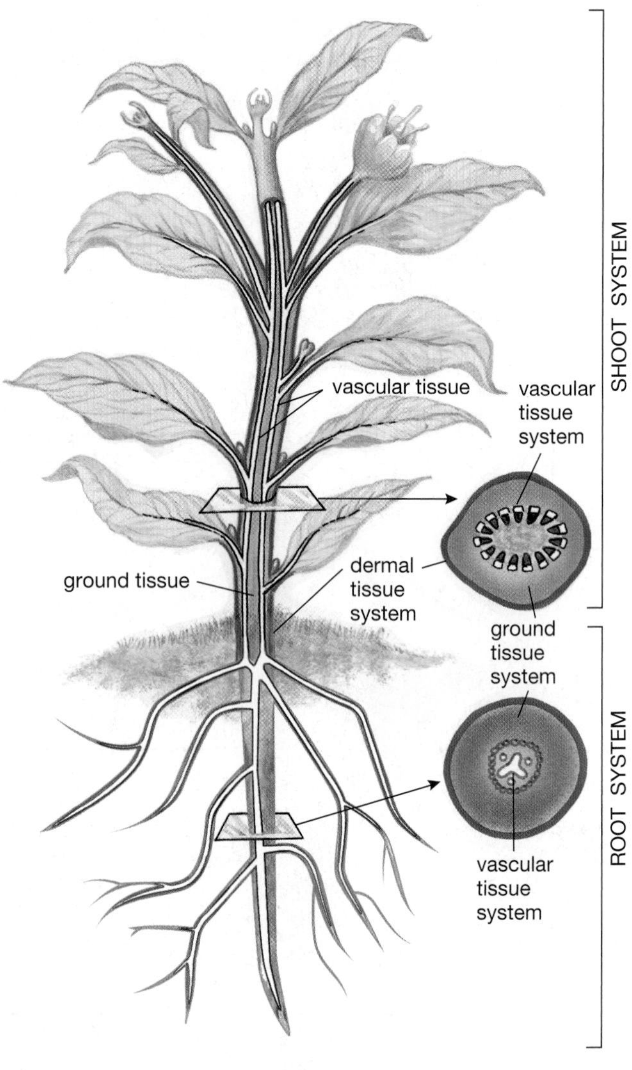

Figure 23-3 The structure of the root and shoot
Both the root and shoot of a flowering plant consist of three tissue systems: dermal, ground, and vascular tissue systems. (The structure of the leaves, vascular tissue system, and roots indicate that our sample plant is a dicot.)

2 What Are the Tissues and Cell Types of Plants?

The major structures of land plants, including roots, stems, and leaves, consist of three *tissue systems*—(1) dermal, (2) ground, and (3) vascular tissue systems—each containing more than one type of tissue. Each tissue, in turn, is composed of one or more specialized cell types (Fig. 23-3). The **dermal tissue system** covers the outer surfaces of the plant body. The **ground tissue system**, which consists of all nondermal and nonvascular tissues, makes up most of the body of young plants. Its functions include photosynthesis, support, and storage. The **vascular tissue system** transports water, minerals, sugars, and plant hormones throughout the plant. Some flowering plants, described as *herbaceous*, are soft-bodied with flexible stems; herbaceous plants include lettuce, beans, and grasses. Such plants usually live only one year. Other plants, such as trees and bushes, are described as *woody*; most are perennial (living many years) and develop hard, thickened, woody stems as a result of secondary growth. As you will see, different types of tissue are present in herbaceous and woody plants.

The Dermal Tissue System Forms the Covering of the Plant Body

The dermal tissue system is the outer covering of the plant body. There are two types of dermal tissue: (1) *epidermal tissue* and (2) *periderm*.

Epidermal tissue forms the **epidermis**, the outermost cell layer that covers the leaves, stems, and roots of all young plants (Fig. 23-4a). Epidermal tissue also covers flowers, seeds, and fruit. In herbaceous plants the epidermis is retained as the outer covering of the entire plant body throughout its life. The epidermal tissue of the aboveground parts of a plant is generally composed of tightly packed thin-walled cells, covered with a waterproof, waxy **cuticle**. Secreted by the epidermal cells, the cuticle reduces the evaporation of water from the plant and helps protect it from the invasion of disease microorganisms. In contrast, the epidermal cells of roots are not covered with cuticle, because a waterproof cuticle would prevent the absorption of water and minerals.

Some epidermal cells produce fine extensions called *hairs.* Many root epidermal cells bear **root hairs**, long projections that greatly increase the absorptive surface area of the root. Epidermal hairs on the stems and leaves of desert plants reduce evaporative water loss by reflecting sunlight and producing an unstirred layer of air near the plant's surface. In contrast, some tropical plants use their hairy leaves to capture and hold water.

Periderm replaces epidermal tissue on the roots and stems of woody plants as they age. This dermal tissue is composed primarily of **cork cells**, which have thick, waterproof walls and are dead at maturity (Fig. 23-4b). Cork cells form the protective outer layers of the bark of trees and woody shrubs and the tough covering of their roots.

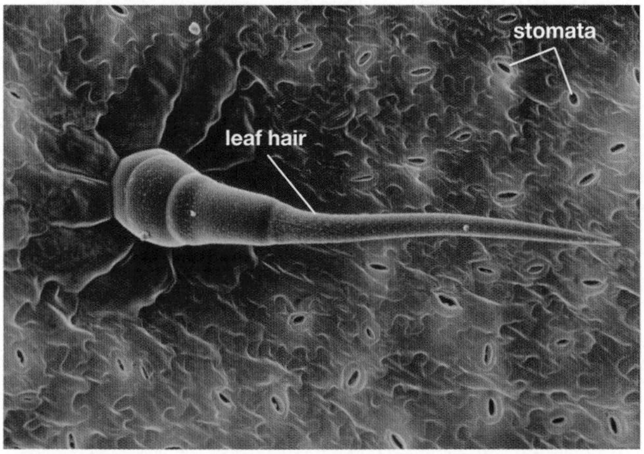

(a)

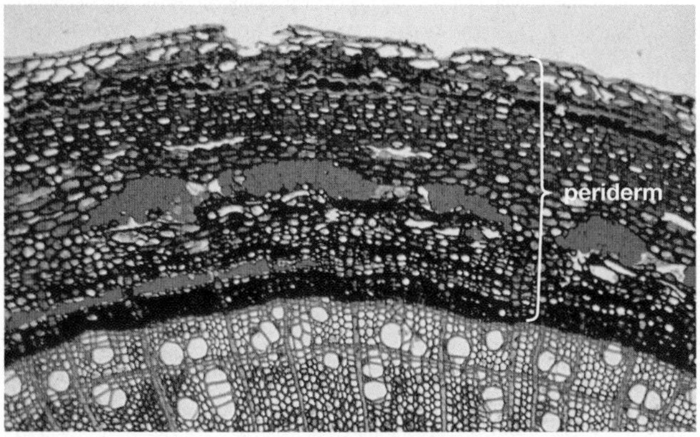

(b)

Figure 23-4 Dermal tissues cover plant surfaces
(a) The epidermis of a young root or shoot is usually a single layer of cells. In shoot epidermis, such as the epidermis of this zinnia leaf, the outer surfaces of the cells are covered with cuticle, a waxy, waterproof coating that reduces the evaporation of water. The "leaf hair" protruding from the epidermis also reduces evaporation by slowing the movement of air across the leaf surface. ***(b)*** Woody stems and roots develop a dense, tough, waterproof periderm, which consists mostly of thick-walled cork cells.

The Ground Tissue System Makes Up Most of the Young Plant Body

The ground tissue system, which makes up the bulk of a young plant, consists of all nondermal and nonvascular tissues. There are three types of ground tissue: (1) *parenchyma*, (2) *collenchyma*, and (3) *sclerenchyma*.

Parenchyma tissue is the most abundant of the ground tissues. Parenchyma cells are thin-walled cells, alive at maturity, that typically carry out most of the metabolic activities of the plant (Fig. 23-5a). Depending on their location within the plant body, parenchyma cells have such diverse functions as photosynthesis, storage of sugars and starches, or secretion of hormones. Under the proper conditions, many parenchyma cells are capable of mitotic cell division. Roots that are adapted for storage, including carrots and sweet potatoes, are packed with parenchyma cells that store carbohydrates, such as starch and sugar. Starch-filled parenchyma cells pack the white potato as well.

Collenchyma tissue consists of elongated, polygonal (many-sided) cells with unevenly thickened cell walls (Fig. 23-5b). Collenchyma cells are alive at maturity but generally cannot divide. Although strong, the cell walls of collenchyma are still somewhat flexible. In herbaceous plants and in the leaf stalks and young growing stems of all plants, collenchyma tissue is an important source of support. The strings in celery stalks, for example, are mostly collenchyma cells in association with vascular tissue.

Sclerenchyma tissue consists of cells with thick, hardened secondary cell walls (located between the outer primary wall and the plasma membrane) reinforced with the stiffening substance *lignin* (Fig. 23-5c). Like collenchyma, sclerenchyma cells support and strengthen the plant body; however, unlike collenchyma, they die after they differentiate. Their hardened cell walls then remain as a source of support. Sclerenchyma tissue can be found in many parts of the plant body, including xylem and phloem (described next). Sclerenchyma cells provide the fibers of hemp and jute, which are used for making rope. Other types of sclerenchyma cells form nut shells, the outer covering of peach pits, and the gritty texture of pears.

The Vascular Tissue System Consists of Xylem and Phloem

The vascular tissue system consists of two complex conducting tissues: (1) *xylem* and (2) *phloem.* The major role of each tissue is the transport of materials. Xylem transports water and minerals up from the roots to the rest of the plant; phloem conveys water, sugars, amino acids, and hormones throughout the plant.

Xylem Conducts Water and Dissolved Minerals from the Roots to the Rest of the Plant

Xylem conducts water and minerals from roots to shoots in tubes that are made from *tracheids* and *vessel elements* (Fig. 23-6). **Tracheids** are thin, tubelike cells with

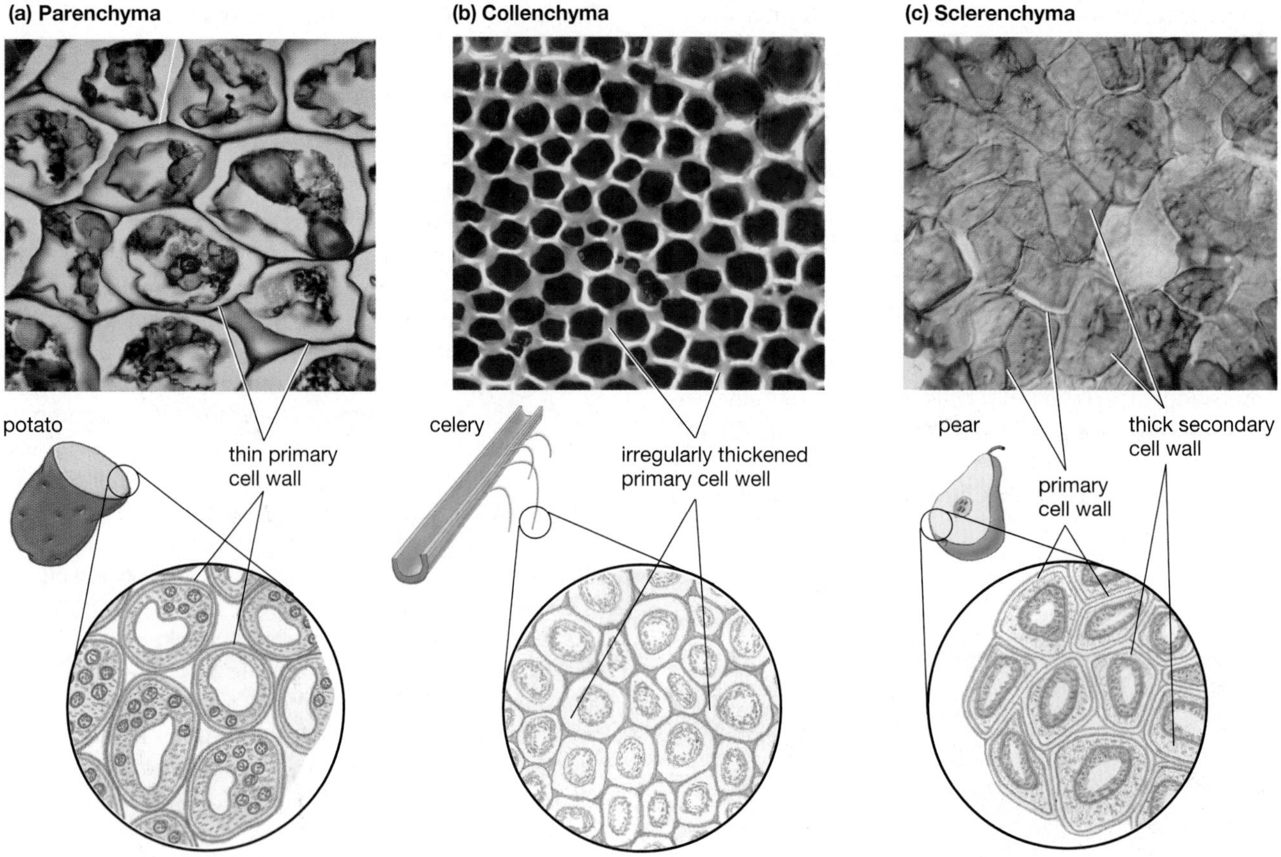

Figure 23-5 The structure of ground tissue
Ground tissue consists of three cell types: parenchyma, collenchyma, and sclerenchyma. ***(a)*** Parenchyma cells are living cells with thin, flexible cell walls. These parenchyma cells are used for starch storage in a potato. ***(b)*** Collenchyma cells, such as these from a celery stalk, are living cells with irregularly thickened, but still somewhat flexible, walls. They help support the plant body. ***(c)*** Sclerenchyma cells, such as these "stone cells" that give the meat of a pear its slightly gritty texture, have thick, rigid secondary cell walls. Sclerenchyma cells also form the shells of nuts.

MediaTutor
23.2 Plant Transport Mechanisms

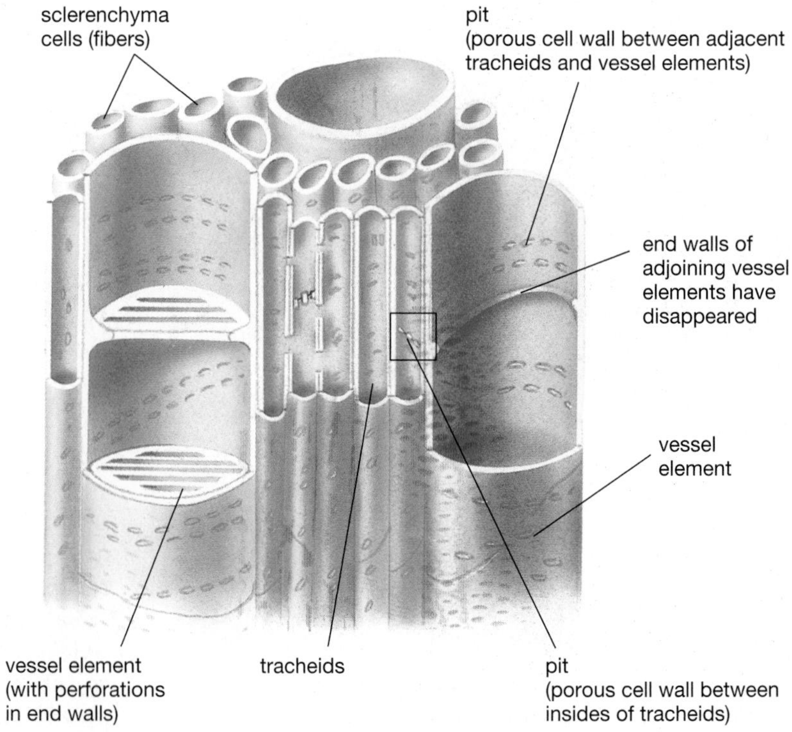

Figure 23-6 The structure of xylem
Xylem is a mixture of cell types, including parenchyma and sclerenchyma fibers (ground tissue) and two types of conducting cells, tracheids and vessel elements. Tracheids are thin with tapered, overlapping ends joined by pits. The pits are not holes; they consist of a water-permeable primary cell wall that separates the interiors of adjoining cells. Vessels consist of vessel elements stacked atop one another. The end walls of some vessel elements are absent, and others have narrow openings that connect adjacent vessel elements. Both tracheids and vessel elements have pits in their side walls, allowing water and dissolved minerals to move sideways between adjacent conducting tubes.

thick cell walls; their slanted ends resemble the tips of hypodermic needles. Tracheids are stacked atop one another with the slanted ends overlapping. The overlapping end walls contain **pits**, porous sections where secondary cell walls are absent. These pits allow water and minerals to pass from one tracheid to the next, or from a tracheid to an adjacent vessel element, by crossing only the thin, water-permeable primary cell wall.

Vessel elements also meet end to end, but they are larger in diameter than tracheids. Their ends may be either flat or overlapping and tapered. Perforations connect adjoining elements. In some cases, the ends of vessel elements completely disintegrate, leaving an open tube (see Fig. 23-6). Thus, vessel elements form wide-diameter, relatively unobstructed pipelines called **vessels** from root to leaf.

Most conifers have only tracheids; flowering plants usually have both tracheids and vessel elements. As these cells differentiate, they develop thick cell walls that help support the weight of the plant. In some trees (pines, for example), the bulk of the tree trunk consists of the thick cell walls of tracheids. The final step in the differentiation of both tracheids and vessel elements is death: The cytoplasm and plasma membrane disintegrate, leaving behind a hollow tube of cell wall.

Phloem Conducts Water, Sugars, Amino Acids, and Hormones Throughout the Plant

Phloem carries water containing dissolved substances synthesized by the plant, including sugars, amino acids, and hormones. Among the supportive sclerenchyma cells of phloem are **sieve tubes**, constructed of a single strand of cells called **sieve-tube elements** (Fig. 23-7). As sieve-tube elements mature, most of their internal contents disintegrate, leaving behind only a thin layer of cytoplasm that lines the plasma membrane. At the ends of sieve-tube elements, where adjacent cells meet, holes form in the primary cell walls, creating structures called **sieve plates**. The interiors of adjacent sieve-tube elements are connected through the openings in the sieve plate. A continuous conducting system is forged by the linking of many sieve-tube elements end to end in this way.

A sieve-tube element has a plasma membrane, a few small mitochondria, and some endoplasmic reticulum. It is therefore considered to be alive, but it generally lacks ribosomes, Golgi complexes, and a nucleus. How, then, can a sieve-tube element remain alive? Each sieve-tube element is nourished by a smaller, adjacent **companion cell**. Companion cells are connected to sieve-tube elements by cytoplasm-filled channels called *plasmodesmata*, described in Chapter 4. The companion cells maintain the integrity of the sieve-tube elements by donating high-energy compounds and perhaps even by repairing the sieve-tube plasma membrane. As we shall see later, companion cells also regulate the movements of sugars into and out of the sieve tubes.

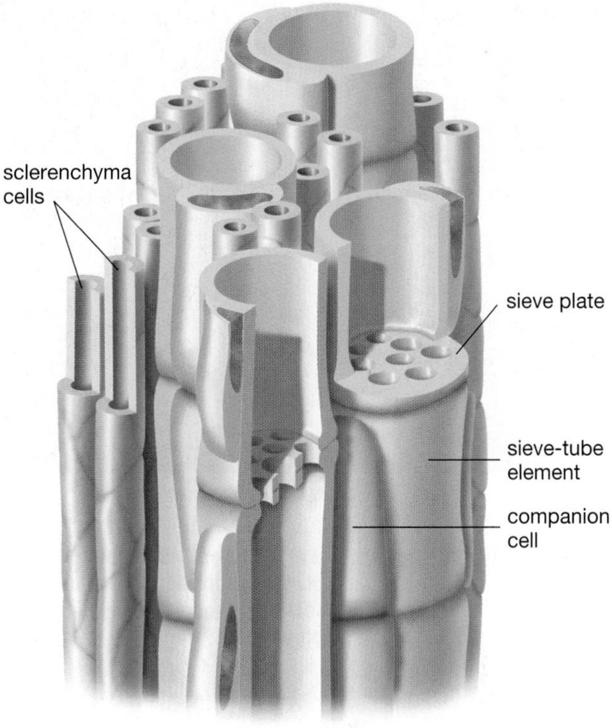

Figure 23-7 The structure of phloem
Phloem is a mixture of cell types, including sclerenchyma, sieve-tube elements, and companion cells. Sieve-tube elements have primary cell walls and a thin layer of cytoplasm around a fluid-filled core. Sieve-tube elements, stacked end to end, form the conducting system of phloem. Where they join, sieve-tube elements form sieve plates, where membrane-lined pores allow fluid to pass from cell to cell. Each sieve-tube element has a companion cell that nourishes it and regulates its function.

3 Roots: Anchorage, Absorption, and Storage

As a seed sprouts, the **primary root**—the first root to develop—grows down into the soil. Many dicots, such as carrots and dandelions, develop a taproot system. A **taproot system** consists of the primary root, which usually becomes longer and stouter with time, and many smaller roots that grow out from the sides of the primary taproot (Fig. 23-8a). In contrast, in monocots such as grasses and palms, the primary root soon dies off, replaced by many new roots that emerge from the base of the stem. These secondary roots are nearly equal in size, forming a **fibrous root system** (Fig. 23-8b).

Primary Growth Causes Roots to Elongate

In young roots of both taproot and fibrous root systems, divisions of the apical meristem give rise to four distinct regions (Fig. 23-9). At the very tip of the root, daughter cells produced on the "soil side" of the apical meristem differentiate into the **root cap**. The root cap protects the apical meristem from being scraped off as the root pushes down between the rocky particles of the soil.

Figure 23-8 Typical root systems in dicots and monocots
(a) Dicots typically have a taproot system, consisting of a long central root with many smaller, secondary roots branching from it. ***(b)*** Monocots normally have a fibrous root system, with many roots of equal size.

Root-cap cells have thick cell walls and secrete a slimy lubricant that helps ease the way between soil particles. Nevertheless, root-cap cells wear away and must be continuously replaced by new cells from the meristem.

Daughter cells produced on the "shoot side" (upper portion) of the apical meristem differentiate into one of three parts: an outer envelope of epidermis; a *vascular cylinder* at the core of the root; and, between the two, a *cortex* (Fig. 23-9).

The Epidermis of the Root Is Very Permeable to Water

The root's outermost covering of cells is the epidermis, which is in contact with the soil and any air or water trapped among the soil particles. The cell walls of the epidermal cells are highly water permeable. Therefore, water can penetrate into the interior of the root either by passing through the membranes of the epidermal cells or by passing between those cells, through the porous cell walls. Many epidermal cells grow root hairs into the surrounding soil (Fig. 23-10). By increasing the surface area, root hairs increase the root's ability to absorb water and minerals. Root hairs may add dozens of square meters of surface area to the roots of even small plants.

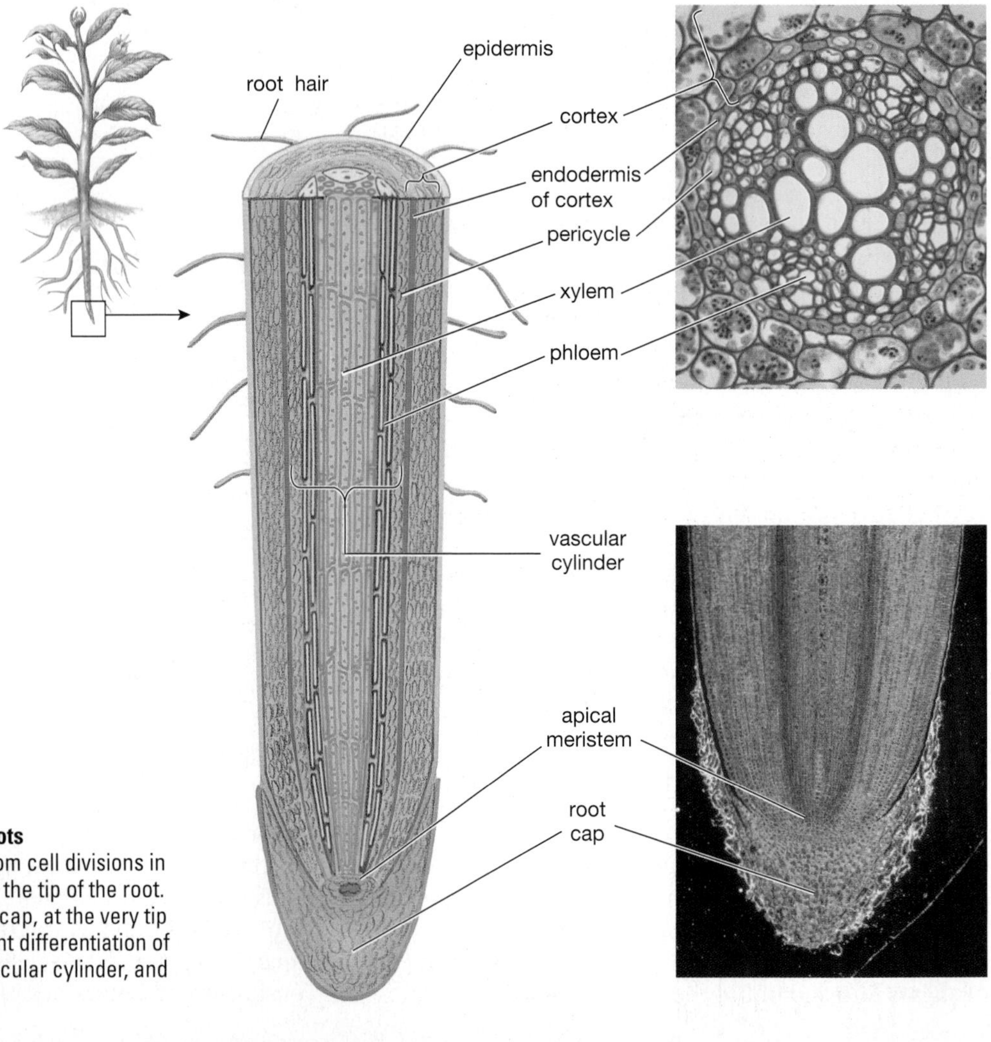

Figure 23-9 Primary growth in roots
Primary growth in roots results from cell divisions in the apical meristem, located near the tip of the root. Four regions are formed: the root cap, at the very tip of the root; and, via the subsequent differentiation of daughter cells, the epidermis, vascular cylinder, and cortex.

Figure 23-10 Root hairs
Root hairs, shown here in a sprouting radish, greatly increase a root's surface area for the absorption of water and minerals from the soil.

Cortex Makes Up Much of the Interior of a Young Root

Cortex occupies most of the inside of a young root. The cortex consists of an outer mass of large, loosely packed parenchyma cells just beneath the epidermis and an inner layer of smaller, close-fitting cells that form a ring called the **endodermis** around the vascular cylinder (see Fig. 23-9). Sugars produced by photosynthesis in the shoot are transported down to the parenchyma cells of the cortex, where they are converted to starch and stored. These cells are particularly abundant in roots specialized for carbohydrate storage, such as the thick roots of carrots and dandelions.

The endodermis is a layer of cells with highly specialized cell walls. At the point of contact between endodermal cells, their cell walls are filled with a waxy material, forming the **Casparian strip**. The Casparian strip resembles the mortar in a brick wall: The waxy waterproofing covers the top, bottom, and ends of the endodermal cells, as mortar surrounds each brick. However, it does not cover the cell surfaces that face the rest of the cortex or those that face the vascular cylinder. Water and dissolved minerals can travel *around* both epidermal and cortex parenchyma cells by moving through their porous cell walls. In contrast, the waxy Casparian strip seals off the vascular cylinder cell walls, forcing water and minerals that enter the vascular cylinder to pass through the living membranes of the endodermal cells (Fig. 23-11). These membranes regulate the types and amounts of materials that the roots can absorb.

The Vascular Cylinder Contains Conducting Tissues

The **vascular cylinder** contains the conducting tissues of xylem and phloem, which transport water and dissolved materials within the plant. The outermost layer of the vascular cylinder, called the **pericycle**, is a remnant of meristem that retains the ability to divide. Under the influence of plant hormones, pericycle cells divide and form the apical meristem of a **branch root**, a root that forms as a branch of an existing root (Fig. 23-12). Branch root development is similar to primary root development except that the branch must first break out

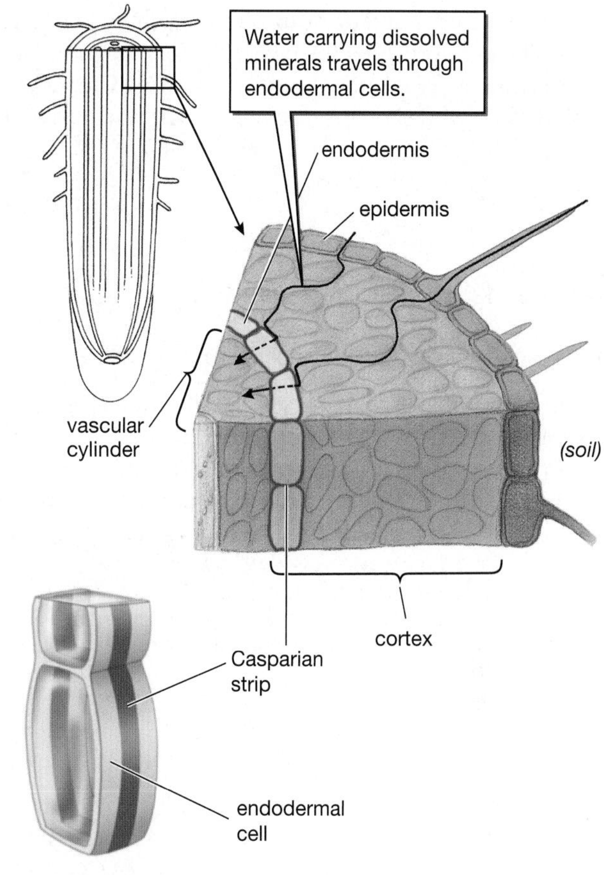

Figure 23-11 The role of the Casparian strip
The Casparian strip is a band of waterproof material in the walls between cells of the endodermis. Arranged like the mortar in a brick wall, the Casparian strip seals off the top, bottom, and ends of the endodermal cells, thereby forcing water to move through the cells rather than between them.

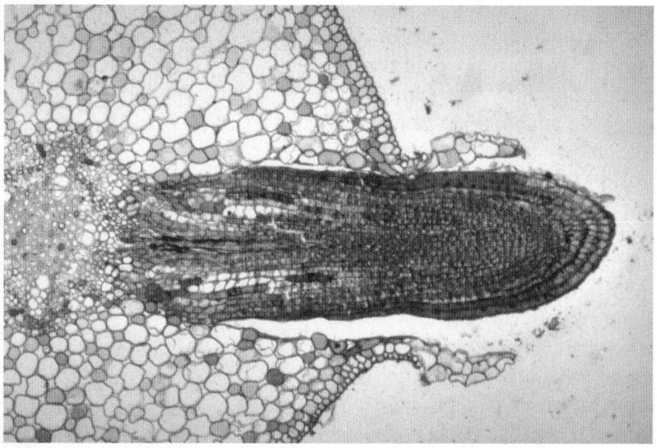

Figure 23-12 Branch roots
Branch roots emerge from the pericycle of a root. The central axis of this branch root is already differentiating into vascular tissue.

through the cortex and epidermis of the primary root. It does so by both crushing the cells that lie in its path and by secreting enzymes that digest them away. The vascular tissues of the branch root connect with the vascular tissues of the primary root.

4 Stems: Reaching for the Light

The Stem Includes Four Types of Tissue

Like roots, stems develop from a small group of actively dividing cells, the apical meristem, which lies at the tip of the young shoot. The apical meristem lies within the **terminal bud**. The terminal bud consists of meristem tissue surrounded by developing leaves called **leaf primordia** (singular, **primordium**) at the tip of the shoot. The daughter cells of the apical meristem differentiate into the specialized cell types of stem, buds, leaves, and flowers.

As the shoot grows, small clusters of meristem cells are "left behind" at the surface of the stem. These meristem cells form the leaf primordia, which develop into the mature leaves unique to each species of plant. The meristem cells also produce **lateral buds**, which, under appropriate conditions, grow into branches. (We will discuss the growth of branches shortly.) Leaf primordia and lateral buds appear at characteristic locations, called **nodes**, on the stem; regions of stem between these nodes are called **internodes** (Fig. 23-13).

Most young stems are composed of four tissues: (1) epidermis (dermal tissue), (2) cortex (ground tissue) , (3) pith (ground tissue), and (4) vascular tissues.

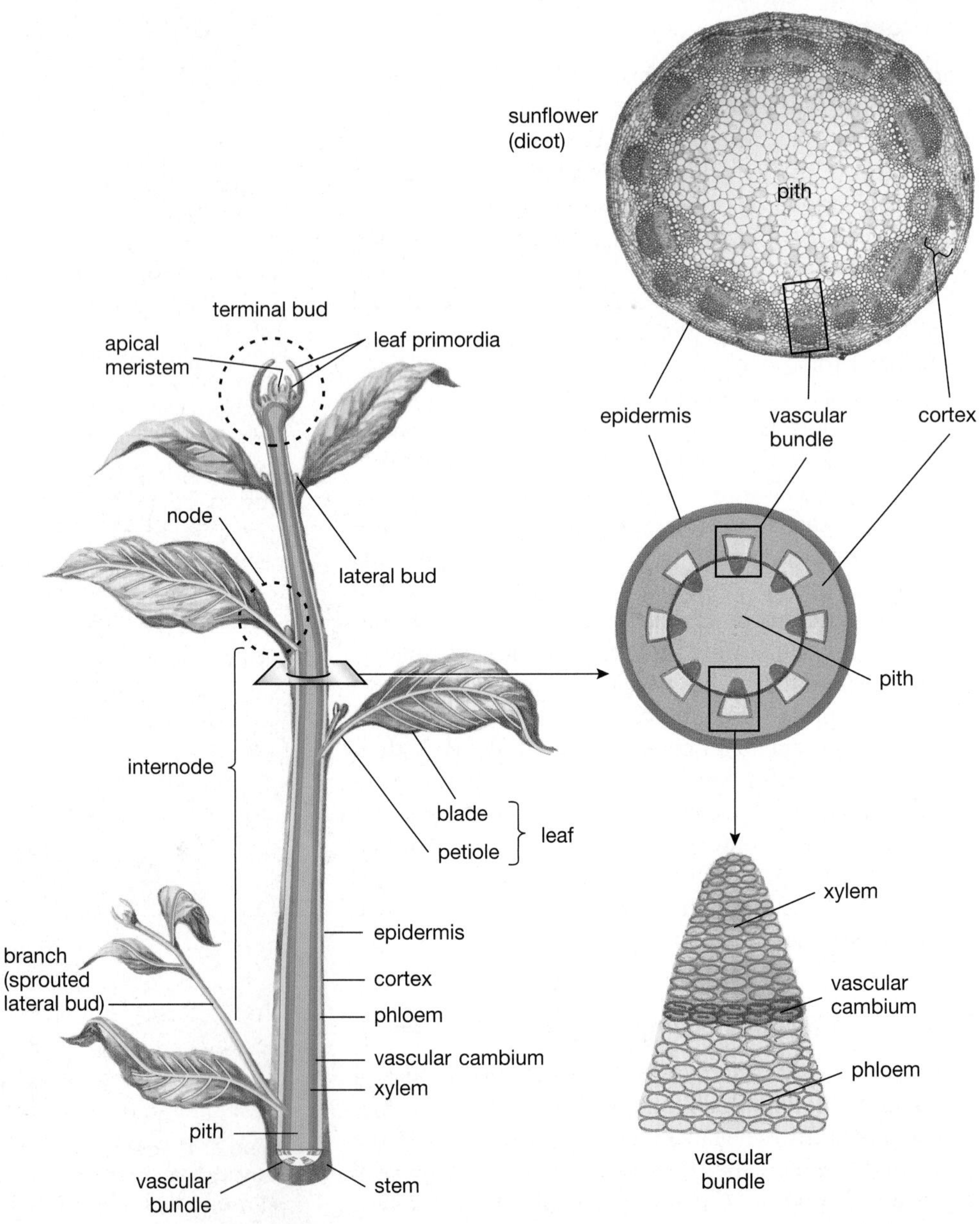

Figure 23-13 The structure of a young dicot stem
At the tip of the stem, the terminal bud includes the apical meristem and several leaf primordia, produced by the meristem. Other daughter cells of the apical meristem differentiate into epidermis, cortex, pith, and vascular tissues. As the young stem grows, the leaf primordia develop into mature leaves. Meanwhile, epidermis, cortex, pith, and vascular cells elongate between the points of attachment of leaf to stem, pushing the leaves apart. A lateral bud, a remnant of meristem tissue, remains between each leaf and the stem. Lateral buds may sprout into branches. Points on the shoot where leaves and lateral buds are located are nodes; the naked stem between nodes is an internode. In cross section, vascular tissue forms a ring of vascular bundles in dicots such as the sunflower shown in the photomicrograph.

As Figure 23-2 illustrates, monocots and dicots differ somewhat in the arrangement of vascular tissues. We will discuss only dicot stems here.

The Epidermis of the Stem Is Specialized to Retard Water Loss While Allowing Carbon Dioxide to Enter

In the stem (and leaves), the epidermis is exposed to dry air, making it a potential pathway for water loss. Epidermal cells of the stem, unlike those of the root, secrete a waxy covering, the cuticle, that reduces evaporation of water. The cuticle also, however, reduces the diffusion of carbon dioxide and oxygen into and out of the plant. Hence the epidermis is commonly perforated with adjustable pores called **stomata** (singular, **stoma**) that regulate this exchange. Stomata are discussed in more detail later in this chapter.

The Cortex and Pith Support the Stem, Store Food, and May Photosynthesize

Cortex (located between the epidermis and vascular tissues) and **pith** (inside the vascular tissues at the center of the stem) are similar in most respects; in fact, in some stems it is difficult to tell where cortex ends and pith begins. Cortex and pith perform three major functions: support, storage, and, in some cases, photosynthesis.

1. *Support*. In very young stems, water filling the central vacuoles of cortex and pith cells causes turgor pressure (see Chapter 5). Turgor pressure stiffens the cells, much as air inflates a tire. If you forget to water your houseplants, their drooping tips show the importance of turgor pressure in keeping young stems erect. Somewhat older stems also have collenchyma or sclerenchyma cells with thickened cell walls, which don't depend on turgor pressure for strength.
2. *Storage*. Parenchyma cells in both cortex and pith convert sugar into starch and store the starch as a food reserve.
3. *Photosynthesis*. In many stems, the outer layers of cortex cells contain chloroplasts and carry out photosynthesis. In some desert plants, such as cacti, the leaves are reduced or absent, and the cortex of the stem is the only green photosynthetic part of the plant.

Vascular Tissues in Stems Transport Water, Dissolved Nutrients, and Hormones

The vascular tissues of stems, like those of roots, transport water, minerals, sugars, and hormones. Vascular tissues are continuous in root, stem, and leaf, interconnecting all the parts of the plant. The **primary xylem** and **primary phloem** found in young stems arise from the apical meristem. In young dicot stems, the primary xylem, **vascular cambium** (meristematic tissue that produces *secondary xylem* and *secondary phloem*), and primary phloem may form concentric cylinders or may appear as a ring of bundles running up the stem, with each bundle containing both phloem and xylem (see Fig. 23-13). Secondary growth in dicot stems, as we shall discuss below, always results in concentric cylinders of xylem and phloem.

Stem Branches Form from Lateral Buds Consisting of Meristem Cells

Branches grow from lateral buds. A lateral bud is a cluster of dormant meristem cells left behind by the apical meristem as the stem grows. Lateral buds are located just above the attachment points of the leaves, at nodes (see Fig. 23-13). When stimulated by the appropriate hormones (as we shall see in Chapter 25), the meristem cells of a lateral bud are activated and the bud sprouts, growing into a branch (Fig. 23-14). As the meristem cells divide, they release hormones that change the developmental fate of the cells between the bud and the vascular tissues of the stem. Parenchyma cells of the cortex differentiate into xylem and phloem, ultimately connecting with the main vascular systems in the stem. As the branch grows, it duplicates the development of the stem: It has an apical meristem at its tip and produces its own leaf primordia and lateral buds as it grows.

Secondary Growth Produces Thicker, Stronger Stems

In conifers and perennial dicots, stems may survive up to hundreds of years, becoming thicker and stronger each year. This secondary growth in stem thickness results from cell division in two lateral meristems: the vascular cambium and *cork cambium* (Fig. 23-15, p. 479).

Vascular Cambium Produces Secondary Xylem and Phloem

The vascular cambium is a cylinder of meristem cells located between the primary xylem and primary phloem. Daughter cells of the vascular cambium produced toward the inside of the stem differentiate into **secondary xylem**; those produced toward the outside of the stem differentiate into **secondary phloem** (see Fig. 23-15). Because the center of the stem is already filled with pith and primary xylem, newly formed secondary xylem pushes the vascular cambium and all outer tissues farther out, increasing the diameter of the stem. This secondary xylem, with its thick cell walls, forms the wood that makes up most of the trunk of a tree. Young xylem, called **sapwood** (located just inside the vascular cambium), transports water and minerals; older xylem, the **heartwood**, contributes only to the strength of the trunk and no longer carries water and solutes. Heartwood serves as a collection site for metabolic wastes of the tree, such as gums, resins, and oils. These wastes increase the density of the heartwood, contribute to its darker color, and help the heartwood resist rotting.

Phloem cells are much weaker than xylem cells. As they die over time, the sieve-tube elements and companion cells are crushed between the hard xylem on the

Figure 23-14 How branches form
Stem branches grow from lateral buds located at the outer surface of a stem. The bud apical meristem generates an outward-growing branch, replicating the pattern of nodes and internodes unique to each type of plant. Meanwhile, cortex cells beneath the sprouting bud differentiate into vascular tissues and connect with the vascular system of the stem.

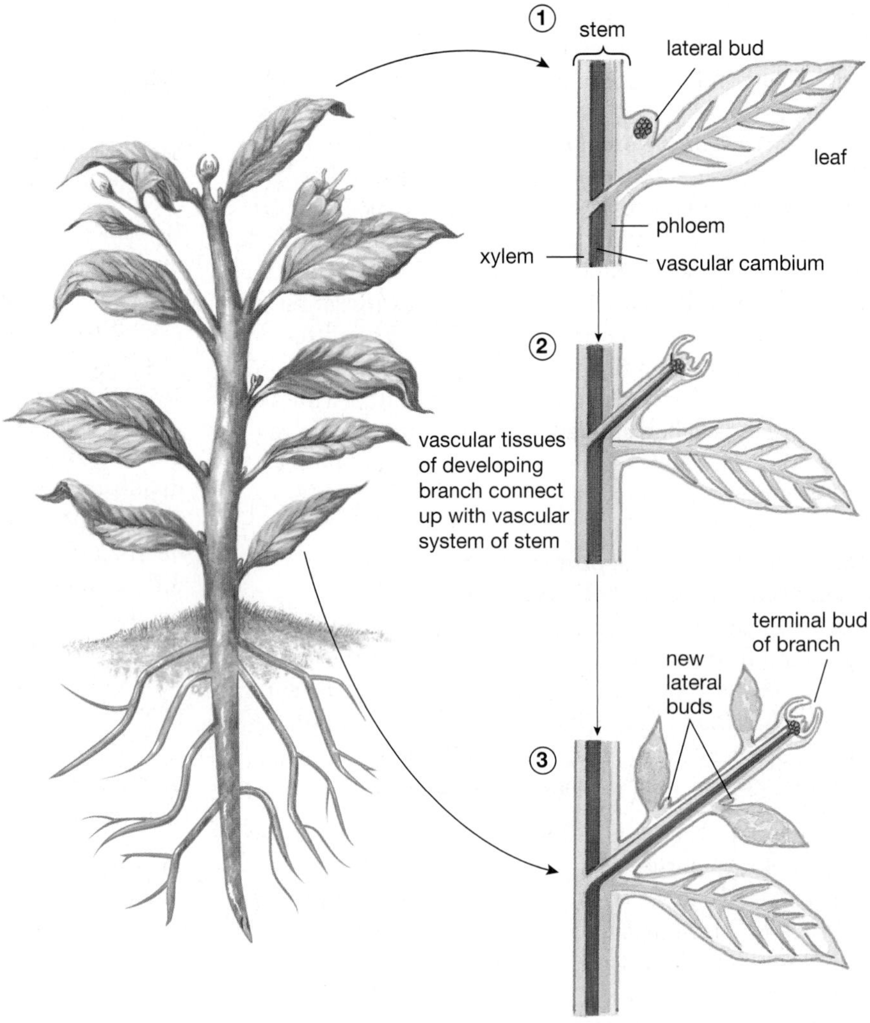

inside of the trunk and the tough cork on the outside (see below). Only a thin strip of recently formed phloem remains alive and functioning.

In trees adapted to temperate latitudes, such as oaks and pines, cell division in the vascular cambium ceases during the cold of winter. In spring the cambium cells divide, forming new xylem and phloem. The young cells grow by absorbing water and swelling, while the newly formed cell walls are still soft. As the cells mature, the cell walls thicken and harden, preventing further growth. Water is readily available in spring; therefore, young xylem cells swell considerably and are large when mature. As summer progresses and water becomes scarcer, new xylem cells absorb less water and consequently are smaller when they mature. As a result, tree trunks in cross section show a pattern of alternating pale regions (large cells formed in spring) and dark r egions (small cells formed in summer), as shown in Figure 23-16 (p. 480). This pattern forms the familiar **annual rings** of growth in temperate trees. You can determine the approximate age of a tree that has been cut by counting the dark growth rings. Scientists can also use the width of each ring to reconstruct past climate. Wet years produce more growth and wider rings. Using this technique and the rings in ancient trees, including a 1000-year-old cypress, researchers have constructed an 800-year record of climate in Virginia. They hypothesize that a 7-year drought from 1606 to 1612, recorded in tree rings, wiped out the colony of Jamestown, which was founded in Virginia in 1607.

Secondary Growth Causes the Epidermis to Be Replaced by Woody Cork

Recall that epidermal cells are mature, differentiated cells that can no longer divide. Therefore, as new secondary xylem and phloem are added each year, enlarging the stem, the epidermis can't expand to keep up with the increasing circumference. The epidermis splits off and dies. Apparently prodded by hormones, some parenchyma cells in the cortex become rejuvenated and

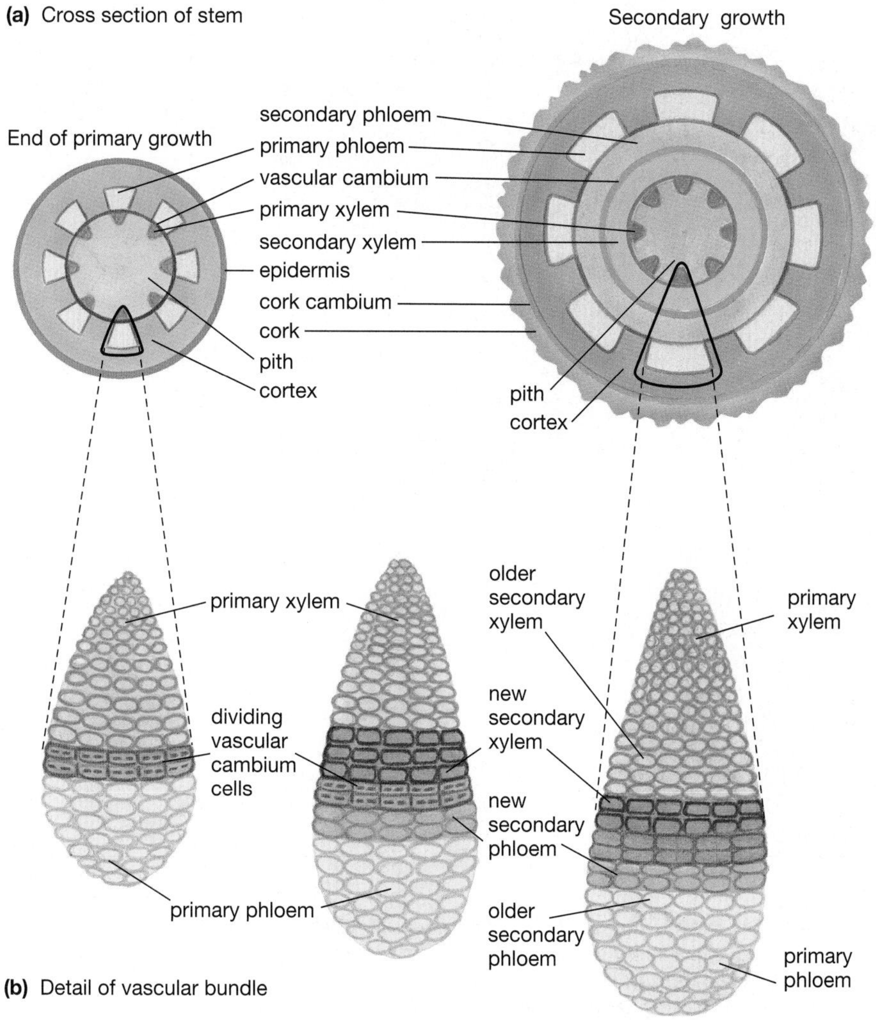

Figure 23-15 Secondary growth in a dicot stem
(a) Cross section of a dicot stem at the end of primary growth (left) and during early secondary growth (right). ***(b)*** Anatomical details of a vascular bundle during secondary growth. A vascular cambium forms between the primary xylem and primary phloem. When vascular cambium cells divide, daughter cells formed on the inside of the vascular cambium differentiate into secondary xylem; cells formed on the outside of the cambium differentiate into secondary phloem. Because xylem and pith already fill the inside of the stem, newly formed secondary xylem forces the cambium, phloem, and all outer tissues farther out, increasing the diameter of the stem. The cork cambium produces cork cells that cover the outside of the stem.

MediaTutor 23.3 Primary and Secondary Growth

form a new lateral meristem, the **cork cambium** (see Fig. 23-15). These cells divide, forming daughter cells toward the outside of the stem. These daughter cells, called *cork cells*, or simply cork, develop tough, waterproof cell walls that protect the trunk both from drying out and from physical damage. Cork cells die as they mature and may form a protective layer a half meter thick in some tree species, such as the fire-resistant sequoia (Fig. 23-17). As the trunk expands from year to year, the outermost layers of cork split apart or peel off, accommodating the growth. Corks used to plug bottles are made from the outermost layer of cork from a certain type of oak, the cork oak, carefully peeled off by harvesters. The cork of the cork oak breaks away from the cork cambium, so the tree is not harmed by stripping it off. The common term **bark** includes all the tissues outside the vascular cambium: phloem, cork cambium, and cork cells. The complete removal of a strip of bark all the way around a tree, called *girdling*, is invariably fatal to a tree because it severs the phloem. With the phloem gone, sugars synthesized in the leaves cannot reach the roots. Deprived of energy, the roots no longer take up water and minerals, causing the tree to die.

5 Leaves: Nature's Solar Collectors

Only structures with chlorophyll are able to carry out photosynthesis and make valuable sugar out of widely available ingredients: sunlight, carbon dioxide, and water. **Leaves** are the major photosynthetic structures of most plants. Water is obtained from the soil and transported to the leaf through the xylem, and carbon dioxide (CO_2) must diffuse into the leaf from the air. An ideal leaf should have a large surface area for gathering light and should be porous to permit CO_2 to enter from the air for photosynthesis. However, a large, porous leaf would also lose large amounts of water by evaporation, so the leaf must be reasonably waterproof as well. The leaves of flowering plants represent an elegant compromise among these conflicting demands (Fig. 23-18, p. 481).

Leaves Have Two Major Parts: Blades and Petioles

A typical angiosperm leaf consists of a broad, flat portion, the **blade**, connected to the stem by a stalk called the **petiole** (see Fig. 23-13). The petiole positions the

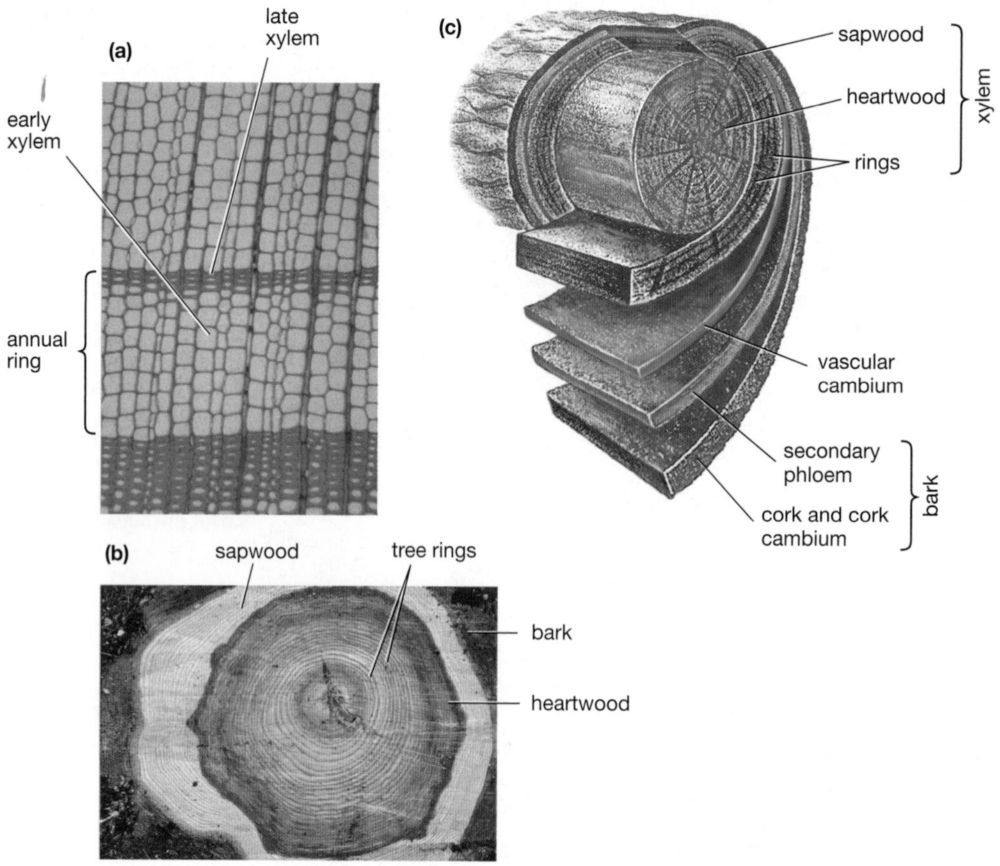

Figure 23-16 How tree rings are formed
Many trees form annual rings of xylem. ***(a)*** As this micrograph shows, secondary xylem cells formed during the wet spring are large, whereas secondary xylem cells formed during the hotter, drier summer are small. ***(b)*** Tree rings are clearly visible in this section of trunk from a larch tree. The ratio of cell wall to "hole" (the now-empty interior of the cell) determines the color of the wood: Early wood, formed during the spring, with lots of hole, is pale; late wood, formed during the summer, with lots of wall, is dark. The water-transporting xylem of sapwood forms a lighter layer inside the bark. Xylem of the older heartwood, where the rings are most easily visible, no longer transports water and minerals. ***(c)*** The layers of a tree trunk.

Figure 23-17 Cork forms the outer layer of bark
An ancient sequoia in the Sierra Nevada of California. The cork cambium of a sequoia produces new layers of cork each year, eventually producing a protective, fire-resistant outer covering half a meter or more thick. This massive cork layer contributes to the sequoia's great longevity; forest fires that kill lesser trees merely burn off a few inches of sequoia cork, leaving the living parts of the tree inside unharmed. The small blackened areas on this cork are from past fires.

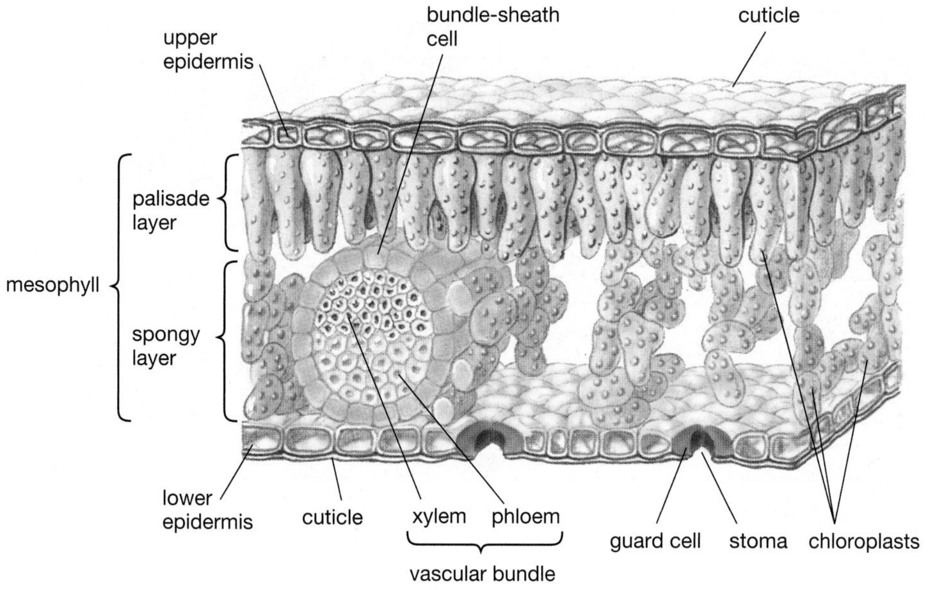

Figure 23-18 The structure of a typical dicot leaf
The cells of the epidermis lack chloroplasts and are transparent, allowing sunlight to penetrate to the chloroplast-containing mesophyll cells beneath. The stomata that pierce the epidermis and the loose, open arrangement of the mesophyll cells ensure that CO_2 can diffuse into the leaf from the air and reach all the photosynthetic cells.

blade in space, usually orienting the leaf for maximum exposure to the sun. Inside the petiole are vascular tissues of xylem and phloem that are continuous with those in the stem, root, and blade. Within the blade, the vascular tissues branch into **vascular bundles**, or **veins**.

The leaf epidermis consists of a layer of nonphotosynthetic, transparent cells that secrete a waxy, waterproof cuticle on their outer surfaces that reduces evaporation. The epidermis and its cuticle are pierced by adjustable pores, the stomata, which regulate the diffusion of CO_2 and water into and out of the leaf. Each stoma is surrounded by two sausage-shaped **guard cells**, which regulate the size of the opening into the interior of the leaf (see Fig. 23-18). Unlike the surrounding epidermal cells, guard cells contain chloroplasts and can carry out photosynthesis. As we shall see later, photosynthesis in the guard cells contributes to their ability to adjust the size of the pore.

Beneath the epidermis lies the loosely packed parenchyma cells of the **mesophyll** ("middle of the leaf"). In many leaves, mesophyll cells are of two types: a layer of columnar **palisade cells** just beneath the upper epidermis and a layer of irregularly shaped **spongy cells** above the lower epidermis. Both palisade and spongy cells contain chloroplasts; these cells perform most of the photosynthesis of the leaf. The openness of the leaf interior (see Fig. 23-18) allows CO_2 to diffuse easily to all the mesophyll cells. Vascular bundles, each containing both xylem and phloem, are embedded within the mesophyll, with fine veins reaching very close to each photosynthetic cell. Thus, each mesophyll cell receives energy from sunlight transmitted through the clear epidermis; carbon dioxide from the air, diffusing through the stomata; and water from the xylem. The sugars it produces are carried away to the rest of the plant by the phloem.

6 How Do Plants Acquire Nutrients?

Nutrients are elements essential to normal life; they differ for different organisms. Plants require relatively large quantities of the following nutrients: carbon (obtained from CO_2), hydrogen (from water), oxygen (from air and water), phosphorus (from phosphate ions in soil), nitrogen (from nitrate and ammonium ions in soil), and magnesium, calcium, and potassium (as ions in soil). Plants also require very small quantities of nutrients such as iron, chlorine, copper, manganese, zinc, boron, and molybdenum. Carbon dioxide and oxygen usually enter a plant by diffusion from the air into leaves, stem, and roots. Roots extract water and all other nutrients, collectively called **minerals**, from the soil.

Roots Acquire Minerals by a Four-Step Process

Soil consists of bits of pulverized rock, air, water, and organic matter (Fig. 23-19). Although both the rock particles and the organic matter contain essential nutrients, only minerals dissolved in the soil water are accessible to roots. The concentration of minerals in the soil water is very low, usually much lower than the concentration within plant cells and fluids. For example, the concentration of potassium (K^+) in root cells is at least 10 times greater than that in soil water, so diffusion cannot move potassium into the root. In general, most minerals are moved into a root against their concentration gradients by active transport. (Recall from Chapter 6 that the movement of molecules from areas of low concentration to areas of high concentration requires energy.) Sugar synthesized in the leaves by photosynthesisis is transported through the phloem to the roots, where mitochondria in root cells use it to produce ATP (adenosine triphosphate) by cellular respiration. Some of this ATP

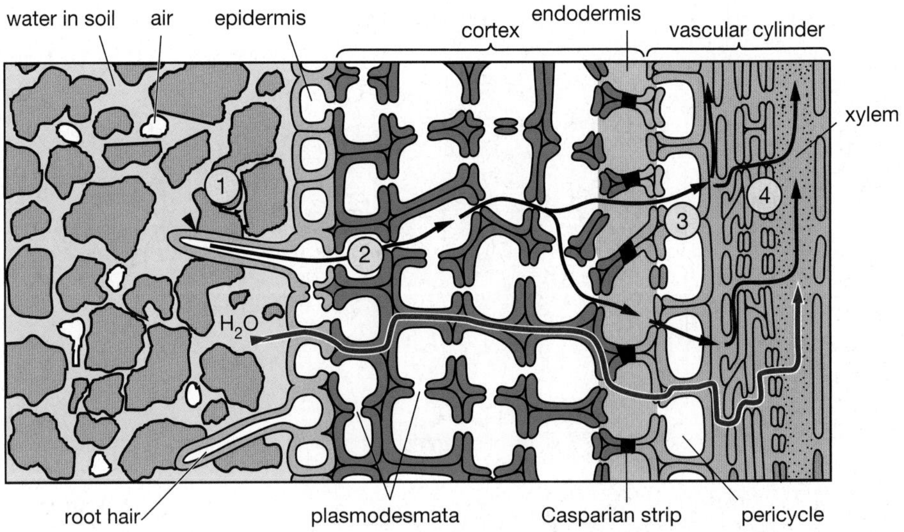

Figure 23-19 Mineral and water uptake by roots
The black line is a pathway for both water and minerals; the blue line is a pathway for water alone. ① Active-transport proteins pump minerals into the root-hair cytoplasm; water follows by osmosis. ② Minerals diffuse inward through plasmodesmata; water follows by osmosis. ③ Pericycle cells actively transport minerals out of their cytoplasm into the extracellular space surrounding the xylem; water follows by osmosis. ④ Entry of the minerals raises the concentration of minerals in the extracellular space, so they diffuse into the xylem cells, drawing in water by osmosis. Water can also move freely (blue path) through the porous cell walls of the epidermis and cortex until it reaches the Casparian strip. There it must move through the interior of the endodermal cells before it enters the vascular cylinder.

drives the active transport of minerals. Because ATP production by mitochondria requires oxygen, soil must have some air spaces within it. Flooding (or overwatering) can kill plants by depriving their roots of oxygen.

Most mineral absorption by roots occurs in a four-step process, as diagrammed in Figure 23-19:

① *Active transport into root hairs.* Root hairs projecting from the epidermal cells provide most of the surface area of the root and are in intimate contact with the soil water. The plasma membranes of the root hairs use the energy of ATP to transport minerals from the soil water, concentrating the minerals in the root hair cytoplasm.

② *Diffusion through cytoplasm to pericycle cells.* The cytoplasm of adjacent living plant cells is interconnected by plasmodesmata. Minerals can diffuse through plasmodesmata from the epidermal cells into the cortex, endodermis, and pericycle cells.

③ *Active transport into the extracellular space of the vascular cylinder.* At the center of the vascular cylinder lies the xylem, into which the minerals must ultimately be transported. The tracheids and vessel elements of xylem are dead, without cytoplasm or plasma membrane—merely an outer skeleton of cell wall shot full of holes (see Fig. 23-6). Any minerals that enter the extracellular space surrounding the xylem can easily diffuse into the xylem cells through the holes in their walls. Therefore, pericycle cells actively transport minerals out of their own cytoplasm into the extracellular space around the xylem.

④ *Diffusion into the xylem.* The active transport of minerals into the extracellular space of the vascular cylinder increases the concentration of minerals in the extracellular space. This high concentration creates a gradient that promotes the diffusion of minerals from the extracellular space into the tracheids and vessel elements of the xylem.

You can now appreciate one of the functions of the waterproof Casparian strip that seals the spaces between the endodermal cells that surround the vascular cylinder. If water and dissolved minerals could flow through the extracellular space *between* endodermal cells, then minerals would leak back out of the extracellular space of the vascular cylinder as fast as they were pumped in. This leakage would waste the energy that was used in actively transporting the minerals into the root and would reduce the concentration gradient that allows the minerals to diffuse into the conducting cells of the xylem. The Casparian strip, however, effectively leakproofs the vascular cylinder, retaining the concentrated mineral solution within the vascular cylinder.

Symbiotic Relationships Help Plants Acquire Nutrients

Many minerals are too scarce in soil water to support plant growth, although plenty of minerals may be bound up in the surrounding rock particles. One nutrient—nitrogen—is almost always in short supply both in rock particles and in soil water. Most plants have evolved beneficial relationships with other organisms that help the plants acquire scarce nutrients such as nitrates and phosphates and minerals. Examples include root–fungus relationships, called *mycorrhizae*, and root–bacteria relationships formed in nodules of legumes.

Fungal Mycorrhizae Help Plants Acquire Minerals

Under normal conditions, water-soluble minerals are released very slowly from rock particles. Furthermore, the chemical forms of the minerals may not be suitable for uptake by the plasma membranes of plant root cells. Most land plants form symbiotic relationships with fungi to form root–fungus complexes called **mycorrhizae** (singular, **mycorrhiza**), which help the plant extract and absorb minerals. Fungal strands intertwine between the root cells and extend out into the soil (Fig. 23-20). In some way that is not yet understood, the fungus renders nutrients, such as phosphorus and certain minerals, accessible for uptake by the roots, perhaps by converting rock-bound minerals into simple soluble compounds that root plasma membranes can transport. The fungus, in return, receives sugars, amino acids, and vitamins from the plant. Both the fungus and the plant can thereby grow in places where neither could survive alone, including deserts and in high-altitude, rocky soils that are low in nutrients.

Recent research has revealed that, in some forests, mycorrhizae form an immense underground web that interlinks trees—even trees of different species. This web of fungi transfers carbon compounds produced by one tree to another. Trees with access to abundant sunlight subsidize their shaded neighbors, with the mycorrhizae (like an underground Robin Hood) transferring photosynthetic products from the rich to the poor. Researchers hypothesize that nutrient transfer among trees by mycorrhizae may be an important factor in the overall health of the forest, which in turn benefits the mycorrhizae.

Bacteria-Filled Nodules on the Roots of Legumes Help Those Plants Acquire Nitrogen

Amino acids, nucleic acids, and chlorophyll all contain nitrogen, so plants need large amounts of this element. Although nitrogen gas (N_2) makes up about 79% of the atmosphere, plants can take up nitrogen only through their roots, in the form of ammonium ion (NH_4^+) or nitrate ion (NO_3^-).

Although N_2 diffuses from the atmosphere into the air spaces in the soil, it cannot be used by plants. Plants don't have the enzymes needed to carry out **nitrogen fixation**, the conversion of N_2 into ammonium or nitrate, although a variety of **nitrogen-fixing bacteria** do have these enzymes. Some of these bacteria are free-living in the soil. However, nitrogen fixation is very costly, energetically speaking, using at least 12 ATPs per ammonium ion synthesized. Consequently, bacteria don't routinely manufacture a lot of extra ammonium and liberate it into the soil.

Some plants, particularly the **legumes** (peas, clover, and soybeans), enter into a mutually beneficial relationship with certain species of nitrogen-fixing bacteria. By secreting chemicals into the soil, legumes attract nitrogen-fixing bacteria to their roots. Once there, the bacteria enter the root hairs. The bacteria then digest channels through the cytoplasm of the epidermal cells and into underlying cortex cells. As both bacteria and their host cortex cells multiply, a **nodule**, or swelling that houses the root–bacteria complexes, is formed (Fig. 23-21). A cooperative relationship develops. The plant transports sugars from its leaves down to the cortex, just as it normally would for storage. The bacteria within the cortex cells take up the sugar and use its energy for all of their metabolic processes, including nitrogen fixation. The bacteria obtain so much energy that they produce more ammonium than they need. The surplus ammonium diffuses into the cytoplasm of their host cells, providing the plant with a steady supply of usable nitrogen. Surplus ammonium also diffuses into the surrounding soil, making it better able to support other types of plants. Farmers plant legumes not only for their commercial value but also to enrich the soil with ammonium for future crops.

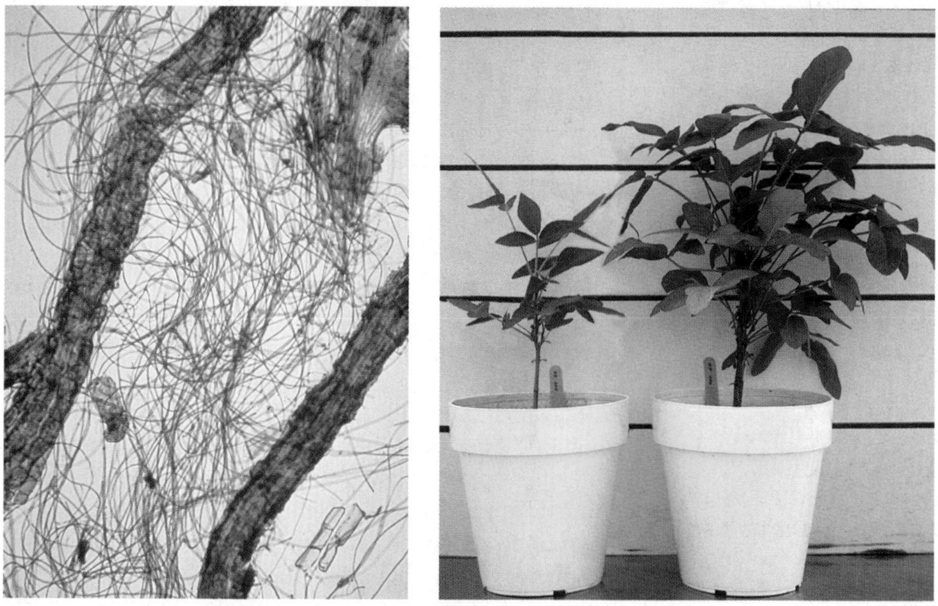

(a) (b)

Figure 23-20 Mycorrhizae, a root–fungus symbiosis
(a) A tangled meshwork of fungal strands surrounds and penetrates into the root. ***(b)*** Seedlings growing under identical conditions with (on the right) and without (left) mycorrhizal fungi show the importance of mycorrhizae in plant nutrition. Plants that participate in this unique relationship tend to grow larger and more vigorously than do those deprived of the fungus.

Figure 23-21 Nitrogen fixation in legumes
A diagram and photograph of root nodules, containing nitrogen-fixing bacteria, in legumes.

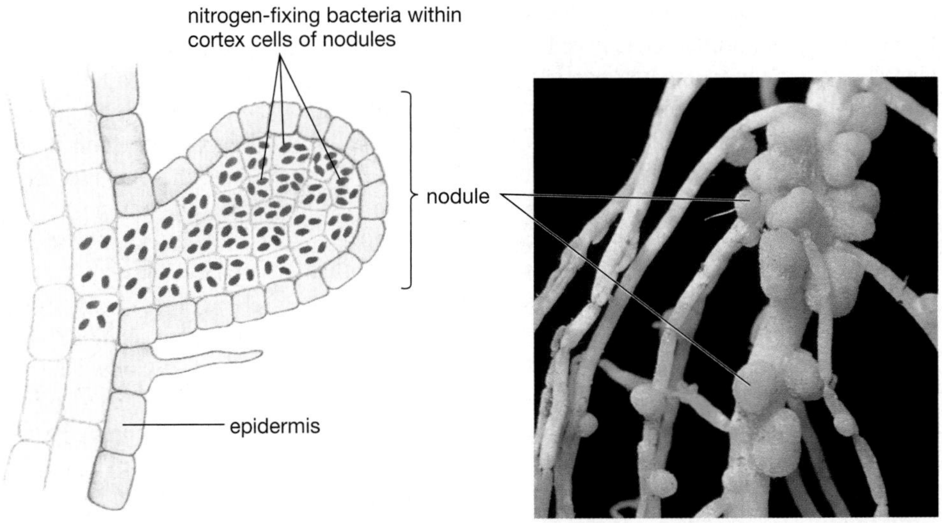

7 How Do Plants Acquire Water and Transport Water and Minerals?

Nearly 99% of the water absorbed by the roots of plants is evaporated through the stomata of leaves and, to a lesser extent, through the stomata of stems in a process called **transpiration**. As you will see in the descriptions that follow, transpiration drives the movement of water through the plant body. To understand how water enters roots, it is first necessary to understand how transpiration, acting in conjunction with the properties of water, can pull water up through the xylem of the roots and stem and into the leaves. Once you understand this process, we will return to the question of how the water enters the roots.

Water Movement in Xylem Is Explained by the Cohesion–Tension Theory

After entering the root xylem, water and minerals still must be moved to the uppermost reaches of the plant. (In redwood trees, the distance may be more than 300 feet!) **Bulk flow** moves fluids up the xylem from root to stem and leaf in land plants. Because minerals are dissolved in water, they are passively carried along as the water flows upward. How do plants overcome the force of gravity and make water flow upward? The *cohesion–tension theory* provides an explanation.

According to the **cohesion–tension theory**, water is pulled up the xylem, powered by transpiration—the evaporation of water from the leaves (Fig. 23-22). As its name suggests, this theory has two essential parts:

1. *Cohesion*. Attraction among water molecules holds water together, forming a solid chainlike column within the xylem tubes.
2. *Tension*. This "water chain" is pulled up the xylem; evaporation provides the necessary energy.

Let's briefly examine both factors.

Hydrogen Bonds Between Water Molecules Produce Cohesion

You may recall from Chapter 2 that water is a polar molecule, with the oxygen carrying a slight negative charge and the hydrogens carrying a slight positive charge. As a result, nearby water molecules attract one another, forming weak hydrogen bonds. Just as individually weak cotton threads together make the strong fabric of your jeans, the network of individually weak hydrogen bonds in water collectively produces a very high cohesion, or tendency to resist being separated. Experiments have demonstrated that the column of water within the xylem is at least as strong—and as unbreakable—as a steel wire of the same diameter. This is the "cohesion" part of the theory: Hydrogen bonds among water molecules provide the cohesion that holds together a chain of water that extends the entire height of the plant within the xylem. Supplementing the cohesion between water molecules is adhesion between water molecules and the walls of xylem. Attraction of water molecules to the cell walls of the thin xylem tubes helps the water creep upward, just as water is pulled upward into a very narrow glass tube.

Transpiration Produces the Tension That Pulls Water Upward

Transpiration provides the force for water movement—the "tension" part of the theory. As a leaf transpires, the concentration of water in the mesophyll drops. This drop causes water to move by osmosis from the xylem into the dehydrating mesophyll cells. Water molecules leaving the xylem are attached to other water molecules in the same xylem tube by hydrogen bonds. Therefore, when one water molecule leaves, it pulls adjacent water molecules up the xylem. As these water molecules move upward, other water molecules farther down the tube move up to replace them. This process continues all the way to the roots, where water in the extracellular space around the xylem is pulled in through the holes in the

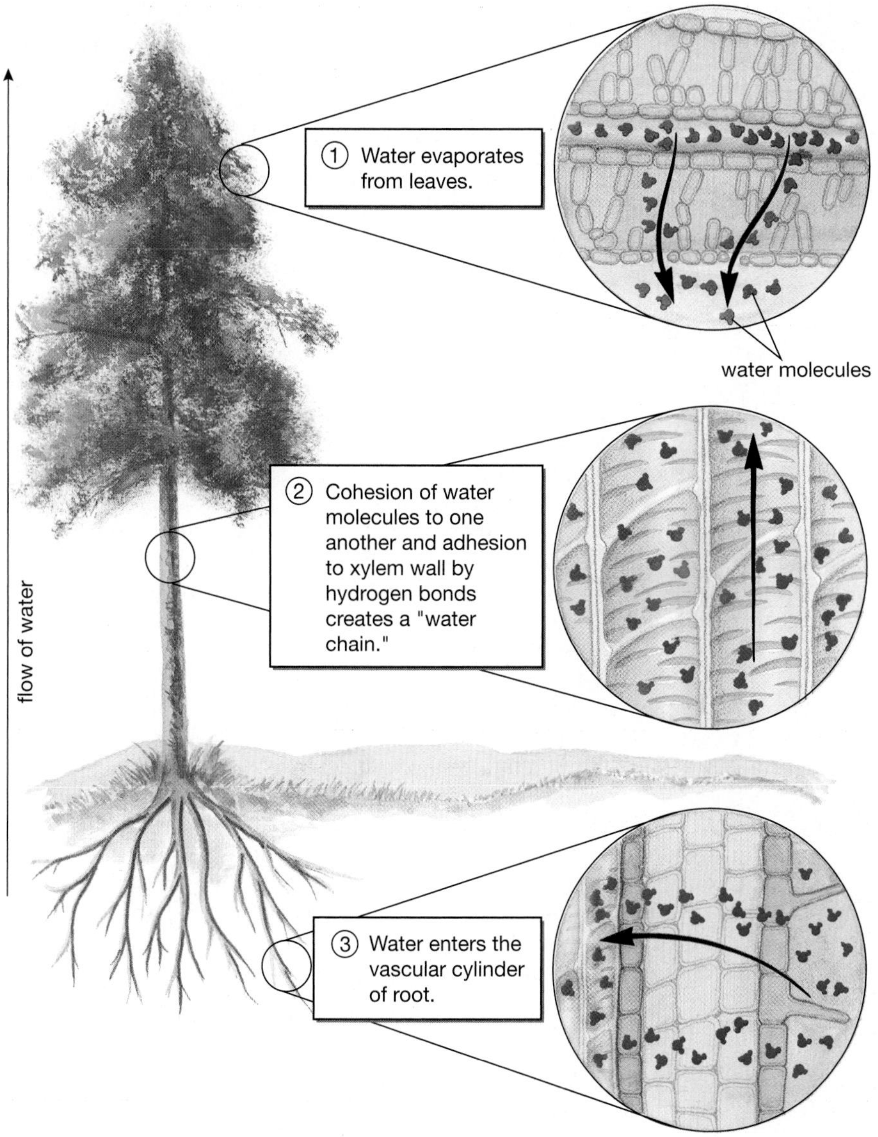

① Water evaporates from leaves.
water molecules
② Cohesion of water molecules to one another and adhesion to xylem wall by hydrogen bonds creates a "water chain."
③ Water enters the vascular cylinder of root.
flow of water

Figure 23-22 The cohesion–tension theory of water flow from root to leaf in xylem ① As water molecules evaporate out of the leaves through transpiration, other water molecules replace them from the xylem of the leaf veins. ② As the top of the "water chain" is pulled up by evaporation, the rest of the chain, all the way down to the roots, comes along as well. ③ As the molecules of the water chain retreat up the xylem in the roots, the decreased water pressure within the root xylem and the surrounding extracellular space causes water to enter from the soil water, thus steadily replenishing the bottom of the chain.

walls of vessel elements and tracheids. This upward and inward movement of water finally causes soil water to move through the endodermal cells into the vascular cylinder by osmosis. The force even extends out into the soil to pull water into the root. The force generated by the evaporation of water from the leaves, transmitted down the xylem to the roots, is so strong that water can be absorbed even from quite dry soils. Can this cohesion–tension theory explain the movement of water from soil to the topmost leaves of giant redwoods? Using a special apparatus, botanists have measured xylem water tensions strong enough to pull water up more than 600 feet (200 meters).

Water Enters Roots Mainly by Pressure Differences Created by Transpiration

Water can take two routes on its way to the vascular cylinder. First, water can follow the minerals that are actively pumped into epidermal cells, moving into the epidermal cells by osmosis (black path in Fig. 23-19). Second, water also moves readily by bulk flow *between* the cells of the epidermis and outer layers of cortex, traveling through their highly porous cell walls, but not penetrating cell membranes until the water reaches the endodermis (blue path in Fig. 23-19). At the endodermis, the waterproof Casparian strip blocks further movement of water through the spaces between cells, so water moves across the membranes of the endodermal cells in both these pathways. After leaving the endodermal cells, water moves into the vascular cylinder, entering the tracheids and vessel elements of the xylem through porous pits in their cell walls.

Movement of water into roots by osmosis due to the chemical gradient of minerals may be an important mechanism in certain plants during times when transpiration is low, such as at night when evaporation from the leaves is reduced. When the transpiration rate is high, such as occurs during the day, the second mechanism dominates. Water moves by bulk flow into the epidermis

Earth Watch
Plants Help Regulate the Distribution of Water

The land teems with a remarkable diversity of plant life. The distribution of plants on Earth is limited by both environmental factors and the adaptations of the plants. Probably the most important environmental factor influencing plant distribution is water: Cacti inhabit deserts because they can withstand drought; orchids and mahogany trees need the frequent drenching rains of the rain forest. However, people often overlook the flip side of this plant–water relationship: Plants, through transpiration, help regulate the amount and distribution of rainfall, soil water, and even river flow.

Consider the Amazon rain forest (Fig. E23-1). An acre of soil supports hundreds of towering trees, each bearing millions of leaves. The surface area of the leaves dwarfs the surface area of the soil, so up to 75% of all the water evaporating from the acre of forest is due to leaf transpiration. This transpiration raises the humidity of the air and causes rain to fall. In fact, about half of the water transpired from the leaves falls again as rain; the overall result being that about one-third of the total rainfall is water recycled by transpiration. Thus, in a very real sense, the high humidity and frequent showers that the rain forest needs to survive are partly *created by the forest itself!* If large areas of rain forest are cut down, less water evaporates in that area, so less rain falls, and new rainforest tree seedlings cannot grow. An entirely different plant community would probably become established on the disturbed land, possibly becoming a *permanent* new community.

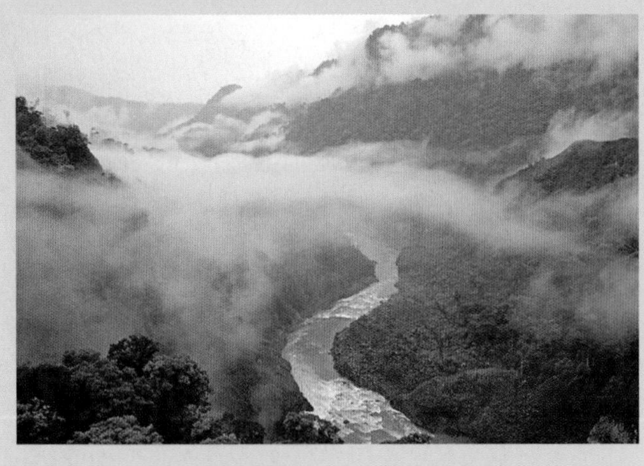

Figure E23-1 The Amazon rain forest
The rainforest community helps mold its own environment.

Plant transpiration might even have a moderating influence on some aspects of the global warming that is being created by increasing atmospheric CO_2 levels, due mostly to burning fossil fuels (such as coal and oil) during industrial activity and to cutting down forests (as we shall see in Chapter 40). According to predictions about global warming, higher CO_2 levels will raise temperatures on Earth, and a warmer planet would increase water evaporation. Increased evaporation, in turn, should lead to drier soils and the expansion of deserts. Stomata, however, open partly in response to low CO_2 levels within the guard cells. Elevated atmospheric CO_2 levels also raise CO_2 within the guard cells and cause partial stomatal closing. Further, a recent study found that plants grown in an atmosphere with high CO_2 levels have fewer stomata per unit area of leaf than do plants grown at low, preindustrial CO_2 levels. If there are fewer stomata, and they are chronically partially closed, then we might expect less transpiration from plants living in a warmer future than from preindustrial plants. Some researchers suggest that, as a result of reduced transpiration, soil moisture and river flow in the southwest United States might actually *increase* in response to increased atmospheric CO_2. These examples show that plants wield an enormous influence on what we often consider to be nonliving aspects of the biosphere, such as humidity, rainfall, soil water, and river flow. The responses of plants to human activities, neither simple nor easily predictable, can have major impacts on ecosystems.

and outer cortex layers of the root between the cells along a pathway created by their porous cell walls and then crosses the endodermal cells by osmosis. The driving force powering this bulk flow of water is lowered water pressure in the vascular cylinder created by water losses due to transpiration from the leaves. Since plants absorb most of their water during periods of high transpiration, this pressure-driven bulk flow mechanism of movement is the primary means by which water enters roots.

Summing Up
Water Transport in Xylem

Transpiration from leaves removes water from the top of a xylem tube. The transpired water is replaced by water from farther down the xylem tube, so water continues to move up the xylem tube by bulk flow. This upward flow removes water from the root xylem and the extracellular space surrounding it, promoting the movement of water by osmosis from the soil into the vascular cylinder of the root. *The flow of water is unidirectional, from root to shoot, because only the shoot can transpire.*

Adjustable Stomata Control the Rate of Transpiration

Although it provides the force that transports water and minerals to the leaves at the top of the plant, transpiration is by far the largest source of water loss—a loss that may threaten the very survival of the plant, especially in hot, dry weather. Most water is transpired through the stomata of the leaves and stem, so you might think that a plant could prevent water loss simply by closing its stomata. Don't forget, however, that photosynthesis requires carbon dioxide from the air, which diffuses into the leaf

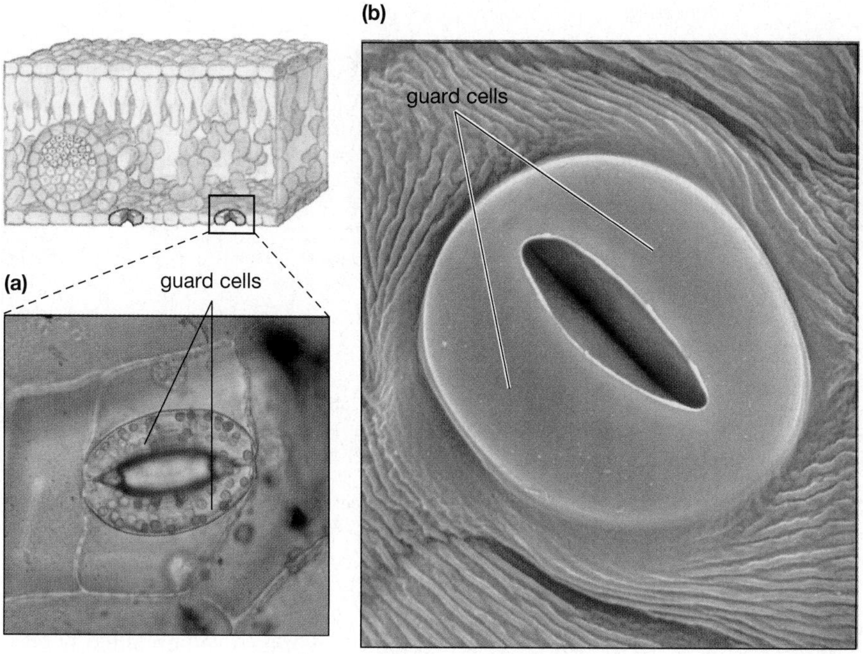

Figure 23-23 Stomata
Stomata seen through ***(a)*** the light microscope and ***(b)*** scanning electron microscope. In the light micrograph, note that the guard cells contain chloroplasts (the green ovals within the cells) but that the other epidermal cells do not.

primarily through open stomata. *Therefore, a plant, by opening and closing its stomata, must achieve a balance between carbon dioxide uptake and water loss.*

A stoma consists of a central opening surrounded by two sausage-shaped, photosynthetic guard cells that regulate the size of the opening (Fig. 23-23). With some exceptions, stomata open during the day, when sunlight allows photosynthesis, and close at night, conserving water. They will also close in the sunlight if the plant is losing too much water. Plants whose leaves are oriented horizontally generally have more stomata on the shaded, lower surface than on the sunny, upper surface, reducing evaporation.

How do plants regulate stomatal opening and closing? Stomata open when the guard cells take up water and elongate, bowing outward and increasing the space between them. Stomata close when guard cells lose water and shrink, reducing the space between them. The entry of water, in turn, is regulated by changes in the potassium content of the guard cells. The plant opens its stomata by actively pumping potassium into the guard cells, causing water to follow by osmosis. When potassium leaves the guard cells, water leaves again by osmosis, and the stomata close.

Several factors regulate the potassium concentration inside guard cells. The three most important factors are availability of both light and carbon dioxide and water levels within the leaf. These factors help the plant achieve a balance between the need to photosynthesize and the need to conserve water:

1. *Light reception.* When light strikes special pigments contained within the guard cells, it triggers a series of reactions that cause potassium to be actively transported into the guard cells. Water follows by osmosis, and the stomata open. At night, when light is not present to activate the pigments, the potassium pumping stops. The "extra" potassium within the guard cells diffuses back out, and the stomata close, conserving water.
2. *Carbon dioxide concentration.* Low CO_2 concentrations (such as those that occur during the day when photosynthesis exceeds cellular respiration) stimulate the active transport of potassium into the guard cells. This transport causes stomata to open and allows CO_2 to diffuse in. At night, cellular respiration in the absence of photosynthesis raises CO_2 levels, halting the inward transport of potassium and allowing the guard cells to close.
3. *Water.* If a leaf loses water faster than it can replace water from the xylem, it begins to wilt, and the mesophyll cells release a hormone called **abscisic acid.** This hormone strongly inhibits the active transport of potassium into the guard cells, overriding the stimulatory effects of light and low CO_2 levels, so potassium pumping stops. As potassium diffuses out of the guard cells, water follows by osmosis, the guard cells shrink, and the stomata close. As you might guess, when your house or garden plants are wilted, they are unable to carry out normal levels of photosynthesis.

8 How Do Plants Transport Sugars?

Sugars synthesized in the leaves must be moved to other parts of the plant. There they nourish nonphotosynthetic structures such as roots or flowers and can be stored in the cortex cells of the root and stem. Sugar transport is the function of phloem.

Botanists studying phloem contents have employed a most unlikely lab assistant: the aphid. *Aphids* are insects

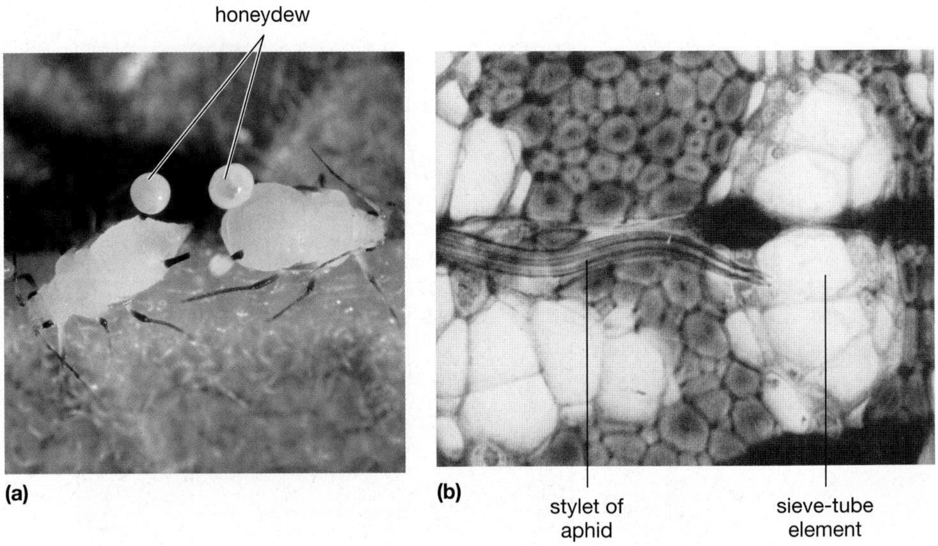

Figure 23-24 Aphids feed on the sugary fluid in phloem sieve tubes ***(a)*** When an aphid pierces a sieve tube, pressure in the tube forces the fluid out of the plant and into the aphid's digestive tract. The pressure can be so great that fluid is forced completely through the aphid and out its anus, as "honeydew." This fluid is collected by certain species of ants that act as "shepherds" to the aphids, defending them from predators in return for a diet of sweet honeydew. ***(b)*** The flexible stylet of an aphid, passing through many layers of cells to penetrate a sieve-tube element.

that feed on the fluid contained in phloem sieve tubes. An aphid inserts its *stylet*, a pointed, hollow tube, through the epidermis and cortex of a young stem into a sieve tube (Fig. 23-24). The aphid can then relax and let the plant do the work. The fluid in the sieve tubes is under pressure and flows through the stylet into the aphid's digestive tract (sometimes with enough pressure to force its way out the other end!). By cutting off the aphid but leaving its stylet in place, botanists have collected sieve-tube fluid and found that it consists mostly of sucrose and water, with the sucrose content ranging from about 10% to almost 25% by weight. How is this concentrated sugar solution moved about the plant?

The Pressure-Flow Theory Explains Sugar Movement in Phloem

The movement of fluid in phloem is directed by sugar production and use. Any structure that actively synthesizes sugar is said to be a **source** away from which phloem fluid will be transported. Conversely, any structure that uses up sugar or converts sugar to starch is said to be a **sink** toward which phloem fluids will flow. A newly forming leaf will be a sink as it develops, with phloem flowing up into it from more mature leaves. When the leaf matures, it will photosynthesize and produce sugar, becoming a source for phloem flow to other newly developing leaves, to flowers or fruits, or to the roots. Therefore, fluid in phloem can move either up or down the plant, depending on the metabolic demands of the various parts of the plant at any given time.

The most widely accepted mechanism of sugar transport in phloem is the **pressure-flow theory** (Fig. 23-25), which relies on differences in hydrostatic pressure (water pressure) to move fluid through sieve tubes. Let's illustrate this theory by following sucrose movements from a mature leaf to a developing fruit:

① *Sucrose source: photosynthesis.* When a leaf is photosynthesizing rapidly, it manufactures lots of glucose, much of which is converted to the larger molecule sucrose.

② *Phloem sieve-tube loading.* Much of this sucrose is actively transported into companion cells of the phloem in the leaf veins. This movement raises the concentration of sucrose within the companion cells, so sucrose then diffuses down its concentration gradient through plasmodesmata into adjacent sieve-tube elements. This diffusion in turn raises the sucrose concentration in the leaf sieve tube.

③ *Osmosis into the leaf sieve tube.* The high sucrose concentration in the leaf sieve tube lowers the water concentration in the sieve tube, causing water to enter the sieve tube by osmosis from nearby xylem. Hydrostatic pressure increases as more water molecules enter the tube.

④ *Sucrose sink: developing fruit.* Meanwhile, some distance away along the same sieve tube, sucrose is actively transported out of sieve-tube elements and companion cells into the cells of a fruit. The concentration of sugar in the fruit is raised, and the concentration of sugar in the sieve tubes is lowered.

⑤ *Osmosis out of the fruit sieve tube.* Water leaves the sieve tube by osmosis and follows the sugar into the fruit. Hydrostatic pressure drops within the tube.

⑥ *Bulk flow, driven by a hydrostatic pressure gradient.* Water moving by osmosis follows the sucrose into the sieve tube near the leaf causes hydrostatic pressure to build up in the leaf portion of the sieve tube. Meanwhile, water entering the fruit by osmosis causes reduced hydrostatic pressure in the sieve tube near the fruit. In response to this pressure gradient, water moves by bulk flow from the leaf portion of the phloem into the fruit portion of the phloem, carrying the dissolved sugar.

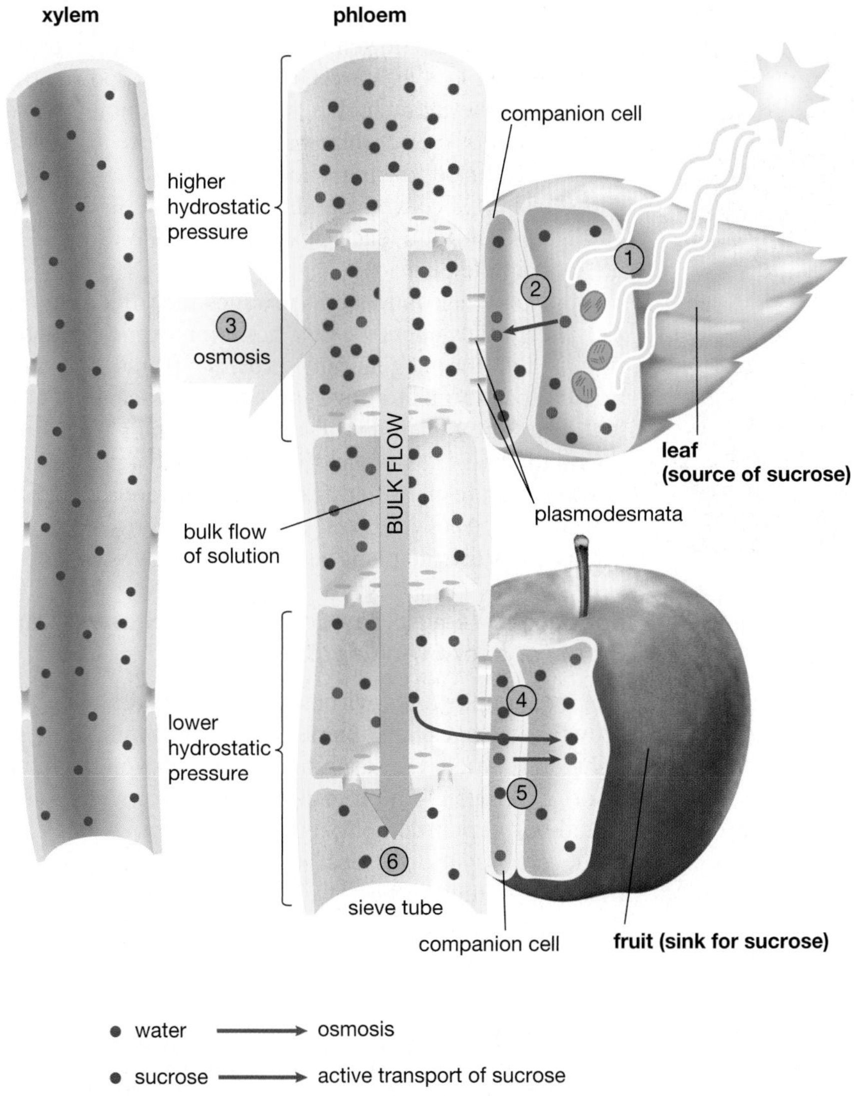

Figure 23-25 The pressure-flow theory This theory relies on differences in hydrostatic pressure to move fluid through phloem sieve tubes. (1) A photosynthesizing leaf manufactures sucrose (red dots), which (2) is actively transported (red arrow) into a nearby companion cell in phloem. The sucrose diffuses into the adjacent sieve-tube element through plasmodesmata, raising the concentration of sucrose in the sieve-tube element. (3) Water (blue dots) leaves nearby xylem and moves into the "leaf end" of the sieve tube by osmosis (blue arrow), raising the hydrostatic pressure as increasing numbers of water molecules enter the fixed volume of the tube. (4) The same sieve tube connects to a developing fruit. At the "fruit end" of the tube, sucrose enters the companion cells by diffusion through plasmodesmata. It is then actively transported out of the companion cells and into the fruit cells. (5) Water moves out of the tube by osmosis, lowering the hydrostatic pressure within the tube. (6) High pressure in the leaf end of the phloem and low pressure in the fruit end cause water, together with any dissolved solutes, to flow in bulk from leaf to fruit.

Evolutionary Connections

What Are Some Special Adaptations of Roots, Stems, and Leaves?

Not all roots are sinuous fibers, not all stems are smooth and upright, and not all leaves are flat and fanlike. Just as evolution has changed the basic shape of the vertebrate forelimb to suit the demands of running, swimming, and flying, so, too, have plant parts become modified in response to environmental demands by the forces of natural selection. You may be surprised to learn that many familiar structures are derived from unlikely parts of a plant.

Although we will highlight unusual adaptations, don't forget that *all plants are adapted to their environments.* The "typical" leaf of an oak or maple is just as much a special adaptation as is a cactus spine or daffodil bulb.

Some Specialized Roots Store Food; Others Photosynthesize

Roots have probably undergone fewer unusual modifications of their basic structure than have either stems or leaves. Some roots have extreme specializations for storage, such as the beet, carrot, or sweet potato. Some of the most bizarre root adaptations occur in certain orchids that grow perched on trees. A few of these aerial orchids have green, photosynthetic roots; in fact, for some orchids, the green roots are the only photosynthetic part of the plant (Fig. 23-26).

Some Specialized Stems Produce New Plants, Store Water or Food, or Produce Thorns or Climbing Tendrils

Many plants have modified stems that perform functions that are very different from the original one of raising leaves up to the light. Strawberries, for example,

Figure 23-26 An adaptation of roots
This *Cattleya* orchid is growing on a tree in the Amazon basin. Its roots hang down below the tree branch.

grow horizontal **runners** that snake out over the soil, sprouting new strawberry plants where nodes touch the soil (Fig. 23-27a). These new plants are connected with the "mother" plant, but once the plantlets form roots, they can live independently.

Some plants, such as the baobab tree (Fig. 23-27b), store water in aboveground stems. Many other plants store carbohydrates in underground stems. The common white potato is actually a storage stem; each eye is a lateral bud, ready to send up a branch next year, using the energy stored as starch in the potato to power the growth of the branch. Irises have horizontal underground stems called **rhizomes**, which store carbohydrates produced during the summer. Irises can be propagated by cutting up the rhizome; if it contains enough stored food, each piece with a node can generate a complete plant.

Many aboveground stems produce modified branches with special functions. One common branch adaptation is the **thorn**, generally growing from the usual branch location just above the site of attachment of a leaf. Hard, pointy thorns discourage animals from dining on the branches. Some of the branches of grapes and Boston ivy are modified into grasping **tendrils**, which coil around trees, trellises, or buildings, providing the otherwise prostrate plant better access to sunlight.

Specialized Leaves May Conserve and Store Water, Store Food, or Even Capture Insects

The most important environmental factors affecting the growth of leaves are temperature and availability of light and water. For example, plants growing on the floor of a tropical rain forest have plenty of water year-round but very little light, owing to the deep shade cast by several layers of trees above them. Consequently, their leaves tend to be extremely large—an adaptation demanded by the low light level and permitted by the abundant water.

(a)

(b)

Figure 23-27 Some adaptations of stems
(a) The beach strawberry can reproduce via runners, horizontal stems that spread out over the surface of the sand. If a node of a runner touches the soil, it will sprout roots and develop into a complete plant. ***(b)*** The baobab tree develops an enormously fat, water-storing trunk. The baobab grows in dry regions; when it rains, it is to the tree's advantage to store all the water it can get. Some baobab trees have trunks so large that they have been hollowed out and used as small houses and, in one case, as a jail!

At the other extreme, deserts receive bright sunlight virtually every day but have limited water and scorching temperatures. Desert plants have evolved two strikingly different adaptations to this situation. Some have very thick leaves with large cells that store water from the infrequent rains against the inevitable long droughts (Fig. 23-28a). Such leaves are covered with a thick cuticle that greatly reduces the evaporation of water. Cacti use the opposite strategy, reducing the leaves to thin spines that protect the plant from herbivores and reduce water loss (Fig. 23-28b). Photosynthesis in cacti occurs in cortex cells of the green, water-storing stems.

Modified leaves in other plants function in ways unrelated to photosynthesis or water conservation. The common edible pea, for example, grasps fences, mailbox posts, or other plants with clinging tendrils. Unlike the tendrils of grapes, which are derived from branches, pea tendrils are slender, supple leaflets. Some plants, such as onions, daffodils, and tulips, use thick, fleshy leaves as storage organs. A daffodil bulb consists of a short stem bearing thick, overlapping leaves that store nutrients during the winter (Fig. 23-28c).

Finally, a few plants have turned the table on the animals and have become predators. For example, both Venus flytraps and sundews (see the opening photo for this chapter; see also Fig. 25-8) have leaves that are modified into snares for trapping unwary insects.

As varied and in some cases bizarre as these leaf specializations are, the most extreme and most important leaf modification is the flower. As we will see in Chapter 24, these "reproductive leaves" enabled flowering plants to become the dominant plants on land.

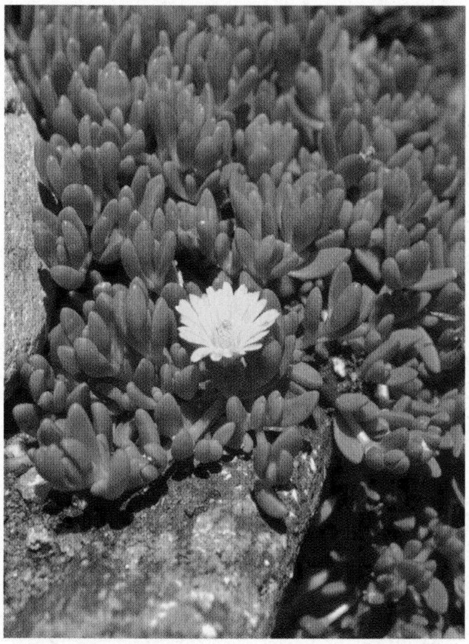

(a)

(b)

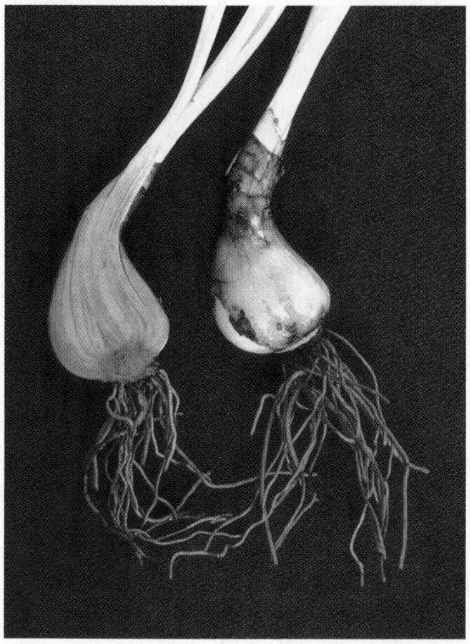

(c)

Figure 23-28 Some adaptations of leaves
(a) Desert plants receive plenty of light but very little water. Some desert plants have evolved fleshy leaves that store water from the occasional rains, just as the baobab tree does in its trunk. ***(b)*** Spines of desert cacti are leaves whose surface area has been minimized, reducing evaporation and protecting the plant from grazing animals. ***(c)*** Daffodil bulbs consist of short central stems surrounded by thick, water- and food-storing leaves. Like other monocots, these bulbs form roots as outgrowths of the base of the stem.

CASE STUDY REVISITED

A Beautiful Death Trap

Why does the sundew trap insects? As you have learned, nitrogen-fixing bacteria in soil are a major source of this crucial nutrient for plants. The soils in bogs are typically very acidic, and acidic conditions are hostile to nitrogen-fixing bacteria. Thus, bog soils tend to be very low in nitrogen and are prime habitat for carnivorous plants such as the sundew. Protein-rich bodies of animals such as insects are excellent sources of nitrogen, if they can only be caught and digested—a formidable challenge for a plant. Sundews meet this challenge using glandular hairs on their leaves. These remarkable structures secrete a complex brew of substances. Sweet, sticky nectar and a glue-like mucilage, like biological flypaper, attract and then entangle unsuspecting insects. Then, several different enzymes in the droplets attack the insects' outer skeleton and inner proteins, breaking them down to their nitrogen-rich components, which are absorbed into the plant's leaves by the same glands that secrete the enzymes. The sundew's hairs actually move toward the insect in response to vibrations caused by its struggles. These vibrations cause special ion channels in the cells of the hairs to open, allowing an influx of charged ions. This electrical signal, by a mechanism that remains somewhat mysterious, causes the hairs to bend.

On a walk, you encounter a plant with leaves shaped like a deep cup. Peering inside, you see a soup of dead insects in various stages of decomposition floating in liquid. Using the flashlight and magnifying glass that you always carry in your pocket on hikes, you examine the inner walls of the cup and see small hairs, all angled in the same direction. In what direction are the hairs pointing, and why? What types of chemicals might you discover in the fluid within the cup if you could analyze them? What do you think is happening in there? In what type of soil is this plant likely to be growing?

Summary of Key Concepts

1) How Are Plant Bodies Organized, and How Do They Grow?

The body of a land plant consists of root and shoot. Roots are normally underground and have six functions: They (1) anchor the plant in the soil; (2) absorb water and minerals from the soil; (3) store surplus photosynthetic products; (4) transport water, minerals, photosynthetic products, and hormones; (5) produce some hormones; and (6) interact with soil fungi and microorganisms that provide nutrients. Shoots are normally located above ground and consist of stem, leaves, buds, and (in season) flowers and fruit. Shoot functions include (1) photosynthesis, (2) transport of materials, (3) reproduction, and (4) hormone synthesis.

Plant bodies are composed of two main classes of cells: meristem cells and differentiated cells. Meristem cells are undifferentiated cells that retain the capacity for mitotic cell division. Differentiated cells arise from divisions of meristem cells, become specialized for particular functions, and usually do not divide. Most meristem cells are located in apical meristems at the tips of roots and shoots and in lateral meristems in the shafts of roots and shoots. Primary growth (growth in length and differentiation of parts) results from the division and differentiation of cells from apical meristems; secondary growth (growth in diameter) results from the division and differentiation of cells from lateral meristems.

2) What Are the Tissues and Cell Types of Plants?

Plant bodies consist of three tissue systems: the dermal, ground, and vascular systems. The dermal tissue system forms the outer covering of the plant body. The dermal tissue system of leaves and of primary roots and stems is normally a single cell layer of epidermis. Dermal tissue after secondary growth is a multilayered covering of cork.

The ground tissue system consists of a variety of cell types, including parenchyma, collenchyma, and sclerenchyma. Most are involved in photosynthesis, support, or storage. Ground tissue makes up most of a young plant during primary growth. During secondary growth of stems and roots, ground tissue becomes an increasingly small part of the plant body.

The vascular tissue system consists of xylem, which transports water and minerals from the roots to the shoots, and phloem, which transports water, sugars, amino acids, and hormones throughout the plant body.

3) Roots: Anchorage, Absorption, and Storage

Primary growth in roots results in a structure consisting of an outer epidermis, an inner vascular cylinder of conducting tissues, and cortex between the two. The apical meristem near the tip of the root is protected by the root cap. Cells of the root epidermis absorb water and minerals from the soil. Root hairs are projections of epidermal cells that increase the surface area for absorption. Most cortex cells store surplus sugars (usually in the form of starch) produced by photosynthesis. The innermost layer of cortex cells is the endodermis, which controls the movement of water and minerals from the soil into the vascular cylinder. The vascular cylinder contains the conducting tissues—xylem and phloem.

4) Stems: Reaching for the Light

Primary growth in dicot stems results in a structure consisting of an outer, waterproof epidermis; supporting and photosynthetic cells of cortex beneath the epidermis; vascular tissues of xylem and phloem; and supporting and storage cells of pith at the center. Leaves and lateral buds are found at nodes along the surface of the stem. Under the proper hormonal conditions, lateral buds may sprout into a branch.

Secondary growth in stems results from cell divisions in the vascular cambium and cork cambium. Vascular cambium produces secondary xylem and secondary phloem, increasing the stem's diameter. Cork cambium produces waterproof cork cells that cover the outside of the stem.

5) **Leaves: Nature's Solar Collectors**

Leaves are the main photosynthetic organs of plants. The blade of a leaf consists of a waterproof outer epidermis surrounding mesophyll cells, which have chloroplasts and which carry out photosynthesis, and vascular bundles of xylem and phloem, which carry water, minerals, and photosynthetic products to and from the leaf. The epidermis is perforated by adjustable pores called *stomata* that regulate the exchange of gases and water.

6) **How Do Plants Acquire Nutrients?**

Most minerals are taken up from the soil water by active transport into the root hairs. These minerals diffuse into the root through plasmodesmata to the pericycle, just inside the vascular cylinder. There they are actively transported into the extracellular space of the vascular cylinder. The minerals diffuse from the extracellular space into the tracheids and vessel elements of xylem.

Many plants have fungi called *mycorrhizae* associated with their roots that help absorb soil nutrients. Nitrogen can be absorbed only as ammonium or nitrate, both of which are scarce in most soils. Legumes have evolved a cooperative relationship with nitrogen-fixing bacteria that invade legume roots. The plant provides the bacteria with sugars, and the bacteria use some of the energy in those sugars to convert atmospheric nitrogen to ammonium, which is then absorbed by the plant.

7) **How Do Plants Acquire Water and Transport Water and Minerals?**

The cohesion–tension theory explains xylem function: The cohesion of water molecules to one another by hydrogen bonds holds together the water within xylem tubes almost as if it were a solid chain. As water molecules evaporate from the leaves during transpiration, the hydrogen bonds pull other water molecules up the xylem to replace them. This movement is transmitted down the xylem to the root, where water loss from the vascular cylinder promotes water movement across the endodermis from the soil water by osmosis. Because the cells of the root epidermis and cortex are loosely packed and have porous walls, water in the soil has a continuous, uninterrupted pathway through the outer layers of the root to the waterproof layer of the Casparian strip between endodermal cells. Both mineral uptake and the upward movement of water in xylem, pulled by transpiration, contribute to a concentration gradient of water across the endodermal cells, with a higher concentration of free water molecules in the extracellular space outside the endodermis than in the extracellular space inside the endodermis. Therefore, water moves by osmosis across the plasma membranes of the endodermal cells into the extracellular space of the vascular cylinder. The water pressure gradient caused by loss of water through transpiration is the primary force drawing water into the root.

8) **How Do Plants Transport Sugars?**

The pressure-flow theory explains sugar transport in phloem: Parts of the plant that synthesize sugar (for example, leaves) export sugar into the sieve tube. Increasing sugar concentrations attract water entry by osmosis, causing high hydrostatic pressure in that part of the phloem. Parts of the plant that consume sugar (for example, fruits) remove sugar from the sieve tube. The loss of sugar causes the loss of water by osmosis, resulting in low hydrostatic pressure. Water and dissolved sugar move by bulk flow in the sieve tube from areas of high to low pressure.

Key Terms

abscisic acid *p. 487*
annual ring *p. 478*
apical meristem *p. 470*
bark *p. 479*
blade *p. 479*
branch root *p. 475*
bulk flow *p. 484*
cambium *p. 470*
Casparian strip *p. 475*
cohesion–tension theory *p. 484*
collenchyma *p. 471*
companion cell *p. 473*
cork cambium *p. 479*
cork cell *p. 471*
cortex *p. 475*
cuticle *p. 470*
dermal tissue system *p. 470*
dicot *p. 468*
differentiated cell *p. 468*
endodermis *p. 475*
epidermal tissue *p. 470*
epidermis *p. 470*
fibrous root system *p. 473*
ground tissue system *p. 470*
guard cell *p. 480*
heartwood *p. 477*
internode *p. 476*
lateral bud *p. 476*
lateral meristem *p. 470*
leaf *p. 479*
leaf primordium *p. 476*
legume *p. 483*
meristem cell *p. 468*
mesophyll *p. 480*
mineral *p. 481*
monocot *p. 468*
mycorrhiza *p. 483*
nitrogen fixation *p. 483*
nitrogen-fixing bacterium *p. 483*
node *p. 476*
nodule *p. 483*
nutrient *p. 481*
palisade cell *p. 480*
parenchyma *p. 471*
pericycle *p. 475*
periderm *p. 471*
petiole *p. 479*
phloem *p. 473*
pit *p. 473*
pith *p. 477*
pressure-flow theory *p. 488*
primary growth *p. 470*
primary phloem *p. 477*
primary root *p. 473*
primary xylem *p. 477*
rhizome *p. 490*
root *p. 468*
root cap *p. 473*
root hair *p. 471*
root system *p. 468*
runner *p. 490*
sapwood *p. 477*
sclerenchyma *p. 471*
secondary growth *p. 470*
secondary phloem *p. 477*
secondary xylem *p. 477*
shoot system *p. 468*
sieve plate *p. 473*
sieve tube *p. 473*
sieve-tube element *p. 473*
sink *p. 488*
source *p. 488*
spongy cell *p. 470*
stem *p. 468*
stoma *p. 477*
taproot system *p. 473*
tendril *p. 490*
terminal bud *p. 476*
thorn *p. 490*
tracheid *p. 471*
transpiration *p. 484*
vascular bundle *p. 480*
vascular cambium *p. 477*
vascular cylinder *p. 475*
vascular tissue system *p. 470*
vein *p. 480*
vessel *p. 473*
vessel element *p. 473*
xylem *p. 471*

Thinking Through the Concepts

Multiple Choice

1. *In a mycorrhizal relationship, a plant root has a symbiotic relationship with*
   a. a fungus that helps obtain minerals from the soil
   b. a fungus that helps in the fixation of nitrogen from the air into a form usable by the plant
   c. bacteria that help obtain minerals from the soil
   d. bacteria that help in the fixation of nitrogen from the air into a form usable by the plant
   e. an alga that helps the plant photosynthesize
2. *Which of the following is involved in the cohesion–tension theory of water movement in plants?*
   a. the presence of hydrogen bonds that hold water molecules together
   b. the attraction of water molecules to the walls of the xylem
   c. the diffusion of water from cells in the root to cells in the shoot
   d. the evaporation of water through the stomata
   e. All of the above are involved in this theory.
3. *You and a friend carve your initials 5 feet above the ground on a tree on campus. The tree is now 40 feet tall. When you come back for your twenty-fifth reunion, the tree will be 100 feet tall. How high above the ground should you look for your initials?*
   a. 5 feet
   b. 40 feet
   c. 60 feet
   d. 95 feet
   e. 100 feet
4. *If you wanted to show a friend how much you had learned in biology class, which of the following would you NOT use to identify a plant as a dicot?*
   a. leaves with veins arranged like a net
   b. hand-shaped leaves
   c. six petals
   d. a taproot
   e. a seed with two cotyledons
5. *When water enters a plant root, it is forced to travel through the endodermal cells by the waxy*
   a. cuticle
   b. epidermis
   c. periderm
   d. xylem
   e. Casparian strip
6. *Which of the following is NOT a special adaptation of certain roots, stems, or leaves?*
   a. roots—photosynthesis
   b. stems—water storage
   c. leaves—defense
   d. roots—prey capture
   e. stems—producing new plants

? Review Questions

1. Describe the locations and functions of the three tissue systems in land plants.
2. Distinguish between primary growth and secondary growth, and describe the cell types involved in each.
3. Distinguish between meristem cells and differentiated cells. Which meristems cause primary growth? Which ones form secondary growth? Where is each type located?
4. Diagram the internal structure of a root after primary growth, labeling and describing the function of epidermis, cortex, endodermis, pericycle, xylem, and phloem. What tissues are located in the vascular cylinder?
5. How do xylem and phloem differ?
6. What are the main functions of roots, stems, and leaves?
7. What types of cells form root hairs? What is the function of root hairs?
8. Diagram the internal structure of leaves. What structures regulate water loss and CO_2 absorption by a leaf?
9. What role does abscisic acid play in controlling the opening and closing of stomata? Describe the daily cycle of the opening and closing of guard cells. How are various environmental conditions involved in this process?

Applying the Concepts

1. A mutant form of aphid, the klutzphid, inserts its stylet into the vessel elements of xylem. What materials are found in the fluids of xylem? Could an aphid live on xylem fluid? Would xylem fluid flow into the aphid? Explain your answer.
2. One of the foremost goals of molecular botanists is to insert the genes for nitrogen fixation, or the ability to enter into symbiotic relationships with nitrogen-fixing bacteria, into crop plants such as corn or wheat (see Chapter 13). Why would the insertion of such genes be useful? What changes in farming practices would this technique allow?
3. We learned in Chapter 2 about the peculiar characteristics of water. Discuss several ways in which the evolution of vascular plants has been greatly influenced by water's special characteristics.
4. A major environmental problem is desertification, in which overgrazing by cattle or other animals results in too few plants in an area. Show how what you know about the movement of water through plants enables you to understand this process, in which there is less water in the atmosphere, less rain, and thus dry, desert-like conditions.
5. The tropical rain forest contains a large number of as-yet unidentified plants, many of which may have uses as medicines or food. If you were given the job of searching a particular portion of the rain forest for useful products, how would you use the information you gained from this chapter to help you narrow your search? What kinds of plant tissues or organs would be most likely to contain such products?

6. The desert tends to have two types of plants with respect to their root systems—small grasses or herbs, and shrubs or small trees. The grasses and herbs typically form fibrous root systems. The shrubs and trees form taproot systems. What advantages can you think of for each system? How does each type of root allow for survival in a desert environment?
7. Grasses (monocots) form their primary meristem near the ground surface rather than at the tips of branches the way dicots do. How does this feature allow you to grow a lawn and mow it every week in the summer? What would happen if you had a dicot lawn and tried to mow it?
8. Discuss the structures and adaptations that might occur in the leaves of plants living in (a) dry, sunny habitats; (b) wet, sunny habitats; (c) dry, shady habitats; and (d) wet, shady habitats. Which of these habitats do you think would be most inhospitable (for example, in which habitat would it be most difficult to design a functioning leaf)?

For More Information

Baskin, Y. "Forests in the Gas." *Discover*, October 1994. As carbon dioxide increases in the atmosphere, the relationships of plants in the natural environment will change.

Day, S. "A Shot in the Arm for Plants." *New Scientist*, January 9, 1993. Plants exposed to insects or disease may produce chemicals that are transported throughout their body to protect them from future attack.

Line, L. "The Return of an American Classic." *Audubon*, September/October 1997. Dutch elm disease has nearly wiped out the majestic American elm tree, but a few hardy survivors with resistance to this fungus are being propagated for sale by the year 2000.

Weiss, R. "When Plants Act Like Animals." *National Wildlife*, December 1994–January 1995. A fun look at plant movements.

Zimmer, C. "The Processing Plant." *Discover*, September 1995. The purple pitcher plant supplements its diet by digesting flies, with the surprising help of insect larvae that thrive in the pitcher plant's "digestive chamber."

Zimmer, C. "The Web Below." *Discover*, November 1997. An underground web of mycorrhizae transfers nutrients between trees and helps maintain forest health.

Answers to Multiple-Choice Questions

1. a 2. e 3. a 4. c 5. e 6. d

MEDIATUTOR
Plant Form and Function

CD Activities

Activity 23.1: Plant Anatomy

Estimated time: 5 minutes

This tutorial will allow the student to move about the plant body, zooming in and out to view and become more familiar with the plant body. Click on the plant structures to view an enlarged illustration of that structure's anatomy.

Activity 23.2: Plant Transport Mechanisms

Estimated time: 5 minutes

In this tutorial you will explore two mechanisms of transport in plants. Section one illustrates the mechanism of water and mineral transport in the xylem, and section two illustrates the mechanism of sugar transport in the phloem.

Activity 23.3: Primary and Secondary Growth

Estimated time: 5 minutes

In this tutorial we will explore the relationship between primary and secondary growth in plants. Plant stems and roots grow in length by primary growth and in width by secondary growth. Secondary growth produces wood and bark. The focus here is on stems; roots show similar development patterns.

Start the MediaTutor Student CD-ROM and enter the activity number in the Quick Search box to be taken directly to that activity.

Web Investigations

Case Study: A Beautiful Death Trap

Estimated time: 15 minutes

Remember the plant Seymour in "The Little Shop of Horrors"? Now, that's a carnivorous plant! Real carnivorous plants may not be quite as flashy, but they have evolved fascinating mechanisms for attracting, catching, and digesting their prey. This exercise takes a brief look at "Seymour's relatives."

Go to http://www.prenhall.com/audesirk6, the Audesirk Companion Web site. Select Chapter 23 and the Web Investigation to begin.

"By the time I get to the wood, I am carrying all manner of seeds hooked in my coat or piercing my socks or sticking by ingenious devices to my shoestrings. I let them ride. After all, who am I to contend against such ingenuity? It is obvious that nature, or some part of it in the shape of these seeds, has intentions beyond this field, and has made plans to travel with me."

Loren Eiseley in* The Immense Journey *(1957)

Poppies blanket California's Antelope Valley after early spring rains. Desert wildflowers bloom and set seed rapidly, taking advantage of the brief spring moisture before the drought and heat of summer. The bright colors, shapes, and scents of flowers attract animal pollinators.

24 Plant Reproduction and Development

AT A GLANCE

Case Study: Walk Through a Meadow

1) **What Are the Features of Plant Life Cycles?**

2) **How Did Flowers Evolve?**
Complete Flowers Have Four Major Parts

3) **How Do Gametophytes Develop in Flowering Plants?**
Pollen Is the Male Gametophyte
The Embryo Sac Is the Female Gametophyte

4) **How Does Pollination Lead to Fertilization?**

5) **How Do Seeds and Fruits Develop?**
The Seed Develops from the Ovule and Embryo Sac
The Fruit Develops from the Ovary Wall
Seed Dormancy Helps Ensure Germination at an Appropriate Time

6) **How Do Seeds Germinate and Grow?**
The Shoot Tip Must Be Protected
Cotyledons Nourish the Sprouting Seed
Controlling the Development of the Seedling

Evolutionary Connections: Adaptations for Pollination and Seed Dispersal

Case Study Revisited: Walk Through a Meadow

CASESTUDY

Walk Through a Meadow

Imagine walking through a wildflower-strewn meadow or amidst the poppy and lupine-carpeted hillsides of the Antelope Valley in southern California shown in the opening photograph of this chapter. You may be tempted to think that these floral displays were created just for human enjoyment. In fact, plants have evolved showy flowers for the birds and bees—and for beetles, moths, and even bats. Perhaps the real wonder is that we, too, are attracted to flowers. Why are they so strikingly and diversely colored? Wouldn't it be simpler and more efficient for the plant if the flowers were green and could photosynthesize?

If you begin sneezing just imagining walking through a flowering or grassy field in summer, you might have a special interest in pollen. What is pollen, and why is the summer air so full of it? If you could return to the Antelope Valley in the fall, where poppy flowers swayed in the spring, you'd find elongated seed pods in various shades of green and brown—some filled with tiny black seeds, others split open and empty. Walk through any wild meadow in fall and your socks will be prickly with small burs and grass seeds that (to paraphrase Loren Eiseley) have made plans to travel with you. Although the plants are dying, you know that the field will bloom again in the spring. How does the cycle of life operate in a plant? Let's find out. ■

1 What Are the Features of Plant Life Cycles?

Do plants engage in sexual activity? Certainly. Many plants can reproduce either sexually or asexually. Asexual reproduction in plants usually involves part of a single plant (say, a stem) giving rise to a new plant. Cells of the parent plant undergo mitosis to produce their offspring asexually. Therefore, these offspring are genetically identical to the parent. In Chapter 23 you encountered several methods of asexual reproduction, including the spreading of runners by strawberries, bulb production by daffodils, and the sprouting of rhizomes by irises. Asexual reproduction is often highly effective, allowing plants to colonize an entire area where the original parent found optimal conditions.

However, if an offspring is genetically identical to its parent, then the offspring is only as well adapted to the environment as its parent. What if the environment changes? Most sexually produced offspring combine genes from both parents, and therefore may be endowed with traits that differ from those of either parent. This new combination of traits may help the offspring cope with changing environments or survive in slightly different habitats. As a result, most organisms, including plants, reproduce sexually, at least some of the time.

During the familiar animal life cycle, animals with diploid ($2n$) cells produce haploid (n) gametes (sperm or eggs) by the process of meiosis. The gametes fuse to form a new diploid cell that develops into the adult organism through repeated mitotic cell divisions. The plant life cycle, however, is a bit more complex. Plants have two distinct, multicellular forms, one diploid and one haploid, that give rise to each other. For this reason, the plant life cycle is called **alternation of generations**: Diploid plants (called *sporophytes*) alternate with haploid plants (called *gametophytes*).

Plants with the most easily visible alternating generations (mosses and ferns, for example) do not produce flowers. To illustrate the alternation of generations, let's examine the life cycle of a fern (Fig. 24-1), starting with the diploid adult form. This stage of the life cycle, the **sporophyte** ("spore plant" in Greek), bears reproductive cells. These cells undergo meiosis to produce haploid cells that are **spores**, not gametes. The difference between spores and gametes is that spores do not fuse together to form a diploid cell. Instead, wind blows a fern spore off the parent leaf, and the spore lands on the soil. There the spore **germinates** (begins to grow and develop), dividing repeatedly by mitosis to form a multicellular, haploid organism. This organism produces gametes and hence is called the **gametophyte** ("gamete plant" in Greek). Because its cells are haploid, the gametophyte can produce sperm and eggs without further meiosis. A single gametophyte usually produces both sperm and eggs, but typically at different times, thereby preventing self-fertilization. Sperm and egg fuse to form a zygote that develops into a new diploid sporophyte plant.

Alternation of generations occurs in all plants. In primitive land plants, including mosses and ferns, the gametophyte is an independent, although generally small, plant. It liberates mobile sperm cells that reach an egg either by swimming through thin films of water that cover adjacent gametophytes or by being splashed by raindrops from one plant to the next. Therefore, ferns and mosses can reproduce only in moist habitats.

However, many terrestrial habitats are relatively dry. In drier habitats a plant can reproduce by surrounding its sperm in a watertight package. This package can be transported to another plant, where the sperm are liberated directly into the egg-bearing structures of the second plant. The seed plants (both nonflowering and flowering plants) do just that. In the flowering plants,

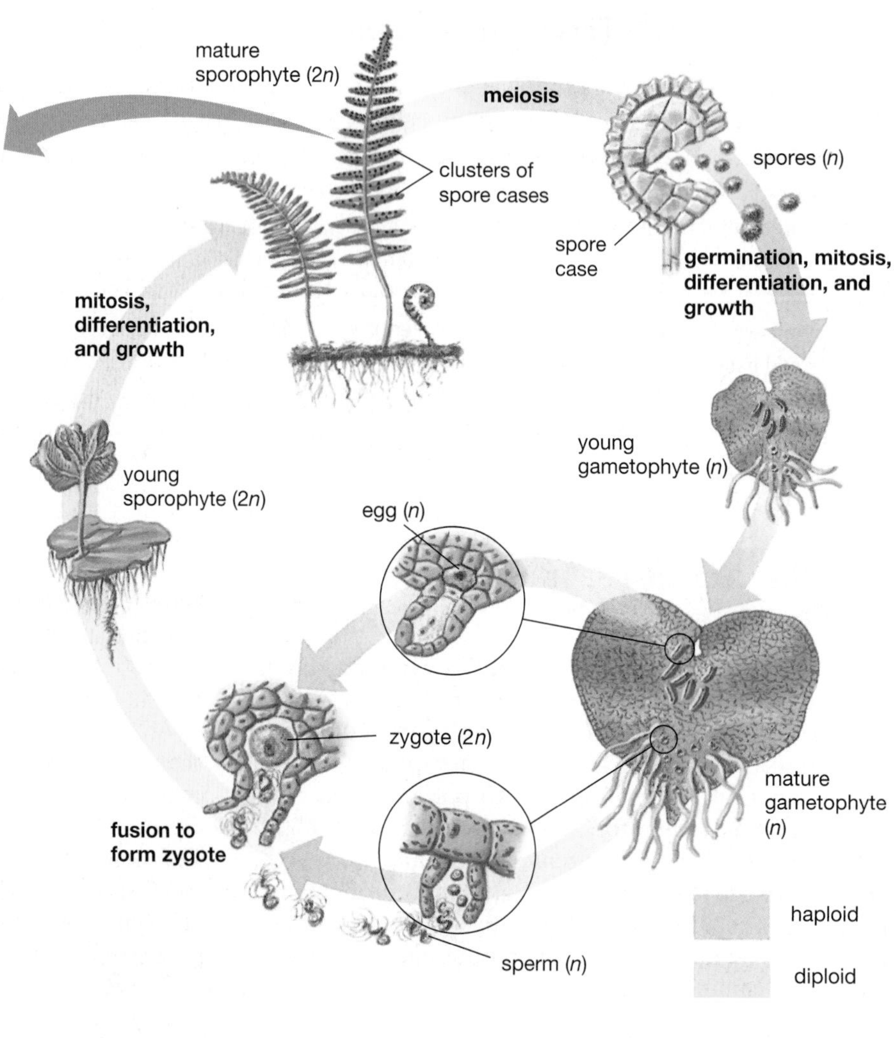

Figure 24-1 The life cycle of a fern—a nonflowering plant
Ferns typify the alternation-of-generations life cycle found in all plants, in which separate multicellular haploid and multicellular diploid organisms occur at different parts of the life cycle. (The text describes the stages of the life cycle.) The letter n refers to the haploid state, and $2n$ to the diploid state. When you see ferns, look on the undersides of their leaves for clusters of brownish sporangia (photo).

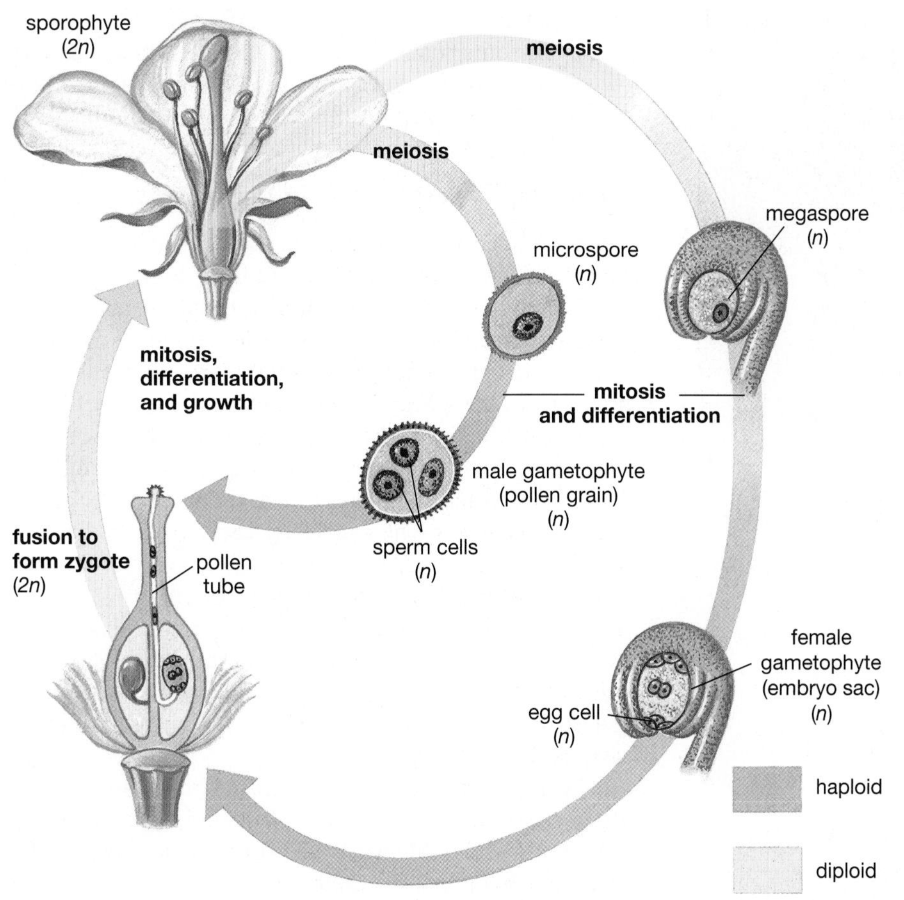

Figure 24-2 The life cycle of a flowering plant
Although this cycle shows the same basic stages as the life cycle of a fern (see Fig. 24-1), the haploid stages are much smaller and cannot live independently of the diploid plant.

two types of spores are formed by meiosis within the flowers borne by the sporophyte generation (Fig. 24-2). These haploid spores develop into gametophytes *within the flower* rather than in the soil. **Flowers** are the reproductive structures of flowering plants. The gametophytes are microscopic and do not live independently of the sporophyte. One type of spore, the *megaspore*, undergoes a few mitotic divisions and develops into the female gametophyte, a small cluster of cells permanently retained within the flower. The other type of spore, the *microspore*, develops into the male gametophyte: a tough, watertight **pollen grain** containing (usually) two sperm. The pollen grain drifts on the wind or is carried by an animal from one flower to another. On the recipient flower, the pollen grain elongates, burrowing through the flower's tissues to the female gametophyte within. This miniature male gametophyte liberates its sperm inside the female gametophyte, where fertilization occurs. The zygote becomes enclosed in a drought-resistant **seed**. The seed, including an embryonic plant and a food reserve within a protective outer coating, may lie *dormant* (it won't germinate) for months or years, waiting for conditions favorable for growth.

In the sections that follow, we examine sexual reproduction in flowering plants, from the evolution of the flower through the formation of the seed and the development of the new seedling.

2 How Did Flowers Evolve?

The flower is actually a sexual display that enhances a plant's reproductive success. By enticing animals to transfer pollen from one plant to another, flowers enable stationary plants to "court" distant members of their own species. This critical advantage has allowed the flowering plants to become the dominant plants on land.

The earliest seed plants were the gymnosperms, represented today mainly by conifers, a group that includes pines, firs, and spruces. As we described in Chapter 21, conifers do not produce flowers; instead, they bear male and female gametophytes on separate cones. During early spring, the small, male cones release millions of pollen grains that float about on breezes (Fig. 24-3). So many grains are floating around that some enter the pollen chambers located on the scales of the female cones, where they are captured by sticky coatings of sugars and resins. The pollen grains germinate and tunnel to the female gametophytes at the base of each cone's scale. Sperm are liberated and fertilize the eggs within a female gametophyte, and a new generation begins.

Clearly, wind pollination is an inefficient operation, because most of the pollen grains are lost. In a world of stationary plants and mobile animals, if a gymnosperm

Figure 24-3 Conifers are wind-pollinated
Even slight breezes blow thick clouds of pollen from ripe male cones. Look for these "soft cones" in clusters near the ends of branches of pine, spruce, and fir trees, often in late spring. The cones disintegrate after releasing their pollen. Woody cones are female.

could entice an animal to carry its pollen from male to female cone, it would greatly enhance its reproductive rate and hence its evolutionary success. As it happens, gymnosperms and insects were poised to establish just such a relationship about 150 million years ago.

Insects, especially beetles, are among the most abundant animals on Earth. They exploit nearly every possible food resource on land, including the reproductive parts of gymnosperms. About 150 million years ago, some beetles fed on both the protein-rich pollen of male cones and the sugar-rich secretions of female cones. Beetles can make quite a mess when they feed, and pollen feeders often wind up with pollen dusted all over their bodies. If the same beetle were to visit one plant and eat pollen, and then wander over to another plant of the same species to dine on the sugary secretions of a female cone, some of the loose pollen would quite likely rub off on the female cone.

The stage was set for the evolution of flowering plants. Efficient pollination by insects requires that a given insect visit several plants of the same species, pollinating them along the way. For the plants, two key adaptations were necessary. First, enough pollen or *nectar* (sugary secretions) must be produced within the reproductive structures so that insects will regularly visit them to feed. Second, the location and richness of these storehouses of pollen and nectar must be advertised to the insects, both to show them where to go and to entice them to specialize on that particular plant species. Any mutation that contributed to these adaptations would enhance the reproductive success of the plant that carried the mutation and would be favored by natural selection. By about 130 million years ago, flowers had evolved with exactly these adaptations. The advantages of flowers are so great that in today's temperate and tropical zones, flowering plants are overwhelmingly dominant, and numerous animals, including bees, moths, butterflies, hummingbirds, and even some mammals, feed at and pollinate flowers (see "Evolutionary Connections: Adaptations for Pollination and Seed Dispersal").

Complete Flowers Have Four Major Parts

We have seen that flowers are the reproductive structures of flowering plants. Evolution commonly produces new structures by modifying old ones, and flower parts are actually highly modified leaves, shaped by mutation and natural selection into a form that enhances pollination. **Complete flowers**, such as those of petunias, roses, and lilies, consist of a central axis on which four successive sets of modified leaves are attached (Fig. 24-4). These modified leaves form the *sepals*, *petals*, *stamens*, and *carpels*. The **sepals** are located at the base of the flower. In dicots, the sepals are typically green and leaflike; in monocots, most sepals resemble the petals (see Fig. 24-4b). In either case, sepals surround and protect the flower bud as the remaining three structures develop. Just above the sepals are the **petals**, which are usually brightly colored and fragrant, advertising the location of the flower.

The male reproductive structures, the **stamens**, are attached just above the petals. Most stamens consist of a long and slender **filament** that supports an **anther**, the structure that produces pollen. The female reproductive structures, the **carpels**, occupy the uppermost position in the flower. An idealized carpel is somewhat vase-shaped, with a sticky **stigma** for catching pollen mounted atop an elongated **style**. The style connects the stigma with the bulbous **ovary**. Inside the ovary are one or more **ovules**, in which the female gametophytes develop. When mature, each ovule will become a seed, and the ovary will develop into a protective, adhesive, or edible enclosure, the **fruit**.

As you may know from your own gardening experience, not all flowers are complete. **Incomplete flowers** lack one or more of the four floral parts. For example, many plants, such as cucumbers and squashes, have separate male and female flowers, which may be borne on the same plant (Fig. 24-5). Alternatively, male and female flowers may be on different plants, as in the American holly: Male flowers lack carpels, and female flowers lack stamens. Incomplete flowers may also lack sepals or petals.

3 How Do Gametophytes Develop in Flowering Plants?

The familiar flowering plant seen in meadows, gardens, and farms is the diploid sporophyte, as Figure 24-2 illustrates. The pollen grain (male) and the **embryo sac** (female) are the haploid gametophytes that develop within

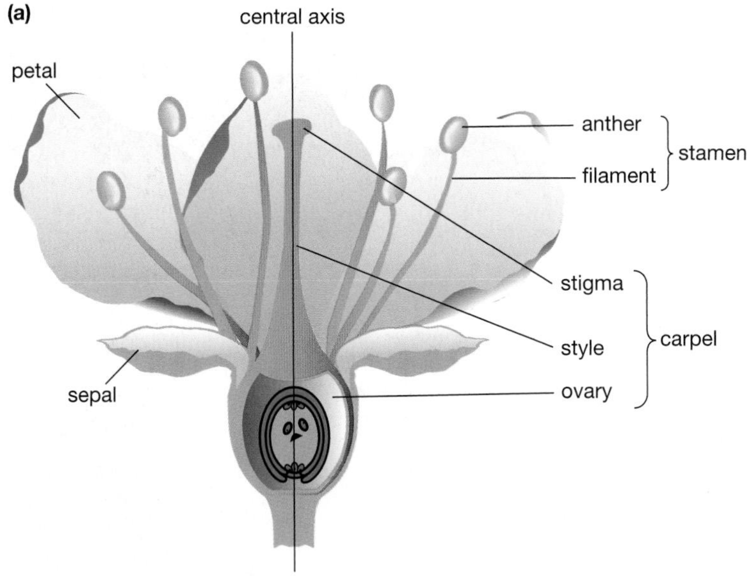

Figure 24-4 A complete flower
(a) A complete flower has four parts: sepals, petals, stamens (the male reproductive structures), and at least one carpel (the female reproductive structure). This drawing shows a complete dicot flower. ***(b)*** The lily is a complete monocot flower, with three sepals (virtually identical to the petals), three petals, six stamens, and three fused carpels. Each stamen consists of a filament bearing an anther at its tip. The carpels consist of an ovary hidden in the base of the flower, with a long style protruding out, ending in a sticky stigma. The anthers are well below the stigma, probably preventing self-pollination: Pollen cannot simply fall from the anther onto the stigma.

MediaTutor
24.1 Flower Structure and Function

Figure 24-5 Some plants have separate male and female flowers
Plants of the squash family, such as these zucchinis, bear separate female (left) and male (right) flowers. Although individual flowers cannot be self-pollinated, the plant could still be self-pollinated if an insect carried pollen from a male flower to a female flower of the same plant. However, each plant initially produces only male flowers, so some cross-pollination between plants that flower at slightly different times is virtually ensured.

the flowers of the sporophyte. Gametophytes of flowering plants are much smaller than those of ferns and mosses and cannot live independently of the sporophyte.

Pollen Is the Male Gametophyte

Each anther consists of four chambers called *pollen sacs* (Fig. 24-6). Within each sac, hundreds to thousands of diploid **microspore mother cells** develop. Each microspore mother cell undergoes meiosis (see Chapter 11) to produce four haploid **microspores.** Each microspore divides once, by mitosis, to produce a male gametophyte, or pollen grain. In many species, each immature pollen grain consists of only two cells: a large **tube cell** and a smaller **generative cell** that resides within the cytoplasm of the tube cell. As the pollen grain matures, the generative cell undergoes mitosis and produces two haploid sperm cells (see Fig. 24-6). A tough surface coat develops around the pollen grain, protecting the cells within during their journey to the carpel (Fig. 24-7).

When the pollen has matured, the pollen sacs of the anther split open. In wind-pollinated flowers, such as those of grasses and oaks, the pollen grains spill out and are carried widely by wind currents; a few of those grains reach and pollinate other flowers of the same

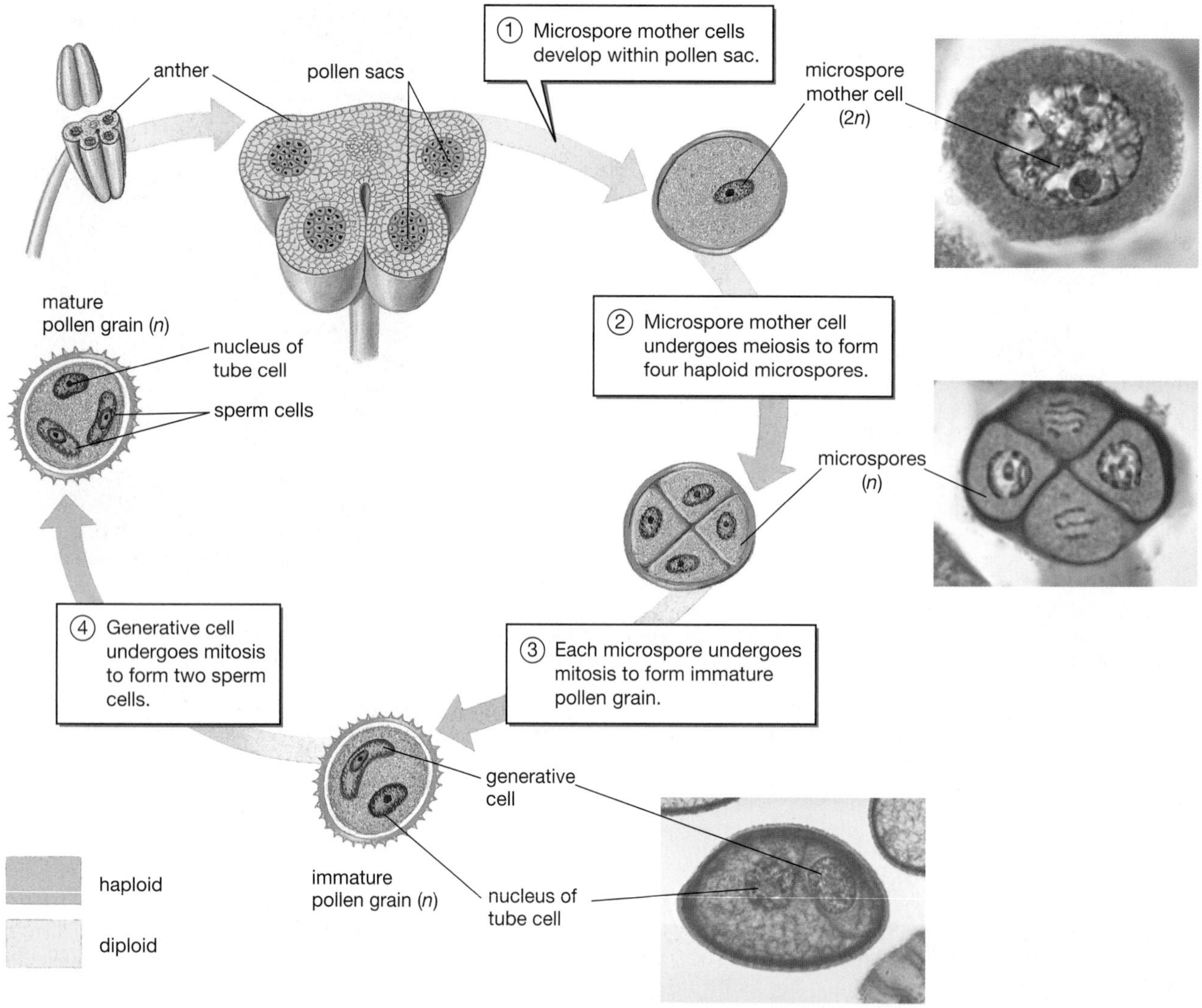
anther
pollen sacs
① Microspore mother cells develop within pollen sac.
microspore mother cell (2n)
mature pollen grain (n)
nucleus of tube cell
sperm cells
② Microspore mother cell undergoes meiosis to form four haploid microspores.
microspores (n)
④ Generative cell undergoes mitosis to form two sperm cells.
③ Each microspore undergoes mitosis to form immature pollen grain.
generative cell
haploid
diploid
immature pollen grain (n)
nucleus of tube cell

Figure 24-6 Pollen development in flowering plants

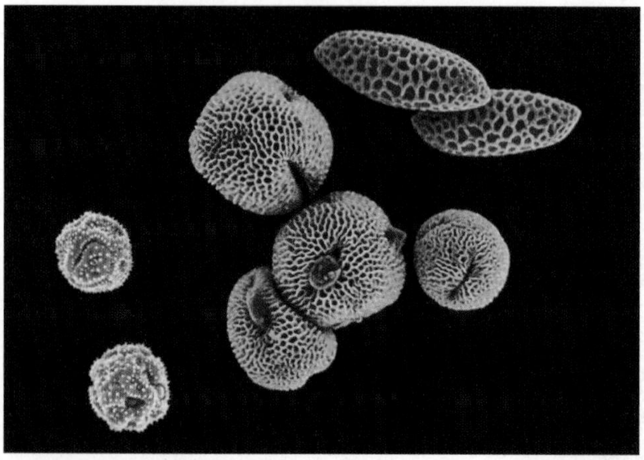

Figure 24-7 Pollen grains
The tough outer coverings of many pollen grains are elaborately sculptured in species-specific shapes and patterns. The pollen grains in this color-enhanced SEM photo are from a geranium (orange), a tiger lily (fuchsia), and a dandelion (yellow).

species. In animal-pollinated flowers, the pollen adheres weakly to the anther case until the pollinator comes along and brushes or picks it off.

The Embryo Sac Is the Female Gametophyte

In an ovary, one or more dome-shaped masses of cells differentiate into ovules. Each ovule consists of protective outer layers of cells called **integuments**, which surround a single, diploid **megaspore mother cell** (Fig. 24-8). That cell divides by meiosis to produce four large haploid **megaspores.** Three megaspores degenerate, and one survives. This remaining megaspore undergoes an unusual set of mitotic divisions. Three nuclear divisions produce a total of eight haploid nuclei. Plasma membranes then divide the cytoplasm into *seven*, not eight, cells: three small cells at each end, each containing one nucleus, and one remaining large cell in the middle containing two nuclei, called **polar nuclei.** This seven-celled

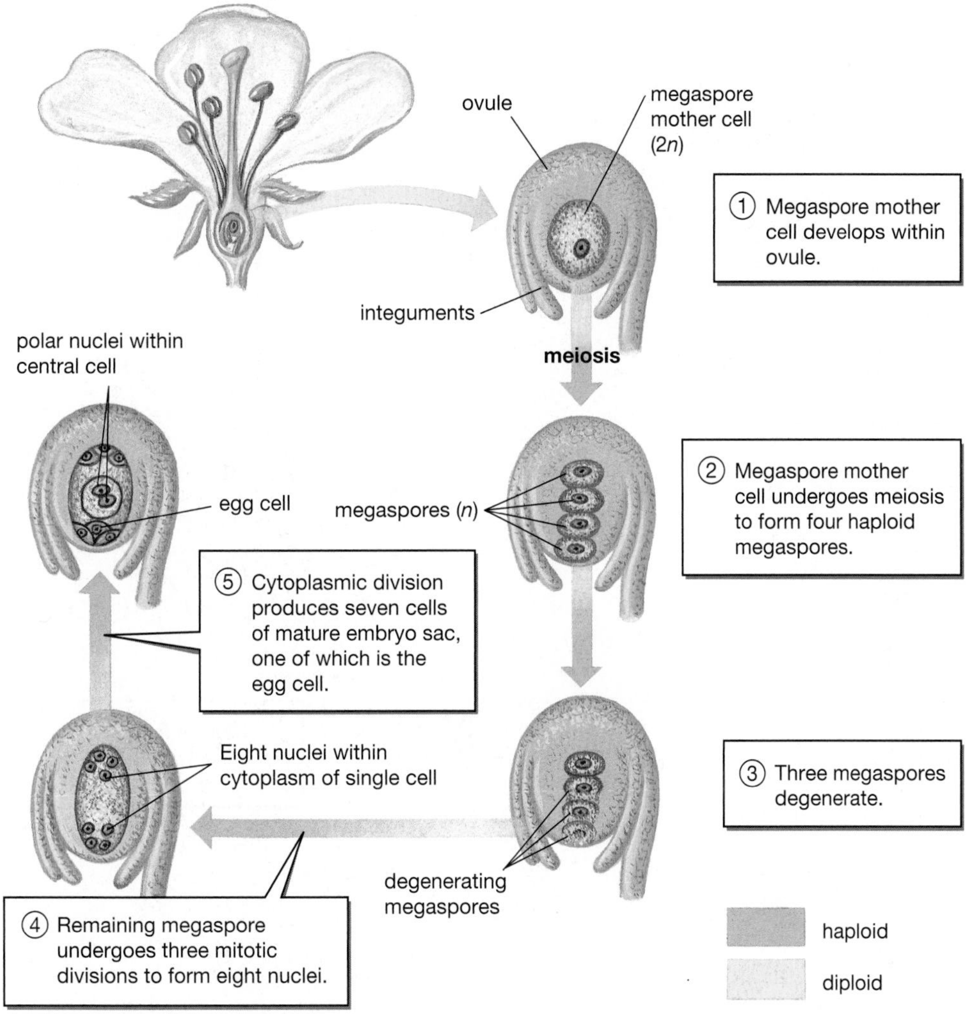
ovule
megaspore mother cell (2n)
① Megaspore mother cell develops within ovule.
integuments
meiosis
polar nuclei within central cell
egg cell
megaspores (n)
② Megaspore mother cell undergoes meiosis to form four haploid megaspores.
⑤ Cytoplasmic division produces seven cells of mature embryo sac, one of which is the egg cell.
Eight nuclei within cytoplasm of single cell
③ Three megaspores degenerate.
degenerating megaspores
④ Remaining megaspore undergoes three mitotic divisions to form eight nuclei.
haploid
diploid

Figure 24-8 Development of the female gametophyte

organism, called the *embryo sac* (because it is where the embryo will develop), is the haploid female gametophyte. The **egg** is the central small cell at the bottom of the embryo sac, located near an opening in the integuments of the ovule.

4 How Does Pollination Lead to Fertilization?

When a pollen grain lands on the stigma of a flower of the same species of plant, a remarkable chain of events occurs (Fig. 24-9). The pollen grain absorbs water from the stigma. The tube cell elongates, growing down the style toward an ovule in the ovary. Meanwhile, the generative cell divides mitotically to form two sperm cells.

If all goes well, the pollen tube reaches the pore in the integument of an ovule and breaks into the embryo sac. The tube's tip ruptures, releasing the two sperm. One sperm fertilizes the egg cell to form the diploid zygote that will develop into the embryo and eventually into a new sporophyte. The second sperm enters the large central cell and its nucleus fuses with *both* polar nuclei, forming a triploid nucleus (having three sets of chromosomes). Through repeated mitotic divisions, this cell will develop into the triploid ($3n$) **endosperm**, a food-storage tissue within the seed. The fusion of the egg with one sperm and the fusion of the polar nuclei with the second sperm are often called **double fertilization**, a process unique to flowering plants. The other five cells of the embryo sac degenerate soon after fertilization.

The distinction between pollination and fertilization is important. **Pollination** occurs when a pollen grain lands on a stigma; **fertilization** is the fusion of sperm and egg. Although pollination is necessary for fertilization, these are two separate events. For example, pollination will not lead to fertilization if the tube cell fails to grow properly, if the embryo sac has no egg, or if a sperm from another pollen grain has already reached the egg.

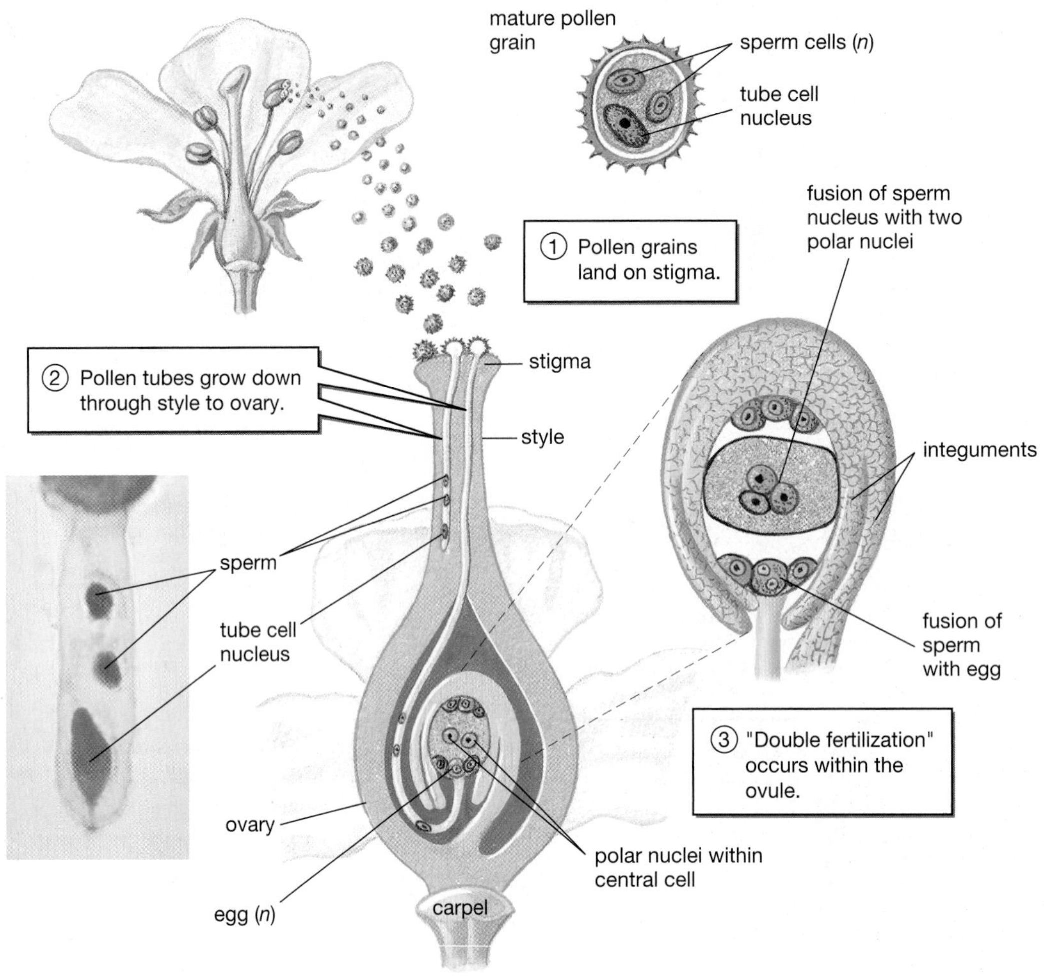

Figure 24-9 Pollination and fertilization of a flower

5 How Do Seeds and Fruits Develop?

Drawing on the resources of the parent plant, the embryo sac and the surrounding integuments of the ovule develop into a seed. The seed is surrounded by the ovary, which develops into a fruit (Fig. 24-10). Having already served their functions of attracting pollinators and producing pollen, petals and stamens shrivel and fall away as the fruit enlarges. So, when you eat a fruit, you are consuming the ripened ovary of a plant.

The Seed Develops from the Ovule and Embryo Sac

The integuments of the ovule develop into the **seed coat**, the outer covering of the seed. As we shall see, in many plants the characteristics of the seed coat play a role in regulating when the seed will germinate. Meanwhile, within the integuments, two distinct developmental processes occur (Fig. 24-11a). First, the triploid endosperm cell divides rapidly. Its daughter cells absorb nutrients from the parent plant, forming a large, food-filled endosperm. Second, the zygote develops into the embryo.

Both dicot and monocot embryos consist of three parts: the shoot, the root, and the **cotyledons**, or *seed leaves*. The cotyledons absorb food molecules from the endosperm and transfer them to other parts of the embryo.

In dicots ("two cotyledons"), the cotyledons usually absorb most of the endosperm during seed development, so the mature seed is virtually filled with embryo (Fig. 24-11b, left). In monocots ("one cotyledon"), the cotyledon absorbs some of the endosperm during seed development, but most of the endosperm remains in the mature seed (Fig. 24-11b, right). Food grains, including wheat, corn, and rice, are monocots. We (like the developing plant) use the stored endosperm as food. In the case of wheat, we grind up the endosperm to make flour and sometimes consume the embryo of the wheat seed in the form of "wheat germ."

The future primary root develops at one end of the embryo. The future shoot, at the other end, is usually divided into two regions at the attachment site of the cotyledons. Below the cotyledons, but above the root, is the **hypocotyl** (*hypo* in Greek means "beneath" or "lower"); above the cotyledons the shoot is called the **epicotyl** (*epi* means "above"). At the tip of the epicotyl lies the apical meristem of the shoot; its daughter cells

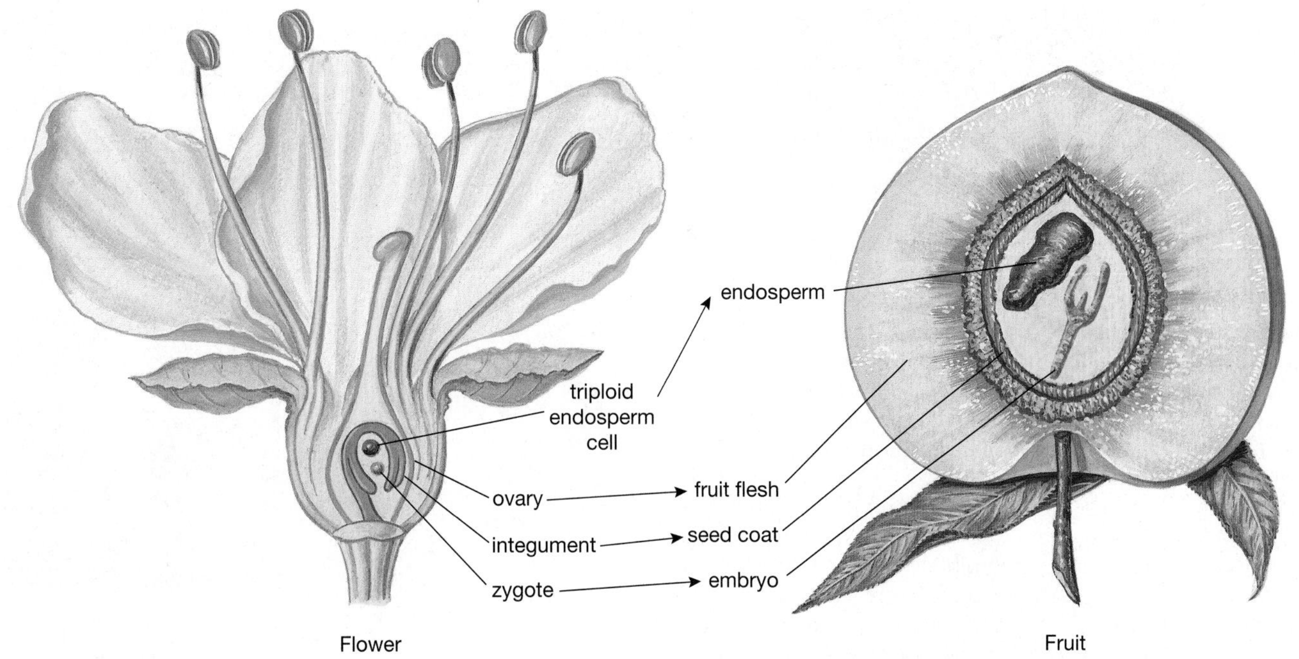

Figure 24-10 Development of fruit and seeds from flower parts
The fruit and seed coat are derived from the parent sporophyte plant. The ovary wall ripens into the fruit flesh, which may be soft and tasty, such as a peach, or variously hard, hooked, or tufted to facilitate dispersal. The integuments of each ovule, which surround the embryo sac, form the seed coat. Within the seed, the triploid endosperm cell divides repeatedly, absorbs nutrients from the parent plant, and becomes the endosperm. The zygote, also within the seed, develops into the embryo.

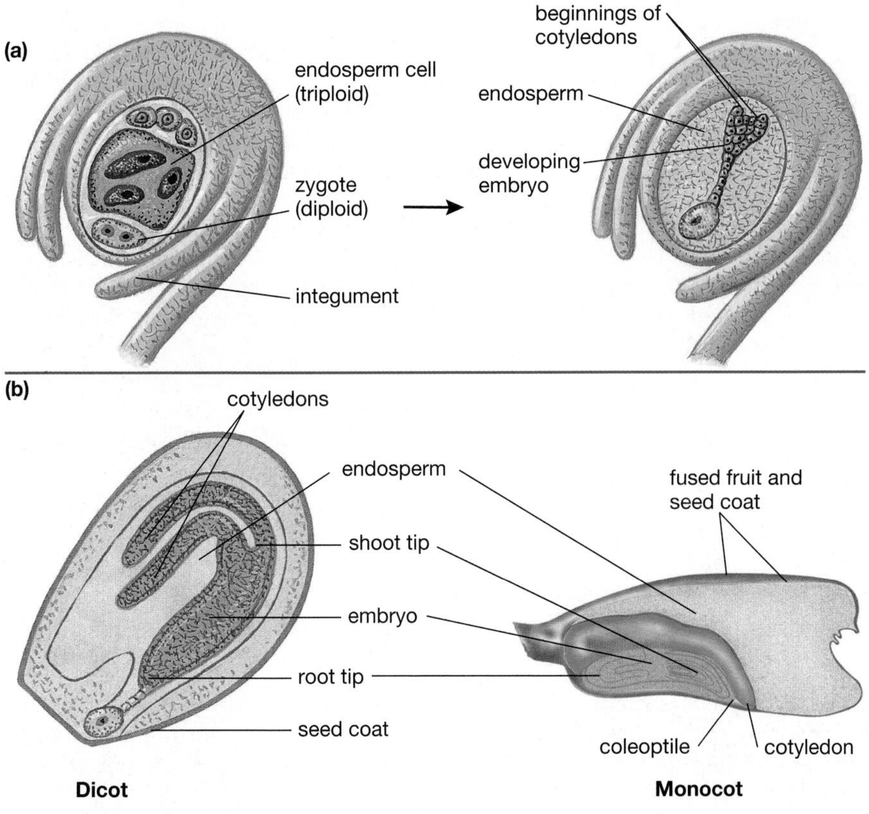

Figure 24-11 Seed development
(a) In a generalized seed, the endosperm develops first, absorbing nutrients from the parent plant. The embryo develops later, absorbing nutrients from the endosperm to fuel its growth. ***(b)*** Monocot and dicot seeds differ in the number of cotyledons and the fate of the endosperm. *(left)* Dicot seeds, such as the shepherd's purse, have two cotyledons, which usually absorb most of the endosperm as the seed develops; hence, the mature seed is mostly cotyledon. *(right)* Monocot seeds, such as the corn kernel, retain a large endosperm. (Cornmeal is the ground-up endosperm of corn seeds.) The embryo produces a single cotyledon. As the seed germinates, the cotyledon absorbs the food reserves of the endosperm and transfers them to the growing embryo.

Earth Watch
On Dodos, Bats, and Disrupted Ecosystems

Flowering plants dominate terrestrial ecosystems largely because of the mutually beneficial relationships they forged with the animals that pollinate their flowers and disperse their seeds. Some plant and animal species, such as the yucca and yucca moths, have become totally dependent on one another. Within complex ecosystems, these mutually beneficial relationships sustain both plant and animal populations and ultimately the ecosystem itself.

On the island of Mauritius in the Indian Ocean, the tambalacoque tree, like most of the native plants on this island, is threatened. Tambalacoque trees produce a large, edible fruit something like a peach, with a pulpy outside surrounding a stone-hard pit. Today, the remaining trees produce healthy fruits that fall to the ground and rapidly rot in the tropical climate. The enclosed seeds are highly susceptible to destruction by fungal and bacterial infections, which are promoted by the rotting fruit. Before humans arrived, the island was home to the dodo (Fig. E24-1a). Early sailors found the large, slow dodos to be easy prey, and by 1681 they had hunted the dodo to extinction. Other native animals, including giant tortoises, large-billed parrots, and the giant skink (a large reptile), were also driven to extinction as humans introduced monkeys, pigs, and deer to Mauritius and destroyed natural habitats by clearing the land for farming. Scientists believe that some of these extinct animals ate the tambalacoque fruit before it had a chance to rot, thoroughly cleaning the seeds and thus protecting them from attack by the fungi that now destroy the seeds. The animals also dispersed the seeds throughout the island, ensuring that some of them reached habitats favorable for germination. As native animals are destroyed, plants that coevolved with them are also threatened, and the natural ecosystem that evolved over millennia collapses.

On the island of Madagascar off the coast of Africa, a burgeoning human population is displacing natural habitats, and lemurs (tree-dwelling primates) are rapidly disappearing. Researchers have identified more than 20 tree species that depend primarily on lemurs for seed dispersal. Where lemurs are disappearing, so are these trees.

In many tropical forests, bats are the most important agents of seed dispersal (Fig. E24-1b). Bats may fly more than 20 miles each night, consuming up to twice their weight in fruit, and defecating the seeds in flight. Biologist Donald Thomas compared germination rates of seeds before and after the seeds' passage through bats' digestive tracts. After passing through a bat, nearly all the seeds germinated; seeds planted directly from fruit, however, had only a 10% germination rate. Today, in the tropical forests of southern Mexico, fruit-eating animals such as monkeys, deer, and tapir have been overhunted. Tropical fruits are rotting on the forest floor or sending up doomed sprouts under the shade of their parents; dispersal has stopped. As Alejandro Estrada of the University of Mexico put it, "The continued existence of tropical forests whose primates and . . . birds and bats have been shot is just as precarious as if their trees had been chain-sawed and bulldozed." The web of interdependent life-forms linked by interactions forged over millennia of coevolution is fragile and easily disrupted. Only by understanding and preserving the complex and crucial interactions among plants and animals can we hope to conserve diverse, functioning ecosystems.

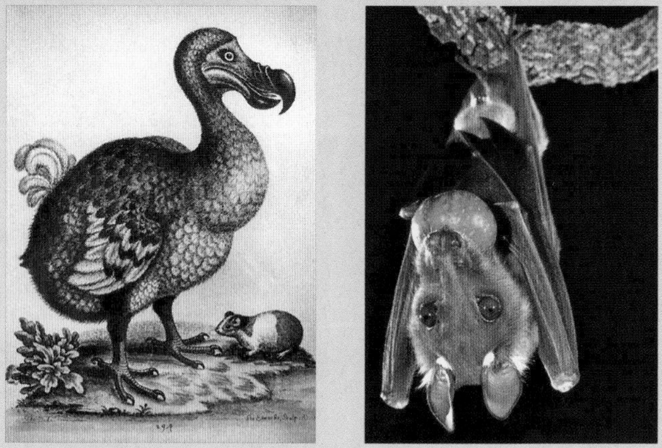

(a) (b)

Figure E24-1 Animal seed dispersers are crucial to some ecosystems
(a) The dodo, driven to extinction in the late 1600s, probably helped disperse and promote germination of the tambalacoque trees of Mauritius. The trees are now seriously threatened and rarely germinate in the wild. **(b)** A Wahlberg's epauleted bat in Kenya eats a ripe fig. Without bats and other seed-dispersing animals, some types of tropical forest communities may not survive.

differentiate into the specialized cell types of stem, leaves, and flowers. One or two developing leaves may already be present.

The Fruit Develops from the Ovary Wall

The ovary wall develops into a fruit (see Fig. 24-10). There are a bewildering variety of fruits, with outer layers that are variously fleshy, hard, winged, or even spiked like a medieval mace. The selective forces favoring the evolution of all fruits, however, are similar: Fruits help disperse the seeds to distant locations away from the parent plant. (See "Earth Watch: On Dodos, Bats, and Disrupted Ecosystems" and "Evolutionary Connections: Adaptations for Pollination and Seed Dispersal.")

Seed Dormancy Helps Ensure Germination at an Appropriate Time

All seeds need warmth and moisture to germinate. But many newly matured seeds will not germinate immediately, even under ideal conditions. Instead, they enter a period of **dormancy**, during which they will not germinate. Dormancy is usually marked by lowered metabolic activity and resistance to adverse environmental conditions.

Seed dormancy solves two problems, one intrinsic to the plant itself and one related to environmental factors. First, if a seed germinated while still enclosed in a fruit and hanging from a tree or vine, it might exhaust its food reserves before it ever touched the ground. Further, seedlings germinating within a fruit that contains many seeds would grow in a dense cluster, competing with one another for nutrients and light. Second, environmental conditions that are suitable for seedling growth (such as adequate moisture and temperatures) may not coincide with seed maturation. Seeds that mature in the late summer in temperate climates, for example, face the harsh winter to come. Spending the winter as a dormant seed is clearly preferable to death by freezing as a tender young sprout. In the warm, moist Tropics, seed dormancy is much less common than in temperate regions because environmental conditions are suitable for germination throughout the year.

Mechanisms for Maintaining and Breaking Dormancy Are Adaptations to Differing Environments

Plants have evolved many mechanisms that produce seed dormancy, and the seeds of each species have their own set of requirements (in addition to adequate moisture and proper temperature) that must be met before germination can occur. Perhaps the three most common requirements are drying, exposure to cold, and disruption of the seed coat:

1. *Drying.* Many seeds must dry out before they are able to germinate. Drying prevents the seed from germinating while it is still within the fruit. Many such seeds are dispersed by animals that eat fruit but cannot digest the seeds, which are therefore excreted in their feces.
2. *Cold.* Seeds of many temperate and arctic plants will not germinate unless those seeds are exposed to prolonged subfreezing temperatures, followed by sufficient warmth and moisture. This requirement ensures that the seeds stay dormant during mild days in autumn and will sprout only after winter yields to spring.
3. *Disruption of the seed coat.* The seed coat itself can be a barrier to seed germination. Many seed coats are impermeable to water and oxygen, or they bind the developing embryo so tightly that growth simply cannot occur; others contain chemicals that inhibit germination (as we shall see in Chapter 25). In deserts, for example, rainfall is spotty and scarce. Years may go by without enough water for plants to germinate, grow, flower, and set more seed. Therefore, the seed must not sprout unless a rainfall is heavy enough to allow the plant to complete its life cycle, because more rain may not fall in time. The seed coats of most desert plants have water-soluble chemicals that inhibit germination. Only a hard rainfall can wash away enough of the inhibitors to allow sprouting.

6 How Do Seeds Germinate and Grow?

During germination, growth, and development of a seed, the formerly dormant embryo resumes growth and emerges from the seed. The embryo absorbs water, which makes it swell and burst its seed coat. The root is usually the first structure to emerge from the seed coat, growing rapidly and absorbing water and minerals from the soil. Much of the water is transported to cells in the shoot. As its cells elongate, the stem lengthens, pushing up through the soil.

The Shoot Tip Must Be Protected

The growing shoot faces a serious difficulty: It must push through the soil without scraping away the apical meristem and tender leaflets at its tip. A root, of course, must always contend with tip abrasion; its apical meristem is protected by a root cap (see Fig. 23-9). Shoots, however, spend most of their time in the air and do not develop permanent protective caps. Instead, germinating shoots have other mechanisms that cope with the abrasion of sprouting. In monocots, the **coleoptile**, a tough sheath, encloses the shoot tip like a glove (Fig. 24-12a). The coleoptile "glove" pushes aside the soil particles as it grows. Once out in the air, the coleoptile tip degenerates, allowing the tender shoot "finger" to emerge. Dicots do not have coleoptiles. Rather, the dicot shoot forms a hook in the hypocotyl or epicotyl (Fig. 24-12b). The bend of the hook, encased in epidermal cells with tough cell walls, leads the way through the soil, clearing the path for the downward-pointing apical meristem with its delicate new leaves.

Cotyledons Nourish the Sprouting Seed

Food stored in the seed provides the energy for sprouting. Recall that the cotyledons of dicots had already absorbed the endosperm while the seed was developing and are now swollen and full of food. In dicots with hypocotyl hooks, such as members of the squash family, the elongating shoot carries the cotyledons out of the soil into the air. These aboveground cotyledons typically become green and photosynthetic and transfer both previously stored food and newly synthesized sugars to

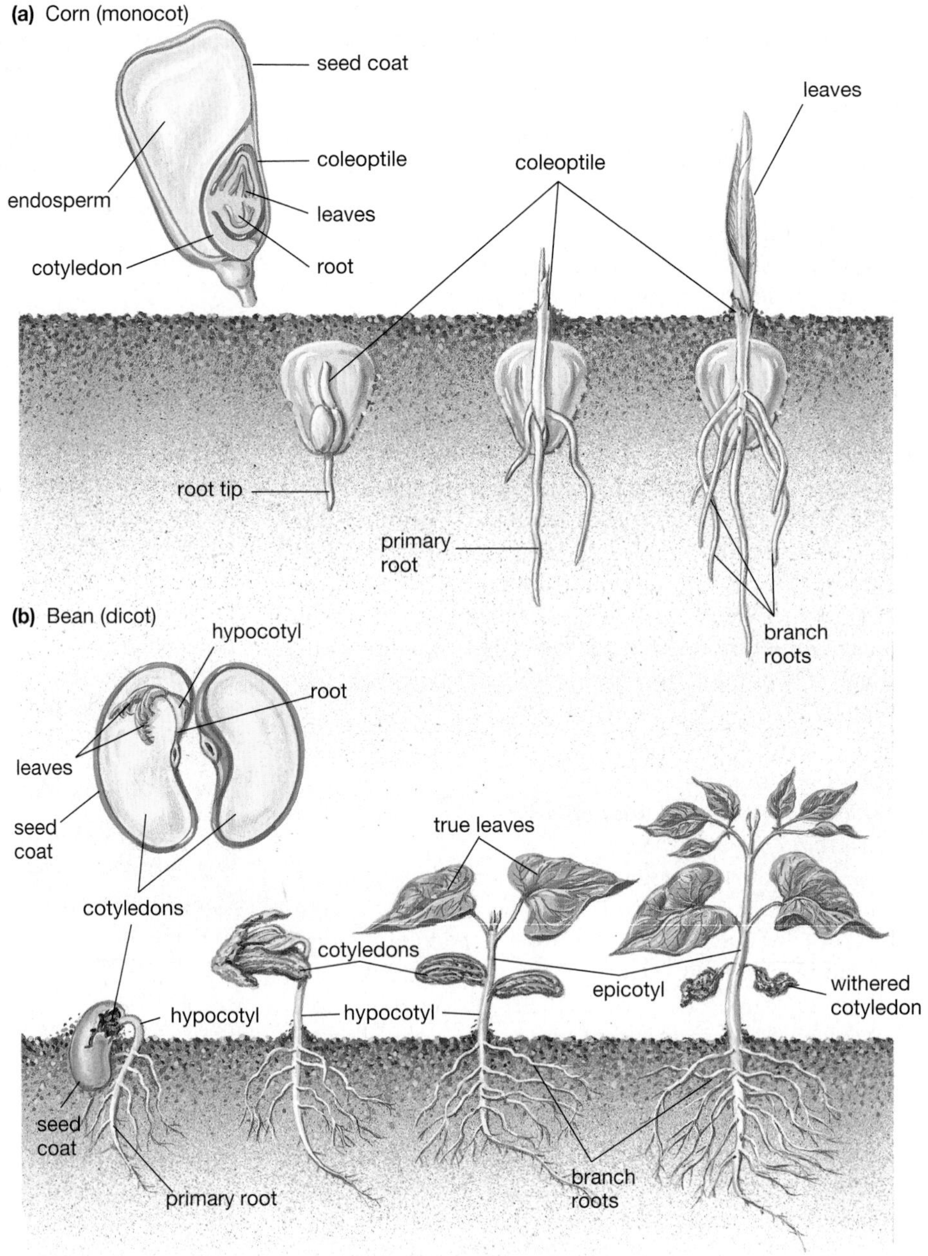

Figure 24-12 Seed germination Seed germination in ***(a)*** corn, a monocot, and ***(b)*** the common bean, a dicot, differs in detail but involves the same three principles: (1) use of stored food to provide energy for seedling growth until photosynthesis can take over; (2) rapid growth and branching of the root to absorb water and nutrients; and (3) protection of the delicate shoot tip as it moves upward through the soil. In monocots (a), the delicate shoot tip is protected within a tough sheath, the coleoptile, which pushes up through the soil when the seed sprouts. In dicots (b), the hypocotyl (shown here) or the epicotyl forms a hook that emerges from the soil first, protecting the shoot tip.

the shoot (Fig. 24-13). In dicots with epicotyl hooks, the cotyledons stay below the ground, shriveling up as the embryo absorbs their stored food. Monocots retain most of their food reserve in the endosperm until germination, when it is digested and absorbed by the cotyledon as the embryo grows. The cotyledon remains below ground in the remnants of the seed.

Controlling the Development of the Seedling

Once out in the air, the shoot rapidly spreads its leaves to the sun. Simultaneously, the root system delves into the soil. The apical meristem cells of shoot and root divide, giving rise to the mature structures discussed in Chapter 23. Eventually this plant, too, will mature, flower, and set seed, renewing the cycle of life. How this cycle is regulated—why shoots grow upward while roots grow downward, and how plants produce flowers at the proper time of year—is the subject of Chapter 25.

Evolutionary Connections

Adaptations for Pollination and Seed Dispersal

Coevolution Matches Plants and Pollinators

Although we have emphasized animal-pollinated flowers, many plants have been enormously successful using the wind to disperse their pollen. This method works well for plants that grow close together, grasses and

Figure 24-13 Cotyledons nourish the developing plant
In the squash family, a hypocotyl hook carries the cotyledons out of the soil. The cotyledons expand into photosynthetic leaves (the pair of smooth oval leaves). The first true leaf (crinkled single leaf) develops a little later. Eventually, the cotyledons shrivel up and die.

many trees, for example. Most wind-pollinated flowers, such as those of grasses and oaks, are inconspicuous and unscented, many of them scarcely more than naked stamens that liberate pollen to the wind (Fig. 24-14). The beautiful flowers so admired by humans, however, are pollinated by animals.

The distinctive shapes, colors, and odors of animal-pollinated flowers as well as the sensory capabilities and lifestyles of their pollinators are an example of **coevolution**—evolution in two species that interact extensively with one another, such that each acts as a major force of natural selection on the other (see Chapter 15). Animal-pollinated flowers must attract useful pollinators and frustrate undesirable visitors who might eat nectar or pollen without fertilizing the flower in return. The animals, in turn, have been under environmental pressures to locate flowers quickly, identify the flowers that can provide adequate nutrition, and extract the nectar or pollen with a minimum expenditure of energy.

Animal-pollinated flowers can be loosely grouped into three categories, depending on the benefits (real or imagined) that they offer to potential pollinators: (1) food, (2) sex, or (3) a nursery.

Some Flowers Provide Food for Pollinators

Many flowers provide food for foraging animals such as beetles, bees, moths and butterflies, or hummingbirds. In return, the animals unwittingly distribute pollen from flower to flower.

We can thank the bees for most of the sweet-smelling flowers, because sweet "flowery" odors attract these pollinators. Bees also have good color vision, but they do not see exactly the same range of colors that humans do (Fig. 24-15a). To attract a bee from afar, bee-pollinated flowers must look brightly colored *to a bee* (Fig. 24-15b). Typically, these flowers are white, blue, yellow, or orange, and many have other markings, such as central spots or lines pointing toward the center, that reflect UV light. Look carefully at the flower colors on the hills shown in our chapter-opening photograph. How are these flowers likely to be pollinated?

Figure 24-14 A wind-pollinated flower
The flowers of grasses and many deciduous trees are wind-pollinated, with anthers (yellow structures hanging beneath the flowers) exposed to the wind. Petals are usually reduced or absent.

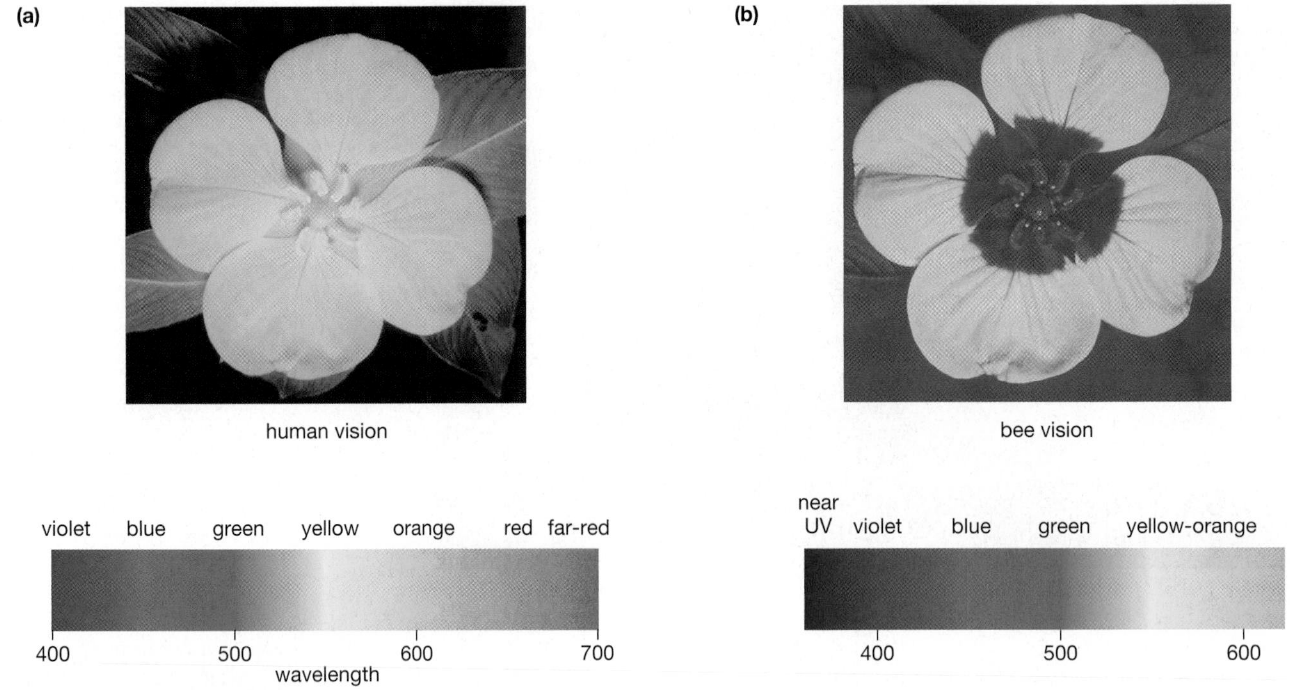

Figure 24-15 Ultraviolet patterns guide bees to nectar
The spectra of color vision for ***(a)*** humans and ***(b)*** bees overlap considerably in the blue, green, and yellow ranges but differ on the edges. Humans are sensitive to red, which bees do not perceive; bees can see UV light, which is invisible to the human eye. Many flowers photographed under ordinary daylight (in part *a*) and under UV light (in part *b*) show striking differences in color patterns. Bees can see the UV patterns that presumably lead them to the nectar- and pollen-containing centers of the flowers.

Bee-pollinated flowers have structural adaptations that help ensure pollen transfer. Many bee-pollinated flowers, such as nasturtiums and foxgloves, produce nectar at the bottom of a tube; in the Scotch broom flower, nectar forms in a crevice between ingeniously evolved petals (Fig. 24-16). Either pollen-laden stamens (in newly opened flowers) or the sticky stigma of the carpel (in older flowers) sticks out of the top of the tube or emerges from confinement when the bee's weight deflects the petals downward. When a bee visits a young flower, the stamens brush pollen onto her back. She may then visit an older flower and repeat her foraging behavior. This time, she leaves pollen behind on the stigma.

Many flowers adapted for moth and butterfly pollinators have tubes containing nectar that accommodate the long tongues of moths and butterflies. Flowers pollinated by night-flying moths open only in the evening; most are white and some give off strong, musky odors that help the moth locate the flower in the dark.

Many beetles and flies prefer to feed on animal material, and beetle-pollinated flowers typically smell like dung or rotting flesh, which attracts these scavenging insects. Their names, including "carrion flower," "skunk cabbage," and "corpse flower" (Fig. 24-17), describe their odor. The corpse flower and its relative, the skunk cabbage common in U.S. swamps, are related. Both share another adaptation: their flowers heat up! By metabolizing stored food, mostly fat, eastern skunk cabbage flowers can reach temperatures up to 25 °C higher than the air around them. The heat probably attracts pollinators and certainly helps broadcast foul-smelling scents. These flowers deceive their pollinators—they smell like nutrient-rich rotting meat, but offer no food at all.

Hummingbirds (see Chapter 8's opening photo) are one of the few important vertebrate pollinators, although several mammals also visit flowers (Fig. 24-18). Since birds have notoriously poor senses of smell, hummingbird-pollinated flowers seldom synthesize fragrant chemicals. However, these flowers often produce more nectar than other flowers do, because hummingbirds need more energy than insects and will favor flowers that provide it. These flowers must protect their large nectar supplies from insects that would become sated on the abundant sugar and fail to transfer pollen to another flower. Adaptations to hummingbird pollination include a deep, tubular shape that matches the long bills and tongues of hummers and no place for insects to land and rest while dining. In addition, most hummingbird-pollinated flowers are red or orange. Red is particularly attractive to hummingbirds but appears gray or black to a bee.

Sexy Deceptions Attract Pollinators

To pollinate their flowers, a few plants, most notably the orchids, take advantage of the mating drive and stereotyped behaviors of male wasps. Some orchid flowers mimic female wasps or bees both in scent (the orchids

(a)

(b)

Figure 24-16 "Pollinating" a pollinator
(a) In Scotch broom flowers, the bee finds nectar near the junction of top and bottom petals. ***(b)*** The bee's weight deflects the bottom petals downward, causing curved, pollen-laden stamens to pop up and cover the bee's hairy back with pollen. The bee will carry the pollen to other Scotch broom flowers, leaving some on ready stigmas.

Figure 24-17 The corpse flower is overwhelming in size and scent
Scavenging insects such as flies and beetles are attracted to the odor of rotting meat produced by some of the largest flowers in the world. To date, only 11 corpse flowers (*Amorphophallus titanum,* a species native to Sumatra, also called "devil's tongue") have ever been coaxed into bloom in the United States, the first of which grew in 1937. Thousands of visitors flocked to see and smell this amazing specimen (nearly 6 feet tall) that bloomed in California's Huntington Botanical Gardens in August 1999. The "bloom" actually contains hundreds of separate male and female flowers, which are pollinated (in Sumatra) by large carrion beetles.

MEDIATUTOR
24.2 Fruit and Seed Structure and Development

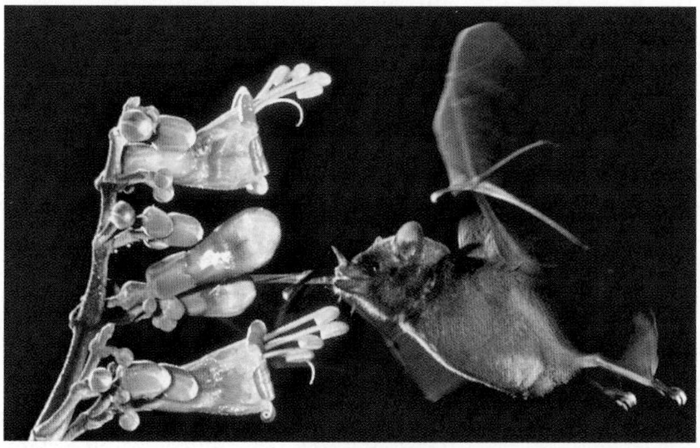

(a)

(b)

Figure 24-18 Some vertebrate pollinators
(a) A tropical bat feeds at a cluster of tubular flowers with protruding stamens and stigma. As the bat hovers before the flower, the top of its head touches either the anthers or stigma or both, thus pollinating the flower. ***(b)*** As the honey possum stuffs its face into this flower, pollen adheres to its muzzle and whiskers. A visit to another flower will transfer the pollen.

Figure 24-19 Sexual deception promotes pollination
This male wasp is trying to copulate with an orchid flower. The result is successful reproduction—for the orchid, not for the wasp!

release a sexual attractant similar to that produced by the female insect) and shape (Fig. 24-19). The males land atop these "fake females" and attempt to copulate, but get only a packet of pollen for their efforts. As they repeat their attempts on other orchids of the same species, the pollen packet is transferred.

Some Plants Provide Nurseries for Pollinators

Perhaps the most elaborate relationships between plants and pollinators occur in a few cases in which insects fertilize a flower and then lay their eggs in the flower's ovary. This arrangement occurs between milkweeds and milkweed bugs, figs and certain wasps, and yuccas and yucca moths (Fig. 24-20). The yucca moth's remarkable behavior results in the pollination of yuccas and a well-stocked pantry for its own offspring. A female moth visits a yucca flower, collects pollen, and rolls it into a compact ball. The moth flies off with the pollen ball to another yucca flower, drills a hole in the ovary wall, and lays its eggs inside the ovary. Then it smears pollen from the pollen ball all over the stigma of the flower, performing this genetically programmed behavior flawlessly! By pollinating the yucca, the moth ensures that the plant will provide a supply of developing seeds for its caterpillar offspring. Because the caterpillars eat only a fraction of the seeds, the yucca also reproduces successfully. The mutual adaptation of yucca and yucca moth is so complete that neither can reproduce without the other.

Fruits Help Disperse Seeds

A plant species will be most successful if its seeds are dispersed far enough so that the young plants don't compete with the parents for light and nutrients. Plants will also be more successful and widespread if its members send seeds to distant habitats. In flowering plants, seed dispersal is the function of fruits. A wide variety of fruits have evolved, each dispersing seeds in a different way.

Figure 24-20 A mutually dependent relationship
(a) Yuccas bloom on the dry plains of eastern Colorado in early summer. ***(b)*** Within many of the yucca flowers, yucca moths carry out their part in this cooperative relationship.

(a) (b)

Figure 24-21 Wind-dispersed fruits
These fruits usually contain only one or two lightweight seeds. Some, such as ***(a)*** dandelions, have filamentous tufts that catch the breezes. Others, such as ***(b)*** maple fruits, are miniature glider–helicopters, silently whirling away from the tree as they fall. To see how the wings aid in seed dispersal, take two maple fruits and pluck the wing off one. Hold both fruits over your head and drop them. The wingless fruit will fall at your feet; the winged one will glide some distance away.

Explosive Fruits Allow Shotgun Dispersal

A few plants develop explosive fruits that eject their seeds meters away from the parent plant. Mistletoes, common parasites of trees, produce fruits that shoot out sticky seeds. If one seed strikes a nearby tree, it sticks to the bark and germinates, sending rootlike fibers into the vascular tissues of its host, from which it draws its nourishment. Because the proper germination site for a mistletoe seed is not the ground but a tree limb, shooting the seeds, not dropping them, is clearly useful.

Lightweight Fruits Allow Wind Dispersal

Dandelions and maples (Fig. 24-21) produce lightweight fruits with surfaces that catch the wind. (Yes, each individual hairy tuft on a dandelion ball is a separate fruit!) Each fruit typically contains a single small seed; having only one seed reduces weight and lets the fruit remain aloft longer. These featherweight fruits aid the seed in traveling away from the parent plant, from a few meters for maples to kilometers for a milkweed or dandelion on a windy day.

Floating Fruits Allow Water Dispersal

Many fruits can float on water for a time and may be dispersed by streams or rivers. The coconut fruit, however, is the ultimate floater. Round, buoyant, and watertight, the coconut drops off its parent palm, rolls to the sea, and floats for weeks or months until it washes ashore on some distant isle (Fig. 24-22). There it germinates, perhaps establishing a new coconut colony on a formerly barren island.

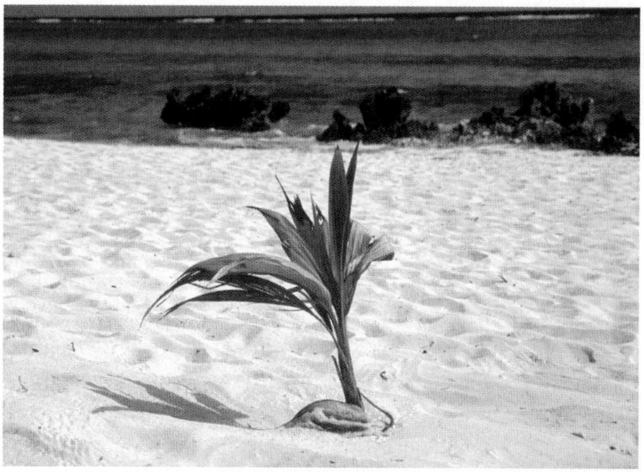

Figure 24-22 Water-dispersed fruit
After a long journey at sea, this coconut was washed high onto a beach by a storm. The large size and massive food reserves of coconuts are probably adaptations required for successful germination and seedling growth on barren, sandy beaches.

Clingy or Tasty Fruits Allow Animal Dispersal

Perhaps the majority of fruits use animals as agents of seed dispersal. Two quite distinct strategies have evolved for dispersal by animals: (1) grabbing an animal as it passes by; or (2) enticing it to eat the fruit but not digest the seeds.

Anyone who takes a long-haired dog on a walk through an abandoned field knows about fruits that hitchhike on animal fur. Burdocks, burr clover, foxtails, and sticktights all develop fruits with prongs, hooks, spines, or adhesive hairs (Fig. 24-23). The parent plants hold these fruits very loosely, so even slight contact with fur pulls the fruit free of the plant and leaves it stuck on the animal. Some of these fruits may fall off the next time the animal brushes against a tree or rock or come out when the animal grooms its fur.

Unlike hitchhiker fruits, edible fruits benefit both animal and plant. The plant stores sugars and tasty flavors

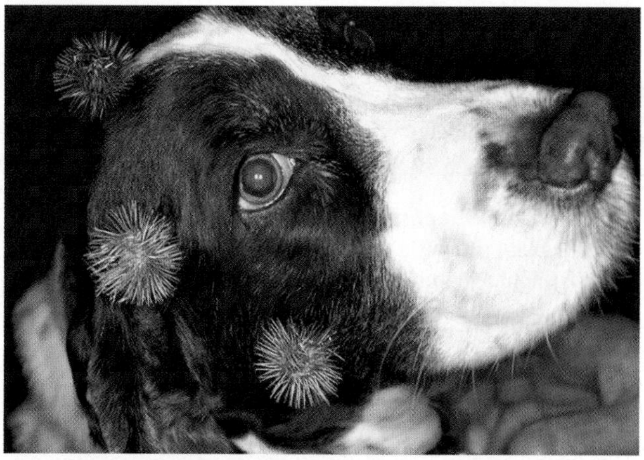

Figure 24-23 The cocklebur fruit uses hooked spines to hitch a ride on furry animals

in a fleshy fruit that surrounds the seeds, enticing hungry animals to eat the fruit (Fig. 24-24). Some fruits, such as peaches and plums, contain large, hard seeds that animals usually do not eat. After consuming the flesh of the fruit, the animals discard the seeds. Other fruits, including blackberries, raspberries, strawberries, and tomatoes, have small seeds that are swallowed along with the fruit flesh. The seeds then pass through the animal's digestive tract without harm. In some cases, passing through an animal's gut may even be essential to seed germination, by scraping or digesting away part of the seed coat. Recently a graduate student in ecology discovered the secret of chili pepper seed dispersal. The burning flavor discourages local mammals, such as desert mice and pack rats, from eating the fruit but does not act as a deterrent to birds that can't taste it. The researcher discovered that the digestive tracts of the potential mammal consumers would destroy the chili seeds, but passage through the digestive tract of birds increases the germination rate by a factor of three compared to seeds that just fall to the ground. Besides transport away from its parent, a seed that is swallowed and excreted benefits in another way: It ends up with its own supply of fertilizer!

Figure 24-24 The colors of ripe fruits attract animals
A bright red raspberry fruit has attracted a resplendent quetzal in Costa Rica. Only ripe fruits with mature seeds inside are sweet and brightly colored. Most unripe fruits are unpalatable: green, hard, and bitter. This too is an evolutionary adaptation. The immature seeds within unripe fruit may not survive passage through an animal's gut.

CASE STUDY REVISITED

Walk Through a Meadow

Hopefully, next time you walk through a meadow or across a deserted lot on which wild plants have been allowed to flourish, you'll see it through different eyes. Flowering grasses don't produce showy flowers, but in the right season you'll see that their "heads" are yellow with pollen. Are there any bees on the grasses? Look closely for pollinators on flowers and notice what colors and shapes of flowers are being visited by bees or butterflies. Find a foraging bee and look at it closely—is there pollen on its body? Are the anthers or the stigma of the flower contacting the bee's body? If so, you're seeing a symbiotic relationship at work! Are there grass seeds or burrs stuck in your socks? They are using you as an agent of seed dispersal. Where will you take them?

Although people may have allergies to many different types of pollen, some of the most common allergies are to the pollen of grasses. Based on what you know about how grasses transfer pollen, devise a hypothesis to explain why grass pollen is such a major contributor to "allergy season."

Summary of Key Concepts

1) What Are the Features of Plant Life Cycles?

The sexual life cycle of plants, called *alternation of generations*, includes both a multicellular diploid form (the sporophyte generation) and a multicellular haploid form (the gametophyte generation).

In seed plants, the gametophyte stage is greatly reduced. The male gametophyte is the pollen grain, a drought-resistant structure that can be carried from plant to plant by wind or animals. The female gametophyte is also reduced and is retained within the body of the sporophyte stage. In this way, seed plants can reproduce independently of liquid water.

2) How Did Flowers Evolve?

Flowering plants evolved from gymnosperms. In gymnosperms, wind blows pollen from male cones to female cones. Wind pollination is inefficient, though. In many habitats, flowering plants enjoy a selective advantage over gymnosperms because many types of flowers attract insects that carry pollen from plant to plant.

Complete flowers consist of four parts: sepals, petals, stamens (male reproductive structures), and carpels (female reproductive structures). The sepals form the outer covering of the flower bud. Most petals (and in some cases, the sepals) are brightly colored and attract pollinators to the flower. The stamen consists of a filament that bears at its tip an anther, in which pollen (the male gametophyte) develops. The carpel consists of the ovary, in which one or more embryo sacs (the female gametophytes) develop, and a style. The style bears at its end a sticky stigma, to which pollen adheres during pollination.

Incomplete flowers lack one or more of the four floral parts.

3) How Do Gametophytes Develop in Flowering Plants?

Pollen develops in the anthers. The diploid microspore mother cell undergoes meiosis to produce four haploid microspores. Each of these divides mitotically to form pollen grains. An immature pollen grain consists of two cells: the tube cell and the generative cell. The generative cell divides once to produce two sperm cells.

The embryo sac develops within the ovules of the ovary. A diploid megaspore mother cell undergoes meiosis to form four haploid megaspores. Three of these degenerate; the fourth undergoes three sets of mitotic divisions to produce the eight nuclei of the embryo sac. These eight nuclei come to reside in only seven cells. One of these cells, with a single nucleus, is the egg cell; another, with two nuclei, is the primary endosperm cell. These two cells are involved in seed formation; the rest of the cells degenerate.

4) How Does Pollination Lead to Fertilization?

Pollination is the transfer of pollen from anther to stigma. When a pollen grain lands on a stigma, its tube cell grows through the style to the embryo sac. The generative cell divides to form two sperm cells that travel down the style within the tube cell, eventually entering the embryo sac. One sperm fuses with the egg to form a diploid zygote, which will give rise to the embryo. The other sperm fuses with the binucleate primary endosperm cell to produce a triploid cell. This cell will give rise to the endosperm, a food-storage tissue within the seed.

5) How Do Seeds and Fruits Develop?

The embryo develops a root, shoot, and cotyledons. Cotyledons digest and absorb food from the endosperm and transfer it to the growing embryo. Monocot embryos have one cotyledon, and dicot embryos have two. The seed is enclosed within a fruit, which develops from the ovary wall. The function of the fruit is to disperse the seeds away from the parent plant. Seeds typically remain dormant for some time after fruit ripening. Environmental conditions involved in the breaking of dormancy include an initial drying, exposure to cold, or disruption of the seed coat.

6) How Do Seeds Germinate and Grow?

Seed germination requires warmth and moisture. Energy for germination comes from food stored in the endosperm; that food is transferred to the embryo by the cotyledons.

Key Terms

alternation of generations *p. 498*
anther *p. 500*
carpel *p. 500*
coevolution *p. 509*
coleoptile *p. 507*
complete flower *p. 500*
cotyledon *p. 504*
dormancy *p. 507*
double fertilization *p. 503*
egg *p. 503*
embryo sac *p. 500*
endosperm *p. 503*
epicotyl *p. 504*
fertilization *p. 503*
filament *p. 500*
flower *p. 499*
fruit *p. 500*
gametophyte *p. 498*
generative cell *p. 501*
germination *p. 498*
hypocotyl *p. 504*
incomplete flower *p. 500*
integument *p. 502*
megaspore *p. 502*
megaspore mother cell *p. 502*
microspore *p. 501*
microspore mother cell *p. 501*
ovary *p. 500*
ovule *p. 500*
petal *p. 500*
polar nucleus *p. 502*
pollen grain *p. 499*
pollination *p. 503*
seed *p. 499*
seed coat *p. 504*
sepal *p. 500*
spore *p. 498*
sporophyte *p. 498*
stamen *p. 500*
stigma *p. 500*
style *p. 500*
tube cell *p. 501*

Thinking Through the Concepts

Multiple Choice

1. *When a bee gets nectar from a flower, the bee picks up pollen from the ____________ and carries it to another flower.*
   a. stigma
   b. ovary
   c. sepals
   d. anthers
   e. filament

2. *In plants, the gametophyte produces eggs and sperm by*
   a. mitosis
   b. meiosis
   c. spore formation
   d. fertilization
   e. germination

3. *If you found a dark, rotten-smelling flower near the ground, it would most likely be pollinated by*
   a. bats
   b. bees
   c. butterflies
   d. moths
   e. beetles
4. *Reproduction in flowering plants is known as double fertilization because*
   a. one tube nucleus fuses with one egg, and one sperm nucleus fuses with another egg
   b. two sperm nuclei fuse with one egg
   c. one sperm fuses with two eggs
   d. one sperm nucleus fuses with one egg, and another sperm nucleus fuses with the two haploid nuclei of the primary endosperm cell
   e. two polar nuclei fuse with one egg nucleus
5. *In a peach, the fruit is derived from the*
   a. wall of the ovary
   b. endosperm
   c. megaspores
   d. sepals
   e. petals
6. *The tassel on top of a corn plant is made up of male flowers that lack female structures. This flower is*
   a. the sporophyte
   b. the gametophyte
   c. incomplete
   d. complete
   e. homologous

? Review Questions

1. Diagram the plant life cycle, comparing ferns with flowering plants. Which stages are haploid, and which are diploid? At which stage are gametes formed?
2. What are the advantages of the reduced gametophyte stages in flowering plants, compared with the more substantial gametophytes of ferns?
3. Diagram a complete flower. Where are the male and female gametophytes formed? What are these gametophytes called?
4. How does an egg develop within an embryo sac? How does this structure allow double fertilization to occur?
5. What does it mean when we say that pollen is the male gametophyte? How is pollen formed?
6. What are the parts of a seed, and how does each part help in the development of a seedling?
7. Describe the characteristics you would expect to find in flowers that are pollinated by the wind, beetles, bees, and hummingbirds, respectively. In each case, explain why.
8. What is the endosperm? From which cell of the embryo sac is it derived? Is endosperm more abundant in the mature seed of a dicot or of a monocot?
9. Describe three mechanisms whereby seed dormancy is broken in different types of seeds. How are these mechanisms related to the normal environment of the plant?
10. How do monocot and dicot seedlings protect the delicate shoot tip during seed germination?
11. Describe three types of fruits and the mechanisms whereby these fruit structures help disperse their seeds.

Applying the Concepts

1. A friend gives you some seeds for you to grow in your yard. When you plant some, nothing happens. What might you try to get the seed to germinate?
2. In areas where farms have been left uncultivated for several years, it is often possible to see certain kinds of trees growing in straight lines, which mark old fences where birds sat and deposited seeds they had eaten. Why are such seeds more likely to germinate than are those of the same species that have not passed through a bird's digestive tract? How might an anthropologist use such lines of trees to study past inhabitants of an area?
3. Charles Darwin once described a flower that produced nectar at the bottom of a tube 25 centimeters (10.5 inches) deep. He predicted that there must be a moth or other animal with a 25-centimeter-long tongue to match; he was right. Such specialization almost certainly means that this particular flower could be pollinated only by that specific moth. What are the advantages and disadvantages of such specialization?
4. Many plants that we call *weeds* were brought from another continent either accidentally or purposefully. In a new environment they have few competitors or animal predators, so they tend to grow in such large amounts that they come to be considered weeds. Think of all the ways you can in which humans become involved in plant dispersal. To what degree do you think humans have changed the distributions of plants? In what ways is this change helpful to humans? In what ways is it a disadvantage?
5. In the Tropics there are a number of plant–animal coevolutionary relationships in which both are dependent on the relationship. In light of the rapid rate of destruction of tropical ecosystems, how does this type of relationship leave both organisms particularly vulnerable to extinction? What political and economic problems might this rapid rate of extinction create?

For More Information

Eiseley, L. "How Flowers Changed the World." *National Wildlife*, April/May 1996. The late philosopher/naturalist Loren Eiseley eloquently explains how the evolution of flowers has changed the history of life on Earth. Beautifully written and lavishly illustrated.

Fleming, T. H. "Cardon and the Night Visitors." *Natural History*, October 1994. A beautifully illustrated look at the way bats pollinate the world's largest cactus and may help determine the numbers of each of the plant's three sexual types.

Handel, S. N., and Beattie, A. J. "Seed Dispersal by Ants." *Scientific American*, August 1990. Many plants depend on ants for seed dispersal, even producing fat deposits on the outside of the seed as a lure for the ants. Dispersal is not the only benefit to the seed: Being "planted" in an anthill seems to provide an ideal environment for germination and growth as well.

Kearns, C. A., and Inouye, D.W. "Pollinators, Flowering Plants, and Conservation Biology." *Bioscience*, May 1997. To preserve plant species, we must preserve their pollinators and the web of life that sustains these pollinators.

Milius, S. "The Science of Big, Weird, Flowers." *Science News*, September 11, 1999. Many giant flowers deceive flies and beetles into pollinating them by smelling like carrion.

Moore, P. D. "The Buzz About Pollination." *Nature*, November 7, 1996. The buzzing of bees may actually shake loose the pollen of certain specially adapted flowers, a kind of "sonication pollination."

Murawski, D. A. "A Taste for Poison." *National Geographic*, December 1993. Certain passion-vine butterflies form cyanide, apparently from eating pollen from the passion vine. Other butterflies mimic the poisonous types.

Seymour, R. S., and Schultz-Motel, P. "Thermoregulating Lotus Flowers." *Nature*, September 26, 1996. The lotus flower generates a significant amount of heat and effectively regulates its own temperature. The heat may serve as an attractant to pollinators.

Simons, P. "An Explosive Start for Plants." *New Scientist*, January 2, 1993. Explosive movements in some plants spread seeds and spores and snap pollen onto bees.

Sunquist, F. "Blessed Are the Fruit Eaters." *International Wildlife*, May/June 1992. Especially in the Tropics, fruit-eating mammals and birds are crucial to the dispersal and germination of seeds. Overhunting threatens not only the animals but also the plants that depend on them.

Answers to Multiple-Choice Questions

1. d 2. a 3. e 4. d 5. a 6. c

MEDIATUTOR
Plant Reproduction and Development

CD Activities

Activity 24.1: Flower Structure and Function

Estimated time: 5 minutes

The following tutorial will help you to explore basic flower structure and function.

Activity 24.2: Fruit and Seed Structure and Development

Estimated time: 5 minutes

In this activity you will first examine the basic structure of fruits and seeds and then explore the process of fruit and seed development.

Start the MediaTutor Student CD-ROM and enter the activity number in the Quick Search box to be taken directly to that activity.

Web Investigations

Case Study: Walk Through a Meadow

Estimated time: 15 minutes

To most people, pollen is just a part of a plant, but for allergy suffers, it's trouble! Why does such a small biological object cause so many problems? The topic here is pollen and pollen allergies.

Go to http://www.prenhall.com/audesirk6, the Audesirk Companion Web site. Select Chapter 24 and the Web Investigation to begin.

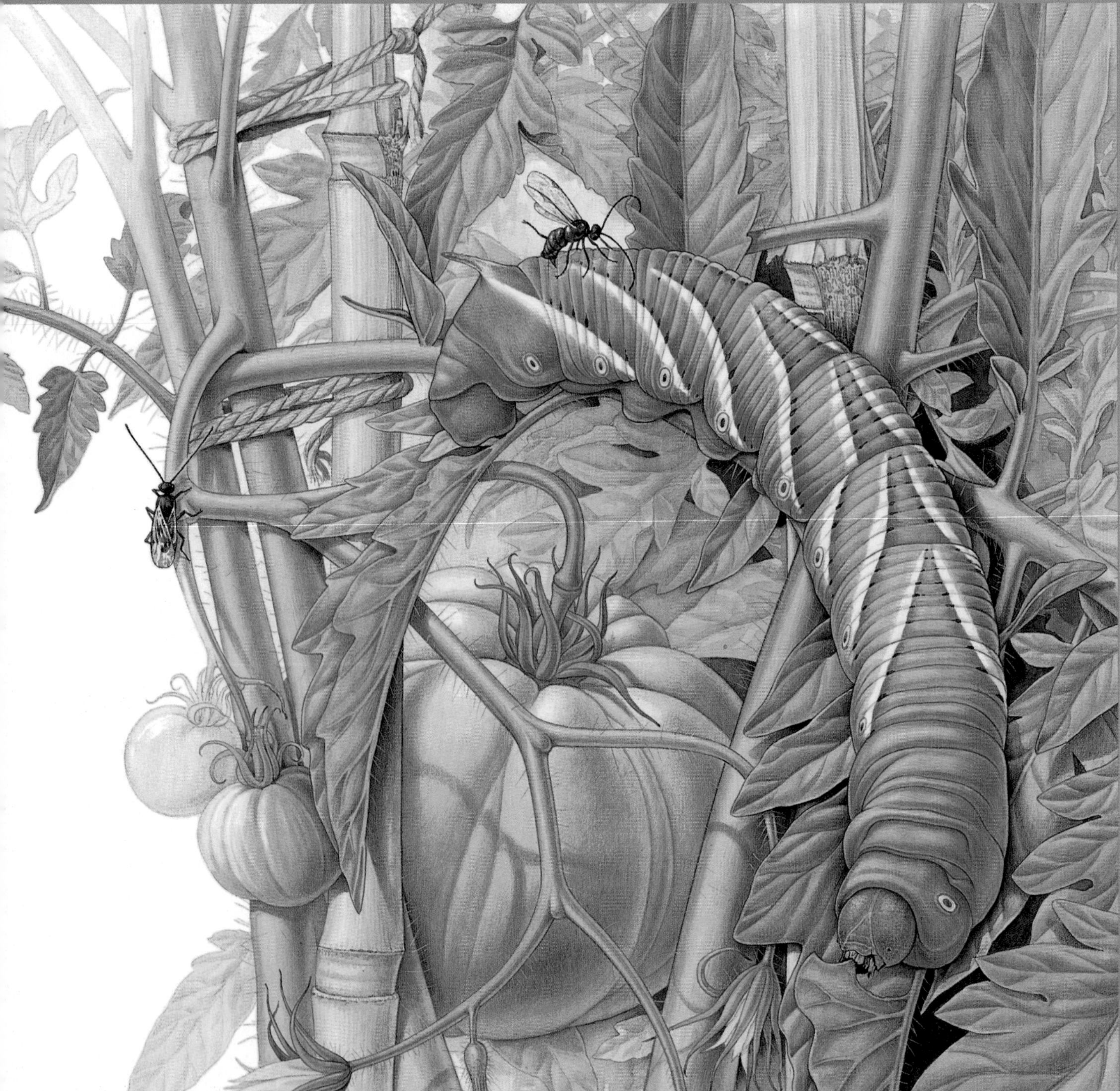

In the battle between plants and their predators, the plants appear helpless, but they actually have subtle ways of defending themselves.

25 Plant Responses to the Environment

AT A GLANCE

Case Study: A Chemical Cry for Help

1) **What Are Plant Hormones, and How Do They Act?**

2) **How Do Hormones Regulate the Plant Life Cycle?**

Abscisic Acid Maintains Seed Dormancy; Gibberellin Stimulates Germination

Auxin Controls the Orientation of the Sprouting Seedling

The Genetically Determined Shape of the Mature Plant Is the Result of Interactions Among Hormones

Daylength Controls Flowering

Hormones Coordinate the Development of Seeds and Fruit

Senescence and Dormancy Prepare the Plant for Winter

Evolutionary Connections: Rapid-Fire Plant Responses

Case Study Revisited: A Chemical Cry for Help

CASESTUDY

A Chemical Cry for Help

The tomato plants in your garden are under attack; some stems are stripped of their leaves, and many remaining leaves seem to have large bites taken out of them. Peering closely at the damage, you are startled to see a green caterpillar, as large as your index finger, voraciously munching. The tomato plant seems helpless against the tomato hornworm (the caterpillar stage of the sphinx moth). But the hundred-million-year war between plants and their parasites and animal predators has led to the evolution of sophisticated plant defenses that surprise people who are accustomed to thinking of most plants as passive, helpless organisms. Researchers studying how plants respond to attack by predators or disease-causing viruses have recently discovered that plants under attack help protect themselves (and even neighboring plants) by releasing volatile chemicals into the air around them. Tomato hornworms, as they prey on plants, are themselves prey for some insects. If you look carefully at enough tomato hornworms, you may spot one with small white cylinders adhering to its skin; these are the cocoons of a parasitic wasp (see the chapter opener illustration). The wasp finds a caterpillar and injects eggs into its body. The eggs hatch and develop into larvae that feed inside the caterpillar, then burrow out through its skin. Here, they attach and form cocoons within which they metamorphose into small wasps. How does the wasp find the caterpillar? Is the plant really a passive victim? ■

1 What Are Plant Hormones, and How Do They Act?

Plant cells, like those of animals, are miniature factories filled with diverse chemicals that allow plants to respond appropriately to their environment. Some convey messages within the plant and others even communicate among individual plants, as you'll see when we revisit "Case Study: A Chemical Cry for Help." Animal physiologists have long recognized that chemicals called **hormones** are produced by cells in one location and transported to other parts of the body, where they exert specific effects. Analogously, plant-regulating chemicals are called **plant hormones**. So far, plant physiologists have identified five major classes of plant hormones: (1) auxins, (2) gibberellins, (3) cytokinins, (4) ethylene, and (5) abscisic acid (Table 25-1). Several other types of hormones are thought to exist, but these are the best known. Each hormone can elicit a variety of responses from plant cells, depending on factors such as the type of target cell, the developmental stage of the plant, the concentration of the hormone, and the presence of other hormones. Further, the roles of some plant hormones vary among plant species. This should not be too surprising; after all, in animals a single hormone can be involved in actions as diverse as a salmon's transition from fresh water to salt water, a frog's metamorphosis from tadpole to adult, and a snake's shedding of its skin

Table 25-1 Hormone Actions in Plants

Hormone	Functions
Abscisic acid	Closing of stomata; seed dormancy; bud dormancy
Auxins	Elongation of cells in coleoptiles and shoots; phototropism; gravitropism in shoots and roots; root growth and branching; apical dominance; development of vascular tissue; fruit development; retarding senescence in leaves and fruit; ethylene production in fruit
Cytokinins	Promotion of sprouting of lateral buds; prevention of leaf senescence; promotion of cell division; stimulation of fruit, endosperm, and embryo development
Ethylene	Ripening of fruit; abscission of fruits, flowers, and leaves; inhibition of stem elongation; formation of hook in dicot seedlings
Gibberellins	Germination of seeds and sprouting of buds; elongation of stems; stimulation of flowering; development of fruit

(see "Evolutionary Connections: The Evolution of Hormones," in Chapter 32).

Auxins promote the elongation of cells in coleoptiles and other parts of the shoot; high concentrations cause cells to elongate (see "Scientific Inquiry: How Were Plant Hormones Discovered?"). In roots, which differ from stems in their response to auxin, low concentrations stimulate elongation, while slightly higher concentrations inhibit elongation. Both light and gravity affect the distribution of auxin in roots and shoots, so auxin plays a major role in both **phototropism** (growth toward light) and **gravitropism** (directional growth with respect to gravity). Auxin also affects many other aspects of plant development, such as the differentiation of conducting tissues (xylem and phloem) and the development of fruits. Auxin may also prevent sprouting by lateral buds. It stimulates root branching and can be used to cause stems of plants to grow roots, which is useful when growing a new plant from a cutting.

Gibberellins are a group of chemically similar molecules that, like auxin, promote the elongation of cells in stems. In some plants, gibberellins stimulate flowering, fruit development, seed germination, and bud sprouting.

Cytokinins promote cell division in many plant tissues; consequently, they stimulate the sprouting of buds and the development of fruit, endosperm, and the embryo. Cytokinins also stimulate plant metabolism, preventing or at least delaying the aging of plant parts, especially leaves.

Ethylene is an unusual plant hormone in that it is a gas at normal environmental temperatures. Ethylene is best known, and most commercially valuable, for its ability to cause fruit to ripen. It also stimulates the separation of cell walls into *abscission layers*, which, we shall soon see, allow leaves, flowers, and fruit to drop off at the appropriate times without harming the plant.

Abscisic acid is a hormone that helps plants withstand unfavorable environmental conditions. As you learned in Chapter 23, it causes stomata to close when water availability is low. It inhibits the activity of gibberellins, thus helping maintain dormancy in buds and seeds at times when germination would be dangerous.

In the rest of this chapter, we describe a year in the life of a plant, illustrating how hormones regulate its growth and development.

2 How Do Hormones Regulate the Plant Life Cycle?

The life cycle of a plant results from a complex interplay between its genetic information and its environment. Hormones mediate many of the genetic determinants of growth and development as well as nearly all responses to environmental factors. At each stage in its life cycle, a plant produces a distinctive set of hormones that interact with one another in directing the growth of the plant body.

Abscisic Acid Maintains Seed Dormancy; Gibberellin Stimulates Germination

As we pointed out in Chapter 24, a seed maturing within a juicy fruit on a warm autumn day has ideal conditions for germination, yet it remains dormant until the following spring. In many seeds, *abscisic acid enforces dormancy*. Abscisic acid slows down the metabolism of the embryo within the seed, preventing its growth. The seeds of some desert plants contain high concentrations of abscisic acid; only a really hard rain can wash out the abscisic acid, freeing the embryo from its inhibitory effects and allowing the seed to germinate. Seeds of most high-latitude plants require a prolonged period of cold, such as occurs during winter, to break dormancy; in these seeds, chilling induces the destruction of abscisic acid.

Germination is stimulated by other hormones, especially gibberellin. The same environmental conditions that cause the breakdown of abscisic acid also promote the synthesis of gibberellin. In germinating seeds, gibberellin initiates the synthesis of enzymes that digest the food reserves of the endosperm and cotyledons, making sugars, lipids, and amino acids available to the growing embryo.

Auxin Controls the Orientation of the Sprouting Seedling

When the growing embryo breaks out of the seed coat, it immediately faces a crucial problem: Which way is up? The roots must burrow downward, while the shoot must grow upward to find the light. Whether the seed was buried by a squirrel or fell randomly to the ground,

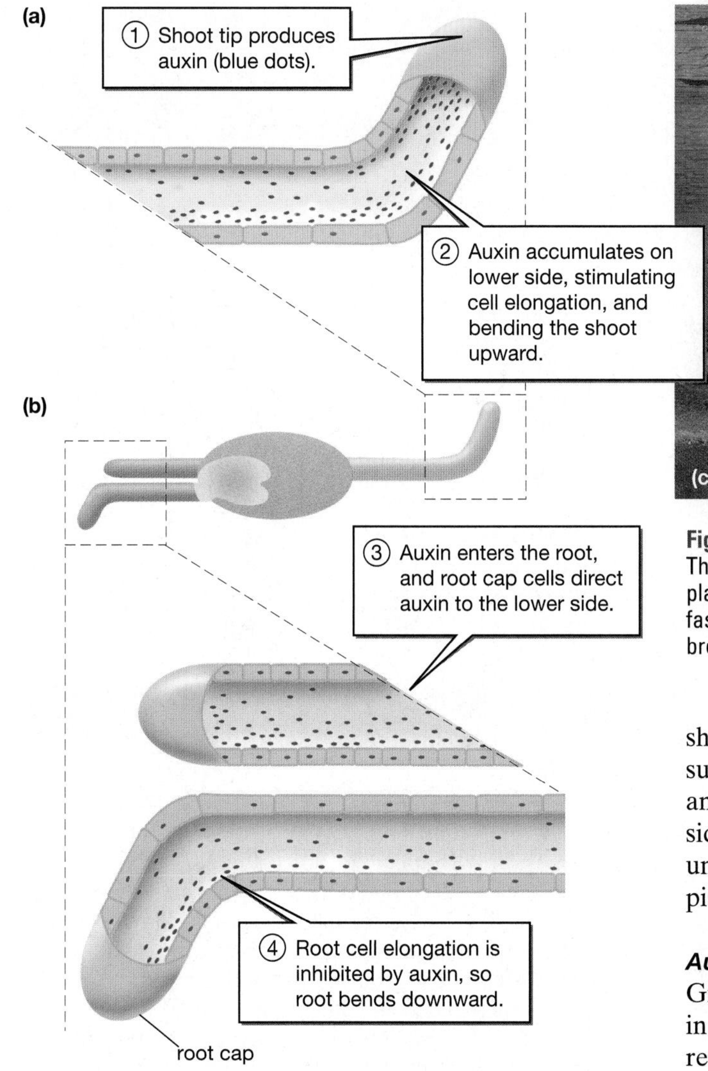
(a)
① Shoot tip produces auxin (blue dots).
② Auxin accumulates on lower side, stimulating cell elongation, and bending the shoot upward.
(b)
③ Auxin enters the root, and root cap cells direct auxin to the lower side.
④ Root cell elongation is inhibited by auxin, so root bends downward.
root cap

(c)

Figure 25-1 The mechanism of gravitropism in shoots and roots The photo shows two impatiens plants; the right-hand one was placed on its side in the dark for 16 hours. In this short time, faster elongation of cells on the lower side of the stem has brought the plant to a nearly vertical orientation.

it faces a high probability of being oriented "upside-down." Auxin apparently controls the responses of both roots and shoots to light and gravity.

Auxin Stimulates Shoot Elongation Away from Gravity and Toward Light

Let's look at the growth of a shoot as it first emerges from the seed, buried underground. Auxin is synthesized in shoot tips, moves down the shaft of the stem, and stimulates cell elongation. If the stem is not exactly vertical, organelles in the cells of the stem detect the direction of gravity and cause auxin to accumulate on the stem's lower side (Fig. 25-1a). Therefore, the lower cells elongate rapidly, forcing the stem to bend upward. When the shoot tip is vertical, the auxin distribution becomes symmetrical. The stem then grows straight up, emerging from the soil into the light (Fig. 25-1c).

Auxin mediates phototropism in addition to gravitropism. Ordinarily, the distribution of auxin caused by light is the same as the distribution caused by gravity, because the direction of brightest light (the sun) is roughly opposite that of gravity. For example, if a young shoot still buried underground is close enough to the surface that some light penetrates down to it, both light and gravity cause auxin to be transported to the lower side of the shoot and promote upward bending. Thus, under normal conditions gravitropism and phototropism add to each other's effects.

Auxin May Control the Direction of Root Growth

Gravitropism in roots is less well understood than it is in stems. According to one model, auxin controls the direction of root growth (Fig. 25-1b). Auxin is transported from the shoot down to the root. If the root is not vertical, the root cap senses the direction of gravity and causes the auxin to accumulate on the lower side of the root. Unlike shoots, in which moderate concentrations of auxin *stimulate* cell elongation, in roots these same concentrations of auxin *inhibit* cell elongation. Therefore, cell elongation in the lower side of the root, where auxin accumulates, is inhibited, whereas cell elongation remains unaffected in the upper side of the root. As a result, the root bends downward. When the root tip points directly downward, the auxin distribution becomes equal on all sides, and the root continues to grow straight downward. Note that *auxin slows down but does not eliminate root cell elongation*: A vertical root continues to grow.

Plants May Sense Gravity by Means of Starch-Filled Plastids

Although auxin plays a major role in the unequal growth rates of cells that cause bending away from gravity (in shoots) and toward gravity (in roots), auxin itself does not detect gravity. Instead, specialized cells in stems and in root caps contain starch-filled plastids. By staining these plastids and observing them under

Scientific Inquiry
How Were Plant Hormones Discovered?

Everyone who keeps houseplants on a windowsill knows that the plants bend toward the window as they grow, in response to the sunlight streaming in. More than a hundred years ago, Charles Darwin and his son, Francis, studied this phenomenon of growth toward the light, or phototropism.

First, the Darwins Determined the Direction of Information Transfer

The Darwins illuminated grass coleoptiles (protective sheaths surrounding monocot seedlings) from various angles. They noted that a region of the coleoptile a few millimeters below the tip bent toward the light, causing the tip to point toward the light source. When they controlled for the presence of light by covering the tip of the coleoptile with an opaque cap, the coleoptile didn't bend.

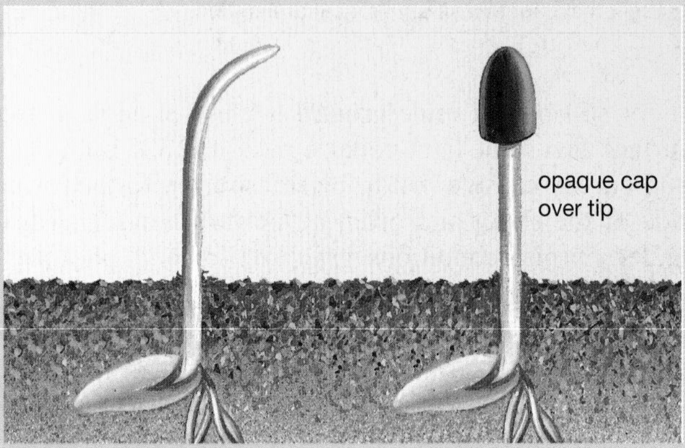

A clear cap allowed the stem to bend, and it still bent if the Darwins covered the bending region below the tip with an opaque sleeve.

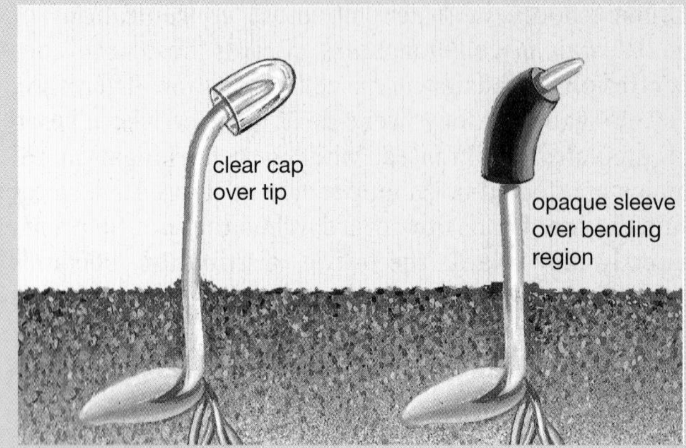

They concluded that (1) the tip of the coleoptile perceives the direction of light and (2) bending occurs farther down the coleoptile; therefore, (3) the tip transmits information about the light direction down to the bending region. How does the coleoptile bend? Although the Darwins didn't know this, the coleoptile bends as a result of unequal elongation of cells on opposite sides of the shaft. The cells on the darker side elongate faster than those facing the light, bending the shaft toward the light. So information transmitted from the tip to the bending region causes unequal cell elongation.

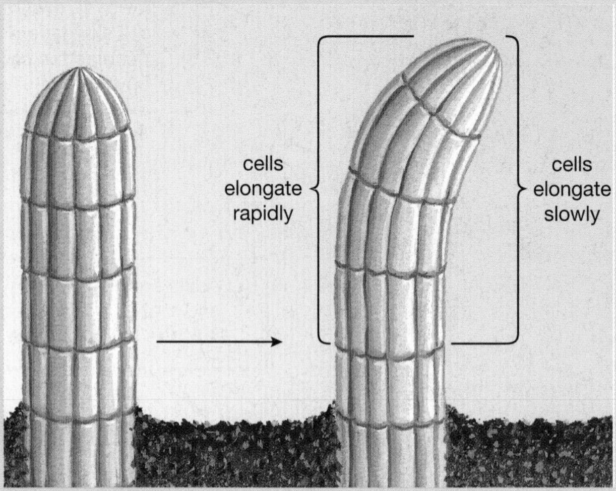

Next, Peter Boysen-Jensen Demonstrated That the Information Is Chemical in Nature

About 30 years after the Darwins' experiments, Peter Boysen-Jensen cut the tips off coleoptiles and found that the remaining stump neither elongated nor bent toward the light. If he replaced the tip and placed the patched-together coleoptile in the dark, it elongated straight up. In the light, it showed normal phototropism. When he inserted a thin layer of porous gelatin that prevented direct contact but permitted diffusion of substances between the severed tip and the stump, he still observed elongation and bending. In contrast, an impenetrable barrier eliminated these responses.

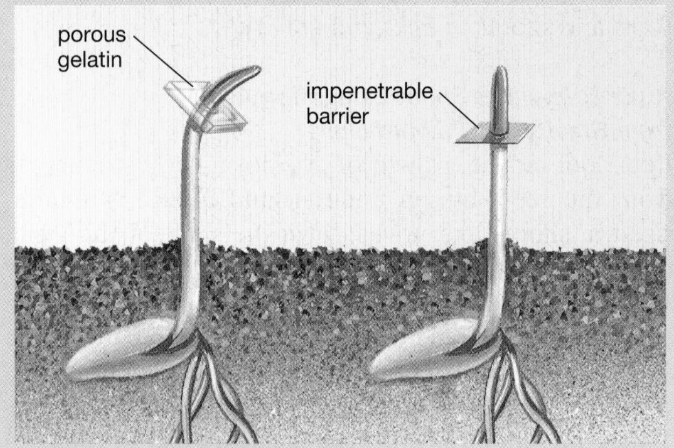

Boysen-Jensen concluded that a chemical is produced in the tip and moves down the shaft, causing cell elongation. In the dark, the chemical that causes the cells to elongate diffuses straight down from the tip and causes the coleoptile to elongate straight up. Presumably, light causes the chemical to become more concentrated on the "shady" far side of the shaft, so cells on the shady side elongate faster than do cells on the "sunny" near side, causing the shaft to bend toward the light.

Finally, the Chemical Auxin Was Identified

The next step was to isolate and identify the chemical. In the 1920s Frits Went devised a way to collect the elongation-promoting chemical. He cut off the tips of oat coleoptiles and placed them on a block of agar (a porous, gelatinous material) for a few hours. Went hypothesized that the chemical would migrate out of the coleoptiles into the agar.

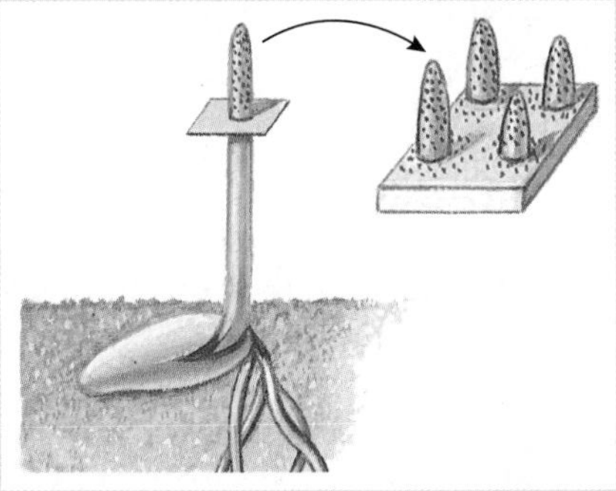

He then cut up the agar, now presumably loaded with the chemical, and placed small pieces on the tops of coleoptile stumps growing in darkness. When he put a piece of agar squarely atop a stump, the stump elongated straight up. All the stump cells received equal amounts of the chemical and elongated at the same rate. If he placed a piece on one side of a cut stump, the stump would invariably bend away from the side with the agar.

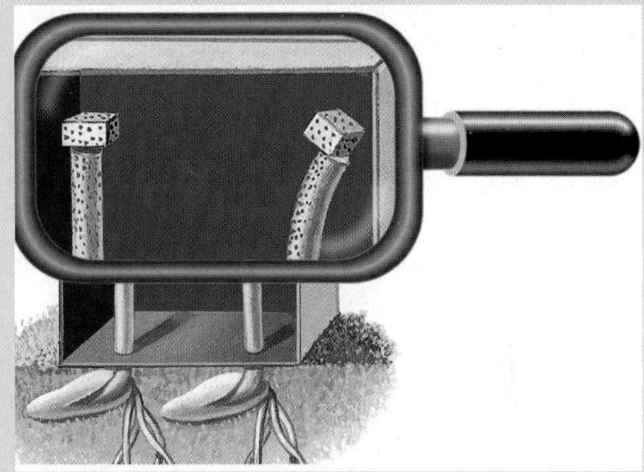

It was apparent that cells on the side under the agar received more of the chemical and were more stimulated to elongate. Went called the chemical *auxin*, from a Greek word meaning "to increase." Kenneth Thimann later purified auxin and determined its molecular structure.

the microscope, plant physiologists have discovered that they settle to the downward side of the cell within minutes when a plant is laid on its side (Fig. 25-2). The time it takes the plastids to settle is similar to the time it takes the root to begin its unequal cell elongation so that it once more heads downward. Exactly how the falling plastids initiate the response to gravity in stems and roots is still under investigation. One hypothesis is that the plastids are enmeshed in fibers of the cytoskeleton (see Chapter 5) that connect them to ion channels. As the plastids are pulled downward by gravity, the ion channels are pulled open, allowing movement of ions, such as calcium, into the cell. Thus, a mechanical stimulus (movement of plasmids in response to gravity) could be converted to a chemical stimulus. This chemical signal might initiate a series of reactions that cause auxin to accumulate on the downward side of the root cap.

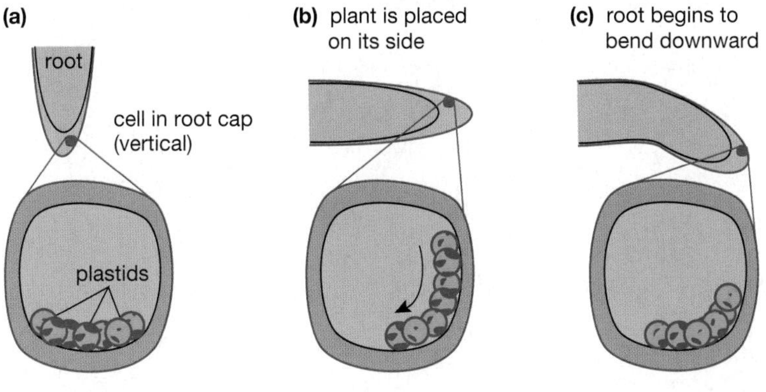

Figure 25-2 Starch-filled plastids allow plants to sense gravity
(a) Normal orientation of starch-filled plastids in a root cap cell. ***(b)*** When the plant is placed on its side (as in Figure 25-1c), the plastids begin tumbling downward, ***(c)*** coming to rest on the new lower surface. This change triggers the bending of the root downward.

The Genetically Determined Shape of the Mature Plant Is the Result of Interactions Among Hormones

As a plant grows, both its root and shoot develop branching patterns that are largely determined by its genetic heritage. For example, the stems of some plants, such as sunflowers, hardly branch at all; others, such as oaks and cottonwoods, branch profusely in seeming confusion; still others branch in a very regular pattern, producing the conical shapes of firs and spruces.

The amount of growth in shoot and root systems must also be kept in balance. The shoot must be large enough to supply the roots with sugars; the roots must be large enough to provide the shoot with water and minerals. Interactions between auxin and cytokinin regulate root and stem branching, thereby regulating the relative sizes of root and shoot systems.

Stem Branching Is Influenced by the Growing Tip of the Shoot

Gardeners know that pinching back the tip of a growing plant causes the plant to become bushier. The botanical explanation for this practice is that the growing tip suppresses the sprouting of lateral buds to form branches, a phenomenon known as **apical dominance**. The control mechanism for the sprouting of lateral buds remains a subject of research. However, some evidence shows that the proper levels and ratios of auxin and cytokinin must be present (Fig. 25-3). Auxin is produced by the shoot tip (its concentration is highest there) and transported down the stem, gradually decreasing in concentration. Cytokinin is produced by the roots (where its concentration is highest) and is transported up the stem. Therefore, the relative concentrations of these two hormones will vary along the length of the stem. Buds at different positions will experience differing hormonal influences.

Auxin by itself appears to inhibit the sprouting of lateral buds, whereas auxin and cytokinin together stimulate bud sprouting. The lateral buds closest to the shoot tip receive a great deal of auxin, probably enough to inhibit their growth, but receive very little cytokinin because they are so far from the roots. Therefore, they remain dormant. Lower buds receive less auxin while receiving much more cytokinin. They are stimulated by optimal concentrations of both hormones, so they sprout (see Fig. 25-3). In many types of plants, this interaction

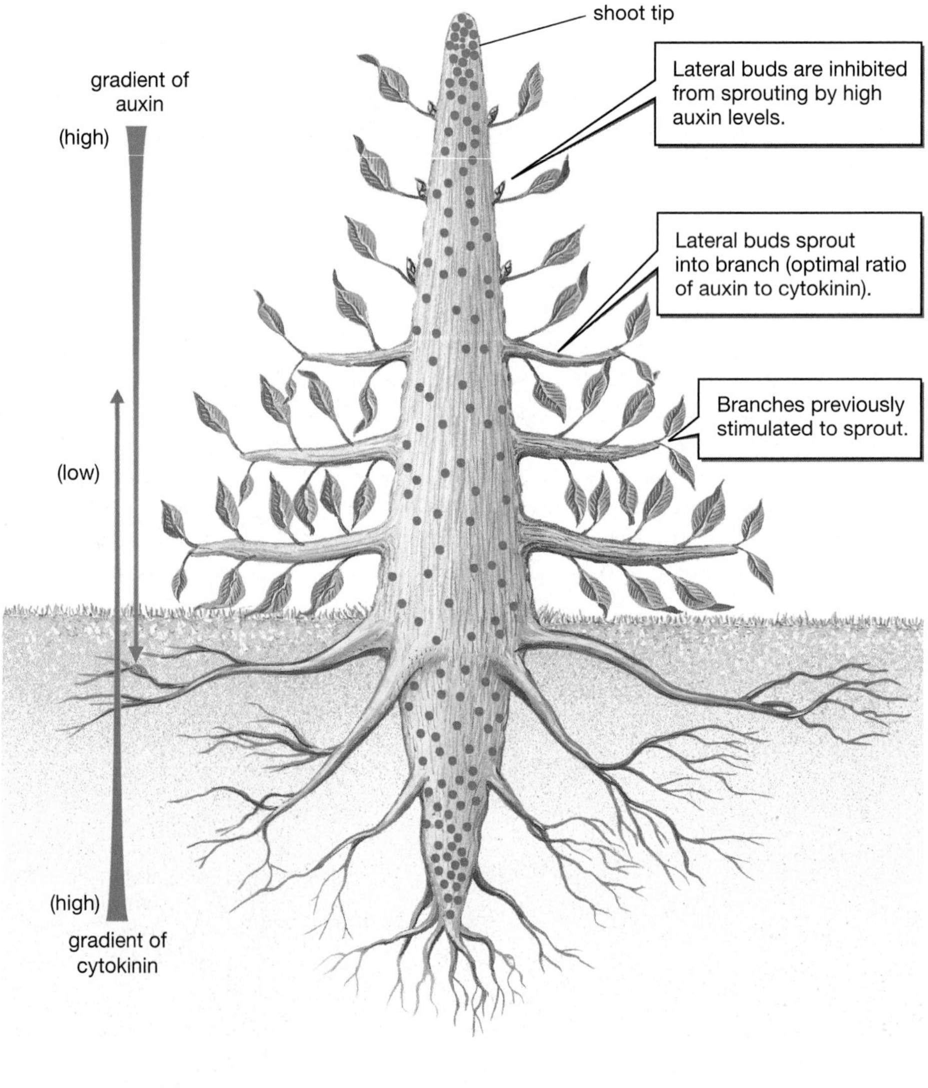
shoot tip
gradient of auxin
(high)
(low)
(high)
gradient of cytokinin
Lateral buds are inhibited from sprouting by high auxin levels.
Lateral buds sprout into branch (optimal ratio of auxin to cytokinin).
Branches previously stimulated to sprout.

Figure 25-3 The role of auxin and cytokinin in lateral bud sprouting A simplified diagram of the interplay of auxin (blue dots) and cytokinin (red dots) in the control of sprouting of lateral buds. Auxin is produced by shoot tips and moves downward; cytokinin is produced by root tips and moves upward.

MEDIATUTOR 25.1 Auxin Action

between auxin and cytokinin produces an orderly progression of bud sprouting from the bottom to the top of the shoot. The exact ratio of cytokinin to auxin that promotes sprouting varies among species.

Auxin Stimulates Root Branching

Even in extremely low concentrations, auxin stimulates the branching of roots. As we described in Chapter 23, branch roots arise from the pericycle layer of the vascular cylinder. Auxin, transported down from the stem, stimulates pericycle cells to divide and form a branch root. Gardeners are also familiar with another effect of auxin: to cause stems to grow roots when the stems are placed in soil. Commercial auxin preparations allow people to produce new plants from a length of stem of a desirable plant.

Gradients of Auxin and Cytokinin Create a Balance Between the Root and Shoot Systems

Through the interaction of auxin and cytokinin, the root and shoot systems regulate each other's growth. This interaction is important, because those systems supply complementary nutrients. An enlarging root system synthesizes large amounts of cytokinin, which stimulates lateral buds to break dormancy and sprout. If the root system isn't keeping up with the growing shoot system, less cytokinin is produced. The sprouting of lateral buds is delayed, slowing the growth of the shoot system. Simultaneously, as the stem grows and branches, it produces lots of auxin, which stimulates root branching and growth. Thus, neither root nor shoot can outgrow the other, and the plant is adequately supplied with all its needs.

Daylength Controls Flowering

Ultimately, the plant matures enough to reproduce. The timing of flowering and seed production is finely tuned to the physiology of the plant and the rigors of its environment. In temperate climates, plants must flower early enough so that their seeds can mature before the killing frosts of autumn. Depending on how quickly the seed and fruit develop, flowering may occur in spring, as it does in oaks; in summer, as in lettuce; or even in autumn, as in asters.

What environmental cues do plants use to determine the season? Most cues, such as temperature or water availability, are quite variable: October can be warm, a late snow may fall in May, or the summer might be unusually cool and wet. *The only reliable cue is daylength*: Longer days always mean that spring and summer are coming; shorter days foretell the onset of autumn and winter.

With respect to flowering, plants are classified as day-neutral, long-day, or short-day (Fig. 25-4). A **day-neutral plant** flowers as soon as it has sufficiently grown and developed, regardless of the length of the day. Day-neutral plants include tomatoes, corn, and snapdragons. A **long-day plant** flowers when the daylength is *longer than some species-specific critical value*. A **short-day plant** flowers when the daylength is *shorter than some species-specific critical value*. Thus, spinach is classified as a long-day plant because it flowers only if the day is *longer than* 13 hours, and cockleburs are short-day plants because they flower only if the day is *shorter than* 15.5 hours. Note that both will flower with 14 hours of light. Spinach, however, will also flower in much longer daylengths (for example, 16 hours of light), but cocklebur will not. Conversely, cocklebur will flower in much shorter daylengths (for example, 12 hours), but spinach will not.

Grafting a branch from a (short-day) cocklebur plant maintained on a short-day schedule onto a cocklebur plant on a long-day schedule will induce flowering. In fact, plant physiologists have been able to induce flowering in the cocklebur by exposing a *single leaf* to short days in a special chamber while the rest of the plant was experiencing long days. Clearly, a signal must be traveling from leaf to bud. Studies on a variety of plants suggest that still-unidentified substances can both trigger and inhibit flowering. These hypothetical substances, which may differ among plant species, are collectively called **florigens** (literally, "flower makers"). Daylength is a crucial stimulus for these substances, but how do plants detect daylength?

Pigments Called Phytochromes Measure Daylength by Resetting the Biological Clock

To measure daylength, a plant needs two things: (1) some sort of metabolic *clock* to measure time (how long it has been light or dark) and (2) a *light-detecting system* to set the clock. Virtually all organisms have an internal **biological clock** that measures time even without environmental cues. However, environmental cues, particularly light, can reset the clock. In most cases, the operating method of the biological clock is poorly understood.

The light-detecting system of plants is a pigment, called **phytochrome** (meaning simply "plant color"), in the leaves. Phytochrome occurs in two interchangeable forms. One form strongly absorbs red light and is called P_r; the other form absorbs far-red light (almost infrared) and is accordingly called P_{fr} (Fig. 25-5). In most plants, P_{fr} is the active form of phytochrome; that is, a suitable concentration of P_{fr} stimulates or inhibits physiological processes, such as flowering or setting the biological clock. P_r, the inactive form, has no effect on these same processes.

Phytochrome flips back and forth from one form to the other when the pigment absorbs light of the appropriate color: Upon absorbing red light, P_r is converted into P_{fr}; upon absorbing far-red light, P_{fr} is transformed back into P_r. Daylight consists of all wavelengths of visible light, including both red and far-red. Therefore, during the day, a

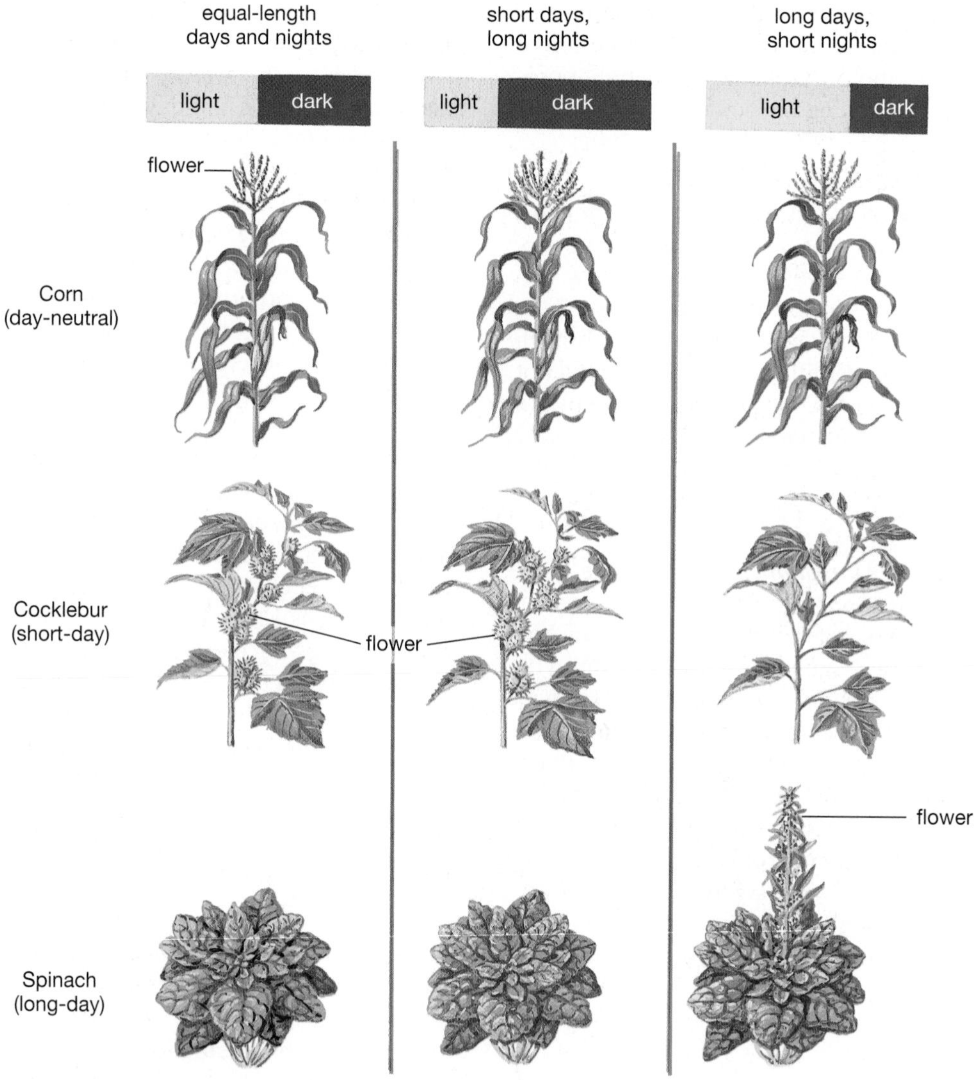

Figure 25-4 The effects of daylength on flowering

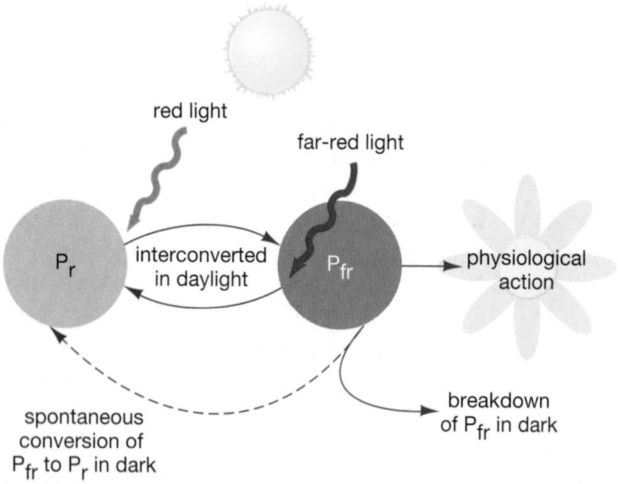

Figure 25-5 The light-sensitive pigment phytochrome
Phytochrome exists in two forms, inactive (P_r) and active (P_{fr}). P_r is converted to P_{fr} by red light. P_{fr} may then participate in physiological responses, be converted to P_r by far-red light, revert spontaneously to P_r, or break down to other, inactive compounds.

MediaTutor 25.2 Is It Time to Grow?

leaf contains both forms of phytochrome. In the dark, P_{fr} rapidly breaks down or reverts to P_r.

Plants seem to use the phytochrome system and their internal biological clocks to detect daylength. Cockleburs, for example, flower under a lighting schedule of 8 hours of light and 16 hours of darkness. However, interrupting the middle of the dark period with just a minute or two of light prevents flowering. Thus, although cockleburs are usually classified as short-day plants, they might more accurately be called "long-night" plants, because what really matters is how long continuous darkness lasts.

The color of the light used for the night flash is also important. A midnight flash of red light inhibits flowering, but a far-red flash allows flowering. This observation, of course, implicates phytochrome in the control of flowering. Scientists are still researching, however, how the response of phytochrome to light determines whether a plant will flower. It seems likely that the biological clock measures the length of the night and that light reception by phytochrome "tells" the clock when sunrise and sunset have occurred, but this is not certain. How phytochromes influence the production of florigens is another area of active research.

Phytochromes Influence Other Responses of Plants to Their Environment

Phytochrome is involved in many plant responses. For example, P_{fr} inhibits the elongation of seedlings. Because P_{fr} breaks down or reverts to P_r in the dark, seedlings germinating in the darkness of the soil contain no P_{fr} and consequently elongate very rapidly, emerging from the soil. Seedlings growing beneath other plants will be exposed largely to far-red light, because the green chlorophyll of the leaves above them absorb most of the red light but transmit the far-red. Far-red light converts P_{fr} to P_r, so shaded seedlings grow rapidly, which may bring them out of the shade. Once out in the sunlight, P_{fr} forms. The P_{fr} slows down elongation, which prevents the seedlings from becoming too spindly.

Other plant responses that are stimulated by P_{fr} include leaf growth, chlorophyll synthesis, and the straightening of the epicotyl or hypocotyl hook of dicot seedlings. As in the case of stem elongation, these responses are adaptations related to burial in the soil or shading by the leaves of other plants. For example, a newly germinating shoot needs to stay in its protective bend while still in the soil (that is, in the dark) and straighten out only in the open air, where sunlight converts P_r to P_{fr}.

Hormones Coordinate the Development of Seeds and Fruit

When a flower is pollinated, auxin or gibberellin released by the pollen stimulates the ovary to begin developing into a fruit. If fertilization also occurs, the developing seeds release still more auxin or gibberellin, or both, into the surrounding ovary tissues. Cells of the ovary multiply and grow larger, commonly storing starches and other food materials, forming a mature fruit. In this way, the plant coordinates the development of seeds and fruit.

Seeds and fruits acquire nutrients for growth and development from their parent plant. If the seed is separated from the parent too soon, it may not complete its development. Not surprisingly, seed maturation and fruit ripening are closely coordinated. Most unripe fruits are inconspicuously colored (normally green like the rest of the plant), hard, bitter, and in some cases even poisonous. As a result, animals seldom eat unripe fruit.

When the seeds mature, the fruit ripens: It becomes sweeter as starches are converted to sugar, softer, and more brightly colored, making it more noticeable and attractive to animals (Fig. 25-6). If you walk in a field in the fall, look around at the wild fruits on the dying plants. A few may be blue, and many are red, such as the "hips" on wild roses. Almost none are green when ripe. Or look around the produce section of your supermarket at all the brightly colored fruits—adapted to attract animal seed dispersers. Interestingly, gibberellin sprayed on fruit such as grapefruit causes the peel to remain tough and green, although the inside continues to ripen. Citrus growers in Florida can now use this technique to discourage fruit flies, which are attracted to the yellow color and must penetrate the ripening peel to lay their eggs.

Ripening is stimulated by ethylene, which also causes the breakdown of green chlorophyll, revealing the attractive pigments that signal a ripe fruit. Ethylene is synthesized by fruit cells in response to a surge of auxin that is released by the seeds (another mechanism by which both seed and fruit development are coordinated). Because ethylene is a gas, a ripe fruit continually leaks ethylene into the air. In nature this probably doesn't make much difference. When you store fruit in a closed container, however, ethylene released from one fruit will hasten ripening in the rest. The discovery of the role of ethylene in ripening revolutionized modern fruit and vegetable marketing. The gibberellin-sprayed green (but ripe) grapefruit described earlier will turn its normal yellow when exposed to ethylene. Bananas grown in Central America can be picked green and tough and shipped to North American markets. By

Figure 25-6 Ripe fruit becomes attractive to animal seed dispersers
The prickly pear cactus fruit is green, hard, and bitter before it ripens, so animals are discouraged from eating it. After the seeds mature, the fruit becomes soft, red, and tasty, attracting animals such as this desert tortoise. The mature seeds are not harmed by the animal's digestive tract and are dispersed in the animal's feces.

exposing them to ethylene at their destination, grocers can market perfectly ripe fruit. Unfortunately, not all fruits seem to ripen properly when separated from the plant. Grapes, strawberries, and tomatoes are not sensitive to ethylene, therefore they must be allowed to ripen on the plant, and shipping them to markets without damage becomes a challenge.

Senescence and Dormancy Prepare the Plant for Winter

The season is autumn and the time has come to let any uneaten fruits drop to the ground. For perennial broadleaf plants, the leaves must be shed as well, because they would be a liability in winter, unable to photosynthesize but still allowing water to evaporate. Leaves, fruits, and flowers undergo a rapid aging called **senescence.** The culmination of senescence is the formation of the **abscission layer** at the base of the petiole (Fig. 25-7). Ethylene stimulates this layer of thin-walled cells to produce an enzyme that digests the cell wall holding a leaf, fruit, or flower to the stem, allowing the leaf or fruit to drop off. Cells interior to the abscission remain intact.

Senescence and abscission are complex processes controlled by several different hormones. In most plants, healthy leaves and developing seeds produce auxin, which in turn helps maintain the health of the leaf or fruit. Simultaneously, the roots synthesize cytokinin, which is transported up the stem and out the branches. Cytokinin also prevents senescence—a leaf plucked from a tree and floated in water in which cytokinin is dissolved stays green for weeks.

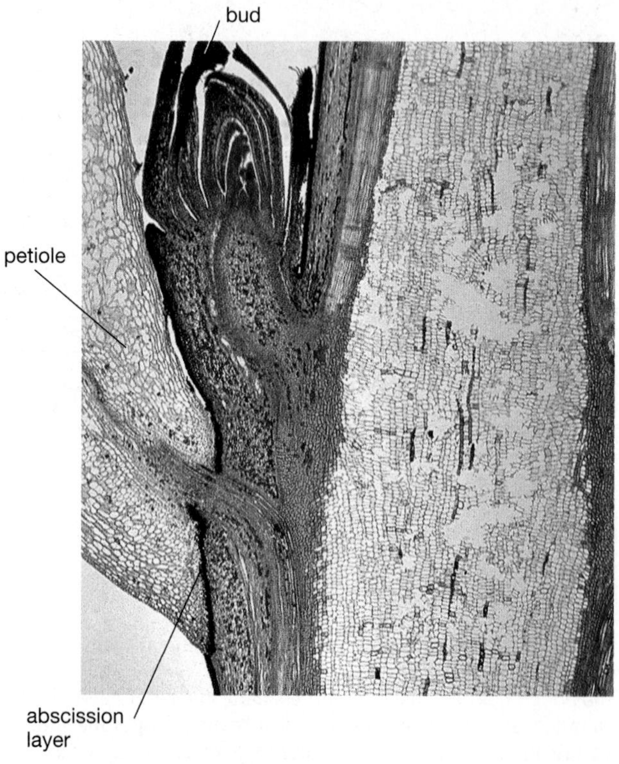

Figure 25-7 The abscission layer
This cross section shows the abscission layer forming at the base of a maple leaf. A new leaf bud is visible above the dying leaf.

As winter approaches, cytokinin production in the roots slows down, and fruits and leaves produce less auxin. Perhaps driven by these hormonal changes, much of the organic material in leaves is broken down to simple molecules that are transported to the roots for winter storage. Meanwhile, ethylene is released by both aging leaves and ripening fruit. Ethylene stimulates the production of enzymes that destroy the cell walls holding together the abscission layers at the base of the petiole. When the abscission layers weaken, leaves and fruits fall from the branches.

Other changes also occur that prepare the plant for winter. Buds, which developed into new leaves and branches during spring and summer, now become tightly wrapped up and dormant, waiting out the winter. Dormancy in buds, as in seeds, is enforced by abscisic acid. Metabolism slows to a crawl, and the plant enters its long winter sleep, waiting for the signals of warmth and longer days in spring before awakening once again.

Evolutionary Connections

Rapid-Fire Plant Responses

All plants are alive, but some are livelier than others. Watch a fly brush against the sensory hairs in a Venus flytrap, and you will see a response that is almost animal-like in its purposefulness and speed of movement (Fig. 25-8). Why is it useful for a Venus flytrap to catch a fly, and how does the plant accomplish this task?

As you learned in Chapter 23, some soils, particularly those of bogs, are nitrogen-deficient. As an evolutionary response to chronic nitrogen shortages, several bog plants, including the Venus flytrap, pitcher plant, and sundew, have resorted to eating animals. By snaring an insect now and then, the plant obtains nitrogen from the chitin (in the exoskeleton), proteins, and nucleic acids of its prey. Although the evolutionary advantage seems clear, the *mechanism* of rapid plant movement is much less obvious. How does the plant perceive the touch of a fly, and how does it move its leaves rapidly enough to catch the fly?

In this chapter you have seen how hormones such as auxin can trigger the expansion of cells, but hormones don't carry signals fast enough to catch a fly. Venus flytraps have evolved a way to transmit signals that resemble the nerve impulse of animals, thus permitting their animal-like predatory behavior. Each of the fringed trapping leaves of a Venus flytrap bears three sensory "hairs" on its inside surface (in Fig. 25-8a, you can see the shadows of these hairs and their small white attachment points). Insects are attracted to nectar secreted by

(a)

(b)

Figure 25-8 A Venus flytrap captures its prey

the leaves. If a foraging insect touches one hair twice in rapid succession or touches two hairs, the hairs initiate a change in electrical potential analogous to the action potential of animal nerve cells (to be discussed in Chapter 33). The electrical potential sets off a rapid chain of events that causes the trap to close (Fig. 25-8b).

In a beautiful set of experiments, botanists Stephen Williams and Alan Bennett found that the flytrap leaf closes because it undergoes *irreversible, differential expansion of cells*. The flytrap leaves can be pictured most simply as two layers of cells, outer and inner epidermis, as illustrated below. The electrical potential triggered by hair movement stimulates the cells of the outer layer to rapidly pump hydrogen ions (H^+) into their cell walls. Enzymes in the cell walls are activated by acidic conditions (created, as you'll remember from Chapter 2, by a high concentration of H^+) and loosen the cellulose fibers of the walls. As the walls weaken, the high osmotic pressure inside causes the cells to absorb water from extracellular fluids, swiftly growing by about 25%. Because the outer layer expands while the inner layer does not, the leaf is pushed closed.

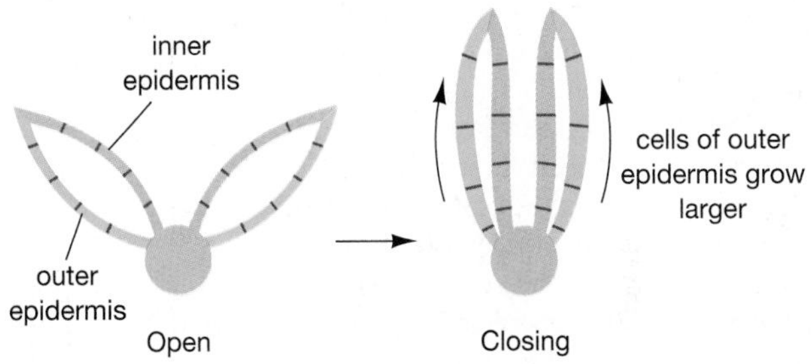

Although reopening the trap takes several hours, the fundamental mechanism is similar: The cells on the inside of the leaf expand, pushing apart the lobes of the trap. So much energy is used up by the hydrogen ion pumps that closing the trap consumes nearly one-third of all the ATP within the entire leaf. It is therefore very important that something digestible actually be in the leaf before it closes the trap! Thus, "teasing" a Venus flytrap by touching its hairs could be very bad for its health.

Although much is known about the mechanisms that produce movement in the Venus flytrap, mysteries still remain. How is touch transformed into an electrical stimulus by the sensory hairs? What is the nature of the electrical potential change? How does the electrical signal cause the cells to begin pumping hydrogen ions? As so often happens in biology, the answer to one question immediately poses several new, usually tougher, questions.

CASE STUDY REVISITED

A Chemical Cry for Help

Recently, researchers have discovered that, when attacked, some plants release a "chemical cry for help." Working with maize, a relative of corn, researchers discovered a sophisticated form of communication between plants and animals: The maize plant calls in a predator to attack a caterpillar that is munching on it. The plant responds to the caterpillar's attack by releasing a mixture of volatile chemicals, which attract a parasitic wasp that lays its eggs in the body of the caterpillar. This "biological time bomb" kills the caterpillar as the larvae hatch and consume their host from the inside out. Scientists discovered that merely tearing the leaves as a caterpillar does will not elicit the chemical alarm signal; the attack on the plant had to come from an actual caterpillar. Recently, the reason was discovered: A chemical called *volicitin* in caterpillar saliva is required to cause the response (Fig. E25-1).

Plants also can defend against diseases. Many plants produce *salicylic acid*, the compound from which aspirin is derived. Ilya Raskin and colleagues at Rutgers University found that tobacco plants infected with a plant virus produce large quantities of salicylic acid. The salicylic acid activates an immune response in the plants, helping them fight off the viral attack. The plant also converts some of the salicylic acid to *methyl salicylate* (the substance used to flavor "wintergreen" candy). This highly volatile compound diffuses into the air from virus-infected plant tissues and is absorbed by other plants nearby. The healthy neighbors reconvert the methyl salicylate to salicylic acid, enhancing their immune defenses and making them better able to resist the viral infection. These findings suggest that future farmers may treat their plants with aspirin or similar substances instead of toxic pesticides to protect them against diseases and plant predators. As Ilya Raskin explains, "Plants can't run away and they can't make . . . noises. But they are wonderful chemists."

Is your tomato plant releasing a cry for help? Assume you work in a lab with a well-equipped greenhouse, healthy tomato plants, tomato hornworms by the dozen, and a supply of the parasitic wasps that attack tomato hornworms. Design a controlled study that will support or refute the hypothesis that tomato plants, like maize, can summon the wasps when attacked by a hornworm. Be sure to control for other types of attack.

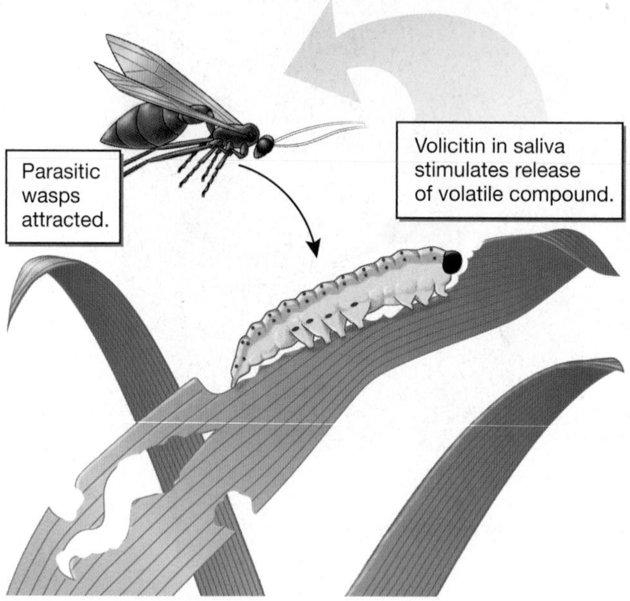
Volicitin in saliva stimulates release of volatile compound.
Parasitic wasps attracted.

Figure E25-1 A chemical cry for help

Summary of Key Concepts

1) What Are Plant Hormones, and How Do They Act?

Plant hormones are chemicals that are produced by cells in one part of a plant body and transported to other parts of the plant, where they exert specific effects. The five major classes of plant hormones are auxins, gibberellins, cytokinins, ethylene, and abscisic acid. The major functions of these hormones are summarized in Table 25-1.

2) How Do Hormones Regulate the Plant Life Cycle?

Dormancy in seeds is enforced by abscisic acid. Falling levels of abscisic acid and rising levels of gibberellin trigger germination. As the seedling grows, it shows differential growth with respect to the direction of light (phototropism) and gravity (gravitropism). Auxin mediates phototropism and gravitropism in shoots and gravitropism in roots. In shoots, auxin stimulates the elongation of cells. In roots, similar concentrations of auxin inhibit elongation. Plants apparently detect gravity by means of organelles called *plastids*.

Branching in stems results from the interplay of two hormones, auxin (produced in shoot tips and transported downward) and cytokinin (synthesized in roots and transported up the shoot). High concentrations of auxin inhibit the growth of lateral buds. An optimum concentration of both auxin and cytokinin stimulates the growth of lateral buds. Auxin also stimulates the growth of branch roots.

The timing of flowering is normally controlled by daylength. Flowering is both stimulated and inhibited by hormones called *florigens*. Plants appear to detect light and darkness by changes in phytochrome, a pigment in the leaves. Plant processes influenced by phytochrome responses to light include flowering, straightening the epicotyl or hypocotyl hook, seedling elongation, leaf growth, and chlorophyll synthesis.

Developing seeds produce auxin, which diffuses into the surrounding ovary tissues and causes the production of a fruit. A surge of auxin as the seed matures stimulates fruit cells to release another hormone, ethylene, which causes the fruit to ripen. Ripening includes the conversion of starches to sugars, softening of the fruit, development of bright colors, and, commonly, the formation of an abscission layer at the base of the petiole.

Several changes prepare perennial plants of temperate zones for winter. Leaves and fruits undergo a rapid aging process called *senescence*, including the formation of an abscission layer. Senescence occurs as a result of a fall in levels of auxin and cytokinin and, perhaps, a rise in ethylene concentrations. Other parts of the plant, including buds, become dormant. Dormancy in buds is enforced by high concentrations of abscisic acid.

Key Terms

abscisic acid *p. 520*
abscission layer *p. 528*
apical dominance *p. 524*
auxin *p. 520*
biological clock *p. 525*
cytokinin *p. 520*
day-neutral plant *p. 525*
ethylene *p. 520*
florigen *p. 525*
gibberellin *p. 520*
gravitropism *p. 520*
hormone *p. 519*
long-day plant *p. 525*
phototropism *p. 520*
phytochrome *p. 525*
plant hormone *p. 519*
senescence *p. 528*
short-day plant *p. 525*

Thinking Through the Concepts

Multiple Choice

1. *If you put an underripe banana in a bag with an apple, the banana will quickly ripen because of the hormone _________ produced by the apple.*
 a. auxin b. cytokinin
 c. gibberellin d. abscisic acid
 e. ethylene
2. *Roots turn downward in a process known as*
 a. phototropism b. abscission
 c. apical dominance d. gravitropism
 e. dormancy
3. *If you grow coleus plants, you will need to cut or pinch off the top bud frequently to keep the plant from becoming tall and spindly. Pinching off this bud will slow the production of _________ by the apical bud and allow the plant to become bushy.*
 a. auxin b. cytokinin
 c. gibberellin d. abscisic acid
 e. ethylene
4. *In your warm house, poinsettias act as short-day plants that turn red when the daylength is less than 12 hours. If you want to get a poinsettia ready for Christmas, which of the following would work best?*
 a. Give it continuous light.
 b. Give it continuous darkness.
 c. Keep it in the dark, but shine a light on it for an hour every 13 hours.
 d. Keep it in the light except for turning off the light once a day for 1 hour.
 e. None of these would get it to turn red.
5. *When a seed is first formed in the fall, it typically will not germinate because _________ must first be washed from the seed by a hard rain or broken down by cycles of freezing and thawing.*
 a. auxin b. cytokinin
 c. gibberellin d. abscisic acid
 e. ethylene
6. *The seed of Question 5 germinates as levels of _________ increase.*
 a. auxin
 b. cytokinin
 c. gibberellin
 d. abscisic acid
 e. ethylene

? Review Questions

1. What did the Darwins, Boysen-Jensen, and Went each contribute to our understanding of phototropism? Do their experiments truly prove that auxin is the hormone controlling phototropism? What other experiments would you like to see?
2. How do hormones interact to cause apical dominance? To control seed dormancy?
3. How can one hormone, an auxin, cause shoots to grow up and roots to grow down?
4. What is the phytochrome system? Why does this chemical exist in two forms? How do the two forms interact to help control the plant life cycle?
5. Which hormones cause fruit development? From where do these hormones come? Which hormone causes fruit ripening?
6. What is a biological clock?
7. Describe the role of phytochrome in stem elongation in seedlings that grow in the shade of other plants. What is the likely adaptive significance of this response?
8. What is apical dominance? How do auxin and cytokinin interact in determining the growth of lateral buds?
9. Which hormone(s) is (are) involved in leaf and fruit drop? In bud dormancy?

Applying the Concepts

1. Suppose you got a job in a greenhouse in which the owner was trying to start the flowering of chrysanthemums (a short-day plant) for Mother's Day. You accidentally turned on the light in the middle of the night. Would you be likely to lose your job? Why or why not? What would happen if you turned on the lights in the day?
2. A student reporting on a project said that one of her seeds did not grow properly because it was planted upside down so that it got confused and tried to grow down. Do you think the teacher accepted this explanation? Why or why not? Which plant hormone or hormones would be involved?
3. Agent Orange, a combination of two synthetic auxins, was used in Vietnam to defoliate the rain forest during the Vietnam War. When they are similar to natural growth hormones, how can synthetic auxins be used to harm or kill plants? What do you think would happen if natural auxins were used in excess quantities on plants?
4. Bean sprouts such as those you might eat in a salad have to be grown in the dark for them to form the long, yellowish stems that you see. We call such stems *etiolated*. If they are grown in the light, they will be short and green. Why do seedlings grow etiolated in the dark? Under what conditions does etiolation occur in nature? How do plant hormones enable these seedlings to form this shape?
5. Suppose that on July 4, you discover that both a short-day plant and a long-day plant have bloomed in your garden. Discuss how it is possible for both to bloom.

For More Information

Farmer, E. E. "New Fatty Acid-Based Signals: A Lesson from the Plant World." *Science*, May 9, 1997. Describes the research leading to the discovery of volicitin, which attracts parasitic wasps to plants under attack by caterpillars.

Mlot, C. "Where There's Smoke, There's Germination." *Science News*, May 31, 1997. Researchers have recently discovered that nitrogen dioxide produced by fires can induce germination in plants that live in ecosystems where fires are common.

Moffatt, A. S. "How Plants Cope with Stress." *Science*, November 1, 1994. A newly discovered hormone, "systemin," similar to animal hormones, enables plants to respond to stress.

Saunders, F. "Keep the Aspirin Flying." *Discover*, January 1998. Describes how plants use methyl salicylate to help nearby plants resist infection.

Answers to Multiple-Choice Questions

1. e 2. d 3. a 4. c 5. d 6. c

MediaTutor
Plant Response to the Environment

CD Activities

Activity 25.1: Auxin Action

Estimated time: 10 minutes

In this simulation you can explore some of the effects of auxin on plant growth and development by altering the source and concentration of auxin in the plant.

Activity 25.2: Is It Time to Grow?

Estimated time: 10 minutes

In this simulation, you will review the conversion of phytochrome by red and far-red light to initiate biological actions. You will be able to explore the effects of different types of light and the length of dark and light periods on different types of plants, and then you will see the response of the plant to the stimulus.

Start the MediaTutor Student CD-ROM and enter the activity number in the Quick Search box to be taken directly to that activity.

Web Investigations

Case Study: A Chemical Cry for Help

Estimated time: 10 minutes

Plants have evolved a number of sophisticated systems to protect themselves from herbivores (plant eaters) and other dangers. This exercise takes a look at some of these defense mechanisms.

Go to http://www.prenhall.com/audesirk6, the Audesirk Companion Web site. Select Chapter 25 and the Web Investigation to begin.

UNIT FIVE

Animal Anatomy and Physiology

The animal body is an exquisite expression of the elegance with which evolution has linked form to function. All the animal body systems work in concert to maintain life.

"As a physiologist I am especially impressed by the common ignorance of bodily organs and their functions. It seems to me now, as it seemed to Robert Boyle nearly 300 years ago, that it is 'highly dishonorable for a Reasonable Soul to live in so Divinely built a Mansion as the Body [he or] she resides in, altogether unacquainted with the exquisite structure of it.'"

W. B. Cannon in* The Way of an Investigator *(1945)

A parachutist challenges his body. Injury disrupts the function of the interrelated systems of the body and threatens its ability to maintain homeostasis.

26 Homeostasis and the Organization of the Animal Body

AT A GLANCE

Case Study: The Limits of Endurance

1) **Homeostasis: How Do Animals Maintain Internal Constancy?**

Negative Feedback Reverses the Effects of Changes

Positive Feedback Drives Events to a Conclusion

The Body's Internal Systems Act in Concert

2) **How Is the Animal Body Organized?**

Animal Tissues Are Composed of Similar Cells That Perform a Specific Function

Organs Include Two or More Interacting Tissue Types

Organ Systems Consist of Two or More Interacting Organs

Case Study Revisited: The Limits of Endurance

CASESTUDY

The Limits of Endurance

On Labor Day of 1999, Wayne Crill climbed to a ledge high on Mount Evans—a 14,000-foot mountain in Colorado—strapped a parachute to his back, and jumped. Sixty-three hours later, after a friend noticed his pickup truck in a parking lot outside the wilderness area with an expired parking ticket, searchers found him lying under his parachute. A helicopter crew rescued his shattered but still-living body. Crill had survived nearly 3 days lying at the base of an enormous boulder at 13,000 feet, experiencing subzero temperatures, the low oxygen conditions of high altitude, dehydration, and massive skeletal injuries, including a broken leg, broken arms and ribs, a broken shoulder blade, fractured skull and facial bones, and a shattered ankle.

The intricate mechanisms of regulation and control that provide animals with internal stability and allowed Crill to survive are among nature's true marvels. Molecules are constantly entering and leaving the body, but physiological systems control this exchange in a way that maintains a relatively constant internal state. The body's cells are thus insulated from the vagaries of the external environment and can go about their biochemical business without interference from drastic fluctuations in their immediate surroundings. Whales can dive from sea level to depths at which the water pressure is tremendous; desert scorpions can survive the daily switch from hot days to cold nights; penguins can stay active on the frigid ice floes of the Antarctic; and (if they are lucky) people sometimes survive days without food, water, or adequate protection against the cold.

How did Crill maintain his body temperature in spite of subzero temperatures? Without food, how did Crill's body obtain energy to carry out the metabolic activities necessary to his survival? To understand more about our resiliency in adverse conditions, we must learn about how our bodies cope with physiological challenges. ■

Source: *Denver Post,* November 21, 1999.

1 Homeostasis: How Do Animals Maintain Internal Constancy?

Whether you dive into a swimming pool, hike in the desert, or swim in the ocean, your cells remain isolated from the outside conditions. They are bathed in extracellular fluid containing a very specific, complex mixture of dissolved substances that must be maintained regardless of conditions on the outside. Many animals have evolved elaborate physiological mechanisms that allow them to maintain precise internal conditions despite a lifetime spent in harsh environments. For example, desert dwellers such as the kangaroo rat have kidneys that allow them to conserve water, while freshwater animals such as the trout or frog excrete copious amounts of water. Ocean-dwelling fish secrete excess salt from their gills. Because the cells of the animal body cannot survive if the conditions of the internal environment deviate from a small range of acceptable states, cells devote a large portion of their energy to actions that serve to keep the cellular environment stable.

This "constancy of the interior milieu" was first recognized by French physiologist Claude Bernard in the mid-nineteenth century. Later, in the 1920s, Walter B. Cannon coined the term **homeostasis** to describe the constancy of the body's internal environment. Although the word *homeostasis* (derived from Greek words meaning "to stay the same") implies a static, unchanging state, the internal environment in fact seethes with activity as the body continuously adjusts to internal and external changes.

The internal state of an animal body is thus better described as a *dynamic equilibrium*. Many physical and chemical changes do occur, but the net result of all this activity is that physical and chemical parameters are kept within the narrow range that cells require to function. These equilibrium conditions are maintained by a host of mechanisms that are collectively known as *feedback systems*. *Negative feedback systems* counteract the effects of changes in the internal environment; less-common *positive feedback systems* reinforce changes when such reinforcement serves a physiological need.

Negative Feedback Reverses the Effects of Changes

The most important mechanism governing homeostasis is **negative feedback**, in which the response to a change is to counteract the change. In other words, an input stimulus causes an output response that "feeds back" to the initial input and decreases its effects. Because the initial change triggers a response that reverses its effects, the overall result is to return the system to its original condition. This kind of feedback is called "negative" because it reverses or negates the initial change.

A familiar example of negative feedback is your home thermostat (Fig. 26-1). In a thermostat, a stimulus (temperature dropping below a *set point*, the thermostat setting) is detected by a thermometer, which signals a control device that switches on a heater. The heater restores the temperature to the set point, and the heater is switched off. Continuously repeated on-off cycles keep your home's temperature near the set point. Note that the thermostat's negative feedback mechanism requires a *control center* with a set point, a *sensor* (the thermometer), and an *effector* (the furnace), which accomplishes the change.

How do people and other "warm-blooded" animals maintain their internal temperature despite extreme fluctuations in the temperature around them, such as those experienced in Colorado at an altitude of 13,000 feet? Maintaining the core body temperature is an important function of homeostatic mechanisms in "warm-blooded" animals that illustrates the concept of negative feedback in physiology. As Wayne Crill lay injured on the mountain, negative feedback kept his body's temperature relatively close to 98.6 °F (37 °C). The set point in this system, which only varies by about 1 °F in healthy humans, is located in a control center in your *hypothalamus*, a region of the brain that controls many homeostatic responses (see Fig. 26-1). Nerve endings in your hypothalamus, abdomen, spinal cord, skin, and large veins act as temperature sensors and transmit this information to the hypothalamus. When your body temperature drops, the hypothalamus activates various effector mechanisms that raise your body temperature, including shivering (which generates heat through muscular activity), constriction of blood supply to the skin (which reduces heat loss), and elevation of metabolic rate (which generates heat). When normal body temperature is restored, sensors signal the hypothalamus to switch off these temperature control mechanisms.

Negative feedback mechanisms abound in physiological systems. In the chapters that follow, you will find many examples of homeostatic control that operate by negative feedback, including those systems regulating blood oxygen content, water balance, blood sugar levels, and many other components of the "internal milieu."

Positive Feedback Drives Events to a Conclusion

When you first think about it, positive feedback is a rather frightening concept. In contrast to a negative feedback system, a change in a **positive feedback** system produces a response that intensifies the original change (see Fig. 26-1). The end result is that change tends to proceed in the same direction as the initial stimulus (rather than reversing to return to a set point). Positive feedback, as you can imagine, tends to create chain reactions that must somehow be controlled. For example, in nuclear fission, each particle that is split from an atom triggers the splitting of another atom, the pieces of which trigger the fission of other atoms, and so on. When controlled, the chain reaction supplies nuclear

MediaTutor 26.1 Feedback Loops and Homeostasis

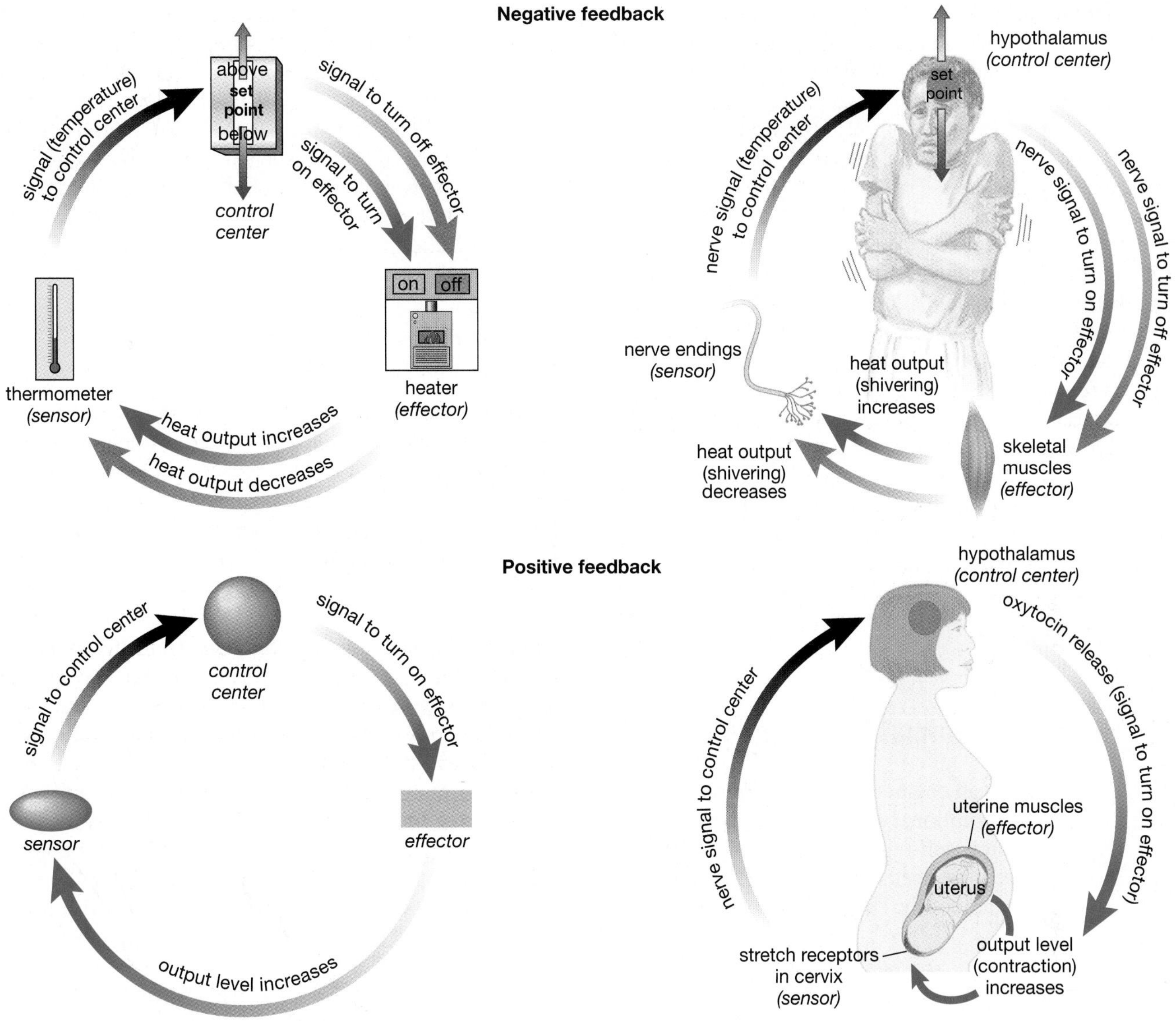

Figure 26-1 Negative and positive feedbacks
(top) Both our homes (left) and our bodies (right) use negative feedback to maintain appropriate temperatures.
(bottom) Positive feedback is seen in this hypothetical model (left) and in childbirth (right).

power. When deliberately set out of control, it produces an atomic explosion. A familiar biological example of positive feedback is population growth (as we shall see in Chapter 38); each offspring gives rise to still more offspring. Ecologist Paul Ehrlich coined the apt expression "population bomb" to describe unchecked population growth.

In physiological systems, events governed by positive feedback mechanisms are generally self-limiting and occur relatively infrequently. Positive feedback occurs, for example, during childbirth (see Fig. 26-1). The early contractions of labor begin to force the baby's head against the cervix, located at the base of the uterus; this pressure causes the cervix to dilate (open). Stretch-receptor neurons in the cervix respond to this expansion by signaling the hypothalamus, which responds by triggering the release of a hormone (oxytocin) that stimulates more and stronger uterine contractions. Stronger contractions create further pressure on the cervix, which in turn prompts the release of more hormones. The feedback cycle is finally terminated by the expulsion of the baby and its placenta.

The Body's Internal Systems Act in Concert

The systems of your body, fortunately, are all "team players" that work together in a coordinated manner to maintain a relatively constant internal environment. The job of regulating a multitude of factors throughout the body cannot be accomplished by a few, independent feedback mechanisms. Instead, numerous mechanisms are constantly at work, responding to various stimuli that continuously change as the animal's activities and external environment change. If all of these control systems worked independently, without regard to the activities of all the other systems, it would not be possible to maintain homeostasis in the body as a whole.

Fortunately, evolution has ensured that the various systems work together. For example, the systems that take substances into the body (for example, the digestive system) act in concert with the systems responsible for transporting substances within the body (the circulatory system) and systems that remove substances from the body (such as the excretory system). This kind of coordinated action is possible because the body has mechanisms for sending signals from one part of the body to another. Each cell in the body is indirectly connected to all of the others by an elaborate network of vessels and nerves that can carry molecules and messages to the appropriate locations. Molecules can be transported to the sites where they are needed and away from locations where their presence is harmful. Messages can be carried from sensors to effectors and back again, allowing feedback mechanisms to maintain homeostasis.

2 How Is the Animal Body Organized?

The animal body is an engineering marvel. From simple, free-living cells, evolutionary change has produced astonishingly complex, self-regulating systems consisting of trillions of specialized cells that accomplish hundreds of functions simultaneously. The parts fit together with a degree of precision and integration of which human engineers can only dream. This exquisite machine is made up of **tissues**, each tissue composed of dozens to billions of structurally similar cells that act in concert to perform a particular function (Fig. 26-2). Tissues are the building blocks of **organs**, such as the stomach, small intestine, kidneys, and urinary bladder. Organs in turn are organized into **organ systems**; for example, the digestive system includes the stomach, small intestine, large intestine, and other organs.

Animal Tissues Are Composed of Similar Cells That Perform a Specific Function

A tissue is composed of cells that are similar in structure and designed to perform a specialized function. Tissue may also include extracellular components produced by these cells, as in the case of cartilage and bone. Here we present a brief overview of the four major categories of animal tissue: (1) epithelial tissue; (2) connective tissue; (3) muscle tissue; and (4) nerve tissue.

Epithelial Tissue Covers the Body and Lines Its Cavities

Epithelial cells are your body's gatekeepers, protecting and regulating the movement of substances into and out of the body. **Epithelial tissues** (also called the *epithelium*) bound to loose connective tissue (described below) form continuous sheets known as **membranes**. Membranes cover the body and line body cavities such as the mouth, the stomach, or the bladder. The epithelial tissue portion of membranes faces either the outside of the body or the inside of a cavity within the

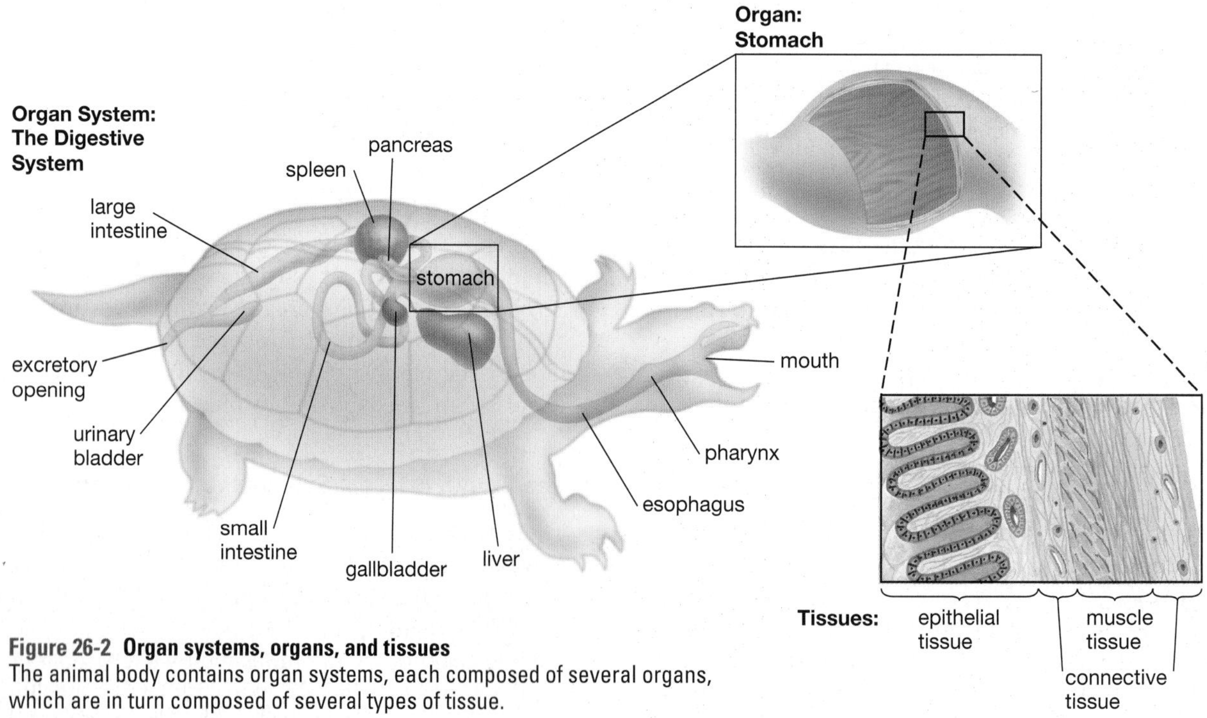

Figure 26-2 Organ systems, organs, and tissues
The animal body contains organ systems, each composed of several organs, which are in turn composed of several types of tissue.

body. The structure of epithelial tissue is adapted to its function. For example, the epithelium that lines the lungs, where exchange of gas molecules takes place, consists of thin, flattened cells arranged in a single layer (Fig. 26-3a). Another form of epithelium consists of elongated cells, often with cilia, and capable of secreting mucus (Fig. 26-3b). This type of epithelium is part of the membrane that lines the trachea that leads to the lungs. Here, the mucus traps dust particles, and the cilia transport them away from the lungs. It also lines tubes of the reproductive organs, where the cilia transport sex cells to their destinations (26-3b).

Membranes create barriers that either resist the movement of substances across them (such as the skin) or that allow the movement of specific substances across them (such as the lining of the small intestine). Epithelial tissues can serve as effective barriers because their cells are packed closely together and connected to one another by several types of junctions (see Chapter 4). No blood vessels penetrate epithelial tissue; it is nourished by diffusion from capillaries (the smallest blood vessels, whose thin walls allow exchange of wastes and nutrients). These capillaries are embedded within the connective tissue that lies beneath the epithelium.

Another important property of epithelial tissues is that they are continuously lost and replaced by mitotic cell division. For example, consider the abuse suffered by the epithelium that lines your mouth. Scalded by coffee and scraped by corn chips, it would be destroyed within a few days if it were not continuously replacing itself. The stomach lining, abraded by food and attacked by acids and protein-digesting enzymes, is completely replaced every 2 to 3 days. Your skin's outer membrane, the epidermis (see Fig. 26-10), is renewed about twice a month.

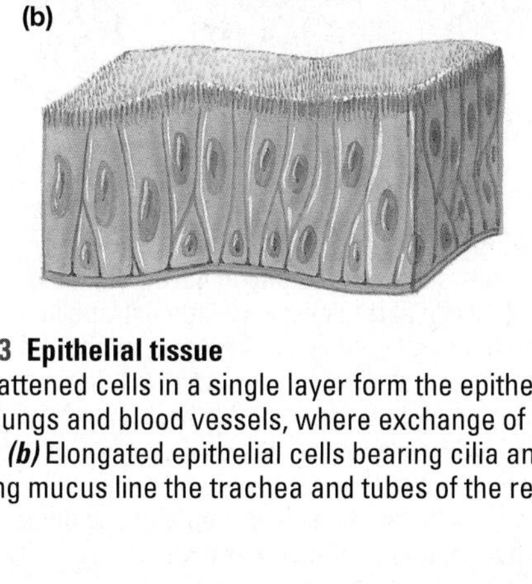

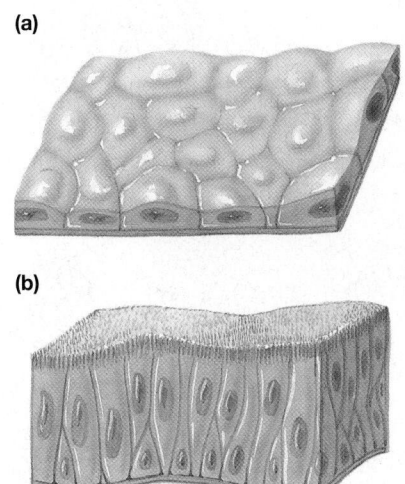

Figure 26-3 Epithelial tissue
(a) Thin, flattened cells in a single layer form the epithelial tissue that lines lungs and blood vessels, where exchange of materials is important. ***(b)*** Elongated epithelial cells bearing cilia and capable of secreting mucus line the trachea and tubes of the reproductive organs.

Some Epithelial Tissues Also Form Glands

During development, some epithelial tissues fold inward; their cells change shape and function to form **glands**, clusters of cells that are specialized to secrete (release) substances. Glands are classified into two broad categories: (1) exocrine glands and (2) endocrine glands. **Exocrine glands** remain connected to the epithelium by a passageway, or *duct.* Examples of exocrine glands are sweat glands and sebaceous (oil-secreting) glands; both types are found in the skin and are derived from skin epithelium (see Fig. 26-10). Exocrine glands called *salivary glands* release saliva into the mouth, and still other exocrine glands line the stomach, where they secrete a protective layer of mucus. **Endocrine glands** become separated from the epithelium that produced them. Most products of endocrine glands are hormones, which are secreted into the extracellular fluid that surrounds the glands and then diffuse into nearby capillaries. Endocrine glands and their hormones are covered in detail in Chapter 32.

Connective Tissues Have Diverse Structures and Functions

Connective tissues serve mainly to support and bind other tissues. Most connective tissues feature cells that are surrounded by large quantities of extracellular substances, typically secreted by the connective tissue cells themselves. Connective tissues can be placed into three main categories: (1) *loose connective tissue*, which combines with epithelial cells to form membranes; (2) *fibrous connective tissues*, which include *tendons* and *ligaments*; and (3) *specialized connective tissues*, which include *cartilage, bone, fat, blood*, and *lymph*. With the exception of blood and lymph, these connective tissues are interwoven with fibrous strands of an extracellular protein called **collagen**, secreted by the cells. Connective tissue underlies all epithelial tissue and contains capillaries and fluid-filled spaces that nourish the epithelium. Underlying the epidermis of the skin, for example, is connective tissue called the **dermis**, which is richly supplied with capillaries (see Fig. 26-10).

Loose connective tissue is characterized by a diffuse network of loosely woven fibers; it serves mainly to bind epithelial cells to underlying tissues and to cushion and support organs. In fibrous connective tissue, collagen fibers are densely packed in an orderly parallel arrangement—a design that gives **tendons** and **ligaments** the strength necessary for their respective functions of attaching muscles to bones and bones to bones. The specialized connective tissues are varied in structure. **Cartilage** is a flexible and resilient form of connective tissue that consists of widely spaced cells surrounded by a thick, nonliving matrix. The matrix is made of collagen

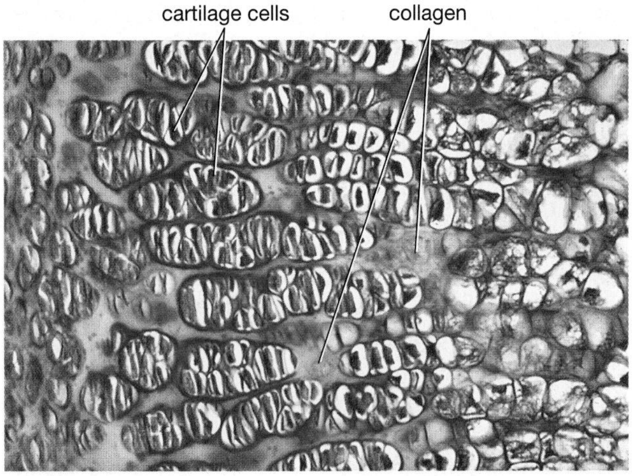

Figure 26-4 Cartilage
This specialized connective tissue supports the body and protects soft tissues. It is more flexible and less rigid than bone. The cells of the cartilage are stained dark purple and surrounded by clear spaces. The homogeneous material stained pale purple is the matrix of collagen secreted by the cartilage cells.

secreted by the cartilage cells (Fig. 26-4). Cartilage covers the ends of bones at joints, provides the supporting framework for the respiratory passages, supports the ear and nose, and forms shock-absorbing pads between the vertebrae. **Bone** (Fig. 26-5) resembles cartilage but is hardened by deposits of calcium phosphate. Bone forms in concentric circles around a central canal, which contains a blood vessel. (We shall discuss cartilage and bone in depth in Chapter 34.) Fat cells, collectively called **adipose tissue** (Fig. 26-6a), are modified for long-term energy storage and helped keep our parachutist alive as he lay stranded without food. Adipose tissue is especially important in the physiology of animals adapted to cold environments, because not only does it store energy, it also serves as insulation (Fig. 26-6b).

Although they are liquids, **blood** and **lymph** are considered connective tissues because they are composed largely of extracellular fluids. The cellular portion of blood consists of red and white blood cells and cell fragments called *platelets* in extracellular fluid called *plasma* (Fig. 26-7). Blood carries dissolved oxygen and other nutrients to cells and carries away carbon dioxide and other cellular waste products. Blood also carries hormonal signals throughout the body. Lymph consists largely of fluid that has leaked out of blood capillaries; it is carried back to the circulatory system within lymphatic vessels, as described in Chapter 27.

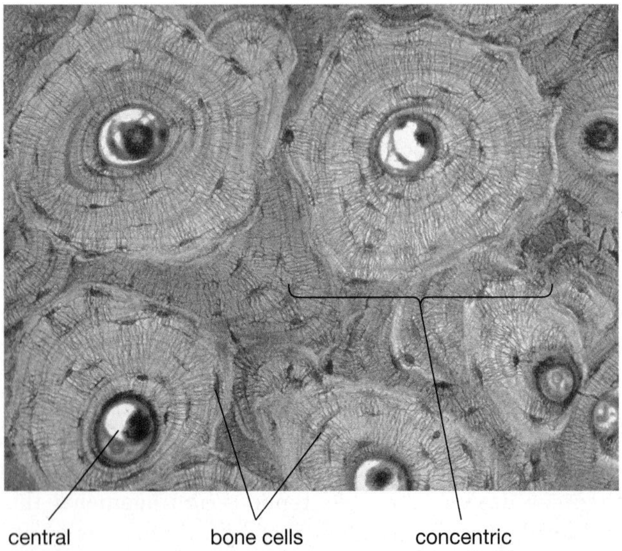

Figure 26-5 Bone consists of cells embedded in a hard matrix
Concentric circles of bone, deposited around a central canal that contains a blood vessel, are clearly visible in this micrograph. Bone cells appear as dark spots trapped in small chambers within the hard matrix that the cells themselves deposit.

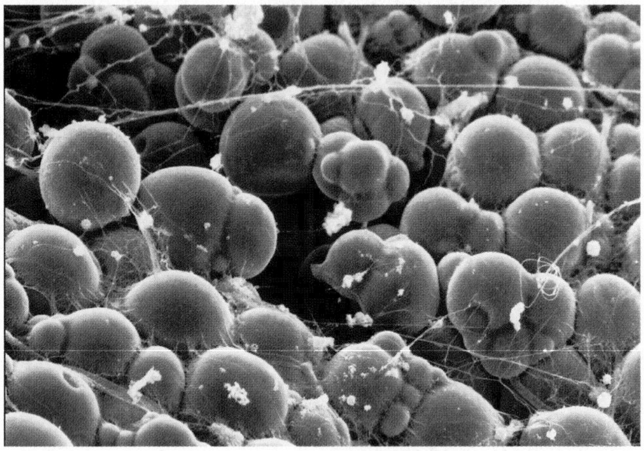

(a)

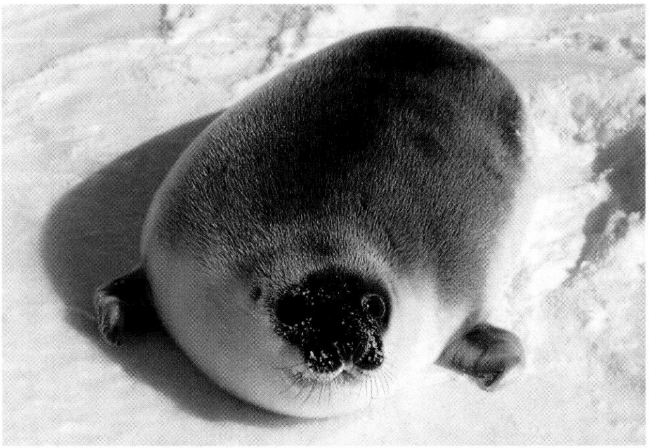

(b)

Figure 26-6 Adipose tissue
(a) Adipose tissue, such as that shown here from a mouse, is made up almost exclusively of fat cells; very little extracellular material is present. Each fat cell contains a droplet of oil that takes up most of the volume of the cell. ***(b)*** A hooded seal at 4 days of age has doubled her birth weight by drinking mother's milk that consists of 61% fat, derived from the mother's own stores of blubber. At 100 pounds, she is almost too fat to move. The fat will feed and insulate the pup as the ice floes break up and she dives into icy water to learn to hunt and feed on her own.

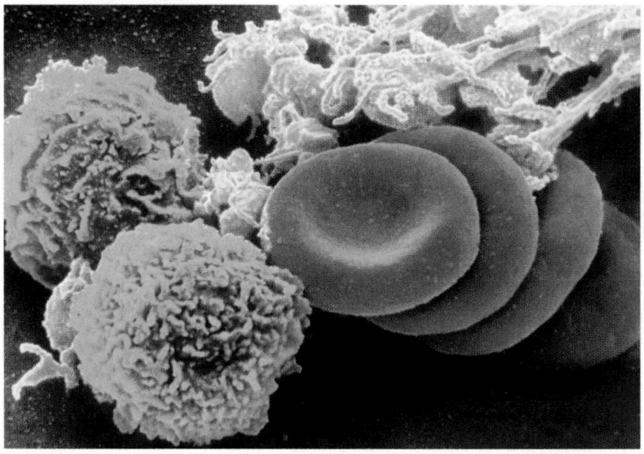

Figure 26-7 Blood
Blood contains three types of cellular components. In this color-enhanced scanning electron micrograph, red blood cells, which carry oxygen to tissues throughout the body, appear as red concave disks (mammalian red blood cells lack nuclei). White blood cells, important in the immune response, appear here as fuzzy blue balls, and platelets, fragments of cells that aid in blood clotting, are colored lavender.

Muscle Tissue Has the Ability to Contract

The long, thin cells of muscle tissue (Fig. 26-8) contract (shorten) when stimulated, then relax passively. There are three types of muscle tissue: (1) skeletal; (2) cardiac; and (3) smooth. **Skeletal muscle** is generally under voluntary, or *conscious,* control. As its name implies, its main function is to move the skeleton, as occurs when you walk or turn the pages of this text. **Cardiac muscle** is located only in the heart. Unlike skeletal muscle, it is spontaneously active and not under conscious control. Cardiac muscle cells are interconnected by gap junctions through which electrical signals spread rapidly through the heart, stimulating the cardiac muscle cells to contract in a coordinated fashion. **Smooth muscle**, so named because it lacks the orderly arrangement of thick and thin filaments seen in cardiac and skeletal muscles, is embedded in the walls of the digestive tract, the uterus, the bladder, and large blood vessels. It produces slow, sustained contractions that are normally involuntary. Muscles and their contraction mechanism are covered in Chapter 34.

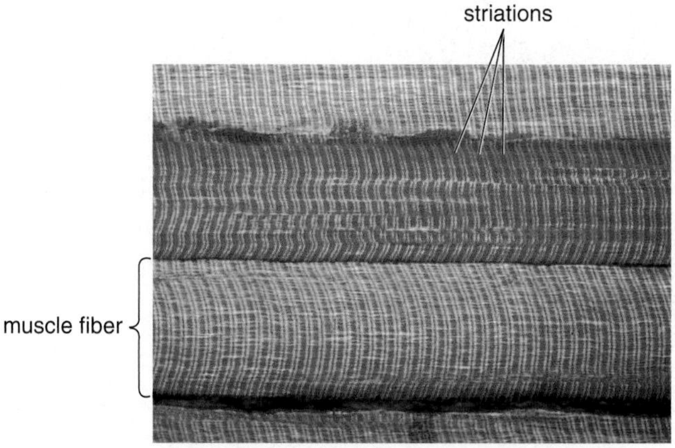

Figure 26-8 Muscle tissue consists of contractile cells called muscle fibers
A regular arrangement of fibrous proteins inside the muscle fibers of skeletal muscle gives this muscle tissue stripes or "striations" when viewed under a microscope. The proteins slide past one another when the muscle is contracted.

Nerve Tissue Is Specialized to Transmit Electrical Signals

You owe your ability to sense and respond to the world to **nerve tissue**, which makes up the brain, the spinal cord, and the nerves that travel from them to all parts of the body. Nerve tissue is composed of two types of cells: (1) neurons and (2) glial cells. **Neurons** are specialized to generate electrical signals and to conduct these signals to other neurons, muscles, or glands. A neuron has four major parts, each with a specialized function (Fig. 26-9). The **dendrites** of a neuron receive signals from other neurons or from the external environment. The **cell body** directs the maintenance and repair of the cell. The **axon** conducts the electrical signal to its target cell, and the **synaptic terminals** transmit the signal to the target cell. **Glial cells** surround, support, and protect neurons and regulate the composition of the extracellular fluid, allowing neurons to function optimally. We shall discuss nerve tissue more fully in Chapter 33.

Organs Include Two or More Interacting Tissue Types

Organs are formed from at least two tissue types that function together. Some organs, such as the skin, include all four of the tissue types described earlier. In this section, we examine the components and functions of a representative animal organ, the skin.

The structure of the skin is, in a general sense, representative of many organs. An outer epithelium is underlain by connective tissue that contains a blood supply, a nerve supply, in some cases muscle, and glandular structures derived from the epithelium. If an organ is hollow, such as the bladder or blood vessels, its interior is also lined with epithelium underlain by connective tissue. Different organs have different types and proportions of glandular, muscular, and nervous tissues.

The Skin Illustrates the Properties of Organs

Although we take our skin for granted, it so important as a barrier against infection and water loss that large-scale destruction of skin, such as by extensive burns, often proves fatal. The **epidermis**, or outer layer of the skin, is a specialized epithelial tissue (Fig. 26-10). It is covered by a protective layer of dead cells produced by underlying living epidermal cells. These dead cells are packed with the protein **keratin**, which helps keep the skin both airtight and relatively waterproof.

Immediately beneath the epidermis lies a layer of connective tissue, the dermis. The loosely packed cells of

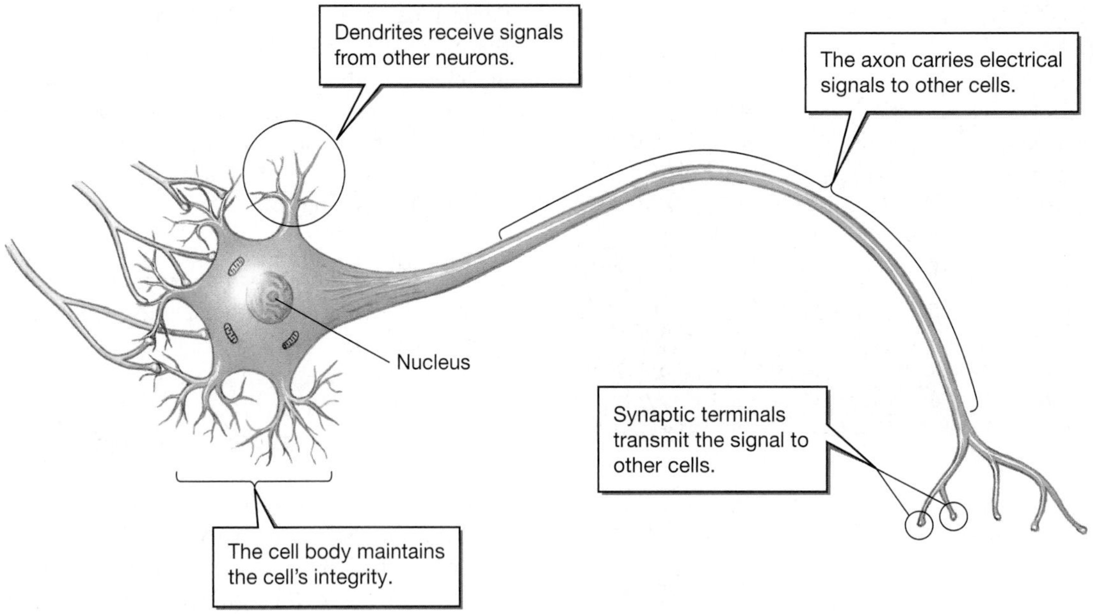

Figure 26-9 A nerve cell
A nerve cell has four major parts, each specialized for a specific function. Nerve cells vary tremendously in shape, depending on their functions in the nervous system.

the dermis are permeated by *arterioles* (small arteries). Arterioles feed blood pumped from the heart into a dense meshwork of capillaries that nourish both the dermal and epidermal tissue and empty into a network of *venules* (small veins) in the dermis. Loss of heat through the skin is precisely regulated by neurons controlling the degree of dilation (expansion) of the arterioles. When cooling is required, the arterioles dilate and flood the capillary beds with blood, thus releasing excess heat; when heat conservation is required, the arterioles supplying the skin capillaries are constricted. Lymph vessels collect and carry off extracellular fluid within the dermis. Various sensory nerve endings responsive to temperature, touch, pressure, vibration, and pain are scattered throughout the dermis and epidermis and provide feedback to the nervous system.

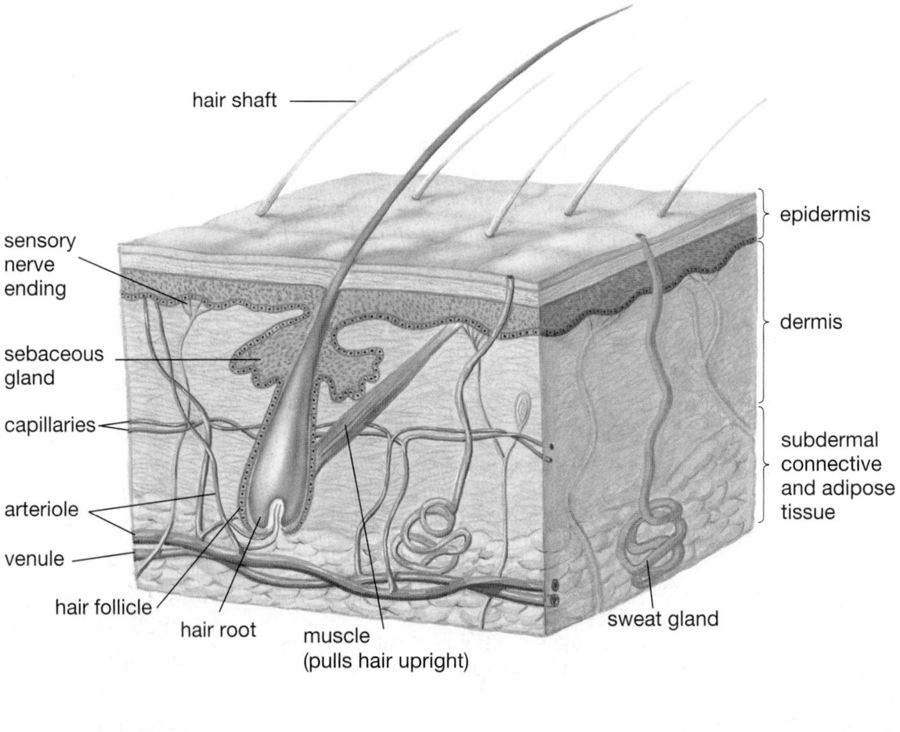

Figure 26-10 Skin
Mammalian skin, a representative organ, in cross section.

The dermis is also packed with glands derived from epithelial tissue. Glands called **hair follicles** produce hair from protein-containing secretions. Sweat glands produce watery secretions that cool the skin and excrete substances such as salts and urea. **Sebaceous glands** secrete an oily substance (*sebum*) that lubricates the epithelium.

In addition to the epithelial, connective, and nerve tissues already mentioned, the skin also contains muscle tissue. Tiny muscles attached to the hair follicles can cause the hairs of the skin to "stand on end" in response to signals from motor neurons. Although this reaction is useless for retaining heat in humans, most mammals are able to increase the thickness of their insulating fur in cold weather by erecting individual hairs.

Organ Systems Consist of Two or More Interacting Organs

Organ systems consist of two or more individual organs (in some cases located in different regions of the body) that work together, performing a common function. An example is the digestive system, in which the mouth, esophagus, stomach, intestines, and other organs that supply digestive enzymes, such as the liver and pancreas, all function together to convert food into nutrient molecules (see Fig. 26-2). The major organ systems of the vertebrate body and their representative organs and functions are listed in Table 26-1. The structure and physiology of these organ systems are the subjects of the following nine chapters.

CASE STUDY REVISITED

The Limits of Endurance

After two months of intensive rehabilitation, Wayne Crill was walking unassisted except for a splint, and debating whether he should continue his death-defying recreational stunts. His survival and his ability to heal and resume a normal, active life are a testimony to the ability of the body to maintain homeostasis and to the marvelous interactions of the body's tissues and organs. How do animal bodies cope with stresses such as dehydration? How do ruptured blood vessels seal themselves off, and how do broken bones heal? How does the body respond to the low-oxygen conditions of high altitude? As you read this unit, you will find answers to these, and many more questions.

Imagine yourself snorkeling in the ocean and encountering a seal. Watching the seal in fascination, you become thoroughly chilled. Describe the homeostatic mechanisms that will help restore your body temperature. Why can the seal remain in the ocean for hours while maintaining its body temperature?

Summary of Key Concepts

1) Homeostasis: How Do Animals Maintain Internal Constancy?

Homeostasis refers to the tendency of many physiological processes to maintain an organism's internal conditions within a narrow range that permits the continuation of life. These conditions are maintained through negative feedback, in which a change triggers a response that counteracts the change and restores conditions to a set point. Temperature regulation as well as many hormone systems use negative feedback to maintain homeostasis. Positive feedback, in which a change initiates events that intensify the change, occurs relatively rarely and is self-limiting. For example, the uterine contractions that lead to childbirth are driven by positive feedback. The animal body contains many feedback mechanisms that work in concert as a coordinated, integrated system.

2) How Is the Animal Body Organized?

The animal body is composed of organ systems consisting of two or more organs. Organs, in turn, are made up of tissues. A tissue is a group of cells and extracellular material that forms a structural and functional unit and is specialized for a specific task. Animal tissues include epithelial, connective, muscle, and nerve tissue.

Epithelial tissue forms membranous coverings over internal and external body surfaces and also gives rise to glands. Connective tissue normally contains considerable extracellular material and includes dermal tissue, bone,

Table 26-1 Major Vertebrate Organ Systems

Organ System	Major Structures	Physiological Role	Organ System	Major Structures	Physiological Role
Circulatory system 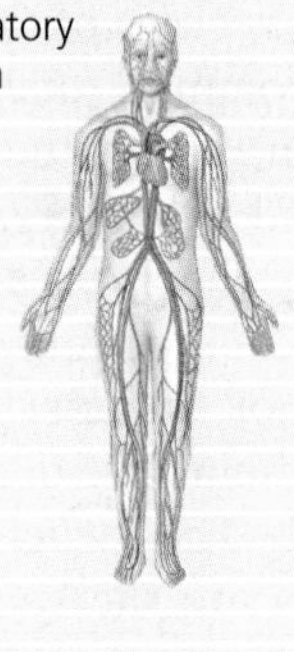	Heart, blood vessels, blood	Transports nutrients, gases, hormones, metabolic wastes; also assists in temperature control	Endocrine system 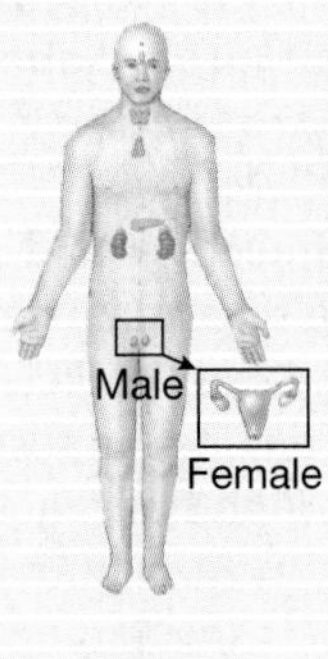	A variety of hormone-secreting glands, including the hypothalamus, pituitary, thyroid, pancreas, and adrenals	Controls physiological processes, typically in conjunction with the nervous system
Lymphatic/immune system 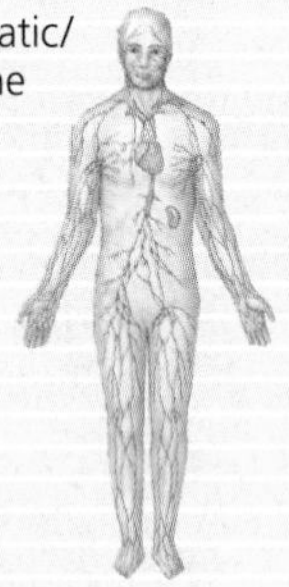	Lymph, lymph nodes and vessels, white blood cells	Carries fat and excess fluids to blood; destroys invading microbes	Nervous system 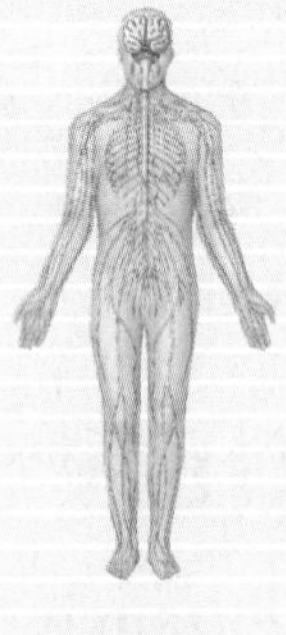	Brain, spinal cord, peripheral nerves	Controls physiological processes in conjunction with the endocrine system; senses the environment, directs behavior
Digestive system 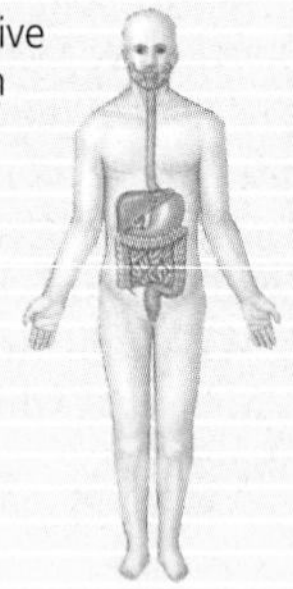	Mouth, esophagus, stomach, small and large intestines, glands producing digestive secretions	Supplies the body with nutrients that provide energy and materials for growth and maintenance	Muscular system 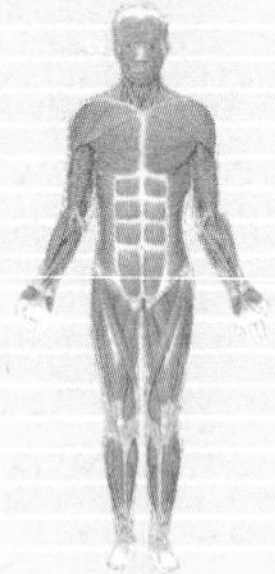	Skeletal muscle Smooth muscle Cardiac muscle	Moves the skeleton Controls movement of substances through hollow organs (digestive tract, large blood vessels) Initiates and implements heart contractions
Excretory system 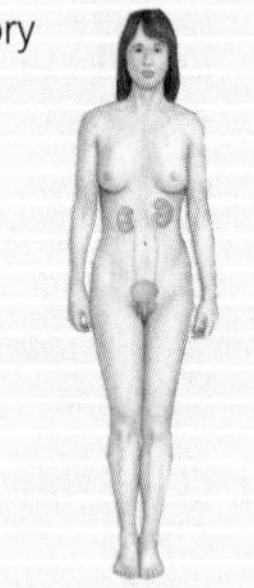	Kidneys, ureters, bladder, urethra	Maintains homeostatic conditions within bloodstream; filters out cellular wastes, certain toxins, and excess water and nutrients	Skeletal system 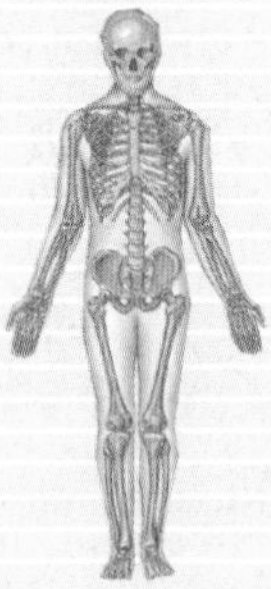	Bones, cartilage, tendons, ligaments	Provides support for the body, attachment sites for muscles, and protection for internal organs
Respiratory system 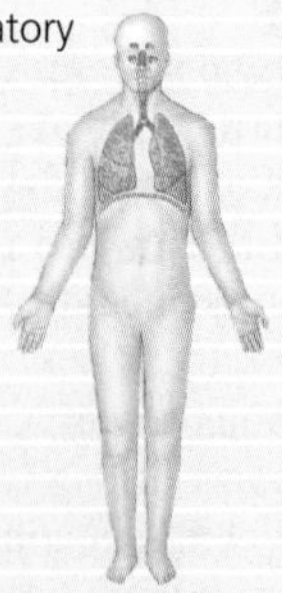	Nose, trachea, lungs (mammals, birds, reptiles, amphibians), gills (fish and some amphibians)	Provides an area for gas exchange between the blood and the environment; allows oxygen acquisition and carbon dioxide elimination	Reproductive system 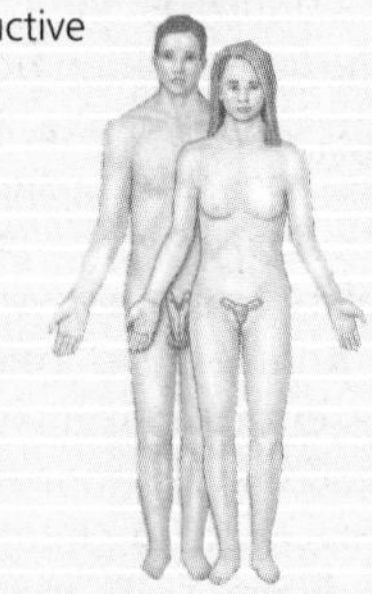	Males: testes, seminal vesicles, penis Female (mammal): ovaries, oviducts, uterus, vagina, mammary glands	Male: produces sperm, inseminates female Female (mammal): Produces egg cells, nurtures developing offspring

cartilage, tendons, ligaments, fat, and blood. Muscle tissue is specialized for movement. There are three types of muscle tissue: skeletal, cardiac, and smooth. Nerve tissue is specialized for the generation and conduction of electrical signals.

Organs include at least two tissue types that function together. Mammalian skin is a representative organ. The epidermis, an epithelial tissue, covers and protects the dermis beneath it. The dermis contains blood and lymph vessels, a variety of glands, and tiny muscles that erect the hairs. Animal organ systems include the digestive, excretory, immune, respiratory, circulatory/lymphatic, nervous, muscular, skeletal, endocrine, and reproductive systems, summarized in Table 26-1.

Key Terms

adipose tissue *p. 540*
axon *p. 541*
blood *p. 540*
bone *p. 540*
cardiac muscle *p. 541*
cartilage *p. 539*
cell body *p. 541*
collagen *p. 539*
connective tissue *p. 539*
dendrite *p. 541*
dermis *p. 539*
endocrine gland *p. 539*
epidermis *p. 541*
epithelial tissue *p. 538*
exocrine gland *p. 539*
gland *p. 539*
glial cell *p. 541*
hair follicle *p. 543*
homeostasis *p. 536*
keratin *p. 541*
ligament *p. 539*
lymph *p. 540*
membrane *p. 538*
negative feedback *p. 536*
nerve tissue *p. 541*
neuron *p. 541*
organ *p. 538*
organ system *p. 538*
positive feedback *p. 536*
sebaceous gland *p. 543*
skeletal muscle *p. 541*
smooth muscle *p. 541*
synaptic terminal *p. 541*
tendon *p. 539*
tissue *p. 538*

Thinking Through the Concepts

Multiple Choice

1. *The skin contains*
   a. epithelial tissue
   b. connective tissue
   c. nerve tissue
   d. muscle tissue
   e. all of the above
2. *Glands that become separated from the epithelium that produced them are called __________ glands.*
   a. sebaceous
   b. sweat
   c. exocrine
   d. endocrine
   e. saliva
3. *Epithelial membranes*
   a. cover the body
   b. line body cavities
   c. may create barriers that alter the movement of certain substances
   d. are continuously replaced by cell division
   e. all of the above
4. *All of the following are examples of connective tissue EXCEPT*
   a. tendons
   b. ligaments
   c. blood
   d. muscle
   e. adipose tissue
5. *Which of the following statements about muscle is true?*
   a. Smooth muscle is important in locomotion.
   b. Skeletal muscle is not under conscious control.
   c. Cardiac muscle utilizes gap junctions.
   d. Smooth muscle is called voluntary muscle.
   e. Smooth muscle moves the skeleton.
6. *All of the following are found in the dermis EXCEPT*
   a. arteries
   b. sensory nerve endings
   c. hair follicles
   d. sebaceous glands
   e. cells packed with keratin

? Review Questions

1. Define *homeostasis,* and explain how negative feedback helps maintain it. Explain one example of homeostasis in the human body.
2. Explain positive feedback, and provide one physiological example. Explain why this type of feedback is relatively rare in physiological processes.
3. Explain why body temperature in humans cannot be maintained at *exactly* 37 °C (98.6 °F) at all times.
4. Describe the structure and functions of epithelial tissue.
5. What property distinguishes connective tissue from all other tissue types? List five types of connective tissue, and briefly describe the function of each type.
6. Describe the skin, a representative organ. Include the various tissues that compose it and the role of each tissue.

Applying the Concepts

1. Why does life on land present more difficulties in maintaining homeostasis than does life in water? What made it evolutionarily advantageous for organisms to colonize dry land?
2. The majority of homeostatic regulatory mechanisms in animals are "autonomic"—that is, not requiring conscious control. Discuss several reasons why this type of regulation is more advantageous to the animal than is conscious regulation of homeostatic controls.
3. Third-degree burns are usually painless. Skin regenerates only from the edges of these wounds. Second-degree burns regenerate from cells located at the burn edges, in hair follicles and in sweat glands. First-degree burns are painful but heal rapidly from undamaged epidermal cells. From this information, draw the depth of first-, second-, and third-degree burns on Figure 26-10.
4. A coroner dictates the following description during an autopsy: "The tissue I am looking at forms part of the fetal skeleton. The extracellular matrix appears transparent. Fibers of collagen are present but are small and evenly dispersed in the extracellular matrix. Chondrocytes appear in tiny spaces, *lacunae,* within the matrix. Blood vessels have not yet penetrated the matrix." What tissue is the coroner describing?
5. Imagine you are a health-care professional teaching a prenatal class for fathers. Design a real-world analogy with sensors, electrical currents, motors, and so on to illustrate feedback relationships involved in the initiation of labor that a layperson could understand.

For More Information

Bruemmer. F. "Five Days with Fat Hoods." *International Wildlife,* January–February 1999. The rapid growth and prodigious fat-storing ability of the hooded seal adapts it to maintaining homeostasis under the extreme conditions of the far north.

Nuland, S. *The Wisdom of the Body.* New York: Alfred A. Knopf, 1997. Human physiology as seen through a surgeon's eyes. A firsthand account of the beauty and power of the body's mechanisms for maintaining homeostasis.

Pool, R. "Saviors." *Discover*, May 1998. Many seriously ill individuals die while waiting for an organ transplant. Thanks to genetic engineering, organ donors of the future may be raised on a farm.

Storey, K. B., and Storey, J. M. "Frozen and Alive." *Scientific American*, December 1990. Some animals have special adaptations that allow them to withstand freezing.

Answers to Multiple-Choice Questions

1.e 2.c 3.e 4.d 5.c 6.e

MEDIATUTOR
Homeostasis and the Organization of the Animal Body

CD Activities

Activity 26.1: Feedback Loops and Homeostasis

Estimated time: 5 minutes

In animals, homeostasis can be maintained through feedback systems. In this tutorial, you will discover the nature of the feedback system and explore some examples.

Start the MediaTutor Student CD-ROM and enter the activity number in the Quick Search box to be taken directly to that activity.

Web Investigations

Case Study: The Limits of Endurance

Estimated time: 15 minutes

The wilderness rescue team has finally located the missing climber at the bottom of the cliff. First step, check the ABCs (Airway, Breathing, Circulation). Next comes treatment for shock and hypothermia. But why aren't they working on the cuts and broken bones yet? Do the Web Investigation to find out.

Go to http://www.prenhall.com/audesirk6, the Audesirk Companion Web site. Select Chapter 26 and the Web Investigation to begin.

In the future, pigs may be raised not only as a source of bacon and ham, but also for hearts that may be transplanted into human recipients. These piglets were cloned by PPL Therapeutics of Scotland from the adult cells of a sow. Born March 5, 2000, and dubbed Millie, Christa, Alexis, Carrel, and Dotcom, the piglets are the first step toward producing genetically engineered animals whose organs might be used for a variety of human organ transplants.

27 Circulation

AT A GLANCE

Case Study: Xenotransplants

1) **What Are the Major Features and Functions of Circulatory Systems?**

Animals Have Two Types of Circulatory Systems

The Vertebrate Circulatory System Has Many Diverse Functions

2) **How Does the Vertebrate Heart Work?**

Increasingly Complex and Efficient Hearts Have Arisen During Vertebrate Evolution

The Vertebrate Heart Consists of Muscular Chambers Whose Contraction Is Controlled by Electrical Impulses

3) **What Is Blood?**

Plasma Is Primarily Water in Which Proteins, Salts, Nutrients, and Wastes Are Dissolved

Red Blood Cells Carry Oxygen from the Lungs to the Tissues

White Blood Cells Help Defend the Body Against Disease

Platelets Are Cell Fragments That Aid in Blood Clotting

4) **What Are the Types and Functions of Blood Vessels?**

Arteries and Arterioles Are Thick-Walled Vessels That Carry Blood Away from the Heart

Capillaries Are Microscopic Vessels That Allow the Blood and Body Cells to Exchange Nutrients and Wastes

Veins and Venules Carry Blood Back to the Heart

Arterioles Control the Distribution of Blood Flow

5) **How Does the Lymphatic System Work with the Circulatory System?**

Lymphatic Vessels Resemble the Veins and Capillaries of the Circulatory System

The Lymphatic System Returns Fluids to the Blood

The Lymphatic System Transports Fats from the Small Intestine to the Blood

The Lymphatic System Helps Defend the Body Against Disease

Case Study Revisited: Xenotransplants

CASE STUDY

Xenotransplants

The patient's heart disease has progressed too far—a transplant is the only option. Working quickly, surgeons remove the heart from a cloned, genetically engineered pig and use it to replace the diseased heart of the human patient.

This scenario could soon become reality. Each year in the U.S., while about 21,000 people receive organ transplants, 62,000 are on a transplant waiting list, and 4000 die while waiting for organ transplants. To many people, the logical solution to this high-stakes crisis is *xenotransplantation*: transplanting an organ from a non-human donor. Today, most governments tightly control xenotransplants, because there are complex ethical and practical questions raised by cross-species transplants. For example, xenotransplants of primate organs might provide a pathway for primate diseases to move into human populations. (The virus that causes AIDS in humans may have originated in wild primates.) Furthermore, sacrificing apes and monkeys for organ transplants is, for many, a disturbing prospect.

In light of these problems, xenotransplant advocates have turned to pigs as the donor animal of choice. Heart valves taken from pigs are now routinely grafted into people whose valves are failing. But when the entire heart needs replacement, the most pressing problem is that the human immune system will attack and destroy ("reject") foreign tissues. The chemical treatment that prevents rejection of isolated pig heart valves would destroy a functioning pig heart. Transplant patients now receive immunosuppressive drugs to prevent them from rejecting the organ from another human. While reducing the chance of rejection, these drugs leave the patient very vulnerable to infections.

Genetic engineering may offer a better solution. PPL Therapeutics, the firm from Edinburgh, Scotland, that brought us the first cloned sheep, have now cloned pigs. Next, they hope to clone pigs whose cells have been genetically engineered to avoid activating the human immune system. Will this work? We may find out soon. ■

1 What Are the Major Features and Functions of Circulatory Systems?

Billions of years ago, the first cells were nurtured by the sea in which they evolved. The sea brought them nutrients, which diffused into the cells, and washed away the wastes, which had diffused out. Today microorganisms and some simple multicellular animals still rely almost exclusively on diffusion for the exchange of nutrients and wastes with the environment. Sponges, for example, circulate seawater through pores in their bodies, bringing the environment close to each cell. As larger, more complex animals evolved, individual cells became increasingly distant from the outside world. The constant demands of a cell require, however, that diffusion distances be kept short so that adequate nutrients reach the cell and that the cell isn't poisoned by its own wastes. With the evolution of the circulatory system, a sort of "internal sea" was created, serving the same purpose as the sea did for the first cells. This internal sea transports a fluid (blood) rich in food and oxygen close to each cell and carries away wastes produced by the cells.

All circulatory systems have three major parts:

1. A fluid, **blood**, that serves as a medium of transport.
2. A system of channels, or **blood vessels**, that conduct the blood throughout the body.
3. A pump, the **heart**, that keeps the blood circulating.

Animals Have Two Types of Circulatory Systems

Animals have one of two major types of circulatory systems: (1) open and (2) closed. **Open circulatory systems** are present in many invertebrates, including arthropods, such as crustaceans, spiders, and insects, and mollusks, such as snails and clams. Animals with an open circulatory system have one or more hearts, a network of blood vessels, and a large open space within the body called a **hemocoel** (Figure 27-1a). The heart pumps blood through vessels that release the blood into the hemocoel. Within the hemocoel (which may occupy 20% to 40% of the body volume), tissues and internal organs are directly bathed in blood. The vessels also deliver blood back to the heart. For example, when the hearts of a grasshopper contract, valves in the hearts are pressed shut, forcing the blood to travel out through the vessels to the hemocoel. When the hearts relax, blood is drawn back into them through openings guarded by valves.

Closed circulatory systems are present in invertebrates, such as the earthworm (Figure 27-1b), and very active mollusks, such as squid and octopuses. Closed circulatory systems are also a characteristic of all vertebrates, including humans. In closed circulatory systems, the blood (whose volume is only 5% to 10% of body volume) is confined to the heart and a continuous series of blood vessels. Closed circulatory systems allow more rapid blood flow, more efficient transport of wastes and nutrients, and higher blood pressure than is possible in open systems. In the earthworm, for example, five contractile vessels serve as hearts. They pump blood through major vessels from which smaller vessels branch.

In this chapter we look at the different parts and functions of the vertebrate circulatory system. We shall focus on the human as a representative vertebrate.

The Vertebrate Circulatory System Has Many Diverse Functions

Your circulatory system supports all the other organ systems in your body. The circulatory systems of humans and other vertebrates perform the following functions:

- Transport oxygen from the lungs or gills to the tissues and transport carbon dioxide from the tissues to the lungs or gills.
- Distribute nutrients from the digestive system to all body cells.
- Transport waste products and toxic substances to the liver (where many of them are detoxified) and to the kidney for excretion.
- Distribute hormones from the glands and organs that produce them to the tissues on which they act.

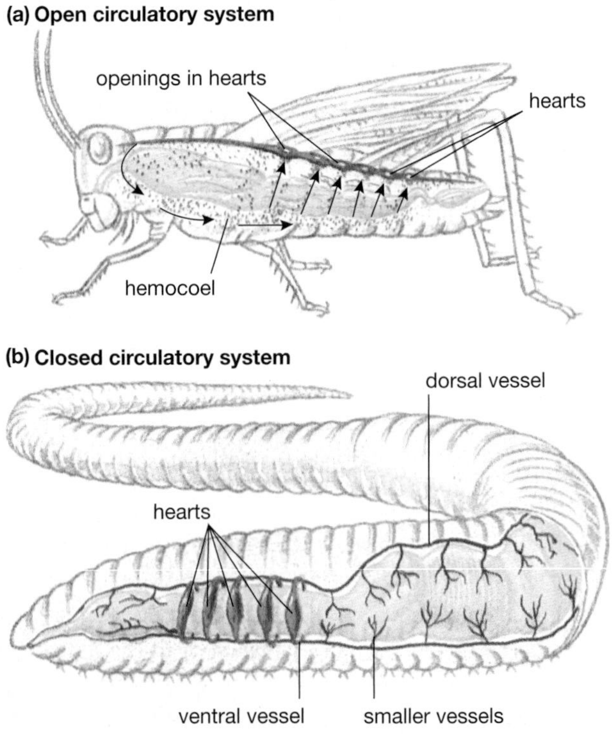

Figure 27-1 Open and closed circulatory systems
(a) In the open circulatory system of insects and other arthropods, a series of hearts pumps blood through vessels into the hemocoel, where blood directly bathes the other organs. When the hearts relax, blood is sucked back into them through openings guarded by one-way valves. When the hearts contract, the valves are pressed shut, forcing the blood to travel out through the vessels and back to the hemocoel. ***(b)*** In a closed circulatory system, blood remains confined to the heart and the blood vessels. In the earthworm, five contractile vessels serve as hearts and pump blood through major ventral and dorsal vessels from which smaller vessels branch.

- Regulate body temperature, which is achieved partly by adjustments in blood flow.
- Prevent blood loss by means of the clotting mechanism.
- Protect the body from bacteria and viruses by circulating antibodies and white blood cells.

In the following sections we examine the three parts of the circulatory system: (1) the heart, (2) the blood, and (3) the vessels, with an emphasis on the human system. Finally, we describe the lymphatic system, which works closely with the circulatory system.

2 How Does the Vertebrate Heart Work?

Increasingly Complex and Efficient Hearts Have Arisen During Vertebrate Evolution

No circulatory system can be constructed without a dependable pump. Blood must be moved through the body continuously throughout an animal's life. The vertebrate heart consists of muscular chambers capable of strong contractions. Chambers called **atria** (singular, **atrium**) collect blood. Atrial contractions send blood into the **ventricles**, chambers whose contractions circulate blood through the body. During the course of vertebrate evolution, the heart has become increasingly complex, with more separation between oxygenated blood (which has picked up oxygen from the lungs or gills) and deoxygenated blood (which, in passing through body tissues, has lost oxygen).

The hearts of fishes, the first vertebrates to evolve, consist of sequential contractile chambers, with a single atrium that empties into a single ventricle (Fig. 27-2a). Blood pumped from the ventricle passes first through the gill *capillaries*, thin-walled vessels where it picks up oxygen and gives off carbon dioxide. Then the blood travels to the rest of the body. In the body capillaries, it delivers oxygen to the tissues and picks up carbon dioxide.

Over evolutionary time, as fish gave rise to amphibians and amphibians to reptiles, a three-chambered heart evolved, consisting of two atria and one ventricle (Fig. 27-2b). In the three-chambered hearts of amphibians and most reptiles, deoxygenated blood from the body is delivered into the right atrium and blood from the lungs into the left atrium. Both atria empty into the single ventricle. Although some mixing occurs there, the deoxygenated blood tends to remain in the right portion of the ventricle and be pumped into vessels that enter the lungs, while most of the oxygenated blood remains in the left portion of the ventricle and is pumped to the rest of the body. The separation is enhanced in reptiles by a partial wall between the right and left portions of the ventricle.

The warm-blooded birds and mammals have high metabolic demands and require more efficient delivery of oxygen to their tissues than do cold-blooded animals. That demand is met by the four-chambered heart (Fig. 27-2c). Separate right and left ventricles isolate oxygenated from deoxygenated blood, ensuring that blood that reaches the tissues has the highest possible oxygen content.

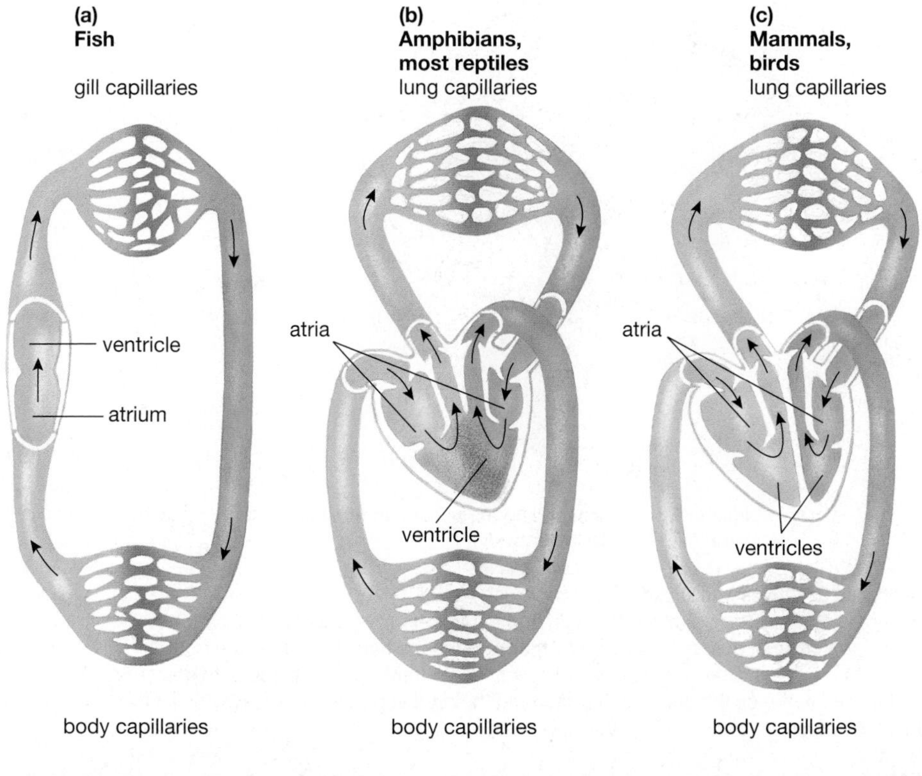

Figure 27-2 The evolution of the vertebrate heart
(a) The earliest vertebrate heart is represented by the two-chambered heart of fishes. ***(b)*** Amphibians and most reptiles have a heart with two atria, from which blood empties into a single ventricle. Many reptiles have a partial wall down the middle of the ventricle. ***(c)*** The hearts of birds and mammals are actually two separate pumps that prevent mixing of oxygenated and deoxygenated blood. Note that in this and in subsequent illustrations, oxygenated blood is depicted as bright red, while deoxgenated blood is colored blue.

The Vertebrate Heart Consists of Muscular Chambers Whose Contraction Is Controlled by Electrical Impulses

Human hearts (and those of other mammals and birds) can be considered as two separate pumps, each with two chambers. In each pump an atrium receives and briefly stores the blood, passing it to a ventricle that propels it through the body (Fig. 27-3). One pump, consisting of the right atrium and right ventricle, deals with deoxygenated blood. The right atrium receives oxygen-depleted blood from the body through a large **vein** (a vessel that carries blood *toward* the heart) called the *superior vena cava*. The right atrium contracts, forcing the blood into the right ventricle. Contraction of the right ventricle sends the oxygen-depleted blood to the lungs via pulmonary **arteries** (vessels that carry blood *away from* the heart). The other pump, consisting of the left atrium and ventricle, deals with oxygenated blood. Oxygen-rich blood from the lungs enters the left atrium through pulmonary veins and is then squeezed into the left ventricle. Strong contractions of the left ventricle, the heart's most muscular chamber, send the oxygenated blood coursing out through a major artery, the *aorta*, to the rest of the body.

The Coordinated Contractions of Atria and Ventricles Produce the Cardiac Cycle

Your heart beats about 100,000 times each day. The alternating contraction and relaxation of its chambers is the **cardiac cycle**. The two atria contract in synchrony, emptying their contents into the ventricles. A fraction of a second later, the two ventricles contract simultaneously, forcing blood into arteries that exit the heart. Both atria and ventricles then relax briefly before the cycle repeats (Fig. 27-4). At a normal resting heart rate, the cardiac cycle lasts just under 1 second. In determining blood pressure (Fig. 27-5), *systolic pressure* (the higher of the two readings) is measured during ventricular contraction and *diastolic pressure* is measured between contractions.

Valves Maintain Directionality of Blood Flow, and Electrical Impulses Coordinate the Sequence of Contractions

Coordinating the activity of the four chambers to maintain blood flow presents challenges. First, when the ventricles contract, the blood must be directed out through the arteries and not back up into the atria. Then, once blood has entered the arteries, it must be prevented from

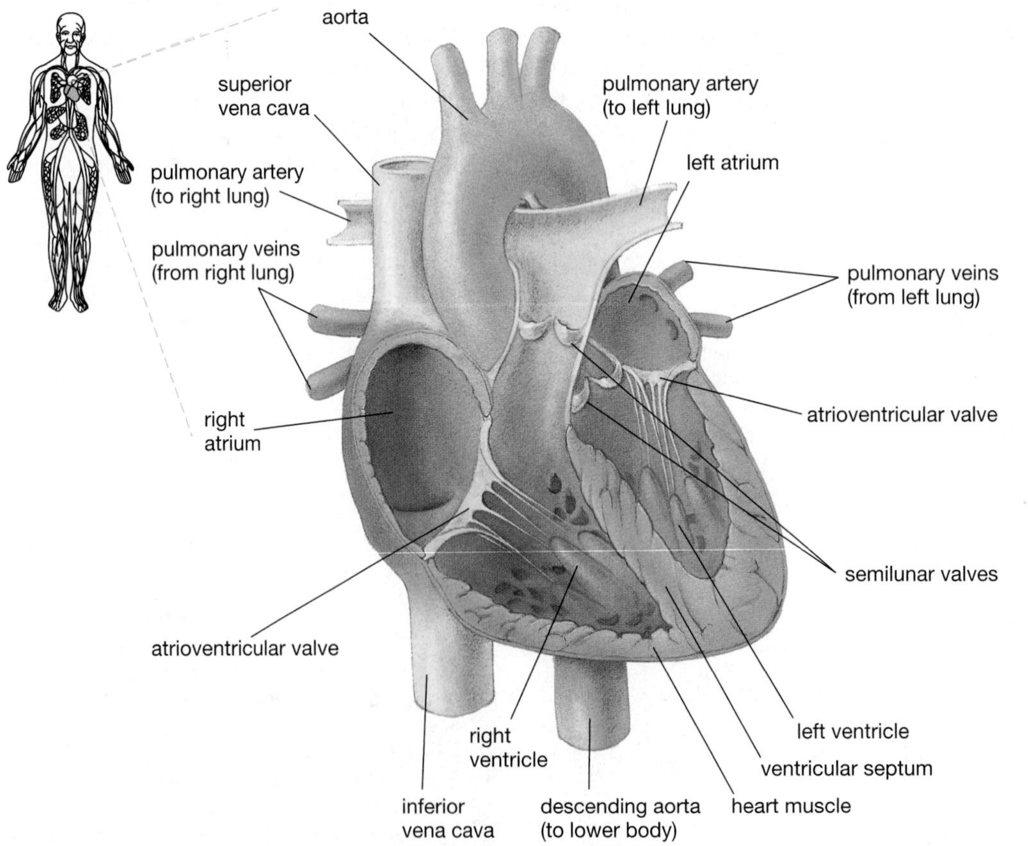

Figure 27-3 The human heart and its valves and vessels
The heart is drawn as if it were in a body facing you, so that right and left appear reversed. Note the thickened walls of the left ventricle, which must pump blood much farther through the body than does the right ventricle, which propels blood to the lungs. One-way valves, called semilunar valves, are located between the aorta and the left ventricle, and between the pulmonary artery and the right ventricle. Atrioventricular valves separate the atria and ventricles.

MediaTutor
27.1 Heart Structure and Function

MEDIATUTOR 27.2 The Cardiac Cycle

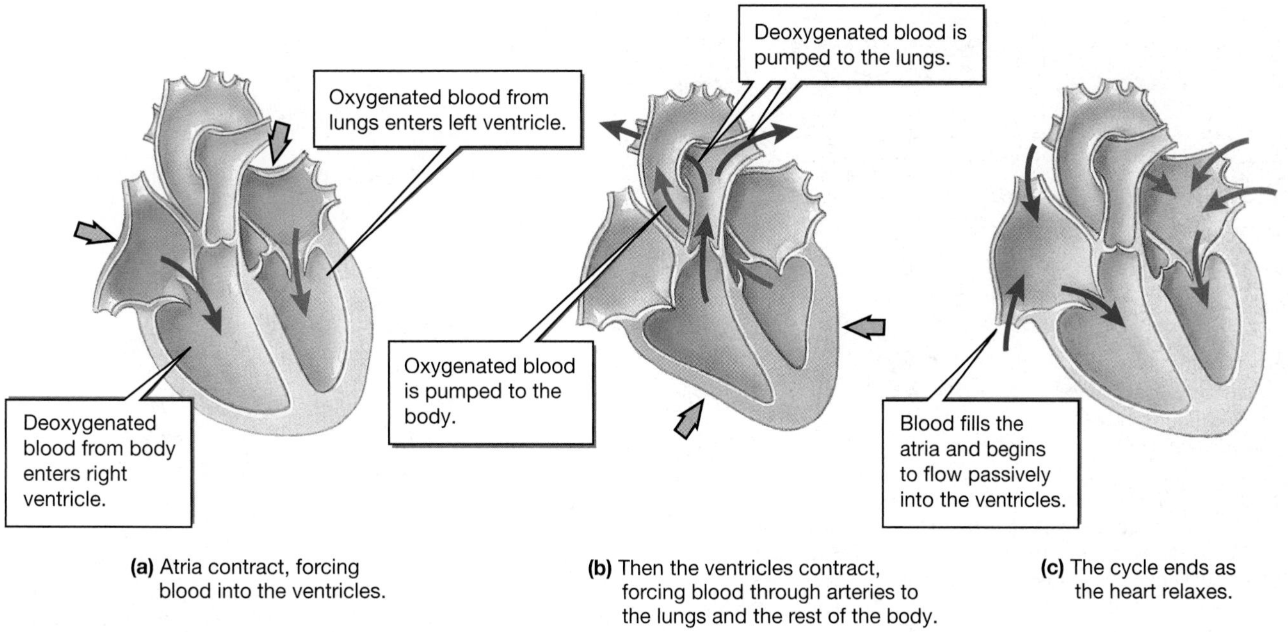
Deoxygenated blood is pumped to the lungs.
Oxygenated blood from lungs enters left ventricle.
Oxygenated blood is pumped to the body.
Deoxygenated blood from body enters right ventricle.
Blood fills the atria and begins to flow passively into the ventricles.
(a) Atria contract, forcing blood into the ventricles.
(b) Then the ventricles contract, forcing blood through arteries to the lungs and the rest of the body.
(c) The cycle ends as the heart relaxes.

Figure 27-4 The cardiac cycle

flowing back as the heart relaxes. These problems are solved by four one-way *valves* (Fig. 27-3; see these valves in action in Fig. 27-4). Pressure in one direction opens them readily, but reverse pressure forces them tightly closed. **Atrioventricular valves** separate the atria from the ventricles. **Semilunar valves** allow blood to enter the pulmonary artery and the aorta when the ventricles contract but prevent it from returning as the ventricles relax.

A second challenge is to create smooth, coordinated contractions of the muscle cells that make up each chamber. Muscle cells produce electrical signals that cause contraction. Individual heart muscle cells communicate directly with one another through gap junctions in their adjacent plasma membranes (Fig. 27-6). These connecting pores, introduced in Chapter 4, allow the electrical signals that cause contraction to pass freely and rapidly between heart cells.

A final challenge is to coordinate contractions of all four chambers. The atria must contract first and empty their contents into the ventricles so that the atria can refill while the ventricles contract. Thus, there must be a delay between the contractions of the atria and those of the ventricles. The contraction of the heart is initiated and coordinated by a **pacemaker**, a cluster of specialized heart muscle cells that produce spontaneous electrical

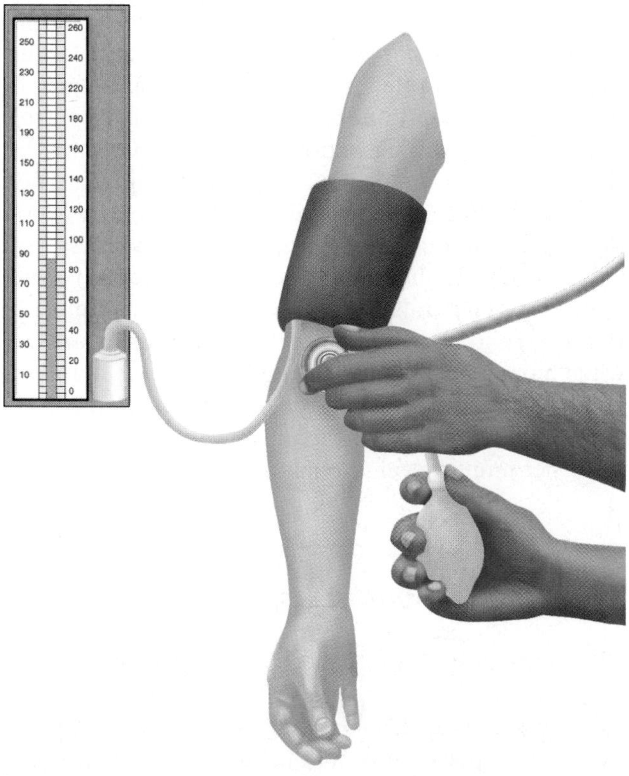

Figure 27-5 Measuring blood pressure
Blood pressure is measured with an inflatable blood pressure cuff and a stethoscope. The cuff is inflated until its pressure closes off the arm's main artery, blood ceases to flow, and no pulse can be detected below the cuff. Then the pressure is gradually reduced. When the pulse is first audible in the artery, the pressure pulses created by the contracting left ventricle are just overcoming the pressure in the cuff and blood is flowing. This is the upper reading: the systolic pressure. Cuff pressure is then further reduced until no pulse is audible, indicating that blood is flowing continuously through the artery and that the pressure between ventricular contractions is just overcoming the cuff pressure. This is the lower reading: the diastolic pressure. The numbers are in millimeters of mercury, a standard measure of pressure also used in barometers.

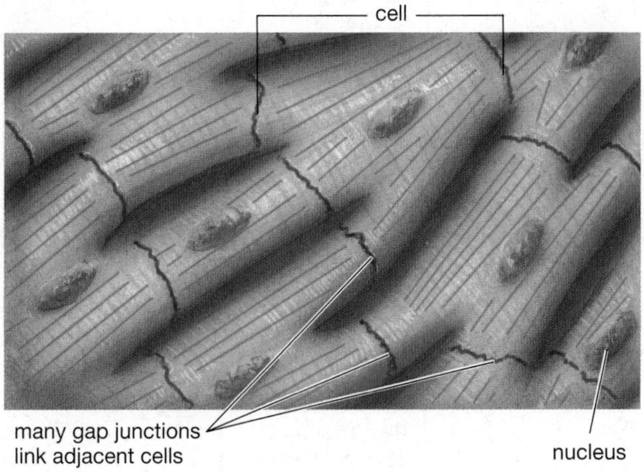

Figure 27-6 The structure of cardiac muscle
Cardiac muscle cells are branched. Adjacent plasma membranes meet in folded areas that are densely packed with gap junctions (pores), which connect the interiors of adjacent cells. This arrangement allows direct transmission of electrical signals between the cells, coordinating their contractions.

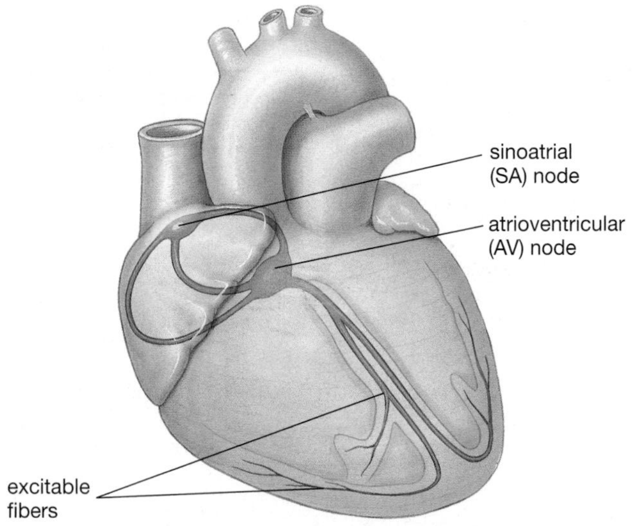

Figure 27-7 The heart's pacemaker and its connections
The sinoatrial (SA) node, a spontaneously active mass of modified muscle fibers in the right atrium, serves as the heart's pacemaker. The signal to contract spreads from the SA node through the muscle fibers of both atria, finally exciting the atrioventricular (AV) node in the lower right atrium. The AV node then transmits the signal to contract through bundles of excitable fibers that stimulate the ventricular muscle.

signals at a regular rate. Although the nervous system can alter the rate of these signals, they are initiated by the pacemaker muscle cells themselves. The heart's primary pacemaker is the **sinoatrial (SA) node**, located in the wall of the right atrium (Fig. 27-7).

From the SA node, an electrical impulse creates a wave of contraction that sweeps through the muscles of the right and left atria, which contract in smooth synchrony. Then the signal reaches a barrier of unexcitable tissue between the atria and the ventricles. There the excitation is channeled through the **atrioventricular (AV) node**, a small mass of specialized muscle cells located in the floor of the right atrium (see Fig. 27-7). The impulse is delayed at the AV node, postponing the ventricular contraction for about 0.1 second after the atria contract. This delay gives the atria time to complete the transfer of blood into the ventricles before ventricular contraction begins. From the AV node, the signal to contract spreads to the base of the two ventricles along tracts of excitable fibers. The impulse then travels rapidly from these fibers through the communicating muscle fibers, causing the ventricles to contract in unison.

A variety of disorders can interfere with the complex series of events that produce smooth, regular heartbeats. When the pacemaker fails, uncoordinated, irregular contractions called **fibrillation** occur. Fibrillation of the ventricles can be fatal, because blood is not pumped out of the heart to the brain and other organs but is merely sloshed around. A defibrillating machine applies a jolt of electricity to the heart, synchronizing the contraction of the ventricular muscle and sometimes allowing the pacemaker to resume its normal coordinating function. Bacterial infections may attack heart valves, allowing blood to flow inappropriately between the chambers. In such cases valve transplants, some using pig heart valves, can be lifesaving. A variety of diseases attack the heart muscle itself. Damaged heart muscle cells die, and the heart walls thin and weaken. Under these conditions, the patient becomes a candidate for a heart transplant.

The Nervous System and Hormones Influence Heart Rate
Your heart rate is finely tuned to whatever you are doing, whether you are running to class or basking in the sun. On its own, the SA node pacemaker would maintain a steady rhythm of about 100 beats per minute. However, nerve impulses and hormones significantly alter heart rate. In a resting individual, activity of the parasympathetic nervous system (which controls body functions during periods of rest; see Chapter 33) slows the heart rate to about 70 beats per minute. When exercise or stress creates a demand for greater blood flow to the muscles, the sympathetic nervous system (which prepares the body for emergency action) accelerates the heart rate. Likewise, the hormone epinephrine (also known as adrenaline) increases the heart rate while mobilizing the entire body for response to threatening or unfamiliar events. When astronauts were landing on the moon, their heart rates were more than 170 beats per minute, even though they were sitting still!

3 What Is Blood?

Blood, which has been called the "river of life," transports dissolved nutrients, gases, hormones, and wastes through the body. It has two major components: (1) a fluid called **plasma** and (2) cellular components (*red* *blood cells*, *white blood cells*, and *platelets*) that are suspended in the plasma. On the average, the cellular components of blood account for 40% to 45% of its volume; the other 55% to 60% is plasma. The average human has 5 to 6 liters of blood, constituting about 8% of the total body weight.

Plasma Is Primarily Water in Which Proteins, Salts, Nutrients, and Wastes Are Dissolved

Water makes up about 90% of the straw-colored plasma. Dissolved in the plasma are proteins, hormones, nutrients (glucose, vitamins, amino acids, lipids), gases (carbon dioxide, oxygen), salts (sodium, calcium, potassium, magnesium), and wastes, such as urea. Plasma proteins are the most abundant of the dissolved substances. The three major plasma proteins are (1) *albumins*, which help maintain the blood's osmotic pressure (which controls the flow of water across plasma membranes); (2) *globulins*, which transport nutrients and play a role in the immune system; and (3) *fibrinogen*, important in blood clotting, which is discussed later in this chapter.

Red Blood Cells Carry Oxygen from the Lungs to the Tissues

The most abundant cells in the blood, not surprisingly, are those that carry oxygen, called red blood cells, or **erythrocytes**. In fact, each cubic millimeter of blood (a small droplet) contains about 5 million erythrocytes, which make up about 99% of all blood cells and constitute about 40% of the total blood volume in females and 45% in males. A red blood cell resembles a ball of clay squeezed between your thumb and forefinger (Fig. 27-8). This shape provides a larger surface area than would a spherical cell of the same volume, and increases the cell's ability to absorb and release oxygen through its plasma membrane.

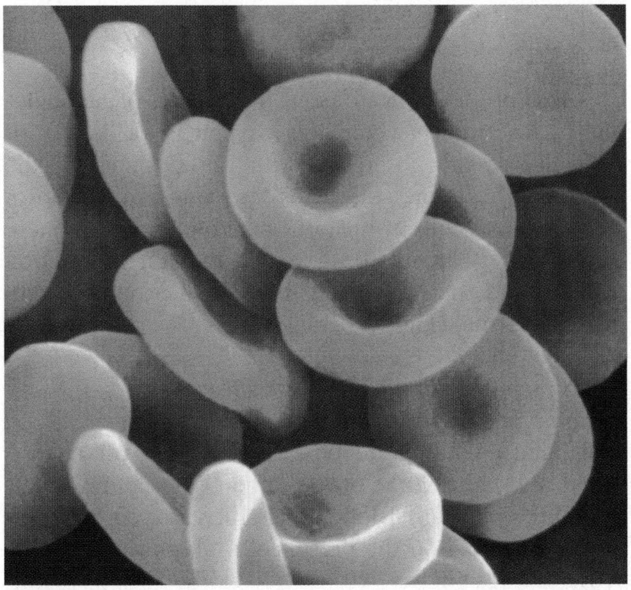

Figure 27-8 Red blood cells
This false-color scanning electron micrograph clearly shows the biconcave disk shape of red blood cells.

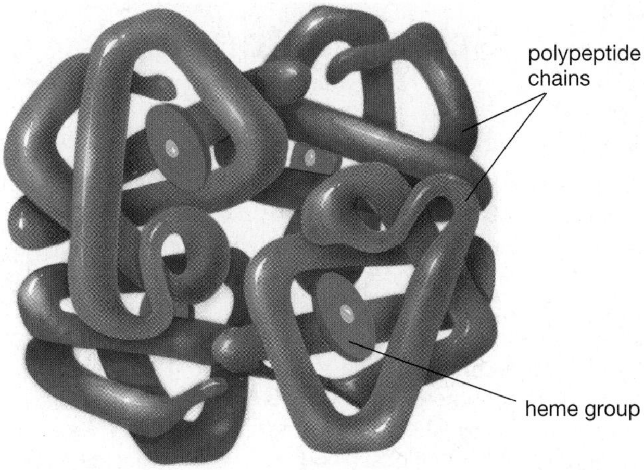

Figure 27-9 Hemoglobin
A molecule of hemoglobin is composed of four polypeptide chains (two pairs of similar chains), each surrounding a heme group. The heme group contains an iron atom and is the site of oxygen binding. When saturated, each hemoglobin molecule can carry four oxygen molecules (eight oxygen atoms).

The red color of erythrocytes is caused by the pigment **hemoglobin** (Fig. 27-9). This large, iron-containing protein accounts for about one-third the weight of each red blood cell and carries about 97% of the blood's oxygen. One hemoglobin molecule can bind and carry up to four molecules of oxygen, permitting blood to hold far more oxygen than would be possible if all the gas were dissolved in plasma. (When it binds oxygen, hemoglobin takes on a cherry-red color; when it loses oxygen it acquires a more bluish tint. Hence the red and blue color conventions in our diagrams of arteries and veins.) Hemoglobin binds loosely to oxygen, picking up oxygen in the capillaries of the lungs, where oxygen concentration is high, and releasing it in other tissues of the body, where the oxygen concentration is lower. After releasing its oxygen, some of the hemoglobin picks up carbon dioxide from the tissues for transport back to the lungs. The ability of blood to transport a large amount of oxygen to tissues is difficult to duplicate with artificial fluids. Nonetheless, because blood, like hearts, can be in short supply, researchers are experimenting with a variety of blood substitutes to use in transfusions (see "Scientific Inquiry: Artificial Blood? Artificial Vessels?"). The role of blood in gas exchange is discussed further in Chapter 28.

Red Blood Cells Have a Relatively Short Life Span

Red blood cells are formed in the *bone marrow*, the soft interior portion of certain bones, including those of the chest, upper arms, upper legs, and hips. During their development, mammalian red blood cells lose their nuclei and, with them, their ability to divide. Without the ability to synthesize cellular materials, their lives are necessarily short; each cell lives only about 120 days. Every

Scientific Inquiry
Artificial Blood? Artificial Vessels?

As some researchers study and debate the issues involved in xenotransplantation of hearts, others work to develop substitutes for the blood and its vessels, often looking to nonhuman sources of materials. Thousands of patients whose vessels are clogged by artherosclerosis would benefit from bypass surgery but do not have vessels suitable for grafting. Researchers have taken collagen from pigs and cows, molded it into tubes, and placed it in a nutrient broth with cells from blood vessels or grafted it directly into the vessels of experimental animals. Under the right conditions, blood vessel cells invade and cover the tubes to form somewhat weak but (at least temporarily) functional vessels. Recently, researchers at Duke University Medical Center induced layers of smooth muscle and endothelial cells from blood vessels to grow on a polymer tube while it was subjected to pressure pulses from a pump, mimicking the conditions under which arteries develop. After the polymer broke down as expected, the remaining artificial vessels were muscular and strong and functioned for about a month when grafted into pigs.

Meanwhile, blood supplies often fail to meet the need for blood transfusions, a problem that authorities predict will worsen as the population ages. Researchers are using two different approaches to make artificial blood with the oxygen-delivering properties of real blood. Some are using perfluorocarbons (PFCs), chemicals that can carry oxygen and carbon dioxide much as blood does (Fig. E27-1). Researchers are trying to create PFC-based blood substitutes that will be nontoxic, will not build up in body tissues, and will exchange appropriate amounts of oxygen and carbon dioxide. Meanwhile, others are working to modify real hemoglobin so it is available in large quantities for transfusion. One approach is to use genetic engineering to induce bacteria to make human hemoglobin. Other scientists are attempting to modify cow's blood so that it will not be rejected by the human immune system or carry harmful diseases to people.

Watch for news of a clinical trial of artificial blood and blood vessels in the next few years. Doctors may routinely use such materials in the future—when you might need them.

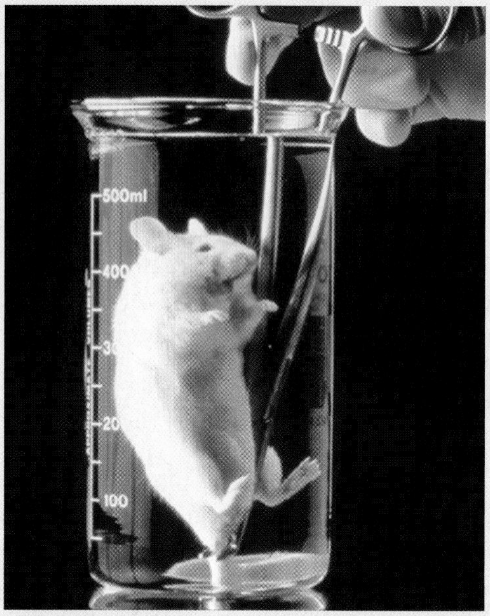

Figure E27-1 Breathing fluid?
After a moment of wide-eyed shock, a mouse submerged in this oxygen-carrying PFC solution begins to breathe the liquid as it would air—with no ill effects. Just as they can deliver oxygen to the lungs of this mouse, PFCs can carry oxygen within the circulatory system, making them promising raw materials for blood substitutes.

second, more than 2 million red blood cells die and are replaced by new ones from the bone marrow. Dead or damaged red blood cells are removed from circulation, primarily in the liver and spleen, and broken down to release their iron. The salvaged iron is carried in the blood to the bone marrow, where it is used to make more hemoglobin and packaged into new red blood cells. Although the recycling process is efficient, small amounts of iron are excreted daily and must be replenished by the diet. Bleeding from injury or menstruation also tends to deplete iron stores.

Negative Feedback Regulates Red Blood Cell Numbers

The number of red blood cells in the blood determines how much oxygen it can carry, and these numbers are maintained by a negative feedback system that involves the hormone **erythropoietin**. Erythropoietin is produced by the kidneys and released into the blood in response to oxygen deficiency. This lack of oxygen may be caused by a loss of blood, insufficient production of hemoglobin, high altitude (where less oxygen is available), or lung disease that interferes with gas exchange in the lungs. The hormone stimulates the rapid production of new red blood cells by the bone marrow. When adequate oxygen levels are restored, erythropoietin production declines and the rate of red blood cell production returns to normal (Fig. 27-10).

Blood Type Is Determined by Specific Proteins on Red Blood Cell Membranes

You inherit your blood type, and if you ever need a transfusion, getting the proper blood type can be a matter of life and death. Blood is classified as type A, B, AB, or O depending on the presence or absence of specific proteins (designated A and B) on the plasma membranes of red blood cells. Because A blood carries antibodies that attack the proteins on B blood (and vice versa), transfusion of B blood into an A blood individual

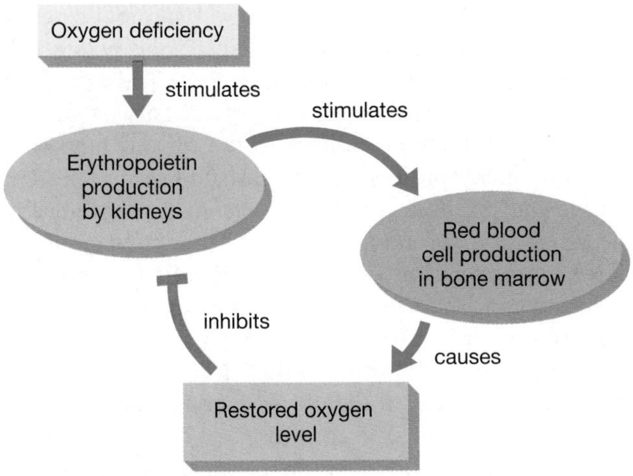

Figure 27-10 Red blood cell regulation by negative feedback

(or vice versa) could be fatal. We discussed the genetics and properties of these major blood types in Chapter 12 (see Table 12-1).

Another type of protein on red blood cells is the **Rh factor**. If it is present, blood is described as *Rh-positive*; if the protein is absent, the blood is *Rh-negative*. The Rh factor is important in transfusions and also in pregnancy. If an Rh-negative woman has children with an Rh-positive man, half of her children are likely to be Rh-positive, because Rh-positive blood is a dominant genetic trait. The first Rh-positive child will trigger antibody production in the mother's blood, usually without noticeable ill effects. Subsequent Rh-positive children, however, could be born with **erythroblastosis fetalis**. In this condition, the mother's antibodies invade the fetus and attack its red blood cells, causing severe anemia in the newborn. Fortunately, this condition can now be prevented by injections of a substance that blocks formation of Rh antibodies by the pregnant woman.

White Blood Cells Help Defend the Body Against Disease

White blood cells are your body's first line of defense when it is invaded by disease-causing microbes. There are five common types of white blood cells, or **leukocytes**, which together make up less than 1% of all blood cells. All white blood cells are derived from cells that originate in bone marrow. Most white blood cells function in some way to protect the body against foreign invaders and use the circulatory system to travel to the site of invasion. Some travel through capillaries to wounds where bacteria have gained entry, then ooze out through narrow openings in the capillary walls. After leaving the capillaries, some change into amoeba-like cells that engulf foreign particles, including bacteria and cancer cells (Fig. 27-11). They typically die in the process, and their dead bodies accumulate and contribute to the white substance called *pus*, seen at infection sites. The **lymphocytes**, described in Chapter 31, are another type of white blood cell. They are responsible for the production of antibodies that help provide immunity against disease. Cells that give rise to lymphocytes migrate from the bone marrow to tissues of the lymphatic system, such as the thymus, spleen, and lymph nodes, described later in this chapter.

Platelets Are Cell Fragments That Aid in Blood Clotting

Platelets, which are crucial to blood clotting, are pieces of large cells called *megakaryocytes*. **Megakaryocytes** remain in the bone marrow, where they pinch off membrane-enclosed chunks of their cytoplasm that we call platelets (Fig. 27-12). The platelets then enter the blood and play a central role in blood clotting. Like red blood cells, platelets lack a nucleus, and their life span is even shorter—about 10 to 12 days.

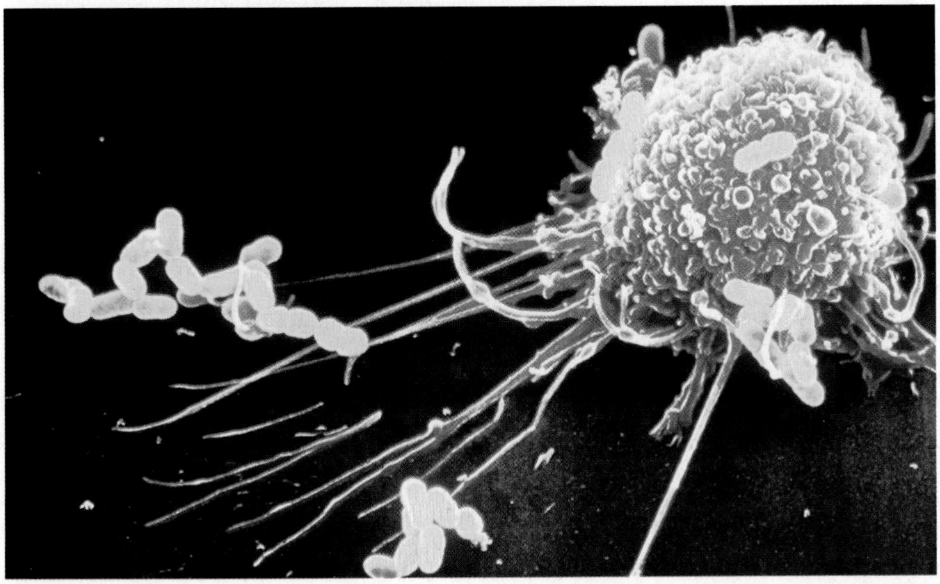

Figure 27-11 A white blood cell attacks bacteria
An amoeba-like white blood cell captures bacteria (in yellow). These bacteria are *Escherichia coli*, intestinal bacteria that can cause disease if they enter the bloodstream.

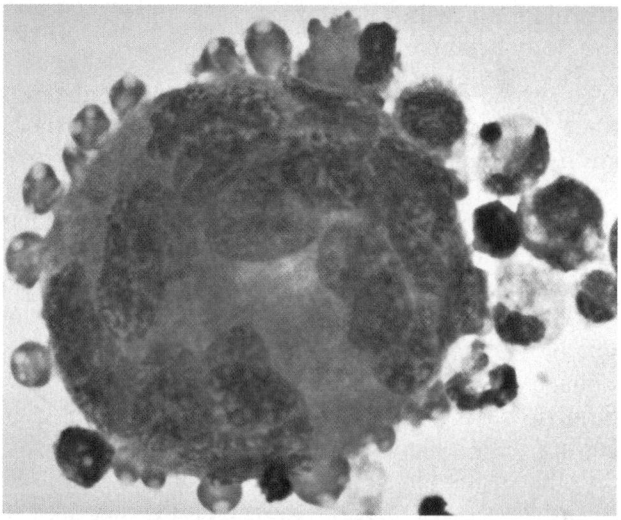

Figure 27-12 The production of platelets
Here a single megakaryocyte, found in bone marrow, is budding off dozens of membrane-enclosed pieces of cytoplasm called *platelets.*

Blood clotting is a complex process that keeps us from bleeding to death from normal wear and tear on the body. Clotting starts when platelets and other factors in the plasma contact an irregular surface, such as a damaged blood vessel. The ruptured surface of an injured blood vessel causes platelets to adhere and partially block the opening. The adhering platelets and the injured tissues initiate a complex sequence of events among circulating plasma proteins. These events result in the production of the enzyme **thrombin**. Thrombin catalyzes the conversion of the plasma protein *fibrinogen* into insoluble, stringlike molecules called **fibrin**. Fibrin molecules adhere to one another, forming a fibrous network (Fig. 27-13a). This protein web immobilizes the fluid portion of the blood, causing it to solidify in much the same way that gelatin does as it cools. The web traps red blood cells, further increasing the density of the clot (Fig. 27-13b). Platelets adhere to the fibrous mass and send out sticky projections that attach to one another. Within half an hour, the platelets contract, pulling the mesh tighter and forcing liquid out. This action creates a denser, stronger clot (on the skin it is called a *scab*) and also constricts the wound, pulling the damaged surfaces closer together in a way that promotes healing.

Despite blood's clotting ability, each year in the U.S. 50,000 people bleed to death from gunshots and other forms of trauma. U.S. Army and Red Cross researchers are now working to develop bandages impregnated with large quantities of thrombin and fibrinogen that will both physically and chemically staunch the flow of blood. Animal tests have been very encouraging, and the product may be available for human use within a few years. To help reduce costs, researchers hope to genetically engineer goats and cows to secrete large quantities of human fibrinogen and other clotting agents into their milk.

4 What Are the Types and Functions of Blood Vessels?

The river of life flows in well-defined channels called *blood vessels*. Some of the major blood vessels of the human circulatory system are diagrammed in Figure 27-14. As it leaves the heart, blood travels from arteries to *arterioles* to capillaries to *venules* to veins, which return it finally to the heart. These vessels are shown in Figure 27-15.

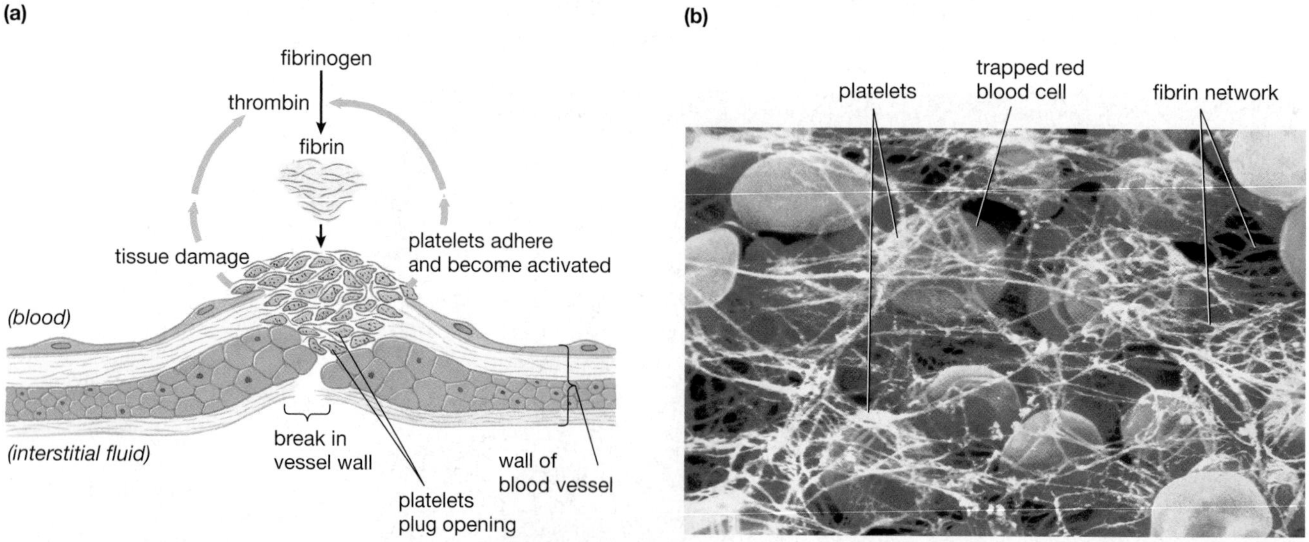

Figure 27-13 Blood clotting
(a) Injured tissue and adhering platelets cause a complex series of biochemical reactions among blood proteins. These reactions produce thrombin, which catalyzes the conversion of fibrinogen to insoluble fibrin strands. ***(b)*** Threadlike fibrin proteins produce a tangled sticky mass that traps red blood cells and eventually forms a clot.

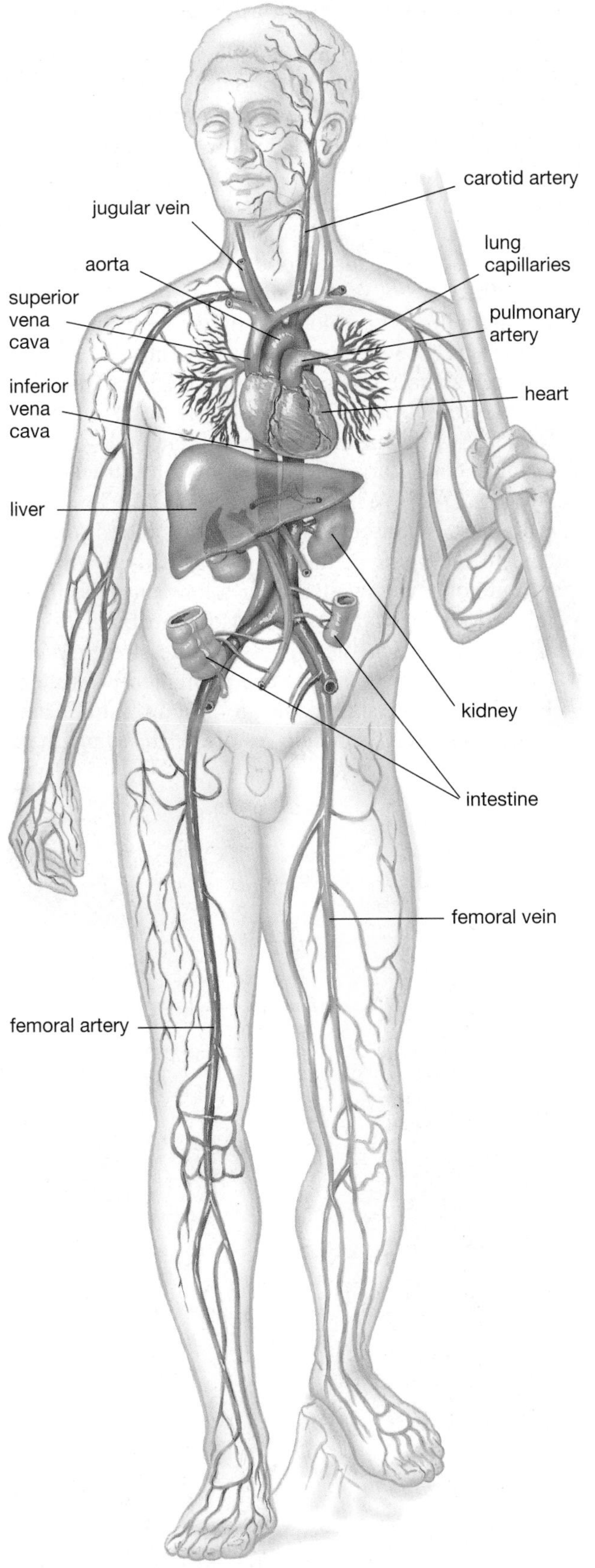

Figure 27-14 The human circulatory system
Most veins (right) carry deoxygenated blood to the heart, and most arteries (left) conduct oxygenated blood away from the heart. The pulmonary veins (carrying oxygenated blood) and arteries (carrying deoxygenated blood) are exceptions. All organs receive blood from arteries, send it back via veins, and are nourished by capillaries (only lung capillaries are illustrated and these are greatly enlarged, since capillaries are microscopic).

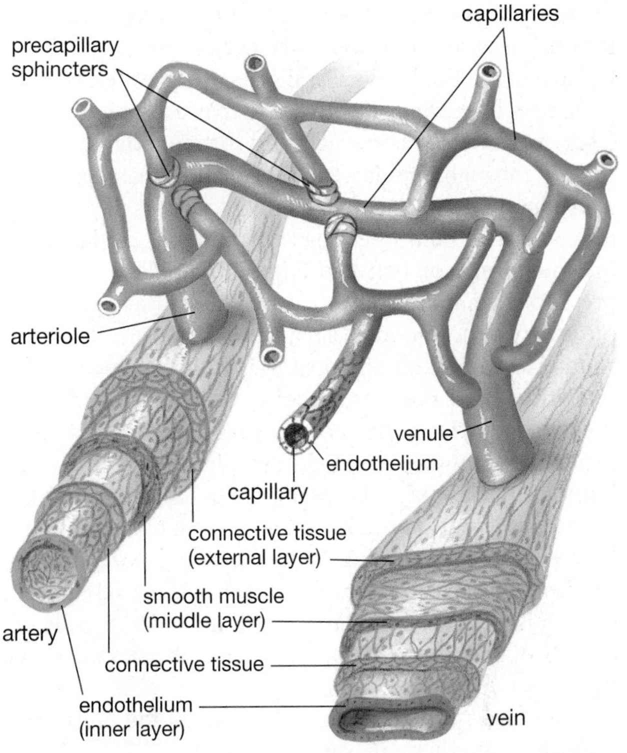

Figure 27-15 Structures and interconnections of blood vessels
Arteries and arterioles are more muscular than are veins and venules. Capillaries have walls only one cell thick. Oxygenated blood moves from arteries to arterioles to capillaries. Capillaries empty deoxygenated blood into venules, which empty into veins. The movement of blood from arterioles into capillaries is regulated by muscular rings called *precapillary sphincters*.

Arteries and Arterioles Are Thick-Walled Vessels That Carry Blood Away from the Heart

Arteries carry blood away from the heart. These vessels have thick walls embedded with smooth muscle and elastic tissue (see Fig. 27-15). With each surge of blood from the ventricles, the arteries expand slightly, like thick-walled balloons. As their elastic walls recoil between heartbeats, the arteries actually help pump the blood and maintain a steady flow through the smaller vessels. Arteries branch into vessels of smaller diameter called **arterioles**, which play a major role in determining how blood is distributed within the body, as described later.

Capillaries Are Microscopic Vessels That Allow the Blood and Body Cells to Exchange Nutrients and Wastes

The entire circulatory system is an elaborate device for allowing each cell of the body to exchange nutrients and wastes by diffusion. The actual process of diffusion occurs in the **capillaries**, the tiniest of all vessels. Here wastes, nutrients, gases, and hormones are exchanged between the blood and the body cells. Capillaries are finely adapted to their role of exchange, with walls that are only one cell thick. Most nutrients, oxygen, and carbon dioxide diffuse readily through capillary plasma membranes. Salts and small charged molecules (including some small proteins) move through fluid-filled spaces within the capillary plasma membrane or between adjacent capillary cells. The pressure within capillaries causes fluid to leak continuously from the blood plasma into the spaces that surround the capillaries and tissues. This fluid, known as the **interstitial fluid**, consists primarily of water in which are dissolved nutrients, hormones, gases, wastes, and small proteins from the blood. Large plasma proteins, red blood cells, and platelets are unable to leave the capillaries because they are too large to fit through plasma membrane channels, but white blood cells can ooze through the openings between capillary cells. The exchange of materials between capillary blood and nearby cells occurs through this interstitial fluid, which bathes nearly all the body's cells.

Capillaries are so narrow that red blood cells must pass through them in single file (Fig. 27-16). Consequently, all the blood is sure to pass very close to the capillary walls, where exchange occurs. In addition, capillaries are so numerous that no body cell is more than 100 micrometers (about as thick as four pages of this book) from a capillary. These factors facilitate the exchange of materials by diffusion. A person has about 50,000 miles (80,600 kilometers) of capillaries, enough to encircle the globe twice! The speed of blood flow drops very quickly as blood moves through this narrow, almost endless capillary network, allowing more time for diffusion to occur.

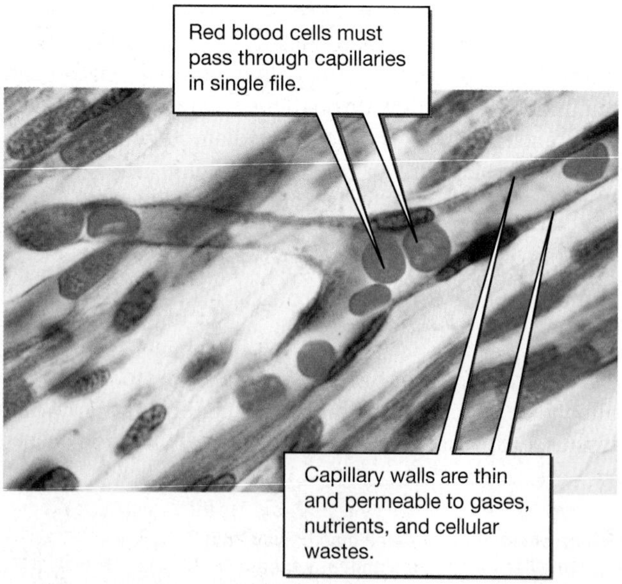
Red blood cells must pass through capillaries in single file.
Capillary walls are thin and permeable to gases, nutrients, and cellular wastes.

Figure 27-16 Red blood cells flow through a capillary

Veins and Venules Carry Blood Back to the Heart

After picking up carbon dioxide and other cellular wastes from cells, capillary blood drains into larger vessels called **venules**, which empty into still larger *veins* (see Fig. 27-15). Veins provide a low-resistance pathway by which blood can return to the heart. The walls of veins are thinner, less muscular, and more expandable than those of arteries, although both contain a layer of smooth muscle. Because blood pressure in the veins is low, contractions of skeletal muscle during exercise and breathing help return blood to the heart by squeezing the veins and forcing blood through them.

When veins are compressed, you might predict that blood would be forced away from the heart as well as toward it. To prevent this, veins are equipped with one-way valves that allow blood to flow only toward the heart (Fig. 27-17). When you sit or stand for long periods, the lack of muscular activity allows blood to accumulate in the veins of the lower legs. This is why you may find your feet swollen after a long airplane flight. Long periods of inactivity can also contribute to varicose veins, in which the valves become stretched and weakened, and the veins become permanently swollen.

If blood pressure should fall—for instance, after extensive bleeding—veins can help restore it. The sympathetic

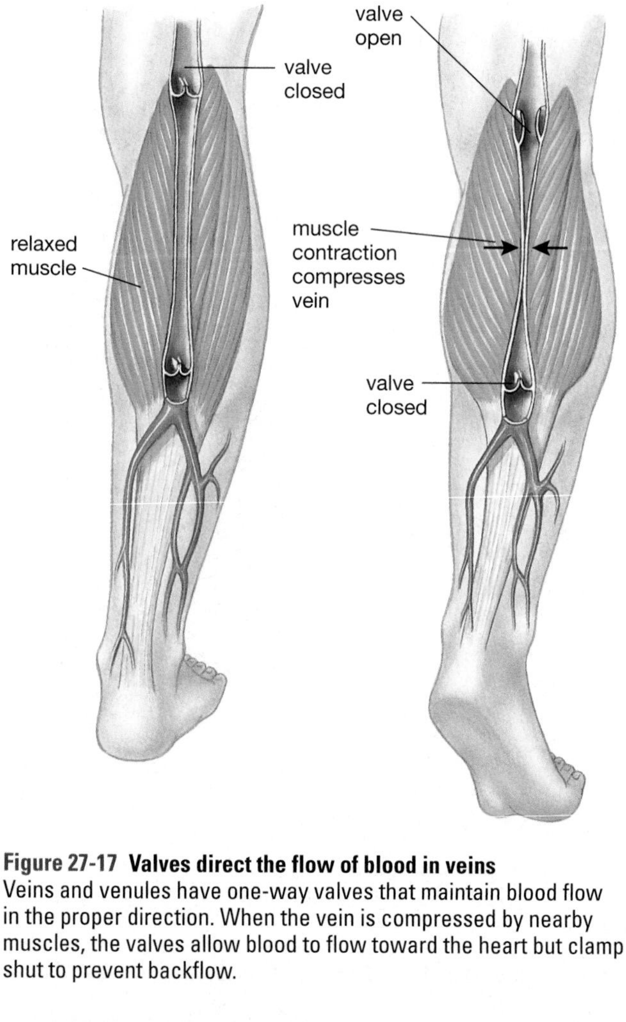
valve open
valve closed
relaxed muscle
muscle contraction compresses vein
valve closed

Figure 27-17 Valves direct the flow of blood in veins
Veins and venules have one-way valves that maintain blood flow in the proper direction. When the vein is compressed by nearby muscles, the valves allow blood to flow toward the heart but clamp shut to prevent backflow.

nervous system (which prepares the body for emergency action) automatically stimulates the contraction of the smooth muscles in the vein walls. This action decreases the internal volume of the veins and raises blood pressure, speeding up the return of blood to the heart.

Arterioles Control the Distribution of Blood Flow

Muscular arteriole walls are under the influence of nerves, hormones, and chemicals produced by nearby tissues. Arterioles therefore contract and relax in response to changing needs of the tissues and organs they supply. As you read in a thriller, ". . . the blood drained from her face as she beheld the gruesome sight . . . ," keep in mind that the heroine is experiencing the constriction of the arterioles that supply her skin with blood. In threatening situations, the sympathetic nervous system stimulates the smooth muscle of the arterioles to contract. This contraction raises blood pressure overall, but selective constriction also redirects blood to the heart and muscles, where it may be needed for vigorous action, and away from the skin, where it is less essential.

On a hot summer day, however, you become flushed as the arterioles in your skin expand and bring more blood to the skin capillaries. Bringing this blood closer to the surface enables your body to dissipate excess heat to the outside and to maintain a proper internal temperature. In contrast, in extremely cold weather, your fingers and toes can become frostbitten because the arterioles that supply blood to the extremities constrict. The blood is shunted to vital organs, such as the heart and brain, which cannot function properly if their temperature drops. By minimizing blood flow to the heat-radiating extremities, your body conserves heat.

The flow of blood in capillaries is regulated by tiny rings of smooth muscle (called **precapillary sphincters**), which surround the junctions between arterioles and capillaries (see Fig. 27-15). These open and close in response to local changes that signal the needs of nearby tissues. For example, the accumulation of carbon dioxide, lactic acid, or other cellular wastes signals the need for increased blood flow to the tissues. These signals cause the precapillary sphincters as well as the muscles in the walls of nearby arterioles to relax, allowing more blood to flow through the capillaries.

What happens when blood vessels rupture, are narrowed by deposits of cholesterol, or are blocked by clots, damming the "river of life"? We explore these questions in "Health Watch: Matters of the Heart."

5 How Does the Lymphatic System Work with the Circulatory System?

The **lymphatic system** consists of a network of lymph capillaries and larger vessels that empty into the circulatory system, numerous small *lymph nodes*, patches of lymphocyte-rich connective tissue (including the *tonsils*), and two additional organs: the *thymus* and the *spleen* (Fig. 27-18). Although not strictly part of the circulatory system, the lymphatic system is closely associated with it. The lymphatic system has several important functions:

1. Removal of excess fluid and dissolved substances that leak from the capillaries.
2. Transport of fats from the small intestine to the bloodstream.
3. Defense of the body by exposing bacteria and viruses to white blood cells.

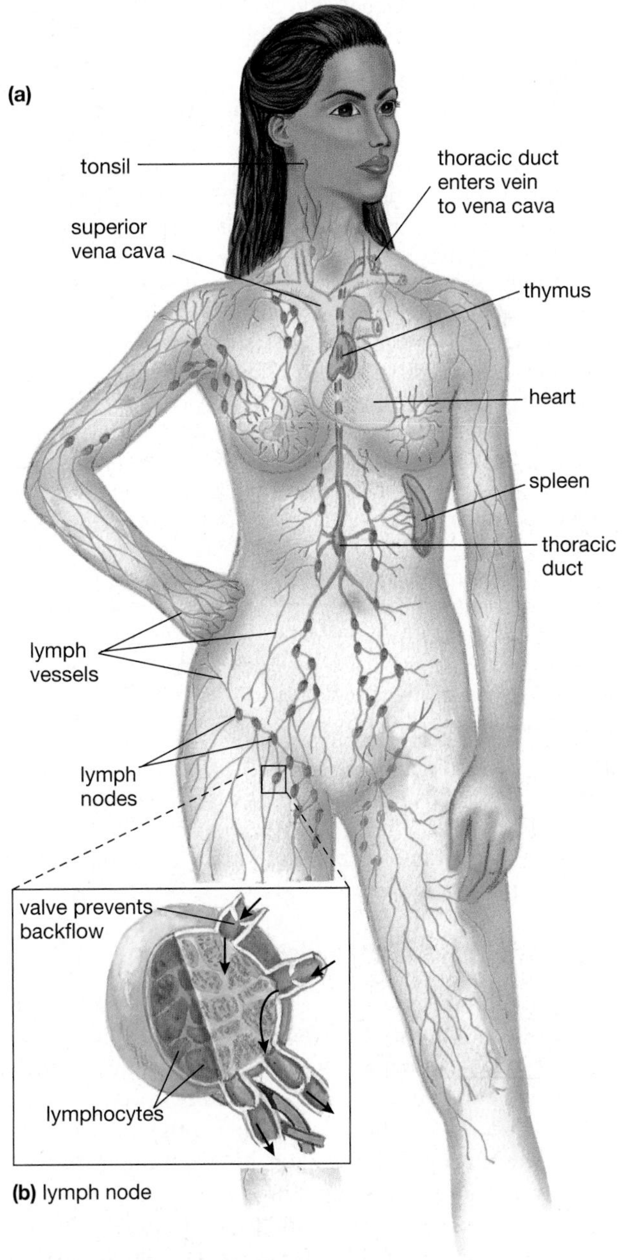

Figure 27-18 The human lymphatic system
(a) Lymph vessels, lymph nodes, and two auxiliary lymph organs, the thymus and spleen. Lymph is returned to the circulatory system by way of the thoracic duct, which empties into the vena cava, a large vein. ***(b)*** A cross section of a lymph node. The node is filled with channels lined with white blood cells (lymphocytes) that attack foreign matter in the lymph.

Health Watch
Matters of the Heart

Disorders of the heart and blood vessels, called *cardiovascular disorders*, are the leading cause of death in the United States, killing nearly 1 million Americans annually, and no wonder. Your heart is expected to contract vigorously more than 2.5 billion times during your lifetime without once stopping to rest, forcing blood through vessels whose total length would encircle the globe twice. Since these vessels may become constricted, weakened, or clogged for a number of reasons, the cardiovascular system is a prime candidate for malfunction.

Hypertension Strains the Heart

High blood pressure, also called **hypertension**, is caused by the constriction of arterioles, causing resistance to blood flow and strain on the heart. For most of the 50 million Americans with this condition, the cause is unknown. An approximate borderline reading for high blood pressure is 140/90. The strain on the heart caused by hypertension may cause it to increase in size, but its own blood supply may not increase proportionately. The heart muscle is then inadequately supplied with blood, especially during exercise. Lack of sufficient oxygen to the heart can cause chest pain called **angina**. High blood pressure also contributes to "hardening of the arteries," or *atherosclerosis*, described below. Hypertension, in conjunction with hardened arteries, can lead to the rupture of an artery and internal bleeding. The rupture of vessels supplying the brain causes **stroke**, in which brain function is lost in the area deprived of blood and of the vital oxygen and nutrients the blood carries.

Mild hypertension is sometimes alleviated by weight reduction, exercise, stress management, and (for some individuals) reduction of dietary salt. For severe cases, doctors may prescribe drugs to reduce fluid in the body, reduce the heart rate, and expand the arteries and arterioles.

Atherosclerosis Obstructs Blood Vessels

Atherosclerosis (derived from the Greek *athero*, meaning "gruel" or "paste," and *scleros*, "hard") causes the walls of the large arteries to thicken and lose their elasticity. This is caused by deposits called **plaques**, which are composed of cholesterol and other fatty substances as well as calcium and fibrin. Plaques are deposited within the wall of the artery between the smooth muscle and the inner lining. A plaque may rupture through the lining into the interior of the vessel, stimulating platelets to adhere to the vessel wall and initiate blood clots. These clots further obstruct the artery and may completely block it (Fig. E27-2). Arterial clots are responsible for the most serious consequences of atherosclerosis: heart attacks and strokes.

About 1.5 million Americans suffer **heart attacks** each year, and about half a million people die from them. A heart attack occurs when one of the coronary arteries is blocked (coronary arteries supply the heart muscle itself; one with blockage is shown in Fig. E27-2a). Deprived of nutrients and oxygen, the heart muscle supplied by the blocked artery rapidly and painfully dies. Although heart attacks are the major cause of death from atherosclerosis, this disease causes plaques and clots to form in arteries throughout the body. If a clot or plaque obstructs an artery that supplies the brain, it can cause a stroke, with results similar to those caused by a ruptured artery.

As with hypertension, the exact cause of atherosclerosis is unclear, but it is promoted by hypertension, cigarette smoking, genetic predisposition, obesity, diabetes, lack of exercise, and high blood levels of a certain type of cholesterol bound to a carrier molecule called *low-density lipoprotein* (LDL). If LDL-bound cholesterol levels are too high, cholesterol can be deposited in arterial walls. In contrast, cholesterol bound to *high-density lipoprotein* (HDL) is metabolized or excreted and hence is often called "good" cholesterol. Cholesterol is discussed in "Health Watch: Cholesterol—Friend and Foe" in Chapter 3.

Traditional treatment of atherosclerosis includes the use of drugs or changes in diet and lifestyle to lower blood pressure and blood cholesterol levels. Plaques are sometimes squashed by inserting a tiny balloon, which is inflated to squash the plaque flat; shaved off the artery wall with a miniature drill; or vaporized with a laser. A new use for a laser beam is *laser*

Lymphatic Vessels Resemble the Veins and Capillaries of the Circulatory System

Like blood capillaries, *lymph capillaries* form a complex network of microscopically narrow, thin-walled vessels into which substances can move readily. In contrast to those of blood capillaries, lymph capillary walls are composed of cells with openings between them that act as one-way valves. These openings allow relatively large particles, along with fluid, to be carried into the lymph capillary. Also unlike blood capillaries, which form a continuous connected network, lymph capillaries dead-end in the body's tissues (Fig. 27-19, p. 564). Materials collected by the lymph capillaries flow into larger lymph vessels. Large lymph vessels have somewhat muscular walls, but, as in blood veins, most of the impetus for lymph flow comes from the contraction of nearby muscles, such as those used in breathing and walking. As in blood veins, the direction of flow is regulated by one-way valves (Fig. 27-20, p. 564).

The Lymphatic System Returns Fluids to the Blood

As described earlier, dissolved substances are exchanged between the capillaries and body cells by means of interstitial fluid (derived from blood plasma), which bathes nearly all the body's cells. In an average person, about 3 liters more fluid leaves the blood capillaries than is reabsorbed by them each day. One function of the lymphatic system is to return this excess fluid and its dissolved proteins and other substances to the blood. As interstitial fluid accumulates, *revascularization*, in which a laser is used to shoot 15 to 30 small (1-mm diameter) channels through the ventricular walls. Blood clots form on the outside of the ventricle, keeping blood from leaking out. As the heart beats, blood from inside the ventricles flows in and out of the channels, supplying oxygen to the ventricular muscle and partially replacing the function of coronary arteries clogged by plaques. *Coronary bypass surgery*, performed on more than 350,000 individuals each year in the United States, consists of bypassing one or more obstructed coronary arteries with a piece of vein (usually obtained from the patient's leg) or with an artery (often from the patient's forearm). Researchers are working to perfect artificial vessels (see "Scientific Inquiry: Artificial Blood? Artificial Vessels?") so the patient's own vessels can be spared.

If a heart attack occurs, rapid treatment can minimize the damage and significantly increase the victim's chances of survival. Blood clots in coronary arteries are commonly dissolved by injecting substances that stimulate the production of an enzyme that breaks down fibrin, the protein that binds the clot together. Although heart disease is still the leading cause of death in the United States, steady progress in treatment has significantly reduced the rate of early deaths from atherosclerosis.

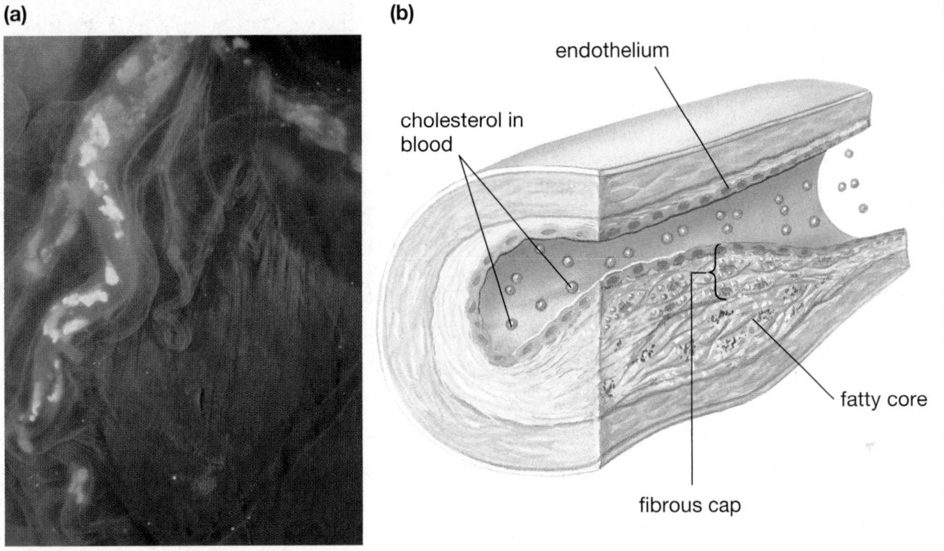

Figure E27-2 Plaques clog arteries
(a) In this remarkable photo of coronary arteries, plaques are seen in glowing yellow. If they block a coronary artery, a heart attack will occur. ***(b)*** Diagrammatic cross section of an artery with a plaque. If the fibrous cap ruptures, a clot will form that can completely obstruct the artery, or the clot can break loose and clog a narrower artery "downstream."

its pressure forces the fluid through the openings in the lymph capillaries (see Fig. 27-19). The lymphatic system transports this fluid, now called **lymph**, back to the circulatory system. The importance of the lymphatic system in returning fluid to the bloodstream is illustrated by the condition known as *elephantiasis* (Fig. 27-21). This disfiguring disorder is caused by a parasitic roundworm that colonizes lymphatic vessels, scarring them and preventing them from draining off excess fluid.

The Lymphatic System Transports Fats from the Small Intestine to the Blood

After a fatty meal, fat globules may make up 1% of the lymphatic fluid. How does this come about? As you will learn in Chapter 29, the small intestine is richly supplied with lymph capillaries. After absorbing digested fats, intestinal cells release fat globules into the interstitial fluid. These globules are too large to diffuse into blood capillaries but can easily move through the openings between lymph capillary cells. Once in the lymph, they are dumped into the vena cava, a large vein that enters the heart.

The Lymphatic System Helps Defend the Body Against Disease

In addition to its other roles, the lymphatic system helps defend the body against foreign invaders, such as bacteria and viruses. In the linings of the respiratory, digestive, and urinary tracts are patches of connective

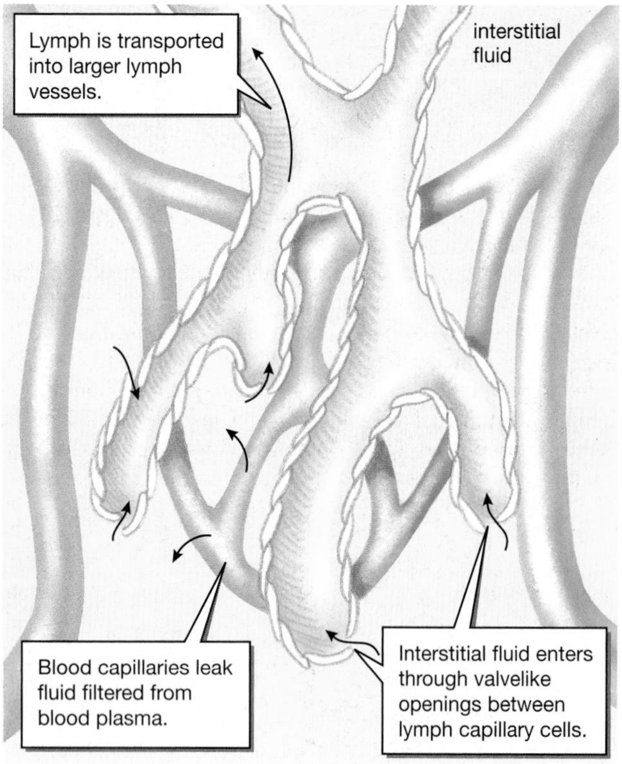

Figure 27-19 Lymph capillary structure
Lymph capillaries end blindly in the body tissues, where pressure from the accumulation of interstitial fluid forces the fluid into the lymph capillaries.

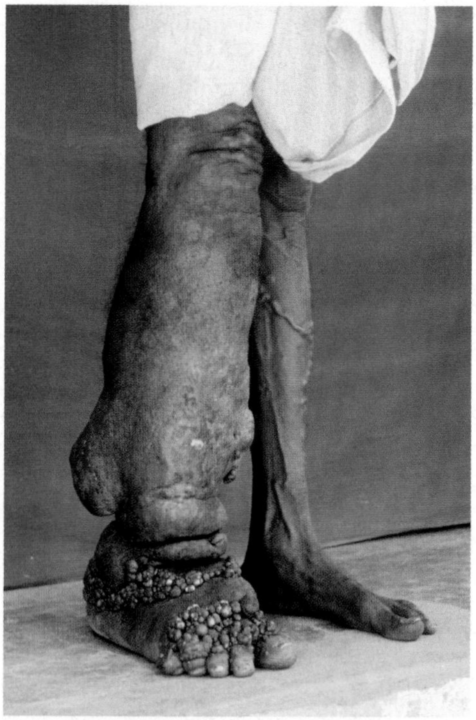

Figure 27-21 Elephantiasis results from blocked lymphatic vessels
When scarring of lymph vessels from infection by a parasitic worm prevents lymph from returning to the bloodstream, the affected area can become massively swollen.

tissue that contain large numbers of lymphocytes. The largest of these patches are the **tonsils**, located in the cavity behind the mouth. The large lymph vessels are interrupted periodically by kidney-bean–shaped structures about 2.5 centimeters (1 inch) long called **lymph nodes** (see Fig. 27-18b). Lymph is forced through channels within the nodes, which are lined with masses of **macrophages**. Lymphocytes are also produced in the lymph nodes. Both macrophages and lymphocytes recognize and destroy foreign particles, such as bacteria and viruses, and are killed in the process. The painful swelling of lymph nodes that accompanies certain diseases (mumps is an extreme example) is largely a result of the accumulation of dead lymphocytes and macrophages and dead virus-infested cells they have engulfed.

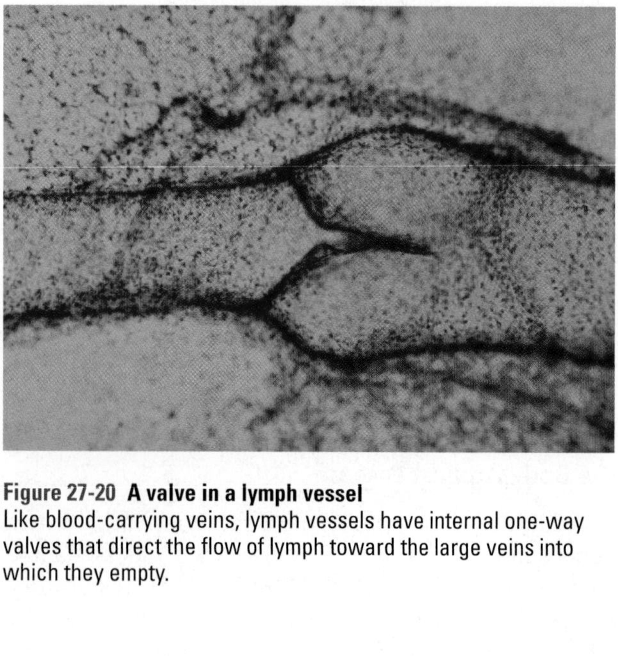

Figure 27-20 A valve in a lymph vessel
Like blood-carrying veins, lymph vessels have internal one-way valves that direct the flow of lymph toward the large veins into which they empty.

The thymus and the spleen are often considered part of the lymphatic system (see Fig. 27-18a). The **thymus**, which is an organ that produces lymphocytes, is located beneath the breastbone slightly above the heart. The thymus is particularly active in infants and young children but decreases in size and importance in early adulthood. The **spleen**, which is another lymphocyte-producing organ, is located in the left side of the abdominal cavity, between the stomach and diaphragm. Just as the lymph nodes filter lymph, the spleen filters blood, exposing it to macrophages and lymphocytes that destroy foreign particles and aged red blood cells.

CASE STUDY REVISITED

Xenotransplants

The promise of organs from genetically engineered pigs and other new developments have raised hope that cross-species transplants will one day make a major contribution to treatment of some of the most devastating human diseases. As you can imagine, however, not everyone is ready to jump on the xenotransplantation bandwagon. Animal-rights activists find xenotransplantation inherently unethical. Some xenotransplantation researchers urge caution as well. Of special concern is the possibility that xenotransplantation could provide a means for disease-causing organisms to jump from other species to humans. The recent discovery that many pigs harbor viruses that have the potential to infect human cells has raised new concerns. Could a virus in pig cells infect an immunosuppressed transplant recipient, and could the infection spread to caregivers and other healthy people? The issues involved in determining how best to proceed remain complex and profound.

Where do you stand on xenotransplantation? Defend your position. If you have objections, describe safeguards that would make the process more acceptable to you. If you or your child, spouse, or parent were dying of a disorder that might be cured by a xenotransplant, would this affect your position?

Summary of Key Concepts

1) What Are the Major Features and Functions of Circulatory Systems?

Circulatory systems transport blood rich in dissolved nutrients and oxygen close to each cell, where nutrients can be released and wastes absorbed by diffusion. All circulatory systems have three major parts: (1) blood, a fluid; (2) vessels, a system of channels to conduct the blood; and (3) a heart, a pump to circulate the blood. Invertebrates have open or closed circulatory systems; nearly all vertebrates have closed systems. In open systems, blood is pumped by a heart into a hemocoel, where it directly bathes internal organs. In closed systems, the blood is confined to the heart and blood vessels.

2) How Does the Vertebrate Heart Work?

Vertebrate circulatory systems transport gases, hormones, and wastes; distribute nutrients; help regulate body temperature; and defend the body against disease.

The vertebrate heart evolved from two chambers in fishes, to three in amphibians and most reptiles, to four in birds and mammals. In the four-chambered heart, blood is pumped separately to the lungs and through the body, maintaining complete separation of oxygenated and deoxygenated blood. Deoxygenated blood is collected from the body in the right atrium and passed to the right ventricle, which pumps it to the lungs. Oxygenated blood from the lungs enters the left atrium, is passed to the left ventricle, and pumped to the rest of the body.

The cardiac cycle consists of two stages: (1) atrial contraction, followed by (2) ventricular contraction. The direction of blood flow is maintained by valves within the heart. The contractions of the heart are initiated and coordinated by the sinoatrial node, the heart's pacemaker. Heart rate can be modified by the nervous system and by hormones such as epinephrine.

3) What Is Blood?

Blood is composed of both fluid and cellular materials. The fluid plasma consists of water that contains proteins, hormones, nutrients, gases, and wastes. Red blood cells, or erythrocytes, are packed with a large iron-containing protein called hemoglobin, which carries oxygen. Their numbers are regulated by the hormone erythropoietin. Proteins on their plasma membrane determine blood type. There are five types of white blood cells, or leukocytes, that fight infection (for more information, see the table on this book's companion Web site, www.prenhall.com/audesirk6). Platelets, which are fragments of megakaryocytes, are important for blood clotting.

4) What Are the Types and Functions of Blood Vessels?

Blood leaving the heart travels (in sequence) through arteries, arterioles, capillaries, venules, veins, and then back to the heart. Each vessel is specialized for its role. Elastic, muscular arteries help pump the blood. The thin-walled capillaries are the sites of exchange of materials between the body cells and the blood. Veins provide a path of low resistance back to the heart, with one-way valves that maintain the direction of blood flow. The distribution of blood is regulated by the constriction and dilation of arterioles under the influence of the sympathetic nervous system and local factors such as the amount of carbon dioxide in the tissues. Local factors also regulate precapillary sphincters, which control blood flow to the capillaries.

5) How Does the Lymphatic System Work with the Circulatory System?

The human lymphatic system consists of lymphatic vessels, tonsils, lymph nodes, and the thymus and spleen. The lymphatic system removes excess interstitial fluid that leaks through blood capillary walls. It transports fats to the bloodstream from the small intestine and fights infection by filtering the lymph through lymph nodes, where white blood cells ingest foreign invaders, such as viruses and bacteria. The thymus, which is most active in young children, produces lymphocytes that function in immunity. The spleen filters blood past macrophages and lymphocytes, which remove bacteria and damaged blood cells.

Key Terms

angina *p. 562*
arteriole *p. 559*
artery *p. 552*
atherosclerosis *p. 562*
atrioventricular (AV) node *p. 554*
atrioventricular valve *p. 553*
atrium *p. 551*
blood *p. 550*
blood clotting *p. 558*
blood vessel *p. 550*
capillary *p. 560*
cardiac cycle *p. 552*
closed circulatory system *p. 550*
erythroblastosis fetalis *p. 557*
erythrocyte *p. 555*
erythropoietin *p. 556*
fibrillation *p. 554*
fibrin *p. 558*
heart *p. 550*
heart attack *p. 562*
hemocoel *p. 550*
hemoglobin *p. 555*
hypertension *p. 562*
interstitial fluid *p. 560*
leukocyte *p. 557*
lymph *p. 563*
lymphatic system *p. 561*
lymph node *p. 564*
lymphocyte *p. 557*
macrophage *p. 564*
megakaryocyte *p. 557*
open circulatory system *p. 550*
pacemaker *p. 553*
plaque *p. 562*
plasma *p. 554*
platelet *p. 557*
precapillary sphincter *p. 561*
Rh factor *p. 557*
semilunar valve *p. 553*
sinoatrial (SA) node *p. 554*
spleen *p. 564*
stroke *p. 562*
thrombin *p. 558*
thymus *p. 564*
tonsil *p. 564*
vein *p. 552*
ventricle *p. 551*
venule *p. 560*

Thinking Through the Concepts

Multiple Choice

1. *Which of the following is NOT an important function of the vertebrate circulatory system?*
   a. transport of nutrients and respiratory gases
   b. regulation of body temperature
   c. protection of the body by circulating antibodies
   d. removal of waste products for excretion from the body
   e. defense against blood loss, through clotting
2. *Which event initiates blood clotting?*
   a. contact with an irregular surface by platelets and other factors in plasma
   b. production of the enzyme thrombin
   c. conversion of fibrinogen into fibrin
   d. conversion of fibrin into fibrinogen
   e. excess flow of blood through a capillary
3. *The sites of exchange of wastes, nutrients, gases, and hormones between the blood and body cells are the*
   a. arteries
   b. arterioles
   c. capillaries
   d. veins
   e. all blood vessels
4. *What produces systolic blood pressure?*
   a. contraction of the right atrium
   b. contraction of the right ventricle
   c. contraction of the left atrium
   d. contraction of the left ventricle
   e. the pause between heartbeats
5. *Which of the following is NOT a component of plasma?*
   a. water
   b. globulins
   c. fibrinogen
   d. albumins
   e. platelets
6. *Lymph most closely resembles which of the following?*
   a. blood
   b. urine
   c. plasma
   d. interstitial fluid
   e. water

? Review Questions

1. Trace the flow of blood through the circulatory system, starting and ending with the right atrium.
2. List three types of blood cells, and describe their principal functions.
3. What are five functions of the vertebrate circulatory system?
4. In what way do veins and lymph vessels resemble one another? Describe how fluid is transported in each of these vessels.
5. Describe three important functions of the lymphatic system.
6. Distinguish among plasma, interstitial fluid, and lymph.
7. Describe veins, capillaries, and arteries, noting their similarities and differences.
8. Trace the evolution of the vertebrate heart from two chambers to four chambers.
9. Explain in detail what causes the vertebrate heart to beat.
10. Describe the cardiac cycle, and relate the contractions of the atria and ventricles to the two readings taken during the measurement of blood pressure.
11. Describe how the number of red blood cells is regulated by a negative feedback system.
12. Describe the formation of an atherosclerotic plaque. What are the risks associated with atherosclerosis?

Applying the Concepts

1. Discuss the steps you can take now and in the future to reduce your risks of developing heart disease.
2. Discuss why a four-chambered heart is much more efficient than a two-chambered heart in delivering oxygenated blood to the various body parts. What evolutionary changes in the lifestyles of organisms selected for the evolution of the four-chambered heart?
3. Heart surgeons have attempted to transplant baboon hearts into humans whose hearts were failing. Discuss the implications of this operation from as many angles as you can think of.
4. Considering the prevalence of cardiovascular disease and the high, increasing costs of treating it, certain treatments may not be available to all who might benefit from them. What factors would you take into account in rationing cardiovascular procedures, such as heart transplants?
5. Joe, a 45-year-old executive of a major corporation, has been diagnosed with mild hypertension. What treatments or lifestyle changes might Joe's physician recommend? If Joe's hypertension becomes more severe, what treatments might Joe's physician use? Should Joe be concerned about mild hypertension? Explain your answer.

For More Information

Gibbons, R., and associates. "Waiting for Organ Transplantation." *Science,* January 14, 2000. The statistics of the need, availability, and waiting time for organ transplantation suggest that new sources of organs are needed.

Hajjar, D. P., and Nicholson, A. C. "Atherosclerosis." *American Scientist,* September–October 1995. Scientists are learning more about the biochemical reactions that occur during plaque formation, an understanding that will lead to more-effective treatments in the future.

Nucci, M. L., and Abuchowski, A. "The Search for Blood Substitutes." *Scientific American*, February 1998. As the authors summarize: "The threat of blood shortages and fears about contamination have hastened attempts to find life-sustaining alternatives."

Radetsky, P. "The Mother of All Blood Cells." *Discover*, March 1995. Discusses the discovery by Irving Weissman of a "stem cell" that gives rise to all other types of blood cells.

Seppa, N. "Secondary Smoke Carries High Price." *Science News*, January 17, 1998. New research suggests that smoking causes atherosclerosis, that the damage persists after a smoker quits, and that exposure to secondhand smoke significantly increases the buildup of plaque in the carotid artery of nonsmokers.

Wu, C. "Engineered Blood Vessel Is Only Human." *Science News*, January 17, 1998. Researchers have constructed an experimental blood vessel composed entirely of human tissue.

Answers to Multiple-Choice Questions

1. d 2. a 3. c 4. d 5. e 6. d

MEDIATUTOR
Circulation

CD Activities

Activity 27.1: Heart Structure and Function

Estimated time: 5 minutes

The vertebrate heart is a complex organ. In this tutorial, you will explore the various structures of the vertebrate heart and the functions of each of those structures.

Activity 27.2: The Cardiac Cycle

Estimated time: 10 minutes

The alternating contraction and relaxation of the heart chambers is called the cardiac cycle. In this tutorial, you will view an animation of this process and explore the various steps involved in the cycle.

Start the MediaTutor Student CD-ROM and enter the activity number in the Quick Search box to be taken directly to that activity.

Web Investigations

Case Study: Xenotransplants

Estimated time: 15 minutes

Sophisticated surgical procedures have become commonplace in current medical practice. Unfortunately, technical expertise is often ahead of biological understanding or careful consideration of ethical issues. One hotly debated issue is xenotransplantation. This exercise examines the question, should xenotransplantation become standard medical practice?

Go to http://www.prenhall.com/audesirk6, the Audesirk Companion Web site. Select Chapter 27 and the Web Investigation to begin.

These students say they will quit smoking later. Only one in five will succeed.

28 Respiration

AT A GLANCE

Case Study: Lives Up in Smoke

1) **Why Exchange Gases?**

2) **What Are Some Evolutionary Adaptations for Gas Exchange?**

Some Animals in Moist Environments Lack Specialized Respiratory Structures

Respiratory Systems Facilitate Gas Exchange by Diffusion

Gills Facilitate Gas Exchange in Aquatic Environments

Terrestrial Animals Have Internal Respiratory Structures

3) **How Does the Human Respiratory System Work?**

The Conducting Portion of the Respiratory System Carries Air to the Lungs

Gas Exchange Occurs in the Alveoli

Oxygen and Carbon Dioxide Are Transported Using Different Mechanisms

Air Is Inhaled Actively and Exhaled Passively

Breathing Rate Is Controlled by the Respiratory Center of the Brain

Case Study Revisited: Lives Up in Smoke

CASESTUDY

Lives Up in Smoke

"I smoke because I like smoking, and when I don't want to anymore, that's when I'll quit," says a 17-year-old high school student. Another student has watched her relatives become addicted to cigarettes but smokes because of peer pressure and to relieve stress. "It scares me because . . . a lot of people get cancer. I don't want to die early," she says, lighting up and inhaling deeply. Alarmingly, as society becomes more aware of the dangers of smoking, teenagers in the U.S. are taking up the habit in larger numbers than ever before. Surveys cited by the American Lung Association indicate that the number of high school seniors who smoke daily rose nearly 19% in the decade between 1988 and 1998. About 40% of high school seniors have lit up within the past month, and more than 19% smoke frequently. Most are aware of the dangers but say they will stop "when the time comes." How likely are these new smokers to quit? What are their chances of dying from smoking? How does smoking interfere with the respiratory system from the bronchi to the bloodstream? Find answers at the end of the chapter and in "Health Watch: Smoking—A Life and Breath Decision." ■

1) Why Exchange Gases?

Late again! Sprinting up two flights of stairs to your classroom, you feel your calves "burning." Remembering Chapter 8, you think "Aha! That's lactic acid building up—my muscle cells are fermenting glucose because they can't get enough oxygen for cellular respiration." As you slip into your seat, quietly panting and feeling your heart pounding, the discomfort eases. Less exertion coupled with rapid breathing ensures that adequate oxygen is now available; the lactic acid is being reconverted to pyruvate and then broken down into carbon dioxide (CO_2) and water, while providing additional energy.

You are experiencing firsthand the relationship between cellular respiration and breathing, which is also called *respiration*. Each cell in your body (and in the bodies of all other organisms) must continuously expend energy to maintain itself. When you call on your muscles to carry you upstairs quickly, the demands are extreme. As cellular respiration converts the energy in nutrients such as sugar into ATP that can be used by body cells, the process requires a steady supply of oxygen and generates carbon dioxide as a waste product. The rapid beating of your heart as you relax after your sprint upstairs is a reminder that the circulatory system works in close harmony with the respiratory system, extracting oxygen from air in your lungs, carrying it within diffusing distance of each cell, then picking up carbon dioxide for release in the lungs.

How, exactly, does breathing support cellular respiration? What does the inside of a lung look like, and how is it adapted for gas exchange? Why are lungs inside our bodies, instead of on the outside, which is already exposed to the air? How do aquatic animals respire? In this chapter we explore the specialized structures of respiratory systems and how they work.

2 What Are Some Evolutionary Adaptations for Gas Exchange?

Gas exchange in all organisms ultimately relies on diffusion. Cellular respiration depletes O_2 and increases CO_2 levels, creating concentration gradients that favor the diffusion of carbon dioxide out of cells and the diffusion of oxygen into them. Although animal respiratory systems are amazingly diverse, they all share three features that facilitate diffusion: (1) The respiratory surface must remain moist, because gases must be dissolved in water when they diffuse into or out of cells. (2) Cells lining respiratory surfaces are very thin, a feature that facilitates diffusion of gases through them. (3) The respiratory system must have a large surface area in contact with the environment to allow adequate gas exchange.

In the following sections we examine some diverse respiratory systems in animals, each shaped by the environment in which it evolved. Notice how each type of system meets the demand for a large, moist surface that allows gas exchange.

Some Animals in Moist Environments Lack Specialized Respiratory Structures

Some animals that live in moist environments are able to exchange gases without specialized respiratory structures. The outside of their bodies provide an adequate surface area for the diffusion of gases. If the body is extremely small and elongated, as in microscopic nematode worms, gases need to diffuse only a short distance to reach all cells of the body. Alternatively, an animal's body may be thin and flattened, producing a large surface area for diffusion. In flatworms, most cells are close to the moist skin through which gases can diffuse (Fig. 28-1a).

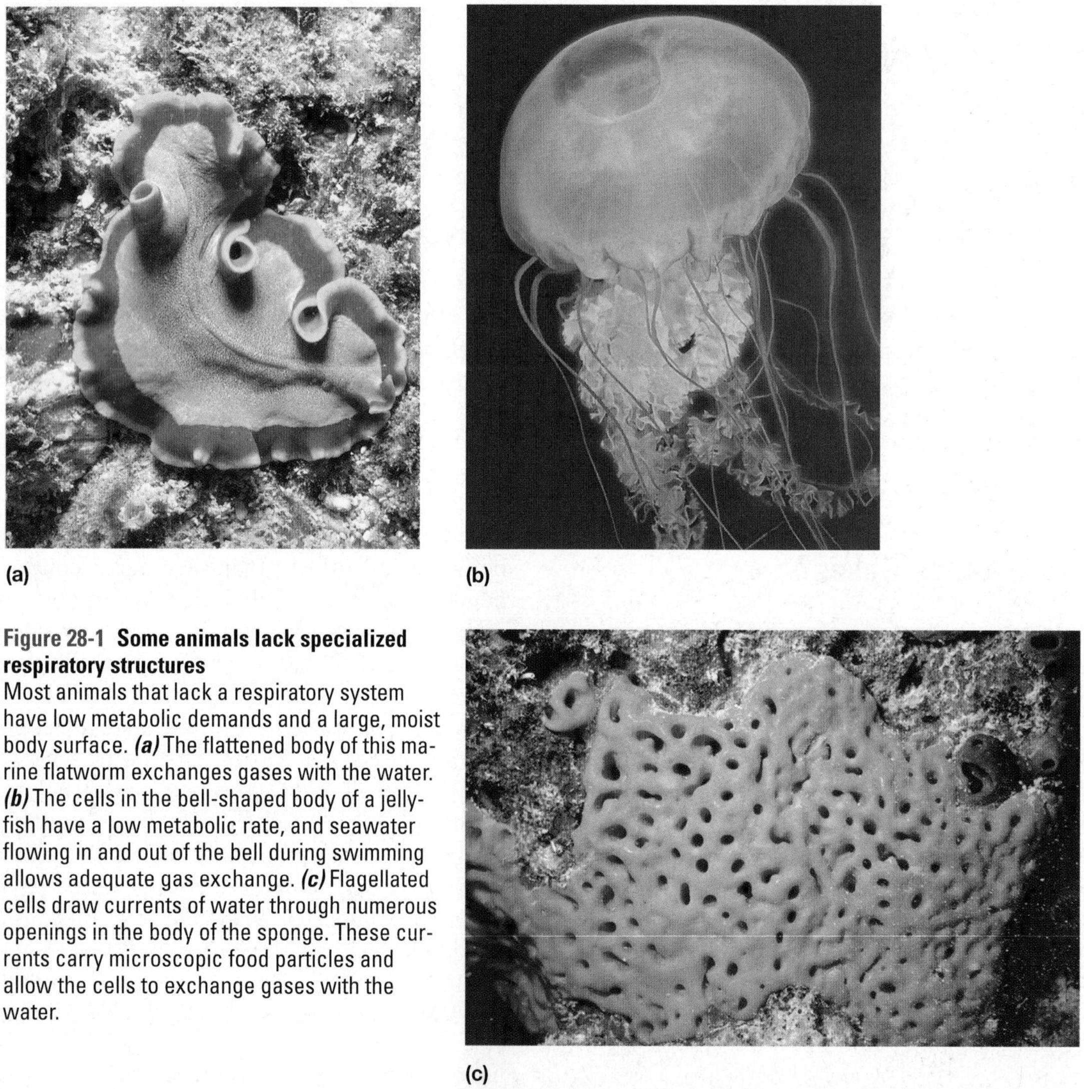

Figure 28-1 Some animals lack specialized respiratory structures
Most animals that lack a respiratory system have low metabolic demands and a large, moist body surface. ***(a)*** The flattened body of this marine flatworm exchanges gases with the water. ***(b)*** The cells in the bell-shaped body of a jellyfish have a low metabolic rate, and seawater flowing in and out of the bell during swimming allows adequate gas exchange. ***(c)*** Flagellated cells draw currents of water through numerous openings in the body of the sponge. These currents carry microscopic food particles and allow the cells to exchange gases with the water.

If energy demands are sufficiently low, the relatively slow rate of gas exchange by diffusion may suffice even for a larger, thicker body. For example, jellyfish can be quite large, but the cells that are far from the surface are relatively inert and require little oxygen (Fig. 28-1b).

Another adaptation for gas exchange is to bring the environment (normally, water) close to all the body cells, allowing direct exchange of gases between the body cells and water. Sponges, for example, circulate seawater through channels within their bodies, where it comes close to all of their cells (Fig. 28-1c).

Some animals combine a large skin surface through which diffusion occurs with a well-developed circulatory system (see Fig. 27-1b). For example, in the earthworm gases diffuse through the moist skin and are distributed throughout the body by an efficient circulatory system. Blood in skin capillaries rapidly carries off oxygen that has diffused through the skin, maintaining a concentration gradient that favors the inward diffusion of oxygen. The worm's elongated shape ensures a large skin surface relative to its internal volume, and the worm's sluggish metabolism demands relatively little oxygen. The skin must stay moist to remain effective as a gas-exchange organ; a dry earthworm will suffocate.

Respiratory Systems Facilitate Gas Exchange by Diffusion

Most animals have evolved specialized respiratory systems that interface closely with their circulatory systems to exchange gases between the cells and the environment. The transfer of gases from the environment to the blood and then to the cells and back again normally occurs in stages that alternate bulk flow with diffusion. During **bulk flow**, fluids or gases move in bulk through relatively large spaces, from areas of higher pressure to areas of lower pressure. Bulk flow contrasts with diffusion, in which molecules move individually from areas of higher concentration to areas of lower concentration (see Chapter 4). In general, gas exchange in respiratory systems occurs in the following stages, illustrated for the mammalian system in Figure 28-2:

① Air or water, containing oxygen, is moved past a respiratory surface by bulk flow, commonly facilitated by muscular breathing movements.

② Oxygen and carbon dioxide are exchanged through the respiratory surface in the lungs by diffusion; oxygen is carried into the capillaries of the circulatory system, and carbon dioxide is removed from the blood.

③ Gases are transported between the respiratory system and the tissues by bulk flow of blood as it is pumped throughout the body by the heart.

④ Gases are exchanged between the tissues and the circulatory system by diffusion. At the tissues, oxygen diffuses out of the capillaries and carbon dioxide diffuses into them along their concentration gradients.

Gills Facilitate Gas Exchange in Aquatic Environments

Gills are the respiratory structures of many aquatic animals. The simplest type of gill, found in certain mollusks and amphibians, consists of numerous projections of the body surface into the surrounding water. In general, gills are elaborately branched or folded to maximize their surface area. In some animals, gill size is determined by the availability of oxygen in the surrounding water. For example, salamanders living in stagnant water (which has little opportunity to mix with air) have larger gills than do those living in well-aerated water. Gills have a dense profusion of capillaries just beneath their delicate outer membrane. These capillaries bring blood close to the surface, where gas exchange occurs.

Fish gills are covered by a protective bony flap, the **operculum**, which keeps the delicate gill membranes from being nibbled off by predators. The operculum

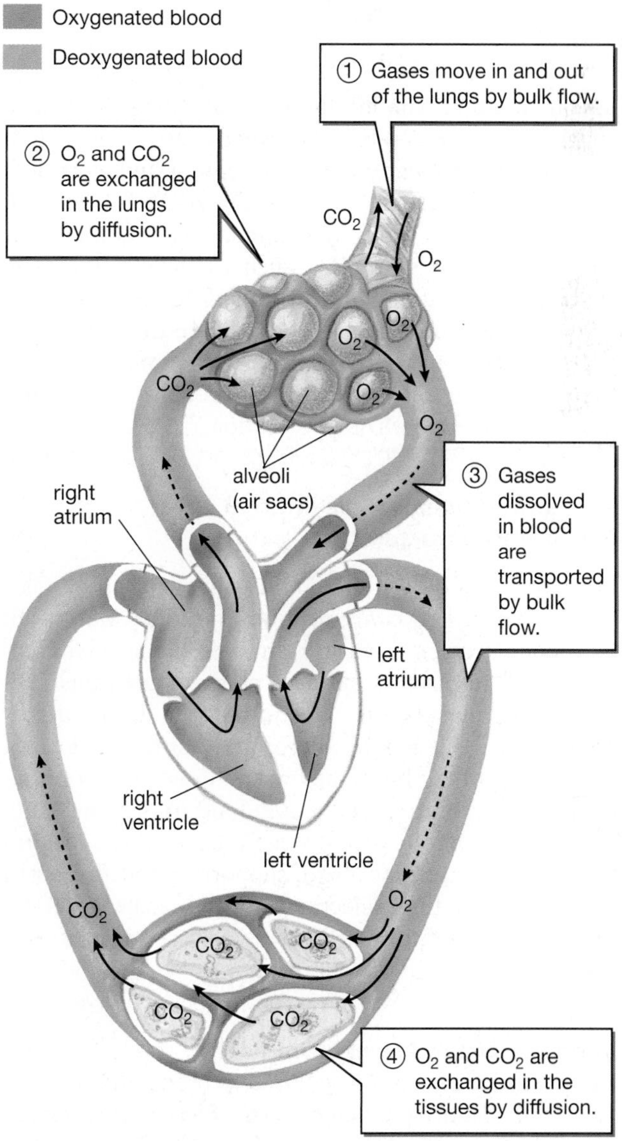

Figure 28-2 An overview of gas exchange

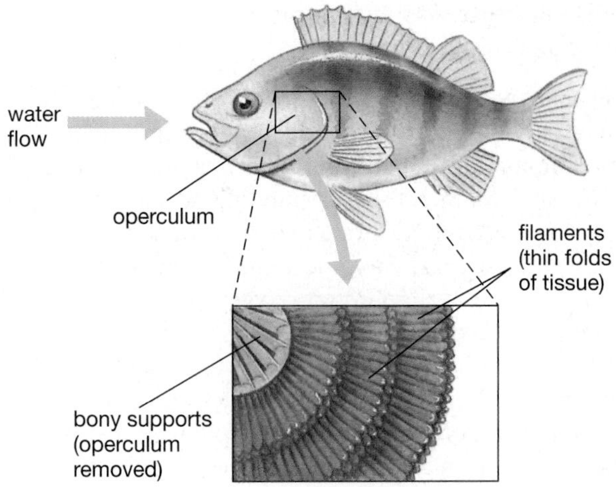

Figure 28-3 Gills exchange gases with water
The gill reaches its greatest complexity in the fish, where it is made of thin folds of tissue called *filaments* and is protected under the operculum, a bony flap. A one-way flow of water is maintained over the gill by the pumping of water through the mouth and out the opercular opening.

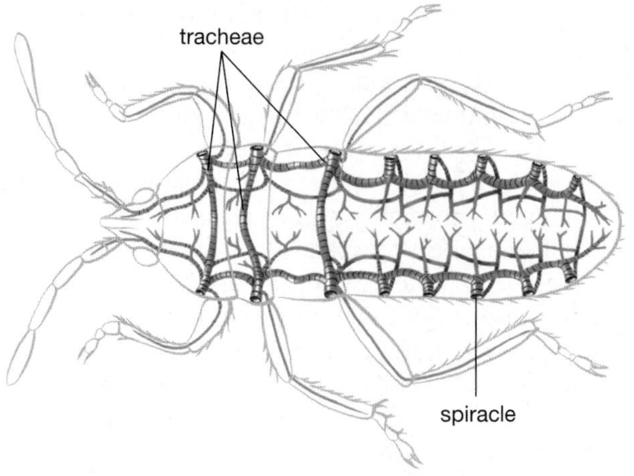

Figure 28-4 Insects breathe via tracheae
The tracheae of insects, such as this beetle, branch intricately throughout the body and open to the air through spiracles in the abdominal wall.

also helps to streamline the body, allowing the fish to swim faster. Fish create a continuous current over their gills by pumping water into their mouths and ejecting it through the opercular openings (Fig. 28-3). They can increase the flow of water by swimming with their mouths open; some fast swimmers, such as tuna and certain sharks, rely exclusively on swimming to ventilate their gills. Gills are useless out of water, because they collapse and dry out in air. As terrestrial animals evolved, therefore, their respiratory organs required both support and protection from desiccation.

Terrestrial Animals Have Internal Respiratory Structures

Terrestrial (land-dwelling) animals live in air that has a far higher oxygen concentration than does water, but extracting oxygen from dry air presents special challenges. All respiratory surfaces must remain moist, because gases must be dissolved in water to diffuse across membranes. But if moist respiratory sufaces were on the outside of the body, they would lose water continuously by evaporation and would tend to dry out. Thus, land animals have evolved structures in which respiratory surfaces are moistened, supported, and protected from drying. Natural selection has produced a variety of terrestrial respiratory structures, including tracheae in insects and lungs in vertebrates.

Insects Respire by Means of Tracheae

Insects use a system of elaborately branching internal tubes called **tracheae** (singular, **trachea**), which convey air directly to the body cells. Reinforced with chitin (which also supports the insect's external skeleton), tracheae subdivide and branch into smaller channels (*tracheoles*) that penetrate the body tissues and allow gas exchange (Fig. 28-4). Each body cell is close to a tracheole, minimizing diffusion distances. Air enters the tracheae through a series of openings called **spiracles**, located along each side of the abdomen. Spiracles have valves that allow them to be opened or closed. Some large insects use muscular pumping movements of the abdomen to enhance air movement through the tracheae.

Most Terrestrial Vertebrates Respire by Means of Lungs

Lungs are chambers containing moist, delicate respiratory surfaces that are protected within the body, where water loss is minimized and the body wall provides support. The first vertebrate lung probably appeared in a freshwater fish and consisted of an outpocketing of the digestive tract. Gas exchange in this simple lung helped the fish survive in stagnant water, in which oxygen is scarce. Amphibians, which straddle the boundary between aquatic and terrestrial life, may use gills in the larval stage and lungs in the more terrestrial adult form. For example, the purely aquatic tadpole exchanges its gills for lungs as it develops into a more terrestrial frog (Fig. 28-5a,b). Frogs and salamanders use their moist skin as a supplemental respiratory surface.

The scales of reptiles (Fig. 28-5c) reduce the loss of water through the skin and allow reptiles to survive in dry environments. But scales also reduce the diffusion of gases through the skin, so the lungs of reptiles are better developed than are those of amphibians.

Birds and mammals are exclusively lung breathers. The bird lung has evolved special adaptations that allow extremely efficient gas exchange, which is necessary to support the enormous energy demands of flight, sometimes at thousands of feet up, where oxygen is very

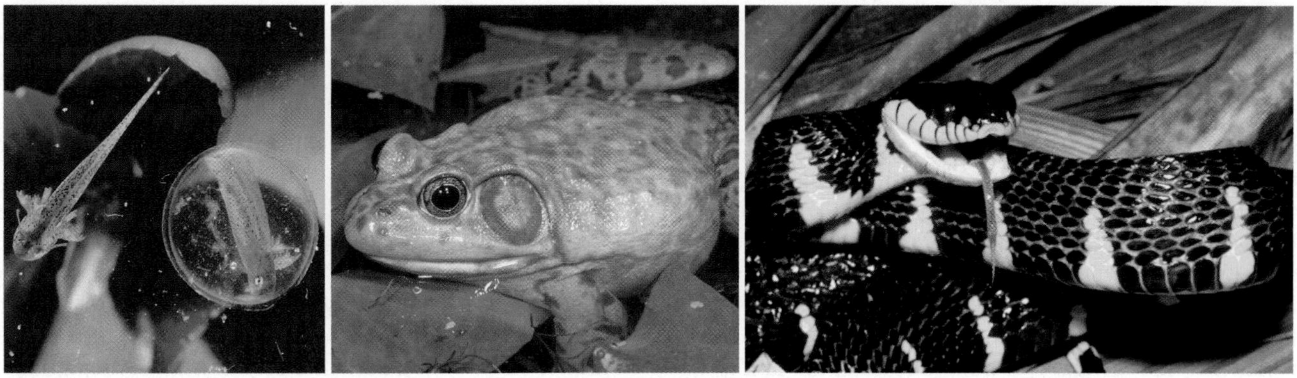

(a) (b) (c)

Figure 28-5 Amphibians and reptiles have different respiratory adaptations
(a) The bullfrog, an amphibian, begins life as a fully aquatic tadpole with feathery external gills that will later become enclosed in a protective chamber. ***(b)*** During metamorphosis into an air-breathing adult frog, the gills are lost and replaced by simple saclike lungs. In both tadpole and adult, gas exchange also occurs by diffusion through the skin, which must be kept moist to function as a respiratory surface. ***(c)*** The fully terrestrial reptile, such as this mangrove snake, is covered with dry scales that restrict gas exchange through the skin. Reptilian lungs are more efficient than are those of amphibians.

scarce. As a bird breathes in, it draws air through its lungs, where oxygen is extracted, and simultaneously pulls air into air sacs, some of which are located beyond the lungs (Fig. 28-6a). As the bird breathes out, oxygenated air from the air sacs is forced back through the lungs, allowing the bird to extract oxygen even as it exhales. In contrast to other vertebrate lungs, which are saclike, bird lungs are filled with hollow, thin-walled tubes called *parabronchi*, which allow air to pass through them in both directions (Fig. 28-6b).

3 How Does the Human Respiratory System Work?

The respiratory system in humans and other lung-breathing vertebrates can be divided into two parts: (1) the **conducting portion** and (2) the **gas-exchange portion**. The conducting portion consists of a series of passageways that carry air into and out of the gas-exchange portion, where gases are exchanged with the blood in tiny sacs in the lungs.

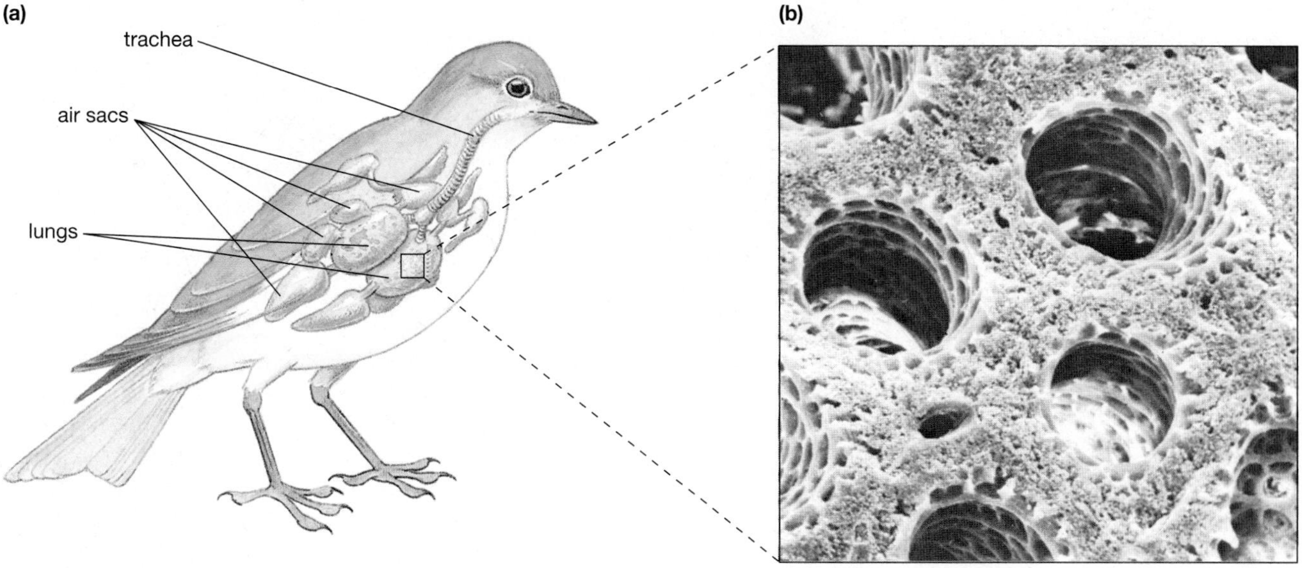

Figure 28-6 The bird respiratory system is extremely efficient
(a) Birds have air sacs in addition to their lungs. When air is inhaled, some air fills the lungs; the rest travels past the lungs to fill air sacs. As air is exhaled, the fresh air that has been temporarily stored in the air sacs fills the lungs on its way out. ***(b)*** Birds use tubular gas-exchange organs called *parabronchi* rather than saclike alveoli. The parabronchi allow the air to flow entirely through the lung to and from the air sacs.

The Conducting Portion of the Respiratory System Carries Air to the Lungs

The conducting portion brings air to the lungs; it also allows you to speak. Air enters through the nose or the mouth, passes through the nasal cavity or oral cavity into a common chamber, the **pharynx**, and then travels through the **larynx** (Fig. 28-7). The opening to the larynx is guarded by the *epiglottis*, a flap of tissue supported by cartilage. During normal breathing, the epiglottis is tilted upward, as shown in Figure 28-7, allowing air to flow freely into the larynx. During swallowing, the epiglottis tilts downward and covers the larynx, directing substances into the esophagus instead. If an individual attempts to inhale and swallow at the same time, this reflex may fail and food can become lodged in the larynx, blocking air from entering the lungs. What should you do if you see this happen? The *Heimlich maneuver* (Fig. 28-8) is easy to perform and has saved countless lives.

Within the larynx are the **vocal cords**, bands of elastic tissue controlled by muscles. Muscular contractions can cause the vocal cords to partially obstruct the opening within the larynx. Exhaled air causes them to vibrate, producing the tones of speech or song. The tones are changed in pitch by stretching the cords and articulated into words by movements of the tongue and lips.

Inhaled air continues past the larynx into the **trachea**, a flexible tube whose walls are reinforced with semicircular bands of stiff cartilage. Within the chest, the trachea splits into two large branches called **bronchi** (singular, **bronchus**), one leading to each lung. Inside the lung, each bronchus branches repeatedly into ever-smaller tubes called **bronchioles**. Bronchioles lead finally to the microscopic **alveoli** (singular, **alveolus**), tiny air sacs where gas exchange occurs (see Fig. 28-7). During its passage through the conducting system, air is warmed and moistened. Much of the dust and bacteria it carries is trapped in mucus secreted by cells that line the respiratory passages. The mucus, with its trapped debris, is continuously swept upward toward the pharynx by cilia that line the bronchioles, bronchi, and trachea. Upon reaching the pharynx, the mucus is coughed up or

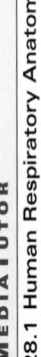

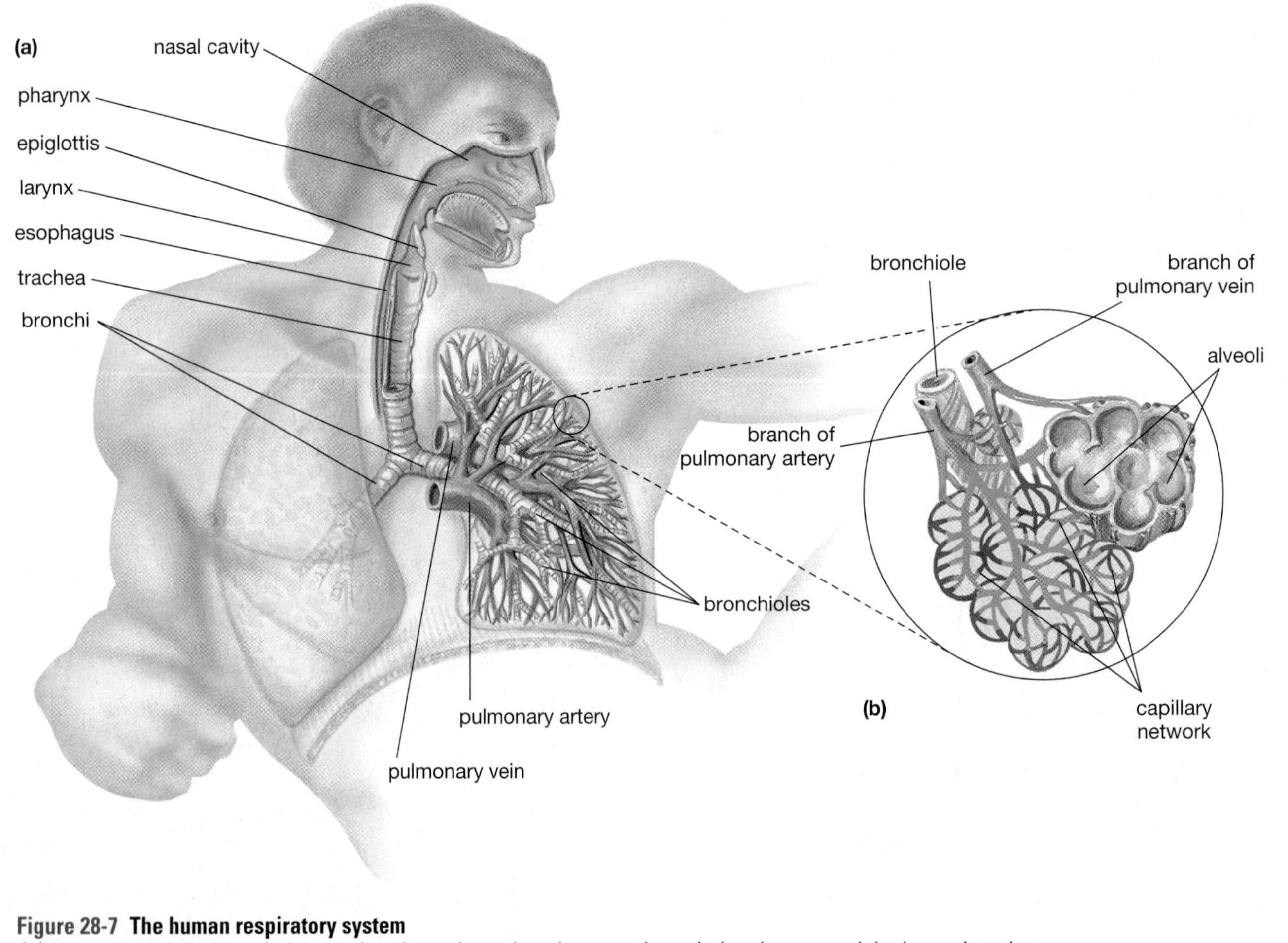

Figure 28-7 The human respiratory system
(a) Air enters mainly through the nasal cavity and mouth and passes through the pharynx and the larynx into the trachea. The epiglottis prevents food from going down the trachea. The trachea splits into two large branches, the bronchi, which lead into the two lungs. The smaller branches of the bronchi, the bronchioles, lead to the microscopic alveoli, which are enmeshed in capillaries, where gas exchange occurs. The pulmonary artery carries deoxygenated blood (in blue) to the lungs; the pulmonary vein carries oxygenated blood (in red) back to the heart. ***(b)*** Close-up of alveoli (their interiors are shown in this cut-away section) and their surrounding capillaries.

MediaTutor 28.1 Human Respiratory Anatomy

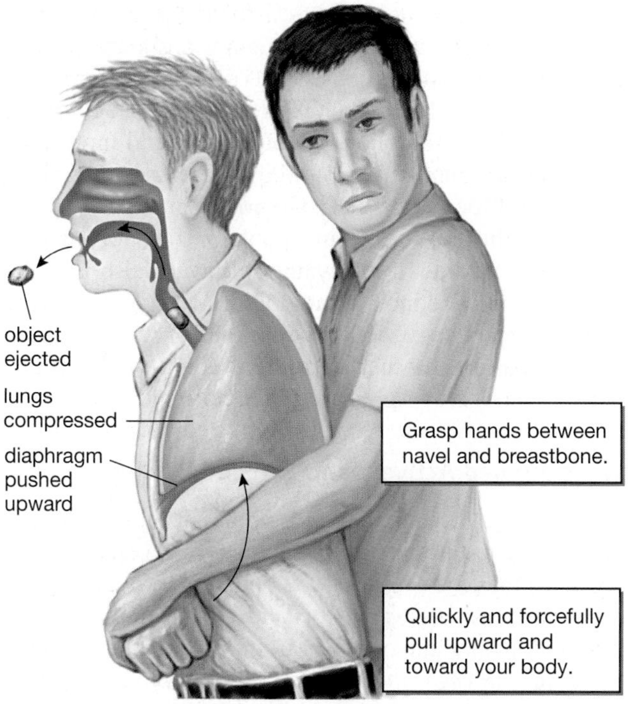

Figure 28-8 The Heimlich maneuver can save lives
If a person has choked on food or other object and is unable to breathe, the Heimlich maneuver will push upward on the victim's diaphragm and force air out of his or her lungs, possibly dislodging the object. Repeat this if necessary.

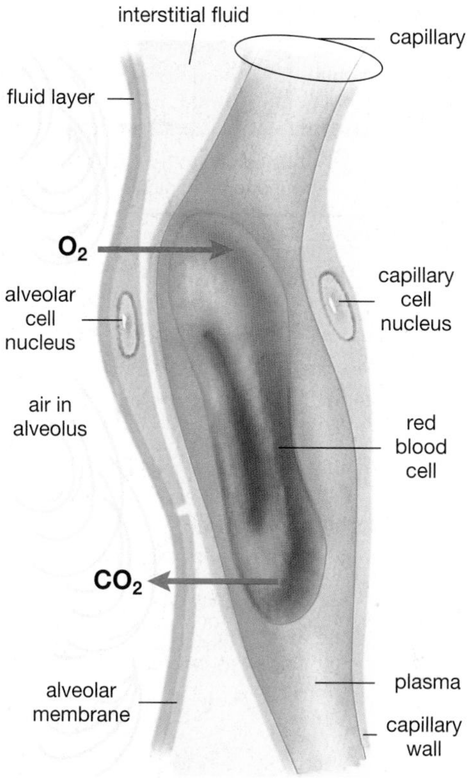

Figure 28-9 Gas exchange between alveoli and capillaries
The alveoli and capillary walls are only one cell thick, very close to one another, and the cells are coated in a thin layer of fluid. This allows gases to dissolve and diffuse easily between the lungs and circulatory system.

MEDIATUTOR 28.2 Gas Exchange

swallowed. Smoking interferes with this cleansing process by paralyzing the cilia (see "Health Watch: Smoking—A Life and Breath Decision").

Gas Exchange Occurs in the Alveoli

The lung has evolved to create an enormous moist surface for gas exchange. The dense network of bronchioles conducts air to tiny structures, the alveoli. Each lung is packed with 1.5 million to 2.5 million alveoli. These microscopic (0.2 millimeter in diameter) chambers give magnified lung tissue the appearance of sponge cake (see Fig. E28-2, p. 579). The thin walls of the alveoli consist of flattened epithelial tissue (see Fig. 26-3a) and the walls of underlying capillaries, fused together by protein filaments. Alveoli resemble tiny bubbles and provide an enormous surface area for diffusion—about 75 square meters (800 square feet), 80 times the total skin surface area of an adult human. The alveoli, which cluster about the end of each bronchiole like a bunch of grapes, are entirely enmeshed in capillaries (see Fig. 28-7b). Because both the alveolar wall and the adjacent capillary walls are only one cell thick, the air is extremely close to the blood in the capillaries. The lung cells remain moist because a thin layer of watery fluid lines each alveolus. Gases dissolve in this fluid and diffuse through the alveolar and capillary membranes (Fig. 28-9).

After the blood circulates through the body tissues, it is pumped to the lungs by the heart. The incoming blood surrounding the alveoli is low in oxygen (because the body cells have used it up) and high in carbon dioxide (which is released by the cells; see Fig. 28-2). Carbon dioxide diffuses out of the blood, where its concentration is high, into the air in the alveoli, where its concentration is lower (see Fig. 28-9). Carbon dioxide concentration in the blood is especially high after heavy exertion, such as when you have sprinted up the stairs on your way to class. In the alveoli of the lungs, oxygen diffuses from the air, where its concentration is high, into the blood, where its concentration is low. Blood from the lungs, now oxygenated and purged of carbon dioxide, returns to the heart, which pumps it to the body tissues. In the tissues, oxygen diffuses into the cells because the concentration of oxygen is lower in the cells than in the blood.

Oxygen and Carbon Dioxide Are Transported Using Different Mechanisms

In the blood, oxygen binds loosely and reversibly with **hemoglobin**, a large, iron-containing protein in the red blood cells, as described in Chapter 27 (see Fig. 27-9). Each hemoglobin molecule can bind up to four oxygen molecules (eight oxygen atoms). Nearly all the oxygen carried by the blood is bound to hemoglobin. By removing oxygen from solution in the plasma,

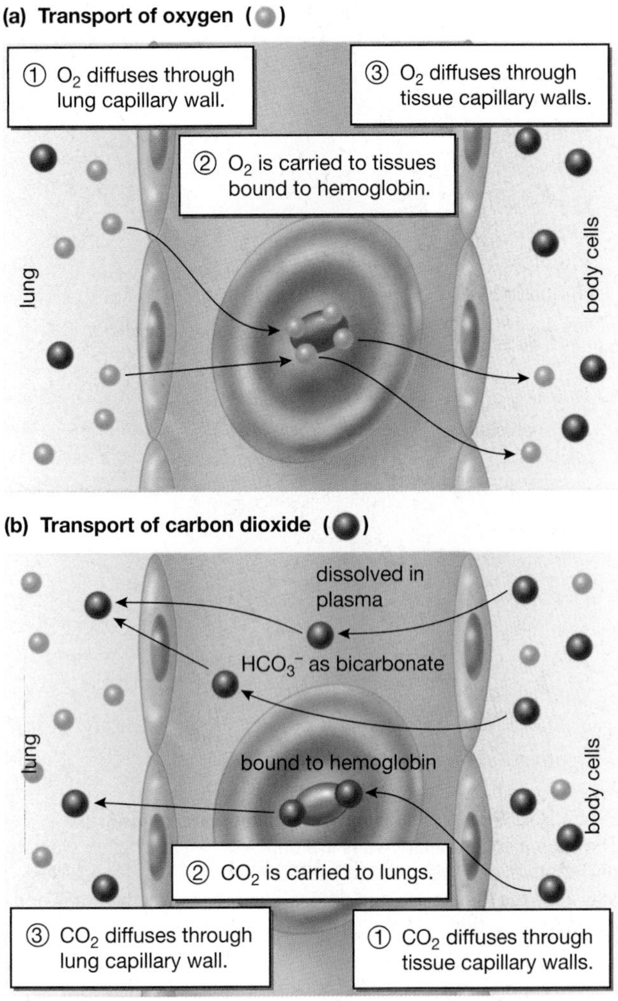
(a) Transport of oxygen
① O_2 diffuses through lung capillary wall.
③ O_2 diffuses through tissue capillary walls.
② O_2 is carried to tissues bound to hemoglobin.
lung
body cells
(b) Transport of carbon dioxide
dissolved in plasma
HCO_3^- as bicarbonate
bound to hemoglobin
lung
body cells
② CO_2 is carried to lungs.
③ CO_2 diffuses through lung capillary wall.
① CO_2 diffuses through tissue capillary walls.

Figure 28-10 The chemistry and mechanism of gas exchange

hemoglobin maintains a concentration gradient that favors the diffusion of oxygen from the air into the blood (Fig. 28-10a). Thanks to hemoglobin, our blood can carry about 70 times as much oxygen as it could if the oxygen were simply dissolved in the plasma.

As hemoglobin binds oxygen, the protein undergoes a slight change in shape, which alters its color. Oxygenated blood is a bright cherry-red; deoxygenated blood is dark maroon-red and appears bluish through the skin.

Carbon dioxide is transported in three different ways. In the presence of an enzyme found in red blood cells (*carbonic anhydrase*), about 70% of the CO_2 reacts with water to form bicarbonate ion (HCO_3^-), which then diffuses into the plasma (Fig. 28-10b). About 20% of the CO_2 binds to hemoglobin (which has released its O_2 to the tissues) for its return trip to the lungs; the remaining 10% stays dissolved in the plasma as CO_2. Both the production of bicarbonate ion and the binding of CO_2 to hemoglobin reduce the concentration of dissolved CO_2 in the blood and increase the gradient for CO_2 to flow from the body cells into the blood.

Carbon monoxide (CO) is a toxic gas produced by combustion such as occurs in engines and furnaces and cigarettes, when the fuel is not completely burned to form CO_2. At high levels, CO is deadly because it "fools" hemoglobin, binding in place of oxygen and more than 200 times as tenaciously. People will die from breathing air with as little as 0.1% CO. Hemoglobin containing CO is also bright red, but it is incapable of transporting oxygen. Most victims of asphyxiation have bluish lips and nail beds, because their hemoglobin is deoxygenated; the lips and nail beds of victims of carbon monoxide poisoning (such as could occur from breathing car exhaust in a closed space) are brighter red than normal.

Air Is Inhaled Actively and Exhaled Passively

Our ability to breathe depends on an airtight chest cavity—if the chest is punctured, the lungs may collapse. Outside the lungs, the chest cavity is bounded by neck muscles and connective tissue on top and by the dome-shaped, muscular **diaphragm** on the bottom. Within the wall of the chest, the *rib cage* surrounds and protects the lungs. Lining the rib cage and surrounding the lungs is a double layer of **pleural membranes**. These membranes contribute to the airtight seal between the lungs and the chest wall.

Breathing occurs in two stages: (1) **inhalation**, when air is actively drawn into the lungs, and (2) **exhalation**, when it is passively expelled from the lungs. Inhalation is accomplished by enlarging the chest cavity. To do so, the diaphragm muscles contract, drawing the diaphragm downward. The rib muscles also contract, lifting the ribs up and outward (Fig. 28-11). When the chest cavity is expanded, the lungs expand with it, because a vacuum holds them tightly against the inner wall of the chest. (If the chest is punctured and air leaks in between the pleural membranes and the lung, the lung will collapse.) As the lungs expand, their increased volume creates a partial vacuum that draws air into the lungs.

Exhalation occurs automatically when the muscles that cause inhalation are relaxed. The relaxed diaphragm domes upward, and the ribs fall down and inward, decreasing the size of the chest cavity and forcing air out of the lungs. More air can be forced out by contracting the abdominal muscles. After exhalation, the lungs still contain air. This air prevents the thin alveoli from collapsing and fills the space within the conducting portion of the respiratory system. A normal breath moves only about 500 milliliters (1 pint) of air into the respiratory system. Of this, only about 350 milliliters reaches the alveoli for gas exchange. Deeper breathing during exercise causes several times this volume to be exchanged.

Breathing Rate Is Controlled by the Respiratory Center of the Brain

Imagine having to think about every breath—fortunately, breathing occurs rhythmically and automatically without conscious thought. But, unlike the heart muscle, the

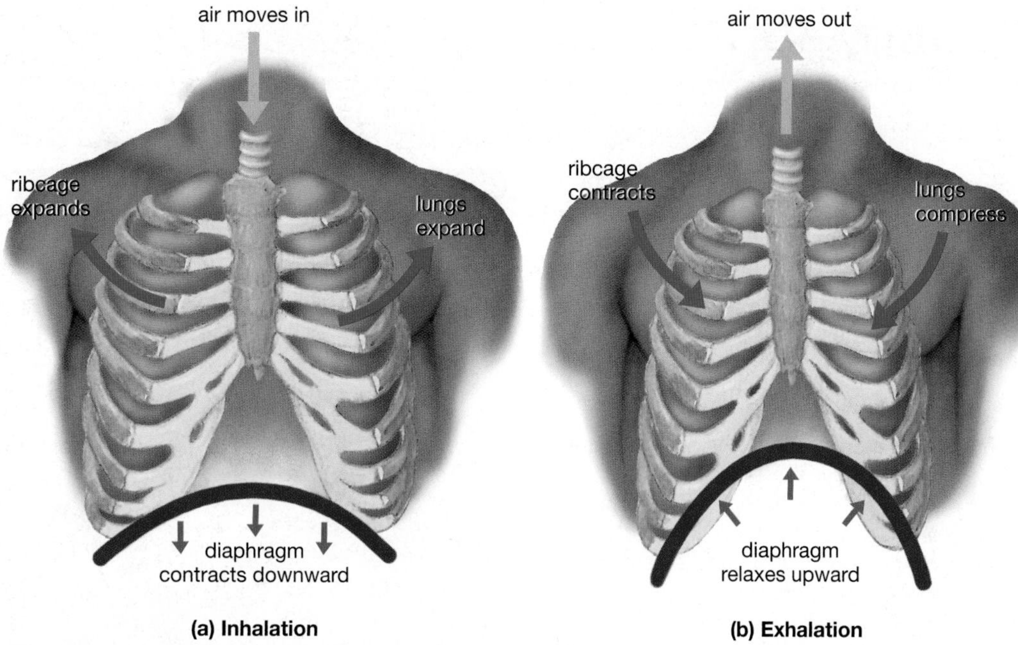

Figure 28-11 The mechanics of breathing
(a) During inhalation, rhythmic nerve impulses from the brain stimulate the diaphragm to contract (pulling it downward) and the muscles surrounding the ribs to contract (moving them up and outward). The result is an increase in the size of the chest cavity, causing air to rush in. ***(b)*** Relaxation of these muscles (exhalation) allows the diaphragm to dome upward and the rib cage to collapse, forcing air out of the lungs.

muscles used in breathing are not self-activating; each contraction is stimulated by impulses from nerve cells. These impulses originate in the **respiratory center**, which is located in the medulla of the brain, just above the spinal cord. Nerve cells in the respiratory center generate cyclic bursts of impulses that cause the alternating contraction and relaxation of the respiratory muscles.

The respiratory center receives input from several sources and adjusts breathing rate and volume to meet the body's changing needs. The respiratory rate is regulated to maintain a constant level of carbon dioxide in the blood, as monitored by carbon dioxide receptors in the medulla. For example, an elevated level of carbon dioxide caused by an increase in cellular activity, such as that experienced by your muscle cells when you run up stairs, signals a need for more oxygen, causing the receptors to stimulate an increase in the rate and depth of breathing. These receptors are extremely sensitive; an increase in carbon dioxide of only 0.3% can double the breathing rate.

The respiratory rate is much less sensitive to changes in oxygen concentration, because normal breathing supplies an overabundance of oxygen. But if blood oxygen levels fall drastically, receptors in the aorta and carotid arteries stimulate the respiratory center. Surprisingly, when you begin strenuous activity, such as running, an increase in breathing rate actually *precedes* any change in blood gas levels. Apparently, when higher brain centers activate muscles during heavy exercise, they simultaneously stimulate the respiratory center to increase breathing rate. Breathing activity is then "fine-tuned" by the receptors that monitor carbon dioxide concentrations.

REVISITED CASE STUDY REVISITED CASE STUDY REVISITED CASE

Lives Up in Smoke

For many smokers, the only time to quit is before they start, because nicotine is a powerfully addictive drug. The percentage of people using nicotine and become addicted is about the same as for cocaine and heroin. Having become regular smokers, only one out of five will successfully quit. Each day in the U.S., 3000 teenagers will begin smoking, and about 1000 of them will eventually die from a smoking-related illness—the legacy of a decision made while they were barely out of childhood.

A school for seventh and eighth graders asks you to develop a lesson plan for keeping their students from taking up smoking. The program you are designing could occupy two school days and could include a field trip. What subjects would you cover, and in what order? Ideally, what "props" would you use? What types of discussion would you lead? What type of field trip might you organize?

Health Watch
Smoking—A Life and Breath Decision

An estimated 430,000 people in the United States die of smoking-related diseases each year, and the cost of health care and loss of productivity due to smoking-related ailments costs the U.S. more than $100 billion annually. The American Cancer Society estimates that nearly 160,000 individuals die of lung cancer annually and that 87% of those deaths are due to smoking. The rest of the smoking-related deaths are from emphysema, chronic bronchitis, heart disease, stroke, and other forms of cancer.

Tobacco smoke has a dramatic impact on the human respiratory tract. As smoke is inhaled, toxic substances such as nicotine and sulfur dioxide paralyze the cilia that line the respiratory tract; a single cigarette can inactivate them for a full hour. Because these ciliary sweepers remove inhaled particles, smoking inhibits them just when they are most needed. The visible portion of cigarette smoke consists of billions of microscopic carbon particles. Adhering to these particles are about 200 different toxic substances, of which more than a dozen are known or probable *carcinogens* (cancer-causing substances). With the cilia out of action, the particles stick to the walls of the respiratory tract and enter the lungs.

Cigarette smoke also impairs the white blood cells that defend the respiratory tract by engulfing foreign particles and bacteria. Consequently, still more bacteria, dust, and smoke particles enter the lungs. In response to the irritation of cigarette smoke, the respiratory tract produces more mucus, another method of trapping foreign particles. But without the cilia to sweep it along, the mucus builds up and can obstruct the airways; the familiar "smoker's cough" is an attempt to expel it. Microscopic smoke particles accumulate in the alveoli over the years until the lungs of a heavy smoker are literally blackened. The longer the delicate tissues of the lungs are exposed to the carcinogens on the trapped particles, the greater the chance that cancer will develop (Fig. E28-1).

Chronic bronchitis and *emphysema* together kill about 96,000 people each year in the United States, and smoking is the leading contributing factor to each of these diseases. **Chronic bronchitis** is a persistent lung infection characterized by coughing, swelling of the lining of the respiratory tract, an increase in mucus production, and a decrease in the number and activity of cilia. The result: a decrease in air flow to the alveoli. **Emphysema** occurs when toxic substances in cigarette smoke cause the body to produce substances that lead to brittle and ruptured alveoli. The lung gradually loses its normal sponge-cake appearance (Fig. E28-2a) and more closely resembles blackened Swiss cheese (Fig. E28-2b). The loss of the alveoli, where gas exchange occurs, leads to oxygen deprivation of all body tissues. In an individual with emphysema, breathing becomes labored and grows increasingly worse until death.

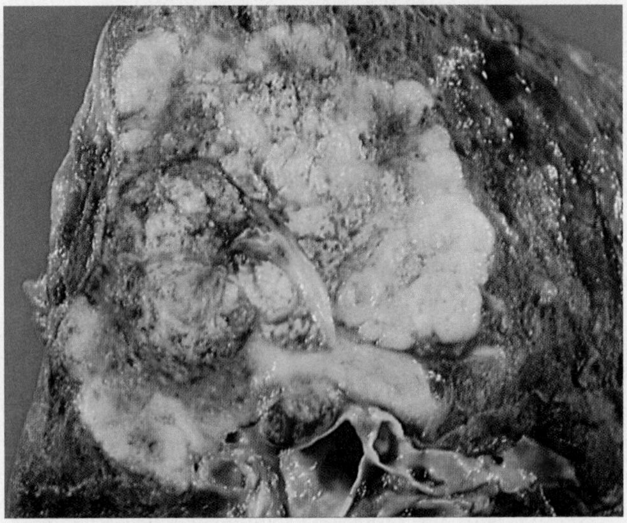

Figure E28-1 Smoking causes lung cancer
A tumor of lung cancer is visible as a large, pale mass; the lung tissue surrounding it is blackened by trapped smoke particles.

Carbon monoxide, present in high levels in cigarette smoke, binds tenaciously to red blood cells in place of oxygen. This binding reduces the blood's oxygen-carrying capacity and thereby increases the work the heart must do. Chronic bronchitis and emphysema compound this problem. Smoking also causes *atherosclerosis*, or thickening of the arterial walls by fatty deposits that can lead to heart attacks (see Chapter 27). As a result, smokers are 70% more likely than nonsmokers to die of heart disease. Smokers' wounds and broken bones take longer to heal, and their skin is more likely to wrinkle prematurely. The carbon monoxide in cigarette smoke may also con-

Summary of Key Concepts

1) Why Exchange Gases?

The respiratory system supports cellular respiration. Oxygen-rich air is inhaled and supplies oxygen to the blood, which carries it to cells throughout the body. Blood also picks up CO_2 (a product of cellular respiration) from body cells and transports it to the lungs, where it is released into the atmosphere.

2) What Are Some Evolutionary Adaptations for Gas Exchange?

Respiration makes possible the exchange of oxygen and carbon dioxide between the body and the environment by diffusion of these gases across a moist surface. In moist environments, animals with very small or flattened bodies may rely exclusively on diffusion through the body surface.

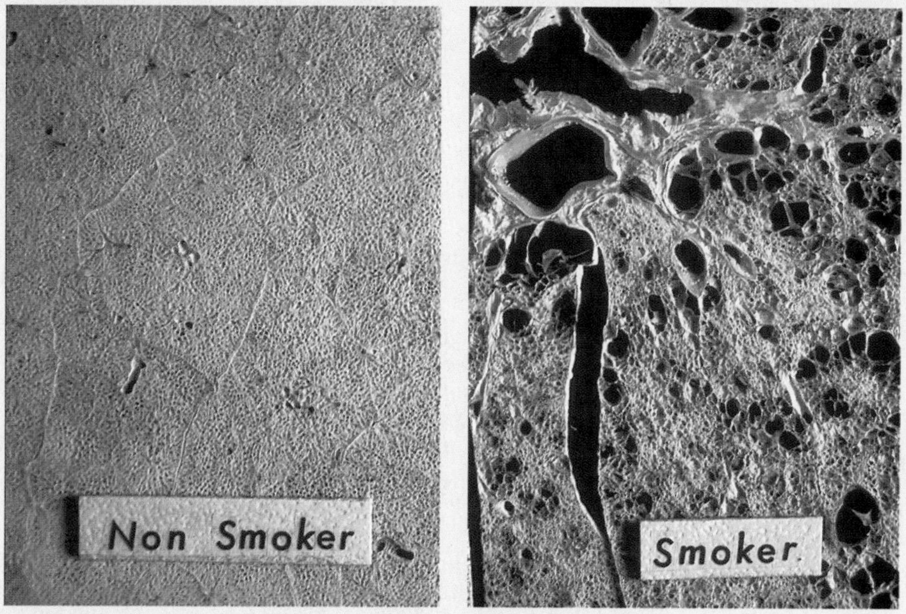

(a) (b)

Figure E28-2 Smoking causes emphysema
(a) Normal lung tissue from a nonsmoker, seen in cross section, has nearly invisible small openings, the alveoli, surrounded by healthy tissue. ***(b)*** The lung of a smoker suffering from emphysema is full of large holes, each caused by the rupture of hundreds of alveoli.

tribute to the reproductive problems experienced by pregnant women who smoke, including an increased incidence of infertility, miscarriage, lower birth weight of their babies, and, for their children, more learning and behavioral problems.

"Passive smoking," or breathing secondhand smoke, poses real health hazards for both children and adults. Researchers have concluded that children whose parents smoke are more likely to contract bronchitis, pneumonia, ear infections, coughs, and colds and to have decreased lung capacity. Children who grow up with smokers are more likely to develop asthma and allergies; for children with asthma, the number and severity of asthma attacks are increased by secondhand smoke. Among adults, studies have concluded that nonsmoking spouses of smokers face a 30% higher risk of both heart attack and lung cancer than do spouses of nonsmokers. A recent study links even relatively infrequent exposure to secondhand smoke with atherosclerosis. Government agencies report that secondhand smoke is responsible for an estimated 3000 lung-cancer deaths and 37,000 deaths from heart disease in nonsmokers in the U.S. each year.

For smokers who quit, healing begins immediately and the chances of heart attack, lung cancer, and numerous other smoking-related illnesses gradually diminish. Atherosclerosis may, however, continue to progress.

Animals with low metabolic demands and/or well-developed circulatory systems may also lack specialized respiratory structures. Larger, more active animals have evolved specialized respiratory systems. Animals in aquatic environments have evolved gills, such as those of fish and many amphibians. On land, moist respiratory surfaces must be protected internally. This need has selected for the evolution of tracheae in insects and lungs in terrestrial vertebrates.

The transfer of gases between respiratory systems and tissues occurs in a series of stages that alternate bulk flow with diffusion. Air or water moves by bulk flow past the respiratory surface, and gases in blood are carried by bulk flow. Gases move by diffusion across membranes between the respiratory system and the capillaries and between the capillaries and the tissues.

3) How Does the Human Respiratory System Work?

The human respiratory system consists of a conducting portion and a gas-exchange portion. Air passes first through the conducting portion, consisting of the nose and mouth, pharynx, larynx, trachea, bronchi, and bronchioles, and then into the gas-exchange portion, composed of alveoli (microscopic sacs). Blood within a dense capillary network surrounding the alveoli releases carbon dioxide and absorbs oxygen from the air.

Most of the oxygen in the blood is bound to hemoglobin within red blood cells. Hemoglobin binds oxygen at concentrations typical of those in alveolar capillaries and releases it at the lower oxygen concentrations of the body tissues. Carbon dioxide diffuses into the blood from the tissues and is transported as bicarbonate, bound to hemoglobin, or dissolved in blood plasma.

Breathing involves actively drawing air into the lungs by contracting the diaphragm and the rib muscles, which expand the chest cavity. Relaxing these muscles causes the chest cavity to collapse, expelling the air. Respiration is controlled by nerve impulses that originate in the medulla's respiratory center. The respiration rate is modified by receptors, such as those in the medulla that monitor carbon dioxide levels in the blood.

Key Terms

alveolus *p. 574*
bronchiole *p. 574*
bronchus *p. 574*
bulk flow *p. 571*
chronic bronchitis *p. 578*
conducting portion *p. 573*
diaphragm *p. 576*
emphysema *p. 578*
exhalation *p. 576*
gas-exchange portion *p. 573*
gill *p. 571*
hemoglobin *p. 575*
inhalation *p. 576*
larynx *p. 574*
lung *p. 572*
operculum *p. 571*
pharynx *p. 574*
pleural membrane *p. 576*
respiratory center *p. 577*
spiracle *p. 572*
trachea (in birds/mammals) *p. 574*
trachea (in insects) *p. 572*
vocal cord *p. 574*

Thinking Through the Concepts

Multiple Choice

1. *With which other system do specialized respiratory systems most closely interface in exchanging gases between the cells and the environment?*
   a. the skin
   b. the excretory system
   c. the circulatory system
   d. the muscular system
   e. the nervous system

2. *The gas-exchange portion of the human respiratory system is the*
   a. larynx b. trachea
   c. bronchi d. pharynx
   e. alveoli

3. *Which of the following types of animal use tracheae for respiration?*
   a. mollusks b. snails
   c. insects d. fish
   e. bookworms

4. *How is most of the oxygen transported in the blood?*
   a. dissolved in plasma
   b. bound to hemoglobin
   c. in the form of CO_2
   d. as bicarbonate
   e. dissolved in water

5. *Which of the following statements regarding cigarette smoking is true?*
   a. Cigarette smoke damages the alveoli.
   b. Cigarette smoke decreases the amount of oxygen in the blood.
   c. Cigarette smoke causes many types of cancer.
   d. Cigarette smoke can lead to heart damage.
   e. all of the above

6. *Which of the following pairs of respiratory adaptations and animals are NOT correct?*
   a. gills: fish b. parabronchi: birds
   c. lungs: mammals d. moist skin: snakes
   e. spiracles: insects

? Review Questions

1. Describe three arthropod respiratory systems and two vertebrate respiratory systems.
2. Trace the route taken by air in the vertebrate respiratory system, listing the structures through which it flows and the point at which gas exchange occurs.
3. Explain some characteristics of animals in moist environments that may supplement respiratory systems or make them unnecessary.
4. How are human respiratory movements initiated? How are they modified, and why are these controls adaptive?
5. What events occur during human inhalation? Exhalation? Which of these is always an active process?
6. Trace the pathway of an oxygen molecule in the human body, starting with the nose and ending with a body cell.
7. Describe the effects of smoking on the human respiratory system.
8. Explain how bulk flow and diffusion interact to promote gas exchange between air and blood and between blood and tissues.
9. Compare carbon dioxide and oxygen transport in the blood. Include the source and destination of each.
10. Explain how the structure and arrangement of alveoli make them well suited for their role in gas exchange.

Applying the Concepts

1. Heart–lung transplants are performed in some cases, but donors are scarce. On the basis of your knowledge of the respiratory and circulatory systems and of lifestyle factors that might damage them, what criteria would you use in selecting a recipient for such a transplant?
2. Nicotine is a drug in tobacco that is responsible for several of the effects that smokers crave. Discuss the advantages and disadvantages of low-nicotine cigarettes.
3. Discuss why a brief exposure to carbon monoxide is much more dangerous than a brief exposure to carbon dioxide.
4. Describe several adaptations that might evolve to help members of a species of mammal respire better if the population began living continuously for many generations at very high altitudes.
5. Mary, a strong-willed 3-year-old, threatens to hold her breath until she dies if she doesn't get her way. Can she carry out her threat? Explain.

For More Information

Gibbs, W. "Breath of Fresh Liquid." *Scientific American,* February 1999. Perfusing the lungs with oxygen-carrying fluid containing perfluorocarbons may help patients with lung disease.

Harding, C. "Going to Extremes." *National Wildlife*, August/September 1993. Describes adaptations that allow diving animals to plunge to enormous depths without running out of oxygen.

Houston, C. "Mountain Sickness." *Scientific American*, October 1992. The mechanisms of potentially fatal altitude sickness are explained.

Seppa, N. "Secondary Smoke Carries High Price." *Science News*, January 17, 1998. New research suggests that smoking causes atherosclerosis, that the damage continues after a smoker quits, and that exposure to secondhand smoke significantly increases the buildup of plaque in the carotid artery of nonsmokers.

Answers to Multiple-Choice Questions

1. c 2. e 3. c 4. b 5. e 6. d

MEDIATUTOR Respiration

CD Activities

Activity 28.1: Human Respiratory Anatomy

Estimated time: 5 minutes

Human respiratory anatomy represents a highly efficient gas-exchange system. In this tutorial, you will explore the various structures of the human respiratory system and the functions of each of those structures.

Activity 28.2: Gas Exchange

Estimated time: 5 minutes

All animals have evolved specialized respiratory surfaces that work together with the circulatory system so that gases can be exchanged between the body's cells and the environment. In this tutorial, you will explore how the human respiratory system accomplishes this important physiological task.

Start the MediaTutor Student CD-ROM and enter the activity number in the Quick Search box to be taken directly to that activity.

Web Investigations

Case Study: Lives Up in Smoke

Estimated time: 10 minutes

Everyone knows that smoking causes cancer, heart disease, emphysema, and other respiratory ailments. Unfortunately, secondhand smoke remains controversial. This exercise surveys the scientific evidence for the harmful effects of environmental tobacco smoke.

Go to http://www.prenhall.com/audesirk6, the Audesirk Companion Web site. Select Chapter 28 and the Web Investigation to begin.

More than 20% of us are obese. As obesity rates climb, researchers seek genetic causes and chemical cures.

29 Nutrition and Digestion

AT A GLANCE

Case Study: Fat in the Family?

1) **What Nutrients Do Animals Need?**

The Primary Sources of Energy Are Carbohydrates and Fats

Lipids Include Fats, Phospholipids, and Cholesterol

Carbohydrates, Including Sugars and Starches, Are a Source of Quick Energy

Proteins, Composed of Amino Acids, Perform a Wide Range of Functions Within the Body

Minerals Are Elements and Small Inorganic Molecules Required by the Body

Vitamins Are Required in Small Amounts and Play Many Roles in Metabolism

Nutritional Guidelines Help People Obtain a Balanced Diet

Are You Too Heavy?

2) **How Is Digestion Accomplished?**

An Overview of Digestion

Digestive Systems Are Adapted to the Lifestyle of Each Animal

3) **How Do Humans and Other Mammals Digest Food?**

The Mechanical and Chemical Breakdown of Food Begins in the Mouth

The Esophagus Conducts Food to the Stomach

Most Digestion Occurs in the Small Intestine

Most Absorption Occurs in the Small Intestine

Water Is Absorbed and Feces Are Formed in the Large Intestine

Digestion Is Controlled by the Nervous System and Hormones

Case Study Revisited: Fat in the Family?

CASE STUDY

Fat in the Family?

"I turn sideways in the mirror and gasp with horror," says Dave Barnett of Dallas. "I stepped on the bathroom scales, and they broke. I couldn't see them over my gut anyway." Obesity is a growing epidemic in the United States. More than half of us are overweight; more than 20% are classified as *obese* (20% above ideal weight). The percentage of overweight people has doubled in the past two decades, and many have given up. "I would rather die than eat fat-free," says Barnett. Exercise? "Forget that. Who has time for it?" Genetic research may fuel this sense of resignation. When an auto parts store employee was fired for "poor job performance" (in spite of an exemplary record) after his weight climbed to 400 pounds, he sued. Upon hearing testimony from an expert medical witness that a person's weight is controlled approximately 80% by genes, a jury agreed that he was a victim of "weight discrimination" and awarded him more than $1 million. The relative contributions of genetics and environment to the weight of any individual is an open question. However, researchers have found at least 130 different genes that in some way contribute to weight. The number of different combinations of these genes that people can inherit is staggering, suggesting thousands of different causes for obesity. "I look at the family pictures of my ancestors, and they are a bunch of pudgy people," Barnett muses. "I'm not going to fight it, because I can't win." People differ genetically in how their bodies respond to exercise, how full they feel after eating, how much fat they store, and how many calories their bodies consume while lying still. Dieters face the dilemma that their bodies react to dieting as if they were facing starvation—their energy consumption plummets. Why do people and other animals have such a strong tendency to store fat? Should overweight people resign themselves to their "genetically fat fate"? ■

1 What Nutrients Do Animals Need?

You might be someone who thinks that if food is nutritious, then it probably doesn't taste very good! In fact, all foods, whether they fall in the "broccoli" or the "chocolate" category, contain nutrients that you need to survive. **Nutrition** is the process of acquiring and processing nutrients into a usable form. Animal **nutrients** are substances that must be supplied in the diet. Nutrients fall into five major categories: (1) lipids, (2) carbohydrates, (3) proteins, (4) minerals, and (5) vitamins. These substances provide the body with its basic needs:

- energy to fuel cellular metabolism and activities, provided by lipids, carbohydrates, and proteins;
- the chemical building blocks, such as amino acids, to construct complex molecules unique to each animal; and
- minerals and vitamins that participate in a variety of metabolic reactions.

The Primary Sources of Energy Are Carbohydrates and Fats

If your cells did not continuously expend energy, they would begin to die within seconds. Cells constantly use energy to maintain their incredible complexity and perform specific functions. Three nutrients provide energy for animals: fats, carbohydrates, and proteins. In a "typical" U.S. diet, fats provide about 38% of the energy; carbohydrates, about 46%, and protein, about 16%. These molecules are broken down during cellular respiration, and the energy derived from them is used to produce adenosine triphosphate (ATP) (see Chapter 8).

The energy in nutrients is measured in calories. A **calorie** is the amount of energy required to raise the temperature of 1 gram of water by 1 degree Celsius. The calorie content of foods is measured in units of 1000 calories (*kilocalories*), also known as **Calories** (with a capital *C*). The "average" human body at rest burns about 1550 Calories per day (somewhat more for males and less for females). Exercise significantly boosts caloric requirements: Well-trained athletes can temporarily raise their calorie consumption from a resting rate of about 1 Calorie per minute to nearly 20 Calories per minute during vigorous exercise (Table 29-1).

Lipids Include Fats, Phospholipids, and Cholesterol

Although in our overweight society they are sometimes viewed as the enemy, fats and other lipids are essential nutrients. Lipids are a diverse group of molecules that generally contain long chains of carbon atoms and are insoluble in water. The principal types of lipids are *fats*, or *triglycerides; phospholipids;* and *cholesterol* (see Chapter 3). Fats are used primarily as a source of energy. Phospholipids are important components of cellular membranes and also provide the insulating covering of neurons (nerve cells). Cholesterol is used in the synthesis of cellular membranes, sex hormones, and bile (which aids in fat breakdown). Some animal species can synthesize all the specialized lipids they need. Others must acquire specific types of lipid building blocks, called **essential fatty acids**, from their food. For example, humans are unable to synthesize linoleic acid, which is required for the synthesis of certain phospholipids, so we need to obtain this essential fatty acid from our diet.

Humans and most other animals store energy primarily as fat. When an animal's diet provides more energy than is expended through metabolic activities, most of the excess carbohydrate, fat, or protein is converted to fat for storage. About 3600 Calories are stored in each pound of fat. Fats have two major advantages as

Table 29-1 Approximate Energy Consumed by a 150-Pound Person for Different Activities

		Time to "Work Off"			
		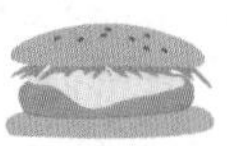			
Activity	**Calories/ hr**	**500 Calories Cheeseburger**	**300 Calories Ice Cream Cone**	**70 Calories Apple**	**40 Calories 1 Cup Broccoli**
Running (6 mph)	700	43 min	26 min	6 min	3 min
Cross-country skiing (moderate)	560	54 min	32 min	7.5 min	4 min
Roller skating	490	1 hr 1 min	37 min	8.6 min	5 min
Bicycling (11 mph)	420	1 hr 11 min	43 min	10 min	6 min
Walking (3 mph)	250	2 hr	1 hr 12 min	17 min	10 min
Frisbee® playing	210	2 hr 23 min	1 hr 26 min	20 min	11 min
Studying	100	5 hr	3 hr	42 min	24 min

energy-storage molecules. First, they are the most concentrated energy source, containing more than twice the energy per unit weight of either carbohydrates or protein (about 9 Calories per gram for fats compared with about 4 per gram for proteins and carbohydrates). Second, lipids are *hydrophobic*—that is, they do not mix with water. Fat deposits, therefore, do not cause any extra accumulation of water in the body. For both these reasons, fats store more calories with less weight than do other molecules. Minimizing weight allows an animal to move faster (important for escaping predators and hunting prey) and to use less energy for movement (important when food supplies are limited). Since people evolved under the same food constraints as other animals, we have a strong tendency to eat when food is available, often in excess of our needs because we may need the energy later. We also have evolved to store food as efficiently as possible—in the form of fat. Only recently have certain societies had access to unlimited food; in this context, our natural tendencies can become liabilities, and we become obese.

In mammals, which maintain an elevated body temperature, fat deposits do double duty by providing insulation as well as storing energy. Fat is typically stored in a layer beneath the skin where it insulates the body, since fat conducts heat at only one-third the rate of other body tissues. Mammals who live near the North or South Pole or in cold ocean waters are particularly dependent on this insulating layer, which reduces the amount of energy they must expend to keep warm (Fig. 29-1).

Carbohydrates, Including Sugars and Starches, Are a Source of Quick Energy

You may have heard of "carbo-loading," such as gorging on pasta, for example. Athletes sometimes do this in an attempt to build up reserves of quick energy for a sporting event. Carbohydrates consist of monosaccharide and disaccharide sugars as well as longer chains of sugars called *polysaccharides* (see Chapter 3). Polysaccharides include starches, the principal energy-storage material of plants; *glycogen,* a short-term energy-storage molecule in animals; and *cellulose,* the major structural component of plant cell walls. During digestion, carbohydrates are broken down into sugars and absorbed. For practical purposes, body cells obtain their energy from a single sugar: glucose. Glucose can be derived from fats, amino acids, and the carbohydrates consumed in the diet.

Animals, including humans, store the carbohydrate **glycogen** (a large, highly branched chain of glucose molecules) in the liver and muscles. Although humans can potentially store hundreds of pounds of fat, we store less than half a pound of glycogen. During exercise, such as running, the body draws on this store of glycogen as a source of quick energy. When the activity is prolonged, as in the case of a marathon runner, the stored glycogen can be totally depleted. The expression "hitting the wall" describes the extreme fatigue that long-distance runners may experience after exhausting their glycogen supply.

Figure 29-1 Fat provides insulation
These walruses can withstand the icy waters of the polar seas because they are insulated with a thick layer of fat beneath the skin.

Proteins, Composed of Amino Acids, Perform a Wide Range of Functions Within the Body

Each day, your body requires 20 to 30 grams of protein, and any excess proteins are broken down into their amino acids and used for energy or stored as fat. Protein breakdown produces the waste product **urea**, which is filtered from the blood by the kidneys. Specialized diets in which protein is the major energy source place extra stress on the kidneys.

Dietary protein serves primarily a source of amino acids to make new molecules (see also Chapter 3). Amino acids are used to synthesize certain hormones, other amino acids, some neurotransmitters (chemicals used in communication between neurons), and new proteins. These proteins have diverse roles in the body, acting as enzymes, receptors on cell membranes, oxygen transport molecules (hemoglobin), structural components (hair and nails), hormones, antibodies, and muscle proteins.

In the digestive tract, dietary protein is broken down into its amino acid subunits. Then, in the body cells, the amino acids are linked in specific sequences to form new proteins. People can synthesize (from other amino acids) 11 of the 20 different amino acids used in proteins. There are 9 others that we cannot synthesize, called **essential amino acids**. Essential amino acids must be supplied by the diet in foods such as meat, milk, eggs, corn, beans, and soybeans. Because many plant proteins are deficient in some of the essential amino acids, individuals on a vegetarian diet must include a variety of plants whose proteins together will provide all 9 of the essential amino acids, or they risk protein deficiency (Table 29-2).

Table 29-2 The Essential Amino Acids and Their Plant Sources

Beans and Other Legumes	Both Legumes and Grains	Corn and Other Grains
Isoleucine	Valine	Tryptophan
Lysine	Histidine	Methionine
	Threonine	
	Phenylalanine	
	Leucine	

Minerals Are Elements and Small Inorganic Molecules Required by the Body

You may have seen a display of **minerals** in a museum—beautiful, brightly colored crystals. Many such substances, which include various elements and inorganic molecules, play crucial roles in animal nutrition (Table 29-3). Minerals must be obtained through the diet, either from food or dissolved in drinking water, because the body cannot manufacture them. Required minerals include calcium, magnesium, and phosphorus, which are major constituents of bones and teeth. Others, such as sodium and potassium, are essential for muscle contraction and the conduction of nerve impulses. Iron is used in the production of hemoglobin, and iodine is found in hormones produced by the thyroid gland. In addition, trace amounts of several other minerals, including zinc, copper, and selenium, are required, typically as parts of enzymes.

Vitamins Are Required in Small Amounts and Play Many Roles in Metabolism

"Take your vitamins!" is a familiar refrain in many households with children. Why take vitamins? **Vitamins** are a diverse group of organic compounds that animals require in small amounts. In general, the body cannot synthesize vitamins (or cannot do so in adequate amounts), so they must be obtained from food. Our modern diet is now so different from the natural diet on which we evolved that many people find vitamin supplements a prudent way to compensate for possible suboptimal vitamin levels. For example, our skin can manufacture some vitamin D when it is exposed to sunlight, but most of us spend so much time indoors that we do not synthesize enough and must augment it from dietary sources or supplements. Some

Table 29-3 Major Mineral Sources and Functions for Humans

Mineral	Dietary Sources	Major Functions in Body	Deficiency Symptoms
Calcium	Milk, cheese, green vegetables, legumes	Bone and tooth formation Blood clotting Nerve impulse transmission	Stunted growth Rickets, osteoporosis Convulsions
Phosphorus	Milk, cheese, meat, poultry, grains	Bone and tooth formation Acid–base balance	Weakness Demineralization of bone Loss of calcium
Potassium	Meats, milk, fruits	Acid–base balance Body water balance Nerve function	Muscular weakness Paralysis
Chlorine	Table salt	Formation of gastric juice Acid–base balance	Muscle cramps Apathy Reduced appetite
Sodium	Table salt	Acid–base balance Body water balance Nerve function	Muscle cramps Apathy Reduced appetite
Magnesium	Whole grains, green leafy vegetables	Activation of enzymes in protein synthesis	Growth failure Behavioral disturbances Weakness, spasms
Iron	Eggs, meats, legumes, whole grains, green vegetables	Constituent of hemoglobin and enzymes involved in energy metabolism	Iron-deficiency anemia (weakness, reduced resistance to infection)
Fluorine	Fluoridated water, tea, seafood	Maintenance of teeth and probably bone structure	High frequency of tooth decay
Zinc	Widely distributed in foods	Constituent of enzymes involved in digestion	Growth failure Small sex glands
Iodine	Seafish and shellfish, dairy products, many vegetables, iodized salt	Constituent of thyroid hormones	Goiter

vitamins, such as C and E, are *antioxidants* in addition to their other functions. As our cells generate and use energy, damaging molecules called free radicals are produced. These molecules react with and can damage DNA, in some cases causing cancer. Free radicals can also promote atherosclerosis, and, over a lifetime, they contribute to the myriad losses in function associated with aging. Antioxidants combine with free radicals to limit their damaging effects. The vitamins considered essential in human nutrition are listed in Table 29-4.

Human vitamins are often grouped into two categories, water soluble and fat soluble. *Water-soluble vitamins* include vitamin C and the 11 compounds that make up the B-vitamin complex. These substances dissolve in the water of the blood plasma and are excreted by the kidneys. They are not stored in the body in any appreciable amounts. Water-soluble vitamins generally work in conjunction with enzymes to promote chemical reactions that supply energy or synthesize materials. Because each vitamin participates in several metabolic

Table 29-4 Vitamin Sources and Functions for Humans

Vitamin	Dietary Sources	Functions in Body	Deficiency Symptoms
Water soluble			
Vitamin B_1 (thiamin)	Milk, meat, bread	Coenzyme in metabolic reactions	Beriberi (muscle weakness, peripheral nerve changes, edema, heart failure)
Vitamin B_2 (riboflavin)	Widely distributed in foods	Constituent of coenzymes in energy metabolism	Reddened lips, cracks at corner of mouth, lesions of eye
Niacin	Liver, lean meats, grains, legumes	Constituent of two coenzymes in energy metabolism	Pellagra (skin and gastrointestinal lesions; nervous, mental disorders)
Vitamin B_6 (pyridoxine)	Meats, vegetables, whole-grain cereals	Coenzyme in amino acid metabolism	Irritability, convulsions, muscular twitching, dermatitis, kidney stones
Pantothenic acid	Milk, meat	Constituent of coenzyme A, with a role in energy metabolism	Fatigue, sleep disturbances, impaired coordination
Folic acid	Legumes, green vegetables, whole wheat	Coenzyme involved in nucleic and amino acid metabolism	Anemia, gastrointestinal disturbances, diarrhea, retarded growth, birth defects
Vitamin B_{12}	Meats, eggs, dairy products	Coenzyme in nucleic acid metabolism	Pernicious anemia, neurological disorders
Biotin	Legumes, vegetables, meats	Coenzymes required for fat synthesis, amino acid metabolism, and glycogen formation	Fatigue, depression, nausea, dermatitis, muscular pains
Choline	Egg yolk, liver, grains, legumes	Constituent of phospholipids, precursor of the neurotransmitter acetylcholine	None reported in humans
Vitamin C (ascorbic acid)	Citrus fruits, tomatoes, green peppers	Maintenance of cartilage, bone, and dentin (hard tissue of teeth); collagen synthesis	Scurvy (degeneration of skin, teeth, blood vessels; epithelial hemorrhages)
Fat soluble			
Vitamin A (retinol)	Beta-carotene in green, yellow, and red vegetables Retinol added to dairy products	Constituent of visual pigment Maintenance of epithelial tissues	Night blindness, permanent blindness
Vitamin D	Cod-liver oil, eggs, dairy products	Promotes bone growth and mineralization Increases calcium absorption	Rickets (bone deformities) in children; skeletal deterioration
Vitamin E (tocopherol)	Seeds, green leafy vegetables, margarines, shortenings	Antioxidant, prevents cellular damage	Possibly anemia
Vitamin K	Green leafy vegetables Product of intestinal bacteria	Important in blood clotting	Bleeding, internal hemorrhages

processes, a deficiency of a single vitamin can have wide-ranging effects (see Table 29-4).

The *fat-soluble* vitamins A, D, E, and K have even more varied roles. Vitamin K, for example, helps regulate blood clotting, and vitamin A is used to produce visual pigment (the light-capturing molecule in the retina of the eye). Fat-soluble vitamins can be stored in body fat and may accumulate in the body over time. For this reason, some fat-soluble vitamins may be toxic if consumed in excessively high doses.

Nutritional Guidelines Help People Obtain a Balanced Diet

Most people in the U.S. are fortunate to live amidst an abundance of food. However, the overwhelming diversity of foods available in a typical U.S. supermarket and the easy availability of "fast food" can lead to poor nutritional choices. To help people make informed choices, the U.S. government has developed recommendations and goals for the average U.S. citizen, which are summarized in Table 29-5.

Further help is provided by the Food Guide Pyramid, designed by the U.S. Department of Agriculture, which illustrates the relative abundances of different food groups in an optimal diet (Fig. 29-2). Another source of information is the nutritional labeling now required on commercially packaged foods. These labels provide complete information about calorie, fiber, fat, sugar, and vitamin content (Fig. 29-3). Some fast-food chains also make fliers available that list nutritional information about their products.

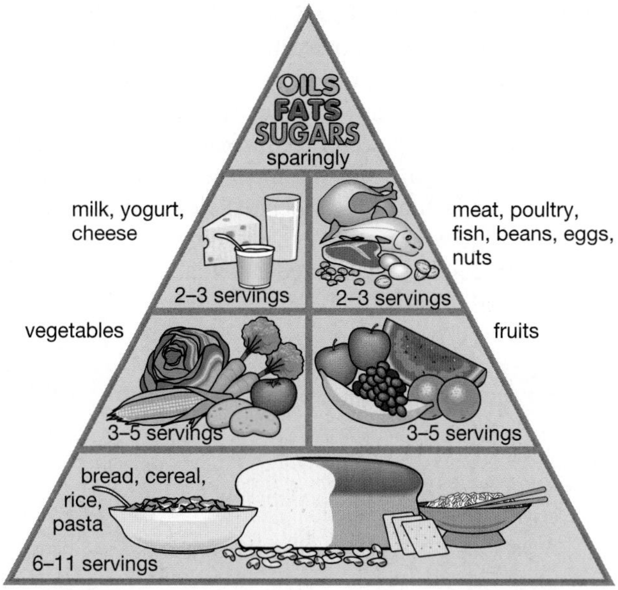

Figure 29-2 The Food Guide Pyramid
Recommended by the U.S. Department of Agriculture, this chart shows suggested daily servings.

Table 29-5 Dietary Changes Advocated by the U.S. Government

Dietary Component	Percentage of Total Daily Energy Intake: Average U.S. Diet	Percentage of Total Daily Energy Intake: Dietary Goals
Carbohydrates	46	58
Lipids	38	30
Proteins	16	12

Summary of Recommendations:

1. Increase consumption of fruits, vegetables, and whole grains.
2. Decrease consumption of refined sugars.
3. Decrease consumption of fats; replace saturated with unsaturated fats.
4. Decrease consumption of animal fats by selecting lean meats, poultry, and fish.
5. Decrease consumption of high-cholesterol foods, such as butter and eggs.
6. Decrease consumption of salt and foods high in salt content.
7. Decrease caloric intake to maintain desirable weight.

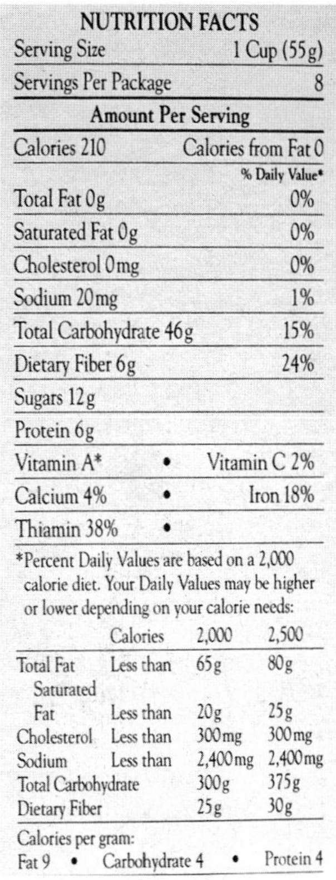
NUTRITION FACTS

Serving Size 1 Cup (55g)
Servings Per Package 8

Amount Per Serving

Calories 210 Calories from Fat 0

% Daily Value*

Total Fat 0g	0%
Saturated Fat 0g	0%
Cholesterol 0mg	0%
Sodium 20mg	1%
Total Carbohydrate 46g	15%
Dietary Fiber 6g	24%
Sugars 12g	
Protein 6g	

Vitamin A* • Vitamin C 2%
Calcium 4% • Iron 18%
Thiamin 38% •

*Percent Daily Values are based on a 2,000 calorie diet. Your Daily Values may be higher or lower depending on your calorie needs:

	Calories	2,000	2,500
Total Fat	Less than	65g	80g
Saturated Fat	Less than	20g	25g
Cholesterol	Less than	300mg	300mg
Sodium	Less than	2,400mg	2,400mg
Total Carbohydrate		300g	375g
Dietary Fiber		25g	30g

Calories per gram:
Fat 9 • Carbohydrate 4 • Protein 4

Figure 29-3 Complete food labels
The U.S. government now requires more complete nutritional labeling of foods, such as in this sample. The weight in grams of various nutrients (such as fat, cholesterol, and sodium) is converted into a percentage of the recommended daily allotment, assuming a 2000-Calorie diet. At the bottom of the label, the total recommended number of grams of these nutrients is listed for a 2000- and a 2500-Calorie diet.

Health Watch
Eating Disorders—Betrayal of the Body

Natural selection has provided animals with strong drives to eat when nutrients are needed (and even when nutrients are not needed, if good food is available). In humans, these natural impulses, so crucial to our health and well-being, can go terribly awry. In recent decades we have seen an increase in the occurrence both of overeating and of *eating disorders*, ailments characterized by the disruption of normal eating behavior.

Eating disorders include two particularly debilitating illnesses, *anorexia nervosa* and *bulimia nervosa*. People with anorexia nervosa experience an intense fear of gaining weight, and they achieve extreme weight loss—often 30% or more below their normal weight—by eating very little food. Victims of this disorder also engage in self-induced vomiting, laxative intake, and excessive exercise. The consequences are disastrous. Anorexics become emaciated, losing both fat and muscle. This emaciation can in turn disrupt cardiac, digestive, endocrine, and reproductive functions. As many as 18% of anorexics die as a result of their disorder. Victims of bulimia nervosa, who maintain a normal weight, engage in binge eating, rapidly consuming huge amounts of food. After bingeing, bulimic individuals typically induce vomiting and may also take laxatives. The repeated vomiting damages the digestive tract and upsets the normal balance of salts in the blood, which can lead to heart disorders.

More than 90% of diagnosed eating disorders occur in females (Fig. E29-1). Among Caucasian females, the disorder usually starts between 12 and 20 years of age, and the incidence of anorexia is about 1%. Scientists don't understand the causes of eating disorders and have not identified a clear genetic link, so suspicion has fallen on environmental factors, especially on societal pressures to be thin, coupled with individual personality traits, such as a high need for achievement and acceptance. Anthropologist Anne Becker of Harvard University studied girls in the Fiji Islands, where 80% of women are overweight by U.S. standards. Shortly after satellite TV introduced U.S. shows glorifying slim actresses, the incidence of vomiting to control weight increased by a factor of five among teenaged Figian girls. Dr. Becker's findings support the concept that social pressures play a major role in eating disorders.

Unfortunately, it is difficult to treat eating disorders. Victims are usually hospitalized to restore nutritional health and counseled, but they often fail to recover completely. Antidepressant drugs are helpful in some cases but are more likely to be effective for bulimia nervosa than for the more serious anorexia nervosa. Hope may be on the horizon, however, because of research into the hormones that control appetite. For example, researchers have discovered a class of hormones known as *orexins* (from the Greek word *orexis*, meaning "appetite"), which, when injected into mice, bind to receptors in the brain and cause food consumption to increase dramatically. Such new discoveries raise hopes that as our understanding of the physiological control of appetite grows, we may yet be able to devise chemical treatments for eating disorders.

Figure E29-1 Eating disorders
The late Princess Diana made headlines when she announced that she had been suffering from bulimia.

Are You Too Heavy?

A simple way to determine whether your weight is likely to pose a health risk is by calculating your **body mass index** (BMI). The BMI takes into account your weight and height to arrive at an estimate of body fat. The original formula is: weight (in kg)/ height^2 (in meters), but you can calculate the same number by multiplying your weight in pounds by 703, then dividing by your height in inches, squared. A BMI between 20 and 24 is considered healthy. Between 25 and 30, your BMI tells you you are overweight, and you are obese if your BMI exceeds 30. Obese people are at greater risk for heart disease, diabetes, stroke, and some forms of cancer.

2 How Is Digestion Accomplished?

An Overview of Digestion

After a meal, you may hear your stomach gurgling and churning; these noises are generated by one of the several phases of digestion. **Digestion** refers to physically grinding up, then chemically breaking down food. The **digestive systems** of animals take in and then digest the complex molecules of their food into simpler molecules that can be absorbed. Material that cannot be used or broken down is then expelled from the body.

Animals eat the bodies of other organisms, bodies that may resist becoming food. The plant body, for example, armors each cell with a wall of indigestible cellulose. Animal bodies may be covered with equally indigestible fur, scales, or feathers. In addition, the complex lipids, proteins, and carbohydrates in food do not occur in a form that can be used directly. The nutrients must be broken down before they can be absorbed and distributed to the cells of the animal that has consumed them, where they recombine in unique ways. Different types of animals meet the challenge of acquiring nutrients with various types of digestive tracts, each finely tuned to a unique diet and lifestyle. Amid this diversity, however, are certain tasks that all digestive systems must accomplish:

1. *Ingestion.* The food must be brought into the digestive tract through an opening, usually called a **mouth**.
2. *Mechanical breakdown.* The food must be physically broken down into smaller pieces. This is accomplished by gizzards or teeth as well as by the churning action of the digestive cavity itself. The particles produced by mechanical breakdown provide a large surface area for attack by digestive enzymes.
3. *Chemical breakdown.* The particles of food must be exposed to digestive enzymes and other digestive fluids that break down large molecules into smaller subunits.
4. *Absorption.* The small molecules must be transported out of the digestive cavity and into cells.
5. *Elimination.* Indigestible materials must be expelled from the body.

In the following section, we will explore briefly some of the diverse mechanisms by which animal digestive systems accomplish these functions.

Digestive Systems Are Adapted to the Lifestyle of Each Animal

Looking at things very simplistically, animals are machines for converting food into more animals. Natural selection has favored adaptations that accomplish this conversion most efficiently. This evolutionary drive has fostered a range of animal behaviors and digestive systems that exploit every conceivable food source.

Digestion Within Single Cells Occurs in the Sponges

Simple digestive systems can be as efficient as more complex ones if a comparatively small amount of energy is used to acquire and digest food. Sponges, for example, are sedentary feeders with no specialized digestive systems; they prosper with a system in which digestion occurs *within* individual cells (Fig. 29-4).

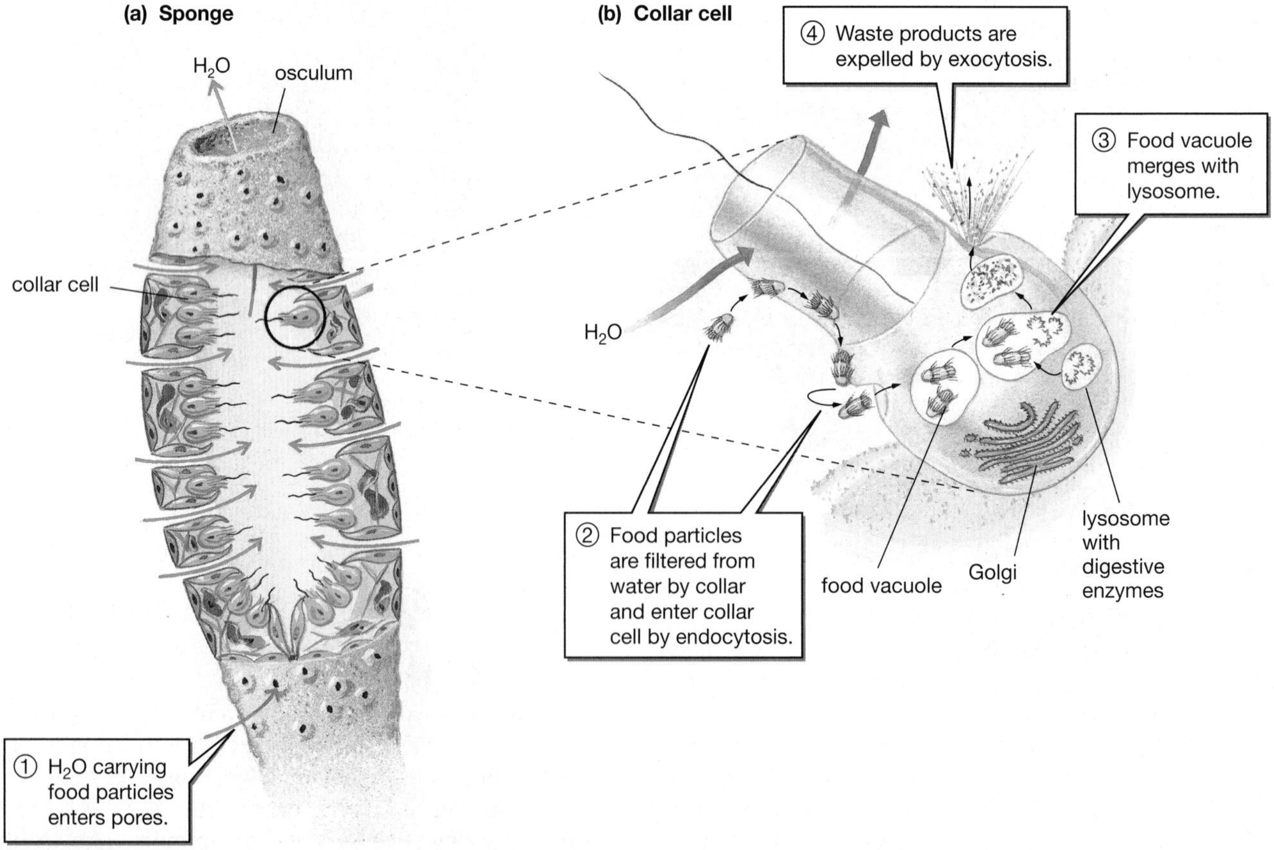
(a) Sponge
H_2O
osculum
collar cell
① H_2O carrying food particles enters pores.
(b) Collar cell
④ Waste products are expelled by exocytosis.
③ Food vacuole merges with lysosome.
H_2O
② Food particles are filtered from water by collar and enter collar cell by endocytosis.
food vacuole
Golgi
lysosome with digestive enzymes

Figure 29-4 Intracellular digestion in a sponge
(a) Internal anatomy of a simple sponge, showing the direction of water flow and the location of the collar cells. ***(b)*** Enlargement of a single collar cell showing digestion of single-celled organisms, which are filtered from the water, trapped on the outside of the collar, engulfed, and digested.

Such **intracellular digestion** occurs after a cell has engulfed microscopic food particles. Once engulfed by a cell, the food is enclosed in a **food vacuole**, a space surrounded by a membrane that serves as a temporary stomach. The vacuole fuses with small packets of digestive enzymes called **lysosomes**, and food is broken down within the vacuole into smaller molecules that can be absorbed into the cell cytoplasm. Undigested remnants remain in the vacuole, which eventually expels its contents outside the cell. Intracellular digestion is seen in single-celled protists and in the simplest animals. Sponges, for example, rely entirely on intracellular digestion. This process limits their menu to microscopic food particles such as protists that are filtered from the surrounding sea by means of the sieve-like collar cells (see Chapter 22).

A Sac with One Opening Forms the Simplest Digestive System

Larger, more complex organisms evolved a chamber within the body where chunks of food could be broken down by enzymes that act outside the cells. This process is called **extracellular digestion**. One of the simplest of these chambers is found in cnidarians, such as sea anemones, hydra, and jellyfish. As we learned in Chapter 22, these animals possess a digestive sac called a **gastrovascular cavity**, with a single opening through which food is ingested and wastes are ejected (Fig. 29-5). Although it is generally referred to as the mouth, this opening also serves as an anus. Food captured by stinging tentacles is escorted into the gastrovascular cavity, where enzymes break it down. Cells lining the cavity absorb the nutrients and engulf small food particles. Further digestion occurs via intracellular digestion. The undigested remains are eventually expelled through the same opening that they entered.

While one meal is being digested, a second cannot be processed efficiently, because the same chamber is used. Thus, this type of digestive system is unsuited to active animals, which require frequent meals, or to animals whose food offers so little nutrition that they must feed continually. The needs of such animals are met by a digestive system that consists of a one-way tube with an opening at each end.

Digestion in a Tube Allows More Frequent Meals

Most animals, from people and other vertebrates to earthworms, mollusks, arthropods, and echinoderms, have a digestive system that is basically a tube through the body, beginning with a mouth and ending with an anus. A tubular digestive tract allows the animal to eat frequently. Further, it consists of a series of specialized regions that process the food in an orderly sequence: first physically grinding it up, then enzymatically breaking it down, then absorbing the small nutrient molecules into body cells. Specialized tubular digestive tracts adapt different types of animals to eat a wide range of foods and to extract the maximum amount of nutrients from them.

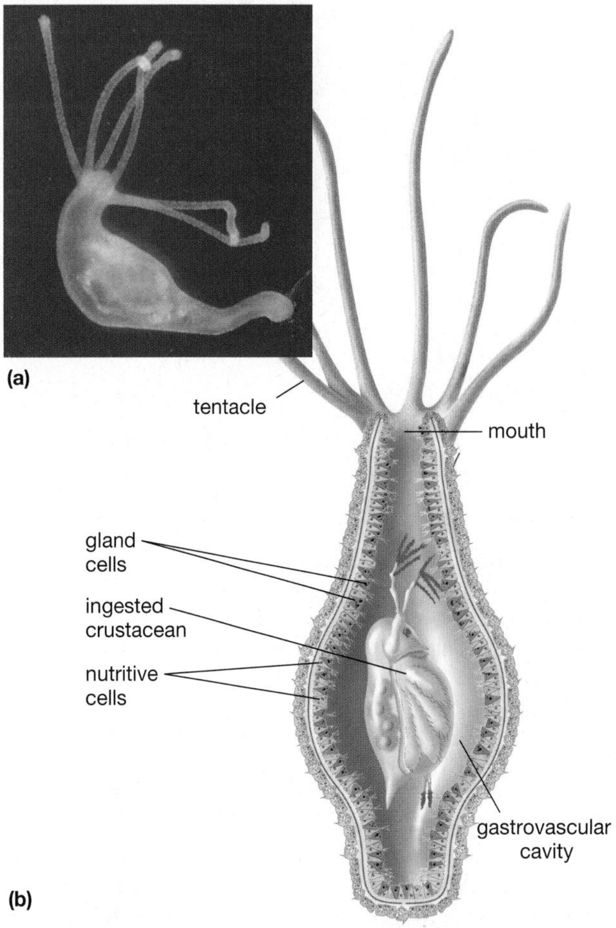

Figure 29-5 Digestion in a sac
(a) A *Hydra* has just captured and ingested a waterflea (*Daphnia,* a microscopic crustacean). ***(b)*** Within the gastrovascular cavity of *Hydra,* gland cells secrete enzymes that digest the prey into smaller particles and nutrients. Elongated cells lining the cavity ingest these particles by intracellular digestion (see Fig. 29-4), and digestion is completed intracellularly. Undigested waste is then expelled through the single opening.

In the earthworm, the tube consists of a series of compartments, each with a specific role in the breakdown of food, which consists of organic material in soil (Fig. 29-6). A tubular digestive system is essential to the earthworm, which continuously ingests soil as it burrows through the earth, passing the soil out at one end while taking it in at the other. The worm's muscular *pharynx* is a cavity that connects the mouth with the *esophagus*, a muscular passageway. Ingested soil and bits of vegetation are passed through the pharynx and the esophagus to the *crop*, a thin-walled storage organ. The crop collects the food and gradually passes it to the *gizzard*. There, bits of sand and the contraction of muscles grind the food into smaller particles. The food then travels to the intestine, where enzymes break it down

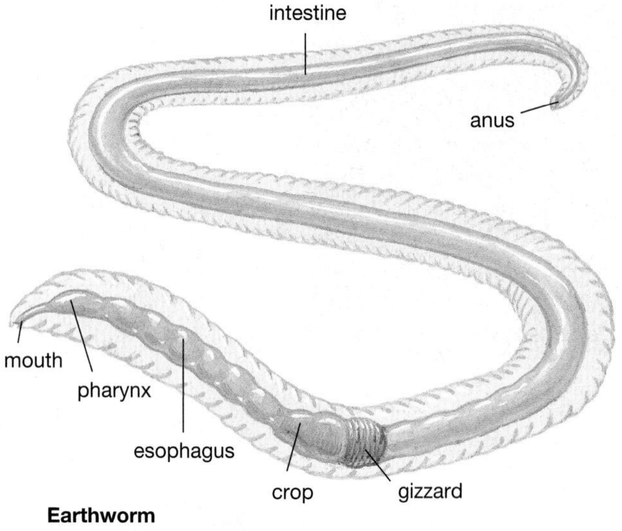

Figure 29-6 Tubular digestive tracts
The earthworm has a one-way digestive system that passes food through a series of compartments, each specialized to play a specific role in breaking down food and absorbing it.

into simple molecules that can be absorbed by the cells lining the intestine.

Like the earthworm, humans and other vertebrates have tubular digestive tracts with several compartments in which food is first physically, then chemically, broken down before being absorbed by individual cells. Animals with tubular digestive systems thus use extracellular digestion to break down their food. Vertebrate digestive tracts are specialized for the particular diet of the animal. For example, birds lack teeth, but some eat seeds, shelled invertebrates, or bony mammalian prey. These birds usually have a large, highly muscular gizzard (Fig. 29-7a) in which these resistant foods are ground up before they are passed to the intestines for further processing. Let's look at the digestive tracts of the cow and the human as well.

Adaptations Help Ruminants Digest Cellulose

If you were restricted to a diet of grass, you would soon starve. Cellulose, like starch, consists of long chains of glucose molecules, but these molecules are linked together in a way that resists the attack of animal digestive enzymes (see Chapter 3). Because cellulose surrounds each plant cell, it is potentially one of the most abundant food energy sources on Earth. *Ruminant* animals—cows, sheep, goats, camels, and hippos, to name a few—have evolved elaborate digestive systems housing microorganisms that can break down cellulose. *Rumination*, or *cud-chewing*, is the process of regurgitating food and rechewing it, one of several adaptations that enable these animals to digest tough plant material. Ruminant stomachs consist of several chambers (Fig. 29-7b). The first two chambers, the *rumen* and the *reticulum*, have evolved into large fermentation vats. There, microorganisms—including many species of bacteria and ciliates—thrive in a mutually beneficial relationship with the ruminant. These microorganisms produce **cellulase**, the enzyme that breaks down cellulose into its component sugars. After it is processed in the rumen, the plant material, now called *cud*, is regurgitated, chewed, and reswallowed. The extra mechanical breakdown exposes more of the cellulose and cell contents to the enzymes of the

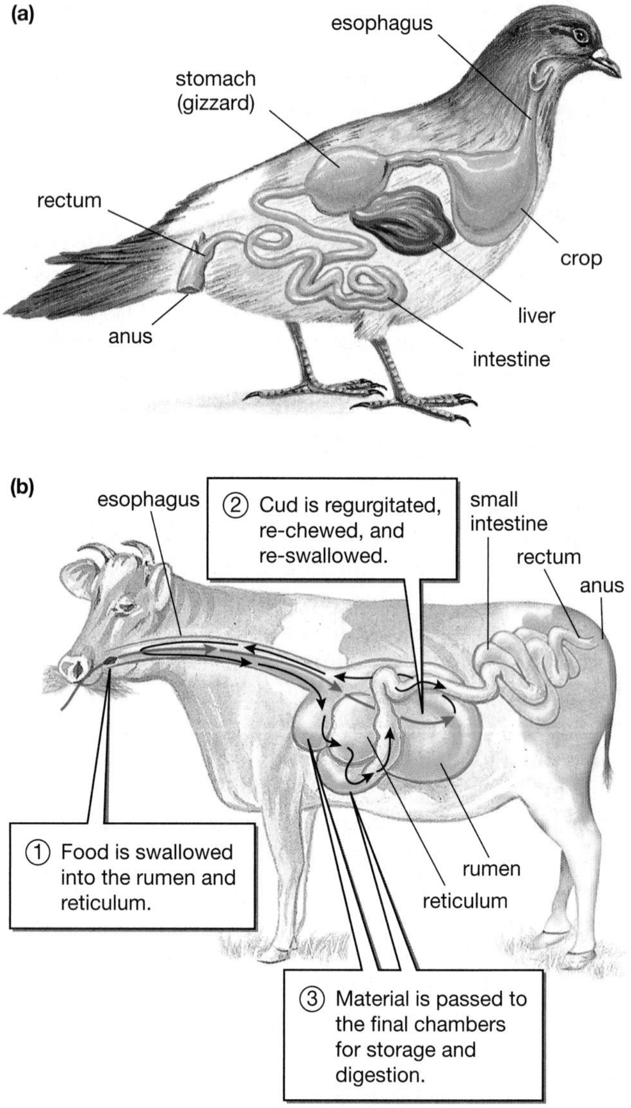

Figure 29-7 Adaptations of vertebrate digestive systems
(a) The digestive system of birds is adapted to the demands of flight. The expandable crop serves as a storage organ, allowing the bird to store food to meet the enormous caloric demands of flight. The gizzard replaces the teeth, using muscular action and small stones that are stored in this organ to break down the hard seeds and insect exoskeletons prevalent in the diet of many birds. Undigested food is expelled. ***(b)*** The stomach of the cow has several chambers. The rumen and reticulum house a flourishing population of microorganisms that digest the cellulose in the cow's vegetarian diet. Arrows trace the path of food through the digestive tract.

microorganisms for further digestion. Gradually the cud is then released into the remaining chambers for further breakdown before it enters the intestine.

3 How Do Humans and Other Mammals Digest Food?

The human digestive system (Fig. 29-8) is adapted for processing a wide variety of different foods. *Herbivores* are animals, such as the ruminants discussed above, that eat only plants; *carnivores* are animals that eat only other animals. In contrast, humans are *omnivores*, adapted to digest all different types of food.

Food travels in a continuous tube from mouth to anus; in the course of this circuitous route, it is subjected to a carefully orchestrated succession of digestive operations. By the time its passage is complete, the food has been chopped, mashed, mixed, churned, and bathed in a series of powerful chemicals. Everything of value has been extracted, and the residue is ejected. This stepwise deconstruction of foodstuffs requires coordinated action from the integrated array of structures that make up the digestive system.

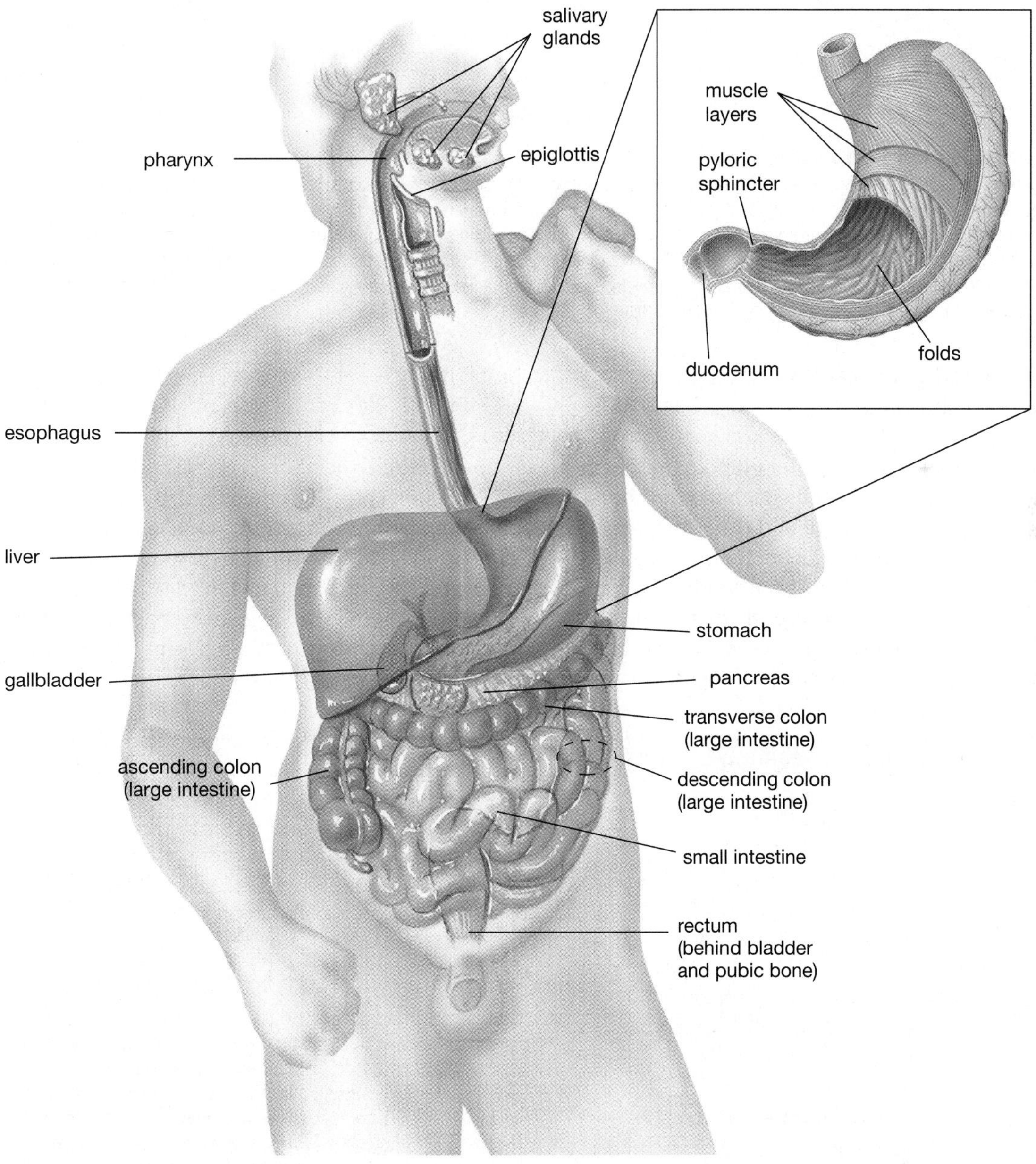

Figure 29-8 The human digestive tract
The digestive system includes both the digestive tube and organs such as the salivary glands, liver, gallbladder, and pancreas, all of which produce and store digestive secretions. The stomach has a muscular wall and folds that allow expansion.

MEDIATUTOR 29.1 The Digestive System

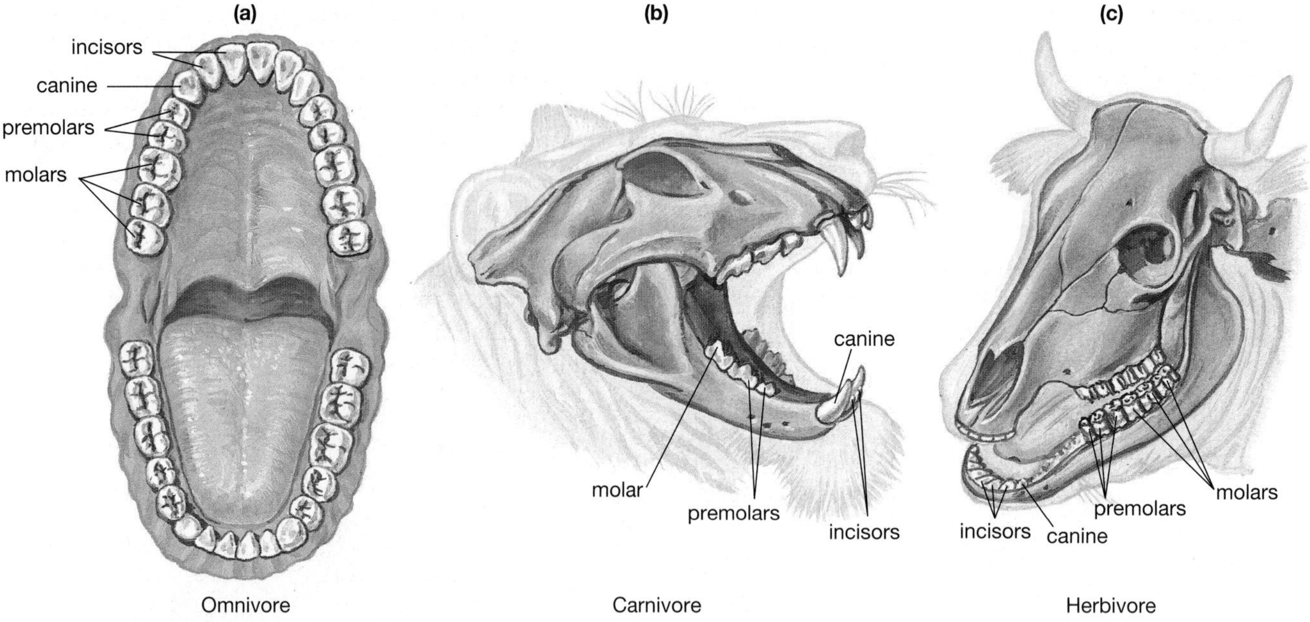

Figure 29-9 Teeth are adapted to different diets
(a) The varied, omnivorous diet of humans has fostered the evolution of an unspecialized dentition that includes flat incisors for biting, pointed canines for tearing, premolars for grinding, and molars for crushing and chewing. ***(b)*** Carnivores (such as the lion) have modest incisors but greatly enlarged canines for stabbing and tearing flesh. Carnivores also have a reduced set of cheek teeth (molars and premolars) that are specialized for shearing through tendon and bone. ***(c)*** Herbivores (such as the cow) have incisors that are specialized for snipping leaves, and their canines have been reduced in size and moved forward to help with that job. The cheek is filled with a full set of wide, flat premolars and molars that grind up the tough, cellulose-containing plant material.

The Mechanical and Chemical Breakdown of Food Begins in the Mouth

You take a bite, your mouth waters, and you begin chewing. This begins both the mechanical and the chemical breakdown of food. In people and other mammals, the mechanical work is done mostly by teeth. *Incisors* at the front of the mouth snip off pieces of food; the pointed *canine* teeth beside them are useful for tearing the pieces apart; and the *premolars* and *molars* at the back of the mouth have flat surfaces for grinding food to a paste. In different species of mammals, teeth are specialized for the diet of the animal (Fig. 29-9). In adult humans, 32 teeth of varying shapes and sizes cut and grind food into small pieces.

While the food is pulverized by the teeth, the first phase of chemical digestion occurs as three pairs of salivary glands pour out saliva in response to the smell, feel, taste, and (if you're hungry) even the thought of food.

Saliva contains the digestive enzyme **amylase**, which begins the breakdown of starches into sugar (Table 29-6). Saliva has other functions as well. It contains a bacteria-

Table 29-6 Digestive Structures and Secretions

Site of Digestion	Secretion	Source of Secretion	Role in Digestion
Mouth	Amylase	Salivary glands	Breaks down starch into disaccharides
	Mucus, water	Salivary glands	Lubricates, dissolves food
Stomach	Hydrochloric acid	Cells lining stomach	Allows pepsin to work, kills bacteria, solubilizes minerals
	Pepsin	Cells lining stomach	Breaks down proteins into large peptides
	Mucus	Cells lining stomach	Protects stomach
Small intestine	Sodium bicarbonate	Pancreas	Neutralizes acidic chyme from stomach
	Amylase	Pancreas	Breaks down starch into disaccharides
	Peptidases	Pancreas	Split large peptides into small peptides
	Trypsin	Pancreas	Breaks down proteins into large peptides
	Chymotrypsin	Pancreas	Breaks down proteins into large peptides
	Lipase	Pancreas	Breaks down lipids into fatty acids and glycerol
	Bile	Liver	Emulsifies lipids
	Peptidases	Cells lining small intestine	Split small peptides into amino acids
	Disaccharidases	Cells lining small intestine	Split disaccharides into monosaccharides

killing enzyme and antibodies that help guard against infection. Saliva also lubricates the food to facilitate swallowing and dissolves some food molecules such as acids and sugars, carrying them to *taste buds* on the tongue. The taste buds are sensory receptors that help identify the type and quality of the food.

With the help of the muscular tongue, the food is manipulated into a mass and pressed backward into the **pharynx**, a muscular cavity connecting the mouth with the esophagus (Fig. 29-10a). Via the *larynx* the pharynx also connects the nose and mouth with the *trachea*, which conducts air to the lungs. This arrangement occasionally causes problems, as anyone who has ever choked on a piece of food can attest. Normally, however, the swallowing reflex (triggered by food entering the pharynx) elevates the larynx so it meets the **epiglottis**, a flap of tissue that blocks off the respiratory passages. Food is thus directed into the esophagus rather than into the trachea (Fig. 29-10b).

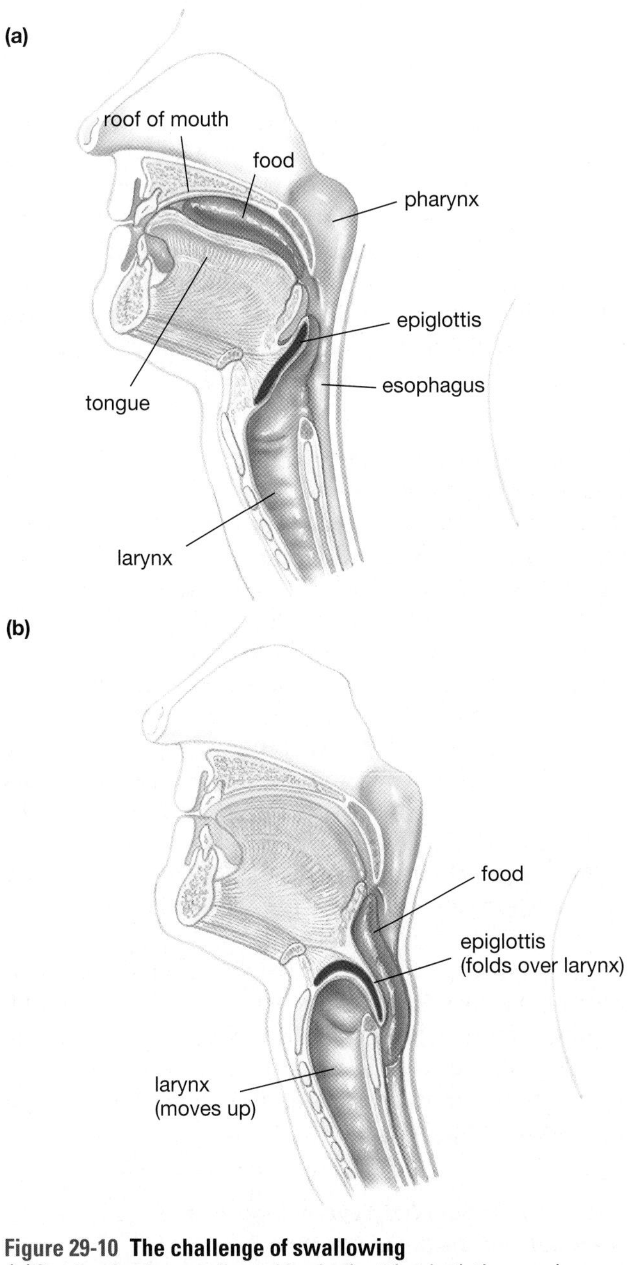

Figure 29-10 The challenge of swallowing
(a) Swallowing is complicated by the fact that both the esophagus (part of the digestive system) and the larynx (part of the respiratory system) open into the pharynx. ***(b)*** During swallowing, the larynx moves upward beneath a small flap of cartilage, the epiglottis. The epiglottis folds down over the larynx, sealing off the opening to the respiratory system and directing food down the esophagus instead.

The Esophagus Conducts Food to the Stomach

Swallowing forces food into the esophagus, a muscular tube that propels food from the mouth to the stomach. Circular muscles surrounding the esophagus contract in sequence above the swallowed food mass, squeezing it down toward the stomach. This muscular action, called **peristalsis**, also occurs in the stomach and intestines, where it helps move food along the digestive tract (Fig. 29-11). Peristalsis is so effective that a person can actually swallow when upside down. Mucus secreted by cells that line the esophagus helps protect it from abrasion and lubricates the food during its passage.

The **stomach** in humans is an expandable muscular sac capable of holding from 2 to 4 liters (as much as a gallon) of food and liquids. Food is retained in the stomach by a ring of circular muscle that separates the lower portion of the stomach from the upper *small intestine.* This muscle, called the **pyloric sphincter**, regulates the passage of food into the small intestine, as described later.

The stomach has three major functions. First, the stomach stores food and releases it gradually into the small intestine at a rate suitable for proper digestion and absorption. Folds in the stomach wall (see Fig. 29-8, inset) increase its capacity, allowing us to eat large, infrequent meals. Carnivores carry this ability to an extreme. A lion, for instance, may consume about 18 kilograms (40 pounds) of meat at one meal, then spend the next few days quietly digesting it. A second function of the stomach is to assist in the mechanical breakdown of food. In addition to peristalsis, its muscular walls undergo a variety of contracting, churning movements that help break apart large pieces of food. The third function of the stomach is the chemical breakdown of food. Glands in the lining of the stomach secrete enzymes and other substances, including gastrin, hydrochloric acid (HCl), pepsinogen, and mucus. **Gastrin** (a hormone) stimulates the secretion of hydrochloric acid by specialized stomach cells. Other cells release *pepsinogen*, an inactive form of the protein-digesting enzyme *pepsin.* Pepsin is a **protease**, an enzyme that helps break proteins into shorter chains of amino acids called *peptides* (see Table 29-6). Pepsin is secreted in an inactive form, pepsinogen, to prevent it from digesting the very cells that produce it. Hydrochloric acid, which gives the fluid in the stomach a pH of 1 to 3, converts the pepsinogen

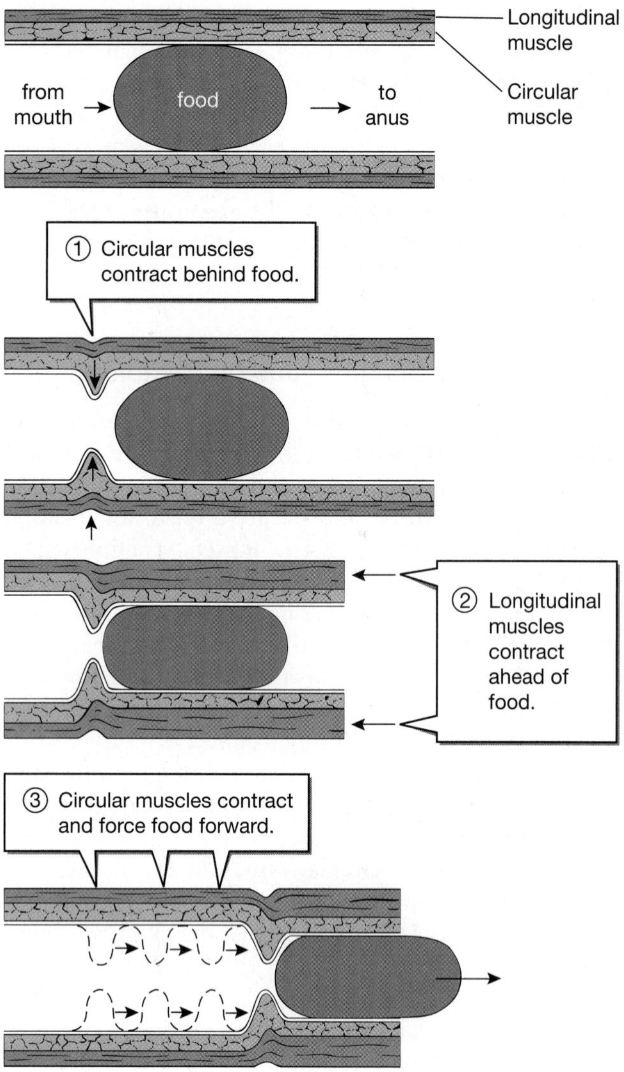

Figure 29-11 Peristalsis
Food is propelled through the digestive system by the rhythmic contractions of circular and longitudinal muscles.

into pepsin, an enzyme that functions best in an acidic environment.

As you may have noticed, the stomach produces all the ingredients necessary to digest itself. Indeed, this is what happens when a person develops ulcers (see "Health Watch: Ulcers: Digesting the Digestive Tract"). However, cells lining the stomach normally produce a large quantity of thick mucus that coats the stomach lining and serves as a barrier to self-digestion. The protection is not perfect, however, and the cells lining the stomach are digested to some extent and must be replaced every few days.

Food in the stomach is gradually converted to a thick, acidic liquid called **chyme**, which consists of partially digested food and digestive secretions. Peristaltic waves then propel the chyme toward the small intestine. The pyloric sphincter allows only about a teaspoon of chyme to be expelled with each contraction, which occurs about every 20 seconds. It takes 2 to 6 hours, depending on the size of the meal, to empty the stomach completely. The continued churning movements of an empty stomach are felt as "hunger pangs."

Only a few substances, including water, some drugs, and alcohol, can enter the bloodstream through the stomach wall. Alcohol that is consumed when the stomach is empty is immediately absorbed into the bloodstream, with strong and rapid effects. Because food in the stomach slows alcohol absorption, the advice "never drink on an empty stomach" is based on sound physiological principles.

Most Digestion Occurs in the Small Intestine

The **small intestine** is narrow (about 1 to 2 inches in diameter in an adult human), but with a length of 10 feet, it is the longest part of the digestive tract. The small intestine functions to digest food into small molecules and to absorb these molecules into the bloodstream. The first role of the small intestine—digestion—is accomplished with the aid of digestive secretions from three sources: (1) the liver, (2) the pancreas, and (3) the cells of the small intestine itself.

The Liver and Gallbladder Provide Bile, Important in Fat Breakdown

The **liver** is perhaps the most versatile organ in the body. Its many functions include storage of fats and carbohydrates for energy, regulation of blood glucose levels, synthesis of blood proteins, storage of iron and certain vitamins, conversion of toxic ammonia (released by the breakdown of amino acids) into urea, and detoxification of other harmful substances such as nicotine and alcohol. The role of the liver in digestion is to produce *bile*, a liquid stored and concentrated in the **gallbladder** and released into the small intestine through a tube called the *bile duct* (see Fig. 29-8).

Bile is a complex mixture composed of **bile salts**, water, other salts, and cholesterol. Bile salts are synthesized in the liver from cholesterol and amino acids. Although they assist in the breakdown of lipids, bile salts are not enzymes. Rather, they act as detergents or emulsifying agents, dispersing globs of fat in the chyme into microscopic particles. These particles expose a large surface area for attack by **lipases**, lipid-digesting enzymes produced by the pancreas.

The Pancreas Supplies Several Digestive Secretions to the Small Intestine

The **pancreas** lies in the loop between the stomach and small intestine (see Fig. 29-8). It consists of two major types of cells. One type produces hormones involved in blood sugar regulation (as we shall see in

Health Watch
Ulcers: Digesting the Digestive Tract

About 1 out of every 10 Americans eventually develops an ulcer. *Ulcers* occur when the mucus barriers of the stomach and upper small intestine break down, and the inner lining and deeper layers of the digestive tract are eroded by acid and the protein-digesting enzyme pepsin. The upper small intestine is the most common site for ulcers (Fig. E29-2), because it receives the highly acidic chyme but is less protected by mucus than the stomach is.

The causes of ulcers are complex and poorly understood. A tendency to develop ulcers can be inherited. Some, but not all, ulcer sufferers secrete excess stomach acid. In some cases, decreased secretion of sodium bicarbonate or mucus by cells that line the stomach and small intestine may contribute to the disease. There is a correlation between the development of ulcers and stress, which may stimulate acid and pepsin secretion. Smoking and chronic consumption of alcohol and aspirin all may reduce the resistance of the digestive tract lining to the effects of pepsin and acids and can aggravate ulcers. In 1983 B. J. Marshall, an Australian physician, identified a bacterium in stomach tissue from ulcer patients. When his claims that the bacterium caused the ulcer were met with skepticism, he swallowed a batch of them and soon developed ulcer symptoms. The bacterium (*Helicobacter pylori*) lives in the digestive tracts of nearly all ulcer patients, and antibiotics that kill this bacterium have been very effective in reducing the recurrence of ulcers. Although an estimated half of the world's population harbors *H. pylori*, the bacterium causes ulcers in only 10% to 20% of those infected with it.

Ulcers can sometimes be relieved by stress-reduction programs, the elimination of smoking and drinking, and dietary changes. Medications for ulcers include antibiotics, antacids, and drugs that decrease stomach acid production.

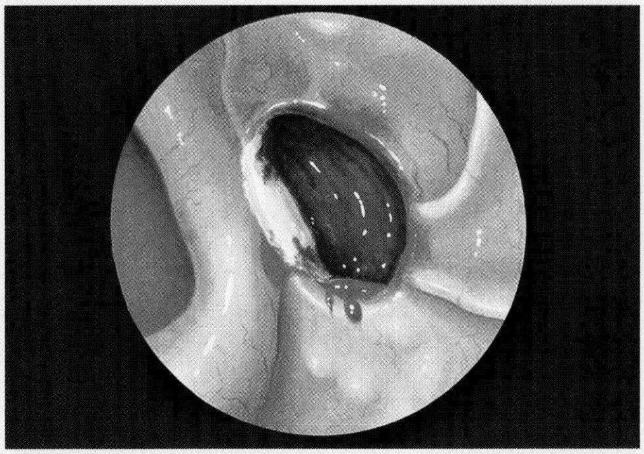

Figure E29-2 A bleeding ulcer in the stomach
An illustration of a bleeding ulcer in the stomach as it would be seen through a fiber-optic viewing device called an endoscope. The stomach lining has been digested away, and blood seeps through the opening.

Chapter 32), and the other produces a digestive secretion called **pancreatic juice**, which is released into the small intestine. Pancreatic juice neutralizes the acidic chyme and digests carbohydrates, lipids, and proteins. About 1 liter (1.06 quarts) of pancreatic juice is released into the small intestine each day. This secretion contains water, sodium bicarbonate, and several digestive enzymes (see Table 29-6). Sodium bicarbonate (the active ingredient in baking soda) neutralizes the acidic chyme in the small intestine, producing a slightly basic pH. Pancreatic digestive enzymes require this basic pH for proper functioning, in contrast to the stomach's digestive enzymes, which require an acidic pH.

The pancreatic digestive enzymes break down three major types of food: (1) amylase breaks down carbohydrates, (2) lipases digest lipids, and (3) several proteases break down proteins and peptides. The pancreatic proteases include *trypsin*, *chymotrypsin*, and *carboxypeptidase*. These proteases are secreted in an inactive form and become activated after they reach the small intestine.

The Intestinal Wall Completes the Digestive Process

The wall of the small intestine is studded with cells that are specialized to complete the digestive process and absorb the small molecules that result. These cells have various enzymes on their external membranes, which form the lining of the small intestine. These enzymes include proteases, which complete the breakdown of peptides into amino acids, and sucrase, lactase, and maltase, which break down disaccharides into monosaccharides (see Chapter 3). Small amounts of lipase digest lipids. Because these enzymes are actually embedded in the membranes of the cells that line the small intestine, this final phase of digestion occurs *as* the nutrient is being absorbed into the cell. Like the stomach, the small intestine is protected from digesting itself by large amounts of mucous secretions from specialized cells in its lining.

Most Absorption Occurs in the Small Intestine

The small intestine is not only the principal site of chemical digestion, it is also the major site of nutrient **absorption** into the blood. The small intestine has

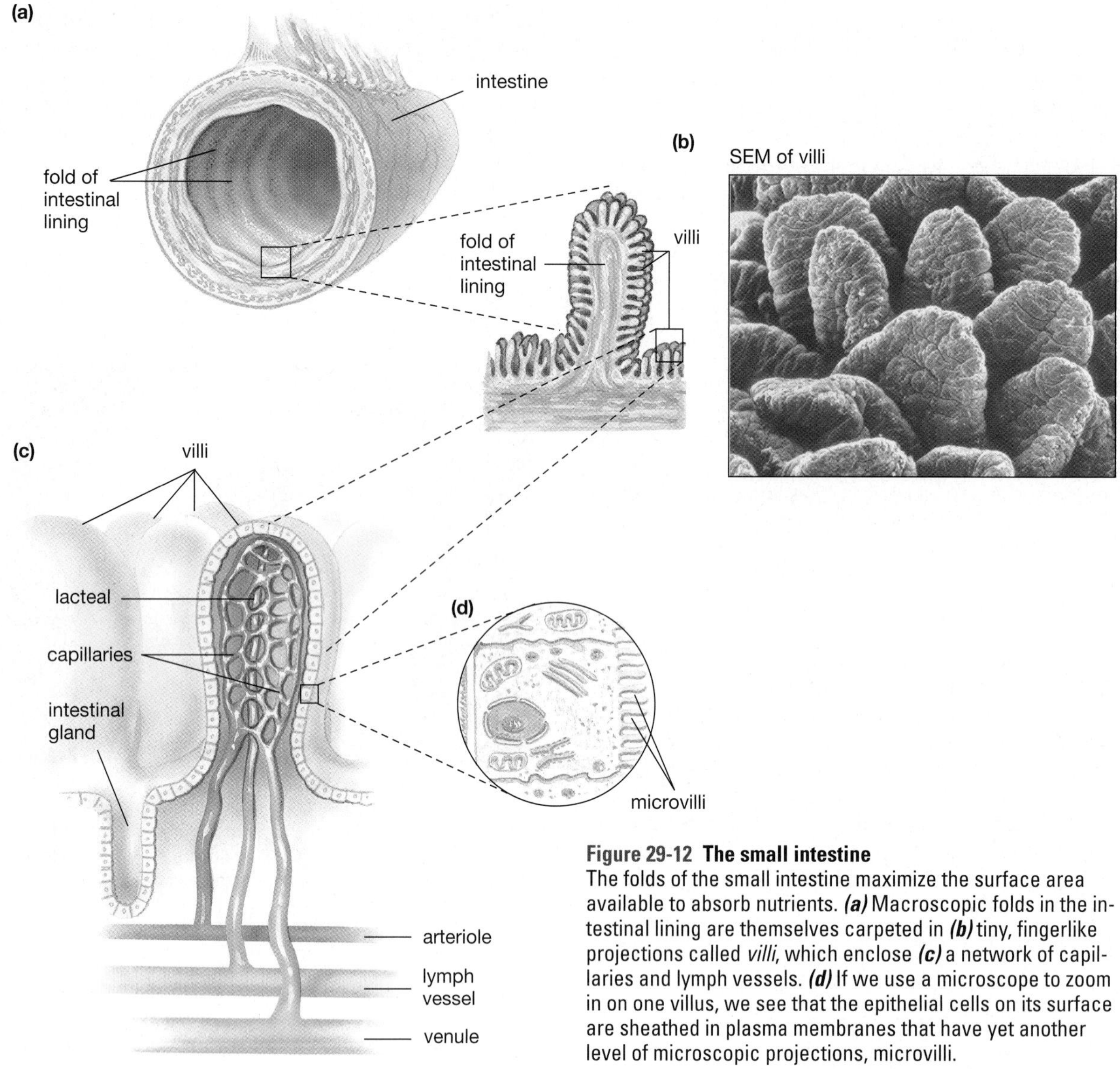

Figure 29-12 The small intestine
The folds of the small intestine maximize the surface area available to absorb nutrients. ***(a)*** Macroscopic folds in the intestinal lining are themselves carpeted in ***(b)*** tiny, fingerlike projections called *villi*, which enclose ***(c)*** a network of capillaries and lymph vessels. ***(d)*** If we use a microscope to zoom in on one villus, we see that the epithelial cells on its surface are sheathed in plasma membranes that have yet another level of microscopic projections, microvilli.

numerous foldings and projections that give it an internal surface area that is 600 times that of a smooth tube of the same length (Fig. 29-12). Minute, fingerlike projections called **villi** (literally, "shaggy hairs"; singular, **villus**) cover the entire folded surface of the intestinal wall. Villi, which range from 0.5 to 1.5 millimeters in length, give the intestinal lining a velvety appearance to the naked eye. They move gently back and forth in the chyme that passes through the intestine. This movement increases the exposure of the villi to the molecules to be digested and absorbed. Further, each individual cell of the villi bears a fringe of microscopic projections called **microvilli**. Taken together, these specializations of the small intestine wall give it a surface area of about 250 square meters (more than 2200 square feet; almost the size of a doubles tennis court).

Unsynchronized contractions of the circular muscles of the intestine, called **segmentation movements**, slosh the chyme back and forth, bringing nutrients into contact with the absorptive surface of the small intestine. When absorption is complete, coordinated peristaltic waves conduct the leftovers into the *large intestine.*

Nutrients absorbed by the small intestine include water, monosaccharides, amino acids and short peptides, fatty acids produced by lipid digestion, vitamins, and minerals. The mechanisms by which this absorption occurs are varied and complex. In most cases, energy is expended to transport nutrients into the intestinal cells. (Water follows by osmosis, as described in Chapter 4.) The nutrients then diffuse out of the intestinal cells into the interstitial fluid, where they then enter the bloodstream.

Each villus of the small intestine is provided with a rich supply of blood capillaries and a single lymph capillary, called a **lacteal**, to carry off the absorbed nutrients and distribute them throughout the body (see Fig. 29-12). Most of the nutrients enter the bloodstream through the capillaries, but fat subunits take a different route. After diffusing into the epithelial cells lining the small intestine, they are resynthesized into fats, combined with other molecules, and then released as droplets into the interstitial fluid (see "A Closer Look: The Fate of Fats" at this text's Web site for Chapter 29). By this means, they enter the lymph vessel and are eventually delivered to the bloodstream when the lymph vessels empty into the veins.

Water Is Absorbed and Feces Are Formed in the Large Intestine

The **large intestine** in an adult human is about 5 feet long and about 3 inches in diameter, wider and shorter than the small intestine. The large intestine has two parts: For most of its length it is called the **colon**, but its final 6-inch compartment is called the **rectum**. Into the large intestine flow the leftovers of digestion: a mixture of water, undigested fats and proteins, and indigestible fibers, such as the cell walls of vegetables and fruits. The large intestine contains a flourishing population of bacteria that live on unabsorbed nutrients. These bacteria earn their keep by synthesizing vitamin B_{12}, thiamin, riboflavin, and, most important, vitamin K, which would otherwise be deficient in a normal diet. Cells lining the large intestine absorb these vitamins as well as leftover water and salts.

After absorption is complete, the result is the semisolid **feces**. Feces consist of indigestible wastes and the dead bodies of bacteria, which account for about one-third of the dry weight of feces. The feces are transported by peristaltic movements until they reach the rectum. Expansion of this chamber stimulates the desire to defecate. Although defecation is a reflex (as any new parent can attest), it is initiated voluntarily after about the age of two.

Digestion Is Controlled by the Nervous System and Hormones

The waiter places a chef salad in front to you, and you hungrily begin to eat. Without you giving it any thought, your body coordinates a complex series of events that convert the salad into nutrients circulating in your blood. As your mouth responds to the first bite, your stomach must be warned that food is on the way. The enzymes of the stomach and small intestine require different environments to function properly (highly acidic in the stomach, slightly basic in the small intestine), and secretions into various parts of the digestive tract correspond with the arrival of food. Not surprisingly, both nerves and hormones coordinate the secretions and activity of the digestive tract (Table 29-7).

The initial phase of digestion is controlled by the nervous system and involves responses to signals that originate in the head. These signals include the sight, smell, taste, and sometimes the thought of food, as well as the muscular activity of chewing. In response to these stimuli, the salivary glands secrete saliva into the mouth while nervous signals to the stomach walls cause the secretion of acid and the hormone gastrin, which stimulates further acid secretion. The concentration of acid is regulated by a negative feedback mechanism (see Chapter 26). When acid levels reach a certain point, they inhibit gastrin secretion, and this inhibits further acid production.

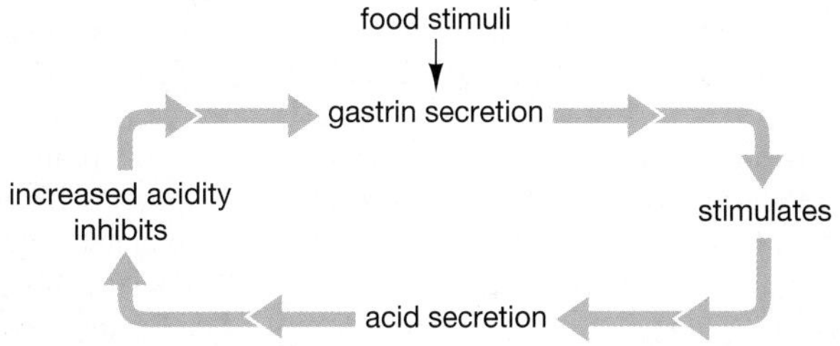

Food arriving in the stomach triggers the second phase of digestion. As the stomach wall is stretched, it produces a large quantity of mucus, which protects the stomach from digesting itself. The acidity of the stomach converts pepsinogen to its active form, pepsin, which begins protein digestion. However, the presence of protein in food tends to reduce the concentration of stomach acid. Thus, as protein is broken down, the acidity drops, and the release of gastrin is no longer inhibited; gastrin is released again and stimulates further acid production.

Table 29-7 Some Important Digestive Hormones

Hormone	Site of Production	Stimulus for Production	Effect
Gastrin	Stomach	Food in mouth Peptides in stomach	Stimulates acid secretion by cells in stomach Distension of stomach
Secretin	Small intestine	Acid in small intestine	Stimulates bicarbonate production by pancreas and liver; increases bile output by liver
Cholecystokinin	Small intestine	Amino acids, fatty acids in small intestine	Stimulates secretion of pancreatic enzymes and release of bile by gallbladder
Gastric inhibitory peptide	Small intestine	Fatty acids and sugars in small intestine	Inhibits stomach movements and release of stomach acid

The cells secreting stomach acid are also activated by the expansion of the stomach and by the presence of peptides produced by protein digestion.

As the liquid chyme is gradually released into the small intestine, its acidity stimulates the release of a second hormone, **secretin**, into the bloodstream by cells of the upper small intestine. Secretin causes the pancreas to pour sodium bicarbonate into the small intestine. Sodium bicarbonate neutralizes the acidity of the incoming chyme, creating an environment in which the pancreatic enzymes can function. A third hormone, **cholecystokinin**, is also produced by cells of the upper small intestine in response to the presence of chyme. This hormone stimulates the pancreas to release various digestive enzymes into the small intestine. It also stimulates the gallbladder to contract, squeezing bile through the bile duct to the small intestine. Bile assists in fat breakdown, as described earlier.

Gastric inhibitory peptide, a hormone secreted by the small intestine in response to the presence of fatty acids and sugars in chyme, inhibits acid production and peristalsis in the stomach. This inhibition slows the rate at which chyme is dumped into the small intestine, providing additional time for digestion and absorption to occur.

CASESTUDYREVISITED

Fat in the Family?

Genes by themselves do not determine our weight. While our genes haven't changed appreciably in the past 20 years, the number of people in the U.S. who are overweight has doubled. Although some people have genetic tendencies that make it easier for them to gain weight and harder to lose it, everyone who is overweight has eaten more calories than his or her body needs. Early in this chapter, we described why animal bodies store fat. Insulated from the forces of natural selection, most people in developed countries live in a fattening environment—delicious, inexpensive, calorie-laden foods beckon us from nearly every street corner. Must we succumb? Researchers at the University of Pittsburgh School of Medicine and the University of Colorado Health Sciences Center are studying a group of 2800 people who were obese but reduced their weight and have maintained a weight loss of at least 30 pounds for more than 5 years. These individuals share certain characteristics: They typically exercise for at least an hour a day, weigh themselves regularly, and get about 24% of their daily calories from fat. Even Dave Barnett is considering some changes: "There must be something between not caring at all and fighting it," he muses. "Maybe it's an alertness. . . . I am learning not to overeat."

Calculate your BMI—are you too heavy? In general, how does your BMI compare with those of your close relatives? Think of ways in which your environment, exercise levels, and eating habits contribute to your weight.

Summary of Key Concepts

1) What Nutrients Do Animals Need?

Each type of animal has specific nutritional requirements. These requirements include molecules that can be broken down to liberate energy, such as lipids, carbohydrates, and protein; chemical building blocks used to construct complex molecules, such as amino acids that can be linked together to form proteins; and minerals and vitamins that facilitate the diverse chemical reactions of metabolism.

2) How Is Digestion Accomplished?

Digestive systems must accomplish five tasks: ingestion, mechanical followed by chemical breakdown of food, absorption, and elimination of wastes. Digestive systems convert the complex molecules of the bodies of other animals or plants that have been eaten into simpler molecules that can be utilized. Animal digestion at its simplest is intracellular, as occurs within the individual cells of a sponge. Extracellular digestion, utilized by all more complex animals, occurs in a body cavity. The simplest form is a saclike gastrovascular cavity in organisms such as flatworms and hydra. Still more complex animals utilize a tubular compartment with specialized chambers where food is processed in a well-defined sequence.

3) How Do Humans and Other Mammals Digest Food?

In humans, digestion begins in the mouth, where food is physically broken down by chewing, and chemical digestion is initiated by saliva. Food is then conducted to the stomach by peristaltic waves of the esophagus. In the acidic environment of the stomach, food is churned into smaller particles, and protein digestion begins. Gradually, the liquefied food, now called chyme, is released to the small intestine. There it is neutralized by sodium bicarbonate from the pancreas. Secretions from the pancreas, liver, and the cells of the intestine complete the breakdown of proteins, fats, and carbohydrates. In the small intestine, the simple molecular products of digestion are absorbed into the bloodstream for distribution to the body cells. The large intestine absorbs the remaining water and converts indigestible material to feces.

Digestion is regulated by the nervous system and hormones. The smell and taste of food and the action of chewing trigger the secretion of saliva in the mouth and the production of gastrin by the stomach. Gastrin stimulates stomach acid production. As chyme enters the small intestine, three additional hormones are produced by intestinal cells: secretin, which causes sodium bicarbonate production to neutralize the acidic chyme; cholecystokinin, which stimulates bile release and causes the pancreas to secrete digestive enzymes into the small intestine; and gastric inhibitory peptide, which inhibits acid production and peristalsis by the stomach. This inhibition slows the movement of food into the intestine.

Key Terms

absorption *p. 597*
amylase *p. 594*
bile *p. 596*
bile salt *p. 596*
body mass index *p. 589*
calorie *p. 584*
Calorie *p. 584*
cellulase *p. 592*
cholecystokinin *p. 600*
chyme *p. 596*
colon *p. 599*
digestion *p. 589*
digestive system *p. 589*
epiglottis *p. 595*
essential amino acid *p. 585*
essential fatty acid *p. 584*
extracellular digestion *p. 591*
feces *p. 599*
food vacuole *p. 591*
gallbladder *p. 596*
gastric inhibitory peptide *p. 600*
gastrin *p. 595*
gastrovascular cavity *p. 591*
glycogen *p. 585*
intracellular digestion *p. 591*
lacteal *p. 598*
large intestine *p. 599*
lipase *p. 596*
liver *p. 596*
lysosome *p. 591*
microvillus *p. 598*
mineral *p. 586*
mouth *p. 590*
nutrients *p. 584*
nutrition *p. 584*
pancreas *p. 596*
pancreatic juice *p. 597*
peristalsis *p. 595*
pharynx (in vertebrates) *p. 595*
protease *p. 595*
pyloric sphincter *p. 595*
rectum *p. 599*
secretin *p. 600*
segmentation movement *p. 598*
small intestine *p. 596*
stomach *p. 595*
urea *p. 585*
villus *p. 598*
vitamin *p. 586*

Thinking Through the Concepts

Multiple Choice

1. *An acidic mixture of partially digested food that moves from the stomach into the small intestine is called*
   a. cholecystokinin
   b. bile
   c. lymph
   d. secretin
   e. chyme
2. *The hormone responsible for stimulating the secretion of hydrochloric acid by stomach cells is*
   a. pepsin
   b. gastrin
   c. cholecystokinin
   d. insulin
   e. secretin
3. *A sudden increase in the amount of secretin circulating in the blood is an indication that food has recently been introduced to the*
   a. mouth
   b. pharynx
   c. stomach
   d. small intestine
   e. large intestine
4. *Humans lack digestive enzymes to attack chitin, a complex polysaccharide in the exoskeleton of lobster and crayfish. We also lack enzymes that degrade*
   a. peptides
   b. plant starch
   c. cellulose
   d. sucrose
   e. lipids
5. *Which of the following is NOT characteristic of bile?*
   a. It is produced in the gallbladder.
   b. It is a mixture of special salts, water, and cholesterol.
   c. It acts as a detergent or emulsifying agent.
   d. It helps expose a large surface area of lipid for attack by lipases.
   e. It works in the small intestine.
6. *"Water-soluble compound that works primarily as an enzyme helper" would be a good definition of*
   a. vitamin C
   b. vitamin A
   c. vitamin B
   d. vitamin E
   e. vitamin D

? Review Questions

1. List four general types of nutrients, and describe the role of each in nutrition.
2. List and describe the function of the three principal secretions of the stomach.
3. List the substances secreted into the small intestine, and describe the origin and function of each.
4. Name and describe the muscular movements that usher food through the human digestive tract.
5. Vitamin C is a vitamin for humans but not for dogs. Certain amino acids are essential for humans but not for plants. Explain.
6. Name four structural or functional adaptations of the human small intestine that ensure good digestion and absorption.
7. Describe protein digestion in the stomach and small intestine.

Applying the Concepts

1. The food label on a soup can shows that the product contains 10 grams of protein, 4 grams of carbohydrate, and 3 grams of fat. How many Calories are in this soup?
2. Stomach ulcers that resist antibiotic therapy are treated with several kinds of drugs. Anticholinergic drugs decrease nerve signals to the stomach walls that are produced by the sight, smell, and taste of food. Antacids neutralize stomach acid. Why would it be inadvisable to take anticholinergics and antacids together?
3. Small birds have high metabolic rates, efficient digestive tracts, and high-calorie diets. Some birds consume an amount of food equivalent to 30% of their body weight every day. They rarely eat leaves or grass but often eat small stones. The bird's small intestine has an attached pancreas and liver. Using this information and Figure 29-7a, explain how a bird's digestive tract is adapted to its lifestyle (foods consumed, flight, habitat, and so on).
4. Control of the human digestive tract involves several feedback loops and messages that coordinate activity in one chamber with those taking place in subsequent chambers. List the coordinating events you discovered in this chapter, in order, beginning with tasting, chewing, and swallowing a piece of meat and ending with residue that enters the large intestine. What turns on and what shuts off each process?
5. Symbiotic protozoa in the digestive tracts of termites produce cellulase that the hosts use. In return, termites provide protozoa with food and shelter. Imagine that the human species is gradually invaded, over many generations, by symbiotic protozoa capable of producing cellulase. What evolutionary adaptive changes in body structure and function might occur simultaneously?
6. Trace a ham and cheese with lettuce sandwich through the human digestive system, discussing what happens to each part of the sandwich as it passes through each region of the digestive tract.
7. One of the common remedies for constipation (difficulty eliminating feces) is a laxative solution that contains magnesium salts. In the large intestine, magnesium salts are absorbed very slowly by the intestinal wall, remaining in the intestinal tract for long periods of time. Thus, the salts affect water movement in the large intestine. On the basis of this information, explain the laxative action of magnesium salts.

For More Information

Blaser, M. "The Bacteria Behind Ulcers." *Scientific American*, February 1996. New discoveries about the causes of ulcers are leading to new treatment strategies.

Diamond, J. "Dining with the Snakes." *Discover*, April 1994. Follow food as it is consumed and digested by a snake.

Martin, R. J., White, B. D., and Hulsey, M. G. "The Regulation of Body Weight." *American Scientist*, November–December 1991. Describes some of the complex mechanisms that control eating, obesity, and the body's energy balance.

Mason, M. "Why Ulcers Run in Families." *Health*, September 1994. The story of the bacterium that causes most ulcers.

Mayfield, E. "A Consumer's Guide to Fats." *FDA Consumer*, May 1994. Commonsense information about fat, cholesterol, diet, and health.

Moog, F. "The Lining of the Small Intestine." *Scientific American*, November 1981. A description of this intricate tissue, which is responsible for absorbing nutrients into the body.

Nuland, S. B. "The Beast in the Belly." *Discover*, February 1995. An interesting medical tale of bacterial disease, digestive enzymes, and food.

Willett, W. C. "Diet and Health: What Should We Eat?" *Science*, April 22, 1994. Summarizes studies that suggest that diet can play a major role in the prevention of disease.

Vogel, S. "Why We Get Fat." *Discover*, April 1999. Americans are getting fatter for many reasons—environmental, genetic, and behavioral.

Answers to Multiple-Choice Questions

1. e 2. b 3. d 4. c 5. a 6. c

MEDIATUTOR
Nutrition and Digestion

CD Activities

Activity 29.1: The Digestive System

Estimated time: 5 minutes

The digestive system involves a multitude of organs and their secretions. This is where the body's food molecules are broken down into their basic building blocks so that they can be absorbed. In this tutorial you will be introduced to the anatomy of the digestive system, and explore the role of the various organs in the digestion of the carbohydrates, lipids, and proteins found in our food.

Start the MediaTutor Student CD-ROM and enter the activity number in the Quick Search box to be taken directly to that activity.

Web Investigations

Case Study: Fat in the Family

Estimated time: 10 minutes

The media messages are simple: "Thin is beautiful," "50 steps to lose weight," "The instant weight loss diet," and so on. Still the number of overweight Americans grows every year. Yet some people can eat several thousand calories a day and never gain a pound. This exercise will examine the roles of heredity, biochemistry, and behavior in obesity.

Go to http://www.prenhall.com/audesirk6, the Audesirk Companion Web site. Select Chapter 29 and the Web Investigation to begin.

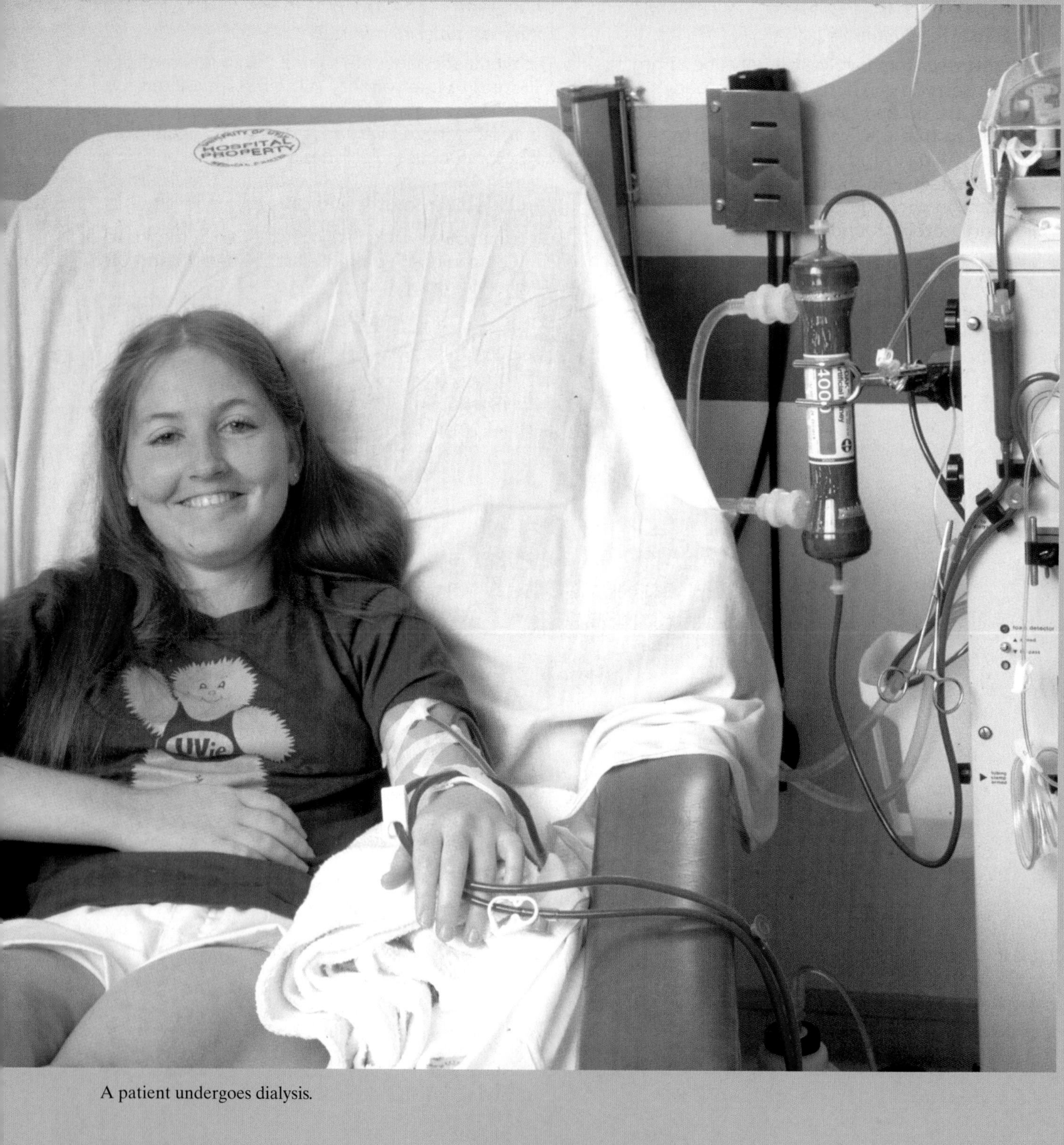

A patient undergoes dialysis.

30 The Urinary System

AT A GLANCE

Case Study: Hemodialysis and Hope

1) **How Does Excretion Occur in Invertebrates?**

Flame Cells Filter Fluids in Flatworms

Nephridia in Earthworms Resemble Parts of the Vertebrate Kidney

2) **What Are the Functions of Vertebrate Urinary Systems?**

3) **How Does the Human Urinary System Function?**

Urine Is Formed in the Kidneys

Blood Is Filtered by the Glomerulus

The Filtrate Is Converted to Urine in the Nephron

The Loop of Henle Allows Urine to Become Concentrated

The Kidneys Are Important Organs of Homeostasis

Case Study Revisited: Hemodialysis and Hope

CASE STUDY

Hemodialysis and Hope

"I have felt convinced since my first days of dialysis that sometime around the year A.D. 2010, some radical new system of dialysis would be invented. Recent miracles with cloning, as performed in Scotland, could mean an unlimited supply of healthy kidneys. Or the brilliance of our medical explorers could very well come up with some new system that I cannot visualize. The bottom line of my personal reaction is that even under the rigid dictates of the present system, a reasonably happy life can be achieved if one is willing to meet the system halfway. . . ."

—James A. Michener

Hemodialysis is the process of filtering blood, using artificial means that replace a failed kidney. James Michener, celebrated author of *Hawaii*, *The Covenant*, *Centennial*, and *The Source*, was on hemodialysis for two years. At the age of 90, facing additional health problems, he chose to withdraw from dialysis, and died about a week later on October 16, 1997, in Austin, Texas.

Tens of thousands of people in the U.S. lose kidney function each year. About 20,000 receive kidney transplants from compatible donors, but at any given time, about 200,000 U.S. citizens are kept alive by hemodialysis. The kidneys are the key component of the urinary system. How accurate was Michener's prediction for the future? What other "medical miracles" are in the works for people with urinary system problems? ■

1 How Does Excretion Occur in Invertebrates?

In this chapter we explore the workings of urinary systems—and discover that they do far more than just produce urine. Urinary systems, also called "excretory systems," serve many crucial functions relating to homeostasis (see Chapter 26), ensuring that the internal environment remains relatively constant. One critical element in homeostasis is water balance. Why? If the volume of water inside body cells fluctuates too much and the chemicals dissolved in internal fluids become too concentrated or too dilute, the chemical equilibrium of the cells will be disrupted, with disastrous consequences for the animal.

As with many physiological systems, the "hardware" that evolution has installed for water regulation serves other crucial purposes as well. Excretory organs not only regulate water balance but also eliminate cellular waste products and ensure that the chemical composition of the blood and extracellular fluid remains within the bounds required for cellular metabolism.

In relatively simple invertebrate animals, excretory systems perform functions similar to those of mammals and other vertebrates, helping to collect and filter body fluids, retaining nutrients, and releasing wastes. Here we describe two of the many types of excretory systems that have evolved in invertebrates.

Flame Cells Filter Fluids in Flatworms

The first specialized excretory structures to arise during the course of animal evolution were most likely **protonephridia**, which consist of a single cell that opens to the environment through a tube. The flatworm, found under rocks in streams, still retains a protonephridial system. The flatworm's excretory system consists of a network of tubes that branch throughout the body (Fig. 30-1). At intervals the tubes end in single-celled bulbs called **flame cells**, which derive their name from the tuft of beating cilia that extends into the hollow bulb. Under the microscope, each beating cluster of cilia resembles a flickering flame. Water and some dissolved wastes are filtered into the bulbs, where the beating cilia produce a current that conducts the fluid through the tubular network. There, more waste products are added and nutrients withdrawn. Eventually, the waste liquid reaches one of numerous pores that release it to the outside. The flattened shape of flatworms also provides a relatively large skin surface through which wastes leave by diffusion.

Nephridia in Earthworms Resemble Parts of the Vertebrate Kidney

Earthworms, mollusks, and several other invertebrates have simple kidney-like structures called **nephridia** (singular, **nephridium**). In the earthworm, fluid fills the body cavity, or coelom, that surrounds the internal organs. This coelomic fluid collects both wastes and nutrients from the blood and tissues. The fluid is conducted into a funnel-shaped opening called the **nephrostome** and swept by cilia along a narrow, twisted tube (Fig. 30-2). There, salts and other dissolved nutrients are absorbed back into the blood, leaving behind water and wastes. The resulting urine is stored in an enlarged bladder-like portion of the nephridium and is then excreted through the **excretory pore**, an opening in the body wall. The earthworm body is composed of repeating segments, nearly every one of which contains its own pair of nephridia.

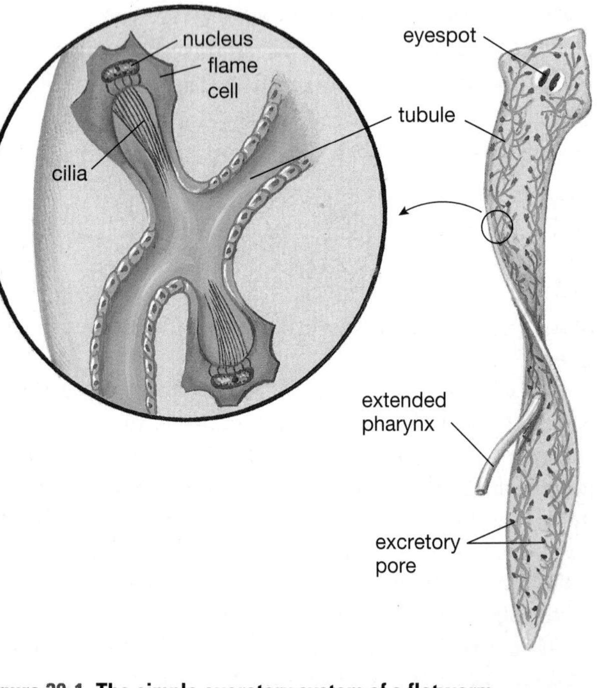

Figure 30-1 The simple excretory system of a flatworm
Hollow flame cells direct excess water and dissolved wastes into a network of tubes. The beating cilia of the flame cells help circulate the fluid to excretory pores.

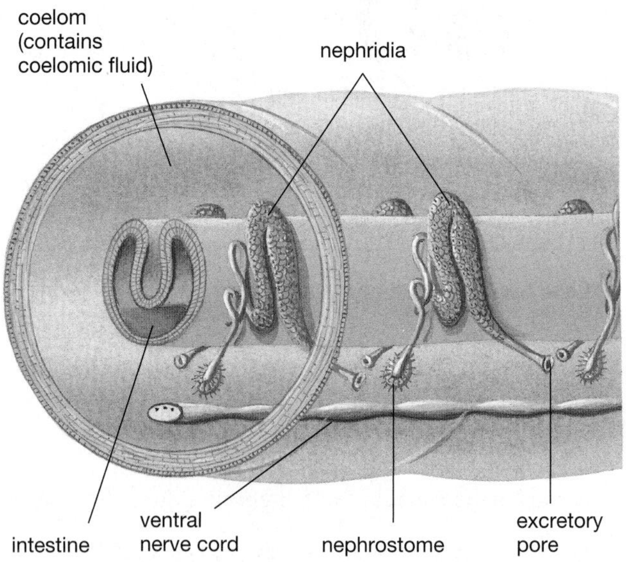

Figure 30-2 The excretory system of the earthworm
This system consists of structures called *nephridia*, one pair per segment. Coelomic fluid is drawn into the nephrostome, and urine is released through the excretory pore. Each nephridium resembles a vertebrate nephron.

2 What Are the Functions of Vertebrate Urinary Systems?

The typical trout urinates more or less constantly, excreting an amount of urine equivalent to its total blood volume every 2 or 3 hours. A kangaroo rat, conversely, might excrete a few milliliters of urine over the course of a day. Both the trout's need to eliminate water and the rat's need to conserve it have the same root cause: The volume of water inside an animal's body must be kept relatively constant. To achieve this constancy, each species has evolved a regulatory mechanism that is effective in its particular environment. For the trout, which spends its life bathed in fresh water, the main regulatory task is getting rid of the water that is driven into its body by osmosis. For the kangaroo rat, which spends its life bathed in dry desert air, the main regulatory task is to prevent water from escaping its body.

As you now know, the urinary system plays a crucial role in **homeostasis**. All of the homeostatic functions of the urinary system in vertebrates are performed as blood is filtered through the kidneys. The **kidneys** are organs in which the fluid portion of the blood is collected. From this fluid, water and important nutrients are reabsorbed into the blood, while toxic substances, cellular waste products, and excess vitamins, salts, hormones, and water are left behind to be excreted as urine. The rest of the urinary system channels and stores urine until it is released from the body, a process called **excretion**. The mammalian urinary system helps maintain homeostasis in several ways, listed below:

1. Regulates blood levels of ions such as sodium, potassium, chloride, and calcium
2. Regulates the water content of the blood
3. Maintains proper pH of the blood
4. Retains important nutrients such as glucose and amino acids in the blood
5. Secretes hormones, such as *erythropoietin,* which stimulates red blood cell production
6. Eliminates cellular waste products such as *urea*

Urea is a product of amino acid metabolism. As you may recall from Chapter 29, the digestive system breaks proteins into their amino acid building blocks, which are then absorbed. When amino acids are taken into cells, some are used directly to synthesize new proteins. Others have their amino ($-NH_2$) groups removed and are then used either as a source of energy or in the synthesis of new molecules. The amino groups are released as **ammonia** (NH_3), which is very toxic. In mammals, ammonia is carried in the blood to the liver, where it is converted to urea, a far less toxic substance (Fig. 30-3). Urea is filtered from the blood by the kidneys and excreted in **urine**, a fluid consisting of water and dissolved wastes and some excess nutrients.

By excreting waste in the form of urea, mammals avoid damage from ammonia's toxicity. Because urea is water soluble, some water must be excreted along with urea, even if water loss is disadvantageous. Birds and reptiles avoid this problem; they excrete the by-products of protein digestion in the form of **uric acid**. Uric acid is not very soluble and can be excreted in crystalline form with very little loss of water.

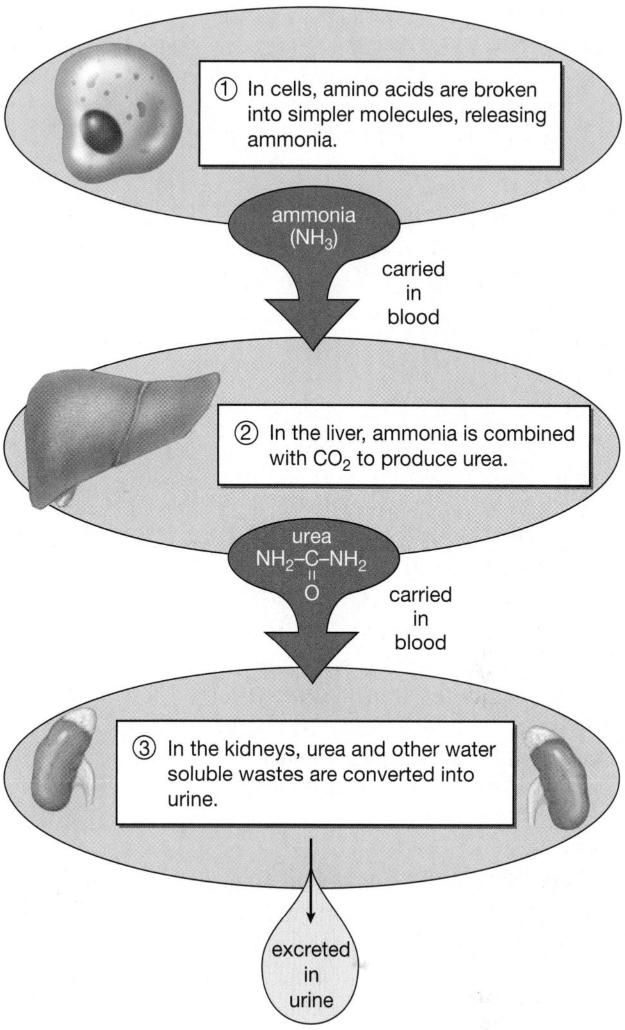

Figure 30-3 A flow diagram showing the formation and excretion of urea

3 How Does the Human Urinary System Function?

The kidneys face a challenge—regulating water, dissolved salts and nutrients, and eliminating dissolved wastes are often in conflict with one another. Excreting dissolved wastes requires excretion of water, but the need to maintain water balance may demand that the water be retained. Furthermore, water containing wastes is very likely also to contain dissolved nutrients and salts that the body cannot afford to lose. How can wastes be excreted without undue loss of water and nutrients? One elegantly evolved solution is found in the mammalian kidneys, complex organs that in some ways resemble dense collections of nephridia, like those found in earthworms. The kidneys are part of a larger group of structures collectively called the *urinary system* (Fig. 30-4). The kidneys produce the urine; other portions of the system transport, store, and eliminate it. In the following sections we examine the major structures of the urinary system, tracing the pathway taken by waste products as they travel through the system. We then explain kidney function in greater detail; finally, we describe the role of the kidneys in homeostasis.

Human kidneys are paired, kidney-bean-shaped organs located on either side of the spinal column and extending slightly above the waist. Each is approximately 5 inches long, 3 inches wide, and 1 inch thick. Blood carrying dissolved cellular wastes enters each kidney through a **renal artery**. After the blood has been filtered,

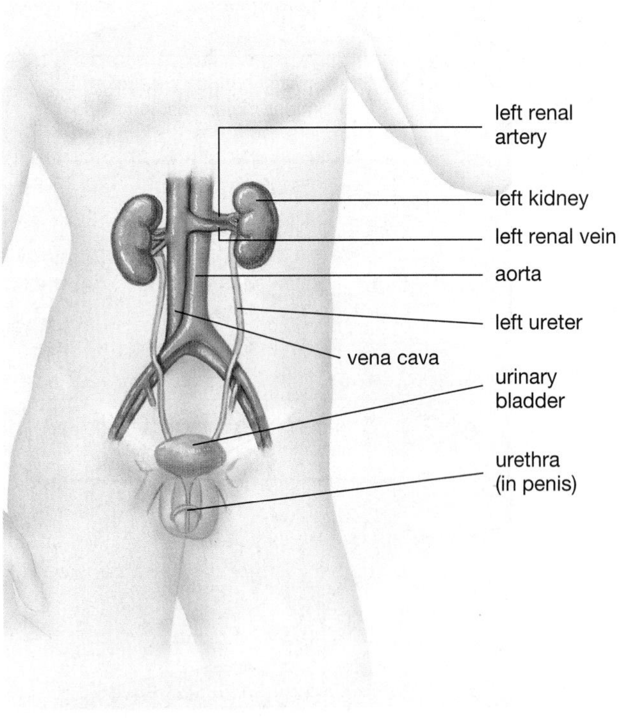

MEDIATUTOR 30.1 Urinary System Anatomy

Figure 30-4 The human urinary system
Diagrammatic view of the human urinary system and its blood supply.

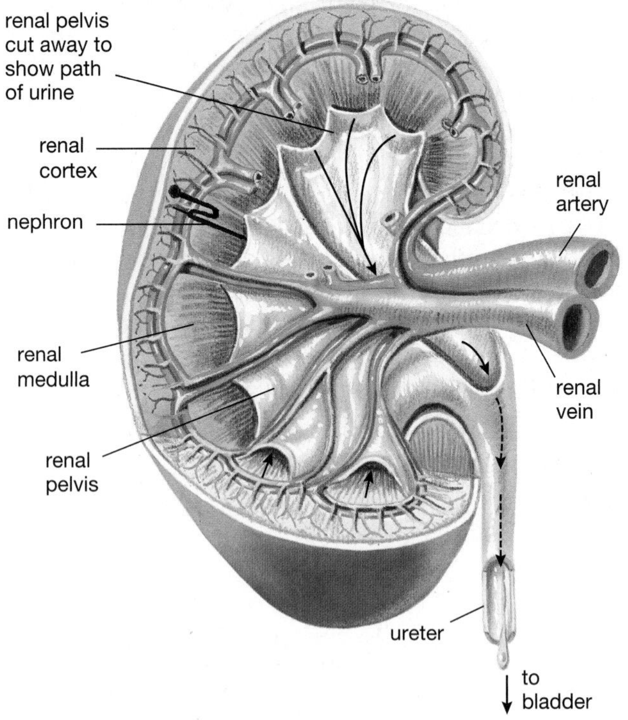

Figure 30-5 Cross section of a kidney
The cross section shows the blood supply and internal structure of a kidney. The renal artery, which brings blood to the kidney, and the renal vein, which carries the filtered blood away, branch extensively within the kidney. The two are joined by a highly permeable capillary network through which substances are exchanged between the blood and the nephrons. A nephron, considerably enlarged, is drawn to show its orientation in the kidney. The renal pelvis is the branched collecting chamber that funnels urine out of the kidney.

it exits through the **renal vein** (Fig. 30-5). Urine leaves each kidney through a narrow, muscular tube called the **ureter**. Using peristaltic contractions, the ureters transport urine to the *urinary bladder,* or simply **bladder**. This hollow, muscular chamber collects and stores the urine.

The walls of the bladder, which contain smooth muscle, are capable of considerable expansion. Urine is retained in the bladder by two sphincter muscles located at its base, just above the junction with the urethra. When the bladder becomes distended, receptors in the walls signal its condition and trigger reflexive contractions. The sphincter nearest the bladder, the *internal sphincter*, is opened during this reflex. However, the lower, or *external*, *sphincter*, is under voluntary control, so the brain can suppress the reflex unless bladder distension becomes acute. The average adult bladder will hold about 500 milliliters (approximately a pint) of urine, but the desire to urinate is triggered by considerably smaller accumulations. Urine completes its journey to the outside through the **urethra**, a single narrow tube about 1.5 inches long in the female and about 8 inches long in the male.

Urine Is Formed in the Kidneys

Each kidney contains a solid outer layer where urine is formed, and a subdivided inner chamber (the *renal pelvis*) that collects urine and funnels it into the ureter (see Fig. 30-5). Microscopic examination of the outer layer of the kidney (consisting of the *renal cortex* overlying the *renal medulla*) reveals an array of tiny individual filters, or **nephrons**, richly supplied with blood vessels. More than 1 million nephrons are packed into the outer layer of each kidney.

The nephron has three major parts: (1) the **glomerulus**, a dense knot of capillaries from which fluid from the blood is collected into (2) a surrounding cuplike structure, called **Bowman's capsule**, and (3), a long, twisted **tubule** (from the Latin, "little tube"). The tubule is further subdivided into first the **proximal tubule**, then the **loop of Henle** (which extends into the renal medulla), and finally the **distal tubule**, which leads to the **collecting duct**, which conducts urine into the renal pelvis (Fig. 30-6; see also Fig. E30-1). In the tubule, nutrients are selectively reabsorbed from the filtered fluid back into the blood, while wastes and some of the water are left behind to form urine. Additional wastes are also secreted into the tubule from the blood. Different portions of the tubule selectively modify the fluid as it travels through. These processes are examined in the following sections and covered in more detail in "A Closer Look: The Nephron and Urine Formation."

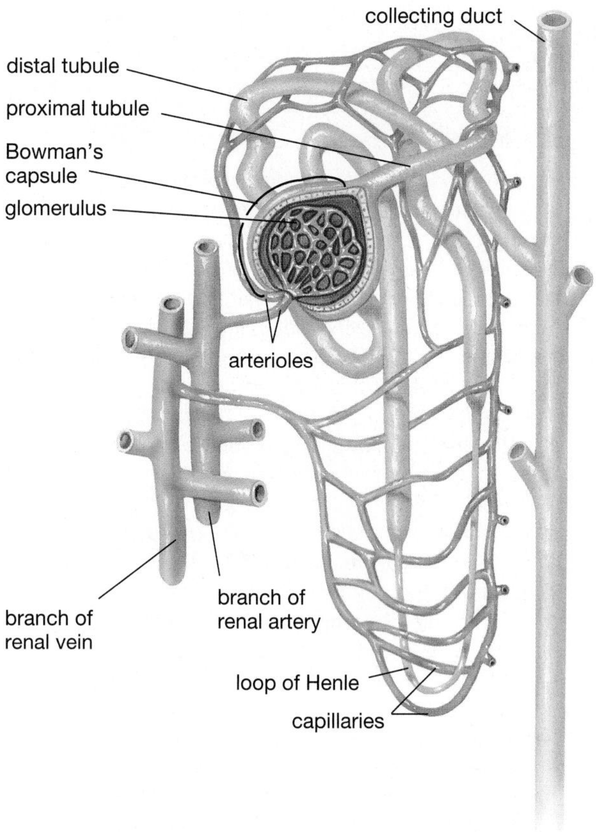

Figure 30-6 An individual nephron and its blood supply

Blood Is Filtered by the Glomerulus

Blood is conducted to each nephron by an arteriole that branches from the renal artery. Within a cup-shaped portion of the nephron—Bowman's capsule—the arteriole branches into numerous microscopic capillaries that form the glomerulus, an intertwined mass (see Fig. 30-6). The walls of the glomerular capillaries are extremely permeable to water and small dissolved molecules, but they prevent the movement of most large proteins, such as the albumin found in blood. Beyond the glomerulus, the capillaries reunite to form an arteriole whose diameter is smaller than that of the incoming arteriole. The differences in diameter between the incoming and outgoing arterioles create pressure within the glomerulus, driving water and many of the dissolved substances from the blood through the capillary walls. This process is called **filtration** (Fig. 30-7, step ①), and the resulting fluid is called the **filtrate.** The watery filtrate, resembling blood plasma without its plasma proteins, is collected in Bowman's capsule for transport through the nephron.

With the filtrate removed, the blood in the arteriole leaving the glomerulus is now very concentrated, having had much of its water removed but retaining substances too large to pass through the glomerular capillary walls, such as blood cells, large proteins, and fat droplets. Beyond the glomerulus, the arteriole branches into smaller, highly porous capillaries. These capillaries surround the tubule, forming intimate contacts with it. At these points of contact, water and nutrients remaining in the filtrate after filtration are reabsorbed and returned to the blood as the filtrate passes through the nephron; in addition, wastes remaining in the blood after filtration are passed into the filtrate for disposal.

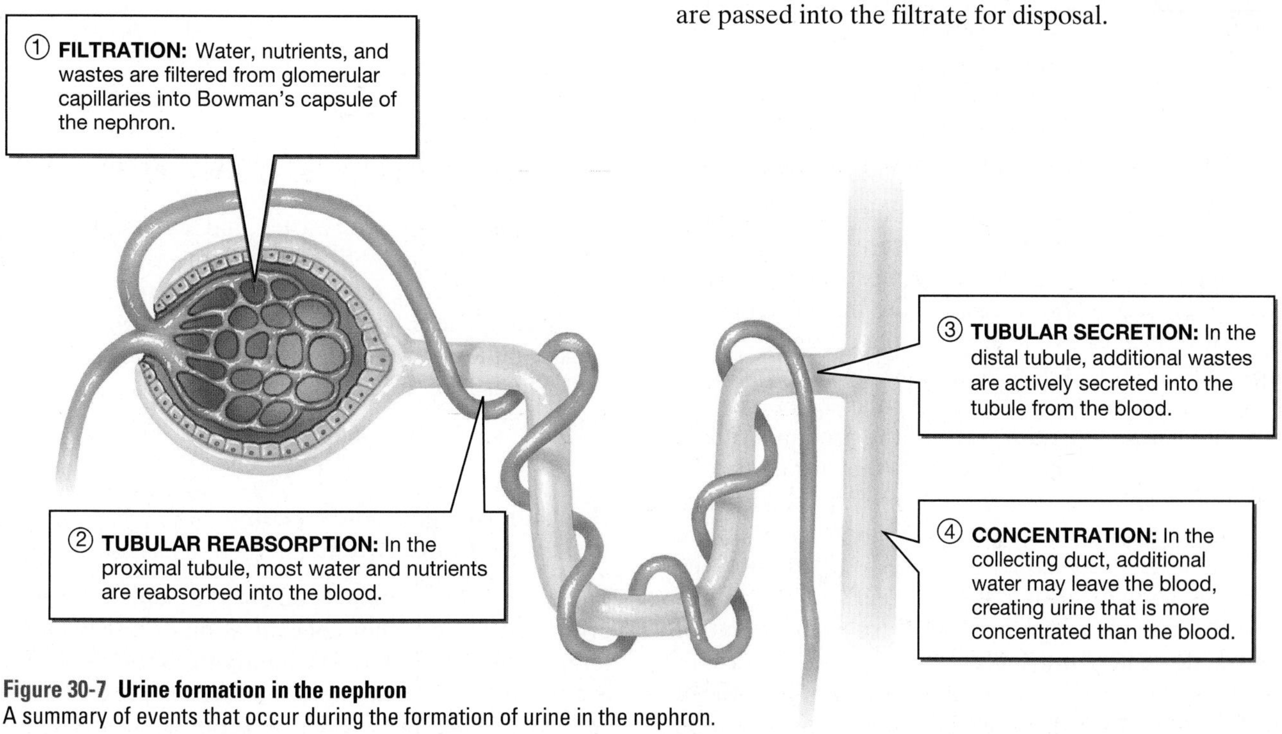

① FILTRATION: Water, nutrients, and wastes are filtered from glomerular capillaries into Bowman's capsule of the nephron.
② TUBULAR REABSORPTION: In the proximal tubule, most water and nutrients are reabsorbed into the blood.
③ TUBULAR SECRETION: In the distal tubule, additional wastes are actively secreted into the tubule from the blood.
④ CONCENTRATION: In the collecting duct, additional water may leave the blood, creating urine that is more concentrated than the blood.

Figure 30-7 Urine formation in the nephron
A summary of events that occur during the formation of urine in the nephron.

A Closer Look
The Nephron and Urine Formation

The complex structure of the nephron is finely adapted to its function. In Figure E30-1, the nephron is presented in diagram form to illustrate the processing that occurs in each part. The circled numbers in the illustration refer to the following descriptions:

① *Filtration.* Water and dissolved substances are forced out of the glomerular capillaries into Bowman's capsule and then funneled into the proximal tubule.

② *Tubular reabsorption.* In the proximal tubule, most of the important nutrients remaining in the filtrate are actively pumped out through the walls of the tubule and are reabsorbed into the blood. These nutrients include about 75% of the salts as well as amino acids, sugars, and vitamins. The proximal tubule is highly permeable to water, so water follows the nutrients, moving by osmosis out of the tubule and back into the blood.

③ The loop of Henle, which is unique to birds and mammals, is essential for urine concentration. It maintains a salt concentration gradient in the extracellular fluid that surrounds it, with the highest concentration at the bottom of the loop. The descending portion of the loop of Henle is very permeable to water but not to salt or other dissolved substances. As the filtrate passes through the descending portion, water leaves by osmosis as the concentration of the surrounding fluid increases.

④ The thin portion of the ascending loop of Henle is relatively impermeable to water and urea but is permeable to salt, which moves out of the filtrate by diffusion. Why? Although the osmotic concentrations inside and outside the tubule are about equal, at this portion of the loop the urea level is higher outside and the salt level is higher inside. Thus, the concentration gradient favors the movement of salt outward. Because water cannot follow it, the filtrate now becomes less concentrated than its surroundings.

⑤ The thick portion of the ascending loop of Henle is also impermeable to water and urea. There, salt is actively pumped out of the filtrate, leaving water and wastes behind.

⑥ The watery filtrate, low in salt but retaining wastes such as urea, now arrives at the distal portion of the tubule, where more salt is pumped out. Because this portion is permeable to water, water follows by osmosis. Tubular secretion occurs throughout the tubule but is especially active in the distal portion. There, substances such as K^+, H^+, NH_3, and some drugs and toxins are actively pumped into the tubule.

⑦ By the time the filtrate reaches the collecting duct, very little salt is left and about 99% of the water has been reabsorbed into the bloodstream. The collecting duct conducts the fluid, now called urine, down through the increasing concentration gradient created by the loop of Henle. The collecting duct is very permeable to water when antidiuretic hormone (ADH) is present, so water moves out by osmosis as the concentration of the external fluid increases. If ADH is absent, the collecting duct remains impermeable to water, and the urine stays dilute and watery.

⑧ The lower portion of the collecting duct is also permeable to urea. Therefore, as the filtrate moves farther down the collecting duct, some urea diffuses out, contributing to the osmotic concentration of the surrounding fluid. As water (when ADH is present) and urea move out, the concentration of dissolved wastes in the urine in the collecting duct approaches equilibrium with the high osmotic concentration of the external fluid.

The Filtrate Is Converted to Urine in the Nephron

Now for the major challenge—the filtrate collected in Bowman's capsule contains a mixture of both wastes and essential nutrients, including most of the blood's vital water. The nephron must restore the nutrients and most of the water to the blood, while retaining wastes for elimination. This is accomplished by two processes: (1) tubular reabsorption and (2) tubular secretion.

In **tubular reabsorption**, cells of the proximal tubule remove water and nutrients from the filtrate within the tubule and pass them back into the blood (Fig. 30-7, step ②). Salts and other nutrients, such as amino acids and glucose, are mostly reabsorbed into the blood by active transport (that is, cells of the tubule expend energy to transport these substances out of the tubule). These nutrients then enter adjacent capillaries by diffusion. Water follows the nutrients out of the tubule by osmosis. Wastes such as urea remain in the tubule and become more concentrated as water leaves.

In **tubular secretion**, wastes and excess substances that were not initially filtered out into Bowman's capsule are removed from the blood for excretion (Fig. 30-7, step ③). These wastes are actively secreted into the distal tubule by tubule cells. Secreted substances include hydrogen and potassium ions, ammonia, and many drugs.

The Loop of Henle Allows Urine to Become Concentrated

The kidneys of mammals and birds are able to produce urine that has a higher concentration of dissolved materials than their blood. The ability to concentrate urine is a result of the structure of both the nephron and the collecting duct into which several nephrons empty

MEDIATUTOR
30.2 Urine Formation

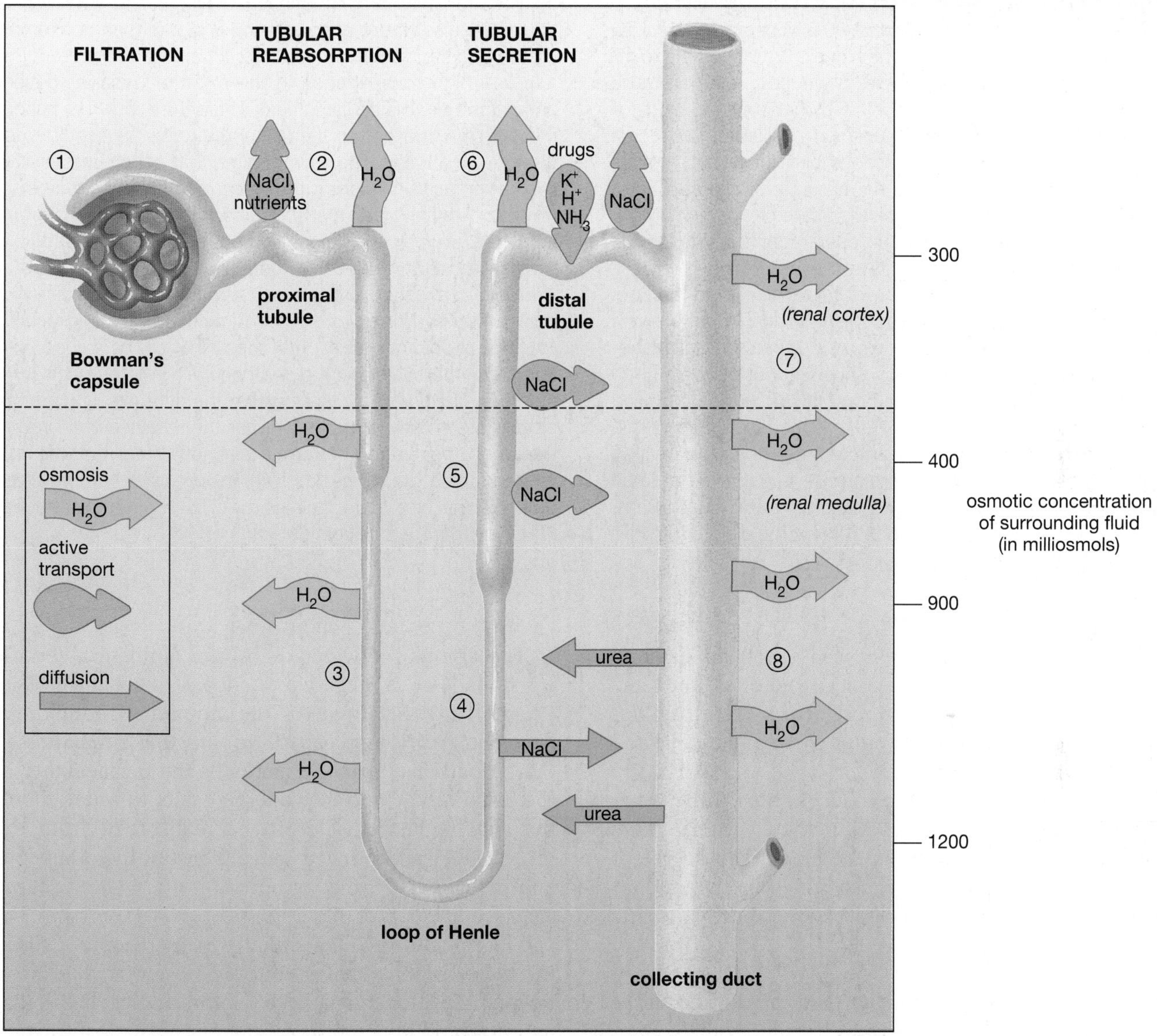

Figure E30-1 Details of urine formation
A single nephron, showing the movement of materials through different regions. The concentration of dissolved substances in the filtrate inside the nephron increases from top to bottom in this diagram. Outside the nephron, darker shades of blue represent higher concentrations of salts in the surrounding fluid. The dashed line marks the boundary between the renal cortex and the renal medulla.

(Fig. 30-7, step ④). Urine can become concentrated because there is an osmotic concentration gradient of salts and urea in the interstitial fluid that surrounds the loop of Henle. This gradient is produced by the loop of Henle within the renal medulla; the longer the loop, the greater the concentration gradient. The most concentrated fluid (with the greatest amount of dissolved substances and the least amount of water), far more concentrated than blood, surrounds the bottom of the loop. The collecting duct passes through this osmotic gradient. As the filtrate passes through the portion of the collecting duct surrounded by the osmotically concentrated fluid, additional water may leave the filtrate by osmosis and be carried off by the surrounding capillaries, while wastes are left behind in the collecting duct. Therefore, as it moves through the collecting duct,

Health Watch

When the Kidneys Collapse

If the kidneys fail, death occurs rapidly, usually within two weeks. Each year in the U.S., about 90,000 people die as a result of urinary system disease. The kidneys are vulnerable to attack from several sources. Diabetes is the most common cause of kidney failure, followed by high blood pressure. Both of these disorders damage the glomerular capillaries. Overdoses of painkilling medicines and infections, particularly by intestinal bacteria that reach the kidney via the urethra, can also harm the kidneys. Kidney failure is typically treated by hemodialysis.

First used in 1945, hemodialysis operates on the simple principle that substances will diffuse from areas of higher concentration to areas of lower concentration across an artificial permeable membrane, a process called *dialysis*. When blood is filtered according to this principle, the process is called **hemodialysis**. During hemodialysis, shown in our opener photo, the patient's blood is diverted from the body and pumped through narrow tubing made of a cellophane membrane suspended in dialyzing fluid. This membrane has pores too small to permit the passage of blood cells and large proteins but large enough to pass small molecules such as water, sugar, salts, amino acids, and urea. The composition of the dialyzing fluid is adjusted to have normal blood levels of salts and nutrients, so only molecules whose concentration is higher than normal will diffuse into the dialyzing fluid. Urea, for example, is present in a relatively high concentration in the blood of a dialysis patient and is absent in the dialyzing fluid, so it diffuses out. The patient must remain attached to the dialysis machine for 4 to 6 hours, three times a week. The average time that people in their 40s can be kept alive on dialysis is 10 years or less, and their lives are far from normal. Their diet and fluid intake must be carefully regulated. Even so, their blood composition fluctuates and toxic substances reach higher than normal levels between sessions. *Peritoneal dialysis* is a less common technique in which dialyzing fluid is pumped or poured through a tube implanted directly into the abdominal cavity. The abdominal cavity is lined with natural membrane called the *peritoneum*. Waste products from blood circulating in capillaries within the peritoneum gradually diffuse into the dialysis fluid, which is then drained out through the tube. A patient can perform this at home, replacing the dialysis fluid about four times daily, or linking his or her implanted tube to a machine that circulates the fluid through the abdominal cavity throughout the night.

the filtrate, now called *urine*, can reach an osmotic equilibrium, becoming as concentrated as the surrounding fluid. Because the rest of the excretory system does not allow water to enter or urea to escape, the urine remains concentrated.

It is important to produce concentrated urine when water is scarce, and to produce dilute, watery urine when there is excess water in the blood. The degree of concentration of the urine is controlled by the amount of *antidiuretic hormone*, to be described shortly.

The Kidneys Are Important Organs of Homeostasis

Each drop of blood in your body passes through a kidney about 350 times a day; thus, the kidney is able to fine-tune the composition of the blood and thereby helps maintain homeostasis. The importance of this task is illustrated by the fact that kidney failure results in death within a short time (see "Health Watch: When the Kidneys Collapse").

The Kidneys Regulate the Water Content of the Blood

One important function of the kidney is to regulate the water content of the blood (see Fig. 30-7). Human kidneys filter out about half a cup of fluid from the blood each minute. Thus, without reabsorption of water, you would produce more than 45 gallons of urine daily! Water reabsorption occurs passively by osmosis as the filtrate travels through the tubule and the collecting duct.

The amount of water reabsorbed into the blood is controlled by a negative feedback mechanism (see Chapter 26) that involves the amount of **antidiuretic hormone** (**ADH**; also called *vasopressin*) circulating in the blood. This hormone increases the permeability of the distal tubule and the collecting duct to water, thereby allowing more water to be reabsorbed from the urine. ADH is produced by cells in the hypothalamus and is released by the posterior pituitary gland (as we shall see in Chapter 32).

The release of ADH is regulated by receptor cells in the hypothalamus that monitor the osmotic concentration of the blood and by receptors in the heart that monitor blood volume. For example, as the lost traveler staggers through the searing desert sun, dehydration occurs. With the loss of water, the osmotic concentration of his blood rises and his blood volume falls, triggering the release of more ADH (Fig. 30-8). The release of ADH increases water reabsorption and produces urine more concentrated than the blood.

In contrast, a partygoer overindulging in liquid refreshment will experience a decrease in blood concentration and an increase in blood volume, and her receptors will cause a decrease in ADH output. Reduced ADH concentration will make the distal tubule and collecting duct less permeable to water. When the ADH level is very low, little water is reabsorbed after the urine leaves the loop of Henle, and the urine produced will be more dilute than the blood. In extreme cases, urine flow may exceed 1 liter (about 1 quart) per hour. As the proper

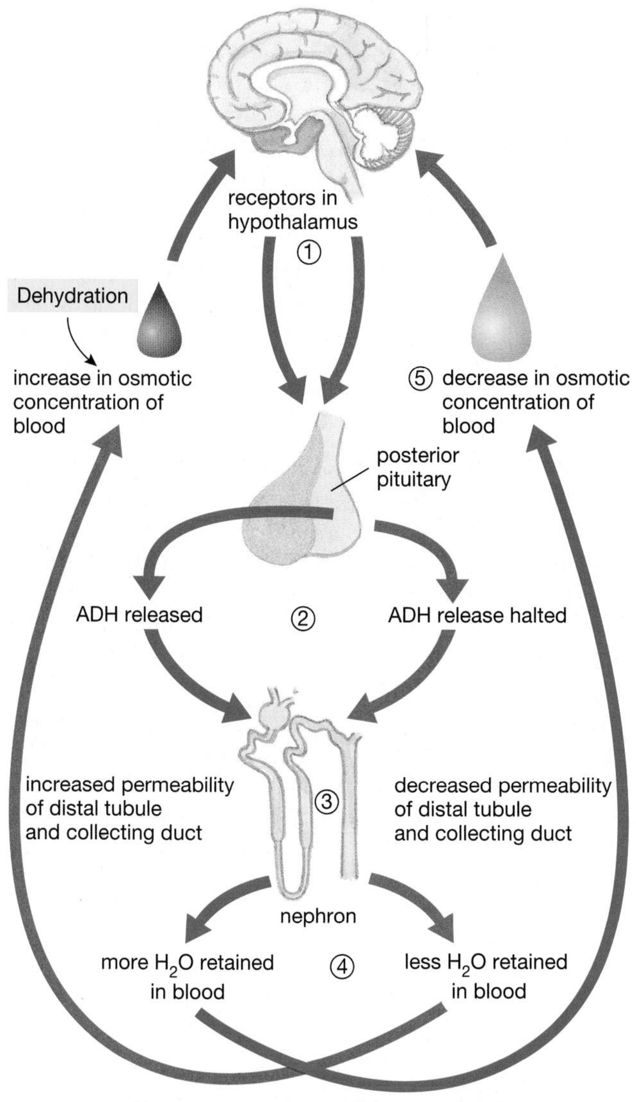

Figure 30-8 Regulation of the water content of the blood
The water content of the blood is regulated in part by a negative feedback mechanism. Sensors in the hypothalamus gauge the concentration of solutes in the blood; deviations from a set point cause the anterior pituitary gland to turn on or turn off a release of the hormone ADH. Any perturbation of the system (for example, by dehydration) triggers a change that will restore blood solute concentration to the appropriate level (red arrows indicate increased osmotic concentration; green arrows indicate decreased concentration).

water level is restored, the increased osmotic concentration of the blood and decreased blood volume will stimulate increased ADH production, which will in turn stimulate increased water reabsorption.

Mammalian Kidneys Are Adapted to Diverse Environments

In addition to regulating water balance by using hormones such as ADH, different types of mammals have kidneys with structures adapted to the availability of water in their natural habitats. Mammals that must conserve water do so by producing urine that is more concentrated than their blood. The degree of concentration that can be achieved is determined by the length of the loop of Henle. The longer the loop, the higher the salt concentration produced in the fluid that surrounds it. Higher salt concentration causes more water to move by osmosis out of the urine as it passes through the collecting duct, resulting in more concentrated urine. As you might predict, animals living in very dry climates have the longest loops of Henle, whereas those living in watery environments have relatively short loops.

The beaver, for example, has only short-looped nephrons and is unable to concentrate its urine to more than twice its plasma concentration. Human kidneys have a mixture of long-looped and short-looped nephrons and can concentrate urine to about four times the plasma concentration. The masters of urine concentration are desert rodents such as kangaroo rats, which can produce urine 14 times more concentrated than their plasma (Fig. 30-9). Kangaroo rats (as you might predict) have only very long-looped nephrons. Because they have an extraordinary ability to conserve water, they can completely dispense with drinking, relying entirely on water derived from their food.

Kidney structure is not the only factor that affects water balance in the mammalian body. The body's water content is also affected by water inputs from drinking and eating, by evaporative water loss through the skin (for example, sweating), and by an animal's behavior. For instance, kangaroo rats have developed their extraordinary ability to conserve water mainly

Figure 30-9 An adaptation to a dry environment
The desert kangaroo rat of the southwestern United States can dispense with drinking partly because its long loops of Henle allow it to produce very concentrated urine.

because their environment affords them little opportunity to drink. They must rely entirely on water derived from their food, and they have developed behaviors that help them both avoid unnecessary water loss and acquire the greatest possible amount of water from their food. Kangaroo rats live in underground burrows and venture above ground only at night. This behavior minimizes their exposure to hot desert temperatures by day, which in turn minimizes the amount of water they lose by sweating. During their nightly sojourns to the surface, kangaroo rats gather seeds to eat. But instead of eating them right away, the rats carry the seeds back to their burrows. Why? Because the air in the underground burrows is far moister than the parched desert air on the surface, and a seed stored underground will absorb precious water before it is eaten. For the kangaroo rat, behaviors that reduce exposure to dry air and dry food make important contributions to maintaining homeostasis.

Kidneys Help Regulate the Blood Pressure and Oxygen Content of the Blood

Two hormones produced by the kidneys are extremely important in regulating blood pressure and the blood's oxygen-carrying capacity. When blood pressure falls, the kidneys release **renin** into the bloodstream. Renin acts as an enzyme, catalyzing the formation of a second hormone, **angiotensin**, from a protein that circulates in the blood. Angiotensin in turn causes arterioles to constrict, elevating blood pressure. The constriction of the arterioles that carry blood to the kidneys also reduces the rate of blood filtration, causing less water to be removed from the blood. Water retention causes an increase in blood volume and, consequently, an increase in blood pressure.

In response to low blood oxygen levels, the kidneys release a second hormone, **erythropoietin**, described in Chapter 27. Erythropoietin travels in the blood to the bone marrow, where it stimulates more rapid production of red blood cells, the role of which is to transport oxygen.

Kidneys Monitor and Regulate Dissolved Substances in the Blood

As the kidney filters the blood, it monitors and regulates blood composition in order to maintain a constant internal environment. Substances the kidney regulates, in addition to water, include nutrients such as glucose, amino acids, vitamins, urea, and a variety of ions, including sodium, potassium, chloride, and sulfate. The kidney maintains a constant blood pH by regulating the amount of hydrogen and sodium bicarbonate ions. This remarkable organ also eliminates potentially harmful substances, including some drugs, food additives, pesticides, and toxic substances from cigarette smoke, such as nicotine.

CASE STUDY REVISITED

Hemodialysis and Hope

What does the future hold for victims of urinary system diseases? At the University of Michigan, Dr. David Humes is working to perfect a "bioartificial kidney." Similar to a hemodialysis machine, Humes's device filters blood through tubes, but these tubes are lined with living kidney cells. Having successfully tested his device on dogs, Humes has applied to the Food and Drug Administration for permission to perform human tests. Eventually, he hopes to use this technology to develop an implantable bioartificial kidney. Meanwhile, at Children's Hospital in Boston, Dr. Anthony Atala has grown bladder muscle and endothelial cells on a plastic framework that biodegrades over time, leaving behind an organ that is difficult to distinguish from a real bladder. For patients who have lost bladders due to infection or cancer, small samples of their own remaining bladder cells might soon be used to grow a new bladder in the lab. Implanted in dogs whose bladders have been removed, these bioartificial bladders develop nerve and blood vessel connections, and have returned normal bladder control and function to the animals. Large medical firms are also continuing work on xenotransplantation—a process that would allow people to receive kidneys from animals such as pigs, genetically engineered to prevent immediate tissue rejection. Michener's prediction that by the year 2010 radical new treatments would be available is almost certainly accurate—we may see them even sooner.

Your teenage son is dying from injuries sustained in a motorcycle accident. You have been asked to allow your son's kidneys to be donated to people who currently rely on dialysis. Explain what benefits these kidneys would confer on a dialysis patient. How could two dialysis patients be helped by your son?

Summary of Key Concepts

1) How Does Excretion Occur in Invertebrates?

The simple excretory system of the flatworm consists of a network of tubules that branch through the body and collect wastes and excess water for excretion through excretory pores. Many of the more complex invertebrates, including earthworms and mollusks, use nephridia, which resemble vertebrate nephrons, and filter the earthworm's coelomic fluid. Wastes and excess water are released through the excretory pore.

2) What Are the Functions of Vertebrate Urinary Systems?

The urinary system plays a crucial role in homeostasis, regulating the water and ion contents of the blood and blood pH. These organs help retain nutrients and eliminate cellular wastes and toxic substances, and they secrete the hormone erythropoietin.

3) How Does the Human Urinary System Function?

The human urinary system consists of kidneys, ureters, bladder, and urethra. Kidneys produce urine, which is conducted by the ureters to the bladder, a storage organ. Distension of the muscular bladder wall triggers urination, during which urine passes out of the body through the urethra.

Each kidney consists of more than a million individual nephrons in an outer layer. Urine formed in the nephrons enters collecting ducts that empty into the renal pelvis, from which it is funneled into the ureter.

Each nephron is served by an arteriole that branches from the renal artery. The arteriole further branches into a mass of porous-walled capillaries called the *glomerulus*. There, water and dissolved substances are filtered from the blood by pressure. The filtrate is collected in the cup-shaped Bowman's capsule and conducted along the tubular portion of the nephron. During tubular reabsorption, nutrients are actively pumped out of the filtrate through the walls of the tubule. Nutrients then enter capillaries that surround the tubule, and water follows by osmosis. Some wastes remain in the filtrate; others are actively pumped into the tubule by tubular secretion. The tubule forms the loop of Henle, which creates a salt concentration gradient surrounding it. After completing its passage through the tubule, the filtrate enters the collecting duct, which passes through the concentration gradient. Final passage of the filtrate through this gradient via the collecting duct allows the concentration of the urine.

The kidneys are important organs of homeostasis. The water content of the blood is regulated by antidiuretic hormone (ADH), produced in the hypothalamus and released by the posterior pituitary gland. Low blood volume and high osmotic concentration of the blood signal dehydration and stimulate the release of ADH into the bloodstream. The ADH increases the permeability to water of the distal tubule and the collecting duct, allowing more water to be reabsorbed into the blood. In addition, the kidneys control blood pH, remove toxic substances, and regulate ions such as sodium, chloride, potassium, and sulfate. Excess glucose, vitamins, and amino acids are also excreted by the kidneys.

Key Terms

ammonia *p. 607*
angiotensin *p. 614*
antidiuretic hormone (ADH) *p. 612*
bladder *p. 608*
Bowman's capsule *p. 608*
collecting duct *p. 608*
distal tubule *p. 608*
erythropoietin *p. 614*
excretion *p. 607*
excretory pore *p. 606*
filtrate *p. 609*
filtration *p. 609*
flame cell *p. 606*
glomerulus *p. 608*
hemodialysis *p. 612*
homeostasis *p. 607*
kidney *p. 607*
loop of Henle *p. 608*
nephridium *p. 606*
nephron *p. 608*
nephrostome *p. 606*
protonephridium *p. 606*
proximal tubule *p. 608*
renal artery *p. 607*
renal vein *p. 608*
renin *p. 614*
tubular reabsorption *p. 610*
tubular secretion *p. 610*
tubule *p. 608*
urea *p. 607*
ureter *p. 608*
urethra *p. 608*
uric acid *p. 607*
urine *p. 607*

Thinking Through the Concepts

Multiple Choice

1. *Which of the following is false?*
   a. Urea is more toxic than ammonia.
   b. Ammonia is converted to urea in the liver.
   c. Ammonia is produced in body cells.
   d. The fluid collected in Bowman's capsule is called the *filtrate*.
   e. *Tubule* means "little tube."

2. *The walls of the ____________ are made more or less permeable to water, depending on the need to conserve water.*
   a. ureter
   b. urethra
   c. proximal tubule
   d. collecting duct
   e. glomerulus

3. *Which of the following conditions will cause a decrease in ADH production?*
   a. dehydration
   b. drinking beer
   c. an increase in osmotic pressure of blood
   d. abnormally high blood sugar
   e. strenuous exercise

4. *The function of the glomerulus and Bowman's capsule of the nephron is to*
   a. reabsorb water into the blood
   b. eliminate ammonia from the body
   c. reabsorb salts and amino acids
   d. filter the blood and capture the filtrate
   e. concentrate the urine

5. *What determines the ability of a mammal to concentrate its urine?*
   a. the number of nephrons
   b. the length of the tubules
   c. the length of the collecting duct
   d. the size of the glomerulus
   e. the length of the loop of Henle

6. *Which of the following processes does NOT occur in the nephron and collecting duct?*
   a. filtration
   b. elimination of urea from the body
   c. reabsorption of nutrients
   d. tubular secretion
   e. concentration of urine

? Review Questions

1. Explain the two major functions of excretory systems.
2. Trace a urea molecule from the bloodstream to the external environment.
3. What is the function of the loop of Henle? The collecting duct? Antidiuretic hormone?
4. Describe and compare the processes of filtration, tubular reabsorption, and tubular secretion.
5. Describe the role of the kidneys as organs of homeostasis.
6. Compare and contrast the excretory systems of humans, earthworms, and flatworms. In what general ways are they similar? Different?

Applying the Concepts

1. Discuss the differences in function of the two major capillary beds in the kidneys: the glomerular capillaries and those surrounding the tubules.
2. Desert animals need to conserve water. These animals have larger kidneys than do animals that live in moist environments and thus need not conserve water. The larger kidneys allow for a greater distance between the glomerulus and the bottom of the loop of Henle. Discuss why this anatomical difference assists water conservation in the desert animals.
3. Some "quick weight loss" diets require the ingestion of much protein-rich food and the elimination of carbohydrates. Two side effects of such diets are increased thirst and increased urination. Explain the connections between the diets and the side effects.
4. Some employers require their employees to submit to urine tests before they can be employed and at random intervals during their employment. Refusal to take the test or failure to "pass" the test could be grounds for termination. What is the purpose of the urine test? What types of employers might find such tests necessary? How do you feel about urine tests for obtaining or keeping a job? Explain your answers.

For More Information

Greenberg, A. (ed.). *Primer of Kidney Disease*. San Diego: Academic Press, 1998. An up-to-date, comprehensive review of kidney diseases, written by a variety of experts.

O'Brien, C. "Lucky Break for Kidney Disease Gene." *Science*, June 24, 1994. Discusses research on a dominant gene that causes kidney disease.

Plafrey, C., and Cossins, A. "Fishy Tales of Kidney Function." *Nature*, September 24, 1994. A description of research findings that have increased our understanding of how fish cope with the physiological challenges posed by watery and/or salty environments.

Taylor, C. R. "The Eland and the Oryx." *Scientific American*, January 1969. An account of the physiological adaptations by which two species of African antelopes cope with the hot, dry environments in which they live.

Answers to Multiple-Choice Questions

1. a 2. d 3. b 4. d 5. e 6. b

MediaTutor
The Urinary System

CD Activities

Activity 30.1: Urinary System Anatomy

Estimated time: 5 minutes

The urinary system plays an important role in maintaining homeostasis. In addition to ridding the body of wastes in the form of urine, this system plays a crucial role in water balance. In this tutorial, you will explore the anatomy of the human urinary system.

Activity 30.2: Urine Formation

Estimated time: 10 minutes

The nephron is the functional unit of the kidney and is responsible for modifying the composition of blood and producing urine. The process of urine formation involves filtration, reabsorption, secretion, and concentration. In this tutorial you will explore these processes in more detail.

Start the MediaTutor Student CD-ROM and enter the activity number in the Quick Search box to be taken directly to that activity.

Web Investigations

Case Study: Hemodialysis and Hope

Estimated time: 10 minutes

A person with end-stage renal disease (kidney failure) has few options. At this time there are only three major treatments, each with considerable side effects, health risks, and costs. Sadly, in many cases, the disease could have been prevented. This exercise looks at causes, treatments, and prevention of end-stage renal disease.

Go to http://www.prenhall.com/audesirk6, the Audesirk Companion Web site. Select Chapter 30 and the Web Investigation to begin.

"If you wish to be well and keep well, take Braggs Vegetable Charcoal and Charcoal Biscuits. Absorbs all impurities in the stomach and bowels, effectively warding off cholera, smallpox, typhoid, and all malignant fevers. Eradicate worms in children. Sweeten the breath."

From an early 1900s newspaper ad

A sneeze propels thousands of microscopic droplets out of the nose and mouth at a rate of 200 miles per hour! There may be as many as 100,000 bacteria released during a single sneeze, and viruses are also spread in this manner.

31 Defenses Against Disease: The Immune Response

AT A GLANCE

Case Study: Fighting the Flu

1) **How Does the Body Defend Against Invasion?**

First, the Skin and Mucous Membranes Form External Barriers to Invasion

Second, Nonspecific Internal Defenses Combat Microbes

Third, the Body Mounts an Immune Response Against Specific Microbes

2) **What Are the Key Characteristics of the Immune Response?**

First, the Immune System Must Recognize the Invader

Second, the Immune System Must Launch an Attack

Third, the Immune System Must Remember Its Past Victories

3) **How Does Medical Care Augment the Immune Response?**

Antibiotics Slow Down Microbial Reproduction

Vaccines Stimulate the Development of Memory Cells

4) **What Happens When the Immune System Malfunctions?**

Allergies Are Misdirected Immune Responses

An Autoimmune Disease Is an Immune Response Against Some of the Body's Own Molecules

An Immune Deficiency Disease Results from the Inability to Mount an Effective Immune Response to Infection

AIDS Is a Devastating Immune Deficiency Disease

Cancer Can Evade or Overwhelm the Immune Response

Case Study Revisited: Fighting the Flu

CASESTUDY

Fighting the Flu

Have you ever been sitting in class when the person next to you lets loose with a juicy, full-bodied sneeze? While drawing in a breath to whisper "Gesundheit," you are unwittingly inhaling thousands of microscopic droplets laden with "germs"; perhaps viruses from the common cold, flu viruses, and certainly bacteria from the flourishing population that grow in everyone's nose and mouth. The symptoms of a cold and of some types of flu—sneezing, coughing, and a runny nose—are all ideally suited to spread the virus. In fact, natural selection favors viruses that cause symptoms that help them spread to other victims—such as you.

After class, your sneezing classmate admits he's been feeling achy and feverish. You grumpily advise him to take some aspirin and go home to bed. Remembering being debilitated by flu last year, you fervently hope you are now immune. Should you go immediately and get a flu shot? Was your suggestion of taking aspirin good advice? How is your body coping with the microbes you have unwittingly inhaled? Look for answers as you read this chapter and rejoin us at the end. ■

1 How Does the Body Defend Against Invasion?

Your body, and indeed the bodies of all forms of life, are under constant attack from other life-forms trying to make a living. Our environment is teeming with viruses, bacteria, fungi, and protists that, should they have the opportunity to take up residence, can harm or even kill; we will refer to this diverse group of parasites as **microbes**. The environment is likewise home to a huge array of toxic substances that can pose a threat to health. Given the diversity and ubiquity of potential threats, it may make more sense to ask, "Why don't we get sick more often?" Among our ancestors, those who best survived and reproduced were those whose bodies best resisted attack by parasites and toxins. These resistant individuals passed on their adaptations to subsequent generations, and eventually to us. On the other side, however, the most evolutionarily successful parasites are those whose attack strategies are most capable of overcoming resistance. As a result, animals and their parasites are engaged in a never-ending, constantly escalating battle in which ever more sophisticated defense systems are challenged by ever more effective tactics for penetrating those defenses. This "evolutionary arms race" has honed our defenses into a stunningly complex system for resisting parasitic invasion. Much of the battle takes place invisibly, at the cellular level. The body mobilizes an army of cells to patrol its territory and seek out and attack invaders. This cellular army, however, faces a complex challenge as it combats diverse and resourceful opponents whose survival depends on gaining access to the stable environment and abundant resources inside the host's body.

The body has three lines of defense against microbial attack: *First*, external barriers to invasion keep microbes out of the body; *second*, nonspecific internal defenses combat invading microbes; and *third*, the immune system, which directs its assault—an *immune response*—against specific microbes (Fig. 31-1).

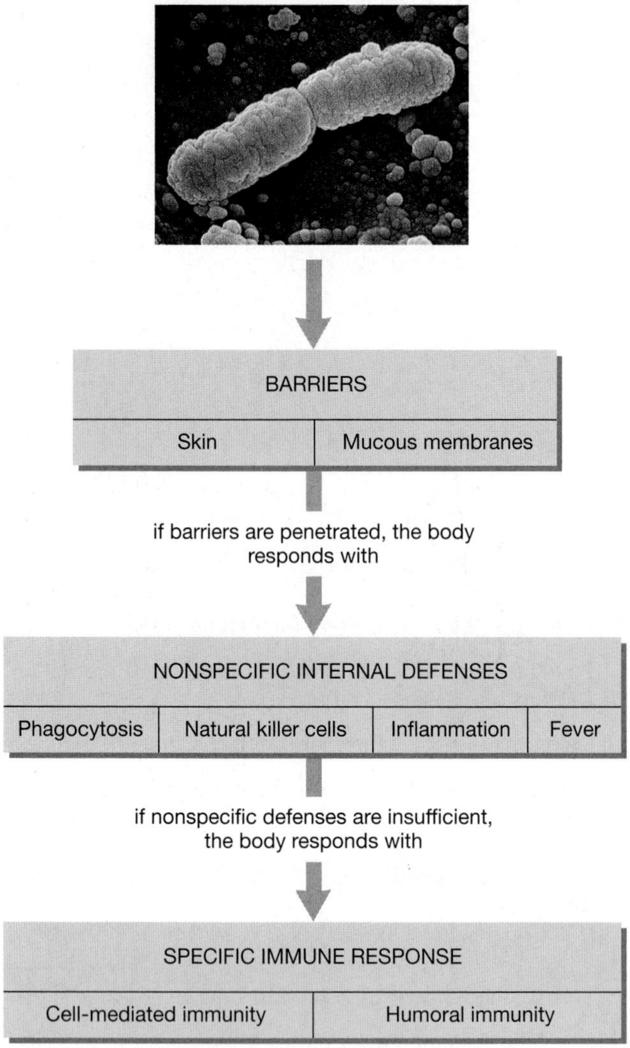

Figure 31-1 Levels of defense against infection

First, the Skin and Mucous Membranes Form External Barriers to Invasion

The best defense strategy is to prevent invaders from entering the body in the first place. In animal bodies, this first line of defense is performed by the two surfaces that are exposed to the environment: (1) the **skin** and (2) the **mucous membranes** of the digestive, respiratory, and urogenital tracts.

Skin Is Both a Barrier to Entry and an Inhospitable Environment for Microbial Growth

The virus and bacteria-laden droplets from your neighbor's sneeze that land on your skin will encounter an outer surface consisting of dry, dead cells filled with horny proteins similar to those in hair and nails. Most of them will not obtain the water and nutrients they need to survive. Those few bacteria and fungi that gain a foothold on skin will usually be shed before they can do harm, because skin cells are constantly sloughed off and replaced from below. Secretions from sweat glands and sebaceous glands also protect the skin. These secretions contain natural antibiotics, such as lactic acid, that inhibit the growth of bacteria and fungi. These multiple defenses make the unbroken skin an extremely effective barrier against microbial invasion.

Antimicrobial Secretions, Mucus, and Ciliary Action Defend the Mucous Membranes Against Microbes

The warm, moist mucous membranes that line the digestive, respiratory, and urogenital tracts are much more hospitable to microbes than is the dry, oily skin, but these membranes nonetheless possess effective defense mechanisms. In particular, the membranes secrete mucus that contains antibacterial enzymes such as lysozyme, which destroys bacterial cell walls. The mucus also physically traps microbes that enter the body

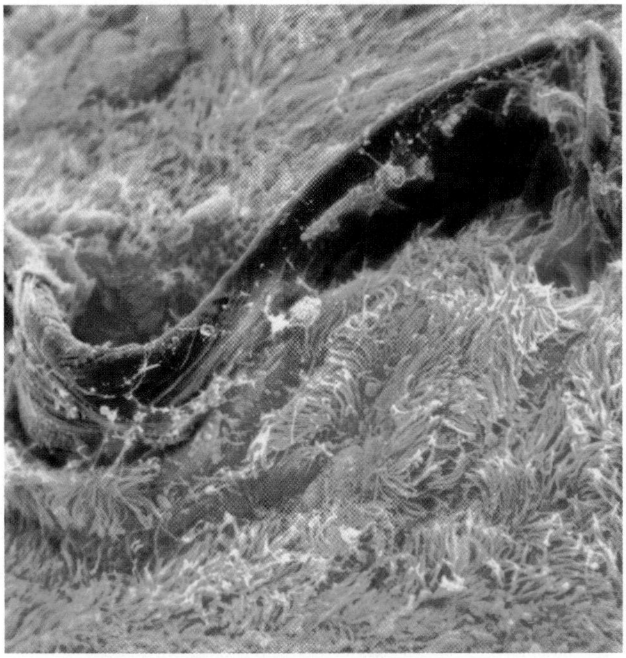

Figure 31-2 The protective function of mucus
Mucus traps microbes and debris in the respiratory tract (a brownish-green strand of dirt is shown caught in mucus atop the orange cilia). The cilia lining the walls of the respiratory tract then sweep both mucus and foreign matter out of the body.

through the nose or mouth (Fig. 31-2). Cilia on the membranes sweep up the mucus, microbes and all, until it is either coughed or sneezed out of the body, or swallowed. If microbes are swallowed, they enter the stomach, where they encounter both extreme acidity (pH 1 to 3) and protein-digesting enzymes, which can kill many types of microbes. Farther along in the digestive tract, the intestine is inhabited by bacteria that are harmless in their intestinal habitat but secrete substances that destroy invading foreign bacteria or fungi. Despite these defenses, many disease organisms manage to enter the body through the mucous membranes. As you inhale the fallout from the sneeze, many viruses enter your lungs; some will probably penetrate these vulnerable membranes. Let's look at what happens next.

Second, Nonspecific Internal Defenses Combat Microbes

Invading parasites that penetrate the first line of defense at the skin and mucous membranes encounter an array of internal defenses. Some of these defenses are nonspecific—that is, they attack a wide variety of microbes rather than targeting specific invaders as the immune response does. Nonspecific defenses fall into three main categories: (1) The body has a standing army of **phagocytic cells**, which directly destroy microbes, and **natural killer cells**, which destroy cells of the body that have been infected by viruses. The steady trickle of microbes that pass through the body's external barriers are mostly mopped up by these cells. (2) An injury, with its combination of tissue damage and relatively massive invasion of microbes, provokes an **inflammatory response**. The inflammatory response simultaneously recruits new members of the army of phagocytic cells and natural killer cells and walls off the injured area, isolating the infected tissue from the rest of the body (Fig. 31-3a,b). (3) If a population of microbes succeeds in establishing a major infection, the body may produce a **fever**, which both slows down microbial reproduction and enhances the body's own fighting abilities.

Phagocytic Cells and Natural Killer Cells Destroy Invading Microbes

The body contains several types of amoeboid white blood cells that can engulf and digest microbes. One important type are the **macrophages** (literally, "big eaters"), white blood cells that crawl around in the blood and extracellular fluid (Fig. 31-3c). Macrophages ingest microbes by phagocytosis; they'll "eat" some of the inhaled bacteria and foreign substances that penetrate the delicate mucous membranes of your lungs. As we shall discuss later, macrophages also play a crucial role in the immune response by acting as *antigen-presenting cells*—"presenting" parts of the microbe to other cells of the immune system.

Natural killer cells are another class of white blood cells. In general, natural killer cells strike at the body's own cells that have been invaded by viruses. Once infected by a virus, cells acquire some viral proteins on their surfaces. Natural killer cells seek out and recognize these proteins and kill the infected cell. Natural killer cells also recognize and kill cancerous cells. Rather than eat their victims, natural killer cells secrete proteins into the plasma membrane of the infected or cancerous cell. These proteins insert themselves and form a ring (much like making a barrel out of individual staves of wood), thereby opening up a large pore in the membrane (Fig. 31-4, p. 623). Killer cells also secrete enzymes that break up some of the molecules of the target cell. Shot full of holes and chewed on by enzymes, the infected or cancerous target cell soon dies.

The Inflammatory Response Defends Against Localized Injury

Have you ever been scratched by a playful but overexuberant cat? The broken skin of your hand and the microbes from the cat's claws set the stage for an infection. But torn skin or mucous membranes cause an inflammatory response (see Fig. 31-3). Damaged cells release the chemical **histamine** into the wounded area. Histamine makes capillary walls leaky and relaxes the smooth muscle that surrounds arterioles, leading to increased blood flow. With extra blood flowing through leaky capillaries, fluid seeps from the capillaries into the tissues around the wound. The wound becomes red,

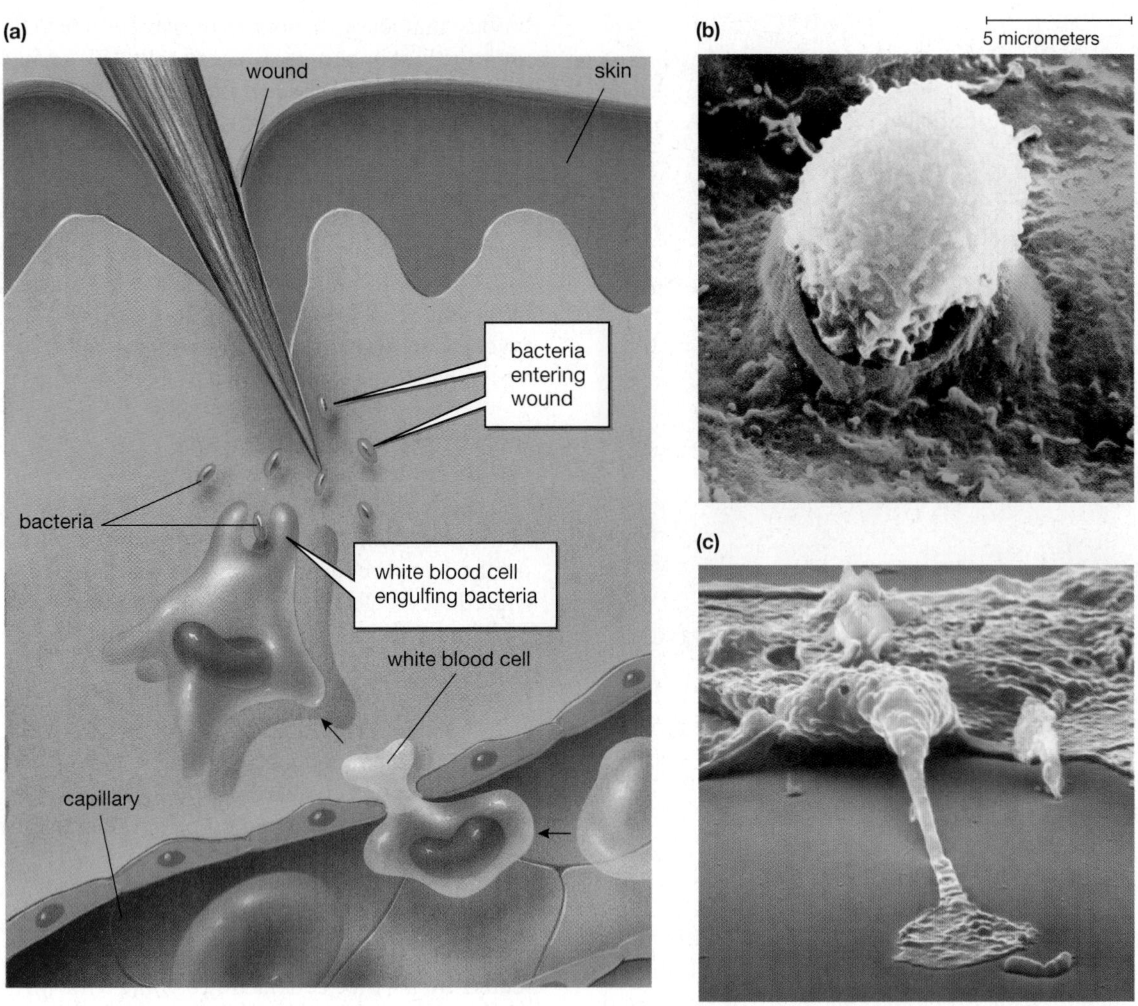

Figure 31-3 The inflammatory response
(a) During an inflammatory response triggered by a cut in the skin, white blood cells such as macrophages squeeze through the gaps and enter the wounded tissue, where they engulf microbes, dirt, and debris from damaged cells.
(b) A micrograph showing a white blood cell leaving a capillary to join the fray against bacteria that have entered a cut.
(c) A macrophage advances toward a rod-shaped bacterium; this photo was taken on an artificial surface for clarity.

swollen, and warm. (*Inflammation* literally means "to set on fire.") Meanwhile, other chemicals released by injured cells initiate blood clotting (see Chapter 27), which blocks damaged blood vessels, preventing microbes from entering the bloodstream, and also seals off the wound from the outside world, limiting the entry of more microbes.

Still other chemicals, some released by wounded cells and others produced by the microbes themselves, attract macrophages and other phagocytic cells to the wound. Some of the phagocytic cells emerge from local tissues; others arrive via the circulatory system and squeeze out through the leaky capillary walls. After entering the wound, the phagocytic cells engulf microbes, dirt, and damaged cells (see Fig. 31-3). Unfortunately, each phagocyte can eat just so many microbes, and then it dies. If tissue damage is too severe or the wound is too dirty, then the phagocytes may be unable to complete the cleanup job. In that case, the fluid around the wound turns into pus, a thick mixture of microbes, tissue debris, and living and dead white blood cells.

Fever Combats Large-Scale Infections

If invaders breach other defenses and mount a large-scale infection, they may trigger a fever. If your sneezing classmate has the flu and you catch it, you will almost certainly develop a fever (a cold will rarely produce a fever). Although high fevers can be dangerous, and even moderate ones can be very unpleasant, fever is actually part of the body's defense against infection. The onset of fever is controlled by the hypothalamus, the part of the brain that contains the temperature-sensing nerve cells that are the body's thermostat. Normally, the thermostat is set at about 37 °C (98.6 °F). Certain macrophages, as they respond to the infection, release hormones called **endogenous pyrogens** ("self-

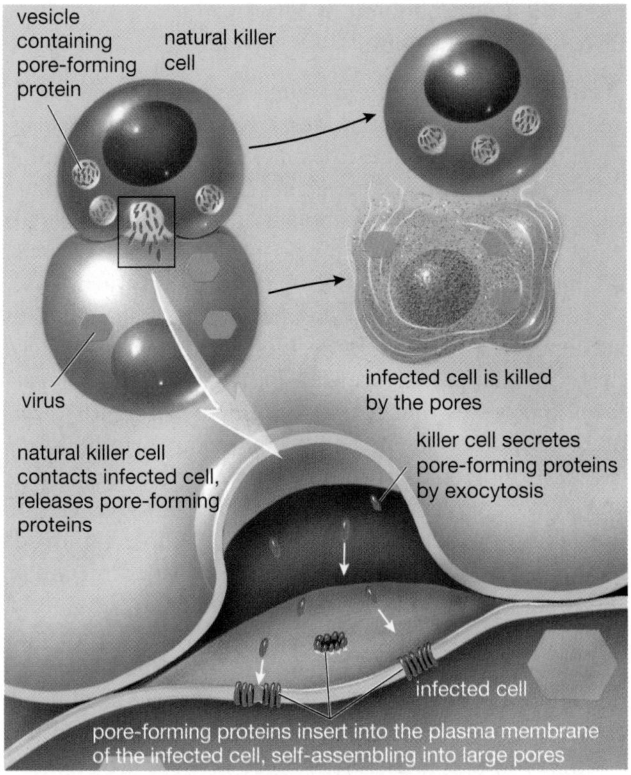
vesicle containing pore-forming protein
natural killer cell
virus
infected cell is killed by the pores
natural killer cell contacts infected cell, releases pore-forming proteins
killer cell secretes pore-forming proteins by exocytosis
infected cell
pore-forming proteins insert into the plasma membrane of the infected cell, self-assembling into large pores

Figure 31-4 How natural killer cells kill their targets
(a) A natural killer cell releases the pore-forming proteins by exocytosis. These proteins assemble into large holes in the target cell's plasma membrane. The cell's contents leak out, and the cell dies. ***(b)*** A close-up view of the killing process. Pore-forming proteins insert themselves into the target cell's plasma membrane and line up next to one another like staves in a barrel, forming large pores.

produced fire-makers"). Pyrogens travel in the bloodstream to the hypothalamus and raise the thermostat's set point, triggering responses that increase body temperature: shivering, increased fat metabolism, constriction of surface blood vessels, and behaviors such as burrowing under blankets. Pyrogens also cause other cells to reduce the concentration of iron in the blood.

Fever both enhances the body's normal defenses and harms invading microbes. In the case of bacterial infection, fever increases the activity of the phagocytic white blood cells that attack bacteria. At the same time, feverish body temperatures of about 102 °F force many bacteria to use more iron for reproduction than is used at the human body's normal temperature, so fever and reduced iron in the blood combine to slow down bacterial reproduction.

Fever also helps fight viral infections by increasing the production of the protein **interferon**. Some types of cells synthesize and release interferon after invasion by viruses, and the interferon travels to other cells and increases their resistance to viral attack. Owing to this benefit of fever, it can be a mistake to attempt to control or reduce mild fevers. In one study, individuals with a viral infection were treated with aspirin (to reduce fever) or a *placebo* (an inactive substance that looks like the real drug, so the subjects don't know whether they have been given the drug or not). The subjects given aspirin had far more viruses in their noses and throats—and, consequently, sneezed and coughed out far more viruses—than did the subjects given the placebo. The immune systems of subjects with fevers lowered by aspirin weren't as effective at controlling infections, and these subjects were much more infectious to other people. Since some fevers are dangerous, and since fevers are almost always a sign of infection, be sure to consult with a health professional when you are dealing with one.

MEDIATUTOR 31.1 Inflammation

Third, the Body Mounts an Immune Response Against Specific Microbes

Phagocytic cells, natural killer cells, the inflammatory response, and fever are all *nonspecific* defenses; their role is to prevent or overcome *any* microbial invasion of the body. Unfortunately, however, these nonspecific defenses are not impregnable. When they fail to do the job, the body's immune system mounts a highly specific *immune response* directed against the particular organism that has successfully invaded the body. The immune system is an unusual "system." Unlike the nervous system, for example, it is not composed of physically attached structures. Instead, as befits its mission of patrolling the entire body for microbial invaders, the **immune system** consists of an army of separate cells. Nevertheless, the army is highly coordinated. This coordination requires complex communications involving hormones, receptors, cells, antigens, and antibodies.

The essential features of the immune response to infection were recognized more than 2000 years ago by Greek historian Thucydides. He observed that occasionally someone would contract a disease, recover, and never catch that particular disease again—the person had become immune. With rare exceptions, however, immunity to one disease confers no protection against other diseases. Thus, the immune system attacks one type of microbe, overcomes it, and provides future protection against that microbe but no others. This is why we refer to the immune response as a defense against the invasion of *specific* microbes or toxins.

2 What Are the Key Characteristics of the Immune Response?

Let's assume that the worst happens and you catch the flu from your neighbor in class. The virus penetrates all your nonspecific defenses and activates your immune system. Your immune system includes about 2 trillion white blood cells called *lymphocytes*. Lymphocytes are distributed throughout the body in the blood and lymph, and many are clustered in specific organs, particularly the thymus, lymph nodes, and spleen. The **immune response** arises from interactions among the various types of lymphocytes and the molecules that they produce, with the result that foreign molecules or microbes are eliminated from the body. The theater of the immune response has a large cast of characters and is difficult to follow without a program. Table 31-1 provides a brief overview of the major cellular actors and their roles.

The key actors in the immune response are two types of lymphocytes, called **B cells** and **T cells** (Table 31-1). Like all white blood cells, B cell lymphocytes and T cell lymphocytes arise from lymphocyte precursor cells in the bone marrow (marrow is found in the interior of certain bones; see Chapter 34). Some of these lymphocyte precursors are released into the bloodstream and travel to the thymus, where they complete their differentiation into T (for thymus) cells. In contrast, B cells differentiate in the bone marrow. The two cell types play quite different roles in the immune response, but immune responses produced by both B cells and T cells consist of the same three fundamental steps. The immune system must *first*, recognize the invader, *second*, launch an attack, and *third*, retain a memory of the invader to ward off future infections.

Table 31-1 The Cells of the Immune Response

Macrophages	White blood cells that both engulf invading microbes and help alert other immune cells to the invasion
Natural killer cells	White blood cells that destroy infected or cancerous cells
B cells	Lymphocytes that produce antibodies
Plasma cells	Offspring of B cells that secrete antibodies into the bloodstream
Memory B cells	Offspring of B cells that provide future immunity against invasion by the same antigen
T cells	A set of lymphocytes that regulate the immune response or kill certain types of cells
Cytotoxic T cells	Offspring of T cells that destroy specific targeted cells, normally either foreign eukaryotic cells, infected body cells, or cancerous body cells
Helper T cells	Offspring of T cells that stimulate immune responses by both B cells and cytotoxic T cells
Suppressor T cells	Offspring of T cells that inhibit immune responses by other lymphocytes after an infection is conquered
Memory T cells	Offspring of T cells that provide future immunity against invasion by the same antigen

First, the Immune System Must Recognize the Invader

To understand how the immune system recognizes invading microbes and initiates a response, we must answer three related questions: (1) How do the immune cells recognize foreign molecules? (2) How can immune cells produce specific responses to so many different types of molecules? (3) How do immune cells avoid mistaking the body's own cells and molecules for invaders?

To answer these questions, first we must explain that immune cells identify molecules or microbes because these invaders either consist of, or bear on their surfaces, molecules that serve as **antigens** (short for "*anti*body response *gen*erating"). In general, only large, complex molecules such as proteins, polysaccharides, and glycoproteins can act as antigens. Antigens can be located on the surfaces of cells (for example, on invading microbes) or dissolved in the blood or extracellular fluid (snake venom and bacterial toxins are dissolved antigens). Next, we must examine the structure and function of two kinds of large proteins: **antibodies**, produced by B cells, and **T-cell receptors**, produced by T cells. Antibody proteins either remain attached to the surfaces of the B cells that produce them or are released by the B cells and become dissolved in the blood plasma. T-cell receptors always remain attached to the surfaces of the T cells that produced them; they are never secreted into the plasma.

Antibodies Recognize and Bind to Foreign Molecules

Antibodies are Y-shaped molecules composed of two pairs of peptide chains: one pair of identical large (heavy) chains and one pair of identical small (light) chains (Fig. 31-5). Both heavy and light chains consist of a **constant region**, which is similar in all antibodies of the same type, and a **variable region**, which differs among individual antibodies. The combination of light and heavy chains results in an antibody with two functional parts: the "arms" and the "stem" of the Y. The variable regions at the tips of an antibody molecule's arms form highly specific binding sites for antigens. These binding sites are a lot like the active sites of enzymes (Chapter 6): Each binding site has a particular shape and electrical charge, so only certain molecules can fit in and bind. The binding sites are so specific that each antibody can bind at most a few types of antigen molecules—perhaps only one.

There are five classes of true antibodies, each serving a different function in the defense of the body. (For more information, please see "A Closer Look: Cellular Communication During the Immune Response" on this text's Web site.) The different types of antibody are distinguished by small differences in the structure of the constant region—that is, the stem. The stem determines the mechanism by which the antibody will act against invaders. For example, the stem of one type of

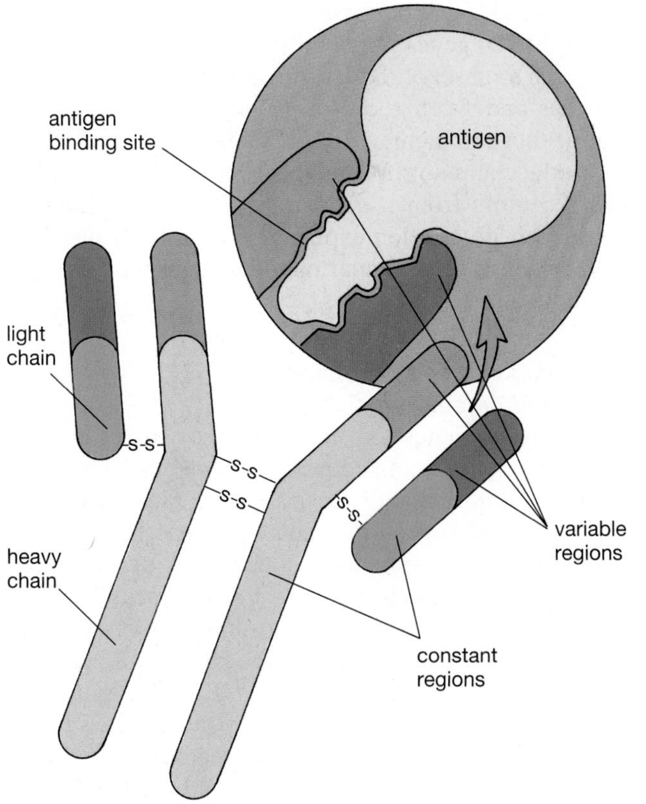

Figure 31-5 Antibody structure
Antibodies are proteins composed of two pairs of peptide chains (light chains and heavy chains) arranged like the letter Y. Constant regions on both chains form the stem of the Y; variable regions on the two chains form a specific binding site at the end of each arm of the Y. Different antibodies have different variable regions, forming unique binding sites. The human body synthesizes millions of distinct antibodies, each binding a different antigen.

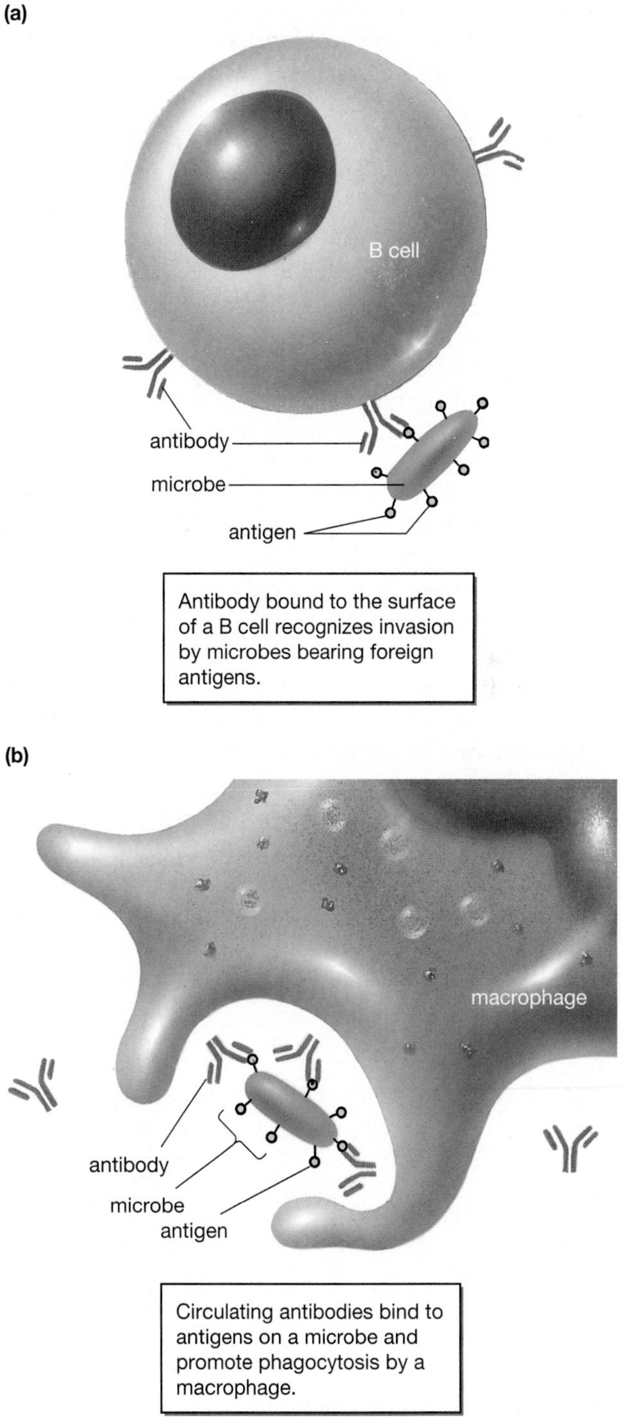

Figure 31-6 Some functions of antibodies

antibody may attach the antibody to the plasma membrane of a B cell, while the stem of another type of antibody may bind to certain proteins in the blood, called **complement** proteins (described later), to promote the destruction of microbes.

When the antibody stem remains attached to the B cell that produced it (Fig. 31-6a), the two arms of the antibody protrude outward from the B cell, sampling the blood and lymph for antigen molecules. When the arms of the antibody encounter an antigen with a compatible chemical structure, they bind to it. This binding triggers a response in the B cell that bears the antibody. Other antibodies circulate in the bloodstream (Fig. 31-6b), where they neutralize poisonous antigens, or destroy microbes that bear antigens, or attract macrophages that eat the antigen-bearing microbes. These functions are described in more detail in the following sections.

T-Cell Receptors Also Recognize and Bind Antigens

T-cell receptors have both differences from and similarities to antibodies. They are found only on the surfaces of T cells; unlike antibodies, none are released into the bloodstream. Although T-cell receptors consist of two peptide chains, these are of about equal size. The ends of the two chains protrude out from the T cell, forming highly specific binding sites for an antigen, just as the ends of the arms of an antibody molecule do. T-cell receptors trigger a response in the T cell only when they encounter antigen molecules borne on the membranes of cancerous or infected cells, or presented on the membranes of macrophages that have ingested the invading microbes. Unlike antibodies, the T-cell receptors do not directly contribute to the destruction of invading microbes or toxic molecules.

The Immune System Can Recognize and Produce Specific Responses to Millions of Types of Molecules

During your lifetime, your body will be challenged by a multitude of different invaders. Your classmates will sneeze cold and flu viruses into the air you breathe. Roadside weeds and trees will release pollen that will find its way to your lungs. Your food may play host to a bacterial population or to toxins released by bacteria. You may drink water that contains the microbes that cause dysentery or the *Giardia* parasite. As you relax outdoors, a tick that carries the bacteria responsible for Lyme disease might bite you. There is no escape from the pervasive and persistent assaults of the invaders. Fortunately, your immune system recognizes and responds to virtually all of the millions of antigens that you might encounter (both natural and artificial), because your immune cells produce millions of different antibodies and T-cell receptors, each capable of binding a different antigen. These amazing recognition abilities are one of the main reasons you don't get sick more often.

The mechanisms by which the immune system recognizes antigens are both complex and fascinating. Antibodies and T-cell receptors are, after all, proteins, and proteins are encoded by genes. It would seem that, to recognize millions of different antigens, each person would have to possess millions of different genes for antibodies and T-cell receptors. But there are probably as few as 35,000 genes in the entire human genome. How, then, can a subset of these genes code for millions of antibodies and T-cell receptors? The answer lies in two distinct but complementary mechanisms that join forces to produce an enormous diversity of antibodies and T-cell receptors from a comparatively small number of genes. Our description explains how antibody diversity develops in B cells; similar phenomena produce a large diversity of T-cell receptors.

Antibodies and T-Cell Receptors Are Pieced Together from Fragments

There are no genes for entire antibody molecules. Instead, the genome includes many genes that encode a variety of different antibody fragments; these fragments can be pieced together in many millions of combinations. As each individual B cell develops, it retains only a few randomly selected fragment genes (including both light and heavy chains and constant and variable regions); the rest are cut out from the DNA as the cell matures (Fig. 31-7). Therefore, each B cell produces a single type of antibody, specified by the chance recombination of variable- and constant-region genes, that is different from the antibody produced by other B cells (except its own daughter cells). The random selection of antibody-fragment genes from a large pool of choices yields a vast number of possible unique combinations. It may be helpful to think of antibody gene formation in terms of card playing. Each B cell is dealt a "hand" of two variable-region genes: one for the light chain and

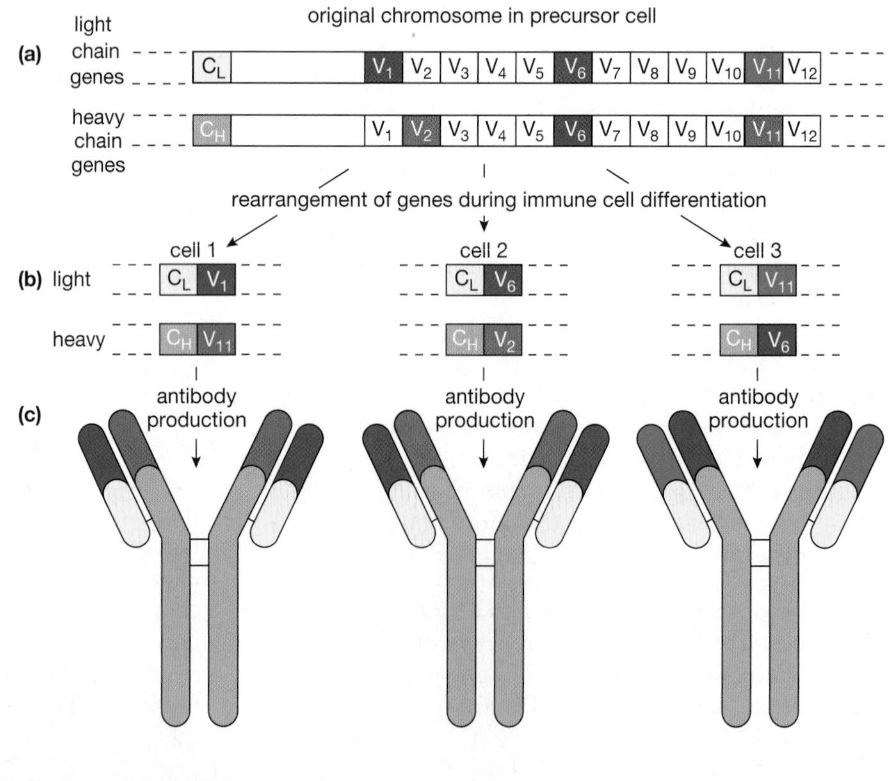

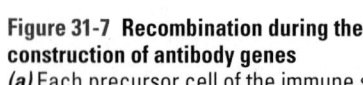

Figure 31-7 Recombination during the construction of antibody genes
(a) Each precursor cell of the immune system contains one or perhaps a few genes for the constant regions (C) of the light (C_L) and heavy (C_H) chains of antibodies and many genes for the variable regions (V_{1-12}). ***(b)*** During the development of each B cell, these genes are rearranged, moving one of the variable-region genes next to a constant-region gene. For each chain, each cell thus generates a "recombined antibody gene" that differs from the recombined antibody genes generated by other B cells. ***(c)*** The different antibodies synthesized by each B cell in (b).

one for the heavy chain, randomly chosen from two large "decks" of genes. With each deck containing hundreds of "cards" (genes), virtually every cell will synthesize its own unique antibody.

The Immune System Does Not Design Antibodies or T-Cell Receptors Expressly to Bind Invading Antigens

The end result of mutation and gene recombination is that each B cell has its own particular antibody genes, different from those of most other B cells. At any time, the human body contains an army of perhaps 100 million different antibodies (and an even larger number of T-cell receptors), so antigens almost always encounter antibodies or receptors that will bind them. It is important to recognize that the immune system does not "design" antibodies and T-cell receptors to fit invading antigens, as a tailor might design custom clothes for the queen of England. Instead, the immune system randomly synthesizes millions of different antibodies and receptors. Like clothes in a department store, the array of antibodies and receptors is simply there, waiting. The fact that virtually every possible invading antigen binds to at least a few antibodies and receptors is due simply to the immense numbers of different antibodies and receptors present in the body. In our clothing analogy, given enough racks of clothing from which to choose, each of us will find something suitable that fits. The binding of antigen to antibody triggers changes in the B cells. These changes usually lead to the destruction of microbes that bear that antigen. We will examine these changes in more detail in a moment.

The Immune System Distinguishes "Self" from "Non-Self"

Why doesn't your own immune system destroy your own cells? The surfaces of the body's own cells bear large proteins and polysaccharides, just as microbes do. Some of these proteins, collectively called the **major histocompatibility complex (MHC)**, are unique to each individual (except identical twins, who have the same genes and hence the same MHC proteins). Because your MHC proteins are different from those of everyone else, they act as antigens in other people's bodies. (That is why transplants are rejected; the recipient's immune system recognizes MHC proteins on the donor's cells as foreign and destroys the transplanted tissue.)

So why don't these "self" antigens arouse your immune system? The key seems to be the continuous presence of the body's antigens while the immune cells mature. As an embryo develops, some differentiating immune cells do indeed produce antibodies or T-cell receptors that can bind the body's own proteins and polysaccharides, treating them as antigens. However, if these *immature* immune cells contact molecules that bind to their antibodies or T-cell receptors, the immune cells are destroyed. In this way, potentially dangerous immune cells, which would attack the body's own cells, are eliminated during immune system development. Thus, the immune system distinguishes "self" from "non-self" by retaining only those immune cells that do not respond to the body's own molecules.

Second, the Immune System Must Launch an Attack

Once your body has been invaded, say by a flu virus, your immune system mounts two types of attack: (1) **Humoral immunity** is provided by B cells and the circulating antibodies they secrete into the bloodstream; invaders are attacked before they can enter body cells. (2) **Cell-mediated immunity** is produced by T cells, which attack invaders that have made their way into body cells. These two types of immune responses require considerable communication among the cell types (as described in "A Closer Look: Cellular Communication During the Immune Response," which appears in Chapter 31 on this text's Web site). Although humoral and cell-mediated immunity are not completely independent, we consider them separately below for ease of understanding.

Humoral Immunity Is Produced by Antibodies Dissolved in Blood

Each B cell bears a specific type of antibody on its surface. When an infection occurs, the antibodies borne by a few B cells are able to bind to antigens on the invader. Antigen–antibody binding causes these B cells to divide rapidly. This process is called **clonal selection** because the resulting population of cells is composed of "clones" (genetically identical to the parent B cells) that have been "selected" to multiply by the presence of particular invading antigens (Fig. 31-8). The daughter cells differentiate into two cell types: (1) **memory B cells** and (2) **plasma cells**. Memory cells do not release antibodies but do play an important role in future immunity to the particular invader that stimulated their production (as we shall soon see). Plasma cells become enlarged and packed with endoplasmic reticulum (Fig. 31-9); these cells churn out huge quantities of their own specific antibodies. These antibodies are released into the bloodstream (hence the name "humoral" immunity; to the ancient Greeks, blood was one of the four "humors," or body fluids). Scientists have exploited the ability of plasma cells to produce large quantities of a specific type of antibody for a variety of medical uses; see "Scientific Inquiry: Monoclonal Antibodies—Designer Drugs Fight Disease."

Because antibodies circulate in the bloodstream, humoral immunity can only defend against invaders that are in the blood or extracellular fluid. Bacteria (most of which never enter the body's cells), bacterial toxins, and some fungi and protists are therefore vulnerable to the humoral immune response. Invaders that penetrate into the body's cells, as viruses do, are safe from antibody attack as long as they are within the cytoplasm of a body cell. Antibodies can attack viruses only during the initial

MEDIATUTOR 31.2 Clonal Selection

Figure 31-8 Clonal selection among B cells by invading antigens

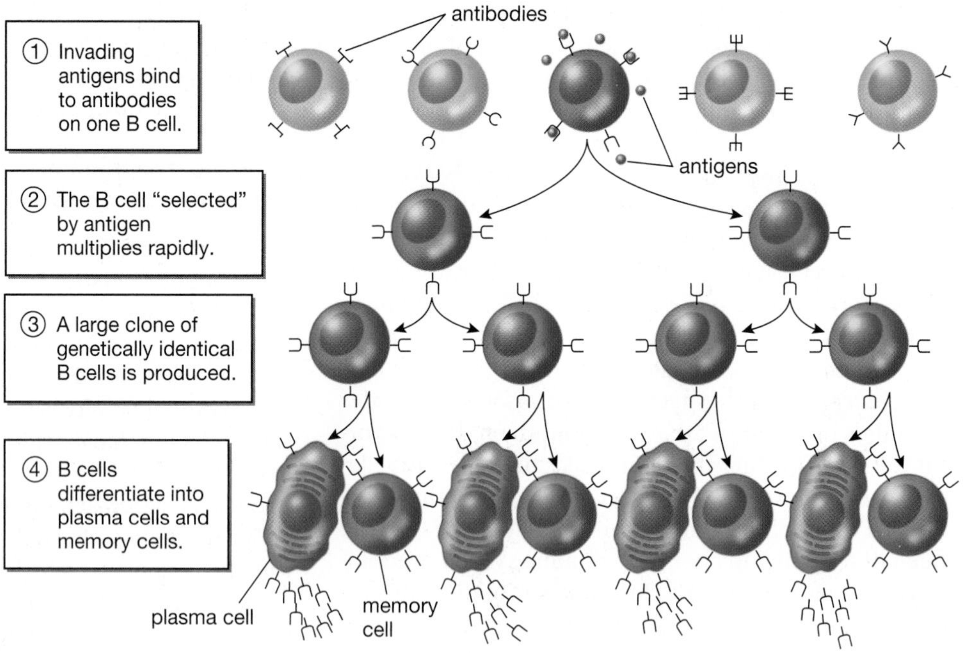

① Invading antigens bind to antibodies on one B cell.
② The B cell "selected" by antigen multiplies rapidly.
③ A large clone of genetically identical B cells is produced.
④ B cells differentiate into plasma cells and memory cells.
antibodies
antigens
plasma cell
memory cell

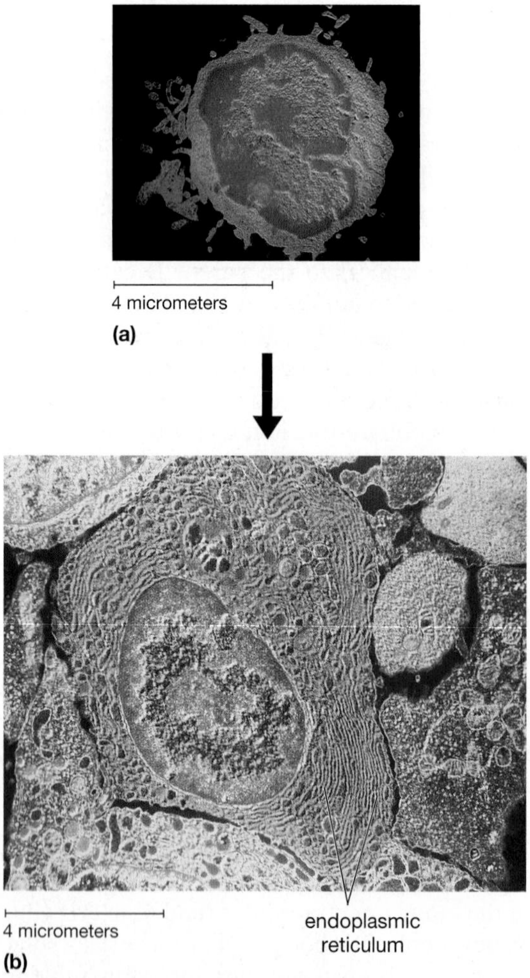

4 micrometers
(a)
4 micrometers
(b)
endoplasmic reticulum

Figure 31-9 A B cell becomes a plasma cell
Colorized micrographs of B cells *(a)* before and *(b)* after conversion to plasma cells. The plasma cell is much larger than the B cell and is virtually filled with rough endoplasmic reticulum that synthesizes antibodies.

infection and when the viruses have finished replicating in one host cell, have ruptured it, and have been released into the body fluids.

Humoral immunity produced by antibodies in the blood may destroy foreign chemicals or microbes in several ways. First, the circulating antibodies can bind to a foreign molecule and render it inactive. Second, the antibodies may coat the surface of an invading cell, or third, each antibody may bind two microbes and cause them to clump together. Both of these tactics encourage macrophages to engulf the microbes. Finally, the antibody–antigen complex on the surface of an invading cell may trigger a series of reactions with a group of blood proteins called the **complement system**. When these complement proteins bind to the antibody stems, a complex series of reactions is produced among the different members of the complement system. The end result is that the proteins attract phagocytic white blood cells to the site, promote phagocytosis of the foreign cells, or in some instances directly destroy the invaders by creating holes in their plasma membranes, much as natural killer cells do (see Fig. 31-4).

Humoral immunity works best against toxic chemicals, or bacterial, fungal, or parasitic microbes. Humoral immunity will also work against viruses as they enter the body or after they are released from infected cells. To fight the viral infection you contract from your sneezing classmate, however, you will also need the help of cell-mediated immune responses, described in the following section.

Cell-Mediated Immunity Is Produced by T Cells

How can the immune system seek out cells of your own body that are cancerous, or harboring viruses? Why are transplanted organs sometimes rejected?

Scientific Inquiry
Monoclonal Antibodies—Designer Drugs Fight Disease

In addition to their crucial role in defending the body, antibodies are valuable for many medical and scientific purposes. To acquire a supply of pure antibody, scientists can inject an animal with the antigen that induces B cells to form plasma cells that churn out the desired antibody. Since normal plasma cells cannot be grown outside the body, scientists induce them to merge with plasma tumor cells to form hybrid cells called **hybridomas**. The hybridomas proliferate under laboratory conditions and produce relatively large quantities of a specific antibody, known as **monoclonal antibodies**. Because a given antibody binds only to one specific type of antigen, antibodies are invaluable tools for finding and/or marking molecules in living animals or in dissected tissues. For example, pregnancy tests typically use an antibody that binds to a hormone released by a developing embryo.

For these "therapeutic antibodies" to reach their full potential in the fight against disease, they must be produced in massive quantities. Modern genetic engineering techniques (see Chapter 13) are being used to introduce the genes for human antibodies into farm animals (engineered cows and goats now secrete antibodies in their milk) and even into corn and other plants. Scientists are experimenting with antibody therapies directed against a variety of invaders and toxins including tooth decay bacteria, herpes viruses, cancer cells, and even addictive drugs. Human clinical trials are now under way that treat cancer victims with cancer cell antibodies that are linked to radioactive particles that they carry directly and specifically to the cancer. In a recent experiment, antibodies to cocaine were used to inactivate the drug in rats, suggesting a possible treatment for addicts. The ability to mass-produce specifically designed antibodies opens up an exciting frontier in the prevention and cure of diseases—using an arsenal derived from the immune system's own weapons.

These are the results of cell-mediated immunity. Four types of T cells contribute to cell-mediated immunity: (1) *cytotoxic T cells*, (2) *helper T cells*, (3) *suppressor T cells*, and (4) *memory T cells.*

Cytotoxic T cells release proteins that disrupt the infected cell's plasma membrane (Fig. 31-10). This attack is activated when T-cell receptors on the cytotoxic T cell's membrane bind to antigens on the surface of an infected cell. In some instances, the proteins released by cytotoxic T cells are similar or identical to those that natural killer cells use to create giant holes in the target cell's membrane. **Helper T cells** also have receptors for microbial antigens presented on the plasma membranes of macrophages that have engulfed the invader or on the surfaces of infected cells. Upon binding an antigen, the helper T cells release hormonelike chemicals that assist other immune cells defend the body. These chemicals stimulate cell division and differentiation in both B cells and cytotoxic T cells that respond to the same microbial invasion. In fact, very little immune response, either cell-mediated or humoral, can occur without the chemical boost provided by helper T cells. That is why AIDS, which destroys helper T cells, is such a deadly disease. **Suppressor T cells** act after an infection has been conquered. They help shut off the immune response in both B and cytotoxic T cells. After an infection is over, some suppressor T cells and helper T cells persist and function as **memory T cells**. Like memory B cells, these help protect the body against future infection. Figure 31-11 compares the humoral immune response with the cell-mediated immune response.

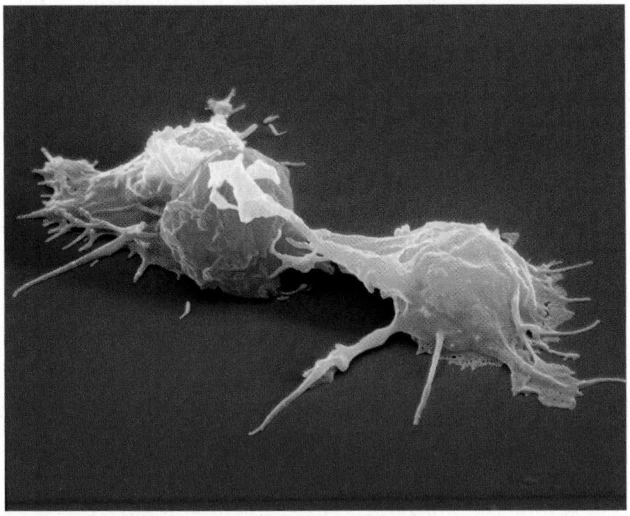

Figure 31-10 Cell-mediated immunity at work
Cytotoxic T cells (yellow) attack a cancer cell (red) in this false-color SEM photo.

Third, the Immune System Must Remember Its Past Victories

After you have recovered from a disease, you will remain immune to that particular strain of microbe for many years, perhaps a lifetime. Retaining immunity is the function of memory cells. Plasma cells and cytotoxic T cells directly fight disease organisms, but they normally live only a few days. Each of these, however, leaves behind hundreds or thousands of memory cells that

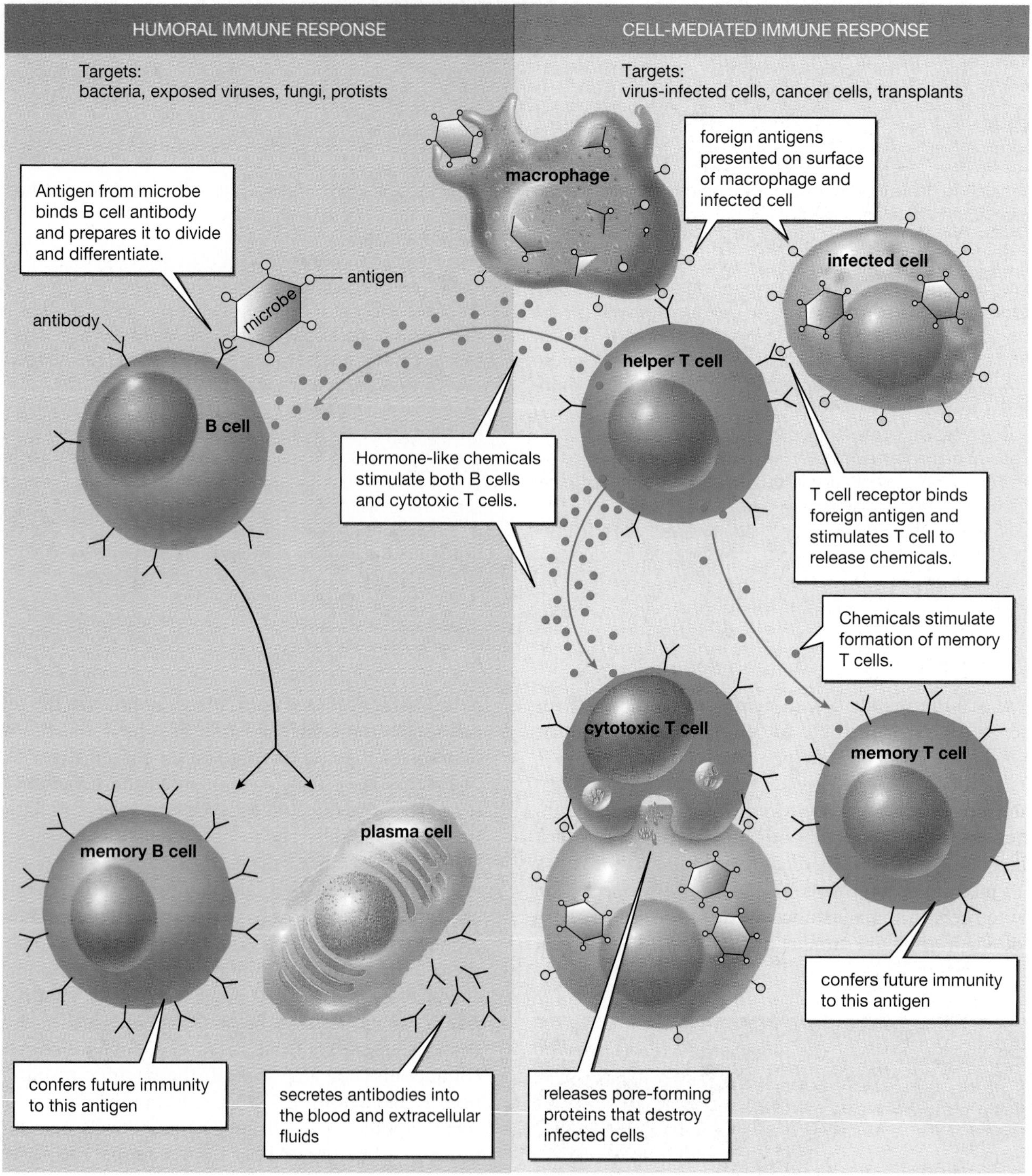
HUMORAL IMMUNE RESPONSE
CELL-MEDIATED IMMUNE RESPONSE
Targets:
bacteria, exposed viruses, fungi, protists
Targets:
virus-infected cells, cancer cells, transplants
macrophage
foreign antigens presented on surface of macrophage and infected cell
Antigen from microbe binds B cell antibody and prepares it to divide and differentiate.
antigen
microbe
antibody
infected cell
helper T cell
B cell
Hormone-like chemicals stimulate both B cells and cytotoxic T cells.
T cell receptor binds foreign antigen and stimulates T cell to release chemicals.
Chemicals stimulate formation of memory T cells.
cytotoxic T cell
memory T cell
memory B cell
plasma cell
confers future immunity to this antigen
confers future immunity to this antigen
secretes antibodies into the blood and extracellular fluids
releases pore-forming proteins that destroy infected cells

Figure 31-11 A summary of humoral and cell-mediated immune responses

may survive for many years. If the body is re-invaded by microbes to which the immune system has already mounted a response, the appropriate memory cells will recognize the invaders. These memory cells will then multiply rapidly and produce a second immune response by generating huge populations of plasma cells and cytotoxic T cells. Because there are so many memory cells primed to respond, the body is able to fend off a second attack before it gains a foothold (Fig. 31-12). Why then, do you seem to keep coming down with colds and flu? The problem, as we describe in "Health Watch: Can We Beat the Flu Bug?" is that this year's colds and flus aren't caused by the same viruses that you fought off last year.

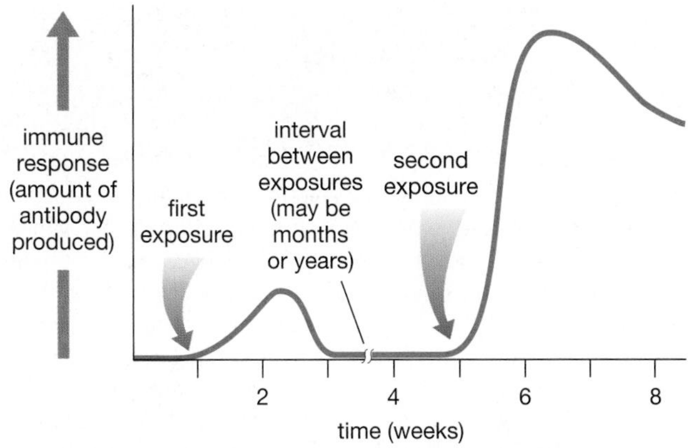

Figure 31-12 Memory cells and the immune response
The immune response to the first exposure to a disease organism is fairly slow and not very large, as B and T cells are selected and multiply. A second exposure activates memory cells formed during the first response, so the second response is both faster and larger.

3 How Does Medical Care Augment the Immune Response?

Antibiotics Slow Down Microbial Reproduction

We may have made the immune system sound invincible, but as you know, this is not the case. If untreated, many diseases kill their victims because, unfortunately, the body provides ideal conditions for the growth and reproduction of disease microbes. The infection thus becomes a race between the invading microbes and the immune response. If the initial infection is massive or if the microbes produce particularly toxic products, the full activation of the immune response may be too late. Further, some microbes have evolved ingenious defenses that help them evade the immune system. These include the microscopic worm responsible for schistosomiasis and the protist responsible for malaria (see "Earth Watch: Dams, Deforestation, and Diseases").

Antibiotics are drugs that help combat infection by slowing down the growth and multiplication of many invaders, including bacteria, fungi, and protists (but not viruses). Although antibiotics usually don't destroy every single microbe, they give the immune system enough time to finish the job. One problem with antibiotics, however, is that they are potent agents of natural selection. The occasional mutant microbe that is resistant to an antibiotic will pass on the gene(s) for resistance to its offspring. The result: Resistant mutants proliferate, whereas susceptible microbes die off. Eventually, many antibiotics become ineffective in treating diseases.

Until very recently, little could be done about viral infections except to treat the symptoms and hope that the immune system would be successful. Now, a new class of drugs called *neuraminidase inhibitors* has entered the physician's arsenal. These anti-viral drugs work by blocking the ability of viruses to enter cells, which is necessary for the viruses to replicate. In the coming years we may see the immune system receiving powerful help from these drugs (see "Health Watch: Can We Beat the Flu Bug?").

Vaccinations Stimulate the Development of Memory Cells

As early as A.D. 1000, people in India, China, and Africa deliberately exposed themselves to mild cases of smallpox to acquire immunity to the disease. In 1798 Edward Jenner discovered that infection with cowpox conferred immunity to smallpox. This discovery initiated the modern practice of *immunization.* In the late 1800s, Louis Pasteur extended the use of immunization to several other diseases by injecting weakened or dead microbes into healthy individuals. The weakened microbes do not cause disease (or at least not a severe case). However, the weakened microbes bear antigens that elicit vigorous immune responses, including an army of memory cells that confer immunity against subsequent exposure to the living, dangerous microbes. These injections of weakened or killed microbes to confer immunity are called **vaccinations**, from the Latin word for "cow," in honor of Jenner's pioneering efforts with cowpox. Today, many diseases, including polio, diphtheria, typhoid fever, and measles, can be controlled through vaccination. Smallpox, one of the deadliest diseases of all, has been completely eradicated since 1980 as a result of a massive vaccination program sponsored by the World Health Organization. Two facilities, one in Siberia and one in Atlanta, legally store what health officials hope are the last smallpox viruses on Earth. One compelling reason to maintain these stocks is as a source of genetic material that will allow scientists to make additional smallpox vaccines. A generation of people has never been vaccinated against smallpox, and some authorities believe that "bioterrorists" may possess their own stocks of smallpox virus.

Through genetic engineering (see Chapter 13), we now enjoy the prospect of manufacturing tailor-made vaccines. One method is to synthesize the antigenic proteins of disease-causing microbes. These antigens can then be used as vaccines without having to raise, isolate, and weaken the disease microbes themselves, or even to inject people with microbes. A vaccine against anthrax, a severe disease of livestock, has been manufactured by this procedure. A second technique is to insert genes that encode the antigens of, for example, herpes into the genome of harmless microbes such as the cowpox virus. These "designer" microbes produce herpes antigens without being able to cause the disease and have been shown to be effective in experiments with animals.

Health Watch
Can We Beat the Flu Bug?

Every winter, a wave of *influenza*, or flu, sweeps across the world. Thousands of the elderly, the newborn, and those already suffering from illness perish, while hundreds of millions more suffer the respiratory distress, fever, and muscle aches of milder cases. Occasionally, devastating flu varieties appear. In the great flu pandemic of 1918, the worldwide toll was 20 million dead in one winter. In 1968 the Hong Kong flu infected 50 million people in the U.S., causing 70,000 deaths in 6 weeks.

Flu Viruses Flu is caused by several viruses that invade the cells of the respiratory tract, turning each cell into a factory for manufacturing new viruses. The outer surface of a flu virus is studded with proteins, some of which are recognized by the immune system as antigens. Most people survive the flu because their immune systems eventually inactivate the viruses or kill off virus-infected body cells. Why then, don't our memory T and B cells make us immune to future outbreaks of flu?

The answer lies in a flu virus's amazing ability to change. Flu virus genes are made up of RNA, which lacks the proofreading mechanisms that reduce mutations in genes made of DNA (see Chapter 10). Therefore, flu RNA genes mutate rapidly. Four or five mutations in the same virus may alter the surface antigens enough that the immune system doesn't fully recognize the virus as the same old flu that was beaten off last year. So your "memory cells" provide only partial protection, and you get the flu again this year.

Deadly New Strains On rare occasions, dramatically new and deadly flu viruses appear, as in the pandemics (worldwide epidemics) of 1918, the Asian flu of 1957, the Hong Kong flu of 1968, and the Russian flu of 1977. In these viruses, entirely new antigens appear suddenly with distinctive structures that the human immune system has never before encountered. In response, the immune system must "start from scratch," selecting out entirely new lines of B cells and T cells to attack the intruder. But in the meantime, the virus multiplies so rapidly that many individuals die or become so weakened that they contract some other fatal disease. Where do the genes that encode these new flu antigens come from? Believe it or not, they come from viruses that infect birds and pigs. The intestinal tracts of birds, especially migratory waterfowl such as ducks, may carry viruses strikingly similar to human flu viruses, although the infected waterfowl show no symptoms. While human flu viruses aren't known to infect birds, *both* human and bird viruses can infect pigs, so occasionally, both viruses will meet within the same pig cell. Once in a great while, new viruses that spring from a doubly infected pig cell end up with a mixture of genes from human and bird viruses (Fig. E31-1). Some of these hybrid viruses combine the worst genes of each (at least from our perspective): From the human virus, the deadly new viruses pick up the genes needed to subvert human cellular metabolism to produce new viruses; from the bird virus, they pick up genes for new surface antigens. The hybrid viruses can move easily from pigs to humans, because pigs live near humans and, like people, pigs cough when they have the flu. Most flu pandemics are thought to originate in China, where crops are grown to feed pigs and ducks, and the feces from the pigs and ducks are used to fertilize fish ponds. This is a very efficient farming practice, but, unfortunately, it also places pigs (ideal mixing vessels for flu viruses) in close proximity to humans and ducks. However, in Hong Kong in 1997, health officials made an alarming discovery—a new and fatal form of flu virus appeared in poultry and could be passed *directly* from the birds to humans. After 6 people died, all the chickens in Hong Kong's markets and farms were slaughtered. Fortunately, this virus had not yet evolved the capacity to spread from person to person; had it acquired this ability, a new and deadly pandemic might have swept the world.

Swatting the Flu Bug Flu viruses mutate so rapidly that they may be impossible to eradicate. Fortunately, we now have two lines of defense. Flu shots are vaccinations, but how can you vaccinate against a mutating virus? Each year, World Health Organization officials collect new samples of mutated flu virus from more than 100 sites throughout the world, and identify three strains they believe are most likely to spread widely. These viruses are allowed to multiply in fertilized chicken eggs, and the viral proteins are then isolated and injected into people to stimulate an immune response that protects against subsequent infections by these strains. Flu shots are often quite effective, especially if the correct strains were selected. For people who

4 What Happens When the Immune System Malfunctions?

Allergies Are Misdirected Immune Responses

As your classmate sneezes nearby, you entertain a fleeting hope that he is merely allergic to something in the air and not coming down with a cold. More than 35 million Americans suffer from **allergies**, adverse reactions to substances that are not harmful in themselves and to which many other people do not respond. Common allergies include those to pollen, dust, mold spores, animal dander, and bee stings. Allergies are actually a form of immune response. A foreign substance, such as a pollen grain, enters the bloodstream and is recognized as an antigen by a particular type of B cell. This B cell proliferates, producing plasma cells that, as a result, pour out antibodies against the pollen antigens. The stems of the antibodies attach to the plasma membranes of **mast cells**, histamine-containing cells located in the respiratory and digestive tracts. Pollen antigens that later encounter and bind to the attached antibodies in the respiratory tract trigger the release of histamine, which (as you learned earlier) causes leaky capillaries and other symptoms of

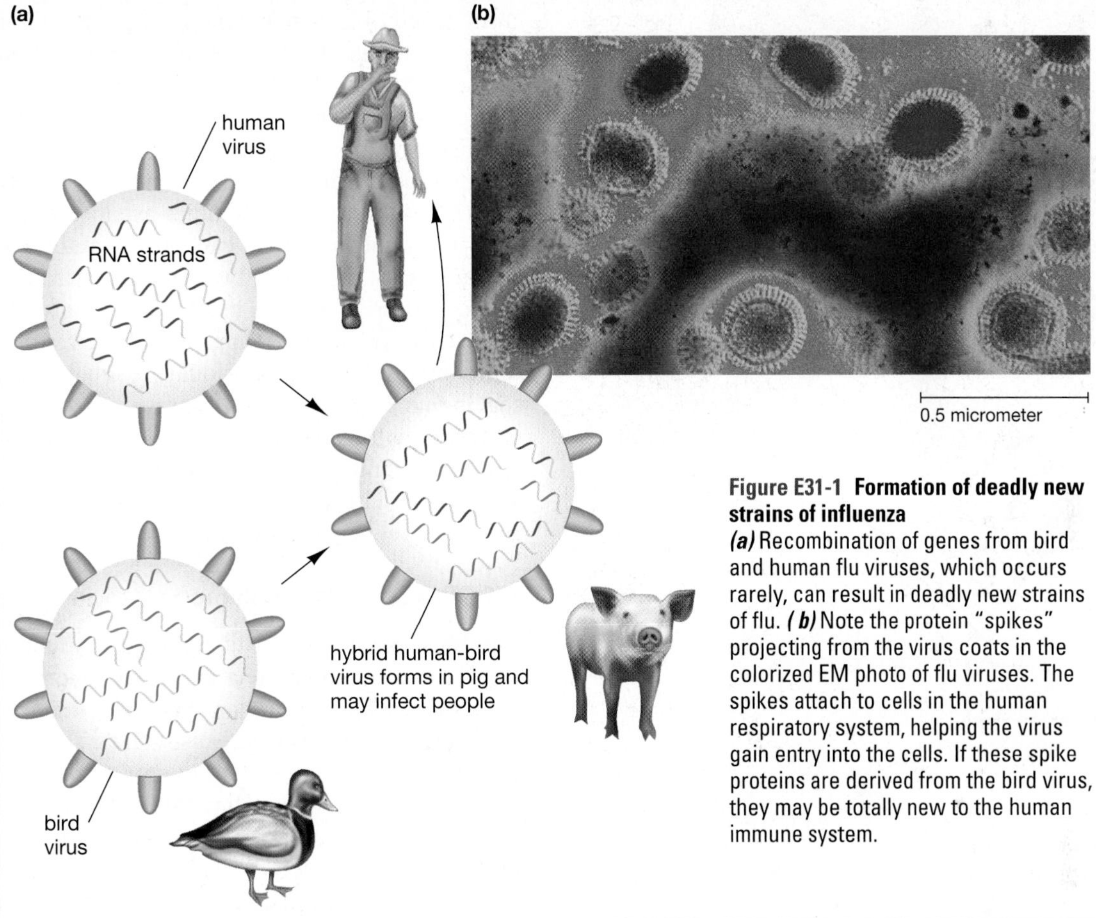

Figure E31-1 Formation of deadly new strains of influenza
(a) Recombination of genes from bird and human flu viruses, which occurs rarely, can result in deadly new strains of flu. ***(b)*** Note the protein "spikes" projecting from the virus coats in the colorized EM photo of flu viruses. The spikes attach to cells in the human respiratory system, helping the virus gain entry into the cells. If these spike proteins are derived from the bird virus, they may be totally new to the human immune system.

contract flu anyway, cleverly designed new drugs, first marketed in the U.S. in 2000, promise to disarm this disease. Certain regions of the spikelike proteins that stud the flu virus cannot mutate without disarming the flu. These crucial sites bind substances on our cell membranes and allow the virus to invade our cells, so they are conserved on all successful flu strains. The new drugs (neuraminidase inhibitors) bind these regions of the spikes tightly and prevent the viruses from entering human cells, the only place where they can multiply. Administered after flu symptoms begin, the drugs hasten recovery. Researchers hope that soon people will be tested for flu virus before symptoms occur; treatment at this early stage might prevent the symptoms entirely. While we may never eradicate the flu virus, it may soon lose its ability to disrupt our lives.

inflammation (Fig. 31-13) . In the respiratory tract, histamine also increases mucus secretion. Because airborne substances such as pollen grains typically enter the nose and throat, the resulting allergic reactions often include the runny nose, sneezing, and congestion of "hay fever." Food allergies cause analogous symptoms, including cramps and diarrhea, in the digestive tract. This type of allergic response might defend the body against parasites, since these same symptoms could help dislodge and expel the parasites. That antibodies may act against pollen as well as disease-causing microbes may be an unfortunate side effect of an otherwise useful immune response.

An Autoimmune Disease Is an Immune Response Against Some of the Body's Own Molecules

Fortunately, our immune systems rarely mistake our own cells for invaders. Occasionally, however, something goes awry, and "anti-self" antibodies are produced. The result is an **autoimmune disease**, in which the immune system attacks some component of one's own body. Some types of anemia, for example, are caused by antibodies that destroy an individual's red blood cells. Many cases of juvenile-onset diabetes occur because

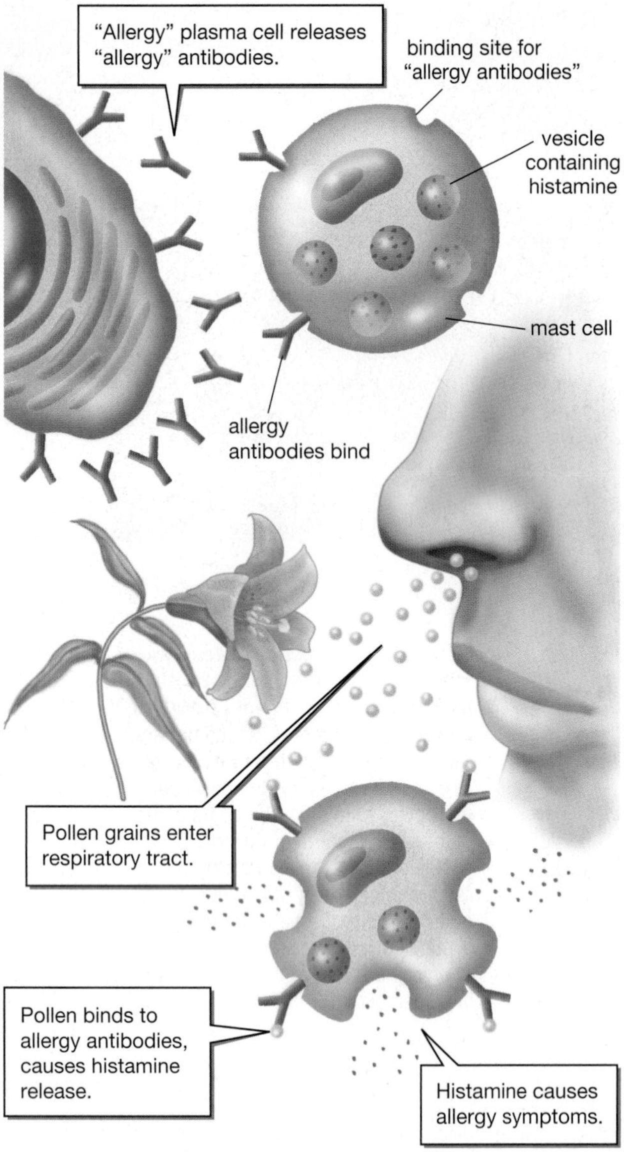
"Allergy" plasma cell releases "allergy" antibodies.
binding site for "allergy antibodies"
vesicle containing histamine
mast cell
allergy antibodies bind
Pollen grains enter respiratory tract.
Pollen binds to allergy antibodies, causes histamine release.
Histamine causes allergy symptoms.

Figure 31-13 Allergic reactions

the immune system attacks the insulin-secreting cells of the pancreas. Multiple sclerosis occurs when immune cells launch a misdirected attack against the insulating fatty sheath that coats parts of neurons in the brain and spinal cord. Rheumatoid arthritis is also caused by inappropriate immune activation. Unfortunately, at present there is no known cure for autoimmune diseases. For some diseases, replacement therapy can alleviate the symptoms—for instance, by administering insulin to diabetics or blood transfusions to anemics. Alternatively, the autoimmune response can be suppressed with drugs. Immune suppression, however, also reduces immune responses to the everyday assaults of disease microbes, so this therapy has major drawbacks.

An Immune Deficiency Disease Results from the Inability to Mount an Effective Immune Response to Infection

David, the "bubble boy," lived all of his short, 12-year life in a germproof "bubble," isolated from contact with every unsterilized object, including other people. On rare occasions, a child like David is born with **severe combined immune deficiency (SCID)**, a defect in which few or no immune cells are formed. A child with SCID may survive the first few months of postnatal life, protected by antibodies acquired from the mother during pregnancy or in her milk. Once these antibodies are lost, however, common bacterial infections can prove fatal. One form of therapy is to transplant bone marrow (from which immune cells arise) from a healthy donor into the child. In some children, marrow transplants have resulted in some antibody production, occasionally enough to confer normal immune responses. In 1990 researchers at the National Institutes of Health began clinical trials of injecting genetically engineered bone marrow cells into children with SCID. The therapy has been somewhat successful but is not yet in widespread use.

AIDS Is a Devastating Immune Deficiency Disease

The most common and widespread immune deficiency disease is **acquired immune deficiency syndrome**, or **AIDS**. Two viruses, named **human immunodeficiency viruses** 1 and 2 (HIV-1 and HIV-2), cause AIDS. These viruses undermine the immune system by infecting and destroying the helper T cells, which stimulate both the cell-mediated and humoral immune responses. AIDS does not kill people directly, but AIDS victims become increasingly susceptible to other diseases as their helper T-cell populations decline. Although it was first recognized in 1981, genetic studies show that the AIDS virus almost certainly arose from viruses that have been infecting chimps in Africa, probably for the past 100,000 years. The infected chimps show no signs of the disease. Preserved tissue samples dating from 1957, taken from a man who lived in Africa, contain fragments of the HIV genome that resemble the common ancestor of three modern strains of HIV and also resemble the ancestral chimp virus, which researchers believed mutated and jumped from chimps to humans between the mid-1940s and early 1950s.

The Human Immunodeficiency Virus Is a Retrovirus That Infects and Destroys Helper T Cells

How does the HIV virus wreak havoc on the human immune system? Both HIV-1 and HIV-2 are **retroviruses**—viruses that have RNA as their genetic material and that reproduce by transcribing that RNA into DNA and then inserting the DNA into the chromosomes of a host cell (retroviruses are described in Chapter 19). HIV consists of an outer envelope, taken from an infected cell's plasma membrane as the virus leaves the cell, and two protein layers, the innermost of which encloses the viral RNA

Earth Watch

Dams, Deforestation, and Diseases

As researchers work in laboratories to devise ways to augment the immune system's fight against diseases, human activities are simultaneously altering the environment in ways that promote the spread of diseases, some of them deadly. A now-classic example is the construction of the Aswan High Dam on the Nile River in Egypt (Fig. E31-2a). Completed in 1968, the dam was built to provide irrigation water and hydroelectric power for Egypt's rapidly expanding population. Officials did not foresee that the stagnant waters of the newly created Lake Nassar and the irrigation ditches fed by the dam would enormously increase the habitat for a snail that carries the *Schistosoma* worm responsible for schistosomiasis—a chronic, debilitating, and potentially fatal disease. Human infection rates in the region around the dam have increased from 5% in 1968 to 77% in 1993. The stagnant waters also created a mosquito-breeding paradise, and several epidemics of the mosquito-transmitted viral infection called *Rift Valley fever* have occurred, starting in 1977; the most recent outbreak was in 1998.

Meanwhile, China is now building the largest dam in the world on the Yangtze River. Twice as tall as the Aswan Dam, the 660-foot-high Three Gorges Dam will create a 400-mile-long reservoir. As occurred in Egypt, experts are predicting an explosion in the population of local schistosomiasis-carrying snails, with major adverse effects on human health.

Throughout the Tropics, rain forests are being felled for agriculture, with devastating consequences—not only for the rare species that inhabit them, but for local human populations as well. As forests are cleared and original habitats destroyed, new types of habitats are created for malaria-carrying mosquitos (Fig. E31-2b). For example, a malaria-carrying *Anopheles* mosquito in Africa prefers to breed in the open areas created by clear-cuts, while the warmer surface temperatures created by removing trees speeds up the life cycle of the mosquito and the development of the malaria parasite. Malaria, which kills more than 1 million people each year, is also spreading in South America as the Brazilian rain forest is cleared.

Why are malaria and schistosomiasis such major health problems and so successful against the human immune system? Malaria is caused by *Plasmodium*, a protist with a complex life cycle (see Chapter 19). Each of the three stages of the life cycle that occur within a person exposes different antigens to the immune system for a relatively short time, so the antigen has changed before the immune system mounts an effective response. Further, two of the stages of the life cycle occur inside body cells: one stage in liver cells, and another stage inside red blood cells, where they are somewhat protected from immune assaults. Finally, the free stage of *Plasmodium* has the ability to shed its surface coat with the attached antibodies, rendering them ineffective. In contrast, schistosomiasis worms can live in an infected person for up to 20 years; the worms actually fool our immune systems by coating themselves with human antigens.

(a)

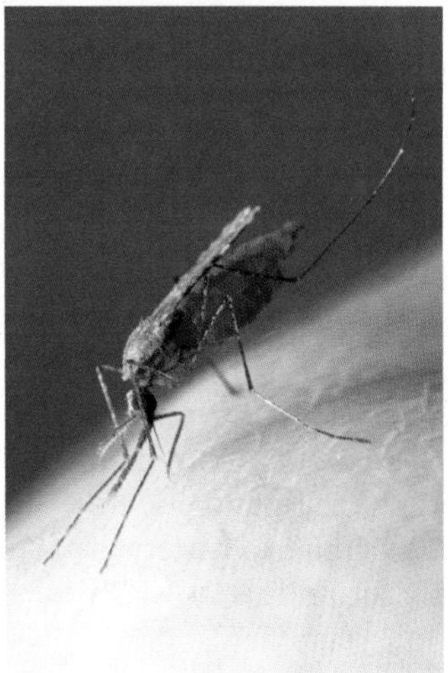

(b)

Figure E31-2 Environmental damage spreads disease
(a) The Aswan High Dam produced Lake Nassar and filled irrigation ditches with stagnant water. The snails that carry the schistosomiasis worm thrive in these quiet waters. ***(b)*** Clear-cutting rain forests has increased the habitat for the *Anopheles* mosquito that carries and spreads the malaria protist.

and an enzyme called **reverse transcriptase** (Fig. 31-14). When HIV attacks a helper T cell, the virus's outer envelope binds to the plasma membrane of the cell and allows the virus to enter the cell. Once the virus is inside, its reverse transcriptase catalyzes the transcription of the virus's RNA genome into DNA (a process known as *reverse transcription*, because it reverses the "normal" DNA-to-RNA transcription in living cells). The resulting DNA then travels to the nucleus and is inserted into the T cell's genome, directing it to make more HIV particles. Early in the infection, the patient may have a fever, rash, muscle aches, headaches, and enlarged lymph nodes, as the immune system fights the infection. After about six months, the rate of viral replication

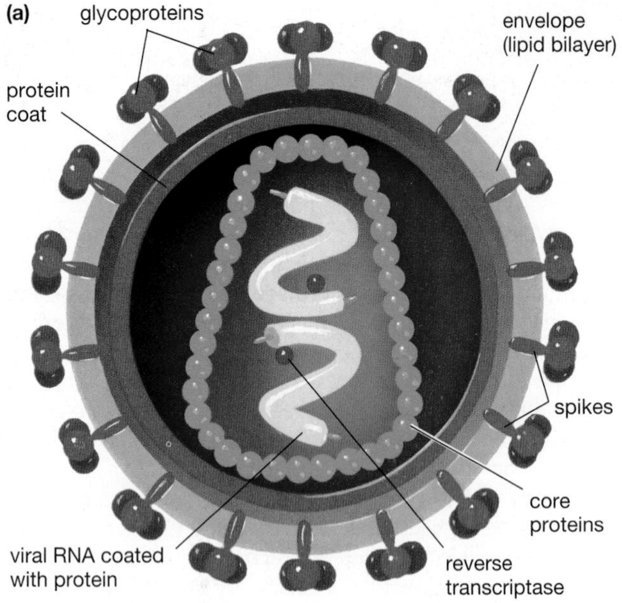

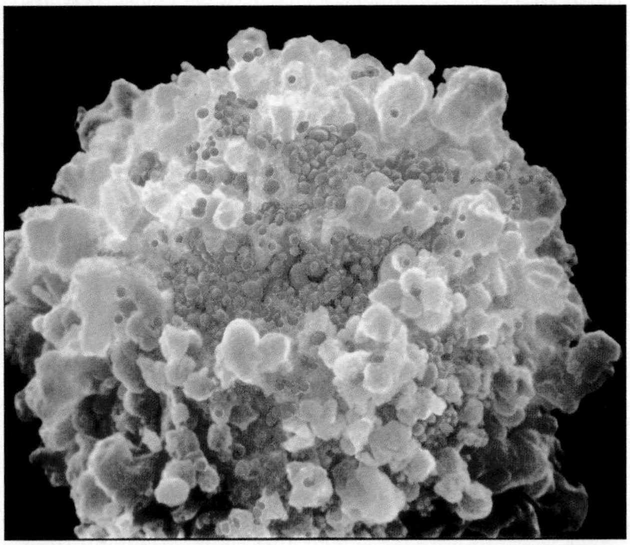
(b)

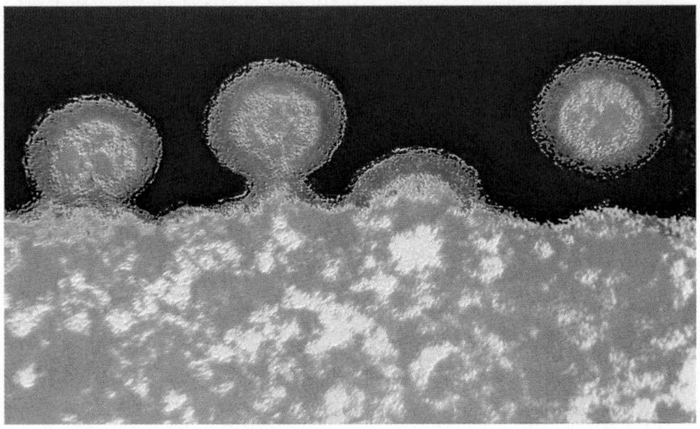
(c)

Figure 31-14 ***(a)*** HIV, the virus that causes AIDS, consists of an outer envelope, a protein coat, and an inner protein capsule, which contains RNA (the genetic material of HIV) and the enzyme reverse transcriptase (which copies the RNA into DNA when the virus infects a cell). Proteins protruding through the envelope attach to the plasma membranes of helper T cells. These proteins are potential targets for AIDS vaccines. ***(b)*** The red specks in this colorized scanning electron micrograph are HIV that have just emerged from the large, green helper T cell. ***(c)*** In this colorized transmission electron micrograph, HIV are emerging from the helper T cell and acquiring an outer envelope of plasma membrane (green) from the infected T cell.

slows and stabilizes at a relatively low level. Enough T cells remain that patients are able to resist disease, and they generally feel quite well. This state may persist for several years. Eventually, however, helper T cell levels begin to drop. When only 200 helper T cells per milliliter of blood remain (one-fourth of normal levels), the patient is described as having AIDS. At this point, in untreated AIDS, HIV levels skyrocket, helper T cell numbers are further reduced, and the patient falls victim to a variety of other infections. The life expectancy for untreated AIDS victims after this occurs is about 1 to 2 years.

The Human Immunodeficiency Virus Is Transmitted by Body Fluids

HIV cannot survive for very long outside the body. The virus can be transmitted only by the direct contact of broken skin or mucous membranes with virus-laden body fluids—including blood, semen, vaginal secretions, and breast milk. HIV infection can be spread by sexual activity, by sharing of needles among intravenous drug users, by blood transfusions (this is rare since it became standard practice to screen all donated blood for anti-HIV antibodies), or from mother to child during pregnancy or childbirth. Nearly all people infected by the HIV virus eventually develop AIDS.

In the early days of the AIDS epidemic in the United States, most individuals with AIDS were homosexual men or intravenous drug users. It is now obvious that AIDS can also be transmitted through heterosexual contact. Heterosexual intercourse is the most common means of infection in Africa and Asia, where the largest number of AIDS victims live.

There Are Partially Effective Treatments, but No Cures, for AIDS

For persons already infected with AIDS, there are two categories of therapy. First, infections that result from the impaired immune system being unable to fight them, called opportunistic infections—such as *Kaposi's sarcoma* (a deadly form of cancer affecting the skin) or *Pneumocystis carinii* pneumonia—can be treated as they would be in any patient. Second, new drugs that target retroviruses can slow the progress of AIDS and improve the quality of life for AIDS patients and help them live longer. These drugs fall into two general classes: (1) *reverse transcriptase inhibitors* and (2) *protease inhibitors.*

Reverse transcriptase inhibiting drugs such as *AZT* and *ddI* inhibit viral reproduction by blocking the production of DNA from RNA. Unfortunately, these drugs are not completely successful, partly because the HIV virus can mutate to forms resistant to the drugs. In some patients they also have severe side effects. Protease inhibitors neutralize an enzyme that helps assemble viruses from the components produced by the host cell's pirated genetic machinery. Often two or more reverse transcriptase inhibitors are given in combination with two or more protease inhibitors, attacking simultaneously at two points in the HIV life cycle. This multi-drug treatment may require patients to take up to 24 pills daily and cost $10,000 to $12,000 each year.

Clearly, the best solution would be to develop an AIDS vaccine. This is a major challenge, since the HIV antibodies that individuals with AIDS produce naturally fail to halt the infection. Therefore, a vaccine would have to evoke a more effective immune response than does the HIV infection itself. Further, HIV has an incredible mutation rate, perhaps a thousand times faster than that of flu viruses, so infected individuals may harbor different strains of HIV. Researchers have recently found that the same individual may harbor different strains of HIV virus in blood and in semen. Nevertheless, trials of AIDS vaccines are under way in several countries, though none yet appear to be particularly promising.

AIDS Is a Widespread, Lethal Disease

We are currently in the midst of a worldwide AIDS epidemic that has killed about 22 million people. UNAIDS, the United Nations organization responsible for tracking AIDS cases, estimates that 36 million people (including 1.4 million children) are infected with HIV and that more than 14,500 new infections and nearly 7000 deaths from AIDS occur daily. Most of those infected (95%) live in the developing world, especially in sub-Saharan Africa, which currently has the highest infection rate. Southern Africa has been especially hard hit; 10% of adults are infected in 10 African nations, and the average life span in these countries has been reduced by at least 10 years due to deaths from AIDS. In the U.S., about 40,000 people are infected with HIV each year; about one-third of these are women. There are about 750,000 AIDS cases in the U.S., and perhaps as many more people harbor the virus but have not developed AIDS symptoms. Here, routes of exposure, in order of frequency, are homosexual contact, intravenous drug use with shared needles, and heterosexual contact. (For more information, please see "A Closer Look: Cellular Communication During the Immune Response," which appears in Chapter 31 on this text's Web site.)

AIDS is one the world's deadliest infectious diseases, rivaling other notorious killers such as malaria and tuberculosis in terms of annual deaths. How concerned should you be about AIDS, personally? Because the disease is deadly and the incidences of other sexually transmitted diseases are on the rise, "safer-sex" practices are important for everyone. Further, health care workers should (and now routinely do) take extra precautions when handling blood products and used needles.

Cancer Can Evade or Overwhelm the Immune Response

Cancer is one of the most dreaded words in the English language, and with good reason. The disease kills more than 500,000 U.S. citizens each year and about 6 million people worldwide. A sobering 40% of U.S. citizens will eventually contract cancer. Advances in detection and treatment have reduced the death rates for some forms of cancer, but the overall cancer death rate has increased since the mid-1970s. Despite decades of intensive research, the quest for a cancer cure remains unfulfilled. Why can't we cure or prevent cancer? We can, after all, prevent smallpox and polio and can cure dozens of other diseases. What makes cancer different?

The usual development of any organ begins with rapid growth during embryonic life, slower growth in childhood, and finally, maintenance of a constant size during adulthood. Individual cells may die and be replaced, but most organs remain about the same size throughout adult life. A **tumor** is a population of cells that has escaped from normal regulatory processes and grows at an abnormal rate. The cells in a *benign tumor*, or *polyp*, normally remain constrained in one area, but the cells in a *malignant tumor* grow uncontrollably and spread to other areas of the body. As a malignant tumor grows, it uses increasing amounts of the body's energy and nutrient supplies and literally squeezes out vital organs nearby. **Cancer** is a disease characterized by the unchecked growth of malignant tumor cells.

Unlike most other diseases, cancer is not a straightforward invasion of the body by a foreign organism. Although some cancers are triggered by viruses, cancer is essentially a failure of the mechanisms that control the growth of the body's own cells—a disease in which the body destroys itself. The mechanisms of regulation of cell division and the genetic mutations that cause cancer are discussed in Chapter 11.

The Immune System Defends Against Cancerous Cells

Cancer cells form in our bodies every day. Even the best of preventive measures cannot eliminate cancer completely. We cannot avoid some carcinogens, such as gamma rays from the sun, radioactivity from the rocks beneath our feet, and naturally produced carcinogens in our food. Fortunately, natural killer cells and cytotoxic T cells screen the body for cancer cells and destroy nearly all of them before they have a chance to proliferate and spread. Cancer cells are, of course, "self" cells (the body's own cells), and the immune system does not respond to "self." How are cancer cells weeded out? It seems likely that the very processes that cause cancer also cause new and slightly different proteins to appear on the surfaces of cancer cells. Cytotoxic T cells encounter these new proteins, recognize them as "non-self" antigens, and destroy the cancer cells (see Fig. 31-10). However, some cancer cells may not be recognized as threats by the immune system. Even tumors that are attacked by the immune system may develop variant cell types that are resistant to immune attack. In other cases, tumors can actively suppress the immune system or can simply grow so fast that the immune response can't keep up. If the immune system is thwarted by any of these means, the tumor grows and spreads. At this point, the individual's health depends on medical treatment. Unfortunately, the available treatments are only partially effective and have serious drawbacks. The rate of cure is increasing, but it is still scarcely a third of all cancers.

Medical Treatments for Cancer Depend on Distinguishing and Selectively Killing Cancerous Cells

The three main forms of treatment are (1) surgery, (2) radiation, and (3) chemotherapy. The surgical removal of the tumor is the first step in the treatment of many cancers. Unfortunately, the surgeon may not be able to see and remove very small patches of cancer cells that may extend from the main tumor. And surgery cannot be used to treat cancer that has begun to spread throughout the body. Alternatively, cancerous cells can be bombarded with radiation in an effort to kill them. Unlike surgery, radiation may destroy even microscopic clusters of cancer cells. Like surgery, however, radiation cannot be used to treat widespread cancers, because to irradiate the whole body would damage a great deal of healthy tissue. Both surgery and radiation therapy can be traumatic and dangerous.

Chemotherapy, or drug treatment, is commonly used to supplement surgery and/or radiation or to treat cancers that cannot be treated with surgery or radiation. Chemotherapy drugs attack the machinery of cell division. Because cancer cells divide much more frequently than do normal cells, the hope is that attacks on dividing cells will selectively kill cancer cells. Unfortunately, other cells of the body divide too, and chemotherapy inevitably kills some healthy cells. It is damage to dividing cells in the hair follicles and intestinal lining that produces the well-known side effects of hair loss, nausea, and vomiting.

A tremendous amount of research has been devoted to the search for cancer treatments that are more effective and have fewer unpleasant side effects. Developing a "cancer vaccine" is a high priority. Other approaches include developing therapies that would stimulate the immune system to attack tumors, and developing antibody molecules that recognize tumor cells and could be used to deliver drugs and other treatments directly to tumor cells without affecting healthy cells (see "Scientific Inquiry: Monoclonal Antibodies—Designer Drugs Fight Disease"). Research and clinical trials continue, however, and cancer patients may one day soon reap the benefits of innovative new treatments.

Without constant surveillance by the immune system, it is unlikely that any of us would survive more than a few years. However, we can reduce our own chances of developing cancer by avoiding cigarette smoke and by eating a diet high in fruits and vegetables and low in red meat and saturated fats. Other behaviors that lower the risk of cancer include avoiding excessive alcohol consumption, using sunscreen, and getting regular exercise.

CASE STUDY REVISITED

Fighting the Flu

Immediately after class, you make your way to the local clinic for a flu shot. From reading this chapter, you know that last year's bout of flu will not protect you from this year's strain. The nurse who administers the flu vaccine points out that it will take a few weeks to activate your immune system and become effective against the virus (see Fig. 31-12), so you realize that today's shot will not protect you from your classmate's sneeze. But since you will undoubtedly be exposed repeatedly during the flu season, the shot seems to be a reasonable precaution. Now you go about your day, hoping that—without any awareness on your part—your own marvelously complex natural defenses will fend off the flu virus before it even gets established. But if you do come down with flu, perhaps your doctor will consider you a candidate for the new neuraminidase inhibitors that will assist your embattled immune system. One way or another, you'll fight it off.

It is likely that neuraminidase inhibitors will be heavily prescribed to fight flu in the coming years. You are aware that both bacteria and viruses mutate frequently and that drug-resistant stains, which arise as a result of these mutations, are a major problem, particularly if drugs are widely prescribed. Based on what you learned about neuraminidase inhibitors and how they work, suggest a reason why it might be difficult for mutations to produce strains of viruses that are resistant to these drugs. Only time will tell if this prediction is correct!

Summary of Key Concepts

1) How Does the Body Defend Against Invasion?

The human body has three lines of defense against invasion by microbes: (1) the barriers of skin and mucous membranes; (2) nonspecific internal defenses, including phagocytosis, killing by natural killer cells, inflammation, and fever; and (3) the immune response. The skin physically blocks the entry of microbes into the body. Skin is also covered with secretions from sweat and sebaceous glands that inhibit bacterial and fungal growth and the entry of microbes into the body. The mucous membranes of the respiratory and digestive tracts secrete antibiotic substances, antibodies, and mucus that traps microbes.

If microbes do enter the body, white blood cells travel to the site of entry and engulf the invading cells. Natural killer cells secrete proteins that kill infected or cancerous cells. Injuries stimulate the inflammatory response, in which chemicals are released that attract phagocytic white blood cells, increase blood flow, and make capillaries leaky. Later, blood clots wall off the injury site. Fever is caused by endogenous pyrogens, chemicals released by white blood cells in response to infection. High temperatures inhibit bacterial growth and accelerate the immune response.

2) What Are the Key Characteristics of the Immune Response?

The immune response involves two types of lymphocytes: B cells and T cells. Plasma cells, which are descendants of B cells, secrete antibodies into the bloodstream, causing humoral immunity. Cytotoxic T cells destroy some microbes, cancer cells, and virus-infected cells on contact, causing cell-mediated immunity. Helper T cells stimulate both the humoral and cell-mediated immune responses. Immune responses have three steps: (1) recognition, (2) attack, and (3) memory.

First, antibodies (on B cells) and T-cell receptors (on T cells) recognize foreign molecules and trigger the immune response. Antibodies are Y-shaped proteins composed of a constant region and a variable region. Antigens are molecules that generate an antibody response. Antibodies both detect and actively destroy antigens. Each B cell synthesizes only one type of antibody, unique to that particular cell and its progeny. The diversity of antibodies arises from gene shuffling and the mutation of antibody genes during B-cell development. Each antibody has specific sites that bind only one or a few types of antigen. Normally, only foreign antigens are recognized by the B cells.

Second, antibodies attack the invaders. Antigens from an invader bind to and activate only those B and T cells with the complementary antibodies or T-cell receptors. In humoral immunity, B cells with the proper antibodies, stimulated by the presence of particular antigens, divide rapidly to produce plasma cells that synthesize massive quantities of the antibody. The circulating antibodies destroy antigens and antigen-bearing microbes by four mechanisms: (1) neutralization, (2) promotion of phagocytosis by white blood cells, (3) agglutination, and (4) complement reactions. In cell-mediated immunity, T cells with the proper receptors bind antigens and divide rapidly. Cytotoxic T cells bind to antigens on microbes, infected cells, or cancer cells and then kill the cells. Helper T cells stimulate, and suppressor T cells turn off, both the B-cell and cytotoxic-T-cell responses.

Finally, some progeny cells of both B and T cells are long-lived memory cells. If the same antigen reappears in the bloodstream, these memory cells are immediately activated, divide rapidly, and cause an immune response that is much faster and more effective than the original response.

3) How Does Medical Care Augment the Immune Response?

Antibiotics kill microbes or slow down their reproduction, thus allowing the immune system more time to respond and exterminate the invaders. Vaccinations are injections of antigens from disease organisms, typically the weakened or dead microbes themselves. An immune response is evoked by the antigens, providing memory and a rapid response should a real infection occur.

4) What Happens When the Immune System Malfunctions?

Allergies are immune responses to normally harmless foreign substances, such as pollen or dust. Mast cells respond to the presence of these substances by releasing histamine, which causes a local inflammatory response. Autoimmune diseases arise when the immune system destroys some of the body's own cells. Immune deficiency diseases occur when the immune system cannot respond strongly enough to ward off normally minor diseases.

Infection with one of two viruses, called human immunodeficiency viruses (HIV) 1 and 2, can lead to AIDS (acquired immune deficiency syndrome). These viruses invade and destroy helper T cells. Without helper T cells to stimulate the immune responses of B cells and cytotoxic T cells, an individual with AIDS is extremely susceptible to a wide assortment of infections, which are eventually fatal.

Cancer is a population of the body's cells that divides without control. Cancerous cells may be recognized as "different" by the immune system and destroyed by natural killer cells and cytotoxic T cells. A few evolve the capacity to evade the immune system, some attack immune cells, and others multiply too fast for the immune system to keep up. In these cases, cancer develops and requires aggressive medical treatment.

Key Terms

acquired immune deficiency syndrome (AIDS) *p. 634*
allergy *p. 632*
antibody *p. 624*
antigen *p. 624*
autoimmune disease *p. 633*
B cell *p. 624*
cancer *p. 637*
cell-mediated immunity *p. 627*
clonal selection *p. 627*
complement *p. 625*
complement system *p. 628*
constant region *p. 624*
cytotoxic T cell *p. 629*
endogenous pyrogen *p. 622*
fever *p. 621*
helper T cell *p. 629*
histamine *p. 621*
human immunodeficiency virus (HIV) *p. 634*

humoral immunity *p. 627*
hybridoma *p. 629*
immune response *p. 624*
immune system *p. 623*
inflammatory response *p. 621*
interferon *p. 623*
macrophage *p. 621*
major histocompatibility complex (MHC) *p. 627*
mast cell *p. 632*
memory B cell *p. 627*
memory T cell *p. 629*
microbe *p. 620*
monoclonal antibody *p. 629*
mucous membrane *p. 620*
natural killer cell *p. 621*
phagocytic cell *p. 621*
plasma cell *p. 627*
retrovirus *p. 634*
reverse transcriptase *p. 635*
severe combined immune deficiency (SCID) *p. 634*
skin *p. 620*
suppressor T cell *p. 629*
T cell *p. 624*
T-cell receptor *p. 624*
tumor *p. 637*
vaccination *p. 631*
variable region *p. 624*

Thinking Through the Concepts

Multiple Choice

1. *In addition to the immune system, we are protected from disease by*
   a. the skin
   b. mucous membranes
   c. natural secretions such as acids, protein-digesting enzymes, and antibiotics
   d. cilia
   e. all of the above

2. *The first line of defense against body cells infected by viruses is produced by*
   a. antibodies
   b. phagocytes
   c. natural killer cells
   d. histamines
   e. natural antibiotics

3. *Fevers*
   a. decrease interferon production
   b. decrease the concentration of iron in the blood
   c. decrease the activity of phagocytes
   d. increase the reproduction rate of invading bacteria
   e. do all of the above

4. *T cells and B cells are*
   a. lymphocytes
   b. macrophages
   c. natural killer cells
   d. red blood cells
   e. phagocytes

5. *During allergic responses*
   a. the foreign substance binds to antibodies on B cells
   b. B cells produce antibodies to the foreign substance
   c. antibodies to the foreign substance bind to that substance and to mast cells
   d. stimulation of mast cells causes them to release histamine
   e. all of the above

6. *What shuts off the immune response in T cells and B cells after an infection has been conquered?*
   a. cytotoxic T cells
   b. histamine
   c. pyrogens
   d. natural killer cells
   e. suppressor T cells

? Review Questions

1. List the human body's three lines of defense against invading microbes. Which are nonspecific (that is, act against all types of invaders), and which are specific (act only against a particular type of invader)? Explain your answer.
2. How do natural killer cells and cytotoxic T cells destroy their targets?
3. Describe humoral immunity and cell-mediated immunity. Include in your answer the types of immune cells involved in each, the location of antibodies and receptors that bind foreign antigens, and the mechanisms by which invading cells are destroyed.
4. How does the immune system construct so many different antibodies?
5. How does the body distinguish "self" from "non-self"?
6. Diagram the structure of an antibody. What parts bind to antigens? Why does each antibody bind only to a specific antigen?
7. What are memory cells? How do they contribute to long-lasting immunity to specific diseases?
8. What is a vaccine? How does it confer immunity to a disease?
9. Compare and contrast an inflammatory response with an allergic reaction from the standpoint of cells involved, substances produced, and symptoms experienced.
10. Distinguish between autoimmune diseases and immune deficiency diseases, and give one example of each.
11. Describe the causes and eventual outcome of AIDS. How do AIDS treatments work? How is the HIV virus spread?
12. Why is cancer sometimes fatal?

Applying the Concepts

1. Why is it essential that antibodies and T-cell receptors bind only relatively large molecules (such as proteins) and not relatively small molecules (such as amino acids)?
2. There are smallpox virus stocks in two laboratories—one in the U.S. and one in Russia. A debate is raging about whether these stocks should be destroyed. In brief, one side argues that having smallpox around is too dangerous. The other side argues that we may be able to learn things from smallpox, answers to questions we don't know enough to ask yet, that may help us conquer future diseases. Do you believe the smallpox stocks should be destroyed? Why or why not?

3. The essay "Health Watch: Can We Beat the Flu Bug?" states that the flu virus is different each year. If that is true, what good is it to get a "flu shot" each winter?
4. Organ transplant patients typically receive the drug cyclosporine. This drug inhibits the production of interleukin-2, a regulatory molecule that stimulates helper T cells to proliferate. How does cyclosporine prevent the rejection of transplanted organs? Some patients who received successful transplants many years ago are now developing various kinds of cancers. Propose a hypothesis to explain this phenomenon.

For More Information

"Defeating AIDS: What Will It Take?" *Scientific American*, July 1998. A series of nine articles covering all aspects of AIDS.

Laver, W. G., Bischofberger, N., and Webster, R. G. "Disarming Flu Viruses." *Scientific American*, January 1999. Presents a clear description of how flu viruses work, how new strains arise, and the development of the new neuraminidase inhibitors that may stop them.

Lichtenstein, L. M. "Allergy and the Immune System." *Scientific American*, September 1993. Allergic responses, evolved as protection against parasites, turn against us when we respond violently to harmless pollen and foods.

Marrack, P., and Kappler, J. W. "How the Immune System Recognizes the Body." *Scientific American*, September 1993. How the immune system's potential weapons against the body are disarmed.

Paul, W. E. "Infectious Diseases and the Immune System." *Scientific American*, September 1993. Microbes and the immune system engage in constant evolutionary warfare.

Steinman, L. "Autoimmune Disease." *Scientific American*, September 1993. An estimated 5% of Americans suffer from some type of autoimmune disorder.

Answers to Multiple-Choice Questions

1. e 2. c 3. b 4. a 5. e 6. e

MEDIATUTOR
Defenses Against Disease: The Immune Response

CD Activities

Activity 31.1: Inflammation

Estimated time: 5 minutes

The immune system includes both your nonspecific and specific defenses to invading organisms. One of the nonspecific defenses that your body uses is an inflammatory response. The symptoms of inflammation (heat, redness, swelling, and pain) help fight off an invasion. In this tutorial you will explore this process in more detail.

Activity 31.2: Clonal Selection

Estimated time: 10 minutes

The production of antibodies by B cells is an important part of your specific immune response. The antibodies that are produced are highly specific for one particular foreign antigen. Before a B cell can begin its production of these important immune proteins, it must be activated by the foreign antigen in a process called *clonal selection*. In this tutorial you will explore this process in more detail.

Start the MediaTutor Student CD-ROM and enter the activity number in the Quick Search box to be taken directly to that activity.

Web Investigations

Case Study: Fighting the Flu

Estimated time: 10 minutes

Fifty years ago, childhood diseases (measles, mumps, chicken pox, and others) were just part of growing up. Everyone knew of people who had died of whooping cough (pertussis) or been crippled by polio. Today, despite extensive vaccination programs, some individuals still get diseases such as polio and rubella. Why?

Go to http://www.prenhall.com/audesirk6, the Audesirk Companion Web site. Select Chapter 31 and the Web Investigation to begin.

The Olympic gold medal awarded to Canada's track superstar Ben Johnson for the 100-meter dash in 1988 was stripped from him after he tested positive for anabolic steroids.

32 Chemical Control of the Animal Body: The Endocrine System

AT A GLANCE

Case Study: Losing on Steroids

1) **What Are the Characteristics of Animal Hormones?**

Hormones Bind to Specific Receptors on Target Cells

Hormone Release Is Regulated by Feedback Mechanisms

2) **What Are the Structures and Hormones of the Mammalian Endocrine System?**

Mammals Have Both Exocrine and Endocrine Glands

The Hypothalamus Controls the Secretions of the Pituitary Gland

The Thyroid and Parathyroid Glands Influence Metabolism and Calcium Levels

The Pancreas Is Both an Exocrine and an Endocrine Gland

The Sex Organs Secrete Steroid Hormones

The Adrenal Glands Have Two Parts That Secrete Different Hormones

Many Types of Cells Produce Prostaglandins

Other Sources of Hormones Include the Pineal Gland, Thymus, Kidneys, Heart, Digestive Tract, and Fat Cells

Evolutionary Connections: The Evolution of Hormones

Case Study Revisited: Losing on Steroids

CASESTUDY

Losing on Steroids

With fractions of a second making the difference between winning and losing, competitive athletes are under enormous pressure to improve their performance—even though they are already pushing the limits of human ability. Many, from world-class athletes to aspiring muscle men on campus to female body builders, turn to "anabolic steroids." Some athletes began to use steroids in the 1950s, and it wasn't until the 1970s that some sports organizations began to ban their use. Despite increasing evidence that steroids pose significant health risks, steroid use has increased, generating a black market in which sales may exceed $400 million annually. Frighteningly, some black-market steroids are manufactured illegally and are contaminated with other harmful substances. *Anabolic* refers to metabolic activities that synthesize molecules. Anabolic steroids resemble the steroid male sex hormone testosterone. For some individuals, anabolic steroids stimulate the body to increase the amount of lean muscle mass. But athletes take dangerous risks to gain this artificial advantage, risks that go well beyond the possible loss of their awards. As you read this chapter, notice how many different effects the same hormone can have throughout the body. Think about what artificially increasing the amount of testosterone-like steroids might do—to men or to women. What risks do steroids pose? We'll revisit this question at the end of the chapter. ■

1 What Are the Characteristics of Animal Hormones?

Hormones are chemical messages produced by specialized cells that have been activated by some environmental or physiological stimulus. Hormones are carried by the circulatory system and influence **target cells**, cells that are specialized to respond to the messages. The changes induced by hormonal messages may be prolonged and irreversible, as in puberty or the metamorphosis of a caterpillar into a butterfly. More typically, the induced changes are transient and reversible and help control and regulate the physiological systems that compose the animal body. In fact, the **endocrine system** (consisting of hormones and the various cells that secrete and receive them) can be viewed as the "postal system of physiology," moving information and instructions between cells that may be some distance apart. To regulate the body requires communication; in animal

MediaTutor 32.1 The Endocrine System Anatomy

Figure 32-1 Major mammalian endocrine glands
The major mammalian endocrine glands are the hypothalamus-pituitary gland complex, the thyroid and parathyroid glands, the pancreas, the sex organs (ovaries in females, testes in males), and the adrenal glands. Other organs that secrete hormones include the pineal gland, thymus, kidneys, heart, and parts of the digestive tract.

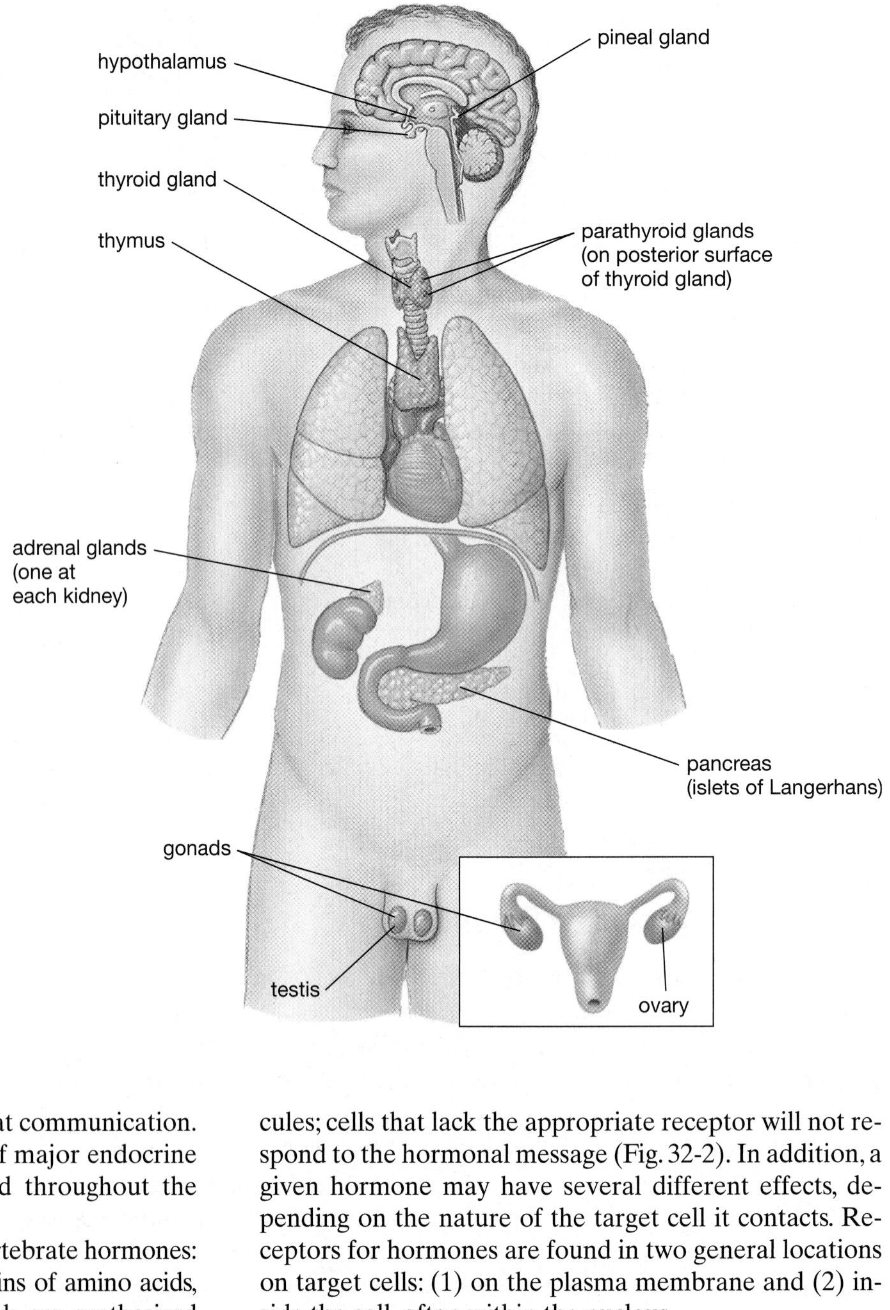

bodies, hormones provide much of that communication. Hormones are released by the cells of major endocrine glands and endocrine organs located throughout the body (Fig. 32-1).

There are four general classes of vertebrate hormones: (1) **peptide hormones**, made from chains of amino acids, (2) *amino acid based hormones,* which are synthesized from single amino acids; (3) **steroid hormones**, which resemble cholesterol, from which most steroid hormones are synthesized; and (4) **prostaglandins**, which are synthesized from fatty acids. For more information about the chemical structures of hormones, see "A Closer Look: The Chemical Diversity of Vertebrate Hormones" in Chapter 32 on this text's Web site.

Hormones Bind to Specific Receptors on Target Cells

Because nearly all cells have a blood supply, a hormone that has been released into the bloodstream will reach nearly every cell of the body. But to exert their precise control, hormones must act only on certain target cells. Target cells have receptors for particular hormone molecules; cells that lack the appropriate receptor will not respond to the hormonal message (Fig. 32-2). In addition, a given hormone may have several different effects, depending on the nature of the target cell it contacts. Receptors for hormones are found in two general locations on target cells: (1) on the plasma membrane and (2) inside the cell, often within the nucleus.

Many peptide and amino acid-based hormones are soluble in water but not in lipid. Hence, these hormones cannot penetrate plasma membranes, which are largely composed of phospholipids. Instead, the hormones react with protein receptors on the target cell's plasma membrane (Fig. 32-3a, p. 646; also see Chapter 4). Typically these "membrane receptors" span the plasma membrane, so a hormone binding to the outside portion of the receptor can cause a shape change on the part of the receptor protein that protrudes into the cell. Membrane receptors generally utilize a **second messenger** molecule, a molecule that transfers the signal from the *first messenger*—the hormone—to other molecules within the cell. In second-messenger systems, a hormone binding to a receptor on the outside of the cell membrane triggers the production of the second mes-

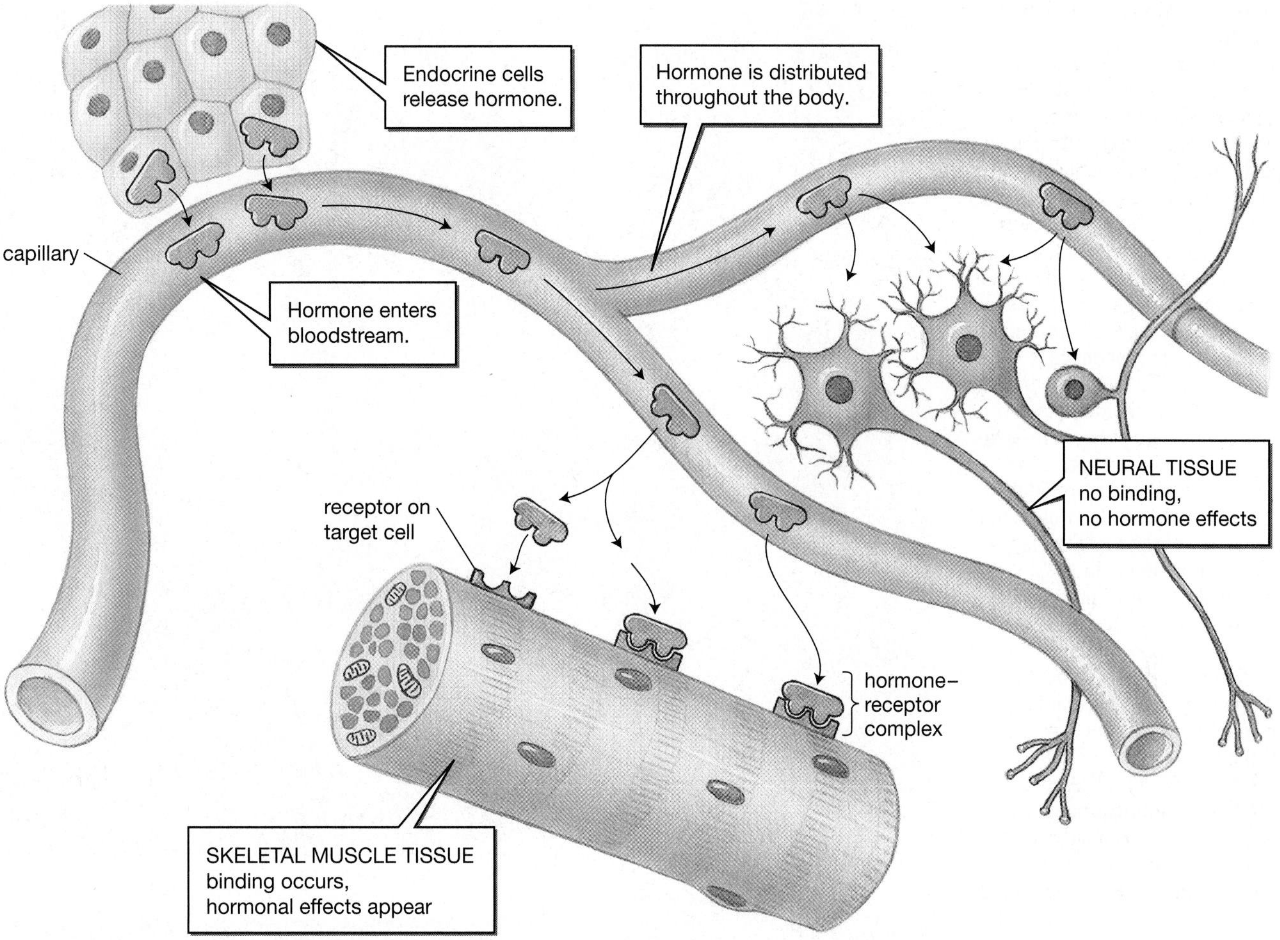
Endocrine cells release hormone.
Hormone is distributed throughout the body.
capillary
Hormone enters bloodstream.
NEURAL TISSUE no binding, no hormone effects
receptor on target cell
hormone–receptor complex
SKELETAL MUSCLE TISSUE binding occurs, hormonal effects appear

Figure 32-2 Endocrine glands, hormones, and target cells
Endocrine glands consist of hormone-producing cells embedded in a network of capillaries. These cells secrete hormones into the extracellular fluid, from which they diffuse into the capillaries. Each hormone is transported around the body by the bloodstream but binds to (and influences) only those cells that contain specific receptors for the hormone. Muscle cells but not neurons have receptors for the particular hormone shown here.

senger, which in turn initiates a cascade of biochemical reactions (Fig. 32-3a). In many cases, the binding of the receptor activates an enzyme that catalyzes the conversion of ATP to **cyclic AMP** (cAMP), a nucleotide that regulates many cellular activities (see Chapter 3). Cyclic AMP acts as a second messenger and initiates a chain of reactions inside the cell. Each reaction in the chain involves an increasing number of molecules, amplifying the original signal. The end result varies with the target cell; channels may be opened in the plasma membrane, or substances may be synthesized or secreted. For example, the hormone epinephrine (also called *adrenaline*) binds to membrane receptors on heart muscle, triggers cAMP formation, and starts a chain of molecular events that keeps calcium channels in the heart muscle open longer. The extra calcium that comes in causes a stronger contraction in the cardiac muscle. (You will learn more about calcium and muscle contraction in Chapter 34). This is one of several ways in which epinephrine helps your body prepare for emergency situations, as described later in this chapter.

In contrast, some hormones, such as the steroid hormones, are lipid soluble and are therefore able to pass through the plasma membrane. Once inside the cell, these hormones bind to receptors inside the cell; typically to receptors in the nucleus (Fig. 32-3b). The receptor–hormone complex then binds to DNA and stimulates particular genes to transcribe messenger RNA, which moves to the cytoplasm and directs the synthesis of proteins that alter the cell's activity. Steroid and thyroid hormones act by stimulating the expression of genes that are otherwise switched off. These hormones may take minutes or even days to exert their full effects (in contrast to the relatively rapid but short-lived effects of hormones that bind to surface receptors). Researchers have also found receptors for steroid hormones on the plasma membrane, which make steroids extremely versatile in their signalling mechanisms. Estrogen is a good example. Although it can bind to membrane receptors, many of estrogen's effects result from its ability to enter the cell nucleus, bind to nuclear receptors, and alter the transcription of genes (see Fig. 32-3b).

Figure 32-3 Modes of action of hormones
(a) ① Peptide and amino-acid hormones, which are not soluble in lipids, bind to a receptor on the outside of the target cell's plasma membrane. ② Hormone-receptor binding triggers the synthesis of cyclic AMP (cAMP). ③ Cyclic AMP activates enzymes that ④ promote specific cellular reactions, producing new products. This cAMP "cascade" may generate a variety of responses, such as an increase in glucose synthesis (induced by epinephrine) and an increase in estrogen synthesis (induced by luteinizing hormone). ***(b)*** ① Lipid-soluble steroid hormones diffuse readily through the plasma membrane into the target cell and into the nucleus, where they combine with a protein receptor molecule. ② The hormone–receptor complex binds to DNA and facilitates ③ the binding of RNA polymerase to promoter sites on specific genes, accelerating ④ the transcription of DNA into messenger RNA (mRNA). ⑤ The mRNA then directs protein synthesis. In hens, for example, estrogen promotes the transcription of the albumin gene, causing the synthesis of albumin (egg white), which is packaged in the egg as a food supply for the developing chick.

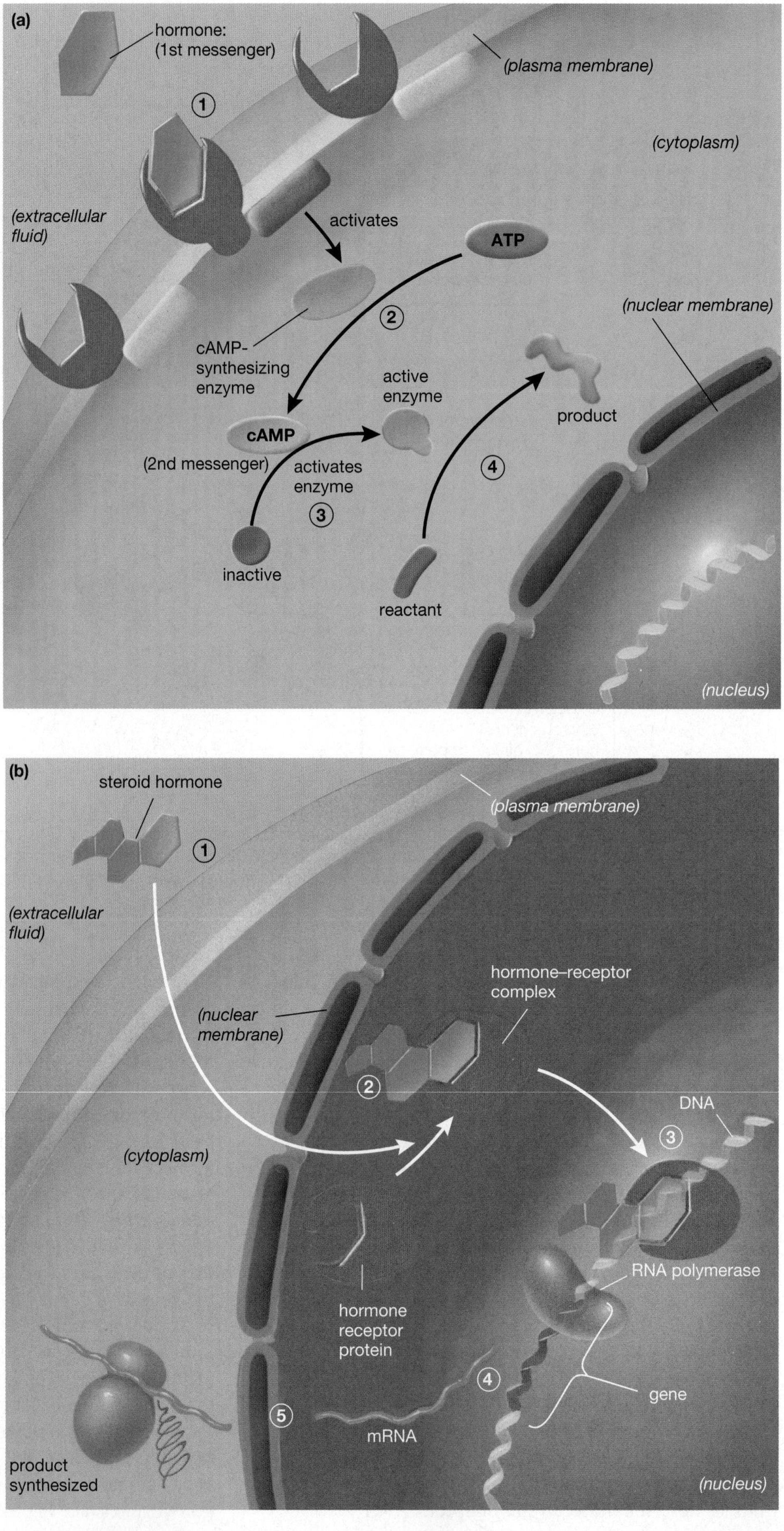

Earth Watch
Endocrine Deception

Human activities have introduced a vast number and quantity of foreign substances into the environment. These substances are now found in small quantities in our water, air, and food, and all of us are exposed to them daily. Some of them have disrupted several aspects of reproductive function in wildlife exposed at high levels. The compounds responsible are extremely diverse in chemical structure and originate from a wide variety of sources, including pesticides (DDT, methoxychlor), plastics (bisphenol A, phthalates), detergents (nonylphenyl), and industrial processes (PCBs). Many such compounds affect the reproductive system of animals. Some, called **environmental estrogens**, mimic the effects of estrogen. Others block the effects of testosterone; still others either mimic or block reproductive hormones, depending on the site of action or the species. Concern about these **endocrine disruptors** is based on studies that have used laboratory animals, cell cultures, exposed wild animals that have been observed in their natural habitats, or pregnant women given the estrogen-like drug DES.

Scientists have identified a wide variety of harmful effects caused by endocrine disruptors; these effects include feminization in males, masculinization in females, reproductive cancers, malformed sex organs, altered blood hormone levels, and reduced fertility. For example, scientists found that male fish living near sewage outlets in England had both male and female sexual features and that their livers were producing an egg yolk protein normally found only in females. Environmental estrogens released from certain detergents and plastics are the prime suspects. When a chemical plant near Lake Apopka in Florida released several known environmental estrogens into the water, wildlife biologists noted an alarming decline in the alligator population of the lake. Scientists investigating the decline found that many eggs were not hatching. In the alligators that did hatch, the researchers found males with higher estrogen and lower testosterone levels than normal, smaller penises, and abnormal testes. Females typically had exceptionally high estrogen levels and abnormal ovaries. Together, these problems severely reduced the alligators' ability to reproduce.

Between 1938 and 1971, women in danger of miscarrying during pregnancy were treated with the estrogen-like substance DES (diethylstilbestrol). Then physicians discovered that the offspring of these women had a higher incidence of both malformations and cancer in their reproductive organs. As a result, the drug was banned in 1971. Some of the most devastating endocrine disruptors, including DDT and PCBs, have been banned in developed countries (although they remain in our air, water, or soil), but many endocrine disruptors are still widely used and are very persistent in the environment. For example, phthalates, which make plastics flexible, are endocrine disruptors that are now among the most common synthetic chemicals in the environment. Infants can be exposed to phthalates as they leach from certain teething toys.

Although high levels of endocrine disruptors are known to be harmful, no one knows what effects long-term, low-level exposure to these substances is having on human and other animal populations, particularly during vulnerable early developmental stages. Important questions in need of further research include: How many of the thousands of industrial chemicals in use act as endocrine disruptors? What are their mechanisms of action? What levels of exposure do various human and wildlife populations experience, and is there a threshold of exposure for toxic effects to occur? Does exposure to multiple endocrine disruptors produce synergistic ill-effects in people or in wildlife? Answers to these questions will allow us to formulate appropriate controls over the release of these chemicals into the environment. Unfortunately, these are difficult and complex questions—and as we seek answers, more and more of these chemicals are entering the environment.

Hormone Release Is Regulated by Feedback Mechanisms

A signal that's always turned on cannot carry an instruction for change. If a hormone is to be useful for physiological control, there must be a way to turn its message on and off. In animals, the switching mechanism usually involves negative feedback: The secretion of a hormone stimulates a response in target cells that inhibits further secretion of the hormone.

Most hormones exert such powerful effects on the body that it would be harmful to have too much hormone working for too long; therefore, control of hormone release by negative feedback is especially important. Suppose you have jogged a few miles on a hot, sunny day and have lost a pint of water through perspiration. In response to the loss of water from your bloodstream, your pituitary gland releases *antidiuretic hormone* (ADH), which causes your kidneys to reabsorb water and to produce a very concentrated urine (see Chapter 30). However, if you arrive home and drink a quart of Gatorade®, you will more than replace the water you lost in sweat. Continued retention of this excess water could raise blood pressure and possibly damage your heart. Negative feedback ensures that ADH secretion is turned off when the water content of your blood returns to normal, and your kidneys begin eliminating the excess water (see Fig. 30-8). Look for more examples of negative feedback as you read through this chapter.

In a few cases, positive feedback controls hormone release. For example, as described in Chapter 26, contractions of the uterus early in childbirth cause the release of the hormone *oxytocin* by the posterior pituitary. Oxytocin stimulates stronger contractions of the uterus, which cause more oxytocin to be released, creating a positive feedback cycle. Simultaneously, oxytocin causes uterine cells to release prostaglandins, which further enhance uterine contractions; this mechanism is another example of positive feedback. Positive feedback systems must be self-limiting. In this case, the birth of the baby eliminates the stretching of the uterus that maintains the positive feedback cycle.

2 What Are the Structures and Hormones of the Mammalian Endocrine System?

Endocrinologists don't fully understand how animal hormones work. New hormones and new roles for known hormones are discovered nearly every year. The key functions of the major endocrine glands and endocrine organs, however, have been known for many years. We shall soon discuss the endocrine functions of the hypothalamus–pituitary complex, the thyroid and parathyroid glands, the pancreas, the sex organs, and the adrenal glands (see Fig. 32-1). Table 32-1 lists these and other glands, their major hormones, and their principal functions.

Mammals Have Both Exocrine and Endocrine Glands

Although this chapter focuses on endocrine glands, you should be aware that glands come in two basic types: (1) exocrine glands and (2) endocrine glands. **Exocrine glands** produce secretions that are released outside the body (*exo* means "out of" in Greek) or into the digestive tract (a hollow tube continuous with the outside world). Exocrine gland secretions are released through tubes or openings called **ducts**. The exocrine glands include the sweat glands and oil-producing (sebaceous) glands of the skin, the tear-producing (lacrimal) glands

Table 32-1 Mammalian Endocrine Glands and Hormones

Endocrine Gland	Hormone	Type of Chemical	Principal Function
Hypothalamus (via posterior pituitary)	Antidiuretic hormone (ADH)	Peptide	Promotes reabsorption of water from kidneys; constricts arterioles
	Oxytocin	Peptide	In females, stimulates contraction of uterine muscles during childbirth, milk ejection, and maternal behaviors; in males, causes sperm ejection
Hypothalamus (to anterior pituitary)	Releasing and inhibiting hormones	Peptides	At least nine hormones; releasing hormones stimulate release of hormones from anterior pituitary; inhibiting hormones inhibit release of hormones from anterior pituitary
Anterior pituitary	Follicle-stimulating hormone (FSH)	Peptide	In females, stimulates growth of follicle, secretion of estrogen, and perhaps ovulation; in males, stimulates spermatogenesis
	Luteinizing hormone (LH)	Peptide	In females, stimulates ovulation, growth of corpus luteum, and secretion of estrogen and progesterone; in males, stimulates secretion of testosterone
	Thyroid-stimulating hormone (TSH)	Peptide	Stimulates thyroid to release thyroxine
	Adrenocorticotropic hormone (ACTH)	Peptide	Stimulates adrenal cortex to release hormones, especially glucocorticoids
	Growth hormone	Peptide	Stimulates growth, protein synthesis, and fat metabolism; inhibits sugar metabolism
	Prolactin	Peptide	Stimulates milk synthesis in and secretion from mammary glands
	Endorphins	Peptides	Reduce perception of pain
	Melanocyte-stimulating hormone (MSH)	Peptide	Promotes synthesis of brown skin pigment, melanin
Thyroid	Thyroxine	Amino acid derivative	Increases metabolic rate of most body cells; increases body temperature; regulates growth and development
	Calcitonin	Peptide	Inhibits release of calcium from bones
Parathyroid	Parathormone	Peptide	Stimulates release of calcium from bones; promotes absorption of calcium by intestines; promotes reabsorption of calcium by kidneys
Pancreas	Insulin	Peptide	Decreases blood glucose levels by increasing uptake of glucose into cells and converting glucose to glycogen, especially in liver; regulates fat metabolism
	Glucagon	Peptide	Converts glycogen to glucose, raising blood glucose levels

MEDIATUTOR 32.2 Modes of Action of Hormones

of the eye, the milk-producing (mammary) glands, as well as glands that produce digestive secretions, such as the salivary glands and some cells of the pancreas.

Endocrine glands, sometimes called *ductless glands*, release their hormones within the body (*endo* means "inside of"). An endocrine gland generally consists of clusters of hormone-producing cells embedded within a network of capillaries. The cells secrete their hormones into the extracellular fluid surrounding the capillaries (see Fig. 32-2). The hormones then enter the capillaries by diffusion and are carried throughout the body by the bloodstream.

The Hypothalamus Controls the Secretions of the Pituitary Gland

If the endocrine system is the body's postal service, then the hypothalamus is the main post office and the pituitary gland is the administrative headquarters. Together, these structures coordinate the action of many key hormonal messaging systems. The **hypothalamus** is a part of the brain that contains clusters of specialized nerve cells called **neurosecretory cells**. Neurosecretory cells synthesize peptide hormones, store them, and release them when stimulated. The **pituitary gland** is a pea-sized gland that dangles from the hypothalamus by a stalk (see Fig. 32-1). Anatomically, the pituitary consists of two distinct parts: (1) the **anterior pituitary** and (2) the **posterior pituitary**. The hypothalamus controls the release of hormones from both parts. The anterior pituitary is a true endocrine gland, composed of several types of hormone-secreting cells enmeshed in a network of capillaries. The posterior pituitary, however, is derived from an outgrowth of the hypothalamus.

Hypothalamic Hormones Control the Anterior Pituitary

Neurosecretory cells of the hypothalamus produce at least nine peptide hormones that regulate the release of hormones from the anterior pituitary. These peptides

Table 32-1 (*Continued*)

Endocrine Gland	Hormone	Type of Chemical	Principal Function
Ovaries[a]	Estrogen	Steroid	Causes development of female secondary sexual characteristics and maturation of eggs; promotes growth of uterine lining
	Progesterone	Steroid	Stimulates development of uterine lining and formation of placenta
Testes[a]	Testosterone	Steroid	Stimulates development of genitalia and male secondary sexual characteristics; stimulates spermatogenesis
Adrenal medulla	Epinephrine (adrenaline) and norepinephrine (noradrenaline)	Amino acid derivative	Increase levels of sugar and fatty acids in blood; increase metabolic rate; increase rate and force of contractions of the heart; constrict some blood vessels
Adrenal cortex	Glucocorticoids	Steroid	Increase blood sugar; regulate sugar, lipid, and fat metabolism; anti-inflammatory effects
	Aldosterone	Steroid	Increases reabsorption of salt in kidney
	Testosterone	Steroid	Causes masculinization of body features, growth
		Other Sources of Hormones	
Pineal gland	Melatonin	Amino acid derivative	Regulates seasonal reproductive cycles and sleep–wake cycles; may regulate onset of puberty
Thymus	Thymosin	Peptide	Stimulates maturation of cells of immune system
Kidney	Renin	Peptide	Acts on blood proteins to produce hormone (angiotensin) that regulates blood pressure
	Erythropoietin	Peptide	Stimulates red blood cell synthesis in bone marrow
Heart	Atrial natriuretic peptide (ANP)	Peptide	Increases salt and water excretion by kidneys; lowers blood pressure
Digestive tract[b]	Secretin, gastrin, cholecystokinin, and others	Peptides	Control secretion of mucus, enzymes, and salts in digestive tract; regulate peristalsis
Fat cells	Leptin	Peptide	Regulates appetite; stimulates immune function; promotes blood vessel growth; required for onset of puberty

[a]See Chapters 35 and 36.

[b]See Chapter 29.

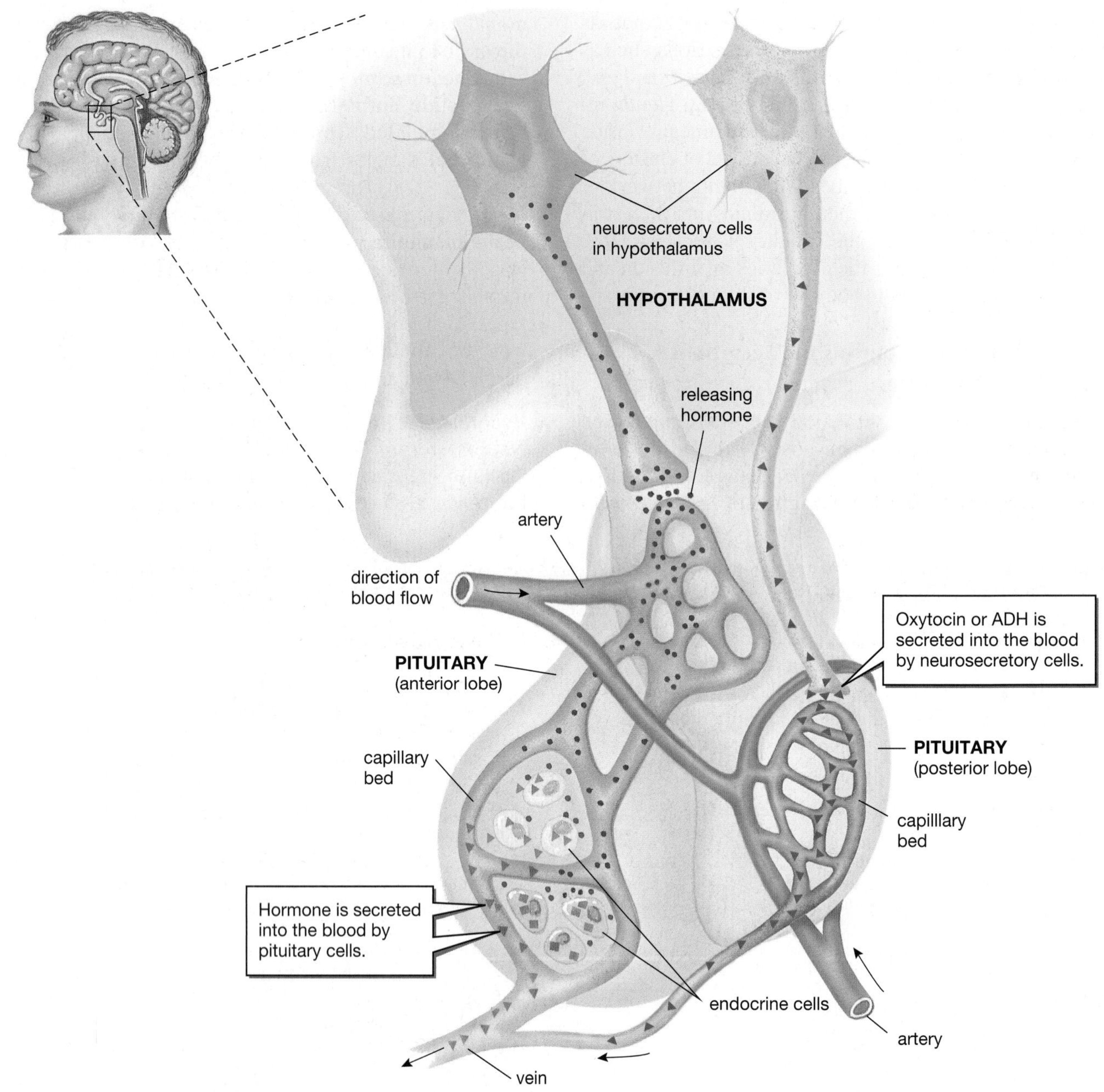

Figure 32-4 The hypothalamus controls the pituitary
Neurosecretory cells of the hypothalamus control hormone release in the anterior lobe of the pituitary by producing releasing hormones (left). These cells secrete their hormones into a capillary network that carries them to the anterior pituitary. There, each releasing hormone stimulates endocrine cells with appropriate receptors to secrete that hormone while leaving other types unaffected. The posterior lobe of the pituitary (right) is an extension of the hypothalamus. Neurosecretory cells in the hypothalamus have cell endings on a capillary bed in the posterior lobe, where the cells release oxytocin or antidiuretic hormone (ADH).

are called **releasing hormones** or **inhibiting hormones**, depending on whether they stimulate or prevent the release of pituitary hormone, respectively (Fig. 32-4). Releasing and inhibiting hormones are synthesized in nerve cells in the hypothalamus, secreted into a capillary bed in the lower portion of the hypothalamus, and travel a short distance through blood vessels to a second capillary bed that surrounds the endocrine cells of the anterior pituitary. There, the releasers and inhibitors diffuse out of the capillaries and influence pituitary hormone secretion.

Because the releasing and inhibiting hormones are secreted very close to the anterior pituitary, they are produced only in minute amounts. Not surprisingly, they were extremely difficult to isolate and study. Andrew Schally and Roger Guillemin, U.S. endocrinologists who shared the Nobel Prize in Medicine in 1977 for characterizing several of these hormones, used the brains of millions of sheep and pigs (obtained from slaughterhouses) to extract enough releasing hormone to analyze.

The Anterior Pituitary Produces and Releases a Variety of Hormones

The anterior pituitary produces several peptide hormones. Four of these regulate hormone production in other endocrine glands. **Follicle-stimulating hormone (FSH)** and **luteinizing hormone (LH)** stimulate the production of sperm and testosterone in males, and of eggs, estrogen, and progesterone in females. We shall discuss the roles of FSH and LH in more detail in Chapter 35. **Thyroid-stimulating hormone (TSH)** stimulates the thyroid gland to release its hormones, and **ACTH**, or **adrenocorticotropic hormone** ("hormone that stimulates the adrenal cortex"), causes the release of hormones from the adrenal cortex. We discuss the effects of thyroid and adrenal cortical hormones later in this chapter.

The remaining hormones of the anterior pituitary do not act on other endocrine glands. **Prolactin**, in conjunction with other hormones, stimulates the development of the mammary glands (which are exocrine glands) during pregnancy. **Endorphins** are anterior pituitary hormones that inhibit the perception of pain by binding to certain receptors in the brain. Some narcotic drugs, including morphine and heroin, mimic endorphins and bind to the same receptors in the brain. **Melanocyte-stimulating hormone (MSH)** stimulates the synthesis of the skin pigment melanin. **Growth hormone** regulates the body's growth by acting on nearly all the body's cells—increasing protein synthesis, fat utilization, and the storage of carbohydrates. During maturation, it has a stimulatory effect on bone growth, which influences the ultimate size of the adult organism. Much of the normal variation in human height is due to differences in the secretion of growth hormone from the anterior pituitary. Too little growth hormone causes some cases of *dwarfism*; too much can cause *gigantism* (Fig. 32-5). Although in adulthood many bones lose their ability to lengthen, growth hormone continues to be secreted throughout life, helping regulate protein, fat, and sugar metabolism.

A major advance in the treatment of pituitary dwarfism occurred when molecular biologists successfully inserted the gene for human growth hormone into bacteria, which churned out large quantities of the substance. Previously, the main commercial source of growth hormone was human cadavers, from which tiny amounts were extracted at great cost. Thanks to the new, cheaper source, many more children with underactive pituitary glands who would previously have been dwarfs can achieve normal height.

Figure 32-5 When the anterior pituitary malfunctions
An improperly functioning anterior pituitary produces either too much or too little growth hormone. Too little results in dwarfism; too much causes gigantism.

The Posterior Pituitary Releases Hormones Produced by Cells in the Hypothalamus

The posterior pituitary contains the endings of two types of neurosecretory cells whose cell bodies are located in the hypothalamus. These neurosecretory cell endings are enmeshed in a capillary bed into which they release hormones to be carried into the bloodstream (see Fig. 32-4). Two peptide hormones are synthesized in the hypothalamus and released from the posterior pituitary: **antidiuretic hormone (ADH)** and **oxytocin**.

Antidiuretic hormone, which literally means "hormone that prevents urination," helps prevent dehydration. As you learned in Chapter 30, by increasing the permeability to water of the collecting ducts of nephrons in the kidney, ADH causes water to be reabsorbed from the urine and retained in the body. Interestingly, alcohol inhibits the release of ADH, greatly increasing urination, so a beer drinker may temporarily lose more fluid than he or she has taken in.

Oxytocin triggers the "milk letdown reflex" in nursing mothers by causing muscle tissue within the breasts (mammary glands) to contract during lactation (breastfeeding). This reflex ejects milk from the saclike milk glands into the nipples (Fig. 32-6). Oxytocin also causes contractions of the muscles of the uterus during childbirth, helping expel the fetus from the womb.

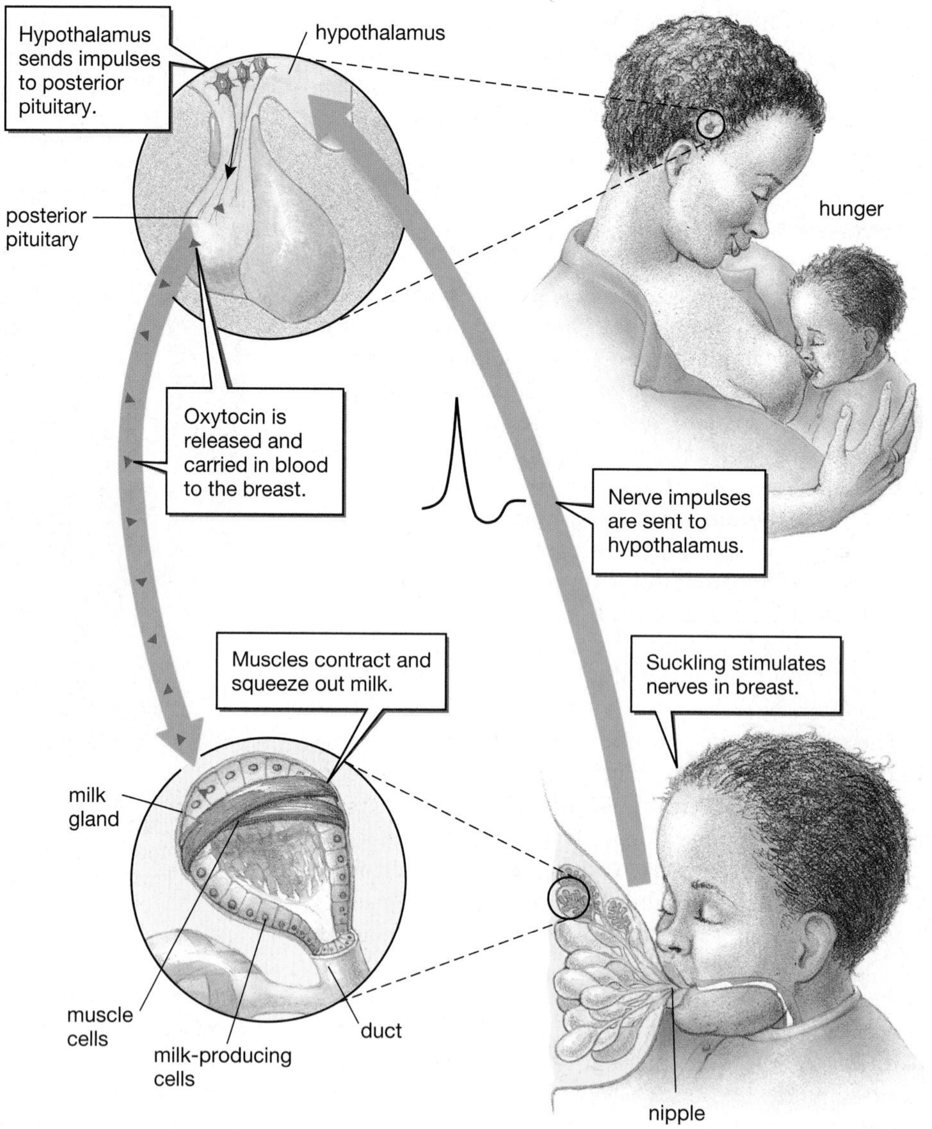

Figure 32-6 Hormones and breastfeeding
The control of milk letdown by oxytocin during breastfeeding is regulated by feedback between a baby and its mother. The mammary gland is an exocrine gland. There, clusters of milk-producing cells surround hollow bulbs, where milk collects in lactating women. The bulbs are surrounded by muscle that can expel the milk through the nipple. Milk is expelled when suckling stimulates nerve endings that send a signal to the mother's hypothalamus, causing the posterior pituitary to secrete oxytocin into the bloodstream. When oxytocin reaches the muscles that surround the milk ducts, it causes them to contract and expel milk through the nipple. This cycle continues until the infant is full and stops suckling. With the nipple no longer being stimulated, oxytocin release stops, the muscles relax, and milk flow ceases.

Recent studies using laboratory animals indicate that oxytocin also has behavioral effects. In rats, for example, oxytocin injections cause virgin females to exhibit maternal behavior such as building a nest, licking pups, and retrieving pups that have strayed. Oxytocin may also have a role in male reproductive behavior. In several types of animals, oxytocin stimulates the contraction of muscles that surround the tubes that conduct sperm from the testes to the penis, causing ejaculation.

The Thyroid and Parathyroid Glands Influence Metabolism and Calcium Levels

Lying at the front of the neck, nestled just below the larynx (Fig. 32-7a), the **thyroid gland** produces two major hormones: (1) thyroxine and (2) calcitonin. **Thyroxine**, often referred to as "thyroid hormone," is an iodine-containing modified amino acid that raises the metabolic rate of most body cells. **Calcitonin** is a peptide important in calcium metabolism.

Thyroxine influences most of the cells in the body, elevating their metabolic rate and stimulating the synthesis of enzymes that break down glucose and provide energy. These cellular effects cause the body to increase oxygen consumption and heart rate. Thyroxine's influence on metabolic rate seems to be at least in part connected to regulating body temperature and response to stress. Exposure to cold, for example, greatly increases thyroid hormone production.

In juvenile animals, including humans, thyroxine helps regulate growth by stimulating both metabolic rate and the development of the nervous system. Undersecretion of thyroid hormone early in life causes *cretinism,* a condition characterized by mental retardation and dwarfism. Fortunately, early diagnosis and thyroxine supplementation can prevent this condition. Conversely, oversecretion of thyroxine in developing vertebrates can trigger precocious development. In 1912, in one of the first demonstrations of hormone action, a physiologist discovered that thyroxine can induce early metamorphosis in

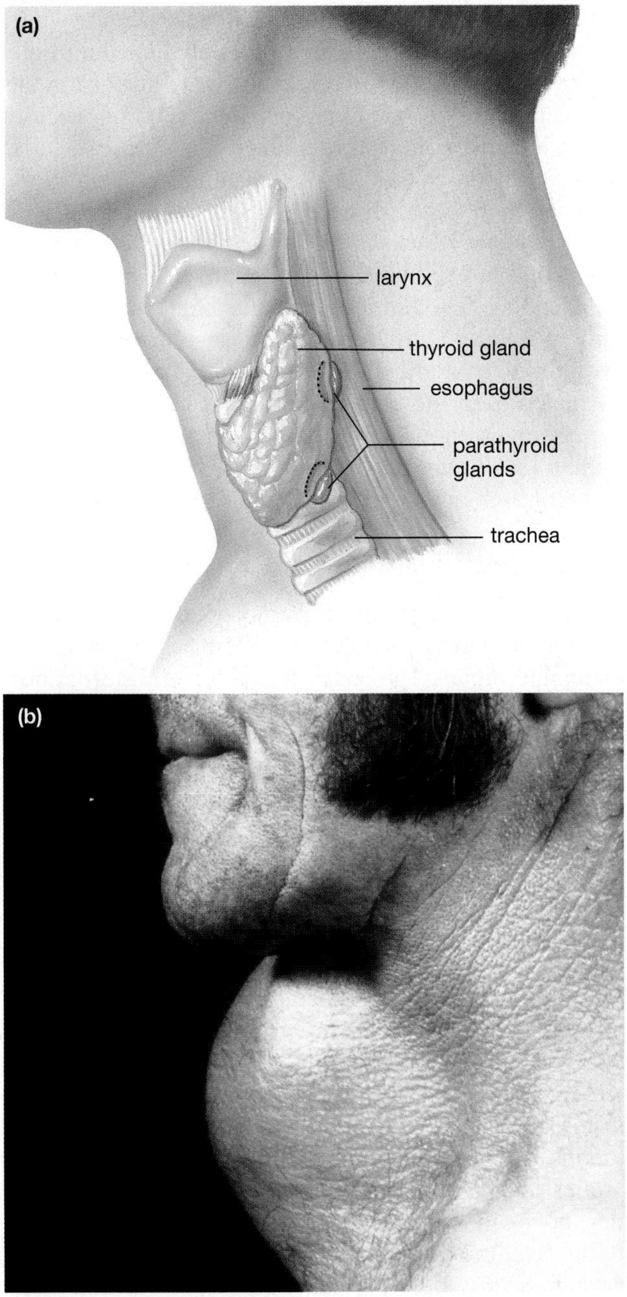

Figure 32-7 The thyroid and parathyroid glands
(a) The thyroid and parathyroid glands wrap around the front of the larynx in the neck. ***(b)*** Goiter, a condition in which the thyroid gland becomes greatly enlarged, is caused by an iodine-deficient diet.

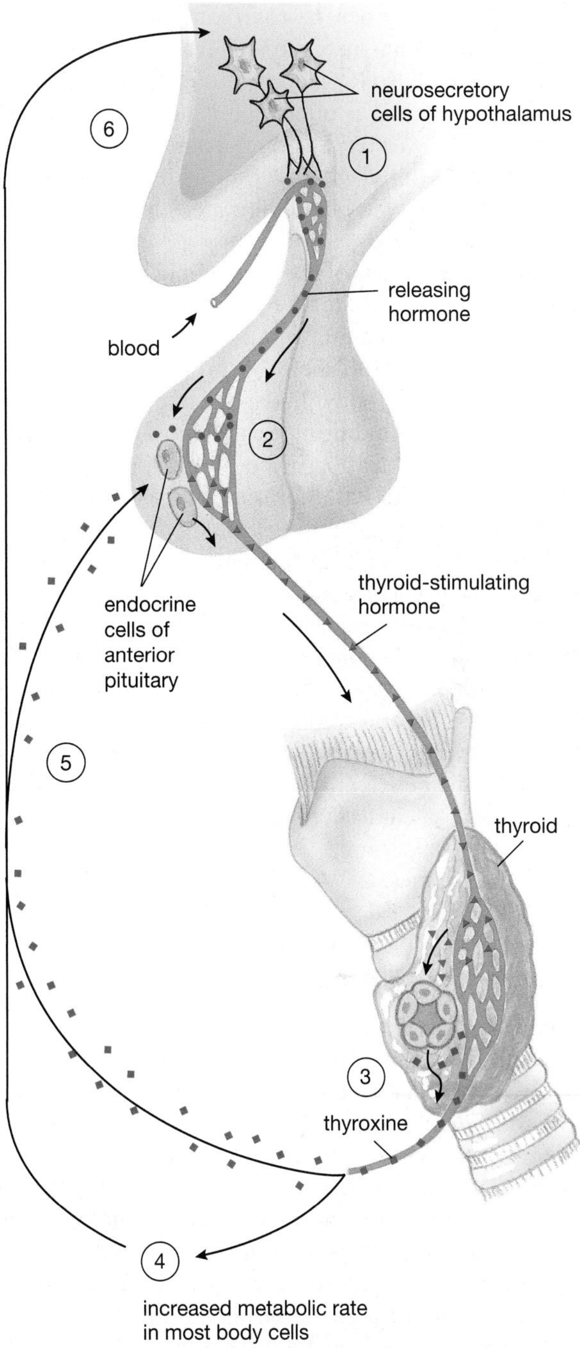

Figure 32-8 Negative feedback in thyroid gland function
① Low body temperature or stress stimulates neurosecretory cells of the hypothalamus to secrete a releasing hormone. ② That hormone triggers the release of thyroid-stimulating hormone (TSH) from the anterior pituitary. ③ TSH stimulates the thyroid gland to release thyroxine. ④ Thyroxine causes an increase in the metabolic activity of most body cells, generating heat. ⑤ Higher thyroxine levels in the blood inhibit the TSH-producing cells. ⑥ Higher body temperature inhibits the hypothalamic cells.

tadpoles (see "Evolutionary Connections: The Evolution of Hormones").

Levels of thyroxine in the bloodstream are finely tuned by negative feedback loops. Thyroxine release is stimulated by thyroid-stimulating hormone (TSH) from the anterior pituitary, which in turn is stimulated by a releasing hormone from the hypothalamus. The amount of TSH released from the pituitary is regulated by thyroxine levels in the blood (Fig. 32-8): High levels of thyroxine inhibit the secretion of both the releasing hormone and TSH, thus inhibiting further release of thyroxine from the thyroid.

A diet deficient in iodine can reduce the production of thyroxine and trigger a feedback mechanism that acts to restore normal hormone levels by dramatically increasing the number of thyroxine-producing cells. This compensating mechanism leads to excessive growth of the thyroid gland; the enlarged gland may bulge from the neck, producing a condition called **goiter** (Fig. 32-7b). Goiter was once common in some regions of the United States, but widespread use of iodized salt has now all but eliminated this condition in developed countries.

The four small disks of the **parathyroid glands** are embedded in the back of the thyroid gland (see Fig. 32-7a). The parathyroids secrete **parathormone**, which, along with calcitonin, controls the concentration of calcium in the blood and other body fluids. Calcium is essential for many processes, including nerve and muscle function, so the calcium concentration in body fluids must be kept within narrow limits. Parathormone and calcitonin regulate calcium absorption and release by the bones, which serve as a bank into which calcium can be deposited or withdrawn as necessary. In response to low blood calcium, the parathyroids release parathormone, which causes the release of calcium from bones. The parathyroids increase in size in pregnant and lactating women, thereby enhancing parathormone output and allowing the mother's body to meet the extra demands for calcium imposed by the developing fetus and, later, milk production. If blood calcium levels become too high, the thyroid releases calcitonin, which inhibits the release of calcium from bones.

The Pancreas Is Both an Exocrine and an Endocrine Gland

The **pancreas** is a gland that produces both exocrine and endocrine secretions. The exocrine portion synthesizes digestive secretions that are released into the *pancreatic duct* and flow into the small intestine (see Chapter 29). The endocrine portion consists of clusters of cells called **islet cells**, which produce peptide hormones. One type of islet cell produces the hormone **insulin**; another type produces the hormone **glucagon**.

Insulin and glucagon work in opposition to regulate carbohydrate and fat metabolism: Insulin reduces the blood glucose level; glucagon increases it (Fig. 32-9). Together the two hormones help keep the blood glucose level nearly constant. When blood glucose rises (for example, after you've eaten), insulin is released. Insulin causes body cells to take up glucose and either metabolize it for energy or convert it to fat or *glycogen* (a starchlike molecule) for storage. When blood glucose levels drop (for example, after you've skipped breakfast or have run a 10-kilometer race), glucagon is released. Glucagon activates an enzyme in the liver that breaks down glycogen (which is stored primarily in the liver), releasing glucose into the blood. Glucagon also promotes lipid breakdown, which releases fatty acids that can be metabolized for energy.

Defects in insulin production, release, or reception by target cells result in **diabetes mellitus**, a condition in which blood glucose levels are high and fluctuate wildly with sugar intake. The lack of functional insulin in diabetics causes the body to rely much more heavily on fats as an energy source, leading to high circulating levels of lipids, including cholesterol. Severe diabetes causes fat deposits in the blood vessels, resulting in high blood pressure and heart disease; diabetes is an important cause of heart attacks in the United States. The fatty deposits in small vessels can also damage the retina of the eye, leading to blindness, and the kidneys, leading to kidney failure. Until recently, the insulin supplements required to treat diabetes were formulated from insulin extracted from the pancreas of cows and pigs, obtained from slaughterhouses. Now, however, large quantities of human insulin are produced by bacteria into which the gene for human insulin has been inserted.

The Sex Organs Secrete Steroid Hormones

The sex organs do far more than produce sperm or eggs. The **testes**, in males, and **ovaries**, in females, are important endocrine organs. The testes secrete several steroid hormones, collectively called **androgens**. The most important of these is **testosterone**. The ovaries secrete two types of steroid hormones: (1) **estrogen** and (2) **progesterone**. The role of the sex hormones in development, the menstrual cycle, and pregnancy is discussed in Chapters 35 and 36.

The sex hormones also play a key role in *puberty*, the phase of life that links maturity and immaturity. Puberty begins when, for reasons not fully understood, the hypothalamus starts to secrete increasing amounts of releasing hormones that in turn stimulate the anterior pituitary to secrete more lutenizing hormone (LH) and follicle-stimulating hormone (FSH) into the bloodstream. Both LH and FSH stimulate target cells in the testes or ovaries to produce higher levels of sex hormones. The resulting increase in circulating sex hormones ultimately affects tissues throughout the body that bear the appropriate receptors. Puberty also leads to the activation of the adult reproductive system, to the new growth underlying secondary sexual characteristics, and to the behavioral changes that make puberty such an interesting time for parents and teenagers.

Mature mammals are capable of reproduction, and puberty encompasses the physiological changes that endow immature individuals with this reproductive capacity. In humans, the development of full reproductive competence is accompanied by a growth spurt and by the development of *secondary sexual characteristics.* Both sexes develop pubic and underarm hair; females develop breasts, while males acquire facial hair, larger muscles, and a prominent larynx ("Adam's apple"), which in turn lowers the voice. Although there is a surge of sex hormone production at puberty, sex hormones are present from the fetal stage onward. They influence brain development in both sexes, and they continue to have an effect on both behavior and cognitive processes throughout life.

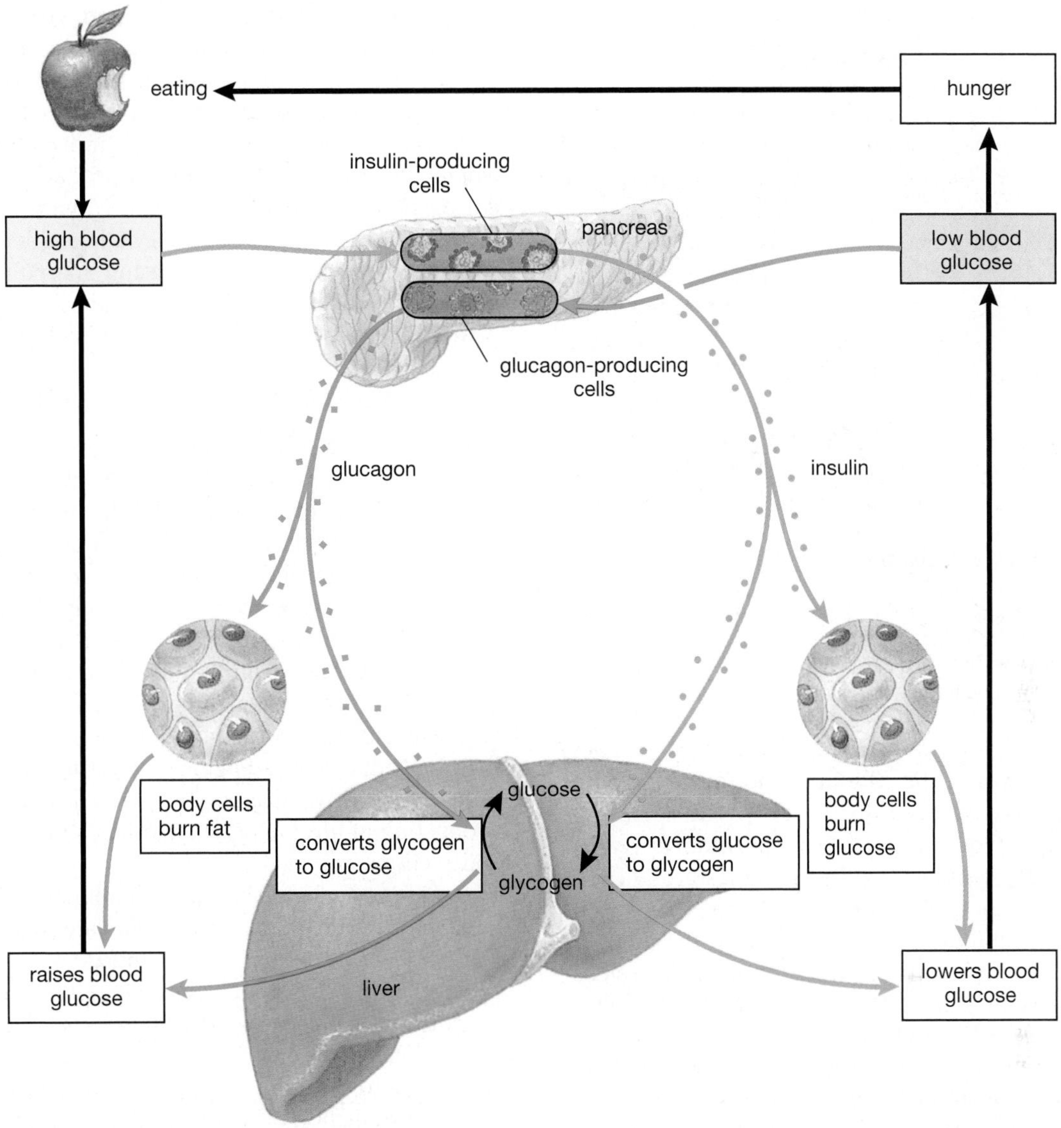

Figure 32-9 The pancreas controls blood glucose levels
The pancreatic islet cells contain two populations of hormone-producing cells: one producing insulin; the other, producing glucagon. These two hormones cooperate in a two-part negative feedback loop to control blood glucose concentrations. High blood glucose stimulates the insulin-producing cells and inhibits the glucagon-producing cells; low blood glucose stimulates the glucagon-producing cells and inhibits the insulin-producing cells. This dual control quickly corrects high or low blood glucose levels.

The Adrenal Glands Have Two Parts That Secrete Different Hormones

Imagine how your body feels when you are startled, afraid, or angry. These physical reactions are the result of hormones produced by the adrenal glands, which act in conjunction with your sympathetic nervous system (see Chapter 33). Like the pituitary gland and pancreas, the **adrenal glands** (Latin for "on the kidney") are two glands in one: (1) the adrenal medulla and (2) the adrenal cortex (Fig. 32-10). The **adrenal medulla** is located in the center of each gland (*medulla* means "marrow" in Latin). It consists of secretory cells derived during development from nervous tissue, and its hormone secretion is controlled directly by the nervous system. The adrenal medulla produces two hormones—**epinephrine** and **norepinephrine** (also called *adrenaline* and *noradrenaline,* respectively)—in response to stress. These hormones, which are amino-acid derivatives, prepare the body for emergency action. They increase the heart and respiratory rates, cause blood glucose levels to rise, and direct blood flow away from the digestive tract and toward the brain and muscles. They also cause the air passages of the lungs to expand, allowing more efficient exchange of gases. For this reason, epinephrine is administered to asthmatics, whose airways become constricted during an attack. The adrenal medulla is activated by the sympathetic nervous system, which prepares the body to respond to emergencies, as we shall describe in Chapter 33.

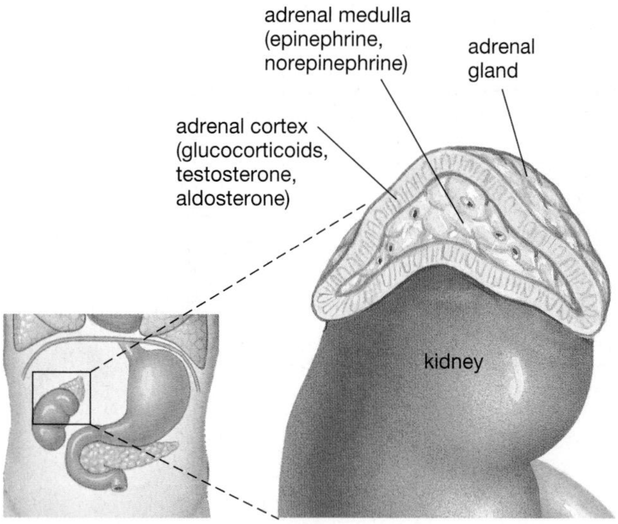

Figure 32-10 The adrenal glands
Atop each kidney sits an adrenal gland, which is a two-part gland composed of very dissimilar cells. The outer cortex consists of ordinary endocrine cells that secrete steroid homones. The inner medulla, derived from nervous tissue during development, secretes epinephrine and norepinephrine.

The outer layer of the adrenal gland forms the **adrenal cortex** (*cortex* is Latin for "bark"). The cortex secretes three types of steroid hormones, called **glucocorticoids**, synthesized from cholesterol. These hormones help control glucose metabolism. Glucocorticoid release is stimulated by ACTH from the anterior pituitary. ACTH release in turn is stimulated by releasing hormones produced by the hypothalamus in response to stressful stimuli, including trauma, infection, or exposure to temperature extremes. In some respects, the glucocorticoids act similarly to glucagon, raising blood glucose levels by stimulating glucose production and promoting the use of fats instead of glucose for energy production.

You may have noticed that many different hormones are involved in glucose metabolism: thyroxine, insulin, glucagon, epinephrine, and the glucocorticoids. Why? The reason can probably be traced to a metabolic requirement of the brain. Although most body cells can produce energy from fats and proteins as well as from carbohydrates, brain cells can burn only glucose. Thus, blood glucose levels cannot be allowed to fall too low, or brain cells rapidly starve, leading to unconsciousness and death.

The adrenal cortex also secretes the hormone **aldosterone**, which regulates the sodium content of the blood. Sodium ions, derived from salt in the diet, are the most abundant positive ions in blood and extracellular fluid. The sodium ion gradient across plasma membranes (high outside, low inside) is a factor in many cellular events, including the production of electrical signals by nerve cells. If blood sodium falls, the adrenal cortex releases aldosterone, which causes the kidneys and sweat glands to retain sodium. Then salt and other sources of dietary sodium, combined with aldosterone-induced sodium conservation, raise blood sodium levels, and shut off further aldosterone secretion.

Finally, the adrenal cortex produces the male sex hormone testosterone, although normally in much smaller amounts than the testes produce. The adrenal cortex produces this hormone in women as well as in men. Tumors of the adrenal medulla can lead to excessive testosterone release, causing masculinization of women. Many of the "bearded ladies" who once appeared in circus sideshows probably had this condition.

Many Types of Cells Produce Prostaglandins

Unlike most other hormones, which are synthesized by a limited number of specialized cells, prostaglandins are produced by many, perhaps all, cells of the body. These hormones are modified fatty acids synthesized by the cell from membrane phospholipids. Several prostaglandins are known, and probably a great many more await discovery. One prostaglandin causes arteries to constrict and stops bleeding from the umbilical cords of newborn infants. Another prostaglandin works in conjunction with oxytocin during labor, stimulating uterine contractions. Prostaglandin-soaked vaginal suppositories are used to induce labor. Menstrual cramps are caused by the overproduction of uterine prostaglandins, stimulating uterine contractions.

Some prostaglandins cause inflammation (such as occurs in arthritic joints) and stimulate pain receptors. Drugs such as aspirin and ibuprofen, which inhibit prostaglandin synthesis, can provide relief by reducing inflammation. Other prostaglandins expand the air passages of the lungs; one day, asthmatics might benefit from research into this effect. Still other prostaglandins stimulate the production of the protective mucus that lines the stomach, a potential boon for ulcer patients. Research on this diverse and potent family of compounds is still in its infancy, but it promises many health benefits in the future.

Other Sources of Hormones Include the Pineal Gland, Thymus, Kidneys, Heart, Digestive Tract, and Fat Cells

The **pineal gland** is located between the two hemispheres of the brain, just above and behind the hypothalamus (see Fig. 32-1). Named for its resemblance to a pine cone, the pineal gland is smaller than a pea. In 1646 philosopher René Descartes described it as "the seat of the rational soul." Since then, scientists have learned more about this organ, but many of its functions are still poorly understood. The pineal produces the hormone **melatonin**, an amino-acid derivative. Melatonin is secreted in a daily rhythm, which in mammals is regulated by the eyes. In some vertebrates, such as the frog, the pineal itself contains photoreceptive cells, and the skull above it is thin, so the pineal can detect sunlight and thus daylength. By responding to daylengths characteristic of different seasons, the pineal appears to regulate the seasonal reproductive cycles of many mammals. Despite years of research, the function of the pineal gland in humans and of melatonin is still unclear. One hypothesis is that the pineal gland and melatonin secretion influence sleep–wake cycles. Melatonin is sold as a sleeping aid, but its use is controversial. Improper pineal function may contribute to the depression that some people experience during the short days of winter.

The **thymus** is located in the chest cavity behind the breastbone, also called the *sternum* (see Fig. 32-1). In addition to producing white blood cells, the thymus produces the hormone **thymosin**, which stimulates the development of specialized white blood cells (T cells) that play an important role in the immune system (see Chapter 31). The thymus is extremely large in infants but, under the influence of sex hormones, begins decreasing in size after puberty.

The kidney, which plays a central role in maintaining body fluid homeostasis, is an important endocrine organ as well. When the oxygen content of the blood drops, the kidney produces the hormone **erythropoietin**, which increases red blood cell production (see Chapter 27). The kidney also produces a second hormone, **renin**, in response to low blood pressure, such as that caused by bleeding. Renin is an enzyme that catalyzes the production of the hormone **angiotensin** from proteins in the blood. Angiotensin raises blood pressure by constricting arterioles. It also stimulates aldosterone release by the adrenal cortex, causing the kidneys to retain sodium, which in turn increases blood volume.

In 1981 a substance that was extracted from atrial tissue of the heart and injected into rats was found to cause an increase in the output of salt and water by the kidneys. Two years later, the active substance **atrial natriuretic peptide (ANP)** was described and its amino acid sequence determined. This peptide is released by cells in the atria when blood volume increases, causing extra distension of the heart. Atrial natriuretic peptide then produces a reduction of blood volume by decreasing the release of both ADH and aldosterone.

The stomach and small intestine produce a variety of peptide hormones that help regulate digestion. These hormones include **gastrin**, **secretin**, and **cholecystokinin**, discussed in Chapter 29.

Can fat be an endocrine organ? In 1995 researchers described the peptide hormone **leptin** (derived from a word meaning "slender"), which is released by adipose (fat) cells. Mice missing the gene for leptin became obese (Fig. 32-11), and leptin injections caused them to lose weight. The researchers hypothesized that adipose tissue, by releasing leptin, tells the body how much fat it has stored and therefore how much to eat. Unfortunately, trials of leptin as a human weight-loss aid have not been encouraging. Many obese people have high levels of leptin but seem to be relatively insensitive to it. However, reasearchers are discovering surprising new functions for

Figure 32-11 Leptin helps regulate body fat
The mouse on the left has been genetically engineered to lack the gene for the hormone leptin.

leptin and are finding leptin receptors in unexpected places, such as blood vessels and white blood cells. Leptin appears to stimulate the growth of new capillaries and to speed wound healing. It also stimulates the immune system and is required for the onset of puberty and the development of secondary sexual characteristics.

Research continues to expand our understanding of the multiple effects of hormones and the wide variety of organs and cells that produce them. In time this understanding will lead to a wealth of new medical treatments. It should also increase our respect for these substances and our recognition that any hormone we take might influence physiological systems throughout our body.

Evolutionary Connections

The Evolution of Hormones

Not long ago, vertebrate endocrine systems were considered unique to our phylum, and the endocrine chemicals were thought to have evolved expressly for their role in vertebrate physiology. In recent years, however, physiologists have discovered that hormones are evolutionarily ancient. Insulin, for example, is found not only in vertebrates but also in protists, fungi, and bacteria, although research has not yet determined the function of insulin in most of those organisms. Protists also manufacture ACTH, even though they have no adrenal glands to stimulate. Yeasts have receptors for estrogen but no ovaries. Thyroid hormones have been found in certain invertebrates, such as worms, insects, and mollusks, as well as in vertebrates. Even among vertebrates, the effects of chemically identical hormones, secreted by the same glands, can vary dramatically from organism to organism. Let's look briefly at the diverse effects that the thyroid hormone thyroxine has on several organisms.

Some fish undergo radical physiological changes during their lifetimes. A salmon, for example, begins life in fresh water, migrates to the ocean, and returns to fresh water to spawn. In the stream where the salmon hatched, fresh water tends to enter the fish's tissues by osmosis; in salt water, the fish tends to lose water, becoming dehydrated. The salmon's migrations, therefore, require complete revamping of salt and water control. In salmon, one of the functions of thyroxine is to produce the metabolic changes necessary to go from life in streams to life in the ocean and back.

In amphibians, thyroxine has the dramatic effect of triggering metamorphosis. In 1912, in one of the first demonstrations of the action of any hormone, tadpoles were fed minced horse thyroid. As a result, the tadpoles metamorphosed prematurely into miniature adult frogs (Fig. 32-12). In high mountain lakes in Mexico, where the water is deficient in the iodine needed to synthesize thyroxine, natural selection has produced one species of salamander that has the ability to reproduce while still in its juvenile form.

Thyroxine regulates the seasonal molting of most vertebrates. From snakes to birds to the family dog, surges of thyroxine stimulate the shedding of skin, feathers, or hair. In humans (who neither migrate regularly, metamorphose, nor molt), thyroxine regulates growth and metabolism.

The use of chemicals to regulate cellular activity is extremely ancient. The diversity of life on Earth rests on a conservative foundation: A relative handful of chemicals coordinate activities within single cells and among groups of cells. Life's diversity originated in part by changing the systems used to deliver the chemicals and by evolving new types of responses. Early in their evolution, animals developed a complement to hormonal communication that provides faster, more precise delivery of chemical messages: the nervous system. As we will explain in the next chapter, the nervous system permits rapid responses to environmental stimuli, flexibility in response options, and ultimately consciousness itself.

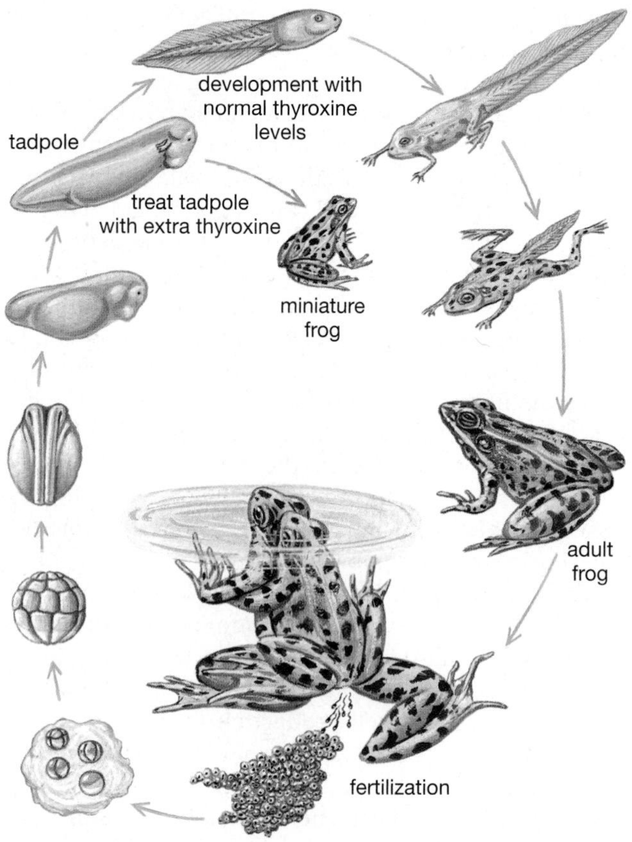

Figure 32-12 Thyroxine controls metamorphosis in amphibians The life cycle of the frog begins with fertilization of the eggs (bottom). The fertilized eggs develop into an aquatic, fishlike tadpole, which grows and ultimately metamorphoses into an adult frog. Metamorphosis is triggered by a surge of thyroxine from the tadpole's thyroid gland. If injected with extra thyroxine, a young tadpole will metamorphose ahead of schedule into a miniature adult frog.

CASE STUDY REVISITED

Losing on Steroids

As with many hormones, anabolic steroids exert their effects throughout the body; a steroid user risks damage to the liver, circulatory system, reproductive system, and immune system. Liver damage can lead to jaundice and liver tumors. Steroid-induced high blood pressure and decreases in the "good" (HDL) form of cholesterol threaten the circulatory system because both of these effects are risk factors for heart attacks and strokes. In males, artificially increased levels of anabolic steroids create a negative feedback effect that can reduce the natural production of testosterone, reduce testicle size and sperm count, and cause the prostate to enlarge and interfere with urination. In females, anabolic steroids can interfere with menstrual periods and cause a more masculine distribution of body hair. For both sexes, anabolic steroids depress the immune system. Aggressiveness and mood swings have also been attributed to steroid use. What unearned glory could possibly be worth such costs?

Many individuals in the world of sports are not surprised when they learn that even elite athletes such as Ben Johnson have used anabolic steroids. The practice is all too widespread, and evading detection has become part of the "game" for some members of this intensely competitive world. Given what you know about world-class sports and what you now know about steroids and other hormones, do you think these substances should be banned? Explain your answer.

Summary of Key Concepts

1) What Are the Characteristics of Animal Hormones?

A hormone is a chemical secreted by cells in one part of the body that is transported in the bloodstream to another part of the body, where it affects the activity of specific target cells. Hormones are synthesized either from amino acids (amino-acid based hormones and peptides) or from lipids (steroids and prostaglandins).

Most hormones act on their target cells in one of two ways: (1) Peptide hormones and amino-acid derivatives bind to receptors on the surfaces of target cells and activate intracellular second messengers, such as cyclic AMP. The second messengers then alter the metabolism of the cell. (2) Steroid hormones can bind to surface receptors or diffuse through the plasma membranes of their target cells and bind with receptors in the cytoplasm. The hormone–receptor complex travels to the nucleus and promotes the transcription of specific genes. Thyroid hormones also penetrate the plasma membrane but diffuse into the nucleus, where they bind to receptors associated with the chromosomes and influence gene transcription.

Hormone action is commonly regulated through negative feedback, a process in which a hormone causes changes that inhibit further secretion of that hormone.

2) What Are the Structures and Hormones of the Mammalian Endocrine System?

Hormones are produced by endocrine glands, which are clusters of cells embedded within a network of capillaries. Hormones are secreted into the extracellular fluid and diffuse into the capillaries. The major endocrine glands of the human body are the hypothalamus–pituitary complex, the thyroid and parathyroid glands, the pancreas, the sex organs, and the adrenal glands. The hormones released by these glands and their actions are summarized in Table 32-1. Prostaglandins, unlike other hormones, are not secreted by discrete glands but are synthesized and released by many cells of the body. Other structures that produce hormones include the pineal gland, thymus, kidneys, heart, stomach and small intestine, and fat cells.

Key Terms

adrenal cortex *p. 656*
adrenal gland *p. 655*
adrenal medulla *p. 655*
adrenocorticotropic hormone (ACTH) *p. 651*
aldosterone *p. 656*
androgen *p. 654*
angiotensin *p. 657*
anterior pituitary *p. 649*
antidiuretic hormone (ADH) *p. 651*
atrial natriuretic peptide (ANP) *p. 657*
calcitonin *p. 652*
cholecystokinin *p. 657*
cyclic AMP *p. 645*
diabetes mellitus *p. 654*
duct *p. 648*
endocrine disruptor *p. 647*
endocrine gland *p. 649*
endocrine system *p. 643*
endorphin *p. 651*
environmental estrogen *p. 647*
epinephrine *p. 655*
erythropoietin *p. 657*
estrogen *p. 654*
exocrine gland *p. 648*
follicle-stimulating hormone (FSH) *p. 651*
gastrin *p. 657*
glucagon *p. 654*
glucocorticoid *p. 656*
goiter *p. 654*
growth hormone *p. 651*
hormone *p. 643*
hypothalamus *p. 649*
inhibiting hormone *p. 650*
insulin *p. 654*
islet cell *p. 654*
leptin *p. 657*
luteinizing hormone (LH) *p. 651*
melanocyte-stimulating hormone (MSH) *p. 651*
melatonin *p. 656*
neurosecretory cell *p. 649*
norepinephrine *p. 655*
ovary *p. 654*
oxytocin *p. 651*
pancreas *p. 654*
parathormone *p. 654*
parathyroid gland *p. 654*
peptide hormone *p. 644*
pineal gland *p. 656*
pituitary gland *p. 649*
posterior pituitary *p. 649*
progesterone *p. 654*
prolactin *p. 651*
prostaglandin *p. 644*
releasing hormone *p. 650*
renin *p. 657*
second messenger *p. 644*
secretin *p. 657*
steroid hormone *p. 644*
target cell *p. 643*
testis *p. 654*
testosterone *p. 654*
thymosin *p. 657*
thymus *p. 657*
thyroid gland *p. 652*
thyroid-stimulating hormone (TSH) *p. 651*
thyroxine *p. 652*

Thinking Through the Concepts

Multiple Choice

1. *Steroid hormones*
   a. alter the activity of genes
   b. trigger rapid, short-term responses in cells
   c. work via second messengers
   d. initiate open channels in plasma membranes
   e. bind to cell-surface receptors

2. *Examples of posterior pituitary hormones are*
   a. FSH and LH
   b. prolactin and parathormone
   c. secretin and cholecystokinin
   d. melatonin and prostaglandin
   e. ADH and oxytocin

3. *Negative feedback to the hypothalamus controls the level of _________ in the blood.*
   a. thyroxine
   b. estrogen
   c. glucocorticoids
   d. insulin
   e. all of the above
4. *The primary targets for FSH are cells in the*
   a. hypothalamus
   b. ovary
   c. thyroid gland
   d. adrenal medulla
   e. pituitary gland
5. *The kidney is a source of*
   a. thyroxine and parathormone
   b. calcitonin and oxytocin
   c. renin and erythropoietin
   d. ANP and epinephrine
   e. glucagon and glucocorticoids
6. *Hormones that are produced by many different types of body cells and cause a variety of localized effects are known as*
   a. peptide hormones
   b. parathormones
   c. releasing hormones
   d. prostaglandins
   e. exocrine hormones

? Review Questions

1. What are the four types of molecules used as hormones in vertebrates? Give an example of each.
2. What is the difference between an endocrine gland and an exocrine gland? Which type releases hormones?
3. When peptide hormones attach to target cell receptors, what cellular events follow? How do steroid hormones behave?
4. Diagram the process of negative feedback, and give an example of it in the control of hormone action.
5. What are the major endocrine glands in the human body, and where are they located?
6. Describe the structure of the hypothalamus–pituitary complex. Which pituitary hormones are neurosecretory? What are their functions?
7. Describe how releasing hormones regulate the secretion of hormones by cells of the anterior pituitary. Name the hormones of the anterior pituitary, and give one function of each.
8. Describe how the hormones of the pancreas act together to regulate the concentration of glucose in the blood.
9. Compare the adrenal cortex and adrenal medulla by answering the following questions: Where are they located within the adrenal gland? What are their embryological origins? Which hormones do they produce? Which organs do their hormones target? What homeostatic processes regulate blood levels of the respective hormones?

Applying the Concepts

1. A student decides to do a science project on the effect of the thyroid gland on frog metamorphosis. She sets up three aquaria with tadpoles. She adds thyroxine to the water of one, the drug thiouracil to a second, and nothing to the third. Thiouracil reacts with thyroxine in tadpoles to produce an ineffective compound. Assuming that the student uses appropriate physiological concentrations, predict what will happen.
2. If you were obese, would you consider injections of leptin? What pros and cons can you think of? Defend your decision.
3. Suggest a hypothesis about the endocrine system to explain why many birds lay their eggs in the spring and why poultry farmers keep lights on at night in their egg-laying operations.
4. Anabolic steroids, used by risk-taking athletes and bodybuilders, are chemically related to testosterone. They increase bone mass and muscle mass and seem to improve athletic performance. But anabolic steroids can cause liver problems, heart attacks, strokes, testicular atrophy, and personality changes in males. Females on anabolic steroids also have liver and circulatory problems. In addition, their voices deepen, their bodies develop more hair, and their menstrual cycles are disturbed. Explain how the same compound can produce different effects in males and females.
5. Some parents who are interested in college sports scholarships for their children are asking physicians to prescribe growth hormone treatments. Farmers also have an economic incentive to treat cows with growth hormone, which can now be produced in large quantities by genetic-engineering techniques. What biological and ethical problems do you foresee for parents, children, physicians, coaches, college scholarship boards, food consumers, farmers, the U.S. Food and Drug Administration, and biotechnology companies?
6. Argue for and against banning or restricting the use of common endocrine disruptors, such as plasticizers and certain pesticides. What compromises can you suggest?

For More Information

Atkinson, M., and MacLaren, N. "What Causes Diabetes?" *Scientific American*, July 1991. An in-depth look at what's behind a common but serious disease.

Beardsley, T. "Melatonin Mania." *Scientific American*, April 1996. The hormone melatonin has been touted as a treatment for ailments ranging from insomnia to cancer. What do scientists think of the evidence?

Berridge, M. J. "The Molecular Basis of Communication Within the Cell." *Scientific American*, October 1985. Both the "classical" cyclic AMP and more-recently discovered second messengers convey information from cell-surface receptors to DNA and cellular metabolism.

Gold, C. "Hormone Hell." *Discover*, September 1996. Environmental pollutants that mimic the chemical activity of hormones may be playing havoc with animal (including human) physiology.

McLachlan, J., and Arnold, S. "Environmental Estrogens." *American Scientist*, September–October 1996. A sober evaluation of the effects of estrogen-like substances in the environment, written by two of the leading researchers in the field.

Raloff, J. "Common Pollutants Undermine Masculinity." *Science News*, April 3, 1999. Several recent research articles report feminizing effects of estrogen disruptors in laboratory animals.

Sapolsky, R. "Testosterone Rules." *Discover*, March 1997. An engaging inquiry into the question of whether testosterone causes aggression.

Snyder, S. H. "The Molecular Basis of Communication Between Cells." *Scientific American*, October 1985. The similarities and differences between neural and hormonal control systems in the body.

Vogel, S. "The Mouse on the Left Needs Leptin." *Discover*, January 1996. The hormone leptin regulates appetite and fat deposition in mice. Can it become a weight-loss treatment for humans?

Answers to Multiple-Choice Questions

1. a 2. e 3. e 4. b 5. c 6. d

MEDIATUTOR
The Endocrine System

CD Activities

Activity 32.1: Endocrine System Anatomy

Estimated time: 10 minutes

The organs of the endocrine system secrete hormones that act as chemical messengers to cells that may be some distance away. The effect of hormones on distant cells may be prolonged and irreversible, or transient and reversible. In this tutorial, you will explore the anatomy of the human endocrine system and the hormones that are produced by the glands that make up this physiological system.

Activity 32.2: Modes of Action of Hormones

Estimated time: 5 minutes

Hormones exert their characteristic effects on target cells by altering the activity of those cells. The way in which this task is accomplished depends on the type of hormone and the type of cell that is responsive to that hormone. In this tutorial, you will explore two different modes of action for hormones through an interactive animation.

Start the MediaTutor Student CD-ROM and enter the activity number in the Quick Search box to be taken directly to that activity.

Web Investigations

Case Study: Losing on Steroids

Estimated time: 10 minutes

For centuries athletes have searched for the performance-enhancing drug that would guarantee victory. The modern pharmacopeia is full of enticing alternatives, but the long-term effects can be devastating. Can a drug-free athlete win in the modern world?

Go to http://www.prenhall.com/audesirk6, the Audesirk Companion Web site. Select Chapter 32 and the Web Investigation to begin.

"Know thyself."

—Inscribed above the entrance to the Temple of Apollo, home of the Oracle of Apollo at Delphi

Christopher Reeve discusses spinal cord injuries at the National Press Club in Washington, D.C., in December 1999.

33 The Nervous System and the Senses

AT A GLANCE

Case Study: From Tragedy to Triumph

1) **What Are the Structures and Functions of Neurons?**

2) **How Is Neural Activity Produced and Transmitted?**

Neurons Create Electrical Signals Across Their Membranes

Neurons Communicate at Synapses

Excitatory or Inhibitory Potentials Are Produced at Synapses and Integrated in the Cell Body

The Nervous System Uses Many Neurotransmitters

3) **What Are Some General Features of Nervous Systems?**

Information Processing Requires Four Basic Operations

Neural Pathways Direct Behavior

Increasingly Complex Nervous Systems Are Increasingly Centralized

4) **How Is the Human Nervous System Organized?**

The Peripheral Nervous System Links the Central Nervous System to the Body

The Central Nervous System Consists of the Spinal Cord and Brain

The Spinal Cord Is a Cable of Axons Protected by the Backbone

The Brain Consists of Many Parts Specialized for Specific Functions

5) **How Does the Brain Produce the Mind?**

The "Left Brain" and "Right Brain" Are Specialized for Different Functions

The Mechanisms of Learning and Memory Are Poorly Understood

Insights on How the Brain Creates the Mind Come from Diverse Sources

6) **How Do Sensory Receptors Work?**

7) **How Is Sound Sensed?**

The Ear Captures, Transmits, and Converts Sound into Electrical Signals

Sound Is Converted into Electrical Signals in the Cochlea

8) **How Is Light Sensed?**

The Compound Eyes of Arthropods Produce a Mosaic Image

The Mammalian Eye Collects, Focuses, and Transduces Light Waves

Binocular Vision Allows Depth Perception

9) **How Are Chemicals Sensed?**

The Ability to Smell Arises from Olfactory Receptors

Taste Receptors Are Located in Clusters on the Tongue

Pain Is a Specialized Chemical Sense

Evolutionary Connections: Uncommon Senses

Case Study Revisited: From Tragedy to Triumph

CASESTUDY

From Tragedy to Triumph

Actor Christopher Reeve's real-life athletic abilities made him well suited for his movie role as Superman. But on Memorial Day in 1995, that phase of his life ended abruptly at an equestrian jumping competition. Racing up to a jump, his horse suddenly balked, throwing Reeve headfirst down onto the jumping rail. The impact shattered the first two vertebrae in his neck and crushed his spinal cord, interrupting the flow of signals from his brain to the rest of his body and from his body to his brain. Unable to move or breathe, Reeve's life was saved by the fast action of paramedics, who began pumping oxygen into his lungs within three minutes of the fall. If one more minute had elapsed, the lack of oxygen could have caused irreversible brain damage.

Completely paralyzed from the neck down, unable even to breathe on his own, Reeve battled depression, finally making a conscious decision to find new ways to be productive and active. In the intervening years, he has become a leader in efforts to raise awareness and funds for research that may help people with spinal cord injuries regain the use of their bodies. Now Christopher Reeve, with a lot of help, is "walking."

How are signals conducted from the brain to and from distant parts of our body? How are the brain and spinal cord protected? Where do the commands to breathe originate? Why doesn't the spinal cord repair itself? What hopes can research offer to individuals with spinal cord injuries? What properties of the spinal cord allow Christopher Reeve to take these steps when his brain can no longer command his muscles? Look for answers as you read this chapter, and rejoin us at the end for an update. ■

1 What Are the Structures and Functions of Neurons?

Our study of the nervous system begins with the individual nerve cell, or **neuron**. As the fundamental unit of the nervous system, each neuron must perform four specialized functions:

1. Receive information from the internal or external environment or from other neurons.
2. Integrate the information it receives and produce an appropriate output signal.
3. Conduct the signal to its terminal ending, which may be some distance away.
4. Transmit the signal to other nerve cells or to glands or muscles.

Although neurons vary enormously in structure, a "typical" vertebrate neuron has four distinct structural regions that carry out the functions listed above. These regions are the *dendrites,* the *cell body,* the *axon,* and the *synaptic terminals* (Fig. 33-1). **Dendrites**, branched tendrils that extend outward from the nerve cell body, respond to signals from other neurons or from the external environment. Dendrites perform the "receive information" function. Their branched form provides a large surface area for receiving signals. In neurons of the brain and spinal cord, dendrites respond to the chemical neurotransmitters released by other neurons. These dendrites have protein receptors in their membranes that bind specific neurotransmitters and, as a result, produce electrical signals. Dendrites of *sensory neurons* have special membrane adaptations that allow them to produce electrical signals in response to specific stimuli from the environment, such as pressure, odorous molecules, light, or heat.

Electrical signals travel down the dendrites and converge on the neuron's **cell body**, which serves as an integration center, the second function. In this role, the cell body adds up the various positive and negative signals from the dendrites. If the sum of all these signals is sufficiently positive, the neuron will produce an **action potential**, the electrical output signal of the neuron. The cell body, containing the usual assortment of organelles, also carries on the routine activities common to most other body cells, such as synthesizing complex molecules and coordinating the metabolic activities of the cell.

In a typical neuron, a long, thin fiber called an **axon** extends outward from the cell body and conducts the electrical signal. Single axons may stretch from your spinal cord to your toes, a distance of about a meter (about 3 feet), making neurons the longest cells in the body. Axons are distribution lines, carrying action potentials from the cell body to the *synaptic terminals*, located at the far end of each axon. Axons are normally bundled together into **nerves**, much like bundles of wires in an electrical cable. However, unlike electrical power distribution cables (in which energy is lost along the way from power station to customer), the plasma membranes of axons conduct action potentials undiminished from the cell body to their synaptic terminals. Some axons are wrapped with insulation called **myelin**, which allows more rapid conduction of the electrical signal. Myelin is formed from non-neuronal cells that wrap around the axon (see Fig. 33-1). In vertebrates, axons bundled into nerves emerge from the brain and spinal cord and extend out to all regions of the body.

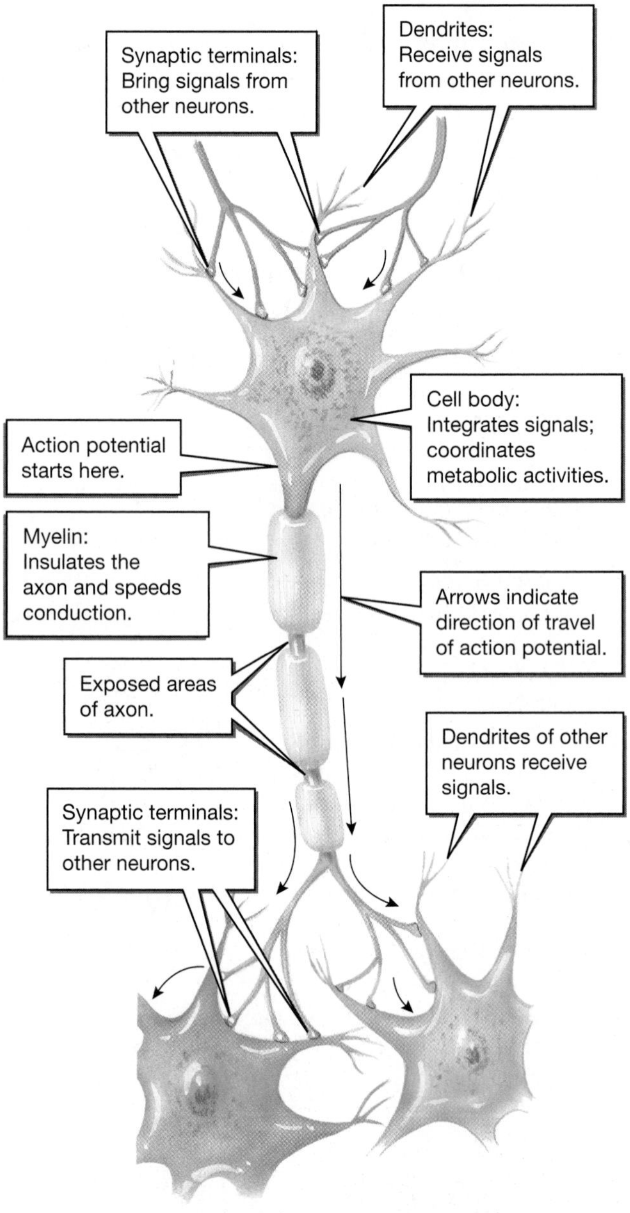

Figure 33-1 A nerve cell, showing its specialized parts and their functions

The transmission of the signal to other cells occurs at **synaptic terminals**, swellings at the branched endings of axons (see Fig. 33-1). Most synaptic terminals contain a **neurotransmitter**, a specific chemical that they release in response to an action potential reaching the terminal. The synaptic terminals of one neuron may communicate with a gland, a muscle cell, or the dendrites or cell body of a second neuron. The output of the first cell becomes the input to the second cell. The site at which synaptic terminals communicate with other cells is called the **synapse**.

2 How Is Neural Activity Produced and Transmitted?

Neurons Create Electrical Signals Across Their Membranes

In the early 1950s, biologists using the giant axon of a squid (a mollusk) developed ways to record electrical events inside individual neurons (Fig. 33-2). The researchers found that unstimulated, inactive neurons maintain a constant electrical difference, or *potential,* across their plasma membranes, similar to that across the poles of a battery. As in a battery, the electrical potential across a neuron membrane stores energy. This potential, called the **resting potential**, is always negative inside the cell and ranges from –40 to –90 millivolts (mV; thousandths of a volt).

If the neuron is stimulated, either naturally or by an experimenter using an electrical current, the negative potential inside the neuron can be made either more or less negative. If the potential is made sufficiently less negative that it reaches a level called **threshold**, an action potential is triggered (Fig. 33-3). During an action potential, the neuron's potential rapidly rises to about +50 mV inside the cell. Action potentials last a few milliseconds (thousandths of a second) before the cell's negative resting potential is restored. The positive charge of the action potential flows rapidly down the axon to the synaptic terminal, where the signal is communicated to another cell at a synapse. In "A Closer Look: Ions and Electrical Signals," we examine these electrical potentials, the language of the nervous system.

Neurons Communicate at Synapses

Once an action potential has been conducted to the synaptic terminal of a neuron, the signal must be transmitted to another cell, such as another neuron, a muscle cell, or a gland cell. This transmission occurs at synapses, and for simplicity we will describe just neuron-to-neuron synapses. The signals transmitted at synapses are called **postsynaptic potentials** (**PSPs**).

At a synaptic terminal, an action potential encounters a synapse, where parts of two neurons are specialized to communicate with one another. A tiny gap separates the synaptic terminal of the first neuron, the

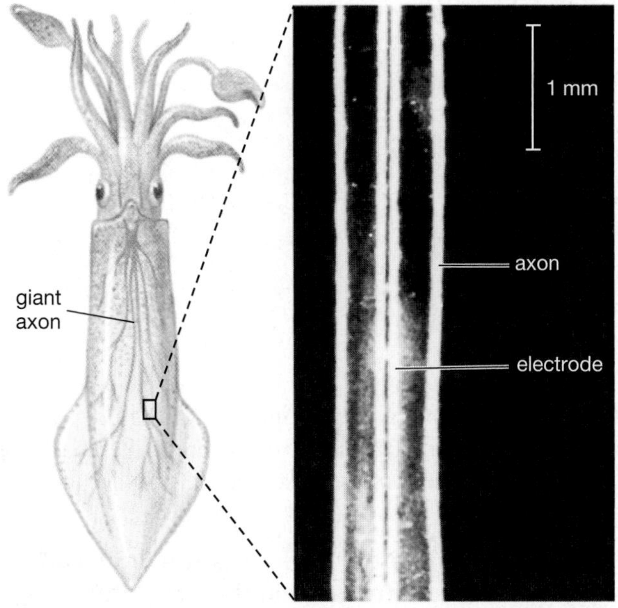

Figure 33-2 The giant axon of the squid
Using the giant axon of the squid, British physiologists Bernard Katz, Alan Hodgkin, and Andrew Huxley pushed electrodes (thin wires or narrow glass tubes filled with conducting solution) inside the axon. These electrodes were connected to voltmeters to record the electrical potential difference between the inside and outside of the axon, about –70 millivolts.

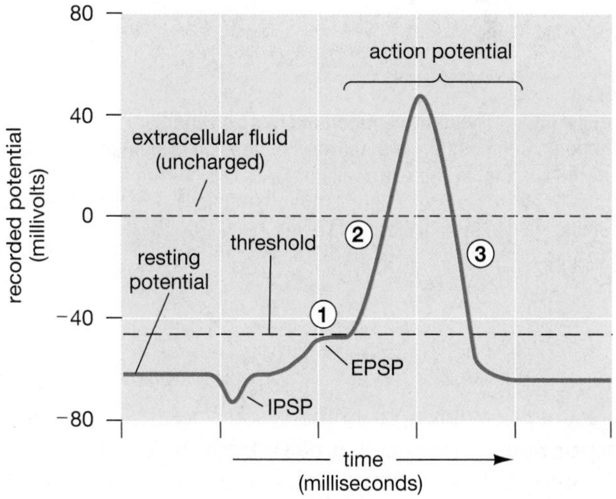

Figure 33-3 The electrical events during an action potential
A recording (by oscilloscope) of the electrical events in a nerve cell. The resting potential is about 60 millivolts negative with respect to the outside. (1) When a PSP stimulates the cell to threshold, membrane channels permeable to sodium open and sodium enters the cell, powered by diffusion and by electrical attraction; (2) the inside of the cell becomes positively charged. (3) Shortly thereafter, other membrane channels permeable to potassium open and potassium leaves, driven by diffusion and electrical repulsion from the now-positive inside of the cell, until the resting potential is reestablished.

MEDIATUTOR 33.1 Ions and Electrical Signals

A Closer Look
Ions and Electrical Signals

How Is the Resting Potential Generated?

The resting potential is based on a balance between chemical and electrical gradients, maintained by active transport and by a membrane that is selectively permeable to specific ions. The ions of the cytoplasm consist mainly of positively charged potassium ions (K^+) and large, negatively charged organic molecules such as proteins, which cannot leave the cell (Fig. E33-1). Outside the cell, the extracellular fluid contains more positively charged sodium ions (Na^+) and negatively charged chloride ions (Cl^-). These concentration differences are maintained by a specialized membrane protein called a *sodium–potassium pump*, which simultaneously pumps K^+ into and Na^+ out of the cell.

In an unstimulated neuron, only K^+ can cross the plasma membrane, by traveling through specific membrane proteins called *potassium channels* (shown in yellow). Although *sodium channels* (shown in purple) are also present, they remain closed in unstimulated neurons. Because the K^+ concentration is higher inside the cell than outside, that ion tends to diffuse out, leaving the negatively charged organic ions behind:

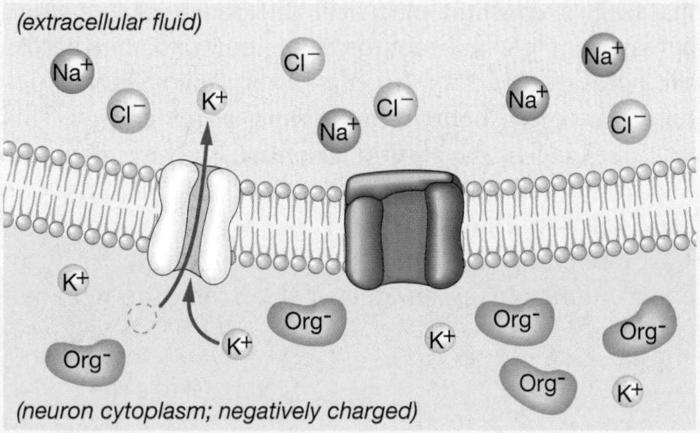

The outward diffusion of K^+ stops when the negative charge (which tends to pull the K^+ back inside the cell) becomes sufficiently large to counteract the diffusion gradient across the membrane. This negative charge at rest is the neuron's resting potential.

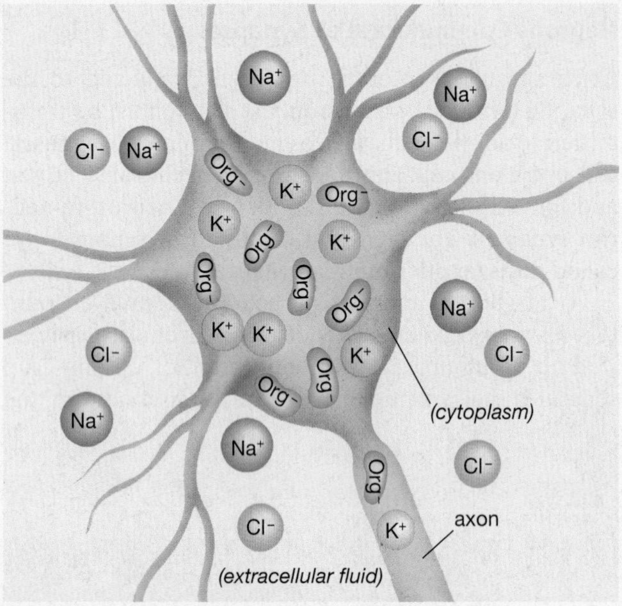

Figure E33-1 The neuron maintains ionic gradients
The ionic composition of a neuron's cytoplasm is significantly different from that of the extracellular fluid. The neuron maintains high concentrations of K^+ and large organic ions (Org^-); the extracellular fluid is high in Na^+ and Cl^-.

presynaptic neuron, from the second neuron, or **postsynaptic neuron** (Fig. 33-4, p. 668). Both the dendrites and the cell bodies of neurons are typically covered with synapses. The synapse includes the synaptic terminal of the presynaptic neuron, the gap, and the specialized membrane of the postsynaptic neuron just across the gap, which contains receptors for neurotransmitters.

When an action potential reaches a synaptic terminal, the inside of the terminal becomes positively charged. This charge triggers storage vesicles in the synaptic terminal to release a chemical neurotransmitter into the gap between the cells. The neurotransmitter molecules rapidly diffuse across the gap and bind briefly to receptors in the membrane of the postsynaptic neuron before diffusing away or being taken back up into the presynaptic neuron (see Fig. 33-4).

Excitatory or Inhibitory Potentials Are Produced at Synapses and Integrated in the Cell Body

Receptor proteins in the postsynaptic membrane bind to a specific type of neurotransmitter. This binding causes specific types of ion channels in the postsynaptic membrane to open, allowing ions to flow across the plasma membrane along their concentration gradients. The flow of ions in the postsynaptic neuron causes a

Action Potentials Can Carry Messages Rapidly over Long Distances

Neuron signals are carried long distances by action potentials. Action potentials occur if the PSPs from other neurons bring the potential inside the neuron (at a specific region where the axon joins the cell body) to threshold. At threshold, Na^+ channels (purple) are triggered to open, allowing a rapid influx of Na^+ (① in the figure below). Soon after the Na^+ channels open, they spontaneously close and additional K^+ channels (orange) are triggered to open by the positive charge inside the axon, allowing more K^+ to flow out of the cell and restore the negative resting potential (②).

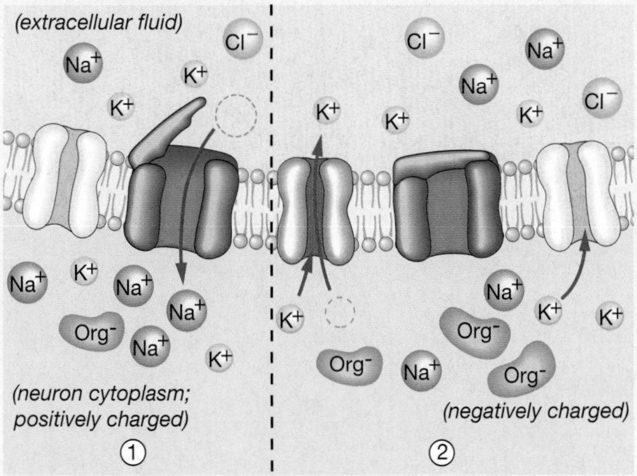

An action potential resembles a fast-moving wave of positive charge that travels, undiminished in size, along the axon to the synaptic terminal. The positive charge carried into the axon by Na^+ causes Na^+ channels farther along the axon to open. More Na^+ can then flow in, more channels further down the axon are opened, and the action potential continues along the axon:

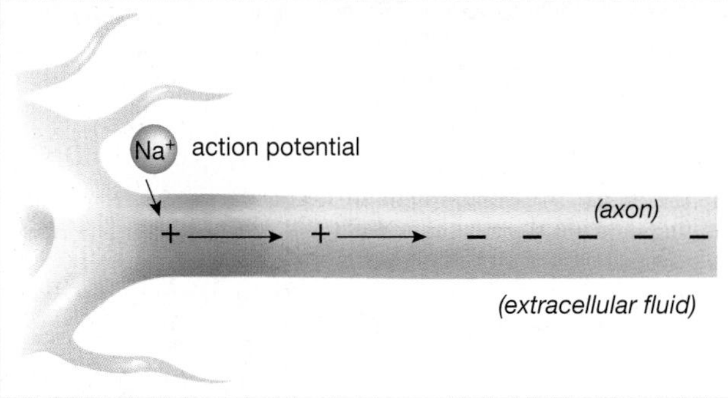

As the wave of positive charges passes a given point along the axon, the resting potential is restored as K^+ flows out:

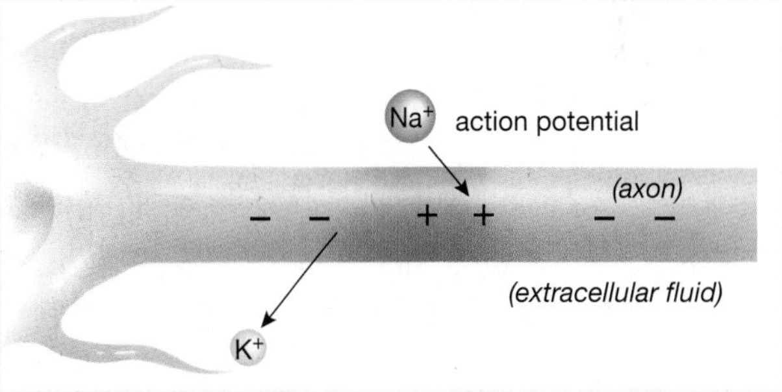

Action potentials are "all-or-none" phenomena. That is, if the neuron does not reach threshold, there will be no action potential, but if threshold is reached, a full-sized action potential will occur and travel the entire length of the axon.

A tiny fraction of the total potassium and sodium in and around each neuron is exchanged during each action potential. The gradients of Na^+ and K^+ are maintained by the sodium–potassium pump.

small, brief change in electrical charge, the postsynaptic potential. Depending on what type of channels are opened and what type of ions flow, PSPs can be either *excitatory* (EPSPs), making the neuron less negative inside and more likely to produce an action potential, or *inhibitory* (IPSPs), making it more negative and less likely to produce an action potential (see Fig. 33-3). Postsynaptic potentials are small, rapidly fading signals, but they travel far enough to reach the cell body. There they determine whether an action potential will be produced. How? The dendrites and cell body of a single neuron often receive EPSPs and IPSPs from the synaptic terminals of thousands of presynaptic neurons. The PSPs that reach the postsynaptic cell body at the same time are "added up," or integrated. If the excitatory and inhibitory potentials, when added together, raise the electrical potential inside the neuron above threshold, the postsynaptic cell will produce an action potential.

The Nervous System Uses Many Neurotransmitters

Over the past few decades, researchers have become increasingly aware that the brain is a teeming cauldron; its neurons synthesize and respond to a vast array of chemicals, including many of the hormones once thought unique to the endocrine system. For example, hormones

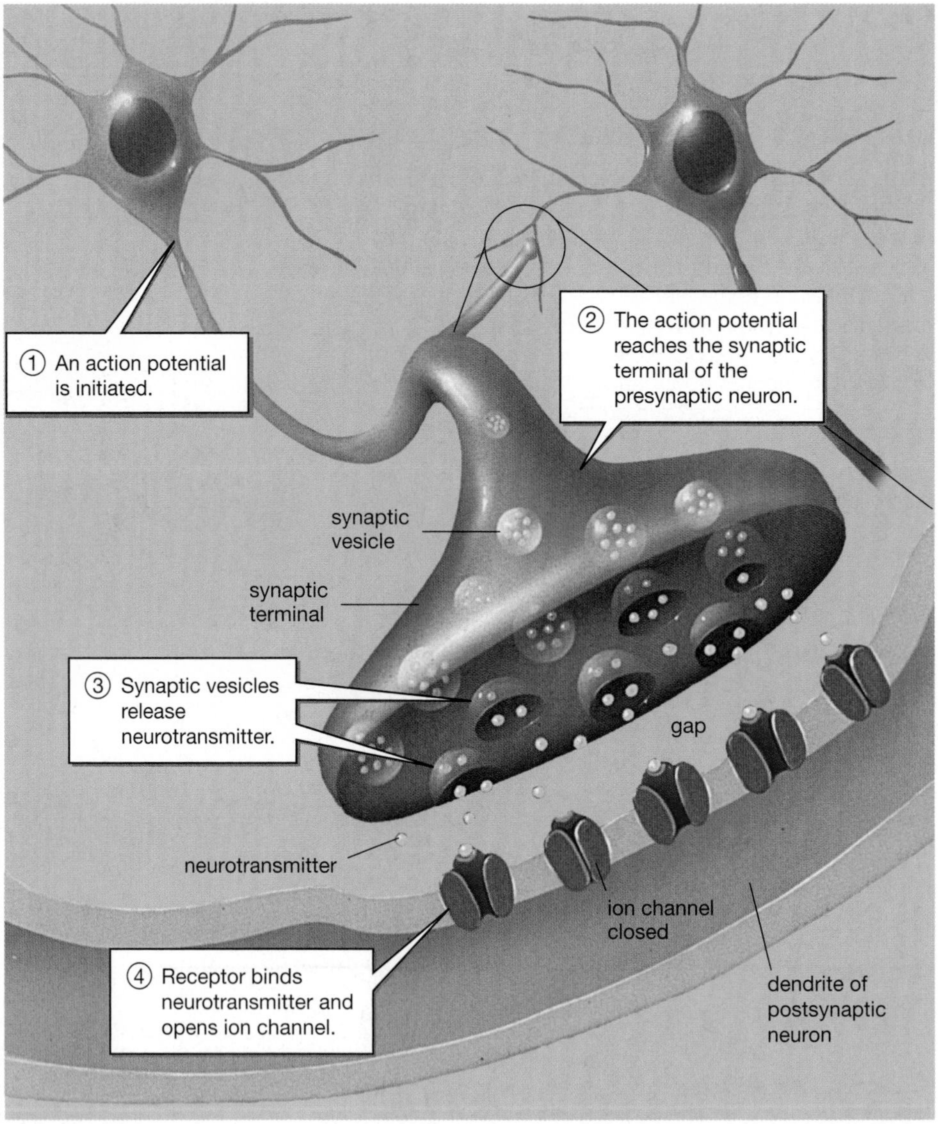
① An action potential is initiated.
② The action potential reaches the synaptic terminal of the presynaptic neuron.
synaptic vesicle
synaptic terminal
③ Synaptic vesicles release neurotransmitter.
gap
neurotransmitter
ion channel closed
④ Receptor binds neurotransmitter and opens ion channel.
dendrite of postsynaptic neuron

Figure 33-4 The structure and operation of the synapse
The synaptic terminal contains numerous vesicles that enclose a neurotransmitter for which the postsynaptic neuron has membrane receptors. When an action potential enters the synaptic terminal of the presynaptic neuron, the vesicles dump their neurotransmitter into the gap between the neurons. The neurotransmitter diffuses rapidly across the space, binds to postsynaptic receptors, and causes ion channels to open. Ions flow through these open channels, causing a postsynaptic potential in the postsynaptic cell.

MediaTutor 33.2 Synapses

that control digestive tract secretions are now known to be synthesized in the brain as well, where they influence appetite. Recently, nitric oxide, a gas that lasts only a few seconds, has been recognized as an important neurotransmitter. At least 50 neurotransmitters have been identified, and the list is growing. In Table 33-1 we list a few well-known neurotransmitters and some of their functions. In "Health Watch: Drugs, Diseases, and Neurotransmitters," we further explore the role of neurotranmitters in addiction and in neurological diseases.

Table 33-1 Some Important Neurotransmitters

Neurotransmitter	Location in Nervous System	Some Functions
Acetylcholine	Motor neuron-to-muscle synapse; autonomic nervous system, brain	Activates skeletal muscles; activates target organs of parasympathetic nervous system
Dopamine	Midbrain	Important in control of movement
Epinephrine (adrenaline)	Sympathetic nervous system	Activates target organs of sympathetic nervous system
Serotonin	Midbrain, pons, and medulla	Influences mood, sleep
Glutamate	Brain and spinal cord	Major excitatory neurotransmitter in CNS
Glycine	Spinal cord	Major inhibitory neurotransmitter in spinal cord
GABA (gamma amino butyric acid)	Throughout brain	Major inhibitory neurotransmitter in brain
Endorphins	Brain and spinal cord	Influence mood, reduce pain sensations
Nitric oxide	Brain	Important in forming memories

Health Watch
Drugs, Diseases, and Neurotransmitters

Chances are, you know someone who is addicted—perhaps to nicotine, to alcohol, or maybe to cocaine. How can these substances exert such profound influence over people's lives? A big part of the answer lies in the effects these drugs have on neurotransmitters and in how the nervous system adapts to those insidious effects. Cocaine is a good example. Synapses in the brain that use the neurotransmitters *dopamine, serotonin,* or *norepinephrine* contribute to our energy level and our overall sense of well-being. Normally, the presynaptic neuron, after releasing one of these neurotransmitters, immediately starts pumping it back in, thus limiting its effects. Researchers have found that cocaine works by blocking this pump mechanism. The result? When a person takes cocaine, these neurotransmitters remain in their synapses much longer and reach higher levels than normal, so their effects are enhanced. The user feels euphoric and energetic. Now the brain attempts to restore the status quo: To reduce the impact of cocaine, the postsynaptic neuron decreases its number of receptors for these neurotransmitters. When fewer receptors are present, the high levels of neurotransmitter caused by cocaine are now *required* for the user to feel normal. When cocaine is withdrawn, the postsynaptic neurons are inadequately stimulated, and the user experiences an emotional "crash" that can be relieved only by more cocaine. Increasing amounts of the drug are required to produce the euphoric effects; the user has become an addict (Fig. E33-2).

Alcohol stimulates receptors for the neurotransmitter GABA (gamma amino butyric acid), enhancing inhibitory neuronal signals, and blocks receptors for glutamate, reducing excitatory signals. When a person drinks frequently, the brain compensates by decreasing GABA receptors and increasing glutamate receptors. How does an alcoholic feel without alcohol? Jittery, nervous, trembling—in short, overstimulated. In extreme cases, withdrawal from alcohol can cause convulsions. Nicotine and other components of cigarette smoke also interfere with normal synaptic transmission, producing a variety of addictive effects. To overcome addictions, drug users must undergo the misery caused by a nervous system that is deprived of a drug to which it has adapted. Although eventually receptors return to normal levels, for unknown reasons, drug cravings often recur periodically.

You may also know someone with Parkinson's or Alzheimer's disease. Both are caused by the death of specific neurons in the brain and the loss of their neurotransmitters, which normally communicate with other neurons. In Parkinson's disease, dopamine-releasing neurons in the midbrain die, interfering with the complex control system that underlies smooth movements. Parkinson's patients experience tremors and have difficulty initiating movement. In Alzheimer's disease, neurons in the temporal lobes that produce the neurotransmitter *acetylcholine* die in large numbers. Memory loss is a prominent symptom of Alzheimer's.

Acetylcholine is also the only neurotransmitter released at the synapses between neurons and skeletal muscles, where it is always excitatory. The drug curare, extracted from a South American plant, blocks acetylcholine receptors. By preventing muscle contraction, curare causes a sometimes deadly paralysis. Some natives of South America have used this drug on arrow tips to stop or kill game animals.

The neurotransmitter serotonin acts in the brain and spinal cord. It can inhibit pain sensory neurons in the spinal cord. Electrical devices that stimulate these neurons may be implanted in individuals with chronic pain. Too little serotonin can cause depression. The antidepressant Prozac® selectively blocks the reuptake of serotonin into the presynaptic neuron, enhancing the neurotransmitter's effects.

The analgesic (pain-relieving) effects of plant-derived opiates, such as morphine, opium, codeine, and heroin, have been recognized for centuries. Since the brain has receptors for these molecules, researchers reasoned that perhaps these plant opiates resemble (then) unknown substances produced by the brain that diminish pain perception. The search for such substances was rewarded in 1975 with the discovery of *opioids* (opiate-like substances); *endorphins* are a group of opioids. Certain opioids suppress pain in times of extreme stress, such as on a battlefield or a football field. Opioids released during strenuous exercise may account for the well-known "runner's high." Some of the analgesic effects of acupuncture are caused by its ability to stimulate the release of opioids.

Figure E33-2
An addict experiences extreme physical and emotional distress without his drug because his nervous system has adapted to it.

3 What Are Some General Features of Nervous Systems?

The individual neuron uses a language of action potentials. Yet somehow this basic language allows even simple animals to perform a variety of complex behaviors. One key to the versatility of the nervous system is the presence of complex networks of neurons. These neural networks range from thousands to billions of cells. As in computers, small, simple elements can perform amazing feats when connected properly.

Information Processing Requires Four Basic Operations

Before we delve into the basic anatomy of nervous systems, we should first examine their operating principles. At a minimum, a nervous system must be able to perform four operations:

1. Determine the type of stimulus.
2. Signal the intensity of a stimulus.
3. Integrate information from many sources.
4. Initiate and direct the response.

Let's examine each of these operations.

The Type of Stimulus Is Distinguished by Wiring Patterns in the Brain

The nervous system must be able to identify the type of stimulus—for example, light, touch, or sound. All action potentials are similar; their properties tell us nothing about the kind of stimulus that elicited them. Instead, the nervous system monitors *which* neurons are producing action potentials. Thus, your brain interprets action potentials that occur in the axons of your optic nerves (originating in the eye and traveling to a specific area of the brain) as the sensation of light, action potentials in olfactory nerves (originating in receptors in the nose and traveling to a different region of the brain) as odors, and so on. In this way you have no trouble distinguishing the action potentials caused by the music from your stereo from those caused by the bitter taste of coffee on your tongue.

However, this genetic wiring may occasionally yield false information. Being poked in the eye may trigger action potentials in the optic nerve, because of slight trauma. Even though the stimulus is mechanical, your brain interprets all action potentials that occur in the optic nerve as light, so you "see stars." For the same reason, a blow to the head can make your ears "ring."

The Intensity of a Stimulus Is Coded by the Frequency of Action Potentials

Because all action potentials are of roughly the same magnitude and duration, no information about the strength, or **intensity**, of a stimulus (for example, the loudness of a sound) can be encoded in a single action potential. Instead, intensity is coded in two other ways (Fig. 33-5). First, intensity can be signaled by the frequency of action potentials in a single neuron. The more intense the stimulus, the faster the neuron produces action potentials, or *fires*. Second, most nervous systems have many neurons that can respond to the same input. Stronger stimuli tend to excite more of these neurons, whereas weaker stimuli excite fewer. Thus, intensity can also be signaled by the number of similar neurons that fire at the same time. The loud wail of a fire alarm activates many of your auditory neurons and causes them to fire action potentials very rapidly.

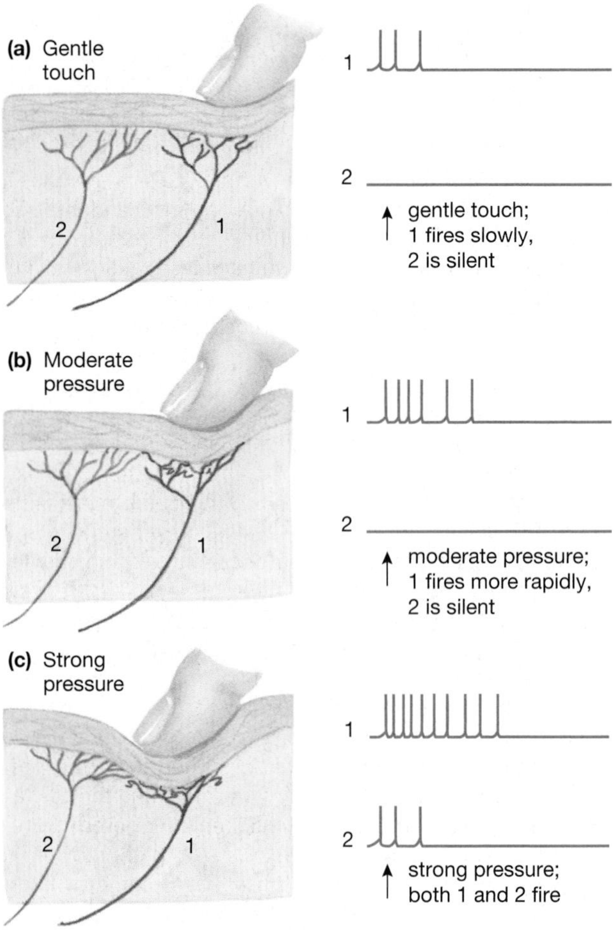

Figure 33-5 Signaling stimulus intensity
The intensity of a stimulus is signaled by the rate at which individual neurons produce action potentials and by the number of neurons that do so. Consider two touch receptors that have endings in adjacent patches of skin. ***(a)*** A gentle touch elicits only a few action potentials from one sensory neuron but none from another. ***(b)*** Moderate pressure stimulates only one receptor, which now fires faster, informing the brain that the touch is more intense than before. ***(c)*** Strong pressure activates both receptors, firing one very fast and the other more slowly, thus signaling that the pressure is very intense and localized over the fastest-firing receptor.

The Nervous System Processes Information from Many Sources Through Convergence

Your brain is continuously bombarded by sensory stimuli that originate both inside and outside the body. The brain must filter all these inputs, determine which ones are important, and decide how to respond. Nervous systems integrate information much as do individual neurons, through **convergence**. In this process, many neurons funnel their signals to fewer neurons. For example, many sensory neurons may converge onto a smaller number of brain cells. The brain cells that might be considered to be "decision-making cells" add up the postsynaptic potentials that result from the synaptic activity of these sensory neurons; depending on their relative strengths (and other internal factors such as hormones or metabolic activity), they produce appropriate outputs.

Divergence of Signals Allows Complex Responses

The output of the decision-making cells is responsible for initiating activity. The actions directed by the brain may involve many parts of the body. These actions require **divergence**, the flow of electrical signals from a relatively small number of decision-making cells onto many different neurons that control the activity of muscles or glands.

Neural Pathways Direct Behavior

Most behaviors are controlled by neuron-to-muscle pathways composed of four elements:

1. **Sensory neurons**, which respond to a stimulus, either internal or external to the body.
2. **Association neurons**, which receive signals from many sources, including sensory neurons, hormones, neurons that store memories, and many others. On the basis of this input, association neurons activate motor neurons.
3. **Motor neurons**, which receive instructions from association neurons and activate muscles or glands.
4. **Effectors**, normally muscles or glands that perform the response directed by the nervous system.

The Simplest Behavior Is the Reflex

The simplest type of behavior in animals is the **reflex**, a largely involuntary movement of a body part in response to a stimulus. Reflexes occur without involving conscious portions of the brain; many occur entirely within the spinal cord and perpiheral neurons. Examples of human reflexes include the familiar knee-jerk and pain-withdrawal reflexes, both of which are produced by neurons in the spinal cord. The pain-withdrawal reflex, which moves a body part away from a painful stimulus such as a thumbtack, uses one of each of the three types of neurons and an effector (see Fig 33-10). Although reflexes of this sort do not require the brain, other pathways inform the brain of pricked fingers and may trigger other, more complex behaviors (cursing, for example!).

Nearly all animals are capable of much more subtle and varied behavior than can be accounted for by simple reflexes. In principle, these more complex behaviors can be organized by *interconnected neural pathways,* in which several types of sensory input (along with memories, hormones, and other factors) converge on a set of association neurons. By integrating the postsynaptic potentials from several sources, the association neurons can "decide" what to do and can stimulate the motor neurons to direct the appropriate activity in muscles and glands.

Increasingly Complex Nervous Systems Are Increasingly Centralized

In all the animal kingdom, there are really only two nervous system designs: (1) a diffuse nervous system, such as that of cnidarians (*Hydra*, *jellyfish*, and their relatives; Fig. 33-6a), and a centralized nervous system,

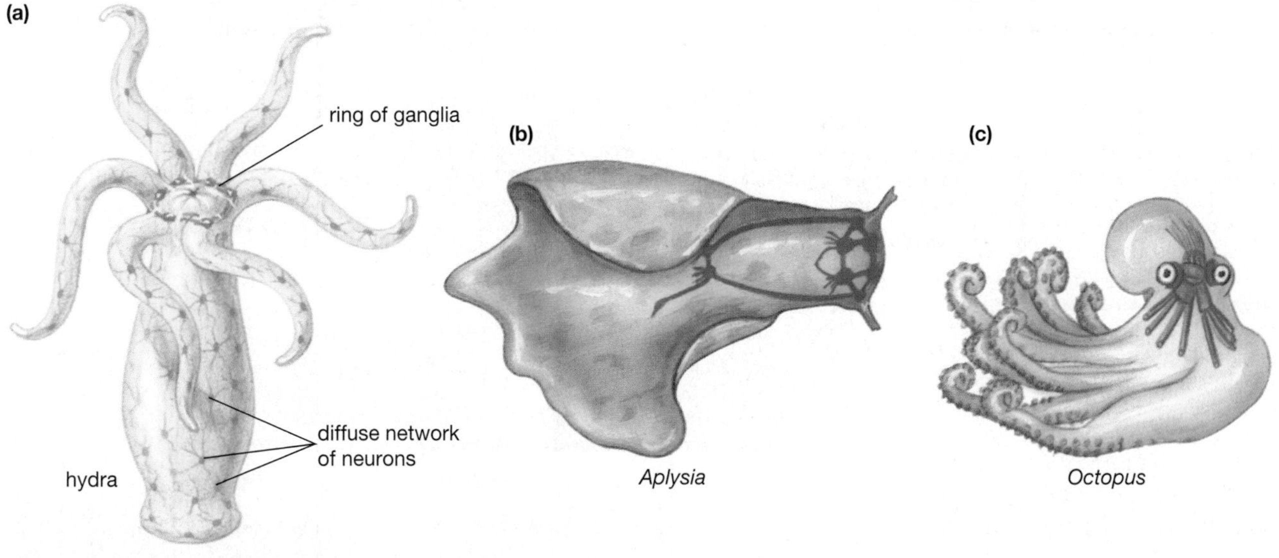

Figure 33-6 Nerve net and cephalization
(a) The diffuse nervous system of *Hydra* contains a few concentrations of neurons, particularly at the bases of tentacles, but no brain. Neural signals are conducted in virtually all directions throughout the body. The trend toward cephalization, increasing concentration of the nervous system in the head, is illustrated by ***(b)*** the shell-less *Aplysia*, which can crawl quite rapidly or even swim. Many of its neurons are aggregated into a brain. ***(c)*** Mollusk mobility and intelligence culminate in *Octopus*, with its large, complex brain and learning capabilities that rival those of some mammals.

found to varying degrees in more complex organisms. Not surprisingly, nervous system design is highly correlated with the animal's lifestyle. The radially symmetrical cnidarians have no "front end," so there has been no evolutionary pressure to concentrate the senses in one place. A *Hydra* sits anchored to the seafloor, and prey or other dangers are equally likely to come from any direction. Cnidarian nervous systems are composed of a network of neurons, often called a **nerve net**, woven through the animal's tissues. Here and there we find a cluster of neurons, called a **ganglion** (plural, **ganglia**), but nothing resembling a real brain.

Almost all other animals are bilaterally symmetrical, with definite head and tail ends. Because the head is usually the first part of the body to encounter food, danger, and potential mates, it is advantageous to have sense organs concentrated there. Sizable ganglia evolved that integrate the information gathered by the senses and initiate appropriate action. Over evolutionary time, the sense organs gathered in the head and the ganglia became centralized into a brain. This trend, called **cephalization**, is clearly seen in different types of mollusks (Fig. 33-6b,c). Cephalization reaches its peak in the vertebrates, in which nearly all the cell bodies of the nervous system are localized in the brain and spinal cord.

4 How Is the Human Nervous System Organized?

The vertebrate nervous system, including that of humans, can be divided into two parts: (1) central and (2) peripheral. Each of these has further subdivisions (Fig. 33-7). The **central nervous system** (**CNS**) consists of a **brain** and a **spinal cord**, which extends down the dorsal part of the torso. The **peripheral nervous system** (**PNS**) consists of nerves that connect the central nervous system to the rest of the body.

The Peripheral Nervous System Links the Central Nervous System to the Body

The peripheral nervous system consists of **peripheral nerves**, which link the brain and spinal cord to the rest of the body, including the muscles, the sensory organs, and the organs of the digestive, respiratory, excretory, and circulatory systems. Within the peripheral nerves are axons of sensory neurons that bring sensory information *to* the central nervous system from all parts of the body. Peripheral nerves also contain the axons of motor neurons that carry signals *from* the central nervous system to the organs and muscles.

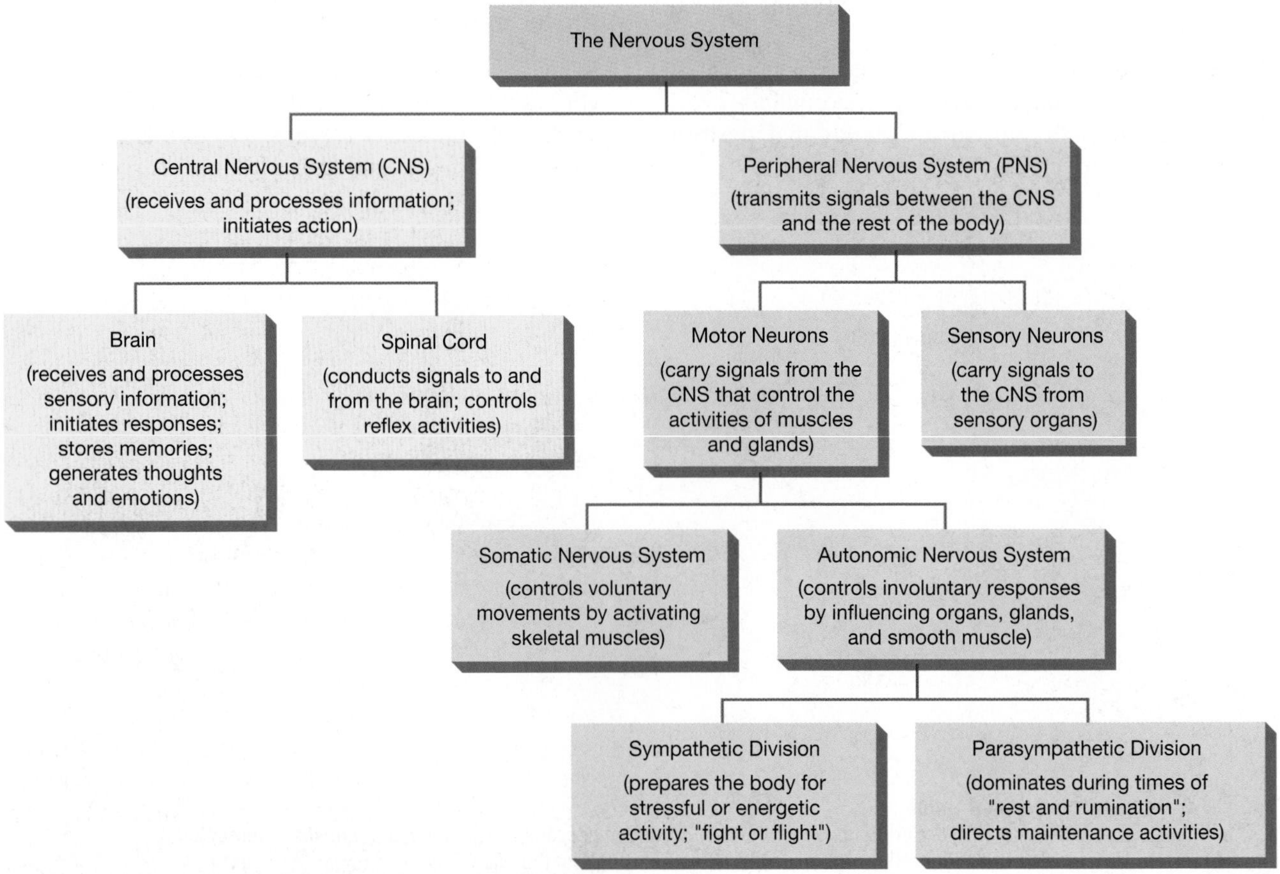

Figure 33-7 The organization and functions of the vertebrate nervous system

The motor portion of the peripheral nervous system can be subdivided into two parts: (1) the **somatic nervous system** and (2) the **autonomic nervous system**. Motor neurons of the somatic nervous system form synapses on skeletal muscles and control voluntary movement. As you take notes, lift a coffee cup, and adjust your stereo, your somatic nervous system is in charge. The cell bodies of somatic motor neurons are located in the *gray matter* of the spinal cord (to be described shortly), and their axons go directly to the muscles they control. (Muscles and their control will be discussed in Chapter 34.)

Motor neurons of the autonomic nervous system control involuntary responses. They form synapses on the heart, smooth muscle, and glands. The autonomic nervous system is controlled both by the *medulla* and the *hypothalamus* of the brain, described later in this chapter. It consists of two divisions: (1) the **sympathetic division** and (2) the **parasympathetic division** (Fig. 33-8). The two divisions of the autonomic nervous system generally make synaptic contacts with the same organs but normally produce opposite effects.

The sympathetic division acts on organs in ways that prepare the body for stressful or highly energetic activity, such as fighting, escaping, or taking an exam. During such "fight-or-flight" activities, the sympathetic nervous system curtails activity of the digestive tract, redirecting some of its blood supply to the muscles of the arms and legs. The heart rate accelerates. The pupils of the eyes open wider, admitting more light, and the air passages in the lungs expand, accommodating more air. These things may happen if you are suddenly called on in class to answer a question—especially if you don't know the answer!

The parasympathetic division, in contrast, dominates during maintenance activities that can be carried on at leisure, often called "rest and rumination." Under its control, the digestive tract becomes active, the heart rate slows, and air passages in the lungs constrict. As you read this text, your parasympathetic nervous system is probably dominating your unconscious functions.

You may notice two differences in the organization of the sympathetic and parasympathetic divisions in Figure 33-8. First, their nerves originate at different levels of the central nervous system. Second, although both divisions use a two-neuron pathway to carry messages to each target organ, in the sympathetic division, the synapse occurs in sympathetic ganglia near the spinal cord. In the parasympathetic division, the synapse occurs in smaller ganglia located on or very near each target organ.

The Central Nervous System Consists of the Spinal Cord and Brain

The spinal cord and brain make up the central nervous system. It is the portion of the nervous system where sensory information is received and processed, thoughts are generated, and responses are directed. The CNS consists primarily of association neurons—somewhere between 10 billion and 100 billion of them!

The brain and spinal cord are physically protected in three ways. The first line of defense is a bony armor, consisting of the *skull,* which surrounds the brain, and the *vertebral column,* which protects the spinal cord. Beneath the bones lies a triple layer of connective tissue called the **meninges** (see Fig. 33-12). Between the layers of the meninges, the **cerebrospinal fluid**, a clear, lymph-like liquid, cushions the brain and spinal cord as it nourishes the cells of the CNS. The delicate cells of the brain are also protected from potentially damaging chemicals that reach the bloodstream because the brain capillaries are far less permeable than capillaries in the rest of the body. This third line of defense is called the **blood–brain barrier**.

The Spinal Cord Is a Cable of Axons Protected by the Backbone

The spinal cord is a neural cable, about as thick as your little finger, that extends from the base of the brain to the lower back. It is protected by the bones of the vertebral column. Although the spinal cord is both strong and flexible, it is no match for the trauma of a headfirst fall off a horse or other types of violent trauma. (To learn what happens when the spinal cord is injured and what treatments are under study, see "Health Watch: Healing the Spinal Cord.")

Between the vertebrae, nerves carrying axons of sensory neurons and motor neurons arise from the dorsal and ventral portions of the spinal cord, respectively, and merge to form the peripheral nerves of the spinal cord (part of the peripheral nervous system; Fig. 33-9, p. 675). In the center of the spinal cord is a butterfly-shaped area of **gray matter**. Gray matter consists of the cell bodies of several types of neurons that control voluntary muscles and the autonomic nervous system, plus neurons that communicate with the brain and other parts of the spinal cord. The gray matter is surrounded by **white matter**, formed of myelin-coated axons of neurons that extend up or down the spinal cord. These axons carry sensory signals from internal organs, muscles, and the skin up to the brain. Axons also extend downward from the brain, carrying signals that direct the motor portions of the peripheral nervous system. Hence the motor neurons of the spinal cord also control the muscles involved in conscious, voluntary activities, such as eating, writing, or playing tennis, that must be directly activated by motor portions of the brain. If the spinal cord is crushed or severed, the body below the brain feels numb, and motor signals from the brain cannot get through to command complex behaviors, even if the motor neurons and peripheral nerves and muscles remain intact.

In addition to relaying neural signals between the brain and the rest of the body, the spinal cord contains

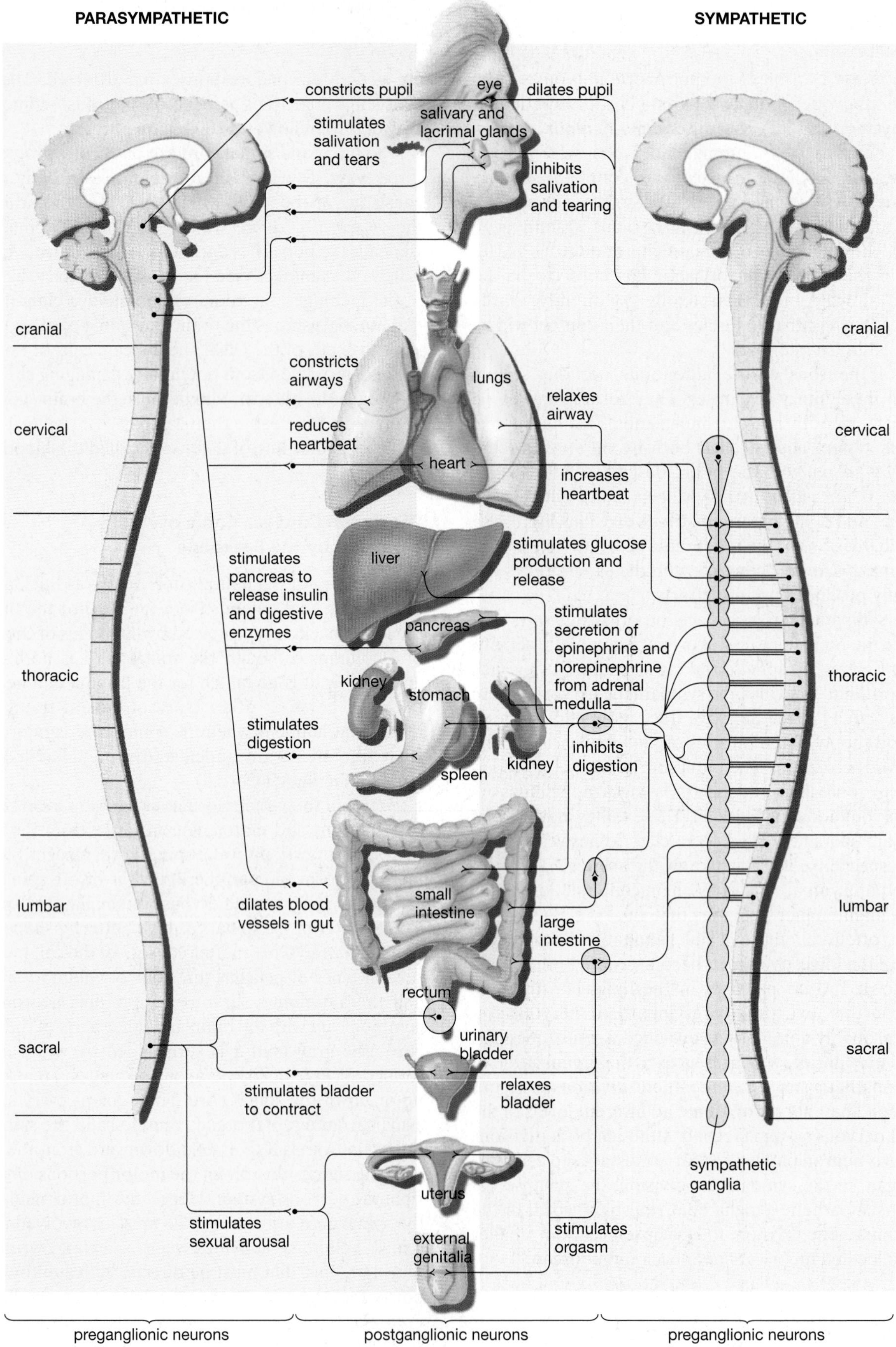

Figure 33-8 The autonomic nervous system
The autonomic nervous system has two divisions: sympathetic and parasympathetic. They supply nerves to many of the same organs but produce opposite effects. Activation of the autonomic nervous system is involuntarily commanded by signals from the hypothalamus, part of the interior of the brain above the spinal cord (see Fig. 33-14).

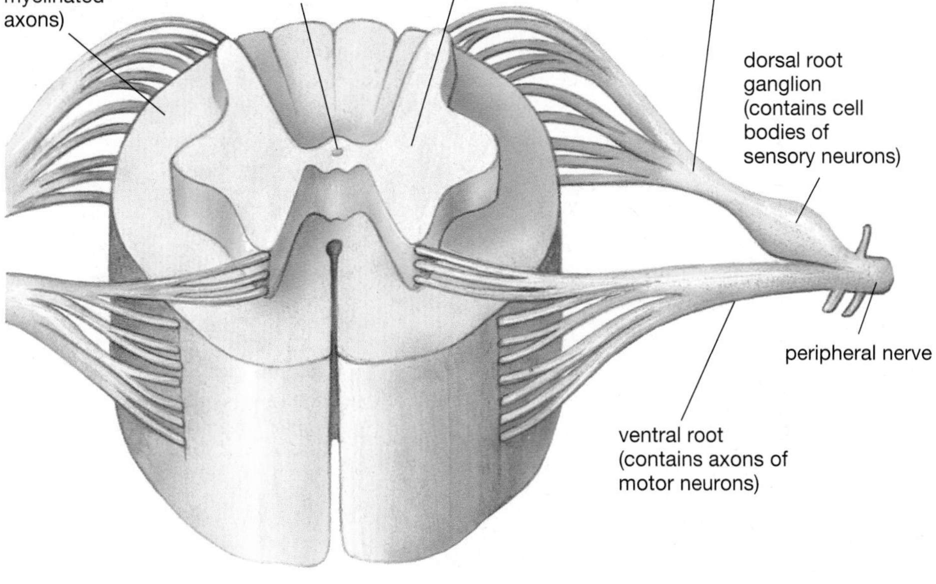

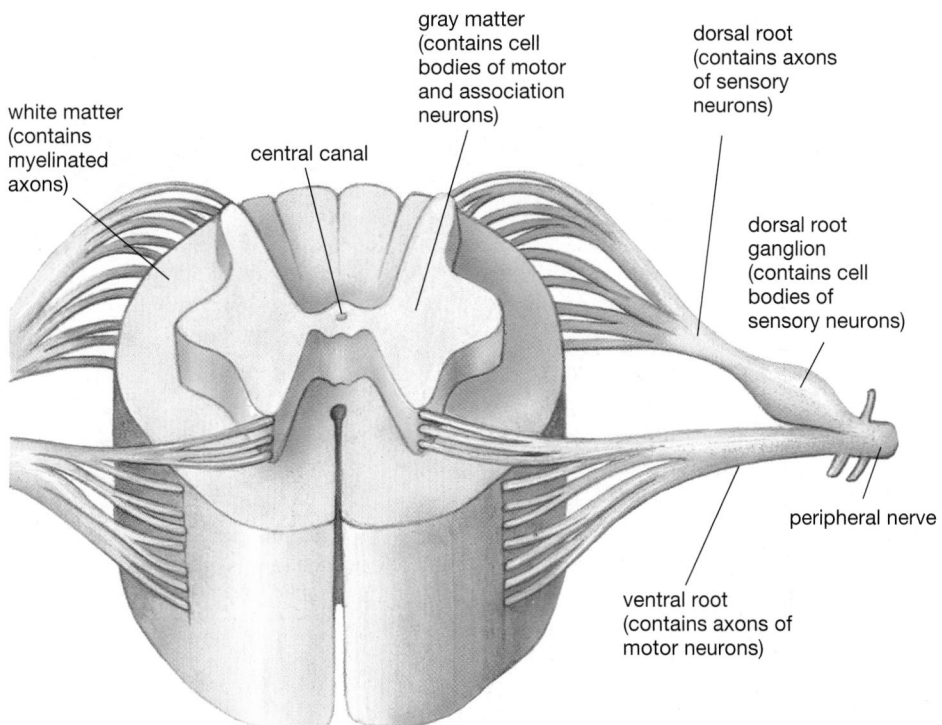

Figure 33-9 The spinal cord
The spinal cord runs from the base of the brain to the lower back and is protected by the vertebrae. Peripheral nerves emerge from between the vertebrae. In cross section, the spinal cord has an outer region of myelinated axons (white matter) that travel to and from the brain and an inner, butterfly-shaped region of dendrites and the cell bodies of association and motor neurons (gray matter). The cell bodies of the sensory neurons are outside the cord in the dorsal root ganglion.

the neural pathways for certain simple behaviors, such as reflexes. Let's examine the simple pain-withdrawal reflex, which involves neurons of both the central and the peripheral nervous systems (Fig. 33-10). The cell bodies of the sensory neurons from the skin (in this case, signaling pain) are located in **dorsal root ganglia**, clusters of neurons on spinal nerves just outside the spinal cord. Both association neuron and motor neuron cell bodies are found in the gray matter in the center of the spinal cord. Association neurons for the pain-withdrawal reflex, for example, not only form synapses on motor neurons but also have axons that extend up to

Health Watch
Healing the Spinal Cord

Why are spinal cord injuries so devastating? If you damage a nerve and lose sensation in part of your hand, chances are excellent that the peripheral nerve will regenerate and feeling will be restored. Unfortunately, the same is not true for injuries to the brain and spinal cord (the CNS). When the spinal cord is crushed, axons are severed. Swelling and inflammation in the cord squashes more axons and cell bodies, adding to the destruction. Damaged cells release substances that are toxic to neighboring cells, further enlarging the neuron death zone. The cut ends of axons that are still attached to living cell bodies will begin to regrow, but their progress is thwarted. They encounter a region of dead tissue and also scar tissue produced by cells that normally surround and support CNS neurons. In addition to creating a physical barrier, this scar tissue secretes proteins that inhibit axon growth. To add insult to injury, even the insulating myelin in the CNS produces proteins that deter axon regeneration.

Research to counteract these setbacks is progressing on several fronts. Medical researchers are investigating the use of drugs that limit swelling and inflammation. Researchers are also working to identify genes that code for the proteins that inhibit axon growth in the CNS, hoping to devise ways to block the action of those proteins or to prevent their synthesis. To help the regenerating axons find their way, scientists are experimenting with grafting tissue, such as myelin from the PNS, to promote and channel the regrowth of axons. Neural transplants are under investigation to replace the dead neurons. These approaches have yielded promising results in laboratory animals. The recent discovery that unspecialized embryonic cells, and even some adult cells, can (under appropriate conditions) be forced to become specialized into neurons has caused great excitement among researchers. Such cells may one day be used as transplants to replace damaged neurons in spinal cords or even in the brain. You'll learn more about these cells, called *stem cells*, in Chapter 36. Finally, researchers are programming computers to stimulate muscles directly with electrical currents, causing the muscles to contract in specific sequences to perform useful functions. A device that stimulates muscles that control the hand will soon be tested on humans. There will be no quick or complete cures, but Christopher Reeve and other individuals with spinal cord injuries have reason to be cautiously optimistic.

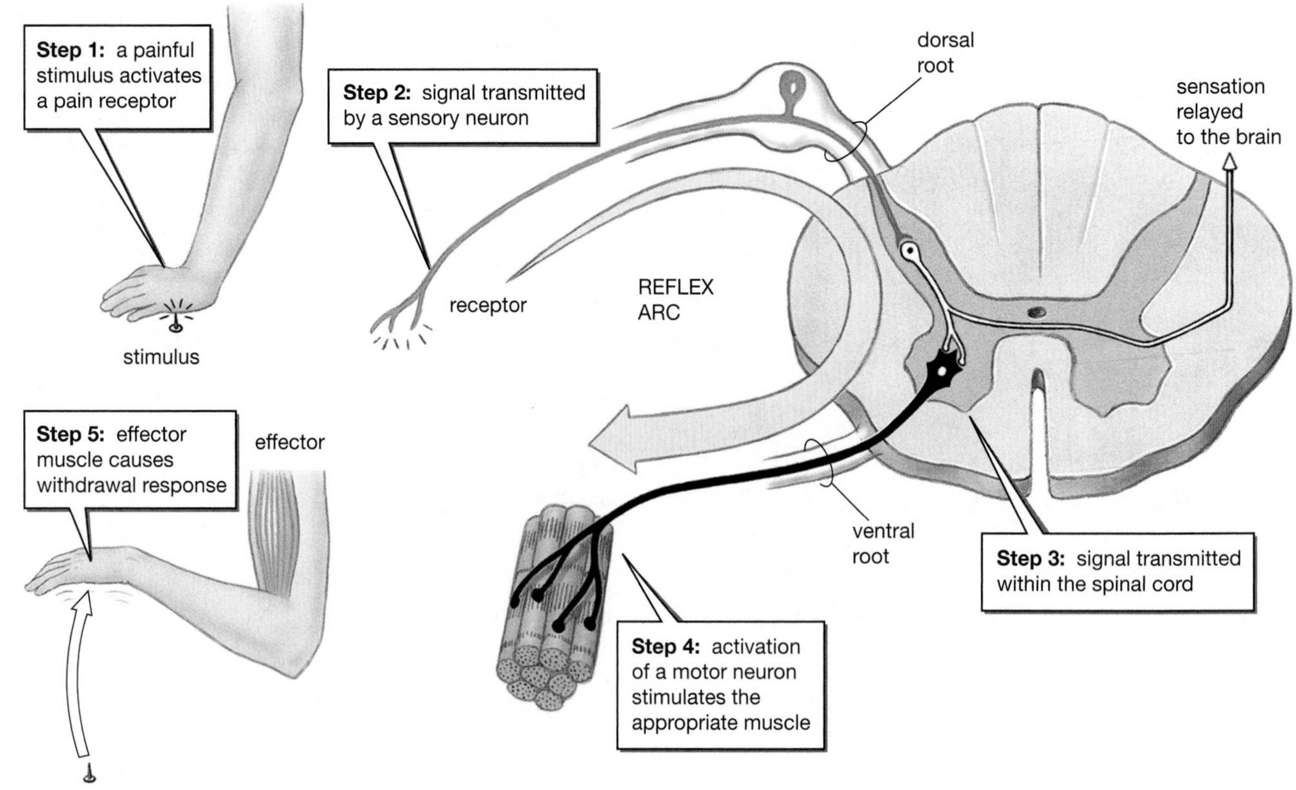
Step 1: a painful stimulus activates a pain receptor
stimulus
Step 2: signal transmitted by a sensory neuron
receptor
REFLEX ARC
dorsal root
sensation relayed to the brain
Step 3: signal transmitted within the spinal cord
ventral root
Step 4: activation of a motor neuron stimulates the appropriate muscle
Step 5: effector muscle causes withdrawal response
effector

Figure 33-10 The pain-withdrawal reflex
This simple reflex circuit includes each of the four elements of a neural pathway. The sensory neuron has pain-sensitive endings in the skin and a long fiber leading to the spinal cord. That neuron stimulates an association neuron in the spinal cord, which in turn stimulates a motor neuron in the cord. The axon of the motor neuron carries action potentials to effectors (muscles), causing them to contract and withdraw the body part from the damaging stimulus. The sensory neuron also makes a synapse on association neurons not involved in the reflex that carry signals to the brain, informing it of the danger.

the brain. Signals carried along these axons alert the brain to the painful event. The brain, in turn, sends impulses down axons in the white matter to cells in the gray matter. These signals can modify spinal reflexes. With sufficient motivation, you can suppress the pain-withdrawal reflex; to rescue a child from a burning crib, for example, you could reach into the flames.

In addition to simple reflexes, the entire "software" for operating some fairly complex activities also resides within the spinal cord. All the neurons and interconnections needed for the basic movements of walking and running, for example, are contained in the spinal cord, allowing Christopher Reeve to "walk" on a treadmill in a harness. The advantage of this semi-independent arrangement between brain and spinal cord is probably an increase in speed and coordination, because messages do not have to travel all the way up to the brain and back down again merely to swing forward one leg (in the case of walking). The brain's role in these "semi-automatic" behaviors is to initiate, guide, and modify the activity of spinal motor neurons, based on conscious decisions (Where are you going? How fast should you walk?). To maintain balance, the brain also uses sensory input from the muscles to command motor neurons to adjust the way the muscles move.

The Brain Consists of Many Parts Specialized for Specific Functions

All vertebrate brains have the same general structure, with major modifications corresponding to lifestyle and intelligence. Embryologically, the vertebrate brain begins as a simple tube that soon develops into three parts: (1) the **hindbrain**, (2) **midbrain**, and (3) **forebrain** (Fig. 33-11a). Scientists believe that in the earliest vertebrates, these three anatomical divisions were also functional divisions: The hindbrain governed automatic behaviors such as breathing and heart rate, the midbrain controlled vision, and the forebrain dealt largely with the sense of smell. In nonmammalian vertebrates, these three divisions remain prominent. However, in adult mammals, and particularly in humans, the brain regions are significantly modified. Some have been reduced in size, and others, especially the forebrain, great-

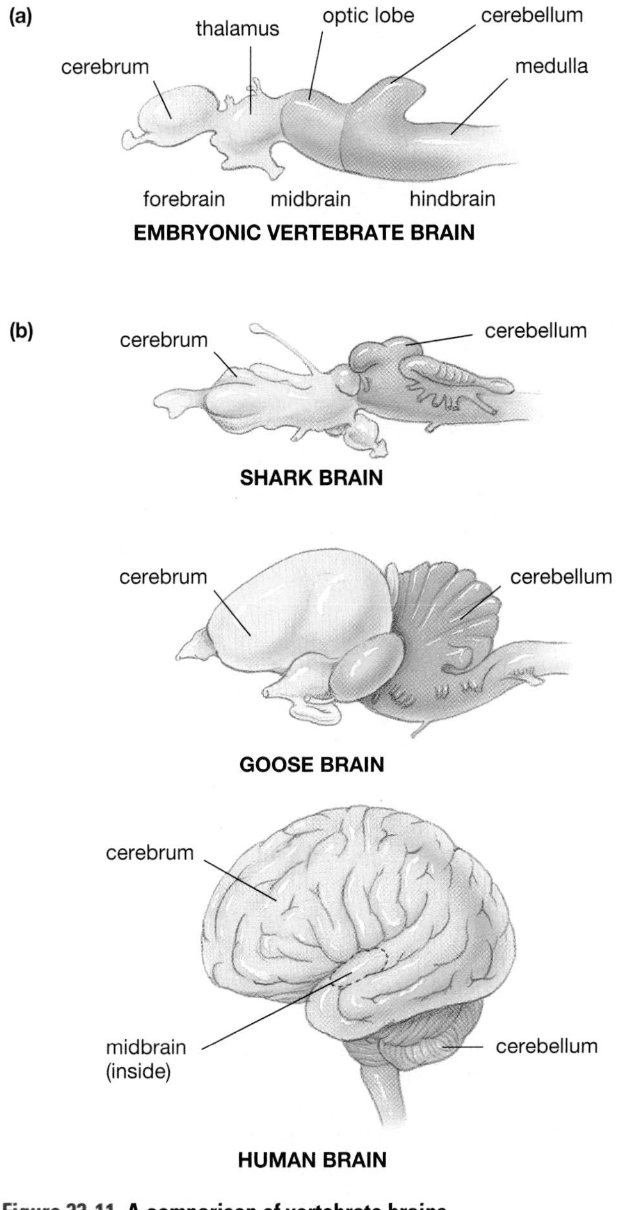

Figure 33-11 A comparison of vertebrate brains
(a) The embryonic vertebrate brain shows three distinct regions: the forebrain, midbrain, and hindbrain. ***(b)*** This basic structure persists in all adult vertebrate brains, but the relative size and importance of the parts vary enormously.

ly enlarged (Fig. 33-11b). A section through the midline of the human brain reveals many of its structural features, as shown in Figure 33-12.

The Hindbrain Includes the Medulla, Pons, and Cerebellum

In humans, the hindbrain is represented by the *medulla,* the *pons,* and the *cerebellum* (see Fig. 33-12). In both structure and function, the **medulla** is very much like an enlarged extension of the spinal cord. Like the spinal cord, the medulla has neuron cell bodies at its center, surrounded by a layer of myelin-covered axons. The medulla controls several automatic functions, such as breathing, heart rate, blood pressure, and swallowing. Certain neurons in the **pons**, located above the medulla, appear to influence transitions between sleep and wakefulness and between stages of sleep. Other neurons influence the rate and pattern of breathing. The **cerebellum** is crucial in coordinating movements of the body. It receives information from command centers in the conscious areas of the brain that control movement and also from position sensors in muscles and joints. By comparing what the command centers order with information from the position sensors, the cerebellum guides smooth, accurate motions and body position. The cerebellum is also involved in learning and memory storage for behaviors. As you take notes, your cerebellum is instructing your brain about the order and timing of muscle movements in your hand. Not surprisingly, the cerebellum is largest in animals whose activities require fine coordination. It is highly developed in birds (see Fig. 33-11b), which engage in the complex activity of flight.

The Midbrain Contains the Reticular Formation

The midbrain is extremely reduced in humans (see Figs. 33-11b and 33-12). It contains an auditory relay center and a center that controls reflex movements of the eyes. Another important relay center, the **reticular formation**, passes through it. The neurons of the reticular formation extend all the way from the central core of the medulla up into lower regions of the forebrain. It receives input from virtually every sense, from every part of the body, and from many areas of the brain as well. The reticular formation plays a role in sleep and arousal, emotion, muscle tone, and certain movements and reflexes. It filters sensory inputs before they reach the conscious regions of the brain, although the selectivity of the filtering seems to be set by higher brain centers, such as those that control conscious thought. Activities of the reticular formation allow you to read and concentrate in the presence of a variety of distracting stimuli, such as the music from your stereo and the smell of coffee. The fact that a mother wakens upon hearing the faint cry of her infant but sleeps through loud traffic noise outside her window testifies to the effectiveness of the reticular formation in screening inputs to the brain.

The Forebrain Includes the Thalamus, Limbic System, and Cerebral Cortex

The forebrain, also called the **cerebrum**, includes the *thalamus* and *limbic system* and the *cerebral cortex.* In mammals, the cerebral cortex is much enlarged compared with that of fish, amphibians, and reptiles. This trend culminates in the human cerebral cortex (see Fig. 33-11b).

The **thalamus** is a kind of complex relay station that channels sensory information from all parts of the body

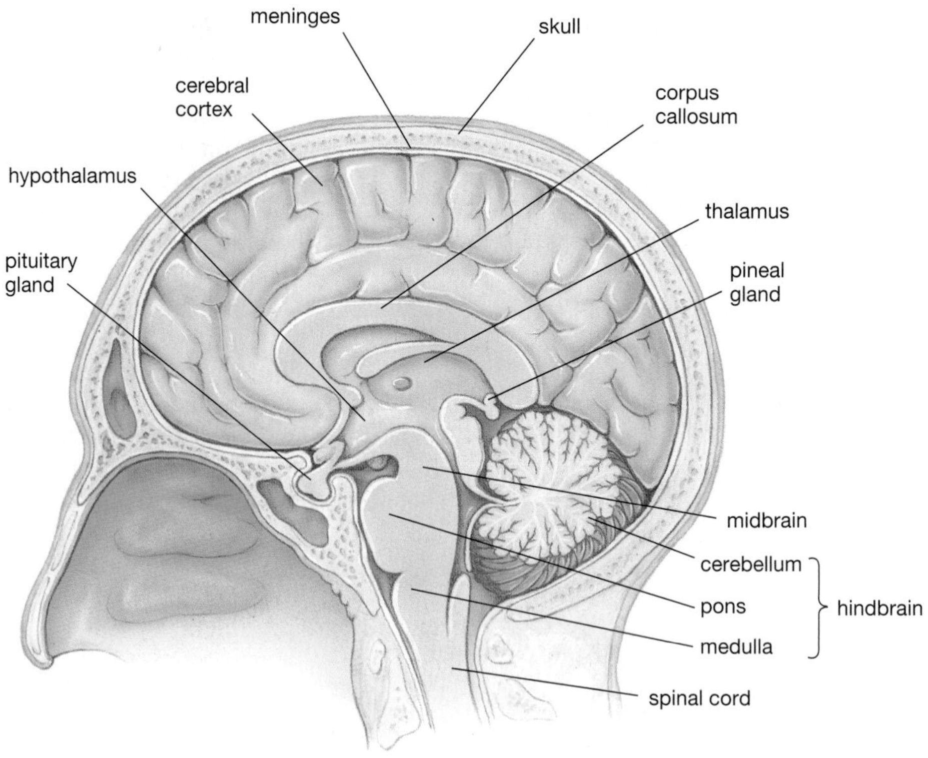

Figure 33-12 The human brain
A section through the midline of the human brain reveals some of its major structures.

to the limbic system and cerebral cortex (Fig. 33-13). Signals from the cerebellum and limbic system back to the cerebral cortex are also channeled through this busy thoroughfare. Anatomically, the **limbic system** is a diverse group of structures located in an arc between the thalamus and cerebral cortex (see Fig. 33-13). These structures work together to produce our most basic and primitive emotions, drives, and behaviors, including fear, rage, tranquility, hunger, thirst, pleasure, and sexual responses. Portions of the limbic system are also important in the formation of memories. The limbic system includes the *hypothalamus*, the *amygdala*, and the *hippocampus* as well as nearby regions of the cerebral cortex.

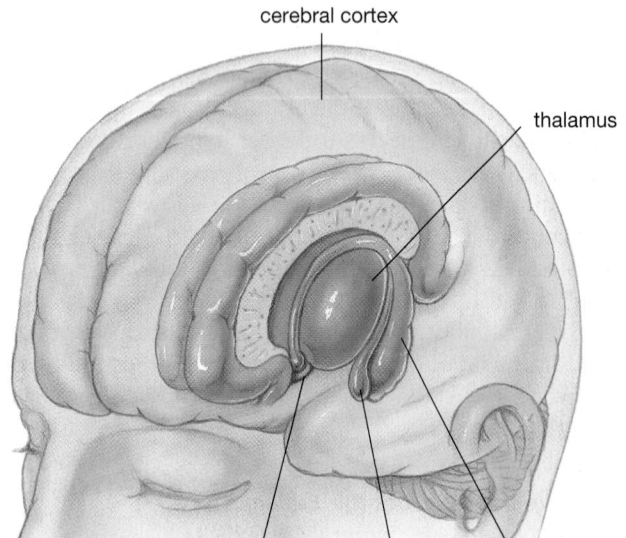

Figure 33-13 The limbic system and thalamus
The limbic system extends through several brain regions. It seems to be the center of most unconscious emotional behaviors, such as love, hate, hunger, sexual responses, and fear. The thalamus is a crucial relay center among the senses, the limbic system, and the cerebral cortex.

The **hypothalamus** (literally, "under the thalamus") contains many clusters of neurons. Some of these neurons are neurosecretory cells that release hormones into the blood; others control the release of a variety of hormones from the pituitary gland (see Chapter 32). Other regions of the hypothalamus direct the activities of the autonomic nervous system. The hypothalamus, through its hormone production and neural connections, acts as a major coordinating center, controlling body temperature, hunger, the menstrual cycle, water balance, and the sleep–wake cycle.

Clusters of neurons in the **amygdala** produce sensations of pleasure, punishment, or sexual arousal when stimulated. Conscious humans whose amygdalas are electrically stimulated have reported feelings of rage or fear. Recent studies have revealed that damage to the amygdala early in life eliminates the ability both to feel fear and to recognize fearful facial expressions in other people.

The shape of the **hippocampus** as it curves around the thalamus inspired its name, which is derived from the Greek word meaning "seahorse." As in the amygdala and hypothalamus, behaviors that reflect a variety of emotions, including rage and sexual arousal, can be elicited by stimulating portions of the hippocampus. The hippocampus also plays an important role in the formation of long-term memory and is thus required for learning, discussed in more detail later in this chapter.

In humans, by far the largest part of the brain is the **cerebral cortex**, the outer layer of the forebrain. The cerebral cortex and underlying parts of the forebrain are divided into two halves, called **cerebral hemispheres**. These halves communicate with each other by means of a large band of axons, the **corpus callosum**. The cerebral cortex is the most sophisticated information-processing center known, but it is also the area of the brain about which scientists know the least. Tens of billions of neurons are packed into this thin surface layer. The cortex is folded into **convolutions**, which greatly increase its area. In the cortex, cell bodies of neurons predominate, giving this outer layer, in preserved specimens, a gray color; hence the term "gray matter." These neurons receive sensory information, process it, store some in memory for future use, direct voluntary movements, and allow us to plan and think in ways we cannot yet understand.

The cerebral cortex is divided into four anatomical regions: the (1) *frontal*, (2) *parietal*, (3) *occipital*, and (4) *temporal lobes* (Fig. 33-14). Functionally, the cortex contains *primary sensory areas,* regions where signals originating in sensory organs such as the eyes and ears are received and converted into subjective impressions—for instance, light and sound. Nearby *association areas* interpret the sounds as speech or music, for

(a)

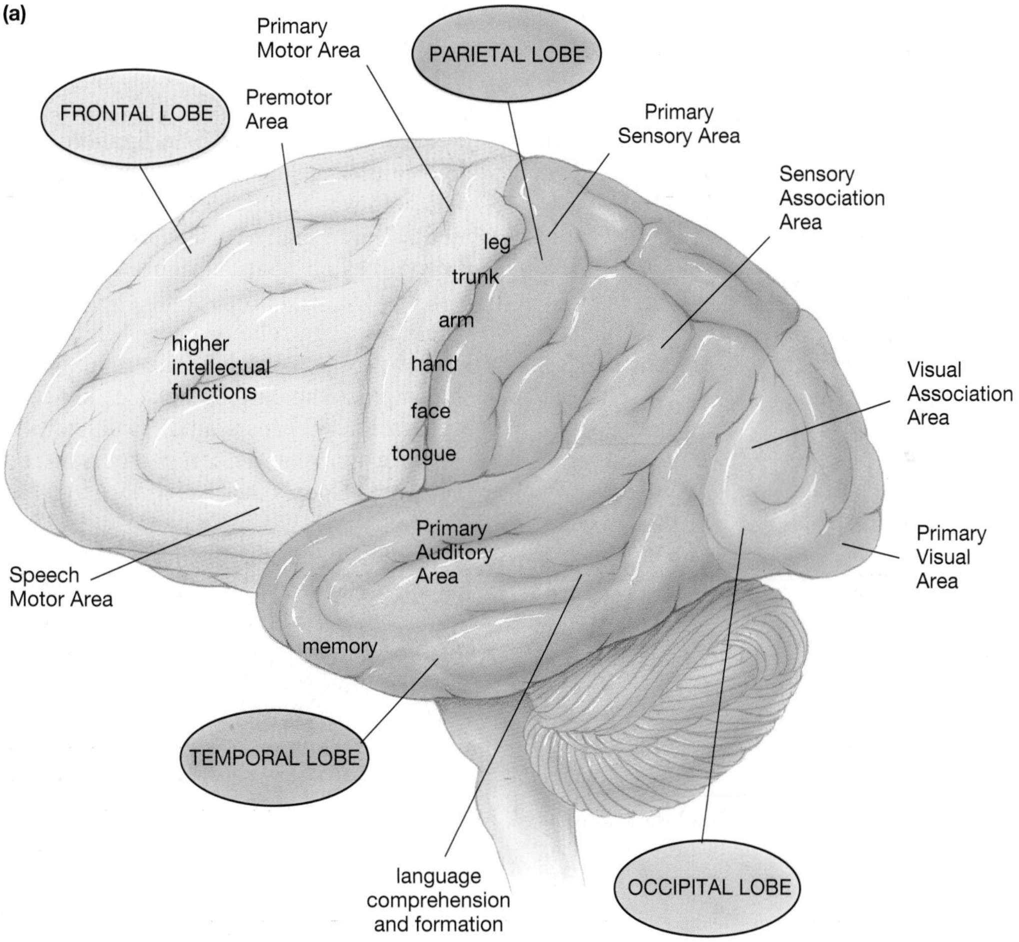

(b)

LEFT HEMISPHERE

1. Controls right side of body
2. Input from right visual field, right ear, left nostril
3. Centers for language, mathematics

RIGHT HEMISPHERE

1. Controls left side of body
2. Input from left visual field, left ear, right nostril
3. Centers for spatial perception, music, creativity

Figure 33-14 The cerebral cortex
(a) Structural (colored) and functional (labeled) regions of the human left cerebral cortex. A map of the right cerebral cortex would be similar, except that speech and language are less well developed there. ***(b)*** The chart shows the distribution of abilities between the two hemispheres.

example, and the visual stimuli as recognizable objects or words on this page. Association areas also link the stimuli with previous memories stored in the cortex and generate commands to produce speech. Research has revealed that the association areas of the brain do not always have the same function in the right and left hemispheres (see Fig. 33-14).

Primary sensory areas in the parietal lobe interpret sensations of touch that originate in all parts of the body, which are "mapped" in an orderly sequence. In an adjacent region of the frontal lobe, *primary motor areas* generate commands for movements in corresponding areas of the body (see Fig. 33-14). These commands control the motor neurons that form synapses with muscles, allowing you to walk to class, throw a Frisbee®, or keyboard a term paper. The association area of the frontal lobe, behind the bones of the forehead, seems to be involved in complex reasoning functions such as decision making, predicting the consequences of actions, controlling aggression, and planning for the future.

Damage to the cortex from trauma, stroke, or a tumor results in specific deficits, such as problems with speech, difficulty reading, or the inability to sense or move specific parts of the body. Most brain cells of adults cannot be replaced, so if a brain region is destroyed, these deficits may be permanent. Fortunately, however, training can sometimes allow undamaged regions of the cortex to take control over and restore some of the lost functions.

5 How Does the Brain Produce the Mind?

Historically, people have always had difficulty reconciling the physical presence of a few pounds of soft material in the skull with the range of thoughts, emotions, and memories of the human mind. This "mind–brain problem" has occupied generations of philosophers and, more recently, neurobiologists. Beginning with observations of individuals with head injuries and progressing to sophisticated surgical, physiological, and biochemical experiments, the outlines of how the brain creates the mind are beginning to emerge. Here, we will be able to touch on only a few of the most fascinating features.

The "Left Brain" and "Right Brain" Are Specialized for Different Functions

Although the cerebrum consists of two extremely similar-looking hemispheres, it has been known since the early 1900s that this symmetry does not extend to brain function. Much of what is known of the differences in hemisphere function comes from studies of accident or stroke victims with localized damage to one hemisphere and studies of patients who have had their corpus callosum (which connects the two hemispheres) severed. This surgical procedure is still performed in certain cases of uncontrollable epilepsy to prevent the spread of seizures through the brain.

Roger Sperry, a neurobiologist working at the California Institute of Technology, studied people whose hemispheres had been surgically separated by cutting the corpus callosum. In his studies, Sperry made use of the knowledge that axons within each optic tract (which are not severed by the surgery) follow a pathway that causes the left half of each visual field to be projected onto the right cerebral hemisphere and the right half to be projected onto the left hemisphere. Through an ingenious device that projected different images onto the left and right visual fields (thus sending different signals to each hemisphere), Sperry and other investigators have gained more insight into the roles of the two hemispheres.

If Sperry projected an image of a nude figure onto just the left visual field, the subjects would blush and smile but would claim to have seen nothing, because the image had reached only the nonverbal right side of the brain! The same figure projected onto the right visual field was readily described verbally. These experiments, begun in the 1950s, and other studies since then, have revealed that in right-handed people, the left hemisphere is dominant in speech, reading, writing, language comprehension, mathematical ability, and logical problem solving. The right side of the brain is superior to the left in musical skills, artistic ability, recognition of faces, spatial visualization, and the ability to recognize and express emotions. For his pioneering work, Sperry was awarded the Nobel Prize for Physiology in 1981.

Recent experiments, however, indicate that the left–right dichotomy is not as rigid as was once believed. An individual who has suffered a stroke that disrupted the blood supply to his or her left hemisphere typically shows symptoms such as loss of speaking ability. Commonly, however, training can partially overcome these speech or reading deficits, even though the left hemisphere itself has not recovered. This fact suggests that the right hemisphere has some latent language capabilities. Interestingly, female stroke victims are more likely to recover some lost abilities than are male stroke victims, and females have a slightly larger corpus callosum. These findings suggest a gender difference in the degree of specialization of the two hemispheres and in the extent of their interconnections. Further evidence of this difference has recently been provided by sensitive techniques that allow imaging of neural activity in the brains of normal subjects while they perform various mental tasks. When subjects were asked to compare word lists for rhyming words, a specific region of the left cortex of male subjects became active, but in females, similar areas in *both* the left and right hemispheres were activated. (Further brain-imaging studies are described in "Scientific Inquiry: Peering into the 'Black Box.'")

Scientific Inquiry

Peering into the "Black Box"

For most of human history, the brain has been seen as a "black box" whose inputs and outputs were observable but whose internal workings were inherently unknowable. New techniques, however, are providing exciting insights into brain function. Two new imaging techniques, *PET* (positron emission tomography, described in Chapter 2) and *fMRI* (functional magnetic resonance imaging), allow researchers to observe the brain in action. Regions of the brain that are most active have higher energy demands; these regions utilize more glucose and attract a greater flow of oxygenated blood than do inactive areas.

In PET scans, scientists inject the subject with a radioactive substance, such as a radioactive form of glucose, and then monitor levels of radioactivity. The level of radioactivity reflects the different metabolic rates, and hence glucose consumption, in various regions throughout the brain. Levels of glucose utilization are translated into colors on cross-sectional images of the brain. By monitoring radioactivity while a specific task is performed, scientists can identify which parts of the brain are most active during that task.

In contrast, fMRI detects differences in the way oxygenated and deoxygenated blood respond to a powerful magnetic field applied in pulses by an enormous electromagnet that surrounds the body. Active brain regions, which receive larger quantities of oxygenated blood than do inactive regions, can be distinguished with fMRI without the use of radioactivity and over much shorter time spans than through PET.

Using fMRI or PET, researchers can observe changes as the brain responds to an odor or to a visual or auditory stimulus or performs a specific reasoning task. Through brain scans, scientists have confirmed that different aspects of the processing of language occur in distinct areas of the cerebral cortex (Fig. E33-3). Using fMRI, researchers analyzed the frontal lobe areas used in generating words in individuals who spoke two languages. In subjects who had grown up speaking two languages, the same region of the frontal lobe was used in speaking each language. In subjects who had learned a second language later in life, different but adjacent frontal lobe areas were activated for the two languages. PET or fMRI scans can also precisely localize damaged portions of the brain, such as the result of a stroke (Fig. E33-4a). We can also observe the contrast between disturbed brain functioning, as in Alzheimer's disease, and normal brain functioning (Fig. E33-4b).

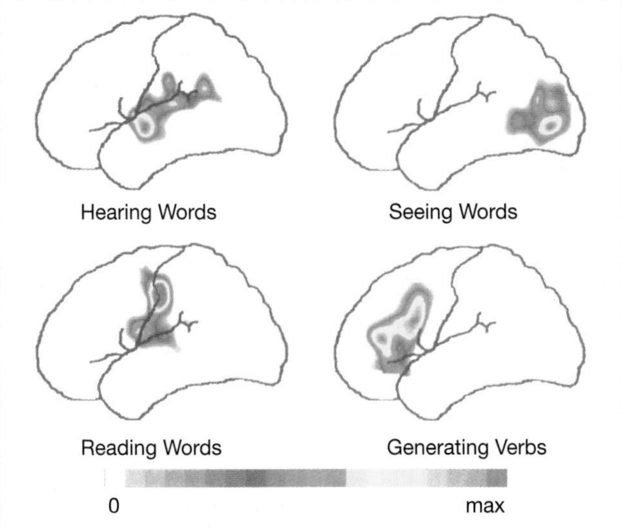

Figure E33-3 Localization of language tasks
Changes in glucose utilization, measured by PET, reveal different cortical regions involved in different language-related tasks, on the basis of research by Dr. Marcus Raichle of the Washington University School of Medicine in St. Louis. The scale ranges from white (lowest) to red (highest).

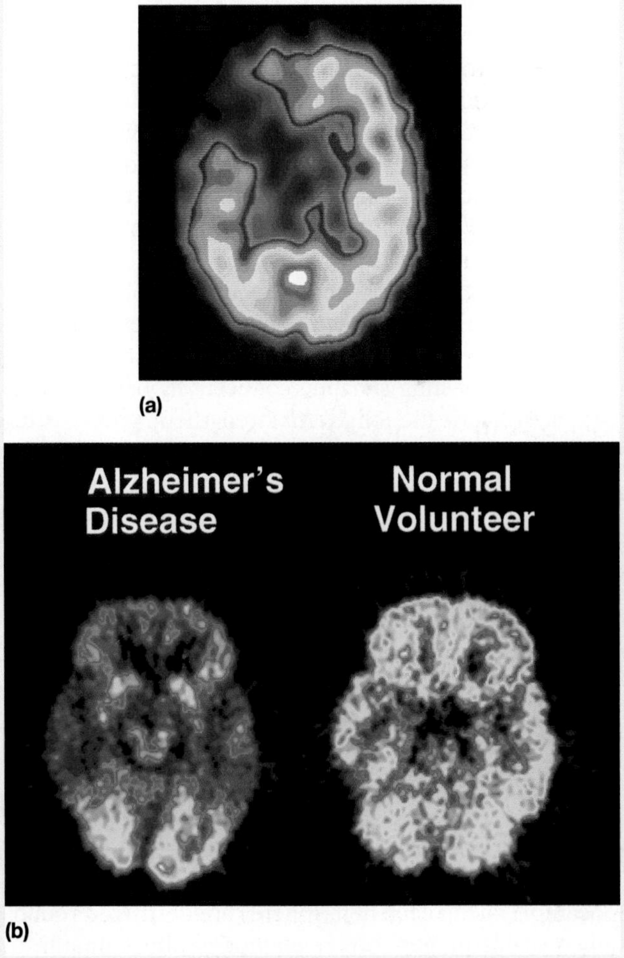

Figure E33-4 PET scans reveal malfunctioning brains
(a) As a result of stroke, part of the brain dies from lack of blood flow. The damaged region can be precisely localized by the lack of neural activity in the upper-left region of this brain scan. ***(b)*** The brain of an individual with Alzheimer's disease is compared with that of a healthy elderly person. The top of each image corresponds to the front of the brain, just behind the forehead. Metabolic activity decreases from red areas to yellow to green and to blue; black areas exhibit little or no metabolic activity.

The Mechanisms of Learning and Memory Are Poorly Understood

Although hypotheses abound as to the cellular mechanisms of learning and memory, we are a long way from understanding these phenomena. In mammals, and particularly in humans, however, we do know a fair amount about two other aspects of learning and memory: the time course of learning and some of the brain sites involved in learning, memory storage, and recall.

Memory May Be Brief or Long Lasting

Experiments show that learning occurs in two phases: (1) an initial **working memory** followed by (2) a **long-term memory**. For example, if you look up a number in the phone book, you will probably remember the number long enough to dial it but will promptly forget it. This is working memory. But if you call the number frequently, eventually you will remember the number more or less permanently. This is long-term memory.

Some working memory seems to be electrical in nature, involving the repeated activity of a particular neural circuit in the brain. As long as the circuit is active, the memory stays. In other cases, working memory involves temporary biochemical changes within neurons of a circuit, with the result that synaptic connections between them are strengthened.

In contrast, long-term memory seems to be structural—the result, perhaps, of persistent changes in the expression of certain genes. It may require the formation of new, long-lasting synaptic connections between specific neurons or the long-term strengthening of existing but weak synaptic connections. Working memory can be converted into long-term memory. This process seems to involve the hippocampus, which is believed to process new memories and then transfer them to the cerebral cortex for permanent storage.

The Temporal Lobes Are Important for Memory

The mechanisms of learning, memory storage, and memory retrieval are subjects of extensive research. There is strong evidence that the hippocampus, deep within the brain's temporal lobe, is involved in learning. For example, intense electrical activity occurs in the hippocampus during learning. Even more striking are the results of hippocampal damage. An individual in whom both hippocampi (one in each hemisphere) are destroyed retains much of his or her former memories but is unable to learn new information after the loss. A patient whose hippocampi and associated brain structures were removed surgically in 1953 in an attempt to control seizures remains unable to recall his address or find his way home after many years at the same residence. He can entertain himself indefinitely by reading the same magazine over and over, and people whom he sees daily require reintroduction at each encounter. This example and others suggest that the hippocampus is responsible for transferring information from working memory to long-term memory.

The temporal lobes of the cerebral hemispheres seem to be important in the *retrieval,* or recall, of long-term memories. In a famous series of experiments in the 1940s, neurosurgeon Wilder Penfield electrically stimulated the temporal lobes of conscious patients undergoing brain surgery. The patients did not merely recall memories but felt that they were experiencing the past events right there in the operating room!

Insights on How the Brain Creates the Mind Come from Diverse Sources

Until about 100 years ago, the mind was more appropriately a subject for philosophers than for scientists, because tools for studying the brain did not yet exist. Through the first half of the twentieth century, the mind was treated by psychologists as a "black box" whose internal workings could be deduced only through the investigation of how past and present experiences were interpreted and influenced behavior. New discoveries, however, are rapidly changing our views of the workings of the brain.

During recent decades, we have begun to understand the neural bases of at least some psychological phenomena. Many forms of mental illness, such as schizophrenia, manic depression, and autism, once thought to be due to childhood trauma or inept parenting, are now recognized as the results of biochemical imbalances or structural abnormalities in the brain. Studies have also revealed a strong heritability factor (and hence, a biological basis) for traits that were once considered entirely learned, such as shyness and alcoholism.

A striking illustration of how the physical structure of the brain is related to personality was unwittingly provided by Phineas Gage in 1848. A railroad construction foreman, Gage was setting an explosive charge when it triggered prematurely. The blast blew a steel rod that weighed 13 pounds and was more than a yard long through his skull, damaging both of his frontal lobes (Fig. 33-15). When he arrived in town, Gage informed the local doctor that "here is business enough for you." Although Gage survived for many years after his accident, his personality changed radically. Before the accident, Gage was conscientious, industrious, and well-liked. After his recovery, he became impetuous, profane, and incapable of working toward a goal. Subsequent research has implicated the frontal lobe in emotional expression, control of aggression, and the ability to work for delayed rewards.

Other sites of damage have revealed additional anatomical specializations. One patient with very localized damage to the left frontal lobe of the cerebral cortex was unable to name fruits and vegetables (although he could name everything else). Describing this patient, one science writer quipped, "Does the brain have a produce section?" Other victims of brain damage

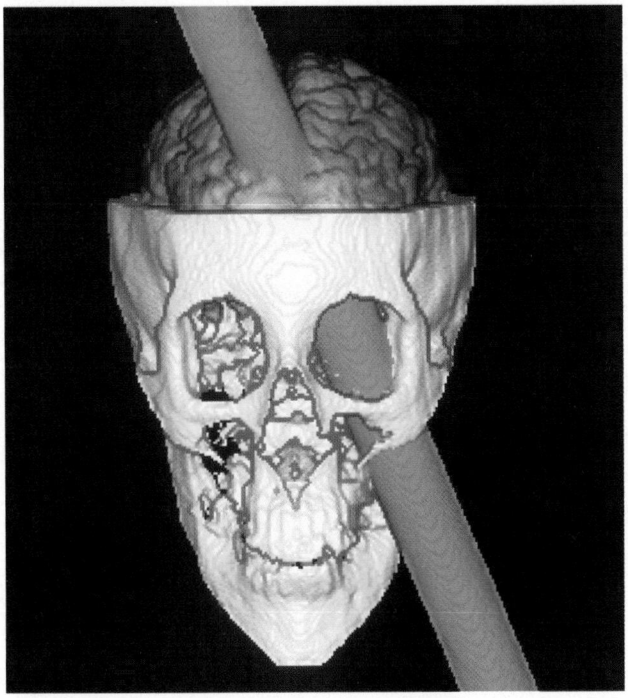

Figure 33-15 A revealing accident
On the basis of studies of the skull of Phineas Gage, scientists have created this computer-generated reconstruction of the path taken by the steel rod that was blown through his head by an explosion. He survived, but the frontal lobe of both cerebral hemispheres was extensively damaged, changing his personality.

have developed a selective inability to recognize faces or, in a recent case, to recognize any object that is *not* a face, suggesting that the brain has a region specialized to recognize faces that is separate from the region that allows it to recognize objects in general.

In the past, much of our understanding of the human mind–brain connection came from the study of victims of brain damage such as that caused by a stroke, trauma, tumor, or surgery. Typically, the exact extent of the damage remained unknown until revealed by autopsy. New techniques, such as the PET and fMRI scans, now permit insight into the functioning of normal, as well as diseased, brains (see "Scientific Inquiry: Peering into the 'Black Box'"). These and increasingly sophisticated techniques of the future will create ever-larger windows into the "black box" that is the human brain and a clearer understanding of how that brain generates the human mind.

6 How Do Sensory Receptors Work?

The word *receptor* is used in several contexts in biology. In the most general sense, a **receptor** is a structure that changes when it is acted on by a stimulus from its surroundings, causing a signal to be produced. A receptor may be a membrane protein that changes configuration when it binds a specific hormone or neurotransmitter, as discussed in previous chapters. Alternatively, as described in this chapter, a **sensory receptor** may be an entire specialized cell (typically, a neuron) that produces an electrical response to particular stimuli—that is, it translates sensory stimuli into the language of the nervous system. All sensory receptors produce electrical signals, but each receptor type is specialized to produce its signal only in response to a particular type of environmental stimulus. Some receptors, called *free nerve endings*, consist of branching dendrites of sensory neurons; other receptors have specialized structures that help them respond to a specific stimulus. Many sensory receptors are clustered into sensory organs, such as the eye, ear, skin, or tongue. Their electrical activity, after being processed by the brain, gives rise to the subjective perceptions of light, sound, touch, and taste that we describe as our "senses."

The stimulation of a sensory receptor causes an electrical signal called a **receptor potential**. Receptor potential amplitude varies with the intensity of the stimulus: The stronger the stimulus, the larger the receptor potential. Sensory receptors of different types have different ways of influencing postsynaptic neurons. In some sensory receptor neurons, a receptor potential will bring the cell above threshold and cause action potentials. In some very small sensory receptors, receptor potentials directly cause neurotransmitters to be released onto postsynaptic neurons, which in turn produce action potentials that travel to the central nervous system. A large positive receptor potential will cause a higher frequency of action potentials; as we learned earlier, that is how the brain interprets intensity. Sensory receptor cells are named after the stimulus to which they respond, as summarized in Table 33-2.

Sensory receptors are our links with the world around us. As Aristotle observed in the fourth century B.C., "Nothing is understood by the intellect which is not first perceived by the senses." In the following sections, we shall focus on some senses that determine the way we as humans perceive the world.

7 How Is Sound Sensed?

Sound is produced by any vibrating object—a drum, a motor, vocal cords, or the speaker of your stereo. These vibrations are transmitted through air and intercepted by our ears, which are elaborate structures that detect the direction, pitch, and intensity of sound.

The Ear Captures, Transmits, and Converts Sound into Electrical Signals

The ear of humans and most other vertebrates consists of three parts: the outer, middle, and inner ear (Fig. 33-16a). The **outer ear** consists of the **external ear** and

Table 33-2 Some Vertebrate Receptor Types

Type of Receptor	Sensory Cell Type	Stimulus	Location
Thermoreceptor	Free nerve ending	Heat, cold	Skin
Mechanoreceptor	Hair cell	Vibration, motion, gravity	Inner ear
	Specialized nerve endings and free nerve endings in skin (Pacinian corpuscle, Merkel's disc)	Vibration, pressure, touch	Skin
	Specialized nerve endings in muscles or joints (muscle spindle, Golgi tendon organ)	Stretch	Muscles, tendons
Photoreceptor	Rod, cone	Light	Retina of eye
Chemoreceptor	Olfactory receptor	Odor (airborne molecules)	Nasal cavity
	Taste receptor	Taste (waterborne molecules)	Tongue
Pain receptor	Free nerve ending	Chemicals released by tissue injury	Widespread in body

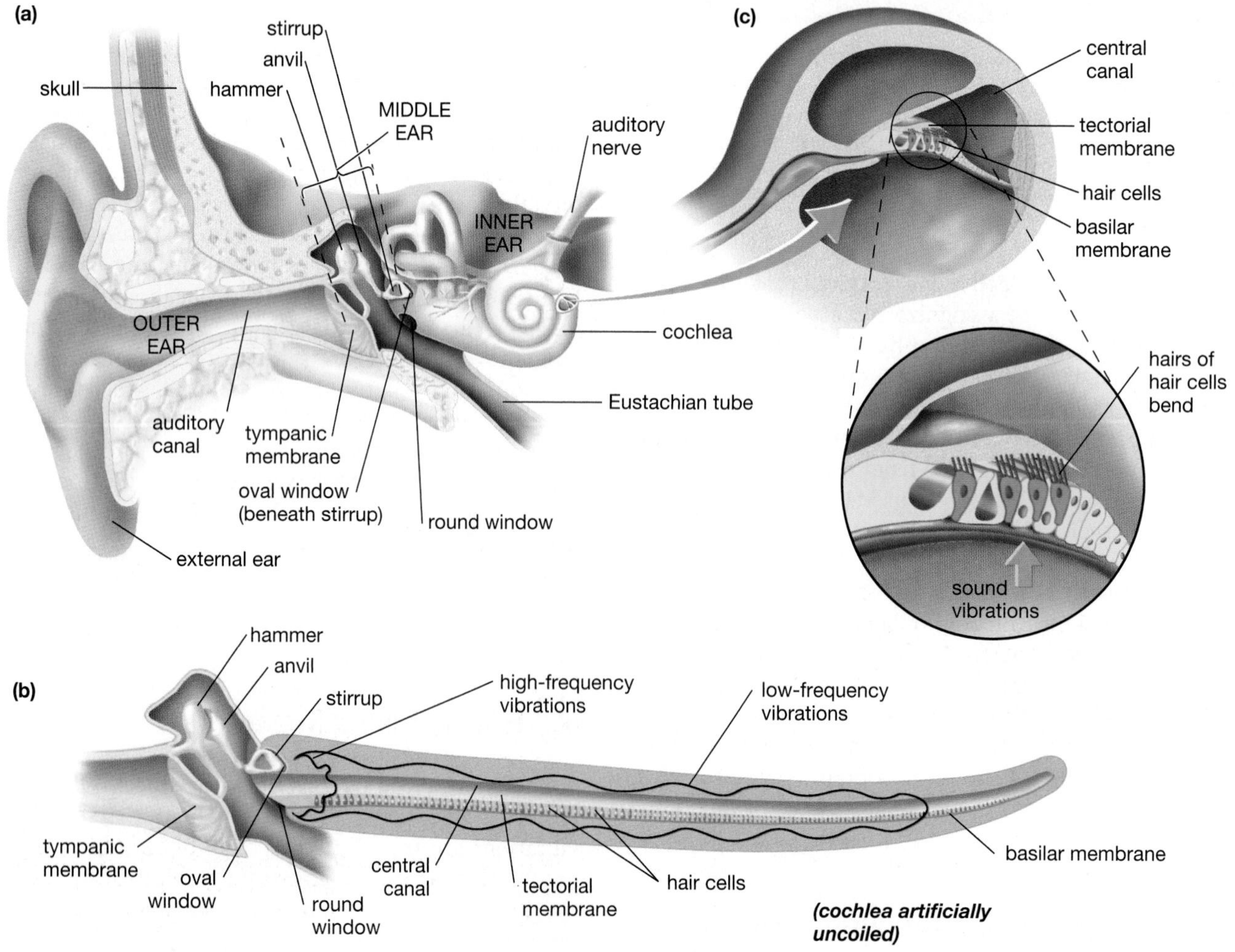

Figure 33-16 The human ear
(a) Overall anatomy of the ear. ***(b)*** Uncoiled, the cochlea consists of an outer tube surrounding the central canal. Different frequencies of sound waves cause the basilar membrane to vibrate at different locations. ***(c)*** The hairs of hair cells span the gap between the membranes in the central canal. Sound vibrations move the membranes relative to one another, bending the hairs and producing a receptor potential in the hair cells.

the **auditory canal**. The external ear, with its fleshy folds, modifies sound waves in ways that the brain uses to determine the location of the sound source. The air-filled auditory canal conducts the sound waves to the **middle ear**, consisting of the **tympanic membrane**, or *eardrum*; three tiny bones called the *hammer*, *anvil*, and *stirrup*; and the **Eustachian tube**. The Eustachian tube connects the middle ear to the pharynx and equalizes the air pressure between the middle ear and the atmosphere.

Within the middle ear, sound vibrates the tympanic membrane, which in turn vibrates the hammer, the anvil, and the stirrup. These bones transmit vibrations to the **inner ear**. The fluid-filled hollow bones of the inner ear form the spiral-shaped **cochlea** ("snail" in Latin) and additional structures that detect head movement and the pull of gravity. The stirrup bone transmits vibrations to the fluid within the cochlea by vibrating a membrane in the cochlea called the *oval window*. The *round window* is a second membrane that allows fluid within the cochlea to shift back and forth as the stirrup bone vibrates the oval window.

Sound Is Converted into Electrical Signals in the Cochlea

If we were to unroll the cochlea, in lengthwise cross section it would consist of two fluid-filled tubes: an outer U-shaped canal and a straight central canal (Fig. 33-16b). The floor of the central canal consists of the **basilar membrane**, on top of which are the receptors, or *hair cells*, which are a type of mechanoreceptor. Protruding into the central canal is the **tectorial membrane**, a gelatinous structure in which the "hairs" of the hair cells are embedded.

How do these structures allow the perception of sound? The oval window passes vibrations to the fluid in the cochlea, which in turn vibrates the basilar membrane, causing it to move up and down. This movement bends the hairs of the hair cells, producing receptor potentials. The hair cells then release neurotransmitters onto neurons of the auditory nerve to the brain. Action potentials triggered in axons of the auditory nerve travel to the brain.

The cochlea also allows us to perceive *loudness* (the magnitude of sound vibrations) and *pitch* (the frequency of sound vibrations). Weak sounds cause small vibrations, which bend the hairs only slightly and result in a low frequency of action potentials in axons of the auditory nerve. Loud sounds cause large vibrations, which cause greater bending of the hairs and a larger receptor potential, producing high-frequency action potentials in the auditory nerve. Loud sounds sustained for a long time can damage the hair cells (Fig. 33-17a), resulting in hearing loss, a fate suffered by many rock musicians and their fans. In fact, many sounds in our everyday environment have the potential to damage hearing, especially if they are prolonged (Fig. 33-17b).

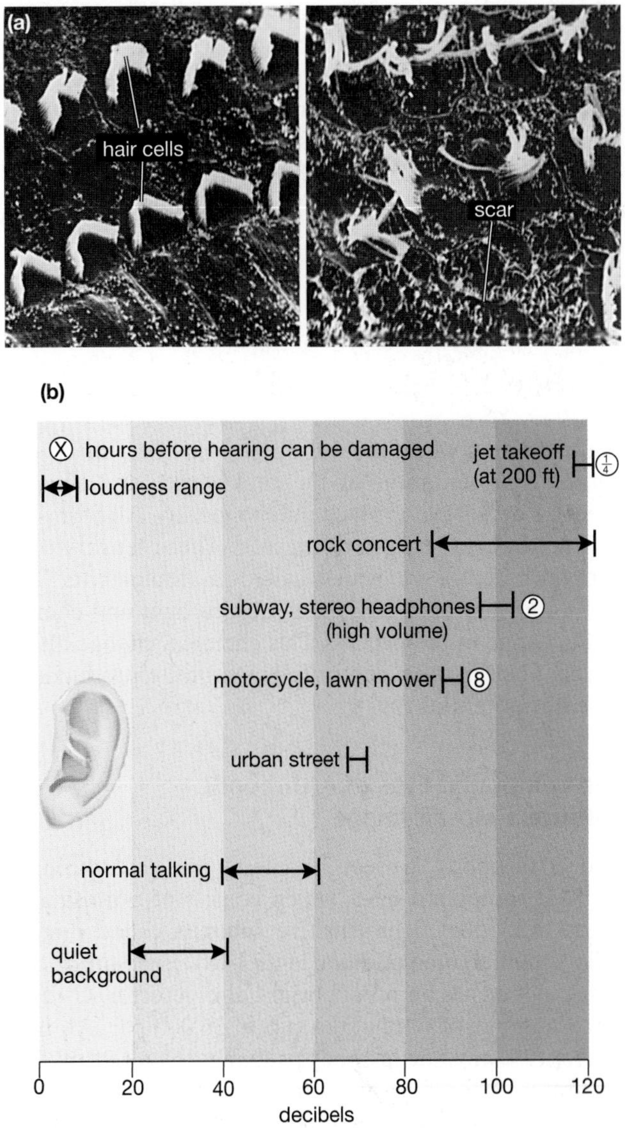

Figure 33-17 Loud sounds can damage hair cells
(a) Scanning electron micrographs show the effect of intense sound on the hair cells of the inner ear. (left) The hairs of hair cells in a normal guinea pig; hairs emerge from each receptor in a V-shaped pattern. (right) After 24-hour exposure to a sound level approached by loud rock music (2000 vibrations per second at 120 decibels), many of the hairs are damaged or missing, leaving "scars." Hair cells in humans do not regenerate, so such hearing loss is permanent. [SEMs by Robert S. Preston, courtesy of Professor J. E. Hawkins, Kresge Hearing Research Institute, University of Michigan Medical School.] ***(b)*** Sound levels of everyday noises and their potential to damage hearing. Sound intensity is measured in *decibels* on a logarithmic scale; a 10-decibel sound is 10 times as loud as a 1-decibel sound, and a 20-decibel sound is 100 times as loud. You feel pain at sound intensities above 120 decibels. [Source: Deafness Research Foundation, National Institute on Deafness and Other Communication Disorders]

The perception of pitch is a little more complex. Humans can detect vibration frequencies from about 30 vibrations per second (very low pitched) to about 20,000 vibrations per second (very high pitched). The basilar membrane is stiff and narrow at the end near the oval window but more flexible and wider near the tip of the cochlea. This progressive change in structure causes each portion of the membrane to vibrate in synchrony with a particular frequency of sound (see Fig. 33-16b). The brain interprets signals from receptors near the oval window as high-pitched sound, whereas signals from receptors farther along the cochlea are interpreted as lower in pitch.

8 How Is Light Sensed?

Animal vision varies in its ability to provide a sharp, accurate representation of the real world, and several types of eyes have evolved independently. All forms of vision, however, use *photoreceptors.* These sensory cells contain receptor molecules called **photopigments** (because they are colored), which absorb light and chemically change in the process. This chemical change alters ion channels in the receptor cell membrane, producing a receptor potential.

The Compound Eyes of Arthropods Produce a Mosaic Image

The arthropods (insects, spiders, and crustaceans) evolved **compound eyes**, which consist of a mosaic of many individual light-sensitive subunits called **ommatidia** (singular, **ommatidium**; Fig. 33-18). Each ommatidium functions as an on/off, bright/dim detector. Using a large number of ommatidia (up to 36,000 per eye in a dragonfly), most arthropods probably see a reasonably accurate, although grainy, image of the world. Compound eyes are excellent at detecting movement, an advantage in avoiding predators and in hunting. In addition, many arthropods, such as bees and butterflies, have good color perception.

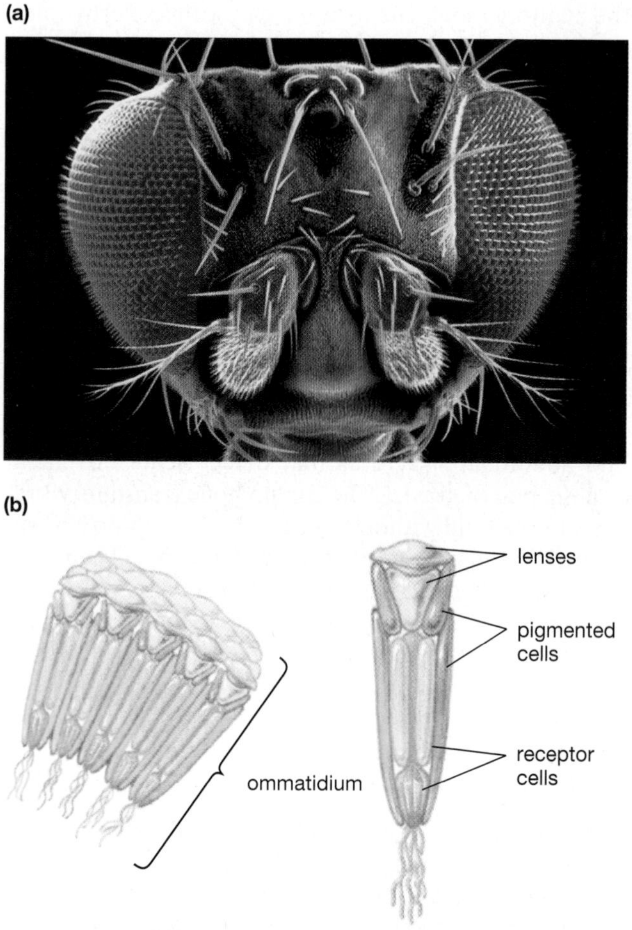

Figure 33-18 Compound eyes
(a) Scanning electron micrograph of the head of a fruit fly, showing a compound eye on each side of the head. ***(b)*** Each eye is made up of numerous light-receptive ommatidia. Within each ommatidium are several receptor cells, capped by a lens. Pigmented cells surrounding each ommatidium prevent the passage of light to adjacent receptors.

The Mammalian Eye Collects, Focuses, and Transduces Light Waves

As you read this, light reflected from the page first encounters the **cornea**, a transparent covering over the front of the eyeball. Behind the cornea, light passes through a chamber filled with a watery fluid called **aqueous humor**, which provides nourishment for both the lens and cornea. The amount of light entering the eye is adjusted by the **iris**, pigmented muscular tissue. The iris regulates the size of the **pupil**, a circular opening in the center of the iris. Light passing through the pupil encounters the **lens**, a structure that resembles a flattened sphere composed of transparent protein fibers. The lens is suspended behind the pupil by muscles that regulate its shape. Behind the lens is a much larger chamber filled with the **vitreous humor**, a clear jellylike substance that helps maintain the shape of the eye (Fig. 33-19a).

After passing through the vitreous humor, light reaches the **retina**, a multilayered sheet of photoreceptors and neurons. There the light energy is converted into electrical nerve impulses that are transmitted to the brain. Behind the retina is the **choroid**, a darkly pigmented tissue. The choroid's rich blood supply helps nourish the cells of the retina. Its dark pigment absorbs stray light whose reflection inside the eyeball would interfere with clear vision. Surrounding the outer portion of the eyeball is the **sclera**, a tough connective tissue layer that is visible as the white of the eye and is continuous with the cornea.

Driving along a country road at night, have you ever been startled by apparently disembodied glowing eyes? In vertebrates (such as deer) that are most active at dusk, the choroid may be modified to reflect light rather than to absorb it. By reflecting light that escaped the photoreceptors during its initial passage, the choroid gives the receptors a second chance to detect it, maximizing the animal's ability to see in dim light. Reflective

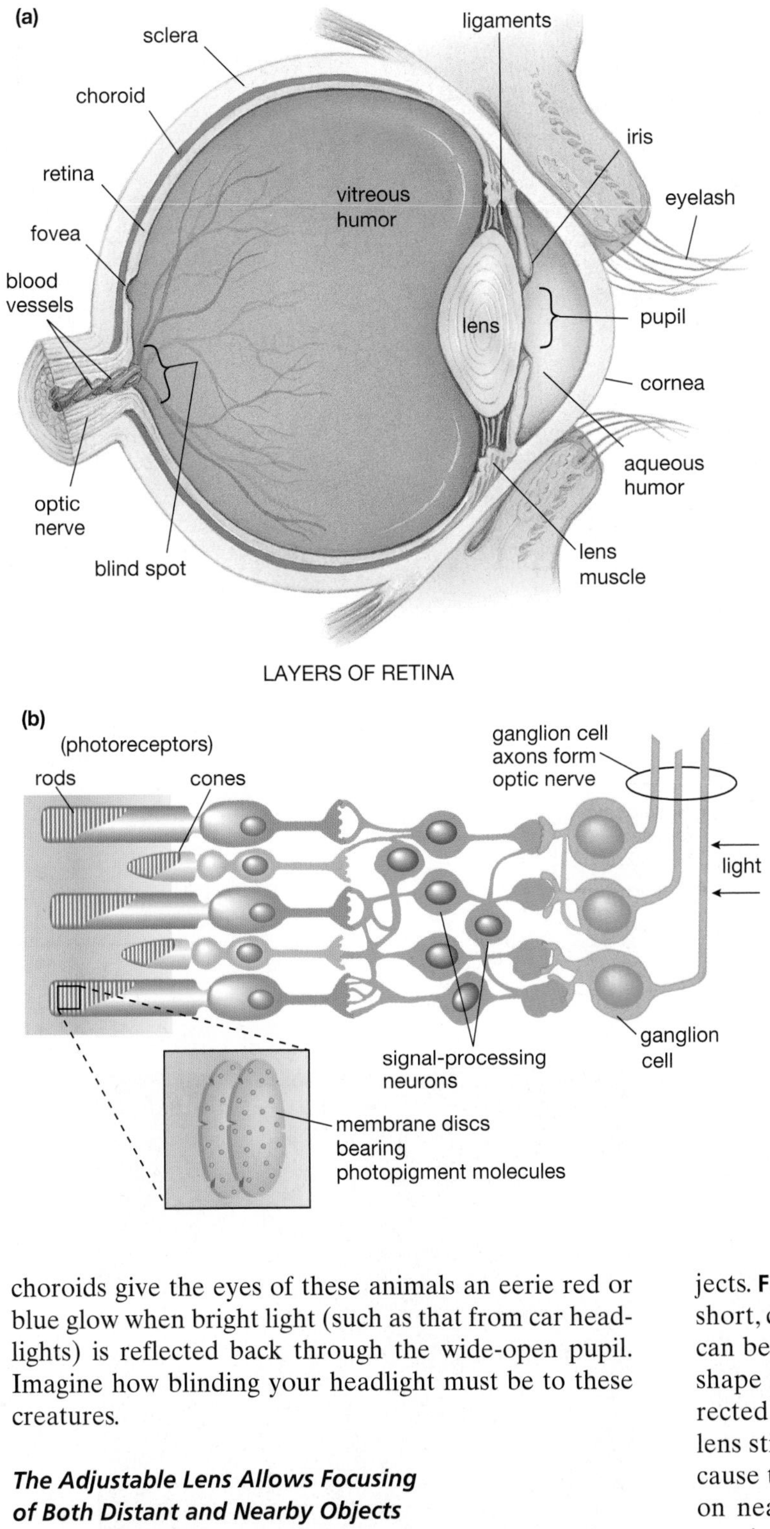

Figure 33-19 The human eye
(a) The anatomy of the human eye. ***(b)*** The human retina has rods and cones (photoreceptors), signal-processing neurons, and ganglion cells. Each rod and cone bears a long extension packed with membranes in which the light-sensitive molecules are embedded.

choroids give the eyes of these animals an eerie red or blue glow when bright light (such as that from car headlights) is reflected back through the wide-open pupil. Imagine how blinding your headlight must be to these creatures.

The Adjustable Lens Allows Focusing of Both Distant and Nearby Objects

The visual image is focused most sharply on a small area of the retina called the **fovea**. Although focusing begins at the cornea, whose rounded contour bends light rays, the lens is responsible for final, sharp focusing. The shape of the lens is adjusted by its encircling muscle. When viewed from the side, the lens is either rounded, to focus on nearby objects, or flattened, to focus on distant objects (Fig. 33-20a).

If your eyeball is even a millimeter too long, you will be **nearsighted**—unable to focus on distant objects. **Farsighted** people, whose eyeballs are slightly too short, cannot focus on nearby objects. These conditions can be corrected by external lenses of the appropriate shape (Fig. 33-20b,c). Nearsightedness can also be corrected by surgery that slightly flattens the cornea. The lens stiffens as humans age, causing farsightedness, because the lens can no longer round up enough to focus on nearby objects. By their mid-forties, most people require glasses for close work such as reading.

Light Striking the Retina Is Captured by Photoreceptors; The Signal Is Processed by Layers of Overlying Neurons

The vertebrate eye provides the sharpest vision in the animal kingdom, even though the complex, multilayered retina is "built backward" from an engineering perspective. The photoreceptors, called **rods** and **cones** after their shapes, have their light-gathering parts farthest from the light, at the rear of the retina (Fig. 33-19b).

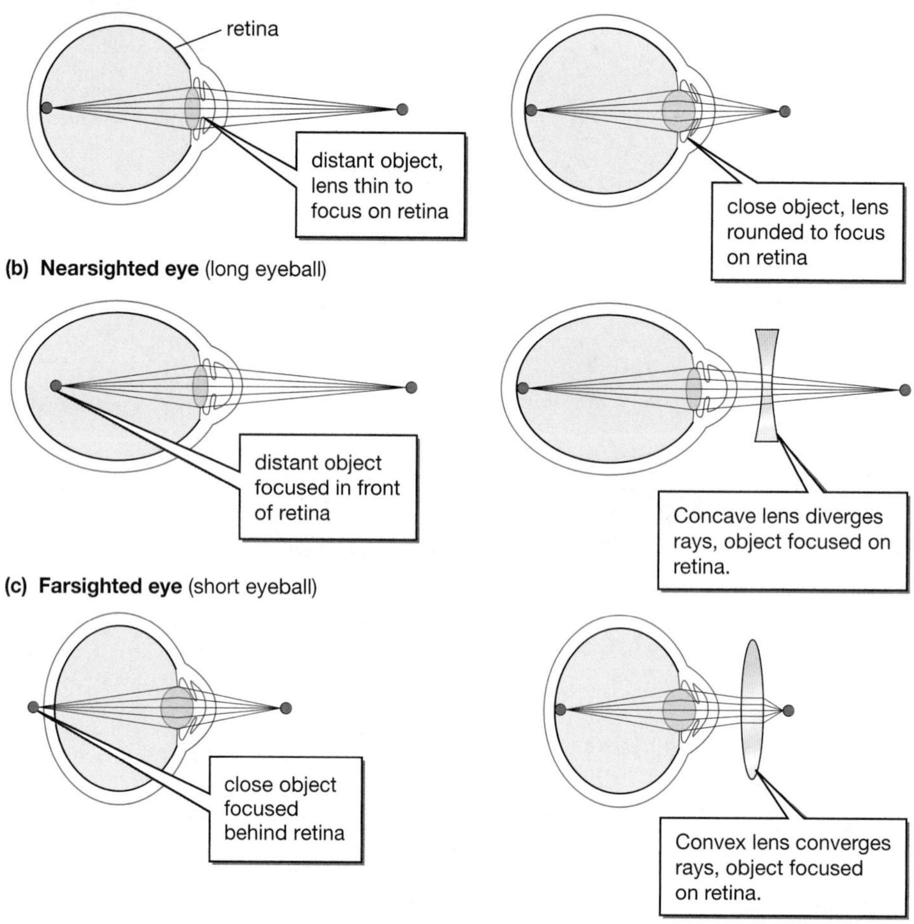

Figure 33-20 Focusing in the human eye
(a) (left) To focus on a distant object, the lens is made thinner, causing relatively little bending of the light rays. (right) To focus on a nearby object, the lens becomes rounder, bending the light rays more sharply. ***(b)*** (left) In a nearsighted eye, light rays focus in front of the retina. (right) Eyeglasses with a concave lens cause the light rays to diverge slightly so that the focal point falls on the retina. ***(c)*** (left) In a farsighted eye, the focal point falls behind the retina. (right) Eyeglasses with a convex lens converge light rays, causing the focal point to fall on the retina.

Between the receptors and incoming light are several layers of neurons that process the signals from the photoreceptors. The retinal layer nearest the vitreous humor consists of **ganglion cells**, whose axons make up the **optic nerve**. Ganglion cell axons must pass back through the retina to reach the brain at a location called the **blind spot** (Fig. 33-21; see also Fig. 33-19a). This area lacks receptors; objects focused there seem to disappear. You can locate your blind spot by closing your left eye and focusing steadily on the star below with your right eye. Start with the book about a foot away and gradually move it closer. The spot will disappear when the image falls on your blind spot.

★ ●

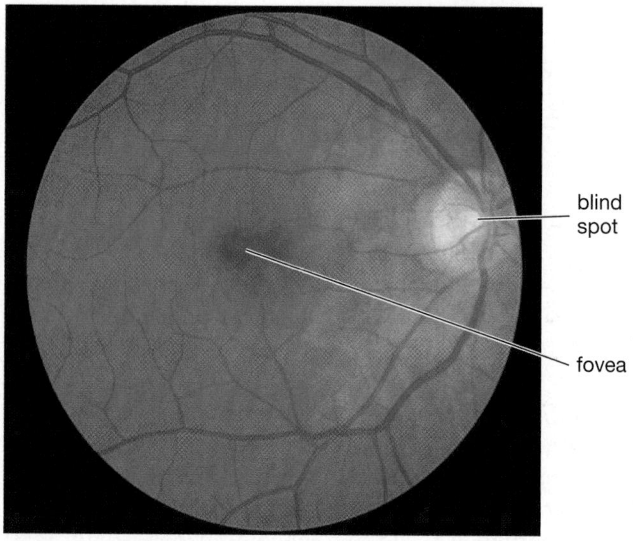

Figure 33-21 The human retina
A portion of the human retina, photographed through the cornea and lens of a living person. The blind spot and fovea are visible. Blood vessels supply oxygen and nutrients; notice that they are dense over the blind spot (where they won't interfere with vision) and scarcer near the fovea.

The receptor potential from the photoreceptors is processed by other retinal neurons in ways that enhance our ability to detect edges, movement, dim light, and changes in light intensity. The much-modified signal from the photoreceptors is finally converted to action potentials, which are carried by the optic nerve to the brain. In the brain, further processing ultimately results in the sensation of vision.

Rods and Cones Differ in Distribution and Light Sensitivity

Photoreception in both rods and cones begins with the absorption of light by photopigment molecules that are embedded in the plasma membranes of the photoreceptors (see Fig. 33-19b). Light hitting a photopigment molecule causes a change in the receptor membrane's permeability to ions, producing a receptor potential in the photoreceptor cell.

Although cones are located throughout the retina, they are concentrated in the fovea, where the lens focuses images most sharply (see Fig. 33-19a). The fovea appears as a depression near the center of the retina, because the layers of signal-processing neurons are pushed aside there while still retaining their synaptic connections. This arrangement allows light to reach the cones of the fovea with relatively little interference.

Human eyes have three varieties of cones, each containing a slightly different photopigment. Each type of photopigment is most strongly stimulated by a particular wavelength of light, corresponding roughly to red, green, or blue. The brain distinguishes color according to the relative intensity of stimulation of different cones. For example, the sensation of yellow is caused by fairly equal stimulation of red and green cones. About 3% of males have difficulty distinguishing red from green, because they possess a defective gene for the red or green photopigment on the X chromosome.

Rods dominate in the peripheral portions of the retina. Rods are far more sensitive to light than are cones and so are largely responsible for our vision in dim light. Unlike cones, rods do not distinguish colors; in moonlight, which is too dim to activate the cones, the world appears in shades of gray.

Not all animals have both rods and cones. Animals that are active almost entirely during the day (certain lizards, for example) may have all-cone retinas, whereas many nocturnal animals (such as the ferret) or those dwelling in dimly lit habitats (such as deep-sea fishes) have mostly or entirely rods.

Binocular Vision Allows Depth Perception

The placement of vertebrate eyes on the head is determined by the lifestyle of the animal. Predators and omnivores, such as humans, have both eyes facing forward (Fig. 33-22a), but most herbivores have one eye on each side of the head (Fig. 33-22b). The forward-facing eyes of predators and omnivores have slightly different but extensively overlapping visual fields. This **binocular vision** allows depth perception and accurate judgment of the size and distance of an object from the eyes. These abilities are important to a cat about to pounce on a mouse or to a monkey leaping from branch to branch.

In contrast, the widely spaced eyes of herbivores have little overlap in their visual fields; accurate depth perception is sacrificed in favor of a nearly 360-degree field of view. This view allows these animals, who are frequently preyed on, to spot a predator approaching from any direction.

(a) (b)

Figure 33-22 Eye position differs in predators and prey
(a) Most predators, such as this owl, and primates have eyes in front; both eyes can be focused on a target, providing binocular vision. ***(b)*** Most herbivorous prey animals, such as rabbits, have eyes at the sides, the better to scan for possible predators.

9 How Are Chemicals Sensed?

Through chemical senses, animals find food, avoid poisonous materials, and may locate homes or find mates. Terrestrial vertebrates have two separate chemical senses: one for airborne molecules, called *smell,* or **olfaction**, and one for chemicals dissolved in water or saliva, the sense of **taste**.

The Ability to Smell Arises from Olfactory Receptors

In humans and most other vertebrates, receptors for olfaction are nerve cells located in a patch of mucus-covered epithelial tissue in the upper portion of each nasal cavity (Fig. 33-23). The human olfactory epithelium is small compared with that of many other mammals (dogs, for example) whose sense of smell is hundreds of times more acute than ours. Olfactory receptors have hairlike dendrites that protrude into the nasal cavity and are embedded in a layer of mucus. Odorous molecules in the air, such as those from your coffee, diffuse into the mucus layer and bind with receptors on the dendrites.

Recent research suggests that there may be 1000 types of receptor proteins embedded in the olfactory dendrites. Each receptor protein is specialized to bind a particular type of molecule and to stimulate the olfactory receptor to send a message to the brain.

Taste Receptors Are Located in Clusters on the Tongue

The human tongue bears about 10,000 **taste buds**, structures embedded in small bumps (called *papillae*) that cover the surface of the tongue (Fig. 33-24a). Each taste bud consists of a cluster of 60 to 80 taste receptor cells surrounded by supporting cells in a small pit. The cells in the pit communicate with the mouth through a taste pore (Fig. 33-24b). Microvilli (thin membrane projections) of taste receptor cells protrude through the pore. Dissolved chemicals enter the pore and bind to receptor molecules on the microvilli, producing a receptor potential.

Four major types of taste receptors have long been known: (1) sweet, (2) sour, (3) salty, and (4) bitter; a fifth type, *umami* (a Japanese word loosely translated as "delicious"), has recently been identified. The umami receptor responds to glutamate, an ingredient in MSG, which is sometimes used as a seasoning and enhances the flavor of foods. Although it was commonly believed that taste buds for specific senses were concentrated on specific areas of the tongue, recent research has shown they are more or less evenly distributed.

We perceive a great variety of tastes as a result of two mechanisms. First, a particular substance may stimulate two or more receptor types to different degrees, making it taste "salty–sweet," for example. Second and more important, a substance being tasted normally releases molecules into the air inside the mouth. These odorous molecules diffuse to the olfactory receptors, which contribute an odor component to the basic flavor. (Recall that the mouth and nasal passages are connected.)

To prove that what we call taste is really mostly smell, try holding your nose (and closing your eyes) while you eat different flavors of jelly beans. The grape, lime, and cherry jelly beans will be indistinguishable sweet, sticky pastes. Likewise, when you have a bad cold, notice how normally tasty foods seem bland and lose much of their appeal and how your coffee tastes merely bitter.

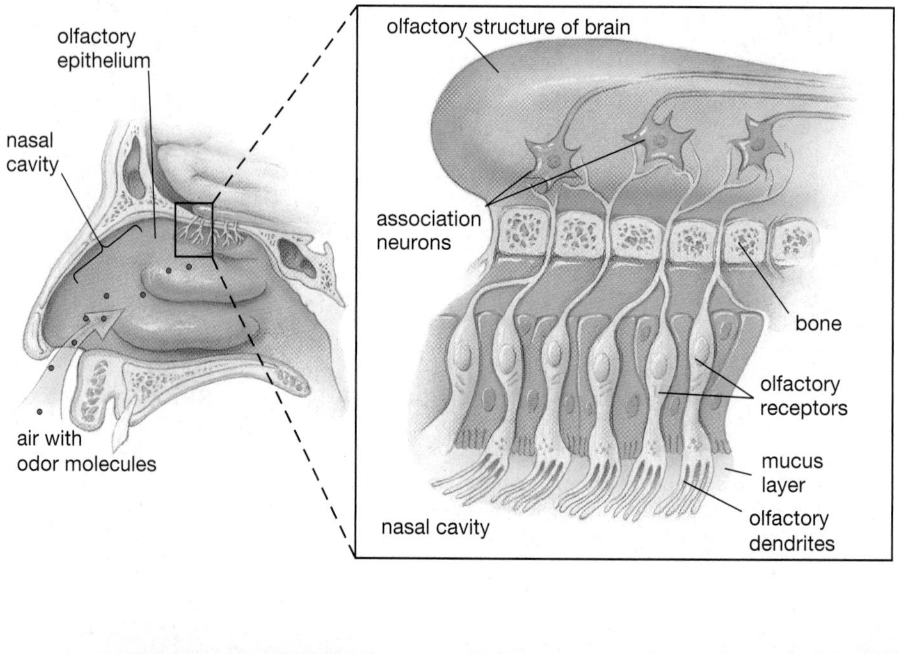

Figure 33-23 Human olfactory receptors The receptors for olfaction in humans are neurons bearing hairlike projections that protrude into the nasal cavity. The projections are embedded in a mucus layer in which odorous molecules dissolve before contacting the receptors.

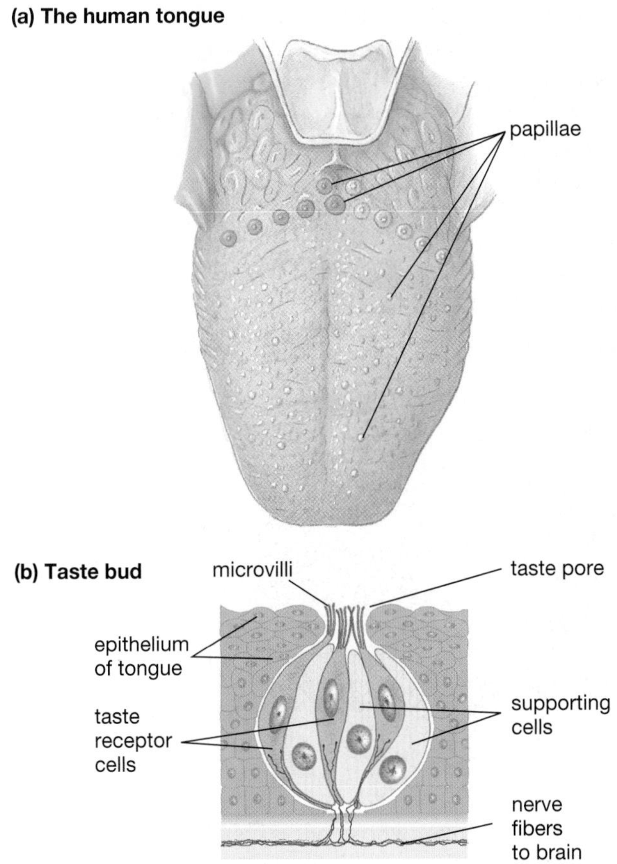

Figure 33-24 Human taste receptors
(a) The human tongue is covered with papillae, bumps in which taste buds are embedded. Small papillae are located on the front two-thirds of the tongue; larger ones with more taste buds are far in the back. ***(b)*** Each taste bud consists of supporting cells surrounding 60 to 80 taste receptor cells, whose microvilli have protein receptors that bind tasty molecules, producing a receptor potential.

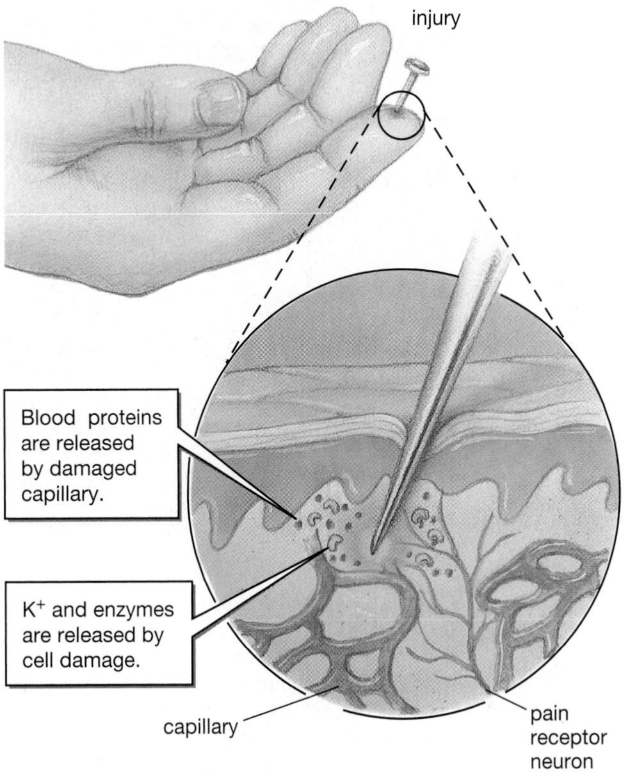

Figure 33-25 Pain perception
Pain perception is a specialized chemical sense. An injury, such as a tack stab, damages both cells and blood vessels. The damaged cells release K^+, which activates pain receptor neurons. Damaged cells also release enzymes that convert certain blood proteins into bradykinin, which also stimulates pain-sensitive neurons.

Pain Is a Specialized Chemical Sense

Whether you burn, cut, or crush a fingertip, you will feel the same sensation: pain. Most pain is produced by tissue damage. Researchers have found that pain perception is actually a special kind of chemical sense (Fig. 33-25).

When cells and capillaries are damaged by a cut or a burn, for example, their contents flow into the extracellular fluid. The cell contents include potassium ions, which stimulate **pain receptors**. Damaged cells also release enzymes that convert certain blood proteins into a chemical called *bradykinin*, another stimulus that activates pain receptors. Each part of the body has a separate set of pain receptor neurons that provide input to particular brain cells. Hence, the brain can identify the location of the pain. Drugs that provide pain relief, such as morphine or Demerol®, block synapses in the pain pathways of the brain or spinal cord. As we mention in "Health Watch: Drugs, Diseases, and Neurotransmitters," the brain can modulate its perception of pain through its own narcotic-like endorphins.

Evolutionary Connections

Uncommon Senses

We have reviewed the "common" senses of sound, sight, smell, taste, and pain. But had this text been written for bats, it undoubtedly would have included a large section on *echolocation* and nearly omitted the coverage of vision! Here we review a few of the "uncommon" senses that have evolved in response to different environments.

Echolocation

Some animals who hunt in darkness or murky water have evolved a type of sonar called **echolocation**, similar to the navigational system used by ships. Using echolocation, bats can navigate and hunt insect prey in total darkness. An echolocating bat emits pulses of noise at ultrasonic frequencies (higher than can be detected by the human ear) that bounce off nearby objects. The patterns of returning sound convey accurate information about the size, shape, surface texture, and location of objects in the environment. Little brown bats can detect wires only 1 millimeter thick from a distance of 2 meters (more than 6 feet). Several adaptations contribute to

(a)

(b)

Figure 33-26 Echolocation
(a) A long-eared bat, showing the enormous size and elaborate folds of the external ears that help it localize returning echoes. ***(b)*** The bottlenose porpoise focuses ultrasonic clicks by using the oil-filled sac in the front of its head.

this remarkable sensitivity. The bat's enormous, elaborately folded outer ears collect the returning echoes and help the bat locate the source of the echoes (Fig. 33-26a). As the bat emits its cry, muscles attached to the bones of the middle ear contract briefly, reducing the bones' vibrations and preventing the bat from being deafened by its own calls. The tympanic membrane and bones of the middle ear are exceptionally light and easily vibrated by the faint returning echoes.

Porpoises produce ultrasonic clicks within their nasal passages and emit the clicks through the front of their head (Fig. 33-26b). There, a large, flexible, oil-filled sac directs the sound forward in a broad beam (for navigation) or a narrow beam (for prey location). An echolocating porpoise can locate a pea-sized object on the floor of its tank and can distinguish among species of fish. Porpoises may also use the narrowly focused beam to stun fish with a blast of sound, making them easier to capture.

Detecting Electrical Fields

Some fish, called weak electric fish, use electrical fields for **electrolocation** in much the same way that bats and porpoises use sound waves for echolocation. These fish produce high-frequency electrical signals from an electric organ just in front of their tails; they detect these signals via *electroreceptor* cells located along both sides of their bodies (Fig. 33-27). Objects near the fish distort the electrical field that surrounds the fish. This distortion is detected by the electroreceptors, which send an altered pattern of action potentials to the brain. The fish uses this information to detect and localize nearby objects. The platypus, whose bill is covered with electroreceptors (as well as sensitive touch receptors), hunts for crayfish and tadpoles at night in murky freshwater ponds and streams.

Detecting Magnetic Fields

Homing pigeons are famous for their ability to fly home after being released some distance away. They can accurately locate their home roost even under cloudy skies in terrain with few landmarks. If a researcher straps a small magnet to a pigeon's back, the bird will find its way home successfully in sunny weather but will lose its

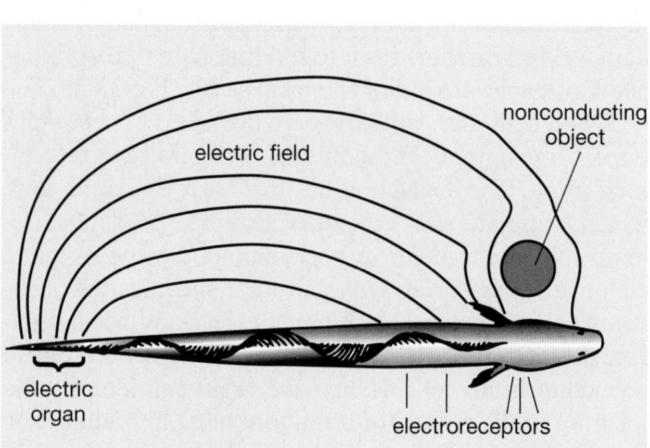

Figure 33-27 Electrolocation
Weak electric fish locate nearby objects by sensing distortions the objects produce in the animal's own electrical field. That field is generated by electric organs near the tail and detected by electroreceptors along the sides of the body.

way under overcast skies. Apparently, pigeons can navigate either by the sun or, if the sun is hidden, by magnetic fields. In cloudy weather, the experimental magnets confuse the pigeon's internal magnetic compass. How do pigeons detect magnetic fields? Pigeons have deposits of magnetite (a magnetic iron compound) just beneath their skull. These deposits may act as a built-in magnet that is used to tell direction.

Eels of eastern North America and western Europe swim out of streams and rivers into the Atlantic Ocean and migrate to the Sargasso Sea (near the West Indies) to spawn. Eels probably also use magnetic fields for navigation. The currents of the Gulf Stream, flowing from the Gulf of Mexico along the east coast of the United States and moving through Earth's magnetic field, generate an electrical field. The electrical field is roughly equivalent to that produced by a 1-volt battery with its poles more than 12 miles (7 km) apart. Researchers have discovered that eels can detect electrical fields as weak as that of a 1-volt battery with poles *more than 5000 kilometers* (3000 miles) apart. For an eel, finding the Gulf Stream must be a piece of cake!

CASE STUDY REVISITED

From Tragedy to Triumph

Whether from auto accidents, bullets, disease, falls, or sporting mishaps, in the U.S. each year nearly 8000 people sustain spinal cord injuries and about 400,000 people are living with the disabling effects of these injuries. Research using laboratory animals has shown that a walking reflex exists within the spinal cord. Recently, patients in experimental programs have demonstrated that a similar walking reflex exists in the human spinal cord as well. With the proper assistance, automatic walking motions, controlled entirely by neurons within the spinal cord, can be activated and improved with practice. Dozens of patients who had completely lost the use of their legs can now temporarily leave their wheelchairs and (grasping a wheeled walker) stand and move about on their own. In 1999 Christopher Reeve, supported by a harness developed to assist astronauts during space travel, began using these spinal "motor programs" to walk on a treadmill. He faces a far greater challenge than some, since he cannot rely on his arms to help support and balance his body. Having directed an HBO movie, played a small role in a recent film, delivered speeches throughout the country, and helped stimulate tremendous support for research since his accident, Reeve hopes to stand and toast everyone on his fiftieth birthday in September 2002. Regardless of whether he will be able do this, he is living testimony to the incredible qualities of the human brain and to the mind that it generates.

Scientists are eagerly searching for ways to promote regeneration of damaged nerves in the CNS, whereas evolution seems to have favored mechanisms that *inhibit* CNS regeneration after the nervous system matures. Speculate on reasons why natural selection has not favored such regrowth. Consider the following as you speculate: First, during development of the nervous system, the billions of neurons in the CNS form trillions of precise synaptic connections that allow the enormously complex information-processing capabilities of the brain. As the nervous system matures, the capability for regrowth of injured axons diminishes in the CNS but not in the PNS. Second, in the wild, are animals with major CNS injuries likely to survive? How does this situation apply to people?

Summary of Key Concepts

1) What Are the Structures and Functions of Neurons?

Nervous systems are composed of billions of individual cells called neurons. A neuron has four major specialized functions, which are reflected in its structure: (1) Dendrites receive information from the environment or from other neurons. (2) The cell body adds together electrical signals from the dendrites and from synapses on the cell body itself and "decides" whether to produce an action potential. (The cell body also coordinates the cell's metabolic activities.) (3) The axon conducts the action potential to its output terminal: the synapse. (4) Synaptic terminals transmit the signal to other nerve cells, to glands, or to muscles.

2) How Is Neural Activity Produced and Transmitted?

An unstimulated neuron maintains a negative resting potential inside the cell. Signals received from other neurons are small, rapidly fading changes in potential called postsynaptic potentials. Inhibitory and excitatory postsynaptic potentials (IPSPs and EPSPs) make the neuron less likely or more likely, respectively, to produce an action potential. If postsynaptic potentials, added together within the cell body, bring the neuron to threshold, an action potential will be triggered. The action potential is a wave of positive charge that travels along the axon to the synaptic terminals, undiminished in magnitude.

A synapse, where two neurons communicate, consists of the synaptic terminal of the presynaptic neuron, a specialized region of the postsynaptic neuron, and the tiny gap between them. Neurotransmitters from the presynaptic neuron, released in response to an action potential, bind to receptors in the postsynaptic cell's plasma membrane, opening ion channels there. Ions flow, producing either an EPSP or an IPSP, depending on the type of ion channels opened.

There are many neurotransmitters. They are being intensively studied to explore their diverse roles in neurological disease, drug addiction, and all aspects of normal nervous system function.

3 What Are Some General Features of Nervous Systems?

Information processing in the nervous system requires four operations. The nervous system must (1) determine the type of stimulus, (2) signal the intensity of the stimulus, (3) integrate information from many sources, and (4) initiate and direct the response. The nervous system collects and processes sensory information from many sources. These sensory stimuli may come together (converge) on fewer neurons whose activity in response to the stimuli determines the action taken. The "decision" to act may then be transmitted to many more neurons (divergence), which direct the activity.

Neural pathways normally have four elements: (1) sensory neurons, (2) association neurons, (3) motor neurons, and (4) effectors. Overall, nervous systems consist of numerous interconnected neural pathways, which may be either diffuse or centralized.

4 How Is the Human Nervous System Organized?

The nervous system of humans and other vertebrates consists of the central nervous system and the peripheral nervous system. The peripheral nervous system is further subdivided into sensory and motor portions. The motor portions consist of the somatic nervous system (which controls voluntary movement) and the autonomic nervous system (which directs involuntary responses).

Within the central nervous system, the spinal cord contains (1) neurons controlling voluntary muscles and the autonomic nervous system and neurons communicating with the brain and other parts of the spinal cord; (2) axons leading to and from the brain; and (3) neural pathways for reflexes and certain simple behaviors. The brain consists of three parts: the hindbrain, midbrain, and forebrain, each further subdivided into distinct regions.

The hindbrain in humans consists of the medulla and pons, which control involuntary functions (such as breathing), and the cerebellum, which coordinates complex motor activities (such as typing). In humans, the small midbrain contains the reticular formation, a filter and relay for sensory stimuli. The forebrain includes the thalamus, a sensory relay station that shuttles information to and from conscious centers in the forebrain. The diverse structures of the limbic system of the forebrain are involved in emotion, learning, and the control of instinctive behaviors such as sex, feeding, and aggression. The cerebral cortex of the forebrain is the center for information processing, memory, and initiation of voluntary actions. It includes primary sensory and motor areas and association areas that analyze sensory information and plan movements.

5 How Does the Brain Produce the Mind?

The cerebral hemispheres are each specialized. In general, the left hemisphere dominates speech, reading, writing, language comprehension, mathematical ability, and logical problem solving. The right hemisphere specializes in recognizing faces and spatial relationships, artistic and musical abilities, and recognition and expression of emotions.

Memory takes two forms. (1) Short-term memory is electrical or chemical. (2) Long-term memory probably involves structural changes that increase the effectiveness of synapses. The hippocampus is an important site for learning and for the transfer of information into long-term memory. The temporal lobes are important for memory.

6 How Do Sensory Receptors Work?

Receptors convert signals from one form to another. Receptor cells are named after the stimulus to which they respond.

7 How Is Sound Sensed?

In the vertebrate ear, air vibrates the tympanic membrane, which transmits vibrations to the bones of the middle ear and then to the oval window of the fluid-filled cochlea. Within the cochlea, vibrations bend the hairs of hair cells, which are receptors located between the basilar and tectorial membranes. This bending produces receptor potentials in the hair cells that cause action potentials in the axons of the auditory nerve, which leads to the brain.

8 How Is Light Sensed?

In the vertebrate eye, light enters the cornea and passes through the pupil to the lens, which focuses an image on the fovea of the retina. Two types of photoreceptor, rods and cones, are located deep in the retina. They produce receptor potentials in response to light. These signals are processed through several layers of neurons in the retina and are translated into action potentials in the optic nerve, which leads to the brain. Rods are more abundant and more light-sensitive than are cones and provide vision in dim light. Cones, which are concentrated in the fovea, provide color vision.

9 How Are Chemicals Sensed?

Terrestrial vertebrates detect chemicals in the external environment either by olfaction or by taste. Each olfactory or taste receptor cell type responds to only one or a few specific types of molecules, allowing discrimination among tastes and odors. Olfactory neurons of vertebrates are located in a tissue that lines the nasal cavity. Taste receptors are located in clusters called taste buds on the tongue. Pain is a special type of chemical sense in which sensory neurons respond to chemicals released by damaged cells.

Key Terms

action potential *p. 664*
amygdala *p. 678*
aqueous humor *p. 686*
association neuron *p. 671*
auditory canal *p. 685*
autonomic nervous system *p. 673*
axon *p. 664*
basilar membrane *p. 685*
binocular vision *p. 689*
blind spot *p. 688*
blood–brain barrier *p. 673*
brain *p. 672*
cell body *p. 664*
central nervous system *p. 672*
cephalization *p. 672*
cerebellum *p. 677*
cerebral cortex *p. 679*
cerebral hemisphere *p. 679*
cerebrospinal fluid *p. 673*
cerebrum *p. 677*
choroid *p. 686*
cochlea *p. 685*
compound eye *p. 686*
cone *p. 687*
convergence *p. 671*
convolution *p. 679*
cornea *p. 686*
corpus callosum *p. 679*
dendrite *p. 664*
divergence *p. 671*
dorsal root ganglion *p. 675*

echolocation *p. 691*
effector *p. 671*
electrolocation *p. 692*
Eustachian tube *p. 685*
external ear *p. 683*
farsighted *p. 687*
forebrain *p. 676*
fovea *p. 687*
ganglion *p. 672*
ganglion cell *p. 688*
gray matter *p. 673*
hindbrain *p. 676*
hippocampus *p. 678*
hypothalamus *p. 678*
inner ear *p. 685*
intensity *p. 670*
iris *p. 686*
lens *p. 686*
limbic system *p. 678*
long-term memory *p. 682*
medulla *p. 677*
meninges *p. 673*
midbrain *p. 676*
middle ear *p. 685*
motor neuron *p. 671*
myelin *p. 664*
nearsighted *p. 687*
nerve *p. 664*
nerve net *p. 672*
neuron *p. 664*
neurotransmitter *p. 664*
olfaction *p. 690*
ommatidium *p. 686*
optic nerve *p. 688*
outer ear *p. 683*
pain receptor *p. 691*
parasympathetic division *p. 673*
peripheral nerve *p. 672*
peripheral nervous system *p. 672*
photopigment *p. 686*
pons *p. 677*
postsynaptic neuron *p. 666*
postsynaptic potential (PSP) *p. 665*
presynaptic neuron *p. 666*
pupil *p. 686*
receptor *p. 683*
receptor potential *p. 683*
reflex *p. 671*
resting potential *p. 665*
reticular formation *p. 677*
retina *p. 686*
rod *p. 687*
sclera *p. 686*
sensory neuron *p. 671*
sensory receptor *p. 683*
somatic nervous system *p. 673*
spinal cord *p. 672*
sympathetic division *p. 673*
synapse *p. 665*
synaptic terminal *p. 664*
taste *p. 690*
taste bud *p. 690*
tectorial membrane *p. 685*
thalamus *p. 677*
threshold *p. 665*
tympanic membrane *p. 685*
vitreous humor *p. 686*
white matter *p. 673*
working memory *p. 682*

Thinking Through the Concepts

Multiple Choice

1. __________ *are integration centers in neurons.*
   a. dendrites b. axons
   c. cell bodies d. ion channels
   e. synapses
2. *The role of the axon is to*
   a. integrate signals from the dendrites
   b. release neurotransmitter
   c. conduct the action potential to the synaptic terminal
   d. synthesize cellular components
   e. stimulate a muscle, gland, or another neuron
3. *Which of the following statements is FALSE?*
   a. The hindbrain contains the medulla, pons, and cerebellum.
   b. The thalamus is an important sensory relay structure.
   c. The cerebral cortex in humans is folded into convolutions.
   d. The hippocampus has an important role in recognizing and experiencing fear.
   e. The hypothalamus coordinates the activities of the autonomic nervous system.
4. *The fovea is*
   a. the blind spot
   b. a clear area in front of the pupil and iris
   c. the tough outer covering of the eyeball
   d. the substance that gives the eyeball its shape
   e. the central focal region of the vertebrate retina
5. *Light entering the eye and striking the choroid would travel first, in order, through the*
   a. lens, vitreous humor, cornea, aqueous humor, and retina
   b. retina, aqueous humor, lens, vitreous humor, and cornea
   c. cornea, aqueous humor, retina, vitreous humor, and lens
   d. cornea, aqueous humor, lens, vitreous humor, and retina
   e. lens, aqueous humor, cornea, vitreous humor, and retina
6. *Sound reception is carried out by hair cells, which are a special type of*
   a. chemoreceptor
   b. photoreceptor
   c. mechanoreceptor
   d. magnetoreceptor
   e. thermoreceptor

? Review Questions

1. List four major parts of a neuron, and explain the specialized function of each part.
2. Diagram a synapse. How are signals transmitted from one neuron to another at a synapse?
3. How does the brain perceive the intensity of a stimulus? The type of stimulus?
4. What are the four elements of a simple nervous pathway? Describe how these elements function in the human pain-withdrawal reflex.
5. Draw a cross section of the spinal cord. What types of neurons are located in the spinal cord? Explain why severing the cord paralyzes the body below the level where it is severed.
6. Describe the functions of the following parts of the human brain: medulla, cerebellum, reticular formation, thalamus, limbic system, and cerebrum.
7. What structure connects the two cerebral hemispheres? Describe the evidence that the two hemispheres are specialized for distinct intellectual functions.
8. Distinguish between long-term memory and working memory.
9. What are the names of the specific receptors used for taste, vision, hearing, smell, and touch?
10. Why are we apparently able to distinguish hundreds of different flavors if we have only five types of taste receptors? How are we able to distinguish so many different odors?

11. Describe the structure and function of the various parts of the human ear by tracing a sound wave from the air outside the ear to the cells that cause action potentials in the auditory nerve.
12. How does the structure of the inner ear allow for the perception of pitch? Of sound intensity?
13. Diagram the overall structure of the human eye. Label the cornea, iris, lens, sclera, retina, and choroid. Describe the function of each structure.
14. How does the lens change shape to allow focusing of distant objects? What defect makes focusing on distant objects impossible, and what is this condition called? What type of lens can be used to correct it, and how does it do so?
15. List the similarities and differences between rods and cones.
16. Distinguish between taste and olfaction.
17. Describe how pain is signaled by tissue damage.

Applying the Concepts

1. Argue for or against the statement "Consciousness by its nature is incomprehensible; the brain will never understand the mind."
2. In Parkinson's disease, which afflicts several million Americans, the cells that produce the neurotransmitter dopamine degenerate in a small part of the brain that is important in the control of movement. Some physicians have reported improvement after injecting cells taken from the same general brain region of an aborted fetus into appropriate parts of the brain of a Parkinson's patient. Discuss this type of surgery from as many viewpoints as possible: ethical, financial, practical, and so on. On the basis of your responses, is fetal transplant surgery the answer to curing Parkinson's disease?
3. If the axons of human spinal cord neurons were unmyelinated, would the spinal cord be larger or smaller? Would you move faster or slower? Explain your answer.
4. What is the adaptive value of reflexes? If Christopher Reeve was accidentally pricked on the toe by a thumbtack, would he withdraw his leg? Explain your answer.
5. Explain the statement "Your sensory perceptions are purely a creation of your brain." Discuss the implications for communicating with other humans, with other animals, and with intelligent life from another universe.
6. Corneal transplants can help restore vision and greatly improve the recipient's quality of life. What properties of the cornea make it an excellent candidate for transplantation? Suggest some ways in which society could improve the availability of corneal and other tissues for transplantation.

For More Information

Angier, N. "Storming the Wall." *Discover*, May 1990. Describes the blood–brain barrier and new methods of penetrating it with drugs.

Axel, R. "The Molecular Logic of Smell." *Scientific American*, October 1995. Describes research that uncovers some of the mechanisms by which the nose and brain decipher scents.

Beardsley, T. "The Machinery of Thought." *Scientific American*, August 1997. Using PET and MRI on monkeys and humans, researchers are learning more about where working memory resides.

Bower, B. "Creatures in the Brain." *Science News*, April 13, 1996. Using imaging techniques, scientists have discovered clues as to how different regions of the brain are specialized for different concepts.

Dolnick, E. "Obsessed." *Health*, September 1994. Researchers have linked obsessive-compulsive disorders to imbalances in neurotransmitters.

Freedman, D. H. "In the Realm of the Chemical." *Discover*, June 1993. Describes how smell and taste help us experience our world.

Gutin, J. C. "Good Vibrations." *Discover*, June 1993. Research uncovers mechanisms of hearing.

Holloway, M. "Rx for Addiction." *Scientific American*, March 1991. By studying drug addiction, neurobiologists are learning more about the brain and new ways to counteract addictions.

Kimura, D. "Sex Differences in the Brain." *Scientific American*, September 1992. Hormonal differences between the sexes influence brain development.

Koretz, J. F., and Handelman, G. H. "How the Human Eye Focuses." *Scientific American*, July 1988. Describes focusing in the human eye with an emphasis on the loss of focusing ability that occurs with age.

McKean, K. "Pain." *Discover*, October 1986. Describes the discovery of bradykinin and its role in pain perception.

Nathans, J. "The Genes for Color Vision." *Scientific American*, February 1989. Good discussion of the physiology and genetics of color blindness.

Raichle, M. E. "Visualizing the Mind." *Scientific American*, April 1994. Brain imaging techniques partially open the "black box" of the mind.

Scientific American special issue on Mind and Brain, September 1992. Articles describe the development of the brain, the biochemical basis of learning, major neurological disorders, and aging effects on the brain.

Answers to Multiple-Choice Questions

1. c 2. c 3. d 4. e 5. d 6. c

MEDIATUTOR
The Nervous System and the Senses

CD Activities

Activity 33.1: Ions and Electrical Signals

Estimated time: 5 minutes

The electrical signal that is generated by a neuron involves a flow of ions across the membrane of the cell. As these ions flow, the electrical potential (or charge) across the membrane of the cell changes. The result is the production of an action potential, which is the electrical output signal of the neuron. In this tutorial, you will explore the events that generate an action potential.

Activity 33.2: Synapses

Estimated time: 5 minutes

Neurons communicate at junctions called synapses. This communication involves the action of a group of chemicals called neurotransmitters. In this tutorial, you will view an animation that simulates the action that occurs at a synapse and the role of the neurotransmitter in this action.

Start the MediaTutor Student CD-ROM and enter the activity number in the Quick Search box to be taken directly to that activity.

Web Investigations

Case Study: From Tragedy to Triumph

Estimated time: 10 minutes

Before WWII, spinal cord injuries were usually lethal. Today many survivors live in wheelchairs, some only with the support of complex and expensive medical apparatus. Scientific researchers are looking for new treatments and developing experimental spinal repair protocols. At this time there is no cure for spinal cord injuries, but there are several promising leads.

Go to http://www.prenhall.com/audesirk6, the Audesirk Companion Web site. Select Chapter 33 and the Web Investigation to begin.

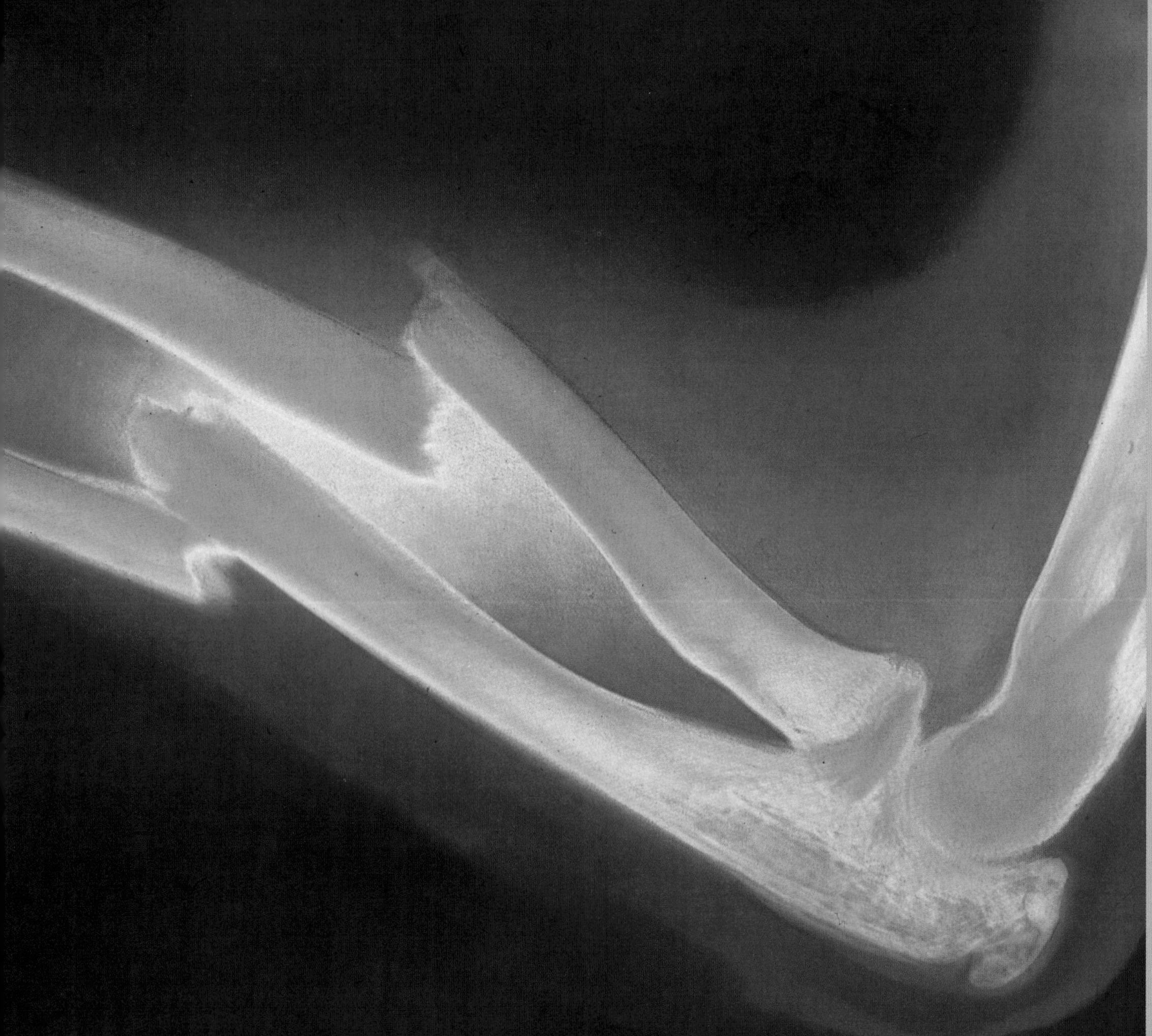

An open, or compound, fracture is seen in this dramatic X-ray of the forearm.

34 Action and Support: The Muscles and Skeleton

AT A GLANCE

Case Study: Healing Broken Bones

1) **How Do Muscles Work?**

Skeletal Muscle Cell Structure and Function Are Closely Linked

Muscle Contraction Results from Thick and Thin Filaments Sliding Past One Another

Cardiac Muscle Powers the Heart

Smooth Muscle Produces Slow, Involuntary Contractions

2) **What Does the Skeleton Do?**

The Vertebrate Skeleton Serves Many Functions

3) **Which Tissues Compose the Vertebrate Skeleton?**

Cartilage Provides Flexible Support and Connections

Bone Provides a Strong, Rigid Framework for the Body

4) **How Does the Body Move?**

Muscles Move the Skeleton Around Flexible Joints

Case Study Revisited: Healing Broken Bones

CASESTUDY

Healing Broken Bones

Ignoring the warning on her stepladder, Margaret steps to the top platform. Paintbrush in hand, she strains to reach the uppermost corner of her bedroom wall—and feels the ladder begin to slip. With nothing to hold onto, she instinctively reaches out her arm to break her fall—but instead, the fall breaks her arm. With a sickening crack, bones shatter into an open, or *compound,* fracture, such as that shown in our opener photo. The bones of the lower arm, the radius and ulna, have both been broken and the jagged point of one bone has pierced the overlying muscle and protrudes out through the skin. Paramedics immobilize the arm in the position they find it and give painkillers as they rush the victim to the hospital. There, after taking X-rays like this one, surgeons work to reposition the bones and to anchor their broken ends in place with screws and metal plates. Now the amazing process of healing begins. Bones seem so "dry" and inert that it may be hard to picture them as living organs, constantly changing, capable of remodeling and repairing themselves. How do bones heal? How do muscles exert the forces that cause the skeleton to move? Will the bones of a person in her sixties heal as rapidly as if she were in her twenties? ■

The system of muscles and skeleton that moves and supports the animal body is an engineering marvel. The flight of a bat, the pounce of a cat, the soaring leap of Michael Jordan, and the simple movements needed to climb a ladder or paint a wall depend on the same humble yet elegant mechanism. The muscle tissue of people and other animals is composed of muscle cells, which perform only one trick: They exert a force by contracting. Under the influence of natural selection, however, this simple unidirectional force has been applied to complex structural elements such as wings, hands, and fins, and its action has come to be coordinated by the nervous system. The resulting capacity for movement gives animals the ability to search for food, seek out new habitats, flee from danger—and, on occasion, to move in ways that we find awe-inspiring.

Muscles and skeletons also perform more mundane, but still crucial, functions. Pumping blood through the circulatory system, moving food through the digestive system, and breathing are some of the essential processes that depend on muscle contraction. Skeletons play an equally essential role by opposing the force of gravity and providing a framework against which muscles exert their forces to move the body. Terrestrial organisms in particular depend on skeletal support to maintain their shapes. Without your skeleton, you'd be a formless, quivering mound of tissue.

1 How Do Muscles Work?

All muscular work requires that muscles alternately contract and lengthen, but muscles are active only during the contraction phase. The lengthening that follows contraction is passive; it occurs when muscles relax and are stretched out by other forces. For example, a relaxed muscle may be lengthened by contractions of opposing muscles, the weight of a limb, or pressure from food on the muscular walls of the stomach.

Animals have a variety of muscle types, each specialized to perform a particular function. Vertebrates have evolved three types of muscle: (1) skeletal, (2) cardiac, and (3) smooth. All work on the same basic principles but differ in function, appearance, and control (Table 34-1).

Table 34-1 Location, Characteristics, and Functions of the Three Muscle Types

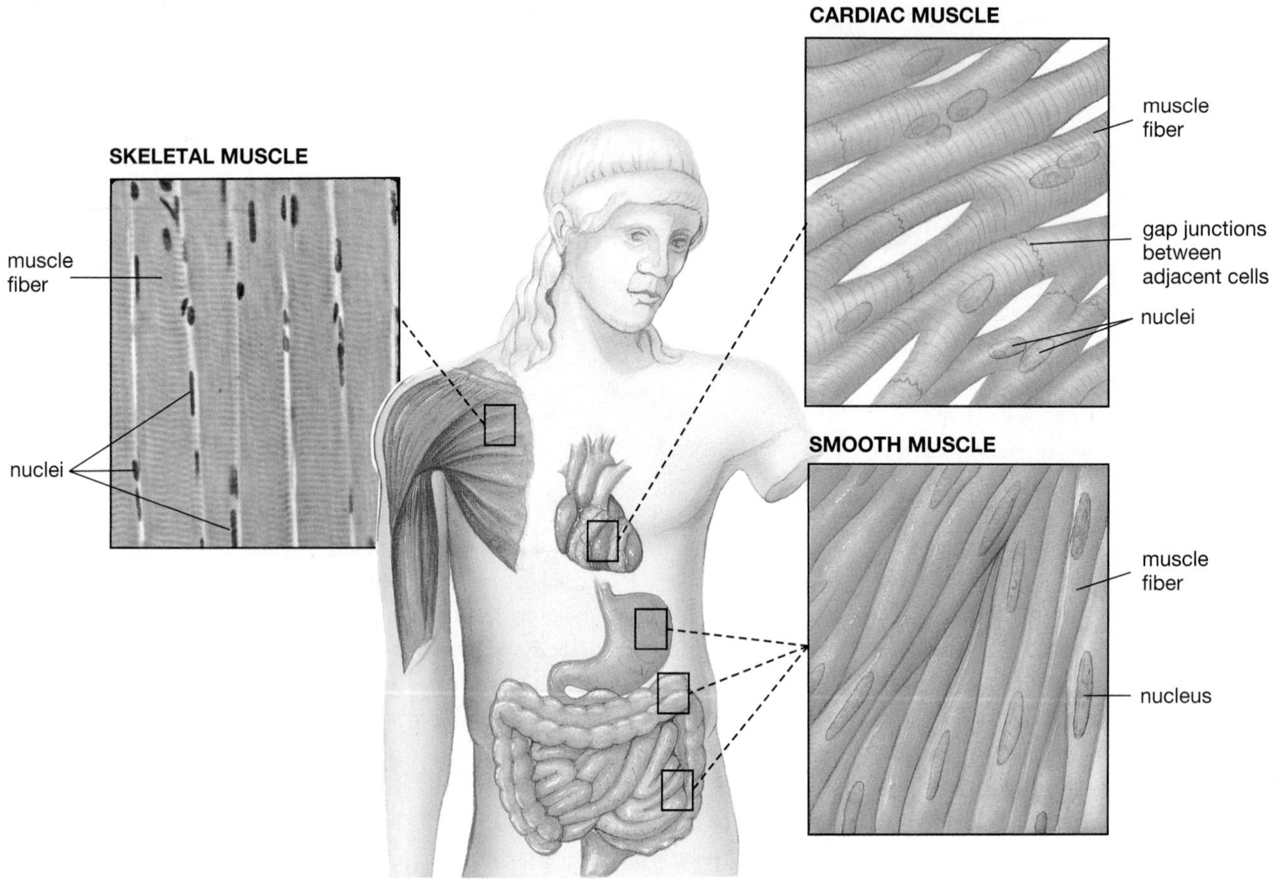

Property	Type of Muscle: Smooth	Cardiac	Skeletal
Muscle appearance	Unstriped	Irregular stripes	Regular stripes
Cell shape	Spindle	Branched	Spindle or cylindrical
Number of nuclei	One per cell	Many per cell	Many per cell
Speed of contraction	Slow	Intermediate	Slow to rapid
Contraction caused by	Spontaneous, stretch, nervous system, hormones	Spontaneous	Nervous system
Function	Controls movement of substances through hollow organs	Pumps blood	Moves the skeleton
Voluntary control	Normally no*	Normally no*	Yes

*Smooth and cardiac muscles normally contract without conscious control. In some cases, however, their contractions may be initiated or modified voluntarily. For example, heart rate can be voluntarily slowed after biofeedback training, and bladder contractions are initiated consciously.

Skeletal muscle, so named because it is used to move the skeleton, is also called **striated muscle**, because it has a striped appearance under the microscope (*striated* means "striped"). Most skeletal muscle is under voluntary, or conscious, control. It can produce contractions ranging from quick twitches (as in blinking) to powerful, sustained tension (as in carrying an armload of textbooks). **Cardiac muscle** is so named because it is located only in the heart. It is spontaneously active, initiating its own contractions, but it is influenced by nerves and hormones. Like skeletal muscle, cardiac muscle has a striped appearance under the microscope. **Smooth muscle**, as its name suggests, lacks the orderly striped appearance of skeletal and cardiac muscles. Smooth muscle lines the walls of the digestive tract and large blood vessels and produces slow, sustained contractions. These contractions are primarily involuntary; that is, they are not under conscious control.

The human body has some 700 different skeletal muscles, which make up 35% to 45% of its total weight. Our discussion of muscles begins with and emphasizes skeletal muscle, followed by a survey of cardiac and smooth muscle.

Skeletal Muscle Cell Structure and Function are Closely Linked

In most eukaryotic cells, motions such as shape changes, movement of organelles, and locomotion depend on microfilaments constructed of the protein **actin** (see Chapter 6) interacting with strands of another protein known as **myosin**. During movements, actin and myosin slide past one another, changing the shape of the cell. This evolutionarily ancient mechanism forms the basis of animal muscle cell function.

Skeletal muscles are attached to the skeleton by cords of connective tissue called **tendons**. Each muscle is also encased in connective tissue and consists of bundles of muscle cells, as well as blood vessels and nerves (Fig. 34-1). Individual muscle cells, called **muscle fibers**, are among the largest cells in the human body. Ranging from 10 to 100 micrometers in diameter (a bit smaller than the period at the end of this sentence), each muscle fiber runs the entire length of the muscle, which may be as long as 35 centimeters (about 14 inches) in a human thigh. Each muscle fiber, in turn, contains many **myofibrils**, contractile cylinders extending from one end of the fiber to the other. Each myofibril is surrounded by **sarcoplasmic reticulum**. Like the endoplasmic reticulum from which it is derived, the sarcoplasmic reticulum consists of flattened, membrane-enclosed compartments (Fig. 34-2a). The fluid within the sarcoplasmic reticulum stores high concentrations of calcium ions, which are crucial in muscle contraction. Surrounding each muscle fiber is a plasma membrane which periodically indents deeply into the muscle fiber. These indentations are called **T tubules**, and they pass very close to portions of the sarcoplasmic reticulum. This arrangement of T tubules and sarcoplasmic reticulum is crucial to the control of muscle contraction, as described later.

Within each myofibril is a beautifully precise arrangement of filaments of actin and myosin. The myofibrils are organized into subunits called **sarcomeres,** which are

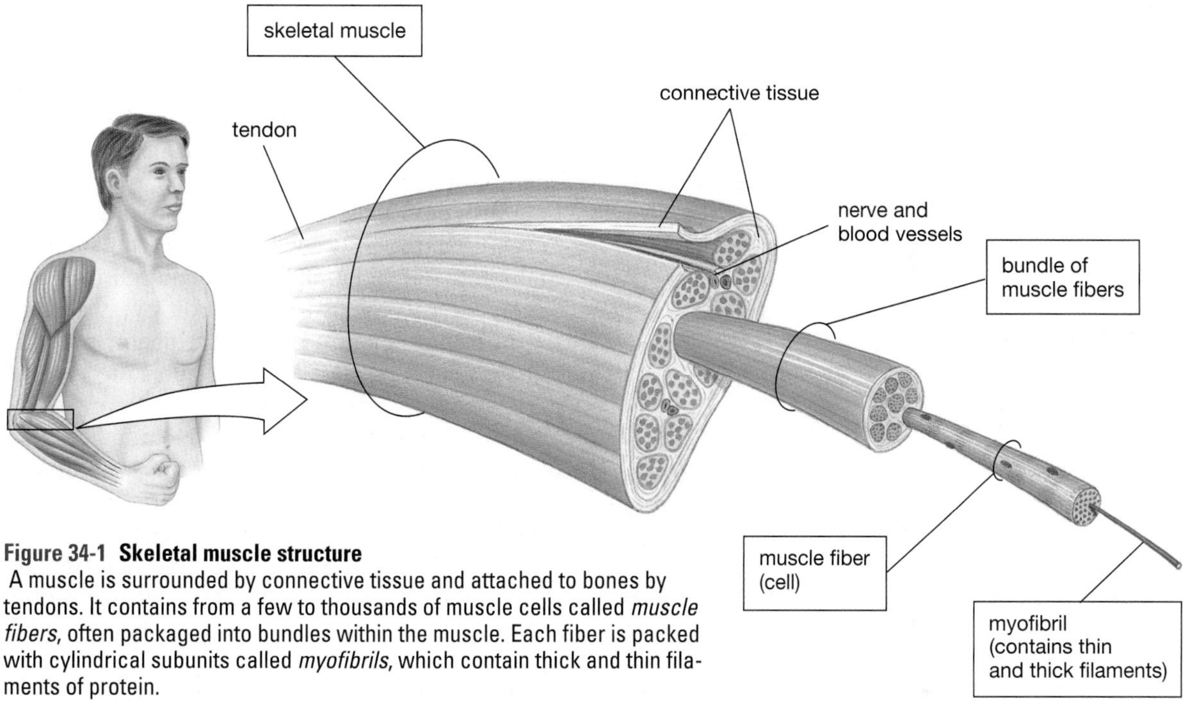

Figure 34-1 Skeletal muscle structure
A muscle is surrounded by connective tissue and attached to bones by tendons. It contains from a few to thousands of muscle cells called *muscle fibers*, often packaged into bundles within the muscle. Each fiber is packed with cylindrical subunits called *myofibrils*, which contain thick and thin filaments of protein.

MEDIATUTOR 34.1 Muscle Structure

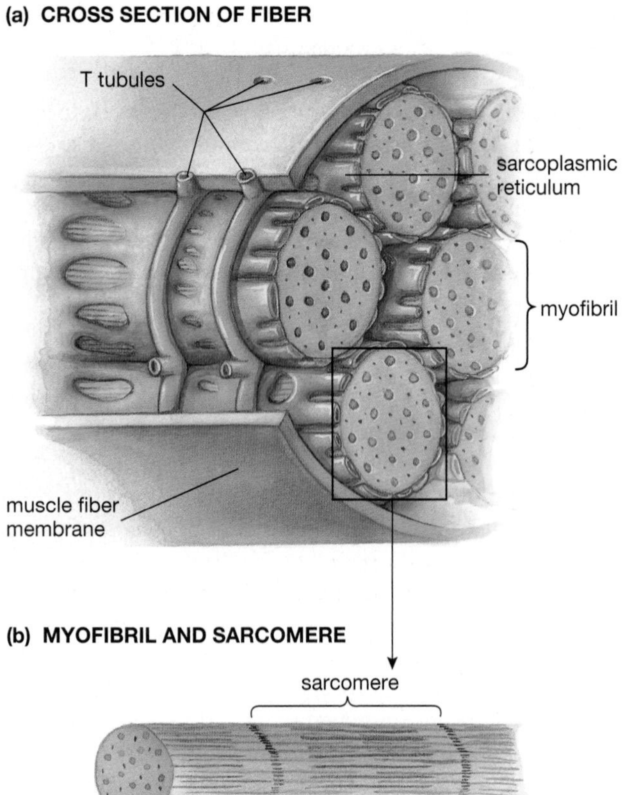

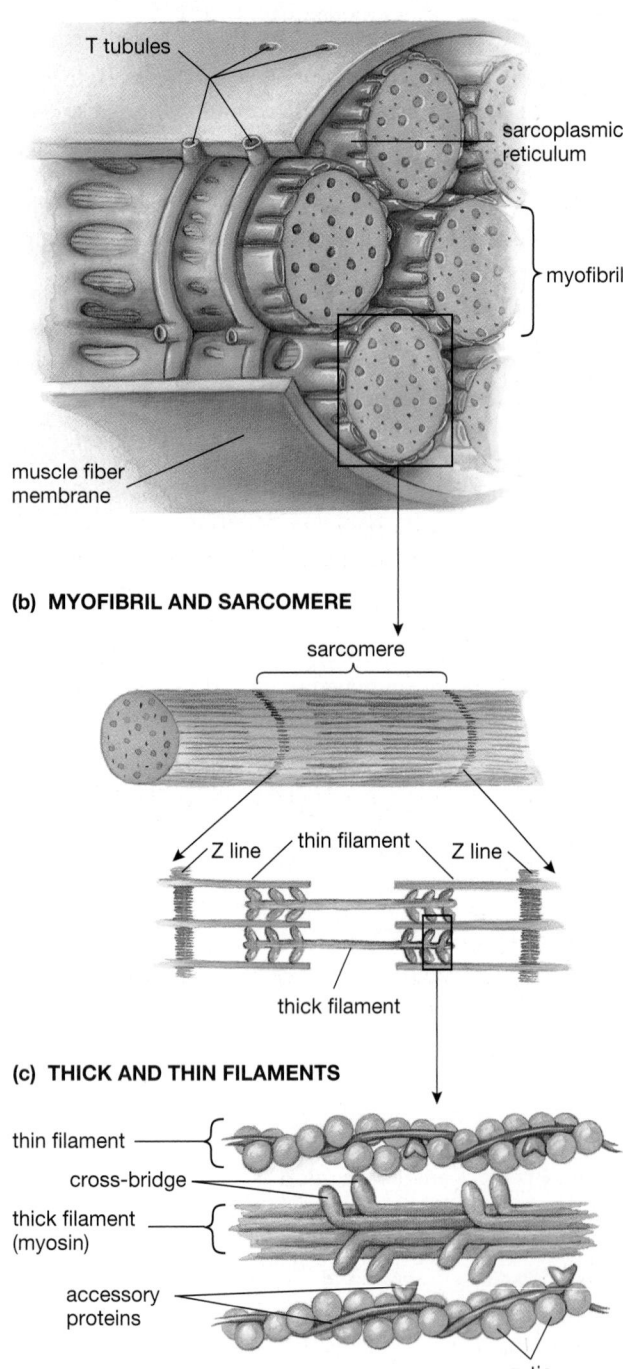

Figure 34-2 A skeletal muscle fiber
(a) Each muscle fiber is surrounded by plasma membrane that burrows inside, forming T tubules. The sarcoplasmic reticulum surrounds each myofibril within the muscle cell. ***(b)*** Each myofibril consists of a series of subunits called *sarcomeres,* attached end to end by proteins called Z lines. ***(c)*** Within each sarcomere are alternating thick and thin filaments, which can be connected by cross-bridges (projections of the myosin molecules that make up the thick filaments).

aligned end to end throughout the length of the myofibril (Fig. 34-2b). Within each sarcomere, actin molecules (in association with two accessory proteins) form the **thin filaments.** Suspended between the thin filaments are **thick filaments.** Thick filaments are composed of myosin protein, which is capable of linking temporarily to the thin filaments using small projections called **cross-bridges** (Fig. 34-2c). The thin filaments are attached to fibrous protein bands called **Z lines**, which separate adjacent sarcomeres. The regular arrangement of thick and thin filaments in all the myofibrils gives the muscle fiber its striped appearance (see Table 34-1).

Muscle Contraction Results from Thick and Thin Filaments Sliding Past One Another

Muscle contraction is controlled by a process that depends on the molecular structure of the thin filament. The actin protein that makes up most of the thin filament is formed from a double chain of subunits, resembling a twisted double strand of pearls. Each of the subunits has a binding site for a myosin cross-bridge. In a relaxed muscle cell, however, these binding sites are blocked by molecules of accessory proteins (see Fig. 34-2c). These accessory proteins prevent the myosin cross-bridges from attaching to the actin of the thin filament.

When a muscle contracts, the accessory proteins of the thin filament are moved aside, exposing the binding sites on the actin. As soon as the sites are exposed, myosin binds the actin, forming cross-bridges. Using energy from the splitting of adenosine triphosphate (ATP), the cross-bridges repeatedly bend, release, and reattach farther along, much like a sailor pulling in an anchor line hand over hand (Fig. 34-3a). The thin filaments are pulled past the thick filaments, shortening the sarcomere and contracting the muscle (Fig. 34-3b). Because the thick and thin filaments slide past one another during contraction, muscle contraction is described as using a *sliding-filament mechanism.*

Skeletal Muscle Contraction Is Controlled by the Nervous System

Motor neurons activate skeletal muscles at specialized synapses called **neuromuscular junctions** (Fig. 34-4). Neuromuscular junctions differ from most other synapses in two important ways: (1) They are always excitatory, never inhibitory. (2) In contrast to the summation of multiple synaptic inputs that is required to generate an action potential in a typical neuron, every action potential in a motor neuron elicits an action potential in a muscle fiber, causing all its sarcomeres to contract. The nervous system controls the strength and degree of muscle contraction by controlling the number of muscle fibers stimulated and the frequency of action potentials in each fiber. A single action potential doesn't cause a muscle cell to contract fully; this requires many action potentials in rapid succession. If rapid firing is prolonged, the muscle produces a sustained maximal contraction, such as you experience when you carry an armful of books like this one.

Most motor neurons form synapses on more than one muscle fiber. The group of fibers on which a single

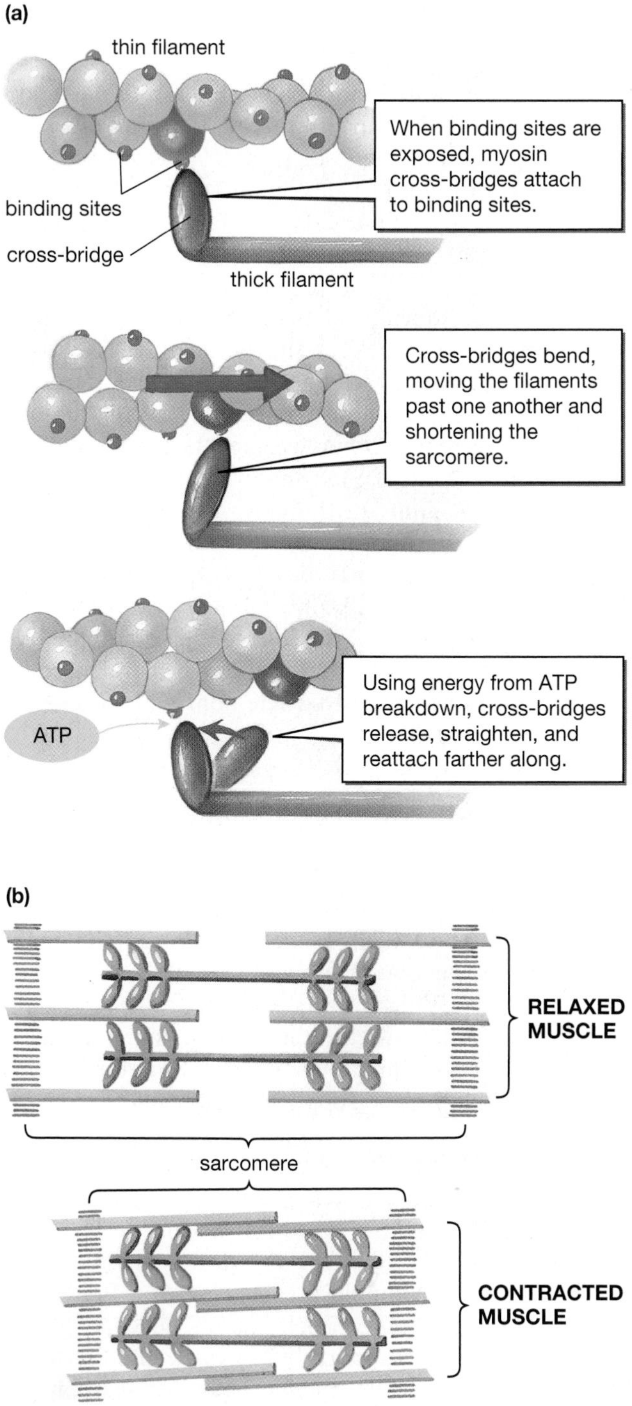

Figure 34-3 Muscle contraction
(a) Repeated cycles of cross-bridge attachment, bending, release, and reattachment result in muscle contraction. ***(b)*** Muscle contraction causes the thick and thin filaments to slide past one another toward the center of each sarcomere, shortening the muscle cell.

motor neuron forms synapses is called a **motor unit**. The number of muscle fibers in a motor unit varies from muscle to muscle. Muscles used for large-scale movement, such as the thigh muscles you'd use when climbing a ladder, or the muscles in your back that help maintain your posture, may have hundreds of muscle

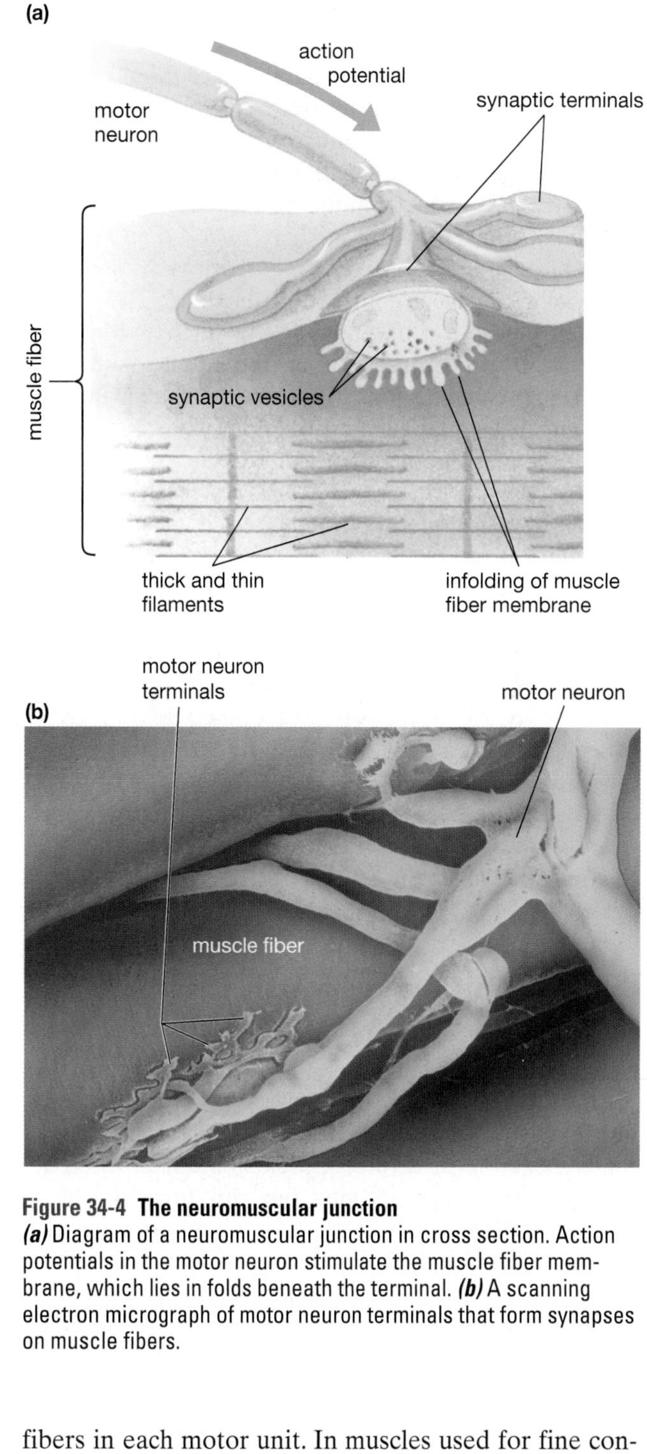

Figure 34-4 The neuromuscular junction
(a) Diagram of a neuromuscular junction in cross section. Action potentials in the motor neuron stimulate the muscle fiber membrane, which lies in folds beneath the terminal. ***(b)*** A scanning electron micrograph of motor neuron terminals that form synapses on muscle fibers.

fibers in each motor unit. In muscles used for fine control of small body parts, such as those of the lips, eyes, and tongue, only a few muscle cells may be stimulated by each motor neuron. Therefore, when a single motor neuron fires an action potential, that action potential can cause the contraction of a few muscle cells or of many, depending on the size of the motor unit.

Muscle Contraction Depends on the Availability of Calcium Ions

An action potential in the muscle cell travels into the cell's interior by passing down the T tubules (see Fig. 34-2a). On reaching the sarcoplasmic reticulum, the

MediaTutor 34.2 Muscle Movement

action potential causes calcium ions to be released from the interior of the sarcoplasmic reticulum and to flow into the cytoplasm surrounding the thick and thin filaments. Once in the cytoplasm, calcium ions bind to the small accessory protein (*troponin*) of the thin filament, changing its shape and pulling the larger protein (*tropomyosin*) away from the binding sites for myosin. As long as these binding sites are exposed and ATP is available, the cross-bridges repeatedly bind, bend, release, and reattach, contracting the muscle fiber. As soon as the action potential is over, active transport proteins in the sarcoplasmic reticulum membrane pump the calcium ions back inside their "holding area" in the sarcoplasmic reticulum. As a result, the calcium leaves the troponin, the accessory proteins return to a configuration that blocks the myosin binding sites, and the muscle fiber relaxes.

You've probably heard of rigor mortis, in which muscles become rigid (without contracting) some hours after death. This occurs because the muscle cells run out of ATP and can no longer pump calcium into the sacrcoplasmic reticulum. As the calcium leaks out, it binds to the accessory protein and allows cross-bridges to form. Without ATP, no contraction occurs, but the calcium remains present so the cross-bridges persist, preventing the muscle from being stretched out. Rigor mortis passes some 15 to 20 hours later, as the muscle cells begin to decompose.

Cardiac Muscle Powers the Heart

Cardiac muscle is located only in the heart. Cardiac muscle, like skeletal muscle, is striated owing to the regular arrangement of sarcomeres with their alternating thick and thin filaments. As with skeletal muscle, the contraction of cardiac muscle is induced when action potentials spread into the cell through the T tubules and cause a release of calcium from the sarcoplasmic reticulum. In contrast to skeletal muscle, calcium also enters the cardiac muscle cytoplasm from the extracellular fluid. Unlike skeletal muscle fibers, which contract in response to action potentials that originate in motor neurons, cardiac muscle fibers can initiate their own contractions. This quality is particularly well developed in the specialized cardiac muscle fibers of the *sinoatrial* (SA) *node,* which serves as the heart's pacemaker (see Chapter 27). Action potentials originating in the pacemaker are spread rapidly throughout the heart by specialized areas in which numerous gap junctions connect the membranes of adjacent muscle cells. Gap junctions allow action potentials to travel from one cell to the next, synchronizing their contractions (see Table 34-1). These connecting areas are absent in skeletal muscle, where each cell is individually stimulated by a branch of a motor neuron.

Smooth Muscle Produces Slow, Involuntary Contractions

Smooth muscle surrounds blood vessels and most hollow organs, including the uterus, bladder, and digestive tract. As its name suggests, smooth muscle lacks the regular arrangement of sarcomeres that characterizes skeletal and cardiac muscles (see Table 34-1). Smooth muscle generally produces either slow, sustained contractions (such as the constriction of arteries to elevate blood pressure during times of stress) or slow, wavelike contractions (such as the peristaltic waves that move food through the digestive tract). Like cardiac muscle cells, most smooth muscle cells are directly connected to one another by gap junctions, allowing synchronized contraction. Smooth muscle lacks sarcoplasmic reticulum; all the calcium needed for contraction flows in from the extracellular fluid during the action potential. Smooth muscle contraction may be initiated by stretching, by hormones, by nervous signals, or by some combination of these stimuli. Although contractions of the bladder can be initiated voluntarily, most smooth muscle contraction is under involuntary control.

2 What Does the Skeleton Do?

For most of us, the word *skeleton* conjures up the image of our own collection of bones, the skull smiling like a Halloween decoration. But skeletons are in fact as diverse as any other structure in the animal kingdom and need not even be made of bone. A **skeleton** can be broadly defined as a supporting framework for the body. Within the animal kingdom, skeletons come in three radically different forms: (1) *hydrostatic skeletons* (made of fluid), (2) *exoskeletons* (on the outside of the animal), and (3) *endoskeletons* (internal).

The **hydrostatic skeletons** of worms, mollusks, and cnidarians are the simplest, consisting of a fluid-filled sac (Fig. 34-5a). Fluid, which cannot be compressed, provides excellent support; but because fluid is formless, these animals rely on two layers of surrounding muscles in the body wall—one circular, the other longitudinal—to determine their shape. The wavelike movements of a burrowing earthworm, alternately extending to stringlike thinness and then fattening as it contracts, provide an excellent illustration of the flexibility of hydrostatic skeletons.

Exoskeletons (literally, "outside skeletons") encase the bodies of arthropods (such as spiders, crustaceans, and insects). Exoskeletons vary tremendously in thickness and rigidity, from the thin flexible covering of many insects and spiders to the armorlike covering of many crustaceans (Fig. 34-5b). All exoskeletons are

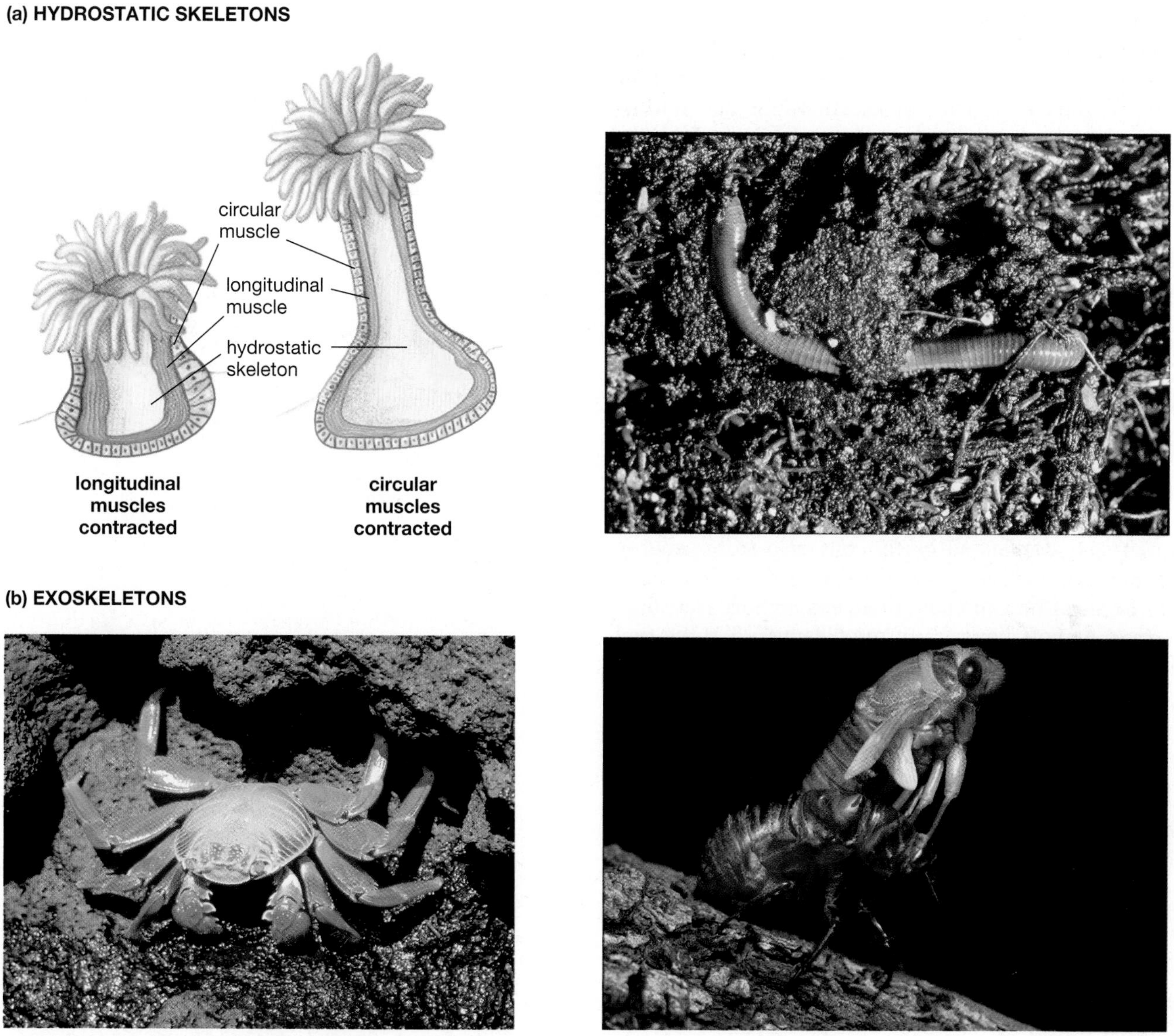

Figure 34-5 Not all skeletons are made of bone
(a) *Hydrostatic skeletons.* The skeleton of cnidarians, such as this sea anemone (art on left), worms such as this earthworm (photo on right), and many mollusks, is essentially a fluid-filled tube with soft walls. The tube is surrounded by two layers of muscles that are perpendicular to one another: The fibers of circular muscles form bands around the circumference of the tube, and the fibers of longitudinal muscles run lengthwise. Because fluids are incompressible and can take any shape, if the circular muscles contract, the animal will become long and thin; if the longitudinal muscles contract, it will become short and fat. ***(b)*** *Exoskeletons.* Arthropods, such as this Sally lightfoot crab (on left) and this cicada (on right, shown molting its exoskeleton), have armorlike skeletons on the outside of their bodies. Joints allow movement, produced by pairs of muscles that span the joint.

thin and flexible at the *joints* (the juncture of two bones), allowing complex and skillful movements such as those of a web-spinning spider. Exoskeletons and hydrostatic skeletons are discussed further in Chapter 22.

Endoskeletons, the internal skeletons of humans and other vertebrates, are found only in echinoderms and chordates (see Chapter 22). Although we think of internal skeletons as "the norm," these are actually the least common type of skeleton in the animal kingdom.

The Vertebrate Skeleton Serves Many Functions

The bony endoskeleton of humans and most vertebrates serves a wide variety of functions:

1. The skeleton provides a rigid framework that supports the body and protects the internal organs. The central nervous system (CNS), for example, is almost completely enclosed within the skull and vertebral column; the rib cage protects the lungs and the heart with its major blood vessels.

2. Vertebrates depend on bones for locomotion. Although muscle contractions provide the power, skeletal elements provide the structures that actually move the animal. Natural selection for efficient locomotion has produced the wonderfully designed wings, limbs, fins, and other complex skeletal structures that allow vertebrates to fly, run, swim, and even slam dunk a basketball.
3. Bones produce red blood cells, white blood cells, and platelets (see Chapter 27). In adults, these cells of the circulatory system are produced by *red bone marrow,* located in porous areas of bone in the sternum (breastbone), ribs, upper arms and legs, and hips.
4. Bone serves as a storage site for calcium and phosphorus. Bone contains 99% of the calcium and 90% of the phosphorus in the human body. It absorbs and releases these minerals as needed, maintaining a constant concentration in the blood. *Yellow bone marrow*, dominated by fat cells, also stores energy reserves.
5. The skeleton even participates in sensory transduction. As you may recall from Chapter 33, three tiny bones of the middle ear (hammer, anvil, and stirrup) transmit the sound vibrations between the eardrum and the cochlea.

The 206 bones of the human skeleton can be placed in two categories: (1) the axial skeleton and (2) the appendicular skeleton. The **axial skeleton**, whose bones form the axis of the body, includes the bones of the head, vertebral column, and rib cage. The **appendicular skeleton**, whose bones form the appendages (extremities) and their attachments to the axial skeleton, includes the pectoral (shoulder) and pelvic (hip) girdles and the bones of the arms, legs, hands, and feet (Fig. 34-6).

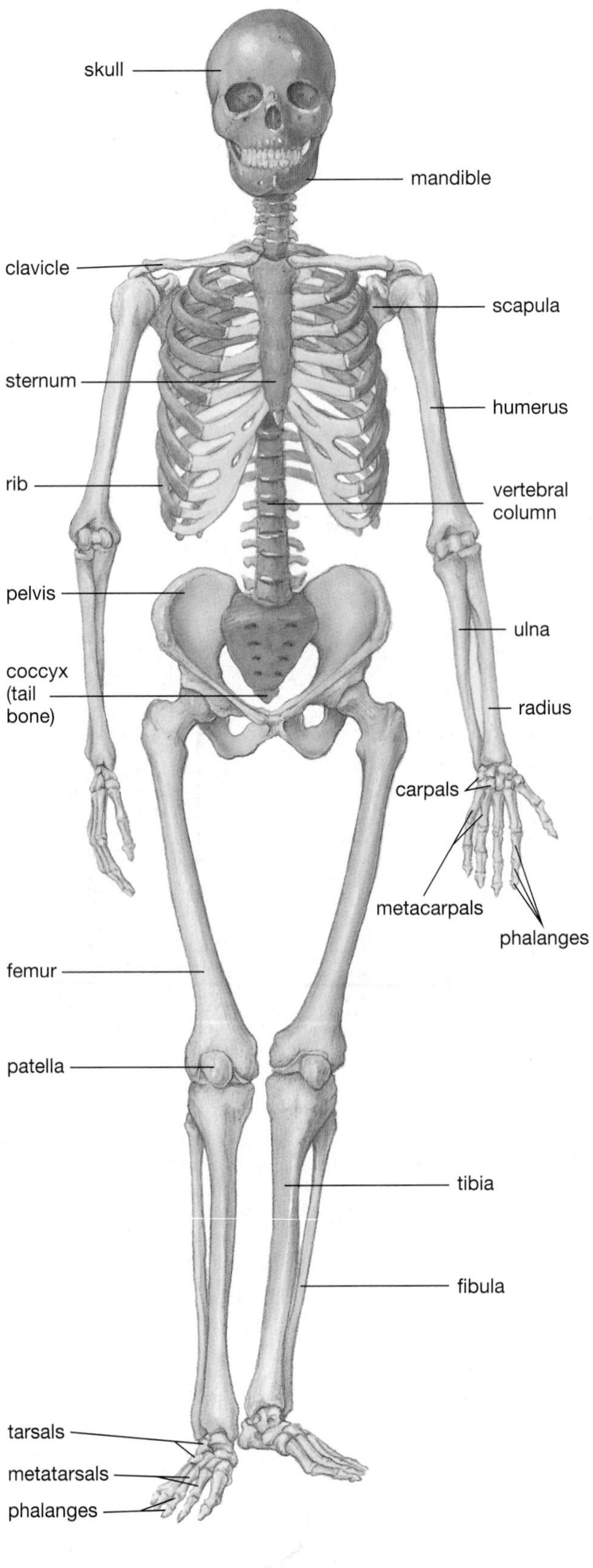

Figure 34-6 The human skeleton
The human skeleton, showing the axial skeleton (tinged in blue-gray) and the appendicular skeleton (bone color).

3 Which Tissues Compose the Vertebrate Skeleton?

The vertebrate skeleton is composed primarily of two types of tissue: *cartilage* and *bone*. Both cartilage and bone are rigid tissues that consist of living cells embedded in a matrix of a protein called **collagen** (see Chapter 26).

Cartilage Provides Flexible Support and Connections

Cartilage plays many roles in the human skeleton. For example, during the development of the embryo, the skeleton is first formed from cartilage that is only later replaced by bone (Fig. 34-7). Cartilage also covers the ends of bones at joints, supports the flexible portion of the nose and external ears, connects the ribs to the sternum (breastbone), and provides the framework for the

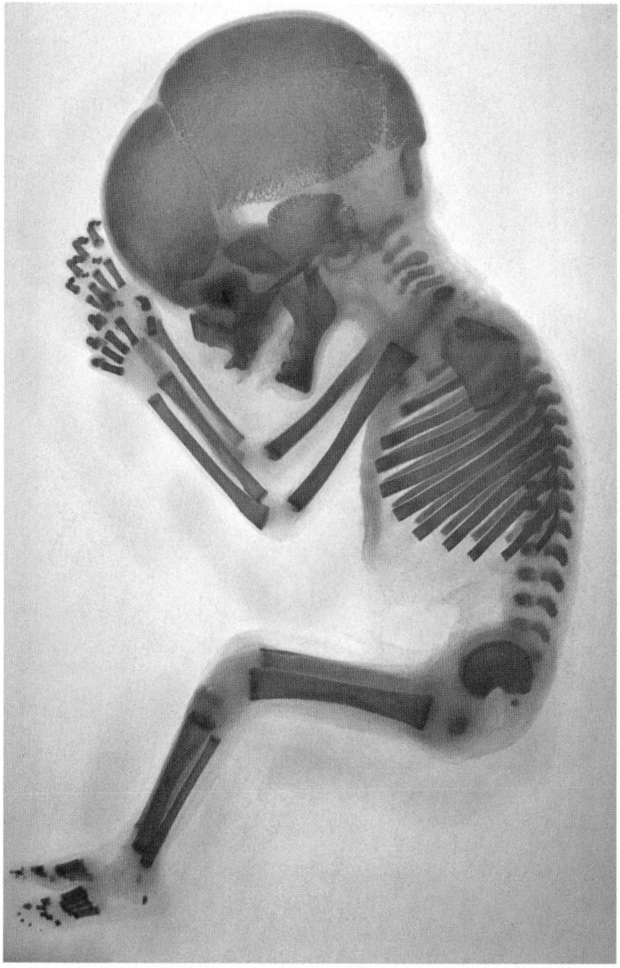

Figure 34-7 Bone replaces cartilage during development
In this 16-week-old human fetus, bone is stained magenta. The clear areas at the wrists, knees, ankles, elbows, and breastbone show where cartilage will later be replaced by bone.

larynx, trachea, and bronchi of the respiratory system. In addition, it forms tough, shock-absorbing pads that cushion the knee joints and that form the **intervertebral discs** between the vertebrae of the backbone.

The living cells of cartilage are called **chondrocytes**. These cells secrete a flexible, elastic, nonliving matrix of collagen that surrounds them and forms the bulk of the cartilage (Fig. 34-8). No blood vessels penetrate cartilage; to exchange wastes and nutrients, chondrocytes must rely on the gradual diffusion of materials through the collagen matrix. As you might predict, cartilage cells have a very slow metabolic rate; damaged cartilage repairs itself very slowly, if at all.

Bone Provides a Strong, Rigid Framework for the Body

Bone is the most rigid form of connective tissue. Although bone resembles cartilage, the collagen fibers of bone are hardened by deposits of the mineral *calcium*

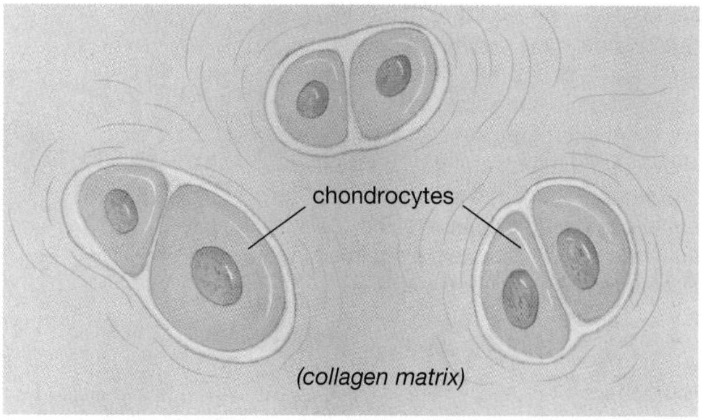

Figure 34-8 Cartilage
In cartilage, the chondrocytes, or cartilage cells, are embedded in an extracellular matrix of the protein collagen, which they secrete.

phosphate. Bones, such as those supporting your arms and legs, consist of a hard outer shell of **compact bone**, with **spongy bone** in the interior (Fig. 34-9). Compact bone is dense and strong and provides an attachment site for muscle. Spongy bone is lightweight, rich in blood vessels, and highly porous. Bone marrow, where blood cells form, is found in cavities of spongy bone. In contrast to cartilage, bone is well supplied with blood capillaries.

Three types of cells are associated with bone: (1) **osteoblasts** (bone-forming cells), (2) **osteocytes** (mature bone cells), and (3) **osteoclasts** (bone-dissolving cells). Early in development, when bone is replacing cartilage, osteoclasts invade and dissolve the cartilage; then osteoblasts replace it with bone.

As bones grow, osteoblasts form a thin layer covering the outside of the bone. The osteoblasts secrete a hardened matrix of bone and gradually become entrapped within it. They then stop secreting matrix and become mature osteocytes. Osteocytes are nourished by nearby capillaries and are connected to other osteocytes by thin extensions that the bone cells send out through narrow channels in the bone. Although unable to produce more bone, osteocytes may secrete substances that control the continuous remodeling of bone.

Bone Remodeling Allows Skeletal Repair and Adaptation to Stresses

Each year 5% to 10% of all the bone in your body is dissolved away and replaced, a process called *bone remodeling.* This process allows your skeleton to alter its shape subtly in response to the demands placed on it. For example, bones that carry heavy loads or are subjected to extra stress become thicker to provide more strength and support. Archeologists excavating the skeletons of individuals buried in volcanic ash at Pompeii were able to identify archers because the bones of their right and left arms differed considerably in thickness. Normal stresses are actually a major factor in maintaining bone

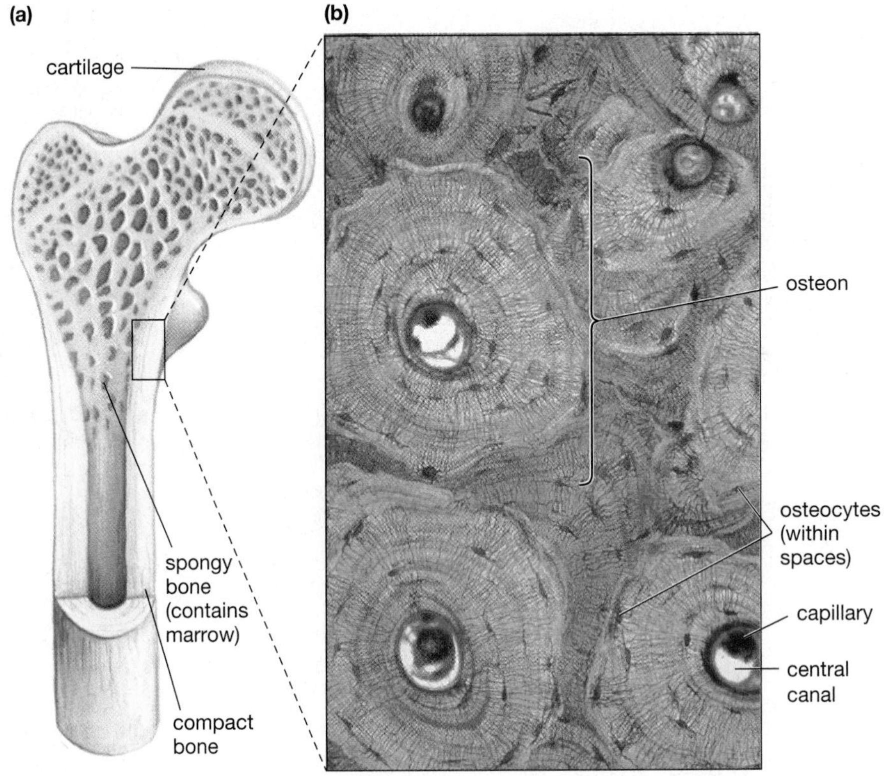

Figure 34-9 The structure of bone ***(a)*** A typical bone, such as those in the arms and legs, is made of an outer layer of compact bone and spongy bone inside. For simplicity, blood vessels are not shown. ***(b)*** Osteons are clearly visible in this micrograph. Each includes a central canal containing a capillary. The capillary nourishes the osteocytes, embedded in the concentric rings of bone material.

strength. The bones of an arm or leg that is immobilized in a cast rapidly lose significant amounts of calcium.

Bone remodeling is the result of the coordinated activity of the osteoclasts that dissolve bone and the osteoblasts that rebuild it. Osteoclasts cling to the bone surface, secreting acids and enzymes that dissolve the hard matrix. Working in small groups, osteoclasts tunnel into the bone, creating channels. These channels are invaded by capillaries and by osteoblasts. The osteoblasts fill the channel with concentric deposits of new bone matrix, leaving only a small opening for the capillary. As a result of this process, hard bone is made up of tightly packed units called **osteons** (also known as *Haversian systems*), each consisting of concentric layers of bone with embedded osteocytes. The concentric deposits surround a *central canal*, through which a capillary passes (see Fig. 34-9). (Osteoclasts and osteocytes also play a crucial role in the repair of bone fractures like the one in our opener photo, described in "Health Watch: How Bones Heal.") The continuous turnover of bone also allows the body to maintain constant levels of calcium in the blood; calcium from bones is retained in the blood if blood calcium levels drop but is returned to bone if blood calcium is adequate or high. This process is regulated by hormones; calcitonin and parathormone cause bones to absorb calcium from the blood and release calcium into the blood, respectively (see Chapter 32). Early in life, the activity of osteoblasts outpaces that of osteoclasts, and bones become larger and thicker as a child grows. In the aging body, however, the balance of power may shift to favor osteoclasts, and bones tend to become more fragile as a result. (See "Health Watch: Osteoporosis—When Bones Become Brittle.") Recently, molecular biologists, puzzling over the function of a gene they had discvered, engineered mice with extra copies of the gene. They were surprised to discover that these mice had unusually thick bones. Further studies revealed that the gene coded for a protein (dubbed "OPG") that binds to and inactivates a second protein (OPG-ligand). This second protein stimulates ostoeclasts that break down bone. Extra OPG protein can shift the balance between bone destruction and bone buildup in favor of regeneration. Some day this may be used by physicians to promote healing of broken bones or to prevent osteoporosis.

4 How Does the Body Move?

In addition to providing support for the body, the skeleton facilitates movement by providing a framework that muscles can move. Movement of the skeleton is accomplished by the action of pairs of **antagonistic muscles**: One muscle actively contracts, causing the other to be passively extended (Fig. 34-10). Antagonistic muscles alter the configuration of the skeleton by causing movement around joints (in exoskeletons and vertebrate endoskeletons) or by altering the shape of the internal fluid (in hydrostatic skeletons; see Fig. 34-5).

Health Watch
How Bones Heal

Bones are strong, but they're not indestructible. All too many of us have experienced the trauma and pain of a broken bone. We cringe at each recollection of the ill-fated mishap, the audible cracking sound, the dawning realization that the painful swelling and oddly bent-looking limb necessitate a trip to the emergency room. There, the attending physician coaxes the damaged bone back into its proper orientation and immobilizes it with a cast or splint. The rest of the healing process is up to the body's own repair mechanisms. Over the next 6 weeks or so, the body orchestrates a systematic restoration of the bone's strength and integrity.

The healing process begins when the break is surrounded by a large blood clot from vessels ruptured during the injury (Fig. E34-1a). Phagocytic cells and osteoclasts in the blood ingest and dissolve cellular debris and bone fragments. The fracture ruptures the *periosteum*, a thin layer of connective tissue that normally surrounds the bone and that is rich in capillaries, osteoblasts, and osteoblast-forming cells. The osteoblasts, in conjunction with cartilage-forming cells, then secrete a *callus*, a porous mass of bone and cartilage that surrounds the break (Fig. E34-1b). The callus replaces the original blood clot and holds the ends of the bones together while remodeling processes re-form the original shape of the bone. Once the callus is in place, osteoclasts, osteoblasts, and capillaries invade it. Nourished by the capillaries, osteoclasts break down cartilage while osteoblasts add new bone (Fig. E34-1c). Finally, osteoclasts remove excess bone, restoring the bone's original shape, but often leaving a slight thickening (Fig. E34-1d).

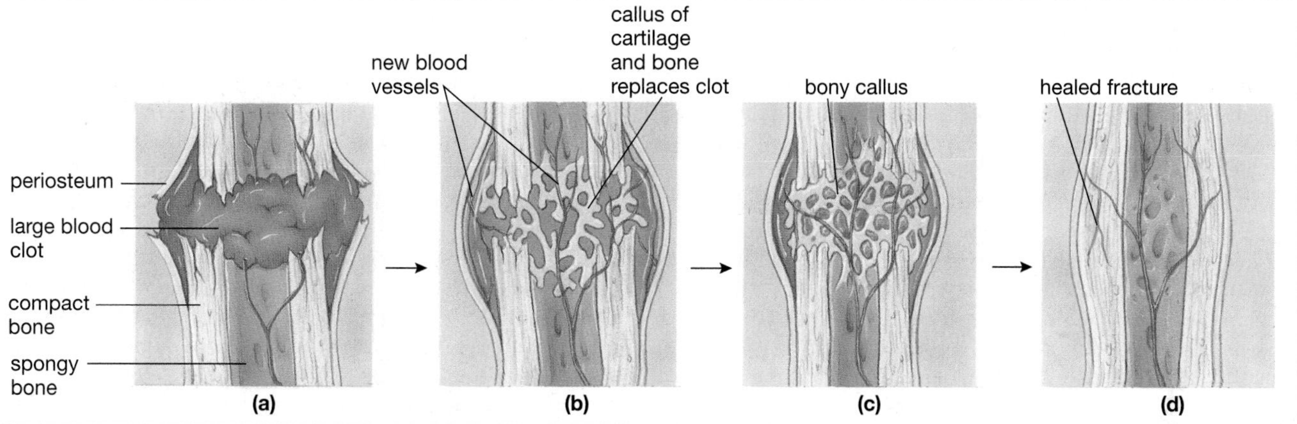

Figure E34-1 The steps in bone repair

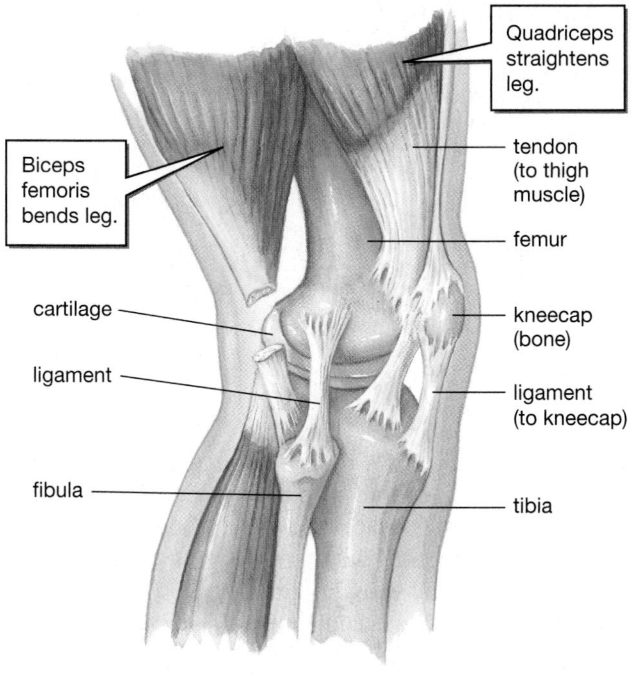

Figure 34-10 A hinge joint
The human knee—a hinge joint—showing antagonistic muscles (here, the biceps femoris and the quadriceps of the thigh), tendons, and ligaments. (The tendon of the biceps femoris has been cut for viewing purposes.) The complexity of this joint, coupled with the extreme stresses placed on it during activities such as jumping, running, or skiing, make it very susceptible to injury.

Health Watch
Osteoporosis—When Bones Become Brittle

As a human grows and matures, bone density increases steadily, reaching a peak at about age 35. After this point, however, the activity of osteoclasts exceeds that of osteoblasts and bone density begins a slow, natural decline. Some bone loss is normal, but in people with the insidious (though preventable) condition known as **osteoporosis** (literally, "porous bones"; Fig. E34-2a), the loss is sufficient to weaken the bones, making them vulnerable to fractures and deformities. As many as 28 million Americans, most of them women past menopause, have osteoporosis. In many cases, the vertebrae of individuals with osteoporosis compress, causing a hunch-backed appearance (Fig. E34-2b). In extreme cases, simple activities such as lifting a shopping bag, opening a window, or sneezing can break a bone. Of women living to age 85, nearly one-third will fracture a hip weakened by osteoporosis. As a result of hip fractures, the elderly can become much less self-sufficient and inactive, and one-quarter of them die within 6 months of complications such as pneumonia.

Women are eight times as likely to suffer from osteoporosis as men. Why? One reason is that, to start with, the bones of women are about 30% less massive than those of men, so they can less afford to lose bone. Another factor is dietary calcium, which tends to be lower in women's diets than in men's. Two-thirds of women between the ages of 18 and 30 get less than the RDA (Recommended Daily Allowance) of calcium, and women tend to consume even less calcium as they become older. As a result, when women reach the age when bone loss begins naturally, their bones may already be more fragile than they should be. Another factor unique to women is the role of the hormone estrogen. In women, estrogen stimulates osteoblasts and helps maintain bone density. After menopause, when estrogen production drops dramatically, women may lose 3% to 5% of their bone mass each year for several years. Half of all women over age 65 are estimated to have some degree of osteoporosis. Alcoholism and smoking also contribute to bone loss and osteoporosis.

Bones thrive on moderate stress, but older people tend to be less active. Being inactive or bedridden (or being weightless, as astronauts have discovered) results in rapid loss of bone calcium. Even in elderly people, weight-bearing exercise such as walking or dancing can reverse bone loss and even increase bone mass.

Fortunately, much of the pain, incapacitation, and expense (estimated at $14 billion per year) caused by fractures due to osteoporosis can be prevented. The best way to prevent osteoporosis is a combination of regular exercise and adequate dietary or supplemental calcium and vitamin D (which is crucial to the proper metabolism of calcium). These steps will ensure that bone mass is as high as possible before natural, age-related losses begin and will also minimize such losses in old age. Some women, in consultation with their physicians, choose hormone-based therapy to help maintain bone density. The hormone calcitonin, administered as a nasal spray, has beneficial effects on bone deposition. A new class of drugs called "selective estrogen receptor modulators," was approved by the FDA in late 1997. These molecules mimic estrogen's effects on the skeletal system while blocking estrogen's effects on breast and uterine tissues. The newly discovered protein OPG is also in clinical trials to prevent bone loss in postmenopausal women. None of these treatments can actually cure osteoporosis, so the best remedy for the disease remains prevention.

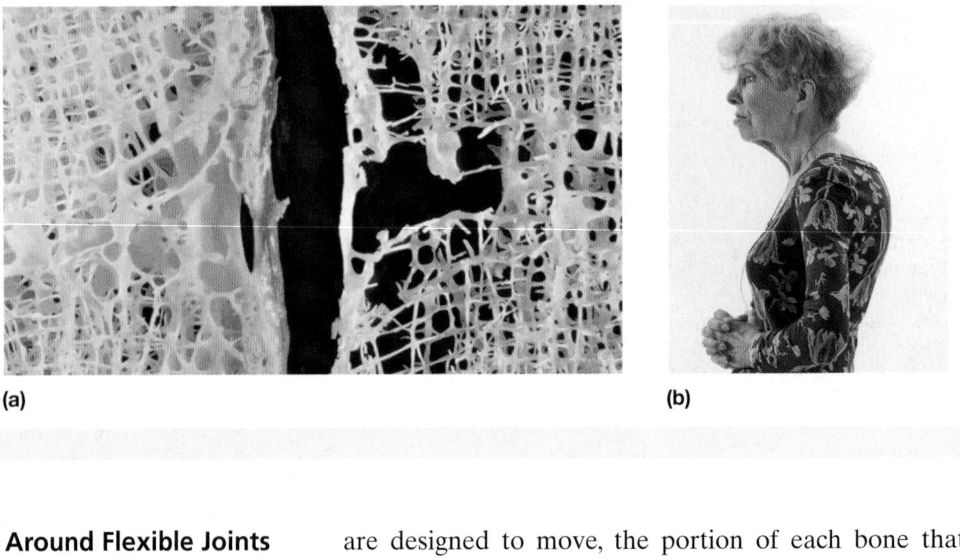
(a) (b)

Figure E34-2 Osteoporosis ***(a)*** Cross section of (left) a normal bone compared with (right) a bone from a woman with osteoporosis. ***(b)*** The devastating effects of osteoporosis extend beyond the obvious deformities, such as a hunch-backed appearance. Its victims are also at high risk for bone fractures.

Muscles Move the Skeleton Around Flexible Joints

In vertebrates, bones act as levers that can be moved by skeletal muscles to which they are attached. The bones typically move around **joints**, the points at which two bones meet. Not all joints are movable, but in those that are designed to move, the portion of each bone that forms the joint is coated with a layer of cartilage, whose smooth, resilient surface allows the bone surfaces to slide past one another during movement. On either side of a joint, skeletal muscles are attached to bones by bands of tough, fibrous connective tissue called **tendons**.

The bones themselves are joined at joints by bands of fibrous connective tissue called **ligaments** (see Fig. 34-10).

Most skeletal muscles are arranged in antagonistic pairs on opposite sides of a joint (see Fig. 34-10). When one of the muscles in a pair contracts, it moves a bone and simultaneously stretches the opposing muscle. In the most common types of joints, skeletal muscles span the joint; their contraction moves the bone on one side of the joint while the bone on the other side of the joint remains in a fixed position. These joints, including the elbows, knees, and finger joints, are called **hinge joints**. Like a hinged door, these joints are movable in only two dimensions.

In hinge joints, pairs of muscles lie in roughly the same plane as the joint, and the members of the muscle pair are known as the **flexor** and the **extensor**. One end of each muscle, called the **origin**, is fixed to a relatively immovable bone on one side of the joint; the other end, the **insertion**, is attached to a mobile bone on the far side of the joint. When the flexor muscle contracts, it bends the joint; when the extensor muscle contracts, it straightens the joint. In Figure 34-10, for example, contraction of the biceps femoris bends the leg at the knee, while contraction of the quadriceps straightens it. Thus, alternate contractions of flexor and extensor muscles cause the movable bone to pivot back and forth at the joint.

Other joints, such as those of the hip and shoulder, are **ball-and-socket joints**, in which the rounded end of one bone fits into a hollow depression in another (Fig. 34-11). Ball-and-socket joints allow movement in several directions: Simply compare the wide-ranging swinging of your arm or upper leg with the limited bending of your knee or elbow. The range of motion in ball-and-socket joints is made possible by at least two pairs of muscles, oriented at right angles to each other, that provide flexibility of movement.

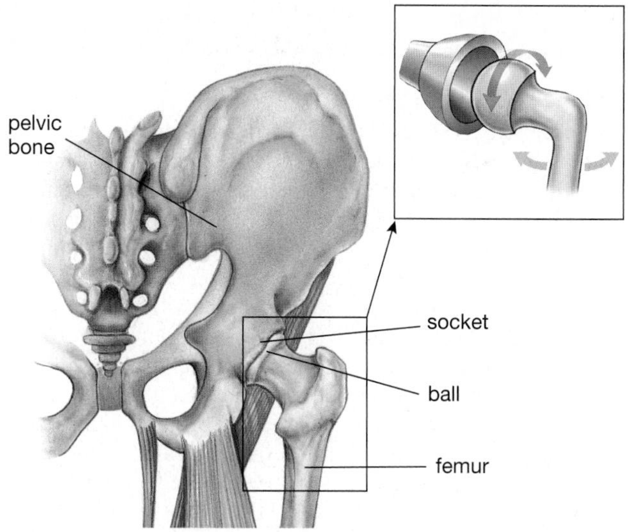

Figure 34-11 A ball-and-socket joint
The human hip—a ball-and-socket joint—consists of a rounded end (the ball; seen at the end of the femur) that fits into a cuplike depression (the socket; seen in the bones of the pelvis). This arrangement permits rotational movement.

CASE STUDY REVISITED

Healing Broken Bones

Margaret is 58 years old. Her physician, looking at X-rays taken through the cast after 6 weeks, is disappointed with the rate at which her bones are healing. She decides to try a new tactic: stimulating the bone to regenerate using an electric field. She has Margaret place her arm inside a device lined with coils of wire. Flipping a switch sends rapid pulses of electric current through the wire, painlessly creating an electromagnetic field around the injured bones. The idea behind this device originated in experiments conducted in the 1950s. These studies revealed that the crystalline structure of the bone itself would produce tiny electric currents when the bones were bent very slightly, as would occur regularly during weight-bearing exercise. The researchers hypothesized that these minuscule electric currents might be responsible for the increase in bone mass that occurs with exercise. This, in turn led to the concept that externally imposed electric currents might stimulate bone growth, a therapy that has been used with some success for more than two decades. After three more weeks of regular electric field treatments, Margaret's physician was satisfied that the bones had knit nicely, and she removed the cast. Brushing away the layers of dead skin that had accumulated during nine weeks under the cast, Margaret gazed with disappointment at a pale, shrunken, weak-looking appendage that she hardly recognized as her arm. Her doctor reassured her, urging her to exercise with hand weights and gradually increase the amount of weights and the number of repetitions.

Based on knowledge gained from this chapter, identify some reasons why Margaret's bones may not have healed as fast as your bones might heal. Explain why her arm looked shrunken, and why her doctor gave her good advice. What might Margaret do in the future to reduce her risk of another fracture?

Summary of Key Concepts

1) How Do Muscles Work?

Vertebrate muscles can only actively contract. Skeletal muscle fibers (muscle cells) consist of subunits—myofibrils—surrounded by the sarcoplasmic reticulum. Sarcomeres within each myofibril are bounded by Z lines and include alternating thick filaments of myosin and thin filaments of actin and two accessory proteins.

When stimulated by motor neurons at neuromuscular junctions, skeletal muscles produce action potentials. Calcium, stored in the sarcoplasmic reticulum, is released when an action potential invades the muscle fiber via the T tubules. Calcium causes the accessory proteins to move off myosin binding sites on the actin molecules of thin filaments. When the binding sites are exposed, myosin cross-bridges bind to them. Using ATP, the cross-bridges bend, release, and reattach, sliding the filaments past each other and shortening the muscle fiber. Both the strength and the degree of muscle contractions are determined by the number of muscle fibers stimulated and the frequency of action potentials in each fiber. Rapid firing can result in maximum contraction.

Cardiac, or heart, muscle also consists of sarcomeres that contain alternating thick and thin filaments. Its cells tend to contract rhythmically and spontaneously, but these contractions are synchronized by electrical signals produced by specialized muscle fibers in the sinoatrial node. Cardiac muscle fibers are interconnected electrically by gap junctions located between cells, allowing coordinated contraction.

Smooth muscle lacks organized sarcomeres, but like cardiac muscle, its cells are electrically coupled by gap junctions. Smooth muscle surrounds hollow organs (uterus, digestive tract, bladder) and blood vessels, producing slow, sustained or rhythmic contractions, which are usually involuntary.

2) What Does the Skeleton Do?

Three types of skeletons are found in animals: Hydrostatic skeletons (in cnidarians, mollusks, and worms) consist of fluid confined in a chamber and surrounded by muscle. Exoskeletons, found in arthropods, are outer coverings with flexible joints. Endoskeletons, including the bony vertebrate skeleton, are found in both echinoderms and chordates.

The vertebrate skeleton provides support for the body, attachment sites for muscles, and protection for internal organs. Red blood cells, white blood cells, and platelets form in the marrow of bones. Bone acts as a storage site for calcium and phosphorus. The axial skeleton includes the skull, vertebral column, and rib cage. The appendicular skeleton consists of the pectoral and pelvic girdles and the bones of the arms, legs, hands, and feet.

3) Which Tissues Compose the Vertebrate Skeleton?

Cartilage is located at the ends of bones and forms pads in the knee joints and the intervertebral discs. It also supports the nose, ears, and respiratory passages. During embryological development, cartilage is the precursor of bone. Cartilage is formed by chondrocytes, which surround themselves with a matrix of fibrous collagen.

Bone is formed by osteoblasts, which secrete a collagen matrix that becomes hardened by calcium phosphate. A typical bone consists of an outer shell of compact, hard bone, to which muscles are attached, and inner spongy bone, which provides a space for bone marrow. Remodeling of bone occurs continuously. Osteoclasts tunnel through the bone by means of acids and enzymes. Nourishing capillaries invade the tunnels, and osteoblasts fill the space with concentric layers of new bone, leaving a small central canal for the capillaries. This process produces osteons. Osteoblasts trapped within the bone are called osteocytes.

4) How Does the Body Move?

Skeletal muscles form antagonistic pairs that move the skeleton. In the vertebrate skeleton, movement occurs around joints, where bones are joined by ligaments. Muscles attach to bones on either side of the joint by tendons. The contraction of one muscle bends the joint and straightens its antagonistic muscle. At hinge joints, muscles are attached to the immovable bone at their origins; their insertions attach to the mobile bone. The contraction of the flexor muscle bends the joint; the contraction of its antagonistic extensor straightens it.

Key Terms

actin *p. 701*
antagonistic muscles *p. 708*
appendicular skeleton *p. 706*
axial skeleton *p. 706*
ball-and-socket joint *p. 711*
cardiac muscle *p. 701*
cartilage *p. 706*
chondrocyte *p. 707*
collagen *p. 706*
compact bone *p. 707*
cross-bridge *p. 702*
endoskeleton *p. 705*
exoskeleton *p. 704*
extensor *p. 711*
flexor *p. 711*
hinge joint *p. 711*
hydrostatic skeleton *p. 704*
insertion *p. 711*
intervertebral disc *p. 707*
joint *p. 710*
ligament *p. 711*
motor unit *p. 703*
muscle fiber *p. 701*
myofibril *p. 701*
myosin *p. 701*
neuromuscular junction *p. 702*
origin *p. 711*
osteoblast *p. 707*
osteoclast *p. 707*
osteocyte *p. 707*
osteon *p. 708*
osteoporosis *p. 710*
sarcomere *p. 701*
sarcoplasmic reticulum *p. 701*
skeletal muscle *p. 701*
skeleton *p. 704*
smooth muscle *p. 701*
spongy bone *p. 707*
striated muscle *p. 701*
tendon *p. 701*
thick filament *p. 702*
thin filament *p. 702*
T tubule *p. 701*
Z line *p. 702*

Thinking Through the Concepts

Multiple Choice

1. *Thin filaments in myofibrils consist of*
   a. actin and accessory proteins
   b. sarcomeres
   c. cross-bridges
   d. Z lines
   e. myosin
2. *The deep infoldings of muscle fiber membranes that conduct action potentials are called*
   a. sarcoplasmic reticula
   b. Z lines
   c. myofilaments
   d. T tubules
   e. sarcomeres
3. *The force of muscle contraction depends on the*
   a. number of muscle fibers stimulated
   b. number of motor units stimulated
   c. frequency of action potentials in each motor unit
   d. frequency of action potentials in each muscle fiber
   e. all of the above
4. *Smooth muscle fibers can be distinguished from striated ones because smooth fibers*
   a. contract more rapidly
   b. lack regular arrangements of sarcomeres
   c. lack gap junctions
   d. contain only actin filaments
   e. contain sarcoplasmic reticulum
5. *Bone-dissolving cells are called*
   a. chondrocytes
   b. osteoblasts
   c. osteoclasts
   d. osteocytes
   e. erythroblasts
6. *Osteons contain all of the following EXCEPT*
   a. blood vessels
   b. intervertebral discs
   c. osteocytes
   d. calcium phosphate crystals
   e. concentric layers of bone

? Review Questions

1. Sketch a relaxed muscle fiber containing a myofibril, sarcomeres, and thick and thin filaments. How would a contracted muscle fiber look by comparison?
2. Describe the process of skeletal muscle contraction, beginning with an action potential in a motor neuron and ending with the relaxation of the muscle. Your answer should include the following words: *neuromuscular junction*, *T tubule*, *sarcoplasmic reticulum*, *calcium*, *thin filaments*, *binding sites*, *thick filaments*, *sarcomere*, *Z line*, and *active transport*.
3. Explain the following two statements: Muscles can only actively contract; muscle fibers lengthen passively.
4. What are the three types of skeletons found in animals? For one of these, describe how the muscles are arranged around the skeleton and how contractions of the muscles result in movement of the skeleton.
5. Compare the structure and function of the following pairs: spongy and compact bone, smooth and striated muscle, and cartilage and bone.
6. Explain the functions of osteoblasts, osteoclasts, and osteocytes.
7. How is cartilage converted to bone during embryonic development? Where is cartilage located in the body, and what functions does it serve?
8. Describe a hinge joint and how it is moved by antagonistic muscles.

Applying the Concepts

1. Discuss some of the problems that would result if the human heart were made of skeletal muscle instead of cardiac muscle.
2. Muscle fibers in individuals with Duchenne muscular dystrophy (DMD) lack a protein called *dystrophin*, which normally helps control calcium release from sarcoplasmic reticulum. Lack of dystrophin leads to a constant leaking of calcium ions, which activates an enzyme that dissolves muscle fibers. The gene that causes DMD is inherited as a sex-linked recessive gene. Women with this gene have a 50% chance of passing the disease to their male children and a 50% chance of passing the gene to their female children. Afflicted children gradually become unable to walk and die of respiratory problems as young adults. Recently tests have been developed that allow a woman to determine if she is a carrier and if her fetus has inherited the gene. What factors would make a woman a candidate for this test? If a woman discovers she carries this gene, what are her options with regard to having children? Discuss the ethical implications of these various options.
3. Myasthenia gravis is caused by the abnormal production of antibodies that bind to acetylcholine receptors on muscle cells and that eventually destroy the receptors. The disease causes muscles to become flaccid, weak, or paralyzed.

Drugs, such as neostigmine, that inhibit the action of acetylcholinesterase (an enzyme that breaks down acetylcholine) are used to treat myasthenia gravis. How does neostigmine restore muscle activity?

4. Some insects would have a tough time flying if one nerve impulse was required for each muscle contraction. Gnats, for example, may beat their wings 1000 times a second. At such high frequencies, contraction is "myogenic," originating from the stretching caused by the contraction of antagonistic muscles. Also, insect flight muscle cells are filled with giant mitochondria. Suggest a mechanism to explain how myogenic contraction works inside cells. Explain the significance of giant mitochondria.

5. Human muscle cells contain a mixture of three types of muscle fibers: slow-twitch oxidative, fast-twitch oxidative, and fast-twitch glycolytic. Slow-twitch muscle cells break down ATP slowly; they contain many mitochondria and large amounts of myoglobin, a dark pigment that acts as a reservoir for oxygen. All fast-twitch muscle cells break down ATP rapidly. Fast-twitch oxidative fibers have moderate amounts of myoglobin and glycogen. Fast-twitch glycolytic fibers have little myoglobin and lots of glycogen. The relative numbers of these fibers in different muscles is under genetic control. Use this information to explain the location of dark and white meat in birds. Predict the relative numbers of slow-twitch and fast-twitch fibers in muscle samples from the thighs of world-class marathon runners and world-class sprinters.

For More Information

Alexander, R. "Muscles Fit for the Job." *New Scientist*, April 15, 1989. Diverse muscles are adapted for different functions.

Huyghe, P. "No Bone Unturned." *Discover*, December 1988. The story of a remarkable forensic anthropologist who uncovers the secrets hidden in unidentified skeletons.

Smith, K. K., and Kier, W. M. "Trunks, Tongues, and Tentacles: Moving with Skeletons of Muscle." *American Scientist*, January–February 1989. In certain organs of many animals, muscles provide support as well as movement.

Stossel, T. P. "The Machinery of Cell Crawling." *Scientific American*, September 1994. Cell movement relies on the orderly assembly and disassembly of scaffold proteins resembling those found in muscle.

Raloff, J. "Medicinal EMFs." *Science News*, November 13, 1999. Describes the use of electromagnetic fields and vibrations to heal bones and possibly help prevent osteoporosis.

Travis, J. "Boning Up." *Science News*, January 15, 2000. Researchers investigate a newly discovered gene whose protein shifts the balance between osteoblasts and osteoclasts.

Answers to Multiple-Choice Questions

1. a 2. d 3. e 4. b 5. c 6. b

MEDIATUTOR
The Muscles and Skeleton

CD Activities

Activity 34.1: Muscle Structure

Estimated time: 10 minutes

Skeletal muscle, also called striated muscle because of its striped appearance under the microscope, is a complex structure that can produce contractions that range from a quick twitch to a powerful, sustained pull. The functional contractile unit of skeletal muscle is a structure called a sarcomere. In this tutorial, you will explore the various levels of structure found within skeletal muscle and begin to learn how the sarcomere functions to produce a contraction.

Activity 34.2: Muscle Movement

Estimated time: 5 minutes

The contraction of skeletal muscle involves the interaction of several proteins found within a sarcomere. This process is described as a sliding-filament mechanism due to the fact that the thin filament slides by the thick filament as the sarcomere shortens. In this tutorial, you will explore the events that occur at the molecular level of muscle contraction.

Start the MediaTutor Student CD-ROM and enter the activity number in the Quick Search box to be taken directly to that activity.

Web Investigations

Case Study: Healing Broken Bones

Estimated time: 15 minutes

There is more to a broken arm than a broken bone. Muscles, blood vessels, skin, and nerves may be injured during the accident. Even the body's own physiological responses may contribute to the final damage toll. Take a closer look at how broken bones heal.

Go to http://www.prenhall.com/audesirk6, the Audesirk Companion Web site. Select Chapter 34 and the Web Investigation to begin.

Newborn seahorses, perfect miniature replicas of their parents, emerge from the male's brood pouch, where they developed and hatched.

35 Animal Reproduction

AT A GLANCE

Case Study: Mr. Mom?

1) **How Do Animals Reproduce?**
Asexual Reproduction Does Not Involve the Fusion of Sperm and Egg
Sexual Reproduction Requires the Union of Sperm and Egg

2) **How Does the Human Reproductive System Work?**
The Male Reproductive Tract Includes the Testes and Accessory Structures
The Female Reproductive Tract Includes the Ovaries and Accessory Structures
Copulation Allows Internal Fertilization

3) **How Can People Limit Fertility?**
Permanent Contraception Can Be Achieved Through Sterilization
There Are Three General Approaches to Temporary Contraception

Case Study Revisited: Mr Mom?

CASESTUDY

Mr. Mom?

The seahorse is an enigma in many ways—a fish with a prehensile tail, a rather rigid body propelled upright through the water by a small dorsal fin, and a tiny mouth through which it seemingly inhales prey such as small shrimp—but the seahorse has more surprises in store. Our opener photo shows their birth—from the male! Scientists have known for decades that the male incubates the eggs, which the female deposits in his brooding pouch. Until recently, they believed that a placenta-like tissue in the pouch provided the developing embryos with nourishment. Dr. Heather Masonjones investigated the intricacies of reproductive behavior in the dwarf seahorse, including the role of the brood pouch. The female seahorse, laden with eggs, approaches the male and initiates an elaborate dance in which the partners approach, quiver, and nod heads before entwining their tails and lining up their bodies face to face. After many attempts, the positioning is perfect and the female inserts a tube for depositing eggs into the pouch in the male's abdomen. As she injects her eggs into the pouch, the male releases a cloud of sperm from an opening just above the pouch. The sperm fertilizes the eggs as they enter, and the male then seals the eggs into his pouch. Here a placenta-like tissue, rich in blood vessels, removes wastes and exchanges gases, but unlike a placenta, it provides no nourishment. Dr. Masonjones found that all the calories the young need to grow, develop, and hatch are contained within the eggs produced by the mother.

The day following birth, the female is ready with a new clutch of eggs to deposit in his pouch, and the cycle begins again—for their seven-month breeding season. This includes perhaps a dozen cycles of breeding, brooding, and birth, producing more than 300 offspring. How do seahorses maintain such synchrony? Why do they produce so many eggs? Why have seahorses evolved such unusual reproductive behavior? Join us at the end of the chapter for some answers. ■

1 How Do Animals Reproduce?

Animals reproduce either sexually or asexually. As you learned in Chapter 11, in **sexual reproduction** an animal produces haploid gametes through meiosis. Two gametes, usually from separate parents, fuse to form a diploid offspring. Because an offspring receives some genes from each of its two parents, its genome is not identical to either parent's genome. In contrast, **asexual reproduction** involves only a single animal that produces offspring through repeated mitoses of cells in some part of its body. The offspring are, therefore, genetically identical to the parent.

Humans reproduce sexually, and we tend to regard sexual reproduction as the normal, best method. In reality, sexual reproduction is a rather inefficient method of producing offspring. Asexual reproduction is far more efficient, because there is no need to find a mate, court, and fend off rivals, and there is no waste of sperm and eggs that never unite to form an offspring. Not surprisingly, many animals reproduce asexually, at least some of the time. Let's begin, then, with a brief survey of asexual reproduction among animals before we move on to sexual reproduction.

Asexual Reproduction Does Not Involve the Fusion of Sperm and Egg

Budding Produces a Miniature Version of the Adult

Many sponges and cnidarians, such as *Hydra* and some sea anemones, reproduce by **budding** (Fig. 35-1). A miniature version of the animal—a **bud**—grows directly on the body of the adult, drawing nourishment from its parent. When it has grown large enough, the bud breaks off and becomes independent.

Fission Followed by Regeneration Can Produce a New Individual

Many animals are capable of **regeneration**, the ability to regrow lost body parts. For example, sea stars will regenerate an arm that is lost to an accident or predator. Even if multiple arms are lost, they can all be replaced as long as most of the central disc remains intact. The regenerative abilities of sea stars do not, however, amount to a reproductive strategy, because in most cases no new individuals are formed.

Nonetheless, some species do use regeneration for true reproduction; they reproduce by **fission**. Several annelid and flatworm species can reproduce by dividing into two or more pieces, each of which regenerates an entire body (Fig. 35-2). A few brittle star species routinely reproduce in a similar fashion; they split apart, and each half regenerates a complete animal. Some coral species can divide lengthwise to produce two smaller but complete individuals. Among cnidarian medusae (that is, cnidarians with bell-shaped bodies; see Chapter 22) a few species can reproduce by a fission process in which the entire bell folds in half and splits the stomach and other organs into two parts and then divides into two individuals.

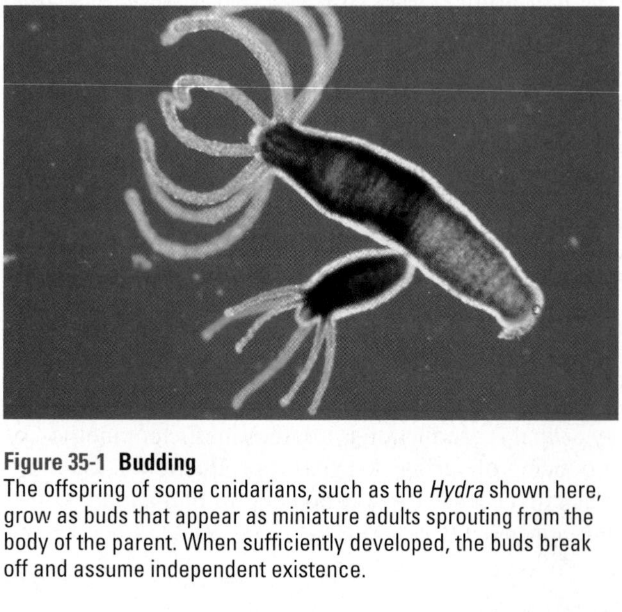

Figure 35-1 Budding
The offspring of some cnidarians, such as the *Hydra* shown here, grow as buds that appear as miniature adults sprouting from the body of the parent. When sufficiently developed, the buds break off and assume independent existence.

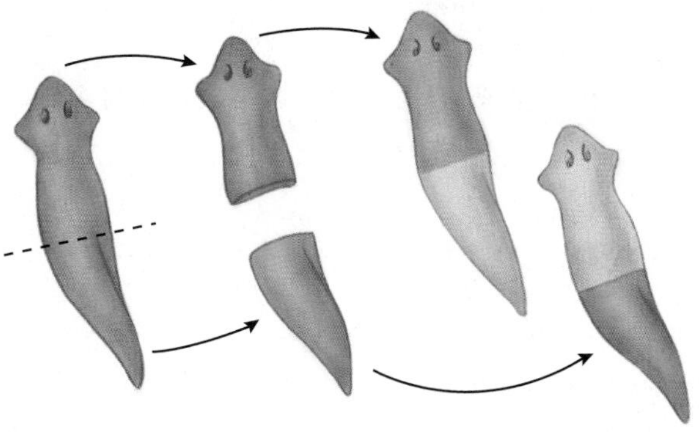

Figure 35-2 Fission followed by regeneration
Some flatworm species reproduce by dividing across the middle. Then each offspring regenerates the missing half of its body.

During Parthenogenesis, Eggs Develop Without Fertilization

The females of some animal species can reproduce by a process known as **parthenogenesis**, in which haploid egg cells develop into adults without being fertilized. **Fertilization** is the union of haploid gametes to form a diploid zygote. Parthenogenetically produced offspring of some species remain haploid. Male honeybees, for example, are haploid, developing from unfertilized eggs; their diploid sisters develop from fertilized eggs. Conversely, some fish, amphibians, and reptiles regain the diploid number of chromosomes in parthenogenetically produced offspring by duplicating all their chromosomes either before or after meiosis. All the resulting offspring are females.

Some species of fish, including relatives of the mollies and platies that are popular in tropical fish stores, and some lizards, such as the whiptail, have done away with males completely. Their populations consist entirely of parthenogenetically reproducing females. Still other animals, such as the aphid, can reproduce either sexually or parthenogenetically, depending on environmental factors such as the season of the year or the availability of food (Fig. 35-3).

Sexual Reproduction Requires the Union of Sperm and Egg

Given the obvious efficiency of asexual reproduction, no one is sure why sex arose. But it did, and sexual reproduction is now a prominent feature in the lives of most animals. It also has an important outcome. The genetic recombination that results from sexual reproduction tends to create novel genotypes—and therefore novel phenotypes—that are an important source of variation on which natural selection may act.

Figure 35-3 A female aphid gives live birth
In spring and early summer, when food is abundant, aphid females reproduce parthenogenetically. In fact, the development of the reproductive tract proceeds so rapidly that females are born pregnant! In fall, they reproduce sexually, as the females mate with males. Aphids have thus evolved the ability to exploit the advantages of both asexual reproduction (rapid population growth during times of abundant food, no energy spent in seeking a mate, no wasted gametes) and sexual reproduction (genetic recombination).

In animals, sexual reproduction occurs when a haploid sperm fertilizes a haploid egg, generating a diploid offspring. In most animal species, an individual is either male or female. These species are termed **dioecious** (Greek for "two houses"). The sexes are defined by the type of gamete that each produces. Females produce **eggs**, which are large, nonmotile cells containing food reserves. Males produce small, motile **sperm**, which have almost no cytoplasm and hence no food reserves.

In **monoecious** ("one house") species, such as earthworms and many snails, single individuals produce both sperm and eggs. Such individuals are commonly called **hermaphrodites** (after Hermaphroditos, a male Greek god whose body was merged with that of a female water nymph, producing a half-male and half-female being). In most hermaphroditic species, reproduction involves a mutual exchange of sperm between individuals. In some hermaphroditic species, however, individuals can fertilize their own eggs if necessary. These animals, including tapeworms and many pond snails, are relatively immobile and may find themselves isolated from other members of their species. Obviously, the ability to fertilize oneself is advantageous under these circumstances.

For dioecious species and for hermaphrodites that cannot self-fertilize, successful reproduction requires that sperm and eggs from different animals be brought together for fertilization. The union of sperm and egg is accomplished in a variety of ways, depending on the mobility of the animals and on whether they breed in water or on land.

External Fertilization Occurs Outside the Parents' Bodies

In **external fertilization**, the union of the sperm and egg takes place outside the bodies of the parents. When animals breed in water, the parents release sperm and eggs into the water, through which the sperm swim to reach an egg. This procedure is called **spawning**. Because sperm and egg are relatively short-lived, spawning animals must synchronize their reproductive behaviors, both *temporally* (male and female spawn at the same time) and *spatially* (male and female spawn in the same place). Synchronization may be achieved by way of signals, behaviors, environmental cues, or some combination of these factors.

Most spawning animals rely on environmental cues to some extent. Breeding usually occurs only during certain seasons of the year, and cues such as seasonal changes in daylength typically stimulate the physiological changes that lead to readiness for breeding. More precise synchrony, however, is required to coordinate the actual release of sperm and egg. Grunion, fish that inhabit coastal waters off southern California, regulate their unusual reproductive rituals by the season, time of day, and phase of the moon (Fig. 35-4a). On fall nights during the highest tides (which occur during a full moon), grunion swim up onto sandy beaches. Writhing masses of males and females release their gametes into the wet sand and then swim back out to sea on the next wave. The fertilized eggs hatch in the warm sand and develop within 2 weeks. Many corals of Australia's Great Barrier Reef also synchronize spawning by the phase of the moon. On the fourth or fifth night after the full moons of November and December, all the corals of a particular species on an entire reef release a blizzard of sperm and eggs into the water (Fig. 35-4b).

Other animals communicate their sexual readiness to one another by sending visual, acoustic, or chemical signals. Chemical signals are especially common among immobile or sluggish invertebrates, such as mussels and sea stars. These animals release chemicals called **pheromones** into the water, where the pheromone can affect the behavior of other animals. Normally, when a female is ready to spawn, she releases eggs and a pheromone into the water. Nearby males, detecting the mating pheromone, quickly release millions of sperm. The sperm themselves are lured by a chemical attractant released by the eggs in some, if not most, animals. Such "egg pheromones," which have been detected in animals as diverse as sea stars and humans, help ensure fertilization.

Synchronized timing alone does not guarantee efficient reproduction. Corals, sea stars, and mussels all waste enormous quantities of sperm and eggs because the gametes are released relatively far apart. In species of mobile animals, both temporal *and* spatial synchrony can be ensured by mating behaviors. Most fish, for example, have some sort of courtship ritual in which the male and female come very close together and release

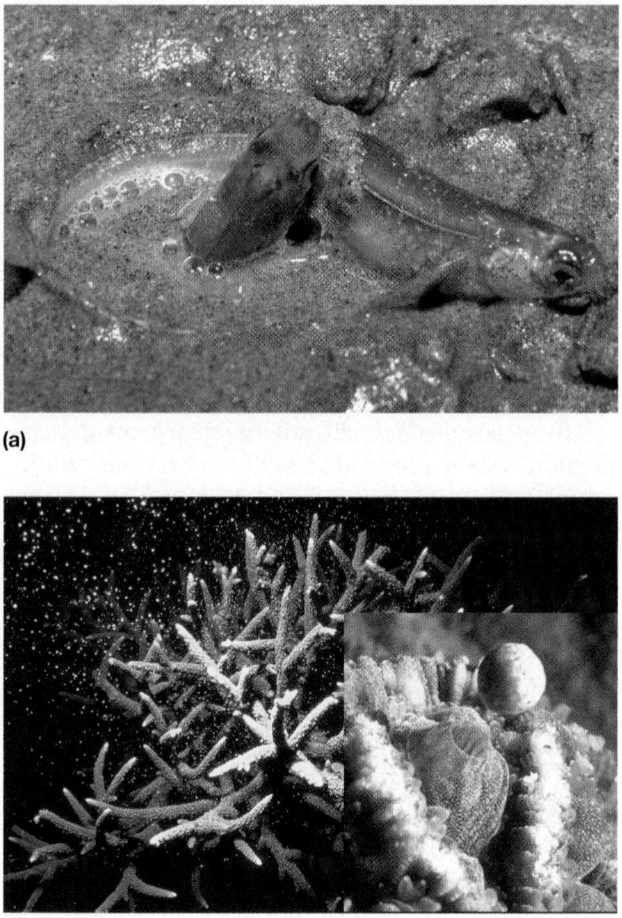

Figure 35-4 **Environmental cues may synchronize spawning**
(a) At the highest tides of fall, grunion swarm ashore on the few undeveloped beaches left in southern California. The fish burrow slightly into the sand and release sperm and eggs. ***(b)*** Along the Great Barrier Reef of Australia, thousands of corals spawn simultaneously, creating this "blizzard" effect. Spawning in these corals is linked to the phase of the moon. (Inset) Close-up of a package of sperm and eggs erupting from a spawning hermaphroditic coral.

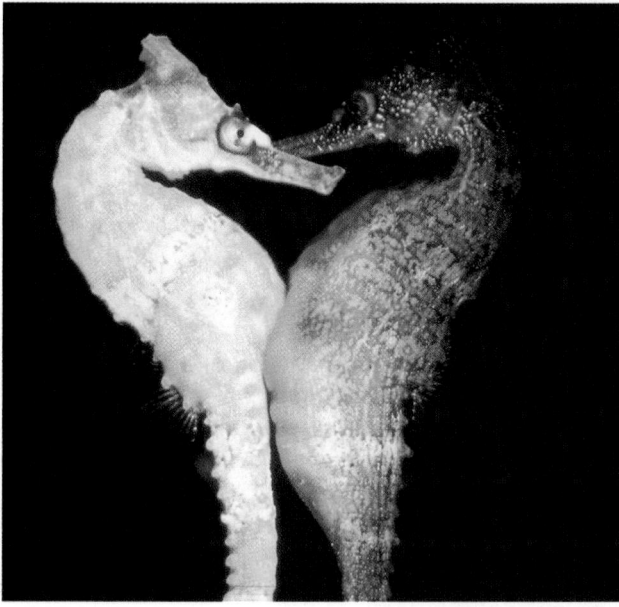

Figure 35-5 **Courtship rituals synchronize release of sperm and eggs**
(a) Courtship rituals among Siamese fighting fish *(Betta splendens)* ensure fertilization of the female's eggs, as male and female curl about one another, releasing sperm and eggs together. The male retrieves the eggs as they fall, spits them into his bubble nest (seen here as bubbles floating on the surface above him), and cares for the offspring during their first few weeks of life. ***(b)*** Spawning in the seahorse requires that the male and female orient their bodies so that the female can deposit her eggs in the male's pouch.

their gametes in the same place and at the same time. The courtship dances of the seahorse and Siamese fighting fish provide exquisite examples (Fig. 35-5). Frogs carry this ritual one step further by assuming a characteristic mating pose called **amplexus** (Fig. 35-6). At the edges of ponds and lakes, the male frog mounts the female and prods the sides of her abdomen. This stimulates her to release eggs, which he fertilizes by releasing a cloud of sperm above them.

Internal Fertilization Occurs Within the Female Body

In **internal fertilization**, sperm are taken into the body of the female, where fertilization occurs. Internal fertilization is an important adaptation to terrestrial life: Sperm must be bathed in fluid until they reach the eggs; on land, this liquid passage is best achieved inside the female's body. Even in aquatic environments, internal fertilization increases the likelihood of successful fertilization, because the sperm and eggs are confined together in a small space rather than relying on encounters within a large volume of water.

Internal fertilization usually occurs by **copulation**, the behavior by which the male deposits sperm directly into the reproductive tract of the female (Fig. 35-7). In a variation of internal fertilization, males of some species package their sperm in a container called a **spermatophore** (Greek for "sperm carrier"). In many spermatophore-producing species, including some scorpions,

Figure 35-6 Golden toads in amplexus
The smaller male clutches the female and stimulates her to release eggs.

grasshoppers, and salamanders, no copulation occurs. The male simply drops a spermatophore on the ground, and if a female finds it, she fertilizes herself by inserting it into her reproductive cavity, where the enclosed sperm are liberated. Other species have evolved more sophisticated ways to transfer a spermatophore. For example, in some scorpion species the male attaches his spermatophore to the ground, then engages his mate in a "push-pull dance," during which he maneuvers her back and forth until she is positioned over the spermatophore. Male octopuses of the genus *Argonauta* use a special tentacle to transfer the spermatophore to the female's mantle cavity. After effecting the transfer, the tentacle breaks off and remains inside the cavity. Before this process was first observed, biologists believed that the detached tentacles inside the females were parasitic worms.

When animals must copulate to reproduce (or when a mating ritual is required for spawning), males may compete for access to copulations with females. This competition has driven the evolution of a wide variety of sexually selected structures and reproductive behaviors (see Chapter 15). One spectacular example of competition for access to mates occurs in the early spring of each year in the woods of western Canada. As the snows melt and the ground warms, male red-sided garter snakes emerge from the underground dens where they have hibernated in groups of thousands. Later the females emerge, and a mating frenzy begins. In a sea of thousands of writhing snake bodies, each female attracts a crowd of dozens or even hundreds of males. Only one will copulate successfully.

Simply depositing sperm into the body of the female, however, does not guarantee fertilization. Fertilization can occur only if a mature egg is released into the female reproductive tract during the limited time when sperm

(a)

(b)

(c)

Figure 35-7 Internal fertilization is essential for reproduction on land
(a) Ladybugs mate on a dandelion flower. ***(b)*** South American tortoises must cope with confining shells. ***(c)*** King penguins mate comfortably in the snow.

are present. Just as with spawning animals, the mating behavior of males and females must be synchronized. Among mammals, for example, copulation usually occurs only at certain seasons of the year or when the female signals readiness to mate. The season or signal typically coincides with **ovulation**, the release of the egg cell from the ovary. Copulation itself triggers ovulation in a few mammals, such as rabbits. An alternative strategy, employed by many female snails and insects, is to store sperm for days, weeks, or even months, thus ensuring a supply of sperm whenever eggs are ready.

2 How Does the Human Reproductive System Work?

Like other mammals, humans have separate sexes and reproduce sexually, with internal fertilization. The **gonads** of mammals are paired organs that produce sex cells—sperm and eggs. Although most mammal species reproduce only during certain seasons of the year and consequently produce sperm and eggs only at that time, human reproduction is not restricted by season. Men produce sperm more or less continuously, and women ovulate (release a mature egg cell) about once a month.

The Male Reproductive Tract Includes the Testes and Accessory Structures

The central structures of the male reproductive tract are the **testes** (singular, **testis**), which are the gonads that produce sperm. The male reproductive system (Table 35-1 and Fig. 35-8) also includes accessory structures that secrete substances that activate and nourish the sperm, store it, and conduct it to the female reproductive tract.

The Testes Are the Site of Sperm Production

The testes, which produce both sperm and male sex hormones, are located in the **scrotum**, a pouch that hangs outside the main body cavity. This location keeps the testes about 4 °C cooler than the core of the body and provides the optimal temperature for sperm development. (Tight jeans push the scrotum up against the body, raising the temperature of the testes. Some researchers believe that wearing tight pants can reduce sperm counts and decrease fertility. This is not, however, a reliable means of birth control!) Coiled, hollow **seminiferous tubules**, in which sperm are produced, nearly fill each testis (Fig. 35-9a). In the spaces between the tubules are the **interstitial cells**, which synthesize the male hormone *testosterone*.

Just inside the wall of each seminiferous tubule lie **spermatogonia** (singular, **spermatogonium**), the diploid cells from which the sperm eventually will arise, and the much larger **Sertoli cells** (Fig. 35-9b). Each time a spermatogonium divides, it can take one of two developmental paths. In the first developmental option, the cell may undergo mitosis. Mitosis ensures that the male has a steady supply of new spermatogonia throughout his life. Alternatively, the spermatogonium may undergo **spermatogenesis**—that is, it may experience a series of developmental events that leads to the production of haploid sperm (Fig. 35-10).

Spermatogenesis begins with the growth and differentiation of spermatogonia into **primary spermatocytes**, which are large diploid cells. The primary spermatocytes then undergo meiosis (see Chapter 11). At the end of meiosis I, each primary spermatocyte gives rise to two haploid **secondary spermatocytes**. Each secondary spermatocyte divides again, during meiosis II, to produce two **spermatids**, for a total of four spermatids per primary spermatocyte. Spermatids undergo

Table 35-1 Structures and Functions of the Human Male Reproductive Tract

Structure	Type of Organ	Function
Testis	Gonad	Produces sperm and testosterone
Epididymis and vas deferens	Ducts	Store sperm; conduct sperm from testes to penis
Urethra	Duct	Conducts semen from vas deferens and urine from urinary bladder to the tip of the penis
Penis	External "appendage"	Deposits sperm in female reproductive tract
Seminal vesicles	Glands	Secrete fluids that contain fructose (energy source) and prostaglandins (possibly cause "upward" contractions of vagina, uterus, and oviducts, assisting sperm transport to oviducts); fluids may wash sperm out of ducts of male reproductive tract into vagina
Prostate	Gland	Secretes fluids that are basic (neutralize acidity of vagina) and contain factors that enhance sperm motility
Bulbourethral glands	Glands	Secrete mucus (may lubricate penis in vagina)

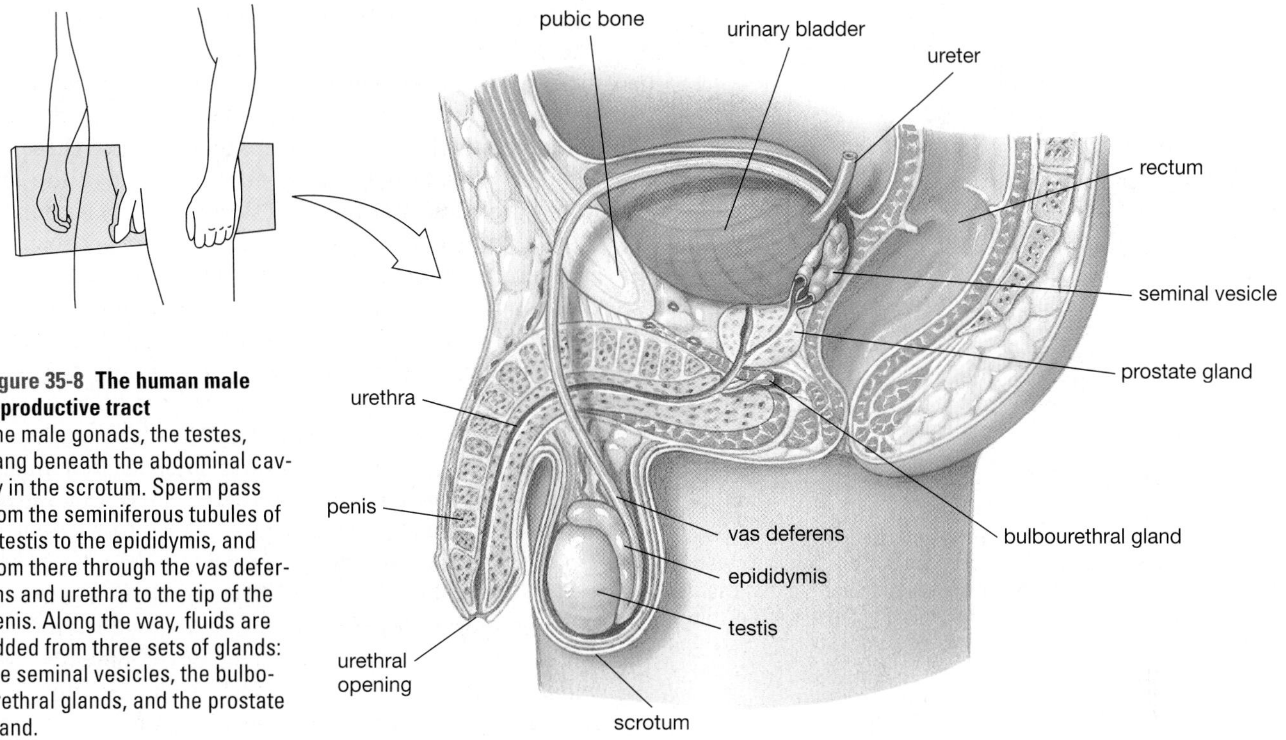

Figure 35-8 The human male reproductive tract
The male gonads, the testes, hang beneath the abdominal cavity in the scrotum. Sperm pass from the seminiferous tubules of a testis to the epididymis, and from there through the vas deferens and urethra to the tip of the penis. Along the way, fluids are added from three sets of glands: the seminal vesicles, the bulbourethral glands, and the prostate gland.

MediaTutor 35.1 Human Reproductive Anatomy

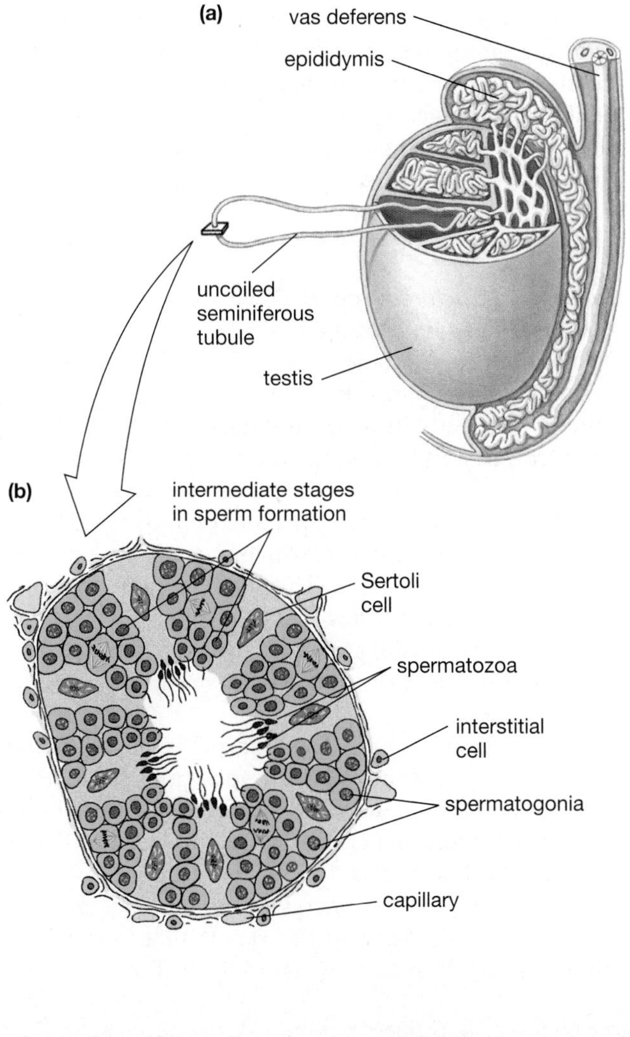

radical rearrangements of their cellular components as they differentiate into sperm (see Fig. 35-10).

Sertoli cells regulate the process of spermatogenesis and nourish the developing sperm. The spermatogonia, spermatocytes, and spermatids are embedded in infoldings of the Sertoli cells. As spermatogenesis proceeds, they migrate up from the outermost edge of the seminiferous tubule to the central cavity of the tubule. The mature sperm (or *spermatozoa*, see Fig. 35-9b) are then liberated into the central cavity.

A human sperm (Fig. 35-11) is unlike any other cell of the body. Most of the cytoplasm disappears, leaving a haploid nucleus nearly filling the head. Atop the

Figure 35-9 The structures involved in spermatogenesis
(a) A section of the testis, showing the location of the seminiferous tubules, epididymis, and vas deferens. ***(b)*** Cross section of a seminiferous tubule. The walls of the seminiferous tubules are lined with Sertoli cells and spermatogonia. As spermatogonia undergo meiosis, the daughter cells move inward, embedded in infoldings of the Sertoli cells. There they differentiate into sperm (spermatazoa), drawing on the Sertoli cells for nourishment. Mature sperm are freed into the central cavity of the tubules for transport to the penis. Testosterone is produced by the interstitial cells in the spaces between tubules.

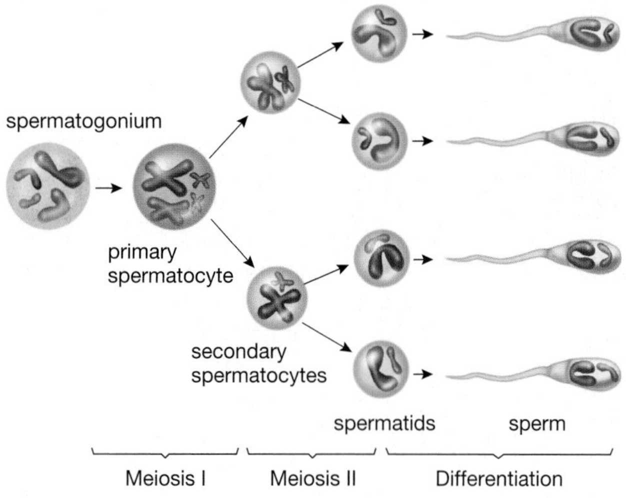

Figure 35-10 Sperm are produced by meiosis
Spermatogenesis is accomplished by meiotic divisions followed by differentiation, producing haploid sperm. Although 4 chromosomes are shown for clarity, in humans the diploid number is 46 and the haploid number is 23.

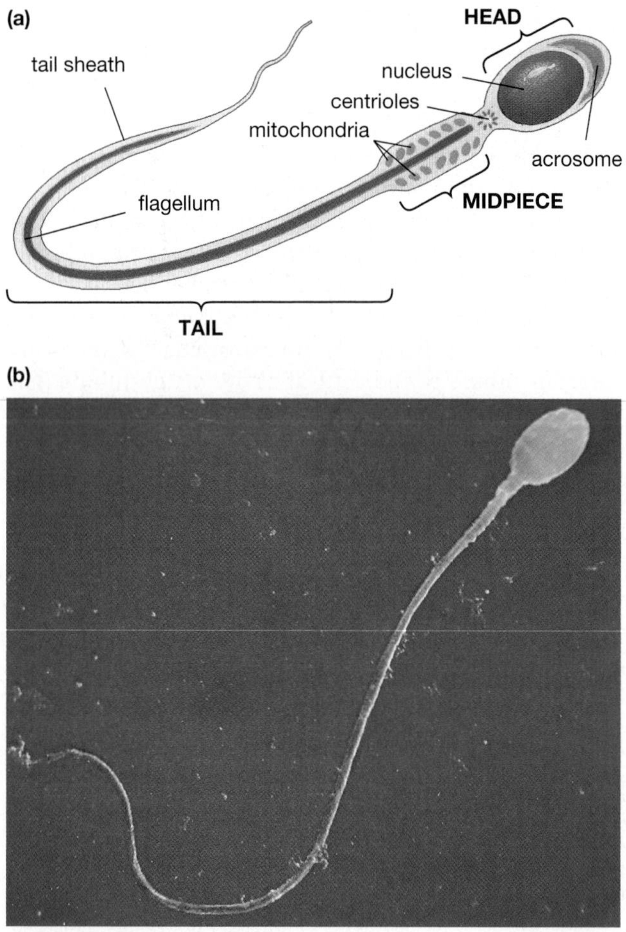

Figure 35-11 A human sperm cell
(a) A mature sperm is a cell equipped with only the essentials: a haploid nucleus containing the male genetic contribution to the future zygote; a lysosome (the *acrosome*) containing enzymes that will digest the barriers surrounding the egg; mitochondria for energy production; and a tail (a long flagellum) for locomotion. ***(b)*** False-color electron micrograph of a human sperm.

nucleus lies a specialized lysosome called the **acrosome**. The acrosome contains enzymes that will be needed to dissolve protective layers around the egg, enabling the sperm to enter and fertilize it. Behind the head is the *midpiece,* which is packed with mitochondria. These organelles provide the energy needed to move the *tail*, which protrudes out the back. Whiplike movements of the tail, which is really a long flagellum, propel the sperm through the female reproductive tract.

In humans and other mammals, immature males do not produce sperm. Spermatogenesis does not begin until *puberty,* a period of rapid growth and transition to sexual maturity. At puberty, the hypothalamus releases **gonadotropin-releasing hormone (GnRH)**, which stimulates the anterior pituitary to produce **luteinizing hormone (LH)** and **follicle-stimulating hormone (FSH)**. Luteinizing hormone stimulates the interstitial cells of the testes to produce the hormone **testosterone** (Fig. 35-12). Testosterone, in combination with FSH, stimulates the Sertoli cells and spermatogonia, causing spermatogenesis.

Testosterone also stimulates the development of *secondary sexual characteristics* (such as the growth of facial hair in males and breast development in females), maintains sexual drive, and is required for successful *intercourse* (a term we will use for human copulation). Sperm, however, are not involved in these functions. Therefore, if FSH release could be suppressed, blocking spermatogenesis, while LH release continues, thereby allowing continued testosterone production, a man would be *infertile* but not *impotent*; in other words, he could not produce sperm but could maintain an erection of the **penis**, the male organ of copulation. Efforts are under way to develop a drug to do just that, as a form of male birth control.

Accessory Structures Produce Semen and Conduct the Sperm Outside the Body

The seminiferous tubules merge to form the **epididymis**, a single continuous, folded tube (see Fig. 35-9a). The epididymis leads into the **vas deferens**, which leaves the scrotum and enters the abdominal cavity. Most of the hundreds of millions of sperm produced each day are stored in the vas deferens and epididymis. The vas deferens joins the **urethra**, which connects the bladder to the tip of the penis. This final common path is shared, at different times, by both urine (during urination) and sperm (during ejaculation—a reflex caused by sexual stimulation that forces sperm out through the penis).

The fluid ejaculated from the penis, called **semen**, consists of sperm mixed with secretions from three glands that empty into the vas deferens or urethra: (1) the **seminal vesicles**, (2) the **prostate gland**, and (3) the **bulbourethral glands**. The secretions activate swimming by the sperm, provide energy for swimming, and neutralize the acidic fluids of the vagina that normally inhibit bacterial growth (see Fig. 35-8 and Table 35-1).

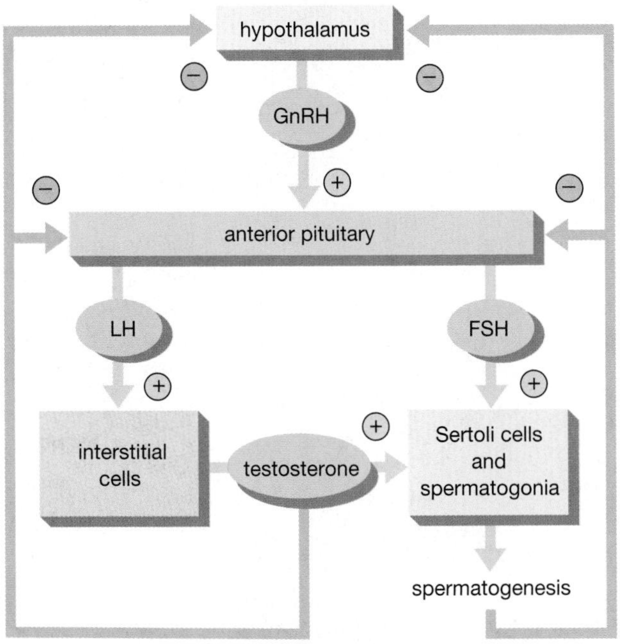

Figure 35-12 Hormonal control of spermatogenesis
GnRH from the hypothalamus stimulates the anterior pituitary to release LH and FSH. LH stimulates the interstitial cells to produce testosterone. Testosterone and FSH stimulate the Sertoli cells and the spermatogonia, causing spermatogenesis. Testosterone and chemicals produced during spermatogenesis inhibit further release of FSH and LH, forming a negative feedback loop that keeps the rate of spermatogenesis and the concentration of testosterone in the blood nearly constant. (+ stimulates; – inhibits)

The Female Reproductive Tract Includes the Ovaries and Accessory Structures

The female reproductive tract is almost entirely contained within the abdominal cavity (Table 35-2 and Fig. 35-13). It consists of paired gonads—the **ovaries** (Fig. 35-14a)—and accessory structures that accept sperm, conduct the sperm to the egg, and nourish the developing **embryo**.

The Ovaries Are the Site of Egg Production

Oogenesis, the formation of egg cells, begins in the developing ovaries of a female **fetus** (an embryo that is sufficiently developed to be recognizably human). This process starts with the formation of precursor egg cells called **oogonia** (singular, **oogonium**). By the end of the third month of fetal development, no oogonia remain, as they have all divided by mitosis and grown into **primary oocytes.** As fetal development continues, meiosis begins in all primary oocytes but is halted at prophase of meiosis I. By birth, a lifetime supply of primary oocytes is already in place, and no new ones will be generated later in life. The ovaries start out with about 2 million primary oocytes. Many primary oocytes die each day, until at puberty (usually 11 to 14 years of age) only about 400,000 remain. That number is more than enough, because only a few oocytes resume meiosis during each month of a woman's reproductive span (from puberty to *menopause* at about age 50).

Surrounding each oocyte is a layer of much smaller cells that both nourish the developing oocyte and secrete female sex hormones. Together, the oocyte and these accessory cells make up a **follicle** (Fig. 35-14b). Approximately once a month during a woman's reproductive years, she undergoes a *menstrual cycle*, which is described later in this chapter. During the menstrual cycle, pituitary hormones stimulate the development of a dozen or more follicles, although normally only one follicle matures completely. At this time, the primary oocyte completes its first meiotic division (which was halted during development), dividing into a single **secondary oocyte** and a **polar body**, which is little more than a discarded set of chromosomes (Fig. 35-15, p. 727). Meanwhile, the small accessory cells of the follicle multiply and secrete the hormone **estrogen**. As the follicle matures, it grows, eventually erupting through the surface of the ovary and releasing the secondary oocyte, a process called *ovulation* (Fig. 35-16, p. 727). The secondary oocyte then travels through the tube leading out of the ovary, called the **uterine tube** or *oviduct.* For convenience, we will refer to the ovulated secondary oocyte as the *egg.* If the egg is fertilized, it may undergo the second meiotic division while in the uterine tube.

Some of the follicle cells accompany the egg, but most remain in the ovary. These cells enlarge and become glandular, forming the **corpus luteum** (see Fig. 35-14b). The corpus luteum secretes both estrogen and a second hormone, **progesterone**. If fertilization does not occur, the corpus luteum breaks down a few days later.

A human male is able to produce large numbers of sperm continuously and can therefore contribute

Table 35-2 Structures and Functions of the Human Female Reproductive Tract

Structure	Type of Organ	Function
Ovary	Gonad	Produces eggs, estrogen, and progesterone
Fimbria	Mouth of duct	Cilia sweep egg into oviduct
Uterine tube	Duct	Conducts egg to uterus; site of fertilization
Uterus	Muscular chamber	Site of development of fetus
Cervix	Connective tissue ring	Closes off lower end of uterus, supports fetus, and prevents foreign material from entering uterus
Vagina	Large "duct"	Receptacle for semen; birth canal

MEDIATUTOR
35.1 Human Reproductive Anatomy

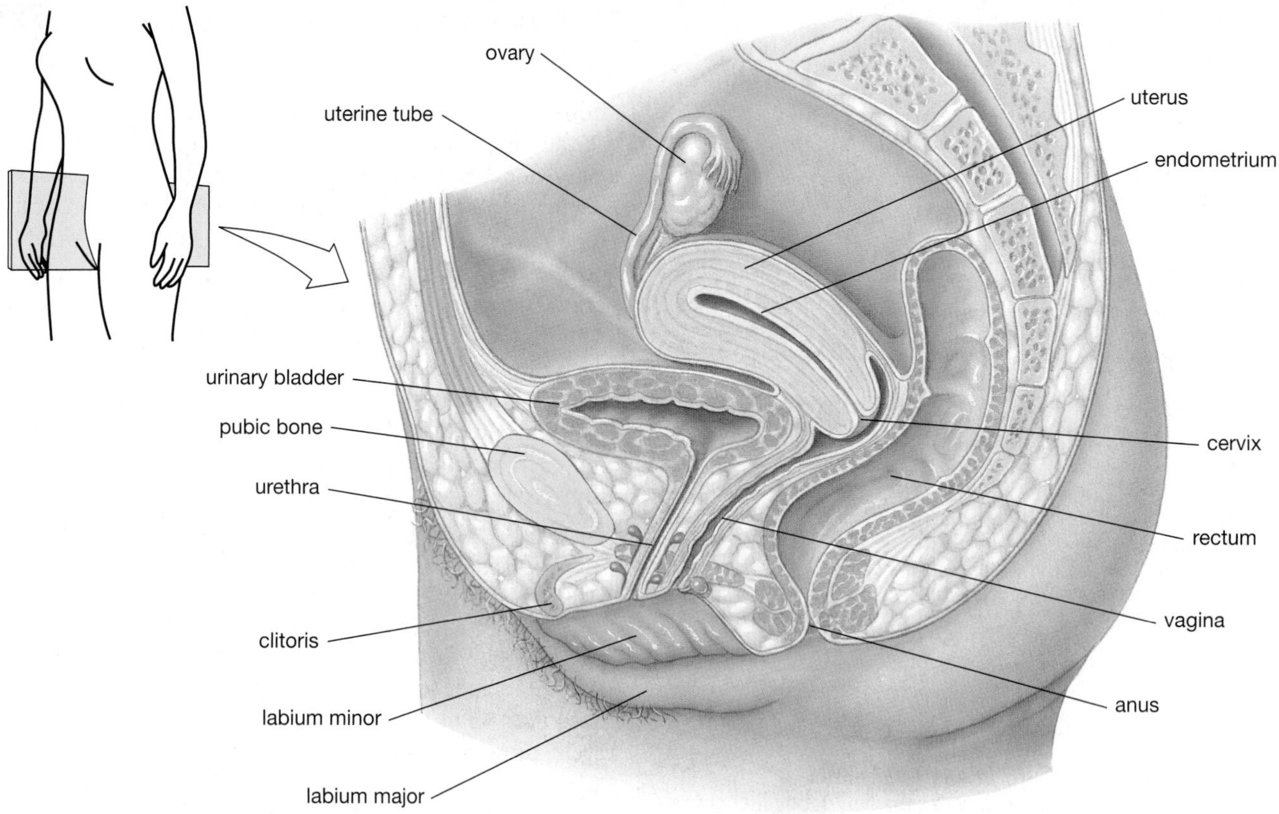

Figure 35-13 The human female reproductive tract
Eggs are produced in the ovaries and swept by cilia into the uterine tube. A male deposits sperm in the vagina, from which they move up through the cervix and uterus into the uterine tube. Sperm and egg normally meet in the uterine tube, where fertilization occurs. The fertilized egg attaches to the lining of the uterus, where the embryo develops.

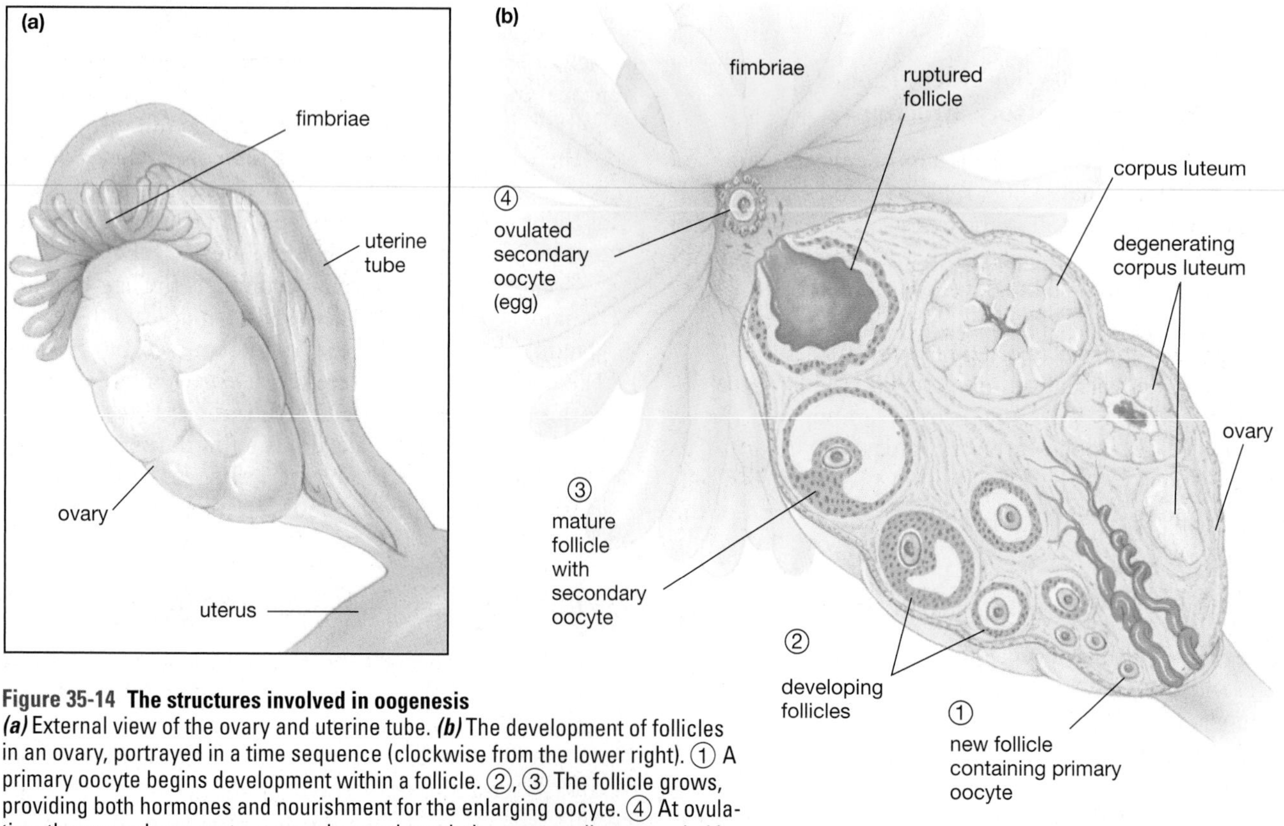

Figure 35-14 The structures involved in oogenesis
(a) External view of the ovary and uterine tube. ***(b)*** The development of follicles in an ovary, portrayed in a time sequence (clockwise from the lower right). ① A primary oocyte begins development within a follicle. ②, ③ The follicle grows, providing both hormones and nourishment for the enlarging oocyte. ④ At ovulation, the secondary oocyte, or egg, bursts through the ovary wall, surrounded by some follicle cells. The remaining follicle cells develop into the corpus luteum, which secretes hormones. If fertilization does not occur, the corpus luteum breaks down after a few days.

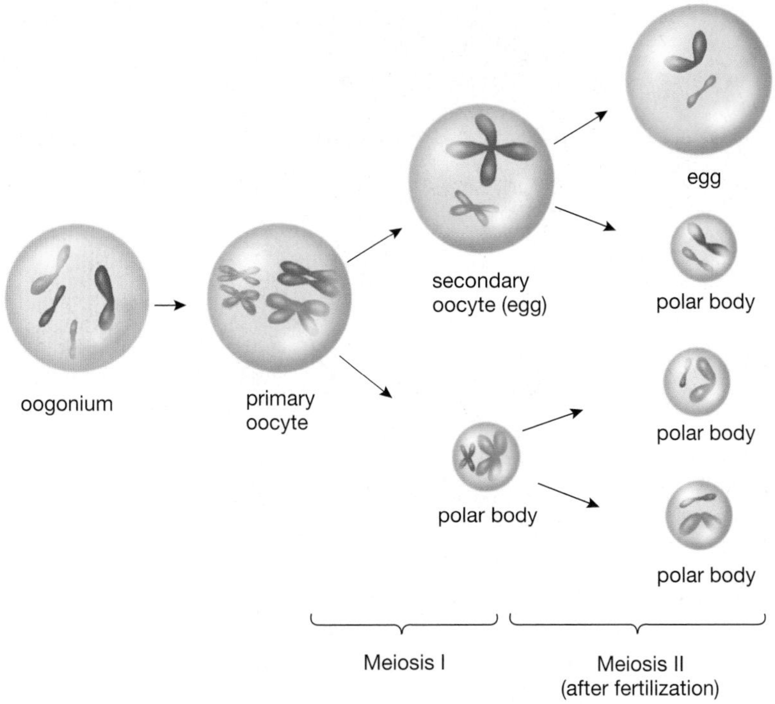

Figure 35-15 Egg cells are formed by meiosis
The cellular stages of oogenesis. The oogonium enlarges to form the primary oocyte. At meiosis I, almost all the cytoplasm is included in one daughter cell, the secondary oocyte. The other daughter cell is a small polar body that contains chromosomes but little cytoplasm. At meiosis II, almost all the cytoplasm of the secondary oocyte is included in the egg, and a second small polar body discards the remaining "extra" chromosomes. The first polar body may also undergo the second meiotic division. In humans, meiosis II does not occur until a sperm penetrates into the egg.

gametes to a fertilization at any time. In contrast, a woman does not produce mature gametes (ovulate) unless her reproductive tract is properly prepared for pregnancy. Ovulation can occur only when the uterus is prepared to receive and nourish a fertilized egg. The necessary coordination of ovulation and uterine preparation is accomplished by a complex **menstrual cycle**, which is regulated by hormonal interactions among the hypothalamus, pituitary gland, and ovary and covered in "A Closer Look: Hormonal Control of the Menstrual Cycle."

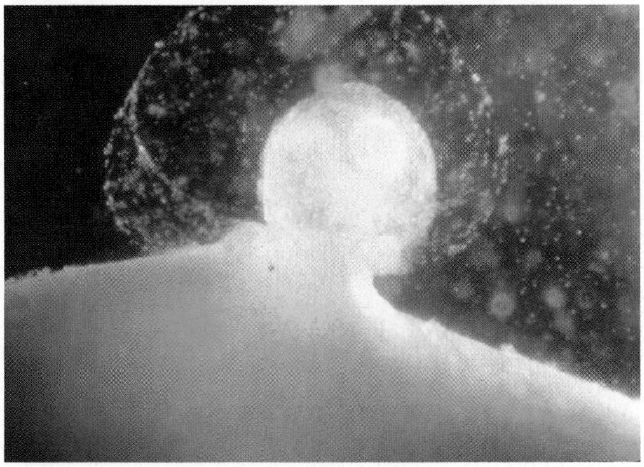

Figure 35-16 A follicle erupts from the ovary
The mature follicle grows so large and is filled with so much fluid that it moves to the surface of the ovary and literally bursts through the ovary wall like a miniature volcano. It then releases the secondary oocyte into the uterine tube.

Accessory Structures Include the Uterine Tubes, Uterus, and Vagina

Each ovary is adjacent to, but not continuous with, a uterine tube (also called the *oviduct*; see Fig. 35-14a). The open end of the uterine tube is fringed with ciliated "fingers" called **fimbriae** (singular, **fimbria**), which nearly surround the ovary. The cilia create a current that sweeps the egg into the mouth of the uterine tube. Fertilization usually occurs in the uterine tube. The **zygote**, as the fertilized egg is now called, is swept down the uterine tube by beating cilia and released into the pear-shaped **uterus**, or *womb*. There it will develop for 9 months. The wall of the uterus has two layers that correspond to its dual functions of (1) nourishment and (2) childbirth. The inner lining, or **endometrium**, is richly supplied with blood vessels. (This lining will form the mother's contribution to the **placenta**, the structure that transfers oxygen, carbon dioxide, nutrients, and wastes between fetus and mother, as we shall see in Chapter 36.) The outer muscular wall of the uterus, the **myometrium**, contracts strongly during delivery, expelling the infant out into the world.

Developing follicles secrete estrogen, which stimulates the uterine lining to grow an extensive network of blood vessels and nutrient-producing glands. After ovulation, estrogen and progesterone released by the corpus luteum promote continued growth of the endometrium. Thus, if an egg is fertilized, it encounters a rich environment for growth. If the egg is not fertilized, however, the corpus luteum disintegrates, estrogen and progesterone levels fall, and the overgrown endometrium disintegrates as well. The uterus contracts, squeezing out the excess endometrial tissue (perhaps causing menstrual cramps in the process). The resulting flow of tissue and

A Closer Look
Hormonal Control of the Menstrual Cycle

We will begin our discussion of the menstrual cycle with the spontaneous release of gonadotropin-releasing hormone (GnRH) by cells in the hypothalamus. This release occurs continuously, unless it is suppressed by other hormones—notably, progesterone. Now follow the descriptions in the diagram by matching the numbers to Figure E35-1.

① GnRH stimulates the anterior pituitary to release FSH (purple line) and LH (blue).

② Both FSH and LH circulate in the bloodstream and initiate the development of several follicles within the ovaries, and the cells of these developing follicles secrete estrogen. Under the combined influences of FSH, LH, and estrogen, the follicles grow during the next 2 weeks. Simultaneously, the primary oocyte within each follicle enlarges, storing both food and regulatory substances (mostly proteins and messenger RNA) that will be needed by the fertilized egg during early development. Only one or, rarely, two follicles complete development each month.

③ As the maturing follicle enlarges, it secretes ever greater amounts of estrogen (red line). This estrogen has three effects. First, it promotes the continued development of the follicle and of the primary oocyte within it (second panel). Second, it stimulates the growth of the endometrium of the uterus (lower panel).

④ High levels of estrogen stimulate both the hypothalamus and pituitary, resulting in a surge of LH and FSH at about the 12th day of the cycle. The function of the peak in FSH concentration is not well understood, but the surge of LH has three important consequences: First, it triggers the resumption of meiosis I in the oocyte, resulting in the formation of the secondary oocyte and the first polar body.

⑤ Second, the LH surge causes the final explosive growth of the follicle, culminating in ovulation, and third, it transforms the remnants of the follicle into the corpus luteum.

⑥ The corpus luteum secretes both estrogen (red line) and progesterone (green).

⑦ The combination of estrogen and progesterone hormones inhibits the hypothalamus and pituitary, shutting down the release of FSH and LH, thereby preventing the development of more follicles. Simultaneously, estrogen and progesterone stimulate further growth of the uterine lining (lower panel), which eventually becomes about 5 millimeters thick.

⑧ If pregnancy does not occur, the corpus luteum starts to disintegrate about 1 week after ovulation. This disintegration is caused by the corpus luteum itself, which secretes the progesterone that in turn shuts down LH secretion. Because the corpus luteum can persist only while it is stimulated by LH (or by a similar hormone released by the developing embryo, described below), it essentially induces its own destruction, in a kind of negative feedback.

⑨ With the corpus luteum gone, estrogen and progesterone levels plummet. Deprived of stimulation by estrogen and progesterone, the endometrium of the uterus also dies, and its blood and tissue are shed. This shedding forms the menstrual flow that begins about the 28th day of the

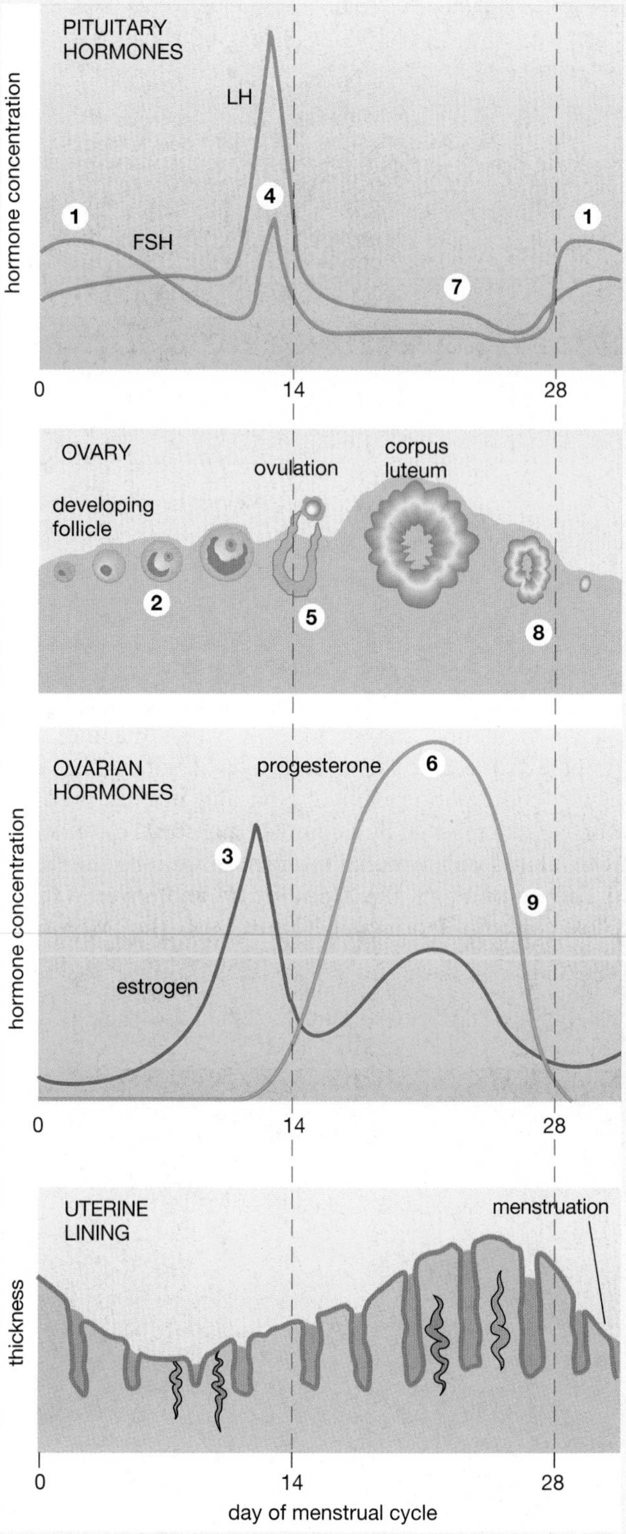

Figure E35-1 Hormonal control of the menstrual cycle
The menstrual cycle is generated by interactions among the hormones of the hypothalamus, the anterior pituitary, and the ovaries. The circled numbers refer to the interactions discussed in the text.

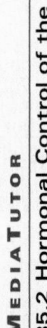

MediaTutor 35.2 Hormonal Control of the Menstrual Cycle

cycle. The reduced level of circulating progesterone no longer inhibits the hypothalamus and pituitary, so the spontaneous release of GnRH from the hypothalamus resumes. The release of GnRH in turn stimulates the release of FSH and LH (back to step ①), initiating the development of a new set of follicles, restarting the cycle.

During pregnancy, the embryo itself prevents these changes from occurring. Shortly after the ball of cells formed by the dividing fertilized egg embeds itself in the endometrium, it starts secreting an LH-like hormone called **chorionic gonadotropin (CG)**. This hormone travels in the bloodstream to the ovary, where it prevents the breakdown of the corpus luteum. The corpus luteum continues to secrete estrogen and progesterone, and the uterine lining continues to grow, nourishing the embryo. The embryo releases so much CG that the hormone is excreted in the mother's urine. In fact, most pregnancy tests use the presence of CG in urine to determine pregnancy.

Although negative feedback regulates the levels of most hormones, the hormones of the menstrual cycle are regulated by both positive and negative feedback. During the first half of the cycle, FSH and LH stimulate estrogen production by the follicles. High levels of estrogen then *stimulate* the mid-cycle surge of FSH and LH release (positive feedback). During the second half of the cycle, estrogen and progesterone together *inhibit* the release of FSH and LH (negative feedback). The early positive feedback causes hormone concentrations to reach high levels, and the later negative feedback shuts the system down again unless pregnancy intervenes.

blood, as a result of the erosion of blood vessels in the endometrium, is called **menstruation** (from the Latin *mensis,* meaning "month").

The outer end of the uterus is nearly closed off by the **cervix**, a ring of connective tissue. The cervix holds a developing baby in the uterus, expanding only at the onset of labor to permit passage of the child. Beyond the cervix is the **vagina**, which opens to the outside. The vagina serves both as the receptacle for the penis during intercourse and as the birth canal (see Fig. 35-13).

Copulation Allows Internal Fertilization

As terrestrial mammals, humans use internal fertilization to deposit sperm into the moist environment of the female's reproductive tract. The penis is inserted into the vagina, where sperm are released during ejaculation. The sperm swim upward in the female reproductive tract, from the vagina through the opening of the cervix into the uterus, and up into the uterine tubes. If the female has ovulated within the past day or so, the sperm will meet an egg in one of the uterine tubes. Only one sperm can succeed in fertilizing the egg and thus begin the development of a new human being.

During Copulation, Sperm Are Deposited in the Female's Vagina

The male role in copulation begins with erection of the penis. Before erection, the penis is relaxed (flaccid), because the arterioles that supply it are constricted, allowing little blood flow (Fig. 35-17a). Under psychological and physical stimulation, the arterioles dilate and blood flows into spaces in the tissue within the penis. As these tissues swell, they squeeze off the veins that drain the penis (Fig. 35-17b). Blood pressure increases, causing an erection. After the penis is inserted into the vagina, movements further stimulate touch receptors on the penis, triggering ejaculation. *Ejaculation* occurs when muscles encircling the epididymis, vas deferens, and urethra contract, forcing semen out of the penis and into the vagina. On average, 3 or 4 milliliters of semen, containing 300 million to 400 million sperm, is ejaculated. *Male orgasm* causes ejaculation and a feeling of intense pleasure and release.

Similar changes occur in the female. Sexual excitement causes increased blood flow to the vagina and to the external parts of the reproductive tract. These external parts consist of the **labia** (singular, **labium**)—paired folds of skin known as the *labia minor* and *labia major*—and the **clitoris**, a rounded projection (see Fig. 35-13). The clitoris, which is derived from the same embryological tissue as the tip of the penis, becomes erect. Stimulation by the penis may result in *female orgasm*, a series of rhythmic contractions of the vagina and uterus accompanied by sensations of pleasure and release. Female orgasm is not necessary for fertilization.

The intimate contact involved in copulation creates a situation in which disease organisms can readily be transmitted. Ever since the "sexual revolution," which began in the 1960s, many individuals have had multiple sexual partners, and the incidence of *sexually transmitted diseases* (STDs) has greatly increased (see "Health Watch: Sexually Transmitted Diseases").

During Fertilization, the Sperm and Egg Nuclei Unite

Neither sperm nor egg lives very long if fertilization does not occur. An unfertilized egg may remain viable for a day, and sperm, under ideal conditions, may live for two. Therefore, fertilization can succeed only if copulation occurs within a couple of days before or after ovulation. You will recall that when it leaves the ovary, the egg is surrounded by follicle cells. These cells, now called the **corona radiata**, form a barrier between the

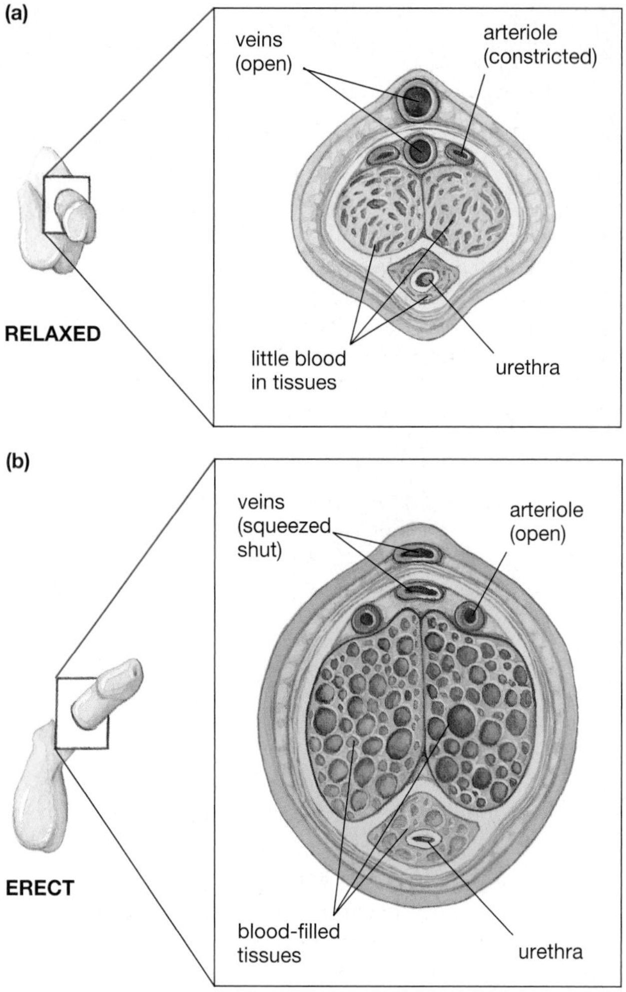

Figure 35-17 Changes in blood flow within the penis cause erection
(a) Normally, smooth muscles encircling the arterioles leading into the penis are contracted, limiting blood flow. ***(b)*** During sexual excitement, these muscles relax, and blood flows into spaces within the penis. The swelling penis squeezes off the veins leaving the penis, thereby increasing the pressure produced by fluids within the penis and causing it to become elongated and firm.

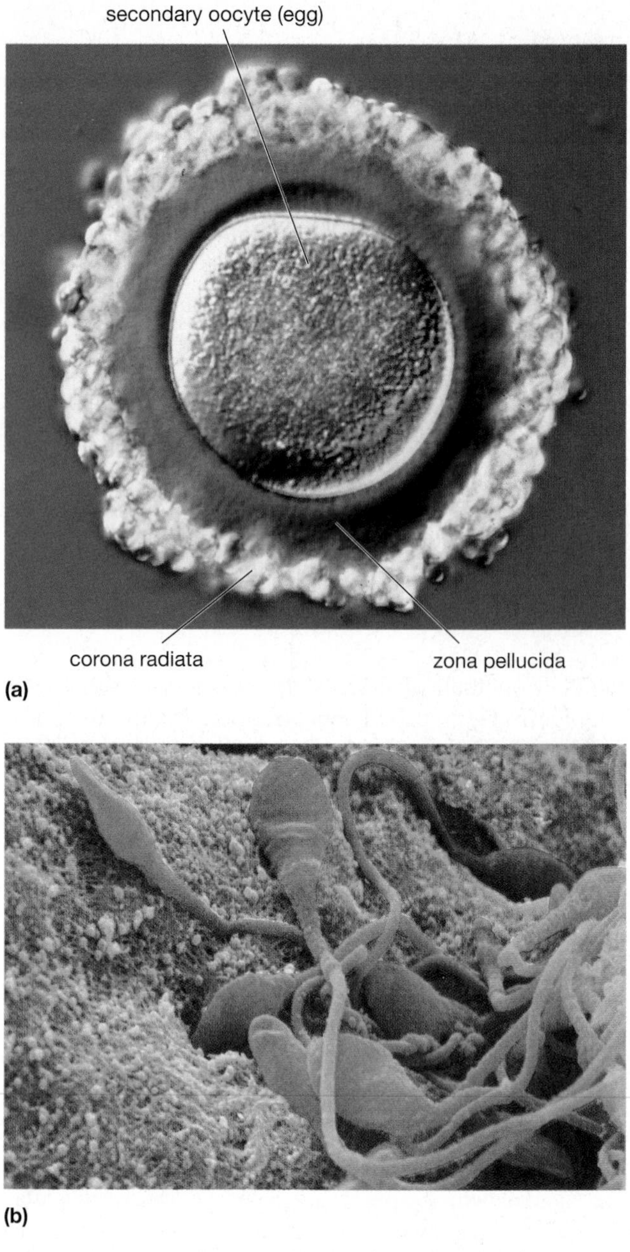

Figure 35-18 The secondary oocyte and fertilization
(a) A human secondary oocyte shortly after ovulation. Sperm must digest their way through the small follicular cells of the corona radiata and the clear zona pellucida to reach the oocyte itself. ***(b)*** Sperm surround the oocyte, attacking its defensive barriers.

sperm and the egg (Fig. 35-18a). A second barrier, the jellylike **zona pellucida** ("clear area"), lies between the corona radiata and the egg. Recent research suggests that the human egg releases a chemical attractant that lures the sperm toward it.

In the uterine tube, hundreds of sperm reach the egg and encircle the corona radiata, each sperm releasing enzymes from its acrosome (Fig. 35-18b). These enzymes weaken both the corona radiata and the zona pellucida, allowing the sperm to wriggle through to the egg. If there aren't enough sperm, an insufficient amount of enzyme is released, and none of the sperm will reach the egg. This may be the reason that natural selection has led to the ejaculation of so many sperm. Perhaps 1 in 100,000 reach the uterine tube, and 1 in 20 of those find the egg, so only a few hundred of the 300 million sperm that were ejaculated join the attack on the barriers around the egg.

When the first sperm finally contacts the surface of the egg, the plasma membranes of egg and sperm fuse, and the sperm's head is drawn into the egg's cytoplasm. As the sperm enters, it triggers two vital changes in the egg: (1) Vesicles near the surface of the egg release chemicals into the zona pellucida, reinforcing it and preventing further sperm from entering the egg. (2) The egg undergoes its second meiotic division, producing a haploid gamete at last. Fertilization occurs as the hap-

Health Watch
Sexally Transmitted Diseases

Sexually transmitted diseases (STDs), caused by viruses, bacteria, protists, or arthropods that infect the sexual organs and reproductive tract, are a serious and growing health problem worldwide. The World Health Organization estimates that there are 250 million new STD cases each year. As the name implies, these diseases are transmitted either exclusively or primarily through sexual contact. Here we discuss some of the more common of these diseases.

Bacterial Infections **Gonorrhea**, an infection of the genital and urinary tract, is a very common STD in the United States, with nearly 2 million cases reported annually. The causative bacterium, *Neisseria gonorrhoeae*, is transmitted almost exclusively by intimate contact. It penetrates the membranes that line the urethra, anus, cervix, uterus, uterine tubes, and throat. In males, inflammation of the urethra results in painful urination and a discharge of pus from the penis. About 10% of infected males and 50% of infected females do not seek treatment, because they have symptoms that are mild or absent, and they become carriers who can readily spread the disease. Gonorrhea can lead to infertility by blocking the uterine tubes with scar tissue. Treatment is by antibiotics. Infants born to infected mothers can acquire the bacterium during delivery. The bacterium attacks the eyes of newborns and was once a major cause of blindness. Today, most newborns are immediately given antibiotic eyedrops preventively to kill the bacterium.

Syphilis is caused by a spiral-shaped bacterium *Treponema pallidum,* which enters the mucous membranes of the genitals, lips, anus, or breasts. Like gonorrhea, it is readily killed by exposure to air and is spread only by intimate contact. Syphilis begins with a sore at the site of infection and can be cured with antibiotics. If untreated, syphilis bacteria spread through the body, multiplying and damaging many organs, including the skin, kidneys, heart, and brain, in some cases with fatal results. Syphilis can be transmitted to the fetus during pregnancy; and the skin, teeth, bones, liver, and central nervous system of such infants may be damaged.

Chlamydia causes inflammation of the urethra in males and the urethra and cervix in females. In many cases there are no obvious symptoms, so the infection goes untreated and is spread. Like the gonorrhea bacterium, the chlamydia bacterium, *Chlamydia trachomatis*, can infect and block the uterine tubes, resulting in sterility. Chlamydial infection can cause eye inflammation in infants born to infected mothers.

Viral Infections **Acquired immune deficiency syndrome**, or **AIDS**, is caused by the HIV virus, discussed in Chapter 31. Because the virus does not survive exposure to air, it is spread primarily by sexual activity and by contaminated blood and needles. The HIV virus attacks the immune system, leaving the victim vulnerable to a variety of infections, which almost invariably prove fatal. Children born to mothers with AIDS can become infected before or during birth. There is no cure, but certain drugs, such as AZT and protease inhibitors, can prolong life.

Genital herpes causes painful blisters on the genitals and surrounding skin and is transmitted primarily when blisters are present. The herpes virus never leaves the body but resides in certain nerve cells, emerging unpredictably, possibly in response to stress. The first outbreak is the most serious; subsequent outbreaks produce fewer blisters and can be quite infrequent. The drug acyclovir, which inhibits viral DNA replication, can reduce the severity of outbreaks. Pregnant women with an active case of genital herpes can transmit the virus to the developing fetus, causing severe mental or physical disability or stillbirth. Herpes can also be transmitted from mother to infant if the infant contacts blisters during childbirth.

Genital warts are growths or bumps that appear on the external genitalia, in or around the vagina or anus, or on the cervix in females, and on the penis, scrotum, groin, or thigh in males. The warts, which are normally painless, are caused by the *human papillomavirus (HPV)*, which is transmitted by skin-to-skin contact during sex. The most common treatment is removal of the warts, typically by freezing the warts with liquid nitrogen. Antiviral drugs can also be effective. Like the herpes virus, the human papillomavirus resides inside cells and is therefore difficult to cure. In women, genital warts are considered to be a risk factor for certain types of cervical cancer.

Protist and Arthropod Infections **Trichomoniasis** is caused by *Trichomonas*, a flagellated protist that colonizes the mucous membranes that line the urinary tract and genitals of both males and females. The symptoms are a discharge caused by inflammation in response to the parasite. The protist is spread by intercourse but can also be acquired through contaminated clothing and toilet articles. Lengthy untreated infections can result in sterility.

Crab lice, also called *pubic lice*, are microscopic arachnids (relatives of spiders) that live and lay their eggs in pubic hair. Their mouthparts are adapted for penetrating the skin and sucking blood and body fluids, a process that causes severe itching. "Crabs" are not only irritating; they can also spread infectious diseases. They can be controlled through careful hygiene and chemical treatments.

loid nuclei of sperm and egg fuse, forming a diploid nucleus that contains all the genes of a new human being.

Anomalies in the male or female reproductive systems can prevent fertilization. For example, a blocked uterine tube can prevent sperm from reaching the egg. Likewise, men who produce fewer than 20 million sperm per milliliter of semen (about one-fifth the normal amount) normally cannot fertilize a woman's egg during intercourse because too few sperm reach the egg. If the sperm are otherwise normal, such men can father children by *artificial insemination*, in which a large quantity of their semen is injected directly into the uterine tube. Today, some couples seek high-technology help in the form of *in vitro* fertilization (see "Scientific Inquiry: High-Tech Reproduction").

3 How Can People Limit Fertility?

Natural selection favors individuals that reproduce successfully and whose offspring survive long enough to reproduce themselves. During most of human evolution, child mortality was high, and natural selection favored people who produced enough children to offset this high mortality rate. Today, however, most humans no longer need to have many children to ensure that a few will survive to adulthood, but we nonetheless retain the reproductive drives with which evolution has endowed us. As a result, each passing week sees nearly 1.6 million new people added to our increasingly overcrowded planet, and the control of birth rates has become an environmental necessity. On the individual level, birth control allows people to plan their families and provide the best opportunities for themselves and their children.

Historically, limiting fertility has not been easy. In the past, some cultures have tried such inventive, if bizarre, techniques as swallowing froth from the mouth of a camel or placing crocodile dung in the vagina. Since the 1970s, however, several effective techniques have been developed for **contraception**, or the prevention of pregnancy. These techniques are described next, and their reliability and some possible side effects are summarized in Table 35-3. The choice of a contraceptive should be made in consultation with a health professional, who can provide more complete information.

Permanent Contraception Can Be Achieved Through Sterilization

In the long run, the most effortless method of contraception is **sterilization**, in which the pathways through which sperm or egg must travel are interrupted (Fig. 35-19). In men, the vas deferens leading from each testis may be severed in an operation called a **vasectomy**. Sperm are still produced, but they cannot reach the penis during ejaculation. The surgery is performed under a local

Table 35-3 Nonpermanent Contraceptive Techniques

Method	Technique and Mechanism	Failure Rate*	Protection from STD
Abstinence	Deciding not to be sexually active.	0%	Protects against STD
Rhythm	Using cervical mucus changes, and body temperature measurements to estimate the time of ovulation and avoiding intercourse during the fertile period.	2% to 20% (almost never done correctly)	No protection
Spermicide	Spermicidal foam (such as non-oxynol-9) is placed in vagina prior to intercourse, forming a chemical barrier to sperm.	6% to 21%	Partial protection; may not protect against HIV
Contraceptive Sponge**	Soft disposable sponge impregnated with spermicide inserted in vagina. Acts as sperm-killing barrier.	10% to 15%	Partial protection
Diaphragm/ Cervical Cap	Flexible reusable domed rubber or rubber-like barriers; spermicide is placed within the dome, and the diaphragm (larger) or cap (smaller) is fitted over the cervix prior to intercourse.	6% to 18%	Partial protection
Condom (male)	Thin, disposable latex sheath placed over penis prior to intercourse. May be lubricated with spermicide. Catches sperm and prevents it from entering vagina.	2% to 12%	Best protection against STD for sexually active people. Should be combined with other methods of birth control for STD protection.
Condom (female)	Lubricated polyurethane pouch inserted into vagina, catches sperm before they enter cervix.	5% to 25%	Partial protection
Birth Control Pill	Pill containing either estrogen and synthetic progesterone (combination pill) or progesterone only (minipill). Must be taken daily. Prevents ovulation.	0.1% to 3%	No protection
Norplant®	Six flexible match-stick-sized tubes of synthetic progesterone are inserted under the skin. Gradual release of hormone prevents ovulation for 5 years.	0.04%	No protection
Depo-Provera®	Injection of synthetic progesterone that blocks ovulation for 3 months. Repeated at 3-month intervals.	0.3%	No protection
IUD (intrauterine device)	Small plastic device treated with hormones or copper and placed in the uterus by physician. Prevents sperm from reaching egg or, if sperm do reach egg, prevents implantation of embryo. The copper IUD may be left in for 8 to 10 years.	0.6% to 2.6%	No protection

*Percentage of women becoming pregnant per year. The low and high numbers, respectively, indicate the differences between consistent and correct use and use in a more typical way that is not always consistent or correct.

**Scheduled to be marketed in the U.S. in 2001.

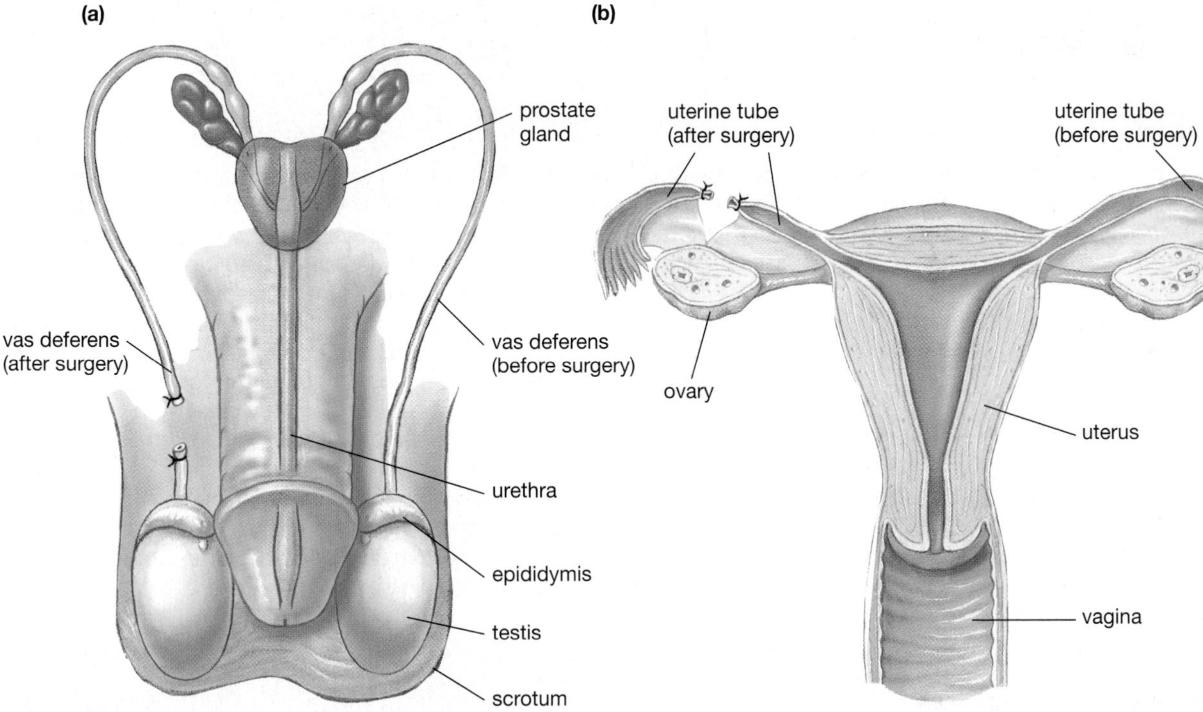

Figure 35-19 Sterilization
(a) Vasectomy in the male involves removing a ½ inch (1 cm) length of the vas deferens and tying off the cut ends. This prevents sperm from leaving the body. ***(b)*** Tubal ligation in the female involves removing a portion of the uterine tube and tying off the cut ends, preventing sperm from reaching the oocyte and preventing the oocyte from reaching the uterus.

anesthetic, and vasectomy has no known effects on health or sexual performance. The slightly more complex operation of **tubal ligation** renders a woman infertile by cutting her uterine tubes. Ovulation still occurs, but sperm cannot travel to the egg; nor can the egg reach the uterus. Sterilization is generally permanent. Sometimes, however, in a delicate and expensive operation, a surgeon can reconnect the vas deferens or uterine tubes.

There Are Three General Approaches to Temporary Contraception

Temporary contraception techniques fall into three general categories: (1) preventing ovulation, (2) preventing sperm and egg from meeting when ovulation does occur, and (3) preventing the implantation of a fertilized egg in the uterus.

Synthetic Hormones Can Prevent Ovulation

As you learned earlier in this chapter, during a normal menstrual cycle, ovulation is triggered by a mid-cycle surge of LH. An obvious way to prevent ovulation is to suppress LH release by providing a continuous supply of estrogen and progesterone. Estrogen and progesterone (generally in synthetic form) are the components of **birth control pills**. "The Pill" is an extremely effective form of birth control, but it must be taken daily, usually for 21 days each menstrual period.

Two long-term contraceptives, Norplant® and Depo-Provera®, have been approved for use in the United States since 1990. Both contain synthetic hormones resembling progesterone that prevent ovulation. Norplant consists of six slim, 1.3-inch-long silicone rubber rods inserted under the skin of the upper arm. The rods provide a gradual, steady diffusion of hormone into the bloodstream for 5 years. In extensive tests, Norplant has proved slightly more effective than the birth control pill. Most women using Norplant become fertile within months after the capsule is removed. Depo-Provera is injected once every 3 months.

Barrier Methods Prevent Sperm from Reaching the Egg

There are several effective *barrier methods,* which prevent the encounter of sperm and egg. The **diaphragm** and **cervical cap** are rubber caps that fit snugly over the cervix, preventing sperm from entering the uterus. In conjunction with a **spermicide**, diaphragms and cervical caps are very effective and have no serious known side effects. Alternatively, the male can wear a **condom** over the penis, preventing sperm from being deposited in the vagina. A female condom is now available that completely lines the vagina. Diaphragms and condoms must be applied shortly before intercourse, typically at a time when the participants would rather be thinking about something else. Furthermore, if the diaphragm or condom has even a small hole, sperm can still enter the uterus and fertilize the egg. However, condoms are

Scientific Inquiry
High-Tech Reproduction

Louise Brown, the first "test-tube baby," was born in England in 1978. Since that well-publicized event, more than 300 centers for *in vitro* fertilization (IVF) have been established in the United States; worldwide, more than 300,000 babies have been born as a result of this process.

IVF is a complex and delicate procedure. First, the woman is given daily injections of drugs, hormones, or both to stimulate multiple ovulations. Surgeons then insert a long, hollow needle into each ripe follicle and suck out the oocyte. Usually, at least four oocytes are incubated in a glass dish to which freshly collected sperm are added (*in vitro* fertilization literally means "fertilization in glass"). In 48 hours, some of the oocytes will have been fertilized, begun dividing, and reached the eight-cell stage. A few of these early embryos are sucked into a tube and expelled very gently into the uterus. (Extra embryos can be frozen for later use in case the first attempt at implantation is unsuccessful.) Transplanting multiple embryos increases the success rate for implantation, but it also increases the probability of multiple births, which are much riskier than single births. The popularity of IVF, which can cost as much as $10,000 per attempt, is a testimony to the strong biological drive to have children. With an average success rate of about 20%, a typical conception via IVF costs more than $40,000.

With a new technique, *intracytoplasmic sperm injection* (ICSI), even men who are unable to produce viable sperm may be able to father their own children. In ICSI, immature sperm (or even sperm precursor cells) are extracted from the testes. Then a micropipette is used to pierce the plasma membrane of an egg cell and inject a single sperm directly into the egg's cytoplasm (Fig. E35-2). In almost all cases, this procedure leads to a successful fertilization. Apparently, sperm are capable of fertilization very early in their developmental process, and further maturation mainly provides them with the ability to swim to the egg and penetrate its protective layers.

IVF is only one of a number of procedures that help formerly childless couples have children. For women who do not ovulate regularly, fertility drugs cause multiple ovulations; as a result of this procedure, the rate of multiple births in the U.S. has quadrupled since 1971 (Fig. E35-3). In the world of assisted reproduction, a widow could be impregnated by her dead husband's sperm stored at subzero temperatures. Recently, twins were born to a woman whose ovaries could not produce eggs. Her children came from donor eggs that had been cryopreserved for two years. A woman can put her uterus up for rent; as a surrogate mother she may bear a child for a woman who has had a hysterectomy or who simply does not want to go through pregnancy. The egg and sperm that produced the fetus developing within the surrogate mother could come from the couple who has hired her, or alternatively, either the egg, or the sperm, or both gametes could come from unrelated individuals. Conceivably, a modern newborn could have up to five "parents."

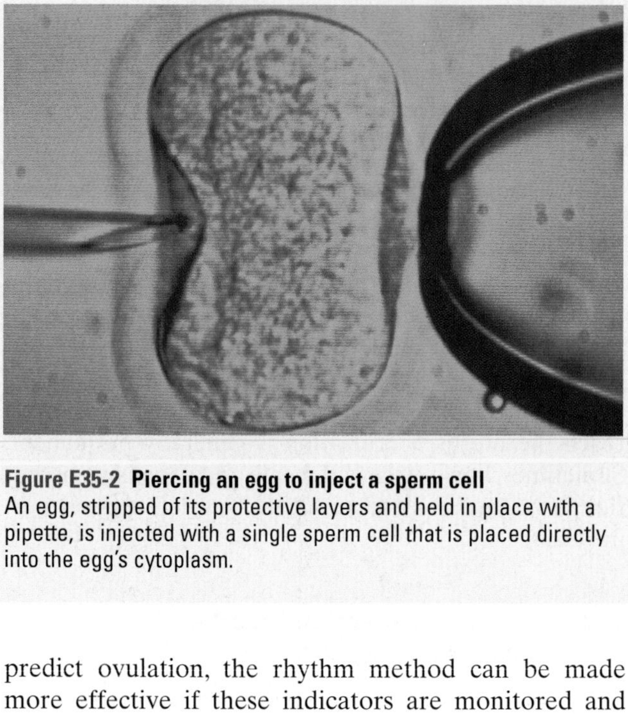

Figure E35-2 Piercing an egg to inject a sperm cell
An egg, stripped of its protective layers and held in place with a pipette, is injected with a single sperm cell that is placed directly into the egg's cytoplasm.

particularly effective against the spread of sexually transmitted diseases.

Other, less effective birth control procedures include the use of spermicides alone, **withdrawal** (the removal of the penis from the vagina just before ejaculation), and **douching** (washing sperm out of the vagina before, it is hoped, they have entered the uterus). Spermicides have some contraceptive effect, but withdrawal and douching are essentially useless. A final method of preventing fertilization is the **rhythm method**: abstinence from intercourse during the ovulatory period of the menstrual cycle. In practice, rhythm normally has a high failure rate, because of lack of discipline on the part of the users or inaccuracies in determining the menstrual cycle, which varies somewhat from month to month. Because a slight rise in body temperature and changes in the discharge of mucus from the cervix can be used to predict ovulation, the rhythm method can be made more effective if these indicators are monitored and sexual activity is regulated accordingly.

Intrauterine Devices and "Morning After" Pills May Prevent Implantation

Even if an egg is fertilized, pregnancy will not occur unless the blastocyst implants in the uterus (see Chapter 36), so contraception can be achieved by preventing implantation. One approach to this kind of contraception is an **intrauterine device (IUD)**, a small copper or plastic loop, squiggle, or shield that is inserted into, and removed from, the uterus by a physician. Highly effective (if it stays in place), the IUD seems to work by preventing implantation of the embryo (in humans, the stages from fertilization through the second month of development) in the uterine lining, although it may also prevent

Figure E35-3 Septuplets
The McCaugheys of Iowa had septuplets in 1997 after Mrs. McCaughey took fertility drugs.

Parents with the money and motivation can now strongly influence the sex of their child. This can be important if the parents are carriers of sex-linked disorders, but some parents are just seeking to balance their families. Researchers have found that sperm carrying an X chromosome have 2.8% more DNA than Y sperm, and this difference can be detected and used to sort the sperm. So far, they have been most successful at increasing the percentage of X sperm in a sample and increasing the probability that a couple will have a girl.

The National Zoo in Washington, D.C., has been working since 1984 to adapt IVF technology to help endangered species reproduce. Zoo officials have developed a mobile IVF laboratory that can travel to zoos where endangered species are housed. A tremendous advantage of IVF is that it will allow sperm from a male of an endangered species to be transported between continents, if necessary, to fertilize an appropriate female. This method eliminates the danger and trauma of transporting the animals themselves. It also overcomes the very real probability that, once together, the animals will refuse to mate. In April 1990, the first "test-tube" tiger was born (Fig. E35-4). Only 200 individuals of this rare Siberian subspecies remain in the wild. The use of IVF increases the chances that we may save this and other precious forms of life on Earth.

Figure E35-4 The world's first test-tube tiger
Born in 1990, this test-tube tiger is the result of the first successful use of IVF in the fight to save endangered species.

sperm from reaching the egg in the first place. A second method of preventing implantation is the "morning after" pill, which contains hormones similar to birth control pills, only in larger dosages. If a woman has intercourse without protection, these pills may be taken within 3 days ("morning after" is a misnomer) to help avoid pregnancy. The pills may stop or delay ovulation or prevent the embryo from implanting. Pregnancy may also be prevented after unprotected intercourse by inserting a copper IUD within 5 days.

Abortion Removes the Embryo from the Uterus

When contraception fails, pregnancy can be terminated by **abortion**. Abortion commonly involves dilating the cervix and removing the embryo and placenta by suction. Most abortions are performed during the first 3 months of pregnancy. Alternatively, abortion can be induced during the first 7 weeks of pregnancy by the drug RU-486 (mifepristone), which binds to progesterone receptors and blocks the actions of progesterone, which is essential to the maintenance of the uterine lining during pregnancy. Under the supervision of a physician, RU-486 is taken as a pill. Two days later, the physician administers a synthetic prostaglandin that induces uterine contractions, causing the embryo to be expelled. The drug RU-486, which has been used extensively abroad, received U.S. FDA approval in 2000. Abortions are potentially more dangerous to a woman's health than the contraceptive techniques described above. Although science can describe fetal development during pregnancy, it cannot provide judgments about when a fetus becomes a "person" or about the relative merits of fetal rights versus maternal rights. Therefore, abortion remains controversial.

CASE STUDY REVISITED

Mr. Mom?

In dwarf seahorses, the day after the young are born the female deposits more ripe eggs in the male's brood pouch. How does the female know she will find a male ready, able, and willing to brood her eggs just as they ripen? Biologist Amanda Vincent, a conservationist and one of the world's foremost researchers on seahorses, discovered the answer. The seahorse couples are faithful for life. The pair bond is renewed each morning with a courtship dance in which the two quiver, entwine tails, and circle around an object such as the sea grass in which many species live. These "married" seahorses show no interest in other sexual partners. From an evolutionary standpoint, this means that the male is guaranteed a steady supply of eggs to fertilize and brood throughout the breeding season, and the female is assured that her eggs will be incubated and protected by a willing male at exactly the time they are ripe. So the adaptation of monogamy enhances their reproductive success.

Why do the males brood the eggs? This frees the female to search more actively for food, which in turn enables the female to produce more eggs. The eggs are produced in large numbers because most of the diminutive offspring die.

Unfortunately the enormous reproductive success and ingenuity of another species—humans—now threatens populations of seahorses worldwide. Dr. Vincent estimates that over the past several years, seahorse populations have plummeted by 15% to 50% depending on the species. Harvested by the millions, seahorse bodies are dried and made into keychains or ground into powder. Others are sold live to aquarium owners, although they rarely survive in aquaria. The majority end up in Chinese pharmacies, where they are used in medicines from aphrodesiacs to stomachache remedies. Further, seahorses live and breed in some of the world's most threatened habitats: coral reefs, mangroves, and sea grass beds. As human activities degrade these areas, more seahorses are lost. In 1996 Dr. Vincent co-founded Project Seahorse. She and other biologists and social workers are educating residents in places such as the Philippines, Vietnam, and Hong Kong, where the seahorse trade flourishes. Project Seahorse is working to find alternatives to seahorse-based medicines and souvenirs and is helping local communities establish seahorse sanctuaries and aquaculture projects to help guarantee the survival of these unique and delightful creatures.

Think back to the description of seahorse reproduction and to the Scientific Inquiry essay on high-tech reproduction. What parallels do you see between the natural way seahorses reproduce and some of the artificial techniques devised by people?

Summary of Key Concepts

1) How Do Animals Reproduce?

Animals reproduce either sexually or asexually. In sexual reproduction, haploid gametes, usually from two separate parents, unite and produce an offspring that is genetically different from either parent. In asexual reproduction, offspring tend to be genetically identical to the parent. Asexual reproduction can occur by budding, fission, or parthenogenesis.

Among animals that engage in sexual reproduction, the female produces large, nonmotile eggs and the male produces small, motile sperm. Animals are either monoecious (a single animal produces both sperm and eggs) or dioecious (a single animal produces one type of gamete). Fertilization, the union of sperm and egg, can occur outside the bodies of the animals (external fertilization) or inside the body of the female (internal fertilization). External fertilization must occur in water so that the sperm can swim to meet the egg. Internal fertilization normally occurs by copulation, in which the male deposits sperm directly into the female's reproductive tract.

2) How Does the Human Reproductive System Work?

The human male reproductive tract consists of paired testes, which produce sperm and testosterone, and accessory structures that conduct the sperm to the female's reproductive tract and secrete fluids that activate swimming by the sperm and provide energy. In human males, spermatogenesis and testosterone production are stimulated by FSH and LH, secreted by the anterior pituitary. Spermatogenesis and testosterone production are nearly continuous, beginning at puberty and lasting until death.

The human female reproductive tract consists of paired ovaries, which produce eggs as well as the hormones estrogen and progesterone, and accessory structures that conduct sperm to the egg and receive and nourish the embryo during prenatal development. In human females, oogenesis, hormone production, and development of the lining of the uterus vary in a monthly menstrual cycle. The cycle is controlled by hormones from the hypothalamus (gonadotropin-releasing), anterior pituitary (FSH and LH), and ovaries (estrogen and progesterone).

During copulation, the male ejaculates semen into the female's vagina. The sperm move through the vagina and uterus into the uterine tube, where fertilization usually takes place. The unfertilized egg is surrounded by two barriers, the corona radiata and the zona pellucida. Enzymes released from the acrosomes in the head of sperm digest these layers, permitting sperm to reach the egg. Only one sperm enters the egg and fertilizes it.

3) How Can People Limit Fertility?

Contraception can be achieved by abstinence or by sterilization: severing the vas deferens in males (vasectomy) or the uterine tubes in females (tubal ligation). Temporary contraception techniques include those that prevent ovulation: birth control pills, Norplant, and Depo-Provera.

Barrier methods, which prevent sperm and egg from meeting, include the diaphragm, the cervical cap, and the condom, accompanied by spermicide. Spermicide alone is less effective. Withdrawal and douching are ineffective techniques. The rhythm method requires abstinence around the time of ovulation. Intrauterine devices may block sperm and also prevent implantation of the early embryo. Abortion causes the expulsion of the developing embryo.

Key Terms

abortion *p. 735*
acquired immune deficiency syndrome (AIDS) *p. 731*
acrosome *p. 724*
amplexus *p. 720*
asexual reproduction *p. 717*
birth control pill *p. 733*
bud *p. 718*
budding *p. 718*
bulbourethral gland *p. 724*
cervical cap *p. 733*
cervix *p. 729*
chlamydia *p. 731*
chorionic gonadotropin (CG) *p. 729*
clitoris *p. 729*
condom *p. 733*
contraception *p. 732*
copulation *p. 720*
corona radiata *p. 729*
corpus luteum *p. 725*
crab lice *p. 731*
diaphragm *p. 733*
dioecious *p. 719*
douching *p. 734*
egg *p. 719*
embryo *p. 725*
endometrium *p. 727*
epididymis *p. 724*
estrogen *p. 725*
external fertilization *p. 719*
fertilization *p. 718*
fetus *p. 725*
fimbria *p. 727*
fission *p. 718*
follicle *p. 725*
follicle-stimulating hormone (FSH) *p. 724*
genital herpes *p. 731*
genital wart *p. 731*
gonad *p. 722*
gonadotropin-releasing hormone (GnRH) *p. 724*
gonorrhea *p. 731*
hermaphrodite *p. 719*
internal fertilization *p. 720*
interstitial cell *p. 722*
intrauterine device (IUD) *p. 734*
labium *p. 729*
luteinizing hormone (LH) *p. 724*
menstrual cycle *p. 727*
menstruation *p. 729*
monoecious *p. 719*
myometrium *p. 727*
oogenesis *p. 725*
oogonium *p. 725*
ovary *p. 724*
ovulation *p. 722*
parthenogenesis *p. 718*
penis *p. 724*
pheromone *p. 719*
placenta *p. 727*
polar body *p. 725*
primary oocyte *p. 725*
primary spermatocyte *p. 722*
progesterone *p. 725*
prostate gland *p. 724*
regeneration *p. 718*
rhythm method *p. 734*
scrotum *p. 722*
secondary oocyte *p. 725*
secondary spermatocyte *p. 722*
semen *p. 724*
seminal vesicle *p. 724*
seminiferous tubule *p. 722*
Sertoli cell *p. 722*
sexually transmitted disease (STD) *p. 731*
sexual reproduction *p. 717*
spawning *p. 719*
sperm *p. 719*
spermatid *p. 722*
spermatogenesis *p. 722*
spermatogonium *p. 722*
spermatophore *p. 720*
spermicide *p. 733*
sterilization *p. 732*
syphilis *p. 731*
testis *p. 722*
testosterone *p. 724*
trichomoniasis *p. 731*
tubal ligation *p. 733*
urethra *p. 724*
uterine tube *p. 725*
uterus *p. 727*
vagina *p. 729*
vas deferens *p. 724*
vasectomy *p. 732*
withdrawal *p. 734*
zona pellucida *p. 730*
zygote *p. 727*

Thinking Through the Concepts

Multiple Choice

1. *Budding and fission are processes used by*
a. dioecious species
b. hermaphroditic organisms
c. organisms requiring new gene combinations for each generation
d. sexually reproducing species
e. asexually reproducing species

2. *All of the following are barrier contraceptive devices EXCEPT the*
a. IUD
b. female condom
c. cervical cap
d. diaphragm
e. male condom

3. *In humans, spermatogenesis yields ______ sperm for each diploid sex cell, and oogenesis yields ______ secondary oocyte(s) for each sex cell.*
a. one, four
b. two, one
c. one, two
d. four, one
e. four, two

4. *Which structure adds the final secretions to semen as it moves out of the male reproductive tract?*
a. epididymis
b. bulbourethral gland
c. seminal vesicle
d. prostate gland
e. interstitial cells

5. *The corpus luteum*
a. accompanies the egg as it enters the oviduct
b. is the fertilized egg
c. secretes prostaglandin
d. forms in the uterus
e. secretes both estrogen and progesterone

6. *The primary hormone that inhibits GnRH is*
a. FSH
b. LH
c. progesterone
d. estrogen
e. a hypothalamic releasing factor

? Review Questions

1. List the advantages and disadvantages of asexual reproduction, sexual reproduction, external fertilization, and internal fertilization, including an example of an animal that uses each type.
2. Compare the structures of the egg and sperm. What structural modifications do sperm have that facilitate movement, energy use, and digestion?
3. What is the role of the corpus luteum in a menstrual cycle? In early pregnancy? What determines its survival after ovulation?
4. Construct a chart of common sexually transmitted diseases. List the disease's name, the cause (organism), symptoms, and treatment.
5. List the structures, in order, through which a sperm passes on its way from the seminiferous tubules of the testis to the uterine tube of the female.
6. Name the three accessory glands of the male reproductive tract. What are the functions of the secretions they produce?
7. Diagram the menstrual cycle, and describe the interactions among hormones secreted by the pituitary gland and ovaries that produce the cycle.

Applying the Concepts

1. Discuss the most effective or appropriate method of birth control for each of the following couples: Couple A, which has intercourse three times a week but never wants to have children; Couple B, which has intercourse once a month and may want to have children someday; and couple C, which has intercourse three times a week and wants to have children someday.
2. *Pelvic endometriosis* is a relatively common disease of women in which bits of the endometrial lining find their way onto abdominal organs and respond in typical ways to hormones during a menstrual cycle. When the uterine lining bleeds during menstruation, so do these implants. Common treatments are oral contraceptives, Danazol™ (a compound that inhibits gonadotropins), and synthetic GnRH analogues that are, paradoxically, powerful inhibitors of FSH and LH. How does each of these compounds provide relief?
3. Would contraceptive drugs that block cell receptors for FSH be useful in males and/or females? Explain. What side effects would such drugs have?
4. Think of all the choices a couple has to obtain a child, including *in vitro* fertilization using the couple's eggs and sperm, *in vitro* fertilization using a donor's sperm or egg, and insemination of a surrogate mother with sperm from the couple's husband. Think of some more. What ethical issues do these various options present?
5. Fertility drugs have greatly increased the incidence of multiple births. When more than two embryos share the uterus, the incidence of premature birth and developmental problems increases substantially. The costs of caring for multiple premature infants is staggering. When fertility drugs produce multiple embryos, the physician can selectively eliminate some of these embryos early in development, so the remaining few have a better chance to develop fully and normally. Given these facts, discuss the ethical implications of taking fertility drugs.

For More Information

Alexander, N. "Future Contraceptives." *Scientific American*, September 1995. A review of the new techniques that may be used for contraception in the twenty-first century.

Christiansen, D. "Male Choice." *Science News*, September 30, 2000. Describes the problems and progress in developing contraceptives for men.

Crews, D. "Animal Sexuality." *Scientific American*, January 1994. Explores the wide range of mechanisms that control male and female sexual development among different types of animals.

Eberhard, W. G. "Runaway Sexual Selection." *Natural History*, December 1987. Interesting article on the evolution of male genitalia.

Fackelmann, K. "It's a Girl." *Science News*, November 28, 1998. Sperm sorting can help parents select the sex of their children, particularly for girls.

Morell, V. "A Clone of One's Own." *Discover*, May 1998. Someday, someone will probably clone a person. What will be the outcome?

Lanza, R. P., Dresser, B. L., and Damiani, P. "Cloning Noah's Ark." *Scientific American*, November 2000. For some endangered species, cloning may offer the best chance for survival.

Riddle, J. M., and Estes, J. W. "Oral Contraceptives in Ancient and Medieval Times." *American Scientist*, May–June 1992. How did women control their fertility before modern medicine stepped in?

Wassarman, P. W. "Fertilization in Mammals." *Scientific American*, December 1988. Describes the events leading to the penetration of the egg by the sperm, with emphasis on the role and composition of the zona pellucida.

Wright, K. "Human in the Age of Mechanical Reproduction." *Discover*, May 1998. The child's question "Where do babies come from?" has become far more difficult to answer as technology has provided more than a dozen ways to have a child.

Answers to Multiple-Choice Questions

1. e 2. a 3. d 4. b 5. e 6. c

MediaTutor
Animal Reproduction

CD Activities

Activity 35.1: Human Reproductive Anatomy

Estimated time: 10 minutes

The human reproductive system is complex and contains a majority of the differences between male and female anatomies. In this tutorial, you will explore the structure and function of the human reproductive system. Your exploration will include a comparison of the male and female anatomies and an introduction to the process of gametogenesis, which is the formation of sperm in the male and an egg in the female.

Activity 35.2: Hormonal Control of the Menstrual Cycle

Estimated time: 10 minutes

The female menstrual cycle is controlled by a variety of complex hormonal interactions that have their effects in both the ovaries and the uterus. The ebb and flow of the various reproductive hormones and the events of the menstrual cycle are depicted over a typical 28-day cycle. These hormonal changes are linked to physical changes occurring in the ovary and the lining of the uterus. In this tutorial, you will explore the events of the menstrual cycle and the function of the various reproductive hormones that are linked to these events.

Start the MediaTutor Student CD-ROM and enter the activity number in the Quick Search box to be taken directly to that activity.

Web Investigations

Case Study: Mr. Mom?

Estimated time: 10 minutes

Male seahorse "mothers" are only one example of a specialized reproductive adaptation optimized for an ecological niche. Why do seahorses, marsupials, and monotremes all have pouches? What other mechanisms do animals use to enhance their reproductive fitness? Take a closer look at some reproductive oddities.

Go to http://www.prenhall.com/audesirk6, the Audesirk Companion Web site. Select Chapter 35 and the Web Investigation to begin.

The pregnant mother's decisions can have a lifelong influence on the health, behavior, and physical attributes of her developing child.

36 Animal Development

AT A GLANCE

Case Study: Far-reaching Choices

1) **How Do Indirect and Direct Development Differ?**
During Indirect Development, Animals Undergo a Radical Change in Body Form
Newborn Animals That Undergo Direct Development Resemble Miniature Adults

2) **How Does Animal Development Proceed?**
Cleavage Begins the Process
Gastrulation Forms Three Tissue Layers
Adult Structures Develop During Organogenesis
Sexual Maturation Is Controlled by Genes and the Environment

3) **How Is Development Controlled?**
Each Cell Contains the Entire Genetic Blueprint for the Organism
Gene Transcription Is Precisely Regulated During Development

4) **How Do Humans Develop?**
During the First 2 Months, Rapid Differentiation and Growth Occur
The Placenta Secretes Hormones and Exchanges Materials Between Mother and Embryo
Growth and Development Continue During the Last 7 Months
Development Culminates in Labor and Delivery
Milk Secretion is Stimulated by the Hormones of Pregnancy
Aging Seems to Be Genetically Programmed

Case Study Revisited: Far-reaching Choices

CASESTUDY

Far-reaching Choices

Maria is excited and nervous as she phones Nicole, a close friend from high school. The two have kept in touch over the past decade or so, and Nicole took part in Maria's recent wedding. "My home pregnancy test showed up blue this morning—I think we did it!" Maria enthuses. Nicole offers her congratulations, sounding temporarily overwhelmed. Then she says, "Let's celebrate! Join me at Sudsy Smith's tonight. Just one drink, and—oh, I'm late for work! How about 5:30?" Recalling the smoky ambiance of Sudsy's and a warning she once noticed on a wine bottle about alcohol and pregnancy, Maria suggests a local coffee shop instead. As she hangs up the phone, she weighs the significance of her new responsibilities. What did the wine bottle say? Retrieving a can of beer from the refrigerator, she reads the small print on the can: "Warning: According to the Surgeon General, women should not drink alcoholic beverages during pregnancy because of the risk of birth defects." As Maria contemplates the seeming miracle by which a single fertilized egg becomes a dazzlingly complex individual, she vows to do everything she can to protect the health of her developing child. Her next phone call is to her doctor to schedule a prenatal visit. As Maria prepares a list of questions for her physician, she also plans a trip to the library to educate herself and her husband.

Although Maria will abstain from all alcohol, the first question on her list is "How does alcohol affect prenatal development? Will children born to mothers who drink suffer serious health problems later on?" As an occasional smoker, her second question is "How does a mother's smoking affect prenatal development?"

As you learn more about the amazing series of events that underlie development, keep Maria's investigation in mind. What questions would you ask your doctor if you were a concerned expectant mother or father? How would you change your life? If you were concerned about a friend who smoked or drank during pregnancy, how would you deal with the issue? ■

How does form arise from formlessness? This question represents one of the central mysteries of biology and is the basis of the burgeoning field of developmental biology. Developmental biologists want to know how a single cell—the zygote formed from the fusion of sperm and egg—transforms itself into a complex organism. As cell division proceeds, some daughter cells become nerves, some become muscles, some become sensory organs, and so on. Because the cells of the embryo proliferate by mitosis, each cell has an identical genome. So what mechanism directs cells to become different specialized types? Perhaps even more puzzling is the question of how animal form arises. How do the cells of, say, a developing dog arrange themselves into the shape of a dog? How do they "know" to arrange themselves as a leg with four toes on a paw and not as a fin, or a wing, or a leg with a hoof, or a tail, or an eye? And once all that is settled, how do the cells age? As you've probably guessed, developmental biology is a field with more questions than answers. Yet, answers are beginning to emerge, and exciting new findings are published almost daily. Let's look at what is known about development and then return to look at a few of the beginnings of answers to these questions.

1 How Do Indirect and Direct Development Differ?

When we think of development, images of a newborn infant may come to mind. Certainly, the proportions are different, but the baby is, in all important ways, a miniature version of an adult person. People, other mammals, and birds and reptiles as well are all born as "miniature adults," a process called **direct development.** For the vast majority of animal species, however, indirect development is the norm.

During Indirect Development, Animals Undergo a Radical Change in Body Form

In **indirect development**, the juvenile animal that hatches from the egg differs significantly from the adult and undergoes radical changes in body form, such as the transformation of a caterpillar into a butterfly. Indirect development occurs in most invertebrates, including insects and echinoderms, and in a few vertebrates—notably, the amphibians. Animals with indirect development typically produce huge numbers of eggs, and each egg has only a small amount of food reserve called **yolk**. The yolk nourishes the developing embryo during a rapid transformation into a small, sexually immature form called a **larva** (Fig. 36-1a,b). Because the yolk is small and the time spent as an embryo is relatively short, indirect development does not place great demands on the mother, and many offspring can be produced.

Some larval animals not only look very different from adult animals but also occupy entirely different habitats. In addition, most larvae feed on different organisms than they will as adults. For instance, the aquatic larva of the dragonfly feeds on aquatic organisms such as tadpoles, but the adult dragonfly, which is terrestrial, feeds on insects (Fig. 36-1b). Eventually, the larvae undergo a revolution in body form, or **metamorphosis**, and become sexually mature adults.

Although we tend to regard the adult form as the "real animal" and larvae as "preparatory stages," most of the life span of some animals, especially insects, is spent as a larva. The adult may live for only a few days, reproducing frantically and in some cases not even eating. The mayfly, for example, metamorphoses from an aquatic larva that may have spent a year or more feeding and growing. Emerging in huge swarms from freshwater streams, ponds, and lakes, adult mayflies live a few hours or, at most, a few days. The sole occupation of the adults is to mate and lay eggs; their fragile dead bodies then accumulate in piles to be swept away by the wind.

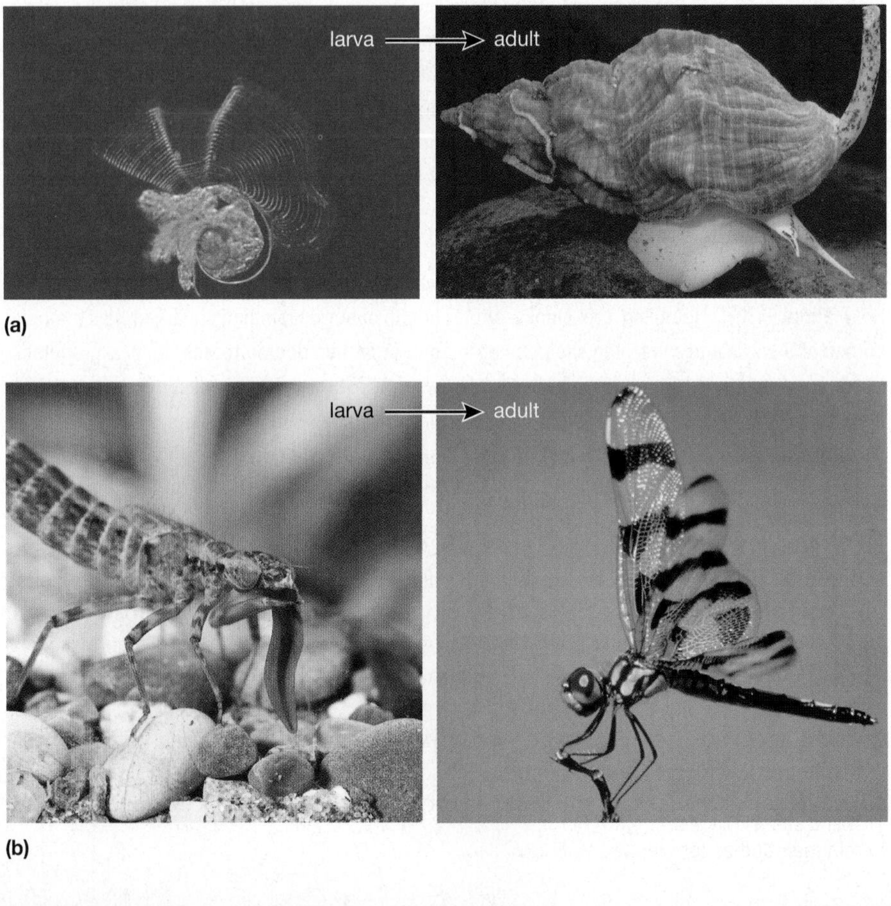

Figure 36-1 Indirect development
Many animals, including ***(a)*** many marine mollusks such as this common whelk, undergo indirect development. The larval stage tends to be very different from the adult in size, appearance, and lifestyle. The molluscan larva is barely visible to the naked eye. ***(b)*** The larval dragonfly is aquatic and feeds on tadpoles and small fish, whereas the adult form is terrestrial and eats other insects.

Newborn Animals That Undergo Direct Development Resemble Miniature Adults

Other animals, including such diverse groups as land snails, reptiles, birds, and mammals, show direct development, in which the newborn animal is a miniature, but sexually immature, version of the adult (Fig. 36-2). As the young animal matures, it may grow much bigger, but it does not radically change its body form.

Juveniles of directly developing species are typically much larger than larvae and consequently need much more nourishment before emerging into the world. Two strategies have evolved that meet the embryo's food requirement. Snails, reptiles, and birds produce eggs that contain relatively large amounts of yolk. An ostrich egg, for example, weighs several pounds. Mammals, some snakes, and a few fish have relatively little yolk in their eggs but instead nourish the developing embryo within the body of the mother. Either way (in contrast to indirect development), providing food for directly developing embryos places great demands on the mother, and relatively few offspring are produced.

Reptiles, Birds, and Mammals Produce Similar Extraembryonic Membranes

Amphibians were the first vertebrates to live on land, but their reproduction remains tied to water, where their eggs are deposited, and their larval offspring grow and metamorphose into adults. For the vertebrates, fully terrestrial

(a) (b) (c) (d)

Figure 36-2 Direct development
The offspring of animals with direct development closely resemble their parents from the moment of birth, except in size. ***(a)*** Land snails, ***(b)*** lizards, and ***(c)*** birds hatch from large, yolk-filled eggs. ***(d)*** Mammalian mothers nourish their young within their bodies for weeks or months.

life was not possible until a final piece of the puzzle evolved: the shelled **amniotic egg**. This innovation, which encases the embryo in a protected, liquid-filled space, arose first in the reptiles and persists today in that group and in its descendants, birds and mammals. It allows these groups to complete their development into the adult form in their own "private pond." The amniote egg is characterized by four membranes, called **extraembryonic membranes**: (1) the *chorion*; (2) the *amnion*; (3) the *allantois*; and (4) the *yolk sac*. The **chorion** lines the shell and exchanges oxygen and carbon dioxide through the shell. The **amnion** encloses the embryo in a watery environment; the **allantois** surrounds wastes; and the **yolk sac** contains the stored food. Although mammalian eggs contain almost no yolk, much of the reptilian genetic program for development still persists, including the four extraembryonic membranes. Table 36-1 compares the structures and functions of these extraembryonic membranes in reptiles and mammals.

2 How Does Animal Development Proceed?

The transformation from fertilized egg—a single cell—to a multicellular, differentiated embryo is a beautiful, nearly magical process that has been carefully described for a number of animals. In textbooks, the changes in the developing embryo are generally depicted in stages, but the stages are just convenient "snapshots" for the purpose of illustration (see "A Closer Look: Stages of Development" in Chapter 36 of this text's Web site). The actual development is a smoothly continuous process. The initial stages of *cleavage, gastrulation, organogenesis,* and *growth* occur during embryonic life, in which nearly all the organs that will be present in the adult are formed. After birth, the animal typically undergoes further growth, achieves *sexual maturity* (at which time the animal may reproduce), *ages,* and finally dies. Let's examine the stages of development.

Table 36-1 Vertebrate Embryonic Membranes

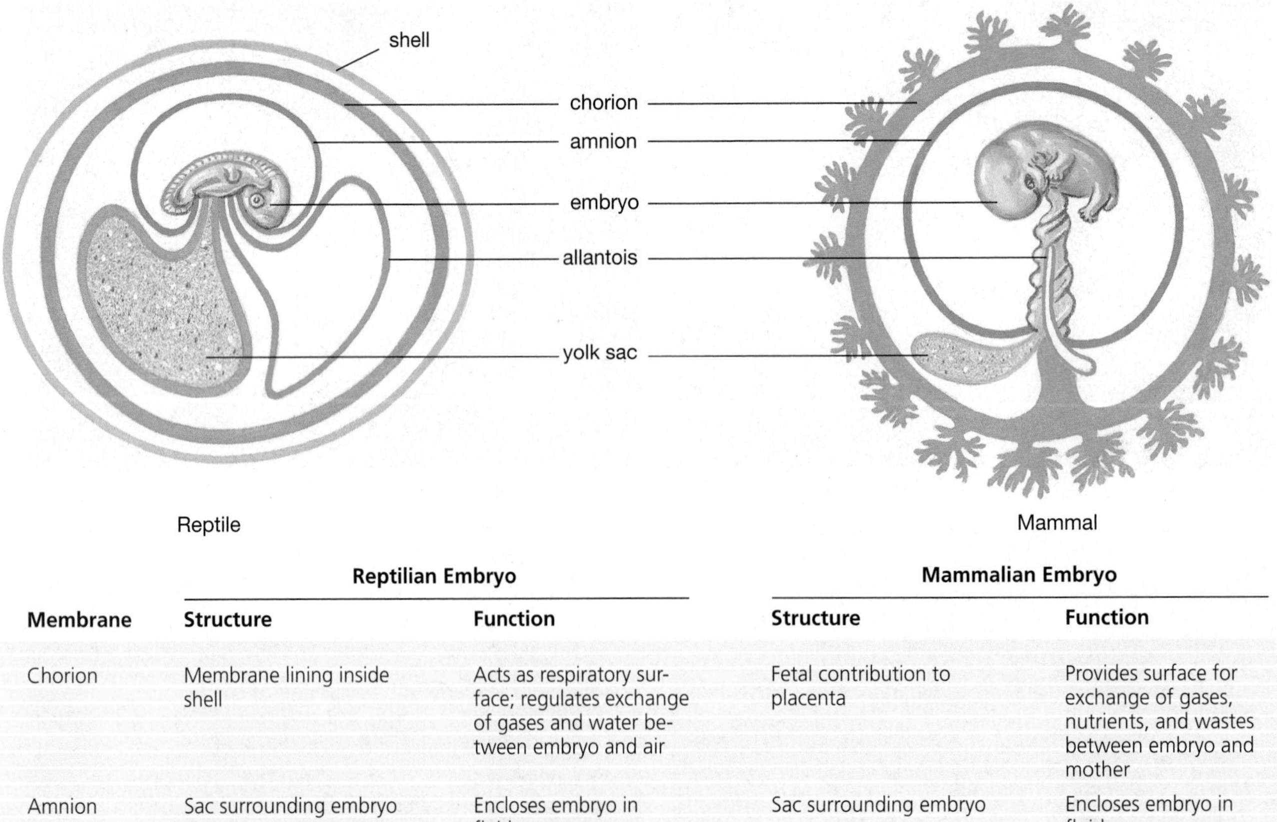

	Reptilian Embryo		Mammalian Embryo	
Membrane	**Structure**	**Function**	**Structure**	**Function**
Chorion	Membrane lining inside shell	Acts as respiratory surface; regulates exchange of gases and water between embryo and air	Fetal contribution to placenta	Provides surface for exchange of gases, nutrients, and wastes between embryo and mother
Amnion	Sac surrounding embryo	Encloses embryo in fluid	Sac surrounding embryo	Encloses embryo in fluid
Allantois	Sac connected to embryonic urinary tract; capillary-rich membrane lining inside of chorion, with blood vessels connecting to embryonic circulation	Stores wastes (especially urine); acts as respiratory surface	Provides blood vessels of umbilical cord	Carries blood between embryo and placenta
Yolk sac	Membrane surrounding yolk	Contains yolk as food; digests yolk and transfers nutrients to embryo; forms part of digestive tract	"Empty" membranous sac	Forms part of digestive tract

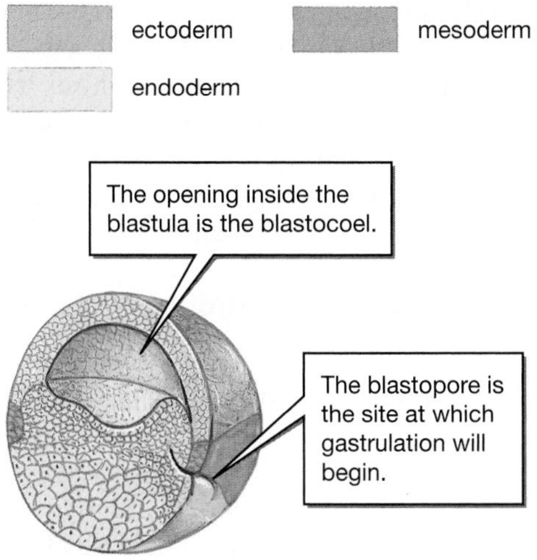

ectoderm
mesoderm
endoderm
The opening inside the blastula is the blastocoel.
The blastopore is the site at which gastrulation will begin.

(a) The blastula just before gastrulation.

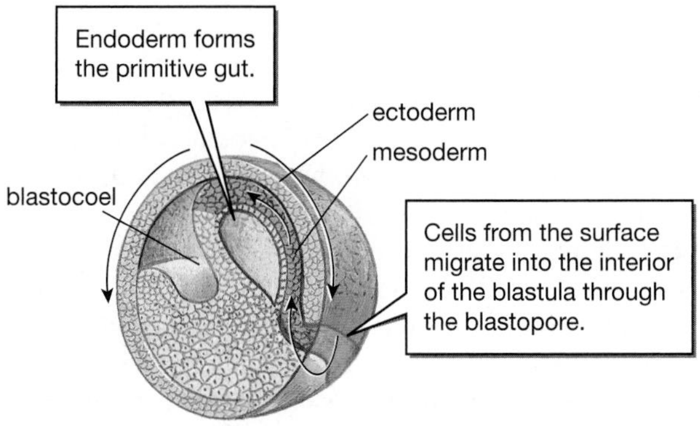

(b) Cells migrate at the start of gastrulation. These cells will form the endoderm and mesoderm layers of the gastrula; the cells remaining on the surface will form ectoderm.

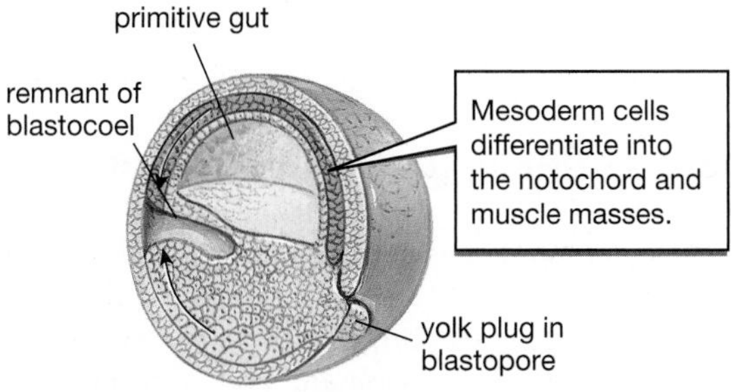

(c) Mesoderm differentiates.

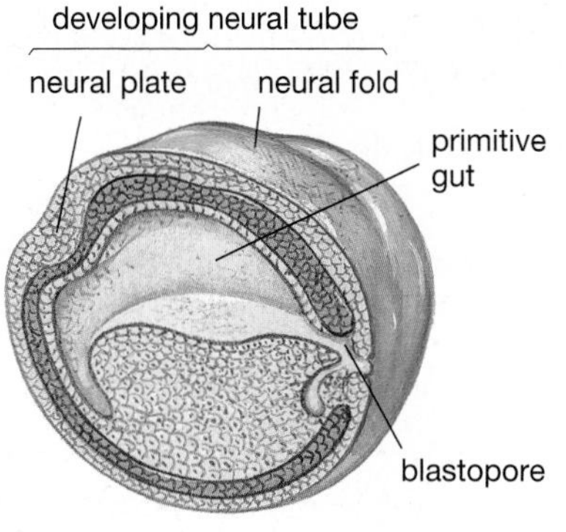

(d) The notochord induces ectoderm cells lying directly above it to form the neural tube.

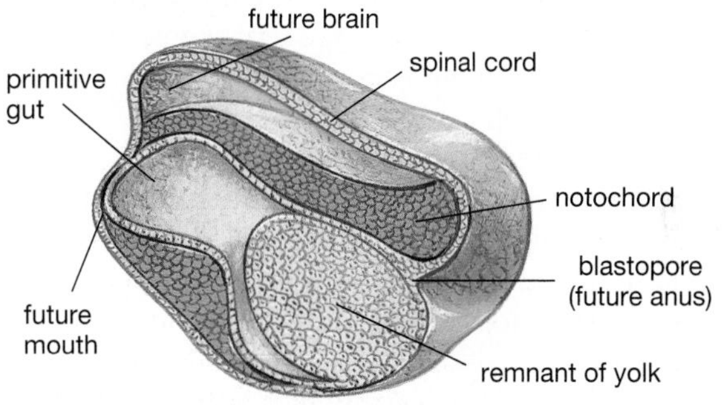

(e) Further development. The neural tube differentiates into brain and spinal cord. A future mouth is produced when the opening formed by the primitive gut breaks through at the end of the embryo opposite the blastopore. The blastopore is the future anus.

Figure 36-3 Gastrulation in the frog

MEDIATUTOR 36.1 Stages of Animal Development

Cleavage Begins the Process

The formation of an embryo begins with **cleavage**, a series of mitotic divisions of the fertilized egg without an increase in size. Cleavage reduces cell size and distributes gene-regulating substances to the newly formed cells. An egg is a very large cell. Unlike most mitotic cell divisions—which divide, grow, duplicate genetic material, then divide again—embryonic cell divisions during cleavage skip the growth phase. Consequently, as cleavage progresses, the available cytoplasm is split up into ever-smaller cells whose sizes approach those of cells in the adult organism. Finally, a solid ball of small cells, the **morula**, is formed. The morula as a whole is about the same size as the fertilized egg. As cleavage continues, a cavity opens within the morula. The cells of the morula then become the outer covering of a hollow (typically spherical) structure, the **blastula**. The space inside the blastula is called the *blastocoel* (Fig. 36-3a), .

The details of cleavage differ by species. The pattern is largely determined by the amount of yolk present, because yolk hinders cytoplasmic division (cytokinesis). The almost yolkless eggs of sea urchins divide symmetrically, but eggs with extremely large yolks, such as a hen's egg, don't divide all the way through. Nevertheless, a hollow blastula is always produced, though in reptiles and birds it is flattened rather than spherical.

Gastrulation Forms Three Tissue Layers

In the next step of development, an indentation called the **blastopore** forms on one side of the blastula. Blastula cells migrate in a continuous sheet in through the

Table 36-2 Derivation of Adult Tissues from Embryonic Cell Layers

Embryonic Layer	Adult Tissue
Ectoderm	Epidermis of skin; lining of mouth and nose; hair; glands of skin (sweat, sebaceous, and mammary glands); nervous system; lens of eye; inner ear
Mesoderm	Dermis of skin; muscle, skeleton; circulatory system; gonads; kidneys; outer layers of digestive and respiratory tracts
Endoderm	Lining of digestive and respiratory tracts; liver; pancreas

blastopore, much as if you punched in an underinflated basketball (Fig. 36-3b), to form three embryonic tissue layers. The enlarging dimple is destined to become the digestive tract; the cells that line its cavity are now called **endoderm** (Greek for "inner skin"). The cells remaining on the outside will form the epidermis of the skin and the nervous system and are called **ectoderm** ("outer skin"). Meanwhile, some cells migrate between the endoderm and ectoderm, forming a third and final layer, the **mesoderm** ("middle skin"). Mesoderm gives rise to muscles, the skeleton (including the **notochord**, a supporting rod found at some stage in all chordates; Fig. 36-3c), and the circulatory system (Table 36-2). This process of cell movement is called **gastrulation**, and the three-layered embryo that results is the **gastrula**.

Adult Structures Develop During Organogenesis

Gradually, the ectoderm, mesoderm, and endoderm rearrange themselves into the organs characteristic of the animal species (see Table 36-2). This process, called **organogenesis**, also usually occurs by induction.

In some cases, adult structures are, in effect, "sculpted" by the death of excess cells produced during embryonic development. Death of some cells is programmed to occur at a precise time during development. At least two mechanisms seem to be at work in different tissues. Some cells die during development unless they receive a "survival signal." Embryonic vertebrates, for example, have far more motor neurons in their spinal cords than do adult animals. Motor neurons are programmed to die unless they successfully form synapses with a skeletal muscle, which releases a chemical that prevents the death of its own motor neuron.

For other cells, the situation is just the reverse: Some cells live unless they receive a "death signal" from other cells of the developing animal. Many embryonic structures disappear during development. For example, all vertebrates pass through developmental stages with tails and webbed hands and feet. In humans, these stages can be seen clearly in the 5-week-old human embryo (see Fig. 36-11a). Two weeks later, the webbing cells have died, revealing separate fingers, and the tail cells are dying, causing the tail to regress (see Fig. 36-11b). In frogs, the tail is lost during metamorphosis from its tadpole larva. Thyroid hormone, which triggers metamorphosis, stimulates cells in the tail to synthesize enzymes that digest the tail away. If the thyroid gland is surgically removed, the frog retains its tail.

Sexual Maturation Is Controlled by Genes and the Environment

Development does not stop at birth; animals continue to change throughout their lives. The period between birth and sexual maturity is one of very active growth in which cells of all types increase in number with a resulting increase in the size of the entire organism. Growth usually slows as the organism becomes sexually mature and stops relatively soon afterward.

Animals are not born with full reproductive capability; they become sexually mature at an age determined by interactions between genetic and environmental factors. Most vertebrates must undergo months to years of genetically regulated growth and development before they are physiologically capable of producing sperm or eggs.

Once the animal has reached the appropriate age, however, the precise onset of sexual maturity must typically await specific environmental stimuli. For example, most songbirds that breed in temperate regions become sexually mature in the spring, stimulated by the increasing daylength. Social factors can also influence maturation in some species. The average age of puberty among women, for instance, has dropped substantially during the past few centuries. Puberty at a younger age is due in part to improved nutrition, but social stimulation by early close contact between the sexes may also be involved.

3 How Is Development Controlled?

Think for a moment about the biological miracle that transformed a single cell—a zygote—into the individual that is you. Biologists use prosaic terms for this incredible series of events: development and differentiation. **Development** is the process by which an organism proceeds from fertilized egg through adulthood. **Differentiation** is the specialization of embryonic cells into different cell types, such as muscle cells, brain cells, and so on. How do cells become differentiated from one another during development? We know that the zygote contains all the genes needed to direct the construction of the entire organism. Are genes lost as cells differentiate?

Each Cell Contains the Entire Genetic Blueprint for the Organism

In the early 1950s American embryologists Thomas King and Robert Briggs began pioneering experiments, later extended by British embryologist John Gurdon. They transplanted the nucleus of a differentiated cell taken from the intestine of a tadpole into an unfertilized frog egg whose nucleus had been removed (Fig. 36-4). In successful experiments, the intestinal nucleus directed the development of a normal tadpole, a feat that would have been impossible if genes were lost during differentiation. The experiment provided strong evidence that each differentiated cell in an animal contains all the genetic information needed for the development of the entire organism. Scientists now know that different types of cells differ not in which genes they *contain* but in which genes they *use*. In other words, cell types differ because different genes are activated, transcribed to messenger RNA, and translated into proteins. Does this experiment sound familiar? Think back to Chapter 13, where you learned that Dolly the sheep (and since then, many other animals) have been "cloned," each from a single adult cell whose nucleus was injected into an egg from which the nucleus had been removed. These frog experiments half a century ago set the stage for the recent breakthrough in mammalian cloning. The knowledge that all cells retain the genetic capacity to produce all of the specialized structures of the adult is now being utilized in another new and exciting way, described in "Scientific Inquiry: The Promise of Stem Cells."

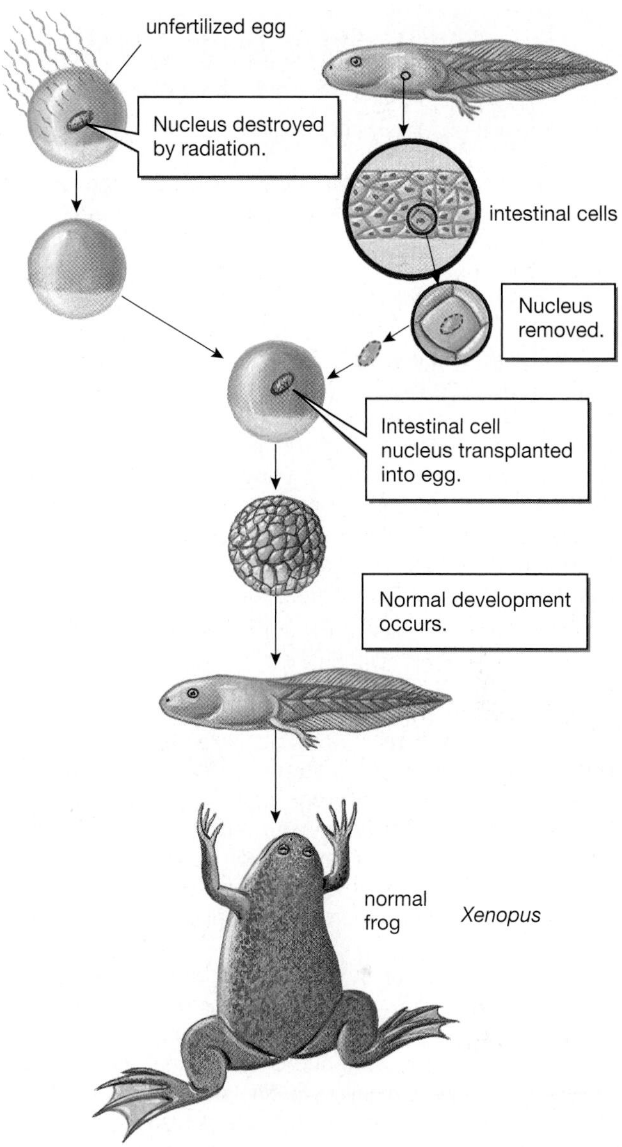

Figure 36-4 Cells retain all of their genes during differentiation This experiment shows that cells do not lose genes as they differentiate. Researchers destroyed the nuclei of unfertilized frog eggs and then transplanted nuclei of intestinal cells from a tadpole into the egg. The resulting egg cells developed into normal tadpoles and eventually adult frogs, demonstrating that the intestinal cells retained all the genes necessary for normal development of an entire organism.

Gene Transcription Is Precisely Regulated During Development

How does a cell "decide" to become bone, muscle, or intestine? In any cell at any time, only a portion of the cell's genes are used, or transcribed. (Recall from Chapter 10 that transcription is the production of messenger RNA, using the gene as a blueprint.) The particular combination of genes that is transcribed in a cell largely determines the shape, structure, and activity of that cell. Thus, differentiation during development is accomplished by selectively activating transcription in different sets of genes. In general, the mechanism by which transcription is controlled involves regulatory molecules, typically proteins or proteins combined with activating substances such as steroid hormones, that travel to the nucleus and bind to the chromosomes (see Chapter 10). These substances bind to specific genes and can either block or promote their transcription.

In some animals, as the egg develops, various gene-regulating substances become concentrated in different places in the egg cytoplasm. Then, as the fertilized egg divides, each of its daughter cells receives different gene-regulating substances (Fig. 36-5), which then influence the fate of the daughter cells.

In mammals, however, gene-regulating substances are present in the fertilized egg but seem to be distributed evenly during early cleavage, producing cells that can each give rise to a complete individual if they are separated. Identical twins can be produced if the two cells formed by the first cleavage are separated. In vertebrates, differentiation occurs as a result of interactions among cells. During later embryonic development (and continuing throughout adult life), cells constantly receive chemical messages, including nutrients, hormones, and neurotransmitters, from other cells of the body. These chemical messages can alter the developmental fate of a cell by altering the transcription of genes and the activity of enzymes within a cell, as described in Chapter 32.

Scientific Inquiry
The Promise of Stem Cells

The ability of a single cell to give rise to the 200 or so types of cells in an adult is one of the wonders of the living world. Scientists have long known that each nucleus contains the entire genetic blueprint for an organism, and that whether a cell becomes muscle or bone or brain is determined by complex factors that shape the environment in which it develops and differentiates. If scientists could harness this potential, they might create new tissues from a person's own donor cells—tissues that wouldn't be rejected by the immune system—to replace those that are damaged or missing. For example, specific types of neurons might be transplanted to replace those that die from Parkinson's disease or to bridge the gap in a crushed spinal cord. Likewise, bone cells might fill in a shattered fracture, or heart muscle cells might replace those killed by a heart attack. In November 1998, James Thompson and co-workers at the University of Wisconsin reported the first success in isolating human embryonic stem cells and inducing them to grow in culture. Transplanted under the skin of mice, the cells differentiated into a wide variety of human tissues. The source of these cells, which are called **embryonic stem cells**, is also a source of controversy. As the name implies, they are derived from embryonic tissue, in this case, leftover human blastostocysts from *in vitro* fertilization clinics. These leftover embryos, each consisting of about 100 cells, are often destroyed by the clinic after a successful pregnancy has been achieved. Other researchers are using tissue from aborted human embryos.

The ability to culture and differentiate human stem cells has generated a flurry of research and tremendous excitement in the scientific community. While researchers of human aging wonder if someday worn-out organs might be replaced, biotech firms are scrambling to find new ways to produce these cells quickly and efficiently for use as tissue transplants—perhaps in the near future. In one company, researchers are implanting the nuclei from human skin cells into cow's eggs and then jolting them with electricity to trick the egg into acting as if fertilization has occurred. If all goes well, the egg may begin dividing and give rise to a blastocyst that could be used as a source of stem cells, which are derived from the embryonic disk inside the blastocyst (Fig. E36-1). The crucial next step is causing these stem cells to differentiate into the desired tissues "on command." Researchers first need to determine what crucial signals in the normal developing embryo carry the message to produce specific tissues. Then they must be able to deliver these messages to cells growing in culture dishes. Alternatively, it might be possible to introduce undifferentiated stem cells into a specific part of the body—say, the brain or the bone marrow—and let the normal chemical environment there guide differentiation.

Early experiments are encouraging. For example, a small remnant of embryonic-type tissue has recently been discovered in the adult brains of humans and other mammals. These *neural stem cells* have promoted limited recovery of nervous system function in rodents with experimentally induced brain and spinal cord injuries. Although these neural stem cells are rare and difficult to extract, researchers have recently discovered that (using appropriate growth factors) they can stimulate the production of neurons from stem cells derived from human bone marrow, and even from human skin. Using stem cells found in such easily accessible tissues from an individual's own body would have tremendous advantages. The ethical concerns that arise from using embryonic stem cells would be circumvented, as would problems of tissue rejection, since the immune system would not attack cells taken from the recipient's own body.

There are still many hurdles to overcome, including finding ways to ensure that transplanted stem cells do not grow out of control and form tumors. But the rewards may be far-reaching. In addition to their tremendous potential for transplantation, stem cells could be cultured in large quantities to screen for drugs that might produce birth defects and to study the incredible complexity of how normal development proceeds.

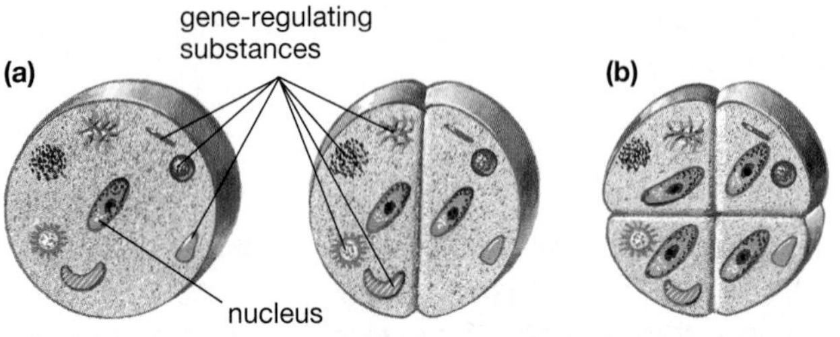

Figure 36-5 Distribution of gene-regulating substances
In some animals, gene-regulating substances are not distributed evenly—nor randomly—during the division of a fertilized egg. ***(a)*** Different gene-regulating substances (represented by the various symbols) are positioned in particular places in the egg's cytoplasm during oogenesis. ***(b)*** As the egg divides, these materials remain in about the same positions, so the daughter cells inherit different substances.

During gastrulation, the developmental fate of most of the embryo's cells is determined by chemical messages received from other cells, a process called **induction**. In amphibian embryos, special cells form at the site of dimpling as the blastula is transformed into the gastrula. This area, called the *dorsal lip* of the blastopore, controls the developmental fate of the cells around it, as Hans Spemann and Hilde Mangold demonstrated in the 1920s (Fig. 36-6a, p. 750). They transplanted the dorsal lip of the blastopore from one embryo to another. The transplanted dorsal lip then induced the cells of the host to form a second embryo, showing that dorsal lip tissue controls differentiation in the surrounding cells. A control experiment can also be performed, transplanting cells from regions of the gastrula other than the dorsal

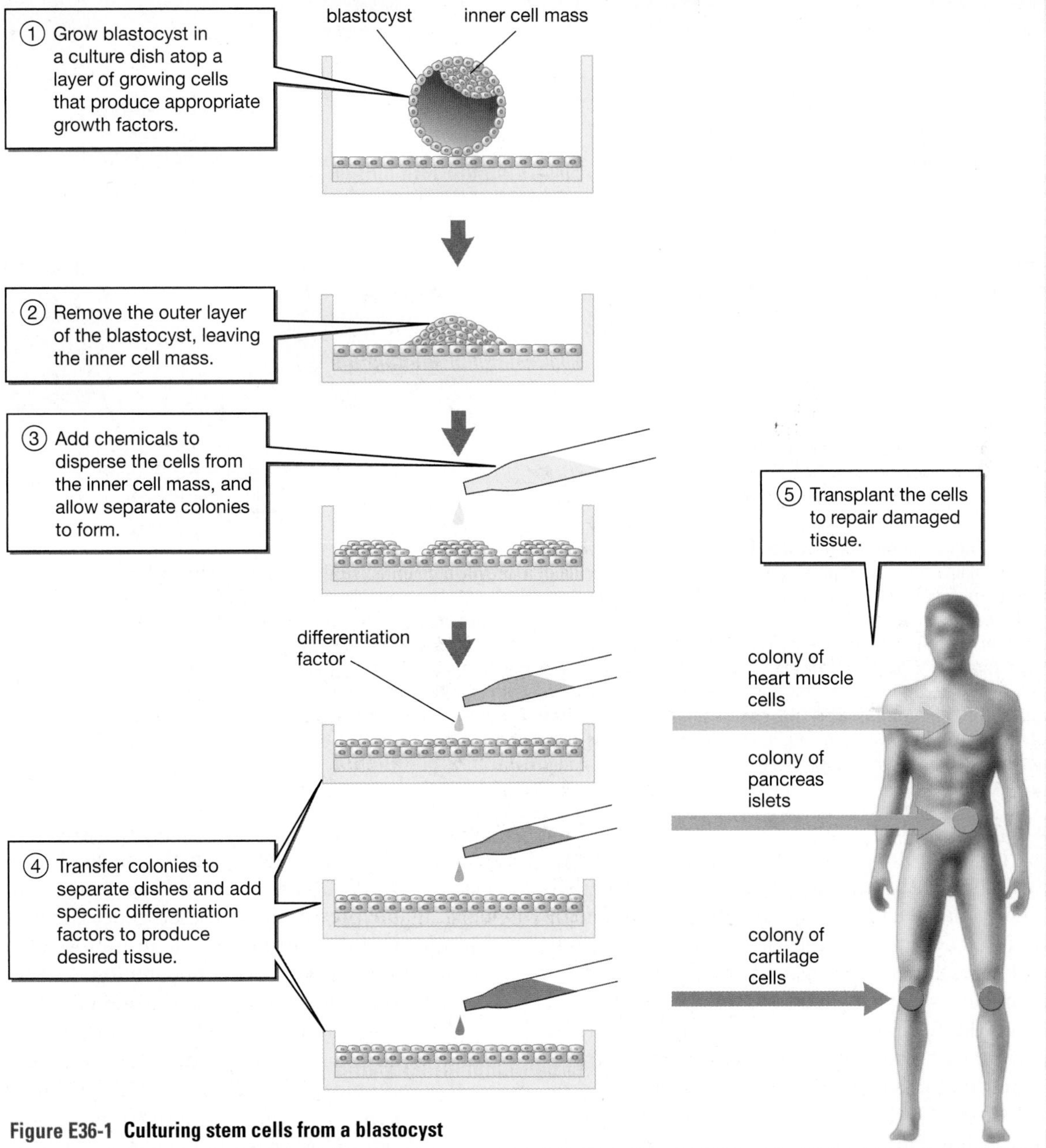

blastocyst
inner cell mass
(1) Grow blastocyst in a culture dish atop a layer of growing cells that produce appropriate growth factors.
(2) Remove the outer layer of the blastocyst, leaving the inner cell mass.
(3) Add chemicals to disperse the cells from the inner cell mass, and allow separate colonies to form.
differentiation factor
(4) Transfer colonies to separate dishes and add specific differentiation factors to produce desired tissue.
(5) Transplant the cells to repair damaged tissue.
colony of heart muscle cells
colony of pancreas islets
colony of cartilage cells

Figure E36-1 Culturing stem cells from a blastocyst

lip of the blastopore (Fig. 36-6b). These cells give rise to tissues appropriate to the region into which they were transplanted rather than the region from which they were taken. These experiments clearly demonstrate that gene expression within differentiating cells (in this case, nonblastopore cells) can be controlled by substances produced by other cells (in this case, cells of the dorsal lip of the blastopore).

Cells migrate within the developing embryo (see Fig. 36-3) and this migration also requires chemical communication among cells. The process by which cells destined to become, for example, the spinal cord, or the muscles of the arm, reach the appropriate positions in the developing embryo is a topic of active research. One mechanism is by the presence of specific surface proteins associated with specific cell types. These proteins recognize

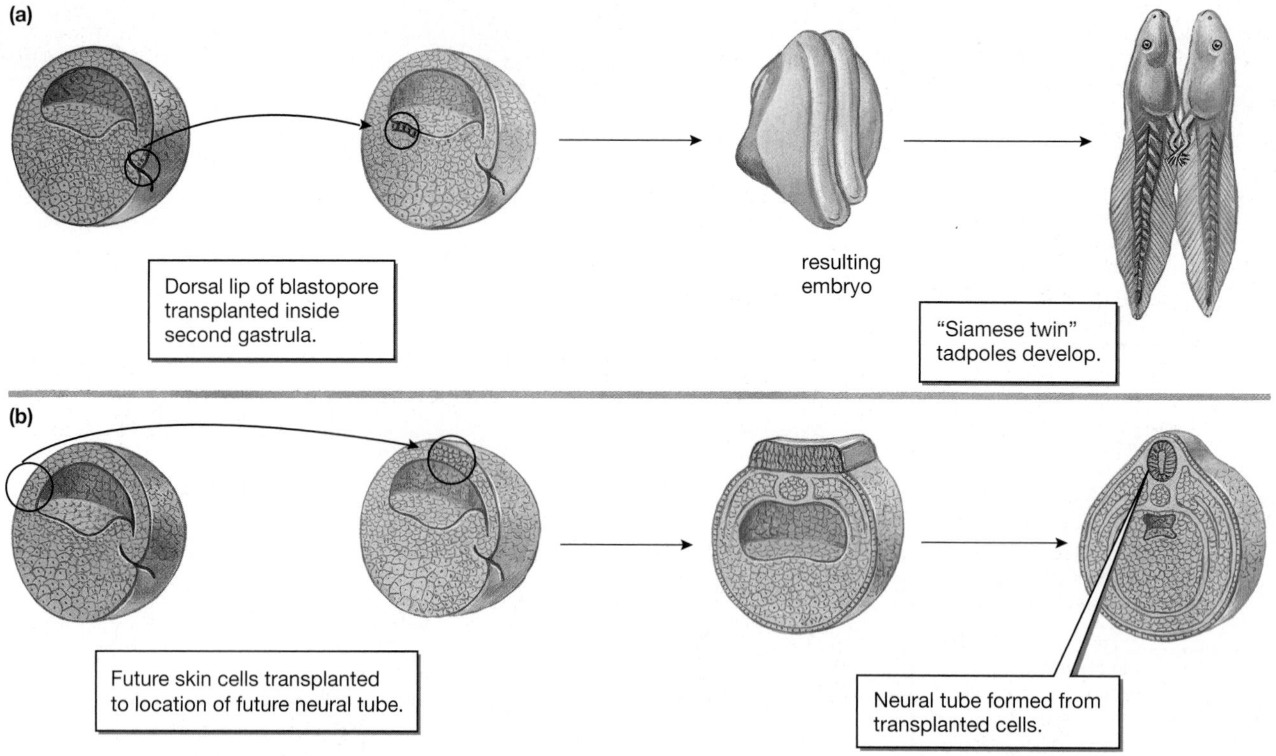
(a)
Dorsal lip of blastopore transplanted inside second gastrula.
resulting embryo
"Siamese twin" tadpoles develop.
(b)
Future skin cells transplanted to location of future neural tube.
Neural tube formed from transplanted cells.

Figure 36-6 Induction and its role in differentiation
(a) The dorsal lip of the blastopore is removed from the gastrula of a heavily pigmented strain of newt and transplanted into a different region of the gastrula from a lightly pigmented strain of newt. The transplanted dorsal lip of the blastopore induced the adjacent regions of the "host" gastrula to form a separate, nearly complete tadpole with the dark pigmentation of the transplanted tissue. This experiment suggests that many cells are "followers," guided along certain developmental paths by their proximity to inducing tissues such as the dorsal lip of the blastopore. ***(b)*** Cells that would normally become skin from the gastrula of a darkly pigmented strain of newt are transplanted to another site on a "host" gastrula of the lighter strain. The host gastrula has induced cells that would normally have formed skin to produce a neural tube. Thus, the developmental fate of follower cells is determined by the region into which they are transplanted, not by the region from which they were removed.

chemical pathways laid out by other cells and cause the cells bearing the protein to migrate along these pathways. Although the exact mechanisms are not fully understood, the production of cell-type specific proteins and of pathways along which these cells migrate in turn depends on our old friend, differential gene expression.

Researchers studying diverse developmental questions, from the formation of segments on a fruit fly to the development of the forelimb in chickens and frogs, have found that chemical gradients of regulatory proteins called *morphogens* commonly specify the fate of cells by causing appropriate genes to be expressed. In other words, during development, cells may take different forms in response to different concentrations of one or a combination of regulatory substances. Thus, a gradient (such as occurs by diffusion from a concentrated source) of a single substance can produce an entire sequence of structural characteristics (such as the sequence of structures from the shoulder to the fingers of a forelimb). The concentrated source of the regulatory chemical is usually a clump of specific cells that act as "signaling centers" that release the protein. As it diffuses from the center, a gradient is formed that affects genes along the gradient in a concentration-dependent manner. For example, during the development of the clawed frog *Xenopus*, a forelimb bud forms from mesoderm that expresses an identified gene. The gene codes for a protein that forms a concentration gradient: high on the "thumb side" of the bud and low on the "pinkie side." As the limb lengthens, this protein forms a second gradient: high at the shoulder and lower toward the hand. Gradients of identical proteins have been found in the forelimbs of developing chick and mouse embryos as well. This topic is explored in more detail in "A Closer Look: Homeobox Genes and the Control of Body Form" in Chapter 36 of this text's Web site.

4 How Do Humans Develop?

Human development is controlled by the same mechanisms that control the development of other animals. In fact, our development strongly reflects our evolutionary heritage. Figure 36-7 summarizes the stages of human embryonic development. You may want to refer to this

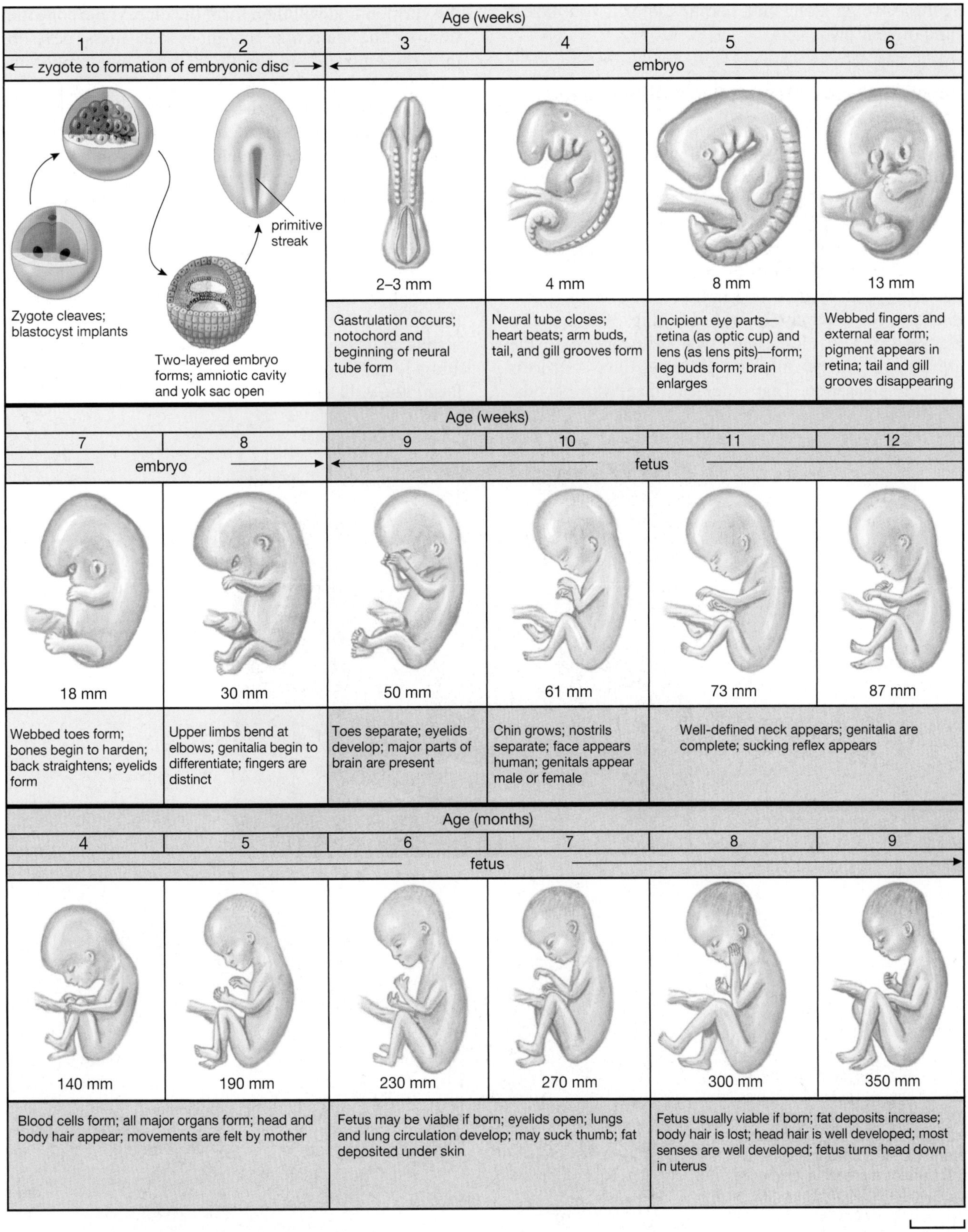

Figure 36-7 Human embryonic development
A calendar of human embryonic development, from blastula to birth.

MediaTutor 36.2 Human Development

figure as we go along and, perhaps, later if you have a child of your own.

During the First 2 Months, Rapid Differentiation and Growth Occur

A human egg is usually fertilized in a woman's oviduct and undergoes a few cleavage divisions on its way to the uterus, a journey that takes about 4 days (Fig. 36-8a,b). By about 1 week after fertilization, the zygote has developed into a hollow ball of cells, known as the **blastocyst** (the mammalian version of a blastula; Fig. 36-8b). The blastocyst consists of a hollow ball of cells with a thicker **inner cell mass** on one side (Fig. 36-9; see also Fig. 36-8b). The sticky outer wall will adhere to the uterus and burrow into the endometrium, a process called **implantation**. That outer cell layer will become the chorion and will then form the embryonic contribution to the placenta; the inner cell mass develops into the embryo and the three other extraembryonic membranes.

After implantation, the inner cell mass grows and splits, forming two fluid-filled sacs that are separated by a double layer of cells called the **embryonic disc** (Fig. 36-9). One sac, bounded by the amnion, forms the amniotic cavity. The amnion eventually grows around the embryo, providing the watery environment needed by all animal embryos. The yolk sac, corresponding to the yolk sac of reptiles and birds, forms the second cavity, although in humans it contains no yolk. At this stage, the embryonic disc consists of an upper layer of future ectoderm cells (on the side facing the amniotic cavity) and a lower layer of future endoderm cells (on the side facing the yolk sac).

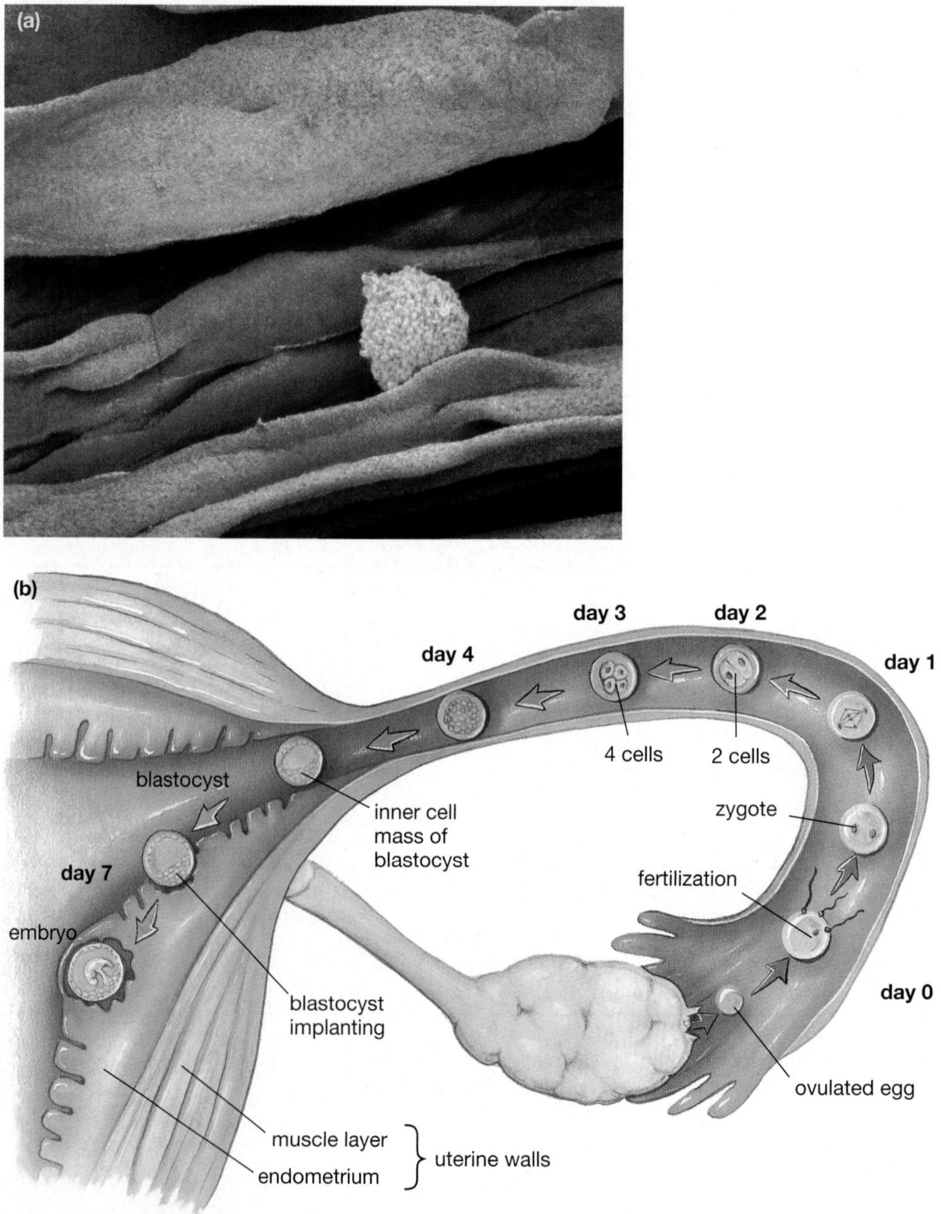

Figure 36-8 The journey of the egg ***(a)*** The egg, surrounded by the cells from the follicle (the corona radiata), travels down the oviduct toward the uterus. It emits chemicals that attract sperm, increasing its chances of being fertilized. ***(b)*** The egg is fertilized in the oviduct and slowly travels down to the uterus. Along the way, the zygote divides a few times, until a hollow blastocyst is formed. The inner cell mass forms the embryo; the surrounding cells adhere to the uterine endometrium, burrow in, and begin forming the placenta.

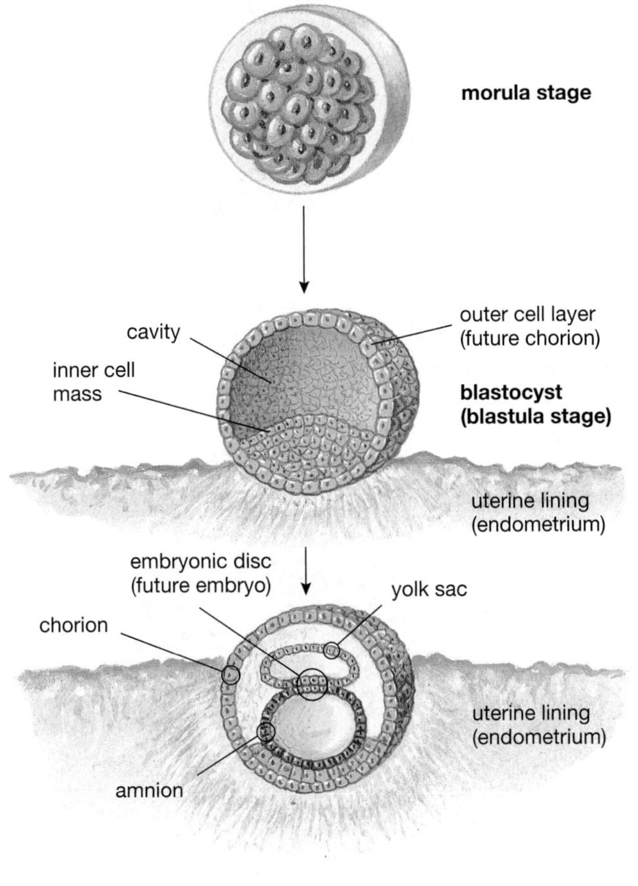

Figure 36-9 Human development during the first and second weeks
As it travels through the oviduct, the fertilized egg undergoes cleavage, forming a morula. In the uterus, the morula becomes a blastocyst (blastula) and implants in the uterine lining. At this stage it consists of an outer layer of cells surrounding an inner cell mass. As it burrows into the uterine lining, the outer cell layer forms the chorion, the embryonic contribution to the placenta. The inner cell mass forms the amnion, yolk sac, and the embryonic disc, which will become the embryo.

Gastrulation begins about the 15th day after fertilization (Fig. 36-10a). The upper and lower layers split apart slightly; a slit, the **primitive streak** (corresponding to the blastopore), appears in the center of the upper layer. Cells of the upper layer migrate through the primitive streak into the interior of the embryo, forming mesoderm. The upper layer is now called *ectoderm* and the lower layer is called *endoderm*. One of the earliest mesoderm structures to develop is the notochord (Fig. 36-10b).

During the third week of development, the embryo and its amniotic sac begin to curl toward the yolk sac (Fig. 36-10c). As the embryo grows, the endoderm pinches, forming a tube that will become the digestive tract (Fig. 36-10d). Simultaneously, the notochord induces the formation of a groove in the overlying ectoderm, which folds inward and then closes over to become the **neural tube**, the forerunner of the brain and spinal cord (see Fig. 36-3d).

By the end of the fourth week, the amnion completely surrounds the embryo. The amnion is punctured only by the *umbilical cord*, which connects the embryo to the placenta.

By the end of the sixth week, the embryo clearly displays its chordate ancestry (see Chapter 22), having developed a notochord, a prominent tail, and *gill grooves* (indentations behind the head that are homologous to the developing gills that many chordates, such as fish, retain as adults; Fig. 36-11a, p. 755). These structures disappear as human development continues. The embryo already has the rudimentary beginnings of the eyes, a beating heart, and separating fingers and toes on its tiny hands and feet (Fig. 36-11b). Especially notable at this stage is the rapid growth of the brain, which is nearly as large as the rest of the body. In fact, many of the structures of the adult brain are already recognizable.

As the second month draws to an end, nearly all the major organs have formed, and the embryo begins to look human (Fig. 36-11c). The gonads appear and develop into testes or ovaries, depending on the presence or absence of the Y chromosome. Sex hormones—either testosterone from the testes or estrogen from the ovaries—are secreted. These hormones will affect the future development of the embryonic organs, including not only the reproductive organs but also certain regions of the brain. After the second month of development, the embryo is called a **fetus**; this designation denotes that the developing being has taken on a generally human appearance.

These first 2 months of pregnancy are times of extremely rapid differentiation and growth for the embryo and also times of considerable danger. Although the fetus is vulnerable throughout development, rapidly developing organs are the most sensitive to environmental insults, such as drugs (including alcohol) or certain medications taken by the mother.

The Placenta Secretes Hormones and Exchanges Materials Between Mother and Embryo

During the first few weeks of pregnancy, embryonic cells burrow into the thickened lining of the uterus and obtain nutrients directly from the nearby cells of the endometrium. Meanwhile, the **placenta**, composed of interlocking tissues of the embryo and the endometrium, begins to form. The outer cells of the embryo form the chorion, which penetrates the endometrium with finger-like projections called **chorionic villi** (singular, **villus**). From this complex interweaving of tissues arises the marvelously intricate placenta. The placenta has two major functions: (1) It secretes hormones, and (2) it allows the selective exchange of materials between the mother and the fetus.

The placenta, as it develops during the first 2 months of pregnancy, begins secreting estrogen and progesterone. Estrogen stimulates the growth of the mother's uterus and mammary glands, whereas progesterone also

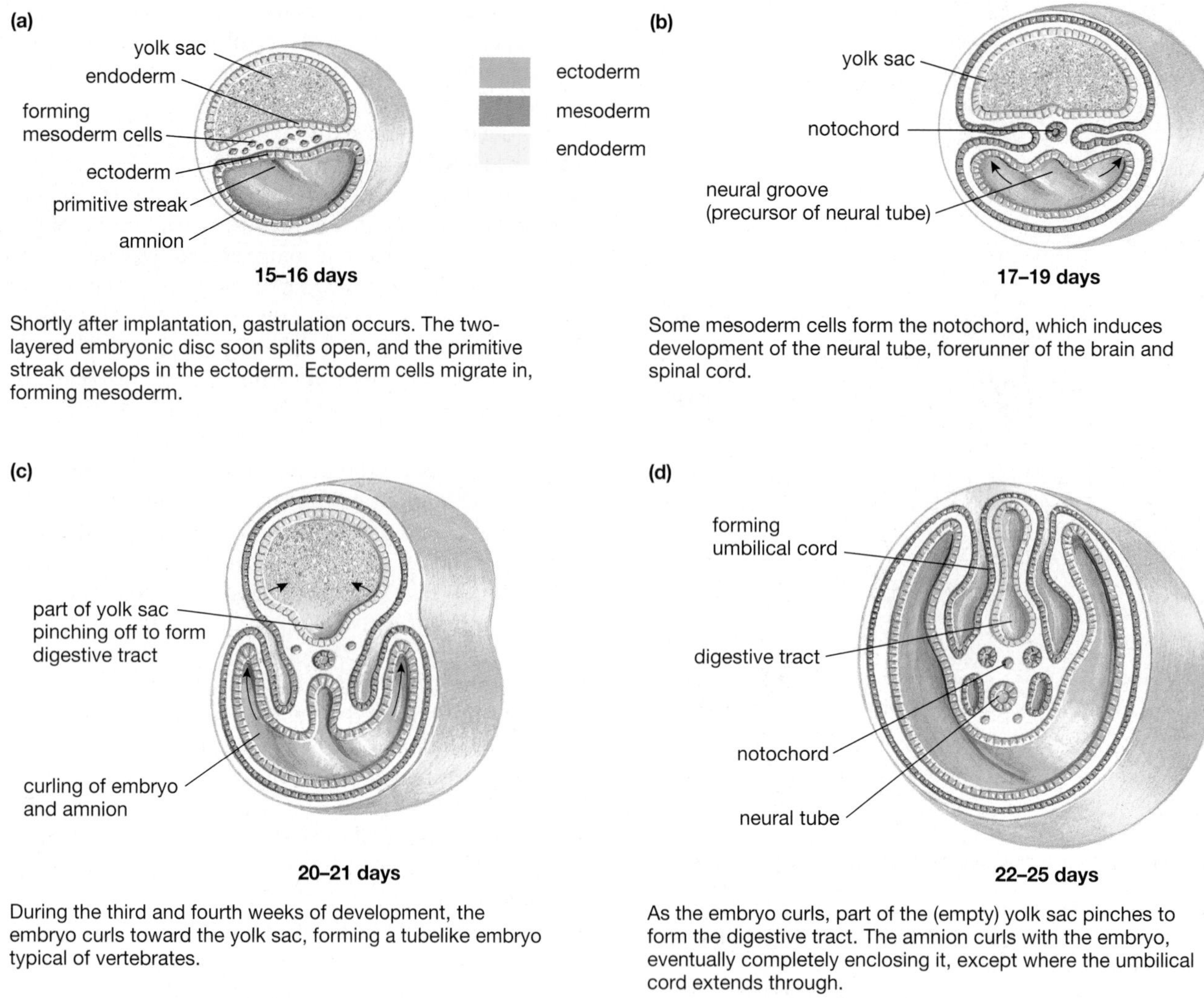
(a)
yolk sac
endoderm
forming mesoderm cells
ectoderm
primitive streak
amnion
15–16 days
Shortly after implantation, gastrulation occurs. The two-layered embryonic disc soon splits open, and the primitive streak develops in the ectoderm. Ectoderm cells migrate in, forming mesoderm.
ectoderm
mesoderm
endoderm
(b)
yolk sac
notochord
neural groove (precursor of neural tube)
17–19 days
Some mesoderm cells form the notochord, which induces development of the neural tube, forerunner of the brain and spinal cord.
(c)
part of yolk sac pinching off to form digestive tract
curling of embryo and amnion
20–21 days
During the third and fourth weeks of development, the embryo curls toward the yolk sac, forming a tubelike embryo typical of vertebrates.
(d)
forming umbilical cord
digestive tract
notochord
neural tube
22–25 days
As the embryo curls, part of the (empty) yolk sac pinches to form the digestive tract. The amnion curls with the embryo, eventually completely enclosing it, except where the umbilical cord extends through.

Figure 36-10 Human development during the third and fourth weeks

stimulates the mammary glands and inhibits premature contractions of the uterus.

The placenta also regulates the exchange of materials between the blood of the mother and the blood of the fetus without allowing the two to mix. The chorionic villi contain a dense network of fetal capillaries and are bathed in pools of maternal blood (Fig. 36-12, p. 756). This arrangement permits many small molecules to diffuse between fetal blood and maternal blood. Oxygen diffuses from maternal blood to fetal blood, and carbon dioxide from fetal blood to maternal blood. Nutrients, some aided by active transport, travel from mother to fetus. Fetal urea diffuses into the mother's blood, to be filtered out by the mother's kidneys.

While allowing exchange by diffusion, the membranes of the capillaries and chorionic villi act as barriers to the passage of some large proteins and most cells. Despite this barrier, some disease-causing organisms and many harmful chemicals, such as alcohol, can penetrate the placental barrier, as described in "Health Watch: The Placenta Provides Only Partial Protection."

Growth and Development Continue During the Last 7 Months

The fetus continues to grow and develop for another 7 months. Although the rest of its body is "catching up" with the head in size, the brain continues to develop rapidly and the head remains disproportionately large. Nearly every nerve cell ever formed during the entire human life span develops during embryonic life, which is one reason that the developing brain is such a sensitive target for drugs ingested during pregnancy. As the brain and spinal cord grow, they begin to generate specific types of behavior. As early as the third month of pregnancy, the fetus begins to move and respond to

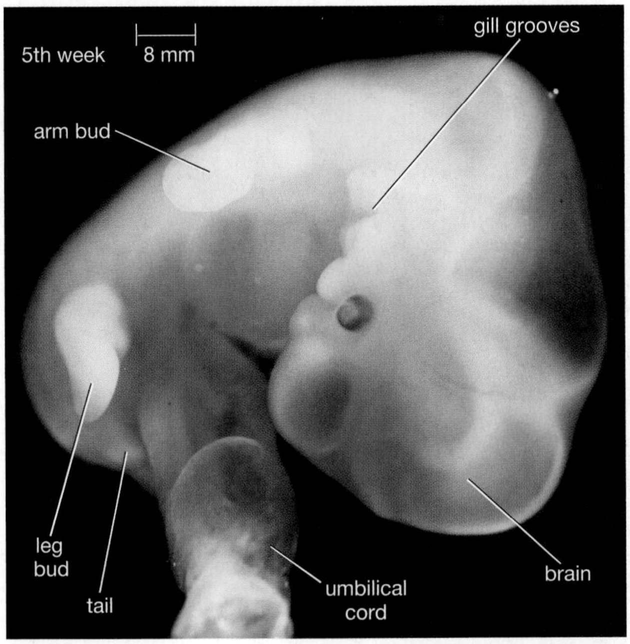

(a)

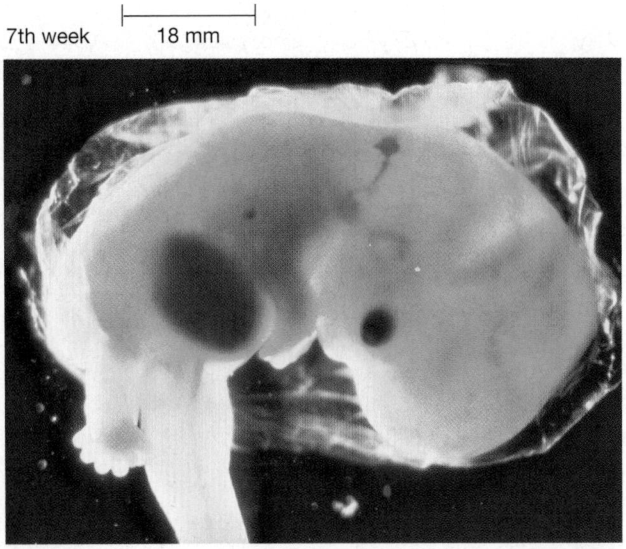

(b)

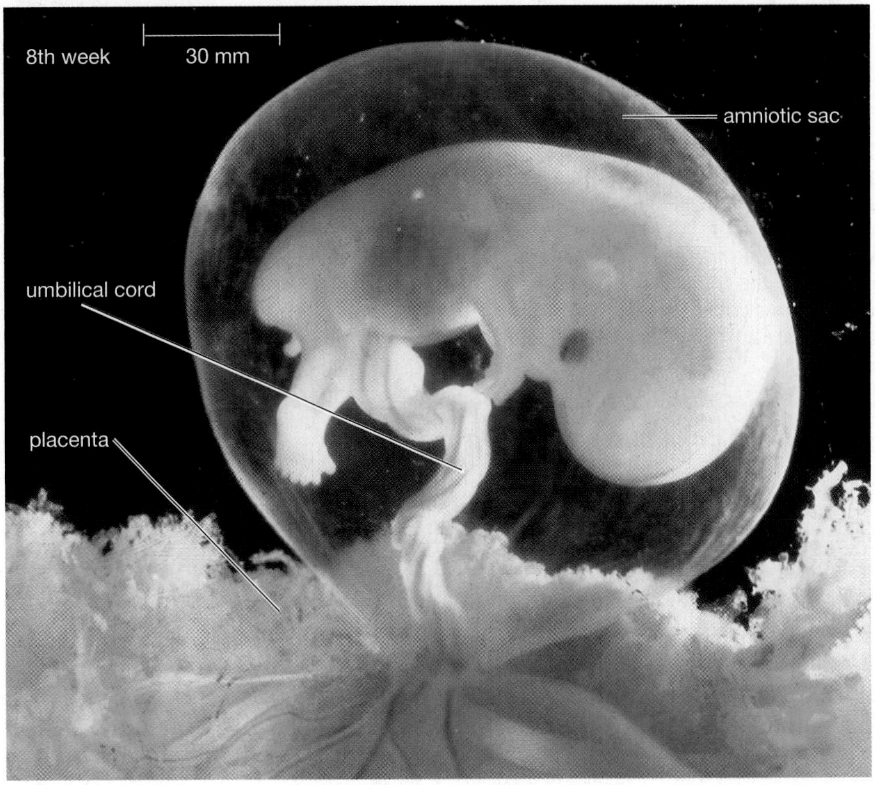

(c)

Figure 36-11 Human development during the fifth through eighth weeks ***(a)*** At the end of the fifth week, the human embryo is about half head. The feet and hands have begun to develop digits. A tail and gill grooves are clearly visible, evidence of our evolutionary relationship to other vertebrates. ***(b)*** By the seventh week, the human form has been more clearly defined by the selective death of cells that form the tail and connect the fingers and toes. The tail has nearly disappeared, and fingers and toes are separated from one another. ***(c)*** At the end of the eighth week, the embryo is clearly human in appearance and is now called a fetus. Most of the major organs of the adult body have begun to develop.

stimuli. Some instinctive behaviors appear, such as sucking, which will have obvious importance soon after birth. Structures that the fetus will need when it emerges from the womb (the mother's uterus), such as the lungs, stomach, intestine, and kidneys, enlarge and become functional, though they will not be used until after birth. Most fetuses 7 months or older can survive outside the womb with medical assistance, but larger and more mature fetuses have a much greater chance of survival.

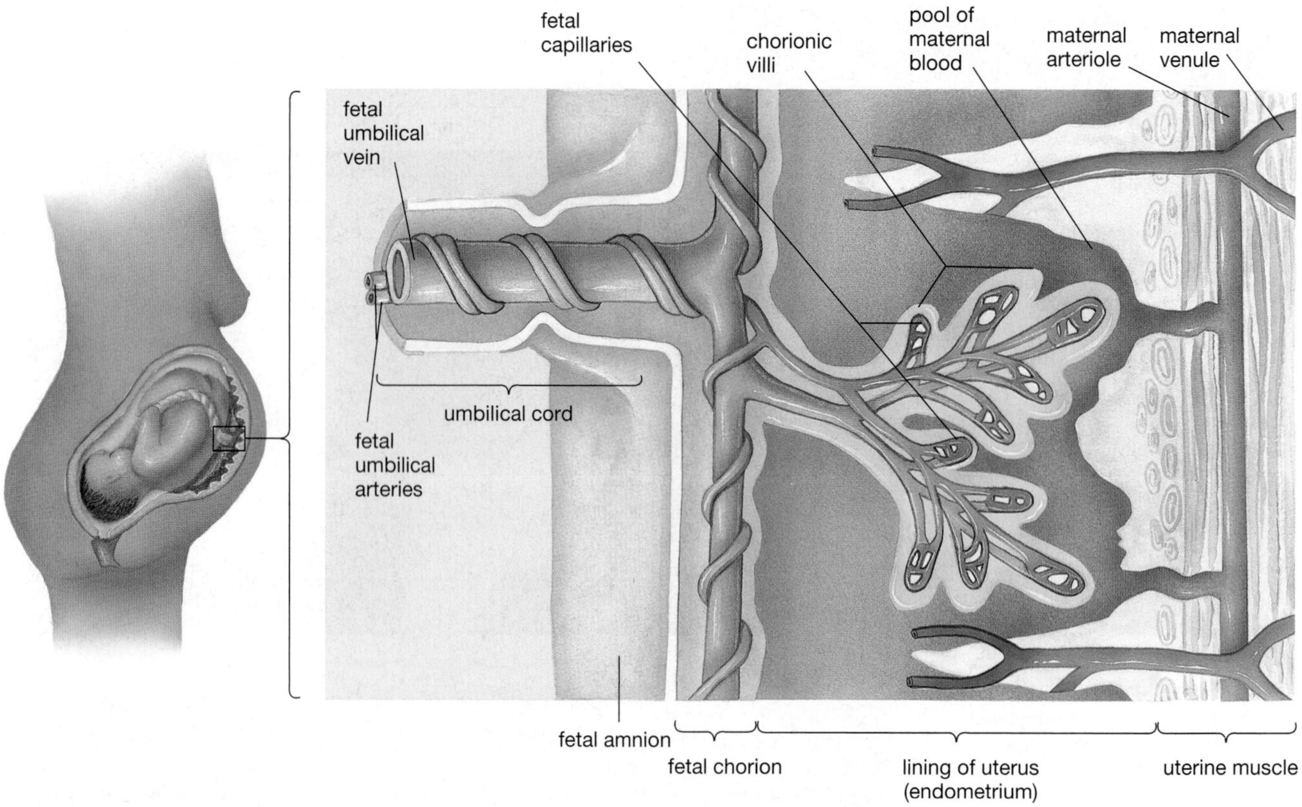

Figure 36-12 The placenta
The placenta is formed from both the chorion of the embryo and the endometrium of the mother. Capillaries of the endometrium break down, releasing blood to form pools within the placenta. Meanwhile, the chorion develops projections (the chorionic villi) that extend into these pools of maternal blood. Blood vessels from the umbilical cord branch extensively within the villi. The resulting structure separates the maternal and fetal blood supplies and generates a large surface area for the diffusion of oxygen, carbon dioxide, nutrients, and wastes between the fetal capillaries and the maternal blood pools. Umbilical arteries carry deoxygenated blood from the fetus to the placenta, and umbilical veins carry oxygenated blood back to the fetus.

Development Culminates in Labor and Delivery

Usually, during the last months of pregnancy, the fetus becomes positioned head downward in the uterus with the crown of the skull resting against (and being held up by) the cervix. Near the end of the ninth month, give or take a few weeks, the process of birth normally begins (Fig. 36-13). Birth is the result of a complex interplay between (1) uterine stretching caused by the growing fetus and (2) fetal and maternal hormones that finally trigger **labor** (contractions of the uterus that result in the birth).

Unlike skeletal muscles, uterine muscles can contract spontaneously, and stretching enhances these contractions. As the baby grows, it stretches the uterine muscles, which contract occasionally weeks before delivery. The final trigger for labor is probably provided by the fetus. The near-term fetus produces steroid hormones that cause increased estrogen and prostaglandin production by the placenta and uterus. These hormones make the uterus even more likely to contract. When the combination of hormones and stretching activate the uterus beyond some critical point, strong contractions begin, signaling the onset of labor. As the contractions proceed, the baby's head pushes against the mother's cervix, making it expand in diameter (dilate). Stretch receptors in the walls of the cervix send signals to the hypothalamus, triggering oxytocin release. Under the dual stimulation of prostaglandin and oxytocin, the uterus contracts even more strongly. This positive feedback cycle is finally halted when the baby emerges from the vagina, or *birth canal.* The infant's head is so large that it can barely fit through the mother's pelvis. The skull is compressed into a slightly conical shape as it passes through the vagina. The infant is in for a rude awakening. The womb was soft, fluid-cushioned, and warm. All of a sudden, the baby must obtain oxygen and eliminate carbon dioxide by breathing. It must regulate its own body temperature, and it must suckle to obtain food.

After a brief rest, uterine contractions resume, causing the uterus to shrink remarkably. During these contractions, the placenta is sheared from the uterus and expelled through the vagina as the *afterbirth.* Further release of prostaglandins in the *umbilical cord,* the stalk

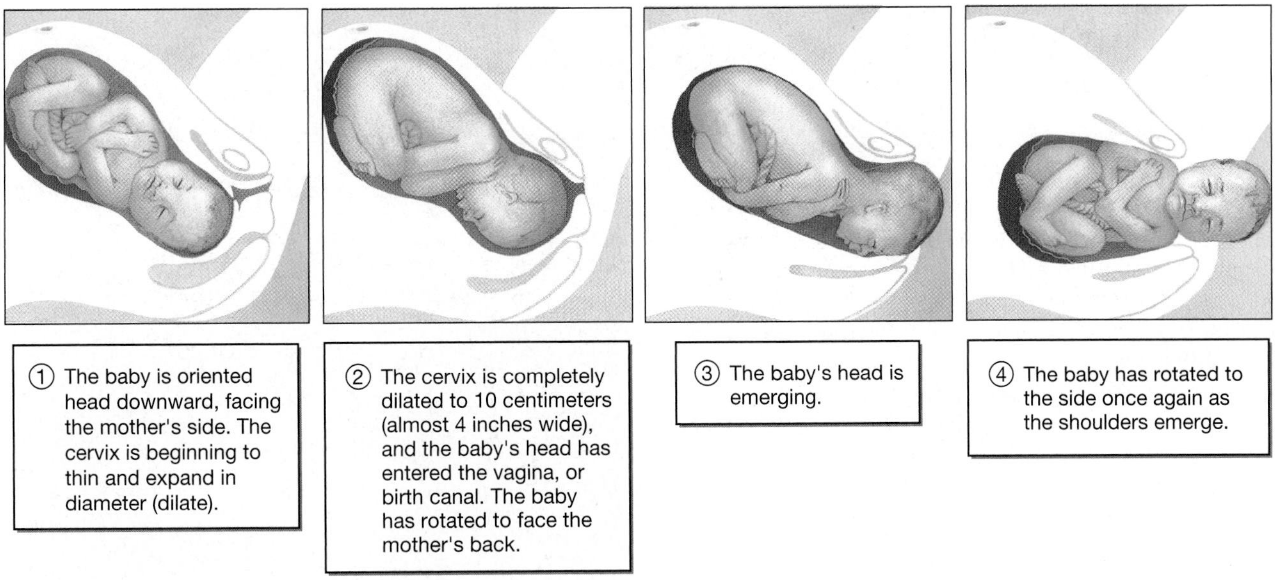
① The baby is oriented head downward, facing the mother's side. The cervix is beginning to thin and expand in diameter (dilate).
② The cervix is completely dilated to 10 centimeters (almost 4 inches wide), and the baby's head has entered the vagina, or birth canal. The baby has rotated to face the mother's back.
③ The baby's head is emerging.
④ The baby has rotated to the side once again as the shoulders emerge.

Figure 36-13 Delivery

that connects the fetus and the placenta, causes the muscles that surround fetal blood vessels in the umbilical cord to contract. This contraction shuts off blood flow. (Tying off the cord is standard practice but is not usually necessary; if it were, other mammals would not survive birth!) A new human being has been born.

Milk Secretion Is Stimulated by Hormones of Pregnancy

As the fetus grows, nourished by nutrients diffusing through the placenta, changes in the mother's breasts prepare her to continue nourishing her child after it is born. When pregnancy occurs, large quantities of estrogen and progesterone (acting together with several other hormones) stimulate **mammary glands**, milk-producing glands in the breasts, to grow, branch, and develop the capacity to secrete milk (Fig. 36-14). The mammary glands are arranged in a circle around the nipple; each gland has a milk duct that leads to the *nipple,* a projection of epithelial tissue. The actual secretion of milk, called **lactation**, is promoted by the pituitary hormone prolactin (see Chapter 32).

The level of prolactin rises steadily from about the fifth week of pregnancy until birth. Immediately after birth, estrogen and progesterone levels plummet, and prolactin takes over, stimulating the production of milk. Milk is released when the infant's suckling stimulates nerve endings in the nipples. The stimulated nerves send a signal to the hypothalamus, triggering an extra surge of prolactin and oxytocin from the pituitary. Oxytocin causes muscles surrounding the mammary glands to contract, ejecting the milk into the ducts that lead to the nipples (see Chapter 32; Fig. 32-6).

During the first few days after birth, the mammary glands secrete a thin, yellowish fluid called **colostrum**. Colostrum is high in protein and contains antibodies that are absorbed directly through the infant's intestine and help protect the newborn against some diseases. Colostrum is gradually replaced by mature milk, which is higher in fat and milk sugar (lactose) and lower in protein.

Aging Seems to Be Genetically Programmed

Most of your cells will function less efficiently, or they will divide more slowly, as you age. Is death, both for cells and for entire organisms, a programmed part of life

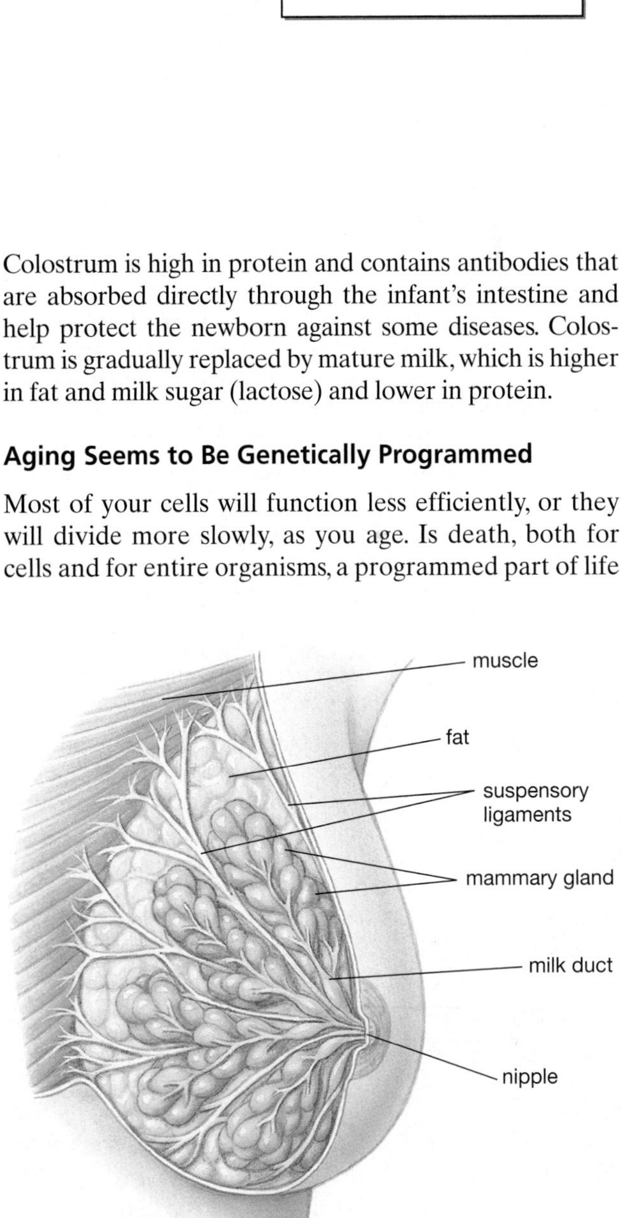

Figure 36-14 The structure of the mammary glands
During pregnancy, both fatty tissue and the milk-secreting glands and ducts increase in size.

Health Watch
The Placenta Provides Only Partial Protection

When your grandparents were bearing children, physicians assumed that the placenta protected the developing fetus from most of the substances in maternal blood that could harm it. We know now that this is far from true. In fact, most drugs (whether medicinal or "recreational"), and even some disease-producing organisms, readily penetrate the placental barrier and affect the fetus.

Infections Can Cross the Placenta The German measles virus can cross the placenta and attack the fetus, causing potentially severe retardation and other defects. As mentioned in Chapter 35, the virus causing genital herpes (during active outbreaks) and the bacterium causing syphilis can cause mental or physical defects in the developing fetus. The virus causing AIDS can also cross the placenta, so some infants are born with this deadly disease.

Drugs Readily Cross the Placenta A tragic example of a drug that crosses the placenta is the tranquilizer *thalidomide,* commonly prescribed in Europe in the early 1960s (Fig. E36-2). Thalidomide's devastating effects on embryos were discovered only when many babies were born with missing or extremely abnormal limbs. In the late 1980s the anti-acne drug Accutane® was found to cause gross deformities in babies born to women using it. Researchers now know that Accutane contains a substance that acts as a morphogen in the developing embryo.

Although these are extreme examples, *any drug, including aspirin, has the potential to harm the fetus, and any woman who thinks she may be pregnant should seek medical advice about any drugs she takes.* The use of so-called recreational drugs, such as heroin and cocaine, has devastating effects on a developing fetus. Many babies of heroin addicts are born addicted. In the U.S., hundreds of thousands of infants have been born in recent years to users of "crack" cocaine. Research suggests that these children may be impaired, both behaviorally and emotionally.

The Effects of Smoking Probably the most common toxic substances to which fetuses are exposed are those in cigarette smoke. Many women who smoke are so addicted that they

Figure E36-2 Drugs interfere with development
Children born to mothers who took the tranquilizer thalidomide during pregnancy cope heroically with missing and deformed limbs. A drug producing no obvious ill effects on the mother can have a dramatic impact on her developing fetus.

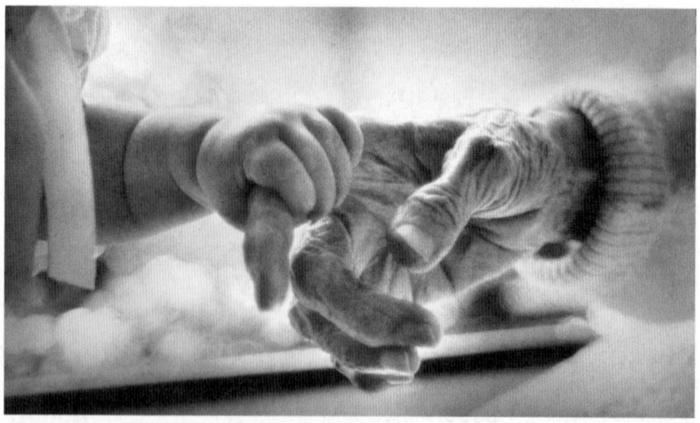

Figure 36-15 Youth vs. age
The contrast between age and youth is dramatically illustrated in this poignant photo.

(Fig. 36-15)? From an evolutionary standpoint, natural selection promotes only those mechanisms that keep an organism alive and healthy during the time that it is producing and nurturing its young. Repair mechanisms that extend longevity past this time are not favored and may even be harmful to the population as a whole. For example, older, nonreproductive individuals might compete with younger ones for limited resources, such as food.

Evidence of programmed cell death is seen in cells grown in dishes in the laboratory. These cells divide a relatively fixed number of times, then stop, and eventually die. The number of cell divisions is at least partly controlled by structures called *telomeres* that cap the ends of chromosomes (see Chapter 11). Telomeres become shorter with every cell cycle, and when they reach don't stop smoking during pregnancy. Consequently, each year in the U.S. nearly a million human embryos are exposed to the poisons in smoke, including nicotine, carbon monoxide, and a host of carcinogens. Women who smoke have a higher incidence of miscarriages than do nonsmokers. Likewise, smokers tend to give birth to smaller infants who have a higher death rate shortly after birth than are infants of nonsmokers. There is evidence that some children born to heavy smokers also suffer behavioral and intellectual impairment. Researchers analyzing the blood of infants born to mothers who smoke found that a potent carcinogen that causes mutations in DNA was passed to the infants through the placenta. Unfortunately, 61% of U.S. women who smoke continue to do so during pregnancy.

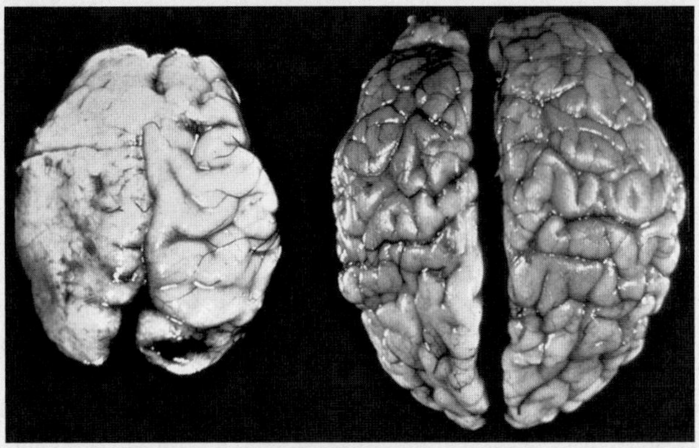

Figure E36-3 Alcohol impairs brain development
(left) The brain of a child with fetal alcohol syndrome and (right) the brain of a normal child of the same age show the devastating effects of alcohol on the developing brain.

Fetal Alcohol Syndrome The effects of alcohol on a developing fetus can be devastating. When a pregnant woman drinks, alcohol in the blood of her unborn child reaches a level as high as that in her own blood. Many children born to alcoholic mothers who drink heavily on a regular basis (4 to 5 drinks per day or more) or go on alcoholic binges exhibit **Fetal Alcohol Syndrome (FAS)**. Such children are mentally retarded and can be hyperactive and irritable. Children afflicted with FAS have small heads, abnormally small, improperly developed brains (Fig. E36-3), facial abnormalities, inhibited growth, and a higher-than-normal incidence of defects of the heart and other organs. The damage is irreversible. Research with rats has recently revealed that even a single 4-hour exposure to blood alcohol levels of 0.2% (such as might result from a single alcoholic binge) during critical stages of brain development (such as occur during the last trimester of human pregnancy) can cause massive death of brain cells.

FAS is the single most common cause of mental retardation in the U.S., with about 7300 FAS infants (1 in every 500) born each year. It is likely that five times as many children suffer from *Fetal Alcohol Effect*, a milder form of FAS that also impairs development. A woman who takes one or two alcoholic drinks a day during the first 3 months of pregnancy also significantly increases her chances of having a miscarriage. *Researchers have established no safe level of alcohol consumption during any phase of pregnancy. The U.S. Surgeon General advises pregnant women and those who are likely to become pregnant to avoid all alcohol consumption.*

In summary, a pregnant woman should assume that whatever drugs she takes will find their way into the bloodstream of her developing infant. Women who are likely to become pregnant need to consider that crucial stages of development may occur before they even realize that they are pregnant and take appropriate precautions. The mother's choices during this critical 9-month period can strongly influence her child's future well-being.

a certain minimum length, cell division ceases. Preventing the telomeres from shortening has extended the life of cells grown in culture dishes. Could it extend our life spans as well? Recent research has also suggested that the maximum life span varies from species to species and that longevity depends on the ability of the cell to repair damage to its DNA. Long-lived cells and long-lived animals have more natural enzymes that protect against DNA damage and are better at repairing damaged DNA than are short-lived cells and short-lived animals. Nevertheless, all normal cells stop dividing and die eventually.

Not all cells, however, are "normal." Cancer cells survive and reproduce indefinitely in cell culture or in the body, which is what makes cancer so devastating. Some lineages of cancer cells have been reproducing in culture for decades. Although scientists are still working to uncover the mechanism that regulates the life span of a cell, cancer cells provide evidence that this mechanism can be bypassed. However, cultured cancer cells frequently mutate and change their characteristics. Can we discover a method of defusing the self-destruct mechanism while retaining proper controls over the cells in our bodies? Would extending the human life span even be desirable, given the health problems that accompany old age and the resource constraints of an overpopulated planet? No one knows, but we can be certain it will not happen soon.

CASE STUDY REVISITED

Far-reaching Choices

Maria's doctor underscores the incredible delicacy of the events of pregnancy and the ease with which they may be disrupted. At two weeks, Maria's child is a small ball of cells beginning to nestle in her endometrium, with complexity emerging inside as the embyonic disk (the future embryo) takes shape. The physician explains that a woman who drinks or smokes during pregnancy is likely to impair the development of her child, in ways that range from subtle to devastating. "The American Medical Society, the Surgeon General of the United States, the Royal Society of Physicians in England, and many other public health organizations advocate *total abstinence from alcohol throughout the duration of the pregnancy.*"

At the library, Maria learns more about the profound consequences of Fetal Alcohol Syndrome (FAS) for children and society. Tragically, FAS, an entirely preventable disorder, is the leading cause of retardation in the U.S. The lifetime cost to society of caring for a single child with Fetal Alcohol Syndrome has been estimated at between \$1.4 and \$4 million and the overall cost to society of FAS at about \$5 million per day, or \$1.9 billion annually. However, the FAS victim's lifelong struggles, the parents' heartbreak, and the loss of a person's unique potential as a fully functional member of society cannot be measured.

Maria's doctor tells her that pregnancy must make her efforts to quit smoking far more urgent, since smoking during pregnancy increases both the chance of miscarriage and of the child having learning and behavioral problems later on. Further, she should avoid smoky places, which will not only tempt her to smoke but will expose her and her fetus to secondhand smoke. Digging a half-empty pack of cigarettes from her purse, Maria tosses it in the wastebasket as she leaves the doctor's office.

Many healthcare providers are not trained to recognize addictions and deal with the problems of addictions and pregnancy; likewise, the availability of treatment for addiction may be limited. How should society deal with the dilemma of pregnant women who are (perhaps unwittingly) damaging their unborn children? Should mothers who cause brain damage by drinking or taking other drugs during pregnancy be charged with child abuse? Based on what you now know about development, how would you deal with a friend who was continuing to smoke, drink, or take other drugs during pregnancy?

Summary of Key Concepts

1) How Do Indirect and Direct Development Differ?

Animals undergo either indirect or direct development. In indirect development, eggs (normally with relatively little yolk) hatch into larval feeding stages, which later undergo metamorphosis to become adults with notably different body forms. In direct development, the newborn animal is sexually immature but otherwise resembles a small adult. Animals with direct development tend to have either large, yolk-filled eggs or nourish the developing embryo within the mother's body. In birds, reptiles, and mammals, extraembryonic membranes (the chorion, amnion, allantois, and yolk sac) encase the embryo in a fluid-filled space and regulate the exchange of nutrients and wastes between the embryo and its environment.

2) How Does Animal Development Proceed?

Animal development occurs in several stages. *Cleavage:* The fertilized egg undergoes cell divisions with little intervening growth, so the egg cytoplasm is partitioned into smaller cells. Cleavage divisions result in the formation of the morula, a solid ball of cells. A cavity then opens up within the morula, forming the blastula, a hollow ball of cells. *Gastrulation:* A dimple forms in the blastula, and cells migrate from the surface into the interior of the ball, eventually forming a three-layered gastrula. The three cell layers of ectoderm, mesoderm, and endoderm give rise to all the adult tissues (see Table 36-2). *Organogenesis:* The cell layers of the gastrula form organs characteristic of the animal species. *Growth* and *Sexual maturation:* The juvenile animal increases in size and achieves sexual maturity. *Aging:* Cells begin to function less efficiently, at least partly as a result of failure to repair their DNA, and eventually the animal dies.

3) How Is Development Controlled?

All the cells of an animal body contain a full set of genetic information, yet cells are specialized for particular functions. During development, cells differentiate by stimulating and repressing the transcription of specific genes. Gene transcription is regulated in two ways: (1) In some animals, gene-regulating substances in the egg cytoplasm are distributed in different proportions to different daughter cells during the first few cleavage divisions. (2) Later in development, certain cells produce chemical messages that induce other cells to differentiate into particular cell types, a process called *induction*. Cells migrate within the developing embryo, a process requiring chemical communication among cells. Surface proteins associated with specific cell types recognize chemical pathways laid out by other cells and cause the cells bearing the protein to migrate along these pathways. Researchers have found that chemical gradients of regulatory proteins called *morphogens* specify the fate of cells by causing appropriate genes to be expressed.

4) How Do Humans Develop?

A fertilized egg (zygote) develops into a hollow blastocyst and implants in the endometrium. The outer wall will become the chorion and will form the embryonic contribution to the placenta; the inner cell mass develops into the embryo and the three other extraembryonic membranes. During gastrulation, cells migrate and differentiate into ectoderm, mesoderm, and endoderm. During the third week, the endoderm forms a tube that will become the digestive tract and the notochord induces the formation of the neural tube, the forerunner of the brain and spinal cord (Fig. 36-3d). By the end of the second month, the major organs have formed, and the embryo appears human and is called a fetus. In the next 7 months, until birth, the fetus continues to grow, and the lungs, stomach, intestine, and kidneys enlarge and become functional. Human embryonic development follows the same principles as the development of other mammals. The stages of human development are summarized in Figure 36-7. During pregnancy, mammary glands in the mother's breasts enlarge under the influence of estrogen, progesterone, and other hormones. After about 9 months, uterine contractions are triggered by a complex interplay of uterine stretch and prostaglandin and oxytocin release. As a result, the uterus expels the baby and then the placenta. After birth, milk secretion is triggered by prolactin and oxytocin, whose release is triggered by suckling.

Aging seems to be genetically programmed, since extending longevity past the time of reproduction and nurturing the young has not been favored by natural selection. Cells grown in the laboratory eventually stop growing and die, a process controlled in part by telomeres that cap the ends of chromosomes.

Key Terms

allantois *p. 744*
amnion *p. 744*
amniotic egg *p. 744*
blastocyst *p. 752*
blastopore *p. 745*
blastula *p. 745*
chorion *p. 744*
chorionic villus *p. 753*
cleavage *p. 745*
colostrum *p. 757*
development *p. 746*
differentiation *p. 746*
direct development *p. 742*
ectoderm *p. 746*
embryonic disc *p. 752*
embryonic stem cells *p. 748*
endoderm *p. 746*
extraembryonic membrane *p. 744*
fetal alcohol syndrome (FAS) *p. 759*
fetus *p. 753*
gastrula *p. 746*
gastrulation *p. 746*
implantation *p. 752*
indirect development *p. 742*
induction *p. 748*
inner cell mass *p. 752*
labor *p. 756*
lactation *p. 757*
larva *p. 742*
mammary glands *p. 757*
mesoderm *p. 746*
metamorphosis *p. 742*
morula *p. 745*
neural tube *p. 753*
notochord *p. 746*
organogenesis *p. 746*
placenta *p. 753*
primitive streak *p. 753*
yolk *p. 742*
yolk sac *p. 744*

Thinking Through the Concepts

Multiple Choice

1. *Cells become differentiated through all of the following events EXCEPT*
a. binding of regulatory molecules to chromosomes
b. chemical messages received from other cells
c. unequal distribution of gene-regulating substances during cleavage
d. transcription of different genes
e. progressive loss of genes as cells divide

2. *Indirect development is characteristic of animals that normally produce*
a. few eggs
b. eggs with large amounts of yolk
c. young that are sexually immature versions of adults
d. all of the above
e. none of the above

3. *In bird eggs, the allantois*
a. exchanges oxygen and carbon dioxide
b. produces the shell
c. stores wastes
d. encloses the embryo in a watery environment
e. contains stored food

4. *The endoderm gives rise to the*
a. lining of the digestive tract
b. epidermis of skin
c. skeletal system
d. muscles
e. nervous system

5. *In human development, ectoderm cells migrate through the primitive streak to form*
a. endoderm
b. mesoderm
c. the chorion
d. the yolk sac
e. the amnion

6. *The process by which a tissue causes another tissue to differentiate is called*
a. gastrulation
b. metamorphosis
c. cleavage
d. induction
e. indirect development

? Review Questions

1. Distinguish between indirect and direct development, and give examples of each.

2. Describe the structure and function of four extraembryonic membranes found in reptiles and birds. Are these four present in placental mammals? In what ways are their roles similar in reptiles and birds *vs.* mammals? How do they differ?

3. What is yolk? How does it influence cleavage?

4. What is gastrulation? Describe gastrulation in frogs and in humans.

5. Name two structures derived from each of the three embryonic tissue layers—endoderm, ectoderm, and mesoderm.
6. How does cell death contribute to development?
7. Describe the process of induction, and give two examples.
8. Define *differentiation*. How do cells differentiate; that is, how is it that adult cells express some but not all the genes of the fertilized egg?
9. What role do gradients of morphogens play in animal development?
10. In humans, where does fertilization occur, and what stages of development occur before the fertilized egg reaches the uterus?
11. Describe how the human blastocyst gives rise to the embryo and its extraembryonic membranes.
12. Explain how the structure of the placenta prevents mixing of fetal and maternal blood while allowing the exchange of substances between the mother and the fetus.
13. Is the placenta an effective barrier against substances that can harm the fetus? Describe two types of harmful agents that can cross the placenta and their effects on the fetus.
14. How do changes in the breast prepare a mother to nurse her newborn? How do hormones influence these changes and stimulate milk production?
15. Describe the events that lead to the expulsion of the baby and the placenta from the uterus. Explain why this is an example of positive feedback.

Applying the Concepts

1. When fetal ectoderm tissue from the mouth of an embryo is transplanted into a region of mesoderm of another embryo of a different species, the mesoderm induces the ectoderm to form mouth structures of the type of organism from which the ectoderm tissue was taken. When ectoderm from a chick is implanted into a mouse embryo, the chicken tissue produces teeth. What does this experiment tell us about why the adage "scarce as hen's teeth" is true? Also, what does this experiment tell us about the evolutionary origin of chickens?
2. On the basis of your knowledge of genetics (Unit II) and evolution (Unit III), explain why the human embryo passes through a developmental stage in which it has gill grooves and a tail.
3. Embryologists have used embryo fusion to produce *tetraparental* (four-parent) mice and have also produced "geeps" from goat and sheep embryos. The resulting bodies are patchworks of cells from both animals. Why does fusion succeed with very early embryos (four-cell to eight-cell stages) and fail when much older embryos are used?
4. If the nuclei of adult cells can be transplanted into eggs from which the nucleus has been removed to produce clones of the parent, is it theoretically possible to produce human clones? Would such clones yield offspring that are *exactly* identical to the parents who supplied the nuclei? Explain.
5. Estimates of the number of women in the U.S. who self-medicate with over-the-counter (OTC) drugs during pregnancy range from 65% to 95%. Evidently, the public doesn't perceive nonprescription substances as drugs. Many subtle disorders, especially of the nervous system, will probably be connected to over-the-counter use in coming years. If you were engaged in pharmaceutical research, which OTC drugs would you focus on as possible culprits? Which months of pregnancy would you study closely? Who would you choose as your study population?

For More Information

Caldwell, M. "How Does a Single Cell Become a Whole Body?" *Discover*, November 1992. Explores the miracles of development in layperson's terms.

DeRobertis, E. M., Oliver, G., and Wright, C. V. E. "Homeobox Genes and the Vertebrate Body Plan." *Scientific American*, July 1990. Genes can be injected into amphibian eggs, allowing direct observation of the effects of specific gene products on development.

Nüsslein-Volhard, C. "Gradients That Organize Embryo Development." *Scientific American*, August 1996. A Nobel laureate describes developmental experiments that have revealed a great deal about how chemical gradients in embryos help guide cell differentiation.

Pedersen, R. A. "Embryonic Stem Cells for Medicine." *Scientific American*, April 1999. Embryonic stem cells hold promise for regenerating damaged tissues.

Rose, M. R. "Can Human Aging Be Postponed?" *Scientific American*, December 1999. Theoretically, yes.

Smith, B. R. "Visualizing Human Embryos." *Scientific American*, March 1999. Magnetic resonance microscopy allows computerized reconstuction of three-dimensional structures within human embryos.

Taubes, G. "Ontongeny Recapitulated." *Discover*, May 1998. Will biotechnology allow us to turn on genes that normally become silent after development has been completed and that will provide the opportunity to regrow limbs or to replace dying brain cells?

Zigova, T., and Sanberg, P. R. "Neural Stem Cells for Brain Repair." *Science and Medicine*, September/October 1999. The recent discovery of stem cells in the adult brain that are capable of giving rise to all types of brain cells suggest that physicians may one day be able to repair damaged portions of the nervous system.

Answers to Multiple-Choice Questions

1. e 2. e 3. c 4. a 5. b 6. d

MediaTutor
Animal Development

CD Activities

Activity 36.1: Stages of Animal Development

Estimated time: 10 minutes

The stages that occur during animal embryonic development are remarkably similar for a wide range of animals, including amphibians, birds and reptiles, and even mammals. One of the characteristics that all animals have in common during embryonic development is the formation of three tissue layers (endoderm, mesoderm, and ectoderm) during gastrulation. In this tutorial, you will explore the various stages in the development of an animal embryo, with special emphasis on the process of gastrulation.

Activity 36.2: Human Development

Estimated time: 10 minutes

Human development is controlled by the same mechanisms that control the development of other animals. This similarity is a product of our evolutionary heritage. Many of the early events of human development resemble those seen in other animals. By the end of the second month, the human embryo becomes a fetus and takes on characteristics that reveal its true identity. In this tutorial, you will explore the key events that occur during human embryonic and fetal development; then you can play an interactive game that will test your knowledge and skill.

Start the MediaTutor Student CD-ROM and enter the activity number in the Quick Search box to be taken directly to that activity.

Web Investigations

Case Study: Far-reaching Choices

Estimated time: 10 minutes

Think about the chemicals you are exposed to every day. Caffeine, alcohol, cigarette smoke, over-the-counter drugs, formaldehyde from carpets, paint fumes, gas. The list goes on and on. Some are generally harmless; others are not. In a few cases, even a short exposure can have devastating effects on a developing fetus. Here are a few things to think about if someone near you is, or plans to become, pregnant.

Go to http://www.prenhall.com/audesirk6, the Audesirk Companion Web site. Select Chapter 36 and the Web Investigation to begin.

"As I watched the geese, it appeared to me as little short of a miracle that a hard, matter-of-fact scientist should have been able to establish a real friendship with the wild, free-living animals, and the realization of this fact made me strangely happy. It made me feel as though man's expulsion from the Garden of Eden had thereby lost some of its bitterness."

Konrad Lorenz in* King Solomon's Ring *(1952)

Both this male scorpionfly and this male human are exceptionally attractive to females of their species. The secret to their sex appeal may be that both have highly symmetrical bodies.

37 Animal Behavior

AT A GLANCE

Case Study: Sex and Symmetry

1) **How Do Innate and Learned Behaviors Differ?**
Innate Behaviors Can Be Performed Without Prior Experience
Learned Behaviors Are Modified by Experience
There Is No Sharp Distinction Between Innate and Learned Behaviors

2) **How Do Animals Communicate?**
Visual Communication Is Most Effective over Short Distances
Communication by Sound Is Effective over Longer Distances
Chemical Messages Persist Longer But Are Hard to Vary
Communication by Touch Helps Establish Social Bonds

3) **How Do Animals Compete for Resources?**
Aggressive Behavior Helps Secure Resources
Dominance Hierarchies Help Manage Aggressive Interactions
Animals May Defend Territories That Contain Resources

4) **How Do Animals Find Mates?**
Vocal and Visual Signals Encode Sex, Species, and Individual Quality
Chemical Signals Bring Mates Together

5) **What Kinds of Societies Do Animals Form?**
Group Living Has Advantages and Disadvantages
Honeybees Form Complex Insect Societies
Bullhead Catfish Form a Simple Vertebrate Society
Naked Mole Rats Form a Complex Vertebrate Society

6) **Can Biology Explain Human Behavior?**
The Behavior of Newborn Infants Has a Large Innate Component
Behaviors Shared by Diverse Cultures May Be Innate
People May Respond to Pheromones
Comparisons of Identical and Fraternal Twins Reveal Genetic Components of Behavior

Evolutionary Connections: Why Do Animals Play?

Case Study Revisited: Sex and Symmetry

CASE STUDY

Sex and Symmetry

For a male Japanese scorpionfly, finding a mate can be a real struggle. Female scorpionflies will mate only with males that can offer a tasty meal (usually a dead insect). Competition for dead insects is fierce, and a male must typically defend an insect from other males. The competition for insects often erupts in bitter combat, characterized by repeated head-butting and grappling with sharp-pointed genital claspers.

In the competition to gain access to mates, then, not all male Japanese scorpionflies can be equally successful. The most successful males must have qualities that help them defeat other males in combat and that are especially attractive to females. Biologist Randy Thornhill has found that one quality in particular accurately predicts the mating success of male scorpionflies: symmetry. In Thornhill's experiments and observations, the most successful males were those whose left and right wings were equal or nearly equal in length. Males in which one wing was longer than the other were less likely to win fights or to copulate; the greater the difference between the two wings, the lower the likelihood of success.

Thornhill's work with scorpionflies led him to wonder if the advantages of symmetry also extend to humans. Working with psychologist Steven Gangestad, he devised some fascinating studies that suggest that male symmetry does indeed play an important role in human sexual relationships. Women find symmetrical men more attractive, and highly symmetrical human males do indeed appear to gain some of the advantages that accrue to symmetrical scorpionflies. The preferences of human females resemble, at least in this one respect, those of female scorpionflies. One could almost say that insects and humans share a standard of beauty. ■

MEDIATUTOR 37.1 Innate Animal Behavior

1 How Do Innate and Learned Behaviors Differ?

Innate Behaviors Can Be Performed Without Prior Experience

Behavior is any observable activity of a living animal. A fruit fly placed in a tunnel with light at one end and darkness at the other will fly toward the light; a hungry newborn human infant, touched on the side of her mouth, will turn her head and attempt to suckle. These are examples of **innate** behaviors. Innate behavior is performed in reasonably complete form even the first time an animal of the right age and motivational state encounters a particular stimulus. (The proper motivational state for feeding, for example, would be hunger.)

Scientists can demonstrate that a behavior is innate by depriving an animal of the opportunity to learn it. For example, wild red squirrels bury extra nuts in the fall for retrieval during the winter. Red squirrels can be raised from birth in a bare cage on a liquid diet, providing them with no experience with nuts, digging, or burying. Presented with nuts for the first time, such a squirrel (after eating several) will carry one to the corner of its cage, then make covering and patting motions with its forefeet. Nut burying is therefore an innate behavior.

Innate behaviors can also be recognized by their occurrence immediately after birth, before there is any opportunity for learning. The cuckoo, for example, lays its eggs in the nest of another bird species, to be raised by the unwitting adoptive parent. Immediately after hatching, the cuckoo chick shoves the nest owner's eggs (or baby birds) out of the nest, eliminating its competitors for food (Fig. 37-1).

Learned Behaviors Are Modified by Experience

Natural selection may favor innate behaviors in many circumstances. For instance, it is clearly to the advantage of a herring gull chick to be able to peck at its parent's bill as soon as possible after hatching, because pecking stimulates the parent to feed the chick. But in other circumstances, rigidly fixed behavior patterns may be less useful. For example, a male red-winged blackbird that copulates with a stuffed female obviously will produce no offspring as a result of that behavior. In many situations, a greater degree of behavioral flexibility is advantageous. The capacity to make *changes* in behavior on the basis of experience is called **learning.** This deceptively simple definition encompasses a vast array of different phenomena. A frog learns to avoid distasteful insects, a baby shrew learns which adult is its mother, a human learns to speak a language, a blackbird learns to use the stars for navigation. Each of the many examples of animal learning represents the outcome of a unique evolutionary history and set of ecological needs, so learning processes are as diverse as animals themselves. Nonetheless, it can be useful to categorize types of learning, as long as we keep in mind that the categories are only rough guides and that many examples of learning will not fit neatly into any category.

(a) (b)

Figure 37-1 Innate behavior
(a) The cuckoo chick, just hours after it hatches, and before its eyes have opened, evicts the eggs of its foster parents from the nest. ***(b)*** The parents, responding to the stimulus of the cuckoo chick's wide-gaping mouth, feed the chick, unaware that it is not related to them.

Habituation Is a Decline in Response to a Repeated Stimulus

A common form of simple learning is **habituation**, defined as a decline in response to a repeated stimulus. The ability to habituate prevents an animal from wasting its energy and attention on irrelevant stimuli. This form of learning is shown by even the simplest animals. For example, a sea anemone will retract its tentacles when touched but gradually stops retracting if touching is continued (Fig. 37-2).

The ability to habituate is clearly adaptive. If a sea anemone withdrew every time it was brushed by a strand of waving seaweed, the animal would waste a great deal of energy, and its retracted posture would prevent it from snaring food. Humans habituate to many stimuli: city dwellers to nighttime traffic sounds and country dwellers to choruses of crickets and tree frogs. Each may initially find the other's habitat unbearably noisy at first but habituates after a time.

Conditioning Is a Learned Association Between a Stimulus and a Response

A more complex form of learning is **trial-and-error learning**, in which animals acquire new and appropriate responses to stimuli through experience. Many animals are faced with naturally occurring rewards and punishments and can learn to modify their responses to them. For example, a hungry toad that captures a bee quickly learns to avoid future encounters with bees (Fig. 37-3). After only one experience with a stung tongue, a toad modifies its response to flying insects so that bees (and even other insects that resemble them) are excluded.

Trial-and-error learning is an important feature in the behavioral development of many animal species and often occurs during play and exploratory behavior (see "Evolutionary Connections: Why Do Animals Play?"). This type of learning plays a key role in human behavior; it allows a child to learn which foods taste good or bad, that a stove can be hot, and not to pull the cat's tail.

Some interesting properties of trial-and-error learning have been revealed by a laboratory technique known as **operant conditioning**. During operant conditioning, an animal learns to perform a behavior (such as pushing a lever or pecking a button) to receive a reward or to avoid punishment. This technique is most closely associated with the American comparative psychologist B. F. Skinner. Skinner designed the "Skinner box," in which an animal is isolated and allowed to train itself. The box might contain a lever that, when pressed, ejects a food pellet. If the animal accidentally bumps the lever, a food reward appears. After a few such occurrences, the animal learns the connection between pressing the lever and receiving food and soon begins to press the lever repeatedly.

Operant conditioning has been used to train animals to perform tasks far more complex than pressing a lever, but perhaps the most interesting revelation fostered by the technique is that species differ in their propensity for learning particular associations. In particular, species seem to be predisposed to learn behaviors that are relevant to their own needs. For example, if a rat is given a distinctively flavored food that has been mixed with a substance that makes the rat sick, the animal will subsequently refuse to eat that food. In contrast, it is very difficult to train a rat to rear up on its hind legs in response to a particular sound or visual cue. The difference can be explained by asking which learning task is more likely to benefit a wild Norway rat (the species from which lab rats are descended). Clearly, an ability to avoid foods that induce illness is beneficial to an animal like the Norway rat, which is well known for eating a tremendous variety of foods. No obvious benefit, however, would arise

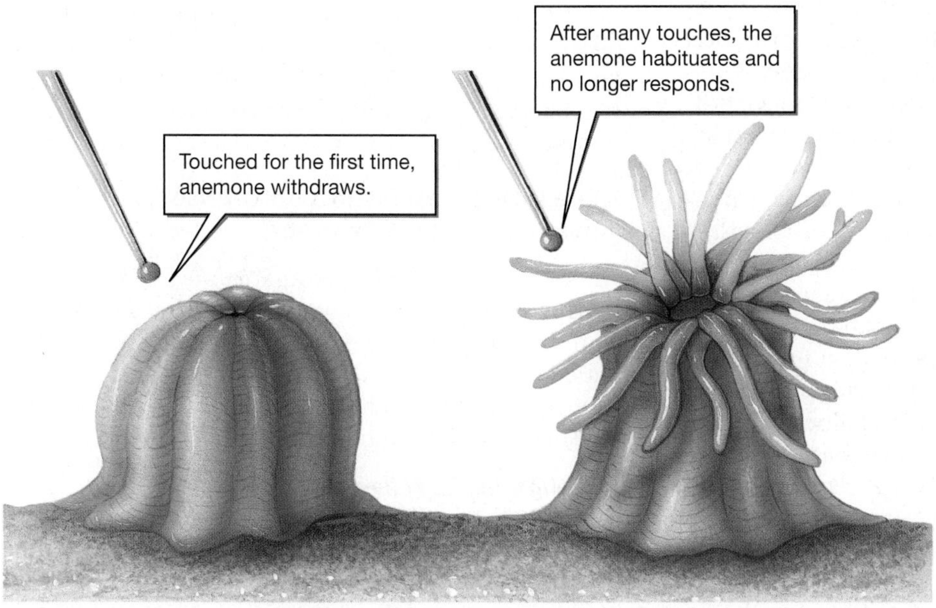
Touched for the first time, anemone withdraws.
After many touches, the anemone habituates and no longer responds.

Figure 37-2 Habituation in a sea anemone

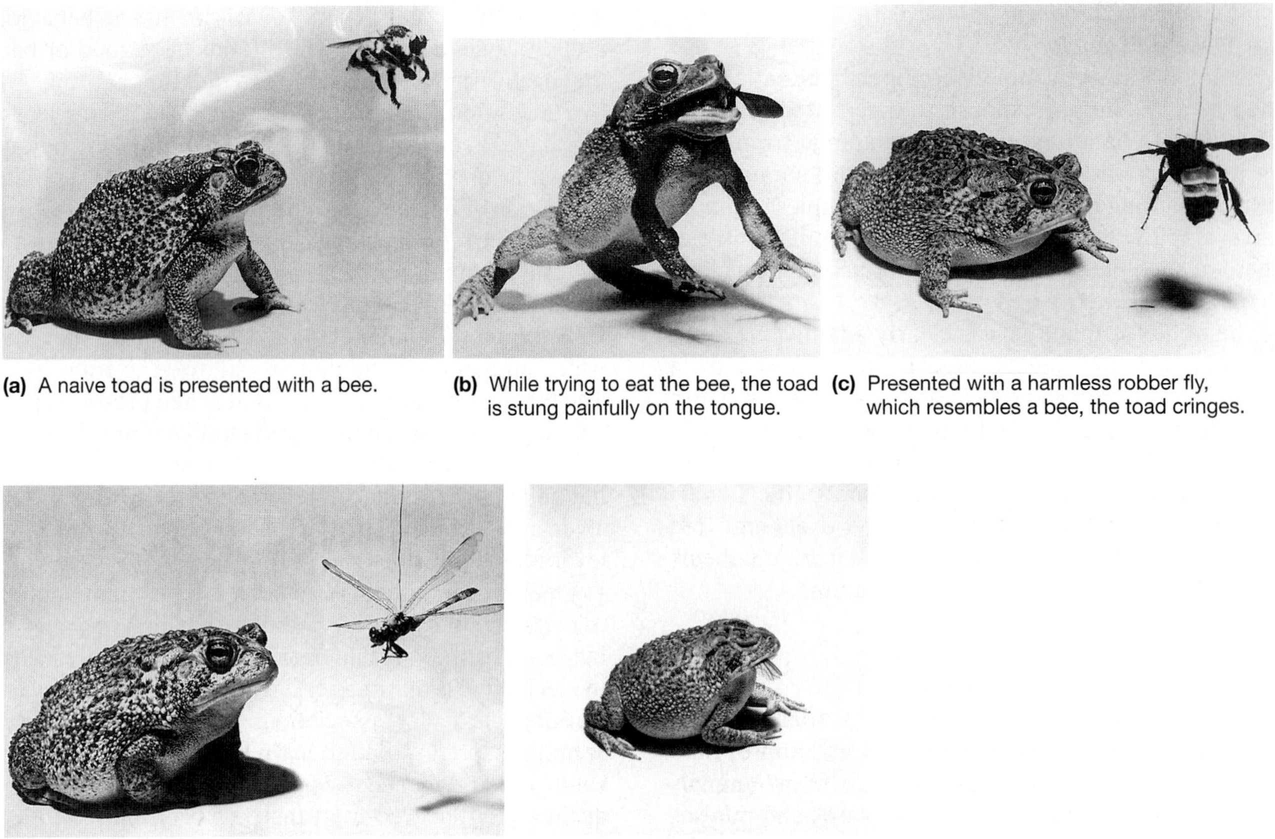

(a) A naive toad is presented with a bee.

(b) While trying to eat the bee, the toad is stung painfully on the tongue.

(c) Presented with a harmless robber fly, which resembles a bee, the toad cringes.

(d) The toad is presented with a dragonfly.

(e) The toad immediately eats the dragonfly, demonstrating that the learned aversion is specific to bees and insect resembling them.

Figure 37-3 Trial-and-error learning in a toad

from learning to stand up in response to a noise. In general, we expect the specific learning abilities of each species to have evolved to support its particular mode of life.

Insight Is Problem Solving Without Trial and Error

In certain situations, animals seem able to solve problems suddenly, without the benefit of prior experience. This kind of sudden problem solving is sometimes called **insight learning**, because it seems at least superficially similar to the process by which humans mentally manipulate concepts to arrive at a solution. We cannot, of course, know for sure if nonhuman animals experience similar mental states when they solve problems.

In 1917 animal behaviorist Wolfgang Kohler showed that a hungry chimpanzee, without any training, would stack boxes to reach a banana suspended from the ceiling (Fig. 37-4). This type of mental problem solving was once believed to be limited to very intelligent types of animals such as primates, but similar abilities may also be present in species that we tend to view as less intelligent. For example, R. Epstein and colleagues performed an experiment that showed that pigeons may be capable of insight learning. In the experiment, pigeons (whose wings had been clipped to prevent flight) were first trained to perform two unrelated tasks in return for food rewards. The tasks were (1) to push a small box around the cage and (2) to peck at a small plastic banana. Later, the trained birds were presented with a novel situation: a plastic banana that hung above their reach in a cage that also contained a small box. Many of the pigeons pushed the box to a position beneath the plastic banana and climbed atop the box to peck the faux fruit. Apparently, a pigeon that has been trained to execute the necessary physical movements can solve the suspended banana problem, just as a chimp can.

There Is No Sharp Distinction Between Innate and Learned Behaviors

Although the terms *innate* and *learned* can be useful tools in describing and understanding behaviors, these words also have the potential to lull us into an oversimplified view of animal behavior. In practice, no behavior is unambiguously instinctive or unequivocally learned; all behaviors are an intimate mixture of the two.

Seemingly Innate Behavior Can Be Modified By Experience

Behaviors that seem to be performed correctly on the first attempt without prior experience can later be modified by experience. For example, a newly hatched her-

Figure 37-4 Insight
(a) A dog, lacking insight in this situation, fails a detour problem on the first try. It will eventually learn by trial and error. ***(b)*** Unable to reach the bananas, the chimpanzee exhibits insight by stacking boxes beneath the bananas to extend its reach.

ring gull chick is able to peck at a red spot on its parent's beak (Fig. 37-5), a stereotyped behavior that causes the parent to regurgitate food for the chick to eat. Biologist Niko Tinbergen studied this pecking behavior and found that the pecking response of very young chicks was triggered by the long, thin shape and red color of the parent's bill. In fact, when Tinbergen offered newly hatched chicks a thin, red rod with white stripes painted on it, they pecked at it more often than at a real beak. Within a few days, however, the chicks learned enough about the appearance of their parents that they began pecking more frequently at models more closely resembling the parents. After 1 week, the young gulls learned enough about how their parents looked to prefer models of their own species to models of a closely related species. Eventually, the young birds learned to beg only from their own parents.

Figure 37-5 Innate behaviors can be modified by experience
A herring gull chick pecks at the red spot on its mother's bill, causing her to regurgitate food. Niko Tinbergen and others have shown that newly hatched chicks will peck instinctively at any object bearing appropriate stimuli, such as a thin stick with a contrasting spot painted on it. As the chick grows older, however, it gradually modifies this innate response until it will peck only at its parent's bill.

Habituation (a decline in response to a repeated stimulus) can also fine-tune an organism's innate responses to environmental stimuli. For example, young birds crouch down when a hawk flies over but ignore harmless birds such as geese. Early observers hypothesized that only the very specific shape of predatory birds provoked crouching. Using an ingenious model, Niko Tinbergen and Konrad Lorenz (two of the founding fathers of **ethology**, the study of animal behavior) tested this hypothesis (Fig. 37-6). When moved in one direction, the model resembled a goose, which the chicks ignored. When its movement was reversed, however, the model resembled a hawk and elicited crouching. Further research revealed that naive chicks instinctively crouch when *any* object moves over their heads. Over time, their response habituates to things that soar by harmlessly and frequently, such as leaves, songbirds, and geese. Predators are much less common, and the novel shape of a hawk still elicits instinctive crouching. Thus, learning modifies the innate response, making it more adaptive.

Learning May Be Governed by Innate Constraints

Learning always occurs within boundaries that help increase the chances that only the appropriate behavior is acquired. For example, young robins routinely learn their songs from the adult robins that they hear but almost never copy the songs of the sparrows, warblers, finches, or other species that they also hear. The innate constraints on learning are perhaps most strikingly

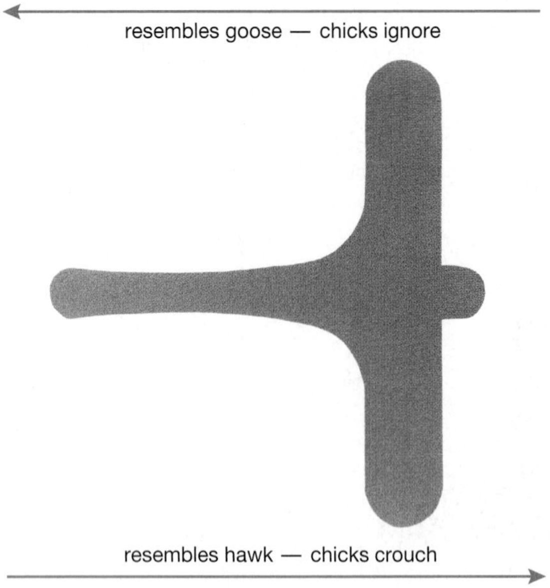

Figure 37-6 Habituation modifies innate responses
The model used by Konrad Lorenz and his student Niko Tinbergen to investigate the response of chicks to the shape of objects flying overhead. The chicks' response depended on which direction the model moved. Moving toward the right, the model resembles a predatory hawk, but when it moves left, it resembles a harmless goose.

Figure 37-7 Konrad Lorenz and imprinting
Konrad Lorenz, known as the "father of ethology," is followed by goslings that imprinted on him shortly after they hatched. They now follow him as they would their mother.

illustrated by **imprinting**, a special form of learning in which learning is rigidly programmed to occur only at a certain critical period of development. A strong association is formed during a particular stage, called a **sensitive period**, in the animal's life. During this stage, the animal is primed to learn a specific type of information, which is then incorporated into a behavior that is not easily altered by further experience. Imprinting is best known in birds such as geese, ducks, and chickens. These birds learn to follow the animal or object that they most frequently encounter during an early sensitive period. In nature, the mother is the object most likely to be nearby during the sensitive period, so the young birds imprint on her. In the laboratory, however, these birds can be forced to imprint on a toy train or other moving object (Fig. 37-7). If given a choice, however, they select an adult of their species.

All Behavior Arises Out of Interaction Between Genes and Environment

Many early ethologists saw innate behaviors as rigidly controlled by genetic factors and viewed learned behaviors as determined exclusively by an animal's environment. Today, however, this false dichotomy has given way to the realization that, just as no behavior is wholly innate or wholly learned, no behavior can be caused strictly by genes or strictly by the environment. Instead, all behavior develops out of an interaction between genes and their environment. The relative contributions of heredity and learning vary among animal species and among behaviors within an individual.

The precise nature of the link among genes, environments, and behaviors is not well understood in most cases. The chain of events between the transcription of genes and the performance of a complex behavior may be so complex that we will never be able to decipher it fully. Nonetheless, a great deal of evidence demonstrates the existence of both genetic and environmental components in the development of even the most complex behaviors. For example, consider the case of bird migration. Even though it is well known that migratory birds must learn by experience how to navigate via celestial cues, this learning process is not the only factor involved.

At the close of the summer, many birds disappear from the habitats where they've spent the breeding season and head for winter quarters hundreds or even thousands of miles to the south. Many of the migrating birds are traveling for the first time, because they were hatched only a few months earlier. Amazingly, these naïve birds depart at the proper time and in the proper direction and locate the proper wintering location, even though they cannot simply follow more-experienced birds (which typically depart a few weeks in advance of the first-year birds). Somehow, these young birds are able to execute a very difficult task the first time they try it. It seems that birds must be born with the ability to migrate; it must be "in their genes." And indeed, birds

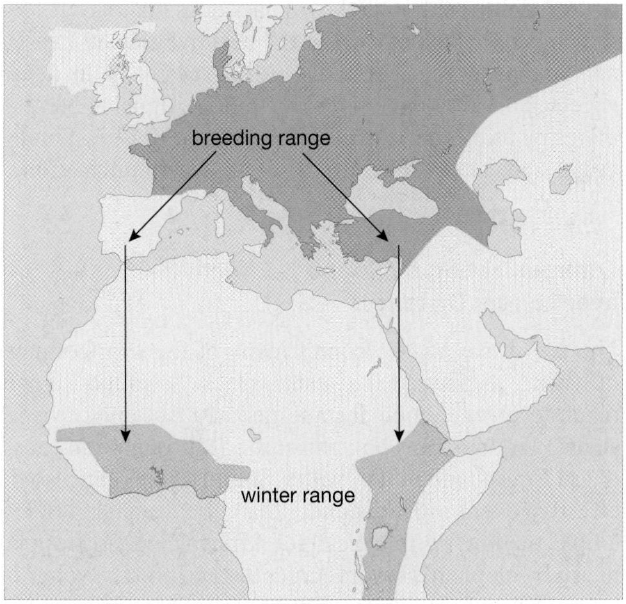

Figure 37-8 Genes influence migratory behavior
Blackcap warblers from western Europe begin their fall migration by flying in a southwesterly direction, but those from eastern Europe fly to the southeast when they begin migrating. If members of the two populations are crossbred in captivity, the hybrid offspring orient in a direction intermediate between the migratory directions of the parents. This result suggests that migratory behavior has a heritable component.

that are experimentally hatched and raised in isolation indoors still orient in the proper migratory direction when autumn comes, apparently without the need for any learning or experience.

The conclusion that birds must have a genetically controlled ability to migrate in the right direction was further supported by hybridization experiments with blackcap warblers. This species breeds in Europe and migrates to Africa, but populations from different areas travel by different routes. Blackcaps from western Europe travel in a southwesterly direction to reach Africa, whereas birds from eastern Europe must travel to the southeast (Fig. 37-8). If birds from the two populations are crossbred in captivity, however, the hybrid offspring will exhibit migratory orientation toward due south, intermediate between that of the two parents. This result suggests that the parents had genes that influenced migratory direction and that the hybrids inherited a mixture of these genes.

2 How Do Animals Communicate?

Animals frequently make information available for sharing. The sounds uttered, movements made, and chemicals emitted by animals can reveal the animals' physical location, level of aggression, readiness to mate, and so on. If revealing such information evokes a response from other individuals, and if that response tends to benefit the sender and the receiver, then a communication channel can form. **Communication** is defined as the production of a signal by one organism that causes another organism to change its behavior in a way beneficial to one or both.

Although animals of different species may communicate (picture a cat, its tail erect and bushy, hissing at a strange dog), most communication occurs between members of the same species. Potential mates must communicate, as must parents and offspring. At the same time, members of the same species compete most directly with one another for food, space, and mates. Communication is often used to resolve such conflicts with minimal damage.

The mechanisms by which animals communicate are astonishingly diverse and use all of the senses. In the following sections, we look at communication by visual displays, sound, chemicals, and touch.

Visual Communication Is Most Effective over Short Distances

Animals with well-developed eyes, from insects to mammals, use visual signals to communicate. Visual signals can be *active,* in which a specific movement (such as baring fangs) or posture (such as lowering the head) conveys a message (Fig. 37-9). Alternatively, visual signals may be *passive,* in which case the size, shape, or color of the animal conveys important information, commonly about its sex and reproductive state. For example, when female mandrills become sexually receptive, they develop a large, brightly colored swelling on their buttocks (Fig. 37-10). Active and passive signals

Figure 37-9 An active visual signal
The wolf signals aggression by lowering its head, ruffling the fur on its neck and along its back, facing its opponent with a direct stare, and exposing its fangs. These signals can vary in intensity, communicating different levels of aggression.

Figure 37-10 A passive visual signal
The female mandrill's colorfully swollen buttocks serve as a passive visual signal that she is fertile and ready to mate.

can be combined, as illustrated by the lizard in Figure 37-11 and the courtship behavior of the three-spined stickleback fish (see Fig. 37-24).

Like all forms of communication, visual signals have both advantages and disadvantages. On the plus side, they are instantaneous, and active signals can be rapidly revised to convey a variety of messages in a short period. Visual communication is quiet and unlikely to alert distant predators, although the signaler does make itself conspicuous to those nearby. On the negative side, visual signals are generally ineffective in darkness and in dense vegetation, although female fireflies signal potential mates by using species-specific patterns of flashes. Finally, visual signals are limited to close-range communication.

Figure 37-11 Active and passive visual signals combined
The South American *Anolis* lizard raises his head high in the air (an active visual signal), revealing a brilliantly colored throat pouch (a passive visual signal) that warns others to keep their distance.

Communication by Sound Is Effective over Longer Distances

The use of sound overcomes many of the shortcomings of visual displays. Like visual displays, sound signals reach receivers almost instantaneously. But unlike visual signals, sound can be transmitted through darkness, dense forests, and murky water. Sound signals can also be effective over longer distances than visual signals. For example, the low, rumbling calls of African elephants can be heard by elephants several miles distant, and the songs of humpback whales are audible hundreds of miles away from the singer. Likewise, the howls of a wolf pack carry for miles on a still night. Even the small kangaroo rat produces a sound (by striking the desert floor with its hind feet) that is audible 150 feet (45 meters) away.

Auditory signals are similar to visual displays in that they can be varied to convey rapidly changing messages. (Think of words and of the emotional nuances conveyed by the human voice during a conversation.) Changes in motivation can be signaled by a change in the loudness or pitch of the sound. An individual can convey different messages by variations in the pattern, volume, and pitch of the sound produced. Ethologist Thomas Struhsaker studied vervet monkeys in Kenya in the 1960s and found that they produced different calls in response to threats from each of their major predators: snakes, leopards, and eagles. In 1980 other researchers reported that the response of other vervet monkeys to each of these calls is appropriate to the particular predator. The "bark" that warns of a leopard or other four-legged carnivore causes monkeys on the ground to take to trees and those in trees to climb higher. The "rraup" call signaling an eagle or other hunting bird causes monkeys on the ground to look upward and take cover, whereas monkeys already in trees drop to the shelter of lower, denser branches. The "chutter" call indicates a snake and causes the monkeys to stand up and search the ground for this slower, lower predator.

The use of sound is by no means limited to birds and mammals. Male crickets produce species-specific songs that attract female crickets of the same species. The annoying whine of the female mosquito as she prepares to bite alerts nearby males that she will soon have the blood meal necessary for laying eggs. Male water striders vibrate their legs, sending species-specific patterns of vibrations through the water, attracting mates and repelling other males (Fig. 37-12). From these rather simple signals to the virtuoso performance of human language, sound is one of the most important forms of communication.

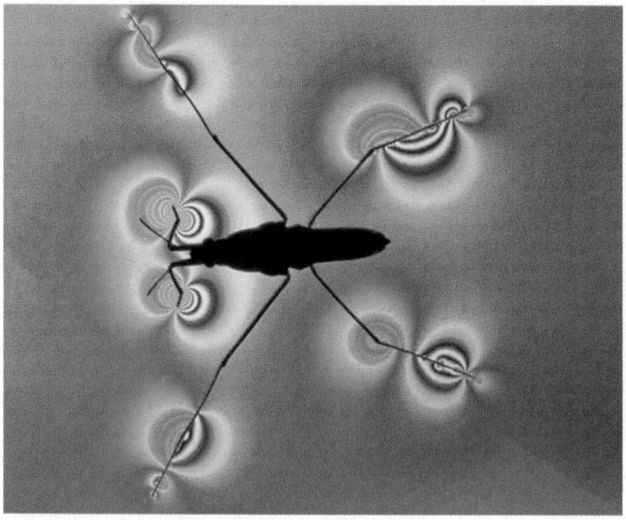

Figure 37-12 Communication by vibration
The light-footed water strider relies on the surface tension of water to support its weight. By vibrating its legs, the water strider sends signals that radiate out over the surface of the water. These vibrations advertise the strider's species and sex to others nearby.

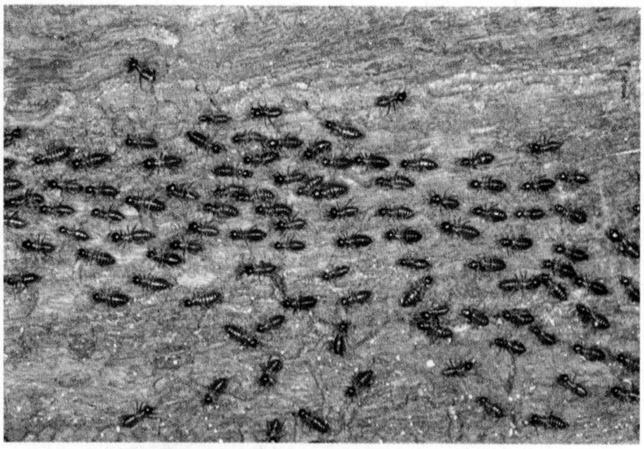

Figure 37-13 Communication by chemical messages
To direct other colony members to a source of food, foraging termites have laid a trail of pheromones, secreted by glands in their abdomens.

Chemical Messages Persist Longer But Are Hard to Vary

Chemical substances produced by an individual that influence the behavior of others of its species are called **pheromones**. Chemicals may carry messages over long distances, and, unlike sound, take very little energy to produce. Pheromones may not even be detected by other species, including predators who might be attracted to visual or auditory signals. Like a signpost, a pheromone persists over time and can convey a message after the animal has departed. Wolf packs, hunting over areas up to 1000 square kilometers (386 square miles), mark the boundaries of their travels with pheromones in urine to warn other packs of their presence. As anyone who has walked a dog can attest, the domesticated dog reveals its wolf ancestry by staking out its neighborhood with urine that carries a chemical message: "I live in this area."

Such communication requires an animal to synthesize as well as respond to a different chemical for each message. As a result, in general, fewer messages are communicated with chemicals than with sight or sound. In addition, pheromone signals lack the diversity and gradation of auditory or visual signals. Nonetheless, chemicals powerfully convey a few simple but critical messages.

Pheromones act in one of two ways. Some cause an immediate, observable behavior in the animal that detects them. They convey messages such as "This area is mine" or "I am ready to mate." For example, foraging ants and termites that discover food lay a trail of pheromones, secreted by abdominal glands, from the food to the nest (Fig. 37-13). Its message: "Follow to find food."

Other pheromones stimulate a physiological change in the animal that detects them; the change is typically in the receiver's reproductive state. For example, the queen honeybee produces a pheromone called **queen substance**, which prevents other females in the hive from becoming sexually mature. Similarly, mature males of some species of mice produce urine that contains a pheromone that influences female reproductive physiology. The pheromone stimulates newly mature females to become fertile and sexually receptive. It will also cause a female mouse that is newly pregnant by another male to abort her litter and become sexually receptive to the new male.

The sex attractant pheromones of some agricultural pests, such as the Japanese beetle and the gypsy moth, have been synthesized. These synthetic pheromones can be used to disrupt mating or to lure these insects into traps. Controlling pests with pheromones has major environmental advantages over conventional pesticides. Pesticides kill beneficial as well as harmful insects and foster the evolution of pesticide-resistant insects. In contrast, each pheromone is specific to a single species, and pheromones do not promote the spread of resistance, because insects resistant to the attraction of their own pheromones do not reproduce successfully.

Communication by Touch Helps Establish Social Bonds

Communication by physical contact between individuals often serves to establish and maintain social bonds among group members. This function is especially apparent in primates, including humans, which have many gestures—including kissing, nuzzling, patting, petting, and grooming—with important social functions (Fig. 37-14a). Touch may even be essential to human wellbeing. Recent research showed that when the limbs of premature human infants were stroked and moved for 45 minutes daily, the infants were more active, responsive, and emotionally stable and gained weight more

Figure 37-14 Communication by touch *(a)* An adult olive baboon grooms a juvenile. Grooming not only reinforces social relationships but also removes debris and parasites from the fur. *(b)* Touch is also important in sexual communication. These land snails (*Helix*) engage in courtship behavior that will culminate in mating.

rapidly than did premature infants who received standard hospital treatment.

Communication by touch is not limited to primates, but is widespread. In many other mammal species, close physical contact helps cement the bond between parent and offspring. Species in which sexual activity is preceded or accompanied by physical contact can be found across the animal kingdom (Fig. 37-14b).

3 How Do Animals Compete for Resources?

The Darwinian contest to survive and reproduce stems from the scarcity of resources relative to the reproductive potential of populations. The resulting competition underlies many of the most frequent types of interactions between animals.

Aggressive Behavior Helps Secure Resources

One of the most obvious manifestations of competition for resources such as food, space, or mates is **aggression**, or antagonistic behavior, between members of the same species. Although the expression "survival of the fittest" evokes images of the strongest animal emerging triumphantly from among the dead bodies of its competitors, in reality most aggressive encounters between members of the same species end without physical damage to the participants. Natural selection has favored the evolution of symbolic displays or rituals for resolving conflicts. During fighting, even the victorious animal can be injured, so serious fighters might not survive to pass on their genes. Aggressive *displays,* in contrast, allow the competitors to assess each other and acknowledge a winner on the basis of its size, strength, and motivation rather than on the wounds it can inflict.

During aggressive displays, animals may exhibit weapons, such as claws and fangs (Fig. 37-15a), and often behave in ways that make them appear larger (Fig. 37-15b). Competitors often stand upright and erect their fur, feathers, ears, or fins (see Fig. 37-9). The displays are typically accompanied by intimidating sounds (growls, croaks, roars, chirps) whose loudness can help decide the winner. Fighting tends to be a last resort when displays fail to resolve the dispute.

In addition to aggressive visual and vocal displays, many animal species engage in ritualized combat. Deadly weapons may clash harmlessly (Fig. 37-16) or may not be used at all. In many cases these encounters involve shoving rather than slashing. The ritual thus allows contestants to assess the strength and the motivation of their rivals, and the loser slinks away in a submissive posture that minimizes the size of its body.

Dominance Hierarchies Help Manage Aggressive Interactions

Aggressive interactions use a lot of energy, can cause injury, and can disrupt other important tasks, such as finding food, watching for predators, or raising young. Thus, there are advantages to resolving conflicts with minimal aggression. In a **dominance hierarchy**, each animal establishes a rank that determines its access to resources. Domestic chickens, after squabbling, sort themselves into a reasonably stable "pecking order." Thereafter, when competition for food occurs, all hens defer to the dominant bird, all but the dominant bird give way to the second, and so on. Conflict is minimized because each bird knows its place. Dominance among male bighorn sheep is reflected in their horn size (Fig. 37-17). In wolf packs, one member of each sex is the dominant, or "alpha," individual to whom all others are subordinate. Although aggressive encounters occur frequently while the dominance hierarchy is being established, after each animal learns its place, disputes are minimized. The dominant individuals obtain most access to the resources needed for reproduction, including food, space, and mates.

(a)

(b)

Figure 37-15 Aggressive displays
(a) Threat display of the male baboon. Despite the potentially lethal fangs so prominently displayed, aggressive encounters between baboons rarely cause injury. ***(b)*** The aggressive display of many male fish (here, striped grunts) includes elevating the fins and flaring the gill covers, thus making the body appear larger.

Perhaps the most thoroughly studied dominance hierarchy is that of chimpanzees. Ethologist Jane Goodall (Fig. 37-18) has devoted more than 30 years to making meticulous observations of chimpanzee behavior in the field at Gombe National Park in Tanzania, and she has described and documented the animals' complex social organization. Chimps live in groups, and dominance hierarchies among males are a key aspect of their social life. Many males devote a significant amount of time to maintaining their position in the hierarchy, largely by way of an aggressive *charging display* in which a male rushes forward, throws rocks, leaps up to shake vegetation, and otherwise seeks to intimidate rival males. The advantages that accrue to dominant males, however, are not clear. According to Goodall, low-ranking males can still secure access to food and successful copulations (albeit not quite as easily as higher-ranking males). In Goodall's view, little evolutionary advantage accrues to a dominant male, and the function of chimpanzee dominance hierarchies remains unexplained.

Figure 37-16 Displays of strength
Ritualized combat of fiddler crabs. Oversized claws, which could severely injure another animal, grasp harmlessly. Eventually one crab, sensing greater vigor in his opponent, typically retreats unharmed.

Figure 37-17 A dominance hierarchy
The dominance hierarchy of the male bighorn sheep is signaled by the size of the horns; these rams increase in status from right to left. These backward-curving horns, clearly not designed to inflict injury, are used in ritualized combat.

Figure 37-18 **Jane Goodall observing chimps at play**

Figure 37-19 **A feeding territory**
Acorn woodpeckers live in communal groups that excavate acorn-sized holes in dead trees, stuffing the holes with green acorns for dining during the lean winter months. The group defends the trees vigorously against other groups of acorn woodpeckers and against acorn-eating birds of other species, such as jays.

Animals May Defend Territories That Contain Resources

In many animal species, competition for resources takes the form of **territoriality**, the defense of an area where important resources are located. The defended resources may include places to mate, raise young, feed, or store food. Territorial animals generally restrict most or all of their activities to the defended area and advertise their presence there. Territories may be defended by males, females, a mated pair, or entire social groups (as in the case of the defense of their nest by social insects). However, territorial behavior is most commonly seen in adult males, and territories are normally defended against members of the same species, who compete most directly for the resources being protected. Territories are as diverse as the animals defending them. For example, a territory can be a tree where a woodpecker stores acorns (Fig. 37-19), small depressions in the sand used as nesting sites by cichlid fish, a hole in the sand used as a home by a crab, or an area of forest providing food for a squirrel.

Acquiring and defending a territory require considerable time and energy, yet territoriality is seen in animals as diverse as worms, arthropods, fish, birds, and mammals. The fact that organisms as unrelated as worms and humans independently evolved similar behavior suggests that territoriality provides some important advantages. Although the particular benefits depend on the species and the type of territory it defends, some broad generalizations are possible. First (as with dominance hierarchies), once a territory is established through aggressive interactions, relative peace prevails as boundaries are recognized and respected. The saying "good fences make good neighbors" also applies to nonhuman territories. One reason is that an animal is highly motivated to defend its territory and will often defeat even larger, stronger animals that attempt to invade it. Conversely, an animal outside its territory is much less secure and more easily defeated. This principle was demonstrated by Niko Tinbergen in an experiment using stickleback fish (Fig. 37-20).

For males of many species, successful territorial defense has a direct impact on reproductive success. In these species, females are attracted to a high-quality breeding territory, which might have features such as large size, abundant food, and secure nesting areas. Males who successfully defend the most desirable territories have the greatest chance of mating and passing on their genes. For example, experiments have shown that male sticklebacks that defend large territories are more successful in attracting mates than are males that defend small territories. Females that select males with the best territories increase their own reproductive success and pass their genetic traits (typically including their mate-selection preferences) to their offspring.

Territories are advertised through sight, sound, and smell. If the territory is small enough, the owner's mere presence, reinforced by aggressive displays at intruders, can be a sufficient defense. A mammal that owns a territory but cannot always be present may use pheromones to scent-mark its territorial boundaries. Male rabbits

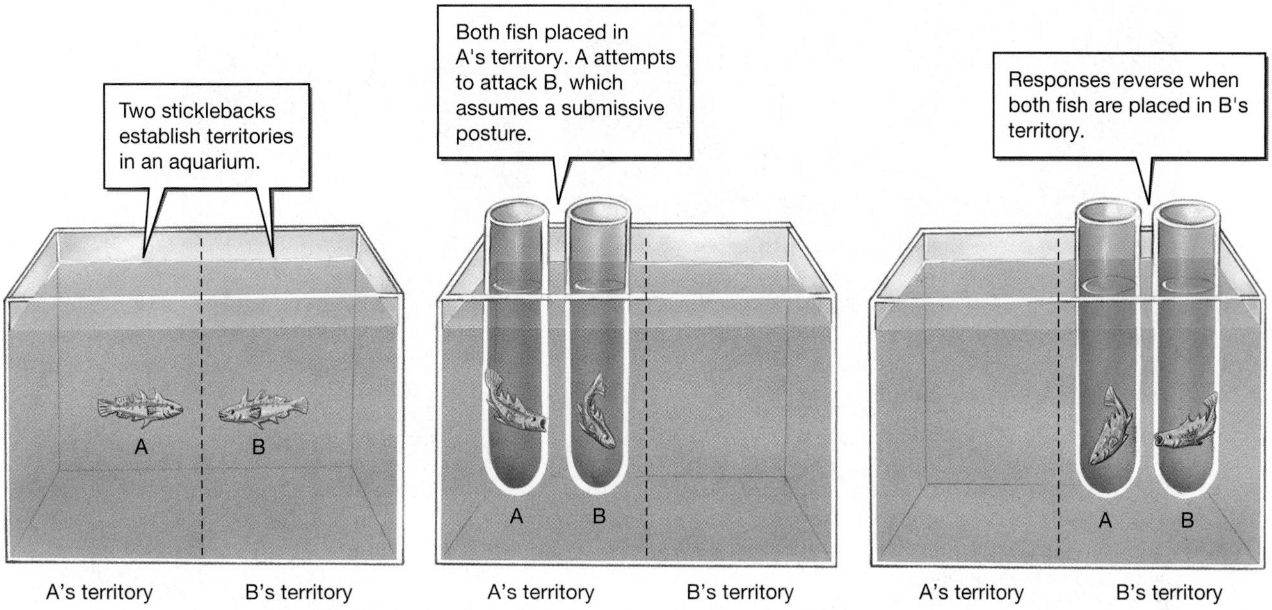

Figure 37-20 Territory ownership and aggression
Niko Tinbergen's experiment demonstrating the effect of territory ownership on aggressive motivation.

use pheromones secreted by glands in their chin and by anal glands to mark their territories. Hamsters rub the areas around their dens with secretions from special glands in their flanks.

Vocal displays are a common form of territorial advertisement. Male sea lions defend a strip of beach by swimming up and down in front of it, calling continuously. Male crickets produce a specific pattern of chirps to warn other males away from their burrows. Birdsong is a striking example of territorial defense. The husky trill of the male seaside sparrow is part of an aggressive display, warning other males to steer clear of his territory (Fig. 37-21). In fact, male sparrows that are unable to sing are unable to defend territories. The importance of singing to seaside sparrows' territorial defense was elegantly demonstrated by ornithologist M. Victoria McDonald, who captured territorial males and performed an operation that left them temporarily unable to sing but still able to utter the other, shorter and quieter signals in their vocal repertoires. The songless males were unable to defend territories or attract mates but regained their lost territories when their songs returned.

Figure 37-21 Defense of a territory by song
A male seaside sparrow announces ownership of his territory.

4 How Do Animals Find Mates?

In many sexually reproducing animal species, mating involves copulation or other close contact between male and female. Before mating can occur, however, animals must identify one another as members of the same species, as members of the opposite sex, and as being sexually receptive. In many species, finding an appropriate potential partner is only the first step, because the male must demonstrate his quality before the female will accept him as a mate. The need to fulfill all of these requirements has resulted in a diverse and fascinating array of courtship behaviors.

Vocal and Visual Signals Encode Sex, Species, and Individual Quality

Individuals that waste energy and gametes by mating with members of the wrong sex or wrong species are at a disadvantage in the contest to reproduce. Thus, natural

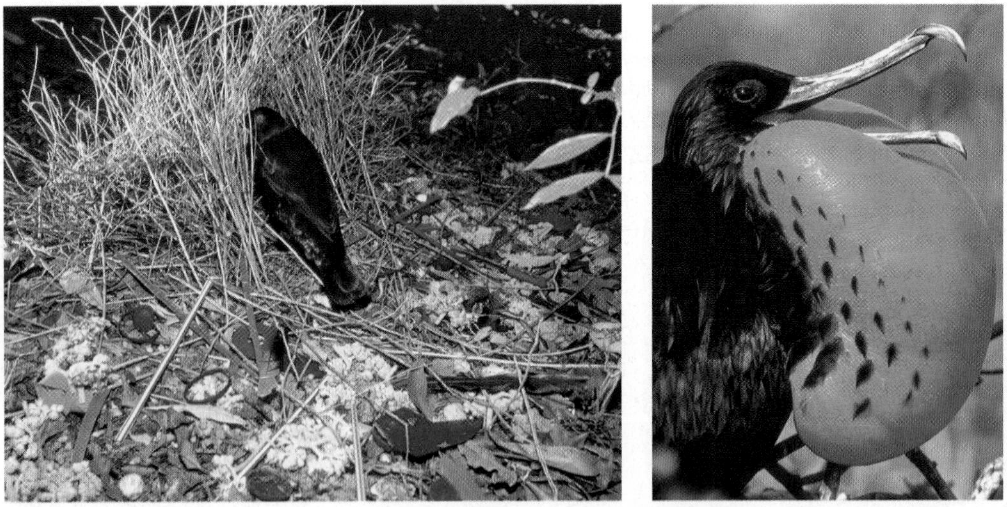

(a) (b)

Figure 37-22 Sexual displays
(a) During courtship, the male satin bowerbird builds a bower out of twigs and decorates it with colorful items that he gathers. ***(b)*** A male frigate bird inflates his scarlet throat pouch to attract passing females.

selection favors behaviors by which animals communicate their sex and species to potential mates. One result of this selection is that many campers have been kept awake by a raucous nighttime chorus of male tree frogs, each singing a species-specific song. Male grasshoppers and crickets also advertise their sex and species by their calls, as does the female mosquito with her high-pitched whine.

The signals that encode sex and species may also be used by potential mates as a basis for comparison among rival suitors. For example, the male bellbird uses its deafening song to defend large territories and attract females from great distances. The females fly from one territory to another, alighting near the male in his tree. The male, beak gaping, leans directly over the flinching female and utters an earsplitting note. The female apparently endures this noise to compare the volumes of the various males, choosing the loudest (who would also be the best defender of a territory) as a mate.

Many species use visual displays for courting. The firefly, for example, flashes a message that identifies its sex and species. Male fence lizards bob their heads in a species-specific rhythm, and females distinguish and prefer the rhythm of their own species. The elaborate construction projects of the male satin bowerbird and the scarlet throat of the male frigate bird serve as flashy advertisements of sex, species, and male quality (Fig. 37-22). Sending these extravagant signals must be risky, as they make it much easier for predators to locate the sender. For males, the added risk is an evolutionary necessity, as females won't mate with males that lack the appropriate signal. Females, in contrast, typically do not need to attract males or assume the risk associated with a conspicuous signal, so in many species females are drab in comparison to males (Fig. 37-23).

The intertwined functions of sex recognition and species recognition, advertisement of individual quality, and synchronization of reproductive behavior commonly require a complex series of signals, both active and passive, by both sexes. Such signals are beautifully illustrated by the complex underwater "ballet" executed by the male and female three-spined stickleback fish (Fig. 37-24).

Chemical Signals Bring Mates Together

Pheromones can play an important role in reproductive behavior. The sexually receptive female silk moth, for example, sits quietly and releases a chemical message so powerful that it can be detected by males up to 3 miles (5 kilometers) away. The exquisitely sensitive and selective receptors on the antennae of the male silk moth respond to just a few molecules of the substance, allowing him to travel upwind along a concentration gradient to find the female (Fig. 37-25a, p. 780).

Figure 37-23 Sexual dimorphism in guppies
As in many animal species, the male guppy (left) is normally much brighter in color than the female.

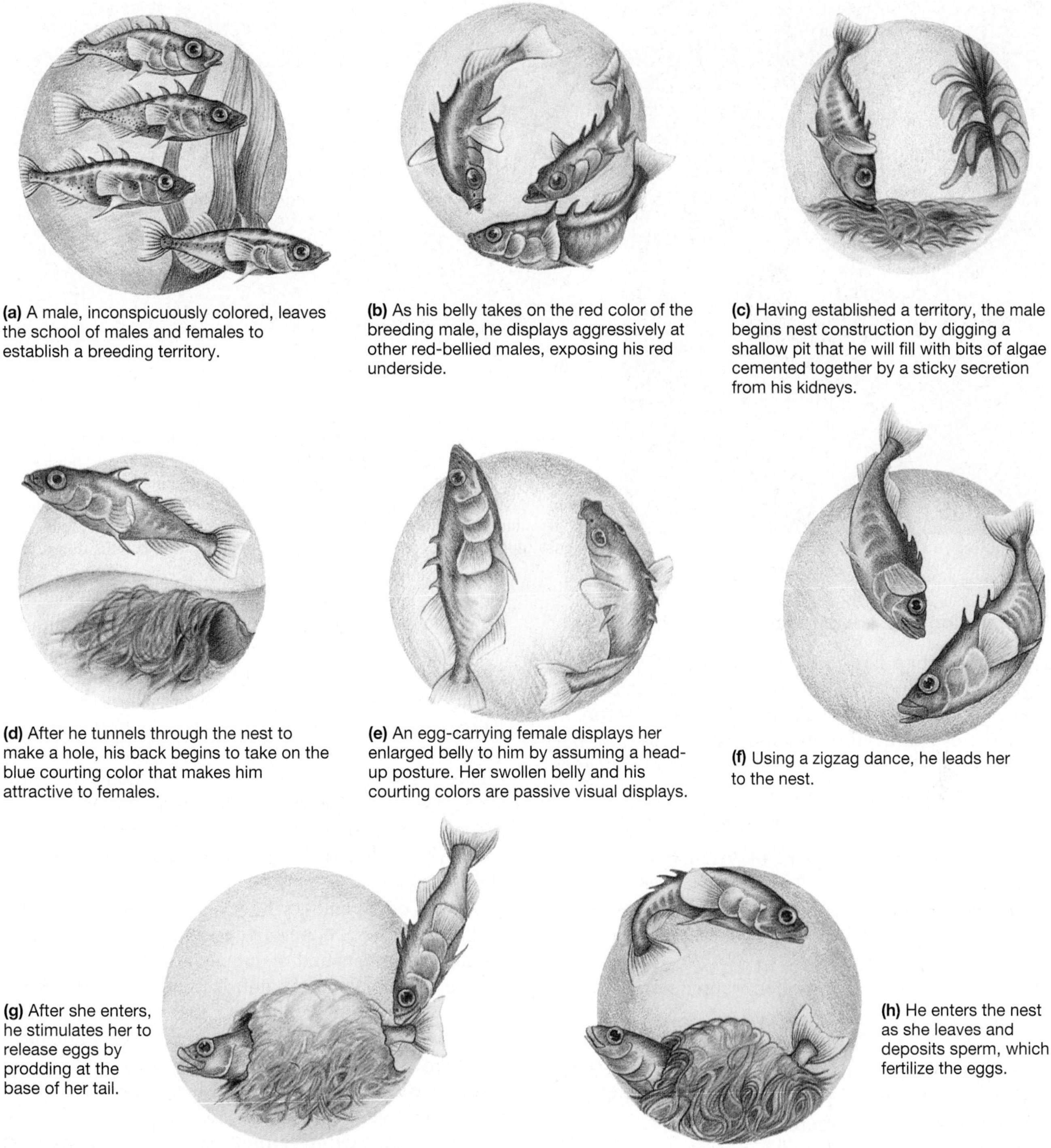

(a) A male, inconspicuously colored, leaves the school of males and females to establish a breeding territory.

(b) As his belly takes on the red color of the breeding male, he displays aggressively at other red-bellied males, exposing his red underside.

(c) Having established a territory, the male begins nest construction by digging a shallow pit that he will fill with bits of algae cemented together by a sticky secretion from his kidneys.

(d) After he tunnels through the nest to make a hole, his back begins to take on the blue courting color that makes him attractive to females.

(e) An egg-carrying female displays her enlarged belly to him by assuming a head-up posture. Her swollen belly and his courting colors are passive visual displays.

(f) Using a zigzag dance, he leads her to the nest.

(g) After she enters, he stimulates her to release eggs by prodding at the base of her tail.

(h) He enters the nest as she leaves and deposits sperm, which fertilize the eggs.

Figure 37-24 Courtship of the three-spined stickleback

Water is an excellent medium for dispersing chemical signals, and fish commonly use a combination of pheromones and elaborate courtship movements to ensure the synchronous release of gametes. Mammals, with their highly developed sense of smell, often rely on pheromones released by the female during her fertile periods to attract males (Fig. 37-25b). The irresistible attraction of a female dog in heat to nearby males is one example; the pheromone in male mouse urine is another.

5 What Kinds of Societies Do Animals Form?

Sociality is a widespread feature of animal life. Most animals have at least a small measure of interaction with other individuals of their species; many spend the bulk of their lives in the company of others; and a few species have developed complex, highly structured societies. Social interaction can be cooperative or competitive and is typically a mixture of the two.

(a)

(b)

Figure 37-25 Pheromone detectors
(a) Male moths find females not by sight but by following airborne pheromones released by females. These odors are sensed by receptors on the male's huge antennae, whose enormous surface area maximizes the chances of detecting the female scent. ***(b)*** When dogs meet, they typically sniff each other near the base of the tail. Scent glands there broadcast information about sex (both type and interest in) and status.

Group Living Has Advantages and Disadvantages

Living in a group has both costs and benefits, and the degree of sociality in a given species depends on the relative importance of the positive and negative factors in that species' particular situation. A species will not evolve social behavior unless the benefits of doing so outweighed the costs.

On the negative side, social animals may encounter:

- Increased competition within the group for limited resources
- Increased risk of infection from contagious diseases
- Increased risk that offspring will be killed by other members of the group
- Increased risk of being spotted by predators

Benefits to social animals include:

- Increased abilities to detect, repel, and confuse predators
- Increased hunting efficiency or increased ability to spot localized food resources
- Advantages resulting from the potential for division of labor within the group
- Increased likelihood of finding mates

The degree to which animals of the same species cooperate varies significantly from one species to the next. Some types of animals, such as the mountain lion, are basically solitary; interactions between adults consist of brief aggressive encounters and mating. Other types of animals cooperate on the basis of changing needs. For example, the coyote is solitary when food is abundant but hunts in packs in the winter when food becomes scarce.

Loose social groupings, such as pods of dolphins, schools of fish, flocks of birds, and herds of musk oxen (Fig. 37-26), provide a variety of benefits. For example, the characteristic spacing of fish in schools or the V-pattern of geese in flight provides a hydrodynamic or aerodynamic advantage for each individual in the group, reducing the energy required for swimming or flying. Some biologists hypothesize that herds of antelope or schools fish confuse predators—their myriad bodies make it difficult for the predator to focus on and pursue a single individual.

At the other end of the social spectrum are a few highly integrated cooperative societies, found primarily among the insects and mammals. As you read the following section, you may notice that some cooperative societies are based on behavior that seems to sacrifice the individual for the good of the group. There are many examples: Young, mature Florida scrub jays may remain at their parents' nest and help them raise subsequent broods instead of breeding; worker ants die in defense of their nest; Belding ground squirrels may sacrifice their own lives to warn the rest of their group of an approaching predator. These behaviors are examples of **altruism**—behavior that decreases the reproductive success of one individual to the benefit of another.

How could such behavior evolve? Why aren't the individuals that perform such self-sacrificing deeds eliminated from the population, taking with them the genes that contribute to this behavior? One possibility is that because closely related individuals are likely to share some genes, an individual can promote the survival of its own genes through behaviors that maximize the survival of its close relatives. This concept is called **kin selection**.

Figure 37-26 Cooperation in loosely organized social groups
A herd of musk oxen functions as a unit when threatened by predators such as wolves. Males form a circle, horns pointed outward, around the females and young.

Kin selection helps explain a variety of altruistic behaviors. Please refer to "Evolutionary Connections: Kin Selection and the Evolution of Altruism" (Chapter 15) for an exploration of the evolutionary basis of the self-sacrificing behaviors that contribute to the success of cooperative societies. In the following section, we present examples of complex societies in insect, fish, and mammal species that vividly illustrate cooperative behavior.

Honeybees Form Complex Insect Societies

Perhaps the most perplexing of all animal societies are those of the bees, ants, and termites. Scientists have long struggled to explain the evolution of a social structure in which most individuals never breed but instead labor slavishly to feed and protect the offspring of a different individual. Whatever its evolutionary explanation, the intricate organization of a social insect colony makes a compelling story. In these communities, the individual is a mere cog in an intricate, smoothly running machine and could not survive by itself.

Individual social insects are born into one of several castes within the society. These castes are groups of similar individuals that perform a specific function. Honeybees emerge from their larval stage into one of three major preordained roles. One role is that of *queen.* Only one queen is tolerated in a hive at any time. Her functions are to produce eggs (up to 1000 per day for a lifetime of 5 to 10 years) and to regulate the lives of the workers. Male bees, called *drones,* serve merely as mates for the queen. Lured by her sex pheromones, drones mate with the queen, perhaps as many as 15 times, during her first week of life. This relatively brief "orgy" supplies her with sperm that will last a lifetime, enough to fertilize more than 3 million eggs. Their sexual chore accomplished, the drones become superfluous and are eventually driven out of the hive or killed.

The hive is run by the third class of bees, sterile female *workers.* A worker's tasks are determined by her age and conditions in the colony. A newly emerged worker starts life as a waitress, carrying food such as honey and pollen to the queen, to other workers, and to developing larvae. As she matures, special glands begin wax production, and she becomes a builder, constructing perfectly hexagonal cells of wax where the queen will deposit her eggs and the larvae will develop. She will also take a shift as maid, cleaning the hive and removing the dead, and as a guard, protecting the hive against intruders. Her final role in life is that of a forager, gathering pollen and nectar, food for the hive. She will spend nearly half of her 2-month life in this role. Acting as a forager scout, she will seek new and rich sources of nectar and, having found one, will return to the hive and communicate its location to other foragers. She communicates by means of the **waggle dance**, an elegant form of symbolic expression (Fig. 37-27). Much of the meaning of the waggle dance was deciphered by Karl von Frisch during 35 years of research beginning in 1915.

Pheromones play a major role in regulating the lives of social insects. Honeybee drones are drawn irresistibly to the queen's sex pheromone (*queen substance*), which she releases during her mating flights. Back at the hive, she uses the same substance to maintain her position as the only fertile female. The queen substance is licked off her body and passed among all the workers, rendering them sterile. The queen's presence and health are signaled by her continuing production of queen substance; a decrease in production (which occurs normally in the spring) alters the behavior of the workers. Almost immediately they begin building extra-large "royal cells." The workers also feed the larvae that develop in these cells a special glandular secretion known as "royal jelly." This unique food alters the development of the growing larvae so that, instead of a worker, a new queen emerges from the royal cell. The old queen will then leave the hive, taking a swarm of workers with her to establish

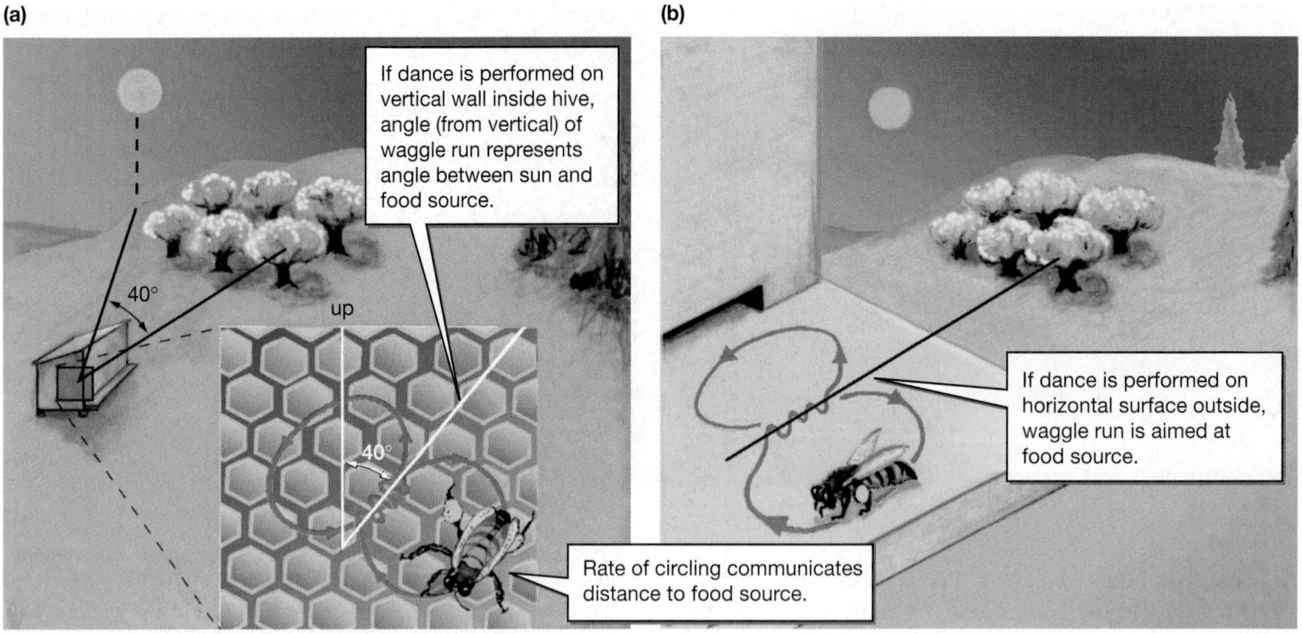
(a)
(b)
If dance is performed on vertical wall inside hive, angle (from vertical) of waggle run represents angle between sun and food source.
40°
up
40°
If dance is performed on horizontal surface outside, waggle run is aimed at food source.
Rate of circling communicates distance to food source.

Figure 37-27 Bee language: the waggle dance
A forager, returning from a rich source of nectar, performs a waggle dance that communicates the distance and direction of the food source, as other foragers crowd around her, touching her with their antennae. The bee moves in a straight line while shaking her abdomen back and forth ("waggling") and buzzing her wings. She repeats this over and over in the same location, circling back in alternating directions.

residence elsewhere. If more than one new queen emerges, a battle to the death ensues, with the victorious queen taking over the hive.

Bullhead Catfish Form a Simple Vertebrate Society

The nervous systems of vertebrates are far more complex than those of insects, and we might therefore expect vertebrate societies to be proportionately more complex. With the exception of human society, however, they are not. Perhaps because the vertebrate brain *is* more complex, vertebrate societies tend to be simpler than are those of the social insects such as honeybees, army ants, and termites. The robotic precision that makes complex insect societies possible is rarely found among vertebrates, which generally exhibit greater behavioral flexibility than do insects and therefore greater behavioral variation among individuals.

The social interactions of bullhead catfish, described by John Todd of the Woods Hole Oceanographic Institution in Massachusetts, provide a fascinating illustration of a vertebrate in which complex social interactions are based almost entirely on pheromones. Todd observed these nocturnal fish in large aquariums in the laboratory. He discovered that when a group was housed together, territories were staked out, and a dominance hierarchy was established, with the dominant fish defending the largest and best-protected area of the tank. Contests between tankmates consisted of open-mouthed aggressive displays. Once a fish became dominant, its aggressive displays caused the subordinate fish to flee. Actual violence occurred only when a stranger was introduced into a tank with an established hierarchy. In this case, the established group exhibited cooperative behavior. The dominant fish allowed others to take refuge in his protected territory, then fought the intruder (Fig. 37-28). When the newcomer was defeated, the dominant fish chased the others back out of his territory.

Todd discovered that blinding the bullheads did not cause any appreciable change in their social interactions. When their sense of smell was temporarily blocked, however, the fish acted like permanent strangers, interacting aggressively for weeks until their sense of smell returned. Both the status of an individual and a change in status were communicated by scent. If a dominant fish was removed from his tank and later returned, in most cases both his territory and his status were remembered and respected by his tankmates. But if he was removed and subjected to defeat in the tank of a more aggressive fish, his pheromones were somehow altered. Upon return to his home tank, he was attacked by his former subordinates.

When many newly caught fish are placed in the same tank, bullheads may form a dense and peaceful community that lacks territories or dominance hierarchies. Todd established such a community in one tank and placed a pair of aggressive rival fish in an adjacent tank. When water was pumped continuously from the "community tank" into the adjacent tank, the aggressive bullheads became peaceful, resuming their fighting only when the flow was stopped. Under the crowded conditions of the community tank, an "anti-aggression pheromone" is apparently produced, minimizing conflict.

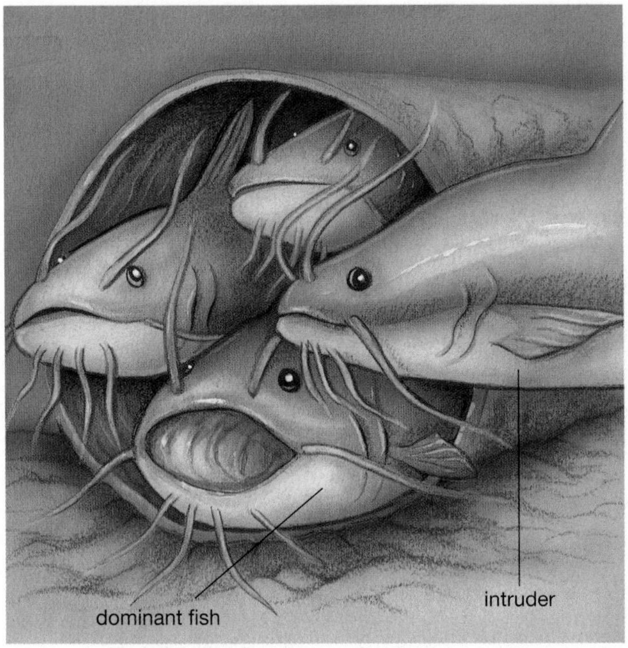

Figure 37-28 Cooperation among bullhead catfish
Three bullhead catfish occupy a section of pipe in a laboratory tank. The dominant fish is normally the exclusive occupant of the pipe, which is part of his territory. When an intruder is introduced into the tank, however, the dominant fish allows two subordinates to seek refuge in the pipe and reacts aggressively to the intruder.

Figure 37-29 The naked mole rat queen
Encountering a lazy worker in one of the colony's underground tunnels, the queen shoves it and knocks it over to stimulate the worker to greater efforts.

Naked Mole Rats Form a Complex Vertebrate Society

Perhaps the most bizarre society among nonhuman mammals is that of the naked mole rat. These nearly blind, nearly hairless relatives of guinea pigs live in large underground colonies in southern Africa and exhibit a form of social organization not unlike that of an ant or termite colony. The colony is dominated by the queen, a single reproducing female to whom all other members are subordinate. The queen is the largest individual in the colony and maintains her status by aggressive behavior, particularly shoving. She prods and shoves lazy workers, stimulating them to become more active (Fig. 37-29). As in honeybee hives, there is a division of labor among the workers, in this case based on size. Small, young rats clean the tunnels, gather food, and tunnel. Tunnelers line up head to tail and pass excavated dirt along the completed tunnel to an opening. Just below the opening, a larger mole rat flings the dirt into the air, adding it to a cone-shaped mound. Biologists observing this behavior from the surface dubbed it "volcanoing." In addition to volcanoing, large mole rats defend the colony against predators and members of other colonies.

If another female begins to become fertile, the queen apparently senses changes in the estrogen levels of the subordinate female's urine. The queen then selectively shoves the would-be breeder, causing stress that prevents her rival from ovulating. Large males are more likely to mate with the queen than are small ones, although all adult males are fertile. When the queen dies, a few of the females gain weight and begin shoving one another. The aggression may escalate until a rival is killed. Ultimately, a single female becomes dominant. Her body lengthens; she assumes the queenship and begins to breed. Litters averaging 14 pups are produced about four times a year. During the first month, the queen nurses her pups, and the workers feed the queen. Then the workers begin feeding the pups solid food.

It is far from clear why naked mole rats, alone among the mammals, have developed a social order in which almost all individuals sacrifice their reproductive potential. One likely contributing factor is that the queen produces the pups that grow up to form her colony, so all colony members are quite closely related. (Helping close relatives to reproduce can be adaptive; see "Evolutionary Connections: Kin Selection and the Evolution of Altruism" in Chapter 15.) Another factor that may have favored the evolution of this communal social structure is the high cost of leaving the colony. It's unlikely that a single departing naked mole rat would be able to construct the elaborate underground burrow that the species inhabits.

6 Can Biology Explain Human Behavior?

Because humans are animals whose behaviors, like those of all other animals, have an evolutionary history, the techniques and concepts of ethology can help us understand and explain human behavior. Human ethology, however, is, and will remain, a less rigorous science than animal ethology. We cannot treat people as laboratory animals, devising experiments that control and manipulate the features that influence our attitudes and actions. In addition, some observers argue that human culture has been freed from the constraints of our evolutionary past for so long that we cannot explain our current behavior in terms of biological evolution. Nevertheless, many scientists have taken an ethological, evolutionary approach to human behavior, and their work has had a major impact on our view of ourselves.

The Behavior of Newborn Infants Has a Large Innate Component

Much of the behavior of very young infants presumably has a large innate component, because infants have had little time for learning to occur. The rhythmic movement of a baby's head in search of the mother's breast is an innate behavior that is expressed in the first days after birth. Sucking, which can be observed even in the human fetus, is also innate (Fig. 37-30). Other behaviors seen in newborns and even premature infants include grasping with the hands and feet and walking movements when the body is supported. Another example is smiling, which can occur soon after birth. Initially, smiling can be induced by almost any object looming over the newborn. This initial indiscriminate response, however, is soon modified by experience. Infants up to 2 months old will smile in response to a stimulus consisting of two dark, eye-sized spots on a light background, which at that stage of development is a more potent stimulus for smiling than is an accurate representation of a human face. But as the child's development continues, learning and further development of the nervous system interact to limit the response to more-correct representations of a face.

One of the most important insights to have emerged from studies of animal learning is that animals of a given species tend to have an inborn predilection for specific types of learning that are important to that species' mode of life. In humans, one such inborn predilection seems to be for the acquisition of language. Young children are able to acquire language rapidly and nearly effortlessly; they typically acquire a vocabulary of 28,000 words before the age of 8. A large body of research suggests that we are born with a brain that is already primed for this early facility with language. For example, the human fetus begins responding to sounds during the third trimester of pregnancy, and researchers have demonstrated that infants are able to distinguish a variety of consonant sounds by 6 weeks after birth. The latter finding stems from an ingenious experiment that took advantage of an infant's ability to make sucking movements in order to gauge the infant's ability to discriminate sounds. In the experiment, a baby responded to the presentation of various consonant sounds by sucking on a pacifier that contained a force transducer to record the sucking rate. When one sound (such as "ba") was presented repeatedly, the infant became habituated and decreased her sucking rate. But when a new sound (such as "pa") was presented, sucking rate increased, revealing that the infant perceived the new sound as different.

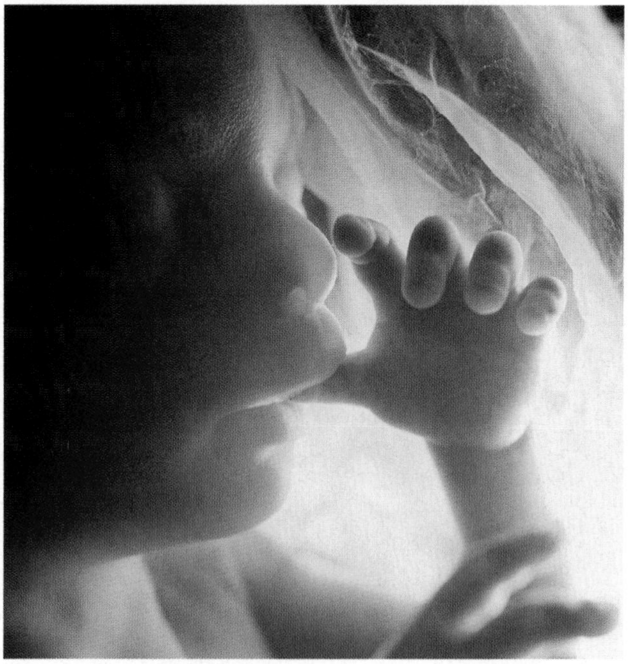

Figure 37-30 A human instinct
Thumb sucking is a difficult habit to discourage in young children, because sucking on appropriately sized objects is an instinctive, food-seeking behavior. This fetus sucks its thumb at about 4½ months of development.

The sucking-measurement technique has also been used to demonstrate that newborns in their first 3 days of life can be conditioned to produce certain rhythms of sucking by using their mother's voice as reinforcement. In an experiment performed by William Fifer of the New York State Psychiatric Institute, infants clearly preferred their own mothers' voices to other female voices, as indicated by their sucking rhythm (Fig. 37-31). The infant's ability to learn his or her mother's voice and respond positively to it within days of birth has strong parallels to imprinting and may help initiate bonding with the mother.

Behaviors Shared by Diverse Cultures May Be Innate

Another way to study the innate bases of human behavior is to compare simple acts performed by people from diverse cultures. This comparative approach, pioneered by ethologist Irenaus Eibl-Eibesfeldt, has revealed several gestures that seem to form a universal, and therefore probably innate, human signaling system. Such gestures include a variety of facial expressions for pleasure, rage, and disdain, as well as greeting movements such as an upraised hand or the "eye flash" (in which the eyes are widely opened and the eyebrows rapidly elevated). The evolution of these "hard-wired"

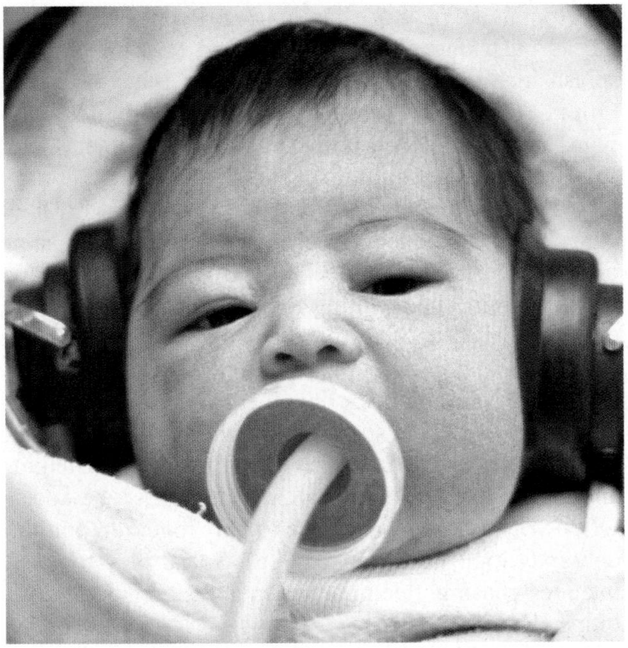

Figure 37-31 Newborns prefer their mother's voices
Using a nipple connected to a computer that plays audio tapes, researcher William Fifer demonstrated that newborns can be conditioned to suck at specific rates for the privilege of listening to their own mothers' voices through headphones. For example, if the infant sucks faster than normal, her mother's voice is played; if she sucks more slowly, another woman's voice is played. Researchers found that infants easily learned and were willing to work hard at this task just to listen to their own mothers' voices, presumably because they had become used to her voice in the womb. This response may be a human version of imprinting.

gestural messages presumably depended on the advantages that accrued to both senders and receivers by sharing information about the emotional state and intentions of the sender. A species-wide method of communication was perhaps especially important before the advent of language, and later remained useful during encounters between individuals who shared no common language.

Complex social behaviors can also be widespread among diverse cultures. For example, the incest taboo (avoidance of mating with close relatives) seems to be universal across human cultures (and even across many species of nonhuman primates). Despite the universality of the taboo, it seems unlikely that something as subtle and complex as a shared belief could be encoded in our genes. Some biologists have suggested that the taboo is instead a cultural expression of an evolved, adaptive behavior. According to this hypothesis, close contact among family members early in life suppresses sexual desire, and this effect arose because of the deleterious consequences of inbreeding (for example, higher incidence of genetic diseases). The hypothesis does not require us to assume an innate social belief, but rather proposes that we inherit a learning program that causes us to undergo a kind of imprinting early in life.

People May Respond to Pheromones

Although the main channels of human communication are visual and auditory, humans also seem to respond to chemical messages. The possible existence of human pheromones was hinted at in the early 1970s, when biologist Martha McClintock found that the menstrual cycles of roommates and close friends tended to become synchronized. McClintock suggested that the synchrony resulted from some chemical signal between the women, but almost 30 years passed before she and her colleagues uncovered more conclusive evidence that a pheromone was at work.

In 1998 McClintock's research group asked nine female volunteers to wear cotton pads in their armpits for eight hours per day over the course of a menstrual cycle. The pads were then disinfected with alcohol, and swabbed above the upper lips of another set of 20 female subjects (who reported that they could detect no odors other than alcohol on the pads). The subjects were exposed to the pads each day for two months, with half the group sniffing secretions from women in the early (pre-ovulation) part of the menstrual cycle, while the other half was exposed to secretions from later in the cycle (post-ovulation). Women exposed to early-cycle secretions had shorter-than-usual menstrual cycles, and women exposed to pads from women who had already ovulated experienced delayed menstruation. It appears that women release different pheromones, with different effects on receivers, at different points in the menstrual cycle.

Although the McClintock experiment offers strong evidence for the existence of human pheromones, little else is known about chemical communication in humans. The actual molecules that caused the effects documented by McClintock remain unknown, as does their function (what benefit would a woman gain by influencing the menstrual cycles of other women?). Receptors for chemical messages have not yet been found in humans, and we don't know if the "menstrual pheromones" are the first known example of an important communication system or merely an isolated case of a vestigial ability. Despite the hopeful advertisements for "sex attraction pheromones" on late-night television, chemical communication in humans is a scientific mystery awaiting a solution.

Comparisons of Identical and Fraternal Twins Reveal Genetic Components of Behavior

Twins present perhaps the best opportunity to examine the hypothesis that differences in human behavior are related to genetic differences. When twins are reared together, environmental influences on behavior are very similar for each member of the pair, so behavioral differences between twins must have a large genetic component. If a particular behavior is heavily influenced by genetic factors, we would expect to find similar expression of that behavior in *identical twins* (which arise from a single fertilized egg and have identical genes) but not in *fraternal twins* (which arise from two individual eggs and are no more similar genetically than are other siblings). Data from twin studies, and from other investigations of the genealogy of behaviors, have tended to confirm the heritability of many human behavioral traits. These studies have documented a significant genetic component for traits such as activity level, alcoholism, sociability, anxiety, intelligence, dominance, and even political attitudes. On the basis of tests designed to measure many aspects of personality, identical twins are about twice as similar in personality as are fraternal twins.

The most fascinating, if perhaps less rigorous, twin findings are based on anecdotal observations of identical twins separated soon after birth, reared in different environments, and reunited for the first time as adults. Identical twins reared apart have been found to be as similar in personality as those reared together, indicating that the differences in their environments had little influence on their personality development. They have been found to share nearly identical taste in jewelry, clothing, humor, food, and names for children and pets. Personal idiosyncrasies such as giggling, nail biting, drinking patterns, hypochondria, and mild phobias are in some cases shared by these unacquainted twins.

The field of human behavioral genetics is controversial, because it challenges the long-held belief that environment is the most important determinant of human behavior. As discussed earlier in this chapter, we now recognize that all behavior has some genetic basis and that complex behavior in nonhuman animals typically combines elements of both innate and learned behaviors. Thus, it seems likely that our own behavior is influenced by both our evolutionary history and our cultural heritage. The debate over the relative importance of heredity and environment in determining human behavior continues and is unlikely ever to be fully resolved. Human ethology is not yet recognized as a rigorous science, and it will always be hampered because we can neither view ourselves with detached objectivity nor treat one another as though we were laboratory rats. Despite these limitations, there is much to be learned about the interaction of learning and innate tendencies in people.

Evolutionary Connections

Why Do Animals Play?

Pigface, a giant 50-year-old African softshell turtle, spends hours each day batting a ball around his aquatic home in the National Zoo in Washington, D.C., to the delight of thousands of visitors and the puzzlement of behavioral biologists. Play has always been somewhat of a mystery. It has been observed in many birds and in most mammals, but, until zookeepers tossed Pigface a ball a few years ago, it had never been seen in animals as evolutionarily ancient as turtles.

Animals at play are fascinating. Pygmy hippopotamuses push one another, shake and toss their heads, splash in the water, and pirouette on their hind legs. Otters delight in elaborate acrobatics. Bottlenose dolphins balance fish on their snouts, throw objects, and carry them in their mouths while swimming. Baby vampire bats chase, wrestle, and slap each other with their wings. Even octopuses have been seen playing a game of catch: pushing an object way from themselves and into a current, then waiting for the object to drift back, only to push it back into the current to start the cycle over again.

Solitary play typically involves a single animal manipulating an object, such as a cat with a ball of yarn, or the dolphin with its fish, or a macaque monkey making and playing with a snowball. Play can also be social. Often young of the same species play together, but parents may join them (Fig. 37-32a). Social play typically includes chasing, fleeing, wrestling, kicking, and gentle biting (Fig. 37-32b,c).

What are the features of play? (1) Play seems to lack any clear immediate function. (2) Play is abandoned in favor of feeding, courtship, and escaping from danger. (3) Young animals play more frequently than do adults. (4) Play typically involves movements borrowed from other behaviors (attacking, fleeing, stalking, and so on). (5) Play uses considerable energy. (6) Play is potentially dangerous. Many young humans and other animals are injured, and some are killed, during play. In addition, play can distract the animal from the presence of danger while making it conspicuous to predators. So, why do animals play?

The only logical conclusion is that play must have survival value and that natural selection has favored those individuals who engage in playful activities. One of the best explanations for the survival value of play is the "practice theory," first proposed by K. Groos in 1898. He suggested that play allows young animals to gain experience in a variety of behaviors that they will use as adults. By performing these acts repeatedly in a lighthearted context, the animal practices skills that will later be important in hunting, fleeing, or social interactions.

More-recent research supports and extends Groos's proposal. Play occurs most intensely early in life when the brain is developing and crucial neural connections are being formed. John Byers, a zoologist at the University of Idaho, has observed that animals with larger brains tend to be more playful than are animals with smaller brains. Because larger brains are generally linked to increased learning ability, this relationship supports the idea that adult skills are learned during juvenile play. Watch children roughhousing or playing tag, and you will see how strength and coordination are fostered by play and how skills are developed that might have helped our hunting ancestors survive. Quiet play with other children, with dolls, blocks, and other toys

(a)

(b)

(c)

Figure 37-32 Young animals at play

helps children prepare to interact socially, nurture their own children, and deal with the physical world.

Although Shakespeare tells us "play needs no excuse," there is good evidence that the tendency to play has evolved as an adaptive behavior in animals capable of learning. Play is quite literally "serious fun"!

CASE STUDY REVISITED

Sex and Symmetry

To assess whether male body symmetry is correlated with male "mating success" in humans, Randy Thornhill, Steven Gangestad, and their colleagues began by measuring symmetry in some young adult males. Each man's degree of symmetry was assessed by measurements of his ear length and the width of his foot, ankle, hand, wrist, elbow, and ear. From these measurements, the researchers derived an index that summarized the degree to which the size of these features differed between the right and left sides of the body.

In one study, the researchers found that, among the males in their sample, the most symmetrical men were judged (by a panel of observers who were unaware of the nature of the study) to be more attractive than other men. A survey of the study subjects revealed that the more symmetrical men also tended to begin having sex earlier in life and to have had a larger number of different sexual partners. Apparently, a man's sexual activity and attractiveness to women are correlated with his symmetry.

Why would male body symmetry affect mating success? The most likely explanation is that symmetry is an indicator of good physical condition. Developmental biologists have long known that disruptions of normal embryological development can cause asymmetrical bodies, so a highly symmetrical body can be an indicator of healthy, normal development. It also may indicate a high-quality genotype that was able to overcome any disturbances (such as diseases or exposure to toxic substances) that may have occurred during development. Females that mate with individuals whose health and vitality are announced by their symmetrical bodies might have offspring that were likewise healthy and vital. Natural selection would thus favor females who chose to mate with symmetrical males, and that kind of mating behavior would come to predominate.

Summary of Key Concepts

1) How Do Innate and Learned Behaviors Differ?

Although all animal behavior is influenced by both genetic and environmental factors, it can be useful to distinguish between behaviors whose development is not highly dependent on external factors and behaviors that require more extensive environmental stimuli in order to develop. Behaviors in the first category are sometimes designated as innate and can be performed properly the first time an animal encounters the appropriate stimulus. Behavior that changes in response to input from an animal's social and physical environment is said to be learned. Learning is especially adaptive in environments that are changing and unpredictable, and learning can modify innate behavior to make it more appropriate. Learning assumes a wide variety of forms, depending on the evolutionary history and the ecological needs of animal species involved.

Among this diverse array of learning methods are imprinting, habituation, conditioning, trial and error, and insight. Imprinting is a special kind of learning that occurs during a limited sensitive period early in life. This form of simple learning typically involves attachment between parent and offspring or learning the features of a future mate. Habituation is the decline in response to a harmless stimulus that is repeated frequently. It commonly modifies innate escape responses or defensive responses. Trial-and-error learning can modify innate behavior or can produce new behavior as a result of rewards and punishments provided by the environment. During operant conditioning, an animal learns to make a new response, such as pressing a button, to obtain a reward or to avoid punishment. Insight, the most complex form of learning, can be considered a form of mental trial-and-error learning. An animal showing insight makes a new and adaptive response to an unfamiliar situation.

Although the distinction between innate and learned behavior is conceptually useful, the distinction is not sharp in naturally occurring behaviors. In virtually all behaviors, learning and instinct interact to produce adaptive behavior. Certain types of learning, such as imprinting, occur instinctively, during a rigidly defined time span. Instinctive responses are typically modified by experience. Learning allows animals to modify these innate responses so that they occur only with appropriate stimuli.

2) How Do Animals Communicate?

Communication, an action by one animal that alters the behavior of another, is the basis of all social behavior. It allows animals of the same species to interact effectively in their quest for mates, food, shelter, and other resources. Animals communicate through visual signals, sound, chemicals (pheromones), and touch.

Visual communication is quiet and can convey subtle, rapidly changing information. Visual signals are active (body movements) or passive (body shape and color). Sound communication can also carry a wide range of rapidly changing information and is effective where vision is impossible. Although sound can attract predators, the animal may remain hidden while communicating.

Some chemical signals cause an immediate, observable behavior in the animal that detects them. Other pheromones alter the physiological state of the recipient; still other pheromones influence the recipient's behavior. Pheromones can be detected after the sender has departed, conveying a message over time. Physical contact reinforces social bonds and helps synchronize mating in a variety of animals, from mammals to snails.

3) How Do Animals Compete for Resources?

Although many competitive interactions are resolved through aggression, serious injuries are rare. Most aggressive encounters are settled by means of displays that communicate the motivation, size, and strength of the combatants.

Some species establish dominance hierarchies that minimize aggression and regulate access to resources. On the basis of initial aggressive encounters, each animal acquires a status in which it defers to more dominant individuals and dominates subordinates. When resources are limited, dominant animals obtain the largest share and are most likely to reproduce.

Territoriality, a behavior in which animals defend areas where important resources are located, also allocates resources and minimizes aggressive encounters. In general, territorial boundaries are respected, and the best-adapted individuals defend the richest territories and produce the most offspring.

4) How Do Animals Find Mates?

Successful reproduction requires that animals recognize the species, sex, and sexual receptivity of potential mates. In many species, animals often must also assess the quality of potential mates. These requirements have resulted in the evolution of sexual displays that use all possible forms of communication.

5) What Kinds of Societies Do Animals Form?

Social living has both advantages and disadvantages, and species show a wide variation in the degree to which their members cooperate. Some species form cooperative societies. The most rigid and highly organized are those of the social insects such as the honeybee, in which the members follow rigidly defined roles throughout life. These roles are maintained through both genetic programming and the influence of certain pheromones. Nonhuman vertebrates also form complex, but normally less rigid, societies, such as those of bullhead catfish. Naked mole rats exhibit the most complex and rigid vertebrate social interactions, resembling insect societies.

6) Can Biology Explain Human Behavior?

The degree to which human behavior is genetically influenced is highly controversial. Because we cannot freely experiment on humans, and because learning plays a major role in nearly all human behavior, investigators must rely on studies of newborn infants and comparative cultural studies, correlations between certain behaviors and physiology (which suggest a role for pheromones), and studies of identical and fraternal twins. Evidence is mounting that our genetic heritage plays a role in personality, intelligence, simple universal gestures, our responses to certain stimuli, and our tendency to learn specific things such as language at particular stages of development.

Key Terms

aggression *p. 774*
altruism *p. 780*
behavior *p. 766*
communication *p. 771*
dominance hierarchy *p. 774*
ethology *p. 769*
habituation *p. 767*
imprinting *p. 770*
innate *p. 766*
insight learning *p. 768*
kin selection *p. 780*
learning *p. 766*
operant conditioning *p. 767*
pheromone *p. 773*
queen substance *p. 773*
sensitive period *p. 770*
territoriality *p. 776*
trial-and-error learning *p. 767*
waggle dance *p. 781*

Thinking Through the Concepts

Multiple Choice

1. *In general, animal behaviors*
   a. are determined mainly by genetic factors
   b. are determined mainly by environmental factors
   c. arise from an interaction between genes and the environment
   d. do not affect an individual's chances of survival

2. *Which of the following is NOT a feature of play?*
   a. Young animals play more often than adults.
   b. It may borrow movements from other behaviors.
   c. It uses considerable energy.
   d. It always has a clear, immediate function.
   e. It is potentially dangerous.

3. *Naked mole rats are unique among mammals, in that*
   a. they live in colonies
   b. most of their life is spent underground
   c. even adults have little body hair
   d. most individuals never reproduce
   e. they use pheromones to communicate

4. *In which of the following ways does an individual benefit from a social grouping with other animals?*
   a. increased ability to detect, repel, or confuse predators
   b. increased hunting efficiency
   c. increased likelihood of finding mates
   d. increased efficiency due to division of labor
   e. all of the above

5. *Which of the following pairs of communication forms and advantages is NOT accurate?*
   a. pheromones, long-lasting
   b. visual displays, instantaneous
   c. sound communication, effective at night
   d. pheromones, convey rapidly changing information
   e. touch, maintains social bonds

6. *In an insect society, such as the honeybee society,*
   a. the division of labor is based on biologically determined castes
   b. all adult members share labor equally
   c. all adult members have the opportunity to reproduce
   d. reproduction is altered seasonally among the adults
   e. the organization of the society is flexible and adaptable

? Review Questions

1. Explain why neither "innate" nor "learned" adequately describes the behavior of any given organism.
2. Explain why animals play. Include the features of play in your answer.
3. List four senses through which animals communicate, and give one example of each form of communication. After each, present both advantages and disadvantages of that form of communication.
4. A bird will ignore a squirrel in its territory but will act aggressively toward a member of its own species. Explain why.
5. Why are most aggressive encounters among members of the same species relatively harmless?
6. Discuss advantages and disadvantages of group living.
7. In what ways do naked mole rat societies resemble those of the honeybee?

Applying the Concepts

1. Male mosquitoes orient toward the high-pitched whine of the female, and female mosquitoes (only females suck blood) are attracted to the warmth, humidity, and carbon dioxide exuded by their prey. Using this information, design a mosquito trap or killer that exploits a mosquito's innate behaviors. Now design one for moths.
2. You raise honeybees but are new at the job. Trying to increase honey production, you introduce several queens into the hive. What is the likely outcome? What different things could you do to increase production?
3. Describe and give an example of a dominance hierarchy. What role does it play in social behavior? Give a human parallel, and describe its role in human society. Are the two roles similar? Why or why not? Repeat this exercise for territorial behavior in humans and in another animal.
4. You are manager of an airport. Planes are being endangered by large numbers of flying birds, which can be sucked into engines, disabling the engines. Without harming the birds, what might you do to discourage them from nesting and flying near the airport and its planes?

For More Information

Brown, S. L. "Animals at Play." *National Geographic*, December 1994. Clearly written description of why animals play, accompanied by beautiful photographs that capture wild animals at play.

Dawkins, R. *The Selfish Gene*, 2nd ed. New York: Oxford University Press, 1989. Exceptionally clear explanation of the evolutionary basis of altruistic behavior, written for both laypeople and scientists.

de Waal, F. "The End of Nature versus Nurture." *Scientific American*, December 1999. An eminent ethologist makes the case for taking a nuanced biological approach to understanding human behavior.

Dugatkin, L., and Godin, J. "How Females Choose Their Mates." *Scientific American*, April 1998. An entertaining look at how biologists study the processes by which females choose among available males.

Gould, J. L., and Gould, C. G. "The Instinct to Learn." *Science 81*, May 1981. Birds, bees, and perhaps even humans are genetically programmed to learn specific things at particular stages in life.

Kirchner, W. H., and Towne, W. F. "The Sensory Basis of the Honeybee Dance Language." *Scientific American*, June 1994. Combines a historical look at the work of Karl von Frisch with a modern update on recent research that has almost fully elucidated the mechanisms of honeybee communication during the waggle dance.

Lorenz, K. *King Solomon's Ring: New Light on Animal Ways.* New York: Thomas Y. Crowell, 1952. Beautifully written and filled with interesting anecdotes; provides important insights into early ethology.

Macdonald, D., and Brown, R. "The Smell of Success." *New Scientist*, May 1985. Describes the amazing diversity of mammalian pheromones.

Phillips, K. "School Riddles." *International Wildlife*, March–April 1995. Explores research on fish schooling behavior and how it helps avoid predation.

Pinker, S. *The Language Instinct.* New York: William Morrow, 1994. An entertaining account of linguists' current understanding of how we develop the ability to use language and how that ability may have evolved.

Seeley, T. D. "The Honeybee Colony as a Superorganism." *American Scientist*, November–December 1989. The honeybee society exemplifies the concept of kin selection in which the society, rather than the individual, serves as a "vehicle for the survival of genes."

Sherman, P., and Alcock, J. *Exploring Animal Behavior.* Sunderland, MA: Sinauer, 1998. A collection of articles in which behavioral biologists describe their research and their conclusions about the mechanisms, function, and evolution of behavior.

Sherman, P. W., Jarvis, J. U. M., and Braude, S. H. "Naked Mole Rats." *Scientific American*, August 1992. Describes the newly investigated society of the naked mole rat, a vertebrate whose social behavior resembles that of some social insects.

Weiner, J. *Time, Love, Memory: A Great Biologist and His Quest for the Origins of Behavior.* New York: Knopf, 2000. A wonderful science writer chronicles the study of the genetics of behavior, especially the work of a great pioneer of the field, Seymour Benzer.

Wilson, E. O. "Empire of the Ants." *Discover*, March 1990. This entertaining article examines the diversity of ant behavior that has contributed to their enormous success.

Answers to Multiple-Choice Questions

1. c 2. d 3. d 4. e 5. d 6. a

MEDIATUTOR
Animal Behavior

CD Activities

Activity 37.1: Innate Animal Behavior

Estimated time: 15 minutes

A behavior is any observable activity of living animal. Each behavior arises through an interaction between an animal's genetic make-up and its environment, but some behaviors appear to be more heavily influenced by genetic factors. These innate behaviors can typically be performed very early in life without the benefit of prior experience. In this tutorial, you will explore the evolutionary significance of an innate behavior that was discovered by the Nobel-Prize-winning scientist Niko Tinbergen. You will have an opportunity to repeat some of Tinbergen's experiments with digger wasps and to draw conclusions based on your observations.

Start the MediaTutor Student CD-ROM and enter the activity number in the Quick Search box to be taken directly to that activity.

Web Investigations

Case Study: Sex and Symmetry

Estimated time: 10 minutes

The research results are in. Women and scorpionflies prefer symmetrical males. What other features do animals use to assess potential mates? How do biologists determine the basis of mate choice in animals?

Go to http://www.prenhall.com/audesirk6, the Audesirk Companion Web site. Select Chapter 37 and the Web Investigation to begin.

UNIT SIX

Ecology

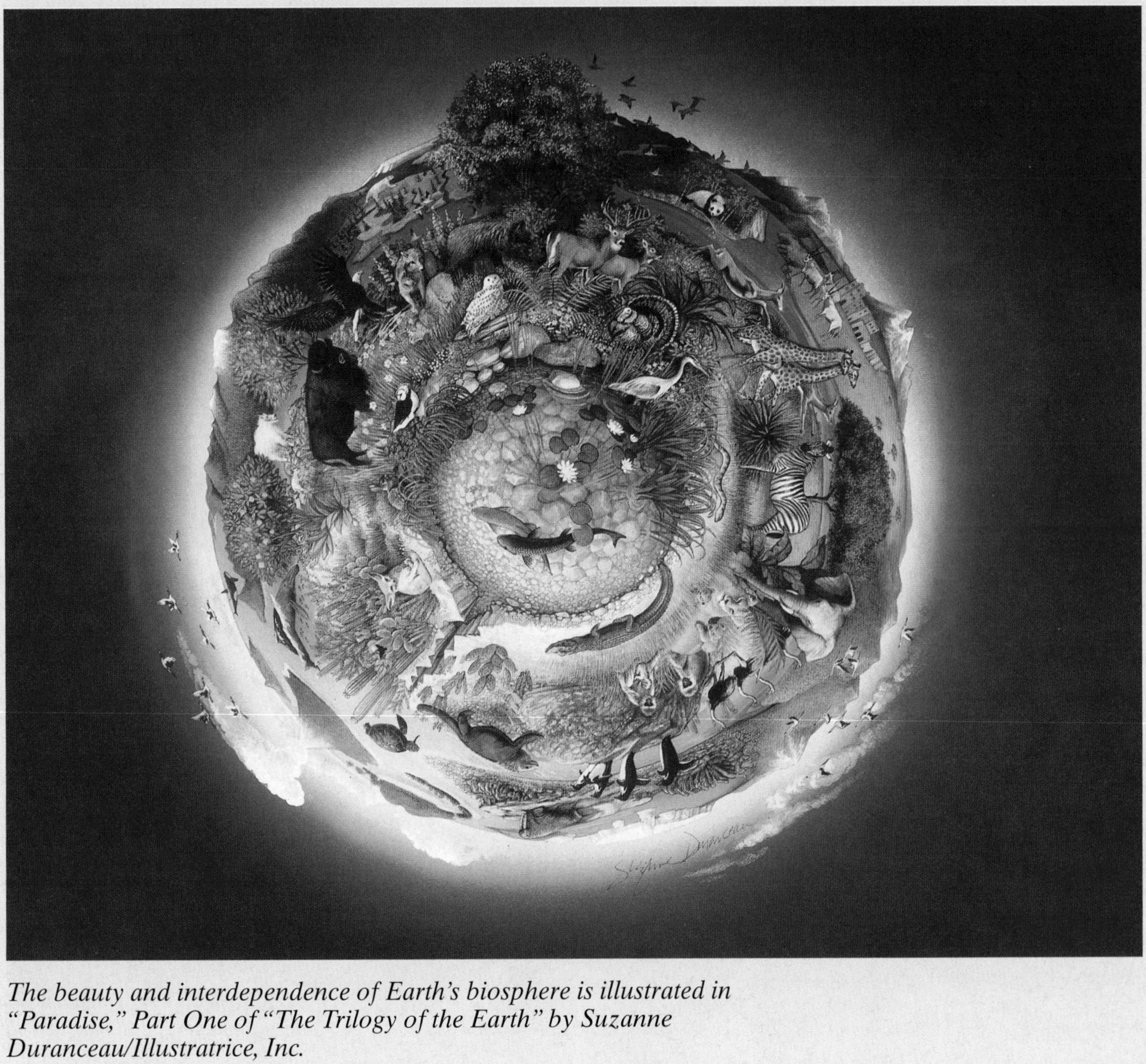

The beauty and interdependence of Earth's biosphere is illustrated in "Paradise," Part One of "The Trilogy of the Earth" by Suzanne Duranceau/Illustratrice, Inc.

"There is no exception to the rule that every organic being naturally increases at so high a rate that, if not destroyed, the Earth would soon be covered by the progeny of a single pair."

Charles Darwin in* On the Origin of Species *(1859)

Populations of white-footed mice, white-tailed deer, gypsy moth caterpillars, deer ticks, and acorns interact in complex ways in the eastern forest ecosystem. This forest is in the Pocono Mountains of Pennsylvania.

38 Population Growth and Regulation

AT A GLANCE

Case Study: Acorns, Mice, Moths, Deer, and Disease

1) **How Do Populations Grow?**
Biotic Potential Can Produce Exponential Growth

2) **How Is Population Growth Regulated?**
Exponential Growth Cannot Continue Indefinitely
Environmental Resistance Limits Population Growth

3) **How Are Populations Distributed in Space and Time?**
Populations Exhibit Differing Spatial Distributions
Survivorship in Populations Follows Three Basic Patterns

4) **How Is the Human Population Changing?**
The Human Population Is Growing Exponentially
Technological Advances Have Increased Earth's Carrying Capacity for Humans
The Age Structure of a Population Predicts Its Future Growth
The U.S. Population Is Growing Rapidly

Case Study Revisited: Acorns, Mice, Moths, Deer, and Disease

CASESTUDY

Acorns, Mice, Moths, Deer, and Disease

In autumn 1995, the hardwood forest in upstate New York was ablaze with fall colors. It was also alive with Girl Scouts industriously strewing acorns on the forest floor. Why bring acorns to an oak forest? The scouts were assisting researchers Clive Jones and Richard Ostfield from the Institute of Ecosystem Studies in Millbrook, New York, who were monitoring several aspects of this forest ecosystem: the availability of acorns and the populations of white-footed mice, gypsy moths, deer, and black-legged deer ticks.

In 1995 the oak trees didn't produce many acorns; in fact, oaks bear large numbers of acorns only every two to five years. White-footed mice feed on acorns; thus, white-footed mouse populations fluctuate dramatically in response to the availability of acorns. The gypsy moth population also undergoes significant fluctuations: Every decade or so, the oak trees are almost stripped of leaves by outbreaks of gypsy moth caterpillars, the larval form of this imported pest. What controls gypsy moth populations? White-footed mice are important because they eat the pupae of the moths. The researchers were testing the hypothesis that an abundant acorn crop should lead to a large mouse population, and therefore a reduction in gypsy moths. By comparing mouse and moth populations in areas of forest artificially enriched with acorns to populations in nearby acorn-poor forests, the researchers could test this link. They also removed mice from selected areas to test the impact of reduced mouse predation on gypsy moths.

Why were the biologists monitoring the population size of the black-legged deer ticks? These ticks spread Lyme disease to people—more than 150,000 cases have been reported since 1980. These parasitic ticks feed on deer as adults but often spend their larval stages sucking the blood of mice. Both deer and mice are attracted to acorns. Blood-sated adult ticks drop from the deer and lay their eggs in the soil. The eggs hatch into larvae, which attach themselves to white-footed mice, nearly all of which carry the bacterium that causes Lyme disease.

By manipulating the numbers of mice and acorns and monitoring the responses of these interacting populations, the researchers made some surprising and potentially important discoveries. What were they? Join us at the end of the chapter. ■

This chapter begins our unit on **ecology** (derived from the Greek word *oikos*, "a place to live"). Ecology refers to the study of interrelationships among living things and their nonliving environment. The environment consists of the **abiotic** (nonliving) component, including soil, water, and weather, and the **biotic** (living) component, including all forms of life. The term **ecosystem** refers both to the nonliving environment and to all the organisms within a defined area, such as our eastern forest. Within an ecosystem, all the interacting populations of organisms—for example, the oaks, mice, deer, ticks, moths, bacteria, and all the other forms of life in the eastern forest—are described as a **community**.

What keeps natural populations from overpopulating their habitat and starving? What happens when different organisms compete for the same type of food, for space, or other resources? Why has the human population continued to expand while other populations have fluctuated, remained stable, or declined? Look for answers to these questions in this chapter as we explore how populations grow and how population growth is controlled. In Chapters 39 through 41, we will proceed from *populations* to levels of increasing complexity: first to *communities* and the interactions within them, then to the organization of ecosystems, and finally to an exploration of the diversity of ecosystems that make up the *biosphere*, which encompasses all life on Earth.

1 How Do Populations Grow?

A **population** consists of all the members of a particular species who live within an ecosystem and can potentially interbreed. The mice, the oaks, the gypsy moths, and the ticks in our case study each constitute a population.

Studies of undisturbed ecosystems show that some populations tend to remain relatively stable in size over time, others fluctuate in a cyclical pattern, and still others vary sporadically in response to complex environmental variables. In contrast to most nonhuman species, the global human population has exhibited steady growth for centuries. Let's examine how and why populations grow and then look at the forces that control this growth.

Three factors determine whether and how much the size of a population changes: (1) births, (2) deaths, and (3) migration. Organisms join a population through birth or **immigration** (migration in) and leave it through death or **emigration** (migration out). A population remains stable if, on the average, as many individuals leave as join. Population growth occurs when the number of births plus immigrants exceeds the number of deaths plus emigrants. Population growth drops when the reverse occurs. A simple equation for the change in population size within a given time period is as follows:

$$\text{(births} - \text{deaths)} + \text{(immigrants} - \text{emigrants)} = \text{change in population size}$$

In many natural populations, organisms moving in and out contribute relatively little to population change, leaving birth and death rates as the primary factors that influence population growth.

The ultimate size of any population (leaving out migration) is the result of a balance between two major opposing factors that determine birth and death rates. The first factor is **biotic potential**, or the maximum rate at which the population could increase, assuming ideal conditions that allow a maximum birth rate and minimum death rate. Opposing this potential for growth are limits set by the living and nonliving environments; collectively, these limits are called **environmental resistance**. Environmental resistance is imposed by the availability of food and space, competition with other organisms, and certain interactions among species, such as predation and parasitism. Natural events such as storms, fires, freezing weather, floods, and droughts also come under this heading. Environmental resistance can both decrease birth rates and increase death rates. For example, a drought might kill plants directly. Drought would also harm animal populations that rely on these plants by reducing reproduction and increasing deaths from starvation. The interaction between biotic potential and environmental resistance usually results in a balance between population size and available resources. To understand how populations grow and how their size is regulated, let's examine each of these forces in more detail.

Biotic Potential Can Produce Exponential Growth

Changes in population size (ignoring migration) are functions of the birth rate, the death rate, and the number of individuals in the original population. Rates of change in population size are often measured as the changes in these variables per individual during a given unit of time. For example, the birth rate may be expressed as the number of births per organism per year.

The **growth rate** (r) of a population is a measure of the change in population size per individual per unit of time. This value is determined by subtracting the death rate (d) from the birth rate (b):

$$\underset{\text{(birth rate)}}{b} - \underset{\text{(death rate)}}{d} = \underset{\text{(growth rate)}}{r}$$

(If the death rate exceeds the birth rate, the growth rate will be negative, and the population will decline; here we will focus on growing populations.) If we wish to calculate the annual growth rate of a human population of 10,000 in which 1500 births and 500 deaths occur each year, we can use this simple equation:

$$\text{birth rate } b - \text{death rate } d = \text{growth rate } r$$

$$1500/10{,}000 - 500/10{,}000 = 0.10, \text{ or } 10\%$$

or

$$0.15 \text{ births per person per year} - 0.05 \text{ deaths per person per year}$$

$$= 0.10, \text{ or } 10\% \text{ increase per person per year} = r$$

To determine the number of individuals added to a population in a given time period, the growth rate (r) is multiplied by the original population size (N):

$$\text{population growth} = rN$$

In this example, population growth (rN) equals $0.10 \times 10{,}000 = 1000$ people in the first year. If this growth rate is constant, then the following year, r must be multiplied by an even larger population size: ($N + rN = 11{,}000$). Thus, in the second year 1100 more individuals are added to the population, which further increases the number of individuals added in the third year, and so on.

This pattern of continuously accelerating increase in population size is **exponential growth**. During exponential growth, the population (over a given time period) grows by a fixed percentage of its size at the beginning of that time period. Thus, an increasing number of individuals is added to the population during each succeeding time period, causing population size to grow at an ever-accelerating pace. The graph of exponential population growth is often called a *J-shaped growth curve*, or a **J-curve**, after its shape (Fig. 38-1). The value *r* allows us to calculate a value called **doubling time**. This is the time it takes a population to double in size at its current rate of growth (*r*). The doubling time, based on the calculation for exponential growth, can be determined by dividing

MEDIATUTOR 38.1 Population Growth and Regulation

(a)

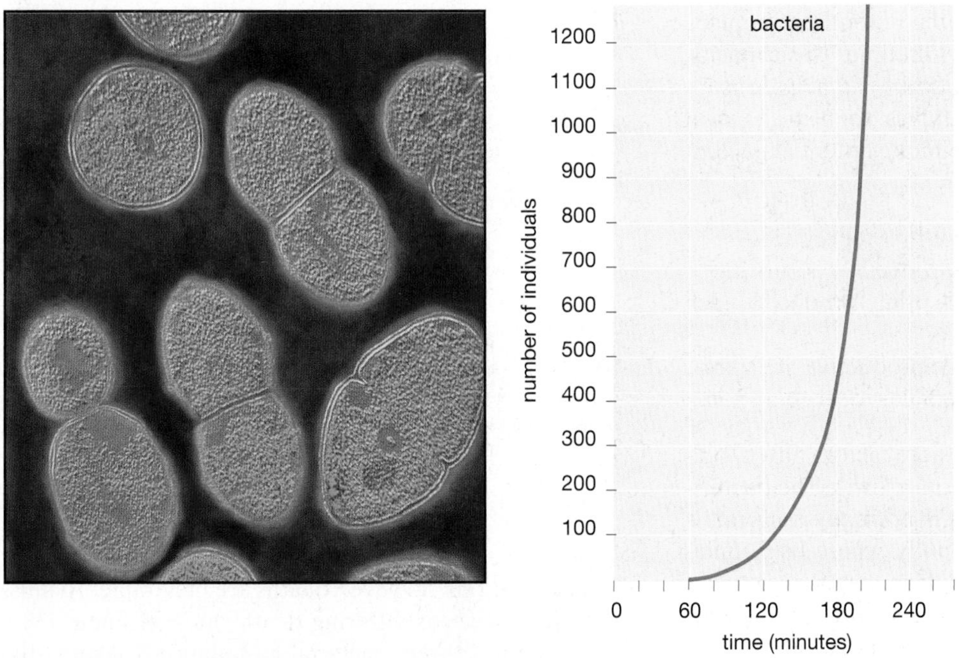

time (minutes)	number of bacteria
0	1
20	2
40	4
60	8
80	16
100	32
120	64
140	128
160	256
180	512
200	1024
220	2048

(b)

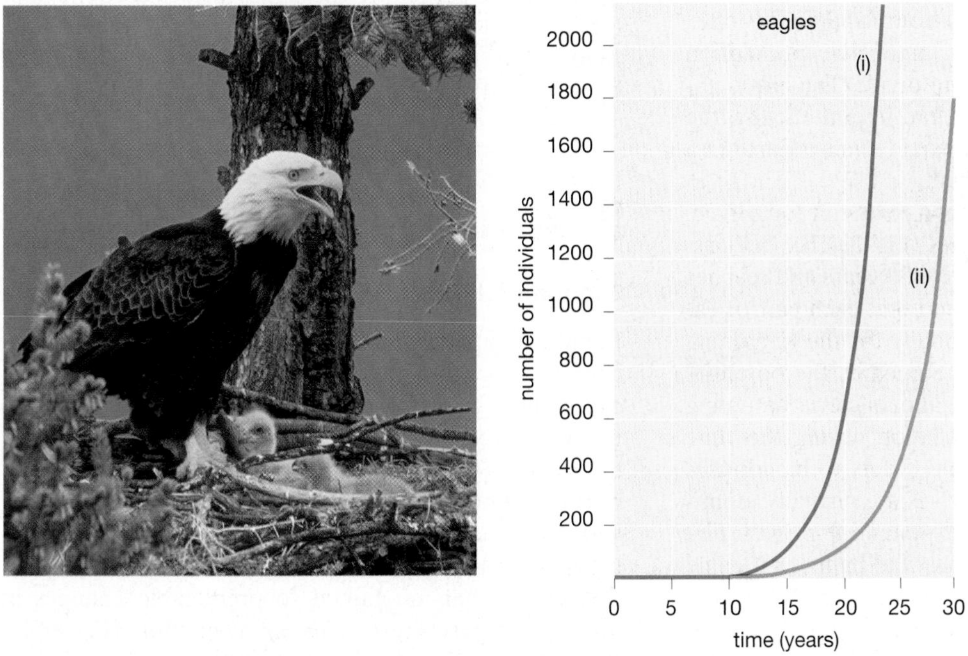

time (years)	number of eagles (i)	number of eagles (ii)
0	2	2
2	2	2
4	4	2
6	8	4
8	14	8
10	28	12
12	52	18
14	100	32
16	190	54
18	362	86
20	630	142
22	1314	238
24	2504	392
26	4770	644
28	9088	1066
30	17314	1764

Figure 38-1 Exponential growth curves
All such curves share a similar J shape; the major difference is the time scale. ***(a)*** Growth of a population of bacteria, starting with a single individual and with a doubling time of 20 minutes. ***(b)*** Growth of a population of eagles, starting with a single pair of hatchlings, for ages at first reproduction of 4 years (red line) and 6 years (green line). Notice in the table that after 26 years, the population of eagles that began reproducing at age 4 is nearly seven times that of the eagles that began reproducing at age 6.

69.3 (about 70) by *r* (with *r* expressed as a percentage). So our hypothetical population has a doubling time of about 70/10 = 7 years.

Exponential population growth occurs whenever births consistently exceed deaths. This occurs if, on the average, each individual produces more than one surviving offspring during its lifetime. Although the number of offspring an individual produces each year varies from millions for an oyster to one or fewer for a human, each organism—whether working alone or as part of a sexually reproducing pair—has the potential to replace itself many times over during its lifetime. This capacity, biotic potential, has evolved because it helps ensure that at least one offspring survives to bear its own young. Several factors influence biotic potential, including the following:

- The age at which the organism first reproduces.
- The frequency with which reproduction occurs.
- The average number of offspring produced each time.
- The length of the organism's reproductive life span.
- The death rate of individuals under ideal conditions.

We will use examples in which these factors differ to illustrate the concept of exponential growth. The bacterium *Staphylococcus* (Fig. 38-1a) is normally a harmless resident in and on the human body, where its population growth is restricted by environmental resistance. But in an ideal culture medium, such as warm custard, where *Staphylococcus* may accidentally be introduced, each bacterial cell can divide every 20 minutes, doubling the population three times each hour (and, in this case, causing food poisoning). The larger the population grows, the more cells there are to divide. The biotic potential of bacteria is so great that, hypothetically, the offspring of a single bacterium would swamp Earth in a layer 7 feet deep within 48 hours!

In contrast, the golden eagle is a relatively long-lived, rather slowly reproducing species (Fig. 38-1b). Let's assume that the golden eagle can live 30 years and reaches sexual maturity at age 4 years, and that each pair of eagles produces two offspring annually for the remaining 26 years (red line). Figure 38-1 compares the potential population growth of eagles to that of bacteria, assuming no deaths occur in either population during the time graphed. Although the time scale differs tremendously, notice that the shapes of the curves are virtually identical—both populations exhibit exponential growth. Figure 38-1b (green line) also shows what happens if eagle reproduction begins at age 6 years instead of 4. Exponential growth still occurs, but the time required to reach a particular size increases considerably. This result has important implications for the human population: Delayed childbearing significantly slows population growth. If each woman bears three children in her early teens, the population will grow much faster than if each woman bears five children but begins at age 30.

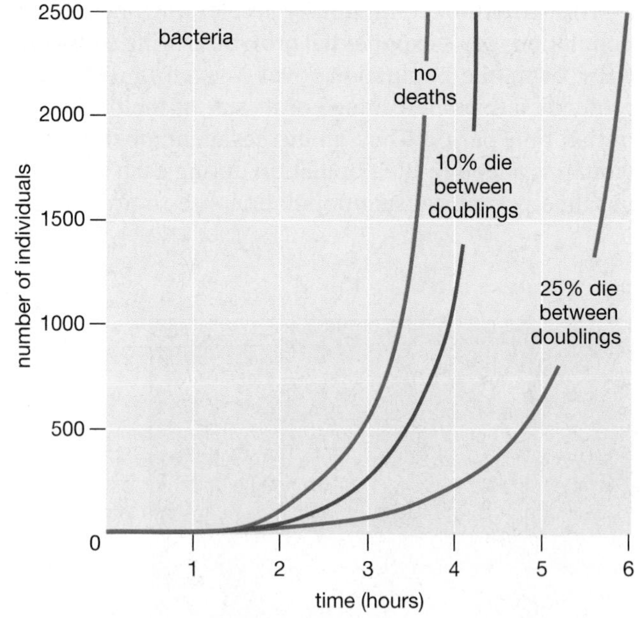

Figure 38-2 The effect of death rates on population growth The graphs assume that a bacterial population doubles every 20 minutes. Notice that the population in which a quarter of the bacteria die every 20 minutes reaches 2500 only 2 hours and 20 minutes later than one in which no deaths occur.

So far we have looked only at birth rates. Even under ideal conditions, however, deaths are inevitable. To illustrate the effect of differing death rates, in Figure 38-2 we compare three bacterial populations. Again, the shapes of the curves are the same. In each case the population eventually approaches infinite size, but the time required to reach any given population size is increased by increased mortality.

2 How Is Population Growth Regulated?

Exponential Growth Cannot Continue Indefinitely

In nature, exponential growth occurs only under special circumstances and only for a limited time. For example, exponential growth is seen in populations that undergo regular cycles, in which rapid population growth is followed by a sudden massive die-off. These **boom-and-bust cycles** occur in a variety of organisms for complex and varied reasons. Many short-lived, rapidly reproducing species, from algae to insects, have seasonal population cycles that are linked to predictable changes in rainfall, temperature, or nutrient availability (Fig. 38-3). In temperate climates, insect populations grow rapidly during the spring and summer and then crash with the killing hard frosts of winter. More complex factors produce roughly 4-year cycles for small rodents such as voles and lemmings, and much longer population cycles in hares, muskrats, and grouse.

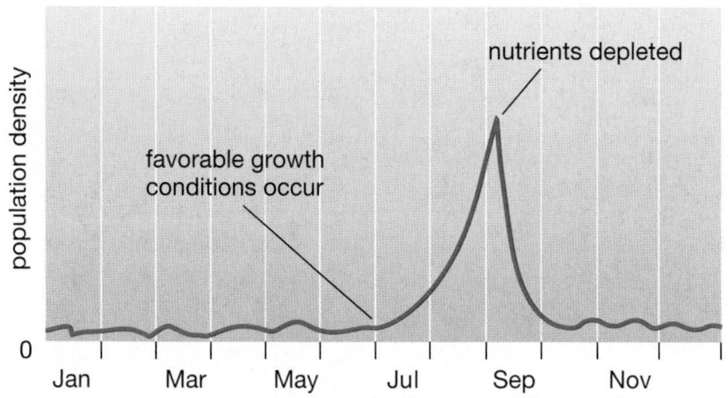

Figure 38-3 A boom-and-bust population cycle
Population density of cyanobacteria (blue-green algae) in an annual boom-and-bust cycle in a lake. Algae survive at a low level through the fall, winter, and spring. Early in July, conditions become favorable for growth, and exponential growth occurs through August. Nutrients soon become depleted, and the population "goes bust."

Lemming populations, for instance, may grow until the rodents overgraze their fragile arctic tundra ecosystem. Lack of food, increasing populations of predators, and social stress caused by overpopulation may all contribute to a sudden high mortality. Many deaths occur as waves of lemmings emigrate from regions of high population density. During these dramatic mass movements, lemmings are easy targets for predators. Many drown; they begin swimming when they encounter a body of water, including the ocean, but cannot make it all the way across. The reduced lemming population eventually contributes to a decline in predator numbers (see "Scientific Inquiry: Cycles in Predator and Prey Populations") and a recovery of the plant community on which the lemmings normally feed. These responses, in turn, set the stage for another round of exponential growth in the lemming population (Fig. 38-4).

In populations that do not show boom-and-bust cycles, exponential growth may occur temporarily under special circumstances—for example, if population-controlling factors, such as predators or parasites, are eliminated or if the food supply is increased. This occurs in white-footed mouse populations when oaks produce large acorn crops. Exponential growth can also occur when individuals invade a new habitat where conditions are favorable and competition is scarce, such as the farm field that is plowed but then abandoned. Invasion of new habitats and exponential growth of the invaders commonly occur when people introduce foreign, or **exotic**, species into ecosystems, in many cases with very damaging results, as illustrated by the gypsy moth's devastation of oak forests. As you will learn in the next section, all populations showing exponential growth must eventually stabilize or crash (decrease in size precipitously).

Environmental Resistance Limits Population Growth

As individuals join the population, competition for resources intensifies. As plants invade the abandoned farm field and their populations grow, competition for space, water, sunlight, and soil nutrients increases until further expansion is impossible. In response to increases in populations of animals such as prairie dogs, predators such as hawks may increase in number or make this newly abundant prey a larger part of their diet. Parasites and diseases spread more readily owing to weakness and crowding, resulting from lack of food or from stress caused by adverse social interactions. Animals might emigrate to establish new populations or die. Consequently, after a period of exponential growth, populations tend to stabilize at or below the maximum number the environment can sustain. The rate of growth gradually declines and reaches a long-term state

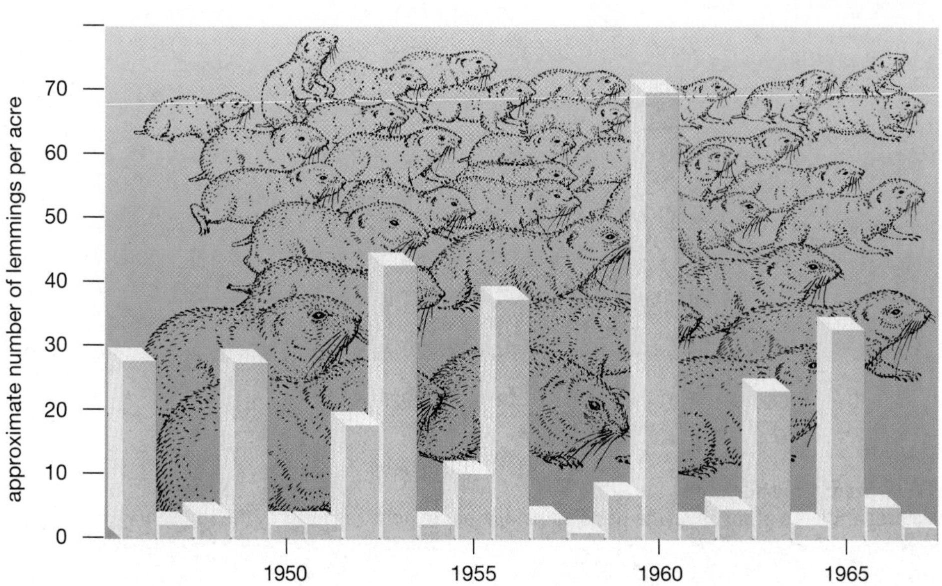

Figure 38-4 Lemming population cycles
Lemming population density follows roughly a 4-year cycle (data from Point Barrow, Alaska).

Scientific Inquiry
Cycles in Predator and Prey Populations

If we assume that prey, such as hares, are eaten exclusively by a particular predator, such as the lynx, it seems logical that both populations might show cyclic changes, with changes in the predator population size lagging behind changes in the prey population size. For example, a large hare population would provide abundant food for lynx and their offspring, which would then survive in large numbers. The increased lynx population would eat more hares, reducing the hare population. With fewer prey, fewer lynx would survive and reproduce, so the lynx population would decline a short time later.

Does this out-of-phase population cycle of predators and prey actually occur in nature? A classic example of such a cycle was demonstrated by using the ingenious method of counting all the pelts of northern Canada lynx and snowshoe hares purchased from trappers by the Hudson Bay Company between 1845 and 1935. The availability of pelts (which presumably reflects population size) showed dramatic, closely linked population cycles of these predators and their prey (Fig. E38-1). Unfortunately, many uncontrolled variables could have influenced the relationship between hares and lynx. One problem is that hare populations sometimes fluctuate even without lynx present, possibly because in the absence of predators hares overshoot their carrying capacity and reduce their food supply. Further, lynx do not feed exclusively on hares but can eat a variety of small mammals. Environmental variables such as exceptionally severe winters could have adversely affected both populations and produced similar cycles. Recently, researchers tested the hare/predator relationship more rigorously by fencing off 1-kilometer-square areas in northern Canada to exclude predators. The crash of the hare population was lessened both by providing extra food and by excluding predators, but by far the greatest success in preventing the crash of the hare population was achieved when the researchers excluded predators *and* provided extra food.

To test the predator–prey cycle hypothesis in an even more controlled manner, investigators turned to laboratory studies on populations of small predators and their prey. The study illustrated in Figure E38-2 involved braconid wasp predators and their bean weevil prey. As predicted, the two populations showed regular cycles, with the predator population rising and falling slightly later than the prey population. The wasps lay their eggs on weevil larvae, which provide food for the newly hatched wasps. A large weevil population ensures a high survival rate for wasp offspring, increasing the predator population. Then, under intense predation pressure, the weevil population plummets, reducing the population size of the next generation of wasps. Reduced predation then allows the weevil population to increase rapidly, and so on.

It is highly unlikely that such a straightforward example will ever be found in nature, but this type of predator–prey interaction clearly can contribute to the fluctuations observed in many natural populations.

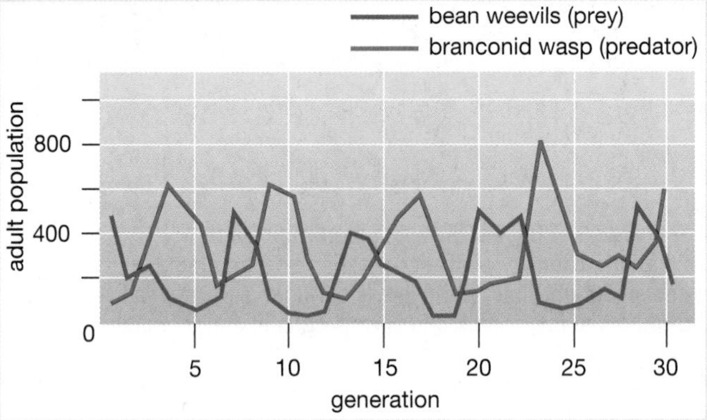

Figure E38-2 Experimental predator–prey cycles
Out-of-phase fluctuations in laboratory populations of the bean weevil and its braconid wasp predator.

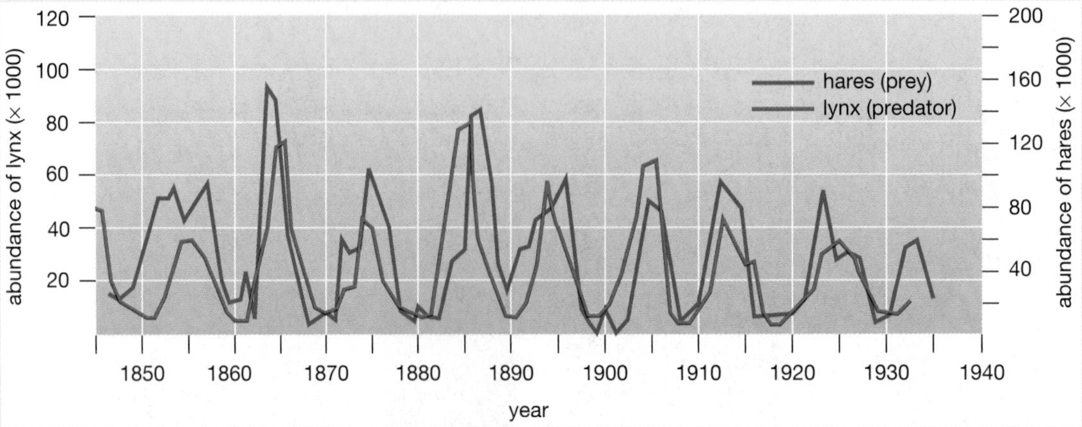

Figure E38-1 Population cycles in predators and prey
Snowshoe hares and their lynx predators are graphed on the basis of the number of pelts received by the Hudson Bay Company.

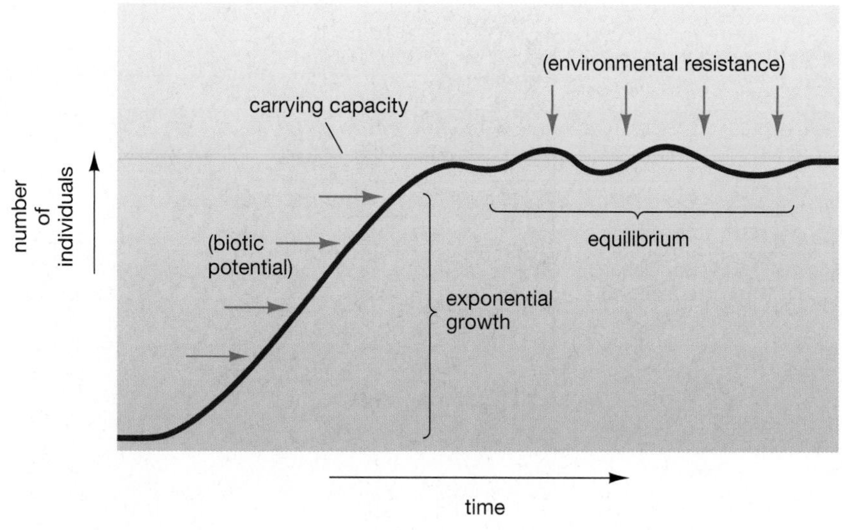

Figure 38-5 The S-curve of population growth The population grows exponentially at first, then fluctuates around the carrying capacity. The growth is driven by biotic potential but levels off owing to environmental resistance.

of equilibrium, fluctuating around a growth rate of zero. In this equilibrium, the birth rate is balanced by the death rate and population size is stabilized. This type of population growth, which is typical of long-lived organisms colonizing a new area, is represented graphically by an S-shaped growth curve, or **S-curve** (Fig. 38-5).

Populations may stabilize at a level at or below the **carrying capacity** of the ecosystem. The carrying capacity is the maximum population size that an ecosystem can support indefinitely. It is determined primarily by the sustained availability of two types of resources. The first type is renewable resources, which are replenished by natural processes. Renewable resources include nutrients, water, and light (the energy source for plants). The second type of resource is nonrenewable: space.

Organisms will starve if demands on renewable resources are too high. If space requirements are exceeded, animals may emigrate, but commonly to less-suitable areas where their death rate will be higher. Reproduction will decline, because animals may not find adequate breeding sites or the seeds of plants may not reach a suitable place to germinate. If a population exceeds its carrying capacity, excess demands on resources may damage ecosystems, reducing *their* carrying capacity. The result is either a population decline until the ecosystem recovers or a permanently reduced population. For example, overgrazing by cattle on dry western grasslands has reduced the grass cover and allowed sagebrush (which cattle will not eat) to thrive. Once established, sagebrush replaces edible grasses and reduces the land's carrying capacity for cattle. Other dramatic cases of overgrazing have occurred when herbivores, such as reindeer, have been introduced onto islands without large predators (Fig. 38-6).

In nature, environmental resistance maintains populations at or below the carrying capacity of their environment. Factors of environmental resistance can be classified into two broad categories. (1) **Density-independent** factors limit population size regardless of the population density (number of individuals per given area). (2) **Density-dependent** factors increase in effectiveness as the population density increases. In the following sections, we will look more closely at these factors and how they control population growth.

Density-Independent Factors Limit Populations Regardless of Their Density

Natural events, including hurricanes, droughts, floods, and fire can have profound effects on local populations; these effects are independent of the population density. Perhaps the most important natural density-independent factor is weather. For example, many insects and annual plant populations are limited in size by the number of individuals that can be produced before the first hard freeze. Such populations typically do not reach their carrying capacity, because density-independent factors intervene first. Weather is largely responsible for the boom-and-bust population cycles described earlier.

Organisms that live several years have evolved various mechanisms to compensate for seasonal changes in weather, thereby circumventing this form of density-independent population control. Many mammals, for example, develop thick coats and store fat for the winter; some also hibernate. Other animals, including many birds, migrate long distances to find food and a hospitable climate. Plants might survive the rigors of winter by entering a period of dormancy, dropping their leaves and drastically slowing their metabolic activities.

Human activities can also limit the growth of natural populations in ways that are independent of population density. Pesticides and pollutants can cause drastic declines in natural populations. Before it was banned in the 1970s, the pesticide DDT had drastically reduced populations of predatory birds, such as eagles, ospreys, and pelicans. Habitat destruction caused by the construction of farms, roads, and housing developments has

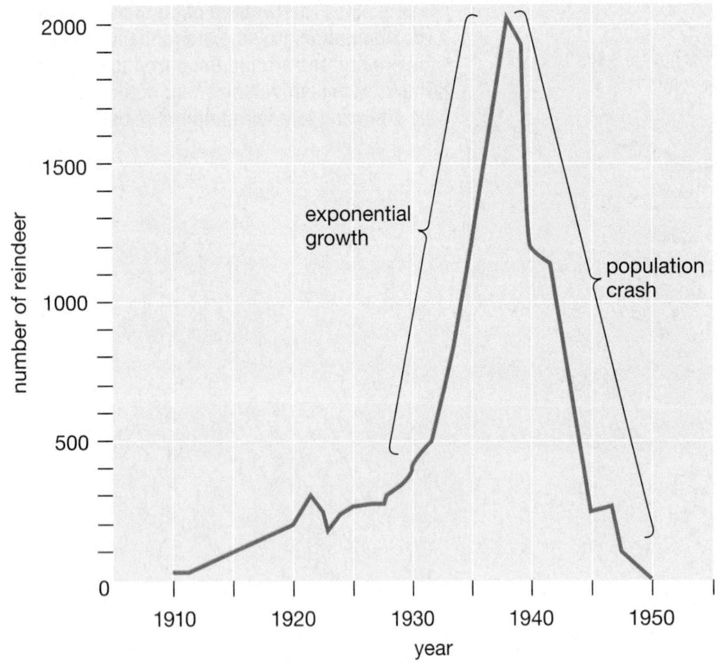

Figure 38-6 The effects of exceeding carrying capacity
Exceeding carrying capacity can damage an ecosystem, reducing its ability to support the population. In 1911, 25 reindeer were introduced onto one of the Pribilof Islands (St. Paul) in the Bering Sea off Alaska. Food was plentiful, and the reindeer encountered no predators on the island. The herd grew exponentially (note the initial J shape) until it reached 2000 reindeer in 1938. At this point, the small island was seriously overgrazed, food was scarce, and the population declined dramatically. By 1950, only eight reindeer survived.

eliminated most of the large prairie dog towns and pushed such diverse species as the American condor and (in China) the giant panda to the brink of extinction.

Density-Dependent Factors Become More Effective as Population Density Increases

For long-lived species, by far the most important elements of environmental resistance are density-dependent factors. Because they become increasingly effective as population density increases, density-dependent factors exert a negative feedback effect on population size. The larger the population grows, the more changes are triggered that counteract this growth. Density-dependent factors include community interactions, such as predation and parasitism, as well as competition within the species or with members of other species. These factors are discussed below and in Chapter 39.

Predators and Parasites Exert Density-Dependent Controls on Populations

Both *predation* and *parasitism* involve one organism feeding on another and harming it in the process. The relationship is called predation when the one organism (the **predator**) kills another (its **prey**; often a smaller organism) in order to eat it. Parasitism occurs when one organism (the **parasite**) lives on another (its **host**; usually a much larger organism) and feeds on the host's body without killing it—at least not immediately. While predators must kill their prey to feed, parasites benefit by having their hosts remain alive.

Predation includes wolves working together to kill an elk (Fig. 38-7), mice munching on acorns or gypsy moth pupae, and the sundew plant digesting an insect (described in Chapter 23). Predation becomes an increasingly important factor in population control as prey populations increase, simply because the more abundant the prey, the more often predators encounter them. Many predators will eat a variety of prey, depending on what is most abundant and easiest to find. Coyotes might eat more mice when the mouse population is high but switch to eating more ground squirrels as the mouse population declines, incidentally allowing the mouse population to recover.

Predators can also exert density-dependent effects by increasing in number as their prey population grows. For instance, predators that feed heavily on lemmings, such as the arctic fox and snowy owl, regulate the number of offspring they produce according to the abundance of lemmings. The snowy owl might produce up to 13 chicks when lemmings are abundant but not reproduce at all in years when lemmings are scarce. In some cases, an increase in predators might cause a crash of the prey population, and a decline in the predator population follows. This pattern can result in out-of-phase **population cycles** of both predators and prey (see "Scientific Inquiry: Cycles in Predator and Prey Populations").

Figure 38-7 Predators help control prey populations
Grey wolves, hunting in a pack, have brought down an elk, who may have been weakened by old age or parasites.

Some predators feed primarily on prey made vulnerable because their populations have exceeded their carrying capacity. Such prey may be weakened by lack of food or may be exposed owing to lack of appropriate shelter. In such cases, predation may maintain prey populations near a density that can be sustained by the resources of the ecosystem. In other cases, predators may maintain their prey at well below carrying capacity. A dramatic example of this phenomenon is the prickly pear cactus, which was introduced into Australia from Latin America. Lacking natural predators, it spread uncontrollably, destroying millions of acres of valuable pasture and range land. Finally, in the 1920s, a cactus moth (a predator of the prickly pear) was imported from Argentina and released to feed on the cacti. Within a few years, the cacti were virtually eliminated. Today the moth continues to maintain its prey at very low population densities, well below the carrying capacity of the ecosystem.

In contrast to a predator, a parasite feeds on a larger organism without killing it immediately or directly. Examples include any organism that causes disease—including some bacteria (such as those causing Lyme disease), fungi, intestinal worms, ticks, and protists such as the malaria parasite. Insects (such as the gypsy moth) that feed on plants without killing them are also parasites. Parasitism is density-dependent. Most parasites have limited motility and spread more readily from host to host at high host-population densities. For example, plant diseases spread readily through acres of densely planted crops, and childhood diseases spread rapidly through schools and day-care centers. Parasites influence population size by weakening their hosts and making them more susceptible to death from other causes. Although parasites do not benefit from the death of their host, their activities may nonetheless kill the host. For example, malaria infects roughly 600 million people each year and kills approximately 2 million annually. Infection by parasites can contribute to deaths by predation by making the host organisms weaker and more vulnerable to predators.

Both predators and parasites can have beneficial effects on the prey population as a whole. Predators, parasites, and their prey evolve together—they *coevolve.* Parasites and predators tend to destroy the least fit of the prey, leaving the better-adapted prey to reproduce. This results in a balance in which the prey population is regulated but not eliminated. The population balance in ecosystems can be destroyed when predators or parasites are introduced into regions where they did not evolve and local prey species have had no opportunity to evolve defenses against them through natural selection. An example is the prickly pear cactus in Australia, mentioned above (this topic will be discussed further in Chapter 39). The smallpox virus, inadvertently carried by traveling Europeans, ravaged the native population of Hawaii, the Amerindians of Argentina, and the Aborigines of Australia. Introduced rats, snakes, and mongooses have exterminated many of Hawaii's native bird populations.

Competition for Resources Helps Control Populations

The resources that determine carrying capacity (space, nutrients, water, light) are often limited relative to the demand for them. In other words, there may not be enough resources to support all the organisms that are produced. Further, use by one individual limits their availability to another. **Competition**, defined as the interaction among individuals who attempt to utilize a limited resource, limits population size in a density-dependent manner. There are two major forms of competition: (1) **interspecific competition** (among individuals of different species; described further in Chapter 39), and (2) **intraspecific competition** (among individuals of the same species). Because the needs of members of the same species for water and nutrients, shelter, breeding sites, light, and other resources are almost identical, intraspecific competition is more intense than interspecific competition.

Organisms have evolved several ways to deal with intraspecific competition. Some organisms, including most plants and many insects, engage in **scramble competition**, a kind of free-for-all with resources as the prize. For example, when a plant disperses its seeds in a small area, hundreds may germinate. However, as they grow, the larger plants begin to shade the smaller ones; those with the most extensive root systems absorb most of the water, and the weaker individuals eventually wither and die.

Many animals (and even a few plants) have evolved **contest competition**, which involves social or chemical interactions used to limit access to important resources (also see Chapter 37). Territorial species—such as wolves, many fish, rabbits, and songbirds—defend an area that contains important resources such as food or nesting sites. When the population begins to exceed the available resources, only the best-adapted individuals are able to defend adequate territories. Those without territories may not reproduce and are easy prey.

As population densities increase and competition becomes more intense, some animals react by emigrating. Large numbers leave their homes to colonize new areas, and many, sometimes most, die in the quest. The mass movements of lemmings, which in some instances end in marches into the sea, are apparently in response to overcrowding. Migrating swarms of locusts plague the African continent, stripping all vegetation in their path (Fig. 38-8).

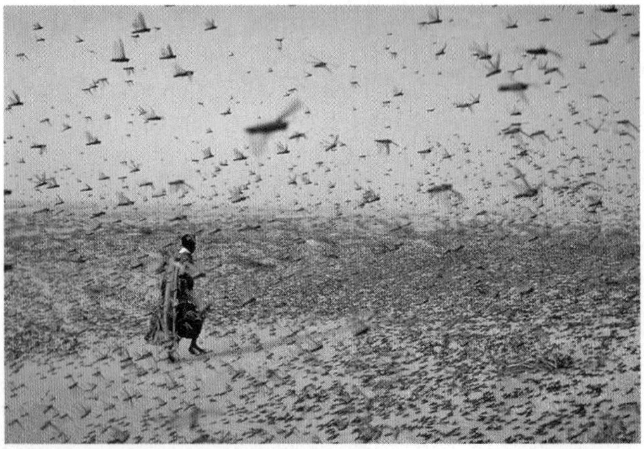

Figure 38-8 Emigration
In response to overcrowding and lack of food, locusts emigrate in swarms, devouring all the vegetation in their path.

Population size is a result of complex interactions between density-independent and density-dependent forms of environmental resistance. For example, a stand of pines weakened by drought (a density-independent factor) may more readily fall victim to the pine bark beetle (a density-dependent parasite). Likewise, a caribou weakened by parasites (density-dependent) is more likely to be killed by an exceptionally cold winter (a density-independent factor).

3 How Are Populations Distributed in Space and Time?

Populations Exhibit Differing Spatial Distributions

Organisms might live in flocks, herds, or pairs or as solitary individuals, or instead they might cluster around resources such as water holes. The spatial pattern in which members of a population are dispersed within a given area is that population's *distribution*. Distribution may vary with time, changing with the breeding season, for example. Ecologists recognize three major types of spatial distribution: *clumped*, *uniform*, and *random* (Fig. 38-9).

There are many populations whose members live in groups and whose distribution can be described as **clumped** (Fig. 38-9a). Examples include family or social groupings, such as elephant herds, wolf packs, prides of lions, flocks of birds, and schools of fish. What are the advantages of clumping? Flocks provide many eyes that can search for localized food, such as a tree full of fruit or a lake with fish. Schooling fish and flocks of birds may confuse predators with their sheer numbers. Some species form temporary groups for mating and caring for their young. Other plant or animal populations cluster, not for social reasons, but because resources are localized. Cottonwood trees, for example, cluster along streams and rivers in grasslands.

Organisms with a **uniform distribution** maintain a relatively constant distance between individuals. This type of distribution is most common among animals that defend territories and exhibit territorial behaviors designed to protect scarce resources. Male Galapagos iguanas establish regularly spaced breeding territories. Shorebirds are also often found in evenly spaced nests, just out of reach of one another. Other territorial species, such as the tawny owl, mate for life and continuously occupy well-defined, relatively uniformly spaced territories. (For animals that remain together to raise their young, spacing often refers to pairs rather than to individuals.) Some plants, such as sage, release chemicals into the soil around them that inhibit germination of other plants and thus space

Figure 38-9 Population distributions
(a) Clumped: a gathering of caterpillars. *(b)* Uniform: creosote bushes in the desert. *(c)* Random: trees and plants in a rain forest.

themselves relatively evenly (Fig. 38-9b). A uniform distribution helps ensure adequate resources for each individual.

Organisms with a **random distribution** are relatively rare. Such individuals do not form social groups. The resources they need are more or less equally available throughout the area they inhabit, and those resources are not scarce enough to require territorial spacing. Trees and other plants in rain forests come close to being randomly distributed (Fig. 38-9c). There are probably no vertebrate species that maintain a random distribution throughout the year, because they must breed—a behavior that makes social interaction inevitable.

Survivorship in Populations Follows Three Basic Patterns

Population patterns can be considered from the perspective of time as well as space. Over time, populations show characteristic patterns of deaths or (more optimistically) survivorship. These patterns, called **survivorship curves**, are revealed when the number of individuals of each age is graphed against their age. Three types of survivorship curve, which can be described as "late loss," "constant loss," and "early loss," according to the part of the life cycle during which most deaths occur, are shown in Figure 38-10.

Late-loss populations produce convex-shaped survivorship curves. Such populations have relatively low infant death rates, and most individuals survive to old age. Late-loss survivorship curves are characteristic of humans and many other large and long-lived animals, such as Dall mountain sheep. These species produce relatively few offspring, which are then protected by the parents.

Populations with constant-loss survivorship curves have a fairly constant death rate; their survivorship graphs appear as a relatively straight line. In these populations, individuals have an equal chance of dying at any time during their life span. This phenomenon is seen in the American robin, the gull, and laboratory populations of organisms that reproduce asexually, such as hydra and bacteria.

Early-loss survivorship produces a *concave* curve and is characteristic of organisms that produce large numbers of offspring. These offspring receive little parental care, being largely left to compete on their own. The death rate is initially very high among the offspring, but those that reach adulthood have a good chance of surviving to old age. Most invertebrates, most plants, and many fish exhibit such early-loss survivorship curves. In some populations of black-tailed deer, 75% of the population dies within the first 10% of its life span, giving this mammalian population an early-loss survivorship curve as well.

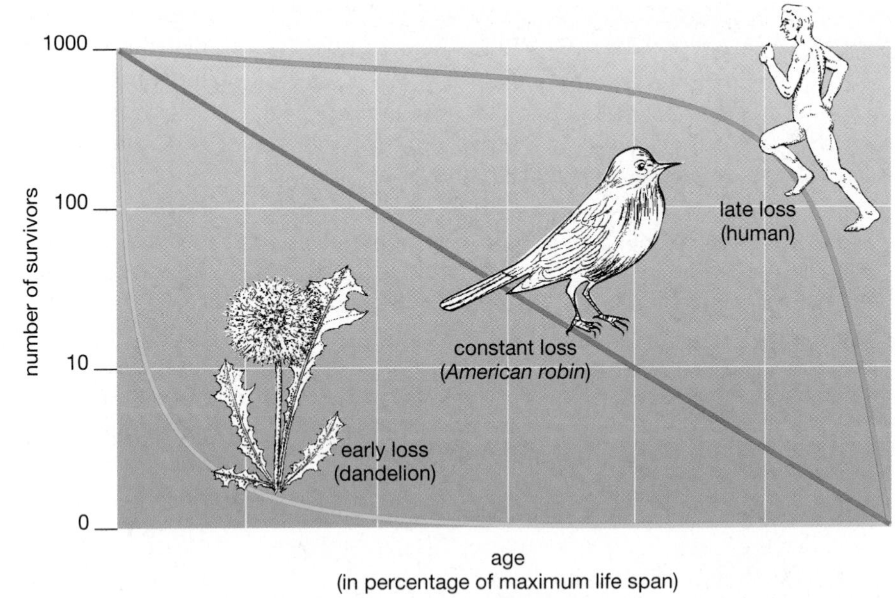

Figure 38-10 Survivorship curves
Three types of survivorship curve are shown. Because the life spans of these organisms differ so much, to compare them on the same graph, the percentages of survivors rather than the ages are used.

4 How Is the Human Population Changing?

The Human Population Is Growing Exponentially

Compare the graph of human population growth in Figure 38-11 with the exponential growth curves in Figure 38-1. The time spans are different, but each has the J-shape characteristic of exponential growth. It took more than 1 million years for the human population to reach 1 billion, but the second billion was added in just 100 years, the third billion in 30 years, the fourth billion in 15 years, and the fifth billion in 12 years. In the year 1750, the human population was 0.75 billion. It doubled in 150 years, reaching 1.6 billion in 1900. In contrast, the 1965 population of 3.5 billion will double to 7 billion in less than 50 years. World population (taking deaths into account) currently grows by about 1.4% or 82 million yearly—more than 224,000 people are added every day, nearly 1.6 million every week! Why hasn't environmen-

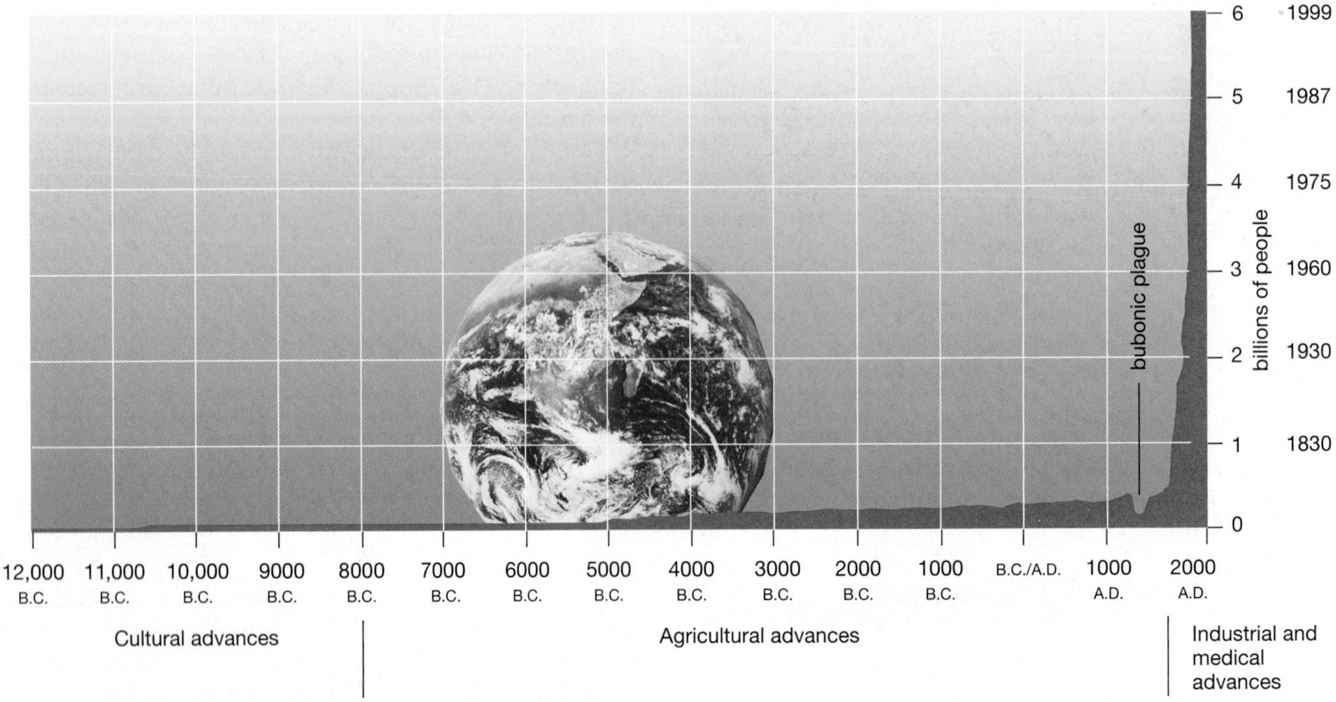

Figure 38-11 Human population growth
The human population from the Stone Age to the present has shown continued exponential growth. Note the dip in the fourteenth century caused by the bubonic plague. Note also the steadily decreasing time intervals at which additional billions are added. (Inset) Earth is an island of life in a sea of emptiness; its space and resources are limited.

MEDIATUTOR 38.2 Population Growth and Regulation in Human Populations tal resistance put an end to our exponential growth? What is the carrying capacity of the world for humans?

Like all populations, humans have encountered environmental resistance, but unlike other populations, we have responded to environmental resistance by devising ways to overcome it. As a result, the human population has grown exponentially for an unprecedented time span. To accommodate our growing numbers, we have altered the face of the globe. Human population growth has been spurred by a series of "revolutions" that conquered various aspects of environmental resistance and increased Earth's carrying capacity for people.

Technological Advances Have Increased Earth's Carrying Capacity for Humans

Primitive peoples produced a *cultural revolution* when they discovered fire, invented tools and weapons, built shelters, and designed protective clothing. Tools and weapons allowed more effective hunting and an increased food supply; shelter and clothing increased the habitable areas of the globe.

Domesticated crops and animals supplanted hunting and gathering by about 8000 B.C. This *agricultural revolution* provided people with a larger, more dependable food supply and further increased Earth's carrying capacity for humans. Increased food resulted in a longer life span and more childbearing years, but a high death rate from disease still restricted the population.

Human population growth continued slowly for thousands of years until the *industrial–medical revolution* began in England in the mid-eighteenth century, spreading through Europe and North America in the nineteenth century. Medical advances dramatically decreased the death rate by reducing environmental resistance from disease. These advances included the discovery of bacteria and their role in infection, leading to the control of bacterial diseases through improved sanitation and the use of antibiotics. Another advance was the discovery of viruses, leading to the development of vaccines for diseases such as smallpox. The revolution continues today as research proceeds on vaccines against major killers such as malaria and AIDS, and sophisticated medical procedures, such as coronary bypass operations and organ transplants, are improved and new procedures are developed to extend the human lifespan.

In developed countries, such as those of Western Europe, the industrial–medical revolution resulted in an initial rise in population due to a decrease in death rates, followed by a decline in birth rates. This decline can be attributed to many factors, including better education, increased availability of contraceptives, a shift to a primarily urban lifestyle, and more career options for women. In developed countries, on the average, populations have more or less stabilized.

In developing countries, such as most of Central and South America, Asia, and Africa (particularly prior to the AIDS epidemic), medical advances have decreased death rates and increased the life span, but birth rates remain high. These countries have not experienced the increase in wealth that was partly responsible for the birth rate decline in developed countries. In developing nations, children may be the only support for elderly parents. In agricultural societies, children are an important source of labor. Social traditions may offer prestige to men who father and to women who bear many children. In Nigeria, the most crowded country in Africa, many men refuse any form of birth control, many women desire large families, and the average woman bears six children. Nigeria is already suffering from loss of forests, the spread of deserts, soil erosion, and water pollution. With nearly half (44%) of its population of 124 million under age 15, continued population growth is a certainty. Although a majority of African nations have concluded that their growth rates are too high and are trying to reduce them, social traditions, lack of education, and lack of access to contraceptives impede progress in curbing population growth.

Of the more than 6 billion people on Earth today, close to 5 billion reside in developing countries. For the year 2050, the United Nations predicts that there will be nearly 9 billion people, with 8 billion of these living in the developing nations.

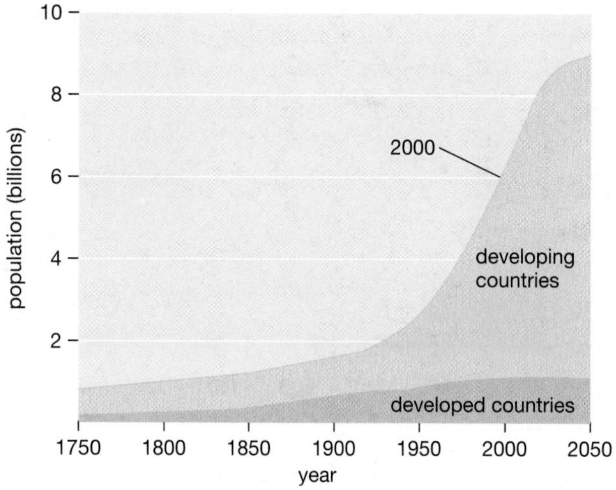

Modified from the *Population Reference Bureau: 1999 World Population Data Sheet.*

The prospects for population stabilization—*zero population growth*—in the near future are nil. In Africa, AIDS infects 25 million people and has taken an enormous toll in human lives, but the United Nations predicts that Africa will contribute 34% to population growth in developing countries until the year 2050. You can see why by examining the age structures of the developing countries and comparing them with countries with stable populations, as we describe next.

The Age Structure of a Population Predicts Its Future Growth

The **age structure** of a population is the distribution of males and females of each age group, which can be shown graphically in an age-structure diagram. The vertical axis of an age-structure diagram shows the age of people in years while the horizontal axis shows numbers of individuals in each age group, with males and females graphed in opposite directions from a central point. All age-structure diagrams rise to a peak at the top, because few people live into their nineties. The shape of the rest of the diagram, however, shows at a glance whether the population is expanding, stable, or shrinking. The shape is determined by the relative numbers of reproductive age adults (ages 15 to 45) and their children (ages 0 to 14). If the adults of reproductive age are having just enough children to replace themselves, the population is said to have **replacement level fertility (RLF)**. The age structure diagram of such a population will have relatively straight sides. If children in the various age classes exceed the number of reproductive-age adults, the population is showing greater than RLF and is expanding as a result. Such expanding populations have a pyramidal shape. In shrinking populations, there are fewer children than reproducing adults and the age-structure diagram narrows at the base.

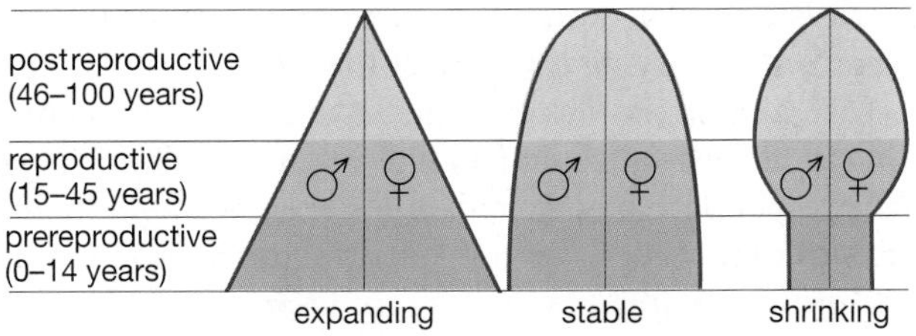

Figure 38-12 shows the average age structures of the populations of developed and developing countries. The outermost boundaries represent the projected population structure for the year 2025; the inner ones are actual values for 1995. Each graph has been divided into three parts to show individuals who are prereproductive (ages 0 to 14), reproductive (15 to 45), and postreproductive (46 and older). In 2000 the developing countries (in Asia, Africa, India, and South and Central America) had an average annual *natural increase* (population growth based on the births minus deaths; this does not

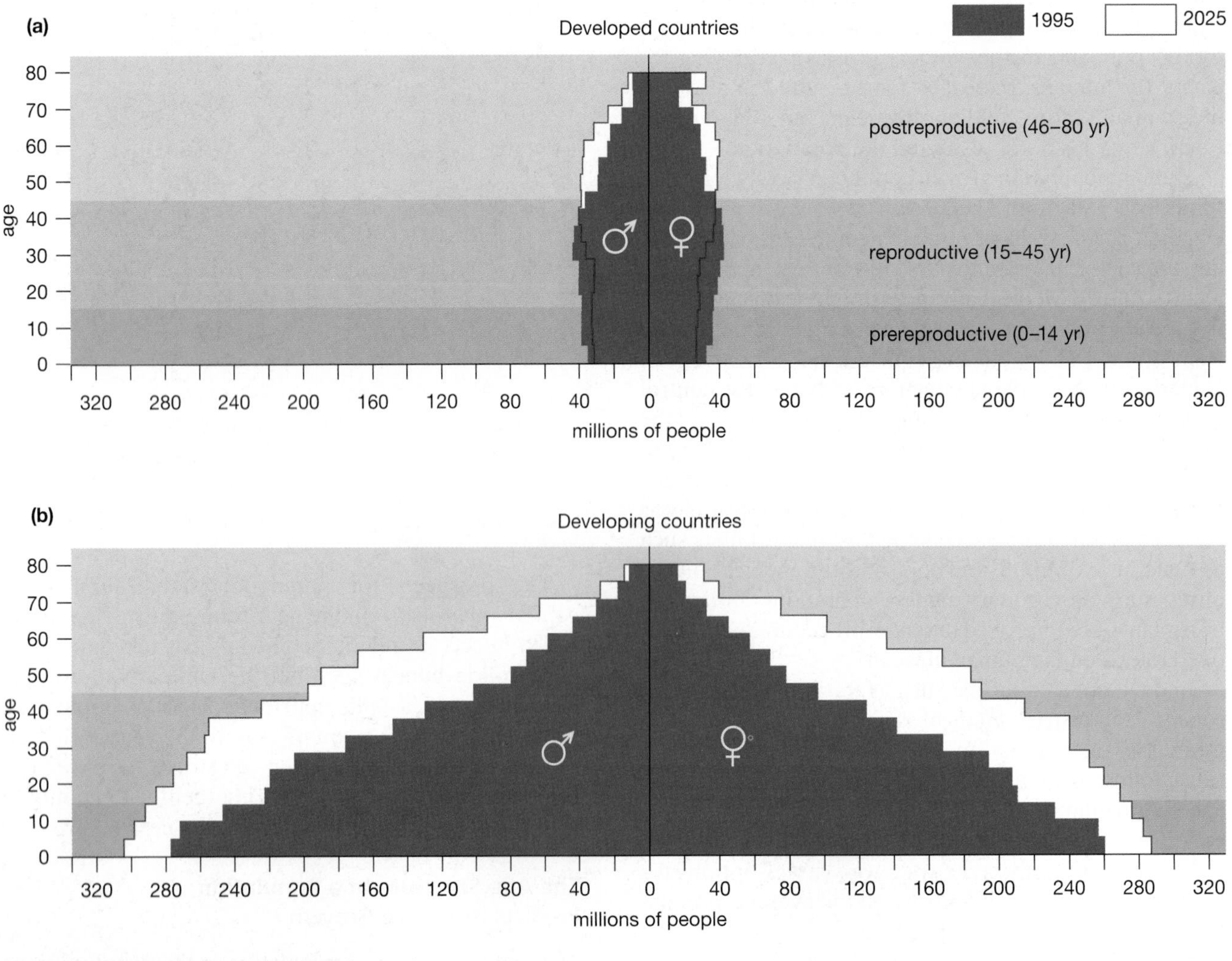

Figure 38-12 Age structures compared
(a) Developed countries and ***(b)*** developing countries. Compare these with the stylized diagrams shown in the text. The outer lines are projections to the year 2025.

include migration) of 1.7% and a doubling time of 42 years. In contrast, developed countries (in North America, Europe, Australia, Japan, and New Zealand) showed an average annual natural increase of 0.1% and a doubling time of 809 years. In the developed countries, projections for the year 2025 are only slightly higher than present levels (see Fig. 38-12a). Because developing countries are home to the vast majority of people, doubling time for the world population as a whole is currently 51 years.

In Europe, the average fertility rate is 1.4—well below RLF—and much of eastern Europe is experiencing a slight decline in population, as poor economic conditions prompt women to delay or forego having children. This raises economic concerns about the availability of future workers and taxpayers to support the increasing percentage of elderly people. Japan, a country the size of Montana, has 127 million people (slightly less than half of the entire U.S. population). Nonetheless, the Japanese fertility rate of 1.3 is causing concern about slowed economic growth and an aging population. The Japanese government provides a variety of subsidies that encourage larger families, and recently a Japanese toy company began offering $10,000 bonuses to their employees for having more than 2 children. It is clear that current economic structures are based on growing populations. Although a reduced and eventually stable population may offer tremendous benefits in the future, the transition can be difficult.

More than 95% of the growth of the human population occurs in the developing countries, where increasing numbers of people enter their reproductive years and give birth to an ever-increasing base of infants. Even if these countries were immediately to reach replacement-level fertility, population growth would continue for decades. This is because the children of the large families of the recent past create a built-in momentum for population growth as they enter their reproductive years and begin having children—even if they have only 2 children each. This momentum fuels China's population growth of 1% annually, even though China's fertility rate of 1.8 is below replacement level. In many African nations, more than 40% of the total population is under 15 years of age.

Study the predicted age structure in 2025 for developing countries. Can you detect a trend that indicates the beginnings of stabilization for these populations? Compare the size of the reproductive population (parents) to the prereproductive population (children). The excess number of children over parents is expected to

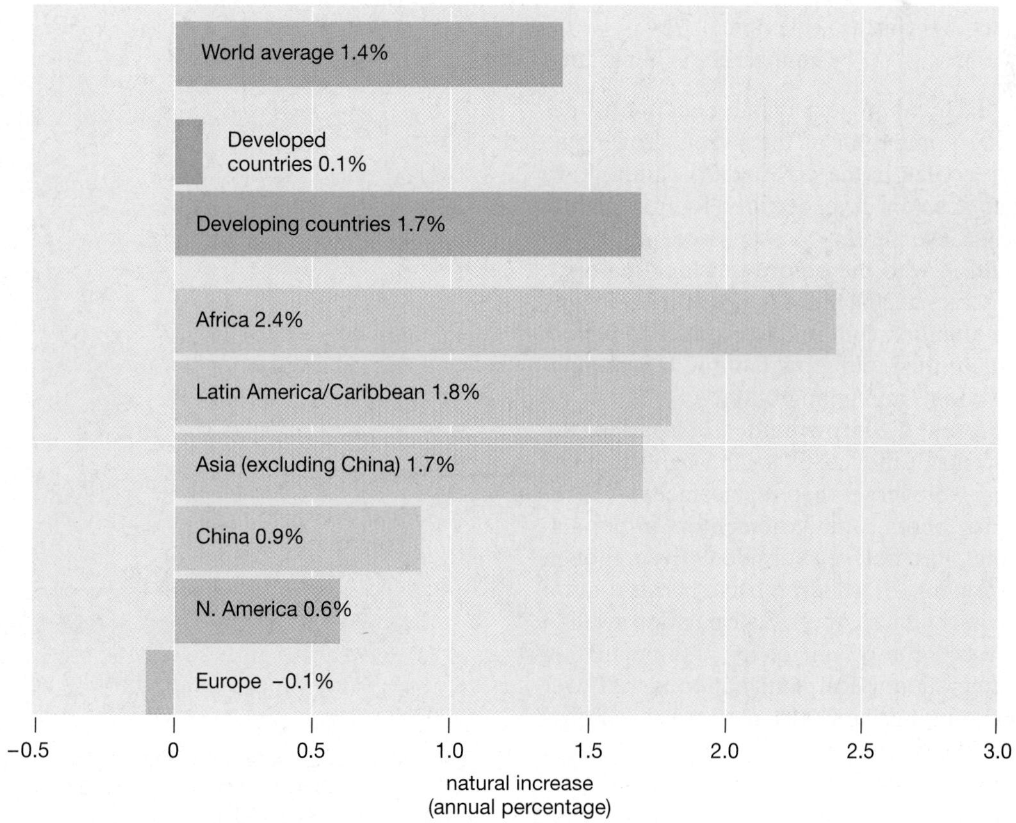

Figure 38-13 Population growth by world regions
Growth rates shown are due to natural increase (births – deaths) expressed as the percentage increase per year for various regions of the world. The bar that represents Europe includes the former Soviet Union. These figures do not include immigration or emigration.

become smaller in 2025 than in the 1990s, indicating that these populations are beginning to move closer to RLF. Nevertheless, as the huge numbers of young people advance up the age pyramid, population growth will continue. The percentage growth per year caused by natural increase for various parts of the world is shown in Figure 38-13.

Ironically, population growth in developing countries helps perpetuate the poverty, lack of access to contraceptives, and lack of education that tend to sustain high birth rates. The relationship of income and education to birth rate has been documented in the United States. Here, on the average, women who do not complete high school have twice as many children as do those with more than 4 years of college.

The U.S. Population Is Growing Rapidly

As shown in Figure 38-14, the U.S. is the fastest growing industrialized country in the world. With a natural increase of 0.6%, the U.S. population grows at six times the average rate of developed countries. Adding immigration to this picture, U.S. annual growth rate approaches 1%. During 1999, for example, the U.S. population increased by almost 2.6 million; the U.S. adds nearly 300 people to its population every hour. The population in 2000 was more than 275 million. Recall the following equation:

$$\text{population change} = (\text{births} - \text{deaths}) + (\text{immigrants} - \text{emigrants})$$

Let's examine each component of the equation to determine why the population of the U.S. is growing so rapidly. If each woman in the U.S. had 2.1 children, we would have replacement-level fertility. RLF is slightly higher than 2 because parents must replace both themselves and children who die before reaching maturity. The U.S. fertility rate in 2000 was 2.1, just at RLF, so why does the U.S. population continue to grow? Two factors are contributing to the rapid growth of the U.S. population: the "baby boom" and immigration.

Part of the current U.S. growth rate is a legacy of the recent past. Parents of the late 1940s through the 1960s had families that were larger than replacement level, resulting in a "baby boom" and a momentum in population growth that has not yet subsided. Even though women are averaging 2.1 children each, because more women are having children, the U.S. population swells.

A second crucial component of the U.S. population equation is immigration. Legal immigration is well over 800,000 per year, causing about 30% of the total U.S. population growth. Illegal immigration adds an estimated 275,000 people each year. The U.S. Census Bureau projects that by the year 2050, there will be 404 million people in the U.S. (about 130 million more than today) and the population will still be growing. Even with RLF, immigration will guarantee continued U.S. population growth.

The rapid growth of the U.S. population has major environmental implications both for us and for the world. The average person in the U.S. uses five times as much energy as the global average, so with less than 5% of the world's population, we account for 25% of world energy use. Because people in developing countries use considerably less energy than the global average, the 2.6 million people added to the U.S. each year contribute more to climate change through the release of CO_2 and other greenhouse gases than do the 32 million people added annually by China and India combined.

When and how will human numbers ever stabilize? How many people can Earth support? There are no certain answers to these questions, but in "Earth Watch: Have We Exceeded Earth's Carrying Capacity?" we explore them in more detail.

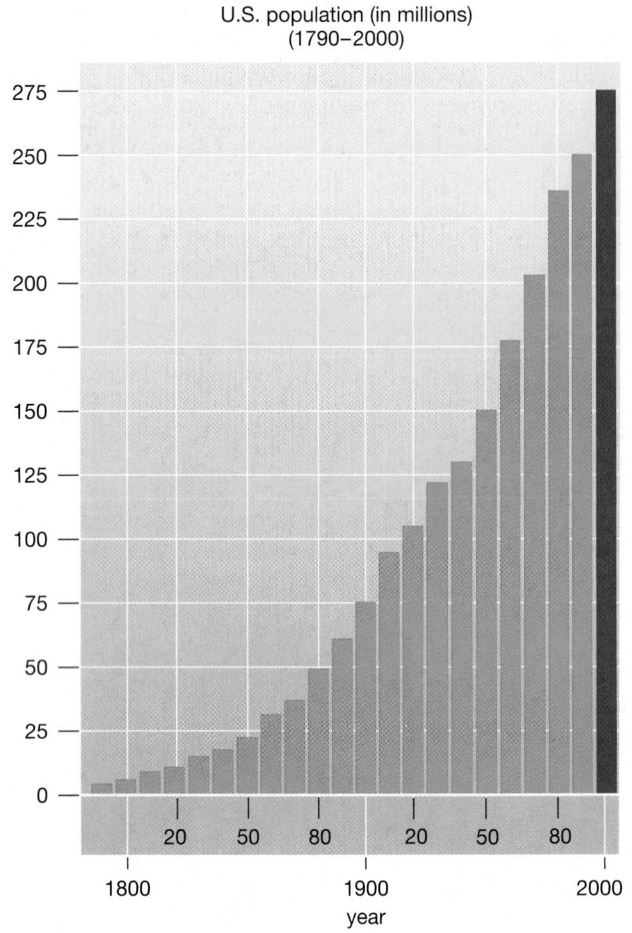

Figure 38-14 U.S. population growth
Since 1790, U.S. population growth has shown the J-shaped curve characteristic of exponential growth.

Earth Watch
Have We Exceeded Earth's Carrying Capacity?

In Côte d'Ivoire, a small country on the west African coast, the government is waging a battle to protect some of its rapidly dwindling tropical rain forest from thousands of illegal hunters, farmers, and loggers. Officials burn the homes of the squatters, who immediately return and rebuild. One illegal resident is Sep Djekoule. "I have 10 children and we must eat," he explains. "The forest is where I can provide for my family, and everybody has that right." His words illustrate the conflict between population growth and environmental protection. A glance at the age structure of developing countries, where most of the world's population resides, shows a tremendous momentum for growth. World population in the year 2010 is predicted to be about 7 billion and growing. A modest United Nations projection is that the human population will reach 8.9 billion by the year 2150, and will still be increasing. How many people can Earth support?

No one knows. Estimates of Earth's carrying capacity for humans have ranged from 3 billion to 44 billion people. Each of these estimates is based on major assumptions regarding technological developments and lifestyles. Earlier we defined carrying capacity as the maximum population that could be indefinitely sustained. This sustainability requires that the ecosystem not be damaged in ways that lower its ability to provide necessary resources. By this definition, we have already exceeded Earth's carrying capacity for humans.

The upper limit of the planet's carrying capacity is determined by the ability of its plant life to harvest the energy from sunlight and produce high-energy molecules that other organisms can use as food. Stanford University biologist Peter Vitousek estimates that human activities have already reduced the productivity of Earth's forests and grasslands by 12%. Each year, overgrazing and deforestation decreases the productivity of land, especially in developing countries (Fig. E38-3). In a world where more than 800 million people are chronically undernourished (Fig. E38-3, inset), two-thirds of the world's agricultural land is suffering moderate to severe erosion. The quest for more agricultural land is leading to deforestation and attempts to farm land that is poorly suited for agriculture. These actions contribute to the destruction of more than 30 million acres of rain forest annually. As a result, some ecologists estimate that we are driving 50 to 150 undescribed species to extinction each day. Each year the U.S. loses nearly half a million acres of prime farmland to development for homes, shopping malls, and roads. Already, most countries must import the grain they need from countries such as the U.S. As our own growing population spreads onto our farmland, our ability to export grain to help sustain other nations will diminish, while their needs increase. Worldwide, the amount of cropland per person has diminished by half in the past 50 years.

Although we take fresh water for granted, in many developing countries, water supplies are badly polluted and underground water supplies, called *aquifers*, are being depleted and not replaced. Both India and China are rapidly depleting aquifers to supply the needs of their growing cities and to irrigate their cropland, a policy that may result in reduced grain harvests in the near future.

The demand for wood in developing countries is far outstripping production, and large areas are being deforested annually. Deforestation in turn causes erosion of precious topsoil, runoff of much-needed fresh water, and the spread of deserts. The total world fish harvest peaked in 1989 and has been declining since, despite increased investments in fishing equipment and fish farming. Almost 70% of commercial ocean fish populations have been fully exploited or overfished, and many fisheries, such as the New England cod harvest, have collapsed because of overfishing. These are clear indications that our present population, at its present level of technology, is "overgrazing" the world ecosystem.

In estimating how many people Earth can—or should—support, we must keep in mind that humans desire more out of life than a minimum caloric intake each day. Do we want to be able to eat meat, drive a car, live in a single-family dwelling, walk in a wilderness, and know that somewhere eagles, pandas, and elephants are living free? For people living at Earth's maximum carrying capacity, these will probably be unattainable luxuries. William Rees of the University of British Columbia has estimated that for all of Earth's 6 billion people to live as people in the U.S. do would require the resources of three additional Earth-sized planets.

Hope for the future lies in using the intelligence that has allowed us to overcome environmental resistance to see signs of overgrazing and to act before we have irrevocably damaged our biosphere. The human population *will* stop its exponential growth. Either we will voluntarily reduce our birth rate, or various forces of environmental resistance such as disease and starvation will dramatically increase human death rates; the choice is ours. Facing the problem of how to limit births is politically and emotionally difficult, but continued failure to do so will be disastrous. Our dignity, our intelligence, and our role as self-appointed stewards of life on Earth demand that we make the decision to halt population growth before we have permanently reduced Earth's ability to support all life, including our own.

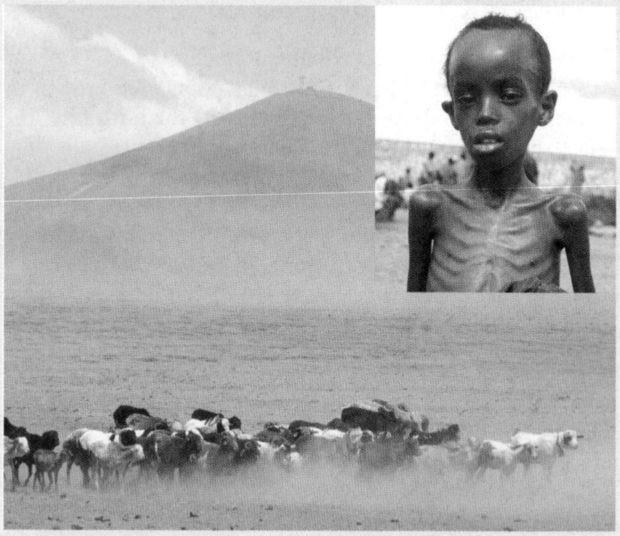

Figure E38-3 Deforestation can lead to the spread of deserts Human activities, including overgrazing livestock, deforestation, and poor agricultural practices, may sometimes convert once-productive land into barren desert. (Inset) An expanding human population, coupled with a loss of productive land, can lead to tragedy.

CASE STUDY REVISITED

Acorns, Mice, Moths, Deer, and Disease

From their carefully designed study, the ecologists were able to conclude that an abundant acorn crop led to an enormous increase in the mouse population. Likewise, in areas where the researchers removed mice, the artificially reduced mouse population allowed far more gypsy moth larvae to survive. This established links among acorns, mice, and moths. Surprisingly, the number of larval ticks increased manyfold in the acorn-enriched area. These were the offspring of parent ticks carried there by deer, which also feed on acorns and were attracted to the bounty. The excess of larval ticks led to a 40% increase in the number of ticks feasting on mice and picking up the Lyme bacterium in the process. Looking back, the researchers noted that in 1994 the oaks had produced a bumper crop of acorns. They predicted that, as a result, 1995 would see extra large populations of mice and tick larvae, and 1996 would see a large population of 1-year-old larval ticks infected with Lyme bacteria. These pinhead-sized "yearling" ticks feed in the summer and pose the greatest threat to people. The prediction was accurate; in 1996 Lyme disease cases in Connecticut doubled compared to 1995. This research makes an important contribution to understanding how populations of oaks, mice, deer, moths, and ticks interact within the forest ecosystem and alter the abundance of these two harmful parasites: gypsy moth caterpillars and the bacterium causing Lyme disease.

A public health official suggests that treating oak trees with hormones that suppress the production of acorns would be a good way to reduce the incidence of Lyme disease. Explain her logic and design a study to test this hypothesis. Now think about how a forester concerned with the health of the oak forest might respond to this suggestion. What would he or she say, and why?

Summary of Key Concepts

1) How Do Populations Grow?

Individuals join populations through birth or immigration and leave through death or emigration. The ultimate size of a stable population is the result of interactions among biotic potential, the maximum possible growth rate, and environmental resistance, which limits population growth.

All organisms have the biotic potential to more than replace themselves over their lifetime, resulting in population growth. Populations tend to grow exponentially, with increasing numbers of individuals added during each successive time period. Populations cannot indefinitely grow exponentially; they either stabilize or undergo periodic boom-and-bust cycles as a result of environmental resistance.

2) How Is Population Growth Regulated?

Environmental resistance restrains population growth by increasing the death rate or decreasing the birth rate. The maximum size at which a population may be sustained indefinitely by an ecosystem is termed the carrying capacity, determined by limited resources such as space, nutrients, and light. Environmental resistance generally maintains populations at or below the carrying capacity.

Population growth is restrained by density-independent forms of environmental resistance (such as weather) and density-dependent forms of resistance (including predation, parasitism, and competition).

3) How Are Populations Distributed in Space and Time?

Populations can be classified into three major types of distribution: clumped, uniform, and random. Clumped distribution may occur for social reasons or around limited resources. Uniform distribution is normally the result of territorial spacing. Random distribution is rare, occurring only when individuals do not interact socially and when resources are abundant and evenly distributed.

Populations show specific survivorship curves that describe the likelihood of survival at any given age. Late-loss (convex) curves are characteristic of long-lived species with few offspring, which receive parental care. Species with constant-loss curves have an equal chance of dying at any age. Early-loss (concave) curves are typical of organisms that produce numerous offspring, most of which die.

4) How Is the Human Population Changing?

The human population has exhibited exponential growth for an unprecedented time by overcoming certain aspects of environmental resistance and increasing Earth's carrying capacity for humans. This has been accomplished by the use of tools, agriculture, industry, and medical advances. Age-structure diagrams depict numbers of males and females in various age groups that a population comprises. Expanding populations have pyramidal age structures, stable populations show rather straight-sided age structures, and shrinking populations are illustrated by age structures that are constricted at the base.

Today, most of the world's people live in developing countries with rapidly expanding populations, where a variety of social and cultural conditions encourage large families. The United States is the fastest growing of the developed countries, owing to high immigration rates and the baby boom of the 1940s through the 1960s. Earth's carrying capacity for humans is unknown, but with a population of more than 6 billion, resources are already too limited for all to be supported at a high standard of living. A steady decline in productive land, fresh water, and in wood and fish harvests suggests that we are already damaging our world ecosystem and decreasing its ability to sustain us.

Key Terms

abiotic *p. 793*
age structure *p. 805*
biotic *p. 793*
biotic potential *p. 794*
boom-and-bust cycle *p. 796*
carrying capacity *p. 799*
clumped distribution *p. 802*
community *p. 793*
competition *p. 801*
contest competition *p. 802*
density-dependent *p. 799*
density-independent *p. 799*
doubling time *p. 795*
ecology *p. 793*
ecosystem *p. 793*
emigration *p. 794*
environmental resistance *p. 794*
exotic *p. 797*
exponential growth *p. 795*
growth rate *p. 794*
host *p. 800*
immigration *p. 794*
interspecific competition *p. 801*
intraspecific competition *p. 802*
J-curve *p. 795*
parasite *p. 800*
population *p. 794*
population cycle *p. 800*
predator *p. 800*
prey *p. 800*
random distribution *p. 803*
replacement-level fertility (RLF) *p. 806*
scramble competition *p. 802*
S-curve *p. 799*
survivorship curve *p. 803*
uniform distribution *p. 802*

Thinking Through the Concepts

Multiple Choice

1. *Which of the following factors is NOT an example of density-dependent environmental resistance?*
   a. weather
   b. competition
   c. predation
   d. parasitism
   e. lack of food
2. *For exponential growth to occur, it is necessary that*
   a. there is no mortality
   b. there are no density-independent limits
   c. the birth rate consistently exceed the death rate
   d. a species is very fast reproducing
   e. the species is an exotic invader in an ecosystem
3. *Which of the following currently contributes most to human population growth within the U.S.?*
   a. the consequences of the baby boom
   b. immigration
   c. a birth rate above RLF
   d. both a and b
   e. all of the above
4. *Which is the most common type of spatial distribution?*
   a. logistic
   b. uniform
   c. random
   d. exponential
   e. clumped
5. *Which continent has the highest rate of natural increase in the human population?*
   a. North America
   b. Africa
   c. Asia
   d. Australia
   e. South America
6. *If a population exceeds its carrying capacity,*
   a. it must immediately crash
   b. it can remain stable at this level indefinitely
   c. it will continue to increase for the indefinite future
   d. it must decline sooner or later
   e. the food supply will increase to support it

? Review Questions

1. Define *biotic potential* and *environmental resistance*.
2. Draw the growth curve of a population before it encounters significant environmental resistance. What is the name of this type of growth, and what is its distinguishing characteristic?
3. Distinguish between density-independent and density-dependent forms of environmental resistance.
4. Describe (or draw a graph illustrating) what is likely to happen to a population that far exceeds the carrying capacity of its ecosystem. Explain your answer.
5. List three density-dependent forms of environmental resistance, and explain why each is density-dependent.
6. Distinguish between populations showing concave and convex survivorship curves. Which is characteristic of people living in the U.S., and why?
7. Given that the U.S. birth rate is currently at replacement-level fertility, why is our population growing?
8. Discuss some reasons why making the transition from a growing to a stable population can be economically difficult.

Applying the Concepts

1. Explain natural selection in terms of biotic potential and environmental resistance.
2. The U.S. has a long history of accepting large numbers of immigrants. Discuss the implications of immigration for population stabilization.
3. What factors encourage rapid population growth in developing countries? What will it take to change this growth?
4. Contrast age structure in rapidly growing versus stable human populations. Why is there a momentum in population growth built into a population that is above RLF?
5. Why is the concept of carrying capacity difficult to apply to human populations? In reference to human population, should the concept be modified to include quality of life?

For More Information

Bender, W., and Smith, M. "Population, Food, and Nutrition." *Population Bulletin*, February 1997. A comprehensive and up-to-date summary of issues regarding food supply and the growing human population.

Bongaarts, J. "Can the Growing Human Population Feed Itself?" *Scientific American*, March 1994. A balanced account of the arguments between optimists and pessimists concerning the future world food supply.

Kates, R. W. "Sustaining Life on the Earth." *Scientific American*, October 1994. Provides a prescription for an environmentally sustainable future.

Korpimaki, E., and Krebs, C. J. "Predation and Population Cycles of Small Mammals." *BioScience*, November 1996. A review of recent studies designed to evaluate cycles of predators and their prey.

Kunzig, R. "Twilight of the Cod." *Discover*, April 1995. Documents the collapse of the cod fishery from Cape Cod to Newfoundland.

Myers, N. "Biotic Holocaust." *International Wildlife*, March–April 1999. Human activities are causing extinction of species unprecedented since the disappearance of the dinosaurs. How can we reverse this trend?

Potts, M. "The Unmet Need for Family Planning." *Scientific American*, January 2000. Reducing population growth and increasing the quality of life requires increased access to contraceptives in developing countries.

Raloff, J. "The Human Numbers Crunch." *Science News*, June 22, 1996. Reviews the challenges for the next 50 years as humans grow in number.

Safina, C. "The World's Imperiled Fish." *Scientific American*, November 1995. The collapse of many fisheries clearly illustrates that wild fish populations cannot sustain themselves against modern fishing techniques.

Sanz, C. "Summer of Danger: Lyme Disease." *Discover*, May 1999. Describes the interrelated populations that contribute to the spread of Lyme disease, and also discusses the new vaccine for Lyme disease.

Answers to Multiple-Choice Questions

1.a 2.c 3.d 4.e 5.b 6.d

MEDIATUTOR
Population Growth and Regulation

CD Activities

Activity 38.1: Population Growth and Regulation

Estimated time: 10 minutes

In this activity you will investigate wolf populations in Yellowstone National Forest to explore the concept of biotic potential, the relationship of predator–prey interactions as they apply to population biology, and the graphical portrayal of population growth over time.

Activity 38.2: Population Growth and Regulation in Human Populations

Estimated time: 5 minutes

In this activity you will run a series of tests to determine how the delayed onset of childbearing by humans affects human population growth.

Start the MediaTutor Student CD-ROM and enter the activity number in the Quick Search box to be taken directly to that activity.

Web Investigations

Case Study: Acorns, Mice, Moths, Deer, and Disease

Estimated time: 15 minutes

It seems like a very straightforward experiment. Add acorns to forest tracts and observe the effects on gypsy moth and mouse populations. But behind this apparently simple protocol lies extensive biological knowledge. This exercise will explore the facts and theories used to design the Lyme disease experiment.

Go to http://www.prenhall.com/audesirk6, the Audesirk Companion Web site. Select Chapter 38 and the Web Investigation to begin.

Workers blast jets of hot water at zebra mussels coating the interior of a Michigan water-treatment plant. (Inset) Zebra mussels blanket a shopping cart that had fallen into a river.

39 Community Interactions

AT A GLANCE

Case Study: Invasion of the Zebra Mussels

1) **Why Are Community Interactions Important?**

2) **What Are the Effects of Competition Among Species?**

The Ecological Niche Defines the Place and Role of Each Species in Its Ecosystem

Adaptations Reduce the Overlap of Ecological Niches Among Coexisting Species

Competition Helps Control Population Size and Distribution

3) **What Are the Results of Interactions Between Predators and Their Prey?**

Predator–Prey Interactions Shape Evolutionary Adaptations

4) **What Is Symbiosis?**

Parasitism Harms, But Does Not Immediately Kill, the Host

In Mutualistic Interactions, Both Species Benefit

5) **How Do Keystone Species Influence Community Structure?**

6) **Succession: How Does a Community Change over Time?**

There Are Two Major Forms of Succession: Primary and Secondary

Succession Also Occurs in Ponds and Lakes

Succession Culminates in the Climax Community

Some ecosystems Are Maintained in a Subclimax State

Case Study Revisited: Invasion of the Zebra Mussels

CASE STUDY

Invasion of the Zebra Mussels

In 1989 residents of Monroe, Michigan, a town on the shores of Lake Erie, suddenly found themselves without water. Their schools, industries, and businesses were closed for two days while workers labored to fix the problem: Zebra mussels had clogged their water-treatment plant. Their problem was not unique; at one treatment plant on Lake Erie, zebra mussel populations reached 600,000 per square yard (see the opening photo). Where did they come from? Sometime in 1985 or 1986, a trading vessel bringing cargo from Europe discharged fresh water into Lake St. Clair, located between Lake Huron and Lake Erie at the border between Ontario and Michigan. The water, used for ballast during the ship's transatlantic voyage, contained stowaways—millions of larvae of the zebra mussel. Although these mollusks are native to the Caspian and Black Seas (large lakes in southeastern Europe), they found ideal conditions here. Now they inhabit all the Great Lakes and the entire Mississippi and Ohio River drainage systems and are spreading throughout eastern waterways. Each year, tens of millions of dollars are expended to keep water pipes clear of the pests. Microscopic mussel larvae can be carried in currents for hundreds of miles. Using sticky threads, the inch-sized adults attach themselves to nearly any underwater surface, including piers, pipes, machinery, underwater debris, boat hulls, and even sand and silt. Since they can survive out of water for days, mussels clinging to small boats may be portaged to other lakes and rivers, where they quickly move in. Since an adult female can produce 50,000 eggs each year, the mussel menace has proven unstoppable. The mussels cover and suffocate other forms of shellfish, threatening many rare varieties with extinction. Why have these invaders been so enormously successful? Will anything control them? Think about the zebra mussel as you read about the community interactions that characterize healthy ecosystems. ■

1 Why Are Community Interactions Important?

An ecological **community** consists of all the interacting populations within an ecosystem; in other words, a community is the *biotic*, or living, component of an ecosystem. In the previous chapter, you learned that community interactions such as predation, parasitism, and competition help limit the size of populations. The interacting web of life that forms a community tends to maintain a balance between resources and the numbers of individuals consuming them. When populations interact with one another, influencing each other's ability to survive and reproduce, they serve as agents of natural selection. For example, in killing prey that are easiest to catch, predators leave behind those individuals with better defenses against predation. These individuals leave the most offspring, and over time their inherited characteristics increase within the prey population. Thus, as community interactions limit population size, they simultaneously shape the bodies and behaviors of the interacting populations. This process by which two interacting species act as agents of natural selection on one another over evolutionary time is called **coevolution**.

You have probably heard the expression "the balance of nature." This balance is the result of community interactions finely tuned to one another over evolutionary time. The sometimes fragile balance can be overturned when organisms are introduced into an ecosystem where they did not evolve, as described in our opening Case Study and in "Earth Watch: Exotic Invaders."

The most important community interactions are competition, predation, parasitism, and mutualism. If we assume that each of these interactions involves two species, the types of interactions can be characterized according to whether each of the two species is harmed or benefits, as shown in Table 39-1. Over evolutionary time, these interactions have shaped the bodies and behaviors of organisms, as described in the following sections.

2 What Are the Effects of Competition Among Species?

Community interactions may be harmful or beneficial to the participating organisms (see Table 39-1). During competition among members of different species, called **interspecific competition**, two or more species attempt to use the same limited resources, particularly food and/or space. In interspecific competition, each species involved is harmed, because access to resources is reduced. The intensity of interspecific competition depends on how similar the requirements of the species are. In other words, the degree of competition is proportional to the amount of overlap in the *ecological niches* of the competing species.

Table 39-1 Interactions Among Organisms

Type of Interaction	Effect on Organism A	Effect on Organism B
Competition between A and B	Harms	Harms
Predation by A on B	Benefits	Harms
Symbiosis		
Parasitism by A on B	Benefits	Harms
Commensalism of A with B	Benefits	No effect
Mutualism between A and B	Benefits	Benefits

The Ecological Niche Defines the Place and Role of Each Species in Its Ecosystem

Although the word *niche* may call to mind a small cubbyhole, in ecology it means much more. Each species occupies a unique **ecological niche** that encompasses all aspects of its way of life. An ecological niche includes the organism's physical home or habitat. The habitat of a white-tailed deer, for example, is the eastern deciduous forest. In addition, the niche includes all the physical environmental factors necessary for survival, such as the range of temperatures under which the organism can survive, the amount of moisture it requires, the pH of the water or soil it may inhabit, the type of soil nutrients required, and the degree of shade it can tolerate. Although different types of organisms share many aspects of their niche with others, no two species ever occupy exactly the same ecological niche.

The ecological niche extends well beyond habitat. It also specifies how the organism gets its supply of energy and materials—what might be called the organism's role or "occupation" within its ecosystem. An organism's predators, prey, and competitors, as well as its behaviors and interactions with other organisms, are considered elements of its niche.

Adaptations Reduce the Overlap of Ecological Niches Among Coexisting Species

Just as no two organisms can occupy exactly the same physical space at the same time, no two species can inhabit exactly the same ecological niche simultaneously and continuously. This important concept, often called the **competitive exclusion principle**, was formulated in 1934 by Russian microbiologist G. F. Gause. If two species with the same niche were placed together and forced to compete for limited resources, inevitably one would outcompete the other, and the less well adapted of the two would die out. Gause used two species of the protist *Paramecium*, *P. aurelia* and *P. caudatum*, to demonstrate this principle. In laboratory flasks, both species thrived on bacteria and fed in the same parts of the flask. Grown separately, each population flourished

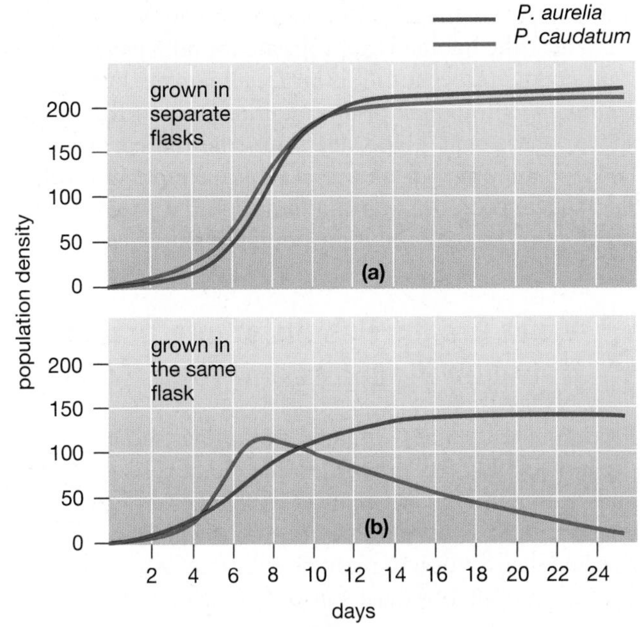

Figure 39-1 Competitive exclusion
(a) Raised separately with a constant food supply, both *Paramecium aurelia* and *P. caudatum* show the S-curve typical of a population that initially grows rapidly, then stabilizes. ***(b)*** Raised together and forced to occupy the same niche, *P. aurelia* consistently outcompetes *P. caudatum* and causes that population to die off. (Modified from G. F. Gause, *The Struggle for Existence*. Baltimore: Williams & Wilkins, 1934.)

(Fig. 39-1a). But when Gause placed the two species together in a flask, one always eliminated, or "competitively excluded," the other (Fig. 39-1b). Gause then repeated the experiment, replacing *P. caudatum* with a different species, *P. bursaria*, which tended to feed in a different part of the flask. In this case, the two species of *Paramecium* were able to coexist indefinitely, because they occupied slightly different niches.

Ecologist R. MacArthur tested Gause's laboratory findings under natural conditions by investigating five species of North American warbler. All these birds hunt for insects and nest in the same type of spruce tree. Although the niches of these birds appear to overlap considerably, MacArthur found that each species concentrates its search in specific areas of the tree, employs different hunting tactics, and nests at slightly different times. By partitioning the resource, the warblers minimize the overlap of their niches and reduce competition among the different species (Fig. 39-2).

As MacArthur discovered, when two species with similar requirements coexist, they typically occupy a smaller niche than either would if it were by itself. This phenomenon, called **resource partitioning**, is an evolutionary adaptation that reduces the harmful effects of interspecific competition. Resource partitioning is the outcome of the coevolution of species with extensive but not total niche overlap. Because natural selection favors individuals with

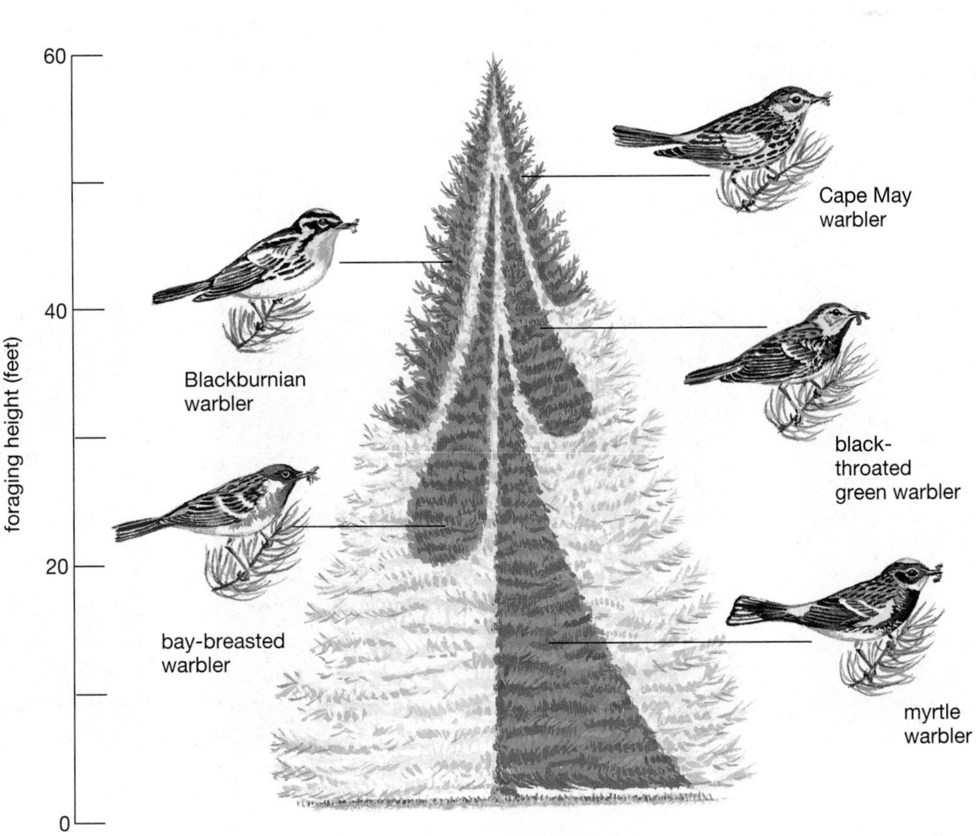

Figure 39-2 Resource partitioning
Each of these five insect-eating species of North American warblers searches for food in different regions of spruce trees.

fewer competitors, over evolutionary time the competing species develop physical and behavioral adaptations that minimize their competitive interactions. A dramatic example of resource partitioning was discovered by Charles Darwin among finches of the Galapagos Islands. The finches had evolved different bill sizes and shapes and different feeding behaviors that reduced the competition among them (described in Chapter 16).

Competition Helps Control Population Size and Distribution

Individuals of the same species have essentially identical requirements for resources and occupy the same ecological niche. For this reason, *intra*specific competition—that is, competition among individuals of the same species—is a major factor in controlling population size. Although natural selection leads to a reduction of niche overlap among individuals of different species, those with similar niches still compete directly for limited resources. This interspecific competition may restrict the size and distribution of the competing populations.

A classic study of the effects of interspecific competition was performed by ecologist J. Connell, using barnacles (barnacles are crustaceans that attach permanently to rocks and other surfaces; gray barnacles coat the rocks in Figure 39-14a). Barnacles of the genus *Chthamalus* share the rocky shores of Scotland with another genus, *Balanus*, and their niches overlap considerably. Both genera live in the **intertidal zone**, an area of the shore that is alternately covered and exposed by the tides. Connell found that *Chthamalus* dominates the upper shore and *Balanus* dominates the lower. When he scraped off *Balanus*, the *Chthamalus* population increased, spreading downward into the area its competitor had once inhabited. Where the habitat is appropriate for both genera, *Balanus* conquers, because it is larger and grows faster. But *Chthamalus* tolerates drier conditions, so on the upper shore, where only high tides submerge the barnacles, it has a competitive advantage. As this example illustrates, interspecific competition can limit both the size and the distribution of the competing populations.

3 What Are the Results of Interactions Between Predators and Their Prey?

Predators eat other organisms, killing them in the process. Ecologists sometimes include herbivorous animals (animals that eat plants) in this general category because herbivores can have a major influence on the size and distribution of plant populations. We will define predation in its broadest sense—to include the grazing cow, the zebra mussel eating microscopic algae, the goby fish eating a zebra mussel, the bat homing in on a moth, and the hawk with a smaller bird in its talons (Fig. 39-3). Most predators are either larger than their prey or hunt collectively, as wolves do when bringing down an elk. Predators are generally less abundant than their prey; you will learn why in the next chapter.

Predator–Prey Interactions Shape Evolutionary Adaptations

To survive, predators must feed and prey must avoid becoming food. Therefore, predator and prey populations exert intense environmental pressure on one another, resulting in coevolution. As prey become more

(a) (b) (c)

Figure 39-3 Forms of predation
(a) A cow grazes on prairie grass. The tough stems of grass have evolved under predation pressure by herbivores. ***(b)*** A long-eared bat uses a sophisticated echolocation system to hunt this moth, which has evolved special sound detectors and behaviors to avoid it. **(c)** The red-tailed hawk prepares to feast on a smaller bird.

difficult to catch, predators must become more adept at hunting. Coevolution has endowed the cheetah with speed and camouflage spots (see Fig. 39-7a) and its zebra prey with speed and camouflage stripes. It has produced the keen eyesight of the hawk, and the poisons and bright colors of the poison arrow frog and the coral snake (see Figs. 39-8 and 39-9a). In the following sections, we examine some of the evolutionary results of predator–prey interactions.

Some Predators and Prey Have Evolved Counteracting Behaviors

Bat and moth adaptations provide an excellent example of how both physical structures and behaviors are molded by coevolution. Most bats are nighttime hunters that navigate and locate prey by echolocation. They emit extremely high-frequency and high-intensity pulses of sound and, by analyzing the returning echoes, create an "image" of their surroundings. Under selection pressure from this specialized prey-location system, certain moths (a favored prey of bats) have evolved simple ears that are particularly sensitive to the frequencies used by echolocating bats. When they hear a bat, these moths take evasive action, flying erratically or dropping to the ground. The bats may counter this defense by switching the frequency of their sound pulses away from the moth's sensitivity range. Some moths have evolved a way to interfere with the bats' echolocation by producing their own high-frequency clicks. In response, when hunting a clicking moth, a bat may turn off its own sound pulses temporarily and zero in on the moth by following the moth's clicks.

Camouflage Conceals Both Predators and Their Prey

An old saying of detective novels is that the best hiding place may be right out in plain sight! Both predators and prey have evolved colors, patterns, and shapes that resemble their surroundings. Such disguises, called **camouflage**, render animals inconspicuous even when they are in plain sight (Fig. 39-4).

Some animals closely resemble specific objects such as leaves, twigs, seaweed, thorns, or even bird droppings (Fig. 39-5a–c). Camouflaged animals tend to remain motionless rather than to flee their predators; a fleeing "bird dropping" would be quite conspicuous! Whereas many camouflaged animals resemble plants, a few types of plants have evolved to resemble rocks, which are ignored by their herbivorous predators (Fig. 39-6).

Predators who ambush are also aided by camouflage. For example, a spotted cheetah becomes inconspicuous in the grass as it watches for grazing mammals. The frogfish closely resembles the algae-covered rocks and algae on which it sits motionless, dangling a small lure from its upper lip (Fig. 39-7). Small fish notice only the lure and are engulfed as they approach it.

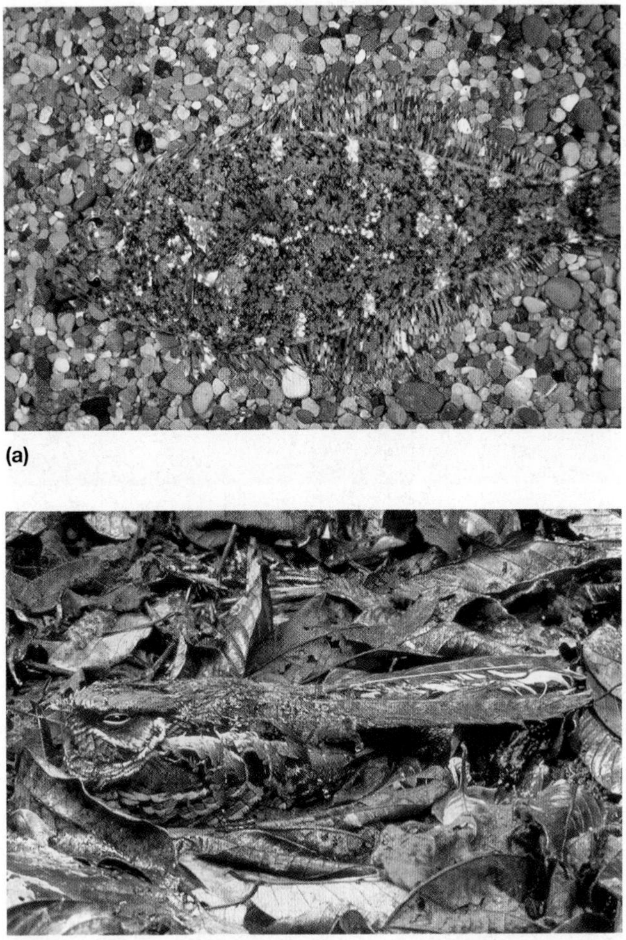

(a)

(b)

Figure 39-4 Camouflage by blending in
(a) The sand dab is a flattened, bottom-dwelling ocean fish with a mottled color that closely resembles the sand on which it rests. ***(b)*** This nightjar on its nest in Belize is barely visible among the surrounding leaf litter.

Bright Colors Often Warn of Danger

Some animals have evolved very differently, exhibiting bright **warning coloration** (Fig. 39-8, p. 821; see also Figs. 39-9, 39-12). These animals are usually distasteful, and many are poisonous, such as the yellow jacket with its bright yellow and black stripes. Because poisoning your predator is small consolation if you have already been eaten, the bright colors declare, "Eat me at your own risk!" A single unpleasant experience is enough to teach predators to avoid these conspicuous prey.

Some Organisms Gain Protection Through Mimicry

Mimicry refers to a situation in which a species evolves to resemble something else, typically another type of organism. For example, once warning coloration evolved, there arose a selective advantage for harmless animals to resemble poisonous ones. The deadly coral snake has brilliant warning coloration, and the harmless mountain king snake avoids predation by resembling the coral

(a)

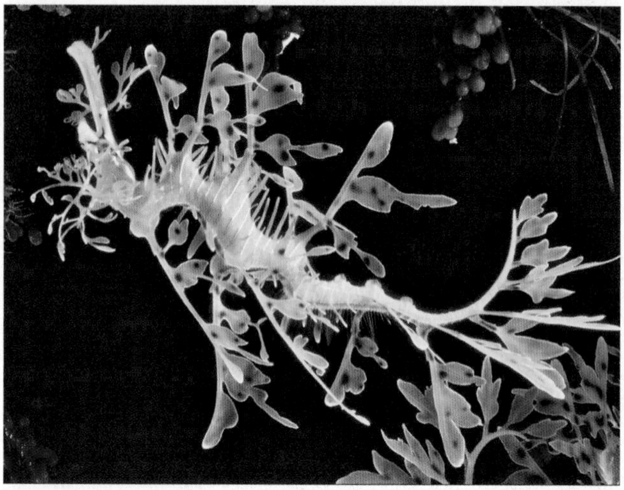

(b)

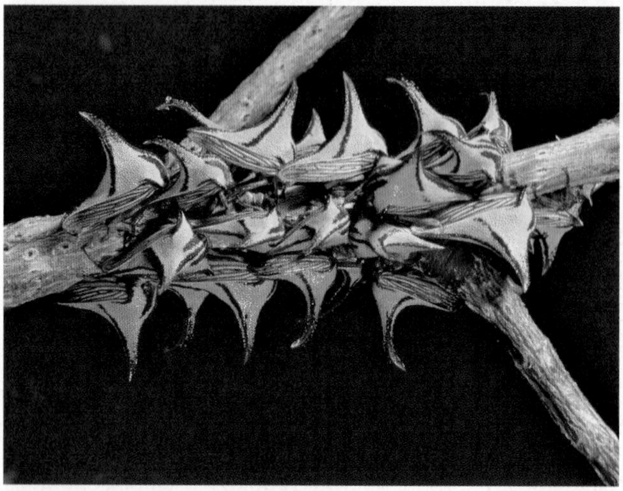

(c)

Figure 39-5 Camouflage by resembling specific objects
Resembling uninteresting parts of the environment allows some animals to avoid predation. ***(a)*** A moth whose color and shape resemble a bird dropping sits motionless on a leaf. ***(b)*** The leafy sea dragon (an Australian "seahorse" fish) has evolved extensions of its body that mimic the algae in which it normally hides. ***(c)*** Florida treehopper insects avoid detection by resembling thorns on a branch.

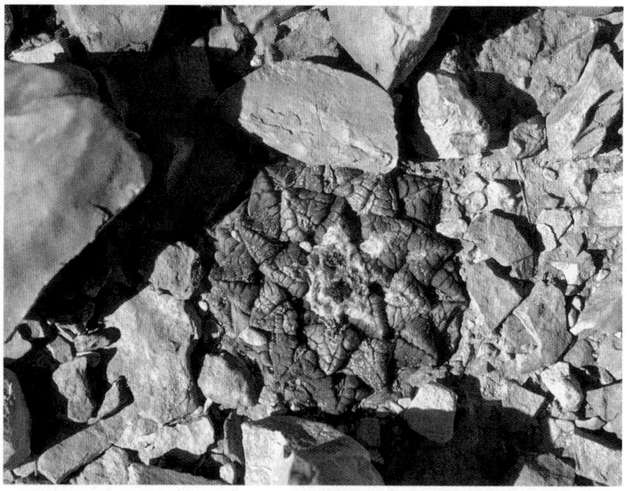

Figure 39-6 A plant that mimics a rock
This cactus of the American Southwest is appropriately called the "living rock cactus."

(a)

(b)

Figure 39-7 Camouflage assists predators
(a) As it waits for prey, a cheetah blends into the background of the grass. ***(b)*** Combining camouflage and aggressive mimicry, a frogfish waits in ambush, its camouflaged body matching the sponge-encrusted rock on which it rests. Above its mouth dangles a lure that closely resembles a small fish. The lure attracts small predators, who will find themselves to be prey.

Figure 39-8 Warning coloration
The South American poison arrow frog, with its poisonous skin, advertises its unpleasant taste with bright and contrasting color patterns.

snake (Fig. 39-9a,b). By resembling each other, two distasteful species may each benefit from predators' painful experience with the other. For example, predators rapidly learn to avoid the conspicuous stripes on bees, hornets, and yellow jackets. The monarch butterfly is poisonous and distasteful, and viceroy butterflies (which researchers have recently discovered are equally distasteful) have wing patterns strikingly similar to those of monarchs (Fig. 39-9c,d). A common color pattern results in faster learning by predators—and less predation on all similarly colored species.

Some predators have evolved **aggressive mimicry**, a "wolf-in-sheep's-clothing" approach, in which they entice their prey to come close by resembling a harmless animal or part of the environment. For example, although the frogfish resembles the algae-covered rocks where it lurks, it dangles a wriggling lure that resembles a small fish just above its mouth (see Fig. 39-7b). A curious fish attracted to the lure is quickly swallowed.

Figure 39-9 Warning mimicry
The warning coloration of ***(a)*** the poisonous coral snake is mimicked by ***(b)*** the harmless mountain king snake. Mutual mimicry of warning coloration by the distasteful monarch butterfly ***(c)*** and the equally distasteful viceroy ***(d)*** protects both species. After attempting to eat either of these butterfly species, a bird will avoid them both.

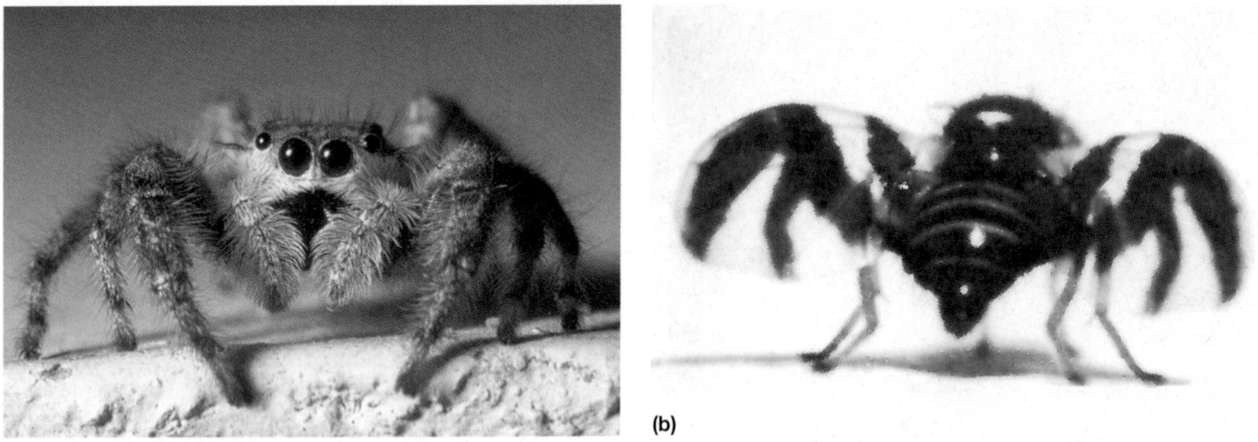

(a) (b)

Figure 39-10 Visual and behavioral mimicry
(a) In response to the approach of a jumping spider, *(b)* the snowberry fly spreads its wings, revealing a pattern that resembles spider legs. The fly enhances the effect by performing a jerky, side-to-side dance that resembles the leg-waving display of a jumping spider defending its territory.

A sophisticated variation on the theme of prey who mimic predators is seen in snowberry flies, which are hunted by territorial jumping spiders. When a fly spots an approaching spider, it spreads its wings, moving them back and forth in a jerky dance. Seeing this display, chances are good that the spider will flee from the harmless fly. Why? Researchers have observed that the markings on the fly's wings closely resemble the legs of another jumping spider. The jerky movements of the fly mimic those of a jumping spider when it drives another spider from its territory (Fig. 39-10). Natural selection has finely tuned both the behavior and the appearance of the fly to avoid predation by jumping spiders.

Certain prey species use another form of mimicry, **startle coloration**. Several insects, and even some vertebrates, such as the false-eyed frog, have evolved patterns of color that closely resemble the eyes of a much larger, and possibly dangerous, animal (Fig. 39-11). If a predator gets close, the prey suddenly flashes its eye-spots, startling the predator and allowing the prey to escape.

Some Predators and Prey Engage in Chemical Warfare

Both predators and prey have evolved a variety of toxic chemicals for attack and defense. The venom of spiders and poisonous snakes, such as the coral snake (see Fig. 39-9), serves both to paralyze prey and to deter its predators. Many plants produce defensive toxins. For example, lupine plants, whose flowers grace both gardens and mountain meadows, produce chemicals called *alkaloids*, which deter attack by the blue butterfly, whose larvae feed on the lupine's buds. In fact, different individuals of the same species of lupine produce different forms of alkaloids, thus making it more difficult for the moths to evolve resistance to it.

(a) (b) (c)

Figure 39-11 Startle coloration
(a) When threatened, the false-eyed frog raises its rump, which resembles the eyes of a larger predator. This startles the real predator, giving the frog a chance to flee. *(b)* The peacock moth from Trinidad is well camouflaged, but should a predator approach, it opens its wings to reveal spots resembling large eyes. *(c)* Predators of this caterpillar larva of the swallowtail butterfly are deterred by its resemblance to a snake. Note that the caterpillar's head is the "snake's nose."

(a)

(b)

Figure 39-12 Chemical warfare
(a) The bombardier beetle sprays a hot toxic brew in response to a leg pinch. ***(b)*** A monarch caterpillar feeds on milkweed that contains a powerful toxin. Why do you think the caterpillar is colored with bright stripes instead of being green?

Certain mollusks, including squid, octopus, and some sea slugs, emit clouds of ink when attacked. These colorful chemical "smoke screens" confuse predators and mask the prey's escape. A dramatic example of chemical defense is seen in the bombardier beetle. In response to the bite of an ant, the beetle releases secretions from special glands into an abdominal chamber. There, enzymes catalyze an explosive chemical reaction that shoots a toxic, boiling-hot spray onto the attacker (Fig. 39-12a).

Plants and Herbivores Have Coevolutionary Adaptations Herbivores don't fit neatly into parasite or predator categories. A grazing horse or cow uproots and kills some grass, but most of the time acts more like a lawn mower, cropping but not killing the plants. Regardless of how we categorize them, herbivores exert strong selective pressure on plants to avoid being eaten. Plants have evolved a variety of chemical adaptations that deter their herbivorous "predators." Many, such as the milkweed, synthesize toxic and distasteful chemicals. As plants evolved toxic chemicals for defense, certain insects evolved increasingly efficient ways to detoxify or even use the chemicals. The result is that nearly every toxic plant is eaten by at least one type of insect. For example, monarch butterflies lay their eggs on milkweed; when their larvae hatch, they consume the toxic plant (Fig. 39-12b). The caterpillars not only tolerate the milkweed poison but also store it in their tissues as a defense against their own predators. The stored toxin is retained in the metamorphosed monarch butterfly (see Fig. 39-9c).

Grasses have evolved tough silicon (glassy) substances in their blades, discouraging all herbivorous predators except those with strong, grinding teeth and powerful jaws, such as the cow in Figure 39-3a. Thus, grazing animals have come under environmental pressure for longer, harder teeth. An example is the coevolution of horses and the grasses they eat. On an evolutionary time scale, grasses evolved tougher blades that reduce predation, and horses evolved longer teeth with thicker enamel coatings that resist wear.

4 What Is Symbiosis?

Symbiosis, which literally means "living together," is defined as a close interaction between organisms of different species for an extended time. Considered in its broadest sense, symbiosis includes parasitism, mutualism, and commensalism. Although one species always benefits in symbiotic relationships, the second species may be unaffected, harmed, or helped (see Table 39-1). **Commensalism** is a relationship in which one species benefits while the other is relatively unaffected. Barnacles that attach themselves to the skin of a whale, for example, get a free ride through nutrient-rich waters without harming the whale. In parasitism and mutualism, in which both organisms are affected, the participants act on each other as strong agents of natural selection. Here we will discuss these two forms of community interaction.

Parasitism Harms, But Does Not Immediately Kill, the Host

In **parasitism**, one organism benefits by feeding on another. **Parasites** live in or on their prey, which are called *hosts*, normally harming or weakening them but not immediately killing them. Although it is sometimes difficult to distinguish clearly between a predator and a parasite, parasites are generally much smaller and more numerous than their hosts. Familiar parasites include tapeworms, fleas, and numerous disease-causing protozoa, bacteria, and viruses. Many parasites, particularly worms and protozoa, have complex life cycles involving two or more hosts (see Chapter 22). There are few parasitic vertebrates; the lamprey eel, which attaches itself to a host fish and sucks its blood, is a rare example.

The variety of infectious bacteria and viruses and the precision of the immune system that counters their attacks are evidence of the powerful forces of coevolution between parasitic microorganisms and their hosts. Consider the malaria parasite, which has provided strong environmental pressure for the defective hemoglobin gene in humans that causes sickle-cell anemia. The parasite can't survive in the affected red blood cells. In certain areas of Africa where malaria is common, up to 20% of the human population carries the sickle-cell gene.

Another example is *Trypanosoma*, a parasitic protozoan that causes both human sleeping sickness and a disease in cattle called *nagana*. African antelope, which coevolved with this parasite, are relatively unaffected by it. Most infected cattle, a more recently introduced species, suffer but survive infection if they have been bred in an infested area for many generations. Newly imported cattle, however, generally die if not treated.

In Mutualistic Interactions, Both Species Benefit

When two species interact in a way that benefits both, the relationship is called **mutualism**. If you see raised, colored patches on rocks, they are probably lichens, a mutualistic association of an alga and a fungus (Fig. 39-13a). The fungus provides support and protection while deriving food from the photosynthetic alga, whose bright colors are light-trapping pigments. The mutualistic interactions between flowering plants and their pollinators are discussed in Chapter 24. Mutualistic associations occur in the digestive tracts of cows and termites, where protists and bacteria find food and shelter while helping their hosts extract nutrients, and in our own intestines, where bacteria synthesize certain vitamins. The nitrogen-fixing bacteria inhabiting special chambers on the roots of legume plants are another important example. These bacteria obtain food and shelter from the plant and in return trap nitrogen in a form the plant can utilize. Some mutualistic partners have coevolved to the extent that neither can survive alone. An example is the ant–acacia mutualism described in "Scientific Inquiry: Ants and Acacias—An Unlikely Match."

Mutualistic relationships involving vertebrates are rare and typically less intimate and extended. The clown fish takes shelter among the venomous tentacles of the anemone, which are harmless to it. The fish derives shelter and protection and, at least occasionally, it brings bits of food to its anemone (Fig. 39-13b).

5 How Do Keystone Species Influence Community Structure?

In some communities, a certain key species, called a **keystone species**, plays a major role in determining community structure—a role that is out of proportion

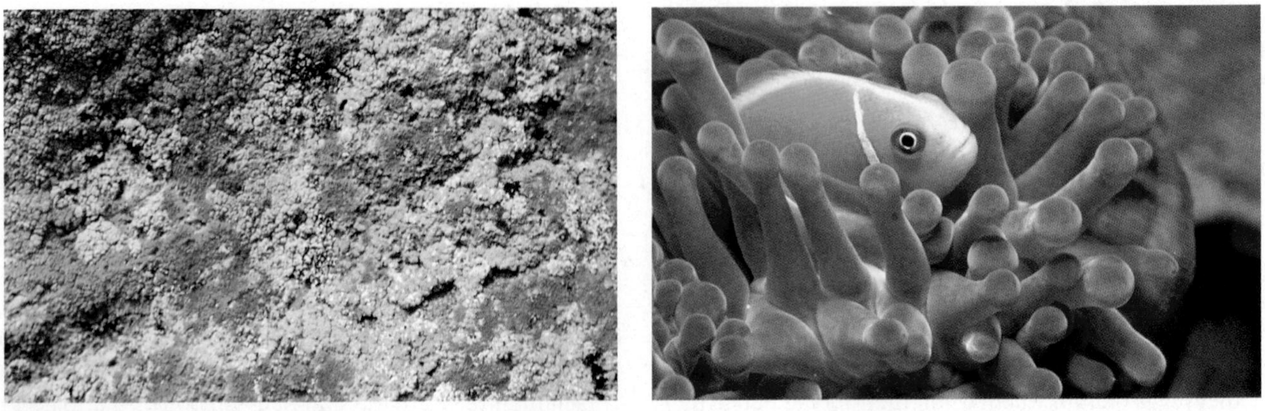

(a) (b)

Figure 39-13 Mutualism
(a) This brightly colored lichen growing on bare rock is a mutualistic relationship between an alga and a fungus. ***(b)*** The clown fish snuggles unharmed among the stinging tentacles of the anemone. The fish obtains protection and sometimes brings food to the anemone in this mutualistic relationship.

Scientific Inquiry

Ants and Acacias—An Unlikely Match

Daniel Janzen, a doctoral student at the University of Pennsylvania, was walking down a road in Veracruz, Mexico, when he saw a flying beetle alight on a thorny tree, only to be driven off by an ant. Looking more closely, he saw that the tree, a bull's-horn acacia, was covered with ants. A large ant colony of the genus *Pseudomyrmex* made its home inside the enlarged thorns of the plant, whose soft pulpy interiors are easily excavated to provide shelter (Fig. E39-1).

To determine how important the ants are to the tree, Janzen began stripping the thorns by hand until he found and removed the thorn that housed the ant queen, thus destroying the colony. Later, he turned to more efficient but dangerous methods, using an insecticide to eliminate all the ants on a large stand of acacias. The acacias were unharmed by the poison, Janzen became ill from it, and all the ants were killed. Within a year of the spraying, Janzen found nearly all the acacia trees dead, consumed by insects and other herbivores and shaded out by competing plants. The ground surrounding the trees, which the ants normally kept trimmed, was overgrown with vegetation. The trees were apparently dependent on their resident ants for survival.

Wondering if the ants could survive off the tree, Janzen painstakingly peeled the ant-inhabited thorns off 100 acacia trees, suffering multiple stings in the process. Janzen housed each ant colony in a jar provided with local non-acacia vegetation and insects for food. All the ant colonies starved. Carefully examining the acacia, he found swollen structures filled with sweet syrup at the base of the leaves and protein-rich capsules on the leaf tips (Fig. E39-1, inset). Together, these materials provide a balanced diet for the ants.

Janzen's experiments strongly suggest that these species of ant and acacia have an obligatory mutualistic relationship: Neither can survive without the other. Of course, further observations were required to support this hypothesis. The fact that the ants starved in Janzen's jars did not rule out that they might survive successfully elsewhere, but, in fact, this species of ant has never been found living independently. Similarly, the bull's-horn acacia has never been found without its resident ant colony. Thus, a chance observation followed by careful research led to the discovery of an important mutualistic association.

Figure E39-1 A mutualistic relationship
Yellow, protein-rich capsules are produced at the tips of certain acacia leaves. These capsules provide food for the resident ants. (Inset) A hole in the enlarged thorn of the bull's-horn acacia provides shelter for members of the ant colony. The ant entering the thorn is carrying a food capsule produced by the acacia. As the ant colony grows, it invades more thorns.

to its abundance in the community. Removal of the keystone species dramatically alters the community. For example, in 1969 R. Paine, an ecologist at the University of Washington, removed predatory starfish *Pisaster* (Fig. 39-14a) from sections of the rocky intertidal coast of Washington. Mussels (bivalve mollusks that are a favored prey of *Pisaster*) became so abundant that they outcompeted algae and other invertebrates that normally coexist in intertidal communities. Another marine invertebrate, the lobster, may be a keystone species off the east coast of Canada. Overfishing of the lobster allowed sea urchins, a prey of lobsters, to increase in numbers. The population explosion of sea urchins nearly eliminated certain types of algae on which the urchins prey, leaving large expanses of bare rock where a diverse community once existed. In the African savanna, the African elephant is a keystone predator. By grazing on small trees and bushes, elephants prevent the encroachment of forests and help maintain the grassland community (Fig. 39-14b).

It is difficult to identify keystone species, because ideally the species should be selectively removed and the community studied for several years before and after its removal. However, many ecological studies

(a)

(b)

Figure 39-14 Keystone species
(a) The starfish *Pisaster* is a keystone species along the rocky coast of the Pacific Northwest. ***(b)*** The elephant is a keystone species on the African savanna.

MEDIATUTOR 39.1 Effects of Keystone Species Extinction

performed since the concept was introduced by Paine provide evidence that keystone species are important in a wide variety of communities. Why is it important to study keystone species? As human activities infringe on natural ecosystems, it becomes increasingly urgent to understand community interactions and to preserve species that are crucial to maintaining the natural community.

6 Succession: How Does a Community Change over Time?

In a mature terrestrial ecosystem, the populations that make up the community interact with one another and with their nonliving environment in intricate ways. But this tangled web of life did not spring fully formed from bare rock or naked soil; rather, it emerged in stages over a long period, a process called succession. **Succession** is a structural change in a community and its nonliving environment over time. It is a kind of "community relay" in which assemblages of plants and animals replace one another in a sequence that is somewhat predictable.

Succession is most easily observed in freshwater and terrestrial ecosystems. Freshwater ponds and lakes tend to undergo a series of changes that, through time, transform them first into marshes and eventually to dry land. Shifting sand dunes are stabilized by creeping plants and may eventually support a forest. Volcanic eruptions may create new islands ripe for colonization. Or, as in the case of Mount St. Helens, they may disrupt previously existing communities, but leave behind a nutrient-rich environment that encourages rapid invasion of new life (Fig. 39-15b). Forest fires, while destroying an existing community, also release nutrients and create conditons that favor rapid succession (Fig. 39-15c).

The precise changes that occur during succession are as diverse as the environments in which succession occurs, but we can recognize certain general stages. In each case, succession is begun by a few hardy invaders called **pioneers**. If undisturbed, succession progresses to a diverse and relatively stable **climax community**, the end point of succession. Our discussion of succession will focus on plant communities.

There Are Two Major Forms of Succession: Primary and Secondary

Succession takes two major forms: primary and secondary. During **primary succession**, a community gradually colonizes bare rock, sand, or a clear glacial pool where there is no trace of a previous community. The generation of a community "from scratch" is a process that typically requires thousands or even tens of thousands of years. During **secondary succession**, a new community develops after an existing ecosystem is disturbed, as in the case of a forest fire or an abandoned farm field. Secondary succession happens much more rapidly than does primary succession, because the previous community has left its mark in the form of soil and seeds. Succession in an abandoned farm field in the southeastern U.S. can reach its climax after just two centuries.

As species gradually replace one another during succession, they interact in various ways. Lichens, pioneer species that grow on bare rock during primary succession, actually change the environment in ways that favor their competitors by liberating nutrients. During secondary succession, most pioneer plants are annuals (grasses, weeds, or wildflowers that live for a single growing season), and they generally produce large numbers of easily dispersed seeds that help them colonize open spaces. These pioneer species are sun-loving and grow rapidly, but they don't compete well against longer-lived (perennial) species that gradually grow larger and shade out the pioneers. In the following examples, we examine these processes in more detail.

Figure 39-15 Succession in progress Pairs of photographs illustrate primary and secondary succession. ***(a)*** Left: The Hawaiian volcano Mount Kilauea has erupted repeatedly since 1983, sending rivers of lava over the surrounding countryside. Right: A pioneer fern takes root on hardened lava. ***(b)*** Left: On May 18, 1980, the explosion of Mount St. Helens in Washington State devastated the pine forest ecosystem on its sides. Right: Twenty years later, life abounds on the once-barren landscape. Because traces of the former ecosystem remained, this is an example of secondary succession. ***(c)*** Left: In the summer of 1988, extensive fires swept through the forests of Yellowstone National Park in Wyoming. Right: Trees and flowering plants sprung up in the sunlight, and wildlife populations are rebounding as secondary succession occurs. (Photo taken in 1998.)

(a)

(b)

(c)

Primary Succession Can Begin on Bare Rock

Figure 39-16 illustrates primary succession on Isle Royale, Michigan, an island in Lake Superior. Bare rock, such as that exposed by a retreating glacier or cooled from molten lava, liberates nutrients such as minerals by *weathering*. In weathering, cracks form as the rock alternately freezes and thaws, contracting and expand. Chemical action such as acid rain further breaks down the surface.

For lichens, weathered rock provides a place to attach where there are no competitors and plenty of sunlight. Lichens can photosynthesize, and they obtain minerals by dissolving some of the rock with an acid they secrete. As the pioneering lichens spread over the rock, drought-resistant, sun-loving mosses begin growing in the cracks. Fortified by nutrients liberated by the lichens, the moss forms a dense mat that traps dust,

MEDIATUTOR 39.2 Primary Succession

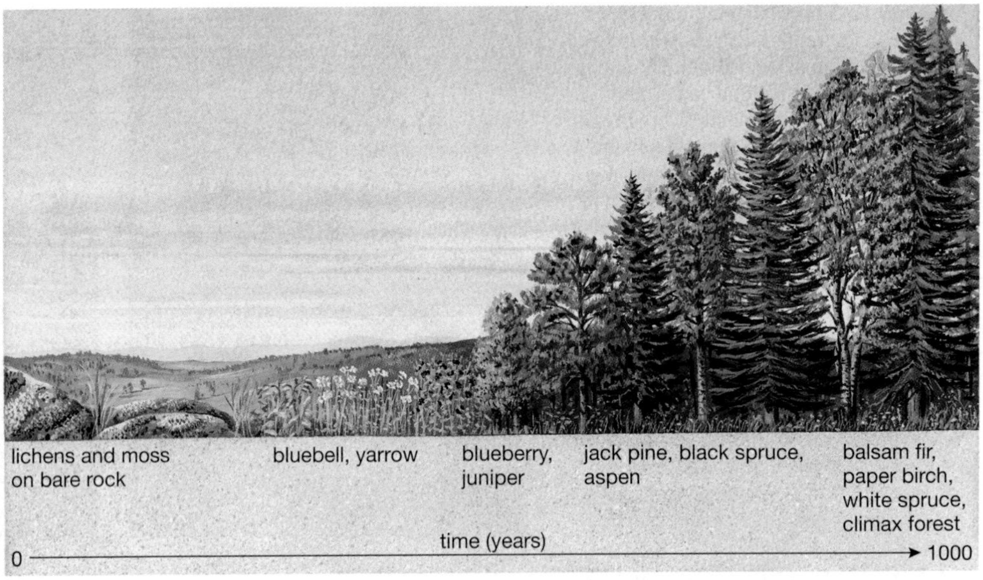

Figure 39-16 Primary succession
Primary succession as it occurs on bare rock in upper Michigan.

tiny rock particles, and bits of organic debris. The death of some of the moss adds to a growing nutrient base, and the moss mat acts like a sponge, trapping moisture. Within the moss, seeds of larger plants, such as bluebell and yarrow, germinate. Later these plants die and their bodies contribute to a growing layer of soil. As woody shrubs such as blueberry and juniper take advantage of the newly formed soil, the moss and lichens may be shaded out and buried by decaying leaves and vegetation. Eventually, trees such as jack pine, blue spruce, and aspen take root in the deeper crevices, and the sun-loving shrubs are shaded out. Within the forest, shade-tolerant seedlings of taller or faster-growing trees, such as balsam fir, paper birch, and white spruce, thrive. In time they tower over and replace the original trees, which are intolerant of shade. After a thousand years or more, a tall climax forest thrives on what was once bare rock.

An Abandoned Farm Will Undergo Secondary Succession
Figure 39-17 illustrates succession on an abandoned southeastern farm. The pioneer species are fast-growing annual weeds such as crabgrass, ragweed, and sorrel, which root in the rich soil already present and thrive in direct sunlight. A few years later, perennial plants such as asters, goldenrod, and broom sedge grass invade, as well as woody shrubs such as blackberry. These plants multiply rapidly and dominate for the next few decades. Eventually, they are replaced by pines and fast-growing deciduous trees, such as tulip poplar and sweet gum, which sprout from windblown seeds. These trees become prominent after about 25 years, and a pine forest

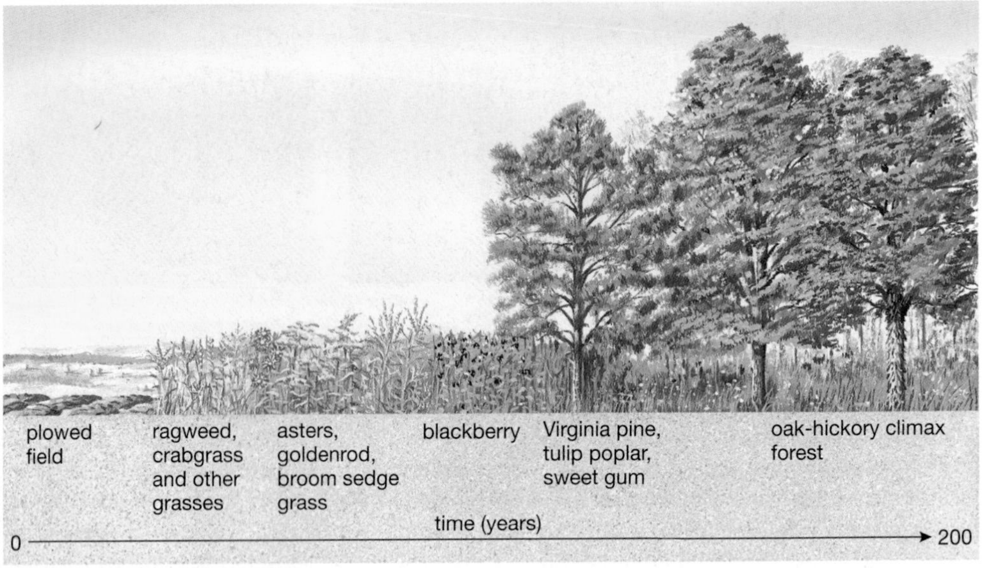

Figure 39-17 Secondary succession
Secondary succession as it occurs on a plowed, abandoned southeastern farm field.

dominates the field for the rest of the first century. Meanwhile, shade-resistant, slow-growing hardwoods such as oak and hickory take root beneath the pines. After the first century, these begin to tower over and shade the pines, which eventually die from lack of sun. A relatively stable climax forest dominated by oak and hickory is present by the end of the second century.

Succession Also Occurs in Ponds and Lakes

In freshwater ponds or lakes, succession occurs both from changes within the pond or lake and as a result of an influx of nutrients from outside the ecosystem. Sediments and nutrients carried in by runoff from the surrounding land have a particularly large impact on small freshwater lakes, ponds, and bogs, which gradually undergo succession to dry land (Fig. 39-18). In forests, meadows may be produced by lakes undergoing succession. As the lake fills in from the edges, grasses colonize the newly exposed soil. As the lake shrinks and the meadow grows, trees will encroach around the meadow's edges. If you visit a forest lake 20 years after your first visit, it probably will be a bit smaller.

Succession Culminates in the Climax Community

Succession ends with a relatively stable climax community, which perpetuates itself if it is not disturbed by external forces (such as fire, invasion of an introduced species, or human activities). The populations within a climax community have ecological niches that allow them to coexist without replacing one another. In general, climax communities have more species and more types of community interactions than do early stages of succession. The plant species that dominate climax communities are generally longer-living and tend to be larger than pioneer species; this trend is particularly evident in ecosystems where forest is the climax community.

In your travels, you have undoubtedly noticed that the type of climax community varies dramatically from one area to the next. For example, if you drive through Colorado, you will see a shortgrass prairie climax community on the eastern plains (in those rare areas where it has not been replaced by farms), pine-spruce forests in the mountains, tundra on their uppermost reaches, and sagebrush-dominated climax community in the western valleys. The exact nature of the climax community is determined by numerous geological and climatic variables, including temperature, rainfall, elevation, latitude, type of rock (which determines the type of nutrients available), exposure to sun and wind, and many more. Natural events such as hurricanes, avalanches, and fires started by lightning may destroy sections of climax forest, reinitiating secondary succession and producing a patchwork of various successional stages within an ecosystem.

In many forests throughout the U.S., rangers are allowing fires set by lightning to run their course, recognizing that this natural process is important for the maintenance of the entire ecosystem. Fires liberate nutrients that are used by plants. Fires also kill some (but usually not all) of the trees they engulf. Sunlight can then reach the forest floor, encouraging the growth of **subclimax** plants, which belong to a successional stage earlier than the climax stage. The combination of climax and subclimax regions within the ecosystem provides habitats for a larger number of species.

Human activities may dramatically alter the climax vegetation. Large stretches of grasslands in the west, for example, are now dominated by sagebrush due to

(a) (b) (c)

Figure 39-18 Succession in a freshwater pond
In small ponds, succession is speeded by an influx of materials from the surroundings. ***(a)*** In this small pond, dissolved minerals carried by runoff from the surroundings support aquatic plants, whose seeds or spores were carried in by the winds or by birds and other animals. ***(b)*** Over time, the decaying bodies of aquatic plants build up soil that provides anchorage for more terrestrial plants. ***(c)*** Finally, the pond is entirely converted to dry land.

Earth Watch
Exotic Invaders

Exotic species, or species introduced into an ecosystem where they did not evolve, sometimes find no predators or parasites in their new environment. The unchecked population growth of these invaders may seriously damage the ecosystem as they displace, outcompete, and prey on native species. Both starlings and English sparrows have spread dramatically since their deliberate introduction into the eastern U.S. in the 1890s. Their success displaces native songbirds, with which they compete for nesting sites. Red fire ants from South America were accidentally introduced into Alabama on shiploads of lumber in the 1930s and have since spread throughout the South, with its hospitable warm, moist climate. Fire ants displace and may kill other insects, birds, and even small mammals. Their mounds can ruin farm fields, and their fiery stings and aggressive temperament can make backyards uninhabitable. Gypsy moths were introduced from Europe in 1866 and still pose a serious threat to fruit and forest trees in North America. Now a new invader, the Asian long-horned beetle, is devouring hardwood trees in New York City and Chicago. The beetle, which officials believe may pose the biggest threat to forests since the gypsy moth, arrived around 1996 in wooden pallets and boxes shipped from China.

Exotic plants also threaten natural communities. In the 1940s a Japanese vine called *kudzu* was planted extensively in the southern U.S. to control erosion. Today kudzu is a major pest, overgrowing and killing trees and underbrush and occasionally engulfing small houses (Fig. E39-2a). The water hyacinth, introduced from South America as an ornamental plant, now clogs about 2 million acres of southern lakes and waterways, slowing boat traffic and displacing natural vegetation (Fig. E39-2b). In the northwestern Mediterranean Sea, an alien algae (*Caulerpa*) recently escaped from a saltwater aquarium and now blankets the seafloor like a carpet of living Astroturf, smothering natural communities. In the summer of 2000, a large patch of the algae was spotted in a lagoon off the California coast near San Diego. The lagoon was quarantined and poisoned with herbicide, and divers and boaters were alerted to keep a lookout for more of the noxious seaweed. Will diligent efforts prevent the spread of *Caulerpa*, or will the coastal marine ecosystems of the U.S. also be smothered? Meanwhile, imported Old World climbing ferns, probably introduced by nurseries, are blanketing treetops throughout southern Florida and rapidly invading the Everglades, killing trees by depriving them of light.

Ecologists estimate that more than 6500 exotic species have established themselves in the U.S., driving 315 native species to the brink of extinction. The cost of attempting to control destructive invaders, when combined with the cost of the damage they do, is estimated at more than $120 billion annually in the U.S. By evading the checks and balances imposed by millennia of coevolution, exotic species are wreaking havoc on natural ecosystems throughout the world. Recently, wildlife officials have made cautious attempts to reestablish these checks and balances by importing predators or parasites to attack selected exotic species. In Australia a dozen European rabbits introduced in the 1840s had exploded in number to 300 million by 1996, wiping out native vegetation and competing for food with native species such as kangaroos. In 1996 officials released a virus deadly to rabbits at hundreds of sites across the continent. This controversial experiment has worked well so far. The virus has not spread to other species, rabbit populations have been reduced by 95% in some areas, and native plants and animals are rebounding. This type of control is fraught with danger, however, because introducing more exotic predators or parasites into an ecosystem can have unpredicted and possibly disastrous consequences for native species. In 1958 a large predatory Florida snail, the rosy wolf snail (Fig. E29-2c), was imported into Hawaii to feed on another exotic, the giant African snail, which was a menace to the native vegetation. The rosy wolf snail is now attacking many species of native snail, threatening them with extinction. Unable to kill the Florida invader without killing native species, Hawaiian officials have resorted to enclosing tracts of land that are home to endangered native snails with corrugated aluminum barriers surrounded by a "moat" of salt crystals to deter the predators. In spite of the risks of importing these biocontrols, there often seems to be little alternative, since poisons intended for exotics will kill natives as well. Biologists are now considering importing sea slugs to feed on *Caulerpa* in the Mediterranean, and U.S. Department of Agriculture scientists are planning to release a fly whose larvae feed selectively on fire ants and are researching possible insects to import that will feed on the Old World climbing fern—hopefully without attacking native plants.

(a) (b) (c)

Figure E39-2 Exotic species
(a) The Japanese vine kudzu will rapidly cover entire trees and houses. ***(b)*** The beauty that became a beast—water hyacinths, originally from South America, today clog waterways in our southern states. ***(c)*** Importing the rosy wolf snail proved disasterous for Hawaii's native snails, one of which is being attacked in this photo.

overgrazing. The grass that usually outcompetes sagebrush is selectively eaten by cattle, allowing the sagebrush to prosper.

Some Ecosystems Are Maintained in a Subclimax State

Some ecosystems are not allowed to reach the climax stage but are maintained in a subclimax stage. The tallgrass prairie that once covered northern Missouri and Illinois is a subclimax of an ecosystem whose climax community is deciduous forest. The prairie was maintained by periodic fires, some set by lightning and others deliberately set by Native Americans to increase grazing land for buffalo. Forest now encroaches, and limited prairie preserves are maintained by carefully managed burning.

Agriculture also depends on the artificial maintenance of carefully selected subclimax communities. Grains are specialized grasses characteristic of the early stages of succession, and much energy goes into preventing competitors (weeds and shrubs) from taking over. The suburban lawn is a painstakingly maintained subclimax ecosystem. Mowing destroys woody invaders, and the selective herbicides that some homeowners use kill pioneers such as crabgrass and dandelions.

To study succession is to study the variations in communities over time. The climax communities that form during succession are strongly influenced by climate and geography—the distribution of ecosystems in *space*. Deserts, grasslands, and deciduous forests are climax communities formed over broad geographical regions with similar environmental conditions. These extensive areas of characteristic plant communities are called **biomes.** In Chapter 41 we will explore some of the great biomes of the world. Although the communities that the various biomes comprise differ radically in the types of populations they support, communities worldwide are structured according to general rules. These principles of ecosystem structure are described in the following chapter.

REVISITED CASE STUDY REVISITED CASESTUDYREVISITEDCASE

Invasion of the Zebra Mussels

About 5 years after the zebra mussel arrived, scientists were pleased to see a native sponge growing on top of zebra mussels. Both sponges and mussels obtain food by filtering water and removing microscopic algae, so these species compete with one another for food. In some study areas, the number of mussels has declined, partly as a result of being smothered by sponges and partly from being eaten by another exotic species: the round goby. In 1990 a biologist at the University of Michigan discovered a round goby in the St. Clair River. The goby probably arrived the same way the mussel did and came from the mussel's home in southeastern Europe. Recognizing one of its favorite prey, the 5-inch-long predator immediately began feasting on small zebra mussels and expanding its range into habitats already invaded by mussels; gobies are now found in all five Great Lakes.

Is this an accidental solution to the mussel problem? Unfortunately not. The gobies ignore the largest zebra mussels, so these continue to spawn. Further, the gobies are not picky eaters. In addition to mussels, they will eat the eggs and young of any other fish in their habitat, including smallmouth bass, walleye, perch, and sculpins. Researchers are now investigating ways to halt the goby's spread toward the Mississippi River, including using a weak but irritating electric current to keep the fish from passing though selected narrow channels along the way. Meanwhile, zebra mussels are still spreading.

Although the round goby was introduced accidently, many exotic predators have been imported to control exotic pests, and some officials have even proposed importing exotic predators to control native pest species, such as grasshoppers. Discuss the implications of this approach for ecological communities and for native species. Describe the types of studies that should be conducted before any new predator is imported.

Summary of Key Concepts

1) Why Are Community Interactions Important?

Community interactions influence population size, and the interacting populations within communities act on one another as agents of natural selection. Thus, community interactions also shape the bodies and behaviors of the interacting populations.

2) What Are the Effects of Competition Among Species?

The ecological niche defines all aspects of a species' habitat and interactions with its living and nonliving environments. Each species occupies a unique ecological niche. Interspecific competition occurs when the niches of two populations within a community overlap. When two species with the same niche are forced under laboratory conditions to occupy the same ecological niche, one species always outcompetes the other. Species within natural communities have evolved in ways that avoid excessive niche overlap, with behavioral and physical adaptations that allow resource partitioning. Interspecific competition limits both the size and the distribution of competing populations.

3) What Are the Results of Interactions Between Predators and Their Prey?

Predators eat other organisms and are generally larger and less abundant than their prey. Predators and prey act as strong agents of selection on one another. Prey animals have evolved a variety of protective colorations that render them either inconspicuous (camouflage) or startling (startle coloration) to their predators. Some prey are poisonous and exhibit warning coloration by which they are readily recognized and avoided. The situation in which an animal has evolved to resemble another is called *mimicry*. Both predators and prey have evolved a variety of toxic chemicals for attack and defense. Plants that are preyed on have evolved elaborate defenses, ranging from poisons to thorns to overall toughness. These defenses, in turn, have selected for predators that can detoxify poisons, ignore thorns, and grind down tough tissues.

4 What Is Symbiosis?

Symbiotic relationships involve two species that interact closely over an extended time. Symbiosis includes parasitism, in which the parasite feeds on a larger, less abundant host, normally harming it but not killing it immediately. In commensal symbiotic relationships, one species benefits, typically by finding food more easily in the presence of the other species, which is not affected by the association. Mutualism benefits both symbiotic species.

5 How Do Keystone Species Influence Community Structure?

Keystone species have a greater influence on community structure than can be predicted by their numbers. Removal of a keystone species will radically alter community structure to an extent that would not be predicted on the basis of the species' abundance.

6 Succession: How Does a Community Change over Time?

Succession is a progressive change over time in the types of populations that comprise a community. Primary succession, which may take thousands of years, occurs where no remnant of a previous community existed (such as on rock scraped bare by a glacier or cooled from molten lava, a sand dune, or in a newly formed glacial lake). Secondary succession occurs much more rapidly, because it builds on the remains of a disrupted community, such as an abandoned field or the aftermath of a fire. Secondary succession on land is initiated by fast-growing, readily dispersing pioneer plants, which are eventually replaced by longer-lived, generally larger and more shade-tolerant species. Uninterrupted succession ends with a climax community, which tends to be self-perpetuating unless acted on by outside forces, such as fire or human activities. Some ecosystems, including tallgrass prairie and farm fields, are maintained in relatively early stages of succession by periodic disruptions.

Key Terms

aggressive mimicry *p. 821*
biome *p. 831*
camouflage *p. 819*
climax community *p. 826*
coevolution *p. 816*
commensalism *p. 823*
community *p. 816*
competitive exclusion principle *p. 816*
ecological niche *p. 816*
exotic species *p. 830*
interspecific competition *p. 816*
intertidal zone *p. 818*
keystone species *p. 824*
mimicry *p. 819*
mutualism *p. 824*
parasite *p. 824*
parasitism *p. 824*
pioneer *p. 826*
primary succession *p. 826*
resource partitioning *p. 817*
secondary succession *p. 826*
startle coloration *p. 822*
subclimax *p. 829*
succession *p. 826*
symbiosis *p. 823*
warning coloration *p. 819*

Thinking Through the Concepts

Multiple Choice

1. *Which of the following is usually NOT true of climax communities?*
   a. They have more species than do pioneer communities.
   b. They have longer-lived species than do early successional stages.
   c. Climax communities are relatively stable.
   d. Some climax communities are maintained by fire.
   e. Climax communities vary with the location of the ecosystem.
2. *What were the differences in the niches of the* Paramecium *in Gause's second experiment that allowed them to coexist?*
   a. food eaten
   b. body size
   c. feeding area
   d. preferred water temperature
   e. preferred pH
3. *Which of the following is a mutualistic relationship?*
   a. flowering plants and their pollinators
   b. a caterpillar eating a tomato plant
   c. bats and moths
   d. monarch and viceroy butterflies
   e. lupines and blue butterflies
4. *What is the function of aggressive mimicry?*
   a. to hide a prey from a predator
   b. to warn a predator that a prey is dangerous
   c. to warn a predator that a prey is distasteful
   d. to keep prey from recognizing a predator
   e. to startle a prey when it sees a predator
5. *What is coevolution?*
   a. two species selecting for traits in each other
   b. individuals of two species living together
   c. the presence of two species in the same community
   d. two species evolving separately through time
   e. individuals of two species learning how to coexist with or to hunt with each other
6. *By the competitive exclusion principle, coexisting species*
   a. can use the same resources
   b. cannot eat exactly the same things
   c. cannot have identical ecological interactions
   d. cannot be exactly the same size
   e. cannot be closely related to one another

? Review Questions

1. Define ecological *community*, and list three important types of community interactions.

2. Describe four very different ways in which specific plants and animals protect themselves against being eaten. In each, describe an adaptation that might evolve in predators of these species that would overcome their defenses.
3. List two important types of symbiosis; define and provide an example of each.
4. Which type of succession would occur on a clear-cut (a region in which all the trees have been removed by logging) in a national forest, and why?
5. List two subclimax and two climax communities. How do they differ?
6. Define *succession*, and explain why it occurs.

Applying the Concepts

1. Herbivorous animals that eat seeds are considered by some ecologists to be predators of plants, and herbivorous animals that eat leaves are considered to be parasites of plants. Discuss the validity of this classification scheme.
2. An interesting interspecific relationship exists between the tarantula spider and the tarantula hawk wasp. This wasp attacks tarantulas, paralyzing them with venom from their stingers. The wasp then lays eggs on the paralyzed spider. The eggs hatch, and the young eat the living, immobilized tissues of the spider. Discuss whether this relationship between spider and wasp exemplifies parasitism or predation.
3. An ecologist visiting an island finds two very closely related species of birds, one of which has a slightly larger bill than the other. Interpret this finding with respect to the competitive exclusion principle and the ecological niche, and explain both concepts.
4. Think about the case of the camouflaged frogfish and its prey. As the frogfish sits camouflaged on the ocean floor, wiggling its lure, a small fish approaches the lure and is eaten, while a very large predatory fish fails to notice the frogfish. Describe all the possible types of community interactions and adaptations these organisms have selected for. Remember that predators can also be prey and that community interactions are complex!
5. Design an experiment to determine whether the kangaroo is a keystone species in the Australian outback.
6. Why is it difficult to study succession? Suggest some ways you would approach this challenge for a few different ecosystems.

For More Information

Amos, W. H. "Hawaii's Volcanic Cradle of Life." *National Geographic*, July 1990. A naturalist explores succession on lava flows.

Enserink, M. "Biological Invaders Sweep In;" Kaiser, J. "Stemming the Tide of Invading Species;" and Malkoff, D. "Fighting Fire with Fire." *Science*, September 17, 1999. A series of articles covering the problems posed by exotic species.

Power, M., et al. "Challenges in the Quest for Keystones." *Bioscience*, September 1996. A comprehensive review of the importance of keystone species and the challenges of studying them.

Answers to Multiple-Choice Questions

1.d 2.c 3.a 4.d 5.a 6.c

MEDIATUTOR
Community Interactions

CD Activities

Activity 39.1: Effects of Keystone Species Extinction

Estimated time: 10 minutes

In this simple simulation you will explore how removing a keystone species, in this case a fruit bat of the family Pteropidae, affects the community of which it is a member.

Activity 39.2: Primary Succession

Estimated time: 10 minutes

Rapid glacial retreat in the last 200 years at Glacier Bay, Alaska, has created a natural laboratory for the study of primary succession. In this tutorial you will explore how succession varies in different sections of the bay and will have an opportunity to view data that indicate how the pattern of invasion by tree species varies among sites.

Start the MediaTutor Student CD-ROM and enter the activity number in the Quick Search box to be taken directly to that activity.

Web Investigations

Case Study: Invasion of the Zebra Mussels

Estimated time: 10 minutes

In these days of routine international commerce, the introduction of harmful, foreign species into ecosystems has become common. Learn more about the current battles against their encroaching ways.

Go to http://www.prenhall.com/audesirk6, the Audesirk Companion Web site. Select Chapter 39 and the Web Investigation to begin.

"A thing is right, when it tends to preserve the integrity, stability, and beauty of the biotic community. It is wrong when it tends otherwise."

Aldo Leopold in **A Sand County Almanac** ***(1949)***

The bald eagle, long a symbol of freedom, is now a symbol of the ability of wildlife to recover from endangered status when it is adequately protected.

40 How Do Ecosystems Work?

AT A GLANCE

Case Study: Flight from Extinction

1) **What Are the Pathways of Energy and Nutrients?**

2) **How Does Energy Flow Through Communities?**

Energy Enters Communities Through Photosynthesis

Energy Is Passed from One Trophic Level to Another

Energy Transfer Through Trophic Levels Is Inefficient

3) **How Do Nutrients Move Within and Among Ecosystems?**

Carbon Cycles Through the Atmosphere, Oceans, and Communities

The Major Reservoir for Nitrogen Is the Atmosphere

The Major Reservoir for Phosphorus Is Rock

Most Water Remains Chemically Unchanged During the Water Cycle

4) **What Is Causing Acid Rain and Global Warming?**

Overloading the Nitrogen and Sulfur Cycles Causes Acid Rain

Interfering with the Carbon Cycle Contributes to Global Warming

Case Study Revisited: Flight from Extinction

CASE STUDY

Flight from Extinction

On July 4, 1999, as the U.S. celebrated the 223rd anniversary of its independence, President Clinton announced plans to officially remove the bald eagle from the "threatened" category of the endangered species list. The eagle's return from the brink of extinction has been impressively rapid—a success story for the U.S. and its national symbol. In 1963 wildlife officials found only 417 breeding pairs of bald eagles in the lower 48 states. Thirty-six years later, they located 5748 pairs and declared the species saved. What threatened this impressive predator? While loss of habitat has caused a substantial decline in eagle populations since the country was founded and eagles numbered in the hundreds of thousands, a pesticide, DDT, finally forced the bald eagle to the brink of extinction. Accumulating in the eagle's bodies, DDT caused female eagles to lay eggs with unusually thin, fragile shells. These eggs were often crushed by the parent birds as they guarded their nests. Because eagles reproduce slowly even under ideal conditions, loss of their young was devastating to the species. Eagle populations were able to recover because DDT was banned in the U.S. in 1972, and eagle habitat was protected under the Endangered Species Act. Why are eagles and other long-lived predators particularly vulnerable to certain pesticides? Do toxic chemicals still threaten wildlife? Is the bald eagle's safety now ensured? We revisit these questions at the end of the chapter. ■

1 What Are the Pathways of Energy and Nutrients?

The activities of life, from the flight of the eagle to active transport of molecules through a cell membrane, are powered by the *energy* of sunlight. The molecules of life are constructed of chemical building blocks that are obtained as *nutrients* from the environment. Solar energy continuously bombards Earth, is used and transformed in the chemical reactions that power life, and is ultimately converted to heat energy that radiates back into space. In contrast, chemical nutrients remain. While they may change in form and distribution, and even be transported among different ecosystems, nutrients do not leave Earth and are constantly recycled. Two basic laws thus underlie ecosystem function: (1) Energy moves through the communities within ecosystems in a continuous one-way flow. Energy needs constant replenishment from an outside source, the sun. (2) Nutrients constantly cycle and recycle within and among ecosystems (Fig. 40-1). These laws shape the complex interactions among populations within ecosystems, and between communities and their abiotic environment.

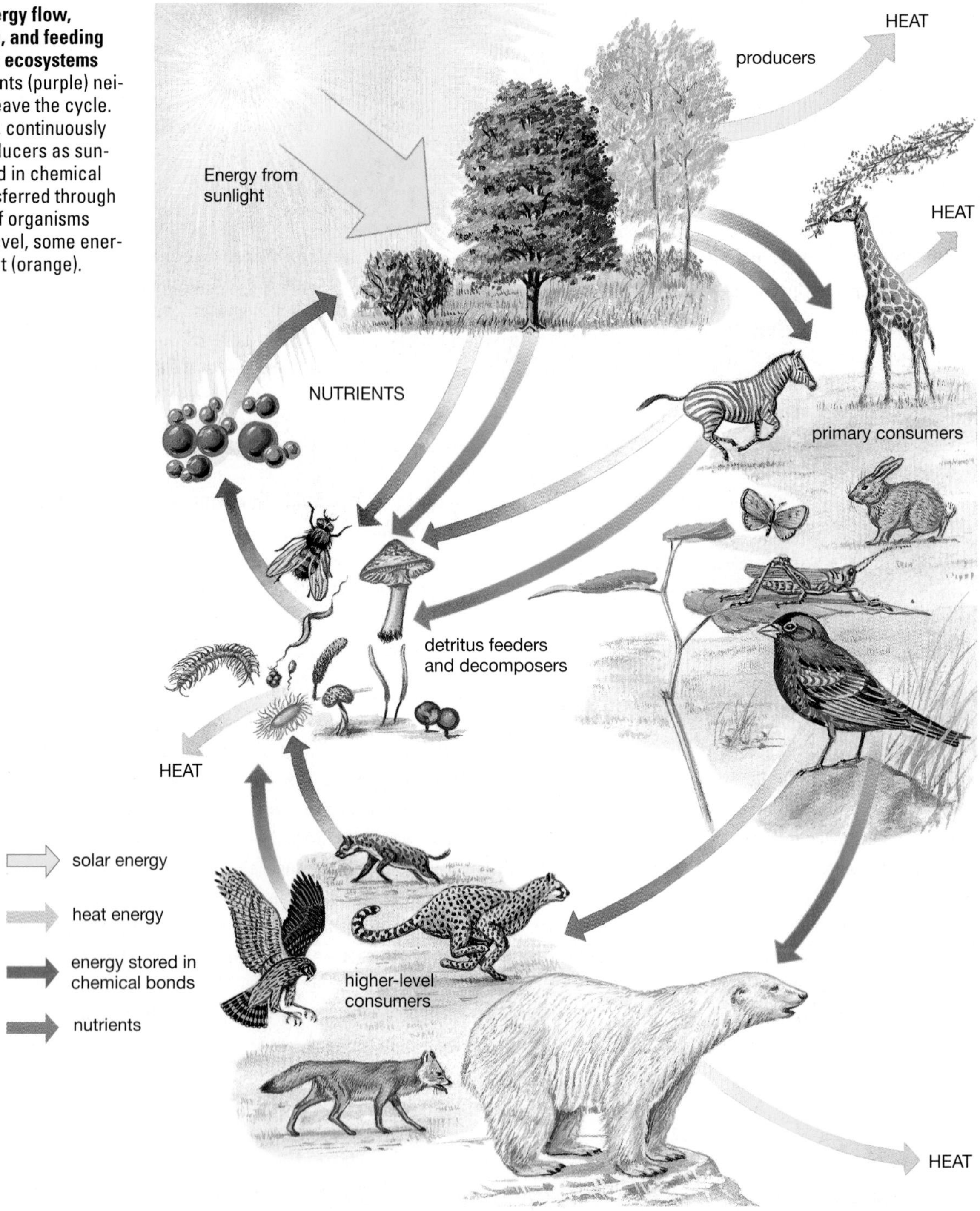

Figure 40-1 Energy flow, nutrient cycling, and feeding relationships in ecosystems Note that nutrients (purple) neither enter nor leave the cycle. Energy (yellow), continuously supplied to producers as sunlight, is captured in chemical bonds and transferred through various levels of organisms (red). At each level, some energy is lost as heat (orange).

2 How Does Energy Flow Through Communities?

Energy Enters Communities Through Photosynthesis

Ninety-three million miles away, the sun fuses hydrogen molecules into helium molecules, releasing tremendous quantities of energy. A tiny fraction of this energy reaches Earth in the form of electromagnetic waves, including heat, light, and ultraviolet energy. Of the energy that reaches Earth, much is reflected by the atmosphere, clouds, and Earth's surface. Still more is absorbed as heat by Earth and its atmosphere, leaving only about 1% to power all life. Of the 1% of the sun's energy that reaches Earth's surface as light, green plants and other photosynthetic organisms capture 3% or less. The teeming life on this planet is thus supported by less than 0.03% of the energy reaching Earth from the sun.

During photosynthesis (described in detail in Chapter 7), pigments such as chlorophyll absorb specific wavelengths of sunlight. This solar energy is then used in reactions that store energy in chemical bonds, producing sugar and other high-energy molecules

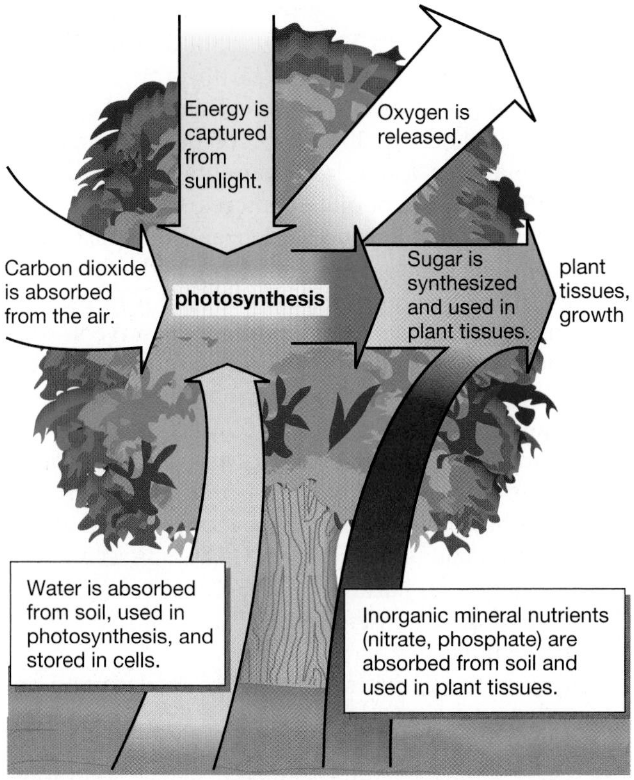

Figure 40-2 Primary productivity: photosynthesis
Photosynthetic organisms capture solar energy. They also remove inorganic nutrients from their nonliving reservoirs and incorporate these nutrients into living tissue. They ultimately provide all the energy and most of the nutrients for organisms in higher trophic levels.

(Fig. 40-2). Photosynthetic organisms, from mighty oaks to the zucchini plants in your garden to single-celled diatoms in the ocean, are called **autotrophs** (Greek, "self-feeders") or **producers**, because they produce food for themselves using nonliving nutrients and sunlight. In doing so, they directly or indirectly produce food for nearly all other forms of life as well. Organisms that cannot photosynthesize, called **heterotrophs** (Greek, "other-feeders") or **consumers**, must acquire energy and many of their nutrients prepackaged in the molecules that compose the bodies of other organisms.

The amount of life that a particular ecosystem can support is determined by the energy captured by the producers in that ecosystem. The energy that photosynthetic organisms store and make available to other members of the community over a given period is called **net primary productivity**. Net primary productivity can be measured in units of energy (calories) stored per unit area during a given timespan. It is also measured as the **biomass**, or dry weight of organic material, in producers that is added to the ecosystem per unit area over a specified time. The productivity of an ecosystem is influenced by many environmental variables, including the amount of nutrients available to the producers, the amount of sunlight reaching them, the availability of water, and the temperature. In the desert, for example, lack of water limits productivity; in the open ocean, light is limited in deep water and nutrients are limited in surface water. When resources are abundant, as in estuaries (where rivers meet the ocean, carrying nutrients washed from the land) and in tropical rain forests, productivity is high. Some average productivities for a variety of ecosystems are shown in Figure 40-3.

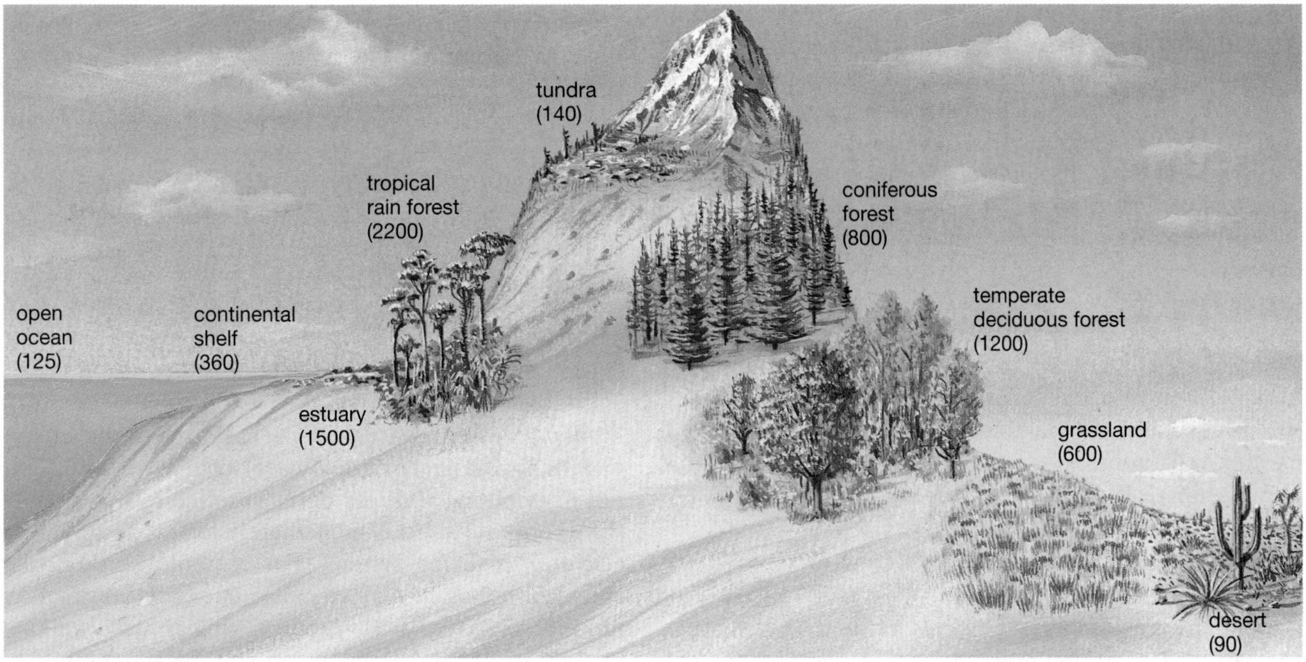

Figure 40-3 Ecosystem productivity compared
Average net primary productivity, in grams of organic material per square meter per year, of some terrestrial and aquatic ecosystems. Notice the enormous differences in productivity among the ecosystems.

Energy Is Passed from One Trophic Level to Another

Energy flows through communities from photosynthetic producers through several levels of consumers. Each category of organism is called a **trophic level** (literally, "feeding level"). The producers, from redwood trees to cyanobacteria, form the first trophic level, obtaining their energy directly from sunlight (see Fig. 40-1). Consumers occupy several trophic levels. Some consumers feed directly and exclusively on producers, the most abundant living energy source in any ecosystem. These **herbivores** (meaning "plant eaters"), ranging from grasshoppers to giraffes, are also called **primary consumers**; they form the second trophic level. **Carnivores** (meaning "meat eaters") such as the spider, eagle, and wolf are predators that feed primarily on primary consumers. Carnivores, also called **secondary consumers**, form the third trophic level. Some carnivores occasionally eat other carnivores. When doing so, they occupy the fourth trophic level, **tertiary consumers**.

Food Chains and Food Webs Describe the Feeding Relationships Within Communities

To illustrate who feeds on whom in a community, it is common to identify a representative of each trophic level that eats a representative of the level below it. This linear feeding relationship is called a **food chain**. As illustrated in Figure 40-4, different ecosystems have radically different food chains.

Natural communities rarely contain well-defined groups of primary, secondary, and tertiary consumers, however. A **food web** shows the many interconnecting food chains in a community, describing the actual feeding relationships within a given community (Fig. 40-5, p. 840). Some animals, such as raccoons, bears, rats, and humans, are **omnivores** (Latin, "eating all")—that is, at different times they act as primary, secondary, and occasionally tertiary (third-level) consumers. Many carnivores will eat either herbivores or other carnivores, thus acting as secondary or tertiary consumers, respectively. An owl, for instance, is a secondary consumer when it eats a mouse, which feeds on plants, but a tertiary consumer when it eats a shrew, which feeds on insects. A shrew that eats a carnivorous insect is a tertiary consumer, and the owl that fed on the shrew is then a quaternary (fourth-level) consumer. When digesting a spider, a carnivorous plant, such as the sundew, can "tangle the web" hopelessly by serving simultaneously as a photosynthetic producer and a secondary consumer!

Detritus Feeders and Decomposers Release Nutrients for Reuse

Among the most important strands in the food web are the *detritus feeders* and *decomposers*. The **detritus feeders** are an army of small and often unnoticed animals and protists that live on the refuse of life: molted exoskeletons, fallen leaves, wastes, and dead bodies. (*Detritus* means "debris.") The network of detritus feeders is extremely complex, including earthworms, mites, protists, centipedes, some insects, a unique land-dwelling crustacean called a pillbug (or "rolypoly"), nematode worms, and even a few large vertebrates such as vultures. With the exception of vultures, these organisms probably abound in your garden and compost heap. They consume dead organic matter, extract some of the energy stored within it, and excrete it in a further decomposed state. Their excretory products serve as food for other detritus feeders and for decomposers. The **decomposers** are primarily fungi and bacteria that digest food outside their bodies by secreting digestive enzymes into the environment. They absorb the nutrients they need and release the remaining nutrients. The black coating or gray fuzz you may notice on the decaying tomatoes and bread crusts in your compost are fungal decomposers hard at work. Thanks to detritus feeders and decomposers, most of the stored energy is eventually utilized.

Through the activities of detritus feeders and decomposers, the bodies and wastes of living organisms are reduced to simple molecules, such as carbon dioxide, water, minerals, and organic molecules, that return to the atmosphere, soil, and water. By liberating nutrients for reuse, detritus feeders and decomposers form a vital link in the nutrient cycles of ecosystems. In some ecosystems, such as deciduous forests, more energy passes through the detritus feeders and decomposers than through the primary, secondary, or tertiary consumers.

What would happen if detritus feeders and decomposers were to disappear? This portion of the food web, although inconspicuous, is absolutely essential to life on Earth. Without them, communities would gradually be smothered by accumulated wastes and dead bodies. The nutrients stored in these bodies would be unavailable to enrich the soil. The quality of the soil would become poorer and poorer until plant life could no longer be sustained. With plants eliminated, energy would cease to enter the community; the higher trophic levels, including humans, would disappear as well.

Energy Transfer Through Trophic Levels Is Inefficient

As we discussed in Chapter 6, a basic law of thermodynamics is that energy use is never completely efficient. For example, as your car burns gasoline, about 75% of the energy released is immediately lost as heat. This is also true in living systems. For example, splitting the chemical bonds of ATP to cause muscular contraction produces heat as a by-product; this is why walking briskly on a cold day will warm you. Small amounts of waste heat are produced by all the biochemical reactions that keep cells alive.

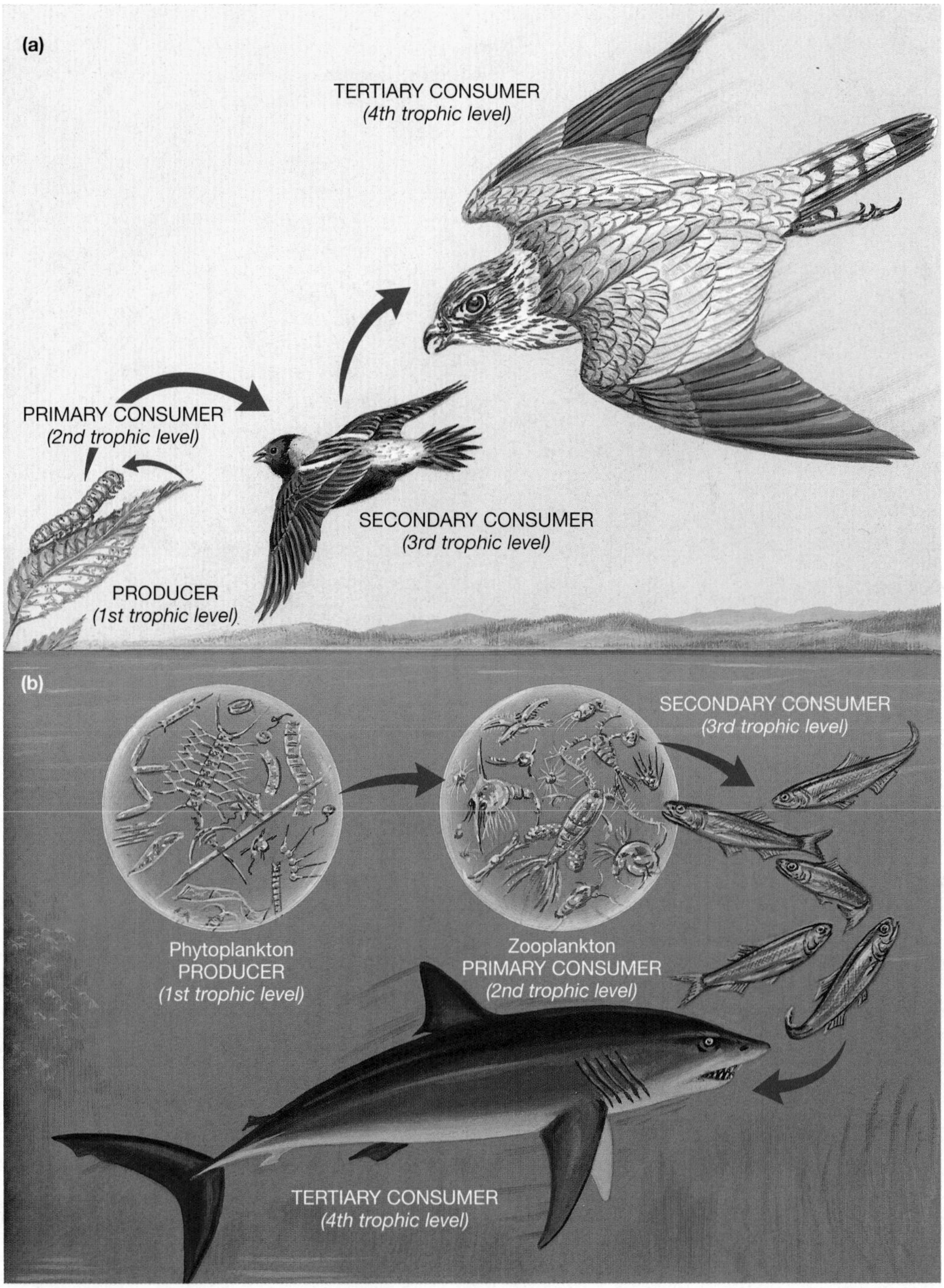

Figure 40-4 Food chains
(a) A simple terrestrial food chain. ***(b)*** A simple marine food chain.

Energy transfer from one trophic level to the next is also quite inefficient. When a caterpillar (a primary consumer) eats the leaves of a tomato plant (a producer), only some of the solar energy originally trapped by the plant is available to the insect. Some energy was used by the plant for growth and maintenance, and more was lost as heat during these processes. Some energy was converted into the chemical bonds of molecules such as cellulose, which the caterpillar cannot digest. Therefore, only a fraction of the energy captured by the first trophic level is available to organisms in the second trophic level. The energy consumed by the caterpillar is in turn partially used to power crawling and the gnashing of mouthparts. Some is used to construct the indigestible exoskeleton, and much is given off as heat. All this energy is unavailable to the songbird in the third trophic level when it eats the caterpillar. The bird loses energy as body heat, uses more in flight, and converts a considerable amount into

MEDIATUTOR
40.1 Building a Food Web

Figure 40-5 A food web
A simple terrestrial food web on a short-grass prairie.

Earth Watch
Food Chains Magnify Toxic Substances

In the 1940s the properties of the new insecticide DDT seemed close to miraculous. In the Tropics, DDT saved millions of lives by killing the mosquitos that spread malaria. Increased crop yields resulting from DDT's destruction of insect pests saved millions more from starvation. DDT is long-lasting, so a single application keeps on killing. Swiss chemist Paul Müller, the discoverer of its properties as a pesticide, was awarded the 1948 Nobel Prize for Medicine and Physiology, and people looked forward to a new age of freedom from insect pests. Little did they realize that the indiscriminate use of this pesticide was unraveling the complex web of life.

For example, in the mid-1950s the World Health Organization sprayed DDT on the island of Borneo to control malaria. A caterpillar that fed on the thatched roofs of houses was relatively unaffected, but a predatory wasp that fed on the caterpillar was destroyed. Thatched roofs collapsed, eaten by the uncontrolled caterpillars. Gecko lizards that ate the poisoned insects accumulated high concentrations of DDT in their bodies. Both they, and the village cats that ate the geckos, died of DDT poisoning. With the cats eliminated, the rat population exploded. The village was then threatened with an outbreak of plague, carried by the uncontrolled rats. The outbreak was avoided by airlifting new cats to the villages.

In the U.S., wildlife biologists during the 1950s and 1960s witnessed an alarming decline in populations of several predatory birds, especially fish-eaters, such as bald eagles, cormorants, ospreys, and brown pelicans. These top predators are never abundant, and the decline pushed some, including the brown pelican and the bald eagle, close to extinction. The aquatic ecosystems supporting these birds had been sprayed with relatively low amounts of DDT to control insects. Scientists were amazed to find concentrations of DDT in the bodies of predatory birds up to *one million* times the concentration present in the water. This led to the discovery of **biological magnification**, the process by which toxic substances accumulate at increasingly high concentrations in progressively higher trophic levels.

The pesticide DDT and many other man-made substances that undergo biological magnification contain chlorine, which tends to stabilize molecules. All such substances have two properties that make them dangerous: (1) Decomposer organisms cannot readily break them down into harmless substances—that is, they are not **biodegradable**; and (2) they are fat soluble, but not water soluble. Thus, they accumulate in the bodies of animals, particularly in fat, rather than being broken down and excreted in the watery urine. Since the transfer of energy from lower to higher trophic levels is extremely inefficient, herbivores must eat large quantities of plant material (which may have been sprayed with pesticides), carnivores must eat many herbivores, and so on. Because these substances remain in the body, a predator accumulates the poison from its prey over many years. Prior to pollution control measures, the Great Lakes built up relatively high levels of bioaccumulating compounds, including DDT and PCBs (chlorinated compounds used in electrical transformers and cables), which continue to affect the health of local wildlife. In the Great Lakes region, populations of fish-eating river otters have declined sharply, and a variety of fish-eating birds, including the newly returned bald eagles, produce deformed offspring or eggs that never hatch (Fig. E40-1).

Figure E40-1 The price of pollution
Deformities such as the twisted beak of this double-crested cormorant from Lake Michigan have been linked to bioaccumulating chemicals. Abnormalities of the reproductive and immune systems are also common in many types of organisms exposed to these pollutants. Predatory animals are especially vulnerable owing to biological magnification.

We must understand the properties of pollutants and the workings of food webs to prevent human health hazards as well as loss of wildlife. Many of the children born to women consuming PCB-contaminated fish from Lake Michigan exhibited slowed intellectual development. Exposure to high levels of pesticides and other persistent pollutants has been linked to certain forms of cancer, infertility, heart disease, and suppressed immune function in people. We often "feed high on the food chain." When we eat tuna or swordfish, for example, we act as a tertiary or even quaternary consumers, so we are vulnerable to bioaccumulating substances. In addition, our long life span provides more time for substances stored in our bodies to accumulate to toxic levels.

Today, although DDT, PCBs, and many other bioaccumulating pollutants are banned in the U.S., they are used in developing countries, persist here from past usage, and are transported worldwide by wind, water, and commerce. Inuit natives living north of the Arctic Circle have high levels of PCBs and chlordane (a banned pesticide similar to DDT) in their bodies from consuming the blubber (fat) of whales and other marine predators. Now there is growing concern that a number of widely used synthetic chemicals share the tendencies of DDT to persist, accumulate, and interfere with hormone functions. These chemicals, called "endocrine disruptors," are released by plastics, detergents, pesticides, and as by-products of manufacturing processes (see "Earth Watch: Endocrine Deception" in Chapter 32).

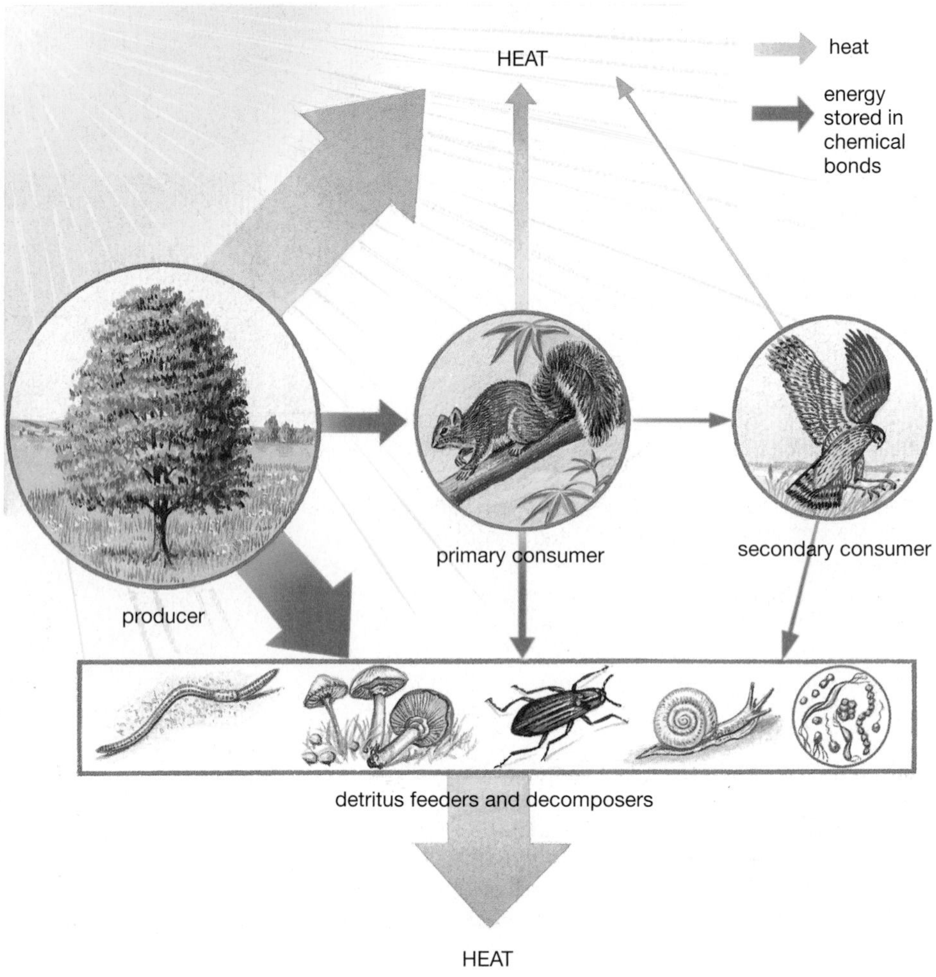

Figure 40-6 Energy transfer and loss
This diagram shows approximate amounts of energy transferred between trophic levels in the form of chemical energy (red) and lost in the form of heat from each trophic level (orange) in a forest community. The width of the arrows is roughly proportional to the quantity of energy transferred or lost.

indigestible feathers, beak, and bone. All this energy will be unavailable to the hawk that catches it. A simplified model of energy flow through the trophic levels in a deciduous forest ecosystem is illustrated in Figure 40-6.

Energy Pyramids Illustrate Energy Transfer Between Trophic Levels

Studies of a variety of communities indicate that the net transfer of energy transfer between trophic levels is roughly 10% efficient, although transfer among levels within different communities varies significantly. This means that, in general, the energy stored in primary consumers (herbivores) is only about 10% of the energy stored in the bodies of producers. In turn, the bodies of secondary consumers possess roughly 10% of the energy stored in primary consumers. In other words, for every 100 calories of solar energy captured by grass, only about 10 calories are converted into herbivores, and only 1 calorie into carnivores. This inefficient energy transfer between trophic levels is called the "10% law." An **energy pyramid**, which shows maximum energy at the base and steadily diminishing amounts at higher levels, illustrates the energy relationships between trophic levels graphically (Fig. 40-7). Ecologists sometimes use biomass as a measure of the energy stored at each trophic level. Because the dry weight of the bodies of organisms at each trophic level is roughly proportional to the amount of energy stored in the organisms at that level, a *biomass pyramid* for a given community often has the same general shape as its energy pyramid.

What does this mean for community structure? If you wander through an undisturbed ecosystem, you will notice that the predominant organisms are plants. Plants have the most energy available to them, because they trap it directly from sunlight. The most abundant animals will be those feeding on plants, and carnivores will be relatively rare. The inefficiency of energy transfer also has important implications for human food production. The lower the trophic level we utilize, the more food energy will be available to us; in other words, far more people can be fed on grain than on meat.

An unfortunate side effect of the inefficiency of energy transfer, coupled with human production and release of toxic chemicals, is that certain persistent toxic chemicals become concentrated in the bodies of carnivores, including people, as described in "Earth Watch: Food Chains Magnify Toxic Substances."

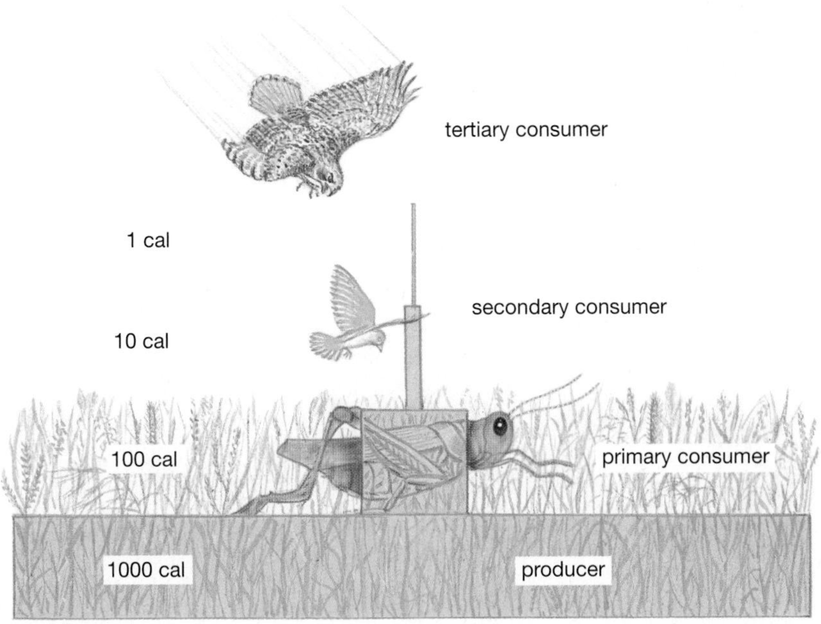

Figure 40-7 An energy pyramid for a prairie ecosystem
Each trophic level from producer to tertiary consumer has less energy stored in it. The width of each rectangle is proportional to the energy stored at that trophic level. A pyramid of biomass for this ecosystem would look quite similar.

3 How Do Nutrients Move Within and Among Ecosystems?

In contrast to the energy of sunlight, nutrients do not flow down onto Earth in a steady stream from above. For practical purposes, the same pool of nutrients has been supporting life for more than 3 billion years. **Nutrients** are elements and small molecules that form all the chemical building blocks of life. Some, called **macronutrients**, are required by organisms in large quantities. These include water, carbon, hydrogen, oxygen, nitrogen, phosphorus, sulfur, and calcium. **Micronutrients**, including zinc, molybdenum, iron, selenium, and iodine, are required only in trace quantities. **Nutrient cycles**, also called **biogeochemical cycles**, describe the pathways these substances follow as they move from the communities to the nonliving portions of ecosystems and then back again to communities.

The sources and storage sites of nutrients are called **reservoirs**. The major reservoirs are generally in the nonliving, or abiotic, environment. For example, carbon has several major reservoirs: It is stored as carbon dioxide gas in the atmosphere, in dissolved form in oceans, and as fossil fuels underground. In the following section we briefly describe the cycles of carbon, nitrogen, phosphorus, and water.

Carbon Cycles Through the Atmosphere, Oceans, and Communities

Chains of carbon atoms form the framework of all organic molecules, the building blocks of life. Carbon enters the living community through capture of carbon dioxide (CO_2) during photosynthesis by producers. On land, producers (such as your garden plants) acquire CO_2 from the atmosphere, where it represents 0.036% of the total gases. Aquatic producers in the ocean, such as seaweeds and diatoms, find abundant CO_2 for photosynthesis dissolved in the water. In fact, far more CO_2 is stored in the oceans than in the atmosphere. Producers return some CO_2 to the atmosphere and ocean during cellular respiration and incorporate the rest into their bodies. Primary consumers, such as cows, shrimp, or tomato hornworms, eat the producers and acquire the carbon stored in their tissues. These herbivores also release some carbon through respiration and store the rest, which is sometimes consumed by organisms in higher trophic levels. All living things eventually die, and their bodies are broken down by detritus feeders and decomposers. Cellular respiration by these organisms, which abound in your compost pile, returns CO_2 to the atmosphere and oceans. Carbon dioxide passes freely between these two great reservoirs (Fig. 40-8).

Some carbon cycles much more slowly. For example, mollusks and marine protists extract CO_2 dissolved in water and combine it with calcium to form calcium carbonate ($CaCO_3$), from which they construct their shells. After death, the shells of these organisms collect in undersea deposits, are buried, and may eventually be converted to limestone. Geological events may expose the limestone, which will dissolve gradually as water runs over it, making the carbon available to living organisms once more.

Another long-term part of the carbon cycle is the production of fossil fuels. **Fossil fuels** are formed from the remains of ancient plants and animals and serve as a third major reservoir for carbon. The carbon found in the organic molecules of the ancient organisms in these deposits is transformed by high temperatures and pressures over millions of years into coal, oil, or natural gas. The energy of prehistoric sunlight is also trapped in fossil fuels. When people burn fossil fuels to utilize this

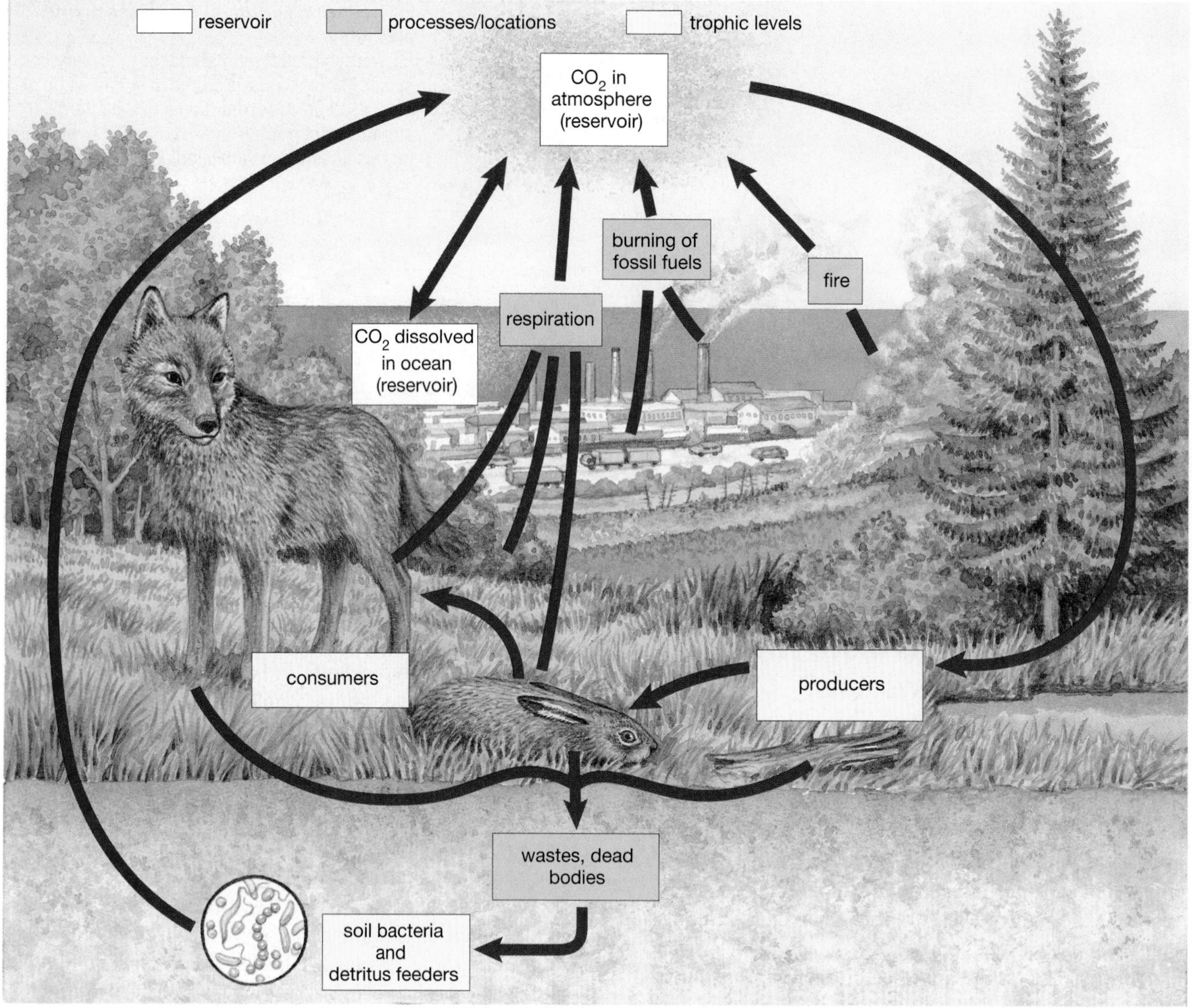

Figure 40-8 The carbon cycle
Carbon is captured from the atmosphere during photosynthesis and passed up through the trophic levels. It is released during respiration from all trophic levels and by the burning of forests and fossil fuels.

stored energy, CO_2 is released into the atmosphere. In addition to the burning of fossil fuels, human activities such as cutting and burning Earth's great forests (where much carbon is stored) are increasing the amount of CO_2 in the atmosphere, as we shall describe later in this chapter.

The Major Reservoir for Nitrogen Is the Atmosphere

The atmosphere contains about 79% nitrogen gas (N_2) and is thus the major reservoir for this important nutrient. Nitrogen is a crucial component of proteins, many vitamins, and the nucleic acids DNA and RNA. Interestingly, neither plants nor animals can extract this gas from the atmosphere. Instead, plants must be supplied with nitrate (NO_3^-) or ammonia (NH_3). But how is atmospheric nitrogen converted to these molecules? Ammonia is synthesized by certain bacteria and cyanobacteria that engage in **nitrogen fixation**, a process that combines nitrogen with hydrogen. Some of these bacteria live in water and soil. Others have entered a symbiotic association with plants called **legumes** (including alfalfa, soybeans, clover, and peas), where they live in special swellings on the roots (see Chapter 19). Decomposer bacteria can also produce ammonia from amino acids and urea found in dead bodies and wastes. Still other bacteria convert ammonia to nitrate.

Nitrogen is combined with oxygen by electrical storms and by burning forests and fossil fuels. Plants incorporate the nitrogen from ammonia and nitrate into amino acids, proteins, nucleic acids, and vitamins. These nitrogen-containing molecules from the plant are eventually consumed, either by primary consumers, detritus feeders, or decomposers. As it is passed through the food web, some of the nitrogen is released in wastes and dead bodies, which decomposer bacteria convert back to nitrate and ammonia. The cycle is balanced by a continuous return of nitrogen to the atmosphere by **denitrifying bacteria**. These residents of wet soil, swamps, and estuaries break down nitrate, releasing nitrogen gas back to the atmosphere (Fig. 40-9).

In human-dominated ecosystems, such as farm fields, gardens, and suburban lawns, ammonia and nitrate are supplied by chemical fertilizers. These fertilizers—like those you sprinkled over your garden before you planted vegetables—are produced by using the energy in fossil fuels to artificially "fix" atmospheric nitrogen. In fact, human activities now dominate the nitrogen cycle, a cause for serious environmental concerns. By burning fossil fuels, by burning forests and draining wetlands that store nitrogen, by cultivating legumes, and by producing fertilizers, humans each year are converting an estimated 140 million metric tons of nitrogen from its gaseous form to ammonia and nitrogen oxides (nitrogen combined with oxygen). Natural processes utilize 90 million to 140 million metric tons of nitrogen yearly. Various types of nitrogen oxides contribute to global warming, destroy the protective ozone layer, and acidify precipitation. Runoff from fertilized farm fields into the Mississippi River is carried to the Gulf of Mexico, where scientists believe it is causing a 7000-square-mile "dead zone" that appears there each summer. When they enter ecosystems, nitrogen oxides may change the composition of plant communities or destroy forests by acid rain, which we shall discuss later in this chapter.

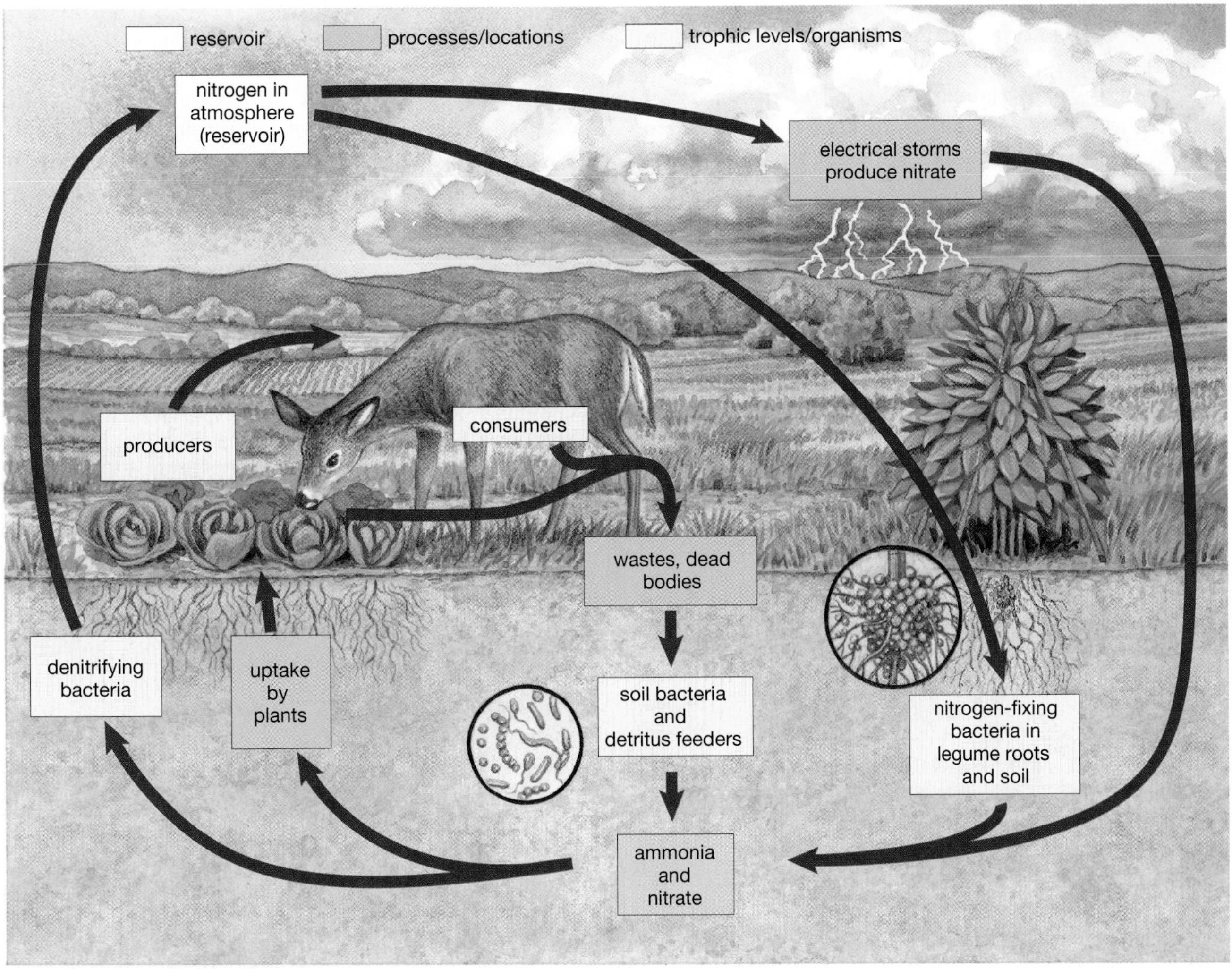

Figure 40-9 The nitrogen cycle
From its reservoir in the atmosphere, nitrogen gas is combined with oxygen to form nitrates by lightning or by burning and then carried to Earth dissolved in rain. Nitrogen-fixing bacteria produce ammonia. Nitrate and ammonia are also synthesized by humans for use in fertilizers. These are absorbed by plants and other producers and incorporated into biological molecules that are passed up through the trophic levels. Nitrate and ammonia are released by excretion or by decomposer bacteria. Other bacteria convert these molecules back to atmospheric nitrogen, completing the cycle.

The Major Reservoir for Phosphorus Is Rock

Phosphorus is a crucial component of biological molecules, including the energy transfer molecules (ATP and NADP), nucleic acids, and the phospholipids of cell membranes. It is also a major component of vertebrate teeth and bones. In contrast to the cycles of carbon and nitrogen, the phosphorus cycle has no atmospheric component. The reservoir of phosphorus in ecosystems is rock, where it is bound to oxygen in the form of phosphate. As phosphate-rich rocks are exposed and eroded, rainwater dissolves the phosphate. Dissolved phosphate is readily absorbed through the roots of plants and by other autotrophs, such as photosynthetic protists and cyanobacteria, where it is incorporated into biological molecules. From these producers, phosphorus is passed through food webs (Fig. 40-10). At each level, excess phosphate is excreted. Ultimately, decomposers return the phosphorus that remains in dead bodies back to the soil and water in the form of phosphate. It may then be reabsorbed by autotrophs, or it may become bound to sediment and eventually reincorporated into rock.

Some of the phosphate dissolved in fresh water is carried to the oceans. Although much of this phosphate ends up in marine sediments, some is absorbed by marine producers and eventually incorporated into the bodies of invertebrates and fish. Some of these, in turn, are consumed by seabirds, which excrete large quantities of phosphorus back onto the land. At one time, the guano (droppings) deposited by seabirds along the western coast of South America was mined, and it provided a major source of the world's phosphorus. Phosphate-rich rock is also mined, and the phosphate is incorporated into fertilizer. Soil that erodes from fertilized fields carries large quantities of phosphates into lakes, streams, and the ocean, where it stimulates the growth of producers. In lakes, phosphorus-rich runoff from land can stimulate such a rich growth of algae and bacteria that the natural community interactions of the lake are disrupted. You'll read more about this in the next chapter.

Most Water Remains Chemically Unchanged During the Water Cycle

The water cycle, or **hydrologic cycle** (Fig. 40-11), differs from most other nutrient cycles in that most water remains in the form of water throughout the cycle and is not used in the synthesis of new molecules. The major reservoir of water is the ocean, which covers about

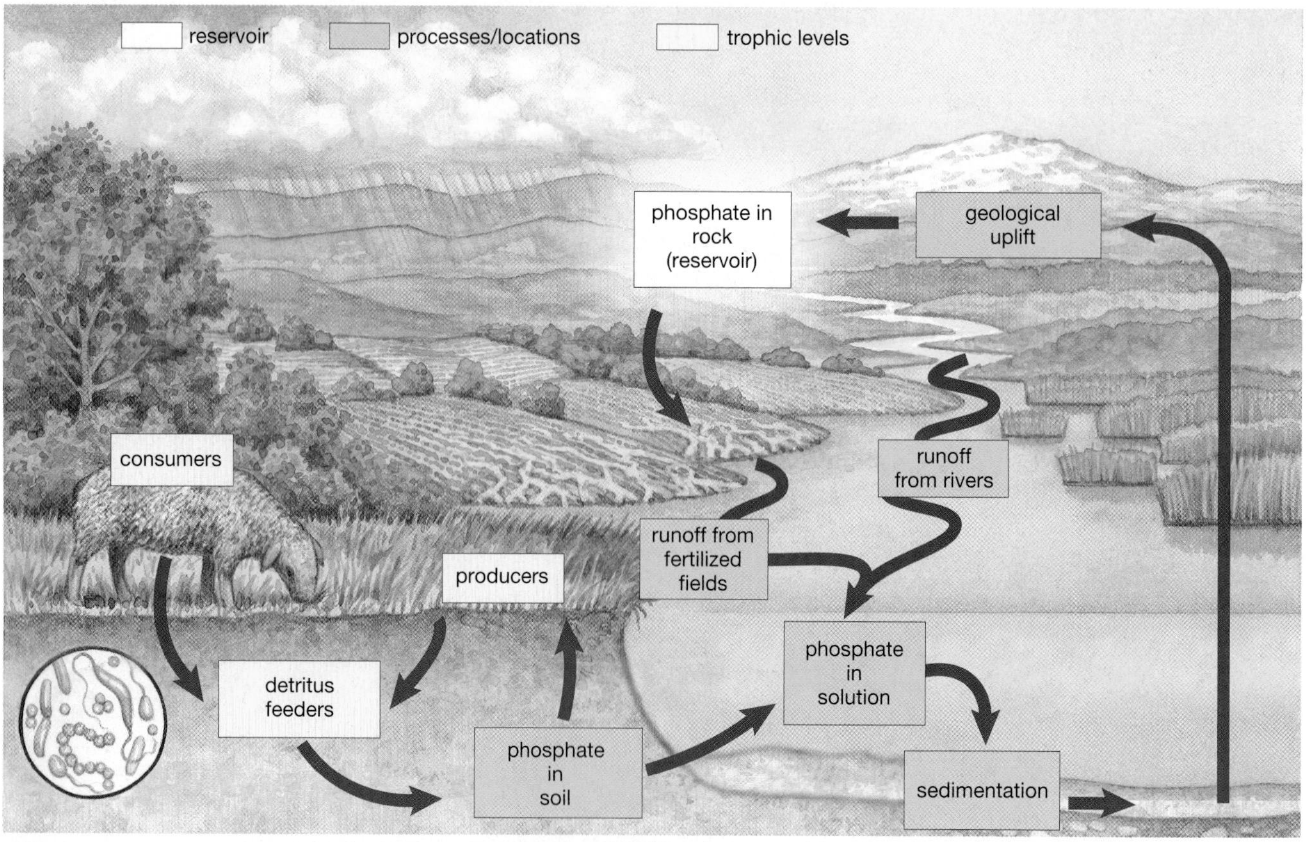

Figure 40-10 The phosphorus cycle
Phosphate dissolves from phosphate-rich rocks or from fertilizers and enters plants and other producers, where it is incorporated into biological molecules. These molecules are passed up through the trophic levels. Phosphate is excreted or returned to the soil and water by decomposer bacteria. It may then be reused by producers or eventually incorporated into rock.

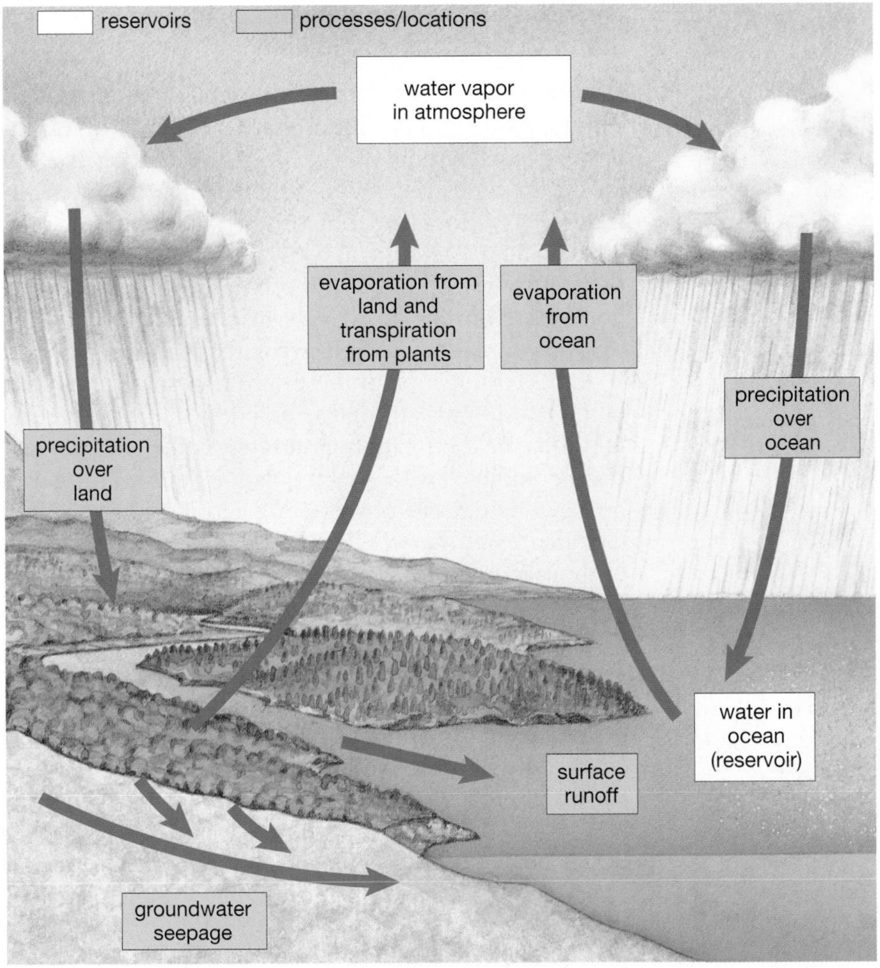

Figure 40-11 The hydrologic cycle
The hydrologic cycle is the simplest of the nutrient cycles. Although water is essential to all life, most of its movement occurs independently of living things.

three-quarters of Earth's surface and contains more than 97% of the available water. The hydrologic cycle is driven by solar energy, which evaporates water, and by gravity, which draws the water back to Earth in the form of precipitation (rain, snow, sleet, and dew). Most evaporation occurs from the oceans, and much water returns directly to them by precipitation. Water falling on land takes various paths. Some water is evaporated from the soil, lakes, and streams. A portion runs off the land back to the oceans, and a small amount enters underground reservoirs. Because the bodies of living things are roughly 70% water, some of the water in the hydrologic cycle enters the living communities of ecosystems. It is absorbed by the roots of plants, and much of this is evaporated back to the atmosphere from their leaves. A small amount is combined with carbon dioxide during photosynthesis to produce high-energy molecules. Eventually these molecules are broken down during cellular respiration, releasing water back to the environment. Consumers get water from their food or by drinking.

As the human population has grown, fresh water has become scarce in many regions of the world where the human demand for it is high. This scarcity limits the ability to grow crops; currently, 40% of the world's food is grown on irrigated cropland. Over the past quarter century farmers have relied increasingly on aquifers—natural underground reservoirs—from which they pump water to irrigate their crops. Unfortunately in many areas of the world, including China, India, Northern Africa, and the western U.S., this groundwater is being "mined"; that is, it is removed faster than it is replenished. The drop in groundwater level is particularly alarming in India, where loss of groundwater places up to 25% of the grain harvest in jeopardy. Contaminated, untreated drinking water is a major problem in developing countries, where 1.2 billion people drink it. The United Nations estimates that impure water causes 80% of the diseases in the developing world and kills 4 to 5 million children each year.

4 What Is Causing Acid Rain and Global Warming?

Many of the environmental problems that beset modern society are the result of human interference in the way ecosystems function. Primitive peoples were sustained solely by the energy flowing from the sun, and

Figure 40-12 A natural substance out of place
This bald eagle was killed after an oil spill off the coast of Alaska.

they produced wastes that were readily taken back into the nutrient cycles. But as the population grew and technology increased, humans began to act more and more independently of these natural processes. We have mined substances, such as lead, arsenic, cadmium, mercury, oil, and uranium, that are foreign to natural ecosystems and toxic to many of the organisms in them (Fig. 40-12). In our factories, we synthesize substances never before found on Earth: pesticides, solvents, and a wide array of other industrial chemicals harmful to many forms of life. The Industrial Revolution, which began in earnest in the mid-nineteenth century, resulted in a tremendous increase in our reliance on energy from fossil fuels (rather than from sunlight) for heat, light, transportation, industry, and even agriculture. Modern industrial and agricultural processes often produce more nutrients than nutrient cycles can efficiently process. In the following sections, we describe two environmental problems of global proportions that are a direct result of human reliance on fossil fuels: acid rain and global warming.

Overloading the Nitrogen and Sulfur Cycles Causes Acid Rain

Each year, the U.S. discharges about 20 million tons of sulfur dioxide into the atmosphere, two-thirds of it from power plants burning coal or oil. The rest is largely a byproduct of industrial boilers, smelters, and refineries. Although volcanoes and hot springs also release sulfur dioxide, human industrial activities account for 90% of the sulfur dioxide in the atmosphere. Twenty-four million tons of nitrogen oxides are also released from vehicles, power plants, and industry in the U.S. annually.

Excess production of these substances was identified in the late 1960s as the cause of a growing environmental threat: *acid rain*, more accurately called **acid deposition**. Combined with water vapor in the atmosphere, nitrogen oxides are converted to nitric acid and sulfur dioxide to sulfuric acid. Days later, and often hundreds or thousands of miles from the source, the acids fall, eating away at statues and buildings (Fig. 40-13), damaging trees and crops, and rendering lakes lifeless. Although both sulfuric and nitric acids form solutions in water vapor, sulfuric acid may also form particles that visibly cloud the air, even under dry conditions. In the U.S., the Northeast, Florida, Mid-Atlantic region, upper Midwest, and the high-elevation West are the most vulnera-

Figure 40-13 Acid deposition is corrosive
This limestone statue at Rheims Cathedral in France is being dissolved by acid deposition.

ble, because the rocks and soils that predominate there are not able to buffer the acidity.

The Acid Rain Program of the U.S. EPA, implemented in 1995, has resulted in substantial reductions in sulfur dioxide emissions from power plants. Air quality has improved in some regions, and there is cautious optimism that damaged ecosystems may begin the slow road to recovery. However, nitrogen oxide emissions remain high despite power plant reductions, partly because of a rise in fuel consumption by automobiles.

Acid Deposition Damages Life in Lakes, Farms, and Forest

In New York's Adirondack Mountains, acid rain has made about 25% of all the lakes and ponds too acidic to support fish. But before the fish die, much of the food web that sustains them is destroyed. Clams, snails, crayfish, and insect larvae die first, then amphibians, and finally fish. The result is a crystal-clear lake—beautiful but dead. The impact is not limited to aquatic organisms. Acid rain also interferes with the growth and yield of many farm crops by leeching out essential nutrients such as calcium and potassium and killing decomposer microorganisms, thus preventing the return of nutrients to the soil. Plants, poisoned and deprived of nutrients, become weak and vulnerable to infection and insect attack. High in the Green Mountains of Vermont, scientists have witnessed the death of about half of the red spruce and beech trees and one-third of the sugar maples since 1965. The snow, rain, and heavy fog that commonly cloak these eastern mountaintops are highly acidic. At a monitoring station atop Mount Mitchell in North Carolina, the pH of fog has been recorded at 2.9, more acidic than vinegar (Fig. 40-14).

Acid deposition increases the exposure of organisms to toxic metals, including aluminum, lead, mercury, and cadmium, which are far more soluble in acidified water than in water of neutral pH. Aluminum dissolved from rock may inhibit plant growth and kill fish. Drinking water in some households has been found dangerously contaminated with lead dissolved by acidic water from lead solder in old pipes. Fish in acidified water have been found to have dangerous levels of mercury in their bodies, which is subject to biological magnification as it is passed through trophic levels (see "Earth Watch: Food Chains Magnify Toxic Substances").

Interfering with the Carbon Cycle Contributes to Global Warming

Between 345 million and 280 million years ago, huge quantities of carbon were diverted from the carbon cycle when, under the warm, wet conditions of the Carboniferous period, the bodies of plants and animals were buried in sediments, escaping decomposition. Over time, heat

Figure 40-14 Acid deposition can destroy forests
Acid rain and fog have destroyed this forest atop Mount Mitchell in North Carolina.

and pressure converted their bodies into fossil fuels, such as coal, oil, and natural gas. Without human intervention, this carbon would have remained untouched. Beginning with the Industrial Revolution, however, we have increasingly relied on the energy stored in these fuels. As we burn them in our power plants, factories, and cars, we release CO_2 into the atmosphere. Since 1850, the CO_2 content of the atmosphere has increased from 280 parts per million (ppm) to 370 ppm, more than 30%, bringing CO_2 concentrations to levels higher than any in more than 400,000 years. That increase is continuing at the rate of 1.5 ppm yearly, primarily as a result of burning fossil fuels.

A second source of added atmospheric CO_2 is global **deforestation**, which eliminates tens of millions of acres of forests annually. Deforestation is occurring principally in the Tropics, where rain forests are rapidly being converted to marginal agricultural land. The carbon stored in the massive trees in these forests returns to the atmosphere after they are cut, primarily through burning.

Greenhouse Gases Trap Heat in the Atmosphere

Carbon dioxide still represents only a tiny fraction of Earth's atmosphere, about 0.036%. However, atmospheric CO_2 acts something like the glass in a greenhouse; it allows solar energy to enter, then absorbs and holds that energy once it has been converted to heat (Fig. 40-15). Several other **greenhouse gases** share this property, including methane, chlorofluorocarbons (CFCs), water vapor, and nitrous oxide. The **greenhouse effect**, the ability of greenhouse gases to trap the sun's energy in a planet's atmosphere as heat, is a natural process. By keeping our atmosphere relatively warm, it allows life on Earth as we know it. However there is very little doubt among scientists that human activities

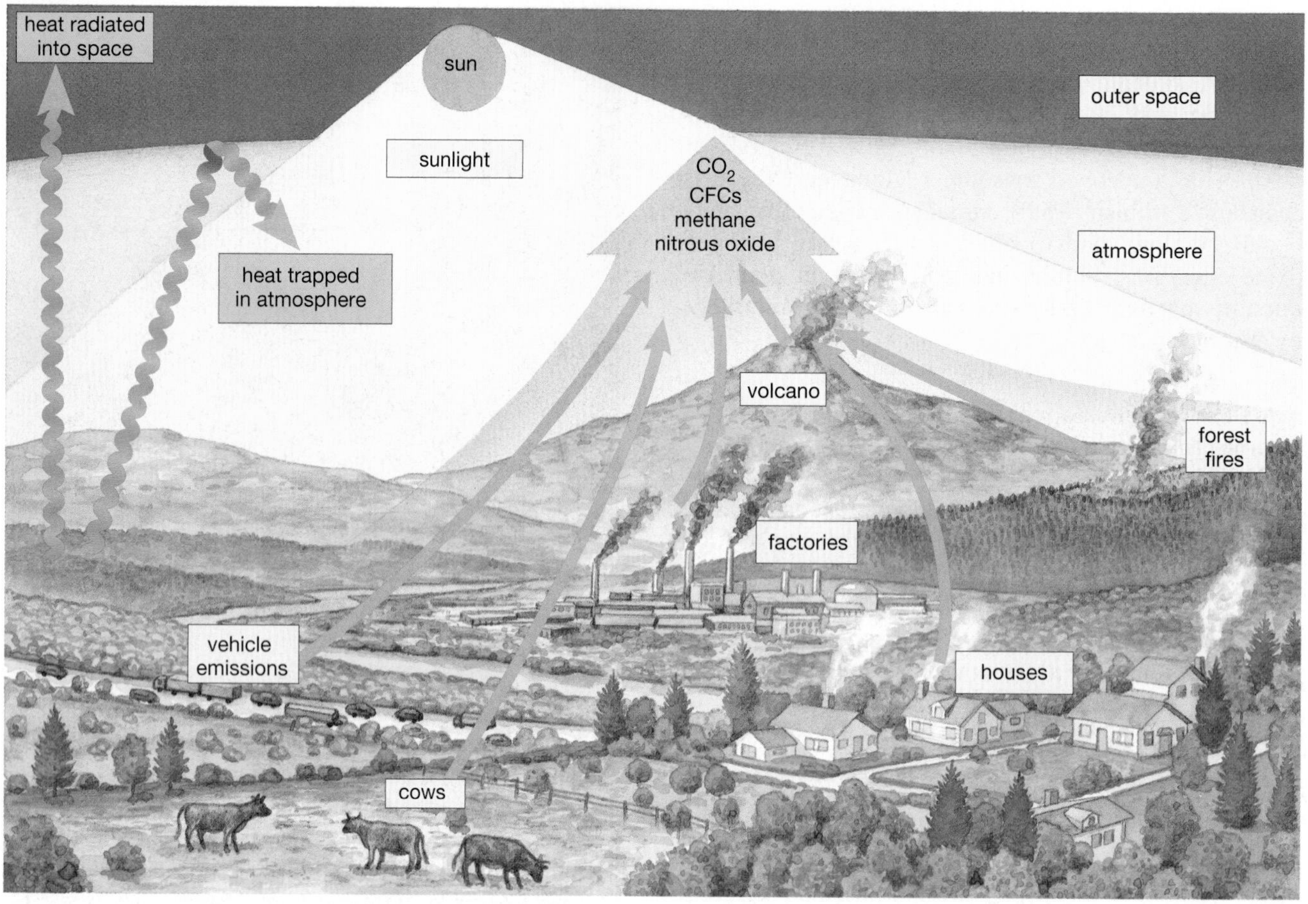

Figure 40-15 Increases in greenhouse gas emissions contribute to global warming
Incoming sunlight warms Earth's surface and is radiated back to the atmosphere. Greenhouse gases absorb some of this heat, trapping it in the atmosphere. Human activities have greatly increased levels of greenhouse gases, resulting in a gradual rise in average global temperatures.

have amplified the natural greenhouse effect, producing a phenomenon called **global warming**.

Historical temperature records have revealed a global temperature increase of 0.5 to 1.0 °C since 1860, paralleling the rise in atmospheric CO_2 (Fig. 40-16). The 1980s were the warmest decade on record, until their record was topped by the 1990s.

Many interacting factors, such as increases in cloud cover caused by increased evaporation, possible increases in net primary productivity due to increased CO_2, and the uncertain capacity of the ocean to absorb CO_2, make future climate prediction difficult and imprecise. However, if greenhouse gas emissions are not curtailed, the United Nations-sponsored Intergovernmental Panel on Climate Change (IPPC) predicts an increase of between 2.5 to 10.4 °F (1.4 to 5.8 °C) in average world surface air temperatures from 1990 to 2100, with temperatures rising more rapidly than in the past 10,000 years. To put this in perspective, average air temperatures during the peak of the last Ice Age (20,000 years ago) were only about 5 °C lower than at present. This extremely rapid temperature rise is of particular concern because it is likely to exceed the rate at which natural selection can provide evolutionary adaptations to the change. Because the temperature change is not distributed evenly worldwide, U.S. temperatures are predicted to increase considerably faster than the global average.

Global Warming May Have Severe Consequences

As geochemist James White at the University of Colorado quips, "If the Earth had an operating manual, the chapter on climate might begin with the caveat that the system has been adjusted at the factory for optimum comfort, so don't touch the dials." The consequences of global warming are only partially predictable and may be severe. As polar ice caps and glaciers melt and ocean waters expand in response to atmospheric warming, sea levels will rise, threatening coastal cities and flooding coastal wetlands.

Even small temperature changes can dramatically alter the paths of major air and ocean currents, altering precipitation patterns in unpredictable ways. Some land might become too hot and dry for agriculture, while other areas might become warmer, wetter, and more productive. Overall, U.S. crop productivity might increase. As the world warms, however, experts predict

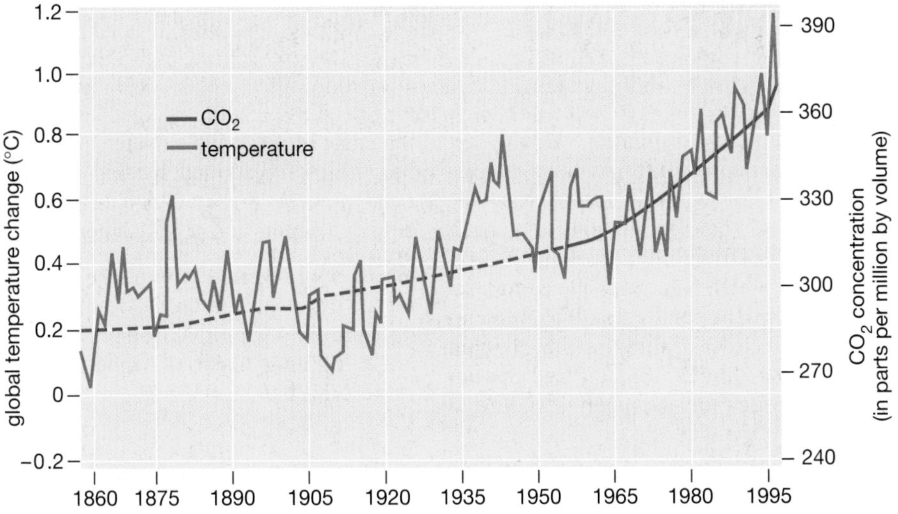

Figure 40-16 Global warming parallels CO_2 increases
The CO_2 concentration of the atmosphere (blue line) has shown a steady increase since 1860. The dashed portion of that curve represents measurements made from air trapped in ice cores; the solid portion reflects direct measurements made at Mauna Loa, Hawaii. Average global temperatures (red line) have also shown a gradual increase, paralleling the increasing atmospheric CO_2. (With thanks to Drs. Kevin Trenberth and Jim Hurrell of the National Center for Atmospheric Research.)

longer, more severe droughts and more extremes in rainfall, leading to more frequent crop failure and flooding. Agricultural disruption could be disastrous for some nations that are already barely able to feed themselves.

The impact of global warming on forests could be profound. While overall forest growth in the U.S. may increase, species distributions will change. For example, sugar maples may disappear from northeastern U.S. forests, while southeastern forests could be partially replaced by grasslands. The IPCC notes that global warming will increase the range of tropical disease-carrying organisms, such as malaria-transmitting mosquitos, with negative consequences for human health. Coral reefs, already stressed by human activities, are likely to suffer further damage from warmer waters.

Biologists worldwide are documenting changes related to warming. In Europe, after analyzing plant data from sites throughout Europe, scientists at the University of Munich, Germany, concluded that the growing season increased by more than 10 days over the past 40 years. Mexican Jays in southern Arizona are nesting 10 days earlier than they did in 1971. Many species of butterflies have shifted their ranges northward. In the Monteverde Cloud Forest Preserve in Costa Rica, rising ocean temperatures have caused more mist-free days, with adverse effects on local amphibians, birds, and reptiles. Some have shifted their range to higher elevations, other populations have disappeared. While each individual report could be due to other factors, the cumulative weight of data from diverse sources worldwide provides strong evidence that warming-related biological changes have begun. These data also provide researchers with a glimpse of what is to come as global warming continues.

Our Decisions Make a Difference

What can we as individuals do to help reduce both acid rain and global warming? The U.S. generates 6.6 tons of greenhouse gases per person each year. With less than 5% of the world's population, the U.S. is responsible for more than 20% of world's greenhouse emissions. A car with a fuel efficiency of 20 miles per gallon releases 1 pound of CO_2 per mile, so we can substantially reduce emissions of CO_2 by using fuel-efficient vehicles, car-pools, and public transportation. As electricity is generated in fossil fuel-fired power plants, so are tremendous quantities of CO_2, sulfur dioxide, and nitrogen oxides. We can support the efforts of utility companies to use renewable energy soures such as wind and solar power. To conserve electricity, we can purchase more efficient appliances, turn off unused computers and lights, and replace incandescent with fluorescent light bulbs. Insulating and weatherproofing our homes, incorporating solar energy features into new homes, and planting deciduous trees near our houses to provide for summer shade and winter sun will significantly reduce fuel consumption while cutting back on heating and air conditioning costs. Recycling is also a tremendous energy saver—for example, 95% of the energy used to produce an aluminum can from raw materials is conserved when the can is recycled. We can also support reforestation efforts to replace trees both in tropical rain forests and in our communities.

Although continued global warming is inevitable, its rate can be reduced by human decisions. Global Earth Summit conferences, such as that held in Kyoto, Japan, in 1997 offer a ray of hope. There, representatives from 160 nations proposed a treaty whose goal is to curb CO_2 emissions. If ratified, the treaty will commit industrialized nations to reducing CO_2 emissions to about 5% below 1990 levels by the year 2012. Developing countries, such as China and India, are not bound by the Kyoto agreement. Many observers are concerned that the benefits of the treaty could be overwhelmed by emissions from such nations, which are home to more than one-third of Earth's people, as they strive to industrialize and increase their standard of living. Although the obstacles are considerable, the Kyoto conference is evidence of a worldwide recognition of global warming and a concern for its consequences, the first steps toward change.

CASESTUDYREVISITED

Flight from Extinction

As you now know, predators such as eagles are relatively rare, because they occupy the highest trophic levels, where food is least abundant. Such predators are also most vulnerable to toxic substances that accumulate in fat and are passed up the food chain. Since DDT was banned in the U.S. in 1972, not only the eagle, but populations of osprey, peregrine falcons, and brown pelicans are also rebounding. Unfortunately, in many developing countries, DDT is still used, particularly to control malaria-carrying mosquitos (some populations of mosquitos are becoming DDT-resistant as a result). Although it remains illegal to hunt bald eagles, as eagles fly off the endangered species list, they lose the habitat protection that the Endangered Species Act provides for endangered wildlife. If the eagle is to survive outside of protected areas, its nests and feeding grounds must now be preserved voluntarily by land developers and private landowners.

While environmental groups are pushing for a worldwide ban on DDT, this ban is opposed by some organizations dedicated to fighting malaria. Argue for and against this ban.

Summary of Key Concepts

1) What Are the Pathways of Energy and Nutrients?

Ecosystems are sustained by a continuous flow of energy from sunlight and a constant recycling of nutrients.

2) How Does Energy Flow Through Communities?

Energy enters the biotic portion of ecosystems when it is harnessed by autotrophs during photosynthesis. Net primary productivity is the amount of energy that autotrophs store in a given unit of area over a given period of time.

Trophic levels describe feeding relationships in ecosystems. Autotrophs are the producers, the lowest trophic level. Herbivores occupy the second level as primary consumers. Carnivores act as secondary consumers when they prey on herbivores and as tertiary or higher-level consumers when they eat other carnivores.

Feeding relationships in which each trophic level is represented by one organism are called *food chains*. In natural ecosystems, feeding relationships are far more complex and are described as food webs. Detritus feeders and decomposers, which digest dead bodies and wastes, use and release the energy stored in these substances and free up nutrients for recycling. In general, only about 10% of the energy captured by organisms at one trophic level is converted to the bodies of organisms in the next highest level. The higher the trophic level, the less energy is available to sustain it. As a result, plants are more abundant than herbivores, and herbivores are more common than carnivores. The storage of energy at each trophic level is illustrated graphically as an energy pyramid. The energy pyramid explains biological magnification, the process by which toxic substances accumulate in increasingly high concentrations in progressively higher trophic levels.

3) How Do Nutrients Move Within and Among Ecosystems?

A nutrient cycle depicts the movement of a particular nutrient from its reservoir (usually in the abiotic, or nonliving, portion of the ecosystem) through the biotic, or living, portion of the ecosystem and back to its reservoir, where it is again available to producers. Carbon reservoirs include the oceans, atmosphere, and fossil fuels. Carbon enters producers through photosynthesis. From autotrophs it is passed through the food web and released to the atmosphere as CO_2 during cellular respiration.

The major reservoir of nitrogen is the atmosphere. Nitrogen gas is converted by bacteria and human industrial processes into ammonia and nitrate, which can be used by plants. Nitrogen passes from producers to consumers and is returned to the environment through excretion and the activities of detritus feeders and decomposers.

The reservoir of phosphorus is in rocks as phosphate, which dissolves in rainwater. Phosphate is absorbed by photosynthetic organisms, then passed through food webs. Some is excreted, and the rest is returned to the soil and water by decomposers. Some is carried to the oceans, where it is deposited in marine sediments. Humans mine phosphate-rich rock to produce fertilizer.

The major reservoir of water is the oceans. Water is evaporated by solar energy and returned to Earth as precipitation. The water flows into lakes and underground reservoirs and in rivers, which flow to the oceans. Water is absorbed directly by plants and animals and is also passed through food webs. A small amount is combined with CO_2 during photosynthesis to form high-energy molecules.

4) What Is Causing Acid Rain and Global Warming?

Environmental problems occur when human activities interfere with the natural functioning of ecosystems. Human industrial processes release toxic substances and produce more nutrients than nutrient cycles can efficiently process. Through massive consumption of fossil fuels, we have increased the flow of energy and disrupted the natural cycles of carbon, sulfur, and nitrogen, causing acid rain and global warming (an amplification of the greenhouse effect).

Key Terms

acid deposition *p. 848*
autotroph *p. 837*
biodegradable *p. 841*
biogeochemical cycle *p. 843*
biological magnification *p. 841*
biomass *p. 837*
carnivore *p. 838*
consumer *p. 837*
decomposer *p. 838*
deforestation *p. 849*
denitrifying bacterium *p. 845*
detritus feeder *p. 838*
energy pyramid *p. 842*
food chain *p. 838*
food web *p. 838*
fossil fuel *p. 843*
global warming *p. 850*
greenhouse effect *p. 849*
greenhouse gas *p. 849*
herbivore *p. 838*
heterotroph *p. 837*
hydrologic cycle *p. 846*
legume *p. 844*
macronutrient *p. 843*
micronutrient *p. 843*
net primary productivity *p. 837*
nitrogen fixation *p. 844*
nutrient *p. 843*
nutrient cycle *p. 843*
omnivore *p. 838*
primary consumer *p. 838*
producer *p. 837*
reservoir *p. 843*
secondary consumer *p. 838*
tertiary consumer *p. 838*
trophic level *p. 838*

Thinking Through the Concepts

Multiple Choice

1. *Why do scientists think that human-induced global warming will be more harmful to plants and animals than were past, natural climate fluctuations?*
   a. because temperatures will change faster
   b. because the temperature changes will be larger
   c. because species now are less adaptable than species in the past
   d. because ecosystems are now more complicated than they used to be
   e. because the temperature changes will last longer
2. *The major source(s) of carbon for living things is (are)*
   a. coal, oil, and natural gas
   b. plants
   c. CO_2 in the atmosphere and oceans
   d. methane in the atmosphere
   e. carbon in animal bodies
3. *Why would you expect there to be a smaller biomass of big predators (lions, leopards, hunting dogs, etc.) than of grazing mammals (gazelles, zebra, elephants, etc.) in the African savanna?*
   a. too little cover for the predators to hide in
   b. the inefficiency of energy transfer between trophic levels
   c. like domestic cats, large predators are susceptible to hair balls
   d. many predators occur in social groups, thus limiting their numbers
   e. because grazers are better adapted to moving long distances, they can better follow the rains
4. *What is the one group of organisms that is able to fix atmospheric nitrogen into forms usable by living organisms?*
   a. plants
   b. fungi
   c. insects
   d. bacteria
   e. viruses
5. *The biological process by which carbon is returned to its reservoir is*
   a. photosynthesis
   b. denitrification
   c. carbon fixation
   d. glycolysis
   e. cellular respiration
6. *As a black widow spider consumes her mate, what is the lowest trophic level she could be occupying?*
   a. third
   b. first
   c. second
   d. fourth
   e. fifth

? Review Questions

1. What makes the flow of energy through ecosystems fundamentally different from the flow of nutrients?
2. What is an autotroph? What trophic level does it occupy, and what is its importance in ecosystems?
3. Define *primary productivity*. Would you predict higher productivity in a farm pond or an alpine lake? Defend your answer.
4. List the first three trophic levels. Among the consumers, which are most abundant? Why would you predict that there will be a greater biomass of plants than herbivores in any ecosystem? Relate your answer to the "10% law."
5. How do food chains and food webs differ? Which is the more accurate representation of actual feeding relationships in ecosystems?
6. Define *detritus feeders* and *decomposers*, and explain their importance in ecosystems.
7. Trace the movement of carbon from its reservoir through the biotic community and back to the reservoir. How have human activities altered the carbon cycle, and what are the implications for future climate?
8. Explain how nitrogen gets from the air to a plant.
9. Trace a phosphorus molecule from a phosphate-rich rock into the DNA of a carnivore. What makes the phosphorus cycle fundamentally different from the carbon and nitrogen cycles?
10. Trace the movement of a water molecule from the moment it leaves the ocean until it eventually reaches a plant root, then a plant stoma, and then makes its way back to the ocean.

Applying the Concepts

1. What could your college or university do to reduce its contribution to acid rain and global warming? Be specific and, if possible, offer practical alternatives to current practices.
2. Relate fossil fuel consumption to (a) the loss of aquatic life in lakes in the Northeast and Canada and (b) the lengthening of the growing season in Europe. Trace each step from the burning of gasoline in a car or power plant to the change in question.
3. Define and give an example of *biological magnification*. What qualities are present in materials that undergo biological magnification? In which trophic level are the problems worst, and why?
4. Discuss the contribution of population growth to (a) acid rain and to (b) the greenhouse effect.
5. Describe what would happen to a population of deer if all predators were removed and hunting was banned. Include effects on vegetation as well as on the deer population itself. Relate your answer to carrying capacity as discussed in Chapter 38.

For More Information

Epstein, P. R. "Is Global Warming Harmful to Health?" *Scientific American*, August 2000. Global warming may increase the incidence of disease as the climate becomes more hospitable for disease-causing organisms and their vectors.

Holmes, B. "Case of the Dwindling Cloud Forest." *International Wildlife*, July–August 2000. The unique and rare animals and plants inhabiting the cloud forest of Monteverde, Costa Rica, are in trouble as their habitat becomes increasingly dry. Global warming may be to blame.

Karl, T. R., Nicholls, N., and Gregory, J. "The Coming Climate." *Scientific American*, May 1997. What changes in weather patterns can we expect in a warmer world?

Mlot, C. "Tallying Nitrogen's Increasing Toll." *Science News*, February 15, 1997. Human activities now dominate the nitrogen cycle, with increasingly disruptive effects.

Post, W. M., and others. "The Global Carbon Cycle." *American Scientist*, July–August 1990. A readable and comprehensive look at the natural carbon cycle and how humans have influenced it.

Wuethrich, B. "How Climate Change Alters Rhythms in the Wild." *Science*, February 4, 2000. The distribution, behavior, and ecology of many animal populations are already being influenced by global warming.

Answers to Multiple-Choice Questions

1. a 2. c 3. b 4. d 5. e 6. d

MEDIATUTOR
How Do Ecosystems Work?

CD Activities

Activity 40.1: Building a Food Web

Estimated time: 10 minutes

In this exercise you will have the opportunity to learn about food webs for a variety of ecosystems. You can also reinforce your understanding of trophic interactions by identifying the roles that organisms play within the ecosystem.

Start the MediaTutor Student CD-ROM and enter the activity number in the Quick Search box to be taken directly to that activity.

Web Investigations

Case Study: Flight from Extinction

Estimated time: 10 minutes

There are thousands of endangered species. Often their favorite habitats are also places where humans would like to work, live, and raise healthy families. The bald eagle nearly became extinct because of the pesticide DDT. Now the number of birds is up, but so is the incidence of malaria. What happens when species conservation and human needs collide?

Go to http://www.prenhall.com/audesirk6, the Audesirk Companion Web site. Select Chapter 40 and the Web Investigation to begin.

"We must consider our planet to be on loan from our children, rather than being a gift from our ancestors."

Gro Harlem Brundtland, former Prime Minister of Norway

Kahindi Samson snares a butterfly. Insets: ***(top)*** A dark blue pansy butterfly. ***(bottom)*** Pupae are identified and sorted for shipment.

41 Earth's Diverse Ecosystems

AT A GLANCE

Case Study: Wings of Hope

1) **What Factors Influence Earth's Climate?**
Both Climate and Weather Are Driven by the Sun
Many Physical Factors Also Influence Climate

2) **What Conditions Does Life Require?**

3) **How Is Life on Land Distributed?**
Terrestrial Biomes Support Characteristic Plant Communities
Rainfall and Temperature Determine What Vegetation a Biome Can Support

4) **How Is Life in Water Distributed?**
Freshwater Lakes Have Distinct Regions of Life
Freshwater Lakes Are Classified According to Their Nutrient Content
Marine Ecosystems Cover Much of Earth
Coastal Waters Support the Most Abundant Marine Life

Case Study Revisited: Wings of Hope

CASESTUDY

Wings of Hope

As a 12-year-old, Kahindi Samson began sneaking into the Arabuko-Sokoke forest, which is designated as protected by the Kenyan government, snaring its endangered antelope and cutting the old growth trees that provide homes for the rare Sokoke Scops owl. Kahindi needed to help support and feed his five younger brothers and sisters. This Kenyan forest is precious—the largest remnant of coastal forest in East Africa, and a final refuge for endangered birds and mammals that have been displaced by the growing human population. But to the farmers on surrounding land, the forest was the enemy, home to elephants and baboons who emerged at night to eat their crops. Most wanted to see the forest felled. Ian Gordon, a butterfly ecologist, watched the poaching and tree-felling with alarm; Arabuko-Sokoke forest is home to 250 species of butterflies. Unwilling to stand by helplessly, Gordon founded Project Kipepeo, meaning "butterfly" in Swahili. His mission was to convince skeptical local farmers to grow butterflies instead of crops. Today, Samson enters the forest with a license and a butterfly net. In a cage outside his home, he places his catch of pregnant female butterflies. As their eggs hatch, Samson fattens the caterpillars on leaves he collects from the forest. Within a month, the caterpillars are ready to form pupae and be shipped to the U.S. and Europe. There they will hatch amidst lush tropical vegetation in butterfly gardens, where they will delight visitors who have never witnessed the splendor of rainforest butterflies. Samson is one of 550 local butterfly workers who now rely on the Arabuko-Sokoke forest for their livelihood and make a far better living than before. "We used to wish the forest would go away," says butterfly farmer Priscilla Kiti, "but these days we earn our living mostly in butterfly farming, so if they cut the forest, things are going to be very difficult." ■

1 What Factors Influence Earth's Climate?

The distribution of life, particularly on land, is dramatically affected by both weather and climate. **Weather** refers to short-term fluctuations in temperature, humidity, cloud cover, wind, and precipitation in a region over periods of hours or days. **Climate**, in contrast, refers to patterns of weather that prevail from year to year and even century to century in a particular region. The amount of sunlight and water and the range of temperatures determine the climate of a given region. Whereas weather affects individual organisms, climate influences and limits the overall distribution of entire species.

Both Climate and Weather Are Driven by the Sun

Both climate and weather are driven by a great thermonuclear engine: the sun. Solar energy reaches Earth in a range of wavelengths, from short, high-energy ultraviolet (UV) rays, through visible light, to the longer infrared wavelengths that produce heat (see Fig. 7-3). The solar energy that reaches Earth drives the wind, the ocean currents, and the global water cycle. Before it

Earth Watch
The Ozone Hole—A Puncture in Our Protective Shield

A small fraction of the radiant energy produced by the sun, called *ultraviolet*, or UV, radiation, is so highly energetic that it can damage biological molecules. In small quantities, UV radiation helps human skin produce vitamin D and causes tanning in fair-skinned people. But in larger doses, UV causes sunburn and premature aging of skin, skin cancer, and cataracts, a condition in which the lens of the eye becomes cloudy.

Fortunately, most of the UV radiation is filtered out by ozone in the stratosphere, a layer of atmosphere extending from 10 to 50 kilometers (6 to 30 miles) above Earth. In pure form, *ozone* (O_3) is a bluish, explosive, and highly poisonous gas. In the stratosphere, the normal concentration of ozone is about 0.1 part per million (ppm), compared with 0.02 ppm in the lower atmosphere. This ozone-enriched layer is called the **ozone layer**. Ultraviolet light striking ozone and oxygen causes reactions that break down as well as regenerate ozone. In the process, the UV radiation is converted to heat, and the overall level of ozone remains reasonably constant—or so it did until humans intervened.

In 1985 British atmospheric scientists published a startling discovery: The springtime levels of stratospheric ozone over Antarctica had declined by more than 40% since 1977. A hole had been punctured in Earth's protective shield. In the *ozone hole* over Antarctica, ozone now dips to about one-third of its pre-depletion levels (Fig. E41-1). Although ozone layer depletion is most severe over Antarctica, the ozone layer is somewhat reduced over most of the world, including nearly all of the continental U.S. Recently, springtime ozone levels over Greenland, Scandinavia, and western Siberia were reduced up to 40%. Evidence of ozone thinning over the Arctic

Figure E41-1 Satellite image of the Antarctic ozone hole
A record-sized ozone hole is seen in purple on this image based on satellite data from September 10, 2000. Notice that the hole extended over the tip of South America, the first time it has exposed a city to dangerous levels of UV radiation. (Image courtesy of NASA.)

reaches Earth's surface, however, sunlight is modified by the atmosphere. A layer relatively rich in ozone (O_3) is located in the middle atmosphere. This *ozone layer* absorbs much of the sun's high-energy UV radiation, which can damage biological molecules (see "Earth Watch: The Ozone Hole—A Puncture in Our Protective Shield"). Dust, water vapor, and clouds scatter light, reflecting some of the energy back into space. Carbon dioxide, water vapor, methane, and other *greenhouse gases* selectively absorb energy at infrared wavelengths, trapping heat in the atmosphere. Human activities have increased levels of greenhouse gases, as described in Chapter 40.

Only about half the solar energy reaching the atmosphere actually strikes Earth's surface. Of this, a small fraction is immediately reflected directly back into space, another small fraction is captured by photosynthetic plants and microorganisms and used to power photosynthesis, and the rest is absorbed as heat. Eventually, nearly all the incoming solar energy is returned to space, either as light or as infrared radiation (heat). The solar energy temporarily absorbed and stored as heat by the atmosphere and by Earth's surface maintains Earth's relative warmth.

Many Physical Factors Also Influence Climate

Many physical factors influence climate. Among the most important are latitude, air currents, ocean currents, and the presence of mountains and irregularly shaped continents.

Latitude Influences the Angle at Which Sunlight Strikes Earth

Latitude, measured in degrees, is the distance north or south from the equator, which is located at 0° latitude. The amount of sunlight that strikes a given area of Earth's surface has a major effect on average yearly temperatures. At the equator, sunlight hits Earth's surface nearly at a right angle, making the weather there consistently warm. Farther north or south, the sun's rays strike Earth's surface at a greater slant. This angle spreads the same amount of sunlight over a larger area, producing lower overall temperatures (Fig. 41-1).

has led researchers to predict that we will soon have an Arctic, as well as an Antarctic, ozone hole.

Satellite data reveal that, since the early 1970s, UV radiation has risen by nearly 7% per decade in the Northern Hemisphere and by almost 10% per decade in the Southern Hemisphere. In New Zealand, researchers have charted a 12% rise during the past decade. Skin cancer rates are predicted to increase as a result. Studies using mice show that exposure to UV light in dosages similar to a day at the beach without sunscreen can depress their immune systems, making them more vulnerable to a variety of diseases. Human health effects are only one cause for concern. Photosynthesis by phytoplankton, the producers for marine ecosystems, is reduced under the ozone hole above Antarctica. Some types of trees and farm crops are also harmed by increased UV radiation.

The decline in the thickness of the ozone layer is caused by increasing levels of chlorofluorocarbons (CFCs), compounds that contain chlorine, fluorine, and carbon. Developed in 1928, these gases were widely used as coolants in refrigerators and air conditioners, aerosol spray propellants, in the production of foam plastic, and as cleansers for electronic parts. These chemicals are very stable and were considered safe. Their stability, however, proved to be a major problem, because they remain chemically unchanged as they gradually rise into the stratosphere. There, under intense bombardment by UV light, the CFCs break down, releasing chlorine atoms. Chlorine catalyzes the breakdown of ozone to oxygen gas (O_2) while remaining unchanged itself. Clouds over the Arctic and Antarctic regions are composed of ice particles that provide a surface on which the reaction can occur.

Fortunately, we have taken the first steps toward "plugging" the ozone hole. In an almost unprecedented example of global cooperation and concern, talks beginning in 1985 led to treaties in 1990 and 1992 in which industrialized nations throughout the world agreed to phase out ozone-depleting chemicals rapidly, eliminating CFCs by 1996. These compounds are no longer used in foam plastic production, and CFC substitutes have been found for use in spray cans, refrigerators, and car air conditioners as well. Ground-level global atmospheric chlorine levels (an indicator of CFC use) peaked in 1994. In 1996 scientists measured a slight reduction in chlorine in the lower regions of the atmosphere. By 1999 they detected chlorine reductions in the stratosphere as well, the result of gradual mixing of air and transport of ground-level air upward. But use of some types of CFCs has increased dramatically in developing countries, and a large reservoir of CFCs remains stored in old refrigeration units. Because these highly stable compounds can persist 50 to 100 years and take a decade or more to ascend into the stratosphere, our current release of CFCs, in conjunction with the millions of tons already released, will continue to erode the protective ozone shield, with possibly severe consequences. For example, in September 2000, the Antarctic ozone hole was larger than ever before, extending over the tip of Chile and exposing residents of the city of Punta Arenas to high levels of UV radiation (see Fig. E41-1). Some scientists are optimistic, however, that the ozone hole may begin to get smaller by the year 2010 and will possibly be eliminated by 2050 if global cooperation continues.

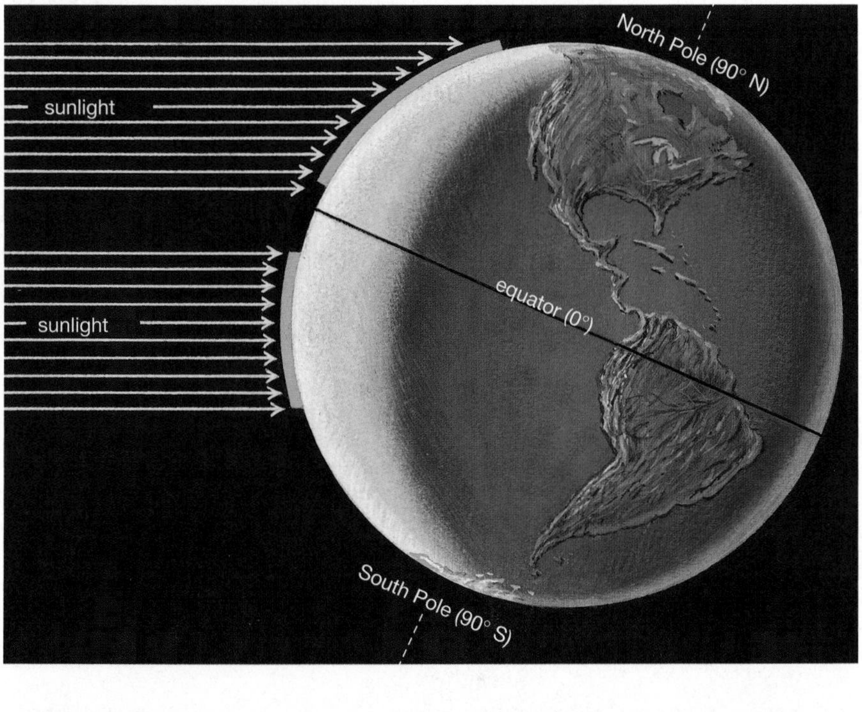

Figure 41-1 Earth's curvature and tilt produce seasons and climate
Near the equator, sunlight falls nearly perpendicularly on Earth's surface, so its warmth is concentrated onto a relatively small area. Toward the poles, the same amount of sunlight falls over a much larger surface area. Temperatures are thus highest at the equator and lowest at the poles. The tilt of Earth on its axis causes seasonal variations in the directness of sunlight. Here, we see winter in the Northern Hemisphere and summer in the Southern Hemisphere.

Earth is tilted on its axis as it makes its yearly trip around the sun. The higher latitudes therefore experience considerable variation in the directness of sunlight throughout the year, resulting in pronounced seasons. For example, when the Northern Hemisphere is tilted toward the sun, it receives sunlight most directly and experiences summer; when that hemisphere is in winter, the Southern Hemisphere is closest to the sun (see Fig. 41-1). The tilting hardly affects the angle of the equator to the sun's rays, so there is little seasonal variation there.

Air Currents Produce Broad Climatic Regions

Air currents are generated by Earth's rotation and by differences in temperature between different air masses. Because warm air is less dense than cold air, as the sun's direct rays fall on the equator, heated air rises there. The warm air near the equator is also laden with water evaporated by solar heat (Fig. 41-2a). As the water-saturated air rises, it cools somewhat. Cool air cannot retain as much moisture as can warm air, so water condenses from the rising air and falls as rain. The direct rays of the sun and the rainfall produced when warm, moist air rises and is cooled create a band around the equator, called the *Tropics*. This region is both the warmest and the wettest on Earth. The cooler dry air then flows north and south from the equator. Near 30° N and 30° S, the cooled air is dense enough to sink. As it sinks, the air is warmed by heat radiated from Earth. By the time it reaches the surface, it is both warm and very dry. Not surprisingly, the major deserts of the world are found at these latitudes (Fig. 41-2a,b). This air then flows back toward the equator. Farther north and south, this general circulation pattern is repeated, dropping moisture at around 60° N and 60° S and creating extremely dry conditions at the North and South Poles.

Ocean Currents Moderate Near-Shore Climates

Ocean currents are driven by Earth's rotation, winds, and the direct heating of the water by the sun. Continents interrupt the currents, breaking them into roughly circular patterns called **gyres**. Gyres rotate clockwise in the Northern Hemisphere and counterclockwise in the Southern Hemisphere (Fig. 41-3). Because water both heats and cools more slowly than does land or air, ocean currents tend to moderate temperature extremes. Coastal areas, then, generally have less-variable climates than do areas near the center of continents. For example, a gyre in the Atlantic Ocean brings warm water (the Gulf Stream) from equatorial regions north along the eastern coast of North America, creating a warmer, moister climate than is found farther inland. It then carries the still-warm water farther north and east, warming the western coast of Europe before returning south.

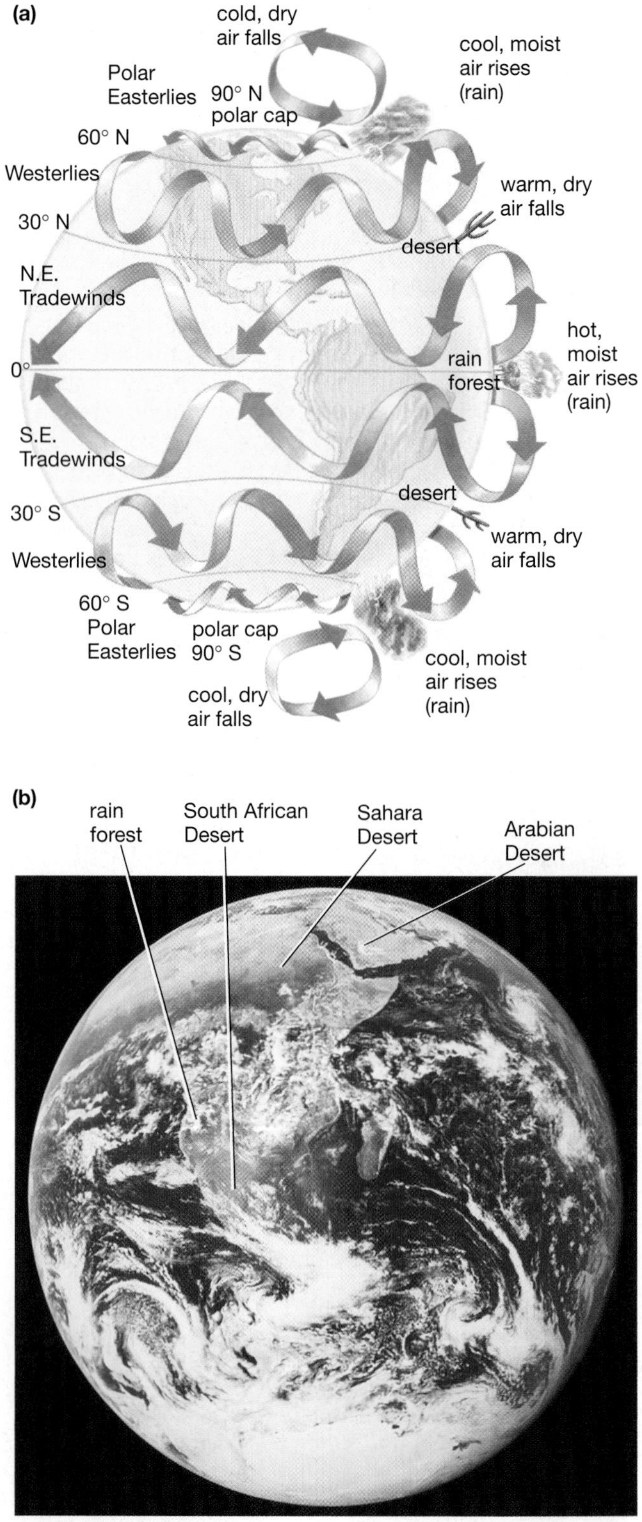

Figure 41-2 Distribution of air currents and climatic regions ***(a)*** Rainfall is determined primarily by the distribution of temperatures and by Earth's rotation. The interaction of these two factors creates air currents that rise and fall predictably with latitude, producing broad climatic regions. ***(b)*** Some of these regions are visible in this photograph of the African continent taken from *Apollo 11.* Along the equator are heavy clouds that drop moisture on the central African rain forests. Note the lack of clouds over the Sahara and Arabian Deserts near 30° N and the South African Desert near 30° S.

MEDIATUTOR 41.1 Climate Builds Biomes

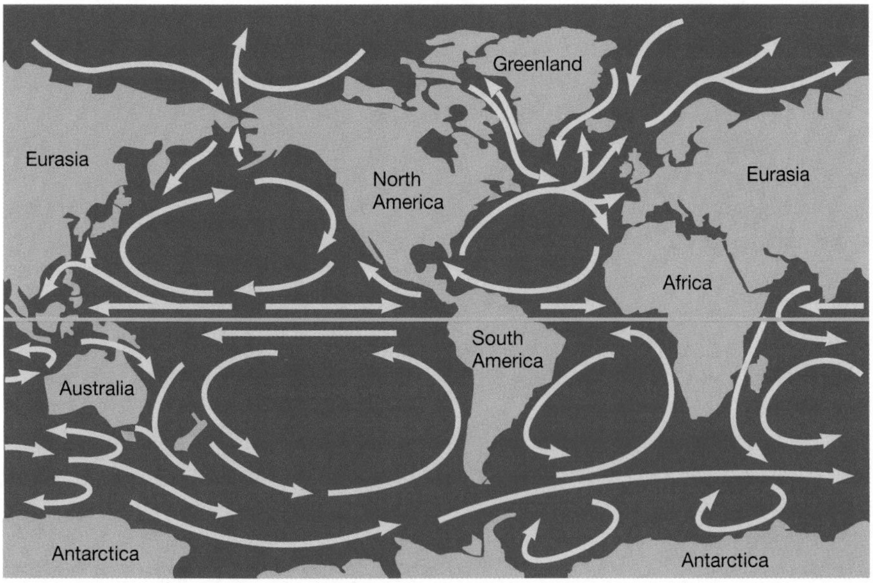

Figure 41-3 Ocean circulation patterns are called gyres
Gyres travel clockwise in the Northern Hemisphere and counterclockwise in the Southern Hemisphere. These currents tend to distribute warmth from the equator to northern and southern coastal areas.

Continents and Mountains Complicate Weather and Climate

If Earth's surface were uniform, climate zones would occur in bands corresponding to latitude. These zones would result from the interaction of temperature and rainfall, which are determined by the rise and fall of air masses (see Fig. 41-2a). The presence of irregularly shaped continents (which heat and cool relatively quickly) amidst oceans (which heat and cool more slowly) alters the flow of wind and water and contributes to the irregular distribution of ecosystems.

Variations in elevation within continents further complicate the situation. As elevation increases, the atmosphere becomes thinner and retains less heat. The temperature drops approximately 3.5 °F (2 °C) for every 1000 feet (305 meters) elevation gain. This characteristic explains why snow-capped mountains are found even in the Tropics (Fig. 41-4).

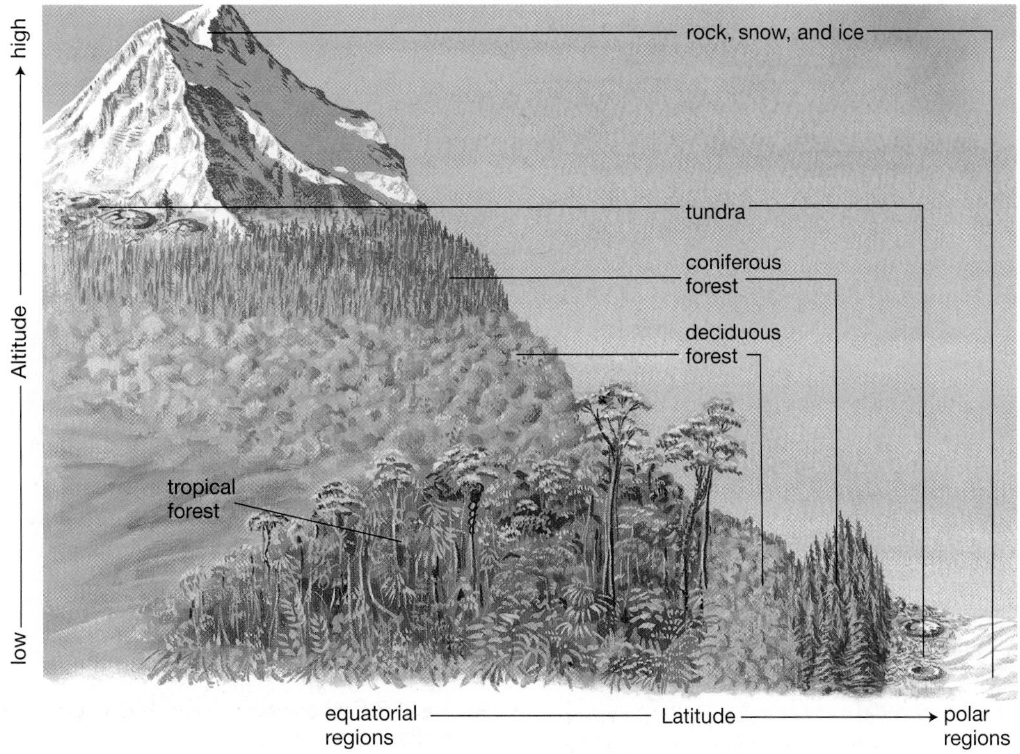

Figure 41-4 Effects of elevation on temperature
Climbing a mountain in some ways is like going north; in both cases, increasingly cool temperatures produce a similar series of biomes.

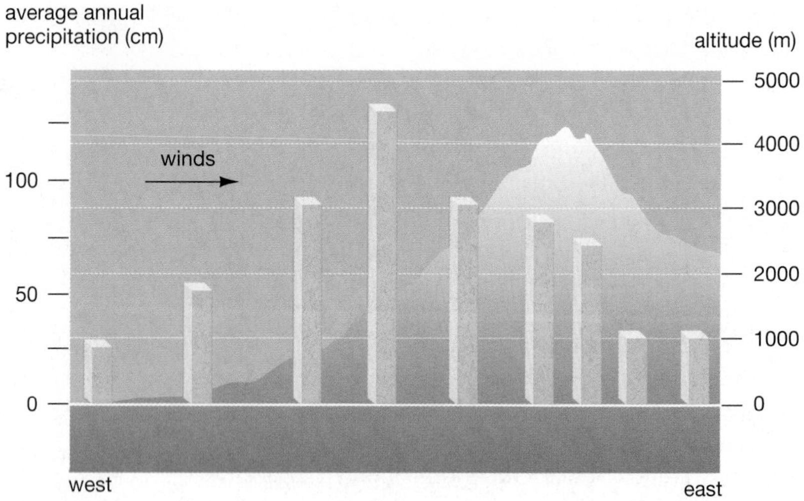

Figure 41-5 The rain shadow of the Sierra Nevada Here, rainfall is graphed against altitude. Westerly (eastward-moving) winds deposit their moisture on the western slope, leaving a desert to the east.

Mountains also modify rainfall patterns. When water-laden air is forced to rise as it meets a mountain, it cools. Cooling reduces the air's ability to retain water, and the water condenses as rain or snow on the windward (near) side of the mountain. The cool, dry air is warmed again as it travels down the far side of the mountain and absorbs water from the land, creating a local dry area called a **rain shadow**. For example, mountain ranges such as the Sierra Nevada of the western U.S. wring the moisture from the westerly winds that come off the Pacific Ocean, leaving deserts in the rain shadow on their eastern sides (Fig. 41-5).

2 What Conditions Does Life Require?

From the lichens on bare rock to the thermophilic (Greek for "heat-loving") algae in the hot springs of Yellowstone National Park to the bacteria thriving under the pressure-cooker conditions of a deep-sea vent, Earth teems with life. Underlying the diversity of habitats is the common ability to provide to varying degrees the four fundamental resources required for life. These resources are:

1. Nutrients from which to construct living tissue
2. Energy to power that construction
3. Liquid water to serve as a medium in which metabolic reactions occur
4. Appropriate temperatures in which to carry out these processes

As we shall see in the following sections, these resources are very unevenly distributed. Their availability limits the types of organisms that can exist within the various terrestrial and aquatic ecosystems on Earth.

Ecosystems are extraordinarily diverse, yet clear patterns exist. The community that is characteristic of each ecosystem is dominated by organisms specifically adapted to particular environmental conditions. Variations in temperature and in the availability of light, water, and nutrients shape the adaptations of organisms that inhabit an ecosystem. The desert community, for example, is dominated by plants adapted to heat and drought. The cacti of the Mojave Desert of the American Southwest are strikingly similar to the euphorbia of the deserts of Africa and of the nearby Canary Islands, although these plants are in separate families and are only distantly related genetically. Their spinelike leaves and thick, green, water-storing stems are adaptations for water conservation (Fig. 41-6). Likewise, the plants of the arctic tundra and those of the alpine tundra of the Rocky Mountains show growth patterns clearly recog-

Figure 41-6 Environmental demands mold physical characteristics
Evolution in response to similar environments has molded the bodies of ***(a)*** American cacti and ***(b)*** Canary Islands euphorbia into nearly identical shapes, although they are in different families. Both of these desert plants have thick, water-conserving stems and leaves reduced to defensive spines that minimize evaporation.

nizable as adaptations to a cold, dry, windy climate. Thus, in regions where the environmental conditions are similar, similar types of organisms are organized into similar types of communities.

3 How Is Life on Land Distributed?

Terrestrial organisms are restricted in their distribution largely by the availability of water and by temperature. Terrestrial ecosystems receive plenty of light, even on an overcast day, and the soil provides abundant nutrients. Water, however, is limited and very unevenly distributed, both in place and in time. Terrestrial organisms must be adapted to obtain water when it is available and to conserve it when it is scarce.

Like water, temperatures favorable to life are very unevenly distributed in place and time. At the South Pole, even in summer, the average temperature is usually well below freezing; not surprisingly, life is scarce there. Places such as central Alaska have favorable temperatures during only the summer, whereas the Tropics have a uniformly warm, moist climate, and life abounds there.

Terrestrial Biomes Support Characteristic Plant Communities

Terrestrial communities are dominated and defined by their plant life. Because plants can't escape from drought, sun, or winter weather, they tend to be precisely adapted to the climate of a particular region. Large land areas with similar environmental conditions and characteristic plant communities are called **biomes** (Fig. 41-7). Biomes are generally named after the major type of vegetation found there. The predominant vegetation of each biome is determined by the complex interplay of rainfall and temperature (Fig. 41-8). These factors determine the available soil moisture needed for plant growth as well as for compensation for water losses through evaporation. In addition to the total amount of

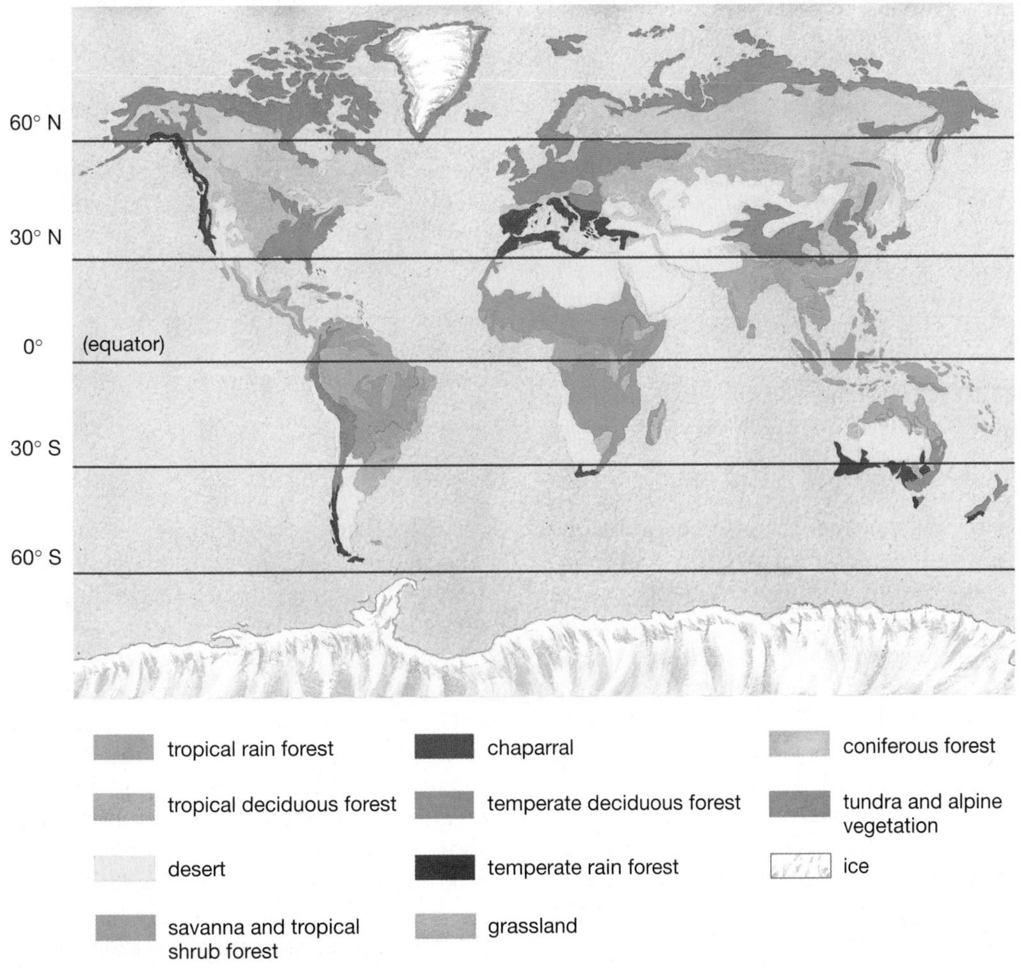

Figure 41-7 The distribution of biomes
Although mountain ranges and the sheer size of the continents complicate the pattern of biomes, note the overall consistencies. Tundras and coniferous forests are in the northernmost parts of the Northern Hemisphere, whereas the deserts of Mexico, the Sahara, Saudi Arabia, South Africa, and Australia are located about 20° to 30° N and S.

MediaTutor 41.2 Biomes of the World

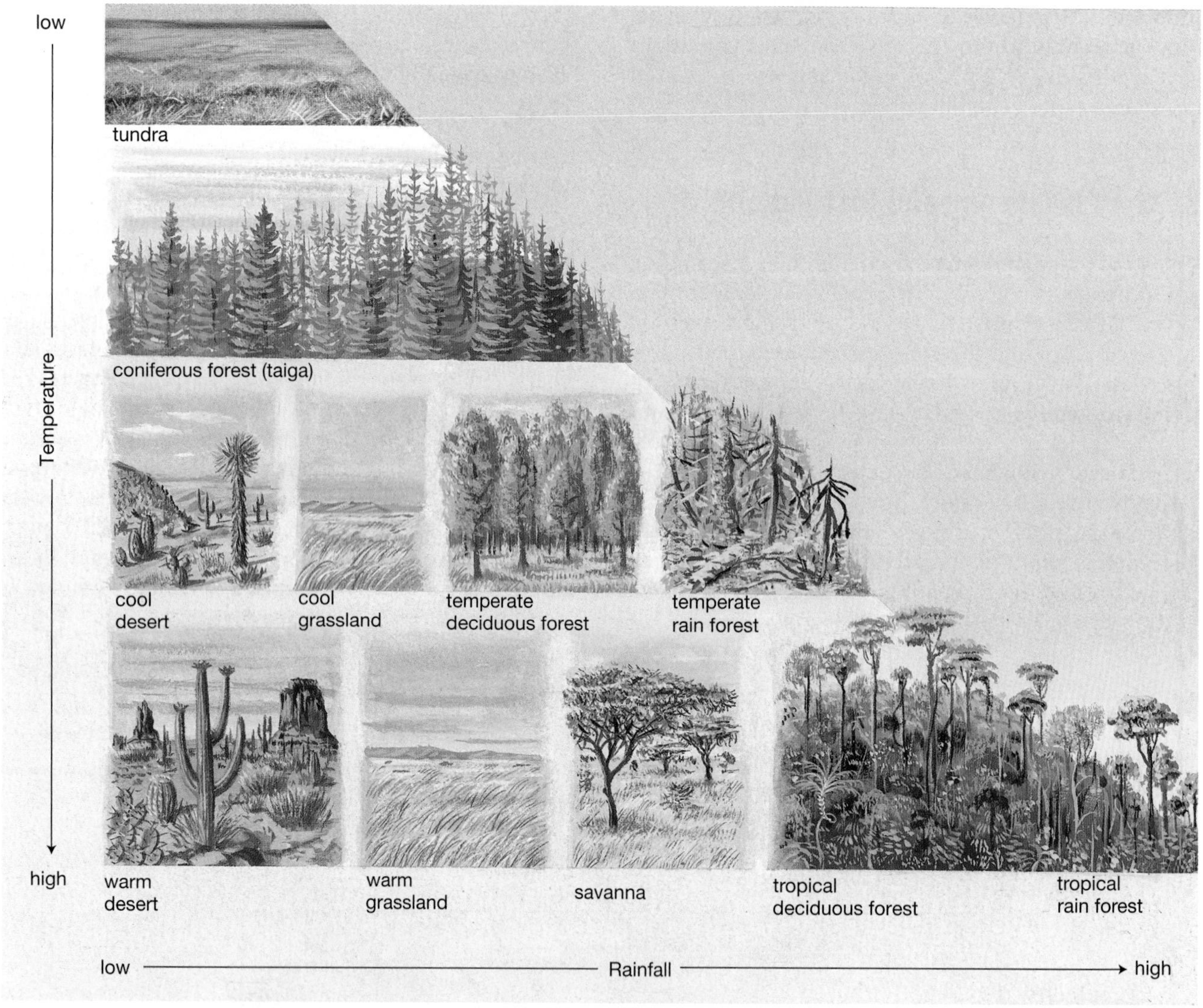

Figure 41-8 Rainfall and temperature influence biome distribution
Together, these factors determine the available soil moisture needed for plant growth.

rainfall and the overall yearly average temperature, the variability of rain and temperature over the year determines which plants can grow in an area. Arctic tundra plants, for example, must be adapted to marshy conditions in the early summer but cold and desertlike conditions for much of the rest of the year, when water is solidly frozen and unavailable. In the following sections we discuss the major biomes, beginning at the equator and working our way poleward. We also discuss some of the effects human activities have on these biomes.

Tropical Rain Forests

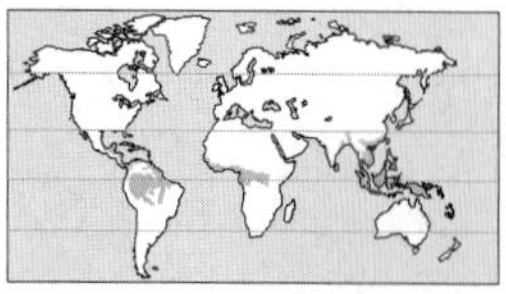

Large areas of South America and Africa lie along the equator. There the temperature averages between 77 °F and 86 °F (25 °C and 30 °C) with little variation, and rainfall ranges from 100 to 160 inches (250 to 400 centimeters) annually. These evenly warm, evenly moist conditions combine to create the most diverse biome on Earth, the **tropical rain forest**, dominated by huge broadleaf evergreen trees (Fig. 41-9).

Although rain forests cover only 6% of Earth's total land area, ecologists estimate that they are home to between 5 million and 8 million species, representing half to two-thirds of the world's total. **Biodiversity** refers to the total number of species within an ecosystem and the resulting complexity of interactions among them; in short, it defines the biological "richness" of an ecosystem. Rain forests have the highest biodiversity of any ecosystem on Earth. For example, a recent survey of a 2.5-acre site in the upper Amazon basin revealed 283 species of trees, most of which were represented by a single individual. In a 5-square-kilometer (about 3-square-mile) tract of rain forest in Peru, scientists counted more than 1300 species of butterfly and 600 species of bird. In

Figure 41-9 The tropical rain forest biome
(a) Towering trees draped with vines reach for the light in the dense tropical rain forest. Because food and light energy are found high in trees, amid their branches dwells the most diverse assortment of animals on Earth, including ***(b)*** a fruit-eating toucan, ***(c)*** a golden-eyed leaf frog, ***(d)*** a tree-climbing orchid, and ***(e)*** a howler monkey.

comparison, the entire United States is home to only 400 butterfly species and 700 bird species.

Tropical rain forests typically have several layers of vegetation. The tallest trees reach 150 feet (50 meters) and tower above the rest of the forest. Below is a fairly continuous canopy of treetops at about 90 to 120 feet (30 to 40 meters). Another layer of shorter trees typically stands below the canopy. Huge woody vines, commonly 328 feet (100 meters) or more in length, grow up the trees, reaching the sunlight far above. These layers of vegetation block out most of the sunlight. Many of the plants that live in the dim green light that filters through to the forest floor have enormous leaves to trap the little available energy.

Because edible plant material close to the ground in tropical rain forests is relatively scarce, much of the animal life, including numerous birds, monkeys, and insects, is *arboreal* (living in the trees). Competition for the nutrients that do reach the ground is intense among both animals and plants. Even such unlikely sources of food as the droppings of monkeys are in great demand. For example, dung beetles feed and lay their eggs on monkey droppings. When ecologists attempted to collect droppings of the South American howler monkeys to find out what the monkeys had been eating, they found themselves in a race with the beetles, hundreds of which would arrive within minutes after a dropping hit the ground!

Almost as soon as bacteria or fungi release any nutrients from dead plants or animals into the soil, rainforest trees and vines absorb the nutrients. This is one of the reasons why, despite the teeming vegetation, agriculture is very risky and normally destructive in rain forests. Virtually all the nutrients in a rain forest are tied up in the vegetation, so the soil is very infertile. If the trees are carried away for lumber, few nutrients remain to support crops. Further, even if the nutrients are released by burning the vegetation, the heavy year-round rainfall quickly dissolves and erodes them away, leaving the soil infertile after a few seasons of cultivation. The exposed soil, which is rich in iron and aluminum, then takes on an impenetrable, bricklike quality as it bakes in the tropical sun. As a result, secondary succession on cleared rainforest land is slowed significantly. Even small forest cuttings take about 70 years to regenerate.

Human Impact Despite their unsuitability for agriculture, rain forests are being felled for lumber or burned down for ranching or agriculture at a rate of at least 38 million acres each year, or more than 70 acres each minute, accounting for 90% of the world's total loss of forests (Fig. 41-10). Rain forest that is left standing is often in fragments too small to allow reproduction of trees and provide adequate habitat for larger animals. Fires in clearcut areas also spread along the ground into adjacent forest, killing young trees and disrupting community interactions. In West Africa, burning the forests is causing acid rain, which is damaging the remaining trees. Further, much of the rainfall in a rain forest comes from water transpired through the forest's leaves. As large tracts of rain forests disappear, the region becomes drier, more stressed, and more susceptible to fires. Researchers estimate that one-quarter of the carbon dioxide released into the atmosphere during the past decade came from cutting and burning tropical rain forests, exacerbating global warming.

At least 40% of the world's rain forests are now gone. Harvard University biologist Edward O. Wilson estimates that the destruction of tropical rain forests may be driving 27,000 species to extinction annually. With the loss of biodiversity, humanity loses access to a wealth of potential drugs and raw materials. For example, in Malaysia, the sap of a rare species of tree yielded a compound called *calanolide A*, which offers promise as an anti-AIDS drug. A synthetic version is now being tested in human clinical trials. A compound newly isolated from an African rain forest leaf fungus may eventually help control diabetes, a health problem that is becoming more common in the U.S.

Although disastrous losses continue, some areas have been set aside as protected preserves, and some reforestation efforts are under way. Local residents are becoming more involved in conservation efforts, as illustrated in our Case Study. In Brazil, a union of people who harvest natural rubber from tree sap is fighting to preserve large tracts of land for rubber production and harvesting fruits and nuts. These efforts are steps toward the ultimate solution, which is tragically slow in coming: sustainable use. Sustainable use means deriving benefits from an ecosystem, whether from tourism or harvesting products, in a way that can be sustained indefinitely without continuing damage to the ecosystem.

Figure 41-10 Fire engulfs a tropical rain forest in Brazil
The cleared area will be converted to ranching or agriculture, doomed to failure because the soil quality is so poor. Spreading fires and the smoke they produce threaten the uncut forests.

Tropical Deciduous Forests

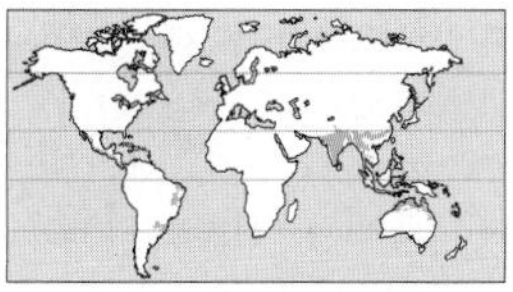

Slightly farther from the equator, the rainfall is not nearly as constant, and there are pronounced wet and dry seasons. In these areas, which include India, much of Southeast Asia, and parts of South and Central America, **tropical deciduous forests** grow. During the dry season, the trees cannot get enough water from the soil to compensate for evaporation from their leaves. As a result, the plants have adapted to the dry season by shedding their leaves, thereby minimizing water loss. If the rains fail to return on schedule, the trees delay the formation of new leaves until the drought passes.

Savanna

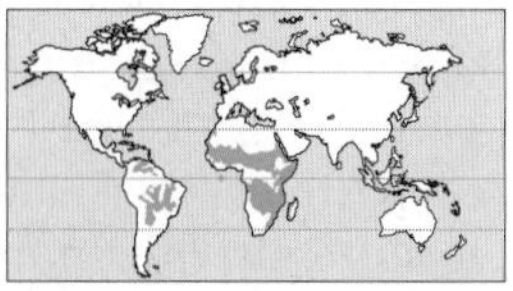

Along the edges of the tropical deciduous forest, the trees gradually become more widely spaced, with grasses growing between them. Eventually, grasses become the dominant vegetation, with only scattered trees and thorny scrub forests here and there; this biome is the **savanna** (Fig. 41-11). Savanna grasslands typically have a rainy season during which virtually all of the year's precipitation falls—12 inches (30 centimeters) or less. When the dry season arrives, it comes with a vengeance. Rain might not fall for months, and the soil becomes hard, dry, and dusty. Grasses are well adapted to this type of climate, growing very rapidly during the rainy season and dying back to drought-resistant roots during the arid times. Only a few specialized trees, such as the thorny acacia or the water-storing baobab, can survive the devastating savanna dry seasons. In areas in which the dry season becomes even more pronounced, virtually no trees can grow, and the savanna imperceptibly grades into tropical grassland.

The African savanna has probably the most diverse and impressive array of large mammals on Earth. These mammals include numerous herbivores such as antelope, wildebeest, water buffalo, elephants, and giraffes and such carnivores as the lion, leopard, hyena, and wild dog.

Human Impact Africa's rapidly expanding human population threatens the wildlife of the savanna. Poaching has driven the black rhinoceros to the brink of extinction (Fig. 41-12) and endangers the African elephant. The abundant grasses that make the savanna a

Figure 41-11 The African savanna
(a) Elephants roam while ***(b)*** a cheetah feasts on its prey. ***(c)*** Large herds of grazing animals, such as zebras, can still be seen on African preserves. The herds of herbivores provide food for the greatest assortment of large carnivores on Earth.

Figure 41-12 Poaching threatens African wildlife
Rhinoceros horns, believed by some to have aphrodisiac properties, fetch staggering prices and encourage poaching. The black rhino is now nearly extinct.

suitable habitat for so much wildlife also make it suitable for grazing domestic cattle. As the human population of East Africa increases, so does the pressure of cattle grazing on the savanna. Fences increasingly disrupt the migration of the great herds of herbivores in search of food and water. Ecologists have discovered that the native herbivores are much more efficient at converting grass into meat than are cattle. Perhaps the future African savanna may support herds of domesticated antelope and other large native grazers in place of cattle.

Deserts

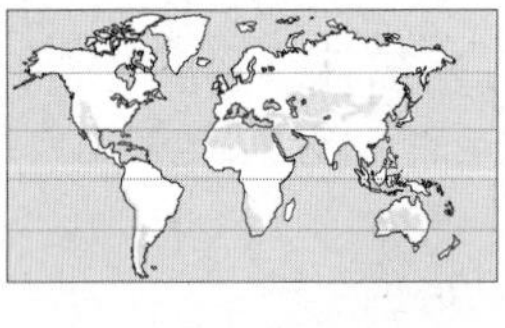

Even drought-resistant grasses need at least 10 to 20 inches (25 to 50 centimeters) of rain a year, depending on its seasonal distribution and the average temperature. When less than 10 inches of rain fall, **desert** biomes result. Although we tend to think of them as hot, deserts are defined by their lack of rain rather than by their temperatures. In the Gobi desert of Asia, for example, temperatures average below freezing for half the year, while summer temperatures average 105 °F to 110 °F (41 °C to 43° C). Desert biomes are found on every continent, typically around 20° to 30° N and S latitude and also in the rain shadows of major mountain ranges. As with all biomes, deserts include a variety of environments. At one extreme are certain areas of the Sahara Desert or Chile, where it virtually never rains and no vegetation grows (Fig. 41-13a). More commonly, deserts are characterized by widely spaced vegetation and large areas of bare ground. The plants tend to be spaced evenly, as if planted by hand (Fig. 41-13b). In many cases, the perennial plants are bushes or cacti with large, shallow root systems. The shallow roots quickly soak up the soil moisture after the infrequent desert storms. The rest of the plant is typically covered with a waterproof, waxy coating to prevent evaporation of precious water. Water is stored in the thick stems of cacti and other succulents. The spines of cacti are leaves modified for protection and water conservation, presenting almost no surface area for evaporation. In many deserts, all the rain falls in just a few storms, and specialized annual wildflowers take advantage of the brief period of moisture to race through germination, growth, flowering, and seed production in a month or less (Fig. 41-14).

The animals of the deserts, like the plants, are specially adapted to survive on little water. Most deserts appear to be almost completely devoid of animal life during summer days, because the animals seek relief from the sun and heat in cool, underground burrows. After dark, when deserts cool down considerably, horned lizards, snakes, and other reptiles emerge to feed, as do mammals such as the kangaroo rat (Fig. 41-13c) and birds

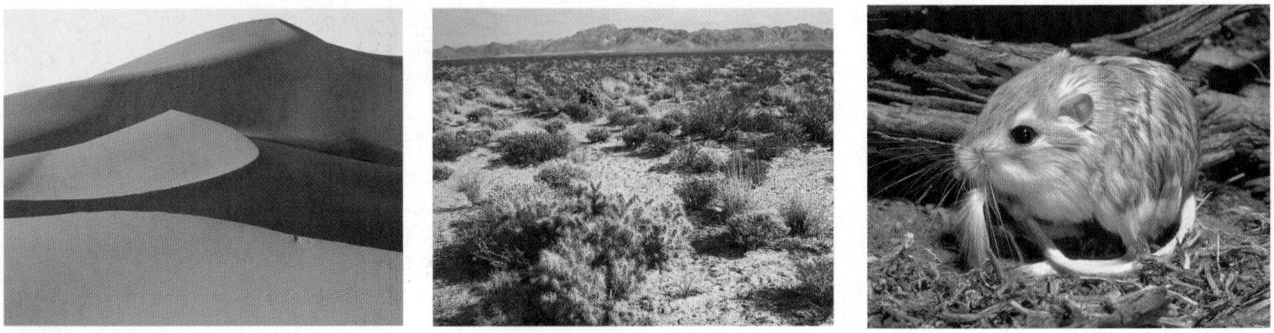

(a) (b) (c)

Figure 41-13 The desert biome
(a) Under the most extreme conditions of heat and drought, deserts can be almost devoid of life, such as these sand dunes of the Sahara Desert in Africa. ***(b)*** Throughout much of Utah and Nevada, the Great Basin Desert presents a monotonous landscape of widely spaced shrubs, such as sagebrush and greasewood. These shrubs often secrete a growth inhibitor from their roots, preventing germination of nearby plants and thus reducing competition for water. ***(c)*** The kangaroo rat is an elusive inhabitant of the deserts of North America.

Figure 41-14 The Sonoran Desert
In spring this Arizona desert is carpeted with wildflowers. Through much of the year, annual wildflower seeds lie dormant, waiting for the spring rains to fall.

such as the burrowing owl. Most of the smaller animals survive without ever drinking, getting all the water they need from their food and from that produced during cellular respiration in their tissues. Larger animals, such as desert bighorn sheep, are dependent on permanent water holes during the driest times of the year.

Human Impact Desert ecosystems are fragile. Ecologists studying the soil of the Mojave desert in southern California recently found treadmarks left by tanks in 1940 when General Patton trained tank crews in preparation for entry into World War II. The desert soil here is stabilized and enriched by microscopic cyanobacteria whose filaments intertwine among sand grains. The tanks, and now numerous off-road vehicles that careen about the desert for recreation, destroy this crucial network. This allows the soil to erode and reduces nutrients available to the desert's slow-growing plants. Ecologists estimate that the desert soil may require hundreds of years to fully recover from heavy vehicle use.

Scientists once believed that human activities were causing a progressive southward spread of the Sahara desert in Africa, a process called **desertification**. Recently published data from satellite photographs taken at intervals since 1980 indicate that the Sahara Desert has spread southward and then has retreated back repeatedly over this 20-year period in accordance with changes in the amount of rainfall. While the dry savanna south of the Sahara Desert has been significantly overgrazed and degraded by a human population that is above its sustainable carrying capacity, the desert biome is not spreading as a result, as was formerly believed.

Chaparral

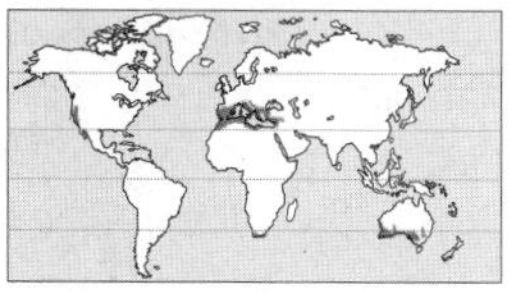

In many coastal regions that border on deserts, such as southern California and much of the Mediterranean, we find a unique type of vegetation called **chaparral**. The annual rainfall in these regions is up to 30 inches, nearly all of which falls during cool, wet winters that alternate with hot, dry summers. The proximity of the sea provides a slightly longer rainy season in the winter and frequent fogs during the spring and fall, reducing evaporation. Chaparral consists of small trees or large bushes such as sages and evergreen oak, with thick waxy or fuzzy leaves that conserve water. These hardy shrubs are also able to withstand the frequent summer fires started by lightning (Fig. 41-15).

Figure 41-15 The chaparral biome
This biome is limited primarily to coastal mountains in dry regions, such as the San Gabriel Mountains in southern California. Chaparral is maintained by frequent fires set by summer lightning. Although the tops of the plants may be burned off, the roots send up new sprouts the following spring.

Grasslands

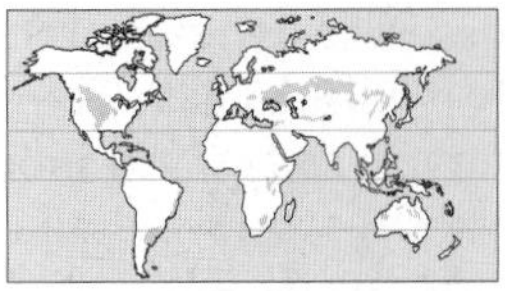

In the temperate regions of North America, deserts occur in the rain shadows east of the mountain ranges, such as the Sierra Nevada and Rocky Mountains. Eastward, as the rainfall gradually increases, the land begins to support more and more grasses, giving rise to the prairies of the Midwest. Most **grassland**, or **prairie**, biomes are located in the centers of continents, where they receive 10 to 30 inches (25 to 75 centimeters) of rain annually (note the large grasslands in the centers of the North American and Eurasian continents). In general, they have a continuous cover of grass and virtually no trees except along the rivers. From the tallgrass prairies of Iowa, Missouri, and Illinois (Fig. 41-16) to the shortgrass prairies of eastern Colorado, Wyoming, and Montana (Fig. 41-17), the North American grassland once stretched across almost half the continent.

Water and fire are the crucial factors in the competition between grasses and trees. The hot, dry summers and frequent droughts of the shortgrass prairies can be tolerated by grass but are fatal to trees. Forests are the climax ecosystems in the more eastern tallgrass prairies, but historically the trees were destroyed by frequent fires, often set by Native Americans to maintain grazing land for the bison. Although the tops of the grasses are destroyed by fire, their root systems usually survive; trees, however, are killed outright. The grasslands of North America once supported huge herds of bison—as many as 60 million in the early nineteenth century. Pronghorn antelope can still be seen in some prairies of the western U.S.; bobcats and coyotes are the major large predators there (a prairie food web is illustrated in Fig. 40-5). Grasses growing and decomposing for thousands of years produced what may be the most fertile soil in the world.

Human Impact When people developed plows that could break through the dense grass turf, this set the stage for converting the midwestern U.S. prairies into the "breadbasket" of North America, so named because enormous quantities of grain are cultivated in its fertile soil. The tallgrass prairie has been converted to agricultural land, except for tiny protected remnants, which are maintained by periodic controlled burning.

On the dry western shortgrass prairie, cattle have replaced the bison and pronghorn antelope. As a result of their overgrazing the grasses, the boundary between the cool deserts and the grassland has commonly been altered in favor of desert plants. Much of the sagebrush desert of the American West is actually overgrazed shortgrass prairie (Fig. 41-18). Cattle prefer grass to sagebrush, so heavy grazing destroys the grass. Consequently, moisture that the grass would have absorbed is left in the soil, encouraging the growth of the woody sagebrush. Thus, the prairie grasses are replaced by plants characteristic of the cool desert.

Figure 41-16 Tallgrass prairie in Missouri
In the central U.S., moisture-bearing winds out of the Gulf of Mexico produce summer rains, allowing a lush growth of tall grasses and wildflowers such as these coneflowers. Periodic fires, now carefully managed, prevent encroachment of forest.

Temperate Deciduous Forests

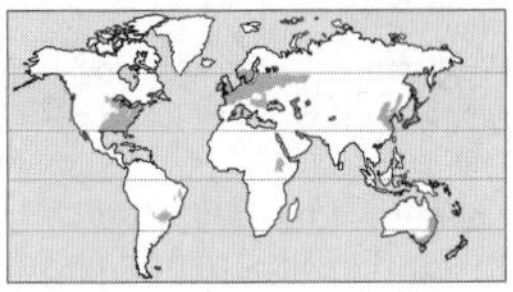

At their eastern edge, the North American grasslands merge into the **temperate deciduous forest** biome, also found in Western Europe and East Asia (Fig. 41-19, p. 872). Higher precipitation occurs there than in the grasslands (30 to 60 inches, or 75 to 150 centimeters), and, in particular, more rain falls during the summer. The soil retains enough moisture for trees to grow, and the resulting forest shades out grasses. In contrast to the tropical forests, the temperate deciduous forest biome has cold winters, usually with at least several hard frosts and often long periods of below-freezing weather. Winter in this biome has an effect on the trees similar to that of the dry season in the tropical deciduous forests: During periods of subfreezing temperatures, liquid water is not available to the trees. To reduce evaporation when water is in short supply, the trees drop their leaves in the fall. They produce leaves again in the spring, when liquid water becomes available. During the brief time in spring when the ground has thawed but the trees have not yet blocked off all the sunlight, abundant wildflowers grace the forest floor.

Insects and other arthropods are numerous and conspicuous in deciduous forests. The decaying leaf litter on the forest floor also provides food and habitat for bacteria, earthworms, fungi, and small plants. Many arthropods

Figure 41-17 Shortgrass prairie
The lands east of the Rocky Mountains receive relatively little rainfall, and ***(a)*** shortgrass prairie results, characterized by low-growing bunch grasses such as buffalo grass and grama grass. ***(b)*** Pronghorn antelope, ***(c)*** prairie dogs, and ***(d)*** protected bison herds occupy this biome, in which ***(e)*** wildflowers such as this coneflower abound.

Figure 41-18 Sagebrush desert or shortgrass prairie?
Biomes are influenced by human activities as well as by temperature, rainfall, and soil. The shortgrass prairie field on the right has been overgrazed by cattle, causing the grasses to be replaced by sagebrush.

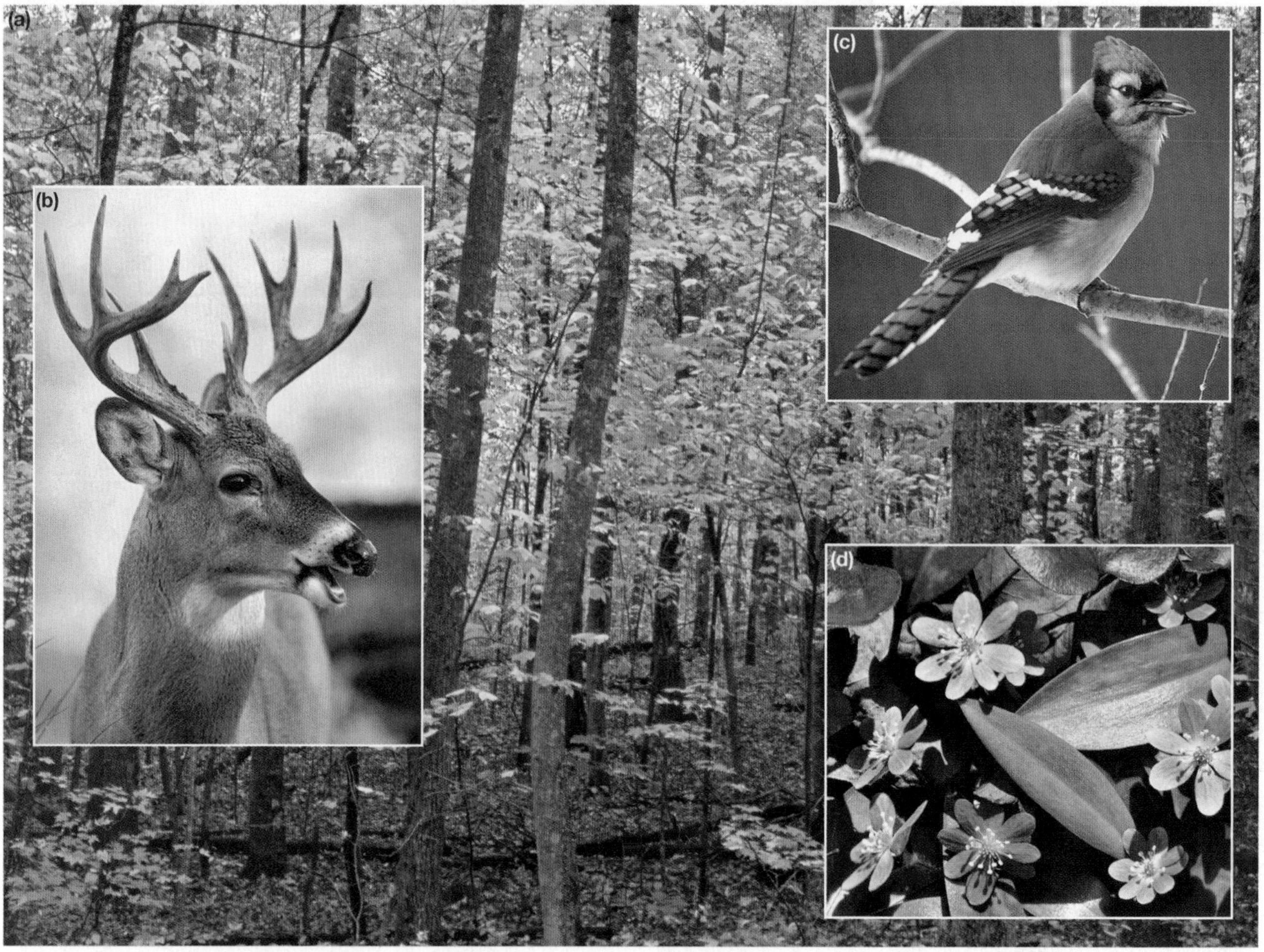

Figure 41-19 The temperate deciduous forest biome
(a) In temperate deciduous forests of the eastern U.S., ***(b)*** the white-tailed deer is the largest herbivore, and ***(c)*** birds such as the blue jay are abundant. ***(d)*** In spring, a profusion of woodland wildflowers (such as these hepaticas) blooms briefly before the trees produce leaves.

feed on these or on each other. A variety of vertebrates, including mice, shrews, squirrels, raccoons, deer, bear, and many species of birds, dwell in the deciduous forests.

Human Impact Large predatory mammals such as black bear, wolves, bobcats, and mountain lions were formerly abundant, but hunting and habitat loss has severely reduced their numbers and effectively eliminated wolves from deciduous forests. In many deciduous forests, deer are plentiful because of lack of natural predators. Clearing for lumber, agriculture, and housing has dramatically reduced deciduous forests in the U.S. from their original extent, and virgin deciduous forests are now almost nonexistent. Over the past 50 years, however, Forest Service data show that forest cover in the U.S. (both evergreen and deciduous) has increased as a result of regrowth of forests on abandoned farms, paper recycling that decreases demand for wood pulp, more efficient lumber milling and tree farming techniques, and the use of alternative building materials.

Temperate Rain Forests

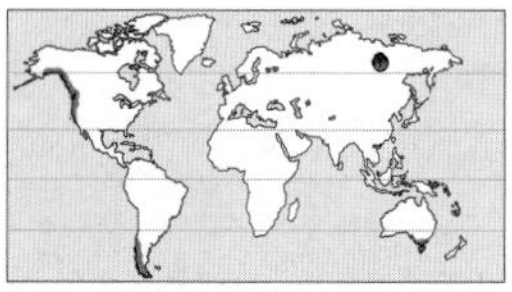

On the U.S. Pacific coast, from the lowlands of the Olympic Peninsula in Washington State to southeast Alaska, lies the **temperate rain forest** biome (Fig. 41-20). Temperate rain forests, which are relatively rare, are also located along the southeastern coast of Australia and the southwestern coast of New Zealand. As in the tropical rain forest, there is no shortage of liquid water year-round. This abundance of water is due to two factors. First, there is a tremendous amount of rain. The Hoh River rain forest in Olympic National Park receives more than 160 inches (400 centimeters) of rain annually, more than 24 inches (60 centimeters) in the month of December alone. Second, the moderating influence of the Pacific Ocean prevents severe frost from occurring along the coast, so the ground seldom freezes and liquid water remains available.

The abundance of water means that the trees have no need to shed their leaves in the fall, and almost all

Figure 41-20 The temperate rain forest biome
(a) The Hoh River temperate rain forest in Olympic National Park. The coniferous trees do not block the light as effectively as do broadleaf trees, so ferns, mosses, and wildflowers grow in the pale green light of the forest floor. ***(b)*** The dead feed the living, as new trees grow from the decay of this fallen giant, called a "nurse log," and ***(c)*** flowering foxglove and ***(d)*** fungi find ideal conditions amid the moist, decaying vegetation.

the trees are evergreens. In contrast to the broadleaf evergreen trees of the Tropics, temperate rain forests are dominated by conifers. The ground and typically the trunks of the trees are covered with mosses and ferns. As in tropical rain forests, so little light reaches the forest floor that tree seedlings usually cannot become established. Whenever one of the forest giants falls, however, it opens up a patch of light, and new seedlings quickly sprout, commonly right atop the fallen log. This event produces a "nurse log" (Fig. 41-20b).

Taiga

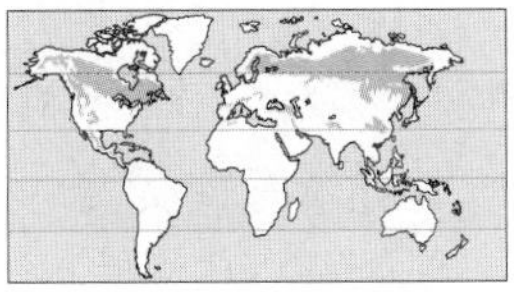

North of the grasslands and temperate forests, the **taiga**, also called the **northern coniferous forest** (Fig. 41-21), stretches horizontally across all of North America and Eurasia, including parts of the northern U.S. and much of southern Canada. Conditions in the taiga are harsher than those in the temperate deciduous forest. In the taiga, the winters are longer and colder, and the growing season is shorter. The few months of warm weather are too short to allow trees the luxury of regrowing leaves in the spring. As a result, the taiga is populated almost entirely by evergreen coniferous trees with narrow, waxy needles. The waxy coating and small surface area of the needles reduce water loss by evaporation during the cold months, and the leaves remain on the trees year-round. Thus, the trees are instantly ready to take advantage of good growing conditions when spring arrives, and they can continue slow growth late into the fall.

Because of the harsh climate in the taiga, the diversity of life there is much lower than in many other biomes. Vast stretches of central Alaska, for example, are covered by a somber forest that consists almost exclusively of black spruce and an occasional birch. Large mammals such as the wood bison, grizzly bear, moose, and wolf, which have mostly been eradicated in the southern regions of their original range, still roam the taiga, as do smaller animals such as the wolverine, fox, snowshoe hare, and deer. Outside of Alaska, where these large mammals remain reasonably abundant, small populations of wolves roam the northern U.S., including Idaho, Montana (where they have been re-introduced into

Figure 41-21 The taiga (or northern coniferous forest) biome
(a) The small needles and pyramidal shape of conifers allow them to shed heavy snows. ***(b)*** Winter is a challenge not only for the trees but also for animals such as this snowshoe hare and the bobcat that preys on it. The hare is also prey for ***(c)*** the great horned owl. Taiga animals face diminished food supply but increased energy requirements during subfreezing weather.

Yellowstone National Park), Michigan, Wisconsin, and Minnesota (which hosts the largest wolf population in the lower 48 states).

Human Impact The taiga is a major source of lumber for construction. *Clear-cutting*, the removal of all the trees in a given area, has destroyed huge expanses of forest, both in Canada and the Pacific Northwest in the U.S. (Fig. 41-22). Owing to the remoteness of the northernmost taiga and the severity of its climate, a greater percentage of the taiga remains in undisturbed condition than any other North American biome except the tundra.

Tundra

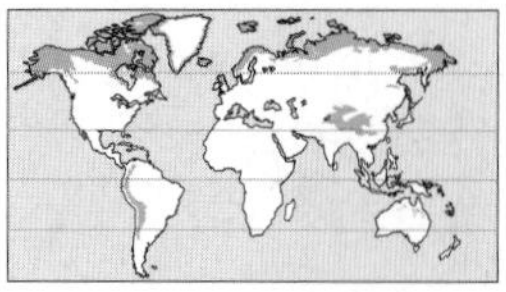

The last biome encountered before we reach the polar ice caps is the arctic **tundra**, a vast treeless region bordering the Arctic Ocean (Fig. 41-23). Conditions in the tundra are severe. Winter temperatures in the arctic tundra often reach –40 °F (–55 °C) or below, winds howl at 30 to

Figure 41-22 Clear-cutting
Clear-cutting, as seen in this Oregon forest, is relatively simple and cheap—but its environmental costs are high. Erosion will diminish the fertility of the soil, slowing new growth. Further, the dense stands of same-age trees that typically regrow are more vulnerable to attack by parasites than a natural stand of trees of various ages would be.

Figure 41-23 The tundra biome
(a) Life on the tundra is adapted to cold. ***(b)*** Plants such as dwarf willows and perennial wildflowers (such as this dwarf clover) grow low to the ground, escaping the chilling tundra wind. Tundra animals, such as ***(c)*** caribou and ***(d)*** arctic foxes, can regulate blood flow in their legs, keeping them just warm enough to prevent frostbite, while preserving precious body heat for the brain and vital organs.

60 miles (50 to 100 kilometers) an hour, and precipitation averages 10 inches (25 centimeters) or less each year, making this a "freezing desert." Even during the summer, the temperatures can drop to freezing, and the growing season may last only a few weeks before a hard frost occurs. Somewhat less cold but similar conditions produce alpine tundra on mountaintops above the altitude where trees can grow.

The cold climate of the arctic tundra results in **permafrost**, a permanently frozen layer of soil typically no more than about 1.5 feet below the surface. As a result, when summer thaws come, the water from melted snow and ice cannot soak into the ground and the tundra becomes a huge marsh. Trees cannot survive in the tundra; the permafrost limits root growth to the topmost meter or so of soil.

Nevertheless, the tundra supports a surprising abundance and variety of life. The ground is carpeted with small perennial flowers and dwarf willows no more than a few centimeters tall and often covered with a large lichen called "reindeer moss," a favorite food of caribou. The standing water provides a superb habitat for mosquitoes. The mosquitoes and other insects provide food for numerous birds, most of which migrate long distances to nest and raise their young during the brief summer feast. The tundra vegetation supports lemmings, which are eaten by wolves, snowy owls, arctic foxes, and even grizzly bears.

Human Impact The tundra is among the most fragile of all the biomes because of its short growing season. A willow 4 inches (10 centimeters) high may have a trunk 3 inches (7 centimeters) in diameter and be 50 years old. Human activities in the tundra leave scars that persist for centuries. Fortunately for the tundra inhabitants, the impact of civilization is localized around oil drilling sites, pipelines, mines, and military bases.

Rainfall and Temperature Determine the Vegetation a Biome Can Support

Terrestrial biomes are greatly influenced by both temperature and rainfall, whose effects interact. Temperature strongly influences the effectiveness of rainfall in providing soil moisture for plants and standing water for animals to drink. The hotter it is, the more rapidly water evaporates, both from the ground and from plants. As a result of this interaction of temperature with rainfall (and to a lesser extent, the distribution of rain throughout the year), areas that receive almost exactly the same rainfall can have startlingly different vegetation, all the way from desert to taiga. Now that you are familiar with biomes, take a trip with us from southern Arizona to central Alaska as we visit ecosystems that each receive about 12 inches (28 centimeters) of rain annually.

The Sonoran Desert near Tucson, Arizona (see Fig. 41-14), has an average annual temperature of 68 °F

(20 °C) and receives about 12 inches (28 centimeters) of rain each year. The landscape is dominated by giant saguaro cactus and low-growing, drought-resistant bushes. Going north for 931 miles (1500 kilometers) will bring you into eastern Montana, where rainfall is about the same, but you will be travelling through shortgrass prairie (see Fig. 41-17), largely because the average temperature is much lower, about 45 °F (7 °C). Much farther north, central Alaska receives about the same annual rainfall (12 inches), yet it is covered with taiga forest (see Fig. 41-21). As a result of the low average annual temperature (about 25 °F, or 24 °C), permafrost underlies much of the ground here. During the summer thaw, the taiga earns its Russian name "swamp forest," although its rainfall is about the same as that of the Sonoran Desert.

4 How Is Life in Water Distributed?

Although thus far this chapter has emphasized terrestrial biomes, the saltwater oceans and seas are the largest ecosystems on Earth, covering about 71% of its surface. Freshwater ecosystems, in contrast, cover less than 1%.

The unique properties of water lend some common features to aquatic ecosystems. *First*, because water is slower to heat and cool than air, temperatures in aquatic ecosystems are more moderate than are those in terrestrial ecosystems. *Second*, although water may appear quite transparent, it absorbs a considerable amount of the light energy that sustains life. Even in the clearest water, the intensity of light decreases rapidly with depth. At depths of 600 feet (200 meters) or more, little light is left to power photosynthesis. If the water is at all cloudy—for example, because of suspended sediment or microorganisms—the depth to which light can penetrate is greatly reduced. *Third*, nutrients in aquatic ecosystems tend to be concentrated near the bottom sediments where light levels are too low to support photosynthesis. This separation of energy and nutrients limits aquatic life. Of the four requirements for life, aquatic ecosystems provide abundant water and appropriate temperatures. Thus, the major factors that determine the quantity and type of life in aquatic ecosystems are the remaining two factors: energy and nutrients.

Although they share some common features, aquatic ecosystems are extremely diverse. Freshwater ecosystems encompass rivers, streams, ponds, lakes, and marshes; marine (saltwater) ecosystems include estuaries, tide pools, coral reefs, the open ocean, and vent communities. In the following sections, we look more closely at some of these important aquatic ecosystems.

Freshwater Lakes Have Distinct Regions of Life

Freshwater lakes vary tremendously in size, depth, and nutrient content. Although each lake is unique, moderate to large lakes in temperate climates share some common features, including distinct zones of life.

Life Zones Are Determined by Access to Light and Nutrients

The distribution of life in lakes depends largely on access to light, to nutrients, and in some cases to a place for attachment (the bottom). The life zones of lakes, then, correspond to specific locations within the lake. We recognize three such zones: (1) the *littoral zone*, (2) the *limnetic zone*, and (3) the *profundal zone* (Fig. 41-24).

Near the shore is the **littoral zone**. In this zone, the water is shallow, and plants find abundant light, anchorage, and adequate nutrients from the bottom sediments. Not surprisingly, littoral-zone communities are the most diverse. Cattails and bulrushes abound nearest the shore, and water lilies and entirely submerged vascular plants and algae may flourish at the deepest reaches of the littoral zone. The plants of the littoral zone trap sediments carried in by streams and by runoff from the surrounding land, increasing the nutrient content in this region. Living among the anchored plants are microscopic organisms called **plankton**. There are two forms of plankton: **phytoplankton** (Greek, "drifting plants"), which includes photosynthetic protists, bacteria, and algae, and **zooplankton** (Greek, "drifting animals"), such as protozoa and tiny crustaceans. The greatest diversity of animals in the lake is also found in this zone. Littoral invertebrate animals include small crustaceans, insect larvae, snails, flatworms, and hydra; littoral vertebrates include frogs, minnows, and aquatic snakes and turtles.

As the water increases in depth farther from shore, plants are unable to anchor to the bottom and still collect enough light for photosynthesis. This open-water area is divided into two regions: the upper limnetic zone and the lower profundal zone (see Fig. 41-24). In the **limnetic zone**, enough light penetrates to support photosynthesis. Here phytoplankton including cyanobacteria (also called *blue-green algae*) serve as producers. These are eaten by protozoa and small crustaceans, which in turn are consumed by fish. Below the limnetic zone lies the **profundal zone**, where light is insufficient to support photosynthesis. This area is nourished mainly by detritus that falls from the littoral and limnetic zones and by incoming sediment. It is inhabited primarily by decomposers and detritus feeders, such as bacteria, snails and insect larvae, and fish that swim freely among the different zones.

Freshwater Lakes Are Classified According to Their Nutrient Content

Although each lake is unique, freshwater lakes can be classified on the basis of their nutrient content as either *oligotrophic* or *eutrophic*.

Oligotrophic (Greek, "poorly fed") **lakes** are very low in nutrients. Many are formed by glaciers that scrape depressions in bare rock, and they are fed by mountain streams carrying little sediment. Because there is little sediment or microscopic life to cloud the water, oligotrophic lakes are clear, and light penetrates deeply.

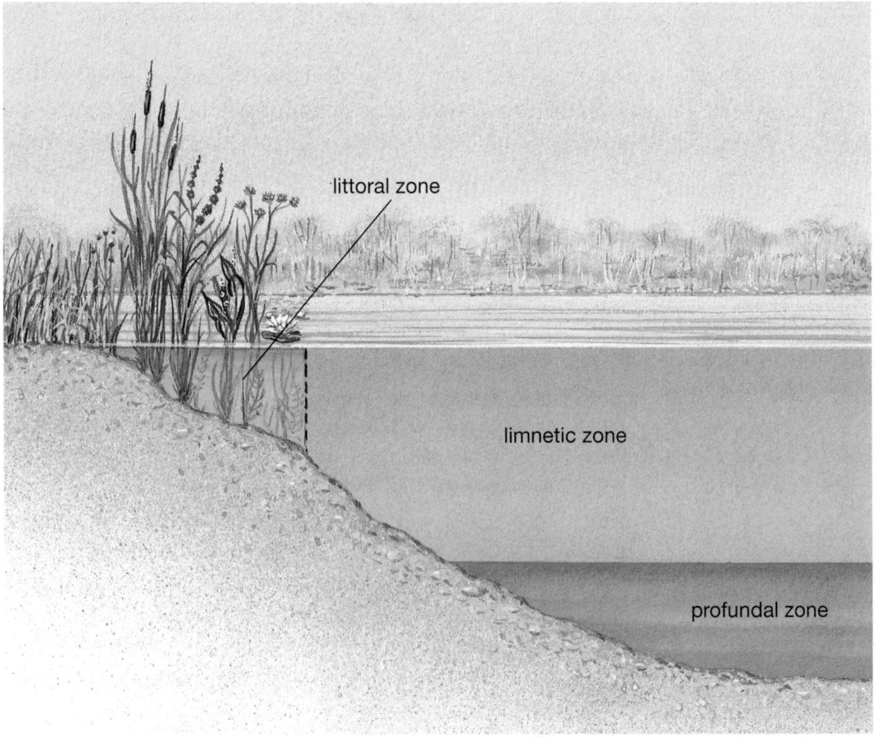

Figure 41-24 Lake life zones
There are three life zones in a "typical" lake: a near-shore littoral zone with rooted plants, an open-water limnetic zone, and a deep, dark profundal zone.

Therefore, photosynthesis (and by extension oxygenation, as oxygen is produced as a by-product of photosynthesis) in deeper water is possible, and the limnetic zone may extend all the way to the bottom. Oxygen-loving fish, such as trout, thrive in oligotrophic lakes.

Eutrophic (Greek, "well fed") **lakes** receive larger inputs of sediments, organic material, and inorganic nutrients (such as phosphorus) from their surroundings, allowing them to support dense communities. They are murkier, both from suspended sediment and dense phytoplankton populations, so the lighted limnetic zone is shallower. Dense "blooms" of algae occur seasonally in the limnetic zone. Their dead bodies fall into the profundal zone, where they are used as food by decomposer organisms. The metabolic activities of these decomposers use oxygen, depleting the oxygen content of the profundal zone in eutrophic lakes.

Lakes are transient ecosystems, although very large lakes may persist for millions of years. Over time, they gradually fill with sediment, undergoing succession to dry land (see Chapter 39). As nutrient-rich sediment accumulates, oligotrophic lakes tend to become eutrophic, a process called *eutrophication*.

Human Impact Human activities can greatly accelerate the process of eutrophication, because nutrients are carried into lakes from farms, feedlots, sewage, and even fertilized suburban lawns. Overenriched lakes become clogged with microorganisms whose dead bodies are attacked by bacteria that deplete the water of oxygen. Normal community interactions are disrupted as organisms in higher trophic levels are smothered. In Seattle, Washington, in the early 1960s local sewage plants released 20 million gallons of effluent into Lake Washington daily. Nutrients from human waste and high-phosphate detergents eutrophied the lake, causing foul odors, murky water, and dead fish. Citizen concern led Seattle to divert sewage effluent out of the lake starting in 1963. By 1975 the lake had fully recovered.

The Great Lakes, whose shores host paper mills and a variety of other polluting industries and whose waters carry vessels from distant countries, have been heavily impacted by human activities. Exotic species such as the zebra mussel are altering their community structures. Imported parasitic sea lampreys and overfishing devastated lake trout populations in the 1940s. Although fishing is now regulated and the lampreys are controlled, the trout still are unable to reproduce because of persistent endocrine-disrupting chemicals (discussed in Chapter 32) that have accumulated in the water and sediments. Recently officials were puzzled to find that the beleaguered trout carried high levels of toxiphene, which was used heavily as a substitute for the pesticide DDT (discussed in Chapter 40). The EPA banned toxiphene in 1982 after discovering that it could cause birth defects and cancer. Some scientists hypothesize that the toxifene is still being carried in the atmosphere and deposited in the Great Lakes from fields in the south, where it was heavily sprayed on cotton 20 years ago.

Acid rain, caused primarily by burning fossil fuels, poses a very different threat, particularly to small freshwater lakes and ponds. In the Adirondack Mountains of New York State, 500 lakes and ponds (almost 20% of the total) have been rendered nearly lifeless by acid rain, and the problem is growing. (Acid rain is discussed in Chapter 40.)

Marine Ecosystems Cover Much of Earth

In the oceans, the upper layer of water to a depth of about 650 feet (200 meters), where the light is strong enough to support photosynthesis, is called the **photic zone**. Below the photic zone lies the **aphotic zone**, where the only energy comes from the excrement and bodies of organisms that sink or swim down there (Fig. 41-25).

As in lakes, most of the nutrients in the oceans are at or near the bottom, where there is not enough light for photosynthesis. Nutrients dissolved in the water of the photic zone are constantly being incorporated into the bodies of living organisms. When these organisms die, some sink into the aphotic zone, providing its organisms with nutrients. If no additional nutrients entered the photic zone, life there would eventually cease.

Fortunately, there are two sources of nutrients to the photic zone: the land, from which rivers constantly remove nutrients and carry them to the oceans, and **upwelling**, an upward flow that brings cold, nutrient-laden water from the ocean depths to the surface. Upwelling occurs along western coastlines, as in California, Peru, and West Africa, where prevailing winds displace surface water, causing it to be replaced by water from below. Upwelling also occurs around Antarctica. Not surprisingly, the major concentrations of life in the oceans are found where abundant light is combined with a source of nutrients, which occurs most commonly in regions of upwelling and in shallow coastal waters.

Coastal Waters Support the Most Abundant Marine Life

The most abundant life in the oceans is found in a narrow strip surrounding Earth's landmasses, where the water is shallow and a steady flow of nutrients washes off the land. Coastal waters consist of (1) the **intertidal zone**, the area that is alternately covered and uncovered by water with the rising and falling of the tides, and (2) the **near-shore zone**, relatively shallow but constantly submerged areas, including bays and coastal wetlands such as salt marshes and **estuaries**, which are wetlands formed where rivers meet the oceans (Fig. 41-26). The near-shore zone is the only part of the ocean where large plants or seaweeds can grow, anchored to the bottom. In addition, the abundance of nutrients and sunlight in this zone promotes the growth of a veritable soup of photosynthetic phytoplankton. Associated with these plants and protists are animals from nearly every phylum: annelid worms, sea anemones, jellyfish, sea urchins, sea stars, mussels,

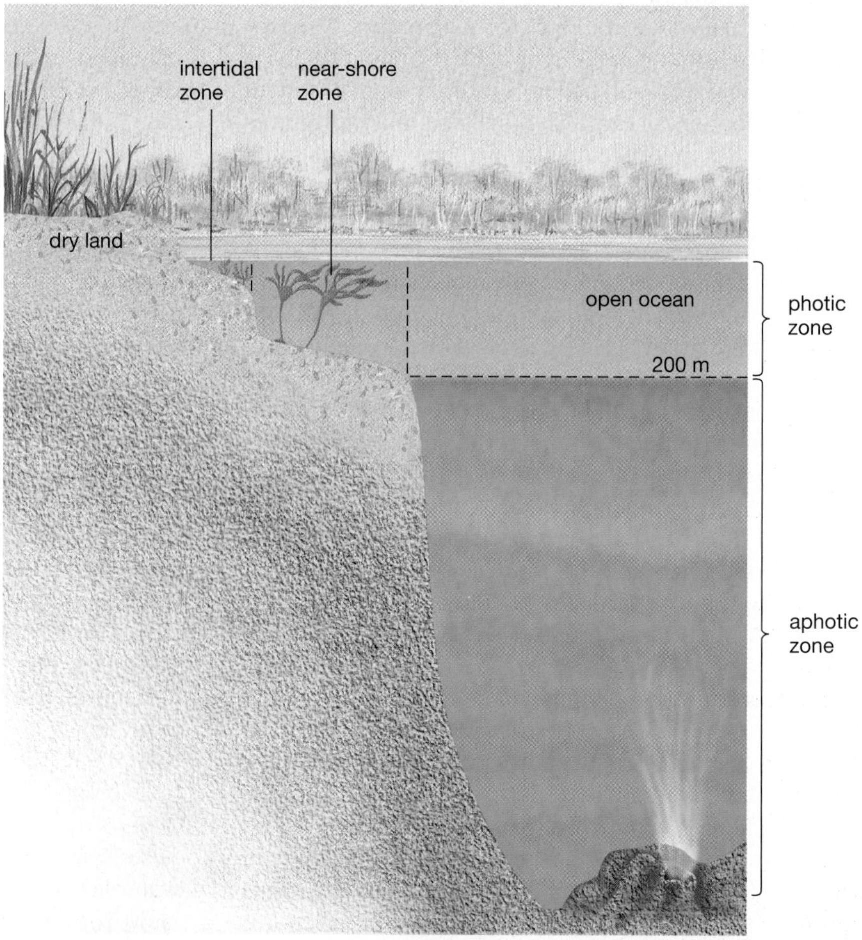

Figure 41-25 Ocean life zones
Photosynthesis can occur only in the upper photic zone, which includes the intertidal and near-shore zones and the upper waters of the open ocean. Life in the aphotic zone relies on energy-rich material that drifts down from the photic zone or (in the unique case of hydrothermal vent communities) on energy that is stored in hydrogen sulfide and trapped by chemosynthesis.

Figure 41-26 Near-shore ecosystems
(a) A salt marsh in the eastern United States. Expanses of shallow water fringed by marsh grass (*Spartina*) provide excellent habitat and breeding grounds for many marine organisms and shorebirds. ***(b)*** Although the shifting sands present a challenge to life, grasses stabilize them, and animals such as (inset) this *Emerita* crab burrow in the sandy intertidal zone. ***(c)*** A rocky intertidal shore in Oregon, where animals and algae grip the rock against the pounding waves and resist drying during low tide. (Inset) Colorful sea stars cling to the rocks surrounded by a seaweed, fucus. ***(d)*** Towering kelp sway through the clear water off southern California, providing the basis for a diverse community of invertebrates, fishes, and (inset) an occasional sea otter.

snails, fish, and sea otters, to name just a few. A large number and variety of organisms live permanently in coastal waters, but many that spend most of their lives in the open ocean come into the coastal waters to reproduce. Bays, salt marshes, and estuaries in particular are the breeding grounds for a wide variety of organisms, such as crabs, shrimp, and an array of fish, including most of our commercially important species. Off the coast of southern California, great undersea forests of kelp provide food and shelter for a rich assemblage of fish and invertebrates that in turn provide food for otters and seals (Fig. 41-26d). The productivity of freshwater and saltwater wetlands is comparable to rain forests. Coral reefs, another unique form of coastal ecosystem, are discussed below.

Human Impact Coastal regions are of great importance not only to the organisms that live or breed there but also to humans who use them for food sources, recreation, mineral and oil extraction, or living places. Like rain forests, wetlands of all types once covered 6% of Earth's surface, but nearly half of them have been dredged or filled in. In the U.S., only about 100 million acres remain out of an original 215 million acres of wetlands. As populations increase in coastal states and resources such as oil become increasingly scarce, the conflict between preservation of coastal wetlands as wildlife and animal habitat and development of these areas for housing, harbors and marinas, and energy extraction will become increasingly intense. Although conservation efforts have slowed the loss of wetlands, the U.S. still loses well over 100,000 acres yearly. Estuaries are also threatened by runoff from farming operations. Pig farms often collect concentrated animal wastes in large holding ponds. When leaks or floods transport this material into local rivers and estuaries, it can be fatal to the estuarian community of phytoplankton, invertebrates, and fish. Because so much of the life of the ocean depends on the well-being of the coastal waters, it is essential to protect these fragile, vital areas.

Coral Reefs

In warm tropical waters, with just the right combination of bottom depth, wave action, and nutrients, specialized algae and corals (types of cnidarians) build reefs from their own calcium carbonate skeletons. **Coral reefs** are most abundant in tropical waters of the Pacific and Indian Oceans, the Caribbean, and the Gulf of Mexico as far north as southern Florida, where the maximum water temperatures range between 72 °F and 82 °F (22 °C and 28 °C).

Reef-building corals (phylum Cnidaria, described in Chapter 22) harbor photosynthetic unicellular algae called *dinoflagellates* within their tissues in a mutualistic relationship. The algae represent up to half the weight of the coral polyp and give the corals their diverse, brilliant colors (Fig. 41-27). These corals thrive within the photic zone at depths of less than 130 feet (40 meters), where light penetrates the clear water and provides energy for photosynthesis. The algae benefit from the high nitrogen, phosphorus, and carbon dioxide levels in the coral tissues. In return, algae provide food for the coral and help produce calcium carbonate, which forms the coral skeleton. The skeletons of corals accumulate over thousands of years, providing an anchoring place for diverse forms of algae, and shelter and food for the most diverse collection of invertebrates and fish in the oceans (Fig. 41-27). In some ways, coral reefs might be considered the ocean equivalent of rain forests, since they are home to more than 93,000 known species with probably ten times that number yet to be identified. The Great Barrier Reef in Australia supports more than 200 species of coral alone, and a single reef may harbor 3000 identified species of fish, invertebrates, and algae.

Figure 41-27 Coral reefs
(a) Coral reefs, composed of the bodies of corals and algae, provide habitat for an extremely diverse community of extravagantly colored animals. ***(b)*** Many fish, including this blue tang, feed on coral (note the bright yellow corals in background). A vast array of invertebrates such as ***(c)*** this sponge and ***(d)*** blue-ringed octopus live among the corals of Australia's Great Barrier Reef. This tiny octopus (6 inches, fully extended) is one of the world's most venomous creatures.

Human Impact Coral reefs are extremely sensitive to many types of disturbance. Anything that diminishes the clarity of the water harms the coral's photosynthetic partners and hinders coral growth. When people farm, log, and develop coastal land, erosion carries silt into the water. As sewage and agricultural runoff pollute coastal waters, eutrophication reduces both sunlight and oxygen. Silt has ruined several reefs near Honolulu, Hawaii. In the Philippines, rain forest logging has dramatically increased erosion; destroying the rain forests is also destroying the local coral reefs.

Overfishing threatens reef communities. In many tropical countries, mollusks, turtles, fish, crustaceans, and the corals themselves are being harvested from reefs faster than they can reproduce. Many are sold to shell enthusiasts and aquarium owners in developed countries. Some collectors use dynamite or poisons to stun the fish before collecting them, leaving most dead. Removing predatory fish and invertebrates from reefs may disrupt the ecological balance of the community, allowing an explosion in populations of coral-eating sea urchins or sea stars.

Throughout the Tropics scientists and concerned citizens have observed a disturbing trend—coral reefs and their communities are becoming "sick." Infections, tumors, and lesions with unknown causes are causing massive mortality of corals and their communities throughout the world. Teams of researchers at Harvard and at the University of New Hampshire who are studying these trends suggest that the declining health of coral reefs results from complex interactions involving both human disturbances and global warming. Higher water temperature combined with runoff from land encourage explosions of microscopic algal populations in nearshore waters. The algae release toxins that damage the health of the entire reef community. Bleaching is another symptom of poor health observed in corals worldwide. When waters become too warm, the corals expel their colorful symbiotic algae, leaving the corals a deathly white. Without their algal partners, the corals will eventually starve. Continued global warming poses a severe threat to the world's coral reefs.

Florida's famous coral reefs are in trouble. Warm water from a nuclear power plant, nutrients from sewage effluent and silt carried in from rivers cloud the water and promote the growth of harmful algae. More than a million divers and snorkelers visit the reefs annually and accidentally (or sometimes deliberately) damage the coral. Boat anchors claw the reefs and boat engines release oil into formerly pristine waters. Infections are increasingly common, and the two warmest decades in recorded history have caused extensive coral bleaching. But there is also some good news—the world's first marine park was established off Florida in 1960, and several other marine sanctuaries have been established in Florida in more recent years. A fishing ban over small coral reef reserves in the Florida Keys in 1997 has already produced larger populations of several important marine species such as lobsters and grouper fish.

As with any ecosystem, both protection and sustainable use are crucial to the survival of these fragile and diverse underwater treasures. Carefully regulated harvesting and tourism produce far greater and longer-lasting economic benefits than do activities that destroy the reefs. Establishing preserves is crucial, but preserves will only be effective if we can prevent human activities on the land from damaging the quality of the water on which reef communities rely.

The Open Ocean

Beyond the coastal regions lie vast areas of the ocean in which the bottom is too deep to allow plants to anchor and still receive enough light to grow. Most life in the open ocean (Fig. 41-28a) is limited to the upper photic zone, where life-forms are **pelagic**—that is, free-swiming or floating—for their entire lives (Fig. 41-28b–d). The food web of the open ocean is dependent on phytoplankton consisting of microscopic photosynthetic protists, mainly diatoms and dinoflagellates (Fig. 41-28e). These organisms are consumed by zooplankton, such as tiny crustaceans that are relatives of crabs and lobsters (Fig. 41-28f). Zooplankton in turn serve as food for larger invertebrates, small fish, and even marine mammals such as the humpback whale (Fig. 41-28d).

One challenge faced by inhabitants of the open ocean is to remain afloat in the photic zone, where sunlight and food are abundant. Many members of the planktonic community have elaborate flotation devices, such as oil droplets in their cells or long projections, to slow their rate of sinking (see Fig. 41-28e). Most fish have swim bladders that can be filled with gas to regulate their buoyancy. Some animals, and even some of the phytoplankton, actively swim to stay in the photic zone. Many small crustaceans migrate to the surface at night to feed, then sink into the dark depths during daylight, thus avoiding visual predators such as fish. The amount of pelagic life varies tremendously from place to place. The blue clarity of tropical waters is a result of a lack of nutrients, which limits the concentration of plankton in the water. Nutrient-rich waters that support a large plankton community are greenish and relatively murky.

Below the photic zone, the only available energy in most regions comes from the excrement and dead bodies that drift down from above. Nevertheless, a surprising quantity and variety of life exist in the aphotic zone, including fishes of bizarre shapes, worms, sea cucumbers, sea stars, and mollusks.

Human Impact Two major threats to the open ocean are pollution and overfishing. Open-ocean pollution takes several forms. Oceangoing vessels such as cruise ships dump millions of plastic containers overboard daily, and plastic six-pack holders, foam cups, and packing material wash and blow off the land, collecting on parts

Figure 41-28 The open ocean
(a) The open ocean supports abundant life in the photic zone, where light is available. ***(b)*** Porpoises skim the surface, ***(c)*** fish such as the blue jack swim, and ***(d)*** rare humpback whales leap clear of the water. ***(e)*** The photosynthetic phytoplankton are the producers on which most other marine life ultimately depends. Phytoplankton are eaten by ***(f)*** zooplankton, represented by this microscopic crustacean, a copepod. The spiny projections on these planktonic creatures help keep them from sinking below the photic zone.

of the ocean surface. The plastic looks like food to unsuspecting sea turtles, gulls, porpoises, seals, and whales, many of which die after trying to consume it. Until 1992 New York City placed its refuse and sewage sludge on barges and towed them out to sea, creating a heavily contaminated area covering 40 square miles of open-ocean floor. The open ocean has served as a dumping ground for radioactive wastes. Oil contaminates the open ocean from many sources, including oil tanker spills, runoff from improper disposal on land, leakage from offshore oil wells, and natural seepage. In the Gulf of Mexico, the Mississippi River deposits nutrient-laden sediment from agricultural runoff onto the seafloor, creating an expanding "dead zone" where oxygen is depleted and the normal marine community is eliminated. The dead zone now covers 7000 square miles during the warmer months of each year.

The increasing demand for fish to feed a growing human population, coupled with increasingly efficient fishing technologies including satellite technology, radar, and sonar, has caused the depletion of many important fisheries. Most of the world's fisheries, despite these technologies, are harvesting fewer fish—evidence that fish are being taken in unsustainable numbers. Many fisheries are heavily subsidized by their governments, a practice that promotes the continuing harvest of a diminishing resource. The cod fishery of the northeastern U.S. has collapsed from overfishing. Natural populations of lobsters, salmon, haddock, king crab, swordfish, and many other types of seafood have also declined dramatically because of overfishing. While dredging for scallops, fishermen often scrape entire marine communities from the seafloor. "Dolphin safe" tuna netting techniques have dramatically reduced accidental dolphin catches but have trapped more sea turtles and sharks.

In 1994 a marine reserve was established off the coast of Massachusetts. Within 5 years, scallop and fish populations were rebounding. Fisheries researchers have now proposed protection of about three dozen key regions off the New England coast to allow fish populations to recover. It has been politically difficult to establish no-fishing zones due to intense opposition from the fishing industry, but this approach holds the most promise for saving endangered fish populations and ultimately fisheries throughout the world. Fish farming, or *aquaculture,* can also help meet the demand for some types of seafood, including shrimp, but fish and shrimp farms must be carefully designed and managed to avoid damaging local ecosystems.

Hydrothermal Vent Communities

In 1977 a new and unusual source of nutrients, forming the basis of a spectacular undersea community, was discovered in the deep ocean. Geologists exploring the Galapagos Rift (an area of the Pacific floor where plates that form Earth's crust are separating) found vents spewing superheated water, black with sulfur and minerals. Surrounding these vents was a rich community of pink fish, blind white crabs, enormous mussels, giant white clams, sea anemones, and giant tube worms (Fig. 41-29). Twenty-two previously undescribed families and 284 new species of organisms have been found in these novel **hydrothermal vent communities**. Scientists have now identified vent communities in many deep-sea areas where tectonic plates are spreading apart and material from Earth's interior is spewing forth to form new crust.

In this unique ecosystem, sulfur bacteria serve as the primary producers. They harvest energy from an unlikely source that is deadly to most other forms of life—hydrogen sulfide spewed from cracks in Earth's crust. This process, called *chemosynthesis*, replaces photosynthesis in these vent communities, which flourish more than a mile below the ocean surface. Bacteria and their close relatives, the archae, proliferate in the hot water surrounding the vents, covering nearby rocks with thick, matlike colonies. These colonies provide the food on which the animals of the vent community thrive. Many vent animals consume the microorganisms directly. Others, such as the giant tube worm (which lacks a digestive tract), harbor the bacteria within special organs in their bodies. In this mutualistic association, the bacteria provide high-energy carbon compounds and the tube worm provides hydrogen sulfide. The worm, which can reach a length of 9 feet, derives its red color from a unique form of hemoglobin that transports hydrogen sulfide, rather than oxygen, to the symbiotic bacteria.

The bacteria and archae that inhabit the vent communities hold the record for survival at high temperatures. One species stops reproducing at water temperatures of 194 °F (90 °C) because it gets too cold; another can survive water temperatures of 248 °F (106 °C; water at this depth can reach temperatures much higher than boiling because of the tremendous pressure). Scientists are investigating how the enzymes and other proteins of these heat-loving microbes can continue to function at temperatures that would destroy the proteins in our bodies.

The world still holds wonders and mysteries for those who seek them. We have only begun to explore the versatility and diversity of life on Earth.

Figure 41-29 Hydrothermal vent communities Located in the ocean depths, vent communities include giant tube worms nearly 4 meters (12 feet) long. Parts of these worms are red with oxygen-trapping hemoglobin. They have no digestive tract but are host to sulfur bacteria, which provide the worms with energy by oxidizing hydrogen sulfide.

Earth Watch
Humans and Ecosystems

The expanding human population has left relatively few ecosystems undisturbed. Human impacts on natural ecosystems are so diverse and wide ranging that they far exceed the scope of this book. However, we can identify some general characteristics of ecosystems dominated by humans and contrast them to the characteristics of undisturbed ecosystems. Listed below are six characteristic differences between these two types of ecosystems and a few ideas for minimizing them.

First, ecosystems dominated by humans tend to be simpler—that is, to have fewer species and fewer community interactions—than do undisturbed ecosystems. Although a city street bustles with life and apparent complexity, count the number of species you encounter in an average block and compare it with the number you encounter while hiking in a wilderness for a similar distance. As humans enter an ecosystem, animals in the highest trophic levels are the first to go. Carnivores are always relatively rare, and their specialized needs are most easily disrupted. Many big carnivores, such as the wolf and mountain lion, require large, undisturbed hunting territories. Humans often selectively destroy large predators, believing them to be a threat to us or our livestock. As a result, even in relatively undisturbed areas many large carnivores have been eliminated. Grizzly bears and wolves no longer roam the southern Rocky Mountains. On the western shortgrass prairie, prairie dog towns fall as human towns and ranches rise (Fig. E41-2). The black-footed ferret, a predator of prairie dogs, faces possible extinction. Farm fields have been deliberately simplified from the original prairie biome to eliminate competition and predation and allow the maximum productivity of a single crop species. Nowhere is the contrast between human and natural ecosystems greater than in the tropical rain forests, whose unparalleled diversity is replaced by failed attempts at farming.

There is probably no practical way to restore to human ecosystems the great diversity found in undisturbed areas, nor is doing so always desirable from a human standpoint. Agriculture, for example, demands a simplified ecosystem (but not all biomes are suited to it). Cities concentrate human activities and culture and may lessen the human impact on the surrounding countryside. However, as we recognize the benefits of artificially simplified ecosystems, we must be aware of the need to preserve intact as many natural communities as possible. The undisturbed forest traps and purifies water and can reduce air pollution. Swamps and estuaries contain a wealth of detritus feeders and decomposers that purify water. Coastal wetlands are breeding places for millions of birds and spawning sites for the majority of our commercially important fish and crustacean species. In undisturbed, diverse ecosystems we find aesthetic pleasure as well as a storehouse of species whose commercial or medicinal values are not yet recognized. For example, nearly half the medicines in use today were originally discovered in plants, and we have examined only a small fraction of existing plants for possible medical uses.

Figure E41-2 Habitat destruction
Loss of habitat due to human activities is a major threat to most of Earth's wildlife.

Second, whereas natural ecosystems run on sunlight, human ecosystems have become dependent on nonrenewable energy from fossil fuels. From the suburbanite pouring gas into a lawn mower to the farmer driving a tractor, managing a simplified ecosystem is an energy-intensive proposition. Energy must be expended to oppose the tendency of the natural system to restore complexity. Fertilizers also require large amounts of energy to produce, and farms must be heavily fertilized because nutrient cycles have been disrupted.

To counteract this trend, some farmers are returning to organic farming. Plant and animal waste and natural nutrient cycles are used to maintain soil fertility. Alternating legume crops, such as soybeans and alfalfa, with other crops helps maintain nitrogen in the soil. Mulching can help retain fertility, reduce water loss, and control weeds. The use of natural insect predators and limited pesticide spraying at critical times during a pest's life cycle can dramatically reduce the need for poisons. Organic farms can be as productive as more conventional farms while using 15% to 50% less energy from fossil fuels per quantity of food produced.

In our homes and commercial buildings, better insulation and increased use of solar heat can result in dramatic energy savings. By increasing our use of renewable energy sources, such as wind and especially sunlight, we can conserve fossil fuels, dramatically reduce pollution, and move human ecosystems a step closer to those that occur naturally.

Third, natural ecosystems recycle nutrients, whereas human ecosystems tend to lose nutrients. Walk around some suburban neighborhoods on trash-pickup day. Grass clippings and leaves are packed in plastic bags to be hauled away. To compensate for this loss of natural nutrients, many suburbanites heavily fertilize their lawns and gardens. A similar trend has occurred in modern farming. The exposed soil is eroded away by wind and water, removing crucial nutrients and requiring large inputs of fertilizer to replace them. Runoff from the field, carrying fertile topsoil and artificial fertilizers, may pollute nearby rivers, streams, and lakes. Pesticides may kill detritus feeders and decomposers, further disrupting natural nutrient cycles. Thus, whereas fertile soil accumulates in many natural ecosystems, it tends to be lost in those dominated by humans. Some 3 billion tons of topsoil are eroded annually from farms in the U.S. The Mississippi River alone carries off 40 tons each hour.

Organic farming can help reverse this trend too. We can apply the same principles to our lawns and gardens by composting organic wastes either in our own gardens or in community compost facilities. Farmers can counteract erosion by using contour planting, in which row crops are oriented so that they slow the flow of water instead of funneling it downslope. Row crops can also be alternated in strips with dense, soil-catching crops such as wheat. Planting rows of trees as windbreaks helps prevent soil loss from blowing wind, and it creates a more diverse ecosystem and a nesting place for insect-eating birds. Farm fields need not be plowed in the fall and left unplanted to erode during the winter; in fact, an increasing number of farmers are planting crops in the stubble from the previous season, reducing erosion.

Fourth, natural ecosystems tend to store water and purify it through biological processes, whereas human ecosystems tend to pollute water and shed it rapidly. A thundershower strikes a forest, an adjacent city, and a farm. The rich soil of the forest sponges up the water, which gradually drains into the ground, filtered by the soil and purified by decomposers that break down organic contaminants. In the nearby city, water pours from sidewalks, rooftops, and streets, picking up soot, silt, oil, heavy metals, and garbage. It races down gutters into storm sewers, and a weakly toxic soup gushes into the nearest stream or river. Farm runoff carries priceless topsoil, expensive fertilizer, and animal manure into rivers and lakes, where these potential resources become pollutants.

Preventing erosion will simultaneously conserve water and reduce water pollution from farm runoff. Manure from cattle feedlots, which is a significant source of both groundwater and surface-water pollution, could be placed on fields, where it would restore needed nutrients.

Although water will continue to run off our cities, the pollutants it carries can be reduced by minimizing our reliance on fossil fuels. We must eliminate leaded gasoline worldwide; U.S. gas is now lead-free, but leaded gas remains available throughout most of the world. We must also tighten standards for emissions from diesel and gasoline engines and smokestacks. Efficient public transportation systems will reduce pollution (and the massive frustration) caused by congested traffic. Insulation will reduce power consumption in our homes and offices, as will increased reliance on solar heat.

Fifth, simple human ecosystems such as farms tend to be unstable, whereas natural ecosystems have many species and tend to remain stable over time. Herbivorous insects are controlled by natural predators, including other insects, birds, and shrews. Insect populations are also limited because their preferred plants are scattered among many other plants rather than growing in a pure stand, as on a farm. On farms, both pest insects and their natural predators are exposed to pesticides. Unfortunately, the pests may develop resistance to the poison, but their predators are killed. In these simplified communities, unfavorable weather conditions or the introduction of an exotic species can be disastrous.

Farmers can counteract this trend by planting smaller fields with a wider variety of crops. Alternating crops helps maintain soil fertility and also helps prevent the proliferation of disease and insect pests that are specialized for a particular crop. Populations of corn borers, for example, will die off during years when the field is planted in alfalfa. The use of biological controls such as natural insect predators and insect diseases can reduce reliance on pesticides.

Finally, human ecosystems are characterized by continuously growing populations, whereas nonhuman populations in natural ecosystems are relatively stable. As our population expands, the spread of human-dominated ecosystems presents a growing threat to the diversity of species and to the delicate balance that has evolved over the 3-billion-year history of life on Earth.

In summary, natural ecosystems tend to be complex, stable, and self-sustaining, powered by solar energy and nourished by recycled nutrients. They provide diverse habitats for wildlife, purify contaminants through the action of decomposers, and build up nutrient-rich soil. Modern human ecosystems are relatively simple and are sustained by large inputs of energy from fossil fuels. They tend to minimize wildlife habitat, to contaminate soil and water, and to lose nutrients and fertile soil. These problems are compounded by continued population growth, which is causing the expansion of human-dominated ecosystems at the expense of undisturbed ones.

As we have pointed out, human ecosystems do not have to be as disruptive and alien to the operation of natural ecosystems as we have allowed them to become. Through understanding, education, commitment, appropriate use of technology, and stabilization of our population, we can reverse many of these destructive trends.

CASE STUDY REVISITED

Wings of Hope

The Arabuko-Sokoke forest remains under siege from squatters who want to clear land and establish homes within its confines. But where the farmers gather their butterflies, the forest suffers far less poaching, because the farmers are now reporting poachers rather than joining them. Over several years of monitoring, the project manager sees no evidence that butterfly populations are being reduced. With full stomachs and money to buy a few minor luxuries, the people can now afford to support the philosophy of one village elder who states, "The forest is here, we found the forest here, and we have to leave it here for our children's generation."

Most conservationists agree that "fencing and fining" is not the way to preserve habitat; the local residents must actively support and participate in its preservation. Design or research other projects that fit the model of sustainable use of rain forests or other endangered ecosystems, such as coral reefs.

Summary of Key Concepts

1) What Factors Influence Earth's Climate?

The availability of sunlight, water, and appropriate temperatures determines the climate of a given region. Sunlight maintains Earth's temperature. Equal amounts of solar energy are spread over a smaller surface at the equator than farther north and south, making the equator relatively warm, whereas higher latitudes have lower overall temperatures. Earth's tilt on its axis causes dramatic seasonal variation at northern and southern latitudes.

The rising of warm air and sinking of cool air in regular patterns from north to south produce areas of low and high moisture. These patterns are modified by the topography of continents and by ocean currents.

2) What Conditions Does Life Require?

The requirements for life on Earth include nutrients, energy, liquid water, and a reasonable temperature. The differences in the form and abundance of living things in various locations on Earth are largely attributable to differences in the interplay of these four factors.

3) How Is Life on Land Distributed?

On land, the crucial limiting factors are temperature and liquid water. Large regions of the continents that have similar climates will have similar vegetation, determined by the interaction of temperature and rainfall or the availability of water. These regions are called biomes.

Tropical forest biomes, located near the equator, are warm and wet, dominated by huge broadleaf evergreen trees. Most nutrients are tied up in vegetation, and most animal life is arboreal. Rain forests, home to at least 50% of all species, are rapidly being cut for agriculture, although the soil is extremely poor.

The African savanna is an extensive grassland with pronounced wet and dry seasons. It is home to the world's most diverse and extensive herds of large mammals.

Most deserts, hot and dry, are located between 20° and 30° north and south latitude and in the rain shadows of mountain ranges. In deserts, plants are widely spaced and have adaptations to conserve water. Animals tend to be small and nocturnal, also adapted to drought.

Chaparral exists in desertlike conditions, which are moderated by their proximity to a coastline, allowing small trees and bushes to thrive. Grasslands, concentrated in the centers of continents, have a continuous grass cover and few trees. They produce the world's richest soils and have largely been converted to agriculture.

Temperate deciduous forests, whose broadleaf trees drop their leaves in winter to conserve moisture, dominate the eastern half of the U.S. and are also found in Western Europe and East Asia. Higher precipitation occurs there than in the grasslands. The wet temperate rain forests, dominated by evergreens, are on the northern Pacific coast of the U.S. The taiga, or northern coniferous forest, covers much of the northern U.S., southern Canada, and northern Eurasia. It is dominated by conifers whose small waxy needles are adapted for water conservation and year-round photosynthesis.

The tundra is a frozen desert where permafrost prevents the growth of trees, and bushes remain stunted. Nonetheless, diverse arrays of animal life and perennial plants flourish in this fragile biome, which is found on mountain peaks and the Arctic.

4) How Is Life in Water Distributed?

Energy and nutrients are the major limiting factors in the distribution and abundance of life in aquatic ecosystems. Nutrients are found in bottom sediments and are washed in from surrounding land, concentrating them near shore and in deep water.

Freshwater lakes have three life zones. The littoral zone, near shore, is rich in energy and nutrients and supports the most diverse community. The limnetic zone is the lighted region of open water where photosynthesis can occur. The profundal zone is the deep water, where light is inadequate for photosynthesis and the community is dominated by heterotrophic organisms. Oligotrophic lakes are clear and low in nutrients, and they support sparse communities. Eutrophic lakes are rich in nutrients and support dense communities. During succession to dry land, lakes tend to go from an oligotrophic to a eutrophic condition.

Most life in the oceans is found in shallow water, where sunlight can penetrate, and is concentrated near the continents and in areas of upwelling, where nutrients are most plentiful. Coastal waters, consisting of the intertidal zone and the near-shore zone, contain the most abundant life. Producers consist of aquatic plants anchored to the bottom and photosynthetic protists called phytoplankton. Coral reefs are confined to warm, shallow seas. The calcium carbonate reefs form a complex habitat supporting the most diverse undersea ecosystem, threatened by silt, overfishing, and global warming.

In the open ocean, most life is found in the photic zone, where light supports phytoplankton. In the lower aphotic zone, life is supported by nutrients that drift down from the photic zone. Many ocean fisheries have been overexploited. Specialized vent communities, supported by chemosynthetic bacteria, thrive at great depths in the superheated waters where Earth's crustal plates are separating.

Key Terms

aphotic zone *p. 878*
biodiversity *p. 864*
biome *p. 863*
chaparral *p. 869*
climate *p. 857*
coral reef *p. 880*
desert *p. 868*
desertification *p. 869*
estuary *p. 878*
eutrophic lake *p. 877*
grassland *p. 870*
gyre *p. 860*
hydrothermal vent community *p. 883*
intertidal zone *p. 878*
limnetic zone *p. 876*
littoral zone *p. 876*
near-shore zone *p. 878*
northern coniferous forest *p. 873*
oligotrophic lake *p. 876*
ozone layer *p. 858*
pelagic *p. 881*
permafrost *p. 875*
photic zone *p. 878*
phytoplankton *p. 876*
plankton *p. 876*
prairie *p. 870*
profundal zone *p. 876*
rain shadow *p. 862*
savanna *p. 867*
taiga *p. 873*
temperate deciduous forest *p. 870*
temperate rain forest *p. 872*
tropical deciduous forest *p. 867*
tropical rain forest *p. 864*
tundra *p. 874*
upwelling *p. 878*
weather *p. 857*
zooplankton *p. 876*

Thinking Through the Concepts

Multiple Choice

1. *Most plant species fit into only a few morphological types. Which type would you think is most restricted by temperature and rainfall?*
a. trees
b. shrubs
c. grasses
d. perennial herbs
e. annual weeds

2. *In which biome is the smallest fraction of carbon and nutrients present in the soil?*
a. tropical rain forest
b. savanna
c. tundra
d. grassland
e. coniferous forest

3. *How do mountain ranges create deserts?*
a. by lifting land up into colder, drier air
b. by completely blocking the flow of air into desert areas, thus preventing clouds from getting there
c. by forcing air to first rise and then fall, thus causing rain on one side of the mountains and desert on the other
d. by causing the global wind patterns that make certain latitudes very dry
e. by causing very steep slopes that are subject to erosion

4. *What is the primary reason that plants from distant, but climatically similar, places commonly look the same?*
a. common ancestry
b. adaptation to the same physical conditions
c. adaptation to similar herbivores
d. continental drift
e. effects of past climate change

5. *Which of these biomes has been increased in area by human activities?*
a. savanna
b. temperate rain forest
c. grassland
d. coniferous forest
e. desert

6. *What biome has the richest soil and has largely been converted to agriculture?*
a. tundra
b. coniferous forest
c. grassland
d. tropical rain forest
e. deciduous forest

? Review Questions

1. Explain how air currents contribute to the formation of the Tropics and the large deserts.

2. What are large, roughly circular ocean currents called? What effect do they have on climate, and where is that effect strongest?

3. What are the four major requirements for life? Which two are most often limiting in terrestrial ecosystems? In ocean ecosystems?

4. Explain why traveling up a mountain takes you through biomes similar to those you would encounter traveling north for a long distance.

5. Where are the nutrients of the tropical forest biome concentrated? Why is life in the tropical rain forest concentrated high above the ground?

6. Explain two undesirable effects of agriculture in the tropical rain forest biome.
7. List some adaptations of (a) desert plants and (b) desert animals to heat and drought.
8. What human activities damage deserts?
9. How are trees of the taiga adapted to a lack of water and a short growing season?
10. How do deciduous and coniferous biomes differ?
11. What single environmental factor best explains why there is shortgrass prairie in Colorado, tallgrass prairie in Illinois, and deciduous forest in Ohio?
12. Where are the world's largest populations of large herbivores and carnivores located?
13. Where is life in the oceans most abundant, and why?
14. Why is the diversity of life so high in coral reefs? What human impacts threaten them?
15. Distinguish among the limnetic, littoral, and profundal zones of lakes in terms of their location and the communities they support.
16. Distinguish between oligotrophic and eutrophic lakes. Describe (a) a natural scenario and (b) a human-created scenario under which an oligotrophic lake might be converted to a eutrophic lake.
17. What is the reason and importance of the spring and fall overturn in temperate lakes?
18. Distinguish between the photic and aphotic zones. How do organisms in the photic zone obtain nutrients? How are nutrients obtained in the aphotic zone?
19. What unusual primary producer forms the basis for hydrothermal vent communities?
20. On the basis of the location of the worst atmospheric ozone depletion, which biomes are likely to be most affected by increased UV penetration?

Applying the Concepts

1. List at least six differences between human-dominated and undisturbed ecosystems, and discuss in some detail how these differences can be minimized.
2. In which terrestrial biome is your college or university located? Discuss similarities and differences between your location and the general description of that biome in the text. If you are living in a city, how has the urban environment modified your interaction with the biome?
3. During the 1960s and 1970s, many parts of the U.S. and Canada banned the use of detergents containing phosphates. Until that time, almost all laundry detergents and many soaps and shampoos had high concentrations of phosphates. What environmental concern do you think prompted these bans, and what ecosystem has benefited most from the bans?
4. Because ozone depletion is expected to get worse, not better, for decades to come, biologists have tried to assess which types of species will be most susceptible to increased UV penetration. Two of the groups that may be most vulnerable—ocean plankton and long-lived birds and mammals—are quite different from each other. Try to think of what makes each of these groups so susceptible to increased UV radiation.
5. Understanding the ways in which the four basic requirements for life determine where different biomes occur can help us predict the consequences of global warming. Global warming is expected to make most areas warmer, but it is also expected to change rainfall in ways that are hard to predict—some areas will get wetter, and others drier. Our ignorance of how rainfall will change is not very important in understanding shifts in more northerly biomes, but it is very important for our understanding of changes in tropical areas. Look at Figure 41-8 and explain why this is true.
6. More-northerly forests are far better able to regenerate after logging than are tropical rain forests. Try to explain why this is true. HINT: The cold soils of northern climates greatly slow down decomposition rates.

For More Information

Burroughs, D. "On the Wings of Hope." *International Wildlife*, July–August 2000. Kenyan butterflies are saving a unique forest and its people, the basis for our Case Study.

Chadwick, D. H. "Blue Refuges." *National Geographic*, March 1998. Beautiful photographs highlight 12 national marine sanctuaries set aside to preserve a variety of marine ecosystems along the coastal U.S.

Helmuth, L. "Can This Swamp Be Saved?" *Science News*, April 17, 1999. Describes the ambitious protection plan for the Florida Everglades.

Hinrichsen, D. "Requiem for Reefs?" *International Wildlife*, March–April 1997. Stunning photographs and a compelling text describe the beauty of Earth's imperiled coral reefs.

Holloway, M. "Sustaining the Amazon." *Scientific American*, July 1993. Describes the threats to the Amazon rain forest and innovative ways to preserve tropical forests and their biodiversity.

Holmes, B. "Case of the Dwindling Cloud Forests." *International Wildlife*, July–August 2000. Global warming may be threatening the unique and diverse cloud forests of Costa Rica.

Kusler, J. A., Mitsch, W. J., and Larson, J. S. "Wetlands." *Scientific American*, January 1994. Discusses the characteristics and values of these threatened but highly important ecosystems.

Mallin, M. A, "Impacts of Industrial Animal Production on Rivers and Estuaries." *American Scientist*, January–February 2000. Lagoons designed to store waste from large numbers of farm animals pose a threat to nearby freshwater ecosystems.

Mares, M. A. "Desert Rodents, Seed Consumption, and Convergence: Evolutionary Shuffling of Adaptations." *BioScience*, Vol. 43, 1993. An example of how animal as well as plant species and communities living in similar climates can show convergence.

Mitchell, J. G. "Our National Forests." *National Geographic*, March 1997. Stunning photographs and engaging text describe the threats to forests in the U.S.

Monks, V. "How Did the Poison Get into the Trout?" *National Wildlife,* August–September 1998. Toxic substances accumulating in the Great Lakes pose a hazard to wildlife and a puzzle to scientists.

Prather, M., Midgley, P., Rowland, F. S., and Storlarski, R. "The Ozone Layer: The Road Not Taken." *Nature*, June 13, 1996. Reviews the progress toward slowing ozone depletion and explores what would have happened had the ozone-protection treaties not been passed.

Toon, B., and Turco, R. P. "Polar Stratospheric Clouds and Ozone Depletion." *Scientific American*, June 1991. Describes the unique chemistry of the Antarctic clouds that causes ozone breakdown.

Tunnicliffe, V. "Hydrothermal-Vent Communities of the Deep Sea." *American Scientist*, July–August 1992. A comprehensive look at the structure and function of undersea vent communities, with excellent illustrations.

Williams, T. "What Good Is a Wetland?" *Audubon*, November–December 1996. In addition to supporting a diversity of wildlife, purifying drinking water, and protecting people from floods, wetlands are beautiful.

Answers to Multiple-Choice Questions

1. a 2. a 3. c 4. b 5. e 6. c

MEDIATUTOR
Earth's Diverse Ecosystems

CD Activities

Activity 41.1: Climate Builds Biomes

Estimated time: 5 minutes

Hadley cells are a major feature of atmospheric circulation that strongly influence climate and community type in the tropics and subtropics. Warm tropical air rises over the equator and produces the wet, warm conditions that support communities like tropical rain forests. The same air circulates poleward and eventually cools enough that it sinks. As it sinks, it warms again. The sinking air is quite dry (since warm air can hold more moisture than cool air can), so it produces the warm, dry conditions responsible for the existence of the major subtropical deserts of the world (such as the Kalahari).

Activity 41.2: Biomes of the World

Estimated time: 10 minutes

In this tutorial you will explore one of the critical factors involved in determining biomes. You will then be able to predict the different characteristics of various biomes, including climatic (abiotic) factors as well as the typical representative plants and animals (biotic factors) of those respective biomes.

Start the MediaTutor Student CD-ROM and enter the activity number in the Quick Search box to be taken directly to that activity.

Web Investigations

Case Study: Wings of Hope

Estimated time: 10 minutes

Is conservation solely for the rich? For many people living at, or near, delicate ecosystems, survival is difficult enough. The Kipepeo Butterfly Project was designed to combine economic development with incentives for conservation practices. The results included a higher standard of living and reduced ecological stress. But developing such programs is difficult. Take a look at the details.

Go to http://www.prenhall.com/audesirk6, the Audesirk Companion Web site. Select Chapter 41 and the Web Investigation to begin.

APPENDIX I
Metric System Conversions

To Convert Metric Units:	Multiply by:	To Get English Equivalent:
	Length	
Centimeters (cm)	0.3937	Inches (in.)
Meters (m)	3.2808	Feet (ft)
Meters (m)	1.0936	Yards (yd)
Kilometers (km)	0.6214	Miles (mi)
	Area	
Square centimeters (cm^2)	0.155	Square inches ($in.^2$)
Square meters (m^2)	10.7639	Square feet (ft^2)
Square meters (m^2)	1.1960	Square yards (yd^2)
Square kilometers (km^2)	0.3831	Square miles (mi^2)
Hectare (ha) (10,000 m^2)	2.4710	Acres (a)
	Volume	
Cubic centimeters (cm^3)	0.06	Cubic inches ($in.^3$)
Cubic meters (m^3)	35.30	Cubic feet (ft^3)
Cubic meters (m^3)	1.3079	Cubic yards (yd^3)
Cubic kilometers (km^3)	0.24	Cubic miles (mi^3)
Liters (L)	1.0567	Quarts (qt), U.S.
Liters (L)	0.26	Gallons (gal), U.S.
	Mass	
Grams (g)	0.03527	Ounces (oz)
Kilograms (kg)	2.2046	Pounds (lb)
Metric ton (tonne) (t)	1.10	Ton (tn), U.S.
	Speed	
Meters/second (mps)	2.24	Miles/hour (mph)
Kilometers/hour (kmph)	0.62	Miles/hour (mph)

To Convert English Units:	Multiply by:	To Get Metric Equivalent:
	Length	
Inches (in.)	2.54	Centimeters (cm)
Feet (ft)	0.3048	Meters (m)
Yards (yd)	0.9144	Meters (m)
Miles (mi)	1.6094	Kilometers (km)
	Area	
Square inches ($in.^2$)	6.45	Square centimeters (cm^2)
Square feet (ft^2)	0.0929	Square meters (m^2)
Square yards (yd^2)	0.8361	Square meters (m^2)
Square miles (mi^2)	2.5900	Square kilometers (km^2)
Acres (a)	0.4047	Hectare (ha) (10,000 m^2)
	Volume	
Cubic inches ($in.^3$)	16.39	Cubic centimeters (cm^3)
Cubic feet (ft^3)	0.028	Cubic meters (m^3)
Cubic yards (yd^3)	0.765	Cubic meters (m^3)
Cubic miles (mi^3)	4.17	Cubic kilometers (km^3)
Quarts (qt), U.S.	0.9463	Liters (L)
Gallons (gal), U.S.	3.8	Liters (L)
	Mass	
Ounces (oz)	28.3495	Grams (g)
Pounds (lb)	0.4536	Kilograms (kg)
Ton (tn), U.S.	0.91	Metric ton (tonne) (t)
	Speed	
Miles/hour (mph)	0.448	Meters/second (mps)
Miles/hour (mph)	1.6094	Kilometers/hour (kmph)

Metric Prefixes

Prefix		Meaning		
giga-	G	10^9	=	1,000,000,000
mega-	M	10^6	=	1,000,000
kilo-	k	10^3	=	1000
hecto-	h	10^2	=	100
deka-	da	10^1	=	10
		10^0	=	1
deci-	d	10^{-1}	=	0.1
centi-	c	10^{-2}	=	0.01
milli-	m	10^{-3}	=	0.001
micro-	μ	10^{-6}	=	0.000001

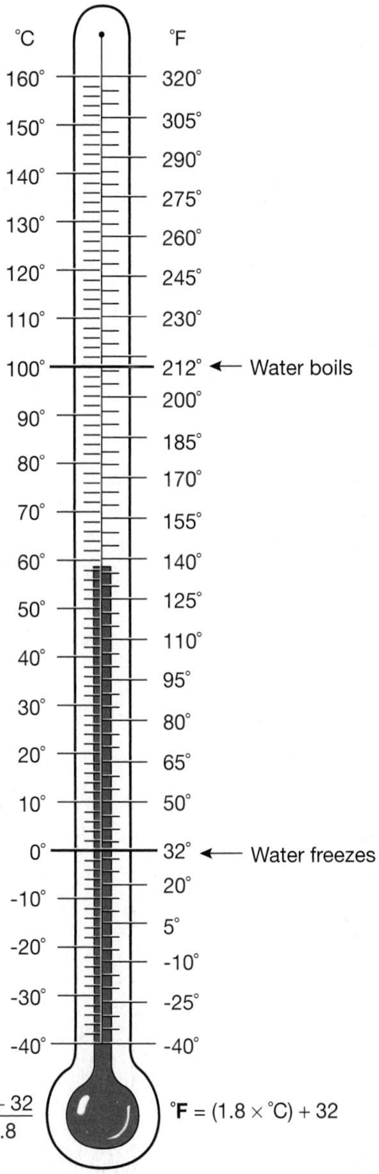

$$°C = \frac{°F - 32}{1.8}$$

$$°F = (1.8 \times °C) + 32$$

APPENDIX II
Classification of Major Groups of Organisms*

Domain	Kingdom	Phylum	Common Name
Bacteria (prokaryotic, peptidoglycan in cell wall)			bacteria
Archaea (prokaryotic, no peptidoglycan in cell wall)			archaeans
Eukarya (eukaryotic)			
	Protista (unicellular)		protists
		Oomycota	egg fungi
		Phaeophyta	brown algae
		Bacillariophyta	diatoms
		Rhodophyta	red algae
		Pyrrophyta	dinoflagellates
		Euglenophyta	euglenoids
		Myxomycota	plasmodial slime molds
		Acrasiomycota	cellular slime molds
		Sarcomastigophora	zooflagellates, amoebae
		Apicomplexa	sporozoans
		Ciliophora	ciliates
		Chlorophyta	green algae
	Fungi (multicellular, heterotrophic, absorb nutrients)		fungi
		Chytridiomycota	chytrids
		Zygomycota	zygote fungi
		Ascomycota	sac fungi
		Basidiomycota	club fungi
	Plantae (multicellular, photosynthetic)		plants
		Bryophyta	liverworts, mosses
		Pteridophyta	ferns
		Coniferophyta	evergreens
		Anthophyta	flowering plants
	Animalia (multicellular, heterotrophic, ingest nutrients)		animals
		Porifera	sponges
		Cnidaria	hydras, sea anemones, jellyfish, corals
		Ctenophora	comb jellies
		Platyhelminthes	flatworms
		Nematoda	roundworms
		Annelida	segmented worms
		Oligochaeta	earthworms
		Polychaeta	tube worms
		Hirudinea	leeches
		Arthropoda	arthropods ("jointed legs")
		Insecta	insects
		Arachnida	spiders, ticks
		Crustacea	crabs, lobsters
		Mollusca	mollusks ("soft-bodied")
		Gastropoda	snails
		Pelecypoda	mussels, clams
		Cephalopoda	squid, octopuses
		Echinodermata	sea stars, sea urchins, sea cucumbers
		Chordata	chordates
		Urochordata	tunicates
		Cephalochordata	lancelets
		Myxini	hagfishes
		Vertebrata	vertebrates
		Pertromyzontiformes	lampreys
		Chondrichthyes	sharks, rays
		Osteichthyes	bony fishes
		Amphibia	frogs, salamanders
		Anapsida	turtles
		Diapsida	
		Archosauria	birds, crocodiles
		Squamata	lizards, snakes
		Mammalia	mammals

*This table lists only those taxonomic categories described in the textbook.

Glossary

abdomen: the body segment at the posterior end of an animal with segmentation; contains most of the digestive structures.

abiotic (ā-bī-ah´-tik): nonliving; the abiotic portion of an ecosystem includes soil, rock, water, and the atmosphere.

abortion: the procedure for terminating pregnancy; the cervix is dilated, and the embryo and placenta are removed.

abscisic acid (ab-sis´-ik): a plant hormone that generally inhibits the action of other hormones, enforcing dormancy in seeds and buds and causing the closing of stomata.

abscission layer: a layer of thin-walled cells, located at the base of the petiole of a leaf, that produces an enzyme that digests the cell wall holding leaf to stem, allowing the leaf to fall off.

absorption: the process by which nutrients are taken into cells.

accessory pigments: colored molecules other than chlorophyll that absorb light energy and pass it to chlorophyll.

acellular slime mold: a type of funguslike protist that forms a multinucleate structure that crawls in amoeboid fashion and ingests decaying organic matter; also called *plasmodial slime mold.*

acetylcholine (ah-sēt´-il-kō´-lēn): a neurotransmitter in the brain and in synapses of motor neurons that innervate skeletal muscles.

acid: a substance that releases hydrogen ions (H^+) into solution; a solution with a pH of less than 7.

acid deposition: the deposition of nitric or sulfuric acid, either dissolved in rain (acid rain) or in the form of dry particles, as a result of the production of nitrogen oxides or sulfur dioxide through burning, primarily of fossil fuels.

acidic: with an H^+ concentration exceeding that of OH^-; releasing H^+.

acquired immune deficiency syndrome (AIDS): an infectious disease caused by the human immunodeficiency virus (HIV); attacks and destroys T cells, thus weakening the immune system.

acrosome (ak´-rō-sōm): a vesicle, located at the tip of an animal sperm, that contains enzymes needed to dissolve protective layers around the egg.

actin (ak´-tin): a major muscle protein whose interactions with myosin produce contraction; found in the thin filaments of the muscle fiber; see also *myosin.*

action potential: a rapid change from a negative to a positive electrical potential in a nerve cell. This signal travels along an axon without a change in intensity.

activation energy: in a chemical reaction, the energy needed to force the electron shells of reactants together, prior to the formation of products.

active site: the region of an enzyme molecule that binds substrates and performs the catalytic function of the enzyme.

active transport: the movement of materials across a membrane through the use of cellular energy, normally against a concentration gradient.

adaptation: a trait that increases the ability of an individual to survive and reproduce compared to individuals without the trait.

adaptive radiation: the rise of many new species in a relatively short time as a result of a single species that invades different habitats and evolves under different environmental pressures in those habitats.

adenine: a nitrogenous base found in both DNA and RNA; abbreviated as *A.*

adenosine diphosphate (a-den´-ō-sēn dī-fos´-fāt; ADP): a molecule composed of the sugar ribose, the base adenine, and two phosphate groups; a component of ATP.

adenosine triphosphate (a-den´-ō-sēn trī-fos´-fāt; ATP): a molecule composed of the sugar ribose, the base adenine, and three phosphate groups; the major energy carrier in cells. The last two phosphate groups are attached by "high-energy" bonds.

adipose tissue (a´-di-pōs): tissue composed of fat cells.

adrenal cortex: the outer part of the adrenal gland, which secretes steroid hormones that regulate metabolism and salt balance.

adrenal gland: a mammalian endocrine gland, adjacent to the kidney; secretes hormones that function in water regulation and in the stress response.

adrenal medulla: the inner part of the adrenal gland, which secretes epinephrine (adrenaline) and norepinephrine (noradrenaline).

adrenocorticotropic hormone (a-drēn-ō-kor-tik-ō-trō´-pik; ACTH): a hormone, secreted by the anterior pituitary, that stimulates the release of hormones by the adrenal glands, especially in response to stress.

aerobic: using oxygen.

age structure: the distribution of males and females in a population according to age groups.

agglutination (a-gloo-tin-ā´-shun): the clumping of foreign substances or microbes, caused by binding with antibodies.

aggression: antagonistic behavior, normally among members of the same species, often resulting from competition for resources.

aggressive mimicry (mim´ik-rē): the evolution of a predatory organism to resemble a harmless animal or part of the environment, thus gaining access to prey.

aldosterone: a hormone, secreted by the adrenal cortex, that helps regulate ion concentration in the blood by stimulating the reabsorption of sodium by the kidneys and sweat glands.

alga (al´-ga; pl., **algae, al´-jē):** any photosynthetic member of the eukaryotic Kingdom Protista.

allantois (al-an-tō´-is): one of the embryonic membranes of reptiles, birds, and mammals; in reptiles and birds, serves as a waste-storage organ; in mammals, forms most of the umbilical cord.

allele (al-ēl´): one of several alternative forms of a particular gene.

allele frequency: for any given gene, the relative proportion of each allele of that gene in a population.

allergy: an inflammatory response produced by the body in response to invasion by foreign materials, such as pollen, that are themselves harmless.

allopatric speciation (al-ō-pat´-rik): speciation that occurs when two populations are separated by a physical barrier that prevents gene flow between them (geographical isolation).

allosteric regulation: the process by which enzyme action is enhanced or inhibited by small organic molecules that act as regulators by binding to the enzyme and altering its active site.

alternation of generations: a life cycle, typical of plants, in which a diploid sporophyte (spore-producing) generation alternates with a haploid gametophyte (gamete-producing) generation.

altruism: a type of behavior that may decrease the reproductive success of the individual performing it but benefits that of other individuals.

alveolus (al-vē´-ō-lus; pl., **alveoli):** a tiny air sac within the lungs, surrounded by capillaries, where gas exchange with the blood occurs.

amino acid: the individual subunit of which proteins are made, composed of a central carbon atom bonded to an amino group ($-NH_2$), a carboxyl group ($-COOH$), a hydrogen atom, and a variable group of atoms denoted by the letter *R.*

ammonia: NH_3; a highly toxic nitrogen-containing waste product of amino acid breakdown. In the mammalian liver, it is converted to urea.

amniocentesis (am-nē-ō-sen-tē´-sis): a procedure for sampling the amniotic fluid surrounding a fetus: A sterile needle is inserted through the abdominal wall, uterus, and amniotic sac of a pregnant woman; 10 to 20 milliliters of amniotic fluid is withdrawn. Various tests may be performed on the fluid and the fetal cells suspended in it to provide information on the developmental and genetic state of the fetus.

amnion (am´-nē-on): one of the embryonic membranes of reptiles, birds, and mammals; encloses a fluid-filled cavity that envelops the embryo.

amniote egg (am-nē-ōt´): the egg of reptiles and birds; contains an amnion that encloses the embryo in a watery environment, allowing the egg to be laid on dry land.
amoeba: a type of animal-like protist that uses a characteristic streaming mode of locomotion by extending a cellular projection called a *pseudopod.*
amoeboid cell: a protist or animal cell that moves by extending a cellular projection called a pseudopod.
amplexus (am-plek´-sus): in amphibians, a form of external fertilization in which the male holds the female during spawning and releases his sperm directly onto her eggs.
ampulla: a muscular bulb that is part of the water-vascular system of echinoderms; controls the movement of tube feet, which are used for locomotion.
amygdala (am-ig´-da-la): part of the forebrain of vertebrates that is involved in the production of appropriate behavioral responses to environmental stimuli.
amylase (am´-i-lās): an enzyme, found in saliva and pancreatic secretions, that catalyzes the breakdown of starch.
anaerobe: an organism whose respiration does not require oxygen.
anaerobic: not using oxygen.
analogous structures: structures that have similar functions and superficially similar appearance but very different anatomies, such as the wings of insects and birds. The similarities are due to similar environmental pressures rather than to common ancestry.
anaphase (an´-a-fāz): in mitosis, the stage in which the sister chromatids of each chromosome separate from one another and are moved to opposite poles of the cell; in meiosis I, the stage in which homologous chromosomes, consisting of two sister chromatids, are separated; in meiosis II, the stage in which the sister chromatids of each chromosome separate from one another and are moved to opposite poles of the cell.
androgen: a male sex hormone.
androgen insensitivity: a rare condition in which an individual with XY chromosomes is female in appearance because the body's cells don't respond to the male hormones that are present.
angina (an-jī´-nuh): chest pain associated with reduced blood flow to the heart muscle, caused by the obstruction of coronary arteries.
angiosperm (an´-jē-ō-sperm): a flowering vascular plant.
angiotensin (an-jē-ō-ten´-sun): a hormone that functions in water regulation in mammals by stimulating physiological changes that increase blood volume and blood pressure.
annual ring: a pattern of alternating light (early) and dark (late) xylem of woody stems and roots, formed as a result of the unequal availability of water in different seasons of the year, normally spring and summer.
antagonistic muscles: a pair of muscles, one of which contracts and in so doing extends the other; an arrangement that makes possible movement of the skeleton at joints.
anterior: the front, forward, or head end of an animal.
anterior pituitary: a lobe of the pituitary gland that produces prolactin and growth hormone as well as hormones that regulate hormone production in other glands.
anther (an´-ther): the uppermost part of the stamen, in which pollen develops.
antheridium (an-ther-id´-ē-um): a structure in which male sex cells are produced, found in the bryophytes and certain seedless vascular plants.
antibiotic resistance: the ability of a mutated pathogen to resist the effects of an antibiotic that normally kills it.
antibody: a protein, produced by cells of the immune system, that combines with a specific antigen and normally facilitates the destruction of the antigen.
anticodon: a sequence of three bases in transfer RNA that is complementary to the three bases of a codon of messenger RNA.
antidiuretic hormone (an-tē-dī-ūr-et´-ik; ADH): a hormone produced by the hypothalamus and released into the bloodstream by the posterior pituitary when blood volume is low; increases the permeability of the distal tubule and the collecting duct to water, allowing more water to be reabsorbed into the bloodstream.
antigen: a complex molecule, normally a protein or polysaccharide, that stimulates the production of a specific antibody.
aphotic zone: the region of the ocean below 200 m, where sunlight does not penetrate.
apical dominance: the phenomenon whereby a growing shoot tip inhibits the sprouting of lateral buds.
apical meristem (āp´-i-kul mer´-i-stem): the cluster of meristematic cells at the tip of a shoot or root (or one of their branches).
appendicular skeleton (ap-pen-dik´-ū-lur): the portion of the skeleton consisting of the bones of the extremities and their attachments to the axial skeleton; the pectoral and pelvic girdles, the arms, legs, hands, and feet.
aqueous humor (ā´-kwē-us): the clear, watery fluid between the cornea and lens of the eye.
Archaea: one of life's three domains; consists of prokaryotes that are only distantly related to members of the domain Bacteria.
archegonium (ar-ke-gō´-nē-um): a structure in which female sex cells are produced; found in the bryophytes and certain seedless vascular plants.
arteriole (ar-tēr´-ē-ōl): a small artery that empties into capillaries. Contraction of the arteriole regulates blood flow to various parts of the body.
artery (ar´-tuh-rē): a vessel with muscular, elastic walls that conducts blood away from the heart.
artificial selection: a selective breeding procedure in which only those individuals with particular traits are chosen as breeders; used mainly to enhance desirable traits in domestic plants and animals; may also be used in evolutionary biology experiments.
ascus (as´-kus): a saclike case in which sexual spores are formed by members of the fungal division Ascomycota.
asexual reproduction: reproduction that does not involve the fusion of haploid sex cells. The parent body may divide and new parts regenerate, or a new, smaller individual may form as an attachment to the parent, to drop off when complete.
association neuron: in a neural network, a nerve cell that is postsynaptic to a sensory neuron and presynaptic to a motor neuron. In actual circuits, there may be many association neurons between individual sensory and motor neurons.
atherosclerosis (ath´-er-ō-skler-ō´-sis): a disease characterized by the obstruction of arteries by cholesterol deposits and thickening of the arterial walls.
atom: the smallest particle of an element that retains the properties of the element.
atomic nucleus: the membrane-bound organelle of eukaryotic cells that contains the cell's genetic material.
atomic number: the number of protons in the nuclei of all atoms of a particular element.
atrial natriuretic peptide (ā´-trē-ul nā-trē-ū-ret´-ik; ANP): a hormone, secreted by cells in the mammalian heart, that reduces blood volume by inhibiting the release of ADH and aldosterone.
atrioventricular (AV) node (ā´-trē-ō-ven-trik´-ū-lar nōd): a specialized mass of muscle at the base of the right atrium through which the electrical activity initiated in the sinoatrial node is transmitted to the ventricles.
atrioventricular valve: a heart valve that separates each atrium from each ventricle, preventing the backflow of blood into the atria during ventricular contraction.
atrium (ā´-trē-um): a chamber of the heart that receives venous blood and passes it to a ventricle.
auditory canal (aw´-di-tor-ē): a canal within the outer ear that conducts sound from the external ear to the tympanic membrane.
auditory nerve: the nerve leading from the mammalian cochlea to the brain, carrying information about sound.
autoimmune disease: a disorder in which the immune system produces antibodies against the body's own cells.
autonomic nervous system: the part of the peripheral nervous system of vertebrates that synapses on glands, internal organs, and smooth muscle and produces largely involuntary responses.
autosome (aw´-tō-sōm): a chromosome that occurs in homologous pairs in both males and females and that does not bear the genes determining sex.
autotroph (aw´-tō-trōf): "self-feeder"; normally, a photosynthetic organism; a producer.
auxin (awk´-sin): a plant hormone that influences many plant functions, including phototropism, apical dominance, and root branching; generally stimulates cell elonga-

tion and, in some cases, cell division and differentiation.

axial skeleton: the skeleton forming the body axis, including the skull, vertebral column, and rib cage.

axon: a long extension of a nerve cell, extending from the cell body to synaptic endings on other nerve cells or on muscles.

bacillus (buh-sil´-us; pl., **bacilli):** a rod-shaped bacterium.

Bacteria: one of life's three domains; consists of prokaryotes that are only distantly related to members of the domain Archaea.

bacterial conjugation: the exchange of genetic material between two bacteria.

bacteriophage (bak-tir´-ē-ō-fāj): a virus specialized to attack bacteria.

bacterium (bak-tir´-ē-um; pl., **bacteria):** an organism consisting of a single prokaryotic cell surrounded by a complex polysaccharide coat.

balanced polymorphism: the prolonged maintenance of two or more alleles in a population, normally because each allele is favored by a separate environmental pressure.

ball-and-socket joint: a joint in which the rounded end of one bone fits into a hollow depression in another, as in the hip; allows movement in several directions.

bark: the outer layer of a woody stem, consisting of phloem, cork cambium, and cork cells.

Barr body: an inactivated X chromosome in cells of female mammals, which have two X chromosomes; normally appears as a dark spot in the nucleus.

basal body: a structure resembling a centriole that produces a cilium or flagellum and anchors this structure within the plasma membrane.

base: (1) a substance capable of combining with and neutralizing H^+ ions in a solution; a solution with a pH of more than 7; (2) in molecular genetics, one of the nitrogen-containing, single- or double-ringed structures that distinguish one nucleotide from another. In DNA, the bases are adenine, guanine, cytosine, and thymine.

basic: with an H^+ concentration less than that of OH^-; combining with H^+.

basidiospore (ba-sid´-ē-ō-spor): a sexual spore formed by members of the fungal division Basidiomycota.

basidium (bas-id´-ē-um): a diploid cell, typically club-shaped, formed by members of the fungal division Basidiomycota; produces basidiospores by meiosis.

basilar membrane (bas´-eh-lar): a membrane in the cochlea that bears hair cells that respond to the vibrations produced by sound.

basophil (bas´-ō-fil): a type of white blood cell that releases both substances that inhibit blood clotting and chemicals that participate in allergic reactions and in responses to tissue damage and microbial invasion.

B cell: a type of lymphocyte that participates in humoral immunity; gives rise to plasma cells, which secrete antibodies into the circulatory system, and to memory cells.

behavior: any observable activity of a living animal.

behavioral isolation: the lack of mating between species of animals that differ substantially in courtship and mating rituals.

bilateral symmetry: a body plan in which only a single plane through the central axis will divide the body into mirror-image halves.

bile (bīl): a liquid secretion, produced by the liver, that is stored in the gallbladder and released into the small intestine during digestion; a complex mixture of bile salts, water, other salts, and cholesterol.

bile salt: a substance that is synthesized in the liver from cholesterol and amino acids and that assists in the breakdown of lipids by dispersing them into small particles on which enzymes can act.

binary fission: the process by which a single bacterium divides in half, producing two identical offspring.

binocular vision: the ability to see objects simultaneously through both eyes, providing greater depth perception and more-accurate judgment of the size and distance of an object from the eyes.

biodegradable: able to be broken down into harmless substances by decomposers.

biodiversity: the total number of species within an ecosystem and the resulting complexity of interactions among them.

biogeochemical cycle: also called a *nutrient cycle*, the process by which a specific nutrient in an ecosystem is transferred between living organisms and the nutrient's reservoir in the nonliving environment.

biological clock: a metabolic timekeeping mechanism found in most organisms, whereby the organism measures the approximate length of a day (24 hours) even without external environmental cues such as light and darkness.

biological magnification: the increasing accumulation of a toxic substance in progressively higher trophic levels.

biomass: the dry weight of organic material in an ecosystem.

biome (bī´-ōm): a terrestrial ecosystem that occupies an extensive geographical area and is characterized by a specific type of plant community: for example, deserts.

biosphere (bī´-ō-sfēr): that part of Earth inhabited by living organisms; includes both living and nonliving components.

biotechnology: any industrial or commercial use or alteration of organisms, cells, or biological molecules to achieve specific practical goals.

biotic (bī-ah´-tik): living.

biotic potential: the maximum rate at which a population could increase, assuming ideal conditions that allow a maximum birth rate and minimum death rate.

birth control pill: a temporary contraceptive method that prevents ovulation by providing a continuing supply of estrogen and progesterone, which in turn suppresses LH release; must be taken daily, normally for 21 days of each menstrual cycle.

bladder: a hollow muscular storage organ for storing urine.

blade: the flat part of a leaf.

blastocyst (blas´-tō-sist): an early stage of human embryonic development, consisting of a hollow ball of cells, enclosing a mass of cells attached to its inner surface, which becomes the embryo.

blastopore: the site at which a blastula indents to form a gastrula.

blastula (blas´-tū-luh): in animals, the embryonic stage attained at the end of cleavage, in which the embryo normally consists of a hollow ball with a wall one or several cell layers thick.

blind spot: the area of the retina at which the axons of the ganglion cell merge to form the optic nerve; the blind spot of the retina.

blood: a fluid consisting of plasma in which blood cells are suspended; carried within the circulatory system.

blood–brain barrier: relatively impermeable capillaries of the brain that protect the cells of the brain from potentially damaging chemicals that reach the bloodstream.

blood clotting: a complex process by which platelets, the protein fibrin, and red blood cells block an irregular surface in or on the body, such as a damaged blood vessel, sealing the wound.

blood vessel: a channel that conducts blood throughout the body.

body mass index (BMI): a number derived from an individual's weight and height used to estimate body fat. The formula is: weight (in kg)/height^2 (in meters^2).

bone: a hard, mineralized connective tissue that is a major component of the vertebrate endoskeleton; provides support and sites for muscle attachment.

book lung: a structure composed of thin layers of tissue, resembling pages in a book, that are enclosed in a chamber and used as a respiratory organ by certain types of arachnids.

boom-and-bust cycle: a population cycle characterized by rapid exponential growth followed by a sudden massive die-off, seen in seasonal species and in some populations of small rodents, such as lemmings.

Bowman's capsule: the cup-shaped portion of the nephron in which blood filtrate is collected from the glomerulus.

bradykinin (brā´-dē-kī´-nin): a chemical, formed during tissue damage, that binds to receptor molecules on pain nerve endings, giving rise to the sensation of pain.

brain: the part of the central nervous system of vertebrates that is enclosed within the skull.

branch root: a root that arises as a branch of a preexisting root, through divisions of pericycle cells and subsequent differentiation of the daughter cells.

bronchiole (bron´-kē-ōl): a narrow tube, formed by repeated branching of the bronchi, that conducts air into the alveoli.

bronchus (bron´-kus): a tube that conducts air from the trachea to each lung.

bryophyte (brī´-ō-fīt): a simple nonvascular plant of the division Bryophyta, including mosses and liverworts.

bud: in animals, a small copy of an adult that develops on the body of the parent and eventually breaks off and becomes independent; in

plants, an embryonic shoot, normally very short and consisting of an apical meristem with several leaf primordia.
budding: asexual reproduction by the growth of a miniature copy, or bud, of the adult animal on the body of the parent. The bud breaks off to begin independent existence.
buffer: a compound that minimizes changes in pH by reversibly taking up or releasing H^+ ions.
bulbourethral gland (bul-bō-ū-rē´-thrul): in male mammals, a gland that secretes a basic, mucus-containing fluid that forms part of the semen.
bulk flow: the movement of many molecules of a gas or fluid in unison from an area of higher pressure to an area of lower pressure.
bundle-sheath cell: one of a group of cells that surround the veins of plants; in C_4 (but not in C_3) plants, bundle-sheath cells contain chloroplasts.
C_3 cycle: the cyclic series of reactions whereby carbon dioxide is fixed into carbohydrates during the light-independent reactions of photosynthesis; also called *Calvin-Benson cycle.*
C_4 pathway: the series of reactions in certain plants that fixes carbon dioxide into oxaloacetic acid, which is later broken down for use in the C_3 cycle of photosynthesis.
calcitonin (kal-si-tōn´-in): a hormone, secreted by the thyroid gland, that inhibits the release of calcium from bone.
calorie (kal´-ō-rē): the amount of energy required to raise the temperature of 1 gram of water by 1 degree Celsius.
Calorie: a unit of energy, in which the energy content of foods is measured; the amount of energy required to raise the temperature of 1 liter of water 1 degree Celsius; also called a *kilocalorie,* equal to 1000 calories.
Calvin-Benson cycle: see *C_3 cycle.*
cambium (kam´-bē-um; pl., **cambia):** a lateral meristem, parallel to the long axis of roots and stems, that causes secondary growth of woody plant stems and roots. See *cork cambium*; *vascular cambium.*
camouflage (cam´-a-flaj): coloration and/or shape that renders an organism inconspicuous in its environment.
cancer: a disease in which some of the body's cells escape from normal regulatory processes and divide without control.
capillary: the smallest type of blood vessel, connecting arterioles with venules. Capillary walls, through which the exchange of nutrients and wastes occurs, are only one cell thick.
capsule: a polysaccharide or protein coating that some disease-causing bacteria secrete outside their cell wall.
carbohydrate: a compound composed of carbon, hydrogen, and oxygen, with the approximate chemical formula $(CH_2O)_n$; includes sugars and starches.
carbon fixation: the initial steps in the C_3 cycle, in which carbon dioxide reacts with ribulose bisphosphate to form a stable organic molecule.
cardiac cycle (kar´-dē-ak): the alternation of contraction and relaxation of the heart chambers.
cardiac muscle (kar´-dē-ak): the specialized muscle of the heart, able to initiate its own contraction, independent of the nervous system.
carnivore (kar´-neh-vor): literally, "meat eater"; a predatory organism that feeds on herbivores or on other carnivores; a secondary (or higher) consumer.
carotenoid (ka-rot´-en-oid): a red, orange, or yellow pigment, found in chloroplasts, that serves as an accessory light-gathering molecule in thylakoid photosystems.
carpel (kar´pel): the female reproductive structure of a flower, composed of stigma, style, and ovary.
carrier: an individual who is heterozygous for a recessive condition; displays the dominant phenotype but can pass on the recessive allele to offspring.
carrier protein: a membrane protein that facilitates the diffusion of specific substances across the membrane. The molecule to be transported binds to the outer surface of the carrier protein; the protein then changes shape, allowing the molecule to move across the membrane through the protein.
carrying capacity: the maximum population size that an ecosystem can support indefinitely; determined primarily by the availability of space, nutrients, water, and light.
cartilage (kar´-teh-lij): a form of connective tissue that forms portions of the skeleton; consists of chondrocytes and their extracellular secretion of collagen; resembles flexible bone.
Casparian strip (kas-par´-ē-un): a waxy, waterproof band, located in the cell walls between endodermal cells in a root, that prevents the movement of water and minerals into and out of the vascular cylinder through the extracellular space.
catalyst (kat´-uh-list): a substance that speeds up a chemical reaction without itself being permanently changed in the process; lowers the activation energy of a reaction.
catastrophism: the hypothesis that Earth has experienced a series of geological catastrophes, probably imposed by a supernatural being that accounts for the multitude of species, both extinct and modern, and preserves creationism.
cell: the smallest unit of life, consisting, at a minimum, of an outer membrane that encloses a watery medium containing organic molecules, including genetic material composed of DNA.
cell body: the part of a nerve cell in which most of the common cellular organelles are located; typically a site of integration of inputs to the nerve cell.
cell cycle: the sequence of events in the life of a cell, from one division to the next.
cell division: splitting of one cell into two; the process of cellular reproduction.
cell-mediated immunity: an immune response in which foreign cells or substances are destroyed by contact with T cells.
cell plate: in plant cell division, a series of vesicles that fuse to form the new plasma membranes and cell wall separating the daughter cells.
cellular respiration: the oxygen-requiring reactions, occurring in mitochondria, that break down the end products of glycolysis into carbon dioxide and water while capturing large amounts of energy as ATP.
cellular slime mold: a funguslike protist consisting of individual amoeboid cells that can aggregate to form a sluglike mass, which in turn forms a fruiting body.
cellulase: an enzyme that catalyzes the breakdown of the carbohydrate cellulose into its component glucose molecules; almost entirely restricted to microorganisms.
cellulose: an insoluble carbohydrate composed of glucose subunits; forms the cell wall of plants.
cell wall: a layer of material, normally made up of cellulose or cellulose-like materials, that is outside the plasma membrane of plants, fungi, bacteria, and some protists.
central nervous system: in vertebrates, the brain and spinal cord.
central vacuole: a large, fluid-filled vacuole occupying most of the volume of many plant cells; performs several functions, including maintaining turgor pressure.
centriole (sen´-trē-ōl): in animal cells, a short, barrel-shaped ring consisting of nine microtubule triplets; a microtubule-containing structure at the base of each cilium and flagellum; gives rise to the microtubules of cilia and flagella and is involved in spindle formation during cell division.
centromere (sen´-trō-mēr): the region of a replicated chromosome at which the sister chromatids are held together until they separate during cell division.
cephalization (sef-ul-ī-zā´-shun): the tendency of sensory organs and nervous tissue to become concentrated in the head region over evolutionary time.
cerebellum (ser-uh-bel´-um): the part of the hindbrain of vertebrates that is concerned with coordinating movements of the body.
cerebral cortex (ser-ē´-brul kor´-tex): a thin layer of neurons on the surface of the vertebrate cerebrum, in which most neural processing and coordination of activity occurs.
cerebral hemisphere: one of two nearly symmetrical halves of the cerebrum, connected by a broad band of axons, the corpus callosum.
cerebrospinal fluid: a clear fluid, produced within the ventricles of the brain, that fills the ventricles and cushions the brain and spinal cord.
cerebrum (ser-ē´-brum): the part of the forebrain of vertebrates that is concerned with sensory processing, the direction of motor output, and the coordination of most bodily activities; consists of two nearly symmetrical halves (the hemispheres) connected by a broad band of axons, the corpus callosum.
cervical cap: a birth control device consisting of a rubber cap that fits over the cervix, preventing sperm form entering the uterus.

cervix (ser´-viks): a ring of connective tissue at the outer end of the uterus, leading into the vagina.
channel protein: a membrane protein that forms a channel or pore completely through the membrane and that is usually permeable to one or to a few water-soluble molecules, especially ions.
chaparral: a biome that is located in coastal regions but has very low annual rainfall.
chemical bond: the force of attraction between neighboring atoms that holds them together in a molecule.
chemical equilibrium: the condition in which the "forward" reaction of reactants to products proceeds at the same rate as the "backward" reaction from products to reactants, so that no net change in chemical composition occurs.
chemical reaction: the process that forms and breaks chemical bonds that hold atoms together.
chemiosmosis (ke-mē-oz-mō´-sis): a process of ATP generation in chloroplasts and mitochondria. The movement of electrons down an electron transport system is used to pump hydrogen ions across a membrane, thereby building up a concentration gradient of hydrogen ions across the membrane; the hydrogen ions diffuse back across the membrane through the pores of ATP-synthesizing enzymes; the energy of their movement down their concentration gradient drives ATP synthesis.
chemoreceptor: a sensory receptor that responds to chemicals from the environment; used in the chemical senses of taste and smell.
chemosynthetic (kēm´-ō-sin-the-tik): capable of oxidizing inorganic molecules to obtain energy.
chemotactic (kēm-ō-tak´-tik): moving toward chemicals given off by food or away from toxic chemicals.
chiasma (kī-as´-muh; pl., chiasmata): a point at which a chromatid of one chromosome crosses with a chromatid of the homologous chromosome during prophase I of meiosis; the site of exchange of chromosomal material between chromosomes.
chitin (kī´-tin): a compound found in the cell walls of fungi and the exoskeletons of insects and some other arthropods; composed of chains of nitrogen-containing, modified glucose molecules.
chlamydia (kla-mid´-ē-uh): a sexually transmitted disease, caused by a bacterium, that causes inflammation of the urethra in males and of the urethra and cervix in females.
chlorophyll (klor´-ō-fil): a pigment found in chloroplasts that captures light energy during photosynthesis; absorbs violet, blue, and red light but reflects green light.
chloroplast (klor´-ō-plast): the organelle in plants and plantlike protists that is the site of photosynthesis; surrounded by a double membrane and containing an extensive internal membrane system that bears chlorophyll.
cholecystokinin (kō´-lē-sis-tō-ki´-nin): a digestive hormone, produced by the small intestine, that stimulates the release of pancreatic enzymes.
chondrocyte (kon´-drō-sīt): a living cell of cartilage. With their extracellular secretions of collagen, chondrocytes form cartilage.
chorion (kor´-ē-on): the outermost embryonic membrane in reptiles, birds, and mammals; in birds and reptiles, functions mostly in gas exchange; in mammals, forms most of the embryonic part of the placenta.
chorionic gonadotropin (CG): a hormone, secreted by the chorion (one of the fetal membranes), that maintains the integrity of the corpus luteum during early pregnancy.
chorionic villus (kor-ē-on-ik; pl., chorionic villi): in mammalian embryos, a fingerlike projection of the chorion that penetrates the uterine lining and forms the embryonic portion of the placenta.
chorionic villus sampling (CVS): a procedure for sampling cells from the chorionic villi produced by a fetus: A tube is inserted into the uterus of a pregnant woman, and a small sample of villi are suctioned off for genetic and biochemical analyses.
choroid (kor´-oid): a darkly pigmented layer of tissue, behind the retina, that contains blood vessels and pigment that absorbs stray light.
chromatid (krō´-ma-tid): one of the two identical strands of DNA and protein that forms a replicated chromosome. The two sister chromatids are joined at the centromere.
chromatin (krō´-ma-tin): the complex of DNA and proteins that makes up eukaryotic chromosomes.
chromosome (krō´-mō-sōm): a single DNA double helix together with proteins that help to organize the DNA.
chronic bronchitis: a persistent lung infection characterized by coughing, swelling of the lining of the respiratory tract, an increase in mucus production, and a decrease in the number and activity of cilia.
chyme (kīm): an acidic, souplike mixture of partially digested food, water, and digestive secretions that is released from the stomach into the small intestine.
ciliate (sil´-ē-et): a protozoan characterized by cilia and by a complex unicellular structure, including harpoonlike organelles called *trichocysts.* Members of the genus *Paramecium* are well-known ciliates.
cilium (sil´-ē-um; pl., cilia): a short, hairlike projection from the surface of certain eukaryotic cells that contains microtubules in a 9 + 2 arrangement. The movement of cilia may propel cells through a fluid medium or move fluids over a stationary surface layer of cells.
circadian rhythm (sir-kā´-dē-un): an event that recurs with a period of about 24 hours, even in the absence of environmental cues.
citric acid cycle: see *Krebs cycle.*
class: the taxonomic category composed of related genera. Closely related classes form a division or phylum.
cleavage: the early cell divisions of embryos, in which little or no growth occurs between divisions; reduces the cell size and distributes gene-regulating substances to the newly formed cell.
climate: patterns of weather that prevail from year to year and even from century to century in a given region.
climax community: a diverse and relatively stable community that forms the endpoint of succession.
clitoris: an external structure of the female reproductive system; composed of erectile tissue; a sensitive point of stimulation during sexual response.
clonal selection: the mechanism by which the immune response gains specificity; an invading antigen elicits a response from only a few lymphocytes, which proliferate to form a clone of cells that attack only the specific antigen that stimulated their production.
clone: offspring that are produced by mitosis and are therefore genetically identical to each other.
cloning: the process of producing many identical copies of a gene; also the production of many genetically identical copies of an organism.
closed circulatory system: the type of circulatory system, found in certain worms and vertebrates, in which the blood is always confined within the heart and vessels.
club fungus: a fungus of the division Basidiomycota, whose members (which include mushrooms, puffballs, and shelf fungi) reproduce by means of basidiospores.
clumped distribution: the distribution characteristic of populations in which individuals are clustered into groups; may be social or based on the need for a localized resource.
cnidocyte (nīd´-ō-sīt): in members of the phylum Cnidaria, a specialized cell that houses a stinging apparatus.
cochlea (kahk´-lē-uh): a coiled, bony, fluid-filled tube found in the mammalian inner ear; contains receptors (hair cells) that respond to the vibration of sound.
codominance: the relation between two alleles of a gene, such that both alleles are phenotypically expressed in heterozygous individuals.
codon: a sequence of three bases of messenger RNA that specifies a particular amino acid to be incorporated into a protein; certain codons also signal the beginning or end of protein synthesis.
coelom (sē´-lōm): a space or cavity that separates the body wall from the inner organs.
coenzyme: an organic molecule that is bound to certain enzymes and is required for the enzymes' proper functioning; typically, a nucleotide bound to a water-soluble vitamin.
coevolution: the evolution of adaptations in two species due to their extensive interactions with one another, such that each species acts as a major force of natural selection on the other.
cohesion: the tendency of the molecules of a substance to stick together.
cohesion–tension theory: a model for the transport of water in xylem, by which water is pulled up the xylem tubes, powered by the force of evaporation of water from the leaves (producing tension) and held together by hydrogen bonds between nearby water molecules (cohesion).

coleoptile (kō-lē-op´-tīl): a protective sheath surrounding the shoot in monocot seeds, allowing the shoot to push aside soil particles as it grows.
collagen (kol´-uh-jen): a fibrous protein in connective tissue such as bone and cartilage.
collar cell: a specialized cell lining the inside channels of sponges. Flagella extend from a sievelike collar, creating a water current that draws microscopic organisms through the collar and into the body, where they become trapped.
collecting duct: a conducting tube, within the kidney, that collects urine from many nephrons and conducts it through the renal medulla into the renal pelvis. Urine may become concentrated in the collecting ducts if ADH is present.
collenchyma (kōl-en´-ki-muh): an elongated, polygonal plant cell type with irregularly thickened primary cell walls that is alive at maturity and that supports the plant body.
colon: the longest part of the large intestine, exclusive of the rectum.
colostrum (kō-los´-trum): a yellowish fluid, high in protein and containing antibodies, that is produced by the mammary glands before milk secretion begins.
commensalism (kum-en´-sal-iz-um): a symbiotic relationship in which one species benefits while another species is neither harmed nor benefited.
communication: the act of producing a signal that causes another animal, normally of the same species, to change its behavior in a way that is beneficial to one or both participants.
community: all the interacting populations within an ecosystem.
compact bone: the hard and strong outer bone; composed of osteons.
companion cell: a cell adjacent to a sieve-tube element in phloem, involved in the control and nutrition of the sieve-tube element.
competition: interaction among individuals who attempt to utilize a resource (for example, food or space) that is limited relative to the demand for it.
competitive exclusion principle: the concept that no two species can simultaneously and continuously occupy the same ecological niche.
competitive inhibition: the process by which two or more molecules that are somewhat similar in structure compete for the active site of an enzyme.
complement: a group of blood-borne proteins that participate in the destruction of foreign cells to which antibodies have bound.
complementary base pair: in nucleic acids, bases that pair by hydrogen bonding. In DNA, adenine is complementary to thymine and guanine is complementary to cytosine; in RNA, adenine is complementary to uracil, and guanine to cytosine.
complement reaction: an interaction among foreign cells, antibodies, and complement proteins that results in the destruction of the foreign cells.
complement system: a series of reactions in which complement proteins bind to antibody stems, attracting to the site phagocytic white blood cells that destroy the invading cell that triggers the reactions.
complete flower: a flower that has all four floral parts (sepals, petals, stamens, and carpels).
compound: a substance whose molecules are formed by different types of atoms; can be broken into its constituent elements by chemical means.
compound eye: a type of eye, found in arthropods, that is composed of numerous independent subunits called *ommatidia.* Each ommatidium apparently contributes a piece of a mosaiclike image perceived by the animal.
concentration: the number of particles of a dissolved substance in a given unit of volume.
concentration gradient: the difference in concentration of a substance between two parts of a fluid or across a barrier such as a membrane.
conclusion: the final operation in the scientific method; a decision made about the validity of a hypothesis on the basis of experimental evidence.
condensation: compaction of eukaryotic chromosomes into discrete units in preparation for mitosis or meiosis.
condom: a contraceptive sheath worn over the penis during intercourse to prevent sperm from being deposited in the vagina.
conducting portion: the portion of the respiratory system in lung-breathing vertebrates that carries air to the lungs.
cone: a cone-shaped photoreceptor cell in the vertebrate retina; not as sensitive to light as are the rods. The three types of cones are most sensitive to different colors of light and provide color vision; see also *rod.*
conifer (kon´-eh-fer): a member of a class of tracheophytes (Coniferophyta) that reproduces by means of seeds formed inside cones and that retains its leaves throughout the year.
connective tissue: a tissue type consisting of diverse tissues, including bone, fat, and blood, that generally contain large amounts of extracellular material.
constant region: the part of an antibody molecule that is similar in all antibodies.
consumer: an organism that eats other organisms; a heterotroph.
contest competition: a mechanism for resolving intraspecific competition by using social or chemical interactions.
contraception: the prevention of pregnancy.
contractile vacuole: a fluid-filled vacuole in certain protists that takes up water from the cytoplasm, contracts, and expels the water outside the cell through a pore in the plasma membrane.
control: that portion of an experiment in which all possible variables are held constant; in contrast to the "experimental" portion, in which a particular variable is altered.
convergence: a condition in which a large number of nerve cells provide input to a smaller number of cells.
convergent evolution: the independent evolution of similar structures among unrelated organisms as a result of similar environmental pressures; see *analogous structures*.
convolution: a folding of the cerebral cortex of the vertebrate brain.
copulation: reproductive behavior in which the penis of the male is inserted into the body of the female, where it releases sperm.
coral reef: a biome created by animals (reef-building corals) and plants in warm tropical waters.
cork cambium: a lateral meristem in woody roots and stems that gives rise to cork cells.
cork cell: a protective cell of the bark of woody stems and roots; at maturity, cork cells are dead, with thick, waterproofed cell walls.
cornea (kor´-nē-uh): the clear outer covering of the eye, in front of the pupil and iris.
corona radiata (kuh-rō´-nuh rā-dē-a´-tuh): the layer of cells surrounding an egg after ovulation.
corpus callosum (kor´pus kal-ō´-sum): the band of axons that connect the two cerebral hemispheres of vertebrates.
corpus luteum (kor´-pus loo´-tē-um): in the mammalian ovary, a structure that is derived from the follicle after ovulation and that secretes the hormones estrogen and progesterone.
cortex: the part of a primary root or stem located between the epidermis and the vascular cylinder.
cotyledon (kot-ul-ē´don): a leaflike structure within a seed that absorbs food molecules from the endosperm and transfers them to the growing embryo; also called *seed leaf.*
coupled reaction: a pair of reactions, one exergonic and one endergonic, that are linked together such that the energy produced by the exergonic reaction provides the energy needed to drive the endergonic reaction.
covalent bond (kō-vā´-lent): a chemical bond between atoms in which electrons are shared.
crab lice: an arthropod parasite that can infest humans; can be transmitted by sexual contact.
creationism: the hypothesis that all species on Earth were created in essentially their present form by a supernatural being and that significant modification of those species—specifically, their transformation into new species—cannot occur by natural processes.
crista (kris´-tuh; pl., **cristae):** a fold in the inner membrane of a mitochondrion.
crop: an organ, found in both earthworms and birds, in which ingested food is temporarily stored before being passed to the gizzard, where it is pulverized.
cross-bridge: in muscles, an extension of myosin that binds to and pulls on actin to produce muscle contraction.
cross-fertilization: the union of sperm and egg from two individuals of the same species.
crossing over: the exchange of corresponding segments of the chromatids of two homologous chromosomes during meiosis.
cultural evolution: changes in the behavior of a population of animals, especially humans, by learning behaviors acquired by members of previous generations.
cuticle (kū´-ti-kul): a waxy or fatty coating on the exposed surfaces of epidermal cells of

many land plants, which aids in the retention of water.

cyanobacterium: a photosynthetic prokaryotic cell that utilizes chlorophyll and releases oxygen as a photosynthetic by-product; sometimes called *blue-green algae*.

cyclic AMP: a cyclic nucleotide, formed within many target cells as a result of the reception of amino acid derivatives or peptide hormones, that causes metabolic changes in the cell; often called a second messenger.

cyclic nucleotide (sik´-lik noo´-klē-ō-tīd): a nucleotide in which the phosphate group is bonded to the sugar at two points, forming a ring; serves as an intracellular messenger.

cyst (sist): an encapsulated resting stage in the life cycle of certain invertebrates, such as parasitic flatworms and roundworms.

cystic fibrosis: an inherited disorder characterized by the buildup of salt in the lungs and the production of thick, sticky mucus that clogs the airways, restricts air exchange, and promotes infection.

cytokinesis (sī-tō-ki-nē´-sis): the division of the cytoplasm and organelles into two daughter cells during cell division; normally occurs during telophase of mitosis.

cytokinin (sī-tō-kī´-nin): a plant hormone that promotes cell division, fruit growth, and the sprouting of lateral buds and prevents the aging of plant parts, especially leaves.

cytoplasm (sī´-tō-plaz-um): the material contained within the plasma membrane of a cell, exclusive of the nucleus.

cytosine: a nitrogenous base found in both DNA and RNA; abbreviated as C.

cytoskeleton: a network of protein fibers in the cytoplasm that gives shape to a cell, holds and moves organelles, and is typically involved in cell movement.

cytotoxic T cell: a type of T cell that, upon contacting foreign cells, directly destroys them.

day-neutral plant: a plant in which flowering occurs as soon as the plant has grown and developed, regardless of daylength.

decomposer: an organism, normally a fungus or bacterium, that digests organic material by secreting digestive enzymes into the environment, in the process liberating nutrients into the environment.

deductive reasoning: the process of generating hypotheses about how a specific experiment or observation will turn out.

deforestation: the excessive cutting of forests, primarily rain forests in the Tropics, to clear space for agriculture.

dehydration synthesis: a chemical reaction in which two molecules are joined by a covalent bond with the simultaneous removal of a hydrogen from one molecule and a hydroxyl group from the other, forming water; the reverse of hydrolysis.

deletion mutation: a mutation in which one or more pairs of nucleotides are removed from a gene.

denature: to disrupt the secondary and/or tertiary structure of a protein while leaving its amino acid sequence intact. Denatured proteins can no longer perform their biological functions.

dendrite (den´-drīt): a branched tendril that extends outward from the cell body of a neuron; specialized to respond to signals from the external environment or from other neurons.

denitrifying bacterium (dē-nī´-treh-fī-ing): a bacterium that breaks down nitrates, releasing nitrogen gas to the atmosphere.

density-dependent: referring to any factor, such as predation, that limits population size more effectively as the population density increases.

density-independent: referring to any factor that limits a population's size and growth regardless of its density.

deoxyribonucleic acid (dē-ox-ē-rī-bō-noo-klā´-ik; DNA): a molecule composed of deoxyribose nucleotides; contains the genetic information of all living cells.

dermal tissue system: a plant tissue system that makes up the outer covering of the plant body.

dermis (dur´-mis): the layer of skin beneath the epidermis; composed of connective tissue and containing blood vessels, muscles, nerve endings, and glands.

desert: a biome in which less than 25 to 50 centimeters (10 to 20 inches) of rain falls each year.

desertification: the spread of deserts by human activities.

desmosome (dez´-mō-sōm): a strong cell-to-cell junction that attaches adjacent cells to one another.

detritus feeder (de-trī´-tus): one of a diverse group of organisms, ranging from worms to vultures, that live off the wastes and dead remains of other organisms.

deuterostome (doo´-ter-ō-stōm): an animal with a mode of embryonic development in which the coelom is derived from outpocketings of the gut; characteristic of echinoderms and chordates.

development: the process by which an organism proceeds from fertilized egg through adulthood to eventual death.

diabetes mellitus (dī-uh-bē´-tēs mel-ī´-tus): a disease characterized by defects in the production, release, or reception of insulin; characterized by high blood glucose levels that fluctuate with sugar intake.

dialysis (dī-al´-i-sis): the passive diffusion of substances across an artificial semipermeable membrane.

diaphragm (dī´-uh-fram): in the respiratory system, a dome-shaped muscle forming the floor of the chest cavity that, when it contracts, pulls itself downward, enlarging the chest cavity and causing air to be drawn into the lungs; in a reproductive sense, a contraceptive rubber cap that fits snugly over the cervix, preventing the sperm from entering the uterus and thereby preventing pregnancy.

diatom (dī´-uh-tom): a protist that includes photosynthetic forms with two-part glassy outer coverings; important photosynthetic organisms in fresh water and salt water.

dicot (dī´-kaht): short for dicotyledon; a type of flowering plant characterized by embryos with two cotyledons, or seed leaves, modified for food storage.

differentially permeable: referring to the ability of some substances to pass through a membrane more readily than can other substances.

differential reproduction: differences in reproductive output among individuals of a population, normally as a result of genetic differences.

differentiated cell: a mature cell specialized for a specific function; in plants, differentiated cells normally do not divide.

differentiation: the process whereby relatively unspecialized cells, especially of embryos, become specialized into particular tissue types.

diffusion: the net movement of particles from a region of high concentration of that particle to a region of low concentration, driven by the concentration gradient; may occur entirely within a fluid or across a barrier such as a membrane.

digestion: the process by which food is physically and chemically broken down into molecules that can be absorbed by cells.

digestive system: a group of organs responsible for ingesting and then digesting food substances into simple molecules that can be absorbed and then expelling undigested wastes from the body.

dinoflagellate (dī-nō-fla´-jel-et): a protist that includes photosynthetic forms in which two flagella project through armorlike plates; abundant in oceans; can reproduce rapidly, causing "red tides."

dioecious (dī-ē´-shus): pertaining to organisms in which male and female gametes are produced by separate individuals rather than in the same individual.

diploid (dip´-loid): referring to a cell with pairs of homologous chromosomes.

direct development: a developmental pathway in which the offspring is born as a miniature version of the adult and does not radically change in body form as it grows and matures.

directional selection: a type of natural selection in which one extreme phenotype is favored over all others.

disaccharide (dī-sak´-uh-rīd): a carbohydrate formed by the covalent bonding of two monosaccharides.

disruptive selection: a type of natural selection in which both extreme phenotypes are favored over the average phenotype.

distal tubule: in the nephrons of the mammalian kidney, the last segment of the renal tubule through which the filtrate passes just before it empties into the collecting duct; a site of selective secretion and reabsorption as water and ions pass between the blood and the filtrate across the tubule membrane.

disulfide bridge: the covalent bond formed between the sulfur atoms of two cysteines in a protein; typically causes the protein to fold by bringing otherwise distant parts of the protein close together.

divergence: a condition in which a small number of nerve cells provide input to a larger number of cells.

divergent evolution: evolutionary change in which the differences between two lineages become more pronounced with the passage of time.

division: the taxonomic category contained within a kingdom and consisting of related classes of plants, fungi, bacteria, or plantlike protists.

DNA–DNA hybridization: a technique by which DNA from two species is separated into single strands and then allowed to re-form; hybrid double-stranded DNA from the two species can occur where the sequence of nucleotides is complementary. The greater the degree of hybridization, the closer the evolutionary relatedness of the two species.

DNA fingerprinting: the use of restriction enzymes to cut DNA segments into a unique set of restriction fragments from one individual that can be distinguished from the restriction fragments of other individuals by gel electrophoresis.

DNA helicase: an enzyme that helps unwind the DNA double helix during DNA replication.

DNA library: a readily accessible, easily duplicable complete set of all the DNA of a particular organism, normally cloned into bacterial plasmids.

DNA ligase: an enzyme that joins the sugars and phosphates in a DNA strand to create a continuous sugar-phosphate backbone.

DNA polymerase: an enzyme that bonds DNA nucleotides together into a continuous strand, using a preexisting DNA strand as a template.

DNA probe: a sequence of nucleotides that is complementary to the nucleotide sequence in a gene under study; used to locate a given gene within a DNA library.

DNA replication: the copying of the double-stranded DNA molecule, producing two identical DNA double helices.

DNA sequencing: the process of determining the chemical composition of a DNA molecule (in particular, the order in which the molecule's constituent nucleic acids are arranged).

domain: the broadest category for classifying organisms; organisms are classified into three domains: Bacteria, Archaea, and Eukarya.

dominance hierarchy: a social arrangement in which a group of animals, usually through aggressive interactions, establishes a rank for some or all of the group members that determines access to resources.

dominant: an allele that can determine the phenotype of heterozygotes completely, such that they are indistinguishable from individuals homozygous for the allele; in the heterozygotes, the expression of the other (recessive) allele is completely masked.

dopamine (dōp´-uh-mēn): a transmitter in the brain whose actions are largely inhibitory. The loss of dopamine-containing neurons causes Parkinson's disease.

dormancy: a state in which an organism does not grow or develop; usually marked by lowered metabolic activity and resistance to adverse environmental conditions.

dorsal (dor´-sul): the top, back, or uppermost surface of an animal oriented with its head forward.

dorsal root ganglion: a ganglion, located on the dorsal (sensory) branch of each spinal nerve, that contains the cell bodies of sensory neurons.

double covalent bond: a covalent bond in which two atoms share two pairs of electrons.

double fertilization: in flowering plants, the fusion of two sperm nuclei with the nuclei of two cells of the female gametophyte. One sperm nucleus fuses with the egg to form the zygote; the second sperm nucleus fuses with the two haploid nuclei of the primary endosperm cell, forming a triploid endosperm cell.

double helix (hē´-liks): the shape of the two-stranded DNA molecule; like a ladder twisted lengthwise into a corkscrew shape.

doubling time: the time it would take a population to double in size at its current growth rate.

douching: washing the vagina; after intercourse, an attempt to wash sperm out of the vagina before they enter the uterus; an ineffective contraceptive method.

Down syndrome: a genetic disorder caused by the presence of three copies of chromosome 21; common characteristics include mental retardation, distinctively shaped eyelids, a small mouth with protruding tongue, heart defects, and low resistance to infectious diseases; also called *trisomy 21*.

duct: a tube or opening through which exocrine secretions are released.

duplicated chromosome: a eukaryotic chromosome following DNA replication; consists of two sister chromatids joined at the centromeres.

echolocation: the use of ultrasonic sounds, which bounce back from nearby objects, to produce an auditory "image" of nearby surroundings; used by bats and porpoises.

ecological isolation: the lack of mating between organisms belonging to different populations that occupy distinct habitats within the same general area.

ecological niche (nitch): the role of a particular species within an ecosystem, including all aspects of its interaction with the living and nonliving environments.

ecology (ē-kol´-uh-jē): the study of the interrelationships of organisms with each other and with their nonliving environment.

ecosystem (ē´kō-sis-tem): all the organisms and their nonliving environment within a defined area.

ectoderm (ek´-tō-derm): the outermost embryonic tissue layer, which gives rise to structures such as hair, the epidermis of the skin, and the nervous system.

effector (ē-fek´-tor): a part of the body (normally a muscle or gland) that carries out responses as directed by the nervous system.

egg: the haploid female gamete, normally large and nonmotile, containing food reserves for the developing embryo.

electrolocation: the production of high-frequency electrical signals from an electric organ in front of the tail of weak electrical fish; used to detect and locate nearly objects.

electron: a subatomic particle, found in an electron shell outside the nucleus of an atom, that bears a unit of negative charge and very little mass.

electron carrier: a molecule that can reversibly gain or lose electrons. Electron carriers generally accept high-energy electrons produced during an exergonic reaction and donate the electrons to acceptor molecules that use the energy to drive endergonic reactions.

electron shell: a region within which electrons orbit that corresponds to a fixed energy level at a given distance from the atomic nucleus of an atom.

electron transport system: a series of electron carrier molecules, found in the thylakoid membranes of chloroplasts and the inner membrane of mitochondria, that extract energy from electrons and generate ATP or other energetic molecules.

element: a substance that cannot be broken down, or converted, to a simpler substance by ordinary chemical means.

embryo: in animals, the stages of development that begin with the fertilization of the egg cell and end with hatching or birth; in mammals in particular, the early stages in which the developing animal does not yet resemble the adult of the species.

embryonic disc: in human embryonic development, the flat, two-layered group of cells that separates the amniotic cavity from the yolk sac.

embryonic stem cell: a cell derived from an early embryo that is capable of differentiating into any of the adult cell types.

embryo sac: the haploid female gametophyte of flowering plants.

emergent property: an intangible attribute that arises as the result of complex ordered interactions among individual parts.

emigration (em-uh-grā´shun): migration of individuals out of an area.

emphysema (em-fuh-sē´-muh): a condition in which the alveoli of the lungs become brittle and rupture, causing decreased area for gas exchange.

endergonic (en-der-gon´-ik): pertaining to a chemical reaction that requires an input of energy to proceed; an "uphill" reaction.

endocrine disruptors: environmental pollutants that interfere with endocrine function, often by disrupting the action of sex hormones.

endocrine gland: a ductless, hormone-producing gland consisting of cells that release their secretions into the extracellular fluid from which the secretions diffuse into nearby capillaries.

endocrine system: an animal's organ system for cell-to-cell communication, composed of hormones and the cells that secrete them and receive them.

endocytosis (en-dō-sī-tō´-sis): the process in which the plasma membrane engulfs extracellular material, forming membrane-bound sacs that enter the cytoplasm and thereby move material into the cell.

endoderm (en´-dō-derm): the innermost embryonic tissue layer, which gives rise to structures such as the lining of the digestive and respiratory tracts.

endodermis (en-dō-der´-mis): the innermost layer of small, close-fitting cells of the cortex of a root that form a ring around the vascular cylinder.
endogenous pyrogen: a chemical, produced by the body, that stimulates the production of a fever.
endometrium (en-dō-mē´-trē-um): the nutritive inner lining of the uterus.
endoplasmic reticulum (ER) (en-dō-plaz´-mik re-tik´-ū-lum): a system of membranous tubes and channels within eukaryotic cells; the site of most protein and lipid syntheses.
endorphin (en-dor´-fin): one of a group of peptide neuromodulators in the vertebrate brain that, by reducing the sensation of pain, mimics some of the actions of opiates.
endoskeleton (en´-dō-skel´-uh-tun): a rigid internal skeleton with flexible joints to allow for movement.
endosperm: a triploid food storage tissue in the seeds of flowering plants that nourishes the developing plant embryo.
endospore: a protective resting structure of some rod-shaped bacteria that withstands unfavorable environmental conditions.
endosymbiont hypothesis: the hypothesis that certain organelles, especially chloroplasts and mitochondria, arose as mutually beneficial associations between the ancestors of eukaryotic cells and captured bacteria that lived within the cytoplasm of the pre-eukaryotic cell.
energy: the capacity to do work.
energy-carrier molecule: a molecule that stores energy in "high-energy" chemical bonds and releases the energy to drive coupled endothermic reactions. In cells, ATP is the most common energy-carrier molecule.
energy level: the specific amount of energy characteristic of a given electron shell in an atom.
energy pyramid: a graphical representation of the energy contained in succeeding trophic levels, with maximum energy at the base (primary producers) and steadily diminishing amounts at higher levels.
entropy (en´-trō-pē): a measure of the amount of randomness and disorder in a system.
environmental estrogens: chemicals in the environment that mimic some of the effects of estrogen in animals.
environmental resistance: any factor that tends to counteract biotic potential, limiting population size.
enzyme (en´zīm): a protein catalyst that speeds up the rate of specific biological reactions.
eosinophil (ē-ō-sin´-ō-fil): a type of white blood cell that converges on parasitic invaders and releases substances to kill them.
epicotyl (ep´-ē-kot-ul): the part of the embryonic shoot located above the cotyledons but below the tip of the shoot.
epidermal tissue: dermal tissue in plants that forms the epidermis, the outermost cell layer that covers young plants.
epidermis (ep-uh-der´-mis): in animals, specialized epithelial tissue that forms the outer layer of skin; in plants, the outermost layer of cells of a leaf, young root, or young stem.
epididymis (e-pi-di´-di-mus): a series of tubes that connect with and receive sperm from the seminiferous tubules of the testis.
epiglottis (ep-eh-glah´-tis): a flap of cartilage in the lower pharynx that covers the opening to the larynx during swallowing; directs food down the esophagus.
epinephrine (ep-i-nef´-rin): a hormone, secreted by the adrenal medulla, that is released in response to stress and that stimulates a variety of responses, including the release of glucose from skeletal muscle and an increase in heart rate.
epithelial cell (eh-puh-thē´-lē-ul): a flattened cell that covers the outer body surfaces of a sponge.
epithelial tissue (eh-puh-thē´-lē-ul): a tissue type that forms membranes that cover the body surface and line body cavities, and that also gives rise to glands.
equilibrium population: a population in which allele frequencies and the distribution of genotypes do not change from generation to generation.
erythroblastosis fetalis (eh-rith´-rō-blas-tō´-sis fē-tal´-is): a condition in which the red blood cells of a newborn Rh-positive baby are attacked by antibodies produced by its Rh-negative mother, causing jaundice and anemia. Retardation and death are possible consequences if treatment is inadequate.
erythrocyte (eh-rith´-rō-sīt): a red blood cell, active in oxygen transport, that contains the red pigment hemoglobin.
erythropoietin (eh-rith´-rō-pō-ē´-tin): a hormone produced by the kidneys in response to oxygen deficiency that stimulates the production of red blood cells by the bone marrow.
esophagus (eh-sof´-eh-gus): a muscular passageway that conducts food from the pharynx to the stomach in humans and other mammals.
essential amino acid: an amino acid that is a required nutrient; the body is unable to manufacture essential amino acids, so they must be supplied in the diet.
essential fatty acid: a fatty acid that is a required nutrient; the body is unable to manufacture essential fatty acids, so they must be supplied in the diet.
estrogen: in vertebrates, a female sex hormone, produced by follicle cells of the ovary, that stimulates follicle development, oogenesis, the development of secondary sex characteristics, and growth of the uterine lining.
estuary: a wetland formed where a river meets the ocean; the salinity there is quite variable but lower than in sea water and higher than in fresh water.
ethology (ē-thol´-ō-jē): the study of animal behavior in natural or near-natural conditions.
ethylene: a plant hormone that promotes the ripening of fruits and the dropping of leaves and fruit.
euglenoid (ū´-gle-noid): a protist characterized by one or more whiplike flagella that are used for locomotion and by a photoreceptor that detects light. Euglenoids are photosynthetic, but if deprived of chlorophyll, some are capable of heterotrophic nutrition.
eukaryotic (ū-kar-ē-ot´-ik): referring to cells of organisms of the domain Eukarya (kingdoms Protista, Fungi, Plantae, and Animalia). Eukaryotic cells have genetic material enclosed within a membrane-bound nucleus and contain other membrane-bound organelles.
Eustachian tube (ū-stā´-shin): a tube connecting the middle ear with the pharynx; allows pressure between the middle ear and the atmosphere to equilibrate.
eutrophic lake: a lake that receives sufficiently large inputs of sediments, organic material, and inorganic nutrients from its surroundings to support dense communities; murky with poor light penetration.
evergreen: a plant that retains green leaves throughout the year.
evolution: the descent of modern organisms with modification from preexisting life-forms; strictly speaking, any change in the proportions of different genotypes in a population from one generation to the next.
excretion: the elimination of waste substances from the body; can occur from the digestive system, skin glands, urinary system, or lungs.
excretory pore: an opening in the body wall of certain invertebrates, such as the earthworm, through which urine is excreted.
exergonic (ex-er-gon´-ik): pertaining to a chemical reaction that liberates energy (either as heat or in the form of increased entropy); a "downhill" reaction.
exhalation: the act of releasing air from the lungs, which results from a relaxation of the respiratory muscles.
exocrine gland: a gland that releases its secretions into ducts that lead to the outside of the body or into the digestive tract.
exocytosis (ex-ō-sī-tō´-sis): the process in which intracellular material is enclosed within a membrane-bound sac that moves to the plasma membrane and fuses with it, releasing the material outside the cell.
exon: a segment of DNA in a eukaryotic gene that codes for amino acids in a protein (see also *intron*).
exoskeleton (ex´-ō-skel´-uh-tun): a rigid external skeleton that supports the body, protects the internal organs, and has flexible joints that allow for movement.
exotic/exotic species: a foreign species introduced into an ecosystem where it did not evolve; such species may flourish and outcompete native species.
experiment: the third operation in the scientific method; the testing of a hypothesis by further observations, leading to a conclusion.
exponential growth: a continuously accelerating increase in population size.
extensor: a muscle that straightens a joint.
external ear: the fleshy portion of the ear that extends outside the skull.
external fertilization: the union of sperm and egg outside the body of either parent.

extinction: the death of all members of a species.
extracellular digestion: the physical and chemical breakdown of food that occurs outside a cell, normally in a digestive cavity.
extraembryonic membrane: in the embryonic development of reptiles, birds, and mammals, either the chorion, amnion, allantois, or yolk sac; functions in gas exchange, provision of the watery environment needed for development, waste storage, and storage of the yolk, respectively.
eyespot: a simple, lensless eye found in various invertebrates, including flatworms and jellyfish. Eyespots can distinguish light from dark and sometimes the direction of light, but they cannot form an image.
facilitated diffusion: the diffusion of molecules across a membrane, assisted by protein pores or carriers embedded in the membrane.
fairy ring: a circular pattern of mushrooms formed when reproductive structures erupt from the underground hyphae of a club fungus that has been growing outward in all directions from its original location.
family: the taxonomic category contained within an order and consisting of related genera.
farsighted: the inability to focus on nearby objects, caused by the eyeball being slightly too short.
fat (molecular): a lipid composed of three saturated fatty acids covalently bonded to glycerol; solid at room temperature.
fat (tissue): adipose tissue; connective tissue that stores the lipid fat; composed of cells packed with triglycerides.
fatty acid: an organic molecule composed of a long chain of carbon atoms, with a carboxylic acid (COOH) group at one end; may be saturated (all single bonds between the carbon atoms) or unsaturated (one or more double bonds between the carbon atoms).
feces: semisolid waste material that remains in the intestine after absorption is complete and is voided through the anus. Feces consist of indigestible wastes and the dead bodies of bacteria.
feedback inhibition: in enzyme-mediated chemical reactions, the condition in which the product of a reaction inhibits one or more of the enzymes involved in synthesizing the product.
fermentation: anaerobic reactions that convert the pyruvic acid produced by glycolysis into lactic acid or alcohol and CO_2.
fertilization: the fusion of male and female haploid gametes, forming a zygote.
fetal alcohol syndrome (FAS): a cluster of symptoms, including retardation and physical abnormalities, that occur in infants born to a mothers who consumed large amounts of alcoholic beverages during pregnancy.
fetus: the later stages of mammalian embryonic development (after the second month for humans), when the developing animal has come to resemble the adult of the species.
fever: an elevation in body temperature caused by chemicals (pyrogens) that are released by white blood cells in response to infection.
fibrillation: rapid, uncoordinated, and ineffective contractions of heart muscle cells.
fibrin (fī´-brin): a clotting protein formed in the blood in response to a wound; binds with other fibrin molecules and provides a matrix around which a blood clot forms.
fibrous root system: a root system, commonly found in monocots, characterized by many roots of approximately the same size arising from the base of the stem.
filament: in flowers, the stalk of a stamen, which bears an anther at its tip.
filtrate: the fluid produced by filtration; in the kidneys, the fluid produced by the filtration of blood through the glomerular capillaries.
filtration: within Bowman's capsule in each nephron of a kidney, the process by which blood is pumped under pressure through permeable capillaries of the glomerulus, forcing out water, dissolved wastes, and nutrients.
fimbria (fim´-brē-uh; pl., **fimbriae):** in female mammals, the ciliated, fingerlike projections of the oviduct that sweep the ovulated egg from the ovary into the oviduct.
first law of thermodynamics: the principle of physics that states that within any isolated system, energy can be neither created nor destroyed but can be converted from one form to another.
fission: asexual reproduction by dividing the body into two smaller, complete organisms.
fitness: the reproductive success of an organism, usually expressed in relation to the average reproductive success of all individuals in the same population.
flagellum (fla-jel´-um; pl., **flagella):** a long, hairlike extension of the plasma membrane; in eukaryotic cells, it contains microtubules arranged in a 9 + 2 pattern. The movement of flagella propel some cells through fluids.
flame cell: in flatworms, a specialized cell, containing beating cilia, that conducts water and wastes through the branching tubes that serve as an excretory system.
flexor: a muscle that flexes (decreases the angle of) a joint.
florigen: one of a group of plant hormones that can both trigger and inhibit flowering; daylength is a stimulus.
flower: the reproductive structure of an angiosperm plant.
fluid: a liquid or gas.
fluid mosaic model: a model of membrane structure; according to this model, membranes are composed of a double layer of phospholipids in which various proteins are embedded. The phospholipid bilayer is a somewhat fluid matrix that allows the movement of proteins within it.
follicle: in the ovary of female mammals, the oocyte and its surrounding accessory cells.
follicle-stimulating hormone (FSH): a hormone, produced by the anterior pituitary, that stimulates spermatogenesis in males and the development of the follicle in females.
food chain: a linear feeding relationship in a community, using a single representative from each of the trophic levels.
food vacuole: a membranous sac, within a single cell, in which food is enclosed. Digestive enzymes are released into the vacuole, where intracellular digestion occurs.
food web: a representation of the complex feeding relationships (in terms of interacting food chains) within a community, including many organisms at various trophic levels, with many of the consumers occupying more than one level simultaneously.
foraminiferan (for-am-i-nif´-er-un): an aquatic (largely marine) protist characterized by a typically elaborate calcium carbonate shell.
forebrain: during development, the anterior portion of the brain. In mammals, the forebrain differentiates into the thalamus, the limbic system, and the cerebrum. In humans, the cerebrum contains about half of all the neurons in the brain.
fossil: the remains of a dead organism, normally preserved in rock; may be petrified bones or wood; shells; impressions of body forms, such as feathers, skin, or leaves; or markings made by organisms, such as footprints.
fossil fuel: a fuel such as coal, oil, and natural gas, derived from the remains of ancient organisms.
founder effect: a type of genetic drift in which an isolated population founded by a small number of individuals may develop allele frequencies that are very different from those of the parent population as a result of chance inclusion of disproportionate numbers of certain alleles in the founders.
fovea (fō´-vē-uh): in the vertebrate retina, the central region on which images are focused; contains closely packed cones.
free-living: not parasitic.
free nerve ending: on some receptor neurons, a finely branched ending that responds to touch and pressure, to heat and cold, or to pain; produces the sensations of itching and tickling.
free nucleotides: nucleotides that have not been joined together to form a DNA or RNA strand.
fruit: in flowering plants, the ripened ovary (plus, in some cases, other parts of the flower), which contains the seeds.
fruiting body: a spore-forming reproductive structure of certain protists, bacteria, and fungi.
functional group: one of several groups of atoms commonly found in an organic molecule, including hydrogen, hydroxyl, amino, carboxyl, and phosphate groups, that determine the characteristics and chemical reactivity of the molecule.
gallbladder: a small sac, next to the liver, in which the bile secreted by the liver is stored and concentrated. Bile is released from the gallbladder to the small intestine through the bile duct.
gamete (gam´-ēt): a haploid sex cell formed in sexually reproducing organisms.
gametic incompatibility: the inability of sperm from one species to fertilize eggs of another species.

gametophyte (ga-mēt´-ō-fīt): the multicellular haploid stage in the life cycle of plants.
ganglion (gang´-lē-un): a cluster of neurons.
ganglion cell: a type of cell, of which the innermost layer of the vertebrate retina is composed, whose axons form the optic nerve.
gap junction: a type of cell-to-cell junction in animals in which channels connect the cytoplasm of adjacent cells.
gas-exchange portion: the portion of the respiratory system in lung-breathing vertebrates where gas is exchanged in the alveoli of the lungs.
gastric inhibitory peptide: a hormone, produced by the small intestine, that inhibits the activity of the stomach.
gastrin: a hormone, produced by the stomach, that stimulates acid secretion in response to the presence of food.
gastrovascular cavity: a saclike chamber with digestive functions, found in simple invertebrates; a single opening serves as both mouth and anus, and the chamber provides direct access of nutrients to the cells.
gastrula (gas´-troo-luh): in animal development, a three-layered embryo with ectoderm, mesoderm, and endoderm cell layers. The endoderm layer normally encloses the primitive gut.
gastrulation (gas-troo-la´-shun): the process whereby a blastula develops into a gastrula, including the formation of endoderm, ectoderm, and mesoderm.
gel electrophoresis: a technique in which molecules (such as DNA fragments) are placed on restricted tracks in a thin sheet of gelatinous material and exposed to an electric field; the molecules then migrate at a rate determined by certain characteristics, such as length.
gene: a unit of heredity that encodes the information needed to specify the amino acid sequence of proteins and hence particular traits; a functional segment of DNA located at a particular place on a chromosome.
gene flow: the movement of alleles from one population to another owing to the migration of individual organisms.
gene pool: the total of all alleles of all genes in a population; for a single gene, the total of all the alleles of that gene that occur in a population.
generative cell: in flowering plants, one of the haploid cells of a pollen grain; undergoes mitosis to form two sperm cells.
genetic code: the collection of codons of mRNA, each of which directs the incorporation of a particular amino acid into a protein during protein synthesis.
genetic drift: a change in the allele frequencies of a small population purely by chance.
genetic engineering: the modification of genetic material to achieve specific goals.
genetic equilibrium: a state in which the allele frequencies and the distribution of genotypes of a population do not change from generation to generation.
genetic recombination: the generation of new combinations of alleles on homologous chromosomes due to the exchange of DNA during crossing over.
genital herpes: a sexually transmitted disease, caused by a virus, that can cause painful blisters on the genitals and surrounding skin.
genital warts: a sexually transmitted disease, caused by a virus, that forms growths or bumps on the external genitalia, in or around the vagina or anus, or on the cervix in females or penis, scrotum, groin, or thigh in males.
genome (jē´-nōm): the entire set of genes carried by a member of any given species.
genotype (jēn´-ō-tīp): the genetic composition of an organism; the actual alleles of each gene carried by the organism.
genus (jē-nus): the taxonomic category contained within a family and consisting of very closely related species.
geographical isolation: the separation of two populations by a physical barrier.
germ layer: a tissue layer formed during early embryonic development.
germination: the growth and development of a seed, spore, or pollen grain.
gibberellin (jib-er-el´-in): a plant hormone that stimulates seed germination, fruit development, and cell division and elongation.
gill: in aquatic animals, a branched tissue richly supplied with capillaries around which water is circulated for gas exchange.
gizzard: a muscular organ, found in earthworms and birds, in which food is mechanically broken down prior to chemical digestion.
gland: a cluster of cells that are specialized to secrete substances such as sweat or hormones.
glial cell: a cell of the nervous system that provides support and insulation for neurons.
global warming: a gradual rise in global atmospheric temperature as a result of an amplification of the natural greenhouse effect due to human activities.
glomerulus (glō-mer´-ū-lus): a dense network of thin-walled capillaries, located within the Bowman's capsule of each nephron of the kidney, where blood pressure forces water and dissolved nutrients through capillary walls for filtration by the nephron.
glucagon (gloo´-ka-gon): a hormone, secreted by the pancreas, that increases blood sugar by stimulating the breakdown of glycogen (to glucose) in the liver.
glucocorticoid (gloo-kō-kor´-tik-oid): a class of hormones, released by the adrenal cortex in response to the presence of ACTH, that make additional energy available to the body by stimulating the synthesis of glucose.
glucose: the most common monosaccharide, with the molecular formula $C_6H_{12}O_6$; most polysaccharides, including cellulose, starch, and glycogen, are made of glucose subunits covalently bonded together.
glycerol (glis´-er-ol): a three-carbon alcohol to which fatty acids are covalently bonded to make fats and oils.
glycogen (glī´-kō-jen): a long, branched polymer of glucose that is stored by animals in the muscles and liver and metabolized as a source of energy.
glycolysis (glī-kol´-i-sis): reactions, carried out in the cytoplasm, that break down glucose into two molecules of pyruvic acid, producing two ATP molecules; does not require oxygen but can proceed when oxygen is present.
glycoprotein: a protein to which a carbohydrate is attached.
goiter: a swelling of the neck caused by iodine deficiency, which affects the functioning of the thyroid gland and its hormones.
Golgi complex (gōl´-jē): a stack of membranous sacs, found in most eukaryotic cells, that is the site of processing and separation of membrane components and secretory materials.
gonad: an organ where reproductive cells are formed; in males, the testes, and in females, the ovaries.
gonadotropin-releasing hormone (GnRH): a hormone produced by the neurosecretory cells of the hypothalamus, which stimulates cells in the anterior pituitary to release FSH and LH. GnRH is involved in the menstrual cycle and in spermatogenesis.
gonorrhea (gon-uh-rē´-uh): a sexually transmitted bacterial infection of the reproductive organs; if untreated, can result in sterility.
gradient: a difference in concentration, pressure, or electrical charge between two regions.
Gram stain: a stain that is selectively taken up by the cell walls of certain types of bacteria (gram-positive bacteria) and rejected by the cell walls of others (gram-negative bacteria); used to distinguish bacteria on the basis of their cell wall construction.
granum (gra´-num; pl., **grana):** a stack of thylakoids in chloroplasts.
grassland: a biome, located in the centers of continents, that supports grasses; also called *prairie*.
gravitropism: growth with respect to the direction of gravity.
gray crescent: in frog embryonic development, an area of intermediate pigmentation in the fertilized egg; contains gene-regulating substances required for the normal development of the tadpole.
gray matter: the outer portion of the brain and inner region of the spinal cord; composed largely of neuron cell bodies, which give this area a gray color.
greenhouse effect: the process in which certain gases such as carbon dioxide and methane trap sunlight energy in a planet's atmosphere as heat; the glass in a greenhouse does the same. The result, global warming, is being enhanced by the production of these gases by humans.
greenhouse gas: a gas, such as carbon dioxide or methane, that traps sunlight energy in a planet's atmosphere as heat; a gas that participates in the greenhouse effect.
ground tissue system: a plant tissue system consisting of parenchyma, collenchyma, and sclerenchyma cells that makes up the bulk of a leaf or young stem, excluding vascular or dermal tissues. Most ground tissue cells function in photosynthesis, support, or carbohydrate storage.

growth hormone: a hormone, released by the anterior pituitary, that stimulates growth, especially of the skeleton.
growth rate: a measure of the change in population size per individual per unit of time.
guanine: a nitrogenous base found in both DNA and RNA; abbreviated as *G*.
guard cell: one of a pair of specialized epidermal cells surrounding the central opening of a stoma of a leaf, which regulates the size of the opening.
gymnosperm (jim´-nō-sperm): a nonflowering seed plant, such as a conifer, cycad, or gingko.
gyre (jīr): a roughly circular pattern of ocean currents, formed because continents interrupt the currents' flow; rotates clockwise in the Northern Hemisphere and counterclockwise in the Southern Hemisphere.
habituation (heh-bich-oo-ā´-shun): simple learning characterized by a decline in response to a harmless, repeated stimulus.
hair cell: the type of receptor cell in the inner ear; bears hairlike projections, the bending of which causes the receptor potential between two membranes.
hair follicle: a gland in the dermis of mammalian skin, formed from epithelial tissue, that produces a hair.
halophile (hā´-lō-fīl): literally, "salt-loving"; a type of archaen that thrives in concentrated salt solutions.
haploid (hap´-loid): referring to a cell that has only one member of each pair of homologous chromosomes.
Hardy-Weinberg principle: a mathematical model proposing that, under certain conditions, the allele frequencies and genotype frequencies in a sexually reproducing population will remain constant over generations.
Haversian system (ha-ver´-sē-un): see *osteon*.
head: the anteriormost segment of an animal with segmentation.
heart: a muscular organ responsible for pumping blood within the circulatory system throughout the body.
heart attack: a severe reduction or blockage of blood flow through a coronary artery, depriving some of the heart muscle of its blood supply.
heartwood: older xylem that contributes to the strength of a tree trunk.
heat of fusion: the energy that must be removed from a compound to transform it from a liquid into a solid at its freezing temperature.
heat of vaporization: the energy that must be supplied to a compound to transform it from a liquid into a gas at its boiling temperature.
heliozoan (hē-lē-ō-zō´-un): an aquatic (largely freshwater) animal-like protist; some have elaborate silica-based shells.
helix (hē´-liks): a coiled, springlike secondary structure of a protein.
helper T cell: a type of T cell that helps other immune cells recognize and act against antigens.
hemocoel (hē´-mō-sēl): a blood cavity within the bodies of certain invertebrates in which blood bathes tissues directly; part of an open circulatory system.
hemodialysis (hē-mō-dī-al´-luh-sis): a procedure that simulates kidney function in individuals with damaged or ineffective kidneys; blood is diverted from the body, artificially filtered, and returned to the body.
hemoglobin (hē´mō-glō-bin): the iron-containing protein that gives red blood cells their color; binds to oxygen in the lungs and releases it to the tissues.
hemophilia: a recessive, sex-linked disease in which the blood fails to clot normally.
herbivore (erb´-i-vor): literally, "plant-eater"; an organism that feeds directly and exclusively on producers; a primary consumer.
hermaphrodite (her-maf´-ruh-dīt´): an organism that possesses both male and female sexual organs.
hermaphroditic (her-maf´-ruh-dit´-ik): possessing both male and female sexual organs. Some hermaphroditic animals can fertilize themselves; others must exchange sex cells with a mate.
heterotroph (het´-er-ō-trōf´): literally, "other-feeder"; an organism that eats other organisms; a consumer.
heterozygous (het-er-ō-zī´-gus): carrying two different alleles of a given gene; also called *hybrid*.
hindbrain: the posterior portion of the brain, containing the medulla, pons, and cerebellum.
hinge joint: a joint at which one bone is moved by muscle and the other bone remains fixed, such as in the knee, elbow, or fingers; allows movement in only two dimensions.
hippocampus (hip-ō-kam´-pus): the part of the forebrain of vertebrates that is important in emotion and especially learning.
histamine: a substance released by certain cells in response to tissue damage and invasion of the body by foreign substances; promotes the dilation of arterioles and the leakiness of capillaries and triggers some of the events of the inflammatory response.
homeobox (hō´-mē-ō-boks): a sequence of DNA coding for special, 60-amino-acid proteins, which activate or inactivate genes that control development; these sequences specify embryonic cell differentiation.
homeostasis (hōm-ē-ō-stā´sis): the maintenance of a relatively constant environment required for the optimal functioning of cells, maintained by the coordinated activity of numerous regulatory mechanisms, including the respiratory, endocrine, circulatory, and excretory systems.
hominid: a human or a prehistoric relative of humans, beginning with the Australopithecines, whose fossils date back at least 4.4 million years.
homologous structures: structures that may differ in function but that have similar anatomy, presumably because the organisms that possess them have descended from common ancestors.
homologue (hō-´mō-log): a chromosome that is similar in appearance and genetic information to another chromosome with which it pairs during meiosis; also called *homologous chromosome*.
homozygous (hō-mō-zī´-gus): carrying two copies of the same allele of a given gene; also called *true-breeding*.
hormone: a chemical that is synthesized by one group of cells, secreted, and then carried in the bloodstream to other cells, whose activity is influenced by reception of the hormone.
host: the prey organism on or in which a parasite lives; is harmed by the relationship.
human immunodeficiency virus (HIV): a pathogenic retrovirus that causes acquired immune deficiency syndrome (AIDS) by attacking and destroying the immune system's T cells.
humoral immunity: an immune response in which foreign substances are inactivated or destroyed by antibodies that circulate in the blood.
Huntington disease: an incurable genetic disorder, caused by a dominant allele, that produces progressive brain deterioration, resulting in the loss of motor coordination, flailing movements, personality disturbances, and eventual death.
hybrid: an organism that is the offspring of parents differing in at least one genetically determined characteristic; also used to refer to the offspring of parents of different species.
hybrid infertility: reduced fertility (typically, complete sterility) in the hybrid offspring of two species.
hybrid inviability: the failure of a hybrid offspring of two species to survive to maturity.
hybridoma: a cell produced by fusing an antibody-producing cell with a myeloma cell; used to produce monoclonal antibodies.
hydrogen bond: the weak attraction between a hydrogen atom that bears a partial positive charge (due to polar covalent bonding with another atom) and another atom, normally oxygen or nitrogen, that bears a partial negative charge; hydrogen bonds may form between atoms of a single molecule or of different molecules.
hydrologic cycle: the water cycle, driven by solar energy; a nutrient cycle in which the main reservoir of water is the ocean and most of the water remains in the form of water throughout the cycle (rather than being used in the synthesis of new molecules).
hydrolysis (hī-drol´-i-sis): the chemical reaction that breaks a covalent bond by means of the addition of hydrogen to the atom on one side of the original bond and a hydroxyl group to the atom on the other side; the reverse of dehydration synthesis.
hydrophilic (hī-drō-fil´-ik): pertaining to a substance that dissolves readily in water or to parts of a large molecule that form hydrogen bonds with water.
hydrophobic (hī-drō-fō´-bik): pertaining to a substance that does not dissolve in water.
hydrophobic interaction: the tendency for hydrophobic molecules to cluster together when immersed in water.
hydrostatic skeleton (hī-drō-stat´-ik): a body type that uses fluid contained in body compartments to provide support and mass against which muscles can contract.

hydrothermal vent community: a community of unusual organisms, living in the deep ocean near hydrothermal vents, that depends on the chemosynthetic activities of sulfur bacteria.

hypertension: arterial blood pressure that is chronically elevated above the normal level.

hypertonic (hī-per-ton´-ik): referring to a solution that has a higher concentration of dissolved particles (and therefore a lower concentration of free water) than has the cytoplasm of a cell.

hypha (hī´-fuh; pl., **hyphae):** a threadlike structure that consists of elongated cells, typically with many haploid nuclei; many hyphae make up the fungal body.

hypocotyl (hī´-pō-kot-ul): the part of the embryonic shoot located below the cotyledons but above the root.

hypothalamus (hī-pō-thal´-a-mus): a region of the brain that controls the secretory activity of the pituitary gland; synthesizes, stores, and releases certain peptide hormones; directs autonomic nervous system responses.

hypothesis (hī-poth´-eh-sis): the second operation in the scientific method; a supposition based on previous observations that is offered as an explanation for the observed phenomenon and is used as the basis for further observations, or experiments.

hypotonic (hī-pō-ton´-ik): referring to a solution that has a lower concentration of dissolved particles (and therefore a higher concentration of free water) than has the cytoplasm of a cell.

immigration (im-uh-grā´-shun): migration of individuals into an area.

immune response: a specific response by the immune system to the invasion of the body by a particular foreign substance or microorganism, characterized by the recognition of the foreign substance by immune cells and its subsequent destruction by antibodies or by cellular attack.

immune system: cells such as macrophages, B cells, and T cells and molecules such as antibodies that work together to combat microbial invasion of the body.

imperfect fungus: a fungus of the division Deuteromycota; no species in this division has been observed to form sexual reproductive structures.

implantation: the process whereby the early embryo embeds itself within the lining of the uterus.

imprinting: the process by which an animal forms an association with another animal or object in the environment during a sensitive period of development.

inclusive fitness: the reproductive success of all organisms that bear a given allele, normally expressed in relation to the average reproductive success of all individuals in the same population; compare with *fitness*.

incomplete dominance: a pattern of inheritance in which the heterozygous phenotype is intermediate between the two homozygous phenotypes.

incomplete flower: a flower that is missing one of the four floral parts (sepals, petals, stamens, or carpels).

independent assortment: see *law of independent assortment*.

indirect development: a developmental pathway in which an offspring goes through radical changes in body form as it matures.

induction: the process by which a group of cells causes other cells to differentiate into a specific tissue type.

inductive reasoning: the process of creating a generalization based on many specific observations that support the generalization, coupled with an absence of observations that contradict it.

inflammatory response: a nonspecific, local response to injury to the body, characterized by the phagocytosis of foreign substances and tissue debris by white blood cells and by the walling off of the injury site by the clotting of fluids that escape from nearby blood vessels.

inhalation: the act of drawing air into the lungs by enlarging the chest cavity.

inheritance: the genetic transmission of characteristics from parent to offspring.

inheritance of acquired characteristics: the hypothesis that organisms' bodies change during their lifetimes by use and disuse and that these changes are inherited by their offspring.

inhibiting hormone: a hormone, secreted by the neurosecretory cells of the hypothalamus, that inhibits the release of specific hormones from the anterior pituitary.

innate (in-āt´): inborn; instinctive; determined by the genetic makeup of the individual.

inner cell mass: in human embryonic development, the cluster of cells, on one side of the blastocyst, that will develop into the embryo.

inner ear: the innermost part of the mammalian ear; composed of the bony, fluid-filled tubes of the cochlea and the vestibular apparatus.

inorganic: describing any molecule that does not contain both carbon and hydrogen.

insertion: the site of attachment of a muscle to the relatively movable bone on one side of a joint.

insertion mutation: a mutation in which one or more pairs of nucleotides are inserted into a gene.

insight learning: a complex form of learning that requires the manipulation of mental concepts to arrive at adaptive behavior.

instinctive: innate; inborn; determined by the genetic makeup of the individual.

insulin: a hormone, secreted by the pancreas, that lowers blood sugar by stimulating the conversion of glucose to glycogen in the liver.

integration: in nerve cells, the process of adding up electrical signals from sensory inputs or other nerve cells to determine the appropriate outputs.

integument (in-teg´-ū-ment): in plants, the outer layers of cells of the ovule that surrounds the embryo sac; develops into the seed coat.

intensity: the strength of stimulation or response.

interferon: a protein released by certain virus-infected cells that increases the resistance of other, uninfected, cells to viral attack.

intermediate filament: part of the cytoskeleton of eukaryotic cells that probably functions mainly for support and is composed of several types of proteins.

intermembrane compartment: the fluid-filled space between the inner and outer membranes of a mitochondrion.

internal fertilization: the union of sperm and egg inside the body of the female.

internode: the part of a stem between two nodes.

interphase: the stage of the cell cycle between cell divisions; the stage in which chromosomes are replicated and other cell functions occur, such as growth, movement, and acquisition of nutrients.

interspecific competition: competition among individuals of different species.

interstitial cell (in-ter-sti´-shul): in the vertebrate testis, a testosterone-producing cell located between the seminiferous tubules.

interstitial fluid (in-ter-sti´-shul): fluid, similar in composition to plasma (except lacking large proteins), that leaks from capillaries and acts as a medium of exchange between the body cells and the capillaries.

intertidal zone: an area of the ocean shore that is alternately covered and exposed by the tides.

intervertebral disc (in-ter-ver-tē´-brul): a pad of cartilage between two vertebrae that acts as a shock absorber.

intracellular digestion: the chemical breakdown of food within single cells.

intraspecific competition: competition among individuals of the same species.

intrauterine device (IUD): a small copper or plastic loop, squiggle, or shield that is inserted in the uterus; a contraceptive method that works by irritating the uterine lining so that it cannot receive the embryo.

intron: a segment of DNA in a eukaryotic gene that does not code for amino acids in a protein.

invertebrate (in-vert´-uh-bret): an animal that never possesses a vertebral column.

ion (ī´-on): a charged atom or molecule; an atom or molecule that has either an excess of electrons (and hence is negatively charged) or has lost electrons (and is positively charged).

ionic bond: a chemical bond formed by the electrical attraction between positively and negatively charged ions.

iris: the pigmented muscular tissue of the vertebrate eye that surrounds and controls the size of the pupil, through which light enters.

islet cell: a cluster of cells in the endocrine portion of the pancreas that produce insulin and glucagon.

isolating mechanism: a morphological, physiological, behavioral, or ecological difference that prevents members of two species from interbreeding.

isotonic (ī-sō-ton´-ik): referring to a solution that has the same concentration of dissolved

particles (and therefore the same concentration of free water) as has the cytoplasm of a cell.
isotope: one of several forms of a single element, the nuclei of which contain the same number of protons but different numbers of neutrons.
J-curve: the J-shaped growth curve of an exponentially growing population in which increasing numbers of individuals join the population during each succeeding time period.
joint: a flexible region between two rigid units of an exoskeleton or endoskeleton, allowing for movement between the units.
karyotype: a preparation showing the number, sizes, and shapes of all chromosomes within a cell and, therefore, within the individual or species from which the cell was obtained.
keratin (ker´-uh-tin): a fibrous protein in hair, nails, and the epidermis of skin.
keystone species: a species whose influence on community structure is greater than its abundance would suggest.
kidney: one of a pair of organs of the excretory system that is located on either side of the spinal column and filters blood, removing wastes and regulating the composition and water content of the blood.
kinetic energy: the energy of movement; includes light, heat, mechanical movement, and electricity.
kinetochore (ki-net´-ō-kor): a protein structure that forms at the centromere regions of chromosomes; attaches the chromosomes to the spindle.
kingdom: the second broadest taxonomic category, contained within a domain and consisting of related phyla or divisions. This textbook recognizes four kingdoms within the domain Eukarya: Protista, Fungi, Plantae, and Animalia.
kin selection: a type of natural selection that favors a certain allele because it increases the survival or reproductive success of relatives that bear the same allele.
Klinefelter syndrome: a set of characteristics typically found in individuals who have two X chromosomes and one Y chromosome; these individuals are phenotypically males but are sterile and have several femalelike traits, including broad hips and partial breast development.
Krebs cycle: a cyclic series of reactions, occurring in the matrix of mitochondria, in which the acetyl groups from the pyruvic acids produced by glycolysis are broken down to CO_2, accompanied by the formation of ATP and electron carriers; also called *citric acid cycle*.
kuru: a degenerative brain disease, first discovered in the cannibalistic Fore tribe of New Guinea, that is caused by a prion.
labium (pl., **labia**): one of a pair of folds of skin of the external structures of the mammalian female reproductive system.
labor: a series of contractions of the uterus that result in birth.
lactation: the secretion of milk from the mammary glands.
lacteal (lak-tēl´): a single lymph capillary that penetrates each villus of the small intestine.
lactose (lak´-tōs): a disaccharide composed of glucose and galactose; found in mammalian milk.
large intestine: the final section of the digestive tract; consists of the colon and the rectum, where feces are formed and stored.
larva (lar´-vuh): an immature form of an organism with indirect development prior to metamorphosis into its adult form; includes the caterpillars of moths and butterflies and the maggots of flies.
larynx (lar´-inks): that portion of the air passage between the pharynx and the trachea; contains the vocal cords.
lateral bud: a cluster of meristematic cells at the node of a stem; under appropriate conditions, it grows into a branch.
lateral meristem: a meristematic tissue that forms cylinders parallel to the long axis of roots and stems; normally located between the primary xylem and primary phloem (vascular cambium) and just outside the phloem (cork cambium); also called *cambium*.
law of independent assortment: the independent inheritance of two or more distinct traits; states that the alleles for one trait may be distributed to the gametes independently of the alleles for other traits.
law of segregation: Gregor Mendel's conclusion that each gamete receives only one of each parent's pair of genes for each trait.
laws of thermodynamics: the physical laws that define the basic properties and behavior of energy.
leaf: an outgrowth of a stem, normally flattened and photosynthetic.
leaf primordium (pri-mor´-dē-um; pl., **primordia):** a cluster of meristem cells, located at the node of a stem, that develops into a leaf.
learning: an adaptive change in behavior as a result of experience.
legume (leg´-ūm): a member of a family of plants characterized by root swellings in which nitrogen-fixing bacteria are housed; includes soybeans, lupines, alfalfa, and clover.
lens: a clear object that bends light rays; in eyes, a flexible or movable structure used to focus light on a layer of photoreceptor cells.
leptin: a peptide hormone. One of the functions of leptin, which is released by fat cells, is to help the body monitor its fat stores and regulate weight.
leukocyte (loo´-kō-sīt): any of the white blood cells circulating in the blood.
lichen (lī´-ken): a symbiotic association between an alga or cyanobacterium and a fungus, resulting in a composite organism.
life cycle: the events in the life of an organism from one generation to the next.
ligament: a tough connective tissue band connecting two bones.
light-dependent reactions: the first stage of photosynthesis, in which the energy of light is captured as ATP and NADPH; occurs in thylakoids of chloroplasts.
light-harvesting complex: in photosystems, the assembly of pigment molecules (chlorophyll and accessory pigments) that absorb light energy and transfer that energy to electrons.
light-independent reactions: the second stage of photosynthesis, in which the energy obtained by the light-dependent reactions is used to fix carbon dioxide into carbohydrates; occurs in the stroma of chloroplasts.
lignin: a hard material that is embedded in the cell walls of vascular plants and provides support in terrestrial species; an early and important adaptation to terrestrial life.
limbic system: a diverse group of brain structures, mostly in the lower forebrain, that includes the thalamus, hypothalamus, amygdala, hippocampus, and parts of the cerebrum and is involved in basic emotions, drives, behaviors, and learning.
limnetic zone: a lake zone in which enough light penetrates to support photosynthesis.
linkage: the inheritance of certain genes as a group because they are parts of the same chromosome. Linked genes do not show independent assortment.
lipase (lī´-pās): an enzyme that catalyzes the breakdown of lipids such as fats.
lipid (li´-pid): one of a number of organic molecules containing large nonpolar regions composed solely of carbon and hydrogen, which make lipids hydrophobic and insoluble in water; includes oils, fats, waxes, phospholipids, and steroids.
littoral zone: a lake zone, near the shore, in which water is shallow and plants find abundant light, anchorage, and adequate nutrients.
liver: an organ with varied functions, including bile production, glycogen storage, and the detoxification of poisons.
locus: the physical location of a gene on a chromosome.
long-day plant: a plant that will flower only if the length of daylight is greater than some species-specific duration.
long-term memory: the second phase of learning; a more-or-less permanent memory formed by a structural change in the brain, brought on by repetition.
loop of Henle (hen´-lē): a specialized portion of the tubule of the nephron in birds and mammals that creates an osmotic concentration gradient in the fluid immediately surrounding it. This gradient in turn makes possible the production of urine more osmotically concentrated than blood plasma.
lung: a paired respiratory organ consisting of inflatable chambers within the chest cavity in which gas exchange occurs.
luteinizing hormone (LH): a hormone, produced by the anterior pituitary, that stimulates testosterone production in males and the development of the follicle, ovulation, and the production of the corpus luteum in females.
lymph (limf): a pale fluid, within the lymphatic system, that is composed primarily of interstitial fluid and lymphocytes.
lymphatic system: a system consisting of lymph vessels, lymph capillaries, lymph nodes, and the thymus and spleen; helps protect the

body against infection, absorbs fats, and returns excess fluid and small proteins to the blood circulatory system.

lymph node: a small structure that filters lymph; contains lymphocytes and macrophages, which inactivate foreign particles such as bacteria.

lymphocyte (lim´-fō-sīt): a type of white blood cell important in the immune response.

lysosome (lī´-sō-sōm): a membrane-bound organelle containing intracellular digestive enzymes.

macronutrient: a nutrient needed in relatively large quantities (often defined as making up more than 0.1% of an organism's body).

macrophage (mak´-rō-fāj): a type of white blood cell that engulfs microbes and destroys them by phagocytosis; also presents microbial antigens to T cells, helping stimulate the immune response.

magnetotactic: able to detect and respond to Earth's magnetic field.

major histocompatibility complex (MHC): proteins, normally located on the surfaces of body cells, that identify the cell as "self"; also important in stimulating and regulating the immune response.

maltose (mal´-tōs): a disaccharide composed of two glucose molecules.

mammary gland (mam´-uh-rē): a milk-producing gland used by female mammals to nourish their young.

mantle (man´-tul): an extension of the body wall in certain invertebrates, such as mollusks; may secrete a shell, protect the gills, and, as in cephalopods, aid in locomotion.

marsupial (mar-soo´-pē-ul): a mammal whose young are born at an extremely immature stage and undergo further development in a pouch while they remain attached to a mammary gland; includes kangaroos, opossums, and koalas.

mass extinction: the extinction of an extraordinarily large number of species in a short period of geologic time. Mass extinctions have recurred periodically throughout the history of life.

mast cell: a cell of the immune system that synthesizes histamine and other molecules used in the body's response to trauma and that are a factor in allergic reactions.

matrix: the fluid contained within the inner membrane of a mitochondrion.

mechanical incompatibility: the inability of male and female organisms to exchange gametes, normally because their reproductive structures are incompatible.

mechanoreceptor: a receptor that responds to mechanical deformation, such as that caused by pressure, touch, or vibration.

medulla (med-ū´-luh): the part of the hindbrain of vertebrates that controls automatic activities such as breathing, swallowing, heart rate, and blood pressure.

medusa (meh-doo´-suh): a bell-shaped, typically free-swimming stage in the life cycle of many cnidarians; includes jellyfish.

megakaryocyte (meg-a-kar´-ē-ō-sīt): a large cell type that remains in the bone marrow, pinching off pieces of itself that then enter the circulation as platelets.

megaspore: a haploid cell formed by meiosis from a diploid megaspore mother cell; through mitosis and differentiation, develops into the female gametophyte.

megaspore mother cell: a diploid cell, within the ovule of a flowering plant, that undergoes meiosis to produce four haploid megaspores.

meiosis (mī-ō´-sis): a type of cell division, used by eukaryotic organisms, in which a diploid cell divides twice to produce four haploid cells.

meiotic cell division: meiosis followed by cytokinesis.

melanocyte-stimulating hormone (me-lan´-ō-sīt): a hormone, released by the anterior pituitary, that regulates the activity of skin pigments in some vertebrates.

melatonin (mel-uh-tōn´-in): a hormone, secreted by the pineal gland, that is involved in the regulation of circadian cycles.

membrane: in multicellular organism, a continuous sheet of epithelial cells that covers the body and lines body cavities; in a cell, a thin sheet of lipids and proteins that surrounds the cell or its organelles, separating them from their surroundings.

memory B cell: a type of white blood cell that is produced as a result of the binding of an antibody on a B cell to an antigen on an invading microorganism. Memory B cells persist in the bloodstream and provide future immunity to invaders bearing that antigen.

memory T cell: a type of white blood cell that is produced as a result of the binding of a receptor on a T cell to an antigen on an invading microorganism. Memory T cells persist in the bloodstream and provide future immunity to invaders bearing that antigen.

meninges (men-in´-jēz): three layers of connective tissue that surround the brain and spinal cord.

menstrual cycle: in human females, a complex 28-day cycle during which hormonal interactions among the hypothalamus, pituitary gland, and ovary coordinate ovulation and the preparation of the uterus to receive and nourish the fertilized egg. If pregnancy does not occur, the uterine lining is shed during menstruation.

menstruation: in human females, the monthly discharge of uterine tissue and blood from the uterus.

meristem cell (mer´-i-stem): an undifferentiated cell that remains capable of cell division throughout the life of a plant.

mesoderm (mēz´-ō-derm): the middle embryonic tissue layer, lying between the endoderm and ectoderm, and normally the last to develop; gives rise to structures such as muscle and skeleton.

mesoglea (mez-ō-glē´-uh): a middle, jellylike layer within the body wall of cnidarians.

mesophyll (mez´-ō-fil): loosely packed parenchyma cells beneath the epidermis of a leaf.

messenger RNA (mRNA): a strand of RNA, complementary to the DNA of a gene, that conveys the genetic information in DNA to the ribosomes to be used during protein synthesis; sequences of three bases (codons) in mRNA specify particular amino acids to be incorporated into a protein.

metabolic pathway: a sequence of chemical reactions within a cell, in which the products of one reaction are the reactants for the next reaction.

metabolism: the sum of all chemical reactions that occur within a single cell or within all the cells of a multicellular organism.

metamorphosis (met-a-mor´-fō-sis): in animals with indirect development, a radical change in body form from larva to sexually mature adult, as seen in amphibians (tadpole to frog) and insects (caterpillar to butterfly).

metaphase (met´-a-fāz): the stage of mitosis in which the chromosomes, attached to spindle fibers at kinetochores, are lined up along the equator of the cell.

methanogen (me-than´-ō-jen): a type of anaerobic archaean capable of converting carbon dioxide to methane.

microbe: a microorganism.

microevolution: change over successive generations in the composition of a population's gene pool.

microfilament: part of the cytoskeleton of eukaryotic cells that is composed of the proteins actin and (in some cases) myosin; functions in the movement of cell organelles and in locomotion by extension of the plasma membrane.

micronutrient: a nutrient needed only in small quantities (often defined as making up less than 0.01% of an organism's body).

microsphere: a small, hollow sphere formed from proteins or proteins complexed with other compounds.

microspore: a haploid cell formed by meiosis from a microspore mother cell; through mitosis and differentiation, develops into the male gametophyte.

microspore mother cell: a diploid cell contained within an anther of a flowering plant, which undergoes meiosis to produce four haploid microspores.

microtubule: a hollow, cylindrical strand, found in eukaryotic cells, that is composed of the protein tubulin; part of the cytoskeleton used in the movement of organelles, cell growth, and the construction of cilia and flagella.

microvillus (mī-krō-vi´-lus; pl., **microvilli):** a microscopic projection of the plasma membrane of each villus; increases the surface area of the villus.

midbrain: during development, the central portion of the brain; contains an important relay center, the reticular formation.

middle ear: the part of the mammalian ear composed of the tympanic membrane, the Eustachian tube, and three bones (hammer, anvil, and stirrup) that transmit vibrations from the auditory canal to the oval window.

middle lamella: a thin layer of sticky polysaccharides, such as pectin, and other carbohydrates that separates and holds together the primary cell walls of adjacent plant cells.

mimicry (mim´-ik-rē): the situation in which a species has evolved to resemble something else—typically another type of organism.
mineral: an inorganic substance, especially one in rocks or soil.
mitochondrion (mī-tō-kon´-drē-un): an organelle, bounded by two membranes, that is the site of the reactions of aerobic metabolism.
mitosis (mī-tō´-sis): a type of nuclear division, used by eukaryotic cells, in which one copy of each chromosome (already duplicated during interphase before mitosis) moves into each of two daughter nuclei; the daughter nuclei are therefore genetically identical to each other.
mitotic cell division: mitosis followed by cytokinesis.
molecule (mol´-e-kūl): a particle composed of one or more atoms held together by chemical bonds; the smallest particle of a compound that displays all the properties of that compound.
molt: to shed an external body covering, such as an exoskeleton, skin, feathers, or fur.
monoclonal antibody: an antibody produced in the laboratory by the cloning of hybridoma cells; each clone of cells produces a single antibody.
monocot: short for monocotyledon; a type of flowering plant characterized by embryos with one seed leaf, or cotyledon.
monoecious (mon-ē´-shus): pertaining to organisms in which male and female gametes are produced in the same individual.
monomer (mo´-nō-mer): a small organic molecule, several of which may be bonded together to form a chain called a *polymer.*
monophyletic: referring to a group of species that contains all the known descendents of an ancestral species.
monosaccharide (mo-nō-sak´-uh-rīd): the basic molecular unit of all carbohydrates, normally composed of a chain of carbon atoms bonded to hydrogen and hydroxyl groups.
monotreme: a mammal that lays eggs; for example, the platypus.
morula (mor´-ū-luh): in animals, an embryonic stage during cleavage, when the embryo consists of a solid ball of cells.
motor neuron: a neuron that receives instructions from the association neurons and activates effector organs, such as muscles or glands.
motor unit: a single motor neuron and all the muscle fibers on which it forms synapses.
mouth: the opening of a tubular digestive system into which food is first introduced.
mucous membrane: the lining of the inside of the respiratory and digestive tracts.
multicellular: many-celled; most members of the kingdoms Fungi, Plantae, and Animalia are multicellular, with intimate cooperation among cells.
multiple alleles: as many as dozens of alleles produced for every gene as a result of different mutations.
muscle fiber: an individual muscle cell.
mutation: a change in the base sequence of DNA in a gene; normally refers to a genetic change significant enough to alter the appearance or function of the organism.
mutualism (mū´-choo-ul-iz-um): a symbiotic relationship in which both participating species benefit.
mycelium (mī-sēl´-ē-um): the body of a fungus, consisting of a mass of hyphae.
mycorrhiza (mī-kō-rī´zuh; pl., **mycorrhizae):** a symbiotic relationship between a fungus and the roots of a land plant that facilitates mineral extraction and absorption.
myelin (mī´-uh-lin): a wrapping of insulating membranes of specialized nonneural cells around the axon of a vertebrate nerve cell; increases the speed of conduction of action potentials.
myofibril (mī-ō-fī´-bril): a cylindrical subunit of a muscle cell, consisting of a series of sarcomeres; surrounded by sarcoplasmic reticulum.
myometrium (mī-ō-mē´-trē-um): the muscular outer layer of the uterus.
myosin (mī´-ō-sin): one of the major proteins of muscle, the interaction of which with the protein actin produces muscle contraction; found in the thick filaments of the muscle fiber; see also *actin.*
natural causality: the scientific principle that natural events occur as a result of preceding natural causes.
natural killer cell: a type of white blood cell that destroys some virus-infected cells and cancerous cells on contact; part of the immune system's nonspecific internal defense against disease.
natural selection: the unequal survival and reproduction of organisms due to environmental forces, resulting in the preservation of favorable adaptations. Usually, natural selection refers specifically to differential survival and reproduction on the basis of genetic differences among individuals.
near-shore zone: the region of coastal water that is relatively shallow but constantly submerged; includes bays and coastal wetlands and can support large plants or seaweeds.
nearsighted: the inability to focus on distant objects caused by an eyeball that is slightly too long.
negative feedback: a situation in which a change initiates a series of events that tend to counteract the change and restore the original state. Negative feedback in physiological systems maintains homeostasis.
nephridium (nef-rid´-ē-um): an excretory organ found in earthworms, mollusks, and certain other invertebrates; somewhat resembles a single vertebrate nephron.
nephron (nef´-ron): the functional unit of the kidney; where blood is filtered and urine formed.
nephrostome (nef´-rō-stōm): the funnel-shaped opening of the nephridium of some invertebrates such as earthworms; coelomic fluid is drawn into the nephrostome for filtration.
nerve: a bundle of axons of nerve cells, bound together in a sheath.
nerve cord: a paired neural structure in most animals that conducts nervous signals to and from the ganglia; in chordates, a nervous structure lying along the dorsal side of the body; also called spinal cord.
nerve net: a simple form of nervous system, consisting of a network of neurons that extend throughout the tissues of an organism such as a cnidarian.
nerve tissue: the tissue that make up the brain, spinal cord, and nerves; consists of neurons and glial cells.
net primary productivity: the energy stored in the autotrophs of an ecosystem over a given time period.
neural tube: a structure, derived from ectoderm during early embryonic development, that later becomes the brain and spinal cord.
neuromuscular junction: the synapse formed between a motor neuron and a muscle fiber.
neuron (noor´-on): a single nerve cell.
neuropeptide: a small protein molecule with neurotransmitter-like actions.
neurosecretory cell: a specialized nerve cell that synthesizes and releases hormones.
neurotransmitter: a chemical that is released by a nerve cell close to a second nerve cell, a muscle, or a gland cell and that influences the activity of the second cell.
neutral mutation: a mutation that has little or no effect on the function of the encoded protein.
neutralization: the process of covering up or inactivating a toxic substance with antibody.
neutron: a subatomic particle that is found in the nuclei of atoms, bears no charge, and has a mass approximately equal to that of a proton.
nitrogen fixation: the process that combines atmospheric nitrogen with hydrogen to form ammonium (NH_4^+).
nitrogen-fixing bacterium: a bacterium that possess the ability to remove nitrogen (N_2) from the atmosphere and combine it with hydrogen to produce ammonium (NH_4^+).
node: in plants, a region of a stem at which leaves and lateral buds are located; in vertebrates, an interruption of the myelin on a myelinated axon, exposing naked membrane at which action potentials are generated.
nodule: a swelling on the root of a legume or other plant that consists of cortex cells inhabited by nitrogen-fixing bacteria.
nondisjunction: an error in meiosis in which chromosomes fail to segregate properly into the daughter cells.
nonpolar covalent bond: a covalent bond with equal sharing of electrons.
norepinephrine (nor-ep-i-nef-rin´): a neurotransmitter, released by neurons of the parasympathetic nervous system, that prepares the body to respond to stressful situations; also called *noradrenaline.*
northern coniferous forest: a biome with long, cold winters and only a few months of warm weather; populated almost entirely by evergreen coniferous trees; also called *taiga.*
notochord (nōt´-ō-kord): a stiff but somewhat flexible, supportive rod found in all members of the phylum Chordata at some stage of development.

nuclear envelope: the double-membrane system surrounding the nucleus of eukaryotic cells; the outer membrane is typically continuous with the endoplasmic reticulum.
nucleic acid (noo-klā´-ik): an organic molecule composed of nucleotide subunits; the two common types of nucleic acids are ribonucleic acid (RNA) and deoxyribonucleic acid (DNA).
nucleoid (noo-klē-oid): the location of the genetic material in prokaryotic cells; not membrane-enclosed.
nucleolus (noo-klē´-ō-lus): the region of the eukaryotic nucleus that is engaged in ribosome synthesis; consists of the genes encoding ribosomal RNA, newly synthesized ribosomal RNA, and ribosomal proteins.
nucleotide: a subunit of which nucleic acids are composed; a phosphate group bonded to a sugar (deoxyribose in DNA), which is in turn bonded to a nitrogen-containing base (adenine, guanine, cytosine, or thymine in DNA). Nucleotides are linked together, forming a strand of nucleic acid, as follows: Bonds between the phosphate of one nucleotide link to the sugar of the next nucleotide.
nucleotide substitution: a mutation that replaces one nucleotide in a DNA molecule with another; for example, a change from an adenine to a guanine.
nucleus (atomic): the central region of an atom, consisting of protons and neutrons.
nucleus (cellular): the membrane-bound organelle of eukaryotic cells that contains the cell's genetic material.
nutrient: a substance acquired from the environment and needed for the survival, growth, and development of an organism.
nutrient cycle: a description of the pathways of a specific nutrient (such as carbon, nitrogen, phosphorus, or water) through the living and nonliving portions of an ecosystem.
nutrition: the process of acquiring nutrients from the environment and, if necessary, processing them into a form that can be used by the body.
observation: the first operation in the scientific method; the noting of a specific phenomenon, leading to the formulation of a hypothesis.
oil: a lipid composed of three fatty acids, some of which are unsaturated, covalently bonded to a molecule of glycerol; liquid at room temperature.
olfaction (ōl-fak´-shun): a chemical sense, the sense of smell; in terrestrial vertebrates, the result of the detection of airborne molecules.
oligotrophic lake: a lake that is very low in nutrients and hence clear with extensive light penetration.
ommatidium (ōm-ma-tid´-ē-um): an individual light-sensitive subunit of a compound eye; consists of a lens and several receptor cells.
omnivore: an organism that consumes both plants and other animals.
one gene, one protein rule: the prem-ise that each gene encodes the information for the synthesis of a single protein.
oogenesis: the process by which egg cells are formed.
oogonium (ō-ō-gō´-nē-um; pl., **oogonia):** in female animals, a diploid cell that gives rise to a primary oocyte.
open circulatory system: a type of circulatory system found in some invertebrates, such as arthropods and mollusks, that includes an open space (the hemocoel) in which blood directly bathes body tissues.
operant conditioning: a laboratory training procedure in which an animal learns to perform a response (such as pressing a lever) through reward or punishment.
operculum: an external flap, supported by bone, that covers and protects the gills of most fish.
opioid (ōp´-ē-oid): one of a group of peptide neuromodulators in the vertebrate brain that mimic some of the actions of opiates (such as opium) and also seem to influence many other processes, including emotion and appetite.
optic nerve: the nerve leading from the eye to the brain, carrying visual information.
order: the taxonomic category contained within a class and consisting of related families.
organ: a structure (such as the liver, kidney, or skin) composed of two or more distinct tissue types that function together.
organelle (or-guh-nel´): a structure, found in the cytoplasm of eukaryotic cells, that performs a specific function; sometimes refers specifically to membrane-bound structures, such as the nucleus or endoplasmic reticulum.
organic/organic molecule: describing a molecule that contains both carbon and hydrogen.
organism (or´-guh-niz-um): an individual living thing.
organogenesis (or-gan-ō-jen´-uh-sis): the process by which the layers of the gastrula (endoderm, ectoderm, mesoderm) rearrange into organs.
organ system: two or more organs that work together to perform a specific function; for example, the digestive system.
origin: the site of attachment of a muscle to the relatively stationary bone on one side of a joint.
osmosis (oz-mō´-sis): the diffusion of water across a differentially permeable membrane, normally down a concentration gradient of free water molecules. Water moves into the solution that has a lower concentration of free water from a solution with the higher concentration of free water.
osmotic pressure: the pressure required to counterbalance the tendency of water to move from a solution with a higher concentration of free water molecules into a solution with a lower concentration of free water molecules.
osteoblast (os´-tē-ō-blast): a cell type that produces bone.
osteoclast (os´-tē-ō-klast): a cell type that dissolves bone.
osteocyte (os´-tē-ō-sīt): a mature bone cell.
osteon: a unit of hard bone consisting of concentric layers of bone matrix, with embedded osteocytes, surrounding a small central canal that contains a capillary.
osteoporosis (os´-tē-ō-por-ō´-sis): a condition in which bones become porous, weak, and easily fractured; most common in elderly women.
outer ear: the outermost part of the mammalian ear, including the external ear and auditory canal leading to the tympanic membrane.
oval window: the membrane-covered entrance to the inner ear.
ovary: in animals, the gonad of females; in flowering plants, a structure at the base of the carpel that contains one or more ovules and develops into the fruit.
oviduct: in mammals, the tube leading from the ovary to the uterus.
ovulation: the release of a secondary oocyte, ready to be fertilized, from the ovary.
ovule: a structure within the ovary of a flower, inside which the female gametophyte develops; after fertilization, develops into the seed.
oxytocin (oks-ē-tō´-sin): a hormone, released by the posterior pituitary, that stimulates the contraction of uterine and mammary gland muscles.
ozone layer: the ozone-enriched layer of the upper atmosphere that filters out some of the sun's ultraviolet radiation.
pacemaker: a cluster of specialized muscle cells in the upper right atrium of the heart that produce spontaneous electrical signals at a regular rate; the sinoatrial node.
pain receptor: a receptor that has extensive areas of membranes studded with special receptor proteins that respond to light or to a chemical.
palisade cell: a columnar mesophyll cell, containing chloroplasts, just beneath the upper epidermis of a leaf.
pancreas (pān´-krē-us): a combined exocrine and endocrine gland located in the abdominal cavity next to the stomach. The endocrine portion secretes the hormones insulin and glucagon, which regulate glucose concentrations in the blood. The exocrine portion secretes enzymes for fat, carbohydrate, and protein digestion into the small intestine and neutralizes the acidic chyme.
pancreatic juice: a mixture of water, sodium bicarbonate, and enzymes released by the pancreas into the small intestine.
parasite (par´-uh-sīt): an organism that lives in or on a larger prey organism, called a *host*, weakening it.
parasitism: a symbiotic relationship in which one organism (commonly smaller and more numerous than its host) benefits by feeding on the other, which is normally harmed but not immediately killed.
parasympathetic division: the division of the autonomic nervous system that produces largely involuntary responses related to the maintenance of normal body functions, such as digestion.
parathormone: a hormone, secreted by the parathyroid gland, that stimulates the release of calcium from bones.
parathyroid gland: one of a set of four small endocrine glands, embedded in the surface of the thyroid gland, that produces parathormone, which (with calcitonin from the thyroid gland) regulates calcium ion concentration in the blood.

parenchyma (par-en´-ki-muh): a plant cell type that is alive at maturity, normally with thin primary cell walls, that carries out most of the metabolism of a plant. Most dividing meristem cells in a plant are parenchyma.

parthenogenesis (par-the-nō-jen´uh-sis): a specialization of sexual reproduction, in which a haploid egg undergoes development without fertilization.

passive transport: the movement of materials across a membrane down a gradient of concentration, pressure, or electrical charge without using cellular energy.

pathogenic (path´-ō-jen-ik): capable of producing disease; refers to an organism with such a capability (a pathogen).

pedigree: a diagram showing genetic relationships among a set of individuals, normally with respect to a specific genetic trait.

pelagic (puh-la´-jik): free-swimming or floating.

penis: an external structure of the male reproductive and urinary systems; serves to deposit sperm into the female reproductive system and delivers urine to the exterior.

peptide (pep´-tīd): a chain composed of two or more amino acids linked together by peptide bonds.

peptide bond: the covalent bond between the amino group's nitrogen of one amino acid and the carboxyl group's carbon of a second amino acid, joining the two amino acids together in a peptide or protein.

peptide hormone: a hormone consisting of a chain of amino acids; includes small proteins that function as hormones.

peptidoglycan (pep-tid-ō-glī´-kan): a component of prokaryotic cell walls that consists of chains of sugars cross-linked by short chains of amino acids called *peptides*.

pericycle (per´-i-sī-kul): the outermost layer of cells of the vascular cylinder of a root.

periderm: the outer cell layers of roots and a stem that have undergone secondary growth, consisting primarily of cork cambium and cork cells.

peripheral nerve: a nerve that links the brain and spinal cord to the rest of the body.

peripheral nervous system: in vertebrates, the part of the nervous system that connects the central nervous system to the rest of the body.

peristalsis: rhythmic coordinated contractions of the smooth muscles of the digestive tract that move substances through the digestive tract.

permafrost: a permanently frozen layer of soil in the arctic tundra that cannot support the growth of trees.

petal: part of a flower, typically brightly colored and fragrant, that attracts potential animal pollinators.

petiole (pet´-ē-ōl): the stalk that connects the blade of a leaf to the stem.

phagocytic cell (fa-gō-sit´-ik): a type of immune system cell that destroys invading microbes by using phagocytosis to engulf and digest the microbes.

phagocytosis (fa-gō-sī-tō´-sis): a type of endocytosis in which extensions of a plasma membrane engulf extracellular particles and transport them into the interior of the cell.

pharyngeal gill slit (far-in´-jē-ul): an opening, located just posterior to the mouth, that connects the digestive tube to the outside environment; present (as some stage of life) in all chordates.

pharynx (far´-inks): in vertebrates, a chamber that is located at the back of the mouth and is shared by the digestive and respiratory systems; in some invertebrates, the portion of the digestive tube just posterior to the mouth.

phenotype (fēn´-ō-tīp): the physical characteristics of an organism; can be defined as outward appearance (such as flower color), as behavior, or in molecular terms (such as glycoproteins on red blood cells).

pheromone (fer´-uh-mōn): a chemical produced by an organism that alters the behavior or physiological state of another member of the same species.

phloem (flō´-um): a conducting tissue of vascular plants that transports a concentrated sugar solution up and down the plant.

phospholipid (fos-fō-li´-pid): a lipid consisting of glycerol bonded to two fatty acids and one phosphate group, which bears another group of atoms, typically charged and containing nitrogen. A double layer of phospholipids is a component of all cellular membranes.

phospholipid bilayer: a double layer of phospholipids that forms the basis of all cellular membranes. The phospholipid heads, which are hydrophilic, face the water of extracellular fluid or the cytoplasm; the tails, which are hydrophobic, are buried in the middle of the bilayer.

photic zone: the region of the ocean where light is strong enough to support photosynthesis.

photon (fō´-ton): the smallest unit of light energy.

photopigment (fō´-tō-pig-ment): a chemical substance in photoreceptor cells that, when struck by light, changes in molecular conformation.

photoreceptor: a receptor cell that responds to light; in vertebrates, rods and cones.

photorespiration: a series of reactions in plants in which O_2 replaces CO_2 during the C_3 cycle, preventing carbon fixation; this wasteful process dominates when C_3 plants are forced to close their stomata to prevent water loss.

photosynthesis: the complete series of chemical reactions in which the energy of light is used to synthesize high-energy organic molecules, normally carbohydrates, from low-energy inorganic molecules, normally carbon dioxide and water.

photosystem: in thylakoid membranes, a light-harvesting complex and its associated electron transport system.

phototactic: capable of detecting and responding to light.

phototropism: growth with respect to the direction of light.

pH scale: a scale, with values from 0 to 14, used for measuring the relative acidity of a solution; at pH 7 a solution is neutral, pH 0 to 7 is acidic, and pH 7 to 14 is basic; each unit on the scale represents a tenfold change in H^+ concentration.

phycocyanin (fī-kō-sī´-uh-nin): a blue or purple pigment that is located in the membranes of chloroplasts and is used as an accessory light-gathering molecule in thylakoid photosystems.

phylogeny (fī-lah´-jen-ē): the evolutionary history of a group of species.

phylum (fī-lum): the taxonomic category of animals and animal-like protists that is contained within a kingdom and consists of related classes.

phytochrome (fī´-tō-krōm): a light-sensitive plant pigment that mediates many plant responses to light, including flowering, stem elongation, and seed germination.

phytoplankton (fī´-tō-plank-ten): photosynthetic protists that are abundant in marine and freshwater environments.

pilus (pil´-us; pl., **pili):** a hairlike projection that is made of protein, located on the surface of certain bacteria, and is typically used to attach a bacterium to another cell.

pineal gland (pī-nē´-al): a small gland within the brain that secretes melatonin; controls the seasonal reproductive cycles of some mammals.

pinocytosis (pi-nō-sī-tō´-sis): the nonselective movement of extracellular fluid, enclosed within a vesicle formed from the plasma membrane, into a cell.

pioneer: an organism that is among the first to colonize an unoccupied habitat in the first stages of succession.

pit: an area in the cell walls between two plant cells in which secondary walls did not form, such that the two cells are separated only by a relatively thin and porous primary cell wall.

pith: cells forming the center of a root or stem.

pituitary gland: an endocrine gland, located at the base of the brain, that produces several hormones, many of which influence the activity of other glands.

placenta (pluh-sen´-tuh): in mammals, a structure formed by a complex interweaving of the uterine lining and the embryonic membranes, especially the chorion; functions in gas, nutrient, and waste exchange between embryonic and maternal circulatory systems and secretes hormones.

placental (pluh-sen´-tul): referring to a mammal, possessing a placenta (that is, species that are not marsupials or monotremes).

plankton: microscopic organisms that live in marine or freshwater environments; includes phytoplankton and zooplankton.

plant hormone: the plant-regulating chemicals auxin, gibberellins, cytokinins, ethylene, and abscisic acid; somewhat resemble animal hormones in that they are chemicals produced by cells in one location that influence the growth or metabolic activity of other cells, typically some distance away in the plant body.

plaque (plak): a deposit of cholesterol and other fatty substances within the wall of an artery.

plasma: the fluid, noncellular portion of the blood.
plasma cell: an antibody-secreting descendant of a B cell.
plasma membrane: the outer membrane of a cell, composed of a bilayer of phospholipids in which proteins are embedded.
plasmid (plaz´-mid): a small, circular piece of DNA located in the cytoplasm of many bacteria; normally does not carry genes required for the normal functioning of the bacterium but may carry genes that assist bacterial survival in certain environments, such as a gene for antibiotic resistance.
plasmodesma (plaz-mō-dez´-muh; pl., **plasmodesmata):** a cell-to-cell junction in plants that connects the cytoplasm of adjacent cells.
plasmodial slime mold: see *acellular slime mold.*
plasmodium (plaz-mō´-dē-um): a sluglike mass of cytoplasm containing thousands of nuclei that are not confined within individual cells.
plastid (plas´-tid): in plant cells, an organelle bounded by two membranes that may be involved in photosynthesis (chloroplasts), pigment storage, or food storage.
platelet (plāt´-let): a cell fragment that is formed from megakaryocytes in bone marrow and lacks a nucleus; circulates in the blood and plays a role in blood clotting.
pleated sheet: a form of secondary structure exhibited by certain proteins, such as silk, in which many protein chains lie side-by-side, with hydrogen bonds holding adjacent chains together.
pleiotropy (plē´-ō-trō-pē): a situation in which a single gene influences more than one phenotypic characteristic.
pleural membrane: a membrane that lines the chest cavity and surrounds the lungs.
point mutation: a mutation in which a single base pair in DNA has been changed.
polar body: in oogenesis, a small cell, containing a nucleus but virtually no cytoplasm, produced by the first meiotic division of the primary oocyte.
polar covalent bond: a covalent bond with unequal sharing of electrons, such that one atom is relatively negative and the other is relatively positive.
polar nucleus: in flowering plants, one of two nuclei in the primary endosperm cell of the female gametophyte; formed by the mitotic division of a megaspore.
pollen/pollen grain: the male gametophyte of a seed plant.
pollination: in flowering plants, when pollen grains land on the stigma of a flower of the same species; in conifers, when pollen grains land within the pollen chamber of a female cone of the same species.
polygenic inheritance: a pattern of inheritance in which the interactions of two or more functionally similar genes determine phenotype.
polymer (pah´-li-mer): a molecule composed of three or more (perhaps thousands) smaller subunits called *monomers*, which may be identical (for example, the glucose monomers of starch) or different (for example, the amino acids of a protein).
polymerase chain reaction (PCR): a method of producing virtually unlimited numbers of copies of a specific piece of DNA, starting with as little as one copy of the desired DNA.
polyp (pah´-lip): the sedentary, vase-shaped stage in the life cycle of many cnidarians; includes hydra and sea anemones.
polypeptide: a short polymer of amino acids; often used as a synonym for protein.
polyploidy (pahl´-ē-ploid-ē): having more than two homologous chromosomes of each type.
polysaccharide (pahl-ē-sak´-uh-rīd): a large carbohydrate molecule composed of branched or unbranched chains of repeating monosaccharide subunits, normally glucose or modified glucose molecules; includes starches, cellulose, and glycogen.
pons: a portion of the hindbrain, just above the medulla, that contains neurons that influence sleep and the rate and pattern of breathing.
population: all the members of a particular species within an ecosystem, found in the same time and place and actually or potentially interbreeding.
population bottleneck: a form of genetic drift in which a population becomes extremely small; may lead to differences in allele frequencies as compared with other populations of the species and to a loss in genetic variability.
population cycle: out-of-phase cyclical patterns of predator and prey populations.
population genetics: the study of the frequency, distribution, and inheritance of alleles in a population.
positive feedback: a situation in which a change initiates events that tend to amplify the original change.
post-anal tail: a tail that extends beyond the anus; exhibited by all chordates at some stage of development.
posterior: the tail, hindmost, or rear end of an animal.
posterior pituitary: a lobe of the pituitary gland that is an outgrowth of the hypothalamus and that releases antidiuretic hormone and oxytocin.
postmating isolating mechanism: any structure, physiological function, or developmental abnormality that prevents organisms of two different populations, once mating has occurred, from producing vigorous, fertile offspring.
postsynaptic neuron: at a synapse, the nerve cell that changes its electrical potential in response to a chemical (the neurotransmitter) released by another (presynaptic) cell.
postsynaptic potential (PSP): an electrical signal produced in a postsynaptic cell by transmission across the synapse; it may be excitatory (EPSP), making the cell more likely to produce an action potential, or inhibitory (IPSP), tending to inhibit an action potential.
potential energy: "stored" energy, normally chemical energy or energy of position within a gravitational field.
prairie: a biome, located in the centers of continents, that supports grasses; also called *grassland.*
preadaptation: a feature evolved under one set of environmental conditions that, purely by chance, helps an organism adapt to new environmental conditions.
prebiotic evolution: evolution before life existed; especially, the abiotic synthesis of organic molecules.
precapillary sphincter (sfink´-ter): a ring of smooth muscle between an arteriole and a capillary that regulates the flow of blood into the capillary bed.
predation (pre-dā´-shun): the act of killing and eating another living organism.
predator: an organism that kills and eats other organisms.
premating isolating mechanism: any structure, physiological function, or behavior that prevents organisms of two different populations from exchanging gametes.
pressure-flow theory: a model for the transport of sugars in phloem, by which the movement of sugars into a phloem sieve tube causes water to enter the tube by osmosis, while the movement of sugars out of another part of the same sieve tube causes water to leave by osmosis; the resulting pressure gradient causes the bulk movement of water and dissolved sugars from the end of the tube into which sugar is transported toward the end of the tube from which sugar is removed.
presynaptic neuron: a nerve cell that releases a chemical (the neurotransmitter) at a synapse, causing changes in the electrical activity of another (postsynaptic) cell.
prey: organisms that are killed and eaten by another organism.
primary cell wall: cellulose and other carbohydrates secreted by a young plant cell between the middle lamella and the plasma membrane.
primary consumer: an organism that feeds on producers; an herbivore.
primary endosperm cell: the central cell of the female gametophyte of a flowering plant, containing the polar nuclei (normally two); after fertilization, undergoes repeated mitotic divisions to produce the endosperm of the seed.
primary growth: growth in length and development of the initial structures of plant roots and shoots, due to the cell division of apical meristems and differentiation of the daughter cells.
primary oocyte (ō´-ō-sīt): a diploid cell, derived from the oogonium by growth and differentiation that undergoes meiosis, producing the egg.
primary phloem: phloem in young stems produced from an apical meristem.
primary root: the first root that develops from a seed.
primary spermatocyte (sper-ma´-tō-sīt): a diploid cell, derived from the spermatogonium by growth and differentiation, that undergoes meiosis, producing four sperm.
primary structure: the amino acid sequence of a protein.

primary succession: succession that occurs in an environment, such as bare rock, in which no trace of a previous community was present.
primary xylem: xylem in young stems produced from an apical meristem.
primate: a mammal characterized by the presence of an opposable thumb, forward-facing eyes, and a well-developed cerebral cortex; includes lemurs, monkeys, apes, and humans.
primitive streak: in reptiles, birds, and mammals, the region of the ectoderm of the two-layered embryonic disc through which cells migrate, forming mesoderm.
prion (prē´-on): a protein that, in mutated form, acts as an infectious agent that causes certain neurodegenerative diseases, including kuru and scrapie.
producer: a photosynthetic organism; an autotroph.
product: an atom or molecule that is formed from reactants in a chemical reaction.
profundal zone: a lake zone in which light is insufficient to support photosynthesis.
progesterone (prō-ge´-ster-ōn): a hormone, produced by the corpus luteum, that promotes the development of the uterine lining in females.
prokaryotic (prō-kar-ē-ot´-ik): referring to cells of the domains Bacteria or Archaea. Prokaryotic cells have genetic material that is not enclosed in a membrane-bound nucleus; they lack other membrane-bound organelles.
prolactin: a hormone, released by the anterior pituitary, that stimulates milk production in human females.
promoter: a specific sequence of DNA to which RNA polymerase binds, initiating gene transcription.
prophase (prō´-fāz): the first stage of mitosis, in which the chromosomes first become visible in the light microscope as thickened, condensed threads and the spindle begins to form; as the spindle is completed, the nuclear envelope breaks apart, and the spindle fibers invade the nuclear region and attach to the kinetochores of the chromosomes. Also, the first stage of meiosis: In meiosis I, the homologous chromosomes pair up and exchange parts at chiasmata; in meiosis II, the spindle re-forms and chromosomes attach to the microtubules.
prostaglandin (pro-stuh-glan´-din): a family of modified fatty acid hormones manufactured by many cells of the body.
prostate gland (pros´-tāt): a gland that produces part of the fluid component of semen; prostatic fluid is basic and contains a chemical that activates sperm movement.
protease (prō´-tē-ās): an enzyme that digests proteins.
protein: polymer of amino acids joined by peptide bonds.
Protista (prō-tis´-tuh): a taxonomic kingdom including unicellular, eukaryotic organisms.
protocell: the hypothetical evolutionary precursor of living cells, consisting of a mixture of organic molecules within a membrane.
proton: a subatomic particle that is found in the nuclei of atoms, bears a unit of positive charge, and has a relatively large mass, roughly equal to the mass of the neutron.
protonephridium (prō-tō-nef-rid´-ē-um; pl., **protonephridia):** an excretory system consisting of tubules that have external opening but lack internal openings; for example, the flame-cell system of flatworms.
protostome (prō´-tō-stōm): an animal with a mode of embryonic development in which the coelom is derived from splits in the mesoderm; characteristic of arthropods, annelids, and mollusks.
protozoan (prō-tuh-zō´-an; pl., **protozoa):** a nonphotosynthetic or animal-like protist.
proximal tubule: in nephrons of the mammalian kidney, the portion of the renal tubule just after the Bowman's capsule; receives filtrate from the capsule and is the site where selective secretion and reabsorption between the filtrate and the blood begins.
pseudocoelom (soo´-dō-sēl´-ōm): "false coelom"; a body cavity that has a different embryological origin than a coelom but serves a similar function; found in roundworms.
pseudoplasmodium (soo´-dō-plaz-mō´-dē-um): an aggregation of individual amoeboid cells that form a sluglike mass.
pseudopod (sood´-ō-pod): an extension of the plasma membrane by which certain cells, such as amoebae, locomote and engulf prey.
Punnett square method: an intuitive way to predict the genotypes and phenotypes of offspring in specific crosses.
pupa: a developmental stage in some insect species in which the organism stops moving and feeding and may be encased in a cocoon; occurs between the larval and the adult phases.
pupil: the adjustable opening in the center of the iris, through which light enters the eye.
pyloric sphincter (pī-lor´-ik sfink´-ter): a circular muscle, located at the base of the stomach, that regulates the passage of chyme into the small intestine.
pyruvate: a three-carbon moelcule that is formed by glycolysis and then used in fermentation or cellular respiration.
quaternary structure (kwat´-er-nuh-rē): the complex three-dimensional structure of a protein composed of more than one peptide chain.
queen substance: a chemical, produced by a queen bee, that can act as both a primer and a pheromone.
radial symmetry: a body plan in which any plane along a central axis will divide the body into approximately mirror-image halves. Cnidarians and many adult echinoderms have radial symmetry.
radioactive: pertaining to an atom with an unstable nucleus that spontaneously disintegrates, with the emission of radiation.
radiolarian (rā-dē-ō-lar´-ē-un): an aquatic protist (largely marine) characterized by typically elaborate silica shells.
radula (ra´-dū-luh): a ribbon of tissue in the mouth of gastropod mollusks; bears numerous teeth on its outer surface and is used to scrape and drag food into the mouth.
rain shadow: a local dry area created by the modification of rainfall patterns by a mountain range.
random distribution: distribution characteristic of populations in which the probability of finding an individual is equal in all parts of an area.
reactant: an atom or molecule that is used up in a chemical reaction to form a product.
reaction center: in the light-harvesting complex of a photosystem, the chlorophyll molecule to which light energy is transferred by the antenna molecules (light-absorbing pigments); the captured energy ejects an electron from the reaction center chlorophyll, and the electron is transferred to the electron transport system.
receptor: a cell that responds to an environmental stimulus (chemicals, sound, light, pH, and so on) by changing its electrical potential; also, a protein molecule in a plasma membrane that binds to another molecule (hormone, neurotransmitter), triggering metabolic or electrical changes in a cell.
receptor-mediated endocytosis: the selective uptake of molecules from the extracellular fluid by binding to a receptor located at a coated pit on the plasma membrane and pinching off the coated pit into a vesicle that moves into the cytoplasm.
receptor potential: an electrical potential change in a receptor cell, produced in response to the reception of an environmental stimulus (chemicals, sound, light, heat, and so on). The size of the receptor potential is proportional to the intensity of the stimulus.
receptor protein: a protein, located on a membrane (or in the cytoplasm), that recognizes and binds to specific molecules. Binding by receptor proteins typically triggers a response by a cell, such as endocytosis, increased metabolic rate, or cell division.
recessive: an allele that is expressed only in homozygotes and is completely masked in heterozygotes.
recognition protein: a protein or glycoprotein protruding from the outside surface of a plasma membrane that identifies a cell as belonging to a particular species, to a specific individual of that species, and in many cases to one specific organ within the individual.
recombinant DNA: DNA that has been altered by the recombination of genes from a different organism, typically from a different species.
recombination: the formation of new combinations of the different alleles of each gene on a chromosome; the result of crossing over.
rectum: the terminal portion of the vertebrate digestive tube, where feces are stored until they can be eliminated.
reflex: a simple, stereotyped movement of part of the body that occurs automatically in response to a stimulus.
regeneration: the regrowth of a body part after loss or damage; also, asexual reproduction by means of the regrowth of an entire body from a fragment.

releasing hormone: a hormone, secreted by the hypothalamus, that causes the release of specific hormones by the anterior pituitary.
renal artery: the artery carrying blood to each kidney.
renal cortex: the outer layer of the kidney; where nephrons are located.
renal medulla: the layer of the kidney just inside the renal cortex; where loops of Henle produce a highly concentrated interstitial fluid, important in the production of concentrated urine.
renal pelvis: the inner chamber of the kidney; where urine from the collecting ducts accumulates before it enters the ureters.
renal vein: the vein carrying cleansed blood away from each kidney.
renin: an enzyme that is released (in mammals) when blood pressure and/or sodium concentration in the blood drops below a set point; initiates a cascade of events that restores blood pressure and sodium concentration.
replacement-level fertility (RLF): the average birthrate at which a reproducing population exactly replaces itself during its lifetime.
replication bubble: the unwound portion of the two parental DNA strands, separated by DNA helicase, in DNA replication.
reproductive isolation: the failure of organisms of one population to breed successfully with members of another; may be due to premating or postmating isolating mechanisms.
reservoir: the major source and storage site of a nutrient in an ecosystem, normally in the abiotic portion.
resource partitioning: the coexistence of two species with similar requirements, each occupying a smaller niche than either would if it were by itself; a means of minimizing their competitive interactions.
respiratory center: a cluster of neurons, located in the medulla of the brain, that sends rhythmic bursts of nerve impulses to the respiratory muscles, resulting in breathing.
resting potential: a negative electrical potential in unstimulated nerve cells.
restriction enzyme: an enzyme, normally isolated from bacteria, that cuts double-stranded DNA at a specific nucleotide sequence; the nucleotide sequence that is cut differs for different restriction enzymes.
restriction fragment: a piece of DNA that has been isolated by cleaving a larger piece of DNA with restriction enzymes.
restriction fragment length polymorphism (RFLP): a difference in the length of restriction fragments, produced by cutting samples of DNA from different individuals of the same species with the same set of restriction enzymes; the result of differences in nucleotide sequences among individuals of the same species.
reticular formation (reh-tik´-ū-lar): a diffuse network of neurons extending from the hindbrain, through the midbrain, and into the lower reaches of the forebrain; involved in filtering sensory input and regulating what information is relayed to conscious brain centers for further attention.
retina (ret´-in-uh): a multilayered sheet of nerve tissue at the rear of camera-type eyes, composed of photoreceptor cells plus associated nerve cells that refine the photoreceptor information and transmit it to the optic nerve.
retrovirus: a virus that uses RNA as its genetic material. When it invades a eukaryotic cell, a retrovirus "reverse transcribes" its RNA into DNA, which then directs the synthesis of more viruses, using the transcription and translation machinery of the cell.
reverse transcriptase: an enzyme found in retroviruses that catalyzes the synthesis of DNA from an RNA template.
Rh factor: a protein on the red blood cells of some people (Rh-positive) but not others (Rh-negative); the exposure of Rh-negative individuals to Rh-positive blood triggers the production of antibodies to Rh-positive blood cells.
rhizoid (rī´-zoid): a rootlike structure found in bryophytes that anchors the plant and absorbs water and nutrients from the soil.
rhizome (rī´-zōm): an undergound stem, usually horizontal, that stores food.
rhythm method: a contraceptive method involving abstinence from intercourse during ovulation.
ribonucleic acid (rī-bō-noo-klā´-ik; RNA): a molecule composed of ribose nucleotides, each of which consists of a phosphate group, the sugar ribose, and one of the bases adenine, cytosine, guanine, or uracil; transfers hereditary instructions from the nucleus to the cytoplasm; also the genetic material of some viruses.
ribosomal RNA (rRNA): a type of RNA that combines with proteins to form ribosomes.
ribosome: an organelle consisting of two subunits, each composed of ribosomal RNA and protein; the site of protein synthesis, during which the sequence of bases of messenger RNA is translated into the sequence of amino acids in a protein.
ribozyme: an RNA molecule that can catalyze certain chemical reactions, especially those involved in the synthesis and processing of RNA itself.
RNA polymerase: in RNA synthesis, an enzyme that catalyzes the bonding of free RNA nucleotides into a continuous strand, using RNA nucleotides that are complementary to those of a strand of DNA.
rod: a rod-shaped photoreceptor cell in the vertebrate retina, sensitive to dim light but not involved in color vision; see also *cone*.
root: the part of the plant body, normally underground, that provides anchorage, absorbs water and dissolved nutrients and transports them to the stem, produces some hormones, and in some plants serves as a storage site for carbohydrates.
root cap: a cluster of cells at the tip of a growing root, derived from the apical meristem; protects the growing tip from damage as it burrows through the soil.
root hair: a fine projection from an epidermal cell of a young root that increases the absorptive surface area of the root.
root system: the part of a plant, normally below ground, that anchors the plant in the soil, absorbs water and minerals, stores food, transports water, minerals, sugars, and hormones, and produces certain hormones.
rough endoplasmic reticulum: endoplasmic reticulum lined on the outside with ribosomes.
runner: a horizontally growing stem that may develop new plants at nodes that touch the soil.
sac fungus: a fungus of the division Ascomycota, whose members form spores in a saclike case called an *ascus*.
sapwood: young xylem that transports water and minerals in a tree trunk.
saprobe (sap´-rōb): an organism that derives its nutrients from the bodies of dead organisms.
sarcodine (sar-kō´-dīn): a nonphotosynthetic protist (protozoan) characterized by the ability to form pseudopodia; some sarcodines, such as amoebae, are naked, whereas others have elaborate shells.
sarcomere (sark´-ō-mēr): the unit of contraction of a muscle fiber; a subunit of the myofibril, consisting of actin and myosin filaments and bounded by Z lines.
sarcoplasmic reticulum (sark´-ō-plas´-mik re-tik´-ū-lum): the specialized endoplasmic reticulum in muscle cells; forms interconnected hollow tubes. The sarcoplasmic reticulum stores calcium ions and releases them into the interior of the muscle cell, initiating contraction.
saturated: referring to a fatty acid with as many hydrogen atoms as possible bonded to the carbon backbone; a fatty acid with no double bonds in its carbon backbone.
savanna: a biome that is dominated by grasses and supports scattered trees and thorny scrub forests; typically has a rainy season in which all the year's precipitation falls.
scientific method: a rigorous procedure for making observations of specific phenomena and searching for the order underlying those phenomena; consists of four operations: observation, hypothesis, experiment, and conclusion.
scientific name: the name of an organism formed from the two smallest major taxonomic categories—the genus and the species.
scientific theory: a general explanation of natural phenomena developed through extensive and reproducible observations; more general and reliable than a hypothesis.
sclera: a tough, white connective tissue layer that covers the outside of the eyeball and forms the white of the eye.
sclerenchyma (skler-en´-ki-muh): a plant cell type with thick, hardened secondary cell walls that normally dies as the last stage of differentiation and both supports and protects the plant body.
scramble competition: a free-for-all scramble for limited resources among individuals of the same species.
scrotum (skrō´-tum): the pouch of skin containing the testes of male mammals.
S-curve: the S-shaped growth curve that describes a population of long-lived organisms introduced into a new area; consists of an initial

period of exponential growth, followed by decreasing growth rate, and, finally, relative stability around a growth rate of zero.
sebaceous gland (se-bā´-shus): a gland in the dermis of skin, formed from epithelial tissue, that produces the oily substance sebum, which lubricates the epidermis.
secondary cell wall: a thick layer of cellulose and other polysaccharides secreted by certain plant cells between the primary cell wall and the plasma membrane.
secondary consumer: an organism that feeds on primary consumers; a carnivore.
secondary growth: growth in the diameter of a stem or root due to cell division in lateral meristems and differentiation of their daughter cells.
secondary oocyte (ō´-ō-sīt): a large haploid cell derived from the first meiotic division of the diploid primary oocyte.
secondary phloem: phloem produced from the cells that arise toward the outside of the vascular cambium.
secondary spermatocyte (sper-ma´-tō-sīt): a large haploid cell derived by meiosis I from the diploid primary spermatocyte.
secondary structure: a repeated, regular structure assumed by protein chains held together by hydrogen bonds; for example, a helix.
secondary succession: succession that occurs after an existing community is disturbed—for example, after a forest fire; much more rapid than primary succession.
secondary xylem: xylem produced from cells that arise at the inside of the vascular cambium.
second law of thermodynamics: the principle of physics that states that any change in an isolated system causes the quantity of concentrated, useful energy to decrease and the amount of randomness and disorder (entropy) to increase.
second messenger: an intracellular chemical, such as cyclic AMP, that is synthesized or released within a cell in response to the binding of a hormone or neurotransmitter (the first messenger) to receptors on the cell surface; brings about specific changes in the metabolism of the cell.
secretin: a hormone, produced by the small intestine, that stimulates the production and release of digestive secretions by the pancreas and liver.
seed: the reproductive structure of a seed plant; protected by a seed coat; contains an embryonic plant and a supply of food for it.
seed coat: the thin, tough, and waterproof outermost covering of a seed, formed from the integuments of the ovule.
segmentation (seg-men-tā´-shun): an animal body plan in which the body is divided into repeated, typically similar units.
segmentation movement: a contraction of the small intestine that results in the mixing of partially digested food and digestive enzymes. Segmentation movements also bring nutrients into contact with the absorptive intestinal wall.
segregation: see *law of segregation*.
self-fertilization: the union of sperm and egg from the same individual.
selfish gene: the concept that genes promote their own survival in individuals through innate self-sacrificing behavior that enhances the survival of others that carry the same genes; helps explain the evolution of altruism.
semen: the sperm-containing fluid produced by the male reproductive tract.
semiconservative replication: the process of replication of the DNA double helix; the two DNA strands separate, and each is used as a template for the synthesis of a complementary DNA strand. Consequently, each daughter double helix consists of one parental strand and one new strand.
semilunar valve: a paired valve between the ventricles of the heart and the pulmonary artery and aorta; prevents the backflow of blood into the ventricles when they relax.
seminal vesicle: in male mammals, a gland that produces a basic, fructose-containing fluid that forms part of the semen.
seminiferous tubule (sem-i-ni´-fer-us): in the vertebrate testis, a series of tubes in which sperm are produced.
senescence: in plants, a specific aging process, typically including deterioration and the dropping of leaves and flowers.
sensitive period: the particular stage in an animal's life during which it imprints.
sensory neuron: a nerve cell that responds to a stimulus from the internal or external environment.
sensory receptor: a cell (typically, a neuron) specialized to respond to particular internal or external environmental stimuli by producing an electrical potential.
sepal (sē´-pul): the set of modified leaves that surround and protect a flower bud, typically opening into green, leaflike structures when the flower blooms.
septum (pl., septa): a partition that separates the fungal hypha into individual cells; pores in septa allow the transfer of materials between cells.
serotonin (ser-uh-tō´-nin): in the central nervous system, a neurotransmitter that is involved in mood, sleep, and the inhibition of pain.
Sertoli cell: in the seminiferous tubule, a large cell that regulates spermatogenesis and nourishes the developing sperm.
sessile (ses´-ul): not free to move about, usually permanently attached to a surface.
severe combined immune deficiency (SCID): a disorder in which no immune cells, or very few, are formed; the immune system is incapable of responding properly to invading disease organisms, and the individual is very vulnerable to common infections.
sex chromosomes: the pair of chromosomes that usually determines the sex of an organism; for example, the X and Y chromosomes in mammals.
sex-linked: referring to a pattern of inheritance characteristic of genes located on one type of sex chromosome (for example, X) and not found on the other type (for example, Y); also called X-linked. In sex-linked inheritance, traits are controlled by genes carried on the X chromosome; females show the dominant trait unless they are homozygous recessive, whereas males express whichever allele is on their single X chromosome.
sexually transmitted disease (STD): a disease that is passed from person to person by sexual contact.
sexual recombination: during sexual reproduction, the formation of new combinations of alleles in offspring as a result of the inheritance of one homologous chromosome from each of two genetically distinct parents.
sexual reproduction: a form of reproduction in which genetic material from two parent organisms is combined in the offspring; normally, two haploid gametes fuse to form a diploid zygote.
sexual selection: a type of natural selection in which the choice of mates by one sex is the selective agent.
shoot system: all the parts of a vascular plant exclusive of the root; normally aboveground, consisting of stem, leaves, buds, and (in season) flowers and fruits; functions include photosynthesis, transport of materials, reproduction, and hormone synthesis.
short-day plant: a plant that will flower only if the length of daylight is shorter than some species-specific duration.
sickle-cell anemia: a recessive disease caused by a single amino acid substitution in the hemoglobin molecule. Sickle-cell hemoglobin molecules tend to cluster together, distorting the shape of red blood cell shape and causing them to break and clog capillaries.
sieve plate: in plants, a structure between two adjacent sieve-tube elements in phloem, where holes formed in the primary cell walls interconnect the cytoplasm of the elements; in echinoderms, the opening through which water enters the water-vascular system.
sieve tube: in phloem, a single strand of sieve-tube elements that transports sugar solutions.
sieve-tube element: one of the cells of a sieve tube, which form the phloem.
simple diffusion: the diffusion of water, dissolved gases, or lipid-soluble molecules through the phospholipid bilayer of a cellular membrane.
single covalent bond: a covalent bond in which two atoms share one pair of electrons.
sink: in plants, any structure that uses up sugars or converts sugars to starch and toward which phloem fluids will flow.
sinoatrial (SA) node (sī´-nō-āt´-rē-ul): a small mass of specialized muscle in the wall of the right atrium; generates electrical signals rhythmically and spontaneously and serves as the heart's pacemaker.
skeletal muscle: the type of muscle that is attached to and moves the skeleton and is under the direct, normally voluntary, control of the nervous system; also called *striated muscle*.
skeleton: a supporting structure for the body, on which muscles act to change the body configuration; may be external or internal.
skin: the tissue that makes up the outer surface of an animal body.

slime layer: a sticky polysaccharide or protein coating that some disease-causing bacteria secrete outside their cell wall; helps the cells aggregate and stick to smooth surfaces.
small intestine: the portion of the digestive tract, located between the stomach and large intestine, in which most digestion and absorption of nutrients occur.
smooth endoplasmic reticulum: endoplasmic reticulum without ribosomes.
smooth muscle: the type of muscle that surrounds hollow organs, such as the digestive tract, bladder, and blood vessels; normally not under voluntary control.
sodium–potassium pump: in nerve cell plasma membranes, a set of active-transport molecules that use the energy of ATP to pump sodium ions out of the cell and potassium ions in, maintaining the concentration gradients of these ions across the membrane.
solvent: a liquid capable of dissolving (uniformly dispersing) other substances in itself.
somatic nervous system: that portion of the peripheral nervous system that controls voluntary movement by activating skeletal muscles.
source: in plants, any structure that actively synthesizes sugar and away from which phloem fluid will be transported.
spawning: a method of external fertilization in which male and female parents shed gametes into the water, and sperm must swim through the water to reach the eggs.
speciation: the process of species formation, in which a single species splits into two or more species.
species (spē´-sēs): the basic unit of taxonomic classification, consisting of a population or series of populations of closely related and similar organisms. In sexually reproducing organisms, a species can be defined as a population or series of populations of organisms that interbreed freely with one another under natural conditions but that do not interbreed with members of other species.
specific heat: the amount of energy required to raise the temperature of 1 gram of a substance by 1 °C.
sperm: the haploid male gamete, normally small, motile, and containing little cytoplasm.
spermatid: a haploid cell derived from the secondary spermatocyte by meiosis II; differentiates into the mature sperm.
spermatogenesis: the process by which sperm cells form.
spermatogonium (pl., **spermatogonia**)**:** a diploid cell, lining the walls of the seminiferous tubules, that gives rise to a primary spermatocyte.
spermatophore: in a variation on internal fertilization in some animals, the males package their sperm in a container that can be inserted into the female reproductive tract.
spermicide: a sperm-killing chemical; used for contraceptive purposes.
spicule (spik´-ūl): a subunit of the endoskeleton of sponges that is made of protein, silica, or calcium carbonate.
spinal cord: the part of the central nervous system of vertebrates that extends from the base of the brain to the hips and is protected by the bones of the vertebral column; contains the cell bodies of motor neurons that form synapses with skeletal muscles, the circuitry for some simple reflex behaviors, and axons that communicate with the brain.
spindle microtubules: microtubules organized in a spindle shape that separate chromosomes during mitosis or meiosis.
spiracle (spi´-ruh-kul): an opening in the abdominal segment of insects through which air enters the tracheae.
spirillum (spi´-ril-um; pl., **spirilla):** a spiral-shaped bacterium.
spleen: an organ of the lymphatic system in which lymphocytes are produced and blood is filtered past lymphocytes and macrophages, which remove foreign particles and aged red blood cells.
spongy bone: porous, lightweight bone tissue in the interior of bones; the location of bone marrow.
spongy cell: an irregularly shaped mesophyll cell, containing chloroplasts, located just above the lower epidermis of a leaf.
spontaneous generation: the proposal that living organisms can arise from nonliving matter.
sporangium (spor-an´-jē-um; pl., **sporangia):** a structure in which spores are produced.
spore: a haploid reproductive cell capable of developing into an adult without fusing with another cell; in the alternation-of-generation life cycle of plants, a haploid cell that is produced by meiosis and then undergoes repeated mitotic divisions and differentiation of daughter cells to produce the gametophyte, a multicellular, haploid organism.
sporophyte (spor´-ō-fīt): the diploid form of a plant that produces haploid, asexual spores through meiosis.
sporozoan (spor-ō-zō´-un): a parasitic protist with a complex life cycle, typically involving more than one host; named for their ability to form infectious spores. A well-known sporozoan (genus *Plasmodium*) causes malaria.
stabilizing selection: a type of natural selection in which those organisms that display extreme phenotypes are selected against.
stamen (stā´-men): the male reproductive structure of a flower, consisting of a filament and an anther, in which pollen grains develop.
starch: a polysaccharide that is composed of branched or unbranched chains or glucose molecules; used by plants as a carbohydrate-storage molecule.
start codon: the first AUG codon in a messenger RNA molecule.
startle coloration: a form of mimicry in which a color pattern (in many cases resembling large eyes) can be displayed suddenly by a prey organism when approached by a predator.
stem: the portion of the plant body, normally located above ground, that bears leaves and reproductive structures such as flowers and fruit.
sterilization: a generally permanent method of contraception in which the pathways through which the sperm (vas deferens) or egg (oviducts) must travel are interrupted; the most common form of contraception.
steroid: see *steroid hormone*.
steroid hormone: a class of hormone whose chemical structure (four fused carbon rings with various functional groups) resembles cholesterol; steroids, which are lipids, are secreted by the ovaries and placenta, the testes, and the adrenal cortex.
stigma (stig´-muh): the pollen-capturing tip of a carpel.
stoma (stō´-muh; pl., **stomata):** an adjustable opening in the epidermis of a leaf, surrounded by a pair of guard cells, that regulates the diffusion of carbon dioxide and water into and out of the leaf.
stomach: the muscular sac between the esophagus and small intestine where food is stored and mechanically broken down and in which protein digestion begins.
stop codon: a codon in messenger RNA that stops protein synthesis and causes the completed protein chain to be released from the ribosome.
strand: a single polymer of nucleotides; DNA is composed of two strands.
striated muscle: see *skeletal muscle*.
stroke: an interruption of blood flow to part of the brain caused by the rupture of an artery or the blocking of an artery by a blood clot. Loss of blood supply leads to rapid death of the area of the brain affected.
stroma (strō´-muh): the semi-fluid material inside chloroplasts in which the grana are embedded.
style: a stalk connecting the stigma of a carpel with the ovary at its base.
subatomic particle: the particles of which atoms are made: electrons, protons, and neutrons.
subclimax: a community in which succession is stopped before the climax community is reached and is maintained by regular disturbances—for example, tallgrass prairie maintained by periodic fires.
substrate: the atoms or molecules that are the reactants for an enzyme-catalyzed chemical reaction.
subunit: a small organic molecule, several of which may be bonded together to form a larger molecule. See also *monomer*.
succession (suk-seh´-shun): a structural change in a community and its nonliving environment over time. Community changes alter the ecosystem in ways that favor competitors, and species replace one another in a somewhat predictable manner until a stable, self-sustaining climax community is reached.
sucrose: a disaccharide composed of glucose and fructose.
sugar: a simple carbohydrate molecule, either a monosaccharide or a disaccharide.
sugar-phosphate backbone: a major feature of DNA structure, formed by attaching the sugar of one nucleotide to the phosphate from the adjacent nucleotide in a DNA strand.
suppressor T cell: a type of T cell that depresses the response of other immune cells to foreign antigens.

surface tension: the property of a liquid to resist penetration by objects at its interface with the air, due to cohesion between molecules of the liquid.
survivorship curve: a curve resulting when the number of individuals of each age in a population is graphed against their age, usually expressed as a percentage of their maximum life span.
symbiosis (sim´-bī-ō´sis): a close interaction between organisms of different species over an extended period. Either or both species may benefit from the association, or (in the case of parasitism) one of the participants is harmed. Symbiosis includes parasitism, mutualism, and commensalism.
symbiotic: referring to an ecological relationship based on symbiosis.
sympathetic division: the division of the autonomic nervous system that produces largely involuntary responses that prepare the body for stressful or highly energetic situations.
sympatric speciation (sim-pat´-rik): speciation that occurs in populations that are not physically divided; normally due to ecological isolation or chromosomal aberrations (such as polyploidy).
synapse (sin´-aps): the site of communication between nerve cells. At a synapse, one cell (presynaptic) normally releases a chemical (the neurotransmitter) that changes the electrical potential of the second (postsynaptic) cell.
synaptic terminal: a swelling at the branched ending of an axon; where the axon forms a synapse.
syphilis (si´-ful-is): a sexually transmitted bacterial infection of the reproductive organs; if untreated, can damage the nervous and circulatory systems.
systematics: the branch of biology concerned with reconstructing phylogenies and with naming and classifying species.
taiga (tī´-guh): a biome with long, cold winters and only a few months of warm weather; dominated by evergreen coniferous trees; also called *northern coniferous forest.*
taproot system: a root system, commonly found in dicots, that consists of a long, thick main root and many smaller lateral roots, all of which grow from the primary root.
target cell: a cell on which a particular hormone exerts its effect.
taste: a chemical sense for substances dissolved in water or saliva; in mammals, perceptions of sweet, sour, bitter, or salt produced by the stimulation of receptors on the tongue.
taste bud: a cluster of taste receptor cells and supporting cells that is located in a small pit beneath the surface of the tongue and that communicates with the mouth through a small pore. The human tongue has about 10,000 taste buds.
taxis (taks´-is; pl., **taxes):** an innate behavior that is a directed movement of an organism toward or away from a stimulus such as heat, light, or gravity.
taxonomy (tax-on´-uh-mē): the science by which organisms are classified into hierarchically arranged categories that reflect their evolutionary relationships.
Tay-Sachs disease: a recessive disease caused by a deficiency in enzymes that regulate lipid breakdown in the brain.
T cell: a type of lymphocyte that recognizes and destroys specific foreign cells or substances or that regulates other cells of the immune system.
T-cell receptor: a protein receptor, located on the surface of a T cell, that binds a specific antigen and triggers the immune response of the T cell.
tectorial membrane (tek-tor´-ē-ul): one of the membranes of the cochlea in which the hairs of the hair cells are embedded. In sound reception, movement of the basilar membrane relative to the tectorial membrane bends the cilia.
telophase (tēl´-ō-fāz): in mitosis, the final stage, in which a nuclear envelope re-forms around each new daughter nucleus, the spindle fibers disappear, and the chromosomes relax from their condensed form; in meiosis I, the stage during which the spindle fibers disappear and the chromosomes normally relax from their condensed form; in meiosis II, the stage during which chromosomes relax into their extended state, nuclear envelopes re-form, and cytokinesis occurs.
temperate deciduous forest: a biome in which winters are cold and summer rainfall is sufficient to allow enough moisture for trees to grow and shade out grasses.
temperate rain forest: a biome in which there is no shortage of liquid water year-round and that is dominated by conifers.
template strand: the strand of the DNA double helix from which RNA is transcribed.
temporal isolation: the inability of organisms to mate if they have significantly different breeding seasons.
tendon: a tough connective tissue band connecting a muscle to a bone.
tendril: a slender outgrowth of a stem that coils about external objects and supports the stem; normally a modified leaf or branch.
tentacle (ten´-te-kul): an elongate, extensible projection of the body of cnidarians and cephalopod mollusks that may be used for grasping, stinging, and immobilizing prey, and for locomotion.
terminal bud: meristem tissue and surrounding leaf primordia that are located at the tip of the plant shoot.
territoriality: the defense of an area in which important resources are located.
tertiary consumer (ter´-shē-er-ē): a carnivore that feeds on other carnivores (secondary consumers).
tertiary structure (ter´-shē-er-ē): the complex three-dimensional structure of a single peptide chain; held in place by disulfide bonds between cysteines.
test cross: a breeding experiment in which an individual showing the dominant phenotype is mated with an individual that is homozygous recessive for the same gene. The ratio of offspring with dominant versus recessive phenotypes can be used to determine the genotype of the phenotypically dominant individual.
testis (pl., **testes):** the gonad of male mammals.
testosterone: in vertebrates, a hormone produced by the interstitial cells of the testis; stimulates spermatogenesis and the development of male secondary sex characteristics.
thalamus: the part of the forebrain that relays sensory information to many parts of the brain.
theory: in science, an explanation for natural events that is based on a large number of observations and is in accord with scientific principles, especially causality.
thermoacidophile (ther-mō-a-sid´-eh-fīl): an archaean that thrives in hot, acidic environments.
thermoreceptor: a sensory receptor that responds to changes in temperature.
thick filament: in the sarcomere, a bundle of myosin that interacts with thin filaments, producing muscle contraction.
thin filament: in the sarcomere, a protein strand that interacts with thick filaments, producing muscle contraction; composed primarily of actin, with accessory proteins.
thorax: the segment between the head and abdomen in animals with segmentation; the segment to which structures used in locomotion are attached.
thorn: a hard, pointed outgrowth of a stem; normally a modified branch.
threshold: the electrical potential (less negative than the resting potential) at which an action potential is triggered.
thrombin: an enzyme produced in the blood as a result of injury to a blood vessel; catalyzes the production of fibrin, a protein that assists in blood clot formation.
thylakoid (thī´-luh-koid): a disk-shaped, membranous sac found in chloroplasts, the membranes of which contain the photosystems and ATP-synthesizing enzymes used in the light-dependent reactions of photosynthesis.
thymine: a nitrogenous base found only in DNA; abbreviated as *T.*
thymosin: a hormone, secreted by the thymus, that stimulates the maturation of cells of the immune system.
thymus (thī´-mus): an organ of the lymphatic system that is located in the upper chest in front of the heart and that secretes thymosin, which stimulates lymphocyte maturation; begins to degenerate at puberty and has little function in the adult.
thyroid gland: an endocrine gland, located in front of the larynx in the neck, that secretes the hormones thyroxine (affecting metabolic rate) and calcitonin (regulating calcium ion concentration in the blood).
thyroid-stimulating hormone (TSH): a hormone, released by the anterior pituitary, that stimulates the thyroid gland to release hormones.
thyroxine (thī-rox´-in): a hormone, secreted by the thyroid gland, that stimulates and regulates metabolism.
tight junction: a type of cell-to-cell junction in animals that prevents the movement of materials through the spaces between cells.

tissue: a group of (normally similar) cells that together carry out a specific function; for example, muscle; may include extracellular material produced by its cells.
tonsil: a patch of lymphatic tissue consisting of connective tissue that contains many lymphocytes; located in the pharynx and throat.
trachea (trā´-kē-uh): in birds and mammals, a rigid but flexible tube, supported by rings of cartilage, that conducts air between the larynx and the bronchi; in insects, an elaborately branching tube that carries air from openings called *spiracles* near each body cell.
tracheid (trā´-kē-id): an elongated xylem cell with tapering ends that contains pits in the cell wall; forms tubes that transport water.
tracheophyte (trā´-kē-ō-fīt): a plant that has conducting vessels; a vascular plant.
transcription: the synthesis of an RNA molecule from a DNA template.
transducer: a device that converts signals from one form to another. Sensory receptors are transducers that convert environmental stimuli, such as heat, light, or vibration, into electrical signals (such as action potentials) recognized by the nervous system.
transfer RNA (tRNA): a type of RNA that binds to a specific amino acid by means of a set of three bases (the anticodon) on the tRNA that are complementary to the mRNA codon for that amino acid; carries its amino acid to a ribosome during protein synthesis, recognizes a codon of mRNA, and positions its amino acid for incorporation into the growing protein chain.
transformation: a method of acquiring new genes, whereby DNA from one bacterium (normally released after the death of the bacterium) becomes incorporated into the DNA of another, living, bacterium.
transgenic: referring to an animal or a plant that expresses DNA derived from another species.
translation: the process whereby the sequence of bases of messenger RNA is converted into the sequence of amino acids of a protein.
transpiration (trans´-per-ā-shun): the evaporation of water through the stomata of a leaf.
transport protein: a protein that regulates the movement of water-soluble molecules through the plasma membrane.
trial-and-error learning: a process by which adaptive responses are learned through rewards or punishments provided by the environment.
trichomoniasis (trik-ō-mō-nī´-uh-sis): a sexually transmitted disease, caused by the protist *Trichomonas*, that causes inflammation of the mucous membranes than line the urinary tract and genitals.
tricuspid valve: the valve between the right ventricle and the right atrium of the heart.
triglyceride (trī-glis´-er-īd): a lipid composed of three fatty-acid molecules bonded to a single glycerol molecule.
triple covalent bond: a covalent bond that occurs when two atoms share three pairs of electrons.
trisomy 21: see *Down syndrome*.
trisomy X: a condition of females who have three X chromosomes instead of the normal two; most such women are phenotypically normal and are fertile.
trophic level: literally, "feeding level"; the categories of organisms in a community, and the position of an organism in a food chain, defined by the organism's source of energy; includes producers, primary consumers, secondary consumers, and so on.
tropical deciduous forest: a biome with pronounced wet and dry seasons and plants that must shed their leaves during the dry season to minimize water loss.
tropical rain forest: a biome with evenly warm, evenly moist conditions; dominated by broadleaf evergreen trees; the most diverse biome.
true-breeding: pertaining to an individual all of whose offspring produced through self-fertilization are identical to the parental type. True-breeding individuals are homozygous for a given trait.
T tubule: a deep infolding of the muscle plasma membrane; conducts the action potential inside a cell.
tubal ligation: a surgical procedure in which a woman's oviducts are cut so that the egg cannot reach the uterus, making her infertile.
tube cell: the outermost cell of a pollen grain; digests a tube through the tissues of the carpel, ultimately penetrating into the female gametophyte.
tube foot: a cylindrical extension of the water-vascular system of echinoderms; used for locomotion, grasping food, and respiration.
tubular reabsorption: the process by which cells of the tubule of the nephron remove water and nutrients from the filtrate within the tubule and return those substances to the blood.
tubular secretion: the process by which cells of the tubule of the nephron remove additional wastes from the blood, actively secreting those wastes into the tubule.
tubule (toob´-ūl): the tubular portion of the nephron; includes a proximal portion, the loop of Henle, and a distal portion. Urine is formed from the blood filtrate as it passes through the tubule.
tumor: a mass that forms in otherwise normal tissue; caused by the uncontrolled growth of cells.
tundra: a biome with severe weather conditions (extreme cold and wind and little rainfall) that cannot support trees.
turgor pressure: pressure developed within a cell (especially the central vacuole of plant cells) as a result of osmotic water entry.
Turner syndrome: a set of characteristics typical of a woman with only one X chromosome: sterile, with a tendency to be very short and to lack normal female secondary sexual characteristics.
tympanic membrane (tim-pan´-ik): the eardrum; a membrane, stretched across the opening of the ear, that transmits vibration of sound waves to bones of the middle ear.
unicellular: single-celled; most members of the domains Bacteria and Archaea and the kingdom Protista are unicellular.
uniform distribution: the distribution characteristic of a population with a relatively regular spacing of individuals, commonly as a result of territorial behavior.
uniformitarianism: the hypothesis that Earth developed gradually through natural processes, similar to those at work today, that occur over long periods of time.
unsaturated: referring to a fatty acid with fewer than the maximum number of hydrogen atoms bonded to its carbon backbone; a fatty acid with one or more double bonds in its carbon backbone.
upwelling: an upward flow that brings cold, nutrient-laden water from the ocean depths to the surface; occurs along western coastlines.
uracil: a nitrogenous base found in RNA; abbreviated as *U*.
urea (ū-rē´-uh): a water-soluble, nitrogen-containing waste product of amino acid breakdown; one of the principal components of mammalian urine.
ureter (ū´-re-ter): a tube that conducts urine from each kidney to the bladder.
urethra (ū-rē´-thruh): the tube leading from the urinary bladder to the outside of the body; in males, the urethra also receives sperm from the vas deferens and conducts both sperm and urine (at different times) to the tip of the penis.
uric acid (ūr´-ik): a nitrogen-containing waste product of amino acid breakdown; a relatively insoluble white crystal excreted by birds, reptiles, and insects.
urine: the fluid produced and excreted by the urinary system of vertebrates; contains water and dissolved wastes, such as urea.
uterine tube: also called the oviduct, the tube leading out of the ovary to the uterus, into which the secondary oocyte (egg cell) is released.
uterus: in female mammals, the part of the reproductive tract that houses the embryo during pregnancy.
vaccination: an injection into the body that contains antigens characteristic of a particular disease organism and that stimulates an immune response.
vacuole (vak´-ū-ōl): a vesicle that is typically large and consists of a single membrane enclosing a fluid-filled space.
vagina: the passageway leading from the outside of a female mammal's body to the cervix of the uterus.
variable: a condition, particularly in a scientific experiment, that is subject to change.
variable region: the part of an antibody molecule that differs among antibodies; the ends of the variable regions of the light and heavy chains form the specific binding site for antigens.
vascular: describing tissues that contain vessels for transporting liquids.
vascular bundle (vas´-kū-lar): a strand of xylem and phloem in leaves and stems; in leaves, commonly called a *vein*.

vascular cambium: a lateral meristem that is located between the xylem and phloem of a woody root or stem and that gives rise to secondary xylem and phloem.
vascular cylinder: the centrally located conducting tissue of a young root, consisting of primary xylem and phloem.
vascular tissue system: a plant tissue system consisting of xylem (which transports water and minerals from root to shoot) and phloem (which transports water and sugars throughout the plant).
vas deferens (vaz de´-fer-enz): the tube connecting the epididymis of the testis with the urethra.
vasectomy: a surgical procedure in which a man's vas deferens are cut, preventing sperm from reaching the penis during ejaculation, thereby making him infertile.
vector: a carrier that introduces foreign genes into cells.
vein: in vertebrates, a large-diameter, thin-walled vessel that carries blood from venules back to the heart; in vascular plants, a vascular bundle, or a strand of xylem and phloem in leaves.
ventral (ven´-trul): the lower side or underside of an animal whose head is oriented forward.
ventricle (ven´-tre-kul): the lower muscular chamber on each side of the heart, which pumps blood out through the arteries. The right ventricle sends blood to the lungs; the left ventricle pumps blood to the rest of the body.
venule (ven´-ūl): a narrow vessel with thin walls that carries blood from capillaries to veins.
vertebral column (ver-tē´-brul): a column of serially arranged skeletal units (the vertebrae) that enclose the nerve cord in vertebrates; the backbone.
vertebrate: an animal that possesses a vertebral column.
vesicle (ves´-i-kul): a small, membrane-bound sac within the cytoplasm.
vessel: a tube of xylem composed of vertically stacked vessel elements with heavily perforated or missing end walls, leaving a continuous, uninterrupted hollow cylinder.
vessel element: one of the cells of a xylem vessel; elongated, dead at maturity, with thick, lignified lateral cell walls for support but with end walls that are either heavily perforated or missing.
vestigial structure (ves-tij´-ē-ul): a structure that serves no apparent purpose but is homologous to functional structures in related organisms and provides evidence of evolution.
villus (vi´-lus; pl., **villi):** a fingerlike projection of the wall of the small intestine that increases the absorptive surface area.
viroid (vī´-roid): a particle of RNA that is capable of infecting a cell and of directing the production of more viroids; responsible for certain plant diseases.
virus (vī´-rus): a noncellular parasitic parti-cle that consists of a protein coat surrounding a strand of genetic material; multiplies only within a cell of a living organism (the host).
vitamin: one of a group of diverse chemicals that must be present in trace amounts in the diet to maintain health; used by the body in conjunction with enzymes in a variety of metabolic reactions.
vitreous humor (vit´-rē-us): a clear, jellylike substance that fills the large chamber of the eye between the lens and the retina.
vocal cord: one of a pair of bands of elastic tissue that extend across the opening of the larynx and produce sound when air is forced between them. Muscles alter the tension on the vocal cords and control the size and shape of the opening, which in turn determines whether sound is produced and what its pitch will be.
waggle dance: a symbolic form of communication used by honeybee foragers to communicate the location of a food source to their hivemates.
warning coloration: bright coloration that warns predators that the potential prey is distasteful or even poisonous.
water mold: a funguslike protist that includes some pathogens, such as the downy mildew, which attacks grapes.
water-vascular system: a system in echinoderms that consists of a series of canals through which seawater is conducted and is used to inflate tube feet for locomotion, grasping food, and respiration.
wax: a lipid composed of fatty acids covalently bonded to long-chain alcohols.
weather: short-term fluctuations in temperature, humidity, cloud cover, wind, and precipitation in a region over periods of hours to days.
Werner syndrome: a rare condition in which a defective gene causes premature aging; caused by a mutation in the gene that codes for DNA replication/repair enzymes.
white matter: the portion of the brain and spinal cord that consists largely of myelin-covered axons and that give these areas a white appearance.
withdrawal: the removal of the penis from the vagina just before ejaculation in an attempt to avoid pregnancy; an ineffective contraceptive method.
working memory: the first phase of learning; short-term memory that is electrical or biochemical in nature.
xylem (zī-lum): a conducting tissue of vascular plants that transports water and minerals from root to shoot.
yolk: protein-rich or lipid-rich substances contained in eggs that provide food for the developing embryo.
yolk sac: one of the embryonic membranes of reptilian, bird, and mammalian embryos; in birds and reptiles, a membrane surrounding the yolk in the egg; in mammals, forms part of the umbilical cord and the digestive tract but is empty.
Z line: a fibrous protein structure to which the thin filaments of skeletal muscle are attached; forms the boundary of a sarcomere.
zona pellucida (pel-oo´-si-duh): a clear, noncellular layer between the corona radiata and the egg.
zooflagellate (zō-ō-fla´-jel-et): a nonphotosynthetic protist that moves by using flagella.
zooplankton: nonphotosynthetic protists that are abundant in marine and freshwater environments.
zoospore (zō´-ō-spor): a nonsexual reproductive cell that swims by using flagella; formed by members of the protistan division Oomycota.
zygospore (zī´-gō-spor): a fungal spore, produced by the division Zygomycota, that is surrounded by a thick, resistant wall and forms from a diploid zygote.
zygote (zī´-gōt): in sexual reproduction, a diploid cell (the fertilized egg) formed by the fusion of two haploid gametes.
zygote fungus: a fungus of the division Zygomycota, which includes the species that cause fruit rot and bread mold.

Photo Credits

Chapter 1

Opener NASA Headquarters **1-1a** Andrew Syred/Science Photo Library/Photo Researchers, Inc. **1-1b** Craig Tuttle/The Stock Market **1-1c** Kim Taylor/Bruce Coleman, Inc. **1-3** Dr. Jeremy Burgess/Science Photo Library/Photo Researchers, Inc. **1-4** William R. Sallaz/Duomo Photography Incorporated **1-5** Kim Taylor/Bruce Coleman, Inc. **1-6** Johnny Johnson/DRK Photo **1-7** Lawrence Livermore National Laboratory/Photo Researchers, Inc. **1-8a** CNRI/Science Photo Library/Photo Researchers, Inc. **1-8b** Dr. M. Rohde, Gesellschaft fur Biotechnologische Forschung/Science Photo Library/Photo Researchers, Inc. **1-8c** Eric V. Grave/Photo Researchers, Inc. **1-8d** Richard L. Carlton/Photo Researchers, Inc. **1-8e** Patti Murray/Animals Animals/Earth Scenes **1-8f** Jeff Rotman/Stone **1-9** John Durham/Science Photo Library/Photo Researchers, Inc. **1-10** Doug Perrine/DRK Photo **1-11** Francois Gohier/Photo Researchers, Inc. **1-13** Teresa and Gerald Audesirk **E1-2** Luiz C. Marigo/Peter Arnold, Inc. **E1-2 inset** Gunter Ziesler/Peter Arnold, Inc.

Unit One

Opener Manfred Kage/Peter Arnold, Inc.

Chapter 2

Opener Chocolate Manufacturers Association **Opener inset** J.-C. Carton/Bruce Coleman, Inc. **2-8a** Stephen Dalton/Photo Researchers, Inc. **2-8b** Teresa and Gerald Audesirk **E2-1c** National Institutes of Health/Science Source/Photo Researchers, Inc.

Chapter 3

Opener Spencer Grant/PhotoEdit **3-2a** Dr. Jeremy Burgess/Science Photo Library/Photo Researchers, Inc. **3-3 (left)** Larry Ulrich/DRK Photo **3-3 (middle)** Dr. Jeremy Burgess/Science Photo Library/Photo Researchers, Inc. **3-3 (right)** Biophoto Associates/Photo Researchers, Inc. **3-4** Richard Kolar/Animals Animals/Earth Scenes **3-5a** Jean-Michel Labat/Auscape International Pty. Ltd. **3-5b** Donald Specker/Animals Animals/Earth Scenes **3-8a** Robert Pearcy/Animals Animals/Earth Scenes **3-8b** Jeff Foott/DRK Photo **3-8c** Nuridsany et Perennou/Photo Researchers, Inc.

Chapter 4

Opener Tom McHugh/Photo Researchers, Inc. **4-5** Dr. Joseph Kurantsin-Mills, The George Washington University Medical Center **4-8** M. M. Perry and A. B. Gilbert, *Journal of Cell Science,* 39:257–272 (1979) **4-9** L. A. Hufnagel, "Ultrastructural Aspects of Chemoreception in Ciliated Protists (Ciliophora)," *Journal of Electron Microscopy Technique,* 1991. Photomicrograph by Jurgen Bohmer and Linda Hufnagel, University of Rhode Island **4-12** Biophoto Associates/Science Source/Photo Researchers, Inc. **4-13** Stephen J. Krasemann/DRK Photo

Chapter 5

Opener NASA/Peter Arnold, Inc. **Opener inset** Oliver Meckes/Photo Researchers, Inc. **5-5b** E. Guth, T. Hashimoto, and S. F. Conti **5-6** Phototake/Carolina Biological Supply Company **5-7** Omikron/Science Source/Photo Researchers, Inc. **5-8** Barry F. King/Biological Photo Service **5-9** D. W. Fawcett/Photo Researchers, Inc. **5-11** Thomas Eisner **5-12** Nigel Cattlin/Holt Studios International/Photo Researchers, Inc. **5-13** Keith R. Porter, University of Pennsylvania **5-14** W. P. Wergin/Biological Photo Service **5-15** Biophoto Associates/Photo Researchers, Inc. **5-17 (left)** William L. Dentler/Biological Photo Service **5-17 (right)** E. de Harven/Photo Researchers, Inc. **5-18a** Ellen R. Dirksen/Visuals Unlimited **5-18b** Yorgos Nikas/Stone **E5-1a (left)** Biophoto Associates/Photo Researchers, Inc. **E5-1a (right)** Cecil Fox/Science Source/Photo Researchers, Inc. **E5-1b** Jean-Claude Revy/Phototake NYC **E5-2a** M. I. Walker/Photo Researchers, Inc. **E5-2b** David M. Phillips/Visuals Unlimited **E5-2c** Manfred Kage/Peter Arnold, Inc. **E5-2d** Kiyoji Tanaka, Osaka University, Osaka, Japan

Chapter 6

Opener Sylvain Grandadam/Stone **6-1** Photograph by Dr. Harold E. Edgerton. © Harold & Esther Edgerton Foundation, 1999, courtesy of Palm Press, Inc.

Chapter 7

Opener Joe Tucciarone **7-2a** Ken W. Davis/Tom Stack & Associates **7-5** Colin Milkins/Oxford Scientific Films/Animals Animals/Earth Scenes

Chapter 8

Opener Wayne Lankinen/DRK Photo **8-3a** Focus on Sports Inc. **8-3b** Teresa Audesirk

Unit Two

Opener USDA–Agricultural Research Service Information Staff

Chapter 9

Opener Christopher Brown/Stock Boston **9-2a** Rosalind Franklin/Photo Researchers, Inc. **9-2b** Cold Spring Harbor Laboratory Archives/Peter Arnold, Inc. **E9-1** A. Barrington Brown/Photo Researchers, Inc.

Chapter 10

Opener Ernest Braun/Stone **10-5** Oscar L. Miller, Jr., University of Virginia **10-9** Frederic Jacana/Photo Researchers, Inc. **E10-2** Howard W. Jones, Jr., M.D., Eastern Virginia Medical School **E10-3** Lippincott Williams & Wilkins (*Medicine,* 1966; 45 (3): 177–221)

Chapter 11

Opener Frank Schneidermeyer/Oxford Scientific Films Ltd. **11-1** CNRI/Science Photo Library/Photo Researchers, Inc. **11-2a** CC Studio/Science Photo Library/Photo Researchers, Inc. **11-2b** Pascal Goetgheluck/Science Photo Library/Photo Researchers, Inc. **11-2c** Neil Harding/Stone **11-2d** G. W. Willis/Biological Photo Service/Stone **11-3a** Michael Abbey/Photo Researchers, Inc. **11-3b** John Durham/Science Photo Library/Photo Researchers, Inc. **11-3c** Carolina Biological Supply Company/Phototake NYC **11-4** Teresa and Gerald Audesirk **11-7** Biophoto Associates/Photo Researchers, Inc. **11-8** CNRI/Science Photo Library/Photo Researchers, Inc. **11-10a–f** Dr. Andrew S. Bajer, University of Oregon **11-12b** T. E. Schroeder/Biological Photo Service **E11-1** Photograph courtesy of The Roslin Institute.

Chapter 12

Opener Tim Davis/Stone **Opener insets** Emmanuel Mignot, M.D., Stanford University School of Medicine **12-1** Archiv/Photo Researchers, Inc. **12-3a** D. Cavagnaro/DRK Photo **12-8** Biophoto Associates/Photo Researchers, Inc. **12-13** Jane Burton/Bruce Coleman, Inc. **12-15a** P. Marazzi/Science Photo Library/Photo Researchers, Inc. **12-15b** K. H. Switak/Photo Researchers, Inc. **12-15c** Gary Retherford/Photo Researchers, Inc. **12-16a** Dennis Kunkel/Phototake NYC **12-16b** Francis Leroy/Biocosmos/Science Photo Library/Photo Researchers, Inc. **12-17a** Hart-Davis/Science Photo Library/Photo Researchers, Inc. **12-18** Gunn and Stewart/Mary Evans/Photo Researchers, Inc. **12-19a** CNRI/Science Photo Library/Photo Researchers, Inc. **12-19b** Lawrence Migdale/Pix **E12-1** Nick Kelsh/Peter Arnold, Inc. **E12-2** National Remote Sensing Centre Ltd./Science Photo Library/Photo Researchers, Inc.

Chapter 13

Opener Penny Tweedie/Stone **13-1a** Stanley N. Cohen/Science Photo Library/Photo Researchers, Inc. **13-4a,b** F. Rudolph Turner, Indiana University **13-7** Keith V. Wood/Science VU/Visuals Unlimited **13-8a** Norm Thomas/Photo Researchers, Inc. **13-8b** Monsanto **13-9a** Derek Bromhall/Oxford Scientific Films Ltd. **13-9b** GenPharm International/Peter Arnold, Inc. **13-9c** Garth Fletcher **13-10** Photo by Ned S. Gilmore/Steven Kurth **13-11a** AP/Wide World Photos **13-12** Bruce Dale/National Geographic Society **E13-1a** Gary Braasch Photography **E13-1b** Joseph R. Newhouse, California University of Pennsylvania

Unit Three

Opener Francois Gohier/Photo Researchers, Inc.

Chapter 14

Opener O. Louis Mazzatenta/NGS Image Collection **14-1** Jim Steinberg/Photo Researchers, Inc. **14-3a** John Cancalosi/DRK Photo **14-3b** David M. Dennis/Tom Stack & Associates **14-3c** Chip Clark **14-7a,b** Stephen Dalton/Photo Researchers, Inc. **14-7c** Douglas T. Cheeseman, Jr./Peter Arnold, Inc. **14-7d** Gerard Lacz/Peter Arnold, Inc. **14-10a–c** Photo Lennart Nilsson/Albert Bonniers Forlag AB **14-11a** Stephen J. Krasemann/DRK Photo **14-11b** Timothy O'Keefe/Tom Stack & Associates **E14-1** Corbis **E14-2a** Frans Lanting/Photo Researchers, Inc. **E14-2b** Christian Grzimek/Okapia/Photo Researchers, Inc.

Chapter 15

Opener David Scharf/Peter Arnold, Inc. **15-3b** Luiz C. Marigo/Peter Arnold, Inc. **15-3c** Y. R. Tymstra/Valan Photos **15-4** Gregory Dimijian/Photo Researchers, Inc. **15-5** Ed Degginger/Color-Pic, Inc. **15-6** W. Perry Conway/Tom Stack & Associates **15-7** D. Cavagnaro/DRK Photo **15-9a** M. P. L. Fogden/Bruce Coleman, Inc. **15-9b** Tim Davis/Photo Researchers, Inc. **15-11** Glen E. Woolfenden/Archbold Biological Station **15-12** Thomas A. Wiewandt **E15-1** Frans Lanting/Minden Pictures

Chapter 16

Opener Alan Rabinowitz **16-1a** Wayne Lankinen/Valan Photos **16-1b** Edgar T. Jones/Bruce Coleman, Inc. **16-3a,b** Pat and Tom Leeson/Photo Researchers, Inc. **16-4** Guy L. Bush, Michigan State University **16-7a,c** Mark Smith/Photo Researchers, Inc. **16-7b** Thomas Kitchin/Tom Stack & Associates **16-8** Tim Laman/NGS Image Collection **16-9** Joy Spurr/Bruce Coleman, Inc. **16-10** Tom McHugh/Steinhart Aquarium/Photo Researchers, Inc. **16-11** The Kern Company

Chapter 17

Opener TSADO/NASA/Tom Stack & Associates **17-3** Sidney Fox/Science VU/Visuals Unlimited **17-5** Michael Abbey/Visuals Unlimited **17-6a** Milwaukee Public Museum, Photograph Collection **17-6b** James L. Amos/Photo Researchers, Inc. **17-6c** Phototake/Carolina Biological Supply Company **17-6d** Douglas Faulkner/Photo Researchers, Inc. **17-7** Illustration by Ludek Pesek/Science Photo Library/Photo Researchers, Inc. **17-8** Terry Whittaker/Photo Researchers, Inc. **17-9** Illustration by Chris Butler/Science Photo Library/Photo Researchers, Inc. **17-12a** Tom McHugh/Chicago Zoological Park/Photo Researchers, Inc. **17-12b** Frans Lanting/Minden Pictures **17-12c** Nancy Adams/Tom Stack & Associates **17-13** Tim D. White/Brill Atlanta **17-15** Neanderthal Museum, Mettmann, Germany **17-16** Jerome Chatin/Liaison Agency, Inc.

Chapter 18

Opener Tom Brakefield/DRK Photo **18-1a** Wayne Lankinen/Bruce Coleman, Inc. **18-1b** M. C. Chamberlain/DRK Photo **18-1c** Maslowski/Photo Researchers, Inc. **18-2a** C. Steven Murphree/Biological Photo Service **18-2b** Dr. Greg Rouse, Department of Invertebrate Zoology, National Museum of Natural History, Smithsonian Institution **18-2c** Dr. Jeremy Burgess/Science Photo Library/Photo Researchers, Inc. **18-4a** Hans Gelderblom/Stone **18-4b** Courtesy W. Jack Jones, reprinted by permission of Springer-Verlag **18-7** Zig Koch/Kino Fotoarquivo/©1992. Reprinted with permission of *Discover* magazine.

Chapter 19

Opener U.S. Army Photo **19-2b** Centers for Disease Control and Prevention/Photo Researchers, Inc. **19-4** Lee Simon/Stammers/Science Photo Library/Photo Researchers, Inc. **19-5** Reproduced by permission from K.-M. Pan et al., *Proceedings of the National Academy of Sciences* 90:1962–1966, Fig. 4c (1993).© National Academy of Sciences, U.S.A. **19-6a** David M. Phillips/Visuals Unlimited **19-6b** Karl O. Stetter, University of Regensburg, Regensburg, Germany **19-6c** CNRI/Science Photo Library/Photo Researchers, Inc. **19-7** Photo Lennart Nilsson/Albert Bonniers Forlag **19-8a** Karl O. Stetter, University of Regensburg, Regensburg, Germany **19-9** A. B. Dowsett/Science Photo Library/Photo Researchers, Inc. **19-10** CNRI/Science Photo Library/Photo Researchers, Inc. **19-11** Dennis Kunkel/Phototake NYC **19-12** Yellowstone National Park **19-13** Biophoto Associates/Photo Researchers, Inc. **19-14** C. P. Vance/Visuals Unlimited **19-15a** Phototake/Carolina Biological Supply Company **19-15b** Eric Grave/Science Source/Photo Researchers, Inc. **19-16** William Merrill, Penn State University **19-17a** P. W. Grace/Science Source/Photo Researchers, Inc. **19-17b** Cabisco/Visuals Unlimited **19-18** Cabisco/Visuals Unlimited **19-19** David M. Phillips/Visuals Unlimited **19-20** Kevin Schafer/Peter Arnold, Inc. **19-21** Stone **19-23** Jeffrey L. Rotman/Jeffrey L. Rotman Photography **19-24a** D. P. Wilson/Eric and David Hosking/Photo Researchers, Inc. **19-24b** Lawrence E. Naylor/Photo Researchers, Inc. **19-25** Harry Rogers/Photo Researchers, Inc. **19-26** Oliver Meckes/Photo Researchers, Inc. **19-27** P. M. Motta and F. M. Magliocca/Science Photo Library/Photo Researchers, Inc. **19-28** M. I. Walker/Photo Researchers, Inc. **19-29** M. I. Walker/Science Source/Photo Researchers, Inc. **19-30a** Ed Degginger/Color-Pic, Inc. **19-30b** Manfred Kage/Peter Arnold, Inc. **19-32** Oliver Meckes & Nicole Ottawa/Eye of Science/Photo Researchers, Inc.

Chapter 20

Opener Hans Reinhard/Bruce Coleman, Inc. **20-1a** Robert and Linda Mitchell/Robert & Linda Mitchell Photography **20-1b** Elmer Koneman/Visuals Unlimited **20-2** Jeff Lepore/Photo Researchers, Inc. **20-3** Thomas J. Volk, University of Wisconsin–La Crosse **20-4b** Breck P. Kent **20-4c** Carolina Biological Supply/Phototake NYC **20-5a** W. K. Fletcher/Photo Researchers, Inc. **20-5b** David Dvorak, Jr. **20-7a** Scott Camazine/Photo Researchers, Inc. **20-7b** David M. Dennis/Tom Stack & Associates **20-7c** Hans Reinhard/Bruce Coleman, Inc. **20-8** Robert & Linda Mitchell Photography **20-10a** Jeff Foott Productions **20-10b** Robert & Linda Mitchell Photography **20-11** Stanley L. Flegler/Visuals Unlimited **20-12** David M. Dennis/Tom Stack & Associates **20-13** Michael Fogden/DRK Photo **20-14** David M. Phillips/Visuals Unlimited **20-15** Teresa and Gerald Audesirk **20-16** M. Viard/Jacana/Photo Researchers, Inc. **20-17** G. L. Barron/Biological Photo Service **20-18** Cabisco/Visuals Unlimited

Chapter 21

Opener Tom and Pat Leeson/Photo Researchers, Inc. **21-3a** John Gerlach/Tom Stack & Associates **21-3b** John Shaw/Tom Stack & Associates **21-4** Carolina Biological Supply Company/Phototake NYC **21-5a** Dwight Kuhn Photography **21-5b** Milton Rand/Tom Stack & Associates **21-5c** Larry Ulrich/DRK Photo **21-6** Milton Rand/Tom Stack & Associates **21-7c** Andy Roberts/Stone **21-7d** John Kaprielian/Photo Researchers, Inc. **21-8a** Maurice Nimmo/A-Z Botanical Collection, Ltd. **21-8b** Teresa and Gerald Audesirk **21-9 (left)** Dr. William M. Harlow/Photo Researchers, Inc. **21-9 (right)** Gilbert S. Grant/Photo Researchers, Inc. **21-10a** Dwight R. Kuhn/Dwight Kuhn Photography **21-10b** David Dare Parker/Auscape International Pty. Ltd. **21-10b inset** Matt Jones/Auscape International Pty. Ltd. **21-10c** Dwight Kuhn Photography **21-10d** Teresa and Gerald Audesirk **21-10e** Larry West/Photo Researchers, Inc. **21-12** Thomas Kitchin/Tom Stack & Associates

Chapter 22

Opener Martin A. Collins, University of Aberdeen, Aberdeen, Scotland, U.K. **22-5a** Larry Lipsky/DRK Photo **22-5b** Brian Parker/Tom Stack & Associates **22-5c** Charles Seaborn/Odyssey Productions **22-6a** Gregory Ochocki/Photo Researchers, Inc. **22-6b** Charles Seaborn/Odyssey Productions **22-6c** Teresa and Gerald Audesirk **22-6d** David B. Fleetham/Innerspace Visions **22-9a** Carolina Biological Supply Company/Phototake NYC **22-10(1)** Martin Rotker/Phototake NYC **22-10(2)** Stanley Flegler/Visuals Unlimited **22-11** Tom E. Adams/Peter Arnold, Inc. **22-12a** Carolina Biological Supply Company/Phototake NYC **22-12b** Howard Shiang, D.V.M., *J. Am. Vet. Med. Assoc.* 163:981, Oct. 1973. **22-14a** Kjell B. Sandved/Butterfly Alphabet, Inc. **22-14b** David L. Bull/David Bull Photography **22-14c** J. H. Robinson/Photo Researchers, Inc. **22-15** Teresa and Gerald Audesirk **22-16** Dwight Kuhn Photography **22-18** David Scharf/Peter Arnold, Inc. **22-19a** Carolina Biological Supply Company/Phototake NYC **22-19b** Peter J. Bryant/Biological Photo Service **22-19c** Stephen Dalton/Photo Researchers, Inc. **22-19d** Werner H. Muller/Peter Arnold, Inc. **22-19e** Stanley Breeden/DRK Photo **22-20a** Dwight Kuhn Photography **22-20b** Tim Flach/Stone **22-20c** Teresa and Gerald Audesirk **22-21a** Tom Branch/Photo Researchers, Inc. **22-21b** Peter J. Bryant/Biological Photo Service **22-21c** Carolina Biological Supply Company/Phototake NYC **22-21d** Alex Kerstitch **22-23a** Ray Coleman/Photo Researchers, Inc. **22-23b** Alex Kerstitch **22-24a** Fred Bavendam/Peter Arnold, Inc. **22-24b** Ed Reschke/Peter Arnold, Inc. **22-25a** Fred Bavendam/Peter Arnold, Inc. **22-25b** Kjell B. Sandved/Photo Researchers, Inc. **22-25c** Alex Kerstitch **22-26a** Teresa and Gerald Audesirk **22-26b** Jeff Foott Productions **22-26c** Chris Newbert/Bruce Coleman, Inc. **22-27b** Michael Male/Photo Researchers, Inc. **22-28** John Giannicchi/Science Source/Photo Researchers, Inc. **22-29b** Tom McHugh/Photo Researchers, Inc. **22-30a** Tom McHugh/Steinhart Aquarium/Photo Researchers, Inc. **22-30b** Tom Stack/Tom Stack & Associates **22-30b inset** Breck P. Kent **22-31a** Jeffrey L. Rotman/Jeffrey L. Rotman Photography **22-31b** David Hall/Photo Researchers, Inc. **22-32a** Peter David/Getty Images/Planet Earth Pictures Ltd. **22-32b** Mike Neumann/Photo Researchers, Inc. **22-32c** Stephen Frink/Stone **22-32d** Peter Scoones/Getty Images/Planet Earth Pictures Ltd. **22-33a** Breck P. Kent/Animals Animals/Earth Scenes **22-33b** Joe McDonald/Tom Stack & Associates **22-33c** Cosmos Blank/National Audubon Society/Photo Researchers, Inc. **22-34a** David G. Barker/Tom Stack & Associates **22-34b** Roger K. Burnard/Biological Photo Service **22-34c** Frans Lanting/Minden Pictures **22-35** Carolina Biological Supply Company/Phototake NYC **22-36a** Walter E. Harvey/Photo Researchers, Inc. **22-36b** Carolina Biological Supply Company/Phototake NYC **22-36c** Ray Ellis/Photo Researchers, Inc. **22-37** Tom McHugh/Photo Researchers, Inc. **22-38a** Flip Nicklin/Minden Pictures **22-38b** Jonathan Watts/Science Photo Library/Photo Researchers, Inc. **22-38c** C. and M. Denis-Huot/Peter Arnold, Inc. **22-38d** S. R. Maglione/Photo Researchers, Inc. **22-39a** Yves Kerban/Jacana/Photo Researchers, Inc. **22-39b** Mark Newman/Auscape International Pty. Ltd. **22-39b inset** D. Parer and E. Parer-Cook/Auscape International Pty. Ltd. **E22-1** Stanley Breeden/DRK Photo

Unit Four

Opener John Shaw/Tom Stack & Associates

Chapter 23

Opener Kim Taylor/Bruce Coleman, Inc. **23-4a** Dr. Jeremy Burgess/Science Photo Library/Photo Researchers, Inc. **23-4b** John D. Cunningham/Visuals Unlimited **23-5a–c** George Wilder/Visuals Unlimited **23-8a** Lynwood M. Chace/Photo Researchers, Inc. **23-8b** Dwight Kuhn Photography **23-9 (top)** Ed Reschke/Peter Arnold, Inc. **23-9 (bottom)** E. R. Degginger/Animals

Animals/Earth Scenes **23-10** Teresa and Gerald Audesirk **23-12** J. Robert Waaland, University of Washington/Biological Photo Service **23-13** Ed Reschke/Peter Arnold, Inc. **23-16a** D. C. Cunningham/Visuals Unlimited **23-16b** Stephen J. Krasemann/Photo Researchers, Inc. **23-17** John D. Cunningham/Visuals Unlimited **23-20a** Biophoto Associates/Science Source/Photo Researchers, Inc. **23-20b** R. Roncadori/Visuals Unlimited **23-21** Hugh Spencer/National Audubon Society/Photo Researchers, Inc. **23-23a** Ray Simon/Photo Researchers, Inc. **23-23b** Dr. Jeremy Burgess/Science Photo Library/Photo Researchers, Inc. **23-24a** E. R. Degginger/Animals Animals/Earth Scenes **23-24b** Dr. Martin H. Zimmermann/Harvard Forest, Harvard University **23-26** Francois Gohier/Photo Researchers, Inc. **23-27a** Kenneth W. Fink/Photo Researchers, Inc. **23-27b** Belinda Wright/DRK Photo **23-28a** Teresa and Gerald Audesirk **23-28b** Carr Clifton/Minden Pictures **23-28c** Teresa and Gerald Audesirk **E23-1** Michael Fogden/DRK Photo

Chapter 24

Opener F. Stuart Westmorland/Photo Researchers, Inc. **24-1** Jeff Foott/Bruce Coleman, Inc. **24-3** Robert & Linda Mitchell Photography **24-4b** James L. Castner **24-5** Teresa and Gerald Audesirk **24-6 (top, middle, bottom)** Carolina Biological Supply Company/Phototake NYC **24-7** Dennis Kunkel/MicroVision **24-9** Carolina Biological Supply Company/Phototake NYC **24-13** Teresa and Gerald Audesirk **24-14** Teresa and Gerald Audesirk **24-15a,b** Thomas Eisner **24-16a,b** Nuridsany et Perennou/The National Audubon Society Collection/Photo Researchers, Inc. **24-17** Huntington Botanical Gardens, San Marino, California **24-18a** Merlin D. Tuttle/Bat Conservation International/Photo Researchers, Inc. **24-18b** Oxford Scientific Films/Animals Animals/Earth Scenes **24-19** Dr. J. A. L. Cooke/Oxford Scientific Films/Animals Animals/Earth Scenes **24-20a** Tom Bean/DRK Photo **24-20b** Michael Fogden/DRK Photo **24-21a** Teresa and Gerald Audesirk **24-21b** Carolina Biological Supply Company/Phototake NYC **24-22** Teresa and Gerald Audesirk **24-23** Scott Camazine/Photo Researchers, Inc. **24-24** Fogden Natural History Photographs **E24-1a** George Holton/Photo Researchers, Inc. **E24-1b** Merlin D. Tuttle/Bat Conservation International/Photo Researchers, Inc.

Chapter 25

Opener Roberto Osti **25-1c** Runk/Schoenberger/Grant Heilman Photography, Inc. **25-6** John Cancalosi/DRK Photo **25-7** Biophoto Associates/Photo Researchers, Inc. **25-8a,b** Carolina Biological Supply Company/Phototake NYC

Unit Five

Opener Randy Wells/Stone

Chapter 26

Opener Karl Weatherly/Corbis **26-4** Biophoto Associates/Photo Researchers, Inc. **26-5** Manfred Kage/Peter Arnold, Inc. **26-6a** John Kennedy/Biological Photo Service/Stone **26-6b** Fred Bruemmer/OKAPIA/Photo Researchers, Inc. **26-7** Yorgos Nikas/Science Photo Library/Photo Researchers, Inc. **26-8** Manfred Kage/Peter Arnold, Inc.

Chapter 27

Opener PPL Therapeutics/AP/Wide World Photos **27-8** Dennis Kunkel/MicroVision **27-11** Photo Lennart Nilsson/Albert Bonniers Forlag **27-12** Kenneth Kaushansky © 1995. Reprinted with permission of *Discover* magazine. **27-13b** CNRI/Science Photo Library/Photo Researchers, Inc. **27-16** Photo Lennart Nilsson/Albert Bonniers Forlag **27-20** John D. Cunningham/Visuals Unlimited **27-21** Chandran/TDR/World Health Organization **E27-1** Gwen Rosenberg/Alliance Pharmaceutical Corp. **E27-2a** Lou Lainey/TimePix

Chapter 28

Opener Spencer Grant/Grantpix/Stock Boston **28-1a** Jeffrey L. Rotman/Jeffrey L. Rotman Photography **28-1b** Brian Parker/Tom Stack & Associates **28-1c** Phil Degginger/Color-Pic, Inc. **28-5a** E. R. Degginger/Bruce Coleman, Inc. **28-5b** Breck P. Kent **28-5c** David M. Dennis/Tom Stack & Associates **28-6b** H. R. Duncker, Department of Anatomy, University of Giessen, Germany **E28-1** Bill Travis, M.D./National Cancer Institute **E28-2** O. Auerbach/Visuals Unlimited

Chapter 29

Opener Barbara Penoya/PhotoDisc, Inc **29-1** Charles Krebs/Stone **29-3** Tobi Zausner **29-5a** E. R. Degginger/Photo Researchers, Inc. **29-12b** Photo Lennart Nilsson/Albert Bonniers Forlag **E29-1** De Keerle-Parker/Liaison Agency, Inc. **E29-2** St. Bartholomew's Hospital/Science Photo Library/Photo Researchers, Inc.

Chapter 30

Opener Dan McCoy/Rainbow **30-9** Allstock/Stone

Chapter 31

Opener Mednet/Phototake NYC **31-1** David Scharf/Peter Arnold, Inc. **31-2** Photo Lennart Nilsson/Albert Bonniers Forlag **31-3b** National Institute for Biological Standards and Control, England/Science Photo Library/Photo Researchers, Inc. **31-3c** Photo Lennart Nilsson/Albert Bonniers Forlag **31-9a** Secchi-Lecaque/Roussel/CNRI/Science Source/Photo Researchers, Inc. **31-9b** CNRI/Science Photo Library/Photo Researchers, Inc. **31-10** Oliver Meckes & Nicole Ottawa/Photo Researchers, Inc. **31-14** National Institute for Biological Standards and Control, U.K./Science Photo Library/Photo Researchers, Inc. **E31-1b** CNRI/Science Photo Library/Photo Researchers, Inc. **E31-2a** Lloyd Cluff/Corbis **E31-2b** Sinclair Stammers/Science Photo Library/Photo Researchers, Inc.

Chapter 32

Opener Pat Benic/Reuters/Corbis **32-5** Feld Entertainment® **32-7b** Biophoto Associates/Photo Researchers, Inc. **32-11** Photo courtesy of The Jackson Laboratory, Bar Harbor, Maine

Chapter 33

Opener Ron Edmonds/AP/Wide World Photos **33-2** The Physiological Society **33-15** Hanna Damasio, M.D. **33-17a** Scanning electron micrographs by Robert S. Preston, courtesy of Professor Joseph E. Hawkins, Jr., Kresge Hearing Research Institute, University of Michigan Medical School. **33-18a** Dennis Kunkel/MicroVision **33-21** Susan Eichorst **33-22a** Frans Lanting/Minden Pictures **33-22b** Larry West/Photo Researchers, Inc. **33-26a** Charles E. Mohr/Photo Researchers, Inc. **33-26b** Brian Parker/Tom Stack & Associates **E33-2** Ken Graham/Bruce Coleman, Inc. **E33-3** Marcus E. Raichle, M.D., Washington University School of Medicine **E33-4a** William J. Powers, M.D., Washington University School of Medicine **E33-4b** Monte S. Buchsbaum, M.D., Mount Sinai School of Medicine, New York, N.Y.

Chapter 34

Opener Dept. of Clinical Radiology, Salisbury District Hospital, Salisbury, U.K./Science Photo Library/Photo Researchers, Inc. **Table 34-1** Eric Grave/Photo Researchers, Inc. **34-4b** Don Fawcett/Science Source/Photo Researchers, Inc. **34-5a** Kent Wood/Photo Researchers, Inc. **34-5b (left)** Lawrence E. Naylor/Photo Researchers, Inc. **34-5b (right)** Jeremy Burgess/Science Photo Library/Photo Researchers, Inc. **34-7** Photo Researchers, Inc. **34-9b** Manfred Kage/Peter Arnold, Inc. **E34-2a** Michael Klein/Peter Arnold, Inc. **E34-2b** Yoav Levy/Phototake NYC

Chapter 35

Opener Paul A. Zahl/Photo Researchers, Inc. **35-1** Biophoto Associates/Photo Researchers, Inc. **35-3** Peter J. Bryant/Biological Photo Service **35-4a** Tom McHugh/Photo Researchers, Inc. **35-4b** Peter Harrison/Teresa and Gerald Audesirk **35-5a** Anne et Jacques Six **35-5b** Rudie Kuiter/Innerspace Visions **35-6** Michael Fogden/Oxford Scientific Films/Animals Animals/Earth Scenes **35-7a** Stephen J. Krasemann/DRK Photo **35-7b** Tom McHugh/Photo Researchers, Inc. **35-7c** Frans Lanting/Minden Pictures **35-11b** CNRI/Science Photo Library/Photo Researchers, Inc. **35-16** C. Edelmann/La Villette/Photo Researchers, Inc. **35-18a** Photo Lennart Nilsson/Albert Bonniers Forlag **35-18b** Andrew Syred/Stone **E35-2** Hank Morgan/Photo Researchers, Inc. **E35-3** Brook Kraft/Corbis/Sygma **E35-4** Juana Anderson/AP/Wide World Photos

Chapter 36

Opener Bob Daemmrich/Bob Daemmrich Photography, Inc. **36-1a (left)** Image Quest 3-D/Natural History Photographic Agency **36-1a (right)** Jim Bain/Natural History Photographic Agency **36-1b (left)** Hans Pfletschinger/Peter Arnold, Inc. **36-1b (right)** Larry West/Photo Researchers, Inc. **36-2a** J. A. L. Cooke/Animals Animals/Earth Scenes **36-2b** K. H. Switak/Photo Researchers, Inc. **36-2c** S. Nielsen/DRK Photo **36-2d** Norbert Rosing/Oxford Scientific Films/Animals Animals/Earth Scenes **36-8a** Photo Lennart Nilsson/Albert Bonniers Forlag **36-11a** Photo Lennart Nilsson/Albert Bonniers Forlag **36-11b** Petit Format/Nestle/Science Source/Photo Researchers, Inc. **36-11c** Photo Lennart Nilsson/Albert Bonniers Forlag **36-15** Jerry Gay/Stone **E36-2** Leonard McCombe/TimePix **E36-3** Sterling K. Clarren, M.D., Children's Hospital & Medical Center, University of Washington School of Medicine

Chapter 37

Opener Reuters NewMedia, Inc./Corbis **Opener inset** Joe McDonald/DRK Photo **37-1** Eric and David Hosking/Frank Lane Picture Agency Limited FLPA **37-3a–e** Boltin Picture Library **37-5** Frans Lanting/Minden Pictures **37-7** Thomas McAvoy/TimePix **37-9** Renee Lynn/Photo Researchers, Inc. **37-10** Ken Cole/Animals Animals/Earth Scenes **37-11** Richard K. LaVal/Animals Animals/Earth Scenes **37-12** Nuridsany et Perennou/Photo Researchers, Inc. **37-13** Robert & Linda Mitchell/Robert & Linda Mitchell Photography **37-14a** Stephen J. Krasemann/Photo Researchers, Inc. **37-14b** Hans Pfletschinger/Peter Arnold, Inc. **37-15a** M. P. Kahl/DRK Photo **37-15b** Marc Chamberlain/Stone **37-16** Ray Dove/Visuals Unlimited **37-17** William Ervin/Natural Imagery **37-18** Michael K. Nichols/National Geographic Image Collection **37-19** John D. Cunningham/

Visuals Unlimited **37-21** Joe McDonald/Visuals Unlimited **37-22a** Norbert Wu/Peter Arnold, Inc. **37-22b** Frans Lanting/Minden Pictures **37-23** Anne et Jacques Six **37-25a** Dwight Kuhn Photography **37-25b** Teresa and Gerald Audesirk **37-26** Fred Bruemmer/Peter Arnold, Inc. **37-29** Raymond A. Mendez/Animals Animals/Earth Scenes **37-30** Photo Lennart Nilsson/Albert Bonniers Forlag **37-31** William P. Fifer, New York State Psychiatric Institute, Columbia University **37-32a** Frans Lanting/Minden Pictures **37-32b** Daniel J. Cox/Liaison Agency, Inc. **37-32c** Michio Hoshino/Minden Pictures

Unit Six

Opener Paradise, Part One of *The Trilogy of the Earth* by Suzanne Duranceau/Illustratrice, Inc.

Chapter 38

Opener Michael P. Gadomski/Photo Researchers, Inc. **Opener inset 1** Ken Wilson; Papilio/Corbis **Opener inset 2** W. Perry Conway/Corbis **38-1a** Prof. S. Cinti/University of Ancona, Italy/CNRI/Phototake NYC **38-1b** Thomas Kitchin/Tom Stack & Associates **38-6** Bryan and Cherry Alexander **38-7** Tom McHugh/Photo Researchers, Inc. **38-8** Hellio-Van Ingen/Auscape International Pty. Ltd. **38-9a** Robert & Linda Mitchell Photography **38-9b** Tom Bean/DRK Photo **38-9c** Gary Braasch Photography **38-11** NASA/Johnson Space Center **E38-3** Frans Lanting/Minden Pictures **E38-3 inset** Andy Holbrooke/The Stock Market

Chapter 39

Opener Ron Peplowski/The Detroit Edison Company **Opener inset** James F. Lubner, University of Wisconsin Sea Grant Advisory Services **39-3a** Henry Ausloos/Animals Animals/Earth Scenes **39-3b** Stephen Dalton/Photo Researchers, Inc. **39-3c** Frans Lanting/Minden Pictures **39-4a** Art Wolfe/Art Wolfe, Inc. **39-4b** Frans Lanting/Minden Pictures **39-5a** Marty Cordano/DRK Photo **39-5b** Paul A. Zahl/Photo Researchers, Inc. **39-5c** Ray Coleman/Photo Researchers, Inc. **39-6** Robert & Linda Mitchell Photography **39-7a** Ferrero-Labat/Auscape International Pty. Ltd. **39-7b** Charles V. Angelo/Photo Researchers, Inc. **39-8** James L. Castner **39-9a** Zig Leszczynski/Animals Animals/Earth Scenes **39-9b** Breck P. Kent **39-9c,d** Ed Degginger/Color-Pic, Inc. **39-10a** Robert P. Carr/Bruce Coleman, Inc. **39-10b** Monica Mather and Bernard Roitberg, Simon Fraser University **39-11a** Zig Leszczynski/Animals Animals/Earth Scenes **39-11b** James L. Castner **39-11c** Jeff Lepore/Photo Researchers, Inc. **39-12a** Dan Aneshausley, Cornell University/Thomas Eisner **39-12b** Frans Lanting/Minden Pictures **39-13a** Teresa Audesirk **39-13b** W. Gregory Brown/Animals Animals/Earth Scenes **39-14a** Jim Zipp/Photo Researchers, Inc. **39-14b** Stephen J. Krasemann/Photo Researchers, Inc. **39-15a (left)** Dennis Oda/Stone **39-15a (right)** Krafft-Explorer/Photo Researchers, Inc. **39-15b** Gary Braasch/Gary Braasch Photography **39-15c (left)** Wendy Shattil/Bob Rozinski/Tom Stack & Associates **39-15c (right)** M. P. Kahl/DRK Photo **39-18a–c** Robert & Linda Mitchell Photography **E39-1** Gilbert Grant/Photo Researchers, Inc. **E39-1 inset** Carol Hughes/Bruce Coleman, Inc. **E39-2a** Chuck Pratt/Bruce Coleman, Inc. **E39-2b** Robert & Linda Mitchell Photography **E39-2c** Jack Jeffrey/Photo Resource Hawaii Stock Photography

Chapter 40

Opener Tom and Pat Leeson/DRK Photo **40-12** Tom Walker/Stock Boston **40-13** William E. Ferguson **40-14** Will McIntyre/Photo Researchers, Inc. **E40-1** Thomas A. Schneider/SchneiderStock Photography

Chapter 41

Opener (and insets) Don L. Boroughs **41-2b** NASA/Johnson Space Center **41-6a** Teresa and Gerald Audesirk **41-6b** Tom McHugh/Photo Researchers, Inc. **41-9a** Loren McIntyre **41-9b** Schafer and Hill/Peter Arnold, Inc. **41-9c** Michael Fogden/Oxford Scientific Films Ltd. **41-9d** Michael Fogden/DRK Photo **41-9e** Frans Lanting/Minden Pictures **41-10** Jacques Jangoux/Peter Arnold, Inc. **41-11a** W. Perry Conway/Aerie Nature Series, Inc. **41-11b** Aubry Lang/Valan Photos **41-11c** Stephen J. Krasemann/DRK Photo **41-12** James Hancock/Photo Researchers, Inc. **41-13a** Don and Pat Valenti/DRK Photo **41-13b** Brian Parker/Tom Stack & Associates **41-13c** Tom McHugh/Photo Researchers, Inc. **41-14** William H. Mullins/Photo Researchers, Inc. **41-15** Brian Parker/Tom Stack & Associates **41-16** Harvey Payne/Harvey Payne **41-17a** Tom Bean/DRK Photo **41-17b** W. Perry Conway/Tom Stack & Associates **41-17c** Nicholas DeVore III/Photographers/Aspen, Inc. **41-17d** Bob Gurr/Valan Photos **41-17e** Jim Brandenburg/Minden Pictures **41-18** Teresa and Gerald Audesirk **41-19a** Gary Braasch Photography **41-19b** Thomas Kitchin/Tom Stack & Associates **41-19c** Stephen J. Krasemann/DRK Photo **41-19d** Joe McDonald/Animals Animals/Earth Scenes **41-20a** Susan G. Drinker/The Stock Market **41-20b** Teresa and Gerald Audesirk **41-20c** Jeff Foott Productions **41-20d** Richard Thom/Tom Stack & Associates **41-21a** Teresa and Gerald Audesirk **41-21b** Marty Stouffer Productions/Animals Animals/Earth Scenes **41-21c** Wm. Bacon III/National Audubon Society/Photo Researchers, Inc. **41-22** Gary Braasch Photography **41-23a** Stephen J. Krasemann/DRK Photo **41-23b** Jeff Foott/Tom Stack & Associates **41-23c** Michael Giannechini/Photo Researchers, Inc. **41-23d** James Simon/Photo Researchers, Inc. **41-26a** T. P. Dickinson/Comstock **41-26b** Teresa and Gerald Audesirk **41-26b inset** Walter Dawn/National Audubon Society/Photo Researchers, Inc. **41-26c** Teresa and Gerald Audesirk **41-26c inset** Thomas Kitchin/Tom Stack & Associates **41-26d** Robert Given, Marymount College **41-26d inset** Stan Wayman/Photo Researchers, Inc. **41-27a** Franklin J. Viola/Viola's Photo Visions, Inc. **41-27b** Alex Kerstitch **41-27c** Charles Seaborn/Odyssey Productions **41-27d** Alex Kerstitch **41-28a** Stephen J. Krasemann/DRK Photo **41-28b** Paul Humann/Jeffrey L. Rotman Photography **41-28c** Fred McConnaughey/Photo Researchers, Inc. **41-28d** James D. Watt/Getty Images/Planet Earth Pictures Ltd. **41-28e** Biophoto Associates/Science Source/Photo Researchers, Inc. **41-28f** Robert Arnold/Getty Images/Planet Earth Pictures Ltd. **41-29** J. Frederick Grassle, Institute of Marine and Coastal Sciences, Rutgers University **E41-1** NASA Headquarters **E41-2** Roy Morsch/The Stock Market

Index

Abdomen, 443
Abiotic environment, 298–99, 793, 843
Abortion, 735
Abscisic acid, 487, 519, 520
Abscission layer, 520, 528
Absorption of nutrient, 597–99
Abstinence, sexual, 732
Acacias, 825
Accessory pigments, 119
Accutane®, 758
Acellular slime molds, 379
Acetylcholine, 669
Acetyl CoA, 139, 140, 141
Acetyl group, 139
Acid rain, 847–49, 877–88
Acid Rain Program, 849
Acid(s), 30. *See also* Amino acids
  fatty, 43, 44, 46, 53, 584
    saturated, 44, 70
    "trans," 46
    unsaturated, 44, 70
  glutamic, 47
Acorn woodpeckers, 776
Acquired immune deficiency syndrome. *See* AIDS
Acridine, 171
Acrosome, 724
Actin, 93, 176, 701
Action potentials, 664, 665, 667
Activation energy, 102, 106–7
Active site, 107
Active transport, 61–62, 65
Adaptations, 6, 13, 298–99
Adaptive radiation, 313–14
Adenine, 51, 151–52
Adeno-associated viruses, 258
Adenosine deaminase (ADA), 257
Adenosine diphosphate (ADP), 104
Adenosine triphosphate. *See* ATP
Adenoviruses, 258
Adhesion, 30
Adipose tissue, 540, 649, 657
Adirondack Mountains, 849, 877–78
ADP, 104
Adrenal cortex, 656
Adrenal glands, 644, 649, 655–56
Adrenaline (epinephrine), 645, 649, 655
Adrenal medulla, 655, 656
Adrenocorticotropic hormone (ACTH), 648, 651, 657
Aerobic metabolism, 90, 329
Aflatoxins, 174, 400–401
African blood lily, 195
African savanna, 867
African sleeping sickness, 384
Afterbirth, 756
Agar, 248, 382
Agassiz, Louis, 271
Age structure of population, 805–8
Aggression, 771, 774–75, 777, 782–83
Aggressive mimicry, 820, 821
Aging, 176–77, 757–59
Agriculture
  biotechnological applications in, 251–55
  fungal impact of, 399–400
  revolution in, 805
Agriculture, U.S. Department of, 251
AIDS, 350, 351, 360, 366, 367, 369, 549, 635–37, 731
  extent of, 637
  HIV and, 635–36
  treatments for, 636–37
Air currents, 860
Alberts, B., 74
Albinism, 226, 227–28
Albumin, 46, 555
Alcohol, 108, 596, 597, 669
  pregnancy and, 741, 759, 760
Alcohol dehydrogenase, 108
Alcoholic fermentation, 137
Alcoholism, 710
Aldosterone, 649, 656
Alexandra, Tsarina, 230
Alexis, Tsarevitch of Russia, 230
Algae, 380–83, 411, 824
  blue-green, 876
  brown, 378, 380, 382, 383
  green, 378, 380, 382–83, 411
  multicellular, 331
  photosynthetic, 397–98
  red, 378, 380, 382
  unicellular. *See* Phytoplankton
Alkaloids, 822
Allantois, 744
Allele frequency, 288
  gene flow and, 289–90
  population bottlenecks and, 292
  small populations and, 290–93
Alleles, 198–99, 212, 288
  codominant, 224
  dominant, 214–16
  multiple, 223–24
  recessive, 214–16
  relationships among genes, chromosomes, and, 212
Allergies, 632–33, 635
Alligators, 456, 459, 647
Allopatric speciation, 309, 310, 318
Allosteric regulation, 108, 109
Allosteric regulatory site, 108
Alpha-1-antitrypsin, 255–57
Altamira, 342
Alternation of generations, 410, 417, 498
Altman, Sidney, 326
Altruism, 300–301, 780
Alvarez, Luis and Walter, 337–38
Alveolus/alveoli, 574, 575
Alzheimer's disease, 669
Amanita, 390, 391, 394, 404
Amatoxins, 404
Amazon rain forest, 486
American alligator, 456
American chestnut tree, 254
Amino acids, 40, 45, 46–48
  as base sequences, 166–67, 173–74
  conduction in plants, 473
  essential, 585–86
  hormones based on, 644
  hydrophilic, 50
  joining of, 47–48
Amino group, 38, 46
Amish founder population, 293
Ammonia, 607
Ammonites, 332
Amniocentesis, 260, 261
Amnion, 260, 455, 457, 744
Amniote egg, 455, 457, 744
Amniotic fluid, 260
Amoeba, 67, 384
Amoebic dysentery, 384
Amoeboid cells, 434
Amorphophallus titanum (corpse flowers), 510, 511
Amphibians, 454–55
  evolution of, 334–35
  gas exchange in, 572, 573
  thyroxine effects on, 658
Amplexus, 720
Ampulla, 450
Amygdala, 678
Amylase, 594, 597
Anabolic steroids, 642, 643, 658
Anaerobic bacteria, 327, 329, 375
Anaerobic metabolism, 90
Analogous structures, 277–78
Anal pore, 385, 386
Anaphase
  in meiosis, 200, 201, 202, 203, 204–5
  in mitosis, 195, 196–97
Anatomy
  evolutionary evidence in, 277–79
  as taxonomic criterion, 353
Androgen insensitivity, 176
Androgen-receptor proteins, 176
Androgens, 176–77, 654
Anemia, 633
Anemone. *See* Sea anemone
Angina, 562
Angiosperms, 416, 418, 419–23, 468. *See also* Flowering plants; Flowers
Angiotensin, 614, 657
Animal cell, cytokinesis in, 197, 198
Animalia. *See* Animal(s)
Animal-like protists. *See* Protozoa
Animal(s), 6, 7, 353, 354, 429–64, 535–47
  adaptations for land living, 334–36
  anatomical features marking evolutionary branch points, 430–32
  arthropods. *See* Arthropods
  body organization of, 538–44
    organs, 538, 541–44
    organ systems, 538, 541–44
    tissues, 538–44
  cannibalistic, 301
  cell number in, 8
  characteristics of, 357
  chordates. *See* Chordates
  cnidarians. *See* Cnidarians
  defining characteristics of, 430
  development. *See* Development
  echinoderms, 430–31, 435, 448–50
  evolution of multicellular, 331–33
  flatworms, 437–40
  genetically engineered, 548
  internal constancy of, 536–38
  major phyla, 432–33
  mollusks. *See* Mollusks
  roundworms, 431, 432, 434, 440–41
  seed dispersal by, 513–14, 527

segmented worms (annelids), 433, 435, 441, 442
sponges, 430, 432–35, 590–91
transgenic, 244, 252–55
Annamite Mountains, 307, 318
Annelids, 433, 435, 441, 442
body cavity in, 431, 432
Annual rings of growth, 478
Anolis lizard, 281, 772
Anopheles, 385, 634
Anorexia nervosa, 588
Antagonistic muscles, 708
Antennae, 445
Antenna molecules, 119
Antennapedia, 249
Anterior (head) end, 431
Anther, 422, 423, 500, 501
Antheridia/antheridium, 413
Anthrax, 365, 387, 631
Antibiotic-resistant bacteria, 286, 287, 294, 302
Antibiotics, 11, 80, 631
Antibodies, 46, 224, 624–28
Anticodon, 170
Antidiuretic hormone (ADH), 610, 612–13, 647, 648, 650, 651
Antigen-presenting cells, 621
Antigens, 624, 632
Antioxidants, 32, 587
Antoinette, Marie, 257
Ants, 780, 825
Anvil, 684, 685
Aorta, 552
Aphids, 487–88, 719
Aphotic zone, 878
Apical dominance, 524
Apical meristems, 470
Aplysia, 671
Apopka, Lake, 647
Appendages, jointed, 443
Appendicular skeleton, 706
Aquaculture, 883
Aqueous humor, 686, 687
Aquifers, 809
Arabuko-Sokoke forest, 857, 886
Arachnids, 433, 435, 445, 446
Archaea, 6, 7, 8, 354, 355, 371, 373, 374, 375, 376, 386
Archaeopteryx, 282, 458
Archegonia/archegonium, 413
Architeuthis (giant squid), 428, 429, 461
Ardipithecus ramidus, 339
Arginine, 47
Argon, 329
Argonauta, 721
Aristelliger, 296
Aristotle, 270, 271, 683
Arteries, 552, 559, 607, 609
Arterioles, 542, 558, 559, 561
Arthritis, 634
Arthrobotrys, 403
Arthropods, 43, 334, 431, 433, 435, 441–45
arachnids, 433, 435, 445, 446
compound eyes of, 686
crustaceans, 433, 435, 445
exoskeletons, 443
gas exchange in, 444
infections from, 731
insects, 432, 433, 435, 444–45, 500, 509–10, 511, 572
segmentation in, 443–44
sensory system of, 444
Artificial insemination, 731
Artificial selection, 280, 281
Artificial sweeteners, 37, 53
Ascomycota, 393, 394–96
Ascorbic acid (vitamin C), 587
Ascus/asci, 394
Asexual reproduction, 187–88, 717, 718
Asian chestnuts, 254
Asian flu, 632
Aspartame (Nutrasweet®), 37, 53
Aspartic acid, 53
Aspen groves, 188
Aspergillus, 400, 401
Aspirin, 597
Association neurons, 671
Assortive mating, 293
Asthma, 579
Aswan High Dam, 634
Atala, Anthony, 614
Atherosclerosis, 562–63, 578–79
Atmosphere, prebiotic evolution and, 324–27
Atomic nucleus, 21, 24
Atomic number, 21–22
Atomic theory, 11
Atom(s), 21–27
composition of, 21–24
defined, 2–4, 21
inert, 24
molecules formed from, 24–27
reactive, 24
stable, 24
ATP, 40, 52, 65, 103–5, 110, 118
active transport and, 65
life span of, 104
synthesis of, 104, 132. *See also* Glycolysis
by chemiosmosis, 119, 121, 140, 141–43
in mitochondria, 90–91, 133
by photosystem II, 119, 121
Atrial natriuretic peptide (ANP), 649, 657
Atrioventricular (AV) node, 554
Atrioventricular valves, 553
Atrium/atria, 551
Auditory canal, 684, 685
Auditory nerve, 684
Audubon's warbler, 38
Australopithecus, 339
Australopithecus afarensis, 341
Australopithecus africanus, 341
Australopithecus anamensis, 341
Australopithecus boisei, 341
Australopithecus robustus, 341
Autoimmune disease, 633–34
Autonomic nervous system, 673, 674
Autosomes, 220, 234
Autotrophs, 8, 116, 837
Auxin, 519, 520–21, 523, 527
discovery of, 523
fruit development and, 527
gradients of, 525
gravitropism and, 521
lateral bud sprouting and, 524
phototropism and, 521, 523
root branching and, 525
seedling orientation and, 520–21
Avery, Oswald, 151
Aves, 455–59
Axial skeleton, 706
Axons, 541, 664, 665, 673–76
damage to, 675
AZT (ziduvidine), 637

Baboon, 289, 774, 775
Baby, test-tube, 734
Bacillus anthracis, 365
Bacillus thuringiensis, 252
Backbone (vertebral column), 451, 673
Bacteria, 1, 6, 7, 75, 354, 355, 370–77, 824
anaerobic, 327, 329, 375
antibiotic-resistant, 286, 287, 294, 302
Archaea vs., 371
binary fission of, 373–74
biotic potential of, 794
capsules around, 371
cell type and number of, 8
cell walls of, 69, 371
chemosynthetic, 375
chemotactic, 372
colonies of, 371
conjugation by, 373–74
cultured, 255
denitrifying, 845
endospores protecting, 372–73
endosymbiotic, 330
features of, 80–81
hardiness of, 95
magnetotactic, 372
nitrogen-fixing, 375–76, 483, 484
pathogenic, 376
photosynthetic, 327, 397–98
phototactic, 372
pili of, 371, 373
on roots, 483
slime layers around, 371
symbiotic, 375
transformed, 150–51
viral infection of, 366, 367
Bacterial cells, 74
Bacterial infections, 731
Bacterial transformation, DNA recombination in, 244–45
Bacteriophages, 171, 366, 367
Baer, Karl von, 279
Balanced polymorphism, 298
Balanus, 818
Bald eagle, 834, 835, 852
Baleen whale, 280
Ball-and-socket joints, 711
Baltimore oriole, 358
Bananas, 527–28
Baobab tree, 490
Bark, 479
Barking deer (giant muntjac), 307, 318
Barnacles, 445, 446, 448
Barnett, Dave, 583, 600
Barr, Murray, 180
Barr body, 180
Barry, Dave, 192
Basal body, 93
Base pairs, 152, 173
Base(s), 30, 151, 166–67, 173–74
anticodon, 170
nitrogen-containing, 51
Base sequences, amino acids as, 166–67
Basidia, 396
Basidiomycota, 384, 393, 394, 396, 399
Basidiospores, 396
Basilar membrane, 684, 685
Basilisk lizard, 29
Bats, 459, 506, 511, 691–92, 819
B cells, 624, 625, 626, 632
clonal selection among, 628
memory, 627
Beach strawberry, 490
Beadle, George, 164, 165
Beagle, 274

Bean, seed germination in, 508
Bears, fat storage in, 44
Beaver, 613
Becker, Anne, 589
Beef fat, 44
Beer, 402
Bees, 509–10, 511, 767, 768, 773
Beetles, 500, 510, 511
Behavior, 765–90
  communication, 771–74
  competition and, 774–77
  cooperative, 781
  defined, 766
  gene-environment interaction in, 770–71
  genetic components of, 785–86
  human, 343–45, 783–86
  innate (instinctive), 766, 768–71, 784–85
  learned, 766–71
    conditioning, 767–68
    habituation, 767, 769
    insight, 768, 769
  play, 786–87
  sexual reproduction and, 777–79
  social, 778–83
    advantages and disadvantages of, 780–81
    in bullhead catfish, 782–83
    in honeybees, 781–82
    in naked mole rats, 783
  territoriality, 776–77
Behavioral isolation, reproductive isolation through, 314, 315–16
Behavioral mimicry, 822
Behavioral specialization, 317
Bellbird, 778
Beltsville Small White turkeys, 205
Benign tumor (polyp), 637
Benirschke, Kurt, 254
Bennett, Alan, 529
Bernard, Claude, 536
Beta-globin protein, 175
Beta-thalassemia, 175
Betta splendens (Siamese fighting fish), 720
Beyer, Peter, 263
Bicarbonate, 576
Bilateral symmetry, 430–32
Bile, 596
Bile duct, 596
Bile salts, 596
Binary fission, 186, 373–74
Binocular vision, 338, 689
Biochemical evidence of evolution, 279
Biodegradability, 375, 841
Biodiversity, 13, 15, 358–60, 864
Biogeochemical cycles (nutrient cycles), 843
Biological clock, 525
Biological magnification, 841
Biological molecules. *See* Molecule(s)
Biological-species concept, 308, 358
Biological weapons, 364, 365, 367, 387
Biology, 8–15
  everyday life illuminated by, 13–15
  evolution as unifying concept of, 12–13
  scope of, 8–12
Biomass, 837
Biomass pyramid, 842
Biomes, 831, 863–75. *See also* Ecosystems
Bioprospecting, 409
Biosphere, 4
Biotechnology, 243–66
  applications of, 249
  defined, 244
  DNA recombination, 244–47
  ethical implications of, 259–62, 748
  gene identification, 248–49
  medical uses of, 255–59
    gene therapy, 257
    genetic screening, 255, 259–61
    Human Genome Project, 259, 356
    knock-out mice, 255
Bio-terrorism, 387
Biotic environment, 299, 793
Biotic potential, 794–96
Biotin, 587
Bipedal locomotion, 339–41, 343
Birch, 421
Birds, 455–59, 743
  digestion in, 592
  evolution from reptiles, 335–36
  extraembryonic membranes in, 743–44
  flu viruses in, 632, 633
  as reptiles, 359
  respiratory system of, 572–73
Birth canal, 729, 756
Birth control pills, 732, 733
Birth rates, 794
Bishop pines, 315
Bison, 871
Bisphosphate, 132
Bivalves, 448
Blackberry, 358
Black bread mold, 394, 395
"Black Death" (plague), 376, 387
Black-faced lion tamarin, 359
Black rhino, 868
Bladder, 608
  bioartificial, 614
  swim, 454
Blades of leaves, 479–80
Blassie, Michael, 257
Blastocoel, 745
Blastocyst, 734, 749, 752
Blastopore, 745, 748–49
Blastula, 745
Blaustein, Andrew, 457
Blind spot (optic disc), 687, 688
Blood, 539, 540, 541, 550, 554–58
  artificial, 556
  dissolved substances in, 614
  fetal cells from maternal, 260
  glomerular filtration of, 609–11
  osmosis and, 64, 65
  oxygen content of, 614
  plasma, 554, 555
  platelets, 540, 555, 557–58
  production of, 706
  red blood cells (erythrocytes), 554, 555–57, 560, 706
  water content of, 612–13
  white blood cells (leukocytes), 555, 557, 578, 706
Blood–brain barrier, 673
Blood clotting, 555, 558
Blood flukes, 439
Blood glucose, pancreatic control of, 655
Blood pressure, 553, 561, 562, 612, 614
Blood types, human, 223–24
Blood vessels, 550, 558–61
  arteries and arterioles, 552, 558, 559, 561, 607, 609
  capillaries, 551, 559–60, 562, 563, 564, 575
  structures and interconnections of, 559
  veins and venules, 552, 558, 559, 560–61
Bluebird, species of, 352
Blue-green algae, 876
Blue jack, 882
Blue jay, 872
Blue-ringed octopus, 880
Blue-spotted sting ray, 453
Boa constrictor, 280
Body cavities, 431–32
Body forms, symmetrical, 430–32
Body mass index (BMI), 589
Body temperatures of birds, 458
Bombardier beetle, 823
Bonds, 22, 24–27
  covalent, 24, 25–27
  in hair, 48
  high-energy, 104
  hydrogen, 24, 26–27, 28, 48
  ionic, 24, 25
  peptide, 48, 164
Bone marrow, 555–57, 706
Bone remodeling, 707–8
Bone(s), 539, 540. *See also* Skeleton
  blood cell production by, 706
  as body framework, 707
  broken, 698, 699, 711
  calcium and phosphorus storage in, 706
  compact, 707
  fossilized, 271
  healing of, 709
  osteoporosis, 710
  spongy, 707
  structure of, 708
Bony fishes, 452, 453–54
Book lungs, 444
Boom-and-bust population cycles, 796–97
Borrelia burgdorferi, 376
Boston ivy, 490
Bottleneck, population, 292–93
Bottlenose dolphins, 786
Botulism, 376
"Bound" water molecules, 63
Bovine spongiform encephalopathy ("mad cow disease"), 370
Bowman's capsule, 608, 609, 610
Boysen-Jensen, Peter, 522–23
Bradykinin, 691
Brain, 672, 676–83
  forebrain (cerebrum), 676, 677–80
  hindbrain, 676, 677
  human, 460, 677, 678
  imaging techniques for, 681
  injuries to, 675
  interpretation of action potentials, 670
  "left" and "right," 680
  mammalian, 460
  midbrain, 676, 677
  mind and, 680–83
  Neanderthal, 341–42
  PET scans in diagnosing disorders of, 23
  respiratory center of, 576–77
  vertebrate, 451
Brain size, 338, 343
Branch root, 475–76
Bray, D., 74
Bread making, 402
Bread mold (*Neurospora*), 164, 165
Breast cancer, genetic tests for, 259–60
Breastfeeding, hormones and, 652
Breathing, cellular respiration and, 569
Breeding, selective, 233, 244
Briggs, Robert, 747
Bronchioles, 574
Bronchitis, chronic, 578
Bronchus/bronchi, 574

Brood pouch, 717
Brown, Louise, 734
Brown algae, 378, 380, 382, 383
Brundtland, Gro Harlem, 856
Bryophytes (nonvascular plants), 413, 414
Bt toxin, 252, 254
Budding, 186, 401, 434, 718
Buds, 468, 528
  lateral, 476, 477, 478, 524
  taste, 595, 690
Buffalo grass plant, 299
Buffer, 30
Bulbourethral glands, 722, 723, 724
Bulimia nervosa, 588
Bulk flow, 484, 488, 571
Bullfrog, 573
Bullhead catfish, 782–83
Bullock's oriole, 358
Bundle-sheath cells, 124
Burs, 423
Butterflies, 445, 510, 856, 857, 886
Butterfly weed, 421
Byers, John, 786

C_3 pathway (Calvin-Benson cycle), 122
C_4 pathway, 123–25
Caco pods, 21
Cacti, 491
Calanolide, 866
Calcitonin, 179, 648, 652, 708, 710
Calcium, 586, 654, 706
Calcium carbonate, 382
Calcium ions, muscle contraction and, 703–4
Calcium phosphate, 707
Calico cat, 180
Callus, 709
Calophyllum trees, 233
calorie (c), 31
Calorie (C), 43, 584
calorie (c), 584
Calvin-Benson cycle (C_3 pathway), 122
Cambia/cambium (lateral meristems), 470
Cambrian period, 332
Camouflage, 819, 820
Cancer, 158
  biodiversity and, 15
  breast, 259–60
  cause of, 637–38
  immune response and, 637–38
  lung, 578, 579
  medical treatments for, 638
  skin, 148, 149, 158, 859
  viruses and, 367
Cancer cells, 759
Candida, 400, 401
Canidae, 359
Canine distemper, 245
Canine teeth, 594
Cannibalism, 301, 370
Cannon, W. B., 534, 536
Capillaries, 559–60
  gas exchange in, 575
  gill, 551
  lymph, 562, 564
Capsules, bacterial, 80, 371
Car, thermodynamics of, 100
Carbohydrates, 37, 39–43, 584, 585
  defined, 39
  disaccharides, 39, 40, 41
  monosaccharides, 39, 40–41
  polysaccharides, 39, 40, 41–43, 585
Carbon, molecular, 38
Carbon-based (organic) molecules, 2, 4, 38, 39
Carbon cycle, 843–44, 849–59
Carbon dioxide, 402
  atmospheric, 486
  C_3 cycle in capture of, 122–23
  global warming and, 850
  photosynthesis and, 124–25
  potassium transport into guard cells and, 487
Carbon dioxide transport, in respiration, 575–76
Carbon fixation, 122–23
Carbonic anhydrase, 576
Carboniferous period, 334
Carbon monoxide, 106, 578–79
Carboxyl group, 38, 43
Carboxypeptidase, 597
Carcinogenic substances, 578
Carcinogens, 637
Cardiac cycle, 552, 553
Cardiac muscles, 541, 554, 700, 701, 704
Cardiovascular disorders, 562–63
Caribou, 69–71
Carnivores, 584, 593, 594, 838, 884
Carotenoids, 119
Carpels, 500, 501
Carrageenan, 382
Carrier proteins, 60, 63
Carriers, heterozygous, 227, 228
Carrion flower, 510
Carrying capacity, 799, 800, 809
Cartilage, 451, 539–40, 706–7
Cartilaginous fishes, 452, 453
Casein, 46
Casparian strip, 475, 482, 485
Cat, calico, 180
Catalysts, activation energy and, 106–7. *See also* Enzymes
Cataplexy, 211
Catastrophism, theory of, 271
Caterpillars, 445
Cats, calico, 180
Cattleya orchid, 490
Caudipteryx, 268, 269
Caulerpa, 830
Causality, natural, 8–9, 10
Cech, Thomas, 326
Cell body of nerve cell, 541, 664
Cell cycle, 191–94
Cell division, 93, 186. *See also* Meiotic cell division; Mitotic cell division
Cell-mediated immunity, 627, 628–29, 630
Cell membrane, 57–73
  diversity of, 69–71
  fluid mosaic model of, 58–60
  transport across, 60–68
    energy-requiring, 61
    passive, 61–63
Cell plate, 187
Cell(s), 75–97. *See also* Eukaryotic cells; Prokaryotic cells; *specific types of cells*
  basic features of, 75–80
  cancer, 759
  connections and communication of, 68–69
  defined, 4
  desmosomes and, 68
  enzyme regulation by, 108
  gap junctions and plasmodesmata and, 68–69
  "information flow" in, 178
  metabolism of, 76, 105–9
  microscopic observations of, 78–79
  nucleus of, 8
  tight junctions and, 68
  types of, 8
Cell theory, 11
Cellular energy, coupled reactions and, 103–5
Cellular reproduction, 185–209. *See also* Meiotic cell division; Mitotic cell division
  asexual, 187–88
  in eukaryotic cells, 191–94
  functions of, 186–88
  sexual, 188
Cellular respiration, 116, 137–43, 569
  chemiosmosis, 140, 141–43
  defined, 137
  electron transport system and, 140–41
  main events of, 138–39
  mitochondrial matrix reactions in, 139–40, 141
  organism function and, 143
Cellular slime molds, 379–80
Cellulase, 592
Cellulose, 39, 40, 41, 42, 585
Cell wall, 42, 69, 70, 80, 82, 83, 85, 371
Centers for Disease Control, 302
Central canal, 684, 708
Central nervous system (CNS), 672–73, 705
Central vacuoles, 65, 82, 83, 89–90
Centrioles, 81, 82, 93, 195
Centromere, 189
Cephalization, 431, 672
Cephalopods, 448, 449
Cerebellum, 677
Cerebral cortex, 679
Cerebral hemispheres, 679
Cerebrospinal fluid, 673
Cerebrum (forebrain), 676, 677–80
Cervical cap, 732, 733
Cervix, 726, 729
Chambered nautilus, 449
Chance mutations, 13
Channel proteins, 59–60, 63
Chaparral, 869
Chargaff, Erwin, 151–52
Charging display, 775
Checkpoints, 194
"Cheddar Man," 257
Cheeses, 401
Cheetahs, 292, 820, 867
Chemical bonds. *See* Bonds
Chemical reactions. *See* Reaction(s)
Chemical senses, 690–91
Chemical warfare, predator–prey, 822–23
Chemiosmosis, 119, 121, 139, 140, 141–43
Chemoreceptor, 684
Chemosynthesis, 883
Chemosynthetic bacteria, 375
Chemotactic bacteria, 372
Chemotherapy, 638
Chestnut blight, 254, 394, 399–400
Chiasmata, 201, 220
Chicxulub crater, 338
Childbirth, 537
Chili pepper, 514
Chimpanzees, 353, 355, 768, 775–76
Chironex fleckeri (sea wasp), 436, 437
Chiszar, David, 185
Chitin, 41, 43, 69, 443
Chlamydia, 731
Chlamydia trachomatis, 731
Chlordane, 843
Chlorella, 331
Chlorine, 25, 586, 859
Chlorofluorocarbons (CFCs), 859
Chlorophyll, 91, 119, 527

Chloroplasts, 82, 83, 90
ATP synthesis in, 119, 121
evolution of, 330, 386
function of, 91
light capture by, 118–19
phagocytosis of, 87
photosynthesis and, 116–17
structure of, 91
Chocolate, 21
Cholecystokinin, 599, 600, 649, 657
Cholera, 376
Cholesterol, 40, 45, 46, 58, 59, 561, 562, 584–85
Choline, 587
Chondrichthyes, 452, 453
Chondrocytes, 707
Chordates, 433, 435, 450–60
amphibians, 334–35, 454–55, 572, 573, 658
birds, 335–36, 455–59, 572–73, 592, 742–43
body cavity in, 432
lancelets, 450–51, 452
mammals, 335–36, 459–60, 572–73
reptiles, 335–36, 455–59, 572, 573, 743–44
tunicates, 450–51, 452
vertebrates, 435, 451–60
Chorion, 744
Chorionic gonadotropin (CG), 729
Chorionic villi, 753
Chorionic villus sampling, 260, 261
Choroid, 686, 687
Chromatids, 155, 189–90, 196, 201–2, 203
Chromatin, 82, 83–84
Chromosome number
errors in, 230–34
reproductive isolation and, 310, 311–13
Chromosome(s), 82, 83, 84, 155
composition of, 150
DNA organization into, 188–91
duplicated, 190
haploid number, 191
homologous. *See* Homologous chromosomes
human and chimp, similarity of, 353, 355
intergenerational passage of, 188–89
during metaphase, 196
during mitosis, 190–91
pairs of, 190–91
during prophase, 194–96
relationships among genes, alleles and, 212
sex, 180, 181, 190, 220, 221, 230–33
structure of, in eukaryotes, 189–91
during telophase, 197
transcriptional regulation of, 180
Chronic bronchitis, 578
Chthamalus, 818
Chyme, 596
Chymotrypsin, 597
Chytrids, 393–94
Cichlid fishes, 314
Cilia, 82, 92, 93–94, 385
Ciliary action, 620–21
Ciliates, 378, 385–86
Circulation, 549–67
blood. *See* Blood
blood vessels, 550, 558–61
arteries and arterioles, 552, 558, 559, 561, 607, 609
capillaries, 551, 559–60, 562, 563, 575
structures and interconnections of, 559
veins and venules, 552, 558, 559, 560–61
of echinoderms, 450
functional diversity of, 550–51
heart (vertebrate), 550, 551–54
chambers of, 552–54
disorders of, 562–63
evolution of, 551
vertebrate, 543
Citrate, 139, 141
Clams, 433, 435, 448
Class, 351
Classification. *See* Systematics
Claviceps purpurea, 401
Clawed frog, 747, 750
Clear-cutting, 874
Cleavage, 745
Climate, 857–62
extinction through changes in, 337
influences on, 857–62
plate tectonics and, 337
prebiotic evolution and, 324–27
sun and, 857–58
Climax community, 826, 829–31
Clinton, Bill, 835
Clitoris, 726, 729
Clonal selection, 627
Clones, 187
Cloning, 192–93, 247, 249
of endangered species, 254
ethical issues in, 261–62
Closed circulatory system, 441, 550
Clostridium, 373
Clostridium botulinum, 376
Clostridium tetani, 376
Club fungi, 393, 394, 396
Club mosses, 413–14
"Clumped" population distribution, 802, 803
Cnidarians, 430, 433, 434, 435–37, 438
body cavity in, 432
budding in, 718
Cnidocytes, 436, 438
Coastal waters, 878–81
Coated pits, 66
Coated vesicle, 67
Cocaine, 669
Cochlea, 685–86
Cockleburs, 525, 526
Cocklebur seed, 513
Coconut fruit, 513
Coconut palm, 418
Codominance, 224
Codons, 167, 168, 173, 175
Coelacanth, 454
Coelom, 431, 432, 441
Coelomates, 431–32
Coenzymes, 52, 109, 139, 140
Coevolution, 299, 508–14, 801, 816, 823
Cohesion of water, 29–30
Cohesion–tension theory, 484–85
Cold, seed germination and, 507
Coleoptiles, 507, 522–23
Collagen, 539–40, 706
Collar cells, 434
Collecting duct, 608, 609
Collenchyma, 471, 472
Colon, 599
Colonies, bacteria, 371
Coloration
startle, 822
warning, 819, 821
Color blindness, 221, 229
Color vision, 338
Colostrum, 757
Combat®, 281
Commensalism, 299, 816, 823
Communication, animal, 771–74
chemical messages, 773
by sound, 772
by touch, 773–74
visual, 771–72
Community(ies), 4, 815–33. *See also* Ecosystems
climax, 826, 829–31
competition in, 816–18
defined, 793
keystone species in, 824–26
predator–prey interactions, 816, 818–23
succession in, 826–31
symbiosis in, 816, 823–24
Compact bone, 707
Companion cell, 473
Competition, 299, 801–2
animal behavior and, 774–77
contest, 802
defined, 801
extinction through, 317
interspecific, 801–2, 816–18
intraspecific, 802, 818
population regulation and, 801–2, 818
scramble, 802
Competitive exclusion principle, 816–17
Competitive inhibition, 108, 109
Complementary base pairs, 152–53, 173
Complementary DNA (cDNA), 167, 250
Complement proteins, 625
Complement system, 628
Complete flowers, 500
Compound, defined, 24
Compound eyes, 444, 686
Compound fracture, 698, 699, 711
Concentration gradients, 60, 63
Concentration of molecules in fluid, 60
Conclusion, 9, 10, 11
Condensation, DNA, 189, 190
Conditioning, 767–68
Condom, 732, 733–34
Conducting portion of respiratory system, 573–75
Cones, 687–89
Conifers (evergreens), 333, 411, 418–19, 499–500
Conjugation
bacterial, 373–74
protistan reproduction by, 377
Connective tissues, 539–40
Connell, J., 818
Conservation of energy, law of, 100
Constant-loss survivorship, 803, 804
Constant region, 624, 625
Consumers (heterotrophs), 8, 837, 838
Contest competition, 802
Continents, weather and climate effects of, 861–62
Continuity of life, 6
Contraception, 732–35
barrier methods of, 733–34
intrauterine devices, 732, 734–35
permanent, 732–33
synthetic hormones, 733–34
Contraceptive sponge, 732
Contractile fibers, 541
Contractile vacuoles, 65, 88, 386
Control center, 536
Controls, experimental, 9, 10
Convergence, 670–71
Convergent evolution, 277–78
Convolutions, 679
Cooperative behavior, 781
Coprolites (fossilized feces), 271, 336
Copulation, 720–22, 729
Coral, 381, 436
Coral reef, 880–81

Corals, 436, 437, 720
Cordyceps, 400
Cork, woody, 478–79, 480
Cork cambium, 479
Cork cells, 471, 479
Corn, 233, 508
Cornea, 686, 687
Corona radiata, 729–30
Coronary bypass surgery, 563
Corpse flowers (*Amorphophallus titanum*), 510, 511
Corpus callosum, 679
Corpus luteum, 725, 726, 728, 729
Correns, Carl, 218
Corroboree toad, 456
Cortex
  adrenal, 656
  cerebral, 679
  renal, 608
  of root, 474
  of stem, 477
Côte d'Ivoire, 809
Cotyledons (seed leaves), 468, 504, 505, 507–8, 509
Coupled reactions, 102–5
Courtship rituals, 79, 315–16, 720, 774
Covalent bonds, 24, 25–27
Cow, 592
Cowpox, 631
Crab lice (pubic lice), 731
Crabs, 445
"Crack" cocaine, 758
Crayfish, 445
Creation, notion of, 271
Creationism, 9
Cretaceous era, 334, 336
Cretinism, 652
Creutzfeldt-Jakob disease, 370
  variant of, 370
Crick, Francis, 11, 152–53, 154, 167, 171
Crill, Wayne, 535, 544
Cristae, 90
Crocodiles, 455, 459
Cro-Magnons, 342–43
Crop (digestive organ), 441, 591
Crops, genetically improved, 233
Cross-bridges, 702, 703
Cross-fertilization, 213, 216–17
Crossing over, 201, 202
  genetic variability and, 205
  linked genes and, 220
Crustaceans, 433, 435, 445
Crust, of Earth, 327
Ctenophores, 430, 433
Cuckoo, 766
Cud, 592
Cud-chewing, 592–93
Cultural evolution, 345
Cultural revolution, 805
Cultured bacteria, 255
Cultures, 11
Curare, 669
Cuticle, 116, 117, 412, 470–71
Cuttlefish, 448
Cuvier, Georges, 271, 307
Cyanide, 143
Cyanobacteria, 330, 375, 876
Cycads, 418, 419
Cyclic adenosine monophosphate (cyclic AMP, cAMP), 40, 52, 645
Cyclic nucleotides, 52
Cysteine, 47
Cystic fibrosis (CF), 251, 255, 257
  genetic tests for, 259–60
Cysts, 384, 438
Cytokinesis
  in animal cell, 197, 198
  during meiotic cell division, 201, 202
  during mitotic cell division, 194, 197–98
  in plant cell, 197, 198
Cytokinins, 519, 520, 524, 525, 528
Cytoplasm, 4, 58, 76, 81
  of eukaryotic cells, 81
  of plant cells, 482
  of prokaryotic cells, 82
Cytosine, 51, 151–52, 158
Cytoskeleton, 81, 82, 91–93
Cytotoxic T cells, 624, 629, 637

Daffodil bulbs, 491
Dall mountain sheep, 803
Dams, 634
Dandelions, 418, 513
Daphnia pulex (waterflea), 2, 446
Darwin, Charles, 12, 273–76, 280, 282, 288, 299, 302, 306, 308, 353, 522, 792
Darwin, Francis, 522
Darwin's finches, 273, 292, 313–14
Daylength, flowering and, 525–27
Day-neutral plant, 525
DDT, 647, 799–800, 835, 841, 852
Death cap mushroom (*Amanita phalloides*), 390, 391
Death rates, 794, 796
Decay, radioactive, 23
Deciduous forests
  temperate, 870–72
  tropical, 867
Decomposers, 838
Deep-sea angler fish, 454
Deep-sea vents, 327, 883
Deforestation, 233, 849
  desertification and, 809
  diseases and, 634
Dehydration synthesis, 39, 41, 48
Delbruck, Max, 154
Deletion mutation, 174
Delivery, 756–57
Demerol®, 691
Denatured proteins, 51
Dendrites, 541, 664
Denitryfing bacteria, 845
Density-dependent factors, 799, 800
Density-independent factors, 799–800
Dental plaque, 371, 372
Deoxyribonucleic acid. *See* DNA
Deoxyribose, 40–41, 51
Department of Energy (DOE), 259
Depo-Provera®, 732, 733
Depth perception, 689
Dermal tissues in plants, 470–71
Dermis, 539
Descartes, René, 656
DES (diethylstilbestrol), 647
Desertification, 809, 869
Deserts, 868–69
"Designer" microbes, 631
DeSilva, Ashanti, 257
Desmosomes, 68
Detritus feeders, 838
Deuteromycota, 393, 396
Deuterostome, 432
Development, 740–63
  aging, 757–59
  cell differentiation, 746–47
  cleavage, 745
  defined, 746
  direct, 742, 743–44
  gastrulation, 745
  human, 750–59
  indirect, 742
  organogenesis, 745
  sexual maturation, 745
Developmental stages, as taxonomic criterion, 353
Devil's Hole pupfish, 316–17
de Vries, Hugo, 218
Diabetes mellitus, 654
  juvenile-onset, 633–34
Dialysis, 605, 612
Diana, Princess, 589
Diaphragm, 576, 732, 733–34
Diastolic pressure, 552
Diatoms, 378, 380, 381
Dicots, 468, 469, 474, 476, 477, 479, 504, 505
Dictyostelium, 379
Dideoxyinosine (ddI), 637
Didinium, 386
Diener, T. O., 369
Dietary pyramid, 588
Diethylstilbestrol (DES), 647
Differentially permeable plasma membrane, 62
Differential reproduction, 296
Differentiated cells, 468–70
Differentiation, 192, 194
  defined, 746
Diffusion, 60–61
  facilitated, 61, 62, 63
  gas exchange by, 571
  simple, 61, 62–63
  transport across cell membranes and, 61–65
Digestion, 589–600
  adaptations in, 590–93
  in annelids, 441
  defined, 589
  extracellular, 591
  human. *See* Human digestion
  intracellular, 590
  in ruminants, 592–93
  in sponges, 590–91
  in vertebrate, 543, 544
Digestive enzymes, 71
Digestive tract, 649
Digestive tube (gut), 431
Dihydroxacetone phosphate, 136
Dimorphism, sexual, 778
Dinoflagellates, 378, 380–81, 880
Dinosaurs, 13, 335, 336
  DNA of, 255
  extinction of, 115, 126
  feathered, 268, 269, 282
Dioecious species, 719
Diploid cells, 190, 191, 199
Direct development, 742, 743–44
Directional selection, 296, 297
Disaccharides, 39, 40, 41
Disease(s)
  dams and, 634
  deforestation and, 634
  from fungi, 399–401
  neurotransmitters and, 669
Displays, sexual, 777–78. *See also* Courtship rituals
Disruptive selection, 296, 297, 298
Distal tubule, 608, 609
Disulfide bridges, 47, 48, 49, 50

Divergence, 671
Diversity of life, 6–8
Division (phylum), 351
Djekoule, Sep, 809
DNA, 6, 40, 41, 52, 82, 148–61
  bases in, 151, 167
  cell structure, function, and reproduction and, 76
  in chromatin, 83–84
  complementary, 167, 250
  condensation of, 189, 190
  in dinosaur, 255
  double helical structure of, 152–53, 189–91
  in eukaryotic cells, 83–84, 188–91
  genetic variation and, 12
  information storage in, 153–54
  "naked," 258
  nucleotides in, 151–52
    order of, 153
  recombinant, 244–47
    in genetic engineering laboratories, 246–47
    library for, 246–47
    in nature, 244–46
  repair enzymes of, 158
  replication of, 154–58, 166
  RNA compared to, 165
  severing with restriction enzymes, 246
  strands (polymers), 152
  viral, 366
  X-ray diffraction pattern of, 152
DNA fingerprinting, 251, 256–57
DNA helicase, 155–57
DNA library, 246–47
DNA ligase, 156–58, 247
DNA polymerase, 156–57, 249–50
DNA probes, 248
DNA sequence, 356–57
  as taxonomic criterion, 353, 356–57
Dobzhansky, Theodosius, 12, 268
Dodo, 506
Dog(s)
  diversity of, 281
  narcoleptic, 210, 211, 235
Dolly (cloned sheep), 192–93
Domains, 6, 351, 354. *See also* Archaea; Bacteria; Eukarya
  of prokaryotic organisms, 355, 356
Dominance, incomplete, 223
Dominance hierarchies, 774–75
Dopamine, 669
Dormancy, plant, 507, 528
  hormonal regulation of, 520
Dorsal lip of blastopore, 748–49, 750
Dorsal root ganglia, 675
Dorsal surface, 430
Double covalent bond, 26
Double fertilization, 503
Double helix, 152, 189–91
Double Helix, The (Watson), 154
Doubling time, 795–96
Douching, 734
Down syndrome (trisomy 21), 180, 234
Downy mildew, 379
Dragonfly, 742
Drones, 781
Drosera rotundifolia (sundew), 466, 467, 491, 492
Drosophila, 220, 221–22
Drug(s)
  crossing placenta, 758
  neurotransmitters and, 669
Drying, seed germination and, 507
Dryopithecines, 338–39
Duckweed, 421
Duct, 539
Ductless glands, 649
Ducts, 648
Duplicated chromosome, 155, 190
Dutch elm disease, 394, 399
Dwarfism, 651, 652
Dwarf seahorses, 716, 717, 736
Dylan, Bob, 400
Dynamic equilibrium, 60, 536
Dysentery, amoebic, 384

Eardrum (tympanic membrane), 684, 685
Early-loss survivorship, 803, 804
Ears, 683–86
Earthstar mushroom, 393
"Earth Summit" (1992), 15
Earthworm, 441, 591–92, 606
Eating disorders, 589
Ebola hemorrhagic fever, 287, 367–68
Echinoderms, 430–31, 433, 435, 448–50
Echolocation, 691–92
E. coli, 186, 246, 252, 366, 373, 374, 557
Ecological isolation, 309–11
  reproductive isolation through, 314, 315
Ecological niches, 816–18
Ecology, 793
  fungi's role in, 402
EcoRI restriction enzyme, 246
Ecosystems, 4, 835–55
  acid rain (acid deposition), 847–49
  chaparral, 869
  climate and weather, 857–62
    influences on, 857–62
    sun and, 857–58
  coastal waters, 878–81
  defined, 793
  deserts, 868–69
  energy flow through, 835–42
    detritus feeders and decomposers, 838
    food chains and food webs, 838, 839, 840
  feeding relationships in, 836, 838
  freshwater lakes, 876–78
  global warming, 486, 849–51
  grasslands (prairie), 799, 809, 840, 843, 870
  human impact on, 866, 867–68, 869, 870, 872, 874, 875, 877–78, 879–80, 881–83, 884–85
  hydrothermal vent communities, 883
  marine, 878
  nutrient pathways through, 835, 836, 843–47
    carbon cycle, 843–44, 849–51
    nitrogen cycle, 844–45, 848–49
    phosphorus cycle, 846
    water cycle (hydrologic cycle), 846–47
  open ocean, 881–83
  rainfall and temperature effects on vegetation, 875–76
  requirements for life, 862–63
  savanna, 867–68
  taiga (northern coniferous forests), 873–74
  temperate deciduous forests, 870–72
  temperate rain forests, 872–73
  tropical deciduous forests, 867
  tropical rain forests, 15, 864–67
  tundra, 874–75
Ectoderm, 430, 746, 753
"Edible" vaccines, 252
Eels, 693
Effectors, 536, 671
Egg(s), 719, 726, 729–30
  amniote, 455, 457, 744
  amphibian, 455, 456, 457
  in flowering plants, 503
  fossilized, 271
  journey of, 752
  reptilian, 335
Ehrlich, Anne, 15
Ehrlich, Paul, 15, 537
Eibl-Eibesfeldt, Irenaus, 784
Eigen, Manfred, 326
Eisely, Loren, 496
Ejaculation, 729
Elastin, 45
Electrical fields, detection of, 692
Electrical signals, ions and, 666–67
Electrolocation, 692
Electron carriers, 105
Electron microscopes, 78, 79
Electrons, 21, 22
Electron shells, 22–24
Electron transport system, 119, 120, 140–41
Electroreceptors, 692
Elements, 2, 21, 22
Elephantiasis, 563, 564
Elephants, 826, 867
Elevation, temperature and, 861
Ellis-van Creveld syndrome, 293
Elongation step
  of transcription, 167, 168, 169
  of translation, 172, 173
Embryological evidence of evolution, 279, 280
Embryonic disc, 752
Embryonic stem cells, 748
Embryo(s), 725, 729, 734, 742
  mammalian, 460
  monocot vs. dicot, 469
Embryo sac, 491–92, 502–3
Emergence of Life on Earth, The (Fry), 327
Emergent properties, 2
Emigration, 794, 802
Emphysema, 578, 579
Endangered species
  cloning of, 254
  habitat destruction and, 295
Endangered Species Act, 295, 835, 852
Endergonic reaction, 101, 102–3
Endocrine disruptors, 647, 843
Endocrine glands, 539, 645
Endocrine system, 643–61. *See also* Hormone(s)
  adrenal glands, 644, 649, 655–56
  heart, 644, 649, 657
  kidneys, 644, 649, 657
  pancreas, 596–97, 644, 648, 654
  parathyroid gland, 644, 648, 652–54
  pineal gland, 644, 649, 656–57
  pituitary gland, 649–51
  sex organs, 644, 649, 654
  thymus, 561, 564, 644, 649, 657
  thyroid gland, 644, 648, 652–54
  vertebrate, 544
Endocytosis, 61, 66–67, 93
Endoderm, 430, 746, 753
Endodermis, 475, 485
Endogenous pyrogens, 622–23
Endometrium, 726, 727, 728
Endoplasmic reticulum (ER), 82, 85–86
  membrane synthesis in and flow through, 87–88
  rough, 81, 83, 85, 87
  smooth, 81, 83, 85, 87
Endorphins, 648, 651, 669
Endoskeleton, 448, 450, 451, 704
Endosperm, 503

Endospores, 95, 372–73
Endosymbiont hypothesis, 90, 330, 386
Endosymbiosis, 90
Energetic molecules, 327
Energy, 5, 99–113. *See also* Metabolism; Photosynthesis
  activation, 102, 106–7
  cellular, 76, 103–5
  in chemical reactions, 101–3
  conservation of, 100
  defined, 100–1
  in ecosystems. *See under* Ecosystems
  extracted from food molecules, 90–91
  kinetic (energy of movement), 100
  kingdoms and acquisition of, 8
  of photon, 118
  potential (stored), 100
  properties of, 100–1
  of sunlight, 101
Energy-carrier molecules, 103–5. *See also* ATP
Energy levels, atomic, 22–24
Energy pyramid, 842–43
Energy-requiring transport, 61
Energy storage, in glucose, 122
Entropy, 101
Environment. *See also* Ecosystems
  abiotic, 298–99, 843
  animal behavior and, 770–71
  biotic, 299
  disease and, 634
  gene expression and, 225–26
  phenotype and, 225–26, 295, 296
  population growth and, 797–802
  recombinants and, 254
  sexual maturation and, 745
  traits and, 288
  transcription and, 176
Environmental estrogens, 647
Environmental Protection Agency, 251
Environmental resistance, 794
Enzymes, 45, 76, 106, 107–9, 110
  antibacterial, 620
  cellular regulation of, 108
  defined, 107
  digestive, 71, 86, 595, 596
  environmental influence on, 108–9
  genes and, 164
  lipid-digesting, 596
  restriction, 246
  structure and function of, 107–8
Epicotyl, 504
Epidermis, 116, 117, 443, 470–71, 541–42
  of leaf, 481, 485
  of root, 474
  of stem, 477, 478–79
Epididymis, 722, 723, 724
Epiglottis, 574, 595
Epinephrine (adrenaline), 645, 649, 655
Epithelial cells, 434
Epithelial tissues (epithelium), 538–39
Epstein, R., 768
Equilibrium
  dynamic, 60, 536
  genetic, 289
Equilibrium population, 289
Equisetum, 414
Erection, 730
Ergot, 401
Erythroblastosis fetalis, 557
Erythrocytes (red blood cells), 554, 555–57, 560, 706
Erythropoietin, 556, 607, 614, 649, 657
Escherichia coli (E. coli), 186, 246, 252, 366, 373, 374, 557
Esophagus, 591, 595–96
Essay on Population (Malthus), 274
Essential amino acids, 585–86
Essential fatty acids, 584
Estrada, Alejandro, 506
Estradiol, 45
Estrogens, 178–80, 645, 646, 649, 654, 725, 728, 753–54
  environmental, 647
Estuaries, 878
Ethanol, 108
Ethical issues in biotechnology, 259–62
  cloning, 261–62
  embryonic stem cells, 748
  genetic screening, 259–60
Ethologists, 770
Ethology, 769
Ethylene, 519, 520, 527–28
Eubacteria, 354
Eucalyptus tree, 421
Euglena, 94, 381–82
Euglenoids, 378, 380, 381–82
Eukarya, 6, 7, 355
  cell type and number of, 8
  characteristics of, 357
Eukaryotic cell cycle, 191–94
Eukaryotic cells, 8, 76, 80
  cellular respiration in, 137
  cytoplasms of, 81
  DNA organization in, 188–91
  evolution of membrane-enclosed organelles and nucleus in, 329–31
  features of, 81–94
    internal membranes, 84–88
    mitochondria, 90–91
    nucleus, 82–84
    organelles, 81–82
    vacuoles, 88–90
  glucose metabolism in, 132
Eukaryotic chromosome, structure of, 189–91
Eukaryotic life cycles, mitosis and meiosis in, 203
Eukaryotic tree of life, 355, 357
Euplotes, 377
Europa, 322, 323, 345
Eustachian tube, 684, 685
Eutrophication, 877
Eutrophic lakes, 877
Everglades kite, 317
Evergreens (conifers), 333, 411, 418–19, 499–500
Evolution, 269–85. *See also* Life on Earth
  capacity for, as characteristic of living things, 6
  causes of, 289–95
    gene flow, 289–90
    mutations, 289, 290
    natural selection. *See* Natural selection
    nonrandom mating, 293–94
    small population size, 290–93
  as change of gene frequencies within population, 288–89, 295
  of chloroplasts, 330, 386
  convergent, 277–78
  cultural, 345
  defined, 269
  evidence of, 270–73, 276–79
    anatomical, 277–79
    biochemical and genetic, 279
    from early explorations, 270
    embryological, 279, 280
    fossil, 270–71, 272, 277
    geological, 271–72
  of heart, 551
  of hormones, 657–58
  of horses, 277
  human. *See* Human evolution
  inheritance and, 288
  of jaws, 452
  Lamarckian mechanism for, 272–73
  of mitochondria, 330, 386
  mutations and, 175
  natural processes underlying, 12–13
  of plant rapid-fire responses, 528–29
  of populations, 288, 290–93
  prebiotic, 324–27
  predator–prey interactions and, 818–23
  present-day, 280–82
  theory of, 6
  as unifying concept of biology, 12–13
Evolutionary tree, 313
Excitatory potential, 666–67
Excretion, 544, 606, 607. *See also* Urinary system
Excretory pore, 606
Exergonic reaction, 101–2, 103
Exhalation, 576
Exocrine glands, 539, 648–49
Exocytosis, 61, 67–68, 87
Exons, 179
Exoskeleton, 41, 334, 443, 704–5
Exotic species, 797, 830
Experience, innate behaviors modified by, 768–69
Experiment, 9, 10, 213
Exponential population growth, 794–96, 803–5
Extensor muscles, 711
External ear, 683–85
External fertilization, 719–20
Extinction (Ehrlich & Ehrlich), 15
Extinctions, 13
  causes of, 316–17
  mass, 336–38
Extracellular digestion, 591
Extracellular fluid, 58
Extraembryonic membranes, 743–44
Exxon Valdez disaster, 375
Eye(s)
  compound, 444
  positioning in predators vs. prey, 689
Eye color, 221–22, 224–25
Eyespot, 382

Facilitated diffusion, 61, 62, 63
FAD, 105
$FADH_2$, 142
Fairy rings, 396, 398
Family, 351
Farming, organic, 884, 885
Farms, 885
  acid deposition damages to, 849
Farsightedness, 687, 688
Fat (adipose) tissue, 540, 649, 657
Fats, 37, 40, 43–45, 134, 539, 584–85
Fat-soluble vitamins, 587, 588
Fatty acids, 43, 44, 46, 53, 584
  saturated, 44, 70
  "trans," 46
  unsaturated, 44, 70
Feathered dinosaur, 268, 269, 282
Feces, 599
  fossilized (coprolites), 271, 336
Feedback inhibition, 108
Feedback systems, 536–37, 556–57, 612
  hormone regulation by, 647, 653

Fermentation, 132, 135–37
alcoholic, 137
of pyruvate, 135, 137
Ferns, 410, 411, 413–14, 417, 498
Fertility
limiting, 732–35
replacement-level (RLF), 806
Fertilization, 503, 729–31, 752
defined, 718
double, 503
external, 719–20
internal, 335, 455, 720–22
in vitro, 731, 734–35
Fetal Alcohol Effect, 759
Fetal alcohol syndrome (FAS), 759, 760
Fetus, 725, 753
Fever, 621, 622–23
Fiber, 41
Fibrillation, 554
Fibrin, 558
Fibrinogen, 555, 558
Fibrous connective tissues, 539
Fibrous root system, 473, 474
Fiddler crabs, 775
Fifer, William, 784, 785
Fig wasp, 315
Filamentous bodies of fungi, 392
Filament(s), 500, 501, 572
intermediate, 92, 93
Filtrate, 609
Filtration, glomerular, 609–11
Fimbria/fimbriae, 726, 727
Firefly, 778
"Fireworm" polychaete, 442
First law of thermodynamics, 100
First messenger, 644
Fir trees, subalpine, 14
Fish(es)
aggressive display in, 775
bony, 452, 453–54
cartilaginous, 452, 453
electrolocation by, 692
genetically engineered, 253
jawless, 451–53
overfishing, 809
thyroxine effects on, 657–58
Fish farming, 883
Fission, 373–74, 718
Fitness, inclusive, 300
Fitness of organism, 295
Fixed behavior patterns, 766
Flagella, 81, 82, 92, 93–94
of dinoflagellates, 380–81
in euglenoids, 381–82
Flagella movement of prokaryotes, 371–72, 373
Flame cells, 606
Flatworms, 433, 434, 437–40, 457, 570
body cavity in, 431, 432
excretory system of, 606
Flavenoids, 32
Flavin adenine dinucleotide (FAD), 105
Fleming, Alexander, 11
"Flesh eating" infection (necrotizing fasciitis), 377
Flexor muscles, 711
Flies, as pollinators, 510, 511
Flight, 444
Florida scrub jay, 300–1, 780
Florida treehopper, 820
Florida tree snail, 447
Florigens, 525
Flowering
daylength and, 525–27
hormonal regulation of, 525–27
Flowering plants, 7, 334, 411, 468
life cycle of, 422
ovary wall of, 506
Flowers, 413, 419, 421, 468, 499, 500–3
coevolution with pollinators, 508–14
complete, 500, 501
gametophyte development in, 499, 500–3
incomplete, 500
as leaf modification, 491
monocot vs. dicot, 469
pollination and fertilization of, 503–4
Flu (influenza), 619, 632–33, 638
Fluid mosaic model of membranes, 58–60
Fluid(s)
amniotic, 260
cerebrospinal, 673
extracellular, 58
interstitial, 560
isotonic, 64
movement of molecules in, 60–61
Flukes, 439
Fluorine, 586
isotope of, 23
fMRI (functional magnetic resonance imaging), 681
Folic acid, 587
Follicle, 725, 726, 728
Follicle-stimulating hormone (FSH), 648, 651, 654, 724, 725, 728, 729
Food allergies, 633
Food and Drug Administration (FDA), 37, 251
Food chains, 838, 839
Food granules, 80
Food molecules, energy extraction from, 90–91
Food vacuole, 67, 87, 591
Food webs, 838, 840
Foose, Thomas, 295
Foot, gastropod, 447
Footprints, fossilized, 271
Foraminiferans, 384, 385
Forebrain (cerebrum), 676, 677–80
Foreman, George, 5
Forensics, DNA fingerprinting in, 251, 256
Forests
deciduous, 867, 870–72
northern coniferous (taiga), 873–74
rain forests, 864–67, 872–73
tropical, 317
Fore tribe, 369–70
Fossil fuels, 843, 848, 884
Fossil record
as evidence of evolution, 270–71, 272, 277–79
primate, 338
sympatric speciation in, 313
Fossil(s)
determining age of, 329
fraudulent, 282
of fungi, 393
living, 13
types of, 271
Founder effect, 292–93
Fovea, 687, 688
Foxgloves, 509
Fracture, compound, 698, 699, 711
Franklin, Rosalind, 152, 154
Fraternal twins, 785–86
Free-living flatworms, 438
Free nerve endings, 683
Free nucleotides, 155
Free radicals, 21, 32, 587
Free ribosome, 81, 83
"Free" water molecules, 63
Freshwater lakes, 876–78
Frigate bird, 458
Frisch, Karl von, 781
Frogfish, 820
Frogs
declining population of, 456–57
gastrulation in, 745
mating pose in, 720, 721
Frontal lobe, 679, 681
Fructose, 39, 40
Fructose bisphosphate, 132, 136
Fruit, 413, 423, 468, 500
development of, 504–7, 527–28
hitchhiker, 513
Fruiting body, 378, 379, 380
Fruits, 512–14
animal-dispersed, 513–14
explosive, 513
water-dispersed, 513
wind-dispersed, 513
Fry, Iris, 327
Fucus, 383
Fumarate, 141
Functional groups, 38
Functional magnetic resonance imaging (fMRI), 681
Fungal pesticides, 400
Fungus/fungi, 6, 7, 353, 354, 391–407, 824
adaptations of, 392–93, 402–3
amphibian decline due to, 456–57
cell number in, 8
cell walls of, 69
characteristics of, 357
chestnut blight, 254
classification of, 393–99
club fungi, 393, 394, 396
imperfect fungi, 393, 396, 402–3
sac fungi, 393, 394–96
symbiotic relationships and, 397–98, 399
zygote fungi, 393, 394
effect on humans, 399–402
as nematode nemesis, 403
reproduction of, 187–88
on roots, 482–83
spores of, 392–93, 402–3
truffles, 402
Funguslike protists, 378–80
Fusion, heat of, 32

G_0 phase, 191, 194
G_1 phase, 191, 194
G_2 phase, 191, 194
G3P (glyceraldehyde-3-phosphate), 122, 132, 136
GABA (gamma amino butyric acid), 669
Gage, Phineas, 682–83
Galactose, 40
Galapagos Islands, 273
Galapagos tortoises, 274–75
Galileo (spacecraft), 323
Gallbladder, 596
Gametes, 199
fusion of, 205
mutations in, 175
spores vs., 498
Gametic incompatibility, 314, 316
Gametophyte, 410, 498, 499
embryo sacs as, 502–3
pollen as, 501–2
Gamma amino butyric acid (GABA), 669
Gangestad, Steven, 765, 787
Ganglion cells, 688

Ganglion/ganglia, 438, 672
dorsal root, 675
Gangrene, 401
Gap junctions, 68–69, 700, 704, 712
Garden spider, 443
Gas exchange, 573, 575, 576. *See also* Respiration
in arthropods, 444
chemistry of, 576
by diffusion, 571
evolutionary adaptations for, 570–73
through gills, 571–72
Gastric brooding frog, 456
Gastric inhibitory peptide, 599, 600
Gastrin, 595, 599, 649, 657
Gastronomy, 401–2
Gastropods, 447
Gastrovascular cavity, 436, 591
Gastrula, 746
Gastrulation, 745, 746
Gatekeeper proteins, 82
Gause, G. F., 816–17
Gel electrophoresis, 248
Gelsinger, Jesse, 257
Gene expression, 176–77, 225–26
Gene flow, 289
evolution through, 289–90
isolation from, speciation and, 308, 309, 310
Gene frequencies, evolution as change of, 288–89, 296
"Gene gun," 251
Gene identification, 248–49
Gene pool, 288
Generative cell, 501
Gene(s), 4, 149–51, 163–83. *See also* DNA; Inheritance
animal behavior and, 770–71
behavior and, 785–86
constancy of, 154–58
enzymes and, 164
fragmented, 179
linkage of, 219–20
locus of, 212
with multiple alleles, 223–24
mutations and, 174–80
proteins and, 164–67
regulation of, 175–80
relationships among alleles, chromosomes, and, 212
sex-linked, 220–22
sexual maturation and, 745
SRY, 225
traits determined by, 288
vectors for human, 258
wild, 233
Gene sequencing, 249–51
Gene therapy, 257–58
Genetic code, 166–67
Genetic disorders, 227–30
albinism, 226, 227–28
Huntington disease, 229, 231, 255, 257
investigation of, 226–27
knock-out mice for modeling, 255
sex-linked, 229–30
sickle-cell anemia, 175, 228, 260–61, 298
Genetic divergence of two populations, 308–9
Genetic drift, 291–92
Genetic engineering, 244, 251–55, 548, 549, 565, 631. *See also* Biotechnology
Genetic equilibrium, 289
Genetic evidence of evolution, 279
Genetic recombination, 220
Genetics. *See also* Inheritance
Mendel and, 212–13
molecular, 353, 356
population, 288
Genetic screening, 255
ethics of, 259–60
prenatal, 260–61
Genetic variability
genetic drift and, 292–93
meiosis and sexual reproduction in producing, 203–5
Genetic variation, 12
Gene transcription, during development, 747–50
Genital herpes, 731
Genital warts, 731
Genome, 251
Genotype, 215, 288
adaptiveness of, 294–95
natural selection and, 294–95, 296
Genus/genera, 351
Geographical isolation of population, 309
reproductive isolation through, 314, 315
Geological evidence, of evolution, 271–72
Germination, 498
Germination, hormonal regulation of, 520
Germ layers, 430
Gestures, universal, 784–85
Giant kelp, 382, 383
Giant muntjac (barking deer), 307, 318
Giant squid (*Architeuthis*), 428, 429, 461
Giant tube worms, 883
Giardia, 384, 626
Gibberellin, 519, 520, 527
Gigantism, 651
Gill capillaries, 551
Gill grooves, 753
Gills, 444, 571–72
Gill slits, pharyngeal, 450
Ginkgo (maidenhair tree), 418, 419
Giraffes, long necks of, 283, 296
Girdling, 479
Gizzard, 441, 591
Gland(s)
bulbourethral, 722, 723, 724
endocrine, 539, 645. *See also* Endocrine system
exocrine, 539, 648–49
mammary, 459
parathyroid, 644, 648, 652–54
prostate, 722, 723, 724
salivary, 539, 594–95
sebaceous, 542, 543
sweat, 539, 542, 543
Glial cells, 541
Global Earth Summit conferences, 851
Global warming, 486, 849–51
Globulins, 555
Glomerulus, 608, 609–11
Glucagon, 648, 654
Glucocorticoids, 649, 656
Glucose, 39, 40, 42, 585
activation of, 132, 133
anaerobic vs. aerobic breakdown of, 135
in blood, pancreatic control of, 655
burning of, 102
energy captured from, 132–37
energy storage in, 122
metabolism of, 132. *See also* Glycolysis
Glutamic acid, 47, 174–75
Glyceraldehyde-3-phosphate (G3P), 122, 132, 136
Glycerol, 43, 45
Glycogen, 40, 41, 585, 654
Glycolipids, 88
Glycolysis, 105, 131–37, 139
defined, 132
NADH from, 133, 136
organism function and, 143
pyruvate from, 132–35
reactions during, 136
Glycoproteins, 59
GnRH (gonadotropin-releasing hormone), 724, 725, 728, 729
Gobea, Andrew, 258
Goiter, 654
Golden eagle, 796
Golden toad, 456
Golgi complex, 81, 82, 83, 86, 87
structure and function of, 86
Gonadotropin-releasing hormone (GnRH), 724, 725, 728, 729
Gonads, 644, 722
Gonorrhea, 371, 376, 731
Goodall, Jane, 775–76
Goose brain, 677
Goose-neck barnacle, 446
Gordon, Ian, 857
Gould, Stephen Jay, 345, 350
Gradient, 60–61
concentration, 60, 63
Gram stain, 371
Grand Canyon, 270
Granum/grana, 91, 117
Grapefruit, 527
Grasping hands, 338
Grasses, 421, 823
Grasslands (prairie), 799, 809, 840, 843, 870
Gravitation, theory of, 12
Gravitropism, 520, 521
Gravity, plants' sense of, 521–23
Gray matter, 673
Gray wolves, 358
Great Barrier Reef, 880
Great Basin Desert, 868
Great horned owl, 874
Green algae, 378, 380, 382–83, 411
Greenhouse effect, 849, 858
Greenhouse gases, 849–50, 858
Green Mountains, 849
Griffith, Frederick, 150
Groos, K., 786
Ground pine (*Lycopodium*), 414
Ground squirrels, 780
Ground tissues, 470, 471, 472
Growth, as property of all organisms, 5–6
Growth hormone, 648, 651
Growth-hormone genes, 253
Growth rate of population, 794
Grunion, 719, 720
Grunts, 775
Guanine, 51, 151–52
Guano, 846
Guard cells, 481, 487
Guillemin, Roger, 650–51
Guppies, 280–81, 778
Gurdon, J.B., 747
Gusella, James, 231
Gut (digestive tube), 431
Gymnosperms, 416, 418–19, 499–500
Gyres, 860, 861

Habitat destruction, 317, 457
Habituation, 767, 769, 770

Hagfishes, 451–52, 453
Hair, 47, 48–49
root, 471, 474, 475, 482
Hair cells, 684, 685
Hair follicles, 542, 543
Haldane, John B. S., 324
Half-life, 329
Hammer, 684, 685
Hands, grasping, 338
Haploid cells, 191, 199
Haploid number, 191
Hardison, Ross, 32
Hardy, Godfrey H., 289
Hardy-Weinberg principle, 289
Haversian systems (osteons), 708
Hay fever, 633
Hcrt2 gene, 235
Head, 443
Heart, 550, 551–54
amphibian, 455
of birds, 458–59
cardiac muscle, 541, 554, 700, 701, 704
chambers of, 552–54
disorders of, 562–63
as endocrine organ, 644, 649, 657
evolution of, 551
reptilian, 455
Heart attacks, 562, 563
Heart rate, influences on, 554
Heartwood, 477
Heartworms, 441
Heat, 110
of fusion, 32
specific, 31
of vaporization, 32
Hecht, M. K., 296
Height, environmental influence on, 226
Heimlich maneuver, 574, 575
Helicobacter pylori, 597
Heliozoans, 384
Helium, 24–25
Helix, 49
Helix (snails), 774
Helper T cells, 624, 629
Heme, 50, 51
Heme group, 555
Hemings, Eston, 257
Hemings, Sally, 257
Hemispheres, cerebral, 679
Hemocoel, 444, 550
Hemodialysis, 605, 612, 614
Hemoglobin, 40, 46, 50, 51, 174–75, 199, 555, 575–76
sickle-cell anemia and, 228
Hemoglobin Machida, 175
Hemophilia, 221, 229–30
Henson, Jim, 377
Herbaceous plants, 470
Herbicide resistance, 254
Herbicides, 252
Herbivores, 593, 594, 838
coevolution of plants and, 823
Hercules beetles, 445
Hereditary Disease Foundation, 231
Hermaphrodites, 440, 719
Hermit crab, 446
Herpes, genital, 731
Herpes viruses, 366, 367, 369
Herpetologists, 456
Herring gull, 766, 769
Heterotrophism, in fungi, 392
Heterotrophs (consumers), 8, 837
Heterozygous carriers, 227, 228
Heterozygous organisms, 212, 214–15, 288
Hierarchies, dominance, 774–75
High-density lipoprotein (HDL), 46, 562
High-energy bonds, 104
Himalayan rabbit, 225–26
Hindbrain, 676, 677
Hinge joints, 709, 711
Hippocampus, 678–79
Histamine, 621, 632–33
Histoplasmosis, 400
HIV, 350, 351, 360, 368, 369, 635–36
cross section of, 367
Hodgkin, Alan, 665
Homeostasis, 2, 4–5, 536
kidneys and, 612–14
urinary system and, 607
Hominids, 338–43
Homo genus, 341
modern humans, 342–43
Neanderthals, 341–42, 343
upright posture and bipedal locomotion, 339–41
Homo erectus, 341, 343, 345
Homo ergaster, 341, 342, 345
Homo habilis, 341, 342, 345
Homo heidelbergensis, 341, 343
Homologous chromosomes, 199, 201–2, 212
dominant and recessive alleles on, 214–16
meiosis I and, 199, 200–3
shuffling of, 203–5
Homologous structures, 278, 279
Homologues, 190
Homo neanderthalensis, 341, 342, 343
Homo rudolfensis, 341, 345
Homo sapiens, 13, 339, 341, 342, 343, 344
Homozygous organisms, 212, 214, 288
Honeybees, 767, 768, 773, 781–82
Honeydew, 488
Honey possum, 511
Hong Kong flu, 632
Hooke, Robert, 78
Hookworm, 440
Hooper, Edward, 360
Hormone(s), 43, 46, 519, 643–48, 667–68. *See also* Endocrine system
adrenocorticotropic (ACTH), 648, 651, 657
antidiuretic (ADH), 610, 612–13, 647, 648, 650, 651
appetite-controlling, 589
breastfeeding and, 652
conduction in plants, 473
contraceptive, 733–34
defined, 643
digestion regulation by, 599–600
evolution of, 657–58
follicle-stimulating (FSH), 648, 651, 654, 724, 725, 728, 729
gene regulation and, 178–80
gonadotropin-releasing (GnRH), 724, 725, 728, 729
growth, 648, 651
heart rate and, 554
human growth, 255
hypothalamic, 649–51
inhibiting, 648, 650
luteinizing (LH), 648, 651, 654, 724, 725, 728, 729
melanocyte-stimulating (MSH), 648, 651
menstrual cycle and, 725–27
modes of action of, 646
peptide, 644, 646
placental, 753–54
plant, 519–32
abscisic acid, 519, 520
abscission and, 528
auxin, 523, 527
classes of, 519
cytokinins, 519, 520, 524, 525, 528
discovery of, 522–23
ethylene, 519, 520, 527–28
florigens, 525
gibberellin, 519, 520, 527
regulation of life cycle, 520–28
root-shoot balance and, 525
seed and fruit development and, 527–28
senescence and, 528
shape at maturity and, 524–25
plant life cycle regulation by, 520–28
of pregnancy, 757
receptor binding by, 644–46
regulation by feedback mechanisms, 647, 653
releasing, 648, 650
steroid, 40, 43, 45, 46, 178–80, 644, 645
thyroid, 657–58
thyroid-stimulating (TSH), 648, 651, 653
transport in plants, 477
Horn, 47
Horses
evolution of, 277
Horsetails, 413–14
Host cell, 245, 246, 366
Hosts, 438, 800, 824
Huitlacoche, 400
Human behavior, 343–45, 783–86
Human brain, 460, 677, 678
Human chromosomes, 353, 355
Human development, 750–59
birth, 756–57
embryonic, 751–54
Human digestion, 593–600
control of, 599–600
esophagus, 595–96
large intestine, 598, 599
mouth, 594–95
small intestine, 594, 595, 596–98
stomach, 594, 595–96
Human evolution, 338–45
behavior, 343–45
binocular vision, 338
brain size, 338, 343
evolutionary tree, 340
grasping hands, 338
hominids, 338–43
Human Genome Project, 259, 356
Human growth hormone, 255
Human immunodeficiency viruses. *See* HIV
Human papillomavirus (HPV), 731
Human reproduction. *See under* Reproduction
Human trachea, 574, 595
Human urinary system, 607–14
bladder, 608
kidneys, 607–8, 609, 610
Humes, David, 614
Hummingbird, 130–31, 143, 458, 510
Humoral immunity, 627–28, 630
Humpback whales, 459, 882
Hunting and scavenging, cooperative, 345
Huntington, George, 231
Huntington disease, 229, 231, 255, 260
Huntington protein, 229
Hutchinson, G. Evelyn, 98, 101
Hutton, James, 271–72
Huxley, Andrew, 665

Hybrid, 212
Hybrid infertility, 314, 316
Hybrid inviability, 314, 316
Hybridomas, 629
Hybrids, 215
Hydra, 187, 188, 434, 435–36, 438, 718
  digestion in, 591
  nervous system of, 671, 672
Hydrochloric acid, 595–96
Hydrogen, 24–25
  covalent bonding of, 26, 27
  as functional group, 38
Hydrogen bonds, 24, 26–27, 28, 48
  cohesion produced by, 484
  between complementary base pairs, 152–53, 155, 158
Hydrogen ions (H^+), 30
Hydroids, 382
Hydrologic (water) cycle, 846–47
Hydrolysis, 39
Hydrophilic head, of phospholipids, 58
Hydrophilicity, 28
Hydrophobic interaction, 24, 28–29
Hydrophobicity, 29, 585
Hydrophobic tail, of phospholipids, 58
Hydrostatic skeleton, 440, 704, 705
Hydrothermal vent communities, 883
Hydroxide ions (OH^-), 30
Hydroxyl group, 38
Hypertension, 562
Hypertonic solutions, 64
Hyphae, 392
Hypocotyl, 504
Hypocotyl hook, 509, 527
Hypocretins (orexins), 235, 589
Hypothalamus, 536, 537, 648, 649–52, 673, 674, 678
Hypothesis, 9, 10
Hypothyroidism, 251
Hypotonic solutions, 64, 65

Ice, 32
"Ice Man," 251
Identical twins, 262, 785–86
Immigration, 794
Immune deficiency disease, 635–37
Immune response, 619–41
  antibodies, 46, 624–25, 629
  cancer and, 637–38
  cell-mediated immunity, 627, 628–29, 630
  characteristics of, 623–31
  defined, 620
  humoral immunity, 627–28, 630
  malfunction of, 632–38
    AIDS, 635–37
    allergies and, 632–33, 635
    autoimmune disease, 633–34
    immune deficiency disease, 635–37
  medical care and, 631
  memory cells and, 627, 629–31
  molecules and cells of, 624
  "self"–"non-self" distinction in, 627
  T-cell receptors, 624, 625–26
  threat recognition in, 624–27
Immunization, 631
Impact winter, 338
Imperfect fungi, 393, 396, 402–3
Implantation, 733, 734–35, 752
Impotence, 724
Imprinting, 770, 784, 785
Incest taboo, 785
Incisors, 594
Inclusive fitness, 300
Incomplete dominance, 223
Incomplete flowers, 500
Incomplete metamorphosis, 445
Independent assortment, 202, 217–18
Indirect development, 742
Induction, 748–50
Industrial–medical revolution, 805
Inert atoms, 24
Infant behavior, 784
Infections
  arthropod, 731
  bacterial, 731
  fever to combat, 621, 622–23
  "flesh eating" (necrotizing fasciitis), 377
  across placenta, 758
  by prions, 369–70
  protist, 731
  vaginal, 400, 401
  viral, 366–68, 731
  by viroids, 369
Infertility, 724
  hybrid, 314, 316
Inflammatory response, 621–22
Influenza (flu), 619, 632–33, 638
Ingestion, 8
Inhalation, 576
Inheritance, 12, 211–41. *See also* Gene(s); Genetics
  of acquired characteristics, 272–73
  chromosome number errors, 230–34
  environmental influence on gene expression, 226–27
  evolution and, 288
  of genes on some chromosomes, 218–20
  genes with multiple alleles, 223–24
  of human disorders, 227–30
    albinism, 226, 227–28
    Huntington disease, 229, 231, 255, 257, 260
    sex-linked, 229–30
    sickle-cell anemia, 228
  incomplete dominance, 223
  of multiple traits on different chromosomes, 217–18
  pleiotropy, 225
  polygenic, 224
  sex-linked, 229
  of sex-linked genes, 220–22
  of single traits, 213–16
    dominant and recessive alleles on homologous chromosomes, 214–16
    prediction of, 216–17
Inhibiting hormones, 648, 650
Inhibition
  competitive, 108, 109
  feedback, 108
Inhibitory potential, 666–67
Initiation complex, 173
Initiation step
  of transcription, 167, 168, 169
  of translation, 172, 173
Innate (instinctive) behavior, 766, 768–71, 784–85
Inner cell mass, 752
Inner compartment (matrix), 91
Inner ear, 684, 685
Inorganic molecules, 38
Insects, 432, 433, 435, 444–45
  as pollinators, 500, 509–10, 511
  societies of, 781–82
  tracheae of, 572
Insemination, artificial, 731
Insertion, 711
Insertion mutations, 171, 174
Insight learning, 768, 769
Instinctive (innate) behavior, 766, 768–71, 784–85
Insulin, 648, 654
  recombinant, 255
Integuments, 502
Intelligence, environmental influence on, 226
Intensity of stimulus, 670
Intercourse, 724. *See also* Copulation
Interferon, 623
Intergovernmental Panel on Climate Change (IPCC), 850, 851
Intermediate filaments, 81, 83, 92, 93
Intermembrane compartment, 91, 137
Internal fertilization, 335, 720–22
Internodes, 476
Interphase, 194
  of daughter cells, 197
  in mitosis, 195
Interspecific competition, 801–2, 816–18
Interstitial cells, 722
Interstitial fluid, 560
Intertidal zone, 818, 878
Intervertebral discs, 707
Intestine
  large, 598, 599
  small, 563, 594, 595, 596–98
Intracellular digestion, 590–91
Intracellular symbiosis, 330, 331
Intracytoplasmic sperm injection (ICSI), 734
Intraspecific competition, 802, 818
Intrauterine devices, 732, 734–35
Introns, 179
Invertebrates, 432
In vitro fertilization, 731, 734–35
Iodine, 586, 653–54
Ionic bonds, 24, 25
Ions, 25
  electrical signals and, 666–67
  of water, 30
IQ level, 226
Iridium, 126
Iris, 686, 687
Iron, 586
Ishihara chart, 229
Isle Royale, Michigan, 826–27
Islet cells, 654
Isocitrate, 141
Isolation of population, 308
  behavioral, 314, 315–16
  ecological, 309–11, 314, 315
  geographical, 309, 314, 315
  reproductive, 310, 311–13, 314–16
  temporal, 314, 315
Isotonic fluids, 64
Isotopes, 22, 23

Janzen, Daniel, 825
Japanese scorpionfly, 765
Jawless fishes, 451–53
Jaws, evolution of, 452
Jeffers, Robinson, 114, 286, 322
Jefferson, Thomas, 257
Jellyfish, 433, 434, 435–36, 437, 571
Jenner, Edward, 631
Johnson, Ben, 642, 658
Jointed appendages, 443
Joints, 705, 709, 711–12
J-shaped population growth curve, 795
June beetle, 445
Jurassic Park scenario, 255
Juvenile-onset diabetes, 633–34

Kangaroo rat, 606, 613–14, 868
Kaposi's sarcoma, 636
Karyotype, human male, 190, 191
Katz, Bernard, 665
Keratin, 40, 45, 48, 49, 541
α-Ketoglutarate, 141
Keystone species, 824–26
Kidneys, 606–10
  adaptations of, 613–14
  bioartificial, 614
  blood oxygen and, 614
  blood pressure and, 614
  blood water content and, 612–13
  collapsed, 612
  as endocrine organ, 644, 649, 657
  homeostasis and, 612–14
  monitoring and regulation of dissolved substances in blood, 614
  nephrons, 608–11
Kinetic energy (energy of movement), 100
Kinetochore, 195–96, 201–2, 203
King, Thomas, 747
Kingdom(s), 6, 351, 353–58. *See also* Animal(s); Fungus/fungi; Plant(s); Protists
  energy acquisition of, 8
King penguins, 721
Kin selection, 300–1, 780–81
Klinefelter syndrome, 232
Knee-jerk reflexes, 671
Knock-out mice, 255
Kohler, Wolfgang, 768
Krebs, Hans, 139
Krebs cycle, 138–39, 140, 141
Kudzu, 830
Kuru, 369–70
Kyoto agreement, 851

Labium/labia, 726, 729
Labor, 727, 756–57
Lactate, 135, 137
Lactation, 757
Lacteal, 598
Lactic acid, 569
Lactoferrin, 253
Lactose, 41
"Ladder of Nature," 270
Lake Maracaibo, Venezuela, 231
Lakes
  acid deposition damages to, 849
  freshwater, 876–78
  succession in, 829
Lamarck, Jean Baptiste, 272–73
Lampreys, 451, 452, 453
Lancelets, 433, 450–51, 452
Language acquisition, 784
Language tasks, localization in brain of, 681
Large intestine, 598, 599
Larva/larvae, 444, 742
Larynx, 574, 595
Lascaux cave, 342, 343
Laser revascularization, 562–63
Late blight, 379
Late-loss survivorship, 803, 804
Lateral bud, 476, 477, 478, 524
Lateral meristems (cambia), 470
Latitude, 858–60
Laws, natural, 9, 11
Lazuli buntings, 299
Leaf primordia, 476
Leafy sea dragon, 820
Leakey, Mary, 339
Learned behavior, 766–71
  conditioning, 767–68
  habituation, 767, 769, 770
Learning, 766
  innate constraints on, 769–70
  insight, 768, 769
  mechanisms of, 682
  trail-and-error, 767, 768
Leaves, 468, 469, 479–81
  as adaptations for photosynthesis, 116–17
  adaptations of, 490–91
  of angiosperms, 423
  growth of, 527
  structure of, 116–17
LeClerc, Georges Louis, 270–71
Leeches, 441, 442
Leeuwenhoek, Anton van, 78
Legumes, 375, 844
  nitrogen fixation in, 483, 484
Lemming populations, 797
Lemur, 339
Lens of eye, 686
Leopard frogs, 456
Leopold, Aldo, 834
Leptin, 649, 657
Leptospirosis interrogans, 372
Leucine, 47
Leukemia, 367
Leukocytes (white blood cells), 555, 557, 578, 706
Lewis, J., 74
LH (luteinizing hormone), 648, 651, 654, 724, 725, 728, 729
Lice, crab (pubic lice), 731
Lichens, 392, 397–98, 399, 824, 826, 827–28
Life
  continuity of, 6
  defined, 2
  diversity of, 6–8
Life cycle
  eukaryotic, 203
Life on Earth, 323–49
  beginning of, 324–27
  earliest organisms, 327–31
  extinctions' role in, 336–38
  hierarchy of structures of, 2–4
  human evolution, 338–45
    behavior, 343–45
    binocular vision, 338
    brain size, 338, 343
    evolutionary tree, 340
    grasping hands, 338
    hominids, 338–43
  invasion of land, 333–36
  multicellularity, 331–33
Ligaments, 539, 711
Light, sensing of, 686–89
  binocular vision, 689
  compound eyes, 686
  mammalian eyes, 686–89
Light-harvesting complex, 119, 120
Light microscopes, 79
Lignin, 412, 471
Lily, 501
Limbic system, 677–78
Limestone, 382
Limnetic zone, 876
Linkage of genes, 219–20
Linné, Carl von (Carolus Linnaeus), 353, 432
Linseed oil, 44
Lion-tail macaque monkey, 339
Lipases, 596, 597
Lipids, 40, 43–45
  defined, 43
  glycolipids, 88
  as nutrients, 584–85
  oils, fats, waxes, 28, 40, 43–45, 134, 539, 584–85
  phospholipids, 40, 45, 46, 58–59, 584–85
  steroids, 40, 43, 45, 46, 178–80, 644
Lipoproteins, 46
Liposomes, as vectors, 258
Littoral zone, 876
Liver, 134, 596
Liver flukes, 439
Liverworts, 410, 411, 413, 414, 415
Livestock, genetically improved, 233, 252–55
Living fossils, 13
Living intermediates, 330
Living rock cactus, 820
Living things, 1–8
  categorizing, 6–8
  characteristics of, 2–6
Lizards, 281, 455, 743, 778
Lobefin fishes, 13, 334–35
Lobster, 445
Localized distribution, extinctions from, 316–17
Locomotion, 332
Locus, 212
Locusts, 445
Long-day plant, 525
Long-term memory, 682
Loop of Henle, 608, 609, 610–12, 613
Loose connective tissue, 539
Lorenz, Konrad, 764, 769, 770
Loudness, 685
Louis XVI, King, 257
Low-density lipoprotein (LDL), 46, 562
LSD, 401
Luciferase, 251, 252
Lung cancer, 578, 579
Lungs, 572–73, 575
  air delivered to, 574–75
  amphibian, 455
  in bony fish, 454
  book, 444
  reptilian, 455
Lupines, 13–15, 822
Luteinizing hormone (LH), 648, 651, 654, 724, 725, 728, 729
Lycopedon giganteum, 398
Lycopodium (ground pine), 414
Lyell, Charles, 271–72
Lyme disease, 376
Lymph, 539, 540, 563
Lymphatic system, 561–64. *See also* Immune response
  defensive function of, 563–64
  fat transport to blood, 563
  fluid return to blood, 562–63
  open and closed circulatory systems, 441, 444, 550
  vertebrate, 544
  vessels, 562
Lymph nodes, 561–62, 564
Lymphocytes, 557, 564, 623–24
Lynx, 798
Lysine, 175
Lysosomes, 81, 82, 84, 87, 591
  function of, 87

MacArthur, R., 817
McCarthy, Joseph, 154
McCarty, Maclyn, 151
McCaughey family, 735

McClintock, Martha, 785
McDonald, M. Victoria, 777
MacLeod, Colin, 151
Macrocystis, 383
Macronutrients, 843
Macrophages, 564, 621, 624
"Mad cow disease" (bovine spongiform encephalopathy), 370
Maggots, 10
Magnesium, 586
Magnetic fields, detection of, 692
Magnetotactic bacteria, 372
Maidenhair tree (ginkgo), 418, 419
Maize plant, 530
Major histocompatibility complex (MHC), 627
Malaria, 385, 631, 634, 801, 824
Malate, 141
Malignant tumor, 637
Malthus, Thomas, 274
Maltose, 41
Mammalian eyes, 686–89
Mammals, 459–60
  evolution from reptiles, 335–36
  extraembryonic membranes in, 743–44
  respiration in, 572–73
  sex determination in, 221
Mammary glands, 459, 757
Mandrill, 771, 772
Mangold, Hilde, 748
Mantle, 447
Maples, 513
Margulis, Lynn, 330
Marine ecosystems, 878
Marshall, B. J., 597
Marsupials, 459, 460
Masonjones, Heather, 717
Mass extinctions, 336–38
Mast cells, 632
Mating
  assortive, 293
  nonrandom, 293–94
Matrix, 91, 137
  mitochondrial, 139–40, 141
Matter, 2
  levels of organization of, 2, 3
  second law of thermodynamics and, 101
Mauritius, 506
Mayr, Ernst, 308–9
Meadowfoam, 233
Measles virus, 366, 367
Mechanical incompatibility, reproductive isolation through, 314, 316
Mechanoreceptor, 684
Medicine, biotechnological applications in, 255–59
  gene therapy, 257–58
  genetic screening, 255, 259–61
  Human Genome Project, 259, 356
  knock-out mice, 255
Medicine, isotopes used in, 23
Medulla
  adrenal, 655, 656
  of brain, 673, 677
  renal, 608
Medusa (jellyfish), 435–36, 437
Megakaryocytes, 557, 558
Megaspore, 499, 502
Megaspore mother cell, 502
Meiotic cell division, 199–205
  cytokinesis in, 201, 202
  genetic variability and, 203–5
  in humans, 189
  in life cycles of organisms, 203
  meiosis, 212
    errors in. *See* Nondisjunction
    genetic variability and, 203–5
    I, 199, 200–3
    II, 201, 203, 205
    spindle microtubules in, 201, 203
Melanin, 227–28
Melanocyte-stimulating hormone (MSH), 648, 651
Melanoma, 149, 158
Melatonin, 649, 656–57
Membrane(s), 538–39
  cellular, 84–88
  extraembryonic, 743–44
  plasma. *See* Plasma membrane
  pleural, 576
  precursors of, 326
Memory, 682
Memory cells, 624, 629–31
  B cells, 627, 628
  T cells, 629
Mendel, Gregor, 212–13, 273
Meninges, 673
Menopause, 725
Menotti-Raymond, Marilyn, 256
Menstrual cycle, 727–29, 785
Menstruation, 729
Mental retardation, 652
Meristem cells, 468–70
Mesoderm, 430, 746
Mesoglea, 435
Mesophyll, 117, 481
Messenger RNA (mRNA), 165–66, 168, 170–73, 174
  codons of, 168
  as DNA probe, 248
  regulation of translation of, 177
  transcription and, 167, 169
Metabolic pathways, 105
Metabolism, 5
  aerobic, 90, 329
  anaerobic, 90
  of cell, 76, 105–9
  of fats, 134
  of glucose, 132. *See also* Glycolysis
  of prokaryotes, 374–75
  of proteins, 134
Metamorphosis, 444–45, 742
Metaphase
  in meiosis, 200, 201–2, 203
  in mitosis, 195, 196
Meteorites, mass extinctions from, 337–38
Methanococcus jannaschi, 355
Methanogens, 375
Methanol, 108
Methionine, 167
Methionine tRNA, 173
Methyl group, 38
Methyl salicylate, 530
MHC (major histocompatibility complex), 627
Mice, knock-out, 255
Michener, James A., 605
Microbes, 365–89, 620
  "designer," 631
  prions, 369–70
  prokaryotes, 370–77
    archaea, 6, 7, 8, 354, 355, 371, 373, 374, 375, 376, 386
    bacteria. *See* Bacteria
    difficulty in classifying, 371
  Protista kingdom, 6, 7, 353, 377–86
    characteristics of, 357
    diversity of, 378
    major groups in, 378
    protozoa, 383–86
    slime molds, 378–80
    unicellular algae. *See* Phytoplankton
    water molds, 378–79
  viroids, 369, 370
  viruses. *See* Virus(es)
Micrococcus, 372
Microfibrils, 48
Microfilaments, 92, 93
Micronutrients, 843
Microorganisms. *See* Microbes
Microscopes, types of, 78–79
Microspheres, 326
Microspore, 499, 501
Microspore mother cells, 501
Microtubules, 81, 83, 92, 93
Microvilli, 598
Midbrain, 676, 677
Middle ear, 684, 685, 706
Middle lamella, 69, 70
Midpiece, 724
Mignot, Dr., 235
Migratory behavior, 143, 771
Milk secretion, 757
Miller, Stanley, 325, 327
Mimicry, 819–22
Mind, brain and, 680–83
Minerals
  as nutrients, 584, 586
  plant transport of, 475, 477, 484–87
  root acquisition of, 481–82
Mistletoes, 513
Mitchell, Peter, 121
Mitochondria, 81, 82, 83, 90–91
  ATP production in, 133
  cellular respiration in, 137, 138
  chemiosmosis in, 141, 142
  electron transport system of, 140–41
  evolution of, 330, 386
  function of, 90
  phagocytosis of, 87
  structure of, 90
Mitochondrial matrix, 139–40, 141
Mitotic cell division, 194
  cytokinesis, 194, 197–98
  functions of, 186–88
  in humans, 189
  mitosis, 194–97
    anaphase, 195, 196–97
    chromosomes during, 190–91
    defined, 194
    life cycle of organisms and, 203
    metaphase, 195, 196
    prophase, 194–96
    spindle microtubules in, 201, 203
    telophase, 195, 197
Mojave desert, 869
Molars, 594
Molds, 400
  black bread, 394, 395
  slime, 378–80
  water, 378–79
Molecular genetics, 353, 356
Molecule(s), 4, 37–55
  antenna, 119
  carbohydrates, 39–43
  carbon in, 38
  defined, 24
  energetic, 327
  energy-carrier, 103–5. *See also* ATP

formed from atoms, 24–27
inorganic, 38
interaction with water, 28–29
lipids, 40, 43–45
nucleic acids, 40, 51–52
organic (carbon-based), 2, 4, 38, 39
proteins, 40, 45–51
defined, 45
formation of, 46–48
functions of, 46, 51
structural, 47
structure of, 48–51
regulator, 108
Mollusks, 431, 433, 435, 445–48, 449
chemical defenses of, 822–23
indirect development in, 742
nervous system of, 671, 672
Molting, 443, 658
Monarch caterpillar, 823
Monera, 353, 354
Monoclonal antibodies, 629
Monocots, 468, 469, 474, 505
Monoecious species, 719
Monomers, 39
Monophyletic groups, 359–60
Monosaccharides, 39, 40–41
Monotremes, 459, 460
Monterey pines, 315
Moray eel, 454
Morels, 396, 401
Morgan, Thomas Hunt, 222
"Morning after" pills, 735
Morphine, 691
Morphogens, 750
Morse code, 153
Morula, 745
Mosquitoes, malaria-carrying, 634
Mosses, 410, 411, 413, 414, 415
Moths, 510, 819, 820
Motor neurons, 668, 671, 672–73, 675–76, 702–3
Motor unit, 703
Mountain king snake, 456
Mountains, weather and climate effects of, 861–62
Mt. Kilauea, 827
Mt. Mitchell, 849
Mt. St. Helens, 826, 827
Mouth, human, 594–95
Movement, energy of (kinetic energy), 100
mRNA. *See* Messenger RNA (mRNA)
Mucous membrane, 620
Mucus, 43, 620–21
Müller, Paul, 841
Mullis, Kary B., 249
Multicellular organisms, 8
Multiple sclerosis, 634
Multiregional hypothesis, 344
Muscle cells, 176
Muscle fibers, 701, 702
Muscle(s), 541, 699–704
antagonistic, 708
cardiac, 541, 554, 700, 701, 704
contraction of, 110, 702–4
extensor, 711
flexor, 711
movement and, 710–11
skeletal, 541, 700, 701–2
smooth, 541, 700, 701, 704
structure of, 701–2
uterine, 756
Muscular dystrophy, 221
Muscular system, vertebrate, 544
Mushrooms, 392, 394
disappearing, 403
fairy rings of, 396, 398
Mussels, 448
Mutations, 6, 116, 164, 174–80, 281–82
aging and, 176–77
chance, 13
defined, 174
evolution and, 175, 289, 290
genetic variability from, 198–99
prion disease inheritance and, 370
protein structure and function and, 174–75
from radiation, 174
in ribozymes, 326
sex and, 176–77
speciation by polyploidy from, 312
types of, 171, 174, 175
viral, 368
Mutualism, 299, 816, 824, 825
Mycelium, fungal, 392
Mycologists, 403
Mycorrhizae/mycorrhiza, 392, 398–99, 400, 482–83
Myelin, 664
Myofibrils, 701
Myometrium, 727
Myosin, 176, 701
Myrtle warbler, 38
Myxini, 451

NAD^+, 105, 135
NADH, 133, 136, 140, 142
$NADP^+$, 135
NADPH, 118, 119, 120
Nagana, 824
"Naked" DNA, 258
Naked mole rats, 783
Narcolepsy, 210, 211, 235
Nasturtiums, 509
National Center for Biotechnology Information, 259
National Institutes of Health (NIH), 259
Natural causality, 8–9, 10
Natural group, 359–60
Natural killer cells, 621, 623, 624, 637
Natural laws, 8–9, 11
Natural selection, 6, 12–13, 273–76
causes of, 298–300
differential reproduction and, 296
directional selection, 296, 297
disruptive selection, 296, 297, 298
evidence for, 279–82
extinction and, 316–17
genotype and, 294–95, 296
individuals and, 294–95
kin selection, 300–1, 780–81
phenotype and, 296
sexual selection, 299–300
stabilizing selection, 296–98
survival of fittest and, 275, 296
Nautilus, chambered, 332
Neanderthals, 341–42, 343
Near-shore zone, 878, 879
Nearsightedness, 687, 688
Necrotizing fasciitis ("flesh eating" infection), 377
Nectar, 500
Negative feedback, 536, 537, 556–57
in thyroid function, 653
water reabsorption in blood and, 612
Neisseria gonorrhoeae, 731
Nematodes, 403, 434
Nematodes. *See* Roundworms
Nephridium/nephridia, 441, 606
Nephrons, 608–11
Nephrostome, 606
Nerve cell. *See* Neuron(s)
Nerve cords, 438, 450
Nerve endings, free, 683
Nerve net, 435, 671, 672
Nerve(s), 541, 542, 664–65
auditory, 684
optic, 687, 688
Nervous system, 662–97
annelid, 441
arthropod, 444
brain, 672, 676–83
forebrain (cerebrum), 676, 677–80
hindbrain, 676, 677
human, 460, 670, 678
imaging techniques for, 681
interpretation of action potentials, 670
"left" and "right," 680
midbrain, 676, 677
mind and, 680–83
Neanderthal, 341–42
PET scans in diagnosing disorders of, 23
respiratory center of, 576–77
vertebrate, 451
central, 672–73, 705
digestion regulation by, 599–600
of echinoderms, 450
general features of, 670–72
heart rate and, 554
mammalian, 460
of mollusks, 447
neurotransmitters, 664–65, 666, 667–69
peripheral, 672–73
spinal cord, 672, 673–76
vertebrate, 544
Nervous tissue, 4
Net primary productivity, 837
Neural pathways, 671
Neural stem cells, 748
Neural transplants, 675
Neural tube, 753
Neuraminidase inhibitors, 631, 633, 638
Neuromuscular junction, 702, 703
Neuron(s), 541, 664–67
association, 671
cell body of, 664
electrical signals by, 665
motor, 668, 671, 672–73, 675–76, 702–3
postganglionic, 674
postsynaptic, 666
preganglionic, 674
presynaptic, 666
sensory, 664, 671
structures and functions of, 664–65
synapses and, 665–67
Neurosecretory cell, 649–50, 678
Neurospora (bread mold), 164, 165
Neurotransmitters, 664–65, 666, 667–69
Neutral mutations, 175
Neutrons, 21, 23
NewLeaf® potatoes, 253
Newt, 750
Niacin, 587
Nicolson, G. L., 58
Nicotinamide adenine dinucleotide (NAD^+), 105, 135
Nicotinamide adenine dinucleotide phosphate (NADPH), 118, 119, 120
Nicotine, 669
Nigeria, 805

Nipple, 757
Nitrates, 375
Nitric oxide, 668
Nitrogen, 483
Nitrogen-containing base, 51
Nitrogen cycle, 844–45, 848–49
Nitrogen fixation, 483, 484, 844
Nitrogen-fixing bacteria, 375–76, 483, 484
Nodes
  lymph, 561–62, 564
  on stem, 476
Nodule, 375, 376, 483, 484
Nondisjunction
  of autosomes, 234
  defined, 230
  of sex chromosomes, 230–33
Nonpolar covalent bond, 27
Nonvascular plants (bryophytes), 413, 414
Norepinephrine (noradrenaline), 649, 655, 669
Norplant®, 732, 733
North Chicago allele of hemoglobin gene, 199
Northern coniferous forests, 873–74
Northern elephant seal, 292
Northern oriole, 358
Norway rat, 767–68
Notochord, 450, 746
Nuclear envelope, 81, 82–83, 87, 197
Nuclear pore, 81, 83
Nucleic acid, 40, 43, 51–52. *See also* DNA; RNA
Nucleoid, 80
Nucleotides, 40, 51, 52, 151–52
  free, 155
  order of, 153
Nucleotide substitutions, 174
Nucleus, 76
Nucleus/nuclei
  atomic, 21, 22, 24
  cell, 81, 82–84
    chromatin, 81, 83–84
    nuclear envelope, 81, 82–83, 87, 197
    nucleolus, 81, 82, 83, 84
  origin of, 330–31
  polar, 502–3
Nutrasweet® (aspartame), 37, 53
Nutrient cycles (biogeochemical cycles), 843
Nutrients, 5, 584
  carbohydrates, 584, 585
  cellular, 76
  from detritus feeders and decomposers, 838
  in ecosystems. *See under* Ecosystems
  energy yield of, 134
  human impact on, 885
  lipids, 584–85
  minerals, 584, 586
  plant acquisition of, 477, 481–84
  proteins, 584, 585–86
  vitamins, 584, 586–88
Nutrition, 583–89
  defined, 584
  eating disorders, 589
  guidelines for, 588

Obesity, 134, 582, 583, 600
Observation, 9, 10
Occipital lobe, 679
Ocean, open, 881–83
Ocean currents, 860–61
Octopods, 433
Octopus, 448, 449, 671, 721, 880
Oils, 28, 40, 43–45, 53
Olestra (Olean®), 37, 53
Olfactory receptors, 690
Oligochaetes, 441
Oligotrophic lakes, 876–77
Olive baboon, 774
Ommatidium/ommatidia, 686
Omnivores, 593, 838
One-gene, one-protein hypothesis, 164
On the Origin of Species (Darwin), 275, 276, 299, 353
Oocytes, 725, 726, 730
Oogenesis, 725–27, 748
Oogonium/oogonia, 725
Oomycetes (water molds), 378–79
Oparin, Alexander, 324
Open circulatory system, 444, 550
Operant conditioning, 767–68
Operculum, 571–72
OPG, 710
Opioids, 669
Opposable thumb, 343
Optic disc (blind spot), 688
Optic nerve, 687, 688
Oral groove, 385, 386
Orangutan, 459
Orbiter spacecraft, 345
Order, 351
Orexins (hypocretins), 235, 589
Organelles, 4, 80, 81–82, 93
Organic (carbon-based) molecules, 2, 4, 38, 39
Organic farming, 884, 885
Organism, 1, 4, 5
Organogenesis, 745, 746
Organs, 4, 432, 538, 541–44
Organ systems, 4, 538
  defined, 543
  in flatworm, 438
  vertebrate, 543
Organ transplants, 549, 565
Orgasm, 729
Origin, 711
Oscula, 433, 435
Osmosis, 61, 63–65
  of water into roots, 485–86
Osteichthyes, 452, 453–54
Osteoblasts, 707, 709
Osteoclasts, 707, 709
Osteocytes, 707
Osteons (Haversian systems), 708
Osteoporosis, 710
Ostrich, 458
Outer ear, 683–85
"Out of Africa" theory, 344
Oval window, 684, 685
Ovary(ies), 644, 649
  of flower, 422, 423, 500, 501
  human, 725–27
  mammalian, 654
Ovary wall of flowering plants, 506
Overfishing, 809
Overgrazing, 871
Overspecialization, extinction from, 316–17
Oviduct, 725, 726
Ovulation, 722, 725, 733
Ovules (immature seeds), 419, 422, 423, 500
Oxaloacetate, 124, 139, 141
Oxidative stress, 21
Oxygen, 24–25, 324, 327, 329
  in blood, 614
  covalent bonding of, 26
Oxygen transport, in respiration, 575–76
Oxytocin, 537, 647, 648, 650, 651–52
Oysters, 448
Ozone layer, 325, 858–59
Pacemaker, cardiac, 553–54
Pacific yew, 15, 233, 408, 409, 424
Pain, 691
Paine, R., 824–25
Pain receptor, 684, 691
Pain-withdrawal reflex, 671, 675–76
Paleontologists, 126
Paleozoic era, 332, 333
Palisade cells, 481
Panama, Isthmus, 317
Pancreas, 596–97, 644, 648, 654
Pancreatic duct, 654
Pancreatic juice, 597
Pantothenic acid, 587
Papillae, 690
Parabronchi, 573
Paramecium, 7, 65, 67, 79, 88, 94, 330, 331, 377, 385–86
Paramecium aurelia, 816–17
Paramecium bursaria, 817
Paramecium caudatum, 816–17
Parasites, 316, 370, 439, 800–1, 824
Parasitism, 299, 800, 816, 824
Parasympathetic division, 672, 673
Parathormone, 648, 654, 708
Parathyroid gland, 644, 648, 652–54
Parenchyma, 471, 472
Parental generation, 213
Parietal lobe, 679
Parkinson's disease, 669
Parthenogenesis, 205, 718
Parthenogeneticity, 358
Particles, subatomic, 4
Passive smoking, 579
Passive transport, 61–63
Pasteur, Louis, 11, 324, 631
Pathogenic bacteria, 376
Patino, Maria Jose Martinez, 176, 180
Pauling, Linus, 154
Pavlov, Ivan, 763
PCBs, 647, 843
Pea, 491
  Mendel's experiments with, 213–18
Peacock, plumage of, 294
Pectin, 69
Pedigrees, 226–27
Pelagic life-forms, 881
Pelomyxa palustris, 330
Pelvis, renal, 608
Penfield, Wilder, 682
Penicillin, 80, 294, 401
Penicillium, 11, 401
Penis, 722, 723, 724, 730
PEP (phosphoenolpyruvate), 124
Pepsin, 109, 595
Pepsinogen, 595
Peptide, 48, 595
Peptide bonds, 48, 164
Peptide hormones, 644, 646
Peptidoglycan, 371
Perception, common, 9
Perfluorocarbons (PFCs), 556
Pericycle, 475, 482
Periderm, 471
Periosteum, 709
Peripheral nervous system, 672–73
Peristalsis, 595, 596
Peritoneal dialysis, 612
Peritoneum, 612
Permafrost, 875
Pest control with pheromones, 773
Pesticides, 647

Pesticides, fungal, 400
Pest resistance in plants, bioengineering, 252–55
Petals, 500, 501
Petioles, 479–80
PET (positron emission tomography), 23, 681
Petromyzontiformes, 451
PFCs (perfluorocarbons), 556
PGA (phosphoglyceric acid), 122
pH, enzyme activity and, 109
Phages. *See* Bacteriophages
Phagocytic cells, 621
Phagocytosis, 66, 67, 68, 87, 620, 621, 625, 628
Pharyngeal gill slits, 450
Pharynx, 441, 574, 591, 595
Phenotype, 215, 288
  environment and, 225–26, 295, 296
  natural selection and, 296
Phenylalanine, 47, 53, 167
Phenylketonuria, 251
Phenylketonurics, 53
Pheromones, 719, 773, 780
  human response to, 785
  insect society and, 781–82
  reproductive behavior and, 778–79
Phloem, 471, 473, 477
  sugar movement in, 488–89
Phosphate group, 38, 51
Phosphatidylcholine, 40
Phosphoenolpyruvate (PEP), 124
Phosphoglyceric acid (PGA), 122
Phospholipases, 71
Phospholipid bilayer, 58–59, 62
Phospholipids, 40, 45, 46, 70, 584–85
  hydrophilic head of, 58
  hydrophobic tail of, 58
Phosphorus, 586, 706
Phosphorus cycle, 846
Photic zone, 878
Photons, 118
Photopigments, 686
Photoreceptor, 684, 686, 688–89
Photorespiration, 124
Photosynthesis, 5, 116–26, 132
  cellular respiration and, 116
  chloroplasts and, 116–17
  as coupled reaction, 103
  as endergonic reaction, 101, 102
  energy flow through community and, 836–37
  equation for, 132
  leaves as adaptations for, 116–17
  by phytoplankton, 859
  reactions for, 117–18
    light-dependent, 117–22, 123
    light-independent, 117–18, 122–23
  site of, 91
  water–CO_2 balance in, 123–24
Photosynthetic bacteria, 327
Photosystems, 119–22
Phototactic bacteria, 372
Phototropism, 520, 522
  auxin and, 521, 523
pH scale, 30, 31
Phthalates, 647
Phycocyanins, 119
Phylogeny, 351
Phylum (division), 351
Physarum, 379
Phytochromes (P_r & P_{fr}), 525–27
Phytoplankton, 380, 859, 876, 881, 882
  diatoms, 378, 380, 381
  dinoflagellates, 378, 380–81
  euglenoids, 378, 380, 381–82
Pigeons, 768
Pigments, accessory, 119
Pigs, 548
  flu viruses in, 632, 633
  genetically engineered, 253
Pili/pilus, 80
  bacterial, 371, 373
Pilobolus, 402
Pine, life cycle of, 420
Pineal gland, 644, 649, 656–57
Pinocytosis, 66
Pioneers, 826
Pisaster, 824–25, 826
Pitch, 685–86
Pith, 477
Pits, 473
Pituitary gland, 649–52
  anterior, 649–51
  posterior, 649, 651–52
Placebo, 623
Placenta, 727, 753–54, 756, 758–59
Placental mammals, 460
Plague ("Black Death"), 376, 387
Planarians, 438
Plankton, 876
Plant cell(s), 42, 83
  central vacuoles in, 89–90
  cytokinesis in, 197, 198
Plantlike protists, 378, 380–83
Plant(s), 6, 7, 353, 354, 409–27, 467–95, 496–517
  adaptations of, 489–91
    for land living, 333–34
  body organization of, 468–70
  C_3 vs. C_4, 124–25
  cell number in, 8
  cell walls of, 69, 70
  characteristics of, 357
  coevolution of herbivores and, 823
  communication through chemicals, 530
  day-neutral, 525
  dormancy, 528
  evolutionary origin of, 410–11
  flowering. *See* Flowering plants; Flowers
  genetically enhanced, 252–55
  growth of, 468–70, 477–79
  herbaceous, 470
  hormones of, 519–32
    abscisic acid, 520
    abscission and, 528
    auxin, 519, 520–21, 523, 527
    classes of, 519
    cytokinins, 519, 520, 524, 525, 528
    discovery of, 522–23
    ethylene, 519, 520, 527–28
    florigens, 525
    gibberellin, 519, 520, 527
    regulation of life cycle, 520–28
    root–shoot balance and, 525
    seed and fruit development and, 527–28
    senescence and, 528
    shape at maturity and, 524–25
  invasion of land, 411–23
    body complexity and, 412
    bryophytes (nonvascular plants), 413, 414
    club mosses, horsetails, and ferns, 411, 413–14, 417
    liverworts and mosses, 410, 411, 413, 414, 415
    protective and reproductive requirements for, 413
    seed plants, 414–18
    tracheophytes (vascular plants), 410, 413–14
  key features of, 410, 411
  leaves of. *See* Leaves
  life cycles of, 497–99
  long-day, 525
  medicinal, 408, 409
  methyl salicylate in, 530
  nutrient acquisition by, 477, 481–84
    minerals, 481–82
    through symbiosis, 482–83
  phloem in, 471, 473, 477–78, 479, 488–89
  plastids in, 82, 83, 91, 92
  polyploidy in, 311
  rapid-fire responses of, 528–29
  recombinant, 254
  reproduction of, 187–88, 410–11. *See also* Flowers; Fruit; Seeds
  roots of, 468, 469, 470, 471, 473–76, 481–82, 489
  salicylic acid in, 530
  seed, 498–99
  sense of gravity of, 521–23
  short-day, 525
  stems of, 476–79, 489–90
    branch formation, 477, 478
    secondary growth in, 477–78, 479
  subclimax, 829
  sugar transport by, 487–89
  tissues of, 470–71
  transgenic, 244, 251–52, 253, 254
  true-breeding, 213
  water acquisition by, 484–87
  water and mineral transport by, 477, 484–87
  water cohesion in, 29
  water distribution in environment and, 486
  woody, 470
  xylem in, 471–73, 477–78, 479, 482, 484–85
Plaques, 46, 562, 563
  dental, 371, 372
Plasma, 540, 554, 555
Plasma cells, 624, 627, 628
Plasma membrane, 4, 57–58, 70, 81, 82, 83, 87, 644. *See also* Cell membrane
  differentially permeable, 62
  function of, 80, 84
Plasmid, 244–45, 251, 374
Plasmodesmata, 68–69, 83, 473
Plasmodium, 379, 385, 634
Plasmopara, 379
Plastids, 82, 83, 91, 92
  starch-filled, 521–23
Platelets, 540, 555, 557–58
Plate tectonics, 336–37
Plato, 270
Platyhelminthes, 433, 434, 437–40
Platypus, 460, 692
Play, 786–87
Pleated sheet, 49, 51
Pleiotropy, 225
Pleural membranes, 576
Pneumocystis carinii pneumonia, 636
Pneumonia, 377, 636
Point mutation, 174
Polar bear, 6
Polar body, 725
Polar covalent bonds, 27
Polar nuclei, 502–3
Pollen, 413, 414–16, 422, 423, 501–2
Pollen antigens, 632
Pollen grains, 334, 499, 500, 502
Pollen sacs, 501
Pollination, 503
  by insects, 500, 509–10, 511
  wind, 499–500, 508, 509

Pollution, 403, 841, 885
  open-ocean, 881–82
Polychaetes, 441, 442
Polygenic inheritance, 224
Polymerase chain reaction (PCR), 249–51, 256, 356
Polymers, 39
  DNA, 152
Polymorphism(s)
  balanced, 298
  restriction fragment length (RFLPs), 248–49, 251, 256
Polyp (benign tumor), 637
Polyp (cnidarian), 435, 437
Polypeptide, 48, 164
Polyploidy, 309, 311–13
Polysaccharides, 39, 40, 41–43, 69, 585
Ponds, succession in, 829
Pongids, 339
Pons, 677
Population bottleneck, 292–93
Population cycles
  boom-and-bust, 796–97
  out-of-phase, 800
  predator–prey, 798
Population genetics, 288
Population(s), 4, 275, 793–813, 885
  carrying capacity and, 799, 800, 809
  changes in human, 803–9
    age structure and, 805–8
    technological advances and, 805
    U.S., 808
  competition and, 818
  defined, 794
  distribution of, 802–3
  equilibrium, 289
  evolution of, 288, 290–93
  genetic divergence of two, 308–9
  growth of, 537, 794–96
    exponential, 794–96, 803–5
    J-curve of, 795
    regulation of, 796–802
    S-curve, 799
  isolation of, 308
    behavioral, 314, 315–16
    ecological, 309–11, 314, 315
    geographical, 309, 314, 315
    reproductive, 310, 311–13, 314–16
  survivorship patterns of, 803, 804
Pore cells, 434
Pores, 82
Porifera, 432–35, 436
Porpoises, 692, 882
Positive feedback, 536–37
Positron emission tomography (PET), 23, 681
Posterior, 431
Postganglionic neurons, 674
Postsynaptic neuron, 666
Postsynaptic potentials, 665
Potassium, 329, 487, 586
Potassium channels, 666
Potato famine, 379
Potential, 665
Potential (stored) energy, 100
Potrykus, Ingo, 263
Power grip, 338
Power stroke of cilia, 94
PPL Therapeutics, 548, 549
Practice theory of play, 786
Prairie dogs, 871
Prairie (grasslands), 799, 809, 840, 843, 870
Praying mantis, 443
Preadaptation, 334
Prebiotic evolution, 324–27
Precambrian era, 327
Precapillary sphincters, 559, 561
Precision grip, 338
Predation, 299
  forms of, 818
  population and, 800–801
Predator–prey interactions, 299, 816, 818–23
  evolutionary adaptations and, 818–23
  population cycles, 798
Preganglionic neurons, 674
Pregnancy, 725, 729, 731, 734–35, 740, 741, 760
  smoking and, 758–59, 760
Premolars, 594
Prenatal genetic screening, 260–61
Pressure, turgor, 89–90
Pressure-flow theory, 488–89
Presynaptic neuron, 666
Prey, 801
Prickly pear cactus, 527, 801
Primary cell wall, 69
Primary consumers, 838
Primary growth, 470, 473–74
Primary oocytes, 725, 726
Primary phloem, 477
Primary root, 473
Primary spermatocytes, 722
Primary structure of protein, 49, 50
Primary succession, 826–28
Primary xylem, 477
Primate evolution, 338
Primer pheromones, 781
Primers, 249
Primitive streak, 753
Primordial soup, 326
Principle, 11
Prions, 369–70
Producers (autotrophs), 8, 116, 837
Products, 101
Profundal zone, 876
Progesterone, 648, 649, 654, 725, 728, 753–54
Project Kipepeo, 857
Project Seahorse, 736
Prokaryotes, 370–77
  archaea, 6, 7, 8, 354, 355, 371, 373, 374, 375, 376, 386
  bacteria. *See* Bacteria
  beneficial functions of, 375–76
  binary fission in, 373–74
  cell wall of, 371
  difficulty in classifying, 371
  domains of, 355, 356
  flagella movement in, 371–72, 373
  habitat specialization of, 374
  metabolisms of, 374–75
Prokaryotic cells, 8, 76, 80, 327
  features of, 80–81
  "flagella" of, 94
Prolactin, 648, 651
Promoter region, 167, 168
Pronghorn antelope, 871
Proofreading, 158
Prophase
  in meiosis, 200–201, 203
  in mitosis, 194–96
Prostaglandin, 644, 648, 656
Prostate gland, 722, 723, 724
Protease, 595, 597
Protease inhibitors, 637
Protein coat, viral, 366
Protein(s), 37, 40, 45–51, 150, 644
  androgen-receptor, 176
  carrier, 60, 63
  in cell membranes, 59–60
  channel, 59–60, 63
  complement, 625
  defined, 45
  denatured, 51
  diversity of, 70–71
  formation of, 46–48
  gatekeeper, 82
  gene identification through, 249
  genes and, 164–67
  life span of, 178
  metabolism of, 134
  modification of, 177–78
  as nutrients, 584, 585–86
  receptor, 60
  recognition, 60
  regulatory, 178–80
  structural, 47
  structure and function of, 46, 48–51, 59–60, 174–75
  synthesis of, 76. *See also* Transcription; Translation
    genetic information for, 164–65, 166, 167
    RNA and, 167–74
    steps in, 166
  therapeutic, 255–57
  transport, 59–60, 63, 65
Protists, 6, 7, 353, 377–86
  animal-like. *See* Protozoa
  characteristics of, 357
  contractile vacuoles in, 88
  diversity of, 378
  infections from, 731
  major groups in, 378
  plantlike, 378, 380–83
  protozoa. *See* Protozoa
  slime molds, 378–80
  unicellular algae. *See* Phytoplankton
  water molds, 378–79
Protocells, 326
Protofibrils, 48
Protonephridia, 606
Protons, 21, 22
Protostome, 432
Protozoa, 378, 383–86
  ciliates, 378, 385–86
  sarcodines, 378, 384
  sporozoans, 378, 384–85
  zooflagellates, 378, 383–84
Proximal tubule, 608, 609
Prozac®, 669
Prusiner, Stanley, 370
Pseudocoelom. *See* Roundworms
Pseudocoelomates, 431–32
Pseudoplasmodium, 379–80
Pseudopods, 66, 67, 379
Psilocin, 404
Psilocybe mushrooms, 391, 404
Psilocybin, 404
Puberty, 654, 724
Pubic lice, 731
Puffball, 398
Pumps (transport proteins), 65
Punnett, R. C., 215
Punnett square method, 215–16
Pupa, 444
Pupil, 686, 687
Pus, 557
Pyloric sphincter, 595, 596

Pyridoxine (vitamin B_6), 587
Pyrogens, endogenous, 622–23
Pyruvate, 132–35
  fermentation of, 135, 137
  in mitochondrial matrix, 139–40

Quaternary structure of proteins, 50
Queen bee, 781
Queen substance, 773, 781
Quetzal, 514

Rabies virus, 366, 367
Radial symmetry, 430–31
Radiation, mutations from, 174
Radiation therapy, 638
Radioactive decay, 23
Radioactive isotopes, 23
Radiolarians, 384, 385
Radiometric dating, 329
Radula, 447
Raff, M., 74
Rainfall, 486, 875–76
Rain forests, 634
  destruction of, 15
  temperate, 872–73
  tropical, 15, 864–67
Rain shadow, 862
Random population distribution, 802–3
Raskin, Ilya, 530
Raspberry fruit, 514
Rasputin, 230
Rats, 767–68
Rays, 453
Reactants, 101
Reaction center, 119
Reaction(s), 25, 99, 101–3
  aerobic, 90
  coupled, 102–5
  endergonic, 101, 102–3
  enzymes and rate of, 107
  exergonic, 101–2, 103
  photosynthetic, 117–18
    light-dependent, 117–22, 123
    light-independent, 117–18, 122–23
  spontaneous, 106
  temperature and, 106
  water in, 39. *See also* Dehydration synthesis; Hydrolysis
Recall, 682
Receptor-mediated endocytosis, 66–67
Receptor potential, 683
Receptor proteins, 60
Receptors
  olfactory, 690
  pain, 691
  sensory, 683, 684
  surface, 645
  taste, 690–91
  T-cell, 624, 625–27
Recognition proteins, 60
Recombinant DNA, 244–47
  DNA library for, 246–47
  in genetic engineering laboratories, 246–47
  in nature, 244–46
Recombinants, environment and, 254
Recombination, 201, 220
  during construction of antibody genes, 626–27
Rectum, 599
Red algae, 378, 380, 382
Red blood cells (erythrocytes), 554, 555–57, 560, 706
Red bone marrow, 706
Red-green color blindness, 229
Redi, Francesco, 10–11, 324
Red-sided garter snakes, 721
Red squirrels, 766
"Red tide," 381
Red-winged blackbird, 766
Red wolf, 358
Redwoods, 29
Reefs, 382
Rees, William, 809
Reeve, Christopher, 662, 663, 675, 693
Reflexes, 671, 676
Regeneration, 718
Regulator molecule, 108
Regulatory proteins, 178–80
Releasing hormones, 648, 650
Renal artery, 607, 609
Renal cortex, 608
Renal medulla, 608
Renal pelvis, 608
Renal vein, 608, 609
Renin, 614, 649, 657
Replacement-level fertility (RLF), 806
Replacement therapy, 634
Replication "bubble," 155–56
Reproduction, 6, 717–39
  in amphibian, 455
  in angiosperms, 419–23
  asexual, 187–88, 717, 718
  cellular. *See* Cellular reproduction
  in conifers, 419
  differential, 296
  in echinoderms, 450
  endocrine disruption and, 647
  in flatworms, 440
  in fungi, 187–88, 392–96
  human, 722–35
    copulation, 721, 729
    female accessory structures, 727–29
    labor and delivery, 727, 756–57
    male accessory structures, 723–24
    menstrual cycle, 725–29
    ovaries, 724–27
    pregnancy, 725, 729, 731, 734–35, 758–59
    testes, 654, 722–24
  limiting fertility, 732–35
  in mollusks, 447
  pregnancy, 740, 741, 760
  sexual. *See* Sexual reproduction
  in sponges, 434–35
  vertebrate, 544
Reproductive isolation between species, 314–16
  postmating mechanisms for, 314, 315, 316
  premating mechanisms for, 314, 315–16
Reproductive isolation of population, 310, 311–13
Reproductive success, 295
Reptiles, 335–36, 455–59
  adaptations to dry land, 335
  bird and mammal evolution from, 335–36
  extraembryonic membranes in, 743–44
  as natural group, 359–60
  scales of, 572, 573
Reservoirs, 843
Resistance, environmental, 794
Resolving power of microscopes, 79
Resource partitioning, 817–18
Respiration, 569–81
  air inhalation and exhalation, 576, 577
  breathing rate, 576–77
  carbon dioxide transport, 575–76
  cellular. *See* Cellular respiration
  conducting portion of, 573–75
  defined, 569
  gas exchange. *See* Gas exchange
  oxygen transport in, 575–76
  vertebrate, 544
Respiratory center of brain, 576–77
Resting potential, 665, 666
Restriction enzymes, 246, 247, 261
Restriction fragment length polymorphisms (RFLPs), 248–49, 251, 256
Restriction fragments, 251
Reticular formation, 677
Retina, 686, 687–89
Retinol (vitamin A), 587
Retroviruses, 258, 369, 635–36
Return stroke of cilia, 94
Reverse transcriptase, 369, 635
Reverse transcriptase inhibitors, 637
Reverse transcription, 635
RFLPs (restriction fragment length polymorphisms), 248–49, 251, 256
Rhagoletis, 309–11, 315
Rheumatoid arthritis, 634
Rh factor, 557
Rhinoceros, 868
Rhizoids, 413
Rhizomes, 490
Rhizopus, 394, 395
Rhythm method, 732, 734
Rib cage, 576
Riboflavin (vitamin B_2), 587
Ribonucleic acid. *See* RNA
Ribose, 40–41, 51
Ribosomal RNA (rRNA), 84, 165–66, 170, 173
Ribosomes, 76, 81, 82, 83, 166, 170, 174
  in bacterial cytoplasm, 80
  free, 81, 83
  on rough endoplasmic reticulum, 85
  synthesis in nucleoli, 85
Ribozymes, 326
Ribulose biphosphate (RuBP), 122
Rice, bioengineered, 242, 243, 263
Rift Valley fever, 634
Rigor mortis, 704
Ripening, ethylene and, 527–28
Rituals
  courtship, 79, 315–16, 720, 774
  Neanderthal, 341–42
RNA, 40, 41, 52, 164. *See also* Messenger RNA (mRNA); Ribosomal RNA (rRNA); Transfer RNA (tRNA)
  archaeal, 371
  in bacterial cytoplasm, 80
  bases in, 167
  DNA compared to, 165
  as first self-reproducing molecule, 326
  flu, 632
  kingdoms and, 354
  protein synthesis and, 165–66, 167–74
  ribosomal, 84
  transfer, 173
  viral, 366
RNA polymerase, 168, 169, 179
Roaches, 281
Roberts, K., 74
Rods, 687–89
Root–bacteria relationship, 483
Root cap, 473–74
Root–fungus relationship (mycorrhizae), 392, 398–99, 400, 482–83
Root hairs, 471, 474, 475, 482
Root nodules, 375, 376

Roots, 468, 473–76
adaptation of, 489, 490
branch, 475–76
epidermis of, 471, 474
gravitropism in, 521
mineral acquisition by, 481–82
monocot vs. dicot, 469
primary, 473
structure of, 470
water entry into, 485–86
Roper, Clyde, 429, 461
Rose aphid, 445
Rosy periwinkle, 15
Rosy wolf snail, 830
Rotifers, 433
Roughage, 41
Rough endoplasmic reticulum, 81, 83, 85, 87
Round goby, 831
Round window, 684, 685
Roundworms, 433, 434, 440–41
body cavity in, 431, 432
fungi as nemesis of, 403
rRNA. *See* Ribosomal RNA (rRNA)
RU-486, 735
Ruby-throated hummingbird, 130–31, 143
Ruminants, 375, 592–93
Runners, 490
energy requirements of, 143
Russian flu, 632
Rusts, 327, 399

Sac fungi, 393, 394–96
"Safe-sex" practices, 637
Sagebrush desert, 868, 870, 871
Sahara Desert, 868, 869
Salamanders, 280, 455, 456
Salicylic acid, 530
Saliva, 594–95
Salivary glands, 539, 594–95
Salmon, 657–58
Salmonella, 387
Salt
crystals, 2
formation of, 25
as preservatives, 109
Samson, Kahindi, 856, 857
Sand tiger shark, 453
San Gabriel mountains, 869
Saola, 306, 307, 318
Saprobes, 392
Sapwood, 477
Sarcodines, 378, 384
Sarcomeres, 701–2, 703
Sarcoplasmic reticulum, 701, 704
Satin bowerbird, 778
Saturated fatty acid, 44, 70
Savanna, 867–68
Scab, 558
Scales of reptiles, 572, 573
Scallops, 448
Scanning electron microscopes (SEMs), 78, 79
Scavenging and hunting, cooperative, 345
Schally, Andrew, 650–51
Schistosoma, 439, 634
Schistosomiasis, 439, 631, 634
Schleiden, Matthias, 78
Schwann, Theodor, 78
Scientific inquiry, 9, 10
Scientific method, 9–11, 216
Scientific name, 352
Scientific theories, 11–12
Sclera, 686, 687
Sclerenchyma, 471, 472
Scorpions, 446
Scotch broom flowers, 509–10, 511
Scouring rushes, 414
Scramble competition, 802
Scrapie, 370
Scrotum, 722, 723
S-curve of population growth, 799
Sea anemone, 433, 434, 435–36, 437, 767, 824
Sea cucumbers, 448–50
Seahorse, 454, 716, 717, 720, 736
Sea lettuce (*Ulva*), 383
Seaside sparrow, 777
Sea slugs, 447
Seasons, 859
Sea squirt, 450, 451
Sea stars, 433, 435, 448–50
Sea urchins, 433, 435, 448–50
Sea wasp (*Chironex fleckeri*), 436, 437
Seaweeds, 380, 382
Sebaceous glands, 539, 542, 543
Sebum, 543
Secondary cell wall, 69
Secondary consumers, 838
Secondary growth, 470, 477–78, 479
Secondary oocyte, 725, 726, 730
Secondary phloem, 477–78, 479
Secondary sexual characteristics, 654, 724
Secondary spermatocytes, 722
Secondary structure of protein, 49, 50, 51
Secondary succession, 826, 828–29
Secondary xylem, 477–78, 479
Secondhand smoke, 579
Second law of thermodynamics, 100–1
Second messenger, 644
Secretin, 599, 600, 649, 657
Secretions, antimicrobial, 620–21
Seed coat, 416, 418, 504, 507
Seed leaves (cotyledons), 468, 504, 505, 507–8, 509
Seedless vascular plants, 410, 413–14
Seedling, development of, 508
Seed plants, 410, 414–18, 498–99
angiosperms, 416, 418, 419–23, 468
gymnosperms, 416, 418–19
Seeds, 413, 416, 499, 504–8
development of, 504–7, 527–28
dispersal of, 512–14, 527
dormancy in, 499, 507, 520
germination of, 507–8
growth of, 507–8
immature (ovules), 412, 422, 423, 500
monocot vs. dicot, 505
Segmentation in arthropods, 443–44
Segmentation movements, 598
Segmented worms. *See* Annelids
Segregation, law of, 214
Selective breeding, 233, 244
Selective estrogen receptor modulators, 710
Self-fertilization, 213
Semen, 724
Semiconservative DNA replication, 155
Semilunar valves, 553
Seminal vesicles, 722, 723, 724
Seminiferous tubules, 722, 723
SEMs (scanning electron microscopes), 78, 79
Senescence, 528
Sense(s), 683–93
of chemicals, 690–91
of light, 686–89
binocular vision, 338, 689
compound eyes, 444, 686
mammalian eyes, 686–89
of sound, 683–86
uncommon, 691–93
Sensitive period, 770
Sensor, 536
Sensory neurons, 664, 671
Sensory receptors, 683, 684
Sensory system of arthropods, 444
Sepals, 500, 501
Septa, fungal, 392
Sequoia, 480
Serotonin, 669
Sertoli cells, 722, 723, 725
Sessile, 431
Sessions, Stanley, 457
Set point, 536
Severe combined immunodeficiency (SCID), 257, 635
Sex
determination of, 163, 220–22
mutations and, 176–77
symmetry and, 764, 765, 787
Sex chromosomes, 180, 181, 190, 220–22, 230–33
X, 180, 181, 220, 221–22
Y, 181, 220, 221–22
Sex-linked genes, 220–22
Sex-linked genetic disorders, 229–30
Sex-linked inheritance, 229
Sex organs, 644, 654
Sex pili, 374
Sexual characteristics, secondary, 654
Sexual deception, pollination and, 510–12
Sexual dimorphism, 778
Sexual displays, 777–78
Sexually transmitted diseases (STDs), 729, 731
Sexual maturation, 746
Sexual reproduction, 188, 717, 718–22, 729–30. *See also* Reproduction
advantages of, 198–99
animal behavior and, 777–79
DNA recombination in, 244
genetic variability and, 203–5
in prokaryotes, 373–74
in protists, 377–78
Sexual selection, 299–300
Sharks, 13, 453, 677
Sheep, cloned, 192–93
Shelf fungi, 398
Shoots, 468
epidermis of, 471
gravitropism in, 521
structure of, 470
Shoot tip, 506
Short-day plant, 525
Shortgrass prairie, 870, 871, 876
Shotgun dispersal of seed, 513
Shrew, 838
Shrimp, 445
Sialia, 352
Siamese fighting fish (*Betta splendens*), 720
Sickle-cell anemia, 228, 251
balanced polymorphism and, 298
cause of, 228
protein mutation and, 175
screening program for, 260–61
Sieve plate, 450, 473
Sieve tubes, 473, 488, 489
Silica, 381
Silk, 40, 47
Silurian period, 332, 334
Simple diffusion, 61, 62–63
Singer, S. J., 58

Single covalent bond, 26
Sink, 488
Sinoatrial (SA) node, 554, 704
Skates, 453
Skeletal muscles, 541, 700, 701–2
Skeleton, 699, 704–8
appendicular, 706
axial, 706
bone healing, 709, 711
composition of, 706–8
defined, 704
endoskeletons, 448, 450, 451, 704
exoskeletons, 41, 334, 443, 704–5
functions of, 704–6
hydrostatic, 440, 704, 705
internal, 333
movement and, 708–11
vertebrate, 544
Skin, 541–44, 620
Skin cancer, 148, 149, 158, 859
Skin color, 225, 226
Skin impression, fossilized, 271
Skinner, B. F., 767
Skinner box, 767
Skull, 673
Skunk cabbage, 510
Sliding-filament mechanism, 702
Slime layers, 80, 371
Slime molds, 378–80
Slugs, 447
Small intestine, 563, 594, 595, 596–97
Smallpox, 287, 631
Smallpox virus, 801
Smell (olfaction), 690
Smiling, 784
Smith, William, 270
Smoking, 568, 569, 578–79, 597, 710
pregnancy and, 758–59, 760
Smooth endoplasmic reticulum, 81, 83, 85, 87
Smooth muscle, 541, 700, 701, 704
Smuts, 399, 400
Snails, 433, 435, 447, 743, 774
Snakes, 456
venom of, 57, 71
Snapdragons, 223
Snowberry flies, 821–22
Snowshoe hares, 798
Social behavior, 778–83
advantages and disadvantages of, 780–81
in bullhead catfish, 782–83
in honeybees, 781–82
in naked mole rats, 783
Sodium, 25, 586
Sodium bicarbonate, 597
Sodium channels, 666
Sodium ions, 656
Sodium–potassium pump, 666
Solution(s), 28
hypertonic, 64
hypotonic, 64, 65
isotonic, 64
Solvent, water as, 28
Somatic nervous system, 673
Sonoran Desert, 869
Sound, 683–86, 772–73
Source, 488
South American tortoises, 721
Sowbug, 446
Space, uniformity in, 9
Spanish shawl sea slugs, 447
Spawning, 719, 720
Specialized connective tissues, 539
Speciation, 308–14
by adaptive radiation, 313–14
allopatric, 309, 310, 318
sympatric, 309–11
by ecological isolation, 309–11
in fossil record, 313
by polyploidy, 311–13
Species, 4, 270, 308–21, 351
"biological species" definition, 358
defined, 308
dioecious, 719
exotic, 797, 830
keystone, 824–26
mass extinctions and diversity of, 336–38
monoecious, 719
reproductive isolation between, 314–16
Specific heat, 31
Spemann, Hans, 748
Sperm, 213, 333, 719, 723, 724, 730, 734
Spermatids, 722–23
Spermatocytes, 722
Spermatogenesis, 722, 723, 725
Spermatogonium/spermatogonia, 722, 723, 725
Spermatophore, 720–21
Spermatozoa, 723
Spermicide, 732, 733
Sperry, Roger, 680
S phase, 194
Sphincter(s)
in bladder, 608
precapillary, 559, 561
pyloric, 595, 596
Spicules, 434
Spiders, 445
Spider web silk, 47
Spinach, 525
Spinal cord, 672, 673–76
healing of, 675
injuries to, 662, 663, 675, 693
Spindle microtubules, 194–96, 201, 203
Spindle poles, 195
Spiracles, 572
Spirogyra, 382–83
Spleen, 561, 564
Splenda® (sucralose), 37, 53
Sponge, contraceptive, 732
Sponges, 432–35, 436, 571, 831, 880
body plan of, 433–34, 436
cellular organization in, 430
digestion in, 590–91
Spongy bone, 707
Spongy cells, 481
Spontaneous generation, 10, 324
Spontaneous reactions, 106
Sporangia, 394, 414
Spores, 392–93, 498, 499
of fungi, 392–93, 402–3
gametes vs., 498
swimming, 393–94
Sporophyte, 410, 498
Sporozoans, 378, 384–85
Sprouting of lateral bud, 524
Squash family, 501, 509
Squid, 435, 448, 449, 665
SRY gene, 181, 225
Stabilizing selection, 296–98
Stamens, 13, 500, 501, 512
Stanford Center for Narcolepsy, 211, 235
Staphylococcus, 796
Staphylococcus aureus, 286, 287
Starch, 39, 40, 41, 42, 585
Starch-filled plastids, 521–23
Start codon, 167
Startle coloration, 822
STDs (sexually transmitted diseases), 729, 731
Stem branching, 524–25
Stem cells, 675, 748
Stems, 468, 476–79, 489–90
adaptations of, 489–90
branch formation, 477, 478
secondary growth in, 477–78, 479
Steps toward Life (Eigen), 326
Sterilization, 732, 733
Steroids, 40, 43, 45, 46, 178–80, 644, 645
anabolic, 642, 643, 658
Stewart, Margaret, 457
Stickleback fish, 776–77, 779
Stigma, 13, 422, 423, 500, 501
Stimuli, responsiveness to, 4
Stirrup, 684, 685
Stomach, human, 594, 595–96
Stomata/stoma, 116, 117, 123–24, 412
of leaves, 481
on stem epidermis, 477
transpiration rate and, 486–87
Stop codon, 167, 173, 175
Stored (potential) energy, 100
Streptococcus, 75, 80, 95, 377
Streptococcus pneumonia, 150
Stress
oxidative, 21
ulcers and, 597
Striated muscles. *See* Skeletal muscles
Striped rabbit, 307, 318
Stroke, 562
Stroma, 91, 117, 118
Structural proteins, 47
Struhsaker, Thomas, 772
Style, 500, 501
Stylet, 488
Subatomic particles, 4
Subclimax plants, 829
Subclimax state, 831
Substrates, 107
Subunits, 39
Succession, 826–31
in climax community, 829–31
in ponds and lakes, 829
primary and secondary, 826–29
Succinate, 141
Sucking behavior, 784, 785
Sucralose (Splenda®), 37, 53
Sucrose, 40, 41, 488–89
Sugar-phosphate backbone, 152
Sugar(s), 37, 39, 585
burning of, 101–2
fat synthesis from, 134
plant transport of, 473, 487–89
Sulfates, 375
Sulfur cycle, 848–49
Sunburn, 158
Sundew (*Drosera rotundifolia*), 466, 467, 491, 492
Supergerms, 287
Superior vena cava, 552
Suppressor T cells, 624, 629
Surface tension, 29
Surgical removal of tumor, 638
Surveyor 3 spacecraft, 75
Survival of the fittest, 275, 296
Survivorship curves, 803, 804
Sverdlovsk, city of, 365
Swallowing, 595
"Swamp gas," 375
Sweat glands, 539, 542, 543

Sweetener, artificial, 37, 53
Swim bladder, 454
Symbiosis, 299, 816, 823–24
in fungi, 397–98, 399
intracellular, 330, 331
plant nutrient acquisition through, 482–83
Symbiotic bacteria, 375
Symbiotic relationship, 330
Symmetrical body forms, 430–32
Symmetry, sex and, 764, 765, 787
Sympathetic division, 673, 674
Sympatric speciation, 309–11
by ecological isolation, 309–11
in fossil record, 313
models of, 310
by polyploidy, 311–13
Synapses, 665–67
excitatory and inhibitory potentials at, 666–67
structure and operation of, 668
Synaptic terminals, 541, 664–65
Syphilis, 376, 731
Systematics, 351–63
biodiversity and, 358–59
defined, 351
taxonomy, 351, 352–53
changes in, 357–58
classification criteria, 353, 356–57
kingdoms, 351, 353–58
origin of, 352–53
Systematists, 351
Systolic pressure, 552

Tadpole, 455
Taiga, 873–74
Tail, 431, 450, 724
Tallgrass prairie, 870
Tambalacoque tree, 506
Tapeworms, 438–39
Taproot system, 473
Tarantula, 446
Target cells, 643, 644–48
Tarsier, 339
Taste buds, 595, 690
Taste receptors, 690–91
Tatum, Edward, 164, 165
Taxis/taxes, 372
Taxol®, 15, 233, 409, 424
Taxonomy, 351, 352–53
changes in, 357–58
classification criteria, 353, 356–57
kingdoms, 351, 353–58
origin of, 352–53
Tay-Sachs disease, 260–61
T-cell leukemia, 367
T-cell receptors, 624, 625–27
T cells, 624, 628–29, 637
memory, 629
suppressor, 629
Tectorial membrane, 684, 685
Teeth, 594
Telomeres, 192, 758–59
Telophase
in meiosis, 200, 201, 202–3
in mitosis, 195, 197
Temperate deciduous forests, 870–72
Temperate rain forests, 872–73
Temperature
elevation and, 861
reactions and, 106, 109
vegetation and, 875–76
water and, 31–32
Template strand of DNA, 168, 170
Temporal isolation, 314, 315
Temporal lobe, 679, 682
TEMs (transmission electron microscopes), 79
Tendons, 539, 701, 710
Tendrils, 489–90, 491
Tentacles, 435
Terminal bud, 476
Termination signal, 168, 169
Termination step
of transcription, 167, 169
of translation, 172, 173
Territoriality, 776–77
Tertiary consumers, 838
Tertiary structure of protein, 49, 50
Test cross, 216
Testes, 644, 649, 654, 722–24
Testosterone, 45, 643, 647, 649, 654, 722, 724
Test-tube baby, 734
Test-tube tiger, 735
Tetanus, 375, 376
Tetrahymena, 187, 188
Tetraploidy, 311–13
Texas bluebonnets, 465
Thalamus, 677–78
Thalidomide, 758
Theories, scientific, 11–12
Thermodynamics, laws of, 100–1
Thermoreceptor, 684
Thiamin (vitamin B_1), 587
Thick filaments, 702, 703
Thimann, Kenneth, 523
Thin filaments, 702, 703
Thomas, Donald, 506
Thomas, Lewis, 56, 377
Thompson, James, 748
Thorax, 443
Thornhill, Randy, 765, 787
Thorns, 489–90
Three Gorges Dam, 634
Threshold, 665
Thrombin, 558
Thucydides, 623
Thumb, opposable, 343
Thumb sucking, 784
Thylakoids, 91, 117, 118, 119
accessory pigments in, 119
structure of, 120
Thymine, 51, 151–52, 158
Thymosin, 649, 657
Thymus, 561, 564, 644, 649, 657
Thyroid gland, 644, 648, 652–54
Thyroid hormones, 657–58
Thyroid-stimulating hormone (TSH), 648, 651, 653
Thyroxine, 648, 652–53, 657–58
Ticks, 446
Tiger, test-tube, 735
Tiger salamanders, 456
Tight junctions, 68
Timber rattlesnake, 184, 185, 205
Time, uniformity in, 9
Tinbergen, Niko, 769, 770, 776–77
Ti plasmid, 251
Tissues, 4, 430–31, 538–44
connective, 539–40
epithelial, 538–39
growth, maintenance and repair of, 186–87
muscle, 541
nerve, 541, 542
plant, 470–71
Toads, 455, 456–57, 767
Tobacco mosaic plant virus, 367
Tobacco plant, 251, 252
Tocopherol (vitamin E), 587
Todd, John, 782
Tomato hornworm, 519
Tonsils, 561, 564
Tool technology, 341, 342, 345
Tooth decay, 371, 372
Tortoises, 455, 456
Trachea/tracheae, 334, 444
human, 574, 595
of insects, 572
Tracheids, 471–73
Tracheoles, 572
Tracheophytes (vascular plants), 410, 413–14
Traits, 213
Transcription, 166, 173–74
complementary DNA strand from, 167
environment and, 176
regulation of, 175–80
reverse, 635
selectiveness of, 168–70
steps in, 167–70
"Trans" fatty acids, 46
Transfer RNA (tRNA), 165–66, 170–74
methionine, 173
Transformation, bacterial, 244–45
Transgenic organisms, 244, 251–55
Translation, 166, 173–74
regulation of, 177
steps in, 170–74
Transmission electron microscopes (TEMs), 79
Transpiration, 484–86
Transport proteins, 59–60, 63, 65
Tree ferns, 414
Tree of life, 355, 357
Tree rings, 478, 480
Treponema pallidum, 731
Trial-and-error learning, 767, 768
Triceratops, 13, 114, 115, 126, 267
Trichinella, 440–41
Trichinosis, 440–41
Trichomonas, 731
Trichomoniasis, 731
Triglyceride, 43
Trilobites, 332
Triple covalent bond, 26
Triploidy, 311–13
Trisomy 21 (Down syndrome), 180, 234
Trisomy X, 232
tRNA. *See* Transfer RNA (tRNA)
Troops, baboon, 289
Trophic levels, 838–42
Tropical deciduous forests, 867
Tropical forests, 317
Tropical rain forests, 15, 864–67
Tropics, 860
Tropomyosin, 704
Troponin, 704
True-breeding plants, 213
Truffles, 402
Tryglyceride, 40
Trypanosoma, 383, 384, 824
Trypsin, 597
Tschermak, Erich, 218
TSH (thyroid-stimulating hormone), 648, 651, 653
T tubules, 701
Tubal ligation, 733
Tube cell, 501

Tube feet, 449, 450
Tuberculosis (TB), 376
"Tube-within-a-tube" body plan, 431
Tubular reabsorption, 609, 610, 611
Tubular secretion, 609, 610, 611
Tubules of nephron, 608
Tubulin, 93
Tumors, 637, 638
Tundra, 874–75
Tunicates, 433, 450–51, 452
Turgor pressure, 89–90
Turner syndrome, 231–32
Turtles, 456
Twins, 262, 785–86
Tympanic membrane (eardrum), 684, 685
Tyndall, John, 324
Tyrannosaurus, 114, 115, 126
Tyrosinase, 227

Ulcers, 597
Ultraviolet (UV) radiation, 158, 457, 858, 859
Ulva (sea lettuce), 383
Umami, 690
Umbilical cord, 753, 756–57
UNAIDS, 637
Unicellular organisms, 8
Uniformitarianism, 272
Uniformity in space and time, 9
Uniform population distribution, 802, 803
United States, population growth in, 808
Unsaturated fatty acid, 44, 70
Upright posture, 339–41
Upwelling, 878
Uracil, 51
Urea, 585, 607
Ureter, 608
Urethra, 722, 723, 724, 726
Urey, Harold, 325
Uric acid, 607
Urinary system, 605–17
  functions of, 606–7
  homeostasis and, 607
  human, 607–14
    bladder, 608
    kidneys, 607–10
  of simple animals, 605–6
  vertebrate, 606–7
Urine, 607, 608, 609–12
Uterine muscles, 756
Uterus (womb), 459, 726, 727
UV (ultraviolet) radiation, 158, 457, 858, 859

Vaccinations, 631
Vaccines, genetically engineered plants as, 252
Vacuole(s), 82, 88–90
  central, 82, 83, 89–90
  contractile, 88
  food, 66, 87, 591
Vagina, 726, 729
  infections of, 400, 401
Valine, 175
Valley fever, 400
Value systems, 9
Valves
  cardiac, 552–54
  venous, 560
Vampire bats, 786
Vaporization, heat of, 32
Variable regions, 624, 625
Variables, 9, 10
Variation, genetic, 12
Vascular bundles (veins), 117, 481
Vascular cambium, 477–78, 479
Vascular cylinder, 474, 482
Vascular plants (tracheophytes), 410, 413–14
Vascular tissues, 333, 470, 471–73
  monocot vs. dicot, 469
  in stems, 477
Vas deferens, 722, 723, 724
Vasectomy, 732–33
Vasopressin. *See* Antidiuretic hormone (ADH)
Vector, 247, 258
Vein(s), 552, 560–61
  of leaves (vascular bundles), 117, 480
  renal, 608, 609
Velociraptors, 336
Venom, snake, 57, 71
Ventral surface, 430
Ventricles, 551
Venules, 558, 560–61
Venus flytrap, 491, 528–29
Vertebral column (backbone), 451, 673
Vertebrates, 432, 433, 435, 451–60
  bony fishes, 452, 453–54
  cartilaginous fishes, 452, 453
  jawless fishes, 451–53
  urinary systems of, 606–7
Vertebrate societies, 782–83
Vesicle, 66, 81, 82, 83, 86
Vessel elements, 471, 473
Vessels, 413, 552
  artificial, 556
Vestigial structures, 278, 280
Vibrio cholerae, 355
Victoria, Queen, 230
Villus/villi, 260, 598
Vincent, Amanda, 736
Viral infections, 731
Virchow, Rudolf, 75, 184
Viroids, 369, 370
Virus(es), 366–70, 824. *See also* HIV
  adeno-associated, 258
  bacteriophages, 171, 366, 367
  DNA recombination through, 245–46
  flu, 632
  infections by, 366–68
  origin of, 370
  retroviruses, 258, 369, 635–36
  smallpox, 801
  structure and replication of, 366, 367, 368–69
Vision, binocular, 338, 689
Visual communication, 771–72
Vitamin A deficiency, 243
Vitamins, 32, 377, 584, 586–88
Vitreous humor, 686, 687
Vocal cords, 574
Volcanic eruptions, 337
Volicitin, 530
Volvox, 383
Vostok, Lake, 345

Waggle dance, 781, 782
Wallaby, 460
Wallace, Alfred Russel, 12, 273, 275, 280, 294
Warblers, 308
Warm-blooded animals, 458–59
Warning coloration, 819, 821
Warts, genital, 731
Wasps, 512, 519
Water, 27–32, 323
  absorption in large intestine, 599
  in blood, 612–13
  "bound" molecules of, 63
  cohesion of, 29–30
  conduction in plants, 473
  covalent bonding in, 26
  electron flow, photosystems and,
  in environment, 486
  "free" molecules of, 63
  ice, 32
  ions of, 30
  molecular interactions of, 28–29
  photosynthesis and, 123–24
  plant acquisition of, 484–87
  plant transport of, 477, 484–87
  pollution of, 885
  potassium transport into guard cells and, 487
  in reactions, 39
  temperature effects of, 31–32
Water dispersal of seed, 513
Waterflea, 2, 446
Water hyacinths, 830
Water (hydrologic) cycle, 846–47
Water molds, 378–79
Water-soluble vitamins, 587–88
Water-vascular system, 449–50
Watson, James, 11, 74, 148, 152–53, 154
Waxes, 40, 43–45
Weapons, biological, 364, 365, 367, 387
Weather, 857–62
Weathering, 827
Weinberg, Wilhelm, 289
Went, Frits, 523
Werner syndrome, 177
Western bluebirds, 299
Western Diamondback rattlesnake, 71
Wexler, Leonore, 231
Wexler, Milton, 231
Wexler, Nancy, 231
Wheeler, Peter, 339–41
White, James, 850
White blood cells (leukocytes), 555, 557, 578, 706
White matter, 673
White-tailed deer, 872
Whittaker, Robert H., 353
Wildflowers, 872, 875
Wild genes, 233
Wilkins, Maurice, 152, 154
Williams, Stephen, 529
Wilmut, Ian, 192
Wilson, E. O., 15, 866
Wilson, H. V., 433
Wind dispersal of seed, 513
Wind pollination, 499–500, 508, 509
Wine, 402
Withdrawal, 734
Woese, Carl, 354
Wolf, 771, 773
Womb (uterus), 459, 726, 727
Woody cork, 478–79, 481
Woody plants, 470
Worker bees, 781
Working memory, 682
Wrasse, 7

X chromosomes, 180, 181, 220, 221–22
Xenopus, 747, 750
Xenotransplantation, 549, 565, 614
Xeroderma pigmentosum, 158
Xylem, 471–73, 477–78
  mineral transport to, 482
  water transport in, 484–86
XYY males, 232

Yucca moths, 512
Yuccas, 512

Zebra mussels, 814, 815, 831
Zero population growth, 805
Ziduvidine (AZT), 637
Zinc, 586
Zinnia, 471
Z lines, 702
Zona pellucida, 730
Zooflagellates, 378, 383–84
Zooplankton, 876, 881, 882
Zooxanthellae, 381
Zucchini, 501
Zygomycota, 393, 394
Zygospores, 394
Zygote, 392, 727
Zygote fungi, 393, 394